CAPILLARY ELECTROPHORESIS METHODS FOR PHARMACEUTICAL ANALYSIS

CAPILLARY ELECTROPHORESIS METHODS FOR PHARMACEUTICAL ANALYSIS

George Lunn

A WILEY-INTERSCIENCE PUBLICATION

JOHN WILEY & SONS, INC.

New York / Chichester / Weinheim / Brisbane / Singapore / Toronto

Library of Congress Cataloging in Publication Data:
Lunn, George.
    Capillary electrophoresis methods for pharmaceutical analysis / by George Lunn
      p. cm.
    "A Wiley-interscience publication."
    Includes bibliographical references.
    ISBN 0-471-33188-0 (cloth : alk. paper)
    1. Capillary electrophoresis. 2. Drugs—Analysis. I. Title.
    [DNLM: 1. Electrophoresis, Capillary—methods. 2. Pharmaceutical Preparations—
analysis. 3. Chemistry, Pharmaceutical—methods. QU 25 L983c 1999]
    RS189.5.C34L86 1999
    615′.1901—dc21
    DNLM/DLC
    for Library of Congress   98-48818
                                                            CIP

CONTENTS

ACKNOWLEDGMENTS

These days the production of a technical book is a complex process requiring the talents of many people. Even by modern standards, however, this work was a particularly involved process. The raw data initially consisted of thousands of individual files, one for each abstract. These files were converted into a database that was used to make the CD for the electronic version as well as prepare the coded report that was used to set the galley and page proofs. The people who worked on this project are as follows:

For Synexchem Consulting Services International LLC: Software Developers Peter Baričič and Martin Mackov; Abstracters Elena Belajová, Research Institute of Foods; Alica Čániová, Slovak Technical University, Department of Analytical Chemistry; Katarina Hroboňová, Slovak Technical University, Department of Analytical Chemistry; Gabriela Krištofiková; Pavol Kubalec, Slovak Technical University, Department of Analytical Chemistry; Gabriela Machalová; Martina Nahálková; Mária Tegzová, Drug Research Institute; and Pavol Valachovič, Drug Research Institute.

For John Wiley: Betty Sun, Senior Editor; Perry King, Associate Editor, Wiley New Media; and John Sollami, Senior Managing Editor, STM Books.

The use of the National Library of Medicine, National Institutes of Health Library, and the FDA Medical Library is greatly appreciated. Although many people have helped with the preparation of this work, the mistakes are my own. I would appreciate hearing from anyone who has corrections, comments, or suggestions. I can be reached at lunng@cder.fda.gov.

The content of this volume does not necessarily reflect the views or policies of the Food and Drug Administration, nor does the mention of trade names, commercial products, or organizations imply endorsement by the U.S. Government.

PREFACE

This book is a collection of procedures for the analysis of a number of pharmaceuticals using capillary electrophoresis (CE). For the purposes of this book we have defined capillary electrophoresis as a procedure involving a narrow-bore tube with an applied voltage. It also includes electrochromatography but not gel electrophoresis or isotachophoresis. For each compound a number of techniques are described in sufficient detail that the analyst can replicate the procedure without reference to the original publication. We have listed procedures for the analysis of drugs used for both medical and veterinary purposes.

Despite the ready availability of laboratory-based literature searching, this resource is not exploited as much as it might be. One reason for this reluctance is, of course, that a computer search merely produces a listing of possibly relevant references. Tedious and time-consuming searches in the library are necessary to find the most relevant reference that can be turned into a practical analytical procedure in the searcher's own laboratory. The reference finally chosen will, naturally, depend on the individual circumstances, such as the matrix in which the drug is present and availability of equipment. This book circumvents this lengthy process by providing a number of abstracted and evaluated procedures for the analysis of each drug. The analyst can rapidly identify a relevant procedure and put it into practice.

In addition to the analytical matrix other factors may be important when choosing an analytical procedure. Accordingly, we have noted such features of the analytical procedures as sensitivity, mode of detection, other compounds that interfere with the analysis, and other drugs that may be determined at the same time, etc.

Readers familiar with our previous publications, *HPLC Methods for Pharmaceutical Analysis* (George Lunn and Norman R. Schmuff, John Wiley, New York, 1997) and *Handbook of Derivatization Reactions for HPLC* (George Lunn and Louise C. Hellwig, John Wiley, New York, 1998), will notice many similarities. The abstract structure is very similar and the philosophy that the procedures should be reproducible without reference to the original literature has been continued. Like the previous volumes, this book is also available on a CD in an electronic form. The software is new and hence the appearance on the computer screen is different but the ease of use has been retained. The CD contains all the data from the above-mentioned pharmaceutical analysis book and the data from the three volumes of HPLC methods that are being published simultaneously with this volume.

ABOUT THIS BOOK

SCOPE

A computer search was used to identify relevant references. The search was conducted using Medline, 1980 to date, but almost all of the references were found in the 1990s. The earliest relevant reference dates to 1987. Since references from the *Journal of Liquid Chromatography* are not included in Medline, these references were manually added to the database.

The search strategy was as follows: electrophoresis (mh) and capillary (tw) or electrokinetic (tw), where tw = text word and mh = MESH heading.

In addition to the Medline search, some journals were routinely surveyed for relevant articles. These journals were:

American Journal of Health-System Pharmacy
Analyst
Analytical Chemistry
Antimicrobial Agents and Chemotherapy
Arzneimittelforschung
Biological and Pharmaceutical Bulletin
Biomedical Chromatography
Biopharmaceutics and Drug Disposition
Chemical and Pharmaceutical Bulletin
Chromatographia
Clinical Chemistry
Clinical Pharmacology and Therapeutics
Drug Metabolism and Disposition
Farmaco
Journal of Analytical Toxicology
Journal of AOAC International
Journal of Chromatographic Science
Journal of Chromatography, Part A and Part B
Journal of Clinical Pharmacology
Journal of Forensic Sciences

Journal of Liquid Chromatography and Related Technology
Journal of Medicinal Chemistry
Journal of Pharmaceutical and Biomedical Analysis
Journal of Pharmaceutical Sciences
Journal of Pharmacology and Experimental Therapeutics
Pharmaceutical Research
Pharmazie
Therapeutic Drug Monitoring
Xenobiotica

Many other journals were consulted when relevant articles were identified by computer searches.

CHAPTER STRUCTURE

Each chapter is headed by the name and structure of the target compound as well as other useful data such as the CAS Registry Number, molecular formula, and molecular weight. In general, the United States Adopted Name (USAN) is used throughout to identify each drug, although we sometimes use a truncated version of this name, e.g., naproxen for naproxen sodium. Exceptions are clavulanate potassium, which is listed in the "Clavulanic Acid" monograph, EDTA, which is used in preference to edetic acid, hyoscyamine, which is listed in the monograph for its racemate atropine, isotretinoin and tretinoin, which are listed in the monograph "Retinoic Acid," levonorgestrel, which is listed in the "Norgestrel" monograph, and some of the steroids, which are listed for convenience in the monograph "Estrogens, conjugated." Names of derivatives, such as esters, which would have different chromatographic properties, are identified by placing the derivative name in parentheses after the migration time.

Reference is also made at the head of each chapter to the relevant abstract in the *Merck Index*. **Note that these numbers refer to the 12th edition of the *Merck Index*.**[1] In addition, the relevant sections in the series *Organic Chemistry of Drug Synthesis*, by Lednicer and Mitscher,[2–6] are referenced. These books give valuable information about the syntheses of various drugs, and this may be helpful in determining impurities, understanding degradation reactions, and so on.

ABSTRACT STRUCTURE

The detailed procedures given normally contain the following sections. Of course, not all papers give full details, so some sections may be missing.

Matrix	Capillary Preparation
Sample Preparation	Running Buffer
Capillary	Injection

Migration Time	Simultaneous
Detector	Also
Internal Standard	Non-interfering
Model (of capillary	Interfering
electrophoresis instrument)	Limit of Detection
Capillary Temperature	Limit of Quantitation
Voltage	Key Words
Current	Reference
Extracted	

ABSTRACT CONVENTIONS

Also	Compounds that can be analyzed at the same time. It is not specified whether or not they interfere but they can be extracted. See also Extracted, Simultaneous.
Capillary	Dimensions are length (cm) $\times$ internal diameter (μm)
Capillary Temperature	If other than ambient (in degrees C)
Current	Can vary (voltage is fixed) unless otherwise stated
Derivatization	Pre-column unless otherwise mentioned (in Key Words)
Detector	Toward the cathode (negative) end of the capillary unless otherwise stated (wavelengths in nm)
Extracted	Compounds that can be extracted from the matrix in question and analyzed at the same time and do not interfere. See also Also, Simultaneous.
Impurities	If method resolves compound and impurities (in Key Words)
Injection	A description of the procedure is provided (1 psi = 6894.76 Pa)
Interfering	Compounds that interfere with the analysis of the target compound. Compounds that interfere with the analysis of the internal standard (IS) are listed under Simultaneous (another IS can always be selected or an external standard procedure can be used).
Matrix	A controlled vocabulary is used (see below)
Metabolites	If method resolves compound and metabolites (in Key Words)
Migration Time	Frequently estimated from a reproduced electropherogram and so the accuracy may not be great (in minutes).
Non-interfering	Compounds that do not interfere with the analysis for various reasons, e.g., they are not extracted, they are not detected.
Simultaneous	Compounds that can be analyzed at the same time and do not interfere. Note that the compound cannot necessarily be extracted from the matrix in question (although it may be). See also Also, Extracted.
Species	If other than human; noun is used instead of adjective, e.g., cow not bovine (in Key Words).
Voltage	Fixed (with varying current) unless otherwise stated.

MATRIX

To help with searching, a controlled vocabulary is used to limit the number of terms in the matrix section. For example, the term "raw materials" is not used, the term "bulk" is used instead. In a number of cases, the matrix is associated with various key words which can be used to narrow the search. For example, the matrix term "blood" has the key words plasma, serum, and whole blood associated with it. Thus, if you are interested in the determination of the drug in blood, in general you should search the matrix field for blood. If, however, you are specifically interested in finding the drug in plasma, you should search the key words field for plasma.

Matrix	Associated Key Words
bile	
blood	plasma, serum, whole blood
bulk	
CSF	
formulations	capsules, creams, injections, ointment, tablets, etc.
microsomal incubations	
milk	
perfusate	
reaction mixtures	
saliva	
solutions	buffer, water
tissue	brain, heart, kidney, liver, muscle, spleen, etc.
urine	

ABBREVIATIONS

BHT	2,6-di-tert-butyl-4-methylphenol, butylated hydroxytoluene
CE	capillary electrophoresis
DMSO	dimethyl sulfoxide
E	electrochemical detection
em	emission wavelength
EtOH	ethanol
ex	excitation wavelength
F	fluorescence detection
FW	formula weight
GPC	gel permeation chromatography
h	hour
HPLC	high-performance liquid chromatography
IS	internal standard
kPa	kiloPascals (1 psi = ca. 6.89476 kPa)
L	liter
LOD	limit of detection or some other description indicating that this is the smallest concentration or quantity that can be detected or analyzed for
LOQ	lower limit of quantitation, either given as such in the paper or taken as the lower limit of the linear quantitation range

M	molar (i.e., moles/L)
MeCN	acetonitrile
MeOH	methanol
min	minutes
mL	milliliter
mM	milli-molar (i.e., milli-moles/L)
MT	migration time
MTBE	methyl tert-butyl ether
nM	nano-molar (i.e., nano-moles/L)
psi	pounds/sq. in. (1 psi = 6.89476 kPa)
s	seconds
SEC	size exclusion chromatography
SFC	supercritical fluid chromatography
SFE	supercritical fluid extraction
SIM	selected-ion monitoring
SPE	solid phase extraction
Temp	temperature
U	units
UV	ultraviolet detection

WORKING PRACTICES

In general, good working practice, e.g., using high-grade materials, is assumed. Solutions containing compounds should be protected from light and silanized glassware should be used unless you have good reason to believe that these precautions are not necessary. A number of excellent texts discuss good working practices and procedures in capillary electrophoresis,[7-13] and these should be consulted. A bibliography of reviews of capillary electrophoresis is appended to this book.

Details of solution preparation are generally not given. It should be remembered that the preparation of a dilute aqueous solution of a relatively water-insoluble compound can frequently be made by dissolving the compound in a small volume of a water-miscible organic solvent and diluting this solution with water.

It is also assumed that safe working practices are observed. **High voltages can be lethal!** Organic solvents should only be evaporated in a properly functioning chemical fume hood, correct protective equipment should be worn when dealing with potentially hazardous biological materials, and waste solutions should be disposed of in accordance with all applicable regulations.

A number of solvents are particularly hazardous. For example, benzene is a human carcinogen;[14] chloroform,[15] dichloromethane,[16] dioxane,[17] and carbon tetrachloride[18] are carcinogenic in experimental animals; and DMF[19] and MTBE[20,21] may be carcinogenic. Organic solvents are, in general, flammable and toxic by inhalation, ingestion, and skin absorption. Sodium azide is carcinogenic and toxic and liberates explosive, volatile, toxic hydrazoic acid with acid. Sodium azide can form explosive heavy metal azides, e.g., with plumbing fixtures, and so should not be discharged down the drain.[22] Disposal procedures have been described for a number of hazardous drugs and reagents,[22] and recent papers describe a procedure for the hydrolysis of acetonitrile in waste solvent to the much less toxic acetic acid and ammonia.[23,24] Recent work has shown that n-hexane is surprisingly toxic.[25]

REFERENCES

1. Budavari, S., Ed., *The Merck Index*, 12th ed., Merck & Co., Whitehouse Station, NJ, 1996.

2. Lednicer, D. and Mitscher, L.A., *The Organic Chemistry of Drug Synthesis*, John Wiley & Sons, New York, 1977.

3. Lednicer, D. and Mitscher, L.A., *The Organic Chemistry of Drug Synthesis*, Vol. 2, John Wiley & Sons, New York, 1980.

4. Lednicer, D. and Mitscher, L.A., *The Organic Chemistry of Drug Synthesis*, Vol. 3, John Wiley & Sons, New York, 1984.

5. Lednicer, D., Mitscher, L.A., and Georg, G.I., *The Organic Chemistry of Drug Synthesis*, Vol. 4, John Wiley & Sons, New York, 1990.

6. Lednicer, D., *The Organic Chemistry of Drug Synthesis*, Vol. 5, John Wiley & Sons, New York, 1995.

7. Landers, J.P., *Handbook of Capillary Electrophoresis*, 2nd ed., CRC Press, Boca Raton, FL, 1996.

8. Lunte, S.M. and Radzik, D.M., *Pharmaceutical and Biomedical Applications of Capillary Electrophoresis*, Elsevier, New York, 1996.

9. Baker, D.R., *Capillary Electrophoresis*, John Wiley & Sons, New York, 1995.

10. Altria, K.D. (Ed.), *Capillary Electrophoresis Guidebook: Principles, Operation, and Applications*, Humana Press, Totowa, NJ, 1995.

11. Righetti, P.G. and Hancock, W. (Eds.), *Capillary Electrophoresis in Analytical Biotechnology*, CRC Press, Boca Raton, FL, 1995.

12. Baker, D., *Capillary Electrophoresis*, Prentice-Hall: Englewood Cliffs, NJ, 1995.

13. Foret, F., Krivankov, L., and Bocek, P., *Capillary Zone Electrophoresis*, VCH, New York, 1994.

14. Lewis, R.J., Sr., *Sax's Dangerous Properties of Industrial Materials*, 8th ed., Van Nostrand-Reinhold, New York, 1992, pp. 356–358.

15. Reference 14, pp. 815–816.

16. Reference 14, pp. 2311–2312.

17. Reference 14, pp. 1449–1450.

18. Reference 14, pp. 701–702.

19. Reference 14, p. 1378.

20. Belpoggi, F., Soffritti, M., and Maltoni, C., Methyl-tertiary-butyl ether (MTBE)—a gasoline additive—causes testicular and lympho-haematopoietic cancers in rats. *Toxicol. Ind. Health*, **1995**, *11*, 119–149.

21. Mehlman, M.A., Dangerous and cancer-causing properties of products and chemicals in the oil refining and petrochemical industry: Part XV. Health hazards and health risks from oxygenated automobile fuels (MTBE): Lessons not heeded. *Int. J. Occup. Med. Toxicol.* **1995**, *4*, 219–236.

22. Lunn, G. and Sansone, E.B., *Destruction of Hazardous Chemicals in the Laboratory*, 2nd ed., John Wiley & Sons, New York, 1994.

23. Gilomen, K., Stauffer, H.P., and Meyer, V.R., Detoxification of acetonitrile–water wastes from liquid chromatography. *Chromatographia*, **1995**, *41*, 488–491.

24. Gilomen, K., Stauffer, H.P., and Meyer, V.R., Management and detoxification of acetonitrile wastes from liquid chromatography. *LC.GC* **1996**, *14*, 56–58.

25. Meyer, V., A safer solvent. *Anal. Chem.* **1997**, *69*, 18A.

CAPILLARY ELECTROPHORESIS METHODS FOR PHARMACEUTICAL ANALYSIS

Acarbose

Molecular formula: $C_{25}H_{43}NO_{18}$
Molecular weight: 645.61
CAS Registry No.: 56180-94-0
Merck Index (12th ed.): 15

SAMPLE
Matrix: urine
Sample preparation: Lyophilize 500 μL urine, add 100 μL 80 mM 7-amino-1,3-naphthalene-disulfonic acid in acetic acid:water 15:85, add 50 μL 900 mM sodium cyanoborohydride in DMSO, vortex, heat at 40° overnight, inject an aliquot. (Prepare reagent solutions just before use.)

CAPILLARY ELECTROPHORESIS
Capillary: 37 cm × 50 μm fused-silica (30 cm to detector) (Grom, Herrenberg, Germany)
Capillary preparation: Between analyses rinse capillary with 100 mM phosphoric acid for 1 min and with running buffer for 1 min.
Capillary temperature: 25
Running buffer: 100 mM Phosphoric acid, adjusted to pH 1.5 with triethylamine
Injection: Pressure injection of sample at 0.5 psi for 5 s followed by an injection of water for 5 s.
Detector: F ex 325 (10 mW Omnichrome He-Cd laser) em 400
Migration time: 10
Voltage: 20 kV
Model: Beckman P/ACE 2100
Limit of detection: 30 ng/mL

OTHER SUBSTANCES
Extracted: metabolites, component 2

KEY WORDS
derivatization; detector is at anode

REFERENCE
Rethfeld,I.; Blaschke,G. Analysis of the antidiabetic drug acarbose by capillary electrophoresis, *J.Chromatogr.B*, **1997**, *700*, 249–253.

Acebutolol

Molecular formula: $C_{18}H_{28}N_2O_4$
Molecular weight: 336.43
CAS Registry No.: 37517-30-9
Merck Index (12th ed.): 16
Lednicer: 2 109

SAMPLE
Matrix: blood
Sample preparation: Hydrolyze 2 mL serum with β-glucuronidase (EC 3.2.1.31, type H-1 from Helix pomatia, 416 800 U/g, Separacor) at 80° for 30 min, cool, add 900 μL MeCN, vortex for 15 min, centrifuge at 2004 g for 10 min, add ephedrine (165 μg/mL), filter (0.5 μm), inject an aliquot of the filtrate.

CAPILLARY ELECTROPHORESIS
Capillary: 58 cm × 50 μm fused-silica (50 cm to detector) (Polymicro Technologies)

Capillary preparation: Before each injection purge capillary with 5% phosphoric acid for 12 s, with water for 30 s, and with running buffer for 10 min.
Capillary temperature: 35
Running buffer: 80 mM pH 6.7 sodium phosphate buffer containing 15 mM cetyltrimethylammonium bromide
Injection: Hydrostatic injection for 20 s
Detector: UV 214
Migration time: 11
Internal standard: ephedrine (17.5)
Voltage: -27 kV
Model: Waters Quanta 4000

OTHER SUBSTANCES
Extracted: alprenolol, atenolol, labetalol, metoprolol, nadolol, oxprenolol, pindolol, propranolol, timolol

KEY WORDS
serum

REFERENCE
Lukkari,P.; Nyman,T.; Riekkola,M.-J. Determination of nine β-blockers in serum by micellar electrokinetic capillary chromatography, *J.Chromatogr.A*, **1994**, *674*, 241–246.

SAMPLE
Matrix: blood, urine
Sample preparation: Urine. Dilute with 2 volumes of water, filter (0.5 μm), inject an aliquot. Serum. Condition a 3 mL Supelclean LC-18 SPE cartridge (Supelco) with MeOH and water. Hydrolyze 900 μL serum with β-glucuronidase (EC 3.2.1.31 type H-1 from Helix pomatia) at 60° with sonication for 30 min, add 500 μL (?) MeOH, centrifuge at 2000 g, add the supernatant to the SPE cartridge, wash with 1 mL water, dry under vacuum, elute with 2 mL MeOH:water 90:10, filter, inject an aliquot.

CAPILLARY ELECTROPHORESIS
Capillary: 68 cm × 50 μm fused-silica (60 cm to detector) (Polymicro Technologies)
Capillary preparation: Purge with running buffer for 2 min before injection.
Running buffer: pH 7.0 Phosphate buffer containing 10 mM N-cetyl-N,N,N-trimethylammonium bromide (Prepare buffer by mixing 100 mM NaH_2PO_4 and 100 mM Na_2HPO_4 to achieve a pH of 7.0.)
Injection: Hydrostatic injection for 30 s
Detector: UV 214
Migration time: 12
Internal standard: 2,6-dimethylphenol (only for urine) (14.5)
Voltage: -26 kV
Current: 97 μA
Model: Waters Quanta 4000

OTHER SUBSTANCES
Extracted: alprenolol, atenolol, labetalol, metoprolol, nadolol, oxprenolol, pindolol, propranolol, timolol

KEY WORDS
serum; comparison with HPLC; SPE

REFERENCE
Lukkari,P.; Sirén,H. Ion-pair chromatography and micellar electrokinetic capillary chromatography in analyzing β-adrenergic blocking agents from human biological fluids, *J.Chromatogr.A*, **1995**, *717*, 211–217.

SAMPLE
Matrix: solutions

CAPILLARY ELECTROPHORESIS
Capillary: 58 cm × 50 μm fused-silica (50 cm to detector)

Running buffer: Isopropanol:buffer 2.5:97.5 (Buffer was 80 mM pH 6.8 Phosphate buffer containing 15 mM cetyltrimethylammonium bromide.)
Injection: Hydrostatic injection for 15 s
Detector: UV 214
Migration time: 11
Voltage: -20 kV
Model: Waters Quanta 4000

OTHER SUBSTANCES
Simultaneous: alprenolol, atenolol, labetalol, metoprolol, nadolol, oxprenolol, pindolol, propranolol, sotalol, timolol

REFERENCE
Lukkari,P.; Vuorela,H.; Riekkola,M.-L. Effects of organic mobile phase modifiers on elution and separation of β-blockers in micellar electrokinetic capillary chromatography, *J.Chromatogr.A*, **1993**, *655*, 317–324.

SAMPLE
Matrix: solutions

CAPILLARY ELECTROPHORESIS
Capillary: 82 cm × 75 μm fused-silica (Waters)
Capillary preparation: Between runs purge capillary with running buffer for 2 min. Purge a new capillary with 100 mM NaOH for 30 min, with water for 30 min, and with running buffer for 30 min.
Capillary temperature: 30
Running buffer: 30 mM pH 7.6 phosphate buffer containing 10 mM cetyltrimethylammonium bromide
Injection: Inject at high pressure for 5 s
Detector: UV 214
Migration time: 5.5
Voltage: 21 kV
Model: Beckman P/ACE System 2000

OTHER SUBSTANCES
Simultaneous: alprenolol, atenolol, nadolol, oxprenolol, pindolol, propranolol, sotalol, timolol

REFERENCE
Lukkari,P.; Ennelin,A.; Sirén,H.; Riekkola,M.-L. Effect of temperature, effective capillary length, and applied voltage on the migration of nine β-blockers in micellar electrokinetic capillary chromatography, *J.Liq.Chromatogr.*, **1993**, *16*, 2069–2079.

SAMPLE
Matrix: solutions

CAPILLARY ELECTROPHORESIS
Capillary: 57 cm × 75 μm fused-silica (50 cm to detector) (Beckman)
Capillary preparation: Before each run rinse capillary with 100 mM NaOH and running buffer.
Capillary temperature: 30
Running buffer: Acetone:100 mM pH 8.1 borate buffer containing 50 mM sodium dodecyl sulfate 15:85 (A) or 100 mM phosphoric acid adjusted to pH 3.1 with triethanolamine (B)
Injection: Pressure injection for 5-10 s.
Detector: UV 200
Migration time: 11 (A), 13 (B)
Voltage: 25 kV
Model: Beckman P/ACE 5510

OTHER SUBSTANCES
Simultaneous: amiodarone, atenolol, bretylium, captopril, diltiazem, disopyramide, lidocaine, lisinopril, metoprolol, nicardipine, nifedipine, phenytoin, pindolol, propafenone, propranolol, quinidine, sotalol, timolol, p-toluenesulfonic acid, verapamil

---

KEY WORDS
only running buffer A separates all compounds

REFERENCE
Bretnall,A.E.; Clarke,G.S. Selectivity of capillary electrophoresis for the analysis of cardiovascular drugs, *J.Chromatogr.A*, **1996**, *745*, 145–154.

SAMPLE
Matrix: solutions

CAPILLARY ELECTROPHORESIS
Capillary: 43 cm × 50 μm fused-silica (36 cm to detector)
Capillary preparation: Before each injection wash with running buffer for 10 min. After each injection wash with water for 2 min. Wash new capillaries with 1 M NaOH at 60° for 50 min, with NaOH solution (?) at 60° for 10 min, and with water at 25°.
Capillary temperature: 25
Running buffer: 320 mM pH 2.0 Citrate buffer
Injection: Hydrodynamic injection at 1.5 psi for 2 s.
Detector: UV 220
Migration time: 10.7
Voltage: 15 kV
Model: Spectra-Physics Model 1000
Limit of detection: 1-18 μg/mL

OTHER SUBSTANCES
Simultaneous: atenolol, labetalol, levobunolol, metoprolol, nadolol, oxprenolol, pindolol, propranolol, timolol

---

REFERENCE
Lin,C.-E.; Chang,C.-C.; Lin,W.-c.; Lin,E.C. Capillary zone electrophoretic separation of β-blockers using citrate buffer at low pH, *J.Chromatogr.A*, **1996**, *753*, 133–138.

SAMPLE
Matrix: solutions
Sample preparation: Inject an aliquot of a solution in running buffer.

CAPILLARY ELECTROPHORESIS
Capillary: 60 cm × 75 μm fused-silica (52.4 cm to detector)
Capillary preparation: After each run flush with 500 mM KOH for 2-3 min then with water.
Running buffer: 10 mM pH 3.8 Phosphate buffer containing 2% sulfated cyclodextrin (ds 7-10)
Injection: Hydrostatic injection.
Detector: UV 214
Migration time: 30.47, 31.23 (enantiomers)
Voltage: 15 kV
Model: Waters Quanta 4000

OTHER SUBSTANCES
Also analyzed: alprenolol, aminoglutethimide, brompheniramine, bupivacaine, bupropion, canadine, carbinoxamine, chloroquine, chlorpheniramine, dimethindene, disopyramide, doxylamine, hydroxychloroquine, idazoxan, isoxsuprine, ketamine, mepenzolate, mepivacaine, methoxyphenamine, mexiletine, midodrine, nefopam, orphenadrine, oxprenolol, oxyphencyclimine, pheniramine, phensuximide, pindolol, piperoxan, terbutaline, tetramisole, tolperisone, tranylcypromine, trihexyphenidyl, trimipramine, verapamil, warfarin

---

KEY WORDS
chiral; detector at anode

REFERENCE
Stalcup,A.M.; Gahm,K.H. Application of sulfated cyclodextrins to chiral separations by capillary zone electrophoresis, *Anal.Chem.*, **1996**, *68*, 1360–1368.

SAMPLE
Matrix: solutions

CAPILLARY ELECTROPHORESIS
Capillary: 67 cm × 50 μm (60 cm to detector)
Capillary preparation: Wash with running buffer for 5 min before each injection. After each injection wash with 1 M NaOH at 60° for 5 min, with 100 mM NaOH at 60° for 10 min, and with water at 25° for 5 min.
Capillary temperature: 25
Running buffer: 70 mM pH 7.0 Sodium phosphate buffer containing 15 mM cetyltriemthylammonium bromide
Injection: Hydrodynamic injection for 1 s.
Detector: UV 220
Migration time: 14
Voltage: 20 kV

OTHER SUBSTANCES
Simultaneous: atenolol, labetalol, levobunolol, metoprolol, nadolol, oxprenolol, pindolol, propranolol, timolol

REFERENCE
Lin,C.-E.; Chen,Y.-C.; Chang,C.-C.; Wang,D.-Z. Migration behavior and selectivity of β-blockers in micellar electrokinetic chromatography. Influence of micelle concentration of cationic surfactants, *J.Chromatogr.A*, **1997**, *775*, 349–357.

SAMPLE
Matrix: solutions

CAPILLARY ELECTROPHORESIS
Capillary: 62.1 cm × 75 μm fused-silica (44.6 cm to detector) (Polymicro Technologies)
Capillary preparation: Coat capillary as follows. Condition capillary with 1 M NaOH for 1 h, rinse with water for 30 min, fill capillary with 1% 3-(trimethoxysilyl)propyl methacrylate (methacryloxypropyltrimethoxysilane) (adjusted to pH 3.5 with acetic acid), allow to react at room temperature for 2 h, rinse with water, fill capillary with 10% 2-aminoethyl methacrylate hydrochloride (Fisher Scientific) solution containing 1-3 mg/mL N,N,N',N'-tetramethylethylenediamine and 1 mg/mL ammonium persulfate, allow to react until polymerization is complete, rinse with water, dry at 30-40° overnight.
Running buffer: 25 mM pH 4.7 Acetate buffer
Injection: Hydrodynamic injection at 5 cm for 5 s.
Detector: UV 210
Migration time: 10.44
Voltage: 15 kV
Model: laboratory constructed

OTHER SUBSTANCES
Simultaneous: atenolol, betaxolol, carteolol, dilevalol, nifenalol, pindolol, propranolol

KEY WORDS
coated capillary; detector at anode

REFERENCE
Liu,Q.; Lin,F.; Hartwick,R.A. Free solution capillary electrophoretic separation of basic proteins and drugs using cationic polymer coated capillaries, *J.Liq.Chromatogr.Rel.Technol.*, **1997**, *20*, 707–718.

SAMPLE
Matrix: solutions

CAPILLARY ELECTROPHORESIS
Capillary: 50.5 cm × 75 μm coated fused-silica (36 cm to detector) (Polymicro Technologies)
Capillary preparation: Coat capillary as follows. Condition capillary with 1 M NaOH for 1 h, rinse with water for 30 min, force 1% methacryloxypropyltrimethoxysilane (adjusted to pH 3.5

with acetic acid) through the capillary using pressure, allow to react at room temperature for 2 h, rinse with water, fill capillary with diallyldimethylammonium chloride solution containing 1 µL/mL N,N,N',N'-tetramethylethylenediamine and 1 mg/mL ammonium persulfate, allow to react until polymerization is complete, rinse with water, dry at 30-40° overnight.
Running buffer: 10 mM pH 5.0 Acetate buffer
Injection: Hydrodynamic injection at 5 cm for 5 s.
Detector: UV 210
Migration time: 9
Voltage: 12 kV
Model: laboratory-constructed

OTHER SUBSTANCES
Simultaneous: atenolol, dilevalol, nifenalol, pindolol

KEY WORDS
coated capillary

REFERENCE
Liu,Q.; Lin,F.; Hartwick,R.A. Poly(diallyldimethylammonium chloride) as a cationic coating for capillary electrophoresis, *J.Chromatogr.Sci.*, **1997**, *35*, 126–130.

SAMPLE
Matrix: solutions

CAPILLARY ELECTROPHORESIS
Capillary: 36 cm × 50 µm fused-silica coated with linear polyacrylamide (31.5 cm to detector) (GL Science)
Capillary preparation: At the beginning and end of each day rinse capillary with capillary wash solution (Bio-Rad Cat. No. 148-5022) at 690 kPa for more than 3 min and with water at 690 kPa for more than 3 min. Coat capillary as follows.
Running buffer: 50 mM pH 6.0 Phosphate buffer
Injection: Before each injection rinse with water at 690 kPa for 30 s, rinse with running buffer at 690 kPa for 30 s, partially fill with separation solution (1 mM α_1-acid glycoprotein (Cohn fraction VI (ICN) in running buffer) at 6.9 kPa for 190 s (27 cm), inject sample at 6.9 kPa for 2 s, electrophorese with running buffer (Note that α_1-acid glycoprotein from other suppliers may provide inferior results).
Detector: UV 210
Migration time: 17, 18 (enantiomers)
Voltage: 12 kV
Model: Bio-rad BioFocus 3000

KEY WORDS
chiral

REFERENCE
Paris,S.; Blaschke,G.; Locher,M.; Borbe,H.O.; Engel,J. Investigation of the stereoselective in vitro metabolism of the chiral antiasthmatic/antiallergenic drug flezelastine by high-performance liquid chromatography and capillary zone electrophoresis, *J.Chromatogr.B*, **1997**, *691*, 463–471.

SAMPLE
Matrix: solutions

CAPILLARY ELECTROPHORESIS
Capillary: 36 cm × 50 µm fused-silica coated with linear polyacrylamide (31.5 cm to detector) (GL Science)
Capillary preparation: At the beginning and end of each day rinse capillary with capillary wash solution (Bio-Rad Cat. No. 148-5022) at 690 kPa for more than 3 min and with water at 690 kPa for more than 3 min. Coat capillary as follows. Treat capillary with 1 M NaOH at room temperature for 1 h, rinse with water, dry by passing nitrogen gas through the capillary at 110° for 6 h. Pass thionyl chloride through the capillary using a suction pump for several min, seal capillary at both ends and heat at 70° for 6 h. Unseal the capillary and fill with 250 mM vinyl magnesium bromide in THF by suction, seal the capillary, heat at 70° for 6 h. Open the

capillary and rinse it with THF for several min, rinse with distilled water, fill the capillary with polymerization solution, heat at 28 ± 2° for 1 h, condition at -100 V/cm for 30 min (Anal. Sci. 1994, 10, 1). (The polymerization solution was 5% acrylamide in water containing 49 mM Tris, 384 mM glycine, and 0.1% sodium dodecyl sulfate, degas in an ultrasonic bath. Add 40 μL 10% N,N,N',N'-tetramethylethylenediamine and 10 μL 10% ammonium persulfate to 5 mL of the degassed solution, mix thoroughly.)
Running buffer: 50 mM pH 6.0 Phosphate buffer
Injection: Before each injection rinse with water at 690 kPa for 30 s, rinse with running buffer at 690 kPa for 30 s, partially fill with separation solution (1 mM α_1-acid glycoprotein (Cohn fraction VI) (ICN) in running buffer) at 6.9 kPa for 190 s (27 cm), inject sample at 6.9 kPa for 2 s, electrophorese with running buffer (Note that α_1-acid glycoprotein from other suppliers may provide inferior results).
Detector: UV 210
Migration time: 17, 18 (enantiomers)
Voltage: 12 kV
Model: Bio-rad BioFocus 3000

KEY WORDS
chiral; coated capillary

REFERENCE
Tanaka,Y.; Terabe,S. Separation of the enantiomers of basic drugs by affinity capillary electrophoresis using a partial filling technique and α_1-acid glycoprotein as chiral selector, *Chromatographia*, **1997**, *44*, 119–128.

SAMPLE
Matrix: solutions

CAPILLARY ELECTROPHORESIS
Capillary: 71.5 cm × 75 μm fused-silica (52.2 cm to detector) (Polymicro Technologies)
Capillary preparation: Rinse capillary with running buffer for 1 min between runs. Rinse new capillaries with 500 mM NaOH for 20 min and with water for 10 min and then condition with running buffer for 1 h.
Running buffer: 50 mM pH 3.0 containing 0.1% guaran (Guaran is the water-soluble fraction of guar gum. Prepare guaran by stirring 1 g guar gum (Sigma) in 100 mL water for 20 min, add 25 mL EtOH dropwise with stirring, let stand overnight, supercentrifuge. Add 30 mL EtOH dropwise with stirring to the centrifugate, centrifuge, collect the guaran as a precipitate, dry in air.)
Injection: Hydrodynamic injection at 5 cm for 5 s.
Detector: UV (wavelengths not given)
Migration time: 16
Voltage: 17.5 kV
Model: laboratory-constructed

OTHER SUBSTANCES
Simultaneous: atenolol, betaxolol, carteolol, dilevalol, nifenalol, pindolol, propranolol

REFERENCE
Liu,Q.; Lin,F.; Hartwick,R.A. Capillary zone electrophoretic separation of basic proteins and drugs using guaran as a buffer modifier, *Chromatographia*, **1998**, *47*, 219–224.

SAMPLE
Matrix: solutions
Sample preparation: Inject an aliquot of a 340 μM solution.

CAPILLARY ELECTROPHORESIS
Capillary: 58 cm × 75 μm fused-silica (50 cm to detector)
Capillary preparation: Rinse with running buffer for 2 min before each run. Before use rinse capillary at 1.36 bar with 100 mM NaOH for 5 min and with water for 10 min.
Capillary temperature: 25
Running buffer: 20 mM pH 7.0 Sodium phosphate buffer (A) or 20 mM pH 7.0 Sodium phosphate buffer containing 10 mM 1-lauroyl-2-hydroxy-sn-glycero-3-phosphocholine (Avanti Polar Lipids, Alabaster AL) (B)

Injection: Hydrodynamic injection at 34 mbar for 1 s followed by 5% MeOH for 5 s.
Detector: UV 200
Migration time: 4.39 min (A); k' 0.12 (B)
Voltage: 15 kV
Current: ca. 50 μA
Model: Beckman P/ACE 2200

OTHER SUBSTANCES
Simultaneous: alprenolol, atenolol, metoprolol, oxprenolol, pindolol, propranolol, timolol

REFERENCE
Masucci,J.A.; Caldwell,G.W.; Foley,J.P. Comparison of the retention behavior of β-blockers using immobilized artificial membrane chromatography and lysophospholipid micellar electrokinetic chromatography, *J.Chromatogr.A*, **1998**, *810*, 95–103.

SAMPLE
Matrix: solutions

CAPILLARY ELECTROPHORESIS
Capillary: 42 cm × 75 μm acrylamide coated (36 cm to detector)
Capillary preparation: Coat column as follows. Adjust the pH of 20 mL water to 3.5 with acetic acid, add 80 μL 3-(trimethoxysilyl)propyl methacrylate (3-methacryloxypropyltrimethoxysilane), mix, suck into capillary, let stand at room temperature for 1 h, remove the solution, wash with water. Fill the capillary with a deaerated 3-4% acrylamide solution containing 1 μL/mL N,N,N',N'-tetramethylethylenediamine and 1 mg/mL potassium persulfate, let stand for 30 min, remove excess solution by aspiration, rinse with water, remove water by aspiration, dry at 35° (J. Chromatogr. 1985, 347, 191).
Capillary temperature: 20
Running buffer: 100 mM pH 6 2-[N-morpholino]ethanesulfonic acid/NaOH buffer
Injection: Inject a 200 mg/mL solution of transferrin in buffer at 30 psi.s then hydrodynamically inject the sample at 10 psi.s. (Buffer was 100 mM pH 6 2-[N-morpholino]ethanesulfonic acid/NaOH buffer. Transferrin was iron-free human serum transferrin (Behring-Werke, Marburg, Germany).)
Detector: UV 215
Migration time: 29.5, 31 (enantiomers)
Voltage: 10 kV
Model: Bio-Rad BioFocus 3000

OTHER SUBSTANCES
Simultaneous: chlophedianol, nylidrin

KEY WORDS
chiral; coated capillary

REFERENCE
Schmid,M.G.; Gübitz,G.; Kilár,F. Stereoselective interaction of drug enantiomers with human serum transferrin in capillary zone electrophoresis (II), *Electrophoresis*, **1998**, *19*, 282–287.

SAMPLE
Matrix: urine
Sample preparation: Dilute urine with 2 volumes of water, add IS, filter (0.5 μm), inject an aliquot.

CAPILLARY ELECTROPHORESIS
Capillary: 68 cm × 50 μm fused-silica (60 cm to detector) (White Associates)
Capillary preparation: Purge capillary with running buffer for 2 min before each injection.
Running buffer: 80 mM pH 7.0 Phosphate buffer containing 10 mM N-cetyl-N,N,N-trimethylammonium bromide
Injection: Hydrostatic injection for 30 s at 10 cm.
Detector: UV 214
Migration time: 12.3
Internal standard: 2,6-dimethylphenol (14.7)

Voltage: -26 kV (at injector end)
Model: Waters Quanta 4000
Limit of detection: 10 μg/mL (S/N 3)

OTHER SUBSTANCES
Extracted: alprenolol, atenolol, labetalol, metoprolol, nadolol, oxprenolol, pindolol, propranolol, timolol
Simultaneous: probenecid
Noninterfering: caffeine

REFERENCE
Lukkari,P.; Sirén,H.; Pantsar,M.; Riekkola,M.-L. Determination of ten β-blockers in urine by micellar electro-kinetic capillary chromatography, *J.Chromatogr.*, **1993**, *632*, 143–148.

SAMPLE
Matrix: urine
Sample preparation: Filter (0.2 μm), inject an aliquot of the filtrate.

CAPILLARY ELECTROPHORESIS
Capillary: 80 cm × 50 μm fused-silica (57 cm to detector)
Capillary preparation: Before each run rinse the capillary with 1 M NaOH for 3 min, with 100 mM NaOH for 3 min, with water for 3 min, and with running buffer for 10 min.
Running buffer: 110 mM Boric acid containing 56 mM NaOH and 44 mM HCl, pH 8
Injection: Vacuum injection for 1 s
Detector: UV 238
Migration time: 4.518
Voltage: 20 kV
Model: Europhor Prime Vision system IV

OTHER SUBSTANCES
Extracted: acetazolamide (UV 222), alprenolol (UV 220), amiloride (UV220), atenolol (UV 228), bendroflumethiazide (UV 220), bumetanide (UV 220), chlorthalidone (UV 220), codeine (UV 220), ethacrynic acid (UV 220), furosemide (UV 232), hydrochlorothiazide (UV 226), methadone (UV 220), metoxiphenamine (UV 220), nadolol (UV 220), oxprenolol (UV 220), propranolol (UV 220), spironolactone (UV 244), triamterene (UV 232), xipamide (UV 234)
Interfering: cocaine (UV 236), norcodeine (UV 220), pentazocine (UV 220)

REFERENCE
Gonzalez,E.; Laserna,J.J. Capillary zone electrophoresis for the rapid screening of banned drugs in sport, *Electrophoresis*, **1994**, *15*, 240–243.

Acenocoumarol

Molecular formula: $C_{19}H_{15}NO_6$
Molecular weight: 353.33
CAS Registry No.: 152-72-7
Merck Index (12th ed.): 29

SAMPLE
Matrix: solutions

CAPILLARY ELECTROPHORESIS
Capillary: 50 cm × 50 μm fused-silica (41 cm to detector) (Grom)
Capillary temperature: 21 ± 1
Running buffer: 50 mM pH 7 Sodium phosphate buffer containing 10 mM carboxymethyl-β-cyclodextrin (Wacker Chemie) (Anode and cathode buffers were the same but contained no cyclodextrin.)

Injection: Hydrostatic injection at 10 cm.
Detector: UV 210
Migration time: 11.8, 12 (enantiomers)
Voltage: 400 V/cm
Model: Grom 100

KEY WORDS
chiral

REFERENCE
Chankvetadze,B.; Endresz,G.; Blaschke,G. Capillary electrophoresis enantioseparation of noncharged and anionic chiral compouds using anionic cyclodextrin derivatives as chiral selectors, *J.Capillary Electrophor.*, **1995**, *2*, 235–240.

SAMPLE
Matrix: solutions
Sample preparation: Inject an aliquot of a 50-100 μM solution.

CAPILLARY ELECTROPHORESIS
Capillary: 35 cm $\times$ 50 μm coated fused-silica (30.5 cm to detector) (Polymicro Technologies)
Capillary preparation: Coat capillary as follows. Adjust the pH of 20 mL water to 3.5 with acetic acid, add 80 μL 3-(trimethoxysilyl)propyl methacrylate (3-methacryloxypropyltrimethoxysilane), mix, suck into capillary, let stand at room temperature for 1 h, remove the solution, wash with water. Fill the capillary with a deaerated 3-4% acrylamide solution containing 1 μL/mL N,N,N',N'-tetramethylethylenediamine and 1 mg/mL potassium persulfate, let stand for 30 min, remove excess solution by aspiration, rinse with water, remove water by aspiration, dry at 35° (J. Chromatogr. 1985, 347, 191).
Capillary temperature: 25
Running buffer: pH 7 Buffer containing 3 mM 6^A-methylamino-β-cyclodextrin (Buffer was 50 mM phosphoric acid containing 50 mM acetic acid and 50 mM boric acid, adjusted to pH 7 with concentrated NaOH solution, diluted with an equal volume of water. Synthesis of 6^A-methylamino-β-cyclodextrin was as follows. Add a solution of 3.65 g p-toluenesulfonyl chloride in 30 mL dry pyridine to 29.60 g β-cyclodextrin stirred at 5° in 300 mL dry pyridine, stir overnight at room temperature, evaporate to dryness under reduced pressure at 40°, add 700 mL diethyl ether to the residue. Collect the precipitate and recrystallize it 3 times from water to obtain mono-(6-O-p-tolylsulfonyl)-β- cyclodextrin in 31% yield (Bull. Chem. Soc. Japan 1978, 51, 3030). Heat 2 g mono-(6-O-p-tolylsulfonyl)-β-cyclodextrin with 35 mL 50% methylamine in MeOH in a sealed tube at 70° for 3 days, purify by chromatography on carboxymethylcellulose with ammonium bicarbonate solution to obtain 6^A-methylamino-β-cyclodextrin (cf. J. Am. Chem. Soc. 1980, 102, 762).)
Injection: Pressure injection at 10 psi.s.
Detector: UV 206
Migration time: 16.5 (R-(+)), 17.5 (S-(-))
Voltage: 15 kV
Current: 54 μA
Model: Biofocus 3000 (Bio-Rad)

OTHER SUBSTANCES
Simultaneous: phenyllactic acid, tiaprofenic acid, warfarin

KEY WORDS
coated capillary; chiral

REFERENCE
Fanali,S.; Camera,E. Use of methylamino-β-cyclodextrin in capillary electrophoresis. Resolution of acidic and basic enantiomers, *Chromatographia*, **1996**, *43*, 247–253.

Acepromazine

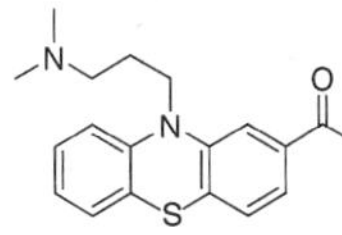

Molecular formula: $C_{19}H_{22}N_2OS$
Molecular weight: 326.46
CAS Registry No.: 61-00-7, 3598-37-6 (maleate)
Merck Index (12th ed.): 32

SAMPLE
Matrix: blood, urine
Sample preparation: Adjust pH of 2 mL urine or plasma to 10.5 with aqueous NaOH, extract gently with chloroform:isopropanol 90:10. Remove the organic layer and evaporate it to dryness under a stream of nitrogen, reconstitute the residue in 50 (urine) or 75 (plasma) μL running buffer, filter, inject an aliquot.

CAPILLARY ELECTROPHORESIS
Capillary: 60 cm × 75 μm AccuSep uncoated silica (52.5 cm to detector) (Waters)
Capillary preparation: At the beginning of each day purge with 500 mM KOH for 5 min, with water for 5 min, and with running buffer for 10 min.
Running buffer: 50 mM NaH_2PO_4 adjusted to pH 2.35 with phosphoric acid
Injection: Hydrostatic injection at 15 cm
Detector: UV 214
Migration time: 7.13
Voltage: 22 kV
Current: 135-145 μA
Model: Waters Quanta 4000

OTHER SUBSTANCES
Extracted: amphetamine, benzocaine, brompheniramine, butacaine, codeine, diazepam, doxapram, lidocaine, medazepam, methamphetamine, methapyrilene, methaqualone, phenmetrazine, procaine, tetrahydrozoline
Simultaneous: meclizine

KEY WORDS
plasma

REFERENCE
Chee,G.L.; Wan,T.S.M. Reproducible and high-speed separation of basic drugs by capillary zone electrophoresis, *J.Chromatogr.*, **1993**, *612*, 172–177.

SAMPLE
Matrix: urine
Sample preparation: 10 mL Urine + 1 mL 5 M NaOH + 10 mL n-hexane, vortex for 1 min, centrifuge at 0° at 3000 g for 5 min, repeat the extraction twice more. Combine the organic layers and add them to 50 μL glacial acetic acid, evaporate to dryness under a stream of nitrogen at 30°, reconstitute the residue in 50 μL 50 mM sodium taurodeoxycholate, filter (0.2 μm PTFE), inject an aliquot.

CAPILLARY ELECTROPHORESIS
Capillary: 100 cm × 50 μm fused-silica (50 cm to detector) (ISCO)
Capillary preparation: Rinse capillary with 20 μL running buffer between injections. Condition a new capillary by filling with 1 M NaOH, let stand for 1 h, fill with 100 mM NaOH, let stand for 1 h, rinse with buffer. Every 20 injections rinse capillary with 200 μL 1 M NaOH, 200 μL water, and 200 μL running buffer, fill with running buffer.
Capillary temperature: 22
Running buffer: 40 mM pH 9.5 Borate buffer containing 10 mM sodium taurodeoxycholate
Injection: Load under vacuum at 7.5 kPa/s.
Detector: UV 240
Migration time: 10
Voltage: 30 kV
Model: ISCO Model 3140 electropherograph

Limit of detection: 10 ng/mL

OTHER SUBSTANCES
Extracted: amiodarone, amitriptyline, azaperone, chlorpromazine, cianopramine, clomipramine, clozapine, desethylamiodarone, desipramine, diclofensine, dothiepin, doxepin, imipramine, isocarboxazid, moclobemide, perphenazine, phenothiazine, pimozide, prochlorperazine, promazine, thioridazine, thiothixene, trifluoperazine, trimipramine

---

KEY WORDS
human; cow; pig; horse; protect from light

REFERENCE
Aumatell,A.; Wells,R.J. Determination of a cardiac antiarrhythmic, tricyclic antipsychotics and antidepressants in human and animal urine by micellar electrokinetic capillary chromatography using a bile salt, *J.Chromatogr.B*, **1995**, *669*, 331–344.

SAMPLE
Matrix: urine
Sample preparation: Dilute 10-fold with water, inject an aliquot.

CAPILLARY ELECTROPHORESIS
Capillary: 70 cm × 50 μm fused-silica (65.4 cm to detector) (Polymicro Technologies)
Capillary temperature: 25
Running buffer: 20 mM pH 5.0 Tris-acetic acid containing 50 μg/mL FC-135 (Fluorad/3M) and 10 mM cetyltrimethylammonium bromide
Injection: Pressure injection at 2 psi.
Detector: UV 240
Migration time: 10.69
Voltage: 20 kV
Model: Bio-Rad BioFocus 3000

OTHER SUBSTANCES
Extracted: chlorpromazine, ethopropazine, fluphenazine, methotrimeprazine, perphenazine, thioridazine, triflupromazine, trimeprazine

---

KEY WORDS
detector at anode

REFERENCE
Muijelsaar,P.G.H.M.; Claessens,H.A.; Cramers,C.A. Determination of structurally related phenothiazines by capillary zone electrophoresis and micellar electrokinetic chromatography, *J.Chromatogr.A*, **1996**, *735*, 395–402.

Acetaminophen

Molecular formula: C₈H₉NO₂
Molecular weight: 151.16
CAS Registry No.: 103-90-2
Merck Index (12th ed.): 45

SAMPLE
Matrix: blood
Sample preparation: Inject directly.

CAPILLARY ELECTROPHORESIS
Capillary: 65 cm × 50 μm untreated fused-silica (50 cm to detector) (SGE)

Capillary preparation: Flush with running buffer every 5 runs. At the end of each day fill capillary with 100 mM KOH, let stand for 30 min, flush with water, let stand for 5 min, fill with running buffer.
Running buffer: pH 8.5 Buffer containing 50 mM sodium dodecyl sulfate (Prepare buffer by mixing 20 mM NaH_2PO_4 solution with 20 mM sodium borate solution to achieve a pH of 8.5.)
Injection: Inject by siphoning at 10 cm for 5-10 s.
Detector: UV 210
Migration time: 7.1
Internal standard: acetaminophen

OTHER SUBSTANCES
Extracted: aspoxicillin

KEY WORDS
plasma; acetaminophen is IS

REFERENCE
Nishi,H.; Fukuyama,T.; Matsuo,M. Separation and determination of aspoxicillin in human plasma by micellar electrokinetic chromatography with direct sample injection, *J.Chromatogr.*, **1990**, *515*, 245–255.

SAMPLE
Matrix: blood
Sample preparation: 100 μL Serum + 150 μL 80 μg/mL 3-isobutyl-1-methylxanthine in MeCN, mix for 30 s, centrifuge at 14000 g for 1 min, inject an aliquot of the supernatant.

CAPILLARY ELECTROPHORESIS
Capillary: 25 cm × 47 μm
Capillary preparation: Wash with running buffer by pressure injection for 1 min after each run. Wash daily with 50 mM phosphoric acid for 2 min, with water for 1 min, with 2 M NaOH for 2 min, with water for 1 min, and with running buffer for 2 min.
Capillary temperature: 24
Running buffer: 220 mM boric acid adjusted to pH 8.8 with 2.5 M NaOH
Injection: Pressure injection for 8 s.
Detector: UV 254
Migration time: 1.95
Internal standard: 3-isobutyl-1-methylxanthine (2.3)
Voltage: 12 kV
Model: Beckman

OTHER SUBSTANCES
Extracted: caffeine, carbamazepine, iohexol, pentobarbital, phenobarbital, phenytoin, salicylic acid, theophylline

KEY WORDS
serum

REFERENCE
Shihabi,Z.K.; Constantinescu,M.S. Iohexol in serum determined by capillary electrophoresis, *Clin.Chem.*, **1992**, *38*, 2117–2120.

SAMPLE
Matrix: blood
Sample preparation: Inject serum directly.

CAPILLARY ELECTROPHORESIS
Capillary: 105 cm × 75 μm fused-silica (68 cm to detector) (Polymicro Technologies)
Capillary preparation: Rinse with running buffer for 10 min between run. At the beginning and end of each day rinse with 100 mM NaOH for 10 min and with water for 10 min.
Running buffer: 10 mM Na_2HPO_4 containing 6 mM sodium borate and 75 mM sodium dodecyl sulfate
Injection: Injection via vacuum suction for 1 s

Detector: UV 215
Migration time: 10.4
Voltage: 23 kV
Current: 62 μA
Model: Europhor Prime Vision IV

OTHER SUBSTANCES
Extracted: ethosuximide, flucytosine, naproxen, primidone, sulfamethoxazole, zomepirac

KEY WORDS
serum

REFERENCE
Schmutz,A.; Thormann,W. Assessment of impact of physico-chemical drug properties on monitoring drug levels
by micellar electrokinetic capillary chromatography with direct serum injection, *Electrophoresis*, **1994**, *15*,
1295–1303.

SAMPLE
Matrix: blood
Sample preparation: Filter (0.45 μm) plasma, inject an aliquot of the filtrate.

CAPILLARY ELECTROPHORESIS
Capillary: 43 cm × 50 μm (22 cm to detector) (Polymicro Technologies)
Capillary preparation: Rinse with running buffer at 68 kPa for 1 min between injections.
Capillary temperature: 30
Running buffer: 60 mM pH 7 Phosphate buffer containing 200 mM sodium dodecyl sulfate
Injection: Vacuum injection at 17 kPa for 1 s
Detector: UV 200
Migration time: 3.2
Voltage: 15 kV
Current: 140 μA
Model: Applied Biosystems 270A-HT

OTHER SUBSTANCES
Extracted: salicylic acid

KEY WORDS
plasma

REFERENCE
Wätzig,H.; Lloyd,D.K. Effect of pH and sodium dodecyl sulfate concentration on the analytical window in the
direct-injection analysis of plasma samples by capillary electrophoresis, *Electrophoresis*, **1995**, *16*, 57–63.

SAMPLE
Matrix: blood
Sample preparation: Filter (0.2 μm) plasma, inject an aliquot of the filtrate.

CAPILLARY ELECTROPHORESIS
Capillary: 42.5 cm × 50 μm fused-silica (34.5 cm to detector) (Polymicro Technologies)
Capillary preparation: After each run rinse capillary at 25° with MeCN:running buffer 50:50
for 15 min then re-equilibrate with running buffer for 20 min. Condition new capillaries with
100 mM NaOH at 50° for 30 min then equilibrate with cathode buffer at 25° for 2 h with
voltage applied.
Capillary temperature: 25
Running buffer: 60 mM pH 10.0 Sodium borate buffer containing 200 mM sodium dodecyl sul-
fate (Cathode buffer was 60 mM pH 10.0 sodium borate buffer prepared by dissolving 3.170 g
boric acid in 100 mL water and adjusting the pH to 10.0 with freshly-prepared 1 M NaOH.)
Injection: Vacuum injection at 0.75 psi for 3 s.
Detector: UV 200
Migration time: 11.8
Voltage: 10 kV

Current: 42 μA
Model: Spectraphoresis 1000 (Thermo Separation Products)

OTHER SUBSTANCES
Extracted: salicylic acid

KEY WORDS
plasma; direct injection

REFERENCE
Kunkel,A.; Gunter,S.; Watzig,H. Determination of pharmaceuticals in plasma by capillary electrophoresis without sample pretreatment reproducibility, limit of quantitation and limit of detection, *Electrophoresis*, **1997**, *18*, 1882–1889.

SAMPLE
Matrix: blood
Sample preparation: Inject an aliquot of plasma directly.

CAPILLARY ELECTROPHORESIS
Capillary: 42.5 cm × 50 μm fused-silica (34.5 cm) (Polymicro Technologies)
Capillary preparation: Between runs rinse with MeCN:running buffer 50:50 for 15 min and with running buffer for 20 min. Condition new capillaries with 100 mM NaOH at 50° for 30 min and equilibrate with cathode buffer at 25° at 10 kV for 2 h. (Cathode buffer was 60 mM pH 10.0 sodium borate buffer.)
Capillary temperature: 25
Running buffer: 60 mM pH 10.0 Sodium borate buffer containing 200 mM sodium dodecyl sulfate
Injection: Vacuum injection at 0.75 psi for 3 s.
Detector: UV 200
Migration time: 11.8
Voltage: 10 kV
Current: 42-44 μA
Model: SpectraPhoresis 1000 (Thermo Separation Products)

OTHER SUBSTANCES
Extracted: salicylic acid

KEY WORDS
plasma; direct injection

REFERENCE
Kunkel,A.; Günter,S.; Wätzig,H. Quantitation of acetaminophen and salicylic acid in plasma using capillary electrophoresis without sample pretreatment. Improvement of precision, *J.Chromatogr.A*, **1997**, *768*, 125–133.

SAMPLE
Matrix: blood, urine
Sample preparation: Filter (Amicon Centrifree Micropartition System) at 1500 g for 20 min, inject an aliquot of the ultrafiltrate.

CAPILLARY ELECTROPHORESIS
Capillary: 90 cm × 75 μm fused-silica (70 cm to detector) (Polymicro Technologies)
Capillary preparation: Between runs rinse capillary with 100 mM NaOH for 5 min and with running buffer for 10 min.
Running buffer: 10 mM pH 9.1 Na_2HPO_4 containing 6 mM sodium borate and 75 mM sodium dodecyl sulfate
Injection: Siphon at 34 cm for 5 s
Detector: UV 195
Migration time: 8.3
Voltage: 20 kV
Current: 80 μA

OTHER SUBSTANCES
Extracted: ethosuximide, phenobarbital, primidone, salicylic acid

KEY WORDS
serum; ultrafiltrate

REFERENCE
Caslavska,J.; Lienhard,S.; Thormann,W. Comparative use of three electrokinetic capillary methods for the determination of drugs in body fluids. Prospects for rapid determination of intoxications, *J.Chromatogr.*, **1993**, *638*, 335–342.

SAMPLE
Matrix: bulk
Sample preparation: Dilute with 400 μg/mL n-propyl p-hydroxybenzoate in running buffer

CAPILLARY ELECTROPHORESIS
Capillary: 27 cm × 50 μm fused-silica (20 cm to detector)
Capillary preparation: Before each run rinse with running buffer at high pressure.
Capillary temperature: 30
Running buffer: MeCN:water 15:85 containing 40 mM sodium dodecyl sulfate, 8.5 mM sodium borate, and 8.5 mM sodium phosphate, pH 8.5
Injection: Inject at high pressure for 1 s
Detector: UV 214
Migration time: 1.0
Internal standard: n-propyl p-hydroxybenzoate (1.8)
Voltage: 20 kV
Model: Beckman P/ACE System 2100

OTHER SUBSTANCES
Simultaneous: acetylcodeine, O-acetylmorphine, aspirin, caffeine, cocaine, codeine, diamorphine, diphenhydramine, hydromorphone, isoniacinamide, lidocaine, methaqualone, morphine, niacinamide, noscapine, papaverine, phenacetin, phenobarbital, phenylpropanolamine, procaine, quinine, salicylic acid, strychnine, thebaine

REFERENCE
Walker,J.A.; Krueger,S.T.; Lurie,I.S.; Marché,H.L.; Newby,N. Analysis of heroin drug seizures by micellar electrokinetic capillary chromatography (MECC), *J.Forensic Sci.*, **1995**, *40*, 6–9.

SAMPLE
Matrix: bulk
Sample preparation: Dissolve in running buffer to a concentration of 1 mg/mL, vortex for 2 min, add an equal volume of 100 μg/mL diphenhydramine in running buffer, mix, inject an aliquot.

CAPILLARY ELECTROPHORESIS
Capillary: 65 cm × 50 μm fused-silica (60 cm to detector)
Capillary preparation: Fill capillary with fresh running buffer before each run. Before use fill with 100 mM NaOH for 20 min, rinse with water, flush with running buffer
Running buffer: MeCN:50 mM 6-aminocaproic acid containing 50 mM 3-N,N-dimethylmyristylammoniopropanesulfonate (MAPS; Fluka) and 5 mM 1-heptanesulfonic acid 10:90, pH adjusted to 4.0 with 1 M phosphoric acid
Injection: Hydrodynamic injection by gravity or pressure.
Detector: UV 214
Migration time: 11.3
Internal standard: diphenhydramine (12)
Voltage: 27 kV
Current: ≤25 μA
Model: Dionex system I

OTHER SUBSTANCES
Simultaneous: allobarbital, barbital, caffeine, codeine, diamorphine, morphine, niacinamide, noscapine, papaverine, phenobarbital, procaine

REFERENCE
Naess,O.; Rasmussen,K.E. Micellar electrokinetic chromatography of charged and neutral drugs in acidic running buffers containing a zwitterionic surfactant, sulfonic acids or sodium dodecyl sulphate. Separation of heroin, basic by-products and adulterants, *J.Chromatogr.A*, **1997**, *760*, 245–251.

SAMPLE
Matrix: bulk
Sample preparation: Dissolve sample in 2 mg/mL histamine diphosphate in 10 mM HCl, inject an aliquot.

CAPILLARY ELECTROPHORESIS
Capillary: 48.5 cm × 50 μm fused-silica (40 cm to detector)
Capillary preparation: Flush capillary with running buffer for 2 min before each run. Replenish buffer in vials after every 10 injections. Condition new capillaries by flushing with 1 M NaOH for 10 min, with 100 mM NaOH for 10 min, with water for 10 min, and with running buffer for 10 min.
Capillary temperature: 40
Running buffer: 75 mM NaH_2PO_4 containing 75 mM sodium borate, adjusted to pH 8.5 with 2 M NaOH
Injection: Pressure injection at 150 mbar for 3 s.
Detector: UV 230
Migration time: 8.2
Internal standard: histamine diphosphate (5)
Voltage: 10 kV
Current: 66-70 μA
Model: Hewlett Packard 3D CE
Limit of detection: 4.1 μg/mL (S/N = 5:1)

OTHER SUBSTANCES
Simultaneous: benzocaine, cis-cinnamoylcocaine, trans-cinnamoylcocaine, cocaine, ephedrine, lidocaine, phenylpropanolamine, procaine, tetracaine

REFERENCE
Krawczeniuk,A.S.; Bravenec,V.A. Quantitative determination of cocaine in illicit powders by free zone capillary electrophoresis, *J.Forensic Sci.*, **1998**, *43*, 738–743.

SAMPLE
Matrix: formulations
Sample preparation: Grind tablet, add 75 mL acidic McOH, sonicate for 5 min, shake for 10 min, make up to 100 mL with acidic MeOH, centrifuge at 2500 g for 10 min. Remove a 5 mL aliquot of the supernatant and add it to 5 mL 10 mM ethyl p-aminobenzoate in MeOH, make up to 50 mL with MeOH, inject an aliquot. (Prepare acidic MeOH by adding 10 mL 100 mM HCl to 1 L MeOH.)

CAPILLARY ELECTROPHORESIS
Capillary: 80 cm × 100 μm fused-silica (50 cm to detector) (Shimadzu)
Running buffer: 20 mM pH 11 Phosphate buffer containing 50 mM sodium dodecyl sulfate
Injection: Injection by siphon at 5 cm for 5 s
Detector: UV 214
Migration time: 12
Internal standard: ethyl p-aminobenzoate (20)
Current: 100 μA (constant current)

OTHER SUBSTANCES
Simultaneous: aspirin, caffeine, o-ethoxybenzamide, salicylamide

KEY WORDS
tablets

REFERENCE
Fujiwara,S.; Honda,S. Determination of ingredients of antipyretic analgesic preparations by micellar electrokinetic capillary chromatography, *Anal.Chem.*, **1987**, *59*, 2773–2776.

SAMPLE
Matrix: formulations
Sample preparation: Grind tablet or powders, add 70 mL MeOH, add 20 mL 5 mg/mL methyl p-hydroxybenzoate in MeOH, sonicate for 10 min, shake for 5 min, make up to 100 mL with water, filter (0.45 μm), inject an aliquot.

CAPILLARY ELECTROPHORESIS
Capillary: 65 cm × 50 μm fused-silica (50 cm to detector) (SGE)
Running buffer: 20 mM NaH_2PO_4 containing 100 mM sodium dodecyl sulfate, adjusted to pH 9.0 with 20 mM sodium tetraborate
Injection: Siphon
Detector: UV 210
Migration time: 7.5
Internal standard: methyl p-hydroxybenzoate
Voltage: +20 kV

OTHER SUBSTANCES
Simultaneous: caffeine, chlorpheniramine, dipyrone (sulpyrin), ethenzamide, guaifenesin, isopropylantipyrine, naproxen, noscapine, phenacetin, tipepidine, trimetoquinol

KEY WORDS
tablets; powders

REFERENCE
Nishi,H.; Fukuyama,T.; Matsuo,M.; Terabe,S. Effect of surfactant structures on the separation of cold medicine ingredients by micellar electrokinetic chromatography, *J.Pharm.Sci.*, **1990**, *79*, 519–523.

SAMPLE
Matrix: formulations
Sample preparation: Grind granules, add 70 mL MeOH, warm at 40° with occasional shaking, cool, add 20 mL 5 mg/mL methyl p-hydroxybenzoate in MeOH, make up to 100 mL with water, filter (0.45 μm), inject an aliquot.

CAPILLARY ELECTROPHORESIS
Capillary: 65 cm × 50 μm fused-silica (SGE)
Running buffer: 20 mM NaH_2PO_4 containing 100 mM sodium cholate, adjusted to pH 9.0 with 20 mM sodium tetraborate
Injection: Siphon for 10 s at 10 cm height
Detector: UV 210
Migration time: 9
Internal standard: methyl p-hydroxybenzoate (16)
Voltage: +20 kV

OTHER SUBSTANCES
Simultaneous: caffeine, chlorpheniramine, dibucaine, dipyrone (sulpyrin), ethenzamide, guaifenesin, isopropylantipyrine, naproxen, noscapine, phenacetin, tipepidine, trimetoquinol, triprolidine

KEY WORDS
granules

REFERENCE
Nishi,H.; Fukuyama,T.; Matsuo,M.; Terabe,S. Separation and determination of the ingredients of a cold medicine by micellar electrokinetic chromatography with bile salts, *J.Chromatogr.*, **1990**, *498*, 313–323.

SAMPLE
Matrix: formulations
Sample preparation: Dilute 10-fold, inject an aliquot.

CAPILLARY ELECTROPHORESIS
Capillary: 62 cm × 50 μm fused silica (50 cm to detector) (Polymicro Technologies)

Capillary temperature: 40
Running buffer: 50 mM pH 2.5 sodium phosphate buffer containing 20 mM β-cyclodextrin and 1 mM hexadecyltrimethylammonium bromide
Injection: Injection by gravity at 10 cm for 5 s
Detector: UV 214
Migration time: 38
Voltage: 20 kV
Current: 36 μA
Model: Laboratory constructed

OTHER SUBSTANCES
Simultaneous: dextromethorphan, doxylamine, pseudoephedrine

KEY WORDS
syrup

REFERENCE
Quang,C.; Khaledi,M.G. Improved chiral separation of basic compounds in capillary electrophoresis using β-cyclodextrin and tetraalkylammonium reagents, *Anal.Chem.*, **1993**, *65*, 3354–3358.

SAMPLE
Matrix: formulations
Sample preparation: Dissolve capsule in 300 mL water, add 50 mL 5 mg/mL 4-hydroxyacetophenone in MeOH, sonicate for 15 min, let stand for 1 h, make up to 500 mL with water, filter (0.45 μm), inject an aliquot.

CAPILLARY ELECTROPHORESIS
Capillary: 57 cm × 50 μm
Capillary preparation: Before each run flush with 100 mM NaOH for 2 min then flush with running buffer for 3 min. Rinse a new capillary with 500 mM NaOH.
Capillary temperature: 40
Running buffer: 40 mM Disodium tetraborate containing 125 mM sodium lauryl sulfate
Injection: Inject sample for 2 s, inject water for 1 s.
Detector: UV 214
Migration time: 11.64
Internal standard: 4-hydroxyacetophenone (17.63)
Voltage: 12 kV
Model: Beckman P/ACE

OTHER SUBSTANCES
Simultaneous: caffeine

KEY WORDS
capsules

REFERENCE
Altria,K.D.; Clayton,N.G.; Hart,M.; Harden,R.C.; Hevizi,J.; Makwana,J.V.; Portsmouth,M.J. An inter-company cross-validation exercise on capillary electrophoresis testing of dose uniformity of paracetamol content in formulations, *Chromatographia*, **1994**, *39*, 180–184.

SAMPLE
Matrix: formulations
Sample preparation: Grind tablet, add MeCN:10 mM HCl 20:80, shake and sonicate for 15 min, centrifuge until a clear solution is obtained. Remove a 25 mL aliquot of the supernatant and add it to 10 mL 500 μg/mL propyl hydroxybenzoate, make up to 50 mL with MeCN:10 mM HCl 20:80, inject an aliquot.

CAPILLARY ELECTROPHORESIS
Capillary: 57 cm × 75 μm fused-silica (Beckman)
Capillary preparation: Pass running buffer through the capillary for at least 4 min before each injection.

Capillary temperature: 25
Running buffer: 20 mM pH 9 Borate buffer containing 25 mM sodium cholate and 50 mM sodium deoxycholate
Injection: Pressure injection for 2 s
Detector: UV 214
Migration time: 6.5
Internal standard: propyl hydroxybenzoate (10.5)
Voltage: 20 kV
Model: Beckman P/ACE System 2100

OTHER SUBSTANCES
Simultaneous: aspirin, caffeine, chlorpheniramine, dextropropoxyphene, salicylic acid

---

KEY WORDS
tablets

REFERENCE
Boonkerd,S.; Lauwers,M.; Detaevernier,M.R.; Michotte,Y. Separation and simultaneous determination of the components in an analgesic tablet formulation by micellar electrokinetic chromatography, *J.Chromatogr.A*, **1995**, *695*, 97–102.

SAMPLE
Matrix: formulations
Sample preparation: Dilute formulation 50-fold, inject an aliquot.

CAPILLARY ELECTROPHORESIS
Capillary: 60 cm × 50 μm (52.5 cm to detector) (AccuSep)
Capillary preparation: Purge capillary with running buffer for 3 min between runs. Purge new capillaries under vacuum with 500 mM NaOH for 10 min, with water for 10 min, and with running buffer for 10 min.
Capillary temperature: 30
Running buffer: 25 mM pH 8.0 Na_2HPO_4/sodium tetraborate containing 50 mM (R)-N-dodecyl-carbonylvaline (Prepare (R)-N-dodecoxycarbonylvaline as follows. Prepare dodecyl chloroformate by reacting 1-dodecanol with 0.33 equivalents of triphosgene in dichloromethane solution in the presence of pyridine. Add dodecyl chloroformate dropwise to (R)-valine in 1 M NaOH solution, filter, wash with hexane, recrystallize from ether/petroleum ether (J. Chromatogr. A 1994, 680, 125).)
Injection: Hydrostatic injection at 10 cm for 15 s.
Detector: UV 214
Migration time: 7.460
Voltage: 15 kV
Model: Waters Quanta 4000E

OTHER SUBSTANCES
Simultaneous: dextromethorphan, doxylamine, pseudoephedrine, saccharin

---

KEY WORDS
not a chiral compound

REFERENCE
Swartz,M.E.; Mazzeo,J.R.; Grover,E.R.; Brown,P.R. Validation of enantiomeric separations by micellar electrokinetic capillary chromatography using synthetic chiral surfactants, *J.Chromatogr.A*, **1996**, *735*, 303–310.

SAMPLE
Matrix: formulations
Sample preparation: Dissolve tablets in water, dilute an aliquot with water, inject an aliquot.

CAPILLARY ELECTROPHORESIS
Capillary: 57 cm × 50 μm fused-silica (50 cm to detector)
Capillary preparation: After each run rinse with 20 mM pH 9.00 borate buffer containing 90 mM sodium dodecyl sulfate for 5 min and with 100 mM NaOH for 5 min.

Running buffer: MeOH:buffer 10:90 (Buffer was 20 mM pH 8.25 borate buffer containing 45 mM sodium dodecyl sulfate.)
Injection: Pressure injection at 3.45 kPa for 3 s.
Detector: UV 214
Migration time: 9
Voltage: 17 kV
Model: Beckman P/ACE 2100

OTHER SUBSTANCES
Simultaneous: caffeine, propyphenazone

KEY WORDS
tablets

REFERENCE
Vogt,C.; Conradi,S.; Rohde,E. Determination of caffeine and other purine compounds in food and pharmaceuticals by micellar electrokinetic chromatography, *J.Chem.Educ.*, **1997**, *74*, 1126–1130.

SAMPLE
Matrix: solutions
Sample preparation: Prepare a 200 µg/mL solution in water, inject an aliquot.

CAPILLARY ELECTROPHORESIS
Capillary: 60 cm × 50 µm
Running buffer: 15 mM pH 11.0 sodium phosphate buffer containing 25 mM sodium dodecyl sulfate
Injection: Hydrodynamic injection at 10 cm for 15 s
Detector: UV 214
Migration time: 2.6
Voltage: 30 kV
Model: Waters Quanta 4000

OTHER SUBSTANCES
Simultaneous: aspirin, caffeine, salicylamide, salicylic acid

REFERENCE
Swartz,M.E. Method development and selectivity control for small molecule pharmaceutical separations by capillary electrophoresis, *J.Liq.Chromatogr.*, **1991**, *14*, 923–938.

SAMPLE
Matrix: solutions
Sample preparation: Prepare a solution in 100 mM cetyltrimethylammonium bromide, inject an aliquot.

CAPILLARY ELECTROPHORESIS
Capillary: 67 cm × 50 µm fused silica capillary (60 cm to detector) (Siemens)
Running buffer: pH 5.0 Acetate buffer containing 10 mM imidazole, 50 µg/mL FC 135 (a fluorinated surfactant, Fluorad/3M), and 100 mM cetyltrimethylammonium bromide
Injection: Pressure injection for 5 s.
Detector: UV 214
Migration time: 6.2
Voltage: 15 kV
Model: Beckman P/ACE System 2000
Limit of detection: 9.87 µg/mL

OTHER SUBSTANCES
Simultaneous: dapsone, dexamethasone, dihydrostreptomycin, kanamycin, sisomycin, tobramycin

KEY WORDS
detector is at anode; indirect UV detection for aminoglycoside antibiotics

REFERENCE

Ackermans,M.T.; Everaerts,F.M.; Beckers,J.L. Determination of aminoglycoside antibiotics in pharmaceuticals by capillary zone electrophoresis with indirect UV detection coupled with micellar electrokinetic capillary chromatography, *J.Chromatogr.*, **1992**, *606*, 229–235.

SAMPLE

Matrix: solutions

CAPILLARY ELECTROPHORESIS

Capillary: 56.9 cm × 75 μm fused-silica (50 cm to detector)
Capillary temperature: 25
Running buffer: MeOH:60 mM pH 8.4 sodium borate buffer containing 60 mM sodium dodecyl sulfate 15:85
Detector: UV 214
Migration time: 8
Voltage: 16 kV
Model: Beckman PACE 2100

OTHER SUBSTANCES

Simultaneous: acetanilide, aspirin, benzamide, caffeine, salicylamide, salicylic acid

REFERENCE

McLaughlin,G.M.; Nolan,J.A.; Lindahl,J.L.; Palmieri,R.H.; Anderson,K.W.; Morris,S.C.; Morrison,J.A.; Bronzert,T.J. Pharmaceutical drug separations by HPCE: Practical guidelines, *J.Liq.Chromatogr.*, **1992**, *15*, 961–1021.

SAMPLE

Matrix: solutions

CAPILLARY ELECTROPHORESIS

Capillary: 65.4 cm × 50 μm fused-silica (56 cm to detector) (Hewlett-Packard)
Capillary temperature: 37
Running buffer: 11.7 mM pH 8.9 Borate buffer containing 8.3 mM phosphate and 50 mM sodium dodecyl sulfate
Injection: Injection at 50 mbar.s
Detector: UV 200
Migration time: 3.544
Voltage: 370 V/cm
Model: Hewlett-Packard G1600A ³ᴰCapillary Electrophoresis system

OTHER SUBSTANCES

Simultaneous: caffeine, guaifenesin, naproxen, noscapine, phenacetin

REFERENCE

Heiger,D.N.; Kaltenbach,P.; Sievet,H.-J. Diode array detection in capillary electrophoresis, *Electrophoresis*, **1994**, *15*, 1234–1247.

SAMPLE

Matrix: solutions
Sample preparation: Inject an aliquot of a solution in MeOH.

CAPILLARY ELECTROPHORESIS

Capillary: 50 cm
Capillary temperature: 25
Running buffer: pH 8.1 Phosphate borate buffer containing 50 mM sodium dodecyl sulfate
Detector: UV 230
Migration time: 2.22
Voltage: 20 kV
Model: Otsuka CAPI-3000

OTHER SUBSTANCES
Simultaneous: caffeine

REFERENCE
Matsuda,R.; Hayashi,Y.; Sasaki,K.; Saito,Y. Deductive prediction of measurement precision and optimization of integration time and wavelength in capillary electrophoresis, *Chromatographia*, **1995**, *41*, 707–714.

SAMPLE
Matrix: solutions

CAPILLARY ELECTROPHORESIS
Capillary: 64.5 cm × 75 μm (56 cm to detector)
Capillary preparation: Between each run replenish inlet and outlet vials, wash capillary with fresh running buffer for 1 min, apply 30 kV for 2 min.
Capillary temperature: 25
Running buffer: 100 mM pH 8.5 Borate buffer
Injection: Pressure injection at 35 mbar for 5 s
Detector: UV 254
Migration time: 2.8
Voltage: 30 kV
Model: Hewlett-Packard HP3D CE

OTHER SUBSTANCES
Simultaneous: aspirin, 2,4-dihydroxybenzoic acid, niacin

REFERENCE
Ross,G.A. Voltage pre-conditioning technique for optimization of migration-time reproducibility in capillary electrophoresis, *J.Chromatogr.A*, **1995**, *718*, 444–447.

SAMPLE
Matrix: solutions

CAPILLARY ELECTROPHORESIS
Capillary: 500 mm × 50 μm
Capillary preparation: Between runs wash capillary for 1 min with 100 mM phosphoric acid and for 1 min with buffer. At the beginning of each day wash the capillary with 2 M NaOH for 3 min, for 3 min with 100 mM phosphoric acid, and for 3 min with buffer.
Capillary temperature: 35
Running buffer: 100 mM boric acid containing 55.4 mM sodium dodecyl sulfate, adjusted to pH 8.4 ± 0.1 with 2 M NaOH
Injection: Pressure injection for 20 s (3% of capillary volume)
Detector: UV 214
Migration time: 4.3
Voltage: 16 kV
Current: 38 μA (fixed)
Model: Beckman Model 2000

OTHER SUBSTANCES
Simultaneous: acetoacetanilide, caffeine, felbamate

REFERENCE
Shihabi,Z.K.; Hinsdale,M.E. Sample matrix effects in micellar electrokinetic capillary electrophoresis, *J.Chromatogr.B*, **1995**, *669*, 75–83.

SAMPLE
Matrix: solutions
Sample preparation: Inject an aliquot of a solution in MeOH:buffer 10:90 (Buffer was 50 mM sodium dodecyl sulfate containing 5 mM boric acid and 5 mM Na_2HPO_4, adjusted to pH 7.4 with concentrated phosphoric acid.)

CAPILLARY ELECTROPHORESIS
Capillary: 380 mm × 50 μm (300 mm to detector)
Capillary preparation: Before use wash capillary with 5 M NaOH for 20 min, wash with 500 mM NaOH for 5 min, wash with 100 mM NaOH for 5 min, and wash with water for 5 min. After the completion of a run flush capillary with 500 mM NaOH for 1 min, with 100 mM NaOH for 1 min, and with water for 1 min. Flush capillary with running buffer for 3 min before a run.
Capillary temperature: 30
Running buffer: Isopropanol:buffer 10:90 (Buffer was 50 mM sodium dodecyl sulfate containing 5 mM boric acid and 5 mM Na_2HPO_4, adjusted to pH 7.4 with concentrated phosphoric acid.)
Injection: Under 50 mbar pressure for 2 s.
Detector: UV 210
Migration time: 1.8
Voltage: 790 V/cm
Model: Hewlett-Packard HP3DCE Model G1602A

OTHER SUBSTANCES
Simultaneous: guaifenesin, phenacetin

REFERENCE
Smith,J.T.; Vinjamoori,D.V. Rapid determination of logarithmic partition coefficients between n-octanol and water using micellar electrokinetic capillary chromatography, *J.Chromatogr.B*, **1995**, *669*, 59–66.

SAMPLE
Matrix: solutions
Sample preparation: Inject an aliquot of an 83 mM solution in MeCN:1% saline 66:34.

CAPILLARY ELECTROPHORESIS
Capillary: 32 cm × 50 μm
Running buffer: 300 mM pH 9.5 Borate buffer
Injection: Hydrodynamic injection for 120 s.
Detector: UV 214
Migration time: 9.6
Voltage: 8 kV
Model: Quanta Model 4000 (Waters)

OTHER SUBSTANCES
Simultaneous: iopamidol, iothalamic acid

REFERENCE
Friedberg,M.A.; Hinsdale,M.; Shihabi,Z.K. Effect of pH and ions in the sample on stacking in capillary electrophoresis, *J.Chromatogr.A*, **1997**, *781*, 35–42.

SAMPLE
Matrix: solutions
Sample preparation: Inject an aliquot of a solution in running buffer.

CAPILLARY ELECTROPHORESIS
Capillary: 57 cm × 50 μm fused-silica (50 cm to detector) (Polymicro Technologies)
Capillary temperature: 25
Running buffer: 25 mM NaH_2PO_4 containing 0.83% poly(sodium 10-undecenyl sulfate), adjusted to pH 7.3 with 12.5 mM sodium tetraborate (Preparation of poly(sodium 10-undecenyl sulfate is as follows. React undecylenyl alcohol (1-undecen-11-ol) with one equivalent of chlorosulfonic acid in diethyl ether under nitrogen at -5° to obtain 10-undecenyl hydrogen sulfate, react with one equivalent of NaOH to obtain sodium 10-undecenyl sulfate (J. Microcolumn Sep. 1996, 8, 115). Dissolve 20 g sodium 10-undecenyl sulfate in 40 mL water (degassed by purging with nitrogen for 24 h), add 600 mg potassium persulfate, stir at 70° under nitrogen for 50 h, cool, add 200 mL cold (0°) EtOH, filter, wash the solid with three 10 mL portions of cold EtOH, dry under vacuum at 65° overnight to obtain poly(sodium 10-undecenyl sulfate) (cf. J. Microcolumn Sep. 1992, 4, 509). Purify by dialysis using dialysis tubing with a 1000 MW cut-off (Spectrum Houston) for 24 h.)
Detector: UV 214

Migration time: 6
Voltage: 16.1 kV
Model: Beckman P/ACE 2000

OTHER SUBSTANCES
Simultaneous: caffeine, ethenzamide, guaifenesin, phenacetin, trimetoquinol

REFERENCE
Palmer,C.P.; Terabe,S. Micelle polymers as pseudostationary phases in MEKC: Chromatographic performance and chemical selectivity, *Anal.Chem.*, **1997**, *69*, 1852–1860.

SAMPLE
Matrix: solutions
Sample preparation: Inject an aliquot of a 10 μg/mL solution

CAPILLARY ELECTROPHORESIS
Capillary: 50 cm × 75 μm silica
Capillary temperature: 25
Running buffer: 20 mM pH 7.0 Phosphate/borate buffer containing 50 mM sodium dodecyl sulfate
Injection: 5 nL
Detector: UV 208
Migration time: 2.4
Voltage: 20 kV
Model: Otsuka CAPI-3000

OTHER SUBSTANCES
Simultaneous: caffeine, ethyl paraben, guaifenesin, methyl paraben, phenacetin, salicylamide

REFERENCE
Poe,R.B.; Hayashi,Y.; Matsuda,R. Precision-optimization of wavelengths in diode-array detection in separation science, *Anal.Sci.*, **1997**, *13*, 951–961.

SAMPLE
Matrix: solutions

CAPILLARY ELECTROPHORESIS
Capillary: 53 cm × 50 μm (44.5 cm to detector) (Polymicro Technologies)
Capillary temperature: 40
Running buffer: 70 mM Boric acid adjusted to pH 9.00 with 1 M NaOH
Injection: Pressure injection at 50 mbar for 2-5 s (2.9-7.2 nL).
Detector: UV 210
Migration time: 2.2
Voltage: 20-25 kV
Model: Hewlett-Packard [3D]CE

OTHER SUBSTANCES
Simultaneous: aspirin, caffeine, norephedrine, salicylic acid

REFERENCE
Thompson,L.; Veening,H.; Strein,T.G. Capillary electrophoresis in the undergraduate instrumental laboratory: Determination of common analgesic formulations, *J.Chem.Educ.*, **1997**, *74*, 1117–1121.

SAMPLE
Matrix: solutions

CAPILLARY ELECTROPHORESIS
Capillary: 57 cm × 50 μm fused-silica (50 cm to detector)
Running buffer: 50 mM pH 8.0 Borate buffer containing 40 mM sodium taurodeoxycholate and 25 mM phosphatidylcholine (Prepare by adding phosphatidylcholine and stirring for 3-6 h until

all cloudiness disappears. Phosphatidylcholine was 95% pure soybean lecithin, Epikuron, Lucas Meyer & Co.)

Injection: Inject a solution of the compound in the running buffer at 20 psi for 1 s, inject MeOH: water 5:95 at 20 psi for 1 s, inject a solution of halofantrine in running buffer at 20 psi for 1 s.

Detector: UV 214
Migration time: k' 0.32
Model: Beckman P/ACE 5000

OTHER SUBSTANCES

Also analyzed: amoxicillin, antipyrine, aspirin, azathioprine, caffeine, captopril, carbamazepine, carprofen, chlorambucil, chlorpheniramine, chlorpromazine, cimetidine, clonidine, codeine, desipramine, diphenhydramine, ephedrine, fenoterol, flufenamic acid, flurbiprofen, haloperidol, hydroxyzine, ibuprofen, imipramine, indomethacin, ketoprofen, lidocaine, melphalan, metoprolol, nabumetone, nadolol, phenobarbital, phenol, promazine, propranolol, pyrilamine, ranitidine, ropinirole, salicylic acid, sulfamethoxazole, testosterone, theophylline, thioridazine, tiaprofenic acid, tolfenamic acid, trifluoperazine, trimethoprim, valproic acid, verapamil

KEY WORDS

comparison with HPLC; k' = (Tr-T0)/(T0(1-Tr/Tm)) where Tr = retention time of analyte; T0 = retention time of water; and Tm = retention time of marker (halofantrine)

REFERENCE

Hanna,M.; de Biasi,V.; Bond,B.; Salter,C.; Hutt,A.J.; Camilleri,P. Estimation of the partitioning characteristics of drugs: A comparison of a large and diverse drug series utilizing chromatographic and electrophoretic methodology, *Anal.Chem.*, **1998**, *70*, 2092–2099.

SAMPLE
Matrix: solutions

CAPILLARY ELECTROPHORESIS
Capillary: 70 cm × 50 μm fused-silica (GL Science, Tokyo)
Capillary preparation: Before each run rinse capillary with running buffer at 94 kPa for 5 min before each run.
Running buffer: 50 mM pH 8.5 ammonium carbonate buffer
Injection: Pressure injection at 5 kPa (50 mbar) for 4 s.
Detector: MS, Perkin-Elmer Sciex API-300 quadrupole, electrospray (ionspray) interface, sheath liquid MeOH:running buffer 50:50 at 2.5 μL/min, ionspray voltage 5 kV, positive ion mode
Migration time: 8.5
Voltage: 20 kV (net voltage across capillary = 15 kV (applied voltage - electrospray voltage))
Model: Hewlett Packard 3D CE

OTHER SUBSTANCES
Simultaneous: ascorbic acid, butylscopolamine bromide, caffeine, ibuprofen, ketoprofen, niacin, niacinamide, riboflavin, thiamine, vitamin B12, warfarin

REFERENCE
Tanaka,Y.; Kishimoto,Y.; Otsuka,K.; Terabe,S. Strategy for selecting separation solutions in capillary electrophoresis-mass spectrometry, *J.Chromatogr.A*, **1998**, *817*, 49–57.

SAMPLE
Matrix: urine
Sample preparation: Filter (Amicon Micron-10) urine while centrifuging at 14000 RCF for 10 min. Dilute the ultrafiltrate 5-10-fold with 5 mM pH 11.3 sodium borate buffer, inject an aliquot.

CAPILLARY ELECTROPHORESIS
Capillary: 47 cm × 50 μm fused-silica (40 cm to detector) (Polymicro Technologies)
Capillary preparation: Before each run rinse capillary with running buffer at 20 psi for 1 min. After each run rinse capillary at 20 psi with 100 mM NaOH for 30 s, with water for 30 s, and with running buffer for 1 min. Before first use rinse capillary at 20 psi with 100 mM NaOH for 20 min, with water for 20 min, and with 50 mM pH 11.3 sodium borate buffer for 20 min.

Capillary temperature: 20
Running buffer: 50 mM pH 11.3 Sodium borate buffer
Injection: Pressure injection of sample at 0.5 psi for 15 s followed by pressure injection of running buffer at 0.5 psi for 1 s.
Detector: UV 254
Migration time: 3.4
Voltage: 25 kV
Model: Beckman P/ACE 2100

OTHER SUBSTANCES
Extracted: iothalamate, sulfamethoxazole

KEY WORDS
ultrafiltrate

REFERENCE
Bergert,J.H.; Liedtke,R.R.; Oda,R.P.; Landers,J.P.; Wilson,D.M. Development of a nonisotopic capillary electrophoresis-based method for measuring glomerular filtration rate, *Electrophoresis*, **1997**, *18*, 1827–1835.

Acetazolamide

Molecular formula: $C_4H_6N_4O_3S_2$
Molecular weight: 222.25
CAS Registry No.: 59-66-5, 1424-27-7 (sodium salt)
Merck Index (12th ed.): 50
Lednicer: 1 249, 1 111

SAMPLE
Matrix: blood, urine
Sample preparation: Condition a 3 mL Supelclean LC-18 SPE cartridge with 3 mL MeOH and 3 mL water. Dilute urine 1:10 with water. Precipitate proteins from serum with MeOH. Add diluted urine or protein supernatant to the SPE cartridge, wash with 3 mL water, elute with 3 mL MeOH, reconstitute to the original volume with 100 mM KOH, inject an aliquot.

CAPILLARY ELECTROPHORESIS
Capillary: 67 cm × 50 μm fused-silica (60 cm to detector) (Polymicro Technologies)
Capillary preparation: Rinse with running buffer for 2 min before run. If necessary, regenerate capillary with 100 mM NaOH for 10 min and with water for 15 min.
Capillary temperature: 20
Running buffer: 60 mM 3-(cyclohexylamino)-1-propanesulfonic acid (CAPS) adjusted to pH 10.6 with 100 mM KOH
Injection: Pressure injection for 5 s
Detector: UV 220
Migration time: 18.5
Voltage: 25 kV
Model: Beckman P/ACE 2000

OTHER SUBSTANCES
Extracted: amiloride, bendroflumethiazide, benzthiazide, bumetanide, caffeine, chlorothiazide, chlorthalidone, clopamide, dichlorphenamide, ethacrynic acid, furosemide, hydrochlorothiazide, metyrapone, probenecid, triamterene, trichlormethiazide

KEY WORDS
serum; SPE

REFERENCE
Jumppanen,J.; Sirén,H.; Riekkola,M.-L. Screening for diuretics in urine and blood serum by capillary zone electrophoresis, *J.Chromatogr.A*, **1993**, *652*, 441–450.

SAMPLE
Matrix: urine
Sample preparation: Filter (0.2 μm), inject an aliquot of the filtrate.

CAPILLARY ELECTROPHORESIS
Capillary: 80 cm × 50 μm fused-silica (57 cm to detector)
Capillary preparation: Before each run rinse the capillary with 1 M NaOH for 3 min, with 100 mM NaOH for 3 min, with water for 3 min, and with running buffer for 10 min.
Running buffer: 110 mM Boric acid containing 56 mM NaOH and 44 mM HCl, pH 8
Injection: Vacuum injection for 1 s
Detector: UV 222
Migration time: 10.798
Voltage: 20 kV
Model: Europhor Prime Vision system IV

OTHER SUBSTANCES
Extracted: acebutolol (UV 238), alprenolol (UV 220), amiloride (UV 220), atenolol (UV 228), bendroflumethiazide (UV 220), bumetanide (UV 220), chlorthalidone (UV 220), cocaine (UV 236), codeine (UV 220), ethacrynic acid (UV 220), furosemide (UV 232), hydrochlorothiazide (UV 226), methadone (UV 220), metoxiphenamine (UV 220), nadolol (UV 220), norcodeine (UV 220), oxprenolol (UV 220), pentazocine (UV 220), propranolol (UV 220), spironolactone (UV 244), triamterene (UV 232), xipamide (UV 234)

REFERENCE
Gonzalez,E.; Laserna,J.J. Capillary zone electrophoresis for the rapid screening of banned drugs in sport, *Electrophoresis*, **1994**, *15*, 240–243.

Acetohexamide

Molecular formula: $C_{15}H_{20}N_2O_4S$
Molecular weight: 324.40
CAS Registry No.: 968-81-0
Merck Index (12th ed.): 59
Lednicer: 1 138

SAMPLE
Matrix: solutions
Sample preparation: Inject a dilute aqueous solution on to the tip of the capillary containing the SPE material at 138 kPa for 12 s, wash for 1 min with running buffer, desorb with elution buffer for 20 s (60 nL), inject running buffer at 3.45 kPa, electrophorese. (Elution buffer was MeCN:50 mM pH 2.5 phosphate buffer 80:20.)

CAPILLARY ELECTROPHORESIS
Capillary: 67 cm × 75 μm with a 0.5 mm long × 0.37 mm diameter column of C18 material (from a J.T. Baker SPE cartridge) at the injection end
Capillary preparation: Before each run rinse capillary with 3 column volumes of elution buffer and 10 column volumes of running buffer. After each run rinse with 500 μL elution buffer.
Capillary temperature: 20
Running buffer: 250 mM pH 8.4 Borate buffer containing 5 mM phosphate
Detector: UV 200
Migration time: 17.5
Voltage: 15 kV
Current: 70 μA
Model: Beckman P/ACE System 5510
Limit of detection: 5 ng/mL

OTHER SUBSTANCES
Simultaneous: chlorpropamide, glipizide, glyburide, tolazamide, tolbutamide

KEY WORDS
SPE

REFERENCE
Strausbauch,M.A.; Xu,S.J.; Ferguson,J.E.; Nunez,M.E.; Machacek,D.; Lawson,G.M.; Wettstein,P.J.; Landers,J.P. Concentration and separation of hypoglycemic drugs using solid-phase extraction-capillary electrophoresis, *J.Chromatogr.A*, **1995**, *717*, 279–291.

SAMPLE
Matrix: solutions
Sample preparation: Inject an aliquot of a solution in MeOH.

CAPILLARY ELECTROPHORESIS
Capillary: 47 cm × 50 μm fused-silica (40 cm to detector)
Capillary preparation: Before each run rinse at 20 psi with 3 column volumes of 100 mM NaOH and 3 column volumes of running buffer. Condition new capillaries by rinsing with 20 column volumes of 1 M NaOH, water, and running buffer
Capillary temperature: 22
Running buffer: 5 mM pH 8.5 Borate buffer containing 5 mM phosphate and 75 mM sodium cholate (Prepare from 500 mM borate buffer (prepared by mixing 125 mM sodium tetraborate and 500 mM boric acid to achieve pH 8.5) and 500 mM Na_2HPO_4 solution with appropriate dilution and the addition of sodium cholate.)
Injection: Pressure injection of sample at 0.5 psi for 1-5 s followed by a pressure injection of running buffer for 1 s.
Detector: UV 200
Migration time: 3.9
Internal standard: N-acetyl-4-(2,3-dichlorophenylureido)benzenesulfonamide (5.5)
Voltage: 25 kV
Current: 61 μA
Model: Beckman P/ACE 5510

OTHER SUBSTANCES
Simultaneous: chlorpropamide, glipizide, glyburide, tolazamide, tolbutamide

REFERENCE
Roche,M.E.; Oda,R.P.; Lawson,G.M.; Landers,J.P. Capillary electrophoretic detection of metabolites in the urine of patients receiving hypoglycemic drug therapy, *Electrophoresis*, **1997**, *18*, 1865–1874.

SAMPLE
Matrix: urine
Sample preparation: Condition a 200 μg Bond Elut C18 SPE cartridge with 1 mL MeCN. 10 mL Urine + 100 μL 10 μg/mL IS, adjust pH to 2.0 with concentrated HCl, add 15 mL dichloromethane, shake vigorously for 2 min, let stand for 5 min. Remove the organic layer and evaporate it to dryness under a stream of nitrogen at 45°, reconstitute the residue in 400 μL MeOH:water 50:50, add to the SPE cartridge, wash with 1.25 mL MeCN:water 10:90, wash with 1 mL MeCN:water 40:60, elute with 1 mL MeCN. Evaporate the eluate to dryness under a stream of nitrogen, reconstitute the residue in 20 μL MeOH, inject an aliquot.

CAPILLARY ELECTROPHORESIS
Capillary: 47 cm × 50 μm fused-silica (40 cm to detector) (Polymicro Technologies)
Capillary preparation: Before each run wash with 3 column volumes of 100 mM NaOH, dip momentarily into water, rinse with 10 column volumes of 50 mM pH 8.5 Sodium borate buffer containing 50 mM Na_2HPO_4 and 75 mM sodium cholate, rinse with 3 column volumes of running buffer. Condition a new capillary by rinsing with 20 column volumes of 100 mM NaOH, 20 column volumes of water, and 20 column volumes of running buffer.
Capillary temperature: 25
Running buffer: 5 mM pH 8.5 Sodium borate buffer containing 5 mM Na_2HPO_4 and 75 mM sodium cholate
Injection: Pressure injection at 0.5 psi for 2 s then a 1 s injection of running buffer.
Detector: UV 200
Migration time: 5.2
Internal standard: N-acetyl-4-(2,3-dichlorophenylureido)benzenesulfonamide (7.4)

Voltage: 25 kV
Current: 61 µA
Model: Beckman P/ACE System 2100 or System 5510
Limit of detection: 50 ng/mL

OTHER SUBSTANCES
Extracted: chlorpropamide, glipizide, glyburide, tolazamide, tolbutamide

KEY WORDS
SPE

REFERENCE
Núñez,M.; Ferguson,J.E.; Machacek,D.; Jacob,G.; Oda,R.P.; Lawson,G.M.; Landers,J.P. Detection of hypoglycemic drugs in human urine using micellar electrokinetic chromatography, *Anal.Chem.*, **1995**, *67*, 3668–3675.

SAMPLE
Matrix: urine
Sample preparation: Inject an aliquot directly.

CAPILLARY ELECTROPHORESIS
Capillary: 37 cm × 50 µm fused-silica (30 cm to detector) (Polymicro Technologies)
Capillary preparation: Every 13 injections rinse capillary with 100 mM NaOH for 1 min and with running buffer for 1 min. Condition a new capillary with 20 column volumes of 100 mM NaOH, 20 column volumes of water, and 20 column volumes of running buffer.
Capillary temperature: 22
Running buffer: 5 mM pH 8.5 Borate buffer containing 5 mM phosphate and 75 mM sodium cholate
Injection: Pressure injection at 0.5 psi for 2 s.
Detector: UV 200
Migration time: 2.3
Internal standard: N-acetyl-4-(2,3-dichlorophenylureido)benzenesulfonamide (3.2)
Voltage: 28 kV
Current: 71 µA
Model: Beckman P/ACE 5510

OTHER SUBSTANCES
Extracted: chlorpropamide, glipizide, glyburide, tolazamide, tolbutamide

KEY WORDS
direct injection

REFERENCE
Roche,M.E.; Oda,R.P.; Machacek,D.; Lawson,G.M.; Landers,J.P. Enhanced throughput with capillary electrophoresis via continuous-sequential sample injection, *Anal.Chem.*, **1997**, *69*, 99–104.

Acetylcysteine

Molecular formula: C_5H_9NO_3S
Molecular weight: 163.20
CAS Registry No.: 616-91-1
Merck Index (12th ed.): 89

SAMPLE
Matrix: blood
Sample preparation: 1 mL Whole blood + 1 mL 10% trichloroacetic acid containing 1 mM disodium EDTA, vortex, centrifuge at 0° at 1850 g for 5 min. Remove a 40 µL aliquot and mix

it with 40 μL 50 μg/mL 2-mercaptoethanol in 2.5 M sodium borate buffer, add 100 μL 2 mg/mL ammonium 7-fluoro-2,1,3-benzoxadiazole-4-sulfonate (SBD-F) in 2.5 M sodium borate buffer, mix, add 220 μL 2.5 M sodium borate buffer, mix vigorously, heat at 60° for 1 h, cool in an ice bath, filter (0.45 μm), inject an aliquot of the filtrate. (Extraction only demonstrated for glutathione.)

CAPILLARY ELECTROPHORESIS
Capillary: 50 cm × 50 μm coated capillary (Bio-Rad ?)
Running buffer: 100 mM pH 2.5 Phosphate buffer
Injection: Electromigration
Detector: F ex 380 em 510
Migration time: 15
Internal standard: 2-mercaptoethanol (13)
Voltage: 12 kV
Model: Bio-Rad HPE 100
Limit of detection: 440 ng/mL

OTHER SUBSTANCES
Extracted: captopril, cysteine, dithiothreitol, glutathione, homocysteine, thiolactic acid

---

KEY WORDS
whole blood; derivatization; detector at anode; coated capillary

---

REFERENCE
Ling,B.L.; Baeyens,W.R.G.; Dewaele,C. Capillary zone electrophoresis with ultraviolet and fluorescence detection for the analysis of thiol. Application to mixtures and blood, *Anal.Chim.Acta*, **1991**, *255*, 283–288.

SAMPLE
Matrix: formulations
Sample preparation: Dissolve formulation in degassed (by sonication) water, cover solution with a 1 mm film of light mineral oil, inject an aliquot.

CAPILLARY ELECTROPHORESIS
Capillary: 30 cm × 50 μm fused-silica (30 cm to detector)
Capillary preparation: Rinse capillary with running buffer before each run. Condition new capillaries with 100 mM NaOH at 50° for 30 min then equilibrate with running buffer for 40 min.
Capillary temperature: 25
Running buffer: 90 mM pH 8.55 Borate buffer containing 5% polyethylene glycol 20000 (Prepare from 329 mg boric acid, 351 mg sodium tetraborate, and 5 g polyethylene glycol 20000 in 100 mL water.)
Injection: Pressure injection at 0.5 psi for 4 s
Detector: UV 214
Migration time: 4.09
Voltage: 20 kV
Model: Beckman P/ACE 2050 or 2100

OTHER SUBSTANCES
Simultaneous: N,N-diacetylcystine

REFERENCE
Wätzig,H. Appropriate calibration functions for capillary electrophoresis. I. Precision and sensitivity using peak areas and heights, *J.Chromatogr.A*, **1995**, *700*, 1–7.

SAMPLE
Matrix: solutions

CAPILLARY ELECTROPHORESIS
Capillary: 52 cm × 75 μm fused-silica (47 cm to detector) (Polymicro Technologies)
Capillary preparation: Prepare capillary by flushing with 500 mM at 5 psi for 3 min, rinse for 3 min with water, repeat cycle twice more, flush for 15 min with running buffer under pressure.

Running buffer: 50 mM boric acid containing 10 mM sodium borate, pH 8.5
Injection: Gravity injection at 5 cm for 10 s
Detector: UV 200
Migration time: 4.6
Internal standard: benzoic acid (5)
Voltage: +18 kV
Model: Dionex CES I

OTHER SUBSTANCES
Simultaneous: degradation products

REFERENCE
Chadwick,R.R.; Hsieh,J.C.; Resham,K.S.; Nelson,R.B. Applications of capillary electrophoresis in the eye-care pharmaceutical industry, *J.Chromatogr.A*, **1994**, *671*, 403–410.

SAMPLE
Matrix: solutions
Sample preparation: 5 mL 10 mM N-acetylcysteine in water + 5 mL 10.1 mM o-phthaldialdehyde in 123 mM pH 10.4 borate buffer, mix, add 5 mL 10.1 mM L-valine in 10 mM HCl, shake, inject an aliquot.

CAPILLARY ELECTROPHORESIS
Capillary: 37 cm × 50 μm fused-silica (30 cm to detector)
Capillary preparation: Condition a new capillary with 100 mM NaOH at 50° for 30 min then equilibrate with running buffer under running conditions for 40 min.
Capillary temperature: 25
Running buffer: 155 mM pH 8.98 Borate buffer containing 5% polyethylene glycol 20000 (Prepare by dissolving 153 mg boric acid, 1.242 mg sodium tetraborate, and 5 g polyethylene glycol 20000 in water, make up to 100 mL with water.)
Injection: Inject 100 mM NaOH for 2 s, inject sample for 4 s, inject 100 mM HCl for 2 s (at 0.5 psi)
Detector: UV 214
Migration time: 3.38 (D), 3.55 (L)
Model: Beckman P/ACE 2050 or 2100
Limit of detection: 0.4% (of major enantiomer)

OTHER SUBSTANCES
Simultaneous: N,N-diacetylcystine

KEY WORDS
derivatization; chiral

REFERENCE
Dette,C.; Watzig,H. Separation of enantiomers of *N*-acetylcysteine by capillary electrophoresis after derivatization by *o*-phthaldialdehyde, *Electrophoresis*, **1994**, *15*, 763–768.

SAMPLE
Matrix: solutions

CAPILLARY ELECTROPHORESIS
Capillary: 50 cm × 75 μm fused-silica (50 cm to detector)
Capillary preparation: Between runs rinse with 100 mM NaOH and water for 2 min and with running buffer for 1 min.
Capillary temperature: 23
Running buffer: 10 mM pH 7.5 Phosphate buffer
Injection: Pressure injection for 5 s.
Detector: UV 214
Migration time: 6.5
Voltage: 20 kV
Model: Beckman P/ACE 2000

OTHER SUBSTANCES
Simultaneous: cysteine, cysteinylglycine, glutamylcysteine, glutathione

KEY WORDS
comparison with HPLC/derivatization

REFERENCE
Ercal,N.; Le,K.; Treeratphan,P.; Matthews,R. Analysis of thiol-containing compounds in biological samples by capillary zone electrophoresis, *Biomed.Chromatogr.*, **1996**, *10*, 15–18.

SAMPLE
Matrix: solutions
Sample preparation: Inject an aliquot of a solution in 3% metaphosphoric acid containing 1 mM EDTA.

CAPILLARY ELECTROPHORESIS
Capillary: 47.5 cm × 50 μm fused-silica (40 cm to detector) (Supelco)
Capillary preparation: Between analyses purge with MeCN:100 mM pH 9.0 borate buffer containing 100 mM sodium dodecyl sulfate 25:75 for 4 min and with running buffer for 6 min. Purge new capillaries under 2.2 kPa vacuum with 500 μM LiOH for 15 min, purge with water, purge with running buffer at 24 kV for 1 h.
Running buffer: MeCN:water:400 mM pH 9.0 sodium borate buffer 20:30:50
Injection: Hydrostatic injection at 10 cm for 40 s.
Detector: UV 185
Migration time: 4.66
Voltage: 26 kV
Current: 85 μA
Model: Waters Quanta 4000

OTHER SUBSTANCES
Simultaneous: glutathione, ascorbic acid

REFERENCE
Davey,M.W.; Bauw,G.; Van Montagu,M. Simultaneous high-performance capillary electrophoresis analysis of the reduced and oxidised forms of ascorbate and glutathione, *J.Chromatogr.B*, **1997**, *697*, 269–276.

Acyclovir

Molecular formula: $C_8H_{11}N_5O_3$
Molecular weight: 225.21
CAS Registry No.: 59277-89-3, 69657-51-8 (sodium salt)
Merck Index (12th ed.): 148
Lednicer: 3 229; 4 31; 4 116; 4 165

SAMPLE
Matrix: formulations
Sample preparation: Ointment. 10 mg Ointment + 2 mL petroleum ether + 1.5 mL buffer + 1.5 mL EtOH, shake for 1 h, remove the aqueous phase, filter the aqueous phase, inject an aliquot of the filtrate. Hydrogel. 10 mg Hydrogel + 2 mL buffer, shake for 1 h, remove the aqueous phase, filter the aqueous phase, inject an aliquot of the filtrate. (Prepare buffer by dissolving 2.48 g boric acid and 800 mg NaOH in 1 L water. Remove a 59 mL aliquot and make it up to 100 mL with 200 mM NaOH so that the final pH is 10.0.)

CAPILLARY ELECTROPHORESIS
Capillary: 64.5 cm × 50 μm fused-silica (56 cm to detector, bubble flow cell) (Hewlett Packard)
Capillary preparation: Before each run flush capillary with 100 mM NaOH for 3 min and with running buffer for 5 min. Wash new capillaries with 1 M NaOH at 40° for 15 min, with water at 40° for 10 min, and with water at 25° for 5 min.

Capillary temperature: 25
Running buffer: Buffer containing 10 mM sodium dodecyl sulfate. (Prepare buffer by dissolving 2.48 g boric acid and 800 mg NaOH in 1 L water. Remove a 59 mL aliquot and make it up to 100 mL with 200 mM NaOH so that the final pH is 10.0.)
Injection: Pressure injection at 200 mbar.s.
Detector: UV 250
Migration time: 3.8
Voltage: 30 kV
Model: Hewlett Packard Model G1600A ^{3D}CE
Limit of detection: 3 µg/mL

OTHER SUBSTANCES
Simultaneous: brivudin

KEY WORDS
ointment; hydrogel

REFERENCE
Neubert,R.H.H.; Mrestani,Y.; Schwarz,M.; Colin,B. Application of micellar electrokinetic chromatography for analyzing antiviral drugs in pharmaceutical semisolid formulations, *J.Pharm.Biomed.Anal.*, **1998**, *16*, 893–897.

SAMPLE
Matrix: solutions

CAPILLARY ELECTROPHORESIS
Capillary: 50 cm × 50 µm fused-silica (38.5 cm to detector) (Yongnian, China)
Capillary preparation: Clean capillary with 100 mM NaOH and water after each run.
Capillary temperature: 24
Running buffer: EtOH:buffer 5:95 (Buffer was 40 mM borax adjusted to pH 2 with phosphoric acid.)
Injection: Electrokinetic injection at 15 kV for 20 s
Detector: UV 260
Migration time: 9
Internal standard: α-amino-5-mercapto-3,4-dithiazole (7)
Voltage: 25 kV
Current: 72 µA
Model: 1229 HPCE (Beijing Institute of New Technology and Application)
Limit of detection: 8.52 µg/mL

OTHER SUBSTANCES
Simultaneous: guanine

KEY WORDS
comparison with HPLC

REFERENCE
Zhang,S.S.; Liu,H.X.; Chen,Y.; Yuan,Z.B. Comparison of high performance capillary electrophoresis and liquid chromatography for the determination of acyclovir and guanine in pharmaceuticals and urine, *Biomed.Chromatogr.*, **1996**, *10*, 256–257.

SAMPLE
Matrix: solutions

CAPILLARY ELECTROPHORESIS
Capillary: 57 cm × 50 µm fused-silica (50 cm to detector) (Composite Metal Services, Worcester, UK)
Capillary temperature: 25
Running buffer: 20 mM pH 2.5 Sodium citrate
Detector: UV 254
Migration time: 10

Voltage: 30 kV
Model: Beckman P/ACE 2210

OTHER SUBSTANCES
Simultaneous: degradation products, guanine

REFERENCE
Assi,K.A.; Altria,K.D.; Clark,B.J. Rapid resolution of drugs and related substances with an eCAP™ polyamine coated capillary, *J.Pharm.Biomed.Anal.*, **1997**, *15*, 1041–1049.

Adenosine

Molecular formula: $C_{10}H_{13}N_5O_4$
Molecular weight: 267.24
CAS Registry No.: 58-61-7
Merck Index (12th ed.): 152

SAMPLE
Matrix: solutions
Sample preparation: Prepare a 2 mM solution in 20 mM pH 4.5 sodium phosphate buffer, add a 20 µL aliquot to 20 µL chloroacetaldehyde solution, heat at 95-100° for 20 min, cool in an ice bath, inject an aliquot. (Chloroacetaldehyde solution was prepared by diluting a 40-45% solution of chloroacetaldehyde in water with 20 mM pH 8.8 sodium phosphate buffer so as to achieve a chloroacetaldehyde concentration of 150 mM, final pH 4.6.)

CAPILLARY ELECTROPHORESIS
Capillary: 60 cm × 25 µm fused-silica (40 cm to detector)
Capillary preparation: Flush with running buffer between runs. Flush with 100 mM NaOH and water before use.
Running buffer: 20 mM pH 8.8 Sodium phosphate buffer
Injection: Siphon at 20.5 cm for 10 s (0.5 nL)
Detector: F ex 325 (He-Cd laser) em 375 (longpass filter) and 400 (bandpass filter)
Migration time: 3
Voltage: 20 kV

OTHER SUBSTANCES
Simultaneous: adenosine triphosphate

KEY WORDS
derivatization

REFERENCE
Tseng,H.C.; Dadoo,R.; Zare,R.N. Selective determination of adenine-containing compounds by capillary electrophoresis with laser-induced fluorescence detection, *Anal.Biochem.*, **1994**, *222*, 55–58.

SAMPLE
Matrix: solutions

CAPILLARY ELECTROPHORESIS
Capillary: 140 cm × 75 µm fused-silica (35 cm to detector) (Polymicro Technologies)
Capillary preparation: Replace running buffer after each run.
Running buffer: 25 mM pH 8.01 Borate buffer
Injection: Siphon at 6 cm for 6 s
Detector: UV 260
Migration time: 4

Voltage: 167 V/cm
Model: Laboratory constructed

OTHER SUBSTANCES
Simultaneous: inosine

REFERENCE
Sun,P.; Hartwick,R.A. On-line kinetic monitoring for biochemical reactions using multi-point detection in high-performance capillary electrophoresis, *J.Chromatogr.A*, **1995**, *695*, 279–285.

SAMPLE
Matrix: solutions
Sample preparation: Inject an aliquot of a solution in running buffer.

CAPILLARY ELECTROPHORESIS
Capillary: 44.7 cm × 75 μm fused-silica (26 cm to detector) (Polymicro Technologies)
Capillary preparation: Replace anodic buffer vial after each run. Every 10-15 runs flush with running buffer and replace cathodic vial. At the start of each day wash capillary with 100 mM NaOH and water.
Running buffer: 100 mM pH 7.5 Sodium phosphate buffer
Injection: Electrokinetic injection at 2 kV for 4 s.
Detector: UV 254
Migration time: 10.47
Voltage: 5 kV
Current: 73 μA
Model: laboratory-constructed

OTHER SUBSTANCES
Simultaneous: inosine

REFERENCE
Saevels,J.; Van Schepdael,A.; Hoogmartens,J. Determination of the kinetic parameters of adensoine deaminase by electrophoretically mediated microanalysis, *Electrophoresis*, **1996**, *17*, 1222–1227.

SAMPLE
Matrix: solutions
Sample preparation: Inject an aliquot of a solution in running buffer.

CAPILLARY ELECTROPHORESIS
Capillary: 57 cm × 75 μm fused-silica (50 cm to detector) (Polymicro Technologies)
Capillary preparation: Rinse with running buffer for 5 min between runs. Condition a new capillary with 100 mM NaOH for 15 min, rinse with water for 5 min, rinse with running buffer for 5 min.
Capillary temperature: 20
Running buffer: 160 mM pH 9.1 Borate buffer containing 60 mM hydroxypropyl-β-cyclodextrin
Injection: Pressure injection at 3.447 kPa for 2 s.
Detector: UV 254
Migration time: 7.4
Voltage: 25 kV
Model: Beckman P/ACE 5500

OTHER SUBSTANCES
Simultaneous: adenosine diphosphate, adenosine monophosphate, adenosine triphosphate, cytidine, cytidine diphosphate, cytidine monophosphate, cytidine triphosphate, guanosine, guanosine diphosphate, guanosine monophosphate, guanosine triphosphate, penciclovir, thymidine, thymidine diphosphate, thymidine monophosphate, thymidine triphosphate, uridine, uridine diphosphate, uridine monophosphate, uridine triphosphate

REFERENCE
Peng,X.; Chen,D.D.Y. Variance contributed by pressure induced injection in capillary electrophoresis, *J.Chromatogr.A*, **1997**, *767*, 205–216.

SAMPLE
Matrix: solutions

CAPILLARY ELECTROPHORESIS
Capillary: 60 cm × 50 μm fused-silica (40 cm to detector) (Polymicro Technologies)
Capillary preparation: Treat new capillary with 1 M NaOH for 30 min, wash with water, wash with MeCN, pass 4 μL/mL methacryloxypropyltrimethoxysilane in MeCN containing 8 μL/mL acetic acid through the capillary for 30 min, let stand for 1.5 h, wash with MeCN. Pass reagent through the capillary and let stand for 1 h, wash with water for 1 h. (Reagent was a saturated solution of m-acrylamidophenylboronic acid in water. To each 1 mL solution 5 μL tetramethyl-ethylenediamine and 5 μL 10% ammonium peroxodisulfate solution were added. Preparation of m-acrylamidophenylboronic acid is as follows. Deareate water by boiling and cooling under vacuum. Stir 1.68 g sodium bicarbonate in 40 mL deaerated water at 0°, add 1.86 g 3-amino-phenylboronic acid in small portions, after the evolution of CO_2 has ceased add 900 μL acryloyl chloride in 50 μL aliquots over 20 min while maintaining the temperature at 0°, cool to -20° for 5 min, filter under suction, wash the solid with a small amount of ice-cold water. Dissolve the solid in ethyl acetate and dry it over anhydrous sodium sulfate overnight, remove the sodium sulfate by filtration, wash the sodium sulfate with a little ethyl acetate, add hexane to the filtrate until no more cloudiness appears, cool at -20° overnight, recrystallize the solid repeatedly from water to obtain m-acrylamidophenylboronic acid as white crystals (Nucleic Acids Research 1985, 13, 6881). (Alternatively it can be purified on a Polyamide CC6 column (Macherey-Nagel) using MeCN:water 10:90 followed by lyophilization of the appropriate fraction.))
Running buffer: 10 mM pH 10.0 Carbonate buffer
Injection: Siphon at 15 cm for 30 s.
Detector: UV 267
Migration time: 5
Voltage: 15 kV

OTHER SUBSTANCES
Simultaneous: deoxyadenosine

KEY WORDS
coated capillary

REFERENCE
Tsukagoshi,K.; Hashimoto,M.; Ichien,K.; Gen,S.; Nakajima,R. Preparation of phenylboronic acid-modified capillary and separation of nucleosides by capillary electrophoresis, *Anal.Sci.*, **1997**, *13*, 485–487.

SAMPLE
Matrix: solutions
Sample preparation: Inject an aliquot of a 50 μg/mL solution in running buffer.

CAPILLARY ELECTROPHORESIS
Capillary: 44 cm × 75 μm fused-silica (36 cm to detector)
Capillary preparation: At the start of each day wash capillary with 100 mM NaOH for 5 min, with water for 5 min, and with running buffer for 5 min. There was no additional washing between runs.
Capillary temperature: 25
Running buffer: 100 mM pH 7.5 Phosphate buffer
Injection: Hydrodynamic injection at 0.75 psi for 1 s.
Detector: UV 255
Migration time: 6.8
Voltage: 10 kV
Model: SpectraPHORESIS 1000

OTHER SUBSTANCES
Simultaneous: inosine, mesityl oxide

REFERENCE
Saevels,J.; Zanoletty Pérez,A.; Salvat Jaumà,A.; Van Schepdael,A.; Hoogmartens,J. Investigation of unexpected migration behavior of adenosine in capillary zone electrophoresis, *Chromatographia*, **1998**, *47*, 225–229.

SAMPLE
Matrix: urine
Sample preparation: Prepare a SPE column by adding 800 μL immobilized boronic acid gel (Pierce, Affipak) to a disposable 3 mL syringe and saturating with 100 mM pH 10 carbonate buffer. Condition a Sep-Pak C18 SPE cartridge with 5 mL MeOH and 5 mL water. Add sodium azide to urine so that the final concentration is 0.03%, centrifuge. Remove a 25 mL aliquot of the supernatant and add it to 25 mL 100 mM pH 7.5 phosphate buffer, mix, cool on ice, add a 10 mL aliquot to the SPE column, wash with 2 mL 25 mM pH 10 carbonate buffer, elute with 9 mL 100 mM HCl. Adjust the pH of the eluate to 7 with 500 mM NaOH, add to the SPE cartridge, wash with 1 mL water, elute with 1.5 mL MeOH:50 mM HCl 25:75. Adjust the pH of the eluate to 6 with concentrated NaOH, lyophilize, resuspend in 1 mL water, add 100 μL 2 M chloroacetaldehyde in water, heat at 90° for 50 min, cool in ice, dilute with an equal volume of 10 mM pH 10 borate buffer containing 0.1% hydroxyethylcellulose, inject a 490 nL aliquot.

CAPILLARY ELECTROPHORESIS
Capillary: 65 cm × 50 μm fused-silica (50 cm to detector) (Polymicro Technologies)
Capillary preparation: Treat new capillaries with 1 M KOH for 15 min, wash with water for 20 min, wash with MeOH for 20 min, pull air through capillary for 30 min, continuously pull 50% trimethylchlorosilane in toluene through the capillary under vacuum for 8 h, rinse with MeOH, wash with water for 20 min, flush with running buffer for 20 min.
Running buffer: 80 mM pH 10 Borate buffer containing 0.1% hydroxyethylcellulose (Terminating electrolyte was the same as running buffer. Leading electrolyte was 180 mM HCl containing 0.1% hydroxyethylcellulose adjusted to pH 7.5 with 1 M Tris.)
Injection: Hydrodynamic injection at 20 cm
Detector: F ex 326 (6 mW He-Cd laser) em 415
Migration time: 20
Voltage: 15 kV
Model: laboratory constructed
Limit of detection: 0.98 nM

KEY WORDS
SPE; derivatization

REFERENCE
Wang,C.-C.; McCann,W.P.; Beale,S.C. Measurement of adenosine by capillary zone electrophoresis with on-column isotachophoretic preconcentration, *J.Chromatogr.B*, **1996**, *676*, 19–28.

SAMPLE
Matrix: urine
Sample preparation: Condition a 140 × 4 SPE column containing 500 mg phenylboronic acid gel (Bio-Rad Affigel 601) with 35 mL 250 mM pH 8.5 ammonium acetate. 10 mL Urine + 300 μL 800 μM 3-deazauridine, mix, add to the SPE column, wash with 20 mL 250 mM ammonium acetate, wash twice with 3 mL portions of MeOH:water 50:50, elute with 25 mL 100 mM formic acid in MeOH:water 50:50 (J. Chromatogr. A 1997, 763, 193), evaporate to dryness under reduced pressure at 39-40°, reconstitute with 1 mL water, inject an aliquot.

CAPILLARY ELECTROPHORESIS
Capillary: 56.5 cm × 50 μm (50 cm to detector) (Grom, Herrenberg, Germany)
Capillary preparation: After each run rinse with water for 100 s, with 100 mM NaOH for 100 s, with water for 100 s, and with running buffer for 120 s.
Running buffer: 50 mM pH 6.7 Phosphate buffer containing 25 mM borate and 300 mM sodium dodecyl sulfate
Injection: Gravity injection at 100 mm for 45 s.
Detector: UV 260
Migration time: 31.06
Internal standard: 3-deazauridine (32.66)
Voltage: 7.5 kV
Current: 41 μA
Model: Dionex CES I
Limit of detection: 2 μM

OTHER SUBSTANCES
Extracted: N[4]-acetylcytidine, cytidine, dihydrouridine, guanosine, inosine, N[6]-methyladenosine, 1-methylguanosine, 2-methylguanosine, 1-methylinosine, 3-methyluridine, 5-methyluridine, pseudouridine, uridine, xanthosine

KEY WORDS
SPE

REFERENCE
Liebich,H.M.; Xu,G.; Di Stefano,C.; Lehmann,R.; Häring,H.U.; Lu,P.; Zhang,Y. Analysis of normal and modified nucleosides in urine by capillary electrophoresis, *Chromatographia*, **1997**, *45*, 396–401.

Adenosine triphosphate

Molecular formula: $C_{10}H_{16}N_5O_{13}P_3$
Molecular weight: 507.18
CAS Registry No.: 56-65-5
Merck Index (12th ed.): 154

SAMPLE
Matrix: blood, solutions
Sample preparation: Place 500 µL plasma or an aqueous solution in a donor compartment separated from 500 µL water in an acceptor compartment by a membrane (30 000 MW cut-off, 16 mm dia; Amicon). The cathode is in the donor compartment and the tip of the capillary is in the acceptor compartment touching the membrane. Electrodialyze at -25 kV for 20 s. When electrodialysis is complete replace the acceptor compartment with a vial of running buffer and electrophorese.

CAPILLARY ELECTROPHORESIS
Capillary: 75 cm × 75 µm fused-silica (50 cm to detector)
Running buffer: 50 mM pH 5.0 Ammonium acetate containing 0.005% hydroxypropylmethyl-cellulose (viscosity 4000 cP for 2% solution; Sigma)
Detector: UV 200 (water), UV 259 (plasma)
Migration time: 6
Voltage: -25 kV
Current: 90 µA
Limit of detection: 300 nM (aqueous solution; UV 200 nm)

KEY WORDS
detector at anode; electrodialysis; plasma

REFERENCE
Buscher,B.A.P.; Tjaden,U.R.; van der Greef,J. On-line electrodialysis-capillary zone electrophoresis of adenosine triphosphate and inositol phosphates, *J.Chromatogr.A*, **1997**, *764*, 135–142.

SAMPLE
Matrix: solutions
Sample preparation: Prepare a 2 mM solution in 20 mM pH 4.5 sodium phosphate buffer, add a 20 µL aliquot to 20 µL chloroacetaldehyde solution, heat at 95-100° for 20 min, cool in an ice bath, inject an aliquot. (Chloroacetaldehyde solution was prepared by diluting a 40-45% solution of chloroacetaldehyde in water with 20 mM pH 8.8 sodium phosphate buffer so as to achieve a chloroacetaldehyde concentration of 150 mM, final pH 4.6.)

CAPILLARY ELECTROPHORESIS
Capillary: 60 cm × 25 µm fused-silica (40 cm to detector)

Capillary preparation: Flush with running buffer between runs. Flush with 100 mM NaOH
and water before use.
Running buffer: 20 mM pH 8.8 Sodium phosphate buffer
Injection: Siphon at 20.5 cm for 10 s (0.5 nL)
Detector: F ex 325 (He-Cd laser) em 375 (longpass filter) and 400 (bandpass filter)
Migration time: 9
Voltage: 20 kV

OTHER SUBSTANCES
Simultaneous: adenosine

KEY WORDS
derivatization

REFERENCE
Tseng,H.C.; Dadoo,R.; Zare,R.N. Selective determination of adenine-containing compounds by capillary electrophoresis with laser-induced fluorescence detection, *Anal.Biochem.*, **1994**, *222*, 55–58.

SAMPLE
Matrix: solutions

CAPILLARY ELECTROPHORESIS
Capillary: 87 cm × 50 μm fused-silica (80 cm to detector) (Composite Metal Services, UK)
Capillary preparation: Before injection flush capillary with running buffer for 5 min. Use fresh
running buffer every 3 runs. Purge new capillaries with 100 mM NaOH for 2 min, with water
for 2 min, and with running buffer for 5 min.
Capillary temperature: 23
Running buffer: 50 mM pH 11.2 3-(Cyclohexylamino)-1-propanesulfonic acid buffer (Prepare
buffer by dissolving 11.2656 g 3-(cyclohexylamino)-1-propanesulfonic acid in water, adjust pH
to 11.2 with 35 mL 1 M NaOH, make up to 1 L with water, sonicate for 10 min, filter (0.45
μm).)
Injection: Hydrostatic injection at 0.5 psi for 10 s
Detector: UV 220
Migration time: 17
Voltage: 22 kV
Model: Beckman P/ACE 2050

OTHER SUBSTANCES
Simultaneous: 2,3-diphenylsuccinic acid, mandelic acid, NADH, NADPH, 1,2-phenylenediacetic
acid, o-phthalic acid

REFERENCE
Sirén,H.; Jumppanen,J.H.; Peltonen,K.; Riekkola,M.-L. Identification of nucleic acids and oligonucleotides by
capillary zone electrophoresis with the four marker technique, *J.Liq.Chromatogr.*, **1995**, *18*, 3577–3589.

SAMPLE
Matrix: solutions
Sample preparation: Inject an aliquot of a solution in running buffer.

CAPILLARY ELECTROPHORESIS
Capillary: 57 cm × 75 μm fused-silica (50 cm to detector) (Polymicro Technologies)
Capillary preparation: Rinse with running buffer for 5 min between runs. Condition a new
capillary with 100 mM NaOH for 15 min, rinse with water for 5 min, rinse with running buffer
for 5 min.
Capillary temperature: 20
Running buffer: 160 mM pH 9.1 Borate buffer containing 60 mM hydroxypropyl-β-cyclodextrin
Injection: Pressure injection at 3.447 kPa for 2 s.
Detector: UV 254
Migration time: 14.3
Voltage: 25 kV
Model: Beckman P/ACE 5500

OTHER SUBSTANCES
Simultaneous: adenosine, adenosine diphosphate, adenosine monophosphate, cytidine, cytidine diphosphate, cytidine monophosphate, cytidine triphosphate, guanosine, guanosine diphosphate, guanosine monophosphate, guanosine triphosphate, penciclovir, thymidine, thymidine diphosphate, thymidine monophosphate, thymidine triphosphate, uridine, uridine diphosphate, uridine monophosphate, uridine triphosphate

REFERENCE
Peng,X.; Chen,D.D.Y. Variance contributed by pressure induced injection in capillary electrophoresis, *J.Chromatogr.A*, **1997**, *767*, 205–216.

Albuterol

Molecular formula: $C_{13}H_{21}NO_3$
Molecular weight: 239.31
CAS Registry No.: 18559-94-9, 51022-70-9 (sulfate)
Merck Index (12th ed.): 217
Lednicer: 2 43

SAMPLE
Matrix: bulk
Sample preparation: Prepare a 1 mg/mL solution in water, inject an aliquot.

CAPILLARY ELECTROPHORESIS
Capillary: 57 cm $\times$ 75 μm (50 cm to detector)
Capillary preparation: Rinse with 500 mM NaOH for 2 min then rinse with 20 mM pH 2.5 sodium citrate for 2 min.
Running buffer: 20 mM pH 2.5 sodium citrate buffer
Injection: Inject for 2 s
Detector: UV 200
Migration time: 9
Voltage: +30 kV
Model: Beckman P/ACE 2000 CE

OTHER SUBSTANCES
Simultaneous: impurities

REFERENCE
Altria,K.D.; Luscombe,D.C.M. Application of capillary electrophoresis as a quantitative identity test for pharmaceuticals employing automated on-column standard addition, *J.Pharm.Biomed.Anal.*, **1993**, *11*, 415–420.

SAMPLE
Matrix: bulk
Sample preparation: Prepare a 100 μg/mL solution, inject an aliquot.

CAPILLARY ELECTROPHORESIS
Capillary: 70 cm $\times$ 75 μm fused-silica
Capillary preparation: Before injection rinse capillary with 100 mM NaOH for 3 min and equilibrate with running buffer for 3 min.
Running buffer: 100 mM NaH_2PO_4 containing 112 mM dimethyl-β-cyclodextrin, adjusted to pH 2.5 with 50 mM citric acid
Injection: Vacuum injection for 2 s
Detector: UV 214
Migration time: 13.2, 13.5
Voltage: 15 kV
Current: about 50 μA
Model: Perkin-Elmer/ABI model 270A-HT

OTHER SUBSTANCES
Simultaneous: impurities

KEY WORDS
chiral

REFERENCE
Rogan,M.M.; Altria,K.D.; Goodall,D.M. Enantiomeric separation of salbutamol and related impurities using
capillary electrophoresis, *Electrophoresis*, **1994**, *15*, 808–817.

SAMPLE
Matrix: formulations
Sample preparation: Dissolve tablets in 500 mL water at 37° (USP apparatus 1, 100 rpm),
remove a 1 mL aliquot, filter, inject an aliquot of the filtrate.

CAPILLARY ELECTROPHORESIS
Capillary: 48.5 cm × 50 μm fused-silica (40 cm to detector)
Capillary preparation: Before each run rinse capillary with 100 mM NaOH and with running
buffer for 2 min.
Capillary temperature: 30
Running buffer: 40 mM Tris containing 20 mM heptakis(2,6-di-O-methyl)-β-cyclodextrin, ad-
justed to pH 2.5 with phosphoric acid
Injection: Hydrodynamic injection at 50 mbar for 5 s.
Detector: UV (wavelength not given)
Migration time: 13.74 (R), 13.98 (S)
Voltage: 15 kV
Model: Hewlett Packard
Limit of quantitation: 1 μg/mL

KEY WORDS
chiral; tablets

REFERENCE
Esquisabel,A.; Hernández,R.M.; Gascón,A.R.; Igartua,M.; Calvo,B.; Pedraz,J.L. Determination of salbutamol
enantiomers by high-performance capillary electrophoresis and its application to dissolution assays,
J.Pharm.Biomed.Anal., **1997**, *16*, 357–366.

SAMPLE
Matrix: solutions

CAPILLARY ELECTROPHORESIS
Capillary: 57 cm × 75 μm fused-silica
Capillary preparation: Rinse with 100 mM NaOH for 1 min and rinse with running buffer for
2 min
Capillary temperature: 25
Running buffer: 25 mM NaH_2PO_4 adjusted to pH 2.4 with concentrated phosphoric acid
Injection: Hydrodynamic injection for 5 s
Detector: UV 200
Migration time: 8.5
Voltage: 30 kV
Model: Beckman P/ACE 2000
Limit of detection: 200 ng/mL

OTHER SUBSTANCES
Simultaneous: impurities

REFERENCE
Altria,K.D. Optimization and improvement of sensitivity in capillary electrophoresis for quantitation of selected
pharmaceuticals, *LC.GC*, **1993**, *11*, 438–442.

SAMPLE
Matrix: solutions

CAPILLARY ELECTROPHORESIS
Capillary: 100 cm $\times$ 50 μm fused-silica (50 cm to detector) (Isco)
Capillary preparation: Flush capillary with 10 μL running buffer between runs. Every 40 sample injections rinse capillary with 200 μL 1 M NaOH, with 200 μL water, and 200 μL running buffer. Before use fill capillary with 1 M NaOH and allow to stand for 1 h, fill with 100 mM NaOH, allow to stand for 1 h, wash with water fill with running buffer.
Capillary temperature: 23
Running buffer: 100 mM citric acid containing 19.27 mM Na_2HPO_4 and 120 mM hydroxypropyl-β-cyclodextrin
Injection: Inject under vacuum at 4.0 kPa.s
Detector: UV 200
Migration time: 25, 25.5 (enantiomers)
Voltage: 30 kV
Model: Isco Model 3140

OTHER SUBSTANCES
Simultaneous: alprenolol, atenolol, cimaterol, clenbuterol, labetalol, nadolol, oxprenolol, pindolol, pirbuterol, propranolol, terbutaline

KEY WORDS
chiral

REFERENCE
Aumatell,A.; Wells,R.J.; Wong,D.K.Y. Enantiomeric differentiation of a wide range of pharmacologically active substances by capillary electrophoresis using modified β-cyclodextrins, *J.Chromatogr.A*, **1994**, *686*, 293–307.

SAMPLE
Matrix: solutions

CAPILLARY ELECTROPHORESIS
Capillary: 74 cm $\times$ 100 μm untreated fused-silica (SGE)
Running buffer: MeOH:50 mM pH 4.8 ammonium acetate 80:20
Injection: Hydrodynamic injection
Detector: MS, Finnigan MAT SSQ 710 single-quadrupole, positive-ion mode, makeup liquid was the same as running buffer (pumped at 1 μL/min), drying gas nitrogen, electrospray tip at ground potential, electrospray counter electrode at -3.8 kV, m/z 240.2
Migration time: 15.3
Voltage: +20 kV

OTHER SUBSTANCES
Simultaneous: clenbuterol, fenoterol, terbutaline

REFERENCE
Lamoree,M.H.; Reinhoud,N.J.; Tjaden,U.R.; Niessen,W.M.A.; van der Greef,J. On-capillary isotachophoresis for loadability enhancement in capillary zone electrophoresis/mass spectrometry of β-agonists, *Biol.Mass.Spectrom.*, **1994**, *23*, 339–345.

SAMPLE
Matrix: solutions
Sample preparation: Inject an aliquot of an aqueous solution.

CAPILLARY ELECTROPHORESIS
Capillary: 37 cm $\times$ 75 μm
Capillary preparation: Before injection rinse with 100 mM NaOH for 1 min and with running buffer for 1 min. Before first use rinse with 100 mM NaOH for 20 min.
Capillary temperature: 30
Running buffer: 25 mM NaH_2PO_4 adjusted to pH 2.3 with concentrated phosphoric acid

Injection: Pressure injection for 5 s
Detector: UV 200
Migration time: 5.856
Internal standard: imidazole (2.722)
Voltage: +13 kV
Model: Beckman P/ACE 2200 or 5100
Limit of quantitation: 1 µg/mL
Limit of detection: 400 ng/mL

OTHER SUBSTANCES
Simultaneous: aspartame, cimetidine

KEY WORDS
robust

REFERENCE
Altria,K.D.; Frake,P.; Gill,I.; Hadgett,T.; Kelly,M.A.; Rudd,D.R. Validated capillary electrophoresis method for the assay of a range of basic drugs, *J.Pharm.Biomed.Anal.*, **1995**, *13*, 951–957.

SAMPLE
Matrix: solutions
Sample preparation: Inject an aliquot of a 10 µg/mL solution in water.

CAPILLARY ELECTROPHORESIS
Capillary: 68.5 × 50 µm fused-silica (60 cm to detector) (Hewlett-Packard)
Capillary preparation: After each run rinse with 100 mM NaOH for 5 min and with running buffer for 5 min. Condition a new capillary with 1 M NaOH at 40° for 20 min and with 100 mM NaOH at 40° for 10 min, rinse with water at 25° for 20 min, and rinse with running buffer.
Capillary temperature: 25
Running buffer: 50 mM pH 8.3 Tris buffer
Injection: Hydrostatic injection at 50 mbar for 2-6 s, flush with buffer for 4 s
Detector: UV 200
Migration time: 3.432
Voltage: 30 kV
Model: Hewlett-packard HP 3DCE

OTHER SUBSTANCES
Simultaneous: cimaterol, clenbuterol, fenoterol, isoxsuprine, metaproterenol, ractopamine, ritodrine, RU 42 173, terbutaline

KEY WORDS
comparison with C18 bonded capillary

REFERENCE
Chevolleau,S.; Tulliez,J. Optimization of the separation of β-agonists by capillary electrophoresis on untreated and C18 bonded silica capillaries, *J.Chromatogr.A*, **1995**, *715*, 345–354.

SAMPLE
Matrix: solutions
Sample preparation: Inject an aliquot of a 100 µg/mL solution in water:running buffer 50:50.

CAPILLARY ELECTROPHORESIS
Capillary: 44.5 cm × 50 µm acrylamide-coated fused-silica (Bio-Rad)
Capillary temperature: 30
Running buffer: 100 mM NaH_2PO_4 containing 15 mM gamma-cyclodextrin, adjusted to pH 2.5 with phosphoric acid
Injection: Electrokinetic injection at 8 kV for 6 s.
Detector: UV 200
Migration time: 9.01
Voltage: 14 kV
Model: Bio-Rad BioFocus 3000

OTHER SUBSTANCES
Also analyzed: alprenolol, atenolol, atropine, baclofen, bamethan, benserazide, biperiden, bisoprolol, bupivacaine, bupranolol, butetamate, carazolol, carbuterol, carvedilol, celiprolol, chloroquine, chlorpheniramine (chlorphenamine), clidinium bromide, clobutinol, disopyramide, dobutamine, flecainide, homatropine, ipratropium bromide, isoproterenol, isothipendyl, ketamine, mefloquine, mequitazine, metaproterenol (orciprenaline), metipranolol, nafronyl (naftidrofuryl), nefopam, ofloxacin, orphenadrine, oxomemazine, oxprenolol, phenoxybenzamine, pholedrine, pindolol, pirbuterol, prilocaine, promethazine, propafenone, propranolol, sotalol, synephrine, terbutaline, tetrahydrozoline (tetryzoline), tocainide, trihexyphenidyl, trimeprazine (alimemazine), trimipramine, tropicamide, verapamil, zopiclone

KEY WORDS
coated capillary; achiral

REFERENCE
Koppenhoefer,B.; Epperlein,U.; Christian,B.; Yibing,J.; Yuying,C.; Bingcheng,L. Separation of enantiomers of drugs by capillary electrophoresis. I. γ-Cyclodextrin as chiral solvating agent, *J Chromatogr.A*, **1995**, *717*, 181–190.

SAMPLE
Matrix: solutions
Sample preparation: Inject an aliquot of a 100 μM solution.

CAPILLARY ELECTROPHORESIS
Capillary: 37 or 47 cm $\times$ 75 μm CElect C8 bonded capillary (effective length 30 or 40 cm) (Supelco)
Capillary preparation: Flush capillary for 2 min with running buffer before injection. Rinse capillary with water at the end of the day.
Running buffer: pH 3.0 Phosphate buffer (I = 0.05) containing 60 mM hydroxypropyl-β-cyclodextrin and 20 mM tetrabutylammonium hydroxide
Injection: Inject at 0.5 psi for 3-5 s.
Detector: UV 214
Migration time: 32, 33 (enantiomers)
Voltage: 15 kV
Current: 40-55 μA
Model: Beckman P/ACE 2100

OTHER SUBSTANCES
Simultaneous: bambuterol, clenbuterol, terbutaline

KEY WORDS
chiral

REFERENCE
Stålberg,O.; Brötell,H.; Westerlund,D. Capillary electrophoretic separation of basic drugs using surface-modified C8 capillaries and derivatized cyclodextrins as structural/chiral selectors, *Chromatographia*, **1995**, *40*, 697–704.

SAMPLE
Matrix: solutions
Sample preparation: Inject an aliquot of a solution in ethyl acetate saturated with 10 mM β-alanine adjusted to pH 5 with 3.3 M acetic acid.

CAPILLARY ELECTROPHORESIS
Capillary: 70 cm $\times$ 100 μm untreated fused-silica (SGE) coupled to 20 cm $\times$ 100 μm untreated fused-silica (BGB Analytik) with a non-conducting coupler
Capillary preparation: Condition capillary daily for 10 min with water, 250 mM KOH, water, and leading buffer
Running buffer: Leading buffer: MeOH:50 mM pH 3.3 ammonium acetate solution 80:20; terminating buffer: MeOH:12 mM β-alanine adjusted to pH 5 with 3.3 M acetic acid 15:85

Injection: Inject terminating buffer at 30 mbar for 18 s to give a 15 mm zone, inject the sample solution at 10 kV for 1 min, inject the sample solution at 10 kV for 9 min with an 8 mbar counterflow so that ethyl acetate does not enter, inject terminating buffer at 15 kV for 1 min with an 8 mbar counterflow, apply 9 kV for 6 s with a 50 mbar counterflow, apply 9 kV with a 23 mbar counterflow until the current reaches 4.0 μA, electrophorese at 21 kV with leading buffer.

Detector: MS, Finnigan MAT, TSQ-70 triple quadrupole, sheath liquid MeOH:50 mM pH 3.3 ammonium acetate solution 80:20 at 1 μL/min, electrospray needle at +3 kV, sampling capillary at ground, sampling capillary 175°, ion source 150°, repeller electrode + 30 V, m/z 240

Migration time: 13.7

Voltage: 9-21 kV

Model: Lauerlabs Prince

Limit of detection: 2 nM

OTHER SUBSTANCES
Simultaneous: clenbuterol, fenoterol, terbutaline

REFERENCE
van der Vlis,E.; Mazereeuw,M.; Tjaden,U.R.; Irth,H.; van der Greef,J. Combined liquid-liquid electroextraction-isotachophoresis for loadability enhancement in capillary zone electrophoresis-mass spectrometry, *J.Chromatogr.A*, **1995**, *712*, 227–234.

SAMPLE
Matrix: solutions

CAPILLARY ELECTROPHORESIS
Capillary: 34 cm × 50 μm (25.5 cm (A) or 8.5 cm (B, C) to detector) (Composite Metal Services, Hallow, UK)

Running buffer: 50 mM pH 2.5 Phosphate buffer (A, B) or 250 mM pH 2.5 phosphate buffer (C)

Injection: Pressure injection at 50 mbar for 5 s.

Detector: UV 200

Migration time: 2.1 (A), 0.9 (B), 3.8 (C)

Voltage: 20 kV (A) or -20 kV (B) or -7.5 kV (C)

Model: Hewlett Packard 3D

OTHER SUBSTANCES
Simultaneous: aminobenzoic acid, imidazole, lamivudine

KEY WORDS
when the distance to the detector was 8.5 cm (B; C) injection was at the normal outlet end and voltage polarity and pressure for injection were reversed.

REFERENCE
Altria,K.D.; Kelly,M.A.; Clark,B.J. The use of a short-end injection procedure to achieve improved performance in capillary electrophoresis, *Chromatographia*, **1996**, *43*, 153–158.

SAMPLE
Matrix: solutions

Sample preparation: Inject an aliquot of a solution in MeOH.

CAPILLARY ELECTROPHORESIS
Capillary: 64.5 cm × 50 μm fused-silica (Polymicro Technologies)

Capillary preparation: Between runs flush with 100 mM NaOH for 2-10 min then with running buffer for 2 min. Before use rinse with 1 M NaOH for 1 h, with 100 mM NaOH for 20 min, with water for 20 min, with 1 M acetic acid in MeOH/MeCN for 10 min, and with running buffer for 10 min.

Capillary temperature: 25

Running buffer: MeCN containing 1 M acetic acid and 60 mM sodium (+)-S-camphorsulfonate (Sodium (+)-S-camphorsulfonate can be prepared by mixing equimolar amounts of (+)-S-camphorsulfonic acid and NaOH in EtOH, collect the precipitate, wash with diethyl ether, dry.)

Injection: Pressure injection of 5 kPa for 3 s.

Detector: UV 214

Migration time: 8, 8.1 (enantiomers)
Voltage: 30 kV
Model: Hewlett-Packard HP^{3D}

KEY WORDS
chiral

REFERENCE
Bjornsdottir,I.; Hansen,S.H.; Terabe,S. Chiral separation in non-aqueous media by capillary electrophoresis using the ion-pair principle, *J.Chromatogr.A*, **1996**, *745*, 37–44.

SAMPLE
Matrix: solutions

CAPILLARY ELECTROPHORESIS
Capillary: 57 cm × 75 μm fused-silica (50 cm to detector)
Capillary preparation: Before each run rinse capillary under pressure with 100 mM NaOH for 5 min then equilibrate with running buffer for 5 min. After each run flush capillary with water for 3 min.
Capillary temperature: 37
Running buffer: 100 mM pH 11 Borate buffer
Injection: Hydrodynamic injection for 3 s.
Detector: UV 280
Migration time: 7
Voltage: 20 kV
Model: Beckman P/ACE 2100

REFERENCE
Lemesle-Lamache,V.; Taverna,M.; Wouessidjewe,D.; Duchêne,D.; Ferrier,D. Determination of the binding constant of salbutamol to unmodified and ethylated cyclodextrins by affinity capillary electrophoresis, *J.Chromatogr.A*, **1996**, *735*, 321–331.

SAMPLE
Matrix: solutions

CAPILLARY ELECTROPHORESIS
Capillary: 52 cm × 75 μm fused-silica (48 cm to detector) (Polymicro Technologies)
Capillary preparation: Before each run purge with running buffer for 3 min. Every 3 runs purge with 100 mM NaOH for 5 min. Purge new capillaries with 1 M NaOH for 20 min and with 100 mM NaOH for 20 min, rinse with running buffer, equilibrate with running buffer at 12 kV for 3 h.
Running buffer: 100 mM CHES (2-(N-cyclohexylamine)ethanesulfonic acid) containing 10 mM triethylamine and 25 mM (R)-dodecoxycarbonylvaline (Waters EnantioSelect (R)-Val-1), pH adjusted to 8.8 with 1 M NaOH
Injection: Hydrostatic injection for 2 s.
Detector: UV 214
Migration time: 9.5, 9.6 (enantiomers)
Voltage: 12 kV
Current: ≤30 μA
Model: Waters Quanta 4000

OTHER SUBSTANCES
Simultaneous: pseudoephedrine, synephrine

KEY WORDS
chiral

REFERENCE
Peterson,A.G.; Ahuja,E.S.; Foley,J.P. Enantiomeric separations of basic pharmaceutical drugs by micellar electrokinetic chromatography using a chiral surfactant, N-dodecoxycarbonylvaline, *J.Chromatogr.B*, **1996**, *683*, 15–28.

SAMPLE
Matrix: solutions

CAPILLARY ELECTROPHORESIS
Capillary: 37 cm × 50 μm eCAP polyamine-coated (30 cm to detector) (Beckman)
Capillary preparation: At the end of each run flush with 1 M NaOH for 2 min, wash with regenerator (Beckman) for 2 min, wash with running buffer for 2 min. Condition a new capillary with running buffer for 2 min and with regenerator for 2 min.
Capillary temperature: 30
Running buffer: 50 mM pH 8 Tris
Detector: UV 214
Migration time: 2
Voltage: 30 kV
Model: Beckman P/ACE 2210

OTHER SUBSTANCES
Simultaneous: aminobenzoic acid, imidazole, lamivudine

KEY WORDS
coated column; comparison with uncoated column

REFERENCE
Assi,K.A.; Altria,K.D.; Clark,B.J. Rapid resolution of drugs and related substances with an eCAP™ polyamine coated capillary, *J.Pharm.Biomed.Anal.*, **1997**, *15*, 1041–1049.

SAMPLE
Matrix: solutions
Sample preparation: Inject an aliquot of a 100 μg/mL solution in running buffer.

CAPILLARY ELECTROPHORESIS
Capillary: 29 cm × 50 μm fused-silica (24.5 cm to detector) (Yongnian Optical Conductive Fiber Plant, China), coated with polyacrylamide
Capillary preparation: No details of the polyacrylamide coating process are provided. However, another paper (LC.GC 1997, 15, 40) by this group indicates that they use the procedure of Hjertén, thus: Adjust the pH of 20 mL water to 3.5 with acetic acid, add 80 μL 3-(trimethoxysilyl)propyl methacrylate (3-methacryloxypropyltrimethoxysilane), mix, suck into capillary, let stand at room temperature for 1 h, remove the solution, wash with water. Fill the capillary with a deaerated 3-4% acrylamide solution containing 1 μL/mL N,N,N',N'-tetramethylethylenediamine and 1 mg/mL potassium persulfate, let stand for 30 min, remove excess solution by aspiration, rinse with water, remove water by aspiration, dry at 35° (J. Chromatogr. 1985, 347, 191).
Capillary temperature: 25
Running buffer: 100 mM NaH_2PO_4 adjusted to pH 2.5
Injection: Electrokinetic injection at 15 kV for 3 s.
Detector: UV 200, UV 210
Migration time: 5.42
Voltage: 15 kV
Model: Bio-Rad BioFocus 3000

OTHER SUBSTANCES
Simultaneous: alprenolol, atenolol, baclofen, bamethan, benproperine, benserazide, bisoprolol, bupranolol, butamirate, butethamate, carbuterol, celiprolol, clenbuterol, clobutinol, dipivefrin, isoproterenol (isoprenaline), metaproterenol (orciprenaline), metipranolol, metoprolol, norfenefrine, ornidazole, oxprenolol, phenylpropanolamine, pholedrine, pirbuterol, prilocaine, procyclidine, sotalol, synephrine, terbutaline, tocainide

KEY WORDS
coated capillary

REFERENCE
Koppenhoefer,B.; Epperlein,U.; Xiaofeng,Z.; Bingcheng,L. Separation of enantiomers of drugs by capillary electrophoresis. Part 4: Hydroxypropyl-γ-cyclodextrin as chiral solvating agent, *Electrophoresis*, **1997**, *18*, 924–930.

SAMPLE
Matrix: solutions
Sample preparation: Inject an aliquot of a 100 μg/mL solution in running buffer.

CAPILLARY ELECTROPHORESIS
Capillary: 30 cm × 50 μm fused-silica (25.5 cm to detector), coated with polyacrylamide
Capillary preparation: Adjust the pH of 20 mL water to 3.5 with acetic acid, add 80 μL 3-(trimethoxysilyl)propyl methacrylate (3-methacryloxypropyltrimethoxysilane), mix, suck into capillary, let stand at room temperature for 1 h, remove the solution, wash with water. Fill the capillary with a deacrated 3-4% acrylamide solution containing 1 μL/mL N,N,N',N'-tetramethylethylenediamine and 1 mg/mL potassium persulfate, let stand for 30 min, remove excess solution by aspiration, rinse with water, remove water by aspiration, dry at 35° (J. Chromatogr. 1985, 347, 191).
Capillary temperature: 25
Running buffer: 100 mM NaH_2PO_4 containing 45 mM heptakis(2,6-di-O- methyl)-β-cyclodextrin, adjusted to pH 2.5 with phosphoric acid
Injection: Electrokinetic injection at 15 kV for 3 s.
Detector: UV 200
Voltage: 15 kV
Model: Bio-Rad BioFocus 3000

KEY WORDS
chiral; coated capillary; comparison with the use of other cyclodextrins; this running buffer gave the greatest enantiomeric separation.; α=1.026

REFERENCE
Lin,B.; Zhu,X.; Koppenhoefer,B.; Epperlein,U. Investigation of 123 chiral drugs by cyclodextrin-modified capillary electrophoresis, *LC.GC*, **1997**, *15*, 40–46.

SAMPLE
Matrix: solutions

CAPILLARY ELECTROPHORESIS
Capillary: 27 cm × 100 μm (20 cm to detector) (Composite Metal Services)
Capillary temperature: 30
Running buffer: 25 mM pH 2.3 NaH_2PO_4
Detector: UV 200
Migration time: 4.84
Internal standard: imidazole (2.46)
Voltage: 7 kV
Model: Beckman

OTHER SUBSTANCES
Simultaneous: alosetron, lamivudine, ranitidine

REFERENCE
Altria,K.D.; Creasey,E.; Howells,J.S. Routine capillary electrophoresis trace level determinations of pharmaceutical and detergent residues on pharmaceutical manufacturing equipment, *J.Liq.Chromatogr. Rel.Technol.*, **1998**, *21*, 1093–1105.

SAMPLE
Matrix: solutions

CAPILLARY ELECTROPHORESIS
Capillary: 48.5 cm × 50 μm fused-silica (40 cm to detector) (Supelco)
Capillary preparation: Condition capillary with running buffer for 3 min before each injection.

Capillary temperature: 15
Running buffer: 25 mM Citric acid adjusted to pH 4.5 with Tris, containing 2% dermatan sulfate
(number average 15270, mass average 22260, z average 30390, polydispersity 1.458) (Opocrin,
Corlo, Italy)
Injection: Pressure injection at 5 kPa for 10 s.
Detector: UV 220
Migration time: 13.6, 14 (enantiomers)
Voltage: 15 kV
Model: Hewlett-Packard ³ᴰCE

OTHER SUBSTANCES
Simultaneous: isoproterenol, metaproterenol, terbutaline

KEY WORDS
chiral

REFERENCE
Gotti,R.; Cavrini,V.; Andrisano,V.; Mascellani,G. Dermatan sulfate as useful chiral selector in capillary electrophoresis, *J.Chromatogr.A,* **1998,** *814,* 205–211.

SAMPLE
Matrix: solutions

CAPILLARY ELECTROPHORESIS
Capillary: 29-36 cm × 50 μm fused-silica (24.5-31.5 cm to detector) (Yongnian Optical Conductive
Fiber Plant, China) coated with polyacrylamide
Capillary preparation: Coat capillary as follows. Adjust the pH of 20 mL water to 3.5 with
acetic acid, add 80 μL 3-(trimethoxysilyl)propyl methacrylate (3-methacryloxypropyltrimethox-
ysilane), mix, suck into capillary, let stand at room temperature for 1 h, remove the solution,
wash with water. Fill the capillary with a deaerated 3-4% acrylamide solution containing 1
μL/mL N,N,N',N'-tetramethylethylenediamine and 1 mg/mL potassium persulfate, let stand
for 30 min, remove excess solution by aspiration, rinse with water, remove water by aspiration,
dry at 35° (J. Chromatogr. 1985, 347, 191).
Capillary temperature: 25
Running buffer: 100 mM pH 2.5 NaH$_2$PO$_4$ (A) or 100 mM pH 2.5 NaH$_2$PO$_4$ containing 45 mM
hydroxypropyl-α-cyclodextrin (Wacker, Munich) (B)
Injection: Electromigration at 15 kV for 3 s.
Detector: UV 200; UV 210
Migration time: 5.42 (A); 6.88 (B) (no separation of enantiomers)
Voltage: 15 kV
Model: Bio-Focus 3000

OTHER SUBSTANCES
Also analyzed: alprenolol, amorolfine, atenolol, atropine, azelastine, baclofen, bamethan, benproperine, benserazide, biperiden, bisoprolol, brompheniramine, bupivacaine, bupranolol, butamirate, butethamate, carazolol, carbuterol, carteolol, carvedilol, celiprolol, chloroquine, chlorpheniramine, chlorphenoxamine, cicletanine, clenbuterol, clidinium bromide, clobutinol,
dimethindene, dipivefrin, disopyramide, dobutamine, doxylamine, fendiline, flecainide, gallopamil, homatropine, ipratropium bromide, isoproterenol (isoprenaline), isothipendyl, ketamine,
meclizine, mefloquine, mepindolol, mequitazine, metaclazepam, metaproterenol (orciprenaline), metipranolol, metoprolol, nafronyl (naftidrofuryl), nefopam, nicardipine, norfenefrine, ofloxacin, ornidazole, orphenadrine, oxomemazine, oxprenolol, oxybutynin, phenoxybenzamine,
phenylpropanolamine, pholedrine, pindolol, pirbuterol, prilocaine, procyclidine, promethazine,
propafenone, propranolol, reproterol, sotalol, sulpride, synephrine, talinolol, terbutaline, tetrahydrozoline (tetryzoline), theodrenaline, tioconazole, tocainide, trihexyphenidyl, trimeprazine (alimemazine), trimipramine, tropicamide, verapamil, zopiclone

KEY WORDS
coated capillary

REFERENCE
Koppenhoefer,B.; Eperlein,U.; Schlunk,R.; Zhu,X.; Lin,B. Separation of enantiomers of drugs by capillary electrophoresis. V. Hydroxypropyl-α-cyclodextrin as chiral solvating agent, *J.Chromatogr.A*, **1998**, *793*, 153–164.

Aldesleukin

Molecular formula: $C_{690}H_{1115}N_{177}O_{203}S_6$
Molecular weight: 15330.91
CAS Registry No.: 110942-02-4

```
                 Pro—Thr—Ser—Ser—Ser—Thr—Lys—Lys—Thr—Gln—Leu—Glu—His—Leu—Leu
                                                                              |
Thr—Leu—Lys—Pro—Asn—Lys—Tyr—Asn—Asn—Ile—Gly—Asn—Leu—Ile—Met—Gln—Leu—Asp—Leu
 |
Arg—Met—Leu—Thr—Phe—Lys—Phe—Tyr—Met—Pro—Lys—Lys—Ala—Thr—Glu—Leu—Lys—His—Leu
                                                                              |
         Leu—Val—Glu—Glu—Leu—Pro—Lys—Leu—Glu—Glu—Glu—Leu—Cys—Gln
          |
         Asn—Leu—Ala—Gln—Ser—Lys—Asn—Pho—His—Leu—Arg—Pro
                                                       |
              Val—Ile—Val—Asn—Ile—Asn—Ser—Ile—Leu—Asp—Arg
                                                             |
              Leu—Glu—Leu—Lys—Gly—Ser—Glu—Thr—Thr—Phe—Met—Cys—Glu
                                                                   |
Ser—Phe—Thr—Ile—Trp—Arg—Asn—Leu—Phe—Glu—Val—Ile—Thr—Ala—Thr—Glu—Asp—Ala—Tyr
 |
Gln—Ser—Ile—Ile—Ser—Thr—Leu—Thr
```

SAMPLE
Matrix: solutions
Sample preparation: Inject an aliquot of a solution in 10 mM pH 7.4 phosphate buffer containing 0/9% NaCl.

CAPILLARY ELECTROPHORESIS
Capillary: 20 cm × 25 μm coated capillary
Running buffer: 100 mM pH 2.5 phosphate buffer
Injection: Electromigration at 8 kV for 8 s.
Detector: UV 200
Migration time: 3-4 (three peaks)
Voltage: 10 kV
Model: Bio-Rad HPE 100

KEY WORDS
recombinant; coated capillary

REFERENCE
Knüver-Hopf,J.; Mohr,H. Differences between natural and recombinant interleukin-2 revealed by gel electrophoresis and capillary electrophoresis, *J.Chromatogr.A*, **1995**, *717*, 71–74.

Allantoin

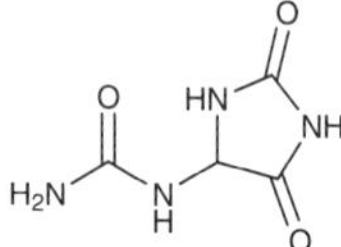

Molecular formula: $C_4H_6N_4O_3$
Molecular weight: 158.12
CAS Registry No.: 97-59-6
Merck Index (12th ed.): 255

SAMPLE
Matrix: urine
Sample preparation: Filter (0.45 μm) urine, dilute 2 (human) or 10 (animal) fold with 15 mM pH 8 sodium tetraborate buffer, inject an aliquot.

CAPILLARY ELECTROPHORESIS
Capillary: 44 cm × 75 μm fused-silica (37 cm to detector) (Composite Metal Services, Hallow UK)
Capillary preparation: Before each run wash capillary with running buffer for 2 min at 20° then fill with running buffer for 2 min. At the start of each day wash capillary with 400 mM NaOH at 40° for 10 min, with running buffer at 40° for 10 min, and with running buffer at 20° for 10 min. At the end of each day wash the capillary with water at 40° for 5 min then leave it in water. Condition new capillaries with 400 mM NaOH at 50° for 30 min.
Capillary temperature: 20
Running buffer: 30 mM pH 9.5 Sodium tetraborate containing 75 mM sodium dodecyl sulfate
Injection: Hydrodynamic injection for 4 s.
Detector: UV 195
Migration time: 4.7
Voltage: 20 kV
Model: TSP CE 2000
Limit of detection: 5 μM

OTHER SUBSTANCES
Extracted: creatinine, hippuric acid, urea, uric acid

KEY WORDS
human; dog; horse; rabbit; rat; mouse

REFERENCE
Alfazema,L.N.; Howells,S.; Perrett,D. Determination of allantoin in biofluids using micellar electrokinetic capillary chromatography, *J.Chromatogr.A*, **1998**, *817*, 345–352.

Allobarbital

Molecular formula: $C_{10}H_{12}N_2O_3$
Molecular weight: 208.22
CAS Registry No.: 52-43-7
Merck Index (12th ed.): 261
Lednicer: 1 269

SAMPLE
Matrix: bulk
Sample preparation: Dissolve in running buffer to a concentration of 1 mg/mL, vortex for 2 min, add an equal volume of 100 μg/mL diphenhydramine in running buffer, mix, inject an aliquot.

CAPILLARY ELECTROPHORESIS
Capillary: 65 cm × 50 μm fused-silica (60 cm to detector)

Capillary preparation: Fill capillary with fresh running buffer before each run. Before use fill
with 100 mM NaOH for 20 min, rinse with water, flush with running buffer
Running buffer: MeCN:50 mM 6-aminocaproic acid containing 50 mM 3-N,N-dimethylmyris-
tylammoniopropanesulfonate (MAPS; Fluka) and 5 mM 1-heptanesulfonic acid 10:90, pH ad-
justed to 4.0 with 1 M phosphoric acid
Injection: Hydrodynamic injection by gravity or pressure.
Detector: UV 214
Migration time: 12.5
Internal standard: diphenhydramine (12)
Voltage: 27 kV
Current: ≤25 μA
Model: Dionex system I

OTHER SUBSTANCES
Simultaneous: acetaminophen, barbital, caffeine, codeine, diamorphine, morphine, niacinamide,
noscapine, papaverine, phenobarbital, procaine

REFERENCE
Naess,O.; Rasmussen,K.E. Micellar electrokinetic chromatography of charged and neutral drugs in acidic running buffers containing a zwitterionic surfactant, sulfonic acids or sodium dodecyl sulphate. Separation of
heroin, basic by-products and adulterants, *J.Chromatogr.A*, **1997**, *760*, 245–251.

Allopurinol

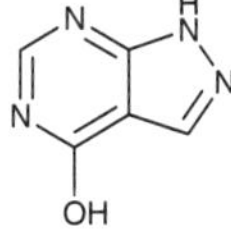

Molecular formula: $C_5H_4N_4O$
Molecular weight: 136.11
CAS Registry No.: 315-30-0
Merck Index (12th ed.): 287
Lednicer: 1 152, 269

SAMPLE
Matrix: blood, hemodialysate fluid
Sample preparation: Serum. 1 mL Serum + 20 μL 100 mM phenyltriethylammonium iodide,
filter (Amicon Centrifree Micropartition System, cut-off 30000) while centrifuging at 311 g for
1 h, inject an aliquot of the ultrafiltrate. Hemodialysate fluid. 9 mL Hemodialysate fluid + 1
mL 20 mM phenyltriethylammonium iodide, inject an aliquot.

CAPILLARY ELECTROPHORESIS
Capillary: 64 cm × 50 μm extended light path capillary (55.8 cm to detector) (Hewlett-Packard)
Capillary temperature: 25
Running buffer: 150 mM Boric acid adjusted to pH 9.0 with 1 M NaOH
Injection: Pressure injection at 50 mbar for 5 s
Detector: UV 210
Migration time: 6.2
Internal standard: phenyltriethylammonium iodide (3.5)
Voltage: 22 kV
Current: 32.5-34.5 μA
Model: Hewlett-Packard [3D]Capillary Electrophoresis System
Limit of detection: 1.2 μM (S/N 3)

OTHER SUBSTANCES
Extracted: ferulic acid, hippuric acid, 2-hydroxyhippuric acid, 4-hydroxyphenylacetic acid, hypoxanthine, indican, indole-3-acetic acid, kynurenic acid, 1-methylniacinamide, niacin, theobromine, theophylline, tryptophan, tyrosine, uracil, uric acid, xanthine

KEY WORDS
serum; ultrafiltrate

REFERENCE

Petucci,C.J.; Kantes,H.L.; Strein,T.G.; Veening,H. Capillary electrophoresis as a clinical tool. Determination of organic anions in normal and uremic serum using photodiode-array detection, *J.Chromatogr.B*, **1995**, *668*, 241–251.

SAMPLE
Matrix: solutions

CAPILLARY ELECTROPHORESIS
Capillary: 50 cm × 50 μm
Capillary temperature: 35
Running buffer: 7 g Boric acid and 7 g sodium carbonate in 1 L water, pH 9.2
Injection: Pressure injection for 20 s
Detector: UV 280
Migration time: 5
Internal standard: iothalamic acid (4.9)
Voltage: 11 kV
Model: Beckman Instruments Model 2000 CE

OTHER SUBSTANCES
Simultaneous: adenine, guanine, hypoxanthine, inosine, oxypurinol, uric acid, xanthine

REFERENCE

Shihabi,Z.K.; Hinsdale,M.E.; Bleyer,A.J. Xanthine analysis in biological fluids by capillary electrophoresis, *J.Chromatogr.B*, **1995**, *669*, 163–169.

Alosetron

Molecular formula: $C_{17}H_{18}N_4O$
Molecular weight: 294.36
CAS Registry No.: 122852-42-0, 122852-69-1 (HCl)

SAMPLE
Matrix: solutions

CAPILLARY ELECTROPHORESIS
Capillary: 27 cm × 100 μm (20 cm to detector) (Composite Metal Services)
Capillary temperature: 30
Running buffer: 25 mM pH 2.3 NaH_2PO_4
Detector: UV 200
Migration time: 5.34
Internal standard: imidazole (2.46)
Voltage: 7 kV
Model: Beckman
Limit of quantitation: 70 ng/mL
Limit of detection: 20 ng/mL

OTHER SUBSTANCES
Simultaneous: albuterol, lamivudine, ranitidine

REFERENCE

Altria,K.D.; Creasey,E.; Howells,J.S. Routine capillary electrophoresis trace level determinations of pharmaceutical and detergent residues on pharmaceutical manufacturing equipment, *J.Liq.Chromatogr.Rel. Technol.*, **1998**, *21*, 1093–1105.

Alprazolam

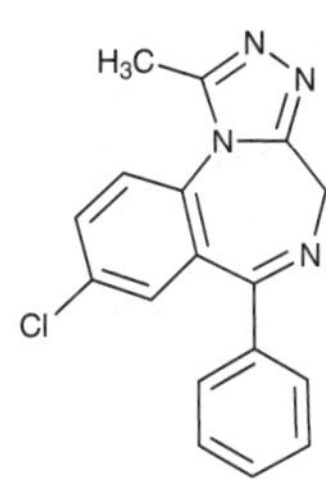

Molecular formula: $C_{17}H_{13}ClN_4$
Molecular weight: 308.77
CAS Registry No.: 28981-97-7
Merck Index (12th ed.): 320
Lednicer: 3 197

SAMPLE
Matrix: solutions

CAPILLARY ELECTROPHORESIS
Capillary: 60 cm × 50 μm fused-silica (40 cm to detector), C18 coated (Supelco)
Capillary preparation: Rinse new capillaries with 250 μL 1 M NaOH, 250 μL 100 mM NaOH, water, MeOH, water, and running buffer then condition with running buffer for at least 15 min. Carry out a similar procedure between runs.
Running buffer: 100 mM pH 8.5 Borate buffer containing 25 mM sodium dodecyl sulfate and 5 M urea
Injection: Electrokinetic injection at 5 kV for 5 s.
Detector: UV 254
Migration time: 25
Voltage: 18 kV
Model: Jasco

OTHER SUBSTANCES
Simultaneous: amobarbital, barbital, bromazepam, clonazepam, clotiazepam, cloxazolam, diazepam, estazolam, etizolam, fludiazepam, flunitrazepam, flurazepam, haloxazolam, medazepam, mephobarbital, metharbital, nimetazepam, nitrazepam, oxazepam, pentobarbital, phenobarbital, secobarbital, triazolam

KEY WORDS
coated capillary

REFERENCE
Jinno,K.; Han,Y.; Nakamura,M Analysis of anxiolytic drugs by capillary electrophoresis with bare and coated capillaries, *J.Capillary Electrophor.*, **1996**, *3*, 139–145.

SAMPLE
Matrix: solutions

CAPILLARY ELECTROPHORESIS
Capillary: 47 cm × 75 μm fused-silica (40 cm to detector) (Beckman)
Capillary preparation: Rinse with running buffer for 1 min before injection. Wash with 100 mM NaOH for 1 min and with water for 1 min between injections.
Capillary temperature: 30
Running buffer: THF:buffer 1:99 (Buffer was 50 mM pH 9.2 borate containing 50 mM sodium dodecyl sulfate, 2 M urea, and 20 mM gamma-cyclodextrin.)
Injection: Pressure injection at 300 kPa for 5 s.
Detector: UV 254
Migration time: 18
Voltage: 15 kV
Model: Beckman P/ACE 2000

OTHER SUBSTANCES
Simultaneous: bromazepam, chlordiazepoxide, clobazam, clonazepam, clorazepate, oxazepam, prazepam, triazolam

REFERENCE
Renou-Gonnord,M.F.; David,K. Optimized micellar electrokinetic chromatographic separation of benzodiazepines, *J.Chromatogr.A*, **1996**, *735*, 249–261.

SAMPLE
Matrix: solutions
Sample preparation: Inject an aliquot of a 100 μg/mL solution in running buffer.

CAPILLARY ELECTROPHORESIS
Capillary: 60 cm × 50 μm acrylamide-coated fused-silica (40 cm to detector) (Supelco)
Capillary preparation: Adjust the pH of 20 mL water to 3.5 with acetic acid, add 80 μL 3-(trimethoxysilyl)propyl methacrylate (3-methacryloxypropyltrimethoxysilane), mix, suck into capillary, let stand at room temperature for 1 h, remove the solution, wash with water. Fill the capillary with a deaerated 3-4% acrylamide solution containing 1 μL/mL N,N,N',N'-tetramethylethylenediamine and 1 mg/mL ammonium persulfate, let stand for 3 h, remove excess solution by aspiration, rinse with water, remove water by aspiration, dry at 35° (cf. J. Chromatogr. 1985, 347, 191).
Running buffer: MeCN:buffer 5:95 (Buffer was 100 mM borate containing 5 M urea and 10 mM sodium dodecyl sulfate, adjusted to pH 8.5 with phosphate.)
Injection: Electrokinetic injection at 5 kV for 5 s.
Detector: UV 254
Migration time: 11.3
Voltage: 18 kV
Model: Jasco Model 870-CE

OTHER SUBSTANCES
Simultaneous: amobarbital, barbital, bromazepam, clonazepam, clotiazepam, cloxazolam, diazepam, estazolam, etizolam, fludiazepam, flunitrazepam, flurazepam, haloxazolam, medazepam, mephobarbital, metharbital, nimetazepam, nitrazepam, oxazepam, pentobarbital, phenobarbital, secobarbital, triazolam

KEY WORDS
coated capillary; detector at anode

REFERENCE
Jinno,K.; Han,Y.; Sawada,H.; Taniguchi,M. Capillary electrophoretic separation of toxic drugs using a polyacrylamide-coated capillary, *Chromatographia*, **1997**, *46*, 309–314.

SAMPLE
Matrix: solutions

CAPILLARY ELECTROPHORESIS
Capillary: 60 cm × 75 μm coated fused-silica (40 cm to detector) (Supelco)
Capillary preparation: Coat column as follows. Adjust the pH of 20 mL water to 3.5 with acetic acid, add 80 μL 3-(trimethoxysilyl)propyl methacrylate (3-methacryloxypropyltrimethoxysilane), mix, suck into capillary, let stand at room temperature for 1 h, remove the solution, wash with water. Fill the capillary with a deaerated 4% acrylamide solution containing 1 mg/mL N,N,N',N'-tetramethylethylenediamine and 1 mg/mL ammonium persulfate, let stand for 3 h, remove excess solution by aspiration, rinse with water, remove water by aspiration, dry at 35° (cf. J. Chromatogr. 1985, 347, 191).
Running buffer: 100 mM pH 8.5 borate buffer containing 10 mM sodium dodecyl sulfate and 5 M urea
Injection: Electromigration at 5 kV for 5 s.
Detector: UV 240, UV 254
Migration time: 8.5
Voltage: 18 kV
Model: Jasco 890-CE

OTHER SUBSTANCES
Simultaneous: clotiazepam, cloxazolam, diazepam, estazolam, etizolam, fludiazepam, flurazepam, haloxazolam, oxazepam

KEY WORDS
injection at cathode; coated capillary

REFERENCE
Jinno,K.; Han,Y.; Sawada,H. Analysis of toxic drugs by capillary electrophoresis using polyacrylamide-coated columns, *Electrophoresis*, **1997**, *18*, 284–286.

Alprenolol

Molecular formula: $C_{15}H_{23}NO_2$
Molecular weight: 249.35
CAS Registry No.: 13655-52-2, 13707-88-5 (HCl)
Merck Index (12th ed.): 321
Lednicer: 1 177

SAMPLE
Matrix: solutions

CAPILLARY ELECTROPHORESIS
Capillary: 52 cm × 75 μm fused-silica (48 cm to detector) (Polymicro Technologies)
Capillary preparation: Before each run purge with running buffer for 3 min. Every 3 runs purge with 100 mM NaOH for 5 min. Purge new capillaries with 1 M NaOH for 20 min and with 100 mM NaOH for 20 min, rinse with running buffer, equilibrate with running buffer at 12 kV for 3 h.
Running buffer: MeCN:buffer 25:75 (Buffer was 25 mM pH 8.8 sodium borate containing 10 mM triethylamine and 25 mM (R)-dodecoxycarbonylvaline (Waters EnantioSelect (R)-Val-1).)
Injection: Hydrostatic injection for 2 s.
Detector: UV 214
Migration time: 28 (R), 29 (S)
Voltage: 12 kV
Current: ≤40 μA
Model: Waters Quanta 4000

OTHER SUBSTANCES
Simultaneous: oxprenolol, propranolol

KEY WORDS
chiral

REFERENCE
Peterson,A.G.; Ahuja,E.S.; Foley,J.P. Enantiomeric separations of basic pharmaceutical drugs by micellar electrokinetic chromatography using a chiral surfactant, N-dodecoxycarbonylvaline, *J.Chromatogr.B*, **1996**, *683*, 15–28.

SAMPLE
Matrix: solutions

CAPILLARY ELECTROPHORESIS
Capillary: 60 cm × 75 μm fused-silica (51 cm to detector) (Supelco)
Capillary preparation: At the end of each day wash capillary with 200 mM NaOH for 10 min and with water for 10 min. Condition new capillaries by washing with 200 mM NaOH for 10 min, with water for 10 min, and with running buffer for 10 min.
Running buffer: 50 mM pH 7.3 Sodium borate buffer containing 3.3% succinyl β-cyclodextrin (Wacker Chemie, Munich, Germany)
Injection: Hydrodynamic injection at 25 mbar for 6 s.
Detector: UV 208
Migration time: 41.02 (first enantiomer)
Voltage: 15 kV
Model: Prince

OTHER SUBSTANCES
Simultaneous: chlorthalidone, ephedrine, etilefrin, homatropine, methoxamine, norephedrine, octopamine, propranolol, synephrine, trihexyphenidyl

KEY WORDS
chiral; $\alpha = 1.0431$

REFERENCE
Schmid,M.G.; Wirnsberger,K.; Gübitz,G. Chiral separation of drug enantiomers by capillary electrophoresis using succinyl-β-cyclodextrin, *Pharmazie*, **1996**, *51*, 852–854.

SAMPLE
Matrix: solutions
Sample preparation: Inject an aliquot of a solution in running buffer.

CAPILLARY ELECTROPHORESIS
Capillary: 60 cm × 75 μm fused-silica (52.4 cm to detector)
Capillary preparation: After each run flush with 500 mM KOH for 2-3 min then with water.
Running buffer: 10 mM pH 3.8 Phosphate buffer containing 2% sulfated cyclodextrin (ds 7-10)
Injection: Hydrostatic injection.
Detector: UV 214
Migration time: 11.62, 12.12 (enantiomers)
Voltage: 15 kV
Model: Waters Quanta 4000

OTHER SUBSTANCES
Also analyzed: acebutolol, aminoglutethimide, brompheniramine, bupivacaine, bupropion, canadine, carbinoxamine, chloroquine, chlorpheniramine, dimethindene, disopyramide, doxylamine, hydroxychloroquine, idazoxan, isoxsuprine, ketamine, mepenzolate, mepivacaine, methoxyphenamine, mexiletine, midodrine, nefopam, orphenadrine, oxprenolol, oxyphencyclimine, pheniramine, phensuximide, pindolol, piperoxan, terbutaline, tetramisole, tolperisone, tranylcypromine, trihexyphenidyl, trimipramine, verapamil, warfarin

KEY WORDS
chiral; detector at anode

REFERENCE
Stalcup,A.M.; Gahm,K.H. Application of sulfated cyclodextrins to chiral separations by capillary zone electrophoresis, *Anal.Chem.*, **1996**, *68*, 1360–1368.

SAMPLE
Matrix: solutions
Sample preparation: Inject an aliquot of a 100 μg/mL solution in running buffer.

CAPILLARY ELECTROPHORESIS
Capillary: 29 cm × 50 μm fused-silica (24.5 cm to detector) (Yongnian Optical Conductive Fiber Plant, China), coated with polyacrylamide
Capillary preparation: No details of the polyacrylamide coating process are provided. However, another paper (LC.GC 1997, 15, 40) by this group indicates that they use the procedure of Hjertén, thus: Adjust the pH of 20 mL water to 3.5 with acetic acid, add 80 μL 3-(trimethoxysilyl)propyl methacrylate (3-methacryloxypropyltrimethoxysilane), mix, suck into capillary, let stand at room temperature for 1 h, remove the solution, wash with water. Fill the capillary with a deaerated 3-4% acrylamide solution containing 1 μL/mL N,N,N',N'-tetramethylethylenediamine and 1 mg/mL potassium persulfate, let stand for 30 min, remove excess solution by aspiration, rinse with water, remove water by aspiration, dry at 35° (J. Chromatogr. 1985, 347, 191).
Capillary temperature: 25
Running buffer: 100 mM NaH_2PO_4 adjusted to pH 2.5 (A) or 100 mM NaH_2PO_4 containing 45 mM hydroxypropyl-gamma-cyclodextrin, adjusted to pH 2.5 (B)
Injection: Electrokinetic injection at 15 kV for 3 s.
Detector: UV 200, UV 210

Migration time: 5.90 (A), 12.18, 12.41 (B, enantiomers)
Voltage: 15 kV
Model: Bio-Rad BioFocus 3000

OTHER SUBSTANCES
Simultaneous: albuterol, atenolol, baclofen, bamethan, benproperine, benserazide, bisoprolol, bupranolol, butamirate, butethamate, carbuterol, celiprolol, clenbuterol, clobutinol, dipivefrin, isoproterenol (isoprenaline), metaproterenol (orciprenaline), metipranolol, metoprolol, norfenefrine, ornidazole, oxprenolol, phenylpropanolamine, pholedrine, pirbuterol, prilocaine, procyclidine, sotalol, synephrine, terbutaline, tocainide

KEY WORDS
chiral; coated capillary

REFERENCE
Koppenhoefer,B.; Epperlein,U.; Xiaofeng,Z.; Bingcheng,L. Separation of enantiomers of drugs by capillary electrophoresis. Part 4: Hydroxypropyl-γ-cyclodextrin as chiral solvating agent, *Electrophoresis*, **1997**, *18*, 924–930.

SAMPLE
Matrix: solutions
Sample preparation: Inject an aliquot of a 100 μg/mL solution in running buffer.

CAPILLARY ELECTROPHORESIS
Capillary: 30 cm × 50 μm fused-silica (25.5 cm to detector), coated with polyacrylamide
Capillary preparation: Adjust the pH of 20 mL water to 3.5 with acetic acid, add 80 μL 3-(trimethoxysilyl)propyl methacrylate (3-methacryloxypropyltrimethoxysilane), mix, suck into capillary, let stand at room temperature for 1 h, remove the solution, wash with water. Fill the capillary with a deaerated 3-4% acrylamide solution containing 1 μL/mL N,N,N',N'-tetramethylethylenediamine and 1 mg/mL potassium persulfate, let stand for 30 min, remove excess solution by aspiration, rinse with water, remove water by aspiration, dry at 35° (J. Chromatogr. 1985, 347, 191).
Capillary temperature: 25
Running buffer: 100 mM NaH_2PO_4 containing 45 mM hydroxypropyl-α-cyclodextrin, adjusted to pH 2.5 with phosphoric acid
Injection: Electrokinetic injection at 15 kV for 3 s.
Detector: UV 200
Voltage: 15 kV
Model: Bio-Rad BioFocus 3000

KEY WORDS
chiral; coated capillary; comparison with the use of other cyclodextrins; this running buffer gave the greatest enantiomeric separation.; α=1.035

REFERENCE
Lin,B.; Zhu,X.; Koppenhoefer,B.; Epperlein,U. Investigation of 123 chiral drugs by cyclodextrin-modified capillary electrophoresis, *LC.GC*, **1997**, *15*, 40–46.

SAMPLE
Matrix: solutions
Sample preparation: Inject an aliquot of a 100 μM solution in water.

CAPILLARY ELECTROPHORESIS
Capillary: 65 cm × 50 μm fused-silica (56.5 cm to detector) (Polymicro Technologies)
Capillary preparation: Before each injection rinse capillary at 930 mbar with 100 mM NaOH for 1 min, with water for 2 min, and with running buffer for 4 min.
Capillary temperature: 20
Running buffer: 50 mM Phosphoric acid containing 4 mM heptakis(2,6-di-O-methyl)-β-cyclodextrin, adjusted to pH 2.5 with triethanolamine
Injection: Pressure injection at 20 mbar for 3 s.
Detector: UV 195
Migration time: 16.5 (R_S = 1.0)

Voltage: 24 kV
Model: Hewlett Packard HP³ᴰ

KEY WORDS
chiral

REFERENCE
Nilsson,S.; Schweitz,L.; Petersson,M. Three approaches to enantiomer separation of β-adrenergic antagonists by capillary electrochromatography, *Electrophoresis*, **1997**, *18*, 884–890.

SAMPLE
Matrix: solutions

CAPILLARY ELECTROPHORESIS
Capillary: 64 cm × 50 μm fused-silica (60 cm to detector) (Polymicro Technologies)
Capillary preparation: Before each run flush capillary with 100 mM NaOH for 3 min and with running buffer for 3 min. Each day flush the capillary with 1 M NaOH for 15 min, with water for 2 min, and with running buffer for 10 min. wash new capillaries with 1 M NaOH for 1 h.
Running buffer: 50 mM pH 9.2 Sodium tetraborate buffer containing 0.5% poly(sodium N-undecylenyl-L-valinevalinate) (Prepare poly(sodium N-undecylenyl-L-valinevalinate) by analogy with the following procedure for poly(sodium N-undecylenyl-D-valinate). Add 30 mmoles undecylenic acid to 30 mmoles N-hydroxysuccinimide in 130 mL dry ethyl acetate, add 30 mmoles dicyclohexylcarbodiimide in 10 mL dry ethyl acetate, let stand at room temperature overnight, filter, recrystallize ester from EtOH. Add 1 mmole undecylenic acid N-hydroxysuccinimide ester in 10 mL THF to 1 mmole L-valine and 1 mmole sodium bicarbonate in 10 mL water, after 16 h acidify to pH 2 with 1 M HCl, remove the organic solvent under reduced pressure, add 50 mL water, filter, dry, recrystallize N-undecylenyl-L-valine from chloroform/petroleum ether (mp 91°) (J.Lipid Res. 1967, 8, 142). Prepare the sodium salt by adding an equimolar amount of sodium bicarbonate in the presence of THF. Polymerize a 100 mM aqueous solution of the sodium salt by irradiating with a ⁶⁰Co gamma source at 70 krad/h for 36-48 h (cf. J.Poly.Sci.: Poly.Lett.Ed. 1979, 17, 749), dialyze against water using a regenerated cellulose membrane (2000 MW cut-off), lyophilize. Dry organic solvents over Type 4A 4-8 mesh molecular sieve.)
Injection: Pressure injection for 2 s.
Detector: UV 214
Migration time: 26.5 (S), 27 (R)
Voltage: +20 kV
Current: 56 μA
Model: Hewlett-Packard ³ᴰCE

OTHER SUBSTANCES
Simultaneous: propranolol

KEY WORDS
chiral

REFERENCE
Shamsi,S.A.; Macossay,J.; Warner,I.M. Improved chiral separations using a polymerized dipeptide anionic chiral surfactant in electrokinetic chromatography: Separations of basic, acidic, and neutral racemates, *Anal.Chem.*, **1997**, *69*, 2980–2987.

SAMPLE
Matrix: solutions

CAPILLARY ELECTROPHORESIS
Capillary: 29-36 cm × 50 μm fused-silica (24.5-31.5 cm to detector) (Yongnian Optical Conductive Fiber Plant, China) coated with polyacrylamide
Capillary preparation: Coat capillary as follows. Adjust the pH of 20 mL water to 3.5 with acetic acid, add 80 μL 3-(trimethoxysilyl)propyl methacrylate (3-methacryloxypropyltrimethoxysilane), mix, suck into capillary, let stand at room temperature for 1 h, remove the solution, wash with water. Fill the capillary with a deaerated 3-4% acrylamide solution containing 1 μL/mL N,N,N',N'-tetramethylethylenediamine and 1 mg/mL potassium persulfate, let stand

for 30 min, remove excess solution by aspiration, rinse with water, remove water by aspiration, dry at 35° (J. Chromatogr. 1985, 347, 191).
Capillary temperature: 25
Running buffer: 100 mM pH 2.5 NaH_2PO_4 (A) or 100 mM pH 2.5 NaH_2PO_4 containing 45 mM hydroxypropyl-α-cyclodextrin (Wacker, Munich) (B)
Injection: Electromigration at 15 kV for 3 s.
Detector: UV 200; UV 210
Migration time: 5.90 (A); 16.64, 17.23 (B) (enantiomers)
Voltage: 15 kV
Model: Bio-Focus 3000

OTHER SUBSTANCES
Also analyzed: albuterol (salbutamol), alprenolol, amorolfine, atenolol, atropine, azelastine, baclofen, bamethan, benproperine, benserazide, biperiden, bisoprolol, brompheniramine, bupivacaine, bupranolol, butamirate, butethamate, carazolol, carbuterol, carteolol, carvedilol, celiprolol, chloroquine, chlorpheniramine, chlorphenoxamine, cicletanine, clenbuterol, clidinium bromide, clobutinol, dimethindene, dipivefrin, disopyramide, dobutamine, doxylamine, fendiline, flecainide, gallopamil, homatropine, ipratropium bromide, isoproterenol (isoprenaline), isothipendyl, ketamine, meclizine, mefloquine, mepindolol, mequitazine, metaclazepam, metaproterenol (orciprenaline), metipranolol, metoprolol, nafronyl (naftidrofuryl), nefopam, nicardipine, norfenefrine, ofloxacin, ornidazole, orphenadrine, oxomemazine, oxprenolol, oxybutynin, phenoxybenzamine, phenylpropanolamine, pholedrine, pindolol, pirbuterol, prilocaine, procyclidine, promethazine, propafenone, propranolol, reproterol, sotalol, sulpride, synephrine, talinolol, terbutaline, tetrahydrozoline (tetryzoline), theodrenaline, tioconazole, tocainide, trihexyphenidyl, trimeprazine (alimemazine), trimipramine, tropicamide, verapamil, zopiclone

KEY WORDS
coated capillary; chiral

REFERENCE
Koppenhoefer,B.; Eperlein,U.; Schlunk,R.; Zhu,X.; Lin,B. Separation of enantiomers of drugs by capillary electrophoresis. V. Hydroxypropyl-α-cyclodextrin as chiral solvating agent, *J.Chromatogr.A*, **1998**, *793*, 153–164.

SAMPLE
Matrix: solutions
Sample preparation: Inject an aliquot of a 340 μM solution.

CAPILLARY ELECTROPHORESIS
Capillary: 58 cm × 75 μm fused-silica (50 cm to detector)
Capillary preparation: Rinse with running buffer for 2 min before each run. Before use rinse capillary at 1.36 bar with 100 mM NaOH for 5 min and with water for 10 min.
Capillary temperature: 25
Running buffer: 20 mM pH 7.0 Sodium phosphate buffer (A) or 20 mM pH 7.0 Sodium phosphate buffer containing 10 mM 1-lauroyl-2-hydroxy-sn-glycero-3-phosphocholine (Avanti Polar Lipids, Alabaster AL) (B)
Injection: Hydrodynamic injection at 34 mbar for 1 s followed by 5% MeOH for 5 s.
Detector: UV 200
Migration time: 3.98 min (A); k' 1.24 (B)
Voltage: 15 kV
Current: ca. 50 μA
Model: Beckman P/ACE 2200

OTHER SUBSTANCES
Simultaneous: acebutolol, atenolol, metoprolol, oxprenolol, pindolol, propranolol, timolol

REFERENCE
Masucci,J.A.; Caldwell,G.W.; Foley,J.P. Comparison of the retention behavior of β-blockers using immobilized artificial membrane chromatography and lysophospholipid micellar electrokinetic chromatography, *J.Chromatogr.A*, **1998**, *810*, 95–103.

Alprostadil

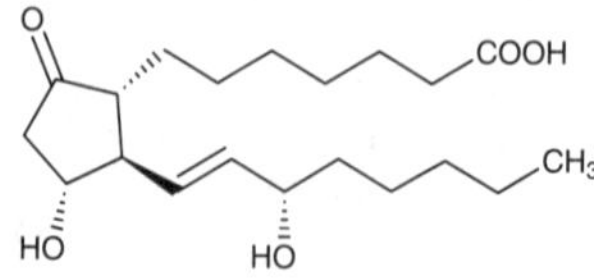

Molecular formula: $C_{20}H_{34}O_5$
Molecular weight: 354.49
CAS Registry No.: 745-65-3
Merck Index (12th ed.): 8063
Lednicer: 3 2

SAMPLE
Matrix: semen
Sample preparation: Mix semen with an equal volume of isopropanol, store at -20°, centrifuge at 14000 rpm, filter (0.22 μm) the supernatant, adjust the pH of the filtrate to 3 with 100 mM HCl, add to a C18 SPE cartridge, wash with 3 mL water, wash with 3 mL hexane, elute with 3 mL methyl acetate. Concentrate the eluate to 300 μL, inject an aliquot.

CAPILLARY ELECTROPHORESIS
Capillary: 47 cm × 50 μm fused-silica (40 cm to detector) (Polymicro Technologies)
Capillary temperature: 30
Running buffer: 38 mM pH 10 Borax/NaOH buffer containing 50 mM sodium dodecyl sulfate
Injection: Pressure injection at 0.5 psi for 3 s.
Detector: UV 200
Migration time: 5.99
Voltage: 20 kV
Model: Beckman P/ACE 2000
Limit of detection: 6.5-22.5 pg

OTHER SUBSTANCES
Extracted: dinoprost, dinoprostone, prostaglandin A_2, prostaglandin B_2, prostaglandin D_2, prostaglandin $F_{1\alpha}$, thromboxane B_2

KEY WORDS
SPE

REFERENCE
Hsieh,Y.-Z.; Kuo,K.-L. Analysis of prostaglandins by micellar electrokinetic capillary chromatography, *J.Chromatogr.A*, **1997**, *761*, 277–284.

Amantadine

Molecular formula: $C_{10}H_{17}N$
Molecular weight: 151.25
CAS Registry No.: 768-94-5, 665-66-7 (HCl)
Merck Index (12th ed.): 389
Lednicer: 2 18

SAMPLE
Matrix: solutions

CAPILLARY ELECTROPHORESIS
Capillary: 44 cm × 75 μm fused-silica (36.5 cm to detector) (Avery Dennison, MA)
Capillary preparation: Before each run wash capillary with running buffer. Before first use wash capillary with 1 M NaOH at 60° for 5 min, with 100 mM NaOH at 60° for 5 min, with water at 30° for 10 min, and with running buffer at 25° for 10 min.
Capillary temperature: 25

Running buffer: MeOH:water 20:80, pH 5.3, containing 5 mM 4-methylbenzylamine (Prepare
   running buffer by dissolving 130 μL 4-methylbenzylamine in 20 mL MeOH, make up to 100
   mL with water, adjust pH to 5.3 with 1 M HCl.)
Injection: Hydrodynamic injection for 10 s.
Detector: UV 210
Migration time: 1.4
Voltage: 30 kV
Model: SpectraPhoresis 2000 (Thermo Bioanalysis)

OTHER SUBSTANCES
Simultaneous: rimantadine

KEY WORDS
indirect UV detection; comparison with derivatization

REFERENCE
Revilla,A.L.; Hamacek,J.; Lubal,P.; Havel,J. Determination of rimantadine in pharmaceutical preparations by
   capillary zone electrophoresis with indirect detection or after derivatization, *Chromatographia*, **1998**, *47*,
   433–439.

SAMPLE
Matrix: urine, blood
Sample preparation: Urine. Adjust urine to pH 11 with NaOH, dilute with an equal volume of
   MeCN, filter (0.45 μm) add a 30 μL aliquot to 20 mg polymeric reagent, heat at 40° for 15 min,
   inject an aliquot of the reaction mixture. Plasma. Lyophilize plasma, reconstitute with 50 mM
   NaOH containing 5 mM sodium dodecyl sulfate, filter (0.45 μm), add a 50 μL aliquot to 30 mg
   polymeric reagent in a Pasteur pipette, heat at 75° for 10 min, wash with 5 mM sodium dodecyl
   sulfate, elute with 200 μL MeCN, dilute the MeCN with 400 μL running buffer, inject an
   aliquot. (Prepare polymeric reagent as follows. Soxhlet extract macroporous polystyrene cross
   linked with divinylbenzene with MeCN for 24 h. Heat 2.5 g resin and 4 g 4-chloro-3-nitroben-
   zoyl chloride in nitrobenzene in the presence of aluminum chloride at 60° for 24 h, pour into
   75 mL DMF, 50 mL concentrated HCl, and 50 g ice. Wash the product with DMF/water, with
   DMF, and with dichloromethane. Add the product to 15 mL trimethylammonium hydroxide,
   15 mL water, and 30 mL dioxane (Caution! Dioxane is a carcinogen!), heat at 90° for 16 h.
   React 1 g of this product with 0.5 g p-nitrobenzoyl chloride and 2.7 mmole pyridine in 20 mL
   dichloromethane at room temperature for 2 h, filter, wash with dichloromethane, with DMF,
   and with THF, dry under vacuum.)

CAPILLARY ELECTROPHORESIS
Capillary: 60 cm × 75 μm (53 cm to detector)
Running buffer: MeCN:buffer 20:80 (Buffer was 25 mM pH 7.2 phosphate buffer.)
Injection: Vacuum inject at 25 kPa
Detector: UV 254
Migration time: 9
Voltage: 20 kV
Current: 70 μA
Model: Isco Model 3140
Limit of quantitation: 10 ppm

KEY WORDS
plasma; cow; derivatization

REFERENCE
Szulc,M.; Krull,I.S. Polymeric reagents for derivatizations in micellar electrokinetic chromatography, *Bio-
   med.Chromatogr.*, **1994**, *8*, 212–218.

Amikacin

Molecular formula: $C_{22}H_{43}N_5O_{13}$
Molecular weight: 585.61
CAS Registry No.: 37517-28-5, 39831-55-5 (sulfate)
Merck Index (12th ed.): 425

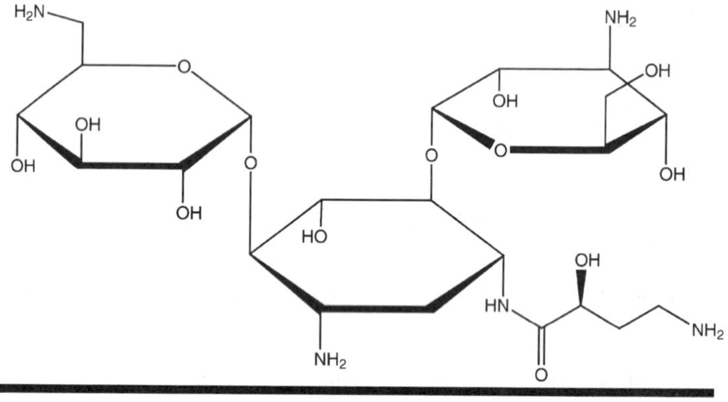

SAMPLE
Matrix: blood
Sample preparation: Evaporate 100 µL of a 1 mg/mL solution of 1-methoxycarbonylindolizine-3,5-dicarbaldehyde in ethyl acetate under reduced pressure into the bottom of a tube. Filter (Ultrafree C3LCC) 200 µL plasma while centrifuging at 4° at 2000 g for 20 min. Add 40 µL reaction buffer to the reaction tube, sonicate for 30 s, add 20 µL plasma ultrafiltrate, mix well, let stand in the dark at room temperature for 15 min, inject an aliquot. (Prepare phosphate-borate buffer by mixing equal volumes of 20 mM NaH_2PO_4 and 20 mM sodium tetraborate, adjust pH to 10 with 1 M NaOH. Prepare reaction buffer by mixing equal volumes of EtOH and phosphate-borate buffer. Prepare reagent (1-methoxycarbonylindolizine-3,5-dicarbalde-hyde) as follows. Reflux 21.4 g 2-pyridinecarboxaldehyde, 24 mL ethylene glycol, 10 g p-tolu-enesulfonic acid, and 300 mL benzene (Caution! Benzene is a carcinogen!) under a Dean-Stark separator for 64 h, pour into concentrated sodium carbonate solution. Remove the organic layer and extract the aqueous layer 4 times with benzene. Combine the organic layers and wash them with water, dry over anhydrous magnesium sulfate, evaporate, distil the residue to give 2-(1,3-dioxolan-2-yl)pyridine (bp 122°/4 mm Hg) (J.Org.Chem. 1963, 28, 83). Reflux 15.1 g 2-(1,3-dioxolan-2-yl)pyridine and 19.5 g tert-butyl bromoacetate in 100 mL dry acetonitrile for 7 h,let stand overnight at room temperature, filter, wash the precipitate with diethyl ether to give 1-(tert-butoxycarbonylmethyl)-2-(1,3-dioxolan-2-yl)pyridinium bromide (mp 110-2° from MeCN). Suspend 51.9 g of this compound in 1. 5 L THF with stirring, add 62.1 g potassium carbonate, add 15.12 g methyl propiolate, stir at room temperature for 9 days, filter, evaporate the filtrate to dryness under reduced pressure, chromatograph the residue on silica gel with hexane:ethyl acetate 20:1-10:1, collect fractions and evaporate to dryness to give methyl 3-tert-butoxycarbonyl-5-(1,3-dioxolan-2-yl)indolizine-1-carboxylate (mp 138-9° from hexane). Reflux 20.82 g of this compound in 600 mL THF and 60 mL 10% HCl for 6 h, concentrate to one quarter of the original volume, add water, extract with chloroform. Wash the chloroform layer with water and dry it over anhydrous sodium sulfate, evaporate, chromatograph on silica gel with hexane:ethyl acetate 10:1 to give 1-methoxycarbonylindolizine-5-carbaldehyde (mp 135-7° from MeOH). Stir 12.18 g of this compound in 116 mL dry DMF at 0° under argon, add 17 mL phosphorus oxychloride, stir at room temperature for 1 h, pour into water, adjust pH to 9.0 with 5% potassium carbonate, extract with chloroform. Wash the organic layer with water and dry it over anhydrous sodium sulfate, concentrate until a precipitate forms, filter to obtain the product, concentrate the filtrate and chromatograph the residue on silica gel with hexane:ethyl acetate 5:1 to obtain more product. The product was 1-methoxycarbonylindolizine-3,5-dicar-baldehyde (mp 164-5° from methyl acetate) (J. Chromatogr. A 1996, 724, 169.)

CAPILLARY ELECTROPHORESIS
Capillary: 75 cm × 50 µm fused-silica
Running buffer: 20 mM pH 7 Phosphate-borate buffer containing 40 mM sodium dodecyl sulfate
Injection: Dynamic compression injection at 50 hPa (sic) for 6 s

Detector: F ex 414 em 482, UV 280
Migration time: 17.6
Voltage: 30 kV
Model: Jasco Model CE-990
Limit of quantitation: 5 µg/mL

KEY WORDS
plasma; ultrafiltrate; derivatization

REFERENCE
Oguri,S.; Miki,Y. Determination of amikacin in human plasma by high-performance capillary electrophoresis with fluorescence detection, *J.Chromatogr.B*, **1996**, *686*, 205–210.

SAMPLE
Matrix: bulk
Sample preparation: Inject an aliquot of a solution in 140 mM sodium tetraborate containing 480 µg/mL sisomicin as IS.

CAPILLARY ELECTROPHORESIS
Capillary: 70 cm × 50 µm uncoated fused silica (45 cm to detector) (Polymicro Technologies or ISCO)
Capillary preparation: Rinse with running buffer for 15 min at the beginning of the day and for 4 min before each injection.
Capillary temperature: 34
Running buffer: 185 mM pH 9 sodium tetraborate
Injection: Vacuum injection at 25.0 kPa
Detector: UV 195
Migration time: 24.6
Internal standard: sisomicin (19.90)
Voltage: +15 kV
Model: ISCO Model 3140 Electropherograph

OTHER SUBSTANCES
Simultaneous: bekanamycin, butirosin, dibekacin, dihydrostreptomycin, gentamicins, kanamycin A, paromomycin, ribostamycin, streptomycin, tobramycin
Noninterfering: neomycin

KEY WORDS
detector is at cathode

REFERENCE
Flurer,C.L. The analysis of aminoglycoside antibiotics by capillary electrophoresis, *J.Pharm.Biomed.Anal.*, **1995**, *13*, 809–816.

SAMPLE
Matrix: solutions
Sample preparation: Prepare a solution in 100 mM cetyltrimethylammonium bromide, inject an aliquot.

CAPILLARY ELECTROPHORESIS
Capillary: 67 cm × 50 µm fused silica capillary (60 cm to detector) (Siemens)
Running buffer: pH 5.0 Acetate buffer containing 10 mM imidazole and 50 µg/mL FC 135 (a fluorinated surfactant, Fluorad/3M)
Injection: Pressure injection for 2 s.
Detector: UV 214
Migration time: 12
Voltage: 12.5 kV
Model: Beckman P/ACE System 2000

OTHER SUBSTANCES
Simultaneous: dihydrostreptomycin, kanamycin, lividomycin, neomycin, paromomycin, sisomycin, streptomycin, tobramycin

Interfering: gentamicin

KEY WORDS
indirect UV detection; detector is at anode

REFERENCE
Ackermans,M.T.; Everaerts,F.M.; Beckers,J.L. Determination of aminoglycoside antibiotics in pharmaceuticals by capillary zone electrophoresis with indirect UV detection coupled with micellar electrokinetic capillary chromatography, *J.Chromatogr.*, **1992**, *606*, 229–235.

SAMPLE
Matrix: solutions
Sample preparation: Inject an aliquot of a 100 μM solution in buffer. (Prepare buffer by mixing equal volumes of 20 mM NaH_2PO_4 and 20 mM sodium tetraborate and adjusting the pH to 10 with 1 M NaOH.)

CAPILLARY ELECTROPHORESIS
Capillary: 110 cm $\times$ 50 μm fused-silica (100 cm to detector)
Capillary preparation: Before injection rinse capillary with running buffer at 200 mbar with no voltage for 3 min and with running buffer at 4 mbar at 20 kV for 30 s.
Capillary temperature: 40
Running buffer: 100 mM pH 10 phosphate borate buffer containing 860 μM 1-methoxycarbonylindolizine-3,5-dicarbaldehyde (Prepare 1-methoxycarbonylindolizine-3,5-dicarbaldehyde as follows. Reflux 21.4 g 2-pyridinecarboxaldehyde, 24 mL ethylene glycol, 10 g p-toluenesulfonic acid, and 300 mL benzene (Caution! Benzene is a carcinogen!) under a Dean-Stark separator for 64 h, pour into concentrated sodium carbonate solution. Remove the organic layer and extract the aqueous layer 4 times with benzene. Combine the organic layers and wash them with water, dry over anhydrous magnesium sulfate, evaporate, distil the residue to give 2-(1,3-dioxolan-2-yl)pyridine (bp 122°/4 mm Hg) (J.Org.Chem. 1963, 28, 83). Reflux 15.1 g 2-(1,3-dioxolan-2-yl)pyridine and 19.5 g tert-butyl bromoacetate in 100 mL dry acetonitrile for 7 h, let stand overnight at room temperature, filter, wash the precipitate with diethyl ether to give 1-(tert-butoxycarbonylmethyl)-2-(1,3-dioxolan-2-yl)pyridinium bromide (mp 110-2° from MeCN). Suspend 51.9 g of this compound in 1. 5 L THF with stirring, add 62.1 g potassium carbonate, add 15.12 g methyl propiolate, stir at room temperature for 9 days, filter, evaporate the filtrate to dryness under reduced pressure, chromatograph the residue on silica gel with hexane:ethyl acetate 20:1-10:1, collect fractions and evaporate to dryness to give methyl 3-tert-butoxycarbonyl-5-(1,3-dioxolan-2-yl)indolizine-1-carboxylate (mp 138-9° from hexane). Reflux 20.82 g of this compound in 600 mL THF and 60 mL 10% HCl for 6 h, concentrate to one quarter of the original volume, add water, extract with chloroform. Wash the chloroform layer with water and dry it over anhydrous sodium sulfate, evaporate, chromatograph on silica gel with hexane:ethyl acetate 10:1 to give 1-methoxycarbonylindolizine-5-carbaldehyde (mp 135-7° from MeOH). Stir 12.18 g of this compound in 116 mL dry DMF at 0° under argon, add 17 mL phosphorus oxychloride, stir at room temperature for 1 h, pour into water, adjust pH to 9.0 with 5% potassium carbonate, extract with chloroform. Wash the organic layer with water and dry it over anhydrous sodium sulfate, concentrate until a precipitate forms, filter to obtain the product, concentrate the filtrate and chromatograph the residue on silica gel with hexane:ethyl acetate 5:1 to obtain more product. The product was 1-methoxycarbonylindolizine-3,5-dicarbaldehyde (mp 164-5° from methyl acetate) (J. Chromatogr. A 1996, 724, 169.))
Injection: Pressure injection at 50 mbar for 6 s (with no applied voltage).
Detector: UV 409
Migration time: 25.5
Voltage: 20 kV
Model: JASCO Model CE-990

OTHER SUBSTANCES
Simultaneous: arbekacin, micromicin, netilmicin

KEY WORDS
derivatization; on-column derivatization

REFERENCE
Oguri,S.; Fujiyoshi,T.; Miki,Y. In-capillary derivatization with 1-methoxycarbonylindolizine-3,5-dicarbaldehyde for high-performance capillary electrophoresis, *Analyst*, **1996**, *121*, 1683–1688.

Amiloride

Molecular formula: $C_6H_8ClN_7O$
Molecular weight: 229.63
CAS Registry No.: 2609-46-3, 2016-88-8 (HCl), 17440-83-4 (HCl dihydrate)
Merck Index (12th ed.): 426
Lednicer: 1 278

SAMPLE
Matrix: blood, urine
Sample preparation: Condition a 3 mL Supelclean LC-18 SPE cartridge with 3 mL MeOH and 3 mL water. Dilute urine 1:10 with water. Precipitate proteins from serum with MeOH. Add diluted urine or protein supernatant to the SPE cartridge, wash with 3 mL water, elute with 3 mL MeOH, reconstitute to the original volume with 100 mM KOH, inject an aliquot.

CAPILLARY ELECTROPHORESIS
Capillary: 67 cm × 50 μm fused-silica (60 cm to detector) (Polymicro Technologies)
Capillary preparation: Rinse with running buffer for 2 min before run. If necessary, regenerate capillary with 100 mM NaOH for 10 min and with water for 15 min.
Capillary temperature: 20
Running buffer: 70 mM pH 4.5 Acetate buffer containing 500 mM betaine
Injection: Pressure injection for 5 s
Detector: UV 215
Migration time: 5.3
Voltage: 25 kV
Model: Beckman P/ACE 2000

OTHER SUBSTANCES
Extracted: metyrapone, triamterene
Noninterfering: caffeine

KEY WORDS
serum; SPE

REFERENCE
Jumppanen,J.; Sirén,H.; Riekkola,M.-L. Screening for diuretics in urine and blood serum by capillary zone electrophoresis, *J.Chromatogr.A*, **1993**, *652*, 441–450.

SAMPLE
Matrix: urine
Sample preparation: Filter (0.2 μm), inject an aliquot of the filtrate.

CAPILLARY ELECTROPHORESIS
Capillary: 80 cm × 50 μm fused-silica (57 cm to detector)
Capillary preparation: Before each run rinse the capillary with 1 M NaOH for 3 min, with 100 mM NaOH for 3 min, with water for 3 min, and with running buffer for 10 min.
Running buffer: 110 mM Boric acid containing 56 mM NaOH and 44 mM HCl, pH 8
Injection: Vacuum injection for 1 s
Detector: UV 220
Migration time: 4.969
Voltage: 20 kV
Model: Europhor Prime Vision system IV

OTHER SUBSTANCES
Extracted: acebutolol (UV 238), acetazolamide (UV 222), alprenolol (UV 220), atenolol (UV 228), bendroflumethiazide (UV 220), bumetanide (UV 220), chlorthalidone (UV 220), cocaine (UV 236), codeine (UV 220), ethacrynic acid (UV 220), furosemide (UV 232), hydrochlorothiazide (UV 226), methadone (UV 220), metoxiphenamine (UV 220), nadolol (UV 220), norcodeine (UV 220), oxprenolol (UV 220), pentazocine (UV 220), propranolol (UV 220), spironolactone (UV 244), triamterene (UV 232), xipamide (UV 234)

REFERENCE

Gonzalez,E.; Laserna,J.J. Capillary zone electrophoresis for the rapid screening of banned drugs in sport, *Electrophoresis*, **1994**, *15*, 240–243.

SAMPLE

Matrix: urine
Sample preparation: Centrifuge urine at 150 g for 10 min, inject an aliquot.

CAPILLARY ELECTROPHORESIS

Capillary: 100 cm × 50 μm (60 cm to detector)
Capillary preparation: Rinse capillary with running buffer for 3 min before each run. Regenerate capillary by flushing with 1 M NaOH for 10 min, with 100 mM NaOH for 10 min, and with water for 15 min.
Running buffer: pH 8 Borate buffer (concentration adjusted to keep current at 42 μA)
Injection: Electrokinetic injection at 10 kV for 10 s (ca. 6 nL).
Detector: F ex 382 em 416
Migration time: 4.0
Voltage: 30 kV
Current: 42 μA
Model: laboratory-constructed
Limit of detection: 2.97 μM

OTHER SUBSTANCES

Extracted: bendroflumethiazide (F ex 272 em 381), bumetanide (F ex 350 em 428), triamterene (F ex 370 em 434)

KEY WORDS

direct injection

REFERENCE

González,E.; Becerra,A.; Laserna,J.J. Direct determination of diuretic drugs in urine by capillary zone electrophoresis using fluorescence detection, *J.Chromatogr.B*, **1996**, *687*, 145–150.

Amino acids

Molecular formula: $C_2H_5NO_2$ (glycine), $C_3H_7NO_2$ (alanine), $C_3H_7NO_2S$ (cysteine), $C_3H_7NO_3$ (serine), $C_4H_7NO_4$ (aspartic acid), $C_4H_8N_2O_3$ (asparagine), $C_4H_9NO_3$ (threonine), $C_5H_{10}N_2O_3$ (glutamine), $C_5H_{11}NO_2$ (valine), $C_5H_{11}NO_2S$ (methionine), $C_5H_9NO_2$ (proline), $C_5H_9NO_3$ (hydroxyproline), $C_5H_9NO_4$ (glutamic acid), $C_6H_{13}NO_2$ (isoleucine), $C_6H_{13}NO_2$ (leucine), $C_6H_{14}N_2O_2$ (lysine), $C_6H_{14}N_4O_2$ (arginine), $C_6H_9N_3O_2$ (histidine), $C_9H_{11}NO_2$ (phenylalanine), $C_9H_{11}NO_3$ (tyrosine), $C_{11}H_{12}N_2O_2$ (tryptophan)

Molecular weight: 75.10 (glycine), 89.10 (alanine), 105.10 (serine), 115.10 (proline), 117.20 (valine), 119.10 (threonine), 121.20 (cysteine), 131.10 (hydroxyproline), 131.20 (isoleucine), 131.20 (leucine), 132.10 (asparagine), 133.10 (aspartic acid), 146.20 (lysine), 146.20 (glutamine), 147.10 (glutamic acid), 149.20 (methionine), 155.20 (histidine), 165.20 (phenylalanine), 174.20 (arginine), 181.20 (tyrosine), 204.20 (tryptophan)

CAS Registry No.: 51-35-4 (hydroxyproline), 52-89-1 (cysteine HCl), 52-90-4 (cysteine), 56-40-6 (glycine), 56-41-7 (alanine), 56-45-1 (serine), 56-84-8 (aspartic acid), 56-85-9 (glutamine), 56-86-0 (glutamic acid), 56-87-1 (lysine), 59-51-8 (DL-methionine), 60-18-4 (tyrosine), 61-90-5 (leucine), 63-68-3 (methionine), 63-91-2 (phenylalanine), 70-47-3 (asparagine), 71-00-1 (histidine), 72-18-4 (valine), 72-19-5 (threonine), 73-22-3 (tryptophan), 73-32-5 (isoleucine), 74-79-3 (arginine), 138-15-8 (glutamic acid HCl), 147-85-3 (proline), 150-30-1 (DL-phenylalanine), 302-72-7 (DL-alanine), 302-84-1 (DL-serine), 312-84-5 (D-serine), 338-69-2 (D-alanine), 348-67-4 (D-methionine), 513-29-1 (glycine sulfate), 585-21-7 (DL-glutamine), 609-36-9 (DL-proline), 617-65-2 (DL-glutamic acid), 673-06-3 (D-phenylalanine), 1119-34-2 (arginine HCl), 1783-96-6 (D-aspartic acid), 2058-58-4 (D-asparagine), 6893-26-1 (D-glutamic acid)

SAMPLE

Matrix: antibiotics

Sample preparation: Hydrolyze antibiotic in 6 M HCl at 110° for 24 h. Dissolve 7.8 mg of the residue in MeCN/water, dilute to 1.7 mg/mL with water. Remove a 50 μL aliquot and mix it with 50 μL 3 mg/mL dansyl chloride in MeCN and 50 μL pH 9.08 borate buffer, let stand at room temperature for 2 h, inject an aliquot.

CAPILLARY ELECTROPHORESIS
Capillary: 43 cm × 75 μm fused-silica (36.2 cm to detector) (Polymicro Technologies)
Capillary preparation: Wash capillary with running buffer between runs. Before use rinse capillary with 1 M NaOH and water then equilibrate with running buffer.
Capillary temperature: 25
Running buffer: 10 mM pH 7.55 Ammonium acetate buffer containing 5 mM aspartame and 2.5 mM copper(II) sulfate pentahydrate
Injection: Pressure injection at 0.5 psi for 2-5 s.
Detector: UV 214
Migration time: 7.2 (D-Trp), 7.4 (L-Trp), 7.6 (D-Val), 7.8 (L-Val), 9.2 (D-Asp), 9.4 (L-Asp), 10.2 (D-Glu), 10.7 (L-Glu)
Voltage: 10 kV
Current: 20 μA
Model: P/ACE Model 2100

KEY WORDS
chiral; derivatization

REFERENCE
Liu,J.; Dabrah,T.T.; Matson,J.A.; Klohr,S.E.; Volk,K.J.; Kerns,E.H.; Lee,M.S. Analysis of amino acid enantiomers derived from antitumor antibiotics using chiral capillary electrophoresis, *J.Pharm.Biomed.Anal.*, **1997**, *16*, 207–214.

SAMPLE
Matrix: beverages
Sample preparation: Filter fruit juice (0.45 μm), inject an aliquot.

CAPILLARY ELECTROPHORESIS
Capillary: 108 cm × 50 μm fused-silica (100 cm to detector) (Polymicro Technologies)
Running buffer: MeCN:buffer 5:95 (Buffer was 10 mM NaH_2PO_4 containing 30 mM octanesulfonic acid, pH 2.36.)
Injection: Hydrostatic injection at 10 cm for 15 s.
Detector: UV 185
Migration time: 15.5 (arginine), 23.5 (alanine), 27 (serine), 28.5 (asparagine), 29.5 (tryptophan), 30.5 (glutamic acid), 31 (phenylalanine), 31.5 (tyrosine), 32 (proline)
Voltage: 30 kV
Model: Waters Quanta 4000
Limit of detection: 0.5-50 ppm

KEY WORDS
underivatized amino acids; fruit juice; orange

REFERENCE
Klampfl,C.W.; Buchberger,W.; Turner,M.; Fritz,J.S. Determination of underivatized amino acids in beverage samples by capillary electrophoresis, *J.Chromatogr.A*, **1998**, *804*, 349–355.

SAMPLE
Matrix: bulk
Sample preparation: Dissolve 3 mg amino acids in 70 μL AccQ.Fluor borate buffer (Waters), add 20 μL 10 mM 6-aminoquinolyl-N-hydroxysuccinimidyl carbamate in MeCN (Waters), vortex for several s, let stand at room temperature for 1 min, heat at 55° for 10 min, inject an aliquot.

CAPILLARY ELECTROPHORESIS
Capillary: 48.5 cm × 50 μm fused-silica (40 cm to detector) (Hewlett-Packard)
Capillary temperature: 20 ± 0.1

Running buffer: 10 mM 1,3-Bis[tris(hydroxymethyl)ethylamino]propane (BTP) containing 5 mM
heptakis(2,6-di-O-methyl)-β-cyclodextrin, adjusted to pH 7.0 with HCl
Detector: UV 245
Migration time: 14.358, 14.933 (alanine enantiomers), 12.740, 13.586 (valine enantiomers)
Voltage: 30 kV
Model: Hewlett-Packard HP-3D

KEY WORDS
derivatization; chiral; discussion of chiral running buffer modifiers

REFERENCE
Cladrowa-Runge,S.; Rizzi,A. Enantioseparation of 6-aminoquinolyl-N-hydroxysuccinimidyl carbamate-deriva-
tized-amino acids by capillary zone electrophoresis using native and substituted β-cyclodextrin as chiral
additives. I. Discussion of optimum separation conditions, *J.Chromatogr.A*, **1997**, *759*, 157–165.

SAMPLE
Matrix: bulk
Sample preparation: Mix 0.1-1 mg amino acids with 100 μL MeOH:water:triethylamine 70:10:
10, add 5-10 μL phenyl isothiocyanate, vortex, heat at 55° for 30 min, evaporate to dryness at
55°, add 100 μL 12.5-25% trifluoroacetic acid in water, heat at 55° for 40 min, extract 3 times
with 1 mL portions of ethyl acetate. Combine the extracts and evaporate them to dryness under
a stream of nitrogen, wash the residue 3 times with 1 mL portions of n-heptane, reconstitute,
inject an aliquot.

CAPILLARY ELECTROPHORESIS
Capillary: 50 cm $\times$ 50 μm (30 cm to detector) (GL Science, Tokyo)
Running buffer: 50 mM pH 3.0 Sodium phosphate buffer containing 25 mM β-escin, 25 mM
digitonin, and 50 mM sodium dodecyl sulfate
Injection: Gravimetric injection at 5 cm for 5-40 s.
Migration time: 18.17, 18.35 (Phe enantiomers), 19.14, 19.29 (Leu enantiomers), 19.27, 19.71
(Ile enantiomers), 22.03, 22.31 (Met enantiomers), 22.53, 22.85 (Tyr enantiomers), 22.74, 23.12
(Val enantiomers), 31.03, 31.63 (Glu enantiomers), 31.14, 31.90 (Ala enantiomers), 33.70, 34.29
(Asp enantiomers), 35.00, 35.62 (Gln enantiomers), 42.87, 43.98 (Thr enantiomers), 47.96, 49.14
(Asn enantiomers), 50.68, 51.78 (Ser enantiomers)
Model: JASCO CE-800

KEY WORDS
derivatization; chiral

REFERENCE
Kurosu,Y.; Murayama,K.; Shindo,N.; Shisa,Y.; Satou,Y.; Ishioka,N. Optical resolution of phenylthiohydantoin-
amino acids by capillary electrophoresis for protein sequencing, *J.Chromatogr.A*, **1997**, *771*, 311–317.

SAMPLE
Matrix: CSF
Sample preparation: 50 μL CSF + 50 μL 210 μM fluorescein isothiocyanate isomer I in acetone,
mix, let stand in the dark for 2 h, dilute 100-1000 times, inject an aliquot.

CAPILLARY ELECTROPHORESIS
Capillary: 75 cm $\times$ 50 μm fused-silica (42 cm to detector) (Polymicro Technologies)
Capillary preparation: Rinse with 100 mM NaOH for 3 min, with water for 2 min, and with
running buffer for 3 min.
Running buffer: 100 mM Boric acid containing 100 mM sodium dodecyl sulfate, adjusted to pH
9.3 with NaOH
Injection: Hydrodynamic injection for 2 s (15 nL)
Detector: F ex 488 (laser)
Migration time: 8.4 (Lys), 8.6 (Arg), 9.1 (ornithine), 9.9 (ammonia), 10.0 (tyramine), 7.2 (Leu),
7.3 (Gln), 7.4 (Tyr), 7.6 (Val), 7.7 (Thr), 7.8 (Phe), 8.2 (Ser), 8.4 (Ala), 8.9 (taurine), 9.2 (Gly),
18.4 (Glu), 19.6 (Asp)
Voltage: +20 kV
Current: 42 μA
Model: TSP Spectraphoresis 100

Limit of detection: <0.2 nM

KEY WORDS
derivatization

REFERENCE
Nouadje,G.; Rubie,H.; Chatelut,E.; Canal,P.; Nertz,M.; Puig,P.; Couderc,F. Child cerebrospinal fluid analysis by capillary electrophoresis and laser-induced fluorescence detection, *J.Chromatogr.A*, **1995**, *717*, 293–298.

SAMPLE
Matrix: dialysate
Sample preparation: Pump dialysate, 7 mM naphthalene-2,3-dicarboxaldehyde in 54:46 MeCN: water, 10 mM fluorescein, and 10 mM NaCN in 50 mM pH 9.3 borate buffer at 1 μL/min each through a 42 cm × 240 μm PEEK capillary, mix this mixture with running buffer pumped at 2-7 μL/min, inject an aliquot electrokinetically.

CAPILLARY ELECTROPHORESIS
Capillary: 30 cm × 25 μm (14 cm to detector) (Polymicro Technologies)
Capillary preparation: Flush with 100 mM NaOH before use.
Running buffer: 100 mM Tris pH 8.65
Injection: Electrokinetic injection
Detector: F ex 442 nm (5-7 mW He-Cd laser)
Migration time: 1.07 (glutamate), 1.17 (aspartate)
Internal standard: fluorescein (0.86)
Voltage: 25 kV
Limit of detection: 30 nM

KEY WORDS
derivatization; rat; on-line derivatization

REFERENCE
Zhou,S.Y.; Zuo,H.; Stobaugh,J.F.; Lunte,C.E.; Lunte,S.M. Continuous in vivo monitoring of amino acid neuro-transmitters by microdialysis sampling with on-line derivatization and capillary electrophoresis separation, *Anal.Chem.*, **1995**, *67*, 594–599.

SAMPLE
Matrix: dialysate
Sample preparation: 10 μL Dialysate + 1 μL 198 μM somatostatin in water + 5 μL 3 mg/mL 3-(4-carboxybenzoyl)-2-quinolinecarboxaldehyde (CBQCA; Molecular Probes, Eugene OR) in MeOH + 2 μL 50 mM KCN in 50 mM pH 9.0 borate buffer, mix, let stand at room temperature for 2 h, inject an aliquot.

CAPILLARY ELECTROPHORESIS
Capillary: 57 cm × 75 μm fused-silica (50 cm to detector) (Polymicro Technologies)
Capillary preparation: Before each injection rinse capillary with 100 mM NaOH at high pressure for 2 min and with running buffer at high pressure for 3 min
Capillary temperature: 25
Running buffer: DMSO:buffer 20:80 (Buffer was 50 mM pH 9.0 Sodium borate buffer containing 30 mM sodium dodecyl sulfate.)
Injection: Pressure injection for 1 s (5.9 nL).
Detector: F ex 488 (4 nW argon ion laser) em 560 ± 40
Migration time: 9.5 (arginine), 14 (glutamine), 14.3 (valine), 14.5 (GABA), 15 (alanine), 15.2 (glycine), 21.6 (glutamic acid), 22.3 (aspartic acid)
Internal standard: somatostatin (3-6) (H-Cys-Lys-Asn-Phe-OH) (10.7)
Voltage: 20 kV
Current: 33 μA
Model: P/ACE 2100
Limit of detection: 0.29-68 nM

OTHER SUBSTANCES
Noninterfering: ammonia, aspargine, citrulline, ethanolamine, histidine, isoleucine, leucine, lysine, phenylalanine, o-phosphoethanolamine, serine, taurine, threonine, tyrosine

KEY WORDS
derivatization; rat

REFERENCE
Bergquist,J.; Vona,M.J.; Stiller,C.-O.; O'Connor,W.T.; Falkenberg,T.; Ekman,R. Capillary electrophoresis with laser-induced fluorescence detection: A sensitive method for monitoring extracellular concentrations of amino acids in the periaqueductal gray matter, *J.Neurosci.Methods*, **1996**, *65*, 33–42.

SAMPLE
Matrix: dialysate
Sample preparation: Dialysate pumped at 79 nL/min mixed in a 50 nL mixing T with reagent pumped at 79 nL/min and the mixture flowed through an 8 cm × 75 μm fused-silica capillary (reaction time 2.2 min). Aliquots were automatically injected (design of injector described in paper). (Reagent was 110 mM o-phthalaldehyde and 220 mM 2-mercaptoethanol in 25 mM pH 9.5 borate buffer.)

CAPILLARY ELECTROPHORESIS
Capillary: 20 cm × 25 μm fused-silica (15 cm to detector) (Polymicro Technologies)
Capillary preparation: At the start of each day rinse with 100 mM NaOH for 10 min, with water for 10 min, and with running buffer for 10 min.
Running buffer: 175 mM 2-(N-Cyclohexylamino)ethanesulfonic acid (CHES) containing 100 mM sodium dodecyl sulfate, adjusted to pH 9.0 with 1 M NaOH
Injection: Electrokinetic injection at -100 V for 2 s.
Detector: F ex 354 (2 mW He-Cd laser, Liconix Model 4210B) em 450
Migration time: 1.8 (glutamine), 2.0 (serine, threonine), 2.2 (tyrosine), 2.3 (alanine, glycine), 2.5 (taurine), 2.55 (valine), 2.6 (glutamic acid), 2.8 (methionine), 2.9 (aspartic acid), 3.2 (leucine), 3.3 (phenylalanine), 3.5 (isoleucine), 4.5 (lysine)
Voltage: 400 V/cm
Limit of detection: 20-40 nM

KEY WORDS
derivatization

REFERENCE
Lada,M.W.; Kennedy,R.T. Quantitative in vivo monitoring of primary amines in rat caudate nucleus using microdialysis coupled by a flow-gated interface to capillary electrophoresis with laser-induced fluorescence detection, *Anal.Chem.*, **1996**, *68*, 2790–2797.

SAMPLE
Matrix: flour
Sample preparation: Hydrolyze 30-40 mg flour with 4 mL 6 M HCl at 120° for 24 h, evaporate to dryness under reduced pressure, reconstitute with 10 mL water. Remove a 2.5 mL aliquot and add it to 2.5 mL 500 mM sodium bicarbonate, add 2.5 mL 20 mM dansyl chloride in acetone, sonicate for 10 min, heat at 50° for 1 h, add 100 μL concentrated ammonia, heat at 50° for 15 min, add to a 30 × 11 glass column of Amberlite CG 50 1, elute twice with 2 mL portions of water, elute with three 5 mL portions of MeOH, combine all the eluates and evaporate them to dryness under reduced pressure, reconstitute with 1 mL MeOH:water 20:80, inject an aliquot.

CAPILLARY ELECTROPHORESIS
Capillary: 57.5 × 50 μm fused-silica (50 cm to detector)
Capillary preparation: Flush capillary with running buffer for 2 min before each run.
Capillary temperature: 10
Running buffer: 20 mM Borax buffer containing 125 mM sodium dodecyl sulfate
Injection: Hydrodynamic injection for 3 s
Detector: UV 214
Migration time: 11.7 (Thr), 11.8 (Ser), 12.1 (Asn), 12.3 (Gln), 12.5 (Ala), 12.7 (Glu), 13.1 (α-ABA), 13.3 (Asp), 13.6 (Gly), 13.8 (Val), 14.3 (cysteic acid), 15.8 (Pro), 16.2 (Met), 17.7 (Ile), 18.5 (Leu), 20.7 (Phe), 21.2 (Trp), 23.3 (Cys, bis derivative), 24.8 (ammonia), 26 (Lys), 26.8 (Arg), 27.5 (Lys, bis derivative), 27.5 (His, bis derivative), 27.9 (Tyr, bis derivative)
Voltage: 25 kV
Model: Thermo Separation Products SpectraPHORESIS 500

KEY WORDS
corn; protect from light; SPE; derivatization

REFERENCE
Skocir,E.; Prosek,M. Determination of amino acid ratios in natural products by micellar electrokinetic chromatography, *Chromatographia*, **1995**, *41*, 638–644.

SAMPLE
Matrix: plant exudate
Sample preparation: 1 μL Plant exudate + 89 μL 100 mM sodium tetraborate + 10 μL reagent, vortex, inject an aliquot within 2 min. (Reagent was 60 mM o-phthalaldehyde in MeCN containing 120 mM 2-mercaptoethanol.)

CAPILLARY ELECTROPHORESIS
Capillary: 95 cm × 50 μm fused-silica (80 cm to detector) (CS-Chromatographie, Langerwehe, Germany)
Running buffer: MeCN:buffer 3:97 (Buffer was 27 mM pH 9.4 borate buffer containing 44 mM sodium dodecyl sulfate.)
Injection: Hydrodynamic injection for 1 s (2.4 nL).
Detector: F 325 (3.5 mW He Cd laser) em >389
Migration time: 17 (Gln), 17.5 (Asn, Thr), 17.8 (Ser), 18.5 (Ala), 19.2 (Val), 19.5 (Met), 21.5 (Ile), 23 (Leu), 27 (Glu), 28 (Asp), 38.5 (Lys)
Model: SpectraPhoresis 100

KEY WORDS
derivatization

REFERENCE
Bazzanella,A.; Lochmann,H.; Mainka,A.; Bächmann,K. Determination of inorganic anions, carboxylic acids and amino acids in plant matrices by capillary zone electrophoresis, *Chromatographia*, **1997**, *45*, 59–62.

SAMPLE
Matrix: protein
Sample preparation: Mix 20 μL 1 mg/mL protein solution and 500 μL concentrated HCl, seal in a tube under vacuum, heat at 110° for 24 h, lyophilize, dissolve the residue in 50 μL water. Add a 2-5 μL aliquot to 2-aminoethanol, add buffer, add 2-5 μL reagent, let stand at room temperature for 1-2 min, inject an aliquot. (Prepare reagent by dissolving 5 mg o-phthalaldehyde in 40 μL MeOH, add 5 μL ethanethiol, add 50 μL buffer. Protect from light, prepare fresh for each use. Buffer was 400 mM boric acid adjusted to pH 9.50 with 1 M KOH.)

CAPILLARY ELECTROPHORESIS
Capillary: 86 cm × 50 μm fused-silica (53 cm to detector) (Polymicro Technologies)
Capillary preparation: Flush with water for 30 min, flush with 1 M KOH for 30 min, flush with 100 mM KOH for 30 min, rinse thoroughly with water, equilibrate with running buffer.
Running buffer: MeOH:THF:buffer 15:2:83 (Buffer was 50 mM pH 9.50 borate buffer containing 50 mM sodium dodecyl sulfate.)
Injection: Hydrodynamic injection for 10 s
Detector: F ex 365 em 418 (cut-off filter)
Migration time: 16.9 (glutamine), 17.3 (threonine), 17.6 (serine), 18.6 (histidine), 19.4 (alanine), 19.7 (glycine), 20.4 (valine), 20.8 (gamma-aminobutyric acid), 21.4 (methionine), 21.8 (taurine), 23.0 (isoleucine), 23.9 (tryptophan), 25.2 (leucine), 27.6 (lysine), 30.2 (glutamic acid), 37.4 (arginine)
Internal standard: 2-aminoethanol (39.8)
Voltage: 23 kV

KEY WORDS
derivatization

REFERENCE
Liu,J.P.; Cobb,K.A.; Novotny,M. Separation of precolumn *ortho*-phthalaldehyde-derivatized amino acids by capillary zone electrophoresis with normal and micellar solutions in the presence of organic modifiers, *J.Chromatogr.*, **1989**, *468*, 55–65.

SAMPLE
Matrix: protein
Sample preparation: Suspend 10 mg protein in 10 mL 6 M HCl, seal under vacuum, heat at 110° for 20 min, evaporate to dryness, reconstitute with 1 mL 100 mM HCl, make up to 10 mL with water, filter, dilute the filtrate 10-fold with water, inject an aliquot.

CAPILLARY ELECTROPHORESIS
Capillary: 100 cm × 50 μm fused-silica (75 cm to detector)
Capillary preparation: Flush with running buffer before each run
Running buffer: 60 mM Sodium tetraborate containing 60 mM NaH_2PO_4, 2 mM o-phthalaldehyde, 2 mM N-acetyl cysteine, and 8 mM β-cyclodextrin, adjusted to pH 10 with 1 M NaOH (Anodic buffer did not contain o-phthalaldehyde or N-acetyl cysteine.)
Injection: Hydrostatic injection at 10 cm for 10 s.
Detector: F ex 340 em 450
Migration time: 11.47 (D-arginine), 11.65 (L-arginine), 15.49 (D-isoleucine), 15.53 (L-phenylalanine), 16.26 (D-leucine), 16.38 (D-phenylalanine), 16.47 (D-valine), 16.71 (L-leucine), 17.40 (L-isoleucine), 17.73 (D-methionine), 18.28 (L-methionine), 18.33 (L-valine), 18.48 (D-histidine), 18.78 (L-histidine), 18.89 (D-lysine), 19.46 (L-lysine), 19.64 (D-threonine), 20.58 (L-alanine), 20.95 (L-threonine), 21.33 (D,L-serine), 21.38 (D-alanine), 38.96 (D-glutamic acid), 39.76 (L-glutamic acid), 40.96 (L-aspartic acid), 43.78 (D-aspartic acid)
Voltage: 25 kV
Model: Jasco Model CE-800
Limit of detection: 2.5-10 μM

OTHER SUBSTANCES
Noninterfering: ammonia, cystine

KEY WORDS
derivatization; on-capillary derivatization; chiral

REFERENCE
Oguri,S.; Yokoi,K.; Motohase,Y. Determination of amino acids by high-performance capillary electrophoresis with on-line mode in-capillary derivatization, *J.Chromatogr.A*, **1997**, *787*, 253–260.

SAMPLE
Matrix: solutions
Sample preparation: Prepare a solution in 5 mM pH 10 carbonate buffer. Add a 2-5 mL aliquot to 20 μL reagent, let stand for 2-4 h in the dark, inject an aliquot. (Reagent was 550 μM fluorescein isothiocyanate isomer 1 in acetone containing a trace of pyridine.)

CAPILLARY ELECTROPHORESIS
Capillary: 99 cm × 50 μm fused-silica
Running buffer: 5 mM pH 10 carbonate buffer
Injection: Electromigration at 2 kV for 10 s
Detector: F ex 488 (1 W Ar laser) em 495 (long pass filter) -560 (short pass filter) (details of detector given in paper)
Migration time: 13.2 (Arg), 14.8 (Lys), 17.7 (Leu), 17.8 (Ile, Trp), 18.0 (Met), 18.2 (Phe, Val, His, Pro), 18.5 (Thr), 18.9 (Ser), 19.0 (Cys), 19.15 (Ala), 19.7 (Gly), 20.5 (Tyr), 24.0 (Glu), 25.0 (Asp)
Voltage: 25 kV
Limit of detection: 0.005-0.086 nM

KEY WORDS
derivatization

REFERENCE
Cheng,Y.-F.; Dovichi,N.J. Subattomole amino acid analysis by capillary zone electrophoresis and laser-induced fluorescence, *Science*, **1988**, *242*, 562–564.

SAMPLE
Matrix: solutions

Sample preparation: 1 mL Amino acid solution in 50 mM pH 8.9 carbonate buffer (chloroform-saturated) + 1 mL 6 mM 4-(dimethylamino)azobenzene-4'-sulfonyl chloride (dabsyl chloride) in acetone, heat at 75° with gentle stirring until the color changes from red to yellow- orange, evaporate to dryness under a stream of nitrogen, reconstitute with 1 mL water, dilute, inject an aliquot.

CAPILLARY ELECTROPHORESIS
Capillary: 115 cm × 50 μm fused-silica (105 cm to detector)
Running buffer: MeCN:20 mM pH 7 phosphate buffer 50:50 containing 5 mM sodium dodecyl sulfate
Injection: Electromigration at 5 kV for 5 s (0.7 nL)
Detector: Thermooptical detection, 458 nm (pump beam, 130 mW Ar laser), 632.8 nm (probe beam, 1 mW He-Ne laser)
Migration time: 22 (arginine), 33 (histidine), 34 (lysine), 38 (cysteine, tyrosine), 38.5 (tryptophan), 39 (proline), 40 (phenylalanine), 40.5 (leucine), 41 (methionine), 41.5 (isoleucine, valine), 42 (tyrosine, serine), 42.5 (alanine), 43.5 (glycine), 66.5 (glutamic acid), 68 (aspartic acid)
Voltage: 27 kV
Limit of detection: 50-500 nM

KEY WORDS
derivatization

REFERENCE
Yu,M.; Dovichi,N.J. Attomole amino acid determination by capillary zone electrophoresis with thermooptical absorbance detection, *Anal.Chem.*, **1989**, *61*, 37–40.

SAMPLE
Matrix: solutions
Sample preparation: 100 μL Solution + 50 μL 4 mM dimethylaminoazobenzene isothiocyanate in acetone, mix, heat at 52 ± 2° for 1 h, evaporate to dryness under reduced pressure at 30°, reconstitute the residue with 40 μL water and 80 μL 6 M HCl:glacial acetic acid 1:2, heat for 50 min, evaporate to dryness under reduced pressure, reconstitute with 500 μL 10 mM pH 2.5 phosphate buffer, inject an aliquot.

CAPILLARY ELECTROPHORESIS
Capillary: 110 cm × 50 μm fused-silica (104.5 cm to detector) (Polymicro Technologies)
Running buffer: MeCN:10 mM pH 2.5 phosphate buffer 40:60 containing 10 mM sodium dodecyl sulfate
Injection: Electrokinetic injection at 20 kV for 5 s
Detector: Thermooptical, pump 442 nm (15 mW He-Cd laser), probe 632.8 nm (2 mW He-Ne laser)
Migration time: 35 (Cis), 36 (His), 36.5 (Lys), 38 (Arg), 42.5 (Lys), 43.5 (Cys), 62 (Asn), 63.5 (Gly), 64 (Thr), 65 (Asp), 66 (Ala), 66.5 (Gln), 67 (Glu, Pro), 68.5 (Val), 70 (Tyr), 71.5 (Met), 73 (Phe), 74 (Trp)

KEY WORDS
derivatization

REFERENCE
Waldron,K.C.; Wu,S.L.; Earle,C.W.; Harke,H.R.; Dovichi,N.J. Capillary zone electrophoresis separation and laser-based detection of both fluorescein thiohydantoin and dimethylaminoazobenzene thiohydantoin derivatives of amino acids, *Electrophoresis*, **1990**, *11*, 777–780.

SAMPLE
Matrix: solutions
Sample preparation: Mix 2-5 μL of a 1-100 μM solution or protein hydrolysate with 10-20 μL 10 mM KCN in water and 5-10 μL 3 mg/mL reagent in MeOH, let stand at room temperature for at least 1 h, inject an aliquot. (Reagent was 3-(4-carboxybenzoyl)-2-quinolinecarboxaldehyde. Synthesis is as follows. Mix 9.69 g p-toluidine, 10.89 g o-nitrobenzaldehyde, and 25 mL EtOH and allow to react for 5 min, filter, wash the solid with EtOH to give 2-nitro-N-(p-tolyl)benzaldimine (Talanta 1989, 36, 321). (More 2-nitro-N-(p-tolyl)benzaldimine can be obtained by adding water to the filtrate.) Add, in portions in a thin stream, a hot solution of 46

g sodium sulfide in 23 mL water and 23 mL EtOH to a stirred refluxing solution of 100 mmole 2-nitro-N-(p-tolyl)benzaldimine in 50 mL EtOH. After a vigorous reaction 2-amino-N-(p-tolyl)benzaldimine crystallizes from the cooling solution (mp 102-103° after recrystallization from dilute MeOH) (Ber. 1943, 76, 1099). Wash 950 mg of a commercial 50% slurry of sodium hydride with pentane, add 7 mL dry THF (distilled from lithium aluminum hydride), add 1.51 g methyl 4-cyanobenzoate in 10 mL THF, add dropwise 1.38 mL acetone (distilled from calcium chloride), reflux for 1.5 h, cool, acidify with 3 M HCl. Remove the organic layer and wash it with brine and sodium bicarbonate solution, dry over anhydrous magnesium sulfate, evaporate to give (4-cyanobenzoyl)acetone. Reflux 433 mg (4-cyanobenzoyl)acetone, 486 mg 2-amino-N-(p-tolyl)benzaldimine, 69 mL piperidine, and 9 mL EtOH:water 95:5 for 18 h, remove volatiles by steam distillation, add the residue to water and dichloromethane. Remove the organic layer and dry it, evaporate to give 3-(4-cyanobenzoyl)-2-methylquinoline. Suspend 547 mg 3-(4-cyanobenzoyl)-2-methylquinoline in 13 mL EtOH:water 95:5, add 500 mg KOH, reflux for 6 h, cool, concentrate, add the residue to ether and water. Remove the aqueous layer and adjust the pH to 5 with tartaric acid, let stand for 15 min, filter, wash the precipitate with water, dry under vacuum to give 3-(4-carboxybenzoyl)-2-methylquinoline. Dissolve 266 mg 3-(4-carboxybenzoyl)-2-methylquinoline in 6 mL acetic acid, add 112 mg selenium dioxide, stir at 80° for 2 h, filter through Celite, wash the precipitate with several volumes of hot MeOH, dilute the filtrate with water, allow to stand, filter, wash the solid with water, dry under vacuum to give 3-(4-carboxybenzoyl)-2-quinolinecarboxaldehyde.)

CAPILLARY ELECTROPHORESIS
Capillary: 104 cm × 50 μm fused-silica (73 cm to detector)
Running buffer: 50 mM pH 7.02 2-[N-[tris(hydroxymethyl)methyl]amino]ethanesulfonic acid (TES) containing 50 mM sodium dodecyl sulfate
Injection: Hydrodynamic or electromigration for 10 s
Detector: F ex 442 (50 mW He-Cd laser) em 560 (bandwith filter)
Migration time: 14 (Arg), 15.2 (Trp), 15.4 (Tyr), 15.6 (His), 15.8 (Met), 16 (Ile), 16.5 (Gln), 16.7 (Asn), 17 (Thr), 18.5 (Phe), 18.8 (Leu), 19.2 (Val), 20.5 (Ser), 21 (Ala), 22.5 (Gly), 24.5 (Glu), 30 (Asp)
Voltage: 25 kV
Current: 14 μA
Limit of detection: 4.6-13.8 amole

KEY WORDS
derivatization

REFERENCE
Liu,J.; Hsieh,Y.-Z.; Wiesler,D.; Novotny,M. Design of 3-(4-carboxybenzoyl)-2-quinolinecarboxaldehyde as a reagent for ultrasensitive determination of primary amines by capillary electrophoresis using laser fluorescence detection, *Anal.Chem.*, **1991**, *63*, 408–412.

SAMPLE
Matrix: solutions
Sample preparation: Prepare dansyl derivatives, inject an aliquot.

CAPILLARY ELECTROPHORESIS
Capillary: 34 cm × 25 μm fused-silica (19 cm to detector) (Polymicro Technologies)
Capillary preparation: Before each run rinse with MeOH, water, and separation buffer
Running buffer: 25 mM pH 2.40 sodium phosphate buffer containing 100 mM Tween 20
Detector: UV 214
Migration time: 6 (Arg), 10 (Asn), 11 (Gln), 11.5 (Ser), 12 (Thr), 13 (Hyp), 15.5 (Glu), 16 (Gly), 16.5 (Ala), 17.5 (Lys, didansyl), 20.5 (Aba), 23 (Pro), 26 (Tyr, O-dansyl derivative), 27 (Val), 32.5 (Nval, Met), 42 (Ile), 45 (Phe), 47 (Leu), 65 (Trp), 67 (Asp), 70 (Cys, didansylcystine)
Voltage: 16 kV

KEY WORDS
derivatization

REFERENCE
Matsubara,N.; Terabe,S. Separation of 24 dansylamino acids by capillary electrophoresis with a non-ionic surfactant, *J.Chromatogr.A*, **1994**, *680*, 311–315.

SAMPLE
Matrix: solutions
Sample preparation: Dissolve 0.25 mmoles leucine in 25 mL MeCN, place under aspirator pressure for 2-3 min to remove traces of water, add 0.5 mmole (+)-O,O'-dibenzoyl-L-tartaric anhydride, stir at 50° for 20 h, evaporate to dryness under aspirator pressure, reconstitute with 3% aqueous ammonia, evaporate to dryness, inject an aliquot of a 100 μM solution in water. (Synthesize (+)-O,O'-dibenzoyl-L-tartaric anhydride as follows. Heat 1 mole L-(+)-tartaric acid and 2.7 mole benzoyl chloride at 150° for 4 h, cool, wash with ligroin, boil with xylene, remove the organic phase, repeat several times, filter, dry in a desiccator.)

CAPILLARY ELECTROPHORESIS
Capillary: 56 cm × 100 μm coated fused-silica (39 cm to detector) (Polymicro Technologies)
Capillary preparation: Coat capillary as follows. Suck 1% 3-(trimethoxysilyl)propylmethacrylate in acetone:water 50:50 into the capillary, after 1 h remove solution, fill capillary with 3% acrylamide (Caution! Acrylamide is a carcinogen!) solution in water containing 0.04% tetramethylethylenediamine and 0.05% ammonium persulfate, after 15-20 min rinse with water, dry by aspiration (Electrophoresis 1989, 10, 23).
Running buffer: 25 mM pH 5.8 phosphate buffer containing x% polyvinylpyrrolidone
Injection: Hydrodynamic injection at 10 cm for 5 s
Detector: UV 233
Migration time: 10.4, 10.7 (Leu, x=0) (enantiomers), 8.5, 8.6 (Val, x=0) (enantiomers), 9.1, 9.3 (Gln, x=0) (enantiomers), 9.2 (D-Thr), 9.4 (L-Thr) (x=3), 11.5 (L-Phe), 12 (D-Phe) (x=3), 14 (L-Trp), 15.7 (D-Trp) (x=2), 11.8 (L-mandelic acid), 12.5 (D-mandelic acid) (x=2)
Voltage: 12 kV
Current: 22 μA
Model: Laboratory constructed

KEY WORDS
chiral; derivatization; anode at detector; coated capillary

REFERENCE
Schützner,W.; Caponecchi,G.; Fanali,S.; Rizzi,A.; Kenndler,E. Improved separation of diastereomeric derivatives of enantiomers by a physical network of linear polyvinylpyrrolidone applied as pseudophase in capillary zone electrophoresis, *Electrophoresis*, **1994**, *15*, 769–773.

SAMPLE
Matrix: solutions
Sample preparation: Dissolve dansylated amino acids in MeOH, dilute with water so that the MeOH concentration does not exceed 7.5%, inject an aliquot.

CAPILLARY ELECTROPHORESIS
Capillary: 57.5 cm × 50 μm fused-silica (50 cm to detector)
Capillary preparation: Rinse with running buffer for 2 min before each run.
Capillary temperature: 10
Running buffer: 20 mM pH 9.2 borax buffer containing 102.5 mM sodium dodecyl sulfate
Injection: Hydrodynamic injection for 3 s
Detector: UV 214
Migration time: 9.21 (Hyp), 10.01 (Thr), 10.21 (Ser), 10.30 (Asn), 10.46 (Gln), 10.61 (Ala), 11.05 (Aba), 11.46 (Gly), 11.72 (Glu), 11.90 (Val), 12.40 (Asp), 12.69 (cysteic), 13.10 (Pro), 13.48 (Met), 14.56 (Ile), 15.21 (Leu), 17.27 (Phe), 17.81 (Trp), 19.78 (Cys, didansyl), 23.02 (Lys), 23.62 (Arg), 24.48 (Lys, didansyl), 24.68 (Tyr, didansyl)
Voltage: 25 kV
Model: SpectraPhoresis 500 (Therm Separation Products)

KEY WORDS
derivatization

REFERENCE
Skocir,E.; Vindevogel,J.; Sandra,P. Separation of 23 dansylated amino acids by micellar electrokinetic chromatography at low temperature, *Chromatographia*, **1994**, *39*, 7–10.

SAMPLE
Matrix: solutions
Sample preparation: 10 μL Amino acid mixture + 10 μL 200 mM pH 8.0 borate buffer + 10 μL 5 mM 9-fluorenylmethyl chloroformate in MeCN, , mix for 1 min, dilute with water, inject an aliquot.

CAPILLARY ELECTROPHORESIS
Capillary: 67 cm × 50 μm fused-silica (60 cm to detector) (Polymicro Technologies)
Capillary preparation: Flush capillary with running buffer for 2 min between runs. Wash new capillaries with 100 mM NaOH for 15 min and with water for 15 min.
Capillary temperature: 20
Running buffer: 5 mM pH 9.2 Sodium borate buffer containing 150 mM sodium dodecyl sulfate and 40 mM gamma-cyclodextrin
Injection: Pressure injection at 0.5 psi for 5 s.
Detector: UV 200
Migration time: 17.7 (D,L-Ser), 18 (D,L-Ala), 19.5 (D-Val), 19.6 (L-val), 21.0 (D-Met), 21.1 (L-Met), 22.5 (D-Leu), 22.6 (L-Leu), 23.4 (D-Phe), L-Phe (23.7), 25.0 (D-Trp), 25.2 (L-Trp)
Voltage: 10 or 20 kV (?)
Model: Beckman P/ACE 2050

KEY WORDS
chiral; derivatization

REFERENCE
Chan,K.C.; Muschik,G.M.; Issaq,H.J. Enantiomeric separation of amino acids using micellar electrokinetic chromatography after pre-column derivatization with the chiral reagent 1-(9-fluorenyl)-ethyl chloroformate, *Electrophoresis*, **1995**, *16*, 504–509.

SAMPLE
Matrix: solutions
Sample preparation: 5 μL 200 μM Amino acid in 200 mM pH 9.0 carbonate buffer + 20 μL 200 mM pH 9.0 carbonate buffer + 24 μL acetone + 1 μL 5 mM fluorescein isothiocyanate isomer 1 (Sigma) in acetone, heat at 40° for 4 h, dilute with water, inject an aliquot.

CAPILLARY ELECTROPHORESIS
Capillary: 57 cm × 50 μm fused-silica (50 cm to detector) (Polymicro Technologies)
Capillary preparation: Before each injection rinse with running buffer for 2 min.
Capillary temperature: 35
Running buffer: 50 mM pH 7.5 Phosphate-borate buffer containing 75 mM sodium dodecyl sulfate
Injection: Pneumatic injection of sample for 2 s then 0.1 s injection of running buffer
Detector: F ex 488 (?) (argon ion laser)
Migration time: 4.0, 4.2 (LYS), 4.5 (ARG), 5.5 (HIS), 5.7 (HYP), 5.78 (TYR), 5.82 (GLN), 5.89 (PRO), 5.91 (LEU, ILE), 5.95 (MET), 6.0 (VAL), 6.05 (ASN), 6.1 (PHE), 6.15 (SER), 6.35 (ALA), 6.52 (GLY), 8.7 (GLU), 9.2 (ASP)
Voltage: 30 kV
Model: Beckman P/ACE System 2100

KEY WORDS
derivatization

REFERENCE
Lalljie,S.P.D.; Sandra,P. MEKC analysis of FITC and DTAF amino acid derivatives with LIF detection, *Chromatographia*, **1995**, *40*, 513–518.

SAMPLE
Matrix: solutions
Sample preparation: Inject an aliquot of a solution of the dansyl derivatives.

CAPILLARY ELECTROPHORESIS
Capillary: 80 cm × 50 μm fused-silica (50 cm to detector) (Polymicro Technologies)

Running buffer: 400 mM pH 7.0 Borate buffer containing 100 mM MEGA 10 (decanoyl-N-methylglucanamide, Calbiochem)
Detector: UV 254
Migration time: 15 (glutamine), 15 (asparagine), 15 (threonine), 15.3 (serine), 15.7 (valine), 16 (methionine), 16.5 (glycine), 16.8 (isoleucine), 17 (leucine), 17.6 (arginine), 18 (phenylalanine), 19 (tryptophan), 21 (glutamic acid), 21.6 (aspartic acid), 22.6 (cysteic acid)
Voltage: 15 kV
Model: Laboratory-constructed

REFERENCE

Mechref,Y.; Smith,J.T.; El Rassi,Z. Micellar electrokinetic capillary chromatography with *in situ* charged micelles. VII. Expanding the utility of alkylglycoside-borate micelles to acidic and neutral pH for capillary electrophoresis of dansyl amino acids and herbicides, *J.Liq.Chromatogr.*, **1995**, *18*, 3769–3786.

SAMPLE

Matrix: solutions
Sample preparation: Evaporate 100 µL of a 1 mg/mL solution of reagent in ethyl acetate into the bottom of a tube using reduced pressure in a desiccator, add 40 µL reaction buffer, add 20 µL of amino acid solution in water containing 2.5 mg/mL 3-nitrophenol, let stand for 15 min in the dark, inject an aliquot. (Prepare reaction buffer by mixing equal volumes of EtOH and phosphate/borate buffer. Prepare phosphate/borate buffer by mixing equal volumes of 20 mM NaH_2PO_4 and 20 mM sodium tetraborate and adjusting the pH to 10 with 1 M NaOH. Prepare reagent (1-methoxycarbonylindolizine-3,5-dicarbaldehyde) as follows. Reflux 21.4 g 2-pyridinecarboxaldehyde, 24 mL ethylene glycol, 10 g p-toluenesulfonic acid, and 300 mL benzene (Caution! Benzene is a carcinogen!) under a Dean-Stark separator for 64 h, pour into concentrated sodium carbonate solution. Remove the organic layer and extract the aqueous layer 4 times with benzene. Combine the organic layers and wash them with water, dry over anhydrous magnesium sulfate, evaporate, distil the residue to give 2-(1,3-dioxolan-2-yl)pyridine (bp 122°/ 4 mm Hg) (J.Org.Chem. 1963, 28, 83). Reflux 15.1 g 2-(1,3-dioxolan-2-yl)pyridine and 19.5 g tert-butyl bromoacetate in 100 mL dry acetonitrile for 7 h, let stand overnight at room temperature, filter, wash the precipitate with diethyl ether to give 1-(tert-butoxycarbonylmethyl)-2-(1,3-dioxolan-2-yl)pyridinium bromide (mp 110-2° from MeCN). Suspend 51.9 g of this compound in 1. 5 L THF with stirring, add 62.1 g potassium carbonate, add 15.12 g methyl propiolate, stir at room temperature for 9 days, filter, evaporate the filtrate to dryness under reduced pressure, chromatograph the residue on silica gel with hexane:ethyl acetate 20:1-10: 1, collect fractions and evaporate to dryness to give methyl 3-tert-butoxycarbonyl-5-(1,3-dioxolan-2-yl)indolizine-1-carboxylate (mp 138-9° from hexane). Reflux 20.82 g of this compound in 600 mL THF and 60 mL 10% HCl for 6 h, concentrate to one quarter of the original volume, add water, extract with chloroform. Wash the chloroform layer with water and dry it over anhydrous sodium sulfate, evaporate, chromatograph on silica gel with hexane:ethyl acetate 10:1 to give 1-methoxycarbonylindolizine-5-carbaldehyde (mp 135-7° from MeOH). Stir 12.18 g of this compound in 116 mL dry DMF at 0° under argon, add 17 mL phosphorus oxychloride, stir at room temperature for 1 h, pour into water, adjust pH to 9.0 with 5% potassium carbonate, extract with chloroform. Wash the organic layer with water and dry it over anhydrous sodium sulfate, concentrate until a precipitate forms, filter to obtain the product, concentrate the filtrate and chromatograph the residue on silica gel with hexane:ethyl acetate 5:1 to obtain more product. The product was 1-methoxycarbonylindolizine-3,5-dicarbaldehyde (mp 164-5° from methyl acetate).)

CAPILLARY ELECTROPHORESIS

Capillary: 60 cm × 50 µm fused-silica (50 cm to detector)
Running buffer: MeOH:buffer 3:97 (Buffer was 40 mM pH 7.0 borate/phosphate buffer containing 20 mM sodium dodecyl sulfate.)
Injection: Hydrostatic injection at 15 cm for 5 s
Detector: UV 280 UV 409 or F ex 282 (or 414) em 482
Migration time: 6.7 (Ser), 6.9 (Cys), 7.1 (Glu), 7.3 (Asp), 7.5 (Gly), 7.7 (Ala), 10.3 (Val), 12.5 (His), 13.2 (Met), 13.8 (Ile), 14.2 (Leu), 15.7 (ammonia), 16.3 (Phe), 17.0 (Lys), 18.0 (Arg), 18.3 (Tyr)
Voltage: 20 kV
Model: JASCO CE-800
Limit of detection: 5 nM

KEY WORDS

derivatization

REFERENCE

Oguri,S.; Uchida,C.; Miyake,Y.; Miki,Y.; Kakehi,K. 1-Methoxycarbonylindolizine-3,5-dicarbaldehyde as a derivatization reagent for amino compounds in high-performance capillary electrophoresis, *Analyst*, **1995**, *120*, 63–68.

SAMPLE

Matrix: solutions

Sample preparation: Prepare derivatives with reaction with a 3-fold excess of (+)-O,O'-dibenzoyl-L-tartaric anhydride in dichloroethane, THF, or acetone in the presence of an excess of trichloroacetic acid. Heat at 50° for several hours (J.Chromatogr. 1984, 316, 605).

CAPILLARY ELECTROPHORESIS

Capillary: 56 cm × 100 μm coated fused-silica (39 cm to detector) (Polymicro Technologies)

Capillary preparation: Coat capillary by filling with a 0.5% solution of gamma- methacryloxypropyltrimethoxysilane in acetone:water 50:50, after 1 h remove solution and fill capillary with 4% acrylamide (Caution! Acrylamide is a carcinogen!) solution containing 0.4 μL/mL tetramethylethylenediamine and 0.5 mg/mL ammonium persulfate, after 15-20 min rinse capillary with water, dry under vacuum (Electrophoresis 1989, 10, 23).

Capillary temperature: ambient

Running buffer: 30 mM NaH_2PO_4 containing 2.5% poly(vinylpyrrolidone) adjusted to pH 5.8 with NaOH

Injection: Hydrodynamic injection at 10 cm for 5 s

Detector: UV 233

Migration time: 9.4 (D-Ser), 9.5 (L-Ser), 9.6 (D-Gln), 9.9 (L-Gln), 10.0 (D-Leu), 10.2 (L- Leu), 11.1 (L-Phe), 11.5 (D-Phe), 14 (L-Trp), 15.5 (D-Trp)

Voltage: 12 kV

Model: laboratory constructed

KEY WORDS

derivatization; chiral; coated capillary

REFERENCE

Schnützner,W.; Fanali,S.; Rizzi,A.; Kenndler,E. Separation of diastereomers by capillary zone electrophoresis with polymer additives: Effect of polymer type and chain length, *Anal.Chem.*, **1995**, *67*, 3866–3870.

SAMPLE

Matrix: solutions

Sample preparation: Mix 100 μL of a 250 μg/mL solution of amino acids in water with 700 μL 200 mM pH 8.8 sodium borate buffer, vortex for 20 s, add 100 μL 3 mg/mL 6-aminoquinoyl-N-hydroxysuccinimidyl carbamate (Waters) in MeCN, vortex for 20 s, dilute with water or running buffer (if desired), inject an aliquot.

CAPILLARY ELECTROPHORESIS

Capillary: 60 cm × 50 μm (52.5 cm to detector) (Waters AccuSep)

Capillary preparation: Rinse with running buffer for 3 min between runs. Rinse new capillaries with 500 mM NaOH for 10 min, with water for 10 min, and with running buffer for 10 min.

Capillary temperature: 30

Running buffer: 25 mM pH 9.0 Na_2HPO_4/sodium tetraborate containing 100 mM (R)-N-dodecoxycarbonylvaline (Prepare (R)-N-dodecoxycarbonylvaline as follows. Prepare dodecyl chloroformate by reacting 1-dodecanol with 0.33 equivalents of triphosgene in dichloromethane solution in the presence of pyridine. Add dodecyl chloroformate dropwise to (R)-valine in 1 M NaOH solution, filter, wash with hexane, recrystallize from ether/petroleum ether (J. Chromatogr. A, 1994, 680, 125).)

Injection: Hydrostatic injection at 10 cm for 20 s

Detector: UV 254

Migration time: 11.8 (d-Ile), 12 (l-Ile), 12.4 (d-Leu), 12.6 (l-Leu), 13.1 (d-Orn), 13.3 (l-Orn), 13.5 (d-Lys), 13.7 (l-Lys), 14.2 (d-Trp), 14.8 (l-Trp), 18.4 (d-Arg), 18.8 (l-Arg)

Voltage: 16 kV

Model: Waters Quanta 4000E

Limit of quantitation: 300 ng/mL

Limit of detection: 100 ng/mL

KEY WORDS
derivatization; chiral

REFERENCE
Swartz,M.E.; Mazzeo,J.R.; Grover,E.R.; Brown,P.R. Separation of amino acid enantiomers by micellar electro-kinetic capillary chromatography using synthetic chiral surfactants, *Anal.Biochem.*, **1995**, *231*, 65–71.

SAMPLE
Matrix: solutions
Sample preparation: Mix 100 μL of a 75 μM amino acid solution with 15 μL buffer and 15 μL reagent, let stand for 6 min, inject an aliquot. (Prepare reagent by dissolving 8 mg o-phthal-dialdehyde and 44 mg O-acetyl-1-thio-β-D-glucopyranose in 1 mL MeOH. Prepare buffer by adjusting 400 mM boric acid to pH 9.5 with ca. 1.2 mL 10 M NaOH.)

CAPILLARY ELECTROPHORESIS
Capillary: 27 cm × 50 μm fused-silica (19 cm to detector) (Polymicro Technologies)
Capillary preparation: Between each run rinse capillary with 100 mM NaOH for 1 min, with water for 1 min, with running buffer for 1 min, and then equilibrate with running buffer by electrokinetic pumping for 2 min. Flush new capillaries with 1 M NaOH for 10 min, rinse with water, and rinse with running buffer.
Running buffer: MeCN:pH 9.55 sodium borate buffer (I=40 mM) containing 45 mM sodium dodecyl sulfate
Injection: Hydrodynamic injection at 0.5 psi for 2 s.
Detector: UV 340 or F ex 350 em 415
Migration time: 1.83 (L-Ser, L-His), 1.90 (L-Tyr, L-Thr), 1.93 (D-Ser, L-ala), 2.0 (D-His), 2.12 (Gly), 2.16 (L-Glu, L-Asp, D-Glu, D-Asp), 2.24 (L-Trp, L-Val, D-Ala, D-Thr), 2.28 (D-Tyr), 2.35 (L-Met), 2.48 (L-Phe), 2.66 (L-Ile), 2.83 (D-Val, D-Met), 2.90 (D-Trp), 3.0 (L-Leu), 3.29 (D-Phe), 3.34 (D-Ile), 3.41 (D-Leu), 3.97 (L-Arg), 4.12 (ammonia), 4.40 (D-Arg), 4.48 (L-Lys), 4.59 (L-Orn), 4.61 (D-Orn), 4.69 (D-Lys)
Voltage: 405 V/cm
Current: ca. 52 μA

KEY WORDS
chiral; derivatization

REFERENCE
Tivesten,A.; Folestad,S. Separation of precolumn-labelled D- and L-amino acids by micellar electrokinetic chromatography with UV and fluorescence detection, *J.Chromatogr.A*, **1995**, *708*, 323–337.

SAMPLE
Matrix: solutions

CAPILLARY ELECTROPHORESIS
Capillary: 50 cm × 50 μm (Polymicro Technologies)
Running buffer: 15 mM pH 10.0 Carbonate buffer containing 5 mM luminol and 25 mM hydrogen peroxide
Injection: Electrokinetic at 21 kV for 2 s
Detector: Chemiluminescence following post-column reaction. The capillary effluent mixed in a PEEK tee connector with 30 μM copper sulfate in 15 mM pH 10.0 carbonate buffer containing 30 μM tartaric acid delivered by gravity (at 40 cm above the buffer reservoir) through a 100 cm × 180 μm ID capillary and this mixture flowed past the detector and through a 30 cm × 180 μm ID capillary to a grounded reservoir. (The copper catalyzes the chemiluminescent reaction between luminol and hydrogen peroxide. The amino acids complex the copper and thus reduce the chemiluminescent light output.)
Migration time: 2 (arginine), 2.6 (proline), 3 (valine), 3.2 (leucine), 3.5 (glutamine), 3.8 (asparagine), 4 (serine), 5.3 (cysteine), 6 (glutamic acid), 6.5 (aspartic acid)
Voltage: 21 kV
Limit of detection: 120-440 fmole (S/N 3)

KEY WORDS
post-column reaction; indirect chemiluminescence detection

REFERENCE

Liao,S.-Y.; Whang,C.-W. Indirect chemiluminescence detection of amino acids separated by capillary electrophoresis, *J.Chromatogr.A*, **1996**, *736*, 247–254.

SAMPLE

Matrix: solutions

Sample preparation: Mix 2 µL of an amino acid solution in 50 mM pH 10.0 borate buffer with 12 µL 50 mM o-phthalaldehyde in 50 mM 50 mM pH 10.0 borate buffer containing 1.5 mM cinnamic acid, let stand for 20 min, inject an aliquot. (The two solutions can also be mixed in the inlet of the capillary and allowed to stand for 20 min before commencing the run. Inject reagent for 3 s, sample for 1 s, and reagent for 3 s.)

CAPILLARY ELECTROPHORESIS

Capillary: 72 cm × 50 µm fused-silica (50 cm to detector) (Polymicro Technologies)
Capillary temperature: 30
Running buffer: 50 mM pH 10.0 Borate buffer
Detector: UV 230
Migration time: 4.4 (Phe), 4.6 (Met), 5 (Ala), 7.5 (Glu)
Internal standard: cinnamic acid (5.5)
Voltage: 30 kV
Model: Perkin Elmer ABI-270A

KEY WORDS

derivatization

REFERENCE

Taga,A.; Honda,S. Derivatization at capillary inlet in high-performance capillary electrophoresis. Its reliability in quantification, *J.Chromatogr.A*, **1996**, *742*, 243–250.

SAMPLE

Matrix: solutions

Sample preparation: Mix 50 µL of a 1 mM amino acid solution in 100 mM pH 9.5 borate buffer with 100 µL 100 mM pH 9.5 borate buffer and 100 µL 8 mM dansyl chloride in MeCN, let stand at room temperature for 1 h, filter 0.45 µm PTFE, inject an aliquot.

CAPILLARY ELECTROPHORESIS

Capillary: 50 cm × 50 µm fused silica (41.5 cm to detector)
Capillary temperature: 25
Running buffer: N-Methylformamide containing 10 mM NaCl and 100 mM β-cyclodextrin
Injection: Pressure injection at 50 mbar for 5 s
Detector: UV 254
Migration time: 12.45 (D-Pro), 23.74 (L-Pro), 12.13 (D-Ala), 12.90 (L-Ala), 12.53 (D-Ser), 13.36 (L-Ser), 11.10 (D-Lys), 11.34 (L-Lys), 12.64 (D-Asn), 15.00 (D-Leu), 15.77 (L-Leu), 12.23 (D-Nval), 12.84 (L-Nval), 14.66 (D-Asp), 15.54 (L-Asp), 12.87 (D-Met), 13.56 (L-Met), 12.99 (D-Thr), 14.17 (L-Thr), 12.38 (D-Val), 13.08 (L-Val)
Voltage: 30 kV
Model: Hewlett-Packard HP3D

KEY WORDS

derivatization; chiral

REFERENCE

Valkó,I.E.; Sirén,H.; Riekkola,M.-L. Chiral separation of dansyl-amino acids in a non-aqueous medium by capillary electrophoresis, *J.Chromatogr.A*, **1996**, *737*, 263–272.

SAMPLE

Matrix: solutions

Sample preparation: Mix 200 µL of a 2 mM amino acid solution in 200 mM pH 9.0 borate buffer with 200 µL 10 mM 2-(9-anthryl)ethyl chloroformate, let stand for 2 min, wash with 500 µL pentane, dilute 10-fold with water, inject an aliquot. (Prepare 2-(9-anthryl)ethyl chloroformate as follows. Stir a solution of 3 g of 9-bromoanthracene in 100 mL ether at 0° under argon or nitrogen, add 9 mL 1.6 M n-butyllithium over 5 min, stir for 30 min, add an ice-cold solution of 3 g ethylene oxide (Caution! Ethylene oxide is a carcinogen!) in 16 mL ether, stir for 1 h, add 70 mL water, add 50 mL ether, remove the organic layer, extract the aqueous layer with 100 mL dichloromethane. Combine the organic layers and wash them with water, dry over anhydrous sodium sulfate, evaporate to dryness, chromatograph on silica gel with dichloromethane to give 2-(9-anthryl)ethanol as pale yellow crystals (mp 106-8°) (J. Org. Chem. 1986, 51, 2956). Stir a solution of 2-(9-anthryl)ethanol in ether in the presence of pyridine (as an HCl scavenger) at 0°, add a solution of phosgene in toluene. 2-(9-anthryl)ethyl chloroformate is obtained as colorless crystals (mp 86-87° from pentane). Protect stock solutions from light and store them in the refrigerator (Anal. Chem. 1991, 63, 292).)

CAPILLARY ELECTROPHORESIS
Capillary: 67 cm × 25 μm (46 cm to detector) (Polymicro Technologies)
Capillary preparation: Before each run flush with MeOH:200 mM NaOH 10:90 for 10 min and with water for 5 min then equilibrate with running buffer for 10 min. Rinse new capillaries with 200 mM NaOH for 2 h.
Capillary temperature: 25
Running buffer: Isopropanol:buffer 15:85 (Buffer was 50 mM pH 7.50 phosphate buffer containing 40 mM sodium dodecyl sulfate and 10 mM gamma-cyclodextrin
Injection: Pressure injection at 50 mbar for 15 s
Detector: UV 256
Migration time: 14.19 (Asp, α = 1.031), 14.23 (Glu, α = 1.030), 15.51 (Ser, α = 1.030), 16.16 (Thr, α = 1.039), 16.41 (Ala, α = 1.022), 19.90 (Val, α = 1.020), 23.70 (Ile, α = 1.053), 26.07 (Leu, α = 1.031), 31.04 (Phe, α = 1.048) [Retention time for the first eluting (L) isomer.]
Voltage: 30 kV
Current: 12 μA
Model: Lauerlabs Prince

KEY WORDS
derivatization; chiral

REFERENCE
Wan,E.; Engström,A.; Blomberg,L.G. Direct chiral separation of amino acids derivatized with 2-(9-anthryl)ethyl chloroformate by capillary electrophoresis using cyclodextrins as chiral selectors. Effect of organic modifiers on resolution and enantiomeric elution order, *J.Chromatogr.A*, **1996**, *731*, 283–292.

SAMPLE
Matrix: solutions
Sample preparation: Prepare dansylated amino acids, inject an aliquot of an aqueous solution.

CAPILLARY ELECTROPHORESIS
Capillary: 50 cm × 75 μm fused-silica (43 cm to detector) (Polymicro Technologies)
Capillary preparation: Between runs purge capillary under vacuum (15 mm Hg) with running buffer. Condition new capillaries by washing them using a 10 cm height differential with MeOH for 1 h then with running buffer for 2 h.
Running buffer: MeOH containing 15 mM tetraethylammonium acetate and 15 mM acetic acid
Injection: Hydrostatic injection at 10 cm for 5 s.
Detector: UV 214
Migration time: 11 (norleucine), 12.2 (glycine), 12.7 (alanine), 13 (threonine), 15 (glutamic acid), 15.4 (isoleucine), 15.9 (leucine), 16.1 (asparagine), 17.2 (proline), 19.5 (hydroxyproline), 22.7 (tyrosine)
Voltage: -20 kV
Model: Waters Quanta 4000

KEY WORDS
derivatization

REFERENCE
Salimi-Moosavi,H.; Cassidy,R.M. Selectivity control in the non-aqueous capillary electrophoretic separation of amino acids, *J.Chromatogr.A*, **1997**, *790*, 185–193.

SAMPLE
Matrix: solutions
Sample preparation: Inject an aliquot of an aqueous solution.

CAPILLARY ELECTROPHORESIS
Capillary: 50 cm × 75 μm fused-silica (43 cm to detector) (Polymicro Technologies)
Capillary preparation: Between runs purge capillary under vacuum (15 mm Hg) with running buffer. Condition new capillaries by washing them using a 10 cm height differential with MeOH for 1 h then with running buffer for 2 h.
Running buffer: MeOH containing 10 mM tetraethylammonium hydroxide
Injection: Hydrostatic injection at 10 cm for 5 s.
Detector: UV 214
Migration time: 7.3 (alanine), 7.5 (lysine), 8.4 (isoleucine), 8.7 (threonine), 10.5 (citrulline)
Voltage: 20 kV
Model: Waters Quanta 4000

KEY WORDS
indirect UV detection

REFERENCE
Salimi-Moosavi,H.; Cassidy,R.M. Selectivity control in the non-aqueous capillary electrophoretic separation of amino acids, *J.Chromatogr.A*, **1997**, *790*, 185–193.

SAMPLE
Matrix: solutions
Sample preparation: Mix a 200 μL aliquot of a solution in 200 mM pH 8.6 borate buffer with enough 10 mM reagent in acetone to give a 10-fold excess of reagent, let stand for 10 min, wash with 500 μL n-hexane, dilute aqueous layer with water so that buffer concentration is <20 mM, inject an aliquot. (The reagent was (+)-APOC ((+)-1-(9-anthryl)-2-propyl chloroformate. Preparation is by analogy with the preparation of 2-(9-anthryl)ethyl chloroformate (Anal. Chem. 1991, 63, 292). Stir a solution of 3 g of 9-bromoanthracene in 100 mL ether at 0° under argon or nitrogen, add 9 mL 1.6 M n-butyllithium over 5 min, stir for 30 min, add an ice-cold solution of 4 g propylene oxide (Caution! Propylene oxide is a carcinogen!) in 16 mL ether, stir for 35 min, add aqueous ammonium chloride, extract the aqueous layer with dichloromethane. Purify by flash chromatography to obtain racemic 2-(9-anthryl)propanol. Prepare the ester by reaction with (-)-camphanic acid chloride and separate the diastereomers by HPLC. Hydrolyze the appropriate diastereomer with acid to obtain (+)-2-(9-anthryl)propanol. Stir a solution of (+)-2-(9-anthryl)propanol in ether in the presence of pyridine (as an HCl scavenger) at 0°, add a solution of phosgene in toluene to obtain (+)-2-(9-anthryl)propyl chloroformate).

CAPILLARY ELECTROPHORESIS
Capillary: 70 cm × 21 μm fused-silica (62 cm to detector) (Polymicro Technologies)
Capillary preparation: Before each run rinse with five column volumes of 100 mM NaOH and five column volumes of water then equilibrate with 10 column volumes of running buffer.
Running buffer: 20 mM pH 9.95 Borax buffer containing 15 mM sodium dodecyl sulfate
Injection: Hydrodynamic injection at 11 cm for 20-80 s.
Detector: UV 256; F ex 351 (argon ion laser) em 412
Migration time: 7.2 (D-Glu), 7.25 (L-Glu), 7.4 (D-Asp), 7.5 (L-Asp), 8.05 (L-Ser, L-Gln), 8.2 (L-Ala), 8.3 (L-Asn). 8.4 (L-Pro, D-Ser, L-Thr), 8.45 (D-Gln), 8.5 (D-Asn), 8.6 (D-Ala), 8.65 (D-Thr), 8.75 (Gly), 9.4 (L-Val), 9.65 (L-Tyr), 9.8 (D-Val, L-norvaline), 10 (L-Met), 10.08 (D-Tyr, D-norvaline), 10.1 (D-Met), 10.25 (D-Pro), 10.5 (L-Ile), 10.75 (D-Ile), 10.95 (L-Leu, D-Leu), 11.65 (L-Phe), 11.85 (D-Phe), 12.15 (L-Trp), 12.3 (D-Trp)
Voltage: 30 kV
Model: laboratory-constructed
Limit of detection: 0.5 amole (F), 2 fmole (UV)

KEY WORDS
chiral; derivatization

REFERENCE
Thorsén,G.; Engstrom,A.; Josefssön,B. Enantiomeric determination of amino compounds with high sensitivity using the chiral reagents (+)- and (-)-1-(9-anthryl)-2-propyl chloroformate, *J.Chromatogr.A*, **1997**, *786*, 347–354.

SAMPLE
Matrix: solutions

Sample preparation: Mix 180 μL of an amino acid solution in buffer with 30 μL 60 mM o-phthaldialdehyde in MeCN containing 120 mM 2,3,4,6-tetra-O-acetyl-1-thio-β-D-glucopyranose, let stand for 6 min, inject an aliquot. (Prepare buffer by adjusting the pH of 400 mM boric acid to 9.5 with 10 M NaOH.)

CAPILLARY ELECTROPHORESIS
Capillary: 56 cm × 50 μm fused-silica (39 cm to detector) (Polymicro Technologies)
Capillary preparation: Between runs rinse capillary with 100 mM NaOH for 1 min, with water for 1 min, and with running buffer for 1 min. Equilibrate with voltage on for 2 min before injection. Rinse new capillaries with 1 M NaOH for 10 min and then with water.
Running buffer: MeCN:buffer 4:96 (Prepare running buffer by dissolving 1.12 moles sodium dodecyl sulfate in a small amount of borate buffer, adding 1 mL MeCN, and making up to 25 mL with borate buffer. Prepare borate buffer by mixing 37.5 mL 40 mM borax solution with 10.6 mL 100 mM NaOH and making up to 100 mL with water, pH 9.50-9.55.)
Injection: Hydrodynamic injection at 80 mbar for 5 s.
Detector: UV 235, UV 340
Migration time: 7 (L-Ala), 8.5 (D-Thr), 9 (L-Met), 11.5 (L-Leu), 13.5 (D-Leu), 15.5 (L-Lys)
Voltage: 15 kV
Current: ca. 30 μA
Model: laboratory-constructed
Limit of detection: 1-2 μM (UV 235), 7-8 μM (UV 340)

KEY WORDS
derivatization; chiral; comparison with other reagents

REFERENCE
Tivesten,A.; Folestad,S. Chiral o-phthaldialdehyde reagents for fluorogenic on-column labeling of D- and L-amino acids in micellar electrokinetic chromatography, *Electrophoresis*, **1997**, *18*, 970–977.

SAMPLE
Matrix: solutions
Sample preparation: Mix 50 μL of a 1 mM solution of the amino acid in pH 9.2 phosphate buffer with 10 μL 2 mg/mL FluoroLink Cy5 Mono Reactive Dye (Amersham) in dried DMF, let stand at room temperature for 1 h, dilute, inject an aliquot.

CAPILLARY ELECTROPHORESIS
Capillary: 60 cm × 50 μm fused-silica (Polymicro Technologies)
Running buffer: 10 mM pH 7.0 Phosphate buffer containing 1% poly(vinylpyrrolidone) (molecular mass 40000; Kishida, Tokyo) and 70 mM gamma-cyclodextrin
Injection: Gravity at 4 cm for 5-10 s.
Detector: F ex 635 (<5 mW diode laser)
Migration time: 18.3, 18.4 (glutamic acid enantiomers), 28.8, 29 (alanine enantiomers), 29.3, 29.5 (valine enantiomers), 32.3, 32.5 (phenylalanine enantiomers), 33.2, 33.5 (tyrosine enantiomers), 34.8, 35.8 (tryptophan enantiomers)
Voltage: 25 kV
Current: 9 μA
Model: laboratory-constructed
Limit of detection: 60 nM (S/N = 3)

KEY WORDS
chiral; derivatization; detector at anode

REFERENCE
Kaneta,T.; Shiba,H.; Imasaka,T. Determination of cyanine-labeled amino acid enantiomers by cyclodextrin-modified capillary gel electrophoresis combined with diode laser fluorescence detection, *J.Chromatogr.A*, **1998**, *805*, 295–300.

SAMPLE
Matrix: tissue
Sample preparation: Heat 10-30 mg tissue protein in 10 mL 6 M HCl at 120° for 20 h, dry with air, reconstitute with water, dry with air, reconstitute with 2 mL water, add 10 μmoles Nor, add 5 μmoles 3,4-dimethoxyphenylammonium chloride. Remove a 50 μL aliquot and add it to

1 mL 25 mM dansyl chloride in MeCN and 1 mL 40 mM pH 9.5 lithium carbonate, let stand in the dark for 2 h, add 100 μL 4% ethylamine, dry with air, reconstitute with MeOH:water 20:80, centrifuge at 2000 g for 2 min, inject an aliquot.

CAPILLARY ELECTROPHORESIS
Capillary: 72 cm × 50 μm fused-silica (52 cm to detector) (J&W Scientific)
Capillary preparation: Before each run wash capillary with 1 M NaOH for 2 min, with water for 2 min, and with running buffer for 5 min.
Capillary temperature: 25
Running buffer: 100 mM pH 8.3 Boric acid containing 150 mM sodium dodecyl sulfate
Injection: Vacuum injection for 1 s (4.5 nL)
Detector: UV 216
Migration time: 18.9 (Thr), 19.1 (Asn), 19.2 (Ser), 19.5 (Gln), 20.2 (Ala), 20.6 (Glu), 21.3 (Asp), 21.6 (Gly), 23.2 (Val), 25.5 (Pro), 26.0 (Met), 28.6 (Ile), 29.7 (Leu), 33.0 (Phe), 33.7 (Trp), 43.6 (Arg), 44.9 (Lys, didansyl), 45.3 (Tryptamine), 46.0 (Tyr, didansyl)
Internal standard: norvaline (25.3)
Voltage: 15 kV
Model: Applied Biosystems Model 270 A-HT
Limit of detection: 0.1-1.3 μM

KEY WORDS
derivatization; skin; mink

REFERENCE
Michaelsen,S.; Moller,P.; Sorensen,H. Analysis of dansyl amino acids in feedstuffs and skin by micellar electrokinetic capillary chromatography, *J.Chromatogr.A*, **1994**, *680*, 299–310.

SAMPLE
Matrix: tissue
Sample preparation: Sonicate (Artix Sonic Dismembrator Model 150, power setting 30) 1-2 mg rat brain tissue and 10 μL EtOH:water 70:30 in an ice bath for 5-10 s, centrifuge at 16000 g for 10 min. 1 μL Supernatant + 1 μL 490 μM α-aminoadipic acid, mix, remove a 1 μL aliquot and add it to 5 μL 20 mM pH 9.0 sodium borate buffer, add 1.5 μL 20 mM NaCN in water, mix, add 1.5 μL 20 mM naphthalene-2,3-dicarboxaldehyde in MeCN, mix thoroughly, let stand at room temperature for 30 min (protect from light), inject an aliquot.

CAPILLARY ELECTROPHORESIS
Capillary: 115 cm × 50 μm fused-silica (95 cm to detector)
Running buffer: 20 mM pH 9.0 sodium borate buffer
Injection: Electrokinetic injection at 5 kV for 12 s
Detector: UV 420
Migration time: 14.0 (Trp), 14.3 (Phe), 14.4 (Tyr), 14.7 (Asn), 15.1 (Ser), 15.4 (Ala), 15.8 (taurine), 16.0 (Gly), 22.7 (Glu), 24.3 (Asp)
Internal standard: α-aminoadipic acid (21.6)
Voltage: 30 kV
Limit of detection: 3.5 μM (Ser), 3.7 μm (Ala), 4.3 μM (Gly), 8.7 μM (Glu), 10.2 μM (Asp)

OTHER SUBSTANCES
Extracted: gamma-aminobutyric acid, dopamine, levodopa, norepinephrine, phospho-ethanolamine

KEY WORDS
rat; brain; derivatization

REFERENCE
Weber,P.L.; O'Shea,T.J.; Lunte,S.M. Separation and quantitation of the amino acid neurotransmitters in rat brain by capillary electrophoresis, *J.Pharm.Biomed.Anal.*, **1994**, *12*, 319–324.

SAMPLE
Matrix: urine
Sample preparation: Filter urine, dilute 10-fold with water, 500 μL diluted urine + 100 μL 400 mM pH 9.3 sodium borate buffer + 400 μL 5 mM 9-fluorenylmethyl chloroformate in MeCN,

mix for 1 min, wash twice with 2 mL pentane, dilute aqueous layer 100-fold with water, inject an aliquot.

CAPILLARY ELECTROPHORESIS
Capillary: 70 cm × 50 μm fused-silica (60 cm to detector) (Polymicro Technologies)
Running buffer: 20 mM pH 9.2 borate buffer containing 25 mM sodium dodecyl sulfate
Injection: Gravity injection at 10 cm for 20 s
Detector: F ex 248 (laser) em 315 (filters ARC No. 310-B-1D (FWHM = 60 nm) bandpass and Melles Griot WG-306 cut-off)
Migration time: 9.7 (alanine), 15.8 (arginine), 11.5 (aspartic acid), 15.6 (cystine), 11.1 (glutamic acid), 9.9 (glycine), 16.1 (histidine), 8.7 (hydroxyproline), 11.8 (isoleucine), 12.4 (leucine), 16.1 (lysine), 11.3 (methionine), 13.5 (phenylalanine), 10.4 (proline), 9.3 (serine), 9.5 (threonine), 10.8 (tyrosine), 10.5 (valine)
Limit of detection: 0.5 nM (S/N 2)

KEY WORDS
derivatization

REFERENCE
Chan,K.C.; Janini,G.M.; Muschik,G.M.; Issaq,H.J. Laser-induced fluorescence detection of 9-fluorenylmethyl chloroformate derivatized amino acids in capillary electrophoresis, *J.Chromatogr.A*, **1993**, *653*, 93–97.

SAMPLE
Matrix: urine
Sample preparation: Inject an aliquot directly (?).

CAPILLARY ELECTROPHORESIS
Capillary: 80 cm × 25 μm fused-silica (Polymicro Technologies)
Running buffer: 50 mM NaOH
Injection: Electromigration at 20 kV for 3 s (1.6 nL).
Detector: E, Bioanalytical Systems Model LC-3, 127 μm Cu wire (construction details in paper) +0.60 V, Ag/AgCl reference electrode, Pt wire counter electrode
Migration time: 10.3 (arginine), 17.3 (tryptophan), 17.6 (lysine), 18.3 (phenylalanine), 18.7 (histidine), 18.9 (glutamine), 19.1 (leucine, isoleucine), 19.3 (methionine), 20.1 (valine), 20.3 (threonine), 20.7 (asparagine), 20.8 (proline), 22.7 (serine), 23.5 (alanine), 28.7 (glycine), 29.8 (cystine), 34.0 (tyrosine) [Some peaks overlap.]
Voltage: 20
Model: laboratory-constructed
Limit of detection: 0.8-640 fmole

OTHER SUBSTANCES
Noninterfering: aspartic acid, cysteine, glutamic acid

REFERENCE
Ye,J.; Baldwin,R.P. Determination of amino acids and peptides by capillary electrophoresis and electrochemical detection at a copper electrode, *Anal.Chem.*, **1994**, *66*, 2669–2674.

SAMPLE
Matrix: urine
Sample preparation: Dilute urine 1:10 with water, inject an aliquot.

CAPILLARY ELECTROPHORESIS
Capillary: 90 cm × 75 μm fused-silica (Polymicro Technologies)
Running buffer: 15 mM pH 9.7 Borate buffer containing 10 mM sodium dodecyl sulfate
Injection: Hydrodynamic injection
Detector: F ex 345 em 455 following post-column reaction. The effluent from the column mixed with the reagent (under 180 mbar pressure) and the mixture flowed through a 22 cm × 50 μm capillary to the detector. (Reagent was 1.5 mL 25 mg/mL o-phthalaldehyde in MeOH, 18 μL mercaptoethanol, 12.5 mL 100 mM pH 10 borax buffer, and 5 mL MeOH made up to 25 mL with water.)
Migration time: 6.7 (Lys), 7 (Arg), 8.2 (Ala), 8.3 (Leu), 8.4(Val), 8.5 (Trp), 9 (Gly), 9.2 (Phe), 9.4 (Met), 10 (Thr), 10.2 (Ser), 13.5 (Glu), 14 (Asp)

Voltage: 20 kV
Model: Lauer labs PRINCE
Limit of detection: 2-4 μM

KEY WORDS
post-column reaction

REFERENCE
Zhu,R.; Kok,W.T. Post-column reaction system for fluorescence detection in capillary electrophoresis, *J.Chromatogr.A*, **1995**, *716*, 123–133.

SAMPLE
Matrix: urine
Sample preparation: Filter (0.45 μm cellulose-acetate), dilute an aliquot of the filtrate with 4 volumes of water, inject an aliquot.

CAPILLARY ELECTROPHORESIS
Capillary: 77 cm × 100 μm fused-silica (Polymicro Technologies)
Capillary preparation: Before use flush capillary with running buffer at 0.5 psi for 3 min, between runs rinse with running buffer for 1.5 min.
Capillary temperature: 25
Running buffer: 5 mM pH 10.7 Carbonate buffer containing 2 mM benzoate and 150 μM myristyltrimethylammonium bromide
Injection: Pressure injection at 0.5 psi for 5 s (57 nL).
Detector: UV 230
Migration time: 9.6 (Asp), 10 (Glu), 10.4 (Cys), 11.9 (uric acid), 14 (Gly), 14.5 (Ser), 15 (Ala), 15.2 (Gln), 16.1 (hydroxyproline), 16.5 (His), 17.2 (Thr), 17.8 (Met), 18 (Val), 18.5 (Orn), 19.3 (Asn), 19.8 (Ile), 20.2 (Leu)
Voltage: 15 kV
Model: Beckman P/ACE 2050
Limit of detection: 65-687 pg

KEY WORDS
indirect UV detection

REFERENCE
Chen,H.; Xu,Y.; Ip,M.P.C. Determination of amino acids in urine by capillary electrophoresis with indirect UV detection, *J.Liq.Chromatogr.Rel.Technol.*, **1997**, *20*, 2475–2493.

Aminobenzoic acid

Molecular formula: $C_7H_7NO_2$
Molecular weight: 137.14
CAS Registry No.: 150-13-0
Merck Index (12th ed.): 443

SAMPLE
Matrix: bulk
Sample preparation: 10-50 mg Bulk compound + 1 mL 1 mg/mL caffeine in 10 mM HCl, make up to 10 mL with 10 mM HCl, sonicate for 2 min, mix thoroughly, filter (0.45 μm cellulose acetate), inject an aliquot.

CAPILLARY ELECTROPHORESIS
Capillary: 75 cm × 75 μm fused-silica (50 cm to the detector) (ISCO)
Capillary preparation: Flush with running buffer for 2 min between analyses. Condition capillary by filling with 1 M NaOH and allowing to stand for 1 h, fill with 100 mM NaOH and allow to stand for 1 h, wash with water, fill with running buffer. Each week wash capillary

with 100 mM HCl for 10 min, with water, with 100 mM NaOH, and with water then fill with running buffer.
Capillary temperature: 30
Running buffer: DMSO:ethanolamine:buffer 11:1:88 (Buffer was 0.92 g cetyltrimethylammonium bromide in 100 mL 10 mM sodium tetraborate, adjust pH to 11.5 with 1 M NaOH.)
Injection: Load under vacuum, vacuum level 2, 10.0 kPa s
Detector: UV 254
Migration time: 5.7
Internal standard: caffeine (6.5)
Voltage: -15 kV
Model: ISCO Model 3140 electropherograph

OTHER SUBSTANCES
Simultaneous: amphetamine, ephedrine, methamphetamine, methylenedioxyamphetamine, methylenedioxymethamphetamine, norephedrine, pseudoephedrine, pseudonorephedrine

REFERENCE
Aumatell,A.; Wells,R.J. Determination of a cardiac antiarrhythmic, tricyclic antipsychotics and antidepressants in human and animal urine by micellar electrokinetic capillary chromatography using a bile salt, *J.Chromatogr.B*, **1995**, *669*, 331–344.

SAMPLE
Matrix: bulk
Sample preparation: 10-50 mg Bulk compound + 1 mL 1 mg/mL caffeine in 10 mM HCl, make up to 10 mL with 10 mM HCl, sonicate for 2 min, mix thoroughly, filter (0.45 μm cellulose acetate), inject an aliquot.

CAPILLARY ELECTROPHORESIS
Capillary: 75 cm × 75 μm fused-silica (50 cm to the detector) (ISCO)
Capillary preparation: Flush with running buffer for 2 min between analyses. Condition capillary by filling with 1 M NaOH and allowing to stand for 1 h, fill with 100 mM NaOH and allow to stand for 1 h, wash with water, fill with running buffer. Each week wash capillary with 100 mM HCl for 10 min, with water, with 100 mM NaOH, and with water then fill with running buffer.
Capillary temperature: 30
Running buffer: DMSO:ethanolamine:buffer 11:1:88 (Buffer was 0.92 g cetyltrimethylammonium bromide in 100 mL 10 mM sodium tetraborate, adjust pH to 11.5 with 1 M NaOH.)
Injection: Load under vacuum, vacuum level 2, 10.0 kPa s
Detector: UV 254
Migration time: 5.7
Internal standard: caffeine (6.5)
Voltage: -15 kV
Model: ISCO Model 3140 electropherograph

OTHER SUBSTANCES
Simultaneous: amphetamine, ephedrine, methamphetamine, methylenedioxyamphetamine, methylenedioxymethamphetamine, norephedrine, pseudoephedrine, pseudonorephedrine

REFERENCE
Trenerry,V.C.; Robertson,J.; Wells,R.J. Analysis of illicit amphetamine seizures by capillary electrophoresis, *J.Chromatogr.A*, **1995**, *708*, 169–176.

SAMPLE
Matrix: cell cultures
Sample preparation: Centrifuge cell culture at 13000 rpm for 10 min. Remove a 900 μL aliquot of the supernatant and add it to 100 μL 200 μg/mL sulfadiazine, mix, inject an aliquot.

CAPILLARY ELECTROPHORESIS
Capillary: 65 cm × 50 μm fused-silica (45 cm to detector)
Capillary preparation: Wash with 100 μL 100 mM NaOH for 1 min, flush with 100 μL running buffer for 1 min, allow to equilibrate under voltage for 5 min.
Running buffer: 50 mM pH 9.1 disodium tetraborate

Injection: Hydrodynamic for 5 s
Detector: UV 270
Migration time: 7.1
Internal standard: sulfadiazine (4.7)
Voltage: 24 kV
Model: Isco Model 3850
Limit of detection: 200 ng/mL (S/N 3)

OTHER SUBSTANCES
Extracted: thymidine, uracil

REFERENCE
Richards,R.M.E.; Xing,D.K.L. Determination of thymidine, uracil and p-aminobenzoic acid in bacteriological cultures by capillary zone electrophoresis, *J.Pharm.Biomed.Anal.*, **1994**, *12*, 1063–1068.

SAMPLE
Matrix: solutions

CAPILLARY ELECTROPHORESIS
Capillary: 34 cm × 50 μm (25.5 cm (A) or 8.5 cm (B, C) to detector) (Composite Metal Services, Hallow, UK)
Running buffer: 50 mM pH 2.5 Phosphate buffer (A, B) or 250 mM pH 2.5 phosphate buffer (C)
Injection: Pressure injection at 50 mbar for 5 s.
Detector: UV 200
Migration time: 1.9 (A), 0.78 (B), 3 (C)
Voltage: 20 kV (A) or -20 kV (B) or -7.5 kV (C)
Model: Hewlett Packard 3D

OTHER SUBSTANCES
Simultaneous: albuterol, imidazole, lamivudine

KEY WORDS
when the distance to the detector was 8.5 cm (B; C) injection was at the normal outlet end and voltage polarity and pressure for injection were reversed.

REFERENCE
Altria,K.D.; Kelly,M.A.; Clark,B.J. The use of a short-end injection procedure to achieve improved performance in capillary electrophoresis, *Chromatographia*, **1996**, *43*, 153–158.

SAMPLE
Matrix: solutions

CAPILLARY ELECTROPHORESIS
Capillary: 27 cm × 75 μm
Capillary preparation: Before each run rinse capillary with 100 mM NaOH for 30 s and with running buffer for 30 s. Condition new capillaries by rinsing with 100 mM NaOH for 20 min.
Capillary temperature: 30
Running buffer: 15 mM Sodium borate
Injection: Pressure injection of sample at 25 mbar for 1 s followed by running buffer at 25 mbar for 1 s.
Detector: UV 200
Migration time: 3.5
Voltage: 6.5 kV
Model: Beckman

OTHER SUBSTANCES
Simultaneous: aspirin, bacitracin, beclomethasone, benzoic acid, ceftizoxime, ceftriaxone, cefuroxime, cephalothin, cromolyn, embonic acid, epoprostenol, glyburide (glibenclamide), levothyroxine, nedocromil, nystatin, omeprazole, prednisolone, warfarin, zidovudine

REFERENCE
Altria,K.D.; Bryant,S.M.; Hadgett,T.A. Validated capillary electrophoresis method for the analysis of a range of acidic drugs and excipients, *J.Pharm.Biomed.Anal.*, **1997**, *15*, 1091–1101.

SAMPLE
Matrix: solutions

CAPILLARY ELECTROPHORESIS
Capillary: 37 cm × 50 μm eCAP polyamine-coated (30 cm to detector) (Beckman)
Capillary preparation: At the end of each run flush with 1 M NaOH for 2 min, wash with regenerator (Beckman) for 2 min, wash with running buffer for 2 min. Condition a new capillary with running buffer for 2 min and with regenerator for 2 min.
Capillary temperature: 30
Running buffer: 50 mM pH 8 Tris
Detector: UV 214
Migration time: 0.8
Voltage: 30 kV
Model: Beckman P/ACE 2210

OTHER SUBSTANCES
Simultaneous: albuterol, imidazole, lamivudine

KEY WORDS
coated column; comparison with uncoated column

REFERENCE
Assi,K.A.; Altria,K.D.; Clark,B.J. Rapid resolution of drugs and related substances with an eCAP™ polyamine coated capillary, *J.Pharm.Biomed.Anal.*, **1997**, *15*, 1041–1049.

Aminoglutethimide

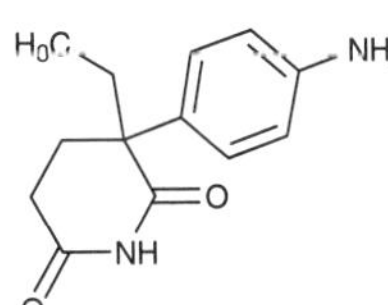

Molecular formula: $C_{13}H_{16}N_2O_2$
Molecular weight: 232.28
CAS Registry No.: 125-84-8
Merck Index (12th ed.): 460
Lednicer: 1 257

SAMPLE
Matrix: solutions
Sample preparation: Prepare a 280 μg/mL solution in running buffer, sonicate for 30 min, inject an aliquot.

CAPILLARY ELECTROPHORESIS
Capillary: 50 cm × 50 μm fused-silica (40 cm to detector) (Polymicro Technologies)
Capillary preparation: Rinse capillary with 10 mM NaOH and water at regular intervals. At the beginning of each day rinse capillary with 100 mM NaOH and water, equilibrate with running buffer at 10 kV for several min.
Running buffer: 10 mM NaH_2PO_4 containing 6 mM sodium borate and 5 mM gamma-cyclodextrin, pH 3
Injection: Hydrostatic injection at 10 cm for 10 s
Detector: UV 205
Migration time: 20.2, 20.7 (enantiomers)
Voltage: 15 kV
Model: Laboratory constructed

KEY WORDS
chiral

REFERENCE

Anigbogu,V.C.; Copper,C.L.; Sepaniak,M.J. Separation of stereoisomers of aminoglutethimide using three capillary electrophoretic techniques, *J.Chromatogr.A*, **1995**, *705*, 343–349.

SAMPLE

Matrix: solutions
Sample preparation: Inject an aliquot of a 0.2-1 µg/mL solution.

CAPILLARY ELECTROPHORESIS

Capillary: 57 cm × 50 µm fused-silica coated with a 0.025 µm film of p-methylbenzoyl cellulose (50 cm to detector) (Polymicro Technologies)
Capillary preparation: Coat capillary as follows. Coat at 0.3 mbar and 35-40° with a filtered 0.2% solution of p-methylbenzoyl cellulose in THF using the static method (further details in K. Grob. Making and Manipulating Capillary Columns for Gas Chromatography, Hüthig, Heidelberg, 1986). Prepare p-methylbenzoyl cellulose as follows. Suspend 100 g of microcrystalline cellulose in a mixture of 1 L pyridine, 420 mL triethylamine, and 2 g dimethylaminopyridine, slowly add 265 mL p-toluoyl chloride (p-methylbenzoyl chloride) at room temperature, stir at 120° under nitrogen for 12 h, cool to RT, treat the solid mass with 4 L MeOH, dissolve in dichloromethane, precipitate twice in EtOH, dry under vacuum at 100° for 2 days (cf. Chirality 1991, 3, 43).
Running buffer: MeCN:40 mM pH 7 phosphate buffer 20:80
Injection: Pressure injection at 35 mbar for 2-5 s.
Detector: UV 214
Migration time: 19.5, 20.5 (enantiomers)
Voltage: 30 kV
Model: Beckman P/ACE 5510

KEY WORDS

electrochromatography; coated capillary; chiral

REFERENCE

Francotte,E.; Jung,M. Enantiomer separation by open-tubular liquid chromatography and electrochromatography in cellulose capillaries, *Chromatographia*, **1996**, *42*, 521–527.

SAMPLE

Matrix: solutions
Sample preparation: Inject an aliquot of a solution in running buffer.

CAPILLARY ELECTROPHORESIS

Capillary: 60 cm × 75 µm fused-silica (52.4 cm to detector)
Capillary preparation: After each run flush with 500 mM KOH for 2-3 min then with water.
Running buffer: 10 mM pH 3.8 Phosphate buffer containing 2% sulfated cyclodextrin (ds 7-10)
Injection: Hydrostatic injection.
Detector: UV 214
Migration time: 8.75, 9.48 (enantiomers)
Voltage: 15 kV
Model: Waters Quanta 4000

OTHER SUBSTANCES

Also analyzed: acebutolol, alprenolol, brompheniramine, bupivacaine, bupropion, canadine, carbinoxamine, chloroquine, chlorpheniramine, dimethindene, disopyramide, doxylamine, hydroxychloroquine, idazoxan, isoxsuprine, ketamine, mepenzolate, mepivacaine, methoxyphenamine, mexiletine, midodrine, nefopam, orphenadrine, oxprenolol, oxyphencyclimine, pheniramine, phensuximide, pindolol, piperoxan, terbutaline, tetramisole, tolperisone, tranylcypromine, trihexyphenidyl, trimipramine, verapamil, warfarin

KEY WORDS

chiral; detector at anode

REFERENCE

Stalcup,A.M.; Gahm,K.H. Application of sulfated cyclodextrins to chiral separations by capillary zone electrophoresis, *Anal.Chem.*, **1996**, *68*, 1360–1368.

SAMPLE
Matrix: solutions
Sample preparation: Inject an aliquot of a 100 µg/mL solution in MeOH:water 10:90.

CAPILLARY ELECTROPHORESIS
Capillary: 60 cm × 50 µm untreated AccuSep (52.5 cm to detector) (Waters)
Capillary preparation: Purge with fresh running buffer for 3 min between runs. Before use purge with vacuum with 500 mM NaOH for 5 min, with water for 10 min, and with running buffer for 10 min.
Capillary temperature: 30
Running buffer: 25 mM pH 9.25 Na_2HPO_4 containing 25 mM sodium tetraborate and 80 mM (S)-dodecoxycarbonylvaline ((S)-DDCV) (Prepare (S)-N-dodecoxycarbonylvaline as follows. Prepare dodecyl chloroformate by reacting 1-dodecanol with 0.33 equivalents of triphosgene in dichloromethane solution in the presence of pyridine. Add dodecyl chloroformate dropwise to (S)-valine in 1 M NaOH solution, filter, wash with hexane, recrystallize from ether/petroleum ether (J.Chromatogr.A 1994, 680, 125).)
Injection: Hydrostatic injection at 10 cm for 15 s.
Detector: UV 214
Migration time: 13.4, 13.8 (enantiomers)
Voltage: 16 kV
Model: Waters Quanta 4000E

OTHER SUBSTANCES
Simultaneous: cyclodexylaminoglutethimide, glutethimide, phenglutarimide, pyridoglutethimide

KEY WORDS
chiral

REFERENCE
Swartz,M.E.; Mazzeo,J.R.; Grover,E.R.; Brown,P.R.; Aboul-Enein,H.Y. Separation of piperidine-2,6-dione drug enantiomers by micellar electrokinetic capillary chromatography using synthetic chiral surfactants, J.Chromatogr.A, **1996**, *724*, 307–316.

Aminohippuric acid

Molecular formula: $C_9H_{10}N_2O_3$
Molecular weight: 194.19
CAS Registry No.: 61-78-9, 94-16-6 (sodium salt)
Merck Index (12th ed.): 462

SAMPLE
Matrix: solutions

CAPILLARY ELECTROPHORESIS
Capillary: 42 cm × 50 µm
Capillary temperature: 24
Running buffer: 175 mM Boric acid adjusted to pH 9.4 with 2 M NaOH
Injection: Pressure injection for 8 s
Detector: UV 254
Migration time: 8.8
Internal standard: 3-isobutyl-1-methylxanthine (7.8)
Voltage: 8 kV
Model: Beckman Model 2000

OTHER SUBSTANCES
Simultaneous: iohexol, iopamidol, iothalamic acid

REFERENCE

Shihabi,Z.K.; Rocco,M.V.; Hinsdale,M.E. Analysis of the contrast agent *iopamidol* in serum by capillary electrophoresis, *J.Liq.Chromatogr.*, **1995**, *18*, 3825–3831.

Aminopyrine

Molecular formula: $C_{13}H_{17}N_3O$
Molecular weight: 231.30
CAS Registry No.: 58-15-1, 94442-12-3 (bicamphorate), 6170-29-2 (HCl)
Merck Index (12th ed.): 495
Lednicer: 1 234

SAMPLE

Matrix: solutions
Sample preparation: Inject an aliquot of a 30-300 µg/mL solution in water.

CAPILLARY ELECTROPHORESIS

Capillary: 36 cm × 50 µm fused-silica (28 cm to detector)
Running buffer: 80 mM Reagent, 120 mM sodium dodecyl sulfate, 900 mM 1-butanol, and 10 mM borate in water where reagent is heptane (A) or octane (B) or 1-butyl chloride (C)
Injection: Hydrostatic injection.
Detector: UV 254
Migration time: 6.366 (A), 6.444 (B), 5.946 (C)
Voltage: 12 kV
Current: ca. 65 µA
Model: Waters Quanta 4000

OTHER SUBSTANCES

Simultaneous: caffeine, phenacetin, phenobarbital

REFERENCE

Fu,X.; Lu,J.; Zhu,A. Microemulsion electrokinetic chromatographic separation of antipyretic analgesic ingredients, *J.Chromatogr.A*, **1996**, *735*, 353–356.

SAMPLE

Matrix: urine
Sample preparation: Dilute urine 1:5 with running buffer, filter (2 µm), inject an aliquot.

CAPILLARY ELECTROPHORESIS

Capillary: 52 cm × 50 µm fused-silica
Running buffer: 4 mM pH 4.82 Phosphate buffer
Injection: Electromigration at 12 kV for 10 s
Detector: E, Ensman EI-400 dual microelectrode potentiostat, electrodes are 30 µm carbon fibers protruding 1-2 mm (Anal.Chem. 1988, 60, 258), electrode potential 0.9 V, pretreat electrode at 1.5 V in pH 5 buffer for 30 s then maintain in buffer for 5 min, Ag/AgCl reference electrode or UV 234
Migration time: 8
Voltage: 12 kV
Limit of detection: 0.40 pg (E), 142 pg (UV)

OTHER SUBSTANCES

Extracted: metabolites, 4-aminoantipyrine

REFERENCE

Zhou,W.; Liu,J.; Wang,E. Determination of aminopyrine and its metabolite by capillary electrophoresis-electrochemical detection, *J.Chromatogr.A*, **1995**, *715*, 355–360.

Amiodarone

Molecular formula: $C_{25}H_{29}I_2NO_3$
Molecular weight: 645.32
CAS Registry No.: 1951-25-3
Merck Index (12th ed.): 504
Lednicer: 4 127, 156

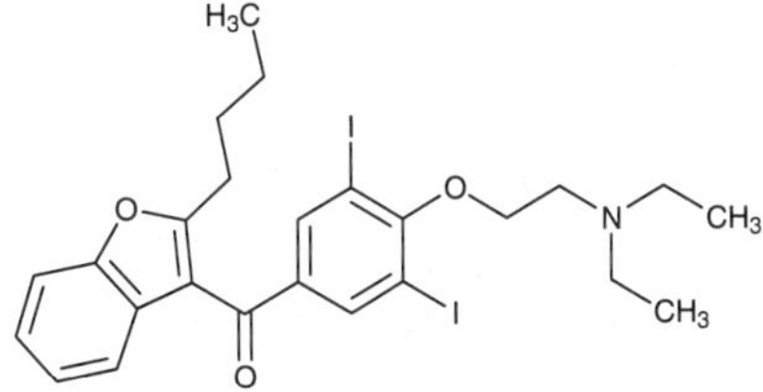

SAMPLE
Matrix: solutions
Sample preparation: Inject an aliquot of a 100 μg/mL solution in 50 mM sodium dodecyl sulfate.

CAPILLARY ELECTROPHORESIS
Capillary: 57 cm × 75 μm fused-silica (50 cm to detector) (Beckman)
Capillary preparation: Rinse capillary with 100 mM NaOH and running buffer between each run.
Capillary temperature: 30
Running buffer: Isopropanol:100 mM pH 8.1 borate buffer containing 50 mM sodium dodecyl sulfate 10 :90
Injection: Pressure injection for 5 s
Detector: UV 200
Migration time: 23.5
Voltage: +25 kV
Model: Beckman P/ACE 5510

OTHER SUBSTANCES
Simultaneous: bretylium, disopyramide, lidocaine, phenytoin, propafenone, quinidine

REFERENCE
Bretnall,A.E.; Clarke,G.S. Investigation and optimisation of the use of organic modifiers in micellar electrokinetic chromatography, *J.Chromatogr.A*, **1995**, *716*, 49–55.

SAMPLE
Matrix: solutions
Sample preparation: Prepare a solution in MeOH, inject an aliquot.

CAPILLARY ELECTROPHORESIS
Capillary: 40 cm × 50 μm fused-silica (35.4 cm to detector) (Polymicro Technologies)
Capillary preparation: Between runs rinse capillary with 100 mM NaOH for 1 min and running buffer for 2 min (at 100 psi).
Capillary temperature: 20
Running buffer: THF:water 60:40 containing 40 mM phosphate, pH 2
Injection: Pressure injection at 5 psi (4 psi.s)
Detector: UV 220
Migration time: 14
Voltage: 20 kV
Current: 7 μA
Model: Bio-Rad BioFocus 3000

OTHER SUBSTANCES
Simultaneous: desethylamiodarone, dextromethorphan, itraconazole, ketoconazole, methadone, naproxen
Noninterfering: caffeine, itraconazole, phenol, theophylline

REFERENCE
Zhang,C.-X.; von Heeren,F.; Thormann,W. Separation of hydrophobic, positively chargeable substances by capillary electrophoresis, *Anal.Chem.*, **1995**, *67*, 2070–2077.

SAMPLE
Matrix: solutions

CAPILLARY ELECTROPHORESIS
Capillary: 57 cm × 75 μm fused-silica (50 cm to detector) (Beckman)
Capillary preparation: Before each run rinse capillary with 100 mM NaOH and running buffer.
Capillary temperature: 30
Running buffer: Acetone:100 mM pH 8.1 borate buffer containing 50 mM sodium dodecyl sulfate 15:85 (A) or isopropanol:100 mM pH 8.1 borate buffer containing 50 mM sodium dodecyl sulfate 10:90 (B)
Injection: Pressure injection for 5-10 s.
Detector: UV 200
Migration time: 21.2 (A), 19 (B)
Voltage: 25 kV
Model: Beckman P/ACE 5510

OTHER SUBSTANCES
Simultaneous: acebutolol, atenolol, bretylium, captopril, diltiazem, disopyramide, lidocaine, lisinopril, metoprolol, nicardipine, nifedipine, phenytoin, pindolol, propafenone, propranolol, quinidine, sotalol, timolol, p-toluenesulfonic acid, verapamil

KEY WORDS
only running buffer A separates all compounds

REFERENCE
Bretnall,A.E.; Clarke,G.S. Selectivity of capillary electrophoresis for the analysis of cardiovascular drugs, *J.Chromatogr.A*, **1996**, *745*, 145–154.

SAMPLE
Matrix: urine
Sample preparation: 10 mL Urine + 1 mL 5 M NaOH + 10 mL n-hexane, vortex for 1 min, centrifuge at 0° at 3000 g for 5 min, repeat the extraction twice more. Combine the organic layers and add them to 50 μL glacial acetic acid, evaporate to dryness under a stream of nitrogen at 30°, reconstitute the residue in 50 μL 50 mM sodium taurodeoxycholate, filter (0.2 μm PTFE), inject an aliquot.

CAPILLARY ELECTROPHORESIS
Capillary: 100 cm × 50 μm fused-silica (50 cm to detector) (ISCO)
Capillary preparation: Rinse capillary with 20 μL running buffer between injections. Condition a new capillary by filling with 1 M NaOH, let stand for 1 h, fill with 100 mM NaOH, let stand for 1 h, rinse with buffer. Every 20 injections rinse capillary with 200 μL 1 M NaOH, 200 μL water, and 200 μL running buffer, fill with running buffer.
Capillary temperature: 22
Running buffer: 40 mM pH 9.5 Borate buffer containing 10 mM sodium taurodeoxycholate
Injection: Load under vacuum at 7.5 kPa/s.
Detector: UV 240
Migration time: 13
Voltage: 30 kV
Model: ISCO Model 3140 electropherograph
Limit of detection: 10 ng/mL

OTHER SUBSTANCES
Extracted: acepromazine, amitriptyline, azaperone, chlorpromazine, cianopramine, clomipramine, clozapine, desethylamiodarone, desipramine, diclofensine, dothiepin, doxepin, imipramine, isocarboxazid, moclobemide, perphenazine, phenothiazine, pimozide, prochlorperazine, promazine, thioridazine, thiothixene, trifluoperazine, trimipramine

KEY WORDS
human; cow; pig; horse; protect from light

REFERENCE
Aumatell,A.; Wells,R.J. Determination of a cardiac antiarrhythmic, tricyclic antipsychotics and antidepressants in human and animal urine by micellar electrokinetic capillary chromatography using a bile salt, *J.Chromatogr.B*, **1995**, *669*, 331–344.

Amitriptyline

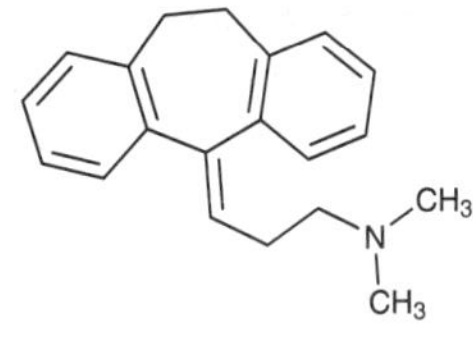

Molecular formula: $C_{20}H_{23}N$
Molecular weight: 277.41
CAS Registry No.: 50-48-6, 549-18-8 (HCl)
Merck Index (12th ed.): 511
Lednicer: 1 151, 404

SAMPLE
Matrix: blood
Sample preparation: 1 mL Serum or plasma + 50 μL 10 μL/mL trimipramine + 1 mL 2 M NaOH + 5 mL hexane:isoamyl alcohol 99:1, vortex vigorously, centrifuge at 2500 g for 5 min. Remove 4 mL of the organic layer and add it to 100 μL 100 mM HCl, extract. Remove 60 μL of the aqueous layer and evaporate it to dryness under reduced pressure, reconstitute with 20 μL water, inject an aliquot.

CAPILLARY ELECTROPHORESIS
Capillary: 71.5 cm × 50 μm fused-silica (50 cm to detector) (Polymicro Technologies)
Capillary preparation: Before each run rinse capillary with 1 M NaOH for 3 min and running buffer for 3 min (use vacuum at 508 mm Hg).
Capillary temperature: 30 or 40
Running buffer: 37.5 mM pH 8.0 Phosphate buffer containing 2 M urea and 25 mM dodecyl-trimethylammonium bromide
Injection: Vacuum injection at 508 mm Hg for 7 s
Detector: UV 254
Migration time: 11.2
Internal standard: trimipramine (11.5)
Voltage: -25 kV
Model: Applied Biosystems Model 270A
Limit of detection: 5-10 ng/mL

OTHER SUBSTANCES
Extracted: desipramine, doxepin, imipramine, nortriptyline

KEY WORDS
serum; plasma

REFERENCE
Lee,K.-J.; Lee,J.J.; Moon,D.C. Determination of tricyclic antidepressants in human plasma by micellar electrokinetic capillary chromatography, *J.Chromatogr.*, **1993**, *616*, 135–143.

SAMPLE
Matrix: solutions
Sample preparation: Inject an aliquot of a 40 μg/mL solution in MeOH.

CAPILLARY ELECTROPHORESIS
Capillary: 64 cm × 50 μm fused silica (55.5 cm to detector) (Polymicro Technologies)
Capillary preparation: Flush with running buffer for 2 min between runs. Before use rinse with 1 M NaOH for 1 h, with 100 mM NaOH for 20 min, with water for 20 min, and with running buffer for 10 min.
Capillary temperature: 25

Running buffer: MeCN:MeOH 50:50 containing 25 mM ammonium acetate and 100 mM sodium acetate
Injection: Pressure injection at 5 kPa for 3 s
Detector: UV 214
Migration time: 8.2
Voltage: 25 kV
Model: Hewlett-Packard HP3D

OTHER SUBSTANCES
Simultaneous: desmethylimipramine, didesmethylimipramine, imipramine, imipramine-N-oxide, litracene, maprotiline, methylimipramine, nortriptyline, protriptyline

REFERENCE
Bjornsdottir,I.; Hansen,S.H. Comparison of separation selectivity in aqueous and non-aqueous capillary electrophoresis, *J.Chromatogr.A*, **1995**, *711*, 313–322.

SAMPLE
Matrix: solutions
Sample preparation: Prepare a 200 µg/mL solution in water, inject an aliquot.

CAPILLARY ELECTROPHORESIS
Capillary: 56 cm × 75 µm fused-silica (56 cm to detector) (Polymicro Technologies)
Capillary temperature: 30
Running buffer: 50 mM pH 4.0 6-Aminocaproic acid containing 25 mM 3-(N,N-dimethylmyristylammonium)propanesulfonate and 15 mM Tween 20
Injection: Pressure injection at 2 kPa for 3 s
Detector: UV 214
Migration time: 23
Voltage: 20 kV
Current: 62 µA
Model: Waters Quanta 4000

OTHER SUBSTANCES
Simultaneous: litracene, maprotiline, nortriptyline, protriptyline

REFERENCE
Hansen,S.H.; Bjornsdottir,I.; Tjornelund,J. Separation of basic drug substances of very similar structure using micellar electrokinetic chromatography, *J.Pharm.Biomed.Anal.*, **1995**, *13*, 489–495.

SAMPLE
Matrix: solutions

CAPILLARY ELECTROPHORESIS
Capillary: 55.5 cm × 50 µm fused-silica (55.5 cm to detector) (Polymicro Technologies)
Capillary temperature: 25
Running buffer: MeCN:MeOH 50:50 containing 25 mM ammonium acetate and 100 mM sodium acetate
Injection: Pressure injection at 5 kPa for 3 s.
Detector: UV 214
Migration time: 8.3
Voltage: 25 kV
Current: 23 µA
Model: Hewlett Packard HP [3D]CE

OTHER SUBSTANCES
Simultaneous: litracene, maprotiline, nortriptyline, protriptyline

REFERENCE
Bjornsdottir,I.; Tjornelund,J.; Hansen,S.H. Nonaqueous capillary electrophoresis in pharmaceutical analysis, *J.Capillary Electrophor.*, **1996**, *3*, 83–87.

SAMPLE
Matrix: solutions

CAPILLARY ELECTROPHORESIS
Capillary: 55 cm × 75 μm fused-silica (40 cm to detector) (Polymicro Technologies)
Running buffer: 10 mM NaOH adjusted to pH 2.75 with 1 M phosphoric acid
Injection: Pressure injection at 20 mbar for 3.6 s.
Detector: UV 211
Migration time: 6.7
Voltage: 15.3 kV
Limit of quantitation: <6 μM

OTHER SUBSTANCES
Simultaneous: metoprolol

KEY WORDS
paper contains details of isotachophoresis pre-concentration.

REFERENCE
Enlund,A.M.; Westerlund,D. Enhancing detectability in CE by combining an isotachophoresis preconcentration with capillary zone electrophoresis in a single capillary, *Chromatographia*, **1997**, *46*, 315–321.

SAMPLE
Matrix: solutions

CAPILLARY ELECTROPHORESIS
Capillary: 70 cm × 50 μm coated fused-silica (55 cm to detector) (Polymicro Technologies)
Capillary preparation: Between runs rinse with water for 2 min. Coat capillaries as follows. Condition with 1 M NaOH for 15 min, rinse with water for 5 min, treat with 30 μL/mL 3-(trimethoxysilyl)propyl methacrylate in acetic acid:water 50:50 for 1 h using house vacuum, rinse with water, fill with polymerization solution, let stand for 1 h, flush, rinse with water. (Prepare polymerization solution by adding 10 μL N,N,N',N'-tetramethylethylenediamine to 10 mL of a degassed 4% solution of acrylamide in water, add 10 μL 10% ammonium persulfate in water.)
Running buffer: 10 mM pH 6.5 Citrate buffer containing 20 mM sodium dodecyl sulfate and 10 mM carboxymethyl-β-cyclodextrin (Cyclodextrin Technologies Development, Gainesville FL)
Injection: Hydrodynamic injection at 15 cm for 8-10 s.
Detector: UV 254
Migration time: 10.9
Voltage: -20 kV
Model: laboratory-constructed

OTHER SUBSTANCES
Simultaneous: carbamazepine, clomipramine, desipramine, imipramine, nortriptyline, opipramol, proptriptyline, trimipramine

KEY WORDS
coated capillary

REFERENCE
Spencer,B.J.; Zhang,W.; Purdy,W.C. Capillary electrophoretic separation of tricyclic antidepressants using charged carboxymethyl-β-cyclodextrin as a buffer additive, *Electrophoresis*, **1997**, *18*, 736–744.

SAMPLE
Matrix: solutions

CAPILLARY ELECTROPHORESIS
Capillary: 57 cm × 50 μm fused-silica (50 cm to detector) (Polymicro Technologies)
Capillary preparation: At the start of each day rinse capillary with water for 5 min, 100 mM NaOH for 30 min, with water for 10 min, and with running buffer for 15 min. Wash new capillaries with 1 M NaOH for 1 h.

Capillary temperature: 23
Running buffer: MeOH:50 mM pH 9.55 sodium phosphate buffer containing 0.06% poly(n-undecyl-α-D-glucopyranoside) (Prepare poly(n-undecyl-α-D-glucopyranoside) as follows. With exclusion of atmospheric moisture stir 40 mL alcohol-free chloroform, 5.2 g yellow mercuric oxide, 0.1 g mercuric bromide, 10 g anhydrous calcium sulfate, and 2.74 g 10-undecen-1-ol then add 10 g acetobromo-α-D-glucose (Fluka), stir at room temperature for 20 h, filter, wash the solid with chloroform. Wash the filtrate with aqueous 1 M KBr until no mercury salts are present in the aqueous layers, wash with later, concentrate. Take up the residue in benzene (Caution! Benzene is a carcinogen!) and filter it through activated silica, crystallize from n-hexane to obtain undecylenyl-2,3,4,6-tetra-O-acetyl-β-D-glucopyranoside. Add 2 mL of a 1 M solution of sodium methoxide in MeOH to a solution of 1 g of undecylenyl-2,3,4,6-tetra-O-acetyl-β-D-glucopyranoside in 50 mL dry MeOH, let stand until the reaction is complete (by TLC) to obtain undecylenyl-β-D-glucopyranoside (cf. Tenside Detergents 1978, 15, 72). Use Dowex 50W (H⁺) resin to deionize the final product in solution. Polymerize the monomer by irradiation of an 870 μM solution in MeOH:water 20:80 with ^{60}Co gamma radiation for 48 h, purify the polymer solution by dialysis using a 1000 Da molecular mass cut-off membrane.)
Injection: Pressure injection at 0.5 psi for 5 s.
Detector: UV 214
Migration time: 13
Voltage: 22.8 kV
Model: Beckman P/ACE 5510

OTHER SUBSTANCES
Simultaneous: desipramine, doxepin, imipramine, nordoxepin, nortriptyline, protriptyline

---

REFERENCE
Harrell,C.W.; Dey,J.; Shamsi,S.A.; Foley,J.P.; Warner,I.M. Enhanced separation of antidepressant drugs using a polymerized nonionic surfactant as a transient capillary coating, *Electrophoresis*, **1998**, *19*, 712–718.

SAMPLE
Matrix: solutions
Sample preparation: Mix 50 μL of a solution in 1% sodium chloride with 100 μL MeCN, vortex for 15 s, centrifuge at 14000 g for 20 s, inject an aliquot.

CAPILLARY ELECTROPHORESIS
Capillary: 40 cm × 50 μm silica
Running buffer: MeCN:isopropanol:buffer 10:10:80 (Buffer was 180 mM triethanolamine, pH adjusted to 8.2.)
Injection: Hydrodynamic injection at low pressure (to fill 6% of the capillary).
Detector: UV 254
Migration time: 4.8
Voltage: 12 kV
Model: Beckman Model 2000 CE

OTHER SUBSTANCES
Simultaneous: imipramine, procainamide, quinine

REFERENCE
Shihabi,Z.K. Stacking of weakly cationic compounds by acetonitrile for capillary electrophoresis, *J.Chromatogr.A*, **1998**, *817*, 25–30.

SAMPLE
Matrix: urine
Sample preparation: 10 mL Urine + 1 mL 5 M NaOH + 10 mL n-hexane, vortex for 1 min, centrifuge at 0° at 3000 g for 5 min. Remove the organic layer and add it to 50 μL glacial acetic acid, evaporate to dryness under a stream of nitrogen at 30°, reconstitute the residue in 50 μL 50 mM sodium taurodeoxycholate, filter (0.2 μm PTFE), inject an aliquot.

CAPILLARY ELECTROPHORESIS
Capillary: 100 cm × 50 μm fused-silica (50 cm to detector) (ISCO)
Capillary preparation: Rinse capillary with 20 μL running buffer between injections. Condition a new capillary by filling with 1 M NaOH, let stand for 1 h, fill with 100 mM NaOH, let stand

for 1 h, rinse with buffer. Every 20 injections rinse capillary with 200 μL 1 M NaOH, 200 μL water, and 200 μL running buffer, fill with running buffer.
Capillary temperature: 22
Running buffer: 40 mM pH 9.5 Borate buffer containing 10 mM sodium taurodeoxycholate
Injection: Load under vacuum at 7.5 kPa/s.
Detector: UV 240
Migration time: 10.5
Voltage: 30 kV
Model: ISCO Model 3140 electropherograph
Limit of detection: 10 ng/mL

OTHER SUBSTANCES
Extracted: acepromazine, amiodarone, azaperone, chlorpromazine, cianopramine, clomipramine, clozapine, desethylamiodarone, desipramine, diclofensine, dothiepin, doxepin, imipramine, isocarboxazid, moclobemide, perphenazine, phenothiazine, pimozide, prochlorperazine, promazine, thioridazine, thiothixene, trifluoperazine, trimipramine

KEY WORDS
human; cow; pig; horse; protect from light

REFERENCE
Aumatell,A.; Wells,R.J. Determination of a cardiac antiarrhythmic, tricyclic antipsychotics and antidepressants in human and animal urine by micellar electrokinetic capillary chromatography using a bile salt, *J.Chromatogr.B*, **1995**, *669*, 331–344.

Amlodipine

Molecular formula: $C_{20}H_{25}ClN_2O_5$
Molecular weight: 408.88
CAS Registry No.: 88150-42-9, 111470-99-6 (besylate), 88150-47-4 (maleate)
Merck Index (12th ed.): 516
Lednicer: 4 108

SAMPLE
Matrix: solutions
Sample preparation: Prepare a 600 μg/mL solution in MeCN:MeOH:water 25:25:50, dilute to 150 μg/mL with 100 mM pH 8.1 borate buffer containing 50 mM sodium dodecyl sulfate, inject an aliquot.

CAPILLARY ELECTROPHORESIS
Capillary: 57 cm $\times$ 75 μm fused-silica (50 cm to detector) (Beckman)
Capillary preparation: Rinse capillary with 100 mM NaOH and running buffer before each run.
Capillary temperature: 30
Running buffer: Acetone:buffer 15:85 (Buffer was 100 mM pH 8.1 borate buffer containing 50 mM sodium dodecyl sulfate.)
Injection: Pressure injection for 5 s
Detector: UV 200
Migration time: 20.5
Voltage: +25 kV
Model: Beckman P/ACE system 5510

OTHER SUBSTANCES
Simultaneous: atenolol, diltiazem, nicardipine, nifedipine, verapamil

REFERENCE
Bretnall,A.E.; Clarke,G.S. Investigation and optimisation of the use of micellar electrokinetic chromatography for the analysis of six cardiovascular drugs, *J.Chromatogr.A*, **1995**, *700*, 173–178.

SAMPLE
Matrix: solutions

CAPILLARY ELECTROPHORESIS
Capillary: 70 cm × 75 μm fused-silica
Capillary preparation: Before each run wash with 100 mM NaOH for 5 min, with water for 5 min, and with running buffer for 5 min.
Capillary temperature: 25
Running buffer: 20 mM pH 9.4 Phosphate/borate buffer containing 1% urea and 10 mM 2-O-carboxymethyl-β-cyclodextrin
Injection: Hydrodynamic injection for 1 s.
Detector: UV 240
Migration time: 15.5, 17 (enantiomers)
Voltage: 20 kV
Model: SpectraPhoresis 500 (Thermo Separation Products)

OTHER SUBSTANCES
Simultaneous: nimodipine, nitrendipine

KEY WORDS
comparison with HPLC; chiral

REFERENCE
Gilar,M.; Uhrová,M.; Tesarová,E. Enantiomer separation of dihydropyridine calcium antagonists with cyclodextrins as chiral selectors: structural correlation, *J.Chromatogr.B*, **1996**, *681*, 133–141.

SAMPLE
Matrix: solutions

CAPILLARY ELECTROPHORESIS
Capillary: 50 cm × 50 μm fused-silica (Bio-Rad)
Running buffer: 20 mM pH 3.2 Tris-HCl buffer containing 18 mM α-cyclodextrin and 0.05% methylcellulose
Injection: Pressure injection at 68.9 kPa.s
Detector: UV 200
Migration time: 11.14 (R-(+)), 11.29 (S-(-))
Voltage: 15 kV
Model: BioFocus 3000 CE

OTHER SUBSTANCES
Simultaneous: benzenesulfonic acid

KEY WORDS
chiral

REFERENCE
Luksa,J.; Josic,D.; Podobnik,B.; Furlan,B.; Kremser,M. Semi-preparative chromatographic purification of the enantiomers of S-(-)-amlodipine and R-(+)-amlodipine, *J.Chromatogr.B*, **1997**, *693*, 367–375.

SAMPLE
Matrix: solutions

CAPILLARY ELECTROPHORESIS
Capillary: 57 cm × 50 μm fused-silica (50 cm to detector) (Beckman)
Capillary preparation: Before each run wash capillary with 100 mM NaOH for 1 min and with running buffer for 3 min. Condition a new capillary with 1 M NaOH for 1 h and with water for 20 min.
Capillary temperature: 17
Running buffer: 20 mM pH 7.0 Sodium phosphate buffer containing 1 mM sulfobutylether β-cyclodextrin

Detector: UV 214
Migration time: 14, 15 (enantiomers)
Voltage: 15 kV
Model: Beckman P/ACE 5510

KEY WORDS
chiral; comparison with other cyclodextrins

REFERENCE
Owens,P.K.; Fell,A.F.; Coleman,M.W.; Kinns,M.; Berridge,J.C. Use of ^{1}H-NMR spectroscopy to determine the enantioselective mechanism of neutral and anionic cyclodextrins by capillary electrophoresis, *J.Pharm.Biomed.Anal.*, **1997**, *15*, 1603–1619.

SAMPLE
Matrix: solutions
Sample preparation: Inject an aliquot of a solution in 10% running buffer.

CAPILLARY ELECTROPHORESIS
Capillary: 57 cm × 50 μm fused-silica (Beckman)
Capillary preparation: Before each run wash capillary with 100 mM NaOH for 1 min and with running buffer for 3 min. Condition new capillaries with 1 M NaOH for 1 h and with water for 20 min.
Capillary temperature: 17
Running buffer: 50 mM pH 3.0 NaH_2PO_4 containing 2.5 mM carboxymethyl-β-cyclodextrin (Wacker)
Detector: UV 214
Migration time: 15 (R), 17 (S)
Voltage: 15 kV
Model: Beckman P/ACE 5510

KEY WORDS
chiral; comparison with HPLC

REFERENCE
Owens,P.K.; Fell,A.F.; Coleman,M.W.; Berridge,J.C. Effect of charged and uncharged chiral additives on the resolution of amlodipine enantiomers in liquid chromatography and capillary electrophoresis, *J.Chromatogr.A*, **1998**, *797*, 187–195.

Amobarbital

Molecular formula: $C_{11}H_{18}N_2O_3$
Molecular weight: 226.28
CAS Registry No.: 57-43-2, 64-43-7 (sodium salt)
Merck Index (12th ed.): 607
Lednicer: 1 268

SAMPLE
Matrix: blood, gastric contents, urine, vitreous humor
Sample preparation: Dilute gastric contents 1:10. 2 mL Serum, blood, vitreous humor, diluted gastric contents, or urine + 2.5 mL water + 200 μL mephobarbital solution, mix, add to TOXI-TUBE B (Toxi-Lab, Irvine CA), rock for 10 min, centrifuge at 2000 rpm for 5 min. Remove the organic layer and evaporate it to dryness under a stream of nitrogen at room temperature, reconstitute the residue in 100 μL running buffer, vortex for 30 s, inject an aliquot. (Mephobarbital solution was 2 mg mephobarbital and 2 drops EtOH, made up to 20 mL with water.)

CAPILLARY ELECTROPHORESIS
Capillary: 60 cm × 75 μm AccuSep (Waters)

Capillary preparation: Purge for 1 min between samples.
Running buffer: MeCN:buffer 15:85 adjusted to pH 8.5 with 1 M phosphoric acid. (Buffer was 3.8 g sodium borate decahydrate, 1 .4 g $NaH_2PO_4.H_2O$, and 28.8 g sodium dodecyl sulfate in 1 L water.)
Injection: Hydrostatic injection for 10 s
Detector: UV 214
Migration time: 9.41
Internal standard: mephobarbital (8.88)
Voltage: 20 kV
Model: Waters Quanta 4000
Limit of detection: 100 ng/mL

OTHER SUBSTANCES
Extracted: butabarbital, butalbital, pentobarbital, phenobarbital, secobarbital

---

KEY WORDS
serum; whole blood

REFERENCE
Ferslew,K.E.; Hagardorn,A.N.; McCormick,W.F. Application of micellar electrokinetic capillary chromatography to forensic analysis of barbiturates in biological fluids, *J.Forensic Sci.*, **1995**, *40*, 245–249.

SAMPLE
Matrix: blood, urine
Sample preparation: Serum, plasma. 200 µL Serum or plasma + 100 µL 1 M HCl + 2 mL chloroform, shake vigorously for 15 min, centrifuge at 500 g for 10 min. Remove the lower organic layer and evaporate it to dryness under a stream of nitrogen at 40°, reconstitute the residue in 200 µL running buffer, shake for 1 min, filter (0.2 µm), inject an aliquot. Urine. Condition a Bond Elut Certify SPE cartridge with 2 mL MeOH and 2 mL 100 mM pH 6 phosphate buffer, do not allow to dry. 5 mL Urine + 2 mL 100 mM pH 6 phosphate buffer, mix, add to the SPE cartridge, wash with 1 mL MeOH:100 mM phosphate buffer 20:80, dry under vacuum for 5 min, wash with 1 mL 1 M acetic acid, dry under vacuum for 10 min, wash with 1 mL hexane, elute with 4 mL dichloromethane. Evaporate the eluate to dryness under a stream of nitrogen at 40°, reconstitute the residue in 100-200 µL running buffer, filter (0.2 µm), inject an aliquot.

CAPILLARY ELECTROPHORESIS
Capillary: 90 cm × 75 µm fused-silica (70 cm to detector) (Polymicro Technologies)
Capillary preparation: Between each run rinse capillary with 100 mM NaOH for 3 min and buffer for 5 min
Running buffer: 15 mM pH 7.8 NaH_2PO_4 containing 9 mM sodium borate and 50 mM sodium dodecyl sulfate
Injection: Siphon at 34 cm for 5 s
Detector: UV 195
Migration time: 12.7
Voltage: 20 kV
Current: 60-63 µA
Model: Laboratory constructed

OTHER SUBSTANCES
Extracted: allobarbital, barbital, butalbital, pentobarbital, phenobarbital, thiopental

---

KEY WORDS
serum; cow; human; plasma; SPE

REFERENCE
Thormann,W.; Meier,P.; Marcolli,C.; Binder,F. Analysis of barbiturates in human serum and urine by high-performance capillary electrophoresis-micellar electrokinetic capillary chromatography with on-column multi-wavelength detection, *J.Chromatogr.*, **1991**, *545*, 445–460.

SAMPLE
Matrix: solutions

CAPILLARY ELECTROPHORESIS
Capillary: 60 cm × 50 μm fused-silica (40 cm to detector), bare (A) or C18 coated (B) (Supelco)
Capillary preparation: Rinse new capillaries with 250 μL 1 M NaOH, 250 μL 100 mM NaOH, water, MeOH, water, and running buffer then condition with running buffer for at least 15 min. Carry out a similar procedure between runs.
Running buffer: 100 mM pH 8.5 Borate buffer containing 25 mM sodium dodecyl sulfate and 5 M urea
Injection: Electrokinetic injection at 5 kV for 5 s.
Detector: UV 254
Migration time: 12 (A), 10.5 (B)
Voltage: 18 kV
Model: Jasco

OTHER SUBSTANCES
Simultaneous: barbital, mephobarbital, metharbital, pentobarbital, phenobarbital, secobarbital

KEY WORDS
coated capillary

REFERENCE
Jinno,K.; Han,Y.; Nakamura,M. Analysis of anxiolytic drugs by capillary electrophoresis with bare and coated capillaries, *J.Capillary Electrophor.*, **1996**, *3*, 139–145.

SAMPLE
Matrix: solutions
Sample preparation: Inject an aliquot of a 100 μg/mL solution in running buffer.

CAPILLARY ELECTROPHORESIS
Capillary: 60 cm × 50 μm acrylamide-coated fused-silica (40 cm to detector) (Supelco)
Capillary preparation: Adjust the pH of 20 mL water to 3.5 with acetic acid, add 80 μL 3-(trimethoxysilyl)propyl methacrylate (3-methacryloxypropyltrimethoxysilane), mix, suck into capillary, let stand at room temperature for 1 h, remove the solution, wash with water. Fill the capillary with a deaerated 3-4% acrylamide solution containing 1 μL/mL N,N,N',N'-tetramethylethylenediamine and 1 mg/mL ammonium persulfate, let stand for 3 h, remove excess solution by aspiration, rinse with water, remove water by aspiration, dry at 35° (cf. J. Chromatogr. 1985, 347, 191).
Running buffer: MeCN:buffer 5:95 (Buffer was 100 mM borate containing 5 M urea and 10 mM sodium dodecyl sulfate, adjusted to pH 8.5 with phosphate.)
Injection: Electrokinetic injection at 5 kV for 5 s.
Detector: UV 254
Migration time: 17
Voltage: 18 kV
Model: Jasco Model 870-CE

OTHER SUBSTANCES
Simultaneous: alprazolam, barbital, bromazepam, clonazepam, clotiazepam, cloxazolam, diazepam, estazolam, etizolam, fludiazepam, flunitrazepam, flurazepam, haloxazolam, medazepam, mephobarbital, metharbital, nimetazepam, nitrazepam, oxazepam, pentobarbital, phenobarbital, secobarbital, triazolam

KEY WORDS
coated capillary; detector at anode

REFERENCE
Jinno,K.; Han,Y.; Sawada,H.; Taniguchi,M. Capillary electrophoretic separation of toxic drugs using a polyacrylamide-coated capillary, *Chromatographia*, **1997**, *46*, 309–314.

SAMPLE
Matrix: solutions

CAPILLARY ELECTROPHORESIS
Capillary: 40 cm × 75 μm coated fused-silica (15 cm to detector) (Supelco)

Capillary preparation: Coat column as follows. Adjust the pH of 20 mL water to 3.5 with acetic acid, add 80 μL 3-(trimethoxysilyl)propyl methacrylate (3-methacryloxypropyltrimethoxysilane), mix, suck into capillary, let stand at room temperature for 1 h, remove the solution, wash with water. Fill the capillary with a deaerated 4% acrylamide solution containing 1 mg/mL N,N,N',N'-tetramethylethylenediamine and 1 mg/mL ammonium persulfate, let stand for 3 h, remove excess solution by aspiration, rinse with water, remove water by aspiration, dry at 35° (cf. J. Chromatogr. 1985, 347, 191).
Running buffer: 100 mM Tris/150 mM boric acid, pH 8.3
Injection: Electromigration at 5 kV for 5 s.
Detector: UV 240, UV 254
Migration time: 6.5
Voltage: 12 kV
Model: Jasco 890-CE

OTHER SUBSTANCES
Simultaneous: barbital, mephobarbital, metharbital, pentobarbital, phenobarbital, secobarbital

---

KEY WORDS
injection at cathode; coated capillary

REFERENCE
Jinno,K.; Han,Y.; Sawada,H. Analysis of toxic drugs by capillary electrophoresis using polyacrylamide-coated columns, *Electrophoresis*, **1997**, *18*, 284–286.

SAMPLE
Matrix: solutions
Sample preparation: Inject an aliquot of a solution in 20 mM pH 5.9 ammonium acetate buffer.

CAPILLARY ELECTROPHORESIS
Capillary: 33 cm × 50 μm fused-silica (20 cm to detector) (Polymicro Technologies)
Running buffer: 10 mM pH 5.9 Ammonium acetate buffer containing 15 mM sodium dodecyl sulfate
Injection: Electrokinetic injection at -7.4 kV for 1 s.
Detector: UV 226, MS, Finnigan MAT TSQ, negative electrospray (details in paper)
Migration time: 9
Voltage: -7.4 kV

OTHER SUBSTANCES
Simultaneous: barbital, butalbital, pentobarbital, secobarbital

---

REFERENCE
Yang,L.; Harrata,A.K.; Lee,C.S. On-line micellar electrokinetic chromatography-electrospray ionization mass spectrometry using anodically migrating micelles, *Anal.Chem.*, **1997**, *69*, 1820–1826.

SAMPLE
Matrix: urine
Sample preparation: Place an extraction rod in 50 μL urine in a 50 mm × 1.5 mm ID length of PTFE tubing for 30 min, place the extraction rod into 5 μL 20-40 mM pH 11.5 phosphate buffer in a 50 mm × 1.2 mm ID length of PTFE tubing for 90 min, inject an aliquot of the buffer. (Prepare the solid-phase extraction rod as follows. Polish a 70 × 1.1 stainless steel rod with emery paper, clean with a Kimwipe and acetone, sonicate in EtOH, sonicate in THF, dip in 3% PVAM in THF momentarily, hold vertically for 1 min, air dry in a hood for at least 5 h, dip in PVC solution momentarily, hold vertically for 1 min, air dry for at least 5 h. The coating is 3 cm long. PVAM was poly(vinyl chloride-co-vinyl acetate-co-maleic acid) consisting of 86% vinyl chloride, 13% vinyl acetate, and 1% maleic acid. Prepare PVC solution by adding very high molecular weight PVC slowly to THF with stirring until the concentration reaches 3.6%, add Santicizer 141 to a concentration of 7.2%. Santicizer 141 (Monsanto) is 92% 2-ethylhexyl diphenyl phosphate, 5% di-2-ethylhexyl phenyl phosphate, and 3% triphenyl phosphate.)

CAPILLARY ELECTROPHORESIS
Capillary: 75 cm × 75 μm fused-silica (50 cm to detector) (Polymicro Technologies)

Running buffer: 50 mM Tris adjusted to pH 7.8 with 3-[N-tris(hydroxymethyl)methylamino]-2-hydroxypropanesulfonic acid (Tapso)
Injection: Pressure injection at 0.5 psi for 4 s.
Detector: UV 230
Migration time: 7.8
Voltage: 35 kV
Current: about 30 μA
Model: Isco 3850
Limit of detection: <1 ppm

OTHER SUBSTANCES
Extracted: aprobarbital, butabarbital, butalbital, mephobarbital, pentobarbital, secobarbital, thiopental
Simultaneous: allobarbital, aspirin, phenobarbital

KEY WORDS
SPE

REFERENCE
Li,S ; Weber,S.G. Determination of barbiturates by solid-phase microextraction and capillary electrophoresis, *Anal.Chem.*, **1997**, *69*, 1217–1222.

Amorolfine

Molecular formula: $C_{21}H_{35}NO$
Molecular weight: 317.52
CAS Registry No.: 78613-35-1, 78613-38-4 (HCl)
Merck Index (12th ed.): 612
Lednicer: 5 99

SAMPLE
Matrix: solutions
Sample preparation: Inject an aliquot of a 100 μg/mL solution in running buffer.

CAPILLARY ELECTROPHORESIS
Capillary: 30 cm × 50 μm fused-silica (25.5 cm to detector) (Yongnian Optical Conductive Fiber Plant, China), coated with polyacrylamide
Capillary preparation: No details of the polyacrylamide coating process are provided. However, another paper (LC.GC 1997, 15, 40) by this group indicates that they use the procedure of Hjertén, thus: Adjust the pH of 20 mL water to 3.5 with acetic acid, add 80 μL 3-(trimethoxysilyl)propyl methacrylate (3-methacryloxypropyltrimethoxysilane), mix, suck into capillary, let stand at room temperature for 1 h, remove the solution, wash with water. Fill the capillary with a deaerated 3-4% acrylamide solution containing 1 μL/mL N,N,N',N'-tetramethylethylenediamine and 1 mg/mL potassium persulfate, let stand for 30 min, remove excess solution by aspiration, rinse with water, remove water by aspiration, dry at 35° (J. Chromatogr. 1985, 347, 191).
Capillary temperature: 25
Running buffer: 100 mM NaH_2PO_4 adjusted to pH 2.5
Injection: Electrokinetic injection at 15 kV for 3 s.
Detector: UV 200, UV 210
Migration time: 5.57
Voltage: 15 kV
Model: Bio-Rad BioFocus 3000

OTHER SUBSTANCES
Simultaneous: brompheniramine, bupivacaine, carteolol, chloroquine, chlorpheniramine, chlorphenoxamine, disopyramide, dobutamine, doxylamine, flecainide, gallopamil, ketamine, mepindolol, orphenadrine, oxybutynin, phenoxybenzamine, pindolol, propafenone, propranolol, sulpiride, talinolol, tropicamide, verapamil

KEY WORDS
coated capillary

REFERENCE
Koppenhoefer,B.; Epperlein,U.; Xiaofeng,Z.; Bingcheng,L. Separation of enantiomers of drugs by capillary electrophoresis. Part 4: Hydroxypropyl-γ-cyclodextrin as chiral solvating agent, *Electrophoresis*, **1997**, *18*, 924–930.

SAMPLE
Matrix: solutions

CAPILLARY ELECTROPHORESIS
Capillary: 29-36 cm × 50 μm fused-silica (24.5-31.5 cm to detector) (Yongnian Optical Conductive Fiber Plant, China) coated with polyacrylamide
Capillary preparation: Coat capillary as follows. Adjust the pH of 20 mL water to 3.5 with acetic acid, add 80 μL 3-(trimethoxysilyl)propyl methacrylate (3-methacryloxypropyltrimethoxysilane), mix, suck into capillary, let stand at room temperature for 1 h, remove the solution, wash with water. Fill the capillary with a deaerated 3-4% acrylamide solution containing 1 μL/mL N,N,N',N'-tetramethylethylenediamine and 1 mg/mL potassium persulfate, let stand for 30 min, remove excess solution by aspiration, rinse with water, remove water by aspiration, dry at 35° (J. Chromatogr. 1985, 347, 191).
Capillary temperature: 25
Running buffer: 100 mM pH 2.5 NaH_2PO_4 (A) or 100 mM pH 2.5 NaH_2PO_4 containing 45 mM hydroxypropyl-α-cyclodextrin (Wacker, Munich) (B)
Injection: Electromigration at 15 kV for 3 s.
Detector: UV 200; UV 210
Migration time: 5.57 (A); 14.37 (B) (no separation of enantiomers)
Voltage: 15 kV
Model: Bio-Focus 3000

OTHER SUBSTANCES
Also analyzed: albuterol (salbutamol), alprenolol, atenolol, atropine, azelastine, baclofen, bamethan, benproperine, benserazide, biperiden, bisoprolol, brompheniramine, bupivacaine, bupranolol, butamirate, butethamate, carazolol, carbuterol, carteolol, carvedilol, celiprolol, chloroquine, chlorpheniramine, chlorphenoxamine, cicletanine, clenbuterol, clidinium bromide, clobutinol, dimethindene, dipivefrin, disopyramide, dobutamine, doxylamine, fendiline, flecainide, gallopamil, homatropine, ipratropium bromide, isoproterenol (isoprenaline), isothipendyl, ketamine, meclizine, mefloquine, mepindolol, mequitazine, metaclazepam, metaproterenol (orciprenaline), metipranolol, metoprolol, nafronyl (naftidrofuryl), nefopam, nicardipine, norfenefrine, ofloxacin, ornidazole, orphenadrine, oxomemazine, oxprenolol, oxybutynin, phenoxybenzamine, phenylpropanolamine, pholedrine, pindolol, pirbuterol, prilocaine, procyclidine, promethazine, propafenone, propranolol, reproterol, sotalol, sulpride, synephrine, talinolol, terbutaline, tetrahydrozoline (tetryzoline), theodrenaline, tioconazole, tocainide, trihexyphenidyl, trimeprazine (alimemazine), trimipramine, tropicamide, verapamil, zopiclone

KEY WORDS
coated capillary

REFERENCE
Koppenhoefer,B.; Eperlein,U.; Schlunk,R.; Zhu,X.; Lin,B. Separation of enantiomers of drugs by capillary electrophoresis. V. Hydroxypropyl-α-cyclodextrin as chiral solvating agent, *J.Chromatogr.A*, **1998**, *793*, 153–164.

Amoxicillin

Molecular formula: $C_{16}H_{19}N_3O_5S$
Molecular weight: 365.41
CAS Registry No.: 26787-78-0 (anhydrous), 61336-70-7 (trihydrate)
Lednicer: 4 179, 180

SAMPLE
Matrix: bulk
Sample preparation: Prepare a solution in pH 6 phosphate buffer, inject an aliquot.

CAPILLARY ELECTROPHORESIS
Capillary: 80 cm $\times$ 50 μm fused silica (75 cm to detector) (Polymicro Technologies or Isco)
Capillary preparation: Rinse with running buffer for 3 min between runs.
Running buffer: 100 mM pH 8 NaH_2PO_4 containing 50 mM sodium borate and 50 mM sodium dodecyl sulfate
Injection: Gravity injection at 10 cm for 15 s
Detector: UV 205
Migration time: 15.63
Voltage: +18 kV
Model: Dionex

OTHER SUBSTANCES
Simultaneous: degradation products, impurities, ampicillin, penicillin G, penicillin V

REFERENCE
Flurer,C.L.; Wolnik,K.A. Chemical profiling of pharmaceuticals by capillary electrophoresis in the determination of drug origin, *J.Chromatogr.A*, **1994**, *674*, 153–163.

SAMPLE
Matrix: bulk
Sample preparation: Inject an aliquot of an aqueous solution.

CAPILLARY ELECTROPHORESIS
Capillary: 44 cm $\times$ 50 μm fused-silica (36 cm to detector) (Polymicro Technologies)
Capillary preparation: Wash capillary with running buffer for 5 min before each run. At the start of each day wash capillary with 100 mM NaOH (at 60° ?) for 5 min and with water at 60° for 5 min. Condition new capillaries with 1 M NaOH (at 60° ?) for 20 min and with water at 60° for 5 min.
Capillary temperature: 25
Running buffer: MeCN:70 mM NaH_2PO_4 5:95 containing 125 mM sodium dodecyl sulfate, adjusted to pH 6.0
Injection: Hydrodynamic injection for 9.9 s.
Detector: UV 230
Migration time: 8
Voltage: 15 kV
Model: Spectraphoresis 500 (Thermo Separation Products, Fremont CA)
Limit of quantitation: 13.3 pg
Limit of detection: 3.3 pg

OTHER SUBSTANCES
Simultaneous: impurities

REFERENCE
Li,Y.M.; Van Schepdael,A.; Zhu,Y.; Roets,E.; Hoogmartens,J. Development and validation of amoxicillin determination by micellar electrokinetic capillary chromatography, *J.Chromatogr.A*, **1998**, *812*, 227–236.

SAMPLE
Matrix: solutions

Sample preparation: Prepare a 0.5-2 mg/mL solution in water, inject an aliquot.

CAPILLARY ELECTROPHORESIS
Capillary: 65 cm × 50 μm fused-silica (50 cm to detector) (Scientific Glass Engineering)
Running buffer: 20 mM NaH_2PO_4 containing 150 mM sodium dodecyl sulfate adjusted to pH 9.0 with 20 mM sodium tetraborate
Injection: Injection by siphon at 5 cm for 5-10 s
Detector: UV 210
Migration time: 8.6
Voltage: 20 kV

OTHER SUBSTANCES
Simultaneous: ampicillin, aspoxicillin, carbenicillin, penicillin G, piperacillin, sulbenicillin
Also analyzed: cefmenoxime, cefminox, cefoperazone, cefotaxime, cefpimizole, cefpiramide, ceftazidime, ceftriaxone

---

REFERENCE
Nishi,H.; Tsumagari,N.; Kakimoto,T.; Terabe,S. Separation of β-lactam antibiotics by micellar electrokinetic chromatography, *J.Chromatogr.*, **1989**, *477*, 259–270.

SAMPLE
Matrix: solutions
Sample preparation: Inject directly.

CAPILLARY ELECTROPHORESIS
Capillary: 65 cm × 50 μm untreated fused-silica (50 cm to detector) (SGE)
Capillary preparation: Flush with running buffer every 5 runs. At the end of each day fill capillary with 100 mM KOH, let stand for 30 min, flush with water, let stand for 5 min, fill with running buffer.
Running buffer: pH 8.5 Buffer containing 100 mM sodium dodecyl sulfate (Prepare buffer by mixing 20 mM NaH_2PO_4 solution with 20 mM sodium borate solution to achieve a pH of 8.5.)
Injection: Inject by siphoning at 10 cm for 5-10 s.
Detector: UV 210
Migration time: 9.3

OTHER SUBSTANCES
Simultaneous: ampicillin, aspoxicillin, carbenicillin, penicillin G, piperacillin, sulbenicillin

---

REFERENCE
Nishi,H.; Fukuyama,T.; Matsuo,M. Separation and determination of aspoxicillin in human plasma by micellar electrokinetic chromatography with direct sample injection, *J.Chromatogr.*, **1990**, *515*, 245–255.

SAMPLE
Matrix: solutions
Sample preparation: Prepare a 120 μg/mL solution in water, inject an aliquot.

CAPILLARY ELECTROPHORESIS
Capillary: 60 cm × 75 μm
Running buffer: 20 mM NaH_2PO_4 containing 50 mM sodium dodecyl sulfate, adjusted to pH 9.0 with sodium tetraborate
Injection: Hydrodynamic injection at 10 cm for 5 s
Detector: UV 214
Migration time: 7.5
Voltage: 18 kV
Model: Waters Quanta 4000

OTHER SUBSTANCES
Simultaneous: 6-aminopenicillanic acid, ampicillin, cloxacillin, dicloxacillin, nafcillin, oxacillin, ticarcillin

REFERENCE
Swartz,M.E. Method development and selectivity control for small molecule pharmaceutical separations by capillary electrophoresis, *J.Liq.Chromatogr.*, **1991**, *14*, 923–938.

SAMPLE
Matrix: solutions
Sample preparation: Prepare a 50 µM solution in water, inject an aliquot.

CAPILLARY ELECTROPHORESIS
Capillary: 57 cm × 50 µm silica (50 cm to detector)
Capillary preparation: Condition with 100 mM NaOH for 5 min, rinse with water for 5 min, fill with running buffer.
Capillary temperature: 20
Running buffer: 10 mM pH 7.42 NaH_2PO_4 containing 30 mM sodium borate and 200 mM sodium dodecyl sulfate
Injection: Inject for 5 s.
Detector: UV 214
Migration time: 9
Voltage: 20 kV
Model: Beckman P/ACE 2100

OTHER SUBSTANCES
Simultaneous: impurities, degradation products, clavulanic acid

REFERENCE
Okafo,G.N.; Camilleri,P. Micellar electrokinetic capillary chromatography of amoxycillin and related molecules, *Analyst*, **1992**, *117*, 1421–1424.

SAMPLE
Matrix: solutions
Sample preparation: Dilute a solution in buffer 10-fold with water, filter, inject an aliquot. (Prepare buffer by dissolving 8.0 g KH_2PO_4 and 2.0 g K_2HPO_4 in 1 L water, adjust pH to 6.0 with 1 M HCl or 1 M NaOH.)

CAPILLARY ELECTROPHORESIS
Capillary: 50 cm × 50 µm fused-silica (Polymicro Technologies)
Capillary preparation: Before each run rinse with water for 80 s, with 1 M NaOH for 80 s, with water for 80 s, and with running buffer for 90 s then the ends are dipped into water
Capillary temperature: 20
Running buffer: 20 mM Sodium tetraborate containing 150 mM sodium dodecyl sulfate, pH 9.20 ± 0.05
Injection: Pressure injection at 5 psi for 6 s (42 nL).
Detector: UV 210
Migration time: 10
Voltage: 14 kV
Model: BioFocus 3000 (Bio-Rad)

OTHER SUBSTANCES
Simultaneous: ampicillin, ceftiofur, cephapirin, cloxacillin, penicillin G

REFERENCE
Cutting,J.H.; Hurlbut,J.A.; Sofos,J.N. Quantitation of penicillin G in medicated premix feeds by micellar electrokinetic capillary chromatography, *J.AOAC Int.*, **1997**, *80*, 951–955.

SAMPLE
Matrix: solutions

CAPILLARY ELECTROPHORESIS
Capillary: 60 cm × 50 µm fused-silica (47 cm to detector) (Polymicro Technologies)
Capillary temperature: 25
Running buffer: 20 mM pH 8.5 Sodium tetraborate containing 100 mM sodium dodecyl sulfate

Injection: Hydrodynamic injection at 50 mbar for 3.6 s (5 nL).
Detector: UV 205
Migration time: 8
Voltage: 22 kV
Model: Crystal 310 (Thermo Unicam)

OTHER SUBSTANCES
Simultaneous: ampicillin, cephapirin, cloxacillin, dicloxacillin, oxacillin, penicillin G, penicillin V, piperacillin, pyrimethamine, sulfacetamide, sulfadimethoxine, sulfaguanidine, sulfamerazine, sulfameter, sulfamethazine, sulfanilamide, sulfanilic acid, sulfapyridine, sulfaquinoxaline, sulfathiazole, sulfisoxazole, trimethoprim

REFERENCE
Hows,M.E.P.; Perrett,D.; Kay,J. Optimization of a simultaneous separation of sulphonamides, dihydrofolate reductase inhibitors and β-lactam antibiotics by capillary electrophoresis, *J.Chromatogr.A*, **1997**, *768*, 97–104.

SAMPLE
Matrix: solutions

CAPILLARY ELECTROPHORESIS
Capillary: 50 cm × 50 μm fused-silica (Polymicro Technologies)
Running buffer: 10 mM pH 7.0 Phosphate buffer
Injection: Electrokinetic injection
Detector: E, Model AFDRE-5 (Pine Instrument, Grove City CA), Au working electrode, Pt auxiliary electrode, Ag/AgCl reference electrode (design of cell described in paper)
Migration time: 10.5
Voltage: 10 kV
Model: laboratory-constructed

OTHER SUBSTANCES
Simultaneous: ampicillin, cloxacillin, penicillin G

REFERENCE
Owens,G.S.; LaCourse,W.R. Pulsed electrochemical detection of thiols and disulfides following capillary electrophoresis, *J.Chromatogr.B*, **1997**, *695*, 15–25.

Amphetamine

Molecular formula: $C_9H_{13}N$
Molecular weight: 135.21
CAS Registry No.: 300-62-9, 139-10-6 (phosphate), 60-13-9 (sulfate), 1407-85-8 (d-form tannate)
Merck Index (12th ed.): 623
Lednicer: 1 37, 70; 2 47

SAMPLE
Matrix: blood, urine
Sample preparation: Adjust pH of 2 mL urine or plasma to 10.5 with aqueous NaOH, extract gently with chloroform:isopropanol 90:10. Remove the organic layer and evaporate it to dryness under a stream of nitrogen, reconstitute the residue in 50 (urine) or 75 (plasma) μL running buffer, filter, inject an aliquot.

CAPILLARY ELECTROPHORESIS
Capillary: 60 cm × 75 μm AccuSep uncoated silica (52.5 cm to detector) (Waters)
Capillary preparation: At the beginning of each day purge with 500 mM KOH for 5 min, with water for 5 min, and with running buffer for 10 min.
Running buffer: 50 mM NaH_2PO_4 adjusted to pH 2.35 with phosphoric acid

Injection: Hydrostatic injection at 15 cm
Detector: UV 214
Migration time: 5.41
Voltage: 22 kV
Current: 135-145 μA
Model: Waters Quanta 4000

OTHER SUBSTANCES
Extracted: acepromazine, benzocaine, brompheniramine, butacaine, codeine, diazepam, doxapram, lidocaine, medazepam, methamphetamine, methapyrilene, methaqualone, phenmetrazine, procaine, tetrahydrozoline
Simultaneous: meclizine

KEY WORDS
plasma

REFERENCE
Chee,G.L.; Wan,T.S.M. Reproducible and high-speed separation of basic drugs by capillary zone electrophoresis, *J.Chromatogr.*, **1993**, *612*, 172–177.

SAMPLE
Matrix: bulk
Sample preparation: Dissolve 4 mg compound in 1 mL MeCN:water 50:50 containing 0.2% triethylamine, vortex for 30 s. Remove a 100 μL aliquot and add it to 100 μL 1.28% 2,3,4,6-tetra-O-acetyl-β-D-glucopyranosyl isothiocyanate in MeCN, vortex for 1 min, let stand for 15 min, make up to 1 mL with 10 mM pH 9.0 phosphate/borate buffer containing 100 mM sodium dodecyl sulfate, vortex for 20 s, filter (Whatman UniPrep), inject an aliquot of the filtrate.

CAPILLARY ELECTROPHORESIS
Capillary: 48 cm × 50 μm fused-silica (26 cm to detector) (Polymicro Technologies)
Capillary preparation: Condition new capillaries with 1 M NaOH for 10 min, with water for 10 min, and with running buffer for 10 min.
Capillary temperature: 30
Running buffer: MeOH:buffer 20:80 (Buffer was 10 mM pH 9.0 phosphate buffer containing 10 mM borate and 100 mM sodium dodecyl sulfate.)
Injection: Vacuum injection for 0.5 s.
Detector: UV 210
Migration time: 21.8 (-), 23.6 (+)
Voltage: 20 kV
Current: 48 μA
Model: Applied Biosystems Model 270A-HT

OTHER SUBSTANCES
Simultaneous: ephedrine, methamphetamine, norpseudoephedrine, phenylpropanolamine (norephedrine), pseudoephedrine

KEY WORDS
derivatization; chiral; comparison with HPLC

REFERENCE
Lurie,I.S. Micellar electrokinetic capillary chromatography of the enantiomers of amphetamine, methamphetamine and their hydroxyphenethylamine precursors, *J.Chromatogr.*, **1992**, *605*, 269–275.

SAMPLE
Matrix: bulk
Sample preparation: 10-50 mg Bulk compound + 1 mL 1 mg/mL caffeine in 10 mM HCl, make up to 10 mL with 10 mM HCl, sonicate for 2 min, mix thoroughly, filter (0.45 μm cellulose acetate), inject an aliquot.

CAPILLARY ELECTROPHORESIS
Capillary: 75 cm × 75 μm fused-silica (50 cm to the detector) (ISCO)

Capillary preparation: Flush with running buffer for 2 min between analyses. Condition capillary by filling with 1 M NaOH and allowing to stand for 1 h, fill with 100 mM NaOH and allow to stand for 1 h, wash with water, fill with running buffer. Each week wash capillary with 100 mM HCl for 10 min, with water, with 100 mM NaOH, and with water then fill with running buffer.
Capillary temperature: 30
Running buffer: DMSO:ethanolamine:buffer 11:1:88 (Buffer was 0.92 g cetyltrimethylammonium bromide in 100 mL 10 mM sodium tetraborate, adjust pH to 11.5 with 1 M NaOH.)
Injection: Load under vacuum, vacuum level 2, 10.0 kPa s
Detector: UV 254
Migration time: 8
Internal standard: caffeine (6.5)
Voltage: -15 kV
Model: ISCO Model 3140 electropherograph

OTHER SUBSTANCES
Simultaneous: p-aminobenzoic acid, ephedrine, methamphetamine, methylenedioxyamphetamine, methylenedioxymethamphetamine, norephedrine, pseudoephedrine, pseudonorephedrine

REFERENCE
Trenerry,V.C.; Robertson,J.; Wells,R.J. Analysis of illicit amphetamine seizures by capillary electrophoresis, *J.Chromatogr.A*, **1995**, *708*, 169–176.

SAMPLE
Matrix: dialysate
Sample preparation: Mix 30 μL dialysate (artificial CSF) with 75 μL buffer and 15 μL 100 μM fluorescein isothiocyanate (isomer I) in acetone, let stand in the dark for 18 h, inject an aliquot.

CAPILLARY ELECTROPHORESIS
Capillary: 30 cm × 20-25 μm fused silica (20 cm to detector) (Polymicro Technologies)
Running buffer: 20 mM carbonate
Injection: Vacuum injection at 19 psi for 0.3 s (0.3 nL)
Detector: F ex 488 (Ar laser) em 520 high-pass filter (with 488 nm notch filter))
Migration time: 3
Voltage: 20 kV
Limit of detection: 3 nM

KEY WORDS
derivatization; rat; brain

REFERENCE
Páez,X.; Rada,P.; Tucci,S.; Rodriguez,N.; Hernández,L. Capillary electrophoresis-laser-induced fluorescence detection of amphetamine in the brain, *J.Chromatogr.A*, **1996**, *735*, 263–269.

SAMPLE
Matrix: formulations
Sample preparation: Add 3 mL buffer to 30 mg capsule contents, sonicate for 30 min, filter (0.2 μm Nylon), inject an aliquot of the filtrate. (Buffer was water acidified to pH 2.0 with concentrated phosphoric acid.)

CAPILLARY ELECTROPHORESIS
Capillary: 900 mm × 50 μm uncoated fused-silica (650 mm to detector)
Capillary preparation: Rinse capillary with buffer for 15 min at the start of each day and for 4 min before each analysis.
Capillary temperature: 31
Running buffer: 70 mM hydroxypropyl-β-cyclodextrin (average molar substitution 0.8, average molecular mass 1500) containing 30 mM tetramethylammonium chloride and 10 mM sodium dodecyl sulfate, pH 2.0
Injection: Vacuum injection at 28 kPa
Detector: UV 210
Migration time: 35.6 (L), 36.7 (D)
Voltage: +28 kV

Model: ISCO Model 3140

OTHER SUBSTANCES
Simultaneous: ephedrine, methamphetamine, methylephedrine, methylpseudoephedrine, nor-ephedrine, norpseudoephedrine, pseudoephedrine

---

KEY WORDS
capsules; chiral

REFERENCE
Flurer,C.L.; Lin,L.A.; Satzger,R.D.; Wolnik,K.A. Determination of ephedrine compounds in nutritional supplements by cyclodextrin-modified capillary electrophoresis, *J.Chromatogr.B*, **1995**, *669*, 133–139.

SAMPLE
Matrix: hair, urine
Sample preparation: Wash 100 mg hair with 0.3% Tween 20 in water, cut in small pieces, add 1 mL 250 mM HCl, heat at 45° overnight, neutralize with NaOH. Add the hair extract or 2 mL urine to a Toxi-tube A, extract the organic layer with 200 μL 10 mM phosphoric acid, inject an aliquot of the aqueous layer.

CAPILLARY ELECTROPHORESIS
Capillary: 45 cm × 50 μm fused-silica (37.5 cm to detector) (Composite Metal Services, Hallow, UK)
Capillary preparation: Condition with running buffer for 10 min before each run. Wash new capillaries with 1 M NaOH for 10 min, with 100 mM NaOH for 10 min, and with water for 10 min, and then condition with running buffer for 20 min.
Running buffer: 100 mM pH 2.5 Potassium phosphate buffer containing 15 mM β-cyclodextrin
Injection: Pressure injection at 35 mbar for 10 s.
Detector: UV 200
Migration time: 19.5, 20 (enantiomers)
Voltage: 10 kV
Model: Beckman P/ACE 2200
Limit of detection: 37 ng/mL

OTHER SUBSTANCES
Extracted: ephedrine, methamphetamine, 3,4-methylenedioxyamphetamine, 3,4-methylenedioxyethylamphetamine, 3,4-methylenedioxymethamphetamine

---

KEY WORDS
chiral; SPE

REFERENCE
Tagliaro,F.; Manetto,G.; Bellini,S.; Scarcella,D.; Smith,F.P.; Marigo,M. Simultaneous chiral separation of 3,4-methylenedioxymethamphetamine (MDMA), 3,4-methylenedioxyamphetamine (MDA), 3,4-methylenedioxyethylamphetamine (MDE), ephedrine, amphetamine and methamphetamine by capillary electrophoresis in uncoated and coated capillaries with native β-cyclodextrin as the chiral selector: Preliminary application to the analysis of urine and hair, *Electrophoresis*, **1998**, *19*, 42–50.

SAMPLE
Matrix: solutions

CAPILLARY ELECTROPHORESIS
Capillary: 100 cm × 50 μm fused-silica (50 cm to detector) (Isco)
Capillary preparation: Flush capillary with 10 μL running buffer between runs. Every 40 sample injections rinse capillary with 200 μL 1 M NaOH, with 200 μL water, and 200 μL running buffer. Before use fill capillary with 1 M NaOH and allow to stand for 1 h, fill with 100 mM NaOH, allow to stand for 1 h, wash with water fill with running buffer.
Capillary temperature: 23
Running buffer: 100 mM citric acid containing 19.27 mM Na_2HPO_4 and 120 mM hydroxypropyl-β-cyclodextrin
Injection: Inject under vacuum at 4.0 kPa.s

Detector: UV 200
Migration time: 29 (S), 30 (R)
Voltage: 30 kV
Model: Isco Model 3140

OTHER SUBSTANCES
Simultaneous: 4-bromo-2,5-dimethoxyamphetamine, ephedrine, epinephrine, methamphetamine, methyldimethoxyethylamphetamine, methyldimethoxyamphetamine, methyldimethoxymethylamphetamine, pseudoephedrine

KEY WORDS
chiral

REFERENCE
Aumatell,A.; Wells,R.J.; Wong,D.K.Y. Enantiomeric differentiation of a wide range of pharmacologically active substances by capillary electrophoresis using modified β-cyclodextrins, *J.Chromatogr.A*, **1994**, *686*, 293–307.

SAMPLE
Matrix: solutions
Sample preparation: Prepare a 50 μg/mL solution in water:buffer:MeOH 90:9:1, inject an aliquot. (Buffer was 25 mM Tris buffer adjusted to pH 2.4 with phosphoric acid.)

CAPILLARY ELECTROPHORESIS
Capillary: 82 cm × 50 μm fused-silica (60 cm to detector)
Capillary preparation: Condition by aspirating with 1 M NaOH for 10 min, with water for 10 min, and with running buffer for 10 min.
Capillary temperature: 30
Running buffer: MeOH:buffer 1.2:98.8 (Buffer was 25 mM Tris buffer adjusted to pH 2.4 with phosphoric acid containing 5 mM heptakis(2,6-di-O-methyl)-β-cyclodextrin and 1 mM sulfobutyl ether β-cyclodextrin (ISCO)
Injection: Vacuum injection for 0.5 s (2 nL)
Detector: UV 210
Migration time: 12.92 (-), 13.28 (+)
Voltage: 30 kV
Model: Applied Biosystems Model 270A-HT

OTHER SUBSTANCES
Simultaneous: cathinone, cocaine, ephedrine, methamphetamine, methcathinone, norephedrine, norpseudoephedrine, propoxyphene, pseudoephedrine

KEY WORDS
chiral

REFERENCE
Lurie,I.S.; Klein,R.F.X.; Dal Cason,T.A.; LeBelle,M.J.; Brenneisen,R.; Weinberger,R.E. Chiral resolution of cationic drugs of forensic interest by capillary electrophoresis with mixtures of neutral and anionic cyclodextrins, *Anal.Chem.*, **1994**, *66*, 4019–4026.

SAMPLE
Matrix: solutions

CAPILLARY ELECTROPHORESIS
Capillary: 48.5 cm × 50 μm fused-silica (40 cm to detector (Hewlett-Packard)
Capillary temperature: 20 ± 0.1
Running buffer: 50 mM NaH_2PO_4 containing 10 mM β-cyclodextrin, adjusted to pH 2.5 with phosphoric acid
Detector: UV 214
Migration time: 8.504, 8.635 (enantiomers)
Voltage: 20 kV
Current: about 30 μA

Model: Hewlett-Packard HP-3D

OTHER SUBSTANCES
Also analyzed: 4-bromo-2,5-dimethoxyamphetamine, 2,5-dimethoxyamphetamine, 2,5-dimethoxymethamphetamine, 4-hydroxyamphetamine, methamphetamine, 4-methoxyamphetamine, 3,4-methylenedioxyethamphetamine (MDEA), 3,4-methylenedioxymethamphetamine (MDMA)

KEY WORDS
chiral; r = 1.015

REFERENCE
Cladrowa-Runge,S.; Hirz,R.; Kenndler,E.; Rizzi,A. Enantiomeric separation of amphetamine related drugs by capillary zone electrophoresis using native and derivatized β-cyclodextrin as chiral additives, *J.Chromatogr.A*, **1995**, *710*, 339–345.

SAMPLE
Matrix: solutions
Sample preparation: Mix a 100 μL aliquot of a 0.1-1 mM solution in 10 mM HCl with 500 μL 100 mM pH 9.5 borate buffer, 100 μL 2.5 mM N-acetyl-L-cysteine in 10 mM HCl, and 200 μL 5 mM o-phthalaldehyde in EtOH, let stand for 10 min, inject an aliquot.

CAPILLARY ELECTROPHORESIS
Capillary: 70 cm × 75 μm fused-silica
Capillary preparation: Before use wash capillary at 60° for 6 min each with water, 1 M NaOH, 100 mM NaOH, water, and running buffer.
Capillary temperature: 25
Running buffer: 100 mM pH 9.5 Borate buffer containing 10 mM sodium dodecyl sulfate
Injection: Hydrodynamic injection for 2 s.
Detector: UV 335
Migration time: 22.9, 23.4 (enantiomers)
Voltage: 20 kV
Model: Spectra Phoresis 1000 (Thermo Separation Products)

OTHER SUBSTANCES
Simultaneous: heptaminol, phenylpropanolamine

KEY WORDS
derivatization; chiral; comparison with HPLC; comparison with other derivatizing reagents

REFERENCE
Leroy,P.; Bellucci,L.; Nicolas,A. Chiral derivatization for separation of racemic amino and thiol drugs by liquid chromatography and capillary electrophoresis, *Chirality*, **1995**, *7*, 235–242.

SAMPLE
Matrix: solutions

CAPILLARY ELECTROPHORESIS
Capillary: 70 cm × 75 μm uncoated fused-silica (55 cm to detector)
Running buffer: 20 mM pH 2.8 Tris-phosphoric acid buffer containing 0.1% hydroxypropylmethylcellulose
Injection: Electrokinetic injection at 3 kV for 12 s
Detector: UV 190
Migration time: 8
Voltage: 21 kV
Current: 32 μA
Model: ATI Unicam Crystal 300

OTHER SUBSTANCES
Simultaneous: methamphetamine, propargylamphetamine, selegiline

---

REFERENCE
Szöko,É.; Magyar,K. Chiral separation of deprenyl and its major metabolites using cyclodextrin-modified capillary zone electrophoresis, *J.Chromatogr.A*, **1995**, *709*, 157–162.

SAMPLE
Matrix: solutions
Sample preparation: Inject an aliquot of a 300 µg/mL solution in water.

CAPILLARY ELECTROPHORESIS
Capillary: 65 cm × 50 µm fused-silica (40 cm to detector) (Isco)
Capillary preparation: Purge with running buffer for 3 min between injections. Purge daily with 1 M NaOH for 3 min, with water for 3 min, and with running buffer for 3 min.
Running buffer: Isopropanol:100 mM pH 7 phosphate buffer containing 25 mM rifamycin B 30: 70
Injection: Electrokinetic injection at 5 kV for 5 s.
Detector: UV 350
Migration time: 36.1 (first enantiomer, R_s 1.1)
Voltage: 8 kV
Model: Isco model 3850

OTHER SUBSTANCES
Simultaneous: alprenolol, epinephrine, metoprolol, norepinephrine, normetanephrine, octapamine, oxprenolol, propranolol

---

KEY WORDS
chiral

REFERENCE
Ward,T.J.; Dann,C.,III; Blaylock,A. Enantiomeric resolution using the macrocyclic antibiotics rifamycin B and rifamycin SV as chiral selectors for capillary electrophoresis, *J.Chromatogr.A*, **1995**, *715*, 337–344.

SAMPLE
Matrix: solutions
Sample preparation: Inject an aliquot of a 1 mg/mL solution in MeOH.

CAPILLARY ELECTROPHORESIS
Capillary: 67 cm × 50 µm (60 cm to detector) (Composite Metal Services, Worcs., UK)
Capillary preparation: Before each run rinse with running buffer for 3 min.
Capillary temperature: 20
Running buffer: 50 mM pH 10.5 Glycine buffer containing 50 mM sodium dodecyl sulfate
Injection: Pressure injection at 30 mbar for 5 s.
Detector: UV 220
Migration time: 17
Voltage: 25 kV
Model: Beckman P/ACE 2050

OTHER SUBSTANCES
Simultaneous: caffeine, codeine, diamorphine, morphine

REFERENCE
Hyötyläinen,T.; Sirén,H.; Riekkola,M.-L. Determination of morphine analogues, caffeine and amphetamine in biological fluids by capillary electrophoresis with the marker technique, *J.Chromatogr.A*, **1996**, *735*, 439–447.

SAMPLE
Matrix: solutions

CAPILLARY ELECTROPHORESIS
Capillary: 58.2 cm × 75 µm fused-silica (50.7 cm to detector) (Polymicro Technologies)

Capillary preparation: Purge with running buffer before each run. At the beginning of each day purge using 50-60 kPa vacuum with 500 mM NaOH for 5 min, with water for 5 min, with MeCN for 5 min, and with running buffer for 5 min.
Running buffer: MeCN:MeOH:acetic acid 49:50:1 containing 20 mM ammonium acetate
Injection: Hydrostatic injection at 10 cm for 5 s.
Detector: UV 214
Migration time: 3.33
Voltage: 25 kV
Model: Waters Quanta 4000

OTHER SUBSTANCES
Simultaneous: benzphetamine, ephedrine, nylidrin, oxymetazoline, phendimetrazine, phenmetrazine, phenylephrine, phenylpropanolamine, xylometazoline

REFERENCE
Leung,G.N.W.; Tang,H.P.O.; Tso,T.S.C.; Wan,T.S.M. Separation of basic drugs with non-aqueous capillary electrophoresis, *J.Chromatogr.A*, **1996**, *738*, 141–154.

SAMPLE
Matrix: solutions
Sample preparation: Inject an aliquot of a 20 μg/mL solution in 12.5 mM pH 9.3 borate buffer.

CAPILLARY ELECTROPHORESIS
Capillary: 66.5 cm × 75 μm fused-silica (41.5 cm to detector) (Polymicro Technologies)
Running buffer: 250 mM pH 9.3 Sodium borate buffer
Injection: Pressure injection at 50 mbar for 15 s.
Detector: UV 195
Migration time: 11
Voltage: 8 kV
Model: Crystal 310

OTHER SUBSTANCES
Simultaneous: chlorpheniramine, ephedrine, MDA, MDMA, methamphetamine, phentermine, phenylpropanolamine, pseudoephedrine
Noninterfering: caffeine

REFERENCE
Lilley,K.A.; Wheat,T.E. Drug identification in biological matrices using capillary electrophoresis and chemometric software, *J.Chromatogr.B*, **1996**, *683*, 67–76.

SAMPLE
Matrix: solutions
Sample preparation: Inject an aliquot of a 5-10 μM solution.

CAPILLARY ELECTROPHORESIS
Capillary: 70 cm × 75 μm fused-silica (55 cm to detector)
Capillary temperature: 21
Running buffer: 20 mM pH 2.7 Tris-phosphate buffer containing 0.5% hydroxypropylmethyl cellulose and 24 mM heptakis(2,6-di-O-methyl)-β-cyclodextrin (Hydroxypropylmethyl cellulose was from Sigma; viscosity of 2% solution = 4000 cP at 25°.)
Injection: Electrokinetic injection at 3 kV for 12 s.
Detector: UV 190
Migration time: 19.5, 19.7 (enantiomers)
Voltage: 21 kV
Model: Crystal 300 (ATI, Unicam)

OTHER SUBSTANCES
Simultaneous: epinephrine, methamphetamine, norepinephrine, selegiline (deprenyl)
Interfering: pseudoephedrine

KEY WORDS
chiral

REFERENCE
Szökö,E.; Gyimesi,J.; Barcza,L.; Magyar,K. Determination of binding constants and the influence of methanol on the separation of drug enantiomers in cyclodextrin modified capillary electrophoresis, *J.Chromatogr.A*, **1996**, *745*, 181–187.

SAMPLE
Matrix: solutions
Sample preparation: Inject an aliquot of a 1-50 µg/mL solution in water.

CAPILLARY ELECTROPHORESIS
Capillary: 55 cm × 50 µm fused-silica (35 cm to detector) (J&W)
Capillary preparation: Flush with running buffer for 5 min before each run. Periodically wash with 100 mM NaOH and flush extensively with running buffer.
Running buffer: MeOH:25 mM pH 9.24 borate buffer 20:80 containing 100 mM sodium dodecyl sulfate (A) or 50 mM pH 2.35 phosphate buffer (B) or 50 mM pH 9.24 borate buffer (C)
Injection: Injection of 5 µL using a split-flow injector ratio of 1:800.
Detector: UV 200
Migration time: 37.63 (A), 8.29 (B), 4.94 (C)
Voltage: 20 kV (A, B) or 12 kV (C)
Current: <60-80 µA
Model: ISCO Model 3850

OTHER SUBSTANCES
Also analyzed: acetylcodeine, barbital, caffeine, cocaine, codeine, diamorphine, diazepam, flunitrazepam, lidocaine, monoacetylmorphine, morphine, nalorphine, narceine, noscapine, papaverine, pentobarbital, procaine, tetracaine, thebaine

KEY WORDS
all compounds were separated with running buffer A; some peaks overlapped with running buffers B and C.

REFERENCE
Tagliaro,F.; Smith,F.P.; Turrina,S.; Equisetto,V.; Marigo,M. Complementary use of capillary zone electrophoresis and micellar electrokinetic capillary chromatography for mutual confirmation of results in forensic drug analysis, *J.Chromatogr.A*, **1996**, *735*, 227–235.

SAMPLE
Matrix: solutions
Sample preparation: Inject an aliquot of a solution in 100 mM sodium phosphate containing 100-300 µg/mL naphazoline.

CAPILLARY ELECTROPHORESIS
Capillary: 67 cm × 50 µm fused-silica (60 cm to detector)
Capillary preparation: Rinse with running buffer for 2 min between sets of analyses.
Capillary temperature: 30
Running buffer: 200 mM pH 4.5 Phosphate buffer
Injection: Injection at high pressure for 2 s.
Detector: UV 210, UV 230
Migration time: 13
Internal standard: naphazoline (14)
Voltage: 20 kV
Model: Beckman P/ACE 5500

OTHER SUBSTANCES
Simultaneous: acetylcodeine, cocaine, codeine, diamorphine, LSD, methadone, methamphetamine, methylenedioxyamphetamine, methylenedioxymethamphetamine, morphine, PCP, psilocyn

REFERENCE
Walker,J.A.; Marché,H.L.; Newby,N.; Bechtold,E.J. A free zone capillary electrophoresis method for the quantitation of common illicit drug samples, *J.Forensic Sci.*, **1996**, *41*, 824–829.

SAMPLE
Matrix: solutions

CAPILLARY ELECTROPHORESIS
Capillary: 102 cm × 75 μm fused-silica (80 cm to detector) (Yongnian Optical Fiber, Hebei, China)
Capillary preparation: Between injections rinse capillary with 100 mM NaOH for 2 min, with water for 2 min, and with running buffer for 5 min. Rinse new capillaries under vacuum with 500 mM NaOH for 30 min and with water for 30 min then equilibrate with running buffer for 10 min.
Running buffer: 50 mM Tris containing 12 mM β-cyclodextrin, adjusted to pH 2.3 with phosphoric acid
Injection: Electrokinetic injection.
Detector: UV 210
Migration time: 27.1, 27.6 (enantiomers)
Voltage: 26.5 kV
Model: laboratory-constructed

OTHER SUBSTANCES
Simultaneous: phenylephrine

KEY WORDS
chiral

REFERENCE
Wang,Z.; Sun,Y.; Sun,Z. Enantiomeric separation of amphetamine and phenylephrine by cyclodextrin-mediated capillary zone electrophoresis, *J.Chromatogr.A*, **1996**, *735*, 295–301.

SAMPLE
Matrix: solutions
Sample preparation: Inject an aliquot of a solution in MeCN:water 1:2.

CAPILLARY ELECTROPHORESIS
Capillary: 64.5 cm × 50 μm fused-silica (56 cm to detector) (Hewlett-Packard)
Capillary preparation: Flush capillary with running buffer for 8 min between runs.
Capillary temperature: 28
Running buffer: 118 mM Tris containing 16 mM hydroxypropyl-β-cyclodextrin, adjusted to pH 3.5 with phosphoric acid
Injection: Pressure injection at 1 bar (ca. 25 nL), ramp to operating voltage over 1 min.
Detector: UV 200
Migration time: 12.5 (l), 12.6 (d)
Voltage: 25 kV
Model: Hewlett-Packard HP[3D]

OTHER SUBSTANCES
Simultaneous: methamphetamine, 3,4-methylenedioxyamphetamine, 3,4-methylenedioxyethylamphetamine, 3,4-methylenedioxymethamphetamine

KEY WORDS
chiral

REFERENCE
Varesio,E.; Gauvrit,J.-Y.; Longeray,R.; Lantéri,P.; Veuthey,J.-L. Central composite design in the chiral analysis of amphetamines by capillary electrophoresis, *Electrophoresis*, **1997**, *18*, 931–937.

SAMPLE
Matrix: solutions

CAPILLARY ELECTROPHORESIS
Capillary: 85 cm × 50 μm fused-silica (Polymicro Technologies)

Capillary preparation: Between runs rinse capillary with running buffer for 5 min. Before use rinse capillary with 1 M NaOH for 10-15 min and with water for 10-15 min.
Running buffer: 25 mM pH 3 Citrate buffer
Injection: Pressure injection at 30 mbar for 12 s.
Detector: MS, laboratory-constructed, electrospray 90°, sheath liquid MeOH:water:acetic acid 80: 20:0.1 at 1 μL/min
Migration time: 9.5
Voltage: 30 kV
Current: 7.8 μA
Model: Crystal CE 300

OTHER SUBSTANCES
Simultaneous: cocaine, diamorphine, methamphetamine, procaine, tetracaine

REFERENCE
Lazar,I.M.; Naisbitt,G.; Lee,M.L. Capillary electrophoresis-time-of-flight mass spectrometry of drugs of abuse, *Analyst*, **1998**, *123*, 1449–1454.

SAMPLE
Matrix: urine
Sample preparation: Condition a Bond Elut Certify SPE cartridge with 2 mL MeOH and 2 mL 100 mM pH 6 phosphate buffer, do not allow to go dry. 5 mL Urine + 2 mL 100 mM pH 6 phosphate buffer, add to the SPE cartridge, wash with 1 mL 1 M acetic acid, wash with 6 mL MeOH, elute with 2 mL ethyl acetate containing 2% ammonium hydroxide. Evaporate the eluate to dryness under a stream of nitrogen at room temperature, reconstitute the residue in 100 μL running buffer, inject an aliquot.

CAPILLARY ELECTROPHORESIS
Capillary: 90 cm × 75 μm fused-silica (70 cm to detector) (Polymicro Technologies)
Capillary preparation: Before each run rinse capillary with 100 mM NaOH for 3 min and with buffer for 5 min.
Running buffer: 10 mM pH 9.1 Na_2HPO_4 containing 6 mM sodium borate and 75 mM sodium dodecyl sulfate
Injection: Gravity injection at 34 cm for 5 s
Detector: UV 195
Migration time: 25
Voltage: 20 kV
Current: 76-80 μA

OTHER SUBSTANCES
Extracted: benzoylecgonine, cocaine, methaqualone, morphine

KEY WORDS
SPE

REFERENCE
Wernly,P.; Thormann,W. Analysis of illicit drugs in human urine by micellar electrokinetic capillary chromatography with on-column fast scanning polychrome absorption detection, *Anal.Chem.*, **1991**, *63*, 2878–2882.

SAMPLE
Matrix: urine
Sample preparation: Condition a Bond Elut Certify SPE cartridge with 2 mL MeOH and 2 mL buffer. 2 mL Urine + 2 mL buffer, vortex for 15 s, add to the SPE cartridge, wash with 1 mL 1 M acetic acid, dry under vacuum for 2 min, elute with 5 mL 2% ammonia in ethyl acetate (freshly prepared). Add 2 drops acetic acid to the eluate and evaporate it to dryness under a stream of nitrogen at room temperature, reconstitute the residue in 1 mL 100 mM HCl, inject an aliquot. (Prepare buffer by dissolving 1.212 g NaH_2PO_4 in water, adjusting to pH 6.0 with 1 M NaOH, and making up to 100 mL with water.)

CAPILLARY ELECTROPHORESIS
Capillary: 64.5 cm × 50 μm fused-silica (56 cm to detector) (Hewlett-Packard)

Capillary preparation: Before each run flush with 100 mM phosphoric acid for 2 min and with running buffer for 5 min. Flush new capillaries with 1 M NaOH for 3 min, with 100 mM NaOH for 5 min, and with water for 10 min.

Capillary temperature: 15

Running buffer: 200 mM pH 2.5 Phosphate buffer containing 20 mM (2-hydroxy)propyl-β-cyclodextrin (Prepare buffer by dissolving 1.153 g 85% phosphoric acid and 1.212 g NaH_2PO_4 in water, adjusting to pH 2.5 with 1 M NaOH, and making up to 100 mL with water. Dissolve 0.3 g (2-hydroxy)propyl-β-cyclodextrin in 10 mL buffer, filter (0.2 μm).)

Injection: Pressure injection at 600 mbar.s (14.3 nL)

Detector: UV 200

Migration time: 15.7 (l), 16 (d)

Internal standard: phenylethylamine (12.5)

Voltage: 25 kV (ramp to 25 kV over 1 min)

Model: Hewlett-Packard HP3D CE

OTHER SUBSTANCES

Extracted: methamphetamine, methylenedioxyamphetamine, methylenedioxyethylamphetamine, methylenedioxymethamphetamine

---

KEY WORDS

chiral; SPE

REFERENCE

Varesio,E.; Veuthey,J.-L. Chiral separation of amphetamines by high-performance capillary electrophoresis, *J.Chromatogr.A*, **1995**, *717*, 219–228.

SAMPLE

Matrix: urine

Sample preparation: 50 μL Urine + 10 μL 210 μM IS + 20 μL concentrated HCl, heat at 80° for 1 h, cool in tap water, add 30 μL 28% ammonia, add 2 mL 100 mM pH 9.0 sodium carbonate/sodium bicarbonate buffer, add to a Sep-Pak C18 SPE cartridge, wash with 5 mL water, elute with 4 mL MeOH, add 400 μL acetic acid to the eluate. Evaporate to dryness, reconstitute the residue in 100 μL 50 mM pH 6.7 sodium borate/NaH_2PO_4 buffer, add 100 μL 2 mM 4-fluoro-7-nitro-2,1,3-benzoxadiazole (NBD-F) in EtOH, mix well, heat at 80° for 10 min, cool to room temperature, add 200 μL 150 mM HCl in EtOH, inject an aliquot.

CAPILLARY ELECTROPHORESIS

Capillary: 57 cm × 75 μm fused-silica (50 cm to detector)

Capillary preparation: Rinse with running buffer for 5 min before each run.

Capillary temperature: 25

Running buffer: 50 mM pH 9.3 sodium borate buffer containing 10 mM sodium dodecyl sulfate

Injection: Pressure injection at 0.5 psi for 1 s (6 nL).

Detector: F ex 488 (4 mW argon ion laser) em 520

Migration time: 34

Internal standard: 1-phenylethylamine (30)

Voltage: 15 kV

Model: Beckman P/ACE 5510

Limit of detection: 517 amole

OTHER SUBSTANCES

Extracted: methamphetamine

KEY WORDS

derivatization; SPE

REFERENCE

Kuroda,N.; Nomura,R.; Al-Dirbashi,O.; Akiyama,S.; Nakashima,K. Determination of methamphetamine and related compounds by capillary electrophoresis with UV and laser-induced fluorescence detection, *J.Chromatogr.A*, **1998**, *798*, 325–334.

SAMPLE

Matrix: urine

Sample preparation: 5 mL Urine + 100 μL 500 μM IS + 1 mL concentrated HCl, heat at 80° for 1 h, cool in tap water, neutralize with 1.5 mL 28% ammonia, add 3.5 mL 500 mM pH 10.5 sodium borate buffer containing 4.5 M NaCl, add 2 mL chloroform:isopropanol 75:25, shake vigorously for 1 min, centrifuge at 1500 g for 10 min, repeat the extraction. Combine the organic layers and wash them with 2.5 mL water, centrifuge at 1500 g for 10 min, add 400 μL acetic acid, evaporate to dryness, reconstitute the residue in 200 μL water, filter (0.45 μm), inject an aliquot.

CAPILLARY ELECTROPHORESIS
Capillary: 50 cm × 75 μm fused-silica (37.5 cm to detector)
Capillary preparation: Rinse with running buffer for 5 min before each run.
Capillary temperature: 25
Running buffer: 50 mM pH 3.0 Sodium acetate/HCl buffer (A) or 50 mM pH 9.3 sodium borate buffer containing 15 mM sodium dodecyl sulfate (B)
Injection: Siphon injection at 20 mM for 20 s (9 nL (A), 17 nL (B))
Detector: UV 210
Migration time: 11 (A), 12.5 (B)
Internal standard: 1-phenylethylamine (10.5 (A), 8.2 (B))
Voltage: 10 kV (A), 15 kV (B)
Model: Otsuka Electronics CAPI-3100A
Limit of detection: 350-580 fmole

OTHER SUBSTANCES
Extracted: methamphetamine

REFERENCE
Kuroda,N.; Nomura,R.; Al-Dirbashi,O.; Akiyama,S.; Nakashima,K. Determination of methamphetamine and related compounds by capillary electrophoresis with UV and laser-induced fluorescence detection, *J.Chromatogr.A*, **1998**, *798*, 325–334.

Ampicillin

Molecular formula: $C_{16}H_{19}N_3O_4S$

Molecular weight: 349.41

CAS Registry No.: 69-53-4 (anhydrous), 32388-53-7 (monohydrate), 23277-71-6 (K salt), 7177-48-2 (trihydrate), 69-52-3 (Na salt)

Merck Index (12th ed.): 628

Lednicer: 1 413, 2 437, 4 179

SAMPLE
Matrix: bulk
Sample preparation: Prepare a solution in pH 6 phosphate buffer, inject an aliquot.

CAPILLARY ELECTROPHORESIS
Capillary: 80 cm × 50 μm fused silica (75 cm to detector) (Polymicro Technologies or Isco)
Capillary preparation: Rinse with running buffer for 3 min between runs.
Running buffer: 100 mM pH 8 NaH_2PO_4 containing 50 mM sodium borate and 50 mM sodium dodecyl sulfate
Injection: Gravity injection at 10 cm for 15 s
Detector: UV 205
Migration time: 17.97
Voltage: +18 kV
Model: Dionex

OTHER SUBSTANCES
Simultaneous: degradation products, impurities, amoxicillin, penicillin G, penicillin V

REFERENCE
Flurer,C.L.; Wolnik,K.A. Chemical profiling of pharmaceuticals by capillary electrophoresis in the determination of drug origin, *J.Chromatogr.A*, **1994**, *674*, 153–163.

SAMPLE
Matrix: solutions
Sample preparation: Prepare a 0.5-2 mg/mL solution in water, inject an aliquot.

CAPILLARY ELECTROPHORESIS
Capillary: 65 cm × 50 μm fused-silica (50 cm to detector) (Scientific Glass Engineering)
Running buffer: 20 mM NaH_2PO_4 containing 150 mM sodium dodecyl sulfate adjusted to pH 9.0 with 20 mM sodium tetraborate
Injection: Injection by siphon at 5 cm for 5-10 s
Detector: UV 210
Migration time: 9.5
Voltage: 20 kV

OTHER SUBSTANCES
Simultaneous: amoxicillin, aspoxicillin, carbenicillin, penicillin G, piperacillin, sulbenicillin
Also analyzed: cefmenoxime, cefminox, cefoperazone, cefotaxime, cefpimizole, cefpiramide, ceftazidime, ceftriaxone

REFERENCE
Nishi,H.; Tsumagari,N.; Kakimoto,T.; Terabe,S. Separation of β-lactam antibiotics by micellar electrokinetic chromatography, *J.Chromatogr.*, **1989**, *477*, 259–270.

SAMPLE
Matrix: solutions
Sample preparation: Inject directly.

CAPILLARY ELECTROPHORESIS
Capillary: 65 cm × 50 μm untreated fused-silica (50 cm to detector) (SGE)
Capillary preparation: Flush with running buffer every 5 runs. At the end of each day fill capillary with 100 mM KOH, let stand for 30 min, flush with water, let stand for 5 min, fill with running buffer.
Running buffer: pH 8.5 Buffer containing 100 mM sodium dodecyl sulfate (Prepare buffer by mixing 20 mM NaH_2PO_4 solution with 20 mM sodium borate solution to achieve a pH of 8.5.)
Injection: Inject by siphoning at 10 cm for 5-10 s.
Detector: UV 210
Migration time: 11.2

OTHER SUBSTANCES
Simultaneous: amoxicillin, aspoxicillin, carbenicillin, penicillin G, piperacillin, sulbenicillin

REFERENCE
Nishi,H.; Fukuyama,T.; Matsuo,M. Separation and determination of aspoxicillin in human plasma by micellar electrokinetic chromatography with direct sample injection, *J.Chromatogr.*, **1990**, *515*, 245–255.

SAMPLE
Matrix: solutions
Sample preparation: Prepare a 120 μg/mL solution in water, inject an aliquot.

CAPILLARY ELECTROPHORESIS
Capillary: 60 cm × 75 μm
Running buffer: 20 mM NaH_2PO_4 containing 50 mM sodium dodecyl sulfate, adjusted to pH 9.0 with sodium tetraborate
Injection: Hydrodynamic injection at 10 cm for 5 s
Detector: UV 214
Migration time: 8.0
Voltage: 18 kV
Model: Waters Quanta 4000

OTHER SUBSTANCES
Simultaneous: 6-aminopenicillanic acid, amoxicillin, cloxacillin, dicloxacillin, nafcillin, oxacillin, ticarcillin

REFERENCE
Swartz,M.E. Method development and selectivity control for small molecule pharmaceutical separations by capillary electrophoresis, *J.Liq.Chromatogr.*, **1991**, *14*, 923–938.

SAMPLE
Matrix: solutions
Sample preparation: Dilute a solution in buffer 10-fold with water, filter, inject an aliquot. (Prepare buffer by dissolving 8.0 g KH_2PO_4 and 2.0 g K_2HPO_4 in 1 L water, adjust pH to 6.0 with 1 M HCl or 1 M NaOH.)

CAPILLARY ELECTROPHORESIS
Capillary: 50 cm × 50 μm fused-silica (Polymicro Technologies)
Capillary preparation: Before each run rinse with water for 80 s, with 1 M NaOH for 80 s, with water for 80 s, and with running buffer for 90 s then the ends are dipped into water
Capillary temperature: 20
Running buffer: 20 mM Sodium tetraborate containing 150 mM sodium dodecyl sulfate, pH 9.20 ± 0.05
Injection: Pressure injection at 5 psi for 6 s (42 nL).
Detector: UV 210
Migration time: 11.7
Voltage: 14 kV
Model: BioFocus 3000 (Bio-Rad)

OTHER SUBSTANCES
Simultaneous: amoxicillin, ceftiofur, cephapirin, cloxacillin, penicillin G

REFERENCE
Cutting,J.H.; Hurlbut,J.A.; Sofos,J.N. Quantitation of penicillin G in medicated premix feeds by micellar electrokinetic capillary chromatography, *J.AOAC Int.*, **1997**, *80*, 951–955.

SAMPLE
Matrix: solutions

CAPILLARY ELECTROPHORESIS
Capillary: 60 cm × 50 μm fused-silica (47 cm to detector) (Polymicro Technologies)
Capillary temperature: 25
Running buffer: 20 mM pH 8.5 Sodium tetraborate containing 100 mM sodium dodecyl sulfate
Injection: Hydrodynamic injection at 50 mbar for 3.6 s (5 nL).
Detector: UV 205
Migration time: 9.3
Voltage: 22 kV
Model: Crystal 310 (Thermo Unicam)

OTHER SUBSTANCES
Simultaneous: amoxicillin, cephapirin, cloxacillin, dicloxacillin, oxacillin, penicillin G, penicillin V, piperacillin, pyrimethamine, sulfacetamide, sulfadimethoxine, sulfaguanidine, sulfamerazine, sulfameter, sulfamethazine, sulfanilamide, sulfanilic acid, sulfapyridine, sulfaquinoxaline, sulfathiazole, sulfisoxazole, trimethoprim

REFERENCE
Hows,M.E.P.; Perrett,D.; Kay,J. Optimization of a simultaneous separation of sulphonamides, dihydrofolate reductase inhibitors and β-lactam antibiotics by capillary electrophoresis, *J.Chromatogr.A*, **1997**, *768*, 97–104.

SAMPLE
Matrix: solutions

CAPILLARY ELECTROPHORESIS
Capillary: 50 cm × 50 μm fused-silica (Polymicro Technologies)
Running buffer: 10 mM pH 7.0 Phosphate buffer
Injection: Electrokinetic injection
Detector: E, Model AFDRE-5 (Pine Instrument, Grove City CA), Au working electrode, Pt auxiliary electrode, Ag/AgCl reference electrode (design of cell described in paper)
Migration time: 11
Voltage: 10 kV
Model: laboratory-constructed

OTHER SUBSTANCES
Simultaneous: amoxicillin, cloxacillin, penicillin G

REFERENCE
Owens,G.S.; LaCourse,W.R. Pulsed electrochemical detection of thiols and disulfides following capillary electrophoresis, *J.Chromatogr.B*, **1997**, *695*, 15–25.

Angiotensin II

Molecular formula: $C_{50}H_{71}N_{13}O_{12}$
Molecular weight: 1046.19
CAS Registry No.: 4474-91-3
Merck Index (12th ed.): 689

SAMPLE
Matrix: blood
Sample preparation: 20 mL Whole blood + 2 mmoles EDTA + 8 mg enalaprilat, centrifuge immediately at 4° at 6000 g for 15 min. Remove the plasma and add HCl to a final concentration of 50 mM, centrifuge at 400000 g for 1 h. Add a 3 mL aliquot of the supernatant to a Sep-Pak C18 SPE cartridge, wash twice with 1 mL portions of MeOH:water 5:95, elute with 2 mL MeOH. Filter (0.45 μm) the eluate and dry it under vacuum, reconstitute with 75 μL 100 mM pH 7.5 phosphate buffer, centrifuge at 400000 g for 1 h, inject an aliquot.

CAPILLARY ELECTROPHORESIS
Capillary: 60 cm × 75 μm fused-silica
Running buffer: 100 mM pH 1.95 Phosphoric acid
Injection: Hydrostatic injection for 4 s (20 nL).
Detector: UV 185
Migration time: 12
Voltage: 10 kV
Model: Waters Quanta 4000
Limit of detection: 1 μM

OTHER SUBSTANCES
Extracted: metabolites, angiotensin I

KEY WORDS
plasma; whole blood; SPE

REFERENCE
Lim,B.C.; Sim,M.K. Determination of angiotensins by capillary electrophoresis, *J.Chromatogr.B*, **1994**, *655*, 127–131.

SAMPLE
Matrix: solutions
Sample preparation: Prepare a 50 μg/mL solution in MeOH:30 mM pH 2.5 NaH_2PO_4 10:90, inject an aliquot.

CAPILLARY ELECTROPHORESIS
Capillary: 44 cm × 75 μm fused-silica
Running buffer: 30 mM pH 2.5 NaH_2PO_4
Injection: Hydrodynamic injection at 20 cm for 5 s
Detector: UV 200
Migration time: 5.5
Voltage: 25 kV
Current: ca. 100 μA

OTHER SUBSTANCES
Simultaneous: oxytocin, bombesin, bradykinin, leucine enkephalin, luteinizing hormone-releasing hormone (LHRH), α-melanocyte stimulating hormone (αMSH), methionine enkephalin, thyrotrophin releasing hormone (TRH)

REFERENCE
Perrett,D.; Birch,A.; Ross,G. Capillary electrophoresis for peptides, including neuropeptides, *Biochem.Soc. Trans.*, **1994**, *22*, 127–131.

SAMPLE
Matrix: solutions
Sample preparation: Mix 10 μL of a 1 mM solution in water with 100 μL 100 mM pH 9.5 carbonate buffer, add 40 μL 1 mg/mL fluorescein isothiocyanate in MeOH, heat at 30-35° for 4 h, dilute with running buffer, inject an aliquot.

CAPILLARY ELECTROPHORESIS
Capillary: 50 cm × 50 μm fused-silica (Polymicro Technologies)
Running buffer: 25 mM pH 8.4 Borate buffer
Injection: Hydrodynamic injection at 20 cm for 5 s.
Detector: F (Wavelengths not given)
Migration time: 5.2
Voltage: 20 kV

OTHER SUBSTANCES
Simultaneous: leucine enkephalin, methionine enkephalin

KEY WORDS
derivatization

REFERENCE
Liu,Y.-M.; Sweedler,J.V. Two-dimensional separations: Capillary electrophoresis coupled to channel gel electrophoresis, *Anal.Chem.*, **1996**, *68*, 3928–3933.

SAMPLE
Matrix: solutions

CAPILLARY ELECTROPHORESIS
Capillary: 100 cm × 50 μm fused-silica
Capillary preparation: Condition capillary with 100 mM KOH for 15 min and with running buffer for 15 min.
Capillary temperature: 22
Running buffer: 0.5% pH 2.5 Phosphoric acid
Injection: Electrokinetic at 5 kV for 45 s.
Detector: UV 214
Migration time: 0.425
Model: Waters Quanta 4000E
Limit of detection: 4.2 μg/mL

OTHER SUBSTANCES
Simultaneous: angiotensin I, angiotensin III

REFERENCE
Prince,S.J.; Allen,L.V. Analysis of benzalkonium chloride and its homologs: HPLC vs HPCE (Abstract APQ 1093), *Pharm.Res.*, **1996**, *13*, S26–S26.

SAMPLE
Matrix: solutions
Sample preparation: Mix 4 μL of a solution in pH 9.0 buffer with 1 μL 1 mM 2-methyl-3-oxo-4-phenyl-2,3-dihydrofuran-2-yl acetate in DMSO, let stand at room temperature for 15 min, inject an aliquot. (Dry DMSO over molecular sieve before use.)

CAPILLARY ELECTROPHORESIS
Capillary: 60 cm × 50 μm coated capillary (45 cm to detector)
Capillary preparation: Coat the capillaries as follows. Treat capillary with 100 mM NaOH for 1 h, rinse with water, rinse with MeOH, fill capillary under nitrogen pressure with 10 μL (gamma-methacryloxypropyl)trimethoxysilane dissolved in dichloromethane containing 20 mM acetic acid, after 1 h rinse with MeOH, rinse with water, pass 4% acrylamide solution containing 1 μL/mL tetramethylethylenediamine and 1 mg/mL ammonium persulfate through the capillary under nitrogen pressure for 30 min (Caution! Acrylamide is a carcinogen!), rinse with water, dry under a stream of nitrogen (Anal. Chem. 1994, 66, 1134).
Running buffer: MeOH:buffer 10:90 (Buffer contained 150 mM Tris and 150 mM boronic acid, pH 8.3.)
Injection: Hydrodynamic injection for 5 s.
Detector: F ex 325 (He-Cd laser) em 500
Migration time: 15.5
Voltage: 25 kV
Current: ca. 5 μA
Model: laboratory-constructed
Limit of detection: <1 pmole

OTHER SUBSTANCES
Simultaneous: leucine enkephalin, methionine enkephalin, Val-Tyr-Val

KEY WORDS
derivatization; coated capillary; comparison with HPLC; comparison with fluorescamine derivatization

REFERENCE
Chen,P.; Novotny,M.V. 2-Methyl-3-oxo-4-phenyl-2,3-dihydrofuran-2-yl acetate: A fluorogenic reagent for detection and analysis of primary amines, *Anal.Chem.*, **1997**, *69*, 2806–2811.

SAMPLE
Matrix: solutions

CAPILLARY ELECTROPHORESIS
Capillary: 80 cm × 75 μm (60 cm to detector) (Bester, Netherlands)
Running buffer: 20 mM Ammonium acetate containing 40 mM acetic acid, pH 4.4
Injection: 77 nL
Detector: UV 214
Migration time: 8
Voltage: 30 kV
Model: Lauerlabs Prince

OTHER SUBSTANCES
Simultaneous: lys-bradykinin

KEY WORDS
also details of MS detection

REFERENCE

Hoitink,M.A.; Hop,E.; Beijnen,J.H.; Kettenes-van den Bosch,J.J.; Underberg,W.J.M. Capillary zone electrophoresis-mass spectrometry as a tool in the stability research of the luteinising hormone-releasing hormone analogue goserelin, *J.Chromatogr.A*, **1997**, *776*, 319–327.

SAMPLE

Matrix: solutions
Sample preparation: Inject an aliquot of a solution in 5-fold diluted running buffer.

CAPILLARY ELECTROPHORESIS

Capillary: 35 cm × 75 μm fused-silica coated with polyacrylamide (Polymicro Technologies)
Capillary preparation: Before each run purge with 100 mM HCl for 2 min, with water for 4 min, and with running buffer for 1 min. (Coat the capillary as follows. Adjust the pH of 20 mL water to 3.5 with acetic acid, add 80 μL 3-(trimethoxysilyl)propyl methacrylate (3-methacryloxypropyltrimethoxysilane), mix, suck into capillary, let stand at room temperature for 1 h, remove the solution, wash with water. Fill the capillary with a deaerated 3-4% acrylamide solution containing 1 μL/mL N,N,N',N'-tetramethylethylenediamine and 1 mg/mL potassium persulfate, let stand for 30 min, remove excess solution by aspiration, rinse with water, remove water by aspiration, dry at 35° (J. Chromatogr. 1985, 347, 191).)
Running buffer: 100 mM pH 2.5 Sodium phosphate buffer
Injection: Pressure injection at 300 mbar.s.
Detector: UV 191
Migration time: 11.5
Voltage: 8 kV
Model: Hewlett-Packard HP³ᴰ CE

OTHER SUBSTANCES

Simultaneous: bombesin, bradykinin, gonadorelin, leucine enkephalin, melanocyte stimulating hormone, methionine enkephalin, oxytocin, protirelin

KEY WORDS

coated capillary

REFERENCE

Zhang,R.; Zhang,H.X.; Eaker,D.; Hjerten,S. The effect of β-cyclodextrins as buffer additives on the (capillary) electrophoretic separation of peptides and proteins, *J.Capillary Electrophor.*, **1997**, *4*, 105–112.

Antipyrine

Molecular formula: $C_{11}H_{12}N_2O$
Molecular weight: 188.23
CAS Registry No.: 60-80-0, 520-07-0 (salicylate)
Merck Index (12th ed.): 757
Lednicer: 1 234

SAMPLE

Matrix: blood
Sample preparation: Inject plasma directly.

CAPILLARY ELECTROPHORESIS

Capillary: 60 cm × 50 μm fused-silica (40 cm to detector) (Scientific Glass Engineering)
Capillary preparation: After each injection flush capillary with 100 mM NaOH.
Capillary temperature: 25
Running buffer: pH 8.0 Phosphate buffer (I = 0.05) containing 10 mM sodium dodecyl sulfate
Injection: Siphon at 5 cm for 10 s
Detector: UV 280
Migration time: 5

Voltage: 15 kV
Current: about 11 μA

OTHER SUBSTANCES
Extracted: cefpiramide

KEY WORDS
plasma; detection at anode

REFERENCE
Nakagawa,T.; Oda,Y.; Shibukawa,A.; Tanaka,H. Separation and determination of cefpiramide in human plasma by electrokinetic chromatography with a micellar solution and an open tubular-fused silica capillary, *Chem.Pharm.Bull.*, **1988**, *36*, 1622–1625.

SAMPLE
Matrix: blood
Sample preparation: Inject directly.

CAPILLARY ELECTROPHORESIS
Capillary: 75 cm × 50 μm fused-silica (55 cm to detector) (SGE)
Capillary preparation: Wash capillary with 1 M NaOH between runs
Capillary temperature: 37
Running buffer: 50 mM pH 8 Phosphate buffer containing 10 mM sodium dodecyl sulfate
Injection: Siphon at 5 cm for 15 s
Detector: UV 280
Migration time: 7.5
Voltage: +15 kV

OTHER SUBSTANCES
Extracted: cefpiramide

KEY WORDS
plasma

REFERENCE
Nakagawa,T.; Oda,Y.; Shibukawa,A.; Fukuda,H.; Tanaka,H. Electrokinetic chromatography for drug analysis. Separation and determination of cefpiramide in human plasma, *Chem.Pharm.Bull.*, **1989**, *37*, 707–711.

SAMPLE
Matrix: blood
Sample preparation: 25 μL Serum + 100 μL 2.5 μg/mL acetaminophen in cold MeCN, vortex for 15 s, centrifuge at 3000 g for 5 min. Remove the supernatant and evaporate it to dryness under a stream of air at room temperature for 2 h, reconstitute the residue in 25 μL water, inject an aliquot.

CAPILLARY ELECTROPHORESIS
Capillary: 70 cm × 75 μm untreated fused silica (65 cm to detector) (Dionex)
Capillary preparation: Before injection rinse capillary with 500 μL 500 mM pH 2.3 phosphoric acid, 500 μL water, 500 μL 500 mM pH 13.4 NaOH, 500 μL water, then fill with running buffer.
Running buffer: 50 mM Boric acid containing 10 mM sodium tetraborate and 50 mM sodium dodecyl sulfate, pH 8.2
Injection: Gravity injection at 100 mm for 10 s (about 13 nL)
Detector: UV 242
Migration time: 7.2
Internal standard: acetaminophen (5.3)
Voltage: 20 kV
Current: 65 μA
Model: Dionex CES-I
Limit of quantitation: 500 ng/mL
Limit of detection: 250 ng/mL

OTHER SUBSTANCES
Simultaneous: sulfanilamide
Noninterfering: benzoic acid, caffeine, furosemide, gentamicin, indocyanine green, ketamine, lidocaine, pentoxifylline, phenacetin, promethazine, riboflavin, sulfacetamide, theobromine, theophylline, triethanolamine, xylazine

KEY WORDS
rat; serum

REFERENCE
Brunner,L.J.; DiPiro,J.T.; Feldman,S. Serum antipyrine concentrations determined by micellar electrokinetic chromatography, *J.Chromatogr.*, **1993**, *622*, 98–102.

SAMPLE
Matrix: solutions

CAPILLARY ELECTROPHORESIS
Capillary: 60 cm × 50 μm fused-silica (25 cm to detector) (Polymicro Technologies)
Capillary preparation: Before use rinse capillary hydrodynamically at 520 kPa with 500 mM pH 13 EDTA for 30 min, with water for 20 min, and with running buffer for 10 min. Fill with running buffer.
Running buffer: 20 mM pH 11.0 Sodium phosphate buffer containing 40 mM sodium dodecyl sulfate
Injection: Vacuum injection of 2 nL.
Detector: UV 274
Migration time: 1.1
Voltage: 29 kV
Model: ISCO Model 3850

OTHER SUBSTANCES
Simultaneous: caffeine, paraxanthine, theobromine, theophylline, 1,3,7-trimethyluric acid

REFERENCE
Zhao,Y.; Lunte,C.E. Determination of caffeine and its metabolites by micellar electrokinetic capillary electrophoresis, *J.Chromatogr.B*, **1997**, *688*, 265–274.

SAMPLE
Matrix: solutions

CAPILLARY ELECTROPHORESIS
Capillary: 57 cm × 50 μm fused-silica (50 cm to detector)
Running buffer: 50 mM pH 8.0 Borate buffer containing 40 mM sodium taurodeoxycholate and 25 mM phosphatidylcholine (Prepare by adding phosphatidylcholine and stirring for 3-6 h until all cloudiness disappears. Phosphatidylcholine was 95% pure soybean lecithin, Epikuron, Lucas Meyer & Co.)
Injection: Inject a solution of the compound in the running buffer at 20 psi for 1 s, inject MeOH: water 5:95 at 20 psi for 1 s, inject a solution of halofantrine in running buffer at 20 psi for 1 s.
Detector: UV 214
Migration time: k' 0.17
Model: Beckman P/ACE 5000

OTHER SUBSTANCES
Also analyzed: acetaminophen, amoxicillin, aspirin, azathioprine, caffeine, captopril, carbamazepine, carprofen, chlorambucil, chlorpheniramine, chlorpromazine, cimetidine, clonidine, codeine, desipramine, diphenhydramine, ephedrine, fenoterol, flufenamic acid, flurbiprofen, haloperidol, hydroxyzine, ibuprofen, imipramine, indomethacin, ketoprofen, lidocaine, melphalan, metoprolol, nabumetone, nadolol, phenobarbital, phenol, promazine, propranolol, pyrilamine, ranitidine, ropinirole, salicylic acid, sulfamethoxazole, testosterone, theophylline, thioridazine, tiaprofenic acid, tolfenamic acid, trifluoperazine, trimethoprim, valproic acid, verapamil

KEY WORDS
comparison with HPLC; k' = (Tr-T0)/(T0(1-Tr/Tm)) where Tr = retention time of analyte; T0 = retention time of water; and Tm = retention time of marker (halofantrine)

REFERENCE
Hanna,M.; de Biasi,V.; Bond,B.; Salter,C.; Hutt,A.J.; Camilleri,P. Estimation of the partitioning characteristics of drugs: A comparison of a large and diverse drug series utilizing chromatographic and electrophoretic methodology, *Anal.Chem.*, **1998**, *70*, 2092–2099.

Aprobarbital

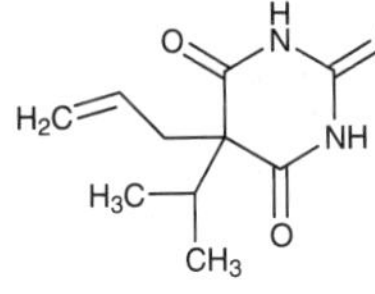

Molecular formula: $C_{10}H_{14}N_2O_3$
Molecular weight: 210.23
CAS Registry No.: 77 02-1
Merck Index (12th ed.): 794
Lednicer: 1 268

SAMPLE
Matrix: blood
Sample preparation: Condition a 1 mL 100 mg Bond Elut C18 SPE cartridge (Varian) with 3 mL MeOH and 3 mL pH 9.0 phosphate buffer, do not allow to dry. Add 1 mL serum to the SPE cartridge, wash with 3 mL buffer, allow to dry for 5 min, elute with 3 mL dichloromethane. Filter (nylon) the eluate and evaporate it to dryness under a stream of nitrogen, reconstitute with 1 mL MeOH:water 30:70, inject an aliquot.

CAPILLARY ELECTROPHORESIS
Capillary: 62 cm × 75 μm fused-silica (52 cm to detector) (Polymicro Technologies)
Capillary preparation: Before each run rinse capillary with 100 mM NaOH for 2 min and with running buffer for 2 min. Condition new capillaries by rinsing with 1 M NaOH for 10 min, with water for 10 min, with 100 mM HCl for 10 min, and with running buffer for 10 min.
Capillary temperature: 25
Running buffer: 50 mM NaH_2PO_4 containing 40 mM hydroxypropyl-gamma-cyclodextrin, adjusted to pH 9.0 with 100 mM NaOH
Injection: Vacuum injection at 0.5 psi for 10 s.
Detector: UV 254
Migration time: 10.86
Internal standard: aprobarbital
Voltage: 15 kV
Current: ca. 100 μA
Model: Beckman P/ACE 5000

OTHER SUBSTANCES
Extracted: pentobarbital
Simultaneous: clonazepam, diazepam, phenytoin

KEY WORDS
aprobarbital is IS; method is chiral for pentobarbital; serum; SPE

REFERENCE
Srinivasan,K.; Bartlett,M.G. Capillary electrophoresis stereoselective determination of *R*-(+)- and *S*-(-)-pentobarbital from serum using hydroxypropyl-γ-cyclodextrin, solid-phase extraction and ultraviolet detection, *J.Chromatogr.B*, **1997**, *703*, 289–294.

SAMPLE
Matrix: blood
Sample preparation: Condition a 1 mL 100 mg Bond Elut C18 SPE cartridge 3 mL MeOH and 3 mL 50 mM pH 9.0 sodium phosphate buffer, do not allow to dry. Add 1 mL serum to the SPE

cartridge, wash with 3 mL 50 mM pH 9.0 sodium phosphate buffer, allow to dry for 5 min, elute with 3 mL dichloromethane. Filter (0.2 μm nylon) the eluate and and evaporate it to dryness under a stream of nitrogen, reconstitute the residue in 1 mL MeOH:water 30:70, inject an aliquot.

CAPILLARY ELECTROPHORESIS

Capillary: 62 cm × 75 μm fused-silica (52 cm to detector) (Polymicro Technologies)
Capillary preparation: Before each run rinse capillary with 100 mM NaOH for 2 min and with running buffer for 2 min. Condition new capillaries by rinsing with 1 M NaOH for 10 min, with water for 10 min, with 100 mM HCl for 10 min, and with running buffer for 10 min.
Capillary temperature: 25
Running buffer: 50 mM pH 9.0 Sodium phosphate buffer containing 40 mM hydroxypropyl-gamma-cyclodextrin (Research Biochemicals International, Natick MA)
Injection: Vacuum injection at 0.5 psi for 10 s.
Detector: UV 254
Migration time: 11.86
Internal standard: aprobarbital
Voltage: 15 kV
Current: 100 μA
Model: Beckman P/ACE 5000

OTHER SUBSTANCES

Extracted: secobarbital
Simultaneous: clonazepam

KEY WORDS

aprobarbital is IS; serum; SPE

REFERENCE

Srinivasan,K.; Zhang,W.; Bartlett,M.G. Rapid simultaneous capillary electrophoretic determination of (*R*)- and (*S*)-secobarbital from serum and prediction of hydroxypropyl-γ-cyclodextrin-secobarbital stereoselective interaction using molecular mechanics simulation, *J.Chromatogr.Sci.*, **1998**, *36*, 85–90.

SAMPLE

Matrix: urine
Sample preparation: Place an extraction rod in 50 μL urine in a 50 mm × 1.5 mm ID length of PTFE tubing for 30 min, place the extraction rod into 5 μL 20-40 mM pH 11.5 phosphate buffer in a 50 mm × 1.2 mm ID length of PTFE tubing for 90 min, inject an aliquot of the buffer. (Prepare the solid-phase extraction rod as follows. Polish a 70 × 1.1 stainless steel rod with emery paper, clean with a Kimwipe and acetone, sonicate in EtOH, sonicate in THF, dip in 3% PVAM in THF momentarily, hold vertically for 1 min, air dry in a hood for at least 5 h, dip in PVC solution momentarily, hold vertically for 1 min, air dry for at least 5 h. The coating is 3 cm long. PVAM was poly(vinyl chloride-co-vinyl acetate-co-maleic acid) consisting of 86% vinyl chloride, 13% vinyl acetate, and 1% maleic acid. Prepare PVC solution by adding very high molecular weight PVC slowly to THF with stirring until the concentration reaches 3.6%, add Santicizer 141 to a concentration of 7.2%. Santicizer 141 (Monsanto) is 92% 2-ethylhexyl diphenyl phosphate, 5% di-2-ethylhexyl phenyl phosphate, and 3% triphenyl phosphate.)

CAPILLARY ELECTROPHORESIS

Capillary: 75 cm × 75 μm fused-silica (50 cm to detector) (Polymicro Technologies)
Running buffer: 50 mM Tris adjusted to pH 7.8 with 3-[N-tris(hydroxymethyl)methylamino]-2-hydroxypropanesulfonic acid (Tapso)
Injection: Pressure injection at 0.5 psi for 4 s.
Detector: UV 230
Migration time: 7.85
Voltage: 35 kV
Current: about 30 μA
Model: Isco 3850
Limit of detection: <1 ppm

OTHER SUBSTANCES

Extracted: amobarbital, butabarbital, butalbital, mephobarbital, pentobarbital, secobarbital, thiopental

Simultaneous: allobarbital, aspirin, phenobarbital

KEY WORDS
SPE

REFERENCE
Li,S.; Weber,S.G. Determination of barbiturates by solid-phase microextraction and capillary electrophoresis, *Anal.Chem.*, **1997**, *69*, 1217–1222.

Aprotinin

Molecular formula: $C_{284}H_{432}N_{84}O_{79}S_7$
Molecular weight: 6511.53
CAS Registry No.: 9087-70-1
Merck Index (12th ed.): 796

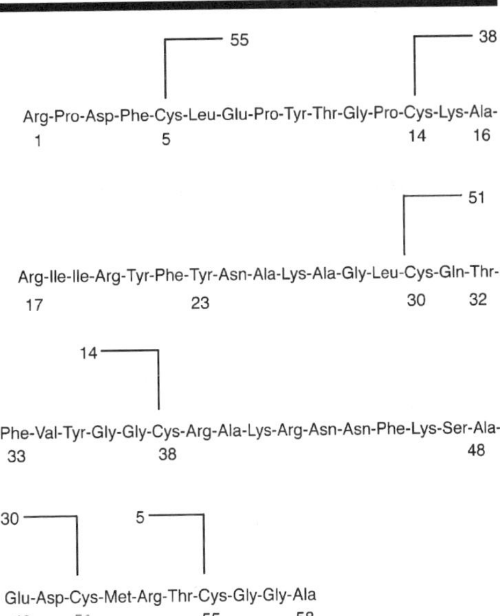

SAMPLE
Matrix: solutions
Sample preparation: Inject an aliquot of a 1 mg/mL solution in water.

CAPILLARY ELECTROPHORESIS
Capillary: 45 cm × 75 μm fused silica (Polymicro Technologies)
Running buffer: 25 mM pH 6.0 Phosphate buffer containing 10 mM sodium salt of phytic acid
Injection: Gravity injection at 10 cm
Detector: UV 214
Migration time: 26
Voltage: 5.5 kV
Model: Waters Quanta 4000 or 4000E

REFERENCE
Okafo,G.N.; Vinther,A.; Kornfelt,T.; Camilleri,P. Effective ion-pairing for the separation of basic proteins in capillary electrophoresis, *Electrophoresis*, **1995**, *16*, 1917–1921.

SAMPLE
Matrix: solutions
Sample preparation: Inject an aliquot of a 1 mg/mL solution in water.

CAPILLARY ELECTROPHORESIS
Capillary: 45 cm × 75 μm fused silica (Polymicro Technologies)
Running buffer: 25 mM pH 6.0 Phosphate buffer containing 10 mM sodium salt of phytic acid
Injection: Gravity injection at 10 cm
Detector: UV 214
Migration time: 26

Voltage: 5.5 kV
Model: Waters Quanta 4000 or 4000E

REFERENCE

Okafo,G.N.; Vinther,A.; Kornfelt,T.; Camilleri,P. Effective ion-pairing for the separation of basic proteins in capillary electrophoresis, *Electrophoresis*, **1995**, *16*, 1917–1921.

Arbekacin

Molecular formula: $C_{22}H_{44}N_6O_{10}$
Molecular weight: 552.63
CAS Registry No.: 51025-84-5
Merck Index (12th ed.): 809

SAMPLE

Matrix: solutions
Sample preparation: Inject an aliquot of a 100 μM solution in buffer. (Prepare buffer by mixing equal volumes of 20 mM NaH_2PO_4 and 20 mM sodium tetraborate and adjusting the pH to 10 with 1 M NaOH.)

CAPILLARY ELECTROPHORESIS

Capillary: 110 cm × 50 μm fused-silica (100 cm to detector)
Capillary preparation: Before injection rinse capillary with running buffer at 200 mbar with no voltage for 3 min and with running buffer at 4 mbar at 20 kV for 30 s.
Capillary temperature: 40
Running buffer: 100 mM pH 10 phosphate borate buffer containing 860 μM 1-methoxycarbonylindolizine-3,5-dicarbaldehyde (Prepare 1-methoxycarbonylindolizine-3,5-dicarbaldehyde as follows. Reflux 21.4 g 2-pyridinecarboxaldehyde, 24 mL ethylene glycol, 10 g p-toluenesulfonic acid, and 300 mL benzene (Caution! Benzene is a carcinogen!) under a Dean-Stark separator for 64 h, pour into concentrated sodium carbonate solution. Remove the organic layer and extract the aqueous layer 4 times with benzene. Combine the organic layers and wash them with water, dry over anhydrous magnesium sulfate, evaporate, distil the residue to give 2-(1,3-dioxolan-2-yl)pyridine (bp 122°/4 mm Hg) (J.Org.Chem. 1963, 28, 83). Reflux 15.1 g 2-(1,3-dioxolan-2-yl)pyridine and 19.5 g tert-butyl bromoacetate in 100 mL dry acetonitrile for 7 h,let stand overnight at room temperature, filter, wash the precipitate with diethyl ether to give 1-(tert-butoxycarbonylmethyl)-2-(1,3-dioxolan-2-yl)pyridinium bromide (mp 110-2° from MeCN). Suspend 51.9 g of this compound in 1. 5 L THF with stirring, add 62.1 g potassium carbonate, add 15.12 g methyl propiolate, stir at room temperature for 9 days, filter, evaporate the filtrate to dryness under reduced pressure, chromatograph the residue on silica gel with hexane:ethyl acetate 20:1-10:1, collect fractions and evaporate to dryness to give methyl 3-tert-butoxycarbonyl-5-(1,3-dioxolan-2-yl)indolizine-1-carboxylate (mp 138-9° from hexane). Reflux 20.82 g of this compound in 600 mL THF and 60 mL 10% HCl for 6 h, concentrate to one quarter of the original volume, add water, extract with chloroform. Wash the chloroform layer with water and dry it over anhydrous sodium sulfate, evaporate, chromatograph on silica gel with hexane:ethyl acetate 10:1 to give 1-methoxycarbonylindolizine-5-carbaldehyde (mp 135-7° from MeOH). Stir 12.18 g of this compound in 116 mL dry DMF at 0° under argon, add 17 mL phosphorus oxychloride, stir at room temperature for 1 h, pour into water, adjust pH to 9.0 with 5% potassium carbonate, extract with chloroform. Wash the organic layer with water and dry it over anhydrous sodium sulfate, concentrate until a precipitate forms, filter to obtain the product, concentrate the filtrate and chromatograph the residue on silica gel with hexane:ethyl acetate 5:1 to obtain more product. The product was 1-methoxycarbonylindolizine-3,5-dicarbaldehyde (mp 164-5° from methyl acetate) (J. Chromatogr. A 1996, 724, 169.))
Injection: Pressure injection at 50 mbar for 6 s (with no applied voltage).
Detector: UV 409
Migration time: 24.5
Voltage: 20 kV
Model: JASCO Model CE-990

OTHER SUBSTANCES

Simultaneous: amikacin, micromicin, netilmicin

KEY WORDS
derivatization; on-column derivatization

REFERENCE
Oguri,S.; Fujiyoshi,T.; Miki,Y. In-capillary derivatization with 1-methoxycarbonylindolizine-3,5-dicarbaldehyde for high-performance capillary electrophoresis, *Analyst*, **1996**, *121*, 1683–1688.

Arotinolol

Molecular formula: $C_{15}H_{21}N_3O_2S_3$
Molecular weight: 371.55
CAS Registry No.: 68377-92-4
Merck Index (12th ed.): 827

SAMPLE
Matrix: solutions

CAPILLARY ELECTROPHORESIS
Capillary: 36 cm × 50 μm polyacrylamide-coated fused-silica (31.5 cm to detector) (Polymicro Technologies)
Capillary preparation: Before each injection rinse with water at 100 psi for 30 s, rinse with running buffer at 100 psi for 30 s, partially fill with separation solution (500 μM ovomucoid in running buffer) at 1 psi for 190 s (27 cm), inject sample, electrophorese with running buffer. Coat capillary with polyacrylamide as follows. Treat with 1 M NaOH for 1 h, rinse with water, dry at 110° by flushing with nitrogen for 6 h, pass thionyl chloride through using suction for 10 min, seal at each end, heat at 70° for 6 h, open, suck 250 mM vinylmagnesium bromide in THF into the capillary, seal at each end, heat at 70° for 6 h, open, rinse with THF for several min, rinse with water, fill by suction with a solution containing 3% acrylamide, 0.1% ammonium persulfate, and 0.1% tetramethylethylenediamine, heat at 28 ± 2° for 1 h, after 1 h wash with water (Biol.Pharm.Bull. 1993, 16, 1185).
Capillary temperature: 25
Running buffer: Isopropanol:50 mM pH 6.0 phosphate buffer 6:94
Injection: Inject at 1 psi for 2 s
Detector: UV 210
Migration time: 12.97, 13.42 (enantiomers)
Voltage: 12 kV
Model: Bio-Rad BioFocus 3000

KEY WORDS
chiral; partial separation zone technique; coated capillary

REFERENCE
Tanaka,Y.; Terabe,S. Partial separation zone technique for the separation of enantiomers by affinity electrokinetic chromatography with proteins as chiral pseudo-stationary phases, *J.Chromatogr.A*, **1995**, *694*, 277–284.

SAMPLE
Matrix: solutions

CAPILLARY ELECTROPHORESIS
Capillary: 36 cm × 50 μm fused-silica coated with linear polyacrylamide (31.5 cm to detector) (GL Science)
Capillary preparation: At the beginning and end of each day rinse capillary with capillary wash solution (Bio-Rad Cat. No. 148-5022) at 690 kPa for more than 3 min and with water at 690 kPa for more than 3 min. Coat capillary as follows. Treat capillary with 1 M NaOH at room temperature for 1 h, rinse with water, dry by passing nitrogen gas through the capillary at 110° for 6 h. Pass thionyl chloride through the capillary using a suction pump for several min, seal capillary at both ends and heat at 70° for 6 h. Unseal the capillary and fill with 250 mM

vinyl magnesium bromide in THF by suction, seal the capillary, heat at 70° for 6 h. Open the capillary and rinse it with THF for several min, rinse with distilled water, fill the capillary with polymerization solution, heat at 28 ± 2° for 1 h, condition at -100 V/cm for 30 min (Anal. Sci. 1994, 10, 1). (The polymerization solution was 5% acrylamide in water containing 49 mM Tris, 384 mM glycine, and 0.1% sodium dodecyl sulfate, degas in an ultrasonic bath. Add 40 μL 10% N,N,N',N'-tetramethylethylenediamine and 10 μL 10% ammonium persulfate to 5 mL of the degassed solution, mix thoroughly.)
Running buffer: Isopropanol:50 mM pH 6.0 Phosphate buffer 10:90
Injection: Before each injection rinse with water at 690 kPa for 30 s, rinse with running buffer at 690 kPa for 30 s, partially fill with separation solution (500 μM α_1-acid glycoprotein (Cohn fraction VI) (ICN) in running buffer) at 6.9 kPa for 190 s (27 cm), inject sample at 6.9 kPa for 2 s, electrophorese with running buffer (Note that α_1-acid glycoprotein from other suppliers may provide inferior results).
Detector: UV 210
Migration time: Resolution of enantiomers 5.9
Voltage: 12 kV
Model: Bio-rad BioFocus 3000

KEY WORDS
chiral

REFERENCE
Tanaka,Y.; Terabe,S. Separation of the enantiomers of basic drugs by affinity capillary electrophoresis using a partial filling technique and α_1-acid glycoprotein as chiral selector, *Chromatographia*, **1997**, *44*, 119–128.

Artesunate

Molecular formula: $C_{19}H_{28}O_8$
Molecular weight: 384.43
CAS Registry No.: 88495-63-0
Merck Index (12th ed.): 857

SAMPLE
Matrix: solutions

CAPILLARY ELECTROPHORESIS
Capillary: 49 cm × 50 μm fused-silica
Running buffer: 40 mM pH 6.8 Lithium perchlorate (degas immediately before use)
Injection: Siphon injection.
Detector: UV 185
Migration time: 1.8
Voltage: 30 kV
Model: Waters Quanta 4000 CE
Limit of quantitation: 25 μM

OTHER SUBSTANCES
Simultaneous: succinic acid
Noninterfering: dihydroartemisinin

REFERENCE
Augustijns,P.; D'Hulst,A.; Van Daele,J.; Kinget,R. Transport of artemisinin and sodium artesunate in Caco-2 intestinal epithelial cells, *J.Pharm.Sci.*, **1996**, *85*, 577–579.

SAMPLE
Matrix: solutions

Sample preparation: Inject an aliquot of a 330 µM solution in 20% HEPES-buffered Hanks' balanced salt solution

---

CAPILLARY ELECTROPHORESIS
Capillary: 47 cm × 50 µm fused-silica
Capillary preparation: Purge capillary with running buffer for 2 min before each run. At the start of each day purge capillary with 500 mM NaOH then with water. Purge with running buffer under vacuum then carry out an electroosmotic purge.
Running buffer: 40 mM pH 6.8 Lithium phosphate
Injection: Siphon injection at 10 cm for 30 s.
Detector: UV 185
Migration time: 2
Voltage: 30 kV
Model: Waters Quanta 4000 CE

OTHER SUBSTANCES
Simultaneous: succinic acid

REFERENCE
D'Hulst,A.; Augustijns,P.; Arens,S.; van Parijs,L.; Colson,S.; Verbeke,N.; Kinget,R. Determination of artesunate by capillary electrophoresis with low UV detection and possible application to analogues, *J.Chromatogr.Sci.*, **1996**, *34*, 276–281.

Ascorbic acid

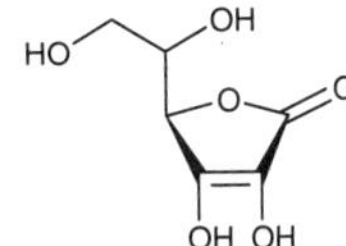

Molecular formula: $C_6H_8O_6$
Molecular weight: 176.13
CAS Registry No.: 50-81-7
Merck Index (12th ed.): 867

SAMPLE
Matrix: beverages
Sample preparation: Wine, beer. 2 mL Wine or 5 mL degassed beer + 0.5 mL 1 mg/mL niacin solution, make up to 10 mL with 2 mg/mL D,L-dithiothreitol solution, inject an aliquot. Fruit juices. 500 µL Fruit juice + 1 mL 300 µg/mL D-erythorbic acid solution, make up to 10 mL with 2 mg/mL D,L-dithiothreitol solution, inject an aliquot.

CAPILLARY ELECTROPHORESIS
Capillary: 76 cm × 75 µm fused silica (50 cm to detector) (ISCO)
Capillary preparation: Flush for 2 min with running buffer before each analysis. Before each batch fill capillary with 100 mM NaOH, let stand for 10 min, wash with water for 5 min, with 100 mM HCl for 5 min, and water for 5 min, fill with running buffer. Each week wash capillary with 100 mM HCl for 10 min, wash with water, fill with buffer.
Capillary temperature: 28
Running buffer: Dissolve 2.16 g sodium deoxycholate in 50 mL 20 mM KH_2PO_4 and 50 mL 20 mM sodium tetraborate, pH 8.6
Injection: Load under vacuum, 2-10 kPa.s
Detector: UV 254
Migration time: 5.5
Internal standard: D-erythorbic acid (5.7), niacin (6.9)
Voltage: +25 kV
Model: ISCO Model 3140 electropherograph

KEY WORDS
beer; wine; fruit juice

REFERENCE
Marshall,P.A.; Trenerry,V.C.; Thompson,C.O. The determination of total ascorbic acid in beers, wines, and fruit drinks by micellar electrokinetic capillary chromatography, *J.Chromatogr.Sci.*, **1995**, *33*, 426–432.

SAMPLE
Matrix: beverages
Sample preparation: Extract 3 g tea leaves with 180 mL boiling water for 5 min, filter (paper), dilute the filtrate 10-fold with 0.1% metaphosphoric acid, filter (0.45 μm), inject an aliquot of the filtrate.

CAPILLARY ELECTROPHORESIS
Capillary: 77 cm $\times$ 50 μm fused-silica (70 cm to detector)
Capillary preparation: Between runs rinse with water, 100 mM HCl, water, 100 mM NaOH, and running buffer (total rinse time 7 min).
Capillary temperature: 23
Running buffer: 20 mM Borax adjusted to pH 8.0 with HCl
Injection: Pressure injection using nitrogen for 5 s.
Detector: UV 200
Migration time: 9.7
Voltage: 30 kV
Model: Beckman P/ACE 5000

OTHER SUBSTANCES
Extracted: caffeine, catechin, epicatechin, epicatechin gallate, epigallocatechin, epigallocatechin gallate, theanine

KEY WORDS
tea; leaves

REFERENCE
Horie,H.; Mukai,T.; Kohata,K. Simultaneous determination of qualitatively important components in green tea infusions using capillary electrophoresis, *J.Chromatogr.A*, **1997**, *758*, 332–335.

SAMPLE
Matrix: beverages

CAPILLARY ELECTROPHORESIS
Capillary: 36 cm $\times$ 50 μm fused-silica (32 cm to detector)
Capillary preparation: Before each run rinse capillary with 1 M NaOH for 2 min and with water for 2 min.
Capillary temperature: 20
Running buffer: 25 mM pH 7.0 Phosphate buffer containing 50 mM borate and 25 mM sodium dodecyl sulfate
Injection: Pressure injection at 350 mbar for 1 s.
Detector: UV 280
Migration time: 3.7
Voltage: 15 kV
Model: Bio-Rad BioFocus 3000
Limit of quantitation: 500 ng/mL

OTHER SUBSTANCES
Simultaneous: caffeine, catechin, catechin gallate, epicatechin, epicatechin gallate, epigallocatechin, epigallocatechin gallate, gallocatechin gallate

KEY WORDS
tea; comparison with HPLC

REFERENCE
Watanabe,T.; Nishiyama,R.; Yamamoto,A.; Nagai,S.; Terabe,S. Simultaneous analysis of individual catechins, caffeine, and ascorbic acid in commercial canned green and black teas by micellar electrokinetic chromatography, *Anal.Sci.*, **1998**, *14*, 435–438.

SAMPLE
Matrix: beverages, blood, urine

Sample preparation: Urine and beverages. Add 100 mg metaphosphoric acid to each 1 mL urine. 200 μL Urine or beverage + 100 μL 100 mg/mL metaphosphoric acid + 100 μL 100 μg/mL isoascorbic acid, vortex for 15 s, filter (0.45 μm) while centrifuging at 5° at 3000 g for 10 min, inject an aliquot of the filtrate.Plasma, serum. 500 μL Plasma or serum + 500 μL 12: trichloroacetic acid, vortex for 30 s, centrifuge at 5° for 5 min, filter the supernatant. Remove a 400 μL aliquot of the filtrate and ad it to 100 μL 81.5 μg/mL isoascorbic acid, inject an aliquot.

CAPILLARY ELECTROPHORESIS
Capillary: 37 cm × 75 μm (30 cm to detector) (Polymicro Technologies)
Capillary preparation: After each run rinse capillary with 100 mM NaOH for 1 min, with water for 1 min, and with buffer for 1 min. Condition new capillary with 1 M NaOH for 10 min, with water for 10 min, and with running buffer for 10 min.
Running buffer: 100 mM pH 8.8 Tricine buffer
Injection: Pressure injection
Detector: UV 254
Migration time: 7.3
Internal standard: isoascorbic acid (7.5)
Voltage: 11 kV
Model: Beckman Model 2100
Limit of detection: 1.6 μg/mL

KEY WORDS
serum; plasma; juice; wine; orange; apple; grapefruit; vegetable; cranberry

REFERENCE
Koh,E.V.; Bissell,M.G.; Ito,R.K. Measurement of vitamin C by capillary electrophoresis in biological fluids and fruit beverages using a stereoisomer as an internal standard, *J.Chromatogr.*, **1993**, *633*, 245–250.

SAMPLE
Matrix: formulations
Sample preparation: Homogenize a multi-vitamin tablet, extract with 100 mM HCl, filter (0.45 μm).

CAPILLARY ELECTROPHORESIS
Capillary: 51 cm × 75 μm fused-silica (33 cm to detector) (Waters)
Running buffer: 20 mM pH 9.0 sodium phosphate buffer
Injection: 3 nL (5 μL with a split ratio of 1:1700)
Detector: UV 254
Migration time: 14
Voltage: 6.0 kV
Model: ISCO Model 3850

OTHER SUBSTANCES
Simultaneous: thiamine (vitamin B1), riboflavin (vitamin B2), pyridoxal, pyridoxamine, pyridoxine

KEY WORDS
tablets

REFERENCE
Huopalahti,R.; Sunell,J. Use of capillary zone electrophoresis in the determination of B vitamins in pharmaceutical products, *J.Chromatogr.*, **1993**, *636*, 133–135.

SAMPLE
Matrix: formulations
Sample preparation: Grind tablet, dissolve in pH 5 citrate buffer, filter (0.2 μm), inject an aliquot.

CAPILLARY ELECTROPHORESIS
Capillary: 48.5 cm × 50 μm fused-silica (40 cm to detector)

Capillary preparation: Between runs flush with 100 mM NaOH for 2 min and with run buffer for 3 min, inject a buffer plug. Before the first use condition capillaries with 1 M NaOH for 10 min, with 100 mM NaOH for 2 min, and with run buffer for 3 min.
Capillary temperature: 25
Running buffer: 20 mM pH 7 sodium phosphate
Injection: Injection at 40 mbar for 4.6 s (post-injection pressure (of run buffer) 40 mbar for 4 s)
Detector: UV 215
Migration time: 12.2
Voltage: 20 kV
Model: Hewlett-Packard

OTHER SUBSTANCES
Simultaneous: folinic acid, niacinamide, niacin, orotic acid, pantothenic acid, pyridoxine, thiamine, vitamin B12 (cyanocobalamin) (UV 360)

KEY WORDS
tablets

REFERENCE
Jegle,U. Separation of water-soluble vitamins via high-performance capillary electrophoresis, *J.Chromatogr.A*, **1993**, *652*, 495–501.

SAMPLE
Matrix: formulations
Sample preparation: Condition a 500 mg Bakerbond C18 SPE cartridge with 3 mL MeOH and 3 mL water:10 mM sodium heptanesulfonate in acetic acid 99:1. Heat 50-100 mg tablet and 40 mL water:10 mM sodium heptanesulfonate in acetic acid 99:1 at 55° with occasional shaking for 5 min, cool, add a 2 mL aliquot to the SPE cartridge, elute with 3.5 mL MeOH, make up the eluate to 10 mL with water, inject an aliquot.

CAPILLARY ELECTROPHORESIS
Capillary: 70 cm × 100 μm
Capillary temperature: 25
Running buffer: MeOH:buffer 10:90 (Buffer was 50 mM pH 8.0 sodium borate containing 22.5 mM sodium dodecyl sulfate.)
Injection: Inject under 3.44 kPa pressure for 5 s (30 nL)
Detector: UV 214
Migration time: 25
Voltage: 16 kV
Current: 33 μA
Model: Beckman P/ACE System 2000

OTHER SUBSTANCES
Simultaneous: niacinamide, pantothenic acid, pyridoxine, riboflavin, thiamine, vitamin B12

KEY WORDS
tablets; SPE

REFERENCE
Dinelli,G.; Bonetti,A. Micellar electrokinetic capillary chromatography analysis of water-soluble vitamins and multi-vitamin integrators, *Electrophoresis*, **1994**, *15*, 1147–1150.

SAMPLE
Matrix: formulations
Sample preparation: Protect from light. Grind tablet, add 50 mL 1% acetic acid, shake at 65° for 10 min, add 5 mL 10 mg/mL niacin in 1% acetic acid, cool to room temperature, make up to 100 mL with 1% acetic acid, filter (0.22 μm cellulose acetate), dilute 4-12-fold, inject an aliquot.

CAPILLARY ELECTROPHORESIS
Capillary: 48.5 cm × 50 μm fused-silica (40 cm to detector) (Hewlett Packard)

Capillary preparation: After each run wash with 100 mM NaOH for 7 min, with water for 1
   min, and with running buffer for 3 min. At the start of each day wash with separation buffer
   for 10 min. Wash new capillaries with 1 M NaOH, 100 mM NaOH, water, and running buffer.
Capillary temperature: 25
Running buffer: 50 mM Borax adjusted to pH 8.5 with boric acid
Injection: Hydrodynamic injection at 5 kPa for 10 s.
Detector: UV 225
Migration time: 6
Internal standard: niacin (7.5)
Voltage: 25 kV
Model: Hewlett Packard ³ᴰCE

OTHER SUBSTANCES
Simultaneous: adenine, biotin, niacinamide, pantothenic acid (UV 215), pyridoxine (UV 215),
   rutin, thiamine, riboflavin

KEY WORDS
tablets

REFERENCE
Fotsing,L.; Fillet,M.; Bechet,I.; Hubert,P.; Crommen,J. Determination of six water-soluble vitamins in a pharmaceutical formulation by capillary electrophoresis, *J.Pharm.Biomed.Anal.*, **1997**, *15*, 1113–1123.

SAMPLE
Matrix: formulations
Sample preparation: Sonicate ground tablet with 50 mL water for 5 min, filter, make up filtrate
   to 100 mL with water, inject an aliquot.

CAPILLARY ELECTROPHORESIS
Capillary: 53 cm × 75 μm fused-silica (45.4 cm to detector) (Waters AccuSep)
Capillary preparation: Before each run purge with 500 mM NaOH for 3 min, with water for 3
   min, and with running buffer for 5 min. Store overnight in 500 mM NaOH, rinse with water
   for 10 min in the morning.
Running buffer: 20 mM pH 7.2 Sodium tetraborate containing 20 mM NaH_2PO_4
Injection: Hydrodynamic injection at 10 cm for 10 s.
Detector: UV 254
Migration time: 2.75
Voltage: 25 kV
Model: Waters Quanta 4000
Limit of detection: 2 μg/mL

OTHER SUBSTANCES
Simultaneous: chlorphenamin males, paracetamolum

KEY WORDS
tablets

REFERENCE
Zhang,Z.; Chen,X.; Hu,Z. Determination of three water-soluble active ingredients in Qiangli Yingqiao containing Vc tablets by capillary zone electrophoresis, *J.Liq.Chromatogr.*, **1997**, *20*, 3245–3255.

SAMPLE
Matrix: formulations
Sample preparation: Grind tablet to a homogeneous powder and dissolve in 2 mL water and
   20 mL EtOH, homogenize (Ultra Turrax), centrifuge. Evaporate the supernatant to dryness
   under reduced pressure, reconstitute the residue with water, inject an aliquot.

CAPILLARY ELECTROPHORESIS
Capillary: 64.5 cm × 50 μm fused-silica (56 cm to detector)
Capillary preparation: Before each run flush capillary at 4 kPa with 1 M NaOH for 2 min and
   with buffer for 5 min. Replace buffer before each analysis.

Capillary temperature: 30
Running buffer: 1-Propanol:buffer 2:98 (Buffer was 100 mM Na_2HPO_4, 500 mM taurine, and 75 mM sodium cholate.)
Injection: Pressure injection at 4 kPa for 1 s.
Detector: UV 214
Migration time: 21.3
Voltage: 17 kV
Model: Hewlett-Packard HP[3D] CE

OTHER SUBSTANCES
Simultaneous: biotin, folic acid, niacinamide, pyridoxamine, pyridoxine, riboflavin, thiamine, vitamin B12

KEY WORDS
tablets

REFERENCE
Buskov,S.; Moller,P.; Sorensen,H.; Sorensen,J.C.; Sorensen,S. Determination of vitamins in food based on supercritical fluid extraction prior to micellar electrokinetic capillary chromatographic analyses of individual vitamins, *J.Chromatogr.A*, **1998**, *802*, 233–241.

SAMPLE
Matrix: formulations, juice
Sample preparation: Dissolve in water, filter (0.2 μm), inject an aliquot. (Add a 50-fold excess of cysteine to solutions containing ascorbic acid.)

CAPILLARY ELECTROPHORESIS
Capillary: 60 cm × 50 μm fused-silica (51.5 cm to detector)
Capillary preparation: Condition with running buffer for 3 min before each run. Condition a new capillary 1 M NaOH for 10 min and with 100 mM NaOH for 2 min.
Capillary temperature: 25
Running buffer: 20 mM pH 8.0 Phosphate buffer
Injection: Pressure injection for 200 mbar.s
Detector: UV 200
Migration time: 7.3
Voltage: -30 kV
Model: Hewlett-Packard 3DCE
Limit of detection: 1.2 μg/mL

OTHER SUBSTANCES
Simultaneous: biotin, niacinamide, pyridoxine, riboflavin phosphate, thiamine

---

KEY WORDS
fruit

REFERENCE
Schiewe,J.; Mrestani,Y.; Neubert,R. Application and optimization of capillary zone electrophoresis in vitamin analysis, *J.Chromatogr.A*, **1995**, *717*, 255–259.

SAMPLE
Matrix: plants
Sample preparation: Freeze and crush 50-200 mg plant tissue in liquid nitrogen, add 400 μL extraction solvent, homogenize thoroughly, let stand on ice until the mixture thaws, add to a 2 mL tube with more extraction solvent, centrifuge at 20800 g for 7 min, remove the supernatant, resuspend the pellet in fresh extraction buffer. Combine and filter (0.22 μm PVDF) the supernatants, inject an aliquot of the filtrate. (Extraction solvent was 3% metaphosphoric acid containing 1 mM EDTA.)

CAPILLARY ELECTROPHORESIS
Capillary: 70 cm × 50 μm fused-silica (62.5 cm to detector) (Supelco)

Capillary preparation: Between analyses purge with MeCN:100 mM pH 9.0 borate buffer containing 100 mM sodium dodecyl sulfate 25:75 for 4 min and with running buffer for 6 min. Purge new capillaries under 2.2 kPa vacuum with 500 μM LiOH for 15 min, purge with water, purge with running buffer at 24 kV for 1 h.
Running buffer: MeCN:water:400 mM pH 9.0 sodium borate buffer 20:30:50
Injection: Hydrostatic injection at 10 cm for 40 s.
Detector: UV 185
Migration time: 13.7
Voltage: 30 kV
Current: 48 μA
Model: Waters Quanta 4000
Limit of detection: 28 μM

OTHER SUBSTANCES
Simultaneous: glutathione

KEY WORDS
leaf

REFERENCE
Davey,M.W.; Bauw,G.; Van Montagu,M. Simultaneous high-performance capillary electrophoresis analysis of the reduced and oxidised forms of ascorbate and glutathione, *J.Chromatogr.B*, **1997**, *697*, 269–276.

SAMPLE
Matrix: solutions
Sample preparation: Prepare a solution in running buffer, inject an aliquot.

CAPILLARY ELECTROPHORESIS
Capillary: 78.6 cm $\times$ 12.7 μm fused-silica (Polymicro Technologies)
Running buffer: 25 mM pH 5.7 2-morpholinoethanesulfonic acid buffer
Injection: Injection by electromigration at 30 kV for 2 s
Detector: E, 0.7 V vs sodium-saturated calomel electrode
Migration time: 17
Voltage: 30 kV

OTHER SUBSTANCES
Simultaneous: 3,4-dihydroxyphenylacetic acid, homovanillic acid, isoproterenol, 4-methylcatechol, norepinephrine, serotonin

REFERENCE
Wallingford,R.A.; Ewing,A.G. Capillary zone electrophoresis with electrochemical detection in 12.7 μm diameter columns, *Anal.Chem.*, **1988**, *60*, 1972–1975.

SAMPLE
Matrix: solutions
Sample preparation: Prepare a 100 μg/mL solution in MeOH:water 50:50, inject an aliquot.

CAPILLARY ELECTROPHORESIS
Capillary: 60 cm $\times$ 75 μm
Running buffer: 20 mM NaH_2PO_4 containing 50 mM sodium dodecyl sulfate, adjusted to pH 9.0 with sodium tetraborate
Injection: Hydrodynamic injection at 10 cm for 5 s
Detector: UV 214
Migration time: 8.7
Voltage: 18 kV
Model: Waters Quanta 4000

OTHER SUBSTANCES
Simultaneous: niacin, pantothenic acid, thiamine, riboflavin, pyridoxine

REFERENCE
Swartz,M.E. Method development and selectivity control for small molecule pharmaceutical separations by capillary electrophoresis, *J.Liq.Chromatogr.*, **1991**, *14*, 923–938.

SAMPLE
Matrix: solutions

CAPILLARY ELECTROPHORESIS
Capillary: 70 cm × 50 μm fused-silica (GL Science, Tokyo)
Capillary preparation: Before each run rinse capillary with running buffer at 94 kPa for 5 min before each run.
Running buffer: 50 mM pH 8.5 ammonium carbonate buffer
Injection: Pressure injection at 5 kPa (50 mbar) for 4 s.
Detector: MS, Perkin-Elmer Sciex API-300 quadrupole, electrospray (ionspray) interface, sheath liquid MeOH:running buffer 50:50 at 2.5 μL/min, ionspray voltage 5 kV, negative ion mode
Migration time: 11
Voltage: 20 kV (net voltage across capillary = 15 kV (applied voltage -electrospray voltage))
Model: Hewlett Packard 3D CE

OTHER SUBSTANCES
Simultaneous: acetaminophen, butylscopolamine bromide, caffeine, ibuprofen, ketoprofen, niacin, niacinamide, riboflavin, thiamine, vitamin B12, warfarin

REFERENCE
Tanaka,Y.; Kishimoto,Y.; Otsuka,K.; Terabe,S. Strategy for selecting separation solutions in capillary electrophoresis-mass spectrometry, *J.Chromatogr.A*, **1998**, *817*, 49–57.

SAMPLE
Matrix: vegetables
Sample preparation: Homogenize 2 g chopped vegetables with 4 mL 2% thiourea in 10 mM HCl in a porcelain mortar, add 14 mL 10 mM HCl, let stand for 15 min, add to a centrifuge tube, wash in with 18 mL water, centrifuge at 1600 g for 5 min, filter (0.45 μm) the supernatant, inject an aliquot of the filtrate.

CAPILLARY ELECTROPHORESIS
Capillary: 72 cm × 50 μm fused-silica (50 cm to detector) (GL Sciences, Tokyo)
Capillary preparation: Before injection fill with running buffer using vacuum for 3 min.
Capillary temperature: 35
Running buffer: 20 mM pH 9.2 Sodium tetraborate
Injection: Vacuum injection for 3 s (12 nL).
Detector: UV 270
Migration time: 9
Voltage: 20 kV
Model: Perkin-Elmer Model 270A
Limit of detection: 350 ng/mL

KEY WORDS
spinach; turnip; perilla; parsley

REFERENCE
Fukushi,K.; Takeda,S.; Wakida,S.; Yamane,M.; Higashi,K.; Hiro,K. Determination of ascorbic acid in vegetables by capillary zone electrophoresis, *J.Chromatogr.A*, **1997**, *772*, 313–320.

Aspartame

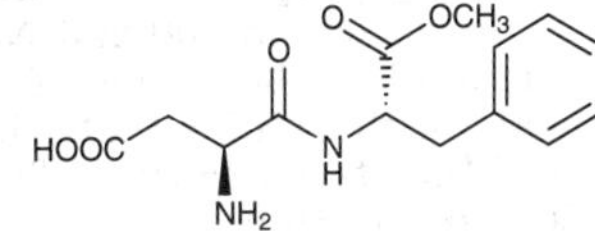

Molecular formula: $C_{14}H_{18}N_2O_5$
Molecular weight: 294.31
CAS Registry No.: 22839-47-0
Merck Index (12th ed.): 874

SAMPLE
Matrix: beverages

CAPILLARY ELECTROPHORESIS
Capillary: 30 cm $\times$ 50 μm
Running buffer: 50 mM pH 2.7 Phosphate buffer
Detector: UV 215
Limit of quantitation: 2 μg/mL
Limit of detection: 1 μg/mL

OTHER SUBSTANCES
Simultaneous: degradation products

REFERENCE
Sabah,S.; Scriba,G.K.E. Determination of aspartame and its degradation and epimerization products by capillary electrophoresis (Abstract APQ 1006), *Pharm.Res.*, **1996**, *13*, S4–S4.

SAMPLE
Matrix: beverages
Sample preparation: Degas the carbonated beverage in an ultrasonic bath, dilute 5-fold with water, filter (0.45 μm), inject an aliquot. Heat coffee to 70° for 10 min, dilute 10-fold with water, filter (0.45 μm), inject an aliquot.

CAPILLARY ELECTROPHORESIS
Capillary: 60 cm $\times$ 75 μm Accusep fused-silica (Waters)
Running buffer: 25 mM pH 9.0 Phosphate buffer containing 25 mM borate
Injection: Hydrostatic injection for 20 s.
Detector: UV 214
Migration time: 10
Voltage: 15 kV
Model: Waters Quanta 4000E
Limit of quantitation: 250 μg/mL

OTHER SUBSTANCES
Simultaneous: degradation products

KEY WORDS
soft drinks; coffee; carbonated beverages

REFERENCE
Aboul-Enein,H.Y.; Bakr,S.A. Comparative study of the separation and determination of aspartame and its decomposition products in bulk material and diet soft drinks by HPLC and CE, *J.Liq.Chromatogr. Rel.Technol.*, **1997**, *20*, 1437–1444.

SAMPLE
Matrix: beverages
Sample preparation: Mix 600 μL beverage with 400 μL water, make up to 5 mL with running buffer, inject an aliquot.

CAPILLARY ELECTROPHORESIS
Capillary: 44 cm $\times$ 50 μm (25 cm to detector) (Polymicro Technologies)

Capillary preparation: Rinse with running buffer for 2 min after each run. After every other run rinse with 100 mM NaOH for 2 min, with water for 2 min, and with running buffer for 2 min. At the beginning of each day rinse with 100 mM NaOH for 5 min, with water for 5 min, and with running buffer for 5 min.
Capillary temperature: 30
Running buffer: 19 mM Tris containing 30 mM phosphoric acid, pH 2.14
Injection: Vacuum injection at 12.5 cm Hg for 3 s.
Detector: UV 211
Migration time: 3.8
Voltage: 30 kV
Model: Model 270A-HT (Perkin Elmer/Applied Biosystems)

OTHER SUBSTANCES
Noninterfering: caffeine

KEY WORDS
soft drinks

REFERENCE
Pesek,J.J.; Matyska,M.T. Determination of aspartame by high-performance capillary electrophoresis, *J.Chromatogr.A*, **1997**, *781*, 423–428.

SAMPLE
Matrix: beverages
Sample preparation: Degas carbonated beverage with sonication under vacuum, mix an aliquot with an equal volume of running buffer, filter (0.45 μm)

CAPILLARY ELECTROPHORESIS
Capillary: 44 cm × 50 μm fused-silica (Polymicro Technologies)
Capillary preparation: Before each injection wash capillary with 2 volumes of fresh running buffer. Condition new capillaries by washing with 1 M NaOH at 60° for 15 min, with 100 mM NaOH at 60° for 15 min, and with water at 25° for 15 min.
Capillary temperature: 35
Running buffer: 20 mM Glycine adjusted to pH 9.0 with NaOH
Injection: Hydrodynamic injection for 1 s.
Detector: UV 215
Migration time: 1.6
Voltage: 20 kV
Model: SpectraPhoresis 1000 CE (ThermoSeparations Products)
Limit of detection: 18 μg/mL

OTHER SUBSTANCES
Simultaneous: degradation products, benzoic acid, caffeine

KEY WORDS
soft drinks

REFERENCE
Walker,J.C.; Zaugg,S.E.; Walker,E.B. Analysis of beverages by capillary electrophoresis, *J.Chromatogr.A*, **1997**, *781*, 481–485.

SAMPLE
Matrix: solutions
Sample preparation: Mix 100 μL of a buffer solution containing aspartame with 200 μL 22 μg/mL salicylic acid in water, inject an aliquot.

CAPILLARY ELECTROPHORESIS
Capillary: 47 cm × 50 μm fused-silica (40 cm to detector) (Beckman)
Capillary preparation: Between analyses flush capillary with 100 mM NaOH for 2 min and with running buffer for 2 min.
Capillary temperature: 20

Running buffer: 50 mM pH 9.35 Borate buffer
Injection: Pressure injection at 0.5 psi for 3 s.
Detector: UV 215
Migration time: 5
Internal standard: salicylic acid (8)
Voltage: 20 kV
Model: Beckman P/ACE 2100

OTHER SUBSTANCES
Simultaneous: degradation products

REFERENCE
Sabah,S.; Scriba,G.K.E. Determination of aspartame and its degradation and epimerization products by capillary electrophoresis, *J.Pharm.Biomed.Anal.*, **1998**, *16*, 1089–1096.

Aspirin

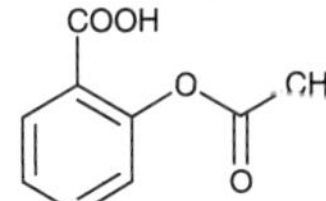

Molecular formula: $C_9H_8O_4$
Molecular weight: 180.16
CAS Registry No.: 50-78-2
Merck Index (12th ed.): 886

SAMPLE
Matrix: blood
Sample preparation: 100 μL Serum + 150 μL 50 μg/mL IS in MeCN, vortex, centrifuge at 13400 g for 2 min, inject an aliquot.

CAPILLARY ELECTROPHORESIS
Capillary: 64.5 cm × 75 μm fused-silica with bubble cell (56 cm to detector) (Hewlett-Packard)
Capillary preparation: Before each run rinse each capillary with 100 mM NaOH for 2 min and with running buffer for 3 min
Capillary temperature: 30
Running buffer: 100 mM Boric acid adjusted to pH 8.8 with 1 M NaOH
Injection: Vacuum injection at 50 mm Hg for 20 s (ca. 100 nL).
Detector: UV 210
Migration time: 5.7
Internal standard: 3-isobutyl-1-methylxanthine (4.3)
Voltage: 25 kV
Model: Hewlett-Packard HP³ᴰ CE

OTHER SUBSTANCES
Extracted: gentisic acid, salicylic acid, salicyluric acid

KEY WORDS
serum

REFERENCE
Goto,Y.; Makino,K.; Kataoka,Y.; Shuto,H.; Oishi,R. Determination of salicylic acid in human serum with capillary zone electrophoresis, *J.Chromatogr.B*, **1998**, *706*, 329–335.

SAMPLE
Matrix: bulk
Sample preparation: Dilute with 400 μg/mL n-propyl p-hydroxybenzoate in running buffer

---

CAPILLARY ELECTROPHORESIS
Capillary: 27 cm × 50 μm fused-silica (20 cm to detector)
Capillary preparation: Before each run rinse with running buffer at high pressure.

Capillary temperature: 30
Running buffer: MeCN:water 15:85 containing 40 mM sodium dodecyl sulfate, 8.5 mM sodium borate, and 8.5 mM sodium phosphate, pH 8.5
Injection: Inject at high pressure for 1 s
Detector: UV 214
Migration time: 1.7
Internal standard: n-propyl p-hydroxybenzoate (1.8)
Voltage: 20 kV
Model: Beckman P/ACE System 2100

OTHER SUBSTANCES

Simultaneous: acetaminophen, acetylcodeine, O-acetylmorphine, caffeine, cocaine, codeine, diamorphine, diphenhydramine, hydromorphone, isoniacinamide, lidocaine, methaqualone, morphine, niacinamide, noscapine, papaverine, phenacetin, phenobarbital, phenylpropanolamine, procaine, quinine, salicylic acid, strychnine, thebaine

REFERENCE

Walker,J.A.; Krueger,S.T.; Lurie,I.S.; Marché,H.L.; Newby,N. Analysis of heroin drug seizures by micellar electrokinetic capillary chromatography (MECC), *J.Forensic Sci.*, **1995**, *40*, 6–9.

SAMPLE

Matrix: formulations
Sample preparation: Grind tablet, add 75 mL acidic MeOH, sonicate for 5 min, shake for 10 min, make up to 100 mL with acidic MeOH, centrifuge at 2500 g for 10 min. Remove a 5 mL aliquot of the supernatant and add it to 5 mL 10 mM ethyl p-aminobenzoate in MeOH, make up to 50 mL with MeOH, inject an aliquot. (Prepare acidic MeOH by adding 10 mL 100 mM HCl to 1 L MeOH.)

CAPILLARY ELECTROPHORESIS

Capillary: 80 cm × 100 μm fused-silica (50 cm to detector) (Shimadzu)
Running buffer: 20 mM pH 11 Phosphate buffer containing 50 mM sodium dodecyl sulfate
Injection: Injection by siphon at 5 cm for 5 s
Detector: UV 214
Migration time: 15
Internal standard: ethyl p-aminobenzoate (20)
Current: 100 μA (constant current)

OTHER SUBSTANCES

Simultaneous: acetaminophen, caffeine, o-ethoxybenzamide, salicylamide

KEY WORDS

tablets

REFERENCE

Fujiwara,S.; Honda,S. Determination of ingredients of antipyretic analgesic preparations by micellar electrokinetic capillary chromatography, *Anal.Chem.*, **1987**, *59*, 2773–2776.

SAMPLE

Matrix: formulations
Sample preparation: Grind tablet, add MeCN:10 mM HCl 20:80, shake and sonicate for 15 min, centrifuge until a clear solution is obtained. Remove a 25 mL aliquot of the supernatant and add it to 10 mL 500 μg/mL propyl hydroxybenzoate, make up to 50 mL with MeCN:10 mM HCl 20:80, inject an aliquot.

CAPILLARY ELECTROPHORESIS

Capillary: 57 cm × 75 μm fused-silica (Beckman)
Capillary preparation: Pass running buffer through the capillary for at least 4 min before each injection.
Capillary temperature: 25
Running buffer: 20 mM pH 9 Borate buffer containing 25 mM sodium cholate and 50 mM sodium deoxycholate
Injection: Pressure injection for 2 s

Detector: UV 214
Migration time: 10
Internal standard: propyl hydroxybenzoate (10.5)
Voltage: 20 kV
Model: Beckman P/ACE System 2100

OTHER SUBSTANCES
Simultaneous: acetaminophen, caffeine, chlorpheniramine, dextropropoxyphene, salicylic acid

KEY WORDS
tablets

REFERENCE
Boonkerd,S.; Lauwers,M.; Detaevernier,M.R.; Michotte,Y. Separation and simultaneous determination of the components in an analgesic tablet formulation by micellar electrokinetic chromatography, *J.Chromatogr.A*, **1995**, *695*, 97–102.

SAMPLE
Matrix: solutions
Sample preparation: Prepare a 200 µg/mL solution in water, inject an aliquot.

CAPILLARY ELECTROPHORESIS
Capillary: 60 cm × 50 µm
Running buffer: 15 mM pH 11.0 sodium phosphate buffer containing 25 mM sodium dodecyl sulfate
Injection: Hydrodynamic injection at 10 cm for 15 s
Detector: UV 214
Migration time: 3.3
Voltage: 30 kV
Model: Waters Quanta 4000

OTHER SUBSTANCES
Simultaneous: acetaminophen, caffeine, salicylamide, salicylic acid

REFERENCE
Swartz,M.E. Method development and selectivity control for small molecule pharmaceutical separations by capillary electrophoresis, *J.Liq.Chromatogr.*, **1991**, *14*, 923–938.

SAMPLE
Matrix: solutions

CAPILLARY ELECTROPHORESIS
Capillary: 56.9 cm × 75 µm fused-silica (50 cm to detector)
Capillary temperature: 25
Running buffer: MeOH:60 mM pH 8.4 sodium borate buffer containing 60 mM sodium dodecyl sulfate 15:85
Detector: UV 214
Migration time: 12.5
Voltage: 16 kV
Model: Beckman PACE 2100

OTHER SUBSTANCES
Simultaneous: acetanilide, acetaminophen, benzamide, caffeine, salicylamide, salicylic acid

REFERENCE
McLaughlin,G.M.; Nolan,J.A.; Lindahl,J.L.; Palmieri,R.H.; Anderson,K.W.; Morris,S.C.; Morrison,J.A.; Bronzert,T.J. Pharmaceutical drug separations by HPCE: Practical guidelines, *J.Liq.Chromatogr.*, **1992**, *15*, 961–1021.

SAMPLE
Matrix: solutions

CAPILLARY ELECTROPHORESIS
Capillary: 64.5 cm × 75 μm (56 cm to detector)
Capillary preparation: Between each run replenish inlet and outlet vials, wash capillary with fresh running buffer for 1 min, apply 30 kV for 2 min.
Capillary temperature: 25
Running buffer: 100 mM pH 8.5 Borate buffer
Injection: Pressure injection at 35 mbar for 5 s
Detector: UV 254
Migration time: 5.0
Voltage: 30 kV
Model: Hewlett-Packard HP3D CE

OTHER SUBSTANCES
Simultaneous: acetaminophen, 2,4-dihydroxybenzoic acid, niacin

---

REFERENCE
Ross,G.A. Voltage pre-conditioning technique for optimization of migration-time reproducibility in capillary electrophoresis, *J.Chromatogr.A*, **1995**, *718*, 444–447.

SAMPLE
Matrix: solutions

CAPILLARY ELECTROPHORESIS
Capillary: 27 cm × 75 μm
Capillary preparation: Before each run rinse capillary with 100 mM NaOH for 30 s and with running buffer for 30 s. Condition new capillaries by rinsing with 100 mM NaOH for 20 min.
Capillary temperature: 30
Running buffer: 15 mM Sodium borate
Injection: Pressure injection of sample at 25 mbar for 1 s followed by running buffer at 25 mbar for 1 s.
Detector: UV 200
Migration time: 3.26
Internal standard: aminobenzoic acid (3.5)
Voltage: 6.5 kV
Model: Beckman

OTHER SUBSTANCES
Simultaneous: bacitracin, beclomethasone, benzoic acid, ceftizoxime, ceftriaxone, cefuroxime, cephalothin, cromolyn, embonic acid, epoprostenol, glyburide (glibenclamide), levothyroxine, nedocromil, nystatin, omeprazole, prednisolone, warfarin, zidovudine

---

REFERENCE
Altria,K.D.; Bryant,S.M.; Hadgett,T.A. Validated capillary electrophoresis method for the analysis of a range of acidic drugs and excipients, *J.Pharm.Biomed.Anal.*, **1997**, *15*, 1091–1101.

SAMPLE
Matrix: solutions
Sample preparation: Inject an aliquot of a 100 μg/mL solution in MeOH:water 10:90.

CAPILLARY ELECTROPHORESIS
Capillary: 27 cm × 75 μm fused-silica (20 cm to detector) (Composite Metal Services, Hallow, UK)
Capillary preparation: Before injection rinse with 100 mM NaOH and running buffer. Condition new capillaries by rinsing with 100 mM NaOH for 20 min.
Capillary temperature: 25
Running buffer: 50 mM pH 7.0 NaH$_2$PO$_4$
Injection: Pressure injection for 3 s.
Detector: UV 214 or 200
Migration time: 2.4
Voltage: 7 kV
Model: Beckman P/ACE 5100

OTHER SUBSTANCES
Simultaneous: benzoic acid, phenylacetic acid

REFERENCE
Kelly,M.A.; Altria,K.D.; Clark,B.J. Approaches used in the reduction of buffer electrolysis effects for routine capillary electrophoresis procedures in pharmaceutical analysis, *J.Chromatogr.A*, **1997**, *768*, 73–80.

SAMPLE
Matrix: solutions

CAPILLARY ELECTROPHORESIS
Capillary: 53 cm × 50 μm (44.5 cm to detector) (Polymicro Technologies)
Capillary temperature: 40
Running buffer: 70 mM Boric acid adjusted to pH 9.00 with 1 M NaOH
Injection: Pressure injection at 50 mbar for 2-5 s (2.9-7.2 nL).
Detector: UV 210
Migration time: 3.3
Voltage: 20-25 kV
Model: Hewlett-Packard [3D]CE

OTHER SUBSTANCES
Simultaneous: acetaminophen, caffeine, norephedrine, salicylic acid

REFERENCE
Thompson,L.; Veening,H.; Strein,T.G. Capillary electrophoresis in the undergraduate instrumental laboratory: Determination of common analgesic formulations, *J.Chem.Educ.*, **1997**, *74*, 1117–1121.

SAMPLE
Matrix: solutions

CAPILLARY ELECTROPHORESIS
Capillary: 57 cm × 50 μm fused-silica.(50 cm to detector)
Running buffer: 50 mM pH 8.0 Borate buffer containing 40 mM sodium taurodeoxycholate and 25 mM phosphatidylcholine (Prepare by adding phosphatidylcholine and stirring for 3-6 h until all cloudiness disappears. Phosphatidylcholine was 95% pure soybean lecithin, Epikuron, Lucas Meyer & Co.)
Injection: Inject a solution of the compound in the running buffer at 20 psi for 1 s, inject MeOH: water 5:95 at 20 psi for 1 s, inject a solution of halofantrine in running buffer at 20 psi for 1 s.
Detector: UV 214
Migration time: k' 2.00
Model: Beckman P/ACE 5000

OTHER SUBSTANCES
Also analyzed: acetaminophen, amoxicillin, antipyrine, azathioprine, caffeine, captopril, carbamazepine, carprofen, chlorambucil, chlorpheniramine, chlorpromazine, cimetidine, clonidine, codeine, desipramine, diphenhydramine, ephedrine, fenoterol, flufenamic acid, flurbiprofen, haloperidol, hydroxyzine, ibuprofen, imipramine, indomethacin, ketoprofen, lidocaine, melphalan, metoprolol, nabumetone, nadolol, phenobarbital, phenol, promazine, propranolol, pyrilamine, ranitidine, ropinirole, salicylic acid, sulfamethoxazole, testosterone, theophylline, thioridazine, tiaprofenic acid, tolfenamic acid, trifluoperazine, trimethoprim, valproic acid, verapamil

KEY WORDS
comparison with HPLC; $k' = (Tr-T0)/(T0(1-Tr/Tm))$ where Tr = retention time of analyte; T0 = retention time of water; and Tm = retention time of marker (halofantrine)

REFERENCE
Hanna,M.; de Biasi,V.; Bond,B.; Salter,C.; Hutt,A.J.; Camilleri,P. Estimation of the partitioning characteristics of drugs: A comparison of a large and diverse drug series utilizing chromatographic and electrophoretic methodology, *Anal.Chem.*, **1998**, *70*, 2092–2099.

Aspoxicillin

Molecular formula: $C_{21}H_{27}N_5O_7S$
Molecular weight: 493.54
CAS Registry No.: 63358-49-6
Merck Index (12th ed.): 887

SAMPLE
Matrix: blood
Sample preparation: Inject directly.

CAPILLARY ELECTROPHORESIS
Capillary: 65 cm × 50 μm untreated fused-silica (50 cm to detector) (SGE)
Capillary preparation: Flush with running buffer every 5 runs. At the end of each day fill capillary with 100 mM KOH, let stand for 30 min, flush with water, let stand for 5 min, fill with running buffer.
Running buffer: pH 8.5 Buffer containing 50 mM sodium dodecyl sulfate (Prepare buffer by mixing 20 mM NaH_2PO_4 solution with 20 mM sodium borate solution to achieve a pH of 8.5.)
Injection: Inject by siphoning at 10 cm for 5-10 s.
Detector: UV 210
Migration time: 7.5
Internal standard: acetaminophen (7.1)
Limit of detection: 1.3 μg/mL (S/N 3)

KEY WORDS
plasma

REFERENCE
Nishi,H.; Fukuyama,T.; Matsuo,M. Separation and determination of aspoxicillin in human plasma by micellar electrokinetic chromatography with direct sample injection, *J.Chromatogr.*, **1990**, *515*, 245–255.

SAMPLE
Matrix: solutions
Sample preparation: Prepare a 0.5-2 mg/mL solution in water, inject an aliquot.

CAPILLARY ELECTROPHORESIS
Capillary: 65 cm × 50 μm fused-silica (50 cm to detector) (Scientific Glass Engineering)
Running buffer: 20 mM NaH_2PO_4 containing 150 mM sodium dodecyl sulfate adjusted to pH 9.0 with 20 mM sodium tetraborate
Injection: Injection by siphon at 5 cm for 5-10 s
Detector: UV 210
Migration time: 8
Voltage: 20 kV

OTHER SUBSTANCES
Simultaneous: amoxicillin, ampicillin, carbenicillin, penicillin G, piperacillin, sulbenicillin
Also analyzed: cefmenoxime, cefminox, cefoperazone, cefotaxime, cefpimizole, cefpiramide, ceftazidime, ceftriaxone

REFERENCE
Nishi,H.; Tsumagari,N.; Kakimoto,T.; Terabe,S. Separation of β-lactam antibiotics by micellar electrokinetic chromatography, *J.Chromatogr.*, **1989**, *477*, 259–270.

Atenolol

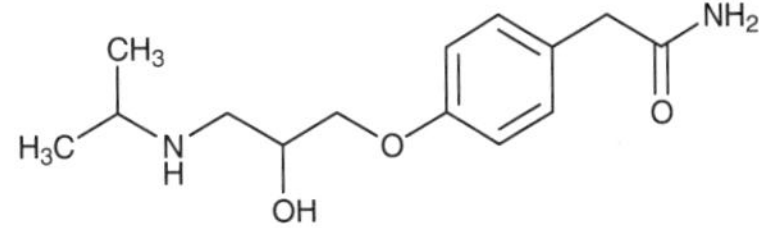

Molecular formula: $C_{14}H_{22}N_2O_3$
Molecular weight: 266.34
CAS Registry No.: 29122-68-7
Merck Index (12th ed.): 892
Lednicer: 2 109

SAMPLE
Matrix: blood
Sample preparation: Hydrolyze 2 mL serum with β-glucuronidase (EC 3.2.1.31, type H-1 from Helix pomatia, 416 800 U/g, Separacor) at 80° for 30 min, cool, add 900 μL MeCN, vortex for 15 min, centrifuge at 2004 g for 10 min, add ephedrine (165 μg/mL), filter (0.5 μm), inject an aliquot of the filtrate.

CAPILLARY ELECTROPHORESIS
Capillary: 58 cm × 50 μm fused-silica (50 cm to detector) (Polymicro Technologies)
Capillary preparation: Before each injection purge capillary with 5% phosphoric acid for 12 s, with water for 30 s, and with running buffer for 10 min.
Capillary temperature: 35
Running buffer: 80 mM pH 6.7 sodium phosphate buffer containing 15 mM cetyltrimethylammonium bromide
Injection: Hydrostatic injection for 20 s
Detector: UV 214
Migration time: 12.5
Internal standard: ephedrine (17.5)
Voltage: -27 kV
Model: Waters Quanta 4000

OTHER SUBSTANCES
Extracted: acebutolol, alprenolol, labetalol, metoprolol, nadolol, oxprenolol, pindolol, propranolol, timolol

KEY WORDS
serum

REFERENCE
Lukkari,P.; Nyman,T.; Riekkola,M.-J. Determination of nine β-blockers in serum by micellar electrokinetic capillary chromatography, *J.Chromatogr.A*, **1904**, *674*, 241–246.

SAMPLE
Matrix: blood
Sample preparation: Filter (1 mL Amicon Centrifree Micropartition System) serum while centrifuging at 1500 g for 20 min, inject an aliquot of the ultrafiltrate.

CAPILLARY ELECTROPHORESIS
Capillary: 92 cm × 75 μm fused-silica (55 cm to detector) (Polymicro Technologies)
Capillary preparation: Rinse with running buffer for 10 min between run. At the beginning and end of each day rinse with 100 mM NaOH for 10 min and with water for 10 min.
Running buffer: Isopropanol:buffer 5:95 (Buffer was 10 mM Na_2HPO_4 containing 6 mM sodium borate and 75 mM sodium dodecyl sulfate.)
Injection: Injection via vacuum suction for 0.7 s
Detector: UV 215
Migration time: 13
Voltage: 25 kV
Current: 82 μA
Model: Europhor Prime Vision IV

OTHER SUBSTANCES
Extracted: carteolol, celiprolol, levobunolol, metoprolol, penbutolol, pindolol, propranolol, timolol

---

KEY WORDS
serum; ultrafiltrate

REFERENCE
Schmutz,A.; Thormann,W. Assessment of impact of physico-chemical drug properties on monitoring drug levels by micellar electrokinetic capillary chromatography with direct serum injection, *Electrophoresis*, **1994**, *15*, 1295–1303.

SAMPLE
Matrix: blood, urine
Sample preparation: Urine. Dilute with 2 volumes of water, filter (0.5 μm), inject an aliquot. Serum. Condition a 3 mL Supelclean LC-18 SPE cartridge (Supelco) with MeOH and water. Hydrolyze 900 μL serum with β-glucuronidase (EC 3.2.1.31 type H-1 from Helix pomatia) at 60° with sonication for 30 min, add 500 μL (?) MeOH, centrifuge at 2000 g, add the supernatant to the SPE cartridge, wash with 1 mL water, dry under vacuum, elute with 2 mL MeOH:water 90:10, filter, inject an aliquot.

CAPILLARY ELECTROPHORESIS
Capillary: 68 cm × 50 μm fused-silica (60 cm to detector) (Polymicro Technologies)
Capillary preparation: Purge with running buffer for 2 min before injection.
Running buffer: pH 7.0 Phosphate buffer containing 10 mM N-cetyl-N,N,N-trimethylammonium bromide (Prepare buffer by mixing 100 mM NaH_2PO_4 and 100 mM Na_2HPO_4 to achieve a pH of 7.0.)
Injection: Hydrostatic injection for 30 s
Detector: UV 214
Migration time: 13.5
Internal standard: 2,6-dimethylphenol (only for urine) (14.5)
Voltage: -26 kV
Current: 97 μA
Model: Waters Quanta 4000

OTHER SUBSTANCES
Extracted: acebutolol, alprenolol, labetalol, metoprolol, nadolol, oxprenolol, pindolol, propranolol, timolol

---

KEY WORDS
serum; comparison with HPLC; SPE

REFERENCE
Lukkari,P.; Sirén,H. Ion-pair chromatography and micellar electrokinetic capillary chromatography in analyzing β-adrenergic blocking agents from human biological fluids, *J.Chromatogr.A*, **1995**, *717*, 211–217.

SAMPLE
Matrix: bulk, formulations
Sample preparation: Grind tablet, add 10 mL MeOH, sonicate for 10 min, filter. Dilute the filtrate with running buffer to give an atenolol concentration of 1 mg/mL. Dilute this solution 20-fold with running buffer, inject an aliquot.

CAPILLARY ELECTROPHORESIS
Capillary: 72 cm × 50 μm fused-silica (50 cm to detector) (Composite Metal Services, Worcester, UK)
Capillary preparation: Before each run wash capillary with 100 mM NaOH for 2 min and with running buffer for 3 min. At the start of each day wash capillary with 1 M NaOH for 20 min and with water for 20 min. At the end of each day wash with 1 M NaOH for 5 min and with water for 5 min. Wash a new capillary with 1 M NaOH for 1 h and then with water for 30-60 min.
Capillary temperature: 25 ± 1
Running buffer: 20 mM pH 9.3 Phosphate buffer containing 20 mM borate
Injection: Hydrodynamic injection for 5 s.
Detector: UV 226
Migration time: 9
Voltage: 15 kV

Model: Applied Biosystems Model 270A
Limit of quantitation: 3.5 µg/mL
Limit of detection: 1 µg/mL

OTHER SUBSTANCES
Simultaneous: impurities

KEY WORDS
tablets

REFERENCE
Shafaati,A.; Clark,B.J. Development and validation of a capillary zone electrophoretic method for the determination of atenolol in presence of its related substances in bulk and tablet dosage form, *J.Pharm.Biomed.Anal.*, **1996**, *14*, 1547–1554.

SAMPLE
Matrix: solutions

CAPILLARY ELECTROPHORESIS
Capillary: 82 cm × 75 µm fused-silica (Waters)
Capillary preparation: Between runs purge capillary with running buffer for 2 min. Purge a new capillary with 100 mM NaOH for 30 min, with water for 30 min, and with running buffer for 30 min.
Capillary temperature: 30
Running buffer: 30 mM pH 7.6 phosphate buffer containing 10 mM cetyltrimethylammonium bromide
Injection: Inject at high pressure for 5 s
Detector: UV 214
Migration time: 7
Voltage: 21 kV
Model: Beckman P/ACE System 2000

OTHER SUBSTANCES
Simultaneous: acebutolol, alprenolol, nadolol, oxprenolol, pindolol, propranolol, sotalol, timolol

---

REFERENCE
Lukkari,P.; Ennelin,A.; Sirén,H.; Riekkola,M.-L. Effect of temperature, effective capillary length, and applied voltage on the migration of nine β-blockers in micellar electrokinetic capillary chromatography, *J.Liq.Chromatogr.*, **1993**, *16*, 2069–2079.

SAMPLE
Matrix: solutions

CAPILLARY ELECTROPHORESIS
Capillary: 58 cm × 50 µm fused-silica (50 cm to detector)
Running buffer: Isopropanol:buffer 2.5:97.5 (Buffer was 80 mM pH 6.8 Phosphate buffer containing 15 mM cetyltrimethylammonium bromide.)
Injection: Hydrostatic injection for 15 s
Detector: UV 214
Migration time: 12
Voltage: -20 kV
Model: Waters Quanta 4000

OTHER SUBSTANCES
Simultaneous: acebutolol, alprenolol, labetalol, metoprolol, nadolol, oxprenolol, pindolol, propranolol, sotalol, timolol

---

REFERENCE
Lukkari,P.; Vuorela,H.; Riekkola,M.-L. Effects of organic mobile phase modifiers on elution and separation of β-blockers in micellar electrokinetic capillary chromatography, *J.Chromatogr.A*, **1993**, *655*, 317–324.

SAMPLE
Matrix: solutions
Sample preparation: Prepare a 10 µg/mL solution in water, inject an aliquot.

CAPILLARY ELECTROPHORESIS
Capillary: 57 cm × 50 µm fused-silica (50 cm to detector) (Beckman)
Capillary temperature: 25
Running buffer: 40 mM Lithium hydroxide containing 37 mM methyl-β-cyclodextrin, adjusted to pH 3.0 with orthophosphoric acid
Injection: Pressure injection for 2 s
Detector: UV 200
Migration time: 24.4, 24.7 (enantiomers)
Voltage: 20 kV
Model: Beckman PACE 2100

OTHER SUBSTANCES
Interfering: oxprenolol

KEY WORDS
chiral

REFERENCE
Wren,S.A.C.; Rowe,R.C. Theoretical aspects of chiral separation in capillary electrophoresis. III. Application to β-blockers, *J.Chromatogr.*, **1993**, *635*, 113–118.

SAMPLE
Matrix: solutions

CAPILLARY ELECTROPHORESIS
Capillary: 100 cm × 50 µm fused-silica (50 cm to detector) (Isco)
Capillary preparation: Flush capillary with 10 µL running buffer between runs. Every 40 sample injections rinse capillary with 200 µL 1 M NaOH, with 200 µL water, and 200 µL running buffer. Before use fill capillary with 1 M NaOH and allow to stand for 1 h, fill with 100 mM NaOH, allow to stand for 1 h, wash with water fill with running buffer.
Capillary temperature: 23
Running buffer: 100 mM citric acid containing 19.27 mM Na_2HPO_4 and 120 mM hydroxypropyl-β-cyclodextrin
Injection: Inject under vacuum at 4.0 kPa.s
Detector: UV 200
Migration time: 33.5 (S), 34 (R)
Voltage: 30 kV
Model: Isco Model 3140

OTHER SUBSTANCES
Simultaneous: albuterol, alprenolol, cimaterol, labetalol, nadolol, oxprenolol, pirbuterol, propranolol, terbutaline
Interfering: clenbuterol, pindolol

KEY WORDS
chiral

REFERENCE
Aumatell,A.; Wells,R.J.; Wong,D.K.Y. Enantiomeric differentiation of a wide range of pharmacologically active substances by capillary electrophoresis using modified β-cyclodextrins, *J.Chromatogr.A*, **1994**, *686*, 293–307.

SAMPLE
Matrix: solutions
Sample preparation: Prepare a 50 µg/mL solution in diluted running buffer, inject an aliquot.

CAPILLARY ELECTROPHORESIS
Capillary: 44 cm × 50 μm fused-silica
Capillary preparation: After each run wash capillary with water for 2 min and with running buffer for 3 min. At the beginning of each day wash capillary with water and running buffer for 5 min. Condition a new capillary with 1 M NaOH, 100 mM NaOH, water, and separation buffer.
Capillary temperature: 15
Running buffer: 100 mM Phosphoric acid containing 30 mM heptakis(2,6-di-O-methyl)-β-cyclodextrin adjusted to pH 3.0 with triethanolamine
Injection: Hydrodynamic injection for 1 s
Detector: UV 210
Voltage: 25 kV
Model: Spectra Physics Spectraphoresis 1000

KEY WORDS
chiral; enantiomer resolution 1.4

REFERENCE
Bechet,I.; Paques,P.; Fillet,M.; Hubert,P.; Crommen,J. Chiral separation of basic drugs by capillary zone electrophoresis with cyclodextrin additives, *Electrophoresis*, **1994**, *15*, 818–823.

SAMPLE
Matrix: solutions

CAPILLARY ELECTROPHORESIS
Capillary: 60 cm × 50 μm AccuSep (52.5 cm to detector) (Waters)
Capillary preparation: Before injection rinse capillary with 100 mM NaOH for 3 min and with running buffer for 3 min. Rinse new capillaries with 500 mM NaOH for 5 min
Running buffer: 50 mM pH 7.0 Na$_2$HPO$_4$ containing 25 mM (S)-N- dodecoxycarbonylvaline (Prepare (S)-N-dodecoxycarbonylvaline as follows. Prepare dodecyl chloroformate by reacting 1-dodecanol with 0.33 equivalents of triphosgene in dichloromethane solution in the presence of pyridine. Add dodecyl chloroformate dropwise to (S)-valine in 1 M NaOH solution, filter, wash with hexane, recrystallize from ether/petroleum ether.)
Injection: Hydrostatic injection 2 s
Detector: UV 214
Voltage: +12 kV
Model: Waters Quanta 4000 or 4000E

OTHER SUBSTANCES
Also analyzed: bupivacaine, ephedrine, homatropine, ketamine, metoprolol, N-methylpseudoephedrine, norephedrine, norphenylephrine, octopamine, pindolol, terbutaline

KEY WORDS
chiral; α = 1.05

REFERENCE
Mazzeo,J.R.; Grover,E.R.; Swartz,M.E.; Petersen,J.S. Novel chiral surfactant for the separation of enantiomers by micellar electrokinetic capillary chromatography, *J.Chromatogr.A*, **1994**, *680*, 125–135.

SAMPLE
Matrix: solutions

CAPILLARY ELECTROPHORESIS
Capillary: 62 cm × 52 μm fused-silica (50 cm to detector) (Polymicro Technologies)
Running buffer: 150 mM pH 2.5 Trimethylammonium phosphate buffer containing 20 mM hydroxypropyl-β-cyclodextrin
Injection: Gravity injection at 10 cm for 5 s
Migration time: 54, 55 (enantiomers)
Voltage: 20 kV
Current: 82 μA
Model: Laboratory constructed

OTHER SUBSTANCES
Simultaneous: nadolol, pindolol

KEY WORDS
chiral

REFERENCE
Quang,C.; Khaledi,M.G. Direct separation of the enantiomers of β-blockers by cyclodextrin-mediated capillary zone electrophoresis, *J.High Res.Chromatogr.*, **1994**, *17*, 99–101.

SAMPLE
Matrix: solutions
Sample preparation: Prepare a 600 μg/mL solution in MeCN:MeOH:water 25:25:50, dilute to 150 μg/mL with 100 mM pH 8.1 borate buffer containing 50 mM sodium dodecyl sulfate, inject an aliquot.

CAPILLARY ELECTROPHORESIS
Capillary: 57 cm × 75 μm fused-silica (50 cm to detector) (Beckman)
Capillary preparation: Rinse capillary with 100 mM NaOH and running buffer before each run.
Capillary temperature: 30
Running buffer: Acetone:buffer 15:85 (Buffer was 100 mM pH 8.1 borate buffer containing 50 mM sodium dodecyl sulfate.)
Injection: Pressure injection for 5 s
Detector: UV 200
Migration time: 6.5
Voltage: +25 kV
Model: Beckman P/ACE system 5510

OTHER SUBSTANCES
Simultaneous: amlodipine, diltiazem, nicardipine, nifedipine, verapamil

REFERENCE
Bretnall,A.E.; Clarke,G.S. Investigation and optimisation of the use of micellar electrokinetic chromatography for the analysis of six cardiovascular drugs, *J.Chromatogr.A*, **1995**, *700*, 173–178.

SAMPLE
Matrix: solutions
Sample preparation: Inject an aliquot of a 100 μg/mL solution in water:running buffer 50:50.

CAPILLARY ELECTROPHORESIS
Capillary: 44.5 cm × 50 μm acrylamide-coated fused-silica (Bio-Rad)
Capillary temperature: 30
Running buffer: 100 mM NaH_2PO_4 containing 15 mM gamma-cyclodextrin, adjusted to pH 2.5 with phosphoric acid
Injection: Electrokinetic injection at 8 kV for 6 s.
Detector: UV 200
Migration time: 9.61
Voltage: 14 kV
Model: Bio-Rad BioFocus 3000

OTHER SUBSTANCES
Also analyzed: albuterol, alprenolol, atropine, baclofen, bamethan, benserazide, biperiden, bisoprolol, bupivacaine, bupranolol, butetamate, carazolol, carbuterol, carvedilol, celiprolol, chloroquine, chlorpheniramine (chlorphenamine), clidinium bromide, clobutinol, disopyramide, dobutamine, flecainide, homatropine, ipratropium bromide, isoproterenol, isothipendyl, ketamine, mefloquine, mequitazine, metaproterenol (orciprenaline), metipranolol, nafronyl (naftidrofuryl), nefopam, ofloxacin, orphenadrine, oxomemazine, oxprenolol, phenoxybenzamine, pholedrine, pindolol, pirbuterol, prilocaine, promethazine, propafenone, propranolol, sotalol, synephrine, terbutaline, tetrahydrozoline (tetryzoline), tocainide, trihexyphenidyl, trimeprazine (alimemazine), trimipramine, tropicamide, verapamil, zopiclone

KEY WORDS
coated capillary; achiral

REFERENCE
Koppenhoefer,B.; Epperlein,U.; Christian,B.; Yibing,J.; Yuying,C.; Bingcheng,L. Separation of enantiomers of drugs by capillary electrophoresis. I. γ-Cyclodextrin as chiral solvating agent, *J.Chromatogr.A*, **1995**, *717*, 181–190.

SAMPLE
Matrix: solutions
Sample preparation: Inject an aliquot of a solution in 25 mM pH 6.8 potassium phosphate buffer.

CAPILLARY ELECTROPHORESIS
Capillary: 23.5 × 75 μm gel-filled fused-silica (16.5 cm to detector) (Polymicro Technologies)
Capillary preparation: Wash a new capillary with 100 μL water, 100 μL 1 M NaOH, 100 μL water, and 100 μL 50 mM pH 5.0 sodium citrate buffer. Pump gel mixture through capillary at 15 μL/min until it is within 1 cm of detector (detection window is filled with buffer, not gel), make sure it contains no bubbles (use a microscope), condition overnight at 2 kV (current limitation 20 μA) at reverse polarity using running buffer. (Gel mixture was 5 parts 20% bovine serum albumin in water, 21 parts 20% cellulase (from Trichoderma reesei) in 50 mM pH 5.0 sodium citrate buffer, and 4 arts 50% glutaraldehyde.)
Running buffer: Isopropanol:50 mM pH 6.8 potassium phosphate buffer 1:99
Injection: Electrokinetic injection at 2 kV for 3 s
Detector: UV 214
Migration time: 6.5, 6.7 (enantiomers)
Voltage: 3.5 kV
Current: 47 μA
Model: Beckman P/ACE 2050

OTHER SUBSTANCES
Simultaneous: metoprolol, pindolol

KEY WORDS
chiral

REFERENCE
Ljungberg,H.; Nilsson,S. Protein-based capillary affinity gel electrophoresis for chiral separation of β-adrenergic blockers, *J.Liq.Chromatogr.*, **1995**, *18*, 3685–3698.

SAMPLE
Matrix: solutions
Sample preparation: Prepare a solution in MeOH/water, inject an aliquot.

CAPILLARY ELECTROPHORESIS
Capillary: 62 cm × 52 μm fused-silica (50 cm to detector) (Polymicro Technologies)
Capillary temperature: 40
Running buffer: 50 mM pH 2.50 Tetrabutylammonium phosphate containing 20 mM hydroxypropyl-β-cyclodextrin
Injection: Siphon injection at 10 cm for 5 s
Detector: UV (wavelength not specified)
Migration time: 18.68, 18.82 (enantiomers)
Voltage: 20 kV
Model: Laboratory constructed

OTHER SUBSTANCES
Simultaneous: alprenolol, chlorpheniramine, epinephrine, isoproterenol, labetalol, norepinephrine, oxprenolol, propranolol

KEY WORDS
chiral

REFERENCE
Quang,C.; Khaledi,M.G. Extending the scope of chiral separation of basic compounds by cyclodextrin-mediated capillary zone electrophoresis, *J.Chromatogr.A*, **1995**, *692*, 253–265.

SAMPLE
Matrix: solutions
Sample preparation: Prepare a 50 µg/mL solution in water, filter, inject an aliquot.

CAPILLARY ELECTROPHORESIS
Capillary: 20 cm × 50 µm fused-silica (Beckman)
Capillary temperature: 25
Running buffer: pH 2.2 or 3.0 Lithium phosphate buffer containing 1.5 mM sodium β-cyclo-dextrin sulfobutyl ether (Applied Biosystems/Perkin Elmer)
Detector: UV 200
Migration time: 7.35 (S-(-)), 7.6 (R-(+))
Voltage: 10 kV
Model: Beckman PACE 2100

KEY WORDS
chiral

REFERENCE
Wren,S.A.C. Chiral separation in capillary electrophoresis, *Electrophoresis*, **1995**, *16*, 2127–2131.

SAMPLE
Matrix: solutions
Sample preparation: Inject an aliquot of a solution in MeOH.

CAPILLARY ELECTROPHORESIS
Capillary: 64.5 cm × 50 µm fused-silica (Polymicro Technologies)
Capillary preparation: Between runs flush with 100 mM NaOH for 2-10 min then with running buffer for 2 min. Before use rinse with 1 M NaOH for 1 h, with 100 mM NaOH for 20 min, with water for 20 min, with 1 M acetic acid in MeOH/MeCN for 10 min, and with running buffer for 10 min.
Capillary temperature: 25
Running buffer: MeCN containing 1 M acetic acid, 200 µM Tween 20, and 30 mM sodium (+)-S-camphorsulfonate (Sodium (+)-S-camphorsulfonate can be prepared by mixing equimolar amounts of (+)-S-camphorsulfonic acid and NaOH in EtOH, collect the precipitate, wash with diethyl ether, dry.)
Injection: Pressure injection of 5 kPa for 3 s.
Detector: UV 214
Migration time: 17.6, 18 (enantiomers)
Voltage: 30 kV
Model: Hewlett-Packard HP[3D]

KEY WORDS
chiral

REFERENCE
Bjornsdottir,I.; Hansen,S.H.; Terabe,S. Chiral separation in non-aqueous media by capillary electrophoresis using the ion-pair principle, *J.Chromatogr.A*, **1996**, *745*, 37–44.

SAMPLE
Matrix: solutions

CAPILLARY ELECTROPHORESIS
Capillary: 57 cm × 75 µm fused-silica (50 cm to detector) (Beckman)
Capillary preparation: Before each run rinse capillary with 100 mM NaOH and running buffer.
Capillary temperature: 30
Running buffer: Acetone:100 mM pH 8.1 borate buffer containing 50 mM sodium dodecyl sulfate 15:85 (A) or 100 mM phosphoric acid adjusted to pH 3.1 with triethanolamine (B)

Injection: Pressure injection for 5-10 s.
Detector: UV 200
Migration time: 6 (A), 10.5 (B)
Voltage: 25 kV
Model: Beckman P/ACE 5510

OTHER SUBSTANCES
Simultaneous: acebutolol, amiodarone, bretylium, captopril, diltiazem, disopyramide, lidocaine, lisinopril, metoprolol, nicardipine, nifedipine, phenytoin, pindolol, propafenone, propranolol, quinidine, sotalol, timolol, p-toluenesulfonic acid, verapamil

KEY WORDS
only running buffer A separates all compounds

REFERENCE
Bretnall,A.E.; Clarke,G.S. Selectivity of capillary electrophoresis for the analysis of cardiovascular drugs, *J.Chromatogr.A*, **1996**, *745*, 145–154.

SAMPLE
Matrix: solutions

CAPILLARY ELECTROPHORESIS
Capillary: 43 cm × 50 μm fused-silica (36 cm to detector)
Capillary preparation: Before each injection wash with running buffer for 10 min. After each injection wash with water for 2 min. Wash new capillaries with 1 M NaOH at 60° for 50 min, with NaOH solution (?) at 60° for 10 min, and with water at 25°.
Capillary temperature: 25
Running buffer: 320 mM pH 2.0 Citrate buffer
Injection: Hydrodynamic injection at 1.5 psi for 2 s.
Detector: UV 220
Migration time: 9.2
Voltage: 15 kV
Model: Spectra-Physics Model 1000
Limit of detection: 1-18 μg/mL

OTHER SUBSTANCES
Simultaneous: acebutolol, labetalol, levobunolol, metoprolol, nadolol, oxprenolol, pindolol, propranolol, timolol

REFERENCE
Lin,C.-E.; Chang,C.-C.; Lin,W.-c.; Lin,E.C. Capillary zone electrophoretic separation of β-blockers using citrate buffer at low pH, *J.Chromatogr.A*, **1996**, *753*, 133–138.

SAMPLE
Matrix: solutions

CAPILLARY ELECTROPHORESIS
Capillary: 52 cm × 75 μm fused-silica (48 cm to detector) (Polymicro Technologies)
Capillary preparation: Before each run purge with running buffer for 3 min. Every 3 runs purge with 100 mM NaOH for 5 min. Purge new capillaries with 1 M NaOH for 20 min and with 100 mM NaOH for 20 min, rinse with running buffer, equilibrate with running buffer at 12 kV for 3 h.
Running buffer: 100 mM CHES (2-(N-cyclohexylamine)ethanesulfonic acid) containing 10 mM triethylamine and 25 mM (R)-dodecoxycarbonylvaline (Waters EnantioSelect (R)-Val-1), pH adjusted to 8.8 with 1 M NaOH
Injection: Hydrostatic injection for 2 s.
Detector: UV 214
Migration time: 8.5 (R), 8.6 (S)
Voltage: 12 kV
Current: ≤30 μA
Model: Waters Quanta 4000

OTHER SUBSTANCES
Simultaneous: ephedrine, methylpseudoephedrine, pindolol

KEY WORDS
chiral

REFERENCE
Peterson,A.G.; Ahuja,E.S.; Foley,J.P. Enantiomeric separations of basic pharmaceutical drugs by micellar electrokinetic chromatography using a chiral surfactant, N-dodecoxycarbonylvaline, *J.Chromatogr.B*, **1996**, *683*,
15–28.

SAMPLE
Matrix: solutions
Sample preparation: Inject an aliquot of a 100 μg/mL solution in running buffer.

CAPILLARY ELECTROPHORESIS
Capillary: 29 cm × 50 μm fused-silica (24.5 cm to detector) (Yongnian Optical Conductive Fiber
Plant, China), coated with polyacrylamide
Capillary preparation: No details of the polyacrylamide coating process are provided. However,
another paper (LC.GC 1997, 15, 40) by this group indicates that they use the procedure of
Hjertén, thus: Adjust the pH of 20 mL water to 3.5 with acetic acid, add 80 μL 3-(trimethoxysilyl)propyl methacrylate (3-methacryloxypropyltrimethoxysilane), mix, suck into capillary, let
stand at room temperature for 1 h, remove the solution, wash with water. Fill the capillary
with a deaerated 3-4% acrylamide solution containing 1 μL/mL N,N,N',N'-tetramethylethyl-
enediamine and 1 mg/mL potassium persulfate, let stand for 30 min, remove excess solution
by aspiration, rinse with water, remove water by aspiration, dry at 35° (J. Chromatogr. 1985,
347, 191).
Capillary temperature: 25
Running buffer: 100 mM NaH$_2$PO$_4$ adjusted to pH 2.5
Injection: Electrokinetic injection at 15 kV for 3 s.
Detector: UV 200, UV 210
Migration time: 6.14
Voltage: 15 kV
Model: Bio-Rad BioFocus 3000

OTHER SUBSTANCES
Simultaneous: albuterol, alprenolol, baclofen, bamethan, benproperine, benserazide, bisoprolol,
bupranolol, butamirate, butethamate, carbuterol, celiprolol, clenbuterol, clobutinol, dipivefrin,
isoproterenol (isoprenaline), metaproterenol (orciprenaline), metipranolol, metoprolol, norfenefrine, ornidazole, oxprenolol, phenylpropanolamine, pholedrine, pirbuterol, prilocaine, procyclidine, sotalol, synephrine, terbutaline, tocainide

KEY WORDS
coated capillary

REFERENCE
Koppenhoefer,B.; Epperlein,U.; Xiaofeng,Z.; Bingcheng,L. Separation of enantiomers of drugs by capillary electrophoresis. Part 4: Hydroxypropyl-γ-cyclodextrin as chiral solvating agent, *Electrophoresis*, **1997**, *18*, 924–
930.

SAMPLE
Matrix: solutions
Sample preparation: Inject an aliquot of a 100 μg/mL solution in running buffer.

CAPILLARY ELECTROPHORESIS
Capillary: 30 cm × 50 μm fused-silica (25.5 cm to detector), coated with polyacrylamide
Capillary preparation: Adjust the pH of 20 mL water to 3.5 with acetic acid, add 80 μL 3-
(trimethoxysilyl)propyl methacrylate (3-methacryloxypropyltrimethoxysilane), mix, suck into
capillary, let stand at room temperature for 1 h, remove the solution, wash with water. Fill the
capillary with a deaerated 3-4% acrylamide solution containing 1 μL/mL N,N,N',N'-tetra-
methylethylenediamine and 1 mg/mL potassium persulfate, let stand for 30 min, remove excess

solution by aspiration, rinse with water, remove water by aspiration, dry at 35° (J. Chromatogr.
1985, 347, 191).
Capillary temperature: 25
Running buffer: 100 mM NaH_2PO_4 containing 45 mM heptakis(2,6-di-O- methyl)-β-cyclodextrin,
adjusted to pH 2.5 with phosphoric acid
Injection: Electrokinetic injection at 15 kV for 3 s.
Detector: UV 200
Voltage: 15 kV
Model: Bio-Rad BioFocus 3000

KEY WORDS
chiral; coated capillary; comparison with the use of other cyclodextrins; this running buffer gave
the greatest enantiomeric separation.; α=1.013

REFERENCE
Lin,B.; Zhu,X.; Koppenhoefer,B.; Epperlein,U. Investigation of 123 chiral drugs by cyclodextrin-modified capillary electrophoresis, *LC.GC*, **1997**, *15*, 40–46.

SAMPLE
Matrix: solutions

CAPILLARY ELECTROPHORESIS
Capillary: 67 cm × 50 μm (60 cm to detector)
Capillary preparation: Wash with running buffer for 5 min before each injection. After each
injection wash with 1 M NaOH at 60° for 5 min, with 100 mM NaOH at 60° for 10 min, and
with water at 25° for 5 min.
Capillary temperature: 25
Running buffer: 70 mM pH 7.0 Sodium phosphate buffer containing 15 mM cetyltriemthylam-
monium bromide
Injection: Hydrodynamic injection for 1 s.
Detector: UV 220
Migration time: 14.8
Voltage: 20 kV

OTHER SUBSTANCES
Simultaneous: acebutolol, labetalol, levobunolol, metoprolol, nadolol, oxprenolol, pindolol, propranolol, timolol

REFERENCE
Lin,C.-E.; Chen,Y.-C.; Chang,C.-C.; Wang,D.-Z. Migration behavior and selectivity of β-blockers in micellar
electrokinetic chromatography. Influence of micelle concentration of cationic surfactants, *J.Chromatogr.A*,
1997, *775*, 349–357.

SAMPLE
Matrix: solutions

CAPILLARY ELECTROPHORESIS
Capillary: 50.5 cm × 75 μm coated fused-silica (36 cm to detector) (Polymicro Technologies)
Capillary preparation: Coat capillary as follows. Condition capillary with 1 M NaOH for 1 h,
rinse with water for 30 min, force 1% methacryloxypropyltrimethoxysilane (adjusted to pH 3.5
with acetic acid) through the capillary using pressure, allow to react at room temperature for
2 h, rinse with water, fill capillary with diallyldimethylammonium chloride solution containing
1 μL/mL N,N,N',N'-tetramethylethylenediamine and 1 mg/mL ammonium persulfate, allow to
react until polymerization is complete, rinse with water, dry at 30-40° overnight.
Running buffer: 10 mM pH 5.0 Acetate buffer
Injection: Hydrodynamic injection at 5 cm for 5 s.
Detector: UV 210
Migration time: 10
Voltage: 12 kV
Model: laboratory-constructed

OTHER SUBSTANCES
Simultaneous: acebutolol, dilevalol, nifenalol, pindolol

KEY WORDS
coated capillary

REFERENCE
Liu,Q.; Lin,F.; Hartwick,R.A. Poly(diallyldimethylammonium chloride) as a cationic coating for capillary electrophoresis, *J.Chromatogr.Sci.*, **1997**, *35*, 126–130.

SAMPLE
Matrix: solutions

CAPILLARY ELECTROPHORESIS
Capillary: 62.1 cm × 75 μm fused-silica (44.6 cm to detector) (Polymicro Technologies)
Capillary preparation: Coat capillary as follows. Condition capillary with 1 M NaOH for 1 h, rinse with water for 30 min, fill capillary with 1% 3-(trimethoxysilyl)propyl methacrylate (methacryloxypropyltrimethoxysilane) (adjusted to pH 3.5 with acetic acid), allow to react at room temperature for 2 h, rinse with water, fill capillary with 10% 2-aminoethyl methacrylate hydrochloride (Fisher Scientific) solution containing 1-3 mg/mL N,N,N',N'-tetramethylethylenediamine and 1 mg/mL ammonium persulfate, allow to react until polymerization is complete, rinse with water, dry at 30-40° overnight.
Running buffer: 25 mM pH 4.7 Acetate buffer
Injection: Hydrodynamic injection at 5 cm for 5 s.
Detector: UV 210
Migration time: 11.57
Voltage: 15 kV
Model: laboratory constructed

OTHER SUBSTANCES
Simultaneous: acebutolol, betaxolol, carteolol, dilevalol, nifenalol, pindolol, propranolol

KEY WORDS
coated capillary; detector at anode

REFERENCE
Liu,Q.; Lin,F.; Hartwick,R.A. Free solution capillary electrophoretic separation of basic proteins and drugs using cationic polymer coated capillaries, *J.Liq.Chromatogr.Rel.Technol.*, **1997**, *20*, 707–718.

SAMPLE
Matrix: solutions
Sample preparation: Inject an aliquot of a 100 μM solution in water.

CAPILLARY ELECTROPHORESIS
Capillary: 65 cm × 50 μm fused-silica (56.5 cm to detector) (Polymicro Technologies)
Capillary preparation: Before each injection rinse capillary at 930 mbar with 100 mM NaOH for 1 min, with water for 2 min, and with running buffer for 4 min.
Capillary temperature: 20
Running buffer: 50 mM Phosphoric acid containing 24 mM heptakis(2,6-di-O-methyl)-β-cyclodextrin, adjusted to pH 2.5 with triethanolamine
Injection: Pressure injection at 20 mbar for 3 s.
Detector: UV 195
Migration time: 17.1 (R_S = 1.2)
Voltage: 24 kV
Model: Hewlett Packard HP³ᴰ

KEY WORDS
chiral

REFERENCE
Nilsson,S.; Schweitz,L.; Petersson,M. Three approaches to enantiomer separation of β-adrenergic antagonists by capillary electrochromatography, *Electrophoresis*, **1997**, *18*, 884–890.

SAMPLE
Matrix: solutions

CAPILLARY ELECTROPHORESIS
Capillary: 100 cm × 50 µm fused-silica (60 cm to detector) (Polymicro Technologies)
Capillary temperature: 30
Running buffer: 50 mM pH 2.5 Sodium phosphate containing 1.5 mM sulfobutyl ether-β-cyclodextrin (average substitution 3.9, MW 1721, Center for Drug Delivery Research, Lawrence KS)
Injection: Pressure injection at 5 psi for 1.0 s.
Detector: UV 215
Migration time: 20.1 (S), 20.7 (R)
Voltage: 25 kV
Current: 27 µA
Model: laboratory-constructed

KEY WORDS
chiral

REFERENCE
Xie,G.-h.; Skanchy,D.J.; Stobaugh,J.F. Chiral separations of enantiomeric pharmaceuticals by capillary electrophoresis using sulphobutyl ether β-cyclodextrin as isomer selector, *Biomed.Chromatogr.*, **1997**, *11*, 193–199.

SAMPLE
Matrix: solutions

CAPILLARY ELECTROPHORESIS
Capillary: 29-36 cm × 50 µm fused-silica (24.5-31.5 cm to detector) (Yongnian Optical Conductive Fiber Plant, China) coated with polyacrylamide
Capillary preparation: Coat capillary as follows. Adjust the pH of 20 mL water to 3.5 with acetic acid, add 80 µL 3-(trimethoxysilyl)propyl methacrylate (3-methacryloxypropyltrimethoxysilane), mix, suck into capillary, let stand at room temperature for 1 h, remove the solution, wash with water. Fill the capillary with a deaerated 3-4% acrylamide solution containing 1 µL/mL N,N,N',N'-tetramethylethylenediamine and 1 mg/mL potassium persulfate, let stand for 30 min, remove excess solution by aspiration, rinse with water, remove water by aspiration, dry at 35° (J. Chromatogr. 1985, 347, 191).
Capillary temperature: 25
Running buffer: 100 mM pH 2.5 NaH_2PO_4 (A) or 100 mM pH 2.5 NaH_2PO_4 containing 45 mM hydroxypropyl-α-cyclodextrin (Wacker, Munich) (B)
Injection: Electromigration at 15 kV for 3 s
Detector: UV 200; UV 210
Migration time: 6.14 (A); 7.67 (B) (no separation of enantiomers)
Voltage: 15 kV
Model: Bio-Focus 3000

OTHER SUBSTANCES
Also analyzed: albuterol (salbutamol), alprenolol, amorolfine, atropine, azelastine, baclofen, bamethan, benproperine, benserazide, biperiden, bisoprolol, brompheniramine, bupivacaine, bupranolol, butamirate, butethamate, carazolol, carbuterol, carteolol, carvedilol, celiprolol, chloroquine, chlorpheniramine, chlorphenoxamine, cicletanine, clenbuterol, clidinium bromide, clobutinol, dimethindene, dipivefrin, disopyramide, dobutamine, doxylamine, fendiline, flecainide, gallopamil, homatropine, ipratropium bromide, isoproterenol (isoprenaline), isothipendyl, ketamine, meclizine, mefloquine, mepindolol, mequitazine, metaclazepam, metaproterenol (orciprenaline), metipranolol, metoprolol, nafronyl (naftidrofuryl), nefopam, nicardipine, norfenefrine, ofloxacin, ornidazole, orphenadrine, oxomemazine, oxprenolol, oxybutynin, phenoxybenzamine, phenylpropanolamine, pholedrine, pindolol, pirbuterol, prilocaine, procyclidine, promethazine, propafenone, propranolol, reproterol, sotalol, sulpride, synephrine, talinolol, terbutaline, tetrahydrozoline (tetryzoline), theodrenaline, tioconazole, tocainide, trihexyphenidyl, trimeprazine (alimemazine), trimipramine, tropicamide, verapamil, zopiclone

---

KEY WORDS
coated capillary

REFERENCE

Koppenhoefer,B.; Eperlein,U.; Schlunk,R.; Zhu,X.; Lin,B. Separation of enantiomers of drugs by capillary electrophoresis. V. Hydroxypropyl-α-cyclodextrin as chiral solvating agent, *J.Chromatogr.A*, **1998**, *793*, 153–164.

SAMPLE

Matrix: solutions

CAPILLARY ELECTROPHORESIS

Capillary: 71.5 cm × 75 μm fused-silica (52.2 cm to detector) (Polymicro Technologies)

Capillary preparation: Rinse capillary with running buffer for 1 min between runs. Rinse new capillaries with 500 mM NaOH for 20 min and with water for 10 min and then condition with running buffer for 1 h.

Running buffer: 50 mM pH 3.0 containing 0.1% guaran (Guaran is the water-soluble fraction of guar gum. Prepare guaran by stirring 1 g guar gum (Sigma) in 100 mL water for 20 min, add 25 mL EtOH dropwise with stirring, let stand overnight, supercentrifuge. Add 30 mL EtOH dropwise with stirring to the centrifugate, centrifuge, collect the guaran as a precipitate, dry in air.)

Injection: Hydrodynamic injection at 5 cm for 5 s.

Detector: UV (wavelengths not given)

Migration time: 13.5

Voltage: 17.5 kV

Model: laboratory-constructed

OTHER SUBSTANCES

Simultaneous: acebutolol, betaxolol, carteolol, dilevalol, nifenalol, pindolol, propranolol

REFERENCE

Liu,Q.; Lin,F.; Hartwick,R.A. Capillary zone electrophoretic separation of basic proteins and drugs using guaran as a buffer modifier, *Chromatographia*, **1998**, *47*, 219–224.

SAMPLE

Matrix: solutions

Sample preparation: Inject an aliquot of a 340 μM solution.

CAPILLARY ELECTROPHORESIS

Capillary: 58 cm × 75 μm fused-silica (50 cm to detector)

Capillary preparation: Rinse with running buffer for 2 min before each run. Before use rinse capillary at 1.36 bar with 100 mM NaOH for 5 min and with water for 10 min.

Capillary temperature: 25

Running buffer: 20 mM pH 7.0 Sodium phosphate buffer (A) or 20 mM pH 7.0 Sodium phosphate buffer containing 10 mM 1-lauroyl-2-hydroxy-sn-glycero-3-phosphocholine (Avanti Polar Lipids, Alabaster AL) (B)

Injection: Hydrodynamic injection at 34 mbar for 1 s followed by 5% MeOH for 5 s.

Detector: UV 200

Migration time: 4.03 min (A); k' 0.0 (B)

Voltage: 15 kV

Current: ca. 50 μA

Model: Beckman P/ACE 2200

OTHER SUBSTANCES

Simultaneous: acebutolol, alprenolol, metoprolol, oxprenolol, pindolol, propranolol, timolol

REFERENCE

Masucci,J.A.; Caldwell,G.W.; Foley,J.P. Comparison of the retention behavior of β-blockers using immobilized artificial membrane chromatography and lysophospholipid micellar electrokinetic chromatography, *J.Chromatogr.A*, **1998**, *810*, 95–103.

SAMPLE

Matrix: urine

Sample preparation: Dilute urine with 2 volumes of water, add IS, filter (0.5 μm), inject an aliquot.

CAPILLARY ELECTROPHORESIS
Capillary: 68 cm × 50 μm fused-silica (60 cm to detector) (White Associates)
Capillary preparation: Purge capillary with running buffer for 2 min before each injection.
Running buffer: 80 mM pH 7.0 Phosphate buffer containing 10 mM N-cetyl-N,N,N-trimethyl-ammonium bromide
Injection: Hydrostatic injection for 30 s at 10 cm.
Detector: UV 214
Migration time: 13.5
Internal standard: 2,6-dimethylphenol (14.7)
Voltage: -26 kV (at injector end)
Model: Waters Quanta 4000
Limit of detection: 10 μg/mL (S/N 3)

OTHER SUBSTANCES
Extracted: acebutolol, alprenolol, labetalol, metoprolol, nadolol, oxprenolol, pindolol, propranolol, timolol
Noninterfering: caffeine
Interfering: probenecid

REFERENCE
Lukkari,P.; Sirén,H.; Pantsar,M.; Riekkola,M.-L. Determination of ten β-blockers in urine by micellar electrokinetic capillary chromatography, *J.Chromatogr.*, **1993**, *632*, 143–148.

SAMPLE
Matrix: urine
Sample preparation: Filter (0.2 μm), inject an aliquot of the filtrate.

CAPILLARY ELECTROPHORESIS
Capillary: 80 cm × 50 μm fused-silica (57 cm to detector)
Capillary preparation: Before each run rinse the capillary with 1 M NaOH for 3 min, with 100 mM NaOH for 3 min, with water for 3 min, and with running buffer for 10 min.
Running buffer: 110 mM Boric acid containing 56 mM NaOH and 44 mM HCl, pH 8
Injection: Vacuum injection for 1 s
Detector: UV 228
Migration time: 4.262
Voltage: 20 kV
Model: Europhor Prime Vision system IV
Limit of detection: 2.12 μM

OTHER SUBSTANCES
Extracted: acebutolol (UV 238), acetazolamide (UV 222), amiloride (UV 220), bendroflumethiazide (UV 220), bumetanide (UV 220), chlorthalidone (UV 220), cocaine (UV 236), codeine (UV 220), ethacrynic acid (UV 220), furosemide (UV 232), hydrochlorothiazide (UV 226), methadone (UV 220), metoxiphenamine (UV 220), nadolol (UV 220), norcodeine (UV 220), oxprenolol (UV 220), pentazocine (UV 220), propranolol (UV 220), spironolactone (UV 244), triamterene (UV 232), xipamide (UV 234)
Interfering: alprenolol (UV 220)

REFERENCE
Gonzalez,E.; Laserna,J.J. Capillary zone electrophoresis for the rapid screening of banned drugs in sport, *Electrophoresis*, **1994**, *15*, 240–243.

SAMPLE
Matrix: urine
Sample preparation: Condition a Bond Elut Certify SP SPE cartridge with 2 mL MeOH and 2 mL water, do not allow to dry. Adjust the pH of 2.5 mL urine to 9 with 500 μL borate buffer, vortex for 5 s, filter (45 μm membrane), add to the SPE cartridge, wash with 2 mL water, wash with 1 mL 100 mM pH 4 acetate buffer, wash with 2 mL MeOH, elute with 2 mL 2% ammonia

in chloroform:isopropanol 80:20. Evaporate the eluate to dryness under a stream of nitrogen at 30°, reconstitute the residue in 25-100 μL running buffer, inject an aliquot.

CAPILLARY ELECTROPHORESIS
Capillary: 78 cm × 75 μm fused-silica (70 cm to detector) (Composite Metal Services, UK)
Capillary preparation: Before each run wash capillary with 100 mM NaOH for 2 min and with running buffer for 3 min. At the start of each day wash with 1 M NaOH for 20 min and with water for 20 min. At the end of each day wash with 1 M NaOH for 5 min and with water for 5 min.
Capillary temperature: 25
Running buffer: pH 9.7 Buffer prepared by mixing equal volumes of 25 mM Na_2HPO_4 and 25 mM sodium tetraborate
Injection: Hydrostatic injection for 25 s.
Detector: UV 214
Migration time: 6
Voltage: 20 kV
Current: 48 μA
Model: Waters Quanta 4000
Limit of quantitation: 100 ng/mL (reconstitute with 25 μL running buffer)

KEY WORDS
SPE

REFERENCE
Maguregui,M.I.; Alonso,R.M.; Jiménez,R.M. Capillary zone electrophoretic method for the quantitative determination of the β-blocker atenolol in human urine, *J.Liq.Chromatogr.*, **1997**, *20*, 3377–3387.

Atosiban

Molecular formula: $C_{43}H_{67}N_{11}O_{12}S_2$
Molecular weight: 994.20
CAS Registry No.: 90779-69-4

SAMPLE
Matrix: solutions
Sample preparation: Inject an aliquot of a 100 μg/mL aqueous solution.

CAPILLARY ELECTROPHORESIS
Capillary: 80 cm × 75 μm fused-silica (60 cm to detector) (Polymicro Technologies)
Capillary preparation: Flush capillary with 1 M NaOH.
Capillary temperature: 30
Running buffer: 5 mM Gamma-cyclodextrin in buffer (Prepare buffer by diluting 10 mL 1 M phosphoric acid to 90 mL with water and adjusting to pH 2.5 with 2 M LiOH, make up to 100 mL with water.)
Injection: 10 nL
Detector: UV 195
Migration time: 21.5
Voltage: 26 kV
Model: Perkin-Elmer Model 270A-HT

OTHER SUBSTANCES
Simultaneous: other diastereomers
Interfering: diastereomers

KEY WORDS
comparison with HPLC

REFERENCE
Oyler,A.R.; Armstrong,B.L.; Cha,J.Y.; Zhou,M.X.; Yang,Q.; Robinson,R.I.; Dunphy,R.; Burinsky,D.J. Hydrophilic interaction chromatography on amino-silica phases complements reversed-phase high-performance liquid chromatography and capillary electrophoresis for peptide analysis, *J.Chromatogr.A*, **1996**, *724*, 378–383.

Atropine

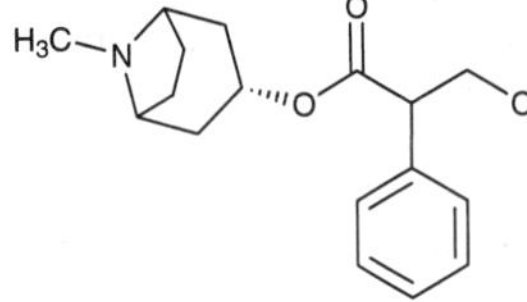

Molecular formula: $C_{17}H_{23}NO_3$
Molecular weight: 289.37
CAS Registry No.: 51-55-8, 52-88-0 (atropine methylnitrate)
Merck Index (12th ed.): 907
Lednicer: 1 35, 71, 93; 2 71

SAMPLE
Matrix: formulations
Sample preparation: Dilute formulation with water to an atropine concentration of about 50 µg/mL.

CAPILLARY ELECTROPHORESIS
Capillary: 64.5 cm × 50 µm fused-silica (56 cm to detector)
Capillary preparation: Between analyses wash capillary with 100 mM NaOH for 2 min, wash with water for 2 min, and equilibrate with running buffer for 3.5 min. At the start of each day flush with 1 M NaOH for 15 min and with water for 10 min.
Capillary temperature: 25
Running buffer: 100 mM pH 7 Tris-phosphate buffer
Injection: Pressure injection at 25 mbar for 20 s (12 nL), ramp voltage at 500 V/s.
Detector: UV 195
Migration time: 3.587
Voltage: 30 kV
Current: 63 µA
Model: Hewlett-Packard HP^{3D}
Limit of quantitation: 3 µg/mL
Limit of detection: 1 µg/mL

OTHER SUBSTANCES
Simultaneous: homatropine, scopolamine

KEY WORDS
ophthalmic solutions

REFERENCE
Cherkaoui,S.; Mateus,L.; Christen,P.; Veuthey,J.-L. Development and validation of a capillary zone electrophoresis method for the determination of atropine, homatropine and scopolamine in ophthalmic solutions, *J.Chromatogr.B*, **1997**, *696*, 283–290.

SAMPLE
Matrix: plants
Sample preparation: Extract 50 mg lyophilized plant material with 5.0 mL 80% MeOH at 60° for 16 h, centrifuge. Remove a 500 µL aliquot of the supernatant and lyophilize it to dryness, make up to 500 µL with running buffer, inject an aliquot.

CAPILLARY ELECTROPHORESIS
Capillary: 67 cm × 75 µm fused-silica (60 cm to detector) (Isco)

Capillary preparation: Between runs rinse the capillary with 100 mM KOH for 1.5 min and with water for 1.5 min then equilibrate with running buffer for 2 min. At the start of each day purge with 100 mM KOH for 10 min, with water for 10 min, and with running buffer for 10 min.
Capillary temperature: 25
Running buffer: 40 mM pH 7.8 Phosphate buffer (Prepare buffer by mixing 40 mM Na_2HPO_4 with 40 mM KH_2PO_4 to achieve a pH of 7.8.)
Injection: Pressure injection at 0.5 psi for 4 s.
Detector: UV 214
Migration time: 5
Voltage: 20 kV
Model: Beckman P/ACE 2200
Limit of quantitation: 5 µg/mL
Limit of detection: 1 µg/mL

OTHER SUBSTANCES
Extracted: norscopolamine, scopolamine, tropic acid

REFERENCE
Eeva,M.; Salo,J.-P.; Oksman-Caldentey,K.-M. Determination of the main tropane alkaloids from transformed *Hyoscyamus muticus* plants by capillary zone electrophoresis, *J.Pharm.Biomed.Anal.*, **1998**, *16*, 717–722.

SAMPLE
Matrix: solutions
Sample preparation: Inject an aliquot of a 100 µg/mL solution in water:running buffer 50:50.

CAPILLARY ELECTROPHORESIS
Capillary: 44.5 cm × 50 µm acrylamide-coated fused-silica (Bio-Rad)
Capillary temperature: 30
Running buffer: 100 mM NaH_2PO_4 containing 15 mM gamma-cyclodextrin, adjusted to pH 2.5 with phosphoric acid
Injection: Electrokinetic injection at 8 kV for 6 s.
Detector: UV 200
Migration time: 11.59
Voltage: 14 kV
Model: Bio-Rad BioFocus 3000

OTHER SUBSTANCES
Also analyzed: albuterol, alprenolol, atenolol, baclofen, bamethan, benserazide, biperiden, bisoprolol, bupivacaine, bupranolol, butetamate, carazolol, carbuterol, carvedilol, celiprolol, chloroquine, chlorpheniramine (chlorphenamine), clidinium bromide, clobutinol, disopyramide, dobutamine, flecainide, homatropine, ipratropium bromide, isoproterenol, isothipendyl, ketamine, mefloquine, mequitazine, metaproterenol (orciprenaline), metipranolol, nafronyl (naftidrofuryl), nefopam, ofloxacin, orphenadrine, oxomemazine, oxprenolol, phenoxybenzamine, pholedrine, pindolol, pirbuterol, prilocaine, promethazine, propafenone, propranolol, sotalol, synephrine, terbutaline, tetrahydrozoline (tetryzoline), tocainide, trihexyphenidyl, trimeprazine (alimemazine), trimipramine, tropicamide, verapamil, zopiclone

KEY WORDS
coated capillary; achiral

REFERENCE
Koppenhoefer,B.; Epperlein,U.; Christian,B.; Yibing,J.; Yuying,C.; Bingcheng,L. Separation of enantiomers of drugs by capillary electrophoresis. I. γ-Cyclodextrin as chiral solvating agent, *J.Chromatogr.A*, **1995**, *717*, 181–190.

SAMPLE
Matrix: solutions
Sample preparation: Inject an aliquot of a solution in MeOH:water 10:90.

CAPILLARY ELECTROPHORESIS
Capillary: 64.5 cm × 75 µm (56 cm to detector)

Capillary preparation: Before each run flush the capillary with 100 mM NaOH for 2 min, with water for 2 min, and equilibrate with running buffer for 3.5 min. At the start of each day wash capillary with 100 mM NaOH for 15 min and with water for 10 min.
Capillary temperature: 25
Running buffer: MeCN:buffer 10:90 (Buffer was 30 mM pH 8.5 phosphate-borate buffer containing 50 mM sodium dodecyl sulfate.)
Injection: Pressure injection at 25 mbar for 10 s (ca. 30 nL), ramp to operating voltage at 500 V/s.
Detector: UV 195
Migration time: 16.5 (hyoscyamine)
Voltage: 30 kV
Model: Hewlett Packard HP 3D

OTHER SUBSTANCES
Simultaneous: apoatropine, homatropine, 6β-hydroxyhyoscyamine, littorine, scopolamine, tropic acid

REFERENCE
Cherkaoui,S.; Mateus,L.; Christen,P.; Veuthey,J.-L. Micellar electrokinetic capillary chromatography for selected tropane alkaloid analysis in plant extract, *Chromatographia*, **1997**, *46*, 351–357.

SAMPLE
Matrix: solutions
Sample preparation: Inject an aliquot of a 100 µg/mL solution in running buffer.

CAPILLARY ELECTROPHORESIS
Capillary: 32 cm × 50 µm fused-silica (27.5 cm to detector) (Yongnian Optical Conductive Fiber Plant, China), coated with polyacrylamide
Capillary preparation: No details of the polyacrylamide coating process are provided. However, another paper (LC.GC 1997, 15, 40) by this group indicates that they use the procedure of Hjertén, thus: Adjust the pH of 20 mL water to 3.5 with acetic acid, add 80 µL 3-(trimethoxysilyl)propyl methacrylate (3-methacryloxypropyltrimethoxysilane), mix, suck into capillary, let stand at room temperature for 1 h, remove the solution, wash with water. Fill the capillary with a deaerated 3-4% acrylamide solution containing 1 µL/mL N,N,N',N'-tetramethylethylenediamine and 1 mg/mL potassium persulfate, let stand for 30 min, remove excess solution by aspiration, rinse with water, remove water by aspiration, dry at 35° (J. Chromatogr. 1985, 347, 191).
Capillary temperature: 25
Running buffer: 100 mM NaH_2PO_4 adjusted to pH 2.5 (A) or 100 mM NaH_2PO_4 containing 45 mM hydroxypropyl-gamma-cyclodextrin, adjusted to pH 2.5 (B)
Injection: Electrokinetic injection at 15 kV for 3 s.
Detector: UV 200, UV 210
Migration time: 5.27, 10.79, 10.84 (B, enantiomers)
Voltage: 15 kV
Model: Bio-Rad BioFocus 3000

OTHER SUBSTANCES
Simultaneous: carazolol, cicletanine, dimethindene, fendiline, homatropine, ipratropium bromide, isothipendyl, mefloquine, metaclazepam, nafronyl (naftidrofuryl), nefopam, nicardipine, promethazine, reproterol, tetrahydrozoline (tetryzoline), theodrenaline, tioconazole, trihexyphenidyl, trimeprazine (alimemazine), trimipramine

KEY WORDS
coated capillary; chiral

REFERENCE
Koppenhoefer,B.; Epperlein,U.; Xiaofeng,Z.; Bingcheng,L. Separation of enantiomers of drugs by capillary electrophoresis. Part 4: Hydroxypropyl-γ-cyclodextrin as chiral solvating agent, *Electrophoresis*, **1997**, *18*, 924–930.

SAMPLE
Matrix: solutions

Sample preparation: Inject an aliquot of a 100 μg/mL solution in running buffer.

CAPILLARY ELECTROPHORESIS
Capillary: 30 cm × 50 μm fused-silica (25.5 cm to detector), coated with polyacrylamide
Capillary preparation: Adjust the pH of 20 mL water to 3.5 with acetic acid, add 80 μL 3-(trimethoxysilyl)propyl methacrylate (3-methacryloxypropyltrimethoxysilane), mix, suck into capillary, let stand at room temperature for 1 h, remove the solution, wash with water. Fill the capillary with a deaerated 3-4% acrylamide solution containing 1 μL/mL N,N,N',N'-tetramethylethylenediamine and 1 mg/mL potassium persulfate, let stand for 30 min, remove excess solution by aspiration, rinse with water, remove water by aspiration, dry at 35° (J. Chromatogr. 1985, 347, 191).
Capillary temperature: 25
Running buffer: 100 mM NaH_2PO_4 containing 45 mM hydroxypropyl-α-cyclodextrin, adjusted to pH 2.5 with phosphoric acid
Injection: Electrokinetic injection at 15 kV for 3 s.
Detector: UV 200
Voltage: 15 kV
Model: Bio-Rad BioFocus 3000

KEY WORDS
chiral; coated capillary; comparison with the use of other cyclodextrins; this running buffer gave the greatest enantiomeric separation.; α=1.013

REFERENCE
Lin,B.; Zhu,X.; Koppenhoefer,B.; Epperlein,U. Investigation of 123 chiral drugs by cyclodextrin-modified capillary electrophoresis, *LC.GC*, **1997**, *15*, 40–46.

SAMPLE
Matrix: solutions

CAPILLARY ELECTROPHORESIS
Capillary: 36 cm × 50 μm fused-silica coated with linear polyacrylamide (31.5 cm to detector) (GL Science)
Capillary preparation: At the beginning and end of each day rinse capillary with capillary wash solution (Bio-Rad Cat. No. 148-5022) at 690 kPa for more than 3 min and with water at 690 kPa for more than 3 min. Coat capillary as follows. Treat capillary with 1 M NaOH at room temperature for 1 h, rinse with water, dry by passing nitrogen gas through the capillary at 110° for 6 h. Pass thionyl chloride through the capillary using a suction pump for several min, seal capillary at both ends and heat at 70° for 6 h. Unseal the capillary and fill with 250 mM vinyl magnesium bromide in THF by suction, seal the capillary, heat at 70° for 6 h. Open the capillary and rinse it with THF for several min, rinse with distilled water, fill the capillary with polymerization solution, heat at 28 ± 2° for 1 h, condition at -100 V/cm for 30 min (Anal. Sci. 1994, 10, 1). (The polymerization solution was 5% acrylamide in water containing 49 mM Tris, 384 mM glycine, and 0.1% sodium dodecyl sulfate, degas in an ultrasonic bath. Add 40 μL 10% N,N,N',N'-tetramethylethylenediamine and 10 μL 10% ammonium persulfate to 5 mL of the degassed solution, mix thoroughly.)
Running buffer: Isopropanol:50 mM pH 6.0 Phosphate buffer 10:90
Injection: Before each injection rinse with water at 690 kPa for 30 s, rinse with running buffer at 690 kPa for 30 s, partially fill with separation solution (750 μM $α_1$-acid glycoprotein (Cohn fraction VI) (ICN) in running buffer) at 6.9 kPa for 190 s (27 cm), inject sample at 6.9 kPa for 2 s, electrophorese with running buffer (Note that $α_1$-acid glycoprotein from other suppliers may provide inferior results).
Detector: UV 210
Migration time: 14, 14.2 (enantiomers)
Voltage: 12 kV
Model: Bio-rad BioFocus 3000

KEY WORDS
chiral; coated capillary

REFERENCE
Tanaka,Y.; Terabe,S. Separation of the enantiomers of basic drugs by affinity capillary electrophoresis using a partial filling technique and $α_1$-acid glycoprotein as chiral selector, *Chromatographia*, **1997**, *44*, 119–128.

SAMPLE
Matrix: solutions

CAPILLARY ELECTROPHORESIS
Capillary: 29-36 cm × 50 μm fused-silica (24.5-31.5 cm to detector) (Yongnian Optical Conductive Fiber Plant, China) coated with polyacrylamide
Capillary preparation: Coat capillary as follows. Adjust the pH of 20 mL water to 3.5 with acetic acid, add 80 μL 3-(trimethoxysilyl)propyl methacrylate (3-methacryloxypropyltrimethoxysilane), mix, suck into capillary, let stand at room temperature for 1 h, remove the solution, wash with water. Fill the capillary with a deaerated 3-4% acrylamide solution containing 1 μL/mL N,N,N',N'-tetramethylethylenediamine and 1 mg/mL potassium persulfate, let stand for 30 min, remove excess solution by aspiration, rinse with water, remove water by aspiration, dry at 35° (J. Chromatogr. 1985, 347, 191).
Capillary temperature: 25
Running buffer: 100 mM pH 2.5 NaH_2PO_4 (A) or 100 mM pH 2.5 NaH_2PO_4 containing 45 mM hydroxypropyl-α-cyclodextrin (Wacker, Munich) (B)
Injection: Electromigration at 15 kV for 3 s.
Detector: UV 200; UV 210
Migration time: 5.27 (A); 8.60, 8.73 (B) (enantiomers)
Voltage: 15 kV
Model: Bio-Focus 3000

OTHER SUBSTANCES
Also analyzed: albuterol (salbutamol), alprenolol, amorolfine, atenolol, azelastine, baclofen, bamethan, benproperine, benserazide, biperiden, bisoprolol, brompheniramine, bupivacaine, bupranolol, butamirate, butethamate, carazolol, carbuterol, carteolol, carvedilol, celiprolol, chloroquine, chlorpheniramine, chlorphenoxamine, cicletanine, clenbuterol, clidinium bromide, clobutinol, dimethindene, dipivefrin, disopyramide, dobutamine, doxylamine, fendiline, flecainide, gallopamil, homatropine, ipratropium bromide, isoproterenol (isoprenaline), isothipendyl, ketamine, meclizine, mefloquine, mepindolol, mequitazine, metaclazepam, metaproterenol (orciprenaline), metipranolol, metoprolol, nafronyl (naftidrofuryl), nefopam, nicardipine, norfenefrine, ofloxacin, ornidazole, orphenadrine, oxomemazine, oxprenolol, oxybutynin, phenoxybenzamine, phenylpropanolamine, pholedrine, pindolol, pirbuterol, prilocaine, procyclidine, promethazine, propafenone, propranolol, reproterol, sotalol, sulpride, synephrine, talinolol, terbutaline, tetrahydrozoline (tetryzoline), theodrenaline, tioconazole, tocainide, trihexyphenidyl, trimeprazine (alimemazine), trimipramine, tropicamide, verapamil, zopiclone

KEY WORDS
coated capillary; chiral

REFERENCE
Koppenhoefer,B.; Eperlein,U.; Schlunk,R.; Zhu,X.; Lin,B. Separation of enantiomers of drugs by capillary electrophoresis. V. Hydroxypropyl-α-cyclodextrin as chiral solvating agent, *J.Chromatogr.A*, **1998**, *793*, 153–164.

SAMPLE
Matrix: solutions
Sample preparation: Inject an aliquot of a solution in MeOH:water 10:90.

CAPILLARY ELECTROPHORESIS
Capillary: 72 cm × 50 μm
Capillary temperature: 30
Running buffer: 20 mM pH 4.0 Sodium acetate buffer containing 5 mM sulfobutylether-β-cyclodextrin
Detector: UV 205
Migration time: 13.5, 14 (enantiomers)
Voltage: 30 kV
Model: Perkin-Elmer ABI 270A-HT

OTHER SUBSTANCES
Simultaneous: methylphenyloxazolidinone, trolox

KEY WORDS
chiral

REFERENCE
Nassar,A.-E.F.; Guarco,F.J.; Gran,D.E.; Stuart,J.D.; Reuter,W.M. Optimizing a method for separating chiral compounds by capillary electrophoresis, *J.Chromatogr.Sci.*, **1998**, *36*, 19–22.

Azaperone

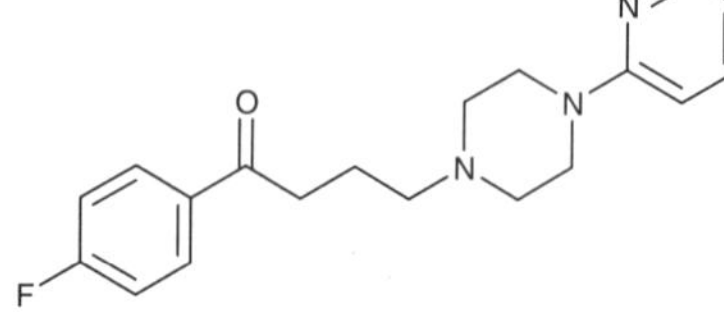

Molecular formula: $C_{19}H_{22}FN_3O$
Molecular weight: 327.40
CAS Registry No.: 1649-18-9
Merck Index (12th ed.): 931
Lednicer: 2 300

SAMPLE
Matrix: urine
Sample preparation: 10 mL Urine + 1 mL 5 M NaOH + 10 mL n-hexane, vortex for 1 min, centrifuge at 0° at 3000 g for 5 min, repeat the extraction twice more. Combine the organic layers and add them to 50 μL glacial acetic acid, evaporate to dryness under a stream of nitrogen at 30°, reconstitute the residue in 50 μL 50 mM sodium taurodeoxycholate, filter (0.2 μm PTFE), inject an aliquot.

CAPILLARY ELECTROPHORESIS
Capillary: 100 cm × 50 μm fused-silica (50 cm to detector) (ISCO)
Capillary preparation: Rinse capillary with 20 μL running buffer between injections. Condition a new capillary by filling with 1 M NaOH, let stand for 1 h, fill with 100 mM NaOH, let stand for 1 h, rinse with buffer. Every 20 injections rinse capillary with 200 μL 1 M NaOH, 200 μL water, and 200 μL running buffer, fill with running buffer.
Capillary temperature: 22
Running buffer: 40 mM pH 9.5 Borate buffer containing 10 mM sodium taurodeoxycholate
Injection: Load under vacuum at 7.5 kPa/s.
Detector: UV 240
Migration time: 5
Voltage: 30 kV
Model: ISCO Model 3140 electropherograph

OTHER SUBSTANCES
Extracted: acepromazine, amiodarone, amitriptyline, chlorpromazine, cianopramine, clomipramine, clozapine, desethylamiodarone, desipramine, diclofensine, dothiepin, doxepin, imipramine, isocarboxazid, moclobemide, perphenazine, phenothiazine, pimozide, prochlorperazine, promazine, thioridazine, thiothixene, trifluoperazine, trimipramine

KEY WORDS
human; cow; pig; horse; protect from light

REFERENCE
Aumatell,A.; Wells,R.J. Determination of a cardiac antiarrhythmic, tricyclic antipsychotics and antidepressants in human and animal urine by micellar electrokinetic capillary chromatography using a bile salt, *J.Chromatogr.B*, **1995**, *669*, 331–344.

Azathioprine

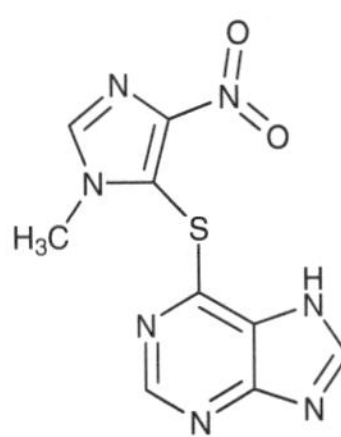

Molecular formula: $C_9H_7N_7O_2S$
Molecular weight: 277.27
CAS Registry No.: 446-86-6
Merck Index (12th ed.): 935
Lednicer: 2 464

SAMPLE
Matrix: solutions

CAPILLARY ELECTROPHORESIS
Capillary: 57 cm × 50 μm fused-silica (50 cm to detector)
Running buffer: 50 mM pH 8.0 Borate buffer containing 40 mM sodium taurodeoxycholate and 25 mM phosphatidylcholine (Prepare by adding phosphatidylcholine and stirring for 3-6 h until all cloudiness disappears. Phosphatidylcholine was 95% pure soybean lecithin, Epikuron, Lucas Meyer & Co.)
Injection: Inject a solution of the compound in the running buffer at 20 psi for 1 s, inject MeOH: water 5:95 at 20 psi for 1 s, inject a solution of halofantrine in running buffer at 20 psi for 1 s.
Detector: UV 214
Migration time: k' 0.83
Model: Beckman P/ACE 5000

OTHER SUBSTANCES
Also analyzed: acetaminophen, amoxicillin, antipyrine, aspirin, caffeine, captopril, carbamazepine, carprofen, chlorambucil, chlorpheniramine, chlorpromazine, cimetidine, clonidine, codeine, desipramine, diphenhydramine, ephedrine, fenoterol, flufenamic acid, flurbiprofen, haloperidol, hydroxyzine, ibuprofen, imipramine, indomethacin, ketoprofen, lidocaine, melphalan, metoprolol, nabumetone, nadolol, phenobarbital, phenol, promazine, propranolol, pyrilamine, ranitidine, ropinirole, salicylic acid, sulfamethoxazole, testosterone, theophylline, thioridazine, tiaprofenic acid, tolfenamic acid, trifluoperazine, trimethoprim, valproic acid, verapamil

KEY WORDS
comparison with HPLC; k' = (Tr-T0)/(T0(1-Tr/Tm)) where Tr = retention time of analyte; T0 = retention time of water; and Tm = retention time of marker (halofantrine)

REFERENCE
Hanna,M.; de Biasi,V.; Bond,B.; Salter,C.; Hutt,A.J.; Camilleri,P. Estimation of the partitioning characteristics of drugs: A comparison of a large and diverse drug series utilizing chromatographic and electrophoretic methodology, *Anal.Chem.*, **1998**, *70*, 2092–2099.

Azelastine

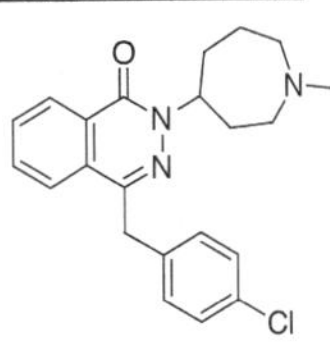

Molecular formula: $C_{22}H_{24}ClN_3O$
Molecular weight: 381.90
CAS Registry No.: 58581-89-8, 79307-93-0 (HCl)
Merck Index (12th ed.): 939
Lednicer: 4 152

SAMPLE
Matrix: solutions
Sample preparation: Inject an aliquot of a 100 μg/mL solution in running buffer.

CAPILLARY ELECTROPHORESIS
Capillary: 36 cm × 50 μm fused-silica (31.5 cm to detector) (Yongnian Optical Conductive Fiber Plant, China), coated with polyacrylamide
Capillary preparation: No details of the polyacrylamide coating process are provided. However, another paper (LC.GC 1997, 15, 40) by this group indicates that they use the procedure of Hjertén, thus: Adjust the pH of 20 mL water to 3.5 with acetic acid, add 80 μL 3-(trimethoxysilyl)propyl methacrylate (3-methacryloxypropyltrimethoxysilane), mix, suck into capillary, let stand at room temperature for 1 h, remove the solution, wash with water. Fill the capillary with a deaerated 3-4% acrylamide solution containing 1 μL/mL N,N,N',N'-tetramethylethylenediamine and 1 mg/mL potassium persulfate, let stand for 30 min, remove excess solution by aspiration, rinse with water, remove water by aspiration, dry at 35° (J. Chromatogr. 1985, 347, 191).
Capillary temperature: 25
Running buffer: 100 mM NaH_2PO_4 adjusted to pH 2.5 (A) or 100 mM NaH_2PO_4 containing 45 mM hydroxypropyl-gamma-cyclodextrin, adjusted to pH 2.5 (B)
Injection: Electrokinetic injection at 15 kV for 3 s.
Detector: UV 200, UV 210
Migration time: 6.68 (A), 11.17, 11.30 (B, enantiomers)
Voltage: 15 kV
Model: Bio-Rad BioFocus 3000

OTHER SUBSTANCES
Simultaneous: biperiden, carvedilol, clidinium bromide, meclizine (meclozine), mequitazine, ofloxacin, zopiclone

KEY WORDS
chiral; coated capillary

REFERENCE
Koppenhoefer,B.; Epperlein,U.; Xiaofeng,Z.; Bingcheng,L. Separation of enantiomers of drugs by capillary electrophoresis. Part 4: Hydroxypropyl-γ-cyclodextrin as chiral solvating agent, *Electrophoresis*, **1997**, *18*, 924–930.

SAMPLE
Matrix: solutions
Sample preparation: Inject an aliquot of a 100 μg/mL solution in running buffer.

CAPILLARY ELECTROPHORESIS
Capillary: 30 cm × 50 μm fused-silica (25.5 cm to detector), coated with polyacrylamide
Capillary preparation: Adjust the pH of 20 mL water to 3.5 with acetic acid, add 80 μL 3-(trimethoxysilyl)propyl methacrylate (3-methacryloxypropyltrimethoxysilane), mix, suck into capillary, let stand at room temperature for 1 h, remove the solution, wash with water. Fill the capillary with a deaerated 3-4% acrylamide solution containing 1 μL/mL N,N,N',N'-tetramethylethylenediamine and 1 mg/mL potassium persulfate, let stand for 30 min, remove excess solution by aspiration, rinse with water, remove water by aspiration, dry at 35° (J. Chromatogr. 1985, 347, 191).
Capillary temperature: 25
Running buffer: 100 mM NaH_2PO_4 containing 45 mM hydroxypropyl-gamma- cyclodextrin, adjusted to pH 2.5 with phosphoric acid
Injection: Electrokinetic injection at 15 kV for 3 s.
Detector: UV 200
Voltage: 15 kV
Model: Bio-Rad BioFocus 3000

KEY WORDS
chiral; coated capillary; comparison with the use of other cyclodextrins; this running buffer gave the greatest enantiomeric separation.; α=1.012

REFERENCE
Lin,B.; Zhu,X.; Koppenhoefer,B.; Epperlein,U. Investigation of 123 chiral drugs by cyclodextrin-modified capillary electrophoresis, *LC.GC*, **1997**, *15*, 40–46.

SAMPLE
Matrix: solutions

CAPILLARY ELECTROPHORESIS
Capillary: 29-36 cm × 50 μm fused-silica (24.5-31.5 cm to detector) (Yongnian Optical Conductive Fiber Plant, China) coated with polyacrylamide
Capillary preparation: Coat capillary as follows. Adjust the pH of 20 mL water to 3.5 with acetic acid, add 80 μL 3-(trimethoxysilyl)propyl methacrylate (3-methacryloxypropyltrimethoxysilane), mix, suck into capillary, let stand at room temperature for 1 h, remove the solution, wash with water. Fill the capillary with a deaerated 3-4% acrylamide solution containing 1 μL/mL N,N,N',N'-tetramethylethylenediamine and 1 mg/mL potassium persulfate, let stand for 30 min, remove excess solution by aspiration, rinse with water, remove water by aspiration, dry at 35° (J. Chromatogr. 1985, 347, 191).
Capillary temperature: 25
Running buffer: 100 mM pH 2.5 NaH_2PO_4 (A) or 100 mM pH 2.5 NaH_2PO_4 containing 45 mM hydroxypropyl-α-cyclodextrin (Wacker, Munich) (B)
Injection: Electromigration at 15 kV for 3 s.
Detector: UV 200; UV 210
Migration time: 6.68 (A); 15.53 (B) (no separation of enantiomers)
Voltage: 15 kV
Model: Bio-Focus 3000

OTHER SUBSTANCES
Also analyzed: albuterol (salbutamol), alprenolol, amorolfine, atenolol, atropine, baclofen, bamethan, benproperine, benserazide, biperiden, bisoprolol, brompheniramine, bupivacaine, bupranolol, butamirate, butethamate, carazolol, carbuterol, carteolol, carvedilol, celiprolol, chloroquine, chlorpheniramine, chlorphenoxamine, cicletanine, clenbuterol, clidinium bromide, clobutinol, dimethindene, dipivefrin, disopyramide, dobutamine, doxylamine, fendiline, flecainide, gallopamil, homatropine, ipratropium bromide, isoproterenol (isoprenaline), isothipendyl, ketamine, meclizine, mefloquine, mepindolol, mequitazine, metaclazepam, metaproterenol (orciprenaline), metipranolol, metoprolol, nafronyl (naftidrofuryl), nefopam, nicardipine, norfenefrine, ofloxacin, ornidazole, orphenadrine, oxomemazine, oxprenolol, oxybutynin, phenoxybenzamine, phenylpropanolamine, pholedrine, pindolol, pirbuterol, prilocaine, procyclidine, promethazine, propafenone, propranolol, reproterol, sotalol, sulpride, synephrine, talinolol, terbutaline, tetrahydrozoline (tetryzoline), theodrenaline, tioconazole, tocainide, trihexyphenidyl, trimeprazine (alimemazine), trimipramine, tropicamide, verapamil, zopiclone

KEY WORDS
coated capillary

REFERENCE
Koppenhoefer,B.; Eperlein,U.; Schlunk,R.; Zhu,X.; Lin,B. Separation of enantiomers of drugs by capillary electrophoresis. V. Hydroxypropyl-α-cyclodextrin as chiral solvating agent, *J.Chromatogr.A*, **1998**, *793*, 153–164.

Bacitracin

Molecular formula: $C_{66}H_{103}N_{17}O_{16}S$ (bacitracin A)
Molecular weight: 1422.71 (bacitracin A)
CAS Registry No.: 1405-87-4, 1405-89-6 (zinc salt),
1405-88-5 (methylenedisalicylate)
Merck Index (12th ed.): 965

SAMPLE
Matrix: solutions

CAPILLARY ELECTROPHORESIS
Capillary: 27 cm × 75 μm
Capillary preparation: Before each run rinse capillary with 100 mM NaOH for 30 s and with
running buffer for 30 s. Condition new capillaries by rinsing with 100 mM NaOH for 20 min.
Capillary temperature: 30
Running buffer: 15 mM Sodium borate
Injection: Pressure injection of sample at 25 mbar for 1 s followed by running buffer at 25 mbar
for 1 s.
Detector: UV 200
Migration time: 2.28
Internal standard: aminobenzoic acid (3.5)
Voltage: 6.5 kV
Model: Beckman

OTHER SUBSTANCES
Simultaneous: aspirin, beclomethasone, benzoic acid, ceftizoxime, ceftriaxone, cefuroxime,
cephalothin, cromolyn, embonic acid, epoprostenol, glyburide (glibenclamide), levothyroxine,
nedocromil, nystatin, omeprazole, prednisolone, warfarin, zidovudine

REFERENCE
Altria,K.D.; Bryant,S.M.; Hadgett,T.A. Validated capillary electrophoresis method for the analysis of a range
of acidic drugs and excipients, *J.Pharm.Biomed.Anal.*, **1997**, *15*, 1091–1101.

Baclofen

Molecular formula: $C_{10}H_{12}ClNO_2$
Molecular weight: 213.66
CAS Registry No.: 1134-47-0
Merck Index (12th ed.): 967
Lednicer: 2 121

SAMPLE
Matrix: solutions
Sample preparation: Inject an aliquot of a 100 μg/mL solution in water:running buffer 50:50.

CAPILLARY ELECTROPHORESIS
Capillary: 44.5 cm × 50 μm acrylamide-coated fused-silica (Bio-Rad)
Capillary temperature: 30
Running buffer: 100 mM NaH_2PO_4 containing 15 mM gamma-cyclodextrin, adjusted to pH 2.5
  with phosphoric acid
Injection: Electrokinetic injection at 8 kV for 6 s.
Detector: UV 200
Migration time: 8.31
Voltage: 14 kV
Model: Bio-Rad BioFocus 3000

OTHER SUBSTANCES
Also analyzed: albuterol, alprenolol, atenolol, atropine, bamethan, benserazide, biperiden, bisoprolol, bupivacaine, bupranolol, butetamate, carazolol, carbuterol, carvedilol, celiprolol, chloroquine, chlorpheniramine (chlorphenamine), clidinium bromide, clobutinol, disopyramide, dobutamine, flecainide, homatropine, ipratropium bromide, isoproterenol, isothipendyl, ketamine,
  mefloquine, mequitazine, metaproterenol (orciprenaline), metipranolol, nafronyl (naftidrofuryl), nefopam, ofloxacin, orphenadrine, oxomemazine, oxprenolol, phenoxybenzamine, pholedrine, pindolol, pirbuterol, prilocaine, promethazine, propafenone, propranolol, sotalol,
  synephrine, terbutaline, tetrahydrozoline (tetryzoline), tocainide, trihexyphenidyl, trimeprazine (alimemazine), trimipramine, tropicamide, verapamil, zopiclone

KEY WORDS
coated capillary; achiral

REFERENCE
Koppenhoefer,B.; Epperlein,U.; Christian,B.; Yibing,J.; Yuying,C.; Bingcheng,L. Separation of enantiomers of
  drugs by capillary electrophoresis. I. γ-Cyclodextrin as chiral solvating agent, *J.Chromatogr.A*, **1995**, *717*,
  181–190.

SAMPLE
Matrix: solutions
Sample preparation: Inject an aliquot of a 100 μg/mL solution in running buffer.

CAPILLARY ELECTROPHORESIS
Capillary: 29 cm × 50 μm fused-silica (24.5 cm to detector) (Yongnian Optical Conductive Fiber
  Plant, China), coated with polyacrylamide
Capillary preparation: No details of the polyacrylamide coating process are provided. However,
  another paper (LC.GC 1997, 15, 40) by this group indicates that they use the procedure of
  Hjertén, thus: Adjust the pH of 20 mL water to 3.5 with acetic acid, add 80 μL 3-(trimethoxysilyl)propyl methacrylate (3-methacryloxypropyltrimethoxysilane), mix, suck into capillary, let
  stand at room temperature for 1 h, remove the solution, wash with water. Fill the capillary
  with a deaerated 3-4% acrylamide solution containing 1 μL/mL N,N,N',N'-tetramethylethyl-
  enediamine and 1 mg/mL potassium persulfate, let stand for 30 min, remove excess solution
  by aspiration, rinse with water, remove water by aspiration, dry at 35° (J. Chromatogr. 1985,
  347, 191).
Capillary temperature: 25
Running buffer: 100 mM NaH_2PO_4 adjusted to pH 2.5
Injection: Electrokinetic injection at 15 kV for 3 s.
Detector: UV 200, UV 210
Migration time: 5.20
Voltage: 15 kV
Model: Bio-Rad BioFocus 3000

OTHER SUBSTANCES
Simultaneous: albuterol, alprenolol, atenolol, bamethan, benproperine, benserazide, bisoprolol, bupranolol, butamirate, butethamate, carbuterol, celiprolol, clenbuterol, clobutinol, dipivefrin, isoproterenol (isoprenaline), metaproterenol (orciprenaline), metipranolol, metoprolol, norfenefrine, ornidazole, oxprenolol, phenylpropanolamine, pholedrine, pirbuterol, prilocaine, procyclidine, sotalol, synephrine, terbutaline, tocainide

KEY WORDS
coated capillary

REFERENCE
Koppenhoefer,B.; Epperlein,U.; Xiaofeng,Z.; Bingcheng,L. Separation of enantiomers of drugs by capillary electrophoresis. Part 4: Hydroxypropyl-γ-cyclodextrin as chiral solvating agent, *Electrophoresis*, **1997**, *18*, 924–930.

SAMPLE
Matrix: solutions
Sample preparation: Inject an aliquot of a 100 μg/mL solution in running buffer.

CAPILLARY ELECTROPHORESIS
Capillary: 30 cm $\times$ 50 μm fused-silica (25.5 cm to detector), coated with polyacrylamide
Capillary preparation: Adjust the pH of 20 mL water to 3.5 with acetic acid, add 80 μL 3-(trimethoxysilyl)propyl methacrylate (3-methacryloxypropyltrimethoxysilane), mix, suck into capillary, let stand at room temperature for 1 h, remove the solution, wash with water. Fill the capillary with a deaerated 3-4% acrylamide solution containing 1 μL/mL N,N,N',N'-tetramethylethylenediamine and 1 mg/mL potassium persulfate, let stand for 30 min, remove excess solution by aspiration, rinse with water, remove water by aspiration, dry at 35° (J. Chromatogr. 1985, 347, 191).
Capillary temperature: 25
Running buffer: 100 mM NaH_2PO_4 containing 45 mM gamma-cyclodextrin, adjusted to pH 2.5 with phosphoric acid
Injection: Electrokinetic injection at 15 kV for 3 s.
Detector: UV 200
Voltage: 15 kV
Model: Bio-Rad BioFocus 3000

KEY WORDS
chiral; coated capillary; comparison with the use of other cyclodextrins; this running buffer gave the greatest enantiomeric separation.; α=1.035

REFERENCE
Lin,B.; Zhu,X.; Koppenhoefer,B.; Epperlein,U. Investigation of 123 chiral drugs by cyclodextrin-modified capillary electrophoresis, *LC.GC*, **1997**, *15*, 40–46.

SAMPLE
Matrix: solutions

CAPILLARY ELECTROPHORESIS
Capillary: 29-36 cm $\times$ 50 μm fused-silica (24.5-31.5 cm to detector) (Yongnian Optical Conductive Fiber Plant, China) coated with polyacrylamide
Capillary preparation: Coat capillary as follows. Adjust the pH of 20 mL water to 3.5 with acetic acid, add 80 μL 3-(trimethoxysilyl)propyl methacrylate (3-methacryloxypropyltrimethoxysilane), mix, suck into capillary, let stand at room temperature for 1 h, remove the solution, wash with water. Fill the capillary with a deaerated 3-4% acrylamide solution containing 1 μL/mL N,N,N',N'-tetramethylethylenediamine and 1 mg/mL potassium persulfate, let stand for 30 min, remove excess solution by aspiration, rinse with water, remove water by aspiration, dry at 35° (J. Chromatogr. 1985, 347, 191).
Capillary temperature: 25
Running buffer: 100 mM pH 2.5 NaH_2PO_4 (A) or 100 mM pH 2.5 NaH_2PO_4 containing 45 mM hydroxypropyl-α-cyclodextrin (Wacker, Munich) (B)
Injection: Electromigration at 15 kV for 3 s.
Detector: UV 200; UV 210

Migration time: 5.20 (A); 16.82, 17.31 (B) (enantiomers)
Voltage: 15 kV
Model: Bio-Focus 3000

OTHER SUBSTANCES
Also analyzed: albuterol (salbutamol), alprenolol, amorolfine, atenolol, atropine, azelastine, bamethan, benproperine, benserazide, biperiden, bisoprolol, brompheniramine, bupivacaine, bupranolol, butamirate, butethamate, carazolol, carbuterol, carteolol, carvedilol, celiprolol, chloroquine, chlorpheniramine, chlorphenoxamine, cicletanine, clenbuterol, clidinium bromide, clobutinol, dimethindene, dipivefrin, disopyramide, dobutamine, doxylamine, fendiline, flecainide, gallopamil, homatropine, ipratropium bromide, isoproterenol (isoprenaline), isothipendyl, ketamine, meclizine, mefloquine, mepindolol, mequitazine, metaclazepam, metaproterenol (orciprenaline), metipranolol, metoprolol, nafronyl (naftidrofuryl), nefopam, nicardipine, norfenefrine, ofloxacin, ornidazole, orphenadrine, oxomemazine, oxprenolol, oxybutynin, phenoxybenzamine, phenylpropanolamine, pholedrine, pindolol, pirbuterol, prilocaine, procyclidine, promethazine, propafenone, propranolol, reproterol, sotalol, sulpride, synephrine, talinolol, terbutaline, tetrahydrozoline (tetryzoline), theodrenaline, tioconazole, tocainide, trihexyphenidyl, trimeprazine (alimemazine), trimipramine, tropicamide, verapamil, zopiclone

KEY WORDS
coated capillary; chiral

REFERENCE
Koppenhoefer,B.; Eperlein,U.; Schlunk,R.; Zhu,X.; Lin,B. Separation of enantiomers of drugs by capillary electrophoresis. V. Hydroxypropyl-α-cyclodextrin as chiral solvating agent, *J.Chromatogr.A*, **1998**, *793*, 153–164.

Bambuterol

Molecular formula: $C_{18}H_{29}N_3O_5$
Molecular weight: 367.45
CAS Registry No.: 81732-65-2, 81732-46-9 (HCl)
Merck Index (12th ed.): 980

SAMPLE
Matrix: blood
Sample preparation: 500 µL Plasma + 5 µL 1 mM physostigmine (esterase inhibitor) in 0.9% NaCl + 100 µL water + 395 µL 300 mM pH 11.1 phosphate buffer, mix. Pump a 900 µL aliquot at 18 µL/min through a dialysis apparatus and dialyze against 36 µL MeOH:1 M phosphoric acid 10:90. Pump the acceptor solution into a vial and rinse out with 14 µL MeOH:1 M phosphoric acid 10:90, add 50 µL MeOH to the dialysate and the rinse, mix, inject an aliquot. (Dialysis apparatus had two 36 µL channels separated by a porous PTFE membrane (Schleicher and Schuell TE35) that was soaked in 6-undecanone for 15 min before use. (Design in paper.) Wash both channels with water before use. After use wash donor channel with 360 µL water and wash acceptor channel with 360 µL MeOH:1 M phosphoric acid 10:90.)

CAPILLARY ELECTROPHORESIS
Capillary: 78 cm × 75 µm fused-silica (70 cm to detector) (Polymicro Technologies)
Capillary preparation: Before each run wash capillary with 4 column volumes of 100 mM NaOH, water, MeOH, and running buffer.
Running buffer: 100 mM pH 2.5 Phosphate buffer containing 3.9 mM dimethyl-β-cyclodextrin
Injection: Fill capillary with 5 mM pH 7.5 phosphate buffer, inject sample at 200 mbar for 2.85 min (3 µL), apply 30 kV with a back-pressure of 68 mbar for 10 min, introduce 100 mM pH 2.5 phosphate buffer into the outlet with a back-pressure of 180 mbar for 1.7 min, apply 30 kV and a back-pressure of 150 mbar with running buffer in inlet and outlet vials for 2 min, lower voltage to 23 kV and continue back-pressure, when current rises (80 µA) switch off back-pressure, run at 23 kv.
Detector: UV 205
Migration time: 16.8, 17.2 (enantiomers)

Voltage: 23 Kv
Model: Prince
Limit of quantitation: 4 nM

KEY WORDS
plasma; dialysis; chiral

REFERENCE
Pálmarsdóttir,S.; Mathiasson,L.; Jönsson,J..; Edholm,L.-E. Determination of a basic drug, bambuterol, in human plasma by capillary electrophoresis using double stacking for large volume injection and supported liquid membranes for sample pretreatment, *J.Chromatogr.B*, **1997**, *688*, 127–134.

SAMPLE
Matrix: solutions
Sample preparation: Prepare a 200 μM solution in water, inject an aliquot.

CAPILLARY ELECTROPHORESIS
Capillary: 57 cm × 75 μm fused-silica (50 cm to detector) (Beckman)
Capillary preparation: Before each analysis wash with three column volumes of 10 mM NaOH, three column volumes of water, and three column volumes of running buffer. At the end of each day wash capillary with 10 column volumes of 10 mM NaOH and 10 column volumes of water. At the beginning of each day wash capillary with 100 column volumes of 10 mM NaOH.
Capillary temperature: 25
Running buffer: 50 mM pH 2.5 Phosphate buffer containing 15 mM heptakis(2,6-di-O-methyl)-β-cyclodextrin
Injection: Injection using pressurized nitrogen for 3 s (about 15 nL)
Detector: UV 214
Migration time: 23.5, 24.5 (enantiomers)
Voltage: 17 kV
Model: Beckman P/ACE 2000

KEY WORDS
chiral

REFERENCE
Pálmarsdóttir,S.; Edholm,L.-E. Capillary zone electrophoresis for separation of drug enantiomers using cyclodextrins as chiral selectors Influence of experimental parameters on separation, *J.Chromatogr.A*, **1994**, *666*, 337–350.

SAMPLE
Matrix: solutions
Sample preparation: Inject an aliquot of a 100 μM solution.

CAPILLARY ELECTROPHORESIS
Capillary: 37 or 47 cm × 75 μm CElect C8 bonded capillary (effective length 30 or 40 cm) (Supelco)
Capillary preparation: Flush capillary for 2 min with running buffer before injection. Rinse capillary with water at the end of the day.
Running buffer: pH 3.0 Phosphate buffer (I = 0.05) containing 60 mM hydroxypropyl-β-cyclodextrin and 20 mM tetrabutylammonium hydroxide
Injection: Inject at 0.5 psi for 3-5 s.
Detector: UV 214
Migration time: 122, 130 (enantiomers)
Voltage: 15 kV
Current: 40-55 μA
Model: Beckman P/ACE 2100

OTHER SUBSTANCES
Simultaneous: albuterol, clenbuterol, terbutaline

KEY WORDS
chiral

REFERENCE
Stålberg,O.; Brötell,H.; Westerlund,D. Capillary electrophoretic separation of basic drugs using surface-modified C8 capillaries and derivatized cyclodextrins as structural/chiral selectors, *Chromatographia*, **1995**, *40*, 697–704.

SAMPLE
Matrix: solutions
Sample preparation: Inject an aliquot of a solution in MeOH:500 μM phosphoric acid 50:50.

CAPILLARY ELECTROPHORESIS
Capillary: 80 cm $\times$ 75 μm fused-silica (70 cm to detector) (Polymicro Technologies)
Capillary preparation: Before each run wash with 4 column volumes of 100 mM NaOH, water, MeOH, and running buffer. [A complex pre-capillary extraction system is described in the paper.]
Running buffer: 100 mM pH 2.5 Phosphate buffer containing 3.9 mM dimethyl-β-cyclodextrin
Detector: UV 205
Migration time: 19.4, 19.7 (enantiomers)
Voltage: 23 kV
Model: Prince

OTHER SUBSTANCES
Simultaneous: physostigmine

KEY WORDS
chiral

REFERENCE
Pálmarsdóttir,S.; Mathiasson,L.; Jönsson,J.Å.; Edholm,L.-E. Micro-CLC as an interface between SLM extraction and CZE for enhancement of sensitivity and selectivity in bioanalysis of drugs, *J.Capillary Electrophor.*, **1996**, *3*, 255–260.

Bamethan

Molecular formula: $C_{12}H_{19}NO_2$
Molecular weight: 209.29
CAS Registry No.: 3703-79-5, 5716-20-1 (sulfate)
Merck Index (12th ed.): 981
Lednicer: 2 39

SAMPLE
Matrix: solutions
Sample preparation: Inject an aliquot of a 100 μg/mL solution in running buffer.

CAPILLARY ELECTROPHORESIS
Capillary: 29 cm $\times$ 50 μm fused-silica (24.5 cm to detector) (Yongnian Optical Conductive Fiber Plant, China), coated with polyacrylamide
Capillary preparation: No details of the polyacrylamide coating process are provided. However, another paper (LC.GC 1997, 15, 40) by this group indicates that they use the procedure of Hjertén, thus: Adjust the pH of 20 mL water to 3.5 with acetic acid, add 80 μL 3-(trimethoxysilyl)propyl methacrylate (3-methacryloxypropyltrimethoxysilane), mix, suck into capillary, let stand at room temperature for 1 h, remove the solution, wash with water. Fill the capillary with a deaerated 3-4% acrylamide solution containing 1 μL/mL N,N,N',N'-tetramethylethylenediamine and 1 mg/mL potassium persulfate, let stand for 30 min, remove excess solution by aspiration, rinse with water, remove water by aspiration, dry at 35° (J. Chromatogr. 1985, 347, 191).
Capillary temperature: 25
Running buffer: 100 mM NaH_2PO_4 adjusted to pH 2.5

Injection: Electrokinetic injection at 15 kV for 3 s.
Detector: UV 200, UV 210
Migration time: 5.42
Voltage: 15 kV
Model: Bio-Rad BioFocus 3000

OTHER SUBSTANCES
Simultaneous: albuterol, alprenolol, atenolol, baclofen, benproperine, benserazide, bisoprolol, bupranolol, butamirate, butethamate, carbuterol, celiprolol, clenbuterol, clobutinol, dipivefrin, isoproterenol (isoprenaline), metaproterenol (orciprenaline), metipranolol, metoprolol, norfenefrine, ornidazole, oxprenolol, phenylpropanolamine, pholedrine, pirbuterol, prilocaine, procyclidine, sotalol, synephrine, terbutaline, tocainide

KEY WORDS
coated capillary

REFERENCE
Koppenhoefer,B.; Epperlein,U.; Xiaofeng,Z.; Bingcheng,L. Separation of enantiomers of drugs by capillary electrophoresis. Part 4: Hydroxypropyl-γ-cyclodextrin as chiral solvating agent, *Electrophoresis*, **1997**, *18*, 924–930.

SAMPLE
Matrix: solutions
Sample preparation: Inject an aliquot of a 100 μg/mL solution in running buffer.

CAPILLARY ELECTROPHORESIS
Capillary: 30 cm $\times$ 50 μm fused-silica (25.5 cm to detector), coated with polyacrylamide
Capillary preparation: Adjust the pH of 20 mL water to 3.5 with acetic acid, add 80 μL 3-(trimethoxysilyl)propyl methacrylate (3-methacryloxypropyltrimethoxysilane), mix, suck into capillary, let stand at room temperature for 1 h, remove the solution, wash with water. Fill the capillary with a deaerated 3-4% acrylamide solution containing 1 μL/mL N,N,N',N'-tetramethylethylenediamine and 1 mg/mL potassium persulfate, let stand for 30 min, remove excess solution by aspiration, rinse with water, remove water by aspiration, dry at 35° (J. Chromatogr. 1985, 347, 191).
Capillary temperature: 25
Running buffer: 100 mM NaH_2PO_4 containing 45 mM hydroxypropyl-β- cyclodextrin, adjusted to pH 2.5 with phosphoric acid
Injection: Electrokinetic injection at 15 kV for 3 s.
Detector: UV 200
Voltage: 15 kV
Model: Bio-Rad BioFocus 3000

KEY WORDS
chiral; coated capillary; comparison with the use of other cyclodextrins; this running buffer gave the greatest enantiomeric separation.; α=1.073

REFERENCE
Lin,B.; Zhu,X.; Koppenhoefer,B.; Epperlein,U. Investigation of 123 chiral drugs by cyclodextrin-modified capillary electrophoresis, *LC.GC*, **1997**, *15*, 40–46.

SAMPLE
Matrix: solutions

CAPILLARY ELECTROPHORESIS
Capillary: 29-36 cm $\times$ 50 μm fused-silica (24.5-31.5 cm to detector) (Yongnian Optical Conductive Fiber Plant, China) coated with polyacrylamide
Capillary preparation: Coat capillary as follows. Adjust the pH of 20 mL water to 3.5 with acetic acid, add 80 μL 3-(trimethoxysilyl)propyl methacrylate (3-methacryloxypropyltrimethoxysilane), mix, suck into capillary, let stand at room temperature for 1 h, remove the solution, wash with water. Fill the capillary with a deaerated 3-4% acrylamide solution containing 1 μL/mL N,N,N',N'-tetramethylethylenediamine and 1 mg/mL potassium persulfate, let stand

for 30 min, remove excess solution by aspiration, rinse with water, remove water by aspiration, dry at 35° (J. Chromatogr. 1985, 347, 191).
Capillary temperature: 25
Running buffer: 100 mM pH 2.5 NaH_2PO_4 (A) or 100 mM pH 2.5 NaH_2PO_4 containing 45 mM hydroxypropyl-α-cyclodextrin (Wacker, Munich) (B)
Injection: Electromigration at 15 kV for 3 s.
Detector: UV 200; UV 210
Migration time: 5.42 (A); 9.76 (B) (no separation of enantiomers)
Voltage: 15 kV
Model: Bio-Focus 3000

OTHER SUBSTANCES
Also analyzed: albuterol (salbutamol), alprenolol, amorolfine, atenolol, atropine, azelastine, baclofen, benproperine, benserazide, biperiden, bisoprolol, brompheniramine, bupivacaine, bupranolol, butamirate, butethamate, carazolol, carbuterol, carteolol, carvedilol, celiprolol, chloroquine, chlorpheniramine, chlorphenoxamine, cicletanine, clenbuterol, clidinium bromide, clobutinol, dimethindene, dipivefrin, disopyramide, dobutamine, doxylamine, fendiline, flecainide, gallopamil, homatropine, ipratropium bromide, isoproterenol (isoprenaline), isothipendyl, ketamine, meclizine, mefloquine, mepindolol, mequitazine, motaclazepam, metaproterenol (orciprenaline), metipranolol, metoprolol, nafronyl (naftidrofuryl), nefopam, nicardipine, norfenefrine, ofloxacin, ornidazole, orphenadrine, oxomemazine, oxprenolol, oxybutynin, phenoxybenzamine, phenylpropanolamine, pholedrine, pindolol, pirbuterol, prilocaine, procyclidine, promethazine, propafenone, propranolol, reproterol, sotalol, sulpride, synephrine, talinolol, terbutaline, tetrahydrozoline (tetryzoline), theodrenaline, tioconazole, tocainide, trihexyphenidyl, trimeprazine (alimemazine), trimipramine, tropicamide, verapamil, zopiclone

KEY WORDS
coated capillary

REFERENCE
Koppenhoefer,B.; Eperlein,U.; Schlunk,R.; Zhu,X.; Lin,B. Separation of enantiomers of drugs by capillary electrophoresis. V. Hydroxypropyl-α-cyclodextrin as chiral solvating agent, *J.Chromatogr.A*, **1998**, *793*, 153–164.

Barbital

Molecular formula: $C_8H_{12}N_2O_3$
Molecular weight: 184.19
CAS Registry No.: 57-44-3, 144-02-5 (sodium salt)
Merck Index (12th ed.): 989
Lednicer: 1 267

SAMPLE
Matrix: bulk
Sample preparation: Dissolve in running buffer to a concentration of 1 mg/mL, vortex for 2 min, add an equal volume of 100 μg/mL diphenhydramine in running buffer, mix, inject an aliquot.

CAPILLARY ELECTROPHORESIS
Capillary: 65 cm × 50 μm fused-silica (60 cm to detector)
Capillary preparation: Fill capillary with fresh running buffer before each run. Before use fill with 100 mM NaOH for 20 min, rinse with water, flush with running buffer
Running buffer: MeCN:50 mM 6-aminocaproic acid containing 50 mM 3-N,N-dimethylmyristylammoniopropanesulfonate (MAPS; Fluka) and 5 mM 1-heptanesulfonic acid 10:90, pH adjusted to 4.0 with 1 M phosphoric acid
Injection: Hydrodynamic injection by gravity or pressure.
Detector: UV 214
Migration time: 11
Internal standard: diphenhydramine (12)

Voltage: 27 kV
Current: ≤25 μA
Model: Dionex system I

OTHER SUBSTANCES
Simultaneous: acetaminophen, allobarbital, caffeine, codeine, diamorphine, morphine, niacinamide, noscapine, papaverine, phenobarbital, procaine

REFERENCE
Naess,O.; Rasmussen,K.E. Micellar electrokinetic chromatography of charged and neutral drugs in acidic running buffers containing a zwitterionic surfactant, sulfonic acids or sodium dodecyl sulphate. Separation of heroin, basic by-products and adulterants, *J.Chromatogr.A*, **1997**, *760*, 245–251.

SAMPLE
Matrix: solutions

CAPILLARY ELECTROPHORESIS
Capillary: 60 cm × 50 μm fused-silica (40 cm to detector), bare (A) or C18 coated (B) (Supelco)
Capillary preparation: Rinse new capillaries with 250 μL 1 M NaOH, 250 μL 100 mM NaOH, water, MeOH, water, and running buffer then condition with running buffer for at least 15 min. Carry out a similar procedure between runs.
Running buffer: 100 mM pH 8.5 Borate buffer containing 25 mM sodium dodecyl sulfate and 5 M urea
Injection: Electrokinetic injection at 5 kV for 5 s.
Detector: UV 254
Migration time: 10 (A), 8.5 (B)
Voltage: 18 kV
Model: Jasco

OTHER SUBSTANCES
Simultaneous: amobarbital, mephobarbital, metharbital, pentobarbital, phenobarbital, secobarbital

KEY WORDS
coated capillary

REFERENCE
Jinno,K.; Han,Y.; Nakamura,M. Analysis of anxiolytic drugs by capillary electrophoresis with bare and coated capillaries, *J.Capillary Electrophor.*, **1996**, *3*, 139–145.

SAMPLE
Matrix: solutions
Sample preparation: Inject an aliquot of a 1-50 μg/mL solution in water.

CAPILLARY ELECTROPHORESIS
Capillary: 55 cm × 50 μm fused-silica (35 cm to detector) (J&W)
Capillary preparation: Flush with running buffer for 5 min before each run. Periodically wash with 100 mM NaOH and flush extensively with running buffer.
Running buffer: MeOH:25 mM pH 9.24 borate buffer 20:80 containing 100 mM sodium dodecyl sulfate (A) or 50 mM pH 2.35 phosphate buffer (B) or 50 mM pH 9.24 borate buffer (C)
Injection: Injection of 5 μL using a split-flow injector ratio of 1:800.
Detector: UV 200
Migration time: 9.70 (A), 39.14 (B), 14.88 (C)
Voltage: 20 kV (A, B) or 12 kV (C)
Current: <60-80 μA
Model: ISCO Model 3850

OTHER SUBSTANCES
Also analyzed: acetylcodeine, amphetamine, caffeine, cocaine, codeine, diamorphine, diazepam, flunitrazepam, lidocaine, monoacetylmorphine, morphine, nalorphine, narceine, noscapine, papaverine, pentobarbital, procaine, tetracaine, thebaine

KEY WORDS
all compounds were separated with running buffer A; some peaks overlapped with running buffers B and C.

REFERENCE
Tagliaro,F.; Smith,F.P.; Turrina,S.; Equisetto,V.; Marigo,M. Complementary use of capillary zone electrophoresis and micellar electrokinetic capillary chromatography for mutual confirmation of results in forensic drug analysis, *J.Chromatogr.A*, **1996**, *735*, 227–235.

SAMPLE
Matrix: solutions

CAPILLARY ELECTROPHORESIS
Capillary: 40 cm × 75 μm coated fused-silica (15 cm to detector) (Supelco)
Capillary preparation: Coat column as follows. Adjust the pH of 20 mL water to 3.5 with acetic acid, add 80 μL 3-(trimethoxysilyl)propyl methacrylate (3-methacryloxypropyltrimethoxysilane), mix, suck into capillary, let stand at room temperature for 1 h, remove the solution, wash with water. Fill the capillary with a deaerated 4% acrylamide solution containing 1 mg/mL N,N,N',N'-tetramethylethylenediamine and 1 mg/mL ammonium persulfate, let stand for 3 h, remove excess solution by aspiration, rinse with water, remove water by aspiration, dry at 35° (cf. J. Chromatogr. 1985, 347, 191).
Running buffer: 100 mM Tris/150 mM boric acid, pH 8.3
Injection: Electromigration at 5 kV for 5 s.
Detector: UV 240, UV 254
Migration time: 5.5
Voltage: 12 kV
Model: Jasco 890-CE

OTHER SUBSTANCES
Simultaneous: amobarbital, mephobarbital, metharbital, pentobarbital, phenobarbital, secobarbital

KEY WORDS
injection at cathode; coated capillary

REFERENCE
Jinno,K.; Han,Y.; Sawada,H. Analysis of toxic drugs by capillary electrophoresis using polyacrylamide-coated columns, *Electrophoresis*, **1997**, *18*, 284–286.

SAMPLE
Matrix: solutions
Sample preparation: Inject an aliquot of a 100 μg/mL solution in running buffer.

CAPILLARY ELECTROPHORESIS
Capillary: 60 cm × 50 μm acrylamide-coated fused-silica (40 cm to detector) (Supelco)
Capillary preparation: Adjust the pH of 20 mL water to 3.5 with acetic acid, add 80 μL 3-(trimethoxysilyl)propyl methacrylate (3-methacryloxypropyltrimethoxysilane), mix, suck into capillary, let stand at room temperature for 1 h, remove the solution, wash with water. Fill the capillary with a deaerated 3-4% acrylamide solution containing 1 μL/mL N,N,N',N'-tetramethylethylenediamine and 1 mg/mL ammonium persulfate, let stand for 3 h, remove excess solution by aspiration, rinse with water, remove water by aspiration, dry at 35° (cf. J. Chromatogr. 1985, 347, 191).
Running buffer: MeCN:buffer 5:95 (Buffer was 100 mM borate containing 5 M urea and 10 mM sodium dodecyl sulfate, adjusted to pH 8.5 with phosphate.)
Injection: Electrokinetic injection at 5 kV for 5 s.
Detector: UV 254
Migration time: 13.6
Voltage: 18 kV
Model: Jasco Model 870-CE

OTHER SUBSTANCES
Simultaneous: alprazolam, amobarbital, bromazepam, clonazepam, clotiazepam, cloxazolam, diazepam, estazolam, etizolam, fludiazepam, flunitrazepam, flurazepam, haloxazolam, medazepam, mephobarbital, metharbital, nimetazepam, nitrazepam, oxazepam, pentobarbital, phenobarbital, secobarbital, triazolam

KEY WORDS
coated capillary; detector at anode

REFERENCE
Jinno,K.; Han,Y.; Sawada,H.; Taniguchi,M. Capillary electrophoretic separation of toxic drugs using a polyacrylamide-coated capillary, *Chromatographia*, **1997**, *46*, 309–314.

SAMPLE
Matrix: solutions
Sample preparation: Inject an aliquot of a solution in 20 mM pH 5.9 ammonium acetate buffer.

CAPILLARY ELECTROPHORESIS
Capillary: 33 cm × 50 μm fused-silica (20 cm to detector) (Polymicro Technologies)
Running buffer: 10 mM pH 5.9 Ammonium acetate buffer containing 15 mM sodium dodecyl sulfate
Injection: Electrokinetic injection at -7.4 kV for 1 s.
Detector: UV 226, MS, Finnigan MAT TSQ, negative electrospray (details in paper)
Migration time: 4.5
Voltage: -7.4 kV

OTHER SUBSTANCES
Simultaneous: amobarbital, butalbital, pentobarbital, secobarbital

REFERENCE
Yang,L.; Harrata,A.K.; Lee,C.S. On-line micellar electrokinetic chromatography-electrospray ionization mass spectrometry using anodically migrating micelles, *Anal.Chem.*, **1997**, *69*, 1820–1826.

Beclomethasone

Molecular formula: $C_{22}H_{29}ClO_5$
Molecular weight: 408.92
CAS Registry No.: 4419-39-0, 5534-09-8
(beclomethasone dipropionate)
Merck Index (12th ed.): 1047

SAMPLE
Matrix: solutions

CAPILLARY ELECTROPHORESIS
Capillary: 27 cm × 75 μm
Capillary preparation: Before each run rinse capillary with 100 mM NaOH for 30 s and with running buffer for 30 s. Condition new capillaries by rinsing with 100 mM NaOH for 20 min.
Capillary temperature: 30
Running buffer: 15 mM Sodium borate
Injection: Pressure injection of sample at 25 mbar for 1 s followed by running buffer at 25 mbar for 1 s.
Detector: UV 200
Migration time: 3.36
Internal standard: aminobenzoic acid (3.5)
Voltage: 6.5 kV
Model: Beckman

OTHER SUBSTANCES
Simultaneous: aspirin, bacitracin, benzoic acid, ceftizoxime, ceftriaxone, cefuroxime, cephalothin, cromolyn, embonic acid, epoprostenol, glyburide (glibenclamide), levothyroxine, nedocromil, nystatin, omeprazole, warfarin, zidovudine
Interfering: prednisolone

REFERENCE
Altria,K.D.; Bryant,S.M.; Hadgett,T.A. Validated capillary electrophoresis method for the analysis of a range of acidic drugs and excipients, *J.Pharm.Biomed.Anal.*, **1997**, *15*, 1091–1101.

Bendroflumethiazide

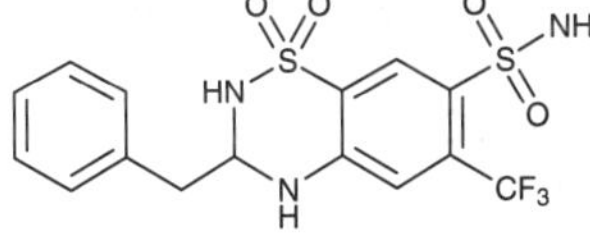

Molecular formula: $C_{15}H_{14}F_3N_3O_4S_2$
Molecular weight: 421.42
CAS Registry No.: 73-48-3
Merck Index (12th ed.): 1064
Lednicer: 2 358

SAMPLE
Matrix: blood, urine
Sample preparation: Condition a 3 mL Supelclean LC-18 SPE cartridge with 3 mL MeOH and 3 mL water. Dilute urine 1:10 with water. Precipitate proteins from serum with MeOH. Add diluted urine or protein supernatant to the SPE cartridge, wash with 3 mL water, elute with 3 mL MeOH, reconstitute to the original volume with 100 mM KOH, inject an aliquot.

CAPILLARY ELECTROPHORESIS
Capillary: 67 cm × 50 μm fused-silica (60 cm to detector) (Polymicro Technologies)
Capillary preparation: Rinse with running buffer for 2 min before run. If necessary, regenerate capillary with 100 mM NaOH for 10 min and with water for 15 min.
Capillary temperature: 20
Running buffer: 60 mM 3-(cyclohexylamino)-1-propanesulfonic acid (CAPS) adjusted to pH 10.6 with 100 mM KOH
Injection: Pressure injection for 5 s
Detector: UV 220
Migration time: 9
Voltage: 25 kV
Model: Beckman P/ACE 2000

OTHER SUBSTANCES
Extracted: acetazolamide, amiloride, benzthiazide, bumetanide, caffeine, chlorothiazide, chlorthalidone, clopamide, dichlorphenamide, ethacrynic acid, furosemide, hydrochlorothiazide, metyrapone, probenecid, triamterene, trichlormethiazide

KEY WORDS
serum; SPE

REFERENCE
Jumppanen,J.; Sirén,H.; Riekkola,M.-L. Screening for diuretics in urine and blood serum by capillary zone electrophoresis, *J.Chromatogr.A*, **1993**, *652*, 441–450.

SAMPLE
Matrix: solutions
Sample preparation: Inject an aliquot of a 100 μg/mL solution in MeOH:running buffer 50:50.

CAPILLARY ELECTROPHORESIS
Capillary: 37 cm × 50 μm fused-silica (30 cm to detector)

Capillary preparation: Between runs rinse capillary with 100 mM KOH for 2 min, with water for 2 min, and with running buffer for 2 min. Rinse new capillaries with 500 mM KOH for 15 min, with water for 10 min, and with running buffer for 15 min.
Capillary temperature: 25
Running buffer: 50 mM pH 7.0 Sodium phosphate buffer containing 25 mM sodium dodecyl sulfate and 2 mM vancomycin
Injection: Hydrostatic injection at 0.5 psi for 1 s
Detector: UV 254
Migration time: 29.2, 29.4 (enantiomers)
Voltage: 5 kV
Model: Beckman P/ACE 2000

OTHER SUBSTANCES
Simultaneous: 5-(4-hydroxyphenyl)-5-phenylhydantoin, warfarin

---

KEY WORDS
chiral

REFERENCE
Armstrong,D.W.; Rundlett,K.L. CE resolution of neutral and anionic racemates with glycopeptide antibiotics and micelles, *J.Liq.Chromatogr.*, **1995**, *18*, 3659–3674.

SAMPLE
Matrix: solutions
Sample preparation: Inject an aliquot of a 200 μg/mL solution in MeCN:water 50:50.

CAPILLARY ELECTROPHORESIS
Capillary: 23 cm × 50 μm 3 μm CEC Hypersil C18
Capillary temperature: 15
Running buffer: Gradient. MeCN:50 mM pH 2.5 Na_2HPO_4 buffer:water 40:20:40 for 6.5 min, 60:20:20 for 10.75 min (step gradient), re-equilibrate at initial conditions for 7.75 min.
Injection: Electrokinetic injection at 5 kV for 15 s.
Detector: UV 210
Migration time: 15.5
Voltage: 30 kV (with 8 bar of pressure at each end of capillary)
Model: Hewlett-Packard HP[3D]

OTHER SUBSTANCES
Simultaneous: bumetanide, chlorothiazide, chlorthalidone, hydrochlorothiazide, hydroflumethiazide

---

KEY WORDS
electrochromatography

REFERENCE
Euerby,M.R.; Gilligan,D.; Johnson,C.M.; Bartle,K.D. Step-gradient capillary electrochromatography, *Analyst*, **1997**, *122*, 1087–1088.

SAMPLE
Matrix: solutions

CAPILLARY ELECTROPHORESIS
Capillary: 40.6 cm × 50 μm fused-silica packed with 3 μm CEC Hypersil ODS (Hypersil)
Running buffer: Gradient. A was 5 mM ammonium acetate in MeCN:water 50:50. B was 5 mM ammonium acetate in MeCN:water 80:20. A:B 100:0 for 3 min, to 0:100 over 0.1 min, maintain at 0:100 (pumped with an HPLC pump at 0.01 mL/min for 3 min then at 0.1 mL/min).
Injection: Inject 5 μL using an HPLC injector.
Detector: MS, VG Biotech Platform, electrospray (details in paper)
Migration time: 24
Voltage: 30 kV

OTHER SUBSTANCES
Simultaneous: epithiazide, hydroflumethiazide, methyclothiazide, metolazone

KEY WORDS
electrochromatography

REFERENCE
Taylor,M.R.; Teale,P. Gradient capillary electrochromatography of drug mixtures with UV and electrospray ionisation mass spectrometric detection, *J.Chromatogr.A*, **1997**, *768*, 89–95.

SAMPLE
Matrix: urine
Sample preparation: Filter (0.2 μm), inject an aliquot of the filtrate.

CAPILLARY ELECTROPHORESIS
Capillary: 80 cm × 50 μm fused-silica (57 cm to detector)
Capillary preparation: Before each run rinse the capillary with 1 M NaOH for 3 min, with 100 mM NaOH for 3 min, with water for 3 min, and with running buffer for 10 min.
Running buffer: 110 mM Boric acid containing 56 mM NaOH and 44 mM HCl, pH 8
Injection: Vacuum injection for 1 s
Detector: UV 220
Migration time: 4.262
Voltage: 20 kV
Model: Europhor Prime Vision system IV

OTHER SUBSTANCES
Extracted: acebutolol (UV 238), acetazolamide (UV 222), alprenolol (UV 220), amiloride (UV 220), atenolol (UV 228), bumetanide (UV 220), chlorthalidone (UV 220), cocaine (UV 236), codeine (UV 220), ethacrynic acid (UV 220), furosemide (UV 232), methadone (UV 220), metoxiphenamine (UV 220), nadolol (UV 220), norcodeine (UV 220), oxprenolol (UV 220), pentazocine (UV 220), propranolol (UV 220), spironolactone (UV 244), triamterene (UV 232), xipamide (UV 234)
Interfering: hydrochlorothiazide (UV 226)

REFERENCE
Gonzalez,E.; Laserna,J.J. Capillary zone electrophoresis for the rapid screening of banned drugs in sport, *Electrophoresis*, **1994**, *15*, 240–243.

SAMPLE
Matrix: urine
Sample preparation: Centrifuge urine at 150 g for 10 min, inject an aliquot.

CAPILLARY ELECTROPHORESIS
Capillary: 100 cm × 50 μm (60 cm to detector)
Capillary preparation: Rinse capillary with running buffer for 3 min before each run. Regenerate capillary by flushing with 1 M NaOH for 10 min, with 100 mM NaOH for 10 min, and with water for 15 min.
Running buffer: pH 8 Borate buffer (concentration adjusted to keep current at 42 μA)
Injection: Electrokinetic injection at 10 kV for 10 s (ca. 6 nL).
Detector: F ex 272 em 381
Migration time: 5.7
Voltage: 30 kV
Current: 42 μA
Model: laboratory-constructed
Limit of detection: 630 nM

OTHER SUBSTANCES
Extracted: amiloride (F ex 382 em 416), bumetanide (F ex 350 em 428), triamterene (F ex 370 em 434)

KEY WORDS
direct injection

REFERENCE

González,E.; Becerra,A.; Laserna,J.J. Direct determination of diuretic drugs in urine by capillary zone electrophoresis using fluorescence detection, *J.Chromatogr.B*, **1996**, *687*, 145–150.

Benproperine

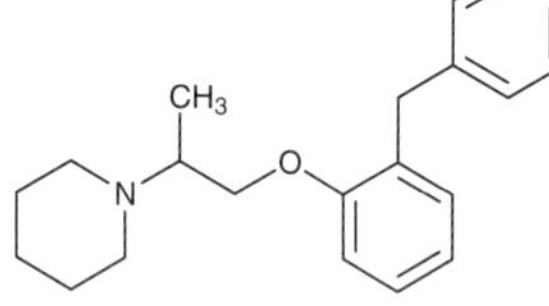

Molecular formula: $C_{21}H_{27}NO$
Molecular weight: 309.45
CAS Registry No.: 2156-27-6, 64238-92-2 (pamoate), 19428-14-9 (trihydrogen phosphate)
Merck Index (12th ed.): 1078

SAMPLE

Matrix: solutions
Sample preparation: Inject an aliquot of a 100 μg/mL solution in running buffer.

CAPILLARY ELECTROPHORESIS

Capillary: 29 cm × 50 μm fused-silica (24.5 cm to detector) (Yongnian Optical Conductive Fiber Plant, China), coated with polyacrylamide
Capillary preparation: No details of the polyacrylamide coating process are provided. However, another paper (LC.GC 1997, 15, 40) by this group indicates that they use the procedure of Hjertén, thus: Adjust the pH of 20 mL water to 3.5 with acetic acid, add 80 μL 3-(trimethoxysilyl)propyl methacrylate (3-methacryloxypropyltrimethoxysilane), mix, suck into capillary, let stand at room temperature for 1 h, remove the solution, wash with water. Fill the capillary with a deaerated 3-4% acrylamide solution containing 1 μL/mL N,N,N',N'-tetramethylethylenediamine and 1 mg/mL potassium persulfate, let stand for 30 min, remove excess solution by aspiration, rinse with water, remove water by aspiration, dry at 35° (J. Chromatogr. 1985, 347, 191).
Capillary temperature: 25
Running buffer: 100 mM NaH_2PO_4 adjusted to pH 2.5 (A) or 100 mM NaH_2PO_4 containing 45 mM hydroxypropyl-gamma-cyclodextrin, adjusted to pH 2.5 (B)
Injection: Electrokinetic injection at 15 kV for 3 s.
Detector: UV 200, UV 210
Migration time: 5.50 (A), 15.73, 16.02 (B, enantiomers)
Voltage: 15 kV
Model: Bio-Rad BioFocus 3000

OTHER SUBSTANCES

Simultaneous: albuterol, alprenolol, atenolol, baclofen, bamethan, benserazide, bisoprolol, bupranolol, butamirate, butethamate, carbuterol, celiprolol, clenbuterol, clobutinol, dipivefrin, isoproterenol (isoprenaline), metaproterenol (orciprenaline), metipranolol, metoprolol, norfenefrine, ornidazole, oxprenolol, phenylpropanolamine, pholedrine, pirbuterol, prilocaine, procyclidine, sotalol, synephrine, terbutaline, tocainide

KEY WORDS

chiral; coated capillary

REFERENCE

Koppenhoefer,B.; Epperlein,U.; Xiaofeng,Z.; Bingcheng,L. Separation of enantiomers of drugs by capillary electrophoresis. Part 4: Hydroxypropyl-γ-cyclodextrin as chiral solvating agent, *Electrophoresis*, **1997**, *18*, 924–930.

SAMPLE

Matrix: solutions
Sample preparation: Inject an aliquot of a 100 μg/mL solution in running buffer.

CAPILLARY ELECTROPHORESIS

Capillary: 30 cm × 50 μm fused-silica (25.5 cm to detector), coated with polyacrylamide

Capillary preparation: Adjust the pH of 20 mL water to 3.5 with acetic acid, add 80 μL 3-(trimethoxysilyl)propyl methacrylate (3-methacryloxypropyltrimethoxysilane), mix, suck into capillary, let stand at room temperature for 1 h, remove the solution, wash with water. Fill the capillary with a deaerated 3-4% acrylamide solution containing 1 μL/mL N,N,N',N'-tetramethylethylenediamine and 1 mg/mL potassium persulfate, let stand for 30 min, remove excess solution by aspiration, rinse with water, remove water by aspiration, dry at 35° (J. Chromatogr. 1985, 347, 191).

Capillary temperature: 25

Running buffer: 100 mM NaH_2PO_4 containing 45 mM hydroxypropyl-gamma- cyclodextrin, adjusted to pH 2.5 with phosphoric acid

Injection: Electrokinetic injection at 15 kV for 3 s.

Detector: UV 200

Voltage: 15 kV

Model: Bio-Rad BioFocus 3000

KEY WORDS

chiral; coated capillary; comparison with the use of other cyclodextrins; this running buffer gave the greatest enantiomeric separation.; α=1.021

REFERENCE

Lin,B.; Zhu,X.; Koppenhoefer,B.; Epperlein,U. Investigation of 123 chiral drugs by cyclodextrin-modified capillary electrophoresis, *LC.GC*, **1997**, *15*, 40–46.

SAMPLE

Matrix: solutions

CAPILLARY ELECTROPHORESIS

Capillary: 29-36 cm × 50 μm fused-silica (24.5-31.5 cm to detector) (Yongnian Optical Conductive Fiber Plant, China) coated with polyacrylamide

Capillary preparation: Coat capillary as follows. Adjust the pH of 20 mL water to 3.5 with acetic acid, add 80 μL 3-(trimethoxysilyl)propyl methacrylate (3-methacryloxypropyltrimethoxysilane), mix, suck into capillary, let stand at room temperature for 1 h, remove the solution, wash with water. Fill the capillary with a deaerated 3-4% acrylamide solution containing 1 μL/mL N,N,N',N'-tetramethylethylenediamine and 1 mg/mL potassium persulfate, let stand for 30 min, remove excess solution by aspiration, rinse with water, remove water by aspiration, dry at 35° (J. Chromatogr. 1985, 347, 191).

Capillary temperature: 25

Running buffer: 100 mM pH 2.5 NaH_2PO_4 (A) or 100 mM pH 2.5 NaH_2PO_4 containing 45 mM hydroxypropyl α-cyclodextrin (Wacker, Munich) (B)

Injection: Electromigration at 15 kV for 3 s.

Detector: UV 200; UV 210

Migration time: 5.50 (A); 10.81, 10.94 (B) (enantiomers)

Voltage: 15 kV

Model: Bio-Focus 3000

OTHER SUBSTANCES

Also analyzed: albuterol (salbutamol), alprenolol, amorolfine, atenolol, atropine, azelastine, baclofen, bamethan, benserazide, biperiden, bisoprolol, brompheniramine, bupivacaine, bupranolol, butamirate, butethamate, carazolol, carbuterol, carteolol, carvedilol, celiprolol, chloroquine, chlorpheniramine, chlorphenoxamine, cicletanine, clenbuterol, clidinium bromide, clobutinol, dimethindene, dipivefrin, disopyramide, dobutamine, doxylamine, fendiline, flecainide, gallopamil, homatropine, ipratropium bromide, isoproterenol (isoprenaline), isothipendyl, ketamine, meclizine, mefloquine, mepindolol, mequitazine, metaclazepam, metaproterenol (orciprenaline), metipranolol, metoprolol, nafronyl (naftidrofuryl), nefopam, nicardipine, norfenefrine, ofloxacin, ornidazole, orphenadrine, oxomemazine, oxprenolol, oxybutynin, phenoxybenzamine, phenylpropanolamine, pholedrine, pindolol, pirbuterol, prilocaine, procyclidine, promethazine, propafenone, propranolol, reproterol, sotalol, sulpride, synephrine, talinolol, terbutaline, tetrahydrozoline (tetryzoline), theodrenaline, tioconazole, tocainide, trihexyphenidyl, trimeprazine (alimemazine), trimipramine, tropicamide, verapamil, zopiclone

---

KEY WORDS

coated capillary; chiral

REFERENCE

Koppenhoefer,B.; Eperlein,U.; Schlunk,R.; Zhu,X.; Lin,B. Separation of enantiomers of drugs by capillary electrophoresis. V. Hydroxypropyl-α-cyclodextrin as chiral solvating agent, *J.Chromatogr.A*, **1998**, *793*, 153–164.

Benserazide

Molecular formula: $C_{10}H_{15}N_3O_5$
Molecular weight: 257.25
CAS Registry No.: 322-35-0, 37270-69-2 (combination with levodopa), 14919-77-8 (HCl)
Merck Index (12th ed.): 1079

SAMPLE

Matrix: solutions
Sample preparation: Inject an aliquot of a 100 μg/mL solution in running buffer.

CAPILLARY ELECTROPHORESIS

Capillary: 29 cm × 50 μm fused-silica (24.5 cm to detector) (Yongnian Optical Conductive Fiber Plant, China), coated with polyacrylamide
Capillary preparation: No details of the polyacrylamide coating process are provided. However, another paper (LC.GC 1997, 15, 40) by this group indicates that they use the procedure of Hjertén, thus: Adjust the pH of 20 mL water to 3.5 with acetic acid, add 80 μL 3-(trimethoxysilyl)propyl methacrylate (3-methacryloxypropyltrimethoxysilane), mix, suck into capillary, let stand at room temperature for 1 h, remove the solution, wash with water. Fill the capillary with a deaerated 3-4% acrylamide solution containing 1 μL/mL N,N,N',N'-tetramethylethylenediamine and 1 mg/mL potassium persulfate, let stand for 30 min, remove excess solution by aspiration, rinse with water, remove water by aspiration, dry at 35° (J. Chromatogr. 1985, 347, 191).
Capillary temperature: 25
Running buffer: 100 mM NaH_2PO_4 adjusted to pH 2.5
Injection: Electrokinetic injection at 15 kV for 3 s.
Detector: UV 200, UV 210
Migration time: 5.20
Voltage: 15 kV
Model: Bio-Rad BioFocus 3000

OTHER SUBSTANCES

Simultaneous: albuterol, alprenolol, atenolol, baclofen, bamethan, benproperine, bisoprolol, bupranolol, butamirate, butethamate, carbuterol, celiprolol, clenbuterol, clobutinol, dipivefrin, isoproterenol (isoprenaline), metaproterenol (orciprenaline), metipranolol, metoprolol, norfenefrine, ornidazole, oxprenolol, phenylpropanolamine, pholedrine, pirbuterol, prilocaine, procyclidine, sotalol, synephrine, terbutaline, tocainide

KEY WORDS

coated capillary

REFERENCE

Koppenhoefer,B.; Epperlein,U.; Xiaofeng,Z.; Bingcheng,L. Separation of enantiomers of drugs by capillary electrophoresis. Part 4: Hydroxypropyl-γ-cyclodextrin as chiral solvating agent, *Electrophoresis*, **1997**, *18*, 924–930.

SAMPLE

Matrix: solutions

CAPILLARY ELECTROPHORESIS

Capillary: 29-36 cm × 50 μm fused-silica (24.5-31.5 cm to detector) (Yongnian Optical Conductive Fiber Plant, China) coated with polyacrylamide

Capillary preparation: Coat capillary as follows. Adjust the pH of 20 mL water to 3.5 with acetic acid, add 80 μL 3-(trimethoxysilyl)propyl methacrylate (3-methacryloxypropyltrimethoxysilane), mix, suck into capillary, let stand at room temperature for 1 h, remove the solution, wash with water. Fill the capillary with a deaerated 3-4% acrylamide solution containing 1 μL/mL N,N,N',N'-tetramethylethylenediamine and 1 mg/mL potassium persulfate, let stand for 30 min, remove excess solution by aspiration, rinse with water, remove water by aspiration, dry at 35° (J. Chromatogr. 1985, 347, 191).
Capillary temperature: 25
Running buffer: 100 mM pH 2.5 NaH_2PO_4 (A) or 100 mM pH 2.5 NaH_2PO_4 containing 45 mM hydroxypropyl-α-cyclodextrin (Wacker, Munich) (B)
Injection: Electromigration at 15 kV for 3 s.
Detector: UV 200; UV 210
Migration time: 5.20 (A); 5.62 (B) (no separation of enantiomers)
Voltage: 15 kV
Model: Bio-Focus 3000

OTHER SUBSTANCES
Also analyzed: albuterol (salbutamol), alprenolol, amorolfine, atenolol, atropine, azelastine, baclofen, bamethan, benproperine, biperiden, bisoprolol, brompheniramine, bupivacaine, bupranolol, butamirate, butethamate, carazolol, carbuterol, carteolol, carvedilol, celiprolol, chloroquine, chlorpheniramine, chlorphenoxamine, cicletanine, clenbuterol, clidinium bromide, clobutinol, dimethindene, dipivefrin, disopyramide, dobutamine, doxylamine, fendiline, flecainide, gallopamil, homatropine, ipratropium bromide, isoproterenol (isoprenaline), isothipendyl, ketamine, meclizine, mefloquine, mepindolol, mequitazine, metaclazepam, metaproterenol (orciprenaline), metipranolol, metoprolol, nafronyl (naftidrofuryl), nefopam, nicardipine, norfenefrine, ofloxacin, ornidazole, orphenadrine, oxomemazine, oxprenolol, oxybutynin, phenoxybenzamine, phenylpropanolamine, pholedrine, pindolol, pirbuterol, prilocaine, procyclidine, promethazine, propafenone, propranolol, reproterol, sotalol, sulpride, synephrine, talinolol, terbutaline, tetrahydrozoline (tetryzoline), theodrenaline, tioconazole, tocainide, trihexyphenidyl, trimeprazine (alimemazine), trimipramine, tropicamide, verapamil, zopiclone

KEY WORDS
coated capillary

REFERENCE
Koppenhoefer,B.; Eperlein,U.; Schlunk,R.; Zhu,X.; Lin,B. Separation of enantiomers of drugs by capillary electrophoresis. V. Hydroxypropyl-α-cyclodextrin as chiral solvating agent, *J.Chromatogr.A*, **1998**, *793*, 153–164.

Benzalkonium chloride

CAS Registry No.: 8001-54-5
Merck Index (12th ed.): 1086

SAMPLE
Matrix: formulations
Sample preparation: Dilute with water containing IS, inject an aliquot.

CAPILLARY ELECTROPHORESIS
Capillary: 37 cm × 75 μm fused-silica (30 cm to detector) (Composite Metal Services, Hallow, UK)
Capillary preparation: Between injections rinse capillary with 100 mM NaOH for 1 min and with running buffer for 1 min.
Capillary temperature: 30
Running buffer: 25 mM NaH_2PO_4 adjusted to pH 2.5 with concentrated phosphoric acid
Injection: Pressure injection for 5 s.
Detector: UV 200
Migration time: 5.3-6 (depending on structure)
Internal standard: imidazole (2.5)

Voltage: 15 kV
Model: Beckman P/ACE 5000
Limit of quantitation: 150 ng/mL
Limit of detection: 50 ng/mL

OTHER SUBSTANCES
Simultaneous: histamine, histidine

REFERENCE
Altria,K.D.; Elgey,J.; Howells,J.S. Validated capillary electrophoretic method for the quantitative analysis of histamine acid phosphate and/or benzalkonium chloride, *J.Chromatogr.B*, **1996**, *686*, 111–117.

SAMPLE
Matrix: formulations
Sample preparation: Mix 2.5 g nasal solution with 500 μL MeOH, add 1 mL 100 mM NaOH, shake by hand for 1 min, make up to 10 mL with MeOH:water 50:50, shake, centrifuge at 3500 rpm for 5 min. Remove the aqueous layer, sonicate for 3 min, inject an aliquot.

CAPILLARY ELECTROPHORESIS
Capillary: 35 cm × 75 μm fused-silica (28.5 cm to detector)
Capillary preparation: Before each run flush capillary with 100 mM NaOH for 3 min and with running buffer for 3 min. Replenish buffer vials before each run.
Running buffer: pH 2.3 Triethylamine/phosphoric acid buffer (Prepare buffer by mixing 2,5 mL 85% orthophosphoric acid and 3 mL triethylamine in water, make up to 500 mL with water.)
Injection: Pressure injection of sample at 35 mbar for 10 s then buffer injection for 2 s.
Detector: UV 215
Migration time: 7 (C12), 8 (C14)
Voltage: 15 kV
Current: 95 μA
Model: HP^{3D}

KEY WORDS
nasal solution

REFERENCE
Jimidar,M.; Beyns,I.; Rome,R.; Peeters,R.; Musch,G. Determination of benzalkonium chloride in drug formulations by capillary electrophoresis (CE), *Biomed.Chromatogr.*, **1998**, *12*, 128–130.

SAMPLE
Matrix: formulations
Sample preparation: Dissolve lozenge in the minimum amount of solvent, sonicate for 30 s, make up to 10 mL with solvent containing IS, filter (0.45 μm), inject an aliquot. (Solvent was MeOH:water 60:40.)

CAPILLARY ELECTROPHORESIS
Capillary: 27.5 × 50 μm fused-silica (20.5 cm to detector)
Capillary preparation: Between runs flush capillary with 100 mM NaOH for 2 min, with water for 1 min, and with running buffer for 3 min. At the end of each day flush with 100 mM NaOH for 5 min and with water for 5 min. At the start of each day flush with 100 mM NaOH for 5 min, with water for 5 min, and with running buffer for 5 min.
Running buffer: MeCN:50 mM pH 3 NaH$_2$PO$_4$ 50:50
Injection: Hydrodynamic injection at 50 mbar for 10 s.
Detector: UV 210
Migration time: 3.7 (C12), 3.9 (C14), 4.1 (C16)
Internal standard: diphenhydramine (3.5)
Voltage: 15 kV
Model: Hewlett-Packard HP3D
Limit of quantitation: 5 μg/mL

KEY WORDS
lozenges

REFERENCE

Taylor,R.B.; Toasaksiri,S.; Reid,R.G. Determination of antibacterial quaternary ammonium compounds in lozenges by capillary electrophoresis, *J.Chromatogr.A*, **1998**, *798*, 335–343.

SAMPLE

Matrix: solutions
Sample preparation: Inject an aliquot of a 10 μM solution in MeOH:water 60:40.

CAPILLARY ELECTROPHORESIS

Capillary: 44 cm × 75 μm fused-silica (37 cm to detector) (Polymicro Technologies)
Capillary preparation: Between runs wash with 100 mM NaOH for 2 min and equilibrate with running buffer for 2 min. Wash a new capillary with 1 M NaOH at 60° for 20 min, with 100 mM NaOH at 60° for 20 min, with water at 60° for 30 min, and with water at 25° for 5 min. Before use wash with 1 M NaOH at 60° for 5 min, with water at 60° for 5 min, and with water at 25° for 5 min.
Capillary temperature: 25
Running buffer: MeCN:20 mM pH 5.0 phosphate buffer 40:60
Injection: Hydrodynamic injection for 2 s
Detector: UV 210
Migration time: 2.85 (C12), 2.92 (C14), 3.0 (C16), 3.5 (C18)
Voltage: 15 kV
Model: Spectra-Physics Model 500

REFERENCE

Lin,C.-E.; Chiou,W.-c.; Lin,W.-c. Separation of alkylbenzyl quaternary ammonium compounds by capillary zone electrophoresis. Effect of organic solvent in sample solution, *J.Chromatogr.A*, **1996**, *723*, 189–195.

Benzocaine

Molecular formula: $C_9H_{11}NO_2$
Molecular weight: 165.19
CAS Registry No.: 94-09-7
Merck Index (12th ed.): 1116
Lednicer: 1 9

SAMPLE

Matrix: blood, urine
Sample preparation: Adjust pH of 2 mL urine or plasma to 10.5 with aqueous NaOH, extract gently with chloroform:isopropanol 90:10. Remove the organic layer and evaporate it to dryness under a stream of nitrogen, reconstitute the residue in 50 (urine) or 75 (plasma) μL running buffer, filter, inject an aliquot.

CAPILLARY ELECTROPHORESIS

Capillary: 60 cm × 75 μm AccuSep uncoated silica (52.5 cm to detector) (Waters)
Capillary preparation: At the beginning of each day purge with 500 mM KOH for 5 min, with water for 5 min, and with running buffer for 10 min.
Running buffer: 50 mM NaH_2PO_4 adjusted to pH 2.35 with phosphoric acid
Injection: Hydrostatic injection at 15 cm
Detector: UV 214
Migration time: 10.10
Voltage: 22 kV
Current: 135-145 μA
Model: Waters Quanta 4000

OTHER SUBSTANCES

Extracted: acepromazine, amphetamine, brompheniramine, butacaine, codeine, diazepam, doxapram, lidocaine, medazepam, methamphetamine, methapyrilene, methaqualone, phenmetrazine, procaine, tetrahydrozoline

Simultaneous: meclizine

KEY WORDS
plasma

REFERENCE
Chee,G.L.; Wan,T.S.M. Reproducible and high-speed separation of basic drugs by capillary zone electrophoresis, *J.Chromatogr.*, **1993**, *612*, 172–177.

SAMPLE
Matrix: bulk
Sample preparation: Prepare a 3 mg/mL solution in 3 mg/mL pholcodine in 10 mM HCl, make up to 25 mL with 10 mM HCl, mix thoroughly, filter (0.45 μm cellulose acetate)

CAPILLARY ELECTROPHORESIS
Capillary: 65 cm $\times$ 75 μm fused-silica (40 cm to detector) (Isco)
Capillary preparation: Flush with running buffer for 2 min between runs. Before each batch of samples fill capillary with 100 mM NaOH and let stand for 10 min, wash with water for 5 min, wash with 100 mM HCl for 5 min, wash with water for 5 min, fill with running buffer (all at vacuum level 4). Clean each week by washing with 100 mM HCl for 10 min, wash with water, fill with running buffer.
Capillary temperature: 30
Running buffer: MeCN:buffer 7.5:92.5. Buffer was 10 mM KH_2PO_4 containing 10 mM sodium tetraborate and 50 mM cetyltrimethylammonium bromide, pH 8.6.)
Injection: Injection by vacuum 5.0 kPa.s (4 nL)
Detector: UV 230
Migration time: 8.9
Internal standard: pholcodine (7.4)
Voltage: -15 kV
Model: Isco Model 3140

OTHER SUBSTANCES
Simultaneous: cocaine, benzoylecgonine, trans-cinnamoylcocaine, lidocaine, procaine, tetracaine

---

REFERENCE
Trenerry,V.C.; Robertson,J.; Wells,R.J. The determination of cocaine and related substances by micellar electrokinetic capillary chromatography, *Electrophoresis*, **1994**, *15*, 103–108.

SAMPLE
Matrix: bulk
Sample preparation: Dissolve sample in 2 mg/mL histamine diphosphate in 10 mM HCl, inject an aliquot.

CAPILLARY ELECTROPHORESIS
Capillary: 48.5 cm $\times$ 50 μm fused-silica (40 cm to detector)
Capillary preparation: Flush capillary with running buffer for 2 min before each run. Replenish buffer in vials after every 10 injections. Condition new capillaries by flushing with 1 M NaOH for 10 min, with 100 mM NaOH for 10 min, with water for 10 min, and with running buffer for 10 min.
Capillary temperature: 40
Running buffer: 75 mM NaH_2PO_4 containing 75 mM sodium borate, adjusted to pH 8.5 with 2 M NaOH
Injection: Pressure injection at 150 mbar for 3 s.
Detector: UV 230
Migration time: 7.5
Internal standard: histamine diphosphate (5)
Voltage: 10 kV
Current: 66-70 μA
Model: Hewlett Packard 3D CE
Limit of detection: 10 μg/mL (S/N = 2:1)

OTHER SUBSTANCES
Simultaneous: acetaminophen, cis-cinnamoylcocaine, trans-cinnamoylcocaine, cocaine, ephedrine, lidocaine, phenylpropanolamine, procaine, tetracaine

REFERENCE
Krawczeniuk,A.S.; Bravenec,V.A. Quantitative determination of cocaine in illicit powders by free zone capillary electrophoresis, *J.Forensic Sci.*, **1998**, *43*, 738–743.

Benzoic acid

Molecular formula: $C_7H_6O_2$
Molecular weight: 122.12
CAS Registry No.: 65-85-0
Merck Index (12th ed.): 1122

SAMPLE
Matrix: beverages
Sample preparation: Degas soft drinks by sonication, dilute 25-fold with water, inject an aliquot.

CAPILLARY ELECTROPHORESIS
Capillary: 60 cm $\times$ 75 μm fused-silica AccuSep (52 cm to detector)
Capillary preparation: Flush with running buffer for 2 min between each run. At the start of each day purge capillary with 500 mM KOH for 5 min, purge with water for 2 min, condition with running buffer for 2 h. At the end of each day flush with 500 mM KOH for 2 min, flush with water for 5 min, and purge with air for 2 min.
Running buffer: pH 11.0 phosphate buffer (μ=0.025)
Injection: Hydrodynamic injection for 30 s
Detector: UV 214
Migration time: 8.7
Voltage: 15 kV
Model: Waters Quanta 4000
Limit of detection: 5 μg/mL

OTHER SUBSTANCES
Simultaneous: aspartame, caffeine

KEY WORDS
soft drinks

REFERENCE
Jimidar,M.; Hamoir,T.P.; Foriers,A.; Massart,D.L. Comparison of capillary zone electrophoresis with high-performance liquid chromatography for the determination of additives in foodstuffs, *J.Chromatogr.*, **1993**, *636*, 179–186.

SAMPLE
Matrix: beverages
Sample preparation: Sonicate to degas beverage. Remove a 3 mL aliquot and add it to 1 mL 300 μg/mL cinchonine in 10 mM HCl, make up to 25 mL with water, inject an aliquot.

CAPILLARY ELECTROPHORESIS
Capillary: 75 cm $\times$ 75 μm fused-silica (66 cm to detector) (Polymicro Technologies)
Capillary preparation: Between runs flush capillary with running buffer for 2 min.
Capillary temperature: 28
Running buffer: MeOH:buffer 15:85 (Prepare buffer by dissolving 920 mg cetyltrimethylammonium bromide in 20 mM sodium tetraborate:20 mM KH_2PO_4 50:50, pH 8.6.)
Injection: Vacuum injection at 10 kPa/s.

Detector: UV 230
Migration time: 9.5
Internal standard: cinchonine (10.5)
Voltage: -25 kV
Model: Hewlett Packard 3D

OTHER SUBSTANCES
Simultaneous: cinchonidine, quinidine, quinine, saccharin, sorbic acid

KEY WORDS
comparison with HPLC

REFERENCE
Trenerry,V.C.; Ward,C.M. The separation and determination of quinine in bitter drinks by micellar electrokinetic capillary chromatography, *J.Capillary Electrophor.*, **1996**, *3*, 271–274.

SAMPLE
Matrix: beverages
Sample preparation: Degas carbonated beverage with sonication under vacuum, mix an aliquot
with an equal volume of running buffer, filter (0.45 μm)

CAPILLARY ELECTROPHORESIS
Capillary: 44 cm × 50 μm fused-silica (Polymicro Technologies)
Capillary preparation: Before each injection wash capillary with 2 volumes of fresh running
buffer. Condition new capillaries by washing with 1 M NaOH at 60° for 15 min, with 100 mM
NaOH at 60° for 15 min, and with water at 25° for 15 min.
Capillary temperature: 35
Running buffer: 20 mM Glycine adjusted to pH 9.0 with NaOH
Injection: Hydrodynamic injection for 1 s.
Detector: UV 215
Migration time: 2
Voltage: 20 kV
Model: SpectraPhoresis 1000 CE (ThermoSeparations Products)
Limit of detection: 4 μg/mL

OTHER SUBSTANCES
Simultaneous: aspartame, caffeine

KEY WORDS
soft drinks

REFERENCE
Walker,J.C.; Zaugg,S.E.; Walker,E.B. Analysis of beverages by capillary electrophoresis, *J.Chromatogr.A*, **1997**,
781, 481–485.

SAMPLE
Matrix: beverages, food
Sample preparation: Dilute sample with water, add dehydroacetic acid to a final concentration
of 10 μg/mL, filter (0.45 μm cellulose acetate), inject an aliquot.

CAPILLARY ELECTROPHORESIS
Capillary: 75 cm × 75 μm fused-silica (50 cm to detector) (Polymicro Technologies)
Capillary preparation: Flush with running buffer for 2 min between runs. Clean each week by
washing with 100 mM NaOH for 10 min and with water for 10 min, fill with running buffer.
Capillary temperature: 27
Running buffer: 10 mM Sodium borate containing 10 mM KH_2PO_4 and 50 mM sodium deoxy-
cholate, pH 8.6
Injection: Injection by vacuum at 20 kPa.s (vacuum level 2)
Detector: UV 220
Migration time: 10.1
Internal standard: dehydroacetic acid (8.4)

Voltage: 20 kV
Model: ISCO Model 3140

OTHER SUBSTANCES
Simultaneous: acesulfame-K, alitame, aspartame, caffeine, dulcin, saccharin, sorbic acid

KEY WORDS
soft drink; cordial; tomato sauce; jam; sweeteners; comparison with HPLC

REFERENCE
Thompson,C.O.; Trenerry,V.C.; Kemmery,B. Micellar electrokinetic capillary chromatographic determination of artificial sweeteners in low-Joule soft drinks and other foods, *J.Chromatogr.A*, **1995**, *694*, 507–514.

SAMPLE
Matrix: bulk
Sample preparation: Heat 100 μg acid with 25 μL 10 mg/mL phenacyl bromide in acetone and 25 μL 10 mg/mL triethylamine in acetone under nitrogen in a capped PTFE tube in a boiling water bath for 5 min, cool, add 40 μL 2 mg/mL acetic acid in acetone, evaporate to dryness under a stream of nitrogen at room temperature , reconstitute with MeOH, inject an aliquot.

CAPILLARY ELECTROPHORESIS
Capillary: 57 cm × 75 μm untreated fused-silica (50 cm to detector) (CElect, Supelco)
Capillary preparation: Between runs wash capillary with water for 1 min, with 1 M NaOH for 7 min, with water for 1 min, with 1 M NaOH for 7 min, with water for 1 min, with MeOH for 1 min, with water for 4 min, and then equilibrate with running buffer for 2 min.
Capillary temperature: 30
Running buffer: 10 mM pH 10.2 borate buffer:heptane:sodium cholate:n-butanol 87.93:0.66: 4.87:6.55 (a microemulsion)
Injection: Hydrodynamic injection at 3.45 kPa for 1 s.
Detector: UV 243
Migration time: 21.5
Voltage: 15 kV
Model: Beckman P/ACE 5500

OTHER SUBSTANCES
Simultaneous: acetic acid, arachidic acid, butyric acid, capric acid, caproic acid, caprylic acid, lauric acid, myristic acid, palmitic acid, stearic acid

KEY WORDS
derivatization

REFERENCE
Miksík,I.; Deyl,Z. Microemulsion electrokinetic chromatography of fatty acids as phenacyl esters, *J.Chromatogr.A*, **1998**, *807*, 111–119.

SAMPLE
Matrix: food
Sample preparation: 100 g Oyster sauce + 30 g NaCl + 100 mL 10% NaOH, shake, make up to 500 mL with saturated NaCl solution, filter. Acidify 100 mL filtrate with concentrated HCl, extract 3 times with 40 mL portions of chloroform, evaporate the extracts to dryness, reconstitute with MeOH, inject an aliquot.

CAPILLARY ELECTROPHORESIS
Capillary: 48 cm × 50 μm fused-silica (Polymicro Technologies)
Running buffer: 50 mM phosphate/50 mM borate buffer containing 10 mM tetrabutylammonium hydrogen sulfate, pH 7.0
Injection: Inject by siphon at 8 cm for 5 s (2 nL)
Detector: UV 190
Migration time: 12
Voltage: 15 kV
Model: Laboratory constructed

Limit of detection: 20 pg

OTHER SUBSTANCES
Extracted: caprylic acid, propionic acid, sorbic acid

KEY WORDS
oyster sauce

REFERENCE
Ng,C.L.; Lee,H.K.; Li,S.F.Y. Analysis of food additives by ion-pairing electrokinetic chromatography, *J.Chromatogr.Sci.*, **1992**, *30*, 167–170.

SAMPLE
Matrix: formulations
Sample preparation: Add capsule or tablet to 10 mL water, sonicate for 30 min, centrifuge, dilute the supernatant with water, inject an aliquot. Dilute liquids with water, inject an aliquot.

CAPILLARY ELECTROPHORESIS
Capillary: 57 cm × 75 μm (50 cm to detector) (Beckman)
Capillary temperature: 25
Running buffer: 10 mM Tris adjusted to pH 5.0 with acetic acid
Injection: Pressure injection for 5 s (about 100 nL)
Detector: UV 214
Migration time: 17.19
Voltage: 12.5 kV
Model: Beckman P/ACE 2000

OTHER SUBSTANCES
Simultaneous: albuterol, m-aminobenzoic acid, aniline, creatinine

KEY WORDS
tablets; capsules; syrup

REFERENCE
Ackermans,M.T.; Beckers,J.L.; Everaerts,F.M.; Seelen,I.G. Comparison of isotachophoresis, capillary zone electrophoresis and high-performance liquid chromatography for the determination of salbutamol, terbutaline sulphate and fenoterol hydrobromide in pharmaceutical dosage forms, *J.Chromatogr.*, **1992**, *590*, 341–353.

SAMPLE
Matrix: formulations
Sample preparation: Dragees. Dissolve dragee in running buffer, add a solution of flurbiprofen in running buffer, filter (paper), dilute with running buffer to a final concentration of 20 μg/mL for ibuprofen and 10 μg/mL for flurbiprofen, inject an aliquot. Suspensions. Dilute 1 mL suspension with running buffer, add a solution of flurbiprofen in running buffer, centrifuge at 3000 rpm for 10 min, dilute the supernatant with running buffer to a final concentration of 50 μg/mL for ibuprofen and 25 μg/mL for flurbiprofen, inject an aliquot.

CAPILLARY ELECTROPHORESIS
Capillary: 60 cm × 75 μm (52.5 cm to detector)
Capillary preparation: Purge with running buffer for 2 min before each injection. Store capillary overnight in water, rinse with 500 mM NaOH, with water, and with running buffer.
Running buffer: 50 mM pH 9.0 Borate buffer containing 40 mM sodium dodecyl sulfate (Mix 2.94 mL 200 mM boric acid, 6.37 mL 75 mM tetraborate solution, and 10 mL 200 mM sodium dodecyl sulfate, make up to 50 mL with water.)
Injection: Hydrodynamic injection at 10 cm for 5 s
Detector: UV 214
Migration time: 7.1
Internal standard: flurbiprofen (6.7)
Voltage: 15 kV
Model: Waters Quanta 4000 CE

OTHER SUBSTANCES
Simultaneous: ibuprofen

KEY WORDS
dragees; suspensions

REFERENCE
Donato,M.G.; Baeyens,W.; Van den Bossche,W.; Sandra,P. The determination of non-steroidal antiinflammatory drugs in pharmaceuticals by capillary zone electrophoresis and micellar electrokinetic capillary chromatography, *J.Pharm.Biomed.Anal.*, **1994**, *12*, 21–26.

SAMPLE
Matrix: solutions

CAPILLARY ELECTROPHORESIS
Capillary: 15 cm × 50 μm fused-silica coated with crosslinked polyacrylamide (13.5 cm to detector) (Scientific Glass Engineering)
Running buffer: 100 mM pH 8.6 Tris-acetic acid containing 10% sodium dodecyl sulfate
Injection: Place a 1-2 mm slug in capillary by capillary action
Detector: UV 220
Migration time: 3.4
Voltage: 2 kV
Current: 70 μA

OTHER SUBSTANCES
Simultaneous: 4-hydroxybenzoic acid, 4-hydroxy-3-methoxybenzoic acid, β-naphthylacetic acid, terephthalic acid

KEY WORDS
coated capillary

REFERENCE
Hjertén,S.; Valtcheva,L.; Elenbring,K.; Eaker,D. High-performance electrophoresis of acidic and basic low-molecular-weight compounds and of proteins in the presence of polymers and neutral surfactants, *J.Liq.Chromatogr.*, **1989**, *12*, 2471–2499.

SAMPLE
Matrix: solutions

CAPILLARY ELECTROPHORESIS
Capillary: 56.5 cm × 50 μm fused-silica (50 cm to detector)
Capillary preparation: Rinse with 1 M NaOH for 30 s between runs. Condition a new capillary with 1 M NaOH for 20 min.
Capillary temperature: 30
Running buffer: 30 mM 2-hydroxybutyric acid adjusted to pH 8.65 with ammonium hydroxide
Injection: Pressure injection 3 s
Detector: UV 200
Migration time: 5
Voltage: 25 kV
Current: 30-35 μA
Model: Beckman P/ACE Model 2100

OTHER SUBSTANCES
Simultaneous: salicylic acid

REFERENCE
Bullock,J.; Strasters,J.; Snider,J. Effect of multiple electrolyte buffers on peak symmetry, resolution, and sensitivity in capillary electrophoresis, *Anal.Chem.*, **1995**, *67*, 3246–3252.

SAMPLE
Matrix: solutions

CAPILLARY ELECTROPHORESIS
Capillary: 43 cm × 75 μM coated capillary (35 cm to detector) (Polymicro Technologies)
Capillary preparation: Treat capillary with 1 M NaOH for 5 h, flush with 100 mM HCl, flush with 100 mM NaOH, after 1 h rinse with water, rinse with THF, pass nitrogen through at 120° for 45 min, pass a 50% solution of gamma- methacryloxypropyltrimethylsilane in THF through under pressure for 20 min, allow to stand for 12 h, flush extensively with THF, flush extensively with water, fill with 3% acrylamide containing 1 μL/mL tetramethylethylenediamine and 1 μL/mL 40% ammonium peroxydisulfate, degas under vacuum (20 mmHg) for 40 min, let stand overnight, empty capillary with a syringe.
Running buffer: 50 mM pH 4.4 Tris-acetate buffer
Injection: Hydrostatic injection
Detector: UV 254
Migration time: 6.5
Voltage: -15 kV
Model: Waters Quanta 4000

OTHER SUBSTANCES
Simultaneous: N-acryloyl-gamma-aminobutyric acid, N-acryloylglycine, caffeic acid, p-hydroxy-cinnamic acid, p-toluenesulfonic acid

KEY WORDS
coated capillary

REFERENCE
Chiari,M.; Kenndler,E. Capillary zone electrophoresis in organic solvents: separation of anions in methanolic buffer solutions, *J.Chromatogr.A*, **1995**, *716*, 303–309.

SAMPLE
Matrix: solutions

CAPILLARY ELECTROPHORESIS
Capillary: 37 cm × 50 μm coated (30 cm to detector) (Polymicro Technologies)
Capillary preparation: The capillary was coated with poly(N-(acryloylaminoethoxyethanol)-β-D-glucoyranoside). Partial details are given in the referenced papers and the references quoted therein.
Running buffer: 50 mM pH 5.5 2-(N-morpholino)ethanesulfonic acid (MES)-Tris buffer containing 1 mM copper sulfate
Injection: Electrokinetic injection for 1 s.
Detector: UV 214
Migration time: 4
Voltage: 15 kV
Model: Waters Quanta 4000

OTHER SUBSTANCES
Simultaneous: N-acryloyl-gamma-aminobutyric acid, N-acryloylglycine, N-acryloyl glycolic acid, caffeic acid, niacin, phthalic acid, p-toluenesulfonic acid

KEY WORDS
reversed polarity; coated capillary

REFERENCE
Chiari,M.; Dell'Orto,N.; Casella,L. Separation of organic acids by capillary zone electrophoresis in buffers containing divalent metal cations, *J.Chromatogr.A*, **1996**, *745*, 93–101.

SAMPLE
Matrix: solutions
Sample preparation: Inject an aliquot of a solution in MeCN:MeOH 50:50 containing 2 mM sodium acetate.

CAPILLARY ELECTROPHORESIS
Capillary: 27 cm × 50 μm fused-silica (20 cm to detector) (Electro-Kinetic Technologies, Edinburgh, UK)

Capillary preparation: Between each injection rinse with 100 mM NaOH for 1 min, with MeOH for 1 min, and with running buffer for 1 min. Rinse new capillaries with 100 mM NaOH then with MeOH for 10 min.
Running buffer: MeCN:MeOH 50:50 containing 10 mM sodium acetate
Injection: Injection for 1 s.
Detector: UV 200
Migration time: 3
Voltage: 25 kV
Model: Beckman P/ACE 5000

OTHER SUBSTANCES
Simultaneous: hydroxynaphthoate, naphthoxyacetic acid, penicillin (sic), troglitazone, warfarin

REFERENCE
Altria,K.D.; Bryant,S.M. Highly selective and efficient separations of a wide range of acidic species in capillary electrophoresis employing non-aqueous media, *Chromatographia*, **1997**, *46*, 122–122.

SAMPLE
Matrix: solutions

CAPILLARY ELECTROPHORESIS
Capillary: 27 cm × 75 μm
Capillary preparation: Before each run rinse capillary with 100 mM NaOH for 30 s and with running buffer for 30 s. Condition new capillaries by rinsing with 100 mM NaOH for 20 min.
Capillary temperature: 30
Running buffer: 15 mM Sodium borate
Injection: Pressure injection of sample at 25 mbar for 1 s followed by running buffer at 25 mbar for 1 s.
Detector: UV 200
Migration time: 3.85
Internal standard: aminobenzoic acid (3.5)
Voltage: 6.5 kV
Model: Beckman

OTHER SUBSTANCES
Simultaneous: aspirin, bacitracin, beclomethasone, ceftizoxime, ceftriaxone, cefuroxime, cephalothin, cromolyn, embonic acid, epoprostenol, glyburide (glibenclamide), levothyroxine, nedocromil, nystatin, omeprazole, prednisolone, warfarin, zidovudine

REFERENCE
Altria,K.D.; Bryant,S.M.; Hadgett,T.A. Validated capillary electrophoresis method for the analysis of a range of acidic drugs and excipients, *J.Pharm.Biomed.Anal.*, **1997**, *15*, 1091–1101.

SAMPLE
Matrix: solutions
Sample preparation: Inject an aliquot of a 100 μg/mL solution in MeOH:water 10:90.

CAPILLARY ELECTROPHORESIS
Capillary: 27 cm × 75 μm fused-silica (20 cm to detector) (Composite Metal Services, Hallow, UK)
Capillary preparation: Before injection rinse with 100 mM NaOH and running buffer. Condition new capillaries by rinsing with 100 mM NaOH for 20 min.
Capillary temperature: 25
Running buffer: 50 mM pH 7.0 NaH_2PO_4
Injection: Pressure injection for 3 s.
Detector: UV 214 or 200
Migration time: 6.2
Voltage: 7 kV
Model: Beckman P/ACE 5100

OTHER SUBSTANCES
Simultaneous: aspirin, phenylacetic acid

REFERENCE
Kelly,M.A.; Altria,K.D.; Clark,B.J. Approaches used in the reduction of buffer electrolysis effects for routine capillary electrophoresis procedures in pharmaceutical analysis, *J.Chromatogr.A*, **1997**, *768*, 73–80.

Benzphetamine

Molecular formula: $C_{17}H_{21}N$
Molecular weight: 239.36
CAS Registry No.: 156-08-1, 5411-22-3 (HCl)
Merck Index (12th ed.): 1151
Lednicer: 1 70

SAMPLE
Matrix: solutions

CAPILLARY ELECTROPHORESIS
Capillary: 58.2 cm × 75 μm fused-silica (50.7 cm to detector) (Polymicro Technologies)
Capillary preparation: Purge with running buffer before each run. At the beginning of each day purge using 50-60 kPa vacuum with 500 mM NaOH for 5 min, with water for 5 min, with MeCN for 5 min, and with running buffer for 5 min.
Running buffer: MeCN:MeOH:acetic acid 49:50:1 containing 20 mM ammonium acetate
Injection: Hydrostatic injection at 10 cm for 5 s.
Detector: UV 214
Migration time: 3.96
Voltage: 25 kV
Model: Waters Quanta 4000

OTHER SUBSTANCES
Simultaneous: amphetamine, ephedrine, nylidrin, oxymetazoline, phendimetrazine, phenmetrazine, phenylephrine, phenylpropanolamine, xylometazoline

REFERENCE
Leung,G.N.W.; Tang,H.P.O.; Tso,T.S.C.; Wan,T.S.M. Separation of basic drugs with non-aqueous capillary electrophoresis, *J.Chromatogr.A*, **1996**, *738*, 141–154.

Benzquinamide

Molecular formula: $C_{22}H_{32}N_2O_5$
Molecular weight: 404.51
CAS Registry No.: 63-12-7
Merck Index (12th ed.): 1154
Lednicer: 1 350

SAMPLE
Matrix: solutions

CAPILLARY ELECTROPHORESIS
Capillary: 59.6 cm × 75 μm fused-silica (52.1 cm to detector) (Polymicro Technologies)
Capillary preparation: Purge with running buffer before each run. At the beginning of each day purge using 50-60 kPa vacuum with 500 mM NaOH for 5 min, with water for 5 min, with MeCN for 5 min, and with running buffer for 5 min.
Running buffer: MeCN:MeOH:acetic acid 49:50:1 containing 20 mM ammonium acetate
Injection: Hydrostatic injection at 10 cm for 5 s.

Detector: UV 214
Migration time: 6.36
Voltage: 30 kV
Model: Waters Quanta 4000

OTHER SUBSTANCES
Simultaneous: deserpidine, ethopropazine, mesoridazine, methotrimeprazine, molindone, promazine, promethazine, reserpine, thioridazine, thiothixene

REFERENCE
Leung,G.N.W.; Tang,H.P.O.; Tso,T.S.C.; Wan,T.S.M. Separation of basic drugs with non-aqueous capillary electrophoresis, *J.Chromatogr.A*, **1996**, *738*, 141–154.

Benzthiazide

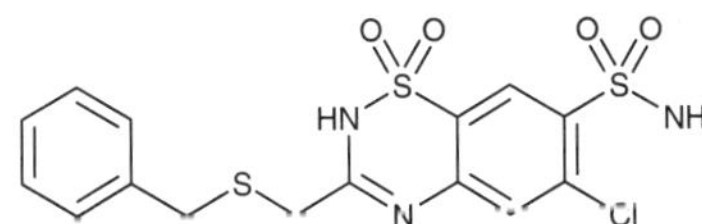

Molecular formula: $C_{15}H_{14}ClN_3O_4S_3$
Molecular weight: 431.94
CAS Registry No.: 91-33-8
Merck Index (12th ed.): 1155

SAMPLE
Matrix: blood, urine
Sample preparation: Condition a 3 mL Supelclean LC-18 SPE cartridge with 3 mL MeOH and 3 mL water. Dilute urine 1:10 with water. Precipitate proteins from serum with MeOH. Add diluted urine or protein supernatant to the SPE cartridge, wash with 3 mL water, elute with 3 mL MeOH, reconstitute to the original volume with 100 mM KOH, inject an aliquot.

CAPILLARY ELECTROPHORESIS
Capillary: 67 cm × 50 µm fused-silica (60 cm to detector) (Polymicro Technologies)
Capillary preparation: Rinse with running buffer for 2 min before run. If necessary, regenerate capillary with 100 mM NaOH for 10 min and with water for 15 min.
Capillary temperature: 20
Running buffer: 60 mM 3-(cyclohexylamino)-1-propanesulfonic acid (CAPS) adjusted to pH 10.6 with 100 mM KOH
Injection: Pressure injection for 5 s
Detector: UV 220
Migration time: 10.8
Voltage: 25 kV
Model: Beckman P/ACE 2000

OTHER SUBSTANCES
Extracted: acetazolamide, amiloride, bendroflumethiazide, bumetanide, caffeine, chlorothiazide, chlorthalidone, clopamide, dichlorphenamide, ethacrynic acid, furosemide, hydrochlorothiazide, metyrapone, probenecid, triamterene, trichlormethiazide

KEY WORDS
serum; SPE

REFERENCE
Jumppanen,J.; Sirén,H.; Riekkola,M.-L. Screening for diuretics in urine and blood serum by capillary zone electrophoresis, *J.Chromatogr.A*, **1993**, *652*, 441–450.

Betamethasone

Molecular formula: $C_{22}H_{29}FO_5$
Molecular weight: 392.47
CAS Registry No.: 378-44-9, 987-24-6 (acetate),
22298-29-9 (benzoate), 5593-20-4 (dipropionate),
151-73-5 (sodium phosphate), 2152-44-5
(17-valerate), 5534-05-4 (acibutate), 360-63-4 (dihydrogen phosphate)
Merck Index (12th ed.): 1226
Lednicer: 1 198

SAMPLE
Matrix: solutions

CAPILLARY ELECTROPHORESIS
Capillary: 65 cm × 50 μm fused-silica (50 cm to detector) (Scientific Glass Engineering)
Capillary preparation: Flush with running buffer for 1 min between runs. Periodically flush
with water for 1 min, sweep with 100 mM KOH, let stand in 100 mM KOH for 30 min, flush
with water until the effluent is neutral to pH paper, fill with running buffer, let stand for 30
min.
Running buffer: 20 mM pH 9.0 Phosphate-borate buffer containing 100 mM sodium cholate
(Buffer was prepared by adjusting pH of 20 mM sodium borate to 9.0 with 20 mM NaH_2PO_4.)
Injection: Injection by siphon at 10 cm for 10 s (about 1 nL)
Detector: UV 210
Migration time: 12.5
Voltage: 20 kV

OTHER SUBSTANCES
Simultaneous: dexamethasone acetate, fluocinolone acetonide, fluocinonide, hydrocortisone acetate, hydrocortisone, triamcinolone, triamcinolone acetonide

REFERENCE
Nishi,H.; Fukuyama,T.; Matsuo,M.; Terabe,S. Separation and determination of lipophilic corticosteroids and
benzothiazepin analogues by micellar electrokinetic chromatography using bile salts, *J.Chromatogr.*, **1990**,
513, 279–295.

SAMPLE
Matrix: solutions
Sample preparation: Prepare a 0.2-1 mg/mL solution in MeOH, inject an aliquot.

CAPILLARY ELECTROPHORESIS
Capillary: 65 cm × 50 μm fused-silica (50 cm to detector)
Running buffer: 20 mM pH 9.0 Phosphate-borate buffer containing 50 mM sodium dodecyl
sulfate, 4 M urea, and 15 mM gamma-cyclodextrin
Detector: UV 220
Migration time: 11.5
Voltage: 20 kV

OTHER SUBSTANCES
Simultaneous: cortisone acetate, dexamethasone acetate, fluocinolone acetonide, fluocinonide,
hydrocortisone, hydrocortisone acetate, triamcinolone acetonide

REFERENCE
Nishi,H.; Matsuo,M. Separation of corticosteroids and aromatic hydrocarbons by cyclodextrin-modified micellar
electrokinetic chromatography, *J.Liq.Chromatogr.*, **1991**, *14*, 973–986.

SAMPLE
Matrix: solutions

Sample preparation: Inject an aliquot of a solution in running buffer.

CAPILLARY ELECTROPHORESIS
Capillary: 57 cm × 75 μm fused-silica (50 cm to detector)
Capillary preparation: Before each run rinse with running buffer for 2 min. After each run rinse capillary with 100 mM NaOH for 2 min and with water for 2 min. Condition new capillaries with 100 mM HCl for 5 min, with 100 mM NaOH for 10 min, and with water for 5 min.
Capillary temperature: 23
Running buffer: MeOH:buffer 10:90 (Buffer was 40 mM pH 9.2 borate buffer containing 20 mM sodium dodecyl sulfate.)
Injection: Low pressure injection for 10 s.
Detector: UV 240
Migration time: 14.2
Voltage: 30 kV
Model: Beckman P/ACE 5510

OTHER SUBSTANCES
Simultaneous: dexamethasone, flumethasone, hydrocortisone, triamcinolone

REFERENCE
Gu,X.; Meleka-Boules,M.; Chen,C.-L. Micellar electrokinetic capillary chromatography combined with immunoaffinity chromatography for identification and determination of dexamethasone and flumethasone in equine urine, *J.Capillary Electrophor.*, **1996**, *3*, 43–49.

SAMPLE
Matrix: solutions

CAPILLARY ELECTROPHORESIS
Capillary: 42 cm × 50 μm fused-silica packed with 3 μm CEC Hypersil ODS (packed length 30 cm, 30.1 cm to detector) (Hypersil)
Running buffer: Gradient. A was 5 mM ammonium acetate in MeCN:water 17:83. B was 5 mM ammonium acetate in MeCN:water 38:62. A:B 100:0 for 3 min, to 0:100 over 15 min, maintain at 0:100 (pumped with an HPLC pump at 0.01 mL/min for 3 min then at 0.1 mL/min).
Injection: Inject 10 μL using an HPLC injector.
Detector: UV 240
Migration time: 35
Voltage: 30 kV

OTHER SUBSTANCES
Simultaneous: adrenosterone, cortisone, dexamethasone, fluocortolone, hydrocortisone, methyl prednisolone, prednisolone, triamcinolone, triamcinolone acetonide

KEY WORDS
electrochromatography

REFERENCE
Taylor,M.R.; Teale,P. Gradient capillary electrochromatography of drug mixtures with UV and electrospray ionisation mass spectrometric detection, *J.Chromatogr.A*, **1997**, *768*, 89–95.

SAMPLE
Matrix: solutions
Sample preparation: Inject an aliquot of a 100 μg/mL solution in running buffer.

CAPILLARY ELECTROPHORESIS
Capillary: 27 cm × 50 μm fused-silica (20 cm to detector) (Composite Metal Services, Hallow, UK) packed with 3 μm Hypersil ODS for 20 cm (details in paper)
Running buffer: MeCN:2 mM pH 7.8 phosphate buffer 80:20
Injection: Electrokinetic injection at 5 kV for 5 s.
Detector: UV 214
Migration time: 7.85 (betamethasone), 12.14 (betamethasone dipropionate), 25.46 (betamethasone-17-valerate)

Voltage: 10 kV
Model: Beckman P/ACE 2050

OTHER SUBSTANCES
Simultaneous: clobetasol butyrate, clobetasone butyrate, fluticasone propionate, hydrocortisone, prednisolone

KEY WORDS
electrochromatography

REFERENCE
Frame,L.A.; Robinson,M.L.; Lough,W.J. Simplification of capillary electrochromatography procedures, *J.Chromatogr.A*, **1998**, *798*, 243–249.

Betaxolol

Molecular formula: $C_{18}H_{29}NO_3$
Molecular weight: 307.43
CAS Registry No.: 63659-18-7, 63659-19-8 (HCl)
Merck Index (12th ed.): 1229
Lednicer: 4 26

SAMPLE
Matrix: solutions

CAPILLARY ELECTROPHORESIS
Capillary: 62.1 cm $\times$ 75 μm fused-silica (44.6 cm to detector) (Polymicro Technologies)
Capillary preparation: Coat capillary as follows. Condition capillary with 1 M NaOH for 1 h, rinse with water for 30 min, fill capillary with 1% 3-(trimethoxysilyl)propyl methacrylate (methacryloxypropyltrimethoxysilane) (adjusted to pH 3.5 with acetic acid), allow to react at room temperature for 2 h, rinse with water, fill capillary with 10% 2-aminoethyl methacrylate hydrochloride (Fisher Scientific) solution containing 1-3 mg/mL N,N,N',N'-tetramethylethyl-enediamine and 1 mg/mL ammonium persulfate, allow to react until polymerization is complete, rinse with water, dry at 30-40° overnight.
Running buffer: 25 mM pH 4.7 Acetate buffer
Injection: Hydrodynamic injection at 5 cm for 5 s.
Detector: UV 210
Migration time: 10.55
Voltage: 15 kV
Model: laboratory constructed

OTHER SUBSTANCES
Simultaneous: acebutolol, atenolol, carteolol, dilevalol, nifenalol, pindolol, propranolol

KEY WORDS
coated capillary; detector at anode

REFERENCE
Liu,Q.; Lin,F.; Hartwick,R.A. Free solution capillary electrophoretic separation of basic proteins and drugs using cationic polymer coated capillaries, *J.Liq.Chromatogr.Rel.Technol.*, **1997**, *20*, 707–718.

SAMPLE
Matrix: solutions

CAPILLARY ELECTROPHORESIS
Capillary: 71.5 cm $\times$ 75 μm fused-silica (52.2 cm to detector) (Polymicro Technologies)

Capillary preparation: Rinse capillary with running buffer for 1 min between runs. Rinse new capillaries with 500 mM NaOH for 20 min and with water for 10 min and then condition with running buffer for 1 h.

Running buffer: 50 mM pH 3.0 containing 0.1% guaran (Guaran is the water-soluble fraction of guar gum. Prepare guaran by stirring 1 g guar gum (Sigma) in 100 mL water for 20 min, add 25 mL EtOH dropwise with stirring, let stand overnight, supercentrifuge. Add 30 mL EtOH dropwise with stirring to the centrifugate, centrifuge, collect the guaran as a precipitate, dry in air.)

Injection: Hydrodynamic injection at 5 cm for 5 s.

Detector: UV (wavelengths not given)

Migration time: 15

Voltage: 17.5 kV

Model: laboratory-constructed

OTHER SUBSTANCES

Simultaneous: acebutolol, atenolol, carteolol, dilevalol, nifenalol, pindolol, propranolol

---

REFERENCE

Liu,Q.; Lin,F.; Hartwick,R.A. Capillary zone electrophoretic separation of basic proteins and drugs using guaran as a buffer modifier, *Chromatographia*, **1998**, *47*, 219–224.

BHT

Molecular formula: $C_{15}H_{24}O$

Molecular weight: 220.35

CAS Registry No.: 128-37-0

Merck Index (12th ed.): 1583

SAMPLE

Matrix: solutions

Sample preparation: Inject an aliquot of a solution in MeOH.

CAPILLARY ELECTROPHORESIS

Capillary: 59.5 × 50 μm fused-silica (51 cm to detector) (Composite Metal services, Hallow, UK)

Capillary preparation: Flush capillary with running buffer for 5.5 min before each injection. At the end of each day flush capillary with 100 mM KOH for 10 min and with water for 10 min. Purge a new capillary with 100 mM KOH for 10 min, with water for 10 min, and with running buffer for 10 min.

Capillary temperature: 23

Running buffer: 30 mM pH 7.00 Phosphate buffer containing 30 mM sodium dodecyl sulfate

Injection: Hydrodynamic injection at 100 mbar

Detector: UV

Migration time: 10.7

Voltage: 22 kV

Current: 32 μA

Model: Hewlett-Packard 3D CE

Limit of detection: 50 ng

OTHER SUBSTANCES

Simultaneous: butylated hydroxyanisole (BHA), n-dodecyl gallate, gallic acid, gallic acid amide, gallic acid trimethyl ether, methyl gallate, n-octyl gallate, n-propyl gallate

---

REFERENCE

Summanen,J.; Vuorela,H.; Hiltunen,R.; Sirén,H.; Riekkola,M.L. Determination of phenolic antioxidants by capillary electrophoresis with ultraviolet detection, *J.Chromatogr.Sci.*, **1995**, *33*, 704–711.

Bifonazole

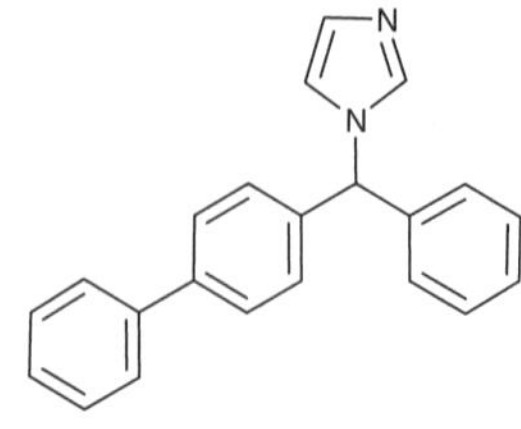

Molecular formula: $C_{22}H_{18}N_2$
Molecular weight: 310.40
CAS Registry No.: 60628-96-8
Merck Index (12th ed.): 1260
Lednicer: 4 93

SAMPLE
Matrix: solutions
Sample preparation: Prepare a 60 µg/mL solution, inject an aliquot.

CAPILLARY ELECTROPHORESIS
Capillary: 61 cm × 50 µm fused-silica (44 cm to detector) (Grom)
Capillary temperature: 21 ± 1
Running buffer: MeOH:buffer 20: 80 (Buffer was 50 mM pH 3.0 (?) phosphate buffer containing 0.1 mM sulfobutyl ether-β-cyclodextrin.)
Injection: Hydrostatic injection at 10 cm for 5 s
Detector: UV 210
Migration time: 15.20, 15.33 (enantiomers)
Voltage: 400 V/cm
Model: Grom 100

OTHER SUBSTANCES
Simultaneous: econazole, enilconazole, ketoconazole, miconazole

KEY WORDS
chiral

REFERENCE
Chankvetadze,B.; Endresz,G.; Blaschke,G. Enantiomeric resolution of chiral imidazole derivatives using capillary electrophoresis with cyclodextrin-type buffer modifiers, *J.Chromatogr.A*, **1995**, *700*, 43–49.

Biotin

Molecular formula: $C_{10}H_{16}N_2O_3S$
Molecular weight: 244.31
CAS Registry No.: 58-85-5
Merck Index (12th ed.): 1272

SAMPLE
Matrix: formulations
Sample preparation: Dissolve in water, inject an aliquot.

CAPILLARY ELECTROPHORESIS
Capillary: 60 cm × 50 µm fused-silica (51.5 cm to detector)
Capillary preparation: Condition capillary with 1 M NaOH for 10 min at the start of each day. Rinse capillary with 100 mM NaOH for 2 min and then with running buffer for 3 min before each run.
Capillary temperature: 25
Running buffer: 20 mM pH 8.0 Phosphate buffer (sonicate for at least 10 min before use)
Injection: Pressure injection at 200 mbar.s.
Migration time: 6
Voltage: 30 kV
Model: Hewlett-Packard [3D]CE

Limit of detection: 3.37 μg/mL

OTHER SUBSTANCES
Simultaneous: niacin, thiamine, ascorbic acid

KEY WORDS
comparison with a spectrophotometric method

REFERENCE
Schiewe,J.; Göbel,S.; Schwarz,M.; Neubert,R. Application of capillary zone electrophoresis for analyzing biotin in pharmaceutical formulations -a comparative study, *J.Pharm.Biomed.Anal.*, **1996**, *14*, 435–439.

SAMPLE
Matrix: formulations
Sample preparation: Protect from light. Grind tablet, add 50 mL 1% acetic acid, shake at 65° for 10 min, add 5 mL 10 mg/mL niacin in 1% acetic acid, cool to room temperature, make up to 100 mL with 1% acetic acid, filter (0.22 μm cellulose acetate), dilute 4-12-fold, inject an aliquot.

CAPILLARY ELECTROPHORESIS
Capillary: 48.5 cm × 50 μm fused-silica (40 cm to detector) (Hewlett Packard)
Capillary preparation: After each run wash with 100 mM NaOH for 7 min, with water for 1 min, and with running buffer for 3 min. At the start of each day wash with separation buffer for 10 min. Wash new capillaries with 1 M NaOH, 100 mM NaOH, water, and running buffer.
Capillary temperature: 25
Running buffer: 50 mM Borax adjusted to pH 8.5 with boric acid
Injection: Hydrodynamic injection at 5 kPa for 10 s.
Detector: UV 225
Migration time: 4.7
Internal standard: niacin (7.5)
Voltage: 25 kV
Model: Hewlett Packard [3D]CE

OTHER SUBSTANCES
Simultaneous: adenine, niacinamide, pantothenic acid (UV 215), pyridoxine (UV 215), rutin, thiamine, riboflavin, ascorbic acid

---

KEY WORDS
tablets

REFERENCE
Fotsing,L.; Fillet,M.; Bechet,I.; Hubert,P.; Crommen,J. Determination of six water-soluble vitamins in a pharmaceutical formulation by capillary electrophoresis, *J.Pharm.Biomed.Anal.*, **1997**, *15*, 1113–1123.

SAMPLE
Matrix: formulations
Sample preparation: Grind tablet to a homogeneous powder and dissolve in 2 mL water and 20 mL EtOH, homogenize (Ultra Turrax), centrifuge. Evaporate the supernatant to dryness under reduced pressure, reconstitute the residue with water, inject an aliquot.

CAPILLARY ELECTROPHORESIS
Capillary: 64.5 cm × 50 μm fused-silica (56 cm to detector)
Capillary preparation: Before each run flush capillary at 4 kPa with 1 M NaOH for 2 min and with buffer for 5 min. Replace buffer before each analysis.
Capillary temperature: 30
Running buffer: 1-Propanol:buffer 2:98 (Buffer was 100 mM Na_2HPO_4, 500 mM taurine, and 75 mM sodium cholate.)
Injection: Pressure injection at 4 kPa for 1 s.
Detector: UV 214
Migration time: 18
Voltage: 17 kV

Model: Hewlett-Packard HP^{3D} CE
Limit of quantitation: 310 μM

OTHER SUBSTANCES
Simultaneous: ascorbic acid, folic acid, niacinamide, pyridoxamine, pyridoxine, riboflavin, thiamine, vitamin B12

KEY WORDS
tablets

REFERENCE
Buskov,S.; Moller,P.; Sorensen,H.; Sorensen,J.C.; Sorensen,S. Determination of vitamins in food based on supercritical fluid extraction prior to micellar electrokinetic capillary chromatographic analyses of individual vitamins, *J.Chromatogr.A*, **1998**, *802*, 233–241.

SAMPLE
Matrix: formulations, juice
Sample preparation: Dissolve in water, filter (0.2 μm), inject an aliquot. (Add a 50-fold excess of cysteine to solutions containing ascorbic acid.)

CAPILLARY ELECTROPHORESIS
Capillary: 60 cm × 50 μm fused-silica (51.5 cm to detector)
Capillary preparation: Condition with running buffer for 3 min before each run. Condition a new capillary 1 M NaOH for 10 min and with 100 mM NaOH for 2 min.
Capillary temperature: 25
Running buffer: 20 mM pH 8.0 Phosphate buffer
Injection: Pressure injection for 200 mbar.s
Detector: UV 200
Migration time: 5.7
Voltage: -30 kV
Model: Hewlett-Packard 3DCE
Limit of detection: 3.4 μg/mL

OTHER SUBSTANCES
Simultaneous: niacinamide, pyridoxine, riboflavin phosphate, thiamine, ascorbic acid

---

KEY WORDS
fruit

REFERENCE
Schiewe,J.; Mrestani,Y.; Neubert,R. Application and optimization of capillary zone electrophoresis in vitamin analysis, *J.Chromatogr.A*, **1995**, *717*, 255–259.

SAMPLE
Matrix: solutions

CAPILLARY ELECTROPHORESIS
Capillary: 70 cm × 50 μm fused-silica (50 cm to detector) (Polymicro Technologies)
Capillary preparation: Before use treat with 100 mM NaOH, water, and running buffer.
Running buffer: 10 mM pH 8.5 Phosphate buffer
Injection: Electrokinetic injection at 21 kV for 2 s.
Detector: UV 200
Migration time: 5.5
Voltage: 21 kV
Model: laboratory-constructed

REFERENCE
Kelly,J.A.; Reddy,K.R.; Lee,C.S. Mechanistic studies of postcapillary affinity detection for capillary zone electrophoresis based on the biotin-streptavidin system, *Anal.Chem.*, **1997**, *69*, 5152–5158.

Biperiden

Molecular formula: $C_{21}H_{29}NO$
Molecular weight: 311.45
CAS Registry No.: 514-65-8, 1235-82-1 (HCl), 7085-45-2 (lactate)
Merck Index (12th ed.): 1274
Lednicer: 1 47

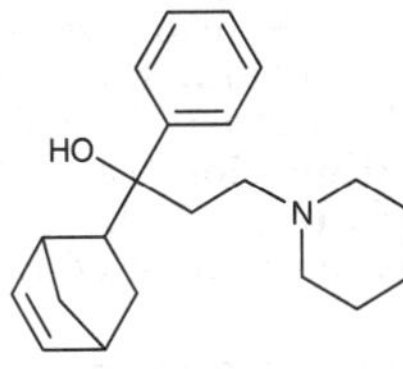

SAMPLE
Matrix: solutions
Sample preparation: Inject an aliquot of a 100 µg/mL solution in water:running buffer 50:50.

CAPILLARY ELECTROPHORESIS
Capillary: 44.5 cm × 50 µm acrylamide-coated fused-silica (Bio-Rad)
Capillary temperature: 30
Running buffer: 100 mM NaH_2PO_4 containing 15 mM gamma-cyclodextrin, adjusted to pH 2.5
  with phosphoric acid
Injection: Electrokinetic injection at 8 kV for 6 s.
Detector: UV 200
Migration time: 14.04
Voltage: 14 kV
Model: Bio-Rad BioFocus 3000

OTHER SUBSTANCES
Also analyzed: albuterol, alprenolol, atenolol, atropine, baclofen, bamethan, benserazide, bisoprolol, bupivacaine, bupranolol, butetamate, carazolol, carbuterol, carvedilol, celiprolol, chloroquine, chlorpheniramine (chlorphenamine), clidinium bromide, clobutinol, disopyramide, dobutamine, flecainide, homatropine, ipratropium bromide, isoproterenol, isothipendyl, ketamine,
  mefloquine, mequitazine, metaproterenol (orciprenaline), metipranolol, nafronyl (naftidrofuryl), nefopam, ofloxacin, orphenadrine, oxomemazine, oxprenolol, phenoxybenzamine, pho-
  ledrinc, pindolol, pirbuterol, prilocaine, promethazine, propafenone, propranolol, sotalol, synephrine, terbutaline, tetrahydrozoline (tetryzoline), tocainide, trihexyphenidyl, trimeprazine
  (alimemazine), trimipramine, tropicamide, verapamil, zopiclone

KEY WORDS
coated capillary; achiral

REFERENCE
Koppenhoefer,B.; Epperlein,U.; Christian,B.; Yibing,J.; Yuying,C.; Bingcheng,L. Separation of enantiomers of
  drugs by capillary electrophoresis. I. γ-Cyclodextrin as chiral solvating agent, *J.Chromatogr.A*, **1995**, *717*,
  181–190.

SAMPLE
Matrix: solutions
Sample preparation: Inject an aliquot of a 100 µg/mL solution in running buffer.

CAPILLARY ELECTROPHORESIS
Capillary: 36 cm × 50 µm fused-silica (31.5 cm to detector) (Yongnian Optical Conductive Fiber
  Plant, China), coated with polyacrylamide
Capillary preparation: No details of the polyacrylamide coating process are provided. However,
  another paper (LC.GC 1997, 15, 40) by this group indicates that they use the procedure of
  Hjertén, thus: Adjust the pH of 20 mL water to 3.5 with acetic acid, add 80 µL 3-(trimethoxysilyl)propyl methacrylate (3-methacryloxypropyltrimethoxysilane), mix, suck into capillary, let
  stand at room temperature for 1 h, remove the solution, wash with water. Fill the capillary
  with a deaerated 3-4% acrylamide solution containing 1 µL/mL N,N,N',N'-tetramethylethyl-
  enediamine and 1 mg/mL potassium persulfate, let stand for 30 min, remove excess solution
  by aspiration, rinse with water, remove water by aspiration, dry at 35° (J. Chromatogr. 1985,
  347, 191).
Capillary temperature: 25

Running buffer: 100 mM NaH_2PO_4 adjusted to pH 2.5 (A) or 100 mM NaH_2PO_4 containing 45 mM hydroxypropyl-gamma-cyclodextrin, adjusted to pH 2.5 (B)
Injection: Electrokinetic injection at 15 kV for 3 s.
Detector: UV 200, UV 210
Migration time: 7.01 (A), 11.09, 11.21 (B, enantiomers)
Voltage: 15 kV
Model: Bio-Rad BioFocus 3000

OTHER SUBSTANCES
Simultaneous: azelastine, carvedilol, clidinium bromide, meclizine (meclozine), mequitazine, ofloxacin, zopiclone

KEY WORDS
chiral; coated capillary

REFERENCE
Koppenhoefer,B.; Epperlein,U.; Xiaofeng,Z.; Bingcheng,L. Separation of enantiomers of drugs by capillary electrophoresis. Part 4: Hydroxypropyl-γ-cyclodextrin as chiral solvating agent, *Electrophoresis*, **1997**, *18*, 924–930.

SAMPLE
Matrix: solutions
Sample preparation: Inject an aliquot of a 100 μg/mL solution in running buffer.

CAPILLARY ELECTROPHORESIS
Capillary: 30 cm × 50 μm fused-silica (25.5 cm to detector), coated with polyacrylamide
Capillary preparation: Adjust the pH of 20 mL water to 3.5 with acetic acid, add 80 μL 3-(trimethoxysilyl)propyl methacrylate (3-methacryloxypropyltrimethoxysilane), mix, suck into capillary, let stand at room temperature for 1 h, remove the solution, wash with water. Fill the capillary with a deaerated 3-4% acrylamide solution containing 1 μL/mL N,N,N',N'-tetramethylethylenediamine and 1 mg/mL potassium persulfate, let stand for 30 min, remove excess solution by aspiration, rinse with water, remove water by aspiration, dry at 35° (J. Chromatogr. 1985, 347, 191).
Capillary temperature: 25
Running buffer: 100 mM NaH_2PO_4 containing 45 mM hydroxypropyl-gamma- cyclodextrin, adjusted to pH 2.5 with phosphoric acid
Injection: Electrokinetic injection at 15 kV for 3 s.
Detector: UV 200
Voltage: 15 kV
Model: Bio-Rad BioFocus 3000

KEY WORDS
chiral; coated capillary; comparison with the use of other cyclodextrins; this running buffer gave the greatest enantiomeric separation.; α=1.011

REFERENCE
Lin,B.; Zhu,X.; Koppenhoefer,B.; Epperlein,U. Investigation of 123 chiral drugs by cyclodextrin-modified capillary electrophoresis, *LC.GC*, **1997**, *15*, 40–46.

SAMPLE
Matrix: solutions

CAPILLARY ELECTROPHORESIS
Capillary: 29-36 cm × 50 μm fused-silica (24.5-31.5 cm to detector) (Yongnian Optical Conductive Fiber Plant, China) coated with polyacrylamide
Capillary preparation: Coat capillary as follows. Adjust the pH of 20 mL water to 3.5 with acetic acid, add 80 μL 3-(trimethoxysilyl)propyl methacrylate (3-methacryloxypropyltrimethoxysilane), mix, suck into capillary, let stand at room temperature for 1 h, remove the solution, wash with water. Fill the capillary with a deaerated 3-4% acrylamide solution containing 1 μL/mL N,N,N',N'-tetramethylethylenediamine and 1 mg/mL potassium persulfate, let stand for 30 min, remove excess solution by aspiration, rinse with water, remove water by aspiration, dry at 35° (J. Chromatogr. 1985, 347, 191).

Capillary temperature: 25
Running buffer: 100 mM pH 2.5 NaH_2PO_4 (A) or 100 mM pH 2.5 NaH_2PO_4 containing 45 mM hydroxypropyl-α-cyclodextrin (Wacker, Munich) (B)
Injection: Electromigration at 15 kV for 3 s.
Detector: UV 200; UV 210
Migration time: 7.01 (A); 14.96 (B) (no separation of enantiomers)
Voltage: 15 kV
Model: Bio-Focus 3000

OTHER SUBSTANCES
Also analyzed: albuterol (salbutamol), alprenolol, amorolfine, atenolol, atropine, azelastine, baclofen, bamethan, benproperine, benserazide, bisoprolol, brompheniramine, bupivacaine, bupranolol, butamirate, butethamate, carazolol, carbuterol, carteolol, carvedilol, celiprolol, chloroquine, chlorpheniramine, chlorphenoxamine, cicletanine, clenbuterol, clidinium bromide, clobutinol, dimethindene, dipivefrin, disopyramide, dobutamine, doxylamine, fendiline, flecainide, gallopamil, homatropine, ipratropium bromide, isoproterenol (isoprenaline), isothipendyl, ketamine, meclizine, mefloquine, mepindolol, mequitazine, metaclazepam, metaproterenol (orciprenaline), metipranolol, metoprolol, nafronyl (naftidrofuryl), nefopam, nicardipine, norfenefrine, ofloxacin, ornidazole, orphenadrine, oxomemazine, oxprenolol, oxybutynin, phenoxybenzamine, phenylpropanolamine, pholedrine, pindolol, pirbuterol, prilocaine, procyclidine, promethazine, propafenone, propranolol, reproterol, sotalol, sulpride, synephrine, talinolol, terbutaline, tetrahydrozoline (tetryzoline), theodrenaline, tioconazole, tocainide, trihexyphenidyl, trimeprazine (alimemazine), trimipramine, tropicamide, verapamil, zopiclone

KEY WORDS
coated capillary

REFERENCE
Koppenhoefer,B.; Eperlein,U.; Schlunk,R.; Zhu,X.; Lin,B. Separation of enantiomers of drugs by capillary electrophoresis. V. Hydroxypropyl-α-cyclodextrin as chiral solvating agent, *J.Chromatogr.A*, **1998**, *793*, 153–164.

Bisoprolol

Molecular formula: $C_{18}H_{31}NO_4$
Molecular weight: 325.45
CAS Registry No.: 66722-44-9, 104344-23-2 (fumarate)
Merck Index (12th ed.): 1336
Lednicer: 4 28

SAMPLE
Matrix: solutions
Sample preparation: Inject an aliquot of a 100 µg/mL solution in water:running buffer 50:50.

CAPILLARY ELECTROPHORESIS
Capillary: 44.5 cm × 50 µm acrylamide-coated fused-silica (Bio-Rad)
Capillary temperature: 30
Running buffer: 100 mM NaH_2PO_4 containing 15 mM gamma-cyclodextrin, adjusted to pH 2.5 with phosphoric acid
Injection: Electrokinetic injection at 8 kV for 6 s.
Detector: UV 200
Migration time: 10.90
Voltage: 14 kV
Model: Bio-Rad BioFocus 3000

OTHER SUBSTANCES
Also analyzed: albuterol, alprenolol, atenolol, atropine, baclofen, bamethan, benserazide, biperiden, bupivacaine, bupranolol, butetamate, carazolol, carbuterol, carvedilol, celiprolol, chloroquine, chlorpheniramine (chlorphenamine), clidinium bromide, clobutinol, disopyramide, dobutamine, flecainide, homatropine, ipratropium bromide, isoproterenol, isothipendyl, ketamine, mefloquine, mequitazine, metaproterenol (orciprenaline), metipranolol, nafronyl (naftidrofuryl), nefopam, ofloxacin, orphenadrine, oxomemazine, oxprenolol, phenoxybenzamine, pholedrine, pindolol, pirbuterol, prilocaine, promethazine, propafenone, propranolol, sotalol, synephrine, terbutaline, tetrahydrozoline (tetryzoline), tocainide, trihexyphenidyl, trimeprazine (alimemazine), trimipramine, tropicamide, verapamil, zopiclone

KEY WORDS
coated capillary; achiral

REFERENCE
Koppenhoefer,B.; Epperlein,U.; Christian,B.; Yibing,J.; Yuying,C.; Bingcheng,L. Separation of enantiomers of drugs by capillary electrophoresis. I. γ-Cyclodextrin as chiral solvating agent, *J.Chromatogr.A*, **1995**, *717*, 181–190.

SAMPLE
Matrix: solutions
Sample preparation: Inject an aliquot of a 100 μg/mL solution in running buffer.

CAPILLARY ELECTROPHORESIS
Capillary: 29 cm × 50 μm fused-silica (24.5 cm to detector) (Yongnian Optical Conductive Fiber Plant, China), coated with polyacrylamide
Capillary preparation: No details of the polyacrylamide coating process are provided. However, another paper (LC.GC 1997, 15, 40) by this group indicates that they use the procedure of Hjertén, thus: Adjust the pH of 20 mL water to 3.5 with acetic acid, add 80 μL 3-(trimethoxysilyl)propyl methacrylate (3-methacryloxypropyltrimethoxysilane), mix, suck into capillary, let stand at room temperature for 1 h, remove the solution, wash with water. Fill the capillary with a deaerated 3-4% acrylamide solution containing 1 μL/mL N,N,N',N'-tetramethylethylenediamine and 1 mg/mL potassium persulfate, let stand for 30 min, remove excess solution by aspiration, rinse with water, remove water by aspiration, dry at 35° (J. Chromatogr. 1985, 347, 191).
Capillary temperature: 25
Running buffer: 100 mM NaH_2PO_4 adjusted to pH 2.5
Injection: Electrokinetic injection at 15 kV for 3 s.
Detector: UV 200, UV 210
Migration time: 6.81
Voltage: 15 kV
Model: Bio-Rad BioFocus 3000

OTHER SUBSTANCES
Simultaneous: albuterol, alprenolol, atenolol, baclofen, bamethan, benproperine, benserazide, bupranolol, butamirate, butethamate, carbuterol, celiprolol, clenbuterol, clobutinol, dipivefrin, isoproterenol (isoprenaline), metaproterenol (orciprenaline), metipranolol, metoprolol, norfenefrine, ornidazole, oxprenolol, phenylpropanolamine, pholedrine, pirbuterol, prilocaine, procyclidine, sotalol, synephrine, terbutaline, tocainide

KEY WORDS
coated capillary

REFERENCE
Koppenhoefer,B.; Epperlein,U.; Xiaofeng,Z.; Bingcheng,L. Separation of enantiomers of drugs by capillary electrophoresis. Part 4: Hydroxypropyl-γ-cyclodextrin as chiral solvating agent, *Electrophoresis*, **1997**, *18*, 924–930.

SAMPLE
Matrix: solutions

CAPILLARY ELECTROPHORESIS

Capillary: 29-36 cm × 50 μm fused-silica (24.5-31.5 cm to detector) (Yongnian Optical Conductive Fiber Plant, China) coated with polyacrylamide

Capillary preparation: Coat capillary as follows. Adjust the pH of 20 mL water to 3.5 with acetic acid, add 80 μL 3-(trimethoxysilyl)propyl methacrylate (3-methacryloxypropyltrimethoxysilane), mix, suck into capillary, let stand at room temperature for 1 h, remove the solution, wash with water. Fill the capillary with a deaerated 3-4% acrylamide solution containing 1 μL/mL N,N,N',N'-tetramethylethylenediamine and 1 mg/mL potassium persulfate, let stand for 30 min, remove excess solution by aspiration, rinse with water, remove water by aspiration, dry at 35° (J. Chromatogr. 1985, 347, 191).

Capillary temperature: 25

Running buffer: 100 mM pH 2.5 NaH_2PO_4 (A) or 100 mM pH 2.5 NaH_2PO_4 containing 45 mM hydroxypropyl-α-cyclodextrin (Wacker, Munich) (B)

Injection: Electromigration at 15 kV for 3 s.

Detector: UV 200; UV 210

Migration time: 6.81 (A); 10.97 (B) (no separation of enantiomers)

Voltage: 15 kV

Model: Bio-Focus 3000

OTHER SUBSTANCES

Also analyzed: albuterol (salbutamol), alprenolol, amorolfine, atenolol, atropine, azelastine, baclofen, bamethan, benproperine, benserazide, biperiden, brompheniramine, bupivacaine, bupranolol, butamirate, butethamate, carazolol, carbuterol, carteolol, carvedilol, celiprolol, chloroquine, chlorpheniramine, chlorphenoxamine, cicletanine, clenbuterol, clidinium bromide, clobutinol, dimethindene, dipivefrin, disopyramide, dobutamine, doxylamine, fendiline, flecainide, gallopamil, homatropine, ipratropium bromide, isoproterenol (isoprenaline), isothipendyl, ketamine, meclizine, mefloquine, mepindolol, mequitazine, metaclazepam, metaproterenol (orciprenaline), metipranolol, metoprolol, nafronyl (naftidrofuryl), nefopam, nicardipine, norfenefrine, ofloxacin, ornidazole, orphenadrine, oxomemazine, oxprenolol, oxybutynin, phenoxybenzamine, phenylpropanolamine, pholedrine, pindolol, pirbuterol, prilocaine, procyclidine, promethazine, propafenone, propranolol, reproterol, sotalol, sulpride, synephrine, talinolol, terbutaline, tetrahydrozoline (tetryzoline), theodrenaline, tioconazole, tocainide, trihexyphenidyl, trimeprazine (alimemazine), trimipramine, tropicamide, verapamil, zopiclone

KEY WORDS

coated capillary

REFERENCE

Koppenhoefer,B.; Eperlein,U.; Schlunk,R.; Zhu,X.; Lin,B. Separation of enantiomers of drugs by capillary electrophoresis. V. Hydroxypropyl-α-cyclodextrin as chiral solvating agent, *J.Chromatogr.A*, **1998**, *793*, 153–164.

Bretylium tosylate

Molecular formula: $C_{18}H_{24}BrNO_3S$

Molecular weight: 414.36

CAS Registry No.: 61-75-6, 59-41-6 (free base)

Merck Index (12th ed.): 1395

Lednicer: 1 55

SAMPLE

Matrix: solutions

Sample preparation: Inject an aliquot of a 100 μg/mL solution in 50 mM sodium dodecyl sulfate.

CAPILLARY ELECTROPHORESIS

Capillary: 57 cm × 75 μm fused-silica (50 cm to detector) (Beckman)

Capillary preparation: Rinse capillary with 100 mM NaOH and running buffer between each run.

Capillary temperature: 30
Running buffer: Isopropanol:100 mM pH 8.1 borate buffer containing 50 mM sodium dodecyl
    sulfate 10:90
Injection: Pressure injection for 5 s
Detector: UV 200
Migration time: 10.2 (tosylate ion at 9.7)
Voltage: +25 kV
Model: Beckman P/ACE 5510

OTHER SUBSTANCES
Simultaneous: amiodarone, disopyramide, lidocaine, phenytoin, propafenone, quinidine

REFERENCE
Bretnall,A.E.; Clarke,G.S. Investigation and optimisation of the use of organic modifiers in micellar electro-
    kinetic chromatography, *J.Chromatogr.A*, **1995**, *716*, 49–55.

SAMPLE
Matrix: solutions

CAPILLARY ELECTROPHORESIS
Capillary: 57 cm × 75 μm fused-silica (50 cm to detector) (Beckman)
Capillary preparation: Before each run rinse capillary with 100 mM NaOH and running buffer.
Capillary temperature: 30
Running buffer: Acetone:100 mM pH 8.1 borate buffer containing 50 mM sodium dodecyl sulfate
    15:85 (A) or isopropanol:100 mM pH 8.1 borate buffer containing 50 mM sodium dodecyl sulfate
    10:90 (B)
Injection: Pressure injection for 5-10 s.
Detector: UV 200
Migration time: 8.5 (A), 13.5 (B)
Voltage: 25 kV
Model: Beckman P/ACE 5510

OTHER SUBSTANCES
Simultaneous: acebutolol, amiodarone, atenolol, captopril, diltiazem, disopyramide, lidocaine,
    lisinopril, metoprolol, nicardipine, nifedipine, phenytoin, pindolol, propafenone, propranolol,
    quinidine, sotalol, timolol, p-toluenesulfonic acid, verapamil

KEY WORDS
only running buffer A separates all compounds

REFERENCE
Bretnall,A.E.; Clarke,G.S. Selectivity of capillary electrophoresis for the analysis of cardiovascular drugs,
    J.Chromatogr.A, **1996**, *745*, 145–154.

Brodimoprim

Molecular formula: $C_{13}H_{15}BrN_4O_2$
Molecular weight: 339.19
CAS Registry No.: 56518-41-3
Merck Index (12th ed.): 1401

SAMPLE
Matrix: solutions
Sample preparation: Prepare a 2.5 μg/mL solution in water, filter (0.45 μm), inject an aliquot.

CAPILLARY ELECTROPHORESIS
Capillary: 71.2 cm × 50 μm fused-silica (52 cm to detector) (Applied Biosystems)

Capillary preparation: Between runs purge the capillary with 100 mM NaOH for 2 min and with running buffer for 2 min. At the start of each day purge capillary with 100 mM NaOH for 2 min, with water for 2 min, and with running buffer for 2 min
Capillary temperature: 30
Running buffer: 250 mM pH 2.1 Phosphate buffer
Injection: Vacuum injection for 10 s (40 nL)
Detector: UV 225
Migration time: 24.375
Voltage: 13 kV
Model: Applied Biosystems Model 270 A

OTHER SUBSTANCES
Simultaneous: aditoprim, diaveridine, metioprim, ormetoprim, pyrimethamine, tetroxoprim, trimethoprim

REFERENCE
Cao,J.; Cross,R.F. The separation of dihydrofolate reductase inhibitors and the determination of pKa,1 values by capillary zone electrophoresis, *J.Chromatogr.A*, **1995**, *695*, 297–308.

Bromazepam

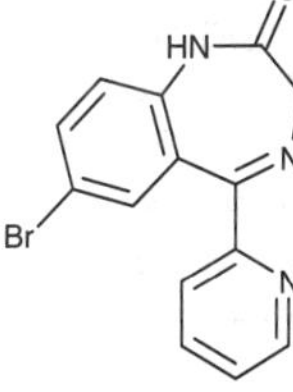

Molecular formula: $C_{14}H_{10}BrN_3O$
Molecular weight: 316.16
CAS Registry No.: 1812-30-2
Merck Index (12th ed.): 1406

SAMPLE
Matrix: blood
Sample preparation: Condition a Sep-Pak C18 SPE cartridge with MeOH, water, and 100 mM pH 9.0 borate buffer. 1 mL Serum + 3 mL 100 mM pH 9.0 borate buffer, mix, add to the SPE cartridge, wash with 3 mL 100 mM pH 9.0 borate buffer, wash with three 2 mL portions of water, wash with 1 mL MeCN:water 5:95, wash with 1 mL hexane, dry under full vacuum. Elute with 4 mL dichloromethane, evaporate the eluate to dryness under a stream of nitrogen, reconstitute with MeOH:water 15:85, filter (0.45 μm), inject an aliquot.

CAPILLARY ELECTROPHORESIS
Capillary: 72 cm × 50 μm fused-silica (50 cm to detector)
Capillary preparation: Before each run wash capillary with 100 mM NaOH for 4 min and with running buffer for 6 min.
Capillary temperature: 25
Running buffer: MeOH:5 mM pH 8.5 phosphate/borate buffer 15:85 containing 50 mM sodium dodecyl sulfate
Injection: Vacuum injection for 4 s.
Detector: UV 200
Migration time: 15
Internal standard: secobarbital (9)
Voltage: 25 kV
Model: Applied Biosystems Model 270A or Beckman P/ACE 5000
Limit of detection: 25 ng/mL

OTHER SUBSTANCES
Extracted: diazepam, estazolam, flurazepam, nitrazepam, triazolam

KEY WORDS
SPE; serum

REFERENCE
Tomita,M.; Okuyama,T. Application of capillary electrophoresis to the simultaneous screening and quantitation of benzodiazepines, *J.Chromatogr.B*, **1996**, *678*, 331–337.

SAMPLE
Matrix: solutions
Sample preparation: Inject an aliquot of a 5 μg/mL solution in running buffer.

CAPILLARY ELECTROPHORESIS
Capillary: 87 cm × 75 μm fused-silica (80 cm to detector) (Beckman)
Capillary preparation: Between runs rinse capillary with running buffer for 2 min then equilibrate for 5 min.
Capillary temperature: 35
Running buffer: 20 mM pH 7 Buffer containing 15 mM sodium cholate and 35 mM sodium deoxycholate (Buffer was 20 mM sodium borate adjusted to pH 7.0 with 20 mM NaH_2PO_4.)
Injection: Pressure injection for 2 s.
Detector: UV 214
Migration time: 14.5
Voltage: 20 kV
Model: Beckman P/ACE 2100

OTHER SUBSTANCES
Simultaneous: chlordiazepoxide, clobazam, clonazepam, diazepam, flunitrazepam, flurazepam, halazepam, lorazepam, lormetazepam, nitrazepam, nordazepam, temazepam

REFERENCE
Boonkerd,S.; Detaevernier,M.R.; Vindevogel,J.; Michotte,Y. Migration behaviour of benzodiazepines in micellar electrokinetic chromatography, *J.Chromatogr.A*, **1996**, *756*, 279–286.

SAMPLE
Matrix: solutions

CAPILLARY ELECTROPHORESIS
Capillary: 60 cm × 50 μm fused-silica (40 cm to detector), C18 coated (Supelco)
Capillary preparation: Rinse new capillaries with 250 μL 1 M NaOH, 250 μL 100 mM NaOH, water, MeOH, water, and running buffer then condition with running buffer for at least 15 min. Carry out a similar procedure between runs.
Running buffer: 100 mM pH 8.5 Borate buffer containing 25 mM sodium dodecyl sulfate and 5 M urea
Injection: Electrokinetic injection at 5 kV for 5 s.
Detector: UV 254
Migration time: 17
Voltage: 18 kV
Model: Jasco

OTHER SUBSTANCES
Simultaneous: alprazolam, amobarbital, barbital, clonazepam, clotiazepam, cloxazolam, diazepam, estazolam, etizolam, fludiazepam, flunitrazepam, flurazepam, haloxazolam, medazepam, mephobarbital, metharbital, nimetazepam, nitrazepam, oxazepam, pentobarbital, phenobarbital, secobarbital, triazolam

KEY WORDS
coated capillary

REFERENCE
Jinno,K.; Han,Y.; Nakamura,M. Analysis of anxiolytic drugs by capillary electrophoresis with bare and coated capillaries, *J.Capillary Electrophor.*, **1996**, *3*, 139–145.

SAMPLE
Matrix: solutions

CAPILLARY ELECTROPHORESIS
Capillary: 47 cm × 75 μm fused-silica (40 cm to detector) (Beckman)
Capillary preparation: Rinse with running buffer for 1 min before injection. Wash with 100 mM NaOH for 1 min and with water for 1 min between injections.

Capillary temperature: 30
Running buffer: THF:buffer 1:99 (Buffer was 50 mM pH 9.2 borate containing 50 mM sodium dodecyl sulfate, 2 M urea, and 20 mM gamma-cyclodextrin.)
Injection: Pressure injection at 300 kPa for 5 s.
Detector: UV 254
Migration time: 12
Voltage: 15 kV
Model: Beckman P/ACE 2000

OTHER SUBSTANCES
Simultaneous: alprazolam, chlordiazepoxide, clobazam, clonazepam, clorazepate, oxazepam, prazepam, triazolam

REFERENCE
Renou-Gonnord,M.F.; David,K. Optimized micellar electrokinetic chromatographic separation of benzodiazepines, *J.Chromatogr.A*, **1996**, *735*, 249–261.

SAMPLE
Matrix: solutions
Sample preparation: Inject an aliquot of a 100 µg/mL solution in running buffer.

CAPILLARY ELECTROPHORESIS
Capillary: 60 cm × 50 µm acrylamide-coated fused-silica (40 cm to detector) (Supelco)
Capillary preparation: Adjust the pH of 20 mL water to 3.5 with acetic acid, add 80 µL 3-(trimethoxysilyl)propyl methacrylate (3-methacryloxypropyltrimethoxysilane), mix, suck into capillary, let stand at room temperature for 1 h, remove the solution, wash with water. Fill the capillary with a deaerated 3-4% acrylamide solution containing 1 µL/mL N,N,N',N'-tetramethylethylenediamine and 1 mg/mL ammonium persulfate, let stand for 3 h, remove excess solution by aspiration, rinse with water, remove water by aspiration, dry at 35° (cf. J. Chromatogr. 1985, 347, 191).
Running buffer: MeCN:buffer 5:95 (Buffer was 100 mM borate containing 5 M urea and 10 mM sodium dodecyl sulfate, adjusted to pH 8.5 with phosphate.)
Injection: Electrokinetic injection at 5 kV for 5 s.
Detector: UV 254
Migration time: 25
Voltage: 18 kV
Model: Jasco Model 870-CE

OTHER SUBSTANCES
Simultaneous: alprazolam, amobarbital, barbital, clonazepam, clotiazepam, cloxazolam, diazepam, estazolam, etizolam, fludiazepam, flunitrazepam, flurazepam, haloxazolam, medazepam, mephobarbital, metharbital, nimetazepam, nitrazepam, oxazepam, pentobarbital, phenobarbital, secobarbital, triazolam

KEY WORDS
coated capillary; detector at anode

REFERENCE
Jinno,K.; Han,Y.; Sawada,H.; Taniguchi,M. Capillary electrophoretic separation of toxic drugs using a polyacrylamide-coated capillary, *Chromatographia*, **1997**, *46*, 309–314.

Brompheniramine

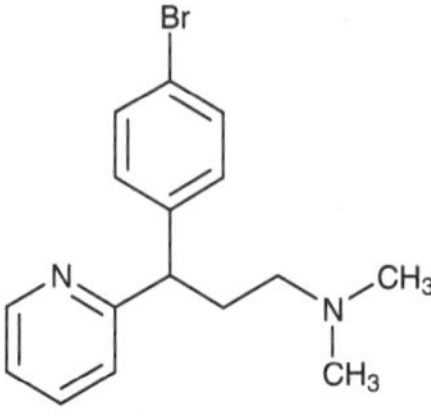

Molecular formula: $C_{16}H_{19}BrN_2$
Molecular weight: 319.24
CAS Registry No.: 86-22-6, 132-21-8 (d-form), 980-71-2 (maleate), 2391-03-9 (d-form maleate)
Merck Index (12th ed.): 1467
Lednicer: 1 77

SAMPLE
Matrix: blood, urine
Sample preparation: Adjust pH of 2 mL urine or plasma to 10.5 with aqueous NaOH, extract gently with chloroform:isopropanol 90:10. Remove the organic layer and evaporate it to dryness under a stream of nitrogen, reconstitute the residue in 50 (urine) or 75 (plasma) μL running buffer, filter, inject an aliquot.

CAPILLARY ELECTROPHORESIS
Capillary: 60 cm × 75 μm AccuSep uncoated silica (52.5 cm to detector) (Waters)
Capillary preparation: At the beginning of each day purge with 500 mM KOH for 5 min, with water for 5 min, and with running buffer for 10 min.
Running buffer: 50 mM NaH_2PO_4 adjusted to pH 2.35 with phosphoric acid
Injection: Hydrostatic injection at 15 cm
Detector: UV 214
Migration time: 4.70
Voltage: 22 kV
Current: 135-145 μA
Model: Waters Quanta 4000

OTHER SUBSTANCES
Extracted: acepromazine, amphetamine, benzocaine, butacaine, codeine, diazepam, doxapram, lidocaine, medazepam, methamphetamine, methapyrilene, methaqualone, phenmetrazine, procaine, tetrahydrozoline
Simultaneous: meclizine

KEY WORDS
plasma

REFERENCE
Chee,G.L.; Wan,T.S.M. Reproducible and high-speed separation of basic drugs by capillary zone electrophoresis, *J.Chromatogr.*, **1993**, *612*, 172–177.

SAMPLE
Matrix: solutions
Sample preparation: Prepare a solution in running buffer, inject an aliquot.

CAPILLARY ELECTROPHORESIS
Capillary: 60 cm × 75 μm fused-silica (52.4 cm to detector)
Capillary preparation: After each run flush with 500 mM KOH for 2-3 min, flush with water, fill with running buffer
Running buffer: 10 mM Na_2HPO_4 containing 2% heparin sodium (MW 10000, 11% S, Scientific Protein Laboratories, Waunakee, WI) adjusted to pH 5 with phosphoric acid
Injection: Hydrostatic injection
Detector: UV 214
Migration time: 17.3 (first (D) enantiomer, R_S 3.3)
Model: Waters Quanta 4000

OTHER SUBSTANCES
Also analyzed: anabasine, bupivacaine, carbinoxamine, chlorcyclizine, chloroquine, chlorpheniramine, dimethindene, doxylamine, enpiroline, halofantrine, hydroxychloroquine, indapamide, mefloquine, nornicotine, pheniramine, primaquine, promethazine, quinacrine, tetramisole

KEY WORDS
chiral

REFERENCE
Stalcup,A.M.; Agyei,N.M. Heparin: a chiral mobile-phase additive for capillary zone electrophoresis, *Anal.Chem.*, **1994**, *66*, 3054–3059.

SAMPLE
Matrix: solutions
Sample preparation: Inject an aliquot of a solution in running buffer.

CAPILLARY ELECTROPHORESIS
Capillary: 60 cm × 75 μm fused-silica (52.4 cm to detector)
Capillary preparation: After each run flush with 500 mM KOH for 2-3 min then with water.
Running buffer: 10 mM pH 3.8 Phosphate buffer containing 2% sulfated cyclodextrin (ds 7-10)
Injection: Hydrostatic injection.
Detector: UV 214
Migration time: 8.26, 8.44 (enantiomers)
Voltage: 15 kV
Model: Waters Quanta 4000

OTHER SUBSTANCES
Also analyzed: acebutolol, alprenolol, aminoglutethimide, bupivacaine, bupropion, canadine, carbinoxamine, chloroquine, chlorpheniramine, dimethindene, disopyramide, doxylamine, hydroxychloroquine, idazoxan, isoxsuprine, ketamine, mepenzolate, mepivacaine, methoxyphenamine, mexiletine, midodrine, nefopam, orphenadrine, oxprenolol, oxyphencyclimine, pheniramine, phensuximide, pindolol, piperoxan, terbutaline, tetramisole, tolperisone, tranylcypromine, trihexyphenidyl, trimipramine, verapamil, warfarin

KEY WORDS
chiral; detector at anode

REFERENCE
Stalcup,A.M.; Gahm,K.H. Application of sulfated cyclodextrins to chiral separations by capillary zone electrophoresis, *Anal.Chem.*, **1996**, *68*, 1360–1368.

SAMPLE
Matrix: solutions
Sample preparation: Inject an aliquot of a 100 μg/mL solution in running buffer.

CAPILLARY ELECTROPHORESIS
Capillary: 30 cm × 50 μm fused-silica (25.5 cm to detector) (Yongnian Optical Conductive Fiber Plant, China), coated with polyacrylamide
Capillary preparation: No details of the polyacrylamide coating process are provided. However, another paper (LC.GC 1997, 15, 40) by this group indicates that they use the procedure of Hjertén, thus: Adjust the pH of 20 mL water to 3.5 with acetic acid, add 80 μL 3-(trimethoxysilyl)propyl methacrylate (3-methacryloxypropyltrimethoxysilane), mix, suck into capillary, let stand at room temperature for 1 h, remove the solution, wash with water. Fill the capillary with a deaerated 3-4% acrylamide solution containing 1 μL/mL N,N,N',N'-tetramethylethylenediamine and 1 mg/mL potassium persulfate, let stand for 30 min, remove excess solution by aspiration, rinse with water, remove water by aspiration, dry at 35° (J. Chromatogr. 1985, 347, 191).
Capillary temperature: 25
Running buffer: 100 mM NaH_2PO_4 adjusted to pH 2.5
Injection: Electrokinetic injection at 15 kV for 3 s.
Detector: UV 200, UV 210
Migration time: 2.71
Voltage: 15 kV
Model: Bio-Rad BioFocus 3000

OTHER SUBSTANCES
Simultaneous: amorolfine, bupivacaine, carteolol, chloroquine, chlorpheniramine, chlorphenoxamine, disopyramide, dobutamine, doxylamine, flecainide, gallopamil, ketamine, mepindolol, orphenadrine, oxybutynin, phenoxybenzamine, pindolol, propafenone, propranolol, sulpiride, talinolol, tropicamide, verapamil

KEY WORDS
coated capillary

REFERENCE
Koppenhoefer,B.; Epperlein,U.; Xiaofeng,Z.; Bingcheng,L. Separation of enantiomers of drugs by capillary electrophoresis. Part 4: Hydroxypropyl-γ-cyclodextrin as chiral solvating agent, *Electrophoresis*, **1997**, *18*, 924–930.

SAMPLE
Matrix: solutions
Sample preparation: Inject an aliquot of a 100 μg/mL solution in running buffer.

CAPILLARY ELECTROPHORESIS
Capillary: 30 cm × 50 μm fused-silica (25.5 cm to detector), coated with polyacrylamide
Capillary preparation: Adjust the pH of 20 mL water to 3.5 with acetic acid, add 80 μL 3-(trimethoxysilyl)propyl methacrylate (3-methacryloxypropyltrimethoxysilane), mix, suck into capillary, let stand at room temperature for 1 h, remove the solution, wash with water. Fill the capillary with a deaerated 3-4% acrylamide solution containing 1 μL/mL N,N,N',N'-tetramethylethylenediamine and 1 mg/mL potassium persulfate, let stand for 30 min, remove excess solution by aspiration, rinse with water, remove water by aspiration, dry at 35° (J. Chromatogr. 1985, 347, 191).
Capillary temperature: 25
Running buffer: 100 mM NaH_2PO_4 containing 45 mM hydroxypropyl-β-cyclodextrin, adjusted to pH 2.5 with phosphoric acid
Injection: Electrokinetic injection at 15 kV for 3 s.
Detector: UV 200
Voltage: 15 kV
Model: Bio-Rad BioFocus 3000

KEY WORDS
chiral; coated capillary; comparison with the use of other cyclodextrins; this running buffer gave the greatest enantiomeric separation.; α=1.019

REFERENCE
Lin,B.; Zhu,X.; Koppenhoefer,B.; Epperlein,U. Investigation of 123 chiral drugs by cyclodextrin-modified capillary electrophoresis, *LC.GC*, **1997**, *15*, 40–46.

SAMPLE
Matrix: solutions

CAPILLARY ELECTROPHORESIS
Capillary: 57 cm × 75 μm fused-silica (34 cm to detector) (Polymicro Technologies)
Capillary preparation: Before each run rinse capillary with 100 mM NaOH for 5 min, with water for 5 min and with running buffer for 5 min. Condition new capillaries with 100 mM NaOH for 10 min, with water for 5 min and with running buffer for 5 min.
Running buffer: 10 mM pH 3.84 Phosphate buffer containing 7 mM sulfated-β-cyclodextrin
Injection: Hydrodynamic injection at 14 cm for 25 s.
Detector: UV 214
Migration time: 7, 9 (enantiomers)
Voltage: 12 kV
Model: laboratory-constructed

OTHER SUBSTANCES
Also analyzed: chloroquine, chlorpheniramine, doxylamine, pheniramine

KEY WORDS
chiral

REFERENCE
Jin,L.J.; Li,S.F.Y. Comparison of chiral recognition capabilities of cyclodextrins for the separation of basic drugs in capillary zone electrophoresis, *J.Chromatogr.B*, **1998**, *708*, 257–266.

SAMPLE
Matrix: solutions

CAPILLARY ELECTROPHORESIS
Capillary: 29-36 cm × 50 μm fused-silica (24.5-31.5 cm to detector) (Yongnian Optical Conductive Fiber Plant, China) coated with polyacrylamide

Capillary preparation: Coat capillary as follows. Adjust the pH of 20 mL water to 3.5 with acetic acid, add 80 μL 3-(trimethoxysilyl)propyl methacrylate (3-methacryloxypropyltrimethoxysilane), mix, suck into capillary, let stand at room temperature for 1 h, remove the solution, wash with water. Fill the capillary with a deaerated 3-4% acrylamide solution containing 1 μL/mL N,N,N',N'-tetramethylethylenediamine and 1 mg/mL potassium persulfate, let stand for 30 min, remove excess solution by aspiration, rinse with water, remove water by aspiration, dry at 35° (J. Chromatogr. 1985, 347, 191).

Capillary temperature: 25

Running buffer: 100 mM pH 2.5 NaH_2PO_4 (A) or 100 mM pH 2.5 NaH_2PO_4 containing 45 mM hydroxypropyl-α-cyclodextrin (Wacker, Munich) (B)

Injection: Electromigration at 15 kV for 3 s.

Detector: UV 200; UV 210

Migration time: 2.71 (A); 8.48, 8.57 (B) (enantiomers)

Voltage: 15 kV

Model: Bio-Focus 3000

OTHER SUBSTANCES
Also analyzed: albuterol (salbutamol), alprenolol, amorolfine, atenolol, atropine, azelastine, baclofen, bamethan, benproperine, benserazide, biperiden, bisoprolol, bupivacaine, bupranolol, butamirate, butethamate, carazolol, carbuterol, carteolol, carvedilol, celiprolol, chloroquine, chlorpheniramine, chlorphenoxamine, cicletanine, clenbuterol, clidinium bromide, clobutinol, dimethindene, dipivefrin, disopyramide, dobutamine, doxylamine, fendiline, flecainide, gallopamil, homatropine, ipratropium bromide, isoproterenol (isoprenaline), isothipendyl, ketamine, meclizine, mefloquine, mepindolol, mequitazine, metaclazepam, metaproterenol (orciprenaline), metipranolol, metoprolol, nafronyl (naftidrofuryl), nefopam, nicardipine, norfenefrine, ofloxacin, ornidazole, orphenadrine, oxomemazine, oxprenolol, oxybutynin, phenoxybenzamine, phenylpropanolamine, pholedrine, pindolol, pirbuterol, prilocaine, procyclidine, promethazine, propafenone, propranolol, reproterol, sotalol, sulpride, synephrine, talinolol, terbutaline, tetrahydrozoline (tetryzoline), theodrenaline, tioconazole, tocainide, trihexyphenidyl, trimeprazine (alimemazine), trimipramine, tropicamide, verapamil, zopiclone

KEY WORDS
coated capillary; chiral

REFERENCE
Koppenhoefer,B.; Eperlein,U.; Schlunk,R.; Zhu,X.; Lin,B. Separation of enantiomers of drugs by capillary electrophoresis. V. Hydroxypropyl-α-cyclodextrin as chiral solvating agent, *J.Chromatogr.A*, **1998**, *793*, 153–164.

Brotizolam

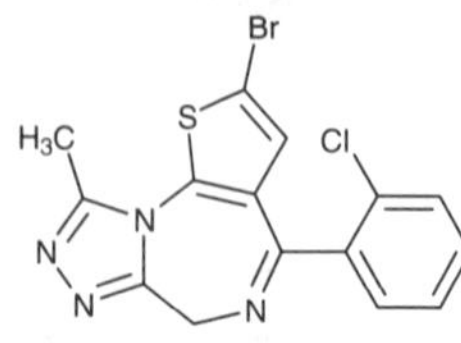

Molecular formula: $C_{15}H_{10}BrClN_4S$
Molecular weight: 393.69
CAS Registry No.: 57801-81-7
Merck Index (12th ed.): 1472
Lednicer: 4 219

SAMPLE
Matrix: solutions
Sample preparation: Prepare a 1-50 μg/mL solution in MeOH:water 20:80, inject an aliquot.

CAPILLARY ELECTROPHORESIS
Capillary: 44 cm × 50 μm fused-silica
Capillary preparation: After each run wash capillary with water for 2 min and buffer for 3 min. At the beginning of each day equilibrate capillary with running buffer for 10 min. Condition a new capillary by washing with 1 M NaOH, 100 mM NaOH, water, and running buffer.
Capillary temperature: 25
Running buffer: MeOH:buffer 20:80 (Buffer was 75 mm pH 9.0 Glycine containing 250 mM triethanolamine and 25 mM sodium dodecyl sulfate.)
Injection: Hydrodynamic injection for 5 s
Detector: UV 235
Migration time: 10.1
Voltage: 25 kV
Current: about 15 μA
Model: Spectraphysics Spectraphoresis 1000 CE
Limit of quantitation: 700 ng/mL
Limit of detection: 200 ng/mL

OTHER SUBSTANCES
Simultaneous: bromazepam, clobazam, clonazepam, lorazepam, lormetazepam, nitrazepam, nordazepam, oxazepam, temazepam

REFERENCE
Bechet,I.; Fillet,M.; Hubert,P.; Crommen,J. Determination of benzodiazepines by micellar electrokinetic chromatography, *Electrophoresis*, **1994**, *15*, 1316–1321.

Buclizine

Molecular formula: $C_{28}H_{33}ClN_2$
Molecular weight: 433.04
CAS Registry No.: 82-95-1, 129-74-8 (HCl)
Merck Index (12th ed.): 1484

SAMPLE
Matrix: solutions

CAPILLARY ELECTROPHORESIS
Capillary: 60 cm × 75 μm fused-silica (52.5 cm to detector) (Polymicro Technologies)
Capillary preparation: Purge with running buffer before each run. At the beginning of each day purge using 50-60 kPa vacuum with 500 mM NaOH for 5 min, with water for 5 min, with MeCN for 5 min, and with running buffer for 5 min.
Running buffer: MeCN:MeOH:acetic acid 49:50:1 containing 20 mM tetrabutylammonium tetrafluoroborate
Injection: Hydrostatic injection at 10 cm for 5 s.

Detector: UV 214
Migration time: 2.68
Voltage: 30 kV
Model: Waters Quanta 4000

OTHER SUBSTANCES
Simultaneous: chlorcyclizine, chlorpheniramine, cyclizine, pheniramine, tripelennamine

REFERENCE
Leung,G.N.W.; Tang,H.P.O.; Tso,T.S.C.; Wan,T.S.M. Separation of basic drugs with non-aqueous capillary electrophoresis, *J.Chromatogr.A*, **1996**, *738*, 141–154.

SAMPLE
Matrix: solutions

CAPILLARY ELECTROPHORESIS
Capillary: 62 cm × 75 μm fused-silica (54.5 cm to detector) (Polymicro Technologies)
Capillary preparation: Purge with running buffer before each run. At the beginning of each day purge using 50-60 kPa vacuum with 500 mM NaOH for 5 min, with water for 5 min, with MeCN for 5 min, and with running buffer for 5 min.
Running buffer: MeCN:MeOH:acetic acid 49:50:1 containing 20 mM ammonium acetate
Injection: Hydrostatic injection at 10 cm for 5 s.
Detector: UV 214
Migration time: 6.05
Voltage: 30 kV
Model: Waters Quanta 4000

OTHER SUBSTANCES
Simultaneous: chlorcyclizine, chlorpheniramine, cyclizine, dimenhydrinate, methaphenilene, pheniramine, promethazine, pyrilamine, pyrrobutamine, tripelennamine

REFERENCE
Leung,G.N.W.; Tang,H.P.O.; Tso,T.S.C.; Wan,T.S.M. Separation of basic drugs with non-aqueous capillary electrophoresis, *J.Chromatogr.A*, **1996**, *738*, 141–154.

Bumetanide

Molecular formula: C$_{17}$H$_{20}$N$_2$O$_5$S
Molecular weight: 364.42
CAS Registry No.: 28395-03-1
Merck Index (12th ed.): 1508
Lednicer: 2 87

SAMPLE
Matrix: blood, urine
Sample preparation: Condition a 3 mL Supelclean LC-18 SPE cartridge with 3 mL MeOH and 3 mL water. Dilute urine 1:10 with water. Precipitate proteins from serum with MeOH. Add diluted urine or protein supernatant to the SPE cartridge, wash with 3 mL water, elute with 3 mL MeOH, reconstitute to the original volume with 100 mM KOH, inject an aliquot.

CAPILLARY ELECTROPHORESIS
Capillary: 67 cm × 50 μm fused-silica (60 cm to detector) (Polymicro Technologies)
Capillary preparation: Rinse with running buffer for 2 min before run. If necessary, regenerate capillary with 100 mM NaOH for 10 min and with water for 15 min.
Capillary temperature: 20
Running buffer: 60 mM 3-(cyclohexylamino)-1-propanesulfonic acid (CAPS) adjusted to pH 10.6 with 100 mM KOH

Injection: Pressure injection for 5 s
Detector: UV 220
Migration time: 8.8
Voltage: 25 kV
Model: Beckman P/ACE 2000

OTHER SUBSTANCES
Extracted: acetazolamide, amiloride, bendroflumethiazide, benzthiazide, caffeine, chlorothiazide, chlorthalidone, clopamide, dichlorphenamide, ethacrynic acid, furosemide, hydrochlorothiazide, metyrapone, probenecid, triamterene, trichlormethiazide

KEY WORDS
serum; SPE

REFERENCE
Jumppanen,J.; Sirén,H.; Riekkola,M.-L. Screening for diuretics in urine and blood serum by capillary zone electrophoresis, *J.Chromatogr.A*, **1993**, *652*, 441–450.

SAMPLE
Matrix: solutions
Sample preparation: Inject an aliquot of a 200 µg/mL solution in MeCN:water 50:50.

CAPILLARY ELECTROPHORESIS
Capillary: 23 cm × 50 µm 3 µm CEC Hypersil C18
Capillary temperature: 15
Running buffer: Gradient. MeCN:50 mM pH 2.5 Na_2HPO_4 buffer:water 40:20:40 for 6.5 min, 60:20:20 for 10.75 min (step gradient), re-equilibrate at initial conditions for 7.75 min.
Injection: Electrokinetic injection at 5 kV for 15 s.
Detector: UV 210
Migration time: 16.5
Voltage: 30 kV (with 8 bar of pressure at each end of capillary)
Model: Hewlett-Packard HP[3D]

OTHER SUBSTANCES
Simultaneous: bendroflumethiazide, chlorothiazide, chlorthalidone, hydrochlorothiazide, hydroflumethiazide

KEY WORDS
electrochromatography

REFERENCE
Euerby,M.R.; Gilligan,D.; Johnson,C.M.; Bartle,K.D. Step-gradient capillary electrochromatography, *Analyst*, **1997**, *122*, 1087–1088.

SAMPLE
Matrix: urine
Sample preparation: Filter (0.2 µm), inject an aliquot of the filtrate.

CAPILLARY ELECTROPHORESIS
Capillary: 80 cm × 50 µm fused-silica (57 cm to detector)
Capillary preparation: Before each run rinse the capillary with 1 M NaOH for 3 min, with 100 mM NaOH for 3 min, with water for 3 min, and with running buffer for 10 min.
Running buffer: 110 mM Boric acid containing 56 mM NaOH and 44 mM HCl, pH 8
Injection: Vacuum injection for 1 s
Detector: UV 220
Migration time: 8.393
Voltage: 20 kV
Model: Europhor Prime Vision system IV
Limit of detection: 1.25 µM

OTHER SUBSTANCES
Extracted: acebutolol (UV 238), acetazolamide (UV 222), alprenolol (UV 220), amiloride (UV 220), atenolol (UV 228), bendroflumethiazide (UV 220), chlorthalidone (UV 220), cocaine (UV 236), codeine (UV 220), ethacrynic acid (UV 220), furosemide (UV 232), hydrochlorothiazide (UV 226), methadone (UV 220), metoxiphenamine (UV 220), nadolol (UV 220), norcodeine (UV 220), oxprenolol (UV 220), pentazocine (UV 220), propranolol (UV 220), spironolactone (UV 244), triamterene (UV 232), xipamide (UV 234)

REFERENCE
Gonzalez,E.; Laserna,J.J. Capillary zone electrophoresis for the rapid screening of banned drugs in sport, *Electrophoresis*, **1994**, *15*, 240–243.

SAMPLE
Matrix: urine
Sample preparation: Centrifuge urine at 150 g for 10 min, inject an aliquot.

CAPILLARY ELECTROPHORESIS
Capillary: 100 cm × 50 μm (60 cm to detector)
Capillary preparation: Rinse capillary with running buffer for 3 min before each run. Regenerate capillary by flushing with 1 M NaOH for 10 min, with 100 mM NaOH for 10 min, and with water for 15 min.
Running buffer: pH 8 Borate buffer (concentration adjusted to keep current at 42 μA)
Injection: Electrokinetic injection at 10 kV for 10 s (ca. 6 nL).
Detector: F ex 350 em 428
Migration time: 7.3
Voltage: 30 kV
Current: 42 μA
Model: laboratory-constructed
Limit of detection: 3.6 μM

OTHER SUBSTANCES
Extracted: amiloride (F ex 382 em 416), bendroflumethiazide (F ex 272 em 381), triamterene (F ex 370 em 434)

KEY WORDS
direct injection

REFERENCE
González,E.; Becerra,A.; Laserna,J.J. Direct determination of diuretic drugs in urine by capillary zone electrophoresis using fluorescence detection, *J.Chromatogr.B*, **1996**, *687*, 145–150.

Bupivacaine

Molecular formula: C₁₈H₂₈N₂O
Molecular weight: 288.43
CAS Registry No.: 2180-92-9, 14252-80-3 (HCl hydrate), 18010-40-7 (HCl)
Merck Index (12th ed.): 1520
Lednicer: 1 17

SAMPLE
Matrix: blood
Sample preparation: 1 mL serum + 30 μL 5.33 mg/mL mepivacaine in MeOH, mix thoroughly, add 200 μL 1 M NaOH, mix for 20 s, add 1.5 mL hexane:diethyl ether 50:50, shake gently horizontally for 5 min, let stand in the refrigerator for 5 min. Remove the 1.3 mL of the organic layer and evaporate it to dryness under a stream of nitrogen, reconstitute the residue in 15 μL MeOH, inject an aliquot.

CAPILLARY ELECTROPHORESIS
Capillary: 64 cm × 50 μm uncoated silica (48 cm to detector) (Polymicro Technologies)
Running buffer: 18 mM pH 2.9 Tris buffer containing 10 mM heptakis(2,6-di-O-methyl)-β-cyclodextrin, 0.1% methylhydroxycellulose 4000, and 0.03 mM hexadecyltrimethylammonium bromide
Injection: Inject hydrodynamically at 13 cm for 15 s.
Detector: UV 220
Migration time: 11.6 (R), 12.0 (S)
Internal standard: mepivacaine (10, 10.4 (enantiomers))
Voltage: 24 kV
Model: Laboratory constructed

OTHER SUBSTANCES
Simultaneous: cimetidine, diltiazem, warfarin
Noninterfering: ibuprofen, indomethacin

KEY WORDS
serum; chiral

REFERENCE
Soini,H.; Riekkola,M.-L.; Novotny,M.V. Chiral separations of basic drugs and quantitation of bupivacaine enantiomers in serum by capillary electrophoresis with modified cyclodextrin buffers, *J.Chromatogr.*, **1992**, *608*, 265–274.

SAMPLE
Matrix: pleural drain fluid
Sample preparation: Dilute pleural drain fluid 10-fold. 1 mL Diluted fluid + 40 μL 300 μg/mL mepivacaine in water + 1 mL 500 mM NaOH + 6 mL hexane, vortex for 30 s, centrifuge at 1500 g for 5 min. Remove the upper organic layer and evaporate it to dryness at 40°, reconstitute the residue in 50 μL running buffer and 50 μL 100 mM HCl, inject an aliquot.

CAPILLARY ELECTROPHORESIS
Capillary: 72 cm × 75 μm fused-silica (72 cm to detector)
Capillary preparation: Before each run rinse capillary with 100 mM NaOH for 2 min, with water for 1 min, and with buffer for 4 min.
Capillary temperature: 30
Running buffer: 45 mM NaH_2PO_4 and 35 mM sodium borate adjusted to pH 8.1 with 100 mM phosphoric acid
Injection: Injection by vacuum suction for 1 s
Detector: UV 200
Migration time: 9.8
Internal standard: mepivacaine (10.3)
Voltage: 19 kV
Current: 96-98 μA
Model: Applied Biosystems Model 270A-HT

OTHER SUBSTANCES
Simultaneous: lidocaine

KEY WORDS
comparison with HPLC

REFERENCE
Wolfisberg,H.; Schmutz,A.; Stotzer,R.; Thormann,W. Assessment of automated capillary electrophoresis for therapeutic and diagnostic drug monitoring: determination of bupivacaine in drain fluid and antipyrine in plasma, *J.Chromatogr.A*, **1993**, *652*, 407–416.

SAMPLE
Matrix: solutions
Sample preparation: Prepare a 50 μg/mL solution in diluted running buffer, inject an aliquot.

CAPILLARY ELECTROPHORESIS
Capillary: 44 cm × 50 μm fused-silica
Capillary preparation: After each run wash capillary with water for 2 min and with running buffer for 3 min. At the beginning of each day wash capillary with water and running buffer for 5 min. Condition a new capillary with 1 M NaOH, 100 mM NaOH, water, and separation buffer.
Capillary temperature: 15
Running buffer: 100 mM Phosphoric acid containing 30 mM heptakis(2,6-di-O-methyl)-β-cyclodextrin adjusted to pH 3.0 with triethanolamine
Injection: Hydrodynamic injection for 1 s
Detector: UV 210
Voltage: 25 kV
Model: Spectra Physics Spectraphoresis 1000

KEY WORDS
chiral; enantiomer resolution 6.0

REFERENCE
Bechet,I.; Paques,P.; Fillet,M.; Hubert,P.; Crommen,J. Chiral separation of basic drugs by capillary zone electrophoresis with cyclodextrin additives, *Electrophoresis*, **1994**, *15*, 818–823.

SAMPLE
Matrix: solutions

CAPILLARY ELECTROPHORESIS
Capillary: 60 cm × 50 μm AccuSep (52.5 cm to detector) (Waters)
Capillary preparation: Before injection rinse capillary with 100 mM NaOH for 3 min and with running buffer for 3 min. Rinse new capillaries with 500 mM NaOH for 5 min
Running buffer: 50 mM pH 7.0 Na$_2$HPO$_4$ containing 25 mM (S)-N- dodecoxycarbonylvaline (Prepare (S)-N-dodecoxycarbonylvaline as follows. Prepare dodecyl chloroformate by reacting 1-dodecanol with 0.33 equivalents of triphosgene in dichloromethane solution in the presence of pyridine. Add dodecyl chloroformate dropwise to (S)-valine in 1 M NaOH solution, filter, wash with hexane, recrystallize from ether/petroleum ether.)
Injection: Hydrostatic injection 2 s
Detector: UV 214
Voltage: +12 kV
Model: Waters Quanta 4000 or 4000E

OTHER SUBSTANCES
Also analyzed: atenolol, ephedrine, homatropine, ketamine, metoprolol, N-methylpseudoephedrine, norephedrine, norphenylephrine, octopamine, pindolol, terbutaline

KEY WORDS
chiral; α = 1.26

REFERENCE
Mazzeo,J.R.; Grover,E.R.; Swartz,M.E.; Petersen,J.S. Novel chiral surfactant for the separation of enantiomers by micellar electrokinetic capillary chromatography, *J.Chromatogr.A*, **1994**, *680*, 125–135.

SAMPLE
Matrix: solutions
Sample preparation: Prepare a solution in running buffer, inject an aliquot.

CAPILLARY ELECTROPHORESIS
Capillary: 60 cm × 75 μm fused-silica (52.4 cm to detector)
Capillary preparation: After each run flush with 500 mM KOH for 2-3 min, flush with water, fill with running buffer
Running buffer: 10 mM Na$_2$HPO$_4$ containing 2% heparin sodium (MW 10000, 11% S, Scientific Protein Laboratories, Waunakee, WI) adjusted to pH 5 with phosphoric acid
Injection: Hydrostatic injection
Detector: UV 214

Migration time: 8.57
Model: Waters Quanta 4000

OTHER SUBSTANCES
Also analyzed: anabasine, brompheniramine, carbinoxamine, chlorcyclizine, chloroquine, chlorpheniramine, dimethindene, doxylamine, enpiroline, halofantrine, hydroxychloroquine, indapamide, mefloquine, nornicotine, pheniramine, primaquine, promethazine, quinacrine, tetramisole

---

REFERENCE
Stalcup,A.M.; Agyei,N.M. Heparin: a chiral mobile-phase additive for capillary zone electrophoresis, *Anal.Chem.*, **1994**, *66*, 3054–3059.

SAMPLE
Matrix: solutions

CAPILLARY ELECTROPHORESIS
Capillary: 40 cm × 50 μm fused-silica (35.5 cm to detector) (Polymicro Technologies)
Capillary preparation: Between runs purge with water for 50 s, with 100 mM NaOH for 50 s, with water for 60 s, and with running buffer for 60 s.
Running buffer: 50 mM pH 6.0 Sodium phosphate buffer containing 6 mg/mL sulfobutyl ether-β-cyclodextrin (Perkin-Elmer)
Injection: Pressure injection at 5 psi
Detector: UV 206
Migration time: 5.1, 5.2 (enantiomers)
Voltage: 15 kV
Model: Bio-Rad Biofocus 3000

KEY WORDS
chiral

REFERENCE
Desiderio,C.; Fanali,S. Use of negatively charged sulfobutyl ether-β-cyclodextrin for enantiomeric separation by capillary electrophoresis, *J.Chromatogr.A*, **1995**, *716*, 183–196.

SAMPLE
Matrix: solutions

CAPILLARY ELECTROPHORESIS
Capillary: 37 cm × 50 μm fused-silica (30 cm to detector) (Microquartz)
Capillary temperature: 20
Running buffer: 200 mM pH 2.7 phosphate buffer containing 200 mg/mL soluble β-cyclodextrin polymer cross-linked with epichlorohydrin (MW 3000-5000) (Cyclolab, Budapest)
Injection: Pressure injection at 0.5 psi for 10 s
Detector: UV 214
Migration time: 20.5, 21 (enantiomers)
Voltage: 15 kV
Model: Beckman P/ACE 2200

KEY WORDS
chiral

REFERENCE
Ingelse,B.A.; Everaerts,F.M.; Desiderio,C.; Fanali,S. Enantiomeric separation by capillary electrophoresis using a soluble neutral β-cyclodextrin polymer, *J.Chromatogr.A*, **1995**, *709*, 89–98.

SAMPLE
Matrix: solutions
Sample preparation: Inject an aliquot of a 100 μg/mL solution in water:running buffer 50:50.

CAPILLARY ELECTROPHORESIS
Capillary: 44.5 cm × 50 μm acrylamide-coated fused-silica (Bio-Rad)
Capillary temperature: 30
Running buffer: 100 mM NaH_2PO_4 containing 15 mM gamma-cyclodextrin, adjusted to pH 2.5 with phosphoric acid
Injection: Electrokinetic injection at 8 kV for 6 s.
Detector: UV 200
Migration time: 8.75
Voltage: 14 kV
Model: Bio-Rad BioFocus 3000

OTHER SUBSTANCES
Also analyzed: albuterol, alprenolol, atenolol, atropine, baclofen, bamethan, benserazide, biperiden, bisoprolol, bupranolol, butetamate, carazolol, carbuterol, carvedilol, celiprolol, chloroquine, chlorpheniramine (chlorphenamine), clidinium bromide, clobutinol, disopyramide, dobutamine, flecainide, homatropine, ipratropium bromide, isoproterenol, isothipendyl, ketamine, mefloquine, mequitazine, metaproterenol (orciprenaline), metipranolol, nafronyl (naftidrofuryl), nefopam, ofloxacin, orphenadrine, oxomemazine, oxprenolol, phenoxybenzamine, pholedrine, pindolol, pirbuterol, prilocaine, promethazine, propafenone, propranolol, sotalol, synephrine, terbutaline, tetrahydrozoline (tetryzoline), tocainide, trihexyphenidyl, trimeprazine (alimemazine), trimipramine, tropicamide, verapamil, zopiclone

KEY WORDS
coated capillary; achiral

REFERENCE
Koppenhoefer,B.; Epperlein,U.; Christian,B.; Yibing,J.; Yuying,C.; Bingcheng,L. Separation of enantiomers of drugs by capillary electrophoresis. I. γ-Cyclodextrin as chiral solvating agent, *J.Chromatogr.A*, **1995**, *717*, 181–190.

SAMPLE
Matrix: solutions
Sample preparation: Inject an aliquot of a 100 μM solution in running buffer diluted 10-fold with water.

CAPILLARY ELECTROPHORESIS
Capillary: 57 cm × 25 μm fused-silica (50 cm to detector) (Polymicro Technologies)
Capillary preparation: Equilibrate with running buffer for 5 min before each analysis. Store in water overnight and rinse with 100 mM NaOH for 5 min at the start of each day and after each change of running buffer. Flush a new capillary with 100 mM HCl for 5 min, with water for 5 min, with 100 mM NaOH for 5 min, with water for 10 min, and with running buffer.
Capillary temperature: 25
Running buffer: pH 3.13 Phosphate buffer (I = 0.02) containing 14 mM taurodeoxycholate and 4.0 mM Brij-35
Injection: Pressure injection at 0.5 psi for 10 s.
Detector: UV 214
Migration time: 36.7 (R), 37.5 (S)
Voltage: 30 kV
Current: 9 μA
Model: Beckman P/ACE 2050

KEY WORDS
chiral

REFERENCE
Amini,A.; Beijersten,I.; Pettersson,C.; Westerlund,D. Enantiomeric separation of local anaesthetic drugs by micellar electrokinetic capillary chromatography with taurodeoxycholate as chiral selector, *J.Chromatogr.A*, **1996**, *737*, 301–313.

SAMPLE
Matrix: solutions
Sample preparation: Inject an aliquot of a 50 μg/mL solution in water.

CAPILLARY ELECTROPHORESIS
Capillary: 48.5 cm × 50 μm fused-silica (40 cm to detector)
Capillary preparation: After each run wash capillary with running buffer for 3 min. At the start of each day wash capillary with running buffer for 10 min. Before use treat new capillaries with 1 M NaOH, 100 mM NaOH, water, and running buffer.
Capillary temperature: 15
Running buffer: 100 mM Phosphoric acid containing 15 mM carboxymethyl-β-cyclodextrin (Cyclolab, Budapest), adjusted to pH 3.0 with 84 mM triethanolamine
Injection: Hydrodynamic injection at 5 kPa for 2 s.
Detector: UV 210
Migration time: 12.68, 12.97 (enantiomers)
Voltage: 25 kV
Model: Hewlett Packard [3D]CE

OTHER SUBSTANCES
Simultaneous: isoproterenol, chlorpheniramine, dimethindene, ephedrine, fenfluramine

KEY WORDS
chiral

REFERENCE
Fillet,M.; Bechet,I.; Hubert,P.; Crommen,J. Resolution improvement by use of carboxymethyl-β-cyclodextrin as chiral additive for the enantiomeric separation of basic drugs by capillary electrophoresis, *J.Pharm.Biomed.Anal.*, **1996**, *14*, 1107–1114.

SAMPLE
Matrix: solutions

CAPILLARY ELECTROPHORESIS
Capillary: 52 cm × 75 μm fused-silica (48 cm to detector) (Polymicro Technologies)
Capillary preparation: Before each run purge with running buffer for 3 min. Every 3 runs purge with 100 mM NaOH for 5 min. Purge new capillaries with 1 M NaOH for 20 min and with 100 mM NaOH for 20 min, rinse with running buffer, equilibrate with running buffer at 12 kV for 3 h.
Running buffer: 100 mM CHES (2-(N-cyclohexylamine)ethanesulfonic acid) containing 10 mM triethylamine and 25 mM (R)-dodecoxycarbonylvaline (Waters EnantioSelect (R)-Val-1), pH adjusted to 8.8 with 1 M NaOH
Injection: Hydrostatic injection for 2 s.
Detector: UV 214
Migration time: 14.4, 14.6 (enantiomers)
Voltage: 12 kV
Current: ≤30 μA
Model: Waters Quanta 4000

OTHER SUBSTANCES
Simultaneous: clenbuterol, disopyramide, metoprolol, norphenylephrine, octopamine

KEY WORDS
chiral

REFERENCE
Peterson,A.G.; Ahuja,E.S.; Foley,J.P. Enantiomeric separations of basic pharmaceutical drugs by micellar electrokinetic chromatography using a chiral surfactant, N-dodecoxycarbonylvaline, *J.Chromatogr.B*, **1996**, *683*, 15–28.

SAMPLE
Matrix: solutions

CAPILLARY ELECTROPHORESIS
Capillary: 80.5 cm × 50 μm fused-silica (72 cm to detector) (Hewlett Packard)

Capillary preparation: Before each run flush with water for 1 min, with 100 mM NaOH for 4
min, with water for 1 min, and with running buffer for 4 min.
Capillary temperature: 30
Running buffer: 100 mM Phosphoric acid adjusted to pH 3.0 with triethanolamine, containing
10 mM heptakis(2,6-di-O-methyl)-β-cyclodextrin
Injection: Pressure injection at 50 mbar for 5 s (5 nL), ramp to operating voltage at 500 V/s.
Detector: UV 206
Migration time: 26.5, 27.8 (enantiomers)
Voltage: 30 kV
Model: Hewlett Packard HP [3D]CE

OTHER SUBSTANCES
Simultaneous: mepivacaine, ropivacaine
Also analyzed: prilocaine

KEY WORDS
chiral

REFERENCE
Sänger-van de Griend,C.E.; Gröningsson,K.; Westerlund,D. Chiral separation of local anaesthetics with capillary electrophoresis. Evaluation of the inclusion complex of the enantiomers with heptakis(2,6-di-O-methyl)-
β-cyclodextrin, *Chromatographia*, **1996**, *42*, 263–268.

SAMPLE
Matrix: solutions

CAPILLARY ELECTROPHORESIS
Capillary: 80.5 cm × 50 μm (72 cm to detector) (Hewlett Packard)
Capillary preparation: Before each run flush with water for 1 min, 100 mM NaOH for 4 min,
with water for 1 min, and with running buffer for 4 min.
Capillary temperature: 30
Running buffer: 100 mM phosphoric acid containing 10 mM heptakis(2,6-di-O-methyl)-β-cy-
clodextrin, adjusted to pH 3.0 with triethanolamine
Injection: Pressure injection at 50 mbar over 5 s (5 nL), ramp to voltage at 500 V/s.
Detector: UV 206
Migration time: 25.4, 27
Voltage: 30 kV
Model: Hewlett Packard HP [3D]CE

OTHER SUBSTANCES
Simultaneous: mepivacaine, ropivacaine

KEY WORDS
chiral

REFERENCE
Sänger-van de Griend,C.E.; Gröningsson,K. Validation of a capillary electrophoresis method for the enantiomeric purity testing of ropivacaine, a new local anaesthetic compound, *J.Pharm.Biomed.Anal.*, **1996**, *14*,
295–304.

SAMPLE
Matrix: solutions
Sample preparation: Inject an aliquot of a solution in running buffer.

CAPILLARY ELECTROPHORESIS
Capillary: 60 cm × 75 μm fused-silica (52.4 cm to detector)
Capillary preparation: After each run flush with 500 mM KOH for 2-3 min then with water.
Running buffer: 10 mM pH 3.8 Phosphate buffer containing 2% sulfated cyclodextrin (ds 7-10)
Injection: Hydrostatic injection.
Detector: UV 214
Migration time: 24.43, 25.33 (enantiomers)

Voltage: 15 kV
Model: Waters Quanta 4000

OTHER SUBSTANCES
Also analyzed: acebutolol, alprenolol, aminoglutethimide, brompheniramine, bupropion, canadine, carbinoxamine, chloroquine, chlorpheniramine, dimethindene, disopyramide, doxylamine, hydroxychloroquine, idazoxan, isoxsuprine, ketamine, mepenzolate, mepivacaine, methoxyphenamine, mexiletine, midodrine, nefopam, orphenadrine, oxprenolol, oxyphencyclimine, pheniramine, phensuximide, pindolol, piperoxan, terbutaline, tetramisole, tolperisone, tranylcypromine, trihexyphenidyl, trimipramine, verapamil, warfarin

KEY WORDS
chiral; detector at anode

REFERENCE
Stalcup,A.M.; Gahm,K.H. Application of sulfated cyclodextrins to chiral separations by capillary zone electrophoresis, *Anal.Chem.*, **1996**, *68*, 1360–1368.

SAMPLE
Matrix: solutions

CAPILLARY ELECTROPHORESIS
Capillary: 47 cm × 50 μm fused-silica (40 cm to detector) (Polymicro Technologies)
Capillary preparation: Before each run flush with running buffer for 2 min, fill with running buffer containing 96 mg/mL 100 mM methyl-β-cyclodextrin at 0.5 psi for 10.0 min. Store capillary in water overnight, rinse with 100 mM NaOH and with water each morning. Rinse a new capillary with 100 mM HCl for 5 min, with water for 5 min, with 100 mM NaOH for 5 min, and with water for 10 min.
Capillary temperature: 25.0
Running buffer: pH 2.90 Phosphate buffer, I = 0.04
Injection: Pressure injection at 0.5 psi for 5 s, ramp to operating voltage at 1.33 kV/s.
Detector: UV 214
Migration time: 13.5 (R), 14.3 (S)
Voltage: 24 kV
Model: Beckman P/ACE 2100

OTHER SUBSTANCES
Simultaneous: mepivacaine

KEY WORDS
chiral

REFERENCE
Amini,A.; Paulsen-Sörman,U. Enantioseparation of local anaesthetic drugs by capillary zone electrophoresis with cyclodextrins as chiral selectors using a partial filling technique, *Electrophoresis*, **1997**, *18*, 1019–1025.

SAMPLE
Matrix: solutions
Sample preparation: Inject an aliquot of a 100 μg/mL solution in running buffer.

CAPILLARY ELECTROPHORESIS
Capillary: 30 cm × 50 μm fused-silica (25.5 cm to detector) (Yongnian Optical Conductive Fiber Plant, China), coated with polyacrylamide
Capillary preparation: No details of the polyacrylamide coating process are provided. However, another paper (LC.GC 1997, 15, 40) by this group indicates that they use the procedure of Hjertén, thus: Adjust the pH of 20 mL water to 3.5 with acetic acid, add 80 μL 3-(trimethoxysilyl)propyl methacrylate (3-methacryloxypropyltrimethoxysilane), mix, suck into capillary, let stand at room temperature for 1 h, remove the solution, wash with water. Fill the capillary with a deaerated 3-4% acrylamide solution containing 1 μL/mL N,N,N',N'-tetramethylethylenediamine and 1 mg/mL potassium persulfate, let stand for 30 min, remove excess solution

by aspiration, rinse with water, remove water by aspiration, dry at 35° (J. Chromatogr. 1985, 347, 191).
Capillary temperature: 25
Running buffer: 100 mM NaH$_2$PO$_4$ adjusted to pH 2.5 (A) or 100 mM NaH$_2$PO$_4$ containing 45 mM hydroxypropyl-gamma-cyclodextrin, adjusted to pH 2.5 (B)
Injection: Electrokinetic injection at 15 kV for 3 s.
Detector: UV 200, UV 210
Migration time: 4.81 (A), 7.31, 7.46 (B, enantiomers)
Voltage: 15 kV
Model: Bio-Rad BioFocus 3000

OTHER SUBSTANCES
Simultaneous: amorolfine, brompheniramine, carteolol, chloroquine, chlorpheniramine, chlorphenoxamine, disopyramide, dobutamine, doxylamine, flecainide, gallopamil, ketamine, mepindolol, orphenadrine, oxybutynin, phenoxybenzamine, pindolol, propafenone, propranolol, sulpiride, talinolol, tropicamide, verapamil

KEY WORDS
chiral; coated capillary

REFERENCE
Koppenhoefer,B.; Epperlein,U.; Xiaofeng,Z.; Bingcheng,L. Separation of enantiomers of drugs by capillary electrophoresis. Part 4: Hydroxypropyl-γ-cyclodextrin as chiral solvating agent, *Electrophoresis*, **1997**, *18*, 924–930.

SAMPLE
Matrix: solutions
Sample preparation: Inject an aliquot of a 100 µg/mL solution in running buffer.

CAPILLARY ELECTROPHORESIS
Capillary: 30 cm × 50 µm fused-silica (25.5 cm to detector), coated with polyacrylamide
Capillary preparation: Adjust the pH of 20 mL water to 3.5 with acetic acid, add 80 µL 3-(trimethoxysilyl)propyl methacrylate (3-methacryloxypropyltrimethoxysilane), mix, suck into capillary, let stand at room temperature for 1 h, remove the solution, wash with water. Fill the capillary with a deaerated 3-4% acrylamide solution containing 1 µL/mL N,N,N',N'-tetramethylethylenediamine and 1 mg/mL potassium persulfate, let stand for 30 min, remove excess solution by aspiration, rinse with water, remove water by aspiration, dry at 35° (J. Chromatogr. 1985, 347, 191).
Capillary temperature: 25
Running buffer: 100 mM NaH$_2$PO$_4$ containing 45 mM heptakis(2,6-di-O-methyl)-β-cyclodextrin, adjusted to pH 2.5 with phosphoric acid
Injection: Electrokinetic injection at 15 kV for 3 s.
Detector: UV 200
Voltage: 15 kV
Model: Bio-Rad BioFocus 3000

KEY WORDS
chiral; coated capillary; comparison with the use of other cyclodextrins; this running buffer gave the greatest enantiomeric separation.; α=1.049

REFERENCE
Lin,B.; Zhu,X.; Koppenhoefer,B.; Epperlein,U. Investigation of 123 chiral drugs by cyclodextrin-modified capillary electrophoresis, *LC.GC*, **1997**, *15*, 40–46.

SAMPLE
Matrix: solutions
Sample preparation: Inject an aliquot of a 1 mM solution in MeOH/water.

CAPILLARY ELECTROPHORESIS
Capillary: 57 cm × 50 µm fused-silica (50 cm to detector) (Polymicro Technologies)
Capillary temperature: 15

Running buffer: 200 mM Sodium phosphate containing 400 mM sodium borate and 150 mM n-octyl-β-D-maltopyranoside (Calbiochem, La Jolla CA), pH 6.5
Injection: Pressure injection at 3.5 kPa.
Detector: UV 230
Migration time: 11.2, 11.4 (enantiomers)
Voltage: 20 kV
Model: Beckman P/ACE 5510

KEY WORDS
chiral

REFERENCE
Mechref,Y.; El Rassi,Z. Comparison of alkylglycoside surfactants in enantioseparation by capillary electrophoresis, *Electrophoresis*, **1997**, *18*, 912–918.

SAMPLE
Matrix: solutions

CAPILLARY ELECTROPHORESIS
Capillary: 36 cm × 50 μm fused-silica coated with linear polyacrylamide (31.5 cm to detector) (GL Science)
Capillary preparation: At the beginning and end of each day rinse capillary with capillary wash solution (Bio-Rad Cat. No. 148-5022) at 690 kPa for more than 3 min and with water at 690 kPa for more than 3 min. Coat capillary as follows. Treat capillary with 1 M NaOH at room temperature for 1 h, rinse with water, dry by passing nitrogen gas through the capillary at 110° for 6 h. Pass thionyl chloride through the capillary using a suction pump for several min, seal capillary at both ends and heat at 70° for 6 h. Unseal the capillary and fill with 250 mM vinyl magnesium bromide in THF by suction, seal the capillary, heat at 70° for 6 h. Open the capillary and rinse it with THF for several min, rinse with distilled water, fill the capillary with polymerization solution, heat at 28 ± 2° for 1 h, condition at -100 V/cm for 30 min (Anal. Sci. 1994, 10, 1). (The polymerization solution was 5% acrylamide in water containing 49 mM Tris, 384 mM glycine, and 0.1% sodium dodecyl sulfate, degas in an ultrasonic bath. Add 40 μL 10% N,N,N',N'-tetramethylethylenediamine and 10 μL 10% ammonium persulfate to 5 mL of the degassed solution, mix thoroughly.)
Running buffer: n-Propanol:50 mM pH 5.0 Phosphate buffer 10:90
Injection: Before each injection rinse with water at 690 kPa for 30 s, rinse with running buffer at 690 kPa for 30 s, partially fill with separation solution (500 μM α_1-acid glycoprotein (Cohn fraction VI) (Sigma) in running buffer) at 6.9 kPa for 190 s (27 cm), inject sample at 6.9 kPa for 2 s, electrophorese with running buffer (Note that α_1-acid glycoprotein from other suppliers may provide inferior results).
Detector: UV 210
Migration time: Resolution of enantiomers 2.1
Voltage: 12 kV
Model: Bio-rad BioFocus 3000

KEY WORDS
chiral; coated capillary

REFERENCE
Tanaka,Y.; Terabe,S. Separation of the enantiomers of basic drugs by affinity capillary electrophoresis using a partial filling technique and α_1-acid glycoprotein as chiral selector, *Chromatographia*, **1997**, *44*, 119–128.

SAMPLE
Matrix: solutions
Sample preparation: Inject an aliquot of a 20 μg/mL solution in buffer with an ionic strength 10 times less than that of the running buffer.

CAPILLARY ELECTROPHORESIS
Capillary: 27 cm × 50 μm fused-silica (20 cm to detector) (Polymicro Technologies)
Capillary preparation: Between runs rinse capillary with running buffer from a different vial for 2 min, equilibrate system for 1 min before analysis. Store capillary in water overnight and rinse with 100 mM NaOH and water each morning. Rinse new capillaries with 100 mM HCl

for 5 min, with water for 5 min, with 100 mM NaOH for 5 min, with water for 10 min, then
with running buffer.
Capillary temperature: 17
Running buffer: pH 2.90 Phosphate buffer (I = 0.04) containing 160 mg/mL 2-hydroxypropyl-β-
cyclodextrin (Sigma) (Prepare buffer by mixing 19.8 mL 1 M NaOH and 23.2 mL 1 M phos-
phoric acid and make up to 500 mL with water.)
Injection: Pressure injection at 3.4 kPa for 5 s (6 nL).
Detector: UV 214
Migration time: 8.8 (R), 9 (S)
Voltage: 15 kV
Current: 20 μA
Model: Beckman P/ACE 2100

OTHER SUBSTANCES
Simultaneous: mepivacaine

KEY WORDS
chiral

REFERENCE
Amini,A.; Sörman,U.P.; Lindgren,B.H.; Westerlund,D. Enantioseparation of anaesthetic drugs by capillary zone
electrophoresis using cyclodextrin-containing background electrolytes, *Electrophoresis*, **1998**, *19*, 731–737.

SAMPLE
Matrix: solutions

CAPILLARY ELECTROPHORESIS
Capillary: 35 cm × 50 μm polyacrylamide-coated fused-silica (30.5 cm to detector) (Composite
Metal Services, UK)
Capillary preparation: Before each run rinse capillary with water for 70 s and with running
buffer for 100 s. (Coat capillary as follows. Adjust the pH of 20 mL water to 3.5 with acetic
acid, add 80 μL 3-(trimethoxysilyl)propyl methacrylate (3-methacryloxypropyltrimethoxysi-
lane), mix, suck into capillary, let stand at room temperature for 1 h, remove the solution,
wash with water. Fill the capillary with a deaerated 3-4% acrylamide solution containing 1
μL/mL N,N,N',N'-tetramethylethylenediamine and 1 mg/mL potassium persulfate, let stand
for 30 min, remove excess solution by aspiration, rinse with water, remove water by aspiration,
dry at 35° (J. Chromatogr. 1985, 347, 191).)
Capillary temperature: 25
Running buffer: Buffer containing 100 mM cyanoethylated-β-cyclodextrin (Cyclolab, Budapest)
(Prepare buffer by adjusting the pH of 50 mM phosphoric acid containing 50 mM acetic acid
and 50 mM boric acid to 2.5 with concentrated NaOH, add the appropriate amount of cyanoe-
thylated-β-cyclodextrin, dilute with an equal volume of water.)
Injection: Pressure injection at 5 psi for 2 s.
Detector: UV 206
Migration time: 11.0 (second enantiomer, α = 1.026)
Voltage: 20 kV
Current: 27-40 μA
Model: Bio-Rad Biofocus 3000

KEY WORDS
chiral; coated capillary

REFERENCE
Aturki,Z.; Desiderio,C.; Mannina,L.; Fanali,S. Chiral separations by capillary zone electrophoresis with the use
of cyanoethylated-β-cyclodextrin as chiral selector, *J.Chromatogr.A*, **1998**, *817*, 91–104.

SAMPLE
Matrix: solutions

CAPILLARY ELECTROPHORESIS
Capillary: 29-36 cm × 50 μm fused-silica (24.5-31.5 cm to detector) (Yongnian Optical Conductive
Fiber Plant, China) coated with polyacrylamide

Capillary preparation: Coat capillary as follows. Adjust the pH of 20 mL water to 3.5 with acetic acid, add 80 µL 3-(trimethoxysilyl)propyl methacrylate (3-methacryloxypropyltrimethoxysilane), mix, suck into capillary, let stand at room temperature for 1 h, remove the solution, wash with water. Fill the capillary with a deaerated 3-4% acrylamide solution containing 1 µL/mL N,N,N',N'-tetramethylethylenediamine and 1 mg/mL potassium persulfate, let stand for 30 min, remove excess solution by aspiration, rinse with water, remove water by aspiration, dry at 35° (J. Chromatogr. 1985, 347, 191).
Capillary temperature: 25
Running buffer: 100 mM pH 2.5 NaH_2PO_4 (A) or 100 mM pH 2.5 NaH_2PO_4 containing 45 mM hydroxypropyl-α-cyclodextrin (Wacker, Munich) (B)
Injection: Electromigration at 15 kV for 3 s.
Detector: UV 200; UV 210
Migration time: 4.81 (A); 7.60, 7.68 (B) (enantiomers)
Voltage: 15 kV
Model: Bio-Focus 3000

OTHER SUBSTANCES
Also analyzed: albuterol (salbutamol), alprenolol, amorolfine, atenolol, atropine, azelastine, baclofen, bamethan, benproperine, benserazide, biperiden, bisoprolol, brompheniramine, bupranolol, butamirate, butethamate, carazolol, carbuterol, carteolol, carvedilol, celiprolol, chloroquine, chlorpheniramine, chlorphenoxamine, cicletanine, clenbuterol, clidinium bromide, clobutinol, dimethindene, dipivefrin, disopyramide, dobutamine, doxylamine, fendiline, flecainide, gallopamil, homatropine, ipratropium bromide, isoproterenol (isoprenaline), isothipendyl, ketamine, meclizine, mefloquine, mepindolol, mequitazine, metaclazepam, metaproterenol (orciprenaline), metipranolol, metoprolol, nafronyl (naftidrofuryl), nefopam, nicardipine, norfenefrine, ofloxacin, ornidazole, orphenadrine, oxomemazine, oxprenolol, oxybutynin, phenoxybenzamine, phenylpropanolamine, pholedrine, pindolol, pirbuterol, prilocaine, procyclidine, promethazine, propafenone, propranolol, reproterol, sotalol, sulpride, synephrine, talinolol, terbutaline, tetrahydrozoline (tetryzoline), theodrenaline, tioconazole, tocainide, trihexyphenidyl, trimeprazine (alimemazine), trimipramine, tropicamide, verapamil, zopiclone

---

KEY WORDS
coated capillary; chiral

REFERENCE
Koppenhoefer,B.; Eperlein,U.; Schlunk,R.; Zhu,X.; Lin,B. Separation of enantiomers of drugs by capillary electrophoresis. V. Hydroxypropyl-α-cyclodextrin as chiral solvating agent, *J.Chromatogr.A*, **1998**, *793*, 153–164.

Bupranolol

Molecular formula: $C_{14}H_{22}ClNO_2$
Molecular weight: 271.79
CAS Registry No.: 14556-46-8, 15148-80-8 (HCl)
Merck Index (12th ed.): 1521

SAMPLE
Matrix: solutions
Sample preparation: Inject an aliquot of a 100 µg/mL solution in running buffer.

CAPILLARY ELECTROPHORESIS
Capillary: 29 cm × 50 µm fused-silica (24.5 cm to detector) (Yongnian Optical Conductive Fiber Plant, China), coated with polyacrylamide
Capillary preparation: No details of the polyacrylamide coating process are provided. However, another paper (LC.GC 1997, 15, 40) by this group indicates that they use the procedure of Hjertén, thus: Adjust the pH of 20 mL water to 3.5 with acetic acid, add 80 µL 3-(trimethoxysilyl)propyl methacrylate (3-methacryloxypropyltrimethoxysilane), mix, suck into capillary, let stand at room temperature for 1 h, remove the solution, wash with water. Fill the capillary with a deaerated 3-4% acrylamide solution containing 1 µL/mL N,N,N',N'-tetramethylethyl-

enediamine and 1 mg/mL potassium persulfate, let stand for 30 min, remove excess solution by aspiration, rinse with water, remove water by aspiration, dry at 35° (J. Chromatogr. 1985, 347, 191).
Capillary temperature: 25
Running buffer: 100 mM NaH_2PO_4 adjusted to pH 2.5
Injection: Electrokinetic injection at 15 kV for 3 s.
Detector: UV 200, UV 210
Migration time: 5.49
Voltage: 15 kV
Model: Bio-Rad BioFocus 3000

OTHER SUBSTANCES
Simultaneous: albuterol, alprenolol, atenolol, baclofen, bamethan, benproperine, benserazide, bisoprolol, butamirate, butethamate, carbuterol, celiprolol, clenbuterol, clobutinol, dipivefrin, isoproterenol (isoprenaline), metaproterenol (orciprenaline), metipranolol, metoprolol, norfenefrine, ornidazole, oxprenolol, phenylpropanolamine, pholedrine, pirbuterol, prilocaine, procyclidine, sotalol, synephrine, terbutaline, tocainide

KEY WORDS
coated capillary

REFERENCE
Koppenhoefer,B.; Epperlein,U.; Xiaofeng,Z.; Bingcheng,L. Separation of enantiomers of drugs by capillary electrophoresis. Part 4: Hydroxypropyl-γ-cyclodextrin as chiral solvating agent, *Electrophoresis*, **1997**, *18*, 924–930.

SAMPLE
Matrix: solutions

CAPILLARY ELECTROPHORESIS
Capillary: 29-36 cm × 50 μm fused-silica (24.5-31.5 cm to detector) (Yongnian Optical Conductive Fiber Plant, China) coated with polyacrylamide
Capillary preparation: Coat capillary as follows. Adjust the pH of 20 mL water to 3.5 with acetic acid, add 80 μL 3-(trimethoxysilyl)propyl methacrylate (3-methacryloxypropyltrimethoxysilane), mix, suck into capillary, let stand at room temperature for 1 h, remove the solution, wash with water. Fill the capillary with a deaerated 3-4% acrylamide solution containing 1 μL/mL N,N,N',N'-tetramethylethylenediamine and 1 mg/mL potassium persulfate, let stand for 30 min, remove excess solution by aspiration, rinse with water, remove water by aspiration, dry at 35° (J. Chromatogr. 1985, 347, 191).
Capillary temperature: 25
Running buffer: 100 mM pH 2.5 NaH_2PO_4 (A) or 100 mM pH 2.5 NaH_2PO_4 containing 45 mM hydroxypropyl-α-cyclodextrin (Wacker, Munich) (B)
Injection: Electromigration at 15 kV for 3 s.
Detector: UV 200; UV 210
Migration time: 5.49 (A); 12.57 (B) (no separation of enantiomers)
Voltage: 15 kV
Model: Bio-Focus 3000

OTHER SUBSTANCES
Also analyzed: albuterol (salbutamol), alprenolol, amorolfine, atenolol, atropine, azelastine, baclofen, bamethan, benproperine, benserazide, biperiden, bisoprolol, brompheniramine, bupivacaine, butamirate, butethamate, carazolol, carbuterol, carteolol, carvedilol, celiprolol, chloroquine, chlorpheniramine, chlorphenoxamine, cicletanine, clenbuterol, clidinium bromide, clobutinol, dimethindene, dipivefrin, disopyramide, dobutamine, doxylamine, fendiline, flecainide, gallopamil, homatropine, ipratropium bromide, isoproterenol (isoprenaline), isothipendyl, ketamine, meclizine, mefloquine, mepindolol, mequitazine, metaclazepam, metaproterenol (orciprenaline), metipranolol, metoprolol, nafronyl (naftidrofuryl), nefopam, nicardipine, norfenefrine, ofloxacin, ornidazole, orphenadrine, oxomemazine, oxprenolol, oxybutynin, phenoxybenzamine, phenylpropanolamine, pholedrine, pindolol, pirbuterol, prilocaine, procyclidine, promethazine, propafenone, propranolol, reproterol, sotalol, sulpride, synephrine, talinolol, terbutaline, tetrahydrozoline (tetryzoline), theodrenaline, tioconazole, tocainide, trihexyphenidyl, trimeprazine (alimemazine), trimipramine, tropicamide, verapamil, zopiclone

KEY WORDS
coated capillary

REFERENCE
Koppenhoefer,B.; Eperlein,U.; Schlunk,R.; Zhu,X.; Lin,B. Separation of enantiomers of drugs by capillary electrophoresis. V. Hydroxypropyl-α-cyclodextrin as chiral solvating agent, *J.Chromatogr.A*, **1998**, *793*, 153–164.

Bupropion

Molecular formula: $C_{13}H_{18}ClNO$
Molecular weight: 239.74
CAS Registry No.: 34911-55-2, 31677-93-7 (HCl)
Merck Index (12th ed.): 1523
Lednicer: 2 124

SAMPLE
Matrix: solutions
Sample preparation: Inject an aliquot of a solution in running buffer.

CAPILLARY ELECTROPHORESIS
Capillary: 60 cm × 75 μm fused-silica (52.4 cm to detector)
Capillary preparation: After each run flush with 500 mM KOH for 2-3 min then with water.
Running buffer: 10 mM pH 3.8 Phosphate buffer containing 2% sulfated cyclodextrin (ds 7-10)
Injection: Hydrostatic injection.
Detector: UV 214
Migration time: 8.34, 8.56 (enantiomers)
Voltage: 15 kV
Model: Waters Quanta 4000

OTHER SUBSTANCES
Also analyzed: acebutolol, alprenolol, aminoglutethimide, brompheniramine, bupivacaine, canadine, carbinoxamine, chloroquine, chlorpheniramine, dimethindene, disopyramide, doxylamine, hydroxychloroquine, idazoxan, isoxsuprine, ketamine, mepenzolate, mepivacaine, methoxyphenamine, mexiletine, midodrine, nefopam, orphenadrine, oxprenolol, oxyphencyclimine, pheniramine, phensuximide, pindolol, piperoxan, terbutaline, tetramisole, tolperisone, tranylcypromine, trihexyphenidyl, trimipramine, verapamil, warfarin

KEY WORDS
chiral; detector at anode

REFERENCE
Stalcup,A.M.; Gahm,K.H. Application of sulfated cyclodextrins to chiral separations by capillary zone electrophoresis, *Anal.Chem.*, **1996**, *68*, 1360–1368.

Buserelin

Molecular formula: $C_{60}H_{86}N_{16}O_{13}$
Molecular weight: 1239.44
CAS Registry No.: 57982-77-1,
68630-75-1 (acetate)
Merck Index (12th ed.): 1527

SAMPLE
Matrix: formulations
Sample preparation: Dissolve 2 rod-shaped implants in 8 drops DMF with sonication, add 8 drops of water, centrifuge at 10000 rpm for 2 min, remove the supernatant, centrifuge repeatedly until clean, inject an aliquot.

CAPILLARY ELECTROPHORESIS
Capillary: 20-30 cm × 25 μm fused-silica (Polymicro Technologies)
Capillary preparation: Before each run rinse capillary with 10 mM NaOH for 1 min and with running buffer for 2 min. Condition new capillaries with 100 mM NaOH for 30 min with heating to 50°, equilibrate with running buffer for 40 min.
Capillary temperature: 30
Running buffer: 60 mM pH 3.0 Phosphate buffer containing 250 mM potassium sulfate (Prepare buffer by dissolving 774 mg 85% phosphoric acid, 7.252 g KH_2PO_4, and 43.56 g potassium sulfate in 1 L water.)
Injection: Electrokinetic injection at 1 kV for 20 s.
Detector: UV 214
Migration time: 15
Voltage: 8 kV
Current: ca. 100 μA
Model: P/ACE 2050 or 2100

OTHER SUBSTANCES
Simultaneous: degradation products

KEY WORDS
implants

REFERENCE
Wätzig,H.; Degenhardt,M. Characterisation of buserelin acetate by capillary electrophoresis, *J.Chromatogr.A*, **1998**, *817*, 239–252.

Buspirone

Molecular formula: $C_{21}H_{31}N_5O_2$
Molecular weight: 385.51
CAS Registry No.: 36505-84-7, 33386-08-2 (HCl)
Merck Index (12th ed.): 1528
Lednicer: 2 300; 4 119

SAMPLE
Matrix: bulk

Sample preparation: Dissolve 100 mg drug in 50 mL running buffer, add IS solution, make up to 100 mL with running buffer, inject an aliquot.

CAPILLARY ELECTROPHORESIS
Capillary: 40 cm × 50 μm fused-silica (Supelco HPE)
Capillary preparation: Between runs wash with 100 mM NaOH for 2 min and with running buffer for 3 min. At the beginning of each day wash capillary with 100 mM NaOH for 30 min.
Capillary temperature: 20
Running buffer: 170 mM Tris adjusted to pH 3 with phosphoric acid
Injection: Hydrodynamic injection for 2 s
Detector: UV 240
Migration time: 6.6
Internal standard: gepirone
Voltage: 20 kV
Current: 83.37 μA
Model: Thermo Separation Products SpectraPHORESIS 1000
Limit of detection: 0.36 ng (S/N 3)

OTHER SUBSTANCES
Simultaneous: ipsapirone, zalospirone

REFERENCE
Quaglia,M.G.; Farina,A.; Bossi,E.; Dell'Aquila,C. Analysis of non-benzodiazepinic anxiolytic agents by capillary zone electrophoresis, *J.Pharm.Biomed.Anal.*, **1995**, *13*, 505–509.

Butabarbital

Molecular formula: $C_{10}H_{15}N_2NaO_3$
Molecular weight: 234.23
CAS Registry No.: 143-81-7, 125-40-6 (free acid)
Merck Index (12th ed.): 1530
Lednicer: 1 268

SAMPLE
Matrix: blood, gastric contents, urine, vitreous humor
Sample preparation: Dilute gastric contents 1:10. 2 mL Serum, blood, vitreous humor, diluted gastric contents, or urine + 2.5 mL water + 200 μL mephobarbital solution, mix, add to TOXI-TUBE B (Toxi-Lab, Irvine CA), rock for 10 min, centrifuge at 2000 rpm for 5 min. Remove the organic layer and evaporate it to dryness under a stream of nitrogen at room temperature, reconstitute the residue in 100 μL running buffer, vortex for 30 s, inject an aliquot. (Mephobarbital solution was 2 mg mephobarbital and 2 drops EtOH, made up to 20 mL with water.)

CAPILLARY ELECTROPHORESIS
Capillary: 60 cm × 75 μm AccuSep (Waters)
Capillary preparation: Purge for 1 min between samples.
Running buffer: MeCN:buffer 15:85 adjusted to pH 8.5 with 1 M phosphoric acid. (Buffer was 3.8 g sodium borate decahydrate, 1 .4 g $NaH_2PO_4.H_2O$, and 28.8 g sodium dodecyl sulfate in 1 L water.)
Injection: Hydrostatic injection for 10 s
Detector: UV 214
Migration time: 8.23
Internal standard: mephobarbital (8.88)
Voltage: 20 kV
Model: Waters Quanta 4000
Limit of detection: 100 ng/mL

OTHER SUBSTANCES
Extracted: amobarbital, butalbital, pentobarbital, phenobarbital, secobarbital

KEY WORDS
serum; whole blood

REFERENCE
Ferslew,K.E.; Hagardorn,A.N.; McCormick,W.F. Application of micellar electrokinetic capillary chromatography to forensic analysis of barbiturates in biological fluids, *J.Forensic Sci.*, **1995**, *40*, 245–249.

SAMPLE
Matrix: solutions

CAPILLARY ELECTROPHORESIS
Capillary: 72 cm × 50 μm silica (50 cm to detector) (Applied Biosystems)
Capillary preparation: Before each run wash capillary with 100 mM NaOH for 3 min, wash with running buffer for 3 min, aspirate neutral marker solution (2 drops DMSO in 10 mL water) for 1 s, place capillary end in buffer vial for 5 s, inject sample. Wash capillary at the beginning of each day by passing 1 M NaOH through for 20 min.
Capillary temperature: 30
Running buffer: MeCN:buffer 20:80 (Buffer was 30 mM pH 9.3 borate buffer containing 30 mM sodium dodecyl sulfate.)
Injection: Inject by applying vacuum for 1 s (2-3 nL)
Detector: UV 200
Migration time: 10.7
Voltage: +30 kV
Current: 41 μA
Model: Applied Biosystems Model 270A

OTHER SUBSTANCES
Simultaneous: heptabarbital, hexobarbital, pentobarbital, phenobarbital

REFERENCE
Evenson,M.A.; Wiktorowicz,J.E. Automated capillary electrophoresis applied to therapeutic drug monitoring, *Clin.Chem.*, **1992**, *38*, 1847–1852.

SAMPLE
Matrix: urine
Sample preparation: Place an extraction rod in 50 μL urine in a 50 mm × 1.5 mm ID length of PTFE tubing for 30 min, place the extraction rod into 5 μL 20-40 mM pH 11.5 phosphate buffer in a 50 mm × 1.2 mm ID length of PTFE tubing for 90 min, inject an aliquot of the buffer. (Prepare the solid-phase extraction rod as follows. Polish a 70 × 1.1 stainless steel rod with emery paper, clean with a Kimwipe and acetone, sonicate in EtOH, sonicate in THF, dip in 3% PVAM in THF momentarily, hold vertically for 1 min, air dry in a hood for at least 5 h, dip in PVC solution momentarily, hold vertically for 1 min, air dry for at least 5 h. The coating is 3 cm long. PVAM was poly(vinyl chloride-co-vinyl acetate-co-maleic acid) consisting of 86% vinyl chloride, 13% vinyl acetate, and 1% maleic acid. Prepare PVC solution by adding very high molecular weight PVC slowly to THF with stirring until the concentration reaches 3.6%, add Santicizer 141 to a concentration of 7.2%. Santicizer 141 (Monsanto) is 92% 2-ethylhexyl diphenyl phosphate, 5% di-2-ethylhexyl phenyl phosphate, and 3% triphenyl phosphate.)

CAPILLARY ELECTROPHORESIS
Capillary: 75 cm × 75 μm fused-silica (50 cm to detector) (Polymicro Technologies)
Running buffer: 50 mM Tris adjusted to pH 7.8 with 3-[N-tris(hydroxymethyl)methylamino]-2-hydroxypropanesulfonic acid (Tapso)
Injection: Pressure injection at 0.5 psi for 4 s.
Detector: UV 230
Migration time: 7.55
Voltage: 35 kV
Current: about 30 μA
Model: Isco 3850
Limit of detection: <1 ppm

OTHER SUBSTANCES
Extracted: amobarbital, aprobarbital, butalbital, mephobarbital, pentobarbital, secobarbital, thiopental
Simultaneous: allobarbital, aspirin, phenobarbital

KEY WORDS
SPE

REFERENCE
Li,S.; Weber,S.G. Determination of barbiturates by solid-phase microextraction and capillary electrophoresis, *Anal.Chem.*, **1997**, *69*, 1217–1222.

Butalbital

Molecular formula: $C_{11}H_{16}N_2O_3$
Molecular weight: 224.26
CAS Registry No.: 77-26-9
Merck Index (12th ed.): 1536
Lednicer: 1 268

SAMPLE
Matrix: blood, gastric contents, urine, vitreous humor
Sample preparation: Dilute gastric contents 1:10. 2 mL Serum, blood, vitreous humor, diluted gastric contents, or urine + 2.5 mL water + 200 µL mephobarbital solution, mix, add to TOXI-TUBE B (Toxi-Lab, Irvine CA), rock for 10 min, centrifuge at 2000 rpm for 5 min. Remove the organic layer and evaporate it to dryness under a stream of nitrogen at room temperature, reconstitute the residue in 100 µL running buffer, vortex for 30 s, inject an aliquot. (Mephobarbital solution was 2 mg mephobarbital and 2 drops EtOH, made up to 20 mL with water.)

CAPILLARY ELECTROPHORESIS
Capillary: 60 cm × 75 µm AccuSep (Waters)
Capillary preparation: Purge for 1 min between samples.
Running buffer: MeCN:buffer 15:85 adjusted to pH 8.5 with 1 M phosphoric acid. (Buffer was 3.8 g sodium borate decahydrate, 1 .4 g $NaH_2PO_4.H_2O$, and 28.8 g sodium dodecyl sulfate in 1 L water.)
Injection: Hydrostatic injection for 10 s
Detector: UV 214
Migration time: 8.01
Internal standard: mephobarbital (8.88)
Voltage: 20 kV
Model: Waters Quanta 4000
Limit of detection: 100 ng/mL

OTHER SUBSTANCES
Extracted: amobarbital, butabarbital, pentobarbital, phenobarbital, secobarbital

KEY WORDS
serum; whole blood

REFERENCE
Ferslew,K.E.; Hagardorn,A.N.; McCormick,W.F. Application of micellar electrokinetic capillary chromatography to forensic analysis of barbiturates in biological fluids, *J.Forensic Sci.*, **1995**, *40*, 245–249.

SAMPLE
Matrix: blood, urine
Sample preparation: Serum, plasma. 200 µL Serum or plasma + 100 µL 1 M HCl + 2 mL chloroform, shake vigorously for 15 min, centrifuge at 500 g for 10 min. Remove the lower

organic layer and evaporate it to dryness under a stream of nitrogen at 40°, reconstitute the residue in 200 µL running buffer, shake for 1 min, filter (0.2 µm), inject an aliquot. Urine. Condition a Bond Elut Certify SPE cartridge with 2 mL MeOH and 2 mL 100 mM pH 6 phosphate buffer, do not allow to dry. 5 mL Urine + 2 mL 100 mM pH 6 phosphate buffer, mix, add to the SPE cartridge, wash with 1 mL MeOH:100 mM phosphate buffer 20:80, dry under vacuum for 5 min, wash with 1 mL 1 M acetic acid, dry under vacuum for 10 min, wash with 1 mL hexane, elute with 4 mL dichloromethane. Evaporate the eluate to dryness under a stream of nitrogen at 40°, reconstitute the residue in 100-200 µL running buffer, filter (0.2 µm), inject an aliquot.

CAPILLARY ELECTROPHORESIS
Capillary: 90 cm × 75 µm fused-silica (70 cm to detector) (Polymicro Technologies)
Capillary preparation: Between each run rinse capillary with 100 mM NaOH for 3 min and buffer for 5 min
Running buffer: 15 mM pH 7.8 NaH$_2$PO$_4$ containing 9 mM sodium borate and 50 mM sodium dodecyl sulfate
Injection: Siphon at 34 cm for 5 s
Detector: UV 195
Migration time: 11.2
Voltage: 20 kV
Current: 60-63 µA
Model: Laboratory constructed

OTHER SUBSTANCES
Extracted: allobarbital, amobarbital, barbital, pentobarbital, phenobarbital, thiopental

KEY WORDS
serum; cow; human; plasma; SPE

REFERENCE
Thormann,W.; Meier,P.; Marcolli,C.; Binder,F. Analysis of barbiturates in human serum and urine by high-performance capillary electrophoresis-micellar electrokinetic capillary chromatography with on-column multi-wavelength detection, *J.Chromatogr.*, **1991**, *545*, 445–460.

SAMPLE
Matrix: solutions

CAPILLARY ELECTROPHORESIS
Capillary: 52 cm × 75 µm fused-silica
Running buffer: 50 mM pH 9.0 Phosphate buffer containing 40 mM hydroxypropyl-gamma-cyclodextrin
Detector: UV 254
Voltage: 15 kV

OTHER SUBSTANCES
Also analyzed: pentobarbital, secobarbital

KEY WORDS
chiral

REFERENCE
Hassan,A.; Alemayehu,B.; Anucha,T.; Chau,S.; Eradiri,O.; Ly,C.; Malaiyandi,P.; Muhuri,G.; Odidi,A.; Odidi,I.; Siwicki,J. Validation of HPLC method for assay and determination of chromatographic purity of diclofenac sodium in extended release formulations (Abstract 4173), *Pharm.Res.*, **1997**, *14*, S687.

SAMPLE
Matrix: solutions
Sample preparation: Inject an aliquot of a solution in 20 mM pH 5.9 ammonium acetate buffer.

CAPILLARY ELECTROPHORESIS
Capillary: 33 cm × 50 µm fused-silica (20 cm to detector) (Polymicro Technologies)

Running buffer: 10 mM pH 5.9 Ammonium acetate buffer containing 15 mM sodium dodecyl
 sulfate
Injection: Electrokinetic injection at -7.4 kV for 1 s.
Detector: UV 226, MS, Finnigan MAT TSQ, negative electrospray (details in paper)
Migration time: 7.5
Voltage: -7.4 kV

OTHER SUBSTANCES
Simultaneous: amobarbital, barbital, pentobarbital, secobarbital

---

REFERENCE
Yang,L.; Harrata,A.K.; Lee,C.S. On-line micellar electrokinetic chromatography-electrospray ionization mass
 spectrometry using anodically migrating micelles, *Anal.Chem.*, **1997**, *69*, 1820–1826.

SAMPLE
Matrix: urine
Sample preparation: Place an extraction rod in 50 µL urine in a 50 mm × 1.5 mm ID length
 of PTFE tubing for 30 min, place the extraction rod into 5 µL 20-40 mM pH 11.5 phosphate
 buffer in a 50 mm × 1.2 mm ID length of PTFE tubing for 90 min, inject an aliquot of the
 buffer. (Prepare the solid-phase extraction rod as follows. Polish a 70 × 1.1 stainless steel rod
 with emery paper, clean with a Kimwipe and acetone, sonicate in EtOH, sonicate in THF, dip
 in 3% PVAM in THF momentarily, hold vertically for 1 min, air dry in a hood for at least 5 h,
 dip in PVC solution momentarily, hold vertically for 1 min, air dry for at least 5 h. The coating
 is 3 cm long. PVAM was poly(vinyl chloride-co-vinyl acetate-co-maleic acid) consisting of 86%
 vinyl chloride, 13% vinyl acetate, and 1% maleic acid. Prepare PVC solution by adding very
 high molecular weight PVC slowly to THF with stirring until the concentration reaches 3.6%,
 add Santicizer 141 to a concentration of 7.2%. Santicizer 141 (Monsanto) is 92% 2-ethylhexyl
 diphenyl phosphate, 5% di-2-ethylhexyl phenyl phosphate, and 3% triphenyl phosphate.)

CAPILLARY ELECTROPHORESIS
Capillary: 75 cm × 75 µm fused-silica (50 cm to detector) (Polymicro Technologies)
Running buffer: 50 mM Tris adjusted to pH 7.8 with 3-[N-tris(hydroxymethyl)methylamino]-2-
 hydroxypropanesulfonic acid (Tapso)
Injection: Pressure injection at 0.5 psi for 4 s.
Detector: UV 230
Migration time: 8.3
Voltage: 35 kV
Current: about 30 µA
Model: Isco 3850
Limit of detection: <1 ppm

OTHER SUBSTANCES
Extracted: amobarbital, aprobarbital, butabarbital, mephobarbital, pentobarbital, secobarbital,
 thiopental
Simultaneous: allobarbital, aspirin, phenobarbital

KEY WORDS
SPE

REFERENCE
Li,S.; Weber,S.G. Determination of barbiturates by solid-phase microextraction and capillary electrophoresis,
 Anal.Chem., **1997**, *69*, 1217–1222.

Butamirate

Molecular formula: $C_{18}H_{29}NO_3$
Molecular weight: 307.43
CAS Registry No.: 18109-80-3, 18109-81-4 (citrate)
Merck Index (12th ed.): 1539

SAMPLE
Matrix: solutions
Sample preparation: Inject an aliquot of a 100 µg/mL solution in running buffer.

CAPILLARY ELECTROPHORESIS
Capillary: 29 cm × 50 µm fused-silica (24.5 cm to detector) (Yongnian Optical Conductive Fiber Plant, China), coated with polyacrylamide
Capillary preparation: No details of the polyacrylamide coating process are provided. However, another paper (LC.GC 1997, 15, 40) by this group indicates that they use the procedure of Hjertén, thus: Adjust the pH of 20 mL water to 3.5 with acetic acid, add 80 µL 3-(trimethoxysilyl)propyl methacrylate (3-methacryloxypropyltrimethoxysilane), mix, suck into capillary, let stand at room temperature for 1 h, remove the solution, wash with water. Fill the capillary with a deaerated 3-4% acrylamide solution containing 1 µL/mL N,N,N',N'-tetramethylethylenediamine and 1 mg/mL potassium persulfate, let stand for 30 min, remove excess solution by aspiration, rinse with water, remove water by aspiration, dry at 35° (J. Chromatogr. 1985, 347, 191).
Capillary temperature: 25
Running buffer: 100 mM NaH_2PO_4 adjusted to pH 2.5
Injection: Electrokinetic injection at 15 kV for 3 s.
Detector: UV 200, UV 210
Migration time: 5.41
Voltage: 15 kV
Model: Bio-Rad BioFocus 3000

OTHER SUBSTANCES
Simultaneous: albuterol, alprenolol, atenolol, baclofen, bamethan, benproperine, benserazide, bisoprolol, bupranolol, butethamate, carbuterol, celiprolol, clenbuterol, clobutinol, dipivefrin, isoproterenol (isoprenaline), metaproterenol (orciprenaline), metipranolol, metoprolol, norfenefrine, ornidazole, oxprenolol, phenylpropanolamine, pholedrine, pirbuterol, prilocaine, procyclidine, sotalol, synephrine, terbutaline, tocainide

KEY WORDS
coated capillary

REFERENCE
Koppenhoefer,B.; Epperlein,U.; Xiaofeng,Z.; Bingcheng,L. Separation of enantiomers of drugs by capillary electrophoresis. Part 4: Hydroxypropyl-γ-cyclodextrin as chiral solvating agent, *Electrophoresis*, **1997**, *18*, 924–930.

SAMPLE
Matrix: solutions

CAPILLARY ELECTROPHORESIS
Capillary: 29-36 cm × 50 µm fused-silica (24.5-31.5 cm to detector) (Yongnian Optical Conductive Fiber Plant, China) coated with polyacrylamide
Capillary preparation: Coat capillary as follows. Adjust the pH of 20 mL water to 3.5 with acetic acid, add 80 µL 3-(trimethoxysilyl)propyl methacrylate (3-methacryloxypropyltrimethoxysilane), mix, suck into capillary, let stand at room temperature for 1 h, remove the solution, wash with water. Fill the capillary with a deaerated 3-4% acrylamide solution containing 1 µL/mL N,N,N',N'-tetramethylethylenediamine and 1 mg/mL potassium persulfate, let stand for 30 min, remove excess solution by aspiration, rinse with water, remove water by aspiration, dry at 35° (J. Chromatogr. 1985, 347, 191).
Capillary temperature: 25

Running buffer: 100 mM pH 2.5 NaH$_2$PO$_4$ (A) or 100 mM pH 2.5 NaH$_2$PO$_4$ containing 45 mM hydroxypropyl-α-cyclodextrin (Wacker, Munich) (B)
Injection: Electromigration at 15 kV for 3 s.
Detector: UV 200; UV 210
Migration time: 5.41 (A); 17.29 (B) (no separation of enantiomers)
Voltage: 15 kV
Model: Bio-Focus 3000

OTHER SUBSTANCES
Also analyzed: albuterol (salbutamol), alprenolol, amorolfine, atenolol, atropine, azelastine, baclofen, bamethan, benproperine, benserazide, biperiden, bisoprolol, brompheniramine, bupivacaine, bupranolol, butethamate, carazolol, carbuterol, carteolol, carvedilol, celiprolol, chloroquine, chlorpheniramine, chlorphenoxamine, cicletanine, clenbuterol, clidinium bromide, clobutinol, dimethindene, dipivefrin, disopyramide, dobutamine, doxylamine, fendiline, flecainide, gallopamil, homatropine, ipratropium bromide, isoproterenol (isoprenaline), isothipendyl, ketamine, meclizine, mefloquine, mepindolol, mequitazine, metaclazepam, metaproterenol (orciprenaline), metipranolol, metoprolol, nafronyl (naftidrofuryl), nefopam, nicardipine, norfenefrine, ofloxacin, ornidazole, orphenadrine, oxomemazine, oxprenolol, oxybutynin, phenoxybenzamine, phenylpropanolamine, pholedrine, pindolol, pirbuterol, prilocaine, procyclidine, promethazine, propafenone, propranolol, reproterol, sotalol, sulpride, synephrine, talinolol, terbutaline, tetrahydrozoline (tetryzoline), theodrenaline, tioconazole, tocainide, trihexyphenidyl, trimeprazine (alimemazine), trimipramine, tropicamide, verapamil, zopiclone

KEY WORDS
coated capillary

REFERENCE
Koppenhoefer,B.; Eperlein,U.; Schlunk,R.; Zhu,X.; Lin,B. Separation of enantiomers of drugs by capillary electrophoresis. V. Hydroxypropyl-α-cyclodextrin as chiral solvating agent, *J.Chromatogr.A*, **1998**, *793*, 153–164.

Butethamate

Molecular formula: C$_{16}$H$_{25}$NO$_2$
Molecular weight: 263.38
CAS Registry No.: 14007-64-8, 3639-12-1 (citrate)
Merck Index (12th ed.): 1551

SAMPLE
Matrix: solutions

CAPILLARY ELECTROPHORESIS
Capillary: 57 cm × 75 μm (50 cm to detector) (Beckman)
Capillary temperature: 35
Running buffer: 100 mM pH 2 Phosphate buffer containing 1.8% β-cyclodextrin and 40 mM (R)-(-)-camphorsulfonic acid
Detector: UV 214
Migration time: 23.68, 23.90 (enantiomers)
Voltage: 12 kV
Model: Beckman P/ACE 2100

OTHER SUBSTANCES
Simultaneous: cyclodrine, cyclopentolate

KEY WORDS
chiral

REFERENCE
Bunke,A.; Jira,T. Enantiomeric separation by capillary electrophoresis using chiral and achiral ion pairing reagents, *Pharmazie*, **1996**, *51*, 479–486.

SAMPLE
Matrix: solutions
Sample preparation: Inject an aliquot of a 100 μg/mL solution in running buffer.

CAPILLARY ELECTROPHORESIS
Capillary: 29 cm × 50 μm fused-silica (24.5 cm to detector) (Yongnian Optical Conductive Fiber Plant, China), coated with polyacrylamide
Capillary preparation: No details of the polyacrylamide coating process are provided. However, another paper (LC.GC 1997, 15, 40) by this group indicates that they use the procedure of Hjertén, thus: Adjust the pH of 20 mL water to 3.5 with acetic acid, add 80 μL 3-(trimethoxysilyl)propyl methacrylate (3-methacryloxypropyltrimethoxysilane), mix, suck into capillary, let stand at room temperature for 1 h, remove the solution, wash with water. Fill the capillary with a deaerated 3-4% acrylamide solution containing 1 μL/mL N,N,N',N'-tetramethylethylenediamine and 1 mg/mL potassium persulfate, let stand for 30 min, remove excess solution by aspiration, rinse with water, remove water by aspiration, dry at 35° (J. Chromatogr. 1985, 347, 191).
Capillary temperature: 25
Running buffer: 100 mM NaH_2PO_4 adjusted to pH 2.5 (A) or 100 mM NaH_2PO_4 containing 45 mM hydroxypropyl-gamma-cyclodextrin, adjusted to pH 2.5 (B)
Injection: Electrokinetic injection at 15 kV for 3 s.
Detector: UV 200, UV 210
Migration time: 5.04 (A), 13.19, 13.40 (B, enantiomers)
Voltage: 15 kV
Model: Bio-Rad BioFocus 3000

OTHER SUBSTANCES
Simultaneous: albuterol, alprenolol, atenolol, baclofen, bamethan, benproperine, benserazide, bisoprolol, bupranolol, butamirate, carbuterol, celiprolol, clenbuterol, clobutinol, dipivefrin, isoproterenol (isoprenaline), metaproterenol (orciprenaline), metipranolol, metoprolol, norfenefrine, ornidazole, oxprenolol, phenylpropanolamine, pholedrine, pirbuterol, prilocaine, procyclidine, sotalol, synephrine, terbutaline, tocainide

KEY WORDS
chiral; coated capillary

REFERENCE
Koppenhoefer,B.; Epperlein,U.; Xiaofeng,Z.; Bingcheng,L. Separation of enantiomers of drugs by capillary electrophoresis. Part 4: Hydroxypropyl-γ-cyclodextrin as chiral solvating agent, *Electrophoresis*, **1997**, *18*, 924–930.

SAMPLE
Matrix: solutions

CAPILLARY ELECTROPHORESIS
Capillary: 29-36 cm × 50 μm fused-silica (24.5-31.5 cm to detector) (Yongnian Optical Conductive Fiber Plant, China) coated with polyacrylamide
Capillary preparation: Coat capillary as follows. Adjust the pH of 20 mL water to 3.5 with acetic acid, add 80 μL 3-(trimethoxysilyl)propyl methacrylate (3-methacryloxypropyltrimethoxysilane), mix, suck into capillary, let stand at room temperature for 1 h, remove the solution, wash with water. Fill the capillary with a deaerated 3-4% acrylamide solution containing 1 μL/mL N,N,N',N'-tetramethylethylenediamine and 1 mg/mL potassium persulfate, let stand for 30 min, remove excess solution by aspiration, rinse with water, remove water by aspiration, dry at 35° (J. Chromatogr. 1985, 347, 191).
Capillary temperature: 25
Running buffer: 100 mM pH 2.5 NaH_2PO_4 (A) or 100 mM pH 2.5 NaH_2PO_4 containing 45 mM hydroxypropyl-α-cyclodextrin (Wacker, Munich) (B)
Injection: Electromigration at 15 kV for 3 s.
Detector: UV 200; UV 210
Migration time: 5.04 (A); 17.08 (B) (no separation of enantiomers)
Voltage: 15 kV
Model: Bio-Focus 3000

OTHER SUBSTANCES
Also analyzed: albuterol (salbutamol), alprenolol, amorolfine, atenolol, atropine, azelastine, baclofen, bamethan, benproperine, benserazide, biperiden, bisoprolol, brompheniramine, bupivacaine, bupranolol, butamirate, carazolol, carbuterol, carteolol, carvedilol, celiprolol, chloroquine, chlorpheniramine, chlorphenoxamine, cicletanine, clenbuterol, clidinium bromide, clobutinol, dimethindene, dipivefrin, disopyramide, dobutamine, doxylamine, fendiline, flecainide, gallopamil, homatropine, ipratropium bromide, isoproterenol (isoprenaline), isothipendyl, ketamine, meclizine, mefloquine, mepindolol, mequitazine, metaclazepam, metaproterenol (orciprenaline), metipranolol, metoprolol, nafronyl (naftidrofuryl), nefopam, nicardipine, norfenefrine, ofloxacin, ornidazole, orphenadrine, oxomemazine, oxprenolol, oxybutynin, phenoxybenzamine, phenylpropanolamine, pholedrine, pindolol, pirbuterol, prilocaine, procyclidine, promethazine, propafenone, propranolol, reproterol, sotalol, sulpride, synephrine, talinolol, terbutaline, tetrahydrozoline (tetryzoline), theodrenaline, tioconazole, tocainide, trihexyphenidyl, trimeprazine (alimemazine), trimipramine, tropicamide, verapamil, zopiclone

KEY WORDS
coated capillary

REFERENCE
Koppenhoefer,B.; Eperlein,U.; Schlunk,R.; Zhu,X.; Lin,B. Separation of enantiomers of drugs by capillary electrophoresis. V. Hydroxypropyl-α-cyclodextrin as chiral solvating agent, *J.Chromatogr.A*, **1998**, *793*, 153–164.

Buthionine sulfoximine

Molecular formula: $C_8H_{18}N_2O_3S$
Molecular weight: 222.31
CAS Registry No.: 5072-26-4
Merck Index (12th ed.): 1556

SAMPLE
Matrix: plasma
Sample preparation: Inject plasma directly.

CAPILLARY ELECTROPHORESIS
Capillary: 72 cm × 50 μm (50 cm to detector) (Polymicro Technologies)
Capillary preparation: Between runs wash capillary with 100 mM NaOH and running buffer.
Capillary temperature: 30
Running buffer: 20 mM pH 6.8 Phosphate buffer containing 170 mM sodium dodecyl sulfate
Injection: Vacuum injection at 17 kPa
Detector: UV 190
Migration time: 11.9 (L,R), 12.1 (L,S)
Voltage: 236 V/cm
Current: 32 μA
Model: Applied Biosystems 270A-HT
Limit of detection: 3.9 μg/mL

KEY WORDS
chiral

REFERENCE
Sandor,V.; Flarakos,T.; Batist,G.; Wainer,I.W.; Lloyd,D.K. Quantitation of the diastereoisomers of L-buthionine-(R,S)-sulfoximine in human plasma: a validated assay by capillary electrophoresis, *J.Chromatogr.B*, **1995**, *673*, 123–131.

Butylscopolammonium bromide

Molecular formula: $C_{21}H_{30}BrNO_4$
Molecular weight: 440.38
CAS Registry No.: 149-64-4
Merck Index (12th ed.): 1624

SAMPLE
Matrix: solutions

CAPILLARY ELECTROPHORESIS
Capillary: 70 cm $\times$ 50 μm fused-silica (GL Science, Tokyo)
Capillary preparation: Before each run rinse capillary with running buffer at 94 kPa for 5 min before each run.
Running buffer: 40 mM pH 6.0 ammonium acetate buffer
Injection: Pressure injection at 5 kPa (50 mbar) for 4 s.
Detector: MS, Perkin-Elmer Sciex API-300 quadrupole, electrospray (ionspray) interface, sheath liquid MeOH:running buffer 50:50 at 2.5 μL/min, ionspray voltage 5 kV, positive ion mode
Migration time: 12
Voltage: 20 kV (net voltage across capillary = 15 kV (applied voltage - electrospray voltage))
Model: Hewlett Packard 3D CE

OTHER SUBSTANCES
Simultaneous: acetaminophen, ascorbic acid, caffeine, ibuprofen, ketoprofen, niacin, niacinamide, riboflavin, thiamine, vitamin B12, warfarin

REFERENCE
Tanaka,Y.; Kishimoto,Y.; Otsuka,K.; Terabe,S. Strategy for selecting separation solutions in capillary electrophoresis-mass spectrometry, *J.Chromatogr.A*, **1998**, *817*, 49–57.

Caffeine

Molecular formula: $C_8H_{10}N_4O_2$
Molecular weight: 194.19
CAS Registry No.: 58-08-2, 5743-12-4 (monohydrate)
Merck Index (12th ed.): 1674
Lednicer: 1 111

SAMPLE
Matrix: beverages
Sample preparation: Extract 3 g tea leaves with 180 mL boiling water for 5 min, filter (paper), dilute the filtrate 10-fold with 0.1% metaphosphoric acid, filter (0.45 μm), inject an aliquot of the filtrate.

CAPILLARY ELECTROPHORESIS
Capillary: 77 cm $\times$ 50 μm fused-silica (70 cm to detector)

Capillary preparation: Between runs rinse with water, 100 mM HCl, water, 100 mM NaOH, and running buffer (total rinse time 7 min).
Capillary temperature: 23
Running buffer: 20 mM Borax adjusted to pH 8.0 with HCl
Injection: Pressure injection using nitrogen for 5 s.
Detector: UV 200
Migration time: 5
Voltage: 30 kV
Model: Beckman P/ACE 5000

OTHER SUBSTANCES
Extracted: catechin, epicatechin, epicatechin gallate, epigallocatechin, epigallocatechin gallate, theanine, ascorbic acid

KEY WORDS
tea; leaves

REFERENCE
Horie,H.; Mukai,T.; Kohata,K. Simultaneous determination of qualitatively important components in green tea infusions using capillary electrophoresis, *J.Chromatogr.A*, **1997**, *758*, 332–335.

SAMPLE
Matrix: beverages
Sample preparation: Degas carbonated beverage with sonication under vacuum, mix an aliquot with an equal volume of running buffer, filter (0.45 μm)

CAPILLARY ELECTROPHORESIS
Capillary: 44 cm × 50 μm fused-silica (Polymicro Technologies)
Capillary preparation: Before each injection wash capillary with 2 volumes of fresh running buffer. Condition new capillaries by washing with 1 M NaOH at 60° for 15 min, with 100 mM NaOH at 60° for 15 min, and with water at 25° for 15 min.
Capillary temperature: 35
Running buffer: 20 mM Glycine adjusted to pH 9.0 with NaOH
Injection: Hydrodynamic injection for 1 s.
Detector: UV 215
Migration time: 1.3
Voltage: 20 kV
Model: SpectraPhoresis 1000 CE (ThermoSeparations Products)
Limit of detection: 1.6 μg/mL

OTHER SUBSTANCES
Simultaneous: aspartame, benzoic acid

KEY WORDS
soft drinks

REFERENCE
Walker,J.C.; Zaugg,S.E.; Walker,E.B. Analysis of beverages by capillary electrophoresis, *J.Chromatogr.A*, **1997**, *781*, 481–485.

SAMPLE
Matrix: beverages

CAPILLARY ELECTROPHORESIS
Capillary: 36 cm × 50 μm fused-silica (32 cm to detector)
Capillary preparation: Before each run rinse capillary with 1 M NaOH for 2 min and with water for 2 min.
Capillary temperature: 20
Running buffer: 25 mM pH 7.0 Phosphate buffer containing 50 mM borate and 25 mM sodium dodecyl sulfate
Injection: Pressure injection at 350 mbar for 1 s.

Detector: UV 280
Migration time: 3.7
Voltage: 15 kV
Model: Bio-Rad BioFocus 3000
Limit of quantitation: 500 ng/mL

OTHER SUBSTANCES
Simultaneous: catechin, catechin gallate, epicatechin, epicatechin gallate, epigallocatechin, epigallocatechin gallate, gallocatechin gallate, vitamin C

KEY WORDS
tea; comparison with HPLC

REFERENCE
Watanabe,T.; Nishiyama,R.; Yamamoto,A.; Nagai,S.; Terabe,S. Simultaneous analysis of individual catechins, caffeine, and ascorbic acid in commercial canned green and black teas by micellar electrokinetic chromatography, *Anal.Sci.*, **1998**, *14*, 435–438.

SAMPLE
Matrix: beverages, blood, formulations
Sample preparation: Beverages. Degas carbonated beverages with argon, filter (0.25 μm), inject an aliquot. Serum. Filter (0.25 μm), inject an aliquot. Alternatively, vortex 600 μL serum with 900 μL ethyl acetate for 40 s, centrifuge. Remove the organic layer, add 500 μL ethyl acetate to the aqueous layer, repeat extraction. Repeat this extraction again. Combine all the organic layers and evaporate them to dryness under a stream of nitrogen, reconstitute with 20 μL water, inject an aliquot. Tablets. Dissolve 1 tablet in 200 mL water, filter (0.25 μm), inject an aliquot.

CAPILLARY ELECTROPHORESIS
Capillary: 60 cm × 50 μm fused-silica (25 cm to detector) (Polymicro Technologies)
Capillary preparation: Before use rinse capillary hydrodynamically at 520 kPa with 500 mM pH 13 EDTA for 30 min, with water for 20 min, and with running buffer for 10 min. Fill with running buffer.
Running buffer: 20 mM pH 11.0 Sodium phosphate buffer containing 40 mM sodium dodecyl sulfate
Injection: Vacuum injection of 2 nL.
Detector: UV 274
Migration time: 0.9
Internal standard: antipyrine (1.1)
Voltage: 29 kV
Model: ISCO Model 3850
Limit of detection: 1.5 μg/mL

OTHER SUBSTANCES
Extracted: paraxanthine, theobromine, theophylline, 1,3,7-trimethyluric acid

KEY WORDS
serum; tablets; coffee; tea; soft drinks

REFERENCE
Zhao,Y.; Lunte,C.E. Determination of caffeine and its metabolites by micellar electrokinetic capillary electrophoresis, *J.Chromatogr.B*, **1997**, *688*, 265–274.

SAMPLE
Matrix: beverages, food
Sample preparation: Dilute sample with water, add dehydroacetic acid to a final concentration of 10 μg/mL, filter (0.45 μm cellulose acetate), inject an aliquot.

CAPILLARY ELECTROPHORESIS
Capillary: 75 cm × 75 μm fused-silica (50 cm to detector) (Polymicro Technologies)

Capillary preparation: Flush with running buffer for 2 min between runs. Clean each week by washing with 100 mM NaOH for 10 min and with water for 10 min, fill with running buffer.
Capillary temperature: 27
Running buffer: 10 mM Sodium borate containing 10 mM KH_2PO_4 and 50 mM sodium deoxycholate, pH 8.6
Injection: Injection by vacuum at 20 kPa.s (vacuum level 2)
Detector: UV 220
Migration time: 5.3
Internal standard: dehydroacetic acid (8.4)
Voltage: 20 kV
Model: ISCO Model 3140

OTHER SUBSTANCES
Simultaneous: acesulfame-K, alitame, aspartame, benzoic acid, dulcin, saccharin, sorbic acid

KEY WORDS
soft drink; cordial; tomato sauce; jam; sweeteners; comparison with HPLC

REFERENCE
Thompson,C.O.; Trenerry,V.C.; Kemmery,B. Micellar electrokinetic capillary chromatographic determination of artificial sweeteners in low-Joule soft drinks and other foods, *J.Chromatogr.A*, **1995**, *694*, 507–514.

SAMPLE
Matrix: blood
Sample preparation: 100 μL Serum or plasma + 900 μL ethyl acetate, vortex for 30 s, centrifuge, remove 500 μL organic layer, repeat procedure twice more. Combine the organic layers and evaporate them to dryness under a stream of nitrogen, reconstitute the residue in 50 μL water, inject an aliquot.

CAPILLARY ELECTROPHORESIS
Capillary: 72 cm × 50 μm fused-silica capillary (50.5 cm to detector) (Applied Biosystems)
Capillary preparation: Before each run rinse capillary with 100 mM NaOH and running buffer at 0.67 bar for 3-5 min.
Capillary temperature: 26.5
Running buffer: 25 mM pH 8.0 Phosphate buffer containing 80 mM sodium dodecyl sulfate
Injection: Vacuum injection at 0.17 bar for 2 s (ca. 8.75 nL)
Detector: UV 274
Migration time: 11
Voltage: 21 kV
Model: Applied Biosystems Model 270A
Limit of detection: 5.7 μM

OTHER SUBSTANCES
Extracted: paraxanthine, theophylline

KEY WORDS
serum; plasma

REFERENCE
Lee,K.J.; Heo,G.S.; Kim,N.J.; Moon,D.C. Separation of theophylline and its analogues by micellar electrokinetic chromatography: application to the determination of theophylline in human plasma, *J.Chromatogr.*, **1992**, *577*, 135–141.

SAMPLE
Matrix: blood
Sample preparation: 100 μL Serum + 150 μL 80 μg/mL 3-isobutyl-1-methylxanthine in MeCN, mix for 30 s, centrifuge at 14000 g for 1 min, inject an aliquot of the supernatant.

CAPILLARY ELECTROPHORESIS
Capillary: 25 cm × 47 μm

Capillary preparation: Wash with running buffer by pressure injection for 1 min after each run. Wash daily with 50 mM phosphoric acid for 2 min, with water for 1 min, with 2 M NaOH for 2 min, with water for 1 min, and with running buffer for 2 min.
Capillary temperature: 24
Running buffer: 220 mM boric acid adjusted to pH 8.8 with 2.5 M NaOH
Injection: Pressure injection for 8 s.
Detector: UV 254
Migration time: 1.7
Internal standard: 3-isobutyl-1-methylxanthine (2.3)
Voltage: 12 kV
Model: Beckman
Limit of quantitation: 10 μg/mL

OTHER SUBSTANCES
Extracted: acetaminophen, carbamazepine, iohexol, pentobarbital, phenobarbital, phenytoin, salicylic acid, theophylline

KEY WORDS
serum; pharmacokinetics

REFERENCE
Shihabi,Z.K.; Constantinescu,M.S. Iohexol in serum determined by capillary electrophoresis, *Clin.Chem.*, **1992**, *38*, 2117–2120.

SAMPLE
Matrix: blood
Sample preparation: Centrifuge at 10000 g for 3 min, inject an aliquot.

CAPILLARY ELECTROPHORESIS
Capillary: 50 cm × 50 μm fused-silica (45 cm to detector) (Polymicro Technologies)
Capillary preparation: Between runs rinse capillary at 100 psi with 100 mM NaOH for 1 min and with running buffer for 2 min.
Capillary temperature: 20
Running buffer: 6 mM Sodium tetraborate containing 10 mM Na_2HPO_4 and 75 mM sodium dodecyl sulfate adjusted to pH 9.6 with 1 M NaOH
Injection: Pressure injection at 5 psi for 0.4 s.
Detector: UV 275
Migration time: 6.3
Voltage: 20 kV
Current: ca. 38 μA
Model: Bio-Rad BioFocus 3000

OTHER SUBSTANCES
Extracted: theophylline

KEY WORDS
serum; direct injection

REFERENCE
Zhang,C.-X.; Thormann,W. Determination of drug levels in human serum by micellar electrokinetic capillary chromatography with direct sample injection using different quantitation strategies, *J.Capillary Electrophor.*, **1994**, *1*, 208–218.

SAMPLE
Matrix: bulk
Sample preparation: Prepare a 10 mg/mL solution in 3 mg/mL pholcodine in 10 mM HCl, dilute 5-fold with 10 mM HCl, filter (0.45 μm cellulose acetate), inject an aliquot.

CAPILLARY ELECTROPHORESIS
Capillary: 72 cm × 75 μm fused-silica (50 cm to detector) (Isco)

Capillary preparation: Flush with running buffer for 2 min between runs. Every 24 h wash capillary with 100 mM HCl for 10 min, wash with water, wash 100 mM NaOH, wash with water, fill with running buffer. Prepare new capillary by filling with 1 M NaOH, let stand for 1 h, fill with 100 mM NaOH, let stand for 1 h, wash with water, fill with running buffer.
Capillary temperature: 30
Running buffer: MeCN:buffer 10:90 (Buffer was 10 mM KH_2PO_4 containing 10 mM sodium tetraborate and 50 mM cetyltrimethylammonium bromide, pH 8.6.)
Injection: Injection by vacuum 5.0 kPa.s (4 nL)
Detector: UV 280
Migration time: 6.3
Internal standard: pholcodine (9.1)
Voltage: 15 kV
Model: Isco Model 3140

OTHER SUBSTANCES
Simultaneous: acetylcodeine, codeine, diamorphine, ethylmorphine, morphine, noscapine, papaverine, strychnine, theophylline

---

REFERENCE
Trenerry,V.C.; Wells,R.J.; Robertson,J. The analysis of illicit heroin seizures by capillary zone electrophoresis, *J.Chromatogr.Sci.*, **1994**, *32*, 1–6.

SAMPLE
Matrix: bulk
Sample preparation: 10-50 mg Bulk compound + 1 mL 1 mg/mL caffeine in 10 mM HCl, make up to 10 mL with 10 mM HCl, sonicate for 2 min, mix thoroughly, filter (0.45 μm cellulose acetate), inject an aliquot.

CAPILLARY ELECTROPHORESIS
Capillary: 75 cm × 75 μm fused-silica (50 cm to the detector) (ISCO)
Capillary preparation: Flush with running buffer for 2 min between analyses. Condition capillary by filling with 1 M NaOH and allowing to stand for 1 h, fill with 100 mM NaOH and allow to stand for 1 h, wash with water, fill with running buffer. Each week wash capillary with 100 mM HCl for 10 min, with water, with 100 mM NaOH, and with water then fill with running buffer.
Capillary temperature: 30
Running buffer: DMSO:ethanolamine:buffer 11:1:88 (Buffer was 0.92 g cetyltrimethylammonium bromide in 100 mL 10 mM sodium tetraborate, adjust pH to 11.5 with 1 M NaOH.)
Injection: Load under vacuum, vacuum level 2, 10.0 kPa s
Detector: UV 254
Migration time: 6.5
Internal standard: caffeine
Voltage: -15 kV
Model: ISCO Model 3140 electropherograph

OTHER SUBSTANCES
Simultaneous: amphetamine, p-aminobenzoic acid, ephedrine, methamphetamine, methylenedioxyamphetamine, methylenedioxymethamphetamine, norephedrine, pseudoephedrine, pseudonorephedrine

---

KEY WORDS
caffeine is IS

REFERENCE
Trenerry,V.C.; Robertson,J.; Wells,R.J. Analysis of illicit amphetamine seizures by capillary electrophoresis, *J.Chromatogr.A*, **1995**, *708*, 169–176.

SAMPLE
Matrix: bulk
Sample preparation: Dilute with 400 μg/mL n-propyl p-hydroxybenzoate in running buffer

---

CAPILLARY ELECTROPHORESIS
Capillary: 27 cm × 50 μm fused-silica (20 cm to detector)

Capillary preparation: Before each run rinse with running buffer at high pressure.
Capillary temperature: 30
Running buffer: MeCN:water 15:85 containing 40 mM sodium dodecyl sulfate, 8.5 mM sodium borate, and 8.5 mM sodium phosphate, pH 8.5
Injection: Inject at high pressure for 1 s
Detector: UV 214
Migration time: 1.1
Internal standard: n-propyl p-hydroxybenzoate (1.8)
Voltage: 20 kV
Model: Beckman P/ACE System 2100

OTHER SUBSTANCES
Simultaneous: acetaminophen, acetylcodeine, O-acetylmorphine, aspirin, cocaine, codeine, diamorphine, diphenhydramine, hydromorphone, isoniacinamide, lidocaine, methaqualone, morphine, niacinamide, noscapine, papaverine, phenacetin, phenobarbital, phenylpropanolamine, procaine, quinine, salicylic acid, strychnine, thebaine

REFERENCE
Walker,J.A.; Krueger,S.T.; Lurie,I.S.; Marché,H.L.; Newby,N. Analysis of heroin drug seizures by micellar electrokinetic capillary chromatography (MECC), *J.Forensic Sci.*, **1995**, *40*, 6–9.

SAMPLE
Matrix: bulk
Sample preparation: Dissolve in running buffer to a concentration of 1 mg/mL, vortex for 2 min, add an equal volume of 100 μg/mL diphenhydramine in running buffer, mix, inject an aliquot.

CAPILLARY ELECTROPHORESIS
Capillary: 65 cm × 50 μm fused-silica (60 cm to detector)
Capillary preparation: Fill capillary with fresh running buffer before each run. Before use fill with 100 mM NaOH for 20 min, rinse with water, flush with running buffer
Running buffer: MeCN:50 mM 6-aminocaproic acid containing 50 mM 3-N,N-dimethylmyristylammoniopropanesulfonate (MAPS; Fluka) and 5 mM 1-heptanesulfonic acid 10:90, pH adjusted to 4.0 with 1 M phosphoric acid
Injection: Hydrodynamic injection by gravity or pressure.
Detector: UV 214
Migration time: 10
Internal standard: diphenhydramine (12)
Voltage: 27 kV
Current: ≤25 μA
Model: Dionex system I

OTHER SUBSTANCES
Simultaneous: acetaminophen, allobarbital, barbital, codeine, diamorphine, morphine, niacinamide, noscapine, papaverine, phenobarbital, procaine

REFERENCE
Naess,O.; Rasmussen,K.E. Micellar electrokinetic chromatography of charged and neutral drugs in acidic running buffers containing a zwitterionic surfactant, sulfonic acids or sodium dodecyl sulphate. Separation of heroin, basic by-products and adulterants, *J.Chromatogr.A*, **1997**, *760*, 245–251.

SAMPLE
Matrix: formulations
Sample preparation: Grind tablet, add 75 mL acidic MeOH, sonicate for 5 min, shake for 10 min, make up to 100 mL with acidic MeOH, centrifuge at 2500 g for 10 min. Remove a 5 mL aliquot of the supernatant and add it to 5 mL 10 mM ethyl p-aminobenzoate in MeOH, make up to 50 mL with MeOH, inject an aliquot. (Prepare acidic MeOH by adding 10 mL 100 mM HCl to 1 L MeOH.)

CAPILLARY ELECTROPHORESIS
Capillary: 80 cm × 100 μm fused-silica (50 cm to detector) (Shimadzu)
Running buffer: 20 mM pH 11 Phosphate buffer containing 50 mM sodium dodecyl sulfate

Injection: Injection by siphon at 5 cm for 5 s
Detector: UV 214
Migration time: 10.5
Internal standard: ethyl p-aminobenzoate (20)
Current: 100 µA (constant current)

OTHER SUBSTANCES
Simultaneous: acetaminophen, aspirin, o-ethoxybenzamide, salicylamide

---

KEY WORDS
tablets

REFERENCE
Fujiwara,S.; Honda,S. Determination of ingredients of antipyretic analgesic preparations by micellar electro-kinetic capillary chromatography, *Anal.Chem.*, **1987**, *59*, 2773–2776.

SAMPLE
Matrix: formulations
Sample preparation: Grind tablet or powders, add 70 mL MeOH, add 20 mL 5 mg/mL methyl p-hydroxybenzoate in MeOH, sonicate for 10 min, shake for 5 min, make up to 100 mL with water, filter (0.45 µm), inject an aliquot.

CAPILLARY ELECTROPHORESIS
Capillary: 65 cm × 50 µm fused-silica (50 cm to detector) (SGE)
Running buffer: 20 mM NaH_2PO_4 containing 100 mM sodium dodecyl sulfate, adjusted to pH 9.0 with 20 mM sodium tetraborate
Injection: Siphon
Detector: UV 210
Migration time: 9
Internal standard: methyl p-hydroxybenzoate
Voltage: +20 kV

OTHER SUBSTANCES
Simultaneous: acetaminophen, chlorpheniramine, dipyrone (sulpyrin), ethenzamide, guaifene-sin, isopropylantipyrine, naproxen, noscapine, phenacetin, tipepidine, trimetoquinol

---

KEY WORDS
tablets; powders

REFERENCE
Nishi,H.; Fukuyama,T.; Matsuo,M.; Terabe,S. Effect of surfactant structures on the separation of cold medicine ingredients by micellar electrokinetic chromatography, *J.Pharm.Sci.*, **1990**, *79*, 519–523.

SAMPLE
Matrix: formulations
Sample preparation: Grind granules, add 70 mL MeOH, warm at 40° with occasional shaking, cool, add 20 mL 5 mg/mL methyl p-hydroxybenzoate in MeOH, make up to 100 mL with water, filter (0.45 µm), inject an aliquot.

CAPILLARY ELECTROPHORESIS
Capillary: 65 cm × 50 µm fused-silica (SGE)
Running buffer: 20 mM NaH_2PO_4 containing 100 mM sodium cholate, adjusted to pH 9.0 with 20 mM sodium tetraborate
Injection: Siphon for 10 s at 10 cm height
Detector: UV 210
Migration time: 8.5
Internal standard: methyl p-hydroxybenzoate (16)
Voltage: +20 kV

OTHER SUBSTANCES
Simultaneous: acetaminophen, chlorpheniramine, dibucaine, dipyrone (sulpyrin), ethenzamide, guaifenesin, isopropylantipyrine, naproxen, noscapine, phenacetin, tipepidine, trimetoquinol, triprolidine

KEY WORDS
granules

REFERENCE
Nishi,H.; Fukuyama,T.; Matsuo,M.; Terabe,S. Separation and determination of the ingredients of a cold medicine by micellar electrokinetic chromatography with bile salts, *J.Chromatogr.*, **1990**, *498*, 313–323.

SAMPLE
Matrix: formulations
Sample preparation: 100 mg Cream + 4 mL MeOH + 4 mL 4 mM caffeine + 32 mL water, heat at 50° until the cream was totally dissolved, cool, filter (Millex-GS), inject an aliquot of the filtrate.

CAPILLARY ELECTROPHORESIS
Capillary: 50 cm × 75 μm fused-silica (38 cm to detector) (SGE)
Running buffer: 10 mM pH 9.5 Borate buffer containing 75 mM sodium dodecyl sulfate
Injection: Hydrodynamic injection at 10 cm for 6 s (ca. 10 nL).
Detector: UV 254
Migration time: 4
Internal standard: caffeine
Voltage: 10 kV
Current: 20 μA
Model: Laboratory constructed

OTHER SUBSTANCES
Extracted: hydroquinone

KEY WORDS
cream; caffeine is IS

REFERENCE
Sakodinskaya,I.K.; Desiderio,C.; Nardi,A.; Fanali,S. Micellar electrokinetic chromatographic study of hydroquinone and some of its ethers. Determination of hydroquinone in skin-toning cream, *J.Chromatogr.*, **1992**, *596*, 95–100.

SAMPLE
Matrix: formulations
Sample preparation: Dissolve capsule in 300 mL water, add 50 mL 5 mg/mL 4-hydroxyacetophenone in MeOH, sonicate for 15 min, let stand for 1 h, make up to 500 mL with water, filter (0.45 μm), inject an aliquot.

CAPILLARY ELECTROPHORESIS
Capillary: 57 cm × 50 μm
Capillary preparation: Before each run flush with 100 mM NaOH for 2 min then flush with running buffer for 3 min. Rinse a new capillary with 500 mM NaOH.
Capillary temperature: 40
Running buffer: 40 mM Disodium tetraborate containing 125 mM sodium lauryl sulfate
Injection: Inject sample for 2 s, inject water for 1 s.
Detector: UV 214
Migration time: 13.93
Internal standard: 4-hydroxyacetophenone (17.63)
Voltage: 12 kV
Model: Beckman P/ACE

OTHER SUBSTANCES
Simultaneous: acetaminophen

KEY WORDS
capsules

REFERENCE
Altria,K.D.; Clayton,N.G.; Hart,M.; Harden,R.C.; Hevizi,J.; Makwana,J.V.; Portsmouth,M.J. An inter-company cross-validation exercise on capillary electrophoresis testing of dose uniformity of paracetamol content in formulations, *Chromatographia*, **1994**, *39*, 180–184.

SAMPLE
Matrix: formulations
Sample preparation: Grind tablet, add MeCN:10 mM HCl 20:80, shake and sonicate for 15 min, centrifuge until a clear solution is obtained. Remove a 25 mL aliquot of the supernatant and add it to 10 mL 500 μg/mL propyl hydroxybenzoate, make up to 50 mL with MeCN:10 mM HCl 20:80, inject an aliquot.

CAPILLARY ELECTROPHORESIS
Capillary: 57 cm × 75 μm fused-silica (Beckman)
Capillary preparation: Pass running buffer through the capillary for at least 4 min before each injection.
Capillary temperature: 25
Running buffer: 20 mM pH 9 Borate buffer containing 25 mM sodium cholate and 50 mM sodium deoxycholate
Injection: Pressure injection for 2 s
Detector: UV 214
Migration time: 6
Internal standard: propyl hydroxybenzoate (10.5)
Voltage: 20 kV
Model: Beckman P/ACE System 2100

OTHER SUBSTANCES
Simultaneous: acetaminophen, aspirin, chlorpheniramine, dextropropoxyphene, salicylic acid

KEY WORDS
tablets

REFERENCE
Boonkerd,S.; Lauwers,M.; Detaevernier,M.R.; Michotte,Y. Separation and simultaneous determination of the components in an analgesic tablet formulation by micellar electrokinetic chromatography, *J.Chromatogr.A*, **1995**, *695*, 97–102.

SAMPLE
Matrix: formulations
Sample preparation: Weigh out amount of powdered tablet corresponding to 50 mg caffeine, add 10 mL running buffer, shake vigorously for 15 min, centrifuge, dilute the supernatant with running buffer, inject an aliquot.

CAPILLARY ELECTROPHORESIS
Capillary: 60 cm × 75 μm fused-silica (Accusep, Waters)
Capillary temperature: 30
Running buffer: 25 mM pH 6.0 Sodium phosphate buffer
Injection: Hydrostatic injection for 10 s.
Detector: UV 214
Migration time: 4.8
Voltage: +20 kV
Model: Waters Quanta 4000E
Limit of quantitation: 2.5 μg/mL

OTHER SUBSTANCES
Simultaneous: ergotamine

KEY WORDS
tablets

REFERENCE
Aboul-Enein,H.Y.; Bakr,S.A. Simultaneous determination of caffeine and ergotamine in pharmaceutical dosage
  formulation by capillary electrophoresis, *J.Liq.Chromatogr.Rel.Technol.*, **1997**, *20*, 47–55.

SAMPLE
Matrix: formulations
Sample preparation: Dissolve tablets in water, dilute an aliquot with water, inject an aliquot.

CAPILLARY ELECTROPHORESIS
Capillary: 57 cm × 50 μm fused-silica (50 cm to detector)
Capillary preparation: After each run rinse with 20 mM pH 9.00 borate buffer containing 90
  mM sodium dodecyl sulfate for 5 min and with 100 mM NaOH for 5 min.
Running buffer: MeOH:buffer 10:90 (Buffer was 10 mM pH 8.25 borate buffer containing 10
  mM phosphate and 45 mM sodium dodecyl sulfate.)
Injection: Pressure injection at 3.45 kPa for 3 s followed by a 2 s injection of 200 μM phenobar-
  bital solution.
Detector: UV 214
Migration time: 9.7
Internal standard: phenobarbital (13.3)
Voltage: 17 kV
Model: Beckman P/ACE 2100

OTHER SUBSTANCES
Simultaneous: theobromine, theophylline

KEY WORDS
tablets

REFERENCE
Vogt,C.; Conradi,S.; Rohde,E. Determination of caffeine and other purine compounds in food and pharmaceuticals by micellar electrokinetic chromatography, *J.Chem.Educ.*, **1997**, *74*, 1126–1130.

SAMPLE
Matrix: solutions
Sample preparation: Prepare a 200 μg/mL solution in water, inject an aliquot.

CAPILLARY ELECTROPHORESIS
Capillary: 60 cm × 50 μm
Running buffer: 15 mM pH 11.0 sodium phosphate buffer containing 25 mM sodium dodecyl
  sulfate
Injection: Hydrodynamic injection at 10 cm for 15 s
Detector: UV 214
Migration time: 2.3
Voltage: 30 kV
Model: Waters Quanta 4000

OTHER SUBSTANCES
Simultaneous: acetaminophen, aspirin, salicylamide, salicylic acid

REFERENCE
Swartz,M.E. Method development and selectivity control for small molecule pharmaceutical separations by
  capillary electrophoresis, *J.Liq.Chromatogr.*, **1991**, *14*, 923–938.

SAMPLE
Matrix: solutions

CAPILLARY ELECTROPHORESIS
Capillary: 72 cm × 50 μm silica (50 cm to detector) (Applied Biosystems)
Capillary preparation: Before each run wash capillary with 100 mM NaOH for 3 min, wash
  with running buffer for 3 min, aspirate neutral marker solution (2 drops DMSO in 10 mL

water) for 1 s, place capillary end in buffer vial for 5 s, inject sample. Wash capillary at the beginning of each day by passing 1 M NaOH through for 20 min.
Capillary temperature: 30
Running buffer: MeCN:buffer 10:90 (Buffer was 30 mM pH 9.3 borate buffer containing 30 mM sodium dodecyl sulfate.)
Injection: Inject by applying vacuum for 1 s (2-3 nL)
Detector: UV 200
Migration time: 5
Voltage: +30 kV
Current: 41 μA
Model: Applied Biosystems Model 270A

OTHER SUBSTANCES
Simultaneous: dyphylline, theobromine, theophylline

REFERENCE
Evenson,M.A.; Wiktorowicz,J.E. Automated capillary electrophoresis applied to therapeutic drug monitoring, *Clin.Chem.*, **1992**, *38*, 1847–1852.

SAMPLE
Matrix: solutions

CAPILLARY ELECTROPHORESIS
Capillary: 56.9 cm × 75 μm fused-silica (50 cm to detector)
Capillary temperature: 25
Running buffer: MeOH:60 mM pH 8.4 sodium borate buffer containing 60 mM sodium dodecyl sulfate 15:85
Detector: UV 214
Migration time: 8.5
Voltage: 16 kV
Model: Beckman PACE 2100

OTHER SUBSTANCES
Simultaneous: acetanilide, acetaminophen, aspirin, benzamide, salicylamide, salicylic acid

REFERENCE
McLaughlin,G.M.; Nolan,J.A.; Lindahl,J.L.; Palmieri,R.H.; Anderson,K.W.; Morris,S.C.; Morrison,J.A.; Bronzert,T.J. Pharmaceutical drug separations by HPCE: Practical guidelines, *J.Liq.Chromatogr.*, **1992**, *15*, 961–1021.

SAMPLE
Matrix: solutions

CAPILLARY ELECTROPHORESIS
Capillary: 65.4 cm × 50 μm fused-silica (56 cm to detector) (Hewlett-Packard)
Capillary temperature: 37
Running buffer: 11.7 mM pH 8.9 Borate buffer containing 8.3 mM phosphate and 50 mM sodium dodecyl sulfate
Injection: Injection at 50 mbar.s
Detector: UV 200
Migration time: 3.755
Voltage: 370 V/cm
Model: Hewlett-Packard G1600A [3D]Capillary Electrophoresis system

OTHER SUBSTANCES
Simultaneous: acetaminophen, guaifenesin, naproxen, noscapine, phenacetin

REFERENCE
Heiger,D.N.; Kaltenbach,P.; Sievet,H.-J. Diode array detection in capillary electrophoresis, *Electrophoresis*, **1994**, *15*, 1234–1247.

SAMPLE
Matrix: solutions

CAPILLARY ELECTROPHORESIS
Capillary: 60 cm × 75 μm fused-silica (53 cm to detector)
Capillary preparation: Rinse capillary with running buffer after each run. At the start of each day rinse with 100 mM NaOH for 5 min and with separation buffer for 10 min. Before storage rinse capillary with water and air dry.
Running buffer: 30 mM pH 9.0 Sodium borate buffer containing 60 mM sodium dodecyl sulfate
Injection: Hydrodynamic injection at 10 cm for 5-10 s
Detector: UV 214
Migration time: 5
Voltage: 20 kV
Model: Waters Quanta 4000

OTHER SUBSTANCES
Simultaneous: ethophylline, methylpropylxanthine, methylxanthine, pentoxifylline, propentofylline, theobromine, theophylline, xanthine

REFERENCE
Korman,M.; Vindevogel,J.; Sandra,P. Application of micellar electrokinetic chromatography to the quality control of pharmaceutical formulations: the analysis of xanthine derivatives, *Electrophoresis*, **1994**, *15*, 1304–1309.

SAMPLE
Matrix: solutions
Sample preparation: Inject an aliquot of a solution in MeOH.

CAPILLARY ELECTROPHORESIS
Capillary: 50 cm
Capillary temperature: 25
Running buffer: pH 8.1 Phosphate borate buffer containing 50 mM sodium dodecyl sulfate
Detector: UV 230
Migration time: 2.75
Voltage: 20 kV
Model: Otsuka CAPI-3000

OTHER SUBSTANCES
Simultaneous: acetaminophen

REFERENCE
Matsuda,R.; Hayashi,Y.; Sasaki,K.; Saito,Y. Deductive prediction of measurement precision and optimization of integration time and wavelength in capillary electrophoresis, *Chromatographia*, **1995**, *41*, 707–714.

SAMPLE
Matrix: solutions

CAPILLARY ELECTROPHORESIS
Capillary: 500 mm × 50 μm
Capillary preparation: Between runs wash capillary for 1 min with 100 mM phosphoric acid and for 1 min with buffer. At the beginning of each day wash the capillary with 2 M NaOH for 3 min, for 3 min with 100 mM phosphoric acid, and for 3 min with buffer.
Capillary temperature: 35
Running buffer: 100 mM boric acid containing 55.4 mM sodium dodecyl sulfate, adjusted to pH 8.4 ± 0.1 with 2 M NaOH
Injection: Pressure injection for 20 s (3% of capillary volume)
Detector: UV 214
Migration time: 4.5
Voltage: 16 kV
Current: 38 μA (fixed)
Model: Beckman Model 2000

OTHER SUBSTANCES
Simultaneous: acetaminophen, acetoacetanilide, felbamate

REFERENCE
Shihabi,Z.K.; Hinsdale,M.E. Sample matrix effects in micellar electrokinetic capillary electrophoresis, *J.Chromatogr.B*, **1995**, *669*, 75–83.

SAMPLE
Matrix: solutions
Sample preparation: Inject an aliquot of a 30-300 µg/mL solution in water.

CAPILLARY ELECTROPHORESIS
Capillary: 36 cm × 50 µm fused-silica (28 cm to detector)
Running buffer: 80 mM Reagent, 120 mM sodium dodecyl sulfate, 900 mM 1-butanol, and 10 mM borate in water where reagent is heptane (A) or octane (B) or 1-butyl chloride (C)
Injection: Hydrostatic injection.
Detector: UV 254
Migration time: 4.447 (A), 4.482 (B), 4.115 (C)
Voltage: 12 kV
Current: ca. 65 µA
Model: Waters Quanta 4000

OTHER SUBSTANCES
Simultaneous: aminopyrine, phenacetin, phenobarbital

REFERENCE
Fu,X.; Lu,J.; Zhu,A. Microemulsion electrokinetic chromatographic separation of antipyretic analgesic ingredients, *J.Chromatogr.A*, **1996**, *735*, 353–356.

SAMPLE
Matrix: solutions
Sample preparation: Inject an aliquot of a 1 mg/mL solution in MeOH.

CAPILLARY ELECTROPHORESIS
Capillary: 67 cm × 50 µm (60 cm to detector) (Composite Metal Services, Worcs., UK)
Capillary preparation: Before each run rinse with running buffer for 3 min.
Capillary temperature: 20
Running buffer: 50 mM pH 10.5 Glycine buffer containing 50 mM sodium dodecyl sulfate
Injection: Pressure injection at 30 mbar for 5 s.
Detector: UV 220
Migration time: 7
Voltage: 25 kV
Model: Beckman P/ACE 2050

OTHER SUBSTANCES
Simultaneous: amphetamine, codeine, diamorphine, morphine

REFERENCE
Hyötyläinen,T.; Sirén,H.; Riekkola,M.-L. Determination of morphine analogues, caffeine and amphetamine in biological fluids by capillary electrophoresis with the marker technique, *J.Chromatogr.A*, **1996**, *735*, 439–447.

SAMPLE
Matrix: solutions
Sample preparation: Inject an aliquot of a 1-50 µg/mL solution in water.

CAPILLARY ELECTROPHORESIS
Capillary: 55 cm × 50 µm fused-silica (35 cm to detector) (J&W)
Capillary preparation: Flush with running buffer for 5 min before each run. Periodically wash with 100 mM NaOH and flush extensively with running buffer.

Running buffer: MeOH:25 mM pH 9.24 borate buffer 20:80 containing 100 mM sodium dodecyl
   sulfate (A) or 50 mM pH 2.35 phosphate buffer (B) or 50 mM pH 9.24 borate buffer (C)
Injection: Injection of 5 μL using a split-flow injector ratio of 1:800.
Detector: UV 200
Migration time: 6.05 (A), 39.14 (B), 7.19 (C)
Voltage: 20 kV (A, B) or 12 kV (C)
Current: <60-80 μA
Model: ISCO Model 3850

OTHER SUBSTANCES
Also analyzed: acetylcodeine, amphetamine, barbital, cocaine, codeine, diamorphine, diazepam,
   flunitrazepam, lidocaine, monoacetylmorphine, morphine, nalorphine, narceine, noscapine, papaverine, pentobarbital, procaine, tetracaine, thebaine

KEY WORDS
all compounds were separated with running buffer A; some peaks overlapped with running buffers
   B and C.

REFERENCE
Tagliaro,F.; Smith,F.P.; Turrina,S.; Equisetto,V.; Marigo,M. Complementary use of capillary zone electrophoresis
   and micellar electrokinetic capillary chromatography for mutual confirmation of results in forensic drug
   analysis, *J.Chromatogr.A*, **1996**, *735*, 227–235.

SAMPLE
Matrix: solutions
Sample preparation: Inject an aliquot of a solution in running buffer.

CAPILLARY ELECTROPHORESIS
Capillary: 57 cm × 50 μm fused-silica (50 cm to detector) (Polymicro Technologies)
Capillary temperature: 25
Running buffer: 25 mM NaH_2PO_4 containing 0.83% poly(sodium 10-undecenyl sulfate), adjusted
   to pH 7.3 with 12.5 mM sodium tetraborate (Preparation of poly(sodium 10-undecenyl sulfate
   is as follows. React undecylenyl alcohol (1-undecen-11-ol) with one equivalent of chlorosulfonic
   acid in diethyl ether under nitrogen at -5° to obtain 10-undecenyl hydrogen sulfate, react with
   one equivalent of NaOH to obtain sodium 10-undecenyl sulfate (J. Microcolumn Sep. 1996, 8,
   115). Dissolve 20 g sodium 10-undecenyl sulfate in 40 mL water (degassed by purging with
   nitrogen for 24 h), add 600 mg potassium persulfate, stir at 70° under nitrogen for 50 h, cool,
   add 200 mL cold (0°) EtOH, filter, wash the solid with three 10 mL portions of cold EtOH, dry
   under vacuum at 65° overnight to obtain poly(sodium 10-undecenyl sulfate) (cf. J. Microcolumn
   Sep. 1992, 4, 509). Purify by dialysis using dialysis tubing with a 1000 MW cut-off (Spectrum
   Houston) for 24 h.)
Detector: UV 214
Migration time: 6.5
Voltage: 16.1 kV
Model: Beckman P/ACE 2000

OTHER SUBSTANCES
Simultaneous: acetaminophen, ethenzamide, guaifenesin, phenacetin, trimetoquinol

REFERENCE
Palmer,C.P.; Terabe,S. Micelle polymers as pseudostationary phases in MEKC: Chromatographic performance
   and chemical selectivity, *Anal.Chem.*, **1997**, *69*, 1852–1860.

SAMPLE
Matrix: solutions
Sample preparation: Inject an aliquot of a 10 μg/mL solution

CAPILLARY ELECTROPHORESIS
Capillary: 50 cm × 75 μm silica
Capillary temperature: 25
Running buffer: 20 mM pH 7.0 Phosphate/borate buffer containing 50 mM sodium dodecyl
   sulfate

Injection: 5 nL
Detector: UV 208
Migration time: 2.65
Voltage: 20 kV
Model: Otsuka CAPI-3000

OTHER SUBSTANCES
Simultaneous: acetaminophen, ethyl paraben, guaifenesin, methyl paraben, phenacetin, salicylamide

---

REFERENCE
Poe,R.B.; Hayashi,Y.; Matsuda,R. Precision-optimization of wavelengths in diode-array detection in separation science, *Anal.Sci.*, **1997**, *13*, 951–961.

SAMPLE
Matrix: solutions

CAPILLARY ELECTROPHORESIS
Capillary: 53 cm × 50 μm (44.5 cm to detector) (Polymicro Technologies)
Capillary temperature: 40
Running buffer: 70 mM Boric acid adjusted to pH 9.00 with 1 M NaOH
Injection: Pressure injection at 50 mbar for 2-5 s (2.9-7.2 nL).
Detector: UV 210
Migration time: 2
Voltage: 20-25 kV
Model: Hewlett-Packard [3D]CE

OTHER SUBSTANCES
Simultaneous: acetaminophen, aspirin, norephedrine, salicylic acid

---

REFERENCE
Thompson,L.; Veening,H.; Strein,T.G. Capillary electrophoresis in the undergraduate instrumental laboratory: Determination of common analgesic formulations, *J.Chem.Educ.*, **1997**, *74*, 1117–1121.

SAMPLE
Matrix: solutions

CAPILLARY ELECTROPHORESIS
Capillary: 57 cm × 50 μm fused-silica (50 cm to detector)
Running buffer: 50 mM pH 8.0 Borate buffer containing 40 mM sodium taurodeoxycholate and 25 mM phosphatidylcholine (Prepare by adding phosphatidylcholine and stirring for 3-6 h until all cloudiness disappears. Phosphatidylcholine was 95% pure soybean lecithin, Epikuron, Lucas Meyer & Co.)
Injection: Inject a solution of the compound in the running buffer at 20 psi for 1 s, inject MeOH: water 5:95 at 20 psi for 1 s, inject a solution of halofantrine in running buffer at 20 psi for 1 s.
Detector: UV 214
Migration time: k' 0.21
Model: Beckman P/ACE 5000

OTHER SUBSTANCES
Also analyzed: acetaminophen, amoxicillin, antipyrine, aspirin, azathioprine, captopril, carbamazepine, carprofen, chlorambucil, chlorpheniramine, chlorpromazine, cimetidine, clonidine, codeine, desipramine, diphenhydramine, ephedrine, fenoterol, flufenamic acid, flurbiprofen, haloperidol, hydroxyzine, ibuprofen, imipramine, indomethacin, ketoprofen, lidocaine, melphalan, metoprolol, nabumetone, nadolol, phenobarbital, phenol, promazine, propranolol, pyrilamine, ranitidine, ropinirole, salicylic acid, sulfamethoxazole, testosterone, theophylline, thioridazine, tiaprofenic acid, tolfenamic acid, trifluoperazine, trimethoprim, valproic acid, verapamil

KEY WORDS
comparison with HPLC; k' = (Tr-T0)/(T0(1-Tr/Tm)) where Tr = retention time of analyte; T0 = retention time of water; and Tm = retention time of marker (halofantrine)

REFERENCE
Hanna,M.; de Biasi,V.; Bond,B.; Salter,C.; Hutt,A.J.; Camilleri,P. Estimation of the partitioning characteristics of drugs: A comparison of a large and diverse drug series utilizing chromatographic and electrophoretic methodology, *Anal.Chem.*, **1998**, *70*, 2092–2099.

SAMPLE
Matrix: solutions

CAPILLARY ELECTROPHORESIS
Capillary: 80 cm × 50 μm fused-silica (Supelco)
Capillary preparation: Before each run rinse capillary with running buffer at 94 kPa for 5 min then with running buffer containing 80 mM sodium dodecyl sulfate at 94 kPa for 2.5 min before each run.
Running buffer: 50 mM pH 8.5 ammonium carbonate buffer
Injection: Pressure injection at 5 kPa (50 mbar) for 8 s into the capillary filled with running buffer containing sodium dodecyl sulfate, electrophoresis is then continued with plain running buffer.
Detector: MS, Perkin-Elmer Sciex API-300 quadrupole, electrospray (ionspray) interface, sheath liquid MeOH:running buffer 50:50 at 4 μL/min, ionspray voltage 5 kV, positive ion mode
Migration time: 15
Voltage: 25 kV (net voltage across capillary = 20 kV (applied voltage - electrospray voltage))
Model: Hewlett Packard 3D CE

OTHER SUBSTANCES
Simultaneous: acetaminophen, acetanilide, guaifenesin, phenacetin, pyridoxine

REFERENCE
Tanaka,Y.; Kishimoto,Y.; Otsuka,K.; Terabe,S. Strategy for selecting separation solutions in capillary electrophoresis-mass spectrometry, *J.Chromatogr.A*, **1998**, *817*, 49–57.

SAMPLE
Matrix: urine
Sample preparation: Prepare a solution in 100 mM NaOH, let stand at room temperature for 15 min, inject an aliquot. (Urine may be cleaned up by HPLC and an appropriate fraction of the eluate lyophilized for analysis by CE (HPLC details in paper).)

CAPILLARY ELECTROPHORESIS
Capillary: 60 cm × 75 μm fused-silica (Polymicro Technologies)
Capillary preparation: Between runs purge capillary with 500 mM KOH for 5 min.
Running buffer: 25 mM pH 11 Sodium acetate containing 12.5 mM Na_2HPO_4 and 50 mM sodium dodecyl sulfate
Injection: Hydrostatic injection for 5-15 s
Detector: UV 280
Migration time: 13
Voltage: 8-11 kV
Model: Quanta 4000 CE (Waters)

OTHER SUBSTANCES
Simultaneous: metabolites

REFERENCE
Rodopoulos,N.; Norman,A. Determination of caffeine and its metabolites in urine by high-performance liquid chromatography and capillary electrophoresis, *Scand.J.Clin.Lab.Invest.*, **1994**, *54*, 305–315.

Calcitonin

Molecular formula: $C_{151}H_{226}N_{40}O_5S_3$
Molecular weight: 2777.92
CAS Registry No.: 9007-12-9, 47931-85-1 (salmon), 21215-62-3 (human)
Merck Index (12th ed.): 1680

SAMPLE
Matrix: solutions
Sample preparation: Inject an aliquot of a 100 µg/mL solution in water.

CAPILLARY ELECTROPHORESIS
Capillary: 25 cm × 25 µm fused-silica (21.6 cm to detector) (Supelco)
Capillary preparation: Before each run rinse capillary with 100 mM NaOH for 2 min and running buffer for 8 min.
Capillary temperature: 25
Running buffer: 100 mM pH 2.5 Phosphate buffer
Injection: Pressure injection.
Detector: UV 200
Migration time: 2.8
Voltage: 30 kV
Model: Bio-Rad BioFocus 3000

KEY WORDS
human; different conformers are seen at different temperatures

REFERENCE
Thunecke,F.; Fischer,G. Separation of cis/trans conformers of human and salmon calcitonin by low temperature capillary electrophoresis, *Electrophoresis*, **1998**, *19*, 288–294.

Canadine

Molecular formula: $C_{20}H_{21}NO_4$
Molecular weight: 339.39
CAS Registry No.: 522-97-4, 2086-96-6 (d-form), 29074-38-2 (dl-form), 5096-57-1 (l-form)
Merck Index (12th ed.): 1785

SAMPLE
Matrix: solutions
Sample preparation: Inject an aliquot of a solution in running buffer.

CAPILLARY ELECTROPHORESIS
Capillary: 60 cm × 75 µm fused-silica (52.4 cm to detector)
Capillary preparation: After each run flush with 500 mM KOH for 2-3 min then with water.
Running buffer: 10 mM pH 3.8 Phosphate buffer containing 2% sulfated cyclodextrin (ds 7-10)
Injection: Hydrostatic injection.
Detector: UV 214
Migration time: 8.32, 8.71 (enantiomers)
Voltage: 15 kV
Model: Waters Quanta 4000

OTHER SUBSTANCES
Also analyzed: acebutolol, alprenolol, aminoglutethimide, brompheniramine, bupivacaine, bupropion, carbinoxamine, chloroquine, chlorpheniramine, dimethindene, disopyramide, doxylamine, hydroxychloroquine, idazoxan, isoxsuprine, ketamine, mepenzolate, mepivacaine, methoxyphenamine, mexiletine, midodrine, nefopam, orphenadrine, oxprenolol, oxyphencyclimine, pheniramine, phensuximide, pindolol, piperoxan, terbutaline, tetramisole, tolperisone, tranylcypromine, trihexyphenidyl, trimipramine, verapamil, warfarin

KEY WORDS
chiral; detector at anode

REFERENCE
Stalcup,A.M.; Gahm,K.H. Application of sulfated cyclodextrins to chiral separations by capillary zone electrophoresis, *Anal.Chem.*, **1996**, *68*, 1360–1368.

Capreomycin

Molecular formula: $C_{25}H_{44}N_{14}O_7$
Molecular weight: 652.71 (capreomycin 1B)
CAS Registry No.: 11003-38-6, 1405-37-4 (sulfate)
Merck Index (12th ed.): 1801

SAMPLE
Matrix: urine
Sample preparation: Inject an aliquot directly (?).

CAPILLARY ELECTROPHORESIS
Capillary: 55.5 cm × 50 μm fused-silica (47.5 cm to detector) (Yongnian, China)
Capillary preparation: After each run rinse capillary with 100 mM NaOH for 5 min and with water for 5 min then equilibrate with running buffer for 10 min.
Capillary temperature: 24 ± 0.2
Running buffer: EtOH:40 mM borax 10:90 adjusted to pH 4.0 with phosphoric acid.
Injection: Electrokinetic at 20 ± 0.3 kV for 20 s.
Detector: UV 280
Migration time: 7
Voltage: 25 ± 0.3 kV
Model: 1229 HPCE
Limit of detection: 150 ng/mL

OTHER SUBSTANCES
Extracted: ofloxacin, pasiniazide

REFERENCE
Zhang,S.S.; Liu,H.X.; Yuan,Z.B.; Yu,C.L. A reproducible, simple and sensitive high-performance capillary electrophoresis method for simultaneous determination of capreomycin, ofloxacin and pasiniazide in urine, *J.Pharm.Biomed.Anal.*, **1998**, *17*, 617–622.

Captan

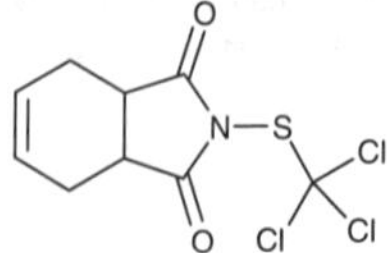

Molecular formula: $C_9H_8Cl_3NO_2S$
Molecular weight: 300.59
CAS Registry No.: 133-06-2
Merck Index (12th ed.): 1815

SAMPLE
Matrix: solutions
Sample preparation: Inject an aliquot of a solution in running buffer.

CAPILLARY ELECTROPHORESIS
Capillary: 60 cm × 50 μm fused-silica (53 cm to detector) (Polymicro Technologies)
Capillary preparation: Condition capillary by flushing with 100 mM NaOH for 20 min, water for 10 min, and running buffer for 10 min then equilibrate under voltage.
Running buffer: 10 mM pH 7.0 Sodium phosphate containing 60 mM sodium dodecyl sulfate
Detector: UV 210
Migration time: 21.5
Voltage: 15 kV
Model: laboratory-constructed

OTHER SUBSTANCES
Simultaneous: acephate, aldicarb, aminocarb, banol, carbaryl, carbendazim, carbofuran, chlorotoluron, dichlorvos, dimethoate, fenobucarb, fensulfothion, iprodione, isofenphos, isoprocarb, malathion, mepronil, methiocarb, napropamide, paclobutrazol, parathion, propoxur, propyzamide, pyriodaphenthion, simazine, simetryn, thiram, triadimefon, tricyclazole
Interfering: pirimicarb

REFERENCE
Wu,Y.S.; Lee,H.K.; Li,S.F.Y. Rapid estimation of octanol-water partition coefficients of pesticides by micellar electrokinetic chromatography, *Electrophoresis*, **1998**, *19*, 1719–1727.

Captopril

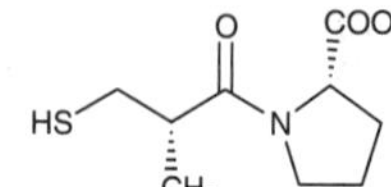

Molecular formula: $C_9H_{15}NO_3S$
Molecular weight: 217.29
CAS Registry No.: 62571-86-2
Merck Index (12th ed.): 1817
Lednicer: 3 128; 4 7, 58, 81, 128

SAMPLE
Matrix: blood
Sample preparation: 1 mL Whole blood + 1 mL 10% trichloroacetic acid containing 1 mM disodium EDTA, vortex, centrifuge at 0° at 1850 g for 5 min. Remove a 40 μL aliquot and mix it with 40 μL 50 μg/mL 2-mercaptoethanol in 2.5 M sodium borate buffer, add 100 μL 2 mg/mL ammonium 7-fluoro-2,1,3-benzoxadiazole-4-sulfonate (SBD-F) in 2.5 M sodium borate buffer, mix, add 220 μL 2.5 M sodium borate buffer, mix vigorously, heat at 60° for 1 h, cool in an ice bath, filter (0.45 μm), inject an aliquot of the filtrate. (Extraction only demonstrated for glutathione.)

CAPILLARY ELECTROPHORESIS
Capillary: 50 cm × 50 μm coated capillary (Bio-Rad ?)
Running buffer: 100 mM pH 2.5 Phosphate buffer
Injection: Electromigration
Detector: F ex 380 em 510

Migration time: 22
Internal standard: 2-mercaptoethanol (13)
Voltage: 12 kV
Model: Bio-Rad HPE 100
Limit of detection: 480 ng/mL

OTHER SUBSTANCES
Extracted: acetylcysteine, cysteine, dithiothreitol, glutathione, homocysteine, thiolactic acid

---

KEY WORDS
whole blood; derivatization; detector at anode; coated capillary

---

REFERENCE
Ling,B.L.; Baeyens,W.R.G.; Dewaele,C. Capillary zone electrophoresis with ultraviolet and fluorescence detection for the analysis of thiol. Application to mixtures and blood, *Anal.Chim.Acta*, **1991**, *255*, 283–288.

SAMPLE
Matrix: solutions

CAPILLARY ELECTROPHORESIS
Capillary: 57 cm × 75 μm fused-silica (50 cm to detector) (Beckman)
Capillary preparation: Before each run rinse capillary with 100 mM NaOH and running buffer.
Capillary temperature: 30
Running buffer: Acetone:100 mM pH 8.1 borate buffer containing 50 mM sodium dodecyl sulfate 15:85 (A) or 50 mM pH 8.2 borate buffer containing 100 mM sodium dodecyl sulfate (B) or 100 mM pH 8.1 borate buffer containing 50 mM sodium dodecyl sulfate (C)
Injection: Pressure injection for 5-10 s.
Detector: UV 200
Migration time: 7 (A), 12 (B), 4.2 (C)
Voltage: 25 kV
Model: Beckman P/ACE 5510

OTHER SUBSTANCES
Simultaneous: acebutolol, amiodarone, atenolol, bretylium, diltiazem, disopyramide, lidocaine, lisinopril, metoprolol, nicardipine, nifedipine, phenytoin, pindolol, propafenone, propranolol, quinidine, sotalol, timolol, p-toluenesulfonic acid, verapamil

---

KEY WORDS
only running buffer A separates all compounds; running buffer C separates degradation products

---

REFERENCE
Bretnall,A.E.; Clarke,G.S. Selectivity of capillary electrophoresis for the analysis of cardiovascular drugs, *J.Chromatogr.A*, **1996**, *745*, 145–154.

SAMPLE
Matrix: solutions

CAPILLARY ELECTROPHORESIS
Capillary: 57 cm × 50 μm fused-silica (50 cm to detector)
Running buffer: 50 mM pH 8.0 Borate buffer containing 40 mM sodium taurodeoxycholate and 25 mM phosphatidylcholine (Prepare by adding phosphatidylcholine and stirring for 3-6 h until all cloudiness disappears. Phosphatidylcholine was 95% pure soybean lecithin, Epikuron, Lucas Meyer & Co.)
Injection: Inject a solution of the compound in the running buffer at 20 psi for 1 s, inject MeOH: water 5:95 at 20 psi for 1 s, inject a solution of halofantrine in running buffer at 20 psi for 1 s.
Detector: UV 214
Migration time: k' 1.55
Model: Beckman P/ACE 5000

OTHER SUBSTANCES
Also analyzed: acetaminophen, amoxicillin, antipyrine, aspirin, azathioprine, caffeine, carbamazepine, carprofen, chlorambucil, chlorpheniramine, chlorpromazine, cimetidine, clonidine, codeine, desipramine, diphenhydramine, ephedrine, fenoterol, flufenamic acid, flurbiprofen, haloperidol, hydroxyzine, ibuprofen, imipramine, indomethacin, ketoprofen, lidocaine, melphalan, metoprolol, nabumetone, nadolol, phenobarbital, phenol, promazine, propranolol, pyrilamine, ranitidine, ropinirole, salicylic acid, sulfamethoxazole, testosterone, theophylline, thioridazine, tiaprofenic acid, tolfenamic acid, trifluoperazine, trimethoprim, valproic acid, verapamil

KEY WORDS
comparison with HPLC; k' = (Tr-T0)/(T0(1-Tr/Tm)) where Tr = retention time of analyte; T0 = retention time of water; and Tm = retention time of marker (halofantrine)

REFERENCE
Hanna,M.; de Biasi,V.; Bond,B.; Salter,C.; Hutt,A.J.; Camilleri,P. Estimation of the partitioning characteristics of drugs: A comparison of a large and diverse drug series utilizing chromatographic and electrophoretic methodology, *Anal.Chem.*, **1998**, *70*, 2092–2099.

Carazolol

Molecular formula: $C_{18}H_{22}N_2O_2$
Molecular weight: 298.38
CAS Registry No.: 57775-29-8
Merck Index (12th ed.): 1822

SAMPLE
Matrix: solutions
Sample preparation: Inject an aliquot of a 100 µg/mL solution in running buffer.

CAPILLARY ELECTROPHORESIS
Capillary: 32 cm × 50 µm fused-silica (27.5 cm to detector) (Yongnian Optical Conductive Fiber Plant, China), coated with polyacrylamide
Capillary preparation: No details of the polyacrylamide coating process are provided. However, another paper (LC.GC 1997, 15, 40) by this group indicates that they use the procedure of Hjertén, thus: Adjust the pH of 20 mL water to 3.5 with acetic acid, add 80 µL 3-(trimethoxysilyl)propyl methacrylate (3-methacryloxypropyltrimethoxysilane), mix, suck into capillary, let stand at room temperature for 1 h, remove the solution, wash with water. Fill the capillary with a deaerated 3-4% acrylamide solution containing 1 µL/mL N,N,N',N'-tetramethylethylenediamine and 1 mg/mL potassium persulfate, let stand for 30 min, remove excess solution by aspiration, rinse with water, remove water by aspiration, dry at 35° (J. Chromatogr. 1985, 347, 191).
Capillary temperature: 25
Running buffer: 100 mM NaH_2PO_4 adjusted to pH 2.5
Injection: Electrokinetic injection at 15 kV for 3 s.
Detector: UV 200, UV 210
Migration time: 5.66
Voltage: 15 kV
Model: Bio-Rad BioFocus 3000

OTHER SUBSTANCES
Simultaneous: atropine, cicletanine, dimethindene, fendiline, homatropine, ipratropium bromide, isothipendyl, mefloquine, metaclazepam, nafronyl (naftidrofuryl), nefopam, nicardipine, promethazine, reproterol, tetrahydrozoline (tetryzoline), theodrenaline, tioconazole, trihexyphenidyl, trimeprazine (alimemazine), trimipramine

KEY WORDS
coated capillary

REFERENCE

Koppenhoefer,B.; Epperlein,U.; Xiaofeng,Z.; Bingcheng,L. Separation of enantiomers of drugs by capillary electrophoresis. Part 4: Hydroxypropyl-γ-cyclodextrin as chiral solvating agent, *Electrophoresis*, **1997**, *18*, 924–930.

SAMPLE

Matrix: solutions
Sample preparation: Inject an aliquot of a 100 μg/mL solution in running buffer.

CAPILLARY ELECTROPHORESIS

Capillary: 30 cm × 50 μm fused-silica (25.5 cm to detector), coated with polyacrylamide
Capillary preparation: Adjust the pH of 20 mL water to 3.5 with acetic acid, add 80 μL 3-(trimethoxysilyl)propyl methacrylate (3-methacryloxypropyltrimethoxysilane), mix, suck into capillary, let stand at room temperature for 1 h, remove the solution, wash with water. Fill the capillary with a deaerated 3-4% acrylamide solution containing 1 μL/mL N,N,N',N'-tetramethylethylenediamine and 1 mg/mL potassium persulfate, let stand for 30 min, remove excess solution by aspiration, rinse with water, remove water by aspiration, dry at 35° (J. Chromatogr. 1985, 347, 191).
Capillary temperature: 25
Running buffer: 100 mM NaH_2PO_4 containing 45 mM hydroxypropyl-α- cyclodextrin, adjusted to pH 2.5 with phosphoric acid
Injection: Electrokinetic injection at 15 kV for 3 s.
Detector: UV 200
Voltage: 15 kV
Model: Bio-Rad BioFocus 3000

KEY WORDS

chiral; coated capillary; comparison with the use of other cyclodextrins; this running buffer gave the greatest enantiomeric separation.; α=1.027

REFERENCE

Lin,B.; Zhu,X.; Koppenhoefer,B.; Epperlein,U. Investigation of 123 chiral drugs by cyclodextrin-modified capillary electrophoresis, *LC.GC*, **1997**, *15*, 40–46.

SAMPLE

Matrix: solutions

CAPILLARY ELECTROPHORESIS

Capillary: 29-36 cm × 50 μm fused-silica (24.5-31.5 cm to detector) (Yongnian Optical Conductive Fiber Plant, China) coated with polyacrylamide
Capillary preparation: Coat capillary as follows. Adjust the pH of 20 mL water to 3.5 with acetic acid, add 80 μL 3-(trimethoxysilyl)propyl methacrylate (3-methacryloxypropyltrimethoxysilane), mix, suck into capillary, let stand at room temperature for 1 h, remove the solution, wash with water. Fill the capillary with a deaerated 3-4% acrylamide solution containing 1 μL/mL N,N,N',N'-tetramethylethylenediamine and 1 mg/mL potassium persulfate, let stand for 30 min, remove excess solution by aspiration, rinse with water, remove water by aspiration, dry at 35° (J. Chromatogr. 1985, 347, 191).
Capillary temperature: 25
Running buffer: 100 mM pH 2.5 NaH_2PO_4 (A) or 100 mM pH 2.5 NaH_2PO_4 containing 45 mM hydroxypropyl-α-cyclodextrin (Wacker, Munich) (B)
Injection: Electromigration at 15 kV for 3 s.
Detector: UV 200; UV 210
Migration time: 5.66 (A); 10.88, 10.99 (B) (enantiomers)
Voltage: 15 kV
Model: Bio-Focus 3000

OTHER SUBSTANCES

Also analyzed: albuterol (salbutamol), alprenolol, amorolfine, atenolol, atropine, azelastine, baclofen, bamethan, benproperine, benserazide, biperiden, bisoprolol, brompheniramine, bupivacaine, bupranolol, butamirate, butethamate, carbuterol, carteolol, carvedilol, celiprolol, chloroquine, chlorpheniramine, chlorphenoxamine, cicletanine, clenbuterol, clidinium bromide, clobutinol, dimethindene, dipivefrin, disopyramide, dobutamine, doxylamine, fendiline, flecainide, gallopamil, homatropine, ipratropium bromide, isoproterenol (isoprenaline), isothipendyl, ketamine, meclizine, mefloquine, mepindolol, mequitazine, metaclazepam, metaproterenol (or-ciprenaline), metipranolol, metoprolol, nafronyl (naftidrofuryl), nefopam, nicardipine, norfe-nefrine, ofloxacin, ornidazole, orphenadrine, oxomemazine, oxprenolol, oxybutynin, phenoxy-benzamine, phenylpropanolamine, pholedrine, pindolol, pirbuterol, prilocaine, procyclidine, promethazine, propafenone, propranolol, reproterol, sotalol, sulpride, synephrine, talinolol, ter-butaline, tetrahydrozoline (tetryzoline), theodrenaline, tioconazole, tocainide, trihexyphenidyl, trimeprazine (alimemazine), trimipramine, tropicamide, verapamil, zopiclone

KEY WORDS
coated capillary; chiral

REFERENCE
Koppenhoefer,B.; Eperlein,U.; Schlunk,R.; Zhu,X.; Lin,B. Separation of enantiomers of drugs by capillary elec-trophoresis. V. Hydroxypropyl-α-cyclodextrin as chiral solvating agent, *J.Chromatogr.A*, **1998**, *793*, 153–164.

Carbamazepine

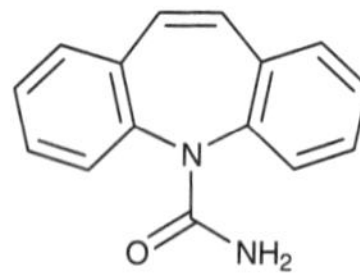

Molecular formula: $C_{15}H_{12}N_2O$
Molecular weight: 236.27
CAS Registry No.: 298-46-4
Merck Index (12th ed.): 1826
Lednicer: 1 403

SAMPLE
Matrix: blood
Sample preparation: 50 µL Serum + 5 mL isotonic saline, add to a Sep-Pak C18 SPE cartridge, wash with 5 mL saline, wash with 5 mL water, wash with 5 mL heptane:chloroform 98:2, wash with 5 mL heptane, wash with 3 mL water, remove all water with a stream of air, elute with 3 mL MeOH (remove all MeOH with a stream of air). Evaporate the eluate to dryness under reduced pressure at 37°, reconstitute with 100 µL water, inject an aliquot.

CAPILLARY ELECTROPHORESIS
Capillary: 72 cm × 50 µm silica (50 cm to detector) (Applied Biosystems)
Capillary preparation: Before each run wash capillary with 100 mM NaOH for 3 min, wash with running buffer for 3 min, aspirate neutral marker solution (2 drops DMSO in 10 mL water) for 1 s, place capillary end in buffer vial for 5 s, inject sample. Wash capillary at the beginning of each day by passing 1 M NaOH through for 20 min.
Capillary temperature: 30
Running buffer: 30 mM pH 9.3 Borate buffer containing 30 mM sodium dodecyl sulfate
Injection: Inject by applying vacuum for 1 s (2-3 nL)
Detector: UV 200
Migration time: 11
Voltage: +30 kV
Current: 72 µA
Model: Applied Biosystems Model 270A

OTHER SUBSTANCES
Extracted: phenobarbital, phenytoin, primidone

KEY WORDS
serum; SPE

REFERENCE
Evenson,M.A.; Wiktorowicz,J.E. Automated capillary electrophoresis applied to therapeutic drug monitoring, *Clin.Chem.*, **1992**, *38*, 1847–1852.

SAMPLE
Matrix: blood
Sample preparation: 100 μL Serum + 150 μL 80 μg/mL 3-isobutyl-1-methylxanthine in MeCN, mix for 30 s, centrifuge at 14000 g for 1 min, inject an aliquot of the supernatant.

CAPILLARY ELECTROPHORESIS
Capillary: 25 cm × 47 μm
Capillary preparation: Wash with running buffer by pressure injection for 1 min after each run. Wash daily with 50 mM phosphoric acid for 2 min, with water for 1 min, with 2 M NaOH for 2 min, with water for 1 min, and with running buffer for 2 min.
Capillary temperature: 24
Running buffer: 220 mM boric acid adjusted to pH 8.8 with 2.5 M NaOH
Injection: Pressure injection for 8 s.
Detector: UV 254
Migration time: 1.6
Internal standard: 3-isobutyl-1-methylxanthine (2.3)
Voltage: 12 kV
Model: Beckman

OTHER SUBSTANCES
Extracted: acetaminophen, caffeine, iohexol, pentobarbital, phenobarbital, phenytoin, salicylic acid, theophylline

KEY WORDS
serum

REFERENCE
Shihabi,Z.K.; Constantinescu,M.S. Iohexol in serum determined by capillary electrophoresis, *Clin.Chem.*, **1992**, *38*, 2117–2120.

SAMPLE
Matrix: blood
Sample preparation: 200 μL Serum + 1 mL 5 μg/mL n-propyl p-hydroxybenzoate in ethyl acetate, vortex for 30 s, centrifuge at 13400 g for 2 min. Remove a 500 μL aliquot of the organic layer and evaporate it to dryness under a stream of nitrogen, reconstitute with 50 μL MeOH: water 5:95, inject an aliquot.

CAPILLARY ELECTROPHORESIS
Capillary: 64.5 cm × 50 μm fused-silica (56 cm to detector) (Hewlett-Packard)
Capillary preparation: Rinse capillary with running buffer at 630 mm Hg for 5 min before each run.
Capillary temperature: 30
Running buffer: 10 mM pH 8.0 Phosphate buffer containing 50 mM sodium dodecyl sulfate
Injection: Vacuum injection at 50 mm Hg for 2 s.
Detector: UV 210
Migration time: 8
Internal standard: n-propyl p-hydroxybenzoate (7.2)
Voltage: 30 kV
Model: Hewlett-Packard HP[3D]
Limit of detection: 600 ng/mL

OTHER SUBSTANCES
Extracted: phenobarbital, phenytoin, primidone, zonisamide
Noninterfering: valproic acid

KEY WORDS
serum; comparison with HPLC

REFERENCE
Makino,K.; Goto,Y.; Sueyasu,M.; Futagami,K.; Kataoka,Y.; Oishi,R. Micellar electrokinetic capillary chromatography for therapeutic drug monitoring of zonisamide, *J.Chromatogr.B*, **1997**, *695*, 417–425.

SAMPLE
Matrix: blood
Sample preparation: 1 mL Serum or plasma + IS, mix, add 6 mL ethyl acetate, vortex, centrifuge at 2000 g for 15 min. Remove 4 mL of the organic layer and evaporate it to dryness under a stream of nitrogen, reconstitute the residue in MeOH:running buffer 5:95, inject an aliquot.

CAPILLARY ELECTROPHORESIS
Capillary: 34 cm × 50 μm silica (25.5 cm to detector) (Microquartz, Munich)
Capillary preparation: Before each run rinse capillary with 100 mM NaOH for 2 min and with running buffer for 5 min.
Capillary temperature: 30
Running buffer: 25 mM pH 9.2 Sodium Tetraborate containing 50 mM sodium dodecyl sulfate
Injection: Pressure injection at 5 kPa for 3 s.
Detector: UV 210
Migration time: 1.8
Internal standard: methylpropylsuccinimide (1)
Voltage: 25 kV
Current: 140 μA
Model: Hewlett-Packard 3D
Limit of detection: 500 nM

OTHER SUBSTANCES
Extracted: metabolites
Simultaneous: ethosuximide, phenytoin, primidone

KEY WORDS
plasma; serum; comparison with HPLC

REFERENCE
Härtter,S.; Jensen,B.; Hiemke,C.; Leal,M.; Weigmann,H.; Unger,K. Micellar electrokinetic capillary chromatography for therapeutic drug monitoring of carbamazepine and its main metabolites, *J.Chromatogr.B*, **1998**, *712*, 253–258.

SAMPLE
Matrix: blood, saliva
Sample preparation: Filter (Amicon Centrifree Micropartition system) serum or saliva at 1500 g for 20 min, inject an aliquot of the ultrafiltrate.

CAPILLARY ELECTROPHORESIS
Capillary: 90 cm × 75 μm fused-silica (70 cm to detector) (Polymicro Technologies)
Capillary preparation: Before each run rinse capillary with 100 mM NaOH for 3 min and with running buffer for 5 min
Running buffer: 10 mM pH 9.1 Na_2HPO_4 containing 6 mM sodium borate and 75 mM sodium dodecyl sulfate
Injection: Injection by siphon at 34 cm for 5 s
Detector: UV 195
Migration time: 25
Voltage: 20 kV
Current: about 80 μA

OTHER SUBSTANCES
Extracted: phenobarbital

KEY WORDS
serum; ultrafiltrate

REFERENCE
Thormann,W.; Lienhard,S.; Wernly,P. Strategies for the monitoring of drugs in body fluids by micellar electrokinetic capillary chromatography, *J.Chromatogr.*, **1993**, *636*, 137–148.

SAMPLE
Matrix: bulk

Sample preparation: Inject an aliquot of a 1 mg/mL solution in 4.767 mg/mL disodium tetra-borate decahydrate in MeOH:water 20:80 containing 14.42 mg/mL sodium dodecyl sulfate.

CAPILLARY ELECTROPHORESIS
Capillary: 42.1 cm × 50 μm fused-silica (34.9 cm to detector) (Polymicro Technologies)
Capillary preparation: Between runs rinse with running buffer for 2 min. Condition new cap-illaries with 1 M NaOH at 30° for 30 min, with 100 mM at 30° for 30 min, with water at 25° for 30 min, and with running buffer at 25° for 1.5 h.
Capillary temperature: 25
Running buffer: 12.5 mM Sodium tetraborate decahydrate in MeOH:water 10:90 (w/w) contain-ing 50 mM sodium dodecyl sulfate
Injection: Hydrodynamic injection at 50 mbar for 5 s.
Detector: UV 210
Migration time: 7.57
Voltage: 25 kV
Current: ca. 45 μA
Model: SpectraPhoresis 1000 (Thermo-Separation Products)

OTHER SUBSTANCES
Simultaneous: impurities

REFERENCE
Degenhardt,M.; Wätzig,H. Purity control of carbamazepine by micellar electrokinetic chromatography, *J.Chromatogr.A*, **1997**, *768*, 113–123.

SAMPLE
Matrix: solutions

CAPILLARY ELECTROPHORESIS
Capillary: 70 cm × 50 μm coated fused-silica (55 cm to detector) (Polymicro Technologies)
Capillary preparation: Between runs rinse with water for 2 min. Coat capillaries as follows. Condition with 1 M NaOH for 15 min, rinse with water for 5 min, treat with 30 μL/mL 3-(trimethoxysilyl)propyl methacrylate in acetic acid:water 50:50 for 1 h using house vacuum, rinse with water, fill with polymerization solution, let stand for 1 h, flush, rinse with water. (Prepare polymerization solution by adding 10 μL N,N,N',N'-tetramethylethylenediamine to 10 mL of a degassed 4% solution of acrylamide in water, add 10 μL 10% ammonium persulfate in water.)
Running buffer: 10 mM pH 6.5 Citrate buffer containing 20 mM sodium dodecyl sulfate and 10 mM carboxymethyl-β-cyclodextrin (Cyclodextrin Technologies Development, Gainesville FL)
Injection: Hydrodynamic injection at 15 cm for 8-10 s.
Detector: UV 254
Migration time: 10.5
Voltage: -20 kV
Model: laboratory-constructed

OTHER SUBSTANCES
Simultaneous: amitriptyline, clomipramine, desipramine, imipramine, nortriptyline, opipramol, proptriptyline, trimipramine

KEY WORDS
coated capillary

REFERENCE
Spencer,B.J.; Zhang,W.; Purdy,W.C. Capillary electrophoretic separation of tricyclic antidepressants using charged carboxymethyl-β-cyclodextrin as a buffer additive, *Electrophoresis*, **1997**, *18*, 736–744.

Carbaryl

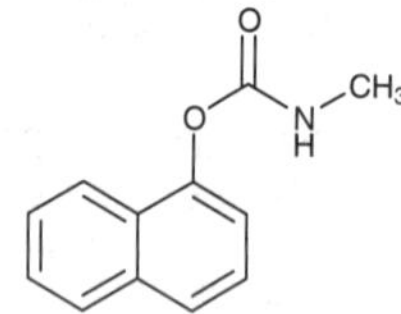

Molecular formula: $C_{12}H_{11}NO_2$
Molecular weight: 201.22
CAS Registry No.: 63-25-2
Merck Index (12th ed.): 1831

SAMPLE
Matrix: plants
Sample preparation: Extract 8 g powdered tobacco at 60° and 100 atmospheres with carbon dioxide:acetone 80:20 for 2 min (restrictor 1.2 m × 100 μm fused-silica; flow rate 160 mL/min), evaporate extract to dryness under a stream of nitrogen, reconstitute with 3 mL acetone, add to a Florisil SPE cartridge (Baker), wash with 20 mL hexane, elute with 30 mL acetone. Evaporate the eluate to dryness under reduced pressure, reconstitute with 500 μL acetone:50 mM pH 8.3 sodium tetraborate buffer 40:60, inject an aliquot.

CAPILLARY ELECTROPHORESIS
Capillary: 60 cm × 75 μm fused-silica
Running buffer: 50 mM pH 8.3 Sodium tetraborate
Injection: Hydrodynamic injection (10 nL).
Detector: UV214
Migration time: 4
Voltage: 20 kV
Model: Waters Quanta 4000
Limit of quantitation: 15 ng/mL

KEY WORDS
tobacco; SFE; SPE

REFERENCE
Lanças,F.M.; Rissato,S.R.; Galhiane,M.S. Off-line SFE-CZE analysis of carbamates residues in tobacco samples, *Chromatographia*, **1996**, *42*, 323–327.

SAMPLE
Matrix: solutions
Sample preparation: Inject an aliquot of a solution in running buffer.

CAPILLARY ELECTROPHORESIS
Capillary: 60 cm × 50 μm fused-silica (53 cm to detector) (Polymicro Technologies)
Capillary preparation: Condition capillary by flushing with 100 mM NaOH for 20 min, water for 10 min, and running buffer for 10 min then equilibrate under voltage.
Running buffer: 10 mM pH 7.0 Sodium phosphate containing 60 mM sodium dodecyl sulfate
Detector: UV 210
Migration time: 21
Voltage: 15 kV
Model: laboratory-constructed

OTHER SUBSTANCES
Simultaneous: acephate, aldicarb, banol, captan, carbendazim, carbofuran, chlorotoluron, dichlorvos, dimethoate, fenobucarb, fensulfothion, iprodione, isofenphos, isoprocarb, malathion, mepronil, methiocarb, napropamide, paclobutrazol, parathion, pirimicarb, propoxur, pyriodaphenthion, simazine, thiram, triadimefon, tricyclazole
Interfering: aminocarb, propyzamide, simetryn

REFERENCE
Wu,Y.S.; Lee,H.K.; Li,S.F.Y. Rapid estimation of octanol-water partition coefficients of pesticides by micellar electrokinetic chromatography, *Electrophoresis*, **1998**, *19*, 1719–1727.

Carbenicillin

Molecular formula: $C_{17}H_{18}N_2O_6S$
Molecular weight: 378.41
CAS Registry No.: 4697-36-3, 4800-94-6 (disodium salt), 17230-86-3 (potassium salt)
Merck Index (12th ed.): 1838
Lednicer: 1 414

SAMPLE
Matrix: solutions
Sample preparation: Prepare a 0.5-2 mg/mL solution in water, inject an aliquot.

CAPILLARY ELECTROPHORESIS
Capillary: 65 cm × 50 μm fused-silica (50 cm to detector) (Scientific Glass Engineering)
Running buffer: 20 mM NaH_2PO_4 containing 150 mM sodium dodecyl sulfate adjusted to pH
 9.0 with 20 mM sodium tetraborate
Injection: Injection by siphon at 5 cm for 5-10 s
Detector: UV 210
Migration time: 12.9
Voltage: 20 kV

OTHER SUBSTANCES
Simultaneous: amoxicillin, ampicillin, aspoxicillin, penicillin G, piperacillin, sulbenicillin
Also analyzed: cefmenoxime, cefminox, cefoperazone, cefotaxime, cefpimizole, cefpiramide, cef-
 tazidime, ceftriaxone

REFERENCE
Nishi,H.; Tsumagari,N.; Kakimoto,T.; Terabe,S. Separation of β-lactam antibiotics by micellar electrokinetic
 chromatography, *J.Chromatogr.*, **1989**, *477*, 259–270.

SAMPLE
Matrix: solutions
Sample preparation: Inject directly.

CAPILLARY ELECTROPHORESIS
Capillary: 65 cm × 50 μm untreated fused-silica (50 cm to detector) (SGE)
Capillary preparation: Flush with running buffer every 5 runs. At the end of each day fill
 capillary with 100 mM KOH, let stand for 30 min, flush with water, let stand for 5 min, fill
 with running buffer.
Running buffer: pH 8.5 Buffer containing 100 mM sodium dodecyl sulfate (Prepare buffer by
 mixing 20 mM NaH_2PO_4 solution with 20 mM sodium borate solution to achieve a pH of 8.5.)
Injection: Inject by siphoning at 10 cm for 5-10 s.
Detector: UV 210
Migration time: 13.8

OTHER SUBSTANCES
Simultaneous: amoxicillin, ampicillin, aspoxicillin, penicillin G, piperacillin, sulbenicillin

REFERENCE
Nishi,H.; Fukuyama,T.; Matsuo,M. Separation and determination of aspoxicillin in human plasma by micellar
 electrokinetic chromatography with direct sample injection, *J.Chromatogr.*, **1990**, *515*, 245–255.

Carbidopa

Molecular formula: $C_{10}H_{14}N_2O_4$
Molecular weight: 226.23
CAS Registry No.: 28860-95-9, 38821-49-7 (monohydrate)
Merck Index (12th ed.): 1843
Lednicer: 2 119

SAMPLE
Matrix: solutions

CAPILLARY ELECTROPHORESIS
Capillary: 23 cm × 25 μm fused-silica (17 cm to detector) (Polymicro Technologies)
Capillary temperature: 16
Running buffer: Buffer containing 20 mM heptakis-(2,3-diacetyl-6-sulfato)-β-cyclodextrin (Regis) and 8% polyethylene glycol 900 (Buffer was phosphoric acid (I = 0.95) adjusted to pH 2.5 with triethylamine. Synthesis of heptakis-(2,3-diacetyl-6-sulfato)-β-cyclodextrin is described in Anal. Chem. 1997, 69, 4226.)
Injection: Pressure injection at 5 psi for 1 s.
Detector: UV 280
Migration time: 6.6 (D), 7.2 (L)
Voltage: 30 kV
Model: Beckman P/ACE 2100

KEY WORDS
chiral

REFERENCE
Vincent,J.B.; Vigh,G. Systematic approach to methods development for the capillary electrophoretic analysis of a minor enantiomer using a single-isomer sulfated cyclodextrin. A case study of L-carbidopa analysis, *J.Chromatogr.A*, **1998**, *817*, 105–111.

Carbinoxamine

Molecular formula: $C_{16}H_{19}ClN_2O$
Molecular weight: 290.79
CAS Registry No.: 486-16-8, 3505-38-2 (maleate), 49746-00-1 (l-form tartrate)
Merck Index (12th ed.): 1845
Lednicer: 1 43

SAMPLE
Matrix: solutions
Sample preparation: Inject an aliquot of a 100 μg/mL solution.

CAPILLARY ELECTROPHORESIS
Capillary: 57 cm × 50 μm fused-silica (50 cm to detector)
Capillary temperature: 20
Running buffer: 100 mM pH 2.5 NaH_2PO_4 containing 1.5 M urea and 30 mM β-cyclodextrin
Injection: Pressure injection for 1 s
Detector: UV 200
Migration time: 14 (D), 14.3 (L)
Voltage: 24 kV
Model: Beckman PACE 2100

KEY WORDS
chiral

REFERENCE
McLaughlin,G.M.; Nolan,J.A.; Lindahl,J.L.; Palmieri,R.H.; Anderson,K.W.; Morris,S.C.; Morrison,J.A.; Bronzert,T.J. Pharmaceutical drug separations by HPCE: Practical guidelines, *J.Liq.Chromatogr.*, **1992**, *15*, 961–1021.

SAMPLE
Matrix: solutions
Sample preparation: Prepare a solution in running buffer, inject an aliquot.

CAPILLARY ELECTROPHORESIS
Capillary: 60 cm × 75 μm fused-silica (52.4 cm to detector)
Capillary preparation: After each run flush with 500 mM KOH for 2-3 min, flush with water, fill with running buffer
Running buffer: 10 mM Na_2HPO_4 containing 2% heparin sodium (MW 10000, 11% S, Scientific Protein Laboratories, Waunakee, WI) adjusted to pH 5 with phosphoric acid
Injection: Hydrostatic injection
Detector: UV 214
Migration time: 14.8 (first enantiomer, R_S 0.5)
Model: Waters Quanta 4000

OTHER SUBSTANCES
Also analyzed: anabasine, brompheniramine, bupivacaine, chlorcyclizine, chloroquine, chlorpheniramine, dimethindene, doxylamine, enpiroline, halofantrine, hydroxychloroquine, indapamide, mefloquine, nornicotine, pheniramine, primaquine, promethazine, quinacrine, tetramisole

KEY WORDS
chiral

REFERENCE
Stalcup,A.M.; Agyei,N.M. Heparin: a chiral mobile-phase additive for capillary zone electrophoresis, *Anal.Chem.*, **1994**, *66*, 3054–3059.

SAMPLE
Matrix: solutions
Sample preparation: Inject an aliquot of a solution in running buffer.

CAPILLARY ELECTROPHORESIS
Capillary: 60 cm × 75 μm fused-silica (52.4 cm to detector)
Capillary preparation: After each run flush with 500 mM KOH for 2-3 min then with water.
Running buffer: 10 mM pH 3.8 Phosphate buffer containing 2% sulfated cyclodextrin (ds 7-10)
Injection: Hydrostatic injection.
Detector: UV 214
Migration time: 8.63, 8.96 (enantiomers)
Voltage: 15 kV
Model: Waters Quanta 4000

OTHER SUBSTANCES
Also analyzed: acebutolol, alprenolol, aminoglutethimide, brompheniramine, bupivacaine, bupropion, canadine, chloroquine, chlorpheniramine, dimethindene, disopyramide, doxylamine, hydroxychloroquine, idazoxan, isoxsuprine, ketamine, mepenzolate, mepivacaine, methoxyphenamine, mexiletine, midodrine, nefopam, orphenadrine, oxprenolol, oxyphencyclimine, pheniramine, phensuximide, pindolol, piperoxan, terbutaline, tetramisole, tolperisone, tranylcypromine, trihexyphenidyl, trimipramine, verapamil, warfarin

KEY WORDS
chiral; detector at anode

REFERENCE
Stalcup,A.M.; Gahm,K.H. Application of sulfated cyclodextrins to chiral separations by capillary zone electrophoresis, *Anal.Chem.*, **1996**, *68*, 1360–1368.

Carboplatin

Molecular formula: $C_6H_{12}N_2O_4Pt$
Molecular weight: 371.25
CAS Registry No.: 41575-94-4
Merck Index (12th ed.): 1870
Lednicer: 4 16

SAMPLE
Matrix: solutions

CAPILLARY ELECTROPHORESIS
Capillary: 75 cm × 75 μm fused-silica (50 cm to detector) (CS Chromatographie Service, Langerwehe, Germany)
Capillary preparation: Purge with running buffer for 2 min before each run.
Running buffer: 50 mM pH 7.0 NaH_2PO_4 containing 25 mM sodium borate and 80 mM sodium dodecyl sulfate
Injection: Hydrostatic injection
Detector: UV 210
Migration time: 9
Voltage: 15 kV
Model: laboratory constructed

OTHER SUBSTANCES
Simultaneous: degradation products, cisplatin, lobaplatin

REFERENCE
Wenclawiak,B.W.; Wollmann,M. Separation of platinum(II) anti-tumour drugs by micellar electrokinetic capillary chromatography, *J.Chromatogr.A*, **1996**, *724*, 317–326.

Carbuterol

Molecular formula: $C_{13}H_{21}N_3O_3$
Molecular weight: 267.33
CAS Registry No.: 34866-47-2, 34866-46-1 (HCl)
Merck Index (12th ed.): 1882
Lednicer: 2 41

SAMPLE
Matrix: solutions
Sample preparation: Inject an aliquot of a 100 μg/mL solution in running buffer.

CAPILLARY ELECTROPHORESIS
Capillary: 29 cm × 50 μm fused-silica (24.5 cm to detector) (Yongnian Optical Conductive Fiber Plant, China), coated with polyacrylamide
Capillary preparation: No details of the polyacrylamide coating process are provided. However, another paper (LC.GC 1997, 15, 40) by this group indicates that they use the procedure of Hjertén, thus: Adjust the pH of 20 mL water to 3.5 with acetic acid, add 80 μL 3-(trimethoxysilyl)propyl methacrylate (3-methacryloxypropyltrimethoxysilane), mix, suck into capillary, let

stand at room temperature for 1 h, remove the solution, wash with water. Fill the capillary with a deaerated 3-4% acrylamide solution containing 1 μL/mL N,N,N',N'-tetramethylethylenediamine and 1 mg/mL potassium persulfate, let stand for 30 min, remove excess solution by aspiration, rinse with water, remove water by aspiration, dry at 35° (J. Chromatogr. 1985, 347, 191).
Capillary temperature: 25
Running buffer: 100 mM NaH_2PO_4 adjusted to pH 2.5
Injection: Electrokinetic injection at 15 kV for 3 s.
Detector: UV 200, UV 210
Migration time: 5.81
Voltage: 15 kV
Model: Bio-Rad BioFocus 3000

OTHER SUBSTANCES
Simultaneous: albuterol, alprenolol, atenolol, baclofen, bamethan, benproperine, benserazide, bisoprolol, bupranolol, butamirate, butethamate, celiprolol, clenbuterol, clobutinol, dipivefrin, isoproterenol (isoprenaline), metaproterenol (orciprenaline), metipranolol, metoprolol, norfenefrine, ornidazole, oxprenolol, phenylpropanolamine, pholedrine, pirbuterol, prilocaine, procyclidine, sotalol, synephrine, terbutaline, tocainide

KEY WORDS
coated capillary

REFERENCE
Koppenhoefer,B.; Epperlein,U.; Xiaofeng,Z.; Bingcheng,L. Separation of enantiomers of drugs by capillary electrophoresis. Part 4: Hydroxypropyl-γ-cyclodextrin as chiral solvating agent, *Electrophoresis*, **1997**, *18*, 924–930.

SAMPLE
Matrix: solutions

CAPILLARY ELECTROPHORESIS
Capillary: 29-36 cm × 50 μm fused-silica (24.5-31.5 cm to detector) (Yongnian Optical Conductive Fiber Plant, China) coated with polyacrylamide
Capillary preparation: Coat capillary as follows. Adjust the pH of 20 mL water to 3.5 with acetic acid, add 80 μL 3-(trimethoxysilyl)propyl methacrylate (3-methacryloxypropyltrimethoxysilane), mix, suck into capillary, let stand at room temperature for 1 h, remove the solution, wash with water. Fill the capillary with a deaerated 3-4% acrylamide solution containing 1 μL/mL N,N,N',N'-tetramethylethylenediamine and 1 mg/mL potassium persulfate, let stand for 30 min, remove excess solution by aspiration, rinse with water, remove water by aspiration, dry at 35° (J. Chromatogr. 1985, 347, 191).
Capillary temperature: 25
Running buffer: 100 mM pH 2.5 NaH_2PO_4 (A) or 100 mM pH 2.5 NaH_2PO_4 containing 45 mM hydroxypropyl-α-cyclodextrin (Wacker, Munich) (B)
Injection: Electromigration at 15 kV for 3 s.
Detector: UV 200; UV 210
Migration time: 5.81 (A); 7.41 (B) (no separation of enantiomers)
Voltage: 15 kV
Model: Bio-Focus 3000

OTHER SUBSTANCES
Also analyzed: albuterol (salbutamol), alprenolol, amorolfine, atenolol, atropine, azelastine, baclofen, bamethan, benproperine, benserazide, biperiden, bisoprolol, brompheniramine, bupivacaine, bupranolol, butamirate, butethamate, carazolol, carteolol, carvedilol, celiprolol, chloroquine, chlorpheniramine, chlorphenoxamine, cicletanine, clenbuterol, clidinium bromide, clobutinol, dimethindene, dipivefrin, disopyramide, dobutamine, doxylamine, fendiline, flecainide, gallopamil, homatropine, ipratropium bromide, isoproterenol (isoprenaline), isothipendyl, ketamine, meclizine, mefloquine, mepindolol, mequitazine, metaclazepam, metaproterenol (orciprenaline), metipranolol, metoprolol, nafronyl (naftidrofuryl), nefopam, nicardipine, norfenefrine, ofloxacin, ornidazole, orphenadrine, oxomemazine, oxprenolol, oxybutynin, phenoxybenzamine, phenylpropanolamine, pholedrine, pindolol, pirbuterol, prilocaine, procyclidine, promethazine, propafenone, propranolol, reproterol, sotalol, sulpride, synephrine, talinolol, terbutaline, tetrahydrozoline (tetryzoline), theodrenaline, tioconazole, tocainide, trihexyphenidyl, trimeprazine (alimemazine), trimipramine, tropicamide, verapamil, zopiclone

KEY WORDS
coated capillary

REFERENCE
Koppenhoefer,B.; Eperlein,U.; Schlunk,R.; Zhu,X.; Lin,B. Separation of enantiomers of drugs by capillary electrophoresis. V. Hydroxypropyl-α-cyclodextrin as chiral solvating agent, *J.Chromatogr.A*, **1998**, *793*, 153–164.

Carnitine

Molecular formula: $C_7H_{15}NO_3$
Molecular weight: 161.20
CAS Registry No.: 461-06-3, 541-15-1 (L-form)
Merck Index (12th ed.): 1898

SAMPLE
Matrix: blood
Sample preparation: 250 μL Plasma + 100 μL 50 mM pH 10.4 carbonate buffer, mix, add 200 μL 15 mM (+)-1-(9-fluorenylethyl)chloroformate in acetone, heat at 45° for 1 h, add 200 μL acetic acid buffer, add 250 μL water, inject an aliquot.

CAPILLARY ELECTROPHORESIS
Capillary: 67 cm × 50 μm CElect H-150 coated capillary (60 cm to detector) (Supelco)
Running buffer: 20 mM pH 4.35 Phosphate buffer containing 20 mM heptakis(2,6-di-O-methyl)-β-cyclodextrin
Injection: Pressure injection for 3 s
Detector: UV 214
Migration time: 13 (D), 14 (L)
Internal standard: 2-amino-4,6-dimethylpyrimidine (12)
Voltage: 17.5 kV
Model: Beckman P/ACE 2100 or 5510
Limit of detection: 20 μM

KEY WORDS
derivatization; chiral; plasma; coated capillary

REFERENCE
Vogt,C.; Kiessig,S. Separation of D/L-carnitine enantiomers by capillary electrophoresis, *J.Chromatogr.A*, **1996**, *745*, 53–60.

SAMPLE
Matrix: blood
Sample preparation: 250 μL Plasma + 100 μL 50 mM pH 10.4 carbonate buffer, mix, add 200 μL 15 mM (+)-1-(9-fluorenylethyl)chloroformate in acetone, heat at 45° for 1 h, add 200 μL acetic acid buffer, add 250 μL water, inject an aliquot (J. Chromatogr. 1996, 745, 53).

CAPILLARY ELECTROPHORESIS
Capillary: 57 cm × 50 μm fused-silica (50 cm to detector) (CS Chromatographie, Langerwehe, Germany)
Capillary preparation: Flush with MeOH:1 M NaOH after each run.
Capillary temperature: 23
Running buffer: MeOH:5% sodium dodecyl sulfate:85% phosphoric acid 88:8:12
Injection: Electrokinetic injection at 10 kV for 120 s.
Detector: F ex 320 (26 mW Omnichrom Series 74 He-Cd laser) em 405
Migration time: 25
Voltage: 20 kV
Model: Beckman P/ACE 2050 or 5510
Limit of detection: <1 μM

OTHER SUBSTANCES
Extracted: acetylcarnitine, dodecylcarnitine, octylcarnitine, palmitoylcarnitine, valerylcarnitine

KEY WORDS
plasma; derivatization

REFERENCE
Kiessig,S.; Vogt,C. Separation of carnitine and acylcarnitines by capillary electrophoresis, *J.Chromatogr.A*, **1997**, *781*, 475–479.

Carprofen

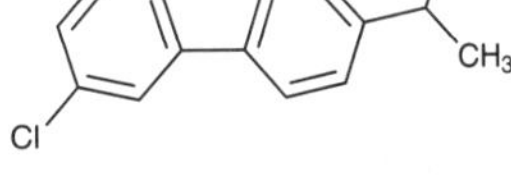

Molecular formula: C$_{15}$H$_{12}$ClNO$_2$
Molecular weight: 273.72
CAS Registry No.: 53716-49-7
Merck Index (12th ed.): 1912
Lednicer: 2 169

SAMPLE
Matrix: solutions
Sample preparation: Inject an aliquot of a 500 μM solution in MeOH:water 50:50.

CAPILLARY ELECTROPHORESIS
Capillary: 60 cm × 50 μm fused-silica (52.5 cm to detector) (Waters)
Capillary preparation: Before each run rinse capillary with 10 mM NaOH for 1 min and with running buffer for 4 min. Condition new capillaries by flushing with 1 M NaOH for 10 min, with water for 5 min, with 6 mM NaOH containing 500 mM NaCl for 10 min, with water for 5 min, and with running buffer for 10 min.
Running buffer: 117.2 mM Formic acid containing 10 mM heptakis(trimethyl-β-cyclodextrin) adjusted to pH 4.0 with 1 M NaOH
Injection: Hydrodynamic injection at 10 cm for 15 s.
Detector: UV 254
Migration time: 17.7, 18.3 (enantiomers)
Voltage: 30 kV
Model: Waters Quanta 4000

KEY WORDS
chiral

REFERENCE
Lelièvre,F.; Gareil,P. Chiral separations of underivatized arylpropionic acids by capillary zone electrophoresis with various cyclodextrins. Acidity and inclusion constant determinations, *J.Chromatogr.A*, **1996**, *735*, 311–320.

SAMPLE
Matrix: solutions

CAPILLARY ELECTROPHORESIS
Capillary: 35 cm × 50 μm coated capillary (30.5 cm to detector) (Composite Metal Services, UK)
Capillary preparation: Before each run purge at high pressure with water for 100 s and with running buffer for 120 s. If vancomycin is used, before each run purge at high pressure with water for 100 s and with running buffer not containing vancomycin for 120 s. Next purge with running buffer containing vancomycin at low pressure (175 psi.s) and inject sample. Coat capillary as follows. Adjust the pH of 20 mL water to 3.5 with acetic acid, add 80 μL 3-(trimethoxysilyl)propyl methacrylate (3-methacryloxypropyltrimethoxysilane), mix, suck into capillary, let stand at room temperature for 1 h, remove the solution, wash with water. Fill the capillary with a deaerated 3-4% acrylamide solution containing 1 μL/mL N,N,N',N'-tetra-

methylethylenediamine and 1 mg/mL potassium persulfate, let stand for 30 min, remove excess solution by aspiration, rinse with water, remove water by aspiration, dry at 35° (J. Chromatogr. 1985, 347, 191).
Capillary temperature: 25
Running buffer: Buffer containing 30 mM heptakis-2,3,6-tri-O-methyl-β-cyclodextrin (A), 5 mM heptamethylamino-β-cyclodextrin (B) or 5 mM vancomycin (C) (Buffer was 50 mM phosphoric acid containing 50 mM acetic acid and 50 mM boric acid, dilute with an equal volume of water, adjust pH to 5 with concentrated NaOH.)
Injection: Pressure injection at 10 psi.
Detector: UV 206
Migration time: 15.7, 17.0 (A); 21.5, 22.7 (B); 8.2, 8.8 (C) (enantiomers)
Voltage: -20 kV
Model: Bio-Rad Biofocus 3000
Limit of detection: 5 μM (A, C), 10 μM (B)

KEY WORDS
chiral; coated capillary; detector at anode

REFERENCE
Fanali,S.; Desiderio,C.; Aturki,Z. Enantiomeric resolution study by capillary electrophoresis. Selection of the appropriate chiral selector, *J.Chromatogr.A*, **1997**, *772*, 185–194.

SAMPLE
Matrix: solutions
Sample preparation: Inject an aliquot of a solution in MeOH:water 10:90.

CAPILLARY ELECTROPHORESIS
Capillary: 44 cm × 50 μm fused-silica (37 cm to detector) (Supelco)
Capillary preparation: Wash with running buffer for 3 min after each injection. Wash with running buffer for 10 min at the end of each day.
Capillary temperature: 25
Running buffer: 100 mM Phosphoric acid containing 5 mM sulfobutyl ether-β-cyclodextrin and 15 mM heptakis(2,3,6-tri-O-methyl)-β-cyclodextrin, adjusted to pH 3.0 with triethanolamine
Injection: Hydrodynamic injection for 5 s (13.3 nL).
Detector: UV 230
Migration time: 11, 14.5 (enantiomers)
Voltage: -25 kV
Current: 60 μA
Model: SpectraPhoresis 1000 CE

KEY WORDS
chiral; detector at anode; resolution (R_s = 30.6)

REFERENCE
Fillet,M.; Hubert,P.; Crommen,J. Enantioseparation of nonsteroidal anti-inflammatory drugs by capillary electrophoresis using mixtures of anionic and uncharged β-cyclodextrins as chiral additives, *Electrophoresis*, **1997**, *18*, 1013–1018.

SAMPLE
Matrix: solutions
Sample preparation: Inject an aliquot of a 500 μM solution in MeOH:water 50:50.

CAPILLARY ELECTROPHORESIS
Capillary: 38.5 × 50 μm fused-silica (30 cm to detector) (Supelco)
Capillary temperature: 25
Running buffer: 40.8 mM pH 2.3 Phosphoric acid containing 20 mM mono-(6-δ-glutamylamino-6-deoxy)-β-cyclodextrin and 10 mM heptakis(2,3,6-tri-O-methyl)-β-cyclodextrin, adjusted to pH 2.3 with ammediol (2-amino-2-methyl-1,3-propanediol)
Injection: Pressure injection at 25 mbar for 4 s.
Detector: UV 254
Migration time: 16.3, 17.3 (enantiomers)
Voltage: 20 kV

Model: Hewlett Packard HP³ᴰ

KEY WORDS
chiral

REFERENCE
Lelièvre,F.; Guelt,C.; Gareil,P.; Bahaddi,Y.; Galons,H. Use of a zwitterionic cyclodextrin as a chiral agent for the separation of enantiomers by capillary electrophoresis, *Electrophoresis*, **1997**, *18*, 891–896.

SAMPLE
Matrix: solutions

CAPILLARY ELECTROPHORESIS
Capillary: 35 cm × 50 μm polyacrylamide-coated fused-silica (30.5 cm to detector) (Composite Metal Services, UK)
Capillary preparation: Before each run rinse capillary with water for 70 s and with running buffer for 100 s. (Coat capillary as follows. Adjust the pH of 20 mL water to 3.5 with acetic acid, add 80 μL 3-(trimethoxysilyl)propyl methacrylate (3-methacryloxypropyltrimethoxysilane), mix, suck into capillary, let stand at room temperature for 1 h, remove the solution, wash with water. Fill the capillary with a deaerated 3-4% acrylamide solution containing 1 μL/mL N,N,N',N'-tetramethylethylenediamine and 1 mg/mL potassium persulfate, let stand for 30 min, remove excess solution by aspiration, rinse with water, remove water by aspiration, dry at 35° (J. Chromatogr. 1985, 347, 191).)
Capillary temperature: 25
Running buffer: Buffer containing 2.5 mM cyanoethylated-β-cyclodextrin (Cyclolab, Budapest) (Prepare buffer by adjusting the pH of 50 mM phosphoric acid containing 50 mM acetic acid and 50 mM boric acid to 5 with concentrated NaOH, add the appropriate amount of cyanoethylated-β-cyclodextrin, dilute with an equal volume of water.)
Injection: Pressure injection at 5 psi for 2 s.
Detector: UV 206
Migration time: 11.7 (second enantiomer, α = 1.029)
Voltage: 20 kV
Current: 27-40 μA
Model: Bio-Rad Biofocus 3000

KEY WORDS
detector at anode; chiral; coated capillary

REFERENCE
Aturki,Z.; Desiderio,C.; Mannina,L.; Fanali,S. Chiral separations by capillary zone electrophoresis with the use of cyanoethylated-β-cyclodextrin as chiral selector, *J.Chromatogr.A*, **1998**, *817*, 91–104.

SAMPLE
Matrix: solutions
Sample preparation: Inject an aliquot of a 50-100 μg/mL solution in 50 mM pH 4.8 ammonium acetate buffer.

CAPILLARY ELECTROPHORESIS
Capillary: 44 cm × 50 μm polyacrylamide coated
Capillary preparation: Adjust the pH of 20 mL water to 3.5 with acetic acid, add 80 μL 3-(trimethoxysilyl)propyl methacrylate (3-methacryloxypropyltrimethoxysilane), mix, suck into capillary, let stand at room temperature for 1 h, remove the solution, wash with water. Fill the capillary with a deaerated 3-4% acrylamide solution containing 1 μL/mL N,N,N',N'-tetramethylethylenediamine and 1 mg/mL potassium persulfate, let stand for 30 min, remove excess solution by aspiration, rinse with water, remove water by aspiration, dry at 35° (J. Chromatogr. 1985, 347, 191).
Running buffer: 50 mM pH 4.8 Ammonium acetate buffer containing 5 mM vancomycin
Injection: Hydrostatic injection at 10 cm for 5-10 s.
Detector: MS, Finnigan LCQ ion trap, electrospray, negative ion mode, probe tip at 2.6 kV, sheath liquid MeOH:water:ammonia 50:48:2 at 6 μL/min (Vancomycin migrates away from the detector.)
Migration time: 9.6, 10 (enantiomers)

Voltage: 20 kV
Model: Grom 100

OTHER SUBSTANCES
Also analyzed: flurbiprofen, ibuprofen, ketoprofen, naproxen

KEY WORDS
chiral; coated capillary

REFERENCE
Fanali,S.; Desiderio,C.; Schulte,G.; Heitmeier,S.; Strickmann,D.; Chankvetadze,B.; Blaschke,G. Chiral capillary
  electrophoresis-electrospray mass spectrometry coupling using vancomycin as chiral selector,
  J.Chromatogr.A, **1998**, *800*, 69–76.

SAMPLE
Matrix: solutions

CAPILLARY ELECTROPHORESIS
Capillary: 57 cm × 50 μm fused-silica (50 cm to detector)
Running buffer: 50 mM pH 8.0 Borate buffer containing 40 mM sodium taurodeoxycholate and
  25 mM phosphatidylcholine (Prepare by adding phosphatidylcholine and stirring for 3-6 h until
  all cloudiness disappears. Phosphatidylcholine was 95% pure soybean lecithin, Epikuron, Lucas
  Meyer & Co.)
Injection: Inject a solution of the compound in the running buffer at 20 psi for 1 s, inject MeOH:
  water 5:95 at 20 psi for 1 s, inject a solution of halofantrine in running buffer at 20 psi for 1
  s.
Detector: UV 214
Migration time: k' 23.44
Model: Beckman P/ACE 5000

OTHER SUBSTANCES
Also analyzed: acetaminophen, amoxicillin, antipyrine, aspirin, azathioprine, caffeine, captopril,
  carbamazepine, chlorambucil, chlorpheniramine, chlorpromazine, cimetidine, clonidine, codeine, desipramine, diphenhydramine, ephedrine, fenoterol, flufenamic acid, flurbiprofen, haloperidol, hydroxyzine, ibuprofen, imipramine, indomethacin, ketoprofen, lidocaine, melphalan,
  metoprolol, nabumetone, nadolol, phenobarbital, phenol, promazine, propranolol, pyrilamine,
  ranitidine, ropinirole, salicylic acid, sulfamethoxazole, testosterone, theophylline, thioridazine,
  tiaprofenic acid, tolfenamic acid, trifluoperazine, trimethoprim, valproic acid, verapamil

---

KEY WORDS
comparison with HPLC; $k' = (Tr\text{-}T0)/(T0(1\text{-}Tr/Tm))$ where Tr = retention time of analyte; T0 =
  retention time of water; and Tm = retention time of marker (halofantrine)

REFERENCE
Hanna,M.; de Biasi,V.; Bond,B.; Salter,C.; Hutt,A.J.; Camilleri,P. Estimation of the partitioning characteristics
  of drugs: A comparison of a large and diverse drug series utilizing chromatographic and electrophoretic
  methodology, *Anal.Chem.*, **1998**, *70*, 2092–2099.

SAMPLE
Matrix: solutions
Sample preparation: Inject an aliquot of a solution in running buffer.

CAPILLARY ELECTROPHORESIS
Capillary: 57 cm × 50 μm fused-silica (Beckman)
Capillary preparation: Rinse with 100 mM NaOH for 5 min, with water for 10 min,
Running buffer: 2-Methoxyethanol:buffer 30:70 (Buffer was 40 mM pH 6 phosphate buffer con-
  taining 500 μM Actaplanin A (Eli Lilly).)
Injection: Pressure injection for 5 s.
Detector: UV 254
Migration time: 42.5, 44.1
Model: Beckman P/ACE 2100

OTHER SUBSTANCES
Also analyzed: flurbiprofen, indoprofen

KEY WORDS
chiral

REFERENCE
Trelli-Seifert,L.A.; Risley,D.S. Capillary electrophoretic enantiomeric separations of nonsteroidal anti-inflammatory compounds using the macrocyclic antibiotic actaplanin A and 2-methoxyethanol, *J.Liq.Chromatogr.Rel.Technol.*, **1998**, *21*, 299–313.

SAMPLE
Matrix: solutions

CAPILLARY ELECTROPHORESIS
Capillary: 42 cm × 50 μm fused-silica (31 cm to detector) (Polymicro Technologies)
Capillary temperature: 30
Running buffer: Formamide containing 3.42% quaternary ammonium-β-cyclodextrin (American Maize Products, Hammond IN), 20 mM ammonium acetate, and 1% acetic acid (pH* = 7.1)
Detector: UV 254
Migration time: 13.91, 14.11 (enantiomers)
Voltage: -30 kV
Model: laboratory-constructed

OTHER SUBSTANCES
Also analyzed: fenoprofen, flurbiprofen, indoprofen, ketoprofen, suprofen

KEY WORDS
chiral

REFERENCE
Wang,F.; Khaledi,M.G. Nonaqueous capillary electrophoresis chiral separations with quaternary ammonium β-cyclodextrin, *J.Chromatogr.A*, **1998**, *817*, 121–128.

Carteolol

Molecular formula: $C_{16}H_{24}N_2O_3$
Molecular weight: 292.38
CAS Registry No.: 51781-06-7, 51781-21-6 (HCl)
Merck Index (12th ed.): 1917
Lednicer: 3 183

SAMPLE
Matrix: blood
Sample preparation: Filter (1 mL Amicon Centrifree Micropartition System) serum while centrifuging at 1500 g for 20 min, inject an aliquot of the ultrafiltrate.

CAPILLARY ELECTROPHORESIS
Capillary: 92 cm × 75 μm fused-silica (55 cm to detector) (Polymicro Technologies)
Capillary preparation: Rinse with running buffer for 10 min between run. At the beginning and end of each day rinse with 100 mM NaOH for 10 min and with water for 10 min.
Running buffer: Isopropanol:buffer 5:95 (Buffer was 10 mM Na_2HPO_4 containing 6 mM sodium borate and 75 mM sodium dodecyl sulfate.)
Injection: Injection via vacuum suction for 0.7 s
Detector: UV 215
Migration time: 19

Voltage: 25 kV
Current: 82 μA
Model: Europhor Prime Vision IV

OTHER SUBSTANCES
Extracted: atenolol, celiprolol, levobunolol, metoprolol, penbutolol, pindolol, propranolol, timolol

KEY WORDS
serum; ultrafiltrate

REFERENCE
Schmutz,A.; Thormann,W. Assessment of impact of physico-chemical drug properties on monitoring drug levels
by micellar electrokinetic capillary chromatography with direct serum injection, *Electrophoresis*, **1994**, *15*,
1295–1303.

SAMPLE
Matrix: solutions
Sample preparation: Inject an aliquot of a 100 μg/mL solution in running buffer.

CAPILLARY ELECTROPHORESIS
Capillary: 30 cm × 50 μm fused-silica (25.5 cm to detector) (Yongnian Optical Conductive Fiber
Plant, China), coated with polyacrylamide
Capillary preparation: No details of the polyacrylamide coating process are provided. However,
another paper (LC.GC 1997, 15, 40) by this group indicates that they use the procedure of
Hjertén, thus: Adjust the pH of 20 mL water to 3.5 with acetic acid, add 80 μL 3-(trimethoxysilyl)propyl methacrylate (3-methacryloxypropyltrimethoxysilane), mix, suck into capillary, let
stand at room temperature for 1 h, remove the solution, wash with water. Fill the capillary
with a deaerated 3-4% acrylamide solution containing 1 μL/mL N,N,N',N'-tetramethylethyl-
enediamine and 1 mg/mL potassium persulfate, let stand for 30 min, remove excess solution
by aspiration, rinse with water, remove water by aspiration, dry at 35° (J. Chromatogr. 1985,
347, 191).
Capillary temperature: 25
Running buffer: 100 mM NaH_2PO_4 adjusted to pH 2.5
Injection: Electrokinetic injection at 15 kV for 3 s.
Detector: UV 200, UV 210
Migration time: 5.10
Voltage: 15 kV
Model: Bio-Rad BioFocus 3000

OTHER SUBSTANCES
Simultaneous: amorolfine, brompheniramine, bupivacaine, chloroquine, chlorpheniramine,
chlorphenoxamine, disopyramide, dobutamine, doxylamine, flecainide, gallopamil, ketamine,
mepindolol, orphenadrine, oxybutynin, phenoxybenzamine, pindolol, propafenone, propranolol,
sulpiride, talinolol, tropicamide, verapamil

KEY WORDS
coated capillary

REFERENCE
Koppenhoefer,B.; Epperlein,U.; Xiaofeng,Z.; Bingcheng,L. Separation of enantiomers of drugs by capillary electrophoresis. Part 4: Hydroxypropyl-γ-cyclodextrin as chiral solvating agent, *Electrophoresis*, **1997**, *18*, 924–
930.

SAMPLE
Matrix: solutions

CAPILLARY ELECTROPHORESIS
Capillary: 62.1 cm × 75 μm fused-silica (44.6 cm to detector) (Polymicro Technologies)
Capillary preparation: Coat capillary as follows. Condition capillary with 1 M NaOH for 1 h,
rinse with water for 30 min, fill capillary with 1% 3-(trimethoxysilyl)propyl methacrylate
(methacryloxypropyltrimethoxysilane) (adjusted to pH 3.5 with acetic acid), allow to react at

room temperature for 2 h, rinse with water, fill capillary with 10% 2-aminoethyl methacrylate hydrochloride (Fisher Scientific) solution containing 1-3 mg/mL N,N,N',N'-tetramethylethylenediamine and 1 mg/mL ammonium persulfate, allow to react until polymerization is complete, rinse with water, dry at 30-40° overnight.
Running buffer: 25 mM pH 4.7 Acetate buffer
Injection: Hydrodynamic injection at 5 cm for 5 s.
Detector: UV 210
Migration time: 11.20
Voltage: 15 kV
Model: laboratory constructed

OTHER SUBSTANCES
Simultaneous: acebutolol, atenolol, betaxolol, dilevalol, nifenalol, pindolol, propranolol

---

KEY WORDS
coated capillary; detector at anode

REFERENCE
Liu,Q.; Lin,F.; Hartwick,R.A. Free solution capillary electrophoretic separation of basic proteins and drugs using cationic polymer coated capillaries, *J.Liq.Chromatogr.Rel.Technol.*, **1997**, *20*, 707–718.

SAMPLE
Matrix: solutions

CAPILLARY ELECTROPHORESIS
Capillary: 29-36 cm × 50 μm fused-silica (24.5-31.5 cm to detector) (Yongnian Optical Conductive Fiber Plant, China) coated with polyacrylamide
Capillary preparation: Coat capillary as follows. Adjust the pH of 20 mL water to 3.5 with acetic acid, add 80 μL 3-(trimethoxysilyl)propyl methacrylate (3-methacryloxypropyltrimethoxysilane), mix, suck into capillary, let stand at room temperature for 1 h, remove the solution, wash with water. Fill the capillary with a deaerated 3-4% acrylamide solution containing 1 μL/mL N,N,N',N'-tetramethylethylenediamine and 1 mg/mL potassium persulfate, let stand for 30 min, remove excess solution by aspiration, rinse with water, remove water by aspiration, dry at 35° (J. Chromatogr. 1985, 347, 191).
Capillary temperature: 25
Running buffer: 100 mM pH 2.5 NaH_2PO_4 (A) or 100 mM pH 2.5 NaH_2PO_4 containing 45 mM hydroxypropyl-α-cyclodextrin (Wacker, Munich) (B)
Injection: Electromigration at 15 kV for 3 s.
Detector: UV 200; UV 210
Migration time: 5.10 (A); 7.82 (B) (no separation of enantiomers)
Voltage: 15 kV
Model: Bio-Focus 3000

OTHER SUBSTANCES
Also analyzed: albuterol (salbutamol), alprenolol, amorolfine, atenolol, atropine, azelastine, baclofen, bamethan, benproperine, benserazide, biperiden, bisoprolol, brompheniramine, bupivacaine, bupranolol, butamirate, butethamate, carazolol, carbuterol, carvedilol, celiprolol, chloroquine, chlorpheniramine, chlorphenoxamine, cicletanine, clenbuterol, clidinium bromide, clobutinol, dimethindene, dipivefrin, disopyramide, dobutamine, doxylamine, fendiline, flecainide, gallopamil, homatropine, ipratropium bromide, isoproterenol (isoprenaline), isothipendyl, ketamine, meclizine, mefloquine, mepindolol, mequitazine, metaclazepam, metaproterenol (orciprenaline), metipranolol, metoprolol, nafronyl (naftidrofuryl), nefopam, nicardipine, norfenefrine, ofloxacin, ornidazole, orphenadrine, oxomemazine, oxprenolol, oxybutynin, phenoxybenzamine, phenylpropanolamine, pholedrine, pindolol, pirbuterol, prilocaine, procyclidine, promethazine, propafenone, propranolol, reproterol, sotalol, sulpride, synephrine, talinolol, terbutaline, tetrahydrozoline (tetryzoline), theodrenaline, tioconazole, tocainide, trihexyphenidyl, trimeprazine (alimemazine), trimipramine, tropicamide, verapamil, zopiclone

---

KEY WORDS
coated capillary

REFERENCE

Koppenhoefer,B.; Eperlein,U.; Schlunk,R.; Zhu,X.; Lin,B. Separation of enantiomers of drugs by capillary electrophoresis. V. Hydroxypropyl-α-cyclodextrin as chiral solvating agent, *J.Chromatogr.A*, **1998**, *793*, 153–164.

SAMPLE
Matrix: solutions

CAPILLARY ELECTROPHORESIS
Capillary: 71.5 cm × 75 μm fused-silica (52.2 cm to detector) (Polymicro Technologies)
Capillary preparation: Rinse capillary with running buffer for 1 min between runs. Rinse new capillaries with 500 mM NaOH for 20 min and with water for 10 min and then condition with running buffer for 1 h.
Running buffer: 50 mM pH 3.0 containing 0.1% guaran (Guaran is the water-soluble fraction of guar gum. Prepare guaran by stirring 1 g guar gum (Sigma) in 100 mL water for 20 min, add 25 mL EtOH dropwise with stirring, let stand overnight, supercentrifuge. Add 30 mL EtOH dropwise with stirring to the centrifugate, centrifuge, collect the guaran as a precipitate, dry in air.)
Injection: Hydrodynamic injection at 5 cm for 5 s.
Detector: UV (wavelengths not given)
Migration time: 14
Voltage: 17.5 kV
Model: laboratory-constructed

OTHER SUBSTANCES
Simultaneous: acebutolol, atenolol, betaxolol, dilevalol, nifenalol, pindolol, propranolol

REFERENCE

Liu,Q.; Lin,F.; Hartwick,R.A. Capillary zone electrophoretic separation of basic proteins and drugs using guaran as a buffer modifier, *Chromatographia*, **1998**, *47*, 219–224.

Carvedilol

Molecular formula: $C_{24}H_{26}N_2O_4$
Molecular weight: 406.48
CAS Registry No.: 72956-09-3
Merck Index (12th ed.): 1924
Lednicer: 5 163

SAMPLE
Matrix: blood
Sample preparation: Extract serum with diethyl ether, evaporate extract to dryness, reconstitute with EtOH, inject an aliquot.

CAPILLARY ELECTROPHORESIS
Capillary: 57 cm × 75 μm fused-silica (50 cm to detector)
Capillary temperature: 20
Running buffer: 25 mM pH 2.5 Sodium phosphate buffer containing 10 mM hydroxypropyl-β-cyclodextrin
Detector: UV 200
Internal standard: l-propranolol
Voltage: 18 kV
Limit of quantitation: 1 μg/mL

KEY WORDS
chiral; serum

REFERENCE

Clohs,L.; McErlane,K. Development of a high performance capillary electrophoresis assay (HPCE) for the determination of carvedilol enantiomers in serum (Abstract APQ 1001), *Pharm.Res.*, **1996**, *13*, S3–S3.

SAMPLE
Matrix: solutions

CAPILLARY ELECTROPHORESIS
Capillary: 64 cm × 50 μm uncoated silica (48 cm to detector) (Polymicro Technologies)
Running buffer: 18 mM pH 2.9 Tris buffer containing 10 mM heptakis(2,6-di-O-methyl)-β-cyclodextrin, 0.1% methylhydroxycellulose 4000, and 0.03 mM hexadecyltrimethylammonium bromide (pH adjusted with phosphoric acid)
Detector: UV 220
Migration time: 20, 21 (enantiomers)
Voltage: 24 kV
Model: Laboratory constructed

OTHER SUBSTANCES
Simultaneous: bupivacaine, mepivacaine

KEY WORDS
chiral

REFERENCE

Soini,H.; Riekkola,M.-L.; Novotny,M.V. Chiral separations of basic drugs and quantitation of bupivacaine enantiomers in serum by capillary electrophoresis with modified cyclodextrin buffers, *J.Chromatogr.*, **1992**, *608*, 265–274.

SAMPLE
Matrix: solutions
Sample preparation: Inject an aliquot of a 100 μg/mL solution in water:running buffer 50:50.

CAPILLARY ELECTROPHORESIS
Capillary: 44.5 cm × 50 μm acrylamide-coated fused-silica (Bio-Rad)
Capillary temperature: 30
Running buffer: 100 mM NaH_2PO_4 containing 15 mM gamma-cyclodextrin, adjusted to pH 2.5 with phosphoric acid
Injection: Electrokinetic injection at 8 kV for 6 s.
Detector: UV 200
Migration time: 14.13
Voltage: 14 kV
Model: Bio-Rad BioFocus 3000

OTHER SUBSTANCES
Also analyzed: albuterol, alprenolol, atenolol, atropine, baclofen, bamethan, benserazide, biperiden, bisoprolol, bupivacaine, bupranolol, butetamate, carazolol, carbuterol, celiprolol, chloroquine, chlorpheniramine (chlorphenamine), clidinium bromide, clobutinol, disopyramide, dobutamine, flecainide, homatropine, ipratropium bromide, isoproterenol, isothipendyl, ketamine, mefloquine, mequitazine, metaproterenol (orciprenaline), metipranolol, nafronyl (naftidrofuryl), nefopam, ofloxacin, orphenadrine, oxomemazine, oxprenolol, phenoxybenzamine, pholedrine, pindolol, pirbuterol, prilocaine, promethazine, propafenone, propranolol, sotalol, synephrine, terbutaline, tetrahydrozoline (tetryzoline), tocainide, trihexyphenidyl, trimeprazine (alimemazine), trimipramine, tropicamide, verapamil, zopiclone

KEY WORDS
coated capillary; achiral

REFERENCE

Koppenhoefer,B.; Epperlein,U.; Christian,B.; Yibing,J.; Yuying,C.; Bingcheng,L. Separation of enantiomers of drugs by capillary electrophoresis. I. γ-Cyclodextrin as chiral solvating agent, *J.Chromatogr.A*, **1995**, *717*, 181–190.

SAMPLE
Matrix: solutions
Sample preparation: Inject an aliquot of a 100 µg/mL solution in water:running buffer 50:50.

CAPILLARY ELECTROPHORESIS
Capillary: 44.5 cm × 50 µm polyacrylamide-coated fused-silica (40 cm to detector) (Bio-Rad)
Capillary temperature: 30
Running buffer: 100 mM NaH_2PO_4 containing 15 mM β-cyclodextrin, adjusted to pH 2.5 with phosphoric acid
Injection: Electrokinetic injection at 8 kV for 6 s.
Detector: UV 200
Migration time: 15.39, 15.75 (enantiomers)
Voltage: 14 kV
Model: Bio-Rad Bio-Focus 3000

OTHER SUBSTANCES
Simultaneous: chlorpheniramine, ketamine, metaproterenol (orciprenaline), tetrahydrozoline, tropicamide, zopiclone
Also analyzed: albuterol (not chiral), atenolol (not chiral), benserazide (not chiral), biperiden (not chiral), bupivacaine (not chiral), butethamate (not chiral), carazolol (not chiral), carbuterol (not chiral), clidinium bromide (not chiral), disopyramide (not chiral), dobutamine (not chiral), flecainide (not chiral), ipratropium (not chiral), isothipendyl (not chiral), mequitazine (not chiral), metipranolol (not chiral), nafronyl (naftidrofuryl) (not chiral), nefopam (not chiral), ofloxacin (not chiral), orphenadrine (not chiral), pindolol (not chiral), pirbuterol (not chiral), prilocaine (not chiral), propafenone (not chiral), sotalol (not chiral), tocainide (not chiral), trimipramine (not chiral)

KEY WORDS
coated capillary; chiral

REFERENCE
Koppenhoefer,B.; Epperlein,U.; Christian,B.; Lin,B.; Ji,Y.; Chen,Y. Separation of enantiomers of drugs by capillary electrophoresis. III. β-cyclodextrin as chiral solvating agent, *J.Chromatogr.A*, **1996**, *735*, 333–343.

SAMPLE
Matrix: solutions
Sample preparation: Inject an aliquot of a 100 µg/mL solution in running buffer.

CAPILLARY ELECTROPHORESIS
Capillary: 36 cm × 50 µm fused-silica (31.5 cm to detector) (Yongnian Optical Conductive Fiber Plant, China), coated with polyacrylamide
Capillary preparation: No details of the polyacrylamide coating process are provided. However, another paper (LC.GC 1997, 15, 40) by this group indicates that they use the procedure of Hjertén, thus: Adjust the pH of 20 mL water to 3.5 with acetic acid, add 80 µL 3-(trimethoxysilyl)propyl methacrylate (3-methacryloxypropyltrimethoxysilane), mix, suck into capillary, let stand at room temperature for 1 h, remove the solution, wash with water. Fill the capillary with a deaerated 3-4% acrylamide solution containing 1 µL/mL N,N,N',N'-tetramethylethylenediamine and 1 mg/mL potassium persulfate, let stand for 30 min, remove excess solution by aspiration, rinse with water, remove water by aspiration, dry at 35° (J. Chromatogr. 1985, 347, 191).
Capillary temperature: 25
Running buffer: 100 mM NaH_2PO_4 adjusted to pH 2.5
Injection: Electrokinetic injection at 15 kV for 3 s.
Detector: UV 200, UV 210
Migration time: 7.12
Voltage: 15 kV
Model: Bio-Rad BioFocus 3000

OTHER SUBSTANCES
Simultaneous: azelastine, biperiden, carvedilol, clidinium bromide, meclizine (meclozine), mequitazine, ofloxacin, zopiclone

KEY WORDS
coated capillary

REFERENCE
Koppenhoefer,B.; Epperlein,U.; Xiaofeng,Z.; Bingcheng,L. Separation of enantiomers of drugs by capillary electrophoresis. Part 4: Hydroxypropyl-γ-cyclodextrin as chiral solvating agent, *Electrophoresis*, **1997**, *18*, 924–930.

SAMPLE
Matrix: solutions
Sample preparation: Inject an aliquot of a 100 μg/mL solution in running buffer.

CAPILLARY ELECTROPHORESIS
Capillary: 30 cm × 50 μm fused-silica (25.5 cm to detector), coated with polyacrylamide
Capillary preparation: Adjust the pH of 20 mL water to 3.5 with acetic acid, add 80 μL 3-(trimethoxysilyl)propyl methacrylate (3-methacryloxypropyltrimethoxysilane), mix, suck into capillary, let stand at room temperature for 1 h, remove the solution, wash with water. Fill the capillary with a deaerated 3 4% acrylamide solution containing 1 μL/mL N,N,N',N'-tetramethylethylenediamine and 1 mg/mL potassium persulfate, let stand for 30 min, remove excess solution by aspiration, rinse with water, remove water by aspiration, dry at 35° (J. Chromatogr. 1985, 347, 191).
Capillary temperature: 25
Running buffer: 100 mM NaH_2PO_4 containing 15 mM β-cyclodextrin, adjusted to pH 2.5 with phosphoric acid
Injection: Electrokinetic injection at 15 kV for 3 s.
Detector: UV 200
Voltage: 15 kV
Model: Bio-Rad BioFocus 3000

KEY WORDS
chiral; coated capillary; comparison with the use of other cyclodextrins; this running buffer gave the greatest enantiomeric separation.; α=1.028

REFERENCE
Lin,B.; Zhu,X.; Koppenhoefer,B.; Epperlein,U. Investigation of 123 chiral drugs by cyclodextrin-modified capillary electrophoresis, *LC.GC*, **1997**, *15*, 40–46.

SAMPLE
Matrix: solutions

CAPILLARY ELECTROPHORESIS
Capillary: 29-36 cm × 50 μm fused-silica (24.5-31.5 cm to detector) (Yongnian Optical Conductive Fiber Plant, China) coated with polyacrylamide
Capillary preparation: Coat capillary as follows. Adjust the pH of 20 mL water to 3.5 with acetic acid, add 80 μL 3-(trimethoxysilyl)propyl methacrylate (3-methacryloxypropyltrimethoxysilane), mix, suck into capillary, let stand at room temperature for 1 h, remove the solution, wash with water. Fill the capillary with a deaerated 3-4% acrylamide solution containing 1 μL/mL N,N,N',N'-tetramethylethylenediamine and 1 mg/mL potassium persulfate, let stand for 30 min, remove excess solution by aspiration, rinse with water, remove water by aspiration, dry at 35° (J. Chromatogr. 1985, 347, 191).
Capillary temperature: 25
Running buffer: 100 mM pH 2.5 NaH_2PO_4 (A) or 100 mM pH 2.5 NaH_2PO_4 containing 45 mM hydroxypropyl-α-cyclodextrin (Wacker, Munich) (B)
Injection: Electromigration at 15 kV for 3 s.
Detector: UV 200; UV 210
Migration time: 7.12 (A); 15.58 (B) (no separation of enantiomers)
Voltage: 15 kV
Model: Bio-Focus 3000

OTHER SUBSTANCES
Also analyzed: albuterol (salbutamol), alprenolol, amorolfine, atenolol, atropine, azelastine, baclofen, bamethan, benproperine, benserazide, biperiden, bisoprolol, brompheniramine, bupivacaine, bupranolol, butamirate, butethamate, carazolol, carbuterol, carteolol, celiprolol, chloroquine, chlorpheniramine, chlorphenoxamine, cicletanine, clenbuterol, clidinium bromide, clobutinol, dimethindene, dipivefrin, disopyramide, dobutamine, doxylamine, fendiline, flecainide, gallopamil, homatropine, ipratropium bromide, isoproterenol (isoprenaline), isothipendyl, ketamine, meclizine, mefloquine, mepindolol, mequitazine, metaclazepam, metaproterenol (orciprenaline), metipranolol, metoprolol, nafronyl (naftidrofuryl), nefopam, nicardipine, norfenefrine, ofloxacin, ornidazole, orphenadrine, oxomemazine, oxprenolol, oxybutynin, phenoxybenzamine, phenylpropanolamine, pholedrine, pindolol, pirbuterol, prilocaine, procyclidine, promethazine, propafenone, propranolol, reproterol, sotalol, sulpride, synephrine, talinolol, terbutaline, tetrahydrozoline (tetryzoline), theodrenaline, tioconazole, tocainide, trihexyphenidyl, trimeprazine (alimemazine), trimipramine, tropicamide, verapamil, zopiclone

KEY WORDS
coated capillary

REFERENCE
Koppenhoefer,B.; Eperlein,U.; Schlunk,R.; Zhu,X.; Lin,B. Separation of enantiomers of drugs by capillary electrophoresis. V. Hydroxypropyl-α-cyclodextrin as chiral solvating agent, *J.Chromatogr.A*, **1998**, *793*, 153–164.

Cefamandole

Molecular formula: $C_{18}H_{18}N_6O_5S_2$
Molecular weight: 462.51
CAS Registry No.: 34444-01-4, 42540-40-9 (nafate), 30034-03-8 (sodium)
Merck Index (12th ed.): 1964
Lednicer: 2 441

SAMPLE
Matrix: formulations
Sample preparation: Dissolve contents of capsule in water so as to achieve a concentration of 100 μg/mL, filter, inject an aliquot.

CAPILLARY ELECTROPHORESIS
Capillary: 72 cm × 50 μm fused-silica (50 cm to detector)
Capillary preparation: Flush with 1 M NaOH for 30 min, rinse with water for 10 min.
Capillary temperature: 30
Running buffer: 20 mM Sodium borate containing 200 mM sodium dodecyl sulfate and 100 mM pentanesulfonic acid
Injection: Hydrostatic injection for 5 s
Detector: UV 210
Migration time: 22
Voltage: 306 V/cm
Model: Applied Biosystems 270A-HT

OTHER SUBSTANCES
Simultaneous: cefuroxime, cephalexin, cephalothin, cephapirin

KEY WORDS
capsules

REFERENCE
Sciacchitano,C.J.; Mopper,B.; Specchio,J.J. Identification and separation of five cephalosporins by micellar electrokinetic chromatography, *J.Chromatogr.B*, **1994**, *657*, 395–399.

Cefixime

Molecular formula: $C_{16}H_{15}N_5O_7S_2$
Molecular weight: 453.46
CAS Registry No.: 79350-37-1
Merck Index (12th ed.): 1975
Lednicer: 4 184,185

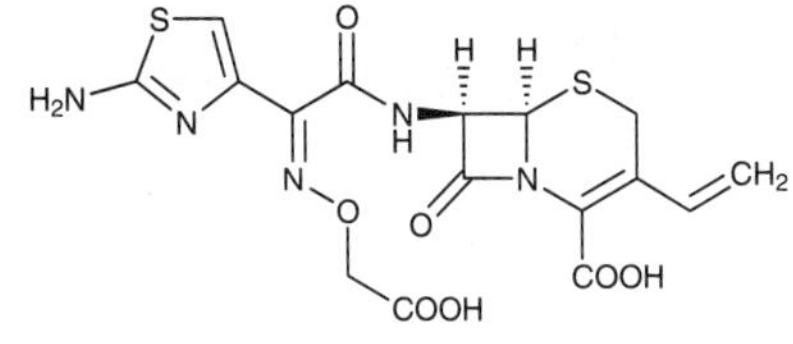

SAMPLE
Matrix: solutions

CAPILLARY ELECTROPHORESIS
Capillary: 75 cm $\times$ 50 μm fused-silica (53 cm to detector)
Running buffer: MeOH:buffer 20:80 (Buffer was 20 mM pH 6.8 phosphate buffer containing 0.15% 3-[(3-cholamidopropyl)dimethylammonio]-1-propanesulfonate.)
Injection: Injection using a weak vacuum for 1.5 s.
Detector: UV 280
Migration time: 25
Voltage: 30 kV
Current: about 30 μA
Model: Applied Biosystems Model 270 A

OTHER SUBSTANCES
Extracted: metabolites

REFERENCE
Honda,S.; Taga,A.; Kakehi,K.; Koda,S.; Okamoto,Y. Determination of cefixime and its metabolites by high-performance capillary electrophoresis, *J.Chromatogr.*, **1992**, *590*, 364–368.

SAMPLE
Matrix: urine

CAPILLARY ELECTROPHORESIS
Capillary: 75 cm $\times$ 50 μm fused-silica (53 cm to detector)
Running buffer: 50 mM pH 6.8 phosphate buffer
Injection: Injection using a weak vacuum for 1.5 s.
Detector: UV 295
Migration time: 20
Internal standard: cinnamic acid (19)
Voltage: 20 kV
Current: about 30 μA
Model: Applied Biosystems Model 270 A
Limit of quantitation: 10 μg/mL

REFERENCE
Honda,S.; Taga,A.; Kakehi,K.; Koda,S.; Okamoto,Y. Determination of cefixime and its metabolites by high-performance capillary electrophoresis, *J.Chromatogr.*, **1992**, *590*, 364–368.

Cefmenoxime

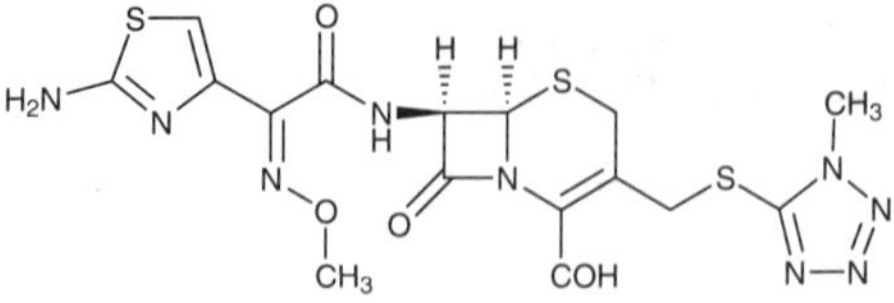

Molecular formula: $C_{16}H_{17}N_9O_5S_3$
Molecular weight: 511.57
CAS Registry No.: 65085-01-0, 75738-58-8 (HCl)
Merck Index (12th ed.): 1976
Lednicer: 4 187

SAMPLE
Matrix: solutions
Sample preparation: Prepare a 0.5-2 mg/mL solution in water, inject an aliquot.

CAPILLARY ELECTROPHORESIS
Capillary: 65 cm × 50 μm fused-silica (50 cm to detector) (Scientific Glass Engineering)
Running buffer: 20 mM NaH_2PO_4 containing 150 mM sodium dodecyl sulfate adjusted to pH 9.0 with 20 mM sodium tetraborate
Injection: Injection by siphon at 5 cm for 5-10 s
Detector: UV 210
Migration time: 8.5
Voltage: 20 kV

OTHER SUBSTANCES
Simultaneous: cefminox, cefoperazone, cefotaxime, cefpimizole, cefpiramide, ceftazidime, ceftriaxone
Also analyzed: amoxicillin, ampicillin, aspoxicillin, carbenicillin, penicillin G, piperacillin, sulbenicillin

REFERENCE
Nishi,H.; Tsumagari,N.; Kakimoto,T.; Terabe,S. Separation of β-lactam antibiotics by micellar electrokinetic chromatography, *J.Chromatogr.*, **1989**, *477*, 259–270.

Cefminox

Molecular formula: $C_{16}H_{21}N_7O_7S_3$
Molecular weight: 519.58
CAS Registry No.: 84305-41-9, 75481-73-1
Merck Index (12th ed.): 1978

SAMPLE
Matrix: solutions
Sample preparation: Prepare a 0.5-2 mg/mL solution in water, inject an aliquot.

CAPILLARY ELECTROPHORESIS
Capillary: 65 cm × 50 μm fused-silica (50 cm to detector) (Scientific Glass Engineering)
Running buffer: 20 mM NaH_2PO_4 containing 150 mM sodium dodecyl sulfate adjusted to pH 9.0 with 20 mM sodium tetraborate
Injection: Injection by siphon at 5 cm for 5-10 s
Detector: UV 210
Migration time: 10.2
Voltage: 20 kV

OTHER SUBSTANCES
Simultaneous: cefmenoxime, cefoperazone, cefotaxime, cefpimizole, cefpiramide, ceftazidime, ceftriaxone

Also analyzed: amoxicillin, ampicillin, aspoxicillin, carbenicillin, penicillin G, piperacillin, sulbenicillin

REFERENCE
Nishi,H.; Tsumagari,N.; Kakimoto,T.; Terabe,S. Separation of β-lactam antibiotics by micellar electrokinetic chromatography, *J.Chromatogr.*, **1989**, *477*, 259–270.

Cefodizime

Molecular formula: $C_{20}H_{20}N_6O_7S_4$
Molecular weight: 584.68
CAS Registry No.: 69739-16-8, 86329-79-5 (disodium salt)
Merck Index (12th ed.): 1979

SAMPLE
Matrix: blood
Sample preparation: Dilute plasma 1:5 with water, inject an aliquot.

CAPILLARY ELECTROPHORESIS
Capillary: 48.5 cm × 50 μm fused-silica (40 cm to detector) (Hewlett-Packard)
Capillary preparation: Before each run flush capillary with 100 mM NaOH for 5 min and with running buffer for 10 min. Condition capillary with 1 M NaOH for 10 min before each run.
Capillary temperature: 25
Running buffer: 20 mM pH 6 Citrate buffer
Injection: Hydrodynamic injection at 50 mbar for 9 s (18.8 nL).
Detector: UV 200
Migration time: 9.3
Voltage: 30 kV
Current: 40-100 μA
Model: Hewlett-Packard Model G1600A [3D]CE
Limit of detection: 4 μg/mL

OTHER SUBSTANCES
Extracted: cefotaxime, cefpirome, cefuroxime

KEY WORDS
plasma

REFERENCE
Mrestani,Y.; Neubert,R.; Schiewe,J.; Härtl,A. Application of capillary zone electrophoresis in cephalosporin analysis, *J.Chromatogr.B*, **1997**, *690*, 321–326.

Cefoperazone

Molecular formula: $C_{25}H_{27}N_9O_8S_2$
Molecular weight: 645.68
CAS Registry No.: 62893-19-0, 62893-20-3 (sodium salt)
Merck Index (12th ed.): 1981
Lednicer: 4 185, 188-190

SAMPLE
Matrix: solutions
Sample preparation: Prepare a 0.5-2 mg/mL solution in water, inject an aliquot.

CAPILLARY ELECTROPHORESIS
Capillary: 65 cm × 50 μm fused-silica (50 cm to detector) (Scientific Glass Engineering)
Running buffer: 20 mM NaH_2PO_4 containing 150 mM sodium dodecyl sulfate adjusted to pH
9.0 with 20 mM sodium tetraborate
Injection: Injection by siphon at 5 cm for 5-10 s
Detector: UV 210
Migration time: 9.5
Voltage: 20 kV

OTHER SUBSTANCES
Simultaneous: cefmenoxime, cefminox, cefotaxime, cefpimizole, cefpiramide, ceftazidime,
ceftriaxone
Also analyzed: amoxicillin, ampicillin, aspoxicillin, carbenicillin, penicillin G, piperacillin,
sulbenicillin

REFERENCE
Nishi,H.; Tsumagari,N.; Kakimoto,T.; Terabe,S. Separation of β-lactam antibiotics by micellar electrokinetic
chromatography, *J.Chromatogr.*, **1989**, *477*, 259–270.

Cefotaxime

Molecular formula: $C_{16}H_{17}N_5O_7S_2$
Molecular weight: 455.47
CAS Registry No.: 63527-52-6, 64485-93-4 (sodium salt)
Merck Index (12th ed.): 1983
Lednicer: 3 216

SAMPLE
Matrix: blood
Sample preparation: 200 μL Plasma + 66 μL water, vortex for 1 min, inject an aliquot.

CAPILLARY ELECTROPHORESIS
Capillary: 57 cm × 75 μm fused-silica (50 cm to detector) (Beckman)
Capillary preparation: Before each run wash with 100 mM NaOH for 2 min, fill with running
buffer for 2 min. At the start of each sequence wash with 100 mM NaOH and with water for
5 min. Before first use rinse with 250 mM NaOH for 30 min and with water for 10 min.
Running buffer: 30 mM NaH_2PO_4 containing 165 mM sodium dodecyl sulfate, adjusted to pH
8.0 with 5 M NaOH
Injection: Hydrodynamic injection of sample for 2 s then hydrodynamic injection of water for
1 s.
Detector: UV 254
Migration time: 11.25
Internal standard: theobromine (11.5)
Voltage: 15 kV (ramp voltage of 37.5 kV/min)
Model: Beckman P/ACE 5500
Limit of quantitation: 2 μg/mL
Limit of detection: 1 μg/mL

OTHER SUBSTANCES
Extracted: metabolites, deacetylcefotaxime

KEY WORDS
plasma

REFERENCE
Castaneda Penalvo,G.; Kelly,M.; Maillols,H.; Fabre,H. Evaluation of capillary zone electrophoresis and micellar
electrokinetic capillary chromatography with direct injection of plasma for the determination of cefotaxime
and its metabolite, *Anal.Chem.*, **1997**, *69*, 1364–1369.

SAMPLE
Matrix: blood
Sample preparation: Dilute plasma 1:5 with water, inject an aliquot.

CAPILLARY ELECTROPHORESIS
Capillary: 48.5 cm × 50 μm fused-silica (40 cm to detector) (Hewlett-Packard)
Capillary preparation: Before each run flush capillary with 100 mM NaOH for 5 min and with running buffer for 10 min. Condition capillary with 1 M NaOH for 10 min before each run.
Capillary temperature: 25
Running buffer: 20 mM pH 6 Citrate buffer
Injection: Hydrodynamic injection at 50 mbar for 9 s (18.8 nL).
Detector: UV 200
Migration time: 4.9
Voltage: 30 kV
Current: 40-100 μA
Model: Hewlett-Packard Model G1600A ^{3D}CE
Limit of detection: 2 μg/mL

OTHER SUBSTANCES
Extracted: cefodizime, cefpirome, cefuroxime

KEY WORDS
plasma

REFERENCE
Mrestani,Y.; Neubert,R.; Schiewe,J.; Härtl,A. Application of capillary zone electrophoresis in cephalosporin analysis, *J.Chromatogr.B*, **1997**, *690*, 321–326.

SAMPLE
Matrix: bulk
Sample preparation: Inject an aliquot of an aqueous solution.

CAPILLARY ELECTROPHORESIS
Capillary: 57 cm × 75 μm fused-silica (50 cm to detector) (Beckman)
Capillary preparation: Rinse at high pressure with running buffer for 3 min before each injection. Before use flush a new capillary with 100 mM NaOH for 30 min and with water for 10 min
Capillary temperature: 22
Running buffer: 30 mM NaH_2PO_4 containing 165 mM sodium dodecyl sulfate, adjusted to pH 7.2 with 5 M NaOH
Injection: Inject at low pressure for 2-5 s then inject running buffer at low pressure for 1 s
Detector: UV 254
Migration time: 12
Voltage: 15 kV (after injection ramp to this value at 37.5 kV/min.)
Current: 148 μA
Model: Beckman P/ACE 5500

OTHER SUBSTANCES
Simultaneous: impurities

REFERENCE
Castaneda Pensalvo,G.; Julien,E.; Fabre,H. Cross validation of capillary electrophoresis and high-performance liquid chromatography for cefotaxime and related impurities, *Chromatographia*, **1996**, *42*, 159–164.

SAMPLE
Matrix: solutions
Sample preparation: Prepare a 0.5-2 mg/mL solution in water, inject an aliquot.

CAPILLARY ELECTROPHORESIS
Capillary: 65 cm × 50 μm fused-silica (50 cm to detector) (Scientific Glass Engineering)

Running buffer: 20 mM NaH_2PO_4 containing 150 mM sodium dodecyl sulfate adjusted to pH 9.0 with 20 mM sodium tetraborate
Injection: Injection by siphon at 5 cm for 5-10 s
Detector: UV 210
Migration time: 8.2
Voltage: 20 kV

OTHER SUBSTANCES
Simultaneous: cefmenoxime, cefminox, cefoperazone, cefpimizole, cefpiramide, ceftazidime, ceftriaxone
Also analyzed: amoxicillin, ampicillin, aspoxicillin, carbenicillin, penicillin G, piperacillin, sulbenicillin

REFERENCE
Nishi,H.; Tsumagari,N.; Kakimoto,T.; Terabe,S. Separation of β-lactam antibiotics by micellar electrokinetic chromatography, *J.Chromatogr.*, **1989**, *477*, 259–270.

SAMPLE
Matrix: solutions
Sample preparation: Inject an aliquot of a 100 µg/mL solution in water.

CAPILLARY ELECTROPHORESIS
Capillary: 57 × 75 µm fused-silica (50 cm to detector) (Beckman)
Capillary preparation: Before each run wash with 250 mM NaOH for 5 min, wash with water for 5 min, fill with running buffer for 2 min. Condition a new capillary with 250 mM NaOH for 30 min and with water for 10 min.
Capillary temperature: about 85 µA
Running buffer: 40 mM pH 8.0 KH_2PO_4
Injection: Inject sample for 5 s then inject running buffer for 1 s
Detector: UV 254
Migration time: 8.1
Voltage: + 15 kV (ramp voltage 37.5 kV/min)
Model: Beckman P/ACE 5500

OTHER SUBSTANCES
Simultaneous: impurities

KEY WORDS
comparison with HPLC

REFERENCE
Fabre,H.; Castaneda Penalvo,G. Capillary electrophoresis as an alternative method for the determination of cefotaxime, *J.Liq.Chromatogr.*, **1995**, *18*, 3877–3887.

Cefpimizole

Molecular formula: $C_{28}H_{26}N_6O_{10}S_2$
Molecular weight: 670.68
CAS Registry No.: 84880-03-5, 85287-61-2 (sodium salt)
Merck Index (12th ed.): 1988
Lednicer: 4 185

SAMPLE
Matrix: solutions
Sample preparation: Prepare a 0.5-2 mg/mL solution in water, inject an aliquot.

CAPILLARY ELECTROPHORESIS
Capillary: 65 cm × 50 μm fused-silica (50 cm to detector) (Scientific Glass Engineering)
Running buffer: 20 mM NaH_2PO_4 containing 150 mM sodium dodecyl sulfate adjusted to pH 9.0 with 20 mM sodium tetraborate
Injection: Injection by siphon at 5 cm for 5-10 s
Detector: UV 210
Migration time: 10.0
Voltage: 20 kV

OTHER SUBSTANCES
Simultaneous: cefmenoxime, cefminox, cefoperazone, cefotaxime, cefpiramide, ceftazidime, ceftriaxone
Also analyzed: amoxicillin, ampicillin, aspoxicillin, carbenicillin, penicillin G, piperacillin, sulbenicillin

REFERENCE
Nishi,H.; Tsumagari,N.; Kakimoto,T.; Terabe,S. Separation of β-lactam antibiotics by micellar electrokinetic chromatography, *J.Chromatogr.*, **1989**, *477*, 259–270.

Cefpiramide

Molecular formula: $C_{25}H_{24}N_8O_7S_2$
Molecular weight: 612.65
CAS Registry No.: 70797-11-4, 74849-93-7 (sodium salt)
Merck Index (12th ed.): 1989
Lednicer: 4 188

SAMPLE
Matrix: blood
Sample preparation: Inject plasma directly.

CAPILLARY ELECTROPHORESIS
Capillary: 60 cm × 50 μm fused-silica (40 cm to detector) (Scientific Glass Engineering)
Capillary preparation: After each injection flush capillary with 100 mM NaOH.
Capillary temperature: 25
Running buffer: pH 8.0 Phosphate buffer (I = 0.05) containing 10 mM sodium dodecyl sulfate
Injection: Siphon at 5 cm for 10 s
Detector: UV 280
Migration time: 5.5
Voltage: 15 kV
Current: about 11 μA
Limit of quantitation: 10 μg/mL
Limit of detection: 5 μg/mL (S/N 3)

OTHER SUBSTANCES
Extracted: antipyrine

KEY WORDS
plasma; detection at anode

REFERENCE
Nakagawa,T.; Oda,Y.; Shibukawa,A.; Tanaka,H. Separation and determination of cefpiramide in human plasma by electrokinetic chromatography with a micellar solution and an open tubular-fused silica capillary, *Chem.Pharm.Bull.*, **1988**, *36*, 1622–1625.

SAMPLE
Matrix: blood

Sample preparation: Inject directly.

CAPILLARY ELECTROPHORESIS
Capillary: 75 cm × 50 μm fused-silica (55 cm to detector) (SGE)
Capillary preparation: Wash capillary with 1 M NaOH between runs
Capillary temperature: 37
Running buffer: 50 mM pH 8 Phosphate buffer containing 10 mM sodium dodecyl sulfate
Injection: Siphon at 5 cm for 15 s
Detector: UV 280
Migration time: 8.5
Voltage: +15 kV
Limit of detection: 5 μg/mL (S/N 3)

OTHER SUBSTANCES
Extracted: antipyrine

KEY WORDS
plasma

REFERENCE
Nakagawa,T.; Oda,Y.; Shibukawa,A.; Fukuda,H.; Tanaka,H. Electrokinetic chromatography for drug analysis. Separation and determination of cefpiramide in human plasma, *Chem.Pharm.Bull.*, **1989**, *37*, 707–711.

SAMPLE
Matrix: solutions
Sample preparation: Prepare a 0.5-2 mg/mL solution in water, inject an aliquot.

CAPILLARY ELECTROPHORESIS
Capillary: 65 cm × 50 μm fused-silica (50 cm to detector) (Scientific Glass Engineering)
Running buffer: 20 mM NaH_2PO_4 containing 150 mM sodium dodecyl sulfate adjusted to pH 9.0 with 20 mM sodium tetraborate
Injection: Injection by siphon at 5 cm for 5-10 s
Detector: UV 210
Migration time: 9.9
Voltage: 20 kV

OTHER SUBSTANCES
Simultaneous: cefmenoxime, cefminox, cefoperazone, cefotaxime, cefpimizole, ceftazidime, ceftriaxone
Also analyzed: amoxicillin, ampicillin, aspoxicillin, carbenicillin, penicillin G, piperacillin, sulbenicillin

REFERENCE
Nishi,H.; Tsumagari,N.; Kakimoto,T.; Terabe,S. Separation of β-lactam antibiotics by micellar electrokinetic chromatography, *J.Chromatogr.*, **1989**, *477*, 259–270.

Cefpirome

Molecular formula: $C_{22}H_{22}N_6O_5S_2$
Molecular weight: 514.59
CAS Registry No.: 84957-29-9, 98753-19-6 (sulfate)
Merck Index (12th ed.): 1990
Lednicer: 5 158

SAMPLE
Matrix: blood
Sample preparation: Dilute plasma 1:5 with water, inject an aliquot.

CAPILLARY ELECTROPHORESIS
Capillary: 48.5 cm × 50 μm fused-silica (40 cm to detector) (Hewlett-Packard)
Capillary preparation: Before each run flush capillary with 100 mM NaOH for 5 min and with running buffer for 10 min. Condition capillary with 1 M NaOH for 10 min before each run.
Capillary temperature: 25
Running buffer: 20 mM pH 6 Citrate buffer
Injection: Hydrodynamic injection at 50 mbar for 9 s (18.8 nL).
Detector: UV 200
Migration time: 2.7
Voltage: 30 kV
Current: 40-100 μA
Model: Hewlett-Packard Model G1600A ^{3D}CE
Limit of detection: 6 μg/mL

OTHER SUBSTANCES
Extracted: cefodizime, cefotaxime, cefuroxime

KEY WORDS
plasma

REFERENCE
Mrestani,Y.; Neubert,R.; Schiewe,J.; Härtl,A. Application of capillary zone electrophoresis in cephalosporin analysis, *J.Chromatogr.B*, **1997**, *690*, 321–326.

Ceftazidime

Molecular formula: $C_{22}H_{22}N_6O_7S_2$
Molecular weight: 546.58
CAS Registry No.: 72558-82-8, 78439-06-2 (pentahydrate)
Merck Index (12th ed.): 1995
Lednicer: 4 192

SAMPLE
Matrix: solutions
Sample preparation: Prepare a 0.5-2 mg/mL solution in water, inject an aliquot.

CAPILLARY ELECTROPHORESIS
Capillary: 65 cm × 50 μm fused-silica (50 cm to detector) (Scientific Glass Engineering)
Running buffer: 20 mM NaH_2PO_4 containing 150 mM sodium dodecyl sulfate adjusted to pH 9.0 with 20 mM sodium tetraborate
Injection: Injection by siphon at 5 cm for 5-10 s
Detector: UV 210
Migration time: 7.9
Voltage: 20 kV

OTHER SUBSTANCES
Simultaneous: cefmenoxime, cefminox, cefoperazone, cefotaxime, cefpimizole, cefpiramide, ceftriaxone
Also analyzed: amoxicillin, ampicillin, aspoxicillin, carbenicillin, penicillin G, piperacillin, sulbenicillin

REFERENCE
Nishi,H.; Tsumagari,N.; Kakimoto,T.; Terabe,S. Separation of β-lactam antibiotics by micellar electrokinetic chromatography, *J.Chromatogr.*, **1989**, *477*, 259–270.

SAMPLE
Matrix: solutions

CAPILLARY ELECTROPHORESIS
Capillary: 57 cm × 75 μm fused-silica
Capillary preparation: Rinse with water and fill with running buffer 5 min prior to injection. Replenish inlet and outlet vials after each run. If necessary because of inconsistent results, rinse with 100 mM NaOH for 15 min, rinse with water for 10 min, rinse with MeOH for 10 min, rinse with water for 10 min, rinse with running buffer for 10 min.
Capillary temperature: 20
Running buffer: 150 mM Sodium dodecyl sulfate containing 20 mM sodium phosphate and 20 mM sodium borate, pH adjusted to 9.0 with 100 mM sodium phosphate
Injection: Pressure injection at 3.45 kPa for 5 s (31 nL).
Detector: UV 274
Migration time: 10
Voltage: 15 kV
Current: about 131 μA
Model: Beckman P/ACE 5500

OTHER SUBSTANCES
Simultaneous: cefoxitin, cefoperazone, cefuroxime, cephalexin, cephradine

KEY WORDS
comparison with HPLC

REFERENCE
Choi,O.-K.; Song,Y.-S. Determination of cefuroxim levels in human serum by micellar electrokinetic capillary chromatography with direct sample injection, *J.Pharm.Biomed.Anal.*, **1997**, *15*, 1265–1270.

SAMPLE
Matrix: solutions
Sample preparation: Inject an aliquot of a 500 μg/mL solution in 50 mM pH 7 buffer.

CAPILLARY ELECTROPHORESIS
Capillary: 48.5 cm × 50 μm fused-silica (40 cm to detector) (Hewlett Packard)
Capillary preparation: Before each injection flush capillary with 100 mM NaOH for 5 min and with running buffer for 5 min. Wash new capillaries with 1 M NaOH at 40° for 15 min, with water at 40° for 10 min, and with water at 25° for 5 min.
Capillary temperature: 25
Running buffer: 50 mM pH 7.0 Phosphate buffer (Prepare buffer by dissolving 5.29 g K_2HPO_4 and 2.61 g KH_2PO_4 in 1 L water.)
Injection: Pressure injection at 50 mbar for 9 s (18.8 nL).
Detector: UV 200
Migration time: 12.5
Voltage: 30 kV
Model: Hewlett Packard Model G1600A [3D]CE

OTHER SUBSTANCES
Simultaneous: diclofenac, dicloxacillin, oxacillin, propranolol, quinine, thiamine

REFERENCE
Mrestani,Y.; Neubert,R.H.H.; Krause,A. Partition behaviour of drugs in microemulsions measured by electrokinetic chromatography, *Pharm.Res.*, **1998**, *15*, 799–801.

Ceftiofur

Molecular formula: $C_{19}H_{17}N_5O_7S_3$
Molecular weight: 523.57
CAS Registry No.: 80370-57-6, 104010-37-9 (sodium salt),
103980-44-5 (HCl)
Merck Index (12th ed.): 1999
Lednicer: 4 187

SAMPLE
Matrix: solutions
Sample preparation: Dilute a solution in buffer 10-fold with water, filter, inject an aliquot.
(Prepare buffer by dissolving 8.0 g KH_2PO_4 and 2.0 g K_2HPO_4 in 1 L water, adjust pH to 6.0
with 1 M HCl or 1 M NaOH.)

CAPILLARY ELECTROPHORESIS
Capillary: 50 cm × 50 μm fused-silica (Polymicro Technologies)
Capillary preparation: Before each run rinse with water for 80 s, with 1 M NaOH for 80 s,
with water for 80 s, and with running buffer for 90 s then the ends are dipped into water
Capillary temperature: 20
Running buffer: 20 mM Sodium tetraborate containing 150 mM sodium dodecyl sulfate, pH
9.20 ± 0.05
Injection: Pressure injection at 5 psi for 6 s (42 nL).
Detector: UV 210
Migration time: 15.5
Voltage: 14 kV
Model: BioFocus 3000 (Bio-Rad)

OTHER SUBSTANCES
Simultaneous: amoxicillin, ampicillin, cephapirin, cloxacillin, penicillin G

REFERENCE
Cutting,J.H.; Hurlbut,J.A.; Sofos,J.N. Quantitation of penicillin G in medicated premix feeds by micellar electrokinetic capillary chromatography, *J.AOAC Int.*, **1997**, *80*, 951–955.

Ceftizoxime

Molecular formula: $C_{13}H_{13}N_5O_5S_2$
Molecular weight: 383.41
CAS Registry No.: 68401-81-0, 68401-82-1 (sodium salt)
Merck Index (12th ed.): 2000
Lednicer: 3 218

SAMPLE
Matrix: solutions

CAPILLARY ELECTROPHORESIS
Capillary: 27 cm × 75 μm
Capillary preparation: Before each run rinse capillary with 100 mM NaOH for 30 s and with
running buffer for 30 s. Condition new capillaries by rinsing with 100 mM NaOH for 20 min.
Capillary temperature: 30
Running buffer: 15 mM Sodium borate
Injection: Pressure injection of sample at 25 mbar for 1 s followed by running buffer at 25 mbar
for 1 s.
Detector: UV 200

Migration time: 2.73
Internal standard: aminobenzoic acid (3.5)
Voltage: 6.5 kV
Model: Beckman

OTHER SUBSTANCES

Simultaneous: aspirin, bacitracin, beclomethasone, benzoic acid, ceftriaxone, cromolyn, embonic
acid, glyburide (glibenclamide), levothyroxine, nedocromil, nystatin, omeprazole, prednisolone,
warfarin, zidovudine
Interfering: cefuroxime, cephalothin, epoprostenol

REFERENCE

Altria,K.D.; Bryant,S.M.; Hadgett,T.A. Validated capillary electrophoresis method for the analysis of a range
of acidic drugs and excipients, *J.Pharm.Biomed.Anal.*, **1997**, *15*, 1091–1101.

Ceftriaxone

Molecular formula: $C_{18}H_{18}N_8O_7S_3$
Molecular weight: 554.59
CAS Registry No.: 73384-59-5, 104376-79-6 (sodium salt)
Merck Index (12th ed.): 2001

SAMPLE

Matrix: solutions
Sample preparation: Prepare a 0.5-2 mg/mL solution in water, inject an aliquot.

CAPILLARY ELECTROPHORESIS

Capillary: 65 cm × 50 μm fused-silica (50 cm to detector) (Scientific Glass Engineering)
Running buffer: 20 mM NaH_2PO_4 containing 150 mM sodium dodecyl sulfate adjusted to pH
9.0 with 20 mM sodium tetraborate
Injection: Injection by siphon at 5 cm for 5-10 s
Detector: UV 210
Migration time: 10.6
Voltage: 20 kV

OTHER SUBSTANCES

Simultaneous: cefmenoxime, cefminox, cefoperazone, cefotaxime, cefpimizole, cefpiramide,
ceftazidime
Also analyzed: amoxicillin, ampicillin, aspoxicillin, carbenicillin, penicillin G, piperacillin,
sulbenicillin

REFERENCE

Nishi,H.; Tsumagari,N.; Kakimoto,T.; Terabe,S. Separation of β-lactam antibiotics by micellar electrokinetic
chromatography, *J.Chromatogr.*, **1989**, *477*, 259–270.

SAMPLE

Matrix: solutions

CAPILLARY ELECTROPHORESIS

Capillary: 27 cm × 75 μm
Capillary preparation: Before each run rinse capillary with 100 mM NaOH for 30 s and with
running buffer for 30 s. Condition new capillaries by rinsing with 100 mM NaOH for 20 min.
Capillary temperature: 30
Running buffer: 15 mM Sodium borate
Injection: Pressure injection of sample at 25 mbar for 1 s followed by running buffer at 25 mbar
for 1 s.
Detector: UV 200

Migration time: 3.22
Internal standard: aminobenzoic acid (3.5)
Voltage: 6.5 kV
Model: Beckman

OTHER SUBSTANCES
Simultaneous: aspirin, bacitracin, beclomethasone, benzoic acid, ceftizoxime, cefuroxime, cephalothin, cromolyn, embonic acid, epoprostenol, glyburide (glibenclamide), levothyroxine, nedocromil, nystatin, omeprazole, prednisolone, warfarin, zidovudine

REFERENCE
Altria,K.D.; Bryant,S.M.; Hadgett,T.A. Validated capillary electrophoresis method for the analysis of a range of acidic drugs and excipients, *J.Pharm.Biomed.Anal.*, **1997**, *15*, 1091–1101.

Cefuroxime

Molecular formula: $C_{16}H_{16}N_4O_8S$
Molecular weight: 424.39
CAS Registry No.: 55268-75-2, 56238-63-2 (sodium salt), 64544-07-6 (axetil), 100680-33-9 (pivoxetil)
Merck Index (12th ed.): 2002
Lednicer: 3 216

SAMPLE
Matrix: blood
Sample preparation: Filter (0.45 μm) serum, inject an aliquot of the filtrate.

CAPILLARY ELECTROPHORESIS
Capillary: 57 cm × 75 μm fused-silica
Capillary preparation: Rinse with water and fill with running buffer 5 min prior to injection. Replenish inlet and outlet vials after each run. If necessary because of inconsistent results, rinse with 100 mM NaOH for 15 min, rinse with water for 10 min, rinse with MeOH for 10 min, rinse with water for 10 min, rinse with running buffer for 10 min.
Capillary temperature: 20
Running buffer: 150 mM Sodium dodecyl sulfate containing 20 mM sodium phosphate and 20 mM sodium borate, pH adjusted to 9.0 with 100 mM sodium phosphate
Injection: Pressure injection at 3.45 kPa for 5 s (31 nL).
Detector: UV 274
Migration time: 11.7
Voltage: 15 kV
Current: about 131 μA
Model: Beckman P/ACE 5500
Limit of detection: 280 nM

OTHER SUBSTANCES
Simultaneous: cefoperazone, cefoxitin, ceftazidime, cephalexin, cephradine

KEY WORDS
serum; comparison with HPLC; pharmacokinetics

REFERENCE
Choi,O.-K.; Song,Y.-S. Determination of cefuroxim levels in human serum by micellar electrokinetic capillary chromatography with direct sample injection, *J.Pharm.Biomed.Anal.*, **1997**, *15*, 1265–1270.

SAMPLE
Matrix: blood
Sample preparation: Dilute plasma 1:5 with water, inject an aliquot.

CAPILLARY ELECTROPHORESIS
Capillary: 48.5 cm × 50 µm fused-silica (40 cm to detector) (Hewlett-Packard)
Capillary preparation: Before each run flush capillary with 100 mM NaOH for 5 min and with running buffer for 10 min. Condition capillary with 1 M NaOH for 10 min before each run.
Capillary temperature: 25
Running buffer: 20 mM pH 6 Citrate buffer
Injection: Hydrodynamic injection at 50 mbar for 9 s (18.8 nL).
Detector: UV 200
Migration time: 4.6
Voltage: 30 kV
Current: 40-100 µA
Model: Hewlett-Packard Model G1600A ^{3D}CE
Limit of detection: 2 µg/mL

OTHER SUBSTANCES
Extracted: cefodizime, cefotaxime, cefpirome

KEY WORDS
plasma

REFERENCE
Mrestani,Y.; Neubert,R.; Schiewe,J.; Härtl,A. Application of capillary zone electrophoresis in cephalosporin analysis, *J.Chromatogr.B*, **1997**, *690*, 321–326.

SAMPLE
Matrix: bulk

CAPILLARY ELECTROPHORESIS
Capillary: 40 cm × 50 µm capillary packed with 3 µm Spherisorb ODS-1 PC12
Capillary temperature: 30
Running buffer: MeCN:10 mM Na_2HPO_4 50:50, pH 9.5
Injection: Electromigration at 20 kV for 12 s.
Detector: UV 276
Migration time: 42, 43 (Z-isomer diastereomers), 47.5, 49 (E-isomer diastereomers)
Voltage: 20 kV (with 500 psi pressure?)
Model: Applied Biosystems ABI 270A

OTHER SUBSTANCES
Simultaneous: impurities

KEY WORDS
electrochromatography

REFERENCE
Smith,N.W.; Evans,M.B. The analysis of pharmaceutical compounds using electrochromatography, *Chromatographia*, **1994**, *38*, 649–657.

SAMPLE
Matrix: formulations
Sample preparation: Dissolve contents of capsule in water so as to achieve a concentration of 100 µg/mL, filter, inject an aliquot.

CAPILLARY ELECTROPHORESIS
Capillary: 72 cm × 50 µm fused-silica (50 cm to detector)
Capillary preparation: Flush with 1 M NaOH for 30 min, rinse with water for 10 min.
Capillary temperature: 30
Running buffer: 20 mM Sodium borate containing 200 mM sodium dodecyl sulfate and 100 mM pentanesulfonic acid
Injection: Hydrostatic injection for 5 s
Detector: UV 210
Migration time: 14

Voltage: 306 V/cm
Model: Applied Biosystems 270A-HT

OTHER SUBSTANCES
Simultaneous: cefamandole, cephalexin, cephalothin, cephapirin

KEY WORDS
capsules

REFERENCE
Sciacchitano,C.J.; Mopper,B.; Specchio,J.J. Identification and separation of five cephalosporins by micellar electrokinetic chromatography, *J.Chromatogr.B*, **1994**, *657*, 395–399.

SAMPLE
Matrix: solutions

CAPILLARY ELECTROPHORESIS
Capillary: 27 cm $\times$ 75 μm
Capillary preparation: Before each run rinse capillary with 100 mM NaOH for 30 s and with running buffer for 30 s. Condition new capillaries by rinsing with 100 mM NaOH for 20 min.
Capillary temperature: 30
Running buffer: 15 mM Sodium borate
Injection: Pressure injection of sample at 25 mbar for 1 s followed by running buffer at 25 mbar for 1 s.
Detector: UV 200
Migration time: 2.70
Internal standard: aminobenzoic acid (3.5)
Voltage: 6.5 kV
Model: Beckman

OTHER SUBSTANCES
Simultaneous: aspirin, bacitracin, beclomethasone, benzoic acid, ceftriaxone, cromolyn, embonic acid, glyburide (glibenclamide), levothyroxine, nedocromil, nystatin, omeprazole, prednisolone, warfarin, zidovudine
Interfering: ceftizoxime, cephalothin, epoprostenol

REFERENCE
Altria,K.D.; Bryant,S.M.; Hadgett,T.A. Validated capillary electrophoresis method for the analysis of a range of acidic drugs and excipients, *J.Pharm.Biomed.Anal.*, **1997**, *15*, 1091–1101.

Celiprolol

Molecular formula: $C_{20}H_{33}N_3O_4$
Molecular weight: 379.50
CAS Registry No.: 56980-93-9, 57470-78-7 (HCl)
Merck Index (12th ed.): 2007
Lednicer: 4 27

SAMPLE
Matrix: blood
Sample preparation: Filter (1 mL Amicon Centrifree Micropartition System) serum while centrifuging at 1500 g for 20 min, inject an aliquot of the ultrafiltrate.

CAPILLARY ELECTROPHORESIS
Capillary: 92 cm $\times$ 75 μm fused-silica (55 cm to detector) (Polymicro Technologies)
Capillary preparation: Rinse with running buffer for 10 min between run. At the beginning and end of each day rinse with 100 mM NaOH for 10 min and with water for 10 min.

Running buffer: Isopropanol:buffer 5:95 (Buffer was 10 mM Na$_2$HPO$_4$ containing 6 mM sodium
  borate and 75 mM sodium dodecyl sulfate.)
Injection: Injection via vacuum suction for 0.7 s
Detector: UV 215
Migration time: 22
Voltage: 25 kV
Current: 82 μA
Model: Europhor Prime Vision IV

OTHER SUBSTANCES
Extracted: atenolol, carteolol, levobunolol, metoprolol, penbutolol, pindolol, propranolol, timolol

KEY WORDS
serum; ultrafiltrate

REFERENCE
Schmutz,A.; Thormann,W. Assessment of impact of physico-chemical drug properties on monitoring drug levels
  by micellar electrokinetic capillary chromatography with direct serum injection, *Electrophoresis*, **1994**, *15*,
  1295–1303.

SAMPLE
Matrix: solutions
Sample preparation: Inject an aliquot of a 100 μg/mL solution in water:running buffer 50:50.

CAPILLARY ELECTROPHORESIS
Capillary: 44.5 cm × 50 μm acrylamide-coated fused-silica (Bio-Rad)
Capillary temperature: 30
Running buffer: 100 mM NaH$_2$PO$_4$ containing 15 mM gamma-cyclodextrin, adjusted to pH 2.5
  with phosphoric acid
Injection: Electrokinetic injection at 8 kV for 6 s.
Detector: UV 200
Migration time: 11.48
Voltage: 14 kV
Model: Bio-Rad BioFocus 3000

OTHER SUBSTANCES
Also analyzed: albuterol, alprenolol, atenolol, atropine, baclofen, bamethan, benserazide, biperiden, bisoprolol, bupivacaine, bupranolol, butetamate, carazolol, carbuterol, carvedilol, chloroquine, chlorpheniramine (chlorphenamine), clidinium bromide, clobutinol, disopyramide, dobutamine, flecainide, homatropine, ipratropium bromide, isoproterenol, isothipendyl, ketamine,
  mefloquine, mequitazine, metaproterenol (orciprenaline), metipranolol, nafronyl (naftidrofuryl), nefopam, ofloxacin, orphenadrine, oxomemazine, oxprenolol, phenoxybenzamine, pholedrine, pindolol, pirbuterol, prilocaine, promethazine, propafenone, propranolol, sotalol, synephrine, terbutaline, tetrahydrozoline (tetryzoline), tocainide, trihexyphenidyl, trimeprazine
  (alimemazine), trimipramine, tropicamide, verapamil, zopiclone

KEY WORDS
coated capillary; achiral

REFERENCE
Koppenhoefer,B.; Epperlein,U.; Christian,B.; Yibing,J.; Yuying,C.; Bingcheng,L. Separation of enantiomers of
  drugs by capillary electrophoresis. I. γ-Cyclodextrin as chiral solvating agent, *J.Chromatogr.A*, **1995**, *717*,
  181–190.

SAMPLE
Matrix: solutions
Sample preparation: Inject an aliquot of a 100 μg/mL solution in running buffer.

CAPILLARY ELECTROPHORESIS
Capillary: 29 cm × 50 μm fused-silica (24.5 cm to detector) (Yongnian Optical Conductive Fiber
  Plant, China), coated with polyacrylamide

Capillary preparation: No details of the polyacrylamide coating process are provided. However, another paper (LC.GC 1997, 15, 40) by this group indicates that they use the procedure of Hjertén, thus: Adjust the pH of 20 mL water to 3.5 with acetic acid, add 80 µL 3-(trimethoxysilyl)propyl methacrylate (3-methacryloxypropyltrimethoxysilane), mix, suck into capillary, let stand at room temperature for 1 h, remove the solution, wash with water. Fill the capillary with a deaerated 3-4% acrylamide solution containing 1 µL/mL N,N,N',N'-tetramethylethylenediamine and 1 mg/mL potassium persulfate, let stand for 30 min, remove excess solution by aspiration, rinse with water, remove water by aspiration, dry at 35° (J. Chromatogr. 1985, 347, 191).
Capillary temperature: 25
Running buffer: 100 mM NaH_2PO_4 adjusted to pH 2.5
Injection: Electrokinetic injection at 15 kV for 3 s.
Detector: UV 200, UV 210
Migration time: 7.05
Voltage: 15 kV
Model: Bio-Rad BioFocus 3000

OTHER SUBSTANCES
Simultaneous: albuterol, alprenolol, atenolol, baclofen, bamethan, benproperine, benserazide, bisoprolol, bupranolol, butamirate, butethamate, carbuterol, clenbuterol, clobutinol, dipivefrin, isoproterenol (isoprenaline), metaproterenol (orciprenaline), metipranolol, metoprolol, norfenefrine, ornidazole, oxprenolol, phenylpropanolamine, pholedrine, pirbuterol, prilocaine, procyclidine, sotalol, synephrine, terbutaline, tocainide

KEY WORDS
coated capillary

REFERENCE
Koppenhoefer,B.; Epperlein,U.; Xiaofeng,Z.; Bingcheng,L. Separation of enantiomers of drugs by capillary electrophoresis. Part 4: Hydroxypropyl-γ-cyclodextrin as chiral solvating agent, *Electrophoresis*, **1997**, *18*, 924–930.

SAMPLE
Matrix: solutions

CAPILLARY ELECTROPHORESIS
Capillary: 29-36 cm × 50 µm fused-silica (24.5-31.5 cm to detector) (Yongnian Optical Conductive Fiber Plant, China) coated with polyacrylamide
Capillary preparation: Coat capillary as follows. Adjust the pH of 20 mL water to 3.5 with acetic acid, add 80 µL 3-(trimethoxysilyl)propyl methacrylate (3-methacryloxypropyltrimethoxysilane), mix, suck into capillary, let stand at room temperature for 1 h, remove the solution, wash with water. Fill the capillary with a deaerated 3-4% acrylamide solution containing 1 µL/mL N,N,N',N'-tetramethylethylenediamine and 1 mg/mL potassium persulfate, let stand for 30 min, remove excess solution by aspiration, rinse with water, remove water by aspiration, dry at 35° (J. Chromatogr. 1985, 347, 191).
Capillary temperature: 25
Running buffer: 100 mM pH 2.5 NaH_2PO_4 (A) or 100 mM pH 2.5 NaH_2PO_4 containing 45 mM hydroxypropyl-α-cyclodextrin (Wacker, Munich) (B)
Injection: Electromigration at 15 kV for 3 s.
Detector: UV 200; UV 210
Migration time: 7.05 (A); 11.52 (B) (no separation of enantiomers)
Voltage: 15 kV
Model: Bio-Focus 3000

OTHER SUBSTANCES
Also analyzed: albuterol (salbutamol), alprenolol, amorolfine, atenolol, atropine, azelastine, baclofen, bamethan, benproperine, benserazide, biperiden, bisoprolol, brompheniramine, bupivacaine, bupranolol, butamirate, butethamate, carazolol, carbuterol, carteolol, carvedilol, chloroquine, chlorpheniramine, chlorphenoxamine, cicletanine, clenbuterol, clidinium bromide, clobutinol, dimethindene, dipivefrin, disopyramide, dobutamine, doxylamine, fendiline, flecainide, gallopamil, homatropine, ipratropium bromide, isoproterenol (isoprenaline), isothipendyl, ketamine, meclizine, mefloquine, mepindolol, mequitazine, metaclazepam, metaproterenol (orciprenaline), metipranolol, metoprolol, nafronyl (naftidrofuryl), nefopam, nicardipine, norfe-

nefrine, ofloxacin, ornidazole, orphenadrine, oxomemazine, oxprenolol, oxybutynin, phenoxybenzamine, phenylpropanolamine, pholedrine, pindolol, pirbuterol, prilocaine, procyclidine, promethazine, propafenone, propranolol, reproterol, sotalol, sulpride, synephrine, talinolol, terbutaline, tetrahydrozoline (tetryzoline), theodrenaline, tioconazole, tocainide, trihexyphenidyl, trimeprazine (alimemazine), trimipramine, tropicamide, verapamil, zopiclone

KEY WORDS
coated capillary

REFERENCE
Koppenhoefer,B.; Eperlein,U.; Schlunk,R.; Zhu,X.; Lin,B. Separation of enantiomers of drugs by capillary electrophoresis. V. Hydroxypropyl-α-cyclodextrin as chiral solvating agent, *J.Chromatogr.A*, **1998**, *793*, 153–164.

Cephalexin

Molecular formula: $C_{16}H_{17}N_3O_4S$
Molecular weight: 347.39
CAS Registry No.: 15686-71-2, 23325-78-2 (monohydrate), 105879-42-3 (HCl)
Merck Index (12th ed.): 2021
Lednicer: 1 417; 2 439; 4 182

SAMPLE
Matrix: bulk
Sample preparation: Prepare a 1 mg/mL solution in water, inject an aliquot.

CAPILLARY ELECTROPHORESIS
Capillary: 57 cm × 75 μm fused-silica (50 cm to detector) (Beckman)
Capillary preparation: Replace running buffer before each run.
Capillary temperature: 25
Running buffer: 2.84 g/L Na_2HPO_4 containing 1.22 g/L boric acid, 14.42 g/L sodium dodecyl sulfate, and 1 g/L Brij 35
Injection: Hydrodynamic injection at 0.5 psi for 5 s (5 nL)
Detector: UV 214
Migration time: 6.5
Voltage: 20 kV
Model: Beckman P/ACE System 2050

OTHER SUBSTANCES
Simultaneous: impurities, cephradine

REFERENCE
Emaldi,P.; Fapanni,S.; Baldini,A. Validation of a capillary electrophoresis method for the determination of cephradine and its related impurities, *J.Chromatogr.A*, **1995**, *711*, 339–346.

SAMPLE
Matrix: formulations
Sample preparation: Dissolve contents of capsule in water so as to achieve a concentration of 100 μg/mL, filter, inject an aliquot.

CAPILLARY ELECTROPHORESIS
Capillary: 72 cm × 50 μm fused-silica (50 cm to detector)
Capillary preparation: Flush with 1 M NaOH for 30 min, rinse with water for 10 min.
Capillary temperature: 30
Running buffer: 20 mM Sodium borate containing 200 mM sodium dodecyl sulfate and 100 mM pentanesulfonic acid
Injection: Hydrostatic injection for 5 s

Detector: UV 210
Migration time: 18.5
Voltage: 306 V/cm
Model: Applied Biosystems 270A-HT

OTHER SUBSTANCES
Simultaneous: cefamandole, cefuroxime, cephalothin, cephapirin

KEY WORDS
capsules

REFERENCE
Sciacchitano,C.J.; Mopper,B.; Specchio,J.J. Identification and separation of five cephalosporins by micellar electrokinetic chromatography, *J.Chromatogr.B*, **1994**, *657*, 395–399.

Cephalothin

Molecular formula: $C_{16}H_{16}N_2O_6S_2$
Molecular weight: 396.44
CAS Registry No.: 153-61-7, 58-71-9 (sodium salt)
Merck Index (12th ed.): 2028
Lednicer: 1 420

SAMPLE
Matrix: formulations
Sample preparation: Dissolve contents of capsule in water so as to achieve a concentration of 100 µg/mL, filter, inject an aliquot.

CAPILLARY ELECTROPHORESIS
Capillary: 72 cm × 50 µm fused-silica (50 cm to detector)
Capillary preparation: Flush with 1 M NaOH for 30 min, rinse with water for 10 min.
Capillary temperature: 30
Running buffer: 20 mM Sodium borate containing 200 mM sodium dodecyl sulfate and 100 mM pentanesulfonic acid
Injection: Hydrostatic injection for 5 s
Detector: UV 210
Migration time: 26
Voltage: 306 V/cm
Model: Applied Biosystems 270A-HT

OTHER SUBSTANCES
Simultaneous: cefamandole, cefuroxime, cephalexin, cephapirin

KEY WORDS
capsules

REFERENCE
Sciacchitano,C.J.; Mopper,B.; Specchio,J.J. Identification and separation of five cephalosporins by micellar electrokinetic chromatography, *J.Chromatogr.B*, **1994**, *657*, 395–399.

SAMPLE
Matrix: solutions

CAPILLARY ELECTROPHORESIS
Capillary: 27 cm × 75 µm
Capillary preparation: Before each run rinse capillary with 100 mM NaOH for 30 s and with running buffer for 30 s. Condition new capillaries by rinsing with 100 mM NaOH for 20 min.

Capillary temperature: 30
Running buffer: 15 mM Sodium borate
Injection: Pressure injection of sample at 25 mbar for 1 s followed by running buffer at 25 mbar for 1 s.
Detector: UV 200
Migration time: 2.70
Internal standard: aminobenzoic acid (3.5)
Voltage: 6.5 kV
Model: Beckman

OTHER SUBSTANCES
Simultaneous: aspirin, bacitracin, beclomethasone, benzoic acid, ceftriaxone, cromolyn, embonic acid, glyburide (glibenclamide), levothyroxine, nedocromil, nystatin, omeprazole, prednisolone, warfarin, zidovudine
Interfering: ceftizoxime, cefuroxime, epoprostenol

REFERENCE
Altria,K.D.; Bryant,S.M.; Hadgett,T.A. Validated capillary electrophoresis method for the analysis of a range of acidic drugs and excipients, *J.Pharm.Biomed.Anal.*, **1997**, *15*, 1091–1101.

Cephapirin sodium

Molecular formula: $C_{17}H_{16}N_3NaO_6S_2$
Molecular weight: 445.45
CAS Registry No.: 24356-60-3, 21595-23-7 (cephapirin)
Merck Index (12th ed.): 2030
Lednicer: 2 441

SAMPLE
Matrix: formulations
Sample preparation: Dissolve contents of capsule in water so as to achieve a concentration of 100 μg/mL, filter, inject an aliquot.

CAPILLARY ELECTROPHORESIS
Capillary: 72 cm × 50 μm fused-silica (50 cm to detector)
Capillary preparation: Flush with 1 M NaOH for 30 min, rinse with water for 10 min.
Capillary temperature: 30
Running buffer: 20 mM Sodium borate containing 200 mM sodium dodecyl sulfate and 100 mM pentanesulfonic acid
Injection: Hydrostatic injection for 5 s
Detector: UV 210
Migration time: 20
Voltage: 306 V/cm
Model: Applied Biosystems 270A-HT

OTHER SUBSTANCES
Simultaneous: cefamandole, cefuroxime, cephalexin, cephalothin

KEY WORDS
capsules

REFERENCE
Sciacchitano,C.J.; Mopper,B.; Specchio,J.J. Identification and separation of five cephalosporins by micellar electrokinetic chromatography, *J.Chromatogr.B*, **1994**, *657*, 395–399.

SAMPLE
Matrix: solutions

Sample preparation: Dilute a solution in buffer 10-fold with water, filter, inject an aliquot. (Prepare buffer by dissolving 8.0 g KH_2PO_4 and 2.0 g K_2HPO_4 in 1 L water, adjust pH to 6.0 with 1 M HCl or 1 M NaOH.)

CAPILLARY ELECTROPHORESIS
Capillary: 50 cm × 50 μm fused-silica (Polymicro Technologies)
Capillary preparation: Before each run rinse with water for 80 s, with 1 M NaOH for 80 s, with water for 80 s, and with running buffer for 90 s then the ends are dipped into water
Capillary temperature: 20
Running buffer: 20 mM Sodium tetraborate containing 150 mM sodium dodecyl sulfate, pH 9.20 ± 0.05
Injection: Pressure injection at 5 psi for 6 s (42 nL).
Detector: UV 210
Migration time: 12
Voltage: 14 kV
Model: BioFocus 3000 (Bio-Rad)

OTHER SUBSTANCES
Simultaneous: amoxicillin, ampicillin, ceftiofur, cloxacillin, penicillin G

REFERENCE
Cutting,J.H.; Hurlbut,J.A.; Sofos,J.N. Quantitation of penicillin G in medicated premix feeds by micellar electrokinetic capillary chromatography, *J.AOAC Int.*, **1997**, *80*, 951–955.

SAMPLE
Matrix: solutions

CAPILLARY ELECTROPHORESIS
Capillary: 60 cm × 50 μm fused-silica (47 cm to detector) (Polymicro Technologies)
Capillary temperature: 25
Running buffer: 20 mM pH 8.5 Sodium tetraborate containing 100 mM sodium dodecyl sulfate
Injection: Hydrodynamic injection at 50 mbar for 3.6 s (5 nL).
Detector: UV 205
Migration time: 9
Voltage: 22 kV
Model: Crystal 310 (Thermo Unicam)

OTHER SUBSTANCES
Simultaneous: amoxicillin, ampicillin, cloxacillin, dicloxacillin, oxacillin, penicillin G, penicillin V, piperacillin, pyrimethamine, sulfacetamide, sulfadimethoxine, sulfaguanidine, sulfamerazine, sulfameter, sulfamethazine, sulfanilamide, sulfanilic acid, sulfapyridine, sulfaquinoxaline, sulfathiazole, sulfisoxazole, trimethoprim

REFERENCE
Hows,M.E.P.; Perrett,D.; Kay,J. Optimization of a simultaneous separation of sulphonamides, dihydrofolate reductase inhibitors and β-lactam antibiotics by capillary electrophoresis, *J.Chromatogr.A*, **1997**, *768*, 97–104.

Cephradine

Molecular formula: $C_{16}H_{19}N_3O_4S$
Molecular weight: 349.41
CAS Registry No.: 38821-53-3, 58456-86-3 (dihydrate), 31828-50-9 (non-stoichiometric hydrate)
Merck Index (12th ed.): 2032
Lednicer: 2 440

SAMPLE
Matrix: bulk
Sample preparation: Prepare a 1 mg/mL solution in water, inject an aliquot.

CAPILLARY ELECTROPHORESIS
Capillary: 57 cm × 75 μm fused-silica (50 cm to detector) (Beckman)
Capillary preparation: Replace running buffer before each run.
Capillary temperature: 25
Running buffer: 2.84 g/L Na_2HPO_4 containing 1.22 g/L boric acid, 14.42 g/L sodium dodecyl
    sulfate, and 1 g/L Brij 35
Injection: Hydrodynamic injection at 0.5 psi for 5 s (5 nL)
Detector: UV 214
Migration time: 6.91
Voltage: 20 kV
Model: Beckman P/ACE System 2050
Limit of quantitation: 1.752 μg/mL
Limit of detection: 526 ng/mL (S/N 3)

OTHER SUBSTANCES
Simultaneous: impurities, cephalexin

REFERENCE
Emaldi,P.; Fapanni,S.; Baldini,A. Validation of a capillary electrophoresis method for the determination of
    cephradine and its related impurities, *J.Chromatogr.A*, **1995**, *711*, 339–346.

Ceronapril

Molecular formula: $C_{21}H_{33}N_2O_6P$
Molecular weight: 440.48
CAS Registry No.: 111223-26-8
Merck Index (12th ed.): 2038
Lednicer: 5 65

SAMPLE
Matrix: solutions

CAPILLARY ELECTROPHORESIS
Capillary: 57 cm × 75 μm fused-silica (50 cm to detector) (Beckman)
Capillary preparation: Before each run rinse capillary with 100 mM NaOH and running buffer.
Capillary temperature: 30
Running buffer: 50 mM pH 8.2 Borate buffer containing 100 mM sodium dodecyl sulfate
Injection: Pressure injection for 5-10 s.
Detector: UV 200
Migration time: 5.3
Voltage: 25 kV
Model: Beckman P/ACE 5510

OTHER SUBSTANCES
Simultaneous: captopril, fosinopril, enalapril, lisinopril, zofenopril

REFERENCE
Bretnall,A.E.; Clarke,G.S. Selectivity of capillary electrophoresis for the analysis of cardiovascular drugs,
    J.Chromatogr.A, **1996**, *745*, 145–154.

Chenodiol

Molecular formula: $C_{24}H_{40}O_4$
Molecular weight: 392.58
CAS Registry No.: 474-25-9
Merck Index (12th ed.): 2096

SAMPLE
Matrix: blood
Sample preparation: Filter (Amicon Centrifree YM) while centrifuging at 2000 rpm for 45 min, inject an aliquot of the ultrafiltrate.

CAPILLARY ELECTROPHORESIS
Capillary: fused-silica (Polymicro Technologies)
Capillary preparation: Before each run flush capillary with 100 mM NaOH for 3 min, with water for 3 min, and with running buffer for 3 min. At the start of each day flush with 1 M NaOH for 30 min, with water for 3 min, and with running buffer for 3 min. Flush new capillaries with 1 M NaOH for 1 h and with water for 10 min.
Capillary temperature: 23
Running buffer: MeOH:100 mM boric acid 75:25 containing 7 mM gamma-cyclodextrin and 5 mM adenosine 5'-monophosphate, pH 7.0
Injection: Pressure injection for 10 s.
Detector: UV 254
Migration time: 24.5
Voltage: 30 kV
Current: 6.8 μA
Model: Beckman P/ACE 5510
Limit of detection: 170-330 μM

OTHER SUBSTANCES
Simultaneous: cholic acid, deoxycholic acid, glycochenodeoxycholic acid, glycocholic acid, glycodeoxycholic acid, glycolithocholic acid, glycoursodeoxycholic acid, lithocholic acid, taurochenodeoxycholic acid, taurocholic acid, taurodeoxycholic acid, taurolithocholic acid, tauroursodeoxycholic acid, ursodiol

KEY WORDS
serum; indirect UV detection; ultrafiltrate

REFERENCE
Yarabe,H.H.; Shamsi,S.A.; Warner,I.M. Capillary zone electrophoresis of bile acids with indirect photometric detection, *Anal.Chem.*, **1998**, *70*, 1412–1418.

SAMPLE
Matrix: solutions
Sample preparation: Inject an aliquot of a 1 mg/mL solution in MeOH.

CAPILLARY ELECTROPHORESIS
Capillary: 50 cm × 50 μm fused-silica (42 cm to detector)
Capillary temperature: 20
Running buffer: 100 mM pH 8.6 Tris containing 50 mM benzoic acid, 0.01% methylcellulose, and 0.5% trimethyl-β-cyclodextrin
Injection: Hydrodynamic injection for 0.5-3 s.
Detector: UV 250
Migration time: 4.3
Internal standard: cholic acid (5.0)
Voltage: 20 kV
Current: 35.4 μA
Model: Spectra PHORESIS 1000 (Thermo Separation Products)

OTHER SUBSTANCES
Simultaneous: deoxycholic acid, lithocholic acid, ursocholic acid, ursodiol
Interfering: 3-hydroxy-7-chetocholanic acid

KEY WORDS
indirect UV

REFERENCE
Quaglia,M.G.; Farina,A.; Bossù,E.; Dell'Aquila,C.; Doldo,A. The indirect UV detection in the analysis of urso-deoxycholic acid and related compounds by HPCE, *J.Pharm.Biomed.Anal.*, **1997**, *16*, 281–285.

Chlophedianol

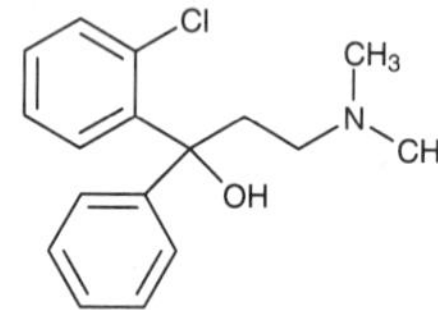

Molecular formula: C₁₇H₂₀ClNO
Molecular weight: 289.80
CAS Registry No.: 791-35-5, 511-13-7 (HCl)
Merck Index (12th ed.): 2106

SAMPLE
Matrix: solutions

CAPILLARY ELECTROPHORESIS
Capillary: 42 cm × 50 μm fused-silica (31 cm to detector) (Polymicro Technologies)
Capillary preparation: Condition capillary by rinsing with running buffer for 10 min, with water for 10 min, with 1 M NaOH for 10 min, with water for 10 min, and with running buffer with the voltage applied for 20 min.
Capillary temperature: 25
Running buffer: Formamide containing 150 mM citric acid, 100 mM Tris, and 100 mM β-cyclodextrin (apparent pH 5.1)
Injection: Gravity injection.
Detector: UV 254
Migration time: 8.73, 8.81 (enantiomers)
Voltage: 30 kV
Model: Beckman P/ACE 5500

OTHER SUBSTANCES
Simultaneous: mianserin, nefopam, propiomazine, trimeprazine, trimipramine, thioridazine

KEY WORDS
chiral

REFERENCE
Wang,F.; Khaledi,M.G. Chiral separations by nonaqueous capillary electrophoresis, *Anal.Chem.*, **1996**, *68*, 3460–3467.

SAMPLE
Matrix: solutions

CAPILLARY ELECTROPHORESIS
Capillary: 42 cm × 75 μm acrylamide coated (36 cm to detector)
Capillary preparation: Coat column as follows. Adjust the pH of 20 mL water to 3.5 with acetic acid, add 80 μL 3-(trimethoxysilyl)propyl methacrylate (3-methacryloxypropyltrimethoxysilane), mix, suck into capillary, let stand at room temperature for 1 h, remove the solution, wash with water. Fill the capillary with a deaerated 3-4% acrylamide solution containing 1 μL/mL N,N,N',N'-tetramethylethylenediamine and 1 mg/mL potassium persulfate, let stand for 30 min, remove excess solution by aspiration, rinse with water, remove water by aspiration, dry at 35° (J. Chromatogr. 1985, 347, 191).

Capillary temperature: 20
Running buffer: 100 mM pH 6 2-[N-morpholino]ethanesulfonic acid/NaOH buffer
Injection: Inject a 200 mg/mL solution of transferrin in buffer at 30 psi.s then hydrodynamically inject the sample at 10 psi.s. (Buffer was 100 mM pH 6 2-[N-morpholino]ethanesulfonic acid/ NaOH buffer. Transferrin was iron-free human serum transferrin (Behring-Werke, Marburg, Germany).)
Detector: UV 215
Migration time: 20.5, 21.5 (enantiomers)
Voltage: 10 kV
Model: Bio-Rad BioFocus 3000

OTHER SUBSTANCES
Simultaneous: acebutolol, nylidrin

KEY WORDS
chiral; coated capillary

REFERENCE
Schmid,M.G.; Gübitz,G.; Kilár,F. Stereoselective interaction of drug enantiomers with human serum transferrin in capillary zone electrophoresis (II), *Electrophoresis*, **1998**, *19*, 282–287.

Chlorambucil

Molecular formula: $C_{14}H_{19}Cl_2NO_2$
Molecular weight: 304.22
CAS Registry No.: 305-03-3
Merck Index (12th ed.): 2116

SAMPLE
Matrix: solutions

CAPILLARY ELECTROPHORESIS
Capillary: 57 cm × 50 μm fused-silica (50 cm to detector)
Running buffer: 50 mM pH 8.0 Borate buffer containing 40 mM sodium taurodeoxycholate and 25 mM phosphatidylcholine (Prepare by adding phosphatidylcholine and stirring for 3-6 h until all cloudiness disappears. Phosphatidylcholine was 95% pure soybean lecithin, Epikuron, Lucas Meyer & Co.)
Injection: Inject a solution of the compound in the running buffer at 20 psi for 1 s, inject MeOH: water 5:95 at 20 psi for 1 s, inject a solution of halofantrine in running buffer at 20 psi for 1 s.
Detector: UV 214
Migration time: k' 3.02
Model: Beckman P/ACE 5000

OTHER SUBSTANCES
Also analyzed: acetaminophen, amoxicillin, antipyrine, aspirin, azathioprine, caffeine, captopril, carbamazepine, carprofen, chlorpheniramine, chlorpromazine, cimetidine, clonidine, codeine, desipramine, diphenhydramine, ephedrine, fenoterol, flufenamic acid, flurbiprofen, haloperidol, hydroxyzine, ibuprofen, imipramine, indomethacin, ketoprofen, lidocaine, melphalan, metoprolol, nabumetone, nadolol, phenobarbital, phenol, promazine, propranolol, pyrilamine, ranitidine, ropinirole, salicylic acid, sulfamethoxazole, testosterone, theophylline, thioridazine, tiaprofenic acid, tolfenamic acid, trifluoperazine, trimethoprim, valproic acid, verapamil

KEY WORDS
comparison with HPLC; k' = (Tr-T0)/(T0(1-Tr/Tm)) where Tr = retention time of analyte; T0 = retention time of water; and Tm = retention time of marker (halofantrine)

REFERENCE
Hanna,M.; de Biasi,V.; Bond,B.; Salter,C.; Hutt,A.J.; Camilleri,P. Estimation of the partitioning characteristics of drugs: A comparison of a large and diverse drug series utilizing chromatographic and electrophoretic methodology, *Anal.Chem.*, **1998**, *70*, 2092–2099.

Chloramphenicol

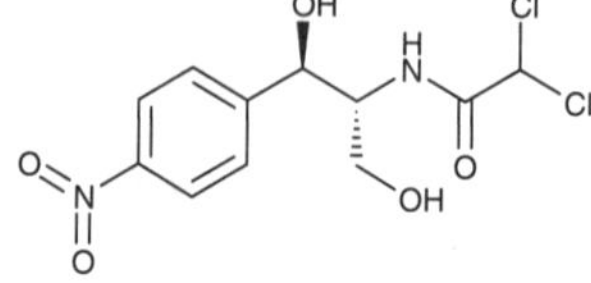

Molecular formula: $C_{11}H_{12}Cl_2N_2O_5$
Molecular weight: 323.13
CAS Registry No.: 56-75-7
Merck Index (12th ed.): 2120

SAMPLE
Matrix: blood
Sample preparation: 500 µL Plasma + 20 µL 100 µg/mL chloramphenicol in water + 3-5 mg K_2HPO_4, mix thoroughly, add 5 mL ethyl acetate, rotate for 1-2 min, centrifuge at 3500 g for 5 min. Remove a 4 mL aliquot of the ethyl acetate layer and evaporate it to dryness under a stream of nitrogen at 40°, reconstitute with 100 µL water, inject an aliquot.

CAPILLARY ELECTROPHORESIS
Capillary: 70 cm × 75 µm fused-silica (62.5 cm to detector) (Yongnian Optical Fiber, Hebei, China)
Capillary preparation: Before each run wash capillary with 100 mM NaOH and water for 1 min, rinse with running buffer for 2 min. Before use condition capillary with 1 M NaOH at 60° for 20 min, with 100 mM NaOH at 60° for 5 min, and with water at 30° for 5 min.
Capillary temperature: 30
Running buffer: MeCN:buffer 10:90 (Buffer was 20 mM pH 9.2 disodium tetraborate containing 40 mM sodium dodecyl sulfate.)
Injection: Hydrodynamic injection for 10 s (40 nL).
Detector: UV 195
Migration time: 11
Internal standard: chloramphenicol
Voltage: 18 kV
Current: 58-69 µA
Model: SpectraPhoresis 1000 with FOCUS detection (Thermo Separation Products)
Limit of quantitation: 100 ng/mL

OTHER SUBSTANCES
Extracted: thiamphenicol

KEY WORDS
plasma; chloramphenicol is IS

REFERENCE
Song,J.-Z.; Wu,X.-J.; Sun,Z.-P.; Tian,S.-J.; Wang,M.-L.; Wang,R.-L. Determination of thiamphenicol in human plasma by micellar electrokinetic capillary chromatography, *J.Chromatogr.B*, **1997**, *692*, 445–451.

SAMPLE
Matrix: reaction mixtures
Sample preparation: Filter (0.22 µm), inject an aliquot.

CAPILLARY ELECTROPHORESIS
Capillary: 57 cm × 50 µm uncoated fused-silica (50 cm to detector)
Capillary preparation: Before each injection rinse with 100 mM NaOH for 1 min and with running buffer for 3 min. Equilibrate new capillary with running buffer overnight.
Capillary temperature: 25
Running buffer: 100 mM pH 8.3 borate buffer
Injection: Pressure injection for 1 s (ca. 1.3 nL/s)

Detector: UV 200
Migration time: 2.3
Voltage: 25 kV
Current: 32 μA
Model: Beckman P/ACE Model 2050

OTHER SUBSTANCES
Simultaneous: acetyl coenzyme A, coenzyme A

REFERENCE
Landers,J.P.; Schuchard,M.D.; Subramaniam,M.; Sismelich,T.P.; Spelsberg,T.C. High-performance capillary electrophoretic analysis of chloramphenicol acetyl transferase activity, *J.Chromatogr.*, **1992**, *603*, 247–257.

SAMPLE
Matrix: solutions

CAPILLARY ELECTROPHORESIS
Capillary: 45 cm × 50 μm fused-silica (37.5 cm to detector) (Polymicro Technologies)
Capillary preparation: Between each run rinse with 100 mM HCl for 2 min and with running buffer for 2 min. At the start of each day condition capillary with 1 M HCl for 10 min. Before the first use rinse capillary with 1 M NaOH for 30 min and with water for 30 min.
Capillary temperature: 25
Running buffer: MeCN:30 mM pH 2.40 ethanesulfonic acid 20:80
Injection: Injection at 4 kV for 3 s.
Detector: UV 214
Migration time: 4.92
Voltage: 20 kV
Model: Waters Quanta 4000E

OTHER SUBSTANCES
Simultaneous: imipramine, lidocaine, phenylpropanolamine, procainamide, quinidine, trimethoprim

REFERENCE
Ding,W.; Fritz,J.S. Separation of neutral compounds and basic drugs by capillary electrophoresis in acidic solution using laurylpoly(oxyethylene) sulfate as an additive, *Anal.Chem.*, **1998**, *70*, 1859–1865.

Chlorcyclizine

Molecular formula: $C_{18}H_{21}ClN_2$
Molecular weight: 300.83
CAS Registry No.: 82-93-9, 1620-21-9 (HCl)
Merck Index (12th ed.): 2128
Lednicer: 1 58

SAMPLE
Matrix: solutions
Sample preparation: Prepare a solution in running buffer, inject an aliquot.

CAPILLARY ELECTROPHORESIS
Capillary: 60 cm × 75 μm fused-silica (52.4 cm to detector)
Capillary preparation: After each run flush with 500 mM KOH for 2-3 min, flush with water, fill with running buffer
Running buffer: 10 mM Na_2HPO_4 containing 2% heparin sodium (MW 10000, 11% S, Scientific Protein Laboratories, Waunakee, WI) adjusted to pH 5 with phosphoric acid
Injection: Hydrostatic injection
Detector: UV 214

Migration time: 16.7
Model: Waters Quanta 4000

OTHER SUBSTANCES
Also analyzed: anabasine, brompheniramine, bupivacaine, carbinoxamine, chloroquine, chlorpheniramine, dimethindene, doxylamine, enpiroline, halofantrine, hydroxychloroquine, indapamide, mefloquine, nornicotine, pheniramine, primaquine, promethazine, quinacrine, tetramisole

---

REFERENCE
Stalcup,A.M.; Agyei,N.M. Heparin: a chiral mobile-phase additive for capillary zone electrophoresis, *Anal.Chem.*, **1994**, *66*, 3054–3059.

SAMPLE
Matrix: solutions

CAPILLARY ELECTROPHORESIS
Capillary: 62 cm × 75 μm fused-silica (54.5 cm to detector) (Polymicro Technologies)
Capillary preparation: Purge with running buffer before each run. At the beginning of each day purge using 50-60 kPa vacuum with 500 mM NaOH for 5 min, with water for 5 min, with MeCN for 5 min, and with running buffer for 5 min.
Running buffer: MeCN:MeOH:acetic acid 49:50:1 containing 20 mM ammonium acetate
Injection: Hydrostatic injection at 10 cm for 5 s.
Detector: UV 214
Migration time: 4.95
Voltage: 30 kV
Model: Waters Quanta 4000

OTHER SUBSTANCES
Simultaneous: buclizine, chlorpheniramine, cyclizine, dimenhydrinate, methaphenilene, pheniramine, promethazine, pyrilamine, pyrrobutamine, tripelennamine

---

REFERENCE
Leung,G.N.W.; Tang,H.P.O.; Tso,T.S.C.; Wan,T.S.M. Separation of basic drugs with non-aqueous capillary electrophoresis, *J.Chromatogr.A*, **1996**, *738*, 141–154.

SAMPLE
Matrix: solutions

CAPILLARY ELECTROPHORESIS
Capillary: 42 cm × 50 μm fused-silica (31 cm to detector) (Polymicro Technologies)
Capillary preparation: Condition capillary by rinsing with running buffer for 10 min, with water for 10 min, with 1 M NaOH for 10 min, with water for 10 min, and with running buffer with the voltage applied for 20 min.
Capillary temperature: 25
Running buffer: Formamide containing 150 mM citric acid, 100 mM Tris, and 100 mM gamma-cyclodextrin (apparent pH 5.1)
Injection: Gravity injection.
Detector: UV 254
Migration time: 11.23, 11.31 (enantiomers)
Voltage: 30 kV
Model: Beckman P/ACE 5500

KEY WORDS
chiral

REFERENCE
Wang,F.; Khaledi,M.G. Chiral separations by nonaqueous capillary electrophoresis, *Anal.Chem.*, **1996**, *68*, 3460–3467.

Chlordiazepoxide

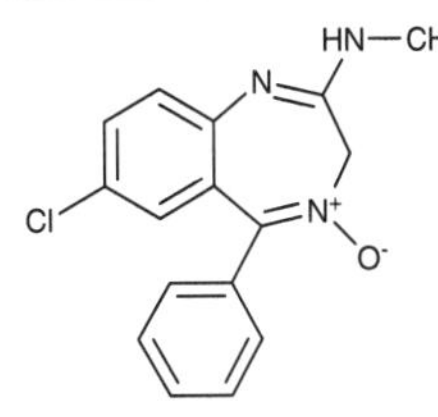

Molecular formula: $C_{16}H_{14}ClN_3O$
Molecular weight: 299.76
CAS Registry No.: 58-25-3, 438-41-5 (HCl)
Merck Index (12th ed.): 2132
Lednicer: 1 365

SAMPLE
Matrix: solutions
Sample preparation: Prepare a solution in running buffer, inject an aliquot.

CAPILLARY ELECTROPHORESIS
Capillary: 100 cm × 75 μm fused-silica (20 cm to UV detector) (Polymicro Technologies)
Capillary preparation: Between runs flush capillaries with 1.5 column volumes of 100 mM NaOH and 9 column volumes of running buffer. Flush new capillaries with 9 column volumes of 1 M NaOH, 3 column volumes of water, 3 column volumes of 100 mM HCl, 3 column volumes of water, and 9 column volumes of running buffer.
Running buffer: MeOH:water 15:85 containing 15 mM ammonium acetate, adjusted to pH 2.5 on trifluoroacetic acid
Injection: Pressure injection at 3.45 kPa (28 nL)
Detector: UV 254 or MS, Sciex TAGA 6000E triple quadrupole, API source, electrospray (ion spray) interface at 4 kV (3 kV to ion-sampling orifice), positive-ion mode, 100 μm dia sampling orifice, nitrogen gas curtain, argon collision gas, m/z 300 [M+H]+
Migration time: 5 (UV), 24.5 (MS)
Voltage: 26 kV
Model: Beckman P/ACE System 2000

OTHER SUBSTANCES
Simultaneous: diazepam, flurazepam, prazepam

REFERENCE
Johansson,I.M.; Pavelka,R.; Henion,J.D. Determination of small drug molecules by capillary electrophoresis-atmospheric pressure ionization mass spectrometry, *J.Chromatogr.*, **1991**, *559*, 515–528.

SAMPLE
Matrix: solutions
Sample preparation: Inject an aliquot of a 5 μg/mL solution in running buffer.

CAPILLARY ELECTROPHORESIS
Capillary: 87 cm × 75 μm fused-silica (80 cm to detector) (Beckman)
Capillary preparation: Between runs rinse capillary with running buffer for 2 min then equilibrate for 5 min.
Capillary temperature: 35
Running buffer: 20 mM pH 7 Buffer containing 15 mM sodium cholate and 35 mM sodium deoxycholate (Buffer was 20 mM sodium borate adjusted to pH 7.0 with 20 mM NaH_2PO_4.)
Injection: Pressure injection for 2 s.
Detector: UV 214
Migration time: 18.7
Voltage: 20 kV
Model: Beckman P/ACE 2100

OTHER SUBSTANCES
Simultaneous: bromazepam, clobazam, clonazepam, diazepam, flunitrazepam, flurazepam, halazepam, lorazepam, lormetazepam, nitrazepam, nordazepam, temazepam

REFERENCE
Boonkerd,S.; Detaevernier,M.R.; Vindevogel,J.; Michotte,Y. Migration behaviour of benzodiazepines in micellar electrokinetic chromatography, *J.Chromatogr.A*, **1996**, *756*, 279–286.

SAMPLE
Matrix: solutions

CAPILLARY ELECTROPHORESIS
Capillary: 47 cm × 75 μm fused-silica (40 cm to detector) (Beckman)
Capillary preparation: Rinse with running buffer for 1 min before injection. Wash with 100 mM NaOH for 1 min and with water for 1 min between injections.
Capillary temperature: 30
Running buffer: THF:buffer 1:99 (Buffer was 50 mM pH 9.2 borate containing 50 mM sodium dodecyl sulfate, 2 M urea, and 20 mM gamma-cyclodextrin.)
Injection: Pressure injection at 300 kPa for 5 s.
Detector: UV 254
Migration time: 14
Voltage: 15 kV
Model: Beckman P/ACE 2000

OTHER SUBSTANCES
Simultaneous: alprazolam, bromazepam, clobazam, clonazepam, clorazepate, oxazepam, prazepam, triazolam

REFERENCE
Renou-Gonnord,M.F.; David,K. Optimized micellar electrokinetic chromatographic separation of benzodiazepines, *J.Chromatogr.A*, **1996**, *735*, 249–261.

Chlormezanone

Molecular formula: $C_{11}H_{12}ClNO_3S$
Molecular weight: 273.74
CAS Registry No.: 80-77-3
Merck Index (12th ed.): 2155

SAMPLE
Matrix: solutions
Sample preparation: Inject an aliquot of a 20 μg/mL solution

CAPILLARY ELECTROPHORESIS
Capillary: 44 cm × 50 μm fused-silica (37 cm to detector) (Supelco)
Capillary preparation: Before each run rinse capillary with running buffer for 3 min. At the start of each day rinse capillary with running buffer for 10 min. Treat new capillaries with 1 M NaOH, 100 mM NaOH, water, and running buffer.
Capillary temperature: 25
Running buffer: 100 mM Phosphoric acid adjusted to pH 5 with triethanolamine containing 10 mM carboxymethyl-β-cyclodextrin (Cyclolab, Budapest) and 10 mM heptakis(2,3,6-tri-O-methyl)-β-cyclodextrin (Sigma)
Injection: Hydrodynamic injection for 5 s (13.3 nL).
Detector: UV 210
Voltage: -25 kV
Model: Spectraphoresis 1000 CE

KEY WORDS
detector at anode; chiral; R_s = 3.3

REFERENCE
Fillet,M.; Fotsing,L.; Crommen,J. Enantioseparation of uncharged compounds by capillary electrophoresis using mixtures of anionic and neutral β-cyclodextrin derivatives, *J.Chromatogr.A*, **1998**, *817*, 113–119.

Chloroquine

Molecular formula: C₁₈H₂₆ClN₃
Molecular weight: 319.88
CAS Registry No.: 54-05-7, 50-63-5 (phosphate), 3545-67-3 (HCl)
Merck Index (12th ed.): 2215
Lednicer: 1 341

SAMPLE
Matrix: formulations
Sample preparation: Dissolve tablet in water, sonicate for 5 min, filter (0.45 μm), dilute the filtrate with water, inject an aliquot.

CAPILLARY ELECTROPHORESIS
Capillary: 57 cm × 75 μm fused-silica (34 cm to detector) (Polymicro Technologies)
Capillary preparation: Before each run rinse capillary with 100 mM NaOH for 5 min, with water for 5 min and with running buffer for 5 min. Condition new capillaries with 100 mM NaOH for 10 min, with water for 5 min and with running buffer for 5 min.
Running buffer: 10 mM pH 3.84 Phosphate buffer containing 7 mM sulfated-β-cyclodextrin
Injection: Hydrodynamic injection at 14 cm for 25 s.
Detector: UV 214
Migration time: 9, 10 (enantiomers)
Voltage: 12 kV
Model: laboratory-constructed

OTHER SUBSTANCES
Also analyzed: brompheniramine, chlorpheniramine, doxylamine, pheniramine

KEY WORDS
chiral; tablets

REFERENCE
Jin,L.J.; Li,S.F.Y. Comparison of chiral recognition capabilities of cyclodextrins for the separation of basic drugs in capillary zone electrophoresis, *J.Chromatogr.B*, **1998**, *708*, 257–266.

SAMPLE
Matrix: solutions
Sample preparation: Prepare a 100-200 ppm solution in MeOH, inject an aliquot.

CAPILLARY ELECTROPHORESIS
Capillary: 45 cm × 50 μm fused-silica (Polymicro Technologies)
Running buffer: 50 mM pH 7.50 Phosphate buffer containing 25 mM borate and 9 mM tetra-butylammonium bromide
Injection: Hydrodynamic injection at 5 cm for 5 s (1 nL)
Detector: UV 240
Migration time: 3.1
Voltage: 15 kV
Current: 39 μA
Model: Laboratory constructed

OTHER SUBSTANCES
Simultaneous: dapsone, primaquine, pyrimethamine, quinacrine, quinine, sulfadiazine

REFERENCE
Ng,C.L.; Toh,Y.L.; Li,S.F.Y.; Lee,H.K. Capillary electrophoresis of biologically important compounds: Optimization of separation conditions by the overlapping resolution mapping scheme, *J.Liq.Chromatogr.*, **1993**, *16*, 3653–3666.

SAMPLE
Matrix: solutions
Sample preparation: Inject an aliquot of a 15 μg/mL solution in water.

CAPILLARY ELECTROPHORESIS
Capillary: 65 cm × 50 μm silica (50 cm to detector)
Capillary preparation: Between injections wash with 100 mM NaOH, rinse with water
Capillary temperature: 20 kV
Running buffer: 25 mM Na_2HPO_4 adjusted to pH 2 with phosphoric acid
Injection: Electrokinetic injection at 5 kV for 10 s
Detector: UV 254
Migration time: 9.4
Model: Isco 3850 Capillary Electropherograph
Limit of detection: 150 ng/mL (40 ng/mL if a water plug is injected first)

OTHER SUBSTANCES
Simultaneous: 4-chlorophenylbiguanide, chlorproguanil, cyclochlorproguanil, cycloguanil, desethylchloroquine, mefloquine, primaquine, proguanil, pyrimethamine, quinine

REFERENCE
Taylor,R.B.; Reid,R.G. Analysis of basic antimalarial drugs by CZE and MEKC. Part 1--Critical factors affecting separation, *J.Pharm.Biomed.Anal.*, **1993**, *11*, 1289–1294.

SAMPLE
Matrix: solutions
Sample preparation: Prepare a solution in running buffer, inject an aliquot.

CAPILLARY ELECTROPHORESIS
Capillary: 60 cm × 75 μm fused-silica (52.4 cm to detector)
Capillary preparation: After each run flush with 500 mM KOH for 2-3 min, flush with water, fill with running buffer
Running buffer: 10 mM Na_2HPO_4 containing 2% heparin sodium (MW 10000, 11% S, Scientific Protein Laboratories, Waunakee, WI) adjusted to pH 5 with phosphoric acid
Injection: Hydrostatic injection
Detector: UV 214
Migration time: 52.8 (first (+) enantiomer, R_s 2.97)
Model: Waters Quanta 4000

OTHER SUBSTANCES
Also analyzed: anabasine, brompheniramine, bupivacaine, carbinoxamine, chlorcyclizine, chlorpheniramine, dimethindene, doxylamine, enpiroline, halofantrine, hydroxychloroquine, indapamide, mefloquine, nornicotine, pheniramine, primaquine, promethazine, quinacrine, tetramisole

KEY WORDS
chiral

REFERENCE
Stalcup,A.M.; Agyei,N.M. Heparin: a chiral mobile-phase additive for capillary zone electrophoresis, *Anal.Chem.*, **1994**, *66*, 3054–3059.

SAMPLE
Matrix: solutions
Sample preparation: Inject an aliquot of a 100 μg/mL solution in water:running buffer 50:50.

CAPILLARY ELECTROPHORESIS
Capillary: 44.5 cm × 50 μm acrylamide-coated fused-silica (Bio-Rad)
Capillary temperature: 30
Running buffer: 100 mM NaH_2PO_4 containing 15 mM gamma-cyclodextrin, adjusted to pH 2.5 with phosphoric acid
Injection: Electrokinetic injection at 8 kV for 6 s.

Detector: UV 200
Migration time: 5.79
Voltage: 14 kV
Model: Bio-Rad BioFocus 3000

OTHER SUBSTANCES
Also analyzed: albuterol, alprenolol, atenolol, atropine, baclofen, bamethan, benserazide, biperiden, bisoprolol, bupivacaine, bupranolol, butetamate, carazolol, carbuterol, carvedilol, celiprolol, chlorpheniramine (chlorphenamine), clidinium bromide, clobutinol, disopyramide, dobutamine, flecainide, homatropine, ipratropium bromide, isoproterenol, isothipendyl, ketamine, mefloquine, mequitazine, metaproterenol (orciprenaline), metipranolol, nafronyl (naftidrofuryl), nefopam, ofloxacin, orphenadrine, oxomemazine, oxprenolol, phenoxybenzamine, pholedrine, pindolol, pirbuterol, prilocaine, promethazine, propafenone, propranolol, sotalol, synephrine, terbutaline, tetrahydrozoline (tetryzoline), tocainide, trihexyphenidyl, trimeprazine (alimemazine), trimipramine, tropicamide, verapamil, zopiclone

KEY WORDS
coated capillary; achiral

REFERENCE
Koppenhoefer,B.; Epperlein,U.; Christian,B.; Yibing,J.; Yuying,C.; Bingcheng,L. Separation of enantiomers of drugs by capillary electrophoresis. I. γ-Cyclodextrin as chiral solvating agent, *J.Chromatogr.A*, **1995**, *717*, 181–190.

SAMPLE
Matrix: solutions
Sample preparation: Inject an aliquot of a solution in running buffer.

CAPILLARY ELECTROPHORESIS
Capillary: 60 cm $\times$ 75 μm fused-silica (52.4 cm to detector)
Capillary preparation: After each run flush with 500 mM KOH for 2-3 min then with water.
Running buffer: 10 mM pH 3.8 Phosphate buffer containing 2% sulfated cyclodextrin (ds 7-10)
Injection: Hydrostatic injection.
Detector: UV 214
Migration time: 10.27, 10.58 (enantiomers)
Voltage: 15 kV
Model: Waters Quanta 4000

OTHER SUBSTANCES
Also analyzed: acebutolol, alprenolol, aminoglutethimide, brompheniramine, bupivacaine, bupropion, canadine, carbinoxamine, chlorpheniramine, dimethindene, disopyramide, doxylamine, hydroxychloroquine, idazoxan, isoxsuprine, ketamine, mepenzolate, mepivacaine, methoxyphenamine, mexiletine, midodrine, nefopam, orphenadrine, oxprenolol, oxyphencyclimine, pheniramine, phensuximide, pindolol, piperoxan, terbutaline, tetramisole, tolperisone, tranylcypromine, trihexyphenidyl, trimipramine, verapamil, warfarin

KEY WORDS
chiral; detector at anode

REFERENCE
Stalcup,A.M.; Gahm,K.H. Application of sulfated cyclodextrins to chiral separations by capillary zone electrophoresis, *Anal.Chem.*, **1996**, *68*, 1360–1368.

SAMPLE
Matrix: solutions
Sample preparation: Inject an aliquot of a 100 μg/mL solution in running buffer.

CAPILLARY ELECTROPHORESIS
Capillary: 30 cm $\times$ 50 μm fused-silica (25.5 cm to detector) (Yongnian Optical Conductive Fiber Plant, China), coated with polyacrylamide

Capillary preparation: No details of the polyacrylamide coating process are provided. However, another paper (LC.GC 1997, 15, 40) by this group indicates that they use the procedure of Hjertén, thus: Adjust the pH of 20 mL water to 3.5 with acetic acid, add 80 μL 3-(trimethoxysilyl)propyl methacrylate (3-methacryloxypropyltrimethoxysilane), mix, suck into capillary, let stand at room temperature for 1 h, remove the solution, wash with water. Fill the capillary with a deaerated 3-4% acrylamide solution containing 1 μL/mL N,N,N',N'-tetramethylethylenediamine and 1 mg/mL potassium persulfate, let stand for 30 min, remove excess solution by aspiration, rinse with water, remove water by aspiration, dry at 35° (J. Chromatogr. 1985, 347, 191).
Capillary temperature: 25
Running buffer: 100 mM NaH_2PO_4 adjusted to pH 2.5
Injection: Electrokinetic injection at 15 kV for 3 s.
Detector: UV 200, UV 210
Migration time: 2.81
Voltage: 15 kV
Model: Bio-Rad BioFocus 3000

OTHER SUBSTANCES
Simultaneous: amorolfine, brompheniramine, bupivacaine, carteolol, chlorpheniramine, chlorphenoxamine, disopyramide, dobutamine, doxylamine, flecainide, gallopamil, ketamine, mepindolol, orphenadrine, oxybutynin, phenoxybenzamine, pindolol, propafenone, propranolol, sulpiride, talinolol, tropicamide, verapamil

KEY WORDS
coated capillary

REFERENCE
Koppenhoefer,B.; Epperlein,U.; Xiaofeng,Z.; Bingcheng,L. Separation of enantiomers of drugs by capillary electrophoresis. Part 4: Hydroxypropyl-γ-cyclodextrin as chiral solvating agent, *Electrophoresis*, **1997**, *18*, 924–930.

SAMPLE
Matrix: solutions

CAPILLARY ELECTROPHORESIS
Capillary: 48 cm × 50 μm fused-silica (30 cm to detector) (Yongnian Optical Fiber, Hebei, China)
Capillary preparation: Between runs rinse with 100 mM NaOH for 2 min, with water for 2 min, and with running buffer for 2 min. Rinse new capillaries with 1 M NaOH overnight, equilibrate with water for 30 min, and equilibrate with running buffer for 30 min.
Running buffer: 50 mM pH 3.0 Sodium phosphate buffer containing 2.5 mM sulfobutyl ether β-cyclodextrin (BioScience Innovations, Lawrence KS)
Injection: Hydrostatic injection at 10 cm for 5 s.
Detector: UV 214
Migration time: 9.8, 11.2 (enantiomers)
Voltage: 20 kV
Model: laboratory-constructed

KEY WORDS
chiral

REFERENCE
Dong,Y.; Huang,A.; Sun,Y.; Sun,Z. Chiral separation of chloroquine and pemoline by capillary zone electrophoresis with sulfobutyl ether β-cyclodextrin as buffer additive, *Chromatographia*, **1998**, *48*, 310–313.

SAMPLE
Matrix: solutions

CAPILLARY ELECTROPHORESIS
Capillary: 29-36 cm × 50 μm fused-silica (24.5-31.5 cm to detector) (Yongnian Optical Conductive Fiber Plant, China) coated with polyacrylamide
Capillary preparation: Coat capillary as follows. Adjust the pH of 20 mL water to 3.5 with acetic acid, add 80 μL 3-(trimethoxysilyl)propyl methacrylate (3-methacryloxypropyltrimethox-

ysilane), mix, suck into capillary, let stand at room temperature for 1 h, remove the solution, wash with water. Fill the capillary with a deaerated 3-4% acrylamide solution containing 1 μL/mL N,N,N',N'-tetramethylethylenediamine and 1 mg/mL potassium persulfate, let stand for 30 min, remove excess solution by aspiration, rinse with water, remove water by aspiration, dry at 35° (J. Chromatogr. 1985, 347, 191).
Capillary temperature: 25
Running buffer: 100 mM pH 2.5 NaH_2PO_4 (A) or 100 mM pH 2.5 NaH_2PO_4 containing 45 mM hydroxypropyl-α-cyclodextrin (Wacker, Munich) (B)
Injection: Electromigration at 15 kV for 3 s.
Detector: UV 200; UV 210
Migration time: 2.81 (A); 5.24 (B) (no separation of enantiomers)
Voltage: 15 kV
Model: Bio-Focus 3000

OTHER SUBSTANCES
Also analyzed: albuterol (salbutamol), alprenolol, amorolfine, atenolol, atropine, azelastine, baclofen, bamethan, benproperine, benserazide, biperiden, bisoprolol, brompheniramine, bupivacaine, bupranolol, butamirate, butethamate, carazolol, carbuterol, carteolol, carvedilol, celiprolol, chlorpheniramine, chlorphenoxamine, ciclotanine, clenbuterol, clidinium bromide, clobutinol, dimethindene, dipivefrin, disopyramide, dobutamine, doxylamine, fendiline, flecainide, gallopamil, homatropine, ipratropium bromide, isoproterenol (isoprenaline), isothipendyl, ketamine, meclizine, mefloquine, mepindolol, mequitazine, metaclazepam, metaproterenol (orciprenaline), metipranolol, metoprolol, nafronyl (naftidrofuryl), nefopam, nicardipine, norfenefrine, ofloxacin, ornidazole, orphenadrine, oxomemazine, oxprenolol, oxybutynin, phenoxybenzamine, phenylpropanolamine, pholedrine, pindolol, pirbuterol, prilocaine, procyclidine, promethazine, propafenone, propranolol, reproterol, sotalol, sulpride, synephrine, talinolol, terbutaline, tetrahydrozoline (tetryzoline), theodrenaline, tioconazole, tocainide, trihexyphenidyl, trimeprazine (alimemazine), trimipramine, tropicamide, verapamil, zopiclone

---

KEY WORDS
coated capillary

REFERENCE
Koppenhoefer,B.; Eperlein,U.; Schlunk,R.; Zhu,X.; Lin,B. Separation of enantiomers of drugs by capillary electrophoresis. V. Hydroxypropyl-α-cyclodextrin as chiral solvating agent, *J.Chromatogr.A*, **1998**, *793*, 153–164.

SAMPLE
Matrix: urine
Sample preparation: Add 1 mL urine containing 2 μg chlorproguanil to a Bond Elut C18 SPE cartridge, wash with pH 2 buffer, wash with water, elute with 1 mL MeOH, inject an aliquot of the eluate. (Samples in saliva and plasma can also be analyzed using elution with perchloric acid.)

CAPILLARY ELECTROPHORESIS
Capillary: 65 cm × 50 μm silica (50 cm to detector)
Capillary preparation: Between injections wash with 100 mM NaOH, rinse with water
Capillary temperature: 20 kV
Running buffer: 25 mM Na_2HPO_4 adjusted to pH 2 with phosphoric acid
Injection: Electrokinetic injection at 5 kV for 10 s
Detector: UV 254
Internal standard: chlorproguanil
Model: Isco 3850 Capillary Electropherograph
Limit of quantitation: 103 ng/mL

OTHER SUBSTANCES
Extracted: 4-chlorophenylbiguanide, cycloguanil, desethylchloroquine, proguanil

---

KEY WORDS
SPE

REFERENCE
Taylor,R.B.; Reid,R.G. Analysis of basic antimalarial drugs by CZE; Part 2. Validation and application to bioanalysis, *J.Pharm.Biomed.Anal.*, **1995**, *13*, 21–26.

Chlorothiazide

Molecular formula: $C_7H_6ClN_3O_4S_2$
Molecular weight: 295.73
CAS Registry No.: 58-94-6, 7085-44-1 (sodium salt)
Merck Index (12th ed.): 2221
Lednicer: 1 321

SAMPLE
Matrix: blood, urine
Sample preparation: Condition a 3 mL Supelclean LC-18 SPE cartridge with 3 mL MeOH and 3 mL water. Dilute urine 1:10 with water. Precipitate proteins from serum with MeOH. Add diluted urine or protein supernatant to the SPE cartridge, wash with 3 mL water, elute with 3 mL MeOH, reconstitute to the original volume with 100 mM KOH, inject an aliquot.

CAPILLARY ELECTROPHORESIS
Capillary: 67 cm × 50 μm fused-silica (60 cm to detector) (Polymicro Technologies)
Capillary preparation: Rinse with running buffer for 2 min before run. If necessary, regenerate capillary with 100 mM NaOH for 10 min and with water for 15 min.
Capillary temperature: 20
Running buffer: 60 mM 3-(cyclohexylamino)-1-propanesulfonic acid (CAPS) adjusted to pH 10.6 with 100 mM KOH
Injection: Pressure injection for 5 s
Detector: UV 220
Migration time: 15.6
Voltage: 25 kV
Model: Beckman P/ACE 2000

OTHER SUBSTANCES
Extracted: acetazolamide, amiloride, bendroflumethiazide, benzthiazide, bumetanide, caffeine, chlorthalidone, clopamide, dichlorphenamide, ethacrynic acid, furosemide, hydrochlorothiazide, metyrapone, probenecid, triamterene, trichlormethiazide

KEY WORDS
serum; SPE

REFERENCE
Jumppanen,J.; Sirén,H.; Riekkola,M.-L. Screening for diuretics in urine and blood serum by capillary zone electrophoresis, *J.Chromatogr.A*, **1993**, *652*, 441–450.

SAMPLE
Matrix: bulk
Sample preparation: Mix with 10 mM pH 2.0 sodium phosphate buffer, add an equal volume of MeCN, sonicate for 10 min, dilute with 8 volumes of 10 mM pH 2.0 sodium phosphate buffer, inject an aliquot.

CAPILLARY ELECTROPHORESIS
Capillary: 70 cm × 100 μm fused-silica (63 cm to detector) (Chrompack or Polymicro Technologies)
Capillary preparation: Before injection fill capillary with buffer for 2 min. At the start of each day wash with 100 mM NaOH for 5 min, with water for 10 min, and with running buffer for 10 min. At the end of each day wash with 100 mM NaOH for 5 min and with water for 10 min.
Capillary temperature: 15

Running buffer: 20 mM pH 9.5 Sodium borate buffer containing 30 mM sodium dodecyl sulfate
Injection: Hydrodynamic injection for 2.5 s.
Detector: UV 225
Migration time: 13.2
Voltage: 20 kV
Model: Spectra Physics SpectraPhoresis 1000

OTHER SUBSTANCES
Simultaneous: impurities, hydrochlorothiazide

KEY WORDS
stability-indicating

REFERENCE
Thomas,B.R.; Fang,X.G.; Chen,X.; Tyrrell,R.J.; Ghodbane,S. Validated micellar electrokinetic chromatography
method for quality control of the drug substances hydrochlorothiazide and chlorothiazide, *J.Chromatogr.B*,
1994, *657*, 383–394.

SAMPLE
Matrix: solutions

CAPILLARY ELECTROPHORESIS
Capillary: 72 cm × 50 μm silica (50 cm to detector) (Applied Biosystems)
Capillary preparation: Before each run wash capillary with 100 mM NaOH for 3 min, wash
with running buffer for 3 min, aspirate neutral marker solution (2 drops DMSO in 10 mL
water) for 1 s, place capillary end in buffer vial for 5 s, inject sample. Wash capillary at the
beginning of each day by passing 1 M NaOH through for 20 min.
Capillary temperature: 30
Running buffer: MeCN:buffer 10:90 (Buffer was 30 mM pH 9.3 borate buffer containing 30 mM
sodium dodecyl sulfate.)
Injection: Inject by applying vacuum for 1 s (2-3 nL)
Detector: UV 200
Migration time: 9.5
Voltage: +30 kV
Current: 56 μA
Model: Applied Biosystems Model 270A

OTHER SUBSTANCES
Simultaneous: furosemide, hydrochlorothiazide, trichlormethiazide

REFERENCE
Evenson,M.A.; Wiktorowicz,J.E. Automated capillary electrophoresis applied to therapeutic drug monitoring,
Clin.Chem., **1992**, *38*, 1847–1852.

SAMPLE
Matrix: solutions
Sample preparation: Inject an aliquot of a 200 μg/mL solution in MeCN:water 50:50.

CAPILLARY ELECTROPHORESIS
Capillary: 23 cm × 50 μm 3 μm CEC Hypersil C18
Capillary temperature: 15
Running buffer: Gradient. MeCN:50 mM pH 2.5 Na$_2$HPO$_4$ buffer:water 40:20:40 for 6.5 min,
60:20:20 for 10.75 min (step gradient), re-equilibrate at initial conditions for 7.75 min.
Injection: Electrokinetic injection at 5 kV for 15 s.
Detector: UV 210
Migration time: 10
Voltage: 30 kV (with 8 bar of pressure at each end of capillary)
Model: Hewlett-Packard HP[3D]

OTHER SUBSTANCES
Simultaneous: bendroflumethiazide, bumetanide, chlorthalidone, hydrochlorothiazide, hydroflumethiazide

KEY WORDS
electrochromatography

REFERENCE
Euerby,M.R.; Gilligan,D.; Johnson,C.M.; Bartle,K.D. Step-gradient capillary electrochromatography, *Analyst*, **1997**, *122*, 1087–1088.

Chlorpheniramine

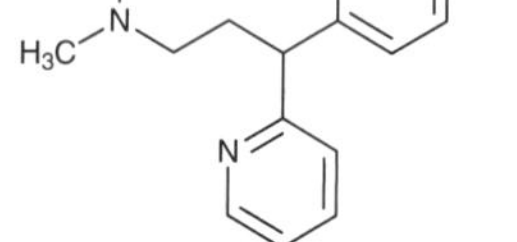

Molecular formula: $C_{16}H_{19}ClN_2$
Molecular weight: 274.79
CAS Registry No.: 132-22-9, 113-92-8 (maleate), 2438-32-6 (maleate), 25523-97-1 (d-form), 2438-32-6 (d-form maleate)
Merck Index (12th ed.): 2232
Lednicer: 1 77

SAMPLE
Matrix: formulations
Sample preparation: Grind tablet or powders, add 70 mL MeOH, add 20 mL 5 mg/mL methyl p-hydroxybenzoate in MeOH, sonicate for 10 min, shake for 5 min, make up to 100 mL with water, filter (0.45 μm), inject an aliquot.

CAPILLARY ELECTROPHORESIS
Capillary: 65 cm × 50 μm fused-silica (50 cm to detector) (SGE)
Running buffer: 20 mM NaH_2PO_4 containing 100 mM sodium dodecyl sulfate, adjusted to pH 9.0 with 20 mM sodium tetraborate
Injection: Siphon
Detector: UV 210
Migration time: 24.5
Internal standard: methyl p-hydroxybenzoate
Voltage: +20 kV

OTHER SUBSTANCES
Simultaneous: acetaminophen, caffeine, dipyrone (sulpyrin), ethenzamide, guaifenesin, isopropylantipyrine, naproxen, noscapine, phenacetin, trimetoquinol
Interfering: tipepidine

KEY WORDS
tablets; powders

REFERENCE
Nishi,H.; Fukuyama,T.; Matsuo,M.; Terabe,S. Effect of surfactant structures on the separation of cold medicine ingredients by micellar electrokinetic chromatography, *J.Pharm.Sci.*, **1990**, *79*, 519–523.

SAMPLE
Matrix: formulations
Sample preparation: Grind granules, add 70 mL MeOH, warm at 40° with occasional shaking, cool, add 20 mL 5 mg/mL methyl p-hydroxybenzoate in MeOH, make up to 100 mL with water, filter (0.45 μm), inject an aliquot.

CAPILLARY ELECTROPHORESIS
Capillary: 65 cm × 50 μm fused-silica (SGE)
Running buffer: 20 mM NaH_2PO_4 containing 100 mM sodium cholate, adjusted to pH 9.0 with 20 mM sodium tetraborate
Injection: Siphon for 10 s at 10 cm height
Detector: UV 210
Migration time: 14

Internal standard: methyl p-hydroxybenzoate (16)
Voltage: +20 kV

OTHER SUBSTANCES
Simultaneous: acetaminophen, caffeine, dibucaine, dipyrone (sulpyrin), ethenzamide, guaifenesin, isopropylantipyrine, naproxen, noscapine, phenacetin, tipepidine, trimetoquinol, triprolidine

KEY WORDS
granules

REFERENCE
Nishi,H.; Fukuyama,T.; Matsuo,M.; Terabe,S. Separation and determination of the ingredients of a cold medicine by micellar electrokinetic chromatography with bile salts, *J.Chromatogr.*, **1990**, *498*, 313–323.

SAMPLE
Matrix: formulations
Sample preparation: Grind tablets, extract twice with MeOH by sonicating at low temperature, centrifuge, filter (0.45 μm) the supernatant, inject an aliquot of the filtrate.

CAPILLARY ELECTROPHORESIS
Capillary: 47 cm × 50 μm (40 cm to detector) (Polymicro Technologies)
Capillary temperature: 25
Running buffer: 50 mM pH 7.5 Borate buffer containing 50 mM phosphate, 10 mM sodium dodecyl sulfate, 10 mM β-cyclodextrin, and 10 mM tetrabutylammonium hydrogen sulfate
Injection: Inject at a pressure of 0.5 psi for 1 s (5.2 nL)
Detector: UV 214
Migration time: 5.3, 5.8 (enantiomers)
Voltage: 21.5 kV
Model: Beckman P/ACE 2000
Limit of detection: 143 pg

OTHER SUBSTANCES
Simultaneous: cyclizine, dimenhydrinate, doxylamine, methapyrilene, pheniramine, promethazine, thonylamine, triprolidine

KEY WORDS
tablets; chiral

REFERENCE
Ong,C.P.; Ng,C.L.; Lee,H.K.; Li,S.F.Y. Determination of antihistamines in pharmaceuticals by capillary electrophoresis, *J.Chromatogr.*, **1991**, *588*, 335–339.

SAMPLE
Matrix: formulations
Sample preparation: Grind tablet, add MeCN:10 mM HCl 20:80, shake and sonicate for 15 min, centrifuge until a clear solution is obtained. Remove a 25 mL aliquot of the supernatant and add it to 10 mL 500 μg/mL propyl hydroxybenzoate, make up to 50 mL with MeCN:10 mM HCl 20:80, inject an aliquot.

CAPILLARY ELECTROPHORESIS
Capillary: 57 cm × 75 μm fused-silica (Beckman)
Capillary preparation: Pass running buffer through the capillary for at least 4 min before each injection.
Capillary temperature: 25
Running buffer: 20 mM pH 9 Borate buffer containing 25 mM sodium cholate and 50 mM sodium deoxycholate
Injection: Pressure injection for 2 s
Detector: UV 214
Migration time: 12.5
Internal standard: propyl hydroxybenzoate (10.5)

Voltage: 20 kV
Model: Beckman P/ACE System 2100

OTHER SUBSTANCES
Simultaneous: acetaminophen, aspirin, caffeine, dextropropoxyphene, salicylic acid

---

KEY WORDS
tablets

REFERENCE
Boonkerd,S.; Lauwers,M.; Detaevernier,M.R.; Michotte,Y. Separation and simultaneous determination of the components in an analgesic tablet formulation by micellar electrokinetic chromatography, *J.Chromatogr.A*, **1995**, *695*, 97–102.

SAMPLE
Matrix: formulations
Sample preparation: Dissolve tablet in water, sonicate for 5 min, filter (0.45 µm), dilute the filtrate with water, inject an aliquot.

CAPILLARY ELECTROPHORESIS
Capillary: 57 cm × 75 µm fused-silica (34 cm to detector) (Polymicro Technologies)
Capillary preparation: Before each run rinse capillary with 100 mM NaOH for 5 min, with water for 5 min and with running buffer for 5 min. Condition new capillaries with 100 mM NaOH for 10 min, with water for 5 min and with running buffer for 5 min.
Running buffer: 10 mM pH 3.84 Phosphate buffer containing 7 mM sulfated-β-cyclodextrin
Injection: Hydrodynamic injection at 14 cm for 25 s.
Detector: UV 214
Migration time: 6, 8 (enantiomers)
Voltage: 12 kV
Model: laboratory-constructed

OTHER SUBSTANCES
Also analyzed: brompheniramine, chloroquine, doxylamine, pheniramine

---

KEY WORDS
chiral; tablets

REFERENCE
Jin,L.J.; Li,S.F.Y. Comparison of chiral recognition capabilities of cyclodextrins for the separation of basic drugs in capillary zone electrophoresis, *J.Chromatogr.B*, **1998**, *708*, 257–266.

SAMPLE
Matrix: solutions

CAPILLARY ELECTROPHORESIS
Capillary: 15 cm × 50 µm fused-silica coated with crosslinked polyacrylamide (13.5 cm to detector) (Scientific Glass Engineering)
Running buffer: 100 mM pH 6.8 Sodium phosphate buffer containing 3% G 3707 (heptaoxyethylene lauryl ether, Atlas Chemie or Sigma P 8800)
Injection: Place a 1-2 mm slug in capillary by capillary action
Detector: UV 205
Migration time: 8
Voltage: 3 kV
Current: 60 µA

OTHER SUBSTANCES
Simultaneous: alprenolol, benzylamine, codeine, ephedrine, protriptyline, terodilin

---

KEY WORDS
coated capillary

REFERENCE
Hjertén,S.; Valtcheva,L.; Elenbring,K.; Eaker,D. High-performance electrophoresis of acidic and basic low-molecular-weight compounds and of proteins in the presence of polymers and neutral surfactants, *J.Liq.Chromatogr.*, **1989**, *12*, 2471–2499.

SAMPLE
Matrix: solutions
Sample preparation: Prepare a 100 μg/mL solution of chlorpheniramine maleate, inject an aliquot.

CAPILLARY ELECTROPHORESIS
Capillary: 26 cm × 50 μm fused-silica (20 cm to detector) (Polymicro Technologies)
Capillary temperature: 25
Running buffer: 50 mM pH 3.0 Phosphate buffer containing 100 mM β-cyclodextrin and 5 M urea
Injection: Pressure injection
Detector: UV 214
Migration time: 7.2 (R), 7.5 (S)
Voltage: 10 kV
Current: 21 μA
Model: Beckman P/ACE System 2000

KEY WORDS
chiral

REFERENCE
Otsuka,K.; Terabe,S. Optical resolution of chlorpheniramine by cyclodextrin added capillary zone electrophoresis and cyclodextrin modified micellar electrokinetic chromatography, *J.Liq.Chromatogr.*, **1993**, *16*, 945–953.

SAMPLE
Matrix: solutions
Sample preparation: Prepare a 50 μg/mL solution in diluted running buffer, inject an aliquot.

CAPILLARY ELECTROPHORESIS
Capillary: 44 cm × 50 μm fused-silica
Capillary preparation: After each run wash capillary with water for 2 min and with running buffer for 3 min. At the beginning of each day wash capillary with water and running buffer for 5 min. Condition a new capillary with 1 M NaOH, 100 mM NaOH, water, and separation buffer.
Capillary temperature: 15
Running buffer: 100 mM Phosphoric acid containing 15 mM cyclodextrin adjusted to pH 3.0 with triethanolamine
Injection: Hydrodynamic injection for 3 s
Detector: UV 210
Migration time: 7.1, 7.4 (enantiomers)
Voltage: 25 kV
Model: Spectra Physics Spectraphoresis 1000

KEY WORDS
chiral

REFERENCE
Bechet,I.; Paques,P.; Fillet,M.; Hubert,P.; Crommen,J. Chiral separation of basic drugs by capillary zone electrophoresis with cyclodextrin additives, *Electrophoresis*, **1994**, *15*, 818–823.

SAMPLE
Matrix: solutions
Sample preparation: Prepare a solution in running buffer, inject an aliquot.

CAPILLARY ELECTROPHORESIS
Capillary: 60 cm × 75 μm fused-silica (52.4 cm to detector)
Capillary preparation: After each run flush with 500 mM KOH for 2-3 min, flush with water, fill with running buffer
Running buffer: 10 mM Na_2HPO_4 containing 2% heparin sodium (MW 10000, 11% S, Scientific Protein Laboratories, Waunakee, WI) adjusted to pH 5 with phosphoric acid
Injection: Hydrostatic injection
Detector: UV 214
Migration time: 18.3 (first enantiomer, R_S 3.5)
Model: Waters Quanta 4000

OTHER SUBSTANCES
Also analyzed: anabasine, brompheniramine, bupivacaine, carbinoxamine, chlorcyclizine, chloroquine, dimethindene, doxylamine, enpiroline, halofantrine, hydroxychloroquine, indapamide, mefloquine, nornicotine, pheniramine, primaquine, promethazine, quinacrine, tetramisole

KEY WORDS
chiral

REFERENCE
Stalcup,A.M.; Agyei,N.M. Heparin: a chiral mobile-phase additive for capillary zone electrophoresis, *Anal.Chem.*, **1994**, *66*, 3054–3059.

SAMPLE
Matrix: solutions
Sample preparation: Inject an aliquot of a 100 μg/mL solution in water:running buffer 50:50.

CAPILLARY ELECTROPHORESIS
Capillary: 44.5 cm × 50 μm acrylamide-coated fused-silica (Bio-Rad)
Capillary temperature: 30
Running buffer: 100 mM NaH_2PO_4 containing 15 mM gamma-cyclodextrin, adjusted to pH 2.5 with phosphoric acid
Injection: Electrokinetic injection at 8 kV for 6 s.
Detector: UV 200
Migration time: 5.85
Voltage: 14 kV
Model: Bio-Rad BioFocus 3000

OTHER SUBSTANCES
Also analyzed: albuterol, alprenolol, atenolol, atropine, baclofen, bamethan, benserazide, biperiden, bisoprolol, bupivacaine, bupranolol, butetamate, carazolol, carbuterol, carvedilol, celiprolol, chloroquine, clidinium bromide, clobutinol, disopyramide, dobutamine, flecainide, homatropine, ipratropium bromide, isoproterenol, isothipendyl, ketamine, mefloquine, mequitazine, metaproterenol (orciprenaline), metipranolol, nafronyl (naftidrofuryl), nefopam, ofloxacin, orphenadrine, oxomemazine, oxprenolol, phenoxybenzamine, pholedrine, pindolol, pirbuterol, prilocaine, promethazine, propafenone, propranolol, sotalol, synephrine, terbutaline, tetrahydrozoline (tetryzoline), tocainide, trihexyphenidyl, trimeprazine (alimemazine), trimipramine, tropicamide, verapamil, zopiclone

KEY WORDS
coated capillary; achiral

REFERENCE
Koppenhoefer,B.; Epperlein,U.; Christian,B.; Yibing,J.; Yuying,C.; Bingcheng,L. Separation of enantiomers of drugs by capillary electrophoresis. I. γ-Cyclodextrin as chiral solvating agent, *J.Chromatogr.A*, **1995**, *717*, 181–190.

SAMPLE
Matrix: solutions
Sample preparation: Prepare a solution in MeOH/water, inject an aliquot.

CAPILLARY ELECTROPHORESIS
Capillary: 62 cm × 52 μm fused-silica (50 cm to detector) (Polymicro Technologies)
Capillary temperature: 40
Running buffer: 50 mM pH 2.50 Tetrabutylammonium phosphate containing 20 mM hydroxy-propyl-β-cyclodextrin
Injection: Siphon injection at 10 cm for 5 s
Detector: UV (wavelength not specified)
Migration time: 10.20, 10.37 (enantiomers)
Voltage: 20 kV
Model: Laboratory constructed

OTHER SUBSTANCES
Simultaneous: alprenolol, atenolol, epinephrine, isoproterenol, labetalol, norepinephrine, oxprenolol, propranolol

KEY WORDS
chiral

REFERENCE
Quang,C.; Khaledi,M.G. Extending the scope of chiral separation of basic compounds by cyclodextrin-mediated capillary zone electrophoresis, *J.Chromatogr.A*, **1995**, *692*, 253–265.

SAMPLE
Matrix: solutions

CAPILLARY ELECTROPHORESIS
Capillary: 36 cm × 50 μm polyacrylamide-coated fused-silica (31.5 cm to detector) (Polymicro Technologies)
Capillary preparation: Before each injection rinse with water at 100 psi for 30 s, rinse with running buffer at 100 psi for 30 s, partially fill with separation solution (500 μM ovomucoid in running buffer) at 1 psi for 190 s (27 cm), inject sample, electrophorese with running buffer. Coat capillary with polyacrylamide as follows. Treat with 1 M NaOH for 1 h, rinse with water, dry at 110° by flushing with nitrogen for 6 h, pass thionyl chloride through using suction for 10 min, seal at each end, heat at 70° for 6 h, open, suck 250 mM vinylmagnesium bromide in THF into the capillary, seal at each end, heat at 70° for 6 h, open, rinse with THF for several min, rinse with water, fill by suction with a solution containing 3% acrylamide, 0.1% ammonium persulfate, and 0.1% tetramethylethylenediamine, heat at 28 ± 2° for 1 h, after 1 h wash with water (Biol.Pharm.Bull. 1993, 16, 1185).
Capillary temperature: 25
Running buffer: 1-Propanol:50 mM pH 6.0 phosphate buffer 8:92
Injection: Inject at 1 psi for 2 s
Detector: UV 210
Migration time: 11.49, 11.83 (enantiomers)
Voltage: 12 kV
Model: Bio-Rad BioFocus 3000

OTHER SUBSTANCES
Also analyzed: primaquine, trimebutine

KEY WORDS
chiral; partial separation zone technique; coated capillary

REFERENCE
Tanaka,Y.; Terabe,S. Partial separation zone technique for the separation of enantiomers by affinity electrokinetic chromatography with proteins as chiral pseudo-stationary phases, *J.Chromatogr.A*, **1995**, *694*, 277–284.

SAMPLE
Matrix: solutions
Sample preparation: Inject an aliquot of a 50-200 μM solution.

CAPILLARY ELECTROPHORESIS
Capillary: 48.5 cm $\times$ 50 μm fused-silica (44.5 cm to detector) (Polymicro Technologies)
Capillary temperature: 25
Running buffer: 100 mM Phosphoric acid containing 7.5 mM 6^A-methylamino-β-cyclodextrin, adjusted to pH 2.5 with tetramethylammonium hydroxide. (Synthesis of 6^A-methylamino-β-cyclodextrin was as follows. Add a solution of 3.65 g p-toluenesulfonyl chloride in 30 mL dry pyridine to 29.60 g β-cyclodextrin stirred at 5° in 300 mL dry pyridine, stir overnight at room temperature, evaporate to dryness under reduced pressure at 40°, add 700 mL diethyl ether to the residue. Collect the precipitate and recrystallize it 3 times from water to obtain mono-(6-O-p-tolylsulfonyl)-β-cyclodextrin in 31% yield (Bull. Chem. Soc. Japan 1978, 51, 3030). Heat 2 g mono-(6-O-p-tolylsulfonyl)-β-cyclodextrin with 35 mL 50% methylamine in MeOH in a sealed tube at 70° for 3 days, purify by chromatography on carboxymethylcellulose with ammonium bicarbonate solution to obtain 6^A-methylamino-β-cyclodextrin (cf. J. Am. Chem. Soc. 1980, 102, 762).
Injection: Pressure injection at 10 psi.s.
Detector: UV 206
Migration time: 19.1 (second enantiomer, R_S = 1.2)
Voltage: 18 kV
Current: 41-48 μA
Model: Biofocus 3000 (Bio-Rad)

OTHER SUBSTANCES
Simultaneous: isoproterenol, ketamine

KEY WORDS
chiral

REFERENCE
Fanali,S.; Camera,E. Use of methylamino-β-cyclodextrin in capillary electrophoresis. Resolution of acidic and basic enantiomers, *Chromatographia*, **1996**, *43*, 247–253.

SAMPLE
Matrix: solutions
Sample preparation: Inject an aliquot of a 50 μg/mL solution in water.

CAPILLARY ELECTROPHORESIS
Capillary: 48.5 cm $\times$ 50 μm fused-silica (40 cm to detector)
Capillary preparation: After each run wash capillary with running buffer for 3 min. At the start of each day wash capillary with running buffer for 10 min. Before use treat new capillaries with 1 M NaOH, 100 mM NaOH, water, and running buffer.
Capillary temperature: 15
Running buffer: 100 mM Phosphoric acid containing 2 mM carboxymethyl-β-cyclodextrin (Cyclolab, Budapest), adjusted to pH 3.0 with 84 mM triethanolamine
Injection: Hydrodynamic injection at 5 kPa for 2 s.
Detector: UV 210
Migration time: 8.55, 10.82 (enantiomers)
Voltage: 25 kV
Model: Hewlett Packard ^{3D}CE

OTHER SUBSTANCES
Simultaneous: fenfluramine, isoproterenol, terbutaline
Interfering: dimethindene, ephedrine

KEY WORDS
chiral

REFERENCE
Fillet,M.; Bechet,I.; Hubert,P.; Crommen,J. Resolution improvement by use of carboxymethyl-β-cyclodextrin as chiral additive for the enantiomeric separation of basic drugs by capillary electrophoresis, *J.Pharm.Biomed.Anal.*, **1996**, *14*, 1107–1114.

SAMPLE
Matrix: solutions
Sample preparation: Inject an aliquot of a 100 μg/mL solution in water:running buffer 50:50.

CAPILLARY ELECTROPHORESIS
Capillary: 44.5 cm × 50 μm polyacrylamide-coated fused-silica (40 cm to detector) (Bio-Rad)
Capillary temperature: 30
Running buffer: 100 mM NaH_2PO_4 containing 15 mM β-cyclodextrin, adjusted to pH 2.5 with phosphoric acid
Injection: Electrokinetic injection at 8 kV for 6 s.
Detector: UV 200
Migration time: 7.75, 7.86 (enantiomers)
Voltage: 14 kV
Model: Bio-Rad Bio-Focus 3000

OTHER SUBSTANCES
Simultaneous: carvedilol, ketamine, metaproterenol (orciprenaline), tetrahydrozoline, tropicamide, zopiclone
Also analyzed: albuterol (not chiral), atenolol (not chiral), benserazide (not chiral), biperiden (not chiral), bupivacaine (not chiral), butethamate (not chiral), carazolol (not chiral), carbuterol (not chiral), clidinium bromide (not chiral), disopyramide (not chiral), dobutamine (not chiral), flecainide (not chiral), ipratropium (not chiral), isothipendyl (not chiral), mequitazine (not chiral), metipranolol (not chiral), nafronyl (naftidrofuryl) (not chiral), nefopam (not chiral), ofloxacin (not chiral), orphenadrine (not chiral), pindolol (not chiral), pirbuterol (not chiral), prilocaine (not chiral), propafenone (not chiral), sotalol (not chiral), tocainide (not chiral), trimipramine (not chiral)

KEY WORDS
coated capillary; chiral

REFERENCE
Koppenhoefer,B.; Epperlein,U.; Christian,B.; Lin,B.; Ji,Y.; Chen,Y. Separation of enantiomers of drugs by capillary electrophoresis. III. β-cyclodextrin as chiral solvating agent, *J.Chromatogr.A*, **1996**, *735*, 333 343.

SAMPLE
Matrix: solutions

CAPILLARY ELECTROPHORESIS
Capillary: 62 cm × 75 μm fused-silica (54.5 cm to detector) (Polymicro Technologies)
Capillary preparation: Purge with running buffer before each run. At the beginning of each day purge using 50-60 kPa vacuum with 500 mM NaOH for 5 min, with water for 5 min, with MeCN for 5 min, and with running buffer for 5 min.
Running buffer: MeCN:MeOH:acetic acid 49:50:1 containing 20 mM ammonium acetate
Injection: Hydrostatic injection at 10 cm for 5 s.
Detector: UV 214
Migration time: 3.93
Voltage: 30 kV
Model: Waters Quanta 4000

OTHER SUBSTANCES
Simultaneous: buclizine, chlorcyclizine, cyclizine, dimenhydrinate, methaphenilene, pheniramine, promethazine, pyrilamine, pyrrobutamine, tripelennamine

REFERENCE
Leung,G.N.W.; Tang,H.P.O.; Tso,T.S.C.; Wan,T.S.M. Separation of basic drugs with non-aqueous capillary electrophoresis, *J.Chromatogr.A*, **1996**, *738*, 141–154.

SAMPLE
Matrix: solutions
Sample preparation: Inject an aliquot of a 20 μg/mL solution in 12.5 mM pH 9.3 borate buffer.

CAPILLARY ELECTROPHORESIS
Capillary: 66.5 cm × 75 μm fused-silica (41.5 cm to detector) (Polymicro Technologies)
Running buffer: 250 mM pH 9.3 Sodium borate buffer
Injection: Pressure injection at 50 mbar for 15 s.
Detector: UV 195
Migration time: 13.3
Voltage: 8 kV
Model: Crystal 310

OTHER SUBSTANCES
Simultaneous: amphetamine, ephedrine, MDA, MDMA, methamphetamine, phentermine, phenylpropanolamine, pseudoephedrine
Noninterfering: caffeine

REFERENCE
Lilley,K.A.; Wheat,T.E. Drug identification in biological matrices using capillary electrophoresis and chemometric software, *J.Chromatogr.B*, **1996**, *683*, 67–76.

SAMPLE
Matrix: solutions
Sample preparation: Inject an aliquot of a solution in running buffer.

CAPILLARY ELECTROPHORESIS
Capillary: 60 cm × 75 μm fused-silica (52.4 cm to detector)
Capillary preparation: After each run flush with 500 mM KOH for 2-3 min then with water.
Running buffer: 10 mM pH 3.8 Phosphate buffer containing 2% sulfated cyclodextrin (ds 7-10)
Injection: Hydrostatic injection.
Detector: UV 214
Migration time: 8.55, 8.80 (enantiomers)
Voltage: 15 kV
Model: Waters Quanta 4000

OTHER SUBSTANCES
Also analyzed: acebutolol, alprenolol, aminoglutethimide, brompheniramine, bupivacaine, bupropion, canadine, carbinoxamine, chloroquine, dimethindene, disopyramide, doxylamine, hydroxychloroquine, idazoxan, isoxsuprine, ketamine, mepenzolate, mepivacaine, methoxyphenamine, mexiletine, midodrine, nefopam, orphenadrine, oxprenolol, oxyphencyclimine, pheniramine, phensuximide, pindolol, piperoxan, terbutaline, tetramisole, tolperisone, tranylcypromine, trihexyphenidyl, trimipramine, verapamil, warfarin

KEY WORDS
chiral; detector at anode

REFERENCE
Stalcup,A.M.; Gahm,K.H. Application of sulfated cyclodextrins to chiral separations by capillary zone electrophoresis, *Anal.Chem.*, **1996**, *68*, 1360–1368.

SAMPLE
Matrix: solutions
Sample preparation: Inject an aliquot of a 100 μg/mL solution in running buffer.

CAPILLARY ELECTROPHORESIS
Capillary: 30 cm × 50 μm fused-silica (25.5 cm to detector) (Yongnian Optical Conductive Fiber Plant, China), coated with polyacrylamide
Capillary preparation: No details of the polyacrylamide coating process are provided. However, another paper (LC.GC 1997, 15, 40) by this group indicates that they use the procedure of Hjertén, thus: Adjust the pH of 20 mL water to 3.5 with acetic acid, add 80 μL 3-(trimethoxysilyl)propyl methacrylate (3-methacryloxypropyltrimethoxysilane), mix, suck into capillary, let stand at room temperature for 1 h, remove the solution, wash with water. Fill the capillary with a deaerated 3-4% acrylamide solution containing 1 μL/mL N,N,N',N'-tetramethylethylenediamine and 1 mg/mL potassium persulfate, let stand for 30 min, remove excess solution

by aspiration, rinse with water, remove water by aspiration, dry at 35° (J. Chromatogr. 1985, 347, 191).
Capillary temperature: 25
Running buffer: 100 mM NaH$_2$PO$_4$ adjusted to pH 2.5
Injection: Electrokinetic injection at 15 kV for 3 s.
Detector: UV 200, UV 210
Migration time: 2.56
Voltage: 15 kV
Model: Bio-Rad BioFocus 3000

OTHER SUBSTANCES
Simultaneous: amorolfine, brompheniramine, bupivacaine, carteolol, chloroquine, chlorphenoxamine, disopyramide, dobutamine, doxylamine, flecainide, gallopamil, ketamine, mepindolol, orphenadrine, oxybutynin, phenoxybenzamine, pindolol, propafenone, propranolol, sulpiride, talinolol, tropicamide, verapamil

KEY WORDS
coated capillary

REFERENCE
Koppenhoefer,B.; Epperlein,U.; Xiaofeng,Z.; Bingcheng,L. Separation of enantiomers of drugs by capillary electrophoresis. Part 4: Hydroxypropyl-γ-cyclodextrin as chiral solvating agent, *Electrophoresis*, **1997**, *18*, 924–930.

SAMPLE
Matrix: solutions
Sample preparation: Inject an aliquot of a 100 μg/mL solution in running buffer.

CAPILLARY ELECTROPHORESIS
Capillary: 30 cm × 50 μm fused-silica (25.5 cm to detector), coated with polyacrylamide
Capillary preparation: Adjust the pH of 20 mL water to 3.5 with acetic acid, add 80 μL 3-(trimethoxysilyl)propyl methacrylate (3-methacryloxypropyltrimethoxysilane), mix, suck into capillary, let stand at room temperature for 1 h, remove the solution, wash with water. Fill the capillary with a deaerated 3-4% acrylamide solution containing 1 μL/mL N,N,N',N'-tetramethylethylenediamine and 1 mg/mL potassium persulfate, let stand for 30 min, remove excess solution by aspiration, rinse with water, remove water by aspiration, dry at 35° (J. Chromatogr. 1985, 347, 191).
Capillary temperature: 25
Running buffer: 100 mM NaH$_2$PO$_4$ containing 45 mM hydroxypropyl-α- cyclodextrin, adjusted to pH 2.5 with phosphoric acid
Injection: Electrokinetic injection at 15 kV for 3 s.
Detector: UV 200
Voltage: 15 kV
Model: Bio-Rad BioFocus 3000

KEY WORDS
chiral; coated capillary; comparison with the use of other cyclodextrins; this running buffer gave the greatest enantiomeric separation.; α=1.021

REFERENCE
Lin,B.; Zhu,X.; Koppenhoefer,B.; Epperlein,U. Investigation of 123 chiral drugs by cyclodextrin-modified capillary electrophoresis, *LC.GC*, **1997**, *15*, 40–46.

SAMPLE
Matrix: solutions

CAPILLARY ELECTROPHORESIS
Capillary: 35 cm × 50 μm polyacrylamide-coated fused-silica (30.5 cm to detector) (Composite Metal Services, UK)
Capillary preparation: Before each run rinse capillary with water for 70 s and with running buffer for 100 s. (Coat capillary as follows. Adjust the pH of 20 mL water to 3.5 with acetic acid, add 80 μL 3-(trimethoxysilyl)propyl methacrylate (3-methacryloxypropyltrimethoxysi-

lane), mix, suck into capillary, let stand at room temperature for 1 h, remove the solution, wash with water. Fill the capillary with a deaerated 3-4% acrylamide solution containing 1 μL/mL N,N,N',N'-tetramethylethylenediamine and 1 mg/mL potassium persulfate, let stand for 30 min, remove excess solution by aspiration, rinse with water, remove water by aspiration, dry at 35° (J. Chromatogr. 1985, 347, 191).)

Capillary temperature: 25

Running buffer: Buffer containing 30 mM cyanoethylated-β-cyclodextrin (Cyclolab, Budapest) (Prepare buffer by adjusting the pH of 50 mM phosphoric acid containing 50 mM acetic acid and 50 mM boric acid to 2.5 with concentrated NaOH, add the appropriate amount of cyanoethylated-β-cyclodextrin, dilute with an equal volume of water.)

Injection: Pressure injection at 5 psi for 2 s.

Detector: UV 206

Migration time: 7.1 (second enantiomer, $\alpha = 1.018$)

Voltage: 20 kV

Current: 27-40 μA

Model: Bio-Rad Biofocus 3000

KEY WORDS
chiral; coated capillary

REFERENCE
Aturki,Z.; Desiderio,C.; Mannina,L.; Fanali,S. Chiral separations by capillary zone electrophoresis with the use of cyanoethylated-β-cyclodextrin as chiral selector, *J.Chromatogr.A*, **1998**, *817*, 91–104.

SAMPLE
Matrix: solutions

Sample preparation: Inject an aliquot of an aqueous solution.

CAPILLARY ELECTROPHORESIS
Capillary: 27 cm × 50 μm fused-silica (20 cm to detector) (Composite Metal Services, Hallow, UK)

Capillary preparation: Before each injection rinse with 100 mM NaOH for 1 min and with running buffer for 1 min.

Capillary temperature: 25

Running buffer: 190 mM pH 2.6 Citrate buffer (Prepare buffer by dissolving 5.52 g trisodium citrate and 21.09 g citric acid in 100 mL water.)

Injection: For 1 s.

Detector: UV 230

Migration time: 19

Internal standard: chlorpheniramine

Voltage: +6 kV

Current: 90 μA

Model: Beckman P/ACE 5100 or 2050

OTHER SUBSTANCES
Simultaneous: ranitidine

KEY WORDS
chlorpheniramine is IS

REFERENCE
Kelly,M.A.; Altria,K.D.; Grace,C.; Clark,B.J. Optimisation, validation and application of a capillary electrophoresis method for the determination of ranitidine hydrochloride and related substances, *J.Chromatogr.A*, **1998**, *798*, 297–306.

SAMPLE
Matrix: solutions

CAPILLARY ELECTROPHORESIS
Capillary: 29-36 cm × 50 μm fused-silica (24.5-31.5 cm to detector) (Yongnian Optical Conductive Fiber Plant, China) coated with polyacrylamide

Capillary preparation: Coat capillary as follows. Adjust the pH of 20 mL water to 3.5 with acetic acid, add 80 µL 3-(trimethoxysilyl)propyl methacrylate (3-methacryloxypropyltrimethoxysilane), mix, suck into capillary, let stand at room temperature for 1 h, remove the solution, wash with water. Fill the capillary with a deaerated 3-4% acrylamide solution containing 1 µL/mL N,N,N',N'-tetramethylethylenediamine and 1 mg/mL potassium persulfate, let stand for 30 min, remove excess solution by aspiration, rinse with water, remove water by aspiration, dry at 35° (J. Chromatogr. 1985, 347, 191).
Capillary temperature: 25
Running buffer: 100 mM pH 2.5 NaH$_2$PO$_4$ (A) or 100 mM pH 2.5 NaH$_2$PO$_4$ containing 45 mM hydroxypropyl-α-cyclodextrin (Wacker, Munich) (B)
Injection: Electromigration at 15 kV for 3 s.
Detector: UV 200; UV 210
Migration time: 2.56 (A); 7.69 (B) (no separation of enantiomers)
Voltage: 15 kV
Model: Bio-Focus 3000

OTHER SUBSTANCES
Also analyzed: albuterol (salbutamol), alprenolol, amorolfine, atenolol, atropine, azelastine, baclofen, bamethan, benproperine, benserazide, biperiden, bisoprolol, brompheniramine, bupivacaine, bupranolol, butamirate, butethamate, carazolol, carbuterol, carteolol, carvedilol, celiprolol, chloroquine, chlorphenoxamine, cicletanine, clenbuterol, clidinium bromide, clobutinol, dimethindene, dipivefrin, disopyramide, dobutamine, doxylamine, fendiline, flecainide, gallopamil, homatropine, ipratropium bromide, isoproterenol (isoprenaline), isothipendyl, ketamine, meclizine, mefloquine, mepindolol, mequitazine, metaclazepam, metaproterenol (orciprenaline), metipranolol, metoprolol, nafronyl (naftidrofuryl), nefopam, nicardipine, norfenefrine, ofloxacin, ornidazole, orphenadrine, oxomemazine, oxprenolol, oxybutynin, phenoxybenzamine, phenylpropanolamine, pholedrine, pindolol, pirbuterol, prilocaine, procyclidine, promethazine, propafenone, propranolol, reproterol, sotalol, sulpride, synephrine, talinolol, terbutaline, tetrahydrozoline (tetryzoline), theodrenaline, tioconazole, tocainide, trihexyphenidyl, trimeprazine (alimemazine), trimipramine, tropicamide, verapamil, zopiclone

---

KEY WORDS
coated capillary

REFERENCE
Koppenhoefer,B.; Eperlein,U.; Schlunk,R.; Zhu,X.; Lin,B. Separation of enantiomers of drugs by capillary electrophoresis. V. Hydroxypropyl-α-cyclodextrin as chiral solvating agent, *J.Chromatogr.A*, **1998**, *793*, 153–164.

SAMPLE
Matrix: solutions

CAPILLARY ELECTROPHORESIS
Capillary: 44 cm × 50 µm fused-silica
Running buffer: 10 mM pH 4.3 acetic acid/ammonium acetate buffer containing 200 µg/mL carboxymethyl ether-β-cyclodextrin (Wacker Chemie, Munich)
Injection: Hydrostatic injection at 10 cm for 5-10 s.
Detector: MS, Finnigan LCQ, electrospray, positive ion mode, ion-spray tip voltage 2.6 kV, sheath liquid MeOH:water:acetic acid 50:49:1 at 6 µL/min, m/z 276
Migration time: 3.7, 4.4 (enantiomers)
Voltage: 20 kV

KEY WORDS
chiral

REFERENCE
Schulte,G.; Heitmeier,S.; Chankvetadze,B.; Blaschke,G. Chiral capillary electrophoresis-electrospray mass spectrometry coupling with charged cyclodextrin derivatives as chiral selectors, *J.Chromatogr.A*, **1998**, *800*, 77–82.

Chlorphenoxamine

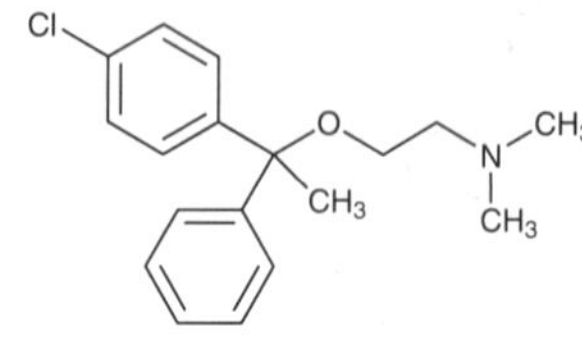

Molecular formula: $C_{18}H_{22}ClNO$
Molecular weight: 303.83
CAS Registry No.: 77-38-3, 562-09-4 (HCl)
Merck Index (12th ed.): 2234
Lednicer: 1 44

SAMPLE
Matrix: solutions
Sample preparation: Inject an aliquot of a 100 µg/mL solution in running buffer.

CAPILLARY ELECTROPHORESIS
Capillary: 30 cm × 50 µm fused-silica (25.5 cm to detector) (Yongnian Optical Conductive Fiber
Plant, China), coated with polyacrylamide
Capillary preparation: No details of the polyacrylamide coating process are provided. However,
another paper (LC.GC 1997, 15, 40) by this group indicates that they use the procedure of
Hjertén, thus: Adjust the pH of 20 mL water to 3.5 with acetic acid, add 80 µL 3-(trimethoxysilyl)propyl methacrylate (3-methacryloxypropyltrimethoxysilane), mix, suck into capillary, let
stand at room temperature for 1 h, remove the solution, wash with water. Fill the capillary
with a deaerated 3-4% acrylamide solution containing 1 µL/mL N,N,N',N'-tetramethylethyl-
enediamine and 1 mg/mL potassium persulfate, let stand for 30 min, remove excess solution
by aspiration, rinse with water, remove water by aspiration, dry at 35° (J. Chromatogr. 1985,
347, 191).
Capillary temperature: 25
Running buffer: 100 mM NaH_2PO_4 adjusted to pH 2.5
Injection: Electrokinetic injection at 15 kV for 3 s.
Detector: UV 200, UV 210
Migration time: 4.47
Voltage: 15 kV
Model: Bio-Rad BioFocus 3000

OTHER SUBSTANCES
Simultaneous: amorolfine, brompheniramine, bupivacaine, carteolol, chloroquine, chlorpheniramine, disopyramide, dobutamine, doxylamine, flecainide, gallopamil, ketamine, mepindolol,
orphenadrine, oxybutynin, phenoxybenzamine, pindolol, propafenone, propranolol, sulpiride,
talinolol, tropicamide, verapamil

KEY WORDS
coated capillary

REFERENCE
Koppenhoefer,B.; Epperlein,U.; Xiaofeng,Z.; Bingcheng,L. Separation of enantiomers of drugs by capillary electrophoresis. Part 4: Hydroxypropyl-γ-cyclodextrin as chiral solvating agent, *Electrophoresis*, **1997**, *18*, 924–
930.

SAMPLE
Matrix: solutions

CAPILLARY ELECTROPHORESIS
Capillary: 29-36 cm × 50 µm fused-silica (24.5-31.5 cm to detector) (Yongnian Optical Conductive
Fiber Plant, China) coated with polyacrylamide
Capillary preparation: Coat capillary as follows. Adjust the pH of 20 mL water to 3.5 with
acetic acid, add 80 µL 3-(trimethoxysilyl)propyl methacrylate (3-methacryloxypropyltrimethox-
ysilane), mix, suck into capillary, let stand at room temperature for 1 h, remove the solution,
wash with water. Fill the capillary with a deaerated 3-4% acrylamide solution containing 1
µL/mL N,N,N',N'-tetramethylethylenediamine and 1 mg/mL potassium persulfate, let stand
for 30 min, remove excess solution by aspiration, rinse with water, remove water by aspiration,
dry at 35° (J. Chromatogr. 1985, 347, 191).

Capillary temperature: 25
Running buffer: 100 mM pH 2.5 NaH$_2$PO$_4$ (A) or 100 mM pH 2.5 NaH$_2$PO$_4$ containing 45 mM hydroxypropyl-α-cyclodextrin (Wacker, Munich) (B)
Injection: Electromigration at 15 kV for 3 s.
Detector: UV 200; UV 210
Migration time: 4.47 (A); 14.93 (B) (no separation of enantiomers)
Voltage: 15 kV
Model: Bio-Focus 3000

OTHER SUBSTANCES
Also analyzed: albuterol (salbutamol), alprenolol, amorolfine, atenolol, atropine, azelastine, baclofen, bamethan, benproperine, benserazide, biperiden, bisoprolol, brompheniramine, bupivacaine, bupranolol, butamirate, butethamate, carazolol, carbuterol, carteolol, carvedilol, celiprolol, chloroquine, chlorpheniramine, cicletanine, clenbuterol, clidinium bromide, clobutinol, dimethindene, dipivefrin, disopyramide, dobutamine, doxylamine, fendiline, flecainide, gallopamil, homatropine, ipratropium bromide, isoproterenol (isoprenaline), isothipendyl, ketamine, meclizine, mefloquine, mepindolol, mequitazine, metaclazepam, metaproterenol (orciprenaline), metipranolol, metoprolol, nafronyl (naftidrofuryl), nefopam, nicardipine, norfenefrine, ofloxacin, ornidazole, orphenadrine, oxomemazine, oxprenolol, oxybutynin, phenoxybenzamine, phenylpropanolamine, pholedrine, pindolol, pirbuterol, prilocaine, procyclidine, promethazine, propafenone, propranolol, reproterol, sotalol, sulpride, synephrine, talinolol, terbutaline, tetrahydrozoline (tetryzoline), theodrenaline, tioconazole, tocainide, trihexyphenidyl, trimeprazine (alimemazine), trimipramine, tropicamide, verapamil, zopiclone

KEY WORDS
coated capillary

REFERENCE
Koppenhoefer,B.; Eperlein,U.; Schlunk,R.; Zhu,X.; Lin,B. Separation of enantiomers of drugs by capillary electrophoresis. V. Hydroxypropyl-α-cyclodextrin as chiral solvating agent, *J.Chromatogr.A*, **1998**, *793*, 153–164.

Chlorpromazine

Molecular formula: C$_{17}$H$_{19}$ClN$_2$S
Molecular weight: 318.87
CAS Registry No.: 50-53-3, 69-09-0 (HCl)
Merck Index (12th ed.): 2238
Lednicer: 1 319

SAMPLE
Matrix: solutions

CAPILLARY ELECTROPHORESIS
Capillary: 57 cm $\times$ 50 μm fused-silica (50 cm to detector)
Running buffer: 50 mM pH 8.0 Borate buffer containing 40 mM sodium taurodeoxycholate and 25 mM phosphatidylcholine (Prepare by adding phosphatidylcholine and stirring for 3-6 h until all cloudiness disappears. Phosphatidylcholine was 95% pure soybean lecithin, Epikuron, Lucas Meyer & Co.)
Injection: Inject a solution of the compound in the running buffer at 20 psi for 1 s, inject MeOH:water 5:95 at 20 psi for 1 s, inject a solution of halofantrine in running buffer at 20 psi for 1 s.
Detector: UV 214
Migration time: k' 144.54
Model: Beckman P/ACE 5000

OTHER SUBSTANCES
Also analyzed: acetaminophen, amoxicillin, antipyrine, aspirin, azathioprine, caffeine, captopril, carbamazepine, carprofen, chlorambucil, chlorpheniramine, cimetidine, clonidine, codeine, desipramine, diphenhydramine, ephedrine, fenoterol, flufenamic acid, flurbiprofen, haloperidol, hydroxyzine, ibuprofen, imipramine, indomethacin, ketoprofen, lidocaine, melphalan, metoprolol, nabumetone, nadolol, phenobarbital, phenol, promazine, propranolol, pyrilamine, ranitidine, ropinirole, salicylic acid, sulfamethoxazole, testosterone, theophylline, thioridazine, tiaprofenic acid, tolfenamic acid, trifluoperazine, trimethoprim, valproic acid, verapamil

KEY WORDS
comparison with HPLC; $k' = (Tr-T0)/(T0(1-Tr/Tm))$ where Tr = retention time of analyte; $T0$ = retention time of water; and Tm = retention time of marker (halofantrine)

REFERENCE
Hanna,M.; de Biasi,V.; Bond,B.; Salter,C.; Hutt,A.J.; Camilleri,P. Estimation of the partitioning characteristics of drugs: A comparison of a large and diverse drug series utilizing chromatographic and electrophoretic methodology, *Anal.Chem.*, **1998**, *70*, 2092–2099.

SAMPLE
Matrix: urine
Sample preparation: 10 mL Urine + 1 mL 5 M NaOH + 10 mL n-hexane, vortex for 1 min, centrifuge at 0° at 3000 g for 5 min. Remove the organic layer and add it to 50 μL glacial acetic acid, evaporate to dryness under a stream of nitrogen at 30°, reconstitute the residue in 50 μL 50 mM sodium taurodeoxycholate, filter (0.2 μm PTFE), inject an aliquot.

CAPILLARY ELECTROPHORESIS
Capillary: 100 cm × 50 μm fused-silica (50 cm to detector) (ISCO)
Capillary preparation: Rinse capillary with 20 μL running buffer between injections. Condition a new capillary by filling with 1 M NaOH, let stand for 1 h, fill with 100 mM NaOH, let stand for 1 h, rinse with buffer. Every 20 injections rinse capillary with 200 μL 1 M NaOH, 200 μL water, and 200 μL running buffer, fill with running buffer.
Capillary temperature: 22
Running buffer: 40 mM pH 9.5 Borate buffer containing 10 mM sodium taurodeoxycholate
Injection: Load under vacuum at 7.5 kPa/s.
Detector: UV 240
Migration time: 11
Voltage: 30 kV
Model: ISCO Model 3140 electropherograph
Limit of detection: 6 ng/mL

OTHER SUBSTANCES
Extracted: acepromazine, amiodarone, amitriptyline, azaperone, cianopramine, clomipramine, clozapine, desethylamiodarone, desipramine, diclofensine, dothiepin, doxepin, imipramine, isocarboxazid, moclobemide, perphenazine, phenothiazine, pimozide, prochlorperazine, promazine, thioridazine, thiothixene, trifluoperazine, trimipramine

KEY WORDS
human; cow; pig; horse; protect from light

REFERENCE
Aumatell,A.; Wells,R.J. Determination of a cardiac antiarrhythmic, tricyclic antipsychotics and antidepressants in human and animal urine by micellar electrokinetic capillary chromatography using a bile salt, *J.Chromatogr.B*, **1995**, *669*, 331–344.

SAMPLE
Matrix: urine
Sample preparation: Dilute 10-fold with water, inject an aliquot.

CAPILLARY ELECTROPHORESIS
Capillary: 70 cm × 50 μm fused-silica (65.4 cm to detector) (Polymicro Technologies)
Capillary temperature: 25

Running buffer: 20 mM pH 5.0 Tris-acetic acid containing 50 µg/mL FC-135 (Fluorad/3M) and 10 mM cetyltrimethylammonium bromide
Injection: Pressure injection at 2 psi.
Detector: UV 240
Migration time: 12.02
Voltage: 20 kV
Model: Bio-Rad BioFocus 3000

OTHER SUBSTANCES
Extracted: acepromazine, ethopropazine, fluphenazine, methotrimeprazine, perphenazine, thioridazine, triflupromazine, trimeprazine

KEY WORDS
detector at anode

REFERENCE
Muijelsaar,P.G.H.M.; Claessens,H.A.; Cramers,C.A. Determination of structurally related phenothiazines by capillary zone electrophoresis and micellar electrokinetic chromatography, *J.Chromatogr.A*, **1996**, *735*, 395–402.

Chlorpropamide

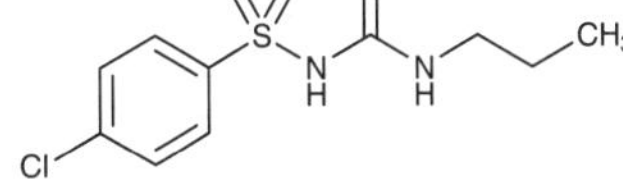

Molecular formula: C$_{10}$H$_{13}$ClN$_2$O$_3$S
Molecular weight: 276.74
CAS Registry No.: 94-20-2
Merck Index (12th ed.): 2239
Lednicer: 1 137

SAMPLE
Matrix: solutions
Sample preparation: Inject a dilute aqueous solution on to the tip of the capillary containing the SPE material at 138 kPa for 12 s, wash for 1 min with running buffer, desorb with elution buffer for 20 s (60 nL), inject running buffer at 3.45 kPa, electrophorese. (Elution buffer was MeCN:50 mM pH 2.5 phosphate buffer 80:20.)

CAPILLARY ELECTROPHORESIS
Capillary: 67 cm × 75 µm with a 0.5 mm long × 0.37 mm diameter column of C18 material (from a J.T. Baker SPE cartridge) at the injection end
Capillary preparation: Before each run rinse capillary with 3 column volumes of elution buffer and 10 column volumes of running buffer. After each run rinse with 500 µL elution buffer.
Capillary temperature: 20
Running buffer: 250 mM pH 8.4 Borate buffer containing 5 mM phosphate
Detector: UV 200
Migration time: 18.5
Voltage: 15 kV
Current: 70 µA
Model: Beckman P/ACE System 5510
Limit of detection: 5 ng/mL

OTHER SUBSTANCES
Simultaneous: acetohexamide, glipizide, glyburide, tolazamide, tolbutamide

KEY WORDS
SPE

REFERENCE
Strausbauch,M.A.; Xu,S.J.; Ferguson,J.E.; Nunez,M.E.; Machacek,D.; Lawson,G.M.; Wettstein,P.J.; Landers,J.P. Concentration and separation of hypoglycemic drugs using solid-phase extraction-capillary electrophoresis, *J.Chromatogr.A*, **1995**, *717*, 279–291.

SAMPLE
Matrix: solutions
Sample preparation: Inject an aliquot of a solution in MeOH.

CAPILLARY ELECTROPHORESIS
Capillary: 47 cm × 50 μm fused-silica (40 cm to detector)
Capillary preparation: Before each run rinse at 20 psi with 3 column volumes of 100 mM NaOH and 3 column volumes of running buffer. Condition new capillaries by rinsing with 20 column volumes of 1 M NaOH, water, and running buffer
Capillary temperature: 22
Running buffer: 5 mM pH 8.5 Borate buffer containing 5 mM phosphate and 75 mM sodium cholate (Prepare from 500 mM borate buffer (prepared by mixing 125 mM sodium tetraborate and 500 mM boric acid to achieve pH 8.5) and 500 mM Na_2HPO_4 solution with appropriate dilution and the addition of sodium cholate.)
Injection: Pressure injection of sample at 0.5 psi for 1-5 s followed by a pressure injection of running buffer for 1 s.
Detector: UV 200
Migration time: 4.1
Internal standard: N-acetyl-4-(2,3-dichlorophenylureido)benzenesulfonamide (5.5)
Voltage: 25 kV
Current: 61 μA
Model: Beckman P/ACE 5510

OTHER SUBSTANCES
Simultaneous: acetohexamide, glipizide, glyburide, tolazamide, tolbutamide

REFERENCE
Roche,M.E.; Oda,R.P.; Lawson,G.M.; Landers,J.P. Capillary electrophoretic detection of metabolites in the urine of patients receiving hypoglycemic drug therapy, *Electrophoresis*, **1997**, *18*, 1865–1874.

SAMPLE
Matrix: urine
Sample preparation: Condition a 200 μg Bond Elut C18 SPE cartridge with 1 mL MeCN. 10 mL Urine + 100 μL 10 μg/mL IS, adjust pH to 2.0 with concentrated HCl, add 15 mL dichloromethane, shake vigorously for 2 min, let stand for 5 min. Remove the organic layer and evaporate it to dryness under a stream of nitrogen at 45°, reconstitute the residue in 400 μL MeOH:water 50:50, add to the SPE cartridge, wash with 1.25 mL MeCN:water 10:90, wash with 1 mL MeCN:water 40:60, elute with 1 mL MeCN. Evaporate the eluate to dryness under a stream of nitrogen, reconstitute the residue in 20 μL MeOH, inject an aliquot.

CAPILLARY ELECTROPHORESIS
Capillary: 47 cm × 50 μm fused-silica (40 cm to detector) (Polymicro Technologies)
Capillary preparation: Before each run wash with 3 column volumes of 100 mM NaOH, dip momentarily into water, rinse with 10 column volumes of 50 mM pH 8.5 Sodium borate buffer containing 50 mM Na_2HPO_4 and 75 mM sodium cholate, rinse with 3 column volumes of running buffer. Condition a new capillary by rinsing with 20 column volumes of 100 mM NaOH, 20 column volumes of water, and 20 column volumes of running buffer.
Capillary temperature: 25
Running buffer: 5 mM pH 8.5 Sodium borate buffer containing 5 mM Na_2HPO_4 and 75 mM sodium cholate
Injection: Pressure injection at 0.5 psi for 2 s then a 1 s injection of running buffer.
Detector: UV 200
Migration time: 5.5
Internal standard: N-acetyl-4-(2,3-dichlorophenylureido)benzenesulfonamide (7.4)
Voltage: 25 kV
Current: 61 μA
Model: Beckman P/ACE System 2100 or System 5510
Limit of detection: 50 ng/mL

OTHER SUBSTANCES
Extracted: acetohexamide, glipizide, glyburide, tolazamide, tolbutamide

KEY WORDS
SPE

REFERENCE
Núñez,M.; Ferguson,J.E.; Machacek,D.; Jacob,G.; Oda,R.P.; Lawson,G.M.; Landers,J.P. Detection of hypoglycemic drugs in human urine using micellar electrokinetic chromatography, *Anal.Chem.*, **1995**, *67*, 3668–
3675.

SAMPLE
Matrix: urine
Sample preparation: Inject an aliquot directly.

CAPILLARY ELECTROPHORESIS
Capillary: 37 cm × 50 μm fused-silica (30 cm to detector) (Polymicro Technologies)
Capillary preparation: Every 13 injections rinse capillary with 100 mM NaOH for 1 min and
with running buffer for 1 min. Condition a new capillary with 20 column volumes of 100 mM
NaOH, 20 column volumes of water, and 20 column volumes of running buffer.
Capillary temperature: 22
Running buffer: 5 mM pH 8.5 Borate buffer containing 5 mM phosphate and 75 mM sodium
cholate
Injection: Pressure injection at 0.5 psi for 2 s.
Detector: UV 200
Migration time: 2.4
Internal standard: N-acetyl-4-(2,3-dichlorophenylureido)benzenesulfonamide (3.2)
Voltage: 28 kV
Current: 71 μA
Model: Beckman P/ACE 5510

OTHER SUBSTANCES
Extracted: acetohexamide, glipizide, glyburide, tolazamide, tolbutamide

KEY WORDS
direct injection

REFERENCE
Roche,M.E.; Oda,R.P.; Machacek,D.; Lawson,G.M.; Landers,J.P. Enhanced throughput with capillary electrophoresis via continuous-sequential sample injection, *Anal.Chem.*, **1997**, *69*, 99–104.

Chlorprothixene

Molecular formula: C$_{18}$H$_{18}$ClNS
Molecular weight: 315.87
CAS Registry No.: 113-59-7
Merck Index (12th ed.): 2241
Lednicer: 1 399

SAMPLE
Matrix: solutions
Sample preparation: Inject an aliquot of a 100 μg/mL solution in MeOH:MeCN 50:50.

CAPILLARY ELECTROPHORESIS
Capillary: 64 cm × 50 μm fused-silica (55.5 cm to detector) (Polymicro Technologies)
Capillary preparation: Between runs flush capillary with running buffer for 2 min. Before use
rinse capillary with 1 M NaOH for 1 h, with 100 mM NaOH for 20 min, and with running
buffer for 10 min.
Capillary temperature: 25
Running buffer: MeCN:MeOH 50:50 containing 50 mM ammonium acetate and 1 M acetic acid

Injection: Pressure injection at 5 kPa for 3 s.
Detector: UV 214
Voltage: 30 kV
Model: Hewlett-Packard ³ᴰCE

KEY WORDS
cis and trans isomers separated; Rs = 3.09

REFERENCE
Hansen,S.H.; Bjornsdottir,I.; Tjornelund,J. Separation of cationic *cis--trans (Z--E)* isomers and diastereoisomers using non-aqueous capillary electrophoresis, *J.Chromatogr.A*, **1997**, *792*, 49–55.

Chlorpyrifos

Molecular formula: $C_9H_{11}Cl_3NO_3PS$
Molecular weight: 350.59
CAS Registry No.: 2921-88-2
Merck Index (12th ed.): 2242

SAMPLE
Matrix: solutions
Sample preparation: Inject an aliquot of a solution in running buffer.

CAPILLARY ELECTROPHORESIS
Capillary: 60 cm × 50 μm fused-silica (53 cm to detector) (Polymicro Technologies)
Capillary preparation: Condition capillary by flushing with 100 mM NaOH for 20 min, water for 10 min, and running buffer for 10 min then equilibrate under voltage.
Running buffer: 10 mM pH 7.0 Sodium phosphate containing 60 mM sodium cholate
Detector: UV 210
Migration time: 13
Voltage: 18 kV
Model: laboratory-constructed

OTHER SUBSTANCES
Simultaneous: acephate, aldicarb, aminocarb, banol, benfluralin, bensulide, buprofezin, captan, carbaryl, carbendazim, carbofuran, chlorotoluron, coumaphos, dichlorvos, dimethoate, EPN, fenchlorphos, fenobucarb, fensulfothion, fenthion, iprodione, isofenphos, isoprocarb, malathion, mepronil, methiocarb, napropamide, oxyfluorfen, paclobutrazol, parathion, pirimicarb, profenophos, propoxur, propyzamide, pyriodaphenthion, simazine, simetryn, thiram, tolclofos-methyl, triadimefon, tricyclazole
Interfering: pendimethalin

REFERENCE
Wu,Y.S.; Lee,H.K.; Li,S.F.Y. Rapid estimation of octanol-water partition coefficients of pesticides by micellar electrokinetic chromatography, *Electrophoresis*, **1998**, *19*, 1719–1727.

Chlortetracycline

Molecular formula: $C_{22}H_{23}ClN_2O_8$
Molecular weight: 478.89
CAS Registry No.: 57-62-5, 64-72-2 (HCl)
Merck Index (12th ed.): 2245
Lednicer: 1 212

SAMPLE
Matrix: blood, milk
Sample preparation: Condition a 3 mL 15 mg SPEC MP1 (Ansys, Irvine CA) SPE cartridge with 1 mL MeOH and 1 mL water. Mix 1 volume milk or plasma with 0.2 volumes 10% trichloroacetic acid in water, centrifuge at 1000 g for 4 min. Add 5 mL supernatant to the SPE cartridge at 2 mL/min, wash with 500 μL water, dry under 40 kPa vacuum for 5 min (more if necessary), elute with 200 μL 50 mM magnesium acetate in N-methylformamide, inject an aliquot of the eluate.

CAPILLARY ELECTROPHORESIS
Capillary: 27 cm × 75 μm fused-silica (20 cm to the detector) (Polymicro Technologies)
Capillary preparation: Flush with running buffer for 30 s before each run. Replace running buffer after each 6 runs. Before long-term storage flush with 100 mM HCl, 100 mM NaOH, water, MeOH, and air. Rinse new capillaries with 1 M NaOH for 1 h, with 100 mM NaOH for 20 min, with water for 20 min, with MeOH for 20 min, and with running buffer for 15 min.
Capillary temperature: 20
Running buffer: 500 mM Magnesium acetate tetrahydrate in N-methylformamide
Injection: Pressure injection at 3.5 kPa for 3 s.
Detector: UV 280, F ex 325 (He-Cd laser) em 514
Migration time: 14
Voltage: 15 kV
Model: Beckman P/ACE 5010
Limit of detection: 25 ng/mL

OTHER SUBSTANCES
Extracted: oxytetracycline, tetracycline

KEY WORDS
plasma; SPE

REFERENCE
Tjornelund,J.; Hansen,S.H. Use of metal complexation in non-aqueous capillary electrophoresis systems for the separation and improved detection of tetracyclines, *J.Chromatogr.A*, **1997**, *779*, 235–243.

SAMPLE
Matrix: blood, milk, urine
Sample preparation: Prepare a column as follows. Swirl Chelating Sepharose Fast Flow resin (Pharmacia) in its bottle, add it to a polypropylene column to give a bed volume of 1.0-1.2 mL, wash 3 times with 2 mL portions of water, wash with 2.5 mL 10 mM copper sulfate, wash with 2 mL water. Condition a PrepSep-C18 SPE cartridge (Fisher Scientific) by allowing 3 mL 4% dimethyldichlorosilane in toluene to drain through it by gravity, after 30 min rinse with 10 mL EtOH, immediately before use wash with 10 mL water. Milk. Centrifuge 10 mL milk at 4° at 2000 g for 20 min, remove the lower layer and add it to 20 mL succinate buffer, centrifuge at 4° at 2000 g for 30 min, pass the supernatant through a Bond Elut SPE cartridge, wash cartridge with 2 mL succinate buffer, collect all the eluate and add it to the column. Wash with 3 mL succinate buffer, wash with 3 mL water, wash with 3 mL MeOH, wash with 3 mL water, elute with 3 mL pH 7 buffer, when the blue band reaches the bottom of the column collect the eluate. Add the eluate to the SPE cartridge, wash with 10 mL water, dry with air, elute with 6 mL EtOH. Evaporate the eluate to dryness under a stream of nitrogen at room temperature, reconstitute with 20 μL running buffer adjusted to pH 7.0 with 5 M HCl, inject an aliquot. Serum. 12 mL Serum + 12 mL PBS, filter (Amicon Centriprep-10) centrifuge at 10° at 2000 g for 20 min, remove the filtrate, centrifuge an additional 4 times, combine the filtrates, add a 20 mL aliquot to the column. Wash with 3 mL succinate buffer, wash with 3 mL water, wash with 3 mL MeOH, wash with 3 mL water, elute with 3 mL pH 7 buffer, when the blue band reaches the bottom of the column collect the eluate. Add the eluate to the SPE cartridge, wash with 10 mL water, dry with air, elute with 6 mL EtOH. Evaporate the eluate to dryness under a stream of nitrogen at room temperature, reconstitute with 20 μL running buffer adjusted to pH 7.0 with 5 M HCl, inject an aliquot. Urine. Centrifuge 10 mL urine at 4° at 2000 g for 15 min, add the supernatant to the column. Wash with 3 mL succinate buffer, wash with 3 mL water, wash with 3 mL MeOH, wash with 3 mL water, elute with 3 mL pH 7 buffer, when the blue band reaches the bottom of the column collect the eluate. Add the eluate to the SPE cartridge, wash with 10 mL water, dry with air, elute with 6 mL EtOH. Evaporate the eluate to dryness under a stream of nitrogen at room temperature, reconstitute with 20 μL running

buffer adjusted to pH 7.0 with 5 M HCl, inject an aliquot. (Prepare succinate buffer by dissolving 11.8 g succinic acid in 980 mL water, adjust pH to 4.0 with 10 M NaOH, make up to 1 L. Prepare PBS by dissolving 20 mG KCl, 8 g NaCl, 1.44 g Na_2HPO_4, and 24 mg KH_2PO_4 in 980 mL water, adjust pH to 7.0 with 1 M HCl, make up to 1 L. Prepare the pH 7 buffer by dissolving 2.31 g citric acid, 7.44 g disodium EDTA dihydrate, and 5.84 g NaCl in 190 mL water, adjust pH to 7.0 with 5 M NaOH, make up to 200 mL.)

CAPILLARY ELECTROPHORESIS

Capillary: 57 cm × 75 μm (50 cm to detector) (Beckman)
Capillary preparation: Before injection rinse with running buffer for 2 min. After each run rinse with NaOH solution for 2 min and with water for 2 min.
Capillary temperature: 23
Running buffer: 50 mM Borate buffer containing 50 mM phosphate and 10 mM sodium dodecyl sulfate (Prepare buffer by dissolving 2.884 g sodium dodecyl sulfate, 19.07 g sodium borate decahydrate, 4.096 g Na_2HPO_4, and 2.537 g NaH_2PO_4 in 980 mL water, adjust pH to 8.5 with 5 M HCl, make up to 1 L.)
Injection: Pressure injection for 5 s (5 nL)
Detector: UV 370
Migration time: 13
Voltage: 15 kV
Model: Beckman P/ACE 5510
Limit of quantitation: 4.2 ppm (milk), 3.1 ppm (serum), 4.7 ppm (urine)
Limit of detection: 2.5 ppm (milk), 2.3 ppm (serum), 2.9 ppm (urine)

OTHER SUBSTANCES
Extracted: doxycycline, oxytetracycline, tetracycline

KEY WORDS
serum; cow; SPE

REFERENCE
Chen,C.-L.; Gu,X. Determination of tetracycline residues in bovine milk, serum, and urine by capillary electrophoresis, *J.AOAC Int.*, **1995**, *78*, 1369–1377.

SAMPLE
Matrix: formulations
Sample preparation: Dissolve capsule contents in 15 mM pH 7.5 sodium phosphate buffer so as to make a 25 mM solution, centrifuge, dilute an aliquot of the supernatant to a concentration of 5 mM with 15 mM pH 7.5 sodium phosphate buffer, inject an aliquot.

CAPILLARY ELECTROPHORESIS
Capillary: 111.9 cm × 75 μm fused-silica (43.4 cm to detector) (Polymicro Technologies)
Capillary preparation: Condition by rinsing with 1 M NaOH for 10 min then running buffer overnight.
Capillary temperature: 25
Running buffer: pH 7.5 Phosphate buffer (buffer concentration 4.3 mM, total sodium concentration 15 mM, ionic strength 18.2 mM)
Injection: Hydrodynamic injection at 2 cm for 1 min
Detector: UV 260
Migration time: 5.5
Current: 20 μA
Limit of detection: 10 μM (S/N 3)

OTHER SUBSTANCES
Simultaneous: demeclocycline, doxycycline, methacycline, minocycline, oxytetracycline, tetracycline

KEY WORDS
capsules

REFERENCE
Tavares,M.F.M.; McGuffin,V.L. Separation and characterization of tetracycline antibiotics by capillary electrophoresis, *J.Chromatogr.A*, **1994**, *686*, 129–142.

SAMPLE
Matrix: solutions
Sample preparation: Inject an aliquot of a 20 µg/mL solution in water.

CAPILLARY ELECTROPHORESIS
Capillary: 20 cm × 25 µm fused-silica (Bio-Rad)
Running buffer: 200 mM pH 2.2 Sodium phosphate buffer containing 0.06% Triton X-100
Injection: Injection by electromigration at 12 kV for 10 s
Detector: UV 265
Migration time: 20
Voltage: 12 kV
Model: Bio-Rad HPE 100

OTHER SUBSTANCES
Simultaneous: oxytetracycline, quatrimycin, tetracycline

REFERENCE
Croubels,S.; Baeyens,W.; Dewaele,C.; Van Petegham,C. Capillary electrophoresis of some tetracycline antibiotics, *J.Chromatogr.A*, **1994**, *673*, 267–274.

SAMPLE
Matrix: solutions

CAPILLARY ELECTROPHORESIS
Capillary: 50 cm × 50 µm (50 cm to detector) (Polymicro Technologies)
Capillary preparation: Purge capillary with 10 mM NaOH for 2 min after each injection.
Running buffer: MeOH:buffer 40:60 (Buffer was 30 mM pH 3.0 citric acid containing 24.5 mM β-alanine.)
Injection: Electrokinetic injection at 10 kV for 5 s.
Detector: UV 254
Migration time: 6.8
Voltage: 30 kV
Model: Perkin Elmer/Applied Biosystems Model 270A-HT

OTHER SUBSTANCES
Simultaneous: doxycycline, meclocycline, methacycline, minocycline, oxytetracycline, tetracycline

REFERENCE
Pesek,J.J.; Matyska,M.T. Separation of tetracyclines by high-performance capillary electrophoresis and capillary electrochromatography, *J.Chromatogr.A*, **1996**, *736*, 313–320.

SAMPLE
Matrix: solutions
Sample preparation: Inject an aliquot of a solution in MeOH.

CAPILLARY ELECTROPHORESIS
Capillary: 64 cm × 50 µm fused-silica (55.5 cm to detector) (Polymicro Technologies)
Capillary preparation: Flush capillary with running buffer for 2 min between runs. Before use rinse capillaries with 1 M NaOH for 1 h, with 100 mM NaOH for 20 min, with water for 20 min, with MeOH for 20 min, and with running buffer for 15 min.
Capillary temperature: 25
Running buffer: MeCN:MeOH:DMF 49:45:6 containing 25 mM ammonium acetate, 10 mM citric acid, and 118 mM methanesulfonic acid
Injection: Pressure injection of sample at 50 mbar for 1 s followed by pressure injection of running buffer at 50 mbar for 3 s.
Detector: UV 254
Migration time: 9.7
Voltage: 25 kV
Model: Hewlett-Packard HP[3D]

OTHER SUBSTANCES
Simultaneous: doxycycline, oxytetracycline, tetracycline

REFERENCE
Tjornelund,J.; Hansen,S.H. Determination of impurities in tetracycline hydrochloride by non-aqueous capillary electrophoresis, *J.Chromatogr.A*, **1996**, *737*, 291–300.

SAMPLE
Matrix: solutions

CAPILLARY ELECTROPHORESIS
Capillary: 43 cm $\times$ 75 μm (36 cm to the detector)
Capillary preparation: Before each run wash capillary with running buffer for 2 min. After each run wash capillary with water at 25° for 5 min, with 100 mM NaOH at 60 ° for 5 min, and with water at 25° for 5 min.
Capillary temperature: 25
Running buffer: 15 mM pH 6.5 Ammonium acetate buffer containing 20 mM sodium dodecyl sulfate and 0.135% Brij 35
Detector: UV 265
Migration time: 5.4
Voltage: 15 kV

OTHER SUBSTANCES
Simultaneous: demeclocycline, doxycycline, minocycline, oxytetracycline, tetracycline

REFERENCE
Chen,Y.-C.; Lin,C.-E. Migration behavior and separation of tetracycline antibiotics by micellar electrokinetic chromatography, *J.Chromatogr.A*, **1998**, *802*, 95–105.

Chlorthalidone

Molecular formula: $C_{14}H_{11}CIN_2O_4S$
Molecular weight: 338.77
CAS Registry No.: 77-36-1
Merck Index (12th ed.): 2246
Lednicer: 1 322

SAMPLE
Matrix: blood, urine
Sample preparation: Condition a 3 mL Supelclean LC-18 SPE cartridge with 3 mL MeOH and 3 mL water. Dilute urine 1:10 with water. Precipitate proteins from serum with MeOH. Add diluted urine or protein supernatant to the SPE cartridge, wash with 3 mL water, elute with 3 mL MeOH, reconstitute to the original volume with 100 mM KOH, inject an aliquot.

CAPILLARY ELECTROPHORESIS
Capillary: 67 cm $\times$ 50 μm fused-silica (60 cm to detector) (Polymicro Technologies)
Capillary preparation: Rinse with running buffer for 2 min before run. If necessary, regenerate capillary with 100 mM NaOH for 10 min and with water for 15 min.
Capillary temperature: 20
Running buffer: 60 mM 3-(cyclohexylamino)-1-propanesulfonic acid (CAPS) adjusted to pH 10.6 with 100 mM KOH
Injection: Pressure injection for 5 s
Detector: UV 220
Migration time: 7.6
Voltage: 25 kV
Model: Beckman P/ACE 2000

OTHER SUBSTANCES
Extracted: acetazolamide, amiloride, bendroflumethiazide, benzthiazide, bumetanide, caffeine, chlorothiazide, clopamide, dichlorphenamide, ethacrynic acid, furosemide, hydrochlorothiazide, metyrapone, probenecid, triamterene, trichlormethiazide

KEY WORDS
serum; SPE

REFERENCE
Jumppanen,J.; Sirén,H.; Riekkola,M.-L. Screening for diuretics in urine and blood serum by capillary zone electrophoresis, *J.Chromatogr.A*, **1993**, *652*, 441–450.

SAMPLE
Matrix: solutions

CAPILLARY ELECTROPHORESIS
Capillary: 60 cm × 75 μm fused-silica (51 cm to detector) (Supelco)
Capillary preparation: At the end of each day wash capillary with 200 mM NaOH for 10 min and with water for 10 min. Condition new capillaries by washing with 200 mM NaOH for 10 min, with water for 10 min, and with running buffer for 10 min.
Running buffer: 50 mM pH 7.3 Sodium borate buffer containing 3.3% succinyl β-cyclodextrin (Wacker Chemie, Munich, Germany)
Injection: Hydrodynamic injection at 25 mbar for 6 s.
Detector: UV 208
Migration time: 40.67, 61.17 (enantiomers)
Voltage: 15 kV
Model: Prince

OTHER SUBSTANCES
Simultaneous: alprenolol, ephedrine, etilefrin, homatropine, methoxamine, norephedrine, octopamine, propranolol, synephrine, trihexyphenidyl

KEY WORDS
chiral

REFERENCE
Schmid,M.G.; Wirnsberger,K.; Gübitz,G. Chiral separation of drug enantiomers by capillary electrophoresis using succinyl-β-cyclodextrin, *Pharmazie*, **1996**, *51*, 852–854

SAMPLE
Matrix: solutions
Sample preparation: Inject an aliquot of a 200 μg/mL solution in MeCN:water 50:50.

CAPILLARY ELECTROPHORESIS
Capillary: 23 cm × 50 μm 3 μm CEC Hypersil C18
Capillary temperature: 15
Running buffer: Gradient. MeCN:50 mM pH 2.5 Na_2HPO_4 buffer:water 40:20:40 for 6.5 min, 60:20:20 for 10.75 min (step gradient), re-equilibrate at initial conditions for 7.75 min.
Injection: Electrokinetic injection at 5 kV for 15 s.
Detector: UV 210
Migration time: 11.5
Voltage: 30 kV (with 8 bar of pressure at each end of capillary)
Model: Hewlett-Packard HP[3D]

OTHER SUBSTANCES
Simultaneous: bendroflumethiazide, bumetanide, chlorothiazide, hydrochlorothiazide, hydroflumethiazide

KEY WORDS
electrochromatography

REFERENCE
Euerby,M.R.; Gilligan,D.; Johnson,C.M.; Bartle,K.D. Step-gradient capillary electrochromatography, *Analyst*, **1997**, *122*, 1087–1088.

SAMPLE
Matrix: solutions

CAPILLARY ELECTROPHORESIS
Capillary: 72 cm × 50 μm coated fused-silica (60 cm to detector) (Ziemer, Mannheim, Germany)
Capillary preparation: Coat capillary as follows. Treat a 1 m length of capillary with 2 mL 1
    M NaOH for 1 h, with 3 mL water for 2 h, and with 2 mL MeOH for 1 h. Push a solution of
    80 μL 3-methacryloyloxypropyltrimethoxysilane in 20 mL water (adjusted to pH 3-5 with acetic
    acid) through with nitrogen at 300 mbar at room temperature for 2 h, rinse with water, fill
    with degassed 4% acrylamide containing 1 μL/mL tetramethylethylenediamine and 1 mg/mL
    sodium persulfate, let stand for 45 min, wash with water, purge with nitrogen, dry at 50°
    overnight.)
Capillary temperature: 25 ± 1
Running buffer: 20 mM pH 7 Phosphate/borate buffer containing 25 mM 6-O-(2-hydroxy-3-tri-
    methylammoniopropyl)-β-cyclodextrin (Preparation of 6-O-(2-hydroxy-3-trimethylammoniopro-
    pyl)-β-cyclodextrin is as follows. Dissolve 3.5 g β-cyclodextrin in 20 mL 10.5 M NaOH, cool to
    0°, add a solution of 6 g glycidyltrimethylammonium chloride (2,3-epoxypropyltrimethylam-
    monium chloride) in 15 mL water dropwise, stir overnight, neutralize (pH meter) with concen-
    trated HCl, evaporate, dissolve the residue in MeOH, filter. Concentrate the filtrate, dissolve
    the white precipitate in water, dialyze 3 times against 1.5 L water to obtain 6-O-(2-hydroxy-3-
    trimethylammoniopropyl)-β-cyclodextrin as a colorless solid (degree of substitution = 0.23).)
Injection: Pressure injection at 40 mbar for 3 s.
Detector: UV 220
Migration time: 72, 78 (enantiomers)
Voltage: 10 kV
Model: Prince (Bischoff, Leonberg, Germany)

KEY WORDS
chiral; coated capillary

REFERENCE
Jakubetz,H.; Juza,M.; Schurig,V. Electrokinetic chromatography employing an anionic and a cationic β-cyclo-
    dextrin derivative, *Electrophoresis*, **1997**, *18*, 897–904.

SAMPLE
Matrix: solutions
Sample preparation: Inject an aliquot of a 500 μM solution in MeCN:water 10:90.

CAPILLARY ELECTROPHORESIS
Capillary: 38.5 cm × 50 μm fused-silica (30 cm to detector) (Supelco)
Capillary temperature: 25
Running buffer: 39.1 mM pH 2.3 Phosphate containing 18 mM ammediol (2-amino-2-methyl-
    1,3-propanediol) and 5 mM mono(6-amino-6-deoxy)-β-cyclodextrin (ionic strength 24 mM)
Injection: Hydrodynamic injection at 25 mbar for 4 s.
Detector: UV 200
Migration time: 13, 14.5 (enantiomers)
Voltage: 20 kV
Model: Hewlett-Packard HP [3D]CE

KEY WORDS
chiral

REFERENCE
Lelièvre,F.; Gareil,P.; Jardy,A. Selectivity in capillary electrophoresis: Application to chiral separations with
    cyclodextrins, *Anal.Chem.*, **1997**, *69*, 385–392.

SAMPLE
Matrix: solutions

Sample preparation: Inject an aliquot of a 20 μg/mL solution

CAPILLARY ELECTROPHORESIS
Capillary: 44 cm × 50 μm fused-silica (37 cm to detector) (Supelco)
Capillary preparation: Before each run rinse capillary with running buffer for 3 min. At the start of each day rinse capillary with running buffer for 10 min. Treat new capillaries with 1 M NaOH, 100 mM NaOH, water, and running buffer.
Capillary temperature: 25
Running buffer: 100 mM Phosphoric acid adjusted to pH 3 with triethanolamine containing 10 mM carboxymethyl-β-cyclodextrin (Cyclolab, Budapest) and 10 mM heptakis(2,3,6-tri-O-methyl)-β-cyclodextrin (Sigma)
Injection: Hydrodynamic injection for 5 s (13.3 nL).
Detector: UV 210
Migration time: 12, 13 (enantiomers)
Voltage: -25 kV
Model: Spectraphoresis 1000 CE

KEY WORDS
detector at anode; chiral; R_s = 2.4

REFERENCE
Fillet,M.; Fotsing,L.; Crommen,J. Enantioseparation of uncharged compounds by capillary electrophoresis using mixtures of anionic and neutral β-cyclodextrin derivatives, *J.Chromatogr.A*, **1998**, *817*, 113–119.

SAMPLE
Matrix: urine
Sample preparation: Filter (0.2 μm), inject an aliquot of the filtrate.

CAPILLARY ELECTROPHORESIS
Capillary: 80 cm × 50 μm fused-silica (57 cm to detector)
Capillary preparation: Before each run rinse the capillary with 1 M NaOH for 3 min, with 100 mM NaOH for 3 min, with water for 3 min, and with running buffer for 10 min.
Running buffer: 110 mM Boric acid containing 56 mM NaOH and 44 mM HCl, pH 8
Injection: Vacuum injection for 1 s
Detector: UV 220
Migration time: 6.017
Voltage: 20 kV
Model: Europhor Prime Vision system IV

OTHER SUBSTANCES
Extracted: acebutolol (UV 238), acetazolamide (UV 222), alprenolol (UV 220), amiloride (UV 220), atenolol (UV 228), bendroflumethiazide (UV 220), bumetanide (UV 220), cocaine (UV 236), codeine (UV 220), ethacrynic acid (UV 220), furosemide (UV 232), hydrochlorothiazide (UV 226), methadone (UV 220), metoxiphenamine (UV 220), nadolol (UV 220), norcodeine (UV 220), oxprenolol (UV 220), pentazocine (UV 220), propranolol (UV 220), spironolactone (UV 244), triamterene (UV 232), xipamide (UV 234)

REFERENCE
Gonzalez,E.; Laserna,J.J. Capillary zone electrophoresis for the rapid screening of banned drugs in sport, *Electrophoresis*, **1994**, *15*, 240–243.

Cholic acid

Molecular formula: $C_{24}H_{40}O_5$
Molecular weight: 408.58
CAS Registry No.: 81-25-4
Merck Index (12th ed.): 2258

SAMPLE
Matrix: blood
Sample preparation: Filter (Amicon Centrifree YM) while centrifuging at 2000 rpm for 45 min, inject an aliquot of the ultrafiltrate.

CAPILLARY ELECTROPHORESIS
Capillary: fused-silica (Polymicro Technologies)
Capillary preparation: Before each run flush capillary with 100 mM NaOH for 3 min, with water for 3 min, and with running buffer for 3 min. At the start of each day flush with 1 M NaOH for 30 min, with water for 3 min, and with running buffer for 3 min. Flush new capillaries with 1 M NaOH for 1 h and with water for 10 min.
Capillary temperature: 23
Running buffer: MeOH:100 mM boric acid 75:25 containing 7 mM gamma-cyclodextrin and 5 mM adenosine 5'-monophosphate, pH 7.0
Injection: Pressure injection for 10 s.
Detector: UV 254
Migration time: 25.8
Voltage: 30 kV
Current: 6.8 µA
Model: Beckman P/ACE 5510
Limit of detection: 170-330 µM

OTHER SUBSTANCES
Simultaneous: chenodiol, deoxycholic acid, glycochenodeoxycholic acid, glycocholic acid, glycodeoxycholic acid, glycolithocholic acid, glycoursodeoxycholic acid, lithocholic acid, taurochenodeoxycholic acid, taurocholic acid, taurodeoxycholic acid, taurolithocholic acid, tauroursodeoxycholic acid, ursodiol

KEY WORDS
serum; indirect UV detection; ultrafiltrate

REFERENCE
Yarabe,H.H.; Shamsi,S.A.; Warner,I.M. Capillary zone electrophoresis of bile acids with indirect photometric detection, *Anal.Chem.*, **1998**, *70*, 1412–1418.

Choline

Molecular formula: $C_5H_{14}NO^+$
Molecular weight: 104.17
CAS Registry No.: 62-49-7, 67-48-1 (chloride), 4201-78-9 (dehydrocholate), 77-91-8 (dihydrogen citrate), 2016-36-6 (salicylate), 4499-40-5 (theophyllinate), 87-67-2 (bitartrate), 507-30-2 (gluconate)
Merck Index (12th ed.): 2259

SAMPLE
Matrix: formulations
Sample preparation: Heat capsule contents or crushed tablet with 40 mL water on a steam bath for 30 min, cool, make up to 50 mL, filter (Whatman No. 4 paper). Remove a 5 mL aliquot of the filtrate and add it to 1 mL 1 mg/mL tetramethylammonium hydroxide, make up to 25 mL with water, inject an aliquot.

CAPILLARY ELECTROPHORESIS
Capillary: 75 cm × 75 µm fused-silica (50 cm to detector) (Polymicro Technologies)
Capillary preparation: Each week wash capillary with 100 mM NaOH for 10 min and water for 5 min, fill with running buffer.
Capillary temperature: 28
Running buffer: 5 mM pH 4.5 1-Methylimidazole buffer (Dissolve 41 mg 1-methylimidazole in 70 mL water, adjust pH to 4.5 with 100 mM HCl, make up to 100 mL. Prepare fresh each day.)
Injection: Vacuum injection at 1-10 kPa.s.

Detector: UV 214
Migration time: 2.5
Internal standard: tetramethylammonium hydroxide (2.3)
Voltage: 30 kV
Model: Isco Model 3140 Electropherograph
Limit of detection: 50 µg/g

KEY WORDS
indirect UV detection; tablets; capsules

REFERENCE
Carter,N.; Trenerry,V.C. The determination of choline in vitamin preparations, infant formula and selected foods by capillary zone electrophoresis with indirect ultraviolet detection, *Electrophoresis*, **1996**, *17*, 1622–1626.

SAMPLE
Matrix: solutions
Sample preparation: Inject an aliquot of a solution containing 100 µg/mL triethylamine as IS.

CAPILLARY ELECTROPHORESIS
Capillary: 80 cm × 50 µm silica (70 cm to detector)
Capillary preparation: Between runs rinse capillary with 100 mM NaOH for 2 min and with running buffer for 3 min.
Capillary temperature: 25
Running buffer: 10 mM Creatinine adjusted to pH 3.6 with acetic acid
Injection: Hydrodynamic injection for 5 s (5 nL).
Detector: UV 200
Migration time: 6.5
Internal standard: triethylamine (7.1)
Voltage: 25 kV
Current: 4.8 µA
Model: Beckman P/ACE 2000
Limit of detection: 5 µg/mL

OTHER SUBSTANCES
Simultaneous: chlormequat, isopropylamine, trimethylamine, trimethylvinylammmonium

KEY WORDS
indirect UV detection

REFERENCE
Wycherley,D.; Rose,M.E.; Giles,K.; Hutton,T.M.; Rimmer,D.A. Capillary electrophoresis with detection by inverse UV spectroscopy and electrospray mass spectrometry for the examination of quaternary ammonium herbicides, *J.Chromatogr.A*, **1996**, *734*, 339–349.

SAMPLE
Matrix: solutions
Sample preparation: Inject an aliquot of a solution in running buffer.

CAPILLARY ELECTROPHORESIS
Capillary: 50 cm × 75 µm fused-silica (50 cm to detector)
Capillary preparation: Between runs rinse capillary with 100 mM NaOH for 1 min and with running buffer for 2 min. At the start of each day rinse capillary with 100 mM NaOH for 10 min, with water for 5 min, and with running buffer for 10 min
Capillary temperature: 25
Running buffer: 5 mM pH 3.2 creatinine
Injection: Hydrodynamic injection for 5 s (ca. 30 nL).
Detector: UV 210
Migration time: 5
Voltage: 25 kV
Model: Beckman P/ACE 5500

OTHER SUBSTANCES
Simultaneous: glycine, lysine

KEY WORDS
indirect UV detection

REFERENCE
Lambert,A.; Colin,J.L.; Leroy,P.; Nicolas,A. Advantages of capillary electrophoresis for determination of choline in pharmaceutical preparations, *Biomed.Chromatogr.*, **1998**, *12*, 181–182.

Chymotrypsin

Molecular formula: indefinite
CAS Registry No.: 9004-07-3
Merck Index (12th ed.): 2320

SAMPLE
Matrix: solutions
Sample preparation: Inject an aliquot of a 1 mg/mL solution in water.

CAPILLARY ELECTROPHORESIS
Capillary: 37 cm × 50 μm fused silica (30 cm to detector) (Polymicro Technologies)
Capillary temperature: 30
Running buffer: 500 mM pH 8.4 borate buffer containing 40 mM sodium salt of phytic acid
Detector: UV 200
Migration time: 5.5
Voltage: 5.0 kV
Model: Beckman P/ACE model 5000

OTHER SUBSTANCES
Simultaneous: cytochrome C, lysozyme, ribonuclease A, trypsinogen
Interfering: Pressure injection for 1 s

REFERENCE
Okafo,G.N.; Vinther,A.; Kornfelt,T.; Camilleri,P. Effective ion-pairing for the separation of basic proteins in capillary electrophoresis, *Electrophoresis*, **1995**, *16*, 1917–1921.

Cicletanine

Molecular formula: $C_{14}H_{12}ClNO_2$
Molecular weight: 261.71
CAS Registry No.: 89943-82-8, 82747-56-6 (HCl)
Merck Index (12th ed.): 2323
Lednicer: 5 143

SAMPLE
Matrix: blood
Sample preparation: 2 mL Plasma + 100 μL 10 μg/mL IS in MeOH:water 10:90 + 7 mL diethyl ether, shake mechanically for 10 min, centrifuge. Remove 5 mL of the organic layer, add 5 mL diethyl ether to the aqueous layer, repeat extraction, remove a 6 mL aliquot of the organic layer. Combine the organic layers and evaporate them to dryness under a stream of nitrogen at 40°, reconstitute the residue in 200 μL MeCN:water 10:90, inject an aliquot.

CAPILLARY ELECTROPHORESIS
Capillary: 57 cm × 75 μm (50 cm to detector) (Beckman)
Capillary preparation: Between injections wash capillary with water, 100 mM NaOH, water, and running buffer for 1 min each
Capillary temperature: 35
Running buffer: MeCN:buffer 10:90 (Buffer was 100 mM pH 8.6 sodium borate buffer containing 110 mM sodium dodecyl sulfate and 25 mM gamma-cyclodextrins (Fluka).)
Injection: Injection by pressure for 10 s (46 nL)
Detector: UV 214
Migration time: 11.7 (S-(+)), 11.9 (R-(-))
Internal standard: (±)-2-methyl-3-hydroxy-4H,5H-5-methyl-(4'-chlorophenyl)isofuropyridine hydrochloride (Expansia, Aramon, France) (11.2, 11.5 (enantiomers))
Voltage: 15 kV
Model: Beckman P/ACE 2000
Limit of detection: 10 ng/mL

KEY WORDS
chiral; plasma; pharmacokinetics

REFERENCE
Prunonosa,J.; Obach,R.; Diez-Cascon,A.; Gouesclou,L. Determination of cicletanine enantiomers in plasma by high-performance capillary electrophoresis, *J.Chromatogr.*, **1992**, *574*, 127–133.

SAMPLE
Matrix: solutions
Sample preparation: Inject an aliquot of a 100 μg/mL solution in running buffer.

CAPILLARY ELECTROPHORESIS
Capillary: 32 cm × 50 μm fused-silica (27.5 cm to detector) (Yongnian Optical Conductive Fiber Plant, China), coated with polyacrylamide
Capillary preparation: No details of the polyacrylamide coating process are provided. However, another paper (LC.GC 1997, 15, 40) by this group indicates that they use the procedure of Hjertén, thus: Adjust the pH of 20 mL water to 3.5 with acetic acid, add 80 μL 3-(trimethoxysilyl)propyl methacrylate (3-methacryloxypropyltrimethoxysilane), mix, suck into capillary, let stand at room temperature for 1 h, remove the solution, wash with water. Fill the capillary with a deaerated 3-4% acrylamide solution containing 1 μL/mL N,N,N',N'-tetramethylethylenediamine and 1 mg/mL potassium persulfate, let stand for 30 min, remove excess solution by aspiration, rinse with water, remove water by aspiration, dry at 35° (J. Chromatogr. 1985, 347, 191).
Capillary temperature: 25
Running buffer: 100 mM NaH_2PO_4 adjusted to pH 2.5
Injection: Electrokinetic injection at 15 kV for 3 s.
Detector: UV 200, UV 210
Migration time: 5.50
Voltage: 15 kV
Model: Bio-Rad BioFocus 3000

OTHER SUBSTANCES
Simultaneous: atropine, carazolol, dimethindene, fendiline, homatropine, ipratropium bromide, isothipendyl, mefloquine, metaclazepam, nafronyl (naftidrofuryl), nefopam, nicardipine, promethazine, reproterol, tetrahydrozoline (tetryzoline), theodrenaline, tioconazole, trihexyphenidyl, trimeprazine (alimemazine), trimipramine

KEY WORDS
coated capillary

REFERENCE
Koppenhoefer,B.; Epperlein,U.; Xiaofeng,Z.; Bingcheng,L. Separation of enantiomers of drugs by capillary electrophoresis. Part 4: Hydroxypropyl-γ-cyclodextrin as chiral solvating agent, *Electrophoresis*, **1997**, *18*, 924–930.

SAMPLE
Matrix: solutions
Sample preparation: Inject an aliquot of a 100 μg/mL solution in running buffer.

CAPILLARY ELECTROPHORESIS
Capillary: 30 cm × 50 μm fused-silica (25.5 cm to detector), coated with polyacrylamide
Capillary preparation: Adjust the pH of 20 mL water to 3.5 with acetic acid, add 80 μL 3-(trimethoxysilyl)propyl methacrylate (3-methacryloxypropyltrimethoxysilane), mix, suck into capillary, let stand at room temperature for 1 h, remove the solution, wash with water. Fill the capillary with a deaerated 3-4% acrylamide solution containing 1 μL/mL N,N,N',N'-tetramethylethylenediamine and 1 mg/mL potassium persulfate, let stand for 30 min, remove excess solution by aspiration, rinse with water, remove water by aspiration, dry at 35° (J. Chromatogr. 1985, 347, 191).
Capillary temperature: 25
Running buffer: 100 mM NaH_2PO_4 containing 45 mM hydroxypropyl-α- cyclodextrin, adjusted to pH 2.5 with phosphoric acid
Injection: Electrokinetic injection at 15 kV for 3 s.
Detector: UV 200
Voltage: 15 kV
Model: Bio-Rad BioFocus 3000

KEY WORDS
chiral; coated capillary; comparison with the use of other cyclodextrins; this running buffer gave the greatest enantiomeric separation.; α=1.026

REFERENCE
Lin,B.; Zhu,X.; Koppenhoefer,B.; Epperlein,U. Investigation of 123 chiral drugs by cyclodextrin-modified capillary electrophoresis, *LC.GC*, **1997**, *15*, 40–46.

SAMPLE
Matrix: solutions

CAPILLARY ELECTROPHORESIS
Capillary: 29-36 cm × 50 μm fused-silica (24.5-31.5 cm to detector) (Yongnian Optical Conductive Fiber Plant, China) coated with polyacrylamide
Capillary preparation: Coat capillary as follows. Adjust the pH of 20 mL water to 3.5 with acetic acid, add 80 μL 3-(trimethoxysilyl)propyl methacrylate (3-methacryloxypropyltrimethoxysilane), mix, suck into capillary, let stand at room temperature for 1 h, remove the solution, wash with water. Fill the capillary with a deaerated 3-4% acrylamide solution containing 1 μL/mL N,N,N',N'-tetramethylethylenediamine and 1 mg/mL potassium persulfate, let stand for 30 min, remove excess solution by aspiration, rinse with water, remove water by aspiration, dry at 35° (J. Chromatogr. 1985, 347, 191).
Capillary temperature: 25
Running buffer: 100 mM pH 2.5 NaH_2PO_4 (A) or 100 mM pH 2.5 NaH_2PO_4 containing 45 mM hydroxypropyl-α-cyclodextrin (Wacker, Munich) (B)
Injection: Electromigration at 15 kV for 3 s.
Detector: UV 200; UV 210
Migration time: 5.50 (A); 14.35, 14.67 (B) (enantiomers)
Voltage: 15 kV
Model: Bio-Focus 3000

OTHER SUBSTANCES
Also analyzed: albuterol (salbutamol), alprenolol, amorolfine, atenolol, atropine, azelastine, baclofen, bamethan, benproperine, benserazide, biperiden, bisoprolol, brompheniramine, bupivacaine, bupranolol, butamirate, butethamate, carazolol, carbuterol, carteolol, carvedilol, celiprolol, chloroquine, chlorpheniramine, chlorphenoxamine, clenbuterol, clidinium bromide, clobutinol, dimethindene, dipivefrin, disopyramide, dobutamine, doxylamine, fendiline, flecainide, gallopamil, homatropine, ipratropium bromide, isoproterenol (isoprenaline), isothipendyl, ketamine, meclizine, mefloquine, mepindolol, mequitazine, metaclazepam, metaproterenol (orciprenaline), metipranolol, metoprolol, nafronyl (naftidrofuryl), nefopam, nicardipine, norfenefrine, ofloxacin, ornidazole, orphenadrine, oxomemazine, oxprenolol, oxybutynin, phenoxybenzamine, phenylpropanolamine, pholedrine, pindolol, pirbuterol, prilocaine, procyclidine,

promethazine, propafenone, propranolol, reproterol, sotalol, sulpride, synephrine, talinolol, terbutaline, tetrahydrozoline (tetryzoline), theodrenaline, tioconazole, tocainide, trihexyphenidyl, trimeprazine (alimemazine), trimipramine, tropicamide, verapamil, zopiclone

KEY WORDS
coated capillary; chiral

REFERENCE
Koppenhoefer,B.; Eperlein,U.; Schlunk,R.; Zhu,X.; Lin,B. Separation of enantiomers of drugs by capillary electrophoresis. V. Hydroxypropyl-α-cyclodextrin as chiral solvating agent, *J.Chromatogr.A*, **1998**, *793*, 153–164.

Cicloprofen

Molecular formula: $C_{16}H_{14}O_2$
Molecular weight: 238.29
CAS Registry No.: 36950-96-6
Lednicer: 2 217

SAMPLE
Matrix: solutions

CAPILLARY ELECTROPHORESIS
Capillary: 35 cm × 50 μm coated capillary (30.5 cm to detector) (Composite Metal Services, UK)
Capillary preparation: Before each run purge at high pressure with water for 100 s and with running buffer for 120 s. If vancomycin is used, before each run purge at high pressure with water for 100 s and with running buffer not containing vancomycin for 120 s. Next purge with running buffer containing vancomycin at low pressure (175 psi.s) and inject sample. Coat capillary as follows. Adjust the pH of 20 mL water to 3.5 with acetic acid, add 80 μL 3-(trimethoxysilyl)propyl methacrylate (3-methacryloxypropyltrimethoxysilane), mix, suck into capillary, let stand at room temperature for 1 h, remove the solution, wash with water. Fill the capillary with a deaerated 3-4% acrylamide solution containing 1 μL/mL N,N,N',N'-tetramethylethylenediamine and 1 mg/mL potassium persulfate, let stand for 30 min, remove excess solution by aspiration, rinse with water, remove water by aspiration, dry at 35° (J. Chromatogr. 1985, 347, 191).
Capillary temperature: 25
Running buffer: Buffer containing 30 mM heptakis-2,3,6-tri-O-methyl-β-cyclodextrin (A), 5 mM heptamethylamino-β-cyclodextrin (B) or 5 mM vancomycin (C) (Buffer was 50 mM phosphoric acid containing 50 mM acetic acid and 50 mM boric acid, dilute with an equal volume of water, adjust pH to 5 with concentrated NaOH.)
Injection: Pressure injection at 10 psi.
Detector: UV 206
Migration time: 21.3, 22.8 (A); 13.8, 14.9 (B); 7.4, 8.3 (C) (enantiomers)
Voltage: -20 kV
Model: Bio-Rad Biofocus 3000
Limit of detection: 5 μM (A, C), 10 μM (B)

KEY WORDS
chiral; coated capillary; detector at anode

REFERENCE
Fanali,S.; Desiderio,C.; Aturki,Z. Enantiomeric resolution study by capillary electrophoresis. Selection of the appropriate chiral selector, *J.Chromatogr.A*, **1997**, *772*, 185–194.

SAMPLE
Matrix: solutions

CAPILLARY ELECTROPHORESIS

Capillary: 35 cm × 50 μm polyacrylamide-coated fused-silica (30.5 cm to detector) (Composite Metal Services, UK)

Capillary preparation: Before each run rinse capillary with water for 70 s and with running buffer for 100 s. (Coat capillary as follows. Adjust the pH of 20 mL water to 3.5 with acetic acid, add 80 μL 3-(trimethoxysilyl)propyl methacrylate (3-methacryloxypropyltrimethoxysilane), mix, suck into capillary, let stand at room temperature for 1 h, remove the solution, wash with water. Fill the capillary with a deaerated 3-4% acrylamide solution containing 1 μL/mL N,N,N',N'-tetramethylethylenediamine and 1 mg/mL potassium persulfate, let stand for 30 min, remove excess solution by aspiration, rinse with water, remove water by aspiration, dry at 35° (J. Chromatogr. 1985, 347, 191).)

Capillary temperature: 25

Running buffer: Buffer containing 2.5 mM cyanoethylated-β-cyclodextrin (Cyclolab, Budapest) (Prepare buffer by adjusting the pH of 50 mM phosphoric acid containing 50 mM acetic acid and 50 mM boric acid to 5 with concentrated NaOH, add the appropriate amount of cyanoethylated-β-cyclodextrin, dilute with an equal volume of water.)

Injection: Pressure injection at 5 psi for 2 s.

Detector: UV 206

Migration time: 13.0 (second enantiomer, $\alpha = 1.033$)

Voltage: 20 kV

Current: 27-40 μA

Model: Bio-Rad Biofocus 3000

KEY WORDS

detector at anode; chiral; coated capillary

REFERENCE

Aturki,Z.; Desiderio,C.; Mannina,L.; Fanali,S. Chiral separations by capillary zone electrophoresis with the use of cyanoethylated-β-cyclodextrin as chiral selector, *J.Chromatogr.A*, **1998**, *817*, 91–104.

Cimetidine

Molecular formula: $C_{10}H_{16}N_6S$
Molecular weight: 252.34
CAS Registry No.: 51481-61-9, 70059-30-2 (HCl)
Merck Index (12th ed.): 2337
Lednicer: 2 253; 4 89, 95, 112

SAMPLE

Matrix: blood

Sample preparation: Condition a 1 mL 100 mg Supelclean LC-18 SPE cartridge with 1 mL MeOH, 1 mL water, and 1 mL 10 mM pH 7 TRIZMA buffer containing 3.2 mM acetic acid. 500 μL Serum + 5 μL 5 mg/mL ranitidine in MeOH + 500 μL 40 mM TRIZMA buffer containing 13 mM acetic acid, add to the bottom of the SPE cartridge, wash with 1 mL water from the bottom of the cartridge, elute downward with 160 μL THF:5 mM pH 7.7 TRIZMA buffer containing 1.7 mM acetic acid 50:50 while applying 150 V using platinum electrodes (negative electrode at bottom), inject an aliquot of the first 50 μL eluate.

CAPILLARY ELECTROPHORESIS

Capillary: 60 cm × 50 μm uncoated fused-silica (45 cm to detector) (Polymicro Technologies)

Capillary preparation: Rinse capillary with buffer between injections. Rinse a new capillary with water for 30 min and with 100 mM NaOH overnight, wash with water, fill with running buffer. recondition capillaries with water for 10 min, with 45% trifluoroacetic acid for 2 h, and water for 10 min.

Running buffer: 9.4 mM pH 6.4 sodium phosphate buffer containing 9.8 mM hexadecyltrimethylammonium bromide and 3.3 mM tris(hydroxymethyl)aminomethane (TRIZMA base)

Injection: Hydrodynamic injection at 13.5 cm for 10-15 s

Detector: UV 228

Migration time: 8.5

Internal standard: ranitidine (10)
Voltage: -20 kV
Model: Laboratory constructed
Limit of quantitation: 233 ng/mL

KEY WORDS
serum; detector is at anode; SPE

REFERENCE
Soini,H.; Tsuda,T.; Novotny,M.V. Electrochromatographic solid-phase extraction for determination of cimetidine in serum by micellar electrokinetic capillary chromatography, *J.Chromatogr.*, **1991**, *559*, 547–558.

SAMPLE
Matrix: blood
Sample preparation: Condition a 1 mL Supelclean LC-18 SPE cartridge (Supelco) with 1 mL water, 1 mL 50 mM pH 8.4 phosphate buffer, 3 mL MeOH, and 5 mL water. 1 mL Plasma + ranitidine, add to the SPE cartridge, wash with 1 mL phosphate buffer, elute with three 1 mL portions of MeOH, evaporate the third eluate to dryness under reduced pressure, reconstitute with 50 µL 10 mM pH 2.0 NaH_2PO_4, inject an aliquot.

CAPILLARY ELECTROPHORESIS
Capillary: 24 cm × 25 µm coated capillary (Bio-Rad)
Capillary preparation: Purge capillary for a few min after each injection.
Capillary temperature: 15
Running buffer: 100 mM pH 2.0 NaH_2PO_4
Injection: Electrokinetic injection at 8 kV for 8 s, positive to negative polarity
Detector: UV 208
Migration time: 4.6
Internal standard: ranitidine (2.9)
Voltage: 15 kV
Model: BioFocus 3000 (Bio-Rad)
Limit of quantitation: 250 ng/mL (S/N 5)
Limit of detection: 100 ng/mL (S/N 3)

OTHER SUBSTANCES
Extracted: metabolites

KEY WORDS
SPE; plasma; coated capillary

REFERENCE
Luksa,J.; Josic,D. Determination of cimetidine in human plasma by free capillary zone electrophoresis, *J.Chromatogr.B*, **1995**, *667*, 321–327.

SAMPLE
Matrix: solutions
Sample preparation: Inject an aliquot of an aqueous solution.

CAPILLARY ELECTROPHORESIS
Capillary: 37 cm × 75 µm
Capillary preparation: Before injection rinse with 100 mM NaOH for 1 min and with running buffer for 1 min. Before first use rinse with 100 mM NaOH for 20 min.
Capillary temperature: 30
Running buffer: 25 mM NaH_2PO_4 adjusted to pH 2.3 with concentrated phosphoric acid
Injection: Pressure injection for 5 s
Detector: UV 200
Migration time: 5.312
Internal standard: imidazole (2.722)
Voltage: +13 kV
Model: Beckman P/ACE 2200 or 5100
Limit of quantitation: 1 µg/mL
Limit of detection: 400 ng/mL

OTHER SUBSTANCES
Simultaneous: albuterol, aspartame

KEY WORDS
robust

REFERENCE
Altria,K.D.; Frake,P.; Gill,I.; Hadgett,T.; Kelly,M.A.; Rudd,D.R. Validated capillary electrophoresis method for the assay of a range of basic drugs, *J.Pharm.Biomed.Anal.*, **1995**, *13*, 951–957.

SAMPLE
Matrix: solutions
Sample preparation: Inject an aliquot of a 100 μg/mL solution in running buffer.

CAPILLARY ELECTROPHORESIS
Capillary: 60 cm × 50 μm fused-silica (44 cm to detector) (Composite Metal Services, Hallow UK)
Capillary preparation: Between runs rinse capillary at 2000 mbar with 100 mM NaOH for 2 min, with water for 0.8 min, and with running buffer for 3 min.
Running buffer: 10 mM pH 3.14 NaH_2PO_4
Injection: Hydrodynamic injection at 25 mbar for 12 s (ca. 7.7 nL).
Detector: UV 230
Migration time: 3.99
Voltage: 30 kV
Model: Thermo-Unicam Crystal CE Model 310

OTHER SUBSTANCES
Simultaneous: impurities

REFERENCE
Ellis,D.R.; Palmer,M.E.; Tetler,L.W.; Eckers,C. Separation of cimetidine and related materials by aqueous and non-aqueous capillary electrophoresis, *J.Chromatogr.A*, **1998**, *808*, 269–275.

SAMPLE
Matrix: solutions

CAPILLARY ELECTROPHORESIS
Capillary: 57 cm × 50 μm fused-silica (50 cm to detector)
Running buffer: 50 mM pH 8.0 Borate buffer containing 40 mM sodium taurodeoxycholate and 25 mM phosphatidylcholine (Prepare by adding phosphatidylcholine and stirring for 3-6 h until all cloudiness disappears. Phosphatidylcholine was 95% pure soybean lecithin, Epikuron, Lucas Meyer & Co.)
Injection: Inject a solution of the compound in the running buffer at 20 psi for 1 s, inject MeOH: water 5:95 at 20 psi for 1 s, inject a solution of halofantrine in running buffer at 20 psi for 1 s.
Detector: UV 214
Migration time: k' 0.50
Model: Beckman P/ACE 5000

OTHER SUBSTANCES
Also analyzed: acetaminophen, amoxicillin, antipyrine, aspirin, azathioprine, caffeine, captopril, carbamazepine, carprofen, chlorambucil, chlorpheniramine, chlorpromazine, clonidine, codeine, desipramine, diphenhydramine, ephedrine, fenoterol, flufenamic acid, flurbiprofen, haloperidol, hydroxyzine, ibuprofen, imipramine, indomethacin, ketoprofen, lidocaine, melphalan, metoprolol, nabumetone, nadolol, phenobarbital, phenol, promazine, propranolol, pyrilamine, ranitidine, ropinirole, salicylic acid, sulfamethoxazole, testosterone, theophylline, thioridazine, tiaprofenic acid, tolfenamic acid, trifluoperazine, trimethoprim, valproic acid, verapamil

KEY WORDS
comparison with HPLC; k' = (Tr-T0)/(T0(1-Tr/Tm)) where Tr = retention time of analyte; T0 = retention time of water; and Tm = retention time of marker (halofantrine)

REFERENCE
Hanna,M.; de Biasi,V.; Bond,B.; Salter,C.; Hutt,A.J.; Camilleri,P. Estimation of the partitioning characteristics of drugs: A comparison of a large and diverse drug series utilizing chromatographic and electrophoretic methodology, *Anal.Chem.*, **1998**, *70*, 2092–2099.

SAMPLE
Matrix: urine
Sample preparation: Centrifuge urine, dilute with 2 volumes of 32 mM pH 4.5 6-aminocaproic acid buffer containing 18 mM adipic acid, inject an aliquot.

CAPILLARY ELECTROPHORESIS
Capillary: 64 cm × 75 μm (48 cm to detector) (Polymicro Technologies)
Running buffer: MeOH:buffer 5:95 (Buffer was 32 mM pH 4.5 6-aminocaproic acid buffer containing 18 mM adipic acid, 5% dextran (MW 18300) and 0.02% polyethylene oxide (MW 300000)
Injection: Hydrodynamic injection at 20 cm
Detector: UV 220
Migration time: 10
Voltage: 22 kV
Model: Laboratory constructed

OTHER SUBSTANCES
Extracted: diltiazem, famotidine, prazosin

REFERENCE
Soini,H.; Riekkola,M.-L.; Novotny,M.V. Mixed polymer networks in the direct analysis of pharmaceuticals in urine by capillary electrophoresis, *J.Chromatogr.A*, **1994**, *680*, 623–634.

Ciprofloxacin

Molecular formula: $C_{17}H_{18}FN_3O_3$
Molecular weight: 331.35
CAS Registry No.: 85721-33-1, 86393-32-0 (hydrochloride monohydrate)
Merck Index (12th ed.): 2374
Lednicer: 4 141

SAMPLE
Matrix: solutions
Sample preparation: Inject an aliquot of a solution in MeOH.

CAPILLARY ELECTROPHORESIS
Capillary: 59 cm × 50 μm fused-silica (43 cm to detector) (Polymicro Technologies)
Capillary preparation: Before each analysis flush with water for 3 min, with 200 mM NaOH for 3 min, with water for 3 min, and with running buffer for 4 min. Flush new capillaries with 1 M NaOH at 2000 mbar for 10 min then with 200 mM NaOH for 10 min.
Capillary temperature: 23
Running buffer: MeCN:buffer 28:72, pH 7.3 (Buffer was 32 mM sodium borate containing 18 mM NaH_2PO_4, 39 mM sodium cholate, and 8 mM sodium heptanesulfonate.)
Injection: Hydrodynamic injection at 40 mbar for 6 s.
Detector: UV 260
Migration time: 4.1
Voltage: 30 kV
Model: Lauer Labs Prince

OTHER SUBSTANCES
Simultaneous: enoxacin, flumequine, lomefloxacin, nalidixic acid, norfloxacin, ofloxacin, oxolinic acid, pefloxacin, pipemidic acid, piromidic acid, rosoxacin, sparfloxacin

REFERENCE
Sun,S.-W.; Chen,L.-Y. Optimization of capillary electrophoretic separation of quinolone antibacterials using the overlapping resolution mapping scheme, *J.Chromatogr.A*, **1997**, *766*, 215–224.

Cisapride

Molecular formula: $C_{23}H_{29}ClFN_3O_4$
Molecular weight: 465.95
CAS Registry No.: 81098-60-4
Merck Index (12th ed.): 2377

SAMPLE
Matrix: solutions
Sample preparation: Inject an aliquot of a solution in MeOH.

CAPILLARY ELECTROPHORESIS
Capillary: 64.5 cm × 50 µm fused-silica (Polymicro Technologies)
Capillary preparation: Between runs flush with 100 mM NaOH for 2-10 min then with running buffer for 2 min. Before use rinse with 1 M NaOH for 1 h, with 100 mM NaOH for 20 min, with water for 20 min, with 1 M acetic acid in MeOH/MeCN for 10 min, and with running buffer for 10 min.
Capillary temperature: 25
Running buffer: MeCN containing 1 M acetic acid, 200 µM Tween 20, and 30 mM sodium (+)-S-camphorsulfonate (Sodium (+)-S-camphorsulfonate can be prepared by mixing equimolar amounts of (+)-S-camphorsulfonic acid and NaOH in EtOH, collect the precipitate, wash with diethyl ether, dry.)
Injection: Pressure injection of 5 kPa for 3 s.
Detector: UV 214
Migration time: 9.5
Voltage: 30 kV
Model: Hewlett-Packard HP³ᴰ

OTHER SUBSTANCES
Simultaneous: impurities

REFERENCE
Bjornsdottir,I.; Hansen,S.H.; Terabe,S. Chiral separation in non-aqueous media by capillary electrophoresis using the ion-pair principle, *J.Chromatogr.A*, **1996**, *745*, 37–44.

Cisplatin

Molecular formula: $Cl_2H_6N_2Pt$
Molecular weight: 300.05
CAS Registry No.: 15663-27-1
Merck Index (12th ed.): 2378
Lednicer: 4 15-17

SAMPLE
Matrix: solutions

CAPILLARY ELECTROPHORESIS
Capillary: 75 cm × 75 µm fused-silica (50 cm to detector) (CS Chromatographie Service, Langerwehe, Germany)

Capillary preparation: Purge with running buffer for 2 min before each run.
Running buffer: 50 mM pH 7.0 NaH_2PO_4 containing 25 mM sodium borate and 100 mM sodium dodecyl sulfate
Injection: Hydrostatic injection
Detector: UV 210
Migration time: 8.3
Voltage: 15 kV
Model: laboratory constructed

OTHER SUBSTANCES
Simultaneous: degradation products, carboplatin, lobaplatin

REFERENCE
Wenclawiak,B.W.; Wollmann,M. Separation of platinum(II) anti-tumour drugs by micellar electrokinetic capillary chromatography, *J.Chromatogr.A*, **1996**, *724*, 317–326.

Clarithromycin

Molecular formula: $C_{38}H_{69}NO_{13}$
Molecular weight: 747.96
CAS Registry No.: 81103-11-9
Merck Index (12th ed.): 2400

SAMPLE
Matrix: solutions
Sample preparation: Inject an aliquot of a solution in MeCN:isopropanol:10 mM pH 7 sodium phosphate buffer 25:25:50.

CAPILLARY ELECTROPHORESIS
Capillary: 68.5 cm × 50 μm fused-silica (60 cm to detector) (Polymicro Technologies)
Capillary preparation: Rinse capillary with MeCN:water 50:50 for 2 min and with running buffer for 4 min before each run.
Capillary temperature: 15
Running buffer: MeCN:80 mM pH 6 phosphate buffer 20:80 containing 30 mM sodium cholate
Injection: Pressure injection at 50 mbar with 7.5 s for sample and 3.7 s for buffer plug
Detector: UV 205
Migration time: 14.4
Voltage: +20 kV
Model: Isco Model 3140 Electropherograph

OTHER SUBSTANCES
Simultaneous: erythromycin, erythromycin estolate, erythromycin ethylsuccinate, oleandomycin, spiramycin, troleandomycin

REFERENCE
Flurer,C.L. Analysis of macrolide antibiotics by capillary electrophoresis, *Electrophoresis*, **1996**, *17*, 359–366.

Clavulanic acid

Molecular formula: $C_8H_9NO_5$
Molecular weight: 199.16
CAS Registry No.: 58001-44-8, 61177-45-5 (K salt)
Merck Index (12th ed.): 2402
Lednicer: 4 180

SAMPLE
Matrix: solutions
Sample preparation: Prepare a 50 μM solution in water, inject an aliquot.

CAPILLARY ELECTROPHORESIS
Capillary: 57 cm × 50 μm silica (50 cm to detector)
Capillary preparation: Condition with 100 mM NaOH for 5 min, rinse with water for 5 min, fill with running buffer.
Capillary temperature: 20
Running buffer: 10 mM pH 7.42 NaH_2PO_4 containing 30 mM sodium borate and 200 mM sodium dodecyl sulfate
Injection: Inject for 5 s.
Detector: UV 214
Migration time: 11
Voltage: 20 kV
Model: Beckman P/ACE 2100

OTHER SUBSTANCES
Simultaneous: amoxicillin

REFERENCE
Okafo,G.N.; Camilleri,P. Micellar electrokinetic capillary chromatography of amoxycillin and related molecules, *Analyst*, **1992**, *117*, 1421–1424.

Clenbuterol

Molecular formula: $C_{12}H_{18}Cl2N_2O$
Molecular weight: 277.19
CAS Registry No.: 37148-27-9, 21898-19-1 (monohydrochloride)
Merck Index (12th ed.): 2407

SAMPLE
Matrix: formulations
Sample preparation: Condition a 13 mM Empore C18 SPE disk (Baker) with 2.5 mL MeOH and 2.5 mL water at 1.5 mL/min. Dissolve tablet in dissolution medium (?). Pass 50 mL through the SPE disk, wash with 2.5 mL water, dry, add 1 mL MeOH and let it soak in for 3 min, elute at 0.5 mL/min, inject an aliquot of the eluate.

CAPILLARY ELECTROPHORESIS
Capillary: 60 cm × 75 μm fused-silica (53 cm to detector) (Waters)
Capillary preparation: Rinse capillary with running buffer for 3 min between runs. At the start of each day rinse capillary with 200 mM KOH for 5 min and with running buffer for 15 min. At the end of each day flush capillary with 200 mM KOH for 2 min then with water.
Capillary temperature: 25
Running buffer: 30 mM pH 6.7 Phosphate buffer (Prepare by dissolving 260 mg KH_2PO_4 and 160 mg Na_2HPO_4 in 100 mL water, adjust pH to 6.7 with phosphoric acid.)
Injection: Hydrodynamic injection at 10 cm for 15 s.

Detector: UV 214
Migration time: 5.2
Voltage: 16 kV
Model: Waters Quanta 4000
Limit of quantitation: 500 ng/mL
Limit of detection: 160 ng/mL

KEY WORDS
tablets; SPE; comparison with HPLC

REFERENCE
Carducci,C.N.; Lucangioli,S.E.; Rodríguez,V.G.; Fernández Otero,G.C. Application of extraction disks in dissolution tests of clenbuterol and levothyroxine tablets by capillary electrophoresis, *J.Chromatogr.A*, **1996**, *730*, 313–319.

SAMPLE
Matrix: solutions

CAPILLARY ELECTROPHORESIS
Capillary: 34 cm × 50 μm (25.5 cm (A) or 8.5 cm (B, C) to detector) (Composite Metal Services, Hallow, UK)
Running buffer: 50 mM pH 2.5 Phosphate buffer containing 25 mM dimethyl-β-cyclodextrin (A, B) or 250 mM pH 2.5 phosphate buffer containing 25 mM dimethyl-β-cyclodextrin (C)
Injection: Pressure injection at 50 mbar for 5 s.
Detector: UV 200
Migration time: 4, 4.3 (A), 1.6, 1.7 (B), 6.6, 7.2 (C) (enantiomers)
Voltage: 20 kV (A) or -20 kV (B) or -7.5 kV (C)
Model: Hewlett Packard 3D

OTHER SUBSTANCES
Simultaneous: picumeterol

KEY WORDS
when the distance to the detector was 8.5 cm (B; C) injection was at the normal outlet end and voltage polarity and pressure for injection were reversed.; chiral

REFERENCE
Altria,K.D.; Kelly,M.A.; Clark,B.J. The use of a short-end injection procedure to achieve improved performance in capillary electrophoresis, *Chromatographia*, **1996**, *43*, 153–158.

SAMPLE
Matrix: solutions

CAPILLARY ELECTROPHORESIS
Capillary: 70 cm × 75 μm fused-silica (63 cm to detector) (Composite Metal Services, Hallow, UK)
Capillary preparation: Wash with running buffer for 5 min before each run. Purge new capillaries with 100 mM NaOH at 60° for 5 min and with water at 60° for 5 min, equilibrate with running buffer for 5 min, apply voltage for 5 min.
Capillary temperature: 20
Running buffer: 50 mM Disodium tetraborate adjusted to pH 2.2 with orthophosphoric acid
Injection: Hydrodynamic (vacuum) injection for 3 s.
Detector: UV (wavelength not given)
Migration time: 15
Voltage: 25 kV
Model: SpectraPhoresis 1000 (Thermo Separation Products)

OTHER SUBSTANCES
Simultaneous: codeine, flurazepam, meperidine (pethidine), noscapine

REFERENCE

McGrath,G.; Smyth,W.F. Large-volume sample stacking of selected drugs of forensic significance by capillary electrophoresis, *J.Chromatogr.B*, **1996**, *681*, 125–131.

SAMPLE

Matrix: solutions

CAPILLARY ELECTROPHORESIS

Capillary: 52 cm × 75 μm fused-silica (48 cm to detector) (Polymicro Technologies)
Capillary preparation: Before each run purge with running buffer for 3 min. Every 3 runs purge with 100 mM NaOH for 5 min. Purge new capillaries with 1 M NaOH for 20 min and with 100 mM NaOH for 20 min, rinse with running buffer, equilibrate with running buffer at 12 kV for 3 h.
Running buffer: 100 mM CHES (2-(N-cyclohexylamine)ethanesulfonic acid) containing 10 mM triethylamine and 25 mM (R)-dodecoxycarbonylvaline (Waters EnantioSelect (R)-Val-1), pH adjusted to 8.8 with 1 M NaOH
Injection: Hydrostatic injection for 2 s.
Detector: UV 214
Migration time: 15, 15.3 (enantiomers)
Voltage: 12 kV
Current: ≤30 μA
Model: Waters Quanta 4000

OTHER SUBSTANCES

Simultaneous: bupivacaine, disopyramide, metoprolol, norphenylephrine, octopamine

KEY WORDS

chiral

REFERENCE

Peterson,A.G.; Ahuja,E.S.; Foley,J.P. Enantiomeric separations of basic pharmaceutical drugs by micellar electrokinetic chromatography using a chiral surfactant, N-dodecoxycarbonylvaline, *J.Chromatogr.B*, **1996**, *683*, 15–28.

SAMPLE

Matrix: solutions

CAPILLARY ELECTROPHORESIS

Capillary: 57 cm × 75 μm fused-silica (50 cm to detector) (Scientific Glass Engineering)
Capillary preparation: Before each injection rinse capillary at 130 kPa with 100 mM HCl for 1 min, with 10 mM KOH for 1 min, and with running buffer for 2 min.
Running buffer: 100 mM pH 8.3 Tris-borate buffer containing 50 μM cetyltrimethylammonium bromide and 0.005% poly(vinyl alcohol) (Hoechst)
Injection: Pressure injection at 3.3 kPa for 10 s.
Detector: UV 214
Migration time: 10
Voltage: 20 kV
Model: Beckman P/ACE 2200

OTHER SUBSTANCES

Simultaneous: histamine

REFERENCE

van der Schans,M.J.; Reijenga,J.C.; Everaerts,F.M. Quality control of histamine and methacholine in diagnostic solutions with capillary electrophoresis, *J.Chromatogr.A*, **1996**, *735*, 387–393.

SAMPLE

Matrix: solutions
Sample preparation: Inject an aliquot of a 100 μg/mL solution in running buffer.

CAPILLARY ELECTROPHORESIS
Capillary: 29 cm × 50 μm fused-silica (24.5 cm to detector) (Yongnian Optical Conductive Fiber Plant, China), coated with polyacrylamide
Capillary preparation: No details of the polyacrylamide coating process are provided. However, another paper (LC.GC 1997, 15, 40) by this group indicates that they use the procedure of Hjertén, thus: Adjust the pH of 20 mL water to 3.5 with acetic acid, add 80 μL 3-(trimethoxysilyl)propyl methacrylate (3-methacryloxypropyltrimethoxysilane), mix, suck into capillary, let stand at room temperature for 1 h, remove the solution, wash with water. Fill the capillary with a deaerated 3-4% acrylamide solution containing 1 μL/mL N,N,N',N'-tetramethylethylenediamine and 1 mg/mL potassium persulfate, let stand for 30 min, remove excess solution by aspiration, rinse with water, remove water by aspiration, dry at 35° (J. Chromatogr. 1985, 347, 191).
Capillary temperature: 25
Running buffer: 100 mM NaH_2PO_4 adjusted to pH 2.5
Injection: Electrokinetic injection at 15 kV for 3 s.
Detector: UV 200, UV 210
Migration time: 6.35
Voltage: 15 kV
Model: Bio-Rad BioFocus 3000

OTHER SUBSTANCES
Simultaneous: albuterol, alprenolol, atenolol, baclofen, bamethan, benproperine, benserazide, bisoprolol, bupranolol, butamirate, butethamate, carbuterol, celiprolol, clobutinol, dipivefrin, isoproterenol (isoprenaline), metaproterenol (orciprenaline), metipranolol, metoprolol, norfenefrine, ornidazole, oxprenolol, phenylpropanolamine, pholedrine, pirbuterol, prilocaine, procyclidine, sotalol, synephrine, terbutaline, tocainide

KEY WORDS
coated capillary

REFERENCE
Koppenhoefer,B.; Epperlein,U.; Xiaofeng,Z.; Bingcheng,L. Separation of enantiomers of drugs by capillary electrophoresis. Part 4: Hydroxypropyl-γ-cyclodextrin as chiral solvating agent, *Electrophoresis*, **1997**, *18*, 924–930.

SAMPLE
Matrix: solutions
Sample preparation: Inject an aliquot of a 100 μg/mL solution in running buffer.

CAPILLARY ELECTROPHORESIS
Capillary: 30 cm × 50 μm fused-silica (25.5 cm to detector), coated with polyacrylamide
Capillary preparation: Adjust the pH of 20 mL water to 3.5 with acetic acid, add 80 μL 3-(trimethoxysilyl)propyl methacrylate (3-methacryloxypropyltrimethoxysilane), mix, suck into capillary, let stand at room temperature for 1 h, remove the solution, wash with water. Fill the capillary with a deaerated 3-4% acrylamide solution containing 1 μL/mL N,N,N',N'-tetramethylethylenediamine and 1 mg/mL potassium persulfate, let stand for 30 min, remove excess solution by aspiration, rinse with water, remove water by aspiration, dry at 35° (J. Chromatogr. 1985, 347, 191).
Capillary temperature: 25
Running buffer: 100 mM NaH_2PO_4 containing 45 mM hydroxypropyl-β- cyclodextrin, adjusted to pH 2.5 with phosphoric acid
Injection: Electrokinetic injection at 15 kV for 3 s.
Detector: UV 200
Voltage: 15 kV
Model: Bio-Rad BioFocus 3000

KEY WORDS
chiral; coated capillary; comparison with the use of other cyclodextrins; this running buffer gave the greatest enantiomeric separation.; α=1.066

REFERENCE
Lin,B.; Zhu,X.; Koppenhoefer,B.; Epperlein,U. Investigation of 123 chiral drugs by cyclodextrin-modified capillary electrophoresis, *LC.GC*, **1997**, *15*, 40–46.

SAMPLE
Matrix: solutions

CAPILLARY ELECTROPHORESIS
Capillary: 35 cm × 50 μm polyacrylamide-coated fused-silica (30.5 cm to detector) (Composite Metal Services, UK)
Capillary preparation: Before each run rinse capillary with water for 70 s and with running buffer for 100 s. (Coat capillary as follows. Adjust the pH of 20 mL water to 3.5 with acetic acid, add 80 μL 3-(trimethoxysilyl)propyl methacrylate (3-methacryloxypropyltrimethoxysilane), mix, suck into capillary, let stand at room temperature for 1 h, remove the solution, wash with water. Fill the capillary with a deaerated 3-4% acrylamide solution containing 1 μL/mL N,N,N',N'-tetramethylethylenediamine and 1 mg/mL potassium persulfate, let stand for 30 min, remove excess solution by aspiration, rinse with water, remove water by aspiration, dry at 35° (J. Chromatogr. 1985, 347, 191).)
Capillary temperature: 25
Running buffer: Buffer containing 100 mM cyanoethylated-β-cyclodextrin (Cyclolab, Budapest) (Prepare buffer by adjusting the pH of 50 mM phosphoric acid containing 50 mM acetic acid and 50 mM boric acid to 2.5 with concentrated NaOH, add the appropriate amount of cyanoethylated-β-cyclodextrin, dilute with an equal volume of water.)
Injection: Pressure injection at 5 psi for 2 s.
Detector: UV 206
Migration time: 16.5 (second enantiomer, α = 1.034)
Voltage: 20 kV
Current: 27-40 μA
Model: Bio-Rad Biofocus 3000

KEY WORDS
chiral; coated capillary

REFERENCE
Aturki,Z.; Desiderio,C.; Mannina,L.; Fanali,S. Chiral separations by capillary zone electrophoresis with the use of cyanoethylated-β-cyclodextrin as chiral selector, *J.Chromatogr.A*, **1998**, *817*, 91–104.

SAMPLE
Matrix: solutions

CAPILLARY ELECTROPHORESIS
Capillary: 29-36 cm × 50 μm fused-silica (24.5-31.5 cm to detector) (Yongnian Optical Conductive Fiber Plant, China) coated with polyacrylamide
Capillary preparation: Coat capillary as follows. Adjust the pH of 20 mL water to 3.5 with acetic acid, add 80 μL 3-(trimethoxysilyl)propyl methacrylate (3-methacryloxypropyltrimethoxysilane), mix, suck into capillary, let stand at room temperature for 1 h, remove the solution, wash with water. Fill the capillary with a deaerated 3-4% acrylamide solution containing 1 μL/mL N,N,N',N'-tetramethylethylenediamine and 1 mg/mL potassium persulfate, let stand for 30 min, remove excess solution by aspiration, rinse with water, remove water by aspiration, dry at 35° (J. Chromatogr. 1985, 347, 191).
Capillary temperature: 25
Running buffer: 100 mM pH 2.5 NaH$_2$PO$_4$ (A) or 100 mM pH 2.5 NaH$_2$PO$_4$ containing 45 mM hydroxypropyl-α-cyclodextrin (Wacker, Munich) (B)
Injection: Electromigration at 15 kV for 3 s.
Detector: UV 200; UV 210
Migration time: 6.35 (A); 7.28, 7.45 (B) (enantiomers)
Voltage: 15 kV
Model: Bio-Focus 3000

OTHER SUBSTANCES
Also analyzed: albuterol (salbutamol), alprenolol, amorolfine, atenolol, atropine, azelastine, baclofen, bamethan, benproperine, benserazide, biperiden, bisoprolol, brompheniramine, bupivacaine, bupranolol, butamirate, butethamate, carazolol, carbuterol, carteolol, carvedilol, celiprolol, chloroquine, chlorpheniramine, chlorphenoxamine, cicletanine, clidinium bromide, clobutinol, dimethindene, dipivefrin, disopyramide, dobutamine, doxylamine, fendiline, flecainide, gallopamil, homatropine, ipratropium bromide, isoproterenol (isoprenaline), isothipendyl,

ketamine, meclizine, mefloquine, mepindolol, mequitazine, metaclazepam, metaproterenol (orciprenaline), metipranolol, metoprolol, nafronyl (naftidrofuryl), nefopam, nicardipine, norfenefrine, ofloxacin, ornidazole, orphenadrine, oxomemazine, oxprenolol, oxybutynin, phenoxybenzamine, phenylpropanolamine, pholedrine, pindolol, pirbuterol, prilocaine, procyclidine, promethazine, propafenone, propranolol, reproterol, sotalol, sulpride, synephrine, talinolol, terbutaline, tetrahydrozoline (tetryzoline), theodrenaline, tioconazole, tocainide, trihexyphenidyl, trimeprazine (alimemazine), trimipramine, tropicamide, verapamil, zopiclone

KEY WORDS
coated capillary; chiral

REFERENCE
Koppenhoefer,B.; Eperlein,U.; Schlunk,R.; Zhu,X.; Lin,B. Separation of enantiomers of drugs by capillary electrophoresis. V. Hydroxypropyl-α-cyclodextrin as chiral solvating agent, *J.Chromatogr.A*, **1998**, *793*, 153–164.

SAMPLE
Matrix: urine
Sample preparation: Mix 5-10 mL urine with 200 μL 5 M NaOH, add 10 mL chloroform, shake vigorously for 10 min. Remove the organic layer and evaporate it to dryness under a stream of nitrogen at 40°, reconstitute the residue in 50 μL water, sonicate for 2 min, inject an aliquot.

CAPILLARY ELECTROPHORESIS
Capillary: 64 cm × 50 μm fused-silica (55 cm to bubble-cell detector)
Capillary preparation: Rinse with running buffer for 2 min before each run. Replace running buffer after every 3 runs.
Capillary temperature: 25
Running buffer: 25 mM pH 6.0 Sodium citrate buffer
Injection: Pressure injection at 50 mbar for 25 s.
Detector: UV 210
Migration time: 5.5
Voltage: 30 kV
Model: Hewlett-Packard HP[3D]
Limit of detection: 0.5 ppb

KEY WORDS
human; cow

REFERENCE
Lalljie,S.P.D.; Vindevogel,J.; Sandra,P. Capillary electrophoresis as a fast screening method for the determination of clenbuterol in urine, *J.Capillary Electrophor.*, **1994**, *1*, 241–245.

Clentiazem

Molecular formula: $C_{22}H_{25}ClN_2O_4S$
Molecular weight: 448.97
CAS Registry No.: 96125-53-0, 96128-92-6 (maleate)
Merck Index (12th ed.): 2408
Lednicer: 5 139

SAMPLE
Matrix: solutions
Sample preparation: Inject an aliquot of a 100 μg/mL solution in water or running buffer.

CAPILLARY ELECTROPHORESIS
Capillary: 47 cm × 75 μm fused-silica (40 cm to detector)

Capillary preparation: Before each run rinse capillary with running buffer for 1-2 min. If peak tailing is observed fill capillary with 500 mM NaOH and let stand for 30 min, wash with water for 3 min, rinse with running buffer for 3 min.
Capillary temperature: 25
Running buffer: 20 mM pH 2.5 Phosphate buffer containing 3% dextrin (Japan Pharmacopeia grade) (Dissolve dextrin in buffer at 90° then cool to room temperature.)
Injection: Pressure injection at 0.5 psi for 2-4 s.
Detector: UV 220
Migration time: 7.795 (S,S), 8.349 (R,R)
Voltage: 20 kV
Model: Beckman P/ACE 5510

KEY WORDS
chiral

REFERENCE
Nishi,H.; Izumoto,S.; Nakamura,K.; Nakai,H.; Sato,T. Dextran and dextrin as chiral selectors in capillary zone electrophoresis, *Chromatographia*, **1996**, *42*, 617–630.

SAMPLE
Matrix: solutions
Sample preparation: Inject an aliquot of a 100 µg/mL solution in MeOH:water 10:90.

CAPILLARY ELECTROPHORESIS
Capillary: 47 cm × 75 µm fused-silica (40 cm to detector)
Capillary preparation: Rinse capillary with running buffer for 1 min before each run. At the end of each day wash with 100 mM NaOH and with water.
Capillary temperature: 20
Running buffer: 20 mM pH 2.9 Phosphate buffer containing 3% chondroitin sulfate C (sodium salt) (Nacalai Tesque, Kyoto)
Injection: Pressure injection at 0.5 psi for 5 s.
Detector: UV 240
Migration time: 18.04, 20.18 (enantiomers)
Voltage: 20 kV
Model: Beckman P/ACE 5510

OTHER SUBSTANCES
Also analyzed: diltiazem, primaquine, propranolol, sulconazole, trimetoquinol, verapamil

KEY WORDS
chiral

REFERENCE
Nishi,H. Enantiomer separation of basic drugs by capillary electrophoresis using ionic and neutral polysaccharides as chiral selectors, *J.Chromatogr.A*, **1996**, *735*, 345–351.

Clidinium bromide

Molecular formula: $C_{22}H_{26}BrNO_3$
Molecular weight: 432.36
CAS Registry No.: 3485-62-9
Merck Index (12th ed.): 2412

SAMPLE
Matrix: bulk
Sample preparation: 100 mg Bulk clidinium bromide + 1 mL 1 mg/mL tetraethylammonium iodide in water, add water, sonicate for <1 min, make up to 10 mL with water, inject an aliquot.

CAPILLARY ELECTROPHORESIS
Capillary: 70 cm × 75 μm
Capillary preparation: Between runs rinse at 10 psi with 1 M NaOH for 5 min, with water for 5 min, and with running buffer for 5 min.
Running buffer: 10 mM NaH_2PO_4 containing 5 mM benzyltrimethylammonium bromide (pH not adjusted)
Injection: Hydrodynamic injection at 5 cm for 10 s (6 nL).
Detector: UV 205
Migration time: 8
Internal standard: tetraethylammonium iodide (used to quantitate impurities) (7.1)
Voltage: 15 kV
Current: 12 μA
Model: Dionex Model CES-1

OTHER SUBSTANCES
Simultaneous: impurities

KEY WORDS
indirect UV detection

REFERENCE
Nickerson,B. The determination of a degradation product in clidinium bromide drug substance by capillary electrophoresis with indirect UV detection, *J.Pharm.Biomed.Anal.*, **1997**, *15*, 965–971.

SAMPLE
Matrix: solutions
Sample preparation: Inject an aliquot of a 100 μg/mL solution in water:running buffer 50:50.

CAPILLARY ELECTROPHORESIS
Capillary: 44.5 cm × 50 μm acrylamide-coated fused-silica (Bio-Rad)
Capillary temperature: 30
Running buffer: 100 mM NaH_2PO_4 containing 15 mM gamma-cyclodextrin, adjusted to pH 2.5 with phosphoric acid
Injection: Electrokinetic injection at 8 kV for 6 s.
Detector: UV 200
Migration time: 11.98, 12.23 (enantiomers)
Voltage: 14 kV
Model: Bio-Rad BioFocus 3000

OTHER SUBSTANCES
Also analyzed: albuterol, alprenolol, atenolol, atropine, baclofen, bamethan, benserazide, biperiden, bisoprolol, bupivacaine, bupranolol, butetamate, carazolol, carbuterol, carvedilol, celiprolol, chloroquine, chlorpheniramine (chlorphenamine), clobutinol, disopyramide, dobutamine, flecainide, homatropine, ipratropium bromide, isoproterenol, isothipendyl, ketamine, mefloquine, mequitazine, metaproterenol (orciprenaline), metipranolol, nafronyl (naftidrofuryl), nefopam, ofloxacin, orphenadrine, oxomemazine, oxprenolol, phenoxybenzamine, pholedrine, pindolol, pirbuterol, prilocaine, promethazine, propafenone, propranolol, sotalol, synephrine, terbutaline, tetrahydrozoline (tetryzoline), tocainide, trihexyphenidyl, trimeprazine (alimemazine), trimipramine, tropicamide, verapamil, zopiclone

KEY WORDS
chiral; coated capillary

REFERENCE
Koppenhoefer,B.; Epperlein,U.; Christian,B.; Yibing,J.; Yuying,C.; Bingcheng,L. Separation of enantiomers of drugs by capillary electrophoresis. I. γ-Cyclodextrin as chiral solvating agent, *J.Chromatogr.A*, **1995**, *717*, 181–190.

SAMPLE
Matrix: solutions
Sample preparation: Inject an aliquot of a 100 μg/mL solution in running buffer:water 50:50.

CAPILLARY ELECTROPHORESIS
Capillary: 44.5 cm × 50 µm polyacrylamide coated fused-silica (40 cm to detector) (Bio-Rad)
Running buffer: 100 mM NaH_2PO_4 containing 15 mM α-cyclodextrin, adjusted to pH 2.5 with
   phosphoric acid
Injection: Electromigration at 8 kV for 6 s
Detector: UV 200
Migration time: 11.58, 11.82 (enantiomers)
Voltage: 14 kV
Model: Bio-Rad Bio-Focus 3000

OTHER SUBSTANCES
Simultaneous: ketamine, orphenadrine, oxomemazine, tetrahydrozoline, tropicamide

KEY WORDS
chiral; α = 1.021; numerous drugs were not separated into their enantiomers under these conditions; coated capillary

REFERENCE
Bingcheng,L.; Yibing,J.; Yoying,C.; Epperlein,U.; Koppenhoefer,B. Separation of drug enantiomers by capillary
   electrophoresis: α-Cyclodextrin as chiral solvating agent, *Chromatographia*, **1996**, *42*, 106–110.

SAMPLE
Matrix: solutions
Sample preparation: Inject an aliquot of a 100 µg/mL solution in running buffer.

CAPILLARY ELECTROPHORESIS
Capillary: 36 cm × 50 µm fused-silica (31.5 cm to detector) (Yongnian Optical Conductive Fiber
   Plant, China), coated with polyacrylamide
Capillary preparation: No details of the polyacrylamide coating process are provided. However,
   another paper (LC.GC 1997, 15, 40) by this group indicates that they use the procedure of
   Hjertén, thus: Adjust the pH of 20 mL water to 3.5 with acetic acid, add 80 µL 3-(trimethoxysilyl)propyl methacrylate (3-methacryloxypropyltrimethoxysilane), mix, suck into capillary, let
   stand at room temperature for 1 h, remove the solution, wash with water. Fill the capillary
   with a deaerated 3-4% acrylamide solution containing 1 µL/mL N,N,N',N'-tetramethylethyl-
   enediamine and 1 mg/mL potassium persulfate, let stand for 30 min, remove excess solution
   by aspiration, rinse with water, remove water by aspiration, dry at 35° (J. Chromatogr. 1985,
   347, 191).
Capillary temperature: 25
Running buffer: 100 mM NaH_2PO_4 adjusted to pH 2.5 (A) or 100 mM NaH_2PO_4 containing 45
   mM hydroxypropyl-gamma-cyclodextrin, adjusted to pH 2.5 (B)
Injection: Electrokinetic injection at 15 kV for 3 s.
Detector: UV 200, UV 210
Migration time: 6.42 (A), 10.65, 10.93 (B, enantiomers)
Voltage: 15 kV
Model: Bio-Rad BioFocus 3000

OTHER SUBSTANCES
Simultaneous: azelastine, biperiden, carvedilol, meclizine (meclozine), mequitazine, ofloxacin,
   zopiclone

KEY WORDS
chiral; coated capillary

REFERENCE
Koppenhoefer,B.; Epperlein,U.; Xiaofeng,Z.; Bingcheng,L. Separation of enantiomers of drugs by capillary electrophoresis. Part 4: Hydroxypropyl-γ-cyclodextrin as chiral solvating agent, *Electrophoresis*, **1997**, *18*, 924–
   930.

SAMPLE
Matrix: solutions
Sample preparation: Inject an aliquot of a 100 µg/mL solution in running buffer.

CAPILLARY ELECTROPHORESIS
Capillary: 30 cm × 50 μm fused-silica (25.5 cm to detector), coated with polyacrylamide
Capillary preparation: Adjust the pH of 20 mL water to 3.5 with acetic acid, add 80 μL 3-(trimethoxysilyl)propyl methacrylate (3-methacryloxypropyltrimethoxysilane), mix, suck into capillary, let stand at room temperature for 1 h, remove the solution, wash with water. Fill the capillary with a deaerated 3-4% acrylamide solution containing 1 μL/mL N,N,N',N'-tetramethylethylenediamine and 1 mg/mL potassium persulfate, let stand for 30 min, remove excess solution by aspiration, rinse with water, remove water by aspiration, dry at 35° (J. Chromatogr. 1985, 347, 191).
Capillary temperature: 25
Running buffer: 100 mM NaH_2PO_4 containing 45 mM hydroxypropyl-gamma- cyclodextrin, adjusted to pH 2.5 with phosphoric acid
Injection: Electrokinetic injection at 15 kV for 3 s.
Detector: UV 200
Voltage: 15 kV
Model: Bio-Rad BioFocus 3000

KEY WORDS
chiral; coated capillary; comparison with the use of other cyclodextrins; this running buffer gave the greatest enantiomeric separation.; α=1.026

REFERENCE
Lin,B.; Zhu,X.; Koppenhoefer,B.; Epperlein,U. Investigation of 123 chiral drugs by cyclodextrin-modified capillary electrophoresis, *LC.GC*, **1997**, *15*, 40–46.

SAMPLE
Matrix: solutions

CAPILLARY ELECTROPHORESIS
Capillary: 29-36 cm × 50 μm fused-silica (24.5-31.5 cm to detector) (Yongnian Optical Conductive Fiber Plant, China) coated with polyacrylamide
Capillary preparation: Coat capillary as follows. Adjust the pH of 20 mL water to 3.5 with acetic acid, add 80 μL 3-(trimethoxysilyl)propyl methacrylate (3-methacryloxypropyltrimethoxysilane), mix, suck into capillary, let stand at room temperature for 1 h, remove the solution, wash with water. Fill the capillary with a deaerated 3-4% acrylamide solution containing 1 μL/mL N,N,N',N'-tetramethylethylenediamine and 1 mg/mL potassium persulfate, let stand for 30 min, remove excess solution by aspiration, rinse with water, remove water by aspiration, dry at 35° (J. Chromatogr. 1985, 347, 191).
Capillary temperature: 25
Running buffer: 100 mM pH 2.5 NaH_2PO_4 (A) or 100 mM pH 2.5 NaH_2PO_4 containing 45 mM hydroxypropyl-α-cyclodextrin (Wacker, Munich) (B)
Injection: Electromigration at 15 kV for 3 s.
Detector: UV 200; UV 210
Migration time: 6.42 (A); 10.59 (B) (no separation of enantiomers)
Voltage: 15 kV
Model: Bio-Focus 3000

OTHER SUBSTANCES
Also analyzed: albuterol (salbutamol), alprenolol, amorolfine, atenolol, atropine, azelastine, baclofen, bamethan, benproperine, benserazide, biperiden, bisoprolol, brompheniramine, bupivacaine, bupranolol, butamirate, butethamate, carazolol, carbuterol, carteolol, carvedilol, celiprolol, chloroquine, chlorpheniramine, chlorphenoxamine, cicletanine, clenbuterol, clobutinol, dimethindene, dipivefrin, disopyramide, dobutamine, doxylamine, fendiline, flecainide, gallopamil, homatropine, ipratropium bromide, isoproterenol (isoprenaline), isothipendyl, ketamine, meclizine, mefloquine, mepindolol, mequitazine, metaclazepam, metaproterenol (orciprenaline), metipranolol, metoprolol, nafronyl (naftidrofuryl), nefopam, nicardipine, norfenefrine, ofloxacin, ornidazole, orphenadrine, oxomemazine, oxprenolol, oxybutynin, phenoxybenzamine, phenylpropanolamine, pholedrine, pindolol, pirbuterol, prilocaine, procyclidine, promethazine, propafenone, propranolol, reproterol, sotalol, sulpride, synephrine, talinolol, terbutaline, tetrahydrozoline (tetryzoline), theodrenaline, tioconazole, tocainide, trihexyphenidyl, trimeprazine (alimemazine), trimipramine, tropicamide, verapamil, zopiclone

KEY WORDS
coated capillary

REFERENCE
Koppenhoefer,B.; Eperlein,U.; Schlunk,R.; Zhu,X.; Lin,B. Separation of enantiomers of drugs by capillary electrophoresis. V. Hydroxypropyl-α-cyclodextrin as chiral solvating agent, *J.Chromatogr.A*, **1998**, *793*, 153–164.

Clindamycin

Molecular formula: $C_{18}H_{33}ClN_2O_5S$
Molecular weight: 424.99
CAS Registry No.: 18323-44-9, 58207-19-5 (HCl monohydrate), 36688-78-5 (palmitate), 25507-04-4 (palmitate HCl), 24729-96-2 (phosphate)
Merck Index (12th ed.): 2414

SAMPLE
Matrix: bulk
Sample preparation: Prepare a solution containing 0.18 mg/mL clindamycin phosphate and 15 mg/mL sucrose, inject an aliquot.

CAPILLARY ELECTROPHORESIS
Capillary: 70 cm × 50 μm fused silica (45 cm to detector) (Polymicro Technologies or Isco)
Capillary preparation: Rinse with running buffer for 3 min between runs.
Capillary temperature: 34
Running buffer: 75 mM pH 9 Sodium borate buffer
Injection: Vacuum injection at 25 kPa s
Detector: UV 195
Migration time: 8.5 (clindamycin), 16 (clindamycin phosphate)
Internal standard: sucrose (9.4)
Voltage: +15 kV
Model: Isco Model 3140 Electropherograph

OTHER SUBSTANCES
Simultaneous: degradation products, impurities, lincomycin

REFERENCE
Flurer,C.L.; Wolnik,K.A. Chemical profiling of pharmaceuticals by capillary electrophoresis in the determination of drug origin, *J.Chromatogr.A*, **1994**, *674*, 153–163.

Clobazam

Molecular formula: $C_{16}H_{13}ClN_2O_2$
Molecular weight: 300.74
CAS Registry No.: 22316-47-8
Merck Index (12th ed.): 2417
Lednicer: 2 406

SAMPLE
Matrix: blood
Sample preparation: 500 μL Serum + 150 ng flunitrazepam + 250 μL 800 mM pH 10.0 N-cyclohexyl-2-hydroxy-3-aminopropanesulfonic acid (CAPSO) buffer + 4 mL n-pentane:ethyl acetate 75:25, vortex for 1 min, centrifuge at 25° at 1500 g for 5 min. Remove the organic layer

and evaporate it to dryness at 28°, reconstitute the residue in 40 μL 50 mM pH 9.5 borate buffer containing 4.5 mM sodium dodecyl sulfate, inject an aliquot.

CAPILLARY ELECTROPHORESIS
Capillary: 47 cm × 50 μm fused-silica (40 cm to detector) (Supelco)
Capillary preparation: Before each run wash with 100 mM NaOH, water, and running buffer.
Capillary temperature: 25
Running buffer: MeCN:50 mM pH 9.5 sodium borate buffer containing 18 mM sodium dodecyl sulfate 14:86
Injection: Pressure injection at 0.5 psi for 10 s.
Detector: UV 214
Migration time: 7.5
Internal standard: flunitrazepam (5.7)
Voltage: 20 kV
Model: Beckman P/ACE 5010
Limit of quantitation: 15 ng/mL

OTHER SUBSTANCES
Extracted: clonazepam, desmethylclobazam, desmethyldiazepam, diazepam, nitrazepam
Noninterfering: carbamazepine, ethosuximide, phenobarbital, phenytoin, primidone, valproic acid, zonisamide

KEY WORDS
serum

REFERENCE
Imazawa,M.; Hatanaka,Y. Micellar electrokinetic capillary chromatography of benzodiazepine antiepileptics and their desmethyl metabolites in blood, *J.Pharm.Biomed.Anal.*, **1997**, *15*, 1503–1508.

SAMPLE
Matrix: solutions
Sample preparation: Inject an aliquot of a 5 μg/mL solution in running buffer.

CAPILLARY ELECTROPHORESIS
Capillary: 87 cm × 75 μm fused-silica (80 cm to detector) (Beckman)
Capillary preparation: Between runs rinse capillary with running buffer for 2 min then equilibrate for 5 min.
Capillary temperature: 35
Running buffer: 20 mM pH 7 Buffer containing 15 mM sodium cholate and 35 mM sodium deoxycholate (Buffer was 20 mM sodium borate adjusted to pH 7.0 with 20 mM NaH_2PO_4.)
Injection: Pressure injection for 2 s.
Detector: UV 214
Migration time: 16
Voltage: 20 kV
Model: Beckman P/ACE 2100

OTHER SUBSTANCES
Simultaneous: bromazepam, chlordiazepoxide, clonazepam, diazepam, flunitrazepam, flurazepam, halazepam, lorazepam, lormetazepam, nitrazepam, nordazepam, temazepam

REFERENCE
Boonkerd,S.; Detaevernier,M.R.; Vindevogel,J.; Michotte,Y. Migration behaviour of benzodiazepines in micellar electrokinetic chromatography, *J.Chromatogr.A*, **1996**, *756*, 279–286.

SAMPLE
Matrix: solutions

CAPILLARY ELECTROPHORESIS
Capillary: 47 cm × 75 μm fused-silica (40 cm to detector) (Beckman)
Capillary preparation: Rinse with running buffer for 1 min before injection. Wash with 100 mM NaOH for 1 min and with water for 1 min between injections.

Capillary temperature: 30
Running buffer: THF:buffer 1:99 (Buffer was 50 mM pH 9.2 borate containing 50 mM sodium dodecyl sulfate, 2 M urea, and 20 mM gamma-cyclodextrin.)
Injection: Pressure injection at 300 kPa for 5 s.
Detector: UV 254
Migration time: 16
Voltage: 15 kV
Model: Beckman P/ACE 2000

OTHER SUBSTANCES
Simultaneous: alprazolam, bromazepam, chlordiazepoxide, clonazepam, clorazepate, oxazepam, prazepam, triazolam

---

REFERENCE
Renou-Gonnord,M.F.; David,K. Optimized micellar electrokinetic chromatographic separation of benzodiazepines, *J.Chromatogr.A*, **1996**, *735*, 249–261.

Clobetasol

Molecular formula: $C_{22}H_{28}ClFO_4$
Molecular weight: 410.91
CAS Registry No.: 25122-41-2, 25122-46-7 (propionate)
Merck Index (12th ed.): 2423
Lednicer: 4 72

SAMPLE
Matrix: solutions
Sample preparation: Inject an aliquot of a 100 μg/mL solution in running buffer.

CAPILLARY ELECTROPHORESIS
Capillary: 27 cm × 50 μm fused-silica (20 cm to detector) (Composite Metal Services, Hallow, UK) packed with 3 μm Hypersil ODS for 20 cm (details in paper)
Running buffer: MeCN:2 mM pH 7.8 phosphate buffer 80:20
Injection: Electrokinetic injection at 5 kV for 5 s.
Detector: UV 214
Migration time: 17.25 (clobetasol butyrate)
Voltage: 10 kV
Model: Beckman P/ACE 2050

OTHER SUBSTANCES
Simultaneous: betamethasone, betamethasone dipropionate, betamethasone-17-valerate, clobetasone butyrate, fluticasone propionate, hydrocortisone, prednisolone

---

KEY WORDS
electrochromatography

REFERENCE
Frame,L.A.; Robinson,M.L.; Lough,W.J. Simplification of capillary electrochromatography procedures, *J.Chromatogr.A*, **1998**, *798*, 243–249.

Clobetasone

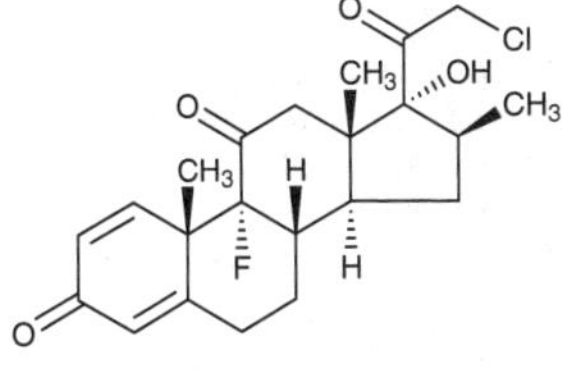

Molecular formula: $C_{22}H_{26}ClFO_4$
Molecular weight: 408.90
CAS Registry No.: 54063-32-0, 25122-57-0 (17-butyrate)
Merck Index (12th ed.): 2424
Lednicer: 4 72

SAMPLE
Matrix: solutions
Sample preparation: Inject an aliquot of a 100 μg/mL solution in running buffer.

CAPILLARY ELECTROPHORESIS
Capillary: 27 cm × 50 μm fused-silica (20 cm to detector) (Composite Metal Services, Hallow, UK) packed with 3 μm Hypersil ODS for 20 cm (details in paper)
Running buffer: MeCN:2 mM pH 7.8 phosphate buffer 80:20
Injection: Electrokinetic injection at 5 kV for 5 s.
Detector: UV 214
Migration time: 23.67 (clobetasone butyrate)
Voltage: 10 kV
Model: Beckman P/ACE 2050

OTHER SUBSTANCES
Simultaneous: betamethasone, betamethasone dipropionate, betamethasone-17-valerate, clobetasol butyrate, fluticasone propionate, hydrocortisone, prednisolone

---

KEY WORDS
electrochromatography

REFERENCE
Frame,L.A.; Robinson,M.L.; Lough,W.J. Simplification of capillary electrochromatography procedures, *J.Chromatogr.A*, **1998**, *798*, 243–249.

Clobutinol

Molecular formula: $C_{14}H_{22}ClNO$
Molecular weight: 255.79
CAS Registry No.: 14860-49-2, 1215-83-4 (HCl)
Merck Index (12th ed.): 2425
Lednicer: 2 121

SAMPLE
Matrix: solutions
Sample preparation: Inject an aliquot of a 100 μg/mL solution in running buffer.

CAPILLARY ELECTROPHORESIS
Capillary: 29 cm × 50 μm fused-silica (24.5 cm to detector) (Yongnian Optical Conductive Fiber Plant, China), coated with polyacrylamide
Capillary preparation: No details of the polyacrylamide coating process are provided. However, another paper (LC.GC 1997, 15, 40) by this group indicates that they use the procedure of Hjertén, thus: Adjust the pH of 20 mL water to 3.5 with acetic acid, add 80 μL 3-(trimethoxysilyl)propyl methacrylate (3-methacryloxypropyltrimethoxysilane), mix, suck into capillary, let stand at room temperature for 1 h, remove the solution, wash with water. Fill the capillary with a deaerated 3-4% acrylamide solution containing 1 μL/mL N,N,N',N'-tetramethylethylenediamine and 1 mg/mL potassium persulfate, let stand for 30 min, remove excess solution

by aspiration, rinse with water, remove water by aspiration, dry at 35° (J. Chromatogr. 1985, 347, 191).
Capillary temperature: 25
Running buffer: 100 mM NaH_2PO_4 adjusted to pH 2.5
Injection: Electrokinetic injection at 15 kV for 3 s.
Detector: UV 200, UV 210
Migration time: 4.85
Voltage: 15 kV
Model: Bio-Rad BioFocus 3000

OTHER SUBSTANCES
Simultaneous: albuterol, alprenolol, atenolol, baclofen, bamethan, benproperine, benserazide, bisoprolol, bupranolol, butamirate, butethamate, carbuterol, celiprolol, clenbuterol, dipivefrin, isoproterenol (isoprenaline), metaproterenol (orciprenaline), metipranolol, metoprolol, norfenefrine, ornidazole, oxprenolol, phenylpropanolamine, pholedrine, pirbuterol, prilocaine, procyclidine, sotalol, synephrine, terbutaline, tocainide

KEY WORDS
coated capillary

REFERENCE
Koppenhoefer,B.; Epperlein,U.; Xiaofeng,Z.; Bingcheng,L. Separation of enantiomers of drugs by capillary electrophoresis. Part 4: Hydroxypropyl-γ-cyclodextrin as chiral solvating agent, *Electrophoresis*, **1997**, *18*, 924–930.

SAMPLE
Matrix: solutions
Sample preparation: Inject an aliquot of a 100 µg/mL solution in running buffer.

CAPILLARY ELECTROPHORESIS
Capillary: 30 cm × 50 µm fused-silica (25.5 cm to detector), coated with polyacrylamide
Capillary preparation: Adjust the pH of 20 mL water to 3.5 with acetic acid, add 80 µL 3-(trimethoxysilyl)propyl methacrylate (3-methacryloxypropyltrimethoxysilane), mix, suck into capillary, let stand at room temperature for 1 h, remove the solution, wash with water. Fill the capillary with a deaerated 3-4% acrylamide solution containing 1 µL/mL N,N,N',N'-tetramethylethylenediamine and 1 mg/mL potassium persulfate, let stand for 30 min, remove excess solution by aspiration, rinse with water, remove water by aspiration, dry at 35° (J. Chromatogr. 1985, 347, 191).
Capillary temperature: 25
Running buffer: 100 mM NaH_2PO_4 containing 45 mM hydroxypropyl-α- cyclodextrin, adjusted to pH 2.5 with phosphoric acid
Injection: Electrokinetic injection at 15 kV for 3 s.
Detector: UV 200
Voltage: 15 kV
Model: Bio-Rad BioFocus 3000

KEY WORDS
chiral; coated capillary; comparison with the use of other cyclodextrins; this running buffer gave the greatest enantiomeric separation.; α=1.013

REFERENCE
Lin,B.; Zhu,X.; Koppenhoefer,B.; Epperlein,U. Investigation of 123 chiral drugs by cyclodextrin-modified capillary electrophoresis, *LC.GC*, **1997**, *15*, 40–46.

SAMPLE
Matrix: solutions

CAPILLARY ELECTROPHORESIS
Capillary: 29-36 cm × 50 µm fused-silica (24.5-31.5 cm to detector) (Yongnian Optical Conductive Fiber Plant, China) coated with polyacrylamide
Capillary preparation: Coat capillary as follows. Adjust the pH of 20 mL water to 3.5 with acetic acid, add 80 µL 3-(trimethoxysilyl)propyl methacrylate (3-methacryloxypropyltrimethox-

ysilane), mix, suck into capillary, let stand at room temperature for 1 h, remove the solution, wash with water. Fill the capillary with a deaerated 3-4% acrylamide solution containing 1 μL/mL N,N,N',N'-tetramethylethylenediamine and 1 mg/mL potassium persulfate, let stand for 30 min, remove excess solution by aspiration, rinse with water, remove water by aspiration, dry at 35° (J. Chromatogr. 1985, 347, 191).
Capillary temperature: 25
Running buffer: 100 mM pH 2.5 NaH_2PO_4 (A) or 100 mM pH 2.5 NaH_2PO_4 containing 45 mM hydroxypropyl-α-cyclodextrin (Wacker, Munich) (B)
Injection: Electromigration at 15 kV for 3 s.
Detector: UV 200; UV 210
Migration time: 4.85 (A); 17.75, 17.98 (B) (enantiomers)
Voltage: 15 kV
Model: Bio-Focus 3000

OTHER SUBSTANCES

Also analyzed: albuterol (salbutamol), alprenolol, amorolfine, atenolol, atropine, azelastine, baclofen, bamethan, benproperine, benserazide, biperiden, bisoprolol, brompheniramine, bupivacaine, bupranolol, butamirate, butethamate, carazolol, carbuterol, carteolol, carvedilol, celiprolol, chloroquine, chlorpheniramine, chlorphenoxamine, cicletanine, clenbuterol, clidinium bromide, dimethindene, dipivefrin, disopyramide, dobutamine, doxylamine, fendiline, flecainide, gallopamil, homatropine, ipratropium bromide, isoproterenol (isoprenaline), isothipendyl, ketamine, meclizine, mefloquine, mepindolol, mequitazine, metaclazepam, metaproterenol (orciprenaline), metipranolol, metoprolol, nafronyl (naftidrofuryl), nefopam, nicardipine, norfenefrine, ofloxacin, ornidazole, orphenadrine, oxomemazine, oxprenolol, oxybutynin, phenoxybenzamine, phenylpropanolamine, pholedrine, pindolol, pirbuterol, prilocaine, procyclidine, promethazine, propafenone, propranolol, reproterol, sotalol, sulpride, synephrine, talinolol, terbutaline, tetrahydrozoline (tetryzoline), theodrenaline, tioconazole, tocainide, trihexyphenidyl, trimeprazine (alimemazine), trimipramine, tropicamide, verapamil, zopiclone

---

KEY WORDS
coated capillary; chiral

REFERENCE
Koppenhoefer,B.; Eperlein,U.; Schlunk,R.; Zhu,X.; Lin,B. Separation of enantiomers of drugs by capillary electrophoresis. V. Hydroxypropyl-α-cyclodextrin as chiral solvating agent, *J.Chromatogr.A*, **1998**, *793*, 153–164.

Clomiphene

Molecular formula: $C_{26}H_{28}ClNO$
Molecular weight: 405.97
CAS Registry No.: 911-45-5, 50-41-9 (citrate)
Merck Index (12th ed.): 2446
Lednicer: 1 105

SAMPLE
Matrix: solutions

CAPILLARY ELECTROPHORESIS
Capillary: 72 cm × 50 μm fused-silica (50 cm to detector)
Capillary preparation: Before each run wash capillary with 100 mM NaOH for 4 min and with running buffer for 4 min.
Capillary temperature: 30
Running buffer: 100 mM pH 2.3 Phosphate buffer containing 5 mM heptakis(2,3,6-tri-O-methyl)-β-cyclodextrin
Injection: Hydrodynamic injection for 1.7 s.
Detector: UV 254
Migration time: 12.7 (zuclomiphene), 14 (enclomiphene)
Voltage: 30 kV

Model: ABI Model 270A

REFERENCE
Bempong,D.K.; Honigberg,I.L. Multivariate analysis of capillary electrophoresis separation conditions for Z-E isomers of clomiphene, *J.Pharm.Biomed.Anal.*, **1996**, *15*, 233–239.

SAMPLE
Matrix: solutions

CAPILLARY ELECTROPHORESIS
Capillary: 33 cm × 50 μm fused-silica (33 cm to detector)(Polymicro Technologies)
Capillary preparation: Between each run wash capillary with 100 mM NaOH for 1 min, with water for 2 min, with 100 mM HCl for 1 min, with water for 2 min, with 80 mM pH 9 phosphate (?) buffer for 2 min, and with running buffer for 2 min. Between each group of experiments wash capillary with 100 mM NaOH for 5 min, with water for 2 min, with 100 mM HCl for 2 min, with water for 2 min, and with 80 mM pH 9 phosphate (?) buffer for 30 min.
Capillary temperature: 20
Running buffer: 80 mM pH 9 Phosphate (?) buffer containing 2 mM dimethyl-β-cyclodextrin and 5 mM carboxymethylated-β-cyclodextrin (Cyclolab, Budapest)
Detector: UV 210
Migration time: 2.05 (E), 2.11 (Z)
Model: Hewlett Packard HP ^{3D}CE

REFERENCE
Juvancz,Z.; Ürmös,I.; Klebovich,I. Capillary electrophoretic separation of clomiphene isomers using various cyclodextrins as additives, *J.Capillary Electrophor.*, **1996**, *3*, 181–189.

SAMPLE
Matrix: solutions
Sample preparation: Inject an aliquot of a 100 μg/mL solution in MeOH:MeCN 50:50.

CAPILLARY ELECTROPHORESIS
Capillary: 64 cm × 50 μm fused-silica (55.5 cm to detector) (Polymicro Technologies)
Capillary preparation: Between runs flush capillary with running buffer for 2 min. Before use rinse capillary with 1 M NaOH for 1 h, with 100 mM NaOH for 20 min, and with running buffer for 10 min.
Capillary temperature: 25
Running buffer: MeCN:MeOH 50:50 containing 50 mM ammonium acetate and 1 M acetic acid
Injection: Pressure injection at 5 kPa for 3 s.
Detector: UV 214
Voltage: 30 kV
Model: Hewlett-Packard ^{3D}CE

KEY WORDS
cis and trans isomers separated; Rs = 1.29

REFERENCE
Hansen,S.H.; Bjornsdottir,I.; Tjornelund,J. Separation of cationic *cis--trans (Z--E)* isomers and diastereoisomers using non-aqueous capillary electrophoresis, *J.Chromatogr.A*, **1997**, *792*, 49–55.

Clomipramine

Molecular formula: $C_{19}H_{23}ClN_2$
Molecular weight: 314.86
CAS Registry No.: 303-49-1, 17231-77-6 (HCl)
Merck Index (12th ed.): 2447

SAMPLE
Matrix: solutions

CAPILLARY ELECTROPHORESIS
Capillary: 70 cm × 50 μm coated fused-silica (55 cm to detector) (Polymicro Technologies)
Capillary preparation: Between runs rinse with water for 2 min. Coat capillaries as follows. Condition with 1 M NaOH for 15 min, rinse with water for 5 min, treat with 30 μL/mL 3-(trimethoxysilyl)propyl methacrylate in acetic acid:water 50:50 for 1 h using house vacuum, rinse with water, fill with polymerization solution, let stand for 1 h, flush, rinse with water. (Prepare polymerization solution by adding 10 μL N,N,N',N'-tetramethylethylenediamine to 10 mL of a degassed 4% solution of acrylamide in water, add 10 μL 10% ammonium persulfate in water.)
Running buffer: 10 mM pH 6.5 Citrate buffer containing 20 mM sodium dodecyl sulfate and 10 mM carboxymethyl-β-cyclodextrin (Cyclodextrin Technologies Development, Gainesville FL)
Injection: Hydrodynamic injection at 15 cm for 8-10 s.
Detector: UV 254
Migration time: 11.4
Voltage: -20 kV
Model: laboratory-constructed

OTHER SUBSTANCES
Simultaneous: amitriptyline, carbamazepine, desipramine, imipramine, nortriptyline, opipramol, proptriptyline, trimipramine

KEY WORDS
coated capillary

REFERENCE
Spencer,B.J.; Zhang,W.; Purdy,W.C. Capillary electrophoretic separation of tricyclic antidepressants using charged carboxymethyl-β-cyclodextrin as a buffer additive, *Electrophoresis*, **1997**, *18*, 736–744.

Clonazepam

Molecular formula: $C_{15}H_{10}ClN_3O_3$
Molecular weight: 315.72
CAS Registry No.: 1622-61-3
Merck Index (12th ed.): 2449

SAMPLE
Matrix: blood
Sample preparation: 500 μL Serum + 150 ng flunitrazepam + 250 μL 800 mM pH 10.0 N-cyclohexyl-2-hydroxy-3-aminopropanesulfonic acid (CAPSO) buffer + 4 mL n-pentane:ethyl acetate 75:25, vortex for 1 min, centrifuge at 25° at 1500 g for 5 min. Remove the organic layer and evaporate it to dryness at 28°, reconstitute the residue in 40 μL 50 mM pH 9.5 borate buffer containing 4.5 mM sodium dodecyl sulfate, inject an aliquot.

CAPILLARY ELECTROPHORESIS
Capillary: 47 cm × 50 μm fused-silica (40 cm to detector) (Supelco)
Capillary preparation: Before each run wash with 100 mM NaOH, water, and running buffer.
Capillary temperature: 25
Running buffer: MeCN:50 mM pH 9.5 sodium borate buffer containing 18 mM sodium dodecyl sulfate 14:86
Injection: Pressure injection at 0.5 psi for 10 s.
Detector: UV 214
Migration time: 6
Internal standard: flunitrazepam (5.7)
Voltage: 20 kV
Model: Beckman P/ACE 5010

Limit of quantitation: 10 ng/mL

OTHER SUBSTANCES
Extracted: clobazam, desmethylclobazam, desmethyldiazepam, diazepam, nitrazepam
Noninterfering: carbamazepine, ethosuximide, phenobarbital, phenytoin, primidone, valproic
acid, zonisamide

KEY WORDS
serum

REFERENCE
Imazawa,M.; Hatanaka,Y. Micellar electrokinetic capillary chromatography of benzodiazepine antiepileptics
and their desmethyl metabolites in blood, *J.Pharm.Biomed.Anal.*, **1997**, *15*, 1503–1508.

SAMPLE
Matrix: solutions
Sample preparation: Prepare a 1-50 μg/mL solution in MeOH:water 20:80, inject an aliquot.

CAPILLARY ELECTROPHORESIS
Capillary: 44 cm × 50 μm fused-silica
Capillary preparation: After each run wash capillary with water for 2 min and buffer for 3
min. At the beginning of each day equilibrate capillary with running buffer for 10 min. Con-
dition a new capillary by washing with 1 M NaOH, 100 mM NaOH, water, and running buffer.
Capillary temperature: 25
Running buffer: MeOH:buffer 20:80 (Buffer was 75 mm pH 9.0 Glycine containing 250 mM
triethanolamine and 25 mM sodium dodecyl sulfate.)
Injection: Hydrodynamic injection for 5 s
Detector: UV 235
Migration time: 7.4
Voltage: 25 kV
Current: about 15 μA
Model: Spectraphysics Spectraphoresis 1000 CE
Limit of quantitation: 700 ng/mL
Limit of detection: 200 ng/mL

OTHER SUBSTANCES
Simultaneous: bromazepam, brotizolam, clobazam, lorazepam, lormetazepam, nitrazepam, nordazepam, oxazepam, temazepam

---

REFERENCE
Bechet,I.; Fillet,M.; Hubert,P.; Crommen,J. Determination of benzodiazepines by micellar electrokinetic chromatography, *Electrophoresis*, **1994**, *15*, 1316–1321.

SAMPLE
Matrix: solutions
Sample preparation: Inject an aliquot of a 5 μg/mL solution in running buffer.

CAPILLARY ELECTROPHORESIS
Capillary: 87 cm × 75 μm fused-silica (80 cm to detector) (Beckman)
Capillary preparation: Between runs rinse capillary with running buffer for 2 min then equil-
ibrate for 5 min.
Capillary temperature: 35
Running buffer: 20 mM pH 7 Buffer containing 15 mM sodium cholate and 35 mM sodium
deoxycholate (Buffer was 20 mM sodium borate adjusted to pH 7.0 with 20 mM NaH_2PO_4.)
Injection: Pressure injection for 2 s.
Detector: UV 214
Migration time: 16.7
Voltage: 20 kV
Model: Beckman P/ACE 2100

OTHER SUBSTANCES
Simultaneous: bromazepam, chlordiazepoxide, clobazam, diazepam, flunitrazepam, flurazepam, halazepam, lorazepam, lormetazepam, nitrazepam, nordazepam, temazepam

REFERENCE
Boonkerd,S.; Detaevernier,M.R.; Vindevogel,J.; Michotte,Y. Migration behaviour of benzodiazepines in micellar electrokinetic chromatography, *J.Chromatogr.A*, **1996**, *756*, 279–286.

SAMPLE
Matrix: solutions

CAPILLARY ELECTROPHORESIS
Capillary: 60 cm × 50 μm fused-silica (40 cm to detector), C18 coated (Supelco)
Capillary preparation: Rinse new capillaries with 250 μL 1 M NaOH, 250 μL 100 mM NaOH, water, MeOH, water, and running buffer then condition with running buffer for at least 15 min. Carry out a similar procedure between runs.
Running buffer: 100 mM pH 8.5 Borate buffer containing 25 mM sodium dodecyl sulfate and 5 M urea
Injection: Electrokinetic injection at 5 kV for 5 s.
Detector: UV 254
Migration time: 19
Voltage: 18 kV
Model: Jasco

OTHER SUBSTANCES
Simultaneous: alprazolam, amobarbital, barbital, bromazepam, clotiazepam, cloxazolam, diazepam, estazolam, etizolam, fludiazepam, flurazepam, haloxazolam, medazepam, mephobarbital, metharbital, nimetazepam, nitrazepam, oxazepam, pentobarbital, phenobarbital, secobarbital, triazolam
Interfering: flunitrazepam

KEY WORDS
coated capillary

REFERENCE
Jinno,K.; Han,Y.; Nakamura,M. Analysis of anxiolytic drugs by capillary electrophoresis with bare and coated capillaries, *J.Capillary Electrophor.*, **1996**, *3*, 139–145.

SAMPLE
Matrix: solutions

CAPILLARY ELECTROPHORESIS
Capillary: 47 cm × 75 μm fused-silica (40 cm to detector) (Beckman)
Capillary preparation: Rinse with running buffer for 1 min before injection. Wash with 100 mM NaOH for 1 min and with water for 1 min between injections.
Capillary temperature: 30
Running buffer: THF:buffer 1:99 (Buffer was 50 mM pH 9.2 borate containing 50 mM sodium dodecyl sulfate, 2 M urea, and 20 mM gamma-cyclodextrin.)
Injection: Pressure injection at 300 kPa for 5 s.
Detector: UV 254
Migration time: 13
Voltage: 15 kV
Model: Beckman P/ACE 2000

OTHER SUBSTANCES
Simultaneous: alprazolam, bromazepam, chlordiazepoxide, clobazam, clorazepate, oxazepam, prazepam, triazolam

REFERENCE
Renou-Gonnord,M.F.; David,K. Optimized micellar electrokinetic chromatographic separation of benzodiazepines, *J.Chromatogr.A*, **1996**, *735*, 249–261.

SAMPLE
Matrix: solutions
Sample preparation: Inject an aliquot of a 100 μg/mL solution in running buffer.

CAPILLARY ELECTROPHORESIS
Capillary: 60 cm × 50 μm acrylamide-coated fused-silica (40 cm to detector) (Supelco)
Capillary preparation: Adjust the pH of 20 mL water to 3.5 with acetic acid, add 80 μL 3-(trimethoxysilyl)propyl methacrylate (3-methacryloxypropyltrimethoxysilane), mix, suck into capillary, let stand at room temperature for 1 h, remove the solution, wash with water. Fill the capillary with a deaerated 3-4% acrylamide solution containing 1 μL/mL N,N,N',N'-tetramethylethylenediamine and 1 mg/mL ammonium persulfate, let stand for 3 h, remove excess solution by aspiration, rinse with water, remove water by aspiration, dry at 35° (cf. J. Chromatogr. 1985, 347, 191).
Running buffer: MeCN:buffer 5:95 (Buffer was 100 mM borate containing 5 M urea and 10 mM sodium dodecyl sulfate, adjusted to pH 8.5 with phosphate.)
Injection: Electrokinetic injection at 5 kV for 5 s.
Detector: UV 254
Migration time: 21.5
Voltage: 18 kV
Model: Jasco Model 870-CE

OTHER SUBSTANCES
Simultaneous: alprazolam, amobarbital, barbital, bromazepam, clotiazepam, cloxazolam, diazepam, estazolam, etizolam, fludiazepam, flunitrazepam, flurazepam, haloxazolam, medazepam, mephobarbital, metharbital, nimetazepam, nitrazepam, oxazepam, pentobarbital, phenobarbital, secobarbital, triazolam

KEY WORDS
coated capillary; detector at anode

REFERENCE
Jinno,K.; Han,Y.; Sawada,H.; Taniguchi,M. Capillary electrophoretic separation of toxic drugs using a polyacrylamide-coated capillary, *Chromatographia*, **1997**, *46*, 309–314.

SAMPLE
Matrix: urine
Sample preparation: Condition a 130 mg Bond Elut Certify SPE cartridge with 2 mL MeOH and 2 mL 100 mM pH 6 phosphate buffer, do not allow to go dry. Mix urine with an equal amount of 200 mM pH 5.4 buffer, add 20 μL β-glucuronidase/arylsulfatase (Helix pomatia, Boehringer Mannheim) for each 1 mL of urine, heat at 37° for 4 h. 5 (or 10) mL Urine + 2 (or 4) mL 100 mM pH 6 phosphate buffer, add to the SPE cartridge, wash with 1 mL MeOH:100 mM phosphate buffer 20:80, wash with 1 mL 1 M acetic acid, wash with 1 mL hexane, suck dry briefly, wash with 4 mL dichloromethane, wash with 5 mL MeOH, elute with 2 mL dichloromethane:isopropanol 80:20 containing 5% ammonium hydroxide. Evaporate the eluate to dryness under a stream of nitrogen at room temperature, reconstitute the residue in 50-100 μL running buffer, inject an aliquot.

CAPILLARY ELECTROPHORESIS
Capillary: 105 cm × 75 μm fused-silica (68 cm to detector)
Capillary preparation: Before each run rinse capillary with 100 mM NaOH for 3.5 min and with buffer for 5 min.
Running buffer: Isopropanol:buffer 5:95 (Buffer was 10 mM pH 9.2-9.3 Na_2HPO_4 containing 6 mM sodium borate and 75 mM sodium dodecyl sulfate.)
Injection: Vacuum injection for 1 s
Detector: UV 235
Migration time: 12
Voltage: 30 kV
Current: 90 μA
Model: Europhor Prime Vision IV

OTHER SUBSTANCES
Extracted: bromazepam, diazepam, flunitrazepam

REFERENCE
Schafroth,M.; Thormann,W.; Allemann,D. Micellar electrokinetic capillary chromatography of benzodiazepines in human urine, *Electrophoresis*, **1994**, *15*, 72–78.

Clopenthixol

Molecular formula: $C_{22}H_{25}ClN_2OS$
Molecular weight: 400.97
CAS Registry No.: 982-24-1, 633-59-0 (2HCl)
Merck Index (12th ed.): 2455
Lednicer: 1 399

SAMPLE
Matrix: solutions
Sample preparation: Inject an aliquot of a 100 μg/mL solution in MeOH:MeCN 50:50.

CAPILLARY ELECTROPHORESIS
Capillary: 64 cm × 50 μm fused-silica (55.5 cm to detector) (Polymicro Technologies)
Capillary preparation: Between runs flush capillary with running buffer for 2 min. Before use rinse capillary with 1 M NaOH for 1 h, with 100 mM NaOH for 20 min, and with running buffer for 10 min.
Capillary temperature: 25
Running buffer: MeCN:MeOH 50:50 containing 25 mM ammonium chloride
Injection: Pressure injection at 5 kPa for 3 s.
Detector: UV 214
Voltage: 30 kV
Model: Hewlett-Packard [3D]CE

KEY WORDS
cis and trans isomers separated; Rs = 2.76

REFERENCE
Hansen,S.H.; Bjornsdottir,I.; Tjornelund,J. Separation of cationic *cis--trans (Z--E)* isomers and diastereoisomers using non-aqueous capillary electrophoresis, *J.Chromatogr.A*, **1997**, *792*, 49–55.

Clorazepate

Molecular formula: $C_{16}H_{13}ClN_2O_4$
Molecular weight: 332.74
CAS Registry No.: 20432-69-3, 57109-90-7 (dipotassium salt), 5991-71-9 (monopotassium slat)
Merck Index (12th ed.): 2465

SAMPLE
Matrix: solutions

CAPILLARY ELECTROPHORESIS
Capillary: 47 cm × 75 μm fused-silica (40 cm to detector) (Beckman)
Capillary preparation: Rinse with running buffer for 1 min before injection. Wash with 100 mM NaOH for 1 min and with water for 1 min between injections.

Capillary temperature: 30
Running buffer: THF:buffer 1:99 (Buffer was 50 mM pH 9.2 borate containing 50 mM sodium dodecyl sulfate, 2 M urea, and 20 mM gamma-cyclodextrin.)
Injection: Pressure injection at 300 kPa for 5 s.
Detector: UV 254
Migration time: 19
Voltage: 15 kV
Model: Beckman P/ACE 2000

OTHER SUBSTANCES
Simultaneous: alprazolam, bromazepam, chlordiazepoxide, clobazam, clonazepam, oxazepam, prazepam, triazolam

REFERENCE
Renou-Gonnord,M.F.; David,K. Optimized micellar electrokinetic chromatographic separation of benzodiazepines, *J.Chromatogr.A*, **1996**, *735*, 249–261.

Clorprenaline

Molecular formula: $C_{11}H_{16}ClNO$
Molecular weight: 213.71
CAS Registry No.: 3811-25-4, 5588-22-7 (HCl monohydrate)
Merck Index (12th ed.): 2470
Lednicer: 2 39

SAMPLE
Matrix: solutions

CAPILLARY ELECTROPHORESIS
Capillary: 36 cm × 50 µm fused-silica coated with linear polyacrylamide (31.5 cm to detector) (GL Science)
Capillary preparation: At the beginning and end of each day rinse capillary with capillary wash solution (Bio-Rad Cat. No. 148-5022) at 690 kPa for more than 3 min and with water at 690 kPa for more than 3 min. Coat capillary as follows. Treat capillary with 1 M NaOH at room temperature for 1 h, rinse with water, dry by passing nitrogen gas through the capillary at 110° for 6 h. Pass thionyl chloride through the capillary using a suction pump for several min, seal capillary at both ends and heat at 70° for 6 h. Unseal the capillary and fill with 250 mM vinyl magnesium bromide in THF by suction, seal the capillary, heat at 70° for 6 h. Open the capillary and rinse it with THF for several min, rinse with distilled water, fill the capillary with polymerization solution, heat at 28 ± 2° for 1 h, condition at -100 V/cm for 30 min (Anal. Sci. 1994, 10, 1). (The polymerization solution was 5% acrylamide in water containing 49 mM Tris, 384 mM glycine, and 0.1% sodium dodecyl sulfate, degas in an ultrasonic bath. Add 40 µL 10% N,N,N',N'-tetramethylethylenediamine and 10 µL 10% ammonium persulfate to 5 mL of the degassed solution, mix thoroughly.)
Running buffer: EtOH:50 mM pH 6.0 Phosphate buffer 10:90
Injection: Before each injection rinse with water at 690 kPa for 30 s, rinse with running buffer at 690 kPa for 30 s, partially fill with separation solution (300 µM α_1-acid glycoprotein (Cohn fraction VI) (Fluka) in running buffer) at 6.9 kPa for 190 s (27 cm), inject sample at 6.9 kPa for 2 s, electrophorese with running buffer (Note that α_1-acid glycoprotein from other suppliers may provide inferior results).
Detector: UV 210
Migration time: Resolution of enantiomers 2.5
Voltage: 12 kV
Model: Bio-rad BioFocus 3000

KEY WORDS
chiral; coated capillary

REFERENCE
Tanaka,Y.; Terabe,S. Separation of the enantiomers of basic drugs by affinity capillary electrophoresis using a partial filling technique and α_1-acid glycoprotein as chiral selector, *Chromatographia*, **1997**, *44*, 119–128.

Clotiazepam

Molecular formula: $C_{16}H_{15}ClN_2OS$
Molecular weight: 318.83
CAS Registry No.: 33671-46-4
Merck Index (12th ed.): 2477

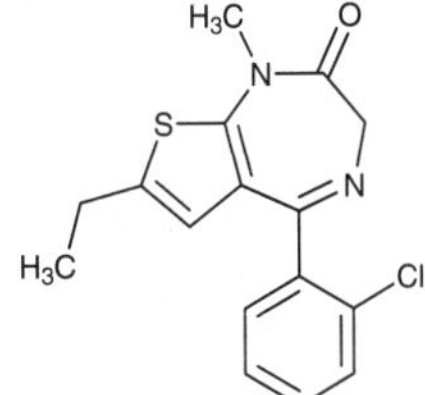

SAMPLE
Matrix: solutions

CAPILLARY ELECTROPHORESIS
Capillary: 60 cm × 50 μm fused-silica (40 cm to detector), C18 coated (Supelco)
Capillary preparation: Rinse new capillaries with 250 μL 1 M NaOH, 250 μL 100 mM NaOH, water, MeOH, water, and running buffer then condition with running buffer for at least 15 min. Carry out a similar procedure between runs.
Running buffer: 100 mM pH 8.5 Borate buffer containing 25 mM sodium dodecyl sulfate and 5 M urea
Injection: Electrokinetic injection at 5 kV for 5 s.
Detector: UV 254
Migration time: 52
Voltage: 18 kV
Model: Jasco

OTHER SUBSTANCES
Simultaneous: alprazolam, amobarbital, barbital, bromazepam, clonazepam, cloxazolam, diazepam, estazolam, etizolam, fludiazepam, flunitrazepam, flurazepam, haloxazolam, medazepam, mephobarbital, metharbital, nimetazepam, nitrazepam, oxazepam, pentobarbital, phenobarbital, secobarbital, triazolam

KEY WORDS
coated capillary

REFERENCE
Jinno,K.; Han,Y.; Nakamura,M. Analysis of anxiolytic drugs by capillary electrophoresis with bare and coated capillaries, *J.Capillary Electrophor.*, **1996**, *3*, 139–145.

SAMPLE
Matrix: solutions

CAPILLARY ELECTROPHORESIS
Capillary: 60 cm × 75 μm coated fused-silica (40 cm to detector) (Supelco)
Capillary preparation: Coat column as follows. Adjust the pH of 20 mL water to 3.5 with acetic acid, add 80 μL 3-(trimethoxysilyl)propyl methacrylate (3-methacryloxypropyltrimethoxysilane), mix, suck into capillary, let stand at room temperature for 1 h, remove the solution, wash with water. Fill the capillary with a deaerated 4% acrylamide solution containing 1 mg/mL N,N,N',N'-tetramethylethylenediamine and 1 mg/mL ammonium persulfate, let stand for 3 h, remove excess solution by aspiration, rinse with water, remove water by aspiration, dry at 35° (cf. J. Chromatogr. 1985, 347, 191).
Running buffer: 100 mM pH 8.5 borate buffer containing 10 mM sodium dodecyl sulfate and 5 M urea
Injection: Electromigration at 5 kV for 5 s.
Detector: UV 240, UV 254
Migration time: 6.8
Voltage: 18 kV

Model: Jasco 890-CE

OTHER SUBSTANCES
Simultaneous: alprazolam, cloxazolam, diazepam, estazolam, etizolam, fludiazepam, flurazepam, haloxazolam, oxazepam

KEY WORDS
injection at cathode; coated capillary

REFERENCE
Jinno,K.; Han,Y.; Sawada,H. Analysis of toxic drugs by capillary electrophoresis using polyacrylamide-coated columns, *Electrophoresis*, **1997**, *18*, 284–286.

SAMPLE
Matrix: solutions
Sample preparation: Inject an aliquot of a 100 μg/mL solution in running buffer.

CAPILLARY ELECTROPHORESIS
Capillary: 60 cm × 50 μm acrylamide-coated fused-silica (40 cm to detector) (Supelco)
Capillary preparation: Adjust the pH of 20 mL water to 3.5 with acetic acid, add 80 μL 3-(trimethoxysilyl)propyl methacrylate (3-methacryloxypropyltrimethoxysilane), mix, suck into capillary, let stand at room temperature for 1 h, remove the solution, wash with water. Fill the capillary with a deaerated 3-4% acrylamide solution containing 1 μL/mL N,N,N',N'-tetramethylethylenediamine and 1 mg/mL ammonium persulfate, let stand for 3 h, remove excess solution by aspiration, rinse with water, remove water by aspiration, dry at 35° (cf. J. Chromatogr. 1985, 347, 191).
Running buffer: MeCN:buffer 5:95 (Buffer was 100 mM borate containing 5 M urea and 10 mM sodium dodecyl sulfate, adjusted to pH 8.5 with phosphate.)
Injection: Electrokinetic injection at 5 kV for 5 s.
Detector: UV 254
Migration time: 8
Voltage: 18 kV
Model: Jasco Model 870-CE

OTHER SUBSTANCES
Simultaneous: alprazolam, amobarbital, barbital, bromazepam, clonazepam, cloxazolam, diazepam, estazolam, etizolam, fludiazepam, flunitrazepam, flurazepam, haloxazolam, medazepam, mephobarbital, metharbital, nimetazepam, nitrazepam, oxazepam, pentobarbital, phenobarbital, secobarbital, triazolam

KEY WORDS
coated capillary; detector at anode

REFERENCE
Jinno,K.; Han,Y.; Sawada,H.; Taniguchi,M. Capillary electrophoretic separation of toxic drugs using a polyacrylamide-coated capillary, *Chromatographia*, **1997**, *46*, 309–314.

Cloxacillin

Molecular formula: C₁₉H₁₈ClN₃O₅S
Molecular weight: 435.89
CAS Registry No.: 61-72-3, 23736-58-5 (benzathine), 7081-44-9 (sodium salt, monohydrate), 642-78-4 (sodium salt)
Merck Index (12th ed.): 2480
Lednicer: 1 413

SAMPLE
Matrix: solutions

Sample preparation: Prepare a 120 µg/mL solution in water, inject an aliquot.

CAPILLARY ELECTROPHORESIS
Capillary: 60 cm × 75 µm
Running buffer: 20 mM NaH_2PO_4 containing 50 mM sodium dodecyl sulfate, adjusted to pH 9.0 with sodium tetraborate
Injection: Hydrodynamic injection at 10 cm for 5 s
Detector: UV 214
Migration time: 10.5
Voltage: 18 kV
Model: Waters Quanta 4000

OTHER SUBSTANCES
Simultaneous: 6-aminopenicillanic acid, amoxicillin, ampicillin, dicloxacillin, nafcillin, oxacillin, ticarcillin

REFERENCE
Swartz,M.E. Method development and selectivity control for small molecule pharmaceutical separations by capillary electrophoresis, *J.Liq.Chromatogr.*, **1991**, *14*, 923–938.

SAMPLE
Matrix: solutions
Sample preparation: Dilute a solution in buffer 10-fold with water, filter, inject an aliquot. (Prepare buffer by dissolving 8.0 g KH_2PO_4 and 2.0 g K_2HPO_4 in 1 L water, adjust pH to 6.0 with 1 M HCl or 1 M NaOH.)

CAPILLARY ELECTROPHORESIS
Capillary: 50 cm × 50 µm fused-silica (Polymicro Technologies)
Capillary preparation: Before each run rinse with water for 80 s, with 1 M NaOH for 80 s, with water for 80 s, and with running buffer for 90 s then the ends are dipped into water
Capillary temperature: 20
Running buffer: 20 mM Sodium tetraborate containing 150 mM sodium dodecyl sulfate, pH 9.20 ± 0.05
Injection: Pressure injection at 5 psi for 6 s (42 nL).
Detector: UV 210
Migration time: 18.3
Voltage: 14 kV
Model: BioFocus 3000 (Bio-Rad)

OTHER SUBSTANCES
Simultaneous: amoxicillin, ampicillin, ceftiofur, cephapirin, penicillin G

REFERENCE
Cutting,J.H.; Hurlbut,J.A.; Sofos,J.N. Quantitation of penicillin G in medicated premix feeds by micellar electrokinetic capillary chromatography, *J.AOAC Int.*, **1997**, *80*, 951–955.

SAMPLE
Matrix: solutions

CAPILLARY ELECTROPHORESIS
Capillary: 60 cm × 50 µm fused-silica (47 cm to detector) (Polymicro Technologies)
Capillary temperature: 25
Running buffer: 20 mM pH 8.5 Sodium tetraborate containing 100 mM sodium dodecyl sulfate
Injection: Hydrodynamic injection at 50 mbar for 3.6 s (5 nL).
Detector: UV 205
Migration time: 11.8
Voltage: 22 kV
Model: Crystal 310 (Thermo Unicam)

OTHER SUBSTANCES
Simultaneous: amoxicillin, ampicillin, cephapirin, dicloxacillin, oxacillin, penicillin G, penicillin V, piperacillin, pyrimethamine, sulfacetamide, sulfadimethoxine, sulfaguanidine, sulfamerazine, sulfameter, sulfamethazine, sulfanilamide, sulfanilic acid, sulfapyridine, sulfaquinoxaline, sulfathiazole, sulfisoxazole, trimethoprim

REFERENCE
Hows,M.E.P.; Perrett,D.; Kay,J. Optimization of a simultaneous separation of sulphonamides, dihydrofolate reductase inhibitors and β-lactam antibiotics by capillary electrophoresis, *J.Chromatogr.A*, **1997**, *768*, 97–104.

SAMPLE
Matrix: solutions

CAPILLARY ELECTROPHORESIS
Capillary: 50 cm × 50 μm fused-silica (Polymicro Technologies)
Running buffer: 10 mM pH 7.0 Phosphate buffer
Injection: Electrokinetic injection
Detector: E, Model AFDRE-5 (Pine Instrument, Grove City CA), Au working electrode, Pt auxiliary electrode, Ag/AgCl reference electrode (design of cell described in paper)
Migration time: 14.3
Voltage: 10 kV
Model: laboratory-constructed

OTHER SUBSTANCES
Simultaneous: amoxicillin, ampicillin, penicillin G

REFERENCE
Owens,G.S.; LaCourse,W.R. Pulsed electrochemical detection of thiols and disulfides following capillary electrophoresis, *J.Chromatogr.B*, **1997**, *695*, 15–25.

Cloxazolam

Molecular formula: $C_{17}H_{14}Cl2N_2O_2$
Molecular weight: 349.22
CAS Registry No.: 24166-13-0
Merck Index (12th ed.): 2481

SAMPLE
Matrix: solutions

CAPILLARY ELECTROPHORESIS
Capillary: 60 cm × 50 μm fused-silica (40 cm to detector), C18 coated (Supelco)
Capillary preparation: Rinse new capillaries with 250 μL 1 M NaOH, 250 μL 100 mM NaOH, water, MeOH, water, and running buffer then condition with running buffer for at least 15 min. Carry out a similar procedure between runs.
Running buffer: 100 mM pH 8.5 Borate buffer containing 25 mM sodium dodecyl sulfate and 5 M urea
Injection: Electrokinetic injection at 5 kV for 5 s.
Detector: UV 254
Migration time: 49
Voltage: 18 kV
Model: Jasco

OTHER SUBSTANCES
Simultaneous: alprazolam, amobarbital, barbital, bromazepam, clonazepam, clotiazepam, diazepam, estazolam, etizolam, fludiazepam, flunitrazepam, flurazepam, haloxazolam, medazepam, mephobarbital, metharbital, nimetazepam, nitrazepam, oxazepam, pentobarbital, phenobarbital, secobarbital, triazolam

KEY WORDS
coated capillary

REFERENCE
Jinno,K.; Han,Y.; Nakamura,M. Analysis of anxiolytic drugs by capillary electrophoresis with bare and coated capillaries, *J.Capillary Electrophor.*, **1996**, *3*, 139–145.

SAMPLE
Matrix: solutions
Sample preparation: Inject an aliquot of a 100 µg/mL solution in running buffer.

CAPILLARY ELECTROPHORESIS
Capillary: 60 cm × 50 µm acrylamide-coated fused-silica (40 cm to detector) (Supelco)
Capillary preparation: Adjust the pH of 20 mL water to 3.5 with acetic acid, add 80 µL 3-(trimethoxysilyl)propyl methacrylate (3-methacryloxypropyltrimethoxysilane), mix, suck into capillary, let stand at room temperature for 1 h, remove the solution, wash with water. Fill the capillary with a deaerated 3-4% acrylamide solution containing 1 µL/mL N,N,N',N'-tetramethylethylenediamine and 1 mg/mL ammonium persulfate, let stand for 3 h, remove excess solution by aspiration, rinse with water, remove water by aspiration, dry at 35° (cf. J. Chromatogr. 1985, 347, 191).
Running buffer: MeCN:buffer 5:95 (Buffer was 100 mM borate containing 5 M urea and 10 mM sodium dodecyl sulfate, adjusted to pH 8.5 with phosphate.)
Injection: Electrokinetic injection at 5 kV for 5 s.
Detector: UV 254
Migration time: 8.5
Voltage: 18 kV
Model: Jasco Model 870-CE

OTHER SUBSTANCES
Simultaneous: alprazolam, amobarbital, barbital, bromazepam, clonazepam, clotiazepam, diazepam, estazolam, etizolam, fludiazepam, flunitrazepam, flurazepam, haloxazolam, medazepam, mephobarbital, metharbital, nimetazepam, nitrazepam, oxazepam, pentobarbital, phenobarbital, secobarbital, triazolam

KEY WORDS
coated capillary; detector at anode

REFERENCE
Jinno,K.; Han,Y.; Sawada,H.; Taniguchi,M. Capillary electrophoretic separation of toxic drugs using a polyacrylamide-coated capillary, *Chromatographia*, **1997**, *46*, 309–314.

SAMPLE
Matrix: solutions

CAPILLARY ELECTROPHORESIS
Capillary: 60 cm × 75 µm coated fused-silica (40 cm to detector) (Supelco)
Capillary preparation: Coat column as follows. Adjust the pH of 20 mL water to 3.5 with acetic acid, add 80 µL 3-(trimethoxysilyl)propyl methacrylate (3-methacryloxypropyltrimethoxysilane), mix, suck into capillary, let stand at room temperature for 1 h, remove the solution, wash with water. Fill the capillary with a deaerated 4% acrylamide solution containing 1 mg/mL N,N,N',N'-tetramethylethylenediamine and 1 mg/mL ammonium persulfate, let stand for 3 h, remove excess solution by aspiration, rinse with water, remove water by aspiration, dry at 35° (cf. J. Chromatogr. 1985, 347, 191).
Running buffer: 100 mM pH 8.5 borate buffer containing 10 mM sodium dodecyl sulfate and 5 M urea

Injection: Electromigration at 5 kV for 5 s.
Detector: UV 240, UV 254
Migration time: 7
Voltage: 18 kV
Model: Jasco 890-CE

OTHER SUBSTANCES
Simultaneous: alprazolam, clotiazepam, diazepam, estazolam, etizolam, fludiazepam, flurazepam, haloxazolam, oxazepam

KEY WORDS
injection at cathode; coated capillary

REFERENCE
Jinno,K.; Han,Y.; Sawada,H. Analysis of toxic drugs by capillary electrophoresis using polyacrylamide-coated columns, *Electrophoresis*, **1997**, *18*, 284–286.

Clozapine

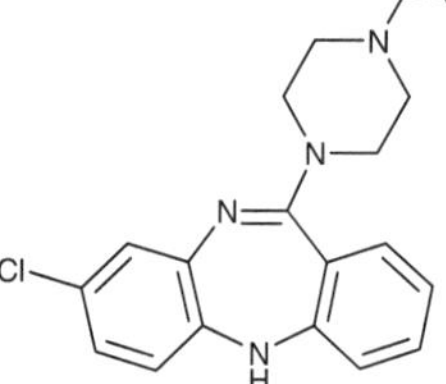

Molecular formula: $C_{18}H_{19}ClN_4$
Molecular weight: 326.83
CAS Registry No.: 5786-21-0
Merck Index (12th ed.): 2484
Lednicer: 2 425, 4 212, 220

SAMPLE
Matrix: urine
Sample preparation: 10 mL Urine + 1 mL 5 M NaOH + 10 mL n-hexane, vortex for 1 min, centrifuge at 0° at 3000 g for 5 min. Remove the organic layer and add it to 50 μL glacial acetic acid, evaporate to dryness under a stream of nitrogen at 30°, reconstitute the residue in 50 μL 50 mM sodium taurodeoxycholate, filter (0.2 μm PTFE), inject an aliquot.

CAPILLARY ELECTROPHORESIS
Capillary: 100 cm × 50 μm fused-silica (50 cm to detector) (ISCO)
Capillary preparation: Rinse capillary with 20 μL running buffer between injections. Condition a new capillary by filling with 1 M NaOH, let stand for 1 h, fill with 100 mM NaOH, let stand for 1 h, rinse with buffer. Every 20 injections rinse capillary with 200 μL 1 M NaOH, 200 μL water, and 200 μL running buffer, fill with running buffer.
Capillary temperature: 22
Running buffer: 40 mM pH 9.5 Borate buffer containing 10 mM sodium taurodeoxycholate
Injection: Load under vacuum at 7.5 kPa/s.
Detector: UV 240
Migration time: 9.5
Voltage: 30 kV
Model: ISCO Model 3140 electropherograph
Limit of detection: 9 ng/mL

OTHER SUBSTANCES
Extracted: acepromazine, amiodarone, amitriptyline, azaperone, chlorpromazine, cianopramine, clomipramine, desethylamiodarone, desipramine, diclofensine, dothiepin, doxepin, imipramine, isocarboxazid, moclobemide, perphenazine, phenothiazine, pimozide, prochlorperazine, promazine, thioridazine, thiothixene, trifluoperazine, trimipramine

KEY WORDS
human; cow; pig; horse; protect from light

REFERENCE
Aumatell,A.; Wells,R.J. Determination of a cardiac antiarrhythmic, tricyclic antipsychotics and antidepressants in human and animal urine by micellar electrokinetic capillary chromatography using a bile salt, *J.Chromatogr.B*, **1995**, *669*, 331–344.

Cocaine

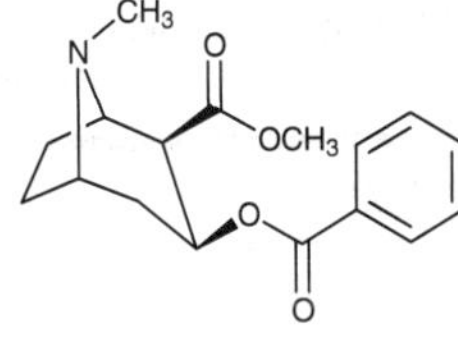

Molecular formula: $C_{17}H_{21}NO_4$
Molecular weight: 303.36
CAS Registry No.: 50-36-2, 53-21-4 (HCl)
Merck Index (12th ed.): 2517

SAMPLE
Matrix: bulk
Sample preparation: Prepare a 3 mg/mL solution in 3 mg/mL pholcodine in 10 mM HCl, make up to 25 mL with 10 mM HCl, mix thoroughly, filter (0.45 μm cellulose acetate)

CAPILLARY ELECTROPHORESIS
Capillary: 65 cm × 75 μm fused-silica (40 cm to detector) (Isco)
Capillary preparation: Flush with running buffer for 2 min between runs. Before each batch of samples fill capillary with 100 mM NaOH and let stand for 10 min, wash with water for 5 min, wash with 100 mM HCl for 5 min, wash with water for 5 min, fill with running buffer (all at vacuum level 4). Clean each week by washing with 100 mM HCl for 10 min, wash with water, fill with running buffer.
Capillary temperature: 30
Running buffer: MeCN:buffer 7.5:92.5. Buffer was 10 mM KH_2PO_4 containing 10 mM sodium tetraborate and 50 mM cetyltrimethylammonium bromide, pH 8.6.)
Injection: Injection by vacuum 5.0 kPa.s (4 nL)
Detector: UV 230
Migration time: 9.6
Internal standard: pholcodine (7.4)
Voltage: -15 kV
Model: Isco Model 3140

OTHER SUBSTANCES
Simultaneous: benzocaine, benzoylecgonine, trans-cinnamoylcocaine, lidocaine, procaine, tetracaine

REFERENCE
Trenerry,V.C.; Robertson,J.; Wells,R.J. The determination of cocaine and related substances by micellar electrokinetic capillary chromatography, *Electrophoresis*, **1994**, *15*, 103–108.

SAMPLE
Matrix: bulk
Sample preparation: Dilute with 400 μg/mL n-propyl p-hydroxybenzoate in running buffer

CAPILLARY ELECTROPHORESIS
Capillary: 27 cm × 50 μm fused-silica (20 cm to detector)
Capillary preparation: Before each run rinse with running buffer at high pressure.
Capillary temperature: 30
Running buffer: MeCN:water 15:85 containing 40 mM sodium dodecyl sulfate, 8.5 mM sodium borate, and 8.5 mM sodium phosphate, pH 8.5
Injection: Inject at high pressure for 1 s
Detector: UV 214
Migration time: 2.7
Internal standard: n-propyl p-hydroxybenzoate (1.8)
Voltage: 20 kV
Model: Beckman P/ACE System 2100

OTHER SUBSTANCES
Simultaneous: acetaminophen, acetylcodeine, O-acetylmorphine, aspirin, caffeine, codeine, diamorphine, diphenhydramine, hydromorphone, isoniacinamide, lidocaine, methaqualone, morphine, niacinamide, noscapine, papaverine, phenacetin, phenobarbital, phenylpropanolamine, procaine, quinine, salicylic acid, strychnine, thebaine

REFERENCE
Walker,J.A.; Krueger,S.T.; Lurie,I.S.; Marché,H.L.; Newby,N. Analysis of heroin drug seizures by micellar electrokinetic capillary chromatography (MECC), *J.Forensic Sci.*, **1995**, *40*, 6–9.

SAMPLE
Matrix: bulk
Sample preparation: Dissolve sample in 2 mg/mL histamine diphosphate in 10 mM HCl, inject an aliquot.

CAPILLARY ELECTROPHORESIS
Capillary: 48.5 cm × 50 μm fused-silica (40 cm to detector)
Capillary preparation: Flush capillary with running buffer for 2 min before each run. Replenish buffer in vials after every 10 injections. Condition new capillaries by flushing with 1 M NaOH for 10 min, with 100 mM NaOH for 10 min, with water for 10 min, and with running buffer for 10 min.
Capillary temperature: 40
Running buffer: 75 mM NaH_2PO_4 containing 75 mM sodium borate, adjusted to pH 8.5 with 2 M NaOH
Injection: Pressure injection at 150 mbar for 3 s.
Detector: UV 230
Migration time: 6
Internal standard: histamine diphosphate (5)
Voltage: 10 kV
Current: 66-70 μA
Model: Hewlett Packard 3D CE
Limit of quantitation: 14 μg/mL
Limit of detection: 2.8 μg/mL (S/N = 8:1)

OTHER SUBSTANCES
Simultaneous: acetaminophen, benzocaine, cis-cinnamoylcocaine, trans-cinnamoylcocaine, ephedrine, lidocaine, phenylpropanolamine, procaine, tetracaine

REFERENCE
Krawczeniuk,A.S.; Bravenec,V.A. Quantitative determination of cocaine in illicit powders by free zone capillary electrophoresis, *J.Forensic Sci.*, **1998**, *43*, 738–743.

SAMPLE
Matrix: hair
Sample preparation: Wash 25-100 mg hair with diethyl ether and 10 mM HCl, add 250 mM HCl, heat at 45° overnight, neutralize with NaOH, extract twice into the organic phase with Toxi-tubes A (Analytical Systems, Laguna Hills, CA). Remove the organic layer and evaporate it to dryness, reconstitute the residue in 20 μL running buffer:water 1:2, inject a 5-10 μL aliquot.

CAPILLARY ELECTROPHORESIS
Capillary: 40 cm × 50 μm bare silica (40 cm to detector) (Isco)
Capillary preparation: After each injection flush with 100 mM NaOH and rinse with running buffer.
Running buffer: 50 mM pH 9.2 borate buffer
Injection: Manual injection of 5-10 μL by syringe through a 1:830 splitter
Detector: UV 238
Migration time: 7.40
Internal standard: tetracaine (7.69)
Voltage: 15 kV
Model: Isco Model 3850
Limit of detection: 150 ng/g (S/N 3)

OTHER SUBSTANCES
Extracted: morphine (UV 214)
Simultaneous: nalorphine (UV 214)
Noninterfering: acetaminophen, amitriptyline, amphetamine, atropine, benzoylecgonine, benztropine, caffeine, carbamazepine, carisoprodol, chlorpheniramine, chlorpromazine, chlorprothixene, cimetidine, cocaine, codeine, dextromethorphan, diazepam, dihydrocodeine, diphenhydramine, diphenoxylate, disopyramide, doxepin, doxylamine, emetine, erythromycin, ethylmorphine, flurazepam, glutethimide, hydrocodone, hydrocortisone, hydromorphone, hydroxyzine, imipramine, lidocaine, loxapine, meperidine, meprobamate, methadone, methamphetamine, methapyrilene, methaqualone, methocarbamol, methylphenidate, naloxone, nicotine, nordiazepam, nortriptyline, orphenadrine, oxycodone, papaverine, pentazocine, phenacetin, phencyclidine, phenmetrazine, phenolphthalein, phentermine, phenylpropanolamine, phenytoin, phetidine, prazepam, procainamide, procaine, propoxyphene, propranolol, protriptyline, pseudoephedrine, pyrilamine, quinine, salicylamide, spironolactone, strychnine, terpin hydrate, thioridazine, thiothixene, triamterene, trifluoperazine, triflupromazine, trihexyphenidyl, trimeprazine, trimethoprim, trimetobenzamide

REFERENCE
Tagliaro,F.; Poiesi,C.; Aiello,R.; Dorizzi,R.; Ghielmi,S.; Marigo,M. Capillary electrophoresis for the investigation of illicit drugs in hair: determination of cocaine and morphine, *J.Chromatogr.*, **1993**, *638*, 303–309.

SAMPLE
Matrix: perfusate
Sample preparation: Inject an aliquot of perfusate (pH 6.5 Ringer's).

CAPILLARY ELECTROPHORESIS
Capillary: 70 cm × 50 μm untreated fused-silica (Polymicro Technologies)
Running buffer: 200 mM sodium tetraborate buffer
Injection: Injection by vacuum for 1 s
Detector: UV 200
Migration time: 7.0
Voltage: 20 kV

OTHER SUBSTANCES
Simultaneous: lidocaine, procaine

REFERENCE
Hernandez,L.; Guzman,N.A.; Hoebel,B.G. Bidirectional microdialysis in vivo shows differential dopaminergic potency of cocaine, procaine and lidocaine in the nucleus accumbens using capillary electrophoresis for calibration of drug outward diffusion, *Psychopharmacology (Berl)*, **1991**, *105*, 264–268.

SAMPLE
Matrix: solutions
Sample preparation: Prepare a 50 μg/mL solution in water:buffer:MeOH 90:9:1, inject an aliquot. (Buffer was 25 mM Tris buffer adjusted to pH 2.4 with phosphoric acid.)

CAPILLARY ELECTROPHORESIS
Capillary: 82 cm × 50 μm fused-silica (60 cm to detector)
Capillary preparation: Condition by aspirating with 1 M NaOH for 10 min, with water for 10 min, and with running buffer for 10 min.
Capillary temperature: 30
Running buffer: MeOH:buffer 1.2:98.8 (Buffer was 25 mM Tris buffer adjusted to pH 2.4 with phosphoric acid containing 5 mM heptakis(2,6-di-O-methyl)-β-cyclodextrin and 0.5 mM sulfobutyl ether β-cyclodextrin (ISCO)
Injection: Vacuum injection for 0.5 s (2 nL)
Detector: UV 210
Migration time: 22.00 (+), 23.30 (–)
Voltage: 30 kV
Model: Applied Biosystems Model 270A-HT

OTHER SUBSTANCES
Simultaneous: amphetamine, cathinone, ephedrine, methamphetamine, methcathinone, nor-ephedrine, norpseudoephedrine, propoxyphene, pseudoephedrine

KEY WORDS
chiral

REFERENCE
Lurie,I.S.; Klein,R.F.X.; Dal Cason,T.A.; LeBelle,M.J.; Brenneisen,R.; Weinberger,R.E. Chiral resolution of cationic drugs of forensic interest by capillary electrophoresis with mixtures of neutral and anionic cyclodextrins, *Anal.Chem.*, **1994**, *66*, 4019–4026.

SAMPLE
Matrix: solutions
Sample preparation: Inject an aliquot of a 1-50 µg/mL solution in water.

CAPILLARY ELECTROPHORESIS
Capillary: 55 cm × 50 µm fused-silica (35 cm to detector) (J&W)
Capillary preparation: Flush with running buffer for 5 min before each run. Periodically wash with 100 mM NaOH and flush extensively with running buffer.
Running buffer: MeOH:25 mM pH 9.24 borate buffer 20:80 containing 100 mM sodium dodecyl sulfate (A) or 50 mM pH 2.35 phosphate buffer (B) or 50 mM pH 9.24 borate buffer (C)
Injection: Injection of 5 µL using a split-flow injector ratio of 1:800.
Detector: UV 200
Migration time: 40.08 (A), 10.23 (B), 6.52 (C)
Voltage: 20 kV (A, B) or 12 kV (C)
Current: <60-80 µA
Model: ISCO Model 3850

OTHER SUBSTANCES
Also analyzed: acetylcodeine, amphetamine, barbital, caffeine, codeine, diamorphine, diazepam, flunitrazepam, lidocaine, monoacetylmorphine, morphine, nalorphine, narceine, noscapine, papaverine, pentobarbital, procaine, tetracaine, thebaine

KEY WORDS
all compounds were separated with running buffer A; some peaks overlapped with running buffers B and C.

REFERENCE
Tagliaro,F.; Smith,F.P.; Turrina,S.; Equisetto,V.; Marigo,M. Complementary use of capillary zone electrophoresis and micellar electrokinetic capillary chromatography for mutual confirmation of results in forensic drug analysis, *J.Chromatogr.A*, **1996**, *735*, 227–235.

SAMPLE
Matrix: solutions
Sample preparation: Inject an aliquot of a solution in 100 mM sodium phosphate containing 100-300 µg/mL naphazoline.

CAPILLARY ELECTROPHORESIS
Capillary: 67 cm × 50 µm fused-silica (60 cm to detector)
Capillary preparation: Rinse with running buffer for 2 min between sets of analyses.
Capillary temperature: 30
Running buffer: 200 mM pH 4.5 Phosphate buffer
Injection: Injection at high pressure for 2 s.
Detector: UV 210, UV 230
Migration time: 15.9
Internal standard: naphazoline (14)
Voltage: 20 kV
Model: Beckman P/ACE 5500

OTHER SUBSTANCES
Simultaneous: acetylcodeine, amphetamine, codeine, diamorphine, LSD, methadone, methamphetamine, methylenedioxyamphetamine, methylenedioxymethamphetamine, morphine, PCP, psilocyn

REFERENCE
Walker,J.A.; Marché,H.L.; Newby,N.; Bechtold,E.J. A free zone capillary electrophoresis method for the quantitation of common illicit drug samples, *J.Forensic Sci.*, **1996**, *41*, 824–829.

SAMPLE
Matrix: solutions

CAPILLARY ELECTROPHORESIS
Capillary: 85 cm × 50 μm fused-silica (Polymicro Technologies)
Capillary preparation: Between runs rinse capillary with running buffer for 5 min. Before use rinse capillary with 1 M NaOH for 10-15 min and with water for 10-15 min.
Running buffer: 25 mM pH 3 Citrate buffer
Injection: Pressure injection at 30 mbar for 12 s.
Detector: MS, laboratory-constructed, electrospray 90°, sheath liquid MeOH:water:acetic acid 80:20:0.1 at 1 μL/min
Migration time: 10.7
Voltage: 30 kV
Current: 7.8 μA
Model: Crystal CE 300

OTHER SUBSTANCES
Simultaneous: amphetamine, diamorphine, methamphetamine, procaine, tetracaine

REFERENCE
Lazar,I.M.; Naisbitt,G.; Lee,M.L. Capillary electrophoresis-time-of-flight mass spectrometry of drugs of abuse, *Analyst*, **1998**, *123*, 1449–1454.

SAMPLE
Matrix: solutions

CAPILLARY ELECTROPHORESIS
Capillary: 20 cm × 75 μm containing 3 μm Micra bare silica (packed for 20 cm; 20 cm to detector) (Unimicro Technologies, Pleasanton CA)
Capillary preparation: Before each run condition capillary at 5 kV until a constant current is achieved. Condition new capillaries with running buffer at 500 psi.
Capillary temperature: 25
Running buffer: MeCN:10 mM pH 8.29 Tris-HCl buffer 80:20
Injection: Electrokinetic injection at 5 kV for 5 s.
Detector: UV 214
Migration time: 5.5
Model: Beckman P/ACE 5510

OTHER SUBSTANCES
Simultaneous: aniline, berberine, codeine, ephedrine, jatrorrhizine, thebaine

KEY WORDS
electrochromatography

REFERENCE
Wei,W.; Luo,G.A.; Hua,G.Y.; Yan,C. Capillary electrochromatographic separation of basic compounds with bare silica as stationary phase, *J.Chromatogr.A*, **1998**, *817*, 65–74.

SAMPLE
Matrix: urine

Sample preparation: Condition a Bond Elut Certify SPE cartridge with 2 mL MeOH and 2 mL 100 mM pH 6 phosphate buffer, do not allow to go dry. 5 mL Urine + 2 mL 100 mM pH 5 phosphate buffer, add to the SPE cartridge, wash with 3 mL water, wash with 3 mL 100 mM HCl, wash with 9 mL MeOH, elute with 2 mL dichloromethane:isopropanol 80:20 containing 2% ammonium hydroxide. Evaporate the eluate to dryness under a stream of nitrogen at room temperature, reconstitute the residue in 100 μL running buffer, inject an aliquot.

CAPILLARY ELECTROPHORESIS
Capillary: 90 cm × 75 μm fused-silica (70 cm to detector) (Polymicro Technologies)
Capillary preparation: Before each run rinse capillary with 100 mM NaOH for 3 min and with buffer for 5 min.
Running buffer: 10 mM pH 9.1 Na_2HPO_4 containing 6 mM sodium borate and 75 mM sodium dodecyl sulfate
Injection: Gravity injection at 34 cm for 5 s
Detector: UV 195
Migration time: 23.7
Voltage: 20 kV
Current: 76-80 μA

OTHER SUBSTANCES
Extracted: benzoylecgonine, codeine, methaqualone, morphine

KEY WORDS
SPE

REFERENCE
Wernly,P.; Thormann,W. Analysis of illicit drugs in human urine by micellar electrokinetic capillary chromatography with on-column fast scanning polychrome absorption detection, *Anal.Chem.*, **1991**, *63*, 2878–2882.

SAMPLE
Matrix: urine
Sample preparation: Filter (0.2 μm), inject an aliquot of the filtrate.

CAPILLARY ELECTROPHORESIS
Capillary: 80 cm × 50 μm fused-silica (57 cm to detector)
Capillary preparation: Before each run rinse the capillary with 1 M NaOH for 3 min, with 100 mM NaOH for 3 min, with water for 3 min, and with running buffer for 10 min.
Running buffer: 110 mM Boric acid containing 56 mM NaOH and 44 mM HCl, pH 8
Injection: Vacuum injection for 1 s
Detector: UV 236
Migration time: 4.524
Voltage: 20 kV
Model: Europhor Prime Vision system IV
Limit of detection: 1.2 μM

OTHER SUBSTANCES
Extracted: acetazolamide (UV 222), alprenolol (UV 220), amiloride (UV220), atenolol (UV 228), bendroflumethiazide (UV 220), bumetanide (UV 220), chlorthalidone (UV 220), codeine (UV 220), ethacrynic acid (UV 220), furosemide (UV 232), hydrochlorothiazide (UV 226), methadone (UV 220), metoxiphenamine (UV 220), nadolol (UV 220), oxprenolol (UV 220), propranolol (UV 220), spironolactone (UV 244), triamterene (UV 232), xipamide (UV 234)
Interfering: acebutolol (UV 238), norcodeine (UV 220), pentazocine (UV 220)

REFERENCE
Gonzalez,E.; Laserna,J.J. Capillary zone electrophoresis for the rapid screening of banned drugs in sport, *Electrophoresis*, **1994**, *15*, 240–243.

Codeine

Molecular formula: $C_{18}H_{21}NO_3$
Molecular weight: 299.37
CAS Registry No.: 76-57-3, 6069-47-8 (monohydrate), 5913-71-3 (acetate), 125-25-7 (HBr), 1422-07-7 (HCl), 6020-73-1 (salicylate), 125-27-9 (methyl bromide), 52-28-8 (phosphate), 41444-62-6 (phosphate hemihydrate), 1420-53-7 (sulfate), 6854-40-6 (sulfate trihydrate)
Merck Index (12th ed.): 2525
Lednicer: 1 287; 2 317

SAMPLE
Matrix: blood, urine
Sample preparation: Adjust pH of 2 mL urine or plasma to 10.5 with aqueous NaOH, extract gently with chloroform:isopropanol 90:10. Remove the organic layer and evaporate it to dryness under a stream of nitrogen, reconstitute the residue in 50 (urine) or 75 (plasma) μL running buffer, filter, inject an aliquot.

CAPILLARY ELECTROPHORESIS
Capillary: 60 cm × 75 μm AccuSep uncoated silica (52.5 cm to detector) (Waters)
Capillary preparation: At the beginning of each day purge with 500 mM KOH for 5 min, with water for 5 min, and with running buffer for 10 min.
Running buffer: 50 mM NaH_2PO_4 adjusted to pH 2.35 with phosphoric acid
Injection: Hydrostatic injection at 15 cm
Detector: UV 214
Migration time: 6.96
Voltage: 22 kV
Current: 135-145 μA
Model: Waters Quanta 4000

OTHER SUBSTANCES
Extracted: acepromazine, amphetamine, benzocaine, brompheniramine, butacaine, diazepam, doxapram, lidocaine, medazepam, methamphetamine, methapyrilene, methaqualone, phenmetrazine, procaine, tetrahydrozoline
Simultaneous: meclizine

KEY WORDS
plasma

REFERENCE
Chee,G.L.; Wan,T.S.M. Reproducible and high-speed separation of basic drugs by capillary zone electrophoresis, *J.Chromatogr.*, **1993**, *612*, 172–177.

SAMPLE
Matrix: bulk
Sample preparation: Prepare a 10 mg/mL solution in 3 mg/mL pholcodine in 10 mM HCl, dilute 5-fold with 10 mM HCl, filter (0.45 μm cellulose acetate), inject an aliquot.

CAPILLARY ELECTROPHORESIS
Capillary: 72 cm × 75 μm fused-silica (50 cm to detector) (Isco)
Capillary preparation: Flush with running buffer for 2 min between runs. Every 24 h wash capillary with 100 mM HCl for 10 min, wash with water, wash 100 mM NaOH, wash with water, fill with running buffer. Prepare new capillary by filling with 1 M NaOH, let stand for 1 h, fill with 100 mM NaOH, let stand for 1 h, wash with water, fill with running buffer.
Capillary temperature: 30
Running buffer: MeCN:buffer 10:90 (Buffer was 10 mM KH_2PO_4 containing 10 mM sodium tetraborate and 50 mM cetyltrimethylammonium bromide, pH 8.6.)
Injection: Injection by vacuum 5.0 kPa.s (4 nL)
Detector: UV 280

Migration time: 10
Internal standard: pholcodine (9.1)
Voltage: 15 kV
Model: Isco Model 3140

OTHER SUBSTANCES
Simultaneous: acetylcodeine, caffeine, diamorphine, ethylmorphine, morphine, noscapine, papaverine, strychnine, theophylline

---

REFERENCE
Trenerry,V.C.; Wells,R.J.; Robertson,J. The analysis of illicit heroin seizures by capillary zone electrophoresis, *J.Chromatogr.Sci.*, **1994**, *32*, 1–6.

SAMPLE
Matrix: bulk
Sample preparation: Dissolve 500 mg crude opium in 5 mL DMSO, make up to 50 mL with MeCN containing 25 mM ammonium acetate and 1 M acetic acid, centrifuge at 18000 g for 2 min. Dilute 1 mL of the supernatant to 20 mL with MeCN containing 25 mM ammonium acetate and 1 M acetic acid, inject an aliquot. Dilute opium tincture 200-fold with water, inject an aliquot.

CAPILLARY ELECTROPHORESIS
Capillary: 64 cm $\times$ 50 μm fused-silica (55.5 cm to detector) (Polymicro Technologies)
Capillary preparation: Flush with running buffer between runs. Before use rinse capillaries with 1 M NaOH for 1 h, with 100 mM NaOH for 20 min, and with running buffer for 10 min
Capillary temperature: 25
Running buffer: MeCN:MeOH 25:75 containing 25 mM ammonium acetate and 1 M acetic acid
Injection: Pressure injection at 5 kPa for 3 s
Detector: UV 214
Migration time: 9
Voltage: 25 kV
Model: Hewlett-Packard HP3D

OTHER SUBSTANCES
Simultaneous: morphine, normorphine, noscapine, papaverine, thebaine

---

KEY WORDS
opium; tincture

REFERENCE
Bjornsdottir,I.; Hansen,S.H. Determination of opium alkaloids in crude opium using non-aqueous capillary electrophoresis, *J.Pharm.Biomed.Anal.*, **1995**, *13*, 1473–1481.

SAMPLE
Matrix: bulk
Sample preparation: Crude preparations. Dissolve in 10 mM HCl containing 400 μg/mL pholcodine, inject an aliquot. Opium. Extract 1 g opium with 20 mL 2.5% acetic acid, make up to 100 mL with water, filter, add 20 mL of the filtrate to 60 mL water, adjust pH to 9.2 with concentrated ammonia solution, extract six times with 50 mL portions of chloroform. Dry the extracts over anhydrous sodium sulfate and evaporate to dryness under reduced pressure, reconstitute with 5 mL 50 mM HCl. Dilute 1 mL to 5 mL with water, add 2 mg pholcodine, inject an aliquot. Poppy straw. 8 g Finely-milled poppy straw + 10 mL water + 2 g calcium hydroxide, shake vigorously for 25 min, filter, dilute 20 mL filtrate with 60 mL water, adjust pH to 9.2 with 10% acetic acid, extract 5 times with 50 mL portions of chloroform:EtOH 75:5. Combine the extracts and wash them twice with 30 mL portions of water, dry over anhydrous sodium sulfate and evaporate to dryness under reduced pressure, reconstitute with 25 mL 50 mM HCl. Dilute 9 mL with 1 mL 4 mg/mL pholcodine, filter, inject an aliquot.

CAPILLARY ELECTROPHORESIS
Capillary: 70 cm $\times$ 50 μm fused-silica (45 cm to detector) (Polymicro Technologies)

Capillary preparation: Flush with running buffer for 2 min between analyses. Replace running buffer after 20 analyses. Each week wash capillary with 100 mM NaOH for 10 min and with water for 10 min then fill with running buffer.
Capillary temperature: 28
Running buffer: DMF:buffer 10:90 [MeCN:buffer 12.5:87.5 may also be used (Electrophoresis 1996, 17, 1361)] (Prepare buffer by dissolving 0.92 g cetyltrimethylammonium bromide in 50 mL 10 mM sodium tetraborate:10 mM KH_2PO_4 50:50 (pH 8.6), mix 2.5 mL DMF and 22.5 mL buffer.)
Injection: Vacuum injection 10 kPa.s (vacuum level 2)
Detector: UV 254
Migration time: 5.8
Internal standard: pholcodine (5.5)
Voltage: -25 kV
Model: Isco Model 3140 Electropherograph

OTHER SUBSTANCES
Simultaneous: cryptopine, morphine, narceine, noscapine, oripavine, papaverine, salutaridine, thebaine

REFERENCE
Trenerry,V.C.; Wells,R.J.; Robertson,J. Determination of morphine and related alkaloids in crude morphine, poppy straw and opium preparations by micellar electrokinetic capillary chromatography, *J.Chromatogr.A*, **1995**, *718*, 217–225.

SAMPLE
Matrix: bulk
Sample preparation: Dilute with 400 µg/mL n-propyl p-hydroxybenzoate in running buffer

CAPILLARY ELECTROPHORESIS
Capillary: 27 cm × 50 µm fused-silica (20 cm to detector)
Capillary preparation: Before each run rinse with running buffer at high pressure.
Capillary temperature: 30
Running buffer: MeCN:water 15:85 containing 40 mM sodium dodecyl sulfate, 8.5 mM sodium borate, and 8.5 mM sodium phosphate, pH 8.5
Injection: Inject at high pressure for 1 s
Detector: UV 214
Migration time: 1.5
Internal standard: n-propyl p-hydroxybenzoate (1.8)
Voltage: 20 kV
Model: Beckman P/ACE System 2100

OTHER SUBSTANCES
Simultaneous: acetaminophen, acetylcodeine, O-acetylmorphine, aspirin, caffeine, cocaine, diamorphine, diphenhydramine, hydromorphone, isoniacinamide, lidocaine, methaqualone, morphine, niacinamide, noscapine, papaverine, phenacetin, phenobarbital, phenylpropanolamine, procaine, quinine, salicylic acid, strychnine, thebaine

REFERENCE
Walker,J.A.; Krueger,S.T.; Lurie,I.S.; Marché,H.L.; Newby,N. Analysis of heroin drug seizures by micellar electrokinetic capillary chromatography (MECC), *J.Forensic Sci.*, **1995**, *40*, 6–9.

SAMPLE
Matrix: bulk
Sample preparation: Dissolve in running buffer to a concentration of 1 mg/mL, vortex for 2 min, add an equal volume of 100 µg/mL diphenhydramine in running buffer, mix, inject an aliquot.

CAPILLARY ELECTROPHORESIS
Capillary: 65 cm × 50 µm fused-silica (60 cm to detector)
Capillary preparation: Fill capillary with fresh running buffer before each run. Before use fill with 100 mM NaOH for 20 min, rinse with water, flush with running buffer

Running buffer: MeCN:50 mM 6-aminocaproic acid containing 50 mM 3-N,N-dimethylmyristylammoniopropanesulfonate (MAPS; Fluka) and 5 mM 1-heptanesulfonic acid 10:90, pH adjusted to 4.0 with 1 M phosphoric acid
Injection: Hydrodynamic injection by gravity or pressure.
Detector: UV 214
Migration time: 5.2
Internal standard: diphenhydramine (12)
Voltage: 27 kV
Current: ≤25 μA
Model: Dionex system I

OTHER SUBSTANCES
Simultaneous: acetaminophen, allobarbital, barbital, caffeine, diamorphine, niacinamide, noscapine, papaverine, phenobarbital, procaine
Interfering: morphine

REFERENCE
Naess,O.; Rasmussen,K.E. Micellar electrokinetic chromatography of charged and neutral drugs in acidic running buffers containing a zwitterionic surfactant, sulfonic acids or sodium dodecyl sulphate. Separation of heroin, basic by-products and adulterants, *J.Chromatogr.A*, **1997**, *760*, 245–251.

SAMPLE
Matrix: formulations
Sample preparation: Dilute formulations with A, inject an aliquot. Dissolve 500 mg crude opium in 5 mL DMSO, make up to 50 mL with 1 M acetic acid, centrifuge. Remove a 1 mL aliquot of the supernatant and make up to 20 mL with water, inject an aliquot.

CAPILLARY ELECTROPHORESIS
Capillary: 55 cm × 50 μm fused-silica (Polymicro Technologies)
Capillary temperature: 30
Running buffer: 50 mM pH 4.0 6-Aminocaproic acid containing 30 mM heptakis(2,6-di-O-methyl-β-cyclodextrin)
Injection: Pressure injection at 5 kPa for 3 s
Detector: UV 214
Migration time: 8.8
Voltage: 30 kV
Model: Hewlett-Packard HP[3D]
Limit of detection: 300 ng/mL

OTHER SUBSTANCES
Simultaneous: morphine, normorphine, noscapine, papaverine, thebaine

KEY WORDS
syrup; tincture; opium

REFERENCE
Bjornsdottir,I.; Hansen,S.H. Determination of opium alkaloids in opium by capillary electrophoresis, *J.Pharm.Biomed.Anal.*, **1995**, *13*, 687–693.

SAMPLE
Matrix: formulations
Sample preparation: Dilute liquid formulation with DMSO:water:1 M acetic acid 0.5:95:4.5, inject an aliquot.

CAPILLARY ELECTROPHORESIS
Capillary: 64 cm × 50 μm fused-silica (55.5 cm to detector) (Polymicro Technologies)
Capillary preparation: Flush with running buffer for 2 min between runs. Rinse new capillaries with 1 M NaOH for 1 h, with 100 mM NaOH for 20 min, with water for 20 min, and with running buffer for 10 min.
Capillary temperature: 30
Running buffer: 50 mM 6-Aminocaproic acid containing 30 mM heptakis(2,6-di-O-methyl)-β-cyclodextrin, adjusted to pH 4.0 with glacial acetic acid

Injection: Pressure injection at 5 kPa for 3 s.
Detector: UV 214
Migration time: 8.8
Voltage: 30 kV
Current: 60 μA
Model: Hewlett-Packard HP[3D]

OTHER SUBSTANCES
Simultaneous: morphine, normorphine, noscapine, papaverine, thebaine

REFERENCE
Bjornsdottir,I.; Hansen,S.H. Comparison of aqueous and non-aqueous capillary electrophoresis for quantitative determination of morphine in pharmaceuticals, *J.Pharm.Biomed.Anal.*, **1997**, *15*, 1083–1089.

SAMPLE
Matrix: formulations
Sample preparation: Dissolve one tablet in 25 mL MeCN:40 mM pH 10 borate buffer 9:91, centrifuge, filter (glass fiber), filter (membrane), dilute filtrate 18-fold with running buffer, inject an aliquot.

CAPILLARY ELECTROPHORESIS
Capillary: 48.5 cm × 50 μm fused-silica (40 cm to detector) (Hewlett Packard)
Capillary preparation: Between runs flush with 100 mM NaOH for 5 min, with water for 2 min, and with running buffer for 5 min.
Capillary temperature: 25
Running buffer: MeCN:buffer 9:91 (Buffer was 40 mM pH 10 borate buffer containing 40 mM sodium dodecyl sulfate.)
Injection: Hydrodynamic injection at 5 kPa for 3 s.
Detector: UV 214
Migration time: 5.77
Voltage: 25 kV
Current: 44 μA
Model: Hewlett Packard [3D]CE

OTHER SUBSTANCES
Simultaneous: impurities, ibuprofen, thebaine

KEY WORDS
tablets

REFERENCE
Persson-Stubberud,K.; Åström,O. Separation of ibuprofen, codeine phosphate, their degradation products and impurities by capillary electrophoresis. 1. Method development and optimization with fractional factorial design, *J.Chromatogr.A*, **1998**, *798*, 307–314.

SAMPLE
Matrix: solutions

CAPILLARY ELECTROPHORESIS
Capillary: 15 cm × 50 μm fused-silica coated with crosslinked polyacrylamide (13.5 cm to detector) (Scientific Glass Engineering)
Running buffer: 100 mM pH 6.8 Sodium phosphate buffer containing 3% G 3707 (heptaoxyethylene lauryl ether, Atlas Chemie or Sigma P 8800)
Injection: Place a 1-2 mm slug in capillary by capillary action
Detector: UV 205
Migration time: 7.2
Voltage: 3 kV
Current: 60 μA

OTHER SUBSTANCES
Simultaneous: alprenolol, benzylamine, chlorpheniramine, ephedrine, protriptyline, terodilin

---

KEY WORDS
coated capillary

REFERENCE
Hjertén,S.; Valtcheva,L.; Elenbring,K.; Eaker,D. High-performance electrophoresis of acidic and basic low-mo-
lecular-weight compounds and of proteins in the presence of polymers and neutral surfactants,
J.Liq.Chromatogr., **1989**, *12*, 2471–2499.

SAMPLE
Matrix: solutions
Sample preparation: Inject an aliquot of a 1 mg/mL solution in MeOH.

CAPILLARY ELECTROPHORESIS
Capillary: 67 cm × 50 μm (60 cm to detector) (Composite Metal Services, Worcs., UK)
Capillary preparation: Before each run rinse with running buffer for 3 min.
Capillary temperature: 20
Running buffer: 50 mM pH 10.5 Glycine buffer containing 50 mM sodium dodecyl sulfate
Injection: Pressure injection at 30 mbar for 5 s.
Detector: UV 220
Migration time: 15
Voltage: 25 kV
Model: Beckman P/ACE 2050

OTHER SUBSTANCES
Simultaneous: amphetamine, caffeine, diamorphine, morphine

REFERENCE
Hyötyläinen,T.; Sirén,H.; Riekkola,M.-L. Determination of morphine analogues, caffeine and amphetamine in
biological fluids by capillary electrophoresis with the marker technique, *J.Chromatogr.A*, **1996**, *735*, 439–
447.

SAMPLE
Matrix: solutions

CAPILLARY ELECTROPHORESIS
Capillary: 70 cm × 75 μm fused-silica (63 cm to detector) (Composite Metal Services, Hallow,
UK)
Capillary preparation: Wash with running buffer for 5 min before each run. Purge new capil-
laries with 100 mM NaOH at 60° for 5 min and with water at 60° for 5 min, equilibrate with
running buffer for 5 min, apply voltage for 5 min.
Capillary temperature: 20
Running buffer: 50 mM Disodium tetraborate adjusted to pH 2.2 with orthophosphoric acid
Injection: Hydrodynamic (vacuum) injection for 3 s.
Detector: UV (wavelength not given)
Migration time: 14
Voltage: 25 kV
Model: SpectraPhoresis 1000 (Thermo Separation Products)

OTHER SUBSTANCES
Simultaneous: clenbuterol, flurazepam, meperidine (pethidine), noscapine

---

REFERENCE
McGrath,G.; Smyth,W.F. Large-volume sample stacking of selected drugs of forensic significance by capillary
electrophoresis, *J.Chromatogr.B*, **1996**, *681*, 125–131.

SAMPLE
Matrix: solutions
Sample preparation: Inject an aliquot of a 1-50 μg/mL solution in water.

CAPILLARY ELECTROPHORESIS
Capillary: 55 cm × 50 μm fused-silica (35 cm to detector) (J&W)

Capillary preparation: Flush with running buffer for 5 min before each run. Periodically wash with 100 mM NaOH and flush extensively with running buffer.
Running buffer: MeOH:25 mM pH 9.24 borate buffer 20:80 containing 100 mM sodium dodecyl sulfate
Injection: Injection of 5 µL using a split-flow injector ratio of 1:800.
Detector: UV 200
Migration time: 21.05
Voltage: 20 kV
Current: <60-80 µA
Model: ISCO Model 3850

OTHER SUBSTANCES
Simultaneous: acetylcodeine, amphetamine, barbital, caffeine, cocaine, diamorphine, diazepam, flunitrazepam, lidocaine, monoacetylmorphine, morphine, nalorphine, narceine, noscapine, papaverine, pentobarbital, procaine, tetracaine, thebaine

REFERENCE
Tagliaro,F.; Smith,F.P.; Turrina,S.; Equisetto,V., Marigo,M. Complementary use of capillary zone electrophoresis and micellar electrokinetic capillary chromatography for mutual confirmation of results in forensic drug analysis, *J.Chromatogr.A*, **1996**, *735*, 227–235.

SAMPLE
Matrix: solutions
Sample preparation: Inject an aliquot of a solution in 100 mM sodium phosphate containing 100-300 µg/mL naphazoline.

CAPILLARY ELECTROPHORESIS
Capillary: 67 cm × 50 µm fused-silica (60 cm to detector)
Capillary preparation: Rinse with running buffer for 2 min between sets of analyses.
Capillary temperature: 30
Running buffer: 200 mM pH 4.5 Phosphate buffer
Injection: Injection at high pressure for 2 s.
Detector: UV 210, UV 230
Migration time: 18
Internal standard: naphazoline (14)
Voltage: 20 kV
Model: Beckman P/ACE 5500

OTHER SUBSTANCES
Simultaneous: acetylcodeine, amphetamine, cocaine, diamorphine, LSD, methadone, methamphetamine, methylenedioxyamphetamine, methylenedioxymethamphetamine, morphine, PCP, psilocyn

REFERENCE
Walker,J.A.; Marché,H.L.; Newby,N.; Bechtold,E.J. A free zone capillary electrophoresis method for the quantitation of common illicit drug samples, *J.Forensic Sci.*, **1996**, *41*, 824–829.

SAMPLE
Matrix: solutions
Sample preparation: Inject an aliquot of a solution in running buffer diluted 10-fold with water.

CAPILLARY ELECTROPHORESIS
Capillary: 70 cm × 75 µm fused-silica (63 cm to detector) (Composite Metal Services, Hallow, UK)
Capillary preparation: Wash with running buffer for 5 min before each injection. Condition new capillaries by purging with 100 mM NaOH at 60° for 5 min and with water at 60° for 5 min. Equilibrate with run buffer for 5 min then apply voltage for 5 min a number of times.
Capillary temperature: 20
Running buffer: 50 mM Disodium tetraborate containing 2 mM cetyltrimethylammonium bromide, adjusted to pH 2.2 with orthophosphoric acid
Injection: Hydrodynamic (vacuum) injection for 30 s then apply + 15 kV (to remove solvent). When the current reaches 95% of its pre-injection level switch polarity to begin the electrophoresis.

Detector: UV 210
Migration time: 31
Voltage: -15 kV
Model: Spectra Phoresis 1000
Limit of detection: 404 nM

OTHER SUBSTANCES
Simultaneous: clenbuterol, flurazepam, meperidine (pethidine), noscapine

---

REFERENCE
Smyth,W.F.; Harland,G.B.; McClean,S.; McGrath,G.; Oxspring,D. Effect of on-capillary large volume sample stacking on limits of detection in the capillary zone electrophoretic determination of selected drugs, dyes and metal chelates, *J.Chromatogr.A*, **1997**, *772*, 161–169.

SAMPLE
Matrix: solutions

CAPILLARY ELECTROPHORESIS
Capillary: 55 cm × 50 μm fused-silica (50 cm to detector) (Polymicro Technologies)
Capillary preparation: Between runs purge with 1 M NaOH for 2 min, with water for 2 min, and with running buffer for 3 min
Capillary temperature: 25
Running buffer: MeCN:buffer 50:50 (Buffer was 100 mM ammonium acetate adjusted to pH 3.1 with acetic acid.)
Injection: Pressure injection at 345 mbar.s.
Detector: UV 224, Finnigan MAT Model 95, electrospray (details in paper)
Migration time: 15
Voltage: 15 kV
Model: Bio-Rad BioFocus 3000

OTHER SUBSTANCES
Simultaneous: morphine, narceine, noscapine, papaverine, thebaine

---

REFERENCE
Unger,M.; Stöckigt,D.; Belder,D.; Stöckigt,J. General approach for the analysis of various alkaloid classes using capillary electrophoresis and capillary electrophoresis-mass spectrometry, *J.Chromatogr.A*, **1997**, *767*, 263–276.

SAMPLE
Matrix: solutions
Sample preparation: Inject an aliquot of a solution in MeOH:0.3 mM phosphoric acid 8:92.

CAPILLARY ELECTROPHORESIS
Capillary: 41 cm × 50 μm fused-silica (22 cm to detector) (Polymicro Technologies)
Capillary preparation: Rinse with running buffer for 10 min using a vacuum of 67.7 kPa.
Capillary temperature: 35
Running buffer: 40 mM pH 10.6 Sodium phosphate buffer containing 80 mM sodium dodecyl sulfate
Injection: Vacuum injection at 16.9 kPa for 1 s.
Detector: UV 210
Migration time: 13
Voltage: 10 kV
Current: 61 μA
Model: Applied Biosystems 270A-HT

OTHER SUBSTANCES
Simultaneous: metabolites, codeine-6-glucuronide, dihydrocodeine, dihydrocodeine-6-glucuronide, dihydromorphine, ethylmorphine, morphine, morphine-3-glucuronide, norcodeine, nordihydrocodeine, nordihydromorphine, normorphine

REFERENCE
Zhang,C.-X.; Thormann,W. Separation of free and glucuronidated opioids by capillary electrophoresis in aqueous, binary and micellar media, *J.Chromatogr.A*, **1997**, *764*, 157–168.

SAMPLE
Matrix: solutions

CAPILLARY ELECTROPHORESIS
Capillary: 20 cm × 75 μm containing 3 μm Micra bare silica (packed for 20 cm; 20 cm to detector) (Unimicro Technologies, Pleasanton CA)
Capillary preparation: Before each run condition capillary at 5 kV until a constant current is achieved. Condition new capillaries with running buffer at 500 psi.
Capillary temperature: 25
Running buffer: MeCN:10 mM pH 8.29 Tris-HCl buffer 80:20
Injection: Electrokinetic injection at 5 kV for 5 s.
Detector: UV 214
Migration time: 12.5
Model: Beckman P/ACE 5510

OTHER SUBSTANCES
Simultaneous: aniline, berberine, cocaine, ephedrine, jatrorrhizine, thebaine

KEY WORDS
electrochromatography

REFERENCE
Wei,W.; Luo,G.A.; Hua,G.Y.; Yan,C. Capillary electrochromatographic separation of basic compounds with bare silica as stationary phase, *J.Chromatogr.A*, **1998**, *817*, 65–74.

SAMPLE
Matrix: solutions
Sample preparation: Inject an aliquot of an aqueous solution.

CAPILLARY ELECTROPHORESIS
Capillary: 41 cm × 50 μm fused-silica (22 cm to detector) (Polymicro Technologies)
Capillary preparation: Flush with running buffer for 5 min between runs.
Running buffer: 40 mM pH 10.7 phosphate buffer containing 70 mM sodium dodecyl sulfate
Injection: Hydrodynamic injection at 5″ Hg for 1 s.
Detector: UV 210
Migration time: 11.2
Internal standard: norcodeine (2)
Voltage: 10 kV
Current: 60 μA
Model: Applied Biosystems 270A-HT

OTHER SUBSTANCES
Simultaneous: codeine-6-glucuronide, dihydrocodeine, dihydrocodeine-6-glucuronide, dihydromorphine, ethylmorphine, morphine, morphine-3-glucuronide, nordihydrocodeine, nordihydromorphine, normorphine

REFERENCE
Zhang,C.-X.; Thormann,W. Head-column field-amplified sample stacking in binary system capillary electrophoresis. 2. Optimization with a preinjection plug and application to micellar electrokinetic chromatography, *Anal.Chem.*, **1998**, *70*, 540–548.

SAMPLE
Matrix: urine
Sample preparation: Condition a Bond Elut Certify SPE cartridge with 2 mL MeOH and 2 mL 100 mM pH 6 phosphate buffer, do not allow to go dry. 5 mL Urine + 2 mL 100 mM pH 5 phosphate buffer, add to the SPE cartridge, wash with 3 mL water, wash with 3 mL 100 mM HCl, wash with 9 mL MeOH, elute with 2 mL dichloromethane:isopropanol 80:20 containing

2% ammonium hydroxide. Evaporate the eluate to dryness under a stream of nitrogen at room temperature, reconstitute the residue in 100 μL running buffer, inject an aliquot.

CAPILLARY ELECTROPHORESIS
Capillary: 90 cm × 75 μm fused-silica (70 cm to detector) (Polymicro Technologies)
Capillary preparation: Before each run rinse capillary with 100 mM NaOH for 3 min and with buffer for 5 min.
Running buffer: 10 mM pH 9.1 Na_2HPO_4 containing 6 mM sodium borate and 75 mM sodium dodecyl sulfate
Injection: Gravity injection at 34 cm for 5 s
Detector: UV 195
Migration time: 22.5
Voltage: 20 kV
Current: 76-80 μA

OTHER SUBSTANCES
Extracted: benzoylecgonine, cocaine, methaqualone, morphine

KEY WORDS
SPE

REFERENCE
Wernly,P.; Thormann,W. Analysis of illicit drugs in human urine by micellar electrokinetic capillary chromatography with on-column fast scanning polychrome absorption detection, *Anal.Chem.*, **1991**, *63*, 2878–2882.

SAMPLE
Matrix: urine
Sample preparation: Filter (0.2 μm), inject an aliquot of the filtrate.

CAPILLARY ELECTROPHORESIS
Capillary: 80 cm × 50 μm fused-silica (57 cm to detector)
Capillary preparation: Before each run rinse the capillary with 1 M NaOH for 3 min, with 100 mM NaOH for 3 min, with water for 3 min, and with running buffer for 10 min.
Running buffer: 110 mM Boric acid containing 56 mM NaOH and 44 mM HCl, pH 8
Injection: Vacuum injection for 1 s
Detector: UV 220
Migration time: 4.820
Voltage: 20 kV
Model: Europhor Prime Vision system IV
Limit of detection: 1.2 μM

OTHER SUBSTANCES
Extracted: acebutolol (UV 238), acetazolamide (UV 222), alprenolol (UV 220), amiloride (UV 220), atenolol (UV 228), bendroflumethiazide (UV 220), bumetanide (UV 220), chlorthalidone (UV 220), cocaine (UV 236), ethacrynic acid (UV 220), furosemide (UV 232), hydrochlorothiazide (UV 226), methadone (UV 220), metoxiphenamine (UV 220), nadolol (UV 220), norcodeine (UV 220), oxprenolol (UV 220), pentazocine (UV 220), propranolol (UV 220), spironolactone (UV 244), triamterene (UV 232), xipamide (UV 234)

REFERENCE
Gonzalez,E.; Laserna,J.J. Capillary zone electrophoresis for the rapid screening of banned drugs in sport, *Electrophoresis*, **1994**, *15*, 240–243.

SAMPLE
Matrix: urine
Sample preparation: Condition a 130 mg Bond Elut Certify SPE cartridge with 2 mL MeOH and 2 mL water. 5 mL Urine + 1 mL concentrated HCl, heat at 120° for 30 min, cool, adjust pH to 7 with about 1.25 mL 10 M KOH, centrifuge at 1500 g for 3 min, add to the SPE cartridge, wash with 2 mL water, 1 mL 100 mM pH 4 acetate buffer, wash with 2 mL MeOH, dry under vacuum, elute with two 750 μL portions of MeOH containing 30% concentrated ammonium hydroxide. Evaporate the eluate to dryness under a stream of air at room temperature, reconstitute the residue in 100 μL running buffer, inject an aliquot.

CAPILLARY ELECTROPHORESIS
Capillary: 70 cm × 75 μm fused-silica (70 cm to detector)
Capillary preparation: Before each run rinse capillary with 100 mM NaOH for 2 min and with buffer for 6 min.
Capillary temperature: 35
Running buffer: 10 mM pH 9.2 Na_2HPO_4 containing 6 mM sodium borate and 75 mM sodium dodecyl sulfate
Injection: Vacuum injection for 0.5 s with electrokinetic injection at 30 kV
Detector: UV 213
Migration time: 14.5
Internal standard: codeine
Voltage: 30 kV
Current: 112-115 μA
Model: Applied Biosystems 270A-HT

OTHER SUBSTANCES
Extracted: dihydrocodeine, dihydromorphine, nordihydrocodeine, nordihydromorphine

KEY WORDS
SPE; codeine is IS

REFERENCE
Hufschmid,E.; Theurillat,R.; Martin,U.; Thormann,W. Exploration of the metabolism of dihydrocodeine via determination of its metabolites in human urine using micellar electrokinetic capillary chromatography, *J.Chromatogr.B*, **1995**, *668*, 159–170.

SAMPLE
Matrix: urine
Sample preparation: Condition a 130 mg Bond Elut Certify SPE cartridge with 2 mL MeOH and 2 mL water. 5 mL Urine + 1 mL concentrated HCl, heat at 120° for 30 min, cool, adjust pH to 7 with about 1.25 mL 10 M KOH, centrifuge at 1500 g for 3 min, add to the SPE cartridge, wash with 2 mL water, 1 mL 100 mM pH 4 acetate buffer, wash with 2 mL MeOH, dry under vacuum, elute with two 750 μL portions of MeOH containing 30% concentrated ammonium hydroxide. Evaporate the eluate to dryness under a stream of nitrogen at room temperature, reconstitute the residue in 100 μL running buffer, inject an aliquot (J. Chromatogr. B 1995, 668, 159).

CAPILLARY ELECTROPHORESIS
Capillary: 70 cm × 75 μm fused-silica (70 cm to detector)
Running buffer: 6 mM pH 9.2 Sodium Tetraborate containing 10 mM Na_2HPO_4 and 75 mM sodium dodecyl sulfate
Injection: Vacuum injection for 0.5 s.
Detector: UV 210
Migration time: 24
Voltage: 25 kV
Current: 68-78 μA
Model: Europhor Prime Vision IV

OTHER SUBSTANCES
Extracted: morphine

KEY WORDS
SPE

REFERENCE
Hufschmid,E.; Theurillat,R.; Wilder-Smith,C.H.; Thormann,W. Characterization of the genetic polymorphism of dihydrocodeine O-demethylation in man via analysis of urinary dihydrocodeine and dihydromorphine by micellar electrokinetic capillary chromatography, *J.Chromatogr.B*, **1996**, *678*, 43–51.

SAMPLE
Matrix: urine

Sample preparation: Condition a 300 mg Bondelut Certify SPE cartridge. 500 μL urine + 500
μL 805 ng/mL levallorphan solution, add to the SPE cartridge, wash, elute with 2 mL dichlo-
romethane:isopropanol:ammonia 80:20:2. Evaporate the eluate to dryness, reconstitute with
100 μL MeOH, dilute to 1 mL with water

CAPILLARY ELECTROPHORESIS
Capillary: 65 cm × 50 μm silica (60 cm to detector)
Capillary preparation: Before each run flush capillary with running buffer at 930 mbar for 2
min
Capillary temperature: 25
Running buffer: 100 mM pH 6 Na_2HPO_4
Injection: Electrokinetic injection at 5 kV for 10 s
Detector: UV 200
Migration time: 10.99
Internal standard: levallorphan (10.18)
Voltage: 20 kV
Model: Hewlett-Packard 3DCE
Limit of detection: 7 ng/mL

OTHER SUBSTANCES
Extracted: diamorphine, dihydrocodeine, 6-monoacetylmorphine, morphine, pholcodine

---

KEY WORDS
SPE

REFERENCE
Taylor,R.B.; Low,A.S.; Reid,R.G. Determination of opiates in urine by capillary electrophoresis, *J.Chromatogr.B*,
1996, *675*, 213–223.

Corticosterone

Molecular formula: $C_{21}H_{30}O_4$
Molecular weight: 346.47
CAS Registry No.: 50-22-6
Merck Index (12th ed.): 2601

SAMPLE
Matrix: solutions

CAPILLARY ELECTROPHORESIS
Capillary: 62 cm × 53 μm fused-silica (50 cm to detector) (Polymicro Technologies)
Capillary preparation: Rinse capillary with running buffer for 1 min between runs.
Running buffer: 50 mM pH 9.0 Phosphate/borate buffer containing 33 mM taurocholate, 33 mM
glycodeoxycholate, and 70 mM butanesulfonate
Detector: UV 254
Migration time: 24.4
Voltage: 15 kV
Model: laboratory-constructed

OTHER SUBSTANCES
Simultaneous: cortisone, cortisone acetate, deoxycorticosterone, fludrocortisone, fludrocortisone
acetate, fluocinolone acetonide, hydrocortisone, hydrocortisone-21-acetate, 6α-methylpredniso-
lone, prednisolone, prednisolone acetate, prednisone, prednisone acetate, progesterone, triamcinolone, triamcinolone acetonide

---

REFERENCE
Bumgarner,J.G.; Khaledi,M.G. Mixed micelles of short chain alkyl surfactants and bile slats in electrokinetic
chromatography: Enhanced separation of corticosteroids, *J.Chromatogr.A*, **1996**, *738*, 275–283.

SAMPLE
Matrix: solutions
Sample preparation: Inject an aliquot of a solution in running buffer.

CAPILLARY ELECTROPHORESIS
Capillary: 60 cm × 50 μm fused-silica (50 cm to detector) (Supelco)
Capillary preparation: Between runs wash capillary with 500 mM NaOH for 4 min.
Running buffer: n-Hexanol:sodium dodecyl sulfate:n-butanol:20 mM pH 10.0 phosphate buffer 0.81:3.31:6.61:89.28
Injection: Electrokinetic injection at 15 kV for 10 s.
Detector: UV 220
Migration time: 23
Voltage: 15 kV
Model: laboratory-constructed
Limit of detection: 1 pmole

OTHER SUBSTANCES
Simultaneous: aldosterone, cortexolone, cortisone, 11-dehydrocorticosterone, deoxycorticosterone, deoxycorticosterone acetate, dexamethasone, hydrocortisone, triamcinolone

REFERENCE
Vomastová,L.; Miksík,I.; Deyl,Z. Microemulsion and micellar electrokinetic chromatography of steroids, *J.Chromatogr.B*, **1996**, *681*, 107–113.

SAMPLE
Matrix: solutions
Sample preparation: Inject an aliquot of a solution in MeOH:water 9.44:91.66 containing 10 mM sodium dodecyl sulfate.

CAPILLARY ELECTROPHORESIS
Capillary: 67 cm × 50 μm fused-silica (60 cm to detector) (Composite Metal Services, Worcs., UK)
Capillary preparation: Before each run rinse capillary with KOH solution for 2 min, with water for 2 min, and with running buffer for 2 min, switch on voltage for 2 min before injection.
Capillary temperature: 25
Running buffer: 49 mM 3-[(1,1-Dimethyl-2-hydroxyethyl)amino]-2-hydroxypropanesulfonic acid (AMPSO) containing 18 mM sodium dodecyl sulfate and 55 mM sodium cholate, adjusted to pH 9.0 with 25% ammonia
Injection: Pressure injection at 35 mbar for 3.5 s.
Detector: UV 260
Migration time: 16.8
Voltage: 20 kV
Current: 33 μA
Model: Beckman P/ACE 2050

OTHER SUBSTANCES
Simultaneous: d-aldosterone, cortisone, 1-dehydroaldosterone, 21-deoxycortisol, hydrocortisone, 17-isoaldosterone

REFERENCE
Wiedmer,S.K.; Jumppanen,J.H.; Haario,H.; Riekkola,M.-L. Optimization of selectivity and resolution in micellar electrokinetic capillary chromatography ith a mixed micellar system of sodium dodecyl sulfate and sodium cholate, *Electrophoresis*, **1996**, *17*, 1931–1937.

SAMPLE
Matrix: solutions

CAPILLARY ELECTROPHORESIS
Capillary: 17.6 cm × 50 μm fused-silica packed with 6 μm Zorbax ODS (9.6 cm to detector) (Quadrex, New Haven CT)

Capillary preparation: Sonicate 10 mg stationary phase in 110 μL acetone and 110 μL toluene for 5 min, pack in the capillary with acetone at 65 MPa for 30 min, wash with acetone at 50 MPa for 30 min, wash with water at 50 MPa, sinter frits while washing with water. Condition with running buffer at 0.15 mL/min for about 15 min.
Capillary temperature: 25
Running buffer: Gradient. A was MeCN:10 mM pH 8.0 borate buffer 65:35. B was MeCN:10 mM pH 8.0 borate buffer 85:15. A:B from 100:0 to 0:100 over 5 min, maintain at 0:100. Pump mobile phase at 0.1 mL/min.
Injection: Electrokinetic injection at 1 kV for 0.5 s.
Detector: UV 205
Migration time: 3
Voltage: 14 kV
Current: 1 μA
Model: Perkin-Elmer/Applied Biosystems ABI 270A-HT

OTHER SUBSTANCES
Simultaneous: androstan-3,17-dione, androsten-3,17-dione, pregnan-3,20-dione, testosterone

KEY WORDS
electrochromatography

REFERENCE
Huber,C.G.; Choudhary,G.; Horváth,C. Capillary electrochromatography with gradient elution, *Anal.Chem.*, **1997**, *69*, 4429–4436.

SAMPLE
Matrix: solutions
Sample preparation: Inject an aliquot of a solution in MeOH:water 30;70.

CAPILLARY ELECTROPHORESIS
Capillary: 83 cm fused-silica (25 cm to detector) (Composite Metal Services)
Capillary temperature: 25
Running buffer: 100 mM ammonium acetate adjusted to pH 9 with ammonia
Injection: Electrokinetic injection of 8.5 mM pH 9 ammonium acetate containing 10 mM sodium dodecyl sulfate and 10 mM sodium cholate at 30 kV for 30 s followed by pressure injection of sample at 50 mbar for 5 s.
Detector: UV 254
Migration time: 7.5
Voltage: 10 kV
Model: Hewlett-Packard 3D

OTHER SUBSTANCES
Simultaneous: cortisone, hydrocortisone

KEY WORDS
MS detector also described

REFERENCE
Wiedmer,S.K.; Jussila,M.; Riekkola,M.-L. On-line partial filling micellar electrokinetic capillary chromatography-electrospray ionization-mass spectrometry of corticosteroids, *Electrophoresis*, **1998**, *19*, 1711–1718.

Corticotropin

CAS Registry No.: 9002-60-2
Merck Index (12th ed.): 136

SAMPLE
Matrix: solutions

Sample preparation: Inject an aliquot of a 200 µg/mL solution in water.

CAPILLARY ELECTROPHORESIS
Capillary: 57 cm × 75 µm fused-silica (50 cm to detector)
Capillary preparation: Before each analysis rinse capillary with 100 mM NaOH for 3 min, with water for 3 min, and with separation buffer for 3 min. At the beginning of each day rinse capillary with 1 M NaOH for 45 min and with water for 45 min.
Capillary temperature: 25
Running buffer: 50 mM Formic acid adjusted to pH 3.80 with β-alanine
Injection: Pressure injection for 5 s (ca. 25 nL)
Detector: UV 214
Migration time: 6.4
Voltage: 25 kV
Model: Beckman P/ACE System 2100

OTHER SUBSTANCES
Simultaneous: degradation products

REFERENCE
Langenhuizen,M.H.J.M.; Janssen,P.S.L. Capillary zone electrophoresis of pharmaceutical peptides, *J.Chromatogr.*, **1993**, *638*, 311–318.

Cortisone

Molecular formula: $C_{21}H_{28}O_5$
Molecular weight: 360.45
CAS Registry No.: 53-06-5, 50-04-4 (acetate), 509-00-2 (21β-cyclopentanepropionate), 508-95-2 (phosphate)
Merck Index (12th ed.): 2602
Lednicer: 1 188, 190

SAMPLE
Matrix: blood
Sample preparation: Condition a C18 SPE cartridge (International Sorbent Technology, Hengoed UK) with three 1 mL portions of MeOH, 1 mL water, and 30 mM pH 7.0 sodium phosphate buffer. 1.5 mL Serum + 1.5 mL water + 10 µL S.H.P. Helix pomatia enzyme (Biosepra, Villneuve-la-Garenne, France), mix, heat at 40° for 5-10 min, add 1 mL metaphosphoric acid, centrifuge (ultracentrifuge) at 3000 rpm for 5 min, add a 2.7 mL aliquot of the supernatant to the SPE cartridge, wash with 1 mL water, elute with 2.5 mL MeOH. Evaporate the eluate to dryness under a stream of nitrogen at 38°, reconstitute with 10 µL MeOH and 90 µL water, inject an aliquot.

CAPILLARY ELECTROPHORESIS
Capillary: 70 cm × 50 µm fused-silica (62.5 cm to detector) (Composite Metal Services, Worcs., UK)
Capillary preparation: Before each run wash capillary with KOH solution for 2 min, with water for 2 min, and with running buffer for 2 min then apply voltage for 2 min. Before first use condition capillary with KOH solution for 20 min and with water for 20 min.
Capillary temperature: 22
Running buffer: 49 mM AMPSO ([(1,1-dimethyl-2-hydroxyethyl)amino]-2-hydroxypropane sulfonic acid) containing 55 mM sodium cholate and 18 mM sodium dodecyl sulfate, pH adjusted to 9.0 with 25% ammonia
Injection: Pressure injection at 50 mbar for 4 s.
Detector: UV 254
Migration time: 11.5
Voltage: 20 kV
Current: 29.5 µA
Model: Hewlett-Packard 3D
Limit of quantitation: 100 ng/mL

OTHER SUBSTANCES
Extracted: dexamethasone, hydrocortisone

KEY WORDS
serum; SPE

REFERENCE
Wiedmer,S.K.; Sirén,H.; Riekkola,M.-L. Determination of serum corticosteroids by mixed micellar electrokinetic capillary chromatography with sodium dodecyl sulfate and sodium cholate, *Electrophoresis*, **1997**, *18*, 1861–1864.

SAMPLE
Matrix: solutions
Sample preparation: Prepare a 0.2-1 mg/mL solution in MeOH, inject an aliquot.

CAPILLARY ELECTROPHORESIS
Capillary: 65 cm $\times$ 50 μm fused-silica (50 cm to detector)
Running buffer: 20 mM pH 9.0 Phosphate-borate buffer containing 50 mM sodium dodecyl sulfate, 4 M urea, and 15 mM gamma-cyclodextrin
Detector: UV 220
Migration time: 17 (cortisone acetate)
Voltage: 20 kV

OTHER SUBSTANCES
Simultaneous: betamethasone, dexamethasone acetate, fluocinolone acetonide, fluocinonide, hydrocortisone, hydrocortisone acetate, triamcinolone acetonide

REFERENCE
Nishi,H.; Matsuo,M. Separation of corticosteroids and aromatic hydrocarbons by cyclodextrin-modified micellar electrokinetic chromatography, *J.Liq.Chromatogr.*, **1991**, *14*, 973–986.

SAMPLE
Matrix: solutions
Sample preparation: Prepare a 10 mg/mL solution in MeOH, inject an aliquot.

CAPILLARY ELECTROPHORESIS
Capillary: 62 cm $\times$ 53 μm fused-silica (50 cm to detector) (Polymicro Technologies)
Capillary preparation: Rinse with running buffer for 1 min between runs.
Running buffer: 50 mM pH 9.0 Phosphate borate buffer containing 50 mM dehydrocholate, 50 mM taurocholate, and 50 mM sodium dodecyl sulfate
Detector: UV 254
Migration time: 17.5 (cortisone), 23.5 (cortisone acetate)
Voltage: 15 kV
Current: 56 μA
Model: Laboratory constructed

OTHER SUBSTANCES
Simultaneous: corticosterone, deoxycorticosterone, fludrocortisone, fludrocortisone acetate, fluocinolone acetonide, hydrocortisone, hydrocortisone-21-acetate, 6α-methylprednisolone, prednisolone, prednisolone acetate, prednisone, prednisone acetate, progesterone, triamcinolone, triamcinolone acetonide

REFERENCE
Bumgarner,J.G.; Khaledi,M.G. Mixed micellar electrokinetic chromatography of corticosteroids, *Electrophoresis*, **1994**, *15*, 1260–1266.

SAMPLE
Matrix: solutions
Sample preparation: Prepare a solution in MeOH:water:100 mM sodium dodecyl sulfate 9.4:80.6:10, inject an aliquot.

CAPILLARY ELECTROPHORESIS
Capillary: 57 cm × 50 μm fused-silica (50 cm to detector) (Polymicro Technologies)
Capillary temperature: 23
Running buffer: 60 mM pH 9.2 Borate buffer containing 10 mM sodium dodecyl sulfate
Injection: Hydrostatic injection at 0.5 psi for 3 s
Detector: UV 260
Migration time: 12.8
Voltage: 22 kV
Current: 48 μA
Model: Beckman 2050 P/ACE

OTHER SUBSTANCES
Simultaneous: aldosterone, corticosterone, 1-dehydroaldosterone, 21-deoxycortisol, hydrocortisone, 17-isoaldosterone

REFERENCE
Jumppanen,J.H.; Wiedmer,S.K.; Sirén,H.; Riekkola,M.L.; Haario,H. Optimized separation of seven corticosteroids by micellar electrokinetic chromatography, *Electrophoresis*, **1994**, *15*, 1267–1272.

SAMPLE
Matrix: solutions

CAPILLARY ELECTROPHORESIS
Capillary: 62 cm × 53 μm fused-silica (50 cm to detector) (Polymicro Technologies)
Capillary preparation: Rinse capillary with running buffer for 1 min between runs.
Running buffer: 50 mM pH 9.0 Phosphate/borate buffer containing 33 mM taurocholate, 33 mM glycodeoxycholate, and 70 mM butanesulfonate
Detector: UV 254
Migration time: 18 (cortisone), 22 (cortisone acetate)
Voltage: 15 kV
Model: laboratory-constructed

OTHER SUBSTANCES
Simultaneous: corticosterone, deoxycorticosterone, fludrocortisone, fludrocortisone acetate, fluocinolone acetonide, hydrocortisone, hydrocortisone-21-acetate, 6α-methylprednisolone, prednisolone, prednisolone acetate, prednisone, prednisone acetate, progesterone, triamcinolone, triamcinolone acetonide

REFERENCE
Bumgarner,J.G.; Khaledi,M.G. Mixed micelles of short chain alkyl surfactants and bile slats in electrokinetic chromatography: Enhanced separation of corticosteroids, *J.Chromatogr.A*, **1996**, *738*, 275–283.

SAMPLE
Matrix: solutions
Sample preparation: Inject an aliquot of a solution in running buffer.

CAPILLARY ELECTROPHORESIS
Capillary: 60 cm × 50 μm fused-silica (50 cm to detector) (Supelco)
Capillary preparation: Between runs wash capillary with 500 mM NaOH for 4 min.
Running buffer: n-Hexanol:sodium dodecyl sulfate:n-butanol:20 mM pH 10.0 phosphate buffer 0.81:3.31:6.61:89.28
Injection: Electrokinetic injection at 15 kV for 10 s.
Detector: UV 220
Migration time: 17
Voltage: 15 kV
Model: laboratory-constructed
Limit of detection: 1 pmole

OTHER SUBSTANCES
Simultaneous: aldosterone, cortexolone, corticosterone, 11-dehydrocorticosterone, deoxycorticosterone, deoxycorticosterone acetate, dexamethasone, hydrocortisone, triamcinolone

REFERENCE
Vomastová,L.; Miksík,I.; Deyl,Z. Microemulsion and micellar electrokinetic chromatography of steroids, *J.Chromatogr.B*, **1996**, *681*, 107–113.

SAMPLE
Matrix: solutions
Sample preparation: Inject an aliquot of a solution in MeOH:water 9.44:91.66 containing 10 mM sodium dodecyl sulfate.

CAPILLARY ELECTROPHORESIS
Capillary: 67 cm × 50 μm fused-silica (60 cm to detector) (Composite Metal Services, Worcs., UK)
Capillary preparation: Before each run rinse capillary with KOH solution for 2 min, with water for 2 min, and with running buffer for 2 min, switch on voltage for 2 min before injection.
Capillary temperature: 25
Running buffer: 49 mM 3-[(1,1-Dimethyl-2-hydroxyethyl)amino]-2-hydroxypropanesulfonic acid (AMPSO) containing 18 mM sodium dodecyl sulfate and 55 mM sodium cholate, adjusted to pH 9.0 with 25% ammonia
Injection: Pressure injection at 35 mbar for 3.5 s.
Detector: UV 260
Migration time: 12
Voltage: 20 kV
Current: 33 μA
Model: Beckman P/ACE 2050

OTHER SUBSTANCES
Simultaneous: d-aldosterone, corticosterone, 1-dehydroaldosterone, 21-deoxycortisol, hydrocortisone, 17-isoaldosterone

REFERENCE
Wiedmer,S.K.; Jumppanen,J.H.; Haario,H.; Riekkola,M.-L. Optimization of selectivity and resolution in micellar electrokinetic capillary chromatography ith a mixed micellar system of sodium dodecyl sulfate and sodium cholate, *Electrophoresis*, **1996**, *17*, 1931–1937.

SAMPLE
Matrix: solutions
Sample preparation: Inject a solution in MeOH:running buffer 10:90.

CAPILLARY ELECTROPHORESIS
Capillary: 50 cm × 50 μm fused-silica (45.4 cm to detector) (Polymicro Technologies)
Capillary temperature: 25
Running buffer: 20 mM pH 7.0 Sodium phosphate buffer containing 50 mM sodium dodecyl sulfate and 10 mM Brij 35
Injection: Pressure injection at 1 psi.s.
Detector: UV 240
Migration time: 5.7
Voltage: 20 kV
Model: Bio-Rad BioFocus 3000

OTHER SUBSTANCES
Simultaneous: corticosterone, hydrocortisone

REFERENCE
Muijselaar,P.G.; Claessens,H.A.; Cramers,C.A. Characterization of pseudostationary phases in micellar electrokinetic chromatography by applying linear solvation energy relationships and retention indexes, *Anal.Chem.*, **1997**, *69*, 1184–1191.

SAMPLE
Matrix: solutions

CAPILLARY ELECTROPHORESIS
Capillary: 42 cm × 50 μm fused-silica packed with 3 μm CEC Hypersil ODS (packed length 30 cm, 30.1 cm to detector) (Hypersil)
Running buffer: Gradient. A was 5 mM ammonium acetate in MeCN:water 17:83. B was 5 mM ammonium acetate in MeCN:water 38:62. A:B 100:0 for 3 min, to 0:100 over 15 min, maintain at 0:100 (pumped with an HPLC pump at 0.01 mL/min for 3 min then at 0.1 mL/min).
Injection: Inject 10 μL using an HPLC injector.
Detector: UV 240
Migration time: 31.3
Voltage: 30 kV

OTHER SUBSTANCES
Simultaneous: adrenosterone, betamethasone, dexamethasone, fluocortolone, hydrocortisone, methylprednisolone, prednisolone, triamcinolone, triamcinolone acetonide

KEY WORDS
electrochromatography

REFERENCE
Taylor,M.R.; Teale,P. Gradient capillary electrochromatography of drug mixtures with UV and electrospray ionisation mass spectrometric detection, *J.Chromatogr.A*, **1997**, *768*, 89–95.

SAMPLE
Matrix: solutions

CAPILLARY ELECTROPHORESIS
Capillary: 67 cm × 50 μm fused-silica (60 cm to detector)
Capillary preparation: Before each run equilibrate with running buffer for 3 min. After each run wash with 100 mM NaOH for 3 min and with water. At the beginning of each day rinse the capillary with 100 mM NaOH for 30 min and with water for 30 min.
Capillary temperature: 25 ± 0.1
Running buffer: 25 mM pH 9 Bicine buffer containing 10 mM sodium dodecyl sulfate and 5 mM N-dodecyl-N,N-dimethyl-3-ammonio-1-propanesulfonate (SB3-12)
Injection: High pressure injection for 5 s
Detector: UV 254
Migration time: 5.9
Voltage: 30 kV
Current: <40 μA
Model: Beckman P/ACE 5010

OTHER SUBSTANCES
Simultaneous: 11-desoxycortisol, 21-desoxycortisol, dimethyltestosterone, hydrocortisone, 17-hydroxyprogesterone, progesterone, testosterone propionate, testosterone

REFERENCE
Valbuena,G.A.; Rao,L.V.; Petersen,J.R.; Okorodudu,A.O.; Bissell,M.G.; Mohammad,A.A. Anionic-zwitterionic mixed micelles in the micellar electrokinetic separation of clinically relevant steroids on a fused-silica capillary, *J.Chromatogr.A*, **1997**, *781*, 467–474.

SAMPLE
Matrix: solutions
Sample preparation: Inject an aliquot of a solution in MeOH:water 30;70.

CAPILLARY ELECTROPHORESIS
Capillary: 83 cm fused-silica (25 cm to detector) (Composite Metal Services)
Capillary temperature: 25
Running buffer: 100 mM ammonium acetate adjusted to pH 9 with ammonia
Injection: Electrokinetic injection of 8.5 mM pH 9 ammonium acetate containing 10 mM sodium dodecyl sulfate and 10 mM sodium cholate at 30 kV for 30 s followed by pressure injection of sample at 50 mbar for 5 s.
Detector: UV 254

Migration time: 6
Voltage: 10 kV
Model: Hewlett-Packard 3D

OTHER SUBSTANCES
Simultaneous: corticosterone, hydrocortisone

KEY WORDS
MS detector also described

REFERENCE
Wiedmer,S.K.; Jussila,M.; Riekkola,M.-L. On-line partial filling micellar electrokinetic capillary chromatography-electrospray ionization-mass spectrometry of corticosteroids, *Electrophoresis*, **1998**, *19*, 1711–1718.

SAMPLE
Matrix: urine
Sample preparation: Acidify 5 mL urine to pH 6, extract twice with 2 mL portions of dichloromethane. Evaporate the extracts to dryness under reduced pressure, reconstitute with 500 µL running buffer, sonicate for 5 min, inject an aliquot.

CAPILLARY ELECTROPHORESIS
Capillary: 57 cm × 50 µm fused-silica (50 cm to detector)
Capillary preparation: Before each run wash capillary with 100 mM HCl for 3 min, wash with 500 mM NaOH, wash with water, and equilibrate with running buffer for 3 min. At the beginning of each day rinse capillary with 500 mM NaOH for 30 min and with water for 30 min.
Capillary temperature: 15 ± 0.1
Running buffer: 10 mM pH 7.4 Sodium phosphate buffer containing 50 mM dodecyltrimethylammonium bromide and 5.2 mM trioctylphosphine oxide
Injection: High-pressure injection for 5 s
Detector: UV 254
Migration time: 29
Voltage: 15 kV
Current: <30 µA
Model: Beckman P/ACE 5010
Limit of quantitation: 500 ng/mL

OTHER SUBSTANCES
Extracted: 17-deoxycorticosterone, dimethyltestosterone, hydrocortisone, testosterone, testosterone propionate

REFERENCE
Abubaker,M.A.; Petersen,J.R.; Bissell,M.G. Micellar electrokinetic capillary chromatographic separation of steroids in urine by trioctylphosphine oxide and cationic surfactant, *J.Chromatogr.B*, **1995**, *674*, 31–38.

SAMPLE
Matrix: urine
Sample preparation: Vortex 5 mL urine and 2 mL hexane for 3 min, centrifuge at 3500 rpm for 3 min, repeat extraction. Combine the organic layers and evaporate them to dryness at room temperature, reconstitute the residue in 500 µL 20 mM pH 2.5 phosphate buffer containing 10 mM sodium dodecyl sulfate, inject an aliquot.

CAPILLARY ELECTROPHORESIS
Capillary: 64.5 cm × 50 µm fused-silica (56 cm to detector) (Polymicro Technologies)
Capillary preparation: Between runs flush capillary with 1 M NaOH for 1 min, with MeOH for 1 min, with 100 mM naOH for 1 min, with water for 2 min, and with running buffer for 2 min. Before use rinse capillary with 1 M NaOH for 10 min, with MeOH for 5 min, with water for 5 min, and with running buffer for 5 min.
Capillary temperature: 30
Running buffer: MeOH:100 mM pH 2 phosphate buffer 20:80 containing 50 mM sodium dodecyl sulfate

Injection: Inject a 5.8 cm water plug at 50 mbar, inject sample using voltage, when the current reaches 90% of current in a capillary filled only with running buffer replace sample with running buffer.
Detector: UV 247
Migration time: 11.5
Voltage: -25 kV
Model: Hewlett Packard 3D
Limit of detection: 9.0 ppb (S/N 3)

OTHER SUBSTANCES
Extracted: hydrocortisone, progesterone, testosterone

KEY WORDS
detector at anode

REFERENCE
Quirino,J.P.; Terabe,S. On-line concentration of neutral analytes for micellar electrokinetic chromatography. 5. Field-enhanced sample injection with reverse migrating micelles, *Anal.Chem.*, **1998**, *70*, 1893 -1901.

Coumachlor

Molecular formula: $C_{19}H_{15}ClO_4$
Molecular weight: 342.78
CAS Registry No.: 81-82-3
Merck Index (12th ed.): 2623

SAMPLE
Matrix: solutions
Sample preparation: Inject an aliquot of a 100-500 µg/mL solution in MeOH.

CAPILLARY ELECTROPHORESIS
Capillary: 55 cm × 50 µm fused-silica (50.5 cm to detector) (Polymicro Technologies)
Capillary preparation: Before each run purge capillary with water and running buffer for 2 min each.
Capillary temperature: 25
Running buffer: 25 mM pH 5.6 NaH_2PO_4 buffer containing 0.5% poly(sodium N-undecylenyl-D-valinate) (Prepare poly(sodium N-undecylenyl-D-valinate) as follows. Add 30 mmoles undecylenic acid to 30 mmoles N-hydroxysuccinimide in 130 mL dry ethyl acetate, add 30 mmoles dicyclohexylcarbodiimide in 10 mL dry ethyl acetate, let stand at room temperature overnight, filter, recrystallize ester from EtOH. Add 1 mmole undecylenic acid N-hydroxysuccinimide ester in 10 mL THF to 1 mmole L-valine and 1 mmole sodium bicarbonate in 10 mL water, after 16 h acidify to pH 2 with 1 M HCl, remove the organic solvent under reduced pressure, add 50 mL water, filter, dry, recrystallize N-undecylenyl-L-valine from chloroform/petroleum ether (mp 91°) (J.Lipid Res. 1967, 8, 142). Prepare the sodium salt by adding an equimolar amount of NaOH in EtOH:water 1:4 (Anal.Chem. 1994, 66, 3773). Polymerize a 50 mM aqueous solution of the sodium salt by irradiating with a ^{60}Co gamma source at 0.16 Mrad/h at 21° (J.Poly.Sci.: Poly.Lett.Ed. 1979, 17, 749), lyophilize, extract with hot EtOH to remove monomer, dry under vacuum. Dry organic solvents over Type 4A 4-8 mesh molecular sieve.)
Injection: Pressure injection at 2 psi.s.
Detector: UV 280
Migration time: 26, 26.5 (enantiomers)
Voltage: +20 kV
Model: Bio-Rad Biofocus 3000

OTHER SUBSTANCES
Simultaneous: warfarin

KEY WORDS
chiral

REFERENCE
Agnew-Heard,K.A.; Sánchez Peña,M.; Shamsi,S.A.; Warner,I.M. Studies of polymerized sodium *N*-undecylenyl-L-valinate in chiral micellar electrokinetic capillary chromatography of neutral, acidic, and basic compounds, *Anal.Chem.*, **1997**, *69*, 958–964.

Coumaphos

Molecular formula: $C_{14}H_{16}ClO_5PS$
Molecular weight: 362.77
CAS Registry No.: 56-72-4
Merck Index (12th ed.): 2626

SAMPLE
Matrix: solutions
Sample preparation: Inject an aliquot of a solution in running buffer.

CAPILLARY ELECTROPHORESIS
Capillary: 60 cm × 50 μm fused-silica (53 cm to detector) (Polymicro Technologies)
Capillary preparation: Condition capillary by flushing with 100 mM NaOH for 20 min, water for 10 min, and running buffer for 10 min then equilibrate under voltage.
Running buffer: 10 mM pH 7.0 Sodium phosphate containing 60 mM sodium cholate
Detector: UV 210
Migration time: 12
Voltage: 18 kV
Model: laboratory-constructed

OTHER SUBSTANCES
Simultaneous: acephate, aldicarb, aminocarb, banol, benfluralin, bensulide, buprofezin, captan, carbaryl, carbendazim, carbofuran, chlorotoluron, chlorpyrifos, dichlorvos, dimethoate, EPN, fenchlorphos, fenobucarb, fensulfothion, fenthion, iprodione, isofenphos, isoprocarb, malathion, mepronil, methiocarb, napropamide, oxyfluorfen, paclobutrazol, parathion, pendimethalin, pirimicarb, propoxur, propyzamide, pyriodaphenthion, simazine, simetryn, thiram, tolclofos-methyl, triadimefon, tricyclazole
Interfering: profenophos

REFERENCE
Wu,Y.S.; Lee,H.K.; Li,S.F.Y. Rapid estimation of octanol-water partition coefficients of pesticides by micellar electrokinetic chromatography, *Electrophoresis*, **1998**, *19*, 1719–1727.

Cromolyn

Molecular formula: $C_{23}H_{16}O_{11}$, $C_{23}H_{14}Na_2O_{11}$ (disodium salt)
Molecular weight: 468.37
CAS Registry No.: 16110-51-3, 15826-37-6 (disodium salt)
Merck Index (12th ed.): 2658
Lednicer: 3 66, 235; 4 44, 137, 150, 205

SAMPLE
Matrix: solutions

CAPILLARY ELECTROPHORESIS
Capillary: 27 cm × 75 μm
Capillary preparation: Before each run rinse capillary with 100 mM NaOH for 30 s and with running buffer for 30 s. Condition new capillaries by rinsing with 100 mM NaOH for 20 min.
Capillary temperature: 30
Running buffer: 15 mM Sodium borate
Injection: Pressure injection of sample at 25 mbar for 1 s followed by running buffer at 25 mbar for 1 s.
Detector: UV 200
Migration time: 3.71
Internal standard: aminobenzoic acid (3.5)
Voltage: 6.5 kV
Model: Beckman

OTHER SUBSTANCES
Simultaneous: aspirin, bacitracin, beclomethasone, benzoic acid, ceftizoxime, ceftriaxone, cefuroxime, cephalothin, embonic acid, epoprostenol, glyburide (glibenclamide), levothyroxine, nedocromil, nystatin, omeprazole, prednisolone, warfarin, zidovudine

REFERENCE
Altria,K.D.; Bryant,S.M.; Hadgett,T.A. Validated capillary electrophoresis method for the analysis of a range of acidic drugs and excipients, *J.Pharm.Biomed.Anal.*, **1997**, *15*, 1091–1101.

Cyclizine

Molecular formula: $C_{18}H_{22}N_2$
Molecular weight: 266.39
CAS Registry No.: 82-92-8, 303-25-3 (HCl), 5897-19-8 (lactate)
Merck Index (12th ed.): 2779
Lednicer: 1 58

SAMPLE
Matrix: formulations
Sample preparation: Grind tablets, extract twice with MeOH by sonicating at low temperature, centrifuge, filter (0.45 μm) the supernatant, inject an aliquot of the filtrate.

CAPILLARY ELECTROPHORESIS
Capillary: 47 cm × 50 μm (40 cm to detector) (Polymicro Technologies)
Capillary temperature: 25
Running buffer: 50 mM pH 7.5 Borate buffer containing 50 mM phosphate, 10 mM sodium dodecyl sulfate, 10 mM β-cyclodextrin, and 10 mM tetrabutylammonium hydrogen sulfate
Injection: Inject at a pressure of 0.5 psi for 1 s (5.2 nL)
Detector: UV 214
Migration time: 4.3
Voltage: 21.5 kV
Model: Beckman P/ACE 2000
Limit of detection: 130 pg

OTHER SUBSTANCES
Simultaneous: chlorpheniramine, dimenhydrinate, doxylamine, methapyrilene, pheniramine, promethazine, thonylamine, triprolidine

KEY WORDS
tablets

REFERENCE
Ong,C.P.; Ng,C.L.; Lee,H.K.; Li,S.F.Y. Determination of antihistamines in pharmaceuticals by capillary electrophoresis, *J.Chromatogr.*, **1991**, *588*, 335–339.

SAMPLE
Matrix: solutions

CAPILLARY ELECTROPHORESIS
Capillary: 62 cm × 75 μm fused-silica (54.5 cm to detector) (Polymicro Technologies)
Capillary preparation: Purge with running buffer before each run. At the beginning of each day purge using 50-60 kPa vacuum with 500 mM NaOH for 5 min, with water for 5 min, with MeCN for 5 min, and with running buffer for 5 min.
Running buffer: MeCN:MeOH:acetic acid 49:50:1 containing 20 mM ammonium acetate
Injection: Hydrostatic injection at 10 cm for 5 s.
Detector: UV 214
Migration time: 4.68
Voltage: 30 kV
Model: Waters Quanta 4000

OTHER SUBSTANCES
Simultaneous: buclizine, chlorcyclizine, chlorpheniramine, dimenhydrinate, methaphenilene, pheniramine, promethazine, pyrilamine, pyrrobutamine, tripelennamine

REFERENCE
Leung,G.N.W.; Tang,H.P.O.; Tso,T.S.C.; Wan,T.S.M. Separation of basic drugs with non-aqueous capillary electrophoresis, *J.Chromatogr.A*, **1996**, *738*, 141–154.

Cyclobarbital

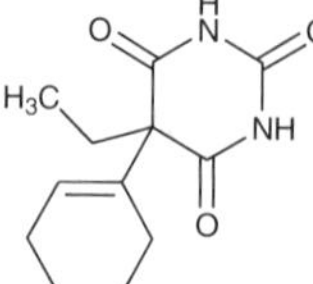

Molecular formula: $C_{12}H_{16}N_2O_3$
Molecular weight: 236.27
CAS Registry No.: 52-31-3, 143-76-0 (calcium salt)
Merck Index (12th ed.): 2780
Lednicer: 1 269

SAMPLE
Matrix: solutions

CAPILLARY ELECTROPHORESIS
Capillary: 57 cm × 50 μm fused-silica (50 cm to detector)
Capillary preparation: After each run rinse with 20 mM pH 9.00 borate buffer containing 90 mM sodium dodecyl sulfate for 5 min and with 100 mM NaOH for 5 min.
Running buffer: 20 mM pH 9.60 Borate buffer containing 10 mM α-cyclodextrin
Injection: Pressure injection at 3.45 kPa for 5 s.
Detector: UV 254
Migration time: 8.3 (S), 8.5 (R)
Voltage: 15 kV
Model: Beckman P/ACE 2100

OTHER SUBSTANCES
Simultaneous: phenobarbital, thiopental

KEY WORDS
chiral

REFERENCE
Conradi,S.; Vogt,C.; Rohde,E. Separation of enantiomeric barbiturates by capillary electrophoresis using a cyclodextrin-containing run buffer, *J.Chem.Educ.*, **1997**, *74*, 1122–1125.

Cyclodrine

Molecular formula: $C_{19}H_{29}NO_3$
Molecular weight: 319.44
CAS Registry No.: 52109-93-0, 78853-39-1 (HCl)
Merck Index (12th ed.): 2788

SAMPLE
Matrix: solutions

CAPILLARY ELECTROPHORESIS
Capillary: 37 cm × 50 μm (30 cm to detector)
Capillary temperature: 25
Running buffer: 50 mM pH 2.5 Phosphate buffer containing 16 mM 2-hydroxy-3-trimethylam-
moniopropyl-β-cyclodextrin
Detector: UV 214
Migration time: 17.56 (first enantiomer; α = 1.128; R_S = 1.720)
Voltage: 20 kV
Model: Beckman P/ACE 2100

OTHER SUBSTANCES
Simultaneous: cyclopentolate, norpseudoephedrine, phendimetrazine, pholedrine, tropicamide

KEY WORDS
chiral

REFERENCE
Bunke,A.; Jira,T. Chiral capillary electrophoresis using a cationic cyclodextrin, *Pharmazie*, **1996**, *51*, 672–673.

SAMPLE
Matrix: solutions

CAPILLARY ELECTROPHORESIS
Capillary: 57 cm × 75 μm (50 cm to detector) (Beckman)
Capillary temperature: 35
Running buffer: 100 mM pH 2 Phosphate buffer containing 1.8% β-cyclodextrin and 40 mM
(R)-(-)-camphorsulfonic acid
Detector: UV 214
Migration time: 25.66, 26.46 (enantiomers)
Voltage: 12 kV
Model: Beckman P/ACE 2100

OTHER SUBSTANCES
Simultaneous: butethamate
Interfering: cyclopentolate

KEY WORDS
chiral

REFERENCE
Bunke,A.; Jira,T. Enantiomeric separation by capillary electrophoresis using chiral and achiral ion pairing
reagents, *Pharmazie*, **1996**, *51*, 479–486.

SAMPLE
Matrix: solutions

CAPILLARY ELECTROPHORESIS
Capillary: 57 cm × 75 μm fused-silica (50 cm to detector) (eCAP, Beckman)

Capillary preparation: Rinse capillary with 100 mM NaOH for 2 min before each analysis.
Running buffer: 100 mM pH 2 Phosphate buffer containing 1.8% β-cyclodextrin and 40 mM
   sodium cyclamate
Injection: Hydrodynamic injection at 0.5 psi for 3 s.
Detector: UV 214
Migration time: 30, 32 (enantiomers)
Voltage: 12 kV
Model: Beckman P/ACE 2100

KEY WORDS
chiral

REFERENCE
Jira,T.; Bunke,A.; Karbaum,A. Use of chiral and achiral ion-pairing reagents in combination with cyclodextrins
   in capillary electrophoresis, *J.Chromatogr.A*, **1998**, *798*, 281–288.

Cyclopentolate

Molecular formula: C$_{17}$H$_{25}$NO3
Molecular weight: 291.39
CAS Registry No.: 512-15-2, 5870-29-1 (HCl)
Merck Index (12th ed.): 2815
Lednicer: 1 92

SAMPLE
Matrix: solutions

CAPILLARY ELECTROPHORESIS
Capillary: 57 cm × 75 μm (50 cm to detector) (Beckman)
Capillary temperature: 35
Running buffer: 100 mM pH 2 Phosphate buffer containing 1.8% β-cyclodextrin and 40 mM
   (R)-(-)-camphorsulfonic acid
Detector: UV 214
Migration time: 25.17, 26.35 (enantiomers)
Voltage: 12 kV
Model: Beckman P/ACE 2100

OTHER SUBSTANCES
Simultaneous: butethamate
Interfering: cyclodrine

KEY WORDS
chiral

REFERENCE
Bunke,A.; Jira,T. Enantiomeric separation by capillary electrophoresis using chiral and achiral ion pairing
   reagents, *Pharmazie*, **1996**, *51*, 479–486.

SAMPLE
Matrix: solutions

CAPILLARY ELECTROPHORESIS
Capillary: 37 cm × 50 μm (30 cm to detector)
Capillary temperature: 25
Running buffer: 50 mM pH 2.5 Phosphate buffer containing 16 mM 2-hydroxy-3-trimethylam-
   moniopropyl-β-cyclodextrin
Detector: UV 214

Migration time: 19.38 (first enantiomer; α = 1.122; R_s = 1.820)
Voltage: 20 kV
Model: Beckman P/ACE 2100

OTHER SUBSTANCES
Simultaneous: cyclodrine, norpseudoephedrine, phendimetrazine, pholedrine, tropicamide

---

KEY WORDS
chiral

REFERENCE
Bunke,A.; Jira,T. Chiral capillary electrophoresis using a cationic cyclodextrin, *Pharmazie*, **1996**, *51*, 672–673.

SAMPLE
Matrix: solutions

CAPILLARY ELECTROPHORESIS
Capillary: 37 cm × 50 μm fused-silica (30 cm to detector) (eCAP, Beckman)
Capillary preparation: Rinse capillary with 100 mM NaOH for 2 min before each analysis.
Capillary temperature: 25
Running buffer: 50 mM pH 5.0 Acetate buffer containing 16 mM 2-hydroxy-3-trimethylammo-
   niopropyl-β-cyclodextrin (Wacker-Chemie, Munich)
Injection: Hydrodynamic injection at 0.5 psi for 3 s.
Detector: UV 214
Migration time: 15, 16.5 (enantiomers)
Voltage: 20 kV
Model: Beckman P/ACE 2100

KEY WORDS
chiral

REFERENCE
Bunke,A.; Jira,T. Use of cationic cyclodextrin for enantioseparation by capillary electrophoresis, *J.Chromatogr.A*, **1998**, *798*, 275–280.

Cytarabine

Molecular formula: $C_9H_{13}N_3O_5$
Molecular weight: 243.22
CAS Registry No.: 147-94-4, 69-74-9 (HCl)
Merck Index (12th ed.): 2853

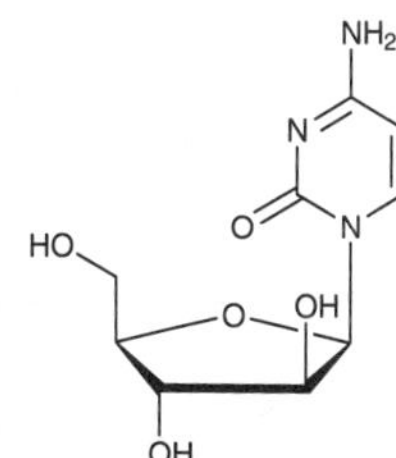

SAMPLE
Matrix: blood
Sample preparation: Condition a 3 mL Bond-Elut SPE cartridge with 3 column volumes of
   MeCN and 2 column volumes of water. Add 200 μL plasma to the SPE cartridge, wash with
   1.2 mL water, suck dry, elute with 1 mL MeCN. Evaporate the eluate to dryness under a stream
   of nitrogen at room temperature, reconstitute the residue in 100 μL water, centrifuge, inject
   an aliquot.

CAPILLARY ELECTROPHORESIS
Capillary: 26 cm × 50 μm fused-silica (Polymicro Technologies)
Capillary preparation: Between injections wash capillary with 100 mM NaOH for 2 min and
   with running buffer for 2 min. At the start of each day wash capillary with 1 M HCl, 1 M
   NaOH, 100 mM NaOH, and water for 5 min each.
Capillary temperature: 30

Running buffer: 40 mM disodium citrate adjusted to pH 2.5 with 1 M HCl
Injection: Pressure injection for 10 s (28 nL)
Detector: UV 280
Migration time: 4.5
Voltage: 8 kV
Current: 90-100 μA
Model: Beckman PACE
Limit of detection: 122 ng/mL (S/N 3)

KEY WORDS
plasma; SPE

REFERENCE
Lloyd,D.K.; Cypess,A.M.; Wainer,I.W. Determination of cytosine-β-D-arabinoside in plasma using capillary electrophoresis, *J.Chromatogr.*, **1991**, *568*, 117–124.

SAMPLE
Matrix: blood
Sample preparation: Inject an aliquot directly.

CAPILLARY ELECTROPHORESIS
Capillary: 80 cm $\times$ 100 μm fused-silica (60 cm to detector) (Polymicro Technologies)
Capillary preparation: Treat before use with 1 M NaOH for 24 h.
Running buffer: 50 mM pH 9 Sodium borate buffer containing 60 mM sodium dodecyl sulfate
Injection: Hydrodynamic injection at 1 cm for 50 s.
Detector: UV 275
Migration time: 11
Voltage: 10 kV
Model: laboratory-constructed
Limit of detection: 800 ng/mL

OTHER SUBSTANCES
Extracted: cytidine

KEY WORDS
serum; direct injection

REFERENCE
Krivánková,L.; Kostálová,A.; Vargas,G.; Havel,J.; Bocek,P. Separation of aracytidine and cytidine by capillary electrophoretic techniques, *Electrophoresis*, **1996**, *17*, 1954–1958.

Dapsone

Molecular formula: $C_{12}H_{12}N_2O_2S$
Molecular weight: 248.31
CAS Registry No.: 80-08-0
Merck Index (12th ed.): 2885
Lednicer: 1 139

SAMPLE
Matrix: solutions
Sample preparation: Prepare a solution in 100 mM cetyltrimethylammonium bromide, inject an aliquot.

CAPILLARY ELECTROPHORESIS
Capillary: 67 cm $\times$ 50 μm fused silica capillary (60 cm to detector) (Siemens)

Running buffer: pH 5.0 Acetate buffer containing 10 mM imidazole, 50 μg/mL FC 135 (a fluorinated surfactant, Fluorad/3M), and 100 mM cetyltrimethylammonium bromide
Injection: Pressure injection for 5 s.
Detector: UV 214
Migration time: 11
Voltage: 15 kV
Model: Beckman P/ACE System 2000
Limit of detection: 29.84 μg/mL

OTHER SUBSTANCES
Simultaneous: acetaminophen, dexamethasone, dihydrostreptomycin, kanamycin, sisomycin, tobramycin

KEY WORDS
detector is at anode; indirect UV detection for aminoglycoside antibiotics

REFERENCE
Ackermans,M.T.; Everaorts,F.M.; Beckers,J.L. Determination of aminoglycoside antibiotics in pharmaceuticals by capillary zone electrophoresis with indirect UV detection coupled with micellar electrokinetic capillary chromatography, *J.Chromatogr.*, **1992**, *606*, 229–235.

SAMPLE
Matrix: solutions
Sample preparation: Prepare a 100-200 ppm solution in MeOH, inject an aliquot.

CAPILLARY ELECTROPHORESIS
Capillary: 45 cm × 50 μm fused-silica (Polymicro Technologies)
Running buffer: 50 mM pH 7.50 Phosphate buffer containing 25 mM borate and 9 mM tetrabutylammonium bromide
Injection: Hydrodynamic injection at 5 cm for 5 s (1 nL)
Detector: UV 240
Migration time: 3.9
Voltage: 15 kV
Current: 39 μA
Model: Laboratory constructed

OTHER SUBSTANCES
Simultaneous: chloroquine, primaquine, pyrimethamine, quinacrine, quinine, sulfadiazine

REFERENCE
Ng,C.L.; Toh,Y.L.; Li,S.F.Y.; Lee,H.K. Capillary electrophoresis of biologically important compounds: Optimization of separation conditions by the overlapping resolution mapping scheme, *J.Liq.Chromatogr.*, **1993**, *16*, 3653–3666.

Daunorubicin

Molecular formula: $C_{27}H_{29}NO_{10}$
Molecular weight: 527.53
CAS Registry No.: 20830-81-3, 23541-50-6 (HCl)
Merck Index (12th ed.): 2890

SAMPLE
Matrix: blood
Sample preparation: 1 mL Plasma + 2 mL chloroform, vortex for 1 min, centrifuge at 1000 g for 10 min. Remove 1.6 mL of the lower organic layer and add it to 100 μL 5 mM pH 2.3

phosphoric acid, vortex for 1 min, centrifuge at 1000 g for 10 min. Remove 50 μL of the upper aqueous layer and add it to 150 μL MeCN, mix, inject an aliquot of this mixture.

CAPILLARY ELECTROPHORESIS
Capillary: 70 cm × 75 μm fused-silica (65 cm to detector) (SGE)
Running buffer: MeCN:100 mM pH 4.2 sodium phosphate buffer 70:30
Injection: Electrokinetic injection at 12 kV for 5 s (14 nL)
Detector: F, laboratory constructed with an 80 mW 476.5 nm argon laser, 595 nm emission filter (10 nm bandwidth)
Migration time: 7.0
Internal standard: daunorubicin
Voltage: 20-25 kV
Current: 35 μA (fixed)
Limit of detection: 50 pg/mL (S/N 3)

OTHER SUBSTANCES
Extracted: doxorubicin, epirubicin

KEY WORDS
plasma; daunorubicin is IS

REFERENCE
Reinhoud,N.J.; Tjaden,U.R.; Irth,H.; van der Greef,J. Bioanalysis of some anthracyclines in human plasma by capillary electrophoresis with laser-induced fluorescence detection, *J.Chromatogr.*, **1992**, *574*, 327–334.

SAMPLE
Matrix: solutions
Sample preparation: Inject an aliquot of a solution in MeCN:water 95:5.

CAPILLARY ELECTROPHORESIS
Capillary: 47 cm × 50 μm fused-silica (40 cm to detector) (Beckman)
Capillary preparation: Between runs rinse capillary with 100 mM NaOH for 1 min and with running buffer for 2 min.
Running buffer: MeCN:buffer 70:30 (Prepare running buffer by adjusting the pH of 100 mM NaH_2PO_4 to 5.0 with 100 mM phosphoric acid, add spermine to a final concentration of 60 μM, add MeCN to a final MeCN concentration of 70%.)
Injection: Electrokinetic injection at 10 kV for 5 s.
Detector: F ex 488 (5 mW Beckman argon ion laser) em 520
Migration time: 4.2
Voltage: 25 kV
Current: 12 μA
Model: Beckman P/ACE 5510

OTHER SUBSTANCES
Simultaneous: idarubicin

REFERENCE
Hempel,G.; Haberland,S.; Schulze-Westhoff,P.; Möhling,N.; Blaschke,G.; Boos,J. Determination of idarubicin and idarubicinol in plasma by capillary electrophoresis, *J.Chromatogr.B*, **1997**, *698*, 287–292.

Debrisoquin

Molecular formula: $C_{10}H_{13}N_3$
Molecular weight: 175.23
CAS Registry No.: 1131-64-2, 581-88-4 (sulfate)
Merck Index (12th ed.): 2901
Lednicer: 2 374

SAMPLE
Matrix: urine
Sample preparation: Condition a 6 mL 200 mg Isolute ENV+ cross-linked polystyrene SPE cartridge (No. IS915-0020-C, International Sorbent Technology, Hengoed, UK) with two 2 mL portions of MeOH and two 2 mL portions of 20 mM pH 6 phosphate buffer, do not allow to dry. Centrifuge urine at 3500 rpm for 5 min. 2 mL Urine+ 2 mL 20 mM pH 6 phosphate buffer, mix, add to the SPE cartridge, wash with 2 mL 50 mM pH 6 phosphate buffer containing 200 mM NaCl, wash with 2 mL MeOH:20 mM pH 6 phosphate buffer 40:60, dry under full vacuum for 10 min, wash with 2 mL ethyl acetate:acetone 50:50, dry under full vacuum for 1 min, wash with 2 mL MeOH, elute with 2 mL MeOH:formic acid 80:20. Evaporate the eluate to dryness under a stream of nitrogen at 37°, reconstitute with 100 μL 4 mM HCl, inject an aliquot.

CAPILLARY ELECTROPHORESIS
Capillary: 60 cm × 50 μm fused-silica (55.4 cm to detector) (Polymicro Technologies)
Capillary preparation: Before each run rinse capillary with 100 mM NaOH for 1 min, with water for 1 min, and with running buffer for 2 min.
Capillary temperature: 20
Running buffer: 50 mM pH 2.5 KH_2PO_4 containing 50 mM heptakis-(2,3,6-tri-O-methyl)-β-cyclodextrin
Injection: Pressure injection at 2 psi.s
Detector: UV 195
Migration time: 17.2
Voltage: 20 kV
Current: 34 μA
Model: Bio-Rad BioFocus 3000
Limit of detection: 150 ng/mL

OTHER SUBSTANCES
Extracted: metabolites, 4-hydroxydebrisoquine (enantiomers are separated under these conditions)

KEY WORDS
SPE

REFERENCE
Lanz,M.; Theurillat,R.; Thormann,W. Characterization of stereoselectivity and genetic polymorphism of the debrisoquine hydroxylation in man via analysis of urinary debrisoquine and 4-hydroxydebrisoquine by capillary electrophoresis, *Electrophoresis*, **1997**, *18*, 1875–1881.

Demeclocycline

Molecular formula: $C_{21}H_{21}ClN_2O_8$
Molecular weight: 464.86
CAS Registry No.: 127-33-3, 64-73-3 (HCl), 13215-10-6 (sesquihydrate)
Merck Index (12th ed.): 2937

SAMPLE
Matrix: bulk

CAPILLARY ELECTROPHORESIS
Capillary: 44 cm × 50 μm fused-silica (38 cm to detector) (Polymicro Technologies)
Capillary temperature: 15
Running buffer: 50 mM pH 12.25 Sodium phosphate buffer containing 0.35% Triton X-100 and 1 mM EDTA
Injection: Hydrodynamic injection for 2 s
Detector: UV 254
Migration time: 12

Voltage: 12 kV
Current: <60 μA
Model: SpectraPhoresis (Thermo Separation Products)
Limit of quantitation: 0.4%
Limit of detection: 0.3%

OTHER SUBSTANCES
Simultaneous: impurities

REFERENCE
Li,Y.M.; Van Schepdael,A.; Roets,E.; Hoogmartens,J. Analysis of demeclocycline by capillary electrophoresis, *J.Chromatogr.A*, **1996**, *740*, 119–123.

SAMPLE
Matrix: bulk
Sample preparation: Inject an aliquot of a solution in running buffer.

CAPILLARY ELECTROPHORESIS
Capillary: 44 cm × 50 μm fused-silica (38 cm to detector) (Polymicro Technologies)
Capillary temperature: 15
Running buffer: 110 mM pH 12.25 Sodium carbonate containing 1 mM EDTA, 0.35% Triton X-100, and 10 mM methyl-β-cyclodextrin
Injection: Hydrodynamic injection for 4 s.
Detector: UV 254
Migration time: 17.5
Voltage: 12 kV
Model: Spectraphoresis 500 (Thermo Separation Products)
Limit of quantitation: 1.5 μg/mL
Limit of detection: 500 ng/mL

OTHER SUBSTANCES
Simultaneous: impurities

REFERENCE
Li,Y.M.; Van Schepdael,A.; Roets,E.; Hoogmartens,J. Optimized methods for capillary electrophoresis of tetracyclines, *J.Pharm.Biomed.Anal.*, **1997**, *15*, 1063–1069.

SAMPLE
Matrix: formulations
Sample preparation: Dissolve capsule contents in 15 mM pH 7.5 sodium phosphate buffer so as to make a 25 mM solution, centrifuge, dilute an aliquot of the supernatant to a concentration of 5 mM with 15 mM pH 7.5 sodium phosphate buffer, inject an aliquot.

CAPILLARY ELECTROPHORESIS
Capillary: 111.9 cm × 75 μm fused-silica (43.4 cm to detector) (Polymicro Technologies)
Capillary preparation: Condition by rinsing with 1 M NaOH for 10 min then running buffer overnight.
Capillary temperature: 25
Running buffer: pH 7.5 Phosphate buffer (buffer concentration 4.3 mM, total sodium concentration 15 mM, ionic strength 18.2 mM)
Injection: Hydrodynamic injection at 2 cm for 1 min
Detector: UV 260
Migration time: 5.6
Current: 20 μA
Limit of detection: 10 μM (S/N 3)

OTHER SUBSTANCES
Simultaneous: chlortetracycline, doxycycline, methacycline, minocycline, oxytetracycline, tetracycline

KEY WORDS
capsules

REFERENCE

Tavares,M.F.M.; McGuffin,V.L. Separation and characterization of tetracycline antibiotics by capillary electrophoresis, *J.Chromatogr.A*, **1994**, *686*, 129–142.

SAMPLE
Matrix: solutions

CAPILLARY ELECTROPHORESIS
Capillary: 43 cm $\times$ 75 μm (36 cm to the detector)
Capillary preparation: Before each run wash capillary with running buffer for 2 min. After each run wash capillary with water at 25° for 5 min, with 100 mM NaOH at 60 ° for 5 min, and with water at 25° for 5 min.
Capillary temperature: 25
Running buffer: 15 mM pH 6.5 Ammonium acetate buffer containing 20 mM sodium dodecyl sulfate and 0.135% Brij 35
Detector: UV 265
Migration time: 4.5
Voltage: 15 kV

OTHER SUBSTANCES
Simultaneous: chlortetracycline, doxycycline, minocycline, oxytetracycline, tetracycline

REFERENCE

Chen,Y.-C.; Lin,C.-E. Migration behavior and separation of tetracycline antibiotics by micellar electrokinetic chromatography, *J.Chromatogr.A*, **1998**, *802*, 95–105.

Denopamine

Molecular formula: $C_{18}H_{23}NO_4$
Molecular weight: 317.38
CAS Registry No.: 7171-90-9
Merck Index (12th ed.): 2943

SAMPLE
Matrix: solutions

CAPILLARY ELECTROPHORESIS
Capillary: 40 cm $\times$ 50 μm fused-silica (25 cm to detector) (SGE)
Running buffer: 25 mM pH 2.7 Phosphate buffer containing 2 M urea and 20 mM heptakis(2,6-di-O-methyl)-β-cyclodextrin
Detector: UV 220
Migration time: 23, 24 (enantiomers)
Voltage: 15 kV
Model: Laboratory constructed

KEY WORDS
chiral

REFERENCE

Nishi,H.; Nakamura,K.; Nakai,H.; Sato,T. Chiral separation of drugs by capillary electrophoresis using β-cyclodextrin polymer, *J.Chromatogr.A*, **1994**, *678*, 333–342.

SAMPLE
Matrix: solutions
Sample preparation: Inject an aliquot of a 100-300 μg/mL solution in water.

CAPILLARY ELECTROPHORESIS
Capillary: 57 cm × 75 μm fused-silica (50 cm to detector)
Capillary temperature: 23
Running buffer: 25 mM pH 2.5 Phosphate buffer containing 2 M urea and 20 mM heptakis(2,6-di-O-methyl)-β-cyclodextrin
Injection: Pressure injection at 0.5 psi for 3-7 s
Detector: UV 220
Migration time: 18.92 (R-(-)), 19.30 (S-(+))
Voltage: 15 kV
Model: Beckman P/ACE 5510

KEY WORDS
chiral

REFERENCE
Nishi,H.; Ishibuchi,K.; Nakamura,K.; Nakai,H.; Sato,T. Enantiomeric separation of denopamine by capillary electrophoresis and high-performance liquid chromatography using cyclodextrins, *J.Pharm.Biomed.Anal.*, **1995**, *13*, 1483–1492.

SAMPLE
Matrix: solutions
Sample preparation: Inject an aliquot of a 100 μg/mL solution in MeOH:water 10:90.

CAPILLARY ELECTROPHORESIS
Capillary: 50 cm × 75 μm fused-silica (37 cm to detector) (Beckman)
Capillary temperature: 23
Running buffer: 20 mM pH 2.5 Phosphate buffer containing 50 mM dimethyl-β-cyclodextrin
Injection: Pressure injection at 0.5 psi for 3-10 s.
Detector: UV 220
Migration time: 11.5 (R), 12 (S)
Voltage: 20 kV
Model: Beckman P/ACE 5510

OTHER SUBSTANCES
Simultaneous: trimetoquinol

KEY WORDS
chiral

REFERENCE
Ishibuchi,K.; Izumoto,S.; Nishi,H.; Sato,T. Enantiomer separation of denopamine by capillary electrophoresis with charged and uncharged cyclodextrins, *Electrophoresis*, **1997**, *18*, 1007–1012.

SAMPLE
Matrix: solutions

CAPILLARY ELECTROPHORESIS
Capillary: 36 cm × 50 μm fused-silica coated with linear polyacrylamide (31.5 cm to detector) (GL Science)
Capillary preparation: At the beginning and end of each day rinse capillary with capillary wash solution (Bio-Rad Cat. No. 148-5022) at 690 kPa for more than 3 min and with water at 690 kPa for more than 3 min. Coat capillary as follows. Treat capillary with 1 M NaOH at room temperature for 1 h, rinse with water, dry by passing nitrogen gas through the capillary at 110° for 6 h. Pass thionyl chloride through the capillary using a suction pump for several min, seal capillary at both ends and heat at 70° for 6 h. Unseal the capillary and fill with 250 mM vinyl magnesium bromide in THF by suction, seal the capillary, heat at 70° for 6 h. Open the capillary and rinse it with THF for several min, rinse with distilled water, fill the capillary with polymerization solution, heat at 28 ± 2° for 1 h, condition at -100 V/cm for 30 min (Anal. Sci. 1994, 10, 1). (The polymerization solution was 5% acrylamide in water containing 49 mM Tris, 384 mM glycine, and 0.1% sodium dodecyl sulfate, degas in an ultrasonic bath. Add 40

μL 10% N,N,N',N'-tetramethylethylenediamine and 10 μL 10% ammonium persulfate to 5 mL of the degassed solution, mix thoroughly.)
Running buffer: MeOH:50 mM pH 6.0 Phosphate buffer 10:90
Injection: Before each injection rinse with water at 690 kPa for 30 s, rinse with running buffer at 690 kPa for 30 s, partially fill with separation solution (500 μM α_1-acid glycoprotein (Cohn fraction VI) (ICN) in running buffer) at 6.9 kPa for 190 s (27 cm), inject sample at 6.9 kPa for 2 s, electrophorese with running buffer (Note that α_1-acid glycoprotein from other suppliers may provide inferior results).
Detector: UV 210
Migration time: Resolution of enantiomers 1.9
Voltage: 12 kV
Model: Bio-rad BioFocus 3000

KEY WORDS
chiral; coated capillary

REFERENCE
Tanaka,Y.; Terabe,S. Separation of the enantiomers of basic drugs by affinity capillary electrophoresis using a partial filling technique and α_1-acid glycoprotein as chiral selector, *Chromatographia*, **1997**, *44*, 119–128.

Deoxycholic acid

Molecular formula: $C_{24}H_{40}O_4$
Molecular weight: 392.58
CAS Registry No.: 83-44-3, 302-95-4 (sodium salt)
Merck Index (12th ed.): 2946

SAMPLE
Matrix: blood
Sample preparation: Filter (Amicon Centrifree YM) while centrifuging at 2000 rpm for 45 min, inject an aliquot of the ultrafiltrate.

CAPILLARY ELECTROPHORESIS
Capillary: fused-silica (Polymicro Technologies)
Capillary preparation: Before each run flush capillary with 100 mM NaOH for 3 min, with water for 3 min, and with running buffer for 3 min. At the start of each day flush with 1 M NaOH for 30 min, with water for 3 min, and with running buffer for 3 min. Flush new capillaries with 1 M NaOH for 1 h and with water for 10 min.
Capillary temperature: 23
Running buffer: MeOH:100 mM boric acid 75:25 containing 7 mM gamma-cyclodextrin and 5 mM adenosine 5'-monophosphate, pH 7.0
Injection: Pressure injection for 10 s.
Detector: UV 254
Migration time: 23
Voltage: 30 kV
Current: 6.8 μA
Model: Beckman P/ACE 5510
Limit of detection: 170-330 μM

OTHER SUBSTANCES
Simultaneous: chenodiol, cholic acid, glycochenodeoxycholic acid, glycocholic acid, glycodeoxycholic acid, glycolithocholic acid, glycoursodeoxycholic acid, lithocholic acid, taurochenodeoxycholic acid, taurocholic acid, taurodeoxycholic acid, taurolithocholic acid, tauroursodeoxycholic acid, ursodiol

KEY WORDS
serum; indirect UV detection; ultrafiltrate

REFERENCE
Yarabe,H.H.; Shamsi,S.A.; Warner,I.M. Capillary zone electrophoresis of bile acids with indirect photometric
detection, *Anal.Chem.*, **1998**, *70*, 1412–1418.

Deserpidine

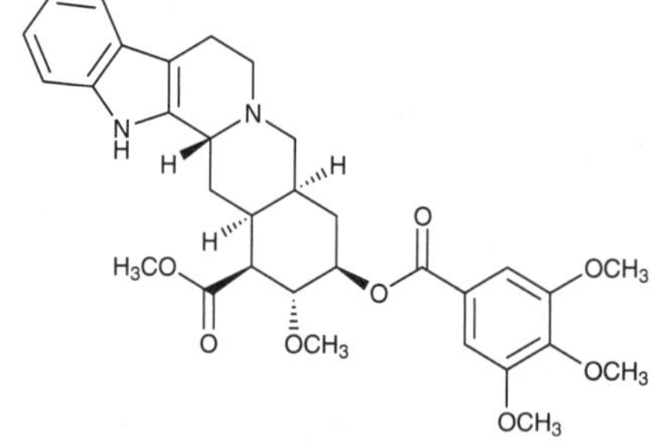

Molecular formula: $C_{32}H_{38}N_2O_8$
Molecular weight: 578.66
CAS Registry No.: 131-01-1
Merck Index (12th ed.): 2964
Lednicer: 1 320

SAMPLE
Matrix: solutions

CAPILLARY ELECTROPHORESIS
Capillary: 59.6 cm × 75 μm fused-silica (52.1 cm to detector) (Polymicro Technologies)
Capillary preparation: Purge with running buffer before each run. At the beginning of each
day purge using 50-60 kPa vacuum with 500 mM NaOH for 5 min, with water for 5 min, with
MeCN for 5 min, and with running buffer for 5 min.
Running buffer: MeCN:MeOH:acetic acid 49:50:1 containing 20 mM ammonium acetate
Injection: Hydrostatic injection at 10 cm for 5 s.
Detector: UV 214
Migration time: 6.14
Voltage: 30 kV
Model: Waters Quanta 4000

OTHER SUBSTANCES
Simultaneous: benzquinamide, ethopropazine, mesoridazine, methotrimeprazine, molindone,
promazine, promethazine, reserpine, thioridazine, thiothixene

REFERENCE
Leung,G.N.W.; Tang,H.P.O.; Tso,T.S.C.; Wan,T.S.M. Separation of basic drugs with non-aqueous capillary electrophoresis, *J.Chromatogr.A*, **1996**, *738*, 141–154.

Desipramine

Molecular formula: $C_{18}H_{22}N_2$
Molecular weight: 266.39
CAS Registry No.: 50-47-5, 58-28-6 (HCl)
Merck Index (12th ed.): 2966
Lednicer: 1 402

SAMPLE
Matrix: blood
Sample preparation: 1 mL Serum or plasma + 50 μL 10 μL/mL trimipramine + 1 mL 2 M
NaOH + 5 mL hexane:isoamyl alcohol 99:1, vortex vigorously, centrifuge at 2500 g for 5 min.
Remove 4 mL of the organic layer and add it to 100 μL 100 mM HCl, extract. Remove 60 μL
of the aqueous layer and evaporate it to dryness under reduced pressure, reconstitute with 20
μL water, inject an aliquot.

CAPILLARY ELECTROPHORESIS
Capillary: 71.5 cm × 50 μm fused-silica (50 cm to detector) (Polymicro Technologies)
Capillary preparation: Before each run rinse capillary with 1 M NaOH for 3 min and running buffer for 3 min (use vacuum at 508 mm Hg).
Capillary temperature: 30 or 40°
Running buffer: 37.5 mM pH 8.0 Phosphate buffer containing 2 M urea and 25 mM dodecyltrimethylammonium bromide
Injection: Vacuum injection at 508 mm Hg for 7 s
Detector: UV 254
Migration time: 9.3
Internal standard: trimipramine (11.5)
Voltage: -25 kV
Model: Applied Biosystems Model 270A
Limit of detection: 5-10 ng/mL

OTHER SUBSTANCES
Extracted: amitriptyline, doxepin, imipramine, nortriptyline

KEY WORDS
serum; plasma

REFERENCE
Lee,K.-J.; Lee,J.J.; Moon,D.C. Determination of tricyclic antidepressants in human plasma by micellar electrokinetic capillary chromatography, *J.Chromatogr.*, **1993**, *616*, 135–143.

SAMPLE
Matrix: solutions

CAPILLARY ELECTROPHORESIS
Capillary: 70 cm × 50 μm coated fused-silica (55 cm to detector) (Polymicro Technologies)
Capillary preparation: Between runs rinse with water for 2 min. Coat capillaries as follows. Condition with 1 M NaOH for 15 min, rinse with water for 5 min, treat with 30 μL/mL 3-(trimethoxysilyl)propyl methacrylate in acetic acid:water 50:50 for 1 h using house vacuum, rinse with water, fill with polymerization solution, let stand for 1 h, flush, rinse with water. (Prepare polymerization solution by adding 10 μL N,N,N',N'-tetramethylethylenediamine to 10 mL of a degassed 4% solution of acrylamide in water, add 10 μL 10% ammonium persulfate in water.)
Running buffer: 10 mM pH 6.5 Citrate buffer containing 20 mM sodium dodecyl sulfate and 10 mM carboxymethyl-β-cyclodextrin (Cyclodextrin Technologies Development, Gainesville FL)
Injection: Hydrodynamic injection at 15 cm for 8-10 s.
Detector: UV 254
Migration time: 10
Voltage: -20 kV
Model: laboratory-constructed

OTHER SUBSTANCES
Simultaneous: amitriptyline, carbamazepine, clomipramine, imipramine, nortriptyline, opipramol, proptriptyline, trimipramine

KEY WORDS
coated capillary

REFERENCE
Spencer,B.J.; Zhang,W.; Purdy,W.C. Capillary electrophoretic separation of tricyclic antidepressants using charged carboxymethyl-β-cyclodextrin as a buffer additive, *Electrophoresis*, **1997**, *18*, 736–744.

SAMPLE
Matrix: solutions

CAPILLARY ELECTROPHORESIS
Capillary: 57 cm × 50 μm fused-silica (50 cm to detector) (Polymicro Technologies)

Capillary preparation: At the start of each day rinse capillary with water for 5 min, 100 mM NaOH for 30 min, with water for 10 min, and with running buffer for 15 min. Wash new capillaries with 1 M NaOH for 1 h.

Capillary temperature: 23

Running buffer: MeOH:50 mM pH 9.55 sodium phosphate buffer containing 0.06% poly(n-undecyl-α-D-glucopyranoside) (Prepare poly(n-undecyl-α-D-glucopyranoside) as follows. With exclusion of atmospheric moisture stir 40 mL alcohol-free chloroform, 5.2 g yellow mercuric oxide, 0.1 g mercuric bromide, 10 g anhydrous calcium sulfate, and 2.74 g 10-undecen-1-ol then add 10 g acetobromo-α-D-glucose (Fluka), stir at room temperature for 20 h, filter, wash the solid with chloroform. Wash the filtrate with aqueous 1 M KBr until no mercury salts are present in the aqueous layers, wash with later, concentrate. Take up the residue in benzene (Caution! Benzene is a carcinogen!) and filter it through activated silica, crystallize from n-hexane to obtain undecylenyl-2,3,4,6-tetra-O-acetyl-β-D-glucopyranoside. Add 2 mL of a 1 M solution of sodium methoxide in MeOH to a solution of 1 g of undecylenyl-2,3,4,6-tetra-O-acetyl-β-D-glucopyranoside in 50 mL dry MeOH, let stand until the reaction is complete (by TLC) to obtain undecylenyl-β-D-glucopyranoside (cf. Tenside Detergents 1978, 15, 72). Use Dowex 50W (H⁺) resin to deionize the final product in solution. Polymerize the monomer by irradiation of an 870 μM solution in MeOH:water 20:80 with ⁶⁰Co gamma radiation for 48 h, purify the polymer solution by dialysis using a 1000 Da molecular mass cut-off membrane.)

Injection: Pressure injection at 0.5 psi for 5 s.

Detector: UV 214

Migration time: 10.5

Voltage: 22.8 kV

Model: Beckman P/ACE 5510

OTHER SUBSTANCES
Simultaneous: amitriptyline, doxepin, imipramine, nordoxepin, nortriptyline, protriptyline

---

REFERENCE
Harrell,C.W.; Dey,J.; Shamsi,S.A.; Foley,J.P.; Warner,I.M. Enhanced separation of antidepressant drugs using a polymerized nonionic surfactant as a transient capillary coating, *Electrophoresis*, **1998**, *19*, 712–718.

SAMPLE
Matrix: urine

Sample preparation: 10 mL Urine + 1 mL 5 M NaOH + 10 mL n-hexane, vortex for 1 min, centrifuge at 0° at 3000 g for 5 min, repeat the extraction twice more. Combine the organic layers and add them to 50 μL glacial acetic acid, evaporate to dryness under a stream of nitrogen at 30°, reconstitute the residue in 50 μL 50 mM sodium taurodeoxycholate, filter (0.2 μm PTFE), inject an aliquot.

CAPILLARY ELECTROPHORESIS
Capillary: 100 cm × 50 μm fused-silica (50 cm to detector) (ISCO)

Capillary preparation: Rinse capillary with 20 μL running buffer between injections. Condition a new capillary by filling with 1 M NaOH, let stand for 1 h, fill with 100 mM NaOH, let stand for 1 h, rinse with buffer. Every 20 injections rinse capillary with 200 μL 1 M NaOH, 200 μL water, and 200 μL running buffer, fill with running buffer.

Capillary temperature: 22

Running buffer: 40 mM pH 9.5 Borate buffer containing 10 mM sodium taurodeoxycholate

Injection: Load under vacuum at 7.5 kPa/s.

Detector: UV 240

Migration time: 6

Voltage: 30 kV

Model: ISCO Model 3140 electropherograph

Limit of detection: 6 ng/mL

OTHER SUBSTANCES
Extracted: acepromazine, amiodarone, amitriptyline, azaperone, chlorpromazine, cianopramine, clomipramine, clozapine, desethylamiodarone, diclofensine, dothiepin, doxepin, imipramine, isocarboxazid, moclobemide, perphenazine, phenothiazine, pimozide, prochlorperazine, promazine, thioridazine, thiothixene, trifluoperazine, trimipramine

---

KEY WORDS
human; cow; pig; horse; protect from light

REFERENCE
Aumatell,A.; Wells,R.J. Determination of a cardiac antiarrhythmic, tricyclic antipsychotics and antidepressants in human and animal urine by micellar electrokinetic capillary chromatography using a bile salt, *J.Chromatogr.B*, **1995**, *669*, 331–344.

Dexamethasone

Molecular formula: $C_{22}H_{29}FO_5$

Molecular weight: 392.47

CAS Registry No.: 50-02-2, 1177-87-3 (acetate), 55812-90-3 (acetate monohydrate), 2392-39-4 (sodium phosphate), 83880-70-0 (acefurate), 55541-30-5 (dipropionate), 312-93-6 (21-dihydrogen phosphate)

Merck Index (12th ed.): 2986

Lednicer: 1 199

SAMPLE

Matrix: blood

Sample preparation: Condition a C18 SPE cartridge (International Sorbent Technology, Hengoed UK) with three 1 mL portions of MeOH, 1 mL water, and 30 mM pH 7.0 sodium phosphate buffer. 1.5 mL Serum + 1.5 mL water + 10 μL S.H.P. Helix pomatia enzyme (Biosepra, Villneuve-la-Garenne, France), mix, heat at 40° for 5-10 min, add 1 mL metaphosphoric acid, centrifuge (ultracentrifuge) at 3000 rpm for 5 min, add a 2.7 mL aliquot of the supernatant to the SPE cartridge, wash with 1 mL water, elute with 2.5 mL MeOH. Evaporate the eluate to dryness under a stream of nitrogen at 38°, reconstitute with 10 μL MeOH and 90 μL water, inject an aliquot.

CAPILLARY ELECTROPHORESIS

Capillary: 70 cm × 50 μm fused-silica (62.5 cm to detector) (Composite Metal Services, Worcs., UK)

Capillary preparation: Before each run wash capillary with KOH solution for 2 min, with water for 2 min, and with running buffer for 2 min then apply voltage for 2 min. Before first use condition capillary with KOH solution for 20 min and with water for 20 min.

Capillary temperature: 22

Running buffer: 49 mM AMPSO ([(1,1-dimethyl-2-hydroxyethyl)amino]-2-hydroxypropane sulfonic acid) containing 55 mM sodium cholate and 18 mM sodium dodecyl sulfate, pH adjusted to 9.0 with 25% ammonia

Injection: Pressure injection at 50 mbar for 4 s.

Detector: UV 254

Migration time: 13.6

Voltage: 20 kV

Current: 29.5 μA

Model: Hewlett-Packard 3D

Limit of quantitation: 100 ng/mL

OTHER SUBSTANCES

Extracted: cortisone, hydrocortisone

KEY WORDS

serum; SPE

REFERENCE
Wiedmer,S.K.; Sirén,H.; Riekkola,M.-L. Determination of serum corticosteroids by mixed micellar electrokinetic capillary chromatography with sodium dodecyl sulfate and sodium cholate, *Electrophoresis*, **1997**, *18*, 1861–1864.

SAMPLE

Matrix: blood, urine

Sample preparation: Condition a 6 mL 300 mg end-capped C8 SPE cartridge (IST, Hengoed UK) with 1 mL MeOH and 1 mL water. Condition a 1 mL 100 mg SAX SPE cartridge (IST, Hengoed UK) with 250 µL MeCN:5 mM ammonium acetate 10:90. Add 2 mL urine to the C8 SPE cartridge, wash with 1 mL water, wash with 2 mL hexane, elute with 1 mL ethyl acetate, elute with 1 mL dichloromethane. Combine the eluates and evaporate them to dryness at 40°, reconstitute with 250 µL MeCN:5 mM ammonium acetate 10:90, add to the SAX SPE cartridge, elute with 150 µL MeCN:5 mM ammonium acetate 10:90, collect all the effluent from the SAX SPE cartridge, mix, inject an aliquot.

CAPILLARY ELECTROPHORESIS

Capillary: 24 cm × 50 µm fused-silica (Composite Metal Services, Hallow UK) slurry packed with 3 µm Apex ODS (Jones Chromatography) to a length of 16 cm (16.1 cm to detector)

Capillary preparation: Slurry pack column with a slurry in MeOH, sonicate during process. Condition with mobile phase at 200 bar for 12 h.

Running buffer: Gradient. MeCN containing 5 mM ammonium acetate:water containing 5 mM ammonium acetate from 9:91 to 80:20 over 5 min, maintain at 80:20 for 5 min, return to initial conditions over 5 min, re-equilibrate for 15 min. Pumped at 0.1 mL/min.

Injection: 10-250 µL

Detector: UV 240

Migration time: 12

Voltage: 25 kV

Limit of detection: 390 ng/mL

OTHER SUBSTANCES

Extracted: fluocortolone, hydrocortisone

KEY WORDS

SPE; horse; electrochromatography

REFERENCE

Taylor,M.R.; Teale,P.; Westwood,S.A.; Perrett,D. Analysis of corticosteroids in biofluids by capillary electrochromatography with gradient elution, *Anal.Chem.*, **1997**, *69*, 2554–2558.

SAMPLE

Matrix: solutions

CAPILLARY ELECTROPHORESIS

Capillary: 65 cm × 50 µm fused-silica (50 cm to detector) (Scientific Glass Engineering)

Capillary preparation: Flush with running buffer for 1 min between runs. Periodically flush with water for 1 min, sweep with 100 mM KOH, let stand in 100 mM KOH for 30 min, flush with water until the effluent is neutral to pH paper, fill with running buffer, let stand for 30 min.

Running buffer: 20 mM pH 9.0 Phosphate-borate buffer containing 100 mM sodium cholate (Buffer was prepared by adjusting pH of 20 mM sodium borate to 9.0 with 20 mM NaH_2PO_4.)

Injection: Injection by siphon at 10 cm for 10 s (about 1 nL)

Detector: UV 210

Migration time: 13 (dexamethasone acetate)

Voltage: 20 kV

OTHER SUBSTANCES

Simultaneous: betamethasone, fluocinolone acetonide, fluocinonide, hydrocortisone acetate, hydrocortisone, triamcinolone, triamcinolone acetonide

REFERENCE

Nishi,H.; Fukuyama,T.; Matsuo,M.; Terabe,S. Separation and determination of lipophilic corticosteroids and benzothiazepin analogues by micellar electrokinetic chromatography using bile salts, *J.Chromatogr.*, **1990**, *513*, 279–295.

SAMPLE

Matrix: solutions

Sample preparation: Prepare a 0.2-1 mg/mL solution in MeOH, inject an aliquot.

CAPILLARY ELECTROPHORESIS
Capillary: 65 cm × 50 μm fused-silica (50 cm to detector)
Running buffer: 20 mM pH 9.0 Phosphate-borate buffer containing 50 mM sodium dodecyl sulfate, 4 M urea, and 15 mM gamma-cyclodextrin
Detector: UV 220
Migration time: 12 (dexamethasone acetate)
Voltage: 20 kV

OTHER SUBSTANCES
Simultaneous: betamethasone, cortisone acetate, fluocinolone acetonide, fluocinonide, hydrocortisone, hydrocortisone acetate, triamcinolone acetonide

REFERENCE
Nishi,H.; Matsuo,M. Separation of corticosteroids and aromatic hydrocarbons by cyclodextrin-modified micellar electrokinetic chromatography, *J.Liq.Chromatogr.*, **1991**, *14*, 973–986.

SAMPLE
Matrix: solutions
Sample preparation: Prepare a solution in 100 mM cetyltrimethylammonium bromide, inject an aliquot.

CAPILLARY ELECTROPHORESIS
Capillary: 67 cm × 50 μm fused silica capillary (60 cm to detector) (Siemens)
Running buffer: pH 5.0 Acetate buffer containing 10 mM imidazole, 50 μg/mL FC 135 (a fluorinated surfactant, Fluorad/3M), and 100 mM cetyltrimethylammonium bromide
Injection: Pressure injection for 5 s.
Detector: UV 214
Migration time: 12.5
Voltage: 15 kV
Model: Beckman P/ACE System 2000

OTHER SUBSTANCES
Simultaneous: acetaminophen, dapsone, dihydrostreptomycin, kanamycin, sisomycin, tobramycin

KEY WORDS
detector is at anode; indirect UV detection for aminoglycoside antibiotics

REFERENCE
Ackermans,M.T.; Everaerts,F.M.; Beckers,J.L. Determination of aminoglycoside antibiotics in pharmaceuticals by capillary zone electrophoresis with indirect UV detection coupled with micellar electrokinetic capillary chromatography, *J.Chromatogr.*, **1992**, *606*, 229–235.

SAMPLE
Matrix: solutions
Sample preparation: Inject an aliquot of a solution in running buffer. (Paper contains details of the preparation of an affinity column for the isolation of this compound from urine prior to capillary electrophoresis analysis.)

CAPILLARY ELECTROPHORESIS
Capillary: 57 cm × 75 μm fused-silica (50 cm to detector)
Capillary preparation: Before each run rinse with running buffer for 2 min. After each run rinse capillary with 100 mM NaOH for 2 min and with water for 2 min. Condition new capillaries with 100 mM HCl for 5 min, with 100 mM NaOH for 10 min, and with water for 5 min.
Capillary temperature: 23
Running buffer: MeOH:buffer 10:90 (Buffer was 40 mM pH 9.2 borate buffer containing 20 mM sodium dodecyl sulfate.)
Injection: Low pressure injection for 10 s.
Detector: UV 240
Migration time: 13.8
Voltage: 30 kV

Model: Beckman P/ACE 5510

OTHER SUBSTANCES
Simultaneous: betamethasone, flumethasone, hydrocortisone, triamcinolone

REFERENCE
Gu,X.; Meleka-Boules,M.; Chen,C.-L. Micellar electrokinetic capillary chromatography combined with immunoaffinity chromatography for identification and determination of dexamethasone and flumethasone in equine urine, *J.Capillary Electrophor.*, **1996**, *3*, 43–49.

SAMPLE
Matrix: solutions
Sample preparation: Inject an aliquot of a solution in running buffer.

CAPILLARY ELECTROPHORESIS
Capillary: 60 cm × 50 μm fused-silica (50 cm to detector) (Supelco)
Capillary preparation: Between runs wash capillary with 500 mM NaOH for 4 min.
Running buffer: n-Hexanol:sodium dodecyl sulfate:n-butanol:20 mM pH 10.0 phosphate buffer 0.81:3.31:6.61:89.28
Injection: Electrokinetic injection at 15 kV for 10 s.
Detector: UV 220
Migration time: 20
Voltage: 15 kV
Model: laboratory-constructed
Limit of detection: 1 pmole

OTHER SUBSTANCES
Simultaneous: aldosterone, cortexolone, corticosterone, cortisone, 11-dehydrocorticosterone, deoxycorticosterone, deoxycorticosterone acetate, hydrocortisone, triamcinolone

REFERENCE
Vomastová,L.; Miksík,I.; Deyl,Z. Microemulsion and micellar electrokinetic chromatography of steroids, *J.Chromatogr.B*, **1996**, *681*, 107–113.

SAMPLE
Matrix: solutions

CAPILLARY ELECTROPHORESIS
Capillary: 33.5 cm × 100 μm fused-silica packed with 3 μm C18 for 25 cm (Innovatech, Stevenage, UK) (Polymicro Technologies)
Capillary preparation: Column packing procedure described in LC.GC 1995, 13, 800.
Capillary temperature: 20
Running buffer: MeCN:25 mM Tris HCl:water 60:20:20, run with 12 bar nitrogen pressure on each side
Injection: Electrokinetic injection at 5 kV for 3 s.
Detector: UV 200
Migration time: 5.2
Voltage: 30 kV
Model: Hewlett-Packard HP[3D]

OTHER SUBSTANCES
Simultaneous: androsterone, dehydrotestosterone, desoxycorticosterone, estrone, ethinyl estradiol, hydrocortisone, prednisone, pregnenolone, testosterone

KEY WORDS
electrochromatography

REFERENCE
Dittmann,M.M.; Rozing,G.P.; Ross,G.; Adam,T.; Unger,K.K. Advances in capillary electrochromatography, *J.Capillary Electrophor.*, **1997**, *4*, 201–212.

SAMPLE
Matrix: solutions

CAPILLARY ELECTROPHORESIS
Capillary: 42 cm × 50 μm fused-silica packed with 3 μm CEC Hypersil ODS (packed length 30 cm, 30.1 cm to detector) (Hypersil)
Running buffer: Gradient. A was 5 mM ammonium acetate in MeCN:water 17:83. B was 5 mM ammonium acetate in MeCN:water 38:62. A:B 100:0 for 3 min, to 0:100 over 15 min, maintain at 0:100 (pumped with an HPLC pump at 0.01 mL/min for 3 min then at 0.1 mL/min).
Injection: Inject 10 μL using an HPLC injector.
Detector: UV 240
Migration time: 35.5
Voltage: 30 kV

OTHER SUBSTANCES
Simultaneous: adrenosterone, betamethasone, cortisone, fluocortolone, hydrocortisone, methylprednisolone, prednisolone, triamcinolone, triamcinolone acetonide

KEY WORDS
electrochromatography

REFERENCE
Taylor,M.R.; Teale,P. Gradient capillary electrochromatography of drug mixtures with UV and electrospray ionisation mass spectrometric detection, *J.Chromatogr.A*, **1997**, *768*, 89–95.

SAMPLE
Matrix: solutions
Sample preparation: Inject an aliquot of a solution in EtOH:water 11.1:78.8 containing 10 mM sodium dodecyl sulfate.

CAPILLARY ELECTROPHORESIS
Capillary: 67 cm × 50 μm fused-silica (60 cm to detector) (Composite Metal Services, Worcestershire UK)
Capillary temperature: 25
Running buffer: 50 mM pH 8.7 [(1,1-Dimethyl-2-hydroxyethyl)amino]-2-hydroxypropanesulfonic acid (AMPSO) containing 20 mM sodium dodecyl sulfate (pH adjusted with 25% ammonia)
Injection: Hydrostatic injection at 0.5 psi for 3.5 s.
Detector: UV 260
Migration time: 11
Voltage: 20 kV
Current: 10.1 μA
Model: Beckman P/ACE 2050

OTHER SUBSTANCES
Simultaneous: 4-androstene-3,17-dione, corticosterone, 11-deoxycortisol, fludrocortisone acetate, hydrocortisone

REFERENCE
Wiedmer,S.K.; Riekkola,M.-L.; Nydén,M.; Söderman,O. Mixed micelles of sodium dodecyl sulfate and sodium cholate: Micellar electrokinetic capillary chromatography and nuclear magnetic resonance spectroscopy, *Anal.Chem.*, **1997**, *69*, 1577–1584.

SAMPLE
Matrix: tears
Sample preparation: 2 μL Tears + 18 μL 50 μg/mL indoprofen in buffer, mix, centrifuge at 700 g for 5 min, inject an aliquot. (Prepare buffer by dissolving 40 mg NaH$_2$PO$_4$, 280 mg Na$_2$HPO$_4$, and 8.5 g NaCl in 1 L water.)

CAPILLARY ELECTROPHORESIS
Capillary: 64.5 cm × 50 μm fused-silica (56 cm to detector) (Hewlett-Packard)

Capillary preparation: Flush with 100 mM sodium dodecyl sulfate for 1 min and with running
  buffer for 3 min between runs. Flush new capillaries with 1 M NaOH for 3 min, with 100 mM
  NaOH for 5 min, and with water for 10 min.
Capillary temperature: 25
Running buffer: 100 mM Sodium tetraborate
Injection: Pressure injection at 5 kPa for 20 s (ca. 24 nL), ramp to running voltage over 1 min.
Detector: UV 242
Migration time: 7.05 (dexamethasone), 15.46 (dexamethasone phosphate)
Internal standard: indoprofen (12.26)
Voltage: 25 kV
Model: Hewlett-Packard HP[3D]
Limit of quantitation: 2 µg/mL
Limit of detection: 500 ng/mL

REFERENCE

Baeyens,V.; Varesio,E.; Veuthey,J.-L.; Gurny,R. Determination of dexamethasone in tears by capillary electrophoresis, *J.Chromatogr.B*, **1997**, *692*, 222–226.

Dextromethorphan

Molecular formula: $C_{18}H_{25}NO$
Molecular weight: 271.40
CAS Registry No.: 125-71-3(dextromethorphan (d-form)), 125-69-9
(dextromethorphan HBr), 6700- 34-1 (dextromethorphan HBr monohydrate),
510-53-2 (racemethorphan)
Merck Index (12th ed.): 8274
Lednicer: 1 293

SAMPLE

Matrix: formulations
Sample preparation: Dilute 10-fold, inject an aliquot.

CAPILLARY ELECTROPHORESIS

Capillary: 62 cm × 50 µm fused silica (50 cm to detector) (Polymicro Technologies)
Capillary temperature: 40
Running buffer: 50 mM pH 2.5 sodium phosphate buffer containing 20 mM β-cyclodextrin and
  1 mM hexadecyltrimethylammonium bromide
Injection: Injection by gravity at 10 cm for 5 s
Detector: UV 214
Migration time: 15
Voltage: 20 kV
Current: 36 µA
Model: Laboratory constructed

OTHER SUBSTANCES

Simultaneous: acetaminophen, doxylamine, pseudoephedrine

KEY WORDS

syrup

REFERENCE

Quang,C.; Khaledi,M.G. Improved chiral separation of basic compounds in capillary electrophoresis using β-
  cyclodextrin and tetraalkylammonium reagents, *Anal.Chem.*, **1993**, *65*, 3354–3358.

SAMPLE

Matrix: formulations
Sample preparation: Dilute formulation 50-fold, inject an aliquot.

CAPILLARY ELECTROPHORESIS
Capillary: 60 cm × 50 μm (52.5 cm to detector) (AccuSep)
Capillary preparation: Purge capillary with running buffer for 3 min between runs. Purge new capillaries under vacuum with 500 mM NaOH for 10 min, with water for 10 min, and with running buffer for 10 min.
Capillary temperature: 30
Running buffer: 25 mM pH 8.0 Na_2HPO_4/sodium tetraborate containing 50 mM (R)-N-dodecyl-carbonylvaline (Prepare (R)-N-dodecoxycarbonylvaline as follows. Prepare dodecyl chloroformate by reacting 1-dodecanol with 0.33 equivalents of triphosgene in dichloromethane solution in the presence of pyridine. Add dodecyl chloroformate dropwise to (R)-valine in 1 M NaOH solution, filter, wash with hexane, recrystallize from ether/petroleum ether (J. Chromatogr. A 1994, 680, 125).)
Injection: Hydrostatic injection at 10 cm for 15 s.
Detector: UV 214
Migration time: 31.966 (dextromethorphan enantiomer)
Voltage: 15 kV
Model: Waters Quanta 4000E

OTHER SUBSTANCES
Simultaneous: acetaminophen, doxylamine, pseudoephedrine, saccharin

REFERENCE
Swartz,M.E.; Mazzeo,J.R.; Grover,E.R.; Brown,P.R. Validation of enantiomeric separations by micellar electrokinetic capillary chromatography using synthetic chiral surfactants, *J.Chromatogr.A*, **1996**, *735*, 303–310.

SAMPLE
Matrix: solutions
Sample preparation: Prepare a solution in MeOH, inject an aliquot.

CAPILLARY ELECTROPHORESIS
Capillary: 55.5 cm × 50 μm fused-silica (44.5 cm to detector) (Polymicro Technologies)
Capillary preparation: Rinse with running buffer at 3 bar for 5 min
Capillary temperature: 35
Running buffer: 40 mM pH 2.2 Sodium phosphate buffer
Injection: Pressure injection at 20 mbar for 6 s
Detector: UV 220
Migration time: 5.5 (dextromethorphan)
Voltage: 20 kV
Current: about 30 μA
Model: Lauerlabs Prince

OTHER SUBSTANCES
Simultaneous: amiodarone, desethylamiodarone, itraconazole, ketoconazole, methadone
Noninterfering: caffeine, naproxen, phenol, theophylline

REFERENCE
Zhang,C.-X.; von Heeren,F.; Thormann,W. Separation of hydrophobic, positively chargeable substances by capillary electrophoresis, *Anal.Chem.*, **1995**, *67*, 2070–2077.

Dextrose

Molecular formula: $C_6H_{12}O_6$
Molecular weight: 180.16
CAS Registry No.: 50-99-7, 77029-61-9 (D-glucopyranose monohydrate), 2280-44-6 (D- glucopyranose), 492-62-5 (α-D-glucopyranose), 492-61-5 (β-D-glucopyranose)
Merck Index (12th ed.): 4467

SAMPLE

Matrix: bulk

Sample preparation: Add 50 μL 500 mM 3-methyl-1-phenyl-2-pyrazolin-5-one in MeOH and 30 μL 500 mM NaOH in water to 10-200 nmoles sugars, heat at 70° for 2 h, cool to room temperature, add 30 μL 500 mM HCl, vortex, wash 4 times with 500 μL aliquots of dichloromethane. Freeze dry the aqueous phase, reconstitute with 200-500 μL water, filter (0.2 μm nylon), inject an aliquot.

CAPILLARY ELECTROPHORESIS

Capillary: 72 cm × 50 μm fused-silica (50 cm to detector) (Polymicro Technologies)

Capillary preparation: Between runs rinse with alkaline rinse solution (NaOH/Na$_3$PO$_4$ in water) for 2 min and with running buffer for 5 min. Flush new capillaries with alkaline rinse solution for 20 min and with running buffer for 20 min.

Capillary temperature: 30

Running buffer: 30 mM Phosphoric acid containing 50 mM sodium dodecyl sulfate, adjusted to pH 7.5 with Tris (High purity sodium dodecyl sulfate (Perkin-Elmer/Applied Biosystems) should be used.)

Injection: Vacuum injection at 16.9 kPa for 3-5 s.

Detector: UV 245

Migration time: 14.5

Voltage: 20 kV

Current: ca. 29 μA

Model: Perkin-Elmer/Applied Biosystems 270A-HT

Limit of detection: 20 fmole

OTHER SUBSTANCES

Simultaneous: N-acetylmannosamine, altrose, cellobiose, 2-deoxy-D-ribose, fucose, galactosamine, galactose, glucosamine, lactose, lyxose, maltose, mannosamine, mannose, rhamnose, xylose

KEY WORDS

derivatization

REFERENCE

Chiesa,C.; Oefner,P.J.; Zieske,L.R.; O'Neill,R.A. Micellar electrokinetic chromatography of monosaccharides derivatized with 1-phenyl-3-methyl-2-pyrazolin-5-one, *J.Capillary Electrophor.*, **1995**, *2*, 175–183.

SAMPLE

Matrix: bulk

Sample preparation: Add the material to 2 μL 200 mM 8-aminopyrene-1,3,6-trisulfonic acid in 15% acetic acid, add 2 μL 1 M sodium cyanoborohydride in THF, heat at 37° overnight, dilute 2500-fold with water, inject an aliquot. (9-Aminopyrene-1,4,6-trisulfonic acid is the same as 8-aminopyrene-1,3,6-trisulfonic acid. It is available from Molecular Probes, Eugene OR. Synthesis is as follows. Rapidly add 300 g anhydrous sodium sulfate to 1300 g concentrated sulfuric acid, cool to 58°, add 202 g finely powdered pyrene over 5 min without cooling, stir for 15 min, cool to 50-55°, add 800 g 65% oleum (with cooling with 12° water) over 20 min, stir for 5 h without cooling, dilute with ice-water, neutralize with calcium carbonate, filter, reduce the filtrate to 10 L, exchange the cation with soda, salt out with 20% sodium chloride to obtain sodium pyrenetetrasulfonate. Heat 61 g sodium pyrenetetrasulfonate with 610 mL 22% aqueous ammonia in a rotating autoclave at 200-210° for 18 h (the pressure rises to 45 atmospheres), remove the excess ammonia by distillation under vacuum, precipitate oxypyrenetrisulfonic acid with a little NaCl, precipitate 9-aminopyrene-1,4,6-trisulfonic acid by saturating with NaCl, recrystallize from dilute NaCl (Liebig's Annalen der Chemie 1939, 540, 189).)

CAPILLARY ELECTROPHORESIS

Capillary: 30 cm × 25 μm fused-silica (30 cm to detector)

Capillary temperature: 20

Running buffer: 25 mM pH 10 Tetraborate buffer

Injection: Pressure injection at 3.45 kPa for 5-10 s.

Detector: F ex 488 (4 mW argon-ion laser) em 520

Migration time: 7.4

Voltage: 750 V/cm

Current: 17 μA

Model: Beckman P/ACE 5500

OTHER SUBSTANCES
Simultaneous: N-acetylgalactosamine, N-acetylglucosamine, fucose, galactose, mannose, xylose

KEY WORDS
derivatization; detector at anode

REFERENCE
Guttman,A.; Brunet,S.; Cooke,N. Capillary electrophoresis fingerprinting of carbohydrates in the biopharmaceutical and food and beverage industries, *LC.GC*, **1996**, *14*, 788–792.

SAMPLE
Matrix: carbohydrates
Sample preparation: Dissolve 1 mg carbohydrate in 100 µL 2 M trifluoroacetic acid, flush with nitrogen, seal tube, heat at 100° for 6 h, evaporate to dryness under reduced pressure in a desiccator containing solid NaOH. Add reagent solution so that the concentration of saccharide is 10-100 mM, vortex gently, heat at 50° for 2 h, inject an aliquot. (Reagent solution was 10 mg/mL sodium cyanoborohydride in MeOH containing 10% 2-aminopyridine and 10% acetic acid.)

CAPILLARY ELECTROPHORESIS
Capillary: 65 cm × 50 µm fused-silica (50 cm to detector) (Scientific Glass Engineering)
Capillary preparation: Rinse capillary with running buffer for 5 min before each run. After 20 runs wash capillary with MeOH for 30 s.
Running buffer: 200 mM pH 10.5 Borate buffer
Injection: Siphon injection at 5 cm for 5 s
Detector: UV 240
Migration time: 13.8
Internal standard: galactose (9)
Voltage: 15 kV

OTHER SUBSTANCES
Extracted: N-acetylgalactosamine, N-acetylglucosamine, arabinose, cinnamic acid, fucose, galacturonic acid, glucuronic acid, lyxose, rhamnose, ribose, xylose

KEY WORDS
derivatization

REFERENCE
Honda,S.; Iwase,S.; Makino,A.; Fujiwara,S. Simultaneous determination of reducing monosaccharides by capillary zone electrophoresis as the borate complexes of N-2-pyridylglycamines, *Anal.Biochem.*, **1989**, *176*, 72–77.

SAMPLE
Matrix: culture media
Sample preparation: Dilute 1:9 with water, inject an aliquot.

CAPILLARY ELECTROPHORESIS
Capillary: 78 cm × 25 µm fused-silica (50 cm to detector) (Polymicro Technologies)
Capillary preparation: After each run flush capillary with running buffer at 668 mbar for 9 min. Fill a new capillary with running buffer for 10 h before use.
Running buffer: 63 mM pH 12.7 LiOH containing 12 mM riboflavin
Injection: Hydrodynamic injection at 167 mbar for 10 s
Detector: UV 267
Migration time: 24
Voltage: +10 kV
Model: ABI Model 270A-HT
Limit of detection: 50 µM

OTHER SUBSTANCES
Simultaneous: fructose, maltose, sucrose

KEY WORDS
indirect UV detection

REFERENCE
Xu,X.; Kok,W.T.; Poppe,H. Sensitive determination of sugars by capillary zone electrophoresis with indirect UV detection under highly alkaline conditions, *J.Chromatogr.A*, **1995**, *716*, 231–240.

SAMPLE
Matrix: juice
Sample preparation: Dilute 300-fold with running buffer, inject an aliquot.

CAPILLARY ELECTROPHORESIS
Capillary: 80 cm × 25 μm fused-silica (Polymicro Technologies)
Running buffer: 50 mM NaOH
Injection: Electromigration at 15 kV for 5 s
Detector: E, Bioanalytical Systems LC-4B, 100 μm Cu-disk electrode (Anal. Chem. 1993, 65, 3525), +0.60 V, Ag/AgCl reference electrode, Pt wire counter electrode
Migration time: 20
Voltage: 15 kV
Model: Laboratory constructed
Limit of detection: 1 fmole

OTHER SUBSTANCES
Simultaneous: fructose, glucitol, sucrose

KEY WORDS
apple juice

REFERENCE
Ye,J.; Baldwin,R.P. Determination of carbohydrates, sugar acids and alditols by capillary electrophoresis and electrochemical detection at a copper electrode, *J.Chromatogr.A*, **1994**, *687*, 141–148.

SAMPLE
Matrix: plant hydrolysate
Sample preparation: Add reagent so that the concentration of saccharide is 2 mg/mL, heat at 50° for 2 h, cool to room temperature, dilute 10-100-fold with MeOH, filter (0.2 μm), inject an aliquot. (Prepare reagent by dissolving 10 mg sodium cyanoborohydride in 1 mL MeOH (?) containing 10% ethyl p-aminobenzoate and 10% acetic acid. The procedure in this paper makes no mention of the use of MeOH as solvent but the paper to which it refers (Chromatographia 1992, 34, 308) does state that MeOH is the solvent.)

CAPILLARY ELECTROPHORESIS
Capillary: 57.5 cm × 50 μm fused-silica (50 cm to detector) (Composite Metal Services, Worcester, UK)
Capillary temperature: 26
Running buffer: MeOH:500 mM pH 10.0 borate buffer 20:80 containing 0.001% hexadimethrine bromide
Injection: Hydrostatic injection at 10 cm for 10 s.
Detector: UV 280
Migration time: 7.3
Voltage: -18 kV
Current: 152 μA
Model: Waters Quanta 4000

OTHER SUBSTANCES
Simultaneous: N-acetyl-D-galactosamine, N-acetyl-D-glucosamine, N-acetylneuraminic acid, arabinose, cellobiose, 2-deoxy-D-ribose, fructose, fucose, galactose, galacturonic acid, gentobiose, glucuronic acid, lactose, lyxose, maltose, maltotetraose, maltotriose, mannose, mannuronic acid, melibiose, rhamnose, ribose, sorbose, xylose

KEY WORDS
derivatization

REFERENCE

Nguyen,D.T.; Lerch,H.; Zemann,A.; Bonn,G. Separation of derivatized carbohydrates by co-electroosmotic capillary electrophoresis, *Chromatographia*, **1997**, *46*, 113–121.

SAMPLE

Matrix: plants
Sample preparation: Stir 40 g dried plant material with 600 mL water at 500 rpm for 24 h, filter (grid mesh 0.3 mm), centrifuge the filtrate at 17700 g for 15 min. Remove an 80 mL aliquot of the supernatant to 400 mL EtOH, let stand at 4° for 48 h, centrifuge at 17700 g for 30 min, discard the supernatant, suspend the pellet in 100 mL ether, evaporate the ether to dryness at 40°. Add 2.5 mL 6 M trifluoroacetic acid to 20 mg of the precipitate, dissolve with gentle swirling, reflux at 110° for 4 h, lyophilize to dryness, add reagent so that the concentration of saccharide is 2 mg/mL, vortex gently, heat at 50° for 2 h, cool to room temperature, dilute 10-100-fold with MeOH, inject an aliquot. (Prepare reagent by dissolving 10 mg sodium cyanoborohydride in 1 mL MeOH containing 10% 2-aminopyridine and 10% acetic acid.)

CAPILLARY ELECTROPHORESIS

Capillary: 72 cm × 50 μm fused-silica (50 cm to detector)
Capillary preparation: Between runs wash capillary with 1 M NaOH for 3-4 min and with 1 mM NaOH for 2 min, equilibrate with running buffer for 3-6 min. Store capillaries in 1 mM NaOH overnight. Flush new capillaries with 1 M NaOH for 1 h and with 1 mM NaOH for 5 min.
Capillary temperature: 30
Running buffer: 150 mM pH 10.5 Borate buffer
Injection: Vacuum injection at 16.9 kPa for 1 s
Detector: UV 237
Migration time: 14.2
Voltage: 20 kV
Model: Applied Biosystems Model 270A

OTHER SUBSTANCES

Extracted: arabinose, 2-deoxy-D-ribose, fucose, galactose, galacturonic acid, glucuronic acid, lyxose, maltose, maltotriose, mannuronic acid, rhamnose, ribose, xylose

KEY WORDS

derivatization

REFERENCE

Oefner,P.J.; Vorndran,A.E.; Grill,E.; Huber,C.; Bonn,G.K. Capillary zone electrophoretic analysis of carbohydrates by direct and indirect UV detection, *Chromatographia*, **1992**, *34*, 308–316.

SAMPLE

Matrix: plants
Sample preparation: Stir 40 g dried plant material with 600 mL water at 500 rpm for 24 h, filter (grid mesh 0.3 mm), centrifuge the filtrate at 17700 g for 15 min. Remove an 80 mL aliquot of the supernatant to 400 mL EtOH, let stand at 4° for 48 h, centrifuge at 17700 g for 30 min, discard the supernatant, suspend the pellet in 100 mL ether, evaporate the ether to dryness at 40°. Add 2.5 mL 6 M trifluoroacetic acid to 20 mg of the precipitate, dissolve with gentle swirling, reflux at 110° for 4 h, lyophilize to dryness, add reagent so that the concentration of saccharide is 2 mg/mL, vortex gently, heat at 50° for 2 h, cool to room temperature, dilute 40-fold with MeOH, inject an aliquot. (Prepare reagent by dissolving 10 mg sodium cyanoborohydride in 1 mL MeOH containing 10% ethyl aminobenzoate and 10% acetic acid.)

CAPILLARY ELECTROPHORESIS

Capillary: 72 cm × 50 μm fused-silica (50 cm to detector)
Capillary preparation: Between runs wash capillary with 1 M NaOH for 4 min and with 1 mM NaOH for 2 min, equilibrate with running buffer for 4 min. Flush new capillaries with 1 M NaOH for 1 h and with 1 mM NaOH for 5 min.
Capillary temperature: 30
Running buffer: 175 mM pH 10.5 Borate buffer
Injection: Vacuum injection at 16.9 kPa for 1 s
Detector: UV 305

Migration time: 10
Voltage: 25 kV
Model: Applied Biosystems Model 270A
Limit of detection: 2 μM

OTHER SUBSTANCES
Extracted: arabinose, cellobiose, 2-deoxy-D-ribose, fucose, galactose, galacturonic acid, glucuronic acid, lactose, maltotriose, mannuronic acid, rhamnose, ribose, xylose

KEY WORDS
derivatization

REFERENCE
Vorndran,A.E.; Grill,E.; Huber,C.; Oefner,P.J.; Bonn,G.K. Capillary zone electrophoresis of aldoses, ketoses and uronic acids derivatized with ethyl p-aminobenzoate, *Chromatographia*, **1992**, *34*, 109–114.

SAMPLE
Matrix: solutions
Sample preparation: Add 50 μL 500 mM 3-methyl-1-phenyl-2-pyrazolin-5-one in MeOH and 50 μL 300 mM NaOH to a dried sample, heat at 70° for 30 min, cool to room temperature, add 50 μL 300 mM HCl, evaporate to dryness under reduced pressure, add 200 μL water, add 200 μL chloroform, shake vigorously. Remove the aqueous layer and evaporate it to dryness, reconstitute with a small volume of MeOH, inject an aliquot.

CAPILLARY ELECTROPHORESIS
Capillary: 78 cm × 50 μm fused-silica (63 cm to detector) (Scientific Glass Engineering)
Capillary preparation: Before each run rinse with 100 mM NaOH and with running buffer. After 10 runs rinse with MeOH.
Running buffer: 200 mM pH 9.5 Borate buffer
Injection: Siphon at 5 cm for 5 s
Detector: UV 245
Migration time: 21.3
Internal standard: amobarbital (17.5)
Voltage: 15 kV

OTHER SUBSTANCES
Simultaneous: allose, altrose, arabinose, galactose, gulose, idose, lyxose, oligoglucans, talose, xylose
Interfering: mannose

KEY WORDS
derivatization

REFERENCE
Honda,S.; Suzuki,S.; Nose,A.; Yamamoto,K.; Kakehi,K. Capillary zone electrophoresis of reducing mono- and oligo-saccharides as the borate complexes of their 3-methyl-1-phenyl-2-pyrazolin-5-one derivatives, *Carbohydrate Res.*, **1991**, *215*, 193–198.

SAMPLE
Matrix: solutions
Sample preparation: Dissolve saccharides in water, add an excess of 2 M ammonium sulfate or 4 M ammonium chloride, add excess 400 mM sodium cyanoborohydride, mix well, heat at 100° for 100-120 min, cool in an ice bath. Mix an aliquot with 10-20 μL 20 mM KCN in water, add 5-10 μL 10 mM reagent in MeOH, let stand at room temperature for 1 h, inject an aliquot. (Reagent was 3-(4-carboxybenzoyl)-2-quinolinecarboxaldehyde. Synthesis is as follows. Mix 9.69 g p-toluidine, 10.89 g o-nitrobenzaldehyde, and 25 mL EtOH and allow to react for 5 min, filter, wash the solid with EtOH to give 2-nitro-N-(p-tolyl)benzaldimine (Talanta 1989, 36, 321). (More 2-nitro-N-(p-tolyl)benzaldimine can be obtained by adding water to the filtrate.) Add, in portions in a thin stream, a hot solution of 46 g sodium sulfide in 23 mL water and 23 mL EtOH to a stirred refluxing solution of 100 mmole 2-nitro-N-(p-tolyl)benzaldimine in 50 mL EtOH. After a vigorous reaction 2-amino-N-(p-tolyl)benzaldimine crystallizes from the cooling solution (mp 102-103° after recrystallization from dilute MeOH) (Ber. 1943, 76, 1099). Wash

950 mg of a commercial 50% slurry of sodium hydride with pentane, add 7 mL dry THF (distilled from lithium aluminum hydride), add 1.51 g methyl 4-cyanobenzoate in 10 mL THF, add dropwise 1.38 mL acetone (distilled from calcium chloride), reflux for 1.5 h, cool, acidify with 3 M HCl. Remove the organic layer and wash it with brine and sodium bicarbonate solution, dry over anhydrous magnesium sulfate, evaporate to give (4-cyanobenzoyl)acetone. Reflux 433 mg (4-cyanobenzoyl)acetone, 486 mg 2-amino-N-(p-tolyl)benzaldimine, 69 mL piperidine, and EtOH:water 95:5 for 18 h, remove volatiles by steam distillation, add the residue to water and dichloromethane. Remove the organic layer and dry it, evaporate to give 3-(4-cyanobenzoyl)-2-methylquinoline. Suspend 547 mg 3-(4-cyanobenzoyl)-2-methylquinoline in 13 mL EtOH:water 95:5, add 500 mg KOH, reflux for 6 h, cool, concentrate, add the residue to ether and water. Remove the aqueous layer and adjust the pH to 5 with tartaric acid, let stand for 15 min, filter, wash the precipitate with water, dry under vacuum to give 3-(4-carboxybenzoyl)-2-methylquinoline. Dissolve 266 mg 3-(4-carboxybenzoyl)-2-methylquinoline in 6 mL acetic acid, add 112 mg selenium dioxide, stir at 80° for 2 h, filter through Celite, wash the precipitate with several volumes of hot MeOH, dilute the filtrate with water, allow to stand, filter, wash the solid with water, dry under vacuum to give 3-(4-carboxybenzoyl)-2-quinolinecarboxaldehyde (Anal. Chem. 1991, 63, 408).)

CAPILLARY ELECTROPHORESIS
Capillary: 88 cm × 50 μm (58 cm to detector) (Polymicro Technologies)
Running buffer: 10 mM pH 9.40 Na_2HPO_4 containing 10 mM sodium tetraborate
Injection: Hydrodynamic injection for 5 s
Detector: F ex 457 (argon laser) em 552
Migration time: 14
Voltage: 20 kV
Current: 12 μA
Model: laboratory constructed
Limit of detection: 1.3 amole

OTHER SUBSTANCES
Simultaneous: erythrose, galactosamine, galactose, galacturonic acid, glucosamine, glucosaminic acid, glucose 6-phosphate, glucuronic acid, mannose, oligosaccharides, ribose, talose

KEY WORDS
derivatization

REFERENCE
Liu,J.; Shirota,O.; Wiesler,D.; Novotny,M. Ultrasensitive fluorometric detection of carbohydrates as derivatives in mixtures separated by capillary electrophoresis, *Proc.Nat.Acad.Sci.USA*, **1991**, *88*, 2302–2306

SAMPLE
Matrix: solutions
Sample preparation: For each 1 mg saccharides add 500 μL reagent, vortex gently, heat at 50° for 2 h, cool to room temperature, dilute 10-100 fold with MeOH, inject an aliquot. (Prepare reagent by dissolving 10 mg sodium cyanoborohydride in MeOH containing 7% p-aminobenzoic acid and 10% acetic acid.)

CAPILLARY ELECTROPHORESIS
Capillary: 72 cm × 50 μm fused-silica (50 cm to detector)
Capillary preparation: Between runs wash capillary with 1 M NaOH for 3 min, wash with 1 mM NaOH for 2 min, equilibrate with running buffer for 3 min. Store capillary in 1 mM NaOH overnight. Flush a new capillary with 1 M NaOH for 1 h then with 1 mM NaOH for 5 min.
Capillary temperature: 30
Running buffer: 150 mM Boric acid adjusted to pH 10.0 with 2 M NaOH
Injection: Vacuum injection at 16.9 kPa for 1 s
Detector: UV 285
Migration time: 11.8
Voltage: 28 kV
Current: 79 μA
Model: Applied Biosystems Model 270A
Limit of detection: 4 μM

OTHER SUBSTANCES
Simultaneous: arabinose, cellobiose, 2-deoxy-D-ribose, fructose, fucose, galactose, galacturonic acid, glucuronic acid, lactose, melibiose, sorbose, xylose

KEY WORDS
derivatization

REFERENCE
Grill,E.; Huber,C.; Oefner,P.; Vorndran,A.; Bonn,G. Capillary zone electrophoresis of p-aminobenzoic acid derivatives of aldoses, ketoses and uronic acids, *Electrophoresis*, **1993**, *14*, 1004–1010.

SAMPLE
Matrix: solutions
Sample preparation: Add reagent to the carbohydrate solution to make a total volume of 2 mL, vortex gently, heat at 90° for 15 min, cool to room temperature, dilute 20-1000-fold with running buffer, inject an aliquot. (Just prior to use prepare reagent by dissolving 10 mg sodium cyanoborohydride in 1 mL 5% acetic acid containing 6% 4-aminobenzonitrile.)

CAPILLARY ELECTROPHORESIS
Capillary: 55 cm $\times$ 50 μm fused-silica (35 cm to detector)
Capillary temperature: 30
Running buffer: 25 mM Tris containing 100 mM sodium dodecyl sulfate, adjusted to pH 7.5 with phosphoric acid
Injection: Vacuum injection at 16.9 kPa for 1 s
Detector: UV 285
Migration time: 3.7
Voltage: 30 kV
Current: 55 μA
Model: Applied Biosystems Model 270A

OTHER SUBSTANCES
Simultaneous: arabinose, cellobiose, fructose, galactose, lactose, lyxose, maltose, maltotriose, mannose, melibiose, ribose, sorbose, xylose

KEY WORDS
derivatization

REFERENCE
Schwaiger,H.; Oefner,P.J.; Huber,C.; Grill,E.; Bonn,G.K. Capillary zone electrophoresis and micellar electrokinetic chromatography of 4-aminobenzonitrile carbohydrate derivatives, *Electrophoresis*, **1994**, *15*, 941–952.

SAMPLE
Matrix: solutions
Sample preparation: Mix 2 μL of a 5 mM solution with 2 μL 100 mM 9-aminopyrene-1,4,6-trisulfonate in 4.2 M acetic acid, add 4 μL 1 M sodium cyanoborohydride in THF, centrifuge, heat at 75° for 1 h, dilute 100-to 1000-fold, inject an aliquot. (9-Aminopyrene-1,4,6-trisulfonic acid is the same as 8-aminopyrene-1,3,6-trisulfonic acid. Synthesis is as follows. Rapidly add 300 g anhydrous sodium sulfate to 1300 g concentrated sulfuric acid, cool to 58°, add 202 g finely powdered pyrene over 5 min without cooling, stir for 15 min, cool to 50-55°, add 800 g 65% oleum (with cooling with 12° water) over 20 min, stir for 5 h without cooling, dilute with ice-water, neutralize with calcium carbonate, filter, reduce the filtrate to 10 L, exchange the cation with soda, salt out with 20% sodium chloride to obtain sodium pyrenetetrasulfonate. Heat 61 g sodium pyrenetetrasulfonate with 610 mL 22% aqueous ammonia in a rotating autoclave at 200-210° for 18 h (the pressure rises to 45 atmospheres), remove the excess ammonia by distillation under vacuum, precipitate oxypyrenetrisulfonic acid with a little NaCl, precipitate 9-aminopyrene-1,4,6-trisulfonic acid by saturating with NaCl, recrystallize from dilute NaCl (Liebig's Annalen der Chemie 1939, 540, 189).)

CAPILLARY ELECTROPHORESIS
Capillary: 27 cm $\times$ 20 μm fused-silica (20 cm to detector) (Polymicro Technologies)

Capillary preparation: Between runs wash capillary with 1 M NaOH at 15 psi for 12 s, wash with water at 15 psi for 12 s, condition with running buffer for 4 min.
Running buffer: 120 mM pH 7.0 MOPS
Detector: F ex 488 (2.5 mW argon ion laser) em 520 ± 9 (narrow-band filter; notch filter at 488 nm)
Migration time: 6.3
Voltage: 25 kV
Current: 19 μA
Model: Beckman P/ACE 2100

OTHER SUBSTANCES
Simultaneous: N-acetylgalactosamine, N-acetylglucosamine, arabinose, fucose, galactose, mannose, rhamnose, ribose, xylose

KEY WORDS
derivatization

REFERENCE
Chen,F.-T.A.; Evangelista,R.A. Analysis of mono- and oligosaccharide isomers derivatized with 9-aminopyrene-1,4,6-trisulfonate by capillary electrophoresis with laser-induced fluorescence, *Anal.Biochem.*, **1995**, *230*, 273–280.

SAMPLE
Matrix: solutions
Sample preparation: Inject an aliquot of an aqueous solution.

CAPILLARY ELECTROPHORESIS
Capillary: 100 cm × 75 μm fused-silica capillaries (Polymicro Technologies)
Capillary preparation: Etch new capillaries with 1 M NaOH for 1 h before use.
Capillary temperature: 28
Running buffer: 100 mM NaOH containing 1 mM NaCl
Injection: Inject at a pressure of 20 mbar for 12 s, ca. 15 nL
Detector: E, cuprous oxide modified carbon working electrode in a 0.5 mm i.d. PEEK tube +0.60 V, stainless steel auxiliary electrode, Ag/AgCl reference electrode. The effluent from the capillary passed through a grounded Pd cylinder and then through an 8 cm × 50 μm coupling capillary to the detector. (Prepare electrode as follows. Push copper wires in to a 4 cm length of 0.5 mm i.d. PEEK tubing so as to leave a 1 mm deep chamber at one end, seal other end with glue. Stir 300 mg conductive carbon cement (Gerhard Neubauer, Münster), 60 mg cuprous oxide (Fluka), and 300 μL acetone until a thick paste forms as the acetone evaporates. Pack paste into the chamber, allow to dry for one week. Use only electrodes with resistance less than 200 Ω. Polish with dry emery paper (grade 2/0, Oakey), 3 μm imperial micro finishing film sheet (3M), and 0.05 μm alumina particles on a Buehler pad, wash with water (Anal. Chim. Acta 1995, 300, 5).)
Migration time: 39
Voltage: 12 kV
Current: 124 μA
Model: Lauer Labs Prince programmable injector and power supply
Limit of detection: 1-2 μM

OTHER SUBSTANCES
Simultaneous: arabinose, fucose, galactose, mannose, raffinose, xylose

REFERENCE
Huang,X.; Kok,W.T. Determination of sugars by capillary electrophoresis with electrochemical detection using cuprous oxide modified electrodes, *J.Chromatogr.A*, **1995**, *707*, 335–342.

SAMPLE
Matrix: solutions
Sample preparation: Evaporate 5 μL of a 1 mM solution in water to dryness under reduced pressure, add 2 μL 100 mM 8-aminopyrene-1,3,6-trisulfonic acid (Lambda Fluoreszenztechnologie, Graz, Austria), add 2 μL 1.8 M citric acid in water, add 2 μL 1 M sodium cyanoborohydride in THF, heat at 75° for 1 h, dilute to 200 μL with water, dilute 25-fold, inject an aliquot.

(9-Aminopyrene-1,4,6-trisulfonic acid is the same as 8-aminopyrene-1,3,6-trisulfonic acid. Synthesis is as follows. Rapidly add 300 g anhydrous sodium sulfate to 1300 g concentrated sulfuric acid, cool to 58°, add 202 g finely powdered pyrene over 5 min without cooling, stir for 15 min, cool to 50-55°, add 800 g 65% oleum (with cooling with 12° water) over 20 min, stir for 5 h without cooling, dilute with ice-water, neutralize with calcium carbonate, filter, reduce the filtrate to 10 L, exchange the cation with soda, salt out with 20% sodium chloride to obtain sodium pyrenetetrasulfonate. Heat 61 g sodium pyrenetetrasulfonate with 610 mL 22% aqueous ammonia in a rotating autoclave at 200-210° for 18 h (the pressure rises to 45 atmospheres), remove the excess ammonia by distillation under vacuum, precipitate oxypyrenetrisulfonic acid with a little NaCl, precipitate 9-aminopyrene-1,4,6-trisulfonic acid by saturating with NaCl, recrystallize from dilute NaCl (Liebig's Annalen der Chemie 1939, 540, 189).)

CAPILLARY ELECTROPHORESIS
Capillary: 25 cm × 19 μm fused-silica
Capillary preparation: Between runs rinse capillary with 1 M NaOH at 15 psi for 12 s and with running buffer at 15 psi for 1.2 min.
Running buffer: 120 mM pH 10.2 Borate buffer
Injection: Pressure injection at 0.5 psi for 20 s
Detector: F ex 488 (laser) em 520
Migration time: 4.4
Voltage: 30 kV
Current: 26 μA
Model: Beckman P/ACE 2100

OTHER SUBSTANCES
Simultaneous: N-acetylgalactosamine, N-acetylglucosamine, fucose, galactose, mannose, xylose

KEY WORDS
derivatization

REFERENCE
Evangelista,R.A.; Guttman,A.; Chen,F.-T.A. Acid-catalyzed reductive amination of aldoses with 8-aminopyrene-1,3,6-trisulfonate, *Electrophoresis*, **1996**, *17*, 347–351.

SAMPLE
Matrix: solutions
Sample preparation: 56 μmoles Dextrose + 100 μL 150 mM 8-aminonaphthalene-1,3,6-trisulfonic acid (Molecular Probes, Eugene OR) in acetic acid:water 15:85 + 100 μL 1 M sodium cyanoborohydride in DMSO, vortex, heat at 40° for 15 h, dilute 1000-fold with water, inject an aliquot.

CAPILLARY ELECTROPHORESIS
Capillary: 27 cm × 50 μm (20.5 cm to detector) (Polymicro Technologies)
Capillary preparation: Between runs flush with 100 mM NaOH for 2 min and with running buffer for 2 min.
Capillary temperature: 25
Running buffer: 150 mM Boric acid adjusted to pH 9.5 with 2 M NaOH
Injection: Hydrodynamic injection
Detector: F ex 325 (2 mW He-Cd laser) em 520 (bandpass filter)
Migration time: 10.4
Voltage: 10 kV
Model: Beckman P/ACE 2100

OTHER SUBSTANCES
Simultaneous: N-acetylgalactosamine, N-acetylglucosamine, fucose, galactose, mannose, xylose

KEY WORDS
derivatization

REFERENCE
Klockow,A.; Amadò,R.; Widmer,H.M.; Paulus,A. The influence of buffer composition on separation efficiency and resolution on capillary electrophoresis of 8-aminonaphthalene-1,3, 6-trisulfonic acid labeled monosaccharides and complex carbohydrates, *Electrophoresis*, **1996**, *17*, 110–119.

SAMPLE
Matrix: solutions
Sample preparation: Inject an aliquot of an aqueous solution.

CAPILLARY ELECTROPHORESIS
Capillary: 57 cm × 25 μm fused-silica (34.5 cm to detector) (Polymicro Technologies)
Capillary preparation: Between injections flush capillary with 100 mM NaOH, with water, and with running buffer. Flush new capillaries with 1 M NaOH.
Running buffer: 50 mM NaOH containing 5 mM tryptophan
Injection: Vacuum injection at 0.5 psi for 1 s
Detector: UV 280
Migration time: 17
Voltage: 7 kV
Model: Isco 3850
Limit of detection: 30 fmole

OTHER SUBSTANCES
Simultaneous: cellobiose, galactose, mannose, melibiose, raffinose, ribose, stachyose, xylose

KEY WORDS
indirect UV detection

REFERENCE
Lu,B.; Westerlund,D. Indirect UV detection of carbohydrates in capillary zone electrophoresis by using tryptophan as a marker, *Electrophoresis*, **1996**, *17*, 325–332.

SAMPLE
Matrix: solutions
Sample preparation: Prepare a solution containing 100 μM monosaccharide, 30 mM 6-aminoquinoline, 5 mM sodium cyanoborohydride, and 150 mM acetic acid, heat at 80° for 45 min (40° for 2 h for oligosaccharides), cool to room temperature, filter, inject an aliquot.

CAPILLARY ELECTROPHORESIS
Capillary: 43 cm × 30 μm fused silica (38 cm to detector) (Skandinaviska GeneTech, Kungsbacka, Sweden)
Running buffer: 420 mM Boric acid adjusted to pH 9 with 1 M NaOH
Injection: Hydrodynamic injection at 75 m for 10 s
Detector: UV 245
Migration time: 8
Current: 55 μA (constant power 1200 mW)
Model: Dionex
Limit of detection: 1 μM

OTHER SUBSTANCES
Simultaneous: arabinose, galactose, galacturonic acid, glucuronic acid, mannose, 4-O-methylglucuronic acid, rhamnose, xylose

KEY WORDS
derivatization

REFERENCE
Rydlund,A.; Dahlman,O. Efficient capillary zone electrophoretic separation of wood-derived neutral and acidic mono- and oligosaccharides, *J.Chromatogr.A*, **1996**, *738*, 129–140.

SAMPLE
Matrix: solutions
Sample preparation: Mix a 20 μL aliquot of a 5 mM solution in EtOH with 90 μL 1% sodium cyanoborohydride in EtOH containing 1-2% acetic acid, add 10 μL benzoic hydrazide in EtOH:water 75:25, heat at 60° for 5 h, cool to room temperature, inject an aliquot. (Adjust the concentration of benzoic hydrazide so that it is present in 20-fold excess for standards and 200-fold excess for hydrolyzed glycoproteins.)

CAPILLARY ELECTROPHORESIS
Capillary: 70 cm × 50 μm fused-silica (52 cm to detector) (Yongnian, Hebei, China)
Capillary preparation: Flush with running buffer before each injection. Flush a new capillary with 1 M NaOH, water, and running buffer.
Capillary temperature: 30
Running buffer: 200 mM pH 10.8 Borate buffer
Injection: Vacuum injection for 8 s.
Detector: UV 220
Migration time: 23
Voltage: 20 kV
Model: Spectra PHORESIS 1000 (Thermo Separation Products)
Limit of detection: 30-40 fmole

OTHER SUBSTANCES
Simultaneous: N-acetylgalactosamine, arabose, fucose, galactose, lyxose, mannose, rhamnose, xylose

KEY WORDS
derivatization

REFERENCE
Lin,Q.; Zhang,R.; Liu,G. Use of benzoyl hydrazine reagent for monosaccharide determination by high performance capillary electrophoresis, *J.Liq.Chromatogr.Rel.Technol.*, **1997**, *20*, 1123–1137.

SAMPLE
Matrix: solutions
Sample preparation: Add 50 μL 500 mM 3-methyl-1-phenyl-2-pyrazolin-5-one in MeOH and 50 μL 300 mM NaOH to 100 nmole saccharides, mix well, heat at 70° for 30 min, cool to room temperature, neutralize with 50 μL 300 mM HCl, evaporate to dryness under reduced pressure, add 200 μL water, add 200 μL chloroform, shake vigorously. Remove the aqueous layer and wash it twice more with chloroform. Inject an aliquot of the aqueous layer.

CAPILLARY ELECTROPHORESIS
Capillary: 58 cm × 50 μm fused-silica (53 cm to detector) (Polymicro Technologies)
Capillary preparation: Before each run flush with 100 mM NaOH for a sufficient time and with running buffer for at least 10 min.
Running buffer: 200 mM pH 9.5 Borate buffer
Injection: Siphon at 10 cm for 10 s.
Detector: UV 254
Migration time: 34
Voltage: 15 kV
Model: Waters Quanta 4000

OTHER SUBSTANCES
Simultaneous: allose, altrose, galactose, gulose, idose, mannose, talose

KEY WORDS
derivatization

REFERENCE
Honda,S.; Togashi,K.; Uegaki,K.; Taga,A. Enhancement of the zone electrophoretic separation of 1-phenyl-5-pyrazolone derivatives of aldoses as borate complexes by concerted ion-interaction electrokinetic chromatography with Polybrene, *J.Chromatogr.A*, **1998**, *805*, 277–284.

SAMPLE
Matrix: tears
Sample preparation: Evaporate 3-5 μL tear fluid to dryness under a stream of nitrogen, reconstitute with 18 μL 25 mM sodium tetraborate solution, add 2 μL 13.4 mM trifluoroacetic acid in water, add 50 μL 2-10 mM dansylhydrazine in EtOH, add 30 μL EtOH, heat at 68° for 15-18 min, store at -20°

CAPILLARY ELECTROPHORESIS
Capillary: 90 cm × 50 μm fused-silica (60 cm to detector) (Polymicro Technologies)
Capillary preparation: Before each injection flush capillary with MeOH for 30 s, with 100 mM NaOH for 30 s, and with running buffer for 30 s. Store overnight in water or running buffer. Flush new capillaries with 100 mM NaOH, water, and running buffer.
Running buffer: 25 mM pH 9.2 sodium tetraborate
Injection: Hydrodynamic injection at 19 cm for 10 s (2 nL)
Detector: F ex 325 (He-Cd laser) em 500 (long-pass filter)
Migration time: 13.2
Voltage: 22 kV
Current: 23 μA
Limit of detection: 50 nM (S/N 3)

OTHER SUBSTANCES
Simultaneous: fructose, fucose, galactose, lactose, maltose, mannose

KEY WORDS
derivatization

REFERENCE
Perez,S.A.; Colón,L.A. Determination of carbohydrates as their dansylhydrazine derivatives by capillary electrophoresis with laser-induced fluorescence detection, *Electrophoresis*, **1996**, *17*, 352–358.

Diamorphine

Molecular formula: $C_{21}H_{23}NO_5$
Molecular weight: 369.42
CAS Registry No.: 561-27-3, 1502-95-0 (hydrochloride)
Merck Index (12th ed.): 3012
Lednicer: 1 288

SAMPLE
Matrix: bulk
Sample preparation: Dissolve in running buffer to a concentration of 1 mg/mL, vortex for 2 min, add an equal volume of 100 μg/mL diphenhydramine in running buffer, mix, inject an aliquot.

CAPILLARY ELECTROPHORESIS
Capillary: 65 cm × 50 μm fused-silica (60 cm to detector)
Capillary preparation: Fill capillary with fresh running buffer before each run. Before use fill with 100 mM NaOH for 20 min, rinse with water, flush with running buffer
Running buffer: MeCN:50 mM 6-aminocaproic acid containing 50 mM 3-N,N-dimethylmyristylammoniopropanesulfonate (MAPS; Fluka) and 5 mM 1-heptanesulfonic acid 10:90, pH adjusted to 4.0 with 1 M phosphoric acid
Injection: Hydrodynamic injection by gravity or pressure.
Detector: UV 214
Migration time: 6.3
Internal standard: diphenhydramine (12)
Voltage: 27 kV
Current: ≤25 μA
Model: Dionex system I

OTHER SUBSTANCES
Simultaneous: acetaminophen, allobarbital, barbital, caffeine, codeine, morphine, niacinamide, noscapine, papaverine, phenobarbital, procaine

REFERENCE

Naess,O.; Rasmussen,K.E. Micellar electrokinetic chromatography of charged and neutral drugs in acidic running buffers containing a zwitterionic surfactant, sulfonic acids or sodium dodecyl sulphate. Separation of heroin, basic by-products and adulterants, *J.Chromatogr.A*, **1997**, *760*, 245–251.

SAMPLE

Matrix: solutions
Sample preparation: Inject an aliquot of a 1 mg/mL solution in MeOH.

CAPILLARY ELECTROPHORESIS

Capillary: 67 cm × 50 μm (60 cm to detector) (Composite Metal Services, Worcs., UK)
Capillary preparation: Before each run rinse with running buffer for 3 min.
Capillary temperature: 20
Running buffer: 50 mM pH 10.5 Glycine buffer containing 50 mM sodium dodecyl sulfate
Injection: Pressure injection at 30 mbar for 5 s.
Detector: UV 220
Migration time: 11
Voltage: 25 kV
Model: Beckman P/ACE 2050

OTHER SUBSTANCES

Simultaneous: amphetamine, caffeine, codeine, morphine

REFERENCE

Hyötyläinen,T.; Sirén,H.; Riekkola,M.-L. Determination of morphine analogues, caffeine and amphetamine in biological fluids by capillary electrophoresis with the marker technique, *J.Chromatogr.A*, **1996**, *735*, 439–447.

SAMPLE

Matrix: solutions
Sample preparation: Inject an aliquot of a 1-50 μg/mL solution in water.

CAPILLARY ELECTROPHORESIS

Capillary: 55 cm × 50 μm fused-silica (35 cm to detector) (J&W)
Capillary preparation: Flush with running buffer for 5 min before each run. Periodically wash with 100 mM NaOH and flush extensively with running buffer.
Running buffer: MeOH:25 mM pH 9.24 borate buffer 20:80 containing 100 mM sodium dodecyl sulfate (A) or 50 mM pH 2.35 phosphate buffer (B) or 50 mM pH 9.24 borate buffer (C)
Injection: Injection of 5 μL using a split-flow injector ratio of 1:800.
Detector: UV 200
Migration time: 28.57 (A), 11.50 (B), 7.07 (C)
Voltage: 20 kV (A, B) or 12 kV (C)
Current: <60-80 μA
Model: ISCO Model 3850

OTHER SUBSTANCES

Also analyzed: acetylcodeine, amphetamine, barbital, caffeine, codeine, cocaine, diazepam, flunitrazepam, lidocaine, monoacetylmorphine, morphine, nalorphine, narceine, noscapine, papaverine, pentobarbital, procaine, tetracaine, thebaine

KEY WORDS

all compounds were separated with running buffer A; some peaks overlapped with running buffers B and C.

REFERENCE

Tagliaro,F.; Smith,F.P.; Turrina,S.; Equisetto,V.; Marigo,M. Complementary use of capillary zone electrophoresis and micellar electrokinetic capillary chromatography for mutual confirmation of results in forensic drug analysis, *J.Chromatogr.A*, **1996**, *735*, 227–235.

SAMPLE

Matrix: solutions

Sample preparation: Inject an aliquot of a solution in 100 mM sodium phosphate containing 100-300 µg/mL naphazoline.

CAPILLARY ELECTROPHORESIS
Capillary: 67 cm × 50 µm fused-silica (60 cm to detector)
Capillary preparation: Rinse with running buffer for 2 min between sets of analyses.
Capillary temperature: 30
Running buffer: 200 mM pH 4.5 Phosphate buffer
Injection: Injection at high pressure for 2 s.
Detector: UV 210, UV 230
Migration time: 19.8
Internal standard: naphazoline (14)
Voltage: 20 kV
Model: Beckman P/ACE 5500

OTHER SUBSTANCES
Simultaneous: acetylcodeine, amphetamine, cocaine, codeine, methadone, methamphetamine, methylenedioxyamphetamine, methylenedioxymethamphetamine, morphine, PCP, psilocyn
Interfering: LSD

REFERENCE
Walker,J.A.; Marché,H.L.; Newby,N.; Bechtold,E.J. A free zone capillary electrophoresis method for the quantitation of common illicit drug samples, *J.Forensic Sci.*, **1996**, *41*, 824–829.

SAMPLE
Matrix: solutions

CAPILLARY ELECTROPHORESIS
Capillary: 85 cm × 50 µm fused-silica (Polymicro Technologies)
Capillary preparation: Between runs rinse capillary with running buffer for 5 min. Before use rinse capillary with 1 M NaOH for 10-15 min and with water for 10-15 min.
Running buffer: 25 mM pH 3 Citrate buffer
Injection: Pressure injection at 30 mbar for 12 s.
Detector: MS, laboratory-constructed, electrospray 90°, sheath liquid MeOH:water:acetic acid 80:20:0.1 at 1 µL/min
Migration time: 11.3
Voltage: 30 kV
Current: 7.8 µA
Model: Crystal CE 300

OTHER SUBSTANCES
Simultaneous: amphetamine, cocaine, methamphetamine, procaine, tetracaine

REFERENCE
Lazar,I.M.; Naisbitt,G.; Lee,M.L. Capillary electrophoresis-time-of-flight mass spectrometry of drugs of abuse, *Analyst*, **1998**, *123*, 1449–1454.

Diazepam

Molecular formula: $C_{16}H_{13}ClN_2O$
Molecular weight: 284.74
CAS Registry No.: 439-14-5
Merck Index (12th ed.): 3042
Lednicer: 1 365; 2 395

SAMPLE
Matrix: blood

Sample preparation: Condition a Sep-Pak C18 SPE cartridge with MeOH, water, and 100 mM pH 9.0 borate buffer. 1 mL Serum + 3 mL 100 mM pH 9.0 borate buffer, mix, add to the SPE cartridge, wash with 3 mL 100 mM pH 9.0 borate buffer, wash with three 2 mL portions of water, wash with 1 mL MeCN:water 5:95, wash with 1 mL hexane, dry under full vacuum. Elute with 4 mL dichloromethane, evaporate the eluate to dryness under a stream of nitrogen, reconstitute with MeOH:water 15:85, filter (0.45 μm), inject an aliquot.

CAPILLARY ELECTROPHORESIS
Capillary: 72 cm × 50 μm fused-silica (50 cm to detector)
Capillary preparation: Before each run wash capillary with 100 mM NaOH for 4 min and with running buffer for 6 min.
Capillary temperature: 25
Running buffer: MeOH:5 mM pH 8.5 phosphate/borate buffer 15:85 containing 50 mM sodium dodecyl sulfate
Injection: Vacuum injection for 4 s.
Detector: UV 200
Migration time: 21.5
Internal standard: secobarbital (9)
Voltage: 25 kV
Model: Applied Biosystems Model 270A or Beckman P/ACE 5000
Limit of detection: 25 ng/mL

OTHER SUBSTANCES
Extracted: bromazepam, estazolam, flurazepam, nitrazepam, triazolam

KEY WORDS
SPE; serum

REFERENCE
Tomita,M.; Okuyama,T. Application of capillary electrophoresis to the simultaneous screening and quantitation of benzodiazepines, *J.Chromatogr.B*, **1996**, *678*, 331–337.

SAMPLE
Matrix: blood
Sample preparation: 500 μL Serum + 150 ng flunitrazepam + 250 μL 800 mM pH 10.0 N-cyclohexyl-2-hydroxy-3-aminopropanesulfonic acid (CAPSO) buffer + 4 mL n-pentane:ethyl acetate 75:25, vortex for 1 min, centrifuge at 25° at 1500 g for 5 min. Remove the organic layer and evaporate it to dryness at 28°, reconstitute the residue in 40 μL 50 mM pH 9.5 borate buffer containing 4.5 mM sodium dodecyl sulfate, inject an aliquot.

CAPILLARY ELECTROPHORESIS
Capillary: 47 cm × 50 μm fused-silica (40 cm to detector) (Supelco)
Capillary preparation: Before each run wash with 100 mM NaOH, water, and running buffer.
Capillary temperature: 25
Running buffer: MeCN:50 mM pH 9.5 sodium borate buffer containing 18 mM sodium dodecyl sulfate 14:86
Injection: Pressure injection at 0.5 psi for 10 s.
Detector: UV 214
Migration time: 9.5
Internal standard: flunitrazepam (5.7)
Voltage: 20 kV
Model: Beckman P/ACE 5010
Limit of quantitation: 20 ng/mL

OTHER SUBSTANCES
Extracted: clobazam, clonazepam, desmethylclobazam, desmethyldiazepam, nitrazepam
Noninterfering: carbamazepine, ethosuximide, phenobarbital, phenytoin, primidone, valproic acid, zonisamide

KEY WORDS
serum

REFERENCE
Imazawa,M.; Hatanaka,Y. Micellar electrokinetic capillary chromatography of benzodiazepine antiepileptics and their desmethyl metabolites in blood, *J.Pharm.Biomed.Anal.*, **1997**, *15*, 1503–1508.

SAMPLE
Matrix: blood, urine
Sample preparation: Adjust pH of 2 mL urine or plasma to 10.5 with aqueous NaOH, extract gently with chloroform:isopropanol 90:10. Remove the organic layer and evaporate it to dryness under a stream of nitrogen, reconstitute the residue in 50 (urine) or 75 (plasma) μL running buffer, filter, inject an aliquot.

CAPILLARY ELECTROPHORESIS
Capillary: 60 cm × 75 μm AccuSep uncoated silica (52.5 cm to detector) (Waters)
Capillary preparation: At the beginning of each day purge with 500 mM KOH for 5 min, with water for 5 min, and with running buffer for 10 min.
Running buffer: 50 mM NaH_2PO_4 adjusted to pH 2.35 with phosphoric acid
Injection: Hydrostatic injection at 15 cm
Detector: UV 214
Migration time: 7.71
Voltage: 22 kV
Current: 135-145 μA
Model: Waters Quanta 4000

OTHER SUBSTANCES
Extracted: acepromazine, amphetamine, benzocaine, brompheniramine, butacaine, codeine, doxapram, lidocaine, medazepam, methamphetamine, methapyrilene, methaqualone, phenmetrazine, procaine, tetrahydrozoline
Simultaneous: meclizine

KEY WORDS
plasma

REFERENCE
Chee,G.L.; Wan,T.S.M. Reproducible and high-speed separation of basic drugs by capillary zone electrophoresis, *J.Chromatogr.*, **1993**, *612*, 172–177.

SAMPLE
Matrix: solutions
Sample preparation: Prepare a solution in running buffer, inject an aliquot.

CAPILLARY ELECTROPHORESIS
Capillary: 100 cm × 75 μm fused-silica (20 cm to UV detector) (Polymicro Technologies)
Capillary preparation: Between runs flush capillaries with 1.5 column volumes of 100 mM NaOH and 9 column volumes of running buffer. Flush new capillaries with 9 column volumes of 1 M NaOH, 3 column volumes of water, 3 column volumes of 100 mM HCl, 3 column volumes of water, and 9 column volumes of running buffer.
Running buffer: MeOH:water 15:85 containing 15 mM ammonium acetate, adjusted to pH 2.5 on trifluoroacetic acid
Injection: Pressure injection at 3.45 kPa (28 nL)
Detector: UV 254 or MS, Sciex TAGA 6000E triple quadrupole, API source, electrospray (ion spray) interface at 4 kV (3 kV to ion-sampling orifice), positive-ion mode, 100 μm dia sampling orifice, nitrogen gas curtain, argon collision gas, m/z 286 [M+H]+
Migration time: 5.8 (UV), 26.5 (MS)
Voltage: 26 kV
Model: Beckman P/ACE System 2000

OTHER SUBSTANCES
Simultaneous: chlordiazepoxide, flurazepam, prazepam

REFERENCE
Johansson,I.M.; Pavelka,R.; Henion,J.D. Determination of small drug molecules by capillary electrophoresis-atmospheric pressure ionization mass spectrometry, *J.Chromatogr.*, **1991**, *559*, 515–528.

SAMPLE
Matrix: solutions

CAPILLARY ELECTROPHORESIS
Capillary: 72 cm × 50 μm silica (50 cm to detector) (Applied Biosystems)
Capillary preparation: Before each run wash capillary with 100 mM NaOH for 3 min, wash with running buffer for 3 min, aspirate neutral marker solution (2 drops DMSO in 10 mL water) for 1 s, place capillary end in buffer vial for 5 s, inject sample. Wash capillary at the beginning of each day by passing 1 M NaOH through for 20 min.
Capillary temperature: 30
Running buffer: MeCN:buffer 10:90 (Buffer was 30 mM pH 9.3 borate buffer containing 30 mM sodium dodecyl sulfate.)
Injection: Inject by applying vacuum for 1 s (2-3 nL)
Detector: UV 200
Migration time: 15
Voltage: +30 kV
Current: 55 μA
Model: Applied Biosystems Model 270A

OTHER SUBSTANCES
Simultaneous: chlordiazepam, desmethylchlordiazepam, desmethyldiazepam, lorazepam, oxazepam, nitrazepam

REFERENCE
Evenson,M.A.; Wiktorowicz,J.E. Automated capillary electrophoresis applied to therapeutic drug monitoring, *Clin.Chem.*, **1992**, *38*, 1847–1852.

SAMPLE
Matrix: solutions
Sample preparation: Inject an aliquot of a 5 μg/mL solution in running buffer.

CAPILLARY ELECTROPHORESIS
Capillary: 87 cm × 75 μm fused-silica (80 cm to detector) (Beckman)
Capillary preparation: Between runs rinse capillary with running buffer for 2 min then equilibrate for 5 min.
Capillary temperature: 35
Running buffer: 20 mM pH 7 Buffer containing 15 mM sodium cholate and 35 mM sodium deoxycholate (Buffer was 20 mM sodium borate adjusted to pH 7.0 with 20 mM NaH_2PO_4.)
Injection: Pressure injection for 2 s.
Detector: UV 214
Migration time: 22.5
Voltage: 20 kV
Model: Beckman P/ACE 2100

OTHER SUBSTANCES
Simultaneous: bromazepam, chlordiazepoxide, clobazam, clonazepam, flunitrazepam, flurazepam, halazepam, lorazepam, lormetazepam, nitrazepam, nordazepam, temazepam

REFERENCE
Boonkerd,S.; Detaevernier,M.R.; Vindevogel,J.; Michotte,Y. Migration behaviour of benzodiazepines in micellar electrokinetic chromatography, *J.Chromatogr.A*, **1996**, *756*, 279–286.

SAMPLE
Matrix: solutions

CAPILLARY ELECTROPHORESIS
Capillary: 60 cm × 50 μm fused-silica (40 cm to detector), C18 coated (Supelco)
Capillary preparation: Rinse new capillaries with 250 μL 1 M NaOH, 250 μL 100 mM NaOH, water, MeOH, water, and running buffer then condition with running buffer for at least 15 min. Carry out a similar procedure between runs.

Running buffer: 100 mM pH 8.5 Borate buffer containing 25 mM sodium dodecyl sulfate and 5 M urea
Injection: Electrokinetic injection at 5 kV for 5 s.
Detector: UV 254
Migration time: 39
Voltage: 18 kV
Model: Jasco

OTHER SUBSTANCES
Simultaneous: alprazolam, amobarbital, barbital, bromazepam, clonazepam, clotiazepam, cloxazolam, estazolam, etizolam, fludiazepam, flunitrazepam, flurazepam, haloxazolam, medazepam, mephobarbital, metharbital, nimetazepam, nitrazepam, oxazepam, pentobarbital, phenobarbital, secobarbital, triazolam

KEY WORDS
coated capillary

REFERENCE
Jinno,K.; Han,Y.; Nakamura,M. Analysis of anxiolytic drugs by capillary electrophoresis with bare and coated capillaries, *J.Capillary Electrophor.*, **1996**, *3*, 139–145.

SAMPLE
Matrix: solutions
Sample preparation: Inject an aliquot of a 1-50 μg/mL solution in water.

CAPILLARY ELECTROPHORESIS
Capillary: 55 cm × 50 μm fused-silica (35 cm to detector) (J&W)
Capillary preparation: Flush with running buffer for 5 min before each run. Periodically wash with 100 mM NaOH and flush extensively with running buffer.
Running buffer: MeOH:25 mM pH 9.24 borate buffer 20:80 containing 100 mM sodium dodecyl sulfate (A) or 50 mM pH 2.35 phosphate buffer (B)
Injection: Injection of 5 μL using a split-flow injector ratio of 1:800.
Detector: UV 200
Migration time: 42.71 (A), 12.11 (B)
Voltage: 20 kV
Current: <60-80 μA
Model: ISCO Model 3850

OTHER SUBSTANCES
Also analyzed: acetylcodeine, amphetamine, barbital, caffeine, codeine, cocaine, diamorphine, flunitrazepam, lidocaine, monoacetylmorphine, morphine, nalorphine, narceine, noscapine, papaverine, pentobarbital, procaine, tetracaine, thebaine

KEY WORDS
all compounds were separated with running buffer A; some peaks overlapped with running buffer B.

REFERENCE
Tagliaro,F.; Smith,F.P.; Turrina,S.; Equisetto,V.; Marigo,M. Complementary use of capillary zone electrophoresis and micellar electrokinetic capillary chromatography for mutual confirmation of results in forensic drug analysis, *J.Chromatogr.A*, **1996**, *735*, 227–235.

SAMPLE
Matrix: solutions

CAPILLARY ELECTROPHORESIS
Capillary: 60 cm × 75 μm coated fused-silica (40 cm to detector) (Supelco)
Capillary preparation: Coat column as follows. Adjust the pH of 20 mL water to 3.5 with acetic acid, add 80 μL 3-(trimethoxysilyl)propyl methacrylate (3-methacryloxypropyltrimethoxysilane), mix, suck into capillary, let stand at room temperature for 1 h, remove the solution, wash with water. Fill the capillary with a deaerated 4% acrylamide solution containing 1 mg/

mL N,N,N',N'-tetramethylethylenediamine and 1 mg/mL ammonium persulfate, let stand for
3 h, remove excess solution by aspiration, rinse with water, remove water by aspiration, dry
at 35° (cf. J. Chromatogr. 1985, 347, 191).
Running buffer: 100 mM pH 8.5 borate buffer containing 10 mM sodium dodecyl sulfate and 5
M urea
Injection: Electromigration at 5 kV for 5 s.
Detector: UV 240, UV 254
Migration time: 7.7
Voltage: 18 kV
Model: Jasco 890-CE

OTHER SUBSTANCES
Simultaneous: alprazolam, clotiazepam, cloxazolam, estazolam, etizolam, fludiazepam, flurazepam, haloxazolam, oxazepam

KEY WORDS
injection at cathode; coated capillary

REFERENCE
Jinno,K.; Han,Y.; Sawada,H. Analysis of toxic drugs by capillary electrophoresis using polyacrylamide-coated
columns, *Electrophoresis*, **1997**, *18*, 284–286.

SAMPLE
Matrix: solutions
Sample preparation: Inject an aliquot of a 100 μg/mL solution in running buffer.

CAPILLARY ELECTROPHORESIS
Capillary: 60 cm × 50 μm acrylamide-coated fused-silica (40 cm to detector) (Supelco)
Capillary preparation: Adjust the pH of 20 mL water to 3.5 with acetic acid, add 80 μL 3-
(trimethoxysilyl)propyl methacrylate (3-methacryloxypropyltrimethoxysilane), mix, suck into
capillary, let stand at room temperature for 1 h, remove the solution, wash with water. Fill the
capillary with a deaerated 3-4% acrylamide solution containing 1 μL/mL N,N,N',N'-tetra-
methylethylenediamine and 1 mg/mL ammonium persulfate, let stand for 3 h, remove excess
solution by aspiration, rinse with water, remove water by aspiration, dry at 35° (cf. J. Chromatogr. 1985, 347, 191).
Running buffer: MeCN:buffer 5:95 (Buffer was 100 mM borate containing 5 M urea and 10 mM
sodium dodecyl sulfate, adjusted to pH 8.5 with phosphate.)
Injection: Electrokinetic injection at 5 kV for 5 s.
Detector: UV 254
Migration time: 9.6
Voltage: 18 kV
Model: Jasco Model 870-CE

OTHER SUBSTANCES
Simultaneous: alprazolam, amobarbital, barbital, bromazepam, clonazepam, clotiazepam, cloxazolam, estazolam, etizolam, fludiazepam, flunitrazepam, flurazepam, haloxazolam, medazepam, mephobarbital, metharbital, nimetazepam, nitrazepam, oxazepam, pentobarbital, phenobarbital, secobarbital, triazolam

KEY WORDS
coated capillary; detector at anode

REFERENCE
Jinno,K.; Han,Y.; Sawada,H.; Taniguchi,M. Capillary electrophoretic separation of toxic drugs using a polyacrylamide-coated capillary, *Chromatographia*, **1997**, *46*, 309–314.

SAMPLE
Matrix: solutions

CAPILLARY ELECTROPHORESIS
Capillary: 53 cm × 50 μm fused-silica packed with 3 μm CEC Hypersil ODS (46 cm packed with
ODS material) (Hypersil)

Running buffer: Gradient. A was 5 mM ammonium acetate in MeCN:water 50:50. B was 5 mM ammonium acetate in MeCN:water 80:20. A:B 100:0 for 3 min, to 0:100 over 0.1 min, maintain at 0:100 (pumped with an HPLC pump at 0.01 mL/min for 3 min then at 0.1 mL/min).
Injection: Inject 10 μL using an HPLC injector.
Detector: MS, VG Biotech Platform, electrospray (details in paper)
Migration time: 25
Voltage: 30 kV

OTHER SUBSTANCES
Simultaneous: nitrazepam

KEY WORDS
electrochromatography

REFERENCE
Taylor,M.R.; Teale,P. Gradient capillary electrochromatography of drug mixtures with UV and electrospray ionisation mass spectrometric detection, *J.Chromatogr.A*, **1997**, *768*, 89–95.

SAMPLE
Matrix: urine
Sample preparation: Condition a 130 mg Bond Elut Certify SPE cartridge with 2 mL MeOH and 2 mL 100 mM pH 6 phosphate buffer, do not allow to go dry. Mix urine with an equal amount of 200 mM pH 5.4 buffer, add 20 μL β-glucuronidase/arylsulfatase (Helix pomatia, Boehringer Mannheim) for each 1 mL of urine, heat at 37° for 4 h. 5 (or 10) mL Urine + 2 (or 4) mL 100 mM pH 6 phosphate buffer, add to the SPE cartridge, wash with 1 mL MeOH:100 mM phosphate buffer 20:80, wash with 1 mL 1 M acetic acid, wash with 1 mL hexane, suck dry briefly, wash with 4 mL dichloromethane, wash with 5 mL MeOH, elute with 2 mL dichloromethane:isopropanol 80:20 containing 5% ammonium hydroxide. Evaporate the eluate to dryness under a stream of nitrogen at room temperature, reconstitute the residue in 50-100 μL running buffer, inject an aliquot.

CAPILLARY ELECTROPHORESIS
Capillary: 105 cm × 75 μm fused-silica (68 cm to detector)
Capillary preparation: Before each run rinse capillary with 100 mM NaOH for 3.5 min and with buffer for 5 min.
Running buffer: Isopropanol:buffer 5:95 (Buffer was 10 mM pH 9.2-9.3 Na_2HPO_4 containing 6 mM sodium borate and 75 mM sodium dodecyl sulfate.)
Injection: Vacuum injection for 1 s
Detector: UV 235
Migration time: 34.3
Voltage: 30 kV
Current: 90 μA
Model: Europhor Prime Vision IV

OTHER SUBSTANCES
Extracted: bromazepam, clonazepam, flunitrazepam

KEY WORDS
SPE

REFERENCE
Schafroth,M.; Thormann,W.; Allemann,D. Micellar electrokinetic capillary chromatography of benzodiazepines in human urine, *Electrophoresis*, **1994**, *15*, 72–78.

Dibucaine

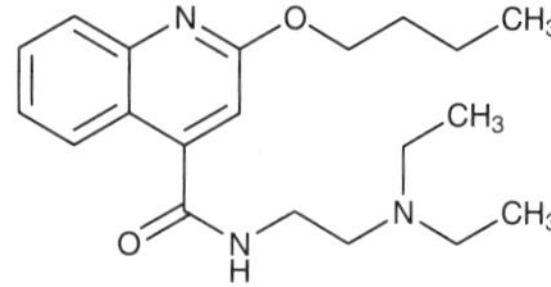

Molecular formula: $C_{20}H_{29}N_3O_2$
Molecular weight: 343.47
CAS Registry No.: 85-79-0, 61-12-1 (HCl)
Merck Index (12th ed.): 3081
Lednicer: 1 15

SAMPLE

Matrix: formulations
Sample preparation: Grind granules, add 70 mL MeOH, warm at 40° with occasional shaking, cool, add 20 mL 5 mg/mL methyl p-hydroxybenzoate in MeOH, make up to 100 mL with water, filter (0.45 μm), inject an aliquot.

CAPILLARY ELECTROPHORESIS

Capillary: 65 cm × 50 μm fused-silica (SGE)
Running buffer: 20 mM NaH_2PO_4 containing 100 mM sodium cholate, adjusted to pH 9.0 with 20 mM sodium tetraborate
Injection: Siphon for 10 s at 10 cm height
Detector: UV 210
Migration time: 16.5
Internal standard: methyl p-hydroxybenzoate (16)
Voltage: +20 kV

OTHER SUBSTANCES

Simultaneous: acetaminophen, caffeine, chlorpheniramine, dipyrone (sulpyrin), ethenzamide, guaifenesin, isopropylantipyrine, naproxen, noscapine, phenacetin, tipepidine, trimetoquinol, triprolidine

KEY WORDS

granules

REFERENCE

Nishi,H.; Fukuyama,T.; Matsuo,M.; Terabe,S. Separation and determination of the ingredients of a cold medicine by micellar electrokinetic chromatography with bile salts, *J.Chromatogr.*, **1990**, *498*, 313–323.

Dichlorphenamide

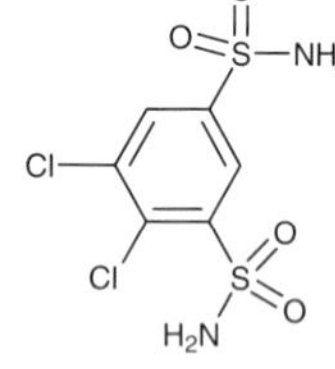

Molecular formula: $C_6H_6Cl_2N_2O_4S_2$
Molecular weight: 305.16
CAS Registry No.: 120-97-8
Merck Index (12th ed.): 3127
Lednicer: 1 133

SAMPLE

Matrix: blood, urine
Sample preparation: Condition a 3 mL Supelclean LC-18 SPE cartridge with 3 mL MeOH and 3 mL water. Dilute urine 1:10 with water. Precipitate proteins from serum with MeOH. Add diluted urine or protein supernatant to the SPE cartridge, wash with 3 mL water, elute with 3 mL MeOH, reconstitute to the original volume with 100 mM KOH, inject an aliquot.

CAPILLARY ELECTROPHORESIS

Capillary: 67 cm × 50 μm fused-silica (60 cm to detector) (Polymicro Technologies)
Capillary preparation: Rinse with running buffer for 2 min before run. If necessary, regenerate capillary with 100 mM NaOH for 10 min and with water for 15 min.

Capillary temperature: 20
Running buffer: 60 mM 3-(cyclohexylamino)-1-propanesulfonic acid (CAPS) adjusted to pH 10.6
  with 100 mM KOH
Injection: Pressure injection for 5 s
Detector: UV 220
Migration time: 14.6
Voltage: 25 kV
Model: Beckman P/ACE 2000

OTHER SUBSTANCES
Extracted: acetazolamide, amiloride, bendroflumethiazide, benzthiazide, bumetanide, caffeine,
  chlorothiazide, chlorthalidone, clopamide, ethacrynic acid, furosemide, hydrochlorothiazide,
  metyrapone, probenecid, triamterene, trichlormethiazide

KEY WORDS
serum; SPE

REFERENCE
Jumppanen,J.; Sirén,H.; Riekkola,M.-L. Screening for diuretics in urine and blood serum by capillary zone
  electrophoresis, *J.Chromatogr.A*, **1993**, *652*, 441–450.

Dichlorvos

Molecular formula: $C_4H_7Cl_2O_4P$
Molecular weight: 220.98
CAS Registry No.: 62-73-7
Merck Index (12th ed.): 3129

SAMPLE
Matrix: solutions
Sample preparation: Inject an aliquot of a solution in running buffer.

CAPILLARY ELECTROPHORESIS
Capillary: 60 cm × 50 μm fused-silica (53 cm to detector) (Polymicro Technologies)
Capillary preparation: Condition capillary by flushing with 100 mM NaOH for 20 min, water
  for 10 min, and running buffer for 10 min then equilibrate under voltage.
Running buffer: 10 mM pH 7.0 Sodium phosphate containing 60 mM sodium dodecyl sulfate
Detector: UV 210
Migration time: 15
Voltage: 15 kV
Model: laboratory-constructed

OTHER SUBSTANCES
Simultaneous: acephate, aldicarb, aminocarb, banol, captan, carbaryl, carbendazim, carbofuran,
  chlorotoluron, dimethoate, fenobucarb, fensulfothion, iprodione, isofenphos, isoprocarb, malathion, mepronil, methiocarb, napropamide, paclobutrazol, parathion, pirimicarb, propoxur, propyzamide, pyriodaphenthion, simazine, simetryn, thiram, triadimefon, tricyclazole

REFERENCE
Wu,Y.S.; Lee,H.K.; Li,S.F.Y. Rapid estimation of octanol-water partition coefficients of pesticides by micellar
  electrokinetic chromatography, *Electrophoresis*, **1998**, *19*, 1719–1727.

Diclofenac sodium

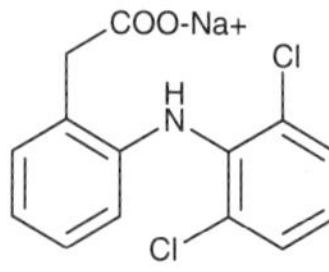

Molecular formula: $C_{14}H_{10}Cl_2NNaO_2$
Molecular weight: 318.13
CAS Registry No.: 15307-79-6
Merck Index (12th ed.): 3132
Lednicer: 2 70

SAMPLE
Matrix: blood
Sample preparation: 0.1-1 mL Plasma + 2 μg IS + 400 μL 1 M HCl + 10 mL diethyl ether, agitate mechanically for 10 min, centrifuge at 3500 g for 10 min. Remove the organic layer and evaporate it to dryness under a stream of nitrogen, reconstitute the residue in MeOH:40 mM NaH_2PO_4 50:50, inject an aliquot.

CAPILLARY ELECTROPHORESIS
Capillary: 58.7 cm × 75 μm fused-silica (50 cm to detector)
Capillary preparation: Wash with running buffer for 2 min before each injection. At the beginning of each day wash capillary with 100 mM NaOH for 10 min.
Capillary temperature: 22
Running buffer: MeOH:buffer 3:97 (Buffer was 40 mM pH 8 NaH_2PO_4 containing 104 mM sodium dodecyl sulfate.)
Injection: Hydrodynamic injection for 2 s
Detector: UV 254
Migration time: 18.5
Internal standard: benzoyl-4-phenyl-2-butyric acid (16.3)
Voltage: 20 kV
Model: Beckman P/ACE 2000
Limit of detection: 670 ng/mL (S/N 5)

OTHER SUBSTANCES
Extracted: diflunisal, etodolac, fenbufen, fenoprofen, flurbiprofen, ibuprofen, indomethacin, ketoprofen, naproxen, niflumic acid, piroxicam, sulindac, tenoxicam, tiaprofenic acid
Noninterfering: acetaminophen, amitriptyline, caffeine, clomipramine, deoxysulindac, desipramine, diazepam, 5'-hydroxytenoxicam, imipramine, maprotiline, nortriptyline, phenobarbital, phenytoin, sulfamethoxazole, theophylline, trimipramine

KEY WORDS
plasma

REFERENCE
Maboundou,C.W.; Paintaud,G.; Bérard,M.; Bechtel,P.R. Separation of fifteen non-steroidal anti-inflammatory drugs using micellar electrokinetic capillary chromatography, *J.Chromatogr.B*, **1994**, *657*, 173–183.

SAMPLE
Matrix: formulations
Sample preparation: Grind tablet, dissolve in running buffer, add a solution of niflumic acid in running buffer, dilute with running buffer to a final concentration of 25 μg/mL for diclofenac and 50 μg/mL for niflumic acid, inject an aliquot.

CAPILLARY ELECTROPHORESIS
Capillary: 60 cm × 75 μm (52.5 cm to detector)
Capillary preparation: Purge with running buffer for 2 min before each injection. Store capillary overnight in water, rinse with 500 mM NaOH, with water, and with running buffer.
Running buffer: 50 mM pH 9.0 Borate buffer containing 40 mM sodium dodecyl sulfate (Mix 2.94 mL 200 mM boric acid, 6.37 mL 75 mM tetraborate solution, and 10 mL 200 mM sodium dodecyl sulfate, make up to 50 mL with water.)
Injection: Hydrodynamic injection at 10 cm for 5 s
Detector: UV 214

Internal standard: niflumic acid
Voltage: 15 kV
Model: Waters Quanta 4000 CE
Limit of quantitation: 5 μg/mL

KEY WORDS
tablets

REFERENCE
Donato,M.G.; Baeyens,W.; Van den Bossche,W.; Sandra,P. The determination of non-steroidal antiinflammatory
drugs in pharmaceuticals by capillary zone electrophoresis and micellar electrokinetic capillary chromatography, *J.Pharm.Biomed.Anal.*, **1994**, *12*, 21–26.

SAMPLE
Matrix: solutions
Sample preparation: Inject an aliquot of a 500 μg/mL solution in 50 mM pH 7 buffer.

CAPILLARY ELECTROPHORESIS
Capillary: 48.5 cm × 50 μm fused-silica (40 cm to detector) (Hewlett Packard)
Capillary preparation: Before each injection flush capillary with 100 mM NaOH for 5 min and
with running buffer for 5 min. Wash new capillaries with 1 M NaOH at 40° for 15 min, with
water at 40° for 10 min, and with water at 25° for 5 min.
Capillary temperature: 25
Running buffer: 50 mM pH 7.0 Phosphate buffer (Prepare buffer by dissolving 5.29 g K_2HPO_4
and 2.61 g KH_2PO_4 in 1 L water.)
Injection: Pressure injection at 50 mbar for 9 s (18.8 nL).
Detector: UV 200
Migration time: 10
Voltage: 30 kV
Model: Hewlett Packard Model G1600A [3D]CE

OTHER SUBSTANCES
Simultaneous: ceftazidime, dicloxacillin, oxacillin, propranolol, quinine, thiamine

REFERENCE
Mrestani,Y.; Noubert,R.H.H.; Krause,A. Partition behaviour of drugs in microemulsions measured by electrokinetic chromatography, *Pharm.Res.*, **1998**, *15*, 799–801.

Dicloxacillin

Molecular formula: $C_{19}H_{17}Cl_2N_3O_5S$
Molecular weight: 470.33
CAS Registry No.: 3116-76-5, 13412-64-1 (sodium salt monohydrate),
343-55-5 (sodium salt)
Merck Index (12th ed.): 3134
Lednicer: 1 413

SAMPLE
Matrix: solutions
Sample preparation: Prepare a 120 μg/mL solution in water, inject an aliquot.

CAPILLARY ELECTROPHORESIS
Capillary: 60 cm × 75 μm
Running buffer: 20 mM NaH_2PO_4 containing 50 mM sodium dodecyl sulfate, adjusted to pH 9.0
with sodium tetraborate
Injection: Hydrodynamic injection at 10 cm for 5 s
Detector: UV 214

Migration time: 13.5
Voltage: 18 kV
Model: Waters Quanta 4000

OTHER SUBSTANCES
Simultaneous: 6-aminopenicillanic acid, amoxicillin, ampicillin, cloxacillin, nafcillin, oxacillin, ticarcillin

---

REFERENCE
Swartz,M.E. Method development and selectivity control for small molecule pharmaceutical separations by capillary electrophoresis, *J.Liq.Chromatogr.*, **1991**, *14*, 923–938.

SAMPLE
Matrix: solutions

CAPILLARY ELECTROPHORESIS
Capillary: 60 cm × 50 μm fused-silica (47 cm to detector) (Polymicro Technologies)
Capillary temperature: 25
Running buffer: 20 mM pH 8.5 Sodium tetraborate containing 100 mM sodium dodecyl sulfate
Injection: Hydrodynamic injection at 50 mbar for 3.6 s (5 nL).
Detector: UV 205
Migration time: 15
Voltage: 22 kV
Model: Crystal 310 (Thermo Unicam)

OTHER SUBSTANCES
Simultaneous: amoxicillin, ampicillin, cephapirin, cloxacillin, oxacillin, penicillin G, penicillin V, piperacillin, pyrimethamine, sulfacetamide, sulfadimethoxine, sulfaguanidine, sulfamerazine, sulfameter, sulfamethazine, sulfanilamide, sulfanilic acid, sulfapyridine, sulfaquinoxaline, sulfathiazole, sulfisoxazole, trimethoprim

---

REFERENCE
Hows,M.E.P.; Perrett,D.; Kay,J. Optimization of a simultaneous separation of sulphonamides, dihydrofolate reductase inhibitors and β-lactam antibiotics by capillary electrophoresis, *J.Chromatogr.A*, **1997**, *768*, 97–104.

SAMPLE
Matrix: solutions
Sample preparation: Inject an aliquot of a 500 μg/mL solution in 50 mM pH 7 buffer.

CAPILLARY ELECTROPHORESIS
Capillary: 48.5 cm × 50 μm fused-silica (40 cm to detector) (Hewlett Packard)
Capillary preparation: Before each injection flush capillary with 100 mM NaOH for 5 min and with running buffer for 5 min. Wash new capillaries with 1 M NaOH at 40° for 15 min, with water at 40° for 10 min, and with water at 25° for 5 min.
Capillary temperature: 25
Running buffer: 50 mM pH 7.0 Phosphate buffer (Prepare buffer by dissolving 5.29 g K_2HPO_4 and 2.61 g KH_2PO_4 in 1 L water.)
Injection: Pressure injection at 50 mbar for 9 s (18.8 nL).
Detector: UV 200
Migration time: 10.4
Voltage: 30 kV
Model: Hewlett Packard Model G1600A [3D]CE

OTHER SUBSTANCES
Simultaneous: ceftazidime, diclofenac, oxacillin, propranolol, quinine, thiamine

---

REFERENCE
Mrestani,Y.; Neubert,R.H.H.; Krause,A. Partition behaviour of drugs in microemulsions measured by electrokinetic chromatography, *Pharm.Res.*, **1998**, *15*, 799–801.

Didanosine

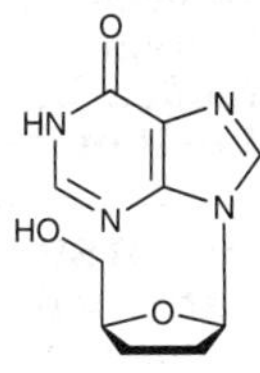

Molecular formula: $C_{10}H_{12}N_4O_3$
Molecular weight: 236.23
CAS Registry No.: 69655-05-6
Merck Index (12th ed.): 3148
Lednicer: 5 146

SAMPLE
Matrix: solutions

CAPILLARY ELECTROPHORESIS
Capillary: 80 cm × 75 µm fused-silica (52 cm to detector) (LC Packing, San Francisco)
Capillary temperature: 20 ± 0.5
Running buffer: 50 mM pH 6.5 Phosphate buffer containing 40 mM sodium dodecyl sulfate
Injection: Positive flow injection.
Detector: UV 260
Migration time: 8.5
Voltage: 20 kV
Current: 69 µA
Model: laboratory-constructed

OTHER SUBSTANCES
Simultaneous: dideoxyadenosine, stavudine, zalcitabine, zidovudine

REFERENCE
Singhal,R.; Xian,J.; Otim,O. Application of spherical and other polymers in capillary zone electrophoresis: separation of antiviral drugs and deoxyribonucleoside phosphates by different principles, *J.Chromatogr.A,* **1996**, *756*, 263–277.

Diflunisal

Molecular formula: $C_{13}H_8F_2O_3$
Molecular weight: 250.20
CAS Registry No.: 22494-42-4
Merck Index (12th ed.): 3190
Lednicer: 2 85

SAMPLE
Matrix: blood
Sample preparation: 0.1-1 mL Plasma + 2 µg IS + 400 µL 1 M HCl + 10 mL diethyl ether, agitate mechanically for 10 min, centrifuge at 3500 g for 10 min. Remove the organic layer and evaporate it to dryness under a stream of nitrogen, reconstitute the residue in MeOH:40 mM NaH_2PO_4 50:50, inject an aliquot.

CAPILLARY ELECTROPHORESIS
Capillary: 58.7 cm × 75 µm fused-silica (50 cm to detector)
Capillary preparation: Wash with running buffer for 2 min before each injection. At the beginning of each day wash capillary with 100 mM NaOH for 10 min.
Capillary temperature: 22
Running buffer: MeOH:buffer 3:97 (Buffer was 40 mM pH 8 NaH_2PO_4 containing 104 mM sodium dodecyl sulfate.)
Injection: Hydrodynamic injection for 2 s
Detector: UV 254
Migration time: 17.6

Internal standard: benzoyl-4-phenyl-2-butyric acid (16.3)
Voltage: 20 kV
Model: Beckman P/ACE 2000
Limit of detection: 330 ng/mL (S/N 5)

OTHER SUBSTANCES
Extracted: diclofenac, etodolac, fenbufen, fenoprofen, flurbiprofen, ibuprofen, indomethacin, ketoprofen, naproxen, niflumic acid, piroxicam, sulindac, tenoxicam, tiaprofenic acid
Noninterfering: acetaminophen, amitriptyline, caffeine, clomipramine, deoxysulindac, desipramine, diazepam, 5'-hydroxytenoxicam, imipramine, maprotiline, nortriptyline, phenobarbital, phenytoin, sulfamethoxazole, theophylline, trimipramine

KEY WORDS
plasma

REFERENCE
Maboundou,C.W.; Paintaud,G.; Bérard,M.; Bechtel,P.R. Separation of fifteen non-steroidal anti-inflammatory drugs using micellar electrokinetic capillary chromatography, *J.Chromatogr.B*, **1994**, *657*, 173–183.

SAMPLE
Matrix: solutions

CAPILLARY ELECTROPHORESIS
Capillary: 42.5 cm × 25 μm
Capillary temperature: 30
Running buffer: 20 mM pH 7.0 Phosphate containing 25 mM sodium dodecyl sulfate
Injection: Vacuum injection for 2 s
Detector: UV 230
Migration time: 4.7
Voltage: 25 kV
Current: 5 μA
Model: Applied Biosystems Model 270A

OTHER SUBSTANCES
Simultaneous: ibuprofen, indomethacin, naproxen, sulindac, tolmetin

REFERENCE
Weinberger,R.; Albin,M. Quantitative micellar electrokinetic capillary chromatography: Linear dynamic range, *J.Liq.Chromatogr.*, **1991**, *14*, 953–972.

Digitoxin

Molecular formula: $C_{41}H_{64}O_{13}$
Molecular weight: 764.95
CAS Registry No.: 71-63-6
Merck Index (12th ed.): 3206

SAMPLE
Matrix: solutions

CAPILLARY ELECTROPHORESIS
Capillary: 80 cm × 50 μm fused-silica (50 cm to detector) (Polymicro Technologies)
Capillary preparation: Between runs flush with 50 μL running buffer. Prepare capillary by flushing with 1 M NaOH for 10 min, with water for 10 min, and with running buffer for 10 min.
Running buffer: 30 mM pH 9.3 sodium borate containing 50 mM sodium dodecyl sulfate and 10 mM gamma-cyclodextrin
Injection: Hydrostatic injection at 100 cm for 30 s
Detector: UV 225
Migration time: 16.5
Voltage: 20 kV
Current: 33 μA
Model: Grom

OTHER SUBSTANCES
Simultaneous: α-acetyldigoxin, digoxin, evatromonoside, β-methyldiginatin, α-methyldigitoxin

---

REFERENCE
Gaus,H.-J.; Treumann,A.; Kreis,W.; Bayer,E. Separation of cardiac glycosides by micellar electrokinetic capillary electrophoresis, *J.Chromatogr.*, **1993**, *635*, 319–327.

Digoxin

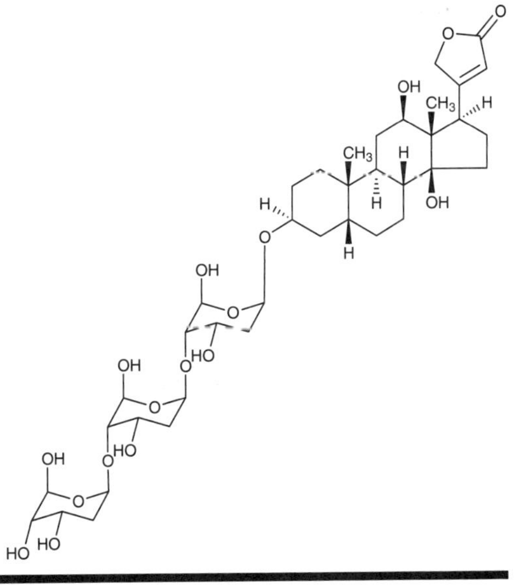

Molecular formula: $C_{41}H_{64}O_{14}$
Molecular weight: 780.95
CAS Registry No.: 20830-75-5
Merck Index (12th ed.): 3210

SAMPLE
Matrix: bulk
Sample preparation: Dissolve compounds in MeOH, dilute with running buffer, inject an aliquot.

CAPILLARY ELECTROPHORESIS
Capillary: 44 cm × 50 μm fused-silica (36 cm to detector)
Capillary preparation: Rinse with running buffer for 2 min before each run. At the end of the day rinse capillary with 1 M NaOH for 2 min and with water for 5 min. Rinse new capillaries with 1 M NaOH for 30 min and with water for 15 min.
Capillary temperature: 25
Running buffer: 30 mM Lithium tetraborate containing 50 mM sodium dodecyl sulfate, adjusted to pH 6.0 with orthophosphoric acid

Detector: UV 220
Migration time: 12
Voltage: 25 kV
Model: SpectraPhoresis 1000 (Thermo Separation Products)

OTHER SUBSTANCES
Simultaneous: acetyldigitoxin, acetyldigoxin, deslanoside

REFERENCE
Debusschere,L.; Demesmay,C.; Rocca,J.L.; Lachatre,G.; Lofti,H. Separation of cardiac glycosides by micellar electrokinetic chromatography and microemulsion electrokinetic chromatography, *J.Chromatogr.A*, **1997**, *779*, 227–233.

SAMPLE
Matrix: solutions

CAPILLARY ELECTROPHORESIS
Capillary: 80 cm × 50 μm fused-silica (50 cm to detector) (Polymicro Technologies)
Capillary preparation: Between runs flush with 50 μL running buffer. Prepare capillary by flushing with 1 M NaOH for 10 min, with water for 10 min, and with running buffer for 10 min.
Running buffer: 30 mM pH 9.3 sodium borate containing 50 mM sodium dodecyl sulfate and 10 mM gamma-cyclodextrin
Injection: Hydrostatic injection at 100 cm for 30 s
Detector: UV 225
Migration time: 10
Voltage: 20 kV
Current: 33 μA
Model: Grom

OTHER SUBSTANCES
Simultaneous: α-acetyldigoxin, digitoxin, evatromonoside, β-methyldiginatin, α-methyldigitoxin

REFERENCE
Gaus,H.-J.; Treumann,A.; Kreis,W.; Bayer,E. Separation of cardiac glycosides by micellar electrokinetic capillary electrophoresis, *J.Chromatogr.*, **1993**, *635*, 319–327.

Dihydrocodeine

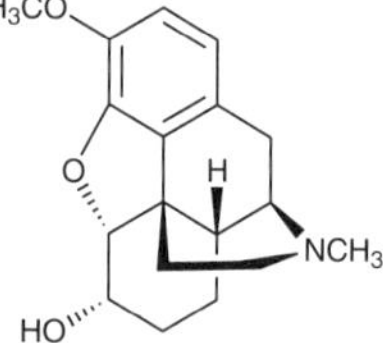

Molecular formula: $C_{18}H_{23}NO_3$
Molecular weight: 301.39
CAS Registry No.: 125-28-0
Merck Index (12th ed.): 3214
Lednicer: 1 288

SAMPLE
Matrix: solutions
Sample preparation: Inject an aliquot of an aqueous solution.

CAPILLARY ELECTROPHORESIS
Capillary: 41 cm × 50 μm fused-silica (22 cm to detector) (Polymicro Technologies)
Capillary preparation: Flush with running buffer for 5 min between runs.
Running buffer: 40 mM pH 10.7 phosphate buffer containing 70 mM sodium dodecyl sulfate
Injection: Hydrodynamic injection at 5" Hg for 1 s.
Detector: UV 210
Migration time: 11.6
Internal standard: norcodeine (2)

Voltage: 10 kV
Current: 60 µA
Model: Applied Biosystems 270A-HT

OTHER SUBSTANCES
Simultaneous: codeine, codeine-6-glucuronide, dihydrocodeine-6-glucuronide, dihydromorphine, ethylmorphine, morphine, morphine-3-glucuronide, nordihydrocodeine, nordihydromorphine, normorphine

REFERENCE
Zhang,C.-X.; Thormann,W. Head-column field-amplified sample stacking in binary system capillary electrophoresis. 2. Optimization with a preinjection plug and application to micellar electrokinetic chromatography, *Anal.Chem.*, **1998**, *70*, 540–548.

SAMPLE
Matrix: urine
Sample preparation: Condition a 130 mg Bond Elut Certify SPE cartridge with 2 mL MeOH and 2 mL water. 5 mL Urine + 1 mL concentrated HCl, heat at 120° for 30 min, cool, adjust pH to 7 with about 1.25 mL 10 M KOH, centrifuge at 1500 g for 3 min, add to the SPE cartridge, wash with 2 mL water, 1 mL 100 mM pH 4 acetate buffer, wash with 2 mL MeOH, dry under vacuum, elute with two 750 µL portions of MeOH containing 30% concentrated ammonium hydroxide. Evaporate the eluate to dryness under a stream of air at room temperature, reconstitute the residue in 100 µL running buffer, inject an aliquot.

CAPILLARY ELECTROPHORESIS
Capillary: 70 cm × 75 µm fused-silica (70 cm to detector)
Capillary preparation: Before each run rinse capillary with 100 mM NaOH for 2 min and with buffer for 6 min.
Capillary temperature: 35
Running buffer: 10 mM pH 9.2 Na_2HPO_4 containing 6 mM sodium borate and 75 mM sodium dodecyl sulfate
Injection: Vacuum injection for 0.5 s with electrokinetic injection at 30 kV
Detector: UV 213
Migration time: 15
Internal standard: codeine (14.5)
Voltage: 30 kV
Current: 112-115 µA
Model: Applied Biosystems 270A-HT
Limit of quantitation: 30 ng/mL (S/N 3)

OTHER SUBSTANCES
Extracted: dihydromorphine, nordihydrocodeine, nordihydromorphine

KEY WORDS
SPE

REFERENCE
Hufschmid,E.; Theurillat,R.; Martin,U.; Thormann,W. Exploration of the metabolism of dihydrocodeine via determination of its metabolites in human urine using micellar electrokinetic capillary chromatography, *J.Chromatogr.B*, **1995**, *668*, 159–170.

Dihydrostreptomycin

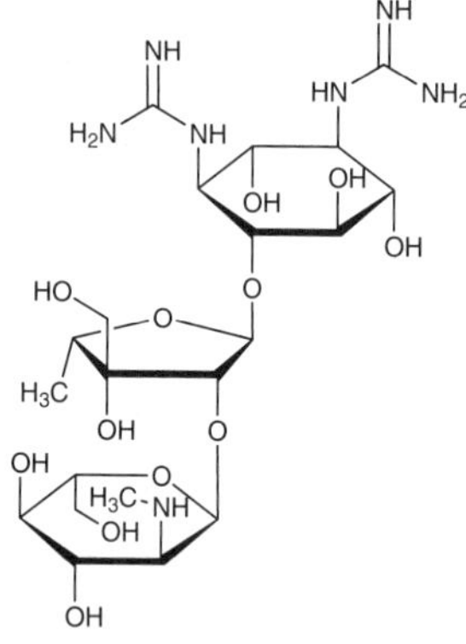

Molecular formula: $C_{21}H_{41}N_7O_{12}$
Molecular weight: 583.60
CAS Registry No.: 128-46-1, 5490-27-7 (sulfate)
Merck Index (12th ed.): 3222

SAMPLE
Matrix: bulk
Sample preparation: Inject an aliquot of an aqueous solution.

CAPILLARY ELECTROPHORESIS
Capillary: 90 cm × 50 μm uncoated fused silica (65 cm to detector) (Polymicro Technologies or ISCO)
Capillary preparation: Rinse with running buffer for 15 min at the beginning of the day and for 4 min before each injection.
Capillary temperature: 34
Running buffer: 160 mM pH 9 sodium tetraborate
Injection: Vacuum injection at 25.0 kPa
Detector: UV 195
Migration time: 15.5
Voltage: +18 kV
Model: ISCO Model 3140 Electropherograph

OTHER SUBSTANCES
Simultaneous: streptomycin

KEY WORDS
detector is at cathode

REFERENCE
Flurer,C.L. The analysis of aminoglycoside antibiotics by capillary electrophoresis, *J.Pharm.Biomed.Anal.*, **1995**, *13*, 809–816.

SAMPLE
Matrix: solutions
Sample preparation: Prepare a solution in 100 mM cetyltrimethylammonium bromide, inject an aliquot.

CAPILLARY ELECTROPHORESIS
Capillary: 67 cm × 50 μm fused silica capillary (60 cm to detector) (Siemens)
Running buffer: pH 5.0 Acetate buffer containing 10 mM imidazole and 50 μg/mL FC 135 (a fluorinated surfactant, Fluorad/3M)
Injection: Pressure injection for 2 s.
Detector: UV 214
Migration time: 9.5
Voltage: 12.5 kV
Model: Beckman P/ACE System 2000
Limit of detection: 23.38 μg/mL

OTHER SUBSTANCES
Simultaneous: amikacin, gentamicin, kanamycin, lividomycin, neomycin, paromomycin, sisomycin, tobramycin

Interfering: streptomycin

KEY WORDS
indirect UV detection; detector is at anode

REFERENCE
Ackermans,M.T.; Everaerts,F.M.; Beckers,J.L. Determination of aminoglycoside antibiotics in pharmaceuticals by capillary zone electrophoresis with indirect UV detection coupled with micellar electrokinetic capillary chromatography, *J.Chromatogr.*, **1992**, *606*, 229–235.

Dilevalol

Molecular formula: $C_{19}H_{24}N_2O_3$
Molecular weight: 328.41
CAS Registry No.: 75659-07-3, 75659-08-4 (HCl)
Merck Index (12th ed.): 3245
Lednicer: 4 20

SAMPLE
Matrix: solutions

CAPILLARY ELECTROPHORESIS
Capillary: 62.1 cm × 75 μm fused-silica (44.6 cm to detector) (Polymicro Technologies)
Capillary preparation: Coat capillary as follows. Condition capillary with 1 M NaOH for 1 h, rinse with water for 30 min, fill capillary with 1% 3-(trimethoxysilyl)propyl methacrylate (methacryloxypropyltrimethoxysilane) (adjusted to pH 3.5 with acetic acid), allow to react at room temperature for 2 h, rinse with water, fill capillary with 10% 2-aminoethyl methacrylate hydrochloride (Fisher Scientific) solution containing 1-3 mg/mL N,N,N',N'-tetramethylethylenediamine and 1 mg/mL ammonium persulfate, allow to react until polymerization is complete, rinse with water, dry at 30-40° overnight.
Running buffer: 25 mM pH 4.7 Acetate buffer
Injection: Hydrodynamic injection at 5 cm for 5 s.
Detector: UV 210
Migration time: 10.80
Voltage: 15 kV
Model: laboratory constructed

OTHER SUBSTANCES
Simultaneous: acebutolol, atenolol, betaxolol, carteolol, nifenalol, pindolol, propranolol

---

KEY WORDS
coated capillary; detector at anode

REFERENCE
Liu,Q.; Lin,F.; Hartwick,R.A. Free solution capillary electrophoretic separation of basic proteins and drugs using cationic polymer coated capillaries, *J.Liq.Chromatogr.Rel.Technol.*, **1997**, *20*, 707–718.

SAMPLE
Matrix: solutions

CAPILLARY ELECTROPHORESIS
Capillary: 50.5 cm × 75 μm coated fused-silica (36 cm to detector) (Polymicro Technologies)
Capillary preparation: Coat capillary as follows. Condition capillary with 1 M NaOH for 1 h, rinse with water for 30 min, force 1% methacryloxypropyltrimethoxysilane (adjusted to pH 3.5 with acetic acid) through the capillary using pressure, allow to react at room temperature for 2 h, rinse with water, fill capillary with diallyldimethylammonium chloride solution containing

1 μL/mL N,N,N',N'-tetramethylethylenediamine and 1 mg/mL ammonium persulfate, allow to react until polymerization is complete, rinse with water, dry at 30-40° overnight.
Running buffer: 10 mM pH 5.0 Acetate buffer
Injection: Hydrodynamic injection at 5 cm for 5 s.
Detector: UV 210
Migration time: 9.3
Voltage: 12 kV
Model: laboratory-constructed

OTHER SUBSTANCES
Simultaneous: acebutolol, atenolol, nifenalol, pindolol

KEY WORDS
coated capillary

REFERENCE
Liu,Q.; Lin,F.; Hartwick,R.A. Poly(diallyldimethylammonium chloride) as a cationic coating for capillary electrophoresis, *J.Chromatogr.Sci.*, **1997**, *35*, 126–130.

SAMPLE
Matrix: solutions

CAPILLARY ELECTROPHORESIS
Capillary: 71.5 cm × 75 μm fused-silica (52.2 cm to detector) (Polymicro Technologies)
Capillary preparation: Rinse capillary with running buffer for 1 min between runs. Rinse new capillaries with 500 mM NaOH for 20 min and with water for 10 min and then condition with running buffer for 1 h.
Running buffer: 50 mM pH 3.0 containing 0.1% guaran (Guaran is the water-soluble fraction of guar gum. Prepare guaran by stirring 1 g guar gum (Sigma) in 100 mL water for 20 min, add 25 mL EtOH dropwise with stirring, let stand overnight, supercentrifuge. Add 30 mL EtOH dropwise with stirring to the centrifugate, centrifuge, collect the guaran as a precipitate, dry in air.)
Injection: Hydrodynamic injection at 5 cm for 5 s.
Detector: UV (wavelengths not given)
Migration time: 14.5
Voltage: 17.5 kV
Model: laboratory-constructed

OTHER SUBSTANCES
Simultaneous: acebutolol, atenolol, betaxolol, carteolol, nifenalol, pindolol, propranolol

REFERENCE
Liu,Q.; Lin,F.; Hartwick,R.A. Capillary zone electrophoretic separation of basic proteins and drugs using guaran as a buffer modifier, *Chromatographia*, **1998**, *47*, 219–224.

Diltiazem

Molecular formula: $C_{22}H_{26}N_2O_4S$
Molecular weight: 414.53
CAS Registry No.: 42399-41-7, 33286-22-5 (HCl)
Merck Index (12th ed.): 3247
Lednicer: 3 198

SAMPLE
Matrix: formulations
Sample preparation: Grind tablet, add 30 mL MeOH, warm at 40° with occasional shaking for 10 min, cool, add 10 mL 3 mg/mL chlorodiltiazem in MeCN, make up to 50 mL with water, filter (0.45 μm), inject an aliquot.

CAPILLARY ELECTROPHORESIS
Capillary: 65 cm × 50 μm fused-silica (50 cm to detector) (Scientific Glass Engineering)
Capillary preparation: Flush with running buffer for 1 min between runs. Periodically flush with water for 1 min, sweep with 100 mM KOH, let stand in 100 mM KOH for 30 min, flush with water until the effluent is neutral to pH paper, fill with running buffer, let stand for 30 min.
Running buffer: 20 mM pH 8.0 Phosphate-borate buffer containing 100 mM sodium cholate (Buffer was prepared by adjusting pH of 20 mM sodium borate to 8.0 with 20 mM NaH_2PO_4.)
Injection: Injection by siphon at 10 cm for 10 s (about 1 nL)
Detector: UV 210
Migration time: 10
Internal standard: chlorodiltiazem (11.5)
Voltage: 20 kV

KEY WORDS
tablets

REFERENCE
Nishi,H.; Fukuyama,T.; Matsuo,M.; Terabe,S. Separation and determination of lipophilic corticosteroids and benzothiazepin analogues by micellar electrokinetic chromatography using bile salts, *J.Chromatogr.*, **1990**, *513*, 279–295.

SAMPLE
Matrix: solutions

CAPILLARY ELECTROPHORESIS
Capillary: 65 cm × 50 μm fused-silica (50 cm to detector) (SGE)
Capillary preparation: Each week fill capillary with 100 mM KOH, let stand for 30 min, flush with water, let stand for 5 min.
Running buffer: 20 mM pH 7.0 Phosphate-borate buffer containing 50 mM sodium taurodeoxycholate
Injection: Inject by siphoning at 10 cm for 5-10 s (about 1 nL)
Detector: UV 210
Migration time: 14.94, 15.05 (enantiomers)
Voltage: 20 kV

OTHER SUBSTANCES
Simultaneous: diltiazem analogs, tetrahydropapaveroline, trimetoquinol

KEY WORDS
chiral

REFERENCE
Nishi,H.; Fukuyama,T.; Matsuo,M.; Terabe,S. Chiral separation of diltiazem, trimetoquinol and related compounds by micellar electrokinetic chromatography with bile salts, *J.Chromatogr.*, **1990**, *515*, 233–243.

SAMPLE
Matrix: solutions
Sample preparation: Prepare a 600 μg/mL solution in MeCN:MeOH:water 25:25:50, dilute to 150 μg/mL with 100 mM pH 8.1 borate buffer containing 50 mM sodium dodecyl sulfate, inject an aliquot.

CAPILLARY ELECTROPHORESIS
Capillary: 57 cm × 75 μm fused-silica (50 cm to detector) (Beckman)
Capillary preparation: Rinse capillary with 100 mM NaOH and running buffer before each run.
Capillary temperature: 30
Running buffer: Acetone:buffer 15:85 (Buffer was 100 mM pH 8.1 borate buffer containing 50 mM sodium dodecyl sulfate.)
Injection: Pressure injection for 5 s
Detector: UV 200

Migration time: 17.5
Voltage: +25 kV
Model: Beckman P/ACE system 5510

OTHER SUBSTANCES
Simultaneous: amlodipine, atenolol, nicardipine, nifedipine, verapamil

REFERENCE
Bretnall,A.E.; Clarke,G.S. Investigation and optimisation of the use of micellar electrokinetic chromatography for the analysis of six cardiovascular drugs, *J.Chromatogr.A*, **1995**, *700*, 173–178.

SAMPLE
Matrix: solutions
Sample preparation: Inject an aliquot of a 100-300 μg/mL solution.

CAPILLARY ELECTROPHORESIS
Capillary: 57 cm × 75 μm fused-silica (50 cm to detector)
Capillary preparation: Rinse capillary with running buffer for 1 min before each run. Wash capillary with 100 mM NaOH and water at the end of each day.
Capillary temperature: 23
Running buffer: 20 mM pH 4.0 Phosphate-borate buffer containing 3% chondroitin sulfate C
Injection: Pressure injection at 0.5 psi for 3-7 s.
Detector: UV 235
Migration time: 8.6 (2S,3R), 8.8 (2R,3S), 9.0 (2S,3S), 9.3 (2R,3R)
Voltage: 20 kV
Model: Beckman P/ACE 5510

OTHER SUBSTANCES
Simultaneous: clentiazem

KEY WORDS
chiral

REFERENCE
Nishi,H.; Terabe,S. Enantiomeric separation of diltiazem, clentiazem, and its related compounds by capillary electrophoresis using polysaccharides, *J.Chromatogr.Sci.*, **1995**, *33*, 698–703.

SAMPLE
Matrix: solutions
Sample preparation: Prepare a 100 μg/mL solution in MeOH:water 10:90, inject an aliquot.

CAPILLARY ELECTROPHORESIS
Capillary: 57 cm × 50 μm fused-silica (50 cm to detector)
Capillary preparation: Rinse with running buffer for 1 min before each run. Wash with 100 mM KOH and water each day.
Capillary temperature: 23
Running buffer: 20 mM pH 2.4 Phosphate borate buffer containing 3% chondroitin sulfate C (sodium salt)
Injection: Pressure injection at 0.5 psi for 5 s
Detector: UV 235
Migration time: 22, 22.5, 24, 26 (enantiomers)
Voltage: 20 kV
Model: Beckman P/ACE 5510
Limit of detection: 0.1% (of major isomer, S/N 3)

KEY WORDS
chiral

REFERENCE
Nishi,H.; Nakamura,K.; Nakai,H.; Sato,T. Enantiomeric separation of drugs by mucopolysaccharide-mediated electrokinetic chromatography, *Anal.Chem.*, **1995**, *67*, 2334–2341.

SAMPLE
Matrix: solutions

CAPILLARY ELECTROPHORESIS
Capillary: 57 cm × 75 μm fused-silica (50 cm to detector) (Beckman)
Capillary preparation: Before each run rinse capillary with 100 mM NaOH and running buffer.
Capillary temperature: 30
Running buffer: Acetone:100 mM pH 8.1 borate buffer containing 50 mM sodium dodecyl sulfate 15:85
Injection: Pressure injection for 5-10 s.
Detector: UV 200
Migration time: 17
Voltage: 25 kV
Model: Beckman P/ACE 5510

OTHER SUBSTANCES
Simultaneous: acebutolol, amiodarone, atenolol, bretylium, captopril, disopyramide, lidocaine, lisinopril, metoprolol, nicardipine, nifedipine, phenytoin, pindolol, propafenone, propranolol, quinidine, sotalol, timolol, p-toluenesulfonic acid, verapamil

REFERENCE
Bretnall,A.E.; Clarke,G.S. Selectivity of capillary electrophoresis for the analysis of cardiovascular drugs, *J.Chromatogr.A*, **1996**, *745*, 145–154.

SAMPLE
Matrix: solutions
Sample preparation: Inject an aliquot of a 100 μg/mL solution in water or running buffer.

CAPILLARY ELECTROPHORESIS
Capillary: 47 cm × 75 μm fused-silica (40 cm to detector)
Capillary preparation: Before each run rinse capillary with running buffer for 1-2 min. If peak tailing is observed fill capillary with 500 mM NaOH and let stand for 30 min, wash with water for 3 min, rinse with running buffer for 3 min.
Capillary temperature: 25
Running buffer: 20 mM pH 2.5 Phosphate buffer containing 3% dextrin (Japan Pharmacopeia grade) (Dissolve dextrin in buffer at 90° then cool to room temperature.)
Injection: Pressure injection at 0.5 psi for 2-4 s.
Detector: UV 220
Migration time: 7.597 (S,S), 8.013 (R,R)
Voltage: 20 kV
Model: Beckman P/ACE 5510

KEY WORDS
chiral

REFERENCE
Nishi,H.; Izumoto,S.; Nakamura,K.; Nakai,H.; Sato,T. Dextran and dextrin as chiral selectors in capillary zone electrophoresis, *Chromatographia*, **1996**, *42*, 617–630.

SAMPLE
Matrix: solutions
Sample preparation: Inject an aliquot of a 100 μg/mL solution in MeOH:water 10:90.

CAPILLARY ELECTROPHORESIS
Capillary: 47 cm × 75 μm fused-silica (40 cm to detector)
Capillary preparation: Rinse capillary with running buffer for 1 min before each run. At the end of each day wash with 100 mM NaOH and with water.
Capillary temperature: 20
Running buffer: 20 mM pH 2.9 Phosphate buffer containing 3% chondroitin sulfate C (sodium salt) (Nacalai Tesque, Kyoto)
Injection: Pressure injection at 0.5 psi for 5 s.

Detector: UV 240
Migration time: 15.81, 17.08 (enantiomers)
Voltage: 20 kV
Model: Beckman P/ACE 5510

OTHER SUBSTANCES
Also analyzed: clentiazem, primaquine, propranolol, sulconazole, trimetoquinol, verapamil

KEY WORDS
chiral

REFERENCE
Nishi,H. Enantiomer separation of basic drugs by capillary electrophoresis using ionic and neutral polysaccharides as chiral selectors, *J.Chromatogr.A*, **1996**, *735*, 345–351.

SAMPLE
Matrix: solutions

CAPILLARY ELECTROPHORESIS
Capillary: 60 cm × 50 μm fused-silica (47 cm to detector) (Polymicro Technologies)
Running buffer: 50 mM pH 3.0 Phosphate buffer containing 10 mg/mL carboxymethyl amylose
Injection: Pressure injection.
Detector: UV 210
Migration time: 16, 17 (enantiomers)
Voltage: 20 kV
Model: Prince

KEY WORDS
chiral

REFERENCE
Chankvetadze,B.; Saito,M.; Yashima,E.; Okamoto,Y. Enantioseparation using selected polysaccharides as chiral buffer additives in capillary electrophoresis, *J.Chromatogr.A*, **1997**, *773*, 331–338.

SAMPLE
Matrix: urine
Sample preparation: Centrifuge urine, dilute with 2 volumes of 32 mM pH 4.5 6-aminocaproic acid buffer containing 18 mM adipic acid, inject an aliquot.

CAPILLARY ELECTROPHORESIS
Capillary: 64 cm × 75 μm (48 cm to detector) (Polymicro Technologies)
Running buffer: MeOH:buffer 5:95 (Buffer was 32 mM pH 4.5 6-aminocaproic acid buffer containing 18 mM adipic acid, 5% dextran (MW 18300) and 0.02% polyethylene oxide (MW 300000)
Injection: Hydrodynamic injection at 20 cm
Detector: UV 220
Migration time: 13
Voltage: 22 kV
Model: Laboratory constructed

OTHER SUBSTANCES
Extracted: degradation products, cimetidine, famotidine, prazosin

REFERENCE
Soini,H.; Riekkola,M.-L.; Novotny,M.V. Mixed polymer networks in the direct analysis of pharmaceuticals in urine by capillary electrophoresis, *J.Chromatogr.A*, **1994**, *680*, 623–634.

Dimenhydrinate

Molecular formula: $C_{24}H_{28}ClN_5O_3$
Molecular weight: 469.97
CAS Registry No.: 523-87-5
Merck Index (12th ed.): 3252

SAMPLE
Matrix: formulations
Sample preparation: Grind tablets, extract twice with MeOH by sonicating at low temperature, centrifuge, filter (0.45 μm) the supernatant, inject an aliquot of the filtrate.

CAPILLARY ELECTROPHORESIS
Capillary: 47 cm × 50 μm (40 cm to detector) (Polymicro Technologies)
Capillary temperature: 25
Running buffer: 50 mM pH 7.5 Borate buffer containing 50 mM phosphate, 10 mM sodium dodecyl sulfate, 10 mM β-cyclodextrin, and 10 mM tetrabutylammonium hydrogen sulfate
Injection: Inject at a pressure of 0.5 psi for 1 s (5.2 nL)
Detector: UV 214
Migration time: 4.2
Voltage: 21.5 kV
Model: Beckman P/ACE 2000
Limit of detection: 133.3 pg

OTHER SUBSTANCES
Simultaneous: chlorpheniramine, cyclizine, doxylamine, methapyrilene, pheniramine, promethazine, thonylamine, triprolidine

KEY WORDS
tablets

REFERENCE
Ong,C.P.; Ng,C.L.; Lee,H.K.; Li,S.F.Y. Determination of antihistamines in pharmaceuticals by capillary electrophoresis, *J.Chromatogr.*, **1991**, *588*, 335–339.

SAMPLE
Matrix: solutions

CAPILLARY ELECTROPHORESIS
Capillary: 62 cm × 75 μm fused-silica (54.5 cm to detector) (Polymicro Technologies)
Capillary preparation: Purge with running buffer before each run. At the beginning of each day purge using 50-60 kPa vacuum with 500 mM NaOH for 5 min, with water for 5 min, with MeCN for 5 min, and with running buffer for 5 min.
Running buffer: MeCN:MeOH:acetic acid 49:50:1 containing 20 mM ammonium acetate
Injection: Hydrostatic injection at 10 cm for 5 s.
Detector: UV 214
Migration time: 3.58
Voltage: 30 kV
Model: Waters Quanta 4000

OTHER SUBSTANCES
Simultaneous: buclizine, chlorcyclizine, chlorpheniramine, cyclizine, methaphenilene, pheniramine, promethazine, pyrilamine, pyrrobutamine, tripelennamine

REFERENCE

Leung,G.N.W.; Tang,H.P.O.; Tso,T.S.C.; Wan,T.S.M. Separation of basic drugs with non-aqueous capillary electrophoresis, *J.Chromatogr.A*, **1996**, *738*, 141–154.

Dimethindene

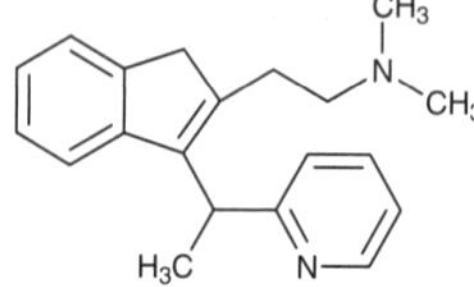

Molecular formula: $C_{20}H_{24}N_2$
Molecular weight: 292.42
CAS Registry No.: 5636-83-9, 3614-69-5 (maleate)
Merck Index (12th ed.): 3265

SAMPLE

Matrix: solutions
Sample preparation: Inject an aliquot of a 50 µg/mL solution in water.

CAPILLARY ELECTROPHORESIS

Capillary: 48.5 cm × 50 µm fused-silica (40 cm to detector)
Capillary preparation: After each run wash capillary with running buffer for 3 min. At the start of each day wash capillary with running buffer for 10 min. Before use treat new capillaries with 1 M NaOH, 100 mM NaOH, water, and running buffer.
Capillary temperature: 15
Running buffer: 100 mM Phosphoric acid containing 2 mM carboxymethyl-β-cyclodextrin (Cyclolab, Budapest), adjusted to pH 3.0 with 84 mM triethanolamine
Injection: Hydrodynamic injection at 5 kPa for 2 s.
Detector: UV 210
Migration time: 7.67, 8.47 (enantiomers)
Voltage: 25 kV
Model: Hewlett Packard ^{3D}CE

OTHER SUBSTANCES

Simultaneous: fenfluramine, isoproterenol, terbutaline
Interfering: chlorpheniramine, ephedrine

KEY WORDS

chiral

REFERENCE

Fillet,M.; Bechet,I.; Hubert,P.; Crommen,J. Resolution improvement by use of carboxymethyl-β-cyclodextrin as chiral additive for the enantiomeric separation of basic drugs by capillary electrophoresis, *J.Pharm.Biomed.Anal.*, **1996**, *14*, 1107–1114.

SAMPLE

Matrix: solutions
Sample preparation: Inject an aliquot of a solution in running buffer.

CAPILLARY ELECTROPHORESIS

Capillary: 60 cm × 75 µm fused-silica (52.4 cm to detector)
Capillary preparation: After each run flush with 500 mM KOH for 2-3 min then with water.
Running buffer: 10 mM pH 3.8 Phosphate buffer containing 2% sulfated cyclodextrin (ds 7-10)
Injection: Hydrostatic injection.
Detector: UV 214
Migration time: 8.87, 9.32 (enantiomers)
Voltage: 15 kV
Model: Waters Quanta 4000

OTHER SUBSTANCES
Also analyzed: acebutolol, alprenolol, aminoglutethimide, brompheniramine, bupivacaine, bupropion, canadine, carbinoxamine, chloroquine, chlorpheniramine, disopyramide, doxylamine, hydroxychloroquine, idazoxan, isoxsuprine, ketamine, mepenzolate, mepivacaine, methoxyphenamine, mexiletine, midodrine, nefopam, orphenadrine, oxprenolol, oxyphencyclimine, pheniramine, phensuximide, pindolol, piperoxan, terbutaline, tetramisole, tolperisone, tranylcypromine, trihexyphenidyl, trimipramine, verapamil, warfarin

KEY WORDS
chiral; detector at anode

REFERENCE
Stalcup,A.M.; Gahm,K.H. Application of sulfated cyclodextrins to chiral separations by capillary zone electrophoresis, *Anal.Chem.*, **1996**, *68*, 1360–1368.

SAMPLE
Matrix: solutions
Sample preparation: Coat capillary as follows. Wash capillary with 1 M NaOH for 30 min and with water for 15 min, flush with a solution of ethylene imine polymer (MW 0.6-1 million, Fluka) in water for 10 min, let stand for 1 h, rinse with water for 15 min, rinse with running buffer for 15 min.

CAPILLARY ELECTROPHORESIS
Capillary: 60 cm × 50 μm fused-silica (Grom, Herrenberg, Germany)
Running buffer: 50 mM pH 3.0 Phosphate buffer containing 5 mg/mL β-cyclodextrin (A) or 50 mM pH 5.3 acetate buffer containing 3 mg/mL β-cyclodextrin (B)
Injection: Hydrostatic injection at 10 cm.
Detector: UV 210
Migration time: 7.5 (S (A)), 8 (R (A)), 27.5 (R (B)), 28.5 (S (B))
Voltage: 24 kV (A), 30 kV (B)
Model: Grom (Herrenberg, Germany)

KEY WORDS
chiral; coated capillary; detector at cathode (A) and detector at anode (B)

REFERENCE
Chankvetadze,B.; Schulte,G.; Blaschke,G. Selected applications of capillaries with dynamic or permanent anodal electroosmotic flow in chiral separations by capillary electrophoresis, *J.Pharm.Biomed.Anal.*, **1997**, *15*, 1577–1584.

SAMPLE
Matrix: solutions
Sample preparation: Inject an aliquot of a 100 μg/mL solution in running buffer.

CAPILLARY ELECTROPHORESIS
Capillary: 32 cm × 50 μm fused-silica (27.5 cm to detector) (Yongnian Optical Conductive Fiber Plant, China), coated with polyacrylamide
Capillary preparation: No details of the polyacrylamide coating process are provided. However, another paper (LC.GC 1997, 15, 40) by this group indicates that they use the procedure of Hjertén, thus: Adjust the pH of 20 mL water to 3.5 with acetic acid, add 80 μL 3-(trimethoxysilyl)propyl methacrylate (3-methacryloxypropyltrimethoxysilane), mix, suck into capillary, let stand at room temperature for 1 h, remove the solution, wash with water. Fill the capillary with a deaerated 3-4% acrylamide solution containing 1 μL/mL N,N,N',N'-tetramethylethylenediamine and 1 mg/mL potassium persulfate, let stand for 30 min, remove excess solution by aspiration, rinse with water, remove water by aspiration, dry at 35° (J. Chromatogr. 1985, 347, 191).
Capillary temperature: 25
Running buffer: 100 mM NaH$_2$PO$_4$ adjusted to pH 2.5
Injection: Electrokinetic injection at 15 kV for 3 s.
Detector: UV 200, UV 210
Migration time: 3.17
Voltage: 15 kV

Model: Bio-Rad BioFocus 3000

OTHER SUBSTANCES
Simultaneous: atropine, carazolol, cicletanine, fendiline, homatropine, ipratropium bromide, isothipendyl, mefloquine, metaclazepam, nafronyl (naftidrofuryl), nefopam, nicardipine, promethazine, reproterol, tetrahydrozoline (tetryzoline), theodrenaline, tioconazole, trihexyphenidyl, trimeprazine (alimemazine), trimipramine

KEY WORDS
coated capillary

REFERENCE
Koppenhoefer,B.; Epperlein,U.; Xiaofeng,Z.; Bingcheng,L. Separation of enantiomers of drugs by capillary electrophoresis. Part 4: Hydroxypropyl-γ-cyclodextrin as chiral solvating agent, *Electrophoresis*, **1997**, *18*, 924–930.

SAMPLE
Matrix: solutions
Sample preparation: Inject an aliquot of a 100 μg/mL solution in running buffer.

CAPILLARY ELECTROPHORESIS
Capillary: 30 cm × 50 μm fused-silica (25.5 cm to detector), coated with polyacrylamide
Capillary preparation: Adjust the pH of 20 mL water to 3.5 with acetic acid, add 80 μL 3-(trimethoxysilyl)propyl methacrylate (3-methacryloxypropyltrimethoxysilane), mix, suck into capillary, let stand at room temperature for 1 h, remove the solution, wash with water. Fill the capillary with a deaerated 3-4% acrylamide solution containing 1 μL/mL N,N,N',N'-tetramethylethylenediamine and 1 mg/mL potassium persulfate, let stand for 30 min, remove excess solution by aspiration, rinse with water, remove water by aspiration, dry at 35° (J. Chromatogr. 1985, 347, 191).
Capillary temperature: 25
Running buffer: 100 mM NaH_2PO_4 containing 30 mM hydroxypropyl-β- cyclodextrin, adjusted to pH 2.5 with phosphoric acid
Injection: Electrokinetic injection at 15 kV for 3 s.
Detector: UV 200
Migration time: 9.9, 10.1 (enantiomers)
Voltage: 15 kV
Model: Bio-Rad BioFocus 3000

OTHER SUBSTANCES
Simultaneous: chlorpheniramine, homatropine, mefloquine, metaclazepam, tetrahydrozoline (tetryzoline), theodrenaline

KEY WORDS
chiral; coated capillary; comparison with the use of other cyclodextrins

REFERENCE
Lin,B.; Zhu,X.; Koppenhoefer,B.; Epperlein,U. Investigation of 123 chiral drugs by cyclodextrin-modified capillary electrophoresis, *LC.GC*, **1997**, *15*, 40–46.

SAMPLE
Matrix: solutions

CAPILLARY ELECTROPHORESIS
Capillary: 60 cm × 50 μm fused-silica (43 cm to detector) (Grom)
Running buffer: 50 mM pH 3.0 Phosphate buffer containing 1 mg/mL carboxymethyl-β-cyclodextrin
Injection: Hydrostatic injection at 10 cm.
Detector: UV (wavelength not given)
Migration time: 4.5 ((R)-(-)), 4.9 ((S)-(+))
Model: Grom 100

KEY WORDS
chiral

REFERENCE
Chankvetadze,B.; Schulte,G.; Bergenthal,D.; Blaschke,G. Comparative capillary electrophoresis and NMR studies of enantioseparation of dimethindene with cyclodextrins, *J.Chromatogr.A*, **1998**, *798*, 315–323.

SAMPLE
Matrix: solutions

CAPILLARY ELECTROPHORESIS
Capillary: 29-36 cm × 50 μm fused-silica (24.5-31.5 cm to detector) (Yongnian Optical Conductive Fiber Plant, China) coated with polyacrylamide
Capillary preparation: Coat capillary as follows. Adjust the pH of 20 mL water to 3.5 with acetic acid, add 80 μL 3-(trimethoxysilyl)propyl methacrylate (3-methacryloxypropyltrimethoxysilane), mix, suck into capillary, let stand at room temperature for 1 h, remove the solution, wash with water. Fill the capillary with a deaerated 3-4% acrylamide solution containing 1 μL/mL N,N,N',N'-tetramethylethylenediamine and 1 mg/mL potassium persulfate, let stand for 30 min, remove excess solution by aspiration, rinse with water, remove water by aspiration, dry at 35° (J. Chromatogr. 1985, 347, 191).
Capillary temperature: 25
Running buffer: 100 mM pH 2.5 NaH$_2$PO$_4$ (A) or 100 mM pH 2.5 NaH$_2$PO$_4$ containing 45 mM hydroxypropyl-α-cyclodextrin (Wacker, Munich) (B)
Injection: Electromigration at 15 kV for 3 s.
Detector: UV 200; UV 210
Migration time: 3.17 (A); 4.94 (B) (no separation of enantiomers)
Voltage: 15 kV
Model: Bio-Focus 3000

OTHER SUBSTANCES
Also analyzed: albuterol (salbutamol), alprenolol, amorolfine, atenolol, atropine, azelastine, baclofen, bamethan, benproperine, benserazide, biperiden, bisoprolol, brompheniramine, bupivacaine, bupranolol, butamirate, butethamate, carazolol, carbuterol, carteolol, carvedilol, celiprolol, chloroquine, chlorpheniramine, chlorphenoxamine, cicletanine, clenbuterol, clidinium bromide, clobutinol, dipivefrin, disopyramide, dobutamine, doxylamine, fendiline, flecainide, gallopamil, homatropine, ipratropium bromide, isoproterenol (isoprenaline), isothipendyl, ketamine, meclizine, mefloquine, mepindolol, mequitazine, metaclazepam, metaproterenol (orciprenaline), metipranolol, metoprolol, nafronyl (naftidrofuryl), nefopam, nicardipine, norfenefrine, ofloxacin, ornidazole, orphenadrine, oxomemazine, oxprenolol, oxybutynin, phenoxybenzamine, phenylpropanolamine, pholedrine, pindolol, pirbuterol, prilocaine, procyclidine, promethazine, propafenone, propranolol, reproterol, sotalol, sulpride, synephrine, talinolol, terbutaline, tetrahydrozoline (tetryzoline), theodrenaline, tioconazole, tocainide, trihexyphenidyl, trimeprazine (alimemazine), trimipramine, tropicamide, verapamil, zopiclone

KEY WORDS
coated capillary

REFERENCE
Koppenhoefer,B.; Eperlein,U.; Schlunk,R.; Zhu,X.; Lin,B. Separation of enantiomers of drugs by capillary electrophoresis. V. Hydroxypropyl-α-cyclodextrin as chiral solvating agent, *J.Chromatogr.A*, **1998**, *793*, 153–164.

SAMPLE
Matrix: solutions

CAPILLARY ELECTROPHORESIS
Capillary: 44 cm × 50 μm fused-silica
Running buffer: 10 mM pH 4.3 acetic acid/ammonium acetate buffer containing 200 μg/mL carboxymethyl ether-β-cyclodextrin (Wacker Chemie, Munich)
Injection: Hydrostatic injection at 10 cm for 5-10 s.
Detector: MS, Finnigan LCQ, electrospray, positive ion mode, ion-spray tip voltage 2.6 kV, sheath liquid MeOH:water:acetic acid 50:49:1 at 6 μL/min, m/z 293

Migration time: 3.8, 4.2 (enantiomers)
Voltage: 20 kV

KEY WORDS
chiral

REFERENCE
Schulte,G.; Heitmeier,S.; Chankvetadze,B.; Blaschke,G. Chiral capillary electrophoresis-electrospray mass
spectrometry coupling with charged cyclodextrin derivatives as chiral selectors, *J.Chromatogr.A*, **1998**, *800*,
77–82.

SAMPLE
Matrix: urine
Sample preparation: Inject an aliquot directly.

CAPILLARY ELECTROPHORESIS
Capillary: 40 cm × 50 μm fused-silica (Grom, Herrenbrerg, Germany)
Capillary temperature: 20
Running buffer: 100 mM pH 3.2 Phosphate buffer
Injection: Low pressure injection for 1 s (ca. 1 nL).
Detector: UV 200
Migration time: 6
Voltage: 400 V/cm
Model: Beckman P/ACE 2100

OTHER SUBSTANCES
Extracted: metabolites

KEY WORDS
comparison with HPLC; rat; direct injection

REFERENCE
Prien,D.; Blaschke,G. Studies of the metabolism of the antihistaminic drug dimethindene by high-performance
liquid chromatography and capillary electrophoresis including enantioselective aspects, *J.Chromatogr.B*,
1997, *688*, 309–318.

Dinoprost

Molecular formula: $C_{20}H_{34}O_5$
Molecular weight: 354.49
CAS Registry No.: 551-11-1, 38562-01-5 (tromethamine salt)
Merck Index (12th ed.): 8065
Lednicer: 1 27

SAMPLE
Matrix: semen
Sample preparation: Mix semen with an equal volume of isopropanol, store at -20°, centrifuge
at 14000 rpm, filter (0.22 μm) the supernatant, adjust the pH of the filtrate to 3 with 100 mM
HCl, add to a C18 SPE cartridge, wash with 3 mL water, wash with 3 mL hexane, elute with
3 mL methyl acetate. Concentrate the eluate to 300 μL, inject an aliquot.

CAPILLARY ELECTROPHORESIS
Capillary: 47 cm × 50 μm fused-silica (40 cm to detector) (Polymicro Technologies)
Capillary temperature: 30
Running buffer: 38 mM pH 10 Borax/NaOH buffer containing 50 mM sodium dodecyl sulfate
Injection: Pressure injection at 0.5 psi for 3 s.
Detector: UV 200

Migration time: 6.25
Voltage: 20 kV
Model: Beckman P/ACE 2000
Limit of detection: 6.5-22.5 pg

OTHER SUBSTANCES
Extracted: alprostadil, dinoprostone, prostaglandin A_2, prostaglandin B_2, prostaglandin D_2, prostaglandin $F_{1\alpha}$, thromboxane B_2

KEY WORDS
SPE

REFERENCE
Hsieh,Y.-Z.; Kuo,K.-L. Analysis of prostaglandins by micellar electrokinetic capillary chromatography, *J.Chromatogr.A*, **1997**, *761*, 277–284.

Dinoprostone

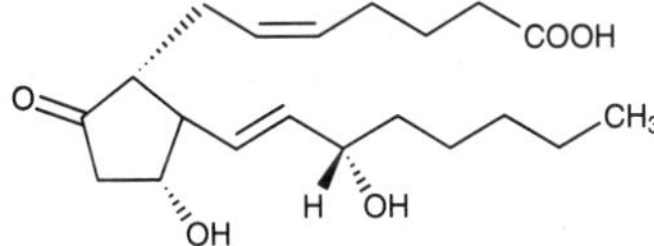

Molecular formula: $C_{20}H_{32}O_5$
Molecular weight: 352.47
CAS Registry No.: 363-24-6
Merck Index (12th ed.): 8064
Lednicer: 1 30

SAMPLE
Matrix: semen
Sample preparation: Mix semen with an equal volume of isopropanol, store at -20°, centrifuge at 14000 rpm, filter (0.22 µm) the supernatant, adjust the pH of the filtrate to 3 with 100 mM HCl, add to a C18 SPE cartridge, wash with 3 mL water, wash with 3 mL hexane, elute with 3 mL methyl acetate. Concentrate the eluate to 300 µL, inject an aliquot.

CAPILLARY ELECTROPHORESIS
Capillary: 47 cm × 50 µm fused-silica (40 cm to detector) (Polymicro Technologies)
Capillary temperature: 30
Running buffer: 38 mM pH 10 Borax/NaOH buffer containing 50 mM sodium dodecyl sulfate
Injection: Pressure injection at 0.5 psi for 3 s.
Detector: UV 200
Migration time: 5.46
Voltage: 20 kV
Model: Beckman P/ACE 2000
Limit of detection: 6.5-22.5 pg

OTHER SUBSTANCES
Extracted: alprostadil, dinoprost, prostaglandin A_2, prostaglandin B_2, prostaglandin D_2, prostaglandin $F_{1\alpha}$, thromboxane B_2

KEY WORDS
SPE

REFERENCE
Hsieh,Y.-Z.; Kuo,K.-L. Analysis of prostaglandins by micellar electrokinetic capillary chromatography, *J.Chromatogr.A*, **1997**, *761*, 277–284.

Diphenhydramine

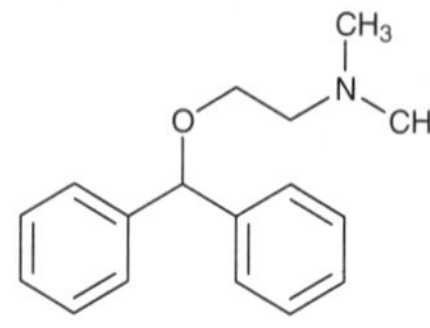

Molecular formula: $C_{17}H_{21}NO$
Molecular weight: 255.36
CAS Registry No.: 58-73-1, 147-24-0 (HCl), 88637-37-0 (citrate)
Merck Index (12th ed.): 3367
Lednicer: 1 41

SAMPLE
Matrix: blood, hair, urine
Sample preparation: Serum. 1 mL Serum + 100 μL MeOH + 200 μL concentrated ammonia adjusted to pH 9-10 + 3 mL n-hexane, shake for 10 min, centrifuge at 3000 g for 10 min. Remove the supernatant and evaporate it to dryness under a stream of nitrogen at 60°, reconstitute with 200 μL MeOH, inject an aliquot. Urine. 1 mL Urine + 100 μL MeOH + 200 μL concentrated ammonia adjusted to pH 9-10 + 3 mL n-hexane, shake for 10 min, centrifuge at 3000 g for 10 min. Remove the supernatant and evaporate it to dryness under a stream of nitrogen at 60°, reconstitute with 1 mL MeOH, inject an aliquot. Hair. Wash hair with MeOH and acetone. 50 mg Washed hair + 4 mL MeOH, sonicate for 5 h, centrifuge at 3000 g for 10 min. Remove the supernatant and evaporate it to dryness under a stream of nitrogen at 60°, reconstitute with 1 mL MeOH, inject an aliquot.

CAPILLARY ELECTROPHORESIS
Capillary: 47 cm × 50 μm fused-silica (40 cm to detector) (Beckman)
Capillary preparation: Before each run rinse with 100 mM NaOH for 2 min and with rinsing buffer for 2 min. Rinse new capillaries with 100 mM NaOH for 20 min. (Rinsing buffer was MeOH:100 mM pH 2.3 phosphate buffer:100 mM pH 2.3 phosphate buffer containing 10 mM heptakis-(2,6-di-O-methyl)-β-cyclodextrin 10:70:20.)
Capillary temperature: 20
Running buffer: MeOH:100 mM pH 2.3 phosphate buffer 10:90
Injection: Electrokinetic injection at 10 kV for 4 s.
Detector: UV 200
Migration time: 19
Internal standard: diphenhydramine
Voltage: 20 kV
Model: Beckman P/ACE 5510

OTHER SUBSTANCES
Extracted: methadone
Simultaneous: aminoflunitrazepam, amphetamine, benzoylecgonine, bromazepam, cocaine, codeine, N-desmethyldiazepam, diazepam, dihydrocodeine, 3,4-methylenedioxyamphetamine, 3,4-methylenedioxyethylamphetamine, 3,4-methylenedioxymethamphetamine, morphine, nicotine
Noninterfering: demethyldiazepam, flunitrazepam, 3-hydroxydiazepam, oxazepam

KEY WORDS
serum; diphenhydramine is IS

REFERENCE
Frost,M.; Köhler,H.; Blaschke,G. Enantioselective determination of methadone and its main metabolite 2-ethylidene-1,5-dimethyl-3,3-diphenylpyrrolidine (EDDP) in serum, urine and hair by capillary electrophoresis, *Electrophoresis*, **1997**, *18*, 1026–1034.

SAMPLE
Matrix: bulk
Sample preparation: Dilute with 400 μg/mL n-propyl p-hydroxybenzoate in running buffer

CAPILLARY ELECTROPHORESIS
Capillary: 27 cm × 50 μm fused-silica (20 cm to detector)
Capillary preparation: Before each run rinse with running buffer at high pressure.

Capillary temperature: 30
Running buffer: MeCN:water 15:85 containing 40 mM sodium dodecyl sulfate, 8.5 mM sodium borate, and 8.5 mM sodium phosphate, pH 8.5
Injection: Inject at high pressure for 1 s
Detector: UV 214
Migration time: 4.4
Internal standard: n-propyl p-hydroxybenzoate (1.8)
Voltage: 20 kV
Model: Beckman P/ACE System 2100

OTHER SUBSTANCES
Simultaneous: acetaminophen, acetylcodeine, O-acetylmorphine, aspirin, caffeine, cocaine, codeine, diamorphine, hydromorphone, isoniacinamide, lidocaine, methaqualone, morphine, niacinamide, noscapine, papaverine, phenacetin, phenobarbital, phenylpropanolamine, procaine, quinine, salicylic acid, strychnine, thebaine

REFERENCE
Walker,J.A.; Krueger,S.T.; Lurie,I.S.; Marché,H.L.; Newby,N. Analysis of heroin drug seizures by micellar electrokinetic capillary chromatography (MECC), *J.Forensic Sci.*, **1995**, *10*, 6–9.

SAMPLE
Matrix: formulations
Sample preparation: Inject an aliquot of a solution in MeOH:water 60:40.

CAPILLARY ELECTROPHORESIS
Capillary: 27.5 × 50 μm fused-silica (20.5 cm to detector)
Capillary preparation: Between runs flush capillary with 100 mM NaOH for 2 min, with water for 1 min, and with running buffer for 3 min. At the end of each day flush with 100 mM NaOH for 5 min and with water for 5 min. At the start of each day flush with 100 mM NaOH for 5 min, with water for 5 min, and with running buffer for 5 min.
Running buffer: MeCN:50 mM pH 3 NaH_2PO_4 50:50
Injection: Hydrodynamic injection at 50 mbar for 10 s.
Detector: UV 210
Migration time: 3.5
Internal standard: diphenhydramine
Voltage: 15 kV
Model: Hewlett-Packard HP3D

OTHER SUBSTANCES
Simultaneous: benzalkonium chloride

KEY WORDS
diphenhydramine is IS

REFERENCE
Taylor,R.B.; Toasaksiri,S.; Reid,R.G. Determination of antibacterial quaternary ammonium compounds in lozenges by capillary electrophoresis, *J.Chromatogr.A*, **1998**, *798*, 335–343.

Diphenoxylate

Molecular formula: $C_{30}H_{32}N_2O_2$
Molecular weight: 452.60
CAS Registry No.: 915-30-0, 3810-80-8 (HCl)
Merck Index (12th ed.): 3371
Lednicer: 1 302

SAMPLE
Matrix: solutions

CAPILLARY ELECTROPHORESIS
Capillary: 59.6 cm × 75 μm fused-silica (52.1 cm to detector) (Polymicro Technologies)
Capillary preparation: Purge with running buffer before each run. At the beginning of each day purge using 50-60 kPa vacuum with 500 mM NaOH for 5 min, with water for 5 min, with MeCN for 5 min, and with running buffer for 5 min.
Running buffer: MeCN:MeOH:acetic acid 49:50:1 containing 20 mM ammonium acetate
Injection: Hydrostatic injection at 10 cm for 5 s.
Detector: UV 214
Migration time: 5.53
Voltage: 30 kV
Model: Waters Quanta 4000

OTHER SUBSTANCES
Simultaneous: ethoheptazine, fentanyl, levallorphan, meperidine, methadone, nalorphine, pentazocine, nikethamide

REFERENCE
Leung,G.N.W.; Tang,H.P.O.; Tso,T.S.C.; Wan,T.S.M. Separation of basic drugs with non-aqueous capillary electrophoresis, *J.Chromatogr.A*, **1996**, *738*, 141–154.

Dipivefrin

Molecular formula: $C_{19}H_{29}NO_5$
Molecular weight: 351.44
CAS Registry No.: 52365-63-6, 64019-93-8 (HCl)
Merck Index (12th ed.): 3400
Lednicer: 3 22

SAMPLE
Matrix: solutions
Sample preparation: Inject an aliquot of a 100 μg/mL solution in running buffer.

CAPILLARY ELECTROPHORESIS
Capillary: 29 cm × 50 μm fused-silica (24.5 cm to detector) (Yongnian Optical Conductive Fiber Plant, China), coated with polyacrylamide
Capillary preparation: No details of the polyacrylamide coating process are provided. However, another paper (LC.GC 1997, 15, 40) by this group indicates that they use the procedure of Hjertén, thus: Adjust the pH of 20 mL water to 3.5 with acetic acid, add 80 μL 3-(trimethoxysilyl)propyl methacrylate (3-methacryloxypropyltrimethoxysilane), mix, suck into capillary, let stand at room temperature for 1 h, remove the solution, wash with water. Fill the capillary with a deaerated 3-4% acrylamide solution containing 1 μL/mL N,N,N',N'-tetramethylethylenediamine and 1 mg/mL potassium persulfate, let stand for 30 min, remove excess solution by aspiration, rinse with water, remove water by aspiration, dry at 35° (J. Chromatogr. 1985, 347, 191).
Capillary temperature: 25
Running buffer: 100 mM NaH_2PO_4 adjusted to pH 2.5
Injection: Electrokinetic injection at 15 kV for 3 s.
Detector: UV 200, UV 210
Migration time: 6.10
Voltage: 15 kV
Model: Bio-Rad BioFocus 3000

OTHER SUBSTANCES
Simultaneous: albuterol, alprenolol, atenolol, baclofen, bamethan, benproperine, benserazide, bisoprolol, bupranolol, butamirate, butethamate, carburerol, celiprolol, clenbuterol, clobutinol, isoproterenol (isoprenaline), metaproterenol (orciprenaline), metipranolol, metoprolol, norfenefrine, ornidazole, oxprenolol, phenylpropanolamine, pholedrine, pirbuterol, prilocaine, procyclidine, sotalol, synephrine, terbutaline, tocainide

KEY WORDS
coated capillary

REFERENCE
Koppenhoefer,B.; Epperlein,U.; Xiaofeng,Z.; Bingcheng,L. Separation of enantiomers of drugs by capillary electrophoresis. Part 4: Hydroxypropyl-γ-cyclodextrin as chiral solvating agent, *Electrophoresis*, **1997**, *18*, 924–930.

SAMPLE
Matrix: solutions

CAPILLARY ELECTROPHORESIS
Capillary: 29-36 cm × 50 μm fused-silica (24.5-31.5 cm to detector) (Yongnian Optical Conductive Fiber Plant, China) coated with polyacrylamide
Capillary preparation: Coat capillary as follows. Adjust the pH of 20 mL water to 3.5 with acetic acid, add 80 μL 3-(trimethoxysilyl)propyl methacrylate (3-methacryloxypropyltrimethoxysilane), mix, suck into capillary, let stand at room temperature for 1 h, remove the solution, wash with water. Fill the capillary with a deaerated 3-4% acrylamide solution containing 1 μL/mL N,N,N',N'-tetramethylethylenediamine and 1 mg/mL potassium persulfate, let stand for 30 min, remove excess solution by aspiration, rinse with water, remove water by aspiration, dry at 35° (J. Chromatogr. 1985, 347, 191).
Capillary temperature: 25
Running buffer: 100 mM pH 2.5 NaH_2PO_4 (A) or 100 mM pH 2.5 NaH_2PO_4 containing 45 mM hydroxypropyl-α-cyclodextrin (Wacker, Munich) (B)
Injection: Electromigration at 15 kV for 3 s.
Detector: UV 200; UV 210
Migration time: 6.10 (A); 10.37 (B) (no separation of enantiomers)
Voltage: 15 kV
Model: Bio-Focus 3000

OTHER SUBSTANCES
Also analyzed: albuterol (salbutamol), alprenolol, amorolfine, atenolol, atropine, azelastine, baclofen, bamethan, benproperine, benserazide, biperiden, bisoprolol, brompheniramine, bupivacaine, bupranolol, butamirate, butethamate, carazolol, carbuterol, carteolol, carvedilol, celiprolol, chloroquine, chlorpheniramine, chlorphenoxamine, cicletanine, clenbuterol, clidinium bromide, clobutinol, dimethindene, disopyramide, dobutamine, doxylamine, fendiline, flecainide, gallopamil, homatropine, ipratropium bromide, isoproterenol (isoprenaline), isothipendyl, ketamine, meclizine, mefloquine, mepindolol, mequitazine, metaclazepam, metaproterenol (orciprenaline), metipranolol, metoprolol, nafronyl (naftidrofuryl), nefopam, nicardipine, norfenefrine, ofloxacin, ornidazole, orphenadrine, oxomemazine, oxprenolol, oxybutynin, phenoxybenzamine, phenylpropanolamine, pholedrine, pindolol, pirbuterol, prilocaine, procyclidine, promethazine, propafenone, propranolol, reproterol, sotalol, sulpride, synephrine, talinolol, terbutaline, tetrahydrozoline (tetryzoline), theodrenaline, tioconazole, tocainide, trihexyphenidyl, trimeprazine (alimemazine), trimipramine, tropicamide, verapamil, zopiclone

KEY WORDS
coated capillary

REFERENCE
Koppenhoefer,B.; Eperlein,U.; Schlunk,R.; Zhu,X.; Lin,B. Separation of enantiomers of drugs by capillary electrophoresis. V. Hydroxypropyl-α-cyclodextrin as chiral solvating agent, *J.Chromatogr.A*, **1998**, *793*, 153–164.

Dipyrone

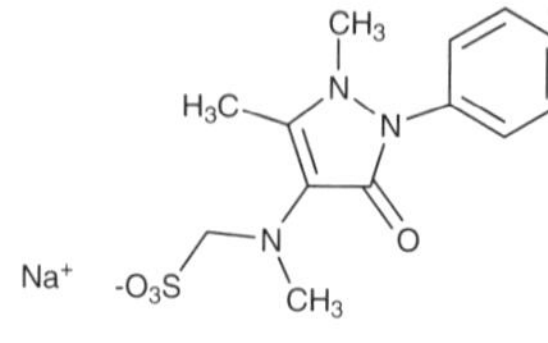

Molecular formula: $C_{13}H_{16}N_3NaO_4S$
Molecular weight: 333.34
CAS Registry No.: 68-89-3, 5907-38-0 (monohydrate)
Merck Index (12th ed.): 3414
Lednicer: 2 262

SAMPLE

Matrix: formulations
Sample preparation: Grind granules, add 70 mL MeOH, warm at 40° with occasional shaking, cool, add 20 mL 5 mg/mL methyl p-hydroxybenzoate in MeOH, make up to 100 mL with water, filter (0.45 μm), inject an aliquot.

CAPILLARY ELECTROPHORESIS

Capillary: 65 cm × 50 μm fused-silica (SGE)
Running buffer: 20 mM NaH_2PO_4 containing 100 mM sodium cholate, adjusted to pH 9.0 with 20 mM sodium tetraborate
Injection: Siphon for 10 s at 10 cm height
Detector: UV 210
Migration time: 12
Internal standard: methyl p-hydroxybenzoate (16)
Voltage: +20 kV

OTHER SUBSTANCES

Simultaneous: acetaminophen, caffeine, chlorpheniramine, dibucaine, ethenzamide, guaifenesin, isopropylantipyrine, naproxen, noscapine, phenacetin, tipepidine, trimetoquinol, triprolidine

KEY WORDS

granules

REFERENCE

Nishi,H.; Fukuyama,T.; Matsuo,M.; Terabe,S. Separation and determination of the ingredients of a cold medicine by micellar electrokinetic chromatography with bile salts, *J.Chromatogr.*, **1990**, *498*, 313–323.

SAMPLE

Matrix: formulations
Sample preparation: Grind tablet or powders, add 70 mL MeOH, add 20 mL 5 mg/mL methyl p-hydroxybenzoate in MeOH, sonicate for 10 min, shake for 5 min, make up to 100 mL with water, filter (0.45 μm), inject an aliquot.

CAPILLARY ELECTROPHORESIS

Capillary: 65 cm × 50 μm fused-silica (50 cm to detector) (SGE)
Running buffer: 20 mM NaH_2PO_4 containing 100 mM sodium dodecyl sulfate, adjusted to pH 9.0 with 20 mM sodium tetraborate
Injection: Siphon
Detector: UV 210
Migration time: 8.5
Internal standard: methyl p-hydroxybenzoate
Voltage: +20 kV

OTHER SUBSTANCES

Simultaneous: acetaminophen, caffeine, chlorpheniramine, ethenzamide, guaifenesin, isopropylantipyrine, naproxen, noscapine, phenacetin, tipepidine, trimetoquinol

KEY WORDS

tablets; powders

REFERENCE
Nishi,H.; Fukuyama,T.; Matsuo,M.; Terabe,S. Effect of surfactant structures on the separation of cold medicine ingredients by micellar electrokinetic chromatography, *J.Pharm.Sci.*, **1990**, *79*, 519–523.

Disopyramide

Molecular formula: $C_{21}H_{29}N_3O$
Molecular weight: 339.48
CAS Registry No.: 3737-09-5, 22059-60-5 (phosphate)
Merck Index (12th ed.): 3424
Lednicer: 2 81

SAMPLE
Matrix: solutions

CAPILLARY ELECTROPHORESIS
Capillary: 72 cm × 50 μm silica (50 cm to detector) (Applied Biosystems)
Capillary preparation: Before each run wash capillary with 100 mM NaOH for 3 min, wash with running buffer for 3 min, aspirate neutral marker solution (2 drops DMSO in 10 mL water) for 1 s, place capillary end in buffer vial for 5 s, inject sample. Wash capillary at the beginning of each day by passing 1 M NaOH through for 20 min.
Capillary temperature: 30
Running buffer: 30 mM pH 9.3 Borate buffer containing 30 mM sodium dodecyl sulfate
Injection: Inject by applying vacuum for 1 s (2-3 nL)
Detector: UV 200
Migration time: 13
Voltage: +30 kV
Current: 64 μA
Model: Applied Biosystems Model 270A

OTHER SUBSTANCES
Simultaneous: N-acetylprocainamide, chlorodisopyramide, procainamide

REFERENCE
Evenson,M.A.; Wiktorowicz,J.E. Automated capillary electrophoresis applied to therapeutic drug monitoring, *Clin.Chem.*, **1992**, *38*, 1847–1852.

SAMPLE
Matrix: solutions
Sample preparation: Inject an aliquot of a 100 μg/mL solution in 50 mM sodium dodecyl sulfate.

CAPILLARY ELECTROPHORESIS
Capillary: 57 cm × 75 μm fused-silica (50 cm to detector) (Beckman)
Capillary preparation: Rinse capillary with 100 mM NaOH and running buffer between each run.
Capillary temperature: 30
Running buffer: Isopropanol:100 mM pH 8.1 borate buffer containing 50 mM sodium dodecyl sulfate 10:90
Injection: Pressure injection for 5 s
Detector: UV 200
Migration time: 10.5
Voltage: +25 kV
Model: Beckman P/ACE 5510

OTHER SUBSTANCES
Simultaneous: amiodarone, bretylium, lidocaine, phenytoin, propafenone, quinidine

REFERENCE

Bretnall,A.E.; Clarke,G.S. Investigation and optimisation of the use of organic modifiers in micellar electrokinetic chromatography, *J.Chromatogr.A*, **1995**, *716*, 49–55.

SAMPLE

Matrix: solutions
Sample preparation: Inject an aliquot of a 100 µg/mL solution in water:running buffer 50:50.

CAPILLARY ELECTROPHORESIS

Capillary: 44.5 cm × 50 µm acrylamide-coated fused-silica (Bio-Rad)
Capillary temperature: 30
Running buffer: 100 mM NaH_2PO_4 containing 15 mM gamma-cyclodextrin, adjusted to pH 2.5 with phosphoric acid
Injection: Electrokinetic injection at 8 kV for 6 s.
Detector: UV 200
Migration time: 8.80
Voltage: 14 kV
Model: Bio-Rad BioFocus 3000

OTHER SUBSTANCES

Also analyzed: albuterol, alprenolol, atenolol, atropine, baclofen, bamethan, benserazide, biperiden, bisoprolol, bupivacaine, bupranolol, butetamate, carazolol, carbuterol, carvedilol, celiprolol, chloroquine, chlorpheniramine (chlorphenamine), clidinium bromide, clobutinol, dobutamine, flecainide, homatropine, ipratropium bromide, isoproterenol, isothipendyl, ketamine, mefloquine, mequitazine, metaproterenol (orciprenaline), metipranolol, nafronyl (naftidrofuryl), nefopam, ofloxacin, orphenadrine, oxomemazine, oxprenolol, phenoxybenzamine, pholedrine, pindolol, pirbuterol, prilocaine, promethazine, propafenone, propranolol, sotalol, synephrine, terbutaline, tetrahydrozoline (tetryzoline), tocainide, trihexyphenidyl, trimeprazine (alimemazine), trimipramine, tropicamide, verapamil, zopiclone

KEY WORDS

coated capillary; achiral

REFERENCE

Koppenhoefer,B.; Epperlein,U.; Christian,B.; Yibing,J.; Yuying,C.; Bingcheng,L. Separation of enantiomers of drugs by capillary electrophoresis. I. γ-Cyclodextrin as chiral solvating agent, *J.Chromatogr.A*, **1995**, *717*, 181–190.

SAMPLE

Matrix: solutions

CAPILLARY ELECTROPHORESIS

Capillary: 57 cm × 75 µm fused-silica (50 cm to detector) (Beckman)
Capillary preparation: Before each run rinse capillary with 100 mM NaOH and running buffer.
Capillary temperature: 30
Running buffer: Acetone:100 mM pH 8.1 borate buffer containing 50 mM sodium dodecyl sulfate 15:85 (A) or isopropanol:100 mM pH 8.1 borate buffer containing 50 mM sodium dodecyl sulfate 10:90 (B)
Injection: Pressure injection for 5-10 s.
Detector: UV 200
Migration time: 8.3 (A), 14.5 (B)
Voltage: 25 kV
Model: Beckman P/ACE 5510

OTHER SUBSTANCES

Simultaneous: acebutolol, amiodarone, atenolol, bretylium, captopril, diltiazem, lidocaine, lisinopril, metoprolol, nicardipine, nifedipine, phenytoin, pindolol, propafenone, propranolol, quinidine, sotalol, timolol, p-toluenesulfonic acid, verapamil

KEY WORDS

only running buffer A separates all compounds

REFERENCE
Bretnall,A.E.; Clarke,G.S. Selectivity of capillary electrophoresis for the analysis of cardiovascular drugs, *J.Chromatogr.A*, **1996**, *745*, 145–154.

SAMPLE
Matrix: solutions

CAPILLARY ELECTROPHORESIS
Capillary: 52 cm × 75 μm fused-silica (48 cm to detector) (Polymicro Technologies)
Capillary preparation: Before each run purge with running buffer for 3 min. Every 3 runs purge with 100 mM NaOH for 5 min. Purge new capillaries with 1 M NaOH for 20 min and with 100 mM NaOH for 20 min, rinse with running buffer, equilibrate with running buffer at 12 kV for 3 h.
Running buffer: 100 mM CHES (2-(N-cyclohexylamine)ethanesulfonic acid) containing 10 mM triethylamine and 25 mM (R)-dodecoxycarbonylvaline (Waters EnantioSelect (R)-Val-1), pH adjusted to 8.8 with 1 M NaOH
Injection: Hydrostatic injection for 2 s.
Detector: UV 214
Migration time: 13.3, 13.4 (enantiomers)
Voltage: 12 kV
Current: ≤30 μA
Model: Waters Quanta 4000

OTHER SUBSTANCES
Simultaneous: bupivacaine, clenbuterol, metoprolol, norphenylephrine, octopamine

KEY WORDS
chiral; separation of enantiomers is poor

REFERENCE
Peterson,A.G.; Ahuja,E.S.; Foley,J.P. Enantiomeric separations of basic pharmaceutical drugs by micellar electrokinetic chromatography using a chiral surfactant, N-dodecoxycarbonylvaline, *J.Chromatogr.B*, **1996**, *683*, 15–28.

SAMPLE
Matrix: solutions
Sample preparation: Inject an aliquot of a solution in running buffer.

CAPILLARY ELECTROPHORESIS
Capillary: 60 cm × 75 μm fused-silica (52.4 cm to detector)
Capillary preparation: After each run flush with 500 mM KOH for 2-3 min then with water.
Running buffer: 10 mM pH 3.8 Phosphate buffer containing 2% sulfated cyclodextrin (ds 7-10)
Injection: Hydrostatic injection.
Detector: UV 214
Migration time: 20.83, 22.64 (enantiomers)
Voltage: 15 kV
Model: Waters Quanta 4000

OTHER SUBSTANCES
Also analyzed: acebutolol, alprenolol, aminoglutethimide, brompheniramine, bupivacaine, bupropion, canadine, carbinoxamine, chloroquine, chlorpheniramine, dimethindene, doxylamine, hydroxychloroquine, idazoxan, isoxsuprine, ketamine, mepenzolate, mepivacaine, methoxyphenamine, mexiletine, midodrine, nefopam, orphenadrine, oxprenolol, oxyphencyclimine, pheniramine, phensuximide, pindolol, piperoxan, terbutaline, tetramisole, tolperisone, tranylcypromine, trihexyphenidyl, trimipramine, verapamil, warfarin

KEY WORDS
chiral; detector at anode

REFERENCE
Stalcup,A.M.; Gahm,K.H. Application of sulfated cyclodextrins to chiral separations by capillary zone electrophoresis, *Anal.Chem.*, **1996**, *68*, 1360–1368.

SAMPLE
Matrix: solutions
Sample preparation: Inject an aliquot of a 10 μg/mL solution of disopyramide in 5 mM pH 3.97 acetate buffer.

CAPILLARY ELECTROPHORESIS
Capillary: 27 cm × 50 μm coated fused-silica (20 cm to detector) (Polymicro Technologies)
Capillary preparation: Between runs wash with 10% acetic acid for 3 min, with 7 M urea in water for 3 min, with 10% sodium dodecyl sulfate in acetic acid:water 10:90 (pH 2.8) for 3 min, with water for 2 min, then with running buffer for 2 min, equilibrate with applied voltage for 1 min, and inject 750 μM α_1-acid glycoprotein (Sigma) in running buffer at 0.5 psi for 1 min. Store in 10% acetic acid overnight, rinse with water each morning. Coat capillary as follows. Wash capillary with chloroform by suction, rinse with acetone, rinse with water, treat with 100 mM NaOH for 5 min, treat with 100 mM HCl for 5 min, rinse with water, rinse with acetone, fill with 5% 3-glycidoxypropyltrimethoxysilane in chloroform, let stand overnight, wash with acetone, drain, fill with 0.25% methyl cellulose (viscosity 7000 cps, Dow Chemical) in water, heat horizontally at 120° for 50 min, flush with acetone, fill with 3% boron trifluoride in chloroform, let stand for 1 h, rinse with acetone, rinse with water, store in 10% acetic acid in water (J. Capillary Electrophoresis 1995, 2, 191).
Capillary temperature: 25.0
Running buffer: 50 mM pH 3.97 Acetate buffer
Injection: Pressure injection at 0.5 psi for 5 s.
Detector: UV 214
Migration time: 5.13, 5.25 (enantiomers)
Voltage: 12 kV
Current: 40 μA
Model: Beckman P/ACE 2050

KEY WORDS
chiral; coated capillary

REFERENCE
Amini,A.; Pettersson,C.; Westerlund,D. Enantioresolution of disopyramide by capillary affinity electrokinetic chromatography with human α1-acid glycoprotein (AGP) as chiral selector applying a partial filling technique, *Electrophoresis*, **1997**, *18*, 950–957.

SAMPLE
Matrix: solutions

CAPILLARY ELECTROPHORESIS
Capillary: 37 cm × 50 μm fused-silica (30 cm to detector) (Polymicro Technologies)
Capillary temperature: 15
Running buffer: 100 mM pH 4.9 Phosphate buffer containing 15 mM carboxymethylated-β-cyclodextrin (substitution 2-3) (Cyclolab, Budapest)
Injection: Hydrodynamic pressure injection for 2 s.
Detector: UV 214
Migration time: 7.61, 7.85 (enantiomers)
Voltage: 30 kV
Model: Beckman P/ACE 2100

KEY WORDS
chiral; comparison with other substituted cyclodextrins

REFERENCE
Juvancz,Z.; Markides,K.E.; Jicsinszky,L. Enantiomer separation of disopyramide with capillary electrophoresis using various cyclodextrins, *Electrophoresis*, **1997**, *18*, 1002–1006.

SAMPLE
Matrix: solutions
Sample preparation: Inject an aliquot of a 100 μg/mL solution in running buffer.

CAPILLARY ELECTROPHORESIS
Capillary: 30 cm × 50 μm fused-silica (25.5 cm to detector) (Yongnian Optical Conductive Fiber Plant, China), coated with polyacrylamide
Capillary preparation: No details of the polyacrylamide coating process are provided. However, another paper (LC.GC 1997, 15, 40) by this group indicates that they use the procedure of Hjertén, thus: Adjust the pH of 20 mL water to 3.5 with acetic acid, add 80 μL 3-(trimethoxysilyl)propyl methacrylate (3-methacryloxypropyltrimethoxysilane), mix, suck into capillary, let stand at room temperature for 1 h, remove the solution, wash with water. Fill the capillary with a deaerated 3-4% acrylamide solution containing 1 μL/mL N,N,N',N'-tetramethylethylenediamine and 1 mg/mL potassium persulfate, let stand for 30 min, remove excess solution by aspiration, rinse with water, remove water by aspiration, dry at 35° (J. Chromatogr. 1985, 347, 191).
Capillary temperature: 25
Running buffer: 100 mM NaH_2PO_4 adjusted to pH 2.5
Injection: Electrokinetic injection at 15 kV for 3 s.
Detector: UV 200, UV 210
Migration time: 3.96
Voltage: 15 kV
Model: Bio-Rad BioFocus 3000

OTHER SUBSTANCES
Simultaneous: amorolfine, brompheniramine, bupivacaine, carteolol, chloroquine, chlorpheniramine, chlorphenoxamine, dobutamine, doxylamine, flecainide, gallopamil, ketamine, mepindolol, orphenadrine, oxybutynin, phenoxybenzamine, pindolol, propafenone, propranolol, sulpiride, talinolol, tropicamide, verapamil

KEY WORDS
coated capillary

REFERENCE
Koppenhoefer,B.; Epperlein,U.; Xiaofeng,Z.; Bingcheng,L. Separation of enantiomers of drugs by capillary electrophoresis. Part 4: Hydroxypropyl-γ-cyclodextrin as chiral solvating agent, *Electrophoresis*, **1997**, *18*, 924–930.

SAMPLE
Matrix: solutions
Sample preparation: Inject an aliquot of a 100 μg/mL solution in running buffer.

CAPILLARY ELECTROPHORESIS
Capillary: 30 cm × 50 μm fused-silica (25.5 cm to detector), coated with polyacrylamide
Capillary preparation: Adjust the pH of 20 mL water to 3.5 with acetic acid, add 80 μL 3-(trimethoxysilyl)propyl methacrylate (3-methacryloxypropyltrimethoxysilane), mix, suck into capillary, let stand at room temperature for 1 h, remove the solution, wash with water. Fill the capillary with a deaerated 3-4% acrylamide solution containing 1 μL/mL N,N,N',N'-tetramethylethylenediamine and 1 mg/mL potassium persulfate, let stand for 30 min, remove excess solution by aspiration, rinse with water, remove water by aspiration, dry at 35° (J. Chromatogr. 1985, 347, 191).
Capillary temperature: 25
Running buffer: 100 mM NaH_2PO_4 containing 45 mM heptakis(2,6-di-O-methyl)-β-cyclodextrin, adjusted to pH 2.5 with phosphoric acid
Injection: Electrokinetic injection at 15 kV for 3 s.
Detector: UV 200
Voltage: 15 kV
Model: Bio-Rad BioFocus 3000

KEY WORDS
chiral; coated capillary; comparison with the use of other cyclodextrins; this running buffer gave the greatest enantiomeric separation.; α=1.017

REFERENCE
Lin,B.; Zhu,X.; Koppenhoefer,B.; Epperlein,U. Investigation of 123 chiral drugs by cyclodextrin-modified capillary electrophoresis, *LC.GC*, **1997**, *15*, 40–46.

SAMPLE
Matrix: solutions

CAPILLARY ELECTROPHORESIS
Capillary: 70 cm × 50 μm fused-silica (50 cm to detector) (Polymicro Technologies)
Capillary temperature: 30
Running buffer: 100 mM pH 4.1 Sodium acetate containing 3.0 mM sulfobutyl ether-β-cyclodextrin (average substitution 3.9, MW 1721, Center for Drug Delivery Research, Lawrence KS)
Injection: Pressure injection at 5 inches Hg for 1.5 s.
Detector: UV 210
Migration time: 9.5, 10 (enantiomers)
Voltage: 15 kV
Current: 26 μA
Model: Perkin Elmer-Applied Biosystems Model 270

KEY WORDS
chiral

REFERENCE
Xie,G.-h.; Skanchy,D.J.; Stobaugh,J.F. Chiral separations of enantiomeric pharmaceuticals by capillary electrophoresis using sulphobutyl ether β-cyclodextrin as isomer selector, *Biomed.Chromatogr.*, **1997**, *11*, 193–199.

SAMPLE
Matrix: solutions

CAPILLARY ELECTROPHORESIS
Capillary: 29-36 cm × 50 μm fused-silica (24.5-31.5 cm to detector) (Yongnian Optical Conductive Fiber Plant, China) coated with polyacrylamide
Capillary preparation: Coat capillary as follows. Adjust the pH of 20 mL water to 3.5 with acetic acid, add 80 μL 3-(trimethoxysilyl)propyl methacrylate (3-methacryloxypropyltrimethoxysilane), mix, suck into capillary, let stand at room temperature for 1 h, remove the solution, wash with water. Fill the capillary with a deaerated 3-4% acrylamide solution containing 1 μL/mL N,N,N',N'-tetramethylethylenediamine and 1 mg/mL potassium persulfate, let stand for 30 min, remove excess solution by aspiration, rinse with water, remove water by aspiration, dry at 35° (J. Chromatogr. 1985, 347, 191).
Capillary temperature: 25
Running buffer: 100 mM pH 2.5 NaH_2PO_4 (A) or 100 mM pH 2.5 NaH_2PO_4 containing 45 mM hydroxypropyl-α-cyclodextrin (Wacker, Munich) (B)
Injection: Electromigration at 15 kV for 3 s.
Detector: UV 200; UV 210
Migration time: 3.96 (A); 6.37 (B) (no separation of enantiomers)
Voltage: 15 kV
Model: Bio-Focus 3000

OTHER SUBSTANCES
Also analyzed: albuterol (salbutamol), alprenolol, amorolfine, atenolol, atropine, azelastine, baclofen, bamethan, benproperine, benserazide, biperiden, bisoprolol, brompheniramine, bupivacaine, bupranolol, butamirate, butethamate, carazolol, carbuterol, carteolol, carvedilol, celiprolol, chloroquine, chlorpheniramine, chlorphenoxamine, cicletanine, clenbuterol, clidinium bromide, clobutinol, dimethindene, dipivefrin, dobutamine, doxylamine, fendiline, flecainide, gallopamil, homatropine, ipratropium bromide, isoproterenol (isoprenaline), isothipendyl, ketamine, meclizine, mefloquine, mepindolol, mequitazine, metaclazepam, metaproterenol (orciprenaline), metipranolol, metoprolol, nafronyl (naftidrofuryl), nefopam, nicardipine, norfenefrine, ofloxacin, ornidazole, orphenadrine, oxomemazine, oxprenolol, oxybutynin, phenoxybenzamine, phenylpropanolamine, pholedrine, pindolol, pirbuterol, prilocaine, procyclidine, promethazine, propafenone, propranolol, reproterol, sotalol, sulpride, synephrine, talinolol, terbutaline, tetrahydrozoline (tetryzoline), theodrenaline, tioconazole, tocainide, trihexyphenidyl, trimeprazine (alimemazine), trimipramine, tropicamide, verapamil, zopiclone

---

KEY WORDS
coated capillary

REFERENCE
Koppenhoefer,B.; Eperlein,U.; Schlunk,R.; Zhu,X.; Lin,B. Separation of enantiomers of drugs by capillary electrophoresis. V. Hydroxypropyl-α-cyclodextrin as chiral solvating agent, *J.Chromatogr.A*, **1998**, *793*, 153–164.

SAMPLE
Matrix: urine
Sample preparation: Inject directly.

CAPILLARY ELECTROPHORESIS
Capillary: 64 cm × 75 μm (48 cm to detector) (Polymicro Technologies)
Running buffer: MeOH:buffer 2:98 (Buffer was 32 mM pH 4.5 6-aminocaproic acid buffer containing 18 mM adipic acid, 4% dextran (MW 515000), 0.04% polyethylene oxide (MW 300000), and 7.5 mM heptakis(2,6-di-O-methyl)-β-cyclodextrin
Injection: Hydrodynamic injection at 20 cm for 9 min
Detector: UV 220
Migration time: 15 (2 peaks seen)
Voltage: 22 kV
Model: Laboratory constructed

OTHER SUBSTANCES
Extracted: famotidine

KEY WORDS
chiral

REFERENCE
Soini,H.; Riekkola,M.-L.; Novotny,M.V. Mixed polymer networks in the direct analysis of pharmaceuticals in urine by capillary electrophoresis, *J.Chromatogr.A*, **1994**, *680*, 623–634.

Dobutamine

Molecular formula: $C_{18}H_{23}NO_3$

Molecular weight: 301.39

CAS Registry No.: 34368-04-2, 49745-95-1 (HCl), 104564-71-8 (lactobionate), 101626-66-8 (tartrate)

Merck Index (12th ed.): 3456

Lednicer: 2 53

SAMPLE
Matrix: solutions
Sample preparation: Inject an aliquot of a 100 μg/mL solution in water:running buffer 50:50.

CAPILLARY ELECTROPHORESIS
Capillary: 44.5 cm × 50 μm acrylamide-coated fused-silica (Bio-Rad)
Capillary temperature: 30
Running buffer: 100 mM NaH_2PO_4 containing 15 mM gamma-cyclodextrin, adjusted to pH 2.5 with phosphoric acid
Injection: Electrokinetic injection at 8 kV for 6 s.
Detector: UV 200
Migration time: 11.12
Voltage: 14 kV
Model: Bio-Rad BioFocus 3000

OTHER SUBSTANCES
Also analyzed: albuterol, alprenolol, atenolol, atropine, baclofen, bamethan, benserazide, biperiden, bisoprolol, bupivacaine, bupranolol, butetamate, carazolol, carbuterol, carvedilol, celiprolol, chloroquine, chlorpheniramine (chlorphenamine), clidinium bromide, clobutinol, disopyramide, flecainide, homatropine, ipratropium bromide, isoproterenol, isothipendyl, ketamine, mefloquine, mequitazine, metaproterenol (orciprenaline), metipranolol, nafronyl (naftidrofuryl), nefopam, ofloxacin, orphenadrine, oxomemazine, oxprenolol, phenoxybenzamine, pholedrine, pindolol, pirbuterol, prilocaine, promethazine, propafenone, propranolol, sotalol, synephrine, terbutaline, tetrahydrozoline (tetryzoline), tocainide, trihexyphenidyl, trimeprazine (alimemazine), trimipramine, tropicamide, verapamil, zopiclone

KEY WORDS
coated capillary; achiral

REFERENCE
Koppenhoefer,B.; Epperlein,U.; Christian,B.; Yibing,J.; Yuying,C.; Bingcheng,L. Separation of enantiomers of drugs by capillary electrophoresis. I. γ-Cyclodextrin as chiral solvating agent, *J.Chromatogr.A*, **1995**, *717*, 181–190.

SAMPLE
Matrix: solutions
Sample preparation: Inject an aliquot of a 100 μg/mL solution in running buffer.

CAPILLARY ELECTROPHORESIS
Capillary: 30 cm × 50 μm fused-silica (25.5 cm to detector) (Yongnian Optical Conductive Fiber Plant, China), coated with polyacrylamide
Capillary preparation: No details of the polyacrylamide coating process are provided. However, another paper (LC.GC 1997, 15, 40) by this group indicates that they use the procedure of Hjertén, thus: Adjust the pH of 20 mL water to 3.5 with acetic acid, add 80 μL 3-(trimethoxysilyl)propyl methacrylate (3-methacryloxypropyltrimethoxysilane), mix, suck into capillary, let stand at room temperature for 1 h, remove the solution, wash with water. Fill the capillary with a deaerated 3-4% acrylamide solution containing 1 μL/mL N,N,N',N'-tetramethylethylenediamine and 1 mg/mL potassium persulfate, let stand for 30 min, remove excess solution by aspiration, rinse with water, remove water by aspiration, dry at 35° (J. Chromatogr. 1985, 347, 191).
Capillary temperature: 25
Running buffer: 100 mM NaH_2PO_4 adjusted to pH 2.5 (A) or 100 mM NaH_2PO_4 containing 45 mM hydroxypropyl-gamma-cyclodextrin, adjusted to pH 2.5 (B)
Injection: Electrokinetic injection at 15 kV for 3 s.
Detector: UV 200, UV 210
Migration time: 5.34 (A), 10.08, 10.17 (B, enantiomers)
Voltage: 15 kV
Model: Bio-Rad BioFocus 3000

OTHER SUBSTANCES
Simultaneous: amorolfine, brompheniramine, bupivacaine, carteolol, chloroquine, chlorpheniramine, chlorphenoxamine, disopyramide, doxylamine, flecainide, gallopamil, ketamine, mepindolol, orphenadrine, oxybutynin, phenoxybenzamine, pindolol, propafenone, propranolol, sulpiride, talinolol, tropicamide, verapamil

KEY WORDS
coated capillary; chiral

REFERENCE
Koppenhoefer,B.; Epperlein,U.; Xiaofeng,Z.; Bingcheng,L. Separation of enantiomers of drugs by capillary electrophoresis. Part 4: Hydroxypropyl-γ-cyclodextrin as chiral solvating agent, *Electrophoresis*, **1997**, *18*, 924–930.

SAMPLE
Matrix: solutions
Sample preparation: Inject an aliquot of a 100 μg/mL solution in running buffer.

CAPILLARY ELECTROPHORESIS
Capillary: 30 cm × 50 μm fused-silica (25.5 cm to detector), coated with polyacrylamide
Capillary preparation: Adjust the pH of 20 mL water to 3.5 with acetic acid, add 80 μL 3-(trimethoxysilyl)propyl methacrylate (3-methacryloxypropyltrimethoxysilane), mix, suck into capillary, let stand at room temperature for 1 h, remove the solution, wash with water. Fill the capillary with a deaerated 3-4% acrylamide solution containing 1 μL/mL N,N,N',N'-tetramethylethylenediamine and 1 mg/mL potassium persulfate, let stand for 30 min, remove excess solution by aspiration, rinse with water, remove water by aspiration, dry at 35° (J. Chromatogr. 1985, 347, 191).
Capillary temperature: 25
Running buffer: 100 mM NaH_2PO_4 containing 45 mM hydroxypropyl-gamma- cyclodextrin, adjusted to pH 2.5 with phosphoric acid
Injection: Electrokinetic injection at 15 kV for 3 s.
Detector: UV 200
Voltage: 15 kV
Model: Bio-Rad BioFocus 3000

KEY WORDS
chiral; coated capillary; comparison with the use of other cyclodextrins; this running buffer gave the greatest enantiomeric separation.; α=1.009

REFERENCE
Lin,B.; Zhu,X.; Koppenhoefer,B.; Epperlein,U. Investigation of 123 chiral drugs by cyclodextrin-modified capillary electrophoresis, *LC.GC*, **1997**, *15*, 40–46.

SAMPLE
Matrix: solutions

CAPILLARY ELECTROPHORESIS
Capillary: 29-36 cm × 50 μm fused-silica (24.5-31.5 cm to detector) (Yongnian Optical Conductive Fiber Plant, China) coated with polyacrylamide
Capillary preparation: Coat capillary as follows. Adjust the pH of 20 mL water to 3.5 with acetic acid, add 80 μL 3-(trimethoxysilyl)propyl methacrylate (3-methacryloxypropyltrimethoxysilane), mix, suck into capillary, let stand at room temperature for 1 h, remove the solution, wash with water. Fill the capillary with a deaerated 3-4% acrylamide solution containing 1 μL/mL N,N,N',N'-tetramethylethylenediamine and 1 mg/mL potassium persulfate, let stand for 30 min, remove excess solution by aspiration, rinse with water, remove water by aspiration, dry at 35° (J. Chromatogr. 1985, 347, 191).
Capillary temperature: 25
Running buffer: 100 mM pH 2.5 NaH_2PO_4 (A) or 100 mM pH 2.5 NaH_2PO_4 containing 45 mM hydroxypropyl-α-cyclodextrin (Wacker, Munich) (B)
Injection: Electromigration at 15 kV for 3 s.
Detector: UV 200; UV 210
Migration time: 5.34 (A); 9.42 (B) (no separation of enantiomers)
Voltage: 15 kV
Model: Bio-Focus 3000

OTHER SUBSTANCES
Also analyzed: albuterol (salbutamol), alprenolol, amorolfine, atenolol, atropine, azelastine, baclofen, bamethan, benproperine, benserazide, biperiden, bisoprolol, brompheniramine, bupivacaine, bupranolol, butamirate, butethamate, carazolol, carbuterol, carteolol, carvedilol, celiprolol, chloroquine, chlorpheniramine, chlorphenoxamine, cicletanine, clenbuterol, clidinium bromide, clobutinol, dimethindene, dipivefrin, disopyramide, doxylamine, fendiline, flecainide, gallopamil, homatropine, ipratropium bromide, isoproterenol (isoprenaline), isothipendyl, ketamine, meclizine, mefloquine, mepindolol, mequitazine, metaclazepam, metaproterenol (orciprenaline), metipranolol, metoprolol, nafronyl (naftidrofuryl), nefopam, nicardipine, norfenefrine, ofloxacin, ornidazole, orphenadrine, oxomemazine, oxprenolol, oxybutynin, phenoxybenzamine, phenylpropanolamine, pholedrine, pindolol, pirbuterol, prilocaine, procyclidine, promethazine, propafenone, propranolol, reproterol, sotalol, sulpride, synephrine, talinolol, terbutaline, tetrahydrozoline (tetryzoline), theodrenaline, tioconazole, tocainide, trihexyphenidyl, trimeprazine (alimemazine), trimipramine, tropicamide, verapamil, zopiclone

KEY WORDS
coated capillary

REFERENCE
Koppenhoefer,B.; Eperlein,U.; Schlunk,R.; Zhu,X.; Lin,B. Separation of enantiomers of drugs by capillary electrophoresis. V. Hydroxypropyl-α-cyclodextrin as chiral solvating agent, *J.Chromatogr.A*, **1998**, *793*, 153–164.

Docetaxel

Molecular formula: $C_{43}H_{53}NO_{14}$
Molecular weight: 807.89
CAS Registry No.: 114977-28-5
Merck Index (12th ed.): 3458

SAMPLE
Matrix: blood, urine
Sample preparation: 500 μL Plasma + 50 μL MeOH + 450 μL water + 4 mL MTBE, shake for 2 min, store at -18° until the bottom layer is frozen. Remove the organic layer and evaporate it to dryness under a stream of nitrogen at 35°, reconstitute with 50 μL running buffer, inject an aliquot. Urine. 200 μL Urine + 50 μL 18.8 μg/mL MeOH + 750 μL water + 4 mL MTBE, shake for 2 min, store at -18° until the bottom layer is frozen. Remove the organic layer and evaporate it to dryness under a stream of nitrogen at 35°, reconstitute with 50 μL running buffer, inject an aliquot.

CAPILLARY ELECTROPHORESIS
Capillary: 67 cm × 50 μm fused-silica (60 cm to detector) (Hewlett-Packard)
Capillary preparation: After each run flush capillary with 100 mM NaOH for 2 min and with running buffer for 3 min.
Running buffer: MeCN:buffer 35:65 (Buffer was 25 mM pH 8.5 Tris-phosphate buffer containing 100 mM sodium dodecyl sulfate.)
Injection: Pressure injection at 0.5 psi for 12 s.
Detector: UV 230
Migration time: 17.5
Voltage: 28 kV
Model: Beckman P/ACE 5510

OTHER SUBSTANCES
Extracted: paclitaxel

KEY WORDS
docetaxel (taxotere) is IS; plasma

REFERENCE
Hempel,G.; Lehmkuhl,D.; Krümpelmann,S.; Blaschke,G.; Boos,J. Determination of paclitaxel in biological fluids by micellar electrokinetic chromatography, *J.Chromatogr.A*, **1996**, *745*, 173–179.

Dopamine

Molecular formula: $C_8H_{11}NO_2$
Molecular weight: 153.18
CAS Registry No.: 51-61-6, 62-31-7 (HCl)
Merck Index (12th ed.): 3479

SAMPLE
Matrix: blood
Sample preparation: Inject an aliquot of 5-fold diluted bovine serum directly.

CAPILLARY ELECTROPHORESIS
Capillary: 50 cm × 50 μm (20 cm to detector) (GL Science)
Capillary preparation: Rinse capillary with running buffer for 5 min before each run. At the beginning of each day rinse capillary with 100 mM NaOH for 10 min and with water for 5 min.
Running buffer: 10 mM pH 2.5 Phosphate buffer containing 25 mM sodium dodecyl sulfate and 12.5 mM Tween 20
Injection: Siphon at 10 cm for 10 s.
Detector: UV 210
Migration time: 6
Voltage: 18 kV
Current: 21-22 μA
Model: Jasco CE-800

OTHER SUBSTANCES
Extracted: epinephrine, levodopa, 3-methoxytyrosine, norepinephrine, phenylalanine, tyrosine

KEY WORDS
serum; cow; detector is at anode

REFERENCE
Esaka,Y.; Tanaka,K.; Uno,B.; Goto,M.; Kano,K. Sodium dodecyl sulfate-Tween 20 mixed micellar electrokinetic chromatography for separation of hydrophobic cations: Application to adrenaline and its precursors, *Anal.Chem.*, **1997**, *69*, 1332–1338.

SAMPLE
Matrix: CSF
Sample preparation: 720 μL CSF + 80 μL 100 nM dihydroxybenzylamine in 175 mM perchloric acid + 160 μL 500 mM pH 8.6 borate buffer + 80 μL 22.5 mM NaCN in water + 40 μL 5 mM naphthalene-2,3-dicarboxaldehyde in MeOH, mix, let stand at room temperature for 5 min, inject an aliquot.

CAPILLARY ELECTROPHORESIS
Capillary: 43 cm × 25 μm fused-silica (23 cm to detector) (Polymicro Technologies)
Capillary temperature: 25.5
Running buffer: 200 mM pH 7.05 ± 0.02 phosphate buffer
Injection: Hydrodynamic injection by vacuum (300 mm Hg) to give 2.5 nL sample injection followed by 60 pL 200 mM orthophosphoric acid
Detector: F ex 442 (He-Cd laser) em 490 (bandpass filter)
Migration time: 1.319
Internal standard: dihydroxybenzylamine (1.346)
Voltage: 29 kV
Model: Zeta Technology IRIS 2000
Limit of detection: 0.086 nM

OTHER SUBSTANCES
Extracted: norepinephrine
Noninterfering: amino acids

KEY WORDS
derivatization

REFERENCE
Bert,L.; Robert,F.; Denoroy,L.; Renaud,B. High-speed separation of subnanomolar concentrations of noradrenaline and dopamine using capillary zone electrophoresis with laser-induced fluorescence detection, *Electrophoresis*, **1996**, *17*, 523–525.

SAMPLE
Matrix: solutions

CAPILLARY ELECTROPHORESIS
Capillary: 87.9 cm × 26 μm fused-silica (Polymicro Technologies)
Running buffer: 20 mM pH 6.05 2-morpholinoethanesulfonic acid buffer
Injection: Inject at 5 kV for 2 s
Detector: E, 0.7 V vs sodium-saturated calomel electrode (details of detector construction in
    paper)
Migration time: 8
Voltage: 25 kV
Current: 2 μA
Limit of detection: 0.2-0.4 fmole (S/N 3)

OTHER SUBSTANCES
Simultaneous: catechol, epinephrine, norepinephrine

REFERENCE
Wallingford,R.A.; Ewing,A.G. Amperometric detection of catechols in capillary zone electrophoresis with normal
    and micellar solutions, *Anal.Chem.*, **1988**, *60*, 258–263.

SAMPLE
Matrix: solutions

CAPILLARY ELECTROPHORESIS
Capillary: 70 cm × 100 μm fused-silica (50 cm to detector) (Gasukuro Kogyo, Tokyo)
Running buffer: 100 mM KOH containing 200 mM boric acid
Injection: Hydrostatic injection for 10 s.
Detector: UV 217
Migration time: 13
Voltage: 9.5 kV
Current: 100 μA

OTHER SUBSTANCES
Simultaneous: epinephrine, isoproterenol, levodopa, metanephrine, norepinephrine, normetanephrine, vanillylmanedlic acid

REFERENCE
Tanaka,S.; Kaneta,T.; Yoshida,H. Separation of catecholamines by capillary zone electrophoresis using complexation with boric acid, *Anal.Sci.*, **1990**, *6*, 467–468.

SAMPLE
Matrix: solutions
Sample preparation: Inject an aliquot of a solution in running buffer.

CAPILLARY ELECTROPHORESIS
Capillary: 44 cm × 5 μm fused-silica (Polymicro Technologies)
Running buffer: Isopropanol:30 mM morpholinoethanesulfonic acid 20:80, pH adjusted to 6.0
Injection: Electromigration at 12 kV for 3 s
Detector: E, Pt electrode 0.700 V, Ag/AgCl reference electrode, details of cell in paper
Migration time: 7
Voltage: 25 kV

OTHER SUBSTANCES
Simultaneous: catechol, epinephrine, isoproterenol, norepinephrine, serotonin

REFERENCE
Chen,M.-C.; Huang,H.-J. An electrochemical cell for end-column amperometric detection in capillary electrophoresis, *Anal.Chem.*, **1995**, *67*, 4010–4014.

SAMPLE
Matrix: solutions
Sample preparation: Inject an aliquot of a solution in Ringer's solution.

CAPILLARY ELECTROPHORESIS
Capillary: 65 cm × 50 μm fused-silica (Polymicro Technologies)
Capillary preparation: Before each run flush capillary with 500 mM disodium EDTA (adjusted to pH 13 with solid NaOH) for 1 min and with running buffer for 1 min.
Running buffer: 100 mM pH 4.75 Lithium acetate containing 0.5 mM EDTA
Injection: Electrokinetic injection for 3 s at 20 kV
Detector: E, Bioanalytical Systems PA-1, 33 μm carbon fiber electrode +0.85 V, Ag/AgCl reference electrode (design of capillary/detector decoupler is given in paper)
Migration time: 4.1
Voltage: 20 kV

OTHER SUBSTANCES
Simultaneous: 3,4-dihydroxybenzylamine, isoproterenol, norepinephrine

REFERENCE
Park,S.; Lunte,C.E. A perfluorosulfonated ionomer end-column electrical decoupler for capillary electrophoresis/electrochemical detection, *Anal.Chem.*, **1995**, *67*, 4366–4370.

SAMPLE
Matrix: solutions

CAPILLARY ELECTROPHORESIS
Capillary: 100 cm × 50 μm fused-silica (Polymicro Technologies)
Running buffer: 20 mM pH 2.5 Sodium citrate
Injection: Electrokinetic injection.
Detector: E, Bioanalytical Systems BAS LC-4C, 25 μm gold wire electrode (design in paper) +800 mV, Pt auxiliary electrode, Ag/AgCl reference electrode
Migration time: 16
Voltage: 30 kV
Limit of quantitation: 500 nM

OTHER SUBSTANCES
Simultaneous: epinephrine, norepinephrine

REFERENCE
Zhong,M.; Lunte,S.M. Integrated on-capillary electrochemical detector for capillary electrophoresis, *Anal.Chem.*, **1996**, *68*, 2488–2493.

SAMPLE
Matrix: solutions

CAPILLARY ELECTROPHORESIS
Capillary: 60 cm × 100 μm (Polymicro Technologies)
Running buffer: 10 mM pH 9.5 Sodium tetraborate buffer
Injection: Electrokinetic injection for 5 s at 15 kV.
Detector: Chemiluminescence. The running buffer passed through a porous cellulose acetate joint (construction details in paper) and mixed with 1 mM potassium permanganate in 1 M sulfuric acid pumped at 2.5 μL/min in an 80 cm × 460 μm PTFE tube.
Migration time: 3.98

Voltage: 15 kV
Current: 56 μA
Model: laboratory constructed
Limit of detection: 100 μM

OTHER SUBSTANCES
Simultaneous: catechol, norepinephrine, serotonin

REFERENCE
Lee,Y.-T.; Whang,C.-W. Off-column chemiluminescence detection in capillary electrophoresis, *J.Chromatogr.A*, **1997**, *771*, 379–384.

SAMPLE
Matrix: solutions

CAPILLARY ELECTROPHORESIS
Capillary: 15 cm × 25 μm fused-silica (Polymicro Technologies)
Running buffer: pH 6.0 Phosphate buffer
Injection: Electrokinetic injection at 10 kV for 5 s.
Detector: E, Bioanalytical Systems Model LC-4B, gold working electrode deposited on the capillary tip (details in paper) +0.60 V, Ag/AgCl reference electrode, Pt wire counter electrode
Migration time: 1
Voltage: 10 kV
Limit of detection: 600 nM

OTHER SUBSTANCES
Simultaneous: catechol

REFERENCE
Voegel,P.D.; Zhou,W.; Baldwin,R.P. Integrated capillary electrophoresis/electrochemical detection with metal film electrodes directly deposited onto the capillary tip, *Anal.Chem.*, **1997**, *69*, 951–957.

SAMPLE
Matrix: solutions

CAPILLARY ELECTROPHORESIS
Capillary: 85 cm × 75 μm fused-silica (Composite Metal Services, Hallow, UK)
Running buffer: pH 10 Borate buffer containing 0.5 mM EDTA (Run with a pressure of 20 mbar at the injection end.)
Injection: Hydrodynamic injection at 40 mbar for 6 s.
Detector: E, Bioanalytical Systems Unijet cell +0.8 V, 1 mm dia. glassy carbon working electrode, Ag/AgCl reference electrode, stainless steel auxiliary electrode, two 16 μm gaskets (design details in paper)
Migration time: 5.8
Model: Prince
Limit of detection: 10-20 nM

OTHER SUBSTANCES
Simultaneous: epinephrine, norepinephrine

REFERENCE
Durgbanshi,A.; Kok,W.T. Capillary electrophoresis and electrochemical detection with a conventional detector cell, *J.Chromatogr.A*, **1998**, *798*, 289–296.

SAMPLE
Matrix: solutions

CAPILLARY ELECTROPHORESIS
Capillary: 40 cm × 25 μm fused-silica (Polymicro Technologies)

Capillary preparation: Before use rinse capillary with 100 mM NaOH for 5 min and with running buffer for 5 min.
Running buffer: 50 mM pH 7.0 Na_2HPO_4/$Na_2B_4O_7$ buffer
Injection: Electrokinetic injection at 5 kV for 10 s.
Detector: E, 25 μm AU electrode, Ag/AgCl reference electrode, stainless steel auxiliary electrode, prescan pulse -500 mV for 30 ms, postscan pulse 1000 mV for 30 s, dc ramp 100-500 mV in 56 ms, square wave 2000 Hz, 50 mV, quantitation method average peak current (first 30% of forward and reverse pulse current response rejected), 70 mV detection bandwith, 15 points included in running average, construction details in paper
Migration time: 7.5
Voltage: 30 kV
Limit of quantitation: 500 nM

OTHER SUBSTANCES
Simultaneous: epinephrine

REFERENCE
Gerhardt,G.C.; Cassidy,R.M.; Baranski,A.S. Square-wave voltammetry detection for capillary electrophoresis, *Anal.Chem.*, **1998**, *70*, 2167–2173.

SAMPLE
Matrix: solutions
Sample preparation: Inject an aliquot of an aqueous solution

CAPILLARY ELECTROPHORESIS
Capillary: 53 cm × 75 μm fused-silica (40.5 cm to detector) (Polymicro Technologies)
Running buffer: 5 mM pH 7.3 NaH_2PO_4 containing 5 mM sodium borate and 20 mM sodium dodecyl sulfate (A) or 10 mM pH 3.1 NaH_2PO_4 (B)
Injection: Electrokinetic injection at 17 kV for 9 s.
Detector: UV 204
Migration time: 6 (A), 7.5 (B)
Voltage: 17 kV
Current: 48 μA (A) or 26 μA (B)
Model: laboratory-constructed
Limit of detection: 820 nM

OTHER SUBSTANCES
Simultaneous: 3,4-dihydroxybenzylamine, epinephrine, norepinephrine

REFERENCE
Shakulashvili,N.; Finkler,C.; Engelhardt,H. Separation of catecholamines and serotonin by micellar electrokinetic chromatography with UV detection, *Chromatographia*, **1998**, *47*, 89–92.

SAMPLE
Matrix: solutions
Sample preparation: Mix 50 μL of a solution in 1% sodium chloride with 100 μL MeCN, vortex for 15 s, centrifuge at 14000 g for 20 s, inject an aliquot.

CAPILLARY ELECTROPHORESIS
Capillary: 40 cm × 50 μm silica
Running buffer: MeCN:buffer 20:80 (Buffer was 80 mM triethanolamine, pH adjusted to 7.2.)
Injection: Hydrodynamic injection at low pressure (to fill 10% of the capillary).
Detector: UV 214
Migration time: 4.5
Voltage: 10 kV
Model: Beckman Model 2000 CE

OTHER SUBSTANCES
Simultaneous: metanephrine, tyramine

REFERENCE

Shihabi,Z.K. Stacking of weakly cationic compounds by acetonitrile for capillary electrophoresis, *J.Chromatogr.A*, **1998**, *817*, 25–30.

SAMPLE

Matrix: solutrions

CAPILLARY ELECTROPHORESIS

Capillary: 50 cm × 75 μm fused-silica (36 cm to detector) (SGE)
Capillary preparation: Fill capillary with 100 mM KOH for 10 min, wash several times with running buffer.
Running buffer: 10 mM Tris adjusted to pH 6.4 with phosphoric acid
Injection: Siphon at 10 cm for 10 s
Detector: UV 206
Migration time: 3.2
Internal standard: dichloroisoproterenol (4.4)
Voltage: 16 kV
Model: Laboratory constructed

OTHER SUBSTANCES

Simultaneous: epinephrine, isoxsuprine, norepinephrine

REFERENCE

Fanali,S.; Cristalli,M.; Nardi,A.; Ossicini,L.; Shukla,S.K. Capillary zone electrophoresis in pharmaceutical analysis, *Farmaco*, **1990**, *45*, 693–702.

SAMPLE

Matrix: tissue
Sample preparation: Sonicate (Artix Sonic Dismembrator Model 150, power setting 30) 1-2 mg rat brain tissue and 10 μL EtOH:water 70:30 in an ice bath for 5-10 s, centrifuge at 16000 g for 10 min. 1 μL Supernatant + 1 μL 490 μM α-aminoadipic acid, mix, remove a 1 μL aliquot and add it to 5 μL 20 mM pH 9.0 sodium borate buffer, add 1.5 μL 20 mM NaCN in water, mix, add 1.5 μL 20 mM naphthalene-2,3-dicarboxaldehyde in MeCN, mix thoroughly, let stand at room temperature for 30 min (protect from light), inject an aliquot.

CAPILLARY ELECTROPHORESIS

Capillary: 115 cm × 50 μm fused-silica (95 cm to detector)
Running buffer: 20 mM pH 9.0 sodium borate buffer
Injection: Electrokinetic injection at 5 kV for 12 s
Detector: UV 420
Migration time: 13.8
Internal standard: α-aminoadipic acid (21.6)
Voltage: 30 kV

OTHER SUBSTANCES

Extracted: amino acids, gamma-aminobutyric acid, levodopa, phosphoethanolamine, norepinephrine, taurine

KEY WORDS

rat; brain; derivatization

REFERENCE

Weber,P.L.; O'Shea,T.J.; Lunte,S.M. Separation and quantitation of the amino acid neurotransmitters in rat brain by capillary electrophoresis, *J.Pharm.Biomed.Anal.*, **1994**, *12*, 319–324.

SAMPLE

Matrix: urine
Sample preparation: Condition a light alumina B SPE cartridge (Waters) with water. Adjust urine to pH 3 with 6 M HCl. 10 mL Acidified urine + 1 mL 200 mM EDTA + 100 μL 500 mM ascorbic acid, adjust pH to 8.5 with NaOH solution, add to the SPE cartridge, wash with two 2 mL portions of water, elute with 1 mL 100 mM HCl, inject an aliquot of the eluate.

CAPILLARY ELECTROPHORESIS
Capillary: 80 cm × 75 μm fused-silica (Composite Metal Services, Worcs., UK)
Running buffer: 30 mM pH 10 Borate buffer containing 0.5 mM EDTA and 1 mM ascorbic acid
Injection: Hydrodynamic injection at 40 mbar for 6 s.
Detector: F ex (Varian Schott glass UG11 filter) em 500 (cut-off filter) [ex 300 em 545] following
post-column reaction. The end of the capillary was connected to a 11.5 cm × 75 μm fused-silica
capillary by means of a porous PTFE sleeve with a narrow gap between the capillaries (de-
scribed in the paper). The detector was 5.5 cm after the junction. The porous sleeve was im-
mersed in a reagent reservoir. The reservoir was grounded and was under 30 mbar air pressure.
(Reagent was 2 mM terbium(III) chloride containing 2 (?) mM EDTA and 800 mM CsCl.)
Migration time: 5.5
Voltage: 30 kV
Model: Prince
Limit of detection: 130 nM

OTHER SUBSTANCES
Extracted: 3,4-dihydroxyphenylacetic acid, 3,4 dihydroxyphenyl glycol, epinephrine, levodopa,
norepinephrine

KEY WORDS
SPE; post-column reaction

REFERENCE
Zhu,R.; Kok,W.T. Determination of catecholamines and related compounds by capillary electrophoresis with
post-column terbium complexation and sensitized luminescence detection, *Anal.Chem.*, **1997**, *69*, 4010–
4016.

Doxapram

Molecular formula: $C_{24}H_{30}N_2O_2$
Molecular weight: 378.51
CAS Registry No.: 309-29-5, 7081-53-0 (HCl monohydrate),
113-07-5 (HCl)
Merck Index (12th ed.): 3488
Lednicer: 2 236

SAMPLE
Matrix: blood, urine
Sample preparation: Adjust pH of 2 mL urine or plasma to 10.5 with aqueous NaOH, extract
gently with chloroform:isopropanol 90:10. Remove the organic layer and evaporate it to dryness
under a stream of nitrogen, reconstitute the residue in 50 (urine) or 75 (plasma) μL running
buffer, filter, inject an aliquot.

CAPILLARY ELECTROPHORESIS
Capillary: 60 cm × 75 μm AccuSep uncoated silica (52.5 cm to detector) (Waters)
Capillary preparation: At the beginning of each day purge with 500 mM KOH for 5 min, with
water for 5 min, and with running buffer for 10 min.
Running buffer: 50 mM NaH_2PO_4 adjusted to pH 2.35 with phosphoric acid
Injection: Hydrostatic injection at 15 cm
Detector: UV 214
Migration time: 8.19
Voltage: 22 kV
Current: 135-145 μA
Model: Waters Quanta 4000

OTHER SUBSTANCES
Extracted: acepromazine, amphetamine, benzocaine, brompheniramine, butacaine, codeine, diazepam, lidocaine, medazepam, methamphetamine, methapyrilene, methaqualone, phenmetrazine, procaine, tetrahydrozoline

Simultaneous: meclizine

KEY WORDS
plasma

REFERENCE
Chee,G.L.; Wan,T.S.M. Reproducible and high-speed separation of basic drugs by capillary zone electrophoresis, *J.Chromatogr.*, **1993**, *612*, 172–177.

Doxepin

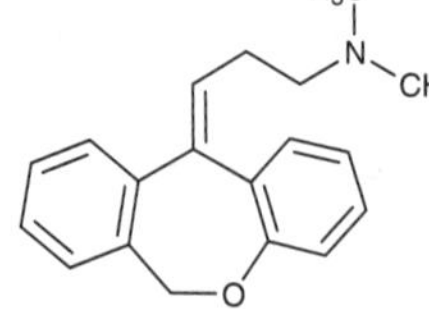

Molecular formula: $C_{19}H_{21}NO$
Molecular weight: 279.38
CAS Registry No.: 1668-19-5, 1229-29-4 (HCl), 4698-39-9 (E), 25127-31-5 (Z), 3607-18-9 (trans HCl)
Merck Index (12th ed.): 3492
Lednicer: 1 404

SAMPLE
Matrix: blood
Sample preparation: 1 mL Serum or plasma + 50 μL 10 μL/mL trimipramine + 1 mL 2 M NaOH + 5 mL hexane:isoamyl alcohol 99:1, vortex vigorously, centrifuge at 2500 g for 5 min. Remove 4 mL of the organic layer and add it to 100 μL 100 mM HCl, extract. Remove 60 μL of the aqueous layer and evaporate it to dryness under reduced pressure, reconstitute with 20 μL water, inject an aliquot.

CAPILLARY ELECTROPHORESIS
Capillary: 71.5 cm × 50 μm fused-silica (50 cm to detector) (Polymicro Technologies)
Capillary preparation: Before each run rinse capillary with 1 M NaOH for 3 min and running buffer for 3 min (use vacuum at 508 mm Hg).
Capillary temperature: 30 or 40
Running buffer: 37.5 mM pH 8.0 Phosphate buffer containing 2 M urea and 25 mM dodecyltrimethylammonium bromide
Injection: Vacuum injection at 508 mm Hg for 7 s
Detector: UV 254
Migration time: 10.5
Internal standard: trimipramine (11.5)
Voltage: -25 kV
Model: Applied Biosystems Model 270A
Limit of detection: 5-10 ng/mL

OTHER SUBSTANCES
Extracted: amitriptyline, desipramine, imipramine, nortriptyline

KEY WORDS
serum; plasma

REFERENCE
Lee,K.-J.; Lee,J.J.; Moon,D.C. Determination of tricyclic antidepressants in human plasma by micellar electrokinetic capillary chromatography, *J.Chromatogr.*, **1993**, *616*, 135–143.

SAMPLE
Matrix: solutions

CAPILLARY ELECTROPHORESIS
Capillary: 57 cm × 50 μm fused-silica (50 cm to detector) (Polymicro Technologies)

Capillary preparation: At the start of each day rinse capillary with water for 5 min, 100 mM NaOH for 30 min, with water for 10 min, and with running buffer for 15 min. Wash new capillaries with 1 M NaOH for 1 h.

Capillary temperature: 23

Running buffer: MeOH:50 mM pH 9.55 sodium phosphate buffer containing 0.06% poly(n-undecyl-α-D-glucopyranoside) (Prepare poly(n-undecyl-α-D-glucopyranoside) as follows. With exclusion of atmospheric moisture stir 40 mL alcohol-free chloroform, 5.2 g yellow mercuric oxide, 0.1 g mercuric bromide, 10 g anhydrous calcium sulfate, and 2.74 g 10-undecen-1-ol then add 10 g acetobromo-α-D-glucose (Fluka), stir at room temperature for 20 h, filter, wash the solid with chloroform. Wash the filtrate with aqueous 1 M KBr until no mercury salts are present in the aqueous layers, wash with later, concentrate. Take up the residue in benzene (Caution! Benzene is a carcinogen!) and filter it through activated silica, crystallize from n-hexane to obtain undecylenyl-2,3,4,6-tetra-O-acetyl-β-D-glucopyranoside. Add 2 mL of a 1 M solution of sodium methoxide in MeOH to a solution of 1 g of undecylenyl-2,3,4,6-tetra-O-acetyl-β-D-glucopyranoside in 50 mL dry MeOH, let stand until the reaction is complete (by TLC) to obtain undecylenyl-β-D-glucopyranoside (cf. Tenside Detergents 1978, 15, 72). Use Dowex 50W (H^+) resin to deionize the final product in solution. Polymerize the monomer by irradiation of an 870 μM solution in MeOH:water 20:80 with ^{60}Co gamma radiation for 48 h, purify the polymer solution by dialysis using a 1000 Da molecular mass cut-off membrane.)

Injection: Pressure injection at 0.5 psi for 5 s.

Detector: UV 214

Migration time: 13.5

Voltage: 22.8 kV

Model: Beckman P/ACE 5510

OTHER SUBSTANCES
Simultaneous: amitriptyline, desipramine, imipramine, nordoxepin, nortriptyline, protriptyline

---

REFERENCE
Harrell,C.W.; Dey,J.; Shamsi,S.A.; Foley,J.P.; Warner,I.M. Enhanced separation of antidepressant drugs using a polymerized nonionic surfactant as a transient capillary coating, *Electrophoresis*, **1998**, *19*, 712–718.

SAMPLE
Matrix: solutions
Sample preparation: Mix 50 μL of a solution in 1% sodium chloride with 100 μL MeCN, vortex for 15 s, centrifuge at 14000 g for 20 s, inject an aliquot.

CAPILLARY ELECTROPHORESIS
Capillary: 40 cm × 50 μm silica
Running buffer: MeCN:buffer 50:50 (Buffer was 60 mM triethanolamine containing 50 mM tricine, pH adjusted to 8.5.)
Injection: Hydrodynamic injection at low pressure for 70 s (to fill 12% of the capillary).
Detector: UV 254
Migration time: 6.5
Voltage: 11 kV
Model: Beckman Model 2000 CE

OTHER SUBSTANCES
Simultaneous: N-acetylprocainamide, quinine

REFERENCE
Shihabi,Z.K. Stacking of weakly cationic compounds by acetonitrile for capillary electrophoresis, *J.Chromatogr.A*, **1998**, *817*, 25–30.

SAMPLE
Matrix: urine
Sample preparation: 10 mL Urine + 1 mL 5 M NaOH + 10 mL n-hexane, vortex for 1 min, centrifuge at 0° at 3000 g for 5 min. Remove the organic layer and add it to 50 μL glacial acetic acid, evaporate to dryness under a stream of nitrogen at 30°, reconstitute the residue in 50 μL 50 mM sodium taurodeoxycholate, filter (0.2 μm PTFE), inject an aliquot.

CAPILLARY ELECTROPHORESIS
Capillary: 100 cm × 50 μm fused-silica (50 cm to detector) (ISCO)

Capillary preparation: Rinse capillary with 20 μL running buffer between injections. Condition a new capillary by filling with 1 M NaOH, let stand for 1 h, fill with 100 mM NaOH, let stand for 1 h, rinse with buffer. Every 20 injections rinse capillary with 200 μL 1 M NaOH, 200 μL water, and 200 μL running buffer, fill with running buffer.
Capillary temperature: 22
Running buffer: 40 mM pH 9.5 Borate buffer containing 10 mM sodium taurodeoxycholate
Injection: Load under vacuum at 7.5 kPa/s.
Detector: UV 240
Migration time: 9
Voltage: 30 kV
Model: ISCO Model 3140 electropherograph
Limit of detection: 4 ng/mL

OTHER SUBSTANCES
Extracted: acepromazine, amiodarone, amitriptyline, azaperone, chlorpromazine, cianopramine, clomipramine, clozapine, desethylamiodarone, desipramine, diclofensine, dothiepin, imipramine, isocarboxazid, moclobemide, perphenazine, phenothiazine, pimozide, prochlorperazine, promazine, thioridazine, thiothixene, trifluoperazine, trimipramine

KEY WORDS
human; cow; pig; horse; protect from light

REFERENCE
Aumatell,A.; Wells,R.J. Determination of a cardiac antiarrhythmic, tricyclic antipsychotics and antidepressants in human and animal urine by micellar electrokinetic capillary chromatography using a bile salt, *J.Chromatogr.B*, **1995**, *669*, 331–344.

Doxorubicin

Molecular formula: $C_{27}H_{29}NO_{11}$
Molecular weight: 543.53
CAS Registry No.: 23214-92-8, 25316-40-9 (HCl)
Merck Index (12th ed.): 3495

SAMPLE
Matrix: blood
Sample preparation: 900 μL Plasma + 100 μL 1 μg/mL daunorubicin + 2 mL chloroform, vortex for 1 min, centrifuge at 1000 g for 10 min. Remove 1.6 mL of the lower organic layer and add it to 100 μL 5 mM pH 2.3 phosphoric acid, vortex for 1 min, centrifuge at 1000 g for 10 min. Remove 50 μL of the upper aqueous layer and add it to 150 μL MeCN, mix, inject an aliquot of this mixture.

CAPILLARY ELECTROPHORESIS
Capillary: 70 cm × 75 μm fused-silica (65 cm to detector) (SGE)
Running buffer: MeCN:100 mM pH 4.2 sodium phosphate buffer 70:30
Injection: Electrokinetic injection at 12 kV for 5 s (14 nL)
Detector: F, laboratory constructed with an 80 mW 476.5 nm argon laser, 595 nm emission filter (10 nm bandwidth)
Migration time: 7.5
Internal standard: daunorubicin (7.0)
Voltage: 20-25 kV
Current: 35 μA (fixed)
Limit of detection: 35 pg/mL (S/N 3)

OTHER SUBSTANCES
Extracted: epirubicin

KEY WORDS
plasma

REFERENCE
Reinhoud,N.J.; Tjaden,U.R.; Irth,H.; van der Greef,J. Bioanalysis of some anthracyclines in human plasma by capillary electrophoresis with laser-induced fluorescence detection, *J.Chromatogr.*, **1992**, *574*, 327–334.

Doxycycline

Molecular formula: $C_{22}H_{24}N_2O_8$
Molecular weight: 444.44
CAS Registry No.: 564-25-0, 17086-28-1 (monohydrate), 24390-14-5 (HCl monohydrate), 83038-87- 3 (fosfatex), 24390-14-5 (hyclate)
Merck Index (12th ed.): 3496

SAMPLE
Matrix: blood, milk, urine
Sample preparation: Prepare a column as follows. Swirl Chelating Sepharose Fast Flow resin (Pharmacia) in its bottle, add it to a polypropylene column to give a bed volume of 1.0-1.2 mL, wash 3 times with 2 mL portions of water, wash with 2.5 mL 10 mM copper sulfate, wash with 2 mL water. Condition a PrepSep-C18 SPE cartridge (Fisher Scientific) by allowing 3 mL 4% dimethyldichlorosilane in toluene to drain through it by gravity, after 30 min rinse with 10 mL EtOH, immediately before use wash with 10 mL water. Milk. Centrifuge 10 mL milk at 4° at 2000 g for 20 min, remove the lower layer and add it to 20 mL succinate buffer, centrifuge at 4° at 2000 g for 30 min, pass the supernatant through a Bond Elut SPE cartridge, wash cartridge with 2 mL succinate buffer, collect all the eluate and add it to the column. Wash with 3 mL succinate buffer, wash with 3 mL water, wash with 3 mL MeOH, wash with 3 mL water, elute with 3 mL pH 7 buffer, when the blue band reaches the bottom of the column collect the eluate. Add the eluate to the SPE cartridge, wash with 10 mL water, dry with air, elute with 6 mL EtOH. Evaporate the eluate to dryness under a stream of nitrogen at room temperature, reconstitute with 20 μL running buffer adjusted to pH 7.0 with 5 M HCl, inject an aliquot. Serum. 12 mL Serum + 12 mL PBS, filter (Amicon Centriprep-10) centrifuge at 10° at 2000 g for 20 min, remove the filtrate, centrifuge an additional 4 times, combine the filtrates, add a 20 mL aliquot to the column. Wash with 3 mL succinate buffer, wash with 3 mL water, wash with 3 mL MeOH, wash with 3 mL water, elute with 3 mL pH 7 buffer, when the blue band reaches the bottom of the column collect the eluate. Add the eluate to the SPE cartridge, wash with 10 mL water, dry with air, elute with 6 mL EtOH. Evaporate the eluate to dryness under a stream of nitrogen at room temperature, reconstitute with 20 μL running buffer adjusted to pH 7.0 with 5 M HCl, inject an aliquot. Urine. Centrifuge 10 mL urine at 4° at 2000 g for 15 min, add the supernatant to the column. Wash with 3 mL succinate buffer, wash with 3 mL water, wash with 3 mL MeOH, wash with 3 mL water, elute with 3 mL pH 7 buffer, when the blue band reaches the bottom of the column collect the eluate. Add the eluate to the SPE cartridge, wash with 10 mL water, dry with air, elute with 6 mL EtOH. Evaporate the eluate to dryness under a stream of nitrogen at room temperature, reconstitute with 20 μL running buffer adjusted to pH 7.0 with 5 M HCl, inject an aliquot. (Prepare succinate buffer by dissolving 11.8 g succinic acid in 980 mL water, adjust pH to 4.0 with 10 M NaOH, make up to 1 L. Prepare PBS by dissolving 20 mG KCl, 8 g NaCl, 1.44 g Na_2HPO_4, and 24 mg KH_2PO_4 in 980 mL water, adjust pH to 7.0 with 1 M HCl, make up to 1 L. Prepare the pH 7 buffer by dissolving 2.31 g citric acid, 7.44 g disodium EDTA dihydrate, and 5.84 g NaCl in 190 mL water, adjust pH to 7.0 with 5 M NaOH, make up to 200 mL.)

CAPILLARY ELECTROPHORESIS
Capillary: 57 cm × 75 μm (50 cm to detector) (Beckman)
Capillary preparation: Before injection rinse with running buffer for 2 min. After each run rinse with NaOH solution for 2 min and with water for 2 min.
Capillary temperature: 23
Running buffer: 50 mM Borate buffer containing 50 mM phosphate and 10 mM sodium dodecyl sulfate (Prepare buffer by dissolving 2.884 g sodium dodecyl sulfate, 19.07 g sodium borate decahydrate, 4.096 g Na_2HPO_4, and 2.537 g NaH_2PO_4 in 980 mL water, adjust pH to 8.5 with 5 M HCl, make up to 1 L.)

Injection: Pressure injection for 5 s (5 nL)
Detector: UV 370
Migration time: 14.4
Voltage: 15 kV
Model: Beckman P/ACE 5510
Limit of quantitation: 4.6 ppm (milk), 3.4 ppm (serum), 5.2 ppm (urine)
Limit of detection: 2.8 ppm (milk), 2.5 ppm (serum), 3.2 ppm (urine)

OTHER SUBSTANCES
Extracted: chlortetracycline, oxytetracycline, tetracycline

KEY WORDS
serum; cow; SPE

REFERENCE
Chen,C.-L.; Gu,X. Determination of tetracycline residues in bovine milk, serum, and urine by capillary electrophoresis, *J.AOAC Int.*, **1995**, *78*, 1369–1377.

SAMPLE
Matrix: bulk
Sample preparation: Inject an aliquot of a 500 μg/mL solution in 10 mM HCl.

CAPILLARY ELECTROPHORESIS
Capillary: 44 cm × 50 μm fused-silica 38 cm to detector (Polymicro Technologies)
Capillary preparation: Wash capillary each day with 100 mM NaOH, 100 mM phosphoric acid, and 20 mM EDTA.
Capillary temperature: 15
Running buffer: 70 mM pH 10.50 Sodium carbonate containing 1 mM EDTA
Injection: Hydrodynamic injection for 4 s
Detector: UV 254
Migration time: 12.5
Voltage: 12 kV
Model: Thermo Separation Products Spectraphoresis 500
Limit of detection: 1 μg/mL

OTHER SUBSTANCES
Simultaneous: impurities, metacycline, oxytetracycline

REFERENCE
Van Schepdael,A.; Kibaya,R.; Roets,E.; Hoogmartens,J. Analysis of doxycycline by capillary electrophoresis, *Chromatographia*, **1995**, *41*, 367–369.

SAMPLE
Matrix: formulations
Sample preparation: Dissolve capsule contents in 15 mM pH 7.5 sodium phosphate buffer so as to make a 25 mM solution, centrifuge, dilute an aliquot of the supernatant to a concentration of 5 mM with 15 mM pH 7.5 sodium phosphate buffer, inject an aliquot.

CAPILLARY ELECTROPHORESIS
Capillary: 111.9 cm × 75 μm fused-silica (43.4 cm to detector) (Polymicro Technologies)
Capillary preparation: Condition by rinsing with 1 M NaOH for 10 min then running buffer overnight.
Capillary temperature: 25
Running buffer: pH 7.5 Phosphate buffer (buffer concentration 4.3 mM, total sodium concentration 15 mM, ionic strength 18.2 mM)
Injection: Hydrodynamic injection at 2 cm for 1 min
Detector: UV 260
Migration time: 5.1
Current: 20 μA
Limit of detection: 10 μM (S/N 3)

OTHER SUBSTANCES
Simultaneous: chlortetracycline, demeclocycline, methacycline, minocycline, oxytetracycline, tetracycline

KEY WORDS
capsules

REFERENCE
Tavares,M.F.M.; McGuffin,V.L. Separation and characterization of tetracycline antibiotics by capillary electrophoresis, *J.Chromatogr.A*, **1994**, *686*, 129–142.

SAMPLE
Matrix: solutions
Sample preparation: Inject an aliquot of a 100 μM solution in buffer. (Prepare buffer by mixing equal volumes of 20 mM NaH_2PO_4 and 20 mM sodium tetraborate and adjusting the pH to 10 with 1 M NaOH.)

CAPILLARY ELECTROPHORESIS
Capillary: 110 cm $\times$ 50 μm fused-silica (100 cm to detector)
Capillary preparation: Before injection rinse capillary with running buffer at 200 mbar with no voltage for 3 min and with running buffer at 4 mbar at 20 kV for 30 s.
Capillary temperature: 40
Running buffer: 100 mM pH 10 phosphate borate buffer containing 860 μM 1-methoxycarbonylindolizine-3,5-dicarbaldehyde (Prepare 1-methoxycarbonylindolizine-3,5-dicarbaldehyde as follows. Reflux 21.4 g 2-pyridinecarboxaldehyde, 24 mL ethylene glycol, 10 g p-toluenesulfonic acid, and 300 mL benzene (Caution! Benzene is a carcinogen!) under a Dean-Stark separator for 64 h, pour into concentrated sodium carbonate solution. Remove the organic layer and extract the aqueous layer 4 times with benzene. Combine the organic layers and wash them with water, dry over anhydrous magnesium sulfate, evaporate, distil the residue to give 2-(1,3-dioxolan-2-yl)pyridine (bp 122°/4 mm Hg) (J.Org.Chem. 1963, 28, 83). Reflux 15.1 g 2-(1,3-dioxolan-2-yl)pyridine and 19.5 g tert-butyl bromoacetate in 100 mL dry acetonitrile for 7 h, let stand overnight at room temperature, filter, wash the precipitate with diethyl ether to give 1-(tert-butoxycarbonylmethyl)-2-(1,3-dioxolan-2-yl)pyridinium bromide (mp 110-2° from MeCN). Suspend 51.9 g of this compound in 1. 5 L THF with stirring, add 62.1 g potassium carbonate, add 15.12 g methyl propiolate, stir at room temperature for 9 days, filter, evaporate the filtrate to dryness under reduced pressure, chromatograph the residue on silica gel with hexane:ethyl acetate 20:1-10:1, collect fractions and evaporate to dryness to give methyl 3-tert-butoxycarbonyl-5-(1,3-dioxolan-2-yl)indolizine-1-carboxylate (mp 138-9° from hexane). Reflux 20.82 g of this compound in 600 mL THF and 60 mL 10% HCl for 6 h, concentrate to one quarter of the original volume, add water, extract with chloroform. Wash the chloroform layer with water and dry it over anhydrous sodium sulfate, evaporate, chromatograph on silica gel with hexane:ethyl acetate 10:1 to give 1-methoxycarbonylindolizine-5-carbaldehyde (mp 135-7° from MeOH). Stir 12.18 g of this compound in 116 mL dry DMF at 0° under argon, add 17 mL phosphorus oxychloride, stir at room temperature for 1 h, pour into water, adjust pH to 9.0 with 5% potassium carbonate, extract with chloroform. Wash the organic layer with water and dry it over anhydrous sodium sulfate, concentrate until a precipitate forms, filter to obtain the product, concentrate the filtrate and chromatograph the residue on silica gel with hexane:ethyl acetate 5:1 to obtain more product. The product was 1-methoxycarbonylindolizine-3,5-dicarbaldehyde (mp 164-5° from methyl acetate) (J. Chromatogr. A 1996, 724, 169.))
Injection: Pressure injection at 50 mbar for 6 s (with no applied voltage).
Detector: UV 409
Migration time: 22
Voltage: 20 kV
Model: JASCO Model CE-990

OTHER SUBSTANCES
Simultaneous: amikacin, arbekacin, netilmicin

KEY WORDS
derivatization; on-column derivatization

REFERENCE
Oguri,S.; Fujiyoshi,T.; Miki,Y. In-capillary derivatization with 1-methoxycarbonylindolizine-3,5-dicarbaldehyde for high-performance capillary electrophoresis, *Analyst*, **1996**, *121*, 1683–1688.

SAMPLE
Matrix: solutions

CAPILLARY ELECTROPHORESIS
Capillary: 50 cm × 50 μm (50 cm to detector) (Polymicro Technologies)
Capillary preparation: Purge capillary with 10 mM NaOH for 2 min after each injection.
Running buffer: MeOH:buffer 40:60 (Buffer was 30 mM pH 3.0 citric acid containing 24.5 mM β-alanine.)
Injection: Electrokinetic injection at 10 kV for 5 s.
Detector: UV 254
Migration time: 7.7
Voltage: 30 kV
Model: Perkin Elmer/Applied Biosystems Model 270A-HT

OTHER SUBSTANCES
Simultaneous: chlortetracycline, meclocycline, minocycline, oxytetracycline, tetracycline
Interfering: methacycline

REFERENCE
Pesek,J.J.; Matyska,M.T. Separation of tetracyclines by high-performance capillary electrophoresis and capillary electrochromatography, *J.Chromatogr.A*, **1996**, *736*, 313–320.

SAMPLE
Matrix: solutions
Sample preparation: Inject an aliquot of a solution in MeOH.

CAPILLARY ELECTROPHORESIS
Capillary: 64 cm × 50 μm fused-silica (55.5 cm to detector) (Polymicro Technologies)
Capillary preparation: Flush capillary with running buffer for 2 min between runs. Before use rinse capillaries with 1 M NaOH for 1 h, with 100 mM NaOH for 20 min, with water for 20 min, with MeOH for 20 min, and with running buffer for 15 min.
Capillary temperature: 25
Running buffer: MeCN:MeOH:DMF 49:45:6 containing 25 mM ammonium acetate, 10 mM citric acid, and 118 mM methanesulfonic acid
Injection: Pressure injection of sample at 50 mbar for 1 s followed by pressure injection of running buffer at 50 mbar for 3 s.
Detector: UV 254
Migration time: 10.5
Voltage: 25 kV
Model: Hewlett-Packard HP[3D]

OTHER SUBSTANCES
Simultaneous: chlortetracycline, oxytetracycline, tetracycline

REFERENCE
Tjornelund,J.; Hansen,S.H. Determination of impurities in tetracycline hydrochloride by non-aqueous capillary electrophoresis, *J.Chromatogr.A*, **1996**, *737*, 291–300.

SAMPLE
Matrix: solutions

CAPILLARY ELECTROPHORESIS
Capillary: 43 cm × 75 μm (36 cm to the detector)
Capillary preparation: Before each run wash capillary with running buffer for 2 min. After each run wash capillary with water at 25° for 5 min, with 100 mM NaOH at 60 ° for 5 min, and with water at 25° for 5 min.
Capillary temperature: 25

Running buffer: 15 mM pH 6.5 Ammonium acetate buffer containing 20 mM sodium dodecyl
  sulfate and 0.135% Brij 35
Detector: UV 265
Migration time: 6
Voltage: 15 kV

OTHER SUBSTANCES
Simultaneous: chlortetracycline, demeclocycline, minocycline, oxytetracycline, tetracycline

REFERENCE
Chen,Y.-C.; Lin,C.-E. Migration behavior and separation of tetracycline antibiotics by micellar electrokinetic
  chromatography, *J.Chromatogr.A*, **1998**, *802*, 95–105.

Doxylamine

Molecular formula: $C_{17}H_{22}N_2O$
Molecular weight: 270.37
CAS Registry No.: 469-21-6, 562-10-7 (succinate)
Merck Index (12th ed.): 3497
Lednicer: 1 44

SAMPLE
Matrix: formulations
Sample preparation: Grind tablets, extract twice with MeOH by sonicating at low temperature,
  centrifuge, filter (0.45 μm) the supernatant, inject an aliquot of the filtrate.

CAPILLARY ELECTROPHORESIS
Capillary: 47 cm × 50 μm (40 cm to detector) (Polymicro Technologies)
Capillary temperature: 25
Running buffer: 50 mM pH 7.5 Borate buffer containing 50 mM phosphate, 10 mM sodium
  dodecyl sulfate, 10 mM β-cyclodextrin, and 10 mM tetrabutylammonium hydrogen sulfate
Injection: Inject at a pressure of 0.5 psi for 1 s (5.2 nL)
Detector: UV 214
Migration time: 3.7
Voltage: 21.5 kV
Model: Beckman P/ACE 2000
Limit of detection: 55.9 pg

OTHER SUBSTANCES
Simultaneous: chlorpheniramine, cyclizine, dimenhydrinate, methapyrilene, pheniramine, promethazine, thonylamine, triprolidine

KEY WORDS
tablets

REFERENCE
Ong,C.P.; Ng,C.L.; Lee,H.K.; Li,S.F.Y. Determination of antihistamines in pharmaceuticals by capillary electrophoresis, *J.Chromatogr.*, **1991**, *588*, 335–339.

SAMPLE
Matrix: formulations
Sample preparation: Dilute 10-fold, inject an aliquot.

CAPILLARY ELECTROPHORESIS
Capillary: 62 cm × 50 μm fused silica (50 cm to detector) (Polymicro Technologies)
Capillary temperature: 40

Running buffer: 50 mM pH 2.5 sodium phosphate buffer containing 20 mM β-cyclodextrin and
1 mM hexadecyltrimethylammonium bromide
Injection: Injection by gravity at 10 cm for 5 s
Detector: UV 214
Migration time: 7 (enantiomers are slightly separated)
Voltage: 20 kV
Current: 36 μA
Model: Laboratory constructed

OTHER SUBSTANCES
Simultaneous: acetaminophen, dextromethorphan, pseudoephedrine

KEY WORDS
syrup; chiral

REFERENCE
Quang,C.; Khaledi,M.G. Improved chiral separation of basic compounds in capillary electrophoresis using β-
cyclodextrin and tetraalkylammonium reagents, *Anal.Chem.*, **1993**, *65*, 3354–3358.

SAMPLE
Matrix: formulations
Sample preparation: Dilute formulation 50-fold, inject an aliquot.

CAPILLARY ELECTROPHORESIS
Capillary: 60 cm × 50 μm (52.5 cm to detector) (AccuSep)
Capillary preparation: Purge capillary with running buffer for 3 min between runs. Purge new
capillaries under vacuum with 500 mM NaOH for 10 min, with water for 10 min, and with
running buffer for 10 min.
Capillary temperature: 30
Running buffer: 25 mM pH 8.0 Na_2HPO_4/sodium tetraborate containing 50 mM (R)-N-dodecyl-
carbonylvaline (Prepare (R)-N-dodecoxycarbonylvaline as follows. Prepare dodecyl chlorofor-
mate by reacting 1-dodecanol with 0.33 equivalents of triphosgene in dichloromethane solution
in the presence of pyridine. Add dodecyl chloroformate dropwise to (R)-valine in 1 M NaOH
solution, filter, wash with hexane, recrystallize from ether/petroleum ether (J. Chromatogr. A
1994, 680, 125).)
Injection: Hydrostatic injection at 10 cm for 15 s.
Detector: UV 214
Migration time: 28.837
Voltage: 15 kV
Model: Waters Quanta 4000E

OTHER SUBSTANCES
Simultaneous: acetaminophen, dextromethorphan, pseudoephedrine, saccharin

KEY WORDS
achiral

REFERENCE
Swartz,M.E.; Mazzeo,J.R.; Grover,E.R.; Brown,P.R. Validation of enantiomeric separations by micellar electrokinetic capillary chromatography using synthetic chiral surfactants, *J.Chromatogr.A*, **1996**, *735*, 303–310.

SAMPLE
Matrix: solutions
Sample preparation: Prepare a solution in running buffer, inject an aliquot.

CAPILLARY ELECTROPHORESIS
Capillary: 60 cm × 75 μm fused-silica (52.4 cm to detector)
Capillary preparation: After each run flush with 500 mM KOH for 2-3 min, flush with water,
fill with running buffer
Running buffer: 10 mM Na_2HPO_4 containing 2% heparin sodium (MW 10000, 11% S, Scientific
Protein Laboratories, Waunakee, WI) adjusted to pH 5 with phosphoric acid

Injection: Hydrostatic injection
Detector: UV 214
Migration time: 17.5 (first enantiomer, R_s 2.4)
Model: Waters Quanta 4000

OTHER SUBSTANCES
Also analyzed: anabasine, brompheniramine, bupivacaine, carbinoxamine, chlorcyclizine, chloroquine, chlorpheniramine, dimethindene, enpiroline, halofantrine, hydroxychloroquine, indapamide, mefloquine, nornicotine, pheniramine, primaquine, promethazine, quinacrine, tetramisole

KEY WORDS
chiral

REFERENCE
Stalcup,A.M.; Agyei,N.M. Heparin: a chiral mobile-phase additive for capillary zone electrophoresis, *Anal.Chem.*, **1994**, *66*, 3054–3059.

SAMPLE
Matrix: solutions
Sample preparation: Prepare a solution in MeOH/water, inject an aliquot.

CAPILLARY ELECTROPHORESIS
Capillary: 62 cm × 52 μm fused-silica (50 cm to detector) (Polymicro Technologies)
Capillary temperature: 40
Running buffer: 50 mM pH 2.50 Tetrabutylammonium phosphate
Injection: Siphon injection at 10 cm for 5 s
Detector: UV (wavelength not specified)
Migration time: 7
Voltage: 20 kV
Current: 29 μA
Model: Laboratory constructed

OTHER SUBSTANCES
Simultaneous: epinephrine, imidazole, isoproterenol, 2-methylphenethylamine, 1-methylphenylpropylamine, metoprolol, nadolol, nicotine, norepinephrine, propranolol, pseudoephedrine

REFERENCE
Quang,C.; Khaledi,M.G. Extending the scope of chiral separation of basic compounds by cyclodextrin-mediated capillary zone electrophoresis, *J.Chromatogr.A*, **1995**, *692*, 253–265.

SAMPLE
Matrix: solutions

CAPILLARY ELECTROPHORESIS
Capillary: 67 cm × 75 μm (60 cm to detector) (Polymicro Technologies)
Capillary preparation: Coat capillary as follows. Treat overnight with 50% 3-(trimethoxysilyl)propyl methacrylate in MeOH, fill capillary with degassed 40 g/L (acrylamide + N,N'-methylenebisacrylamide) in 100 mM pH 8.2 Tris-borate buffer containing 2 mM EDTA, 5 μL/L 10% ammonium persulfate, and 5 μL/mL N,N,N',N'-tetramethylethylenediamine, after polymerization force out gel leaving coating sticking to wall (Caution! Acrylamide is a carcinogen!), rinse with water, remove water by aspiration, dry at 35°.
Running buffer: 20 mM pH 2.5 Citric acid containing 2% carboxymethyl β-cyclodextrin
Detector: UV 214
Migration time: 9.7, 10 (enantiomers)
Voltage: 240 V/cm
Model: Beckman P/ACE 2000

OTHER SUBSTANCES
Simultaneous: arterenol, dimethindene, ephedrine, pindolol, propranolol

KEY WORDS
chiral; coated capillary

REFERENCE
Schmitt,T.; Engelhardt,H. Optimization of enantiomeric separations in capillary electrophoresis by reversal of the migration order and using different cyclodextrins, *J.Chromatogr.A*, **1995**, *697*, 561–570.

SAMPLE
Matrix: solutions
Sample preparation: Inject an aliquot of a solution in running buffer.

CAPILLARY ELECTROPHORESIS
Capillary: 60 cm × 75 μm fused-silica (52.4 cm to detector)
Capillary preparation: After each run flush with 500 mM KOH for 2-3 min then with water.
Running buffer: 10 mM pH 3.8 Phosphate buffer containing 2% sulfated cyclodextrin (ds 7-10)
Injection: Hydrostatic injection.
Detector: UV 214
Migration time: 8.88, 8.97 (enantiomers)
Voltage: 15 kV
Model: Waters Quanta 4000

OTHER SUBSTANCES
Also analyzed: acebutolol, alprenolol, aminoglutethimide, brompheniramine, bupivacaine, bupropion, canadine, carbinoxamine, chloroquine, chlorpheniramine, dimethindene, disopyramide, hydroxychloroquine, idazoxan, isoxsuprine, ketamine, mepenzolate, mepivacaine, methoxyphenamine, mexiletine, midodrine, nefopam, orphenadrine, oxprenolol, oxyphencyclimine, pheniramine, phensuximide, pindolol, piperoxan, terbutaline, tetramisole, tolperisone, tranylcypromine, trihexyphenidyl, trimipramine, verapamil, warfarin

KEY WORDS
chiral; detector at anode

REFERENCE
Stalcup,A.M.; Gahm,K.H. Application of sulfated cyclodextrins to chiral separations by capillary zone electrophoresis, *Anal.Chem.*, **1996**, *68*, 1360–1368.

SAMPLE
Matrix: solutions
Sample preparation: Inject an aliquot of a 100 μg/mL solution in running buffer.

CAPILLARY ELECTROPHORESIS
Capillary: 30 cm × 50 μm fused-silica (25.5 cm to detector) (Yongnian Optical Conductive Fiber Plant, China), coated with polyacrylamide
Capillary preparation: No details of the polyacrylamide coating process are provided. However, another paper (LC.GC 1997, 15, 40) by this group indicates that they use the procedure of Hjertén, thus: Adjust the pH of 20 mL water to 3.5 with acetic acid, add 80 μL 3-(trimethoxysilyl)propyl methacrylate (3-methacryloxypropyltrimethoxysilane), mix, suck into capillary, let stand at room temperature for 1 h, remove the solution, wash with water. Fill the capillary with a deaerated 3-4% acrylamide solution containing 1 μL/mL N,N,N',N'-tetramethylethylenediamine and 1 mg/mL potassium persulfate, let stand for 30 min, remove excess solution by aspiration, rinse with water, remove water by aspiration, dry at 35° (J. Chromatogr. 1985, 347, 191).
Capillary temperature: 25
Running buffer: 100 mM NaH_2PO_4 adjusted to pH 2.5
Injection: Electrokinetic injection at 15 kV for 3 s.
Detector: UV 200, UV 210
Migration time: 2.53
Voltage: 15 kV
Model: Bio-Rad BioFocus 3000

OTHER SUBSTANCES
Simultaneous: amorolfine, brompheniramine, bupivacaine, carteolol, chloroquine, chlorpheniramine, chlorphenoxamine, disopyramide, dobutamine, flecainide, gallopamil, ketamine, mepindolol, orphenadrine, oxybutynin, phenoxybenzamine, pindolol, propafenone, propranolol,
sulpiride, talinolol, tropicamide, verapamil

KEY WORDS
coated capillary

REFERENCE
Koppenhoefer,B.; Epperlein,U.; Xiaofeng,Z.; Bingcheng,L. Separation of enantiomers of drugs by capillary electrophoresis. Part 4: Hydroxypropyl-γ-cyclodextrin as chiral solvating agent, *Electrophoresis*, **1997**, *18*, 924–
930.

SAMPLE
Matrix: solutions
Sample preparation: Inject an aliquot of a 100 μg/mL solution in running buffer.

CAPILLARY ELECTROPHORESIS
Capillary: 30 cm × 50 μm fused-silica (25.5 cm to detector), coated with polyacrylamide
Capillary preparation: Adjust the pH of 20 mL water to 3.5 with acetic acid, add 80 μL 3-
(trimethoxysilyl)propyl methacrylate (3-methacryloxypropyltrimethoxysilane), mix, suck into
capillary, let stand at room temperature for 1 h, remove the solution, wash with water. Fill the
capillary with a deaerated 3-4% acrylamide solution containing 1 μL/mL N,N,N',N'-tetra-
methylethylenediamine and 1 mg/mL potassium persulfate, let stand for 30 min, remove excess
solution by aspiration, rinse with water, remove water by aspiration, dry at 35° (J. Chromatogr.
1985, 347, 191).
Capillary temperature: 25
Running buffer: 100 mM NaH_2PO_4 containing 45 mM hydroxypropyl-α- cyclodextrin, adjusted
to pH 2.5 with phosphoric acid
Injection: Electrokinetic injection at 15 kV for 3 s.
Detector: UV 200
Voltage: 15 kV
Model: Bio-Rad BioFocus 3000

KEY WORDS
chiral; coated capillary; comparison with the use of other cyclodextrins; this running buffer gave
the greatest enantiomeric separation.; α=1.035

REFERENCE
Lin,B.; Zhu,X.; Koppenhoefer,B.; Epperlein,U. Investigation of 123 chiral drugs by cyclodextrin-modified capillary electrophoresis, *LC.GC*, **1997**, *15*, 40–46.

SAMPLE
Matrix: solutions

CAPILLARY ELECTROPHORESIS
Capillary: 57 cm × 75 μm fused-silica (34 cm to detector) (Polymicro Technologies)
Capillary preparation: Before each run rinse capillary with 100 mM NaOH for 5 min, with
water for 5 min and with running buffer for 5 min. Condition new capillaries with 100 mM
NaOH for 10 min, with water for 5 min and with running buffer for 5 min.
Running buffer: 10 mM pH 3.84 Phosphate buffer containing 7 mM sulfated-β-cyclodextrin
Injection: Hydrodynamic injection at 14 cm for 25 s.
Detector: UV 214
Migration time: 7.5, 8 (enantiomers)
Voltage: 12 kV
Model: laboratory-constructed

OTHER SUBSTANCES
Also analyzed: brompheniramine, chloroquine, chlorpheniramine, pheniramine

KEY WORDS
chiral

REFERENCE
Jin,L.J.; Li,S.F.Y. Comparison of chiral recognition capabilities of cyclodextrins for the separation of basic drugs in capillary zone electrophoresis, *J.Chromatogr.B*, **1998**, *708*, 257–266.

SAMPLE
Matrix: solutions

CAPILLARY ELECTROPHORESIS
Capillary: 29-36 cm $\times$ 50 μm fused-silica (24.5-31.5 cm to detector) (Yongnian Optical Conductive Fiber Plant, China) coated with polyacrylamide
Capillary preparation: Coat capillary as follows. Adjust the pH of 20 mL water to 3.5 with acetic acid, add 80 μL 3-(trimethoxysilyl)propyl methacrylate (3-methacryloxypropyltrimethoxysilane), mix, suck into capillary, let stand at room temperature for 1 h, remove the solution, wash with water. Fill the capillary with a deaerated 3-4% acrylamide solution containing 1 μL/mL N,N,N',N'-tetramethylethylenediamine and 1 mg/mL potassium persulfate, let stand for 30 min, remove excess solution by aspiration, rinse with water, remove water by aspiration, dry at 35° (J. Chromatogr. 1985, 347, 191).
Capillary temperature: 25
Running buffer: 100 mM pH 2.5 NaH_2PO_4 (A) or 100 mM pH 2.5 NaH_2PO_4 containing 45 mM hydroxypropyl-α-cyclodextrin (Wacker, Munich) (B)
Injection: Electromigration at 15 kV for 3 s.
Detector: UV 200; UV 210
Migration time: 2.53 (A); 4.58, 4.74 (B) (enantiomers)
Voltage: 15 kV
Model: Bio-Focus 3000

OTHER SUBSTANCES
Also analyzed: albuterol (salbutamol), alprenolol, amorolfine, atenolol, atropine, azelastine, baclofen, bamethan, benproperine, benserazide, biperiden, bisoprolol, brompheniramine, bupivacaine, bupranolol, butamirate, butethamate, carazolol, carbuterol, carteolol, carvedilol, celiprolol, chloroquine, chlorpheniramine, chlorphenoxamine, cicletanine, clenbuterol, clidinium bromide, clobutinol, dimethindene, dipivefrin, disopyramide, dobutamine, fendiline, flecainide, gallopamil, homatropine, ipratropium bromide, isoproterenol (isoprenaline), isothipendyl, ketamine, meclizine, mefloquine, mepindolol, mequitazine, metaclazepam, metaproterenol (orciprenaline), metipranolol, metoprolol, nafronyl (naftidrofuryl), nefopam, nicardipine, norfenefrine, ofloxacin, ornidazole, orphenadrine, oxomemazine, oxprenolol, oxybutynin, phenoxybenzamine, phenylpropanolamine, pholedrine, pindolol, pirbuterol, prilocaine, procyclidine, promethazine, propafenone, propranolol, reproterol, sotalol, sulpride, synephrine, talinolol, terbutaline, tetrahydrozoline (tetryzoline), theodrenaline, tioconazole, tocainide, trihexyphenidyl, trimeprazine (alimemazine), trimipramine, tropicamide, verapamil, zopiclone

KEY WORDS
coated capillary; chiral

REFERENCE
Koppenhoefer,B.; Eperlein,U.; Schlunk,R.; Zhu,X.; Lin,B. Separation of enantiomers of drugs by capillary electrophoresis. V. Hydroxypropyl-α-cyclodextrin as chiral solvating agent, *J.Chromatogr.A*, **1998**, *793*, 153–164.

Dyphylline

Molecular formula: $C_{10}H_{14}N_4O_4$
Molecular weight: 254.25
CAS Registry No.: 479-18-5
Merck Index (12th ed.): 3529

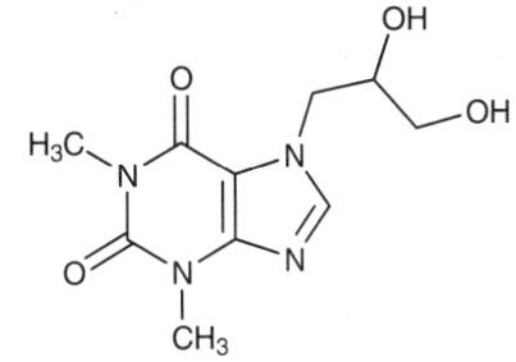

SAMPLE
Matrix: solutions

CAPILLARY ELECTROPHORESIS
Capillary: 72 cm × 50 μm silica (50 cm to detector) (Applied Biosystems)
Capillary preparation: Before each run wash capillary with 100 mM NaOH for 3 min, wash with running buffer for 3 min, aspirate neutral marker solution (2 drops DMSO in 10 mL water) for 1 s, place capillary end in buffer vial for 5 s, inject sample. Wash capillary at the beginning of each day by passing 1 M NaOH through for 20 min.
Capillary temperature: 30
Running buffer: MeCN:buffer 10:90 (Buffer was 30 mM pH 9.3 borate buffer containing 30 mM sodium dodecyl sulfate.)
Injection: Inject by applying vacuum for 1 s (2-3 nL)
Detector: UV 200
Migration time: 6.1
Voltage: +30 kV
Current: 41 μA
Model: Applied Biosystems Model 270A

OTHER SUBSTANCES
Simultaneous: caffeine, theobromine, theophylline

REFERENCE
Evenson,M.A.; Wiktorowicz,J.E. Automated capillary electrophoresis applied to therapeutic drug monitoring, *Clin.Chem.*, **1992**, *38*, 1847–1852.

Econazole

Molecular formula: $C_{18}H_{15}Cl_3N_2O$
Molecular weight: 381.69
CAS Registry No.: 27220-47-9, 68797-31-9 (nitrate)
Merck Index (12th ed.): 3550
Lednicer: 2 249

SAMPLE
Matrix: solutions
Sample preparation: Prepare a 60 μg/mL solution, inject an aliquot.

CAPILLARY ELECTROPHORESIS
Capillary: 61 cm × 50 μm fused-silica (44 cm to detector) (Grom)
Capillary temperature: 21 ± 1
Running buffer: MeOH:buffer 20: 80 (Buffer was 50 mM pH 3.0 (?) phosphate buffer containing 0.1 mM sulfobutyl ether-β-cyclodextrin.)
Injection: Hydrostatic injection at 10 cm for 5 s
Detector: UV 210
Migration time: 12.92, 13.69 (enantiomers)
Voltage: 400 V/cm
Model: Grom 100

OTHER SUBSTANCES
Simultaneous: bifonazole, enilconazole, ketoconazole, miconazole

KEY WORDS
chiral

REFERENCE
Chankvetadze,B.; Endresz,G.; Blaschke,G. Enantiomeric resolution of chiral imidazole derivatives using capillary electrophoresis with cyclodextrin-type buffer modifiers, *J.Chromatogr.A*, **1995**, *700*, 43–49.

EDTA

Molecular formula: $C_{10}H_{16}N_2O_8$
Molecular weight: 292.25
CAS Registry No.: 60-00-4, 150-38-9 (tri Na salt), 64-02-8 (tetra Na salt), 6381-92-6 (di Na salt dihydrate), 139-33-3 (di Na salt), 25102-12-9 (di K salt dihydrate), 2001- 94-7 (di K salt), 58167-76-3 (di K salt monohydrate), 23411-34-9 (Ca di Na salt hydrate), 62-33-9 (Ca di Na salt)
Merck Index (12th ed.): 3559

SAMPLE
Matrix: solutions
Sample preparation: Inject an aliquot of a 500 µM solution.

CAPILLARY ELECTROPHORESIS
Capillary: 50 cm × 75 µm fused-silica (40 cm to detector) (Polymicro Technologies)
Capillary preparation: Flush capillary with running buffer for 1 min. At the start of each day flush capillary with 100 mM NaOH for 10 min, with water for 5 min, and with running buffer for 10 min. When not in use flush capillary with water for 10 min before storage.
Capillary temperature: 25 ± 1
Running buffer: 10 mM pH 3.50 Disodium adenosine 5'-phosphate containing 50 µM cetyltrimethylammonium bromide
Injection: Hydrostatic injection at 4.5 cm for 15 s
Detector: UV 260
Migration time: 7
Voltage: 15 kV
Model: laboratory constructed

OTHER SUBSTANCES
Simultaneous: citric acid, diethylenetriaminepentaacetic acid (DTPA), nitrilotriacetic acid (NTA), orthophosphate, pyrophosphate, tripolyphosphate

KEY WORDS
indirect UV detection; detection at anode

REFERENCE
Wang,T.; Li,S.F.Y. Separation of polyphosphates and polycarboxylates by capillary electrophoresis in a carrier electrolyte containing adenosine 5'-triphosphate and cetyltrimethylammonium bromide with indirect UV detection, *J.Chromatogr.A*, **1996**, *723*, 197–205.

SAMPLE
Matrix: solutions

CAPILLARY ELECTROPHORESIS
Capillary: 68 cm × 75 µm fused-silica (60 cm to detector) (Polymicro Technologies)
Capillary preparation: Between runs wash with 100 mM KOH then water.
Capillary temperature: 25

Running buffer: 25 mM pH 7 Phosphate buffer containing 500 μM tetradecyltrimethylammonium bromide
Injection: Hydrostatic injection at 10 cm for 30 s.
Detector: UV 185
Migration time: 5.1
Voltage: 19 kV
Current: 50 μA (constant)
Model: Waters Quanta 4000E
Limit of detection: 50 μM

OTHER SUBSTANCES
Simultaneous: acetic acid, diethylenetriaminepentaacetic acid (DTPA), ethylenediamine triacetic acid, ethylenediaminomonoacetic acid, ethylenediimino-N,N'-diacetic acid, formic acid, glycine, glyoxylic acid, N-(2-hydroxyethyl)ethylenediaminetriacetic acid, N-(2-hydroxyethyl) iminodiacetic acid, iminodiacetic acid, methyliminodiacetic acid, nitrilotriacetic acid, oxalic acid

KEY WORDS
detector at anode; constant current

REFERENCE
Bürgisser,C.S.; Stone,A.T. Determination of EDTA, NTA, and other amino carboxylic acids and their Co(II) and Co(III) complexes by capillary electrophoresis, *Environ.Sci.Technol.*, **1997**, *31*, 2656–2664.

SAMPLE
Matrix: solutions
Sample preparation: Mix the test solution with an aqueous solution containing an equimolar amount of lutetium trichloride, inject an aliquot.

CAPILLARY ELECTROPHORESIS
Capillary: 47 cm × 74 μm fused-silica (40 cm to detector) (Polymicro Technologies)
Capillary preparation: Before each run rinse capillary with water for 1 min and with running buffer for 2 min. After each run rinse with 100 mM KOH at high pressure for 2 min. (Rinse new capillaries with 100 mM KOH at high pressure (20 psi) for 10 min, with water at high pressure for 5 min, and with running buffer for 10 min.)
Capillary temperature: 50
Running buffer: 10 mM pH 11 phosphate buffer containing 1 mM 8-hydroxyquinoline-5-sulfonic acid
Injection: Pressure injection at 0.5 psi for 3 s
Detector: F ex 325 (He-Cd laser) em 520 ± 20 (filter)
Migration time: 4.5
Voltage: 15 kV
Model: Beckman P/ACE 2100
Limit of detection: 7 ng/mL

OTHER SUBSTANCES
Simultaneous: N-(hydroxyethyl)ethylenediaminetriacetic acid, HEDTA, trans-1,2-diaminocyclohexane-N,N,N',N'-tetraacetic acid, CDTA

KEY WORDS
derivatization; complexation

REFERENCE
Ye,L.; Wong,J.E.; Lucy,C.A. Determination of aminopolycarboxylate ligands using 8-hydroxyquinoline-5-sulfonic acid-based ternary complexes in capillary zone electrophoresis with laser-induced fluorescence, *Anal.Chem.*, **1997**, *69*, 1837–1843.

SAMPLE
Matrix: water
Sample preparation: Condition a SPEC strong anion-exchange SPE extraction disc (Ansys, Irvine CA) with 200 μL MeOH and 200 μL water adjusted to pH 3 with formic acid. If necessary adjust pH of water to 7-9 with ammonium hydroxide. 5 mL Water + 100 μL 10 mM nickel nitrate, vortex for 1 min, adjust pH to 3 ± 0.2 with about 12 μL 9% formic acid, vortex for 1

min, push through the extraction disc at about 1 mL/min, wash with 1 mL water adjusted to pH 3 with formic acid, wash with 2 mL water, wash with 1 mL MeOH, elute with 300 μL MeOH:water 5:95 containing 50 mM trifluoroacetic acid and 1 mM bromothymol blue. Evaporate the eluate to dryness under a stream of nitrogen at 80°, reconstitute the residue in 100 μL 0.1% ammonium hydroxide, evaporate to dryness, reconstitute with 30 μL water, inject an aliquot. (The bromothymol blue displaces the Ni-EDTA complex from the extraction disc.)

CAPILLARY ELECTROPHORESIS
Capillary: 60 cm × 50 μm CElect-Amine coated capillary (Supelco)
Running buffer: 30 mM Ammonium formate adjusted to pH 3 with formic acid
Injection: Pressure injection of water at 50 mb for 2 s then electrokinetic injection of the sample at -20 kV for 30 s followed by ramping to -30 kV plus 50 mb inlet pressure over 1 min.
Detector: MS, PE Sciex API 300 triple quadrupole, makeup liquid MeOH:water 95:5 containing 10 mM ammonium formate at 10 μL/min, electrospray ionization -4 kV, collision gas nitrogen, selective reaction monitoring m/z 347 then m/z 329
Migration time: 5
Internal standard: [$^{13}C_4$]EDTA
Voltage: -30 kV
Current: about 32 μA
Model: Hewlett-Packard 3D HPCE
Limit of quantitation: 300 pg/mL
Limit of detection: 150 pg/mL

KEY WORDS
derivatization; complexation; SPE; coated capillary

REFERENCE
Sheppard,R.L.; Henion,J. Determination of ethylenediaminetetraacetic acid as the nickel chelate in environmental water by solid-phase extraction and capillary electrophoresis/tandem mass spectrometry, *Electrophoresis*, **1997**, *18*, 287–291.

Eflornithine

Molecular formula: $C_6H_{12}F_2N_2O_2$
Molecular weight: 182.17
CAS Registry No.: 67037-37-0, 68278-23-9 (HCl), 96020-91-6 (HCl monohydrate)
Merck Index (12th ed.): 3564
Lednicer: 4 2

SAMPLE
Matrix: dialysate
Sample preparation: 4 μL Dialysate + 2 μL 50 mM pH 10 borate buffer + 2 μL 15 mM NaCN in water + 2 μL 15 mM naphthalene-2,3-dicarboxaldehyde in MeCN, let stand for 10 min (color changes from colorless to bright yellow), inject an aliquot.

CAPILLARY ELECTROPHORESIS
Capillary: 75 cm × 50 μm fused-silica (50 cm to detector) (Polymicro Technologies)
Running buffer: 100 mM pH 6.5 Phosphate buffer
Injection: Vacuum injection for 4 s (3.7 nL)
Detector: UV 254
Migration time: 11
Voltage: 15 kV
Current: 49 μA
Model: ISCO Model 3850
Limit of detection: 5 μM

KEY WORDS
rat; derivatization; pharmacokinetics

REFERENCE

Hu,T.; Zuo,H.; Riley,C.M.; Stobaugh,J.F.; Lunte,S.M. Determination of α-difluoromethylornithine in blood by microdialysis sampling and capillary electrophoresis with UV detection, *J.Chromatogr.A*, **1995**, *716*, 381–388.

Enalapril

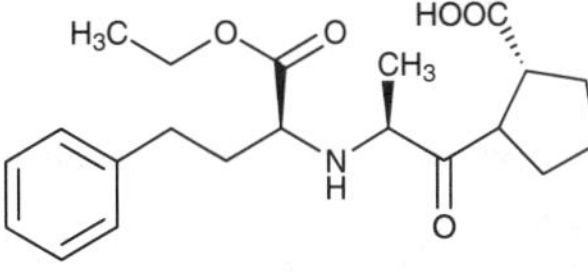

Molecular formula: $C_{20}H_{28}N_2O_5$
Molecular weight: 376.45
CAS Registry No.: 75847-73-3, 76095-16-4 (maleate)
Merck Index (12th ed.): 3605
Lednicer: 4 58, 81-84

SAMPLE

Matrix: formulations
Sample preparation: Sonicate and shake tablets for 30 min with MeOH:10 mM pH 8.5 sodium borate 50:50, inject an aliquot.

CAPILLARY ELECTROPHORESIS

Capillary: 70 cm × 100 μm (63 cm to detector) (Chrompack)
Capillary preparation: Wash capillary with running buffer for 2 min between runs. Wash with running buffer at 60° for 10 min at the start of each sequence. Wash with 100 mM NaOH for 5 min and with water for 5 min at the beginning and end of each day.
Capillary temperature: 50
Running buffer: 20 mM pH 8.5 sodium tetraborate containing 2% Brij 35 and 20 mM sodium dodecyl sulfate
Injection: Hydrodynamic for 2.5 s
Detector: UV 200
Migration time: 6.3
Voltage: 15 kV
Current: about 45 μA
Model: Spectra Physics SpectraPhoresis 1000
Limit of quantitation: 1 μg/mL
Limit of detection: 100 ng/mL (S/N 3)

OTHER SUBSTANCES

Simultaneous: impurities, enalaprilat

KEY WORDS

tablets

REFERENCE

Thomas,B.R.; Ghodbane,S. Evaluation of a mixed micellar electrokinetic capillary electrophoresis method for validated pharmaceutical quality control, *J.Liq.Chromatogr.*, **1993**, *16*, 1983–2006.

SAMPLE

Matrix: solutions

CAPILLARY ELECTROPHORESIS

Capillary: 57 cm × 75 μm fused-silica (50 cm to detector) (Beckman)
Capillary preparation: Before each run rinse capillary with 100 mM NaOH and running buffer.
Capillary temperature: 30
Running buffer: 50 mM pH 8.2 Borate buffer containing 100 mM sodium dodecyl sulfate
Injection: Pressure injection for 5-10 s.
Detector: UV 200
Migration time: 7
Voltage: 25 kV

Model: Beckman P/ACE 5510

OTHER SUBSTANCES
Simultaneous: captopril, ceronapril, fosinopril, lisinopril, zofenopril

REFERENCE
Bretnall,A.E.; Clarke,G.S. Selectivity of capillary electrophoresis for the analysis of cardiovascular drugs, *J.Chromatogr.A*, **1996**, *745*, 145–154.

Enalaprilat

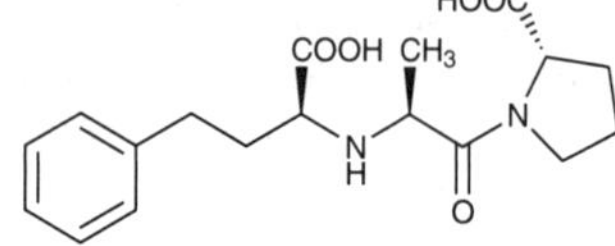

Molecular formula: $C_{18}H_{24}N_2O_5.2H_2O$
Molecular weight: 384.43
CAS Registry No.: 76420-72-9, 84680-54-6 (dihydrate)
Merck Index (12th ed.): 3606

SAMPLE
Matrix: formulations
Sample preparation: Sonicate and shake tablets for 30 min with MeOH:10 mM pH 8.5 sodium borate 50:50, inject an aliquot.

CAPILLARY ELECTROPHORESIS
Capillary: 70 cm × 100 μm (63 cm to detector) (Chrompack)
Capillary preparation: Wash capillary with running buffer for 2 min between runs. Wash with running buffer at 60° for 10 min at the start of each sequence. Wash with 100 mM NaOH for 5 min and with water for 5 min at the beginning and end of each day.
Capillary temperature: 50
Running buffer: 20 mM pH 8.5 sodium tetraborate containing 2% Brij 35 and 20 mM sodium dodecyl sulfate
Injection: Hydrodynamic for 2.5 s
Detector: UV 200
Migration time: 7.8
Voltage: 15 kV
Current: about 45 μA
Model: Spectra Physics SpectraPhoresis 1000
Limit of quantitation: 1 μg/mL
Limit of detection: 100 ng/mL (S/N 3)

OTHER SUBSTANCES
Simultaneous: impurities, enalapril

KEY WORDS
tablets

REFERENCE
Thomas,B.R.; Ghodbane,S. Evaluation of a mixed micellar electrokinetic capillary electrophoresis method for validated pharmaceutical quality control, *J.Liq.Chromatogr.*, **1993**, *16*, 1983–2006.

Enoxacin

Molecular formula: $C_{15}H_{17}FN_4O_3$
Molecular weight: 320.32
CAS Registry No.: 74011-58-8
Merck Index (12th ed.): 3625

SAMPLE
Matrix: solutions
Sample preparation: Inject an aliquot of a solution in MeOH.

CAPILLARY ELECTROPHORESIS
Capillary: 59 cm × 50 μm fused-silica (43 cm to detector) (Polymicro Technologies)
Capillary preparation: Before each analysis flush with water for 3 min, with 200 mM NaOH for 3 min, with water for 3 min, and with running buffer for 4 min. Flush new capillaries with 1 M NaOH at 2000 mbar for 10 min then with 200 mM NaOH for 10 min.
Capillary temperature: 23
Running buffer: MeCN:buffer 28:72, pH 7.3 (Buffer was 32 mM sodium borate containing 18 mM NaH_2PO_4, 39 mM sodium cholate, and 8 mM sodium heptanesulfonate.)
Injection: Hydrodynamic injection at 40 mbar for 6 s.
Detector: UV 260
Migration time: 4.15
Voltage: 30 kV
Model: Lauer Labs Prince

OTHER SUBSTANCES
Simultaneous: ciprofloxacin, flumequine, lomefloxacin, nalidixic acid, norfloxacin, ofloxacin, oxolinic acid, pefloxacin, pipemidic acid, piromidic acid, rosoxacin, sparfloxacin

REFERENCE
Sun,S.-W.; Chen,L.-Y. Optimization of capillary electrophoretic separation of quinolone antibacterials using the overlapping resolution mapping scheme, *J.Chromatogr.A*, **1997**, *766*, 215–224.

Eperisone

Molecular formula: $C_{17}H_{25}NO$
Molecular weight: 259.39
CAS Registry No.: 64840-90-0
Merck Index (12th ed.): 3642

SAMPLE
Matrix: solutions

CAPILLARY ELECTROPHORESIS
Capillary: 36 cm × 50 μm fused-silica coated with linear polyacrylamide (31.5 cm to detector) (GL Science)
Capillary preparation: At the beginning and end of each day rinse capillary with capillary wash solution (Bio-Rad Cat. No. 148-5022) at 690 kPa for more than 3 min and with water at 690 kPa for more than 3 min. Coat capillary as follows. Treat capillary with 1 M NaOH at room temperature for 1 h, rinse with water, dry by passing nitrogen gas through the capillary at 110° for 6 h. Pass thionyl chloride through the capillary using a suction pump for several min, seal capillary at both ends and heat at 70° for 6 h. Unseal the capillary and fill with 250 mM vinyl magnesium bromide in THF by suction, seal the capillary, heat at 70° for 6 h. Open the capillary and rinse it with THF for several min, rinse with distilled water, fill the capillary with polymerization solution, heat at 28 ± 2° for 1 h, condition at -100 V/cm for 30 min (Anal. Sci. 1994, 10, 1). (The polymerization solution was 5% acrylamide in water containing 49 mM Tris, 384 mM glycine, and 0.1% sodium dodecyl sulfate, degas in an ultrasonic bath. Add 40 μL 10% N,N,N',N'-tetramethylethylenediamine and 10 μL 10% ammonium persulfate to 5 mL of the degassed solution, mix thoroughly.)
Running buffer: n-Propanol:50 mM pH 6.0 Phosphate buffer 10:90
Injection: Before each injection rinse with water at 690 kPa for 30 s, rinse with running buffer at 690 kPa for 30 s, partially fill with separation solution (500 μM α_1-acid glycoprotein (Cohn fraction VI) (Fluka) in running buffer) at 6.9 kPa for 190 s (27 cm), inject sample at 6.9 kPa for 2 s, electrophorese with running buffer (Note that α_1-acid glycoprotein from other suppliers may provide inferior results).

Detector: UV 210
Migration time: Resolution of enantiomers 4.8
Voltage: 12 kV
Model: Bio-rad BioFocus 3000

KEY WORDS
chiral; coated capillary

REFERENCE
Tanaka,Y.; Terabe,S. Separation of the enantiomers of basic drugs by affinity capillary electrophoresis using a partial filling technique and α_1-acid glycoprotein as chiral selector, *Chromatographia*, **1997**, *44*, 119–128.

Ephedrine

Molecular formula: $C_{10}H_{15}NO$
Molecular weight: 165.24
CAS Registry No.: 90-81-3 (DL), 134-71-4 (DL HCl), 50-98-6 (L HCl), 134-72-5 (L sulfate), 299-42-3 (-), 50906-05-3 ((-) hemihydrate)
Merck Index (12th ed.): 3645
Lednicer: 1 66

SAMPLE
Matrix: blood
Sample preparation: 500 µL Plasma + 3 mL MeCN, agitate on a rotary mixer for 10 min, centrifuge at 2000 g for 10 min. Remove the supernatant and evaporate it to dryness under a stream of nitrogen, reconstitute the residue in 500 µL running buffer, centrifuge at 2000 g for 54 min, inject an aliquot.

CAPILLARY ELECTROPHORESIS
Capillary: 58 cm × 50 µm fused-silica (51 cm to detector)
Capillary preparation: Between analyses flush with 100 mM HCl for 30 s, with 100 mM NaOH for 2 min, with water for 2 min, and with running buffer for 2 min. Before first use rinse capillary with 100 mM NaOH for 20 min, with water for 10 min, and with running buffer for 10 min.
Capillary temperature: 25
Running buffer: 10 mM pH 9.36 Borax containing 40 mM α-cyclodextrin
Injection: Pressure injection under argon at 0.5 psi for 7 s.
Detector: UV 200
Migration time: 4.5
Internal standard: ephedrine
Voltage: 25 kV
Current: 16 µA
Model: Beckman P/ACE 2100

OTHER SUBSTANCES
Extracted: piracetam

KEY WORDS
plasma; ephedrine is IS

REFERENCE
Lamparczyk,H.; Kowalski,P.; Rajzer,D.; Nowakowska,J. Determination of piracetam in human plasma by capillary electrophoresis, *J.Chromatogr.B*, **1997**, *692*, 483–487.

SAMPLE
Matrix: bulk

Sample preparation: Dissolve 4 mg compound in 1 mL MeCN:water 50:50 containing 0.2% triethylamine, vortex for 30 s. Remove a 100 μL aliquot and add it to 100 μL 1.28% 2,3,4,6-tetra-O-acetyl-β-D-glucopyranosyl isothiocyanate in MeCN, vortex for 1 min, let stand for 15 min, make up to 1 mL with 10 mM pH 9.0 phosphate/borate buffer containing 100 mM sodium dodecyl sulfate, vortex for 20 s, filter (Whatman UniPrep), inject an aliquot of the filtrate.

CAPILLARY ELECTROPHORESIS
Capillary: 48 cm × 50 μm fused-silica (26 cm to detector) (Polymicro Technologies)
Capillary preparation: Condition new capillaries with 1 M NaOH for 10 min, with water for 10 min, and with running buffer for 10 min.
Capillary temperature: 30
Running buffer: MeOH:buffer 20:80 (Buffer was 10 mM pH 9.0 phosphate buffer containing 10 mM borate and 100 mM sodium dodecyl sulfate.)
Injection: Vacuum injection for 0.5 s.
Detector: UV 210
Migration time: 9.4 (-), 11.6 (+)
Voltage: 20 kV
Current: 48 μA
Model: Applied Biosystems Model 270A-HT

OTHER SUBSTANCES
Simultaneous: amphetamine, methamphetamine, norpseudoephedrine, phenylpropanolamine (norephedrine), pseudoephedrine

KEY WORDS
derivatization; chiral; comparison with HPLC

REFERENCE
Lurie,I.S. Micellar electrokinetic capillary chromatography of the enantiomers of amphetamine, methamphetamine and their hydroxyphenethylamine precursors, *J.Chromatogr.*, **1992**, *605*, 269–275.

SAMPLE
Matrix: bulk
Sample preparation: 10-50 mg Bulk compound + 1 mL 1 mg/mL caffeine in 10 mM HCl, make up to 10 mL with 10 mM HCl, sonicate for 2 min, mix thoroughly, filter (0.45 μm cellulose acetate), inject an aliquot.

CAPILLARY ELECTROPHORESIS
Capillary: 75 cm × 75 μm fused-silica (50 cm to the detector) (ISCO)
Capillary preparation: Flush with running buffer for 2 min between analyses. Condition capillary by filling with 1 M NaOH and allowing to stand for 1 h, fill with 100 mM NaOH and allow to stand for 1 h, wash with water, fill with running buffer. Each week wash capillary with 100 mM HCl for 10 min, with water, with 100 mM NaOH, and with water then fill with running buffer.
Capillary temperature: 30
Running buffer: DMSO:ethanolamine:buffer 11:1:88 (Buffer was 0.92 g cetyltrimethylammonium bromide in 100 mL 10 mM sodium tetraborate, adjust pH to 11.5 with 1 M NaOH.)
Injection: Load under vacuum, vacuum level 2, 10.0 kPa s
Detector: UV 254
Migration time: 7.3
Internal standard: caffeine (6.5)
Voltage: -15 kV
Model: ISCO Model 3140 electropherograph

OTHER SUBSTANCES
Simultaneous: amphetamine, p-aminobenzoic acid, methamphetamine, methylenedioxyamphetamine, methylenedioxymethamphetamine, norephedrine, pseudoephedrine, pseudonorephedrine

REFERENCE
Trenerry,V.C.; Robertson,J.; Wells,R.J. Analysis of illicit amphetamine seizures by capillary electrophoresis, *J.Chromatogr.A*, **1995**, *708*, 169–176.

SAMPLE
Matrix: bulk
Sample preparation: Dissolve sample in 2 mg/mL histamine diphosphate in 10 mM HCl, inject an aliquot.

CAPILLARY ELECTROPHORESIS
Capillary: 48.5 cm × 50 μm fused-silica (40 cm to detector)
Capillary preparation: Flush capillary with running buffer for 2 min before each run. Replenish buffer in vials after every 10 injections. Condition new capillaries by flushing with 1 M NaOH for 10 min, with 100 mM NaOH for 10 min, with water for 10 min, and with running buffer for 10 min.
Capillary temperature: 40
Running buffer: 75 mM NaH_2PO_4 containing 75 mM sodium borate, adjusted to pH 8.5 with 2 M NaOH
Injection: Pressure injection at 150 mbar for 3 s.
Detector: UV 230
Migration time: 5.5
Internal standard: histamine diphosphate (5)
Voltage: 10 kV
Current: 66-70 μA
Model: Hewlett Packard 3D CE
Limit of detection: 34 μg/mL (S/N = 2:1)

OTHER SUBSTANCES
Simultaneous: acetaminophen, benzocaine, cis-cinnamoylcocaine, trans-cinnamoylcocaine, cocaine, lidocaine, phenylpropanolamine, procaine, tetracaine

REFERENCE
Krawczeniuk,A.S.; Bravenec,V.A. Quantitative determination of cocaine in illicit powders by free zone capillary electrophoresis, *J.Forensic Sci.*, **1998**, *43*, 738–743.

SAMPLE
Matrix: formulations

CAPILLARY ELECTROPHORESIS
Capillary: 60 cm × 75 μm fused-silica (53 cm to detector)
Capillary preparation: Rinse capillary with running buffer after each run. At the start of each day rinse with 100 mM NaOH for 5 min and with separation buffer for 10 min. Before storage rinse capillary with water and air dry.
Running buffer: 30 mM pH 8.8 Sodium borate buffer containing 40 mM sodium dodecyl sulfate
Injection: Hydrodynamic injection at 10 cm for 5-10 s
Detector: UV 214
Migration time: 8.3
Voltage: 20 kV
Model: Waters Quanta 4000

OTHER SUBSTANCES
Simultaneous: ethophylline, mebromphenhydramine, theophylline

KEY WORDS
tablets

REFERENCE
Korman,M.; Vindevogel,J.; Sandra,P. Application of micellar electrokinetic chromatography to the quality control of pharmaceutical formulations: the analysis of xanthine derivatives, *Electrophoresis*, **1994**, *15*, 1304–1309.

SAMPLE
Matrix: formulations

Sample preparation: Add 3 mL buffer to 30 mg capsule contents, sonicate for 30 min, filter (0.2 μm Nylon), inject an aliquot of the filtrate. (Buffer was water acidified to pH 2.0 with concentrated phosphoric acid.)

CAPILLARY ELECTROPHORESIS

Capillary: 900 mm × 50 μm uncoated fused-silica (650 mm to detector)
Capillary preparation: Rinse capillary with buffer for 15 min at the start of each day and for 4 min before each analysis.
Capillary temperature: 31
Running buffer: 70 mM hydroxypropyl-β-cyclodextrin (average molar substitution 0.8, average molecular mass 1500) containing 30 mM tetramethylammonium chloride and 10 mM sodium dodecyl sulfate, pH 2.0
Injection: Vacuum injection at 28 kPa
Detector: UV 210
Migration time: 33.7 (+), 34.1 (-)
Voltage: +28 kV
Model: ISCO Model 3140
Limit of detection: 4 μg/mL

OTHER SUBSTANCES

Simultaneous: amphetamine, methamphetamine, methylephedrine, methylpseudoephedrine, norephedrine, norpseudoephedrine, pseudoephedrine

---

KEY WORDS

capsules; chiral

REFERENCE

Flurer,C.L.; Lin,L.A.; Satzger,R.D.; Wolnik,K.A. Determination of ephedrine compounds in nutritional supplements by cyclodextrin-modified capillary electrophoresis, *J.Chromatogr.B*, **1995**, *669*, 133–139.

SAMPLE

Matrix: hair, urine
Sample preparation: Wash 100 mg hair with 0.3% Tween 20 in water, cut in small pieces, add 1 mL 250 mM HCl, heat at 45° overnight, neutralize with NaOH. Add the hair extract or 2 mL urine to a Toxi-tube A, extract the organic layer with 200 μL 10 mM phosphoric acid, inject an aliquot of the aqueous layer.

CAPILLARY ELECTROPHORESIS

Capillary: 45 cm × 50 μm fused-silica (37.5 cm to detector) (Composite Metal Services, Hallow, UK)
Capillary preparation: Condition with running buffer for 10 min before each run. Wash new capillaries with 1 M NaOH for 10 min, with 100 mM NaOH for 10 min, and with water for 10 min, and then condition with running buffer for 20 min.
Running buffer: 100 mM pH 2.5 Potassium phosphate buffer containing 15 mM β-cyclodextrin
Injection: Pressure injection at 35 mbar for 10 s.
Detector: UV 200
Migration time: 19, 19.3 (enantiomers)
Voltage: 10 kV
Model: Beckman P/ACE 2200
Limit of detection: 75 ng/mL

OTHER SUBSTANCES

Extracted: amphetamine, methamphetamine, 3,4-methylenedioxyamphetamine, 3,4-methylenedioxyethylamphetamine, 3,4-methylenedioxymethamphetamine

---

KEY WORDS

chiral; SPE

REFERENCE
Tagliaro,F.; Manetto,G.; Bellini,S.; Scarcella,D.; Smith,F.P.; Marigo,M. Simultaneous chiral separation of 3,4-methylenedioxymethamphetamine (MDMA), 3,4-methylenedioxyamphetamine (MDA), 3,4-methylenedioxyethylamphetamine (MDE), ephedrine, amphetamine and methamphetamine by capillary electrophoresis in uncoated and coated capillaries with native β-cyclodextrin as the chiral selector: Preliminary application to the analysis of urine and hair, *Electrophoresis*, **1998**, *19*, 42–50.

SAMPLE
Matrix: solutions

CAPILLARY ELECTROPHORESIS
Capillary: 15 cm × 50 μm fused-silica coated with crosslinked polyacrylamide (13.5 cm to detector) (Scientific Glass Engineering)
Running buffer: 100 mM pH 6.8 Sodium phosphate buffer containing 3% G 3707 (heptaoxyethylene lauryl ether, Atlas Chemie or Sigma P 8800)
Injection: Place a 1-2 mm slug in capillary by capillary action
Detector: UV 205
Migration time: 5
Voltage: 3 kV
Current: 60 μA

OTHER SUBSTANCES
Simultaneous: alprenolol, benzylamine, chlorpheniramine, codeine, protriptyline, terodilin

KEY WORDS
coated capillary

REFERENCE
Hjertén,S.; Valtcheva,L.; Elenbring,K.; Eaker,D. High-performance electrophoresis of acidic and basic low-molecular-weight compounds and of proteins in the presence of polymers and neutral surfactants, *J.Liq.Chromatogr.*, **1989**, *12*, 2471–2499.

SAMPLE
Matrix: solutions
Sample preparation: Prepare a 100 μg/mL solution in MeOH:water 50:50, inject an aliquot.

CAPILLARY ELECTROPHORESIS
Capillary: 35 cm × 50 μm
Running buffer: MeOH:buffer 20:80 (Buffer was 25 mM Tris containing 15 mM heptakis(di-O-methyl)-β-cyclodextrin, adjusted to pH 2.5 with phosphoric acid.)
Injection: Hydrodynamic injection at 10 cm for 10 s
Detector: UV 214
Migration time: 3.0 (-), 5.2 (+)
Voltage: 18 kV
Model: Waters Quanta 4000

KEY WORDS
chiral

REFERENCE
Swartz,M.E. Method development and selectivity control for small molecule pharmaceutical separations by capillary electrophoresis, *J.Liq.Chromatogr.*, **1991**, *14*, 923–938.

SAMPLE
Matrix: solutions

CAPILLARY ELECTROPHORESIS
Capillary: 62 cm × 50 μm fused silica (50 cm to detector) (Polymicro Technologies)
Capillary temperature: 40

Running buffer: 150 mM pH 2.5 tetrabutylammonium phosphate buffer containing 20 mM β-cyclodextrin
Injection: Injection by gravity at 10 cm for 5 s
Detector: UV 214
Migration time: 21.6 (1R,2S-(-)), 22.2 (1S,2R-(+))
Voltage: 20 kV
Current: 56 μA
Model: Laboratory constructed

OTHER SUBSTANCES
Simultaneous: norephedrine, pseudoephedrine

KEY WORDS
chiral

REFERENCE
Quang,C.; Khaledi,M.G. Improved chiral separation of basic compounds in capillary electrophoresis using β-cyclodextrin and tetraalkylammonium reagents, *Anal.Chem.*, **1993**, *65*, 3354–3358.

SAMPLE
Matrix: solutions

CAPILLARY ELECTROPHORESIS
Capillary: 100 cm × 50 μm fused-silica (50 cm to detector) (Isco)
Capillary preparation: Flush capillary with 10 μL running buffer between runs. Every 40 sample injections rinse capillary with 200 μL 1 M NaOH, with 200 μL water, and 200 μL running buffer. Before use fill capillary with 1 M NaOH and allow to stand for 1 h, fill with 100 mM NaOH, allow to stand for 1 h, wash with water fill with running buffer.
Capillary temperature: 23
Running buffer: 100 mM citric acid containing 19.27 mM Na_2HPO_4 and 120 mM hydroxypropyl-β-cyclodextrin
Injection: Inject under vacuum at 4.0 kPa.s
Detector: UV 200
Migration time: 28.5 (S), 29 (R)
Voltage: 30 kV
Model: Isco Model 3140

OTHER SUBSTANCES
Simultaneous: amphetamine, 4-bromo-2,5-dimethoxyamphetamine, epinephrine, methamphetamine, methyldimethoxyethylamphetamine, methyldimethoxyamphetamine, methyldimethoxymethylamphetamine, pseudoephedrine

---

KEY WORDS
chiral

REFERENCE
Aumatell,A.; Wells,R.J.; Wong,D.K.Y. Enantiomeric differentiation of a wide range of pharmacologically active substances by capillary electrophoresis using modified β-cyclodextrins, *J.Chromatogr.A*, **1994**, *686*, 293–307.

SAMPLE
Matrix: solutions
Sample preparation: Prepare a 50 μg/mL solution in diluted running buffer, inject an aliquot.

CAPILLARY ELECTROPHORESIS
Capillary: 44 cm × 50 μm fused-silica
Capillary preparation: After each run wash capillary with water for 2 min and with running buffer for 3 min. At the beginning of each day wash capillary with water and running buffer for 5 min. Condition a new capillary with 1 M NaOH, 100 mM NaOH, water, and separation buffer.
Capillary temperature: 15

Running buffer: MeOH:buffer 30:70 (Buffer was 100 mM Phosphoric acid containing 30 mM heptakis(2,6-di-O-methyl)-β-cyclodextrin adjusted to pH 3.0 with triethanolamine.)
Injection: Hydrodynamic injection for 1 s
Detector: UV 210
Voltage: 25 kV
Model: Spectra Physics Spectraphoresis 1000

KEY WORDS
chiral; enantiomer resolution 2.7

REFERENCE
Bechet,I.; Paques,P.; Fillet,M.; Hubert,P.; Crommen,J. Chiral separation of basic drugs by capillary zone electrophoresis with cyclodextrin additives, *Electrophoresis*, **1994**, *15*, 818–823.

SAMPLE
Matrix: solutions

CAPILLARY ELECTROPHORESIS
Capillary: 50 cm × 50 μm fused-silica (50 cm to detector) (Polymicro Technologies)
Capillary preparation: Rinse capillary with running buffer for 1 min between runs. Flush a new capillary with 100 mM NaOH for 30 min, flush for 20 min with running buffer, equilibrate with running buffer for 20 min at 15 kV.
Capillary temperature: 26
Running buffer: 20 mM pH 10.0 Borate buffer containing 40 mM tetrakis[6-O-(4-sulfobutyl)]-β-cyclodextrin sodium salt (Center of Drug Delivery Research, Higuchi Bioscience Center, University of Kansas, Lawrence KS)
Injection: Pressure injection at 0.5 psi for 4 s
Detector: UV 214
Migration time: 15.83 (-), 16.04 (+)
Voltage: 15 kV
Model: Beckman P/ACE 2000

OTHER SUBSTANCES
Simultaneous: methylephedrine, methylpseudoephedrine, pseudoephedrine
Interfering: (+)-norephedrine (interferes with (-) ephedrine)

KEY WORDS
chiral

REFERENCE
Dette,C.; Ebel,S.; Terabe,S. Neutral and anionic cyclodextrins in capillary zone electrophoresis: enantiomeric separation of ephedrine and related compounds, *Electrophoresis*, **1994**, *15*, 799–803.

SAMPLE
Matrix: solutions

CAPILLARY ELECTROPHORESIS
Capillary: 58 cm × 50 μm fused-silica (50 cm to detector) (Polymicro Technologies)
Capillary preparation: Before each injection purge capillary with 5% phosphoric acid for 12 s, with water for 30 s, and with running buffer for 10 min.
Capillary temperature: 35
Running buffer: 80 mM pH 6.7 sodium phosphate buffer containing 15 mM cetyltrimethylammonium bromide
Injection: Hydrostatic injection for 20 s
Detector: UV 214
Migration time: 17.5
Voltage: -27 kV
Model: Waters Quanta 4000

OTHER SUBSTANCES
Simultaneous: acebutolol, alprenolol, atenolol, labetalol, metoprolol, nadolol, oxprenolol, pindolol, propranolol, timolol

REFERENCE
Lukkari,P.; Nyman,T.; Riekkola,M.-J. Determination of nine β-blockers in serum by micellar electrokinetic capillary chromatography, *J.Chromatogr.A*, **1994**, *674*, 241–246.

SAMPLE
Matrix: solutions
Sample preparation: Prepare a 50 μg/mL solution in water:buffer:MeOH 90:9:1, inject an aliquot. (Buffer was 25 mM Tris buffer adjusted to pH 2.4 with phosphoric acid.)

CAPILLARY ELECTROPHORESIS
Capillary: 82 cm × 50 μm fused-silica (60 cm to detector)
Capillary preparation: Condition by aspirating with 1 M NaOH for 10 min, with water for 10 min, and with running buffer for 10 min.
Capillary temperature: 30
Running buffer: MeOH:buffer 1.2:98.8 (Buffer was 25 mM Tris buffer adjusted to pH 2.4 with phosphoric acid containing 5 mM heptakis(2,6-di-O-methyl)-β-cyclodextrin and 1 mM sulfobutyl ether β-cyclodextrin (ISCO)
Injection: Vacuum injection for 0.5 s (2 nL)
Detector: UV 210
Migration time: 11.93 (+), 12.10 (-)
Voltage: 30 kV
Model: Applied Biosystems Model 270A-HT

OTHER SUBSTANCES
Simultaneous: amphetamine, cathinone, cocaine, methamphetamine, methcathinone, norephedrine, norpseudoephedrine, propoxyphene, pseudoephedrine

KEY WORDS
chiral

REFERENCE
Lurie,I.S.; Klein,R.F.X.; Dal Cason,T.A.; LeBelle,M.J.; Brenneisen,R.; Weinberger,R.E. Chiral resolution of cationic drugs of forensic interest by capillary electrophoresis with mixtures of neutral and anionic cyclodextrins, *Anal.Chem.*, **1994**, *66*, 4019–4026.

SAMPLE
Matrix: solutions

CAPILLARY ELECTROPHORESIS
Capillary: 60 cm × 50 μm AccuSep (52.5 cm to detector) (Waters)
Capillary preparation: Before injection rinse capillary with 100 mM NaOH for 3 min and with running buffer for 3 min. Rinse new capillaries with 500 mM NaOH for 5 min
Running buffer: 50 mM pH 7.0 Na$_2$HPO$_4$ containing 25 mM (S)-N- dodecoxycarbonylvaline (Prepare (S)-N-dodecoxycarbonylvaline as follows. Prepare dodecyl chloroformate by reacting 1-dodecanol with 0.33 equivalents of triphosgene in dichloromethane solution in the presence of pyridine. Add dodecyl chloroformate dropwise to (S)-valine in 1 M NaOH solution, filter, wash with hexane, recrystallize from ether/petroleum ether.)
Injection: Hydrostatic injection 2 s
Detector: UV 214
Voltage: +12 kV
Model: Waters Quanta 4000 or 4000E

OTHER SUBSTANCES
Also analyzed: atenolol, bupivacaine, homatropine, ketamine, metoprolol, N-methylpseudoephedrine, norephedrine, norphenylephrine, octopamine, pindolol, terbutaline

KEY WORDS
chiral; α = 1.14

REFERENCE
Mazzeo,J.R.; Grover,E.R.; Swartz,M.E.; Petersen,J.S. Novel chiral surfactant for the separation of enantiomers by micellar electrokinetic capillary chromatography, *J.Chromatogr.A*, **1994**, *680*, 125–135.

SAMPLE
Matrix: solutions
Sample preparation: Prepare a 200 μM solution in water, inject an aliquot.

CAPILLARY ELECTROPHORESIS
Capillary: 57 cm × 75 μm fused-silica (50 cm to detector) (Beckman)
Capillary preparation: Before each analysis wash with three column volumes of 10 mM NaOH, three column volumes of water, and three column volumes of running buffer. At the end of each day wash capillary with 10 column volumes of 10 mM NaOH and 10 column volumes of water. At the beginning of each day wash capillary with 100 column volumes of 10 mM NaOH.
Capillary temperature: 25
Running buffer: 50 mM pH 2.5 Phosphate buffer containing 15 mM heptakis(2,6-di-O-methyl)-β-cyclodextrin
Injection: Injection using pressurized nitrogen for 3 s (about 15 nL)
Detector: UV 214
Migration time: 22.5, 23 (enantiomers)
Voltage: 11 kV
Model: Beckman P/ACE 2000

KEY WORDS
chiral

REFERENCE
Pálmarsdóttir,S.; Edholm,L.-E. Capillary zone electrophoresis for separation of drug enantiomers using cyclo-dextrins as chiral selectors Influence of experimental parameters on separation, *J.Chromatogr.A*, **1994**, *666*, 337–350.

SAMPLE
Matrix: solutions

CAPILLARY ELECTROPHORESIS
Capillary: 57 cm × 50 μm fused-silica (50 cm to detector) (Polymicro Technologies)
Capillary temperature: 40
Running buffer: MeCN:buffer 10:90 (Buffer was 50 mM pH 11.0 phosphate/carbonate buffer containing 40 mM sodium dodecyl sulfate.)
Detector: UV 214
Migration time: 6
Voltage: 18 kV
Current: <60 μA
Model: Beckman P/ACE 2000

OTHER SUBSTANCES
Simultaneous: nicotine

REFERENCE
Quang,C.; Strasters,J.K.; Khaledi,M.G. Computer-assisted modeling, prediction, and multifactor optimization in micellar electrokinetic chromatography of ionizable compounds, *Anal.Chem.*, **1994**, *66*, 1646–1653.

SAMPLE
Matrix: solutions

CAPILLARY ELECTROPHORESIS
Capillary: 67 cm × 75 μm (60 cm to detector) (Polymicro Technologies)
Capillary preparation: Coat capillary as follows. Treat overnight with 50% 3-(trimethoxysi-lyl)propyl methacrylate in MeOH, fill capillary with degassed 40 g/L (acrylamide + N,N'-meth-ylenebisacrylamide) in 100 mM pH 8.2 Tris-borate buffer containing 2 mM EDTA, 5 μL/L 10% ammonium persulfate, and 5 μL/mL N,N,N',N'-tetramethylethylenediamine, after polymeri-zation force out gel leaving coating sticking to wall (Caution! Acrylamide is a carcinogen!), rinse with water, remove water by aspiration, dry at 35°.
Running buffer: 20 mM pH 2.5 Citric acid containing 2% carboxymethyl β-cyclodextrin
Detector: UV 214

Migration time: 12 (+), 12.4 (-)
Voltage: 240 V/cm
Model: Beckman P/ACE 2000

OTHER SUBSTANCES
Simultaneous: arterenol, dimethindene, doxylamine, pindolol, propranolol

KEY WORDS
chiral; coated capillary

REFERENCE
Cao,J.; Cross,R.F. The separation of dihydrofolate reductase inhibitors and the determination of pKa,1 values by capillary zone electrophoresis, *J.Chromatogr.A*, **1995**, *695*, 297–308.

SAMPLE
Matrix: solutions
Sample preparation: Prepare a solution in MeOH/water, inject an aliquot.

CAPILLARY ELECTROPHORESIS
Capillary: 62 cm × 52 μm fused-silica (50 cm to detector) (Polymicro Technologies)
Capillary temperature: 40
Running buffer: 50 mM pH 2.50 Tetrabutylammonium phosphate containing 20 mM dimethyl-β-cyclodextrin
Injection: Siphon injection at 10 cm for 5 s
Detector: UV (wavelength not specified)
Migration time: 12.97, 13.16 (enantiomers)
Voltage: 20 kV
Model: Laboratory constructed

KEY WORDS
chiral

REFERENCE
Quang,C.; Khaledi,M.G. Extending the scope of chiral separation of basic compounds by cyclodextrin-mediated capillary zone electrophoresis, *J.Chromatogr.A*, **1995**, *692*, 253–265.

SAMPLE
Matrix: solutions

CAPILLARY ELECTROPHORESIS
Capillary: 67 cm × 75 μm (60 cm to detector) (Polymicro Technologies)
Capillary preparation: Coat capillary as follows. Treat overnight with 50% 3-(trimethoxysilyl)propyl methacrylate in MeOH, fill capillary with degassed 40 g/L (acrylamide + N,N'-methylenebisacrylamide) in 100 mM pH 8.2 Tris-borate buffer containing 2 mM EDTA, 5 μL/L 10% ammonium persulfate, and 5 μL/mL N,N,N',N'-tetramethylethylenediamine, after polymerization force out gel leaving coating sticking to wall (Caution! Acrylamide is a carcinogen!), rinse with water, remove water by aspiration, dry at 35°.
Running buffer: 20 mM pH 2.5 Citric acid containing 2% carboxymethyl β-cyclodextrin
Detector: UV 214
Migration time: 12 (+), 12.4 (-)
Voltage: 240 V/cm
Model: Beckman P/ACE 2000

OTHER SUBSTANCES
Simultaneous: arterenol, dimethindene, doxylamine, pindolol, propranolol

KEY WORDS
chiral; coated capillary

REFERENCE
Schmitt,T.; Engelhardt,H. Optimization of enantiomeric separations in capillary electrophoresis by reversal of the migration order and using different cyclodextrins, *J.Chromatogr.A*, **1995**, *697*, 561–570.

SAMPLE
Matrix: solutions
Sample preparation: Inject an aliquot of an aqueous solution.

CAPILLARY ELECTROPHORESIS
Capillary: 120 cm × 75 μm fused-silica
Capillary preparation: Rinse with Bio-Rad Capillary wash Solution (20 mM phosphoric acid) and water, equilibrate with running buffer.
Running buffer: 10 mM pH 3.0 Tris-formic acid buffer containing 20 mM heptakis(2,6-di-O-methyl)-β-cyclodextrin
Injection: Electrokinetic injection at 10 kV or pressure injection for 5-10 s
Detector: MS, Sciex TAGA 6000E, API III, nebulizing gas nitrogen at 35 psi, collision gas argon, ion spray tip voltage 3.5-4.5 kV, sheath liquid 2-4 μL/min (MeOH:water 50:50 to 90:10 with or without 2 mM ammonium acetate), m/z 166
Migration time: 22.4 (-), 22.5 (+)
Voltage: 30 kV
Current: 20 μA
Model: Beckman P/ACE Model 2050

KEY WORDS
chiral

REFERENCE
Sheppard,R.L.; Tong,X.; Cai,J.; Henion,J.D. Chiral separation and detection of terbutaline and ephedrine by capillary electrophoresis coupled with ion spray mass spectrometry, *Anal.Chem.*, **1995**, *67*, 2054–2058.

SAMPLE
Matrix: solutions
Sample preparation: Inject an aliquot of a 50 μg/mL solution in water.

CAPILLARY ELECTROPHORESIS
Capillary: 48.5 cm × 50 μm fused-silica (40 cm to detector)
Capillary preparation: After each run wash capillary with running buffer for 3 min. At the start of each day wash capillary with running buffer for 10 min. Before use treat new capillaries with 1 M NaOH, 100 mM NaOH, water, and running buffer.
Capillary temperature: 15
Running buffer: 100 mM Phosphoric acid containing 15 mM carboxymethyl-β-cyclodextrin (Cyclolab, Budapest), adjusted to pH 3.0 with 84 mM triethanolamine
Injection: Hydrodynamic injection at 5 kPa for 2 s.
Detector: UV 210
Migration time: 15.89, 17.16 (enantiomers)
Voltage: 25 kV
Model: Hewlett Packard ³ᴰCE

OTHER SUBSTANCES
Simultaneous: bupivacaine, chlorpheniramine, dimethindene, fenfluramine, isoproterenol

KEY WORDS
chiral

REFERENCE
Fillet,M.; Bechet,I.; Hubert,P.; Crommen,J. Resolution improvement by use of carboxymethyl-β-cyclodextrin as chiral additive for the enantiomeric separation of basic drugs by capillary electrophoresis, *J.Pharm.Biomed.Anal.*, **1996**, *14*, 1107–1114.

SAMPLE
Matrix: solutions

CAPILLARY ELECTROPHORESIS
Capillary: 58.2 cm × 75 μm fused-silica (50.7 cm to detector) (Polymicro Technologies)

Capillary preparation: Purge with running buffer before each run. At the beginning of each day purge using 50-60 kPa vacuum with 500 mM NaOH for 5 min, with water for 5 min, with MeCN for 5 min, and with running buffer for 5 min.
Running buffer: MeCN:MeOH:acetic acid 49:50:1 containing 20 mM ammonium acetate
Injection: Hydrostatic injection at 10 cm for 5 s.
Detector: UV 214
Migration time: 3.65
Voltage: 25 kV
Model: Waters Quanta 4000

OTHER SUBSTANCES
Simultaneous: amphetamine, benzphetamine, nylidrin, oxymetazoline, phendimetrazine, phenmetrazine, phenylephrine, phenylpropanolamine, xylometazoline

REFERENCE
Leung,G.N.W.; Tang,H.P.O.; Tso,T.S.C.; Wan,T.S.M. Separation of basic drugs with non-aqueous capillary electrophoresis, *J.Chromatogr.A*, **1996**, *738*, 141–154.

SAMPLE
Matrix: solutions
Sample preparation: Inject an aliquot of a 20 μg/mL solution in 12.5 mM pH 9.3 borate buffer.

CAPILLARY ELECTROPHORESIS
Capillary: 66.5 cm $\times$ 75 μm fused-silica (41.5 cm to detector) (Polymicro Technologies)
Running buffer: 250 mM pH 9.3 Sodium borate buffer
Injection: Pressure injection at 50 mbar for 15 s.
Detector: UV 195
Migration time: 12.5
Voltage: 8 kV
Model: Crystal 310

OTHER SUBSTANCES
Simultaneous: amphetamine, chlorpheniramine, MDA, MDMA, methamphetamine, phentermine, phenylpropanolamine, pseudoephedrine
Noninterfering: caffeine

REFERENCE
Lilley,K.A.; Wheat,T.E. Drug identification in biological matrices using capillary electrophoresis and chemometric software, *J.Chromatogr.B*, **1996**, *683*, 67–76.

SAMPLE
Matrix: solutions

CAPILLARY ELECTROPHORESIS
Capillary: 52 cm $\times$ 75 μm fused-silica (48 cm to detector) (Polymicro Technologies)
Capillary preparation: Before each run purge with running buffer for 3 min. Every 3 runs purge with 100 mM NaOH for 5 min. Purge new capillaries with 1 M NaOH for 20 min and with 100 mM NaOH for 20 min, rinse with running buffer, equilibrate with running buffer at 12 kV for 3 h.
Running buffer: 100 mM CHES (2-(N-cyclohexylamine)ethanesulfonic acid) containing 10 mM triethylamine and 25 mM (R)-dodecoxycarbonylvaline (Waters EnantioSelect (R)-Val-1), pH adjusted to 8.8 with 1 M NaOH
Injection: Hydrostatic injection for 2 s.
Detector: UV 214
Migration time: 10.1 (1S,2R), 10.2 (1R,2S)
Voltage: 12 kV
Current: $\leq$30 μA
Model: Waters Quanta 4000

OTHER SUBSTANCES
Simultaneous: atenolol, methylpseudoephedrine, pindolol

KEY WORDS
chiral

REFERENCE
Peterson,A.G.; Ahuja,E.S.; Foley,J.P. Enantiomeric separations of basic pharmaceutical drugs by micellar electrokinetic chromatography using a chiral surfactant, N-dodecoxycarbonylvaline, *J.Chromatogr.B*, **1996**, *683*, 15–28.

SAMPLE
Matrix: solutions

CAPILLARY ELECTROPHORESIS
Capillary: 60 cm × 75 μm fused-silica (51 cm to detector) (Supelco)
Capillary preparation: At the end of each day wash capillary with 200 mM NaOH for 10 min and with water for 10 min. Condition new capillaries by washing with 200 mM NaOH for 10 min, with water for 10 min, and with running buffer for 10 min.
Running buffer: 50 mM pH 7.3 Sodium borate buffer containing 3.3% succinyl β-cyclodextrin (Wacker Chemie, Munich, Germany)
Injection: Hydrodynamic injection at 25 mbar for 6 s.
Detector: UV 208
Migration time: 55.24 (first enantiomer)
Voltage: 15 kV
Model: Prince

OTHER SUBSTANCES
Simultaneous: alprenolol, chlorthalidone, etilefrin, homatropine, methoxamine, norephedrine, octopamine, propranolol, synephrine, trihexyphenidyl

KEY WORDS
chiral; α = 1.1462

REFERENCE
Schmid,M.G.; Wirnsberger,K.; Gübitz,G. Chiral separation of drug enantiomers by capillary electrophoresis using succinyl-β-cyclodextrin, *Pharmazie*, **1996**, *51*, 852–854.

SAMPLE
Matrix: solutions

CAPILLARY ELECTROPHORESIS
Capillary: 37 cm × 50 μm fused-silica (30 cm to detector)
Capillary temperature: 50
Running buffer: MeOH:100 mM pH 2.5 Tris/phosphate buffer 20:80 containing 300 mg/mL epichlorohydrin-β-cyclodextrin polymer (MW 3000-5000, Cyclolab, Budapest)
Injection: Pressure injection at 0.5 psi for 8 s.
Detector: UV 214
Migration time: 10.8 (-), 11 (+)
Voltage: 14.8 kV
Model: Beckman P/ACE 2200

OTHER SUBSTANCES
Simultaneous: methamphetamine, selegiline

KEY WORDS
chiral

REFERENCE
Sevcik,J.; Stránsky,Z.; Ingelse,B.A.; Lemr,K. Capillary electrophoretic enantioseparation of selegiline, methamphetamine and ephedrine using a neutral β-cyclodextrin epichlorhydrin polymer, *J.Pharm.Biomed.Anal.*, **1996**, *14*, 1089–1094.

SAMPLE
Matrix: solutions
Sample preparation: Inject an aliquot of a 100 μg/mL solution in water.

CAPILLARY ELECTROPHORESIS
Capillary: 60 cm × 50 μm (52.5 cm to detector) (AccuSep)
Capillary preparation: Purge capillary with running buffer for 3 min between runs. Purge new capillaries under vacuum with 500 mM NaOH for 10 min, with water for 10 min, and with running buffer for 10 min.
Capillary temperature: 30
Running buffer: 25 mM pH 8.0 Na_2HPO_4/sodium tetraborate containing 50 mM (R)-N-dodecylcarbonylvaline (Prepare (R)-N-dodecoxycarbonylvaline as follows. Prepare dodecyl chloroformate by reacting 1-dodecanol with 0.33 equivalents of triphosgene in dichloromethane solution in the presence of pyridine. Add dodecyl chloroformate dropwise to (R)-valine in 1 M NaOH solution, filter, wash with hexane, recrystallize from ether/petroleum ether (J. Chromatogr. A 1994, 680, 125).)
Injection: Hydrostatic injection at 10 cm for 15 s.
Detector: UV 214
Migration time: 22.6 (+), 23.3 (-)
Voltage: 15 kV
Model: Waters Quanta 4000E
Limit of detection: 0.5% ((+) enantiomer of (-) enantiomer)

KEY WORDS
chiral

REFERENCE
Swartz,M.E.; Mazzeo,J.R.; Grover,E.R.; Brown,P.R. Validation of enantiomeric separations by micellar electrokinetic capillary chromatography using synthetic chiral surfactants, *J.Chromatogr.A*, **1996**, *735*, 303–310.

SAMPLE
Matrix: solutions

CAPILLARY ELECTROPHORESIS
Capillary: 52-55 cm × 75 μm fused-silica (45-48 cm to detector) (Polymicro Technologies)
Capillary preparation: After each run purge with running buffer for 3 min. After every 5 runs purge with 100 mM LiOH. Purge new capillaries with 1 M LiOH for 20 min, purge with 100 mM LiOH for 20 min, rinse with running buffer, equilibrate with running buffer with voltage applied for 3 h.
Running buffer: Buffer (Prepare a 100 mM CHES (2-[N-cyclohexylamine]ethanesulfonic acid) solution containing 25 mM (S)-N-dodecoxycarbonylvaline (EnantioSelect (S)-Val-1, Waters), adjust pH to 7 with 1 M LiOH, add triethylamine to a final concentration of 10 mM, adjust pH to 8.8 with 1 M LiOH. Preparation of (S)-N-dodecoxycarbonylvaline is as follows. Prepare dodecyl chloroformate by reacting 1-dodecanol with 0.33 equivalents of triphosgene in dichloromethane solution in the presence of pyridine. Add dodecyl chloroformate dropwise to (S)-valine in 1 M NaOH solution, filter, wash with hexane, recrystallize from ether/petroleum ether (J. Chromatogr. 1994, 680, 125).)
Injection: Hydrostatic injection for 2 s.
Detector: UV 214
Migration time: 12.5, 13 (enantiomers)
Voltage: 12 kV
Model: Waters Quanta 4000 CE

OTHER SUBSTANCES
Simultaneous: N-methylpseudoephedrine, metoprolol, norphenylephrine, synephrine

KEY WORDS
chiral

REFERENCE
Peterson,A.G.; Foley,J.P. Influence of the inorganic counterion on the chiral micellar electrokinetic separation of basic drugs using the surfactant N-dodecoxycarbonylvaline, *J.Chromatogr.B*, **1997**, *695*, 131–145.

SAMPLE
Matrix: solutions

CAPILLARY ELECTROPHORESIS
Capillary: 35 cm × 50 μm polyacrylamide-coated fused-silica (30.5 cm to detector) (Composite Metal Services, UK)
Capillary preparation: Before each run rinse capillary with water for 70 s and with running buffer for 100 s. (Coat capillary as follows. Adjust the pH of 20 mL water to 3.5 with acetic acid, add 80 μL 3-(trimethoxysilyl)propyl methacrylate (3-methacryloxypropyltrimethoxysilane), mix, suck into capillary, let stand at room temperature for 1 h, remove the solution, wash with water. Fill the capillary with a deaerated 3-4% acrylamide solution containing 1 μL/mL N,N,N',N'-tetramethylethylenediamine and 1 mg/mL potassium persulfate, let stand for 30 min, remove excess solution by aspiration, rinse with water, remove water by aspiration, dry at 35° (J. Chromatogr. 1985, 347, 191).)
Capillary temperature: 25
Running buffer: Buffer containing 150 mM cyanoethylated-β-cyclodextrin (Cyclolab, Budapest) (Prepare buffer by adjusting the pH of 50 mM phosphoric acid containing 50 mM acetic acid and 50 mM boric acid to 2.5 with concentrated NaOH, add the appropriate amount of cyanoethylated-β-cyclodextrin, dilute with an equal volume of water.)
Injection: Pressure injection at 5 psi for 2 s.
Detector: UV 206
Migration time: 14.4 (second enantiomer, α = 1.014)
Voltage: 20 kV
Current: 27-40 μA
Model: Bio-Rad Biofocus 3000

KEY WORDS
chiral; coated capillary

REFERENCE
Aturki,Z.; Desiderio,C.; Mannina,L.; Fanali,S. Chiral separations by capillary zone electrophoresis with the use of cyanoethylated-β-cyclodextrin as chiral selector, *J.Chromatogr.A*, **1998**, *817*, 91–104.

SAMPLE
Matrix: solutions

CAPILLARY ELECTROPHORESIS
Capillary: 57 cm × 50 μm fused-silica (50 cm to detector)
Running buffer: 50 mM pH 8.0 Borate buffer containing 40 mM sodium taurodeoxycholate and 25 mM phosphatidylcholine (Prepare by adding phosphatidylcholine and stirring for 3-6 h until all cloudiness disappears. Phosphatidylcholine was 95% pure soybean lecithin, Epikuron, Lucas Meyer & Co.)
Injection: Inject a solution of the compound in the running buffer at 20 psi for 1 s, inject MeOH: water 5:95 at 20 psi for 1 s, inject a solution of halofantrine in running buffer at 20 psi for 1 s.
Detector: UV 214
Migration time: k' 1.12
Model: Beckman P/ACE 5000

OTHER SUBSTANCES
Also analyzed: acetaminophen, amoxicillin, antipyrine, aspirin, azathioprine, caffeine, captopril, carbamazepine, carprofen, chlorambucil, chlorpheniramine, chlorpromazine, cimetidine, clonidine, codeine, desipramine, diphenhydramine, fenoterol, flufenamic acid, flurbiprofen, haloperidol, hydroxyzine, ibuprofen, imipramine, indomethacin, ketoprofen, lidocaine, melphalan, metoprolol, nabumetone, nadolol, phenobarbital, phenol, promazine, propranolol, pyrilamine, ranitidine, ropinirole, salicylic acid, sulfamethoxazole, testosterone, theophylline, thioridazine, tiaprofenic acid, tolfenamic acid, trifluoperazine, trimethoprim, valproic acid, verapamil

---

KEY WORDS
comparison with HPLC; k' = (Tr-T0)/(T0(1-Tr/Tm)) where Tr = retention time of analyte; T0 = retention time of water; and Tm = retention time of marker (halofantrine)

REFERENCE
Hanna,M.; de Biasi,V.; Bond,B.; Salter,C.; Hutt,A.J.; Camilleri,P. Estimation of the partitioning characteristics of drugs: A comparison of a large and diverse drug series utilizing chromatographic and electrophoretic methodology, *Anal.Chem.*, **1998**, *70*, 2092–2099.

SAMPLE
Matrix: solutions

CAPILLARY ELECTROPHORESIS
Capillary: 72 cm × 50 μm
Capillary temperature: 30
Running buffer: 20 mM pH 2.5 Tris-phosphate buffer containing 1.5 mM sulfobutylether-β-cyclodextrin
Detector: UV 205
Migration time: 25 (+), 24 (-)
Voltage: 30 kV
Model: Perkin-Elmer ABI 270A-HT

OTHER SUBSTANCES
Simultaneous: pseudoephedrine

KEY WORDS
chiral

REFERENCE
Nassar,A.-E.F.; Guarco,F.J.; Gran,D.E.; Stuart,J.D.; Reuter,W.M. Optimizing a method for separating chiral compounds by capillary electrophoresis, *J.Chromatogr.Sci.*, **1998**, *36*, 19–22.

SAMPLE
Matrix: solutions

CAPILLARY ELECTROPHORESIS
Capillary: 20 cm × 75 μm containing 3 μm Micra bare silica (packed for 20 cm; 20 cm to detector) (Unimicro Technologies, Pleasanton CA)
Capillary preparation: Before each run condition capillary at 5 kV until a constant current is achieved. Condition new capillaries with running buffer at 500 psi.
Capillary temperature: 25
Running buffer: MeCN:10 mM pH 8.29 Tris-HCl buffer 80:20
Injection: Electrokinetic injection at 5 kV for 5 s.
Detector: UV 214
Migration time: 9
Model: Beckman P/ACE 5510

OTHER SUBSTANCES
Simultaneous: aniline, berberine, cocaine, codeine, jatrorrhizine, thebaine

KEY WORDS
electrochromatography

REFERENCE
Wei,W.; Luo,G.A.; Hua,G.Y.; Yan,C. Capillary electrochromatographic separation of basic compounds with bare silica as stationary phase, *J.Chromatogr.A*, **1998**, *817*, 65–74.

SAMPLE
Matrix: urine
Sample preparation: Dilute 3 volumes of urine with 10 volumes of water, inject an aliquot.

CAPILLARY ELECTROPHORESIS
Capillary: 100 cm × 75 μm untreated silica (65 cm to detector) (Supelco)

Capillary preparation: Flush (under pressure) with 100 mM NaOH for 2 min, with water for 2 min, and with running buffer for 2 min.
Running buffer: MeCN:50 mM pH 9.5 phosphate buffer 1:99
Injection: Inject hydrodynamically using a vacuum for 2 s
Detector: UV 210
Migration time: 12.5
Voltage: 12.5 kV
Current: 34 μA
Model: Europhor
Limit of detection: 2.6 μg/mL (S/N 10)

REFERENCE

Chicharro,M.; Zapardiel,A.; Bermejo,E.; Perez,J.A.; Hernández,L. Direct determination of ephedrine and norephedrine in human urine by capillary zone electrophoresis, *J.Chromatogr.*, **1993**, *622*, 103–108.

SAMPLE
Matrix: urine
Sample preparation: Dilute 2 mL urine to 10 mL with water, inject an aliquot.

CAPILLARY ELECTROPHORESIS
Capillary: 100 cm × 75 μm (65 cm to detector) (Supelco)
Capillary preparation: Between runs rinse capillary with 100 mM NaOH for 2 min, with water for 2 min, and with running buffer for 2 min.
Running buffer: 40 mM phosphoric acid and 10 mM boric acid adjusted to pH with 9.7 with 1 M NaOH
Injection: Hydrodynamic injection for 2 s
Detector: UV 210
Migration time: 7.2
Voltage: 20 kV
Model: Europhor
Limit of quantitation: 1.4 μg/mL
Limit of detection: LOD 400 ng/mL (S/N 3)

OTHER SUBSTANCES
Extracted: epinephrine, methylephedrine, methylpseudoephedrine, norephedrine, norpseudoephedrine, pseudoephedrine

REFERENCE

Chicarro,M.; Zapardiel,A.; Bermejo,E.; Pérez-López,J.A.; Hernández,L. Direct determination of ephedrine alkaloids and epinephrine in human urine by capillary electrophoresis, *J.Liq.Chromatogr.*, **1995**, *18*, 1363–1381.

Epinastine

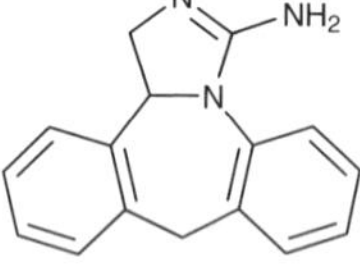

Molecular formula: $C_{16}H_{15}N_3$
Molecular weight: 249.32
CAS Registry No.: 80012-43-7 (HCl)
Merck Index (12th ed.): 3655

SAMPLE
Matrix: solutions

CAPILLARY ELECTROPHORESIS
Capillary: 36 cm × 50 μm polyacrylamide-coated fused-silica (31.5 cm to detector) (Polymicro Technologies)
Capillary preparation: Before each injection rinse with water at 100 psi for 30 s, rinse with running buffer at 100 psi for 30 s, partially fill with separation solution (750 μM bovine serum albumin in running buffer) at 1 psi for 190 s (27 cm), inject sample, electrophorese with running

buffer. Coat capillary with polyacrylamide as follows. Treat with 1 M NaOH for 1 h, rinse with water, dry at 110° by flushing with nitrogen for 6 h, pass thionyl chloride through using suction for 10 min, seal at each end, heat at 70° for 6 h, open, suck 250 mM vinylmagnesium bromide in THF into the capillary, seal at each end, heat at 70° for 6 h, open, rinse with THF for several min, rinse with water, fill by suction with a solution containing 3% acrylamide, 0.1% ammonium persulfate, and 0.1% tetramethylethylenediamine, heat at 28 ± 2° for 1 h, after 1 h wash with water (Biol.Pharm.Bull. 1993, 16, 1185).

Capillary temperature: 25
Running buffer: 50 mM pH 6.0 Phosphate buffer
Injection: Inject at 1 psi for 2 s
Detector: UV 210
Migration time: 10.20 (+), 10.86 (−)
Voltage: 12 kV
Model: Bio-Rad BioFocus 3000
Limit of quantitation: 10 µg/mL

KEY WORDS
chiral; partial separation zone technique; coated capillary

REFERENCE
Tanaka,Y.; Terabe,S. Partial separation zone technique for the separation of enantiomers by affinity electrokinetic chromatography with proteins as chiral pseudo-stationary phases, *J.Chromatogr.A*, **1995**, *694*, 277–284.

SAMPLE
Matrix: solutions

CAPILLARY ELECTROPHORESIS
Capillary: 36 cm × 50 µm fused-silica coated with linear polyacrylamide (31.5 cm to detector) (GL Science)
Capillary preparation: At the beginning and end of each day rinse capillary with capillary wash solution (Bio-Rad Cat. No. 148-5022) at 690 kPa for more than 3 min and with water at 690 kPa for more than 3 min. Coat capillary as follows. Treat capillary with 1 M NaOH at room temperature for 1 h, rinse with water, dry by passing nitrogen gas through the capillary at 110° for 6 h. Pass thionyl chloride through the capillary using a suction pump for several min, seal capillary at both ends and heat at 70° for 6 h. Unseal the capillary and fill with 250 mM vinyl magnesium bromide in THF by suction, seal the capillary, heat at 70° for 6 h. Open the capillary and rinse it with THF for several min, rinse with distilled water, fill the capillary with polymerization solution, heat at 28 ± 2° for 1 h, condition at -100 V/cm for 30 min (Anal. Sci. 1994, 10, 1). (The polymerization solution was 5% acrylamide in water containing 49 mM Tris, 384 mM glycine, and 0.1% sodium dodecyl sulfate, degas in an ultrasonic bath. Add 40 µL 10% N,N,N',N'-tetramethylethylenediamine and 10 µL 10% ammonium persulfate to 5 mL of the degassed solution, mix thoroughly.)
Running buffer: n-Propanol:50 mM pH 6.0 Phosphate buffer 10:90
Injection: Before each injection rinse with water at 690 kPa for 30 s, rinse with running buffer at 690 kPa for 30 s, partially fill with separation solution (100 µM α₁-acid glycoprotein (Cohn fraction VI) (Sigma) in running buffer) at 6.9 kPa for 190 s (27 cm), inject sample at 6.9 kPa for 2 s, electrophorese with running buffer (Note that α₁-acid glycoprotein from other suppliers may provide inferior results).
Detector: UV 210
Migration time: Resolution of enantiomers 2.1
Voltage: 12 kV
Model: Bio-rad BioFocus 3000

KEY WORDS
chiral; coated capillary

REFERENCE
Tanaka,Y.; Terabe,S. Separation of the enantiomers of basic drugs by affinity capillary electrophoresis using a partial filling technique and α₁-acid glycoprotein as chiral selector, *Chromatographia*, **1997**, *44*, 119–128.

Epinephrine

Molecular formula: $C_9H_{13}NO_3$
Molecular weight: 183.21
CAS Registry No.: 51-43-4 (-), 51-42-3 ((-) bitartrate), 329-65-7 (racemic)
Merck Index (12th ed.): 3656
Lednicer: 1 95

SAMPLE
Matrix: blood
Sample preparation: Inject an aliquot of 5-fold diluted bovine serum directly.

CAPILLARY ELECTROPHORESIS
Capillary: 50 cm × 50 μm (20 cm to detector) (GL Science)
Capillary preparation: Rinse capillary with running buffer for 5 min before each run. At the beginning of each day rinse capillary with 100 mM NaOH for 10 min and with water for 5 min.
Running buffer: 10 mM pH 2.5 Phosphate buffer containing 25 mM sodium dodecyl sulfate and 12.5 mM Tween 20
Injection: Siphon at 10 cm for 10 s.
Detector: UV 210
Migration time: 13
Voltage: 18 kV
Current: 21-22 μA
Model: Jasco CE-800

OTHER SUBSTANCES
Extracted: dopamine, levodopa, 3-methoxytyrosine, norepinephrine, phenylalanine, tyrosine

KEY WORDS
serum; cow; detector is at anode

REFERENCE
Esaka,Y.; Tanaka,K.; Uno,B.; Goto,M.; Kano,K. Sodium dodecyl sulfate-Tween 20 mixed micellar electrokinetic chromatography for separation of hydrophobic cations: Application to adrenaline and its precursors, *Anal.Chem.*, **1997**, *69*, 1332–1338.

SAMPLE
Matrix: cells
Sample preparation: 5 μL Cell suspension + 200 μL water, inject an aliquot (cells lyse on contact with running buffer).

CAPILLARY ELECTROPHORESIS
Capillary: 65 × 16 μm cm fused-silica (45 cm to detector) (Polymicro Technologies)
Capillary preparation: Every few runs flush capillary with running buffer, equilibrate for 5 min.
Running buffer: 100 mM pH 2.3 Citric acid
Injection: Hydrodynamic
Detector: F ex 275 (Ar laser (or frequency doubled Kr laser at 284 nm))
Migration time: 3.75
Voltage: 30 kV
Model: Laboratory constructed

OTHER SUBSTANCES
Extracted: norepinephrine

REFERENCE
Chang,H.-T.; Yeung,E.S. Determination of catecholamines in single adrenal medullary cells by capillary electrophoresis and laser-induced native fluorescence, *Anal.Chem.*, **1995**, *67*, 1079–1083.

SAMPLE
Matrix: formulations
Sample preparation: Dilute formulation 50-fold with water. Remove a 9.8 mL aliquot and add
it to 200 μL 1 mM (-)-isoproterenol, make up to 20 mL with water, inject an aliquot.

CAPILLARY ELECTROPHORESIS
Capillary: 20 cm $\times$ 25 μm coated fused-silica (Bio-Rad 148-3002)
Running buffer: 100 mM pH 2.5 NaH_2PO_4 buffer containing 20 mM di-O-methyl-β-cyclodextrin
Injection: Injection by electromigration at 7 kV for 7 s
Detector: UV 206
Migration time: 3.7 (-), 4 (+)
Internal standard: (-)-isoproterenol (4.5)
Voltage: 8 kV
Current: 14.6 μA
Model: Bio-Rad HPE 100

KEY WORDS
injections; chiral; coated capillary

REFERENCE
Fanali,S.; Bocek,P. Enantiomer resolution by using capillary zone electrophoresis: resolution of racemic tryptophan and determination of the enantiomer composition of commercial pharmaceutical epinephrine, *Electrophoresis*, **1990**, *11*, 757–760.

SAMPLE
Matrix: formulations
Sample preparation: Dilute formulation to an epinephrine concentration of 25 ppm with 10
mM HCl containing 100 mM l-pseudoephedrine, inject an aliquot.

CAPILLARY ELECTROPHORESIS
Capillary: 50 cm $\times$ 75 μm unmodified fused silica (45 cm to detector) (Polymicro Technologies)
Capillary preparation: Rinse with running buffer after each injection. At the beginning of each
day rinse the capillary twice for 3 min with 100 mM phosphoric acid and twice with 500 mM
NaOH, rinse entire system four times with water and five times with running buffer.
Running buffer: 10 mM Tris buffer containing 18 mM heptakis-(2,6-di-O-methyl)-β-cyclodextrin
adjusted to pH 2.4 with concentrated phosphoric acid
Injection: Gravity injection for 10 s at 50 mm.
Detector: UV 206
Migration time: 10.20 (l), 10.51 (d)
Internal standard: pseudoephedrine (11.75)
Voltage: 15 kV
Current: 30-40 μA
Model: Dionex CES

KEY WORDS
chiral

REFERENCE
Peterson,T.E.; Trowbridge,D. Quantitation of *l*-epinephrine and determination of the *d-/l*-epinephrine enantiomer ratio in a pharmaceutical formulation by capillary electrophoresis, *J.Chromatogr.*, **1992**, *603*, 298–
301.

SAMPLE
Matrix: formulations
Sample preparation: Inject an aliquot of the injection directly.

CAPILLARY ELECTROPHORESIS
Capillary: 57 cm $\times$ 75 μm fused-silica (Polymicro Technologies)
Capillary preparation: Before each run rinse capillary at high pressure with 100 mM NaOH
for 1.5 min then with running buffer for 4 min. Rinse new capillaries under high pressure with

1 M NaOH for 5 min and with running buffer for 10 min then equilibrate overnight with running buffer.
Capillary temperature: 25
Running buffer: 160 mM pH 10.1 Borate buffer containing 1 mM EDTA
Injection: Pressure injection at 0.5 psi for 25 s (150 nL).
Detector: UV 280
Migration time: 12
Voltage: 15 kV
Model: Beckman P/ACE 5500
Limit of detection: 90 ng/mL

OTHER SUBSTANCES
Noninterfering: methylparaben, xylocaine
Interfering: lidocaine

KEY WORDS
injections

REFERENCE
Britz-McKibbin,P.; Kranack,A.R.; Paprica,A.; Chen,D.D.Y. Quantitative assay for epinephrine in dental anesthetic solutions by capillary electrophoresis, *Analyst*, **1998**, *123*, 1461–1463.

SAMPLE
Matrix: solutions

CAPILLARY ELECTROPHORESIS
Capillary: 87.9 cm × 26 μm fused-silica (Polymicro Technologies)
Running buffer: 20 mM pH 6.05 2-morpholinoethanesulfonic acid buffer
Injection: Inject at 5 kV for 2 s
Detector: E, 0.7 V vs sodium-saturated calomel electrode (details of detector construction in paper)
Migration time: 8.4
Voltage: 25 kV
Current: 2 μA
Limit of detection: 0.2-0.4 fmole (S/N 3)

OTHER SUBSTANCES
Simultaneous: catechol, dopamine, norepinephrine

REFERENCE
Wallingford,R.A.; Ewing,A.G. Amperometric detection of catechols in capillary zone electrophoresis with normal and micellar solutions, *Anal.Chem.*, **1988**, *60*, 258–263.

SAMPLE
Matrix: solutions

CAPILLARY ELECTROPHORESIS
Capillary: 70 cm × 100 μm fused-silica (50 cm to detector) (Gasukuro Kogyo, Tokyo)
Running buffer: 100 mM KOH containing 200 mM boric acid
Injection: Hydrostatic injection for 10 s.
Detector: UV 217
Migration time: 14
Voltage: 9.5 kV
Current: 100 μA

OTHER SUBSTANCES
Simultaneous: dopamine, isoproterenol, levodopa, metanephrine, norepinephrine, normetanephrine, vanillylmandelic acid

REFERENCE
Tanaka,S.; Kaneta,T.; Yoshida,H. Separation of catecholamines by capillary zone electrophoresis using complexation with boric acid, *Anal.Sci.*, **1990**, *6*, 467–468.

SAMPLE
Matrix: solutions

CAPILLARY ELECTROPHORESIS
Capillary: 57 cm × 50 μm fused-silica (50 cm to detector)
Capillary temperature: 25
Running buffer: 100 mM pH 2.5 NaH$_2$PO$_4$ buffer containing 1.5 M urea and 30 mM heptakis
(heptakis(2,6-di-O-methyl)-β-cyclodextrin or heptakis(2,3,6-tri-O-methyl)-β-cyclodextrin (?))
Injection: Pressure injection for 6 s
Detector: UV 200
Migration time: 15.5 (D), 16 (L)
Voltage: 15 kV
Model: Beckman PACE 2100

KEY WORDS
chiral

REFERENCE
McLaughlin,G.M.; Nolan,J.A.; Lindahl,J.L.; Palmieri,R.H.; Anderson,K.W.; Morris,S.C.; Morrison,J.A.; Bronzert,T.J. Pharmaceutical drug separations by HPCE: Practical guidelines, *J.Liq.Chromatogr.*, **1992**, *15*, 961–
1021.

SAMPLE
Matrix: solutions

CAPILLARY ELECTROPHORESIS
Capillary: 100 cm × 50 μm fused-silica (50 cm to detector) (Isco)
Capillary preparation: Flush capillary with 10 μL running buffer between runs. Every 40 sam-
ple injections rinse capillary with 200 μL 1 M NaOH, with 200 μL water, and 200 μL running
buffer. Before use fill capillary with 1 M NaOH and allow to stand for 1 h, fill with 100 mM
NaOH, allow to stand for 1 h, wash with water fill with running buffer.
Capillary temperature: 23
Running buffer: 100 mM citric acid containing 19.27 mM Na$_2$HPO$_4$ and 120 mM hydroxypropyl-
β-cyclodextrin
Injection: Inject under vacuum at 4.0 kPa.s
Detector: UV 200
Migration time: 27 (S), 28.5 (R)
Voltage: 30 kV
Model: Isco Model 3140

OTHER SUBSTANCES
Simultaneous: amphetamine, 4-bromo-2,5-dimethoxyamphetamine, ephedrine, methamphetamine, methyldimethoxyethylamphetamine, methyldimethoxyamphetamine, methyldimethoxymethylamphetamine, pseudoephedrine

KEY WORDS
chiral

REFERENCE
Aumatell,A.; Wells,R.J.; Wong,D.K.Y. Enantiomeric differentiation of a wide range of pharmacologically active
substances by capillary electrophoresis using modified β-cyclodextrins, *J.Chromatogr.A*, **1994**, *686*, 293–
307.

SAMPLE
Matrix: solutions
Sample preparation: Inject an aliquot of a solution in running buffer.

CAPILLARY ELECTROPHORESIS
Capillary: 44 cm × 5 μm fused-silica (Polymicro Technologies)
Running buffer: Isopropanol:30 mM morpholinoethanesulfonic acid 20:80, pH adjusted to 6.0
Injection: Electromigration at 12 kV for 3 s

Detector: E, Pt electrode 0.700 V, Ag/AgCl reference electrode, details of cell in paper
Migration time: 7.5
Voltage: 25 kV

OTHER SUBSTANCES
Simultaneous: catechol, dopamine, isoproterenol, norepinephrine, serotonin

REFERENCE
Chen,M.-C.; Huang,H.-J. An electrochemical cell for end-column amperometric detection in capillary electrophoresis, *Anal.Chem.*, **1995**, *67*, 4010–4014.

SAMPLE
Matrix: solutions
Sample preparation: Prepare a solution in MeOH/water, inject an aliquot.

CAPILLARY ELECTROPHORESIS
Capillary: 62 cm × 52 μm fused-silica (50 cm to detector) (Polymicro Technologies)
Capillary temperature: 40
Running buffer: 50 mM pH 2.50 Tetrabutylammonium phosphate
Injection: Siphon injection at 10 cm for 5 s
Detector: UV (wavelength not specified)
Migration time: 10.5
Voltage: 20 kV
Current: 29 μA
Model: Laboratory constructed

OTHER SUBSTANCES
Simultaneous: doxylamine, imidazole, isoproterenol, 2-methylphenethylamine, 1-methylphenylpropylamine, metoprolol, nadolol, nicotine, norepinephrine, propranolol, pseudoephedrine

REFERENCE
Quang,C.; Khaledi,M.G. Extending the scope of chiral separation of basic compounds by cyclodextrin-mediated capillary zone electrophoresis, *J.Chromatogr.A*, **1995**, *692*, 253–265.

SAMPLE
Matrix: solutions
Sample preparation: Inject an aliquot of a 300 μg/mL solution in water.

CAPILLARY ELECTROPHORESIS
Capillary: 65 cm × 50 μm fused-silica (40 cm to detector) (Isco)
Capillary preparation: Purge with running buffer for 3 min between injections. Purge daily with 1 M NaOH for 3 min, with water for 3 min, and with running buffer for 3 min.
Running buffer: Isopropanol:100 mM pH 7 phosphate buffer containing 25 mM rifamycin B 30:70
Injection: Electrokinetic injection at 5 kV for 5 s.
Detector: UV 350
Migration time: 48, 50 (enantiomers)
Voltage: 8 kV
Model: Isco model 3850

OTHER SUBSTANCES
Simultaneous: alprenolol, amphetamine, metoprolol, norepinephrine, normetanephrine, octapamine, oxprenolol, propranolol

KEY WORDS
chiral

REFERENCE
Ward,T.J.; Dann,C.,III; Blaylock,A. Enantiomeric resolution using the macrocyclic antibiotics rifamycin B and rifamycin SV as chiral selectors for capillary electrophoresis, *J.Chromatogr.A*, **1995**, *715*, 337–344.

SAMPLE
Matrix: solutions

CAPILLARY ELECTROPHORESIS
Capillary: 65 cm × 50 μm coated fused-silica (65 cm to detector) (Polymicro Technologies)
Capillary preparation: Rinse with running buffer for 2 min between runs. Coat capillary as
   follows. Flush capillary with 1 M NaOH for 1 h, rinse with water for 1 h, dry at 180° with a
   flow of nitrogen overnight, fill the capillary with reagent using a syringe, let stand for 10 min,
   force out excess reagent using pressure, dry at 90° with a flow of nitrogen overnight, wash with
   acetone for 30 min, wash with MeOH for 30 min, rinse with water for 20 min, equilibrate with
   running buffer for 1 h before use. (Prepare reagent by mixing 500 μL 3-aminopropyltriethox-
   ysilane, 375 μL EtOH, 54 μL water, and 100 μL 1.2 m HCl, stir at room temperature for 10 h
   before use.)
Capillary temperature: 22
Running buffer: 10 mM pH 3.4 Phosphate buffer
Injection: Hydrodynamic injection at 10 cm for 5 s.
Detector: UV 214
Migration time: 4.7
Voltage: -24 kV
Model: Beckman P/ACE Model 2200

OTHER SUBSTANCES
Simultaneous: norepinephrine

KEY WORDS
coated capillary

REFERENCE
Guo,Y.; Imahori,G.A.; Colón,L.A. Hydrolytically stable amino-silica glass coating material for manipulation of
   the electroosmotic flow in capillary electrophoresis, *J.Chromatogr.A*, **1996**, *744*, 17–29.

SAMPLE
Matrix: solutions
Sample preparation: Inject an aliquot of a 5-10 μM solution.

CAPILLARY ELECTROPHORESIS
Capillary: 70 cm × 75 μm fused-silica (55 cm to detector)
Capillary temperature: 21
Running buffer: 20 mM pH 2.7 Tris-phosphate buffer containing 0.5% hydroxypropylmethyl
   cellulose and 24 mM heptakis(2,6-di-O-methyl)-β-cyclodextrin (Hydroxypropylmethyl cellulose
   was from Sigma; viscosity of 2% solution = 4000 cP at 25°.)
Injection: Electrokinetic injection at 3 kV for 12 s.
Detector: UV 190
Migration time: 18.2, 18.7 (enantiomers)
Voltage: 21 kV
Model: Crystal 300 (ATI, Unicam)

OTHER SUBSTANCES
Simultaneous: amphetamine, methamphetamine, norepinephrine, selegiline (deprenyl)
Interfering: pseudoephedrine

KEY WORDS
chiral

REFERENCE
Szökö,E.; Gyimesi,J.; Barcza,L.; Magyar,K. Determination of binding constants and the influence of methanol
   on the separation of drug enantiomers in cyclodextrin modified capillary electrophoresis, *J.Chromatogr.A*,
   1996, *745*, 181–187.

SAMPLE
Matrix: solutions

CAPILLARY ELECTROPHORESIS
Capillary: 100 cm × 50 μm fused-silica (Polymicro Technologies)
Running buffer: 20 mM pH 2.5 Sodium citrate
Injection: Electrokinetic injection.
Detector: E, Bioanalytical Systems BAS LC-4C, 25 μm gold wire electrode (design in paper) +800 mV, Pt auxiliary electrode, Ag/AgCl reference electrode
Migration time: 17.7
Voltage: 30 kV
Limit of quantitation: 500 nM

OTHER SUBSTANCES
Simultaneous: dopamine, norepinephrine

REFERENCE
Zhong,M.; Lunte,S.M. Integrated on-capillary electrochemical detector for capillary electrophoresis, *Anal.Chem.*, **1996**, *68*, 2488–2493.

SAMPLE
Matrix: solutions

CAPILLARY ELECTROPHORESIS
Capillary: 37 cm × 50 μm fused-silica (30 cm to detector) (Quadrex, New Haven CT)
Capillary temperature: -16
Running buffer: MeOH:water 10:90 containing 20 mM heptakis(2,6-di-O-methyl)-β-cyclodextrin, 5 M urea, and 150 mM sodium phosphate, pH 2.5
Injection: Injection for 6 s
Detector: UV 214
Migration time: 14, 15 (enantiomers)
Voltage: 30 kV
Current: 45 μA
Model: Beckman P/ACE 2210

OTHER SUBSTANCES
Simultaneous: β-hydroxyphenethylamine, isoproterenol, norepinephrine, octopamine

KEY WORDS
chiral

REFERENCE
Ma,S.; Horváth,C. Capillary zone electrophoresis at subzero temperatures. II: Chiral separation of biogenic amines, *Electrophoresis*, **1997**, *18*, 873–883.

SAMPLE
Matrix: solutions

CAPILLARY ELECTROPHORESIS
Capillary: 35 cm × 50 μm polyacrylamide-coated fused-silica (30.5 cm to detector) (Composite Metal Services, UK)
Capillary preparation: Before each run rinse capillary with water for 70 s and with running buffer for 100 s. (Coat capillary as follows. Adjust the pH of 20 mL water to 3.5 with acetic acid, add 80 μL 3-(trimethoxysilyl)propyl methacrylate (3-methacryloxypropyltrimethoxysilane), mix, suck into capillary, let stand at room temperature for 1 h, remove the solution, wash with water. Fill the capillary with a deaerated 3-4% acrylamide solution containing 1 μL/mL N,N,N',N'-tetramethylethylenediamine and 1 mg/mL potassium persulfate, let stand for 30 min, remove excess solution by aspiration, rinse with water, remove water by aspiration, dry at 35° (J. Chromatogr. 1985, 347, 191).)
Capillary temperature: 25
Running buffer: Buffer containing 100 mM cyanoethylated-β-cyclodextrin (Cyclolab, Budapest) (Prepare buffer by adjusting the pH of 50 mM phosphoric acid containing 50 mM acetic acid and 50 mM boric acid to 2.5 with concentrated NaOH, add the appropriate amount of cyanoethylated-β-cyclodextrin, dilute with an equal volume of water.)

Injection: Pressure injection at 5 psi for 2 s.
Detector: UV 206
Migration time: 11.4 (second enantiomer, α = 1.056)
Voltage: 20 kV
Current: 27-40 μA
Model: Bio-Rad Biofocus 3000

KEY WORDS
chiral; coated capillary

REFERENCE
Aturki,Z.; Desiderio,C.; Mannina,L.; Fanali,S. Chiral separations by capillary zone electrophoresis with the use of cyanoethylated-β-cyclodextrin as chiral selector, *J.Chromatogr.A*, **1998**, *817*, 91–104.

SAMPLE
Matrix: solutions

CAPILLARY ELECTROPHORESIS
Capillary: 85 cm $\times$ 75 μm fused-silica (Composite Metal Services, Hallow, UK)
Running buffer: pH 10 Borate buffer containing 0.5 mM EDTA (Run with a pressure of 20 mbar at the injection end.)
Injection: Hydrodynamic injection at 40 mbar for 6 s.
Detector: E, Bioanalytical Systems Unijet cell +0.8 V, 1 mm dia. glassy carbon working electrode, Ag/AgCl reference electrode, stainless steel auxiliary electrode, two 16 μm gaskets (design details in paper)
Migration time: 6.1
Model: Prince
Limit of detection: 10-20 nM

OTHER SUBSTANCES
Simultaneous: dopamine, norepinephrine

REFERENCE
Durgbanshi,A.; Kok,W.T. Capillary electrophoresis and electrochemical detection with a conventional detector cell, *J.Chromatogr.A*, **1998**, *798*, 289–296.

SAMPLE
Matrix: solutions

CAPILLARY ELECTROPHORESIS
Capillary: 40 cm $\times$ 25 μm fused-silica (Polymicro Technologies)
Capillary preparation: Before use rinse capillary with 100 mM NaOH for 5 min and with running buffer for 5 min.
Running buffer: 50 mM pH 7.0 $Na_2HPO_4/Na_2B_4O_7$ buffer
Injection: Electrokinetic injection at 5 kV for 10 s.
Detector: E, 25 μm AU electrode, Ag/AgCl reference electrode, stainless steel auxiliary electrode, prescan pulse -500 mV for 30 ms, postscan pulse 1000 mV for 30 s, dc ramp 100-500 mV in 56 ms, square wave 2000 Hz, 50 mV, quantitation method average peak current (first 30% of forward and reverse pulse current response rejected), 70 mV detection bandwith, 15 points included in running average, construction details in paper
Migration time: 8.5
Voltage: 30 kV
Limit of quantitation: 500 nM

OTHER SUBSTANCES
Simultaneous: dopamine

REFERENCE
Gerhardt,G.C.; Cassidy,R.M.; Baranski,A.S. Square-wave voltammetry detection for capillary electrophoresis, *Anal.Chem.*, **1998**, *70*, 2167–2173.

SAMPLE
Matrix: solutions
Sample preparation: Inject an aliquot of an aqueous solution

CAPILLARY ELECTROPHORESIS
Capillary: 53 cm × 75 μm fused-silica (40.5 cm to detector) (Polymicro Technologies)
Running buffer: 5 mM pH 7.3 NaH_2PO_4 containing 5 mM sodium borate and 20 mM sodium dodecyl sulfate (A) or 10 mM pH 3.1 NaH_2PO_4 (B)
Injection: Electrokinetic injection at 17 kV for 9 s.
Detector: UV 204
Migration time: 4 (A), 9 (B)
Voltage: 17 kV
Current: 48 μA (A) or 26 μA (B)
Model: laboratory-constructed
Limit of detection: 680 nM

OTHER SUBSTANCES
Simultaneous: 3,4-dihydroxybenzylamine, dopamine, norepinephrine

REFERENCE
Shakulashvili,N.; Finkler,C.; Engelhardt,H. Separation of catecholamines and serotonin by micellar electrokinetic chromatography with UV detection, *Chromatographia*, **1998**, *47*, 89–92.

SAMPLE
Matrix: solutrions

CAPILLARY ELECTROPHORESIS
Capillary: 50 cm × 75 μm fused-silica (36 cm to detector) (SGE)
Capillary preparation: Fill capillary with 100 mM KOH for 10 min, wash several times with running buffer.
Running buffer: 10 mM Tris adjusted to pH 6.4 with phosphoric acid
Injection: Siphon at 10 cm for 10 s
Detector: UV 206
Migration time: 3.4
Internal standard: dichloroisoproterenol (4.4)
Voltage: 16 kV
Model: Laboratory constructed

OTHER SUBSTANCES
Simultaneous: dopamine, isoxsuprine, norepinephrine

REFERENCE
Fanali,S.; Cristalli,M.; Nardi,A.; Ossicini,L.; Shukla,S.K. Capillary zone electrophoresis in pharmaceutical analysis, *Farmaco*, **1990**, *45*, 693–702.

SAMPLE
Matrix: urine
Sample preparation: Dilute 2 mL urine to 10 mL with water, inject an aliquot.

CAPILLARY ELECTROPHORESIS
Capillary: 100 cm × 75 μm (65 cm to detector) (Supelco)
Capillary preparation: Between runs rinse capillary with 100 mM NaOH for 2 min, with water for 2 min, and with running buffer for 2 min.
Running buffer: 40 mM phosphoric acid and 10 mM boric acid adjusted to pH with 9.7 with 1 M NaOH
Injection: Hydrodynamic injection for 2 s
Detector: UV 210
Migration time: 7.8
Voltage: 20 kV
Model: Europhor
Limit of quantitation: 1 μg/mL

Limit of detection: LOD 300 ng/mL (S/N 3)

OTHER SUBSTANCES
Extracted: ephedrine, methylephedrine, methylpseudoephedrine, norephedrine, norpseudoephedrine, pseudoephedrine

REFERENCE
Chicarro,M.; Zapardiel,A.; Bermejo,E.; Pérez-López,J.A.; Hernández,L. Direct determination of ephedrine alkaloids and epinephrine in human urine by capillary electrophoresis, *J.Liq.Chromatogr.*, **1995**, *18*, 1363–1381.

SAMPLE
Matrix: urine
Sample preparation: Condition a light alumina B SPE cartridge (Waters) with water. Adjust urine to pH 3 with 6 M HCl. 10 mL Acidified urine + 1 mL 200 mM EDTA + 100 μL 500 mM ascorbic acid, adjust pH to 8.5 with NaOH solution, add to the SPE cartridge, wash with two 2 mL portions of water, elute with 1 mL 100 mM HCl, inject an aliquot of the eluate.

CAPILLARY ELECTROPHORESIS
Capillary: 80 cm × 75 μm fused-silica (Composite Metal Services, Worcs., UK)
Running buffer: 30 mM pH 10 Borate buffer containing 0.5 mM EDTA and 1 mM ascorbic acid
Injection: Hydrodynamic injection at 40 mbar for 6 s.
Detector: F ex (Varian Schott glass UG11 filter) em 500 (cut-off filter) [ex 300 em 545] following post-column reaction. The end of the capillary was connected to a 11.5 cm × 75 μm fused-silica capillary by means of a porous PTFE sleeve with a narrow gap between the capillaries (described in the paper). The detector was 5.5 cm after the junction. The porous sleeve was immersed in a reagent reservoir. The reservoir was grounded and was under 30 mbar air pressure. (Reagent was 2 mM terbium(III) chloride containing 2 (?) mM EDTA and 800 mM CsCl.)
Migration time: 6
Voltage: 30 kV
Model: Prince
Limit of detection: 70 nM

OTHER SUBSTANCES
Extracted: 3,4-dihydroxyphenylacetic acid, 3,4-dihydroxyphenyl glycol, dopamine, levodopa, norepinephrine

KEY WORDS
SPE; post-column reaction

REFERENCE
Zhu,R.; Kok,W.T. Determination of catecholamines and related compounds by capillary electrophoresis with post-column terbium complexation and sensitized luminescence detection, *Anal.Chem.*, **1997**, *69*, 4010–4016.

Epirubicin

Molecular formula: $C_{27}H_{29}NO_{11}$
Molecular weight: 543.53
CAS Registry No.: 56420-45-2, 56390-09-1 (HCl)
Merck Index (12th ed.): 3660

SAMPLE
Matrix: blood
Sample preparation: 900 μL Plasma + 100 μL 1 μg/mL daunorubicin + 2 mL chloroform, vortex for 1 min, centrifuge at 1000 g for 10 min. Remove 1.6 mL of the lower organic layer and add it to 100 μL 5 mM pH 2.3 phosphoric acid, vortex for 1 min, centrifuge at 1000 g for 10 min. Remove 50 μL of the upper aqueous layer and add it to 150 μL MeCN, mix, inject an aliquot of this mixture.

CAPILLARY ELECTROPHORESIS
Capillary: 70 cm × 75 μm fused-silica (65 cm to detector) (SGE)
Running buffer: MeCN:100 mM pH 4.2 sodium phosphate buffer 70:30
Injection: Electrokinetic injection at 12 kV for 5 s (14 nL)
Detector: F, laboratory constructed with an 80 mW 476.5 nm argon laser, 595 nm emission filter (10 nm bandwidth)
Migration time: 7.3
Internal standard: daunorubicin (7.0)
Voltage: 20-25 kV
Current: 35 μA (fixed)
Limit of detection: 70 pg/mL (S/N 3)

OTHER SUBSTANCES
Extracted: doxorubicin

KEY WORDS
plasma

REFERENCE
Reinhoud,N.J.; Tjaden,U.R.; Irth,H.; van der Greef,J. Bioanalysis of some anthracyclines in human plasma by capillary electrophoresis with laser-induced fluorescence detection, *J.Chromatogr.*, **1992**, *574*, 327–334.

Epithiazide

Molecular formula: $C_{10}H_{11}ClF_3N_3O_4S_3$
Molecular weight: 425.86
CAS Registry No.: 1764-85-8
Merck Index (12th ed.): 3661
Lednicer: 1 359

SAMPLE
Matrix: solutions

CAPILLARY ELECTROPHORESIS
Capillary: 40.6 cm × 50 μm fused-silica packed with 3 μm CEC Hypersil ODS (Hypersil)
Running buffer: Gradient. A was 5 mM ammonium acetate in MeCN:water 50:50. B was 5 mM ammonium acetate in MeCN:water 80:20. A:B 100:0 for 3 min, to 0:100 over 0.1 min, maintain at 0:100 (pumped with an HPLC pump at 0.01 mL/min for 3 min then at 0.1 mL/min).
Injection: Inject 5 μL using an HPLC injector.

Detector: MS, VG Biotech Platform, electrospray (details in paper)
Migration time: 28
Voltage: 30 kV

OTHER SUBSTANCES
Simultaneous: bendroflumethiazide, hydroflumethiazide, methyclothiazide, metolazone

KEY WORDS
electrochromatography

REFERENCE
Taylor,M.R.; Teale,P. Gradient capillary electrochromatography of drug mixtures with UV and electrospray
  ionisation mass spectrometric detection, *J.Chromatogr.A*, **1997**, *768*, 89–95.

Epoetin

Molecular formula: $C_{809}H_{1301}N_{229}O_{240}S_5$
Molecular weight: 30400 ± 400
CAS Registry No.: 113427-24-0 (alfa), 122312-54-3 (beta)
Merck Index (12th ed.): 3729

```
H—Ala—Pro—Pro—Arg—Leu—Ile—Cys—Asp—Ser—Arg—Val—Leu—Glu—Arg—Tyr
                          |                                     |
Ala—Cys—Gly—Thr—Thr—Ile—Asn—Glu—Ala—Glu—Lys—Ala—Glu—Leu—Leu
  |   |
Glu—His—Cys—Ser—Leu—Asn—Glu—Asn—Ile—Thr—Val—Pro—Asp—Thr—Lys
                                                              |
Ala—Gln—Gln—Gly—Val—Glu—Met—Arg—Lys—Trp—Ala—Tyr—Phe—Asn—Val
  |
Val—Glu—Val—Trp—Gln—Gly—Leu—Ala—Leu—Leu—Ser—Glu—Ala—Val—Leu
                                                              |
Pro—Glu—Trp—Pro—Gln—Ser—Ser—Asn—Val—Leu—Leu    Ala—Gln—Gly—Arg
  |
Leu—Gln—Leu—His—Val—Asp—Lys—Ala—Val—Ser—Gly—Leu—Arg—Ser—Leu
                                                              |
Ser—Ile—Ala—Glu—Lys—Gln—Ala—Gly—Leu—Ala—Arg—Leu—Leu—Thr—Thr
  |
Pro—Pro—Asp—Ala—Ala—Ser—Ala—Ala—Pro—Leu—Arg—Thr—Ile—Thr—Ala
                                                              |
Arg—Leu—Phe—Asn—Ser—Tyr—Val—Arg—Phe—Leu—Lys—Arg—Phe—Thr—Asp
  |
Gly—Lys—Leu—Lys—Leu—Leu—Tyr—Thr—Gly—Glu—Ala—Cys—Arg—Thr—Gly—Asp—OH
```

SAMPLE
Matrix: formulations
Sample preparation: Dilute with water, inject an aliquot.

CAPILLARY ELECTROPHORESIS
Capillary: 47 cm × 50 µm (40 cm to detector)

Capillary temperature: 20
Running buffer: 200 mM pH 4.0 Sodium phosphate buffer containing 1 mM nickel chloride
Injection: Hydrodynamic injection at 0.5 psi of nitrogen for 8 s.
Detector: UV 200
Migration time: 31-34
Voltage: 8 kV
Model: Beckman P/ACE 5500
Limit of quantitation: 30 μg/mL
Limit of detection: 10 μg/mL

OTHER SUBSTANCES
Noninterfering: human serum albumin

KEY WORDS
detector at anode; human; recombinant

REFERENCE
Bietlot,H.P.; Girard,M. Analysis of recombinant human erythropoietin in drug formulations by high-performance capillary electrophoresis, *J.Chromatogr.A*, **1997**, *759*, 177–184.

SAMPLE
Matrix: solutions
Sample preparation: Inject an aliquot of a 1 mg/mL solution.

CAPILLARY ELECTROPHORESIS
Capillary: 50 cm $\times$ 75 μm uncoated (Polymicro Technologies)
Capillary preparation: Rinse with 100 mM NaOH for 3 min, with water for 3 min, and with
running buffer for 3 min before each run.
Running buffer: 10 mM Tricine containing 10 mM NaCl, 2.5 mM 1,2-diaminobutane, and 7 M
urea, pH 6.2
Injection: Inject with low pressure for 5 s
Detector: UV 214
Migration time: 30-34 (multiple peaks)
Voltage: 10 kV
Model: Beckman PACE

REFERENCE
Watson,E.; Yao,F. Capillary electrophoretic separation of human recombinant erythropoietin (r-HuEPO) glycoforms, *Anal.Biochem.*, **1993**, *210*, 389–393.

Epoprostenol

Molecular formula: $C_{20}H_{32}O_5$
Molecular weight: 352.47
CAS Registry No.: 35121-78-9, 61849-14-7 (sodium salt)
Merck Index (12th ed.): 8061
Lednicer: 3 10

SAMPLE
Matrix: solutions

CAPILLARY ELECTROPHORESIS
Capillary: 27 cm $\times$ 75 μm
Capillary preparation: Before each run rinse capillary with 100 mM NaOH for 30 s and with
running buffer for 30 s. Condition new capillaries by rinsing with 100 mM NaOH for 20 min.
Capillary temperature: 30
Running buffer: 15 mM Sodium borate

Injection: Pressure injection of sample at 25 mbar for 1 s followed by running buffer at 25 mbar for 1 s.
Detector: UV 200
Migration time: 2.70
Internal standard: aminobenzoic acid (3.5)
Voltage: 6.5 kV
Model: Beckman

OTHER SUBSTANCES
Simultaneous: aspirin, bacitracin, beclomethasone, benzoic acid, ceftriaxone, cromolyn, embonic acid, glyburide (glibenclamide), levothyroxine, nedocromil, nystatin, omeprazole, prednisolone, warfarin, zidovudine
Interfering: ceftizoxime, cefuroxime, cephalothin

REFERENCE
Altria,K.D.; Bryant,S.M.; Hadgett,T.A Validated capillary electrophoresis method for the analysis of a range of acidic drugs and excipients, *J.Pharm.Biomed.Anal.*, **1997**, *15*, 1091–1101.

Equilenin

Molecular formula: $C_{18}H_{18}O_2$
Molecular weight: 266.34
CAS Registry No.: 517-09-9
Merck Index (12th ed.): 3675

SAMPLE
Matrix: urine
Sample preparation: Condition a 3 mL 200 mg Bond Elut C18 SPE cartridge with 3 mL water, 3 mL MeCN, and 3 mL water. 3 mL Urine + 450 µL concentrated HCl, heat at 100° for 1 h, cool to room temperature, add 10 mL dichloromethane, vortex for 2 min, centrifuge at 4° at 1500 g for 5 min. Remove the organic phase and add it to 1 mL 100 mM NaOH containing 100 mM sodium bicarbonate, vortex, centrifuge at 1500 g for 3 min, discard the aqueous phase. Wash the organic phase with 1 mL 8% sodium bicarbonate then 1 mL water. Remove the organic layer and evaporate it to dryness under a stream of nitrogen at 45°, reconstitute the residue in 400 µL MeOH:water 50:50, add to the SPE cartridge, wash with 2 mL water, wash with 2 mL MeCN:water 22:78, elute with 1 mL MeCN:water 30:70 (retain this eluate), wash with 1 mL MeCN:water 40:60, elute with 1 mL MeCN:water 55:45. Evaporate the eluates (i.e., those obtained with 30% and 55% MeCN) to dryness under a stream of nitrogen at 45°, reconstitute with 15 µL MeOH, inject immediately.

CAPILLARY ELECTROPHORESIS
Capillary: 47 cm × 50 µm bare fused silica (40 cm to the detector) (Polymicro Technologies, Phoenix, AR)
Capillary preparation: After each run rinse with 100 mM NaOH for 1 min, with water for 1 min, and with running buffer for 20 min.
Capillary temperature: 25
Running buffer: MeOH:buffer 20:80 pH 8.86 (Buffer was 5 mM borate containing 5 mM phosphate and 75 mM cholate.)
Detector: UV 200
Migration time: 10.85
Internal standard: d-equilenin (10.85)
Voltage: 28 kV
Model: Beckman P/ACE 5510

OTHER SUBSTANCES
Extracted: estradiol, estriol, estrone

KEY WORDS
outlet is at cathode; SPE; equilenin is IS

REFERENCE

Ji,A.J.; Nunez,M.F.; Machacek,D.; Ferguson,J.E.; Iossi,M.F.; Kao,P.C.; Landers,J.P. Separation of urinary estrogens by micellar electrokinetic chromatography, *J.Chromatogr.B*, **1995**, *669*, 15–26.

Ergonovine

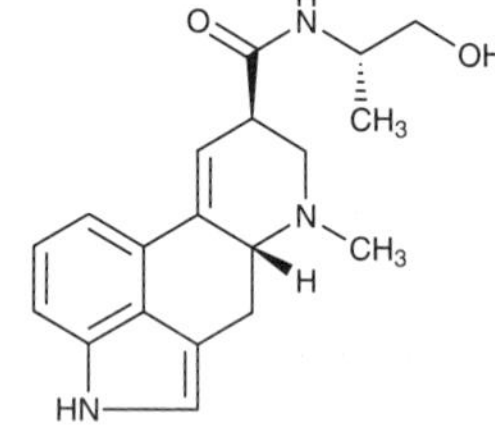

Molecular formula: $C_{19}H_{23}N_3O_2$
Molecular weight: 325.41
CAS Registry No.: 60-79-7, 129-51-1 (maleate)
Merck Index (12th ed.): 3694

SAMPLE

Matrix: solutions
Sample preparation: Inject an aliquot of a 5 µg/mL solution.

CAPILLARY ELECTROPHORESIS

Running buffer: 100 mM pH 2.5 Phosphate buffer
Injection: Electromigration at 7 kV for 7 s
Detector: UV 206
Migration time: 6.4
Voltage: 7 kV
Current: 12 µA (constant current)
Model: Bio-Rad HPE 100

OTHER SUBSTANCES

Simultaneous: ergonovinine, ergotamine, ergotaminine
Also analyzed: isolysergic acid, lisuride, meluol, nicergoline, terguride

REFERENCE

Fanali,S.; Flieger,M.; Steinerova,N.; Nardi,A. Use of cyclodextrins for the enantioselective separation of ergot alkaloids by capillary zone electrophoresis, *Electrophoresis*, **1992**, *13*, 39–43.

Ergotamine

Molecular formula: $C_{33}H_{35}N_5O_5$
Molecular weight: 581.67
CAS Registry No.: 113-15-5, 379-79-3 (tartrate)
Merck Index (12th ed.): 3703

SAMPLE

Matrix: formulations
Sample preparation: Weigh out amount of powdered tablet corresponding to 50 mg caffeine, add 10 mL running buffer, shake vigorously for 15 min, centrifuge, dilute the supernatant with running buffer, inject an aliquot.

CAPILLARY ELECTROPHORESIS

Capillary: 60 cm × 75 µm fused-silica (Accusep, Waters)
Capillary temperature: 30
Running buffer: 25 mM pH 6.0 Sodium phosphate buffer

Injection: Hydrostatic injection for 10 s.
Detector: UV 214
Migration time: 4.2
Voltage: +20 kV
Model: Waters Quanta 4000E
Limit of quantitation: 10 µg/mL

OTHER SUBSTANCES
Simultaneous: caffeine

KEY WORDS
tablets

REFERENCE
Aboul-Enein,H.Y.; Bakr,S.A. Simultaneous determination of caffeine and ergotamine in pharmaceutical dosage
formulation by capillary electrophoresis, *J.Liq.Chromatogr.Rel.Technol.*, **1997**, *20*, 47–55.

SAMPLE
Matrix: solutions
Sample preparation: Inject an aliquot of a 5 µg/mL solution.

CAPILLARY ELECTROPHORESIS
Running buffer: 100 mM pH 2.5 Phosphate buffer
Injection: Electromigration at 7 kV for 7 s
Detector: UV 206
Migration time: 4.5
Voltage: 7 kV
Current: 12 µA (constant current)
Model: Bio-Rad HPE 100

OTHER SUBSTANCES
Simultaneous: ergonovine, ergonovinine, ergotaminine
Also analyzed: isolysergic acid, lisuride, meluol, nicergoline, terguride

REFERENCE
Fanali,S.; Flieger,M.; Steinerova,N.; Nardi,A. Use of cyclodextrins for the enantioselective separation of ergot
alkaloids by capillary zone electrophoresis, *Electrophoresis*, **1992**, *13*, 39–43.

SAMPLE
Matrix: solutions
Sample preparation: Pass 20 mL of a solution in water (?) through an Empore C18 SPE disc.
Wash with 2.5 mL water, add 1 mL MeCN:25 mM pH 3.0 phosphate buffer 65:35, let soak for
3 min, elute, inject an aliquot of the eluate.

CAPILLARY ELECTROPHORESIS
Capillary: 35 cm × 75 µm fused-silica (27 cm to detector) (Waters)
Capillary preparation: Before each run rinse with running buffer for 3 min. At the start of
each day flush with 100 mM KOH for 5 min and with water for 10 min then condition with
running buffer for 20 min. At the end of each day wash with 100 mM KOH and water.
Capillary temperature: 25
Running buffer: 25 mM KH_2PO_4 adjusted to pH 3.0 with dilute phosphoric acid
Injection: Hydrodynamic injection at 10 cm for 8 s (17.4 nL).
Detector: UV 254
Migration time: 4.2
Voltage: 15 kV
Model: Waters Quanta 4000
Limit of quantitation: 2.5 µg/mL
Limit of detection: 800 ng/mL

OTHER SUBSTANCES
Noninterfering: excipients

KEY WORDS
comparison with HPLC; SPE

REFERENCE
Lucangioli,S.E.; Rodriguez,V.G.; Fernandez Otero,G.C.; Vizioli,N.M.; Carducci,C.N. Development and validation of capillary electrophoresis methods for pharmaceutical dissolution assays, *J.Capillary Electrophor.*, **1997**, *4*, 27–31.

Erythromycin

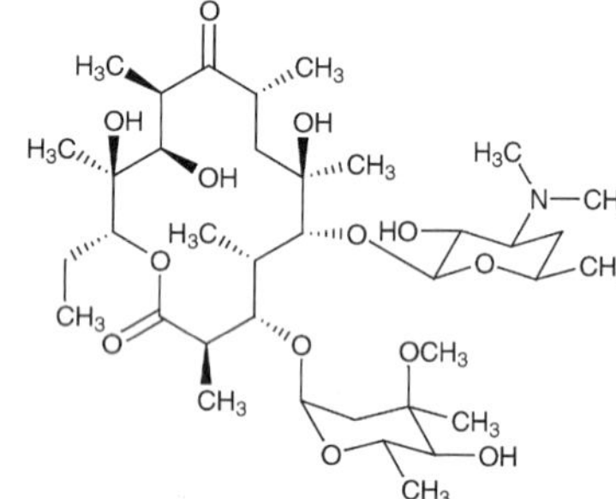

Molecular formula: $C_{37}H_{67}NO_{13}$
Molecular weight: 733.94
CAS Registry No.: 114-07-8, 41342-53-4 (ethylsuccinate), 96128-89-1 (acistrate), 3521-62-8 (estolate), 304-63-2 (gluheptonate), 23067-13-2 (gluheptonate), 3847-29-8 (lactobionate), 134-36-1 (propionate), 643-22-1 (stearate), 84252-03-9 (stinoprate)
Merck Index (12th ed.): 3720

SAMPLE
Matrix: bulk
Sample preparation: Prepare a 4 mg/mL solution in isopropanol:water 50:50, inject an aliquot.

CAPILLARY ELECTROPHORESIS
Capillary: 90 cm × 50 μm fused silica (65 cm to detector) (Polymicro Technologies or Isco)
Capillary preparation: Rinse with running buffer for 4 min between runs.
Capillary temperature: 34
Running buffer: MeOH:buffer 5:95, pH 9 (Buffer was 60 mM sodium borate containing 10 mM methyl-β-cyclodextrin.)
Injection: Vacuum injection at 25 kPa s
Detector: UV 205
Migration time: 12
Voltage: +22 kV
Model: Isco Model 3140 Electropherograph

OTHER SUBSTANCES
Simultaneous: impurities, degradation products

REFERENCE
Flurer,C.L.; Wolnik,K.A. Chemical profiling of pharmaceuticals by capillary electrophoresis in the determination of drug origin, *J.Chromatogr.A*, **1994**, *674*, 153–163.

SAMPLE
Matrix: solutions

CAPILLARY ELECTROPHORESIS
Capillary: 90 cm × 50 μm untreated fused-silica (Polymicro Technologies)
Running buffer: Trisma buffer pH 7.2 (about 7 g Trisma HCl and 0.7 g Trisma base in 1 L water)
Injection: 3 s vacuum injection
Detector: UV 200 or MS, SCIEX API III triple quadrupole, API source, positive ion analysis, make-up flow 0.2% aqueous formic acid, nebulizing gas air, target gas argon, collision energy 20-35 eV
Migration time: 6.7
Voltage: 24.4 kV
Model: Applied Biosystems Model 270A

OTHER SUBSTANCES
Simultaneous: sulfaisomidine, sulfamethazine, trimethoprim
Interfering: ormetoprim

REFERENCE
Pleasance,S.; Thibault,P.; Kelly,J. Comparison of liquid-junction and coaxial interfaces for capillary electrophoresis-mass spectrometry with application to compounds of concern to the aquaculture industry, *J.Chromatogr.*, **1992**, *591*, 325–339.

SAMPLE
Matrix: solutions
Sample preparation: Inject an aliquot of a solution in MeCN:isopropanol:10 mM pH 7 sodium phosphate buffer 25:25:50.

CAPILLARY ELECTROPHORESIS
Capillary: 68.5 cm × 50 μm fused-silica (60 cm to detector) (Polymicro Technologies)
Capillary preparation: Rinse capillary with MeCN:water 50:50 for 2 min and with running buffer for 4 min before each run.
Capillary temperature: 15
Running buffer: MeCN:80 mM pH 6 phosphate buffer 20:80 containing 30 mM sodium cholate
Injection: Pressure injection at 50 mbar with 7.5 s for sample and 3.7 s for buffer plug
Detector: UV 205
Migration time: 14 (erythromycin), 15 (erythromycin estolate), 15.4 (erythromycin ethylsuccinate)
Voltage: +20 kV
Model: Isco Model 3140 Electropherograph

OTHER SUBSTANCES
Simultaneous: clarithromycin, oleandomycin, spiramycin, troleandomycin

REFERENCE
Flurer,C.L. Analysis of macrolide antibiotics by capillary electrophoresis, *Electrophoresis*, **1996**, *17*, 359–366.

SAMPLE
Matrix: solutions
Sample preparation: Inject an aliquot of a solution in MeCN:isopropanol:water 25:25:50.

CAPILLARY ELECTROPHORESIS
Capillary: 68.5 cm × 50 μm fused-silica (60 cm to detector) (Polymicro Technologies)
Capillary preparation: Rinse capillary with running buffer for 4 min before each run.
Capillary temperature: 15
Running buffer: MeCN:35 mM pH 6 phosphate buffer 65:35
Injection: Pressure injection at 50 mbar with 7.5 s for sample and 3.7 s for buffer plug
Detector: UV 205
Migration time: 10.83 (erythromycin), 12.59 (erythromycin ethylsuccinate)
Voltage: +20 kV
Model: Isco Model 3140 Electropherograph

OTHER SUBSTANCES
Simultaneous: clarithromycin, oleandomycin, spiramycin, troleandomycin

REFERENCE
Flurer,C.L. Analysis of macrolide antibiotics by capillary electrophoresis, *Electrophoresis*, **1996**, *17*, 359–366.

SAMPLE
Matrix: solutions
Sample preparation: Inject an aliquot of a 500 μg/mL solution in MeOH:water 10:80.

CAPILLARY ELECTROPHORESIS
Capillary: 114.5 cm × 75 μm fused-silica (75 cm)

Capillary preparation: Between runs rinse capillary with running buffer for 3 min. Before use flush capillary with 100 mM NaOH for 10 min, with water for 10 min, and equilibrate with running buffer for 15 min. Condition a new capillary at 2000 mbar with 1 M NaOH for 30 min, 100 mM NaOH for 30 min, and water for 30 min.
Capillary temperature: 25
Running buffer: MeOH:75 mM pH 7.5 phosphate buffer 50:50
Injection: Pressure injection at 50 mbar for 5 s, ramp to operating voltage at 6 kV/s.
Detector: UV 200
Migration time: 25
Voltage: 25 kV
Model: Prince

OTHER SUBSTANCES
Simultaneous: josamycin, oleandomycin

REFERENCE
Lalloo,A.K.; Chataraj,S.C.; Kanfer,I. Development of a capillary electrophoretic method for the separation of the macrolide antibiotics, erythromycin, josamycin and oleandomycin, *J.Chromatogr.B*, **1997**, *704*, 333–341.

SAMPLE
Matrix: solutions
Sample preparation: Inject an aliquot of a 500 µg/mL solution in MeCN:water 20:80.

CAPILLARY ELECTROPHORESIS
Capillary: 144.5 cm × 50 µm fused-silica (75 cm to detector) (Polymicro Technologies)
Capillary preparation: Between runs rinse capillary with running buffer for 3 min. Before use flush capillary with 100 mM NaOH for 10 min, with water for 10 min, and equilibrate with running buffer for 15 min. Condition a new capillary at 2000 mbar with 1 M NaOH for 30 min, 100 mM NaOH for 30 min, and water for 30 min.
Capillary temperature: 30
Running buffer: EtOH:150 mM pH 7.5 phosphate buffer 35:65
Injection: Pressure injection at 100 mbar for 5 s, ramp to operating voltage at 6 kV/s. (Perform two blank injections of sample solvent before each experimental set.)
Detector: UV 200
Migration time: 39.3 (erythromycin B), 39.7 (erythromycin A), 40.7 (erythromycin C)
Internal standard: josamycin, phenylpropanolamine
Voltage: 30 kV
Model: Prince
Limit of quantitation: 40 µg/mL
Limit of detection: 30 µg/mL

OTHER SUBSTANCES
Simultaneous: impurities, anhydroerythromycin, N-demethylerythromycin, erythromycin enol ether

REFERENCE
Lalloo,A.K.; Kanfer,I. Determination of erythromycin and related substances by capillary electrophoresis, *J.Chromatogr.B*, **1997**, *704*, 343–350.

Estazolam

Molecular formula: $C_{16}H_{11}ClN_4$
Molecular weight: 294.74
CAS Registry No.: 29975-16-4
Merck Index (12th ed.): 3744

SAMPLE
Matrix: blood

Sample preparation: Condition a Sep-Pak C18 SPE cartridge with MeOH, water, and 100 mM
   pH 9.0 borate buffer. 1 mL Serum + 3 mL 100 mM pH 9.0 borate buffer, mix, add to the SPE
   cartridge, wash with 3 mL 100 mM pH 9.0 borate buffer, wash with three 2 mL portions of
   water, wash with 1 mL MeCN:water 5:95, wash with 1 mL hexane, dry under full vacuum.
   Elute with 4 mL dichloromethane, evaporate the eluate to dryness under a stream of nitrogen,
   reconstitute with MeOH:water 15:85, filter (0.45 μm), inject an aliquot.

CAPILLARY ELECTROPHORESIS
Capillary: 72 cm × 50 μm fused-silica (50 cm to detector)
Capillary preparation: Before each run wash capillary with 100 mM NaOH for 4 min and with
   running buffer for 6 min.
Capillary temperature: 25
Running buffer: MeOH:5 mM pH 8.5 phosphate/borate buffer 15:85 containing 50 mM sodium
   dodecyl sulfate
Injection: Vacuum injection for 4 s.
Detector: UV 200
Migration time: 20
Internal standard: secobarbital (9)
Voltage: 25 kV
Model: Applied Biosystems Model 270A or Beckman P/ACE 5000
Limit of detection: 25 ng/mL

OTHER SUBSTANCES
Extracted: bromazepam, diazepam, flurazepam, nitrazepam, triazolam

KEY WORDS
SPE; serum

REFERENCE
Tomita,M.; Okuyama,T. Application of capillary electrophoresis to the simultaneous screening and quantitation
   of benzodiazepines, *J.Chromatogr.B*, **1996**, *678*, 331–337.

SAMPLE
Matrix: solutions

CAPILLARY ELECTROPHORESIS
Capillary: 60 cm × 50 μm fused-silica (40 cm to detector), C18 coated (Supelco)
Capillary preparation: Rinse new capillaries with 250 μL 1 M NaOII, 250 μL 100 mM NaOH,
   water, MeOH, water, and running buffer then condition with running buffer for at least 15
   min. Carry out a similar procedure between runs.
Running buffer: 100 mM pH 8.5 Borate buffer containing 25 mM sodium dodecyl sulfate and 5
   M urea
Injection: Electrokinetic injection at 5 kV for 5 s.
Detector: UV 254
Migration time: 31
Voltage: 18 kV
Model: Jasco

OTHER SUBSTANCES
Simultaneous: alprazolam, amobarbital, barbital, bromazepam, clonazepam, clotiazepam, cloxazolam, diazepam, etizolam, fludiazepam, flunitrazepam, flurazepam, haloxazolam, medazepam, mephobarbital, metharbital, nimetazepam, nitrazepam, oxazepam, pentobarbital, phenobarbital, secobarbital, triazolam

KEY WORDS
coated capillary

REFERENCE
Jinno,K.; Han,Y.; Nakamura,M. Analysis of anxiolytic drugs by capillary electrophoresis with bare and coated
   capillaries, *J.Capillary Electrophor.*, **1996**, *3*, 139–145.

SAMPLE
Matrix: solutions

CAPILLARY ELECTROPHORESIS
Capillary: 60 cm × 75 μm coated fused-silica (40 cm to detector) (Supelco)
Capillary preparation: Coat column as follows. Adjust the pH of 20 mL water to 3.5 with acetic acid, add 80 μL 3-(trimethoxysilyl)propyl methacrylate (3-methacryloxypropyltrimethoxysilane), mix, suck into capillary, let stand at room temperature for 1 h, remove the solution, wash with water. Fill the capillary with a deaerated 4% acrylamide solution containing 1 mg/mL N,N,N',N'-tetramethylethylenediamine and 1 mg/mL ammonium persulfate, let stand for 3 h, remove excess solution by aspiration, rinse with water, remove water by aspiration, dry at 35° (cf. J. Chromatogr. 1985, 347, 191).
Running buffer: 100 mM pH 8.5 borate buffer containing 10 mM sodium dodecyl sulfate and 5 M urea
Injection: Electromigration at 5 kV for 5 s.
Detector: UV 240, UV 254
Migration time: 9
Voltage: 18 kV
Model: Jasco 890-CE

OTHER SUBSTANCES
Simultaneous: alprazolam, clotiazepam, cloxazolam, diazepam, etizolam, fludiazepam, flurazepam, haloxazolam, oxazepam

---

KEY WORDS
injection at cathode; coated capillary

REFERENCE
Jinno,K.; Han,Y.; Sawada,H. Analysis of toxic drugs by capillary electrophoresis using polyacrylamide-coated columns, *Electrophoresis*, **1997**, *18*, 284–286.

SAMPLE
Matrix: solutions
Sample preparation: Inject an aliquot of a 100 μg/mL solution in running buffer.

CAPILLARY ELECTROPHORESIS
Capillary: 60 cm × 50 μm acrylamide-coated fused-silica (40 cm to detector) (Supelco)
Capillary preparation: Adjust the pH of 20 mL water to 3.5 with acetic acid, add 80 μL 3-(trimethoxysilyl)propyl methacrylate (3-methacryloxypropyltrimethoxysilane), mix, suck into capillary, let stand at room temperature for 1 h, remove the solution, wash with water. Fill the capillary with a deaerated 3-4% acrylamide solution containing 1 μL/mL N,N,N',N'-tetramethylethylenediamine and 1 mg/mL ammonium persulfate, let stand for 3 h, remove excess solution by aspiration, rinse with water, remove water by aspiration, dry at 35° (cf. J. Chromatogr. 1985, 347, 191).
Running buffer: MeCN:buffer 5:95 (Buffer was 100 mM borate containing 5 M urea and 10 mM sodium dodecyl sulfate, adjusted to pH 8.5 with phosphate.)
Injection: Electrokinetic injection at 5 kV for 5 s.
Detector: UV 254
Migration time: 13.2
Voltage: 18 kV
Model: Jasco Model 870-CE

OTHER SUBSTANCES
Simultaneous: alprazolam, amobarbital, barbital, bromazepam, clonazepam, clotiazepam, cloxazolam, diazepam, etizolam, fludiazepam, flunitrazepam, flurazepam, haloxazolam, medazepam, mephobarbital, metharbital, nimetazepam, nitrazepam, oxazepam, pentobarbital, phenobarbital, secobarbital, triazolam

---

KEY WORDS
coated capillary; detector at anode

REFERENCE
Jinno,K.; Han,Y.; Sawada,H.; Taniguchi,M. Capillary electrophoretic separation of toxic drugs using a poly-acrylamide-coated capillary, *Chromatographia*, **1997**, *46*, 309–314.

Estradiol

Molecular formula: $C_{18}H_{24}O_2$
Molecular weight: 272.39
CAS Registry No.: 50-28-2, 113-38-2 (dipropionate), 979-32-8 (valerate), 57-91-0 (α- estradiol), 50-50-0 (benzoate), 313-06-4 (cypionate), 4956-37-0
(enanthate), 3571-53-7 (undecylenate)
Merck Index (12th ed.): 3746
Lednicer: 1 162; 2 136

SAMPLE
Matrix: solutions

CAPILLARY ELECTROPHORESIS
Capillary: 70 cm × 75 μm fused-silica (45 cm to detector) (Polymicro Technologies)
Capillary preparation: Treat with 100 mM NaOH and rinse with water before use.
Capillary temperature: 20
Running buffer: MeOH:buffer 20:80 (Buffer was 100 mM pH 11.5 cyclohexylamino-1-propane-sulfonic acid.)
Injection: Injection by vacuum (21 nL)
Detector: UV 237
Migration time: 19
Voltage: 357 V/cm
Model: Isco Model 3850 or Model 3140

OTHER SUBSTANCES
Simultaneous: estriol, estrone

REFERENCE
Potter,K.J.; Allington,R.J.; Algaier,J. Separation of estrogens and rodenticides using capillary electrophoresis with aqueous-methanolic buffers, *J.Chromatogr.A*, **1993**, *652*, 427–429.

SAMPLE
Matrix: solutions

CAPILLARY ELECTROPHORESIS
Capillary: 57 cm × 75 μm fused-silica (50 cm to detector)
Capillary preparation: Before each run rinse with running buffer for 2 min. After each run rinse capillary with 100 mM NaOH for 2 min and with water for 2 min. Condition new capillaries with 100 mM HCl for 5 min, with 100 mM NaOH for 10 min, and with water for 5 min.
Capillary temperature: 23
Running buffer: EtOH:buffer 2:98 (Buffer was 50 mM pH 9.2 borate buffer containing 75 mM sodium dodecyl sulfate.)
Injection: Low pressure injection for 5 s.
Detector: UV 214
Migration time: 54
Voltage: 20 kV
Model: Beckman P/ACE 5510

OTHER SUBSTANCES
Simultaneous: metabolites, estriol, estrone, hydrocortisone, progesterone, testosterone

REFERENCE
Gu,X.; Meleka-Boules,M.; Chen,C.-L. Micellar electrokinetic capillary chromatography combined with immunoaffinity chromatography for identification and determination of dexamethasone and flumethasone in equine urine, *J.Capillary Electrophor.*, **1996**, *3*, 43–49.

SAMPLE
Matrix: solutions

CAPILLARY ELECTROPHORESIS
Capillary: 57 cm × 50 μm fused-silica (50 cm to detector) (Polymicro Technologies)
Capillary preparation: After each run rinse capillaries with 100 mM NaOH for 10 min and with water for 5 min. Fill with running buffer and apply voltage for 10 min. Rinse new capillaries with 1 M HCl for 5 min, with 100 mM NaOH for 10 min, and with water for 5 min.
Capillary temperature: 25
Running buffer: MeOH:buffer 20:80 (Buffer was 38 mM pH 7.4 phosphate buffer containing 40 mM sodium cholate.)
Injection: Pressure injection for 5 s (3.1 nL).
Detector: UV (wavelength not given)
Migration time: 18.91
Voltage: 15 kV
Model: Beckman P/ACE 5500

OTHER SUBSTANCES
Simultaneous: aldosterone, dehydroepiandrosterone, hydrocortisone, 17α-hydroxyprogesterone, progesterone

REFERENCE
Hsiao,M.-W.; Lin,S.-T. Separation of steroids by micellar electrokinetic capillary chromatography with sodium cholate, *J.Chromatogr.Sci.*, **1997**, *35*, 259–264.

SAMPLE
Matrix: solutions
Sample preparation: Inject an aliquot of a sample in 50% diluted running buffer.

CAPILLARY ELECTROPHORESIS
Capillary: 63.5 cm × 50 μm fused-silica (55 cm to detector) (Polymicro Technologies)
Capillary preparation: Between runs flush capillary with 1 M NaOH for 1 min, with MeOH for 1 min, with 100 mM NaOH for 1 min, with water for 2 min, and with running buffer for 2 min. At the start of each day wash capillary with 1 M NaOH for 10 min, with MeOH for 5 min, with 100 mM NaOH for 5 min, with water for 5 min, and with running buffer for 5 min.
Capillary temperature: 20
Running buffer: MeOH:buffer 20:80 containing 0.75% butyl acrylate-butyl methacrylate-methacrylic acid copolymers, sodium salt (Dai-ichi Kogyo Seiyaku, Kyoto) (Buffer was 50 mM NaH$_2$PO$_4$ containing 100 mM sodium tetraborate, pH 8.5.)
Injection: Flush capillary at high pressure with water for 1 min or until filled, inject sample by electrokinetic injection at -20 kV until current reaches 97-99% of that of a capillary filled with running buffer, turn off voltage, replace injection vial with a vial of running buffer, electrophorese.
Detector: UV 200
Migration time: 16
Voltage: 20 kV
Model: Hewlett-Packard 3D

OTHER SUBSTANCES
Simultaneous: estriol

REFERENCE
Quirino,J.P.; Terabe,S. On-line concentration of neutral analytes for micellar electrokinetic chromatography. IV. Field-enhanced sample injection, *J.Chromatogr.A*, **1998**, *798*, 251–257.

SAMPLE
Matrix: solutions

CAPILLARY ELECTROPHORESIS
Capillary: 32 cm × 50 μm fused-silica packed for 24 cm with 1.8 μm Zorbax ODS (Polymicro Technologies) (Column preparation described in paper.)
Capillary temperature: 25
Running buffer: MeCN:0.8 mM sodium tetraborate 80:20 containing 5 mM sodium dodecyl sulfate
Injection: Electrokinetic injection at 20 kV for 1 s.
Detector: UV 254
Migration time: 2.7
Voltage: 25 kV
Model: Hewlett-Packard HP [3D]CE

OTHER SUBSTANCES
Simultaneous: estriol, estrone, hydrocortisone, methyltestosterone, 4-pregnen-20α-ol-3-one, progesterone, testosterone

KEY WORDS
electrochromatography

REFERENCE
Seifar,R.M; Kraak,J.C.; Kok,W.T.; Poppe,H. Capillary electrochromatography with 1.8-μm ODS-modified porous silica particles, *J.Chromatogr.A*, **1998**, *808*, 71–77.

SAMPLE
Matrix: urine
Sample preparation: Condition a 3 mL 200 mg Bond Elut C18 SPE cartridge with 3 mL water, 3 mL MeCN, and 3 mL water. 3 mL Urine + 450 μL concentrated HCl, heat at 100° for 1 h, cool to room temperature, add 50 μL 100 μg/mL d-equilenin in MeOH, add 10 mL dichloromethane, vortex for 2 min, centrifuge at 4° at 1500 g for 5 min. Remove the organic phase and add it to 1 mL 100 mM NaOH containing 100 mM sodium bicarbonate, vortex, centrifuge at 1500 g for 3 min, discard the aqueous phase. Wash the organic phase with 1 mL 8% sodium bicarbonate then 1 mL water. Remove the organic layer and evaporate it to dryness under a stream of nitrogen at 45°, reconstitute the residue in 400 μL MeOH:water 50:50, add to the SPE cartridge, wash with 2 mL water, wash with 2 mL MeCN:water 22:78, elute with 1 mL MeCN:water 30:70 (retain this eluate), wash with 1 mL MeCN:water 40:60, elute with 1 mL MeCN:water 55:45. Evaporate the eluates (i.e., those obtained with 30% and 55% MeCN) to dryness under a stream of nitrogen at 45°, reconstitute with 15 μL MeOH, inject immediately.

CAPILLARY ELECTROPHORESIS
Capillary: 47 cm × 50 μm bare fused silica (40 cm to the detector) (Polymicro Technologies, Phoenix, AR)
Capillary preparation: After each run rinse with 100 mM NaOH for 1 min, with water for 1 min, and with running buffer for 20 min.
Capillary temperature: 25
Running buffer: MeOH:buffer 20:80 pH 8.86 (Buffer was 5 mM borate containing 5 mM phosphate and 75 mM cholate.)
Detector: UV 200
Migration time: 11.17
Internal standard: d-equilenin (10.85)
Voltage: 28 kV
Model: Beckman P/ACE 5510

OTHER SUBSTANCES
Extracted: estriol, estrone

KEY WORDS
outlet is at cathode; SPE

REFERENCE
Ji,A.J.; Nunez,M.F.; Machacek,D.; Ferguson,J.E.; Iossi,M.F.; Kao,P.C.; Landers,J.P. Separation of urinary estrogens by micellar electrokinetic chromatography, *J.Chromatogr.B*, **1995**, *669*, 15–26.

Estrogens, conjugated

Molecular formula: $C_{18}H_{18}O_2$ (equilenin), $C_{18}H_{22}O_2$ (17α-dihydroequilin), $C_{18}H_{20}O_2$ (equilin), $C_{18}H_{22}O_2$ (estrone), $C_{18}H_{24}O_2$ (estradiol)

Molecular weight: 266.34 (equilenin), 272.40 (estradiol), 268.30 (equilin), 270.40 (17α-dihydroequilin), 270.40 (estrone)

CAS Registry No.: 474-86-2 (equilin), 50-28-2 (estradiol), 57-91-0 (α-estradiol), 517-09-9 (equilenin), 53-16-7 (estrone), 338-67-5 (estrone sodium sulfate), 481-97-0 (estrone hydrogen sulfate)

Merck Index (12th ed.): 3216 (17α-dihydroequilin), 3675 (equilenin), 3676 (equilin), 3746 (estradiol), 3751 (estrone)

Lednicer: 1 156 (estrone); 1 162 (estradiol); 2 136 (estradiol)

SAMPLE
Matrix: amniotic fluid, urine
Sample preparation: Dilute urine 1:50 with running buffer, inject an aliquot. Dilute amniotic fluid 1:25 with running buffer, inject an aliquot.

CAPILLARY ELECTROPHORESIS
Capillary: 50 cm $\times$ 50 μm fused-silica (31 cm to detector)
Running buffer: 200 mM pH 10.2 Borate/NaOH buffer
Injection: Electromigration of 2-4 nL
Detector: UV 210
Migration time: 16.6 (estrone sulfate), 16.2 (equilin sulfate), 17.5 (equilenin sulfate)
Voltage: 20 kV
Model: ISCO Model 3850

REFERENCE
Touchstone,J.C.; Levin,S.S. Capillary electrophoresis of steroid conjugates, *J.Liq.Chromatogr.*, **1994**, *17*, 3925–3931.

SAMPLE
Matrix: formulations
Sample preparation: Remove sugar coating on tablets with a moist paper towel. Grind to pass a 60 mesh sieve, weigh out an amount containing 25 mg estrogens, add 20 mL MeOH, sonicate, centrifuge. Add the supernatant to 20 mL water, add 4 mL concentrated HCl, reflux for 30 min, cool to room temperature, extract three times with 10 mL portions of chloroform. Combine the extracts and wash them with 5 mL water, pass the organic layer through a short column of anhydrous sodium sulfate, evaporate to dryness under reduced pressure, take up the residue in 20 mL MeOH, make up to 25 mL with water, inject an aliquot.

CAPILLARY ELECTROPHORESIS
Capillary: 48.5 cm $\times$ 50 μm fused-silica (40 cm to detector) (Hewlett-Packard)
Capillary preparation: Before each run flush capillary with 100 mM NaOH for 2 min and with running buffer for 5 min.
Capillary temperature: 25
Running buffer: 20 mM pH 8 Sodium borate/sodium phosphate containing 15 mM β-cyclodextrin and 75 mM sodium cholate
Injection: Pressure injection at 50 mbar for 1-2 s.
Detector: UV 210
Migration time: 7 (estriol), 8 (17β-dihydroequilin), 8.7 (17α-dihydroequilin), 9.3 (17β-estradiol), 10 (17α-estradiol), 10.4 (17β-dihydroequilenin), 10.5 (17α-dihydroequilenin), 12 (equilin), 13 (estrone), 15.5 (equilenin)
Voltage: 20 kV
Model: Hewlett-Packard [3D]CE

KEY WORDS
tablets

REFERENCE
Poole,S.K.; Poole,C.F. Separation of pharmaceutically important estrogens by micellar electrokinetic chromatography, *J.Chromatogr.A*, **1996**, *749*, 247–255.

Estrone

Molecular formula: $C_{18}H_{22}O_2$
Molecular weight: 270.37
CAS Registry No.: 53-16-7, 338-67-5 (sodium sulfate), 481-97-0 (hydrogen sulfate)
Merck Index (12th ed.): 3751
Lednicer: 1 156

SAMPLE
Matrix: solutions

CAPILLARY ELECTROPHORESIS
Capillary: 70 cm × 75 μm fused-silica (45 cm to detector) (Polymicro Technologies)
Capillary preparation: Treat with 100 mM NaOH and rinse with water before use.
Capillary temperature: 20
Running buffer: MeOH:buffer 20:80 (Buffer was 100 mM pH 11.5 cyclohexylamino-1-propane-sulfonic acid.)
Injection: Injection by vacuum (21 nL)
Detector: UV 237
Migration time: 20
Voltage: 357 V/cm
Model: Isco Model 3850 or Model 3140

OTHER SUBSTANCES
Simultaneous: β-estradiol, estriol

REFERENCE
Potter,K.J.; Allington,R.J.; Algaier,J. Separation of estrogens and rodenticides using capillary electrophoresis with aqueous-methanolic buffers, *J.Chromatogr.A*, **1993**, *652*, 427–429.

SAMPLE
Matrix: solutions

CAPILLARY ELECTROPHORESIS
Capillary: 57 cm × 75 μm fused-silica (50 cm to detector)
Capillary preparation: Before each run rinse with running buffer for 2 min. After each run rinse capillary with 100 mM NaOH for 2 min and with water for 2 min. Condition new capillaries with 100 mM HCl for 5 min, with 100 mM NaOH for 10 min, and with water for 5 min.
Capillary temperature: 23
Running buffer: EtOH:buffer 2:98 (Buffer was 50 mM pH 9.2 borate buffer containing 75 mM sodium dodecyl sulfate.)
Injection: Low pressure injection for 5 s.
Detector: UV 214
Migration time: 51
Voltage: 20 kV
Model: Beckman P/ACE 5510

OTHER SUBSTANCES
Simultaneous: metabolites, estradiol, estriol, hydrocortisone, progesterone, testosterone

REFERENCE

Gu,X.; Meleka-Boules,M.; Chen,C.-L. Micellar electrokinetic capillary chromatography combined with immunoaffinity chromatography for identification and determination of dexamethasone and flumethasone in equine urine, *J.Capillary Electrophor.*, **1996**, *3*, 43–49.

SAMPLE
Matrix: solutions

CAPILLARY ELECTROPHORESIS
Capillary: 33.5 cm × 100 μm fused-silica packed with 3 μm C18 for 25 cm (Innovatech, Stevenage, UK) (Polymicro Technologies)
Capillary preparation: Column packing procedure described in LC.GC 1995, 13, 800.
Capillary temperature: 20
Running buffer: MeCN:25 mM Tris HCl:water 60:20:20, run with 12 bar nitrogen pressure on each side
Injection: Electrokinetic injection at 5 kV for 3 s.
Detector: UV 200
Migration time: 6.5
Voltage: 30 kV
Model: Hewlett-Packard HP[3D]

OTHER SUBSTANCES
Simultaneous: androsterone, dehydrotestosterone, desoxycorticosterone, dexamethasone, ethinyl estradiol, hydrocortisone, prednisone, pregnenolone, testosterone

KEY WORDS
electrochromatography

REFERENCE

Dittmann,M.M.; Rozing,G.P.; Ross,G.; Adam,T.; Unger,K.K. Advances in capillary electrochromatography, *J.Capillary Electrophor.*, **1997**, *4*, 201–212.

SAMPLE
Matrix: solutions

CAPILLARY ELECTROPHORESIS
Capillary: 32 cm × 50 μm fused-silica packed for 24 cm with 1.8 μm Zorbax ODS (Polymicro Technologies) (Column preparation described in paper.)
Capillary temperature: 25
Running buffer: MeCN:0.8 mM sodium tetraborate 80:20 containing 5 mM sodium dodecyl sulfate
Injection: Electrokinetic injection at 20 kV for 1 s.
Detector: UV 254
Migration time: 2.8
Voltage: 25 kV
Model: Hewlett-Packard HP [3D]CE

OTHER SUBSTANCES
Simultaneous: estradiol, estriol, hydrocortisone, methyltestosterone, 4-pregnen-20α-ol-3-one, progesterone, testosterone

KEY WORDS
electrochromatography

REFERENCE

Seifar,R.M; Kraak,J.C.; Kok,W.T.; Poppe,H. Capillary electrochromatography with 1.8-μm ODS-modified porous silica particles, *J.Chromatogr.A*, **1998**, *808*, 71–77.

SAMPLE
Matrix: urine

Sample preparation: Condition a 3 mL 200 mg Bond Elut C18 SPE cartridge with 3 mL water, 3 mL MeCN, and 3 mL water. 3 mL Urine + 450 µL concentrated HCl, heat at 100° for 1 h, cool to room temperature, add 50 µL 100 µg/mL d-equilenin in MeOH, add 10 mL dichloromethane, vortex for 2 min, centrifuge at 4° at 1500 g for 5 min. Remove the organic phase and add it to 1 mL 100 mM NaOH containing 100 mM sodium bicarbonate, vortex, centrifuge at 1500 g for 3 min, discard the aqueous phase. Wash the organic phase with 1 mL 8% sodium bicarbonate then 1 mL water. Remove the organic layer and evaporate it to dryness under a stream of nitrogen at 45°, reconstitute the residue in 400 µL MeOH:water 50:50, add to the SPE cartridge, wash with 2 mL water, wash with 2 mL MeCN:water 22:78, elute with 1 mL MeCN:water 30:70 (retain this eluate), wash with 1 mL MeCN:water 40:60, elute with 1 mL MeCN:water 55:45. Evaporate the eluates (i.e., those obtained with 30% and 55% MeCN) to dryness under a stream of nitrogen at 45°, reconstitute with 15 µL MeOH, inject immediately.

CAPILLARY ELECTROPHORESIS
Capillary: 47 cm × 50 µm bare fused silica (40 cm to the detector) (Polymicro Technologies, Phoenix, AR)
Capillary preparation: After each run rinse with 100 mM NaOH for 1 min, with water for 1 min, and with running buffer for 20 min.
Capillary temperature: 25
Running buffer: MeOH:buffer 20:80 pH 8.86 (Buffer was 5 mM borate containing 5 mM phosphate and 75 mM cholate.)
Detector: UV 200
Migration time: 10.35
Internal standard: d-equilenin (10.85)
Voltage: 28 kV
Model: Beckman P/ACE 5510

OTHER SUBSTANCES
Extracted: estradiol, estriol

KEY WORDS
outlet is at cathode; SPE

REFERENCE
Ji,A.J.; Nunez,M.F.; Machacek,D.; Ferguson,J.E.; Iossi,M.F.; Kao,P.C.; Landers,J.P. Separation of urinary estrogens by micellar electrokinetic chromatography, *J.Chromatogr.B*, **1995**, *669*, 15–26.

Ethacrynic acid

Molecular formula: $C_{13}H_{12}Cl_2O_4$
Molecular weight: 303.14
CAS Registry No.: 58-54-8, 6500-81-8 (sodium salt)
Merck Index (12th ed.): 3761
Lednicer: 1 120

SAMPLE
Matrix: blood, urine
Sample preparation: Condition a 3 mL Supelclean LC-18 SPE cartridge with 3 mL MeOH and 3 mL water. Dilute urine 1:10 with water. Precipitate proteins from serum with MeOH. Add diluted urine or protein supernatant to the SPE cartridge, wash with 3 mL water, elute with 3 mL MeOH, reconstitute to the original volume with 100 mM KOH, inject an aliquot.

CAPILLARY ELECTROPHORESIS
Capillary: 67 cm × 50 µm fused-silica (60 cm to detector) (Polymicro Technologies)
Capillary preparation: Rinse with running buffer for 2 min before run. If necessary, regenerate capillary with 100 mM NaOH for 10 min and with water for 15 min.
Capillary temperature: 20

Running buffer: 60 mM 3-(cyclohexylamino)-1-propanesulfonic acid (CAPS) adjusted to pH 10.6
   with 100 mM KOH
Injection: Pressure injection for 5 s
Detector: UV 220
Migration time: 8
Voltage: 25 kV
Model: Beckman P/ACE 2000

OTHER SUBSTANCES
Extracted: acetazolamide, amiloride, bendroflumethiazide, benzthiazide, bumetanide, caffeine,
   chlorothiazide, chlorthalidone, clopamide, dichlorphenamide, furosemide, hydrochlorothiazide,
   metyrapone, triamterene, trichlormethiazide
Interfering: probenecid

KEY WORDS
serum; SPE

REFERENCE
Jumppanen,J.; Sirén,H.; Riekkola,M.-L. Screening for diuretics in urine and blood serum by capillary zone
   electrophoresis, *J.Chromatogr.A*, **1993**, *652*, 441–450.

SAMPLE
Matrix: urine
Sample preparation: Filter (0.2 μm), inject an aliquot of the filtrate.

CAPILLARY ELECTROPHORESIS
Capillary: 80 cm × 50 μm fused-silica (57 cm to detector)
Capillary preparation: Before each run rinse the capillary with 1 M NaOH for 3 min, with 100
   mM NaOH for 3 min, with water for 3 min, and with running buffer for 10 min.
Running buffer: 110 mM Boric acid containing 56 mM NaOH and 44 mM HCl, pH 8
Injection: Vacuum injection for 1 s
Detector: UV 220
Migration time: 8.964
Voltage: 20 kV
Model: Europhor Prime Vision system IV

OTHER SUBSTANCES
Extracted: acebutolol (UV 238), acetazolamide (UV 222), alprenolol (UV 220), amiloride (UV
   220), atenolol (UV 228), bendroflumethiazide (UV 220), bumetanide (UV 220), chlorthalidone
   (UV 220), cocaine (UV 236), codeine (UV 220), furosemide (UV 232), hydrochlorothiazide (UV
   226), methadone (UV 220), metoxiphenamine (UV 220), nadolol (UV 220), norcodeine (UV 220),
   oxprenolol (UV 220), pentazocine (UV 220), propranolol (UV 220), spironolactone (UV 244),
   triamterene (UV 232), xipamide (UV 234)

REFERENCE
Gonzalez,E.; Laserna,J.J. Capillary zone electrophoresis for the rapid screening of banned drugs in sport,
   Electrophoresis, **1994**, *15*, 240–243.

Ethenzamide

Molecular formula: $C_9H_{11}NO_2$
Molecular weight: 165.19
CAS Registry No.: 938-73-8
Merck Index (12th ed.): 3776

SAMPLE
Matrix: solutions

Sample preparation: Inject an aliquot of a solution in running buffer.

CAPILLARY ELECTROPHORESIS
Capillary: 57 cm × 50 μm fused-silica (50 cm to detector) (Polymicro Technologies)
Capillary temperature: 25
Running buffer: 25 mM NaH_2PO_4 containing 0.83% poly(sodium 10-undecenyl sulfate), adjusted
   to pH 7.3 with 12.5 mM sodium tetraborate (Preparation of poly(sodium 10-undecenyl sulfate
   is as follows. React undecylenyl alcohol (1-undecen-11-ol) with one equivalent of chlorosulfonic
   acid in diethyl ether under nitrogen at -5° to obtain 10-undecenyl hydrogen sulfate, react with
   one equivalent of NaOH to obtain sodium 10-undecenyl sulfate (J. Microcolumn Sep. 1996, 8,
   115). Dissolve 20 g sodium 10-undecenyl sulfate in 40 mL water (degassed by purging with
   nitrogen for 24 h), add 600 mg potassium persulfate, stir at 70° under nitrogen for 50 h, cool,
   add 200 mL cold (0°) EtOH, filter, wash the solid with three 10 mL portions of cold EtOH, dry
   under vacuum at 65° overnight to obtain poly(sodium 10-undecenyl sulfate) (cf. J. Microcolumn
   Sep. 1992, 4, 509). Purify by dialysis using dialysis tubing with a 1000 MW cut-off (Spectrum
   Houston) for 24 h.)
Detector: UV 214
Migration time: 9
Voltage: 16.1 kV
Model: Beckman P/ACE 2000

OTHER SUBSTANCES
Simultaneous: acetaminophen, caffeine, guaifenesin, trimetoquinol
Interfering: phenacetin

REFERENCE
Palmer,C.P.; Terabe,S. Micelle polymers as pseudostationary phases in MEKC: Chromatographic performance
   and chemical selectivity, *Anal.Chem.*, **1997**, *69*, 1852–1860.

Ethinyl estradiol

Molecular formula: $C_{20}H_{24}O_2$
Molecular weight: 296.41
CAS Registry No.: 57-63-6
Merck Index (12th ed.): 3780
Lednicer: 1 162

SAMPLE
Matrix: solutions

CAPILLARY ELECTROPHORESIS
Capillary: 48.5 cm × 50 μm fused-silica (40 cm to detector) (Hewlett-Packard)
Capillary preparation: Before each run flush capillary with 100 mM NaOH for 2 min and with
   running buffer for 5 min.
Capillary temperature: 25
Running buffer: MeOH:buffer 10:90 (Buffer was 20 mM pH 8 Sodium borate/sodium phosphate
   buffer containing 50 mM sodium cholate.)
Injection: Pressure injection at 50 mbar for 1-2 s.
Detector: UV 210
Migration time: 10.7
Voltage: 20 kV
Model: Hewlett-Packard ³ᴰCE

OTHER SUBSTANCES
Simultaneous: ethynodiol diacetate, mestranol, norethindrone, norethindrone acetate, norethynodrel, norgestrel

REFERENCE
Poole,S.K.; Poole,C.F. Separation of pharmaceutically important estrogens by micellar electrokinetic chromatography, *J.Chromatogr.A*, **1996**, *749*, 247–255.

SAMPLE
Matrix: solutions

CAPILLARY ELECTROPHORESIS
Capillary: 33.5 cm × 100 μm fused-silica packed with 3 μm C18 for 25 cm (Innovatech, Stevenage, UK) (Polymicro Technologies)
Capillary preparation: Column packing procedure described in LC.GC 1995, 13, 800.
Capillary temperature: 20
Running buffer: MeCN:25 mM Tris HCl:water 60:20:20, run with 12 bar nitrogen pressure on each side
Injection: Electrokinetic injection at 5 kV for 3 s.
Detector: UV 200
Migration time: 5.7
Voltage: 30 kV
Model: Hewlett-Packard HP³ᴰ

OTHER SUBSTANCES
Simultaneous: androsterone, dehydrotestosterone, desoxycorticosterone, dexamethasone, estrone, hydrocortisone, prednisone, pregnenolone, testosterone

KEY WORDS
electrochromatography

REFERENCE
Dittmann,M.M.; Rozing,G.P.; Ross,G.; Adam,T.; Unger,K.K. Advances in capillary electrochromatography, *J.Capillary Electrophor.*, **1997**, *4*, 201–212.

Ethoheptazine

Molecular formula: C₁₆H₂₃NO₂
Molecular weight: 261.36
CAS Registry No.: 77-15-6, 6700-56-7 (citrate)
Merck Index (12th ed.): 3789
Lednicer: 1 303

SAMPLE
Matrix: solutions

CAPILLARY ELECTROPHORESIS
Capillary: 59.6 cm × 75 μm fused-silica (52.1 cm to detector) (Polymicro Technologies)
Capillary preparation: Purge with running buffer before each run. At the beginning of each day purge using 50-60 kPa vacuum with 500 mM NaOH for 5 min, with water for 5 min, with MeCN for 5 min, and with running buffer for 5 min.
Running buffer: MeCN:MeOH:acetic acid 49:50:1 containing 20 mM ammonium acetate
Injection: Hydrostatic injection at 10 cm for 5 s.
Detector: UV 214
Migration time: 3.47
Voltage: 30 kV
Model: Waters Quanta 4000

OTHER SUBSTANCES
Simultaneous: diphenoxylate, fentanyl, levallorphan, meperidine, methadone, nalorphine, pentazocine, nikethamide

REFERENCE
Leung,G.N.W.; Tang,H.P.O.; Tso,T.S.C.; Wan,T.S.M. Separation of basic drugs with non-aqueous capillary electrophoresis, *J.Chromatogr.A*, **1996**, *738*, 141–154.

Ethopropazine

Molecular formula: C$_{19}$H$_{24}$N$_2$S
Molecular weight: 312.48
CAS Registry No.: 522-00-9, 1094-08-2 (HCl)
Merck Index (12th ed.): 3793
Lednicer: 1 373

SAMPLE
Matrix: solutions

CAPILLARY ELECTROPHORESIS
Capillary: 59.6 cm × 75 μm fused-silica (52.1 cm to detector) (Polymicro Technologies)
Capillary preparation: Purge with running buffer before each run. At the beginning of each day purge using 50-60 kPa vacuum with 500 mM NaOH for 5 min, with water for 5 min, with MeCN for 5 min, and with running buffer for 5 min.
Running buffer: MeCN:MeOH:acetic acid 49:50:1 containing 20 mM ammonium acetate
Injection: Hydrostatic injection at 10 cm for 5 s.
Detector: UV 214
Migration time: 3.09
Voltage: 30 kV
Model: Waters Quanta 4000

OTHER SUBSTANCES
Simultaneous: benzquinamide, deserpidine, mesoridazine, methotrimeprazine, molindone, promazine, promethazine, reserpine, thioridazine, thiothixene

REFERENCE
Leung,G.N.W.; Tang,H.P.O.; Tso,T.S.C.; Wan,T.S.M. Separation of basic drugs with non-aqueous capillary electrophoresis, *J.Chromatogr.A*, **1996**, *738*, 141–154.

SAMPLE
Matrix: solutions

CAPILLARY ELECTROPHORESIS
Capillary: 42 cm × 50 μm fused-silica (31 cm to detector) (Polymicro Technologies)
Capillary preparation: Condition capillary by rinsing with running buffer for 10 min, with water for 10 min, with 1 M NaOH for 10 min, with water for 10 min, and with running buffer with the voltage applied for 20 min.
Capillary temperature: 25
Running buffer: Formamide containing 150 mM citric acid, 100 mM Tris, and 100 mM hydroxypropyl-β-cyclodextrin (apparent pH 5.1)
Injection: Gravity injection.
Detector: UV 254
Migration time: 9.92, 9.98 (enantiomers)
Voltage: 30 kV
Model: Beckman P/ACE 5500

KEY WORDS
chiral

REFERENCE
Wang,F.; Khaledi,M.G. Chiral separations by nonaqueous capillary electrophoresis, *Anal.Chem.*, **1996**, *68*, 3460–3467.

SAMPLE
Matrix: urine
Sample preparation: Dilute 10-fold with water, inject an aliquot.

CAPILLARY ELECTROPHORESIS
Capillary: 70 cm × 50 μm fused-silica (65.4 cm to detector) (Polymicro Technologies)
Capillary temperature: 25
Running buffer: 20 mM pH 5.0 Tris-acetic acid containing 50 μg/mL FC-135 (Fluorad/3M) and 10 mM cetyltrimethylammonium bromide
Injection: Pressure injection at 2 psi.
Detector: UV 240
Migration time: 10.53
Voltage: 20 kV
Model: Bio-Rad BioFocus 3000

OTHER SUBSTANCES
Extracted: acepromazine, chlorpromazine, fluphenazine, methotrimeprazine, perphenazine, thioridazine, triflupromazine, trimeprazine

KEY WORDS
detector at anode

REFERENCE
Muijelsaar,P.G.H.M.; Claessens,H.A.; Cramers,C.A. Determination of structurally related phenothiazines by capillary zone electrophoresis and micellar electrokinetic chromatography, *J.Chromatogr.A*, **1996**, *735*, 395–402.

Ethosuximide

Molecular formula: $C_7H_{11}NO_2$
Molecular weight: 141.17
CAS Registry No.: 77-67-8
Merck Index (12th ed.): 3794
Lednicer: 1 228

SAMPLE
Matrix: blood
Sample preparation: Vortex serum for 20 s, filter (0.45 μm), inject an aliquot.

CAPILLARY ELECTROPHORESIS
Capillary: 92 cm × 75 μm (68 cm to detector)
Capillary preparation: Before each run rinse capillary with 100 mM NaOH for 3 min and with running buffer for 5 min.
Capillary temperature: 35
Running buffer: 10 mM pH 9.2 Na_2HPO_4 containing 6 mM sodium borate and 75 mM sodium dodecyl sulfate
Injection: Vacuum suction for 1 s
Detector: UV 220, UV 245
Migration time: 11.3
Voltage: 20 kV
Current: 59 μA
Model: Applied Biosystems model 270A-HT

OTHER SUBSTANCES
Extracted: phenobarbital, primidone

KEY WORDS
serum

REFERENCE
Schmutz,A.; Thormann,W. Determination of phenobarbital, ethosuximide, and primidone in human serum by micellar electrokinetic capillary chromatography with direct sample injection, *Ther.Drug Monit.*, **1993**, *15*, 310–316.

SAMPLE
Matrix: blood
Sample preparation: Inject serum directly.

CAPILLARY ELECTROPHORESIS
Capillary: 105 cm × 75 μm fused-silica (68 cm to detector) (Polymicro Technologies)
Capillary preparation: Rinse with running buffer for 10 min between run. At the beginning and end of each day rinse with 100 mM NaOH for 10 min and with water for 10 min.
Running buffer: 10 mM Na_2HPO_4 containing 6 mM sodium borate and 75 mM sodium dodecyl sulfate
Injection: Injection via vacuum suction for 1 s
Detector: UV 215
Migration time: 11.4
Voltage: 23 kV
Current: 62 μA
Model: Europhor Prime Vision IV

OTHER SUBSTANCES
Extracted: acetaminophen, flucytosine, naproxen, primidone, sulfamethoxazole, zomepirac

KEY WORDS
serum

REFERENCE
Schmutz,A.; Thormann,W. Assessment of impact of physico-chemical drug properties on monitoring drug levels by micellar electrokinetic capillary chromatography with direct serum injection, *Electrophoresis*, **1994**, *15*, 1295–1303.

SAMPLE
Matrix: blood
Sample preparation: Vortex serum for 15 s, filter (0.45 μm), add sodium dodecyl sulfate to a concentration of 100 mM, let stand for 1 h, filter (Amicon Centrifree Micropartition) while centrifuging at 1500 g for 20 min, inject an aliquot of the ultrafiltrate.

CAPILLARY ELECTROPHORESIS
Capillary: 63 cm × 75 μm fused-silica (63 cm to detector)
Capillary preparation: Before each run rinse capillary with 100 mM NaOH for 3 min and with running buffer for 5 min.
Capillary temperature: 35
Running buffer: 10 mM pH 9.2 Na_2HPO_4 containing 6 mM sodium borate and 100 mM sodium dodecyl sulfate
Injection: Vacuum suction for 1 s
Detector: UV 195
Migration time: 9.2
Voltage: 25 kV
Current: 80 μA
Model: Applied Biosystems model 270A-HT

OTHER SUBSTANCES
Extracted: phenobarbital, phenytoin

KEY WORDS
serum; ultrafiltrate

REFERENCE

Schmutz,A.; Thormann,W. Factors affecting the determination of drugs and endogenous low molecular mass compounds in human serum by micellar electrokinetic capillary chromatography with direct sample injection, *Electrophoresis*, **1994**, *15*, 51–61.

SAMPLE

Matrix: blood, urine
Sample preparation: Filter (Amicon Centrifree Micropartition System) at 1500 g for 20 min, inject an aliquot of the ultrafiltrate.

CAPILLARY ELECTROPHORESIS

Capillary: 90 cm × 75 μm fused-silica (70 cm to detector) (Polymicro Technologies)
Capillary preparation: Between runs rinse capillary with 100 mM NaOH for 5 min and with running buffer for 10 min.
Running buffer: 10 mM pH 9.1 Na_2HPO_4 containing 6 mM sodium borate and 75 mM sodium dodecyl sulfate
Injection: Siphon at 34 cm for 5 s
Detector: UV 195
Migration time: 9.5
Voltage: 20 kV
Current: 80 μA

OTHER SUBSTANCES

Extracted: acetaminophen, phenobarbital, primidone, salicylic acid

KEY WORDS

serum; ultrafiltrate

REFERENCE

Caslavska,J.; Lienhard,S.; Thormann,W. Comparative use of three electrokinetic capillary methods for the determination of drugs in body fluids. Prospects for rapid determination of intoxications, *J.Chromatogr.*, **1993**, *638*, 335–342.

Ethynodiol

Molecular formula: $C_{20}H_{28}O_2$
Molecular weight: 300.44
CAS Registry No.: 1231-93-2, 297-76-7 (diacetate)
Merck Index (12th ed.): 3905
Lednicer: 1 165

SAMPLE

Matrix: solutions

CAPILLARY ELECTROPHORESIS

Capillary: 48.5 cm × 50 μm fused-silica (40 cm to detector) (Hewlett-Packard)
Capillary preparation: Before each run flush capillary with 100 mM NaOH for 2 min and with running buffer for 5 min.
Capillary temperature: 25
Running buffer: MeOH:buffer 10:90 (Buffer was 20 mM pH 8 Sodium borate/sodium phosphate buffer containing 50 mM sodium cholate.)
Injection: Pressure injection at 50 mbar for 1-2 s.
Detector: UV 210
Migration time: 13
Voltage: 20 kV
Model: Hewlett-Packard ^{3D}CE

OTHER SUBSTANCES
Simultaneous: ethinyl estradiol, mestranol, norethindrone, norethindrone acetate, norethynodrel, norgestrel

REFERENCE
Poole,S.K.; Poole,C.F. Separation of pharmaceutically important estrogens by micellar electrokinetic chromatography, *J.Chromatogr.A*, **1996**, *749*, 247–255.

Etilefrin

Molecular formula: $C_{10}H_{15}NO_2$
Molecular weight: 181.23
CAS Registry No.: 709-55-7, 10128-36-6 (dl-form), 534-87-2 (dl-form HCl)
Merck Index (12th ed.): 3911

SAMPLE
Matrix: solutions

CAPILLARY ELECTROPHORESIS
Capillary: 60 cm × 75 μm fused-silica (51 cm to detector) (Supelco)
Capillary preparation: At the end of each day wash capillary with 200 mM NaOH for 10 min and with water for 10 min. Condition new capillaries by washing with 200 mM NaOH for 10 min, with water for 10 min, and with running buffer for 10 min.
Running buffer: 50 mM pH 7.3 Sodium borate buffer containing 3.3% succinyl β-cyclodextrin (Wacker Chemie, Munich, Germany)
Injection: Hydrodynamic injection at 25 mbar for 6 s.
Detector: UV 208
Migration time: 72.79 (first enantiomer)
Voltage: 15 kV
Model: Prince

OTHER SUBSTANCES
Simultaneous: alprenolol, chlorthalidone, ephedrine, homatropine, methoxamine, norephedrine, octopamine, propranolol, synephrine, trihexyphenidyl

---

KEY WORDS
chiral; $\alpha = 1.0764$

REFERENCE
Schmid,M.G.; Wirnsberger,K.; Gübitz,G. Chiral separation of drug enantiomers by capillary electrophoresis using succinyl-β-cyclodextrin, *Pharmazie*, **1996**, *51*, 852–854.

SAMPLE
Matrix: solutions

CAPILLARY ELECTROPHORESIS
Capillary: 36 cm × 50 μm fused-silica coated with linear polyacrylamide (31.5 cm to detector) (GL Science)
Capillary preparation: At the beginning and end of each day rinse capillary with capillary wash solution (Bio-Rad Cat. No. 148-5022) at 690 kPa for more than 3 min and with water at 690 kPa for more than 3 min. Coat capillary as follows. Treat capillary with 1 M NaOH at room temperature for 1 h, rinse with water, dry by passing nitrogen gas through the capillary at 110° for 6 h. Pass thionyl chloride through the capillary using a suction pump for several min, seal capillary at both ends and heat at 70° for 6 h. Unseal the capillary and fill with 250 mM vinyl magnesium bromide in THF by suction, seal the capillary, heat at 70° for 6 h. Open the capillary and rinse it with THF for several min, rinse with distilled water, fill the capillary with polymerization solution, heat at 28 ± 2° for 1 h, condition at -100 V/cm for 30 min (Anal.

Sci. 1994, 10, 1). (The polymerization solution was 5% acrylamide in water containing 49 mM Tris, 384 mM glycine, and 0.1% sodium dodecyl sulfate, degas in an ultrasonic bath. Add 40 µL 10% N,N,N',N'-tetramethylethylenediamine and 10 µL 10% ammonium persulfate to 5 mL of the degassed solution, mix thoroughly.)
Running buffer: 50 mM pH 6.0 Phosphate buffer
Injection: Before each injection rinse with water at 690 kPa for 30 s, rinse with running buffer at 690 kPa for 30 s, partially fill with separation solution (500 µM α_1-acid glycoprotein (Cohn fraction VI) (ICN) in running buffer) at 6.9 kPa for 190 s (27 cm), inject sample at 6.9 kPa for 2 s, electrophorese with running buffer (Note that α_1-acid glycoprotein from other suppliers may provide inferior results).
Detector: UV 210
Migration time: Resolution of enantiomers 2.6
Voltage: 12 kV
Model: Bio-rad BioFocus 3000

KEY WORDS
chiral; coated capillary

REFERENCE
Tanaka,Y.; Terabe,S. Separation of the enantiomers of basic drugs by affinity capillary electrophoresis using a partial filling technique and α_1-acid glycoprotein as chiral selector, *Chromatographia*, **1997**, *44*, 119–128.

SAMPLE
Matrix: solutions

CAPILLARY ELECTROPHORESIS
Capillary: 35 cm × 50 µm polyacrylamide-coated fused-silica (30.5 cm to detector) (Composite Metal Services, UK)
Capillary preparation: Before each run rinse capillary with water for 70 s and with running buffer for 100 s. (Coat capillary as follows. Adjust the pH of 20 mL water to 3.5 with acetic acid, add 80 µL 3-(trimethoxysilyl)propyl methacrylate (3-methacryloxypropyltrimethoxysilane), mix, suck into capillary, let stand at room temperature for 1 h, remove the solution, wash with water. Fill the capillary with a deaerated 3-4% acrylamide solution containing 1 µL/mL N,N,N',N'-tetramethylethylenediamine and 1 mg/mL potassium persulfate, let stand for 30 min, remove excess solution by aspiration, rinse with water, remove water by aspiration, dry at 35° (J. Chromatogr. 1985, 347, 191).)
Capillary temperature: 25
Running buffer: Buffer containing 150 mM cyanoethylated-β-cyclodextrin (Cyclolab, Budapest) (Prepare buffer by adjusting the pH of 50 mM phosphoric acid containing 50 mM acetic acid and 50 mM boric acid to 2.5 with concentrated NaOH, add the appropriate amount of cyanoethylated-β-cyclodextrin, dilute with an equal volume of water.)
Injection: Pressure injection at 5 psi for 2 s.
Detector: UV 206
Migration time: 25.9 (second enantiomer, α = 1.068)
Voltage: 20 kV
Current: 27-40 µA
Model: Bio-Rad Biofocus 3000

KEY WORDS
chiral; coated capillary

REFERENCE
Aturki,Z.; Desiderio,C.; Mannina,L.; Fanali,S. Chiral separations by capillary zone electrophoresis with the use of cyanoethylated-β-cyclodextrin as chiral selector, *J.Chromatogr.A*, **1998**, *817*, 91–104.

SAMPLE
Matrix: solutions

CAPILLARY ELECTROPHORESIS
Capillary: 44 cm × 50 µm fused-silica
Running buffer: 10 mM pH 4.3 acetic acid/ammonium acetate buffer containing 3 mg/mL carboxymethyl ether-β-cyclodextrin (Wacker Chemie, Munich)

Injection: Hydrostatic injection at 10 cm for 5-10 s.
Detector: MS, Finnigan LCQ, electrospray, positive ion mode, ion-spray tip voltage 2.6 kV, sheath liquid MeOH:water:acetic acid 50:49:1 at 6 µL/min, m/z 182
Migration time: 4.8, 5.1 (enantiomers)
Voltage: 20 kV

KEY WORDS
chiral

REFERENCE
Schulte,G.; Heitmeier,S.; Chankvetadze,B.; Blaschke,G. Chiral capillary electrophoresis-electrospray mass spectrometry coupling with charged cyclodextrin derivatives as chiral selectors, *J.Chromatogr.A*, **1998**, *800*, 77–82.

Etizolam

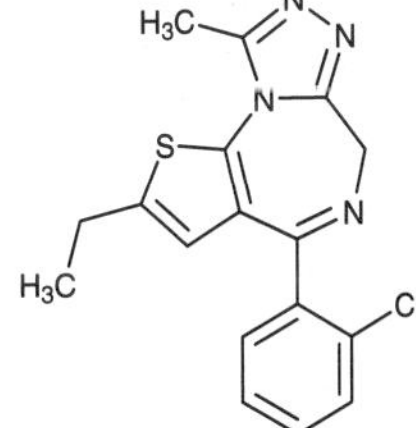

Molecular formula: $C_{17}H_{15}ClN_4S$
Molecular weight: 342.85
CAS Registry No.: 40054-69-1
Merck Index (12th ed.): 3919

SAMPLE
Matrix: solutions

CAPILLARY ELECTROPHORESIS
Capillary: 60 cm × 50 µm fused-silica (40 cm to detector), C18 coated (Supelco)
Capillary preparation: Rinse new capillaries with 250 µL 1 M NaOH, 250 µL 100 mM NaOH, water, MeOH, water, and running buffer then condition with running buffer for at least 15 min. Carry out a similar procedure between runs.
Running buffer: 100 mM pH 8.5 Borate buffer containing 25 mM sodium dodecyl sulfate and 5 M urea
Injection: Electrokinetic injection at 5 kV for 5 s.
Detector: UV 254
Migration time: 47
Voltage: 18 kV
Model: Jasco

OTHER SUBSTANCES
Simultaneous: alprazolam, amobarbital, barbital, bromazepam, clonazepam, clotiazepam, cloxazolam, diazepam, estazolam, fludiazepam, flunitrazepam, flurazepam, haloxazolam, medazepam, mephobarbital, metharbital, nimetazepam, nitrazepam, oxazepam, pentobarbital, phenobarbital, secobarbital, triazolam

KEY WORDS
coated capillary

REFERENCE
Jinno,K.; Han,Y.; Nakamura,M. Analysis of anxiolytic drugs by capillary electrophoresis with bare and coated capillaries, *J.Capillary Electrophor.*, **1996**, *3*, 139–145.

SAMPLE
Matrix: solutions
Sample preparation: Inject an aliquot of a 100 µg/mL solution in running buffer.

CAPILLARY ELECTROPHORESIS
Capillary: 60 cm × 50 µm acrylamide-coated fused-silica (40 cm to detector) (Supelco)

Capillary preparation: Adjust the pH of 20 mL water to 3.5 with acetic acid, add 80 µL 3-(trimethoxysilyl)propyl methacrylate (3-methacryloxypropyltrimethoxysilane), mix, suck into capillary, let stand at room temperature for 1 h, remove the solution, wash with water. Fill the capillary with a deaerated 3-4% acrylamide solution containing 1 µL/mL N,N,N',N'-tetramethylethylenediamine and 1 mg/mL ammonium persulfate, let stand for 3 h, remove excess solution by aspiration, rinse with water, remove water by aspiration, dry at 35° (cf. J. Chromatogr. 1985, 347, 191).

Running buffer: MeCN:buffer 5:95 (Buffer was 100 mM borate containing 5 M urea and 10 mM sodium dodecyl sulfate, adjusted to pH 8.5 with phosphate.)

Injection: Electrokinetic injection at 5 kV for 5 s.

Detector: UV 254

Migration time: 9.5

Voltage: 18 kV

Model: Jasco Model 870-CE

OTHER SUBSTANCES

Simultaneous: alprazolam, amobarbital, barbital, bromazepam, clonazepam, clotiazepam, cloxazolam, diazepam, estazolam, fludiazepam, flunitrazepam, flurazepam, haloxazolam, medazepam, mephobarbital, metharbital, nimetazepam, nitrazepam, oxazepam, pentobarbital, phenobarbital, secobarbital, triazolam

KEY WORDS

coated capillary; detector at anode

REFERENCE

Jinno,K.; Han,Y.; Sawada,H.; Taniguchi,M. Capillary electrophoretic separation of toxic drugs using a polyacrylamide-coated capillary, *Chromatographia*, **1997**, *46*, 309–314.

SAMPLE

Matrix: solutions

CAPILLARY ELECTROPHORESIS

Capillary: 60 cm × 75 µm coated fused-silica (40 cm to detector) (Supelco)

Capillary preparation: Coat column as follows. Adjust the pH of 20 mL water to 3.5 with acetic acid, add 80 µL 3-(trimethoxysilyl)propyl methacrylate (3-methacryloxypropyltrimethoxysilane), mix, suck into capillary, let stand at room temperature for 1 h, remove the solution, wash with water. Fill the capillary with a deaerated 4% acrylamide solution containing 1 mg/mL N,N,N',N'-tetramethylethylenediamine and 1 mg/mL ammonium persulfate, let stand for 3 h, remove excess solution by aspiration, rinse with water, remove water by aspiration, dry at 35° (cf. J. Chromatogr. 1985, 347, 191).

Running buffer: 100 mM pH 8.5 borate buffer containing 10 mM sodium dodecyl sulfate and 5 M urea

Injection: Electromigration at 5 kV for 5 s.

Detector: UV 240, UV 254

Migration time: 7.2

Voltage: 18 kV

Model: Jasco 890-CE

OTHER SUBSTANCES

Simultaneous: alprazolam, clotiazepam, cloxazolam, diazepam, estazolam, fludiazepam, flurazepam, haloxazolam, oxazepam

KEY WORDS

injection at cathode; coated capillary

REFERENCE

Jinno,K.; Han,Y.; Sawada,H. Analysis of toxic drugs by capillary electrophoresis using polyacrylamide-coated columns, *Electrophoresis*, **1997**, *18*, 284–286.

Etodolac

Molecular formula: C$_{17}$H$_{21}$NO$_3$
Molecular weight: 287.36
CAS Registry No.: 41340-25-4
Merck Index (12th ed.): 3920

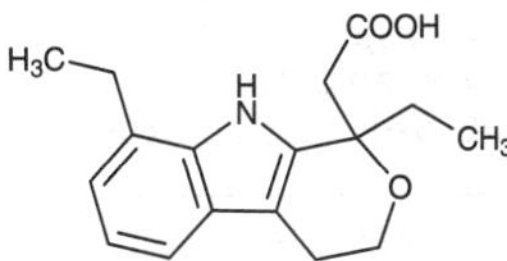

SAMPLE

Matrix: blood
Sample preparation: 0.1-1 mL Plasma + 2 μg IS + 400 μL 1 M HCl + 10 mL diethyl ether, agitate mechanically for 10 min, centrifuge at 3500 g for 10 min. Remove the organic layer and evaporate it to dryness under a stream of nitrogen, reconstitute the residue in MeOH:40 mM NaH$_2$PO$_4$ 50:50, inject an aliquot.

CAPILLARY ELECTROPHORESIS

Capillary: 58.7 cm × 75 μm fused-silica (50 cm to detector)
Capillary preparation: Wash with running buffer for 2 min before each injection. At the beginning of each day wash capillary with 100 mM NaOH for 10 min.
Capillary temperature: 22
Running buffer: MeOH:buffer 3:97 (Buffer was 40 mM pH 8 NaH$_2$PO$_4$ containing 104 mM sodium dodecyl sulfate.)
Injection: Hydrodynamic injection for 2 s
Detector: UV 254
Migration time: 18.1
Internal standard: benzoyl-4-phenyl-2-butyric acid (16.3)
Voltage: 20 kV
Model: Beckman P/ACE 2000
Limit of detection: 670 ng/mL (S/N 5)

OTHER SUBSTANCES

Extracted: diclofenac, diflunisal, fenbufen, fenoprofen, flurbiprofen, ibuprofen, indomethacin, ketoprofen, naproxen, niflumic acid, piroxicam, sulindac, tenoxicam, tiaprofenic acid
Noninterfering: acetaminophen, amitriptyline, caffeine, clomipramine, deoxysulindac, desipramine, diazepam, 5'-hydroxytenoxicam, imipramine, maprotiline, nortriptyline, phenobarbital, phenytoin, sulfamethoxazole, theophylline, trimipramine

KEY WORDS

plasma

REFERENCE

Maboundou,C.W.; Paintaud,G.; Bérard,M.; Bechtel,P.R. Separation of fifteen non-steroidal anti-inflammatory drugs using micellar electrokinetic capillary chromatography, *J.Chromatogr.B*, **1994**, *657*, 173–183.

SAMPLE

Matrix: blood
Sample preparation: 0.5 mL Serum + 200 μL 2 M HCl + 5 mL diethyl ether, tumble 10 min on a rotary mixer. Remove organic layer and evaporate it to dryness under vacuum centrifugation. Reconstitute residue in 500 μL running buffer, inject an aliquot (J.Chromatogr. 1992, 578, 251).

CAPILLARY ELECTROPHORESIS

Capillary: 50 cm × 75 μm fused-silica (30 cm to detector) (Polymicro Technologies)
Capillary preparation: Rinse with running buffer for 10 min between run. At the beginning and end of each day rinse with 100 mM NaOH for 10 min and with water for 10 min.
Capillary temperature: 35
Running buffer: 10 mM Na$_2$HPO$_4$ containing 6 mM sodium borate and 75 mM sodium dodecyl sulfate
Injection: Injection via vacuum suction for 1 s
Detector: UV 225
Migration time: 5

Internal standard: etodolac
Voltage: 13 kV
Current: 84 µA
Model: Applied Biosystems model 270A-HT

OTHER SUBSTANCES
Extracted: naproxen

KEY WORDS
serum; etodolac is IS

REFERENCE
Schmutz,A.; Thormann,W. Assessment of impact of physico-chemical drug properties on monitoring drug levels
by micellar electrokinetic capillary chromatography with direct serum injection, *Electrophoresis*, **1994**, *15*,
1295–1303.

SAMPLE
Matrix: solutions
Sample preparation: Prepare a 10 µg/mL solution in MeOH, inject an aliquot.

CAPILLARY ELECTROPHORESIS
Capillary: 97 cm × 50 µm fused-silica (80 cm to detector) (Chrompack), coat with Chirasil-Dex
(J.High Res.Chromatogr. 1991, 14, 58)
Capillary preparation: Rinse with water then operating buffer between each run. Condition
coated columns with running buffer for half a day.
Running buffer: 20 mM pH 7.0 Borate-phosphate buffer
Injection: Hydrostatic injection for 5 s
Detector: UV 220
Migration time: 22.5, 26.0 (enantiomers)
Voltage: 30 kV
Model: Kapillar-Elektrophorese-System 100 (Grom)

KEY WORDS
chiral; coated capillary

REFERENCE
Mayer,S.; Schurig,V. Enantiomer separation by electrochromatography in open tubular columns coated with
Chirasil-Dex, *J.Liq.Chromatogr.*, **1993**, *16*, 915–931.

SAMPLE
Matrix: solutions
Sample preparation: Inject an aliquot of a 50-100 µg/mL solution in 50 mM pH 4.8 ammonium
acetate buffer.

CAPILLARY ELECTROPHORESIS
Capillary: 44 cm × 50 µm polyacrylamide coated
Capillary preparation: Adjust the pH of 20 mL water to 3.5 with acetic acid, add 80 µL 3-
(trimethoxysilyl)propyl methacrylate (3-methacryloxypropyltrimethoxysilane), mix, suck into
capillary, let stand at room temperature for 1 h, remove the solution, wash with water. Fill the
capillary with a deaerated 3-4% acrylamide solution containing 1 µL/mL N,N,N',N'-tetra-
methylethylenediamine and 1 mg/mL potassium persulfate, let stand for 30 min, remove excess
solution by aspiration, rinse with water, remove water by aspiration, dry at 35° (J. Chromatogr.
1985, 347, 191).
Running buffer: 50 mM pH 4.8 Ammonium acetate buffer containing 5 mM vancomycin
Injection: Hydrostatic injection at 10 cm for 5-10 s.
Detector: MS, Finnigan LCQ ion trap, electrospray, negative ion mode, probe tip at 2.6 kV,
sheath liquid MeOH:water:ammonia 50:48:2 at 6 µL/min, m/z 286 (Vancomycin migrates away
from the detector.)
Migration time: 11, 12 (enantiomers)
Voltage: 15 kV
Model: Grom 100

KEY WORDS
chiral; coated capillary

REFERENCE
Fanali,S.; Desiderio,C.; Schulte,G.; Heitmeier,S.; Strickmann,D.; Chankvetadze,B.; Blaschke,G. Chiral capillary
electrophoresis-electrospray mass spectrometry coupling using vancomycin as chiral selector,
J.Chromatogr.A, **1998**, *800*, 69–76.

Famotidine

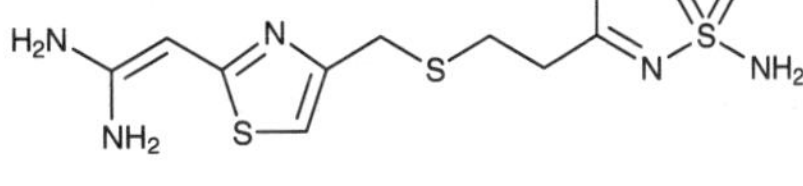

Molecular formula: $C_8H_{15}N_7O_2S_3$
Molecular weight: 337.45
CAS Registry No.: 76824-35-6
Merck Index (12th ed.): 3972
Lednicer: 2 37

SAMPLE
Matrix: urine
Sample preparation: Centrifuge urine, dilute with 2 volumes of 32 mM pH 4.5 6-aminocaproic
acid buffer containing 18 mM adipic acid, inject an aliquot.

CAPILLARY ELECTROPHORESIS
Capillary: 64 cm × 75 μm (48 cm to detector) (Polymicro Technologies)
Running buffer: MeOH:buffer 5:95 (Buffer was 32 mM pH 4.5 6-aminocaproic acid buffer con-
taining 18 mM adipic acid, 5% dextran (MW 18300) and 0.02% polyethylene oxide (MW 300000)
Injection: Hydrodynamic injection at 20 cm
Detector: UV 220
Migration time: 11
Voltage: 22 kV
Model: Laboratory constructed

OTHER SUBSTANCES
Extracted: degradation products, cimetidine, diltiazem, prazosin

REFERENCE
Soini,H.; Riekkola,M.-L.; Novotny,M.V. Mixed polymer networks in the direct analysis of pharmaceuticals in
urine by capillary electrophoresis, *J.Chromatogr.A*, **1994**, *680*, 623–634.

Felbamate

Molecular formula: $C_{11}H_{14}N_2O_4$
Molecular weight: 238.24
CAS Registry No.: 25451-15-4
Merck Index (12th ed.): 3988

SAMPLE
Matrix: blood
Sample preparation: Inject serum for 20 s.

CAPILLARY ELECTROPHORESIS
Capillary: 500 mm × 50 μm

Capillary preparation: Between runs wash capillary for 1 min with 100 mM phosphoric acid
and for 1 min with buffer. At the beginning of each day wash the capillary with 2 M NaOH for
3 min, for 3 min with 100 mM phosphoric acid, and for 3 min with buffer.
Capillary temperature: 35
Running buffer: 100 mM boric acid containing 55.4 mM sodium dodecyl sulfate, adjusted to pH
8.4 ± 0.1 with 2 M NaOH
Injection: Pressure injection for 20 s (3% of capillary volume)
Detector: UV 214
Migration time: 5.5
Voltage: 16 kV
Current: 38 μA (fixed)
Model: Beckman Model 2000

OTHER SUBSTANCES
Simultaneous: acetaminophen, acetoacetanilide, caffeine

KEY WORDS
serum

REFERENCE
Shihabi,Z.K.; Hinsdale,M.E. Sample matrix effects in micellar electrokinetic capillary electrophoresis,
J.Chromatogr.B, **1995**, *669*, 75–83.

Fendiline

Molecular formula: $C_{23}H_{25}N$
Molecular weight: 315.46
CAS Registry No.: 13042-18-7, 13636-18-5 (HCl)
Merck Index (12th ed.): 4011

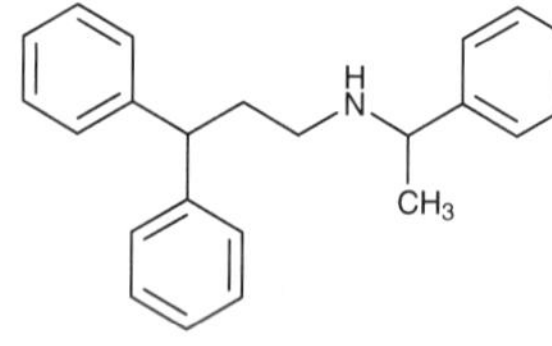

SAMPLE
Matrix: solutions
Sample preparation: Inject an aliquot of a 100 μg/mL solution in running buffer.

CAPILLARY ELECTROPHORESIS
Capillary: 32 cm × 50 μm fused-silica (27.5 cm to detector) (Yongnian Optical Conductive Fiber
Plant, China), coated with polyacrylamide
Capillary preparation: No details of the polyacrylamide coating process are provided. However,
another paper (LC.GC 1997, 15, 40) by this group indicates that they use the procedure of
Hjertén, thus: Adjust the pH of 20 mL water to 3.5 with acetic acid, add 80 μL 3-(trimethoxysilyl)propyl methacrylate (3-methacryloxypropyltrimethoxysilane), mix, suck into capillary, let
stand at room temperature for 1 h, remove the solution, wash with water. Fill the capillary
with a deaerated 3-4% acrylamide solution containing 1 μL/mL N,N,N',N'-tetramethylethyl-
enediamine and 1 mg/mL potassium persulfate, let stand for 30 min, remove excess solution
by aspiration, rinse with water, remove water by aspiration, dry at 35° (J. Chromatogr. 1985,
347, 191).
Capillary temperature: 25
Running buffer: 100 mM NaH_2PO_4 adjusted to pH 2.5 (A) or 100 mM NaH_2PO_4 containing 45
mM hydroxypropyl-gamma-cyclodextrin, adjusted to pH 2.5 (B)
Injection: Electrokinetic injection at 15 kV for 3 s.
Detector: UV 200, UV 210
Migration time: 6.05 (A), 14.92, 15.09 (B, enantiomers)
Voltage: 15 kV
Model: Bio-Rad BioFocus 3000

OTHER SUBSTANCES
Simultaneous: atropine, carazolol, cicletanine, dimethindene, homatropine, ipratropium bromide, isothipendyl, mefloquine, metaclazepam, nafronyl (naftidrofuryl), nefopam, nicardipine,
promethazine, reproterol, tetrahydrozoline (tetryzoline), theodrenaline, tioconazole, trihexyphenidyl, trimeprazine (alimemazine), trimipramine

KEY WORDS
coated capillary; chiral

REFERENCE
Koppenhoefer,B.; Epperlein,U.; Xiaofeng,Z.; Bingcheng,L. Separation of enantiomers of drugs by capillary electrophoresis. Part 4: Hydroxypropyl-γ-cyclodextrin as chiral solvating agent, *Electrophoresis*, **1997**, *18*, 924–930.

SAMPLE
Matrix: solutions
Sample preparation: Inject an aliquot of a 100 μg/mL solution in running buffer.

CAPILLARY ELECTROPHORESIS
Capillary: 30 cm × 50 μm fused-silica (25.5 cm to detector), coated with polyacrylamide
Capillary preparation: Adjust the pH of 20 mL water to 3.5 with acetic acid, add 80 μL 3-(trimethoxysilyl)propyl methacrylate (3-methacryloxypropyltrimethoxysilane), mix, suck into capillary, let stand at room temperature for 1 h, remove the solution, wash with water. Fill the capillary with a deaerated 3-4% acrylamide solution containing 1 μL/mL N,N,N',N'-tetramethylethylenediamine and 1 mg/mL potassium persulfate, let stand for 30 min, remove excess solution by aspiration, rinse with water, remove water by aspiration, dry at 35° (J. Chromatogr. 1985, 347, 191).
Capillary temperature: 25
Running buffer: 100 mM NaH_2PO_4 containing 45 mM hydroxypropyl-gamma- cyclodextrin, adjusted to pH 2.5 with phosphoric acid
Injection: Electrokinetic injection at 15 kV for 3 s.
Detector: UV 200
Voltage: 15 kV
Model: Bio-Rad BioFocus 3000

KEY WORDS
chiral; coated capillary; comparison with the use of other cyclodextrins; this running buffer gave the greatest enantiomeric separation.; α=1.013

REFERENCE
Lin,B.; Zhu,X.; Koppenhoefer,B.; Epperlein,U. Investigation of 123 chiral drugs by cyclodextrin-modified capillary electrophoresis, *LC.GC*, **1997**, *15*, 40–46.

SAMPLE
Matrix: solutions

CAPILLARY ELECTROPHORESIS
Capillary: 29-36 cm × 50 μm fused-silica (24.5-31.5 cm to detector) (Yongnian Optical Conductive Fiber Plant, China) coated with polyacrylamide
Capillary preparation: Coat capillary as follows. Adjust the pH of 20 mL water to 3.5 with acetic acid, add 80 μL 3-(trimethoxysilyl)propyl methacrylate (3-methacryloxypropyltrimethoxysilane), mix, suck into capillary, let stand at room temperature for 1 h, remove the solution, wash with water. Fill the capillary with a deaerated 3-4% acrylamide solution containing 1 μL/mL N,N,N',N'-tetramethylethylenediamine and 1 mg/mL potassium persulfate, let stand for 30 min, remove excess solution by aspiration, rinse with water, remove water by aspiration, dry at 35° (J. Chromatogr. 1985, 347, 191).
Capillary temperature: 25
Running buffer: 100 mM pH 2.5 NaH_2PO_4 (A) or 100 mM pH 2.5 NaH_2PO_4 containing 45 mM hydroxypropyl-α-cyclodextrin (Wacker, Munich) (B)
Injection: Electromigration at 15 kV for 3 s.
Detector: UV 200; UV 210
Migration time: 6.05 (A); 11.77 (B) (no separation of enantiomers)
Voltage: 15 kV
Model: Bio-Focus 3000

OTHER SUBSTANCES
Also analyzed: albuterol (salbutamol), alprenolol, amorolfine, atenolol, atropine, azelastine, baclofen, bamethan, benproperine, benserazide, biperiden, bisoprolol, brompheniramine, bupivacaine, bupranolol, butamirate, butethamate, carazolol, carbuterol, carteolol, carvedilol, celiprolol, chloroquine, chlorpheniramine, chlorphenoxamine, cicletanine, clenbuterol, clidinium bromide, clobutinol, dimethindene, dipivefrin, disopyramide, dobutamine, doxylamine, flecainide, gallopamil, homatropine, ipratropium bromide, isoproterenol (isoprenaline), isothipendyl, ketamine, meclizine, mefloquine, mepindolol, mequitazine, metaclazepam, metaproterenol (orciprenaline), metipranolol, metoprolol, nafronyl (naftidrofuryl), nefopam, nicardipine, norfenefrine, ofloxacin, ornidazole, orphenadrine, oxomemazine, oxprenolol, oxybutynin, phenoxybenzamine, phenylpropanolamine, pholedrine, pindolol, pirbuterol, prilocaine, procyclidine, promethazine, propafenone, propranolol, reproterol, sotalol, sulpride, synephrine, talinolol, terbutaline, tetrahydrozoline (tetryzoline), theodrenaline, tioconazole, tocainide, trihexyphenidyl, trimeprazine (alimemazine), trimipramine, tropicamide, verapamil, zopiclone

KEY WORDS
coated capillary

REFERENCE
Koppenhoefer,B.; Eperlein,U.; Schlunk,R.; Zhu,X.; Lin,B. Separation of enantiomers of drugs by capillary electrophoresis. V. Hydroxypropyl-α-cyclodextrin as chiral solvating agent, *J.Chromatogr.A*, **1998**, *793*, 153–164.

Fenfluramine

Molecular formula: $C_{12}H_{16}F_3N$
Molecular weight: 231.26
CAS Registry No.: 458-24-2, 404-82-0 (HCl)
Merck Index (12th ed.): 4015
Lednicer: 1 70

SAMPLE
Matrix: formulations
Sample preparation: Add 98 mg capsule contents to 5 mL 10 mM pH 2.5 Tris/phosphate buffer, stir at room temperature for 15 min, filter (0.45 μm), make up the filtrate to 10 mL, dilute 40 times with 10 mM pH 2.5 Tris/phosphate buffer, inject an aliquot.

CAPILLARY ELECTROPHORESIS
Capillary: 48.5 cm × 50 μm fused-silica (40 cm to detector) (Hewlett-Packard)
Capillary preparation: Before each run wash with water for 4 min and with running buffer for 10 min. Rinse a new capillary with 500 mM NaOH at 60° for 15 min and wash with water for 15 min.
Capillary temperature: 15
Running buffer: Buffer containing 30 mM gamma-cyclodextrin and 40 mM 2,3,6-tri-O-methyl-β-cyclodextrin (Buffer was 100 mM phosphoric acid adjusted to pH 2.5 with Tris.)
Injection: Pressure injection at 50 mbar for 8 s
Detector: UV 205
Migration time: 14.2 (d), 14.5 (l)
Voltage: 22 kV
Current: 26-35 μA
Model: Hewlett-Packard HP3D
Limit of detection: 1.8 μM

OTHER SUBSTANCES
Simultaneous: impurities

KEY WORDS
capsules; chiral

REFERENCE

Porrà,R.; Quaglia,M.G.; Fanali,S. Determination of fenfluramine enantiomers in pharmaceutical formulations by capillary zone electrophoresis, *Chromatographia*, **1995**, *41*, 383–388.

SAMPLE

Matrix: solutions
Sample preparation: Prepare a 50 μg/mL solution in diluted running buffer, inject an aliquot.

CAPILLARY ELECTROPHORESIS

Capillary: 44 cm × 50 μm fused-silica
Capillary preparation: After each run wash capillary with water for 2 min and with running buffer for 3 min. At the beginning of each day wash capillary with water and running buffer for 5 min. Condition a new capillary with 1 M NaOH, 100 mM NaOH, water, and separation buffer.
Capillary temperature: 15
Running buffer: MeOH:buffer 30:70 (Buffer was 100 mM Phosphoric acid containing 15 mM heptakis(2,6-di-O-methyl)-β-cyclodextrin adjusted to pH 3.0 with triethanolamine.)
Injection: Hydrodynamic injection for 1 s
Detector: UV 210
Voltage: 25 kV
Model: Spectra Physics Spectraphoresis 1000

KEY WORDS

chiral; enantiomer resolution 1.7

REFERENCE

Bechet,I.; Paques,P.; Fillet,M.; Hubert,P.; Crommen,J. Chiral separation of basic drugs by capillary zone electrophoresis with cyclodextrin additives, *Electrophoresis*, **1994**, *15*, 818–823.

SAMPLE

Matrix: solutions
Sample preparation: Inject an aliquot of a 50 μg/mL solution in water.

CAPILLARY ELECTROPHORESIS

Capillary: 57 cm × 75 μm fused-silica (50 cm to detector) (Beckman)
Capillary preparation: Between runs flush capillary with buffer for 2 min then equilibrate for 5 min.
Capillary temperature: 20
Running buffer: MeOH:buffer 40:60 (Buffer was 100 mM orthophosphoric acid containing 10 mM heptakis(2,6-di-O-methyl)-β-cyclodextrin, pH adjusted to 3 with Tris.)
Injection: Pressure injection for 1 s.
Detector: UV 214
Migration time: 18 (S-(+)), 18.5 (R-(-))
Internal standard: d-chlorpheniramine maleate (19.2)
Voltage: 25 kV
Model: Beckman P/ACE 2100

KEY WORDS

chiral

REFERENCE

Boomkerd,S.; Detaevernier,M.R.; Heyden,Y.V.; Vindevogel,J.; Michotte,Y. Determination of the enantiomeric purity of dexfenfluramine by capillary electrophoresis: use of a Plackett-Burman design for the optimization of the separation, *J.Chromatogr.A*, **1996**, *736*, 281–289.

SAMPLE

Matrix: solutions
Sample preparation: Inject an aliquot of a 50 μg/mL solution in water.

CAPILLARY ELECTROPHORESIS

Capillary: 48.5 cm × 50 μm fused-silica (40 cm to detector)

Capillary preparation: After each run wash capillary with running buffer for 3 min. At the start of each day wash capillary with running buffer for 10 min. Before use treat new capillaries with 1 M NaOH, 100 mM NaOH, water, and running buffer.
Capillary temperature: 15
Running buffer: 100 mM Phosphoric acid containing 15 mM carboxymethyl-β-cyclodextrin (Cyclolab, Budapest), adjusted to pH 3.0 with 84 mM triethanolamine
Injection: Hydrodynamic injection at 5 kPa for 2 s.
Detector: UV 210
Migration time: 41.76, 43.38 (enantiomers)
Voltage: 25 kV
Model: Hewlett Packard ^{3D}CE

OTHER SUBSTANCES
Simultaneous: bupivacaine, chlorpheniramine, dimethindene, ephedrine, isoproterenol

---

KEY WORDS
chiral

REFERENCE
Fillet,M.; Bechet,I.; Hubert,P.; Crommen,J. Resolution improvement by use of carboxymethyl-β-cyclodextrin as chiral additive for the enantiomeric separation of basic drugs by capillary electrophoresis, *J.Pharm.Biomed.Anal.*, **1996**, *14*, 1107–1114.

Fenoprofen

Molecular formula: $C_{15}H_{14}O_3$
Molecular weight: 242.27
CAS Registry No.: 31879-05-7, 34507-40-5 (calcium salt), 53746-45-5 (calcium salt dihydrate)
Merck Index (12th ed.): 4021
Lednicer: 2 67

SAMPLE
Matrix: blood
Sample preparation: 0.1-1 mL Plasma + 2 μg IS + 400 μL 1 M HCl + 10 mL diethyl ether, agitate mechanically for 10 min, centrifuge at 3500 g for 10 min. Remove the organic layer and evaporate it to dryness under a stream of nitrogen, reconstitute the residue in MeOH:40 mM NaH_2PO_4 50:50, inject an aliquot.

CAPILLARY ELECTROPHORESIS
Capillary: 58.7 cm $\times$ 75 μm fused-silica (50 cm to detector)
Capillary preparation: Wash with running buffer for 2 min before each injection. At the beginning of each day wash capillary with 100 mM NaOH for 10 min.
Capillary temperature: 22
Running buffer: MeOH:buffer 3:97 (Buffer was 40 mM pH 8 NaH_2PO_4 containing 104 mM sodium dodecyl sulfate.)
Injection: Hydrodynamic injection for 2 s
Detector: UV 254
Migration time: 14.2
Internal standard: benzoyl-4-phenyl-2-butyric acid (16.3)
Voltage: 20 kV
Model: Beckman P/ACE 2000
Limit of detection: 1 μg/mL (S/N 5)

OTHER SUBSTANCES
Extracted: diclofenac, diflunisal, etodolac, fenbufen, flurbiprofen, ibuprofen, indomethacin, ketoprofen, naproxen, niflumic acid, piroxicam, sulindac, tenoxicam, tiaprofenic acid
Noninterfering: acetaminophen, amitriptyline, caffeine, clomipramine, deoxysulindac, desipramine, diazepam, 5'-hydroxytenoxicam, imipramine, maprotiline, nortriptyline, phenobarbital, phenytoin, sulfamethoxazole, theophylline, trimipramine

KEY WORDS
plasma

REFERENCE
Maboundou,C.W.; Paintaud,G.; Bérard,M.; Bechtel,P.R. Separation of fifteen non-steroidal anti-inflammatory drugs using micellar electrokinetic capillary chromatography, *J.Chromatogr.B*, **1994**, *657*, 173–183.

SAMPLE
Matrix: formulations
Sample preparation: Stir formulation in MeOH for 10 min, filter, dilute with 5 mM pH 5 morpholinoethanesulfonic acid, inject an aliquot.

CAPILLARY ELECTROPHORESIS
Capillary: 35 cm × 50 μm fused-silica (31.5 cm to detector) (Polymicro Technologies)
Capillary temperature: 25
Running buffer: 100 mM Morpholinoethanesulfonic acid containing 30 mM heptakis-2,3,6-tri-O-methyl-β-cyclodextrin, adjusted to pH 5 with NaOH
Injection: Pressure injection at 10 psi
Detector: UV 206
Migration time: 21.8, 25.3 (enantiomers)
Voltage: 20 kV
Current: 6.6 μA
Model: Biofocus 3000 (Bio-Rad)

OTHER SUBSTANCES
Simultaneous: flurbiprofen, ibuprofen, ketoprofen

KEY WORDS
chiral

REFERENCE
Fanali,S.; Aturki,Z. Use of cyclodextrins in capillary electrophoresis for the chiral resolution of some 2-arylpropionic acid non-steroidal anti-inflammatory drugs, *J.Chromatogr.A*, **1995**, *694*, 297–305.

SAMPLE
Matrix: solutions

CAPILLARY ELECTROPHORESIS
Capillary: 60 cm × 50 μm fused-silica
Capillary preparation: Before each run purge with running buffer for 3 min. At the beginning of each day vacuum purge with 500 mM NaOH, vacuum purge with water, perform an electroosmotic purge, vacuum purge with running buffer.
Running buffer: 10 mM pH 7 Tris-phosphate buffer containing 2.5% Glucidex2 (maltodextrin preparation, Roquette, Lestrem, France)
Injection: Injection by siphon at 10 cm
Detector: UV 185
Migration time: 4.7, 4.75 (enantiomers)
Voltage: 30 kV
Model: Waters Quanta 4000 CE

OTHER SUBSTANCES
Simultaneous: ibuprofen, indoprofen, suprofen

KEY WORDS
chiral

REFERENCE
D'Hulst,A.; Verbeke,N. Quantitation in chiral capillary electrophoresis: theoretical and practical considerations, *Electrophoresis*, **1994**, *15*, 854–863.

SAMPLE
Matrix: solutions
Sample preparation: Inject an aliquot of a 100 µg/mL solution.

CAPILLARY ELECTROPHORESIS
Capillary: 32.5 cm × 50 µm fused-silica (25 cm to detector) (Quadrex, New Haven, CT)
Capillary preparation: Condition new capillaries with 100 mM KOH for 10 min, purge with water for 5 min, equilibrate with running buffer for 5 min.
Running buffer: 100 mM pH 6.0 Phosphate buffer containing 2 mM vancomycin (A) or ristocetin (B) or teicoplanin (C)
Injection: Hydrostatic injection for 3 s.
Detector: UV 254, UV 280
Migration time: 27.5, 29.6 (A) or 16.1, 16.5 (B) or 16.3, 16.7 (C) (enantiomers)
Voltage: 5 kV
Model: Waters Quanta 4000

OTHER SUBSTANCES
Also analyzed: ketoprofen, methotrexate

KEY WORDS
chiral

REFERENCE
Gasper,M.P.; Berthod,A.; Nair,U.B.; Armstrong,D.W. Comparison and modeling study of vancomycin, ristocetin A, and teicoplanin for CE enantioseparations, *Anal.Chem.*, **1996**, *68*, 2501–2514.

SAMPLE
Matrix: solutions
Sample preparation: Inject an aliquot of a solution in MeOH:water 10:90.

CAPILLARY ELECTROPHORESIS
Capillary: 44 cm × 50 µm fused-silica (37 cm to detector) (Supelco)
Capillary preparation: Wash with running buffer for 3 min after each injection. Wash with running buffer for 10 min at the end of each day.
Capillary temperature: 25
Running buffer: 100 mM Phosphoric acid containing 5 mM sulfobutyl ether-β-cyclodextrin and 10 mM heptakis(2,3,6-tri-O-methyl)-β-cyclodextrin, adjusted to pH 3.0 with triethanolamine
Injection: Hydrodynamic injection for 5 s (13.3 nL).
Detector: UV 230
Migration time: 5, 5.2 (enantiomers)
Voltage: -25 kV
Current: 60 µA
Model: SpectraPhoresis 1000 CE

KEY WORDS
chiral; detector at anode; resolution (R_s = 16.3)

REFERENCE
Fillet,M.; Hubert,P.; Crommen,J. Enantioseparation of nonsteroidal anti-inflammatory drugs by capillary electrophoresis using mixtures of anionic and uncharged β-cyclodextrins as chiral additives, *Electrophoresis*, **1997**, *18*, 1013–1018.

SAMPLE
Matrix: solutions
Sample preparation: Inject an aliquot of a solution in MeOH:water 50:50.

CAPILLARY ELECTROPHORESIS
Capillary: 57 cm × 75 µm fused-silica (50 cm to detector) (Polymicro Technologies)
Capillary preparation: Rinse capillary with running buffer for 2 min between samples. At the start of each day wash capillary with 1 M NaOH for 10 min. Wash new capillaries with 1 M NaOH for 1 h.

Capillary temperature: 23 ± 0.1
Running buffer: 50 mM NaH_2PO_4 containing 3 mM heptakis(6-hydroxyethylamino-6-deoxy-β-cyclodextrin), adjusted to pH 5.0 with 500 mM NaOH (Prepare heptakis(6-hydroxyethylamino-6-deoxy-β-cyclodextrin) as follows. Add 4.32 g β-cyclodextrin to a stirred solution of 4 mL bromine and 21 g triphenylphosphine in 80 mL DMF and the mixture stirred at 80° for 15 h, concentrate under reduced pressure to half volume, adjust pH to 9-10 by adding 30 mL 3 M sodium methoxide in MeOH with simultaneous cooling. Keep at room temperature for 30 min then pour into 1.5 L ice water, filter to obtain heptakis(6-bromo-6-deoxy-β-cyclodextrin) (Ang. Chem. Int. Ed. Engl. 1991, 30, 78). Dissolve heptakis(6-bromo-6-deoxy-β-cyclodextrin) in ethanolamine at 65°, heat at 65° for 48 h. Remove the solvent by evaporation under reduced pressure, dissolve the residue in hot MeOH, slowly add the methanolic solution to reagent-grade acetone, collect the precipitate by filtration. Dissolve the precipitate in water, treat with a basic ion-exchange resin, lyophilize to obtain heptakis(6-hydroxyethylamino-6-deoxy-β-cyclodextrin).)
Detector: UV 214
Migration time: 18, 19 (enantiomers)
Voltage: -15 kV
Model: Beckman P/ACE 5510

OTHER SUBSTANCES
Simultaneous: flurbiprofen, ibuprofen (peak shape very poor)

KEY WORDS
chiral

REFERENCE
O'Keeffe,F.; Shamsi,S.A.; Darcy,R.; Schwinté,P.; Warner,I.M. A persubstituted cationic β-cyclodextrin for chiral separations, *Anal.Chem.*, **1997**, *69*, 4773–4782.

SAMPLE
Matrix: solutions

CAPILLARY ELECTROPHORESIS
Capillary: 35 cm × 50 μm polyacrylamide-coated fused-silica (30.5 cm to detector) (Composite Metal Services, UK)
Capillary preparation: Before each run rinse capillary with water for 70 s and with running buffer for 100 s. (Coat capillary as follows. Adjust the pH of 20 mL water to 3.5 with acetic acid, add 80 μL 3-(trimethoxysilyl)propyl methacrylate (3-methacryloxypropyltrimethoxysilane), mix, suck into capillary, let stand at room temperature for 1 h, remove the solution, wash with water. Fill the capillary with a deaerated 3-4% acrylamide solution containing 1 μL/mL N,N,N',N'-tetramethylethylenediamine and 1 mg/mL potassium persulfate, let stand for 30 min, remove excess solution by aspiration, rinse with water, remove water by aspiration, dry at 35° (J. Chromatogr. 1985, 347, 191).)
Capillary temperature: 25
Running buffer: Buffer containing 10 mM cyanoethylated-β-cyclodextrin (Cyclolab, Budapest) (Prepare buffer by adjusting the pH of 50 mM phosphoric acid containing 50 mM acetic acid and 50 mM boric acid to 5 with concentrated NaOH, add the appropriate amount of cyanoethylated-β-cyclodextrin, dilute with an equal volume of water.)
Injection: Pressure injection at 5 psi for 2 s.
Detector: UV 206
Migration time: 14.5 (second enantiomer, α = 1.022)
Voltage: 20 kV
Current: 27-40 μA
Model: Bio-Rad Biofocus 3000

KEY WORDS
detector at anode; chiral; coated capillary

REFERENCE
Aturki,Z.; Desiderio,C.; Mannina,L.; Fanali,S. Chiral separations by capillary zone electrophoresis with the use of cyanoethylated-β-cyclodextrin as chiral selector, *J.Chromatogr.A*, **1998**, *817*, 91–104.

SAMPLE
Matrix: solutions

CAPILLARY ELECTROPHORESIS
Capillary: 64.5 cm × 50 μm fused-silica (56 cm to detector) (Hewlett-Packard)
Capillary preparation: Before each run flush with 100 mM NaOH for 3 min and with running buffer at 8 min then equilibrate at 20 kV for 6 min. At the start of each batch flush capillary with 1 M NaOH for 10 min, with 100 mM NaOH for 10 min, and with running buffer for 15 min then equilibrate at 20 kV for 20 min.
Capillary temperature: 35
Running buffer: 20 mM Phosphoric acid containing 20 mM triethanolamine and 50 mM 2,3,6-tri-O-methyl-β-cyclodextrin, adjusted to pH 5.0 with NaOH
Injection: Pressure injection at 75 mbar (5 nL).
Detector: UV 253
Migration time: 13.2, 13.8 (enantiomers)
Voltage: 20 kV
Current: 8.1 μA
Model: Hewlett-Packard ^{3D}CE

OTHER SUBSTANCES
Simultaneous: ibuprofen, ketoprofen

KEY WORDS
chiral

REFERENCE
Blanco,M.; Coello,J.; Iturriaga,H.; Maspoch,S.; Pérez-Maseda,C. Separation of profen enantiomers by capillary electrophoresis using cyclodextrins as chiral selectors, *J.Chromatogr.A*, **1998**, *793*, 165–175.

SAMPLE
Matrix: solutions
Sample preparation: Inject an aliquot of a 100 μg/mL solution in MeOH:water 50:50.

CAPILLARY ELECTROPHORESIS
Capillary: 47 cm × 75 μm fused-silica (40 cm to detector) (Polymicro Technologies)
Capillary preparation: Rinse capillary with 100 mM NaOH for 1 min, with water for 1.5 min, and with running buffer for 2 min between samples. Wash new capillaries with 1 M NaOH for 1 h, with 100 mM NaOH for 30 min, and with water for 30 min.
Running buffer: 50 mM NaH_2PO_4 containing 3 mM heptakis(6-methoxyethylamine-6-deoxy-β-cyclodextrin), adjusted to pH 6 (Prepare heptakis(6-methoxyethylamine-6-deoxy-β-cyclodextrin) as follows. Add 4.32 g β-cyclodextrin to a stirred solution of 4 mL bromine and 21 g triphenylphosphine in 80 mL DMF and the mixture stirred at 80°C for 15 h, concentrate under reduced pressure to half volume, adjust pH to 9-10 by adding 30 mL 3 M sodium methoxide in MeOH with simultaneous cooling. Keep at room temperature for 30 min then pour into 1.5 L ice water, filter to obtain heptakis(6-bromo-6-deoxy-β-cyclodextrin) (Ang. Chem. Int. Ed. Engl. 1991, 30, 78). Dissolve heptakis(6-bromo-6-deoxy-β-cyclodextrin) in methoxyethylamine, heat at 65°C for 48 h. Remove the solvent by evaporation under reduced pressure, dissolve the residue in MeOH at 55°C, slowly add the methanolic solution to reagent-grade acetone with stirring, collect the precipitate by filtration. Dissolve the precipitate in water, treat with a basic ion-exchange resin. Lyophilize the precipitate to obtain heptakis(6-methoxyethylamine-6-deoxy-β-cyclodextrin).)
Injection: Pressure injection at 85 kPa.s.
Detector: UV 214
Migration time: 22, 32 (enantiomers)
Voltage: -30 kV
Current: 25 μA
Model: Beckman P/ACE 5510

OTHER SUBSTANCES
Simultaneous: flurbiprofen, ketoprofen

KEY WORDS
chiral

REFERENCE
Haynes,J.L.,III; Shamsi,S.A.; O'Keefe,F.; Darcey,R.; Warner,I.M. Cationic β-cyclodextrin derivative for chiral
separations, *J.Chromatogr.A*, **1998**, *803*, 261–271.

SAMPLE
Matrix: solutions
Sample preparation: Inject an aliquot of a solution in running buffer.

CAPILLARY ELECTROPHORESIS
Capillary: 57 cm × 50 μm fused-silica (Beckman)
Capillary preparation: Rinse with 100 mM NaOH for 5 min, with water for 10 min,
Running buffer: 2-Methoxyethanol:buffer 15:85 (Buffer was 40 mM pH 6 phosphate buffer con-
taining 500 μM Actaplanin A (Eli Lilly).)
Injection: Pressure injection for 5 s.
Detector: UV 254
Migration time: 18.9, 20.3
Model: Beckman P/ACE 2100

OTHER SUBSTANCES
Simultaneous: ketoprofen

KEY WORDS
chiral

REFERENCE
Trelli-Seifert,L.A.; Risley,D.S. Capillary electrophoretic enantiomeric separations of nonsteroidal anti-inflam-
matory compounds using the macrocyclic antibiotic actaplanin A and 2-methoxyethanol,
J.Liq.Chromatogr.Rel.Technol., **1998**, *21*, 299–313.

SAMPLE
Matrix: solutions

CAPILLARY ELECTROPHORESIS
Capillary: 42 cm × 50 μm fused-silica (31 cm to detector) (Polymicro Technologies)
Capillary temperature: 30
Running buffer: Formamide containing 3.42% quaternary ammonium-β-cyclodextrin (American
Maize Products, Hammond IN), 20 mM ammonium acetate, and 1% acetic acid (pH* = 7.1)
Detector: UV 254
Migration time: 14.39, 14.81 (enantiomers)
Voltage: -30 kV
Model: laboratory-constructed

OTHER SUBSTANCES
Also analyzed: carprofen, flurbiprofen, indoprofen, ketoprofen, suprofen

KEY WORDS
chiral

REFERENCE
Wang,F.; Khaledi,M.G. Nonaqueous capillary electrophoresis chiral separations with quaternary ammonium β-
cyclodextrin, *J.Chromatogr.A*, **1998**, *817*, 121–128.

Fenoterol

Molecular formula: $C_{17}H_{21}NO_4$
Molecular weight: 303.36
CAS Registry No.: 13392-18-2, 1944-12-3 (HBr), 1944-10-1 (HBr)
Merck Index (12th ed.): 4022
Lednicer: 2 38

SAMPLE
Matrix: solutions

CAPILLARY ELECTROPHORESIS
Capillary: 36 cm × 50 μm fused-silica coated with linear polyacrylamide (31.5 cm to detector) (GL Science)
Capillary preparation: At the beginning and end of each day rinse capillary with capillary wash solution (Bio-Rad Cat. No. 148-5022) at 690 kPa for more than 3 min and with water at 690 kPa for more than 3 min. Coat capillary as follows. Treat capillary with 1 M NaOH at room temperature for 1 h, rinse with water, dry by passing nitrogen gas through the capillary at 110° for 6 h. Pass thionyl chloride through the capillary using a suction pump for several min, seal capillary at both ends and heat at 70° for 6 h. Unseal the capillary and fill with 250 mM vinyl magnesium bromide in THF by suction, seal the capillary, heat at 70° for 6 h. Open the capillary and rinse it with THF for several min, rinse with distilled water, fill the capillary with polymerization solution, heat at 28 ± 2° for 1 h, condition at -100 V/cm for 30 min (Anal. Sci. 1994, 10, 1). (The polymerization solution was 5% acrylamide in water containing 49 mM Tris, 384 mM glycine, and 0.1% sodium dodecyl sulfate, degas in an ultrasonic bath. Add 40 μL 10% N,N,N',N'-tetramethylethylenediamine and 10 μL 10% ammonium persulfate to 5 mL of the degassed solution, mix thoroughly.)
Running buffer: MeOH:50 mM pH 6.0 Phosphate buffer 10:90
Injection: Before each injection rinse with water at 690 kPa for 30 s, rinse with running buffer at 690 kPa for 30 s, partially fill with separation solution (500 μM α_1-acid glycoprotein (Cohn fraction VI) (ICN) in running buffer) at 6.9 kPa for 190 s (27 cm), inject sample at 6.9 kPa for 2 s, electrophorese with running buffer (Note that α_1-acid glycoprotein from other suppliers may provide inferior results).
Detector: UV 210
Migration time: Resolution of enantiomers 3.1
Voltage: 12 kV
Model: Bio-rad BioFocus 3000

KEY WORDS
chiral; coated capillary

REFERENCE
Tanaka,Y.; Terabe,S. Separation of the enantiomers of basic drugs by affinity capillary electrophoresis using a partial filling technique and α_1-acid glycoprotein as chiral selector, *Chromatographia*, **1997**, *44*, 119–128.

SAMPLE
Matrix: solutions

CAPILLARY ELECTROPHORESIS
Capillary: 57 cm × 50 μm fused-silica (50 cm to detector)
Running buffer: 50 mM pH 8.0 Borate buffer containing 40 mM sodium taurodeoxycholate and 25 mM phosphatidylcholine (Prepare by adding phosphatidylcholine and stirring for 3-6 h until all cloudiness disappears. Phosphatidylcholine was 95% pure soybean lecithin, Epikuron, Lucas Meyer & Co.)
Injection: Inject a solution of the compound in the running buffer at 20 psi for 1 s, inject MeOH:water 5:95 at 20 psi for 1 s, inject a solution of halofantrine in running buffer at 20 psi for 1 s.
Detector: UV 214
Migration time: k' 5.13
Model: Beckman P/ACE 5000

OTHER SUBSTANCES
Also analyzed: acetaminophen, amoxicillin, antipyrine, aspirin, azathioprine, caffeine, captopril, carbamazepine, carprofen, chlorambucil, chlorpheniramine, chlorpromazine, cimetidine, clonidine, codeine, desipramine, diphenhydramine, ephedrine, flufenamic acid, flurbiprofen, haloperidol, hydroxyzine, ibuprofen, imipramine, indomethacin, ketoprofen, lidocaine, melphalan, metoprolol, nabumetone, nadolol, phenobarbital, phenol, promazine, propranolol, pyrilamine, ranitidine, ropinirole, salicylic acid, sulfamethoxazole, testosterone, theophylline, thioridazine, tiaprofenic acid, tolfenamic acid, trifluoperazine, trimethoprim, valproic acid, verapamil

KEY WORDS
comparison with HPLC; $k' = (Tr-T0)/(T0(1-Tr/Tm))$ where Tr = retention time of analyte; $T0$ = retention time of water; and Tm = retention time of marker (halofantrine)

REFERENCE
Hanna,M.; de Biasi,V.; Bond,B.; Salter,C.; Hutt,A.J.; Camilleri,P. Estimation of the partitioning characteristics of drugs: A comparison of a large and diverse drug series utilizing chromatographic and electrophoretic methodology, *Anal.Chem.*, **1998**, *70*, 2092–2099.

Fentanyl

Molecular formula: $C_{22}H_{28}N_2O$
Molecular weight: 336.48
CAS Registry No.: 437-38-7, 990-73-8 (citrate)
Merck Index (12th ed.): 4043
Lednicer: 1 299

SAMPLE
Matrix: solutions

CAPILLARY ELECTROPHORESIS
Capillary: 59.6 cm $\times$ 75 μm fused-silica (52.1 cm to detector) (Polymicro Technologies)
Capillary preparation: Purge with running buffer before each run. At the beginning of each day purge using 50-60 kPa vacuum with 500 mM NaOH for 5 min, with water for 5 min, with MeCN for 5 min, and with running buffer for 5 min.
Running buffer: MeCN:MeOH:acetic acid 49:50:1 containing 20 mM ammonium acetate
Injection: Hydrostatic injection at 10 cm for 5 s.
Detector: UV 214
Migration time: 4.15
Voltage: 30 kV
Model: Waters Quanta 4000

OTHER SUBSTANCES
Simultaneous: diphenoxylate, ethoheptazine, levallorphan, meperidine, methadone, nalorphine, pentazocine, nikethamide

REFERENCE
Leung,G.N.W.; Tang,H.P.O.; Tso,T.S.C.; Wan,T.S.M. Separation of basic drugs with non-aqueous capillary electrophoresis, *J.Chromatogr.A*, **1996**, *738*, 141–154.

Flecainide

Molecular formula: $C_{17}H_{20}F_6N_2O_3$
Molecular weight: 414.35
CAS Registry No.: 54143-55-4, 54143-56-5 (acetate)
Merck Index (12th ed.): 4136
Lednicer: 3 59

SAMPLE
Matrix: solutions
Sample preparation: Inject an aliquot of a 100 µg/mL solution in water:running buffer 50:50.

CAPILLARY ELECTROPHORESIS
Capillary: 44.5 cm × 50 µm acrylamide-coated fused-silica (Bio-Rad)
Capillary temperature: 30
Running buffer: 100 mM NaH_2PO_4 containing 15 mM gamma-cyclodextrin, adjusted to pH 2.5
   with phosphoric acid
Injection: Electrokinetic injection at 8 kV for 6 s.
Detector: UV 200
Migration time: 9.99
Voltage: 14 kV
Model: Bio-Rad BioFocus 3000

OTHER SUBSTANCES
Also analyzed: albuterol, alprenolol, atenolol, atropine, baclofen, bamethan, benserazide, biperiden, bisoprolol, bupivacaine, bupranolol, butetamate, carazolol, carbuterol, carvedilol, celiprolol, chloroquine, chlorpheniramine (chlorphenamine), clidinium bromide, clobutinol, disopyramide, dobutamine, homatropine, ipratropium bromide, isoproterenol, isothipendyl,
   ketamine, mefloquine, mequitazine, metaproterenol (orciprenaline), metipranolol, nafronyl
   (naftidrofuryl), nefopam, ofloxacin, orphenadrine, oxomemazine, oxprenolol, phenoxybenzamine, pholedrine, pindolol, pirbuterol, prilocaine, promethazine, propafenone, propranolol, sotalol, synephrine, terbutaline, tetrahydrozoline (tetryzoline), tocainide, trihexyphenidyl, trimeprazine (alimemazine), trimipramine, tropicamide, verapamil, zopiclone

KEY WORDS
coated capillary; achiral

REFERENCE
Koppenhoefer,B.; Epperlein,U.; Christian,B.; Yibing,J.; Yuying,C.; Bingcheng,L. Separation of enantiomers of
   drugs by capillary electrophoresis. I. γ-Cyclodextrin as chiral solvating agent, *J.Chromatogr.A*, **1995**, *717*,
   181–190.

SAMPLE
Matrix: solutions
Sample preparation: Inject an aliquot of a 100 µg/mL solution in running buffer.

CAPILLARY ELECTROPHORESIS
Capillary: 30 cm × 50 µm fused-silica (25.5 cm to detector) (Yongnian Optical Conductive Fiber
   Plant, China), coated with polyacrylamide
Capillary preparation: No details of the polyacrylamide coating process are provided. However,
   another paper (LC.GC 1997, 15, 40) by this group indicates that they use the procedure of
   Hjertén, thus: Adjust the pH of 20 mL water to 3.5 with acetic acid, add 80 µL 3-(trimethoxysilyl)propyl methacrylate (3-methacryloxypropyltrimethoxysilane), mix, suck into capillary, let
   stand at room temperature for 1 h, remove the solution, wash with water. Fill the capillary
   with a deaerated 3-4% acrylamide solution containing 1 µL/mL N,N,N',N'-tetramethylethyl-
   enediamine and 1 mg/mL potassium persulfate, let stand for 30 min, remove excess solution
   by aspiration, rinse with water, remove water by aspiration, dry at 35° (J. Chromatogr. 1985,
   347, 191).
Capillary temperature: 25

Running buffer: 100 mM NaH_2PO_4 adjusted to pH 2.5
Injection: Electrokinetic injection at 15 kV for 3 s.
Detector: UV 200, UV 210
Migration time: 5.27
Voltage: 15 kV
Model: Bio-Rad BioFocus 3000

OTHER SUBSTANCES
Simultaneous: amorolfine, brompheniramine, bupivacaine, carteolol, chloroquine, chlorpheniramine, chlorphenoxamine, disopyramide, dobutamine, doxylamine, gallopamil, ketamine, mepindolol, orphenadrine, oxybutynin, phenoxybenzamine, pindolol, propafenone, propranolol,
sulpiride, talinolol, tropicamide, verapamil

KEY WORDS
coated capillary

REFERENCE
Koppenhoefer,B.; Epperlein,U.; Xiaofeng,Z.; Bingcheng,L. Separation of enantiomers of drugs by capillary electrophoresis. Part 4: Hydroxypropyl-γ-cyclodextrin as chiral solvating agent, *Electrophoresis*, **1997**, *18*, 924–
930.

SAMPLE
Matrix: solutions

CAPILLARY ELECTROPHORESIS
Capillary: 29-36 cm × 50 μm fused-silica (24.5-31.5 cm to detector) (Yongnian Optical Conductive
Fiber Plant, China) coated with polyacrylamide
Capillary preparation: Coat capillary as follows. Adjust the pH of 20 mL water to 3.5 with
acetic acid, add 80 μL 3-(trimethoxysilyl)propyl methacrylate (3-methacryloxypropyltrimethox-
ysilane), mix, suck into capillary, let stand at room temperature for 1 h, remove the solution,
wash with water. Fill the capillary with a deaerated 3-4% acrylamide solution containing 1
μL/mL N,N,N',N'-tetramethylethylenediamine and 1 mg/mL potassium persulfate, let stand
for 30 min, remove excess solution by aspiration, rinse with water, remove water by aspiration,
dry at 35° (J. Chromatogr. 1985, 347, 191).
Capillary temperature: 25
Running buffer: 100 mM pH 2.5 NaH_2PO_4 (A) or 100 mM pH 2.5 NaH_2PO_4 containing 45 mM
hydroxypropyl-α-cyclodextrin (Wacker, Munich) (B)
Injection: Electromigration at 15 kV for 3 s.
Detector: UV 200; UV 210
Migration time: 5.27 (A); 12.19 (B) (no separation of enantiomers)
Voltage: 15 kV
Model: Bio-Focus 3000

OTHER SUBSTANCES
Also analyzed: albuterol (salbutamol), alprenolol, amorolfine, atenolol, atropine, azelastine, baclofen, bamethan, benproperine, benserazide, biperiden, bisoprolol, brompheniramine, bupivacaine, bupranolol, butamirate, butethamate, carazolol, carbuterol, carteolol, carvedilol, celiprolol, chloroquine, chlorpheniramine, chlorphenoxamine, cicletanine, clenbuterol, clidinium
bromide, clobutinol, dimethindene, dipivefrin, disopyramide, dobutamine, doxylamine, fendiline, gallopamil, homatropine, ipratropium bromide, isoproterenol (isoprenaline), isothipendyl,
ketamine, meclizine, mefloquine, mepindolol, mequitazine, metaclazepam, metaproterenol (orciprenaline), metipranolol, metoprolol, nafronyl (naftidrofuryl), nefopam, nicardipine, norfenefrine, ofloxacin, ornidazole, orphenadrine, oxomemazine, oxprenolol, oxybutynin, phenoxybenzamine, phenylpropanolamine, pholedrine, pindolol, pirbuterol, prilocaine, procyclidine,
promethazine, propafenone, propranolol, reproterol, sotalol, sulpride, synephrine, talinolol, terbutaline, tetrahydrozoline (tetryzoline), theodrenaline, tioconazole, tocainide, trihexyphenidyl,
trimeprazine (alimemazine), trimipramine, tropicamide, verapamil, zopiclone

KEY WORDS
coated capillary

REFERENCE

Koppenhoefer,B.; Eperlein,U.; Schlunk,R.; Zhu,X.; Lin,B. Separation of enantiomers of drugs by capillary electrophoresis. V. Hydroxypropyl-α-cyclodextrin as chiral solvating agent, *J.Chromatogr.A*, **1998**, *793*, 153–164.

Fleroxacin

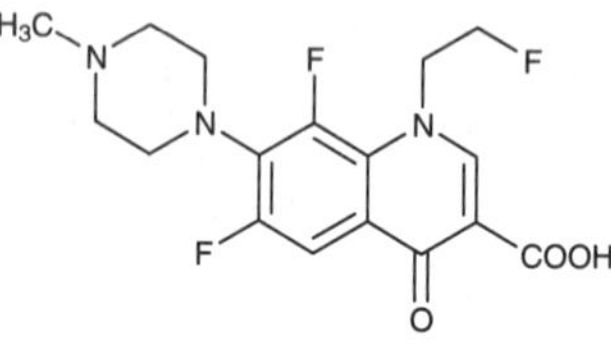

Molecular formula: $C_{17}H_{18}F_3N_3O_3$
Molecular weight: 369.34
CAS Registry No.: 79660-72-3
Merck Index (12th ed.): 4137
Lednicer: 5 125

SAMPLE

Matrix: bulk
Sample preparation: Inject an aliquot of an aqueous solution.

CAPILLARY ELECTROPHORESIS

Capillary: 70 cm × 75 μm fused-silica
Capillary preparation: Rinse capillary with running buffer for 3 min before each run. Rinse reservoirs with running buffer for 6 s before each run.
Running buffer: 50 mM Boric acid containing 10 mM sodium borate, pH 8.6
Injection: Hydrodynamic injection at 50 mm for 10 s
Migration time: 6.73
Voltage: 20 kV
Current: 21 μA
Model: Dionex Capillary Electrophoresis System I

OTHER SUBSTANCES

Simultaneous: Ro 23-9424, Ro 19-4885

REFERENCE

Nickerson,B.; Cunningham,B.; Scypinski,S. The use of capillary electrophoresis to monitor the stability of a dual-action cephalosporin in solution, *J.Pharm.Biomed.Anal.*, **1996**, *14*, 73–83.

Fluconazole

Molecular formula: $C_{13}H_{12}F_2N_6O$
Molecular weight: 306.27
CAS Registry No.: 86386-73-4
Merck Index (12th ed.): 4158

SAMPLE

Matrix: blood
Sample preparation: Inject nL quantities of plasma directly or perform a preliminary cleanup using a C18 SPE cartridge.

CAPILLARY ELECTROPHORESIS

Capillary: 500-700 mm × 75 μm
Running buffer: pH 9.2 Phosphate/borate buffer containing 75 mM sodium dodecyl sulfate
Detector: UV 190, UV 200
Migration time: <11
Internal standard: 1,7-dimethylxanthine

Voltage: 250 V/cm
Limit of detection: 5000 µg/mL (direct injection), 200 ng/mL (SPE)

KEY WORDS
plasma; SPE

REFERENCE
von Heeren,F.; Tanner,R.; Stotzer,R.; Thormann,W. Determination of the antimycotic drug fluconazole in human plasma by micellar electrokinetic capillary chromatography (Abstract 118), *Ther.Drug Monit.*, **1995**, *17*, 412.

Flucytosine

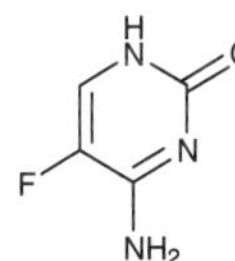

Molecular formula: $C_4H_4FN_3O$
Molecular weight: 129.09
CAS Registry No.: 2022-85-7
Merck Index (12th ed.): 4161

SAMPLE
Matrix: blood
Sample preparation: Inject serum directly.

CAPILLARY ELECTROPHORESIS
Capillary: 105 cm × 75 µm fused-silica (68 cm to detector) (Polymicro Technologies)
Capillary preparation: Rinse with running buffer for 10 min between run. At the beginning and end of each day rinse with 100 mM NaOH for 10 min and with water for 10 min.
Running buffer: 10 mM Na_2HPO_4 containing 6 mM sodium borate and 75 mM sodium dodecyl sulfate
Injection: Injection via vacuum suction for 1 s
Detector: UV 215
Migration time: 8.6
Voltage: 23 kV
Current: 62 µA
Model: Europhor Prime Vision IV

OTHER SUBSTANCES
Extracted: acetaminophen, ethosuximide, naproxen, primidone, sulfamethoxazole, zomepirac

KEY WORDS
serum

REFERENCE
Schmutz,A.; Thormann,W. Assessment of impact of physico-chemical drug properties on monitoring drug levels by micellar electrokinetic capillary chromatography with direct serum injection, *Electrophoresis*, **1994**, *15*, 1295–1303.

SAMPLE
Matrix: blood
Sample preparation: Vortex serum for 20 s, inject directly.

CAPILLARY ELECTROPHORESIS
Capillary: 50 cm × 50 µm fused-silica (45 cm to detector) (Bio-Rad)
Capillary preparation: Between runs rinse capillary with 100 mM NaOH for 40 s and with running buffer for 100 s. Replace buffer vial every 10 runs.
Capillary temperature: 20
Running buffer: 10 mM pH 9.2 Na_2HPO_4 containing 6 mM sodium tetraborate and 75 mM sodium dodecyl sulfate
Injection: Pressure injection 2 psi.s

Detector: UV 210
Migration time: 3.3
Voltage: 25 kV
Current: 48 μA
Model: BioFocus 3000 (Bio-Rad)
Limit of detection: 2 μg/mL (S/N 3)

OTHER SUBSTANCES
Simultaneous: amphotericin B

KEY WORDS
serum

REFERENCE
Schmutz,A.; Thormann,W. Rapid determination of the antimycotic drug flucytosine in human serum by micellar electrokinetic capillary chromatography with direct sample injection, *Ther.Drug Monit.*, **1994**, *16*, 483–490.

Fludiazepam

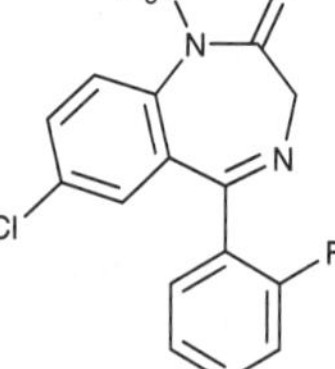

Molecular formula: $C_{16}H_{12}ClFN_2O$
Molecular weight: 302.74
CAS Registry No.: 3900-31-0
Merck Index (12th ed.): 4164

SAMPLE
Matrix: solutions

CAPILLARY ELECTROPHORESIS
Capillary: 60 cm × 50 μm fused-silica (40 cm to detector), C18 coated (Supelco)
Capillary preparation: Rinse new capillaries with 250 μL 1 M NaOH, 250 μL 100 mM NaOH, water, MeOH, water, and running buffer then condition with running buffer for at least 15 min. Carry out a similar procedure between runs.
Running buffer: 100 mM pH 8.5 Borate buffer containing 25 mM sodium dodecyl sulfate and 5 M urea
Injection: Electrokinetic injection at 5 kV for 5 s.
Detector: UV 254
Migration time: 38
Voltage: 18 kV
Model: Jasco

OTHER SUBSTANCES
Simultaneous: alprazolam, amobarbital, barbital, bromazepam, clonazepam, clotiazepam, cloxazolam, diazepam, estazolam, etizolam, flunitrazepam, flurazepam, haloxazolam, medazepam, mephobarbital, metharbital, nimetazepam, nitrazepam, oxazepam, pentobarbital, phenobarbital, secobarbital, triazolam

KEY WORDS
coated capillary

REFERENCE
Jinno,K.; Han,Y.; Nakamura,M. Analysis of anxiolytic drugs by capillary electrophoresis with bare and coated capillaries, *J.Capillary Electrophor.*, **1996**, *3*, 139–145.

SAMPLE
Matrix: solutions

CAPILLARY ELECTROPHORESIS
Capillary: 60 cm × 75 μm coated fused-silica (40 cm to detector) (Supelco)
Capillary preparation: Coat column as follows. Adjust the pH of 20 mL water to 3.5 with acetic acid, add 80 μL 3-(trimethoxysilyl)propyl methacrylate (3-methacryloxypropyltrimethoxysilane), mix, suck into capillary, let stand at room temperature for 1 h, remove the solution, wash with water. Fill the capillary with a deaerated 4% acrylamide solution containing 1 mg/mL N,N,N',N'-tetramethylethylenediamine and 1 mg/mL ammonium persulfate, let stand for 3 h, remove excess solution by aspiration, rinse with water, remove water by aspiration, dry at 35° (cf. J. Chromatogr. 1985, 347, 191).
Running buffer: 100 mM pH 8.5 borate buffer containing 10 mM sodium dodecyl sulfate and 5 M urea
Injection: Electromigration at 5 kV for 5 s.
Detector: UV 240, UV 254
Migration time: 8.2
Voltage: 18 kV
Model: Jasco 890-CE

OTHER SUBSTANCES
Simultaneous: alprazolam, clotiazepam, cloxazolam, diazepam, estazolam, etizolam, flurazepam, haloxazolam, oxazepam

KEY WORDS
injection at cathode; coated capillary

REFERENCE
Jinno,K.; Han,Y.; Sawada,H. Analysis of toxic drugs by capillary electrophoresis using polyacrylamide-coated columns, *Electrophoresis*, **1997**, *18*, 284–286.

SAMPLE
Matrix: solutions
Sample preparation: Inject an aliquot of a 100 μg/mL solution in running buffer.

CAPILLARY ELECTROPHORESIS
Capillary: 60 cm × 50 μm acrylamide-coated fused-silica (40 cm to detector) (Supelco)
Capillary preparation: Adjust the pH of 20 mL water to 3.5 with acetic acid, add 80 μL 3-(trimethoxysilyl)propyl methacrylate (3-methacryloxypropyltrimethoxysilane), mix, suck into capillary, let stand at room temperature for 1 h, remove the solution, wash with water. Fill the capillary with a deaerated 3-4% acrylamide solution containing 1 μL/mL N,N,N',N'-tetramethylethylenediamine and 1 mg/mL ammonium persulfate, let stand for 3 h, remove excess solution by aspiration, rinse with water, remove water by aspiration, dry at 35° (cf. J. Chromatogr. 1985, 347, 191).
Running buffer: MeCN:buffer 5:95 (Buffer was 100 mM borate containing 5 M urea and 10 mM sodium dodecyl sulfate, adjusted to pH 8.5 with phosphate.)
Injection: Electrokinetic injection at 5 kV for 5 s.
Detector: UV 254
Migration time: 10.8
Voltage: 18 kV
Model: Jasco Model 870-CE

OTHER SUBSTANCES
Simultaneous: alprazolam, amobarbital, barbital, bromazepam, clonazepam, clotiazepam, cloxazolam, diazepam, estazolam, etizolam, flunitrazepam, flurazepam, haloxazolam, medazepam, mephobarbital, metharbital, nimetazepam, nitrazepam, oxazepam, pentobarbital, phenobarbital, secobarbital, triazolam

KEY WORDS
coated capillary; detector at anode

REFERENCE
Jinno,K.; Han,Y.; Sawada,H.; Taniguchi,M. Capillary electrophoretic separation of toxic drugs using a polyacrylamide-coated capillary, *Chromatographia*, **1997**, *46*, 309–314.

Fludrocortisone

Molecular formula: $C_{21}H_{29}FO_5$
Molecular weight: 380.46
CAS Registry No.: 127-31-1, 514-36-3 (acetate)
Merck Index (12th ed.): 4166
Lednicer: 1 192

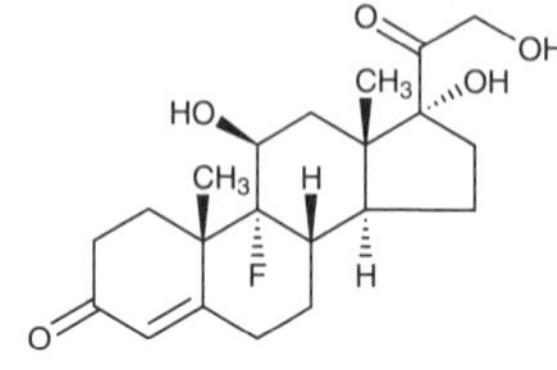

SAMPLE
Matrix: solutions
Sample preparation: Prepare a 10 mg/mL solution in MeOH, inject an aliquot.

CAPILLARY ELECTROPHORESIS
Capillary: 62 cm × 53 μm fused-silica (50 cm to detector) (Polymicro Technologies)
Capillary preparation: Rinse with running buffer for 1 min between runs.
Running buffer: 50 mM pH 9.0 Phosphate borate buffer containing 50 mM dehydrocholate, 50 mM taurocholate, and 50 mM sodium dodecyl sulfate
Detector: UV 254
Migration time: 19.5 (fludrocortisone), 23 (fludrocortisone acetate)
Voltage: 15 kV
Current: 56 μA
Model: Laboratory constructed

OTHER SUBSTANCES
Simultaneous: corticosterone, cortisone, cortisone acetate, deoxycorticosterone, fluocinolone acetonide, hydrocortisone, hydrocortisone-21-acetate, 6α-methylprednisolone, prednisolone, prednisolone acetate, prednisone, prednisone acetate, progesterone, triamcinolone, triamcinolone acetonide

REFERENCE
Bumgarner,J.G.; Khaledi,M.G. Mixed micellar electrokinetic chromatography of corticosteroids, *Electrophoresis*, **1994**, *15*, 1260–1266.

SAMPLE
Matrix: solutions

CAPILLARY ELECTROPHORESIS
Capillary: 62 cm × 53 μm fused-silica (50 cm to detector) (Polymicro Technologies)
Capillary preparation: Rinse capillary with running buffer for 1 min between runs.
Running buffer: 50 mM pH 9.0 Phosphate/borate buffer containing 33 mM taurocholate, 33 mM glycodeoxycholate, and 70 mM butanesulfonate
Detector: UV 254
Migration time: 20 (fludrocortisone), 22.5 (fludrocortisone acetate)
Voltage: 15 kV
Model: laboratory-constructed

OTHER SUBSTANCES
Simultaneous: corticosterone, cortisone, cortisone acetate, deoxycorticosterone, fluocinolone acetonide, hydrocortisone, hydrocortisone-21-acetate, 6α-methylprednisolone, prednisolone, prednisolone acetate, prednisone, prednisone acetate, progesterone, triamcinolone, triamcinolone acetonide

REFERENCE
Bumgarner,J.G.; Khaledi,M.G. Mixed micelles of short chain alkyl surfactants and bile slats in electrokinetic chromatography: Enhanced separation of corticosteroids, *J.Chromatogr.A*, **1996**, *738*, 275–283.

SAMPLE
Matrix: solutions

Sample preparation: Inject an aliquot of a solution in EtOH:water 11.1:78.8 containing 10 mM sodium dodecyl sulfate.

CAPILLARY ELECTROPHORESIS
Capillary: 67 cm × 50 μm fused-silica (60 cm to detector) (Composite Metal Services, Worcestershire UK)
Capillary temperature: 25
Running buffer: 50 mM pH 8.7 [(1,1-Dimethyl-2-hydroxyethyl)amino]-2-hydroxypropanesulfonic acid (AMPSO) containing 20 mM sodium dodecyl sulfate (pH adjusted with 25% ammonia)
Injection: Hydrostatic injection at 0.5 psi for 3.5 s.
Detector: UV 260
Migration time: 11.3
Voltage: 20 kV
Current: 10.1 μA
Model: Beckman P/ACE 2050

OTHER SUBSTANCES
Simultaneous: 4-androstene-3,17-dione,   corticosterone,   11-deoxycortisol,   dexamethasone, hydrocortisone

REFERENCE
Wiedmer,S.K.; Riekkola,M.-L.; Nydén,M.; Söderman,O. Mixed micelles of sodium dodecyl sulfate and sodium cholate: Micellar electrokinetic capillary chromatography and nuclear magnetic resonance spectroscopy, *Anal.Chem.*, **1997**, *69*, 1577–1584.

Flufenamic acid

Molecular formula: $C_{14}H_{10}F_3NO_2$
Molecular weight: 281.23
CAS Registry No.: 530-78-9
Merck Index (12th ed.): 4167
Lednicer: 1 110

SAMPLE
Matrix: urine
Sample preparation: Dilute 10-fold (with water ?), inject an aliquot

CAPILLARY ELECTROPHORESIS
Capillary: 57 cm × 50 μm fused-silica (50 cm to detector) (Beckman)
Capillary preparation: Between runs rinse capillary with running buffer for 2 min. Before first use flush capillary with freshly prepared 100 mM NaOH for 5 min, rinse with buffer, equilibrate with buffer at 12 kV for 15 min.
Running buffer: MeCN:buffer 10:90 (Buffer was 30 mM pH 3.0 phosphate buffer containing 2 mM β-cyclodextrin.)
Injection: Pressure injection at 0.5 psi for 4-10 s.
Detector: UV 285
Migration time: 10.2
Voltage: 12 kV
Current: 90-92 μA
Model: Beckman P/ACE 5500
Limit of detection: 0.3 ng/mL

OTHER SUBSTANCES
Extracted: meclofenamic acid, mefenamic acid

REFERENCE
Pérez-Ruiz,T.; Martínez-Lozano,C.; Sanz,A.; Bravo,E. Determination of flufenamic, meclofenamic and mefenamic acids by capillary electrophoresis using β-cyclodextrin, *J.Chromatogr.B*, **1998**, *708*, 249–256.

Flumequine

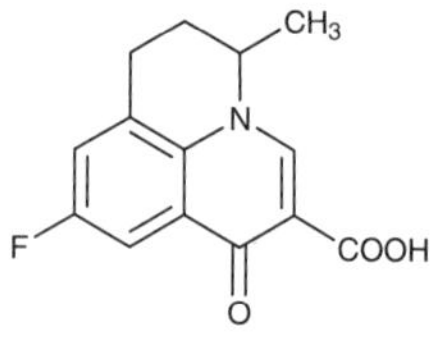

Molecular formula: $C_{14}H_{12}FNO_3$
Molecular weight: 261.25
CAS Registry No.: 42835-25-6
Merck Index (12th ed.): 4172
Lednicer: 3 186

SAMPLE

Matrix: solutions
Sample preparation: Inject an aliquot of a solution in MeOH.

CAPILLARY ELECTROPHORESIS

Capillary: 59 cm × 50 μm fused-silica (43 cm to detector) (Polymicro Technologies)
Capillary preparation: Before each analysis flush with water for 3 min, with 200 mM NaOH for 3 min, with water for 3 min, and with running buffer for 4 min. Flush new capillaries with 1 M NaOH at 2000 mbar for 10 min then with 200 mM NaOH for 10 min.
Capillary temperature: 23
Running buffer: MeCN:buffer 28:72, pH 7.3 (Buffer was 32 mM sodium borate containing 18 mM NaH_2PO_4, 39 mM sodium cholate, and 8 mM sodium heptanesulfonate.)
Injection: Hydrodynamic injection at 40 mbar for 6 s.
Detector: UV 260
Migration time: 5.85
Voltage: 30 kV
Model: Lauer Labs Prince

OTHER SUBSTANCES

Simultaneous: ciprofloxacin, enoxacin, lomefloxacin, nalidixic acid, norfloxacin, ofloxacin, oxolinic acid, pefloxacin, pipemidic acid, piromidic acid, rosoxacin, sparfloxacin

REFERENCE

Sun,S.-W.; Chen,L.-Y. Optimization of capillary electrophoretic separation of quinolone antibacterials using the overlapping resolution mapping scheme, *J.Chromatogr.A*, **1997**, *766*, 215–224.

Flumethasone

Molecular formula: $C_{22}H_{28}F_2O_5$
Molecular weight: 410.46
CAS Registry No.: 2135-17-3, 2002-29-1 (pivalate)
Merck Index (12th ed.): 4173
Lednicer: 1 200

SAMPLE

Matrix: solutions
Sample preparation: Inject an aliquot of a solution in running buffer. (Paper contains details of the preparation of an affinity column for the isolation of this compound from urine prior to capillary electrophoresis analysis.)

CAPILLARY ELECTROPHORESIS

Capillary: 57 cm × 75 μm fused-silica (50 cm to detector)
Capillary preparation: Before each run rinse with running buffer for 2 min. After each run rinse capillary with 100 mM NaOH for 2 min and with water for 2 min. Condition new capillaries with 100 mM HCl for 5 min, with 100 mM NaOH for 10 min, and with water for 5 min.
Capillary temperature: 23

Running buffer: MeOH:buffer 10:90 (Buffer was 40 mM pH 9.2 borate buffer containing 20 mM sodium dodecyl sulfate.)
Injection: Low pressure injection for 10 s.
Detector: UV 240
Migration time: 11
Voltage: 30 kV
Model: Beckman P/ACE 5510

OTHER SUBSTANCES
Simultaneous: betamethasone, dexamethasone, hydrocortisone, triamcinolone

REFERENCE
Gu,X.; Meleka-Boules,M.; Chen,C.-L. Micellar electrokinetic capillary chromatography combined with immunoaffinity chromatography for identification and determination of dexamethasone and flumethasone in equine urine, *J.Capillary Electrophor.*, **1996**, *3*, 43–49.

Flunitrazepam

Molecular formula: $C_{16}H_{12}FN_3O_3$
Molecular weight: 313.29
CAS Registry No.: 1622-62-4
Merck Index (12th ed.): 4181
Lednicer: 2 406

SAMPLE
Matrix: blood
Sample preparation: 500 µL Serum + 250 µL 800 mM pH 10.0 N-cyclohexyl-2-hydroxy-3-aminopropanesulfonic acid (CAPSO) buffer + 4 mL n-pentane:ethyl acetate 75:25, vortex for 1 min, centrifuge at 25° at 1500 g for 5 min. Remove the organic layer and evaporate it to dryness at 28°, reconstitute the residue in 40 µL 50 mM pH 9.5 borate buffer containing 4.5 mM sodium dodecyl sulfate, inject an aliquot.

CAPILLARY ELECTROPHORESIS
Capillary: 47 cm × 50 µm fused-silica (40 cm to detector) (Supelco)
Capillary preparation: Before each run wash with 100 mM NaOH, water, and running buffer.
Capillary temperature: 25
Running buffer: MeCN:50 mM pH 9.5 sodium borate buffer containing 18 mM sodium dodecyl sulfate 14:86
Injection: Pressure injection at 0.5 psi for 10 s.
Detector: UV 214
Migration time: 5.7
Internal standard: flunitrazepam
Voltage: 20 kV
Model: Beckman P/ACE 5010

OTHER SUBSTANCES
Extracted: clobazam, clonazepam, desmethylclobazam, desmethyldiazepam, diazepam, nitrazepam
Noninterfering: carbamazepine, ethosuximide, phenobarbital, phenytoin, primidone, valproic acid, zonisamide

KEY WORDS
serum; flunitrazepam is IS

REFERENCE
Imazawa,M.; Hatanaka,Y. Micellar electrokinetic capillary chromatography of benzodiazepine antiepileptics and their desmethyl metabolites in blood, *J.Pharm.Biomed.Anal.*, **1997**, *15*, 1503–1508.

SAMPLE
Matrix: solutions
Sample preparation: Inject an aliquot of a 5 μg/mL solution in running buffer.

CAPILLARY ELECTROPHORESIS
Capillary: 87 cm × 75 μm fused-silica (80 cm to detector) (Beckman)
Capillary preparation: Between runs rinse capillary with running buffer for 2 min then equilibrate for 5 min.
Capillary temperature: 35
Running buffer: 20 mM pH 7 Buffer containing 15 mM sodium cholate and 35 mM sodium deoxycholate (Buffer was 20 mM sodium borate adjusted to pH 7.0 with 20 mM NaH_2PO_4.)
Injection: Pressure injection for 2 s.
Detector: UV 214
Migration time: 15.5
Voltage: 20 kV
Model: Beckman P/ACE 2100

OTHER SUBSTANCES
Simultaneous: bromazepam, chlordiazepoxide, clobazam, clonazepam, diazepam, flurazepam, halazepam, lorazepam, lormetazepam, nitrazepam, nordazepam, temazepam

REFERENCE
Boonkerd,S.; Detaevernier,M.R.; Vindevogel,J.; Michotte,Y. Migration behaviour of benzodiazepines in micellar electrokinetic chromatography, *J.Chromatogr.A*, **1996**, *756*, 279–286.

SAMPLE
Matrix: solutions

CAPILLARY ELECTROPHORESIS
Capillary: 60 cm × 50 μm fused-silica (40 cm to detector), C18 coated (Supelco)
Capillary preparation: Rinse new capillaries with 250 μL 1 M NaOH, 250 μL 100 mM NaOH, water, MeOH, water, and running buffer then condition with running buffer for at least 15 min. Carry out a similar procedure between runs.
Running buffer: 100 mM pH 8.5 Borate buffer containing 25 mM sodium dodecyl sulfate and 5 M urea
Injection: Electrokinetic injection at 5 kV for 5 s.
Detector: UV 254
Migration time: 19
Voltage: 18 kV
Model: Jasco

OTHER SUBSTANCES
Simultaneous: alprazolam, amobarbital, barbital, bromazepam, clotiazepam, cloxazolam, diazepam, estazolam, etizolam, fludiazepam, flurazepam, haloxazolam, medazepam, mephobarbital, metharbital, nimetazepam, nitrazepam, oxazepam, pentobarbital, phenobarbital, secobarbital, triazolam
Interfering: clonazepam

KEY WORDS
coated capillary

REFERENCE
Jinno,K.; Han,Y.; Nakamura,M. Analysis of anxiolytic drugs by capillary electrophoresis with bare and coated capillaries, *J.Capillary Electrophor.*, **1996**, *3*, 139–145.

SAMPLE
Matrix: solutions
Sample preparation: Inject an aliquot of a 1-50 μg/mL solution in water.

CAPILLARY ELECTROPHORESIS
Capillary: 55 cm × 50 μm fused-silica (35 cm to detector) (J&W)

Capillary preparation: Flush with running buffer for 5 min before each run. Periodically wash with 100 mM NaOH and flush extensively with running buffer.
Running buffer: MeOH:25 mM pH 9.24 borate buffer 20:80 containing 100 mM sodium dodecyl sulfate (A) or 50 mM pH 2.35 phosphate buffer (B) or 50 mM pH 9.24 borate buffer (C)
Injection: Injection of 5 μL using a split-flow injector ratio of 1:800.
Detector: UV 200
Migration time: 29.36 (A), 28.67 (B), 7.19 (C)
Voltage: 20 kV (A, B) or 12 kV (C)
Current: <60-80 μA
Model: ISCO Model 3850

OTHER SUBSTANCES
Also analyzed: acetylcodeine, amphetamine, barbital, caffeine, codeine, cocaine, diamorphine, diazepam, lidocaine, monoacetylmorphine, morphine, nalorphine, narceine, noscapine, papaverine, pentobarbital, procaine, tetracaine, thebaine

KEY WORDS
all compounds were separated with running buffer A; some peaks overlapped with running buffers B and C.

REFERENCE
Tagliaro,F.; Smith,F.P.; Turrina,S.; Equisetto,V.; Marigo,M. Complementary use of capillary zone electrophoresis and micellar electrokinetic capillary chromatography for mutual confirmation of results in forensic drug analysis, *J.Chromatogr.A*, **1996**, *735*, 227–235.

SAMPLE
Matrix: urine
Sample preparation: Place a silica fiber coated with an 85 μm-thick layer of polyacrylate (Supelco) in 10 mL urine, stir at 60° for 2 h, place the fiber in a capillary tube containing 20 μL MeCN, let stand at room temperature for 30 min, inject an aliquot of the solution. (Desorb a new fiber several times with MeCN before use.)

CAPILLARY ELECTROPHORESIS
Capillary: 60 cm × 50 μm acrylamide-coated fused-silica (40 cm to detector) (Supelco)
Capillary preparation: Adjust the pH of 20 mL water to 3.5 with acetic acid, add 80 μL 3-(trimethoxysilyl)propyl methacrylate (3-methacryloxypropyltrimethoxysilane), mix, suck into capillary, let stand at room temperature for 1 h, remove the solution, wash with water. Fill the capillary with a deaerated 3-4% acrylamide solution containing 1 μL/mL N,N,N',N'-tetramethylethylenediamine and 1 mg/mL ammonium persulfate, let stand for 3 h, remove excess solution by aspiration, rinse with water, remove water by aspiration, dry at 35° (cf. J. Chromatogr. 1985, 347, 191).
Running buffer: MeCN:buffer 5:95 (Buffer was 100 mM borate containing 5 M urea and 10 mM sodium dodecyl sulfate, adjusted to pH 8.5 with phosphate.)
Injection: Electrokinetic injection at 5 kV for 5 s.
Detector: UV 254
Migration time: 22.5
Voltage: 18 kV
Model: Jasco Model 870-CE
Limit of detection: <1 ppm

OTHER SUBSTANCES
Extracted: nitrazepam, triazolam
Simultaneous: alprazolam, amobarbital, barbital, bromazepam, clonazepam, clotiazepam, cloxazolam, diazepam, estazolam, etizolam, fludiazepam, flurazepam, haloxazolam, medazepam, mephobarbital, metharbital, nimetazepam, oxazepam, pentobarbital, phenobarbital, secobarbital

KEY WORDS
coated capillary; detector at anode; SPE

REFERENCE
Jinno,K.; Han,Y.; Sawada,H.; Taniguchi,M. Capillary electrophoretic separation of toxic drugs using a polyacrylamide-coated capillary, *Chromatographia*, **1997**, *46*, 309–314.

Flunixin

Molecular formula: $C_{14}H_{11}F_3N_2O_2$
Molecular weight: 296.25
CAS Registry No.: 38677-85-9, 42461-84-7 (meglumine salt)
Merck Index (12th ed.): 4182
Lednicer: 2 281

SAMPLE
Matrix: blood, urine
Sample preparation: Condition a Bond Elut Certify C18 SPE cartridge with 5 mL EtOH and 10 mL water. 2 mL Serum or urine + 4 mL buffer, mix, add to the SPE cartridge, wash with 5 mL buffer, wash with 1 mL water, dry under vacuum (15 psi) for 2 min, elute with 5 mL dichloromethane. Evaporate the eluate at 56°, reconstitute with 500 μL EtOH, evaporate to dryness under a stream of nitrogen, reconstitute with sample solvent, inject an aliquot. (Buffer was 1 M acetic acid containing 1 M sodium acetate, pH adjusted to 4.5 with 6 M HCl. Sample solvent was EtOH:50 mM acetic acid containing 75 mM sodium dodecyl sulfate 2:98, adjusted to pH 4.5 with 1 M NaOH.)

CAPILLARY ELECTROPHORESIS
Capillary: 57 cm × 75 μm silica (50 cm to detector) (J&W)
Capillary preparation: Rinse with running buffer for 2 min before each injection and with 10 mM NaOH for 2 min and with water for 2 min after each injection. At the start of each day rinse with 1 M HCl for 5 min, with 10 mM NaOH for 10 min, and with water for 5 min.
Capillary temperature: 23
Running buffer: EtOH:buffer 2:98 (Buffer was 50 mM pH 9.2 sodium borate containing 75 mM sodium dodecyl sulfate.)
Injection: Pressure injection at 0.5 psi for 10 s.
Detector: UV 286
Migration time: 15.7
Voltage: 20 kV
Model: Beckman P/ACE 5510
Limit of quantitation: 5.57 ng/mL (serum), 33.1 ng/mL (urine)
Limit of detection: 3.35 ng/mL (serum), 16.9 ng/mL (urine)

KEY WORDS
horse; serum; SPE

REFERENCE
Gu,X.; Meleka-Boules,M.; Chen,C.-L.; Ceska,D.M.; Tiffany,D.M. Determination of flunixin in equine urine and serum by capillary electrophoresis, *J.Chromatogr.B*, **1997**, *692*, 187–198.

Fluocinolone acetonide

Molecular formula: $C_{24}H_{30}F_2O_6$
Molecular weight: 452.50
CAS Registry No.: 67-73-2
Merck Index (12th ed.): 4185
Lednicer: 1 202; 3 94

SAMPLE
Matrix: solutions

CAPILLARY ELECTROPHORESIS
Capillary: 65 cm × 50 μm fused-silica (50 cm to detector) (Scientific Glass Engineering)

Capillary preparation: Flush with running buffer for 1 min between runs. Periodically flush with water for 1 min, sweep with 100 mM KOH, let stand in 100 mM KOH for 30 min, flush with water until the effluent is neutral to pH paper, fill with running buffer, let stand for 30 min.

Running buffer: 20 mM pH 9.0 Phosphate-borate buffer containing 100 mM sodium cholate (Buffer was prepared by adjusting pH of 20 mM sodium borate to 9.0 with 20 mM NaH_2PO_4.)

Injection: Injection by siphon at 10 cm for 10 s (about 1 nL)

Detector: UV 210

Migration time: 14.3 (fluocinolone acetonide)

Voltage: 20 kV

OTHER SUBSTANCES

Simultaneous: betamethasone, dexamethasone acetate, fluocinonide, hydrocortisone acetate, hydrocortisone, triamcinolone, triamcinolone acetonide

REFERENCE

Nishi,H.; Fukuyama,T.; Matsuo,M.; Terabe,S. Separation and determination of lipophilic corticosteroids and benzothiazepin analogues by micellar electrokinetic chromatography using bile salts, *J.Chromatogr.*, **1990**, *513*, 279–295.

SAMPLE

Matrix: solutions

Sample preparation: Prepare a 0.2-1 mg/mL solution in MeOH, inject an aliquot.

CAPILLARY ELECTROPHORESIS

Capillary: 65 cm × 50 μm fused-silica (50 cm to detector)

Running buffer: 20 mM pH 9.0 Phosphate-borate buffer containing 50 mM sodium dodecyl sulfate, 4 M urea, and 15 mM gamma-cyclodextrin

Detector: UV 220

Migration time: 11 (fluocinolone acetonide)

Voltage: 20 kV

OTHER SUBSTANCES

Simultaneous: betamethasone, cortisone acetate, dexamethasone acetate, fluocinonide, hydrocortisone, hydrocortisone acetate, triamcinolone acetonide

REFERENCE

Nishi,H.; Matsuo,M. Separation of corticosteroids and aromatic hydrocarbons by cyclodextrin-modified micellar electrokinetic chromatography, *J.Liq.Chromatogr.*, **1991**, *14*, 973–986.

SAMPLE

Matrix: solutions

Sample preparation: Prepare a 10 mg/mL solution in MeOH, inject an aliquot.

CAPILLARY ELECTROPHORESIS

Capillary: 62 cm × 53 μm fused-silica (50 cm to detector) (Polymicro Technologies)

Capillary preparation: Rinse with running buffer for 1 min between runs.

Running buffer: 50 mM pH 9.0 Phosphate borate buffer containing 50 mM dehydrocholate, 50 mM taurocholate, and 50 mM sodium dodecyl sulfate

Detector: UV 254

Migration time: 29 (fluocinolone acetonide)

Voltage: 15 kV

Current: 56 μA

Model: Laboratory constructed

OTHER SUBSTANCES

Simultaneous: corticosterone, cortisone, cortisone acetate, deoxycorticosterone, fludrocortisone, fludrocortisone acetate, hydrocortisone, hydrocortisone-21-acetate, 6α-methylprednisolone, prednisolone, prednisolone acetate, prednisone, prednisone acetate, progesterone, triamcinolone, triamcinolone acetonide

REFERENCE

Bumgarner,J.G.; Khaledi,M.G. Mixed micellar electrokinetic chromatography of corticosteroids, *Electrophoresis*, **1994**, *15*, 1260–1266.

SAMPLE
Matrix: solutions

CAPILLARY ELECTROPHORESIS
Capillary: 62 cm × 53 μm fused-silica (50 cm to detector) (Polymicro Technologies)
Capillary preparation: Rinse capillary with running buffer for 1 min between runs.
Running buffer: 50 mM pH 9.0 Phosphate/borate buffer containing 33 mM taurocholate, 33 mM glycodeoxycholate, and 70 mM butanesulfonate
Detector: UV 254
Migration time: 29 (fluocinolone acetonide)
Voltage: 15 kV
Model: laboratory-constructed

OTHER SUBSTANCES
Simultaneous: corticosterone, cortisone, cortisone acetate, deoxycorticosterone, fludrocortisone, fludrocortisone acetate, hydrocortisone, hydrocortisone-21-acetate, 6α-methylprednisolone, prednisolone, prednisolone acetate, prednisone, prednisone acetate, progesterone, triamcinolone, triamcinolone acetonide

REFERENCE

Bumgarner,J.G.; Khaledi,M.G. Mixed micelles of short chain alkyl surfactants and bile slats in electrokinetic chromatography: Enhanced separation of corticosteroids, *J.Chromatogr.A*, **1996**, *738*, 275–283.

Fluocinonide

Molecular formula: $C_{26}H_{32}F_2O_7$
Molecular weight: 494.53
CAS Registry No.: 356-12-7
Merck Index (12th ed.): 4186

SAMPLE
Matrix: formulations
Sample preparation: 1 g Cream + 5 mL MeOH + 5 mL 200 μg/mL IS in MeOH, warm at 40° for 5 min, cool, filter (0.45 μm). Remove a 1 mL aliquot and add it to 100 μL water, filter (0.45 μm), inject an aliquot.

CAPILLARY ELECTROPHORESIS
Capillary: 65 cm × 50 μm fused-silica (50 cm to detector) (Scientific Glass Engineering)
Capillary preparation: Flush with running buffer for 1 min between runs. Periodically flush with water for 1 min, sweep with 100 mM KOH, let stand in 100 mM KOH for 30 min, flush with water until the effluent is neutral to pH paper, fill with running buffer, let stand for 30 min.
Running buffer: 20 mM pH 9.0 Phosphate-borate buffer containing 100 mM sodium cholate (Buffer was prepared by adjusting pH of 20 mM sodium borate to 9.0 with 20 mM NaH_2PO_4.)
Injection: Injection by siphon at 10 cm for 10 s (about 1 nL)
Detector: UV 210
Migration time: 12.5
Internal standard: 3-(acetyloxy)-chloro-5-[2-(benzoylamino)ethyl]-2,3-dihydro-2-(4-methoxyphenyl)-1,5-benzothiazepin-4(5H)-one acetate (ester) (13.5)
Voltage: 20 kV

OTHER SUBSTANCES
Simultaneous: betamethasone, dexamethasone acetate, fluocinolone acetonide, hydrocortisone acetate, hydrocortisone, triamcinolone, triamcinolone acetonide

KEY WORDS
cream

REFERENCE
Nishi,H.; Fukuyama,T.; Matsuo,M.; Terabe,S. Separation and determination of lipophilic corticosteroids and benzothiazepin analogues by micellar electrokinetic chromatography using bile salts, *J.Chromatogr.*, **1990**, *513*, 279–295.

SAMPLE
Matrix: solutions
Sample preparation: Prepare a 0.2-1 mg/mL solution in MeOH, inject an aliquot.

CAPILLARY ELECTROPHORESIS
Capillary: 65 cm × 50 μm fused-silica (50 cm to detector)
Running buffer: 20 mM pH 9.0 Phosphate-borate buffer containing 50 mM sodium dodecyl sulfate, 4 M urea, and 15 mM gamma-cyclodextrin
Detector: UV 220
Migration time: 14.3
Voltage: 20 kV

OTHER SUBSTANCES
Simultaneous: betamethasone, cortisone acetate, dexamethasone acetate, fluocinolone acetonide, hydrocortisone, hydrocortisone acetate, triamcinolone acetonide

REFERENCE
Nishi,H.; Matsuo,M. Separation of corticosteroids and aromatic hydrocarbons by cyclodextrin-modified micellar electrokinetic chromatography, *J.Liq.Chromatogr.*, **1991**, *14*, 973–986.

Fluocortolone

Molecular formula: $C_{22}H_{29}FO_4$
Molecular weight: 376.47
CAS Registry No.: 152-97-6, 303-40-2 (21-hexanoate)
Merck Index (12th ed.): 4188

SAMPLE
Matrix: urine
Sample preparation: Condition a 6 mL 300 mg end-capped C8 SPE cartridge (IST, Hengoed UK) with 1 mL MeOH and 1 mL water. Condition a 1 mL 100 mg SAX SPE cartridge (IST, Hengoed UK) with 250 μL MeCN:5 mM ammonium acetate 10:90. Add 2 mL urine to the C8 SPE cartridge, wash with 1 mL water, wash with 2 mL hexane, elute with 1 mL ethyl acetate, elute with 1 mL dichloromethane. Combine the eluates and evaporate them to dryness at 40°, reconstitute with 250 μL MeCN:5 mM ammonium acetate 10:90, add to the SAX SPE cartridge, elute with 150 μL MeCN:5 mM ammonium acetate 10:90, collect all the effluent from the SAX SPE cartridge, mix, inject an aliquot.

CAPILLARY ELECTROPHORESIS
Capillary: 24 cm × 50 μm fused-silica (Composite Metal Services, Hallow UK) slurry packed with 3 μm Apex ODS (Jones Chromatography) to a length of 16 cm (16.1 cm to detector)
Capillary preparation: Slurry pack column with a slurry in MeOH, sonicate during process. Condition with mobile phase at 200 bar for 12 h.
Running buffer: Gradient. MeCN containing 5 mM ammonium acetate:water containing 5 mM ammonium acetate from 9:91 to 80:20 over 5 min, maintain at 80:20 for 5 min, return to initial conditions over 5 min, re-equilibrate for 15 min. Pumped at 0.1 mL/min.

Injection: 10-250 μL
Detector: UV 240
Migration time: 12.5
Voltage: 25 kV
Limit of detection: 390 ng/mL

OTHER SUBSTANCES
Extracted: dexamethasone, hydrocortisone

KEY WORDS
SPE; horse; electrochromatography

REFERENCE
Taylor,M.R.; Teale,P.; Westwood,S.A.; Perrett,D. Analysis of corticosteroids in biofluids by capillary electrochromatography with gradient elution, *Anal.Chem.*, **1997**, *69*, 2554–2558.

Fluoxetine

Molecular formula: $C_{17}H_{18}F_3NO$
Molecular weight: 309.33
CAS Registry No.: 54910-89-3, 59333-67-4 (HCl)
Merck Index (12th ed.): 4222
Lednicer: 3 32

SAMPLE
Matrix: solutions

CAPILLARY ELECTROPHORESIS
Capillary: 50 cm $\times$ 30 μm uncoated silica (34 cm to detector) (Polymicro Technologies)
Running buffer: 20 mM pH 2.7 Tris buffer containing 12 mM heptakis(2,3,6-tri-O-methyl)-β-cyclodextrin, 0.1% methylhydroxycellulose 1000, and 0.05 mM hexadecyltrimethylammonium bromide
Detector: UV 220
Migration time: 8.5, 9.5 (enantiomers)
Voltage: 25 kV
Model: Laboratory constructed

OTHER SUBSTANCES
Simultaneous: verapamil

KEY WORDS
chiral

REFERENCE
Soini,H.; Riekkola,M.-L.; Novotny,M.V. Chiral separations of basic drugs and quantitation of bupivacaine enantiomers in serum by capillary electrophoresis with modified cyclodextrin buffers, *J.Chromatogr.*, **1992**, *608*, 265–274.

SAMPLE
Matrix: solutions
Sample preparation: Inject an aliquot of an aqueous solution.

CAPILLARY ELECTROPHORESIS
Capillary: 49 cm $\times$ 50 μm fused-silica
Capillary preparation: Purge with running buffer for 3 min before each run. At the start of each day purge under vacuum with 500 mM NaOH then with water. Finally use an electroosmotic purge with running buffer then a vacuum purge with running buffer.

Running buffer: 50 mM L-(+)-tartrate containing 2.5% Glucidex6 maltodextrin (Roquette, Lestrem, France), adjusted to pH 3.25 with triethylamine
Injection: Gravity siphon at 10 cm for 30 s.
Detector: UV 214
Migration time: 8, 8.3 (enantiomers)
Voltage: 30 kV
Model: Waters Quanta 4000 CE

OTHER SUBSTANCES
Simultaneous: mianserin, sulpride (not chiral), sultopride (not chiral), viloxacine (not chiral)

KEY WORDS
chiral

REFERENCE
D'Hulst,A.; Verbeke,N. Chiral analysis of basic drugs by oligosaccharide-mediated capillary electrophoresis, *J.Chromatogr.A*, **1996**, *735*, 283–293.

Flupentixol

Molecular formula: $C_{23}H_{25}F_3N_2OS$
Molecular weight: 434.53
CAS Registry No.: 2709-56-0, 30909-51-4 (decanoate), 2413-38-9 (2.HCl)
Merck Index (12th ed.): 4224

SAMPLE
Matrix: solutions
Sample preparation: Inject an aliquot of a 100 µg/mL solution in MeOH:MeCN 50:50.

CAPILLARY ELECTROPHORESIS
Capillary: 64 cm × 50 µm fused-silica (55.5 cm to detector) (Polymicro Technologies)
Capillary preparation: Between runs flush capillary with running buffer for 2 min. Before use rinse capillary with 1 M NaOH for 1 h, with 100 mM NaOH for 20 min, and with running buffer for 10 min.
Capillary temperature: 25
Running buffer: MeCN:MeOH 50:50 containing 25 mM ammonium chloride
Injection: Pressure injection at 5 kPa for 3 s.
Detector: UV 214
Voltage: 30 kV
Model: Hewlett-Packard [3D]CE

KEY WORDS
cis and trans isomers separated; Rs = 3.32

REFERENCE
Hansen,S.H.; Bjornsdottir,I.; Tjornelund,J. Separation of cationic *cis--trans (Z--E)* isomers and diastereoisomers using non-aqueous capillary electrophoresis, *J.Chromatogr.A*, **1997**, *792*, 49–55.

Fluphenazine

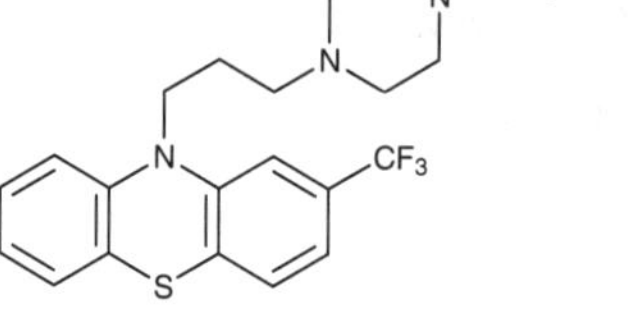

Molecular formula: $C_{22}H_{26}F_3N_3OS$
Molecular weight: 437.53
CAS Registry No.: 69-23-8, 146-56-5 (di HCl), 2746-81-8 (enanthate)
Merck Index (12th ed.): 4226
Lednicer: 1 383

SAMPLE
Matrix: urine
Sample preparation: Dilute 10-fold with water, inject an aliquot.

CAPILLARY ELECTROPHORESIS
Capillary: 70 cm × 50 μm fused-silica (65.4 cm to detector) (Polymicro Technologies)
Capillary temperature: 25
Running buffer: 20 mM pH 5.0 Tris-acetic acid containing 50 μg/mL FC-135 (Fluorad/3M) and 10 mM cetyltrimethylammonium bromide
Injection: Pressure injection at 2 psi.
Detector: UV 240
Migration time: 13.88
Voltage: 20 kV
Model: Bio-Rad BioFocus 3000
Limit of detection: 2.18 μg/mL

OTHER SUBSTANCES
Extracted: acepromazine, chlorpromazine, ethopropazine, methotrimeprazine, perphenazine, thioridazine, triflupromazine, trimeprazine

KEY WORDS
detector at anode

REFERENCE
Muijelsaar,P.G.H.M.; Claessens,H.A.; Cramers,C.A. Determination of structurally related phenothiazines by capillary zone electrophoresis and micellar electrokinetic chromatography, *J.Chromatogr.A*, **1996**, *735*, 395–402.

Flurazepam

Molecular formula: $C_{21}H_{23}ClFN_3O$
Molecular weight: 387.88
CAS Registry No.: 17617-23-1, 1172-18-5 (HCl)
Merck Index (12th ed.): 4233

SAMPLE
Matrix: blood
Sample preparation: Condition a Sep-Pak C18 SPE cartridge with MeOH, water, and 100 mM pH 9.0 borate buffer. 1 mL Serum + 3 mL 100 mM pH 9.0 borate buffer, mix, add to the SPE cartridge, wash with 3 mL 100 mM pH 9.0 borate buffer, wash with three 2 mL portions of water, wash with 1 mL MeCN:water 5:95, wash with 1 mL hexane, dry under full vacuum. Elute with 4 mL dichloromethane, evaporate the eluate to dryness under a stream of nitrogen, reconstitute with MeOH:water 15:85, filter (0.45 μm), inject an aliquot.

CAPILLARY ELECTROPHORESIS
Capillary: 72 cm × 50 μm fused-silica (50 cm to detector)
Capillary preparation: Before each run wash capillary with 100 mM NaOH for 4 min and with running buffer for 6 min.
Capillary temperature: 25
Running buffer: MeOH:5 mM pH 8.5 phosphate/borate buffer 15:85 containing 50 mM sodium dodecyl sulfate
Injection: Vacuum injection for 4 s.
Detector: UV 200
Migration time: 24
Internal standard: secobarbital (9)
Voltage: 25 kV
Model: Applied Biosystems Model 270A or Beckman P/ACE 5000
Limit of detection: 200 ng/mL

OTHER SUBSTANCES
Extracted: bromazepam, diazepam, estazolam, nitrazepam, triazolam

KEY WORDS
SPE; serum

REFERENCE
Tomita,M.; Okuyama,T. Application of capillary electrophoresis to the simultaneous screening and quantitation of benzodiazepines, *J.Chromatogr.B*, **1996**, *678*, 331–337.

SAMPLE
Matrix: solutions
Sample preparation: Prepare a solution in running buffer, inject an aliquot.

CAPILLARY ELECTROPHORESIS
Capillary: 100 cm × 75 μm fused-silica (20 cm to UV detector) (Polymicro Technologies)
Capillary preparation: Between runs flush capillaries with 1.5 column volumes of 100 mM NaOH and 9 column volumes of running buffer. Flush new capillaries with 9 column volumes of 1 M NaOH, 3 column volumes of water, 3 column volumes of 100 mM HCl, 3 column volumes of water, and 9 column volumes of running buffer.
Running buffer: MeOH:water 15:85 containing 15 mM ammonium acetate, adjusted to pH 2.5 on trifluoroacetic acid
Injection: Pressure injection at 3.45 kPa (28 nL)
Detector: UV 254 or MS, Sciex TAGA 6000E triple quadrupole, API source, electrospray (ion spray) interface at 4 kV (3 kV to ion-sampling orifice), positive-ion mode, 100 μm dia sampling orifice, nitrogen gas curtain, argon collision gas, m/z 388 [M+H]+
Migration time: 5.3 (UV), 25 (MS)
Voltage: 26 kV
Model: Beckman P/ACE System 2000

OTHER SUBSTANCES
Simultaneous: chlordiazepoxide, diazepam, prazepam

REFERENCE
Johansson,I.M.; Pavelka,R.; Henion,J.D. Determination of small drug molecules by capillary electrophoresis-atmospheric pressure ionization mass spectrometry, *J.Chromatogr.*, **1991**, *559*, 515–528.

SAMPLE
Matrix: solutions
Sample preparation: Inject an aliquot of a 5 μg/mL solution in running buffer.

CAPILLARY ELECTROPHORESIS
Capillary: 87 cm × 75 μm fused-silica (80 cm to detector) (Beckman)
Capillary preparation: Between runs rinse capillary with running buffer for 2 min then equilibrate for 5 min.
Capillary temperature: 35

Running buffer: 20 mM pH 7 Buffer containing 15 mM sodium cholate and 35 mM sodium deoxycholate (Buffer was 20 mM sodium borate adjusted to pH 7.0 with 20 mM NaH_2PO_4.)
Injection: Pressure injection for 2 s.
Detector: UV 214
Migration time: 22
Voltage: 20 kV
Model: Beckman P/ACE 2100

OTHER SUBSTANCES
Simultaneous: bromazepam, chlordiazepoxide, clobazam, clonazepam, diazepam, flunitrazepam, halazepam, lorazepam, lormetazepam, nitrazepam, nordazepam, temazepam

REFERENCE
Boonkerd,S.; Detaevernier,M.R.; Vindevogel,J.; Michotte,Y. Migration behaviour of benzodiazepines in micellar electrokinetic chromatography, *J.Chromatogr.A*, **1996**, *756*, 279–286.

SAMPLE
Matrix: solutions

CAPILLARY ELECTROPHORESIS
Capillary: 60 cm × 50 μm fused-silica (40 cm to detector), C18 coated (Supelco)
Capillary preparation: Rinse new capillaries with 250 μL 1 M NaOH, 250 μL 100 mM NaOH, water, MeOH, water, and running buffer then condition with running buffer for at least 15 min. Carry out a similar procedure between runs.
Running buffer: 100 mM pH 8.5 Borate buffer containing 25 mM sodium dodecyl sulfate and 5 M urea
Injection: Electrokinetic injection at 5 kV for 5 s.
Detector: UV 254
Migration time: 66
Voltage: 18 kV
Model: Jasco

OTHER SUBSTANCES
Simultaneous: alprazolam, amobarbital, barbital, bromazepam, clonazepam, clotiazepam, cloxazolam, diazepam, estazolam, etizolam, fludiazepam, flunitrazepam, haloxazolam, medazepam, mephobarbital, metharbital, nimetazepam, nitrazepam, oxazepam, pentobarbital, phenobarbital, secobarbital, triazolam

KEY WORDS
coated capillary

REFERENCE
Jinno,K.; Han,Y.; Nakamura,M. Analysis of anxiolytic drugs by capillary electrophoresis with bare and coated capillaries, *J.Capillary Electrophor.*, **1996**, *3*, 139–145.

SAMPLE
Matrix: solutions

CAPILLARY ELECTROPHORESIS
Capillary: 70 cm × 75 μm fused-silica (63 cm to detector) (Composite Metal Services, Hallow, UK)
Capillary preparation: Wash with running buffer for 5 min before each run. Purge new capillaries with 100 mM NaOH at 60° for 5 min and with water at 60° for 5 min, equilibrate with running buffer for 5 min, apply voltage for 5 min.
Capillary temperature: 20
Running buffer: 50 mM Disodium tetraborate adjusted to pH 2.2 with orthophosphoric acid
Injection: Hydrodynamic (vacuum) injection for 3 s.
Detector: UV (wavelength not given)
Migration time: 14.3
Voltage: 25 kV
Model: SpectraPhoresis 1000 (Thermo Separation Products)

OTHER SUBSTANCES
Simultaneous: clenbuterol, codeine, meperidine (pethidine), noscapine

REFERENCE
McGrath,G.; Smyth,W.F. Large-volume sample stacking of selected drugs of forensic significance by capillary electrophoresis, *J.Chromatogr.B*, **1996**, *681*, 125–131.

SAMPLE
Matrix: solutions
Sample preparation: Inject an aliquot of a 100 μg/mL solution in running buffer.

CAPILLARY ELECTROPHORESIS
Capillary: 60 cm × 50 μm acrylamide-coated fused-silica (40 cm to detector) (Supelco)
Capillary preparation: Adjust the pH of 20 mL water to 3.5 with acetic acid, add 80 μL 3-(trimethoxysilyl)propyl methacrylate (3-methacryloxypropyltrimethoxysilane), mix, suck into capillary, let stand at room temperature for 1 h, remove the solution, wash with water. Fill the capillary with a deaerated 3-4% acrylamide solution containing 1 μL/mL N,N,N',N'-tetramethylethylenediamine and 1 mg/mL ammonium persulfate, let stand for 3 h, remove excess solution by aspiration, rinse with water, remove water by aspiration, dry at 35° (cf. J. Chromatogr. 1985, 347, 191).
Running buffer: MeCN:buffer 5:95 (Buffer was 100 mM borate containing 5 M urea and 10 mM sodium dodecyl sulfate, adjusted to pH 8.5 with phosphate.)
Injection: Electrokinetic injection at 5 kV for 5 s.
Detector: UV 254
Migration time: 6.5
Voltage: 18 kV
Model: Jasco Model 870-CE

OTHER SUBSTANCES
Simultaneous: alprazolam, amobarbital, barbital, bromazepam, clonazepam, clotiazepam, cloxazolam, diazepam, estazolam, etizolam, fludiazepam, flunitrazepam, haloxazolam, medazepam, mephobarbital, metharbital, nimetazepam, nitrazepam, oxazepam, pentobarbital, phenobarbital, secobarbital, triazolam

KEY WORDS
coated capillary; detector at anode

REFERENCE
Jinno,K.; Han,Y.; Sawada,H.; Taniguchi,M. Capillary electrophoretic separation of toxic drugs using a polyacrylamide-coated capillary, *Chromatographia*, **1997**, *46*, 309–314.

SAMPLE
Matrix: solutions

CAPILLARY ELECTROPHORESIS
Capillary: 60 cm × 75 μm coated fused-silica (40 cm to detector) (Supelco)
Capillary preparation: Coat column as follows. Adjust the pH of 20 mL water to 3.5 with acetic acid, add 80 μL 3-(trimethoxysilyl)propyl methacrylate (3-methacryloxypropyltrimethoxysilane), mix, suck into capillary, let stand at room temperature for 1 h, remove the solution, wash with water. Fill the capillary with a deaerated 4% acrylamide solution containing 1 mg/mL N,N,N',N'-tetramethylethylenediamine and 1 mg/mL ammonium persulfate, let stand for 3 h, remove excess solution by aspiration, rinse with water, remove water by aspiration, dry at 35° (cf. J. Chromatogr. 1985, 347, 191).
Running buffer: 100 mM pH 8.5 borate buffer containing 10 mM sodium dodecyl sulfate and 5 M urea
Injection: Electromigration at 5 kV for 5 s.
Detector: UV 240, UV 254
Migration time: 6.5
Voltage: 18 kV
Model: Jasco 890-CE

OTHER SUBSTANCES
Simultaneous: alprazolam, clotiazepam, cloxazolam, diazepam, estazolam, etizolam, fludiazepam, haloxazolam, oxazepam

KEY WORDS
injection at cathode; coated capillary

REFERENCE
Jinno,K.; Han,Y.; Sawada,H. Analysis of toxic drugs by capillary electrophoresis using polyacrylamide-coated columns, *Electrophoresis*, **1997**, *18*, 284–286.

SAMPLE
Matrix: solutions
Sample preparation: Inject an aliquot of a solution in running buffer diluted 10-fold with water.

CAPILLARY ELECTROPHORESIS
Capillary: 70 cm × 75 μm fused-silica (63 cm to detector) (Composite Metal Services, Hallow, UK)
Capillary preparation: Wash with running buffer for 5 min before each injection. Condition new capillaries by purging with 100 mM NaOH at 60° for 5 min and with water at 60° for 5 min. Equilibrate with run buffer for 5 min then apply voltage for 5 min a number of times.
Capillary temperature: 20
Running buffer: 50 mM Disodium tetraborate containing 2 mM cetyltrimethylammonium bromide, adjusted to pH 2.2 with orthophosphoric acid
Injection: Hydrodynamic (vacuum) injection for 30 s then apply +15 kV (to remove solvent). When the current reaches 95% of its pre-injection level switch polarity to begin the electrophoresis.
Detector: UV 210
Migration time: 30
Voltage: -15 kV
Model: Spectra Phoresis 1000
Limit of detection: 465 nM

OTHER SUBSTANCES
Simultaneous: clenbuterol, codeine, meperidine (pethidine), noscapine

REFERENCE
Smyth,W.F.; Harland,G.B.; McClean,S.; McGrath,G.; Oxspring,D. Effect of on-capillary large volume sample stacking on limits of detection in the capillary zone electrophoretic determination of selected drugs, dyes and metal chelates, *J.Chromatogr.A*, **1997**, *772*, 161–169.

Flurbiprofen

Molecular formula: $C_{15}H_{13}FO_2$
Molecular weight: 244.27
CAS Registry No.: 5104-49-4
Merck Index (12th ed.): 4234
Lednicer: 1 86

SAMPLE
Matrix: blood
Sample preparation: 0.1-1 mL Plasma + 2 μg IS + 400 μL 1 M HCl + 10 mL diethyl ether, agitate mechanically for 10 min, centrifuge at 3500 g for 10 min. Remove the organic layer and evaporate it to dryness under a stream of nitrogen, reconstitute the residue in MeOH:40 mM NaH_2PO_4 50:50, inject an aliquot.

CAPILLARY ELECTROPHORESIS
Capillary: 58.7 cm × 75 μm fused-silica (50 cm to detector)

Capillary preparation: Wash with running buffer for 2 min before each injection. At the beginning of each day wash capillary with 100 mM NaOH for 10 min.
Capillary temperature: 22
Running buffer: MeOH:buffer 3:97 (Buffer was 40 mM pH 8 NaH_2PO_4 containing 104 mM sodium dodecyl sulfate.)
Injection: Hydrodynamic injection for 2 s
Detector: UV 254
Migration time: 16.7
Internal standard: benzoyl-4-phenyl-2-butyric acid (16.3)
Voltage: 20 kV
Model: Beckman P/ACE 2000
Limit of detection: 130 ng/mL (S/N 5)

OTHER SUBSTANCES
Extracted: diclofenac, diflunisal, etodolac, fenbufen, fenoprofen, ibuprofen, indomethacin, ketoprofen, naproxen, niflumic acid, piroxicam, sulindac, tenoxicam, tiaprofenic acid
Noninterfering: acetaminophen, amitriptyline, caffeine, clomipramine, deoxysulindac, desipramine, diazepam, 5'-hydroxytenoxicam, imipramine, maprotiline, nortriptyline, phenobarbital, phenytoin, sulfamethoxazole, theophylline, trimipramine

KEY WORDS
plasma

REFERENCE
Maboundou,C.W.; Paintaud,G.; Bérard,M.; Bechtel,P.R. Separation of fifteen non-steroidal anti-inflammatory drugs using micellar electrokinetic capillary chromatography, *J.Chromatogr.B*, **1994**, *657*, 173–183.

SAMPLE
Matrix: formulations
Sample preparation: Dragees. Dissolve dragee in running buffer, add a solution of flurbiprofen in running buffer, filter (paper), dilute with running buffer to a final concentration of 20 µg/mL for ibuprofen and 10 µg/mL for flurbiprofen, inject an aliquot. Suspensions. Dilute 1 mL suspension with running buffer, add a solution of flurbiprofen in running buffer, centrifuge at 3000 rpm for 10 min, dilute the supernatant with running buffer to a final concentration of 50 µg/mL for ibuprofen and 25 µg/mL for flurbiprofen, inject an aliquot.

CAPILLARY ELECTROPHORESIS
Capillary: 60 cm $\times$ 75 µm (52.5 cm to detector)
Capillary preparation: Purge with running buffer for 2 min before each injection. Store capillary overnight in water, rinse with 500 mM NaOH, with water, and with running buffer.
Running buffer: 50 mM pH 9.0 Borate buffer containing 40 mM sodium dodecyl sulfate (Mix 2.94 mL 200 mM boric acid, 6.37 mL 75 mM tetraborate solution, and 10 mL 200 mM sodium dodecyl sulfate, make up to 50 mL with water.)
Injection: Hydrodynamic injection at 10 cm for 5 s
Detector: UV 214
Migration time: 6.7
Internal standard: flurbiprofen
Voltage: 15 kV
Model: Waters Quanta 4000 CE

OTHER SUBSTANCES
Simultaneous: benzoic acid, ibuprofen

KEY WORDS
dragees; suspensions; flurbiprofen is IS

REFERENCE
Donato,M.G.; Baeyens,W.; Van den Bossche,W.; Sandra,P. The determination of non-steroidal antiinflammatory drugs in pharmaceuticals by capillary zone electrophoresis and micellar electrokinetic capillary chromatography, *J.Pharm.Biomed.Anal.*, **1994**, *12*, 21–26.

SAMPLE
Matrix: formulations

Sample preparation: Stir formulation in MeOH for 10 min, filter, dilute with 5 mM pH 5 morpholinoethanesulfonic acid, inject an aliquot.

CAPILLARY ELECTROPHORESIS
Capillary: 35 cm × 50 μm fused-silica (31.5 cm to detector) (Polymicro Technologies)
Capillary temperature: 25
Running buffer: 100 mM Morpholinoethanesulfonic acid containing 30 mM heptakis-2,3,6-tri-O-methyl-β-cyclodextrin, adjusted to pH 5 with NaOH
Injection: Pressure injection at 10 psi
Detector: UV 206
Migration time: 20, 21 (enantiomers)
Voltage: 20 kV
Current: 6.6 μA
Model: Biofocus 3000 (Bio-Rad)

OTHER SUBSTANCES
Simultaneous: fenoprofen, ibuprofen, ketoprofen

KEY WORDS
chiral

REFERENCE
Fanali,S.; Aturki,Z. Use of cyclodextrins in capillary electrophoresis for the chiral resolution of some 2-aryl-propionic acid non-steroidal anti-inflammatory drugs, *J.Chromatogr.A*, **1995**, *694*, 297–305.

SAMPLE
Matrix: solutions
Sample preparation: Dissolve in MeCN, dilute with water to a concentration of 25 μg/mL, inject an aliquot.

CAPILLARY ELECTROPHORESIS
Capillary: 60 cm × 75 μm fused-silica (52.5 cm to detector)
Running buffer: MeCN:30 mM pH 7.0 phosphate buffer 20:80
Injection: Hydrostatic injection at 9.8 cm for 5 s
Detector: UV 214
Migration time: 20.55
Model: Waters Quanta 4000

OTHER SUBSTANCES
Simultaneous: acemetacin, alclofenac, fenbufen, ibuprofen, indomethacin, ketoprofen, lonazolac, naproxen, niflumic acid, piroxicam, tenoxicam, tiaprofenic acid, tolmetin, voltaren

REFERENCE
Donato,M.G.; Van den Eeckhout,E.; Van den Bossche,W.; Sandra,P. Capillary zone electrophoresis and micellar electrokinetic capillary chromatography of some non-steroidal antiinflammatory drugs (NSAIDs), *J.Pharm.Biomed.Anal.*, **1993**, *11*, 197–201.

SAMPLE
Matrix: solutions
Sample preparation: Prepare a 10 μg/mL solution in MeOH, inject an aliquot.

CAPILLARY ELECTROPHORESIS
Capillary: 97 cm × 50 μm fused-silica (80 cm to detector) (Chrompack), coat with Chirasil-Dex (J.High Res.Chromatogr. 1991, 14, 58)
Capillary preparation: Rinse with water then operating buffer between each run. Condition coated columns with running buffer for half a day.
Running buffer: 20 mM pH 7.0 Borate-phosphate buffer
Injection: Hydrostatic injection for 5 s
Detector: UV 220
Migration time: 22.5, 23.5 (enantiomers)
Voltage: 30 kV

Model: Kapillar-Elektrophorese-System 100 (Grom)

OTHER SUBSTANCES
Simultaneous: cicloprofen, ibuprofen

KEY WORDS
chiral

REFERENCE
Mayer,S.; Schurig,V. Enantiomer separation by electrochromatography in open tubular columns coated with Chirasil-Dex, *J.Liq.Chromatogr.*, **1993**, *16*, 915–931.

SAMPLE
Matrix: solutions
Sample preparation: Prepare an aqueous solution, filter (0.45 μm), inject an aliquot.

CAPILLARY ELECTROPHORESIS
Capillary: 36 cm × 50 μm (31.5 cm to detector)
Capillary preparation: Before each run rinse with water for 1 min and with running buffer for 1 min.
Capillary temperature: 25
Running buffer: EtOH:buffer 10:90 (Buffer was 50 mM pH 6.0 phosphate buffer containing 25 μm avidin.)
Injection: Inject at 1 psi pressure for 2 s
Detector: UV 250
Migration time: 16, 17.3 (enantiomers)
Voltage: -12 kV
Model: BioFocus 3000 (Bio-Rad)

KEY WORDS
chiral

REFERENCE
Tanaka,Y.; Matsubara,N.; Terabe,S. Separation of enantiomers by affinity electrokinetic chromatography using avidin, *Electrophoresis*, **1994**, *15*, 848–853.

SAMPLE
Matrix: solutions
Sample preparation: Inject an aliquot of a 50 μM solution in MeOH:water 40:60.

CAPILLARY ELECTROPHORESIS
Capillary temperature: 25
Running buffer: MeOH:water 40:60 containing 75 mM glycine and 15 mM hydroxypropyl-β-cyclodextrin adjusted to pH 9.1 with triethanolamine
Injection: Hydrodynamic injection for 2 s
Detector: UV 280
Migration time: 13.3
Voltage: 15 kV

OTHER SUBSTANCES
Simultaneous: alclofenac, bufexamac, carprofen, flufenamic acid, indomethacin, ketoprofen, naproxen, niflumic acid, piroxicam, sulindac, tiaprofenic acid

---

REFERENCE
Bechet,I.; Fillet,M.; Hubert,P.; Crommen,J. Improvement of achiral resolution in capillary zone electrophoresis by use of cyclodextrin additives, *Biomed.Chromatogr.*, **1995**, *9*, 267–268.

SAMPLE
Matrix: solutions

Sample preparation: Prepare a 100 µg/mL solution in 50 mM pH 7.0 phosphate buffer, inject an aliquot.

CAPILLARY ELECTROPHORESIS
Capillary: 37 cm × 50 µm (30 cm to detector)
Capillary preparation: Between runs rinse capillary with 100 mM KOH for 2 min, with water for 2 min, and with running buffer for 2 min. At the beginning of each day treat capillary with 500 mM KOH for 5 min, flush with water for 10 min, and flush with running buffer for 10 min.
Capillary temperature: 20
Running buffer: 50 mM pH 7.0 Phosphate buffer containing 2 mM vancomycin and 30 mM sodium dodecyl sulfate
Injection: Pressure injection at 0.5 psi for 1 s
Detector: UV 254
Migration time: 25, 26 (enantiomers)
Voltage: +5 kV
Model: Beckman P/ACE 2000

OTHER SUBSTANCES
Simultaneous: indoprofen, ketoprofen

KEY WORDS
chiral

REFERENCE
Rundlett,K.L.; Armstrong,D.W. Effect of micelles and mixed micelles on efficiency and selectivity of antibiotic based capillary electrophoretic enantioseparations, *Anal.Chem.*, **1995**, *67*, 2088–2095.

SAMPLE
Matrix: solutions
Sample preparation: Inject an aliquot of a 500 µM solution in MeOH:water 50:50.

CAPILLARY ELECTROPHORESIS
Capillary: 60 cm × 50 µm fused-silica (52.5 cm to detector) (Waters)
Capillary preparation: Before each run rinse capillary with 10 mM NaOH for 1 min and with running buffer for 4 min. Condition new capillaries by flushing with 1 M NaOH for 10 min, with water for 5 min, with 6 mM NaOH containing 500 mM NaCl for 10 min, with water for 5 min, and with running buffer for 10 min.
Running buffer: Formic acid containing 10 mM heptakis(trimethyl-β-cyclodextrin) adjusted to pH 4.0 with 1 M NaOH (ionic strength, I = 75 mM)
Injection: Hydrodynamic injection at 10 cm for 15 s.
Detector: UV 254
Voltage: 30 kV
Model: Waters Quanta 4000

KEY WORDS
chiral; R_S = 0.7; electroosmotic mobility = 11.2 × 10^{-5} $cm^2V^{-1}s^{-1}$

REFERENCE
Lelièvre,F.; Gareil,P. Chiral separations of underivatized arylpropionic acids by capillary zone electrophoresis with various cyclodextrins. Acidity and inclusion constant determinations, *J.Chromatogr.A*, **1996**, *735*, 311–320.

SAMPLE
Matrix: solutions
Sample preparation: Inject an aliquot of a 100 µg/mL solution in water or running buffer.

CAPILLARY ELECTROPHORESIS
Capillary: 47 cm × 75 µm fused-silica (40 cm to detector)
Capillary preparation: Before each run rinse capillary with running buffer for 1-2 min. If peak tailing is observed fill capillary with 500 mM NaOH and let stand for 30 min, wash with water for 3 min, rinse with running buffer for 3 min.

Capillary temperature: 20
Running buffer: 20 mM pH 7.0 Phosphate buffer containing 6% dextrin (Japan Pharmacopeia
  grade) (Dissolve dextrin in buffer at 90° then cool to room temperature.)
Injection: Pressure injection at 0.5 psi for 2-4 s.
Detector: UV 220
Migration time: 10.357, 10.615 (enantiomers)
Voltage: 20 kV
Model: Beckman P/ACE 5510

KEY WORDS
chiral

REFERENCE
Nishi,H.; Izumoto,S.; Nakamura,K.; Nakai,H.; Sato,T. Dextran and dextrin as chiral selectors in capillary zone
  electrophoresis, *Chromatographia*, **1996**, *42*, 617–630.

SAMPLE
Matrix: solutions

CAPILLARY ELECTROPHORESIS
Capillary: 37 cm × 50 μm fused-silica (30 cm to detector) (Polymicro Technologies)
Capillary preparation: Purge with 3 volumes of running buffer between runs. Before use wash
  with 10 volumes of 100 mM NaOH.
Capillary temperature: 25 ± 0.1
Running buffer: 50 mM pH 7.0 phosphate buffer containing 5 mM vancomycin
Injection: Pressure injection at 0.5 psi for 5 s
Detector: UV 254
Migration time: 11.5, 13.8 (enantiomers)
Voltage: 20 kV
Model: Beckman P/ACe 2000

KEY WORDS
chiral

REFERENCE
Strege,M.A.; Huff,B.E.; Risley,D.S. Evaluation of macrocyclic antibiotic A82846B as a chiral selector for capillary electrophoresis separations, *LC.GC*, **1996**, *14*, 144–150.

SAMPLE
Matrix: solutions

CAPILLARY ELECTROPHORESIS
Capillary: 35 cm × 50 μm coated capillary (30.5 cm to detector) (Composite Metal Services, UK)
Capillary preparation: Before each run purge at high pressure with water for 100 s and with
  running buffer for 120 s. If vancomycin is used, before each run purge at high pressure with
  water for 100 s and with running buffer not containing vancomycin for 120 s. Next purge with
  running buffer containing vancomycin at low pressure (175 psi.s) and inject sample. Coat capillary as follows. Adjust the pH of 20 mL water to 3.5 with acetic acid, add 80 μL 3-(trimethoxysilyl)propyl methacrylate (3-methacryloxypropyltrimethoxysilane), mix, suck into capillary, let stand at room temperature for 1 h, remove the solution, wash with water. Fill the
  capillary with a deaerated 3-4% acrylamide solution containing 1 μL/mL N,N,N',N'-tetra-
  methylethylenediamine and 1 mg/mL potassium persulfate, let stand for 30 min, remove excess
  solution by aspiration, rinse with water, remove water by aspiration, dry at 35° (J. Chromatogr.
  1985, 347, 191).
Capillary temperature: 25
Running buffer: Buffer containing 30 mM heptakis-2,3,6-tri-O-methyl-β-cyclodextrin (A), 5 mM
  heptamethylamino-β-cyclodextrin (B) or 5 mM vancomycin (C) (Buffer was 50 mM phosphoric
  acid containing 50 mM acetic acid and 50 mM boric acid, dilute with an equal volume of water,
  adjust pH to 5 with concentrated NaOH.)
Injection: Pressure injection at 10 psi.
Detector: UV 206
Migration time: 20.2, 21.2 (A); 16.7, 18.8 (B); 6.8, 7.7 (C)

Voltage: -20 kV
Model: Bio-Rad Biofocus 3000
Limit of detection: 5 μM (A, C), 10 μM (B)

KEY WORDS
chiral; coated capillary; detector at anode

REFERENCE
Fanali,S.; Desiderio,C.; Aturki,Z. Enantiomeric resolution study by capillary electrophoresis. Selection of the appropriate chiral selector, *J.Chromatogr.A*, **1997**, *772*, 185–194.

SAMPLE
Matrix: solutions
Sample preparation: Inject an aliquot of a solution in MeOH:water 10:90.

CAPILLARY ELECTROPHORESIS
Capillary: 44 cm $\times$ 50 μm fused-silica (37 cm to detector) (Supelco)
Capillary preparation: Wash with running buffer for 3 min after each injection. Wash with running buffer for 10 min at the end of each day.
Capillary temperature: 25
Running buffer: 100 mM Phosphoric acid containing 5 mM sulfobutyl ether-β-cyclodextrin and 30 mM heptakis(2,3,6-tri-O-methyl)-β-cyclodextrin, adjusted to pH 3.0 with triethanolamine
Injection: Hydrodynamic injection for 5 s (13.3 nL).
Detector: UV 230
Voltage: -25 kV
Current: 60 μA
Model: SpectraPhoresis 1000 CE

KEY WORDS
chiral; detector at anode; resolution (R_s = 16.3)

REFERENCE
Fillet,M.; Hubert,P.; Crommen,J. Enantioseparation of nonsteroidal anti-inflammatory drugs by capillary electrophoresis using mixtures of anionic and uncharged β-cyclodextrins as chiral additives, *Electrophoresis*, **1997**, *18*, 1013–1018.

SAMPLE
Matrix: solutions
Sample preparation: Inject an aliquot of a solution in MeOH:water 50:50.

CAPILLARY ELECTROPHORESIS
Capillary: 57 cm $\times$ 75 μm fused-silica (50 cm to detector) (Polymicro Technologies)
Capillary preparation: Rinse capillary with running buffer for 2 min between samples. At the start of each day wash capillary with 1 M NaOH for 10 min. Wash new capillaries with 1 M NaOH for 1 h.
Capillary temperature: 23 $\pm$ 0.1
Running buffer: 50 mM NaH$_2$PO$_4$ containing 3 mM heptakis(6-hydroxyethylamino-6-deoxy-β-cyclodextrin), adjusted to pH 5.0 with 500 mM NaOH (Prepare heptakis(6-hydroxyethylamino-6-deoxy-β-cyclodextrin) as follows. Add 4.32 g β-cyclodextrin to a stirred solution of 4 mL bromine and 21 g triphenylphosphine in 80 mL DMF and the mixture stirred at 80° for 15 h, concentrate under reduced pressure to half volume, adjust pH to 9-10 by adding 30 mL 3 M sodium methoxide in MeOH with simultaneous cooling. Keep at room temperature for 30 min then pour into 1.5 L ice water, filter to obtain heptakis(6-bromo-6-deoxy-β-cyclodextrin) (Ang. Chem. Int. Ed. Engl. 1991, 30, 78). Dissolve heptakis(6-bromo-6-deoxy-β-cyclodextrin) in ethanolamine at 65°, heat at 65° for 48 h. Remove the solvent by evaporation under reduced pressure, dissolve the residue in hot MeOH, slowly add the methanolic solution to reagent-grade acetone, collect the precipitate by filtration. Dissolve the precipitate in water, treat with a basic ion-exchange resin, lyophilize to obtain heptakis(6-hydroxyethylamino-6-deoxy-β-cyclodextrin).)
Detector: UV 214
Migration time: 19.5, 20.5 (enantiomers)
Voltage: -15 kV

Model: Beckman P/ACE 5510

OTHER SUBSTANCES
Simultaneous: fenoprofen, ibuprofen (peak shape very poor)

KEY WORDS
chiral

REFERENCE
O'Keeffe,F.; Shamsi,S.A.; Darcy,R.; Schwinté,P.; Warner,I.M. A persubstituted cationic β-cyclodextrin for chiral separations, *Anal.Chem.*, **1997**, *69*, 4773–4782.

SAMPLE
Matrix: solutions

CAPILLARY ELECTROPHORESIS
Capillary: 37 cm × 50 μm eCAP bare silica (30 cm to detector) (Beckman)
Capillary temperature: 25
Running buffer: MeOH:100 mM pH 9.2 borate buffer 15:85 containing 4 mM LY307599 (Eli Lilly)
Injection: Pressure injection at 0.5 psi for 5 s.
Detector: UV 254
Migration time: 37, 39 (enantiomers)
Voltage: 10 kV
Model: Beckman P/ACE 2000

KEY WORDS
chiral

REFERENCE
Sharp,V.S.; Risley,D.S.; McCarthy,S.; Huff,B.E.; Strege,M.A. Evaluation of a new macrocyclic antibiotic as a chiral selector for use in capillary electrophoresis, *J.Liq.Chromatogr.Rel.Technol.*, **1997**, *20*, 887–898.

SAMPLE
Matrix: solutions
Sample preparation: Inject an aliquot of a 50-100 μg/mL solution in 50 mM pH 4.8 ammonium acetate buffer.

CAPILLARY ELECTROPHORESIS
Capillary: 44 cm × 50 μm polyacrylamide coated
Capillary preparation: Adjust the pH of 20 mL water to 3.5 with acetic acid, add 80 μL 3-(trimethoxysilyl)propyl methacrylate (3-methacryloxypropyltrimethoxysilane), mix, suck into capillary, let stand at room temperature for 1 h, remove the solution, wash with water. Fill the capillary with a deaerated 3-4% acrylamide solution containing 1 μL/mL N,N,N',N'-tetramethylethylenediamine and 1 mg/mL potassium persulfate, let stand for 30 min, remove excess solution by aspiration, rinse with water, remove water by aspiration, dry at 35° (J. Chromatogr. 1985, 347, 191).
Running buffer: 50 mM pH 4.8 Ammonium acetate buffer containing 5 mM vancomycin
Injection: Hydrostatic injection at 10 cm for 5-10 s.
Detector: MS, Finnigan LCQ ion trap, electrospray, negative ion mode, probe tip at 2.6 kV, sheath liquid MeOH:water:ammonia 50:48:2 at 6 μL/min (Vancomycin migrates away from the detector.)
Migration time: 10.4, 11.6 (enantiomers)
Voltage: 20 kV
Model: Grom 100

OTHER SUBSTANCES
Also analyzed: carprofen, ibuprofen, ketoprofen, naproxen

KEY WORDS
chiral

REFERENCE

Fanali,S.; Desiderio,C.; Schulte,G.; Heitmeier,S.; Strickmann,D.; Chankvetadze,B.; Blaschke,G. Chiral capillary electrophoresis-electrospray mass spectrometry coupling using vancomycin as chiral selector, *J.Chromatogr.A*, **1998**, *800*, 69–76.

SAMPLE

Matrix: solutions

CAPILLARY ELECTROPHORESIS

Capillary: 57 cm × 50 μm fused-silica (50 cm to detector)

Running buffer: 50 mM pH 8.0 Borate buffer containing 40 mM sodium taurodeoxycholate and 25 mM phosphatidylcholine (Prepare by adding phosphatidylcholine and stirring for 3-6 h until all cloudiness disappears. Phosphatidylcholine was 95% pure soybean lecithin, Epikuron, Lucas Meyer & Co.)

Injection: Inject a solution of the compound in the running buffer at 20 psi for 1 s, inject MeOH: water 5:95 at 20 psi for 1 s, inject a solution of halofantrine in running buffer at 20 psi for 1 s.

Detector: UV 214

Migration time: k' 3.24

Model: Beckman P/ACE 5000

OTHER SUBSTANCES

Also analyzed: acetaminophen, amoxicillin, antipyrine, aspirin, azathioprine, caffeine, captopril, carbamazepine, carprofen, chlorambucil, chlorpheniramine, chlorpromazine, cimetidine, clonidine, codeine, desipramine, diphenhydramine, ephedrine, fenoterol, flufenamic acid, haloperidol, hydroxyzine, imipramine, indomethacin, ketoprofen, lidocaine, melphalan, metoprolol, nabumetone, nadolol, phenobarbital, phenol, promazine, propranolol, pyrilamine, ranitidine, ropinirole, salicylic acid, sulfamethoxazole, testosterone, theophylline, thioridazine, tiaprofenic acid, tolfenamic acid, trifluoperazine, trimethoprim, valproic acid, verapamil

KEY WORDS

comparison with HPLC; k' = (Tr-T0)/(T0(1-Tr/Tm)) where Tr = retention time of analyte; T0 = retention time of water; and Tm = retention time of marker (halofantrine)

REFERENCE

Hanna,M.; de Biasi,V.; Bond,B.; Salter,C.; Hutt,A.J.; Camilleri,P. Estimation of the partitioning characteristics of drugs: A comparison of a large and diverse drug series utilizing chromatographic and electrophoretic methodology, *Anal.Chem.*, **1998**, *70*, 2092–2099.

SAMPLE

Matrix: solutions

Sample preparation: Inject an aliquot of a 100 μg/mL solution in MeOH:water 50:50.

CAPILLARY ELECTROPHORESIS

Capillary: 47 cm × 75 μm fused-silica (40 cm to detector) (Polymicro Technologies)

Capillary preparation: Rinse capillary with 100 mM NaOH for 1 min, with water for 1.5 min, and with running buffer for 2 min between samples. Wash new capillaries with 1 M NaOH for 1 h, with 100 mM NaOH for 30 min, and with water for 30 min.

Running buffer: 50 mM NaH_2PO_4 containing 3 mM heptakis(6-methoxyethylamine-6-deoxy-β-cyclodextrin), adjusted to pH 6 (Prepare heptakis(6-methoxyethylamine-6-deoxy-β-cyclodextrin) as follows. Add 4.32 g β-cyclodextrin to a stirred solution of 4 mL bromine and 21 g triphenylphosphine in 80 mL DMF and the mixture stirred at 80°C for 15 h, concentrate under reduced pressure to half volume, adjust pH to 9-10 by adding 30 mL 3 M sodium methoxide in MeOH with simultaneous cooling. Keep at room temperature for 30 min then pour into 1.5 L ice water, filter to obtain heptakis(6-bromo-6-deoxy-β-cyclodextrin) (Ang. Chem. Int. Ed. Engl. 1991, 30, 78). Dissolve heptakis(6-bromo-6-deoxy-β-cyclodextrin) in methoxyethylamine, heat at 65°C for 48 h. Remove the solvent by evaporation under reduced pressure, dissolve the residue in MeOH at 55°C, slowly add the methanolic solution to reagent-grade acetone with stirring, collect the precipitate by filtration. Dissolve the precipitate in water, treat with a basic ion-exchange resin. Lyophilize the precipitate to obtain heptakis(6-methoxyethylamine-6-deoxy-β-cyclodextrin).)

Injection: Pressure injection at 85 kPa.s.

Detector: UV 214
Migration time: 34, 36 (enantiomers)
Voltage: -30 kV
Current: 25 µA
Model: Beckman P/ACE 5510

OTHER SUBSTANCES
Simultaneous: fenoprofen, ketoprofen

KEY WORDS
chiral

REFERENCE
Haynes,J.L.,III; Shamsi,S.A.; O'Keefe,F.; Darcey,R.; Warner,I.M. Cationic β-cyclodextrin derivative for chiral separations, *J.Chromatogr.A*, **1998**, *803*, 261–271.

SAMPLE
Matrix: solutions
Sample preparation: Inject an aliquot of a 100 µg/mL solution in water.

CAPILLARY ELECTROPHORESIS
Capillary: 27 cm × 50 µm eCAP neutral capillary (20 cm to detector) (Beckman)
Running buffer: 100 mM pH 6.0 Sodium phosphate buffer containing 2 mM A82846B (Eli Lilly)
Injection: Fill capillary with running buffer for 5 min, apply -10 kV for 5 min, hydrodynamic injection at 0.5 psi for 2 s.
Detector: UV 254
Migration time: 5.4, 6.4 (enantiomers)
Model: Beckman P/ACE 2000

OTHER SUBSTANCES
Also analyzed: ketoprofen, suprofen

KEY WORDS
chiral; detector at anode

REFERENCE
Reilly,J.; Risley,D.S. The separation of enantiomers by countercurrent capillary electrophoresis using the macrocyclic antibiotic A82846B, *LC.GC*, **1998**, *16*, 170–178.

SAMPLE
Matrix: solutions
Sample preparation: Inject an aliquot of a solution in running buffer.

CAPILLARY ELECTROPHORESIS
Capillary: 57 cm × 50 µm fused-silica (Beckman)
Capillary preparation: Rinse with 100 mM NaOH for 5 min, with water for 10 min,
Running buffer: 2-Methoxyethanol:buffer 30:70 (Buffer was 40 mM pH 6 phosphate buffer containing 500 µM Actaplanin A (Eli Lilly).)
Injection: Pressure injection for 5 s.
Detector: UV 254
Migration time: 44.3, 47.0
Model: Beckman P/ACE 2100

OTHER SUBSTANCES
Also analyzed: carprofen, indoprofen

KEY WORDS
chiral

REFERENCE
Trelli-Seifert,L.A.; Risley,D.S. Capillary electrophoretic enantiomeric separations of nonsteroidal anti-inflammatory compounds using the macrocyclic antibiotic actaplanin A and 2-methoxyethanol, *J.Liq.Chromatogr.Rel.Technol.*, **1998**, *21*, 299–313.

SAMPLE
Matrix: solutions

CAPILLARY ELECTROPHORESIS
Capillary: 42 cm × 50 μm fused-silica (31 cm to detector) (Polymicro Technologies)
Capillary temperature: 30
Running buffer: Formamide containing 3.42% quaternary ammonium-β-cyclodextrin (American Maize Products, Hammond IN), 20 mM ammonium acetate, and 1% acetic acid (pH* = 7.1)
Detector: UV 254
Migration time: 14.63, 14.73 (enantiomers)
Voltage: -30 kV
Model: laboratory-constructed

OTHER SUBSTANCES
Also analyzed: carprofen, fenoprofen, indoprofen, ketoprofen, suprofen

KEY WORDS
chiral

REFERENCE
Wang,F.; Khaledi,M.G. Nonaqueous capillary electrophoresis chiral separations with quaternary ammonium β-cyclodextrin, *J.Chromatogr.A*, **1998**, *817*, 121–128.

Fluticasone propionate

Molecular formula: C$_{25}$H$_{31}$F$_3$O$_5$S
Molecular weight: 500.58
CAS Registry No.: 80474-14-2, 90566-53-3 (free acid)
Merck Index (12th ed.): 4244
Lednicer: 4 75

SAMPLE
Matrix: bulk

CAPILLARY ELECTROPHORESIS
Capillary: 60 cm × 50 μm capillary packed with 3 μm Spherisorb ODS-1
Capillary temperature: 30
Running buffer: MeCN:5 mM pH 9.0 borate 80:20
Injection: Electromigration at 30 kV for 24 s.
Detector: UV 238
Migration time: 22
Voltage: 30 kV (with 500 psi pressure?)
Model: Applied Biosystems ABI 270A

OTHER SUBSTANCES
Simultaneous: impurities

KEY WORDS
electrochromatography

REFERENCE

Smith,N.W.; Evans,M.B. The analysis of pharmaceutical compounds using electrochromatography, *Chromatographia*, **1994**, *38*, 649–657.

SAMPLE

Matrix: solutions
Sample preparation: Inject an aliquot of a 100 μg/mL solution in running buffer.

CAPILLARY ELECTROPHORESIS

Capillary: 27 cm × 50 μm fused-silica (20 cm to detector) (Composite Metal Services, Hallow, UK) packed with 3 μm Hypersil ODS for 20 cm (details in paper)
Running buffer: MeCN:2 mM pH 7.8 phosphate buffer 80:20
Injection: Electrokinetic injection at 5 kV for 5 s.
Detector: UV 214
Migration time: 19.99 (fluticasone propionate)
Voltage: 10 kV
Model: Beckman P/ACE 2050

OTHER SUBSTANCES

Simultaneous: betamethasone, betamethasone dipropionate, betamethasone-17-valerate, clobetasol butyrate, clobetasone butyrate, hydrocortisone, prednisolone

KEY WORDS

electrochromatography

REFERENCE

Frame,L.A.; Robinson,M.L.; Lough,W.J. Simplification of capillary electrochromatography procedures, *J.Chromatogr.A*, **1998**, *798*, 243–249.

Fluvoxamine

Molecular formula: $C_{15}H_{21}F_3N_2O_2$
Molecular weight: 318.34
CAS Registry No.: 54739-18-3
Merck Index (12th ed.): 4251

SAMPLE

Matrix: solutions

CAPILLARY ELECTROPHORESIS

Capillary: 50 cm × 50 μm fused-silica (41.5 cm to detector)
Capillary preparation: Flush capillary with running buffer for 2 min between runs. Rinse new capillaries with 1 M NaOH for 15 min, with water for 15 min, and with running buffer for 30 min.
Capillary temperature: 30
Running buffer: 25 mM pH 7.0 Phosphate buffer containing 10 mM cetyltrimethylammonium bromide
Injection: Pressure injection at 50 mbar for 2 s.
Detector: UV (Wavelength not given)
Migration time: 6.1 (Z), 6.4 (E)
Voltage: 30 kV
Model: Hewlett-Packard HP^{3D} CE

OTHER SUBSTANCES

Simultaneous: impurities

REFERENCE
Hilhorst,M.J.; Somsen,G.W.; de Jong,G.J. Choice of capillary electrophoresis systems for the impurity profiling of drugs, *J.Pharm.Biomed.Anal.*, **1998**, *16*, 1251–1260.

Folic acid

Molecular formula: $C_{19}H_{19}N_7O_6$
Molecular weight: 441.40
CAS Registry No.: 59-30-3
Merck Index (12th ed.): 4253

SAMPLE
Matrix: formulations
Sample preparation: Grind tablet to a homogeneous powder and dissolve in 2 mL water and 20 mL EtOH, homogenize (Ultra Turrax), centrifuge. Evaporate the supernatant to dryness under reduced pressure, reconstitute the residue with water, inject an aliquot.

CAPILLARY ELECTROPHORESIS
Capillary: 64.5 cm $\times$ 50 μm fused-silica (56 cm to detector)
Capillary preparation: Before each run flush capillary at 4 kPa with 1 M NaOH for 2 min and with buffer for 5 min. Replace buffer before each analysis.
Capillary temperature: 30
Running buffer: 1-Propanol:buffer 2:98 (Buffer was 100 mM Na_2HPO_4, 500 mM taurine, and 75 mM sodium cholate.)
Injection: Pressure injection at 4 kPa for 1 s.
Detector: UV 214
Migration time: 25
Voltage: 17 kV
Model: Hewlett-Packard HP^{3D} CE
Limit of quantitation: 30 μM

OTHER SUBSTANCES
Simultaneous: ascorbic acid, biotin, niacinamide, pyridoxamine, pyridoxine, riboflavin, thiamine, vitamin B12

KEY WORDS
tablets

REFERENCE
Buskov,S.; Moller,P.; Sorensen,H.; Sorensen,J.C.; Sorensen,S. Determination of vitamins in food based on supercritical fluid extraction prior to micellar electrokinetic capillary chromatographic analyses of individual vitamins, *J.Chromatogr.A*, **1998**, *802*, 233–241.

Formoterol

Molecular formula: $C_{19}H_{24}N_2O_4$
Molecular weight: 344.41
CAS Registry No.: 73573-87-2, 43229-80-7 (fumarate)
Merck Index (12th ed.): 4272

SAMPLE
Matrix: solutions
Sample preparation: Add 15 mL 8.57 mg/mL fluorescein isothiocyanate in EtOH dropwise to 103 mg formoterol and 500 μL triethylamine in 15 mL EtOH, stir at 40° for 4 h, evaporate,

purify by TLC on silica gel using chloroform:ethyl acetate:MeOH 25:37.5:37.5, inject an aliquot of a 0.1 ng/mL solution.

CAPILLARY ELECTROPHORESIS
Capillary: 57 cm × 75 μm fused-silica (50 cm to detector) (Supelco)
Capillary preparation: Before analysis rinse capillary with 100 mM NaOH for 2 min, fill with running buffer.
Capillary temperature: 25
Running buffer: MeOH:67 mM pH 8 phosphate buffer 10:90 containing 20 mM heptakis(2,3,6-tri-O-methyl-β-cyclodextrin) and 40 mM octanesulfonic acid
Injection: Pressure injection at 0.3 psi for 5 s (5 nL)
Detector: F ex 488 (4 mW argon-ion laser), em 520 (band-pass filter)
Migration time: 27, 28 (enantiomers)
Model: Beckman PACE 2100
Limit of detection: 0.01 ng/mL

KEY WORDS
derivatization; chiral

REFERENCE
Cherkaoui,S.; Faupel,M.; Francotte,E. Separation of formoterol enantiomers and detection of zeptomolar amounts by capillary electrophoresis using laser-induced fluorescence, *J.Chromatogr.A*, **1995**, *715*, 159–165.

SAMPLE
Matrix: solutions

CAPILLARY ELECTROPHORESIS
Capillary: 47 cm × 50 μm fused-silica (40 cm to detector)
Capillary preparation: After each run rinse with 100 mM NaOH for 2-5 min, with water for 2 min, and with running buffer for 2 min
Running buffer: 80 mM Tris containing 6% dextrin 10 sulfopropyl ether, adjusted to pH 4 with 1 M phosphoric acid (Prepare dextrin 10 sulfopropyl ether as follows. Dissolve 10 g dextrin 10 in 50 mL water, stir vigorously, add 8 g 25% aqueous NaOH dropwise at room temperature over 2.5 h, at the same time add 5 g propane sultone (Caution! Propane sultone is a carcinogen!) in three portions, add 10 mL water, adjust pH to 6 with 4 M HCl, remove inorganic salts by 3-fold membrane filtration in a stirred cell (Spec, Basel; exclusion size 1000 Da; initial volume 150 mL; final volume 20 mL; cell pressure 4 bar), concentrate in a rotary evaporator, dry at 0.1 torr and 60° overnight, to gibe dextrin 10 sulfopropyl ether as a white powder (yield 5 g).)
Detector: UV 214
Migration time: 15.5, 15.9 (enantiomers)
Voltage: 30 kV
Model: Beckman P/ACE 5000

OTHER SUBSTANCES
Simultaneous: Tröger's base

KEY WORDS
chiral

REFERENCE
Jung,M.; Börnsen,K.O.; Francotte,E. Dextrin sulfopropyl ether: A novel anionic chiral buffer additive for enantiomer separation by electrokinetic chromatography, *Electrophoresis*, **1996**, *17*, 130–136.

Fosinopril

Molecular formula: $C_{30}H_{46}NO_7P$
Molecular weight: 563.67
CAS Registry No.: 98048-97-6, 97825-24-6, 88889-14-9
(sodium salt)
Merck Index (12th ed.): 4282
Lednicer: 5 66

SAMPLE
Matrix: solutions

CAPILLARY ELECTROPHORESIS
Capillary: 57 cm × 75 μm fused-silica (50 cm to detector) (Beckman)
Capillary preparation: Before each run rinse capillary with 100 mM NaOH and running buffer.
Capillary temperature: 30
Running buffer: 50 mM pH 8.2 Borate buffer containing 100 mM sodium dodecyl sulfate
Injection: Pressure injection for 5-10 s.
Detector: UV 200
Migration time: 12.7
Voltage: 25 kV
Model: Beckman P/ACE 5510

OTHER SUBSTANCES
Simultaneous: captopril, ceronapril, enalapril, lisinopril, zofenopril

REFERENCE
Bretnall,A.E.; Clarke,G.S. Selectivity of capillary electrophoresis for the analysis of cardiovascular drugs, *J.Chromatogr.A*, **1996**, *745*, 145–154.

Furosemide

Molecular formula: $C_{12}H_{11}ClN_2O_5S$
Molecular weight: 330.75
CAS Registry No.: 54-31-9
Merck Index (12th ed.): 4331
Lednicer: 1 134; 2 87

SAMPLE
Matrix: blood, urine
Sample preparation: Condition a 3 mL Supelclean LC-18 SPE cartridge with 3 mL MeOH and 3 mL water. Dilute urine 1:10 with water. Precipitate proteins from serum with MeOH. Add diluted urine or protein supernatant to the SPE cartridge, wash with 3 mL water, elute with 3 mL MeOH, reconstitute to the original volume with 100 mM KOH, inject an aliquot.

CAPILLARY ELECTROPHORESIS
Capillary: 67 cm × 50 μm fused-silica (60 cm to detector) (Polymicro Technologies)
Capillary preparation: Rinse with running buffer for 2 min before run. If necessary, regenerate capillary with 100 mM NaOH for 10 min and with water for 15 min.
Capillary temperature: 20
Running buffer: 60 mM 3-(cyclohexylamino)-1-propanesulfonic acid (CAPS) adjusted to pH 10.6 with 100 mM KOH
Injection: Pressure injection for 5 s
Detector: UV 220
Migration time: 9.4

Voltage: 25 kV
Model: Beckman P/ACE 2000

OTHER SUBSTANCES
Extracted: acetazolamide, amiloride, bendroflumethiazide, benzthiazide, bumetanide, caffeine, chlorothiazide, chlorthalidone, clopamide, dichlorphenamide, ethacrynic acid, hydrochlorothiazide, metyrapone, probenecid, triamterene, trichlormethiazide

KEY WORDS
serum; SPE

REFERENCE
Jumppanen,J.; Sirén,H.; Riekkola,M.-L. Screening for diuretics in urine and blood serum by capillary zone electrophoresis, *J.Chromatogr.A*, **1993**, *652*, 441–450.

SAMPLE
Matrix: formulations
Sample preparation: Add 2 mL injection to 2 mL IS solution, make up to 10 mL with water, inject an aliquot.

CAPILLARY ELECTROPHORESIS
Capillary: 50 cm × 50 μm fused-silica (Bio-Rad)
Capillary temperature: 25
Running buffer: 20 mM pH 8 phosphate buffer containing 30 mM sodium dodecyl sulfate
Injection: Injection by pressure 3 psi.sec (4.2 nL)
Detector: UV 230
Migration time: 6
Internal standard: hydroquinone monobenzyl ether (12)
Voltage: 17 kV
Current: 31 μA
Model: Biofocus 3000 (Bio-Rad)

OTHER SUBSTANCES
Simultaneous: degradation products, aminophylline, methylprednisolone-21-hemisuccinate, prednisolone

KEY WORDS
injections; saline

REFERENCE
Quaglia,M.G.; Bossù,E.; Desiderio,C.; Fanali,S. Use of capillary electrophoresis for testing the stability of a drugs mixture in perfusional solution, *Farmaco*, **1994**, *49*, 403–406.

SAMPLE
Matrix: solutions

CAPILLARY ELECTROPHORESIS
Capillary: 72 cm × 50 μm silica (50 cm to detector) (Applied Biosystems)
Capillary preparation: Before each run wash capillary with 100 mM NaOH for 3 min, wash with running buffer for 3 min, aspirate neutral marker solution (2 drops DMSO in 10 mL water) for 1 s, place capillary end in buffer vial for 5 s, inject sample. Wash capillary at the beginning of each day by passing 1 M NaOH through for 20 min.
Capillary temperature: 30
Running buffer: MeCN:buffer 10:90 (Buffer was 30 mM pH 9.3 borate buffer containing 30 mM sodium dodecyl sulfate.)
Injection: Inject by applying vacuum for 1 s (2-3 nL)
Detector: UV 200
Migration time: 5.8
Voltage: +30 kV
Current: 56 μA
Model: Applied Biosystems Model 270A

OTHER SUBSTANCES
Simultaneous: chlorothiazide, hydrochlorothiazide, trichlormethiazide

REFERENCE
Evenson,M.A.; Wiktorowicz,J.E. Automated capillary electrophoresis applied to therapeutic drug monitoring, *Clin.Chem.*, **1992**, *38*, 1847–1852.

SAMPLE
Matrix: solutions
Sample preparation: Inject an aliquot of a 4 μg/mL solution in running buffer.

CAPILLARY ELECTROPHORESIS
Capillary: 42 cm × 50 μm fused-silica (34 cm to detector) (Supelco)
Capillary preparation: Before each run wash with water for 10 min and with running buffer for 3 min. At the start of each day wash with 100 mM NaOH for 30 min.
Capillary temperature: 35
Running buffer: 70 mM pH 7.4 Sodium phosphate buffer
Injection: Vacuum injection for 2.5 s (8.8 nL).
Detector: UV 260
Migration time: 7.5
Voltage: 10 kV
Current: 55 μA
Model: Spectraphoresis 1000 (Thermo Separation Products)

REFERENCE
Quaglia,M.G.; Bossu,E.; Dell'Aquila,C.; Guidotti,M. Determination of the binding of a β_2-blocker drug, frusemide and ceftriaxone to serum proteins by capillary zone electrophoresis, *J.Pharm.Biomed.Anal.*, **1997**, *15*, 1033–1039.

SAMPLE
Matrix: urine
Sample preparation: Filter (0.2 μm), inject an aliquot of the filtrate.

CAPILLARY ELECTROPHORESIS
Capillary: 80 cm × 50 μm fused-silica (57 cm to detector)
Capillary preparation: Before each run rinse the capillary with 1 M NaOH for 3 min, with 100 mM NaOH for 3 min, with water for 3 min, and with running buffer for 10 min.
Running buffer: 110 mM Boric acid containing 56 mM NaOH and 44 mM HCl, pH 8
Injection: Vacuum injection for 1 s
Detector: UV 232
Migration time: 8.767
Voltage: 20 kV
Model: Europhor Prime Vision system IV
Limit of detection: 443 nM

OTHER SUBSTANCES
Extracted: acebutolol (UV 238), acetazolamide (UV 222), alprenolol (UV 220), amiloride (UV 220), atenolol (UV 228), bendroflumethiazide (UV 220), bumetanide (UV 220), chlorthalidone (UV 220), cocaine (UV 236), codeine (UV 220), ethacrynic acid (UV 220), hydrochlorothiazide (UV 226), methadone (UV 220), metoxiphenamine (UV 220), nadolol (UV 220), norcodeine (UV 220), oxprenolol (UV 220), pentazocine (UV 220), propranolol (UV 220), spironolactone (UV 244), triamterene (UV 232), xipamide (UV 234)

REFERENCE
Gonzalez,E.; Laserna,J.J. Capillary zone electrophoresis for the rapid screening of banned drugs in sport, *Electrophoresis*, **1994**, *15*, 240–243.

SAMPLE
Matrix: urine
Sample preparation: 5 mL Urine + 120 μL 1 M phosphoric acid + 10 mL chloroform, shake by hand for 2 min. Remove the organic layer and evaporate it to dryness under a stream of

nitrogen at 45°, reconstitute with 100 μL 20 mM sodium dodecyl sulfate in water, sonicate for 1 min, inject an aliquot.

CAPILLARY ELECTROPHORESIS
Capillary: 56 cm × 50 μm fused-silica (48 cm to detector)
Capillary preparation: After each run rinse capillary with running buffer for 2 min.
Capillary temperature: 25
Running buffer: 20 mM pH 9 Sodium borate buffer containing 150 mM sodium dodecyl sulfate (Prepare by mixing 20 mM sodium tetraborate containing 150 mM sodium dodecyl sulfate with 20 mM boric acid containing 150 mM sodium dodecyl sulfate so as to achieve a pH of 9.)
Injection: Hydrodynamic injection at 50 mbar for 10 s.
Detector: UV 230
Migration time: 4.3
Voltage: 30 kV
Model: Hewlett-Packard HP[3D]
Limit of quantitation: 25 ng/mL

OTHER SUBSTANCES
Extracted: piretanide
Simultaneous: ethacrynic acid, torsemide

REFERENCE
Lalljie,S.P.D.; Begoña Barroso,M.; Steenackers,D.; Alonso,R.M.; Jiménez,R.M.; Sandra,P. Micellar electrokinetic chromatography as a fast screening method for the determination of the doping agents furosemide and piretanide in urine, *J.Chromatogr.B*, **1997**, *688*, 71–78.

Gabapentin

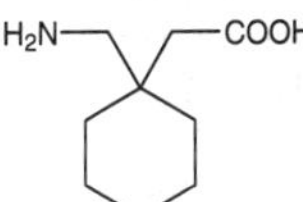

Molecular formula: $C_9H_{17}NO_2$
Molecular weight: 171.24
CAS Registry No.: 60142-96-3
Merck Index (12th ed.): 4343

SAMPLE
Matrix: blood
Sample preparation: 50 μL Serum + 200 μL 2.5 mg/mL fluorescamine in MeCN (prepare fresh daily), vortex for 1 min, centrifuge at 14000 g for 30 s, inject an aliquot of the supernatant.

CAPILLARY ELECTROPHORESIS
Capillary: 60 cm × 50 μm
Capillary preparation: After each run wash capillary with 2 M NaOH for 2 min then with buffer. Condition capillary each morning with 2 M NaOH for 7 min and buffer for 7 min.
Capillary temperature: 24
Running buffer: 200 mM Boric acid containing 26 mM K_2HPO_4
Injection: Pressure injection for 15 s
Detector: UV 200
Migration time: 11
Current: 35 μA
Model: Beckman Model 2000
Limit of detection: 1 μg/mL

OTHER SUBSTANCES
Noninterfering: acetaminophen, N-acetylprocainamide, amino acids, carbamazepine, disopyramide, GABA, gentamicin, lidocaine, phenytoin, primidone, procainamide, quinidine, salicylic acid, theophylline, tobramycin, valproic acid, vancomycin

KEY WORDS
serum; derivatization

REFERENCE
Garcia,L.L.; Shihabi,Z.K.; Oles,K. Determination of gabapentin in serum by capillary electrophoresis, *J.Chromatogr.B*, **1995**, *669*, 157–162.

Gallopamil

Molecular formula: $C_{28}H_{40}N_2O_5$
Molecular weight: 484.64
CAS Registry No.: 16662-47-8
Merck Index (12th ed.): 4369

SAMPLE
Matrix: blood
Sample preparation: 1 mL Plasma + 1 mL 100 mM NaOH + 5 mL hexane:isopropanol 90:10, vortex for 30 s, centrifuge at 1500 g for 10 min. Remove the organic layer and evaporate it to dryness under a stream of nitrogen, reconstitute the residue in 100 μL MeOH:water 25:75, centrifuge, inject an aliquot.

CAPILLARY ELECTROPHORESIS
Capillary: 18 cm × 75 μm fused-silica (18 cm to detector)
Capillary preparation: Before each analysis wash with 100 mM NaOH for 1 min, wash with buffer for 1 min, fill with running buffer for 2 min. (Buffer was 4 mL/L phosphoric acid in water adjusted to pH 2.5 with 1 M NaOH.)
Capillary temperature: 15
Running buffer: 60 mM pH 2.5 Phosphate buffer containing trimethyl-β-cyclodextrin
Injection: Injection by electromigration at 10 kV for 7 s
Detector: UV 200
Migration time: 8
Internal standard: gallopamil
Voltage: 12 kV
Current: 90-100 μA
Model: Spectra-Physics Phoresis 1000

OTHER SUBSTANCES
Extracted: norverapamil, verapamil

KEY WORDS
plasma; gallopamil is IS

REFERENCE
Dethy,J.-M.; De Broux,S.; Lesne,M.; Longstreth,J.; Gilbert,P. Stereoselective determination of verapamil and norverapamil by capillary electrophoresis, *J.Chromatogr.B*, **1994**, *654*, 121–127.

SAMPLE
Matrix: solutions
Sample preparation: Prepare a 50 μg/mL solution in diluted running buffer, inject an aliquot.

CAPILLARY ELECTROPHORESIS
Capillary: 44 cm × 50 μm fused-silica
Capillary preparation: After each run wash capillary with water for 2 min and with running buffer for 3 min. At the beginning of each day wash capillary with water and running buffer for 5 min. Condition a new capillary with 1 M NaOH, 100 mM NaOH, water, and separation buffer.
Capillary temperature: 15
Running buffer: 100 mM Phosphoric acid containing 20 mM hydroxypropyl-β-cyclodextrin adjusted to pH 3.0 with triethanolamine
Injection: Hydrodynamic injection for 1 s
Detector: UV 210

Voltage: 25 kV
Model: Spectra Physics Spectraphoresis 1000

KEY WORDS
chiral; enantiomer resolution 1.3

REFERENCE
Bechet,I.; Paques,P.; Fillet,M.; Hubert,P.; Crommen,J. Chiral separation of basic drugs by capillary zone electrophoresis with cyclodextrin additives, *Electrophoresis*, **1994**, *15*, 818–823.

SAMPLE
Matrix: solutions
Sample preparation: Inject an aliquot of a solution in MeOH.

CAPILLARY ELECTROPHORESIS
Capillary: 75 cm × 50 μm fused-silica (65 cm to detector) (Supelco)
Capillary preparation: Before each run rinse capillary with 100 mM NaOH for 2 min and with running buffer for 2 min.
Running buffer: MeOH:16 mM NaCl 25:75 containing 50 mM polyoxyethylene-4-dodecyl ether (Brij 30) (mole fraction 0.3)
Injection: Hydrodynamic injection at 15 cm for 1-6 s
Detector: UV 210
Migration time: 21.4 (-), 21.6 (+)
Voltage: 20 kV
Model: laboratory constructed

OTHER SUBSTANCES
Simultaneous: norverapamil, verapamil

KEY WORDS
chiral

REFERENCE
Clothier,J.G.,Jr.; Tomellini,S.A. Chiral separation of verapamil and related compounds using micellar electrokinetic capillary chromatography with mixed micelles of bile salt and polyoxyethylene ethers, *J.Chromatogr.A*, **1996**, *723*, 179–187.

SAMPLE
Matrix: solutions
Sample preparation: Inject an aliquot of a 100 μg/mL solution in running buffer.

CAPILLARY ELECTROPHORESIS
Capillary: 30 cm × 50 μm fused-silica (25.5 cm to detector) (Yongnian Optical Conductive Fiber Plant, China), coated with polyacrylamide
Capillary preparation: No details of the polyacrylamide coating process are provided. However, another paper (LC.GC 1997, 15, 40) by this group indicates that they use the procedure of Hjertén, thus: Adjust the pH of 20 mL water to 3.5 with acetic acid, add 80 μL 3-(trimethoxysilyl)propyl methacrylate (3-methacryloxypropyltrimethoxysilane), mix, suck into capillary, let stand at room temperature for 1 h, remove the solution, wash with water. Fill the capillary with a deaerated 3-4% acrylamide solution containing 1 μL/mL N,N,N',N'-tetramethylethylenediamine and 1 mg/mL potassium persulfate, let stand for 30 min, remove excess solution by aspiration, rinse with water, remove water by aspiration, dry at 35° (J. Chromatogr. 1985, 347, 191).
Capillary temperature: 25
Running buffer: 100 mM NaH_2PO_4 adjusted to pH 2.5
Injection: Electrokinetic injection at 15 kV for 3 s.
Detector: UV 200, UV 210
Migration time: 6.27
Voltage: 15 kV
Model: Bio-Rad BioFocus 3000

OTHER SUBSTANCES
Simultaneous: amorolfine, brompheniramine, bupivacaine, carteolol, chloroquine, chlorpheniramine, chlorphenoxamine, disopyramide, dobutamine, doxylamine, flecainide, ketamine, mepindolol, orphenadrine, oxybutynin, phenoxybenzamine, pindolol, propafenone, propranolol, sulpiride, talinolol, tropicamide, verapamil

KEY WORDS
coated capillary

REFERENCE
Koppenhoefer,B.; Epperlein,U.; Xiaofeng,Z.; Bingcheng,L. Separation of enantiomers of drugs by capillary electrophoresis. Part 4: Hydroxypropyl-γ-cyclodextrin as chiral solvating agent, *Electrophoresis*, **1997**, *18*, 924–930.

SAMPLE
Matrix: solutions

CAPILLARY ELECTROPHORESIS
Capillary: 29-36 cm × 50 μm fused-silica (24.5-31.5 cm to detector) (Yongnian Optical Conductive Fiber Plant, China) coated with polyacrylamide
Capillary preparation: Coat capillary as follows. Adjust the pH of 20 mL water to 3.5 with acetic acid, add 80 μL 3-(trimethoxysilyl)propyl methacrylate (3-methacryloxypropyltrimethoxysilane), mix, suck into capillary, let stand at room temperature for 1 h, remove the solution, wash with water. Fill the capillary with a deaerated 3-4% acrylamide solution containing 1 μL/mL N,N,N',N'-tetramethylethylenediamine and 1 mg/mL potassium persulfate, let stand for 30 min, remove excess solution by aspiration, rinse with water, remove water by aspiration, dry at 35° (J. Chromatogr. 1985, 347, 191).
Capillary temperature: 25
Running buffer: 100 mM pH 2.5 NaH_2PO_4 (A) or 100 mM pH 2.5 NaH_2PO_4 containing 45 mM hydroxypropyl-α-cyclodextrin (Wacker, Munich) (B)
Injection: Electromigration at 15 kV for 3 s.
Detector: UV 200; UV 210
Migration time: 6.27 (A); 10.64 (B) (no separation of enantiomers)
Voltage: 15 kV
Model: Bio-Focus 3000

OTHER SUBSTANCES
Also analyzed: albuterol (salbutamol), alprenolol, amorolfine, atenolol, atropine, azelastine, baclofen, bamethan, benproperine, benserazide, biperiden, bisoprolol, brompheniramine, bupivacaine, bupranolol, butamirate, butethamate, carazolol, carbuterol, carteolol, carvedilol, celiprolol, chloroquine, chlorpheniramine, chlorphenoxamine, cicletanine, clenbuterol, clidinium bromide, clobutinol, dimethindene, dipivefrin, disopyramide, dobutamine, doxylamine, fendiline, flecainide, homatropine, ipratropium bromide, isoproterenol (isoprenaline), isothipendyl, ketamine, meclizine, mefloquine, mepindolol, mequitazine, metaclazepam, metaproterenol (orciprenaline), metipranolol, metoprolol, nafronyl (naftidrofuryl), nefopam, nicardipine, norfenefrine, ofloxacin, ornidazole, orphenadrine, oxomemazine, oxprenolol, oxybutynin, phenoxybenzamine, phenylpropanolamine, pholedrine, pindolol, pirbuterol, prilocaine, procyclidine, promethazine, propafenone, propranolol, reproterol, sotalol, sulpride, synephrine, talinolol, terbutaline, tetrahydrozoline (tetryzoline), theodrenaline, tioconazole, tocainide, trihexyphenidyl, trimeprazine (alimemazine), trimipramine, tropicamide, verapamil, zopiclone

KEY WORDS
coated capillary

REFERENCE
Koppenhoefer,B.; Eperlein,U.; Schlunk,R.; Zhu,X.; Lin,B. Separation of enantiomers of drugs by capillary electrophoresis. V. Hydroxypropyl-α-cyclodextrin as chiral solvating agent, *J.Chromatogr.A*, **1998**, *793*, 153–164.

Gentamicin

Molecular formula: $C_{19}H_{39}N_5O_7$
Molecular weight: 449.55
CAS Registry No.: 1403-66-3, 1405-41-0 (sulfate)
Merck Index (12th ed.): 4398

Gentamicin C_1 $R_1 = R_2 = CH_3$

Gentamicin C_2 $R_1 = CH_3$, $R_2 = H$

Gentamicin C_{1a} $R_1 = R_2 = H$

SAMPLE
Matrix: bulk
Sample preparation: Inject an aliquot of a solution in 140 mM sodium tetraborate containing 480 µg/mL sisomicin as IS.

CAPILLARY ELECTROPHORESIS
Capillary: 70 cm × 50 µm uncoated fused silica (45 cm to detector) (Polymicro Technologies or ISCO)
Capillary preparation: Rinse with running buffer for 15 min at the beginning of the day and for 4 min before each injection.
Capillary temperature: 34
Running buffer: 185 mM pH 9 sodium tetraborate
Injection: Vacuum injection at 25.0 kPa
Detector: UV 195
Migration time: 17.88 (C_1), 18.88 (C_{1a}), 19.22 (C_2 and C_{2a})
Internal standard: sisomicin (19.90)
Voltage: +15 kV
Model: ISCO Model 3140 Electropherograph

OTHER SUBSTANCES
Simultaneous: amikacin, bekanamycin, butirosin, dibekacin, dihydrostreptomycin, kanamycin A, paromomycin, ribostamycin, streptomycin, tobramycin
Noninterfering: neomycin

KEY WORDS
detector is at cathode

REFERENCE
Flurer,C.L. The analysis of aminoglycoside antibiotics by capillary electrophoresis, *J.Pharm.Biomed.Anal.*, **1995**, *13*, 809–816.

SAMPLE
Matrix: formulations
Sample preparation: Dilute injection with water to a gentamicin concentration of 2 mg/mL, inject an aliquot.

CAPILLARY ELECTROPHORESIS
Capillary: 70 cm × 50 µm fused-silica (45 cm to detector) (Isco)
Capillary preparation: Rinse with running buffer for 3 min before each analysis.
Capillary temperature: 34
Running buffer: 150 mM pH 9.4 borate buffer
Injection: Vacuum injection at 25 kPa.s
Detector: UV 195

Migration time: 8.08 (C_1), 8.51 (C_{1a}), 8.64 (C_2, C_{2a})
Voltage: +15 kV
Model: Isco Model 3140

KEY WORDS
injections

REFERENCE
Flurer,C.L.; Wolnik,K.A. Quantitation of gentamicin sulfate in injectable solutions by capillary electrophoresis, *J.Chromatogr.A*, **1994**, *663*, 259–263.

SAMPLE
Matrix: solutions
Sample preparation: Prepare a solution in 100 mM cetyltrimethylammonium bromide, inject an aliquot.

CAPILLARY ELECTROPHORESIS
Capillary: 67 cm × 50 μm fused silica capillary (60 cm to detector) (Siemens)
Running buffer: pH 5.0 Acetate buffer containing 10 mM imidazole and 50 μg/mL FC 135 (a fluorinated surfactant, Fluorad/3M)
Injection: Pressure injection for 2 s.
Detector: UV 214
Migration time: 12
Voltage: 12.5 kV
Model: Beckman P/ACE System 2000

OTHER SUBSTANCES
Simultaneous: dihydrostreptomycin, kanamycin, lividomycin, neomycin, paromomycin, sisomycin, streptomycin, tobramycin
Interfering: amikacin

KEY WORDS
indirect UV detection; detector is at anode

REFERENCE
Ackermans,M.T.; Everaerts,F.M.; Beckers,J.L. Determination of aminoglycoside antibiotics in pharmaceuticals by capillary zone electrophoresis with indirect UV detection coupled with micellar electrokinetic capillary chromatography, *J.Chromatogr.*, **1992**, *606*, 229–235.

Glipizide

Molecular formula: $C_{21}H_{27}N_5O_4S$
Molecular weight: 445.54
CAS Registry No.: 29094-61-9
Merck Index (12th ed.): 4451
Lednicer: 2 117

SAMPLE
Matrix: solutions
Sample preparation: Inject a dilute aqueous solution on to the tip of the capillary containing the SPE material at 138 kPa for 12 s, wash for 1 min with running buffer, desorb with elution buffer for 20 s (60 nL), inject running buffer at 3.45 kPa, electrophorese. (Elution buffer was MeCN:50 mM pH 2.5 phosphate buffer 80:20.)

CAPILLARY ELECTROPHORESIS
Capillary: 67 cm × 75 μm with a 0.5 mm long × 0.37 mm diameter column of C18 material (from a J.T. Baker SPE cartridge) at the injection end

Capillary preparation: Before each run rinse capillary with 3 column volumes of elution buffer and 10 column volumes of running buffer. After each run rinse with 500 μL elution buffer.
Capillary temperature: 20
Running buffer: 250 mM pH 8.4 Borate buffer containing 5 mM phosphate
Detector: UV 200
Migration time: 15.5
Voltage: 15 kV
Current: 70 μA
Model: Beckman P/ACE System 5510
Limit of detection: 5 ng/mL

OTHER SUBSTANCES
Simultaneous: acetohexamide, chlorpropamide, glyburide, tolazamide, tolbutamide

KEY WORDS
SPE

REFERENCE
Strausbauch,M.A.; Xu,S.J.; Ferguson,J.E.; Nunez,M.E.; Machacek,D.; Lawson,G.M.; Wettstein,P.J.; Landers,J.P. Concentration and separation of hypoglycemic drugs using solid-phase extraction-capillary electrophoresis, *J.Chromatogr.A*, **1995**, *717*, 279–291.

SAMPLE
Matrix: urine
Sample preparation: Condition a 200 μg Bond Elut C18 SPE cartridge with 1 mL MeCN. 10 mL Urine + 100 μL 10 μg/mL IS, adjust pH to 2.0 with concentrated HCl, add 15 mL dichloromethane, shake vigorously for 2 min, let stand for 5 min. Remove the organic layer and evaporate it to dryness under a stream of nitrogen at 45°, reconstitute the residue in 400 μL MeOH:water 50:50, add to the SPE cartridge, wash with 1.25 mL MeCN:water 10:90, wash with 1 mL MeCN:water 40:60, elute with 1 mL MeCN. Evaporate the eluate to dryness under a stream of nitrogen, reconstitute the residue in 20 μL MeOH, inject an aliquot.

CAPILLARY ELECTROPHORESIS
Capillary: 47 cm × 50 μm fused-silica (40 cm to detector) (Polymicro Technologies)
Capillary preparation: Before each run wash with 3 column volumes of 100 mM NaOH, dip momentarily into water, rinse with 10 column volumes of 50 mM pH 8.5 Sodium borate buffer containing 50 mM Na₂HPO₄ and 75 mM sodium cholate, rinse with 3 column volumes of running buffer. Condition a new capillary by rinsing with 20 column volumes of 100 mM NaOH, 20 column volumes of water, and 20 column volumes of running buffer.
Capillary temperature: 25
Running buffer: 5 mM pH 8.5 Sodium borate buffer containing 5 mM Na$_2$HPO$_4$ and 75 mM sodium cholate
Injection: Pressure injection at 0.5 psi for 2 s then a 1 s injection of running buffer.
Detector: UV 200
Migration time: 5
Internal standard: N-acetyl-4-(2,3-dichlorophenylureido)benzenesulfonamide (7.4)
Voltage: 25 kV
Current: 61 μA
Model: Beckman P/ACE System 2100 or System 5510
Limit of detection: 50 ng/mL

OTHER SUBSTANCES
Extracted: acetohexamide, chlorpropamide, glyburide, tolazamide, tolbutamide

KEY WORDS
SPE

REFERENCE
Núñez,M.; Ferguson,J.E.; Machacek,D.; Jacob,G.; Oda,R.P.; Lawson,G.M.; Landers,J.P. Detection of hypoglycemic drugs in human urine using micellar electrokinetic chromatography, *Anal.Chem.*, **1995**, *67*, 3668–3675.

SAMPLE
Matrix: urine
Sample preparation: Inject an aliquot directly.

CAPILLARY ELECTROPHORESIS
Capillary: 37 cm × 50 μm fused-silica (30 cm to detector) (Polymicro Technologies)
Capillary preparation: Every 13 injections rinse capillary with 100 mM NaOH for 1 min and with running buffer for 1 min. Condition a new capillary with 20 column volumes of 100 mM NaOH, 20 column volumes of water, and 20 column volumes of running buffer.
Capillary temperature: 22
Running buffer: 5 mM pH 8.5 Borate buffer containing 5 mM phosphate and 75 mM sodium cholate
Injection: Pressure injection at 0.5 psi for 2 s.
Detector: UV 200
Migration time: 2.2
Internal standard: N-acetyl-4-(2,3-dichlorophenylureido)benzenesulfonamide (3.2)
Voltage: 28 kV
Current: 71 μA
Model: Beckman P/ACE 5510

OTHER SUBSTANCES
Extracted: acetohexamide, chlorpropamide, glyburide, tolazamide, tolbutamide

KEY WORDS
direct injection

REFERENCE
Roche,M.E.; Oda,R.P.; Machacek,D.; Lawson,G.M.; Landers,J.P. Enhanced throughput with capillary electrophoresis via continuous-sequential sample injection, *Anal.Chem.*, **1997**, *69*, 99–104.

SAMPLE
Matrix: urine
Sample preparation: Condition a 200 μg (sic) Bond Elut C18 SPE cartridge with 1 mL MeCN. 10 mL Urine + 100 μL 10 μg/mL IS, mix, adjust pH to 2.0 with concentrated HCl, add 15 mL dichloromethane, shake vigorously for 2 min, let stand for 5 min. Remove the lower organic layer and evaporate it to dryness under a stream of nitrogen at 45°, reconstitute with 400 μL MeOH:water 50:50, add to the SPE cartridge, wash with 1.25 mL MeCN:water 10:90, wash with 1 mL MeCN:water 40:60, elute with 1 mL MeCN, evaporate the eluate to dryness under a stream of nitrogen, reconstitute with 30 μL MeOH, inject an aliquot.

CAPILLARY ELECTROPHORESIS
Capillary: 47 cm × 50 μm fused-silica (40 cm to detector)
Capillary preparation: Before each run rinse at 20 psi with 3 column volumes of 100 mM NaOH and 3 column volumes of running buffer. Condition new capillaries by rinsing with 20 column volumes of 1 M NaOH, water, and running buffer
Capillary temperature: 22
Running buffer: 5 mM pH 8.5 Borate buffer containing 5 mM phosphate and 75 mM sodium cholate (Prepare from 500 mM borate buffer (prepared by mixing 125 mM sodium tetraborate and 500 mM boric acid to achieve pH 8.5) and 500 mM Na$_2$HPO$_4$ solution with appropriate dilution and the addition of sodium cholate.)
Injection: Pressure injection of sample at 0.5 psi for 1-5 s followed by a pressure injection of running buffer for 1 s.
Detector: UV 200
Migration time: 3.7
Internal standard: N-acetyl-4-(2,3-dichlorophenylureido)benzenesulfonamide (5.5)
Voltage: 25 kV
Current: 61 μA
Model: Beckman P/ACE 5510

OTHER SUBSTANCES
Extracted: glyburide
Simultaneous: acetohexamide, chlorpropamide, tolazamide, tolbutamide

KEY WORDS
SPE

REFERENCE
Roche,M.E.; Oda,R.P.; Lawson,G.M.; Landers,J.P. Capillary electrophoretic detection of metabolites in the urine of patients receiving hypoglycemic drug therapy, *Electrophoresis*, **1997**, *18*, 1865–1874.

Glutethimide

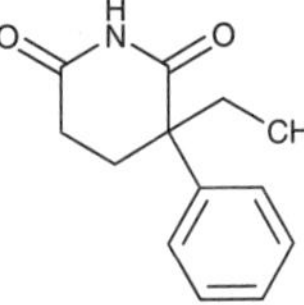

Molecular formula: $C_{13}H_{15}NO_2$
Molecular weight: 217.27
CAS Registry No.: 77-21-4
Merck Index (12th ed.): 4485
Lednicer: 1 257

SAMPLE
Matrix: solutions
Sample preparation: Inject an aliquot of a 300 µg/mL solution in water.

CAPILLARY ELECTROPHORESIS
Capillary: 65 cm × 50 µm fused-silica (40 cm to detector) (Isco)
Capillary preparation: Purge with running buffer for 3 min between injections. Purge daily with 1 M NaOH for 3 min, with water for 3 min, and with running buffer for 3 min.
Running buffer: Isopropanol:100 mM pH 7 phosphate buffer containing 25 mM rifamycin SV 30:70
Injection: Electrokinetic injection at 5 kV for 5 s.
Detector: UV 350
Migration time: 34.8 (first enantiomer, R_S 4.0)
Voltage: 8 kV
Model: Isco model 3850

OTHER SUBSTANCES
Simultaneous: dansyl aspartic acid, hexobarbital

KEY WORDS
chiral

REFERENCE
Ward,T.J.; Dann,C.,III; Blaylock,A. Enantiomeric resolution using the macrocyclic antibiotics rifamycin B and rifamycin SV as chiral selectors for capillary electrophoresis, *J.Chromatogr.A*, **1995**, *715*, 337–344.

SAMPLE
Matrix: solutions
Sample preparation: Inject an aliquot of a 0.2-1 µg/mL solution.

CAPILLARY ELECTROPHORESIS
Capillary: 57 cm × 50 µm fused-silica coated with a 0.025 µm film of p-methylbenzoyl cellulose (50 cm to detector) (Polymicro Technologies)
Capillary preparation: Coat capillary as follows. Coat at 0.3 mbar and 35-40° with a filtered 0.2% solution of p-methylbenzoyl cellulose in THF using the static method (further details in K. Grob. Making and Manipulating Capillary Columns for Gas Chromatography, Hüthig, Heidelberg, 1986). Prepare p-methylbenzoyl cellulose as follows. Suspend 100 g of microcrystalline cellulose in a mixture of 1 L pyridine, 420 mL triethylamine, and 2 g dimethylaminopyridine, slowly add 265 mL p-toluoyl chloride (p-methylbenzoyl chloride) at room temperature, stir at 120° under nitrogen for 12 h, cool to RT, treat the solid mass with 4 L MeOH, dissolve in dichloromethane, precipitate twice in EtOH, dry under vacuum at 100° for 2 days (cf. Chirality 1991, 3, 43).

Running buffer: MeCN:40 mM pH 7 phosphate buffer 25:75
Injection: Pressure injection at 35 mbar for 2-5 s.
Detector: UV 214
Migration time: 23, 24 (enantiomers)
Voltage: 30 kV
Model: Beckman P/ACE 5510

KEY WORDS
electrochromatography; coated capillary; chiral

REFERENCE
Francotte,E.; Jung,M. Enantiomer separation by open-tubular liquid chromatography and electrochromatography in cellulose capillaries, *Chromatographia*, **1996**, *42*, 521–527.

SAMPLE
Matrix: solutions
Sample preparation: Inject an aliquot of a 100 µg/mL solution in MeOH:water 10:90.

CAPILLARY ELECTROPHORESIS
Capillary: 60 cm × 50 µm untreated AccuSep (52.5 cm to detector) (Waters)
Capillary preparation: Purge with fresh running buffer for 3 min between runs. Before use purge with vacuum with 500 mM NaOH for 5 min, with water for 10 min, and with running buffer for 10 min.
Capillary temperature: 30
Running buffer: 25 mM pH 8.80 Na_2HPO_4 containing 25 mM sodium tetraborate and 75 mM (S)-dodecoxycarbonylproline ((S)-DDCP) ((S)-N-dodecoxycarbonylproline is presumably prepared in analagous fashion to (S)-dodecoxycarbonylvaline. Prepare dodecyl chloroformate by reacting 1-dodecanol with 0.33 equivalents of triphosgene in dichloromethane solution in the presence of pyridine. Add dodecyl chloroformate dropwise to (S)-proline in 1 M NaOH solution, filter, wash with hexane, recrystallize from ether/petroleum ether (J.Chromatogr.A 1994, 680, 125).)
Injection: Hydrostatic injection at 10 cm for 5 s.
Detector: UV 214
Migration time: 34.7, 35.3 (enantiomers)
Voltage: 16 kV
Model: Waters Quanta 4000E

KEY WORDS
chiral

REFERENCE
Swartz,M.E.; Mazzeo,J.R.; Grover,E.R.; Brown,P.R.; Aboul-Enein,H.Y. Separation of piperidine-2,6-dione drug enantiomers by micellar electrokinetic capillary chromatography using synthetic chiral surfactants, *J.Chromatogr.A*, **1996**, *724*, 307–316.

Glyburide

Molecular formula: $C_{23}H_{28}ClN_3O_5S$
Molecular weight: 494.01
CAS Registry No.: 10238-21-8
Merck Index (12th ed.): 4486
Lednicer: 2 139

SAMPLE
Matrix: solutions

Sample preparation: Inject a dilute aqueous solution on to the tip of the capillary containing the SPE material at 138 kPa for 12 s, wash for 1 min with running buffer, desorb with elution buffer for 20 s (60 nL), inject running buffer at 3.45 kPa, electrophorese. (Elution buffer was MeCN:50 mM pH 2.5 phosphate buffer 80:20.)

CAPILLARY ELECTROPHORESIS
Capillary: 67 cm × 75 µm with a 0.5 mm long × 0.37 mm diameter column of C18 material (from a J.T. Baker SPE cartridge) at the injection end
Capillary preparation: Before each run rinse capillary with 3 column volumes of elution buffer and 10 column volumes of running buffer. After each run rinse with 500 µL elution buffer.
Capillary temperature: 20
Running buffer: 250 mM pH 8.4 Borate buffer containing 5 mM phosphate
Detector: UV 200
Migration time: 16
Voltage: 15 kV
Current: 70 µA
Model: Beckman P/ACE System 5510
Limit of detection: 5 ng/mL

OTHER SUBSTANCES
Simultaneous: acetohexamide, chlorpropamide, glipizide, tolazamide, tolbutamide

KEY WORDS
SPE

REFERENCE
Strausbauch,M.A.; Xu,S.J.; Ferguson,J.E.; Nunez,M.E.; Machacek,D.; Lawson,G.M.; Wettstein,P.J.; Landers,J.P. Concentration and separation of hypoglycemic drugs using solid-phase extraction-capillary electrophoresis, *J.Chromatogr.A*, **1995**, *717*, 279–291.

SAMPLE
Matrix: solutions

CAPILLARY ELECTROPHORESIS
Capillary: 27 cm × 75 µm
Capillary preparation: Before each run rinse capillary with 100 mM NaOH for 30 s and with running buffer for 30 s. Condition new capillaries by rinsing with 100 mM NaOH for 20 min.
Capillary temperature: 30
Running buffer: 15 mM Sodium borate
Injection: Pressure injection of sample at 25 mbar for 1 s followed by running buffer at 25 mbar for 1 s.
Detector: UV 200
Migration time: 2.49
Internal standard: aminobenzoic acid (3.5)
Voltage: 6.5 kV
Model: Beckman

OTHER SUBSTANCES
Simultaneous: aspirin, bacitracin, beclomethasone, benzoic acid, ceftizoxime, ceftriaxone, cefuroxime, cephalothin, cromolyn, embonic acid, epoprostenol, levothyroxine, nedocromil, nystatin, omeprazole, prednisolone, warfarin, zidovudine

REFERENCE
Altria,K.D.; Bryant,S.M.; Hadgett,T.A. Validated capillary electrophoresis method for the analysis of a range of acidic drugs and excipients, *J.Pharm.Biomed.Anal.*, **1997**, *15*, 1091–1101.

SAMPLE
Matrix: urine
Sample preparation: Condition a 200 µg (sic) Bond Elut C18 SPE cartridge with 1 mL MeCN. 10 mL Urine + 100 µL 10 µg/mL IS, mix, adjust pH to 2.0 with concentrated HCl, add 15 mL dichloromethane, shake vigorously for 2 min, let stand for 5 min. Remove the lower organic layer and evaporate it to dryness under a stream of nitrogen at 45°, reconstitute with 400 µL

MeOH:water 50:50, add to the SPE cartridge, wash with 1.25 mL MeCN:water 10:90, wash with 1 mL MeCN:water 40:60, elute with 1 mL MeCN, evaporate the eluate to dryness under a stream of nitrogen, reconstitute with 30 µL MeOH, inject an aliquot.

CAPILLARY ELECTROPHORESIS
Capillary: 47 cm × 50 µm fused-silica (40 cm to detector)
Capillary preparation: Before each run rinse at 20 psi with 3 column volumes of 100 mM NaOH and 3 column volumes of running buffer. Condition new capillaries by rinsing with 20 column volumes of 1 M NaOH, water, and running buffer
Capillary temperature: 22
Running buffer: 5 mM pH 8.5 Borate buffer containing 5 mM phosphate and 75 mM sodium cholate (Prepare from 500 mM borate buffer (prepared by mixing 125 mM sodium tetraborate and 500 mM boric acid to achieve pH 8.5) and 500 mM Na_2HPO_4 solution with appropriate dilution and the addition of sodium cholate.)
Injection: Pressure injection of sample at 0.5 psi for 1-5 s followed by a pressure injection of running buffer for 1 s.
Detector: UV 200
Migration time: 4.7
Internal standard: N-acetyl-4-(2,3-dichlorophenylureido)benzenesulfonamide (5.5)
Voltage: 25 kV
Current: 61 µA
Model: Beckman P/ACE 5510

OTHER SUBSTANCES
Extracted: metabolites, glipizide
Simultaneous: acetohexamide, chlorpropamide, tolazamide, tolbutamide

KEY WORDS
SPE

REFERENCE
Roche,M.E.; Oda,R.P.; Lawson,G.M.; Landers,J.P. Capillary electrophoretic detection of metabolites in the urine of patients receiving hypoglycemic drug therapy, *Electrophoresis*, **1997**, *18*, 1865–1874.

SAMPLE
Matrix: urine
Sample preparation: Inject an aliquot directly.

CAPILLARY ELECTROPHORESIS
Capillary: 37 cm × 50 µm fused-silica (30 cm to detector) (Polymicro Technologies)
Capillary preparation: Every 13 injections rinse capillary with 100 mM NaOH for 1 min and with running buffer for 1 min. Condition a new capillary with 20 column volumes of 100 mM NaOH, 20 column volumes of water, and 20 column volumes of running buffer.
Capillary temperature: 22
Running buffer: 5 mM pH 8.5 Borate buffer containing 5 mM phosphate and 75 mM sodium cholate
Injection: Pressure injection at 0.5 psi for 2 s.
Detector: UV 200
Migration time: 2.8
Internal standard: N-acetyl-4-(2,3-dichlorophenylureido)benzenesulfonamide (3.2)
Voltage: 28 kV
Current: 71 µA
Model: Beckman P/ACE 5510

OTHER SUBSTANCES
Extracted: acetohexamide, chlorpropamide, glipizide, tolazamide, tolbutamide

KEY WORDS
direct injection

REFERENCE
Roche,M.E.; Oda,R.P.; Machacek,D.; Lawson,G.M.; Landers,J.P. Enhanced throughput with capillary electrophoresis via continuous-sequential sample injection, *Anal.Chem.*, **1997**, *69*, 99–104.

Glycerin

HO⟍⟋⟍⟋OH
    OH

Molecular formula: $C_3H_8O_3$
Molecular weight: 92.09
CAS Registry No.: 56-81-5
Merck Index (12th ed.): 4493

SAMPLE
Matrix: solutions

CAPILLARY ELECTROPHORESIS
Capillary: 80 cm × 25 μm fused-silica (Polymicro Technologies)
Running buffer: 250 mM NaOH
Injection: Electromigration at 15 kV for 3 s
Detector: E, Bioanalytical Systems LC-4B, 100 μm Cu-disk electrode (Anal. Chem. 1993, 65, 3525), +0.60 V, Ag/AgCl reference electrode, Pt wire counter electrode
Migration time: 30
Voltage: 15 kV
Model: Laboratory constructed
Limit of detection: 2.4 fmole

OTHER SUBSTANCES
Simultaneous: adonitol, erythritol, ethylene glycol, galactitol, glucitol, inositol, mannitol

REFERENCE
Ye,J.; Baldwin,R.P. Determination of carbohydrates, sugar acids and alditols by capillary electrophoresis and electrochemical detection at a copper electrode, *J.Chromatogr.A*, **1994**, *687*, 141–148.

SAMPLE
Matrix: solutions
Sample preparation: Inject an aliquot of an aqueous solution.

CAPILLARY ELECTROPHORESIS
Capillary: 100 cm × 75 μm fused-silica capillaries (Polymicro Technologies)
Capillary preparation: Etch new capillaries with 1 M NaOH for 1 h before use.
Capillary temperature: 28
Running buffer: 50 mM NaOH containing 1 mM NaCl
Injection: Inject at a pressure of 20 mbar for 12 s, ca. 15 nL
Detector: E, cuprous oxide modified carbon working electrode in a 0.5 mm i.d. PEEK tube +0.60 V, stainless steel auxiliary electrode, Ag/AgCl reference electrode. The effluent from the capillary passed through a grounded Pd cylinder and then through an 8 cm × 50 μm coupling capillary to the detector. (Prepare electrode as follows. Push copper wires in to a 4 cm length of 0.5 mm i.d. PEEK tubing so as to leave a 1 mm deep chamber at one end, seal other end with glue. Stir 300 mg conductive carbon cement (Gerhard Neubauer, Münster), 60 mg cuprous oxide (Fluka), and 300 μL acetone until a thick paste forms as the acetone evaporates. Pack paste into the chamber, allow to dry for one week. Use only electrodes with resistance less than 200 Ω. Polish with dry emery paper (grade 2/0, Oakey), 3 μm imperial micro finishing film sheet (3M), and 0.05 μm alumina particles on a Buehler pad, wash with water (Anal. Chim. Acta 1995, 300, 5).)
Migration time: 22
Voltage: 12 kV
Current: 124 μA
Model: Lauer Labs Prince programmable injector and power supply
Limit of detection: 1-2 μM

OTHER SUBSTANCES
Simultaneous: arabinose, dextrose, fucose, galactose, mannose, raffinose, xylose

REFERENCE

Huang,X.; Kok,W.T. Determination of sugars by capillary electrophoresis with electrochemical detection using cuprous oxide modified electrodes, *J.Chromatogr.A*, **1995**, *707*, 335–342.

Gonadorelin

Molecular formula: $C_{55}H_{75}N_{17}O_{13}$
Molecular weight: 1182.31
CAS Registry No.: 33515-09-2, 51952-41-1 (HCl), 52699-48-6 (sulfate)
Merck Index (12th ed.): 5500

SAMPLE

Matrix: solutions
Sample preparation: Inject an aliquot of a solution in 5-fold diluted running buffer.

CAPILLARY ELECTROPHORESIS

Capillary: 35 cm × 75 µm fused-silica coated with polyacrylamide (Polymicro Technologies)
Capillary preparation: Before each run purge with 100 mM HCl for 2 min, with water for 4 min, and with running buffer for 1 min. (Coat the capillary as follows. Adjust the pH of 20 mL water to 3.5 with acetic acid, add 80 µL 3-(trimethoxysilyl)propyl methacrylate (3-methacryl-oxypropyltrimethoxysilane), mix, suck into capillary, let stand at room temperature for 1 h, remove the solution, wash with water. Fill the capillary with a deaerated 3-4% acrylamide solution containing 1 µL/mL N,N,N',N'-tetramethylethylenediamine and 1 mg/mL potassium persulfate, let stand for 30 min, remove excess solution by aspiration, rinse with water, remove water by aspiration, dry at 35° (J. Chromatogr. 1985, 347, 191).)
Running buffer: 100 mM pH 2.5 Sodium phosphate buffer
Injection: Pressure injection at 300 mbar.s.
Detector: UV 191
Migration time: 14
Voltage: 8 kV
Model: Hewlett-Packard HP³ᴰ CE

OTHER SUBSTANCES

Simultaneous: angiotensin II, bombesin, bradykinin, leucine enkephalin, melanocyte stimulating hormone, methionine enkephalin, oxytocin, protirelin

KEY WORDS

coated capillary

REFERENCE

Zhang,R.; Zhang,H.X.; Eaker,D.; Hjerten,S. The effect of β-cyclodextrins as buffer additives on the (capillary) electrophoretic separation of peptides and proteins, *J.Capillary Electrophor.*, **1997**, *4*, 105–112.

Goserelin

Molecular formula: $C_{59}H_{84}N_{18}O_{14}$
Molecular weight: 1269.43
CAS Registry No.: 65807-02-5
Merck Index (12th ed.): 4547

SAMPLE
Matrix: solutions

CAPILLARY ELECTROPHORESIS
Capillary: 80 cm × 75 μm (60 cm to detector) (Bester, Netherlands)
Running buffer: Water:glacial acetic acid 90:10 (pH 2.3)
Injection: 77 nL
Detector: UV 214
Migration time: 19
Voltage: 30 kV
Model: Lauerlabs Prince

KEY WORDS
also details of MS detection

REFERENCE
Hoitink,M.A.; Hop,F.; Beijnen,J.H.; Kettenes-van den Bosch,J.J.; Underberg,W.J.M. Capillary zone electrophoresis-mass spectrometry as a tool in the stability research of the luteinising hormone-releasing hormone analogue goserelin, *J.Chromatogr.A*, **1997**, *776*, 319–327.

Guaifenesin

Molecular formula: $C_{10}H_{14}O_4$
Molecular weight: 198.22
CAS Registry No.: 93-14-1
Merck Index (12th ed.): 4582
Lednicer: 1 118

SAMPLE
Matrix: formulations
Sample preparation: Grind tablet or powders, add 70 mL MeOH, add 20 mL 5 mg/mL methyl p-hydroxybenzoate in MeOH, sonicate for 10 min, shake for 5 min, make up to 100 mL with water, filter (0.45 μm), inject an aliquot.

CAPILLARY ELECTROPHORESIS
Capillary: 65 cm × 50 μm fused-silica (50 cm to detector) (SGE)
Running buffer: 20 mM NaH_2PO_4 containing 100 mM sodium dodecyl sulfate, adjusted to pH 9.0 with 20 mM sodium tetraborate

Injection: Siphon
Detector: UV 210
Migration time: 13
Internal standard: methyl p-hydroxybenzoate
Voltage: +20 kV

OTHER SUBSTANCES
Simultaneous: acetaminophen, caffeine, chlorpheniramine, dipyrone (sulpyrin), ethenzamide, isopropylantipyrine, naproxen, noscapine, phenacetin, tipepidine, trimetoquinol

KEY WORDS
tablets; powders

REFERENCE
Nishi,H.; Fukuyama,T.; Matsuo,M.; Terabe,S. Effect of surfactant structures on the separation of cold medicine ingredients by micellar electrokinetic chromatography, *J.Pharm.Sci.*, **1990**, *79*, 519–523.

SAMPLE
Matrix: formulations
Sample preparation: Grind granules, add 70 mL MeOH, warm at 40° with occasional shaking, cool, add 20 mL 5 mg/mL methyl p-hydroxybenzoate in MeOH, make up to 100 mL with water, filter (0.45 μm), inject an aliquot.

CAPILLARY ELECTROPHORESIS
Capillary: 65 cm × 50 μm fused-silica (SGE)
Running buffer: 20 mM NaH_2PO_4 containing 100 mM sodium cholate, adjusted to pH 9.0 with 20 mM sodium tetraborate
Injection: Siphon for 10 s at 10 cm height
Detector: UV 210
Migration time: 9
Internal standard: methyl p-hydroxybenzoate (16)
Voltage: +20 kV

OTHER SUBSTANCES
Simultaneous: acetaminophen, caffeine, chlorpheniramine, dibucaine, dipyrone (sulpyrin), ethenzamide, isopropylantipyrine, naproxen, noscapine, phenacetin, tipepidine, trimetoquinol, triprolidine

KEY WORDS
granules

REFERENCE
Nishi,H.; Fukuyama,T.; Matsuo,M.; Terabe,S. Separation and determination of the ingredients of a cold medicine by micellar electrokinetic chromatography with bile salts, *J.Chromatogr.*, **1990**, *498*, 313–323.

SAMPLE
Matrix: solutions

CAPILLARY ELECTROPHORESIS
Capillary: 65.4 cm × 50 μm fused-silica (56 cm to detector) (Hewlett-Packard)
Capillary temperature: 37
Running buffer: 11.7 mM pH 8.9 Borate buffer containing 8.3 mM phosphate and 50 mM sodium dodecyl sulfate
Injection: Injection at 50 mbar.s
Detector: UV 200
Migration time: 4.9
Voltage: 370 V/cm
Model: Hewlett-Packard G1600A [3D]Capillary Electrophoresis system

OTHER SUBSTANCES
Simultaneous: acetaminophen, caffeine, naproxen, noscapine, phenacetin

REFERENCE
Heiger,D.N.; Kaltenbach,P.; Sievet,H.-J. Diode array detection in capillary electrophoresis, *Electrophoresis*, **1994**, *15*, 1234–1247.

SAMPLE
Matrix: solutions
Sample preparation: Inject an aliquot of a solution in MeOH:buffer 10:90 (Buffer was 50 mM sodium dodecyl sulfate containing 5 mM boric acid and 5 mM Na_2HPO_4, adjusted to pH 7.4 with concentrated phosphoric acid.)

CAPILLARY ELECTROPHORESIS
Capillary: 380 mm $\times$ 50 μm (300 mm to detector)
Capillary preparation: Before use wash capillary with 5 M NaOH for 20 min, wash with 500 mM NaOH for 5 min, wash with 100 mM NaOH for 5 min, and wash with water for 5 min. After the completion of a run flush capillary with 500 mM NaOH for 1 min, with 100 mM NaOH for 1 min, and with water for 1 min. Flush capillary with running buffer for 3 min before a run.
Capillary temperature: 30
Running buffer: Isopropanol:buffer 10:90 (Buffer was 50 mM sodium dodecyl sulfate containing 5 mM boric acid and 5 mM Na_2HPO_4, adjusted to pH 7.4 with concentrated phosphoric acid.)
Injection: Under 50 mbar pressure for 2 s.
Detector: UV 210
Migration time: 2.6
Voltage: 790 V/cm
Model: Hewlett-Packard HP3DCE Model G1602A

OTHER SUBSTANCES
Simultaneous: acetaminophen, phenacetin

REFERENCE
Smith,J.T.; Vinjamoori,D.V. Rapid determination of logarithmic partition coefficients between n-octanol and water using micellar electrokinetic capillary chromatography, *J.Chromatogr.B*, **1995**, *669*, 59–66.

SAMPLE
Matrix: solutions
Sample preparation: Inject an aliquot of a solution in running buffer.

CAPILLARY ELECTROPHORESIS
Capillary: 57 cm $\times$ 50 μm fused-silica (50 cm to detector) (Polymicro Technologies)
Capillary temperature: 25
Running buffer: 25 mM NaH_2PO_4 containing 0.83% poly(sodium 10-undecenyl sulfate), adjusted to pH 7.3 with 12.5 mM sodium tetraborate (Preparation of poly(sodium 10-undecenyl sulfate is as follows. React undecylenyl alcohol (1-undecen-11-ol) with one equivalent of chlorosulfonic acid in diethyl ether under nitrogen at -5° to obtain 10-undecenyl hydrogen sulfate, react with one equivalent of NaOH to obtain sodium 10-undecenyl sulfate (J. Microcolumn Sep. 1996, 8, 115). Dissolve 20 g sodium 10-undecenyl sulfate in 40 mL water (degassed by purging with nitrogen for 24 h), add 600 mg potassium persulfate, stir at 70° under nitrogen for 50 h, cool, add 200 mL cold (0°) EtOH, filter, wash the solid with three 10 mL portions of cold EtOH, dry under vacuum at 65° overnight to obtain poly(sodium 10-undecenyl sulfate) (cf. J. Microcolumn Sep. 1992, 4, 509). Purify by dialysis using dialysis tubing with a 1000 MW cut-off (Spectrum Houston) for 24 h.)
Detector: UV 214
Migration time: 7.5
Voltage: 16.1 kV
Model: Beckman P/ACE 2000

OTHER SUBSTANCES
Simultaneous: acetaminophen, caffeine, ethenzamide, phenacetin, trimetoquinol

REFERENCE
Palmer,C.P.; Terabe,S. Micelle polymers as pseudostationary phases in MEKC: Chromatographic performance and chemical selectivity, *Anal.Chem.*, **1997**, *69*, 1852–1860.

SAMPLE
Matrix: solutions
Sample preparation: Inject an aliquot of a 10 µg/mL solution

CAPILLARY ELECTROPHORESIS
Capillary: 50 cm × 75 µm silica
Capillary temperature: 25
Running buffer: 20 mM pH 7.0 Phosphate/borate buffer containing 50 mM sodium dodecyl sulfate
Injection: 5 nL
Detector: UV 208
Migration time: 3.5
Voltage: 20 kV
Model: Otsuka CAPI-3000

OTHER SUBSTANCES
Simultaneous: acetaminophen, ethyl paraben, caffeine, methyl paraben, phenacetin, salicylamide

REFERENCE
Poe,R.B.; Hayashi,Y.; Matsuda,R. Precision-optimization of wavelengths in diode-array detection in separation science, *Anal.Sci.*, **1997**, *13*, 951–961.

SAMPLE
Matrix: solutions

CAPILLARY ELECTROPHORESIS
Capillary: 80 cm × 50 µm fused-silica (Supelco)
Capillary preparation: Before each run rinse capillary with running buffer at 94 kPa for 5 min then with running buffer containing 80 mM sodium dodecyl sulfate at 94 kPa for 2.5 min before each run.
Running buffer: 50 mM pH 8.5 ammonium carbonate buffer
Injection: Pressure injection at 5 kPa (50 mbar) for 8 s into the capillary filled with running buffer containing sodium dodecyl sulfate, electrophoresis is then continued with plain running buffer.
Detector: MS, Perkin-Elmer Sciex API-300 quadrupole, electrospray (ionspray) interface, sheath liquid MeOH:running buffer 50:50 at 4 µL/min, ionspray voltage 5 kV, positive ion mode
Migration time: 23
Voltage: 25 kV (net voltage across capillary = 20 kV (applied voltage - electrospray voltage))
Model: Hewlett Packard 3D CE

OTHER SUBSTANCES
Simultaneous: acetaminophen, acetanilide, caffeine, phenacetin, pyridoxine

REFERENCE
Tanaka,Y.; Kishimoto,Y.; Otsuka,K.; Terabe,S. Strategy for selecting separation solutions in capillary electrophoresis-mass spectrometry, *J.Chromatogr.A*, **1998**, *817*, 49–57.

Halazepam

Molecular formula: $C_{17}H_{12}ClF_3N_2O$
Molecular weight: 352.74
CAS Registry No.: 23092-17-3
Merck Index (12th ed.): 4619

SAMPLE
Matrix: solutions

Sample preparation: Inject an aliquot of a 5 µg/mL solution in running buffer.

CAPILLARY ELECTROPHORESIS
Capillary: 87 cm × 75 µm fused-silica (80 cm to detector) (Beckman)
Capillary preparation: Between runs rinse capillary with running buffer for 2 min then equilibrate for 5 min.
Capillary temperature: 35
Running buffer: 20 mM pH 7 Buffer containing 15 mM sodium cholate and 35 mM sodium deoxycholate (Buffer was 20 mM sodium borate adjusted to pH 7.0 with 20 mM NaH_2PO_4.)
Injection: Pressure injection for 2 s.
Detector: UV 214
Migration time: 24
Voltage: 20 kV
Model: Beckman P/ACE 2100

OTHER SUBSTANCES
Simultaneous: bromazepam, chlordiazepoxide, clobazam, clonazepam, diazepam, flunitrazepam, flurazepam, lorazepam, lormetazepam, nitrazepam, nordazepam, temazepam

REFERENCE
Boonkerd,S.; Detaevernier,M.R.; Vindevogel,J.; Michotte,Y. Migration behaviour of benzodiazepines in micellar electrokinetic chromatography, *J.Chromatogr.A*, **1996**, *756*, 279–286.

Halofantrine

Molecular formula: $C_{26}H_{30}Cl_2F_3NO$
Molecular weight: 500.43
CAS Registry No.: 69756-53-2, 36167-63-2 (HCl)
Merck Index (12th ed.): 4626
Lednicer: 3 76

SAMPLE
Matrix: solutions
Sample preparation: Prepare a solution in running buffer, inject an aliquot.

CAPILLARY ELECTROPHORESIS
Capillary: 60 cm × 75 µm fused-silica (52.4 cm to detector)
Capillary preparation: After each run flush with 500 mM KOH for 2-3 min, flush with water, fill with running buffer
Running buffer: 10 mM Na_2HPO_4 containing 2% heparin sodium (MW 10000, 11% S, Scientific Protein Laboratories, Waunakee, WI) adjusted to pH 5 with phosphoric acid
Injection: Hydrostatic injection
Detector: UV 214
Migration time: 27.1
Model: Waters Quanta 4000

OTHER SUBSTANCES
Also analyzed: anabasine, brompheniramine, bupivacaine, carbinoxamine, chlorcyclizine, chloroquine, chlorpheniramine, dimethindene, doxylamine, enpiroline, hydroxychloroquine, indapamide, mefloquine, nornicotine, pheniramine, primaquine, promethazine, quinacrine, tetramisole

REFERENCE
Stalcup,A.M.; Agyei,N.M. Heparin: a chiral mobile-phase additive for capillary zone electrophoresis, *Anal.Chem.*, **1994**, *66*, 3054–3059.

Haloperidol

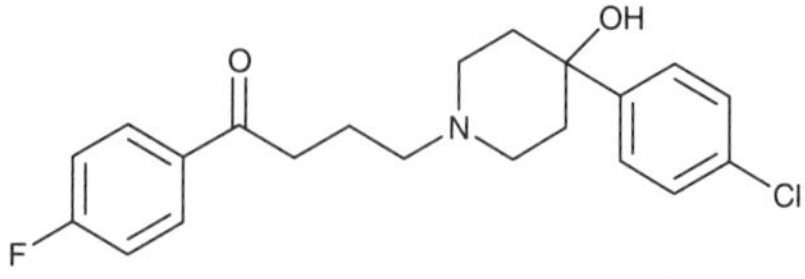

Molecular formula: $C_{21}H_{23}CIFNO_2$
Molecular weight: 375.87
CAS Registry No.: 52-86-8, 74050-97-8 (decanoate)
Merck Index (12th ed.): 4629
Lednicer: 1 306

SAMPLE

Matrix: microsomal incubations
Sample preparation: Condition a Sep-Pak C18 SPE cartridge with 4 mL MeOH and 4 mL water. 2 mL Microsomal incubation + 200 mg zinc sulfate, centrifuge at 1200 g 20 min, add the supernatant to the SPE cartridge, wash with 4 mL water, elute with 4 mL MeOH. Evaporate the eluate to dryness under a stream of nitrogen at 45°, reconstitute the residue in 200 μL MeOH, inject an aliquot.

CAPILLARY ELECTROPHORESIS

Capillary: 65 or 100 cm × 50 μm uncoated
Capillary preparation: Rinse capillary with 10 column volumes of running buffer between injections. Before use rinse capillary with 20 column volumes of 100 mM NaOH, 20 column volumes of water, and 10 column volumes of buffer.
Capillary temperature: 25
Running buffer: MeOH:50 mM ammonium acetate:glacial acetic acid 10:89:1, pH 4.1
Injection: Pressure injection for 15 s.
Detector: MS, Finnigan MAT 900, EB configuration, PATRIC focal plane detector, Analytica electrospray ion source, positive ion mode, needle assembly (end of capillary with coating removed) at ground potential, liquid sheath electrode of isopropanol:water:acetic acid 60:40:1 at 3 μL/min, ESI drying gas nitrogen at 140° and 3.6 L/min, ESI voltage -3400 V, m/z 386 (MH+)
Migration time: 33.5
Voltage: 15 kV
Model: Beckman P/ACE 2100 Model CE
Limit of detection: 100 fmole

OTHER SUBSTANCES

Extracted: metabolites

KEY WORDS

mouse; guinea pig; liver; SPE

REFERENCE

Tomlinson,A.J.; Benson,L.M.; Johnson,K.L.; Naylor,S. Investigation of the metabolic fate of the neuroleptic drug haloperidol by capillary electrophoresis-electrospray ionization mass spectrometry, *J.Chromatogr.*, **1993**, *621*, 239–248.

SAMPLE

Matrix: microsomal incubations
Sample preparation: Condition a Sep-Pak C18 SPE cartridge with 4 mL MeOH and 4 mL water. 2 mL Microsomal incubation + 200 mg zinc sulfate, centrifuge at 1200 g for 20 min, add the supernatant to the SPE cartridge, wash with 4 mL water, elute with 4 mL MeOH. Evaporate the eluate to dryness under a stream of nitrogen at 45°, reconstitute the residue in 200 μL MeOH, inject an aliquot.

CAPILLARY ELECTROPHORESIS

Capillary: 57 cm × 75 μm (50 cm to detector)
Capillary preparation: Between runs wash capillary with 10 column volumes of 100 mM NaOH, with 10 column volumes of water, and with 10 column volumes of running buffer. Before use wash capillary with 20 column volumes of 100 mM NaOH, with 20 column volumes of water, and with 10 column volumes of running buffer.
Capillary temperature: 25

Running buffer: MeOH:50 mM ammonium acetate:1% aqueous acetic acid 10:89:1, pH 4.1
Injection: Pressure injection for 1 s (10 nL)
Detector: UV 214
Migration time: 3.9 (haloperidol), 4.3 (reduced haloperidol)
Voltage: 30 kV
Current: 140 µA
Model: Beckman P/ACE 2100
Limit of detection: 15 µM

OTHER SUBSTANCES
Extracted: metabolites

KEY WORDS
mouse; liver; guinea pig; comparison with HPLC; SPE

REFERENCE
Tomlinson,A.J.; Benson,L.M.; Landers,J.P.; Scanlan,G.F.; Fang,J.; Gorrod,J.W.; Naylor,S. Investigation of the metabolism of the neuroleptic drug haloperidol by capillary electrophoresis, *J.Chromatogr.A*, **1993**, *652*, 417–426.

SAMPLE
Matrix: microsomal incubations, urine
Sample preparation: Microsomal incubations. Condition a Sep-Pak C18 SPE cartridge with 4 mL MeOH and 4 mL water. 2 mL Microsomal incubation + 200 mg zinc sulfate, centrifuge at 1200 g for 20 min, add to the SPE cartridge, wash with 4 mL water, elute with 4 mL MeOH. Evaporate the eluate to dryness under a stream of nitrogen at 45°, reconstitute the residue in 200 µL MeOH, inject an aliquot. Urine. Condition a Sep-Pak C18 SPE cartridge with 4 mL MeOH and 4 mL water. Add 3 mL urine to the SPE cartridge, wash with 4 mL water, elute with 4 mL MeOH. Evaporate the eluate to dryness under a stream of nitrogen at 45°, reconstitute the residue in 10 µL MeOH:water 50:50, inject an aliquot.

CAPILLARY ELECTROPHORESIS
Capillary: 57 cm × 50 µm (50 cm to detector) (UV) or 65 cm × 50 µm (MS) (Beckman)
Capillary preparation: Between analyses wash with 10 column volumes of running buffer. Before use rinse with 20 column volumes of 1 M NaOH, rinse with 20 column volumes of water, and rinse with 10 column volumes of running buffer.
Capillary temperature: 25
Running buffer: MeOH:50 mM ammonium acetate:glacial acetic acid 10:89:1, pH 4.1
Injection: Pressure injection for 1 s
Detector: UV 214 or MS, Finnigan MAT 900, EB configuration, PATRIC focal plane detector, Analytica modified electrospray ion source, positive ion mode, needle assembly at ground potential, liquid sheath electrode isopropanol:water:acetic acid at 3 µL/min, ESI drying gas at 140° and 3.6 L/min, ESI voltage -3400 V, m/z 376
Migration time: 9 (UV), 20 (MS)
Voltage: 20 kV (UV), 15 kV (MS)
Model: Beckman P/ACE 2100

OTHER SUBSTANCES
Extracted: metabolites

KEY WORDS
guinea pig; liver; SPE

REFERENCE
Tomlinson,A.J.; Benson,L.M.; Johnson,K.L.; Naylor,S. Investigation of drug metabolism using capillary electrophoresis with photodiode array detection and online mass spectrometry equipped with an array detector, *Electrophoresis*, **1994**, *15*, 62–71.

SAMPLE
Matrix: solutions
Sample preparation: Prepare a solution in MeOH, inject an aliquot.

CAPILLARY ELECTROPHORESIS
Capillary: 65 cm × 50 μm (Beckman)
Capillary preparation: Between analyses rinse with 5 column volumes of running buffer. Before use rinse capillary with 20 column volumes of 100 mM NaOH, 20 column volumes of water, and 10 column volumes of running buffer.
Capillary temperature: 25
Running buffer: MeOH:50 mM ammonium acetate:acetic acid 10:89:1
Injection: Inject by pressure for 1 s
Detector: MS, Finnigan MAT 900, EB configuration, electrospray ion source, positive-ion mode, nitrogen drying gas at 140° and 3.6 L/min, ESI voltage -3400 V, sheath liquid isopropanol:water:acetic acid 60:40:1, m/z 376
Migration time: 18
Voltage: 15 kV
Model: Beckman P/ACE 2100

REFERENCE
Tomlinson,A.J.; Benson,L.M.; Naylor,S. Effects of organic solvent in the CE and on-line CE-MS analysis of drug metabolite mixtures, *Am.Lab.*, **1994**, *26(9)*, 26–36.

SAMPLE
Matrix: solutions

CAPILLARY ELECTROPHORESIS
Capillary: 57 cm × 50 μm fused-silica
Capillary preparation: Between analyses rinse with 2 capillary volumes of MeOH and 2 capillary volumes of running buffer. Rinse new capillaries with 10 capillary volumes of 1 M NaOH, 1 M HCl, water, MeOH, and running buffer.
Capillary temperature: 25
Running buffer: MeOH:5 mM ammonium acetate:acetic acid 50:49:1
Injection: Pressure injection for 1-10 s.
Detector: UV 214
Migration time: 13
Voltage: 30 kV
Model: Beckman P/ACE 2100

REFERENCE
Naylor,S.; Benson,L.M.; Tomlinson,A.J. Application of capillary electrophoresis and related techniques to drug metabolism studies, *J.Chromatogr.A*, **1996**, *735*, 415–438.

SAMPLE
Matrix: solutions
Sample preparation: Inject an aliquot of a solution in 40 mM pH 2.5 phosphate buffer.

CAPILLARY ELECTROPHORESIS
Capillary: 37 cm × 50 μm neutral-coated capillary (30 cm to detector) (Beckman)
Running buffer: 40 mM pH 2.5 Phosphate buffer containing 10 mM heptakis(2,6-di-O-methyl)-β-cyclodextrin
Injection: Pressure injection for 1 s (1.78 nL).
Detector: UV 200
Migration time: 9.67 (haloperidol), 9.88, 10.05 (reduced haloperidol enantiomers)
Voltage: 22 kV
Model: Beckman P/ACE 2000

KEY WORDS
chiral; coated capillary

REFERENCE
Wu,H.L.; Otsuka,K.; Terabe,S. Chiral separation of reduced haloperidol by capillary zone electrophoresis with heptakis (2,6-di-O-methyl)-β-cyclodextrin, *J.Liq.Chromatogr.Rel.Technol.*, **1996**, *19*, 1567–1577.

Haloxazolam

Molecular formula: $C_{17}H_{14}BrFN_2O_2$
Molecular weight: 377.21
CAS Registry No.: 59128-97-1
Merck Index (12th ed.): 4635

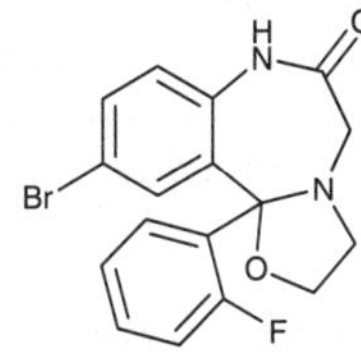

SAMPLE
Matrix: solutions

CAPILLARY ELECTROPHORESIS
Capillary: 60 cm × 50 μm fused-silica (40 cm to detector), C18 coated (Supelco)
Capillary preparation: Rinse new capillaries with 250 μL 1 M NaOH, 250 μL 100 mM NaOH, water, MeOH, water, and running buffer then condition with running buffer for at least 15 min. Carry out a similar procedure between runs.
Running buffer: 100 mM pH 8.5 Borate buffer containing 25 mM sodium dodecyl sulfate and 5 M urea
Injection: Electrokinetic injection at 5 kV for 5 s.
Detector: UV 254
Migration time: 40
Voltage: 18 kV
Model: Jasco

OTHER SUBSTANCES
Simultaneous: alprazolam, amobarbital, barbital, bromazepam, clonazepam, clotiazepam, cloxazolam, diazepam, estazolam, etizolam, fludiazepam, flunitrazepam, flurazepam, medazepam, mephobarbital, metharbital, nimetazepam, nitrazepam, oxazepam, pentobarbital, phenobarbital, secobarbital, triazolam

KEY WORDS
coated capillary

REFERENCE
Jinno,K.; IIan,Y.; Nakamura,M. Analysis of anxiolytic drugs by capillary electrophoresis with bare and coated capillaries, *J.Capillary Electrophor.*, **1996**, *3*, 139–145.

SAMPLE
Matrix: solutions

CAPILLARY ELECTROPHORESIS
Capillary: 60 cm × 75 μm coated fused-silica (40 cm to detector) (Supelco)
Capillary preparation: Coat column as follows. Adjust the pH of 20 mL water to 3.5 with acetic acid, add 80 μL 3-(trimethoxysilyl)propyl methacrylate (3-methacryloxypropyltrimethoxysilane), mix, suck into capillary, let stand at room temperature for 1 h, remove the solution, wash with water. Fill the capillary with a deaerated 4% acrylamide solution containing 1 mg/mL N,N,N',N'-tetramethylethylenediamine and 1 mg/mL ammonium persulfate, let stand for 3 h, remove excess solution by aspiration, rinse with water, remove water by aspiration, dry at 35° (cf. J. Chromatogr. 1985, 347, 191).
Running buffer: 100 mM pH 8.5 borate buffer containing 10 mM sodium dodecyl sulfate and 5 M urea
Injection: Electromigration at 5 kV for 5 s.
Detector: UV 240, UV 254
Migration time: 8
Voltage: 18 kV
Model: Jasco 890-CE

OTHER SUBSTANCES
Simultaneous: alprazolam, clotiazepam, cloxazolam, diazepam, estazolam, etizolam, fludiazepam, flurazepam, oxazepam

KEY WORDS
injection at cathode; coated capillary

REFERENCE
Jinno,K.; Han,Y.; Sawada,H. Analysis of toxic drugs by capillary electrophoresis using polyacrylamide-coated columns, *Electrophoresis*, **1997**, *18*, 284–286.

SAMPLE
Matrix: solutions
Sample preparation: Inject an aliquot of a 100 μg/mL solution in running buffer.

CAPILLARY ELECTROPHORESIS
Capillary: 60 cm × 50 μm acrylamide-coated fused-silica (40 cm to detector) (Supelco)
Capillary preparation: Adjust the pH of 20 mL water to 3.5 with acetic acid, add 80 μL 3-(trimethoxysilyl)propyl methacrylate (3-methacryloxypropyltrimethoxysilane), mix, suck into capillary, let stand at room temperature for 1 h, remove the solution, wash with water. Fill the capillary with a deaerated 3-4% acrylamide solution containing 1 μL/mL N,N,N',N'-tetramethylethylenediamine and 1 mg/mL ammonium persulfate, let stand for 3 h, remove excess solution by aspiration, rinse with water, remove water by aspiration, dry at 35° (cf. J. Chromatogr. 1985, 347, 191).
Running buffer: MeCN:buffer 5:95 (Buffer was 100 mM borate containing 5 M urea and 10 mM sodium dodecyl sulfate, adjusted to pH 8.5 with phosphate.)
Injection: Electrokinetic injection at 5 kV for 5 s.
Detector: UV 254
Migration time: 15.2
Voltage: 18 kV
Model: Jasco Model 870-CE

OTHER SUBSTANCES
Simultaneous: alprazolam, amobarbital, barbital, bromazepam, clonazepam, clotiazepam, cloxazolam, diazepam, estazolam, etizolam, fludiazepam, flunitrazepam, flurazepam, medazepam, mephobarbital, metharbital, nimetazepam, nitrazepam, oxazepam, pentobarbital, phenobarbital, secobarbital, triazolam

KEY WORDS
coated capillary; detector at anode

REFERENCE
Jinno,K.; Han,Y.; Sawada,H.; Taniguchi,M. Capillary electrophoretic separation of toxic drugs using a polyacrylamide-coated capillary, *Chromatographia*, **1997**, *46*, 309–314.

Heparin

Molecular weight: 6000-30000
CAS Registry No.: 9005-49-6, 9041-08-1 (Na salt)
Merck Index (12th ed.): 4685

SAMPLE
Matrix: bulk

Sample preparation: Dissolve 5 mg heparin and 10 mg 1-maltoheptaosyl-1,5-diaminonaphthalene in 25 μL 10 mM phosphoric acid, add 5 μL 2 M sodium cyanoborohydride, heat at 65° for 3 h, make up to 500 μL with water, inject an aliquot. (Synthesis of 1-maltoheptaosyl-1,5-diaminonaphthalene is as follows. Heat 500 mg maltoheptaose and 362.5 mg 1,5-diaminonaphthalene in 10 mL EtOH:20 mM phosphoric acid 50:50 at 85° for 30 min, add sodium cyanoborohydride to a final concentration of 300 mM, heat at 85° for 3 h, remove the solvent by evaporation under a stream of nitrogen, centrifuge, purify the supernatant three times by precipitation from 90% acetone with drying under nitrogen, store at 4°.)

CAPILLARY ELECTROPHORESIS
Capillary: 60 cm × 50 μm (50 cm to detector) (Polymicro Technologies)
Capillary preparation: Coat capillary as follows. Adjust the pH of 20 mL water to 3.5 with acetic acid, add 80 μL 3-(trimethoxysilyl)propyl methacrylate (3-methacryloxypropyltrimethoxysilane), mix, suck into capillary, let stand at room temperature for 1 h, remove the solution, wash with water. Fill the capillary with a deaerated 3-4% acrylamide solution containing 1 μL/mL N,N,N',N'-tetramethylethylenediamine and 1 mg/mL potassium persulfate, let stand for 30 min, remove excess solution by aspiration, rinse with water, remove water by aspiration, dry at 35° (J. Chromatogr. 1985, 347, 191).
Running buffer: 10 mM pH 3.1 Sodium citrate
Detector: F ex 325 (He-Cd laser) em >360 (Cut-off filter)
Migration time: 15-42 (depending on structure)
Voltage: 21 kV
Model: laboratory-constructed

KEY WORDS
derivatization; coated capillary

REFERENCE
Sudor,J.; Novotny,M.V. End-label free-solution electrophoresis of the low molecular weight heparins, *Anal.Chem.*, **1997**, *69*, 3199–3204.

SAMPLE
Matrix: solutions
Sample preparation: Inject an aliquot of a solution in 20 mM pH 7.0 acetate buffer.

CAPILLARY ELECTROPHORESIS
Capillary: 57 cm × 75 μm fused-silica (50 cm to detector)
Capillary preparation: Before each run wash capillary with 100 mM NaOH for 1 min and with running buffer for 4 min.
Capillary temperature: 25
Running buffer: 15 mM pH 3.50 Sodium phosphate buffer
Injection: Pressure injection for 25 s (15 nL)
Detector: UV 231
Migration time: 3-15 (depending on structure and degree of sulfation of disaccharide)
Voltage: 20 kV
Model: Beckman P/ACE system 5510

KEY WORDS
detector is at anode

REFERENCE
Karamanos,N.K.; Vanky,P.; Tzanakakis,G.N.; Hjerpe,A. High-performance capillary electrophoresis method to characterize heparin and heparin sulfate disaccharides, *Electrophoresis*, **1996**, *17*, 391–395.

SAMPLE
Matrix: solutions

CAPILLARY ELECTROPHORESIS
Capillary: 57 cm × 50 μm fused-silica (50 cm to detector) (Beckman)
Capillary preparation: Before each run condition capillary for 2 min at reversed polarity and 2 min at normal polarity. After each run wash with 1 M NaOH for 2 min. Condition new

capillaries by washing with 500 mM NaOH for 20 min and with water for 20 min then rinsing with running buffer for 30 min.
Capillary temperature: 25 ± 1
Running buffer: 50 mM pH 10.4 Sodium borate containing 10 mM boric acid and 200 mM triethylamine
Injection: Pressure injection at 0.5 psi for 5 s (45 nL).
Detector: UV 214
Migration time: 5-27 (depending on structure)
Voltage: 30 kV
Model: Beckman P/ACE 2100

REFERENCE

Scapol,L.; Marchi,E.; Viscomi,G.C. Capillary electrophoresis of heparin and dermatan sulfate unsaturated disaccharides with triethylamine and acetonitrile as electrolyte additives, *J.Chromatogr.A*, **1996**, *735*, 367–374.

Heptaminol

Molecular formula: $C_8H_{19}NO$
Molecular weight: 145.25
CAS Registry No.: 372-66-7, 57249-13-5 (5'-adenylate), 543-15-7 (HCl)
Merck Index (12th ed.): 4691

SAMPLE

Matrix: solutions
Sample preparation: Mix a 100 μL aliquot of a 0.1-1 mM solution in 10 mM HCl with 500 μL 100 mM pH 9.5 borate buffer, 100 μL 2.5 mM N-acetyl-L-cysteine in 10 mM HCl, and 200 μL 5 mM o-phthalaldehyde in EtOH, let stand for 10 min, inject an aliquot.

CAPILLARY ELECTROPHORESIS

Capillary: 70 cm × 75 μm fused-silica
Capillary preparation: Before use wash capillary at 60° for 6 min each with water, 1 M NaOH, 100 mM NaOH, water, and running buffer.
Capillary temperature: 25
Running buffer: 100 mM pH 9.5 Borate buffer containing 10 mM sodium dodecyl sulfate
Injection: Hydrodynamic injection for 2 s.
Detector: UV 335
Migration time: 18.2, 18.7 (enantiomers)
Voltage: 20 kV
Model: Spectra Phoresis 1000 (Thermo Separation Products)

OTHER SUBSTANCES

Simultaneous: amphetamine, phenylpropanolamine

KEY WORDS

derivatization; chiral; comparison with HPLC; comparison with other derivatizing reagents

REFERENCE

Leroy,P.; Bellucci,L.; Nicolas,A. Chiral derivatization for separation of racemic amino and thiol drugs by liquid chromatography and capillary electrophoresis, *Chirality*, **1995**, *7*, 235–242.

Hexobarbital

Molecular formula: $C_{12}H_{16}N_2O_3$
Molecular weight: 236.27
CAS Registry No.: 56-29-1, 50-09-9 (sodium salt)
Merck Index (12th ed.): 4742
Lednicer: 1 273

SAMPLE
Matrix: solutions
Sample preparation: Inject an aliquot of a 20 μg/mL solution

CAPILLARY ELECTROPHORESIS
Capillary: 44 cm × 50 μm fused-silica (37 cm to detector) (Supelco)
Capillary preparation: Before each run rinse capillary with running buffer for 3 min. At the
start of each day rinse capillary with running buffer for 10 min. Treat new capillaries with 1
M NaOH, 100 mM NaOH, water, and running buffer.
Capillary temperature: 25
Running buffer: 100 mM Phosphoric acid adjusted to pH 5 with triethanolamine containing 10
mM carboxymethyl-β-cyclodextrin (Cyclolab, Budapest) and 10 mM heptakis(2,6-di-O-methyl)-
β-cyclodextrin (Sigma)
Injection: Hydrodynamic injection for 5 s (13.3 nL).
Detector: UV 210
Voltage: -25 kV
Model: Spectraphoresis 1000 CE

KEY WORDS
detector at anode; chiral; R_s = 3.2

REFERENCE
Fillet,M.; Fotsing,L.; Crommen,J. Enantioseparation of uncharged compounds by capillary electrophoresis using mixtures of anionic and neutral β-cyclodextrin derivatives, *J.Chromatogr.A*, **1998**, *817*, 113–119.

SAMPLE
Matrix: solutions

CAPILLARY ELECTROPHORESIS
Capillary: 95 cm × 50 μm fused-silica coated with Chirasil-Dex (80 cm to detector)
Capillary preparation: Conduct all reactions under nitrogen. Dissolve 3.395 g dry β-cyclodextrin in 75 mL anhydrous DMSO, add 600 mg powdered NaOH, stir vigorously at room temperature for 1 h, add 1.78 mL 5-bromo-1-pentene in 10 mL anhydrous DMSO dropwise, stir at
room temperature for 48 h, filter. Concentrate the filtrate almost to dryness at 0.01 torr and
60°, dilute with not more than 10 mL MeOH, precipitate with 200 mL diethyl ether to obtain
5-pent-1-enylated β-cyclodextrin, dry at 0.01 torr and 60°. Wash 3.024 g NaH (55-60% in paraffin) repeatedly with n-hexane to remove the paraffin. Place in an ice-cooled flask and add 1.5
g 5-pent-1-enylated β-cyclodextrin in 50 mL anhydrous DMF (Caution! Vigorous evolution of
hydrogen). When the reaction has stopped add 5.9 mL MeI at a bath temperature of 20°, stir
for 30 min, add 1.5 g 5-pent-1-enylated β-cyclodextrin in 50 mL anhydrous DMF, add 5.9 mL
MeI (?), stir for 1 h. Carefully decant from unreacted sodium hydride and pour into 200 mL
water, extract three times with 70 mL portions of chloroform. Combine the chloroform layers
and wash them 3 times with 15 mL portions of water, dry over anhydrous sodium sulfate,
concentrate, purify the residue by chromatography on a 200 × 50 column of Sephadex LH 20
with dichloromethane:methanol 2:1. Collect the yellow fraction and repeat the purification until
the dry solid is white, dry at 0.01 torr at 40° over phosphorus pentoxide for 72 h. This is
permethylated-5-pent-1-enyl β-cyclodextrin. Reflux 3.0 g (1 mmole) poly(dimethylsiloxane) containing 5% Si-H groups, 0.72 g permethylated-5-pent-1-enyl β-cyclodextrin, and 100 mL dry
toluene. Add a few drops of a semi-concentrated solution of hydrogen hexachloropalatinate(IV)
(H_2PtCl_6) in anhydrous THF at 150 min intervals. After 24 h evaporate to dryness in a rotary
evaporator under reduced pressure, take up the residue in anhydrous MeOH. Separate the
turbid MeOH phase from a blackish phase and evaporate it. Extract the residue with petroleum

ether, filter, evaporate to dryness to obtain Chirasil-Dex (J. High Res. Chromatogr. 1990, 13, 713). Heat the capillary at 250° with a low hydrogen flow for 2.5 h, coat with 0.4% Chirasil-Dex in diethyl ether. Heat the capillary at 190° with a reduced flow of hydrogen for 24 h, rinse with MeOH, rinse with dichloromethane, rinse with diethyl ether, rinse with 2 mL water under pressure for 1 h, rinse with 4 mL buffer under pressure for 2 h. (Dry β-cyclodextrin under reduced pressure over phosphorus pentoxide at 80° for 48 h.)
Capillary temperature: 25
Running buffer: 20 mM pH 7 borate/phosphate buffer
Injection: Pressure injection at 40 mbar for 3 s.
Detector: UV 220
Migration time: 10, 18 (enantiomers)
Voltage: 30 kV
Model: Prince

KEY WORDS
chiral; coated capillary

REFERENCE
Jakubetz,H.; Juza,M.; Schurig,V. Dual chiral recognition system involving cyclodextrin derivatives in capillary electrophoresis. II. Enhancement of enantioselectivity, *Electrophoresis*, **1998**, *19*, 738–744.

Hirudin
CAS Registry No.: 8001-27-2
Merck Index (12th ed.): 4754

SAMPLE
Matrix: solutions

CAPILLARY ELECTROPHORESIS
Capillary: 117 cm × 50 μm fused-silica (110 cm to detector) (BGB Analytik AG, Rothenfluh, Switzerland)
Capillary temperature: 25
Running buffer: 20 mM pH 8.3 Tricine containing 10 mM sodium tetraborate and ca. 270 μM 1,4-diaminobutane
Injection: Pressure injection at 138 kPa for 5 s.
Detector: UV 200
Migration time: 28
Voltage: 28 kV
Current: 8 μA
Model: Beckman P/ACE 5010
Limit of quantitation: 50 μg/mL

OTHER SUBSTANCES
Simultaneous: degradation products

REFERENCE
Gietz,U.; Alder,R.; Langguth,P.; Arvinte,T.; Merkle,H.P. Chemical degradation kinetics of recombinant hirudin (HV1) in aqueous solution: Effect of pH, *Pharm.Res.*, **1998**, *15*, 1456–1462.

Histamine

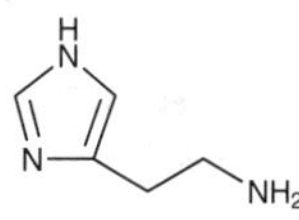

Molecular formula: $C_5H_9N_3$
Molecular weight: 111.15
CAS Registry No.: 51-45-6, 51-74-1 (phosphate)
Merck Index (12th ed.): 4756

SAMPLE

Matrix: cheese
Sample preparation: Homogenize 4 parts cheese with 3 parts 100 mM HCl, suspend 230 mg paste in 5 mL 100 mM HCl, centrifuge, extract the residue twice with 5 mL portions of 100 mM HCl. Filter the supernatants and neutralize them with 200 mM sodium carbonate solution. Remove a 1 mL aliquot and add it to 1 mL 83 μg/mL fluorescein isothiocyanate in acetone, let stand in the dark for 4 h (J.Chromatogr. 1991, 559, 183), dilute 100 000-fold, inject an aliquot.

CAPILLARY ELECTROPHORESIS

Capillary: 80 cm × 50 μm fused-silica (50 cm to detector) (Polymicro Technologies)
Capillary preparation: Rinse with 100 mM NaOH for 3 min, with water for 3 min, and with running buffer for 3 min.
Running buffer: 100 mM Boric acid containing 100 mM sodium dodecyl sulfate, pH adjusted to 9.2 with NaOH
Injection: Hydrodynamic injection for 2 s (17.5 nL)
Detector: F ex 488 (7 mW argon ion laser)
Migration time: 9.65
Voltage: 24 kV
Model: TSP Spectra-Phoresis 100
Limit of detection: 0.1 nM

OTHER SUBSTANCES

Extracted: amino acids, cadaverine, β-phenylethylamine, putrescine, tryptamine, tyramine

KEY WORDS

derivatization

REFERENCE

Nouadje,G.; Nertz,M.; Verdeguer,P.; Couderc,F. Ball-lens laser-induced fluorescence detector as an easy-to-use highly sensitive detector for capillary electrophoresis. Application to the identification of biogenic amines in dairy products, *J.Chromatogr.A*, **1995**, *717*, 335–343.

SAMPLE

Matrix: food
Sample preparation: Homogenize 5 g fish with 20 mL water and 10 mL HCl, dilute to 50 mL with water. Remove a 5 mL aliquot and adjust pH to 6-7 with 1 M NaOH, add 25 mL buffer, add to a 100 × 10 column of 100-200 mesh Type I Amberlite CG-50 made up in buffer, wash with 80 mL buffer:water 25:75, elute with 1 M HCl. Collect 20 mL eluate and neutralize with 1 M NaOH, make up to 50 mL with water, inject an aliquot. (Buffer was 100 mM pH 5.6 sodium acetate.)

CAPILLARY ELECTROPHORESIS

Capillary: 75 cm × 50 μm fused-silica (50 cm to detector)
Capillary preparation: Rinse with running buffer before each run.
Running buffer: 20 mM Sodium tetraborate containing 20 mM NaH_2PO_4, 20 mM sodium dodecyl sulfate, 2 mM o-phthalaldehyde, and 2 mM N-acetylcysteine, pH adjusted to 10 with 1 M NaOH (Anode buffer was 60 mM pH 10 phosphate/borate buffer.)
Injection: Hydrostatic injection at 10 cm for 10 s.
Detector: F ex 340 em 450; UV 340
Migration time: 12.7
Voltage: 25 kV
Model: Jasco CE-800
Limit of quantitation: 20 μM

Limit of detection: 1 μM

OTHER SUBSTANCES
Extracted: cadaverine, spermidine (UV only), tyramine

KEY WORDS
fish; derivatization; on-capillary derivatization; SPE

REFERENCE
Oguri,S.; Watanabe,S.; Abe,S. Determination of histamine and some other amines by high-performance capillary electrophoresis with on-line mode in-capillary derivatization, *J.Chromatogr.A*, **1997**, *790*, 177–183.

SAMPLE
Matrix: formulations
Sample preparation: Dilute with water containing IS, inject an aliquot.

CAPILLARY ELECTROPHORESIS
Capillary: 37 cm × 75 μm fused-silica (30 cm to detector) (Composite Metal Services, Hallow, UK)
Capillary preparation: Between injections rinse capillary with 100 mM NaOH for 1 min and with running buffer for 1 min.
Capillary temperature: 30
Running buffer: 25 mM NaH_2PO_4 adjusted to pH 2.5 with concentrated phosphoric acid
Injection: Pressure injection for 5 s.
Detector: UV 200
Migration time: 2.1
Internal standard: imidazole (2.5)
Voltage: 15 kV
Model: Beckman P/ACE 5000
Limit of quantitation: 150 ng/mL
Limit of detection: 50 ng/mL

OTHER SUBSTANCES
Simultaneous: benzalkonium chloride, histidine

REFERENCE
Altria,K.D.; Elgey,J.; Howells,J.S. Validated capillary electrophoretic method for the quantitative analysis of histamine acid phosphate and/or benzalkonium chloride, *J.Chromatogr.B*, **1996**, *686*, 111–117.

SAMPLE
Matrix: solutions

CAPILLARY ELECTROPHORESIS
Capillary: 44 cm × 50 μm
Capillary preparation: Wash with running buffer for 3 min before each injection. After each injection wash with water for 2 min, 100 mM NaOH for 3 min, and water for 2 min. Wash new capillaries at 60° with 1 M NaOH for 2 h and wash at 25° with water for 30 min.
Capillary temperature: 25
Running buffer: MeCN:100 mM pH 7.5 ammonium acetate buffer 30:70
Injection: Hydrodynamic injection for 1 s.
Detector: UV 215
Migration time: 5.14
Voltage: 10 kV
Current: <100 μA
Model: Spectra-Physics Model 1000

OTHER SUBSTANCES
Simultaneous: benzylamine, 2-phenylethylamine, serotonin, tryptamine, tyramine

---

REFERENCE
Lin,W.-c.; Lin,C.-E.; Lin,E.-C. Capillary zone electrophoretic separation of biogenic amines: influence of organic modifier, *J.Chromatogr.A*, **1996**, *755*, 142–146.

SAMPLE
Matrix: solutions

CAPILLARY ELECTROPHORESIS
Capillary: 57 cm × 75 μm fused-silica (50 cm to detector) (Scientific Glass Engineering)
Capillary preparation: Before each injection rinse capillary at 130 kPa with 100 mM HCl for 1 min, with 10 mM KOH for 1 min, and with running buffer for 2 min.
Running buffer: 100 mM pH 8.3 Tris-borate buffer containing 50 μM cetyltrimethylammonium bromide and 0.005% poly(vinyl alcohol) (Hoechst)
Injection: Pressure injection at 3.3 kPa for 10 s.
Detector: UV 214
Migration time: 7.5
Internal standard: clenbuterol (10)
Voltage: 20 kV
Model: Beckman P/ACE 2200

REFERENCE
van der Schans,M.J.; Reijenga,J.C.; Everaerts,F.M. Quality control of histamine and methacholine in diagnostic solutions with capillary electrophoresis, *J.Chromatogr.A*, **1996**, *735*, 387 393.

SAMPLE
Matrix: tissue
Sample preparation: Blend 19 g fish with 40-50 mL MeOH:water 50:50 at low speed for 30 s and at high speed for 2 min, warm at 60° for 15-20 min, cool to room temperature, make up to 100 mL with MeOH:water 50:50, filter, inject an aliquot.

CAPILLARY ELECTROPHORESIS
Capillary: 72 cm × 50 μm fused-silica
Capillary preparation: Before each injection rinse with 100 mM NaOH for 2 min, rinse with water for 3 min, and equilibrate with running buffer for 5 min.
Capillary temperature: 35
Running buffer: 20 mM pH 2.5 Citrate buffer
Injection: Inject using vacuum for 5 s (17.5 nL)
Detector: UV 210
Migration time: 3
Voltage: 375 V/cm
Model: Applied Biosystems Model 270A-HT
Limit of quantitation: 0.5 ppm

KEY WORDS
fish

REFERENCE
Mopper,B.; Sciacchitano,C.J. Capillary zone electrophoretic determination of histamine in fish, *J.AOAC Int.*, **1994**, *77*, 881–884.

SAMPLE
Matrix: tissue
Sample preparation: Shake 20 g fish muscle with 100-150 mL 600 mM perchloric acid for 1 h, filter (paper), wash through with perchloric acid, make up to 200 mL. Dilute a 1-5 mL aliquot to 5 mL with perchloric acid, add 125 μL 400 mg/mL 1,7-heptanediamine, add 1 mL 9.8 M NaOH, vortex briefly, add 100 μL benzoyl chloride, shake for 2.5 min, sonicate for 15-10 min, add 2.5 g NaCl, shake for 1 min, extract twice with 3 mL portions of diethyl ether. Combine the organic layers and evaporate them to dryness under a stream of air, reconstitute the residue in 400 μL MeOH:water 50:50, inject an aliquot.

CAPILLARY ELECTROPHORESIS
Capillary: 43 cm × 75 μm fused-silica (36 cm to detector) (CElect FS75, Supelco)
Capillary preparation: Wash capillary with running buffer for 3 min before each injection. After each injection wash capillary with water for 3 min, with 100 mM NaOH for 3 min, and with water for 3 min. Wash new capillaries with 1 M NaOH at 30° for 1 h and with water at 30° for 1 h.

Capillary temperature: 30 ± 0.01
Running buffer: MeOH:15 mM sodium tetraborate 25:75 containing 40 mM sodium dodecyl
   sulfate, pH 9.45
Injection: Vacuum injection for 2.5 s (15 nL).
Detector: UV 200
Migration time: 27.88
Internal standard: 1,7-heptanediamine (15.97)
Voltage: 15 kV
Current: 48 μA
Model: Spectraphoresis 2000 (Thermo Separation Products, Fremont CA)
Limit of detection: 2.1 μg/g

OTHER SUBSTANCES
Extracted: cadaverine, putrescine, spermidine, spermine, tryptamine, tyramine

KEY WORDS
derivatization; fish; muscle

REFERENCE
Krízek,M.; Pelikánová,T. Determination of seven biogenic amines in foods by micellar electrokinetic capillary
   chromatography, *J.Chromatogr.A*, **1998**, *815*, 243–250.

Homatropine

Molecular formula: $C_{16}H_{21}NO_3$
Molecular weight: 275.35
CAS Registry No.: 87-00-3, 51-56-9 (HBr)
Merck Index (12th ed.): 4766

SAMPLE
Matrix: formulations
Sample preparation: Dilute formulation with water to a homatropine concentration of about
   50 μg/mL.

CAPILLARY ELECTROPHORESIS
Capillary: 64.5 cm × 50 μm fused-silica (56 cm to detector)
Capillary preparation: Between analyses wash capillary with 100 mM NaOH for 2 min, wash
   with water for 2 min, and equilibrate with running buffer for 3.5 min. At the start of each day
   flush with 1 M NaOH for 15 min and with water for 10 min.
Capillary temperature: 25
Running buffer: 100 mM pH 7 Tris-phosphate buffer
Injection: Pressure injection at 25 mbar for 20 s (12 nL), ramp voltage at 500 V/s.
Detector: UV 195
Migration time: 3.564
Voltage: 30 kV
Current: 63 μA
Model: Hewlett-Packard HP[3D]
Limit of quantitation: 3 μg/mL
Limit of detection: 1 μg/mL

OTHER SUBSTANCES
Simultaneous: atropine, scopolamine

KEY WORDS
ophthalmic solutions

REFERENCE

Cherkaoui,S.; Mateus,L.; Christen,P.; Veuthey,J.-L. Development and validation of a capillary zone electrophoresis method for the determination of atropine, homatropine and scopalamine in ophthalmic solutions, *J.Chromatogr.B*, **1997**, *696*, 283–290.

SAMPLE
Matrix: solutions

CAPILLARY ELECTROPHORESIS
Capillary: 60 cm × 50 μm AccuSep (52.5 cm to detector) (Waters)
Capillary preparation: Before injection rinse capillary with 100 mM NaOH for 3 min and with running buffer for 3 min. Rinse new capillaries with 500 mM NaOH for 5 min
Running buffer: 50 mM pH 7.0 Na_2HPO_4 containing 25 mM (S)-N- dodecoxycarbonylvaline (Prepare (S)-N-dodecoxycarbonylvaline as follows. Prepare dodecyl chloroformate by reacting 1-dodecanol with 0.33 equivalents of triphosgene in dichloromethane solution in the presence of pyridine. Add dodecyl chloroformate dropwise to (S)-valine in 1 M NaOH solution, filter, wash with hexane, recrystallize from ether/petroleum ether.)
Injection: Hydrostatic injection 2 s
Detector: UV 214
Voltage: +12 kV
Model: Waters Quanta 4000 or 4000E

OTHER SUBSTANCES
Also analyzed: atenolol, bupivacaine, ephedrine, ketamine, metoprolol, N-methylpseudoephedrine, norephedrine, norphenylephrine, octopamine, pindolol, terbutaline

KEY WORDS
chiral; $\alpha = 1.03$

REFERENCE

Mazzeo,J.R.; Grover,E.R.; Swartz,M.E.; Petersen,J.S. Novel chiral surfactant for the separation of enantiomers by micellar electrokinetic capillary chromatography, *J.Chromatogr.A*, **1994**, *680*, 125–135.

SAMPLE
Matrix: solutions
Sample preparation: Inject an aliquot of a 100 μg/mL solution in water:running buffer 50:50.

CAPILLARY ELECTROPHORESIS
Capillary: 44.5 cm × 50 μm acrylamide-coated fused-silica (Bio-Rad)
Capillary temperature: 30
Running buffer: 100 mM NaH_2PO_4 containing 15 mM gamma-cyclodextrin, adjusted to pH 2.5 with phosphoric acid
Injection: Electrokinetic injection at 8 kV for 6 s.
Detector: UV 200
Migration time: 8.54
Voltage: 14 kV
Model: Bio-Rad BioFocus 3000

OTHER SUBSTANCES
Also analyzed: albuterol, alprenolol, atenolol, atropine, baclofen, bamethan, benserazide, biperiden, bisoprolol, bupivacaine, bupranolol, butetamate, carazolol, carbuterol, carvedilol, celiprolol, chloroquine, chlorpheniramine (chlorphenamine), clidinium bromide, clobutinol, disopyramide, dobutamine, flecainide, ipratropium bromide, isoproterenol, isothipendyl, ketamine, mefloquine, mequitazine, metaproterenol (orciprenaline), metipranolol, nafronyl (naftidrofuryl), nefopam, ofloxacin, orphenadrine, oxomemazine, oxprenolol, phenoxybenzamine, pholedrine, pindolol, pirbuterol, prilocaine, promethazine, propafenone, propranolol, sotalol, synephrine, terbutaline, tetrahydrozoline (tetryzoline), tocainide, trihexyphenidyl, trimeprazine (alimemazine), trimipramine, tropicamide, verapamil, zopiclone

KEY WORDS
coated capillary; achiral

REFERENCE
Koppenhoefer,B.; Epperlein,U.; Christian,B.; Yibing,J.; Yuying,C.; Bingcheng,L. Separation of enantiomers of drugs by capillary electrophoresis. I. γ-Cyclodextrin as chiral solvating agent, *J.Chromatogr.A*, **1995**, *717*, 181–190.

SAMPLE
Matrix: solutions

CAPILLARY ELECTROPHORESIS
Capillary: 60 cm × 75 μm fused-silica (51 cm to detector) (Supelco)
Capillary preparation: At the end of each day wash capillary with 200 mM NaOH for 10 min and with water for 10 min. Condition new capillaries by washing with 200 mM NaOH for 10 min, with water for 10 min, and with running buffer for 10 min.
Running buffer: 50 mM pH 7.3 Sodium borate buffer containing 3.3% succinyl β-cyclodextrin (Wacker Chemie, Munich, Germany)
Injection: Hydrodynamic injection at 25 mbar for 6 s.
Detector: UV 208
Migration time: 38.73, 40.76 (enantiomers)
Voltage: 15 kV
Model: Prince

OTHER SUBSTANCES
Simultaneous: alprenolol, chlorthalidone, ephedrine, etilefrin, methoxamine, norephedrine, octopamine, propranolol, synephrine, trihexyphenidyl

KEY WORDS
chiral

REFERENCE
Schmid,M.G.; Wirnsberger,K.; Gübitz,G. Chiral separation of drug enantiomers by capillary electrophoresis using succinyl-β-cyclodextrin, *Pharmazie*, **1996**, *51*, 852–854.

SAMPLE
Matrix: solutions
Sample preparation: Inject an aliquot of a solution in MeOH:water 10:90.

CAPILLARY ELECTROPHORESIS
Capillary: 64.5 cm × 75 μm (56 cm to detector)
Capillary preparation: Before each run flush the capillary with 100 mM NaOH for 2 min, with water for 2 min, and equilibrate with running buffer for 3.5 min. At the start of each day wash capillary with 100 mM NaOH for 15 min and with water for 10 min.
Capillary temperature: 25
Running buffer: MeCN:buffer 10:90 (Buffer was 30 mM pH 8.5 phosphate-borate buffer containing 50 mM sodium dodecyl sulfate.)
Injection: Pressure injection at 25 mbar for 10 s (ca. 30 nL), ramp to operating voltage at 500 V/s.
Detector: UV 195
Migration time: 14
Voltage: 30 kV
Model: Hewlett Packard HP 3D

OTHER SUBSTANCES
Simultaneous: apoatropine, 6β-hydroxyhyoscyamine, hyoscyamine, littorine, scopolamine, tropic acid

REFERENCE
Cherkaoui,S.; Mateus,L.; Christen,P.; Veuthey,J.-L. Micellar electrokinetic capillary chromatography for selected tropane alkaloid analysis in plant extract, *Chromatographia*, **1997**, *46*, 351–357.

SAMPLE
Matrix: solutions

Sample preparation: Inject an aliquot of a 100 µg/mL solution in running buffer.

CAPILLARY ELECTROPHORESIS
Capillary: 32 cm × 50 µm fused-silica (27.5 cm to detector) (Yongnian Optical Conductive Fiber Plant, China), coated with polyacrylamide
Capillary preparation: No details of the polyacrylamide coating process are provided. However, another paper (LC.GC 1997, 15, 40) by this group indicates that they use the procedure of Hjertén, thus: Adjust the pH of 20 mL water to 3.5 with acetic acid, add 80 µL 3-(trimethoxysilyl)propyl methacrylate (3-methacryloxypropyltrimethoxysilane), mix, suck into capillary, let stand at room temperature for 1 h, remove the solution, wash with water. Fill the capillary with a deaerated 3-4% acrylamide solution containing 1 µL/mL N,N,N',N'-tetramethylethylenediamine and 1 mg/mL potassium persulfate, let stand for 30 min, remove excess solution by aspiration, rinse with water, remove water by aspiration, dry at 35° (J. Chromatogr. 1985, 347, 191).
Capillary temperature: 25
Running buffer: 100 mM NaH_2PO_4 adjusted to pH 2.5 (A) or 100 mM NaH_2PO_4 containing 45 mM hydroxypropyl-gamma-cyclodextrin, adjusted to pH 2.5 (B)
Injection: Electrokinetic injection at 15 kV for 3 s.
Detector: UV 200, UV 210
Migration time: 5.05 (A), 8.70, 8.80 (B, enantiomers)
Voltage: 15 kV
Model: Bio-Rad BioFocus 3000

OTHER SUBSTANCES
Simultaneous: atropine, carazolol, cicletanine, dimethindene, fendiline, ipratropium bromide, isothipendyl, mefloquine, metaclazepam, nafronyl (naftidrofuryl), nefopam, nicardipine, promethazine, reproterol, tetrahydrozoline (tetryzoline), theodrenaline, tioconazole, trihexyphenidyl, trimeprazine (alimemazine), trimipramine

---

KEY WORDS
coated capillary; chiral

REFERENCE
Koppenhoefer,B.; Epperlein,U.; Xiaofeng,Z.; Bingcheng,L. Separation of enantiomers of drugs by capillary electrophoresis. Part 4: Hydroxypropyl-γ-cyclodextrin as chiral solvating agent, *Electrophoresis*, **1997**, *18*, 924–930.

SAMPLE
Matrix: solutions
Sample preparation: Inject an aliquot of a 100 µg/mL solution in running buffer.

CAPILLARY ELECTROPHORESIS
Capillary: 30 cm × 50 µm fused-silica (25.5 cm to detector), coated with polyacrylamide
Capillary preparation: Adjust the pH of 20 mL water to 3.5 with acetic acid, add 80 µL 3-(trimethoxysilyl)propyl methacrylate (3-methacryloxypropyltrimethoxysilane), mix, suck into capillary, let stand at room temperature for 1 h, remove the solution, wash with water. Fill the capillary with a deaerated 3-4% acrylamide solution containing 1 µL/mL N,N,N',N'-tetramethylethylenediamine and 1 mg/mL potassium persulfate, let stand for 30 min, remove excess solution by aspiration, rinse with water, remove water by aspiration, dry at 35° (J. Chromatogr. 1985, 347, 191).
Capillary temperature: 25
Running buffer: 100 mM NaH_2PO_4 containing 30 mM hydroxypropyl-β-cyclodextrin, adjusted to pH 2.5 with phosphoric acid
Injection: Electrokinetic injection at 15 kV for 3 s.
Detector: UV 200
Migration time: 17.5, 18.9 (enantiomers)
Voltage: 15 kV
Model: Bio-Rad BioFocus 3000

OTHER SUBSTANCES
Simultaneous: chlorpheniramine, dimethindene, mefloquine, metaclazepam, tetrahydrozoline (tetryzoline), theodrenaline

KEY WORDS

chiral; coated capillary; comparison with the use of other cyclodextrins

REFERENCE

Lin,B.; Zhu,X.; Koppenhoefer,B.; Epperlein,U. Investigation of 123 chiral drugs by cyclodextrin-modified capillary electrophoresis, *LC.GC*, **1997**, *15*, 40–46.

SAMPLE

Matrix: solutions

CAPILLARY ELECTROPHORESIS

Capillary: 36 cm × 50 μm fused-silica coated with linear polyacrylamide (31.5 cm to detector) (GL Science)

Capillary preparation: At the beginning and end of each day rinse capillary with capillary wash solution (Bio-Rad Cat. No. 148-5022) at 690 kPa for more than 3 min and with water at 690 kPa for more than 3 min. Coat capillary as follows. Treat capillary with 1 M NaOH at room temperature for 1 h, rinse with water, dry by passing nitrogen gas through the capillary at 110° for 6 h. Pass thionyl chloride through the capillary using a suction pump for several min, seal capillary at both ends and heat at 70° for 6 h. Unseal the capillary and fill with 250 mM vinyl magnesium bromide in THF by suction, seal the capillary, heat at 70° for 6 h. Open the capillary and rinse it with THF for several min, rinse with distilled water, fill the capillary with polymerization solution, heat at 28 ± 2° for 1 h, condition at -100 V/cm for 30 min (Anal. Sci. 1994, 10, 1). (The polymerization solution was 5% acrylamide in water containing 49 mM Tris, 384 mM glycine, and 0.1% sodium dodecyl sulfate, degas in an ultrasonic bath. Add 40 μL 10% N,N,N',N'-tetramethylethylenediamine and 10 μL 10% ammonium persulfate to 5 mL of the degassed solution, mix thoroughly.)

Running buffer: n-Propanol:50 mM pH 5.0 Phosphate buffer 10:90

Injection: Before each injection rinse with water at 690 kPa for 30 s, rinse with running buffer at 690 kPa for 30 s, partially fill with separation solution (50 μM α_1-acid glycoprotein (Cohn fraction VI) (Sigma) in running buffer) at 6.9 kPa for 190 s (27 cm), inject sample at 6.9 kPa for 2 s, electrophorese with running buffer (Note that α_1-acid glycoprotein from other suppliers may provide inferior results).

Detector: UV 210

Migration time: Resolution of enantiomers 3.0

Voltage: 12 kV

Model: Bio-rad BioFocus 3000

KEY WORDS

chiral; coated capillary

REFERENCE

Tanaka,Y.; Terabe,S. Separation of the enantiomers of basic drugs by affinity capillary electrophoresis using a partial filling technique and α_1-acid glycoprotein as chiral selector, *Chromatographia*, **1997**, *44*, 119–128.

SAMPLE

Matrix: solutions

CAPILLARY ELECTROPHORESIS

Capillary: 29-36 cm × 50 μm fused-silica (24.5-31.5 cm to detector) (Yongnian Optical Conductive Fiber Plant, China) coated with polyacrylamide

Capillary preparation: Coat capillary as follows. Adjust the pH of 20 mL water to 3.5 with acetic acid, add 80 μL 3-(trimethoxysilyl)propyl methacrylate (3-methacryloxypropyltrimethoxysilane), mix, suck into capillary, let stand at room temperature for 1 h, remove the solution, wash with water. Fill the capillary with a deaerated 3-4% acrylamide solution containing 1 μL/mL N,N,N',N'-tetramethylethylenediamine and 1 mg/mL potassium persulfate, let stand for 30 min, remove excess solution by aspiration, rinse with water, remove water by aspiration, dry at 35° (J. Chromatogr. 1985, 347, 191).

Capillary temperature: 25

Running buffer: 100 mM pH 2.5 NaH_2PO_4 (A) or 100 mM pH 2.5 NaH_2PO_4 containing 45 mM hydroxypropyl-α-cyclodextrin (Wacker, Munich) (B)

Injection: Electromigration at 15 kV for 3 s.

Detector: UV 200; UV 210

Migration time: 5.05 (A); 7.67, 7.91 (B) (enantiomers)
Voltage: 15 kV
Model: Bio-Focus 3000

OTHER SUBSTANCES
Also analyzed: albuterol (salbutamol), alprenolol, amorolfine, atenolol, atropine, azelastine, baclofen, bamethan, benproperine, benserazide, biperiden, bisoprolol, brompheniramine, bupivacaine, bupranolol, butamirate, butethamate, carazolol, carbuterol, carteolol, carvedilol, celiprolol, chloroquine, chlorpheniramine, chlorphenoxamine, cicletanine, clenbuterol, clidinium bromide, clobutinol, dimethindene, dipivefrin, disopyramide, dobutamine, doxylamine, fendiline, flecainide, gallopamil, ipratropium bromide, isoproterenol (isoprenaline), isothipendyl, ketamine, meclizine, mefloquine, mepindolol, mequitazine, metaclazepam, metaproterenol (orciprenaline), metipranolol, metoprolol, nafronyl (naftidrofuryl), nefopam, nicardipine, norfenefrine, ofloxacin, ornidazole, orphenadrine, oxomemazine, oxprenolol, oxybutynin, phenoxybenzamine, phenylpropanolamine, pholedrine, pindolol, pirbuterol, prilocaine, procyclidine, promethazine, propafenone, propranolol, reproterol, sotalol, sulpride, synephrine, talinolol, terbutaline, tetrahydrozoline (tetryzoline), theodrenaline, tioconazole, tocainide, trihexyphenidyl, trimeprazine (alimemazine), trimipramine, tropicamide, verapamil, zopiclone

KEY WORDS
coated capillary; chiral

REFERENCE
Koppenhoefer,B.; Eperlein,U.; Schlunk,R.; Zhu,X.; Lin,B. Separation of enantiomers of drugs by capillary electrophoresis. V. Hydroxypropyl-α-cyclodextrin as chiral solvating agent, *J.Chromatogr.A*, **1998**, *793*, 153–164.

Hyaluronic acid

CAS Registry No.: 9004-61-9
Merck Index (12th ed.): 4793

SAMPLE
Matrix: bulk
Sample preparation: Dissolve 250 μg in 1 mL water. Remove a 50 μL aliquot and mix it with 25 μL 0.1% IS in water, inject an aliquot.

CAPILLARY ELECTROPHORESIS
Capillary: 58 cm × 75 μm fused-silica (50 cm to detector) (Waters)
Capillary preparation: Before use condition with 100 mM NaOH for 5 min, with water for 3 min, and with running buffer for 5 min.
Capillary temperature: 25
Running buffer: 50 mM pH 4.0 Phosphate buffer
Injection: Hydrostatic injection at 10 cm for 30 s.
Detector: UV 185
Migration time: 15
Internal standard: naphthalene-1,3,6-trisulfonic acid, trisodium salt (4.70)
Voltage: 20 kV
Model: Waters Quanta 4000E

Limit of detection: 1 μg/mL

KEY WORDS
detector at anode

REFERENCE
Hayase,S.; Oda,Y.; Honda,S.; Kakehi,K. High-performance capillary electrophoresis of hyaluronic acid: determination of its amount and molecular mass, *J.Chromatogr.A*, **1997**, *768*, 295–305.

SAMPLE
Matrix: bulk
Sample preparation: 2 mg Hyaluronic acid + 30 μL 20 mM 8-aminopyrene-1,3,6-trisulfonic acid (Molecular Probes, Eugene OR) in 3% acetic acid, mix, add 1 μL 1 M sodium cyanoborohydride in water, heat at 90° for 1 h, cool, make up to 500 μL with water, inject an aliquot.

CAPILLARY ELECTROPHORESIS
Capillary: 50 cm × 50 μm fused-silica (50 cm to detector) (Polymicro Technologies)
Capillary preparation: Treat capillary with 1 M NaOH for 1 h, rinse with water for 15 min, rinse with MeOH for 15 min. Fill with a solution of 30 μL 3-(trimethoxysilyl)propyl methacrylate (3-methacryloxypropyltrimethoxysilane) in 1 mL dichloromethane, let stand at room temperature for 1 h under nitrogen pressure, remove the solution, rinse with MeOH, rinse with water. Pass a 4% acrylamide solution containing 2 μL/mL N,N,N',N'-tetramethylethylenediamine and 2 mg/mL ammonium persulfate through the capillary under nitrogen pressure for 1 h, rinse with water, dry under a stream of nitrogen.
Running buffer: Polyacrylamide:buffer 5:95 (Buffer was 25 mM citric acid containing 12.5 mM Tris, pH 3.0. Polyacrylamide (MW 700 000-1 000 000, 10% aqueous solution) from Polysciences Inc., Warrington PA.)
Detector: F ex 488 (5 mW argon ion laser) em 514
Migration time: 10-100 (depending on degree of polymerization)
Voltage: -403 V/cm
Current: 10 μA
Model: laboratory-constructed

KEY WORDS
derivatization; coated capillary

REFERENCE
Hong,M.; Sudor,J.; Stefansson,M.; Novotny,M.V. High-resolution studies of hyaluronic acid mixtures through capillary gel electrophoresis, *Anal.Chem.*, **1998**, *70*, 568–573.

SAMPLE
Matrix: synovial fluid
Sample preparation: 200 μL Synovial fluid + 200 μL buffer + 100 μL enzyme, heat at 37° for 72 h, centrifuge at 15000 g for 5 min. Remove a 10 μL aliquot of the supernatant and add it to 5 μL IS solution, inject an aliquot. (Prepare buffer by mixing 8 mL 80 mM NaH_2PO_4 containing 0.05% sodium azide with 2 mL 80 mM Na_2HPO_4 containing 0.05% sodium azide, adjust pH to 6.3. The enzyme contained 5.4 mg/mL sheep testicular hyaluronidase (EC 2.2.1.35, 300 U/mg, Sigma) in 0.05% sodium azide solution. The IS solution contained 1.05 mg/mL Omnipaque in water. Omnipaque is a solution of iohexol and contains 240 mg iodine/mL. Store in the dark.)

CAPILLARY ELECTROPHORESIS
Capillary: 58.7 cm × 75 μm (50 cm to detector)
Capillary preparation: Before each injection wash capillary with 100 mM NaOH for 2 min, with 100 mM HCl for 2 min, with water for 2 min, and with running buffer for 2 min. At the beginning of each sequence of operations pressure wash capillary with 100 mM NaOH for 4 min and with water for 4 min.
Capillary temperature: 30
Running buffer: 50 mM pH 9 Na_2HPO_4 containing 10 mM sodium tetraborate and 40 mM sodium dodecyl sulfate
Injection: Pressure injection using nitrogen at 0.5 psi for 5 s
Detector: UV 200

Migration time: 10.5 (tetrasaccharide), 11.1 (hexasaccharide)
Internal standard: iohexol (7.3)
Voltage: 15 kV
Model: Beckman P/ACE system 2050
Limit of detection: 50 µg/mL

REFERENCE
Grimshaw,J.; Trocha-Grimshaw,J.; Fisher,W.; Rice,A.; Smith,S.; Spedding,P.; Duffy,J.; Mollan,R. Quantitative analysis of hyaluronan in human synovial using capillary electrophoresis, *Electrophoresis*, **1996**, *17*, 396–400.

Hydrochlorothiazide

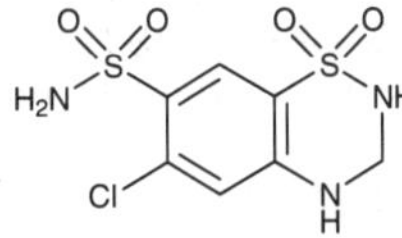

Molecular formula: $C_7H_8ClN_3O_4S_2$
Molecular weight: 297.74
CAS Registry No.: 58-93-5
Merck Index (12th ed.): 4822
Lednicer: 1 358

SAMPLE
Matrix: blood, urine
Sample preparation: Condition a 3 mL Supelclean LC-18 SPE cartridge with 3 mL MeOH and 3 mL water. Dilute urine 1:10 with water. Precipitate proteins from serum with MeOH. Add diluted urine or protein supernatant to the SPE cartridge, wash with 3 mL water, elute with 3 mL MeOH, reconstitute to the original volume with 100 mM KOH, inject an aliquot.

CAPILLARY ELECTROPHORESIS
Capillary: 67 cm × 50 µm fused-silica (60 cm to detector) (Polymicro Technologies)
Capillary preparation: Rinse with running buffer for 2 min before run. If necessary, regenerate capillary with 100 mM NaOH for 10 min and with water for 15 min.
Capillary temperature: 20
Running buffer: 60 mM 3-(cyclohexylamino)-1-propanesulfonic acid (CAPS) adjusted to pH 10.6 with 100 mM KOH
Injection: Pressure injection for 5 s
Detector: UV 220
Migration time: 11.2
Voltage: 25 kV
Model: Beckman P/ACE 2000
Limit of detection: 100 ng/mL (S/N 3, 30 s injection)

OTHER SUBSTANCES
Extracted: acetazolamide, amiloride, bendroflumethiazide, benzthiazide, bumetanide, caffeine, chlorothiazide, chlorthalidone, clopamide, dichlorphenamide, ethacrynic acid, furosemide, metyrapone, probenecid, triamterene, trichlormethiazide

KEY WORDS
serum; SPE

REFERENCE
Jumppanen,J.; Sirén,H.; Riekkola,M.-L. Screening for diuretics in urine and blood serum by capillary zone electrophoresis, *J.Chromatogr.A*, **1993**, *652*, 441–450.

SAMPLE
Matrix: bulk
Sample preparation: Dissolve in MeOH, sonicate for 5 min, dilute with 9 volumes 10 mM pH 3.0 sodium phosphate buffer, inject an aliquot.

CAPILLARY ELECTROPHORESIS
Capillary: 70 cm × 100 μm fused-silica (63 cm to detector) (Chrompack or Polymicro
 Technologies)
Capillary preparation: Before injection fill capillary with buffer for 2 min. At the start of each
 day wash with 100 mM NaOH for 5 min, with water for 10 min, and with running buffer for
 10 min. At the end of each day wash with 100 mM NaOH for 5 min and with water for 10 min.
Capillary temperature: 15
Running buffer: 20 mM pH 9.5 Sodium borate buffer containing 30 mM sodium dodecyl sulfate
Injection: Hydrodynamic injection for 2.5 s.
Detector: UV 225
Migration time: 8
Voltage: 20 kV
Model: Spectra Physics SpectraPhoresis 1000

OTHER SUBSTANCES
Simultaneous: impurities, chlorothiazide

KEY WORDS
stability-indicating

REFERENCE
Thomas,B.R.; Fang,X.G.; Chen,X.; Tyrrell,R.J.; Ghodbane,S. Validated micellar electrokinetic chromatography
 method for quality control of the drug substances hydrochlorothiazide and chlorothiazide, *J.Chromatogr.B*,
 1994, *657*, 383–394.

SAMPLE
Matrix: solutions

CAPILLARY ELECTROPHORESIS
Capillary: 72 cm × 50 μm silica (50 cm to detector) (Applied Biosystems)
Capillary preparation: Before each run wash capillary with 100 mM NaOH for 3 min, wash
 with running buffer for 3 min, aspirate neutral marker solution (2 drops DMSO in 10 mL
 water) for 1 s, place capillary end in buffer vial for 5 s, inject sample. Wash capillary at the
 beginning of each day by passing 1 M NaOH through for 20 min.
Capillary temperature: 30
Running buffer: MeCN:buffer 10:90 (Buffer was 30 mM pH 9.3 borate buffer containing 30 mM
 sodium dodecyl sulfate.)
Injection: Inject by applying vacuum for 1 s (2-3 nL)
Detector: UV 200
Migration time: 5.3
Voltage: +30 kV
Current: 56 μA
Model: Applied Biosystems Model 270A

OTHER SUBSTANCES
Simultaneous: chlorothiazide, furosemide, trichlormethiazide

REFERENCE
Evenson,M.A.; Wiktorowicz,J.E. Automated capillary electrophoresis applied to therapeutic drug monitoring,
 Clin.Chem., **1992**, *38*, 1847–1852.

SAMPLE
Matrix: solutions
Sample preparation: Inject an aliquot of a 200 μg/mL solution in MeCN:water 50:50.

CAPILLARY ELECTROPHORESIS
Capillary: 23 cm × 50 μm 3 μm CEC Hypersil C18
Capillary temperature: 15
Running buffer: Gradient. MeCN:50 mM pH 2.5 Na$_2$HPO$_4$ buffer:water 40:20:40 for 6.5 min,
 60:20:20 for 10.75 min (step gradient), re-equilibrate at initial conditions for 7.75 min.
Injection: Electrokinetic injection at 5 kV for 15 s.

Detector: UV 210
Migration time: 10.5
Voltage: 30 kV (with 8 bar of pressure at each end of capillary)
Model: Hewlett-Packard HP^{3D}

OTHER SUBSTANCES
Simultaneous: bendroflumethiazide, bumetanide, chlorothiazide, chlorthalidone, hydroflumethiazide

KEY WORDS
electrochromatography

REFERENCE
Euerby,M.R.; Gilligan,D.; Johnson,C.M.; Bartle,K.D. Step-gradient capillary electrochromatography, *Analyst*, **1997**, *122*, 1087–1088.

SAMPLE
Matrix: urine
Sample preparation: Filter (0.2 μm), inject an aliquot of the filtrate.

CAPILLARY ELECTROPHORESIS
Capillary: 80 cm × 50 μm fused-silica (57 cm to detector)
Capillary preparation: Before each run rinse the capillary with 1 M NaOH for 3 min, with 100 mM NaOH for 3 min, with water for 3 min, and with running buffer for 10 min.
Running buffer: 110 mM Boric acid containing 56 mM NaOH and 44 mM HCl, pH 8
Injection: Vacuum injection for 1 s
Detector: UV 226
Migration time: 6.100
Voltage: 20 kV
Model: Europhor Prime Vision system IV

OTHER SUBSTANCES
Extracted: acebutolol (UV 238), acetazolamide (UV 222), alprenolol (UV 220), amiloride (UV 220), atenolol (UV 228), bumetanide (UV 220), chlorthalidone (UV 220), cocaine (UV 236), codeine (UV 220), ethacrynic acid (UV 220), furosemide (UV 232), methadone (UV 220), metoxiphenamine (UV 220), nadolol (UV 220), norcodeine (UV 220), oxprenolol (UV 220), pentazocine (UV 220), propranolol (UV 220), spironolactone (UV 244), triamterene (UV 232), xipamide (UV 234)
Interfering: bendroflumethiazide (UV 220)

REFERENCE
Gonzalez,E.; Laserna,J.J. Capillary zone electrophoresis for the rapid screening of banned drugs in sport, *Electrophoresis*, **1994**, *15*, 240–243.

Hydrocortisone

Molecular formula: C$_{21}$H$_{30}$O$_5$
Molecular weight: 362.47
CAS Registry No.: 50-23-7, 13609-67-1 (butyrate), 57524-89-7 (valerate), 50-03-3 (acetate), 3863-59-0 (phosphate), 6000-74-4 (sodium phosphate), 125-04-2 (21-sodium succinate), 508-96-3 (tebutate), 74050-20-7 (aceponate), 72590-77-3 (buteprate), 508-99-6 (cypionate), 83784-20-7 (hemisuccinate monohydrate), 2203-97-6 (hemisuccinate), 2203-97-6 (succinate), 5752489-7 (valerate)
Merck Index (12th ed.): 4828
Lednicer: 1 190

SAMPLE
Matrix: blood

Sample preparation: 100 μL Serum + 900 μL 50 mM pH 4.5 acetate buffer, vortex for 1 min, filter (Amicon MPS-1 micropartition system) while centrifuging, inject an aliquot of the ultrafiltrate.

CAPILLARY ELECTROPHORESIS
Capillary: 37 cm × 50 μm eCAP (37 cm to detector) (Beckman)
Capillary preparation: At the beginning of each day rinse capillary with 100 mM HCl then rinse with running buffer for 10 min.
Capillary temperature: 16 ± 0.1
Running buffer: MeCN:buffer 20:80 (Buffer was 20 mM pH 6.0 4-morpholinopropanesulfonic acid containing 100 mM sodium dodecyl sulfate.)
Injection: High pressure injection for 20 s
Detector: UV 254
Migration time: 5.12
Voltage: 15 kV
Current: <50 μA
Model: Beckman P/ACE 5010
Limit of detection: 1 μg/mL (S/N 3)

OTHER SUBSTANCES
Extracted: 11-deoxycortisol, 21-deoxycortisol, 17-hydroxyprogesterone

---

KEY WORDS
serum

REFERENCE
Abubaker,M.A.; Bissell,M.G.; Petersen,J.R. Micellar electrokinetic capillary chromatography to separate steroids that are increased in congenital adrenal hyperplasia, *Clin.Chem.*, **1995**, *41*, 1369–1370.

SAMPLE
Matrix: blood
Sample preparation: Treat 200 μL serum with 1% sodium salicylate for 1 h, add to a 20 × 9 ASPEC pak octadecylsilane SPE cartridge ((M&S Instruments Trading Inc., Osaka), wash with 20 mL water, elute with 2 mL MeOH. Evaporate the eluate to dryness under educed pressure, reconstitute with 20 μL running buffer, filter (0.5 μm), inject an aliquot.

CAPILLARY ELECTROPHORESIS
Capillary: 37.5 cm × 75 μm fused-silica
Capillary preparation: Change the running buffer for each run.
Capillary temperature: 25
Running buffer: MeCN:buffer 16:84 (Buffer was 20 mM pH 8.0 sodium tetraborate/potassium phosphate buffer containing 50 mM sodium dodecyl sulfate.)
Injection: Pressure injection at 0.05 kg/cm^2 for 2 s
Detector: UV 205
Migration time: 13
Current: 60 μA (constant current)
Model: Ohtsuka Electronic model CAPI-3000

OTHER SUBSTANCES
Extracted: hydroxypregnenolone, hydroxyprogesterone, nandrolone, pregnenolone

---

KEY WORDS
serum; SPE

REFERENCE
Kobayashi,Y.; Matsui,J.; Watanabe,F. Simultaneous separation of free and conjugated steroids by micellar electrokinetic chromatography and its clinical application, *Biol.Pharm.Bull.*, **1995**, *18*, 1614–1616.

SAMPLE
Matrix: blood

Sample preparation: Condition a C18 SPE cartridge (International Sorbent Technology, Hengoed UK) with three 1 mL portions of MeOH, 1 mL water, and 30 mM pH 7.0 sodium phosphate buffer. 1.5 mL Serum + 1.5 mL water + 10 μL S.H.P. Helix pomatia enzyme (Biosepra, Villneuve-la-Garenne, France), mix, heat at 40° for 5-10 min, add 1 mL metaphosphoric acid, centrifuge (ultracentrifuge) at 3000 rpm for 5 min, add a 2.7 mL aliquot of the supernatant to the SPE cartridge, wash with 1 mL water, elute with 2.5 mL MeOH. Evaporate the eluate to dryness under a stream of nitrogen at 38°, reconstitute with 10 μL MeOH and 90 μL water, inject an aliquot.

CAPILLARY ELECTROPHORESIS
Capillary: 70 cm × 50 μm fused-silica (62.5 cm to detector) (Composite Metal Services, Worcs., UK)
Capillary preparation: Before each run wash capillary with KOH solution for 2 min, with water for 2 min, and with running buffer for 2 min then apply voltage for 2 min. Before first use condition capillary with KOH solution for 20 min and with water for 20 min.
Capillary temperature: 22
Running buffer: 49 mM AMPSO ([(1,1-dimethyl-2-hydroxyethyl)amino]-2-hydroxypropane sulfonic acid) containing 55 mM sodium cholate and 18 mM sodium dodecyl sulfate, pH adjusted to 9.0 with 25% ammonia
Injection: Pressure injection at 50 mbar for 4 s.
Detector: UV 254
Migration time: 13.3
Voltage: 20 kV
Current: 29.5 μA
Model: Hewlett-Packard 3D
Limit of quantitation: 100 ng/mL

OTHER SUBSTANCES
Extracted: cortisone, dexamethasone

KEY WORDS
serum; SPE

REFERENCE
Wiedmer,S.K.; Sirén,H.; Riekkola,M.-L. Determination of serum corticosteroids by mixed micellar electrokinetic capillary chromatography with sodium dodecyl sulfate and sodium cholate, *Electrophoresis*, **1997**, *18*, 1861–1864.

SAMPLE
Matrix: blood, urine
Sample preparation: Condition a 6 mL 300 mg end-capped C8 SPE cartridge (IST, Hengoed UK) with 1 mL MeOH and 1 mL water. Condition a 1 mL 100 mg SAX SPE cartridge (IST, Hengoed UK) with 250 μL MeCN:5 mM ammonium acetate 10:90. Add 2 mL urine to the C8 SPE cartridge, wash with 1 mL water, wash with 2 mL hexane, elute with 1 mL ethyl acetate, elute with 1 mL dichloromethane. Combine the eluates and evaporate them to dryness at 40°, reconstitute with 250 μL MeCN:5 mM ammonium acetate 10:90, add to the SAX SPE cartridge, elute with 150 μL MeCN:5 mM ammonium acetate 10:90, collect all the effluent from the SAX SPE cartridge, mix, inject an aliquot.

CAPILLARY ELECTROPHORESIS
Capillary: 24 cm × 50 μm fused-silica (Composite Metal Services, Hallow UK) slurry packed with 3 μm Apex ODS (Jones Chromatography) to a length of 16 cm (16.1 cm to detector)
Capillary preparation: Slurry pack column with a slurry in MeOH, sonicate during process. Condition with mobile phase at 200 bar for 12 h.
Running buffer: Gradient. MeCN containing 5 mM ammonium acetate:water containing 5 mM ammonium acetate from 9:91 to 80:20 over 5 min, maintain at 80:20 for 5 min, return to initial conditions over 5 min, re-equilibrate for 15 min. Pumped at 0.1 mL/min.
Injection: 10-250 μL
Detector: UV 240
Migration time: 11.5
Voltage: 25 kV
Limit of detection: 390 ng/mL

OTHER SUBSTANCES
Extracted: dexamethasone, fluocortolone

KEY WORDS
SPE; horse; electrochromatography

REFERENCE
Taylor,M.R.; Teale,P.; Westwood,S.A.; Perrett,D. Analysis of corticosteroids in biofluids by capillary electrochromatography with gradient elution, *Anal.Chem.*, **1997**, *69*, 2554–2558.

SAMPLE
Matrix: solutions

CAPILLARY ELECTROPHORESIS
Capillary: 65 cm × 50 μm fused-silica (50 cm to detector) (Scientific Glass Engineering)
Capillary preparation: Flush with running buffer for 1 min between runs. Periodically flush with water for 1 min, sweep with 100 mM KOH, let stand in 100 mM KOH for 30 min, flush with water until the effluent is neutral to pH paper, fill with running buffer, let stand for 30 min.
Running buffer: 20 mM pH 9.0 Phosphate-borate buffer containing 100 mM sodium cholate (Buffer was prepared by adjusting pH of 20 mM sodium borate to 9.0 with 20 mM NaH_2PO_4.)
Injection: Injection by siphon at 10 cm for 10 s (about 1 nL)
Detector: UV 210
Migration time: 12 (hydrocortisone), 12.9 (hydrocortisone acetate)
Voltage: 20 kV

OTHER SUBSTANCES
Simultaneous: betamethasone, dexamethasone acetate, fluocinonide, fluocinolone acetonide, triamcinolone, triamcinolone acetonide

REFERENCE
Nishi,H.; Fukuyama,T.; Matsuo,M.; Terabe,S. Separation and determination of lipophilic corticosteroids and benzothiazepin analogues by micellar electrokinetic chromatography using bile salts, *J.Chromatogr.*, **1990**, *513*, 279–295.

SAMPLE
Matrix: solutions
Sample preparation: Prepare a 0.2-1 mg/mL solution in MeOH, inject an aliquot.

CAPILLARY ELECTROPHORESIS
Capillary: 65 cm × 50 μm fused-silica (50 cm to detector)
Running buffer: 20 mM pH 9.0 Phosphate-borate buffer containing 50 mM sodium dodecyl sulfate, 4 M urea, and 15 mM gamma-cyclodextrin
Detector: UV 220
Migration time: 13.2 (hydrocortisone), 19 (hydrocortisone acetate)
Voltage: 20 kV

OTHER SUBSTANCES
Simultaneous: betamethasone, cortisone acetate, dexamethasone acetate, fluocinolone acetonide, fluocinonide, triamcinolone acetonide

REFERENCE
Nishi,H.; Matsuo,M. Separation of corticosteroids and aromatic hydrocarbons by cyclodextrin-modified micellar electrokinetic chromatography, *J.Liq.Chromatogr.*, **1991**, *14*, 973–986.

SAMPLE
Matrix: solutions
Sample preparation: Prepare a 10 mg/mL solution in MeOH, inject an aliquot.

CAPILLARY ELECTROPHORESIS
Capillary: 62 cm × 53 μm fused-silica (50 cm to detector) (Polymicro Technologies)

Capillary preparation: Rinse with running buffer for 1 min between runs.
Running buffer: 50 mM pH 9.0 Phosphate borate buffer containing 50 mM dehydrocholate, 50 mM taurocholate, and 50 mM sodium dodecyl sulfate
Detector: UV 254
Migration time: 21.5 (hydrocortisone), 24.5 (hydrocortisone-21-acetate)
Voltage: 15 kV
Current: 56 μA
Model: Laboratory constructed

OTHER SUBSTANCES
Simultaneous: corticosterone, cortisone, cortisone acetate, deoxycorticosterone, fludrocortisone, fludrocortisone acetate, fluocinolone acetonide, 6α-methylprednisolone, prednisolone, prednisolone acetate, prednisone, prednisone acetate, progesterone, triamcinolone, triamcinolone acetonide

REFERENCE
Bumgarner,J.G.; Khaledi,M.G. Mixed micellar electrokinetic chromatography of corticosteroids, *Electrophoresis*, **1994**, *15*, 1260–1266.

SAMPLE
Matrix: solutions
Sample preparation: Prepare a solution in MeOH:water:100 mM sodium dodecyl sulfate 9.4: 80.6:10, inject an aliquot.

CAPILLARY ELECTROPHORESIS
Capillary: 57 cm × 50 μm fused-silica (50 cm to detector) (Polymicro Technologies)
Capillary temperature: 23
Running buffer: 60 mM pH 9.2 Borate buffer containing 10 mM sodium dodecyl sulfate
Injection: Hydrostatic injection at 0.5 psi for 3 s
Detector: UV 260
Migration time: 14.6
Voltage: 22 kV
Current: 48 μA
Model: Beckman 2050 P/ACE

OTHER SUBSTANCES
Simultaneous: aldosterone, corticosterone, cortisone, 1-dehydroaldosterone, 21-deoxycortisol, 17-isoaldosterone

REFERENCE
Jumppanen,J.H.; Wiedmer,S.K.; Sirén,H.; Riekkola,M.L.; Haario,H. Optimized separation of seven corticosteroids by micellar electrokinetic chromatography, *Electrophoresis*, **1994**, *15*, 1267–1272.

SAMPLE
Matrix: solutions

CAPILLARY ELECTROPHORESIS
Capillary: 62 cm × 53 μm fused-silica (50 cm to detector) (Polymicro Technologies)
Capillary preparation: Rinse capillary with running buffer for 1 min between runs.
Running buffer: 50 mM pH 9.0 Phosphate/borate buffer containing 33 mM taurocholate, 33 mM glycodeoxycholate, and 70 mM butanesulfonate
Detector: UV 254
Migration time: 20.5 (hydrocortisone), 23 (hydrocortisone-21-acetate)
Voltage: 15 kV
Model: laboratory-constructed

OTHER SUBSTANCES
Simultaneous: corticosterone, cortisone, cortisone acetate, deoxycorticosterone, fludrocortisone, fludrocortisone acetate, fluocinolone acetonide, 6α-methylprednisolone, prednisolone, prednisolone acetate, prednisone, prednisone acetate, progesterone, triamcinolone, triamcinolone acetonide

REFERENCE
Bumgarner,J.G.; Khaledi,M.G. Mixed micelles of short chain alkyl surfactants and bile slats in electrokinetic chromatography: Enhanced separation of corticosteroids, *J.Chromatogr.A*, **1996**, *738*, 275–283.

SAMPLE
Matrix: solutions
Sample preparation: Inject an aliquot of a solution in running buffer.

CAPILLARY ELECTROPHORESIS
Capillary: 57 cm × 75 μm fused-silica (50 cm to detector)
Capillary preparation: Before each run rinse with running buffer for 2 min. After each run rinse capillary with 100 mM NaOH for 2 min and with water for 2 min. Condition new capillaries with 100 mM HCl for 5 min, with 100 mM NaOH for 10 min, and with water for 5 min.
Capillary temperature: 23
Running buffer: MeOH:buffer 10:90 (Buffer was 40 mM pH 9.2 borate buffer containing 20 mM sodium dodecyl sulfate.)
Injection: Low pressure injection for 10 s.
Detector: UV 240
Migration time: 10
Voltage: 30 kV
Model: Beckman P/ACE 5510

OTHER SUBSTANCES
Simultaneous: betamethasone, dexamethasone, flumethasone, triamcinolone

REFERENCE
Gu,X.; Meleka-Boules,M.; Chen,C.-L. Micellar electrokinetic capillary chromatography combined with immunoaffinity chromatography for identification and determination of dexamethasone and flumethasone in equine urine, *J.Capillary Electrophor.*, **1996**, *3*, 43–49.

SAMPLE
Matrix: solutions
Sample preparation: Inject an aliquot of a solution in running buffer.

CAPILLARY ELECTROPHORESIS
Capillary: 60 cm × 50 μm fused-silica (50 cm to detector) (Supelco)
Capillary preparation: Between runs wash capillary with 500 mM NaOH for 4 min.
Running buffer: n-Hexanol:sodium dodecyl sulfate:n-butanol:20 mM pH 10.0 phosphate buffer 0.81:3.31:6.61:89.28
Injection: Electrokinetic injection at 15 kV for 10 s.
Detector: UV 220
Migration time: 18
Voltage: 15 kV
Model: laboratory-constructed
Limit of detection: 1 pmole

OTHER SUBSTANCES
Simultaneous: aldosterone, cortexolone, corticosterone, cortisone, 11-dehydrocorticosterone, deoxycorticosterone, deoxycorticosterone acetate, dexamethasone, triamcinolone

REFERENCE
Vomastová,L.; Miksík,I.; Deyl,Z. Microemulsion and micellar electrokinetic chromatography of steroids, *J.Chromatogr.B*, **1996**, *681*, 107–113.

SAMPLE
Matrix: solutions
Sample preparation: Inject an aliquot of a solution in MeOH:water 9.44:91.66 containing 10 mM sodium dodecyl sulfate.

CAPILLARY ELECTROPHORESIS
Capillary: 67 cm × 50 μm fused-silica (60 cm to detector) (Composite Metal Services, Worcs., UK)

Capillary preparation: Before each run rinse capillary with KOH solution for 2 min, with water for 2 min, and with running buffer for 2 min, switch on voltage for 2 min before injection.
Capillary temperature: 25
Running buffer: 49 mM 3-[(1,1-Dimethyl-2-hydroxyethyl)amino]-2-hydroxypropanesulfonic acid (AMPSO) containing 18 mM sodium dodecyl sulfate and 55 mM sodium cholate, adjusted to pH 9.0 with 25% ammonia
Injection: Pressure injection at 35 mbar for 3.5 s.
Detector: UV 260
Migration time: 14
Voltage: 20 kV
Current: 33 μA
Model: Beckman P/ACE 2050

OTHER SUBSTANCES
Simultaneous: d-aldosterone, corticosterone, cortisone, 1-dehydroaldosterone, 21-deoxycortisol, 17-isoaldosterone

REFERENCE
Wiedmer,S.K.; Jumppanen,J.H.; Haario,H.; Riekkola,M.-L. Optimization of selectivity and resolution in micellar electrokinetic capillary chromatography ith a mixed micellar system of sodium dodecyl sulfate and sodium cholate, *Electrophoresis*, **1996**, *17*, 1931–1937.

SAMPLE
Matrix: solutions

CAPILLARY ELECTROPHORESIS
Capillary: 33.5 cm × 100 μm fused-silica packed with 3 μm C18 for 25 cm (Innovatech, Stevenage, UK) (Polymicro Technologies)
Capillary preparation: Column packing procedure described in LC.GC 1995, 13, 800.
Capillary temperature: 20
Running buffer: MeCN:25 mM Tris HCl:water 60:20:20, run with 12 bar nitrogen pressure on each side
Injection: Electrokinetic injection at 5 kV for 3 s.
Detector: UV 200
Migration time: 4.6
Voltage: 30 kV
Model: Hewlett-Packard HP[3D]

OTHER SUBSTANCES
Simultaneous: androsterone, dehydrotestosterone, desoxycorticosterone, dexamethasone, estrone, ethinyl estradiol, prednisone, pregnenolone, testosterone

KEY WORDS
electrochromatography

REFERENCE
Dittmann,M.M.; Rozing,G.P.; Ross,G.; Adam,T.; Unger,K.K. Advances in capillary electrochromatography, *J.Capillary Electrophor.*, **1997**, *4*, 201–212.

SAMPLE
Matrix: solutions

CAPILLARY ELECTROPHORESIS
Capillary: 57 cm × 50 μm fused-silica (50 cm to detector) (Polymicro Technologies)
Capillary preparation: After each run rinse capillaries with 100 mM NaOH for 10 min and with water for 5 min. Fill with running buffer and apply voltage for 10 min. Rinse new capillaries with 1 M HCl for 5 min, with 100 mM NaOH for 10 min, and with water for 5 min.
Capillary temperature: 25
Running buffer: MeOH:buffer 20:80 (Buffer was 38 mM pH 7.4 phosphate buffer containing 40 mM sodium cholate.)
Injection: Pressure injection for 5 s (3.1 nL).
Detector: UV (wavelength not given)

Migration time: 12.57
Voltage: 15 kV
Model: Beckman P/ACE 5500

OTHER SUBSTANCES
Simultaneous: aldosterone, dehydroepiandrosterone, estradiol, 17α-hydroxyprogesterone, progesterone

---

REFERENCE
Hsiao,M.-W.; Lin,S.-T. Separation of steroids by micellar electrokinetic capillary chromatography with sodium cholate, *J.Chromatogr.Sci.*, **1997**, *35*, 259–264.

SAMPLE
Matrix: solutions
Sample preparation: Inject a solution in MeOH:running buffer 10:90.

CAPILLARY ELECTROPHORESIS
Capillary: 50 cm × 50 μm fused-silica (45.4 cm to detector) (Polymicro Technologies)
Capillary temperature: 25
Running buffer: 20 mM pH 7.0 Sodium phosphate buffer containing 50 mM sodium dodecyl sulfate and 10 mM Brij 35
Injection: Pressure injection at 1 psi.s.
Detector: UV 240
Migration time: 6
Voltage: 20 kV
Model: Bio-Rad BioFocus 3000

OTHER SUBSTANCES
Simultaneous: corticosterone, cortisone

REFERENCE
Muijselaar,P.G.; Claessens,H.A.; Cramers,C.A. Characterization of pseudostationary phases in micellar electrokinetic chromatography by applying linear solvation energy relationships and retention indexes, *Anal.Chem.*, **1997**, *69*, 1184–1191.

SAMPLE
Matrix: solutions

CAPILLARY ELECTROPHORESIS
Capillary: 32 cm × 100 μm fused-silica (Polymicro Technologies) packed for 24 cm with 1.5 μm Chromspher ODS (Chrompack)
Capillary preparation: Sonicate 30 mg 1.5 μm Chromspher ODS in 2 mL hexane:isopropanol 50:50 for 1 h, pump the slurry into the capillary at 600 bar using hexane:isopropanol 90:10 for 4 h, release the pressure over 1 min, wash with acetone, wash with water at 600 bar. Fabrication of frits is described in the paper.
Capillary temperature: 25
Running buffer: MeCN:water 60:40 containing 5 mM sodium dodecyl sulfate and 1.6 mM sodium tetraborate, pH 9.25
Injection: Electrokinetic injection at 10 kV.
Detector: UV (wavelength not given)
Migration time: 2.6
Voltage: 20 kV
Model: Hewlett Packard HP [3D]CE

OTHER SUBSTANCES
Simultaneous: 17α-methyltestosterone, progesterone, testosterone

---

KEY WORDS
electrochromatography

REFERENCE
Seifar,R.M.; Kok,W.T.; Kraak,J.C.; Poppe,H. Capillary electrochromatography with 1.5 μm ODS-modified non-porous silica spheres, *Chromatographia*, **1997**, *46*, 131–136.

SAMPLE
Matrix: solutions

CAPILLARY ELECTROPHORESIS
Capillary: 42 cm × 50 μm fused-silica packed with 3 μm CEC Hypersil ODS (packed length 30 cm, 30.1 cm to detector) (Hypersil)
Running buffer: Gradient. A was 5 mM ammonium acetate in MeCN:water 17:83. B was 5 mM ammonium acetate in MeCN:water 38:62. A:B 100:0 for 3 min, to 0:100 over 15 min, maintain at 0:100 (pumped with an HPLC pump at 0.01 mL/min for 3 min then at 0.1 mL/min).
Injection: Inject 10 μL using an HPLC injector.
Detector: UV 240
Migration time: 31
Voltage: 30 kV

OTHER SUBSTANCES
Simultaneous: adrenosterone, betamethasone, cortisone, fluocortolone, dexamethasone, methylprednisolone, triamcinolone, triamcinolone acetonide
Interfering: prednisolone

KEY WORDS
electrochromatography

REFERENCE
Taylor,M.R.; Teale,P. Gradient capillary electrochromatography of drug mixtures with UV and electrospray ionisation mass spectrometric detection, *J.Chromatogr.A*, **1997**, *768*, 89–95.

SAMPLE
Matrix: solutions

CAPILLARY ELECTROPHORESIS
Capillary: 67 cm × 50 μm fused-silica (60 cm to detector)
Capillary preparation: Before each run equilibrate with running buffer for 3 min. After each run wash with 100 mM NaOH for 3 min and with water. At the beginning of each day rinse the capillary with 100 mM NaOH for 30 min and with water for 30 min.
Capillary temperature: 25 ± 0.1
Running buffer: 25 mM pH 9 Bicine buffer containing 10 mM sodium dodecyl sulfate and 5 mM N-dodecyl-N,N-dimethyl-3-ammonio-1-propanesulfonate (SB3-12)
Injection: High pressure injection for 5 s
Detector: UV 254
Migration time: 6.2
Voltage: 30 kV
Current: <40 μA
Model: Beckman P/ACE 5010

OTHER SUBSTANCES
Simultaneous: cortisone, 11-desoxycortisol, 21-desoxycortisol, dimethyltestosterone, 17-hydroxyprogesterone, progesterone, testosterone propionate, testosterone

REFERENCE
Valbuena,G.A.; Rao,L.V.; Petersen,J.R.; Okorodudu,A.O.; Bissell,M.G.; Mohammad,A.A. Anionic-zwitterionic mixed micelles in the micellar electrokinetic separation of clinically relevant steroids on a fused-silica capillary, *J.Chromatogr.A*, **1997**, *781*, 467–474.

SAMPLE
Matrix: solutions
Sample preparation: Inject an aliquot of a solution in EtOH:water 11.1:78.8 containing 10 mM sodium dodecyl sulfate.

CAPILLARY ELECTROPHORESIS
Capillary: 67 cm × 50 μm fused-silica (60 cm to detector) (Composite Metal Services, Worcestershire UK)
Capillary temperature: 25
Running buffer: 50 mM pH 8.7 [(1,1-Dimethyl-2-hydroxyethyl)amino]-2-hydroxypropanesulfonic acid (AMPSO) containing 20 mM sodium dodecyl sulfate (pH adjusted with 25% ammonia)
Injection: Hydrostatic injection at 0.5 psi for 3.5 s.
Detector: UV 260
Migration time: 10.5
Voltage: 20 kV
Current: 10.1 μA
Model: Beckman P/ACE 2050

OTHER SUBSTANCES
Simultaneous: 4-androstene-3,17-dione, corticosterone, 11-deoxycortisol, dexamethasone, fludrocortisone acetate

REFERENCE
Wiedmer,S.K.; Riekkola,M.-L.; Nydén,M.; Söderman,O. Mixed micelles of sodium dodecyl sulfate and sodium cholate: Micellar electrokinetic capillary chromatography and nuclear magnetic resonance spectroscopy, *Anal.Chem.*, **1997**, *69*, 1577–1584.

SAMPLE
Matrix: solutions
Sample preparation: Inject an aliquot of a 100 μg/mL solution in running buffer.

CAPILLARY ELECTROPHORESIS
Capillary: 27 cm × 50 μm fused-silica (20 cm to detector) (Composite Metal Services, Hallow, UK) packed with 3 μm Hypersil ODS for 20 cm (details in paper)
Running buffer: MeCN:2 mM pH 7.8 phosphate buffer 80:20
Injection: Electrokinetic injection at 5 kV for 5 s.
Detector: UV 214
Migration time: 6.29
Voltage: 10 kV
Model: Beckman P/ACE 2050

OTHER SUBSTANCES
Simultaneous: betamethasone, betamethasone dipropionate, betamethasone-17-valerate, clobetasol butyrate, clobetasone butyrate, fluticasone propionate, prednisolone

KEY WORDS
electrochromatography

REFERENCE
Frame,L.A.; Robinson,M.L.; Lough,W.J. Simplification of capillary electrochromatography procedures, *J.Chromatogr.A*, **1998**, *798*, 243–249.

SAMPLE
Matrix: solutions

CAPILLARY ELECTROPHORESIS
Capillary: 32 cm × 50 μm fused-silica packed for 24 cm with 1.8 μm Zorbax ODS (Polymicro Technologies) (Column preparation described in paper.)
Capillary temperature: 25
Running buffer: MeCN:0.8 mM sodium tetraborate 80:20 containing 5 mM sodium dodecyl sulfate
Injection: Electrokinetic injection at 20 kV for 1 s.
Detector: UV 254
Migration time: 2.4
Voltage: 25 kV
Model: Hewlett-Packard HP ^{3D}CE

OTHER SUBSTANCES
Simultaneous: estradiol, estriol, estrone, methyltestosterone, 4-pregnen-20α-ol-3-one, progesterone, testosterone

KEY WORDS
electrochromatography

REFERENCE
Seifar,R.M; Kraak,J.C.; Kok,W.T.; Poppe,H. Capillary electrochromatography with 1.8-μm ODS-modified porous silica particles, *J.Chromatogr.A*, **1998**, *808*, 71–77.

SAMPLE
Matrix: solutions
Sample preparation: Inject an aliquot of a solution in MeOH:water 30;70.

CAPILLARY ELECTROPHORESIS
Capillary: 83 cm fused-silica (25 cm to detector) (Composite Metal Services)
Capillary temperature: 25
Running buffer: 100 mM ammonium acetate adjusted to pH 9 with ammonia
Injection: Electrokinetic injection of 8.5 mM pH 9 ammonium acetate containing 10 mM sodium dodecyl sulfate and 10 mM sodium cholate at 30 kV for 30 s followed by pressure injection of sample at 50 mbar for 5 s.
Detector: UV 254
Migration time: 6.5
Voltage: 10 kV
Model: Hewlett-Packard 3D

OTHER SUBSTANCES
Simultaneous: corticosterone, cortisone

KEY WORDS
MS detector also described

REFERENCE
Wiedmer,S.K.; Jussila,M.; Riekkola,M.-L. On-line partial filling micellar electrokinetic capillary chromatography-electrospray ionization-mass spectrometry of corticosteroids, *Electrophoresis*, **1998**, *19*, 1711–1718.

SAMPLE
Matrix: urine
Sample preparation: Acidify 5 mL urine to pH 6, extract twice with 2 mL portions of dichloromethane. Evaporate the extracts to dryness under reduced pressure, reconstitute with 500 μL running buffer, sonicate for 5 min, inject an aliquot.

CAPILLARY ELECTROPHORESIS
Capillary: 57 cm × 50 μm fused-silica (50 cm to detector)
Capillary preparation: Before each run wash capillary with 100 mM HCl for 3 min, wash with 500 mM NaOH, wash with water, and equilibrate with running buffer for 3 min. At the beginning of each day rinse capillary with 500 mM NaOH for 30 min and with water for 30 min.
Capillary temperature: 15 ± 0.1
Running buffer: 10 mM pH 7.4 Sodium phosphate buffer containing 50 mM dodecyltrimethylammonium bromide and 5.2 mM trioctylphosphine oxide
Injection: High-pressure injection for 5 s
Detector: UV 254
Migration time: 30
Voltage: 15 kV
Current: <30 μA
Model: Beckman P/ACE 5010
Limit of quantitation: 500 ng/mL

OTHER SUBSTANCES
Extracted: cortisone, 17-deoxycorticosterone, dimethyltestosterone, testosterone, testosterone propionate

REFERENCE
Abubaker,M.A.; Petersen,J.R.; Bissell,M.G. Micellar electrokinetic capillary chromatographic separation of steroids in urine by trioctylphosphine oxide and cationic surfactant, *J.Chromatogr.B*, **1995**, *674*, 31–38.

SAMPLE
Matrix: urine
Sample preparation: Condition a 3M Empore C18 SPE disc with 250 μL water and 1 mL water. Mix 10 mL urine with 2.76 nmoles corticosterone, pass through the SPE disc under vacuum, wash with two 1 mL portions of acetone:water 10:90, wash with 1 mL water, elute with 80 μL MeCN, elute with 320 μL 10 mM sodium dodecyl sulfate, mix the eluate, inject an aliquot.

CAPILLARY ELECTROPHORESIS
Capillary: 57 cm × 75 μm fused-silica (Supelco)
Capillary temperature: 16 ± 0.1
Running buffer: MeCN:buffer 20:80 (Buffer was 15 mM pH 2.5 phosphate buffer containing 100 mM sodium dodecyl sulfate.)
Injection: High pressure injection for 20 s.
Detector: UV 254
Migration time: 9.4
Internal standard: corticosterone (8.2)
Voltage: 10 kV
Model: Beckman P/ACE 5010
Limit of detection: 138 nM

KEY WORDS
SPE

REFERENCE
Rao,L.V.; Petersen,J.R.; Bissell,M.G.; Okorodudu,A.O.; Mohammad,A.A. Specific determination of urinary free cortisol by solid-phase microparticle extraction capillary electrophoresis with fused silica capillaries, *Clin.Chem.*, **1997**, *43*, 1801–1803.

SAMPLE
Matrix: urine
Sample preparation: Vortex 5 mL urine and 2 mL hexane for 3 min, centrifuge at 3500 rpm for 3 min, repeat extraction. Combine the organic layers and evaporate them to dryness at room temperature, reconstitute the residue in 500 μL 20 mM pH 2.5 phosphate buffer containing 10 mM sodium dodecyl sulfate, inject an aliquot.

CAPILLARY ELECTROPHORESIS
Capillary: 64.5 cm × 50 μm fused-silica (56 cm to detector) (Polymicro Technologies)
Capillary preparation: Between runs flush capillary with 1 M NaOH for 1 min, with MeOH for 1 min, with 100 mM naOH for 1 min, with water for 2 min, and with running buffer for 2 min. Before use rinse capillary with 1 M NaOH for 10 min, with MeOH for 5 min, with water for 5 min, and with running buffer for 5 min.
Capillary temperature: 30
Running buffer: MeOH:100 mM pH 2 phosphate buffer 20:80 containing 50 mM sodium dodecyl sulfate
Injection: Inject a 5.8 cm water plug at 50 mbar, inject sample using voltage, when the current reaches 90% of current in a capillary filled only with running buffer replace sample with running buffer.
Detector: UV 247
Migration time: 11
Voltage: -25 kV
Model: Hewlett Packard 3D
Limit of detection: 8.0 ppb (S/N 3)

OTHER SUBSTANCES
Extracted: cortisone, progesterone, testosterone

KEY WORDS
detector at anode

REFERENCE
Quirino,J.P.; Terabe,S. On-line concentration of neutral analytes for micellar electrokinetic chromatography. 5. Field-enhanced sample injection with reverse migrating micelles, *Anal.Chem.*, **1998**, *70*, 1893–1901.

Hydroflumethiazide

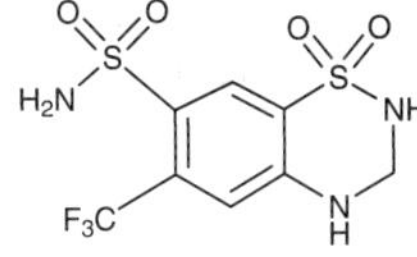

Molecular formula: $C_8H_8F_3N_3O_4S_2$
Molecular weight: 331.30
CAS Registry No.: 135-09-1
Merck Index (12th ed.): 4830
Lednicer: 1 358

SAMPLE
Matrix: solutions
Sample preparation: Inject an aliquot of a 200 μg/mL solution in MeCN:water 50:50.

CAPILLARY ELECTROPHORESIS
Capillary: 23 cm × 50 μm 3 μm CEC Hypersil C18
Capillary temperature: 15
Running buffer: Gradient. MeCN:50 mM pH 2.5 Na_2HPO_4 buffer:water 40:20:40 for 6.5 min, 60:20:20 for 10.75 min (step gradient), re-equilibrate at initial conditions for 7.75 min.
Injection: Electrokinetic injection at 5 kV for 15 s.
Detector: UV 210
Migration time: 12
Voltage: 30 kV (with 8 bar of pressure at each end of capillary)
Model: Hewlett-Packard HP³ᴰ

OTHER SUBSTANCES
Simultaneous: bendroflumethiazide, bumetanide, chlorothiazide, chlorthalidone, hydrochlorothiazide

KEY WORDS
electrochromatography

REFERENCE
Euerby,M.R.; Gilligan,D.; Johnson,C.M.; Bartle,K.D. Step-gradient capillary electrochromatography, *Analyst*, **1997**, *122*, 1087–1088.

SAMPLE
Matrix: solutions

CAPILLARY ELECTROPHORESIS
Capillary: 40.6 cm × 50 μm fused-silica packed with 3 μm CEC Hypersil ODS (Hypersil)
Running buffer: Gradient. A was 5 mM ammonium acetate in MeCN:water 50:50. B was 5 mM ammonium acetate in MeCN:water 80:20. A:B 100:0 for 3 min, to 0:100 over 0.1 min, maintain at 0:100 (pumped with an HPLC pump at 0.01 mL/min for 3 min then at 0.1 mL/min).
Injection: Inject 5 μL using an HPLC injector.
Detector: MS, VG Biotech Platform, electrospray (details in paper)
Migration time: 24
Voltage: 30 kV

OTHER SUBSTANCES
Simultaneous: bendroflumethiazide, epithiazide, methyclothiazide, metolazone

KEY WORDS
electrochromatography

REFERENCE

Taylor,M.R.; Teale,P. Gradient capillary electrochromatography of drug mixtures with UV and electrospray ionisation mass spectrometric detection, *J.Chromatogr.A*, **1997**, *768*, 89–95.

Hydromorphone

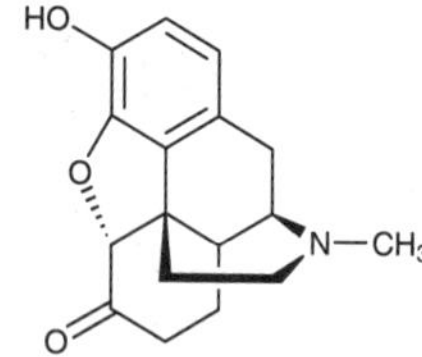

Molecular formula: $C_{17}H_{19}NO_3$
Molecular weight: 285.34
CAS Registry No.: 466-99-9, 71-68-1 (HCl)
Merck Index (12th ed.): 4847
Lednicer: 1 288

SAMPLE

Matrix: bulk
Sample preparation: Dilute with 400 µg/mL n-propyl p-hydroxybenzoate in running buffer

CAPILLARY ELECTROPHORESIS

Capillary: 27 cm × 50 µm fused-silica (20 cm to detector)
Capillary preparation: Before each run rinse with running buffer at high pressure.
Capillary temperature: 30
Running buffer: MeCN:water 15:85 containing 40 mM sodium dodecyl sulfate, 8.5 mM sodium borate, and 8.5 mM sodium phosphate, pH 8.5
Injection: Inject at high pressure for 1 s
Detector: UV 214
Migration time: 1.3
Internal standard: n-propyl p-hydroxybenzoate (1.8)
Voltage: 20 kV
Model: Beckman P/ACE System 2100

OTHER SUBSTANCES

Simultaneous: acetaminophen, acetylcodeine, O-acetylmorphine, aspirin, caffeine, cocaine, codeine, diamorphine, diphenhydramine, isoniacinamide, lidocaine, methaqualone, morphine, niacinamide, noscapine, papaverine, phenacetin, phenobarbital, phenylpropanolamine, procaine, quinine, salicylic acid, strychnine, thebaine

REFERENCE

Walker,J.A.; Krueger,S.T.; Lurie,I.S.; Marché,H.L.; Newby,N. Analysis of heroin drug seizures by micellar electrokinetic capillary chromatography (MECC), *J.Forensic Sci.*, **1995**, *40*, 6–9.

Hydroquinone

Molecular formula: $C_6H_6O_2$
Molecular weight: 110.11
CAS Registry No.: 123-31-9
Merck Index (12th ed.): 4853

SAMPLE

Matrix: formulations
Sample preparation: 100 mg Cream + 4 mL MeOH + 4 mL 4 mM caffeine + 32 mL water, heat at 50° until the cream was totally dissolved, cool, filter (Millex-GS), inject an aliquot of the filtrate.

CAPILLARY ELECTROPHORESIS

Capillary: 50 cm × 75 µm fused-silica (38 cm to detector) (SGE)

Running buffer: 10 mM pH 9.5 Borate buffer containing 75 mM sodium dodecyl sulfate
Injection: Hydrodynamic injection at 10 cm for 6 s (ca. 10 nL).
Detector: UV 254
Migration time: 3.5
Internal standard: caffeine (4)
Voltage: 10 kV
Current: 20 μA
Model: Laboratory constructed
Limit of quantitation: 50 μM

KEY WORDS
cream

REFERENCE
Sakodinskaya,I.K.; Desiderio,C.; Nardi,A.; Fanali,S. Micellar electrokinetic chromatographic study of hydroquinone and some of its ethers. Determination of hydroquinone in skin-toning cream, *J.Chromatogr.*, **1992**, *596*, 95–100.

SAMPLE
Matrix: solutions

CAPILLARY ELECTROPHORESIS
Capillary: 5.5 cm × 50 μm fused-silica (Polymicro Technologies)
Running buffer: 10 mM pH 10 CAPS (3-(cyclohexylamino)-1-propanesulfonic acid)
Injection: Electrokinetic injection.
Detector: E, Bioanalytical Systems BAS LC-4C, 25 μm gold wire electrode (design in paper) +800 mV, Pt auxiliary electrode, Ag/AgCl reference electrode
Migration time: 0.3
Voltage: 5.5 kV

OTHER SUBSTANCES
Simultaneous: dopamine

REFERENCE
Zhong,M.; Lunte,S.M. Integrated on-capillary electrochemical detector for capillary electrophoresis, *Anal.Chem.*, **1996**, *68*, 2488–2493.

Hydroxychloroquine

Molecular formula: $C_{18}H_{26}ClN_3O$
Molecular weight: 335.88
CAS Registry No.: 118-42-3, 747-36-4 (sulfate)
Merck Index (12th ed.): 4863
Lednicer: 1 342

SAMPLE
Matrix: solutions
Sample preparation: Prepare a solution in running buffer, inject an aliquot.

CAPILLARY ELECTROPHORESIS
Capillary: 60 cm × 75 μm fused-silica (52.4 cm to detector)
Capillary preparation: After each run flush with 500 mM KOH for 2-3 min, flush with water, fill with running buffer
Running buffer: 10 mM Na_2HPO_4 containing 2% heparin sodium (MW 10000, 11% S, Scientific Protein Laboratories, Waunakee, WI) adjusted to pH 5 with phosphoric acid
Injection: Hydrostatic injection
Detector: UV 214

Migration time: 54.0 (first enantiomer (+), R_S 2.31)
Model: Waters Quanta 4000

OTHER SUBSTANCES
Also analyzed: anabasine, brompheniramine, bupivacaine, carbinoxamine, chlorcyclizine, chloroquine, chlorpheniramine, dimethindene, doxylamine, enpiroline, halofantrine, indapamide, mefloquine, nornicotine, pheniramine, primaquine, promethazine, quinacrine, tetramisole

KEY WORDS
chiral

REFERENCE
Stalcup,A.M.; Agyei,N.M. Heparin: a chiral mobile-phase additive for capillary zone electrophoresis, *Anal.Chem.*, **1994**, *66*, 3054–3059.

SAMPLE
Matrix: solutions
Sample preparation: Inject an aliquot of a solution in running buffer.

CAPILLARY ELECTROPHORESIS
Capillary: 60 cm $\times$ 75 μm fused-silica (52.4 cm to detector)
Capillary preparation: After each run flush with 500 mM KOH for 2-3 min then with water.
Running buffer: 10 mM pH 3.8 Phosphate buffer containing 2% sulfated cyclodextrin (ds 7-10)
Injection: Hydrostatic injection.
Detector: UV 214
Migration time: 10.38, 10.75 (enantiomers)
Voltage: 15 kV
Model: Waters Quanta 4000

OTHER SUBSTANCES
Also analyzed: acebutolol, alprenolol, aminoglutethimide, brompheniramine, bupivacaine, bupropion, canadine, carbinoxamine, chloroquine, chlorpheniramine, dimethindene, disopyramide, doxylamine, idazoxan, isoxsuprine, ketamine, mepenzolate, mepivacaine, methoxyphenamine, mexiletine, midodrine, nefopam, orphenadrine, oxprenolol, oxyphencyclimine, pheniramine, phensuximide, pindolol, piperoxan, terbutaline, tetramisole, tolperisone, tranylcypromine, trihexyphenidyl, trimipramine, verapamil, warfarin

KEY WORDS
chiral; detector at anode

REFERENCE
Stalcup,A.M.; Gahm,K.H. Application of sulfated cyclodextrins to chiral separations by capillary zone electrophoresis, *Anal.Chem.*, **1996**, *68*, 1360–1368.

Hydroxyprogesterone

Molecular formula: $C_{21}H_{30}O_3$
Molecular weight: 330.47
CAS Registry No.: 68-96-2
Merck Index (12th ed.): 4886
Lednicer: 1 176

SAMPLE
Matrix: blood
Sample preparation: 100 μL Serum + 900 μL 50 mM pH 4.5 acetate buffer, vortex for 1 min, filter (Amicon MPS-1 micropartition system) while centrifuging, inject an aliquot of the ultrafiltrate.

CAPILLARY ELECTROPHORESIS
Capillary: 37 cm × 50 μm eCAP (37 cm to detector) (Beckman)
Capillary preparation: At the beginning of each day rinse capillary with 100 mM HCl then rinse with running buffer for 10 min.
Capillary temperature: 16 ± 0.1
Running buffer: MeCN:buffer 20:80 (Buffer was 20 mM pH 6.0 4-morpholinopropanesulfonic acid containing 100 mM sodium dodecyl sulfate.)
Injection: High pressure injection for 20 s
Detector: UV 254
Migration time: 4.62
Voltage: 15 kV
Current: <50 μA
Model: Beckman P/ACE 5010
Limit of detection: 1 μg/mL (S/N 3)

OTHER SUBSTANCES
Extracted: 11-deoxycortisol, 21-deoxycortisol, hydrocortisone

KEY WORDS
serum

REFERENCE
Abubaker,M.A.; Bissell,M.G.; Petersen,J.R. Micellar electrokinetic capillary chromatography to separate steroids that are increased in congenital adrenal hyperplasia, *Clin.Chem.*, **1995**, *41*, 1369–1370.

SAMPLE
Matrix: blood
Sample preparation: Treat 200 μL serum with 1% sodium salicylate for 1 h, add to a 20 × 9 ASPEC pak octadecylsilane SPE cartridge ((M&S Instruments Trading Inc., Osaka), wash with 20 mL water, elute with 2 mL MeOH. Evaporate the eluate to dryness under educed pressure, reconstitute with 20 μL running buffer, filter (0.5 μm), inject an aliquot.

CAPILLARY ELECTROPHORESIS
Capillary: 37.5 cm × 75 μm fused-silica
Capillary preparation: Change the running buffer for each run.
Capillary temperature: 25
Running buffer: MeCN:buffer 16:84 (Buffer was 20 mM pH 8.0 sodium tetraborate/potassium phosphate buffer containing 50 mM sodium dodecyl sulfate.)
Injection: Pressure injection at 0.05 kg/cm^2 for 2 s
Detector: UV 205
Migration time: 25
Current: 60 μA (constant current)
Model: Ohtsuka Electronic model CAPI-3000

OTHER SUBSTANCES
Extracted: hydrocortisone, hydroxypregnenolone, nandrolone, pregnenolone

KEY WORDS
serum; SPE

REFERENCE
Kobayashi,Y.; Matsui,J.; Watanabe,F. Simultaneous separation of free and conjugated steroids by micellar electrokinetic chromatography and its clinical application, *Biol.Pharm.Bull.*, **1995**, *18*, 1614–1616.

SAMPLE
Matrix: blood
Sample preparation: Condition a 1 mL 100 mg Bond Elut C18 SPE cartridge with 1 mL MeOH and 1 mL water. Add 500 μL plasma to the cartridge, wash with 2 mL water, wash with 2 mL MeOH:water 20:80, elute with two 500 μL portions of MeOH. Evaporate the eluate to dryness under a stream of nitrogen, reconstitute the residue in 50 μL MeCN:MeOH:20 mM pH 8 Tris-HCl buffer 22.5:22.5:45, inject an aliquot.

CAPILLARY ELECTROPHORESIS
Capillary: 50 cm × 100 µm fused-silica (Polymicro Technologies) packed with 3 µm ODS Hypersil for 20 cm (20 cm to detector) (details in paper)
Running buffer: MeCN:MeOH:20 mM pH 8 Tris-HCl buffer 37.5:37.5:25
Injection: Electrokinetic injection at 15 kV for 5 s (25 nL). (Injector modified for injection from small volume samples; details in paper.)
Detector: UV 240
Migration time: 8.5 (17α), 11.5 (20α)
Internal standard: norethindrone
Voltage: 25 kV (?)
Model: Isco Model 3850

OTHER SUBSTANCES
Extracted: androstenedione, progesterone, testosterone

KEY WORDS
electrochromatography; comparison with HPLC; plasma; SPE

REFERENCE
Stead,D.A.; Reid,R.G.; Taylor,R.B. Capillary electrochromatography of steroids. Increased sensitivity by on-line concentration and comparison with high-performance liquid chromatography, *J.Chromatogr.A*, **1998**, *798*, 259–267.

SAMPLE
Matrix: solutions

CAPILLARY ELECTROPHORESIS
Capillary: 57 cm × 50 µm fused-silica (50 cm to detector) (Polymicro Technologies)
Capillary preparation: After each run rinse capillaries with 100 mM NaOH for 10 min and with water for 5 min. Fill with running buffer and apply voltage for 10 min. Rinse new capillaries with 1 M HCl for 5 min, with 100 mM NaOH for 10 min, and with water for 5 min.
Capillary temperature: 25
Running buffer: MeOH:buffer 20:80 (Buffer was 38 mM pH 7.4 phosphate buffer containing 40 mM sodium cholate.)
Injection: Pressure injection for 5 s (3.1 nL).
Detector: UV (wavelength not given)
Migration time: 14.34
Voltage: 15 kV
Model: Beckman P/ACE 5500

OTHER SUBSTANCES
Simultaneous: aldosterone, dehydroepiandrosterone, estradiol, hydrocortisone, progesterone

REFERENCE
Hsiao,M.-W.; Lin,S.-T. Separation of steroids by micellar electrokinetic capillary chromatography with sodium cholate, *J.Chromatogr.Sci.*, **1997**, *35*, 259–264.

SAMPLE
Matrix: solutions

CAPILLARY ELECTROPHORESIS
Capillary: 67 cm × 50 µm fused-silica (60 cm to detector)
Capillary preparation: Before each run equilibrate with running buffer for 3 min. After each run wash with 100 mM NaOH for 3 min and with water. At the beginning of each day rinse the capillary with 100 mM NaOH for 30 min and with water for 30 min.
Capillary temperature: 25 ± 0.1
Running buffer: 25 mM pH 9 Bicine buffer containing 10 mM sodium dodecyl sulfate and 5 mM N-dodecyl-N,N-dimethyl-3-ammonio-1-propanesulfonate (SB3-12)
Injection: High pressure injection for 5 s
Detector: UV 254
Migration time: 8.07

Voltage: 30 kV
Current: <40 μA
Model: Beckman P/ACE 5010

OTHER SUBSTANCES
Simultaneous: cortisone, 11-desoxycortisol, 21-desoxycortisol, dimethyltestosterone, hydrocortisone, progesterone, testosterone propionate, testosterone

REFERENCE
Valbuena,G.A.; Rao,L.V.; Petersen,J.R.; Okorodudu,A.O.; Bissell,M.G.; Mohammad,A.A. Anionic-zwitterionic mixed micelles in the micellar electrokinetic separation of clinically relevant steroids on a fused-silica capillary, *J.Chromatogr.A*, **1997**, *781*, 467–474.

Hydroxyzine

Molecular formula: $C_{21}H_{27}ClN_2O_2$
Molecular weight: 374.91
CAS Registry No.: 68-88-2, 2192-20-3 (di HCl), 10246-75-0 (pamoate)
Merck Index (12th ed.): 4897
Lednicer: 1 59, 4 118

SAMPLE
Matrix: solutions

CAPILLARY ELECTROPHORESIS
Capillary: 57 cm × 50 μm fused-silica (50 cm to detector)
Running buffer: 50 mM pH 8.0 Borate buffer containing 40 mM sodium taurodeoxycholate and 25 mM phosphatidylcholine (Prepare by adding phosphatidylcholine and stirring for 3-6 h until all cloudiness disappears. Phosphatidylcholine was 95% pure soybean lecithin, Epikuron, Lucas Meyer & Co.)
Injection: Inject a solution of the compound in the running buffer at 20 psi for 1 s, inject MeOH:water 5:95 at 20 psi for 1 s, inject a solution of halofantrine in running buffer at 20 psi for 1 s.
Detector: UV 214
Migration time: k' 77.62
Model: Beckman P/ACE 5000

OTHER SUBSTANCES
Also analyzed: acetaminophen, amoxicillin, antipyrine, aspirin, azathioprine, caffeine, captopril, carbamazepine, carprofen, chlorambucil, chlorpheniramine, chlorpromazine, cimetidine, clonidine, codeine, desipramine, diphenhydramine, ephedrine, fenoterol, flufenamic acid, flurbiprofen, haloperidol, ibuprofen, imipramine, indomethacin, ketoprofen, lidocaine, melphalan, metoprolol, nabumetone, nadolol, phenobarbital, phenol, promazine, propranolol, pyrilamine, ranitidine, ropinirole, salicylic acid, sulfamethoxazole, testosterone, theophylline, thioridazine, tiaprofenic acid, tolfenamic acid, trifluoperazine, trimethoprim, valproic acid, verapamil

KEY WORDS
comparison with HPLC; k' = (Tr-T0)/(T0(1-Tr/Tm)) where Tr = retention time of analyte; T0 = retention time of water; and Tm = retention time of marker (halofantrine)

REFERENCE
Hanna,M.; de Biasi,V.; Bond,B.; Salter,C.; Hutt,A.J.; Camilleri,P. Estimation of the partitioning characteristics of drugs: A comparison of a large and diverse drug series utilizing chromatographic and electrophoretic methodology, *Anal.Chem.*, **1998**, *70*, 2092–2099.

Ibuprofen

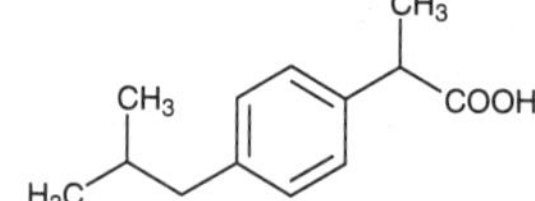

Molecular formula: $C_{13}H_{18}O_2$
Molecular weight: 206.28
CAS Registry No.: 15687-27-1, 58560-75-1 ($\pm$ mixture), 61054-06-6 (Al salt), 112017-99-9 (piconol)
Merck Index (12th ed.): 4925
Lednicer: 1 86; 2 218, 356

SAMPLE
Matrix: blood
Sample preparation: 0.1-1 mL Plasma + 2 µg IS + 400 µL 1 M HCl + 10 mL diethyl ether, agitate mechanically for 10 min, centrifuge at 3500 g for 10 min. Remove the organic layer and evaporate it to dryness under a stream of nitrogen, reconstitute the residue in MeOH:40 mM NaH_2PO_4 50:50, inject an aliquot.

CAPILLARY ELECTROPHORESIS
Capillary: 58.7 cm $\times$ 75 µm fused-silica (50 cm to detector)
Capillary preparation: Wash with running buffer for 2 min before each injection. At the beginning of each day wash capillary with 100 mM NaOH for 10 min.
Capillary temperature: 22
Running buffer: MeOH:buffer 3:97 (Buffer was 40 mM pH 8 NaH_2PO_4 containing 104 mM sodium dodecyl sulfate.)
Injection: Hydrodynamic injection for 2 s
Detector: UV 254
Migration time: 14.5
Internal standard: benzoyl-4-phenyl-2-butyric acid (16.3)
Voltage: 20 kV
Model: Beckman P/ACE 2000
Limit of detection: 10 µg/mL (S/N 5)

OTHER SUBSTANCES
Extracted: diclofenac, diflunisal, etodolac, fenbufen, fenoprofen, flurbiprofen, indomethacin, ketoprofen, naproxen, niflumic acid, piroxicam, sulindac, tenoxicam, tiaprofenic acid
Noninterfering: acetaminophen, amitriptyline, caffeine, clomipramine, deoxysulindac, desipramine, diazepam, 5'-hydroxytenoxicam, imipramine, maprotiline, nortriptyline, phenobarbital, phenytoin, sulfamethoxazole, theophylline, trimipramine

KEY WORDS
plasma

REFERENCE
Maboundou,C.W.; Paintaud,G.; Bérard,M.; Bechtel,P.R. Separation of fifteen non-steroidal anti-inflammatory drugs using micellar electrokinetic capillary chromatography, *J.Chromatogr.B*, **1994**, *657*, 173–183.

SAMPLE
Matrix: blood
Sample preparation: 50 µL Serum + 100 µL 30 µg/mL IS in MeCN, vortex for 30 s, centrifuge at 14000 g for 1 min, inject an aliquot of the supernatant.

CAPILLARY ELECTROPHORESIS
Capillary: 60 cm $\times$ 50 µm
Capillary preparation: Between runs rinse capillary with 200 mM NaOH for 2 min, with water for 1 min, and with running buffer for 2 min. Wash new capillaries with 200 mM NaOH for 30 min and with water for 5 min.
Capillary temperature: 24
Running buffer: 160 mM Boric acid adjusted to pH 8.5 with 2 M NaOH
Injection: Pressure injection for 20 s.
Detector: UV 214
Migration time: 5.6

Internal standard: 3-isobutyl-1-methylxanthine (4.8)
Voltage: 17 kV
Model: Beckman P/ACE 2000
Limit of detection: 8 μg/m

OTHER SUBSTANCES
Simultaneous: acetaminophen, daypro, pentobarbital, phenobarbital, phenytoin, salicylic acid,
theophylline
Noninterfering: bilirubin, uric acid

KEY WORDS
serum

REFERENCE
Shihabi,Z.K.; Hinsdale,M.E. Analysis of ibuprofen in serum by capillary electrophoresis, *J.Chromatogr.B*, **1996**,
683, 115–118.

SAMPLE
Matrix: blood
Sample preparation: 50 μL Plasma + 1 mL 30 μg/mL oxaprozin in MeCN, mix vigorously for
30 s, centrifuge at 14000 g for 1 min, inject an aliquot of the supernatant.

CAPILLARY ELECTROPHORESIS
Capillary: 66 cm × 50 μm fused-silica (62 cm to detector) (Polymicro Technologies)
Capillary preparation: After each run rinse capillary at 100 psi with water for 1 min, with 100
mM NaOH for 1 min, and with running buffer for 3 min.
Capillary temperature: 20
Running buffer: 250 mM Boric acid adjusted to pH 8.5 with 1 M NaOH
Injection: Pressure injection at 5 psi for 0.6 s (5 nL).
Detector: UV 200
Migration time: 9.7
Internal standard: oxaprozin (9)
Voltage: 25 kV
Model: Bio-Rad BioFocus 3000
Limit of quantitation: 1 μg/mL
Limit of detection: 300 ng/mL

KEY WORDS
plasma; rat; comparison with HPLC

REFERENCE
Kang,S.H.; Chang,S.-Y.; Do,K.-C.; Chi,S.-C.; Chung,D.S. High-performance liquid chromatography with a col-
umn-switching system and capillary electrophoresis for the determination of ibuprofen in plasma,
J.Chromatogr.B, **1998**, *712*, 153–160.

SAMPLE
Matrix: formulations
Sample preparation: Dragees. Dissolve dragee in running buffer, add a solution of flurbiprofen
in running buffer, filter (paper), dilute with running buffer to a final concentration of 20 μg/
mL for ibuprofen and 10 μg/mL for flurbiprofen, inject an aliquot. Suspensions. Dilute 1 mL
suspension with running buffer, add a solution of flurbiprofen in running buffer, centrifuge at
3000 rpm for 10 min, dilute the supernatant with running buffer to a final concentration of 50
μg/mL for ibuprofen and 25 μg/mL for flurbiprofen, inject an aliquot.

CAPILLARY ELECTROPHORESIS
Capillary: 60 cm × 75 μm (52.5 cm to detector)
Capillary preparation: Purge with running buffer for 2 min before each injection. Store capil-
lary overnight in water, rinse with 500 mM NaOH, with water, and with running buffer.
Running buffer: 50 mM pH 9.0 Borate buffer containing 40 mM sodium dodecyl sulfate (Mix
2.94 mL 200 mM boric acid, 6.37 mL 75 mM tetraborate solution, and 10 mL 200 mM sodium
dodecyl sulfate, make up to 50 mL with water.)
Injection: Hydrodynamic injection at 10 cm for 5 s

Detector: UV 214
Migration time: 6.4
Internal standard: flurbiprofen (6.7)
Voltage: 15 kV
Model: Waters Quanta 4000 CE
Limit of quantitation: 5 μg/mL

OTHER SUBSTANCES
Simultaneous: benzoic acid

KEY WORDS
dragees; suspensions

REFERENCE
Donato,M.G.; Baeyens,W.; Van den Bossche,W.; Sandra,P. The determination of non-steroidal antiinflammatory drugs in pharmaceuticals by capillary zone electrophoresis and micellar electrokinetic capillary chromatography, *J.Pharm.Biomed.Anal.*, **1994**, *12*, 21–26.

SAMPLE
Matrix: formulations
Sample preparation: Stir formulation in MeOH for 10 min, filter, dilute with 5 mM pH 5 morpholinoethanesulfonic acid, inject an aliquot.

CAPILLARY ELECTROPHORESIS
Capillary: 35 cm × 50 μm fused-silica (31.5 cm to detector) (Polymicro Technologies)
Capillary temperature: 25
Running buffer: 100 mM Morpholinoethanesulfonic acid containing 30 mM heptakis-2,3,6-tri-O-methyl-β-cyclodextrin, adjusted to pH 5 with NaOH
Injection: Pressure injection at 10 psi
Detector: UV 206
Migration time: 29, 33 (enantiomers)
Voltage: 20 kV
Current: 6.6 μA
Model: Biofocus 3000 (Bio-Rad)

OTHER SUBSTANCES
Simultaneous: fenoprofen, flurbiprofen, ketoprofen

KEY WORDS
chiral

REFERENCE
Fanali,S.; Aturki,Z. Use of cyclodextrins in capillary electrophoresis for the chiral resolution of some 2-arylpropionic acid non-steroidal anti-inflammatory drugs, *J.Chromatogr.A*, **1995**, *694*, 297–305.

SAMPLE
Matrix: formulations
Sample preparation: Dissolve one tablet in 25 mL MeCN:40 mM pH 10 borate buffer 9:91, centrifuge, filter (glass fiber), filter (membrane), dilute filtrate 18-fold with running buffer, inject an aliquot.

CAPILLARY ELECTROPHORESIS
Capillary: 48.5 cm × 50 μm fused-silica (40 cm to detector) (Hewlett Packard)
Capillary preparation: Between runs flush with 100 mM NaOH for 5 min, with water for 2 min, and with running buffer for 5 min.
Capillary temperature: 25
Running buffer: MeCN:buffer 9:91 (Buffer was 40 mM pH 10 borate buffer containing 40 mM sodium dodecyl sulfate.)
Injection: Hydrodynamic injection at 5 kPa for 3 s.
Detector: UV 214
Migration time: 4.45

Voltage: 25 kV
Current: 44 μA
Model: Hewlett Packard [3D]CE

OTHER SUBSTANCES
Simultaneous: impurities, codeine, thebaine

KEY WORDS
tablets

REFERENCE
Persson-Stubberud,K.; Åström,O. Separation of ibuprofen, codeine phosphate, their degradation products and impurities by capillary electrophoresis. 1. Method development and optimization with fractional factorial design, *J.Chromatogr.A*, **1998**, *798*, 307–314.

SAMPLE
Matrix: solutions

CAPILLARY ELECTROPHORESIS
Capillary: 42.5 cm × 25 μm
Capillary temperature: 30
Running buffer: 20 mM pH 7.0 Phosphate containing 25 mM sodium dodecyl sulfate
Injection: Vacuum injection for 2 s
Detector: UV 230
Migration time: 4.3
Voltage: 25 kV
Current: 5 μA
Model: Applied Biosystems Model 270A

OTHER SUBSTANCES
Simultaneous: diflunisal, indomethacin, naproxen, sulindac, tolmetin

---

REFERENCE
Weinberger,R.; Albin,M. Quantitative micellar electrokinetic capillary chromatography: Linear dynamic range, *J.Liq.Chromatogr.*, **1991**, *14*, 953–972.

SAMPLE
Matrix: solutions
Sample preparation: Dissolve in MeCN, dilute with water to a concentration of 25 μg/mL, inject an aliquot.

CAPILLARY ELECTROPHORESIS
Capillary: 60 cm × 75 μm fused-silica (52.5 cm to detector)
Running buffer: MeCN:30 mM pH 7.0 phosphate buffer 20:80
Injection: Hydrostatic injection at 9.8 cm for 5 s
Detector: UV 214
Migration time: 21.20
Model: Waters Quanta 4000

OTHER SUBSTANCES
Simultaneous: acemetacin, alclofenac, fenbufen, flurbiprofen, indomethacin, ketoprofen, lonazolac, naproxen, piroxicam, tenoxicam, tiaprofenic acid, tolmetin, voltaren
Interfering: niflumic acid

REFERENCE
Donato,M.G.; Van den Eeckhout,E.; Van den Bossche,W.; Sandra,P. Capillary zone electrophoresis and micellar electrokinetic capillary chromatography of some non-steroidal antiinflammatory drugs (NSAIDs), *J.Pharm.Biomed.Anal.*, **1993**, *11*, 197–201.

SAMPLE
Matrix: solutions

Sample preparation: Prepare a 10 μg/mL solution in MeOH, inject an aliquot.

CAPILLARY ELECTROPHORESIS
Capillary: 97 cm × 50 μm fused-silica (80 cm to detector) (Chrompack), coat with Chirasil-Dex
(J.High Res.Chromatogr. 1991, 14, 58)
Capillary preparation: Rinse with water then operating buffer between each run. Condition
coated columns with running buffer for half a day.
Running buffer: 20 mM pH 7.0 Borate-phosphate buffer
Injection: Hydrostatic injection for 5 s
Detector: UV 220
Migration time: 21, 22 (enantiomers)
Voltage: 30 kV
Model: Kapillar-Elektrophorese-System 100 (Grom)

OTHER SUBSTANCES
Simultaneous: cicloprofen, flurbiprofen

KEY WORDS
chiral; coated capillary

REFERENCE
Mayer,S.; Schurig,V. Enantiomer separation by electrochromatography in open tubular columns coated with
Chirasil-Dex, *J.Liq.Chromatogr.*, **1993**, *16*, 915–931.

SAMPLE
Matrix: solutions

CAPILLARY ELECTROPHORESIS
Capillary: 60 cm × 50 μm fused-silica
Capillary preparation: Before each run purge with running buffer for 3 min. At the beginning
of each day vacuum purge with 500 mM NaOH, vacuum purge with water, perform an elec-
troosmotic purge, vacuum purge with running buffer.
Running buffer: 10 mM pH 7 Tris-phosphate buffer containing 2.5% Glucidex2 (maltodextrin
preparation, Roquette, Lestrem, France)
Injection: Injection by siphon at 10 cm
Detector: UV 185
Migration time: 5.1, 5.2 (enantiomers)
Voltage: 30 kV
Model: Waters Quanta 4000 CE

OTHER SUBSTANCES
Simultaneous: fenoprofen
Interfering: indoprofen, suprofen

KEY WORDS
chiral

REFERENCE
D'Hulst,A.; Verbeke,N. Quantitation in chiral capillary electrophoresis: theoretical and practical considerations,
Electrophoresis, **1994**, *15*, 854–863.

SAMPLE
Matrix: solutions
Sample preparation: Prepare an aqueous solution, filter (0.45 μm), inject an aliquot.

CAPILLARY ELECTROPHORESIS
Capillary: 36 cm × 50 μm (31.5 cm to detector)
Capillary preparation: Before each run rinse with water for 1 min and with running buffer for
1 min.
Capillary temperature: 25

Running buffer: Isopropanol:buffer 10:90 (Buffer was 50 mM pH 6.0 phosphate buffer containing 25 μm avidin.)
Injection: Inject at 1 psi pressure for 2 s
Detector: UV 230
Migration time: 11.6, 12 (enantiomers)
Voltage: -12 kV
Model: BioFocus 3000 (Bio-Rad)

KEY WORDS
chiral

REFERENCE
Tanaka,Y.; Matsubara,N.; Terabe,S. Separation of enantiomers by affinity electrokinetic chromatography using avidin, *Electrophoresis*, **1994**, *15*, 848–853.

SAMPLE
Matrix: solutions
Sample preparation: Inject an aliquot of a 100 μg/mL solution.

CAPILLARY ELECTROPHORESIS
Capillary: 27 cm × 50 μm neutrally coated (20 cm to detector)
Capillary preparation: Before each run rinse capillary with 100 mM HCl for 30 s, with water for 2 min, and with running buffer for 2 min
Capillary temperature: 20
Running buffer: 200 mM pH 4.6 2-(N-morpholino)ethanesulfonic acid buffer containing 15 mM β-cyclodextrin
Injection: Pressure injection at 0.5 psi for 5 s
Detector: UV 230
Migration time: 9.9, 10.1 (enantiomers)
Voltage: 500 V/cm
Current: 13 μA
Model: Beckman P/ACE system 5500

KEY WORDS
chiral; detector is at anode; coated capillary

REFERENCE
Guttman,A. Novel separation scheme for capillary electrophoresis of enantiomers, *Electrophoresis*, **1995**, *16*, 1900–1905.

SAMPLE
Matrix: solutions
Sample preparation: Inject an aliquot of a 100 μg/mL solution in water or running buffer.

CAPILLARY ELECTROPHORESIS
Capillary: 47 cm × 75 μm fused-silica (40 cm to detector)
Capillary preparation: Before each run rinse capillary with running buffer for 1-2 min. If peak tailing is observed fill capillary with 500 mM NaOH and let stand for 30 min, wash with water for 3 min, rinse with running buffer for 3 min.
Capillary temperature: 20
Running buffer: 20 mM pH 7.0 Phosphate buffer containing 6% dextrin (Japan Pharmacopeia grade) (Dissolve dextrin in buffer at 90° then cool to room temperature.)
Injection: Pressure injection at 0.5 psi for 2-4 s.
Detector: UV 220
Migration time: 12.566, 13.055 (enantiomers)
Voltage: 20 kV
Model: Beckman P/ACE 5510

KEY WORDS
chiral

REFERENCE

Nishi,H.; Izumoto,S.; Nakamura,K.; Nakai,H.; Sato,T. Dextran and dextrin as chiral selectors in capillary zone electrophoresis, *Chromatographia*, **1996**, *42*, 617–630.

SAMPLE
Matrix: solutions

CAPILLARY ELECTROPHORESIS
Capillary: 35 cm × 50 µm coated capillary (30.5 cm to detector) (Composite Metal Services, UK)
Capillary preparation: Before each run purge at high pressure with water for 100 s and with running buffer for 120 s. If vancomycin is used, before each run purge at high pressure with water for 100 s and with running buffer not containing vancomycin for 120 s. Next purge with running buffer containing vancomycin at low pressure (175 psi.s) and inject sample. Coat capillary as follows. Adjust the pH of 20 mL water to 3.5 with acetic acid, add 80 µL 3-(trimethoxysilyl)propyl methacrylate (3-methacryloxypropyltrimethoxysilane), mix, suck into capillary, let stand at room temperature for 1 h, remove the solution, wash with water. Fill the capillary with a deaerated 3-4% acrylamide solution containing 1 µL/mL N,N,N',N'-tetramethylethylenediamine and 1 mg/mL potassium persulfate, let stand for 30 min, remove excess solution by aspiration, rinse with water, remove water by aspiration, dry at 35° (J. Chromatogr. 1985, 347, 191).
Capillary temperature: 25
Running buffer: Buffer containing 30 mM heptakis-2,3,6-tri-O-methyl-β-cyclodextrin (A) or 5 mM vancomycin (B) (Buffer was 50 mM phosphoric acid containing 50 mM acetic acid and 50 mM boric acid, dilute with an equal volume of water, adjust pH to 5 with concentrated NaOH.)
Injection: Pressure injection at 10 psi.
Detector: UV 206
Migration time: 31.0, 34.9 (A) or 7.3, 7.8 (B)
Voltage: -20 kV
Model: Bio-Rad Biofocus 3000
Limit of detection: 5 µM

KEY WORDS
chiral; coated capillary; detector at anode

REFERENCE

Fanali,S.; Desiderio,C.; Aturki,Z. Enantiomeric resolution study by capillary electrophoresis. Selection of the appropriate chiral selector, *J.Chromatogr.A*, **1997**, *772*, 185–194.

SAMPLE
Matrix: solutions
Sample preparation: Inject an aliquot of a solution in MeOH:water 10:90.

CAPILLARY ELECTROPHORESIS
Capillary: 44 cm × 50 µm fused-silica (37 cm to detector) (Supelco)
Capillary preparation: Wash with running buffer for 3 min after each injection. Wash with running buffer for 10 min at the end of each day.
Capillary temperature: 25
Running buffer: 100 mM Phosphoric acid containing 5 mM sulfobutyl ether-β-cyclodextrin and 30 mM heptakis(2,3,6-tri-O-methyl)-β-cyclodextrin, adjusted to pH 3.0 with triethanolamine
Injection: Hydrodynamic injection for 5 s (13.3 nL).
Detector: UV 210
Voltage: -25 kV
Current: 60 µA
Model: SpectraPhoresis 1000 CE

KEY WORDS
chiral; detector at anode; resolution (R_s = 5.6)

REFERENCE

Fillet,M.; Hubert,P.; Crommen,J. Enantioseparation of nonsteroidal anti-inflammatory drugs by capillary electrophoresis using mixtures of anionic and uncharged β-cyclodextrins as chiral additives, *Electrophoresis*, **1997**, *18*, 1013–1018.

SAMPLE
Matrix: solutions

CAPILLARY ELECTROPHORESIS
Capillary: 35 cm × 50 μm polyacrylamide-coated fused-silica (30.5 cm to detector) (Composite Metal Services, UK)
Capillary preparation: Before each run rinse capillary with water for 70 s and with running buffer for 100 s. (Coat capillary as follows. Adjust the pH of 20 mL water to 3.5 with acetic acid, add 80 μL 3-(trimethoxysilyl)propyl methacrylate (3-methacryloxypropyltrimethoxysilane), mix, suck into capillary, let stand at room temperature for 1 h, remove the solution, wash with water. Fill the capillary with a deaerated 3-4% acrylamide solution containing 1 μL/mL N,N,N',N'-tetramethylethylenediamine and 1 mg/mL potassium persulfate, let stand for 30 min, remove excess solution by aspiration, rinse with water, remove water by aspiration, dry at 35° (J. Chromatogr. 1985, 347, 191).)
Capillary temperature: 25
Running buffer: Buffer containing 2.5 mM cyanoethylated-β-cyclodextrin (Cyclolab, Budapest) (Prepare buffer by adjusting the pH of 50 mM phosphoric acid containing 50 mM acetic acid and 50 mM boric acid to 5 with concentrated NaOH, add the appropriate amount of cyanoethylated-β-cyclodextrin, dilute with an equal volume of water.)
Injection: Pressure injection at 5 psi for 2 s.
Detector: UV 206
Migration time: 14.2 (second enantiomer, α = 1.017)
Voltage: 20 kV
Current: 27-40 μA
Model: Bio-Rad Biofocus 3000

KEY WORDS
detector at anode; chiral; coated capillary

REFERENCE
Aturki,Z.; Desiderio,C.; Mannina,L.; Fanali,S. Chiral separations by capillary zone electrophoresis with the use of cyanoethylated-β-cyclodextrin as chiral selector, *J.Chromatogr.A*, **1998**, *817*, 91–104.

SAMPLE
Matrix: solutions

CAPILLARY ELECTROPHORESIS
Capillary: 64.5 cm × 50 μm fused-silica (56 cm to detector) (Hewlett-Packard)
Capillary preparation: Before each run flush with 100 mM NaOH for 3 min and with running buffer at 8 min then equilibrate at 20 kV for 6 min. At the start of each batch flush capillary with 1 M NaOH for 10 min, with 100 mM NaOH for 10 min, and with running buffer for 15 min then equilibrate at 20 kV for 20 min.
Capillary temperature: 35
Running buffer: 20 mM Phosphoric acid containing 20 mM triethanolamine and 50 mM 2,3,6-tri-O-methyl-β-cyclodextrin, adjusted to pH 5.0 with NaOH
Injection: Pressure injection at 75 mbar (5 nL).
Detector: UV 253
Migration time: 13.2, 13.8 (enantiomers)
Voltage: 20 kV
Current: 8.1 μA
Model: Hewlett-Packard [3D]CE

OTHER SUBSTANCES
Simultaneous: fenoprofen, ketoprofen

KEY WORDS
chiral

REFERENCE
Blanco,M.; Coello,J.; Iturriaga,H.; Maspoch,S.; Pérez-Maseda,C. Separation of profen enantiomers by capillary electrophoresis using cyclodextrins as chiral selectors, *J.Chromatogr.A*, **1998**, *793*, 165–175.

SAMPLE
Matrix: solutions
Sample preparation: Inject an aliquot of a 50-100 μg/mL solution in 50 mM pH 4.8 ammonium acetate buffer.

CAPILLARY ELECTROPHORESIS
Capillary: 44 cm × 50 μm polyacrylamide coated
Capillary preparation: Adjust the pH of 20 mL water to 3.5 with acetic acid, add 80 μL 3-(trimethoxysilyl)propyl methacrylate (3-methacryloxypropyltrimethoxysilane), mix, suck into capillary, let stand at room temperature for 1 h, remove the solution, wash with water. Fill the capillary with a deaerated 3-4% acrylamide solution containing 1 μL/mL N,N,N',N'-tetramethylethylenediamine and 1 mg/mL potassium persulfate, let stand for 30 min, remove excess solution by aspiration, rinse with water, remove water by aspiration, dry at 35° (J. Chromatogr. 1985, 347, 191).
Running buffer: 50 mM pH 4.8 Ammonium acetate buffer containing 5 mM vancomycin
Injection: Hydrostatic injection at 10 cm for 5-10 s.
Detector: MS, Finnigan LCQ ion trap, electrospray, negative ion mode, probe tip at 2.6 kV, sheath liquid MeOH:water:ammonia 50:48:2 at 6 μL/min, m/z 205 (Vancomycin migrates away from the detector.)
Migration time: 12.63, 13.36 (enantiomers)
Voltage: 20 kV
Model: Grom 100

OTHER SUBSTANCES
Also analyzed: carprofen, flurbiprofen, ketoprofen, naproxen

KEY WORDS
chiral; coated capillary

REFERENCE
Fanali,S.; Desiderio,C.; Schulte,G.; Heitmeier,S.; Strickmann,D.; Chankvetadze,B.; Blaschke,G. Chiral capillary electrophoresis-electrospray mass spectrometry coupling using vancomycin as chiral selector, *J.Chromatogr.A*, **1998**, *800*, 69–76.

SAMPLE
Matrix: solutions

CAPILLARY ELECTROPHORESIS
Capillary: 57 cm × 50 μm fused-silica (50 cm to detector)
Running buffer: 50 mM pH 8.0 Borate buffer containing 40 mM sodium taurodeoxycholate and 25 mM phosphatidylcholine (Prepare by adding phosphatidylcholine and stirring for 3-6 h until all cloudiness disappears. Phosphatidylcholine was 95% pure soybean lecithin, Epikuron, Lucas Meyer & Co.)
Injection: Inject a solution of the compound in the running buffer at 20 psi for 1 s, inject MeOH:water 5:95 at 20 psi for 1 s, inject a solution of halofantrine in running buffer at 20 psi for 1 s.
Detector: UV 214
Migration time: k' 2.00
Model: Beckman P/ACE 5000

OTHER SUBSTANCES
Also analyzed: acetaminophen, amoxicillin, antipyrine, aspirin, azathioprine, caffeine, captopril, carbamazepine, carprofen, chlorambucil, chlorpheniramine, chlorpromazine, cimetidine, clonidine, codeine, desipramine, diphenhydramine, ephedrine, fenoterol, flufenamic acid, flurbiprofen, haloperidol, hydroxyzine, imipramine, indomethacin, ketoprofen, lidocaine, melphalan, metoprolol, nabumetone, nadolol, phenobarbital, phenol, promazine, propranolol, pyrilamine, ranitidine, ropinirole, salicylic acid, sulfamethoxazole, testosterone, theophylline, thioridazine, tiaprofenic acid, tolfenamic acid, trifluoperazine, trimethoprim, valproic acid, verapamil

KEY WORDS
comparison with HPLC; k' = (Tr-T0)/(T0(1-Tr/Tm)) where Tr = retention time of analyte; T0 = retention time of water; and Tm = retention time of marker (halofantrine)

REFERENCE
Hanna,M.; de Biasi,V.; Bond,B.; Salter,C.; Hutt,A.J.; Camilleri,P. Estimation of the partitioning characteristics of drugs: A comparison of a large and diverse drug series utilizing chromatographic and electrophoretic methodology, *Anal.Chem.*, **1998**, *70*, 2092–2099.

SAMPLE
Matrix: solutions

CAPILLARY ELECTROPHORESIS
Capillary: 70 cm × 50 μm fused-silica (GL Science, Tokyo)
Capillary preparation: Before each run rinse capillary with running buffer at 94 kPa for 5 min before each run.
Running buffer: 50 mM pH 8.5 ammonium carbonate buffer
Injection: Pressure injection at 5 kPa (50 mbar) for 4 s.
Detector: MS, Perkin-Elmer Sciex API-300 quadrupole, electrospray (ionspray) interface, sheath liquid MeOH:running buffer 50:50 at 2.5 μL/min, ionspray voltage 5 kV, negative ion mode
Migration time: 10.6
Voltage: 20 kV (net voltage across capillary = 15 kV (applied voltage - electrospray voltage))
Model: Hewlett Packard 3D CE

OTHER SUBSTANCES
Simultaneous: acetaminophen, ascorbic acid, butylscopolamine bromide, caffeine, ketoprofen, niacin, niacinamide, riboflavin, thiamine, vitamin B12, warfarin

REFERENCE
Tanaka,Y.; Kishimoto,Y.; Otsuka,K.; Terabe,S. Strategy for selecting separation solutions in capillary electrophoresis-mass spectrometry, *J.Chromatogr.A*, **1998**, *817*, 49–57.

SAMPLE
Matrix: urine
Sample preparation: Condition a 100 mg Bond Elut C18 SPE cartridge with two 1 mL portions of MeCN and two 1 mL portions of 1% acetic acid. Add 1 mL urine to the SPE cartridge, wash with 500 μL MeCN:1% acetic acid 10:90, elute with 500 μL MeCN. Evaporate the eluate to dryness, reconstitute the residue in 500 μL MeCN:water 25:75, inject an aliquot.

CAPILLARY ELECTROPHORESIS
Capillary: 64.5 cm × 50 μm fused-silica (56 cm to detector) (Polymicro Technologies)
Capillary preparation: Between runs flush capillary with running buffer for 3 min. Before use rinse capillary with 1 M NaOH for 1 h, with 100 mM NaOH for 20 min, with water for 20 min, and with running buffer for 1 h.
Capillary temperature: 25
Running buffer: 100 mM 2-[N-morpholino]ethanesulfonic acid containing 0.01% hexadimethrine bromide, 10% dextrin 10 (Fluka), and 20 mM heptakis(2,3,6-tri-O-methyl)-β-cyclodextrin, adjusted to pH 5.26 with 1 M NaOH
Injection: Pressure injection at 5 kPa for 3 s.
Detector: UV 214
Migration time: 24.5 (S), 25.5 (R)
Voltage: 30 kV
Model: Hewlett-Packard HP[3D]

OTHER SUBSTANCES
Extracted: metabolites

KEY WORDS
SPE; chiral

REFERENCE
Bjornsdottir,I.; Kepp,D.R.; Tjornelund,J.; Hansen,S.H. Separation of the enantiomers of ibuprofen and its major phase I metabolites in urine using capillary electrophoresis, *Electrophoresis*, **1998**, *19*, 455–460.

Idarubicin

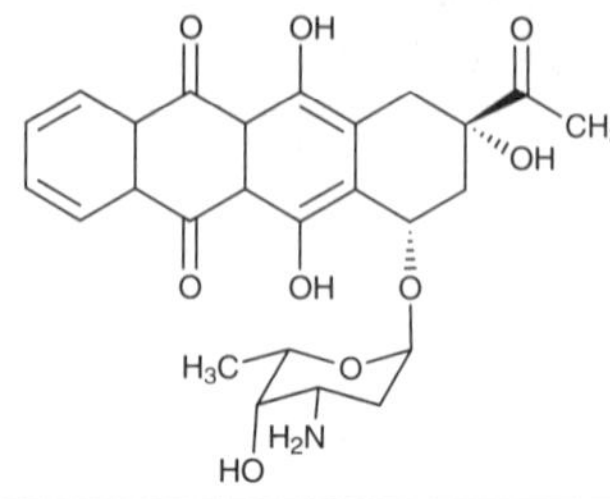

Molecular formula: $C_{26}H_{27}NO_9$
Molecular weight: 497.50
CAS Registry No.: 58957-92-9, 57852-57-0 (HCl)
Merck Index (12th ed.): 4931
Lednicer: 5 47

SAMPLE

Matrix: blood
Sample preparation: 100 μL Plasma + 100 μL 100 mM pH 7.4 sodium phosphate buffer + 50 μL 20 ng/mL daunorubicin in MeCN + 1 mL chloroform, vortex for 1 min, centrifuge at 1500 g for 5 min. Remove 800 μL of the organic layer and evaporate it to dryness under a stream of nitrogen at 35°, reconstitute the residue in 50 μL MeCN:water 95:5, inject an aliquot.

CAPILLARY ELECTROPHORESIS

Capillary: 47 cm × 50 μm fused-silica (40 cm to detector) (Beckman)
Capillary preparation: Between runs rinse capillary with 100 mM NaOH for 1 min and with running buffer for 2 min.
Running buffer: MeCN:buffer 70:30 (Prepare running buffer by adjusting the pH of 100 mM NaH_2PO_4 to 5.0 with 100 mM phosphoric acid, add spermine to a final concentration of 60 μM, add MeCN to a final MeCN concentration of 70%.)
Injection: Electrokinetic injection at 10 kV for 5 s.
Detector: F ex 488 (5 mW Beckman argon ion laser) em 520
Migration time: 4
Internal standard: daunorubicin (4.2)
Voltage: 25 kV
Current: 12 μA
Model: Beckman P/ACE 5510
Limit of quantitation: 0.5 ng/mL

OTHER SUBSTANCES

Extracted: metabolites, idarubicinol

KEY WORDS

plasma; pharmacokinetics

REFERENCE

Hempel,G.; Haberland,S.; Schulze-Westhoff,P.; Möhling,N.; Blaschke,G.; Boos,J. Determination of idarubicin and idarubicinol in plasma by capillary electrophoresis, *J.Chromatogr.B*, **1997**, *698*, 287–292.

Idazoxan

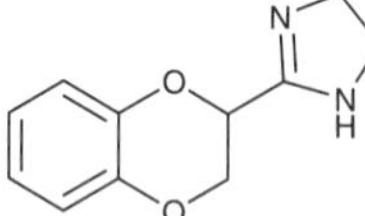

Molecular formula: $C_{11}H_{12}N_2O_2$
Molecular weight: 204.23
CAS Registry No.: 79944-58-4

SAMPLE

Matrix: solutions
Sample preparation: Inject an aliquot of a solution in running buffer.

CAPILLARY ELECTROPHORESIS

Capillary: 60 cm × 75 μm fused-silica (52.4 cm to detector)
Capillary preparation: After each run flush with 500 mM KOH for 2-3 min then with water.
Running buffer: 10 mM pH 3.8 Phosphate buffer containing 2% sulfated cyclodextrin (ds 7-10)

Injection: Hydrostatic injection.
Detector: UV 214
Migration time: 8.83, 9.15 (enantiomers)
Voltage: 15 kV
Model: Waters Quanta 4000

OTHER SUBSTANCES
Also analyzed: acebutolol, alprenolol, aminoglutethimide, brompheniramine, bupivacaine, bupropion, canadine, carbinoxamine, chloroquine, chlorpheniramine, dimethindene, disopyramide, doxylamine, hydroxychloroquine, isoxsuprine, ketamine, mepenzolate, mepivacaine, methoxyphenamine, mexiletine, midodrine, nefopam, orphenadrine, oxprenolol, oxyphencyclimine, pheniramine, phensuximide, pindolol, piperoxan, terbutaline, tetramisole, tolperisone, tranylcypromine, trihexyphenidyl, trimipramine, verapamil, warfarin

KEY WORDS
chiral; detector at anode

REFERENCE
Stalcup,A.M.; Gahm,K.H. Application of sulfated cyclodextrins to chiral separations by capillary zone electrophoresis, *Anal.Chem.*, **1996**, *68*, 1360–1368.

Imipramine

Molecular formula: $C_{19}H_{24}N_2$
Molecular weight: 280.41
CAS Registry No.: 50-49-7, 113-52-0 (HCl)
Merck Index (12th ed.): 4955
Lednicer: 1 401; 2 420; 3 32; 4 146 201 203

SAMPLE
Matrix: blood
Sample preparation: 1 mL Serum or plasma + 50 µL 10 µL/mL trimipramine + 1 mL 2 M NaOH + 5 mL hexane:isoamyl alcohol 99:1, vortex vigorously, centrifuge at 2500 g for 5 min. Remove 4 mL of the organic layer and add it to 100 µL 100 mM HCl, extract. Remove 60 µL of the aqueous layer and evaporate it to dryness under reduced pressure, reconstitute with 20 µL water, inject an aliquot.

CAPILLARY ELECTROPHORESIS
Capillary: 71.5 cm × 50 µm fused-silica (50 cm to detector) (Polymicro Technologies)
Capillary preparation: Before each run rinse capillary with 1 M NaOH for 3 min and running buffer for 3 min (use vacuum at 508 mm Hg).
Capillary temperature: 30 or 40
Running buffer: 37.5 mM pH 8.0 Phosphate buffer containing 2 M urea and 25 mM dodecyltrimethylammonium bromide
Injection: Vacuum injection at 508 mm Hg for 7 s
Detector: UV 254
Migration time: 10.7
Internal standard: trimipramine (11.5)
Voltage: -25 kV
Model: Applied Biosystems Model 270A
Limit of detection: 5-10 ng/mL

OTHER SUBSTANCES
Extracted: amitriptyline, desipramine, doxepin, nortriptyline

KEY WORDS
serum; plasma

REFERENCE
Lee,K.-J.; Lee,J.J.; Moon,D.C. Determination of tricyclic antidepressants in human plasma by micellar electro-kinetic capillary chromatography, *J.Chromatogr.*, **1993**, *616*, 135–143.

SAMPLE
Matrix: solutions
Sample preparation: Inject an aliquot of a 40 μg/mL solution in MeOH.

CAPILLARY ELECTROPHORESIS
Capillary: 64 cm × 50 μm fused silica (55.5 cm to detector) (Polymicro Technologies)
Capillary preparation: Flush with running buffer for 2 min between runs. Before use rinse with 1 M NaOH for 1 h, with 100 mM NaOH for 20 min, with water for 20 min, and with running buffer for 10 min.
Capillary temperature: 25
Running buffer: MeCN:MeOH 50:50 containing 25 mM ammonium acetate and 100 mM sodium acetate
Injection: Pressure injection at 5 kPa for 3 s
Detector: UV 214
Migration time: 8
Voltage: 25 kV
Model: Hewlett-Packard HP3D

OTHER SUBSTANCES
Simultaneous: amitriptyline, desmethylimipramine, didesmethylimipramine, imipramine-N-oxide, litracene, maprotiline, methylimipramine, nortriptyline, protriptyline

REFERENCE
Bjornsdottir,I.; Hansen,S.H. Comparison of separation selectivity in aqueous and non-aqueous capillary electrophoresis, *J.Chromatogr.A*, **1995**, *711*, 313–322.

SAMPLE
Matrix: solutions
Sample preparation: Prepare a 200 μg/mL solution in water, inject an aliquot.

CAPILLARY ELECTROPHORESIS
Capillary: 56 cm × 75 μm fused-silica (56 cm to detector) (Polymicro Technologies)
Capillary temperature: 30
Running buffer: 50 mM pH 4.0 6-Aminocaproic acid containing 25 mM 3-(N,N-dimethyl-myristylammonium)propanesulfonate
Injection: Pressure injection at 2 kPa for 3 s
Detector: UV 214
Migration time: 15
Voltage: 20 kV
Current: 62 μA
Model: Waters Quanta 4000

OTHER SUBSTANCES
Simultaneous: degradation products, impurities

REFERENCE
Hansen,S.H.; Bjornsdottir,I.; Tjornelund,J. Separation of basic drug substances of very similar structure using micellar electrokinetic chromatography, *J.Pharm.Biomed.Anal.*, **1995**, *13*, 489–495.

SAMPLE
Matrix: solutions

CAPILLARY ELECTROPHORESIS
Capillary: 55.5 cm × 50 μm fused-silica (55.5 cm to detector) (Polymicro Technologies)
Capillary temperature: 25
Running buffer: MeCN containing 25 mM ammonium acetate and 1 M acetic acid
Injection: Pressure injection at 5 kPa for 3 s.

Detector: UV 214
Migration time: 4.5
Voltage: 25 kV
Current: 7 μA
Model: Hewlett Packard HP ^{3D}CE

OTHER SUBSTANCES
Simultaneous: degradation products

REFERENCE
Bjornsdottir,I.; Tjornelund,J.; Hansen,S.H. Nonaqueous capillary electrophoresis in pharmaceutical analysis, *J.Capillary Electrophor.*, **1996**, *3*, 83–87.

SAMPLE
Matrix: solutions

CAPILLARY ELECTROPHORESIS
Capillary: 70 cm × 50 μm coated fused-silica (55 cm to detector) (Polymicro Technologies)
Capillary preparation: Between runs rinse with water for 2 min. Coat capillaries as follows. Condition with 1 M NaOH for 15 min, rinse with water for 5 min, treat with 30 μL/mL 3-(trimethoxysilyl)propyl methacrylate in acetic acid:water 50:50 for 1 h using house vacuum, rinse with water, fill with polymerization solution, let stand for 1 h, flush, rinse with water. (Prepare polymerization solution by adding 10 μL N,N,N',N'-tetramethylethylenediamine to 10 mL of a degassed 4% solution of acrylamide in water, add 10 μL 10% ammonium persulfate in water.)
Running buffer: 10 mM pH 6.5 Citrate buffer containing 20 mM sodium dodecyl sulfate and 10 mM carboxymethyl-β-cyclodextrin (Cyclodextrin Technologies Development, Gainesville FL)
Injection: Hydrodynamic injection at 15 cm for 8-10 s.
Detector: UV 254
Migration time: 9.8
Voltage: 20 kV
Model: laboratory-constructed

OTHER SUBSTANCES
Simultaneous: amitriptyline, carbamazepine, clomipramine, desipramine, nortriptyline, opipramol, proptriptyline, trimipramine

KEY WORDS
coated capillary

REFERENCE
Spencer,B.J.; Zhang,W.; Purdy,W.C. Capillary electrophoretic separation of tricyclic antidepressants using charged carboxymethyl-β-cyclodextrin as a buffer additive, *Electrophoresis*, **1997**, *18*, 736–744.

SAMPLE
Matrix: solutions

CAPILLARY ELECTROPHORESIS
Capillary: 45 cm × 50 μm fused-silica (37.5 cm to detector) (Polymicro Technologies)
Capillary preparation: Between each run rinse with 100 mM HCl for 2 min and with running buffer for 2 min. At the start of each day condition capillary with 1 M HCl for 10 min. Before the first use rinse capillary with 1 M NaOH for 30 min and with water for 30 min.
Capillary temperature: 25
Running buffer: MeCN:30 mM pH 2.40 ethanesulfonic acid 20:80
Injection: Injection at 4 kV for 3 s.
Detector: UV 214
Migration time: 5.35
Voltage: 20 kV
Model: Waters Quanta 4000E

OTHER SUBSTANCES
Simultaneous: chloramphenicol, lidocaine, phenylpropanolamine, procainamide, quinidine, trimethoprim

REFERENCE
Ding,W.; Fritz,J.S. Separation of neutral compounds and basic drugs by capillary electrophoresis in acidic solution using laurylpoly(oxyethylene) sulfate as an additive, *Anal.Chem.*, **1998**, *70*, 1859–1865.

SAMPLE
Matrix: solutions

CAPILLARY ELECTROPHORESIS
Capillary: 57 cm × 50 μm fused-silica (50 cm to detector) (Polymicro Technologies)
Capillary preparation: At the start of each day rinse capillary with water for 5 min, 100 mM NaOH for 30 min, with water for 10 min, and with running buffer for 15 min. Wash new capillaries with 1 M NaOH for 1 h.
Capillary temperature: 23
Running buffer: MeOH:50 mM pH 9.55 sodium phosphate buffer containing 0.06% poly(n-undecyl-α-D-glucopyranoside) (Prepare poly(n-undecyl-α-D-glucopyranoside) as follows. With exclusion of atmospheric moisture stir 40 mL alcohol-free chloroform, 5.2 g yellow mercuric oxide, 0.1 g mercuric bromide, 10 g anhydrous calcium sulfate, and 2.74 g 10-undecen-1-ol then add 10 g acetobromo-α-D-glucose (Fluka), stir at room temperature for 20 h, filter, wash the solid with chloroform. Wash the filtrate with aqueous 1 M KBr until no mercury salts are present in the aqueous layers, wash with later, concentrate. Take up the residue in benzene (Caution! Benzene is a carcinogen!) and filter it through activated silica, crystallize from n-hexane to obtain undecylenyl-2,3,4,6-tetra-O-acetyl-β-D-glucopyranoside. Add 2 mL of a 1 M solution of sodium methoxide in MeOH to a solution of 1 g of undecylenyl-2,3,4,6-tetra-O-acetyl-β-D-glucopyranoside in 50 mL dry MeOH, let stand until the reaction is complete (by TLC) to obtain undecylenyl-β-D-glucopyranoside (cf. Tenside Detergents 1978, 15, 72). Use Dowex 50W (H⁺) resin to deionize the final product in solution. Polymerize the monomer by irradiation of an 870 μM solution in MeOH:water 20:80 with ⁶⁰Co gamma radiation for 48 h, purify the polymer solution by dialysis using a 1000 Da molecular mass cut-off membrane.)
Injection: Pressure injection at 0.5 psi for 5 s.
Detector: UV 214
Migration time: 12.5
Voltage: 22.8 kV
Model: Beckman P/ACE 5510

OTHER SUBSTANCES
Simultaneous: amitriptyline, desipramine, doxepin, nordoxepin, nortriptyline, protriptyline

REFERENCE
Harrell,C.W.; Dey,J.; Shamsi,S.A.; Foley,J.P.; Warner,I.M. Enhanced separation of antidepressant drugs using a polymerized nonionic surfactant as a transient capillary coating, *Electrophoresis*, **1998**, *19*, 712–718.

SAMPLE
Matrix: solutions
Sample preparation: Mix 50 μL of a solution in 1% sodium chloride with 100 μL MeCN, vortex for 15 s, centrifuge at 14000 g for 20 s, inject an aliquot.

CAPILLARY ELECTROPHORESIS
Capillary: 40 cm × 50 μm silica
Running buffer: MeCN:isopropanol:buffer 10:10:80 (Buffer was 180 mM triethanolamine, pH adjusted to 8.2.)
Injection: Hydrodynamic injection at low pressure (to fill 6% of the capillary).
Detector: UV 254
Migration time: 5
Voltage: 12 kV
Model: Beckman Model 2000 CE

OTHER SUBSTANCES
Simultaneous: amitriptyline, procainamide, quinine

REFERENCE
Shihabi,Z.K. Stacking of weakly cationic compounds by acetonitrile for capillary electrophoresis, *J.Chromatogr.A*, **1998**, *817*, 25–30.

SAMPLE
Matrix: urine
Sample preparation: 10 mL Urine + 1 mL 5 M NaOH + 10 mL n-hexane, vortex for 1 min, centrifuge at 0° at 3000 g for 5 min. Remove the organic layer and add it to 50 µL glacial acetic acid, evaporate to dryness under a stream of nitrogen at 30°, reconstitute the residue in 50 µL 50 mM sodium taurodeoxycholate, filter (0.2 µm PTFE), inject an aliquot.

CAPILLARY ELECTROPHORESIS
Capillary: 100 cm × 50 µm fused-silica (50 cm to detector) (ISCO)
Capillary preparation: Rinse capillary with 20 µL running buffer between injections. Condition a new capillary by filling with 1 M NaOH, let stand for 1 h, fill with 100 mM NaOH, let stand for 1 h, rinse with buffer. Every 20 injections rinse capillary with 200 µL 1 M NaOH, 200 µL water, and 200 µL running buffer, fill with running buffer.
Capillary temperature: 22
Running buffer: 40 mM pH 9.5 Borate buffer containing 10 mM sodium taurodeoxycholate
Injection: Load under vacuum at 7.5 kPa/s.
Detector: UV 240
Migration time: 9
Voltage: 30 kV
Model: ISCO Model 3140 electropherograph
Limit of detection: 7 ng/mL

OTHER SUBSTANCES
Extracted: acepromazine, amiodarone, amitriptyline, azaperone, chlorpromazine, cianopramine, clomipramine, clozapine, desethylamiodarone, desipramine, diclofensine, dothiepin, doxepin, isocarboxazid, moclobemide, perphenazine, phenothiazine, pimozide, prochlorperazine, promazine, thioridazine, thiothixene, trifluoperazine, trimipramine

KEY WORDS
human; cow; pig; horse; protect from light

REFERENCE
Aumatell,A.; Wells,R.J. Determination of a cardiac antiarrhythmic, tricyclic antipsychotics and antidepressants in human and animal urine by micellar electrokinetic capillary chromatography using a bile salt, *J.Chromatogr.B*, **1995**, *669*, 331–344.

Indapamide

Molecular formula: $C_{16}H_{16}ClN_3O_3S$
Molecular weight: 365.84
CAS Registry No.: 26807-65-8
Merck Index (12th ed.): 4969
Lednicer: 2 349

SAMPLE
Matrix: solutions
Sample preparation: Prepare a solution in running buffer, inject an aliquot.

CAPILLARY ELECTROPHORESIS
Capillary: 60 cm × 75 µm fused-silica (52.4 cm to detector)
Capillary preparation: After each run flush with 500 mM KOH for 2-3 min, flush with water, fill with running buffer

Running buffer: 10 mM Na_2HPO_4 containing 2% heparin sodium (MW 10000, 11% S, Scientific
Protein Laboratories, Waunakee, WI) adjusted to pH 5 with phosphoric acid
Injection: Hydrostatic injection
Detector: UV 214
Migration time: 10.4
Model: Waters Quanta 4000

OTHER SUBSTANCES
Also analyzed: anabasine, brompheniramine, bupivacaine, carbinoxamine, chlorcyclizine, chloroquine, chlorpheniramine, dimethindene, doxylamine, enpiroline, halofantrine, hydroxychloroquine, mefloquine, nornicotine, pheniramine, primaquine, promethazine, quinacrine, tetramisole

---

REFERENCE
Stalcup,A.M.; Agyei,N.M. Heparin: a chiral mobile-phase additive for capillary zone electrophoresis,
Anal.Chem., **1994**, *66*, 3054–3059.

SAMPLE
Matrix: solutions
Sample preparation: Prepare a 500 µg/mL solution in MeOH, inject an aliquot.

CAPILLARY ELECTROPHORESIS
Capillary: 60 cm × 75 µm fused-silica (52.4 cm to detector)
Capillary preparation: Before each run rinse capillary with 100 mM KOH for 2 min then with
running buffer for 2 min.
Running buffer: 10 mM Na_2HPO_4 containing 4% sulfated β-cyclodextrin (β-cyclodextrin, sul-
fated) adjusted to pH 7.0 with phosphoric acid
Injection: Hydrostatic injection for 2 s
Detector: UV 214
Migration time: 24.08, 24.96 (enantiomers)
Voltage: 8 kV
Model: Waters Quanta 4000

KEY WORDS
chiral

REFERENCE
Wu,W.; Stalcup,A.M. Capillary electrophoretic chiral separations using a sulfated β-cyclodextrin-containing
electrolyte, *J.Liq.Chromatogr.*, **1995**, *18*, 1289–1315.

Indomethacin

Molecular formula: $C_{19}H_{16}ClNO_4$
Molecular weight: 357.79
CAS Registry No.: 53-86-1, 74252-25-8 (sodium salt trihydrate)
Merck Index (12th ed.): 4998
Lednicer: 1 318; 2 345; 3 165

SAMPLE
Matrix: blood
Sample preparation: 0.1-1 mL Plasma + 2 µg IS + 400 µL 1 M HCl + 10 mL diethyl ether,
agitate mechanically for 10 min, centrifuge at 3500 g for 10 min. Remove the organic layer
and evaporate it to dryness under a stream of nitrogen, reconstitute the residue in MeOH:40
mM NaH_2PO_4 50:50, inject an aliquot.

CAPILLARY ELECTROPHORESIS
Capillary: 58.7 cm × 75 µm fused-silica (50 cm to detector)

Capillary preparation: Wash with running buffer for 2 min before each injection. At the beginning of each day wash capillary with 100 mM NaOH for 10 min.
Capillary temperature: 22
Running buffer: MeOH:buffer 3:97 (Buffer was 40 mM pH 8 NaH_2PO_4 containing 104 mM sodium dodecyl sulfate.)
Injection: Hydrodynamic injection for 2 s
Detector: UV 254
Migration time: 30.5
Internal standard: benzoyl-4-phenyl-2-butyric acid (16.3)
Voltage: 20 kV
Model: Beckman P/ACE 2000
Limit of detection: 670 ng/mL (S/N 5)

OTHER SUBSTANCES
Extracted: diclofenac, diflunisal, etodolac, fenbufen, fenoprofen, flurbiprofen, ibuprofen, ketoprofen, naproxen, niflumic acid, piroxicam, sulindac, tenoxicam, tiaprofenic acid
Noninterfering: acetaminophen, amitriptyline, caffeine, clomipramine, deoxysulindac, desipramine, diazepam, 5'-hydroxytenoxicam, imipramine, maprotiline, nortriptyline, phenobarbital, phenytoin, sulfamethoxazole, theophylline, trimipramine

KEY WORDS
plasma

REFERENCE
Maboundou,C.W.; Paintaud,G.; Bérard,M.; Bechtel,P.R. Separation of fifteen non-steroidal anti-inflammatory drugs using micellar electrokinetic capillary chromatography, *J.Chromatogr.B*, **1994**, *657*, 173–183.

SAMPLE
Matrix: formulations
Sample preparation: Dissolve suppository in running buffer at 36°, add a solution of acemetacin in running buffer, dilute with running buffer to a final concentration of 20 μg/mL for indomethacin and 20 μg/mL for acemetacin, inject an aliquot.

CAPILLARY ELECTROPHORESIS
Capillary: 60 cm × 75 μm (52.5 cm to detector)
Capillary preparation: Purge with running buffer for 2 min before each injection. Store capillary overnight in water, rinse with 500 mM NaOH, with water, and with running buffer.
Running buffer: 50 mM pH 9.0 Borate buffer containing 40 mM sodium dodecyl sulfate (Mix 2.94 mL 200 mM boric acid, 6.37 mL 75 mM tetraborate solution, and 10 mL 200 mM sodium dodecyl sulfate, make up to 50 mL with water.)
Injection: Hydrodynamic injection at 10 cm for 5 s
Detector: UV 214
Internal standard: acemetacin
Voltage: 15 kV
Model: Waters Quanta 4000 CE
Limit of quantitation: 5 μg/mL

KEY WORDS
suppositories

REFERENCE
Donato,M.G.; Baeyens,W.; Van den Bossche,W.; Sandra,P. The determination of non-steroidal antiinflammatory drugs in pharmaceuticals by capillary zone electrophoresis and micellar electrokinetic capillary chromatography, *J.Pharm.Biomed.Anal.*, **1994**, *12*, 21–26.

SAMPLE
Matrix: solutions

CAPILLARY ELECTROPHORESIS
Capillary: 42.5 cm × 25 μm
Capillary temperature: 30
Running buffer: 20 mM pH 7.0 Phosphate containing 25 mM sodium dodecyl sulfate

Injection: Vacuum injection for 2 s
Detector: UV 230
Migration time: 5.4
Voltage: 25 kV
Current: 5 μA
Model: Applied Biosystems Model 270A

OTHER SUBSTANCES
Simultaneous: diflunisal, ibuprofen, naproxen, sulindac, tolmetin

---

REFERENCE
Weinberger,R.; Albin,M. Quantitative micellar electrokinetic capillary chromatography: Linear dynamic range, *J.Liq.Chromatogr.*, **1991**, *14*, 953–972.

SAMPLE
Matrix: solutions
Sample preparation: Dissolve in MeCN, dilute with water to a concentration of 25 μg/mL, inject an aliquot.

CAPILLARY ELECTROPHORESIS
Capillary: 60 cm $\times$ 75 μm fused-silica (52.5 cm to detector)
Running buffer: MeCN:30 mM pH 7.0 phosphate buffer 20:80
Injection: Hydrostatic injection at 9.8 cm for 5 s
Detector: UV 214
Migration time: 17.38
Model: Waters Quanta 4000

OTHER SUBSTANCES
Simultaneous: acemetacin, alclofenac, fenbufen, flurbiprofen, ibuprofen, ketoprofen, lonazolac, naproxen, niflumic acid, piroxicam, tenoxicam, tiaprofenic acid, tolmetin, voltaren

---

REFERENCE
Donato,M.G.; Van den Eeckhout,E.; Van den Bossche,W.; Sandra,P. Capillary zone electrophoresis and micellar electrokinetic capillary chromatography of some non-steroidal antiinflammatory drugs (NSAIDs), *J.Pharm.Biomed.Anal.*, **1993**, *11*, 197–201.

SAMPLE
Matrix: solutions
Sample preparation: Inject an aliquot of a 50 μM solution in MeOH:water 40:60.

CAPILLARY ELECTROPHORESIS
Capillary temperature: 25
Running buffer: MeOH:water 40:60 containing 75 mM glycine and 15 mM hydroxypropyl-β-cyclodextrin adjusted to pH 9.1 with triethanolamine
Injection: Hydrodynamic injection for 2 s
Detector: UV 280
Migration time: 16.3
Voltage: 15 kV

OTHER SUBSTANCES
Simultaneous: alclofenac, bufexamac, carprofen, flufenamic acid, flurbiprofen, ketoprofen, naproxen, niflumic acid, piroxicam, sulindac, tiaprofenic acid

---

REFERENCE
Bechet,I.; Fillet,M.; Hubert,P.; Crommen,J. Improvement of achiral resolution in capillary zone electrophoresis by use of cyclodextrin additives, *Biomed.Chromatogr.*, **1995**, *9*, 267–268.

SAMPLE
Matrix: solutions
Sample preparation: Prepare a 20 μg/mL solution in 7.5 mM glycine adjusted to pH 8.0 with triethanolamine, inject an aliquot.

CAPILLARY ELECTROPHORESIS
Capillary: 44 cm × 50 μm fused-silica (37 cm to detector) (Supelco)
Capillary preparation: After each injection wash capillary with water for 2 min and with running buffer for 3 min. At the beginning of each day wash with running buffer for 10 min. Condition a new capillary with 1 M NaOH, 100 mM NaOH, water, and running buffer.
Capillary temperature: 25
Running buffer: 75 mM Glycine adjusted to pH 9.1 with triethanolamine
Injection: Hydrodynamic injection for 5 s
Detector: UV 280
Migration time: 5.0
Voltage: 15 kV
Model: Spectra-physics Spectraphoresis 1000 CE

OTHER SUBSTANCES
Simultaneous: alclofenac, piroxicam, sulindac, tiaprofenic acid

REFERENCE
Bechet,I.; Fillet,M.; Hubert,P.; Crommen,J. Quantitative analysis of non-steroidal anti-inflammatory drugs by capillary zone electrophoresis, *J.Pharm.Biomed.Anal.*, **1995**, *13*, 497–503.

SAMPLE
Matrix: solutions

CAPILLARY ELECTROPHORESIS
Capillary: 57 cm × 50 μm fused-silica (50 cm to detector)
Running buffer: 50 mM pH 8.0 Borate buffer containing 40 mM sodium taurodeoxycholate and 25 mM phosphatidylcholine (Prepare by adding phosphatidylcholine and stirring for 3-6 h until all cloudiness disappears. Phosphatidylcholine was 95% pure soybean lecithin, Epikuron, Lucas Meyer & Co.)
Injection: Inject a solution of the compound in the running buffer at 20 psi for 1 s, inject MeOH: water 5:95 at 20 psi for 1 s, inject a solution of halofantrine in running buffer at 20 psi for 1 s.
Detector: UV 214
Migration time: k' 6.61
Model: Beckman P/ACE 5000

OTHER SUBSTANCES
Also analyzed: acetaminophen, amoxicillin, antipyrine, aspirin, azathioprine, caffeine, captopril, carbamazepine, carprofen, chlorambucil, chlorpheniramine, chlorpromazine, cimetidine, clonidine, codeine, desipramine, diphenhydramine, ephedrine, fenoterol, flufenamic acid, flurbiprofen, haloperidol, hydroxyzine, ibuprofen, imipramine, ketoprofen, lidocaine, melphalan, metoprolol, nabumetone, nadolol, phenobarbital, phenol, promazine, propranolol, pyrilamine, ranitidine, ropinirole, salicylic acid, sulfamethoxazole, testosterone, theophylline, thioridazine, tiaprofenic acid, tolfenamic acid, trifluoperazine, trimethoprim, valproic acid, verapamil

KEY WORDS
comparison with HPLC; k' = (Tr-T0)/(T0(1-Tr/Tm)) where Tr = retention time of analyte; T0 = retention time of water; and Tm = retention time of marker (halofantrine)

REFERENCE
Hanna,M.; de Biasi,V.; Bond,B.; Salter,C.; Hutt,A.J.; Camilleri,P. Estimation of the partitioning characteristics of drugs: A comparison of a large and diverse drug series utilizing chromatographic and electrophoretic methodology, *Anal.Chem.*, **1998**, *70*, 2092–2099.

Indoprofen

Molecular formula: $C_{17}H_{15}NO_3$
Molecular weight: 281.31
CAS Registry No.: 31842-01-0, 53086-14-9 ((-) form), 53086-13-8 ((+) form)
Merck Index (12th ed.): 4999
Lednicer: 3 171

SAMPLE
Matrix: solutions
Sample preparation: Inject an aliquot of a 100 µg/mL solution.

CAPILLARY ELECTROPHORESIS
Capillary: 27 cm × 50 µm fused-silica (20 cm to detector)
Capillary preparation: Between runs rinse capillary with 500 mM KOH for 3 min, with water
for 2 min, and with running buffer for 2 min. Condition new capillaries with 100 mM KOH for
10 min, purge with water for 5 min, equilibrate with running buffer for 5 min.
Capillary temperature: 20
Running buffer: 100 mM pH 6.0 Phosphate buffer containing 2 mM ristocetin A and 25 mM
sodium dodecyl sulfate
Injection: Pressure injection at 0.5 psi for 1-3 s.
Detector: UV 254
Migration time: 34, 37 (enantiomers)
Voltage: 5 kV
Model: Beckman P/ACE 2000

OTHER SUBSTANCES
Simultaneous: ketoprofen

KEY WORDS
chiral

REFERENCE
Gasper,M.P.; Berthod,A.; Nair,U.B.; Armstrong,D.W. Comparison and modeling study of vancomycin, ristocetin
A, and teicoplanin for CE enantioseparations, *Anal.Chem.*, **1996**, *68*, 2501–2514.

SAMPLE
Matrix: solutions
Sample preparation: Inject an aliquot of a 500 µM solution in MeOH:water 50:50.

CAPILLARY ELECTROPHORESIS
Capillary: 60 cm × 50 µm fused-silica (52.5 cm to detector) (Waters)
Capillary preparation: Before each run rinse capillary with 10 mM NaOH for 1 min and with
running buffer for 4 min. Condition new capillaries by flushing with 1 M NaOH for 10 min,
with water for 5 min, with 6 mM NaOH containing 500 mM NaCl for 10 min, with water for
5 min, and with running buffer for 10 min.
Running buffer: Formic acid containing 10 mM heptakis(trimethyl-β-cyclodextrin) adjusted to
pH 4.0 with 1 M NaOH (ionic strength, I = 35 mM)
Injection: Hydrodynamic injection at 10 cm for 15 s.
Detector: UV 254
Voltage: 30 kV
Model: Waters Quanta 4000

KEY WORDS
chiral; R_S = 2.4; electroosmotic mobility = 8.7×10^{-5} cm^{2}V^{-1}s^{-1}

REFERENCE
Lelièvre,F.; Gareil,P. Chiral separations of underivatized arylpropionic acids by capillary zone electrophoresis
with various cyclodextrins. Acidity and inclusion constant determinations, *J.Chromatogr.A*, **1996**, *735*, 311–
320.

SAMPLE
Matrix: solutions

CAPILLARY ELECTROPHORESIS
Capillary: 35 cm × 50 μm coated capillary (30.5 cm to detector) (Composite Metal Services, UK)
Capillary preparation: Before each run purge at high pressure with water for 100 s and with
running buffer for 120 s. If vancomycin is used, before each run purge at high pressure with
water for 100 s and with running buffer not containing vancomycin for 120 s. Next purge with
running buffer containing vancomycin at low pressure (175 psi.s) and inject sample. Coat capillary as follows. Adjust the pH of 20 mL water to 3.5 with acetic acid, add 80 μL 3-(trimethoxysilyl)propyl methacrylate (3-methacryloxypropyltrimethoxysilane), mix, suck into capillary, let stand at room temperature for 1 h, remove the solution, wash with water. Fill the
capillary with a deaerated 3-4% acrylamide solution containing 1 μL/mL N,N,N',N'-tetra-
methylethylenediamine and 1 mg/mL potassium persulfate, let stand for 30 min, remove excess
solution by aspiration, rinse with water, remove water by aspiration, dry at 35° (J. Chromatogr.
1985, 347, 191).
Capillary temperature: 25
Running buffer: Buffer containing 30 mM heptakis-2,3,6-tri-O-methyl-β-cyclodextrin (A), 5 mM
heptamethylamino β-cyclodextrin (B) or 5 mM vancomycin (C) (Buffer was 50 mM phosphoric
acid containing 50 mM acetic acid and 50 mM boric acid, dilute with an equal volume of water,
adjust pH to 5 with concentrated NaOH.)
Injection: Pressure injection at 10 psi.
Detector: UV 206
Migration time: 13.4, 13.8 (A); 12.8, 13.5 (B); 8.5, 8.8 (C)
Voltage: -20 kV
Model: Bio-Rad Biofocus 3000
Limit of detection: 1 μM (C), 5 μM (A), 10 μM (B)

KEY WORDS
chiral; coated capillary; detector at anode

REFERENCE
Fanali,S.; Desiderio,C.; Aturki,Z. Enantiomeric resolution study by capillary electrophoresis. Selection of the
appropriate chiral selector, *J.Chromatogr.A*, **1997**, *772*, 185–194.

SAMPLE
Matrix: solutions
Sample preparation: Inject an aliquot of a solution in MeOH:water 10:90.

CAPILLARY ELECTROPHORESIS
Capillary: 44 cm × 50 μm fused-silica (37 cm to detector) (Supelco)
Capillary preparation: Wash with running buffer for 3 min after each injection. Wash with
running buffer for 10 min at the end of each day.
Capillary temperature: 25
Running buffer: 100 mM Phosphoric acid containing 5 mM sulfobutyl ether-β-cyclodextrin and
40 mM heptakis(2,3,6-tri-O-methyl)-β-cyclodextrin, adjusted to pH 3.0 with triethanolamine
Injection: Hydrodynamic injection for 5 s (13.3 nL).
Detector: UV 280
Voltage: -25 kV
Current: 60 μA
Model: SpectraPhoresis 1000 CE

KEY WORDS
chiral; detector at anode; resolution (R_s = 4.9)

REFERENCE
Fillet,M.; Hubert,P.; Crommen,J. Enantioseparation of nonsteroidal anti-inflammatory drugs by capillary electrophoresis using mixtures of anionic and uncharged β-cyclodextrins as chiral additives, *Electrophoresis*,
1997, *18*, 1013–1018.

SAMPLE
Matrix: solutions

CAPILLARY ELECTROPHORESIS

Capillary: 35 cm × 50 μm polyacrylamide-coated fused-silica (30.5 cm to detector) (Composite Metal Services, UK)

Capillary preparation: Before each run rinse capillary with water for 70 s and with running buffer for 100 s. (Coat capillary as follows. Adjust the pH of 20 mL water to 3.5 with acetic acid, add 80 μL 3-(trimethoxysilyl)propyl methacrylate (3-methacryloxypropyltrimethoxysilane), mix, suck into capillary, let stand at room temperature for 1 h, remove the solution, wash with water. Fill the capillary with a deaerated 3-4% acrylamide solution containing 1 μL/mL N,N,N',N'-tetramethylethylenediamine and 1 mg/mL potassium persulfate, let stand for 30 min, remove excess solution by aspiration, rinse with water, remove water by aspiration, dry at 35° (J. Chromatogr. 1985, 347, 191).)

Capillary temperature: 25

Running buffer: Buffer containing 30 mM cyanoethylated-β-cyclodextrin (Cyclolab, Budapest) (Prepare buffer by adjusting the pH of 50 mM phosphoric acid containing 50 mM acetic acid and 50 mM boric acid to 5 with concentrated NaOH, add the appropriate amount of cyanoethylated-β-cyclodextrin, dilute with an equal volume of water.)

Injection: Pressure injection at 5 psi for 2 s.

Detector: UV 206

Migration time: 18.9 (second enantiomer, $\alpha = 1.027$)

Voltage: 20 kV

Current: 27-40 μA

Model: Bio-Rad Biofocus 3000

KEY WORDS

detector at anode; chiral; coated capillary

REFERENCE

Aturki,Z.; Desiderio,C.; Mannina,L.; Fanali,S. Chiral separations by capillary zone electrophoresis with the use of cyanoethylated-β-cyclodextrin as chiral selector, *J.Chromatogr.A*, **1998**, *817*, 91–104.

SAMPLE

Matrix: solutions

Sample preparation: Inject an aliquot of a 100 μg/mL solution in MeOH:water 50:50.

CAPILLARY ELECTROPHORESIS

Capillary: 47 cm × 75 μm fused-silica (40 cm to detector) (Polymicro Technologies)

Capillary preparation: Rinse capillary with 100 mM NaOH for 1 min, with water for 1.5 min, and with running buffer for 2 min between samples. Wash new capillaries with 1 M NaOH for 1 h, with 100 mM NaOH for 30 min, and with water for 30 min.

Running buffer: 50 mM NaH_2PO_4 containing 20 mM heptakis(6-methoxyethylamine-6-deoxy-β-cyclodextrin), adjusted to pH 5 (Prepare heptakis(6-methoxyethylamine-6-deoxy-β-cyclodextrin) as follows. Add 4.32 g β-cyclodextrin to a stirred solution of 4 mL bromine and 21 g triphenylphosphine in 80 mL DMF and the mixture stirred at 80°C for 15 h, concentrate under reduced pressure to half volume, adjust pH to 9-10 by adding 30 mL 3 M sodium methoxide in MeOH with simultaneous cooling. Keep at room temperature for 30 min then pour into 1.5 L ice water, filter to obtain heptakis(6-bromo-6-deoxy-β-cyclodextrin) (Ang. Chem. Int. Ed. Engl. 1991, 30, 78). Dissolve heptakis(6-bromo-6-deoxy-β-cyclodextrin) in methoxyethylamine, heat at 65°C for 48 h. Remove the solvent by evaporation under reduced pressure, dissolve the residue in MeOH at 55°C, slowly add the methanolic solution to reagent-grade acetone with stirring, collect the precipitate by filtration. Dissolve the precipitate in water, treat with a basic ion-exchange resin. Lyophilize the precipitate to obtain heptakis(6-methoxyethylamine-6-deoxy-β-cyclodextrin).)

Injection: Pressure injection at 85 kPa.s.

Detector: UV 214

Migration time: 58, 68 (enantiomers)

Voltage: -15 kV

Current: 30 μA

Model: Beckman P/ACE 5510

KEY WORDS

chiral; peak shape is poor

REFERENCE
Haynes,J.L.,III; Shamsi,S.A.; O'Keefe,F.; Darcey,R.; Warner,I.M. Cationic β-cyclodextrin derivative for chiral separations, *J.Chromatogr.A*, **1998**, *803*, 261–271.

SAMPLE
Matrix: solutions
Sample preparation: Inject an aliquot of a solution in running buffer.

CAPILLARY ELECTROPHORESIS
Capillary: 57 cm × 50 μm fused-silica (Beckman)
Capillary preparation: Rinse with 100 mM NaOH for 5 min, with water for 10 min,
Running buffer: 2-Methoxyethanol:buffer 30:70 (Buffer was 40 mM pH 6 phosphate buffer containing 500 μM Actaplanin A (Eli Lilly).)
Injection: Pressure injection for 5 s.
Detector: UV 254
Migration time: 40.0, 42.2 (enantiomers)
Model: Beckman P/ACE 2100

OTHER SUBSTANCES
Also analyzed: carprofen, flurbiprofen

KEY WORDS
chiral

REFERENCE
Trelli-Seifert,L.A.; Risley,D.S. Capillary electrophoretic enantiomeric separations of nonsteroidal anti-inflammatory compounds using the macrocyclic antibiotic actaplanin A and 2-methoxyethanol, *J.Liq.Chromatogr.Rel.Technol.*, **1998**, *21*, 299–313.

SAMPLE
Matrix: solutions

CAPILLARY ELECTROPHORESIS
Capillary: 42 cm × 50 μm fused-silica (31 cm to detector) (Polymicro Technologies)
Capillary temperature: 30
Running buffer: Formamide containing 3.42% quaternary ammonium-β-cyclodextrin (American Maize Products, Hammond IN), 20 mM ammonium acetate, and 1% acetic acid (pH* = 7.1)
Detector: UV 254
Migration time: 13.61, 13.67 (enantiomers)
Voltage: -30 kV
Model: laboratory-constructed

OTHER SUBSTANCES
Also analyzed: carprofen, fenoprofen, flurbiprofen, ketoprofen, suprofen

KEY WORDS
chiral

REFERENCE
Wang,F.; Khaledi,M.G. Nonaqueous capillary electrophoresis chiral separations with quaternary ammonium β-cyclodextrin, *J.Chromatogr.A*, **1998**, *817*, 121–128.

Insulin

Molecular formula: $C_{258}H_{383}N_{65}O_{77}S_6$ (human)
Molecular weight: 5807.60 (human)
CAS Registry No.: 9004-10-8 (injection), 8049-62-5 (zinc suspension), 11061-68-0 (human), 12584-58-6 (pig), 11070-73-8 (cow), 9004-14-2 (neutral insulin), 8049-62-5 (isophane insulin), 9004-17-5 (protamine zinc suspension)
Merck Index (12th ed.): 5011

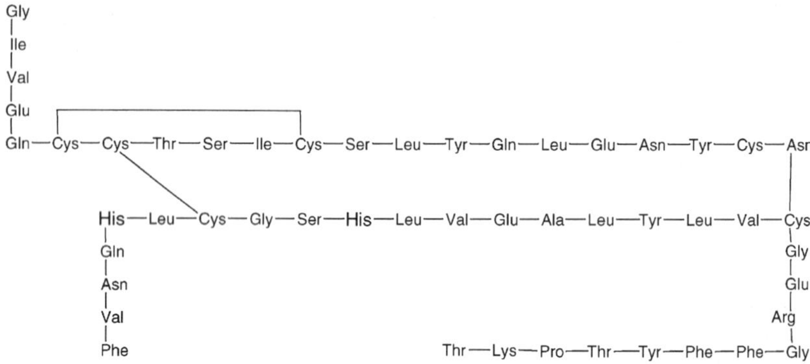

SAMPLE
Matrix: blood
Sample preparation: 200 µL Serum + 200 µL EtOH, vortex, centrifuge. Remove the supernatant and concentrate to 20 µL under a stream of nitrogen, inject an aliquot.

CAPILLARY ELECTROPHORESIS
Capillary: 57 cm × 75 µm uncoated fused-silica (50 cm to detector)
Capillary preparation: Before each analysis wash with 100 mM NaOH for 2 min and twice with water for 2 min, condition with running buffer for 5 min.
Capillary temperature: 23
Running buffer: 20 mM pH 8.5 Tricine containing 40 mM NaCl
Injection: 11.8 nL
Detector: UV 200
Migration time: 10.4
Current: 70 µA
Model: Beckman P/ACE System 2000

KEY WORDS
serum

REFERENCE
Arcelloni,C.; Fermo,I.; Banfi,G.; Pontiroli,A.E.; Paroni,R. Capillary electrophoresis for protein analysis: separation of human growth hormone and human insulin molecular forms, *Anal.Biochem.*, **1993**, *212*, 160–167.

SAMPLE
Matrix: solutions
Sample preparation: Prepare a 1 mg/mL solution in water, inject an aliquot.

CAPILLARY ELECTROPHORESIS
Capillary: 95.5 cm × 50 µm fused-silica (81.5 cm to detector) (Polymicro Technologies)
Capillary preparation: Replace buffer in capillary after each run.
Capillary temperature: 24.4 ± 0.1
Running buffer: 10 mM Tricine containing 5.8 mM morpholine and 20 mM NaCl, adjusted to pH 8 with 1 M NaOH
Injection: Inject using an 11.4 cm Hg vacuum for 0.9 s (2.5 nL)
Detector: UV 200

Migration time: 9.5
Voltage: 300 V/cm
Current: about 20 µA

OTHER SUBSTANCES
Simultaneous: degradation products

KEY WORDS
recombinant

REFERENCE
Nielsen,R.G.; Sittampalam,G.S.; Rickard,E.C. Capillary zone electrophoresis of insulin and growth hormone, *Anal.Biochem.*, **1989**, *177*, 20–26.

SAMPLE
Matrix: solutions
Sample preparation: Prepare a solution in running buffer, centrifuge at 5000 g for 10 min, inject an aliquot.

CAPILLARY ELECTROPHORESIS
Capillary: 52 cm × 100 µm fused-silica uncoated (Beckman)
Capillary preparation: Clean periodically by rinsing with potassium dichromate.
Running buffer: 20 mM Na_2HPO_4 adjusted to pH 11.2 with 100 mM NaOH
Injection: Inject at +5 kV for 0.1 s
Detector: UV 214
Migration time: 6.6
Voltage: + 20 kV
Model: Applied Biosystems Model 270A

OTHER SUBSTANCES
Simultaneous: degradation products, proinsulin

REFERENCE
Koulich,D.M.; Nazimov,I.V.; Maltsev,K.V.; Wulfson,A.N. Recombinant human insulin VI. Determination of recombinant human proinsulin by capillary zone electrophoresis: optimization of efficiency and selectivity, *J.Chromatogr.B*, **1994**, *662*, 357–362.

SAMPLE
Matrix: solutions
Sample preparation: Inject an aliquot of an aqueous solution

CAPILLARY ELECTROPHORESIS
Capillary: 50 µm fused-silica (Polymicro Technologies)
Capillary preparation: Before each injection flush capillary with 100 mM NaOH for 5 min, with water for 5 min, and with running buffer for 5 min.
Capillary temperature: 25 ± 2
Running buffer: 25 mM pH 7.4 Tris buffer containing 192 mM glycine
Injection: Vacuum injection (8 nL)
Detector: UV 200
Migration time: 3
Voltage: 30 kV
Model: Isco Model 3140 or Beckman P/ACE 5500

OTHER SUBSTANCES
Simultaneous: acetylated insulins

REFERENCE
Gao,J.; Mrksich,M.; Gomez,F.A.; Whitesides,G.M. Using capillary electrophoresis to follow the acetylation of the amino groups of insulin and to estimate their basicities, *Anal.Chem.*, **1995**, *67*, 3093–3100.

SAMPLE
Matrix: solutions

CAPILLARY ELECTROPHORESIS
Capillary: 50 cm × 50 μm fused-silica (37.5 cm to detector)
Capillary temperature: 25
Running buffer: 100 mM pH 8.1 Tricine buffer containing 20 mM NaCl, 1 M urea, and 450 mM ethyleneglycol
Injection: Hydrostatic injection at 1 cm for 10 s.
Detector: UV 214
Migration time: 14.5
Voltage: 7 kV
Model: Photal CAPI-3000 (Otsuka Electronics)

OTHER SUBSTANCES
Simultaneous: degradation products

REFERENCE
Yomota,C.; Matsumoto,Y; Okada,S.; Hayashi,Y.; Matsuda,R. Discrimination limit for purity test of human insulin by capillary electrophoresis, *J.Chromatogr.B*, **1997**, *703*, 139–145.

SAMPLE
Matrix: tissue
Sample preparation: Homogenize 300 mg pancreatic tissue in 1 mL 1% sodium chloride, centrifuge at 14000 g for 1 min. Remove a 200 μL aliquot of the supernatant and mix it with 400 μL MeCN, centrifuge. Remove the supernatant and evaporate it to dryness under a stream of air, reconstitute the residue in a mixture of 50 μL saline and 100 μL MeCN, inject an aliquot.

CAPILLARY ELECTROPHORESIS
Capillary: 42 cm × 50 μm
Running buffer: MeCN:buffer 10:90 (Buffer was 660 mM 2[N-cyclohexylamino]ethanesulfonic acid (CHES) containing 45 mM acetic acid, adjusted to pH 7.8 with triethanolamine.)
Injection: Hydrodynamic injection for 30 s.
Detector: UV 214
Migration time: 7.5
Voltage: 16.5 kV
Model: Beckman Model 2000 CE

KEY WORDS
pancreas

REFERENCE
Shihabi,Z.K.; Friedberg,M. Insulin stacking for capillary electrophoresis, *J.Chromatogr.A*, **1998**, *807*, 129–133.

Interferon

Molecular formula: $C_{860}H_{1353}N_{227}O_{255}S_9$

Molecular weight: 19241.17

CAS Registry No.: 76543-88-9 (α2A), 99210-65-8 (α2B), 98059-61-1 (γ1B)

Merck Index (12th ed.): 5015

Lednicer: 4 1

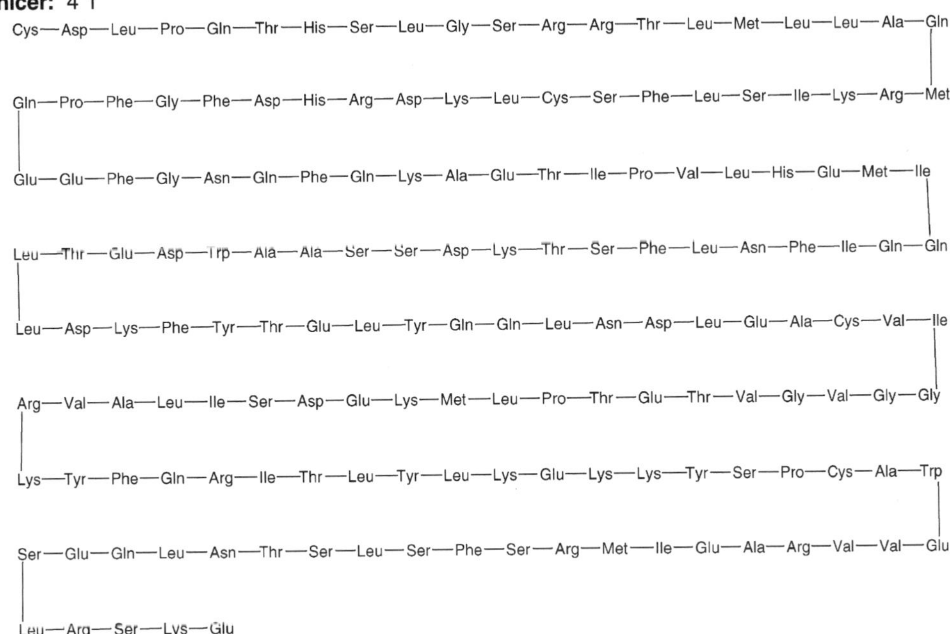

SAMPLE

Matrix: bulk

Sample preparation: Prepare a solution in 100 mM pH 9.0 borax buffer. While continuously and vigorously vortexing add 30 µL 3 mg/mL fluorescamine in acetone containing 20 µL/mL (?) pyridine to 70 µL sample, vortex for 2 min, inject an aliquot.

CAPILLARY ELECTROPHORESIS

Capillary: 57 cm $\times$ 75 µm (50 cm to the detector)

Capillary preparation: At the end of each run wash capillary with 2 M NaOH, 100 mM NaOH, water, and running buffer.

Capillary temperature: 25

Running buffer: 50 mM pH 7.0 Sodium phosphate buffer containing 50 mM LiCl

Injection: Pressure injection at 0.5 psi for 4 s (24 nL)

Detector: UV 280

Migration time: 23 (recombinant human leukocyte-A interferon)

Voltage: 12 kV

Model: Beckman P/ACE System 2000

OTHER SUBSTANCES

Simultaneous: acetone, ammonia

KEY WORDS

recombinant; derivatization

REFERENCE

Guzman,N.A.; Moschera,J.; Bailey,C.A.; Iqbal,K.; Malick,A.W. Assay of protein drug substances present in solution mixtures by fluorescamine derivatization and capillary electrophoresis, *J.Chromatogr.*, **1992**, *598*, 123–131.

SAMPLE

Matrix: solutions
Sample preparation: Inject an aliquot of a solution in 50 mM pH 8.5 borate buffer containing 50 mM sodium dodecyl sulfate

CAPILLARY ELECTROPHORESIS

Capillary: 57 cm × 50 μm fused-silica (50 cm to detector)
Capillary preparation: Before each run rinse with 100 mM NaOH for 5 min, with water for 5 min, and with running buffer for 5 min. Prepare capillaries for use by rinsing with 100 mM NaOH for 10 min, with water for 5 min, with 100 mM pH 8.5 borate buffer for 1 h, with 100 mM NaOH for 10 min, and water for 10 min, then equilibrate with running buffer for 1 h.
Capillary temperature: 25
Running buffer: 400 mM pH 8.5 Borate buffer containing 100 mM sodium dodecyl sulfate
Injection: Electrokinetic injection at 22 kV for 5 s
Detector: UV 200
Migration time: 18 (PG1), 23 (PG2), 34 (PG3)
Voltage: 22 kV
Current: 120 μA
Model: Beckman P/ACE 2100

KEY WORDS

interferon-γ

REFERENCE

James,D.C.; Freedman,R.B.; Hoare,M.; Jenkins,N. High-resolution separation of recombinant human interferon-γ glycoforms by micellar electrokinetic capillary chromatography, *Anal.Biochem.*, **1994**, *222*, 315–322.

SAMPLE

Matrix: solutions
Sample preparation: Inject an aliquot of a 100-500 μg/mL solution in water.

CAPILLARY ELECTROPHORESIS

Capillary: 80 cm × 50 μm fused-silica (50 cm to detector) (ISCO)
Capillary preparation: Rinse with running buffer before each analysis. After each buffer change rinse with water, 100 mM NaOH, and water. Store in 100 mM NaOH overnight.
Running buffer: 200 mM pH 6.1 Phosphate buffer
Injection: Manually inject 4 nL (8 μL split 1:1930)
Detector: UV 200
Migration time: 23-38 (multiple peaks)
Voltage: 12 kV
Current: 42 μA
Model: ISCO Model 3850

REFERENCE

Craig,S.J.; Ashton,D.S.; Beddell,C.; Valko,K. The effect of operating parameters on the analysis of a human α-interferon by capillary zone electrophoresis, *J.Liq.Chromatogr.*, **1995**, *18*, 3629–3641.

Iohexol

Molecular formula: $C_{19}H_{26}I_3N_3O_9$
Molecular weight: 821.14
CAS Registry No.: 66108-95-0
Merck Index (12th ed.): 5068

SAMPLE
Matrix: blood
Sample preparation: 100 μL Serum + 150 μL MeCN, mix for 15 s, centrifuge at 15000 g for 1 min, inject an aliquot of the supernatant.

CAPILLARY ELECTROPHORESIS
Capillary: 25 cm × 50 μm
Capillary preparation: Wash with running buffer for 1 min after each run.
Capillary temperature: 24
Running buffer: 300 mM boric acid adjusted to pH 8.5 with NaOH
Injection: Pressure injection for 15 s.
Detector: UV 254
Migration time: 2
Voltage: 13 kV
Model: Beckman

OTHER SUBSTANCES
Extracted: phenobarbital, phenytoin, theophylline

KEY WORDS
serum

REFERENCE
Shihabi,Z.K. Sample matrix effects in capillary electrophoresis. II. Acetonitrile deproteinization, *J.Chromatogr.A*, **1993**, *652*, 471–475.

SAMPLE
Matrix: solutions

CAPILLARY ELECTROPHORESIS
Capillary: 72 cm × 50 μm fused-silica (50 cm to detector) (Applied Biosystems)
Capillary preparation: Before each injection flush with 100 mM NaOH for 2 min then with running buffer for 2 min.
Capillary temperature: 25
Running buffer: MeOH:50 mM pH 9.2 borate buffer 20:80
Injection: Vacuum injection for 2 s
Detector: UV 245
Voltage: 20 kV
Model: Applied Biosystems Model 270A-HT

OTHER SUBSTANCES
Simultaneous: iopamidol, iopentol, iosemide, ioversol

REFERENCE
Thanh,H.H. Migration behavior of triiodinated X-ray contrast media containing diol groups as borate complexes in capillary electrophoresis, *J.Chromatogr.A*, **1994**, *678*, 343–350.

SAMPLE
Matrix: solutions

CAPILLARY ELECTROPHORESIS
Capillary: 42 cm $\times$ 50 μm
Capillary temperature: 24
Running buffer: 175 mM Boric acid adjusted to pH 9.4 with 2 M NaOH
Injection: Pressure injection for 8 s
Detector: UV 254
Migration time: 6.4
Internal standard: 3-isobutyl-1-methylxanthine (7.8)
Voltage: 8 kV
Model: Beckman Model 2000

OTHER SUBSTANCES
Simultaneous: p-aminohippuric acid, iopamidol, iothalamic acid

REFERENCE
Shihabi,Z.K.; Rocco,M.V.; Hinsdale,M.E. Analysis of the contrast agent *iopamidol* in serum by capillary electrophoresis, *J.Liq.Chromatogr.*, **1995**, *18*, 3825–3831.

SAMPLE
Matrix: solutions

CAPILLARY ELECTROPHORESIS
Capillary: 58.7 cm $\times$ 75 μm (50 cm to detector)
Capillary preparation: Before each injection wash capillary with 100 mM NaOH for 2 min, with 100 mM HCl for 2 min, with water for 2 min, and with running buffer for 2 min. At the beginning of each sequence of operations pressure wash capillary with 100 mM NaOH for 4 min and with water for 4 min.
Capillary temperature: 30
Running buffer: 50 mM pH 9 Na_2HPO_4 containing 10 mM sodium tetraborate and 40 mM sodium dodecyl sulfate
Injection: Pressure injection using nitrogen at 0.5 psi for 5 s
Detector: UV 200
Migration time: 7.3
Internal standard: iohexol (7.3)
Voltage: 15 kV
Model: Beckman P/ACE system 2050

OTHER SUBSTANCES
Simultaneous: hyaluronic acid tetrasaccharide, hyaluronic acid hexasaccharide

REFERENCE
Grimshaw,J.; Trocha-Grimshaw,J.; Fisher,W.; Rice,A.; Smith,S.; Spedding,P.; Duffy,J.; Mollan,R. Quantitative analysis of hyaluronan in human synovial using capillary electrophoresis, *Electrophoresis*, **1996**, *17*, 396–400.

Iopamidol

Molecular formula: $C_{17}H_{22}I_3N_3O_8$
Molecular weight: 777.09
CAS Registry No.: 60166-93-0
Merck Index (12th ed.): 5071

SAMPLE
Matrix: blood
Sample preparation: 50 μL Serum + 100 μL 80 μg/mL 3-isobutyl-1-methylxanthine in MeCN, mix for 15 s, centrifuge at 15000 rpm for 30 s, inject an aliquot of the supernatant.

CAPILLARY ELECTROPHORESIS
Capillary: 42 cm × 50 μm
Capillary temperature: 24
Running buffer: 175 mM Boric acid adjusted to pH 9.4 with 2 M NaOH
Injection: Pressure injection for 8 s
Detector: UV 254
Migration time: 6.3
Internal standard: 3-isobutyl-1-methylxanthine (8.0)
Voltage: 8 kV
Model: Beckman Model 2000
Limit of detection: 5 μg/mL

OTHER SUBSTANCES
Simultaneous: p-aminohippuric acid, iohexol, iothalamic acid

KEY WORDS
serum

REFERENCE
Shihabi,Z.K.; Rocco,M.V.; Hinsdale,M.E. Analysis of the contrast agent *iopamidol* in serum by capillary electrophoresis, *J.Liq.Chromatogr.*, **1995**, *18*, 3825–3831.

SAMPLE
Matrix: solutions

CAPILLARY ELECTROPHORESIS
Capillary: 72 cm × 50 μm fused-silica (50 cm to detector) (Applied Biosystems)
Capillary preparation: Before each injection flush with 100 mM NaOH for 2 min then with running buffer for 2 min.
Capillary temperature: 25
Running buffer: MeOH:50 mM pH 9.2 borate buffer 20:80
Injection: Vacuum injection for 2 s
Detector: UV 245
Voltage: 20 kV
Model: Applied Biosystems Model 270A-HT

OTHER SUBSTANCES
Simultaneous: iohexol, iopentol, iosemide, ioversol

REFERENCE
Thanh,H.H. Migration behavior of triiodinated X-ray contrast media containing diol groups as borate complexes in capillary electrophoresis, *J.Chromatogr.A*, **1994**, *678*, 343–350.

SAMPLE
Matrix: solutions
Sample preparation: Inject an aliquot of an 83 mM solution in MeCN:1% saline 66:34.

CAPILLARY ELECTROPHORESIS
Capillary: 32 cm × 50 μm
Running buffer: 300 mM pH 9.5 Borate buffer
Injection: Hydrodynamic injection for 120 s.
Detector: UV 214
Migration time: 10.9
Voltage: 8 kV
Model: Quanta Model 4000 (Waters)

OTHER SUBSTANCES
Simultaneous: acetaminophen, iothalamic acid

REFERENCE
Friedberg,M.A.; Hinsdale,M.; Shihabi,Z.K. Effect of pH and ions in the sample on stacking in capillary electrophoresis, *J.Chromatogr.A*, **1997**, *781*, 35–42.

Iopanoic acid

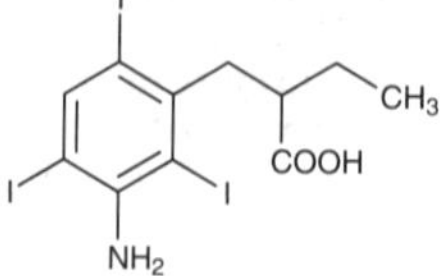

Molecular formula: $C_{11}H_{12}I_3NO_2$
Molecular weight: 570.93
CAS Registry No.: 96-83-3
Merck Index (12th ed.): 5072

SAMPLE
Matrix: solutions
Sample preparation: Inject an aliquot of a solution in 10 mM NaOH.

CAPILLARY ELECTROPHORESIS
Capillary: 50 cm $\times$ 50 μm
Running buffer: MeOH:buffer:polyvinyl alcohol 10:89.5:0.5 (Buffer was 500 mM pH 8.5 sodium borate buffer containing 40 mM hydroxypropyl-β-cyclodextrin.)
Injection: 40 psi.s
Detector: UV 210
Migration time: 12.8, 13.8 (enantiomers)
Voltage: 15 kV

KEY WORDS
chiral

REFERENCE
Yang,Y.; Botha,A. Chiral separation of a low aqueous soluble compound, (R,S)-iopanoic acid by capillary electrophoresis (Abstract APQ 1109), *Pharm.Res.*, **1996**, *13*, S30.

Iothalamic acid

Molecular formula: $C_{11}H_9I_3N_2O_4$
Molecular weight: 613.92
CAS Registry No.: 2276-90-6, 13087-53-1 (meglumine salt), 1225-20-3 (Na salt)
Merck Index (12th ed.): 5080

SAMPLE
Matrix: blood, urine
Sample preparation: Filter (Amicon Micron-10) plasma or urine while centrifuging at 14000 RCF for 10 min. Inject an aliquot of the plasma ultrafiltrate. Dilute the urine ultrafiltrate 5-10-fold with 5 mM pH 11.3 sodium borate buffer, inject an aliquot.

CAPILLARY ELECTROPHORESIS
Capillary: 47 cm $\times$ 50 μm fused-silica (40 cm to detector) (Polymicro Technologies)
Capillary preparation: Before each run rinse capillary with running buffer at 20 psi for 1 min. After each run rinse capillary at 20 psi with 100 mM NaOH for 30 s, with water for 30 s, and with running buffer for 1 min. Before first use rinse capillary at 20 psi with 100 mM NaOH for 20 min, with water for 20 min, and with 50 mM pH 11.3 sodium borate buffer for 20 min.
Capillary temperature: 20
Running buffer: 50 mM pH 11.3 Sodium borate buffer
Injection: Pressure injection of sample at 0.5 psi for 15 s followed by pressure injection of running buffer at 0.5 psi for 1 s.
Detector: UV 254
Migration time: 3.8
Voltage: 25 kV
Model: Beckman P/ACE 2100

OTHER SUBSTANCES
Extracted: acetaminophen (only demonstrated in urine), sulfamethoxazole (only demonstrated in urine)

KEY WORDS
plasma; ultrafiltrate

REFERENCE
Bergert,J.H.; Liedtke,R.R.; Oda,R.P.; Landers,J.P.; Wilson,D.M. Development of a nonisotopic capillary electrophoresis-based method for measuring glomerular filtration rate, *Electrophoresis*, **1997**, *18*, 1827–1835.

SAMPLE
Matrix: solutions

CAPILLARY ELECTROPHORESIS
Capillary: 50 cm × 50 μm
Capillary temperature: 35
Running buffer: 7 g Boric acid and 7 g sodium carbonate in 1 L water, pH 9.2
Injection: Pressure injection for 20 s
Detector: UV 280
Migration time: 4.9
Voltage: 11 kV
Model: Beckman Instruments Model 2000 CE

OTHER SUBSTANCES
Simultaneous: adenine, allopurinol, guanine, hypoxanthine, inosine, oxypurinol, uric acid, xanthine

REFERENCE
Shihabi,Z.K.; Hinsdale,M.E.; Bleyer,A.J. Xanthine analysis in biological fluids by capillary electrophoresis, *J.Chromatogr.B*, **1995**, *669*, 163–169.

SAMPLE
Matrix: solutions

CAPILLARY ELECTROPHORESIS
Capillary: 42 cm × 50 μm
Capillary temperature: 24
Running buffer: 175 mM Boric acid adjusted to pH 9.4 with 2 M NaOH
Injection: Pressure injection for 8 s
Detector: UV 254
Migration time: 7.3
Internal standard: 3-isobutyl-1-methylxanthine (7.8)
Voltage: 8 kV
Model: Beckman Model 2000

OTHER SUBSTANCES
Simultaneous: p-aminohippuric acid, iohexol, iopamidol

REFERENCE
Shihabi,Z.K.; Rocco,M.V.; Hinsdale,M.E. Analysis of the contrast agent *iopamidol* in serum by capillary electrophoresis, *J.Liq.Chromatogr.*, **1995**, *18*, 3825–3831.

SAMPLE
Matrix: solutions

CAPILLARY ELECTROPHORESIS
Capillary: 42 cm × 50 μm (35.5 cm to detector)
Capillary preparation: After each run wash with 200 mM NaOH for 1 min and filled with running buffer for 1 min.

Capillary temperature: 24
Running buffer: 175 mM Boric acid adjusted to pH 8.4 with 4 M ammonium hydroxide
Detector: UV 214
Migration time: 3
Voltage: 15 kV
Model: Beckman P/ACE 2000

REFERENCE
Shihabi,Z.K.; Kute,T.E. Analysis of cathepsin D from breast tissues by capillary electrophoresis, *J.Chromatogr.B*, **1996**, *683*, 125–131.

SAMPLE
Matrix: solutions
Sample preparation: Inject an aliquot of an 83 mM solution in MeCN:1% saline 66:34.

CAPILLARY ELECTROPHORESIS
Capillary: 32 cm × 50 μm
Running buffer: 300 mM pH 9.5 Borate buffer
Injection: Hydrodynamic injection for 120 s.
Detector: UV 214
Migration time: 11.5
Voltage: 8 kV
Model: Quanta Model 4000 (Waters)

OTHER SUBSTANCES
Simultaneous: acetaminophen, iopamidol

REFERENCE
Friedberg,M.A.; Hinsdale,M.; Shihabi,Z.K. Effect of pH and ions in the sample on stacking in capillary electrophoresis, *J.Chromatogr.A*, **1997**, *781*, 35–42.

SAMPLE
Matrix: solutions
Sample preparation: Mix 50 μL of a solution in 1% sodium chloride with 100 μL MeCN, vortex for 15 s, centrifuge at 14000 g for 20 s, inject an aliquot.

CAPILLARY ELECTROPHORESIS
Capillary: 40 cm × 50 μm silica
Running buffer: MeCN:buffer 10:90 (Buffer was 160 mM triethanolamine containing 50 mM tricine, pH adjusted to 8.6.)
Injection: Hydrodynamic injection at low pressure for 70 s (to fill 12% of the capillary).
Detector: UV 254
Migration time: 11.5
Voltage: 14 kV
Model: Beckman Model 2000 CE

OTHER SUBSTANCES
Simultaneous: N-acetylprocainamide, doxepin, quinine, theophylline

REFERENCE
Shihabi,Z.K. Stacking of weakly cationic compounds by acetonitrile for capillary electrophoresis, *J.Chromatogr.A*, **1998**, *817*, 25–30.

Ipratropium bromide

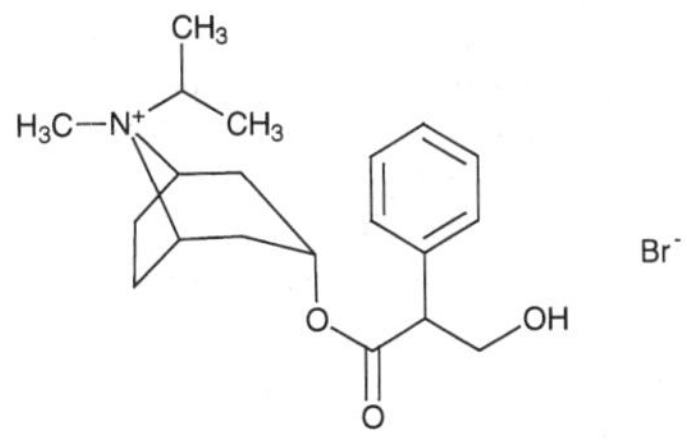

Molecular formula: $C_{20}H_{30}BrNO_3$
Molecular weight: 412.37
CAS Registry No.: 22254-24-6, 66985-17-9 (monohydrate)
Merck Index (12th ed.): 5089
Lednicer: 3 160

SAMPLE
Matrix: solutions
Sample preparation: Inject an aliquot of a 100 μg/mL solution in water:running buffer 50:50.

CAPILLARY ELECTROPHORESIS
Capillary: 44.5 cm × 50 μm acrylamide-coated fused-silica (Bio-Rad)
Capillary temperature: 30
Running buffer: 100 mM NaH_2PO_4 containing 15 mM gamma cyclodextrin, adjusted to pH 2.5
   with phosphoric acid
Injection: Electrokinetic injection at 8 kV for 6 s.
Detector: UV 200
Migration time: 10.97
Voltage: 14 kV
Model: Bio-Rad BioFocus 3000

OTHER SUBSTANCES
Also analyzed: albuterol, alprenolol, atenolol, atropine, baclofen, bamethan, benserazide, biperiden, bisoprolol, bupivacaine, bupranolol, butetamate, carazolol, carbuterol, carvedilol, celiprolol, chloroquine, chlorpheniramine (chlorphenamine), clidinium bromide, clobutinol, disopyramide, dobutamine, flecainide, homatropine, isoproterenol, isothipendyl, ketamine, mefloquine, mequitazine, metaproterenol (orciprenaline), metipranolol, nafronyl (naftidrofuryl), nefopam, ofloxacin, orphenadrine, oxomemazine, oxprenolol, phenoxybenzamine, pholedrine, pindolol, pirbuterol, prilocaine, promethazine, propafenone, propranolol, sotalol, synephrine, terbutaline, tetrahydrozoline (tetryzoline), tocainide, trihexyphenidyl, trimeprazine (alimemazine), trimipramine, tropicamide, verapamil, zopiclone

KEY WORDS
coated capillary; achiral

REFERENCE
Koppenhoefer,B.; Epperlein,U.; Christian,B.; Yibing,J.; Yuying,C.; Bingcheng,L. Separation of enantiomers of
   drugs by capillary electrophoresis. I. γ-Cyclodextrin as chiral solvating agent, *J.Chromatogr.A*, **1995**, *717*,
   181–190.

SAMPLE
Matrix: solutions
Sample preparation: Inject an aliquot of a 100 μg/mL solution in running buffer.

CAPILLARY ELECTROPHORESIS
Capillary: 32 cm × 50 μm fused-silica (27.5 cm to detector) (Yongnian Optical Conductive Fiber
   Plant, China), coated with polyacrylamide
Capillary preparation: No details of the polyacrylamide coating process are provided. However,
   another paper (LC.GC 1997, 15, 40) by this group indicates that they use the procedure of
   Hjertén, thus: Adjust the pH of 20 mL water to 3.5 with acetic acid, add 80 μL 3-(trimethoxysilyl)propyl methacrylate (3-methacryloxypropyltrimethoxysilane), mix, suck into capillary, let
   stand at room temperature for 1 h, remove the solution, wash with water. Fill the capillary
   with a deaerated 3-4% acrylamide solution containing 1 μL/mL N,N,N',N'-tetramethylethylenediamine and 1 mg/mL potassium persulfate, let stand for 30 min, remove excess solution
   by aspiration, rinse with water, remove water by aspiration, dry at 35° (J. Chromatogr. 1985,
   347, 191).
Capillary temperature: 25

Running buffer: 100 mM NaH_2PO_4 adjusted to pH 2.5 (A) or 100 mM NaH_2PO_4 containing 45 mM hydroxypropyl-gamma-cyclodextrin, adjusted to pH 2.5 (B)
Injection: Electrokinetic injection at 15 kV for 3 s.
Detector: UV 200, UV 210
Migration time: 5.41 (A), 10.10, 10.21 (B, enantiomers)
Voltage: 15 kV
Model: Bio-Rad BioFocus 3000

OTHER SUBSTANCES
Simultaneous: atropine, carazolol, cicletanine, dimethindene, fendiline, homatropine, isothipendyl, mefloquine, metaclazepam, nafronyl (naftidrofuryl), nefopam, nicardipine, promethazine, reproterol, tetrahydrozoline (tetryzoline), theodrenaline, tioconazole, trihexyphenidyl, trimeprazine (alimemazine), trimipramine

KEY WORDS
coated capillary; chiral

REFERENCE
Koppenhoefer,B.; Epperlein,U.; Xiaofeng,Z.; Bingcheng,L. Separation of enantiomers of drugs by capillary electrophoresis. Part 4: Hydroxypropyl-γ-cyclodextrin as chiral solvating agent, *Electrophoresis*, **1997**, *18*, 924–930.

SAMPLE
Matrix: solutions
Sample preparation: Inject an aliquot of a 100 μg/mL solution in running buffer.

CAPILLARY ELECTROPHORESIS
Capillary: 30 cm × 50 μm fused-silica (25.5 cm to detector), coated with polyacrylamide
Capillary preparation: Adjust the pH of 20 mL water to 3.5 with acetic acid, add 80 μL 3-(trimethoxysilyl)propyl methacrylate (3-methacryloxypropyltrimethoxysilane), mix, suck into capillary, let stand at room temperature for 1 h, remove the solution, wash with water. Fill the capillary with a deaerated 3-4% acrylamide solution containing 1 μL/mL N,N,N',N'-tetramethylethylenediamine and 1 mg/mL potassium persulfate, let stand for 30 min, remove excess solution by aspiration, rinse with water, remove water by aspiration, dry at 35° (J. Chromatogr. 1985, 347, 191).
Capillary temperature: 25
Running buffer: 100 mM NaH_2PO_4 containing 45 mM hydroxypropyl-β- cyclodextrin, adjusted to pH 2.5 with phosphoric acid
Injection: Electrokinetic injection at 15 kV for 3 s.
Detector: UV 200
Voltage: 15 kV
Model: Bio-Rad BioFocus 3000

KEY WORDS
chiral; coated capillary; comparison with the use of other cyclodextrins; this running buffer gave the greatest enantiomeric separation.; α=1.010

REFERENCE
Lin,B.; Zhu,X.; Koppenhoefer,B.; Epperlein,U. Investigation of 123 chiral drugs by cyclodextrin-modified capillary electrophoresis, *LC.GC*, **1997**, *15*, 40–46.

SAMPLE
Matrix: solutions

CAPILLARY ELECTROPHORESIS
Capillary: 29-36 cm × 50 μm fused-silica (24.5-31.5 cm to detector) (Yongnian Optical Conductive Fiber Plant, China) coated with polyacrylamide
Capillary preparation: Coat capillary as follows. Adjust the pH of 20 mL water to 3.5 with acetic acid, add 80 μL 3-(trimethoxysilyl)propyl methacrylate (3-methacryloxypropyltrimethoxysilane), mix, suck into capillary, let stand at room temperature for 1 h, remove the solution, wash with water. Fill the capillary with a deaerated 3-4% acrylamide solution containing 1 μL/mL N,N,N',N'-tetramethylethylenediamine and 1 mg/mL potassium persulfate, let stand

for 30 min, remove excess solution by aspiration, rinse with water, remove water by aspiration, dry at 35° (J. Chromatogr. 1985, 347, 191).
Capillary temperature: 25
Running buffer: 100 mM pH 2.5 NaH_2PO_4 (A) or 100 mM pH 2.5 NaH_2PO_4 containing 45 mM hydroxypropyl-α-cyclodextrin (Wacker, Munich) (B)
Injection: Electromigration at 15 kV for 3 s.
Detector: UV 200; UV 210
Migration time: 5.41 (A); 8.96 (B) (no separation of enantiomers)
Voltage: 15 kV
Model: Bio-Focus 3000

OTHER SUBSTANCES
Also analyzed: albuterol (salbutamol), alprenolol, amorolfine, atenolol, atropine, azelastine, baclofen, bamethan, benproperine, benserazide, biperiden, bisoprolol, brompheniramine, bupivacaine, bupranolol, butamirate, butethamate, carazolol, carbuterol, carteolol, carvedilol, celiprolol, chloroquine, chlorpheniramine, chlorphenoxamine, cicletanine, clenbuterol, clidinium bromide, clobutinol, dimethindene, dipivefrin, disopyramide, dobutamine, doxylamine, fendiline, flecainide, gallopamil, homatropine, isoproterenol (isoprenaline), isothipendyl, ketamine, meclizine, mefloquine, mepindolol, mequitazine, metaclazepam, metaproterenol (orciprenaline), mctipranolol, mctoprolol, nafronyl (naftidrofuryl), nefopam, nicardipine, norfenefrine, ofloxacin, ornidazole, orphenadrine, oxomemazine, oxprenolol, oxybutynin, phenoxybenzamine, phenylpropanolamine, pholedrine, pindolol, pirbuterol, prilocaine, procyclidine, promethazine, propafenone, propranolol, reproterol, sotalol, sulpride, synephrine, talinolol, terbutaline, tetrahydrozoline (tetryzoline), theodrenaline, tioconazole, tocainide, trihexyphenidyl, trimeprazine (alimemazine), trimipramine, tropicamide, verapamil, zopiclone

---

KEY WORDS
coated capillary

REFERENCE
Koppenhoefer,B.; Eperlein,U.; Schlunk,R.; Zhu,X.; Lin,B. Separation of enantiomers of drugs by capillary electrophoresis. V. Hydroxypropyl-α-cyclodextrin as chiral solvating agent, *J.Chromatogr.A*, **1998**, *793*, 153–164.

Isocarboxazid

Molecular formula: $C_{12}H_{13}N_3O_2$
Molecular weight: 231.25
CAS Registry No.: 59-63-2
Merck Index (12th ed.): 5172
Lednicer: 1 233

SAMPLE
Matrix: urine
Sample preparation: 10 mL Urine + 1 mL 5 M NaOH + 10 mL n-hexane, vortex for 1 min, centrifuge at 0° at 3000 g for 5 min. Remove the organic layer and add it to 50 μL glacial acetic acid, evaporate to dryness under a stream of nitrogen at 30°, reconstitute the residue in 50 μL 50 mM sodium taurodeoxycholate, filter (0.2 μm PTFE), inject an aliquot.

CAPILLARY ELECTROPHORESIS
Capillary: 100 cm × 50 μm fused-silica (50 cm to detector) (ISCO)
Capillary preparation: Rinse capillary with 20 μL running buffer between injections. Condition a new capillary by filling with 1 M NaOH, let stand for 1 h, fill with 100 mM NaOH, let stand for 1 h, rinse with buffer. Every 20 injections rinse capillary with 200 μL 1 M NaOH, 200 μL water, and 200 μL running buffer, fill with running buffer.
Capillary temperature: 22
Running buffer: 40 mM pH 9.5 Borate buffer containing 10 mM sodium taurodeoxycholate
Injection: Load under vacuum at 7.5 kPa/s.
Detector: UV 240

Migration time: 10.5
Voltage: 30 kV
Model: ISCO Model 3140 electropherograph
Limit of detection: 86 ng/mL

OTHER SUBSTANCES
Extracted: acepromazine, amiodarone, amitriptyline, azaperone, chlorpromazine, cianopramine, clomipramine, clozapine, desethylamiodarone, desipramine, diclofensine, dothiepin, doxepin, imipramine, moclobemide, perphenazine, phenothiazine, pimozide, prochlorperazine, promazine, thioridazine, thiothixene, trifluoperazine, trimipramine

KEY WORDS
human; cow; pig; horse; protect from light

REFERENCE
Aumatell,A.; Wells,R.J. Determination of a cardiac antiarrhythmic, tricyclic antipsychotics and antidepressants in human and animal urine by micellar electrokinetic capillary chromatography using a bile salt, *J.Chromatogr.B*, **1995**, *669*, 331–344.

Isoniazid

Molecular formula: $C_6H_7N_3O$
Molecular weight: 137.14
CAS Registry No.: 54-85-3
Merck Index (12th ed.): 5203
Lednicer: 1 254

SAMPLE
Matrix: solutions
Sample preparation: Inject an aliquot of an aqueous solution.

CAPILLARY ELECTROPHORESIS
Capillary: 70 cm × 50 μm fused-silica (50 cm to detector) (GL Science)
Running buffer: 20 mM pH 9.1 Borate buffer containing 10 mM KOH and 150 mM sodium dodecyl sulfate
Injection: Hydrostatic injection at 4 cm
Detector: UV 185, UV 210
Migration time: 8
Voltage: 15 kV
Current: 30 μA

OTHER SUBSTANCES
Simultaneous: 3-acetylpyridine, 6-aminoniacinamide, ethyl nicotinate, N-methylniacinamide, niacin, niacinamide, β-picoline, pyridine, pyridine-3-aldehyde, 3-pyridinemethanol, pyridine-3-sulfonic acid, thioniacinamide

REFERENCE
Tanaka,S.; Kodama,K.; Kaneta,T.; Nakamura,H. Migration behavior of niacin derivatives in capillary electrophoresis, *J.Chromatogr.A*, **1995**, *718*, 233–237.

SAMPLE
Matrix: urine
Sample preparation: Dilute urine 2-fold, inject an aliquot.

CAPILLARY ELECTROPHORESIS
Capillary: 60 cm × 75 μm fused-silica (Yongnian Optical Fiber, Hebei, China)

Capillary preparation: Between runs rinse capillary with 100 mM NaOH for 3 min, water for 3 min, and running buffer for 5 min.

Running buffer: 10 mM pH 5.52 Sodium phosphate buffer containing 10 mM sodium dodecyl sulfate

Injection: Electrokinetic injection at 15 kV for 3 s.

Detector: E, Pd-modified carbon fiber electrode (details in paper) 0.5 V, Ag/AgCl reference electrode

Migration time: 10

Voltage: 15 kV

Model: laboratory-constructed

Limit of detection: 5 μM

OTHER SUBSTANCES

Extracted: hydrazine, methylhydrazine

REFERENCE

Liu,J.; Zhou,W.; You,T.; Li,F.; Wang,E.; Dong,S. Detection of hydrazine, methylhydrazine, and isoniazid by capillary electrophoresis with a palladium-modified microdisk array electrode, *Anal.Chem.*, **1996**, *68*, 3350–3353.

Isoproterenol

Molecular formula: $C_{11}H_{17}NO_3$

Molecular weight: 211.26

CAS Registry No.: 7683-59-2, 51-30-9 (HCl), 6700-39-6 (sulfate dihydrate), 299-95-6 (sulfate)

Merck Index (12th ed.): 5236

SAMPLE

Matrix: dialysate

Sample preparation: Inject an aliquot of dialysate (Ringer's).

CAPILLARY ELECTROPHORESIS

Capillary: 65 cm × 50 μm fused-silica (40 cm to detector) (Polymicro Technologies)

Capillary preparation: Between injections flush capillary with 500 μL water, 500 μL 3% v/v Microcleaning solution (International Products, Trenton, NJ), 500 μL water, and 500 μL running buffer.

Running buffer: 250 mM pH 4.75 Sodium acetate buffer containing 100 mg/mL methyl-O-β-cyclodextrin

Injection: Vacuum injection for 8 s.

Detector: UV 220

Migration time: 10.5 (-), 11 (+)

Voltage: 20 kV

Model: Isco 3850

Limit of detection: 2.8 μg/mL

KEY WORDS

chiral

REFERENCE

Hadwiger,M.E.; Torchia,S.R.; Park,S.; Biggin,M.E.; Lunte,C.E. Optimization of the separation and detection of the enantiomers of isoproterenol in microdialysis samples by cyclodextrin-modified capillary electrophoresis using electrochemical detection, *J.Chromatogr.B*, **1996**, *681*, 241–249.

SAMPLE

Matrix: solutions

Sample preparation: Prepare a solution in running buffer, inject an aliquot.

CAPILLARY ELECTROPHORESIS
Capillary: 78.6 cm × 12.7 μm fused-silica (Polymicro Technologies)
Running buffer: 25 mM pH 5.7 2-morpholinoethanesulfonic acid buffer
Injection: Injection by electromigration at 30 kV for 2 s
Detector: E, 0.7 V vs sodium-saturated calomel electrode
Migration time: 5.5
Voltage: 30 kV

OTHER SUBSTANCES
Simultaneous: 3,4-dihydroxyphenylacetic acid, homovanillic acid, 4-methylcatechol, norepinephrine, serotonin, ascorbic acid

REFERENCE
Wallingford,R.A.; Ewing,A.G. Capillary zone electrophoresis with electrochemical detection in 12.7 μm diameter columns, *Anal.Chem.*, **1988**, *60*, 1972–1975.

SAMPLE
Matrix: solutions

CAPILLARY ELECTROPHORESIS
Capillary: 70 cm × 100 μm fused-silica (50 cm to detector) (Gasukuro Kogyo, Tokyo)
Running buffer: 100 mM KOH containing 200 mM boric acid
Injection: Hydrostatic injection for 10 s.
Detector: UV 217
Migration time: 13.5
Voltage: 9.5 kV
Current: 100 μA

OTHER SUBSTANCES
Simultaneous: dopamine, epinephrine, levodopa, metanephrine, norepinephrine, normetanephrine, vanillylmanedlic acid

REFERENCE
Tanaka,S.; Kaneta,T.; Yoshida,H. Separation of catecholamines by capillary zone electrophoresis using complexation with boric acid, *Anal.Sci.*, **1990**, *6*, 467–468.

SAMPLE
Matrix: solutions

CAPILLARY ELECTROPHORESIS
Capillary: 62 cm × 50 μm fused silica (50 cm to detector) (Polymicro Technologies)
Capillary temperature: 40
Running buffer: 150 mM pH 2.5 tetrabutylammonium phosphate buffer containing 35 mM β-cyclodextrin
Injection: Injection by gravity at 10 cm for 5 s
Detector: UV 214
Migration time: 27.2, 28.8 (enantiomers)
Voltage: 20 kV
Current: 56 μA
Model: Laboratory constructed

OTHER SUBSTANCES
Simultaneous: propranolol

KEY WORDS
chiral

REFERENCE
Quang,C.; Khaledi,M.G. Improved chiral separation of basic compounds in capillary electrophoresis using β-cyclodextrin and tetraalkylammonium reagents, *Anal.Chem.*, **1993**, *65*, 3354–3358.

SAMPLE
Matrix: solutions

Sample preparation: Prepare a 50 µg/mL solution in diluted running buffer, inject an aliquot.

CAPILLARY ELECTROPHORESIS
Capillary: 44 cm × 50 µm fused-silica
Capillary preparation: After each run wash capillary with water for 2 min and with running buffer for 3 min. At the beginning of each day wash capillary with water and running buffer for 5 min. Condition a new capillary with 1 M NaOH, 100 mM NaOH, water, and separation buffer.
Capillary temperature: 15
Running buffer: 100 mM Phosphoric acid containing 15 mM heptakis(2,6-di-O-methyl)-β-cyclodextrin adjusted to pH 3.0 with triethanolamine
Injection: Hydrodynamic injection for 3 s
Detector: UV 210
Migration time: 9.3, 10.0 (enantiomers)
Voltage: 25 kV
Model: Spectra Physics Spectraphoresis 1000

KEY WORDS
chiral

REFERENCE
Bechet,I.; Paques,P.; Fillet,M.; Hubert,P.; Crommen,J. Chiral separation of basic drugs by capillary zone electrophoresis with cyclodextrin additives, *Electrophoresis*, **1994**, *15*, 818–823.

SAMPLE
Matrix: solutions

CAPILLARY ELECTROPHORESIS
Capillary: 41 cm × 50 µm fused-silica (41 cm to detector) (Grom)
Capillary temperature: 22 ± 1
Running buffer: pH 3.10 50 mM Phosphate buffer containing 1 mM β-cyclodextrin sulfobutyl ether (Center for Drug Delivery Research,University of Kansas, Lawrence KS)
Injection: Hydrostatic injection at 10 cm for 30 s
Detector: UV 210
Migration time: 8.82, 8.94 (enantiomers)
Voltage: 340 V/cm
Model: Grom system 100

KEY WORDS
chiral

REFERENCE
Chankvetadze,B.; Endresz,G.; Blaschke,G. About some aspects of the use of charged cyclodextrins for capillary electrophoresis enantioseparation, *Electrophoresis*, **1994**, *15*, 804–807.

SAMPLE
Matrix: solutions

CAPILLARY ELECTROPHORESIS
Capillary: 62 cm × 52 µm fused-silica (50 cm to detector) (Polymicro Technologies)
Running buffer: 70 mM pH 2.5 Trimethylammonium phosphate buffer containing 20 mM hydroxypropyl-β-cyclodextrin
Injection: Gravity injection at 10 cm for 5 s
Migration time: 20, 21 (enantiomers)
Voltage: 20 kV
Current: 43 µA
Model: Laboratory constructed

OTHER SUBSTANCES
Simultaneous: acebutolol, alprenolol, atenolol, labetalol, nadolol, pindolol, propranolol

KEY WORDS
chiral

REFERENCE
Quang,C.; Khaledi,M.G. Direct separation of the enantiomers of β-blockers by cyclodextrin-mediated capillary zone electrophoresis, *J.High Res.Chromatogr.*, **1994**, *17*, 99–101.

SAMPLE
Matrix: solutions
Sample preparation: Inject an aliquot of a 10 μg/mL solution of the compound in water.

CAPILLARY ELECTROPHORESIS
Capillary: 27 cm × 50 μm neutrally-coated capillary (7 cm to detector) (Beckman)
Capillary preparation: Before each analysis rinse capillary with 100 mM HCl for 30 s, with water for 2 min, and with running buffer for 2 min
Capillary temperature: 15 ± 0.1
Running buffer: 25 mM pH 2.5 Phosphate buffer containing 30 mM dimethyl-β-cyclodextrin
Injection: Electrokinetic injection at 10 kV for 2 s
Detector: UV 200
Migration time: 0.66, 0.68 (enantiomers)
Voltage: 30 kV
Model: Beckman P/ACE system 5500

OTHER SUBSTANCES
Simultaneous: metaproterenol

KEY WORDS
chiral; coated capillary

REFERENCE
Aumatell,A.; Guttman,A. Ultra-fast chiral separation of basic drugs by capillary electrophoresis, *J.Chromatogr.A*, **1995**, *717*, 229–234.

SAMPLE
Matrix: solutions
Sample preparation: Inject an aliquot of a solution in running buffer.

CAPILLARY ELECTROPHORESIS
Capillary: 44 cm × 5 μm fused-silica (Polymicro Technologies)
Running buffer: Isopropanol:30 mM morpholinoethanesulfonic acid 20:80, pH adjusted to 6.0
Injection: Electromigration at 12 kV for 3 s
Detector: E, Pt electrode 0.700 V, Ag/AgCl reference electrode, details of cell in paper
Migration time: 8
Voltage: 25 kV

OTHER SUBSTANCES
Simultaneous: catechol, dopamine, epinephrine, norepinephrine, serotonin

REFERENCE
Chen,M.-C.; Huang,H.-J. An electrochemical cell for end-column amperometric detection in capillary electrophoresis, *Anal.Chem.*, **1995**, *67*, 4010–4014.

SAMPLE
Matrix: solutions

CAPILLARY ELECTROPHORESIS
Capillary: 56 cm × 50 μm fused-silica (36 cm to detector) (Supelco)
Running buffer: 25 mM pH 2.7 phosphate buffer containing 50 mg/mL soluble β-cyclodextrin polymer cross-linked with epichlorohydrin (MW 3000-5000) (Cyclolab, Budapest)
Injection: Hydrodynamic injection at 20 cm for 10 s

Detector: UV 206
Migration time: 15.0 (-), 15.3 (+)
Voltage: 12 kV
Model: Laboratory made

OTHER SUBSTANCES
Simultaneous: norepinephrine, propranolol

KEY WORDS
chiral

REFERENCE
Ingelse,B.A.; Everaerts,F.M.; Desiderio,C.; Fanali,S. Enantiomeric separation by capillary electrophoresis using a soluble neutral β-cyclodextrin polymer, *J.Chromatogr.A*, **1995**, *709*, 89–98.

SAMPLE
Matrix: solutions
Sample preparation: Inject an aliquot of a 100 μg/mL solution in water:running buffer 50:50.

CAPILLARY ELECTROPHORESIS
Capillary: 44.5 cm × 50 μm acrylamide-coated fused-silica (Bio-Rad)
Capillary temperature: 30
Running buffer: 100 mM NaH_2PO_4 containing 15 mM gamma-cyclodextrin, adjusted to pH 2.5 with phosphoric acid
Injection: Electrokinetic injection at 8 kV for 6 s.
Detector: UV 200
Migration time: 8.50
Voltage: 14 kV
Model: Bio-Rad BioFocus 3000

OTHER SUBSTANCES
Also analyzed: albuterol, alprenolol, atenolol, atropine, baclofen, bamethan, benserazide, biperiden, bisoprolol, bupivacaine, bupranolol, butetamate, carazolol, carbuterol, carvedilol, celiprolol, chloroquine, chlorpheniramine (chlorphenamine), clidinium bromide, clobutinol, disopyramide, dobutamine, flecainide, homatropine, ipratropium bromide, isothipendyl, ketamine, mefloquine, mequitazine, metaproterenol (orciprenaline), metipranolol, nafronyl (naftidrofuryl), nefopam, ofloxacin, orphenadrine, oxomemazine, oxprenolol, phenoxybenzamine, pholedrine, pindolol, pirbuterol, prilocaine, promethazine, propafenone, propranolol, sotalol, synephrine, terbutaline, tetrahydrozoline (tetryzoline), tocainide, trihexyphenidyl, trimoprazine (alimemazine), trimipramine, tropicamide, verapamil, zopiclone

---

KEY WORDS
coated capillary; achiral

REFERENCE
Koppenhoefer,B.; Epperlein,U.; Christian,B.; Yibing,J.; Yuying,C.; Bingcheng,L. Separation of enantiomers of drugs by capillary electrophoresis. I. γ-Cyclodextrin as chiral solvating agent, *J.Chromatogr.A*, **1995**, *717*, 181–190.

SAMPLE
Matrix: solutions
Sample preparation: Inject an aliquot of a solution in 8 mM disodium EDTA containing 0.1 mM sodium bisulfite.

CAPILLARY ELECTROPHORESIS
Capillary: 65 cm × 50 μm fused-silica (Polymicro Technologies)
Capillary preparation: Before each run flush capillary with 500 mM disodium EDTA (adjusted to pH 13 with solid NaOH) for 1 min and with running buffer for 1 min.
Running buffer: 25 mM pH 6.0 Sodium phosphate buffer
Injection: Pressure injection at 2 psi for 2 s

Detector: E, Bioanalytical Systems PA-1, 33 μm carbon fiber electrode +0.85 V, Ag/AgCl reference
   electrode (design of capillary/detector decoupler is given in paper)
Migration time: 2.5
Voltage: 28 kV

OTHER SUBSTANCES
Simultaneous: caffeic acid, catechol, chlorogenic acid, 3,4-dihydroxybenzylamine, gentisic acid,
   protocatechuic acid

REFERENCE
Park,S.; Lunte,C.E. A perfluorosulfonated ionomer end-column electrical decoupler for capillary electrophoresis/
   electrochemical detection, *Anal.Chem.*, **1995**, *67*, 4366–4370.

SAMPLE
Matrix: solutions
Sample preparation: Prepare a solution in MeOH/water, inject an aliquot.

CAPILLARY ELECTROPHORESIS
Capillary: 62 cm × 52 μm fused-silica (50 cm to detector) (Polymicro Technologies)
Capillary temperature: 40
Running buffer: 50 mM pH 2.50 Tetrabutylammonium phosphate containing 20 mM hydroxy-
   propyl-β-cyclodextrin
Injection: Siphon injection at 10 cm for 5 s
Detector: UV (wavelength not specified)
Migration time: 15.64, 16.12 (enantiomers)
Voltage: 20 kV
Model: Laboratory constructed

OTHER SUBSTANCES
Simultaneous: alprenolol, atenolol, chlorpheniramine, epinephrine, labetalol, norepinephrine,
   oxprenolol, propranolol

KEY WORDS
chiral

REFERENCE
Quang,C.; Khaledi,M.G. Extending the scope of chiral separation of basic compounds by cyclodextrin-mediated
   capillary zone electrophoresis, *J.Chromatogr.A*, **1995**, *692*, 253–265.

SAMPLE
Matrix: solutions
Sample preparation: Prepare a solution in MeOH/water, inject an aliquot.

CAPILLARY ELECTROPHORESIS
Capillary: 62 cm × 52 μm fused-silica (50 cm to detector) (Polymicro Technologies)
Capillary temperature: 40
Running buffer: 50 mM pH 2.50 Tetrabutylammonium phosphate
Injection: Siphon injection at 10 cm for 5 s
Detector: UV (wavelength not specified)
Migration time: 12
Voltage: 20 kV
Current: 29 μA
Model: Laboratory constructed

OTHER SUBSTANCES
Simultaneous: doxylamine, epinephrine, imidazole, 2-methylphenethylamine, 1-methylphenyl-
   propylamine, metoprolol, nadolol, nicotine, norepinephrine, propranolol, pseudoephedrine

REFERENCE
Quang,C.; Khaledi,M.G. Extending the scope of chiral separation of basic compounds by cyclodextrin-mediated
   capillary zone electrophoresis, *J.Chromatogr.A*, **1995**, *692*, 253–265.

SAMPLE
Matrix: solutions
Sample preparation: Inject an aliquot of an aqueous solution.

CAPILLARY ELECTROPHORESIS
Capillary: 65 cm × 50 μm fused-silica (40 cm to detector) (Isco)
Capillary preparation: Purge with running buffer for 3 min between injections. Purge daily with 1 M NaOH for 3 min, with water for 3 min, and with running buffer for 3 min.
Running buffer: Isopropanol:100 mM pH 7 phosphate buffer containing 25 mM rifamycin B 30: 70
Injection: Electrokinetic injection at 5 kV for 8 s.
Detector: UV 350
Migration time: 29 (-), 30 (+)
Voltage: 8 kV
Model: Isco model 3850

KEY WORDS
chiral; indirect UV detection

REFERENCE
Ward,T.J.; Dann,C.,III; Blaylock,A. Enantiomeric resolution using the macrocyclic antibiotics rifamycin B and rifamycin SV as chiral selectors for capillary electrophoresis, *J.Chromatogr.A*, **1995**, *715*, 337–344.

SAMPLE
Matrix: solutions
Sample preparation: Inject an aliquot of a 50-200 μM solution.

CAPILLARY ELECTROPHORESIS
Capillary: 48.5 cm × 50 μm fused-silica (44.5 cm to detector) (Polymicro Technologies)
Capillary temperature: 25
Running buffer: 100 mM Phosphoric acid containing 7.5 mM 6^A-methylamino-β-cyclodextrin, adjusted to pH 2.5 with tetramethylammonium hydroxide. (Synthesis of 6^A-methylamino-β-cyclodextrin was as follows. Add a solution of 3.65 g p-toluenesulfonyl chloride in 30 mL dry pyridine to 29.60 g β-cyclodextrin stirred at 5° in 300 mL dry pyridine, stir overnight at room temperature, evaporate to dryness under reduced pressure at 40°, add 700 mL diethyl ether to the residue. Collect the precipitate and recrystallize it 3 times from water to obtain mono-(6-O-p-tolylsulfonyl)-β-cyclodextrin in 31% yield (Bull. Chem. Soc. Japan 1978, 51, 3030). Heat 2 g mono-(6-O-p-tolylsulfonyl)-β-cyclodextrin with 35 mL 50% methylamine in MeOH in a sealed tube at 70° for 3 days, purify by chromatography on carboxymethylcellulose with ammonium bicarbonate solution to obtain 6^A-methylamino-β-cyclodextrin (cf. J. Am. Chem. Soc. 1980, 102, 762).
Injection: Pressure injection at 10 psi.s.
Detector: UV 206
Migration time: 35.8 (second enantiomer, R_S = 0.7)
Voltage: 18 kV
Current: 41-48 μA
Model: Biofocus 3000 (Bio-Rad)

OTHER SUBSTANCES
Simultaneous: chlorpheniramine, ketamine

KEY WORDS
chiral

REFERENCE
Fanali,S.; Camera,E. Use of methylamino-β-cyclodextrin in capillary electrophoresis. Resolution of acidic and basic enantiomers, *Chromatographia*, **1996**, *43*, 247–253.

SAMPLE
Matrix: solutions
Sample preparation: Inject an aliquot of a 50 μg/mL solution in water.

CAPILLARY ELECTROPHORESIS
Capillary: 48.5 cm × 50 μm fused-silica (40 cm to detector)
Capillary preparation: After each run wash capillary with running buffer for 3 min. At the start of each day wash capillary with running buffer for 10 min. Before use treat new capillaries with 1 M NaOH, 100 mM NaOH, water, and running buffer.
Capillary temperature: 15
Running buffer: 100 mM Phosphoric acid containing 15 mM carboxymethyl-β-cyclodextrin (Cyclolab, Budapest), adjusted to pH 3.0 with 84 mM triethanolamine
Injection: Hydrodynamic injection at 5 kPa for 2 s.
Detector: UV 210
Migration time: 16.62, 17.69 (enantiomers)
Voltage: 25 kV
Model: Hewlett Packard [3D]CE

OTHER SUBSTANCES
Simultaneous: bupivacaine, chlorpheniramine, dimethindene, ephedrine, fenfluramine

KEY WORDS
chiral

REFERENCE
Fillet,M.; Bechet,I.; Hubert,P.; Crommen,J. Resolution improvement by use of carboxymethyl-β-cyclodextrin as chiral additive for the enantiomeric separation of basic drugs by capillary electrophoresis, *J.Pharm.Biomed.Anal.*, **1996**, *14*, 1107–1114.

SAMPLE
Matrix: solutions
Sample preparation: Inject an aliquot of a 100 μg/mL solution in running buffer.

CAPILLARY ELECTROPHORESIS
Capillary: 29 cm × 50 μm fused-silica (24.5 cm to detector) (Yongnian Optical Conductive Fiber Plant, China), coated with polyacrylamide
Capillary preparation: No details of the polyacrylamide coating process are provided. However, another paper (LC.GC 1997, 15, 40) by this group indicates that they use the procedure of Hjertén, thus: Adjust the pH of 20 mL water to 3.5 with acetic acid, add 80 μL 3-(trimethoxysilyl)propyl methacrylate (3-methacryloxypropyltrimethoxysilane), mix, suck into capillary, let stand at room temperature for 1 h, remove the solution, wash with water. Fill the capillary with a deaerated 3-4% acrylamide solution containing 1 μL/mL N,N,N',N'-tetramethylethylenediamine and 1 mg/mL potassium persulfate, let stand for 30 min, remove excess solution by aspiration, rinse with water, remove water by aspiration, dry at 35° (J. Chromatogr. 1985, 347, 191).
Capillary temperature: 25
Running buffer: 100 mM NaH_2PO_4 adjusted to pH 2.5
Injection: Electrokinetic injection at 15 kV for 3 s.
Detector: UV 200, UV 210
Migration time: 5.11
Voltage: 15 kV
Model: Bio-Rad BioFocus 3000

OTHER SUBSTANCES
Simultaneous: albuterol, alprenolol, atenolol, baclofen, bamethan, benproperine, benserazide, bisoprolol, bupranolol, butamirate, butethamate, carbuterol, celiprolol, clenbuterol, clobutinol, dipivefrin, metaproterenol (orciprenaline), metipranolol, metoprolol, norfenefrine, ornidazole, oxprenolol, phenylpropanolamine, pholedrine, pirbuterol, prilocaine, procyclidine, sotalol, synephrine, terbutaline, tocainide

KEY WORDS
coated capillary

REFERENCE
Koppenhoefer,B.; Epperlein,U.; Xiaofeng,Z.; Bingcheng,L. Separation of enantiomers of drugs by capillary electrophoresis. Part 4: Hydroxypropyl-γ-cyclodextrin as chiral solvating agent, *Electrophoresis*, **1997**, *18*, 924–930.

SAMPLE
Matrix: solutions
Sample preparation: Inject an aliquot of a 100 μg/mL solution in running buffer.

CAPILLARY ELECTROPHORESIS
Capillary: 30 cm × 50 μm fused-silica (25.5 cm to detector), coated with polyacrylamide
Capillary preparation: Adjust the pH of 20 mL water to 3.5 with acetic acid, add 80 μL 3-(trimethoxysilyl)propyl methacrylate (3-methacryloxypropyltrimethoxysilane), mix, suck into capillary, let stand at room temperature for 1 h, remove the solution, wash with water. Fill the capillary with a deaerated 3-4% acrylamide solution containing 1 μL/mL N,N,N',N'-tetramethylethylenediamine and 1 mg/mL potassium persulfate, let stand for 30 min, remove excess solution by aspiration, rinse with water, remove water by aspiration, dry at 35° (J. Chromatogr. 1985, 347, 191).
Capillary temperature: 25
Running buffer: 100 mM NaH_2PO_4 containing 45 mM hydroxypropyl-β- cyclodextrin, adjusted to pH 2.5 with phosphoric acid
Injection: Electrokinetic injection at 15 kV for 3 s.
Detector: UV 200
Voltage: 15 kV
Model: Bio-Rad BioFocus 3000

KEY WORDS
chiral; coated capillary; comparison with the use of other cyclodextrins; this running buffer gave the greatest enantiomeric separation.; α=1.063

REFERENCE
Lin,B.; Zhu,X.; Koppenhoefer,B.; Epperlein,U. Investigation of 123 chiral drugs by cyclodextrin-modified capillary electrophoresis, *LC.GC*, **1997**, *15*, 40–46.

SAMPLE
Matrix: solutions

CAPILLARY ELECTROPHORESIS
Capillary: 37 cm × 50 μm fused-silica (30 cm to detector) (Quadrex, New Haven CT)
Capillary temperature: -16
Running buffer: MeOH:water 10:90 containing 20 mM heptakis(2,6-di-O-methyl)-β-cyclodextrin, 5 M urea, and 150 mM sodium phosphate, pH 2.5
Injection: Injection for 6 s
Detector: UV 214
Migration time: 16, 17 (enantiomers)
Voltage: 30 kV
Current: 45 μA
Model: Beckman P/ACE 2210

OTHER SUBSTANCES
Simultaneous: epinephrine, β-hydroxyphenethylamine, norepinephrine, octopamine

KEY WORDS
chiral

REFERENCE
Ma,S.; Horváth,C. Capillary zone electrophoresis at subzero temperatures. II: Chiral separation of biogenic amines, *Electrophoresis*, **1997**, *18*, 873–883.

SAMPLE
Matrix: solutions

CAPILLARY ELECTROPHORESIS
Capillary: 35 cm × 50 μm polyacrylamide-coated fused-silica (30.5 cm to detector) (Composite Metal Services, UK)

Capillary preparation: Before each run rinse capillary with water for 70 s and with running buffer for 100 s. (Coat capillary as follows. Adjust the pH of 20 mL water to 3.5 with acetic acid, add 80 µL 3-(trimethoxysilyl)propyl methacrylate (3-methacryloxypropyltrimethoxysilane), mix, suck into capillary, let stand at room temperature for 1 h, remove the solution, wash with water. Fill the capillary with a deaerated 3-4% acrylamide solution containing 1 µL/mL N,N,N',N'-tetramethylethylenediamine and 1 mg/mL potassium persulfate, let stand for 30 min, remove excess solution by aspiration, rinse with water, remove water by aspiration, dry at 35° (J. Chromatogr. 1985, 347, 191).)

Capillary temperature: 25

Running buffer: Buffer containing 150 mM cyanoethylated-β-cyclodextrin (Cyclolab, Budapest) (Prepare buffer by adjusting the pH of 50 mM phosphoric acid containing 50 mM acetic acid and 50 mM boric acid to 2.5 with concentrated NaOH, add the appropriate amount of cyanoethylated-β-cyclodextrin, dilute with an equal volume of water.)

Injection: Pressure injection at 5 psi for 2 s.

Detector: UV 206

Migration time: 19.8 (second enantiomer, α = 1.059)

Voltage: 20 kV

Current: 27-40 µA

Model: Bio-Rad Biofocus 3000

KEY WORDS
chiral; coated capillary

REFERENCE
Aturki,Z.; Desiderio,C.; Mannina,L.; Fanali,S. Chiral separations by capillary zone electrophoresis with the use of cyanoethylated-β-cyclodextrin as chiral selector, *J.Chromatogr.A*, **1998**, *817*, 91–104.

SAMPLE
Matrix: solutions

CAPILLARY ELECTROPHORESIS
Capillary: 48.5 cm × 50 µm fused-silica (40 cm to detector) (Supelco)

Capillary preparation: Condition capillary with running buffer for 3 min before each injection.

Capillary temperature: 15

Running buffer: 25 mM Citric acid adjusted to pH 4.5 with Tris, containing 2% dermatan sulfate (number average 15270, mass average 22260, z average 30390, polydispersity 1.458) (Opocrin, Corlo, Italy)

Injection: Pressure injection at 5 kPa for 10 s.

Detector: UV 220

Migration time: 21. 21.4 (enantiomers)

Voltage: 15 kV

Model: Hewlett-Packard ³ᴰCE

OTHER SUBSTANCES
Simultaneous: albuterol

Also analyzed: metaproterenol, terbutaline

KEY WORDS
chiral

REFERENCE
Gotti,R.; Cavrini,V.; Andrisano,V.; Mascellani,G. Dermatan sulfate as useful chiral selector in capillary electrophoresis, *J.Chromatogr.A*, **1998**, *814*, 205–211.

SAMPLE
Matrix: solutions

CAPILLARY ELECTROPHORESIS
Capillary: 29-36 cm × 50 µm fused-silica (24.5-31.5 cm to detector) (Yongnian Optical Conductive Fiber Plant, China) coated with polyacrylamide

Capillary preparation: Coat capillary as follows. Adjust the pH of 20 mL water to 3.5 with acetic acid, add 80 µL 3-(trimethoxysilyl)propyl methacrylate (3-methacryloxypropyltrimethox-

ysilane), mix, suck into capillary, let stand at room temperature for 1 h, remove the solution, wash with water. Fill the capillary with a deaerated 3-4% acrylamide solution containing 1 μL/mL N,N,N',N'-tetramethylethylenediamine and 1 mg/mL potassium persulfate, let stand for 30 min, remove excess solution by aspiration, rinse with water, remove water by aspiration, dry at 35° (J. Chromatogr. 1985, 347, 191).
Capillary temperature: 25
Running buffer: 100 mM pH 2.5 NaH$_2$PO$_4$ (A) or 100 mM pH 2.5 NaH$_2$PO$_4$ containing 45 mM hydroxypropyl-α-cyclodextrin (Wacker, Munich) (B)
Injection: Electromigration at 15 kV for 3 s.
Detector: UV 200; UV 210
Migration time: 5.11 (A); 7.18 (B) (no separation of enantiomers)
Voltage: 15 kV
Model: Bio-Focus 3000

OTHER SUBSTANCES
Also analyzed: albuterol (salbutamol), alprenolol, amorolfine, atenolol, atropine, azelastine, baclofen, bamethan, benproperine, benserazide, biperiden, bisoprolol, brompheniramine, bupivacaine, bupranolol, butamirate, butethamate, carazolol, carbuterol, carteolol, carvedilol, celiprolol, chloroquine, chlorpheniramine, chlorphenoxamine, cicletanine, clenbuterol, clidinium bromide, clobutinol, dimethindene, dipivefrin, disopyramide, dobutamine, doxylamine, fendiline, flecainide, gallopamil, homatropine, ipratropium bromide, isothipendyl, ketamine, meclizine, mefloquine, mepindolol, mequitazine, metaclazepam, metaproterenol (orciprenaline), metipranolol, metoprolol, nafronyl (naftidrofuryl), nefopam, nicardipine, norfenefrine, ofloxacin, ornidazole, orphenadrine, oxomemazine, oxprenolol, oxybutynin, phenoxybenzamine, phenylpropanolamine, pholedrine, pindolol, pirbuterol, prilocaine, procyclidine, promethazine, propafenone, propranolol, reproterol, sotalol, sulpride, synephrine, talinolol, terbutaline, tetrahydrozoline (tetryzoline), theodrenaline, tioconazole, tocainide, trihexyphenidyl, trimeprazine (alimemazine), trimipramine, tropicamide, verapamil, zopiclone

---

KEY WORDS
coated capillary

REFERENCE
Koppenhoefer,B.; Eperlein,U.; Schlunk,R.; Zhu,X.; Lin,B. Separation of enantiomers of drugs by capillary electrophoresis. V. Hydroxypropyl-α-cyclodextrin as chiral solvating agent, *J.Chromatogr.A*, **1998**, *793*, 153–164.

Isothipendyl

Molecular formula: C$_{16}$H$_{19}$N$_3$S
Molecular weight: 285.41
CAS Registry No.: 482-15-5, 1225-60-1 (HCl)
Merck Index (12th ed.): 5248
Lednicer: 1 430

SAMPLE
Matrix: solutions
Sample preparation: Inject an aliquot of a 100 μg/mL solution in running buffer.

CAPILLARY ELECTROPHORESIS
Capillary: 32 cm × 50 μm fused-silica (27.5 cm to detector) (Yongnian Optical Conductive Fiber Plant, China), coated with polyacrylamide
Capillary preparation: No details of the polyacrylamide coating process are provided. However, another paper (LC.GC 1997, 15, 40) by this group indicates that they use the procedure of Hjertén, thus: Adjust the pH of 20 mL water to 3.5 with acetic acid, add 80 μL 3-(trimethoxysilyl)propyl methacrylate (3-methacryloxypropyltrimethoxysilane), mix, suck into capillary, let stand at room temperature for 1 h, remove the solution, wash with water. Fill the capillary with a deaerated 3-4% acrylamide solution containing 1 μL/mL N,N,N',N'-tetramethylethylenediamine and 1 mg/mL potassium persulfate, let stand for 30 min, remove excess solution

by aspiration, rinse with water, remove water by aspiration, dry at 35° (J. Chromatogr. 1985, 347, 191).
Capillary temperature: 25
Running buffer: 100 mM NaH_2PO_4 adjusted to pH 2.5 (A) or 100 mM NaH_2PO_4 containing 45 mM hydroxypropyl-gamma-cyclodextrin, adjusted to pH 2.5 (B)
Injection: Electrokinetic injection at 15 kV for 3 s.
Detector: UV 200, UV 210
Migration time: 4.80 (A), 11.54, 11.75 (B, enantiomers)
Voltage: 15 kV
Model: Bio-Rad BioFocus 3000

OTHER SUBSTANCES
Simultaneous: atropine, carazolol, cicletanine, dimethindene, fendiline, homatropine, ipratropium bromide, mefloquine, metaclazepam, nafronyl (naftidrofuryl), nefopam, nicardipine, promethazine, reproterol, tetrahydrozoline (tetryzoline), theodrenaline, tioconazole, trihexyphenidyl, trimeprazine (alimemazine), trimipramine

KEY WORDS
coated capillary; chiral

REFERENCE
Koppenhoefer,B.; Epperlein,U.; Xiaofeng,Z.; Bingcheng,L. Separation of enantiomers of drugs by capillary electrophoresis. Part 4: Hydroxypropyl-γ-cyclodextrin as chiral solvating agent, *Electrophoresis*, **1997**, *18*, 924–930.

SAMPLE
Matrix: solutions
Sample preparation: Inject an aliquot of a 100 μg/mL solution in running buffer.

CAPILLARY ELECTROPHORESIS
Capillary: 30 cm × 50 μm fused-silica (25.5 cm to detector), coated with polyacrylamide
Capillary preparation: Adjust the pH of 20 mL water to 3.5 with acetic acid, add 80 μL 3-(trimethoxysilyl)propyl methacrylate (3-methacryloxypropyltrimethoxysilane), mix, suck into capillary, let stand at room temperature for 1 h, remove the solution, wash with water. Fill the capillary with a deaerated 3-4% acrylamide solution containing 1 μL/mL N,N,N',N'-tetramethylethylenediamine and 1 mg/mL potassium persulfate, let stand for 30 min, remove excess solution by aspiration, rinse with water, remove water by aspiration, dry at 35° (J. Chromatogr. 1985, 347, 191).
Capillary temperature: 25
Running buffer: 100 mM NaH_2PO_4 containing 45 mM α-cyclodextrin, adjusted to pH 2.5 with phosphoric acid
Injection: Electrokinetic injection at 15 kV for 3 s.
Detector: UV 200
Voltage: 15 kV
Model: Bio-Rad BioFocus 3000

KEY WORDS
chiral; coated capillary; comparison with the use of other cyclodextrins; this running buffer gave the greatest enantiomeric separation.; α=1.019

REFERENCE
Lin,B.; Zhu,X.; Koppenhoefer,B.; Epperlein,U. Investigation of 123 chiral drugs by cyclodextrin-modified capillary electrophoresis, *LC.GC*, **1997**, *15*, 40–46.

SAMPLE
Matrix: solutions

CAPILLARY ELECTROPHORESIS
Capillary: 29-36 cm × 50 μm fused-silica (24.5-31.5 cm to detector) (Yongnian Optical Conductive Fiber Plant, China) coated with polyacrylamide
Capillary preparation: Coat capillary as follows. Adjust the pH of 20 mL water to 3.5 with acetic acid, add 80 μL 3-(trimethoxysilyl)propyl methacrylate (3-methacryloxypropyltrimethox-

ysilane), mix, suck into capillary, let stand at room temperature for 1 h, remove the solution, wash with water. Fill the capillary with a deaerated 3-4% acrylamide solution containing 1 μL/mL N,N,N',N'-tetramethylethylenediamine and 1 mg/mL potassium persulfate, let stand for 30 min, remove excess solution by aspiration, rinse with water, remove water by aspiration, dry at 35° (J. Chromatogr. 1985, 347, 191).
Capillary temperature: 25
Running buffer: 100 mM pH 2.5 NaH_2PO_4 (A) or 100 mM pH 2.5 NaH_2PO_4 containing 45 mM hydroxypropyl-α-cyclodextrin (Wacker, Munich) (B)
Injection: Electromigration at 15 kV for 3 s.
Detector: UV 200; UV 210
Migration time: 4.80 (A); 11.04 (B) (no separation of enantiomers)
Voltage: 15 kV
Model: Bio-Focus 3000

OTHER SUBSTANCES
Also analyzed: albuterol (salbutamol), alprenolol, amorolfine, atenolol, atropine, azelastine, baclofen, bamethan, benproperine, benserazide, biperiden, bisoprolol, brompheniramine, bupivacaine, bupranolol, butamirate, butethamate, carazolol, carbuterol, carteolol, carvedilol, celiprolol, chloroquine, chlorpheniramine, chlorphenoxamine, cicletanine, clenbuterol, clidinium bromide, clobutinol, dimethindene, dipivefrin, disopyramide, dobutamine, doxylamine, fendiline, flecainide, gallopamil, homatropine, ipratropium bromide, isoproterenol (isoprenaline), ketamine, meclizine, mefloquine, mepindolol, mequitazine, metaclazepam, metaproterenol (orciprenaline), metipranolol, metoprolol, nafronyl (naftidrofuryl), nefopam, nicardipine, norfenefrine, ofloxacin, ornidazole, orphenadrine, oxomemazine, oxprenolol, oxybutynin, phenoxybenzamine, phenylpropanolamine, pholedrine, pindolol, pirbuterol, prilocaine, procyclidine, promethazine, propafenone, propranolol, reproterol, sotalol, sulpride, synephrine, talinolol, terbutaline, tetrahydrozoline (tetryzoline), theodrenaline, tioconazole, tocainide, trihexyphenidyl, trimeprazine (alimemazine), trimipramine, tropicamide, verapamil, zopiclone

---

KEY WORDS
coated capillary

REFERENCE
Koppenhoefer,B.; Eperlein,U.; Schlunk,R.; Zhu,X.; Lin,B. Separation of enantiomers of drugs by capillary electrophoresis. V. Hydroxypropyl-α-cyclodextrin as chiral solvating agent, *J.Chromatogr.A*, **1998**, *793*, 153–164.

Isoxsuprine

Molecular formula: $C_{18}H_{23}NO_3$
Molecular weight: 301.39
CAS Registry No.: 395-28-8, 579-56-6 (HCl)
Merck Index (12th ed.): 5259
Lednicer: 1 69

SAMPLE
Matrix: formulations
Sample preparation: Tablets. Powder tablet, dissolve in 500 mL 10 μg/mL dichloroisoproterenol in water, stir at room temperature, filter, inject an aliquot of the filtrate. Drops. Dilute 1 mL drops five fold with water, dilute 1 mL of this solution to 100 mL with 10 μg/mL dichloroisoproterenol in water, inject an aliquot.

CAPILLARY ELECTROPHORESIS
Capillary: 50 cm × 75 μm fused-silica (36 cm to detector) (SGE)
Capillary preparation: Fill capillary with 100 mM KOH for 10 min, wash several times with running buffer.
Running buffer: 10 mM Tris adjusted to pH 6.4 with phosphoric acid
Injection: Siphon at 10 cm for 10 s
Detector: UV 206

Migration time: 4.8
Internal standard: dichloroisoproterenol (4.4)
Voltage: 12 kV
Model: Laboratory constructed

OTHER SUBSTANCES
Simultaneous: dopamine, epinephrine, norepinephrine

KEY WORDS
tablets; drops

REFERENCE
Fanali,S.; Cristalli,M.; Nardi,A.; Ossicini,L.; Shukla,S.K. Capillary zone electrophoresis in pharmaceutical analysis, *Farmaco*, **1990**, *45*, 693–702.

SAMPLE
Matrix: solutions
Sample preparation: Inject an aliquot of a 10 μg/mL solution in water.

CAPILLARY ELECTROPHORESIS
Capillary: 68.5 × 50 μm fused-silica (60 cm to detector) (Hewlett-Packard)
Capillary preparation: After each run rinse with 100 mM NaOH for 5 min and with running buffer for 5 min. Condition a new capillary with 1 M NaOH at 40° for 20 min and with 100 mM NaOH at 40° for 10 min, rinse with water at 25° for 20 min, and rinse with running buffer.
Capillary temperature: 25
Running buffer: 50 mM pH 8.3 Tris buffer
Injection: Hydrostatic injection at 50 mbar for 2-6 s, flush with buffer for 4 s
Detector: UV 200
Migration time: 3.941
Voltage: 30 kV
Model: Hewlett-packard HP 3DCE

OTHER SUBSTANCES
Simultaneous: albuterol, cimaterol, clenbuterol, metaproterenol, ractopamine, ritodrine, RU 42 173, terbutaline
Interfering: fenoterol

KEY WORDS
comparison with C18 bonded capillary

REFERENCE
Chevolleau,S.; Tulliez,J. Optimization of the separation of β-agonists by capillary electrophoresis on untreated and C18 bonded silica capillaries, *J.Chromatogr.A*, **1995**, *715*, 345–354.

SAMPLE
Matrix: solutions
Sample preparation: Inject an aliquot of a solution in running buffer.

CAPILLARY ELECTROPHORESIS
Capillary: 60 cm × 75 μm fused-silica (52.4 cm to detector)
Capillary preparation: After each run flush with 500 mM KOH for 2-3 min then with water.
Running buffer: 10 mM pH 3.8 Phosphate buffer containing 2% sulfated cyclodextrin (ds 7-10)
Injection: Hydrostatic injection.
Detector: UV 214
Migration time: 10.24, 10.37 (enantiomers)
Voltage: 15 kV
Model: Waters Quanta 4000

OTHER SUBSTANCES
Also analyzed: acebutolol, alprenolol, aminoglutethimide, brompheniramine, bupivacaine, bupropion, canadine, carbinoxamine, chloroquine, chlorpheniramine, dimethindene, disopyramide, doxylamine, hydroxychloroquine, idazoxan, ketamine, mepenzolate, mepivacaine, methoxyphenamine, mexiletine, midodrine, nefopam, orphenadrine, oxprenolol, oxyphencyclimine, pheniramine, phensuximide, pindolol, piperoxan, terbutaline, tetramisole, tolperisone, tranylcypromine, trihexyphenidyl, trimipramine, verapamil, warfarin

KEY WORDS
chiral; detector at anode

REFERENCE
Stalcup,A.M.; Gahm,K.H. Application of sulfated cyclodextrins to chiral separations by capillary zone electrophoresis, *Anal.Chem.*, **1996**, *68*, 1360–1368.

Itraconazole

Molecular formula: $C_{35}H_{38}Cl_2N_8O_4$
Molecular weight: 705.64
CAS Registry No.: 84625-61-6
Merck Index (12th ed.): 5262

SAMPLE
Matrix: solutions
Sample preparation: Prepare a solution in MeOH, inject an aliquot.

CAPILLARY ELECTROPHORESIS
Capillary: 55.5 cm × 50 μm fused-silica (44.5 cm to detector) (Polymicro Technologies)
Capillary preparation: Rinse with running buffer at 3 bar for 5 min
Capillary temperature: 35
Running buffer: 40 mM pH 2.2 Sodium phosphate buffer
Injection: Pressure injection at 20 mbar for 6 s
Detector: UV 220
Migration time: 8.5
Voltage: 20 kV
Current: about 30 μA
Model: Lauerlabs Prince

OTHER SUBSTANCES
Simultaneous: amiodarone, desethylamiodarone, dextromethorphan, ketoconazole, methadone
Noninterfering: caffeine, naproxen, phenol, theophylline

REFERENCE
Zhang,C.-X.; von Heeren,F.; Thormann,W. Separation of hydrophobic, positively chargeable substances by capillary electrophoresis, *Anal.Chem.*, **1995**, *67*, 2070–2077.

Josamycin

Molecular formula: $C_{42}H_{69}NO_{15}$
Molecular weight: 828.01
CAS Registry No.: 16846-24-5,
51016-68-3 (propionate)
Merck Index (12th ed.): 5280

SAMPLE
Matrix: solutions
Sample preparation: Inject an aliquot of a 500 µg/mL solution in MeOH:water 10:80.

CAPILLARY ELECTROPHORESIS
Capillary: 114.5 cm × 75 µm fused-silica (75 cm)
Capillary preparation: Between runs rinse capillary with running buffer for 3 min. Before use flush capillary with 100 mM NaOH for 10 min, with water for 10 min, and equilibrate with running buffer for 15 min. Condition a new capillary at 2000 mbar with 1 M NaOH for 30 min, 100 mM NaOH for 30 min, and water for 30 min.
Capillary temperature: 25
Running buffer: MeOH:75 mM pH 7.5 phosphate buffer 50:50
Injection: Pressure injection at 50 mbar for 5 s, ramp to operating voltage at 6 kV/s.
Detector: UV 200
Migration time: 30
Voltage: 25 kV
Model: Prince

OTHER SUBSTANCES
Simultaneous: erythromycin, oleandomycin

REFERENCE
Lalloo,A.K.; Chataraj,S.C.; Kanfer,I. Development of a capillary electrophoretic method for the separation of the macrolide antibiotics, erythromycin, josamycin and oleandomycin, *J.Chromatogr.B*, **1997**, *704*, 333–341.

Kanamycin

Molecular formula: $C_{18}H_{36}N_4O_{11}$ (A)
Molecular weight: 484.50 (A)
CAS Registry No.: 8063-07-8, 25389-94-0 (A sulfate), 59-01-8 (A),
4696-78-8 (B)
Merck Index (12th ed.): 5293

	R	R'
Kanamycin A	NH_2	OH
Kanamycin B	NH_2	NH_2
Kanamycin C	OH	NH_2

SAMPLE
Matrix: bulk
Sample preparation: Inject an aliquot of a solution in 140 mM sodium tetraborate containing 480 µg/mL sisomicin as IS.

CAPILLARY ELECTROPHORESIS
Capillary: 70 cm × 50 µm uncoated fused silica (45 cm to detector) (Polymicro Technologies or ISCO)

Capillary preparation: Rinse with running buffer for 15 min at the beginning of the day and for 4 min before each injection.
Capillary temperature: 34
Running buffer: 185 mM pH 9 sodium tetraborate
Injection: Vacuum injection at 25.0 kPa
Detector: UV 195
Migration time: 25.7
Internal standard: sisomicin (19.90)
Voltage: +15 kV
Model: ISCO Model 3140 Electropherograph

OTHER SUBSTANCES
Simultaneous: amikacin, bekanamycin, butirosin, dibekacin, dihydrostreptomycin, gentamicins, paromomycin, ribostamycin, streptomycin, tobramycin
Noninterfering: neomycin

KEY WORDS
detector is at cathode

REFERENCE
Flurer,C.L. The analysis of aminoglycoside antibiotics by capillary electrophoresis, *J.Pharm.Biomed.Anal.*, **1995**, *13*, 809–816.

Ketamine

Molecular formula: $C_{13}H_{16}CINO$
Molecular weight: 237.73
CAS Registry No.: 6740-88-1, 1867-66-9 (HCl)
Merck Index (12th ed.): 5306
Lednicer: 1 57

SAMPLE
Matrix: solutions

CAPILLARY ELECTROPHORESIS
Capillary: 60 cm × 50 μm AccuSep (52.5 cm to detector) (Waters)
Capillary preparation: Before injection rinse capillary with 100 mM NaOH for 3 min and with running buffer for 3 min. Rinse new capillaries with 500 mM NaOH for 5 min
Running buffer: 50 mM pH 7.0 Na_2HPO_4 containing 25 mM (S)-N- dodecoxycarbonylvaline (Prepare (S)-N-dodecoxycarbonylvaline as follows. Prepare dodecyl chloroformate by reacting 1-dodecanol with 0.33 equivalents of triphosgene in dichloromethane solution in the presence of pyridine. Add dodecyl chloroformate dropwise to (S)-valine in 1 M NaOH solution, filter, wash with hexane, recrystallize from ether/petroleum ether.)
Injection: Hydrostatic injection 2 s
Detector: UV 214
Voltage: +12 kV
Model: Waters Quanta 4000 or 4000E

OTHER SUBSTANCES
Also analyzed: atenolol, bupivacaine, ephedrine, homatropine, metoprolol, N-methylpseudoephedrine, norephedrine, norphenylephrine, octopamine, pindolol, terbutaline

KEY WORDS
chiral; α = 1.06

REFERENCE
Mazzeo,J.R.; Grover,E.R.; Swartz,M.E.; Petersen,J.S. Novel chiral surfactant for the separation of enantiomers by micellar electrokinetic capillary chromatography, *J.Chromatogr.A*, **1994**, *680*, 125–135.

SAMPLE
Matrix: solutions
Sample preparation: Inject an aliquot of a 100 µg/mL solution in water:running buffer 50:50.

CAPILLARY ELECTROPHORESIS
Capillary: 44.5 cm × 50 µm acrylamide-coated fused-silica (Bio-Rad)
Capillary temperature: 30
Running buffer: 100 mM NaH_2PO_4 containing 15 mM gamma-cyclodextrin, adjusted to pH 2.5 with phosphoric acid
Injection: Electrokinetic injection at 8 kV for 6 s.
Detector: UV 200
Migration time: 10.01
Voltage: 14 kV
Model: Bio-Rad BioFocus 3000

OTHER SUBSTANCES
Also analyzed: albuterol, alprenolol, atenolol, atropine, baclofen, bamethan, benserazide, biperiden, bisoprolol, bupivacaine, bupranolol, butetamate, carazolol, carbuterol, carvedilol, celiprolol, chloroquine, chlorpheniramine (chlorphenamine), clidinium bromide, clobutinol, disopyramide, dobutamine, flecainide, homatropine, ipratropium bromide, isoproterenol, isothipendyl, mefloquine, mequitazine, metaproterenol (orciprenaline), metipranolol, nafronyl (naftidrofuryl), nefopam, ofloxacin, orphenadrine, oxomemazine, oxprenolol, phenoxybenzamine, pholedrine, pindolol, pirbuterol, prilocaine, promethazine, propafenone, propranolol, sotalol, synephrine, terbutaline, tetrahydrozoline (tetryzoline), tocainide, trihexyphenidyl, trimeprazine (alimemazine), trimipramine, tropicamide, verapamil, zopiclone

KEY WORDS
coated capillary; achiral

REFERENCE
Koppenhoefer,B.; Epperlein,U.; Christian,B.; Yibing,J.; Yuying,C.; Bingcheng,L. Separation of enantiomers of drugs by capillary electrophoresis. I. γ-Cyclodextrin as chiral solvating agent, *J.Chromatogr.A*, **1995**, *717*, 181–190.

SAMPLE
Matrix: solutions
Sample preparation: Inject an aliquot of a 100 µg/mL solution in running buffer:water 50:50.

CAPILLARY ELECTROPHORESIS
Capillary: 44.5 cm × 50 µm polyacrylamide coated fused-silica (40 cm to detector) (Bio-Rad)
Running buffer: 100 mM NaH_2PO_4 containing 15 mM α-cyclodextrin, adjusted to pH 2.5 with phosphoric acid
Injection: Electromigration at 8 kV for 6 s
Detector: UV 200
Migration time: 12.54, 12.65 (enantiomers)
Voltage: 14 kV
Model: Bio-Rad Bio-Focus 3000

OTHER SUBSTANCES
Simultaneous: clidinium bromide, orphenadrine, oxomemazine, tetrahydrozoline, tropicamide

KEY WORDS
chiral; α = 1.008; numerous drugs were not separated into their enantiomers under these conditions; coated capillary

REFERENCE
Bingcheng,L.; Yibing,J.; Yoying,C.; Epperlein,U.; Koppenhoefer,B. Separation of drug enantiomers by capillary electrophoresis: α-Cyclodextrin as chiral solvating agent, *Chromatographia*, **1996**, *42*, 106–110.

SAMPLE
Matrix: solutions

Sample preparation: Inject an aliquot of a 50-200 μM solution.

CAPILLARY ELECTROPHORESIS
Capillary: 48.5 cm × 50 μm fused-silica (44.5 cm to detector) (Polymicro Technologies)
Capillary temperature: 25
Running buffer: 100 mM Phosphoric acid containing 7.5 mM 6^A-methylamino-β-cyclodextrin, adjusted to pH 2.5 with tetramethylammonium hydroxide. (Synthesis of 6^A-methylamino-β-cyclodextrin was as follows. Add a solution of 3.65 g p-toluenesulfonyl chloride in 30 mL dry pyridine to 29.60 g β-cyclodextrin stirred at 5° in 300 mL dry pyridine, stir overnight at room temperature, evaporate to dryness under reduced pressure at 40°, add 700 mL diethyl ether to the residue. Collect the precipitate and recrystallize it 3 times from water to obtain mono-(6-O-p-tolylsulfonyl)-β-cyclodextrin in 31% yield (Bull. Chem. Soc. Japan 1978, 51, 3030). Heat 2 g mono-(6-O-p-tolylsulfonyl)-β-cyclodextrin with 35 mL 50% methylamine in MeOH in a sealed tube at 70° for 3 days, purify by chromatography on carboxymethylcellulose with ammonium bicarbonate solution to obtain 6^A-methylamino-β-cyclodextrin (cf. J. Am. Chem. Soc. 1980, 102, 762).
Injection: Pressure injection at 10 psi.s.
Detector: UV 206
Migration time: 34.3 (second enantiomer, R_S = 1.0)
Voltage: 18 kV
Current: 41-48 μA
Model: Biofocus 3000 (Bio-Rad)

OTHER SUBSTANCES
Simultaneous: chlorpheniramine, isoproterenol

KEY WORDS
chiral

REFERENCE
Fanali,S.; Camera,E. Use of methylamino-β-cyclodextrin in capillary electrophoresis. Resolution of acidic and basic enantiomers, *Chromatographia*, **1996**, *43*, 247–253.

SAMPLE
Matrix: solutions
Sample preparation: Inject an aliquot of a 100 μg/mL solution in water:running buffer 50:50.

CAPILLARY ELECTROPHORESIS
Capillary: 44.5 cm × 50 μm polyacrylamide-coated fused-silica (40 cm to detector) (Bio-Rad)
Capillary temperature: 30
Running buffer: 100 mM NaH$_2$PO$_4$ containing 15 mM β-cyclodextrin, adjusted to pH 2.5 with phosphoric acid
Injection: Electrokinetic injection at 8 kV for 6 s.
Detector: UV 200
Migration time: 8.44, 8.52 (enantiomers)
Voltage: 14 kV
Model: Bio-Rad Bio-Focus 3000

OTHER SUBSTANCES
Simultaneous: carvedilol, chlorpheniramine, metaproterenol (orciprenaline), tetrahydrozoline, tropicamide, zopiclone
Also analyzed: albuterol (not chiral), atenolol (not chiral), benserazide (not chiral), biperiden (not chiral), bupivacaine (not chiral), butethamate (not chiral), carazolol (not chiral), carbuterol (not chiral), clidinium bromide (not chiral), disopyramide (not chiral), dobutamine (not chiral), flecainide (not chiral), ipratropium (not chiral), isothipendyl (not chiral), mequitazine (not chiral), metipranolol (not chiral), nafronyl (naftidrofuryl) (not chiral), nefopam (not chiral), ofloxacin (not chiral), orphenadrine (not chiral), pindolol (not chiral), pirbuterol (not chiral), prilocaine (not chiral), propafenone (not chiral), sotalol (not chiral), tocainide (not chiral), trimipramine (not chiral)

KEY WORDS
coated capillary; chiral

REFERENCE
Koppenhoefer,B.; Epperlein,U.; Christian,B.; Lin,B.; Ji,Y.; Chen,Y. Separation of enantiomers of drugs by capillary electrophoresis. III. β-cyclodextrin as chiral solvating agent, *J.Chromatogr.A*, **1996**, *735*, 333–343.

SAMPLE
Matrix: solutions
Sample preparation: Inject an aliquot of a solution in running buffer.

CAPILLARY ELECTROPHORESIS
Capillary: 60 cm × 75 μm fused-silica (52.4 cm to detector)
Capillary preparation: After each run flush with 500 mM KOH for 2-3 min then with water.
Running buffer: 10 mM pH 3.8 Phosphate buffer containing 2% sulfated cyclodextrin (ds 7-10)
Injection: Hydrostatic injection.
Detector: UV 214
Migration time: 13.08, 13.62 (enantiomers)
Voltage: 15 kV
Model: Waters Quanta 4000

OTHER SUBSTANCES
Also analyzed: acebutolol, alprenolol, aminoglutethimide, brompheniramine, bupivacaine, bupropion, canadine, carbinoxamine, chloroquine, chlorpheniramine, dimethindene, disopyramide, doxylamine, hydroxychloroquine, idazoxan, isoxsuprine, mepenzolate, mepivacaine, methoxyphenamine, mexiletine, midodrine, nefopam, orphenadrine, oxprenolol, oxyphencyclimine, pheniramine, phensuximide, pindolol, piperoxan, terbutaline, tetramisole, tolperisone, tranylcypromine, trihexyphenidyl, trimipramine, verapamil, warfarin

KEY WORDS
chiral; detector at anode

REFERENCE
Stalcup,A.M.; Gahm,K.H. Application of sulfated cyclodextrins to chiral separations by capillary zone electrophoresis, *Anal.Chem.*, **1996**, *68*, 1360–1368.

SAMPLE
Matrix: solutions
Sample preparation: Inject an aliquot of a 100 μg/mL solution in running buffer.

CAPILLARY ELECTROPHORESIS
Capillary: 30 cm × 50 μm fused-silica (25.5 cm to detector) (Yongnian Optical Conductive Fiber Plant, China), coated with polyacrylamide
Capillary preparation: No details of the polyacrylamide coating process are provided. However, another paper (LC.GC 1997, 15, 40) by this group indicates that they use the procedure of Hjertén, thus: Adjust the pH of 20 mL water to 3.5 with acetic acid, add 80 μL 3-(trimethoxysilyl)propyl methacrylate (3-methacryloxypropyltrimethoxysilane), mix, suck into capillary, let stand at room temperature for 1 h, remove the solution, wash with water. Fill the capillary with a deaerated 3-4% acrylamide solution containing 1 μL/mL N,N,N',N'-tetramethylethylenediamine and 1 mg/mL potassium persulfate, let stand for 30 min, remove excess solution by aspiration, rinse with water, remove water by aspiration, dry at 35° (J. Chromatogr. 1985, 347, 191).
Capillary temperature: 25
Running buffer: 100 mM NaH$_2$PO$_4$ adjusted to pH 2.5 (A) or 100 mM NaH$_2$PO$_4$ containing 45 mM hydroxypropyl-gamma-cyclodextrin, adjusted to pH 2.5 (B)
Injection: Electrokinetic injection at 15 kV for 3 s.
Detector: UV 200, UV 210
Migration time: 3.76 (A), 7.42, 7.55 (B, enantiomers)
Voltage: 15 kV
Model: Bio-Rad BioFocus 3000

OTHER SUBSTANCES
Simultaneous: amorolfine, brompheniramine, bupivacaine, carteolol, chloroquine, chlorpheniramine, chlorphenoxamine, disopyramide, dobutamine, doxylamine, flecainide, gallopamil, mepindolol, orphenadrine, oxybutynin, phenoxybenzamine, pindolol, propafenone, propranolol, sulpiride, talinolol, tropicamide, verapamil

KEY WORDS
coated capillary; chiral

REFERENCE
Koppenhoefer,B.; Epperlein,U.; Xiaofeng,Z.; Bingcheng,L. Separation of enantiomers of drugs by capillary electrophoresis. Part 4: Hydroxypropyl-γ-cyclodextrin as chiral solvating agent, *Electrophoresis*, **1997**, *18*, 924–930.

SAMPLE
Matrix: solutions
Sample preparation: Inject an aliquot of a 100 μg/mL solution in running buffer.

CAPILLARY ELECTROPHORESIS
Capillary: 30 cm $\times$ 50 μm fused-silica (25.5 cm to detector), coated with polyacrylamide
Capillary preparation: Adjust the pH of 20 mL water to 3.5 with acetic acid, add 80 μL 3-(trimethoxysilyl)propyl methacrylate (3-methacryloxypropyltrimethoxysilane), mix, suck into capillary, let stand at room temperature for 1 h, remove the solution, wash with water. Fill the capillary with a deaerated 3-4% acrylamide solution containing 1 μL/mL N,N,N',N'-tetramethylethylenediamine and 1 mg/mL potassium persulfate, let stand for 30 min, remove excess solution by aspiration, rinse with water, remove water by aspiration, dry at 35° (J. Chromatogr. 1985, 347, 191).
Capillary temperature: 25
Running buffer: 100 mM NaH_2PO_4 containing 45 mM hydroxypropyl-α-cyclodextrin, adjusted to pH 2.5 with phosphoric acid
Injection: Electrokinetic injection at 15 kV for 3 s.
Detector: UV 200
Voltage: 15 kV
Model: Bio-Rad BioFocus 3000

KEY WORDS
chiral; coated capillary; comparison with the use of other cyclodextrins; this running buffer gave the greatest enantiomeric separation.; α=1.032

REFERENCE
Lin,B.; Zhu,X.; Koppenhoefer,B.; Epperlein,U. Investigation of 123 chiral drugs by cyclodextrin-modified capillary electrophoresis, *LC.GC*, **1997**, *15*, 40–46.

SAMPLE
Matrix: solutions

CAPILLARY ELECTROPHORESIS
Capillary: 36 cm $\times$ 50 μm fused-silica coated with linear polyacrylamide (31.5 cm to detector) (GL Science)
Capillary preparation: At the beginning and end of each day rinse capillary with capillary wash solution (Bio-Rad Cat. No. 148-5022) at 690 kPa for more than 3 min and with water at 690 kPa for more than 3 min. Coat capillary as follows. Treat capillary with 1 M NaOH at room temperature for 1 h, rinse with water, dry by passing nitrogen gas through the capillary at 110° for 6 h. Pass thionyl chloride through the capillary using a suction pump for several min, seal capillary at both ends and heat at 70° for 6 h. Unseal the capillary and fill with 250 mM vinyl magnesium bromide in THF by suction, seal the capillary, heat at 70° for 6 h. Open the capillary and rinse it with THF for several min, rinse with distilled water, fill the capillary with polymerization solution, heat at 28 $\pm$ 2° for 1 h, condition at -100 V/cm for 30 min (Anal. Sci. 1994, 10, 1). (The polymerization solution was 5% acrylamide in water containing 49 mM Tris, 384 mM glycine, and 0.1% sodium dodecyl sulfate, degas in an ultrasonic bath. Add 40 μL 10% N,N,N',N'-tetramethylethylenediamine and 10 μL 10% ammonium persulfate to 5 mL of the degassed solution, mix thoroughly.)
Running buffer: 50 mM pH 6.0 Phosphate buffer
Injection: Before each injection rinse with water at 690 kPa for 30 s, rinse with running buffer at 690 kPa for 30 s, partially fill with separation solution (200 μM α_1-acid glycoprotein (Cohn fraction VI) (Sigma) in running buffer) at 6.9 kPa for 190 s (27 cm), inject sample at 6.9 kPa for 2 s, electrophorese with running buffer (Note that α_1-acid glycoprotein from other suppliers may provide inferior results).

Detector: UV 210
Migration time: 8.6, 9.0 (enantiomers)
Voltage: 12 kV
Model: Bio-Rad BioFocus 3000

KEY WORDS
chiral; coated capillary

REFERENCE
Tanaka,Y.; Terabe,S. Separation of the enantiomers of basic drugs by affinity capillary electrophoresis using a partial filling technique and α_1-acid glycoprotein as chiral selector, *Chromatographia*, **1997**, *44*, 119–128.

SAMPLE
Matrix: solutions
Sample preparation: Inject an aliquot of a 20 µg/mL solution in buffer with an ionic strength 10 times less than that of the running buffer.

CAPILLARY ELECTROPHORESIS
Capillary: 57 cm × 50 µm fused-silica (50 cm to detector) (Polymicro Technologies)
Capillary preparation: Between runs rinse capillary with running buffer from a different vial for 2 min, equilibrate system for 1 min before analysis. Store capillary in water overnight and rinse with 100 mM NaOH and water each morning. Rinse new capillaries with 100 mM HCl for 5 min, with water for 5 min, with 100 mM NaOH for 5 min, with water for 10 min, then with running buffer.
Capillary temperature: 15
Running buffer: pH 2.90 Phosphate buffer (I = 0.04) containing 50 mM α-cyclodextrin (Prepare buffer by mixing 19.8 mL 1 M NaOH and 23.2 mL 1 M phosphoric acid and make up to 500 mL with water.)
Injection: Pressure injection at 3.4 kPa for 5 s (6 nL).
Detector: UV 214
Migration time: 21, 21.3 (enantiomers)
Voltage: 21 kV
Current: 17 µA
Model: Beckman P/ACE 2100

OTHER SUBSTANCES
Simultaneous: prilocaine

KEY WORDS
chiral

REFERENCE
Amini,A.; Sörman,U.P.; Lindgren,B.H.; Westerlund,D. Enantioseparation of anaesthetic drugs by capillary zone electrophoresis using cyclodextrin-containing background electrolytes, *Electrophoresis*, **1998**, *19*, 731–737.

SAMPLE
Matrix: solutions

CAPILLARY ELECTROPHORESIS
Capillary: 29-36 cm × 50 µm fused-silica (24.5-31.5 cm to detector) (Yongnian Optical Conductive Fiber Plant, China) coated with polyacrylamide
Capillary preparation: Coat capillary as follows. Adjust the pH of 20 mL water to 3.5 with acetic acid, add 80 µL 3-(trimethoxysilyl)propyl methacrylate (3-methacryloxypropyltrimethoxysilane), mix, suck into capillary, let stand at room temperature for 1 h, remove the solution, wash with water. Fill the capillary with a deaerated 3-4% acrylamide solution containing 1 µL/mL N,N,N',N'-tetramethylethylenediamine and 1 mg/mL potassium persulfate, let stand for 30 min, remove excess solution by aspiration, rinse with water, remove water by aspiration, dry at 35° (J. Chromatogr. 1985, 347, 191).
Capillary temperature: 25
Running buffer: 100 mM pH 2.5 NaH$_2$PO$_4$ (A) or 100 mM pH 2.5 NaH$_2$PO$_4$ containing 45 mM hydroxypropyl-α-cyclodextrin (Wacker, Munich) (B)
Injection: Electromigration at 15 kV for 3 s.

Detector: UV 200; UV 210
Migration time: 3.76 (A); 7.13, 7.36 (B) (enantiomers)
Voltage: 15 kV
Model: Bio-Focus 3000

OTHER SUBSTANCES
Also analyzed: albuterol (salbutamol), alprenolol, amorolfine, atenolol, atropine, azelastine, baclofen, bamethan, benproperine, benserazide, biperiden, bisoprolol, brompheniramine, bupivacaine, bupranolol, butamirate, butethamate, carazolol, carbuterol, carteolol, carvedilol, celiprolol, chloroquine, chlorpheniramine, chlorphenoxamine, cicletanine, clenbuterol, clidinium bromide, clobutinol, dimethindene, dipivefrin, disopyramide, dobutamine, doxylamine, fendiline, flecainide, gallopamil, homatropine, ipratropium bromide, isoproterenol (isoprenaline), isothipendyl, meclizine, mefloquine, mepindolol, mequitazine, metaclazepam, metaproterenol (orciprenaline), metipranolol, metoprolol, nafronyl (naftidrofuryl), nefopam, nicardipine, norfenefrine, ofloxacin, ornidazole, orphenadrine, oxomemazine, oxprenolol, oxybutynin, phenoxybenzamine, phenylpropanolamine, pholedrine, pindolol, pirbuterol, prilocaine, procyclidine, promethazine, propafenone, propranolol, reproterol, sotalol, sulpride, synephrine, talinolol, terbutaline, tetrahydrozoline (tetryzoline), theodrenaline, tioconazole, tocainide, trihexyphenidyl, trimeprazine (alimemazine), trimipramine, tropicamide, verapamil, zopiclone

KEY WORDS
coated capillary; chiral

REFERENCE
Koppenhoefer,B.; Eperlein,U.; Schlunk,R.; Zhu,X.; Lin,B. Separation of enantiomers of drugs by capillary electrophoresis. V. Hydroxypropyl-α-cyclodextrin as chiral solvating agent, *J.Chromatogr.A*, **1998**, *793*, 153–164.

SAMPLE
Matrix: solutions

CAPILLARY ELECTROPHORESIS
Capillary: 100 cm × 100 μm fused-silica (Polymicro Technologies)
Capillary preparation: Between runs wash capillary with running buffer for 10 min. Replace the electrolyte every other run. Wash new capillaries with water for 20 min and with running buffer for 20 min.
Running buffer: MeOH:water 80:20 containing 15 mM dimethyl-β-cyclodextrin, 5 mM ammonium acetate and 800 mM acetic acid
Injection: Gravity injection at 15 cm for 2 s.
Detector: MS, Vestec 201, electrospray voltage 2.48 kV, sheath liquid MeOH:acetic acid 99.1 flowing at 2.5 μL/min, m/z 238 (design of interface in J. Chromatogr. 1993, 639, 303)
Migration time: 27.5, 28.6 (enantiomers)
Voltage: 25 kV
Current: 16 μA
Model: Dionex CE System I
Limit of detection: 10 μM

KEY WORDS
chiral

REFERENCE
Lu,W.; Cole,R.B. Determination of chiral pharmaceutical compounds, terbutaline, ketamine and propranolol, by on-line capillary electrophoresis-electrospray mass spectrometry, *J.Chromatogr.B*, **1998**, *714*, 69–75.

Ketoconazole

Molecular formula: $C_{26}H_{28}Cl_2N_4O_4$
Molecular weight: 531.44
CAS Registry No.: 65277-42-1
Merck Index (12th ed.): 5313
Lednicer: 3 132

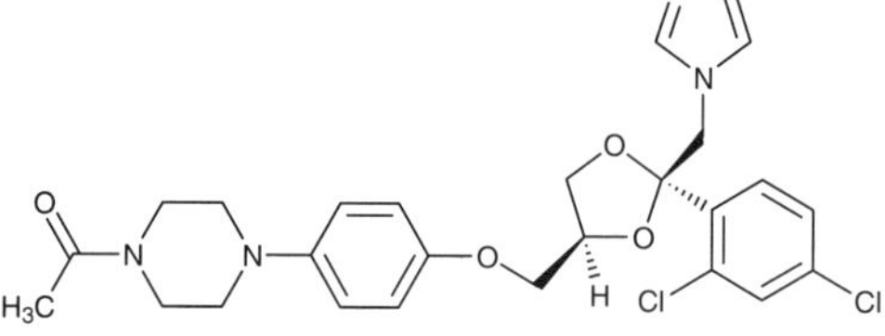

SAMPLE
Matrix: solutions
Sample preparation: Prepare a 60 μg/mL solution, inject an aliquot.

CAPILLARY ELECTROPHORESIS
Capillary: 61 cm × 50 μm fused-silica (44 cm to detector) (Grom)
Capillary temperature: 21 ± 1
Running buffer: MeOH:buffer 20:80 (Buffer was 50 mM pH 3.0 (?) phosphate buffer containing 0.1 mM sulfobutyl ether-β-cyclodextrin.)
Injection: Hydrostatic injection at 10 cm for 5 s
Detector: UV 210
Migration time: 12.01, 12.62 (enantiomers)
Voltage: 400 V/cm
Model: Grom 100

OTHER SUBSTANCES
Simultaneous: bifonazole, econazole, enilconazole, miconazole

KEY WORDS
chiral

REFERENCE
Chankvetadze,B.; Endresz,G.; Blaschke,G. Enantiomeric resolution of chiral imidazole derivatives using capillary electrophoresis with cyclodextrin-type buffer modifiers, *J.Chromatogr.A*, **1995**, *700*, 43–49.

SAMPLE
Matrix: solutions
Sample preparation: Prepare a solution in MeOH, inject an aliquot.

CAPILLARY ELECTROPHORESIS
Capillary: 55.5 cm × 50 μm fused-silica (44.5 cm to detector) (Polymicro Technologies)
Capillary preparation: Rinse with running buffer at 3 bar for 5 min
Capillary temperature: 35
Running buffer: 40 mM pH 2.2 Sodium phosphate buffer
Injection: Pressure injection at 20 mbar for 6 s
Detector: UV 220
Migration time: 5
Voltage: 20 kV
Current: about 30 μA
Model: Lauerlabs Prince

OTHER SUBSTANCES
Simultaneous: amiodarone, desethylamiodarone, dextromethorphan, itraconazole, methadone
Noninterfering: caffeine, naproxen, phenol, theophylline

REFERENCE
Zhang,C.-X.; von Heeren,F.; Thormann,W. Separation of hydrophobic, positively chargeable substances by capillary electrophoresis, *Anal.Chem.*, **1995**, *67*, 2070–2077.

Ketoprofen

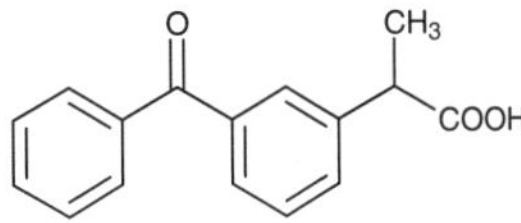

Molecular formula: $C_{16}H_{14}O_3$
Molecular weight: 254.29
CAS Registry No.: 22071-15-4
Merck Index (12th ed.): 5316
Lednicer: 2 64

SAMPLE
Matrix: blood
Sample preparation: 0.1-1 mL Plasma + 2 μg IS + 400 μL 1 M HCl + 10 mL diethyl ether, agitate mechanically for 10 min, centrifuge at 3500 g for 10 min. Remove the organic layer and evaporate it to dryness under a stream of nitrogen, reconstitute the residue in MeOH:40 mM NaH_2PO_4 50:50, inject an aliquot.

CAPILLARY ELECTROPHORESIS
Capillary: 58.7 cm × 75 μm fused-silica (50 cm to detector)
Capillary preparation: Wash with running buffer for 2 min before each injection. At the beginning of each day wash capillary with 100 mM NaOH for 10 min.
Capillary temperature: 22
Running buffer: MeOH:buffer 3:97 (Buffer was 40 mM pH 8 NaH_2PO_4 containing 104 mM sodium dodecyl sulfate.)
Injection: Hydrodynamic injection for 2 s
Detector: UV 254
Migration time: 13.4
Internal standard: benzoyl-4-phenyl-2-butyric acid (16.3)
Voltage: 20 kV
Model: Beckman P/ACE 2000
Limit of detection: 130 ng/mL (S/N 5)

OTHER SUBSTANCES
Extracted: diclofenac, diflunisal, etodolac, fenbufen, fenoprofen, flurbiprofen, ibuprofen, indomethacin, naproxen, niflumic acid, piroxicam, sulindac, tenoxicam, tiaprofenic acid
Noninterfering: acetaminophen, amitriptyline, caffeine, clomipramine, deoxysulindac, desipramine, diazepam, 5'-hydroxytenoxicam, imipramine, maprotiline, nortriptyline, phenobarbital, phenytoin, sulfamethoxazole, theophylline, trimipramine

KEY WORDS
plasma

REFERENCE
Maboundou,C.W.; Paintaud,G.; Bérard,M.; Bechtel,P.R. Separation of fifteen non-steroidal anti-inflammatory drugs using micellar electrokinetic capillary chromatography, *J.Chromatogr.B*, **1994**, *657*, 173–183.

SAMPLE
Matrix: formulations
Sample preparation: Stir formulation in MeOH for 10 min, filter, dilute with 5 mM pH 5 morpholinoethanesulfonic acid, inject an aliquot.

CAPILLARY ELECTROPHORESIS
Capillary: 35 cm × 50 μm fused-silica (31.5 cm to detector) (Polymicro Technologies)
Capillary temperature: 25
Running buffer: 100 mM Morpholinoethanesulfonic acid containing 30 mM heptakis-2,3,6-tri-O-methyl-β-cyclodextrin, adjusted to pH 5 with NaOH
Injection: Pressure injection at 10 psi
Detector: UV 206
Migration time: 11.5, 12 (enantiomers)
Voltage: 20 kV
Current: 6.6 μA

Model: Biofocus 3000 (Bio-Rad)

OTHER SUBSTANCES
Simultaneous: fenoprofen, flurbiprofen, ibuprofen

KEY WORDS
chiral

REFERENCE
Fanali,S.; Aturki,Z. Use of cyclodextrins in capillary electrophoresis for the chiral resolution of some 2-aryl-propionic acid non-steroidal anti-inflammatory drugs, *J.Chromatogr.A*, **1995**, *694*, 297–305.

SAMPLE
Matrix: formulations
Sample preparation: Gel. Suspend 800 mg gel in 30 mL MeOH, heat at 50° for 10 min, sonicate for 5 min, make up to 50 mL with EtOH, centrifuge at 3000 rpm for 20 min. Remove a 10 mL aliquot and make it up to 50 mL with water, filter, inject an aliquot of the filtrate. Capsules, tablets. Grind several tablets or the contents of a few capsules to a fine powder, weigh out 160 mg, suspend in 30 mL MeOH, sonicate for 5 min, make up to 50 mL with MeOH, centrifuge at 3000 rpm for 5 min. Remove a 5 (capsules) or 10 (tablets) mL aliquot of the supernatant and make it up to 50 mL with water, filter, inject an aliquot of the filtrate.

CAPILLARY ELECTROPHORESIS
Capillary: 64.5 cm × 50 μm fused-silica (56 cm to detector) (Hewlett-Packard)
Capillary temperature: 35
Running buffer: 20 mM pH 5 Phosphate buffer containing 20 mM triethanolamine and 50 mM heptakis-tri-O-methyl-β-cyclodextrin
Injection: Pressure injection at 50 mbar for 3-12 s (6.5-26 nL).
Detector: UV (wavelength not given)
Migration time: 14.7 (R-(-)), 15.2 (S-(+))
Voltage: 20 kV
Current: 8.1 μA
Model: Hewlett-Packard [3D]CE
Limit of quantitation: 1.6 μM
Limit of detection: 700 nM

KEY WORDS
tablets; capsules; gel; chiral

REFERENCE
Blanco,M.; Coello,J.; Iturriaga,H.; Maspoch,S.; Pérez-Maseda,C. Chiral and nonchiral determination of keto-profen in pharmaceuticals by capillary zone electrophoresis, *J.Chromatogr.A*, **1998**, *799*, 301–307.

SAMPLE
Matrix: solutions
Sample preparation: Dissolve in MeCN, dilute with water to a concentration of 25 μg/mL, inject an aliquot.

CAPILLARY ELECTROPHORESIS
Capillary: 60 cm × 75 μm fused-silica (52.5 cm to detector)
Running buffer: MeCN:30 mM pH 7.0 phosphate buffer 20:80
Injection: Hydrostatic injection at 9.8 cm for 5 s
Detector: UV 214
Migration time: 19.71
Model: Waters Quanta 4000

OTHER SUBSTANCES
Simultaneous: acemetacin, alclofenac, flurbiprofen, ibuprofen, indomethacin, lonazolac, naproxen, niflumic acid, piroxicam, tenoxicam, tiaprofenic acid, tolmetin, voltaren
Interfering: fenbufen

REFERENCE
Donato,M.G.; Van den Eeckhout,E.; Van den Bossche,W.; Sandra,P. Capillary zone electrophoresis and micellar electrokinetic capillary chromatography of some non-steroidal antiinflammatory drugs (NSAIDs), *J.Pharm.Biomed.Anal.*, **1993**, *11*, 197–201.

SAMPLE
Matrix: solutions
Sample preparation: Prepare an aqueous solution, filter (0.45 µm), inject an aliquot.

CAPILLARY ELECTROPHORESIS
Capillary: 36 cm × 50 µm (31.5 cm to detector)
Capillary preparation: Before each run rinse with water for 1 min and with running buffer for 1 min.
Capillary temperature: 25
Running buffer: Isopropanol:buffer 10:90 (Buffer was 50 mM pH 6.0 phosphate buffer containing 25 µm avidin.)
Injection: Inject at 1 psi pressure for 2 s
Detector: UV 270
Migration time: 16.8, 17.7 (enantiomers)
Voltage: -12 kV
Model: BioFocus 3000 (Bio-Rad)

KEY WORDS
chiral

REFERENCE
Tanaka,Y.; Matsubara,N.; Terabe,S. Separation of enantiomers by affinity electrokinetic chromatography using avidin, *Electrophoresis*, **1994**, *15*, 848–853.

SAMPLE
Matrix: solutions
Sample preparation: Inject an aliquot of a 50 µM solution in MeOH:water 40:60.

CAPILLARY ELECTROPHORESIS
Capillary temperature: 25
Running buffer: MeOH:water 40:60 containing 75 mM glycine and 15 mM hydroxypropyl-β-cyclodextrin adjusted to pH 9.1 with triethanolamine
Injection: Hydrodynamic injection for 2 s
Detector: UV 280
Migration time: 17.4
Voltage: 15 kV

OTHER SUBSTANCES
Simultaneous: alclofenac, bufexamac, carprofen, flufenamic acid, flurbiprofen, indomethacin, niflumic acid, piroxicam, sulindac, tiaprofenic acid
Interfering: naproxen

REFERENCE
Bechet,I.; Fillet,M.; Hubert,P.; Crommen,J. Improvement of achiral resolution in capillary zone electrophoresis by use of cyclodextrin additives, *Biomed.Chromatogr.*, **1995**, *9*, 267–268.

SAMPLE
Matrix: solutions
Sample preparation: Prepare a 100 µg/mL solution in 50 mM pH 7.0 phosphate buffer, inject an aliquot.

CAPILLARY ELECTROPHORESIS
Capillary: 37 cm × 50 µm (30 cm to detector)
Capillary preparation: Between runs rinse capillary with 100 mM KOH for 2 min, with water for 2 min, and with running buffer for 2 min. At the beginning of each day treat capillary with

500 mM KOH for 5 min, flush with water for 10 min, and flush with running buffer for 10 min.
Capillary temperature: 20
Running buffer: 50 mM pH 7.0 Phosphate buffer containing 2 mM vancomycin and 30 mM sodium dodecyl sulfate
Injection: Pressure injection at 0.5 psi for 1 s
Detector: UV 254
Migration time: 22, 23 (enantiomers)
Voltage: +5 kV
Model: Beckman P/ACE 2000

OTHER SUBSTANCES
Simultaneous: flurbiprofen, indoprofen

KEY WORDS
chiral

REFERENCE
Rundlett,K.L.; Armstrong,D.W. Effect of micelles and mixed micelles on efficiency and selectivity of antibiotic based capillary electrophoretic enantioseparations, *Anal.Chem.*, **1995**, *67*, 2088–2095.

SAMPLE
Matrix: solutions
Sample preparation: Inject an aliquot of a 100 µg/mL solution.

CAPILLARY ELECTROPHORESIS
Capillary: 32.5 cm × 50 µm fused-silica (25 cm to detector) (Quadrex, New Haven, CT)
Capillary preparation: Condition new capillaries with 100 mM KOH for 10 min, purge with water for 5 min, equilibrate with running buffer for 5 min.
Running buffer: 100 mM pH 6.0 Phosphate buffer containing 2 mM vancomycin (A) or ristocetin (B) or teicoplanin (C)
Injection: Hydrostatic injection for 3 s.
Detector: UV 254, UV 280
Migration time: 22.1, 26.3 (A) or 12.3, 14.3 (B) or 12.5, 13.0 (C) (enantiomers)
Voltage: 5 kV
Model: Waters Quanta 4000

OTHER SUBSTANCES
Also analyzed: fenoprofen, methotrexate

KEY WORDS
chiral

REFERENCE
Gasper,M.P.; Berthod,A.; Nair,U.B.; Armstrong,D.W. Comparison and modeling study of vancomycin, ristocetin A, and teicoplanin for CE enantioseparations, *Anal.Chem.*, **1996**, *68*, 2501–2514.

SAMPLE
Matrix: solutions
Sample preparation: Inject an aliquot of a 500 µM solution in MeOH:water 50:50.

CAPILLARY ELECTROPHORESIS
Capillary: 60 cm × 50 µm fused-silica (52.5 cm to detector) (Waters)
Capillary preparation: Before each run rinse capillary with 10 mM NaOH for 1 min and with running buffer for 4 min. Condition new capillaries by flushing with 1 M NaOH for 10 min, with water for 5 min, with 6 mM NaOH containing 500 mM NaCl for 10 min, with water for 5 min, and with running buffer for 10 min.
Running buffer: Formic acid containing 10 mM heptakis(trimethyl-β-cyclodextrin) adjusted to pH 4.0 with 1 M NaOH (ionic strength, I = 35 mM)
Injection: Hydrodynamic injection at 10 cm for 15 s.
Detector: UV 254

Voltage: 30 kV
Model: Waters Quanta 4000

KEY WORDS
chiral; R_S = 1.5; electroosmotic mobility = 8.7 $\times$ 10^{-5} $cm^2V^{-1}s^{-1}$

REFERENCE
Lelièvre,F.; Gareil,P. Chiral separations of underivatized arylpropionic acids by capillary zone electrophoresis with various cyclodextrins. Acidity and inclusion constant determinations, *J.Chromatogr.A*, **1996**, *735*, 311–320.

SAMPLE
Matrix: solutions

CAPILLARY ELECTROPHORESIS
Capillary: 35 cm $\times$ 50 μm coated capillary (30.5 cm to detector) (Composite Metal Services, UK)
Capillary preparation: Before each run purge at high pressure with water for 100 s and with running buffer for 120 s. If vancomycin is used, before each run purge at high pressure with water for 100 s and with running buffer not containing vancomycin for 120 s. Next purge with running buffer containing vancomycin at low pressure (175 psi.s) and inject sample. Coat capillary as follows. Adjust the pH of 20 mL water to 3.5 with acetic acid, add 80 μL 3-(trimethoxysilyl)propyl methacrylate (3-methacryloxypropyltrimethoxysilane), mix, suck into capillary, let stand at room temperature for 1 h, remove the solution, wash with water. Fill the capillary with a deaerated 3-4% acrylamide solution containing 1 μL/mL N,N,N',N'-tetramethylethylenediamine and 1 mg/mL potassium persulfate, let stand for 30 min, remove excess solution by aspiration, rinse with water, remove water by aspiration, dry at 35° (J. Chromatogr. 1985, 347, 191).
Capillary temperature: 25
Running buffer: Buffer containing 30 mM heptakis-2,3,6-tri-O-methyl-β-cyclodextrin (A), 5 mM heptamethylamino-β-cyclodextrin (B) or 5 mM vancomycin (C) (Buffer was 50 mM phosphoric acid containing 50 mM acetic acid and 50 mM boric acid, dilute with an equal volume of water, adjust pH to 5 with concentrated NaOH.)
Injection: Pressure injection at 10 psi.
Detector: UV 206
Migration time: 11.2, 11.6 (A); 12.1, 12.4 (B); 7.4, 9.6 (C) (enantiomers)
Voltage: -20 kV
Model: Bio-Rad Biofocus 3000
Limit of detection: 5 μM (A, C), 10 μM (B)

KEY WORDS
chiral; coated capillary; detector at anode

REFERENCE
Fanali,S.; Desiderio,C.; Aturki,Z. Enantiomeric resolution study by capillary electrophoresis. Selection of the appropriate chiral selector, *J.Chromatogr.A*, **1997**, *772*, 185–194.

SAMPLE
Matrix: solutions
Sample preparation: Inject an aliquot of a solution in MeOH:water 10:90.

CAPILLARY ELECTROPHORESIS
Capillary: 44 cm $\times$ 50 μm fused-silica (37 cm to detector) (Supelco)
Capillary preparation: Wash with running buffer for 3 min after each injection. Wash with running buffer for 10 min at the end of each day.
Capillary temperature: 25
Running buffer: 100 mM Phosphoric acid containing 5 mM sulfobutyl ether-β-cyclodextrin and 30 mM heptakis(2,3,6-tri-O-methyl)-β-cyclodextrin, adjusted to pH 3.0 with triethanolamine
Injection: Hydrodynamic injection for 5 s (13.3 nL).
Detector: UV 280
Voltage: -25 kV
Current: 60 μA
Model: SpectraPhoresis 1000 CE

KEY WORDS

chiral; detector at anode; resolution (R_s = 11.3)

REFERENCE

Fillet,M.; Hubert,P.; Crommen,J. Enantioseparation of nonsteroidal anti-inflammatory drugs by capillary electrophoresis using mixtures of anionic and uncharged β-cyclodextrins as chiral additives, *Electrophoresis*, **1997**, *18*, 1013–1018.

SAMPLE

Matrix: solutions

Sample preparation: Inject an aliquot of a solution in MeOH:water 50:50.

CAPILLARY ELECTROPHORESIS

Capillary: 57 cm × 75 μm fused-silica (50 cm to detector) (Polymicro Technologies)

Capillary preparation: Rinse capillary with running buffer for 2 min between samples. At the start of each day wash capillary with 1 M NaOH for 10 min. Wash new capillaries with 1 M NaOH for 1 h.

Capillary temperature: 23 ± 0.1

Running buffer: 50 mM Na_2HPO_4 containing 20 mM heptakis(6-hydroxyethylamino-6-deoxy-β-cyclodextrin), adjusted to pH 7.0 with 100 mM HCl (Prepare heptakis(6-hydroxyethylamino-6-deoxy-β-cyclodextrin) as follows. Add 4.32 g β-cyclodextrin to a stirred solution of 4 mL bromine and 21 g triphenylphosphine in 80 mL DMF and the mixture stirred at 80° for 15 h, concentrate under reduced pressure to half volume, adjust pH to 9-10 by adding 30 mL 3 M sodium methoxide in MeOH with simultaneous cooling. Keep at room temperature for 30 min then pour into 1.5 L ice water, filter to obtain heptakis(6-bromo-6-deoxy-β-cyclodextrin) (Ang. Chem. Int. Ed. Engl. 1991, 30, 78). Dissolve heptakis(6-bromo-6-deoxy-β-cyclodextrin) in ethanolamine at 65°, heat at 65° for 48 h. Remove the solvent by evaporation under reduced pressure, dissolve the residue in hot MeOH, slowly add the methanolic solution to reagent-grade acetone, collect the precipitate by filtration. Dissolve the precipitate in water, treat with a basic ion-exchange resin, lyophilize to obtain heptakis(6-hydroxyethylamino-6-deoxy-β-cyclodextrin).)

Detector: UV 214

Migration time: 18.4, 18.6 (enantiomers)

Voltage: -15 kV

Model: Beckman P/ACE 5510

OTHER SUBSTANCES

Simultaneous: fenoprofen, ibuprofen (peak shape very poor)

KEY WORDS

chiral

REFERENCE

O'Keeffe,F.; Shamsi,S.A.; Darcy,R.; Schwinté,P.; Warner,I.M. A persubstituted cationic β-cyclodextrin for chiral separations, *Anal.Chem.*, **1997**, *69*, 4773–4782.

SAMPLE

Matrix: solutions

CAPILLARY ELECTROPHORESIS

Capillary: 64.5 cm × 50 μm fused-silica (56 cm to detector) (Hewlett-Packard)

Capillary preparation: Before each run flush with 100 mM NaOH for 3 min and with running buffer at 8 min then equilibrate at 20 kV for 6 min. At the start of each batch flush capillary with 1 M NaOH for 10 min, with 100 mM NaOH for 10 min, and with running buffer for 15 min then equilibrate at 20 kV for 20 min.

Capillary temperature: 35

Running buffer: 20 mM Phosphoric acid containing 20 mM triethanolamine and 50 mM 2,3,6-tri-O-methyl-β-cyclodextrin, adjusted to pH 5.0 with NaOH

Injection: Pressure injection at 75 mbar (5 nL).

Detector: UV 253

Migration time: 13.2, 13.8 (enantiomers)

Voltage: 20 kV

Current: 8.1 μA

Model: Hewlett-Packard ³ᴰCE

OTHER SUBSTANCES
Simultaneous: fenoprofen, ibuprofen

KEY WORDS
chiral

REFERENCE
Blanco,M.; Coello,J.; Iturriaga,H.; Maspoch,S.; Pérez-Maseda,C. Separation of profen enantiomers by capillary electrophoresis using cyclodextrins as chiral selectors, *J.Chromatogr.A*, **1998**, *793*, 165–175.

SAMPLE
Matrix: solutions
Sample preparation: Inject an aliquot of a 50-100 μg/mL solution in 50 mM pH 4.8 ammonium acetate buffer.

CAPILLARY ELECTROPHORESIS
Capillary: 44 cm × 50 μm polyacrylamide coated
Capillary preparation: Adjust the pH of 20 mL water to 3.5 with acetic acid, add 80 μL 3-(trimethoxysilyl)propyl methacrylate (3-methacryloxypropyltrimethoxysilane), mix, suck into capillary, let stand at room temperature for 1 h, remove the solution, wash with water. Fill the capillary with a deaerated 3-4% acrylamide solution containing 1 μL/mL N,N,N',N'-tetramethylethylenediamine and 1 mg/mL potassium persulfate, let stand for 30 min, remove excess solution by aspiration, rinse with water, remove water by aspiration, dry at 35° (J. Chromatogr. 1985, 347, 191).
Running buffer: 50 mM pH 4.8 Ammonium acetate buffer containing 5 mM vancomycin
Injection: Hydrostatic injection at 10 cm for 5-10 s.
Detector: MS, Finnigan LCQ ion trap, electrospray, negative ion mode, probe tip at 2.6 kV, sheath liquid MeOH:water:ammonia 50:48:2 at 6 μL/min (Vancomycin migrates away from the detector.)
Migration time: 10, 12 (enantiomers)
Voltage: 20 kV
Model: Grom 100

OTHER SUBSTANCES
Also analyzed: carprofen, flurbiprofen, ibuprofen, naproxen

KEY WORDS
chiral; coated capillary

REFERENCE
Fanali,S.; Desiderio,C.; Schulte,G.; Heitmeier,S.; Strickmann,D.; Chankvetadze,B.; Blaschke,G. Chiral capillary electrophoresis-electrospray mass spectrometry coupling using vancomycin as chiral selector, *J.Chromatogr.A*, **1998**, *800*, 69–76.

SAMPLE
Matrix: solutions

CAPILLARY ELECTROPHORESIS
Capillary: 57 cm × 50 μm fused-silica (50 cm to detector)
Running buffer: 50 mM pH 8.0 Borate buffer containing 40 mM sodium taurodeoxycholate and 25 mM phosphatidylcholine (Prepare by adding phosphatidylcholine and stirring for 3-6 h until all cloudiness disappears. Phosphatidylcholine was 95% pure soybean lecithin, Epikuron, Lucas Meyer & Co.)
Injection: Inject a solution of the compound in the running buffer at 20 psi for 1 s, inject MeOH:water 5:95 at 20 psi for 1 s, inject a solution of halofantrine in running buffer at 20 psi for 1 s.
Detector: UV 214
Migration time: k' 1.51
Model: Beckman P/ACE 5000

OTHER SUBSTANCES
Also analyzed: acetaminophen, amoxicillin, antipyrine, aspirin, azathioprine, caffeine, captopril, carbamazepine, carprofen, chlorambucil, chlorpheniramine, chlorpromazine, cimetidine, clonidine, codeine, desipramine, diphenhydramine, ephedrine, fenoterol, flufenamic acid, flurbiprofen, haloperidol, hydroxyzine, ibuprofen, imipramine, indomethacin, lidocaine, melphalan, metoprolol, nabumetone, nadolol, phenobarbital, phenol, promazine, propranolol, pyrilamine, ranitidine, ropinirole, salicylic acid, sulfamethoxazole, testosterone, theophylline, thioridazine, tiaprofenic acid, tolfenamic acid, trifluoperazine, trimethoprim, valproic acid, verapamil

KEY WORDS
comparison with HPLC; k' = (Tr-T0)/(T0(1-Tr/Tm)) where Tr = retention time of analyte; T0 = retention time of water; and Tm = retention time of marker (halofantrine)

REFERENCE
Hanna,M.; de Biasi,V.; Bond,B.; Salter,C.; Hutt,A.J.; Camilleri,P. Estimation of the partitioning characteristics of drugs: A comparison of a large and diverse drug series utilizing chromatographic and electrophoretic methodology, *Anal.Chem.*, **1998**, *70*, 2092–2099.

SAMPLE
Matrix: solutions
Sample preparation: Inject an aliquot of a 100 μg/mL solution in MeOH:water 50:50.

CAPILLARY ELECTROPHORESIS
Capillary: 47 cm $\times$ 75 μm fused-silica (40 cm to detector) (Polymicro Technologies)
Capillary preparation: Rinse capillary with 100 mM NaOH for 1 min, with water for 1.5 min, and with running buffer for 2 min between samples. Wash new capillaries with 1 M NaOH for 1 h, with 100 mM NaOH for 30 min, and with water for 30 min.
Running buffer: 50 mM NaH_2PO_4 containing 3 mM heptakis(6-methoxyethylamine-6-deoxy-β-cyclodextrin), adjusted to pH 6 (Prepare heptakis(6-methoxyethylamine-6-deoxy-β-cyclodextrin) as follows. Add 4.32 g β-cyclodextrin to a stirred solution of 4 mL bromine and 21 g triphenylphosphine in 80 mL DMF and the mixture stirred at 80°C for 15 h, concentrate under reduced pressure to half volume, adjust pH to 9-10 by adding 30 mL 3 M sodium methoxide in MeOH with simultaneous cooling. Keep at room temperature for 30 min then pour into 1.5 L ice water, filter to obtain heptakis(6-bromo-6-deoxy-β-cyclodextrin) (Ang. Chem. Int. Ed. Engl. 1991, 30, 78). Dissolve heptakis(6-bromo-6-deoxy-β-cyclodextrin) in methoxyethylamine, heat at 65°C for 48 h. Remove the solvent by evaporation under reduced pressure, dissolve the residue in MeOH at 55°C, slowly add the methanolic solution to reagent-grade acetone with stirring, collect the precipitate by filtration. Dissolve the precipitate in water, treat with a basic ion-exchange resin. Lyophilize the precipitate to obtain heptakis(6-methoxyethylamine-6-deoxy-β-cyclodextrin).)
Injection: Pressure injection at 85 kPa.s.
Detector: UV 214
Migration time: 13, 14 (enantiomers)
Voltage: -30 kV
Current: 25 μA
Model: Beckman P/ACE 5510

OTHER SUBSTANCES
Simultaneous: fenoprofen, flurbiprofen

KEY WORDS
chiral

REFERENCE
Haynes,J.L.,III; Shamsi,S.A.; O'Keefe,F.; Darcey,R.; Warner,I.M. Cationic β-cyclodextrin derivative for chiral separations, *J.Chromatogr.A*, **1998**, *803*, 261–271.

SAMPLE
Matrix: solutions

CAPILLARY ELECTROPHORESIS
Capillary: 47 cm $\times$ 50 μm fused-silica (40 cm to detector)

Capillary preparation: Before each run rinse capillary with water for 2 min, with 100 mM HCl
for 2 min, with water for 2 min, and with 500 mM KOH for 2 h
Capillary temperature: 25
Running buffer: MeOH:50 mM pH 7.0 phosphate buffer containing 15 mM d-(+)-tubocurarine
chloride (Caution! Tubocurarine is highly toxic!)
Injection: Pressure injection at 0.5 psi for 1 s.
Detector: UV 254
Migration time: 17.54, 17.91 (enantiomers)
Voltage: -15 kV
Model: Beckman P/ACE 2000

OTHER SUBSTANCES
Simultaneous: proglumide

KEY WORDS
detector at anode; chiral

REFERENCE
Nair,U.B.; Armstrong,D.W.; Hinze,W.L. Characterization and evaluation of d-(+)-tubocurarine chloride as a
chiral selector for capillary electrophoretic enantioseparations, *Anal.Chem.*, **1998**, *70*, 1059–1065.

SAMPLE
Matrix: solutions
Sample preparation: Inject an aliquot of a 100 µg/mL solution in water.

CAPILLARY ELECTROPHORESIS
Capillary: 27 cm × 50 µm eCAP neutral capillary (20 cm to detector) (Beckman)
Running buffer: 100 mM pH 6.0 Sodium phosphate buffer containing 2 mM A82846B (Eli Lilly)
Injection: Fill capillary with running buffer for 5 min, apply -10 kV for 5 min, hydrodynamic
injection at 0.5 psi for 2 s.
Detector: UV 254
Migration time: 6.9, 8.2 (enantiomers)
Model: Beckman P/ACE 2000

OTHER SUBSTANCES
Also analyzed: flurbiprofen, suprofen

KEY WORDS
chiral; detector at anode

REFERENCE
Reilly,J.; Risley,D.S. The separation of enantiomers by countercurrent capillary electrophoresis using the
macrocyclic antibiotic A82846B, *LC.GC*, **1998**, *16*, 170–178.

SAMPLE
Matrix: solutions

CAPILLARY ELECTROPHORESIS
Capillary: 70 cm × 50 µm fused-silica (GL Science, Tokyo)
Capillary preparation: Before each run rinse capillary with running buffer at 94 kPa for 5 min
before each run.
Running buffer: 50 mM pH 8.5 ammonium carbonate buffer
Injection: Pressure injection at 5 kPa (50 mbar) for 4 s.
Detector: MS, Perkin-Elmer Sciex API-300 quadrupole, electrospray (ionspray) interface, sheath
liquid MeOH:running buffer 50:50 at 2.5 µL/min, ionspray voltage 5 kV, negative ion mode
Migration time: 10.4
Voltage: 20 kV (net voltage across capillary = 15 kV (applied voltage - electrospray voltage))
Model: Hewlett Packard 3D CE

OTHER SUBSTANCES
Simultaneous: acetaminophen, ascorbic acid, butylscopolamine bromide, caffeine, ibuprofen, niacin, niacinamide, riboflavin, thiamine, vitamin B12, warfarin

REFERENCE
Tanaka,Y.; Kishimoto,Y.; Otsuka,K.; Terabe,S. Strategy for selecting separation solutions in capillary electrophoresis-mass spectrometry, *J.Chromatogr.A*, **1998**, *817*, 49–57.

SAMPLE
Matrix: solutions
Sample preparation: Inject an aliquot of a solution in running buffer.

CAPILLARY ELECTROPHORESIS
Capillary: 57 cm × 50 μm fused-silica (Beckman)
Capillary preparation: Rinse with 100 mM NaOH for 5 min, with water for 10 min,
Running buffer: 2-Methoxyethanol:buffer 15:85 (Buffer was 40 mM pH 6 phosphate buffer containing 500 μM Actaplanin A (Eli Lilly).)
Injection: Pressure injection for 5 s.
Detector: UV 254
Migration time: 21.4, 22.2
Model: Beckman P/ACE 2100

OTHER SUBSTANCES
Simultaneous: fenoprofen

KEY WORDS
chiral

REFERENCE
Trelli-Seifert,L.A.; Risley,D.S. Capillary electrophoretic enantiomeric separations of nonsteroidal anti-inflammatory compounds using the macrocyclic antibiotic actaplanin A and 2-methoxyethanol, *J.Liq.Chromatogr.Rel.Technol.*, **1998**, *21*, 299–313.

SAMPLE
Matrix: solutions

CAPILLARY ELECTROPHORESIS
Capillary: 42 cm × 50 μm fused-silica (31 cm to detector) (Polymicro Technologies)
Capillary temperature: 30
Running buffer: Formamide containing 3.42% quaternary ammonium-β-cyclodextrin (American Maize Products, Hammond IN), 20 mM ammonium acetate, and 1% acetic acid (pH* = 7.1)
Detector: UV 254
Migration time: 11.55, 11.66 (enantiomers)
Voltage: -30 kV
Model: laboratory-constructed

OTHER SUBSTANCES
Also analyzed: carprofen, fenoprofen, flurbiprofen, indoprofen, suprofen

KEY WORDS
chiral

REFERENCE
Wang,F.; Khaledi,M.G. Nonaqueous capillary electrophoresis chiral separations with quaternary ammonium β-cyclodextrin, *J.Chromatogr.A*, **1998**, *817*, 121–128.

Labetalol

Molecular formula: C$_{19}$H$_{24}$N$_2$O$_3$
Molecular weight: 328.41
CAS Registry No.: 36894-69-6, 32780-64-6 (HCl)
Merck Index (12th ed.): 5341
Lednicer: 3 24; 4 20

SAMPLE
Matrix: blood
Sample preparation: Hydrolyze 2 mL serum with β-glucuronidase (EC 3.2.1.31, type H-1 from Helix pomatia, 416 800 U/g, Separacor) at 80° for 30 min, cool, add 900 μL MeCN, vortex for 15 min, centrifuge at 2004 g for 10 min, add ephedrine (165 μg/mL), filter (0.5 μm), inject an aliquot of the filtrate.

CAPILLARY ELECTROPHORESIS
Capillary: 58 cm × 50 μm fused-silica (50 cm to detector) (Polymicro Technologies)
Capillary preparation: Before each injection purge capillary with 5% phosphoric acid for 12 s, with water for 30 s, and with running buffer for 10 min.
Capillary temperature: 35
Running buffer: 80 mM pH 6.7 sodium phosphate buffer containing 15 mM cetyltrimethylammonium bromide
Injection: Hydrostatic injection for 20 s
Detector: UV 214
Migration time: 16.5
Internal standard: ephedrine (17.5)
Voltage: -27 kV
Model: Waters Quanta 4000

OTHER SUBSTANCES
Extracted: acebutolol, alprenolol, atenolol, metoprolol, nadolol, oxprenolol, pindolol, timolol
Interfering: propranolol

KEY WORDS
serum

REFERENCE
Lukkari,P.; Nyman,T.; Riekkola,M.-J. Determination of nine β-blockers in serum by micellar electrokinetic capillary chromatography, *J.Chromatogr.A*, **1994**, *674*, 241–246.

SAMPLE
Matrix: blood, urine
Sample preparation: Urine. Dilute with 2 volumes of water, filter (0.5 μm), inject an aliquot. Serum. Condition a 3 mL Supelclean LC-18 SPE cartridge (Supelco) with MeOH and water. Hydrolyze 900 μL serum with β-glucuronidase (EC 3.2.1.31 type H-1 from Helix pomatia) at 60° with sonication for 30 min, add 500 μL (?) MeOH, centrifuge at 2000 g, add the supernatant to the SPE cartridge, wash with 1 mL water, dry under vacuum, elute with 2 mL MeOH:water 90:10, filter, inject an aliquot.

CAPILLARY ELECTROPHORESIS
Capillary: 68 cm × 50 μm fused-silica (60 cm to detector) (Polymicro Technologies)
Capillary preparation: Purge with running buffer for 2 min before injection.
Running buffer: pH 7.0 Phosphate buffer containing 10 mM N-cetyl-N,N,N-trimethylammonium bromide (Prepare buffer by mixing 100 mM NaH$_2$PO$_4$ and 100 mM Na$_2$HPO$_4$ to achieve a pH of 7.0.)
Injection: Hydrostatic injection for 30 s
Detector: UV 214
Migration time: 17
Internal standard: 2,6-dimethylphenol (only for urine) (14.5)

Voltage: -26 kV
Current: 97 µA
Model: Waters Quanta 4000

OTHER SUBSTANCES
Extracted: acebutolol, alprenolol, atenolol, metoprolol, nadolol, oxprenolol, pindolol, propranolol, timolol

KEY WORDS
serum; comparison with HPLC; SPE

REFERENCE
Lukkari,P.; Sirén,H. Ion-pair chromatography and micellar electrokinetic capillary chromatography in analyzing β-adrenergic blocking agents from human biological fluids, *J.Chromatogr.A*, **1995**, *717*, 211–217.

SAMPLE
Matrix: solutions

CAPILLARY ELECTROPHORESIS
Capillary: 58 cm × 50 µm fused-silica (50 cm to detector)
Running buffer: Isopropanol:buffer 2.5:97.5 (Buffer was 80 mM pH 6.8 Phosphate buffer containing 15 mM cetyltrimethylammonium bromide.)
Injection: Hydrostatic injection for 15 s
Detector: UV 214
Migration time: 15
Voltage: -20 kV
Model: Waters Quanta 4000

OTHER SUBSTANCES
Simultaneous: acebutolol, alprenolol, atenolol, metoprolol, nadolol, oxprenolol, pindolol, propranolol, sotalol, timolol

REFERENCE
Lukkari,P.; Vuorela,H.; Riekkola,M.-L. Effects of organic mobile phase modifiers on elution and separation of β-blockers in micellar electrokinetic capillary chromatography, *J.Chromatogr.A*, **1993**, *655*, 317–324.

SAMPLE
Matrix: solutions
Sample preparation: Prepare a 1 mg/mL solution in 40 mM pH 5.1 sodium phosphate buffer.

CAPILLARY ELECTROPHORESIS
Capillary: 11.5 cm × 75 µm fused silica tubing (8.5 cm to the detector) (Polymicro Technologies, Phoenix, AZ) coated on the inner surface with non-cross-linked polyacrylamide. Coat as follows. Mix 80 µL gamma-methacryloxypropyltrimethoxysilane with 20 mL water adjusted to pH 3.5 with acetic acid, suck into capillary, let stand for 1 h, remove solution, wash with water fill capillary with a 3 or 4% solution of acrylamide containing 1 mL/L N,N,N',N'-tetramethylethylenediamine and 1 g/L potassium persulfate, let stand for 30 min, remove excess solution, wash capillary with water, suck capillary dry, dry at 35° (J. Chromatogr. 1985, 347, 191).
Running buffer: Isopropanol:400 mM pH 5.1 sodium phosphate buffer 30:70
Injection: Fill 7 cm of tube with running buffer, fill rest of tube with 40 mg/mL cellobiohydrolase I in buffer, add a 2-3 mm plug of agarose by pressing end of tube into agarose gel, inject sample at 1000 V for 20 s. (Buffer was 400 mM pH 5.1 sodium phosphate buffer. Prepare agarose gel by boiling 20 mg agarose in 1 mL buffer, cool, allow to gel.)
Detector: UV 220
Migration time: 19, 22 (RR, SS); 40, 85 (RS, SR)
Voltage: 1000 V
Model: Laboratory constructed

OTHER SUBSTANCES
Also analyzed: alprenolol, metoprolol, propranolol

KEY WORDS
detector is at cathode; chiral; coated capillary

REFERENCE
Valtcheva,L.; Mohammad,J.; Pettersson,G.; Hjertén,S. Chiral separation of β-blockers by high-performance capillary electrophoresis based on non-immobilized cellulase as enantioselective protein, *J.Chromatogr.*, **1993**, *638*, 263–267.

SAMPLE
Matrix: solutions

CAPILLARY ELECTROPHORESIS
Capillary: 100 cm × 50 μm fused-silica (50 cm to detector) (Isco)
Capillary preparation: Flush capillary with 10 μL running buffer between runs. Every 40 sample injections rinse capillary with 200 μL 1 M NaOH, with 200 μL water, and 200 μL running buffer. Before use fill capillary with 1 M NaOH and allow to stand for 1 h, fill with 100 mM NaOII, allow to otand for 1 h, wash with water fill with running buffer.
Capillary temperature: 23
Running buffer: 100 mM citric acid containing 19.27 mM Na_2HPO_4 and 120 mM hydroxypropyl-β-cyclodextrin
Injection: Inject under vacuum at 4.0 kPa.s
Detector: UV 200
Migration time: 45, 45.5 (enantiomers)
Voltage: 30 kV
Model: Isco Model 3140

OTHER SUBSTANCES
Simultaneous: albuterol, alprenolol, atenolol, cimaterol, clenbuterol, nadolol, oxprenolol, pindolol, pirbuterol, propranolol, terbutaline

KEY WORDS
chiral

REFERENCE
Aumatell,A.; Wells,R.J.; Wong,D.K.Y. Enantiomeric differentiation of a wide range of pharmacologically active substances by capillary electrophoresis using modified β-cyclodextrins, *J.Chromatogr.A*, **1994**, *686*, 293–307.

SAMPLE
Matrix: solutions

CAPILLARY ELECTROPHORESIS
Capillary: 62 cm × 52 μm fused-silica (50 cm to detector) (Polymicro Technologies)
Running buffer: 70 mM pH 2.5 Trimethylammonium phosphate buffer containing 20 mM hydroxypropyl-β-cyclodextrin
Injection: Gravity injection at 10 cm for 5 s
Migration time: 46, 47.5 (enantiomers)
Voltage: 20 kV
Current: 43 μA
Model: Laboratory constructed

OTHER SUBSTANCES
Simultaneous: acebutolol, alprenolol, atenolol, isoproterenol, nadolol, pindolol, propranolol

KEY WORDS
chiral

REFERENCE
Quang,C.; Khaledi,M.G. Direct separation of the enantiomers of β-blockers by cyclodextrin-mediated capillary zone electrophoresis, *J.High Res.Chromatogr.*, **1994**, *17*, 99–101.

SAMPLE
Matrix: solutions
Sample preparation: Prepare a solution in MeOH/water, inject an aliquot.

CAPILLARY ELECTROPHORESIS
Capillary: 62 cm × 52 μm fused-silica (50 cm to detector) (Polymicro Technologies)
Capillary temperature: 40
Running buffer: 50 mM pH 2.50 Tetrabutylammonium phosphate containing 20 mM hydroxy-propyl-β-cyclodextrin
Injection: Siphon injection at 10 cm for 5 s
Detector: UV (wavelength not specified)
Migration time: 28.67, 29.10 (enantiomers)
Voltage: 20 kV
Model: Laboratory constructed

OTHER SUBSTANCES
Simultaneous: alprenolol, atenolol, chlorpheniramine, epinephrine, isoproterenol, norepinephrine, oxprenolol, propranolol

KEY WORDS
chiral

REFERENCE
Quang,C.; Khaledi,M.G. Extending the scope of chiral separation of basic compounds by cyclodextrin-mediated capillary zone electrophoresis, *J.Chromatogr.A*, **1995**, *692*, 253–265.

SAMPLE
Matrix: solutions

CAPILLARY ELECTROPHORESIS
Capillary: 57 cm × 75 μm fused-silica (50 cm to detector) (Beckman)
Capillary preparation: Before each run rinse capillary with 100 mM NaOH and running buffer.
Capillary temperature: 30
Running buffer: 100 mM Phosphoric acid adjusted to pH 3.1 with triethanolamine
Injection: Pressure injection for 5-10 s.
Detector: UV 200
Migration time: 12.3
Voltage: 25 kV
Model: Beckman P/ACE 5510

OTHER SUBSTANCES
Simultaneous: acebutolol, atenolol, metoprolol, nadolol, oxprenolol, pindolol, propranolol, sotalol, timolol

REFERENCE
Bretnall,A.E.; Clarke,G.S. Selectivity of capillary electrophoresis for the analysis of cardiovascular drugs, *J.Chromatogr.A*, **1996**, *745*, 145–154.

SAMPLE
Matrix: solutions

CAPILLARY ELECTROPHORESIS
Capillary: 43 cm × 50 μm fused-silica (36 cm to detector)
Capillary preparation: Before each injection wash with running buffer for 10 min. After each injection wash with water for 2 min. Wash new capillaries with 1 M NaOH at 60° for 50 min, with NaOH solution (?) at 60° for 10 min, and with water at 25°.
Capillary temperature: 25
Running buffer: 320 mM pH 2.0 Citrate buffer
Injection: Hydrodynamic injection at 1.5 psi for 2 s.
Detector: UV 220
Migration time: 10.5

Voltage: 15 kV
Model: Spectra-Physics Model 1000
Limit of detection: 1-18 μg/mL

OTHER SUBSTANCES
Simultaneous: acebutolol, atenolol, levobunolol, metoprolol, nadolol, oxprenolol, pindolol, propranolol, timolol

REFERENCE
Lin,C.-E.; Chang,C.-C.; Lin,W.-c.; Lin,E.C. Capillary zone electrophoretic separation of β-blockers using citrate buffer at low pH, *J.Chromatogr.A*, **1996**, *753*, 133–138.

SAMPLE
Matrix: solutions

CAPILLARY ELECTROPHORESIS
Capillary: 67 cm × 50 μm (60 cm to detector)
Capillary preparation: Wash with running buffer for 5 min before each injection. After each injection wash with 1 M NaOH at 60° for 5 min, with 100 mM NaOH at 60° for 10 min, and with water at 25° for 5 min.
Capillary temperature: 25
Running buffer: 70 mM pH 7.0 Sodium phosphate buffer containing 15 mM cetyltriemthylammonium bromide
Injection: Hydrodynamic injection for 1 s.
Detector: UV 220
Migration time: 20
Voltage: 20 kV

OTHER SUBSTANCES
Simultaneous: acebutolol, atenolol, levobunolol, metoprolol, nadolol, oxprenolol, pindolol, propranolol, timolol

REFERENCE
Lin,C.-E.; Chen,Y.-C.; Chang,C.-C.; Wang,D.-Z. Migration behavior and selectivity of β-blockers in micellar electrokinetic chromatography. Influence of micelle concentration of cationic surfactants, *J.Chromatogr.A*, **1997**, *775*, 349–357.

SAMPLE
Matrix: urine
Sample preparation: Dilute urine with 2 volumes of water, add IS, filter (0.5 μm), inject an aliquot.

CAPILLARY ELECTROPHORESIS
Capillary: 68 cm × 50 μm fused-silica (60 cm to detector) (White Associates)
Capillary preparation: Purge capillary with running buffer for 2 min before each injection.
Running buffer: 80 mM pH 7.0 Phosphate buffer containing 10 mM N-cetyl-N,N,N-trimethylammonium bromide
Injection: Hydrostatic injection for 30 s at 10 cm.
Detector: UV 214
Migration time: 17.6
Internal standard: 2,6-dimethylphenol (14.7)
Voltage: -26 kV (at injector end)
Model: Waters Quanta 4000
Limit of detection: 10 μg/mL (S/N 3)

OTHER SUBSTANCES
Extracted: acebutolol, alprenolol, atenolol, metoprolol, nadolol, oxprenolol, pindolol, propranolol, timolol
Simultaneous: probenecid
Noninterfering: caffeine

REFERENCE

Lukkari,P.; Sirén,H.; Pantsar,M.; Riekkola,M.-L. Determination of ten β-blockers in urine by micellar electro-kinetic capillary chromatography, *J.Chromatogr.*, **1993**, *632*, 143–148.

Lamivudine

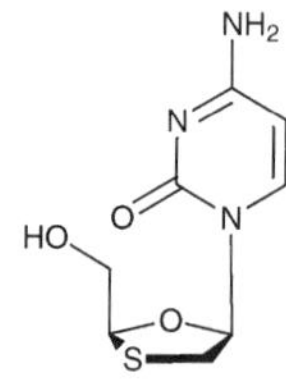

Molecular formula: $C_8H_{11}N_3O_3S$
Molecular weight: 229.26
CAS Registry No.: 134678-17-4
Merck Index (12th ed.): 5365
Lednicer: 5 99

SAMPLE

Matrix: enzyme incubations
Sample preparation: Filter (Millipore Ultrafree MC, 10 kDa cut-off) enzyme reactions involving immobilized cytidine deaminase. Dilute an aliquot of the ultrafiltrate 10-fold with water, filter (0.45 μm), inject an aliquot of the filtrate.

CAPILLARY ELECTROPHORESIS

Capillary: 57 cm × 75 μm fused-silica
Capillary preparation: Before each injection rinse capillary with 500 mM NaOH for 3 min and equilibrate with running buffer for 3 min.
Running buffer: pH 2.3 Sodium phosphate buffer containing 27 mM dimethyl-β-cyclodextrin
Injection: Pressure injection for 5 s.
Detector: UV 214
Migration time: 17 (-), 18 (+)
Voltage: 15 kV
Model: Beckman P/ACE 2000

OTHER SUBSTANCES

Simultaneous: uridine analog

KEY WORDS

chiral

REFERENCE

Rogan,M.M.; Drake,C.; Goodall,D.M.; Altria,K.D. Enantioselective enzymatic biotransformation of 2'-deoxy-3'-thiacytidine (BCH 189) monitored by capillary electrophoresis, *Anal.Biochem.*, **1993**, *208*, 343–347.

SAMPLE

Matrix: solutions

CAPILLARY ELECTROPHORESIS

Capillary: 34 cm × 50 μm (25.5 cm (A) or 8.5 cm (B, C) to detector) (Composite Metal Services, Hallow, UK)
Running buffer: 50 mM pH 2.5 Phosphate buffer (A, B) or 250 mM pH 2.5 phosphate buffer (C)
Injection: Pressure injection at 50 mbar for 5 s.
Detector: UV 200
Migration time: 1.95 (A), 0.8 (B), 3.3 (C)
Voltage: 20 kV (A) or -20 kV (B) or -7.5 kV (C)
Model: Hewlett Packard 3D

OTHER SUBSTANCES

Simultaneous: albuterol, aminobenzoic acid, imidazole

KEY WORDS

when the distance to the detector was 8.5 cm (B; C) injection was at the normal outlet end and voltage polarity and pressure for injection were reversed.

REFERENCE
Altria,K.D.; Kelly,M.A.; Clark,B.J. The use of a short-end injection procedure to achieve improved performance in capillary electrophoresis, *Chromatographia*, **1996**, *43*, 153–158.

SAMPLE
Matrix: solutions

CAPILLARY ELECTROPHORESIS
Capillary: 37 cm × 50 μm eCAP polyamine-coated (30 cm to detector) (Beckman)
Capillary preparation: At the end of each run flush with 1 M NaOH for 2 min, wash with regenerator (Beckman) for 2 min, wash with running buffer for 2 min. Condition a new capillary with running buffer for 2 min and with regenerator for 2 min.
Capillary temperature: 30
Running buffer: 50 mM pH 8 Tris
Detector: UV 214
Migration time: 1.3
Voltage: 30 kV
Model: Beckman P/ACE 2210

OTHER SUBSTANCES
Simultaneous: albuterol, aminobenzoic acid, imidazole

KEY WORDS
coated capillary; comparison with uncoated capillary

REFERENCE
Assi,K.A.; Altria,K.D.; Clark,B.J. Rapid resolution of drugs and related substances with an eCAP™ polyamine coated capillary, *J.Pharm.Biomed.Anal.*, **1997**, *15*, 1041–1049.

SAMPLE
Matrix: solutions

CAPILLARY ELECTROPHORESIS
Capillary: 27 cm × 100 μm (20 cm to detector) (Composite Metal Services)
Capillary temperature: 30
Running buffer: 25 mM pH 2.3 NaH_2PO_4
Detector: UV 200
Migration time: 4.54
Internal standard: imidazole (2.46)
Voltage: 7 kV
Model: Beckman

OTHER SUBSTANCES
Simultaneous: albuterol, alosetron, ranitidine

REFERENCE
Altria,K.D.; Creasey,E.; Howells,J.S. Routine capillary electrophoresis trace level determinations of pharmaceutical and detergent residues on pharmaceutical manufacturing equipment, *J.Liq.Chromatogr.Rel.Technol.*, **1998**, *21*, 1093–1105.

Lamotrigine

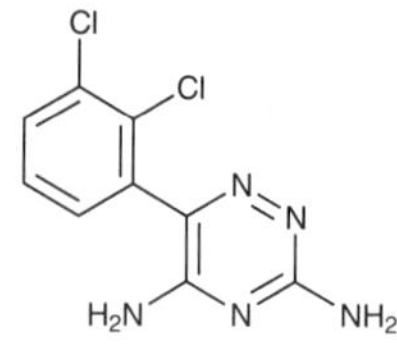

Molecular formula: $C_9H_7Cl_2N_5$
Molecular weight: 256.09
CAS Registry No.: 84057-84-1
Merck Index (12th ed.): 5367
Lednicer: 4 120

SAMPLE
Matrix: blood
Sample preparation: 50 μL Serum + 100 μL 40 μg/mL tyramine in MeCN, mix, centrifuge at 14000 g for 30 s. Remove the supernatant and mix it with 100 μL 900 mM acetic acid, inject an aliquot.

CAPILLARY ELECTROPHORESIS
Capillary: 42 cm × 50 μm
Capillary preparation: After each run wash with 200 mM NaOH for 1 min and with running buffer for 1 min.
Capillary temperature: 24
Running buffer: 130 mM Sodium acetate adjusted to pH 4.8 with 10 mM acetic acid
Injection: Pressure injection for 25 s.
Detector: UV 214
Migration time: 3.3
Internal standard: tyramine (2.6)
Voltage: 10 kV
Model: Beckman P/ACE 2000
Limit of quantitation: 500 ng/mL
Limit of detection: 300 ng/mL

KEY WORDS
comparison with HPLC; serum

REFERENCE
Shihabi,Z.K.; Oles,K.S. Serum lamotrigine analysis by capillary electrophoresis, *J.Chromatogr.B*, **1996**, *683*, 119–123.

Lansoprazole

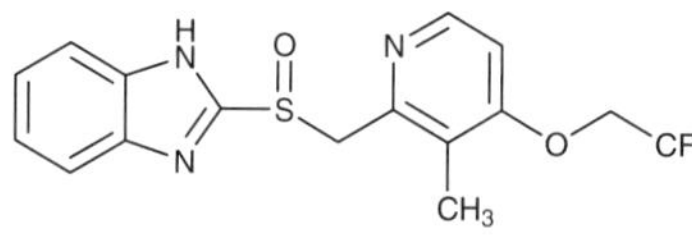

Molecular formula: $C_{16}H_{14}F_3N_3O_2S$
Molecular weight: 369.37
CAS Registry No.: 103577-45-3
Merck Index (12th ed.): 5373
Lednicer: 5 115

SAMPLE
Matrix: solutions
Sample preparation: Inject an aliquot of a solution in water.

CAPILLARY ELECTROPHORESIS
Capillary: 66 cm × 50 μm fused-silica (50 cm to detector)
Capillary preparation: After each analysis rinse with 100 mM NaOH for 3 min and with running buffer for 2 min.
Running buffer: n-Propanol:10 mM pH 7.4 potassium phosphate buffer containing 40 μM bovine serum albumin 7:93
Injection: Electrokinetic injection at 5-8 kV for 7 s.

Detector: UV 290
Migration time: 12.8, 13.3 (enantiomers)
Voltage: about 300 V/cm
Current: 35 μA (constant current)
Model: Grom CE-System 100
Limit of detection: 40 μg/mL

KEY WORDS
chiral

REFERENCE
Eberle,D.; Hummel,R.P.; Kuhn,R. Chiral resolution of pantoprazole sodium and related sulfoxides by complex formation with bovine serum albumin in capillary electrophoresis, *J.Chromatogr.A*, **1997**, *759*, 185–192.

Leucovorin

Molecular formula: $C_{20}H_{23}N_7O_7$
Molecular weight: 473.45
CAS Registry No.: 58-05-9, 1492-18-8 (Ca salt), 6035-45-6 (Ca salt pentahydrate)
Merck Index (12th ed.): 4254

SAMPLE
Matrix: solutions

CAPILLARY ELECTROPHORESIS
Capillary: 95 cm × 75 μm fused-silica (Polymicro Technologies)
Running buffer: 20 mM pH 7.0 Phosphate buffer containing 1 mg/mL bovine serum albumin
Detector: UV 280
Migration time: 19.5 (6R), 20.5 (6S)
Voltage: 325 V/cm
Model: Laboratory constructed

KEY WORDS
chiral

REFERENCE
Barker,G.E.; Russo,P.; Hartwick,R.A. Chiral separation of leucovorin with bovine serum albumin using affinity capillary electrophoresis, *Anal.Chem.*, **1992**, *64*, 3024–3028.

Levallorphan

Molecular formula: $C_{19}H_{25}NO$
Molecular weight: 283.41
CAS Registry No.: 152-02-3, 71-82-9 (tartrate)
Merck Index (12th ed.): 5485

SAMPLE
Matrix: solutions

CAPILLARY ELECTROPHORESIS
Capillary: 59.6 cm × 75 μm fused-silica (52.1 cm to detector) (Polymicro Technologies)

Capillary preparation: Purge with running buffer before each run. At the beginning of each day purge using 50-60 kPa vacuum with 500 mM NaOH for 5 min, with water for 5 min, with MeCN for 5 min, and with running buffer for 5 min.
Running buffer: MeCN:MeOH:acetic acid 49:50:1 containing 20 mM ammonium acetate
Injection: Hydrostatic injection at 10 cm for 5 s.
Detector: UV 214
Migration time: 3.59
Voltage: 30 kV
Model: Waters Quanta 4000

OTHER SUBSTANCES
Simultaneous: diphenoxylate, ethoheptazine, fentanyl, meperidine, methadone, nalorphine, pentazocine, nikethamide

REFERENCE
Leung,G.N.W.; Tang,H.P.O.; Tso,T.S.C.; Wan,T.S.M. Separation of basic drugs with non-aqueous capillary electrophoresis, *J.Chromatogr.A*, **1996**, *738*, 141–154.

Levamisole

Molecular formula: $C_{11}H_{12}N_2S$
Molecular weight: 204.30
CAS Registry No.: 14769-73-4, 16595-80-5 (HCl), 5036-02-2 (racemic), 5086-74-8 (racemic HCl)
Merck Index (12th ed.): 5486
Lednicer: 4 217

SAMPLE
Matrix: solutions
Sample preparation: Inject an aliquot of a solution in running buffer.

CAPILLARY ELECTROPHORESIS
Capillary: 60 cm × 75 μm fused-silica (52.4 cm to detector)
Capillary preparation: After each run flush with 500 mM KOH for 2-3 min then with water.
Running buffer: 10 mM pH 3.8 Phosphate buffer containing 2% sulfated cyclodextrin (ds 7-10)
Injection: Hydrostatic injection.
Detector: UV 214
Migration time: 11.35, 11.98 (enantiomers)
Voltage: 15 kV
Model: Waters Quanta 4000

OTHER SUBSTANCES
Also analyzed: acebutolol, alprenolol, aminoglutethimide, brompheniramine, bupivacaine, bupropion, canadine, carbinoxamine, chloroquine, chlorpheniramine, dimethindene, disopyramide, doxylamine, hydroxychloroquine, idazoxan, isoxsuprine, ketamine, mepenzolate, mepivacaine, methoxyphenamine, mexiletine, midodrine, nefopam, orphenadrine, oxprenolol, oxyphencyclimine, pheniramine, phensuximide, pindolol, piperoxan, terbutaline, tolperisone, tranylcypromine, trihexyphenidyl, trimipramine, verapamil, warfarin

KEY WORDS
chiral; detector at anode

REFERENCE
Stalcup,A.M.; Gahm,K.H. Application of sulfated cyclodextrins to chiral separations by capillary zone electrophoresis, *Anal.Chem.*, **1996**, *68*, 1360–1368.

Levobunolol

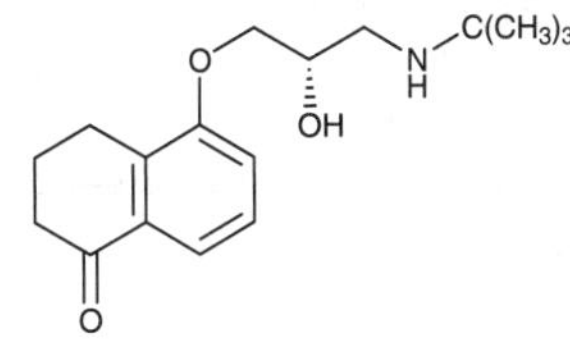

Molecular formula: $C_{17}H_{25}NO_3$
Molecular weight: 291.39
CAS Registry No.: 47141-42-4, 27912-14-7 (HCl)
Merck Index (12th ed.): 5488
Lednicer: 2 110, 215, 280

SAMPLE
Matrix: blood
Sample preparation: Filter (1 mL Amicon Centrifree Micropartition System) serum while centrifuging at 1500 g for 20 min, inject an aliquot of the ultrafiltrate.

CAPILLARY ELECTROPHORESIS
Capillary: 92 cm × 75 μm fused-silica (55 cm to detector) (Polymicro Technologies)
Capillary preparation: Rinse with running buffer for 10 min between run. At the beginning and end of each day rinse with 100 mM NaOH for 10 min and with water for 10 min.
Running buffer: Isopropanol:buffer 5:95 (Buffer was 10 mM Na_2HPO_4 containing 6 mM sodium borate and 75 mM sodium dodecyl sulfate.)
Injection: Injection via vacuum suction for 0.7 s
Detector: UV 215
Migration time: 22.5
Voltage: 25 kV
Current: 82 μA
Model: Europhor Prime Vision IV

OTHER SUBSTANCES
Extracted: atenolol, carteolol, celiprolol, metoprolol, penbutolol, pindolol, propranolol, timolol

KEY WORDS
serum; ultrafiltrate

REFERENCE
Schmutz,A.; Thormann,W. Assessment of impact of physico-chemical drug properties on monitoring drug levels by micellar electrokinetic capillary chromatography with direct serum injection, *Electrophoresis*, **1994**, *15*, 1295–1303.

SAMPLE
Matrix: solutions

CAPILLARY ELECTROPHORESIS
Capillary: 43 cm × 50 μm fused-silica (36 cm to detector)
Capillary preparation: Before each injection wash with running buffer for 10 min. After each injection wash with water for 2 min. Wash new capillaries with 1 M NaOH at 60° for 50 min, with NaOH solution (?) at 60° for 10 min, and with water at 25°.
Capillary temperature: 25
Running buffer: 320 mM pH 2.0 Citrate buffer
Injection: Hydrodynamic injection at 1.5 psi for 2 s.
Detector: UV 220
Migration time: 9.8
Voltage: 15 kV
Model: Spectra-Physics Model 1000
Limit of detection: 1-18 μg/mL

OTHER SUBSTANCES
Simultaneous: acebutolol, atenolol, labetalol, metoprolol, nadolol, oxprenolol, pindolol, propranolol, timolol

REFERENCE

Lin,C.-E.; Chang,C.-C.; Lin,W.-c.; Lin,E.C. Capillary zone electrophoretic separation of β-blockers using citrate buffer at low pH, *J.Chromatogr.A*, **1996**, *753*, 133–138.

SAMPLE

Matrix: solutions

CAPILLARY ELECTROPHORESIS

Capillary: 67 cm × 50 μm (60 cm to detector)
Capillary preparation: Wash with running buffer for 5 min before each injection. After each injection wash with 1 M NaOH at 60° for 5 min, with 100 mM NaOH at 60° for 10 min, and with water at 25° for 5 min.
Capillary temperature: 25
Running buffer: 70 mM pH 7.0 Sodium phosphate buffer containing 15 mM cetyltriemthylammonium bromide
Injection: Hydrodynamic injection for 1 s.
Detector: UV 220
Migration time: 15.2
Voltage: 20 kV

OTHER SUBSTANCES

Simultaneous: acebutolol, atenolol, labetalol, metoprolol, nadolol, oxprenolol, pindolol, propranolol, timolol

REFERENCE

Lin,C.-E.; Chen,Y.-C.; Chang,C.-C.; Wang,D.-Z. Migration behavior and selectivity of β-blockers in micellar electrokinetic chromatography. Influence of micelle concentration of cationic surfactants, *J.Chromatogr.A*, **1997**, *775*, 349–357.

Levodopa

Molecular formula: $C_9H_{11}NO_4$
Molecular weight: 197.19
CAS Registry No.: 59-92-7
Merck Index (12th ed.): 5490

SAMPLE

Matrix: blood
Sample preparation: Inject an aliquot of 5-fold diluted bovine serum directly.

CAPILLARY ELECTROPHORESIS

Capillary: 50 cm × 50 μm (20 cm to detector) (GL Science)
Capillary preparation: Rinse capillary with running buffer for 5 min before each run. At the beginning of each day rinse capillary with 100 mM NaOH for 10 min and with water for 5 min.
Running buffer: 10 mM pH 2.5 Phosphate buffer containing 25 mM sodium dodecyl sulfate and 12.5 mM Tween 20
Injection: Siphon at 10 cm for 10 s.
Detector: UV 210
Migration time: 11
Voltage: 18 kV
Current: 21-22 μA
Model: Jasco CE-800

OTHER SUBSTANCES

Extracted: dopamine, epinephrine, 3-methoxytyrosine, norepinephrine, phenylalanine, tyrosine

KEY WORDS
serum; cow; detector is at anode

REFERENCE
Esaka,Y.; Tanaka,K.; Uno,B.; Goto,M.; Kano,K. Sodium dodecyl sulfate-Tween 20 mixed micellar electrokinetic chromatography for separation of hydrophobic cations: Application to adrenaline and its precursors, *Anal.Chem.*, **1997**, *69*, 1332–1338.

SAMPLE
Matrix: solutions

CAPILLARY ELECTROPHORESIS
Capillary: 70 cm × 100 μm fused-silica (50 cm to detector) (Gasukuro Kogyo, Tokyo)
Running buffer: 100 mM KOH containing 200 mM boric acid
Injection: Hydrostatic injection for 10 s.
Detector: UV 217
Migration time: 27
Voltage: 9.5 kV
Current: 100 μA

OTHER SUBSTANCES
Simultaneous: dopamine, epinephrine, isoproterenol, metanephrine, norepinephrine, normetanephrine, vanillylmanedlic acid

REFERENCE
Tanaka,S.; Kaneta,T.; Yoshida,H. Separation of catecholamines by capillary zone electrophoresis using complexation with boric acid, *Anal.Sci.*, **1990**, *6*, 467–468.

SAMPLE
Matrix: solutions

CAPILLARY ELECTROPHORESIS
Capillary: 50 cm × 75 μm fused-silica
Capillary temperature: 25
Running buffer: 30 mM [18]-crown-6 tetracarboxylic acid (Merck), pH 2.2
Injection: Hydrodynamic injection for 1 s
Detector: UV 254
Migration time: 23.51
Voltage: 15 kV
Model: Beckman P/ACE 2000

OTHER SUBSTANCES
Simultaneous: D-dopa, D,L-phenylalanine, D,L-tryptophan, D,L-tyrosine

KEY WORDS
chiral

REFERENCE
Kuhn,R.; Stoecklin,F.; Erni,F. Chiral separations by host-guest complexation with cyclodextrin and crown ether in capillary zone electrophoresis, *Chromatographia*, **1992**, *33*, 32–36.

SAMPLE
Matrix: solutions
Sample preparation: Prepare a solution in running buffer or 50 mM HCl, inject an aliquot.

CAPILLARY ELECTROPHORESIS
Capillary: 57 cm × 75 μm fused-silica (50 cm to detector) (Polymicro)
Capillary temperature: 25
Running buffer: 10 mM pH 2.2 Tris buffer containing 10 mM 18-crown-6-tetracarboxylic acid (Merck)

Injection: Hydrodynamic injection for 1 s
Detector: UV 254
Migration time: 29.18 (first eluted enantiomer)
Voltage: 263 V/cm
Model: Beckman P/ACE 2000 or 2100

KEY WORDS
chiral; $\alpha = 1.053$

REFERENCE
Kuhn,R.; Wagner,J.; Walbroehl,Y.; Bereuter,T. Potential and limitations of an optically active crown ether for chiral separation in capillary zone electrophoresis, *Electrophoresis*, **1994**, *15*, 828–834.

SAMPLE
Matrix: solutions
Sample preparation: Inject an aliquot of a solution in running buffer.

CAPILLARY ELECTROPHORESIS
Capillary: 56 cm $\times$ 75 μm fused-silica (36 cm to detector) (Otsuka Electronics, Osaka)
Running buffer: Formamide containing 25 mM (+)-18-crown-6 tetracarboxylic acid (Merck, Darmstadt, Germany) and 2.5 mM tetra-n-butylammonium perchlorate
Injection: Gravity injection at 10 cm for 5 s.
Detector: UV 260-300
Migration time: 23.53, 24.18 (enantiomers)
Voltage: 20 kV
Model: Jasco CE-800

KEY WORDS
chiral

REFERENCE
Mori,Y.; Ueno,K.; Umeda,T. Enantiomeric separations of primary amino compounds by non-aqueous capillary zone electrophoresis with a chiral crown ether, *J.Chromatogr.A*, **1997**, *757*, 328–332.

SAMPLE
Matrix: solutions
Sample preparation: Inject an aliquot of a 100 μg/mL solution in MeOH:water 10:90.

CAPILLARY ELECTROPHORESIS
Capillary: 37 cm $\times$ 75 μm
Capillary temperature: 15
Running buffer: 20 mM pH 2.06 Tris-phosphoric acid containing 10 mM 18-crown-6 tetracarboxylic acid (Merck, Darmstadt, Germany)
Injection: Pressure injection at 0.5 psi for 2-9 s.
Detector: UV 210, UV 235
Migration time: 10.94 (L), 11.28 (D) (enantiomers)
Voltage: 20 kV
Model: Beckman P/ACE 5510

OTHER SUBSTANCES
Also analyzed: aminoglutethimide, baclofen, mexiletine, norephedrine, norepinephrine, octopamine, primaquine

---

KEY WORDS
chiral; comparison with HPLC

REFERENCE
Nishi,H.; Nakamura,K.; Nakai,H.; Sato,T. Separation of enantiomers and isomers of amino compounds by capillary electrophoresis and high-performance liquid chromatography utilizing crown ethers, *J.Chromatogr.A*, **1997**, *757*, 225–235.

SAMPLE
Matrix: solutions

CAPILLARY ELECTROPHORESIS
Capillary: 27 cm × 50 μm fused-silica (20 cm to detector) (Beckman)
Capillary preparation: Before each run rinse with 100 mM NaOH for 2 min and with running buffer for 3-4 min.
Capillary temperature: 20
Running buffer: 25 mM pH 9.05 Sodium tetraborate buffer
Injection: Pressure injection for 5 s.
Detector: UV 285
Migration time: 1.25
Voltage: 25 kV
Model: Beckman P/ACE 5000

OTHER SUBSTANCES
Simultaneous: dopachrome, dopaquinone, leucodopachrome, topaquinone

REFERENCE
Robinson,G.M.; Smyth,M.R. Simultaneous determination of products and intermediates of L-dopa oxidation using capillary electrophoresis with diode-array detection, *Analyst*, **1997**, *122*, 797–802.

SAMPLE
Matrix: solutions

CAPILLARY ELECTROPHORESIS
Capillary: 27 cm × 75 μm fused-silica (20 cm to detector) (Polymicro Technologies)
Capillary preparation: Before each injection rinse capillary with 400 mM KOH for 2 min, with water for 2 min, and with running buffer for 3 min.
Capillary temperature: 20
Running buffer: 25 mM pH 2.4 Phosphate buffer containing 1 mM starburst dendrimer PAMAM G0.5 (Aldrich)
Injection: Hydrodynamic injection at high pressure.
Detector: UV 214
Migration time: 17
Voltage: 10 kV
Model: Beckman P/ACE 5050

OTHER SUBSTANCES
Simultaneous: homophenylalanine, methyldopa, phenylalanine, phenylglycine, tyrosine

REFERENCE
Gao,H.; Carlson,J.; Stalcup,A.M.; Heineman,W.R. Separation of aromatic acids, DOPA, and methyl dopa by capillary electrophoresis with dendrimers as buffer additives, *J.Chromatogr.Sci.*, **1998**, *36*, 146–154.

SAMPLE
Matrix: tissue
Sample preparation: Sonicate (Artix Sonic Dismembrator Model 150, power setting 30) 1-2 mg rat brain tissue and 10 μL EtOH:water 70:30 in an ice bath for 5-10 s, centrifuge at 16000 g for 10 min. 1 μL Supernatant + 1 μL 490 μM α-aminoadipic acid, mix, remove a 1 μL aliquot and add it to 5 μL 20 mM pH 9.0 sodium borate buffer, add 1.5 μL 20 mM NaCN in water, mix, add 1.5 μL 20 mM naphthalene-2,3-dicarboxaldehyde in MeCN, mix thoroughly, let stand at room temperature for 30 min (protect from light), inject an aliquot.

CAPILLARY ELECTROPHORESIS
Capillary: 115 cm × 50 μm fused-silica (95 cm to detector)
Running buffer: 20 mM pH 9.0 sodium borate buffer
Injection: Electrokinetic injection at 5 kV for 12 s
Detector: UV 420
Migration time: 19.6
Internal standard: α-aminoadipic acid (21.6)

Voltage: 30 kV

OTHER SUBSTANCES
Extracted: amino acids, gamma-aminobutyric acid, dopamine, phosphoethanolamine, norepinephrine, taurine

---

KEY WORDS
rat; brain; derivatization

REFERENCE
Weber,P.L.; O'Shea,T.J.; Lunte,S.M. Separation and quantitation of the amino acid neurotransmitters in rat brain by capillary electrophoresis, *J.Pharm.Biomed.Anal.*, **1994**, *12*, 319–324.

SAMPLE
Matrix: urine
Sample preparation: Condition a light alumina B SPE cartridge (Waters) with water. Adjust urine to pH 3 with 6 M HCl. 10 mL Acidified urine + 1 mL 200 mM EDTA + 100 µL 500 mM ascorbic acid, adjust pH to 8.5 with NaOH solution, add to the SPE cartridge, wash with two 2 mL portions of water, elute with 1 mL 100 mM HCl, inject an aliquot of the eluate.

CAPILLARY ELECTROPHORESIS
Capillary: 80 cm × 75 µm fused-silica (Composite Metal Services, Worcs., UK)
Running buffer: 30 mM pH 10 Borate buffer containing 0.5 mM EDTA and 1 mM ascorbic acid
Injection: Hydrodynamic injection at 40 mbar for 6 s.
Detector: F ex (Varian Schott glass UG11 filter) em 500 (cut-off filter) [ex 300 em 545] following post-column reaction. The end of the capillary was connected to a 11.5 cm × 75 µm fused-silica capillary by means of a porous PTFE sleeve with a narrow gap between the capillaries (described in the paper). The detector was 5.5 cm after the junction. The porous sleeve was immersed in a reagent reservoir. The reservoir was grounded and was under 30 mbar air pressure. (Reagent was 2 mM terbium(III) chloride containing 2 (?) mM EDTA and 800 mM CsCl.)
Migration time: 8.7
Voltage: 30 kV
Model: Prince
Limit of detection: 90 nM

OTHER SUBSTANCES
Extracted: 3,4-dihydroxyphenylacetic acid, 3,4-dihydroxyphenyl glycol, dopamine, epinephrine, norepinephrine

---

KEY WORDS
SPE; post-column reaction

REFERENCE
Zhu,R.; Kok,W.T. Determination of catecholamines and related compounds by capillary electrophoresis with post-column terbium complexation and sensitized luminescence detection, *Anal.Chem.*, **1997**, *69*, 4010–4016.

Levorphanol

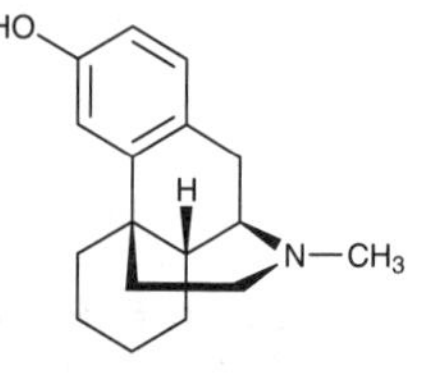

Molecular formula: $C_{17}H_{23}NO$
Molecular weight: 257.38
CAS Registry No.: 77-07-6, 5985-38-6 (tartrate dihydrate), 125-72-4 (tartrate)
Merck Index (12th ed.): 5496
Lednicer: 1 293

SAMPLE
Matrix: solutions

CAPILLARY ELECTROPHORESIS
Capillary: 57 cm × 75 μm fused-silica (Polymicro Technologies)
Capillary preparation: Before each run rinse capillary with running buffer for 2 min. After each run rinse capillary with 100 mM NaOH then with water.
Capillary temperature: 25
Running buffer: 22.5 mM Na_2HPO_4 containing 10 mM sodium tetraborate, pH 9.1-9.2
Injection: Pressure injection at 0.5 psi for 5 s (30 nL).
Detector: UV 214
Migration time: 8.5
Voltage: 15 kv
Model: Beckman P/ACE 2200

OTHER SUBSTANCES
Simultaneous: honkiol, magnolol

REFERENCE
Chou,C.Y.C.; Tsai,T.H.; Lin,M.F.; Chen,C.F. Simultaneous determination of honokiol and magnolol in magnolia officinalis by capillary zone electrophoresis, *J.Liq.Chromatogr.Rel.Technol.*, **1996**, *19*, 1909–1915.

SAMPLE
Matrix: urine
Sample preparation: Condition a 1 g BakerBond SPE Octadecyl HC SPE cartridge with 12 mL MeOH, 12 mL water, and 12 mL 15 mM pH 9.5 borate buffer. 5 mL Urine + 2.5 μg ethylmorphine + 1 mL concentrated HCl, heat at 120° for 1 h, cool to 4° in an ice bath, neutralize with 3 mL 5 M NaOH, cool, extract 3 times with 10 mL aliquots of diethyl ether. Combine the organic layers and add 20 μL glacial acetic acid, evaporate to dryness under a stream of nitrogen at 60°, reconstitute the residue in 5 mL 15 mM pH 9.5 sodium tetraborate buffer, sonicate at 70° for 3 min, cool to 25°, add to the SPE cartridge, wash with 4 mL 15 mM pH 9.5 borate buffer, elute with 15 mL MeOH:water:glacial acetic acid 80:19:1, evaporate to dryness at 115° under a stream of nitrogen, reconstitute with 100 μL running buffer, sonicate, inject an aliquot.

CAPILLARY ELECTROPHORESIS
Capillary: 100 cm × 50 μm fused-silica (50 cm to detector) (Isco)
Capillary preparation: At least every 14 samples rinse with 200 μL 1 M NaOH, 200 μL water, and 200 μL running buffer. Before use fill capillary with 1 M NaOH and let stand for 1 h, fill with 100 mM NaOH and let stand for 1 h, wash with water, fill with running buffer.
Capillary temperature: 30
Running buffer: Isopropanol:buffer 20:80 (Buffer 50 mM sodium tetraborate containing 50 mM sodium dodecyl sulfate and 60 mM β-cyclodextrin, adjusted to pH 9.05 with 50 mM HCl.)
Injection: Inject under vacuum at 2200 kPa/s
Detector: UV 200
Migration time: 22
Internal standard: ethylmorphine (25)
Voltage: 30 kV
Model: Isco Model 3140
Limit of detection: 20 ppb

OTHER SUBSTANCES
Extracted: dextromethorphan, dextrophan, levomethorphan

KEY WORDS
SPE; chiral

REFERENCE
Aumatell,A.; Wells,R.J. Chiral differentiation of the optical isomers of racemethorphan and racemorphan in urine by capillary zone electrophoresis, *J.Chromatogr.Sci.*, **1993**, *31*, 502–508.

SAMPLE
Matrix: urine

Sample preparation: 1 mL Urine + 1 mL 200 mM pH 5.4 sodium acetate buffer + 20 µL β-glucuronidase/arylsulfatase (Helix pomatia, Boehringer Mannheim), heat at 37° overnight, inject directly or dilute with water before injection.

CAPILLARY ELECTROPHORESIS
Capillary: 105 cm × 75 µm fused-silica (68-70 cm to detector)
Capillary preparation: Before each injection rinse capillary with 100 mM NaOH for 3-5 min and with running buffer for 5-8 min.
Running buffer: 140 mM pH 9.4 sodium borate buffer
Injection: Vacuum injection for 1 s
Detector: UV 200
Migration time: 20.5
Voltage: 14 kV
Current: 117 µA
Model: Europhor Prime Vision IV
Limit of detection: 500 mg/mL

OTHER SUBSTANCES
Extracted: dextromethorphan

REFERENCE
Caslavska,J.; Hufschmid,E.; Theurillat,R.; Desiderio,C.; Wolfisberg,H.; Thormann,W. Screening for hydroxylation and acetylation polymorphisms in man via simultaneous analysis of urinary metabolites of mephenytoin, dextromethorphan and caffeine by capillary electrophoretic procedures, *J.Chromatogr.B*, **1994**, *656*, 219–231.

Levothyroxine sodium

Molecular formula: $C_{15}H_{10}I_4NNaO_4$
Molecular weight: 798.86
CAS Registry No.: 55-03-8, 25416-65-3 (hydrate), 51-48-9 (free acid)
Merck Index (12th ed.): 5497
Lednicer: 1 97

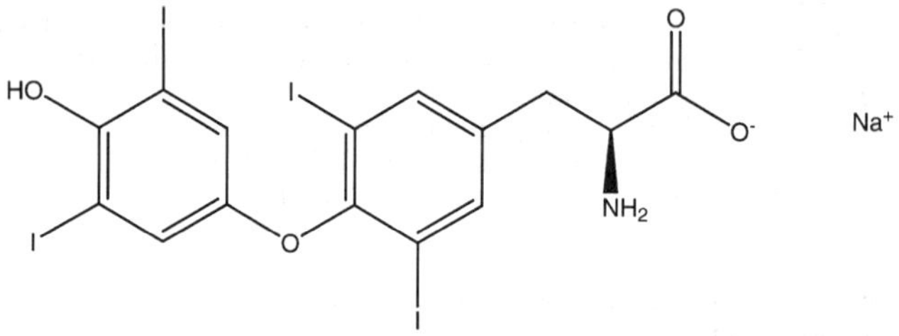

SAMPLE
Matrix: solutions
Sample preparation: Inject an aliquot of a 10 µg/mL solution in pH 7.4 phosphate buffer.

CAPILLARY ELECTROPHORESIS
Capillary: 27 cm × 75 µm uncoated silica (20 cm to detector)
Capillary preparation: Equilibrate overnight with running buffer.
Running buffer: 100 mM pH 2.5 phosphate buffer
Injection: Pressure injection for 4 s (20 nL)
Detector: UV 214
Migration time: 9.2
Model: Beckman 2100 P/ACE
Limit of detection: 500 ng/mL

OTHER SUBSTANCES
Simultaneous: 3,5-diiodo-L-thyronine, liothyronine

KEY WORDS
protect from light; analysis at constant power (2.0 W)

REFERENCE
Dalal,P.S.; Albuquerque,P.; Bhagat,H.R. Development and standardization of levothyroxine analysis by high-performance capillary electrophoresis, *Anal.Biochem.*, **1993**, *211*, 34–36.

SAMPLE
Matrix: solutions

CAPILLARY ELECTROPHORESIS
Capillary: 27 cm × 75 μm
Capillary preparation: Before each run rinse capillary with 100 mM NaOH for 30 s and with running buffer for 30 s. Condition new capillaries by rinsing with 100 mM NaOH for 20 min.
Capillary temperature: 30
Running buffer: 15 mM Sodium borate
Injection: Pressure injection of sample at 25 mbar for 1 s followed by running buffer at 25 mbar for 1 s.
Detector: UV 200
Migration time: 3.12
Internal standard: aminobenzoic acid (3.5)
Voltage: 6.5 kV
Model: Beckman

OTHER SUBSTANCES
Simultaneous: aspirin, bacitracin, beclomethasone, benzoic acid, ceftizoxime, ceftriaxone, cefuroxime, cephalothin, cromolyn, embonic acid, epoprostenol, glyburide (glibenclamide), levothyroxine, nedocromil, nystatin, omeprazole, prednisolone, warfarin, zidovudine

REFERENCE
Altria,K.D.; Bryant,S.M.; Hadgett,T.A. Validated capillary electrophoresis method for the analysis of a range of acidic drugs and excipients, *J.Pharm.Biomed.Anal.*, **1997**, *15*, 1091–1101.

Lidocaine

Molecular formula: $C_{14}H_{22}N_2O$
Molecular weight: 234.34
CAS Registry No.: 137-58-6, 6108-05-0 (HCl monohydrate), 73-78-9 (HCl)
Merck Index (12th ed.): 5505
Lednicer: 1 16

SAMPLE
Matrix: blood, urine
Sample preparation: Adjust pH of 2 mL urine or plasma to 10.5 with aqueous NaOH, extract gently with chloroform:isopropanol 90:10. Remove the organic layer and evaporate it to dryness under a stream of nitrogen, reconstitute the residue in 50 (urine) or 75 (plasma) μL running buffer, filter, inject an aliquot.

CAPILLARY ELECTROPHORESIS
Capillary: 60 cm × 75 μm AccuSep uncoated silica (52.5 cm to detector) (Waters)
Capillary preparation: At the beginning of each day purge with 500 mM KOH for 5 min, with water for 5 min, and with running buffer for 10 min.
Running buffer: 50 mM NaH_2PO_4 adjusted to pH 2.35 with phosphoric acid
Injection: Hydrostatic injection at 15 cm
Detector: UV 214
Migration time: 6.81
Voltage: 22 kV

Current: 135-145 μA
Model: Waters Quanta 4000

OTHER SUBSTANCES
Extracted: acepromazine, amphetamine, benzocaine, brompheniramine, butacaine, codeine, diazepam, doxapram, medazepam, methamphetamine, methapyrilene, methaqualone, phenmetrazine, procaine, tetrahydrozoline
Simultaneous: meclizine

KEY WORDS
plasma

REFERENCE
Chee,G.L.; Wan,T.S.M. Reproducible and high-speed separation of basic drugs by capillary zone electrophoresis, *J.Chromatogr.*, **1993**, *612*, 172–177.

SAMPLE
Matrix: bulk
Sample preparation: Prepare a 3 mg/mL solution in 3 mg/mL pholcodine in 10 mM HCl, make up to 25 mL with 10 mM HCl, mix thoroughly, filter (0.45 μm cellulose acetate)

CAPILLARY ELECTROPHORESIS
Capillary: 65 cm × 75 μm fused-silica (40 cm to detector) (Isco)
Capillary preparation: Flush with running buffer for 2 min between runs. Before each batch of samples fill capillary with 100 mM NaOH and let stand for 10 min, wash with water for 5 min, wash with 100 mM HCl for 5 min, wash with water for 5 min, fill with running buffer (all at vacuum level 4). Clean each week by washing with 100 mM HCl for 10 min, wash with water, fill with running buffer.
Capillary temperature: 30
Running buffer: MeCN:buffer 7.5:92.5. Buffer was 10 mM KH_2PO_4 containing 10 mM sodium tetraborate and 50 mM cetyltrimethylammonium bromide, pH 8.6.)
Injection: Injection by vacuum 5.0 kPa.s (4 nL)
Detector: UV 230
Migration time: 9.4
Internal standard: pholcodine (7.4)
Voltage: -15 kV
Model: Isco Model 3140

OTHER SUBSTANCES
Simultaneous: cocaine, benzocaine, benzoylecgonine, trans-cinnamoylcocaine, procaine, tetracaine

REFERENCE
Trenerry,V.C.; Robertson,J.; Wells,R.J. The determination of cocaine and related substances by micellar electrokinetic capillary chromatography, *Electrophoresis*, **1994**, *15*, 103–108.

SAMPLE
Matrix: bulk
Sample preparation: Dilute with 400 μg/mL n-propyl p-hydroxybenzoate in running buffer

CAPILLARY ELECTROPHORESIS
Capillary: 27 cm × 50 μm fused-silica (20 cm to detector)
Capillary preparation: Before each run rinse with running buffer at high pressure.
Capillary temperature: 30
Running buffer: MeCN:water 15:85 containing 40 mM sodium dodecyl sulfate, 8.5 mM sodium borate, and 8.5 mM sodium phosphate, pH 8.5
Injection: Inject at high pressure for 1 s
Detector: UV 214
Migration time: 2.1
Internal standard: n-propyl p-hydroxybenzoate (1.8)
Voltage: 20 kV
Model: Beckman P/ACE System 2100

OTHER SUBSTANCES
Simultaneous: acetaminophen, acetylcodeine, O-acetylmorphine, aspirin, caffeine, cocaine, codeine, diamorphine, diphenhydramine, hydromorphone, isoniacinamide, methaqualone, morphine, niacinamide, noscapine, papaverine, phenacetin, phenobarbital, phenylpropanolamine, procaine, quinine, salicylic acid, strychnine, thebaine

REFERENCE
Walker,J.A.; Krueger,S.T.; Lurie,I.S.; Marché,H.L.; Newby,N. Analysis of heroin drug seizures by micellar electrokinetic capillary chromatography (MECC), *J.Forensic Sci.*, **1995**, *40*, 6–9.

SAMPLE
Matrix: bulk
Sample preparation: Dissolve sample in 2 mg/mL histamine diphosphate in 10 mM HCl, inject an aliquot.

CAPILLARY ELECTROPHORESIS
Capillary: 48.5 cm × 50 μm fused-silica (40 cm to detector)
Capillary preparation: Flush capillary with running buffer for 2 min before each run. Replenish buffer in vials after every 10 injections. Condition new capillaries by flushing with 1 M NaOH for 10 min, with 100 mM NaOH for 10 min, with water for 10 min, and with running buffer for 10 min.
Capillary temperature: 40
Running buffer: 75 mM NaH_2PO_4 containing 75 mM sodium borate, adjusted to pH 8.5 with 2 M NaOH
Injection: Pressure injection at 150 mbar for 3 s.
Detector: UV 230
Migration time: 7
Internal standard: histamine diphosphate (5)
Voltage: 10 kV
Current: 66-70 μA
Model: Hewlett Packard 3D CE
Limit of detection: 5 μg/mL (S/N = 4:1)

OTHER SUBSTANCES
Simultaneous: acetaminophen, benzocaine, cis-cinnamoylcocaine, trans-cinnamoylcocaine, cocaine, ephedrine, phenylpropanolamine, procaine, tetracaine

REFERENCE
Krawczeniuk,A.S.; Bravenec,V.A. Quantitative determination of cocaine in illicit powders by free zone capillary electrophoresis, *J.Forensic Sci.*, **1998**, *43*, 738–743.

SAMPLE
Matrix: perfusate
Sample preparation: Inject an aliquot of perfusate (pH 6.5 Ringer's).

CAPILLARY ELECTROPHORESIS
Capillary: 70 cm × 50 μm untreated fused-silica (Polymicro Technologies)
Running buffer: 200 mM sodium tetraborate buffer
Injection: Injection by vacuum for 1 s
Detector: UV 200
Migration time: 7.6
Voltage: 20 kV

OTHER SUBSTANCES
Simultaneous: cocaine, procaine

REFERENCE
Hernandez,L.; Guzman,N.A.; Hoebel,B.G. Bidirectional microdialysis in vivo shows differential dopaminergic potency of cocaine, procaine and lidocaine in the nucleus accumbens using capillary electrophoresis for calibration of drug outward diffusion, *Psychopharmacology (Berl)*, **1991**, *105*, 264–268.

SAMPLE
Matrix: solutions
Sample preparation: Inject an aliquot of a 100 μg/mL solution in 50 mM sodium dodecyl sulfate.

CAPILLARY ELECTROPHORESIS
Capillary: 57 cm × 75 μm fused-silica (50 cm to detector) (Beckman)
Capillary preparation: Rinse capillary with 100 mM NaOH and running buffer between each run.
Capillary temperature: 30
Running buffer: Isopropanol:100 mM pH 8.1 borate buffer containing 50 mM sodium dodecyl sulfate 10:90
Injection: Pressure injection for 5 s
Detector: UV 200
Migration time: 9.3
Voltage: +25 kV
Model: Beckman P/ACE 5510

OTHER SUBSTANCES
Simultaneous: amiodarone, bretylium, disopyramide, phenytoin, propafenone, quinidine

REFERENCE
Bretnall,A.E.; Clarke,G.S. Investigation and optimisation of the use of organic modifiers in micellar electro-kinetic chromatography, *J.Chromatogr.A*, **1995**, *716*, 49–55.

SAMPLE
Matrix: solutions

CAPILLARY ELECTROPHORESIS
Capillary: 57 cm × 75 μm fused-silica (50 cm to detector) (Beckman)
Capillary preparation: Before each run rinse capillary with 100 mM NaOH and running buffer.
Capillary temperature: 30
Running buffer: Acetone:100 mM pH 8.1 borate buffer containing 50 mM sodium dodecyl sulfate 15:85 (A) or isopropanol:100 mM pH 8.1 borate buffer containing 50 mM sodium dodecyl sulfate 10:90 (B)
Injection: Pressure injection for 5-10 s.
Detector: UV 200
Migration time: 8 (A), 11.5 (B)
Voltage: 25 kV
Model: Beckman P/ACE 5510

OTHER SUBSTANCES
Simultaneous: acebutolol, amiodarone, atenolol, bretylium, captopril, diltiazem, disopyramide, lisinopril, metoprolol, nicardipine, nifedipine, phenytoin, pindolol, propafenone, propranolol, quinidine, sotalol, timolol, p-toluenesulfonic acid, verapamil

KEY WORDS
only running buffer A separates all compounds

REFERENCE
Bretnall,A.E.; Clarke,G.S. Selectivity of capillary electrophoresis for the analysis of cardiovascular drugs, *J.Chromatogr.A*, **1996**, *745*, 145–154.

SAMPLE
Matrix: solutions
Sample preparation: Inject an aliquot of a 1-50 μg/mL solution in water.

CAPILLARY ELECTROPHORESIS
Capillary: 55 cm × 50 μm fused-silica (35 cm to detector) (J&W)
Capillary preparation: Flush with running buffer for 5 min before each run. Periodically wash with 100 mM NaOH and flush extensively with running buffer.

Running buffer: MeOH:25 mM pH 9.24 borate buffer 20:80 containing 100 mM sodium dodecyl
   sulfate (A) or 50 mM pH 2.35 phosphate buffer (B) or 50 mM pH 9.24 borate buffer (C)
Injection: Injection of 5 μL using a split-flow injector ratio of 1:800.
Detector: UV 200
Migration time: 27.02 (A), 10.36 (B), 7.07 (C)
Voltage: 20 kV (A, B) or 12 kV (C)
Current: <60-80 μA
Model: ISCO Model 3850

OTHER SUBSTANCES
Also analyzed: acetylcodeine, amphetamine, barbital, caffeine, codeine, cocaine, diamorphine,
   diazepam, flunitrazepam, monoacetylmorphine, morphine, nalorphine, narceine, noscapine, papaverine, pentobarbital, procaine, tetracaine, thebaine

KEY WORDS
all compounds were separated with running buffer A; some peaks overlapped with running buffers
   B and C.

REFERENCE
Tagliaro,F.; Smith,F.P.; Turrina,S.; Equisetto,V.; Marigo,M. Complementary use of capillary zone electrophoresis
   and micellar electrokinetic capillary chromatography for mutual confirmation of results in forensic drug
   analysis, *J.Chromatogr.A*, **1996**, *735*, 227–235.

SAMPLE
Matrix: solutions

CAPILLARY ELECTROPHORESIS
Capillary: 45 cm × 50 μm fused-silica (37.5 cm to detector) (Polymicro Technologies)
Capillary preparation: Between each run rinse with 100 mM HCl for 2 min and with running
   buffer for 2 min. At the start of each day condition capillary with 1 M HCl for 10 min. Before
   the first use rinse capillary with 1 M NaOH for 30 min and with water for 30 min.
Capillary temperature: 25
Running buffer: MeCN:30 mM pH 2.40 ethanesulfonic acid 20:80
Injection: Injection at 4 kV for 3 s.
Detector: UV 214
Migration time: 5.22
Voltage: 20 kV
Model: Waters Quanta 4000E

OTHER SUBSTANCES
Simultaneous: chloramphenicol, imipramine, phenylpropanolamine, procainamide, quinidine,
   trimethoprim

REFERENCE
Ding,W.; Fritz,J.S. Separation of neutral compounds and basic drugs by capillary electrophoresis in acidic
   solution using laurylpoly(oxyethylene) sulfate as an additive, *Anal.Chem.*, **1998**, *70*, 1859–1865.

Lincomycin

Molecular formula: $C_{18}H_{34}N_2O_6S$
Molecular weight: 406.54
CAS Registry No.: 154-21-2, 7179-49-9 (HCl monohydrate), 859-18-7 (HCl)
Merck Index (12th ed.): 5525

SAMPLE
Matrix: bulk

Sample preparation: Prepare a solution in 15 mg/mL sucrose, inject an aliquot.

CAPILLARY ELECTROPHORESIS
Capillary: 70 cm × 50 μm fused silica (45 cm to detector) (Polymicro Technologies or Isco)
Capillary preparation: Rinse with running buffer for 3 min between runs.
Capillary temperature: 34
Running buffer: 75 mM pH 9 Sodium borate buffer
Injection: Vacuum injection at 25 kPa s
Detector: UV 195
Migration time: 8.7
Internal standard: sucrose (9.4)
Voltage: +15 kV
Model: Isco Model 3140 Electropherograph

OTHER SUBSTANCES
Simultaneous: clindamycin, clindamycin phosphate

REFERENCE
Flurer,C.L.; Wolnik,K.A. Chemical profiling of pharmaceuticals by capillary electrophoresis in the determination of drug origin, *J.Chromatogr.A*, **1994**, *674*, 153–163.

Liothyronine

Molecular formula: $C_{15}H_{12}I_3NO_4$
Molecular weight: 650.98
CAS Registry No.: 6893-02-3, 66-06-1 (Na salt)
Merck Index (12th ed.): 5535
Lednicer: 1 97

SAMPLE
Matrix: solutions
Sample preparation: Inject an aliquot of a 10 μg/mL solution in pH 7.4 phosphate buffer.

CAPILLARY ELECTROPHORESIS
Capillary: 27 cm × 75 μm uncoated silica (20 cm to detector)
Capillary preparation: Equilibrate overnight with running buffer.
Running buffer: 100 mM pH 2.5 phosphate buffer
Injection: Pressure injection for 4 s (20 nL)
Detector: UV 214
Migration time: 7.5
Model: Beckman 2100 P/ACE

OTHER SUBSTANCES
Simultaneous: 3,5-diiodo-L-thyronine, levothyroxine

KEY WORDS
protect from light; analysis at constant power (2.0 W)

REFERENCE
Dalal,P.S.; Albuquerque,P.; Bhagat,H.R. Development and standardization of levothyroxine analysis by high-performance capillary electrophoresis, *Anal.Biochem.*, **1993**, *211*, 34–36.

Lisinopril

Molecular formula: $C_{21}H_{31}N_3O_5$
Molecular weight: 405.49
CAS Registry No.: 76547-98-3 (anhydrous), 83915-83-7 (dihydrate)
Merck Index (12th ed.): 5540
Lednicer: 4 83

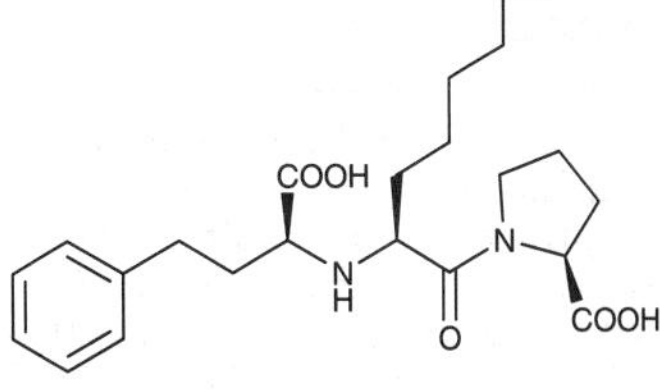

SAMPLE
Matrix: solutions
Sample preparation: Prepare a 100 µg/mL solution in water, inject an aliquot.

CAPILLARY ELECTROPHORESIS
Capillary: 72 cm × 50 µm fused-silica (50 cm to dctcctor)
Capillary preparation: Flush with running buffer between injections. Prepare capillaries by washing with 100 mM NaOH, water, and running buffer.
Capillary temperature: 30
Running buffer: MeOH:25 mM phosphate buffer containing 50 mM sodium cholate 55:45, pH 9.55
Injection: Hydrodynamic injection at 12.5 cm Hg for 2 s
Detector: UV 210
Migration time: 22.91
Voltage: 30 kV
Model: Applied Biosystems Model 270-HT

OTHER SUBSTANCES
Simultaneous: RSS diastereomer

REFERENCE
Qin,X.-Z.; Nguyen,D.-S.T.; Ip,D.P. Separation of lisinopril and its RSS diastereoisomer by micellar electrokinetic chromatography, *J.Liq.Chromatogr.*, **1993**, *16*, 3713–3734.

SAMPLE
Matrix: solutions

CAPILLARY ELECTROPHORESIS
Capillary: 70 cm × 100 µm fused-silica (63 cm to detector) (Chrompack or Polymicro Technologies)
Capillary preparation: Before injection fill capillary with buffer for 2 min. At the start of each day wash with 100 mM NaOH for 5 min, with water for 10 min, and with running buffer for 10 min. At the end of each day wash with 100 mM NaOH for 5 min and with water for 10 min.
Capillary temperature: 15
Running buffer: 20 mM pH 9.5 Sodium borate buffer containing 20 mM sodium dodecyl sulfate
Injection: Hydrodynamic injection for 2.5 s.
Detector: UV 210
Migration time: 8
Voltage: 20 kV
Model: Spectra Physics SpectraPhoresis 1000

OTHER SUBSTANCES
Simultaneous: chlorothiazide, hydrochlorothiazide

REFERENCE
Thomas,B.R.; Fang,X.G.; Chen,X.; Tyrrell,R.J.; Ghodbane,S. Validated micellar electrokinetic chromatography method for quality control of the drug substances hydrochlorothiazide and chlorothiazide, *J.Chromatogr.B*, **1994**, *657*, 383–394.

SAMPLE
Matrix: solutions

CAPILLARY ELECTROPHORESIS
Capillary: 57 cm × 75 μm fused-silica (50 cm to detector) (Beckman)
Capillary preparation: Before each run rinse capillary with 100 mM NaOH and running buffer.
Capillary temperature: 30
Running buffer: Acetone:100 mM pH 8.1 borate buffer containing 50 mM sodium dodecyl sulfate 15:85 (A) or 50 mM pH 8.2 borate buffer containing 100 mM sodium dodecyl sulfate (B)
Injection: Pressure injection for 5-10 s.
Detector: UV 200
Migration time: 4.5 (A), 5 (B)
Voltage: 25 kV
Model: Beckman P/ACE 5510

OTHER SUBSTANCES
Simultaneous: acebutolol, amiodarone, atenolol, bretylium, captopril, diltiazem, disopyramide, lidocaine, metoprolol, nicardipine, nifedipine, phenytoin, pindolol, propafenone, propranolol, quinidine, sotalol, timolol, p-toluenesulfonic acid, verapamil

KEY WORDS
only running buffer A separates all compounds

REFERENCE
Bretnall,A.E.; Clarke,G.S. Selectivity of capillary electrophoresis for the analysis of cardiovascular drugs, *J.Chromatogr.A*, **1996**, *745*, 145–154.

Litracen

Molecular formula: $C_{20}H_{23}N$
Molecular weight: 277.41
CAS Registry No.: 5118-30-9

SAMPLE
Matrix: solutions

CAPILLARY ELECTROPHORESIS
Capillary: 55.5 cm × 50 μm fused-silica (55.5 cm to detector) (Polymicro Technologies)
Capillary temperature: 25
Running buffer: MeCN:MeOH 50:50 containing 25 mM ammonium acetate and 100 mM sodium acetate
Injection: Pressure injection at 5 kPa for 3 s.
Detector: UV 214
Migration time: 7.1
Voltage: 25 kV
Current: 23 μA
Model: Hewlett Packard HP [3D]CE

OTHER SUBSTANCES
Simultaneous: amitriptyline, maprotiline, nortriptyline, protriptyline

REFERENCE
Bjornsdottir,I.; Tjornelund,J.; Hansen,S.H. Nonaqueous capillary electrophoresis in pharmaceutical analysis, *J.Capillary Electrophor.*, **1996**, *3*, 83–87.

Lomefloxacin

Molecular formula: $C_{17}H_{19}F_2N_3O_3$
Molecular weight: 351.35
CAS Registry No.: 98079-51-7, 98079-52-8 (HCl)
Merck Index (12th ed.): 5592
Lednicer: 5 125

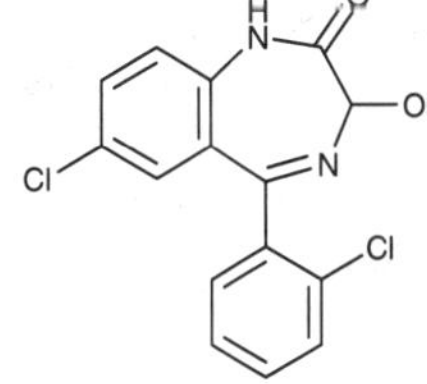

SAMPLE
Matrix: solutions
Sample preparation: Inject an aliquot of a solution in MeOH.

CAPILLARY ELECTROPHORESIS
Capillary: 59 cm × 50 µm fused-silica (43 cm to detector) (Polymicro Technologies)
Capillary preparation: Before each analysis flush with water for 3 min, with 200 mM NaOH
   for 3 min, with water for 3 min, and with running buffer for 4 min. Flush new capillaries with
   1 M NaOH at 2000 mbar for 10 min then with 200 mM NaOH for 10 min.
Capillary temperature: 23
Running buffer: MeCN:buffer 28:72, pH 7.3 (Buffer was 32 mM sodium borate containing 18
   mM NaH_2PO_4, 39 mM sodium cholate, and 8 mM sodium heptanesulfonate.)
Injection: Hydrodynamic injection at 40 mbar for 6 s.
Detector: UV 260
Migration time: 4
Voltage: 30 kV
Model: Lauer Labs Prince

OTHER SUBSTANCES
Simultaneous: ciprofloxacin, enoxacin, flumequine, nalidixic acid, norfloxacin, ofloxacin, oxolinic
   acid, pefloxacin, pipemidic acid, piromidic acid, rosoxacin, sparfloxacin

REFERENCE
Sun,S.-W.; Chen,L.-Y. Optimization of capillary electrophoretic separation of quinolone antibacterials using the
   overlapping resolution mapping scheme, *J.Chromatogr.A*, **1997**, *766*, 215–224.

Lorazepam

Molecular formula: $C_{15}H_{10}Cl_2N_2O_2$
Molecular weight: 321.16
CAS Registry No.: 846-49-1
Merck Index (12th ed.): 5609
Lednicer: 1 368

SAMPLE
Matrix: solutions

CAPILLARY ELECTROPHORESIS
Capillary: 72 cm × 50 µm silica (50 cm to detector) (Applied Biosystems)
Capillary preparation: Before each run wash capillary with 100 mM NaOH for 3 min, wash
   with running buffer for 3 min, aspirate neutral marker solution (2 drops DMSO in 10 mL
   water) for 1 s, place capillary end in buffer vial for 5 s, inject sample. Wash capillary at the
   beginning of each day by passing 1 M NaOH through for 20 min.
Capillary temperature: 30
Running buffer: MeCN:buffer 10:90 (Buffer was 30 mM pH 9.3 borate buffer containing 30 mM
   sodium dodecyl sulfate.)
Injection: Inject by applying vacuum for 1 s (2-3 nL)

Detector: UV 200
Migration time: 14.8
Voltage: +30 kV
Current: 55 μA
Model: Applied Biosystems Model 270A

OTHER SUBSTANCES
Simultaneous: chlordiazepam, desmethylchlordiazepam, desmethyldiazepam, diazepam, oxazepam, nitrazepam

REFERENCE
Evenson,M.A.; Wiktorowicz,J.E. Automated capillary electrophoresis applied to therapeutic drug monitoring, *Clin.Chem.*, **1992**, *38*, 1847–1852.

SAMPLE
Matrix: solutions
Sample preparation: Prepare a 1-50 μg/mL solution in MeOH:water 20:80, inject an aliquot.

CAPILLARY ELECTROPHORESIS
Capillary: 44 cm × 50 μm fused-silica
Capillary preparation: After each run wash capillary with water for 2 min and buffer for 3 min. At the beginning of each day equilibrate capillary with running buffer for 10 min. Condition a new capillary by washing with 1 M NaOH, 100 mM NaOH, water, and running buffer.
Capillary temperature: 25
Running buffer: MeOH:buffer 20:80 (Buffer was 75 mm pH 9.0 Glycine containing 250 mM triethanolamine and 25 mM sodium dodecyl sulfate.)
Injection: Hydrodynamic injection for 5 s
Detector: UV 235
Migration time: 10.6
Voltage: 25 kV
Current: about 15 μA
Model: Spectraphysics Spectraphoresis 1000 CE
Limit of quantitation: 700 ng/mL
Limit of detection: 200 ng/mL

OTHER SUBSTANCES
Simultaneous: bromazepam, brotizolam, clobazam, clonazepam, lormetazepam, nitrazepam, nordazepam, oxazepam, temazepam

REFERENCE
Bechet,I.; Fillet,M.; Hubert,P.; Crommen,J. Determination of benzodiazepines by micellar electrokinetic chromatography, *Electrophoresis*, **1994**, *15*, 1316–1321.

SAMPLE
Matrix: solutions

CAPILLARY ELECTROPHORESIS
Capillary: 57 cm × 75 μm fused-silica (50 cm to detector)
Capillary preparation: Before each run flush with running buffer for 2 min, equilibrate for 5 min.
Capillary temperature: 33
Running buffer: MeCN:buffer 10:90 (Buffer was 20 mM pH 10.2 borate containing 50 mM sodium cholate.)
Injection: Pressure injection for 2 s
Detector: UV 214
Migration time: 7.2
Voltage: 20 kV
Model: Beckman P/ACE 2100

OTHER SUBSTANCES
Simultaneous: lormetazepam, oxazepam, temazepam

REFERENCE

Boonkerd,S.; Detaevernier,M.R.; Michotte,Y.; Vindevogel,J. Suppression of chiral recognition of 3-hydroxy-1,4-benzodiazepines during micellar electrokinetic capillary chromatography with bile salts, *J.Chromatogr.A*, **1995**, *704*, 238–241.

SAMPLE

Matrix: solutions
Sample preparation: Inject an aliquot of a 5 μg/mL solution in running buffer.

CAPILLARY ELECTROPHORESIS

Capillary: 87 cm × 75 μm fused-silica (80 cm to detector) (Beckman)
Capillary preparation: Between runs rinse capillary with running buffer for 2 min then equilibrate for 5 min.
Capillary temperature: 35
Running buffer: 20 mM pH 7 Buffer containing 15 mM sodium cholate and 35 mM sodium deoxycholate (Buffer was 20 mM sodium borate adjusted to pH 7.0 with 20 mM NaH_2PO_4.)
Injection: Pressure injection for 2 s.
Detector: UV 214
Migration time: 20.5
Voltage: 20 kV
Model: Beckman P/ACE 2100

OTHER SUBSTANCES

Simultaneous: bromazepam, chlordiazepoxide, clobazam, clonazepam, diazepam, flunitrazepam, flurazepam, halazepam, lormetazepam, nitrazepam, nordazepam, temazepam

REFERENCE

Boonkerd,S.; Detaevernier,M.R.; Vindevogel,J.; Michotte,Y. Migration behaviour of benzodiazepines in micellar electrokinetic chromatography, *J.Chromatogr.A*, **1996**, *756*, 279–286.

SAMPLE

Matrix: urine
Sample preparation: Condition a 130 mg Bond Elut Certify SPE cartridge with 2 mL MeOH and 2 mL 100 mM pH 6 phosphate buffer, do not allow to go dry. Mix urine with an equal amount of 200 mM pH 5.4 buffer, add 20 μL β-glucuronidase/arylsulfatase (Helix pomatia, Boehringer Mannheim) for each 1 mL of urine, heat at 37° for 4 h. 5 (or 10) mL Urine + 2 (or 4) mL 100 mM pH 6 phosphate buffer, add to the SPE cartridge, wash with 1 mL MeOH:100 mM phosphate buffer 20:80, wash with 1 mL 1 M acetic acid, wash with 1 mL hexane, suck dry briefly, elute with 4 mL dichloromethane. Evaporate the eluate to dryness under a stream of nitrogen at room temperature, reconstitute the residue in 50-100 μL running buffer, inject an aliquot.

CAPILLARY ELECTROPHORESIS

Capillary: 105 cm × 75 μm fused-silica (68 cm to detector)
Capillary preparation: Before each run rinse capillary with 100 mM NaOH for 3.5 min and with buffer for 5 min.
Running buffer: Isopropanol:buffer 5:95 (Buffer was 10 mM pH 9.2-9.3 Na_2HPO_4 containing 6 mM sodium borate and 75 mM sodium dodecyl sulfate.)
Injection: Vacuum injection for 1 s
Detector: UV 235
Migration time: 33
Voltage: 30 kV
Current: 90 μA
Model: Europhor Prime Vision IV

OTHER SUBSTANCES

Extracted: oxazepam

KEY WORDS

SPE

REFERENCE

Schafroth,M.; Thormann,W.; Allemann,D. Micellar electrokinetic capillary chromatography of benzodiazepines in human urine, *Electrophoresis*, **1994**, *15*, 72–78.

Lypressin

Molecular formula: $C_{46}H_{65}N_{13}O_{12}S_2$
Molecular weight: 1056.23
CAS Registry No.: 50-57-7
Merck Index (12th ed.): 5661

SAMPLE

Matrix: solutions

CAPILLARY ELECTROPHORESIS

Capillary: 50 cm × 50 μm uncoated fused-silica
Running buffer: 20 mM pH 2.2 Phosphate buffer containing 90 mM hexanesulfonic acid
Injection: Electrokinetic injection at 8 kV for 10 s
Detector: UV 200
Voltage: 9 kV
Model: Bio-Rad HPE 100

OTHER SUBSTANCES

Simultaneous: felypressin, ornipressin, oxytocin

REFERENCE

Adhiambo Hagono,P.; Baeyens,W.R.G.; Van Der Weken,G. Capillary electrophoretic separation of some pharmaceutically important polypeptides, *Biomed.Chromatogr.*, **1995**, *9*, 291.

Lysozyme

Molecular weight: ca. 14400
CAS Registry No.: 9001-63-2
Merck Index (12th ed.): 5671

CAPILLARY ELECTROPHORESIS

Capillary: 66 cm × 50 μm fused-silica (44 cm to detector) (Herbei Yongnian Optical Fibre Factory, Herbei, China)
Capillary preparation: Flush with running buffer for 3 min between runs. Condition new capillaries with 1 M NaOH for 30 min, wash with water for 30 min, evaporate water at 100° under a 400 kPa pressure of nitrogen for 1 h, coat with gamma-glycidopropyltrimethoxysilane, heat at 90° for 1 h, aspirate reagent through the capillary, heat at 30° under 200 psi of nitrogen for 30 min, heat at 80° for 2 h, heat at 150° for 1 h, heat at 180° for 3 h, wash with water for 24 h. (Reagent was acetone containing 500 mg/mL bisphenol epoxy resins (Beijing Institute of Aeronautical Materials) and 140 mg/mL 4,4'-diaminodiphenylmethane.)
Running buffer: 70 mM pH 7.0 Phosphate buffer
Injection: Hydrodynamic injection at 10 cm for 5 s
Detector: UV 214

Migration time: 6
Voltage: 19 kV

OTHER SUBSTANCES
Simultaneous: chymotrypsinogen, cytochrome C, ribonuclease A

KEY WORDS
coated capillary

REFERENCE
Liu,Y.; Fu,R.; Gu,J. Epoxy resin coatings for capillary zone electrophoretic separation of basic proteins, *J.Chromatogr.A*, **1996**, *723*, 157–167.

SAMPLE
Matrix: solutions

CAPILLARY ELECTROPHORESIS
Capillary: 60 cm × 50 μm fused-silica (50 cm to detector)
Capillary preparation: Flush a new capillary with 400 mM NaOH for 0.5-1 h, flush with water until the effluent is neutral, flush with running buffer for 20 min.
Running buffer: 10 mM pH 4 Phosphate buffer containing 25 μg/mL zwitterionic fluorosurfactant $F(CF_2CF_2)_{3-8}CH_2CH(OCOCH_3)CH_2N^+(CH_3)_2CH_2CO_2^-$ (DuPont) and 50 μg/mL FC 134 (a cationic surfactant, $C_8F_{17}O_2NH(CH_2)_3N^+(CH_3)_3$ I⁻, 3M)
Injection: Electromigration at 10 kV for 10 s
Detector: UV 210
Migration time: 7
Voltage: -20 kV
Model: Laboratory constructed

OTHER SUBSTANCES
Simultaneous: myoglobin, ribonuclease

REFERENCE
Emmer,Å.; Roeraade,J. Performance of zwitterionic and cationic fluorosurfactants as buffer additives for capillary electrophoresis of proteins, *J.Liq.Chromatogr.*, **1994**, *17*, 3831–3846.

SAMPLE
Matrix: solutions
Sample preparation: Prepare a 1-3 mg/mL solution in water, inject an aliquot.

CAPILLARY ELECTROPHORESIS
Capillary: 37 cm × 75 μm fused-silica (30 cm to detector) (Quadrex)
Capillary preparation: Rinse with running buffer for 3 min between runs. Renew running buffer after 5 or 6 runs. Flush a new capillary with 500 mM NaOH for 30 min, with water for 10 min, with 100 mM HCl for 30 min, with 500 mM NaOH for 30 min, and with water for 10 min, then rinse with running buffer.
Capillary temperature: 25
Running buffer: 50 mM pH 2.5 phosphate buffer containing 60 mM triethylamine (ionic strength 123 mM)
Injection: Pressure injection of 0.5 psi for 1 s (9 nL)
Detector: UV 214
Migration time: 15
Voltage: 10 kV
Current: 160 μA
Model: Beckman P/ACE Model 2100

OTHER SUBSTANCES
Simultaneous: α-chymotrypsinogen, cytochrome c, ribonuclease A

REFERENCE
Corradini,D.; Cannarsa,G.; Fabbri,E.; Corradini,C. Effects of alkylamines on electroosmotic flow and protein migration behaviour in capillary electrophoresis, *J.Chromatogr.A*, **1995**, *709*, 127–134.

SAMPLE
Matrix: solutions
Sample preparation: Prepare a 300 µg/mL solution in water, inject an aliquot.

CAPILLARY ELECTROPHORESIS
Capillary: 50 cm × 50 µm coated fused-silica (30 cm to detector) (GL Sciences)
Capillary preparation: Coat capillary as follows. Treat with 1 M NaOH for 1 h, rinse with
water, dry at 110° by flushing with nitrogen for 6 h, pass thionyl chloride through using suction
for 10 min, seal at each end, heat at 70° for 6 h, open, suck 250 mM vinylmagnesium bromide
in THF into the capillary, seal at each end, heat at 70° for 6 h, open, rinse with THF for several
min, rinse with water, fill by suction with a solution containing 3% acrylamide, 0.1% ammo-
nium persulfate, and 0.1% tetramethylethylenediamine, heat at 28 ± 2° for 1 h, after 1 h wash
with water.
Running buffer: 50 mM pH 4.6 sodium phosphate buffer
Injection: Electrokinetic injection at 100 V/cm for 1 s
Detector: UV 214
Migration time: 50 mM pH 4.6 sodium phosphate
Internal standard: 5
Voltage: 400 V/cm
Model: Jasco CE-800

OTHER SUBSTANCES
Simultaneous: cytochrome, trypsinogen

KEY WORDS
coated capillary

REFERENCE
Nakatani,M.; Shibukawa,A.; Nakagawa,T. Chemical stability of polyacrylamide-coating on fused silica capillary,
Electrophoresis, **1995**, *16*, 1451–1456.

SAMPLE
Matrix: solutions
Sample preparation: Inject an aliquot of a 1 mg/mL solution in water.

CAPILLARY ELECTROPHORESIS
Capillary: 37 cm × 50 µm fused silica (30 cm to detector) (Polymicro Technologies)
Capillary temperature: 30
Running buffer: 500 mM pH 8.4 borate buffer containing 40 mM sodium salt of phytic acid
Injection: Pressure injection for 1 s
Detector: UV 200
Migration time: 4.5
Voltage: 5.0 kV
Model: Beckman P/ACE model 5000

OTHER SUBSTANCES
Simultaneous: α-chymotrypsin, cytochrome C, ribonuclease A, trypsinogen

REFERENCE
Okafo,G.N.; Vinther,A.; Kornfelt,T.; Camilleri,P. Effective ion-pairing for the separation of basic proteins in
capillary electrophoresis, *Electrophoresis*, **1995**, *16*, 1917–1921.

SAMPLE
Matrix: solutions
Sample preparation: Inject an aliquot of an aqueous solution.

CAPILLARY ELECTROPHORESIS
Capillary: 57 cm × 75 µm fused-silica (50 cm to detector) (Composite Metal Services, Worcester,
UK)
Capillary preparation: Between injections rinse with 100 mM NaOH for 3 min and with run-
ning buffer for 3 min.

Capillary temperature: 10
Running buffer: 50 mM pH 7 2-[N-morpholino]ethanesulfonic acid buffer containing 10 mM cetyl trimethylammonium bromide
Injection: Pressure injection at 0.5 psi nitrogen for 2 s.
Detector: UV 210
Migration time: 23
Voltage: -10 kV
Model: Beckman P/ACE 5510

OTHER SUBSTANCES
Simultaneous: α-chymotrypsinogen, cytochrome c, ribonuclease A, trypsinogen

REFERENCE
Cifuentes,A.; Rodríguez,M.A.; Garcia-Montelongo,F.J. Separation of basic proteins in free solution capillary electrophoresis: effect of additive, temperature and voltage, *J.Chromatogr.A*, **1996**, *742*, 257–266.

SAMPLE
Matrix: solutions

CAPILLARY ELECTROPHORESIS
Capillary: 60 cm × 50 μm coated fused-silica (45 cm to detector) (Polymicro Technologies)
Capillary preparation: Rinse with running buffer for 2 min between runs. Coat capillary as follows. Flush capillary with 1 M NaOH for 1 h, rinse with water for 1 h, dry at 180° with a flow of nitrogen overnight, fill the capillary with reagent using a syringe, let stand for 10 min, force out excess reagent using pressure, dry at 90° with a flow of nitrogen overnight, wash with acetone for 30 min, wash with MeOH for 30 min, rinse with water for 20 min, equilibrate with running buffer for 1 h before use. (Prepare reagent by mixing 500 μL 3-aminopropyltriethoxysilane, 375 μL EtOH, 54 μL water, and 100 μL 1.2 m HCl, stir at room temperature for 10 h before use.)
Capillary temperature: 22
Running buffer: 40 mM pH 3.9 Tris-HCl
Injection: Hydrodynamic injection at 5 cm for 2 s.
Detector: UV 210
Migration time: 18
Voltage: -10 kV
Model: Beckman P/ACE Model 2200

OTHER SUBSTANCES
Simultaneous: ribonuclease A, trypsinogen

KEY WORDS
coated capillary

REFERENCE
Guo,Y.; Imahori,G.A.; Colón,L.A. Hydrolytically stable amino-silica glass coating material for manipulation of the electroosmotic flow in capillary electrophoresis, *J.Chromatogr.A*, **1996**, *744*, 17–29.

SAMPLE
Matrix: solutions
Sample preparation: Inject an aliquot of a 1-3 mg/mL solution in running buffer:water 1:5.

CAPILLARY ELECTROPHORESIS
Capillary: 75 μm ID CElect-H275 C18 (Supelco)
Capillary preparation: Before injection rinse with 0.5% Brij-35 in water for 1 h.
Capillary temperature: 30
Running buffer: 10 mM pH 6.0 Sodium phosphate buffer containing 0.001: Brij-35
Injection: Hydrodynamic injection for 2 s
Detector: UV 214
Migration time: 10
Voltage: 30 kV
Model: ABI Model 270A-HT or Beckman Model P/ACE 2100

OTHER SUBSTANCES
Simultaneous: α-chymotrypsinogen A, ribonuclease

REFERENCE
Huang,M.; Mitchell,D.; Bigelow,M. Highly efficient protein separations in high-performance capillary electrophoresis, using hydrophilic coatings on polysiloxane-bonded columns, *J.Chromatogr.B*, **1996**, *677*, 77–84.

SAMPLE
Matrix: solutions
Sample preparation: Inject an aliquot of an aqueous solution.

CAPILLARY ELECTROPHORESIS
Capillary: 100 cm × 75 μm fused-silica (60 cm to detector) (Polymicro Technologies)
Capillary preparation: Condition with two column volumes of water, 1 M NaOH, water, 100 mM HCl, water, and running buffer.
Capillary temperature: 30 ± 0.5
Running buffer: 30 mM Ammonium formate containing 30 mM triethylamine, adjusted to pH 2.5 with trifluoroacetic acid
Injection: Vacuum injection of 60 nL.
Detector: UV 200
Migration time: 13
Voltage: 27 kV
Current: 98 μA
Model: ISCO Model 3140

OTHER SUBSTANCES
Simultaneous: α-lactalbumin, myoglobin, ribonuclease A, trypsinogen

REFERENCE
Lee,H.G.; Desiderio,D.M. Optimization of the capillary zone electrophoresis loading limit and resolution of proteins, using triethylamine, ammonium formate and acidic pH, *J.Chromatogr.B*, **1997**, *691*, 67–75.

SAMPLE
Matrix: solutions

CAPILLARY ELECTROPHORESIS
Capillary: 47 cm × 75 μm silica (40 cm to detector) (Polymicro Technologies)
Capillary preparation: Rinse at 20 psi with 100 mM NaOH for 1 min, with water for 2 min, and with running buffer for 2 min. Treat new capillaries with 100 mM NaOH for 10 min.
Capillary temperature: 25
Running buffer: 10 mM pH 7.2 Sodium phosphate buffer adjusted to a total ionic strength of 50 mM with NaCl, containing 2 mM Rewoteric AM CAS U (average MW 450) (Witco) (Rewoteric AM CAS U is a zwitterionic surfactant.)
Injection: Hydrodynamic injection at 0.5 psi for 3 s.
Detector: UV 214
Migration time: 11
Voltage: 15 kV
Model: Beckman P/ACE 2100

OTHER SUBSTANCES
Simultaneous: α-chymotrypsinogen

REFERENCE
Yeung,K.K.-C.; Lucy,C.A. Suppression of electroosmotic flow and prevention of wall adsorption in capillary zone electrophoresis using zwitterionic surfactants, *Anal.Chem.*, **1997**, *69*, 3435–3441.

Mannitol

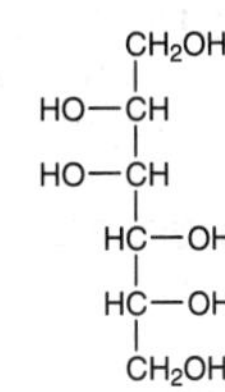

Molecular formula: $C_6H_{14}O_6$
Molecular weight: 182.17
CAS Registry No.: 69-65-8
Merck Index (12th ed.): 5788

SAMPLE
Matrix: solutions

CAPILLARY ELECTROPHORESIS
Capillary: 80 cm × 25 μm fused-silica (Polymicro Technologies)
Running buffer: 250 mM NaOH
Injection: Electromigration at 15 kV for 3 s
Detector: E, Bioanalytical Systems LC-4B, 100 μm Cu-disk electrode (Anal. Chem. 1993, 65, 3525), +0.60 V, Ag/AgCl reference electrode, Pt wire counter electrode
Migration time: 40
Voltage: 15 kV
Model: Laboratory constructed
Limit of detection: 4 fmole

OTHER SUBSTANCES
Simultaneous: adonitol, erythritol, ethylene glycol, galactitol, glucitol, glycerin, inositol

REFERENCE
Ye,J.; Baldwin,R.P. Determination of carbohydrates, sugar acids and alditols by capillary electrophoresis and electrochemical detection at a copper electrode, *J.Chromatogr.A*, **1994**, *687*, 141–148.

Maprotiline

Molecular formula: $C_{20}H_{23}N$
Molecular weight: 277.41
CAS Registry No.: 10262-69-8, 10347-81-6 (HCl)
Merck Index (12th ed.): 5792
Lednicer: 2 220

SAMPLE
Matrix: solutions
Sample preparation: Inject an aliquot of a 40 μg/mL solution in MeOH.

CAPILLARY ELECTROPHORESIS
Capillary: 64 cm × 50 μm fused silica (55.5 cm to detector) (Polymicro Technologies)
Capillary preparation: Flush with running buffer for 2 min between runs. Before use rinse with 1 M NaOH for 1 h, with 100 mM NaOH for 20 min, with water for 20 min, and with running buffer for 10 min.
Capillary temperature: 25
Running buffer: MeCN:MeOH 50:50 containing 25 mM ammonium acetate and 100 mM sodium acetate
Injection: Pressure injection at 5 kPa for 3 s
Detector: UV 214
Migration time: 7
Voltage: 25 kV
Model: Hewlett-Packard HP3D

OTHER SUBSTANCES
Simultaneous: amitriptyline, desmethylimipramine, didesmethylimipramine, imipramine, imipramine-N-oxide, litracene, methylimipramine, nortriptyline, protriptyline

REFERENCE
Bjornsdottir,I.; Hansen,S.H. Comparison of separation selectivity in aqueous and non-aqueous capillary electrophoresis, *J.Chromatogr.A*, **1995**, *711*, 313–322.

SAMPLE
Matrix: solutions
Sample preparation: Prepare a 200 µg/mL solution in water, inject an aliquot.

CAPILLARY ELECTROPHORESIS
Capillary: 56 cm × 75 µm fused-silica (56 cm to detector) (Polymicro Technologies)
Capillary temperature: 30
Running buffer: 50 mM pH 4.0 6-Aminocaproic acid containing 25 mM 3-(N,N-dimethylmyristylammonium)propanesulfonate and 15 mM Tween 20
Injection: Pressure injection at 2 kPa for 3 s
Detector: UV 214
Migration time: 32
Voltage: 20 kV
Current: 62 µA
Model: Waters Quanta 4000

OTHER SUBSTANCES
Simultaneous: amitriptyline, litracene, nortriptyline, protriptyline

REFERENCE
Hansen,S.H.; Bjornsdottir,I.; Tjornelund,J. Separation of basic drug substances of very similar structure using micellar electrokinetic chromatography, *J.Pharm.Biomed.Anal.*, **1995**, *13*, 489–495.

SAMPLE
Matrix: solutions

CAPILLARY ELECTROPHORESIS
Capillary: 55.5 cm × 50 µm fused-silica (55.5 cm to detector) (Polymicro Technologies)
Capillary temperature: 25
Running buffer: MeCN:MeOH 50:50 containing 25 mM ammonium acetate and 100 mM sodium acetate
Injection: Pressure injection at 5 kPa for 3 s.
Detector: UV 214
Migration time: 7
Voltage: 25 kV
Current: 23 µA
Model: Hewlett Packard HP ^{3D}CE

OTHER SUBSTANCES
Simultaneous: amitriptyline, litracene, nortriptyline, protriptyline

REFERENCE
Bjornsdottir,I.; Tjornelund,J.; Hansen,S.H. Nonaqueous capillary electrophoresis in pharmaceutical analysis, *J.Capillary Electrophor.*, **1996**, *3*, 83–87.

Meclizine

Molecular formula: $C_{25}H_{27}ClN_2$
Molecular weight: 390.96
CAS Registry No.: 569-65-3, 31884-77-2
(di HCl monohydrate), 1104-22-9 (di HCl)
Merck Index (12th ed.): 5817
Lednicer: 1 59

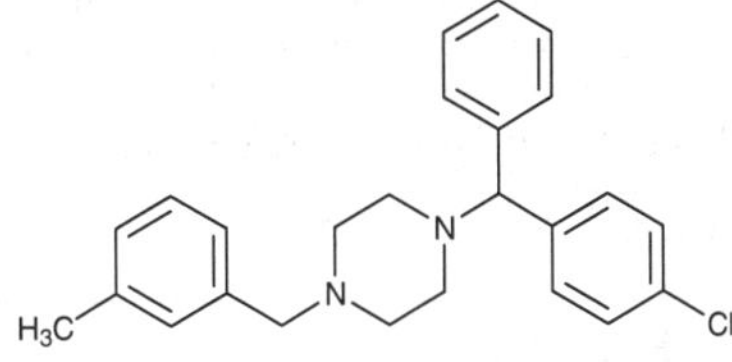

SAMPLE
Matrix: solutions
Sample preparation: Inject an aliquot of a 100 μg/mL solution in running buffer.

CAPILLARY ELECTROPHORESIS
Capillary: 36 cm × 50 μm fused-silica (31.5 cm to detector) (Yongnian Optical Conductive Fiber Plant, China), coated with polyacrylamide
Capillary preparation: No details of the polyacrylamide coating process are provided. However, another paper (LC.GC 1997, 15, 40) by this group indicates that they use the procedure of Hjertén, thus: Adjust the pH of 20 mL water to 3.5 with acetic acid, add 80 μL 3-(trimethoxysilyl)propyl methacrylate (3-methacryloxypropyltrimethoxysilane), mix, suck into capillary, let stand at room temperature for 1 h, remove the solution, wash with water. Fill the capillary with a deaerated 3-4% acrylamide solution containing 1 μL/mL N,N,N',N'-tetramethylethylenediamine and 1 mg/mL potassium persulfate, let stand for 30 min, remove excess solution by aspiration, rinse with water, remove water by aspiration, dry at 35° (J. Chromatogr. 1985, 347, 191).
Capillary temperature: 25
Running buffer: 100 mM NaH_2PO_4 adjusted to pH 2.5
Injection: Electrokinetic injection at 15 kV for 3 s.
Detector: UV 200, UV 210
Migration time: 6.62
Voltage: 15 kV
Model: Bio-Rad BioFocus 3000

OTHER SUBSTANCES
Simultaneous: azelastine, biperiden, carvedilol, clidinium bromide, mequitazine, ofloxacin, zopiclone

KEY WORDS
coated capillary

REFERENCE
Koppenhoefer,B.; Epperlein,U.; Xiaofeng,Z.; Bingcheng,L. Separation of enantiomers of drugs by capillary electrophoresis. Part 4: Hydroxypropyl-γ-cyclodextrin as chiral solvating agent, *Electrophoresis*, **1997**, *18*, 924–930.

SAMPLE
Matrix: solutions

CAPILLARY ELECTROPHORESIS
Capillary: 29-36 cm × 50 μm fused-silica (24.5-31.5 cm to detector) (Yongnian Optical Conductive Fiber Plant, China) coated with polyacrylamide
Capillary preparation: Coat capillary as follows. Adjust the pH of 20 mL water to 3.5 with acetic acid, add 80 μL 3-(trimethoxysilyl)propyl methacrylate (3-methacryloxypropyltrimethoxysilane), mix, suck into capillary, let stand at room temperature for 1 h, remove the solution, wash with water. Fill the capillary with a deaerated 3-4% acrylamide solution containing 1 μL/mL N,N,N',N'-tetramethylethylenediamine and 1 mg/mL potassium persulfate, let stand for 30 min, remove excess solution by aspiration, rinse with water, remove water by aspiration, dry at 35° (J. Chromatogr. 1985, 347, 191).
Capillary temperature: 25

Running buffer: 100 mM pH 2.5 NaH_2PO_4 (A) or 100 mM pH 2.5 NaH_2PO_4 containing 45 mM
   hydroxypropyl-α-cyclodextrin (Wacker, Munich) (B)
Injection: Electromigration at 15 kV for 3 s.
Detector: UV 200; UV 210
Migration time: 6.62 (A); 17.82 (B) (no separation of enantiomers)
Voltage: 15 kV
Model: Bio-Focus 3000

OTHER SUBSTANCES

Also analyzed: albuterol (salbutamol), alprenolol, amorolfine, atenolol, atropine, azelastine, baclofen, bamethan, benproperine, benserazide, biperiden, bisoprolol, brompheniramine, bupivacaine, bupranolol, butamirate, butethamate, carazolol, carbuterol, carteolol, carvedilol, celiprolol, chloroquine, chlorpheniramine, chlorphenoxamine, cicletanine, clenbuterol, clidinium
bromide, clobutinol, dimethindene, dipivefrin, disopyramide, dobutamine, doxylamine, fendiline, flecainide, gallopamil, homatropine, ipratropium bromide, isoproterenol (isoprenaline),
isothipendyl, ketamine, mefloquine, mepindolol, mequitazine, metaclazepam, metaproterenol
(orciprenaline), metipranolol, metoprolol, nafronyl (naftidrofuryl), nefopam, nicardipine, norfenefrine, ofloxacin, ornidazole, orphenadrine, oxomemazine, oxprenolol, oxybutynin, phenoxybenzamine, phenylpropanolamine, pholedrine, pindolol, pirbuterol, prilocaine, procyclidine,
promethazine, propafenone, propranolol, reproterol, sotalol, sulpride, synephrine, talinolol, terbutaline, tetrahydrozoline (tetryzoline), theodrenaline, tioconazole, tocainide, trihexyphenidyl,
trimeprazine (alimemazine), trimipramine, tropicamide, verapamil, zopiclone

KEY WORDS
coated capillary

REFERENCE
Koppenhoefer,B.; Eperlein,U.; Schlunk,R.; Zhu,X.; Lin,B. Separation of enantiomers of drugs by capillary electrophoresis. V. Hydroxypropyl-α-cyclodextrin as chiral solvating agent, *J.Chromatogr.A*, **1998**, *793*, 153–
164.

SAMPLE
Matrix: urine
Sample preparation: Filter urine, inject an aliquot.

CAPILLARY ELECTROPHORESIS
Capillary: 60 cm × 75 μm AccuSep uncoated silica (52.5 cm to detector) (Waters)
Capillary preparation: At the beginning of each day purge with 500 mM KOH for 5 min, with
   water for 5 min, and with running buffer for 10 min.
Running buffer: 50 mM NaH_2PO_4 adjusted to pH 2.35 with phosphoric acid
Injection: Hydrostatic injection at 15 cm
Detector: UV 214
Migration time: 7.48
Voltage: 22 kV
Current: 135-145 μA
Model: Waters Quanta 4000

OTHER SUBSTANCES
Extracted: acepromazine, amphetamine, benzocaine, brompheniramine, butacaine, codeine, diazepam, doxapram, lidocaine, medazepam, methamphetamine, methapyrilene, methaqualone,
phenmetrazine, procaine, tetrahydrozoline

REFERENCE
Chee,G.L.; Wan,T.S.M. Reproducible and high-speed separation of basic drugs by capillary zone electrophoresis,
J.Chromatogr., **1993**, *612*, 172–177.

Meclocycline

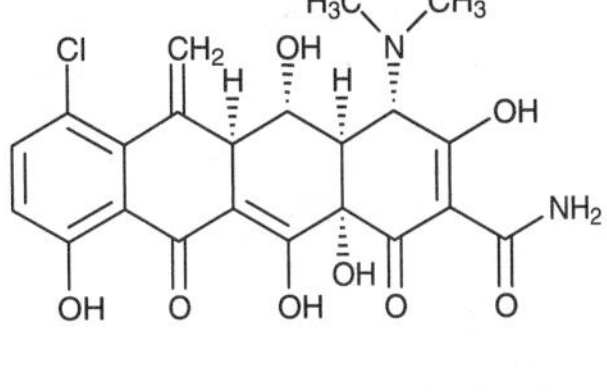

Molecular formula: $C_{22}H_{21}ClN_2O_8$
Molecular weight: 476.87
CAS Registry No.: 2013-58-3, 73816-42-9 (sulfosalicylate)
Merck Index (12th ed.): 5818

SAMPLE
Matrix: solutions

CAPILLARY ELECTROPHORESIS
Capillary: 50 cm × 50 μm (50 cm to detector) (Polymicro Technologies)
Capillary preparation: Purge capillary with 10 mM NaOH for 2 min after each injection.
Running buffer: MeOH:buffer 40:60 (Buffer was 30 mM pH 3.0 citric acid containing 24.5 mM β-alanine.)
Injection: Electrokinetic injection at 10 kV for 5 s.
Detector: UV 254
Migration time: 8.2
Voltage: 30 kV
Model: Perkin Elmer/Applied Biosystems Model 270A-HT

OTHER SUBSTANCES
Simultaneous: chlortetracycline, doxycycline, methacycline, minocycline, oxytetracycline, tetracycline

REFERENCE
Pesek,J.J.; Matyska,M.T. Separation of tetracyclines by high-performance capillary electrophoresis and capillary electrochromatography, *J.Chromatogr.A*, **1996**, *736*, 313–320.

Meclofenamic acid

Molecular formula: $C_{14}H_{11}Cl_2NO_2$
Molecular weight: 296.15
CAS Registry No.: 644-62-2, 6385-02-0 (Na salt)
Merck Index (12th ed.): 5819
Lednicer: 1 110

SAMPLE
Matrix: urine
Sample preparation: Condition a 300 μg Isolute HAX (Jones Chromatography) SPE cartridge with 2 mL MeOH and 2 mL 0.1% trifluoroacetic acid in water. 1 mL Urine + 5 mL water, mix, adjust pH to 3.0 with HCl, add to the SPE cartridge, wash with 2 mL 0.1% trifluoroacetic acid in water, wash with 2 mL 10 mM pH 8 ammonium acetate buffer, wash with 5 mL MeOH, elute with 2 mL 0.1% trifluoroacetic acid in MeCN. Evaporate the eluate to dryness under reduced pressure, reconstitute with 50 μL MeCN:running buffer:water 10:10:100, inject an aliquot.

CAPILLARY ELECTROPHORESIS
Capillary: 48.5 cm × 50 μm fused-silica (40 cm to detector) (Composite Metal Services, Hallow, UK)
Capillary preparation: Flush with running buffer for 1-4 min between runs. Flush new capillaries under 900 mbar pressure with 1 M NaOH for 10 min, with 100 mM NaOH for 5 min, and with running buffer for 5 min.
Capillary temperature: 25

Running buffer: MeOH:buffer 10:90 (Buffer was 70 mM pH 9.3 borate buffer containing 70 mM sodium dodecyl sulfate.)
Injection: Pressure injection of sample at 50 mbar for 3 s and of running buffer at 50 mbar for 1 s.
Detector: UV 240
Migration time: 5.8
Voltage: 22.8 kV
Model: Hewlett-Packard HP ^{3D}CE

OTHER SUBSTANCES
Extracted: metabolites

KEY WORDS
SPE; horse

REFERENCE
Taylor,M.R.; Westwood,S.A.; Perrett,D. Determination of Phase II drug metabolites in equine urine by micellar electrokinetic capillary chromatography, *J.Chromatogr.A*, **1996**, *745*, 155–163.

SAMPLE
Matrix: urine
Sample preparation: Dilute 10-fold (with water ?), inject an aliquot

---

CAPILLARY ELECTROPHORESIS
Capillary: 57 cm × 50 μm fused-silica (50 cm to detector) (Beckman)
Capillary preparation: Between runs rinse capillary with running buffer for 2 min. Before first use flush capillary with freshly prepared 100 mM NaOH for 5 min, rinse with buffer, equilibrate with buffer at 12 kV for 15 min.
Running buffer: MeCN:buffer 10:90 (Buffer was 30 mM pH 3.0 phosphate buffer containing 2 mM β-cyclodextrin.)
Injection: Pressure injection at 0.5 psi for 4-10 s.
Detector: UV 285
Migration time: 10.7
Voltage: 12 kV
Current: 90-92 μA
Model: Beckman P/ACE 5500
Limit of detection: 0.5 ng/mL

OTHER SUBSTANCES
Extracted: flufenamic acid, mefenamic acid

REFERENCE
Pérez-Ruiz,T.; Martínez-Lozano,C.; Sanz,A.; Bravo,E. Determination of flufenamic, meclofenamic and mefenamic acids by capillary electrophoresis using β-cyclodextrin, *J.Chromatogr.B*, **1998**, *708*, 249–256.

Medazepam

Molecular formula: $C_{16}H_{15}ClN_2$
Molecular weight: 270.76
CAS Registry No.: 2898-12-6, 2898-11-5 (HCl)
Merck Index (12th ed.): 5829
Lednicer: 1 368

SAMPLE
Matrix: solutions

CAPILLARY ELECTROPHORESIS
Capillary: 60 cm × 50 μm fused-silica (40 cm to detector), C18 coated (Supelco)

Capillary preparation: Rinse new capillaries with 250 μL 1 M NaOH, 250 μL 100 mM NaOH, water, MeOH, water, and running buffer then condition with running buffer for at least 15 min. Carry out a similar procedure between runs.

Running buffer: 100 mM pH 8.5 Borate buffer containing 25 mM sodium dodecyl sulfate and 5 M urea

Injection: Electrokinetic injection at 5 kV for 5 s.

Detector: UV 254

Migration time: 53

Voltage: 18 kV

Model: Jasco

OTHER SUBSTANCES

Simultaneous: alprazolam, amobarbital, barbital, bromazepam, clonazepam, clotiazepam, cloxazolam, diazepam, estazolam, etizolam, fludiazepam, flunitrazepam, flurazepam, haloxazolam, mephobarbital, metharbital, nimetazepam, nitrazepam, oxazepam, pentobarbital, phenobarbital, secobarbital, triazolam

KEY WORDS

coated capillary

REFERENCE

Jinno,K.; Han,Y.; Nakamura,M. Analysis of anxiolytic drugs by capillary electrophoresis with bare and coated capillaries, *J.Capillary Electrophor.*, **1996**, *3*, 139–145.

SAMPLE

Matrix: solutions

Sample preparation: Inject an aliquot of a 100 μg/mL solution in running buffer.

CAPILLARY ELECTROPHORESIS

Capillary: 60 cm × 50 μm acrylamide-coated fused-silica (40 cm to detector) (Supelco)

Capillary preparation: Adjust the pH of 20 mL water to 3.5 with acetic acid, add 80 μL 3-(trimethoxysilyl)propyl methacrylate (3-methacryloxypropyltrimethoxysilane), mix, suck into capillary, let stand at room temperature for 1 h, remove the solution, wash with water. Fill the capillary with a deaerated 3-4% acrylamide solution containing 1 μL/mL N,N,N',N'-tetra-methylethylenediamine and 1 mg/mL ammonium persulfate, let stand for 3 h, remove excess solution by aspiration, rinse with water, remove water by aspiration, dry at 35° (cf. J. Chromatogr. 1985, 347, 191).

Running buffer: MeCN:buffer 5:95 (Buffer was 100 mM borate containing 5 M urea and 10 mM sodium dodecyl sulfate, adjusted to pH 8.5 with phosphate.)

Injection: Electrokinetic injection at 5 kV for 5 s.

Detector: UV 254

Migration time: 7.5

Voltage: 18 kV

Model: Jasco Model 870-CE

OTHER SUBSTANCES

Simultaneous: alprazolam, amobarbital, barbital, bromazepam, clonazepam, clotiazepam, cloxazolam, diazepam, estazolam, etizolam, fludiazepam, flunitrazepam, flurazepam, haloxazolam, mephobarbital, metharbital, nimetazepam, nitrazepam, oxazepam, pentobarbital, phenobarbital, secobarbital, triazolam

KEY WORDS

coated capillary; detector at anode

REFERENCE

Jinno,K.; Han,Y.; Sawada,H.; Taniguchi,M. Capillary electrophoretic separation of toxic drugs using a polyacrylamide-coated capillary, *Chromatographia*, **1997**, *46*, 309–314.

Mefenamic acid

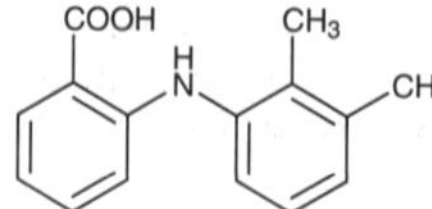

Molecular formula: $C_{15}H_{15}NO_2$
Molecular weight: 241.29
CAS Registry No.: 61-68-7
Merck Index (12th ed.): 5842
Lednicer: 1 10

SAMPLE
Matrix: urine
Sample preparation: Dilute 10-fold (with water ?), inject an aliquot

CAPILLARY ELECTROPHORESIS
Capillary: 57 cm × 50 μm fused-silica (50 cm to detector) (Beckman)
Capillary preparation: Between runs rinse capillary with running buffer for 2 min. Before first use flush capillary with freshly prepared 100 mM NaOH for 5 min, rinse with buffer, equilibrate with buffer at 12 kV for 15 min.
Running buffer: MeCN:buffer 10:90 (Buffer was 30 mM pH 3.0 phosphate buffer containing 2 mM β-cyclodextrin.)
Injection: Pressure injection at 0.5 psi for 4-10 s.
Detector: UV 285
Migration time: 10.8
Voltage: 12 kV
Current: 90-92 μA
Model: Beckman P/ACE 5500
Limit of detection: 0.4 ng/mL

OTHER SUBSTANCES
Extracted: flufenamic acid, meclofenamic acid

REFERENCE
Pérez-Ruiz,T.; Martínez-Lozano,C.; Sanz,A.; Bravo,E. Determination of flufenamic, meclofenamic and mefenamic acids by capillary electrophoresis using β-cyclodextrin, *J.Chromatogr.B*, **1998**, *708*, 249–256.

Mefloquine hydrochloride

Molecular formula: $C_{17}H_{17}ClF_6N_2O$
Molecular weight: 414.78
CAS Registry No.: 51773-92-3, 53230-10-7 (free base)
Merck Index (12th ed.): 5845

SAMPLE
Matrix: formulations
Sample preparation: Dissolve ground tablet containing 41.477 mg mefloquine in 100 mL water, stir for 30 min, dilute 5 times, inject an aliquot.

CAPILLARY ELECTROPHORESIS
Capillary: 48.5 cm × 50 μm fused-silica (39.5 cm to detector) (Hewlett-Packard)
Capillary preparation: Before each run rinse with water for 1 min, with 100 mM NaOH for 1 min, with water for 1 min, and with running buffer for 1 min. Condition new capillaries with 1 M NaOH for 30 min then flush with water.
Capillary temperature: 25
Running buffer: 100 mM Phosphoric acid containing 2.5 mM heptakis(2,6-di-O-methyl)-β-cyclodextrin, adjusted to pH 2.5 with concentrated NaOH
Injection: Pressure injection at 20 mbar for 10 s.

Detector: UV 223
Migration time: 4.995, 5.459 (enantiomers)
Voltage: 30 kV
Model: Hewlett-Packard HP^{3D}
Limit of detection: 1 μM

OTHER SUBSTANCES
Simultaneous: threo diastereomers

KEY WORDS
chiral; tablets

REFERENCE
Fanali,S.; Camera,E. Use of cyclodextrins in the capillary electrophoretic separation of *erythro-* and *threo-*mefloquine enantiomers, *J.Chromatogr.A*, **1996**, *745*, 17–23.

SAMPLE
Matrix: solutions
Sample preparation: Inject an aliquot of a 15 μg/mL solution in water.

CAPILLARY ELECTROPHORESIS
Capillary: 65 cm × 50 μm silica (50 cm to detector)
Capillary preparation: Between injections wash with 100 mM NaOH, rinse with water
Capillary temperature: 20 kV
Running buffer: 25 mM Na$_2$HPO$_4$ adjusted to pH 2 with phosphoric acid
Injection: Electrokinetic injection at 5 kV for 10 s
Detector: UV 254
Migration time: 16
Model: Isco 3850 Capillary Electropherograph
Limit of detection: 150 ng/mL (40 ng/mL if a water plug is injected first)

OTHER SUBSTANCES
Simultaneous: 4-chlorophenylbiguanide, chloroquine, chlorproguanil, cyclochlorproguanil, cyclo-guanil, desethylchloroquine, primaquine, proguanil, pyrimethamine, quinine

---

REFERENCE
Taylor,R.B.; Reid,R.G. Analysis of basic antimalarial drugs by CZE and MEKC. Part 1--Critical factors affecting separation, *J.Pharm.Biomed.Anal.*, **1993**, *11*, 1289–1294.

SAMPLE
Matrix: solutions

CAPILLARY ELECTROPHORESIS
Capillary: 41 cm × 50 μm fused-silica (41 cm to detector) (Grom)
Capillary temperature: 22 ± 1
Running buffer: pH 3.10 50 mM Phosphate buffer containing 80 μM β-cyclodextrin sulfobutyl ether (Center for Drug Delivery Research,University of Kansas, Lawrence KS)
Injection: Hydrostatic injection at 10 cm for 30 s
Detector: UV 210
Migration time: 8.65, 8.89 (enantiomers)
Voltage: 340 V/cm
Model: Grom system 100

KEY WORDS
chiral

REFERENCE
Chankvetadze,B.; Endresz,G.; Blaschke,G. About some aspects of the use of charged cyclodextrins for capillary electrophoresis enantioseparation, *Electrophoresis*, **1994**, *15*, 804–807.

SAMPLE
Matrix: solutions
Sample preparation: Prepare a solution in running buffer, inject an aliquot.

CAPILLARY ELECTROPHORESIS
Capillary: 60 cm $\times$ 75 μm fused-silica (52.4 cm to detector)
Capillary preparation: After each run flush with 500 mM KOH for 2-3 min, flush with water, fill with running buffer
Running buffer: 10 mM Na_2HPO_4 containing 2% heparin sodium (MW 10000, 11% S, Scientific Protein Laboratories, Waunakee, WI) adjusted to pH 5 with phosphoric acid
Injection: Hydrostatic injection
Detector: UV 214
Migration time: 12.1 (first enantiomer, R_S 0.82)
Model: Waters Quanta 4000

OTHER SUBSTANCES
Also analyzed: anabasine, brompheniramine, bupivacaine, carbinoxamine, chlorcyclizine, chloroquine, chlorpheniramine, dimethindene, doxylamine, enpiroline, halofantrine, hydroxychloroquine, indapamide, nornicotine, pheniramine, primaquine, promethazine, quinacrine, tetramisole

KEY WORDS
chiral

REFERENCE
Stalcup,A.M.; Agyei,N.M. Heparin: a chiral mobile-phase additive for capillary zone electrophoresis, *Anal.Chem.*, **1994**, *66*, 3054–3059.

SAMPLE
Matrix: solutions
Sample preparation: Inject an aliquot of a 100 μg/mL solution in water:running buffer 50:50.

CAPILLARY ELECTROPHORESIS
Capillary: 44.5 cm $\times$ 50 μm acrylamide-coated fused-silica (Bio-Rad)
Capillary temperature: 30
Running buffer: 100 mM NaH_2PO_4 containing 15 mM gamma-cyclodextrin, adjusted to pH 2.5 with phosphoric acid
Injection: Electrokinetic injection at 8 kV for 6 s.
Detector: UV 200
Migration time: 10.10
Voltage: 14 kV
Model: Bio-Rad BioFocus 3000

OTHER SUBSTANCES
Also analyzed: albuterol, alprenolol, atenolol, atropine, baclofen, bamethan, benserazide, biperiden, bisoprolol, bupivacaine, bupranolol, butetamate, carazolol, carbuterol, carvedilol, celiprolol, chloroquine, chlorpheniramine (chlorphenamine), clidinium bromide, clobutinol, disopyramide, dobutamine, flecainide, homatropine, ipratropium bromide, isoproterenol, isothipendyl, ketamine, mequitazine, metaproterenol (orciprenaline), metipranolol, nafronyl (naftidrofuryl), nefopam, ofloxacin, orphenadrine, oxomemazine, oxprenolol, phenoxybenzamine, pholedrine, pindolol, pirbuterol, prilocaine, promethazine, propafenone, propranolol, sotalol, synephrine, terbutaline, tetrahydrozoline (tetryzoline), tocainide, trihexyphenidyl, trimeprazine (alimemazine), trimipramine, tropicamide, verapamil, zopiclone

KEY WORDS
coated capillary; achiral

REFERENCE
Koppenhoefer,B.; Epperlein,U.; Christian,B.; Yibing,J.; Yuying,C.; Bingcheng,L. Separation of enantiomers of drugs by capillary electrophoresis. I. γ-Cyclodextrin as chiral solvating agent, *J.Chromatogr.A*, **1995**, *717*, 181–190.

SAMPLE
Matrix: solutions
Sample preparation: Inject an aliquot of a 100 μg/mL solution in running buffer.

CAPILLARY ELECTROPHORESIS
Capillary: 32 cm × 50 μm fused-silica (27.5 cm to detector) (Yongnian Optical Conductive Fiber Plant, China), coated with polyacrylamide
Capillary preparation: No details of the polyacrylamide coating process are provided. However, another paper (LC.GC 1997, 15, 40) by this group indicates that they use the procedure of Hjertén, thus: Adjust the pH of 20 mL water to 3.5 with acetic acid, add 80 μL 3-(trimethoxysilyl)propyl methacrylate (3-methacryloxypropyltrimethoxysilane), mix, suck into capillary, let stand at room temperature for 1 h, remove the solution, wash with water. Fill the capillary with a deaerated 3-4% acrylamide solution containing 1 μL/mL N,N,N',N'-tetramethylethylenediamine and 1 mg/mL potassium persulfate, let stand for 30 min, remove excess solution by aspiration, rinse with water, remove water by aspiration, dry at 35° (J. Chromatogr. 1985, 347, 191).
Capillary temperature: 25
Running buffer: 100 mM NaH_2PO_4 adjusted to pH 2.5 (A) or 100 mM NaH_2PO_4 containing 45 mM hydroxypropyl-gamma-cyclodextrin, adjusted to pII 2.5 (B)
Injection: Electrokinetic injection at 15 kV for 3 s.
Detector: UV 200, UV 210
Migration time: 6.31 (A), 11.77, 12.16 (B, enantiomers)
Voltage: 15 kV
Model: Bio-Rad BioFocus 3000

OTHER SUBSTANCES
Simultaneous: atropine, carazolol, cicletanine, dimethindene, fendiline, homatropine, ipratropium bromide, isothipendyl, metaclazepam, nafronyl (naftidrofuryl), nefopam, nicardipine, promethazine, reproterol, tetrahydrozoline (tetryzoline), theodrenaline, tioconazole, trihexyphenidyl, trimeprazine (alimemazine), trimipramine

KEY WORDS
coated capillary; chiral

REFERENCE
Koppenhoefer,B.; Epperlein,U.; Xiaofeng,Z.; Bingcheng,L. Separation of enantiomers of drugs by capillary electrophoresis. Part 4: Hydroxypropyl-γ-cyclodextrin as chiral solvating agent, *Electrophoresis*, **1997**, *18*, 924–930.

SAMPLE
Matrix: solutions
Sample preparation: Inject an aliquot of a 100 μg/mL solution in running buffer.

CAPILLARY ELECTROPHORESIS
Capillary: 30 cm × 50 μm fused-silica (25.5 cm to detector), coated with polyacrylamide
Capillary preparation: Adjust the pH of 20 mL water to 3.5 with acetic acid, add 80 μL 3-(trimethoxysilyl)propyl methacrylate (3-methacryloxypropyltrimethoxysilane), mix, suck into capillary, let stand at room temperature for 1 h, remove the solution, wash with water. Fill the capillary with a deaerated 3-4% acrylamide solution containing 1 μL/mL N,N,N',N'-tetramethylethylenediamine and 1 mg/mL potassium persulfate, let stand for 30 min, remove excess solution by aspiration, rinse with water, remove water by aspiration, dry at 35° (J. Chromatogr. 1985, 347, 191).
Capillary temperature: 25
Running buffer: 100 mM NaH_2PO_4 containing 30 mM hydroxypropyl-β- cyclodextrin, adjusted to pH 2.5 with phosphoric acid
Injection: Electrokinetic injection at 15 kV for 3 s.
Detector: UV 200
Migration time: 23.2, 26 (enantiomers)
Voltage: 15 kV
Model: Bio-Rad BioFocus 3000

OTHER SUBSTANCES
Simultaneous: chlorpheniramine, dimethindene, homatropine, metaclazepam, tetrahydrozoline (tetryzoline), theodrenaline

KEY WORDS
chiral; coated capillary; comparison with the use of other cyclodextrins

REFERENCE
Lin,B.; Zhu,X.; Koppenhoefer,B.; Epperlein,U. Investigation of 123 chiral drugs by cyclodextrin-modified capillary electrophoresis, *LC.GC*, **1997**, *15*, 40–46.

SAMPLE
Matrix: solutions

CAPILLARY ELECTROPHORESIS
Capillary: 29-36 cm × 50 μm fused-silica (24.5-31.5 cm to detector) (Yongnian Optical Conductive Fiber Plant, China) coated with polyacrylamide
Capillary preparation: Coat capillary as follows. Adjust the pH of 20 mL water to 3.5 with acetic acid, add 80 μL 3-(trimethoxysilyl)propyl methacrylate (3-methacryloxypropyltrimethoxysilane), mix, suck into capillary, let stand at room temperature for 1 h, remove the solution, wash with water. Fill the capillary with a deaerated 3-4% acrylamide solution containing 1 μL/mL N,N,N',N'-tetramethylethylenediamine and 1 mg/mL potassium persulfate, let stand for 30 min, remove excess solution by aspiration, rinse with water, remove water by aspiration, dry at 35° (J. Chromatogr. 1985, 347, 191).
Capillary temperature: 25
Running buffer: 100 mM pH 2.5 NaH_2PO_4 (A) or 100 mM pH 2.5 NaH_2PO_4 containing 45 mM hydroxypropyl-α-cyclodextrin (Wacker, Munich) (B)
Injection: Electromigration at 15 kV for 3 s.
Detector: UV 200; UV 210
Migration time: 6.31 (A); 11.65, 13.39 (B) (enantiomers)
Voltage: 15 kV
Model: Bio-Focus 3000

OTHER SUBSTANCES
Also analyzed: albuterol (salbutamol), alprenolol, amorolfine, atenolol, atropine, azelastine, baclofen, bamethan, benproperine, benserazide, biperiden, bisoprolol, brompheniramine, bupivacaine, bupranolol, butamirate, butethamate, carazolol, carbuterol, carteolol, carvedilol, celiprolol, chloroquine, chlorpheniramine, chlorphenoxamine, cicletanine, clenbuterol, clidinium bromide, clobutinol, dimethindene, dipivefrin, disopyramide, dobutamine, doxylamine, fendiline, flecainide, gallopamil, homatropine, ipratropium bromide, isoproterenol (isoprenaline), isothipendyl, ketamine, meclizine, mepindolol, mequitazine, metaclazepam, metaproterenol (orciprenaline), metipranolol, metoprolol, nafronyl (naftidrofuryl), nefopam, nicardipine, norfenefrine, ofloxacin, ornidazole, orphenadrine, oxomemazine, oxprenolol, oxybutynin, phenoxybenzamine, phenylpropanolamine, pholedrine, pindolol, pirbuterol, prilocaine, procyclidine, promethazine, propafenone, propranolol, reproterol, sotalol, sulpride, synephrine, talinolol, terbutaline, tetrahydrozoline (tetryzoline), theodrenaline, tioconazole, tocainide, trihexyphenidyl, trimeprazine (alimemazine), trimipramine, tropicamide, verapamil, zopiclone

KEY WORDS
coated capillary; chiral

REFERENCE
Koppenhoefer,B.; Eperlein,U.; Schlunk,R.; Zhu,X.; Lin,B. Separation of enantiomers of drugs by capillary electrophoresis. V. Hydroxypropyl-α-cyclodextrin as chiral solvating agent, *J.Chromatogr.A*, **1998**, *793*, 153–164.

Melphalan

Molecular formula: $C_{13}H_{18}Cl_2N_2O_2$
Molecular weight: 305.20
CAS Registry No.: 148-82-3
Merck Index (12th ed.): 5871
Lednicer: 2 120

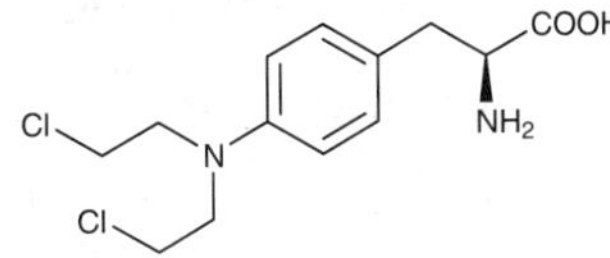

SAMPLE
Matrix: solutions

CAPILLARY ELECTROPHORESIS
Capillary: 57 cm $\times$ 50 μm fused-silica (50 cm to detector)
Running buffer: 50 mM pH 8.0 Borate buffer containing 40 mM sodium taurodeoxycholate and
   25 mM phosphatidylcholine (Prepare by adding phosphatidylcholine and stirring for 3-6 h until
   all cloudiness disappears. Phosphatidylcholine was 95% pure soybean lecithin, Epikuron, Lucas
   Meyer & Co.)
Injection: Inject a solution of the compound in the running buffer at 20 psi for 1 s, inject MeOH:
   water 5:95 at 20 psi for 1 s, inject a solution of halofantrine in running buffer at 20 psi for
   1 s.
Detector: UV 214
Migration time: k' 0.18
Model: Beckman P/ACE 5000

OTHER SUBSTANCES
Also analyzed: acetaminophen, amoxicillin, antipyrine, aspirin, azathioprine, caffeine, captopril,
   carbamazepine, carprofen, chlorambucil, chlorpheniramine, chlorpromazine, cimetidine, clonidine, codeine, desipramine, diphenhydramine, ephedrine, fenoterol, flufenamic acid, flurbiprofen, haloperidol, hydroxyzine, ibuprofen, imipramine, indomethacin, ketoprofen, lidocaine,
   metoprolol, nabumetone, nadolol, phenobarbital, phenol, promazine, propranolol, pyrilamine,
   ranitidine, ropinirole, salicylic acid, sulfamethoxazole, testosterone, theophylline, thioridazine,
   tiaprofenic acid, tolfenamic acid, trifluoperazine, trimethoprim, valproic acid, verapamil

KEY WORDS
comparison with HPLC; k' = (Tr-T0)/(T0(1-Tr/Tm)) where Tr = retention time of analyte; T0 =
retention time of water; and Tm = retention time of marker (halofantrine)

REFERENCE
Hanna,M.; de Biasi,V.; Bond,B.; Salter,C.; Hutt,A.J.; Camilleri,P. Estimation of the partitioning characteristics
   of drugs: A comparison of a large and diverse drug series utilizing chromatographic and electrophoretic
   methodology, *Anal.Chem.*, **1998**, *70*, 2092–2099.

Mepenzolate bromide

Molecular formula: $C_{21}H_{26}BrNO_3$
Molecular weight: 420.35
CAS Registry No.: 76-90-4
Merck Index (12th ed.): 5893

SAMPLE
Matrix: solutions
Sample preparation: Inject an aliquot of a solution in MeOH:10 mM pH 4.5 ammonium acetate
   buffer containing 25 mM sodium dodecylsulfate 40:60.

CAPILLARY ELECTROPHORESIS
Capillary: 57.5 × 75 μm fused-silica (SGE) and 32.5 × 75 μm fused-silica (SGE) coupled with a non-conducting coupler which allowed buffer to pass
Running buffer: MeOH:10 mM pH 4.5 ammonium acetate buffer containing 25 mM sodium dodecylsulfate 40:60 for the first capillary, MeOH:50 mM pH 4.5 ammonium acetate buffer 40: 60 for the second capillary
Injection: Pressure injection of about 25 nL
Detector: MS, Finnigan MAT SSQ 710 single-quadrupole, electrospray, positive-ion mode, sheath liquid was MeOH:100 mM pH 4.5 ammonium acetate buffer 80:20 at 1 μL/min, nitrogen drying gas, electrospray tip at ground, electrospray counter-electrode at -4 kV
Migration time: 20
Voltage: 25 kV for first capillary, 15 kV for second capillary

OTHER SUBSTANCES
Simultaneous: pipenzolate

KEY WORDS
9 min in first capillary; 3 min transfer; then electrophorese in second capillary

REFERENCE
Lamoree,M.H.; Tjaden,U.R.; van der Greef,J. On-line coupling of micellar electrokinetic chromatography to electrospray mass spectrometry, *J.Chromatogr.A*, **1995**, *712*, 219–225.

SAMPLE
Matrix: solutions
Sample preparation: Inject an aliquot of a solution in running buffer.

CAPILLARY ELECTROPHORESIS
Capillary: 60 cm × 75 μm fused-silica (52.4 cm to detector)
Capillary preparation: After each run flush with 500 mM KOH for 2-3 min then with water.
Running buffer: 10 mM pH 3.8 Phosphate buffer containing 2% sulfated cyclodextrin (ds 7-10)
Injection: Hydrostatic injection.
Detector: UV 214
Migration time: 9.74, 9.98 (enantiomers)
Voltage: 15 kV
Model: Waters Quanta 4000

OTHER SUBSTANCES
Also analyzed: acebutolol, alprenolol, aminoglutethimide, brompheniramine, bupivacaine, bupropion, canadine, carbinoxamine, chloroquine, chlorpheniramine, dimethindene, disopyramide, doxylamine, hydroxychloroquine, idazoxan, isoxsuprine, ketamine, mepivacaine, methoxyphenamine, mexiletine, midodrine, nefopam, orphenadrine, oxprenolol, oxyphencyclimine, pheniramine, phensuximide, pindolol, piperoxan, terbutaline, tetramisole, tolperisone, tranylcypromine, trihexyphenidyl, trimipramine, verapamil, warfarin

KEY WORDS
chiral; detector at anode

REFERENCE
Stalcup,A.M.; Gahm,K.H. Application of sulfated cyclodextrins to chiral separations by capillary zone electrophoresis, *Anal.Chem.*, **1996**, *68*, 1360–1368.

Meperidine

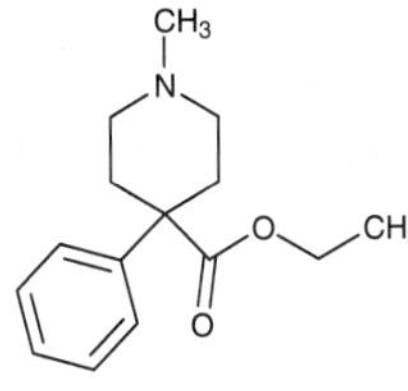

Molecular formula: $C_{15}H_{21}NO_2$
Molecular weight: 247.34
CAS Registry No.: 57-42-1, 50-13-5 (HCl)
Merck Index (12th ed.): 5894
Lednicer: 1 300

SAMPLE
Matrix: solutions

CAPILLARY ELECTROPHORESIS
Capillary: 59.6 cm × 75 μm fused-silica (52.1 cm to detector) (Polymicro Technologies)
Capillary preparation: Purge with running buffer before each run. At the beginning of each day purge using 50-60 kPa vacuum with 500 mM NaOH for 5 min, with water for 5 min, with MeCN for 5 min, and with running buffer for 5 min.
Running buffer: MeCN:MeOH:acetic acid 49:50:1 containing 20 mM ammonium acetate
Injection: Hydrostatic injection at 10 cm for 5 s.
Detector: UV 214
Migration time: 3.84
Voltage: 30 kV
Model: Waters Quanta 4000

OTHER SUBSTANCES
Simultaneous: diphenoxylate, ethoheptazine, fentanyl, levallorphan, methadone, nalorphine, pentazocine, nikethamide

REFERENCE
Leung,G.N.W.; Tang,H.P.O.; Tso,T.S.C.; Wan,T.S.M. Separation of basic drugs with non-aqueous capillary electrophoresis, *J.Chromatogr.A*, **1996**, *738*, 141–154.

SAMPLE
Matrix: solutions

CAPILLARY ELECTROPHORESIS
Capillary: 70 cm × 75 μm fused-silica (63 cm to detector) (Composite Metal Services, Hallow, UK)
Capillary preparation: Wash with running buffer for 5 min before each run. Purge new capillaries with 100 mM NaOH at 60° for 5 min and with water at 60° for 5 min, equilibrate with running buffer for 5 min, apply voltage for 5 min.
Capillary temperature: 20
Running buffer: 50 mM Disodium tetraborate adjusted to pH 2.2 with orthophosphoric acid
Injection: Hydrodynamic (vacuum) injection for 3 s.
Detector: UV (wavelength not given)
Migration time: 12
Voltage: 25 kV
Model: SpectraPhoresis 1000 (Thermo Separation Products)

OTHER SUBSTANCES
Simultaneous: clenbuterol, codeine, flurazepam, noscapine

REFERENCE
McGrath,G.; Smyth,W.F. Large-volume sample stacking of selected drugs of forensic significance by capillary electrophoresis, *J.Chromatogr.B*, **1996**, *681*, 125–131.

SAMPLE
Matrix: solutions
Sample preparation: Inject an aliquot of a solution in running buffer diluted 10-fold with water.

CAPILLARY ELECTROPHORESIS
Capillary: 70 cm × 75 μm fused-silica (63 cm to detector) (Composite Metal Services, Hallow, UK)
Capillary preparation: Wash with running buffer for 5 min before each injection. Condition new capillaries by purging with 100 mM NaOH at 60° for 5 min and with water at 60° for 5 min. Equilibrate with run buffer for 5 min then apply voltage for 5 min a number of times.
Capillary temperature: 20
Running buffer: 50 mM Disodium tetraborate containing 2 mM cetyltrimethylammonium bromide, adjusted to pH 2.2 with orthophosphoric acid
Injection: Hydrodynamic (vacuum) injection for 30 s then apply + 15 kV (to remove solvent). When the current reaches 95% of its pre-injection level switch polarity to begin the electrophoresis.
Detector: UV 210
Migration time: 35
Voltage: -15 kV
Model: Spectra Phoresis 1000
Limit of detection: 405 nM

OTHER SUBSTANCES
Simultaneous: clenbuterol, codeine, flurazepam, noscapine

REFERENCE
Smyth,W.F.; Harland,G.B.; McClean,S.; McGrath,G.; Oxspring,D. Effect of on-capillary large volume sample stacking on limits of detection in the capillary zone electrophoretic determination of selected drugs, dyes and metal chelates, *J.Chromatogr.A*, **1997**, *772*, 161–169.

SAMPLE
Matrix: solutions

CAPILLARY ELECTROPHORESIS
Capillary: 65 cm × 50 μm polyacrylamide coated (Bio-Rad)
Capillary preparation: Between runs rinse capillary with 50 mM pH 2.5 phosphate buffer for 2 min, with 100 mM NaOH for 2 min, with water for 1 min, with washing buffer for 2 min, and with running buffer for 2 min. (The washing buffer was the same as the running buffer but it was contained in a separate vial to minimize contamination.)
Capillary temperature: 15
Running buffer: 50 mM pH 2.5 Phosphate buffer containing 75 mM methyl-β-cyclodextrin, 220 mM urea, and 15 mM trimethylamine
Injection: Electrokinetic injection at 5 kV for 18 s.
Detector: UV 200
Migration time: 28.9
Voltage: 25 kV
Model: Bio-Rad BioFocus 3000

OTHER SUBSTANCES
Simultaneous: morphine, tramadol

KEY WORDS
achiral except for tramadol; coated capillary

REFERENCE
Chan,E.C.Y.; Ho,P.C. Enantiomeric separation of tramadol hydrochloride and its metabolites by cyclodextrin-mediated capillary zone electrophoresis, *J.Chromatogr.B*, **1998**, *707*, 287–294.

Mephenytoin

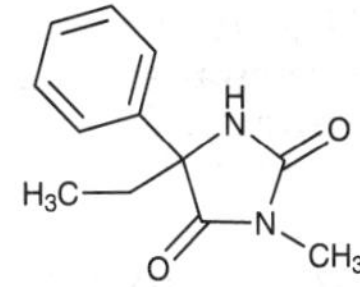

Molecular formula: $C_{12}H_{14}N_2O_2$
Molecular weight: 218.26
CAS Registry No.: 50-12-4
Merck Index (12th ed.): 5898

SAMPLE
Matrix: solutions
Sample preparation: Inject an aliquot of a 20 µg/mL solution

CAPILLARY ELECTROPHORESIS
Capillary: 44 cm × 50 µm fused-silica (37 cm to detector) (Supelco)
Capillary preparation: Before each run rinse capillary with running buffer for 3 min. At the start of each day rinse capillary with running buffer for 10 min. Treat new capillaries with 1 M NaOH, 100 mM NaOH, water, and running buffer.
Capillary temperature: 25
Running buffer: 100 mM Phosphoric acid adjusted to pH 3 with triethanolamine containing 10 mM carboxymethyl-β-cyclodextrin (Cyclolab, Budapest) and 10 mM heptakis(2,3,6-tri-O-methyl)-β-cyclodextrin (Sigma)
Injection: Hydrodynamic injection for 5 s (13.3 nL).
Detector: UV 210
Migration time: 9, 10 (enantiomers)
Voltage: -25 kV
Model: Spectraphoresis 1000 CE

KEY WORDS
detector at anode; chiral; R_s = 5.5

REFERENCE
Fillet,M.; Fotsing,L.; Crommen,J. Enantioseparation of uncharged compounds by capillary electrophoresis using mixtures of anionic and neutral β-cyclodextrin derivatives, *J.Chromatogr.A*, **1998**, *817*, 113–119.

Mephobarbital

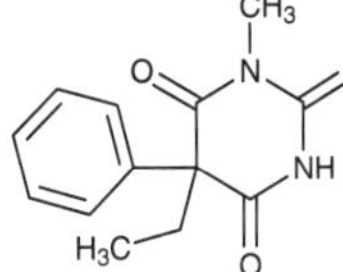

Molecular formula: $C_{13}H_{14}N_2O_3$
Molecular weight: 246.27
CAS Registry No.: 115-38-8
Merck Index (12th ed.): 5899

SAMPLE
Matrix: blood, gastric contents, urine, vitreous humor
Sample preparation: Dilute gastric contents 1:10. 2 mL Serum, blood, vitreous humor, diluted gastric contents, or urine + 2.5 mL water, mix, add to TOXI-TUBE B (Toxi-Lab, Irvine CA), rock for 10 min, centrifuge at 2000 rpm for 5 min. Remove the organic layer and evaporate it to dryness under a stream of nitrogen at room temperature, reconstitute the residue in 100 µL running buffer, vortex for 30 s, inject an aliquot.

CAPILLARY ELECTROPHORESIS
Capillary: 60 cm × 75 µm AccuSep (Waters)
Capillary preparation: Purge for 1 min between samples.
Running buffer: MeCN:buffer 15:85 adjusted to pH 8.5 with 1 M phosphoric acid. (Buffer was 3.8 g sodium borate decahydrate, 1.4 g $NaH_2PO_4.H_2O$, and 28.8 g sodium dodecyl sulfate in 1 L water.)
Injection: Hydrostatic injection for 10 s

Detector: UV 214
Migration time: 8.88
Internal standard: mephobarbital
Voltage: 20 kV
Model: Waters Quanta 4000

OTHER SUBSTANCES
Extracted: amobarbital, butabarbital, butalbital, pentobarbital, phenobarbital, secobarbital

KEY WORDS
serum; whole blood; mephobarbital is IS

REFERENCE
Ferslew,K.E.; Hagardorn,A.N.; McCormick,W.F. Application of micellar electrokinetic capillary chromatography to forensic analysis of barbiturates in biological fluids, *J.Forensic Sci.*, **1995**, *40*, 245–249.

SAMPLE
Matrix: solutions
Sample preparation: Inject an aliquot of a 0.2-1 μg/mL solution.

CAPILLARY ELECTROPHORESIS
Capillary: 57 cm × 50 μm fused-silica coated with a 0.025 μm film of p-methylbenzoyl cellulose (50 cm to detector) (Polymicro Technologies)
Capillary preparation: Coat capillary as follows. Coat at 0.3 mbar and 35-40° with a filtered 0.2% solution of p-methylbenzoyl cellulose in THF using the static method (further details in K. Grob. Making and Manipulating Capillary Columns for Gas Chromatography, Hüthig, Heidelberg, 1986). Prepare p-methylbenzoyl cellulose as follows. Suspend 100 g of microcrystalline cellulose in a mixture of 1 L pyridine, 420 mL triethylamine, and 2 g dimethylaminopyridine, slowly add 265 mL p-toluoyl chloride (p-methylbenzoyl chloride) at room temperature, stir at 120° under nitrogen for 12 h, cool to RT, treat the solid mass with 4 L MeOH, dissolve in dichloromethane, precipitate twice in EtOH, dry under vacuum at 100° for 2 days (cf. Chirality 1991, 3, 43).
Running buffer: MeCN:40 mM pH 7 phosphate buffer 25:75
Injection: Pressure injection at 35 mbar for 2-5 s.
Detector: UV 214
Migration time: 40, 43 (enantiomers)
Voltage: 30 kV
Model: Beckman P/ACE 5510

KEY WORDS
electrochromatography; coated capillary; chiral

REFERENCE
Francotte,E.; Jung,M. Enantiomer separation by open-tubular liquid chromatography and electrochromatography in cellulose capillaries, *Chromatographia*, **1996**, *42*, 521–527.

SAMPLE
Matrix: solutions

CAPILLARY ELECTROPHORESIS
Capillary: 60 cm × 50 μm fused-silica (40 cm to detector), bare (A) or C18 coated (B) (Supelco)
Capillary preparation: Rinse new capillaries with 250 μL 1 M NaOH, 250 μL 100 mM NaOH, water, MeOH, water, and running buffer then condition with running buffer for at least 15 min. Carry out a similar procedure between runs.
Running buffer: 100 mM pH 8.5 Borate buffer containing 25 mM sodium dodecyl sulfate and 5 M urea
Injection: Electrokinetic injection at 5 kV for 5 s.
Detector: UV 254
Migration time: 11 (A), 10 (B)
Voltage: 18 kV
Model: Jasco

OTHER SUBSTANCES
Simultaneous: amobarbital, barbital, metharbital, pentobarbital, phenobarbital, secobarbital

KEY WORDS
coated capillary

REFERENCE
Jinno,K.; Han,Y.; Nakamura,M. Analysis of anxiolytic drugs by capillary electrophoresis with bare and coated capillaries, *J.Capillary Electrophor.*, **1996**, *3*, 139–145.

SAMPLE
Matrix: solutions

CAPILLARY ELECTROPHORESIS
Capillary: 40 cm × 75 μm coated fused-silica (15 cm to detector) (Supelco)
Capillary preparation: Coat column as follows. Adjust the pH of 20 mL water to 3.5 with acetic acid, add 80 μL 3-(trimethoxysilyl)propyl methacrylate (3-methacryloxypropyltrimethoxysilane), mix, suck into capillary, let stand at room temperature for 1 h, remove the solution, wash with water. Fill the capillary with a deaerated 4% acrylamide solution containing 1 mg/mL N,N,N',N'-tetramethylethylenediamine and 1 mg/mL ammonium persulfate, let stand for 3 h, remove excess solution by aspiration, rinse with water, remove water by aspiration, dry at 35° (cf. J. Chromatogr. 1985, 347, 191).
Running buffer: 100 mM Tris/150 mM boric acid, pH 8.3
Injection: Electromigration at 5 kV for 5 s.
Detector: UV 240, UV 254
Migration time: 5.75
Voltage: 12 kV
Model: Jasco 890-CE

OTHER SUBSTANCES
Simultaneous: amobarbital, barbital, metharbital, pentobarbital, phenobarbital, secobarbital

KEY WORDS
injection at cathode; coated capillary

REFERENCE
Jinno,K.; Han,Y.; Sawada,H. Analysis of toxic drugs by capillary electrophoresis using polyacrylamide-coated columns, *Electrophoresis*, **1997**, *18*, 284–286.

SAMPLE
Matrix: solutions
Sample preparation: Inject an aliquot of a 100 μg/mL solution in running buffer.

CAPILLARY ELECTROPHORESIS
Capillary: 60 cm × 50 μm acrylamide-coated fused-silica (40 cm to detector) (Supelco)
Capillary preparation: Adjust the pH of 20 mL water to 3.5 with acetic acid, add 80 μL 3-(trimethoxysilyl)propyl methacrylate (3-methacryloxypropyltrimethoxysilane), mix, suck into capillary, let stand at room temperature for 1 h, remove the solution, wash with water. Fill the capillary with a deaerated 3-4% acrylamide solution containing 1 μL/mL N,N,N',N'-tetramethylethylenediamine and 1 mg/mL ammonium persulfate, let stand for 3 h, remove excess solution by aspiration, rinse with water, remove water by aspiration, dry at 35° (cf. J. Chromatogr. 1985, 347, 191).
Running buffer: MeCN:buffer 5:95 (Buffer was 100 mM borate containing 5 M urea and 10 mM sodium dodecyl sulfate, adjusted to pH 8.5 with phosphate.)
Injection: Electrokinetic injection at 5 kV for 5 s.
Detector: UV 254
Migration time: 14.8
Voltage: 18 kV
Model: Jasco Model 870-CE

OTHER SUBSTANCES
Simultaneous: alprazolam, amobarbital, barbital, bromazepam, clonazepam, clotiazepam, cloxazolam, diazepam, estazolam, etizolam, fludiazepam, flunitrazepam, flurazepam, haloxazolam, medazepam, metharbital, nimetazepam, nitrazepam, oxazepam, pentobarbital, phenobarbital, secobarbital, triazolam

KEY WORDS
coated capillary; detector at anode

REFERENCE
Jinno,K.; Han,Y.; Sawada,H.; Taniguchi,M. Capillary electrophoretic separation of toxic drugs using a polyacrylamide-coated capillary, *Chromatographia*, **1997**, *46*, 309–314.

SAMPLE
Matrix: solutions

CAPILLARY ELECTROPHORESIS
Capillary: 70 cm × 50 μm fused-silica (50 cm to detector) (Polymicro Technologies)
Capillary temperature: 30
Running buffer: 35 mM pH 5.4 Sodium phosphate containing 2.5 mM sulfobutyl ether-β-cyclodextrin (average substitution 3.9, MW 1721, Center for Drug Delivery Research, Lawrence KS)
Injection: Pressure injection at 5 inches Hg for 2.5 s.
Detector: UV 210
Migration time: 15, 15.5 (enantiomers)
Voltage: 20 kV
Current: 34 μA
Model: Perkin Elmer-Applied Biosystems Model 270

KEY WORDS
chiral

REFERENCE
Xie,G.-h.; Skanchy,D.J.; Stobaugh,J.F. Chiral separations of enantiomeric pharmaceuticals by capillary electrophoresis using sulphobutyl ether β-cyclodextrin as isomer selector, *Biomed.Chromatogr.*, **1997**, *11*, 193–199.

SAMPLE
Matrix: solutions
Sample preparation: Inject an aliquot of a 20 μg/mL solution

CAPILLARY ELECTROPHORESIS
Capillary: 44 cm × 50 μm fused-silica (37 cm to detector) (Supelco)
Capillary preparation: Before each run rinse capillary with running buffer for 3 min. At the start of each day rinse capillary with running buffer for 10 min. Treat new capillaries with 1 M NaOH, 100 mM NaOH, water, and running buffer.
Capillary temperature: 25
Running buffer: 100 mM Phosphoric acid adjusted to pH 5 with triethanolamine containing 10 mM carboxymethyl-β-cyclodextrin (Cyclolab, Budapest) and 50 mM heptakis(2,3,6-tri-O-methyl)-β-cyclodextrin (Sigma)
Injection: Hydrodynamic injection for 5 s (13.3 nL).
Detector: UV 210
Voltage: -25 kV
Model: Spectraphoresis 1000 CE

KEY WORDS
detector at anode; chiral; R_s = 5.1

REFERENCE
Fillet,M.; Fotsing,L.; Crommen,J. Enantioseparation of uncharged compounds by capillary electrophoresis using mixtures of anionic and neutral β-cyclodextrin derivatives, *J.Chromatogr.A*, **1998**, *817*, 113–119.

SAMPLE
Matrix: urine
Sample preparation: Place an extraction rod in 50 μL urine in a 50 mm × 1.5 mm ID length
of PTFE tubing for 30 min, place the extraction rod into 5 μL 20-40 mM pH 11.5 phosphate
buffer in a 50 mm × 1.2 mm ID length of PTFE tubing for 90 min, inject an aliquot of the
buffer. (Prepare the solid-phase extraction rod as follows. Polish a 70 × 1.1 stainless steel rod
with emery paper, clean with a Kimwipe and acetone, sonicate in EtOH, sonicate in THF, dip
in 3% PVAM in THF momentarily, hold vertically for 1 min, air dry in a hood for at least 5 h,
dip in PVC solution momentarily, hold vertically for 1 min, air dry for at least 5 h. The coating
is 3 cm long. PVAM was poly(vinyl chloride-co-vinyl acetate-co-maleic acid) consisting of 86%
vinyl chloride, 13% vinyl acetate, and 1% maleic acid. Prepare PVC solution by adding very
high molecular weight PVC slowly to THF with stirring until the concentration reaches 3.6%,
add Santicizer 141 to a concentration of 7.2%. Santicizer 141 (Monsanto) is 92% 2-ethylhexyl
diphenyl phosphate, 5% di-2-ethylhexyl phenyl phosphate, and 3% triphenyl phosphate.)

CAPILLARY ELECTROPHORESIS
Capillary: 75 cm × 75 μm fused-silica (50 cm to detector) (Polymicro Technologies)
Running buffer: 50 mM Tris adjusted to pH 7.8 with 3-[N-tris(hydroxymethyl)methylamino]-2-
hydroxypropanesulfonic acid (Tapso)
Injection: Pressure injection at 0.5 psi for 4 s.
Detector: UV 230
Migration time: 8.1
Voltage: 35 kV
Current: about 30 μA
Model: Isco 3850
Limit of detection: <1 ppm

OTHER SUBSTANCES
Extracted: amobarbital, aprobarbital, butabarbital, butalbital, pentobarbital, secobarbital,
thiopental
Simultaneous: allobarbital, aspirin, phenobarbital

KEY WORDS
SPE

REFERENCE
Li,S.; Weber,S.G. Determination of barbiturates by solid-phase microextraction and capillary electrophoresis,
Anal.Chem., **1997**, *69*, 1217–1222.

Mepindolol

Molecular formula: $C_{15}H_{22}N_2O_2$
Molecular weight: 262.35
CAS Registry No.: 23694-81-7, 56396-94-2 (sulfate salt)
Merck Index (12th ed.): 5901

SAMPLE
Matrix: solutions
Sample preparation: Inject an aliquot of a 100 μg/mL solution in running buffer.

CAPILLARY ELECTROPHORESIS
Capillary: 30 cm × 50 μm fused-silica (25.5 cm to detector) (Yongnian Optical Conductive Fiber
Plant, China), coated with polyacrylamide
Capillary preparation: No details of the polyacrylamide coating process are provided. However,
another paper (LC.GC 1997, 15, 40) by this group indicates that they use the procedure of
Hjertén, thus: Adjust the pH of 20 mL water to 3.5 with acetic acid, add 80 μL 3-(trimethox-
ysilyl)propyl methacrylate (3-methacryloxypropyltrimethoxysilane), mix, suck into capillary, let
stand at room temperature for 1 h, remove the solution, wash with water. Fill the capillary

with a deaerated 3-4% acrylamide solution containing 1 μL/mL N,N,N',N'-tetramethylethyl-enediamine and 1 mg/mL potassium persulfate, let stand for 30 min, remove excess solution by aspiration, rinse with water, remove water by aspiration, dry at 35° (J. Chromatogr. 1985, 347, 191).
Capillary temperature: 25
Running buffer: 100 mM NaH$_2$PO$_4$ adjusted to pH 2.5
Injection: Electrokinetic injection at 15 kV for 3 s.
Detector: UV 200, UV 210
Migration time: 4.68
Voltage: 15 kV
Model: Bio-Rad BioFocus 3000

OTHER SUBSTANCES

Simultaneous: amorolfine, brompheniramine, bupivacaine, carteolol, chloroquine, chlorpheni-ramine, chlorphenoxamine, disopyramide, dobutamine, doxylamine, flecainide, gallopamil, ke-tamine, orphenadrine, oxybutynin, phenoxybenzamine, pindolol, propafenone, propranolol, sul-piride, talinolol, tropicamide, verapamil

KEY WORDS

coated capillary

REFERENCE

Koppenhoefer,B.; Epperlein,U.; Xiaofeng,Z.; Bingcheng,L. Separation of enantiomers of drugs by capillary elec-trophoresis. Part 4: Hydroxypropyl-γ-cyclodextrin as chiral solvating agent, *Electrophoresis*, **1997**, *18*, 924–930.

SAMPLE

Matrix: solutions
Sample preparation: Inject an aliquot of a 100 μg/mL solution in running buffer.

CAPILLARY ELECTROPHORESIS

Capillary: 30 cm × 50 μm fused-silica (25.5 cm to detector), coated with polyacrylamide
Capillary preparation: Adjust the pH of 20 mL water to 3.5 with acetic acid, add 80 μL 3-(trimethoxysilyl)propyl methacrylate (3-methacryloxypropyltrimethoxysilane), mix, suck into capillary, let stand at room temperature for 1 h, remove the solution, wash with water. Fill the capillary with a deaerated 3-4% acrylamide solution containing 1 μL/mL N,N,N',N'-tetra-methylethylenediamine and 1 mg/mL potassium persulfate, let stand for 30 min, remove excess solution by aspiration, rinse with water, remove water by aspiration, dry at 35° (J. Chromatogr. 1985, 347, 191).
Capillary temperature: 25
Running buffer: 100 mM NaH$_2$PO$_4$ containing 45 mM heptakis(2,6-di-O-methyl)-β-cyclodextrin, adjusted to pH 2.5 with phosphoric acid
Injection: Electrokinetic injection at 15 kV for 3 s.
Detector: UV 200
Voltage: 15 kV
Model: Bio-Rad BioFocus 3000

KEY WORDS

chiral; coated capillary; comparison with the use of other cyclodextrins; this running buffer gave the greatest enantiomeric separation.; α=1.018

REFERENCE

Lin,B.; Zhu,X.; Koppenhoefer,B.; Epperlein,U. Investigation of 123 chiral drugs by cyclodextrin-modified cap-illary electrophoresis, *LC.GC*, **1997**, *15*, 40–46.

SAMPLE

Matrix: solutions

CAPILLARY ELECTROPHORESIS

Capillary: 29-36 cm × 50 μm fused-silica (24.5-31.5 cm to detector) (Yongnian Optical Conductive Fiber Plant, China) coated with polyacrylamide

Capillary preparation: Coat capillary as follows. Adjust the pH of 20 mL water to 3.5 with acetic acid, add 80 μL 3-(trimethoxysilyl)propyl methacrylate (3-methacryloxypropyltrimethoxysilane), mix, suck into capillary, let stand at room temperature for 1 h, remove the solution, wash with water. Fill the capillary with a deaerated 3-4% acrylamide solution containing 1 μL/mL N,N,N',N'-tetramethylethylenediamine and 1 mg/mL potassium persulfate, let stand for 30 min, remove excess solution by aspiration, rinse with water, remove water by aspiration, dry at 35° (J. Chromatogr. 1985, 347, 191).

Capillary temperature: 25

Running buffer: 100 mM pH 2.5 NaH_2PO_4 (A) or 100 mM pH 2.5 NaH_2PO_4 containing 45 mM hydroxypropyl-α-cyclodextrin (Wacker, Munich) (B)

Injection: Electromigration at 15 kV for 3 s.

Detector: UV 200; UV 210

Migration time: 4.68 (A); 8.14, 8.26 (B) (enantiomers)

Voltage: 15 kV

Model: Bio-Focus 3000

OTHER SUBSTANCES

Also analyzed: albuterol (salbutamol), alprenolol, amorolfine, atenolol, atropine, azelastine, baclofen, bamethan, benproperine, benserazide, biperiden, bisoprolol, brompheniramine, bupivacaine, bupranolol, butamirate, butethamate, carazolol, carbuterol, carteolol, carvedilol, celiprolol, chloroquine, chlorpheniramine, chlorphenoxamine, cicletanine, clenbuterol, clidinium bromide, clobutinol, dimethindene, dipivefrin, disopyramide, dobutamine, doxylamine, fendiline, flecainide, gallopamil, homatropine, ipratropium bromide, isoproterenol (isoprenaline), isothipendyl, ketamine, meclizine, mefloquine, mequitazine, metaclazepam, metaproterenol (orciprenaline), metipranolol, metoprolol, nafronyl (naftidrofuryl), nefopam, nicardipine, norfenefrine, ofloxacin, ornidazole, orphenadrine, oxomemazine, oxprenolol, oxybutynin, phenoxybenzamine, phenylpropanolamine, pholedrine, pindolol, pirbuterol, prilocaine, procyclidine, promethazine, propafenone, propranolol, reproterol, sotalol, sulpride, synephrine, talinolol, terbutaline, tetrahydrozoline (tetryzoline), theodrenaline, tioconazole, tocainide, trihexyphenidyl, trimeprazine (alimemazine), trimipramine, tropicamide, verapamil, zopiclone

KEY WORDS

coated capillary; chiral

REFERENCE

Koppenhoefer,B.; Eperlein,U.; Schlunk,R.; Zhu,X.; Lin,B. Separation of enantiomers of drugs by capillary electrophoresis. V. Hydroxypropyl-α-cyclodextrin as chiral solvating agent, *J.Chromatogr.A*, **1998**, *793*, 153–164.

Mepivacaine

Molecular formula: $C_{15}H_{22}N_2O$

Molecular weight: 246.35

CAS Registry No.: 96-88-8, 1722-62-9 (HCl)

Merck Index (12th ed.): 5905

Lednicer: 1 17

SAMPLE

Matrix: blood

Sample preparation: 1 mL serum + 30 μL 5.33 mg/mL mepivacaine in MeOH, mix thoroughly, add 200 μL 1 M NaOH, mix for 20 s, add 1.5 mL hexane:diethyl ether 50:50, shake gently horizontally for 5 min, let stand in the refrigerator for 5 min. Remove the 1.3 mL of the organic layer and evaporate it to dryness under a stream of nitrogen, reconstitute the residue in 15 μL MeOH, inject an aliquot.

CAPILLARY ELECTROPHORESIS

Capillary: 64 cm × 50 μm uncoated silica (48 cm to detector) (Polymicro Technologies)

Running buffer: 18 mM pH 2.9 Tris buffer containing 10 mM heptakis(2,6-di-O-methyl)-β-cy-
clodextrin, 0.1% methylhydroxycellulose 4000, and 0.03 mM hexadecyltrimethylammonium
bromide
Injection: Inject hydrodynamically at 13 cm for 15 s.
Detector: UV 220
Migration time: 10, 10.4 (enantiomers)
Internal standard: mepivacaine
Voltage: 24 kV
Model: Laboratory constructed

OTHER SUBSTANCES
Extracted: bupivacaine
Simultaneous: cimetidine, diltiazem, warfarin
Noninterfering: ibuprofen, indomethacin

KEY WORDS
serum; chiral; mepivacaine is IS

REFERENCE
Soini,H.; Riekkola,M.-L.; Novotny,M.V. Chiral separations of basic drugs and quantitation of bupivacaine enantiomers in serum by capillary electrophoresis with modified cyclodextrin buffers, *J.Chromatogr.*, **1992**,
608, 265–274.

SAMPLE
Matrix: blood
Sample preparation: 1 mL Serum + 40 μL 10 μg/mL (R)-prilocaine hydrochloride + 100 μL 1
M NaOH + 6 mL diethyl ether:ethyl acetate 50:50, vortex for 2 min, centrifuge at 4000 rpm
for 5 min. Filter (0.22 μm) the supernatant and evaporate it to dryness under a stream of
nitrogen, reconstitute the residue in 100 μL water, inject an aliquot.

CAPILLARY ELECTROPHORESIS
Capillary: 72 cm × 50 μm fused-silica (50 cm to detector) (Polymicro Technologies)
Capillary preparation: After each run rinse with 100 mM NaOH for 2 min and with running
buffer for 3 min. Condition a new capillary by rinsing with 1 M NaOH for 10 min and with
water for 10 min.
Capillary temperature: 35
Running buffer: 100 mM NaH_2PO_4 containing 20 mM heptakis(2,6-di-O-methyl)-β-cyclodextrin
and 30 nM (sic) hexadecyltrimethylammonium bromide, adjusted to pH 2.5 with 100 mM phos-
phoric acid
Injection: Vacuum injection for 20 s.
Detector: UV 215
Migration time: 14.9 (R-(+)), 15.4 (S-(-))
Internal standard: (R)-prilocaine (17.9)
Voltage: 25 kV
Model: ABI Model 270A (Applied Biosystems)
Limit of quantitation: 200 ng/mL
Limit of detection: 150 ng/mL

KEY WORDS
serum; chiral

REFERENCE
Siluveru,M.; Stewart,J.T. HPCE determination of *R*(+) and *S*(-) mepivacaine in human serum using a derivatized cyclodextrin and ultraviolet detection, *J.Pharm.Biomed.Anal.*, **1997**, *15*, 1751–1756.

SAMPLE
Matrix: formulations
Sample preparation: If necessary dilute with water, inject an aliquot.

CAPILLARY ELECTROPHORESIS
Capillary: 80.5 cm × 50 μm (72 cm to detector) (Hewlett Packard)

Capillary preparation: Before each run flush with water for 1 min, with 100 mM NaOH for 4 min, with water for 1 min, and with running buffer for 4 min.
Capillary temperature: 30
Running buffer: 100 mM Phosphoric acid containing 10 mM heptakis(2,6-di-O-methyl)-β-cyclodextrin (Sigma), adjusted to pH 3.0 with triethanolamine.
Injection: Pressure injection at 50 mbar for 5 s (5 nL), ramp to operating voltage at 500 V/s.
Detector: UV 206
Migration time: 21.8
Voltage: 30 kV
Model: Hewlett Packard HP ^{3D}CE

OTHER SUBSTANCES
Simultaneous: ropivacaine

KEY WORDS
injections

REFERENCE
Sänger-van de Griend,C.E.; Wahlström,H.; Gröningsson,K.; Widahl-Näsman,M. A chiral capillary electrophoresis method for ropivacaine hydrochloride in pharmaceutical formulations: validation and comparison with chiral liquid chromatography, *J.Pharm.Biomed.Anal.*, **1997**, *15*, 1051–1061.

SAMPLE
Matrix: pleural drain fluid
Sample preparation: Dilute pleural drain fluid 10-fold. 1 mL Diluted fluid + 1 mL 500 mM NaOH + 6 mL hexane, vortex for 30 s, centrifuge at 1500 g for 5 min. Remove the upper organic layer and evaporate it to dryness at 40°, reconstitute the residue in 50 μL running buffer and 50 μL 100 mM HCl, inject an aliquot.

CAPILLARY ELECTROPHORESIS
Capillary: 72 cm × 75 μm fused-silica (72 cm to detector)
Capillary preparation: Before each run rinse capillary with 100 mM NaOH for 2 min, with water for 1 min, and with buffer for 4 min.
Capillary temperature: 30
Running buffer: 45 mM NaH_2PO_4 and 35 mM sodium borate adjusted to pH 8.1 with 100 mM phosphoric acid
Injection: Injection by vacuum suction for 1 s
Detector: UV 200
Migration time: 10.3
Internal standard: mepivacaine
Voltage: 19 kV
Current: 96-98 μA
Model: Applied Biosystems Model 270A-HT

OTHER SUBSTANCES
Extracted: bupivacaine
Simultaneous: lidocaine

KEY WORDS
mepivacaine is IS; comparison with HPLC

REFERENCE
Wolfisberg,H.; Schmutz,A.; Stotzer,R.; Thormann,W. Assessment of automated capillary electrophoresis for therapeutic and diagnostic drug monitoring: determination of bupivacaine in drain fluid and antipyrine in plasma, *J.Chromatogr.A*, **1993**, *652*, 407–416.

SAMPLE
Matrix: solutions
Sample preparation: Inject an aliquot of a 100 μM solution in running buffer diluted 10-fold with water.

CAPILLARY ELECTROPHORESIS
Capillary: 57 cm × 25 μm fused-silica (50 cm to detector) (Polymicro Technologies)
Capillary preparation: Equilibrate with running buffer for 5 min before each analysis. Store in water overnight and rinse with 100 mM NaOH for 5 min at the start of each day and after each change of running buffer. Flush a new capillary with 100 mM HCl for 5 min, with water for 5 min, with 100 mM NaOH for 5 min, with water for 10 min, and with running buffer.
Capillary temperature: 25
Running buffer: pH 3.13 Phosphate buffer (I = 0.02) containing 14 mM taurodeoxycholate and 5.0 mM Brij-35
Injection: Pressure injection at 0.5 psi for 10 s.
Detector: UV 214
Migration time: 13 (R), 13.3 (S)
Voltage: 30 kV
Current: 9 μA
Model: Beckman P/ACE 2050

OTHER SUBSTANCES
Simultaneous: prilocaine

KEY WORDS
chiral

REFERENCE
Amini,A.; Beijersten,I.; Pettersson,C.; Westerlund,D. Enantiomeric separation of local anaesthetic drugs by micellar electrokinetic capillary chromatography with taurodeoxycholate as chiral selector, *J.Chromatogr.A*, **1996**, *737*, 301–313.

SAMPLE
Matrix: solutions

CAPILLARY ELECTROPHORESIS
Capillary: 80.5 cm × 50 μm fused-silica (72 cm to detector) (Hewlett Packard)
Capillary preparation: Before each run flush with water for 1 min, with 100 mM NaOH for 4 min, with water for 1 min, and with running buffer for 4 min.
Capillary temperature: 30
Running buffer: 100 mM Phosphoric acid adjusted to pH 3.0 with triethanolamine, containing 10 mM heptakis(2,6-di-O-methyl)-β-cyclodextrin
Injection: Pressure injection at 50 mbar for 5 s (5 nL), ramp to operating voltage at 500 V/s.
Detector: UV 206
Migration time: 23, 23.5 (enantiomers)
Voltage: 30 kV
Model: Hewlett Packard HP [3D]CE

OTHER SUBSTANCES
Simultaneous: bupivacaine, ropivacaine
Also analyzed: prilocaine

KEY WORDS
chiral

REFERENCE
Sänger-van de Griend,C.E.; Gröningsson,K.; Westerlund,D. Chiral separation of local anaesthetics with capillary electrophoresis. Evaluation of the inclusion complex of the enantiomers with heptakis(2,6-di-O-methyl)-β-cyclodextrin, *Chromatographia*, **1996**, *42*, 263–268.

SAMPLE
Matrix: solutions

CAPILLARY ELECTROPHORESIS
Capillary: 80.5 cm × 50 μm (72 cm to detector) (Hewlett Packard)

Capillary preparation: Before each run flush with water for 1 min, 100 mM NaOH for 4 min, with water for 1 min, and with running buffer for 4 min.
Capillary temperature: 30
Running buffer: 100 mM phosphoric acid containing 10 mM heptakis(2,6-di-O-methyl)-β-cyclodextrin, adjusted to pH 3.0 with triethanolamine
Injection: Pressure injection at 50 mbar over 5 s (5 nL), ramp to voltage at 500 V/s.
Detector: UV 206
Migration time: 22, 22.5 (enantiomers)
Voltage: 30 kV
Model: Hewlett Packard HP ^{3D}CE

OTHER SUBSTANCES
Simultaneous: bupivacaine, ropivacaine

KEY WORDS
chiral

REFERENCE
Sänger-van de Griend,C.E.; Gröningsson,K. Validation of a capillary electrophoresis method for the enantiomeric purity testing of ropivacaine, a new local anaesthetic compound, *J.Pharm.Biomed.Anal.*, **1996**, *14*, 295–304.

SAMPLE
Matrix: solutions
Sample preparation: Inject an aliquot of a solution in running buffer.

CAPILLARY ELECTROPHORESIS
Capillary: 60 cm × 75 μm fused-silica (52.4 cm to detector)
Capillary preparation: After each run flush with 500 mM KOH for 2-3 min then with water.
Running buffer: 10 mM pH 3.8 Phosphate buffer containing 2% sulfated cyclodextrin (ds 7-10)
Injection: Hydrostatic injection.
Detector: UV 214
Migration time: 19.79, 21.43 (enantiomers)
Voltage: 15 kV
Model: Waters Quanta 4000

OTHER SUBSTANCES
Also analyzed: acebutolol, alprenolol, aminoglutethimide, brompheniramine, bupivacaine, bupropion, canadine, carbinoxamine, chloroquine, chlorpheniramine, dimethindene, disopyramide, doxylamine, hydroxychloroquine, idazoxan, isoxsuprine, ketamine, mepenzolate, methoxyphenamine, mexiletine, midodrine, nefopam, orphenadrine, oxprenolol, oxyphencyclimine, pheniramine, phensuximide, pindolol, piperoxan, terbutaline, tetramisole, tolperisone, tranylcypromine, trihexyphenidyl, trimipramine, verapamil, warfarin

KEY WORDS
chiral; detector at anode

REFERENCE
Stalcup,A.M.; Gahm,K.H. Application of sulfated cyclodextrins to chiral separations by capillary zone electrophoresis, *Anal.Chem.*, **1996**, *68*, 1360–1368.

SAMPLE
Matrix: solutions

CAPILLARY ELECTROPHORESIS
Capillary: 47 cm × 50 μm fused-silica (40 cm to detector) (Polymicro Technologies)
Capillary preparation: Before each run flush with running buffer for 2 min, fill with running buffer containing 96 mg/mL 100 mM methyl-β-cyclodextrin at 0.5 psi for 10.0 min. Store capillary in water overnight, rinse with 100 mM NaOH and with water each morning. Rinse a new capillary with 100 mM HCl for 5 min, with water for 5 min, with 100 mM NaOH for 5 min, and with water for 10 min.

Capillary temperature: 25.0
Running buffer: pH 2.90 Phosphate buffer, I = 0.04
Injection: Pressure injection at 0.5 psi for 5 s, ramp to operating voltage at 1.33 kV/s.
Detector: UV 214
Migration time: 12 (R), 12.4 (S)
Voltage: 24 kV
Model: Beckman P/ACE 2100

OTHER SUBSTANCES
Simultaneous: bupivacaine

KEY WORDS
chiral

REFERENCE
Amini,A.; Paulsen-Sörman,U. Enantioseparation of local anaesthetic drugs by capillary zone electrophoresis with cyclodextrins as chiral selectors using a partial filling technique, *Electrophoresis*, **1997**, *18*, 1019–1025.

SAMPLE
Matrix: solutions
Sample preparation: Inject an aliquot of a 20 μg/mL solution in buffer with an ionic strength 10 times less than that of the running buffer.

CAPILLARY ELECTROPHORESIS
Capillary: 27 cm × 50 μm fused-silica (20 cm to detector) (Polymicro Technologies)
Capillary preparation: Between runs rinse capillary with running buffer from a different vial for 2 min, equilibrate system for 1 min before analysis. Store capillary in water overnight and rinse with 100 mM NaOH and water each morning. Rinse new capillaries with 100 mM HCl for 5 min, with water for 5 min, with 100 mM NaOH for 5 min, with water for 10 min, then with running buffer.
Capillary temperature: 17
Running buffer: pH 2.90 Phosphate buffer (I = 0.04) containing 160 mg/mL 2-hydroxypropyl-β-cyclodextrin (Sigma) (Prepare buffer by mixing 19.8 mL 1 M NaOH and 23.2 mL 1 M phosphoric acid and make up to 500 mL with water.)
Injection: Pressure injection at 3.4 kPa for 5 s (6 nL).
Detector: UV 214
Migration time: 7.8 (R), 8 (S)
Voltage: 15 kV
Current: 20 μA
Model: Beckman P/ACE 2100

OTHER SUBSTANCES
Simultaneous: bupivacaine

KEY WORDS
chiral

REFERENCE
Amini,A.; Sörman,U.P.; Lindgren,B.H.; Westerlund,D. Enantioseparation of anaesthetic drugs by capillary zone electrophoresis using cyclodextrin-containing background electrolytes, *Electrophoresis*, **1998**, *19*, 731–737.

Mequitazine

Molecular formula: $C_{20}H_{22}N_2S$
Molecular weight: 322.47
CAS Registry No.: 29216-28-2
Merck Index (12th ed.): 5911

SAMPLE
Matrix: solutions
Sample preparation: Inject an aliquot of a 100 μg/mL solution in running buffer.

CAPILLARY ELECTROPHORESIS
Capillary: 36 cm × 50 μm fused-silica (31.5 cm to detector) (Yongnian Optical Conductive Fiber Plant, China), coated with polyacrylamide
Capillary preparation: No details of the polyacrylamide coating process are provided. However, another paper (LC.GC 1997, 15, 40) by this group indicates that they use the procedure of Hjertén, thus: Adjust the pH of 20 mL water to 3.5 with acetic acid, add 80 μL 3-(trimethoxysilyl)propyl methacrylate (3-methacryloxypropyltrimethoxysilane), mix, suck into capillary, let stand at room temperature for 1 h, remove the solution, wash with water. Fill the capillary with a deaerated 3-4% acrylamide solution containing 1 μL/mL N,N,N',N'-tetramethylethylenediamine and 1 mg/mL potassium persulfate, let stand for 30 min, remove excess solution by aspiration, rinse with water, remove water by aspiration, dry at 35° (J. Chromatogr. 1985, 347, 191).
Capillary temperature: 25
Running buffer: 100 mM NaH_2PO_4 adjusted to pH 2.5
Injection: Electrokinetic injection at 15 kV for 3 s.
Detector: UV 200, UV 210
Migration time: 6.16
Voltage: 15 kV
Model: Bio-Rad BioFocus 3000

OTHER SUBSTANCES
Simultaneous: azelastine, biperiden, carvedilol, clidinium bromide, meclizine (meclozine), ofloxacin, zopiclone

KEY WORDS
coated capillary

REFERENCE
Koppenhoefer,B.; Epperlein,U.; Xiaofeng,Z.; Bingcheng,L. Separation of enantiomers of drugs by capillary electrophoresis. Part 4: Hydroxypropyl-γ-cyclodextrin as chiral solvating agent, *Electrophoresis*, **1997**, *18*, 924–930.

SAMPLE
Matrix: solutions
Sample preparation: Inject an aliquot of a 100 μg/mL solution in running buffer.

CAPILLARY ELECTROPHORESIS
Capillary: 30 cm × 50 μm fused-silica (25.5 cm to detector), coated with polyacrylamide
Capillary preparation: Adjust the pH of 20 mL water to 3.5 with acetic acid, add 80 μL 3-(trimethoxysilyl)propyl methacrylate (3-methacryloxypropyltrimethoxysilane), mix, suck into capillary, let stand at room temperature for 1 h, remove the solution, wash with water. Fill the capillary with a deaerated 3-4% acrylamide solution containing 1 μL/mL N,N,N',N'-tetramethylethylenediamine and 1 mg/mL potassium persulfate, let stand for 30 min, remove excess solution by aspiration, rinse with water, remove water by aspiration, dry at 35° (J. Chromatogr. 1985, 347, 191).
Capillary temperature: 25
Running buffer: 100 mM NaH_2PO_4 containing 45 mM α-cyclodextrin, adjusted to pH 2.5 with phosphoric acid
Injection: Electrokinetic injection at 15 kV for 3 s.
Detector: UV 200
Voltage: 15 kV
Model: Bio-Rad BioFocus 3000

KEY WORDS
chiral; coated capillary; comparison with the use of other cyclodextrins; this running buffer gave the greatest enantiomeric separation.; α=1.018

REFERENCE

Lin,B.; Zhu,X.; Koppenhoefer,B.; Epperlein,U. Investigation of 123 chiral drugs by cyclodextrin-modified capillary electrophoresis, *LC.GC*, **1997**, *15*, 40–46.

SAMPLE
Matrix: solutions

CAPILLARY ELECTROPHORESIS

Capillary: 29-36 cm × 50 μm fused-silica (24.5-31.5 cm to detector) (Yongnian Optical Conductive Fiber Plant, China) coated with polyacrylamide

Capillary preparation: Coat capillary as follows. Adjust the pH of 20 mL water to 3.5 with acetic acid, add 80 μL 3-(trimethoxysilyl)propyl methacrylate (3-methacryloxypropyltrimethoxysilane), mix, suck into capillary, let stand at room temperature for 1 h, remove the solution, wash with water. Fill the capillary with a deaerated 3-4% acrylamide solution containing 1 μL/mL N,N,N',N'-tetramethylethylenediamine and 1 mg/mL potassium persulfate, let stand for 30 min, remove excess solution by aspiration, rinse with water, remove water by aspiration, dry at 35° (J. Chromatogr. 1985, 347, 191).

Capillary temperature: 25

Running buffer: 100 mM pH 2.5 NaH_2PO_4 (A) or 100 mM pH 2.5 NaH_2PO_4 containing 45 mM hydroxypropyl-α-cyclodextrin (Wacker, Munich) (B)

Injection: Electromigration at 15 kV for 3 s.

Detector: UV 200; UV 210

Migration time: 6.16 (A); 14.22 (B) (no separation of enantiomers)

Voltage: 15 kV

Model: Bio-Focus 3000

OTHER SUBSTANCES

Also analyzed: albuterol (salbutamol), alprenolol, amorolfine, atenolol, atropine, azelastine, baclofen, bamethan, benproperine, benserazide, biperiden, bisoprolol, brompheniramine, bupivacaine, bupranolol, butamirate, butethamate, carazolol, carbuterol, carteolol, carvedilol, celiprolol, chloroquine, chlorpheniramine, chlorphenoxamine, cicletanine, clenbuterol, clidinium bromide, clobutinol, dimethindene, dipivefrin, disopyramide, dobutamine, doxylamine, fendiline, flecainide, gallopamil, homatropine, ipratropium bromide, isoproterenol (isoprenaline), isothipendyl, ketamine, meclizine, mefloquine, mepindolol, metaclazepam, metaproterenol (orciprenaline), metipranolol, metoprolol, nafronyl (naftidrofuryl), nefopam, nicardipine, norfenefrine, ofloxacin, ornidazole, orphenadrine, oxomemazine, oxprenolol, oxybutynin, phenoxybenzamine, phenylpropanolamine, pholedrine, pindolol, pirbuterol, prilocaine, procyclidine, promethazine, propafenone, propranolol, reproterol, sotalol, sulpride, synephrine, talinolol, terbutaline, tetrahydrozoline (tetryzoline), theodrenaline, tioconazole, tocainide, trihexyphenidyl, trimeprazine (alimemazine), trimipramine, tropicamide, verapamil, zopiclone

KEY WORDS
coated capillary

REFERENCE

Koppenhoefer,B.; Eperlein,U.; Schlunk,R.; Zhu,X.; Lin,B. Separation of enantiomers of drugs by capillary electrophoresis. V. Hydroxypropyl-α-cyclodextrin as chiral solvating agent, *J.Chromatogr.A*, **1998**, *793*, 153–164.

Mercaptobenzothiazole

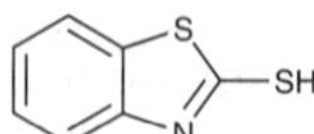

Molecular formula: $C_7H_5NS_2$
Molecular weight: 167.26
CAS Registry No.: 149-30-4
Merck Index (12th ed.): 5916

SAMPLE
Matrix: solutions

Sample preparation: Inject an aliquot of a solution in MeOH.

CAPILLARY ELECTROPHORESIS
Capillary: 43 cm × 50 μm fused-silica (36 cm to detector)
Capillary preparation: Flush capillary with running buffer for 5 min before each run. Wash new capillaries with 1 M NaOH at 60° for 1 h, with water at 60° for 10 min, and with water at 25° for 10 min.
Capillary temperature: 25
Running buffer: 10 mM pH 11.06 Phytic acid (Prepare by mixing a 100 mM solution of phytic acid with a 100 mM solution of the dodecasodium salt of phytic acid to achieve pH 11.06, dilute to 10 mM.)
Injection: Hydrodynamic injection for 1 s.
Detector: UV 200
Migration time: 3
Voltage: 25 kV
Model: Spectra-Physics Model 1000

OTHER SUBSTANCES
Simultaneous: benzotriazole, dimethylbenzotriazole, 5-tolyltriazole

REFERENCE
Lin,C.-E.; Chen,C.-C. Migration behavior and separation of aromatic triazole and thiazole compounds by capillary electrophoresis, *J.Chromatogr.A*, **1996**, *731*, 299–303.

Mesalamine

Molecular formula: $C_7H_7NO_3$
Molecular weight: 153.14
CAS Registry No.: 89-57-6
Merck Index (12th ed.): 5964

SAMPLE
Matrix: solutions

CAPILLARY ELECTROPHORESIS
Capillary: 65.5 cm × 50 μm fused-silica (57.3 cm to detector) (Yongnian Optical Factory, China)
Capillary preparation: After each run rinse capillary with 100 mM NaOH for 5 min and with water for 5 min then re-equilibrate with running buffer for 10 min.
Capillary temperature: 22 ± 0.5
Running buffer: EtOH:buffer 10:90 (Buffer was 40 mM borax containing 10 mM β-cyclodextrin,adjusted to pH 7.0 with phosphoric acid.)
Injection: Electromigration at 20 kV for 20 s.
Detector: UV 280
Migration time: 14.5
Voltage: 20 ± 0.3
Model: 1229 HPCE (Beijing)
Limit of detection: 5.41 μg/mL

OTHER SUBSTANCES
Simultaneous: p-aminophenol, benzeneazosalicylic acid, salicylic acid, zinc 5-aminosalicylate

KEY WORDS
comparison with HPLC

REFERENCE
Zhang,S.S.; Liu,H.X.; Yuan,Z.B. Comparison of high-performance capillary electrophoresis and liquid chromatography on analysis of zinc 5-aminosalicylate dihydrate and related materials, *J.Chromatogr.B*, **1998**, *705*, 165–170.

SAMPLE
Matrix: urine
Sample preparation: Inject a sample directly.

CAPILLARY ELECTROPHORESIS
Capillary: 44 cm × 50 μm fused-silica (36.5 cm to detector)
Capillary preparation: Before each run wash with 100 mM NaOH for 1 min, with water for 1 min, and with running buffer for 2 min.
Capillary temperature: 30
Running buffer: 75 mM pH 7 Sodium phosphate buffer
Injection: Hydrodynamic injection for 1 s.
Detector: UV 305
Migration time: <10
Voltage: 13.8 kV
Current: 70 μA
Model: SpectraPhoresis 1000 (Thermo Separation Products)

OTHER SUBSTANCES
Extracted: metabolites

KEY WORDS
direct injection

REFERENCE
Cummins,C.L.; O'Neil,W.M.; Lloyd,D.K.; Wainer,I.W. CZE assay of *p*-aminosalicylic acid and its N-acetyl metabolite in urine (Abstract APQ 1011), *Pharm.Res.*, **1996**, *13*, S5.

SAMPLE
Matrix: urine
Sample preparation: Inject an aliquot directly.

CAPILLARY ELECTROPHORESIS
Capillary: 45 cm fused-silica (45 cm to detector)
Capillary preparation: Before each run equilibrate with running buffer for 3 min. After each run wash with 100 mM NaOH then water for 3 min. Condition new capillaries by rinsing with 100 mM NaOH for 30 min and with water for 30 min.
Capillary temperature: 16 ± 0.1
Running buffer: 20 mM Sodium carbonate containing 2 mM terbium chloride and 2 mM EDTA, pH adjusted to 11 by 100 mM NaOH (The terbium absorbs the laser energy and transfers it to the analyte which then fluoresces.)
Injection: Injection at high pressure for 5 s.
Detector: F ex 325 (Omnichrome Series 39 He-Cd laser) em 547
Migration time: 7
Voltage: 20 kV
Current: <100 μA
Model: Beckman P/ACE 5500
Limit of detection: 100 nM

OTHER SUBSTANCES
Extracted: gentisic acid, salicylic acid, salicyluric acid

KEY WORDS
direct injection; resonance energy transfer

REFERENCE
Petersen,J.R.; Bissell,M.G.; Mohammad,A.A. Laser induced resonance energy transfer -a novel approach towards achieving high sensitivity in capillary electrophoresis. I. Clinical diagnostic application, *J.Chromatogr.A*, **1996**, *744*, 37–44.

SAMPLE
Matrix: urine

Sample preparation: Centrifuge if necessary, inject an aliquot.

CAPILLARY ELECTROPHORESIS
Capillary: 44 cm × 50 μm (36.5 cm to detector) (Polymicro Technologies)
Capillary preparation: Between each run condition capillary with 100 mM NaOH for 1 min, with water for 1 min, and with running buffer for 2 min. At the start of each day wash capillary with 1 M HCl at 30° for 5 min, with 1 M HCl at 60° for 5 min, with 100 mM NaOH at 60° for 5 min, and with water at 30° for 5 min.
Capillary temperature: 30
Running buffer: 75 mM pH 7.0 Sodium phosphate buffer
Injection: Hydrodynamic injection with 10.3 kPa vacuum for 1 s.
Detector: UV 254 for 5 min then UV 305
Migration time: 8.7
Voltage: 13.8 kV
Current: ca. 70 μA
Model: SpectraPhoresis 1000 (Thermo Separation Products)

OTHER SUBSTANCES
Extracted: metabolites, N-acetyl-p-aminosalicylic acid

KEY WORDS
direct injection

REFERENCE
Cummins,C.L.; O'Neil,W.M.; Soo,E.C.; Lloyd,D.K.; Wainer,I.W. Determination of p-aminosalicylic acid and its N-acetylated metabolite in human urine by capillary zone electrophoresis as a measure of in vivo N-acetyltransferase 1 activity, *J.Chromatogr.B*, **1997**, *697*, 283–288.

Mesoridazine

Molecular formula: $C_{21}H_{26}N_2OS_2$
Molecular weight: 386.58
CAS Registry No.: 5588-33-0, 32672-69-8 (besylate)
Merck Index (12th ed.): 5970
Lednicer: 1 389

SAMPLE
Matrix: solutions

CAPILLARY ELECTROPHORESIS
Capillary: 59.6 cm × 75 μm fused-silica (52.1 cm to detector) (Polymicro Technologies)
Capillary preparation: Purge with running buffer before each run. At the beginning of each day purge using 50-60 kPa vacuum with 500 mM NaOH for 5 min, with water for 5 min, with MeCN for 5 min, and with running buffer for 5 min.
Running buffer: MeCN:MeOH:acetic acid 49:50:1 containing 20 mM ammonium acetate
Injection: Hydrostatic injection at 10 cm for 5 s.
Detector: UV 214
Migration time: 3.82
Voltage: 30 kV
Model: Waters Quanta 4000

OTHER SUBSTANCES
Simultaneous: benzquinamide, deserpidine, ethopropazine, methotrimeprazine, molindone, promazine, promethazine, reserpine, thioridazine, thiothixene

REFERENCE
Leung,G.N.W.; Tang,H.P.O.; Tso,T.S.C.; Wan,T.S.M. Separation of basic drugs with non-aqueous capillary electrophoresis, *J.Chromatogr.A*, **1996**, *738*, 141–154.

Mestranol

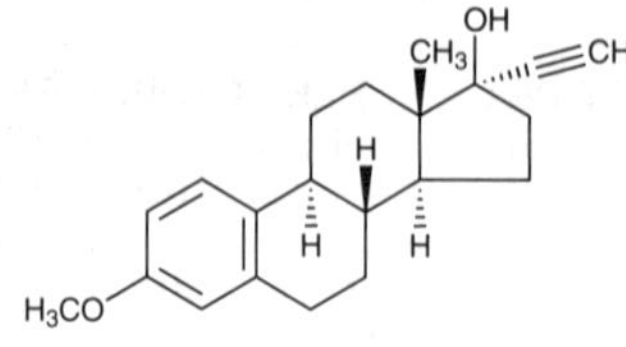

Molecular formula: $C_{21}H_{26}O_2$
Molecular weight: 310.44
CAS Registry No.: 72-33-3
Merck Index (12th ed.): 5976
Lednicer: 1 162

SAMPLE
Matrix: solutions

CAPILLARY ELECTROPHORESIS
Capillary: 48.5 cm × 50 μm fused-silica (40 cm to detector) (Hewlett-Packard)
Capillary preparation: Before each run flush capillary with 100 mM NaOH for 2 min and with running buffer for 5 min.
Capillary temperature: 25
Running buffer: MeOH:buffer 10:90 (Buffer was 20 mM pH 8 Sodium borate/sodium phosphate buffer containing 50 mM sodium cholate.)
Injection: Pressure injection at 50 mbar for 1-2 s.
Detector: UV 210
Migration time: 12.5
Voltage: 20 kV
Model: Hewlett-Packard ^{3D}CE

OTHER SUBSTANCES
Simultaneous: ethinyl estradiol, ethynodiol diacetate, norethindrone, norethindrone acetate, norethynodrel, norgestrel

REFERENCE
Poole,S.K.; Poole,C.F. Separation of pharmaceutically important estrogens by micellar electrokinetic chromatography, *J.Chromatogr.A*, **1996**, *749*, 247–255.

Metaclazepam

Molecular formula: $C_{18}H_{18}BrClN_2O$
Molecular weight: 393.71
CAS Registry No.: 84031-17-4, 65517-27-3
Merck Index (12th ed.): 5980

SAMPLE
Matrix: solutions
Sample preparation: Inject an aliquot of a 100 μg/mL solution in running buffer.

CAPILLARY ELECTROPHORESIS
Capillary: 32 cm × 50 μm fused-silica (27.5 cm to detector) (Yongnian Optical Conductive Fiber Plant, China), coated with polyacrylamide
Capillary preparation: No details of the polyacrylamide coating process are provided. However, another paper (LC.GC 1997, 15, 40) by this group indicates that they use the procedure of Hjertén, thus: Adjust the pH of 20 mL water to 3.5 with acetic acid, add 80 μL 3-(trimethoxysilyl)propyl methacrylate (3-methacryloxypropyltrimethoxysilane), mix, suck into capillary, let stand at room temperature for 1 h, remove the solution, wash with water. Fill the capillary with a deaerated 3-4% acrylamide solution containing 1 μL/mL N,N,N',N'-tetramethylethylenediamine and 1 mg/mL potassium persulfate, let stand for 30 min, remove excess solution by aspiration, rinse with water, remove water by aspiration, dry at 35° (J. Chromatogr. 1985, 347, 191).

Capillary temperature: 25
Running buffer: 100 mM NaH_2PO_4 adjusted to pH 2.5 (A) or 100 mM NaH_2PO_4 containing 45 mM hydroxypropyl-gamma-cyclodextrin, adjusted to pH 2.5 (B)
Injection: Electrokinetic injection at 15 kV for 3 s.
Detector: UV 200, UV 210
Migration time: 5.67 (A), 9.61, 9.97 (B, enantiomers)
Voltage: 15 kV
Model: Bio-Rad BioFocus 3000

OTHER SUBSTANCES
Simultaneous: atropine, carazolol, cicletanine, dimethindene, fendiline, homatropine, ipratropium bromide, isothipendyl, mefloquine, nafronyl (naftidrofuryl), nefopam, nicardipine, promethazine, reproterol, tetrahydrozoline (tetryzoline), theodrenaline, tioconazole, trihexyphenidyl, trimeprazine (alimemazine), trimipramine

KEY WORDS
coated capillary; chiral

REFERENCE
Koppenhoefer,B.; Epperlein,U.; Xiaofeng,Z.; Bingcheng,L. Separation of enantiomers of drugs by capillary electrophoresis. Part 4: Hydroxypropyl-γ-cyclodextrin as chiral solvating agent, *Electrophoresis*, **1997**, *18*, 924–930.

SAMPLE
Matrix: solutions
Sample preparation: Inject an aliquot of a 100 µg/mL solution in running buffer.

CAPILLARY ELECTROPHORESIS
Capillary: 30 cm × 50 µm fused-silica (25.5 cm to detector), coated with polyacrylamide
Capillary preparation: Adjust the pH of 20 mL water to 3.5 with acetic acid, add 80 µL 3-(trimethoxysilyl)propyl methacrylate (3-methacryloxypropyltrimethoxysilane), mix, suck into capillary, let stand at room temperature for 1 h, remove the solution, wash with water. Fill the capillary with a deaerated 3-4% acrylamide solution containing 1 µL/mL N,N,N',N'-tetramethylethylenediamine and 1 mg/mL potassium persulfate, let stand for 30 min, remove excess solution by aspiration, rinse with water, remove water by aspiration, dry at 35° (J. Chromatogr. 1985, 347, 191).
Capillary temperature: 25
Running buffer: 100 mM NaH_2PO_4 containing 30 mM hydroxypropyl-β- cyclodextrin, adjusted to pH 2.5 with phosphoric acid
Injection: Electrokinetic injection at 15 kV for 3 s.
Detector: UV 200
Migration time: 21.9, 22.2 (enantiomers)
Voltage: 15 kV
Model: Bio-Rad BioFocus 3000

OTHER SUBSTANCES
Simultaneous: chlorpheniramine, dimethindene, homatropine, mefloquine, tetrahydrozoline (tetryzoline), theodrenaline

KEY WORDS
chiral; coated capillary; comparison with the use of other cyclodextrins

REFERENCE
Lin,B.; Zhu,X.; Koppenhoefer,B.; Epperlein,U. Investigation of 123 chiral drugs by cyclodextrin-modified capillary electrophoresis, *LC.GC*, **1997**, *15*, 40–46.

SAMPLE
Matrix: solutions

CAPILLARY ELECTROPHORESIS
Capillary: 29-36 cm × 50 µm fused-silica (24.5-31.5 cm to detector) (Yongnian Optical Conductive Fiber Plant, China) coated with polyacrylamide

Capillary preparation: Coat capillary as follows. Adjust the pH of 20 mL water to 3.5 with acetic acid, add 80 µL 3-(trimethoxysilyl)propyl methacrylate (3-methacryloxypropyltrimethoxysilane), mix, suck into capillary, let stand at room temperature for 1 h, remove the solution, wash with water. Fill the capillary with a deaerated 3-4% acrylamide solution containing 1 µL/mL N,N,N',N'-tetramethylethylenediamine and 1 mg/mL potassium persulfate, let stand for 30 min, remove excess solution by aspiration, rinse with water, remove water by aspiration, dry at 35° (J. Chromatogr. 1985, 347, 191).
Capillary temperature: 25
Running buffer: 100 mM pH 2.5 NaH$_2$PO$_4$ (A) or 100 mM pH 2.5 NaH$_2$PO$_4$ containing 45 mM hydroxypropyl-α-cyclodextrin (Wacker, Munich) (B)
Injection: Electromigration at 15 kV for 3 s.
Detector: UV 200; UV 210
Migration time: 5.67 (A); 9.78, 10.32 (B) (enantiomers)
Voltage: 15 kV
Model: Bio-Focus 3000

OTHER SUBSTANCES
Also analyzed: albuterol (salbutamol), alprenolol, amorolfine, atenolol, atropine, azelastine, baclofen, bamethan, benproperine, benserazide, biperiden, bisoprolol, brompheniramine, bupivacaine, bupranolol, butamirate, butethamate, carazolol, carbuterol, carteolol, carvedilol, celiprolol, chloroquine, chlorpheniramine, chlorphenoxamine, cicletanine, clenbuterol, clidinium bromide, clobutinol, dimethindene, dipivefrin, disopyramide, dobutamine, doxylamine, fendiline, flecainide, gallopamil, homatropine, ipratropium bromide, isoproterenol (isoprenaline), isothipendyl, ketamine, meclizine, mefloquine, mepindolol, mequitazine, metaproterenol (orciprenaline), metipranolol, metoprolol, nafronyl (naftidrofuryl), nefopam, nicardipine, norfenefrine, ofloxacin, ornidazole, orphenadrine, oxomemazine, oxprenolol, oxybutynin, phenoxybenzamine, phenylpropanolamine, pholedrine, pindolol, pirbuterol, prilocaine, procyclidine, promethazine, propafenone, propranolol, reproterol, sotalol, sulpride, synephrine, talinolol, terbutaline, tetrahydrozoline (tetryzoline), theodrenaline, tioconazole, tocainide, trihexyphenidyl, trimeprazine (alimemazine), trimipramine, tropicamide, verapamil, zopiclone

---

KEY WORDS
coated capillary; chiral

REFERENCE
Koppenhoefer,B.; Eperlein,U.; Schlunk,R.; Zhu,X.; Lin,B. Separation of enantiomers of drugs by capillary electrophoresis. V. Hydroxypropyl-α-cyclodextrin as chiral solvating agent, *J.Chromatogr.A*, **1998**, *793*, 153–164.

Metaproterenol

Molecular formula: C$_{11}$H$_{17}$NO$_3$
Molecular weight: 211.26
CAS Registry No.: 586-06-1, 5874-97-5 (sulfate)
Merck Index (12th ed.): 5992
Lednicer: 1 64

SAMPLE
Matrix: solutions
Sample preparation: Inject an aliquot of a 10 µg/mL solution of the compound in water.

CAPILLARY ELECTROPHORESIS
Capillary: 27 cm × 50 µm neutrally-coated capillary (7 cm to detector) (Beckman)
Capillary preparation: Before each analysis rinse capillary with 100 mM HCl for 30 s, with water for 2 min, and with running buffer for 2 min
Capillary temperature: 15 ± 0.1
Running buffer: 25 mM pH 2.5 Phosphate buffer containing 30 mM dimethyl-β-cyclodextrin
Injection: Electrokinetic injection at 10 kV for 2 s
Detector: UV 200

Migration time: 0.73, 0.76 (enantiomers)
Voltage: 30 kV
Model: Beckman P/ACE system 5500

OTHER SUBSTANCES
Simultaneous: isoproterenol

KEY WORDS
chiral; coated capillary

REFERENCE
Aumatell,A.; Guttman,A. Ultra-fast chiral separation of basic drugs by capillary electrophoresis, *J.Chromatogr.A*, **1995**, *717*, 229–234.

SAMPLE
Matrix: solutions
Sample preparation: Inject an aliquot of a 10 μg/mL solution in water.

CAPILLARY ELECTROPHORESIS
Capillary: 68.5 × 50 μm fused-silica (60 cm to detector) (Hewlett-Packard)
Capillary preparation: After each run rinse with 100 mM NaOH for 5 min and with running buffer for 5 min. Condition a new capillary with 1 M NaOH at 40° for 20 min and with 100 mM NaOH at 40° for 10 min, rinse with water at 25° for 20 min, and rinse with running buffer.
Capillary temperature: 25
Running buffer: 50 mM pH 8.3 Tris buffer
Injection: Hydrostatic injection at 50 mbar for 2-6 s, flush with buffer for 4 s
Detector: UV 200
Migration time: 3.520
Voltage: 30 kV
Model: Hewlett-packard HP 3DCE

OTHER SUBSTANCES
Simultaneous: albuterol, cimaterol, clenbuterol, fenoterol, isoxsuprine, ractopamine, ritodrine, RU 42 173, terbutaline

KEY WORDS
comparison with C18 bonded capillary

REFERENCE
Chevolleau,S.; Tulliez,J. Optimization of the separation of β-agonists by capillary electrophoresis on untreated and C18 bonded silica capillaries, *J.Chromatogr.A*, **1995**, *715*, 345–354.

SAMPLE
Matrix: solutions
Sample preparation: Inject an aliquot of a 100 μg/mL solution.

CAPILLARY ELECTROPHORESIS
Capillary: 27 cm × 50 μm neutrally coated (20 cm to detector)
Capillary preparation: Before each run rinse capillary with 100 mM HCl for 30 s, with water for 2 min, and with running buffer for 2 min
Capillary temperature: 20
Running buffer: 25 mM pH 2.5 Phosphate buffer containing 10 mM dimethyl-β-cyclodextrin
Injection: Pressure injection at 0.5 psi for 5 s
Detector: UV 214
Migration time: 5, 5.5 (enantiomers)
Voltage: 500 V/cm
Current: 28 μA
Model: Beckman P/ACE system 5500

KEY WORDS
chiral; coated capillary

REFERENCE

Guttman,A. Novel separation scheme for capillary electrophoresis of enantiomers, *Electrophoresis*, **1995**, *16*, 1900–1905.

SAMPLE

Matrix: solutions
Sample preparation: Inject an aliquot of a 100 µg/mL solution in water:running buffer 50:50.

CAPILLARY ELECTROPHORESIS

Capillary: 44.5 cm × 50 µm acrylamide-coated fused-silica (Bio-Rad)
Capillary temperature: 30
Running buffer: 100 mM NaH_2PO_4 containing 15 mM gamma-cyclodextrin, adjusted to pH 2.5 with phosphoric acid
Injection: Electrokinetic injection at 8 kV for 6 s.
Detector: UV 200
Migration time: 9.45
Voltage: 14 kV
Model: Bio-Rad BioFocus 3000

OTHER SUBSTANCES

Also analyzed: albuterol, alprenolol, atenolol, atropine, baclofen, bamethan, benserazide, biperiden, bisoprolol, bupivacaine, bupranolol, butetamate, carazolol, carbuterol, carvedilol, celiprolol, chloroquine, chlorpheniramine (chlorphenamine), clidinium bromide, clobutinol, disopyramide, dobutamine, flecainide, homatropine, ipratropium bromide, isoproterenol, isothipendyl, ketamine, mefloquine, mequitazine, metipranolol, nafronyl (naftidrofuryl), nefopam, ofloxacin, orphenadrine, oxomemazine, oxprenolol, phenoxybenzamine, pholedrine, pindolol, pirbuterol, prilocaine, promethazine, propafenone, propranolol, sotalol, synephrine, terbutaline, tetrahydrozoline (tetryzoline), tocainide, trihexyphenidyl, trimeprazine (alimemazine), trimipramine, tropicamide, verapamil, zopiclone

KEY WORDS

coated capillary; achiral

REFERENCE

Koppenhoefer,B.; Epperlein,U.; Christian,B.; Yibing,J.; Yuying,C.; Bingcheng,L. Separation of enantiomers of drugs by capillary electrophoresis. I. γ-Cyclodextrin as chiral solvating agent, *J.Chromatogr.A*, **1995**, *717*, 181–190.

SAMPLE

Matrix: solutions
Sample preparation: Inject an aliquot of a 100 µg/mL solution in water:running buffer 50:50.

CAPILLARY ELECTROPHORESIS

Capillary: 44.5 cm × 50 µm polyacrylamide-coated fused-silica (40 cm to detector) (Bio-Rad)
Capillary temperature: 30
Running buffer: 100 mM NaH_2PO_4 containing 15 mM β-cyclodextrin, adjusted to pH 2.5 with phosphoric acid
Injection: Electrokinetic injection at 8 kV for 6 s.
Detector: UV 200
Migration time: 9.25, 9.39 (enantiomers)
Voltage: 14 kV
Model: Bio-Rad Bio-Focus 3000

OTHER SUBSTANCES

Simultaneous: carvedilol, chlorpheniramine, ketamine, tetrahydrozoline, tropicamide, zopiclone
Also analyzed: albuterol (not chiral), atenolol (not chiral), benserazide (not chiral), biperiden (not chiral), bupivacaine (not chiral), butethamate (not chiral), carazolol (not chiral), carbuterol (not chiral), clidinium bromide (not chiral), disopyramide (not chiral), dobutamine (not chiral), flecainide (not chiral), ipratropium (not chiral), isothipendyl (not chiral), mequitazine (not chiral), metipranolol (not chiral), nafronyl (naftidrofuryl) (not chiral), nefopam (not chiral), ofloxacin (not chiral), orphenadrine (not chiral), pindolol (not chiral), pirbuterol (not chiral), prilo-

caine (not chiral), propafenone (not chiral), sotalol (not chiral), tocainide (not chiral), trimipramine (not chiral)

KEY WORDS
coated capillary; chiral

REFERENCE
Koppenhoefer,B.; Epperlein,U.; Christian,B.; Lin,B.; Ji,Y.; Chen,Y. Separation of enantiomers of drugs by capillary electrophoresis. III. β-cyclodextrin as chiral solvating agent, *J.Chromatogr.A*, **1996**, *735*, 333–343.

SAMPLE
Matrix: solutions
Sample preparation: Inject an aliquot of a 100 μg/mL solution in running buffer.

CAPILLARY ELECTROPHORESIS
Capillary: 29 cm × 50 μm fused-silica (24.5 cm to detector) (Yongnian Optical Conductive Fiber Plant, China), coated with polyacrylamide
Capillary preparation: No details of the polyacrylamide coating process are provided. However, another paper (LC.GC 1997, 15, 40) by this group indicates that they use the procedure of Hjertén, thus: Adjust the pH of 20 mL water to 3.5 with acetic acid, add 80 μL 3-(trimethoxysilyl)propyl methacrylate (3-methacryloxypropyltrimethoxysilane), mix, suck into capillary, let stand at room temperature for 1 h, remove the solution, wash with water. Fill the capillary with a deaerated 3-4% acrylamide solution containing 1 μL/mL N,N,N',N'-tetramethylethylenediamine and 1 mg/mL potassium persulfate, let stand for 30 min, remove excess solution by aspiration, rinse with water, remove water by aspiration, dry at 35° (J. Chromatogr. 1985, 347, 191).
Capillary temperature: 25
Running buffer: 100 mM NaH$_2$PO$_4$ adjusted to pH 2.5
Injection: Electrokinetic injection at 15 kV for 3 s.
Detector: UV 200, UV 210
Migration time: 5.17
Voltage: 15 kV
Model: Bio-Rad BioFocus 3000

OTHER SUBSTANCES
Simultaneous: albuterol, alprenolol, atenolol, baclofen, bamethan, benproperine, benserazide, bisoprolol, bupranolol, butamirate, butethamate, carbuterol, celiprolol, clenbuterol, clobutinol, dipivefrin, isoproterenol (isoprenaline), metipranolol, metoprolol, norfenefrine, ornidazole, oxprenolol, phenylpropanolamine, pholedrine, pirbuterol, prilocaine, procyclidine, sotalol, synephrine, terbutaline, tocainide

KEY WORDS
coated capillary

REFERENCE
Koppenhoefer,B.; Epperlein,U.; Xiaofeng,Z.; Bingcheng,L. Separation of enantiomers of drugs by capillary electrophoresis. Part 4: Hydroxypropyl-γ-cyclodextrin as chiral solvating agent, *Electrophoresis*, **1997**, *18*, 924–930.

SAMPLE
Matrix: solutions
Sample preparation: Inject an aliquot of a 100 μg/mL solution in running buffer.

CAPILLARY ELECTROPHORESIS
Capillary: 30 cm × 50 μm fused-silica (25.5 cm to detector), coated with polyacrylamide
Capillary preparation: Adjust the pH of 20 mL water to 3.5 with acetic acid, add 80 μL 3-(trimethoxysilyl)propyl methacrylate (3-methacryloxypropyltrimethoxysilane), mix, suck into capillary, let stand at room temperature for 1 h, remove the solution, wash with water. Fill the capillary with a deaerated 3-4% acrylamide solution containing 1 μL/mL N,N,N',N'-tetramethylethylenediamine and 1 mg/mL potassium persulfate, let stand for 30 min, remove excess solution by aspiration, rinse with water, remove water by aspiration, dry at 35° (J. Chromatogr. 1985, 347, 191).

Capillary temperature: 25
Running buffer: 100 mM NaH_2PO_4 containing 45 mM hydroxypropyl-β- cyclodextrin, adjusted
to pH 2.5 with phosphoric acid
Injection: Electrokinetic injection at 15 kV for 3 s.
Detector: UV 200
Voltage: 15 kV
Model: Bio-Rad BioFocus 3000

KEY WORDS
chiral; coated capillary; comparison with the use of other cyclodextrins; this running buffer gave
the greatest enantiomeric separation.; α=1.059

REFERENCE
Lin,B.; Zhu,X.; Koppenhoefer,B.; Epperlein,U. Investigation of 123 chiral drugs by cyclodextrin-modified capillary electrophoresis, *LC.GC*, **1997**, *15*, 40–46.

SAMPLE
Matrix: solutions

CAPILLARY ELECTROPHORESIS
Capillary: 48.5 cm × 50 μm fused-silica (40 cm to detector) (Supelco)
Capillary preparation: Condition capillary with running buffer for 3 min before each injection.
Capillary temperature: 15
Running buffer: 25 mM Citric acid adjusted to pH 4.5 with Tris, containing 2% dermatan sulfate
(number average 15270, mass average 22260, z average 30390, polydispersity 1.458) (Opocrin,
Corlo, Italy)
Injection: Pressure injection at 5 kPa for 10 s.
Detector: UV 220
Migration time: 19.8, 20.6 (enantiomers)
Voltage: 15 kV
Model: Hewlett-Packard ^{3D}CE

OTHER SUBSTANCES
Simultaneous: albuterol
Also analyzed: metaproterenol, terbutaline

KEY WORDS
chiral

REFERENCE
Gotti,R.; Cavrini,V.; Andrisano,V.; Mascellani,G. Dermatan sulfate as useful chiral selector in capillary electrophoresis, *J.Chromatogr.A*, **1998**, *814*, 205–211.

SAMPLE
Matrix: solutions

CAPILLARY ELECTROPHORESIS
Capillary: 29-36 cm × 50 μm fused-silica (24.5-31.5 cm to detector) (Yongnian Optical Conductive
Fiber Plant, China) coated with polyacrylamide
Capillary preparation: Coat capillary as follows. Adjust the pH of 20 mL water to 3.5 with
acetic acid, add 80 μL 3-(trimethoxysilyl)propyl methacrylate (3-methacryloxypropyltrimethox-
ysilane), mix, suck into capillary, let stand at room temperature for 1 h, remove the solution,
wash with water. Fill the capillary with a deaerated 3-4% acrylamide solution containing 1
μL/mL N,N,N',N'-tetramethylethylenediamine and 1 mg/mL potassium persulfate, let stand
for 30 min, remove excess solution by aspiration, rinse with water, remove water by aspiration,
dry at 35° (J. Chromatogr. 1985, 347, 191).
Capillary temperature: 25
Running buffer: 100 mM pH 2.5 NaH_2PO_4 (A) or 100 mM pH 2.5 NaH_2PO_4 containing 45 mM
hydroxypropyl-α-cyclodextrin (Wacker, Munich) (B)
Injection: Electromigration at 15 kV for 3 s.
Detector: UV 200; UV 210
Migration time: 5.17 (A); 6.97 (B) (no separation of enantiomers)

Voltage: 15 kV
Model: Bio-Focus 3000

OTHER SUBSTANCES
Also analyzed: albuterol (salbutamol), alprenolol, amorolfine, atenolol, atropine, azelastine, baclofen, bamethan, benproperine, benserazide, biperiden, bisoprolol, brompheniramine, bupivacaine, bupranolol, butamirate, butethamate, carazolol, carbuterol, carteolol, carvedilol, celiprolol, chloroquine, chlorpheniramine, chlorphenoxamine, cicletanine, clenbuterol, clidinium bromide, clobutinol, dimethindene, dipivefrin, disopyramide, dobutamine, doxylamine, fendiline, flecainide, gallopamil, homatropine, ipratropium bromide, isoproterenol (isoprenaline), isothipendyl, ketamine, meclizine, mefloquine, mepindolol, mequitazine, metaclazepam, metipranolol, metoprolol, nafronyl (naftidrofuryl), nefopam, nicardipine, norfenefrine, ofloxacin, ornidazole, orphenadrine, oxomemazine, oxprenolol, oxybutynin, phenoxybenzamine, phenylpropanolamine, pholedrine, pindolol, pirbuterol, prilocaine, procyclidine, promethazine, propafenone, propranolol, reproterol, sotalol, sulpride, synephrine, talinolol, terbutaline, tetrahydrozoline (tetryzoline), theodrenaline, tioconazole, tocainide, trihexyphenidyl, trimeprazine (alimemazine), trimipramine, tropicamide, verapamil, zopiclone

KEY WORDS
coated capillary

REFERENCE
Koppenhoefer,B.; Eperlein,U.; Schlunk,R.; Zhu,X.; Lin,B. Separation of enantiomers of drugs by capillary electrophoresis. V. Hydroxypropyl-α-cyclodextrin as chiral solvating agent, *J.Chromatogr.A*, **1998**, *793*, 153–164.

Metformin

Molecular formula: $C_4H_{11}N_5$
Molecular weight: 129.17
CAS Registry No.: 657-24-9
Merck Index (12th ed.): 6001

SAMPLE
Matrix: blood
Sample preparation: 100 μL Plasma + 5 μL 40 μg/mL phenformin in water + 300 μL McCN, mix thoroughly, centrifuge at 12000 g for 1 min. Remove the supernatant and evaporate it to dryness under reduced pressure at 70°, reconstitute the residue in 50 μL pH 7.8 phosphate buffer, add 5 μL bromothymol blue solution, mix, extract with 1 mL chloroform for 1 min, centrifuge at 12000 g for 2 min, repeat extraction. Combine the organic layers and evaporate them to dryness under a stream of nitrogen at 40°, reconstitute the residue in 100 μL 200 μM phosphoric acid, inject an aliquot. (The analyte forms an ion pair with bromothymol blue and this analyte is extracted. Prepare bromothymol blue solution by dissolving 62 mg bromothymol blue in 2 mL 100 mM NaOH, make up top 10 mL with water, sonicate, adjust to pH 7.8 with concentrated HCl or NaOH. Prepare the pH 7.8 phosphate buffer by dissolving 5.52 g $NaH_2PO_4.H_2O$ in 21 mL 2 M NaOH, adjust to pH 7.8 with saturated NaOH, make up to 100 mL with water.)

CAPILLARY ELECTROPHORESIS
Capillary: 40 cm × 50 μm fused-silica (32.5 cm to detector) (Polymicro Technologies)
Capillary preparation: After each run rinse capillary with 100 mM NaOH for 1 min, with water for 1 min, then with running buffer for 5 min. Condition a new capillary with 1 M NaOH at 60° for 20 min, with 100 mM NaOH at 60° for 5 min, and with water at 20° for 5 min.
Capillary temperature: 20
Running buffer: 100 mM pH 2.5 Phosphate buffer
Injection: Electrokinetic injection at 10 kV for 10 s.
Detector: UV 195
Migration time: 4
Internal standard: phenformin (3)

Voltage: 20 kV
Current: 68 μA
Model: SpectraPhoresis 1000
Limit of quantitation: 250 ng/mL
Limit of detection: 100 ng/mL

KEY WORDS
plasma

REFERENCE
Song,J.-Z.; Chen,H.-F.; Tian,S.-J.; Sun,Z.-P. Determination of metformin in plasma by capillary electrophoresis using field-amplified sample stacking technique, *J.Chromatogr.B*, **1998**, *708*, 277–283.

Methacholine chloride

Molecular formula: $C_8H_{18}CINO_2$
Molecular weight: 195.69
CAS Registry No.: 62-51-1
Merck Index (12th ed.): 6003

SAMPLE
Matrix: solutions
Sample preparation: Inject an aliquot of a solution in 100 μg/mL KCl.

CAPILLARY ELECTROPHORESIS
Capillary: 87 cm × 75 μm fused-silica (80 cm to detector) (Scientific Glass Engineering)
Capillary preparation: Before each injection rinse capillary at 130 kPa with 100 mM HCl for 1 min, with 10 mM KOH for 1 min, and with running buffer for 2 min.
Running buffer: 10 mM pH 4.85 Creatinine hydrochloride buffer
Injection: Pressure injection at 3.3 kPa for 10 s.
Detector: UV 230
Migration time: 8
Internal standard: potassium (5)
Voltage: 20 kV
Model: Beckman P/ACE 2200

OTHER SUBSTANCES
Simultaneous: degradation products

KEY WORDS
indirect UV detection

REFERENCE
van der Schans,M.J.; Reijenga,J.C.; Everaerts,F.M. Quality control of histamine and methacholine in diagnostic solutions with capillary electrophoresis, *J.Chromatogr.A*, **1996**, *735*, 387–393.

Methacycline

Molecular formula: $C_{22}H_{22}N_2O_8$
Molecular weight: 442.43
Merck Index (12th ed.): 6007
Lednicer: 2 227

SAMPLE
Matrix: solutions

CAPILLARY ELECTROPHORESIS
Capillary: 50 cm × 50 μm (50 cm to detector) (Polymicro Technologies)
Capillary preparation: Purge capillary with 10 mM NaOH for 2 min after each injection.
Running buffer: MeOH:buffer 40:60 (Buffer was 30 mM pH 3.0 citric acid containing 24.5 mM β-alanine.)
Injection: Electrokinetic injection at 10 kV for 5 s.
Detector: UV 254
Migration time: 7.7
Voltage: 30 kV
Model: Perkin Elmer/Applied Biosystems Model 270A-HT

OTHER SUBSTANCES
Simultaneous: chlortetracycline, meclocycline, minocycline, oxytetracycline, tetracycline
Interfering: doxycycline

REFERENCE
Pesek,J.J.; Matyska,M.T. Separation of tetracyclines by high-performance capillary electrophoresis and capillary electrochromatography, *J.Chromatogr.A*, **1996**, *736*, 313–320.

Methadone

Molecular formula: $C_{21}H_{27}NO$
Molecular weight: 309.45
CAS Registry No.: 76-99-3, 1095-90-5 (HCl)
Merck Index (12th ed.): 6008

SAMPLE
Matrix: blood, hair, urine
Sample preparation: Serum. 1 mL Serum + 100 μL 2 μg/mL diphenhydramine in MeOH + 200 μL concentrated ammonia adjusted to pH 9-10 + 3 mL n-hexane, shake for 10 min, centrifuge at 3000 g for 10 min. Remove the supernatant and evaporate it to dryness under a stream of nitrogen at 60°, reconstitute with 200 μL MeOH, inject an aliquot. Urine. 1 mL Urine + 100 μL 30 μg/mL diphenhydramine in MeOH + 200 μL concentrated ammonia adjusted to pH 9-10 + 3 mL n-hexane, shake for 10 min, centrifuge at 3000 g for 10 min. Remove the supernatant and evaporate it to dryness under a stream of nitrogen at 60°, reconstitute with 1 mL MeOH, inject an aliquot. Hair. Wash hair with MeOH and acetone. 50 mg Washed hair + 200 ng diphenhydramine + 4 mL MeOH, sonicate for 5 h, centrifuge at 3000 g for 10 min. Remove the supernatant and evaporate it to dryness under a stream of nitrogen at 60°, reconstitute with 1 mL MeOH, inject an aliquot.

CAPILLARY ELECTROPHORESIS
Capillary: 47 cm × 50 μm fused-silica (40 cm to detector) (Beckman)
Capillary preparation: Before each run rinse with 100 mM NaOH for 2 min and with rinsing buffer for 2 min. Rinse new capillaries with 100 mM NaOH for 20 min. (Rinsing buffer was

MeOH:100 mM pH 2.3 phosphate buffer:100 mM pH 2.3 phosphate buffer containing 10 mM heptakis-(2,6-di-O-methyl)-β-cyclodextrin 10:70:20.)
Capillary temperature: 20
Running buffer: MeOH:100 mM pH 2.3 phosphate buffer 10:90
Injection: Electrokinetic injection at 10 kV for 4 s.
Detector: UV 200
Migration time: 15.5 (R), 16.5 (S)
Internal standard: diphenhydramine (19)
Voltage: 20 kV
Model: Beckman P/ACE 5510
Limit of quantitation: 5 ng/mL

OTHER SUBSTANCES
Extracted: metabolites
Simultaneous: aminoflunitrazepam, amphetamine, benzoylecgonine, bromazepam, cocaine, codeine, N-desmethyldiazepam, diazepam, dihydrocodeine, 3,4-methylenedioxyamphetamine, 3,4-methylenedioxyethylamphetamine, 3,4-methylenedioxymethamphetamine, morphine, nicotine
Noninterfering: demethyldiazepam, flunitrazepam, 3-hydroxydiazepam, oxazepam

KEY WORDS
serum; chiral

REFERENCE
Frost,M.; Köhler,H.; Blaschke,G. Enantioselective determination of methadone and its main metabolite 2-ethylidene-1,5-dimethyl-3,3-diphenylpyrrolidine (EDDP) in serum, urine and hair by capillary electrophoresis, *Electrophoresis*, **1997**, *18*, 1026–1034.

SAMPLE
Matrix: solutions
Sample preparation: Prepare a solution in MeOH, inject an aliquot.

CAPILLARY ELECTROPHORESIS
Capillary: 55.5 cm × 50 μm fused-silica (44.5 cm to detector) (Polymicro Technologies)
Capillary preparation: Rinse with running buffer at 3 bar for 5 min
Capillary temperature: 35
Running buffer: 40 mM pH 2.2 Sodium phosphate buffer
Injection: Pressure injection at 20 mbar for 6 s
Detector: UV 220
Migration time: 6
Voltage: 20 kV
Current: about 30 μA
Model: Lauerlabs Prince

OTHER SUBSTANCES
Simultaneous: amiodarone, desethylamiodarone, dextromethorphan, itraconazole, ketoconazole
Noninterfering: caffeine, naproxen, phenol, theophylline

REFERENCE
Zhang,C.-X.; von Heeren,F.; Thormann,W. Separation of hydrophobic, positively chargeable substances by capillary electrophoresis, *Anal.Chem.*, **1995**, *67*, 2070–2077.

SAMPLE
Matrix: solutions

CAPILLARY ELECTROPHORESIS
Capillary: 59.6 cm × 75 μm fused-silica (52.1 cm to detector) (Polymicro Technologies)
Capillary preparation: Purge with running buffer before each run. At the beginning of each day purge using 50-60 kPa vacuum with 500 mM NaOH for 5 min, with water for 5 min, with MeCN for 5 min, and with running buffer for 5 min.
Running buffer: MeCN:MeOH:acetic acid 49:50:1 containing 20 mM ammonium acetate
Injection: Hydrostatic injection at 10 cm for 5 s.

Detector: UV 214
Migration time: 2.32
Voltage: 30 kV
Model: Waters Quanta 4000

OTHER SUBSTANCES
Simultaneous: diphenoxylate, ethoheptazine, fentanyl, levallorphan, meperidine, nalorphine, pentazocine, nikethamide

REFERENCE
Leung,G.N.W.; Tang,H.P.O.; Tso,T.S.C.; Wan,T.S.M. Separation of basic drugs with non-aqueous capillary electrophoresis, *J.Chromatogr.A*, **1996**, *738*, 141–154.

SAMPLE
Matrix: solutions
Sample preparation: Inject an aliquot of a solution in 100 mM sodium phosphate containing 100-300 μg/mL naphazoline.

CAPILLARY ELECTROPHORESIS
Capillary: 67 cm × 50 μm fused-silica (60 cm to detector)
Capillary preparation: Rinse with running buffer for 2 min between sets of analyses.
Capillary temperature: 30
Running buffer: 200 mM pH 4.5 Phosphate buffer
Injection: Injection at high pressure for 2 s.
Detector: UV 210, UV 230
Migration time: 17.5
Internal standard: naphazoline (14)
Voltage: 20 kV
Model: Beckman P/ACE 5500

OTHER SUBSTANCES
Simultaneous: acetylcodeine, amphetamine, cocaine, codeine, diamorphine, LSD, methamphetamine, methylenedioxyamphetamine, methylenedioxymethamphetamine, morphine, PCP, psilocyn

REFERENCE
Walker,J.A.; Marché,H.L.; Newby,N.; Bechtold,E.J. A free zone capillary electrophoresis method for the quantitation of common illicit drug samples, *J.Forensic Sci.*, **1996**, *41*, 824–829.

SAMPLE
Matrix: urine
Sample preparation: Filter (0.2 μm), inject an aliquot of the filtrate.

CAPILLARY ELECTROPHORESIS
Capillary: 80 cm × 50 μm fused-silica (57 cm to detector)
Capillary preparation: Before each run rinse the capillary with 1 M NaOH for 3 min, with 100 mM NaOH for 3 min, with water for 3 min, and with running buffer for 10 min.
Running buffer: 110 mM Boric acid containing 56 mM NaOH and 44 mM HCl, pH 8
Injection: Vacuum injection for 1 s
Detector: UV 220
Migration time: 4.426
Voltage: 20 kV
Model: Europhor Prime Vision system IV
Limit of detection: 4.68 μM

OTHER SUBSTANCES
Extracted: acebutolol (UV 238), acetazolamide (UV 222), alprenolol (UV 220), amiloride (UV 220), atenolol (UV 228), bendroflumethiazide (UV 220), bumetanide (UV 220), chlorthalidone (UV 220), cocaine (UV 236), codeine (UV 220), ethacrynic acid (UV 220), furosemide (UV 232), hydrochlorothiazide (UV 226), metoxiphenamine (UV 220), nadolol (UV 220), norcodeine (UV 220), oxprenolol (UV 220), pentazocine (UV 220), propranolol (UV 220), spironolactone (UV 244), triamterene (UV 232), xipamide (UV 234)

REFERENCE
Gonzalez,E.; Laserna,J.J. Capillary zone electrophoresis for the rapid screening of banned drugs in sport, *Electrophoresis*, **1994**, *15*, 240–243.

SAMPLE
Matrix: urine
Sample preparation: Inject directly. Alternatively, condition a Bond Elut Certify SPE cartridge with 2 mL MeOH and 2 mL 100 mM pH 6 phosphate buffer, do not allow to dry. 2 mL Urine + 2 mL 100 mM pH 6 phosphate buffer, mix, add to the SPE cartridge, wash with 1 mL MeOH: 100 mM phosphate buffer 20:80, dry under full vacuum for 5 min, wash with 1 mL 1 M acetic acid, dry under full vacuum for 10 min, wash with 1 mL hexane, dry under full vacuum for 2 min, wash with 4 mL dichloromethane, wash with 6 mL MeOH, elute with 2 mL dichloromethane:isopropanol 80:20 containing 5% concentrated ammonium hydroxide solution. Evaporate the eluate to dryness under a stream of air at room temperature, reconstitute with 50 μL diluted running buffer or water, inject an aliquot.

CAPILLARY ELECTROPHORESIS
Capillary: 90 cm × 75 μm fused-silica (70 cm to detector) (Polymicro Technologies)
Capillary preparation: Before each injection rinse with 100 mM NaOH for 5 min and with running buffer for 5 min.
Running buffer: 50 mM pH 9.3 tetraborate buffer
Detector: UV 195
Migration time: 7.3
Voltage: 20 kV
Current: about 90 μA
Model: Laboratory constructed
Limit of detection: 2 μg/mL (direct), 20-50 ng/mL (SPE)

OTHER SUBSTANCES
Extracted: metabolites
Simultaneous: amphetamine, methamphetamine
Noninterfering: benzodiazepines, benzoylecgonine, codeine, diphenhydramine, morphine

KEY WORDS
SPE

REFERENCE
Molteni,S.; Caslavska,J.; Alleman,D.; Thormann,W. Determination of methadone and its primary metabolite in human urine by capillary electrophoretic techniques, *J.Chromatogr.B*, **1994**, *658*, 355–367.

SAMPLE
Matrix: urine
Sample preparation: Condition a Bond Elut Certify SPE cartridge with 2 mL MeOH and 2 mL 100 mM pH 6 phosphate buffer, do not allow to dry. Filter (0.45 μm) urine. Mix 1 mL filtered urine with 2 mL 100 mM pH 6 phosphate buffer, add to the SPE cartridge over at least 2 min, wash with 1 mL MeOH:100 mM pH 6 phosphate buffer 20:80, wash with 1 mL 1 M acetic acid, wash with 1 mL hexane, dry under full vacuum for 2 min, wash with 4 mL dichloromethane, wash with 6 mL MeOH, elute with 2 mL dichloromethane:isopropanol:concentrated ammonium hydroxide 80:20:5. Evaporate the eluate to dryness under a stream of nitrogen at 37°, reconstitute with 500 μL water, sonicate for 10 min, inject an aliquot.

CAPILLARY ELECTROPHORESIS
Capillary: 60 cm × 50 μm fused-silica (55.4 cm to detector) (Polymicro Technologies)
Capillary preparation: Before each run rinse capillary with 100 mM NaOH for 1 min and running buffer for 2 min.
Capillary temperature: 20
Running buffer: 100 mM pH 3 KH_2PO_4 containing hydroxypropyl-β-cyclodextrin
Injection: Pressure injection at 2 psi s.
Detector: UV 195
Migration time: 23 (R-(-)), 25 (S-(+))
Internal standard: codeine (17)
Voltage: 19 kV

Current: ca. 60 μA
Model: BioFocus 3000 (Bio-Rad)
Limit of detection: 100 ng/mL

OTHER SUBSTANCES
Extracted: metabolites

KEY WORDS
SPE; chiral

REFERENCE
Lanz,M.; Thormann,W. Characterization of the stereoselective metabolism of methadone and its primary metabolite via cyclodextrin capillary electrophoretic determination of their urinary enantiomers, *Electrophoresis*, **1996**, *17*, 1945–1949.

SAMPLE
Matrix: urine
Sample preparation: Inject an aliquot directly.

CAPILLARY ELECTROPHORESIS
Capillary: 75 cm × 50 μm fused-silica (70.4 cm to detector) (Polymicro Technologies)
Capillary preparation: Before each run rinse capillary with 100 mM NaOH for 1 min, with water for 1 min, and with running buffer for 2 min.
Capillary temperature: 20
Running buffer: 20 mM Ammonium acetate containing 20 mM acetic acid, pH 4.6 (Outlet buffer was 1% acetic acid in MeOH:water 60:40.)
Injection: Pressure injection at 2 psi.s.
Detector: UV 195
Migration time: 10.7
Voltage: 20 kV
Current: ca. 11 μA
Model: Bio-Rad BioFocus 3000

OTHER SUBSTANCES
Simultaneous: metabolites, 2-ethylidene-1,5-dimethyl-3,3-diphenylpyrrolidine

---

KEY WORDS
direct injection

REFERENCE
Thormann,W.; Lanz,M.; Caslavska,J.; Siegenthaler,P.; Portmann,R. Screening for urinary methadone by capillary electrophoretic immunoassays and confirmation by capillary electrophoresis-mass spectrometry, *Electrophoresis*, **1998**, *19*, 57–65.

Methamphetamine

Molecular formula: $C_{10}H_{15}N$
Molecular weight: 149.24
CAS Registry No.: 537-46-2
Merck Index (12th ed.): 6015
Lednicer: 1 37

SAMPLE

Matrix: blood, urine
Sample preparation: Adjust pH of 2 mL urine or plasma to 10.5 with aqueous NaOH, extract gently with chloroform:isopropanol 90:10. Remove the organic layer and evaporate it to dryness under a stream of nitrogen, reconstitute the residue in 50 (urine) or 75 (plasma) μL running buffer, filter, inject an aliquot.

CAPILLARY ELECTROPHORESIS

Capillary: 60 cm × 75 μm AccuSep uncoated silica (52.5 cm to detector) (Waters)
Capillary preparation: At the beginning of each day purge with 500 mM KOH for 5 min, with water for 5 min, and with running buffer for 10 min.
Running buffer: 50 mM NaH_2PO_4 adjusted to pH 2.35 with phosphoric acid
Injection: Hydrostatic injection at 15 cm
Detector: UV 214
Migration time: 5.56
Voltage: 22 kV
Current: 135-145 μA
Model: Waters Quanta 4000

OTHER SUBSTANCES

Extracted: acepromazine, amphetamine, benzocaine, brompheniramine, butacaine, codeine, diazepam, doxapram, lidocaine, medazepam, methapyrilene, methaqualone, phenmetrazine, procaine, tetrahydrozoline
Simultaneous: meclizine

KEY WORDS

plasma

REFERENCE

Chee,G.L.; Wan,T.S.M. Reproducible and high-speed separation of basic drugs by capillary zone electrophoresis, *J.Chromatogr.*, **1993**, *612*, 172–177.

SAMPLE

Matrix: bulk
Sample preparation: Dissolve 4 mg compound in 1 mL MeCN:water 50:50 containing 0.2% triethylamine, vortex for 30 s. Remove a 100 μL aliquot and add it to 100 μL 1.28% 2,3,4,6-tetra-O-acetyl-β-D-glucopyranosyl isothiocyanate in MeCN, vortex for 1 min, let stand for 15 min, make up to 1 mL with 10 mM pH 9.0 phosphate/borate buffer containing 100 mM sodium dodecyl sulfate, vortex for 20 s, filter (Whatman UniPrep), inject an aliquot of the filtrate.

CAPILLARY ELECTROPHORESIS

Capillary: 48 cm × 50 μm fused-silica (26 cm to detector) (Polymicro Technologies)
Capillary preparation: Condition new capillaries with 1 M NaOH for 10 min, with water for 10 min, and with running buffer for 10 min.
Capillary temperature: 30
Running buffer: MeOH:buffer 20:80 (Buffer was 10 mM pH 9.0 phosphate buffer containing 10 mM borate and 100 mM sodium dodecyl sulfate.)
Injection: Vacuum injection for 0.5 s.
Detector: UV 210
Migration time: 21 (-), 2.8 (+)
Voltage: 20 kV

Current: 48 μA
Model: Applied Biosystems Model 270A-HT

OTHER SUBSTANCES
Simultaneous: amphetamine, ephedrine, norpseudoephedrine, phenylpropanolamine (norephedrine), pseudoephedrine

---

KEY WORDS
derivatization; chiral; comparison with HPLC

REFERENCE
Lurie,I.S. Micellar electrokinetic capillary chromatography of the enantiomers of amphetamine, methamphetamine and their hydroxyphenethylamine precursors, *J.Chromatogr.*, **1992**, *605*, 269–275.

SAMPLE
Matrix: bulk
Sample preparation: 10 50 mg Bulk compound + 1 mL 1 mg/mL caffeine in 10 mM HCl, make up to 10 mL with 10 mM HCl, sonicate for 2 min, mix thoroughly, filter (0.45 μm cellulose acetate), inject an aliquot.

CAPILLARY ELECTROPHORESIS
Capillary: 75 cm × 75 μm fused-silica (50 cm to the detector) (ISCO)
Capillary preparation: Flush with running buffer for 2 min between analyses. Condition capillary by filling with 1 M NaOH and allowing to stand for 1 h, fill with 100 mM NaOH and allow to stand for 1 h, wash with water, fill with running buffer. Each week wash capillary with 100 mM HCl for 10 min, with water, with 100 mM NaOH, and with water then fill with running buffer.
Capillary temperature: 30
Running buffer: DMSO:ethanolamine:buffer 11:1:88 (Buffer was 0.92 g cetyltrimethylammonium bromide in 100 mL 10 mM sodium tetraborate, adjust pH to 11.5 with 1 M NaOH.)
Injection: Load under vacuum, vacuum level 2, 10.0 kPa s
Detector: UV 254
Migration time: 9.2
Internal standard: caffeine (6.5)
Voltage: -15 kV
Model: ISCO Model 3140 electropherograph

OTHER SUBSTANCES
Simultaneous: amphetamine, p-aminobenzoic acid, ephedrine, methylenedioxyamphetamine, methylenedioxymethamphetamine, norephedrine, pseudoephedrine, pseudonorephedrine

---

REFERENCE
Trenerry,V.C.; Robertson,J.; Wells,R.J. Analysis of illicit amphetamine seizures by capillary electrophoresis, *J.Chromatogr.A*, **1995**, *708*, 169–176.

SAMPLE
Matrix: formulations
Sample preparation: Add 3 mL buffer to 30 mg capsule contents, sonicate for 30 min, filter (0.2 μm Nylon), inject an aliquot of the filtrate. (Buffer was water acidified to pH 2.0 with concentrated phosphoric acid.)

CAPILLARY ELECTROPHORESIS
Capillary: 90 cm × 50 μm uncoated fused-silica (65 cm to detector)
Capillary preparation: Rinse capillary with buffer for 15 min at the start of each day and for 4 min before each analysis.
Capillary temperature: 31
Running buffer: 70 mM hydroxypropyl-β-cyclodextrin (average molar substitution 0.8, average molecular mass 1500) containing 30 mM tetramethylammonium chloride and 10 mM sodium dodecyl sulfate, pH 2.0
Injection: Vacuum injection at 28 kPa
Detector: UV 210
Migration time: 38.3 (L), 40.0 (D)

Voltage: +28 kV
Model: ISCO Model 3140

OTHER SUBSTANCES
Simultaneous: amphetamine, ephedrine, methylephedrine, methylpseudoephedrine, norephedrine, norpseudoephedrine, pseudoephedrine

---

KEY WORDS
capsules; chiral

REFERENCE
Flurer,C.L.; Lin,L.A.; Satzger,R.D.; Wolnik,K.A. Determination of ephedrine compounds in nutritional supplements by cyclodextrin-modified capillary electrophoresis, *J.Chromatogr.B*, **1995**, *669*, 133–139.

SAMPLE
Matrix: hair, urine
Sample preparation: Wash 100 mg hair with 0.3% Tween 20 in water, cut in small pieces, add 1 mL 250 mM HCl, heat at 45° overnight, neutralize with NaOH. Add the hair extract or 2 mL urine to a Toxi-tube A, extract the organic layer with 200 μL 10 mM phosphoric acid, inject an aliquot of the aqueous layer.

CAPILLARY ELECTROPHORESIS
Capillary: 45 cm × 50 μm fused-silica (37.5 cm to detector) (Composite Metal Services, Hallow, UK)
Capillary preparation: Condition with running buffer for 10 min before each run. Wash new capillaries with 1 M NaOH for 10 min, with 100 mM NaOH for 10 min, and with water for 10 min, and then condition with running buffer for 20 min.
Running buffer: 100 mM pH 2.5 Potassium phosphate buffer containing 15 mM β-cyclodextrin
Injection: Pressure injection at 35 mbar for 10 s.
Detector: UV 200
Migration time: 21, 21.5 (enantiomers)
Voltage: 10 kV
Model: Beckman P/ACE 2200
Limit of detection: 75 ng/mL

OTHER SUBSTANCES
Extracted: amphetamine, ephedrine, 3,4-methylenedioxyamphetamine, 3,4-methylenedioxyethylamphetamine, 3,4-methylenedioxymethamphetamine

---

KEY WORDS
chiral

REFERENCE
Tagliaro,F.; Manetto,G.; Bellini,S.; Scarcella,D.; Smith,F.P.; Marigo,M. Simultaneous chiral separation of 3,4-methylenedioxymethamphetamine (MDMA), 3,4-methylenedioxyamphetamine (MDA), 3,4-methylenedioxyethylamphetamine (MDE), ephedrine, amphetamine and methamphetamine by capillary electrophoresis in uncoated and coated capillaries with native β-cyclodextrin as the chiral selector: Preliminary application to the analysis of urine and hair, *Electrophoresis*, **1998**, *19*, 42–50.

SAMPLE
Matrix: solutions

CAPILLARY ELECTROPHORESIS
Capillary: 100 cm × 50 μm fused-silica (50 cm to detector) (Isco)
Capillary preparation: Flush capillary with 10 μL running buffer between runs. Every 40 sample injections rinse capillary with 200 μL 1 M NaOH, with 200 μL water, and 200 μL running buffer. Before use fill capillary with 1 M NaOH and allow to stand for 1 h, fill with 100 mM NaOH, allow to stand for 1 h, wash with water fill with running buffer.
Capillary temperature: 23
Running buffer: 100 mM citric acid containing 19.27 mM Na_2HPO_4 and 120 mM hydroxypropyl-β-cyclodextrin

Injection: Inject under vacuum at 4.0 kPa.s
Detector: UV 200
Migration time: 30.5 (S), 31 (R)
Voltage: 30 kV
Model: Isco Model 3140

OTHER SUBSTANCES
Simultaneous: amphetamine, 4-bromo-2,5-dimethoxyamphetamine, ephedrine, epinephrine, methyldimethoxyethylamphetamine, methyldimethoxyamphetamine, methyldimethoxymethylamphetamine, pseudoephedrine

KEY WORDS
chiral

REFERENCE
Aumatell,A.; Wells,R.J.; Wong,D.K.Y. Enantiomeric differentiation of a wide range of pharmacologically active substances by capillary electrophoresis using modified β-cyclodextrins, *J.Chromatogr.A*, **1994**, *686*, 293–307.

SAMPLE
Matrix: solutions
Sample preparation: Prepare a 50 μg/mL solution in water:buffer:MeOH 90:9:1, inject an aliquot. (Buffer was 25 mM Tris buffer adjusted to pH 2.4 with phosphoric acid.)

CAPILLARY ELECTROPHORESIS
Capillary: 82 cm × 50 μm fused-silica (60 cm to detector)
Capillary preparation: Condition by aspirating with 1 M NaOH for 10 min, with water for 10 min, and with running buffer for 10 min.
Capillary temperature: 30
Running buffer: MeOH:buffer 1.2:98.8 (Buffer was 25 mM Tris buffer adjusted to pH 2.4 with phosphoric acid containing 5 mM heptakis(2,6-di-O-methyl)-β-cyclodextrin and 1 mM sulfobutyl ether β-cyclodextrin (ISCO)
Injection: Vacuum injection for 0.5 s (2 nL)
Detector: UV 210
Migration time: 13.99 (-), 14.40 (+)
Voltage: 30 kV
Model: Applied Biosystems Model 270A-HT

OTHER SUBSTANCES
Simultaneous: amphetamine, cathinone, cocaine, ephedrine, methcathinone, norephedrine, norpseudoephedrine, propoxyphene, pseudoephedrine

KEY WORDS
chiral

REFERENCE
Lurie,I.S.; Klein,R.F.X.; Dal Cason,T.A.; LeBelle,M.J.; Brenneisen,R.; Weinberger,R.E. Chiral resolution of cationic drugs of forensic interest by capillary electrophoresis with mixtures of neutral and anionic cyclodextrins, *Anal.Chem.*, **1994**, *66*, 4019–4026.

SAMPLE
Matrix: solutions

CAPILLARY ELECTROPHORESIS
Capillary: 48.5 cm × 50 μm fused-silica (40 cm to detector (Hewlett-Packard)
Capillary temperature: 20 ± 0.1
Running buffer: 50 mM NaH_2PO_4 containing 10 mM β-cyclodextrin, adjusted to pH 2.5 with phosphoric acid
Detector: UV 214
Migration time: 8.292, 8.412 (enantiomers)
Voltage: 20 kV

Current: about 30 μA
Model: Hewlett-Packard HP-3D

OTHER SUBSTANCES
Also analyzed: amphetamine, 4-bromo-2,5-dimethoxyamphetamine, 2,5-dimethoxyamphetamine, 2,5-dimethoxymethamphetamine, 4-hydroxyamphetamine, 4-methoxyamphetamine, 3,4-methylenedioxyethamphetamine (MDEA), 3,4-methylenedioxymethamphetamine (MDMA)

KEY WORDS
chiral; r = 1.015

REFERENCE
Cladrowa-Runge,S.; Hirz,R.; Kenndler,E.; Rizzi,A. Enantiomeric separation of amphetamine related drugs by capillary zone electrophoresis using native and derivatized β-cyclodextrin as chiral additives, *J.Chromatogr.A*, **1995**, *710*, 339–345.

SAMPLE
Matrix: solutions

CAPILLARY ELECTROPHORESIS
Capillary: 70 cm × 75 μm uncoated fused-silica (55 cm to detector)
Running buffer: 20 mM pH 2.8 Tris-phosphoric acid buffer containing 0.1% hydroxypropylmethylcellulose and 6 mM heptakis-(2,6-di-O-methyl)-β-cyclodextrin
Injection: Electrokinetic injection at 3 kV for 12 s
Detector: UV 190
Migration time: 12.5 (-), 12.7 (+)
Voltage: 21 kV
Current: 32 μA
Model: ATI Unicam Crystal 300

OTHER SUBSTANCES
Simultaneous: amphetamine, selegiline

KEY WORDS
chiral

REFERENCE
Szöko,É.; Magyar,K. Chiral separation of deprenyl and its major metabolites using cyclodextrin-modified capillary zone electrophoresis, *J.Chromatogr.A*, **1995**, *709*, 157–162.

SAMPLE
Matrix: solutions

CAPILLARY ELECTROPHORESIS
Capillary: 70 cm × 75 μm uncoated fused-silica (55 cm to detector)
Running buffer: 20 mM pH 2.8 Tris-phosphoric acid buffer containing 0.1% hydroxypropylmethylcellulose
Injection: Electrokinetic injection at 3 kV for 12 s
Detector: UV 190
Migration time: 8.3
Voltage: 21 kV
Current: 32 μA
Model: ATI Unicam Crystal 300

OTHER SUBSTANCES
Simultaneous: amphetamine, propargylamphetamine, selegiline

REFERENCE
Szöko,É.; Magyar,K. Chiral separation of deprenyl and its major metabolites using cyclodextrin-modified capillary zone electrophoresis, *J.Chromatogr.A*, **1995**, *709*, 157–162.

SAMPLE
Matrix: solutions
Sample preparation: Inject an aliquot of a 20 μg/mL solution in 12.5 mM pH 9.3 borate buffer.

CAPILLARY ELECTROPHORESIS
Capillary: 66.5 cm × 75 μm fused-silica (41.5 cm to detector) (Polymicro Technologies)
Running buffer: 250 mM pH 9.3 Sodium borate buffer
Injection: Pressure injection at 50 mbar for 15 s.
Detector: UV 195
Migration time: 10.5
Voltage: 8 kV
Model: Crystal 310

OTHER SUBSTANCES
Simultaneous: amphetamine, chlorpheniramine, ephedrine, MDA, MDMA, phentermine, phenylpropanolamine, pseudoephedrine
Noninterfering: caffeine

REFERENCE
Lilley,K.A.; Wheat,T.E. Drug identification in biological matrices using capillary electrophoresis and chemometric software, *J.Chromatogr.B*, **1996**, *683*, 67–76.

SAMPLE
Matrix: solutions

CAPILLARY ELECTROPHORESIS
Capillary: 37 cm × 50 μm fused-silica (30 cm to detector)
Capillary temperature: 20
Running buffer: 100 mM pH 2.5 Tris/phosphate buffer containing 200 mg/mL epichlorohydrin-β-cyclodextrin polymer (MW 3000-5000, Cyclolab, Budapest)
Injection: Pressure injection at 0.5 psi for 8 s.
Detector: UV 214
Migration time: 23 (-), 24 (+)
Voltage: 14.8 kV
Model: Beckman P/ACE 2200

OTHER SUBSTANCES
Simultaneous: ephedrine (not chiral), selegiline

KEY WORDS
chiral

REFERENCE
Sevcik,J.; Stránsky,Z.; Ingelse,B.A.; Lemr,K. Capillary electrophoretic enantioseparation of selegiline, methamphetamine and ephedrine using a neutral β-cyclodextrin epichlorhydrin polymer, *J.Pharm.Biomed.Anal.*, **1996**, *14*, 1089–1094.

SAMPLE
Matrix: solutions
Sample preparation: Inject an aliquot of a 5-10 μM solution.

CAPILLARY ELECTROPHORESIS
Capillary: 70 cm × 75 μm fused-silica (55 cm to detector)
Capillary temperature: 21
Running buffer: 20 mM pH 2.7 Tris-phosphate buffer containing 0.5% hydroxypropylmethyl cellulose and 24 mM heptakis(2,6-di-O-methyl)-β-cyclodextrin (Hydroxypropylmethyl cellulose was from Sigma; viscosity of 2% solution = 4000 cP at 25°.)
Injection: Electrokinetic injection at 3 kV for 12 s.
Detector: UV 190
Migration time: 20.9, 21.3 (enantiomers)
Voltage: 21 kV

Model: Crystal 300 (ATI, Unicam)

OTHER SUBSTANCES
Simultaneous: amphetamine,   epinephrine,   norepinephrine,   pseudoephedrine,   selegiline
(deprenyl)

KEY WORDS
chiral

REFERENCE
Szökö,E.; Gyimesi,J.; Barcza,L.; Magyar,K. Determination of binding constants and the influence of methanol
on the separation of drug enantiomers in cyclodextrin modified capillary electrophoresis, *J.Chromatogr.A*,
1996, *745*, 181–187.

SAMPLE
Matrix: solutions
Sample preparation: Inject an aliquot of a solution in 100 mM sodium phosphate containing
100-300 µg/mL naphazoline.

CAPILLARY ELECTROPHORESIS
Capillary: 67 cm × 50 µm fused-silica (60 cm to detector)
Capillary preparation: Rinse with running buffer for 2 min between sets of analyses.
Capillary temperature: 30
Running buffer: 200 mM pH 4.5 Phosphate buffer
Injection: Injection at high pressure for 2 s.
Detector: UV 210, UV 230
Migration time: 13.3
Internal standard: naphazoline (14)
Voltage: 20 kV
Model: Beckman P/ACE 5500

OTHER SUBSTANCES
Simultaneous: acetylcodeine, amphetamine, cocaine, codeine, diamorphine, LSD, methadone,
methylenedioxyamphetamine, methylenedioxymethamphetamine, morphine, PCP, psilocyn

REFERENCE
Walker,J.A.; Marché,H.L.; Newby,N.; Bechtold,E.J. A free zone capillary electrophoresis method for the quantitation of common illicit drug samples, *J.Forensic Sci.*, **1996**, *41*, 824–829.

SAMPLE
Matrix: solutions
Sample preparation: Inject an aliquot of a solution in MeCN:water 1:2.

CAPILLARY ELECTROPHORESIS
Capillary: 64.5 cm × 50 µm fused-silica (56 cm to detector) (Hewlett-Packard)
Capillary preparation: Flush capillary with running buffer for 8 min between runs.
Capillary temperature: 28
Running buffer: 118 mM Tris containing 16 mM hydroxypropyl-β-cyclodextrin, adjusted to pH
3.5 with phosphoric acid
Injection: Pressure injection at 1 bar (ca. 25 nL), ramp to operating voltage over 1 min.
Detector: UV 200
Migration time: 13 (l), 13.2 (d)
Voltage: 25 kV
Model: Hewlett-Packard HP[3D]

OTHER SUBSTANCES
Simultaneous: amphetamine, 3,4-methylenedioxyamphetamine, 3,4-methylenedioxyethylam-
phetamine, 3,4-methylenedioxymethamphetamine

KEY WORDS
chiral

REFERENCE
Varesio,E.; Gauvrit,J.-Y.; Longeray,R.; Lantéri,P.; Veuthey,J.-L. Central composite design in the chiral analysis of amphetamines by capillary electrophoresis, *Electrophoresis*, **1997**, *18*, 931–937.

SAMPLE
Matrix: solutions

CAPILLARY ELECTROPHORESIS
Capillary: 85 cm × 50 μm fused-silica (Polymicro Technologies)
Capillary preparation: Between runs rinse capillary with running buffer for 5 min. Before use rinse capillary with 1 M NaOH for 10-15 min and with water for 10-15 min.
Running buffer: 25 mM pH 3 Citrate buffer
Injection: Pressure injection at 30 mbar for 12 s.
Detector: MS, laboratory-constructed, electrospray 90°, sheath liquid MeOH:water:acetic acid 80:20:0.1 at 1 μL/min
Migration time: 9.7
Voltage: 30 kV
Current: 7.8 μA
Model: Crystal CE 300

OTHER SUBSTANCES
Simultaneous: amphetamine, cocaine, diamorphine, procaine, tetracaine

REFERENCE
Lazar,I.M.; Naisbitt,G.; Lee,M.L. Capillary electrophoresis-time-of-flight mass spectrometry of drugs of abuse, *Analyst*, **1998**, *123*, 1449–1454.

SAMPLE
Matrix: urine
Sample preparation: Condition a Bond Elut Certify SPE cartridge with 2 mL MeOH and 2 mL buffer. 2 mL Urine + 2 mL buffer, vortex for 15 s, add to the SPE cartridge, wash with 1 mL 1 M acetic acid, dry under vacuum for 2 min, elute with 5 mL 2% ammonia in ethyl acetate (freshly prepared). Add 2 drops acetic acid to the eluate and evaporate it to dryness under a stream of nitrogen at room temperature, reconstitute the residue in 1 mL 100 mM HCl, inject an aliquot. (Prepare buffer by dissolving 1.212 g NaH$_2$PO$_4$ in water, adjusting to pH 6.0 with 1 M NaOH, and making up to 100 mL with water.)

CAPILLARY ELECTROPHORESIS
Capillary: 64.5 cm × 50 μm fused-silica (56 cm to detector) (Hewlett-Packard)
Capillary preparation: Before each run flush with 100 mM phosphoric acid for 2 min and with running buffer for 5 min. Flush new capillaries with 1 M NaOH for 3 min, with 100 mM NaOH for 5 min, and with water for 10 min.
Capillary temperature: 15
Running buffer: 200 mM pH 2.5 Phosphate buffer containing 20 mM (2-hydroxy)propyl-β-cyclodextrin (Prepare buffer by dissolving 1.153 g 85% phosphoric acid and 1.212 g NaH$_2$PO$_4$ in water, adjusting to pH 2.5 with 1 M NaOH, and making up to 100 mL with water. Dissolve 0.3 g (2-hydroxy)propyl-β-cyclodextrin in 10 mL buffer, filter (0.2 μm).)
Injection: Pressure injection at 600 mbar.s (14.3 nL)
Detector: UV 200
Migration time: 16.5 (l), 17.0 (d)
Internal standard: phenylethylamine (12.5)
Voltage: 25 kV (ramp to 25 kV over 1 min)
Model: Hewlett-Packard HP3D CE

OTHER SUBSTANCES
Extracted: amphetamine, methylenedioxyamphetamine, methylenedioxyethylamphetamine, methylenedioxymethamphetamine

KEY WORDS
chiral; SPE

REFERENCE
Varesio,E.; Veuthey,J.-L. Chiral separation of amphetamines by high-performance capillary electrophoresis, *J.Chromatogr.A*, **1995**, *717*, 219–228.

SAMPLE
Matrix: urine
Sample preparation: 5 mL Urine + 100 μL 500 μM IS + 1 mL concentrated HCl, heat at 80° for 1 h, cool in tap water, neutralize with 1.5 mL 28% ammonia, add 3.5 mL 500 mM pH 10.5 sodium borate buffer containing 4.5 M NaCl, add 2 mL chloroform:isopropanol 75:25, shake vigorously for 1 min, centrifuge at 1500 g for 10 min, repeat the extraction. Combine the organic layers and wash them with 2.5 mL water, centrifuge at 1500 g for 10 min, add 400 μL acetic acid, evaporate to dryness, reconstitute the residue in 200 μL water, filter (0.45 μm), inject an aliquot.

CAPILLARY ELECTROPHORESIS
Capillary: 50 cm × 75 μm fused-silica (37.5 cm to detector)
Capillary preparation: Rinse with running buffer for 5 min before each run.
Capillary temperature: 25
Running buffer: 50 mM pH 3.0 Sodium acetate/HCl buffer (A) or 50 mM pH 9.3 sodium borate buffer containing 15 mM sodium dodecyl sulfate (B)
Injection: Siphon injection at 20 mM for 20 s (9 nL (A), 17 nL (B))
Detector: UV 210
Migration time: 11.4 (A), 13.6 (B)
Internal standard: 1-phenylethylamine (10.5 (A), 8.2 (B))
Voltage: 10 kV (A), 15 kV (B)
Model: Otsuka Electronics CAPI-3100A
Limit of detection: 350-580 fmole

OTHER SUBSTANCES
Extracted: amphetamine

REFERENCE
Kuroda,N.; Nomura,R.; Al-Dirbashi,O.; Akiyama,S.; Nakashima,K. Determination of methamphetamine and related compounds by capillary electrophoresis with UV and laser-induced fluorescence detection, *J.Chromatogr.A*, **1998**, *798*, 325–334.

Methaphenilene

Molecular formula: $C_{15}H_{20}N_2S$
Molecular weight: 260.40
CAS Registry No.: 493-78-7, 7084-07-3 (HCl)
Merck Index (12th ed.): 6026

SAMPLE
Matrix: solutions

CAPILLARY ELECTROPHORESIS
Capillary: 62 cm × 75 μm fused-silica (54.5 cm to detector) (Polymicro Technologies)
Capillary preparation: Purge with running buffer before each run. At the beginning of each day purge using 50-60 kPa vacuum with 500 mM NaOH for 5 min, with water for 5 min, with MeCN for 5 min, and with running buffer for 5 min.
Running buffer: MeCN:MeOH:acetic acid 49:50:1 containing 20 mM ammonium acetate
Injection: Hydrostatic injection at 10 cm for 5 s.
Detector: UV 214
Migration time: 4.46
Voltage: 30 kV
Model: Waters Quanta 4000

OTHER SUBSTANCES
Simultaneous: buclizine, chlorcyclizine, chlorpheniramine, cyclizine, dimenhydrinate, pheniramine, promethazine, pyrilamine, pyrrobutamine, tripelennamine

REFERENCE
Leung,G.N.W.; Tang,H.P.O.; Tso,T.S.C.; Wan,T.S.M. Separation of basic drugs with non-aqueous capillary electrophoresis, *J.Chromatogr.A*, **1996**, *738*, 141–154.

Methaqualone

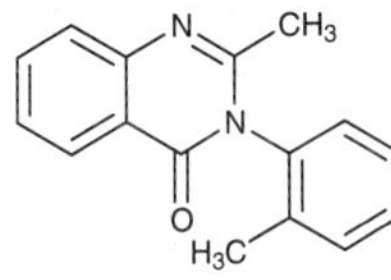

Molecular formula: $C_{16}H_{14}N_2O$
Molecular weight: 250.30
CAS Registry No.: 72-44-6, 340-56-7 (HCl)
Merck Index (12th ed.): 6028
Lednicer: 1 353

SAMPLE
Matrix: solutions

CAPILLARY ELECTROPHORESIS
Capillary: 57 cm × 50 μm (50 cm to detector) (Beckman)
Capillary temperature: 25
Running buffer: 150 mM pH 9.4 Borate buffer containing 20 mM (-)-N-dodecyl-N-methylephed-
rinium bromide
Detector: UV 280
Migration time: 66, 68 (enantiomers)
Voltage: -12 kV
Model: Beckman P/ACE 2100

KEY WORDS
chiral

REFERENCE
Bunke,A.; Jira,T.; Beyrich,T. (-)-N-Dodecyl-N-methylephedrinium bromide as chiral selector in capillary electrophoresis, *Pharmazie*, **1997**, *52*, 762–764.

Metharbital

Molecular formula: $C_9H_{14}N_2O_3$
Molecular weight: 198.22
CAS Registry No.: 50-11-3
Merck Index (12th ed.): 6029
Lednicer: 1 273

SAMPLE
Matrix: solutions

CAPILLARY ELECTROPHORESIS
Capillary: 60 cm × 50 μm fused-silica (40 cm to detector), bare (A) or C18 coated (B) (Supelco)
Capillary preparation: Rinse new capillaries with 250 μL 1 M NaOH, 250 μL 100 mM NaOH,
water, MeOH, water, and running buffer then condition with running buffer for at least 15
min. Carry out a similar procedure between runs.

Running buffer: 100 mM pH 8.5 Borate buffer containing 25 mM sodium dodecyl sulfate and 5 M urea
Injection: Electrokinetic injection at 5 kV for 5 s.
Detector: UV 254
Migration time: 9.2 (A), 8 (B)
Voltage: 18 kV
Model: Jasco

OTHER SUBSTANCES
Simultaneous: amobarbital, barbital, mephobarbital, pentobarbital, phenobarbital, secobarbital

KEY WORDS
coated capillary

REFERENCE
Jinno,K.; Han,Y.; Nakamura,M. Analysis of anxiolytic drugs by capillary electrophoresis with bare and coated capillaries, *J.Capillary Electrophor.*, **1996**, *3*, 139–145.

SAMPLE
Matrix: solutions

CAPILLARY ELECTROPHORESIS
Capillary: 40 cm × 75 μm coated fused-silica (15 cm to detector) (Supelco)
Capillary preparation: Coat column as follows. Adjust the pH of 20 mL water to 3.5 with acetic acid, add 80 μL 3-(trimethoxysilyl)propyl methacrylate (3-methacryloxypropyltrimethoxysilane), mix, suck into capillary, let stand at room temperature for 1 h, remove the solution, wash with water. Fill the capillary with a deaerated 4% acrylamide solution containing 1 mg/mL N,N,N',N'-tetramethylethylenediamine and 1 mg/mL ammonium persulfate, let stand for 3 h, remove excess solution by aspiration, rinse with water, remove water by aspiration, dry at 35° (cf. J. Chromatogr. 1985, 347, 191).
Running buffer: 100 mM Tris/150 mM boric acid, pH 8.3
Injection: Electromigration at 5 kV for 5 s.
Detector: UV 240, UV 254
Migration time: 9.3
Voltage: 12 kV
Model: Jasco 890-CE

OTHER SUBSTANCES
Simultaneous: amobarbital, barbital, mephobarbital, pentobarbital, phenobarbital, secobarbital

KEY WORDS
injection at cathode; coated capillary

REFERENCE
Jinno,K.; Han,Y.; Sawada,H. Analysis of toxic drugs by capillary electrophoresis using polyacrylamide-coated columns, *Electrophoresis*, **1997**, *18*, 284–286.

SAMPLE
Matrix: solutions
Sample preparation: Inject an aliquot of a 100 μg/mL solution in running buffer.

CAPILLARY ELECTROPHORESIS
Capillary: 60 cm × 50 μm acrylamide-coated fused-silica (40 cm to detector) (Supelco)
Capillary preparation: Adjust the pH of 20 mL water to 3.5 with acetic acid, add 80 μL 3-(trimethoxysilyl)propyl methacrylate (3-methacryloxypropyltrimethoxysilane), mix, suck into capillary, let stand at room temperature for 1 h, remove the solution, wash with water. Fill the capillary with a deaerated 3-4% acrylamide solution containing 1 μL/mL N,N,N',N'-tetramethylethylenediamine and 1 mg/mL ammonium persulfate, let stand for 3 h, remove excess solution by aspiration, rinse with water, remove water by aspiration, dry at 35° (cf. J. Chromatogr. 1985, 347, 191).

Running buffer: MeCN:buffer 5:95 (Buffer was 100 mM borate containing 5 M urea and 10 mM sodium dodecyl sulfate, adjusted to pH 8.5 with phosphate.)
Injection: Electrokinetic injection at 5 kV for 5 s.
Detector: UV 254
Migration time: 19.5
Voltage: 18 kV
Model: Jasco Model 870-CE

OTHER SUBSTANCES
Simultaneous: alprazolam, amobarbital, barbital, bromazepam, clonazepam, clotiazepam, cloxazolam, diazepam, estazolam, etizolam, fludiazepam, flunitrazepam, flurazepam, haloxazolam, medazepam, mephobarbital, nimetazepam, nitrazepam, oxazepam, pentobarbital, phenobarbital, secobarbital, triazolam

KEY WORDS
coated capillary; detector at anode

REFERENCE
Jinno,K.; Han,Y.; Sawada,H.; Taniguchi,M. Capillary electrophoretic separation of toxic drugs using a polyacrylamide-coated capillary, *Chromatographia*, **1997**, *46*, 309–314.

Methotrexate

Molecular formula: $C_{20}H_{22}N_8O_5$
Molecular weight: 454.46
CAS Registry No.: 59-05-2
Merck Index (12th ed.): 6065

SAMPLE
Matrix: blood
Sample preparation: Condition a Sep-Pak C18 SPE cartridge with 10 mL MeOH and 10 mL 200 mM pH 5.10 sodium acetate. 500 μL Serum + 1.5 mL 200 mM pH 5.10 sodium acetate, mix, add to the SPE cartridge, wash with 10 mL water, elute with 2 mL MeOH. Evaporate the eluate to dryness under reduced pressure, reconstitute the residue with 1 M pH 4.00 2-(N-morpholino)ethanesulfonic acid buffer. Remove a 100 μL aliquot and add it to 75 μL 0.35% potassium permanganate, vortex, let stand for 1 min, add 20 μL 3% hydrogen peroxide, dilute with running buffer, inject an aliquot.

CAPILLARY ELECTROPHORESIS
Capillary: 100 cm × 75 μm fused-silica (75 cm to detector) (Scientific Glass Engineering)
Running buffer: MeOH:buffer 30:70 (Buffer was 5 mM 2-(N-morpholino)ethanesulfonic acid containing 16 mM sodium sulfate, adjusted to pH 6.7 with 1 M NaCl.)
Injection: Electrokinetic injection at 10 kV for 10 s
Detector: F ex 325 (17 mW He-Cd laser) em 450
Migration time: 15.5
Voltage: 25 kV
Current: about 40 μA
Limit of detection: 50 nM

OTHER SUBSTANCES
Extracted: metabolites, 7-hydroxymethotrexate

KEY WORDS
serum; SPE

REFERENCE
Roach,M.C.; Gozel,P.; Zare,R.N. Determination of methotrexate and its major metabolite, 7-hydroxymethotrexate, using capillary zone electrophoresis and laser-induced fluorescence detection, *J.Chromatogr.*, **1988**, *426*, 129–140.

SAMPLE
Matrix: solutions
Sample preparation: Inject an aliquot of a 100 µg/mL solution.

CAPILLARY ELECTROPHORESIS
Capillary: 32.5 cm × 50 µm fused-silica (25 cm to detector) (Quadrex, New Haven, CT)
Capillary preparation: Condition new capillaries with 100 mM KOH for 10 min, purge with water for 5 min, equilibrate with running buffer for 5 min.
Running buffer: 100 mM pH 6.0 Phosphate buffer containing 2 mM vancomycin (A) or ristocetin (B) or teicoplanin (C)
Injection: Hydrostatic injection for 3 s.
Detector: UV 254, UV 280
Migration time: 22.0, 25.0 (A) or 12.5, 16.5 (B) or 14.0, 19.5 (C) (enantiomers)
Voltage: 5 kV
Model: Waters Quanta 4000

OTHER SUBSTANCES
Also analyzed: fenoprofen, ketoprofen

KEY WORDS
chiral

REFERENCE
Gasper,M.P.; Berthod,A.; Nair,U.B.; Armstrong,D.W. Comparison and modeling study of vancomycin, ristocetin A, and teicoplanin for CE enantioseparations, *Anal.Chem.*, **1996**, *68*, 2501–2514.

SAMPLE
Matrix: solutions

CAPILLARY ELECTROPHORESIS
Capillary: 47 cm × 50 µm fused-silica (40 cm to detector)
Capillary preparation: Before each run rinse capillary with water for 2 min, with 100 mM HCl for 2 min, with water for 2 min, and with 500 mM KOH for 2 h
Capillary temperature: 25
Running buffer: MeOH:50 mM pH 6.0 phosphate buffer containing 15 mM d-(+)-tubocurarine chloride (Caution! Tubocurarine is highly toxic!)
Injection: Pressure injection at 0.5 psi for 1 s.
Detector: UV 254
Migration time: 20.92, 21.60 (enantiomers)
Voltage: -15 kV
Model: Beckman P/ACE 2000

KEY WORDS
detector at anode; chiral

REFERENCE
Nair,U.B.; Armstrong,D.W.; Hinze,W.L. Characterization and evaluation of d-(+)-tubocurarine chloride as a chiral selector for capillary electrophoretic enantioseparations, *Anal.Chem.*, **1998**, *70*, 1059–1065.

Methotrimeprazine

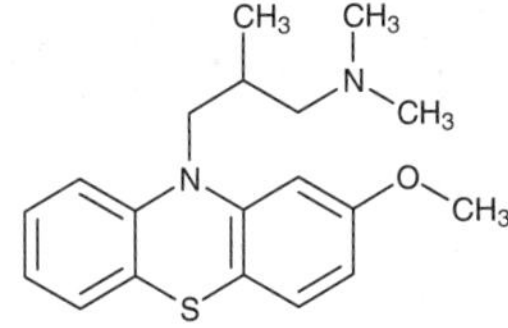

Molecular formula: $C_{19}H_{24}N_2OS$
Molecular weight: 328.48
CAS Registry No.: 60-91-1
Merck Index (12th ed.): 6066

SAMPLE
Matrix: solutions

CAPILLARY ELECTROPHORESIS
Capillary: 59.6 cm × 75 µm fused-silica (52.1 cm to detector) (Polymicro Technologies)
Capillary preparation: Purge with running buffer before each run. At the beginning of each day purge using 50-60 kPa vacuum with 500 mM NaOH for 5 min, with water for 5 min, with MeCN for 5 min, and with running buffer for 5 min.
Running buffer: MeCN:MeOH:acetic acid 49:50:1 containing 20 mM ammonium acetate
Injection: Hydrostatic injection at 10 cm for 5 s.
Detector: UV 214
Migration time: 3.55
Voltage: 30 kV
Model: Waters Quanta 4000

OTHER SUBSTANCES
Simultaneous: benzquinamide, deserpidine, ethopropazine, mesoridazine, molindone, promazine, promethazine, reserpine, thioridazine, thiothixene

REFERENCE
Leung,G.N.W.; Tang,H.P.O.; Tso,T.S.C.; Wan,T.S.M. Separation of basic drugs with non-aqueous capillary electrophoresis, *J.Chromatogr.A*, **1996**, *738*, 141–154.

SAMPLE
Matrix: urine
Sample preparation: Dilute 10-fold with water, inject an aliquot.

CAPILLARY ELECTROPHORESIS
Capillary: 70 cm × 50 µm fused-silica (65.4 cm to detector) (Polymicro Technologies)
Capillary temperature: 25
Running buffer: 20 mM pH 5.0 Tris-acetic acid containing 50 µg/mL FC-135 (Fluorad/3M) and 10 mM cetyltrimethylammonium bromide
Injection: Pressure injection at 2 psi.
Detector: UV 240
Migration time: 11.45
Voltage: 20 kV
Model: Bio-Rad BioFocus 3000

OTHER SUBSTANCES
Extracted: acepromazine, chlorpromazine, ethopropazine, fluphenazine, perphenazine, thioridazine, triflupromazine, trimeprazine

KEY WORDS
detector at anode

REFERENCE
Muijelsaar,P.G.H.M.; Claessens,H.A.; Cramers,C.A. Determination of structurally related phenothiazines by capillary zone electrophoresis and micellar electrokinetic chromatography, *J.Chromatogr.A*, **1996**, *735*, 395–402.

Methoxamine

Molecular formula: $C_{11}H_{17}NO_3$
Molecular weight: 211.26
CAS Registry No.: 390-28-3, 61-16-5 (HCl)
Merck Index (12th ed.): 6067

SAMPLE
Matrix: solutions

CAPILLARY ELECTROPHORESIS
Capillary: 60 cm × 75 μm fused-silica (51 cm to detector) (Supelco)
Capillary preparation: At the end of each day wash capillary with 200 mM NaOH for 10 min and with water for 10 min. Condition new capillaries by washing with 200 mM NaOH for 10 min, with water for 10 min, and with running buffer for 10 min.
Running buffer: 50 mM pH 7.3 Sodium borate buffer containing 3.3% succinyl β-cyclodextrin (Wacker Chemie, Munich, Germany)
Injection: Hydrodynamic injection at 25 mbar for 6 s.
Detector: UV 208
Migration time: 31.05 (first enantiomer)
Voltage: 15 kV
Model: Prince

OTHER SUBSTANCES
Simultaneous: alprenolol, chlorthalidone, ephedrine, etilefrin, homatropine, norephedrine, octopamine, propranolol, synephrine, trihexyphenidyl

KEY WORDS
chiral; $\alpha = 1.0431$

REFERENCE
Schmid,M.G.; Wirnsberger,K.; Gübitz,G. Chiral separation of drug enantiomers by capillary electrophoresis using succinyl-β-cyclodextrin, *Pharmazie*, **1996**, *51*, 852–854.

Methoxyphenamine

Molecular formula: $C_{11}H_{17}NO$
Molecular weight: 179.26
CAS Registry No.: 93-30-1, 5588-10-3 (HCl)
Merck Index (12th ed.): 6077

SAMPLE
Matrix: solutions
Sample preparation: Inject an aliquot of a solution in running buffer.

CAPILLARY ELECTROPHORESIS
Capillary: 60 cm × 75 μm fused-silica (52.4 cm to detector)
Capillary preparation: After each run flush with 500 mM KOH for 2-3 min then with water.
Running buffer: 10 mM pH 3.8 Phosphate buffer containing 2% sulfated cyclodextrin (ds 7-10)
Injection: Hydrostatic injection.
Detector: UV 214
Migration time: 17.16, 22.81 (enantiomers)
Voltage: 15 kV
Model: Waters Quanta 4000

OTHER SUBSTANCES
Also analyzed: acebutolol, alprenolol, aminoglutethimide, brompheniramine, bupivacaine, bupropion, canadine, carbinoxamine, chloroquine, chlorpheniramine, dimethindene, disopyramide, doxylamine, hydroxychloroquine, idazoxan, isoxsuprine, ketamine, mepenzolate, mepivacaine, mexiletine, midodrine, nefopam, orphenadrine, oxprenolol, oxyphencyclimine, pheniramine, phensuximide, pindolol, piperoxan, terbutaline, tetramisole, tolperisone, tranylcypromine, trihexyphenidyl, trimipramine, verapamil, warfarin

KEY WORDS
chiral; detector at anode

REFERENCE
Stalcup,A.M.; Gahm,K.H. Application of sulfated cyclodextrins to chiral separations by capillary zone electrophoresis, *Anal.Chem.*, **1996**, *68*, 1360–1368.

Methyclothiazide

Molecular formula: $C_9H_{11}Cl_2N_3O_4S_2$
Molecular weight: 360.24
CAS Registry No.: 135-07-9
Merck Index (12th ed.): 6086
Lednicer: 1 360

SAMPLE
Matrix: solutions

CAPILLARY ELECTROPHORESIS
Capillary: 40.6 cm × 50 μm fused-silica packed with 3 μm CEC Hypersil ODS (Hypersil)
Running buffer: Gradient. A was 5 mM ammonium acetate in MeCN:water 50:50. B was 5 mM ammonium acetate in MeCN:water 80:20. A:B 100:0 for 3 min, to 0:100 over 0.1 min, maintain at 0:100 (pumped with an HPLC pump at 0.01 mL/min for 3 min then at 0.1 mL/min).
Injection: Inject 5 μL using an HPLC injector.
Detector: MS, VG Biotech Platform, electrospray (details in paper)
Migration time: 27
Voltage: 30 kV

OTHER SUBSTANCES
Simultaneous: bendroflumethiazide, epithiazide, hydroflumethiazide, metolazone

KEY WORDS
electrochromatography

REFERENCE
Taylor,M.R.; Teale,P. Gradient capillary electrochromatography of drug mixtures with UV and electrospray ionisation mass spectrometric detection, *J.Chromatogr.A*, **1997**, *768*, 89–95.

Methyldopa

Molecular formula: C₁₀H₁₃NO₄
Molecular weight: 211.22
CAS Registry No.: 555-30-6, 41372-08-1 (sesquihydrate)
Merck Index (12th ed.): 6132
Lednicer: 1 95

SAMPLE
Matrix: solutions

CAPILLARY ELECTROPHORESIS
Capillary: 27 cm × 75 μm fused-silica (20 cm to detector) (Polymicro Technologies)
Capillary preparation: Before each injection rinse capillary with 400 mM KOH for 2 min, with water for 2 min, and with running buffer for 3 min.
Capillary temperature: 20
Running buffer: 25 mM pH 2.4 Phosphate buffer containing 1 mM starburst dendrimer PAMAM G0.5 (Aldrich)
Injection: Hydrodynamic injection at high pressure.
Detector: UV 214
Migration time: 18
Voltage: 10 kV
Model: Beckman P/ACE 5050

OTHER SUBSTANCES
Simultaneous: homophenylalanine, levodopa, phenylalanine, phenylglycine, tyrosine

REFERENCE
Gao,H.; Carlson,J.; Stalcup,A.M.; Heineman,W.R. Separation of aromatic acids, DOPA, and methyl dopa by capillary electrophoresis with dendrimers as buffer additives, *J.Chromatogr.Sci.*, **1998**, *36*, 146–154.

Methylene blue

Molecular formula: C₁₆H₁₈ClN₃S
Molecular weight: 319.86
CAS Registry No.: 61-73-4, 7220-79-3 (trihydrate)
Merck Index (12th ed.): 6137

SAMPLE
Matrix: solutions

CAPILLARY ELECTROPHORESIS
Capillary: 100 cm × 75 μm fused-silica (75 cm to detector) (J&W, Folsom CA)
Capillary preparation: Flush capillary under pressure with 100 mM NaOH for 10 min before each run.
Running buffer: 20 mM pH 3.4 Sodium phosphate buffer
Injection: Electrokinetic injection (10 nL).
Detector: F ex 655 (29 mW semiconductor laser, Hamamatsu LDH-065/PLP-01) em 700 (design of detector given in paper)
Migration time: 12
Voltage: 28 kV

KEY WORDS
peak shape is poor

REFERENCE
Kawazumi,H.; Song,J.M.; Inoue,T.; Ogawa,T. Application of an avalanche photodiode in a near-Geiger operation as a fluorescence detector for capillary electrophoresis, *J.Chromatogr.A*, **1996**, *744*, 31–36.

SAMPLE
Matrix: solutions

CAPILLARY ELECTROPHORESIS
Capillary: 100 cm × 75 μm fused-silica (J & W)
Capillary preparation: After each run rinse capillary with 100 mM NaOH and running buffer for about 10 min.
Running buffer: 20 mM pH 3.4 Sodium phosphate buffer
Injection: Electrokinetic injection at 28 kV for 2 s (10 nL).
Detector: F ex 655 (Hamamtsu Photonics LHD-065 pulsed semiconductor laser, peak power 29 mW, average power 21 μW at 10 MHz) em 700
Migration time: 13.5
Voltage: 28 kV
Limit of detection: 14 fmole

OTHER SUBSTANCES
Simultaneous: HIDC, oxazine 725, rhodamine 700

REFERENCE
Song,J.M.; Kawazumi,H.; Inoue,T.; Ogawa,T. Time-resolved photon-counting fluorimetry with a semiconductor laser for capillary electrophoresis, *J.Chromatogr.A*, **1996**, *727*, 330–333.

Methylphenidate

Molecular formula: $C_{14}H_{19}NO_2$
Molecular weight: 233.31
CAS Registry No.: 113-45-1, 298-59-9 (HCl)
Merck Index (12th ed.): 6186
Lednicer: 1 88

SAMPLE
Matrix: urine
Sample preparation: Perform at 0°. 4 mL Urine + 1 mL saturated sodium tetraborate (1 M, pH 9.3) + 40 μL 10 μg/mL IS in MeCN, add 4 mL cyclohexane, rotate at 40 rpm for 10 min, centrifuge at 268 g for 10 min. Remove the organic layer and evaporate it to dryness under a stream of nitrogen in a siliconized polypropylene tube at room temperature, reconstitute the residue in 200 μL water.

CAPILLARY ELECTROPHORESIS
Capillary: 65 cm × 50 μm fused-silica (Polymicro Technologies)
Running buffer: 40 mM Ammonium acetate adjusted to pH 9.0 with 30% ammonium hydroxide
Injection: Electrokinetic injection at 20 kV for 20 s (20 kV at anode, 0 kV at cathode).
Detector: MS, Finnigan LCQ ion-trap, electrospray, positive ion mode, sheath gas pressure 20 psi, sheath liquid MeCN at 7 μL/min, electrospray needle at 4 kV, heated capillary at 110°C, SRM (details in paper), m/z 234
Migration time: 5.3
Internal standard: methylphenidate-d$_3$
Voltage: 16 kV (20 kV to anode, 4 kV to cathode)
Model: Hewlett-Packard ^{3D}CE
Limit of quantitation: 1.5 ng/mL

KEY WORDS
pharmacokinetics

REFERENCE

Bach,G.A.; Henion,J. Quantitative capillary electrophoresis-ion-trap mass spectrometry determination of methylphenidate in human urine, *J.Chromatogr.B*, **1998**, *707*, 275–285.

Methylprednisolone

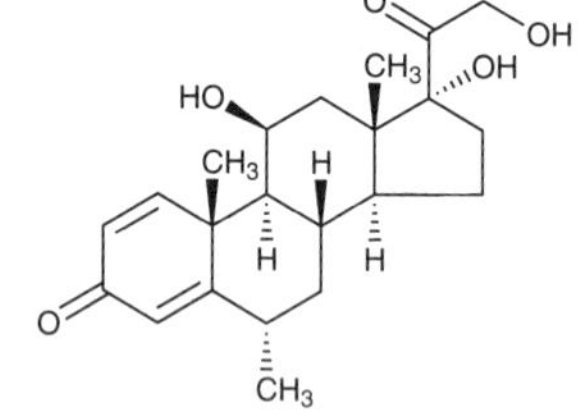

Molecular formula: C$_{22}$H$_{30}$O$_5$

Molecular weight: 374.48

CAS Registry No.: 83-43-2, 53-36-1 (acetate), 5015-36-1 (sodium phosphate), 2375-03-3 (sodium succinate), 2921-57-5 (hemisuccinate), 90350-40-6 (suleptanate)

Merck Index (12th ed.): 6189

Lednicer: 1 193, 196

SAMPLE

Matrix: formulations

Sample preparation: Add 2 mL injection to 2 mL IS solution, make up to 10 mL with water, inject an aliquot.

CAPILLARY ELECTROPHORESIS

Capillary: 50 cm × 50 μm fused-silica (Bio-Rad)

Capillary temperature: 25

Running buffer: 20 mM pH 8 phosphate buffer containing 30 mM sodium dodecyl sulfate

Injection: Injection by pressure 3 psi.sec (4.2 nL)

Detector: UV 230

Migration time: 9 (methylprednisolone-21-hemisuccinate)

Internal standard: hydroquinone monobenzyl ether (12)

Voltage: 17 kV

Current: 31 μA

Model: Biofocus 3000 (Bio-Rad)

OTHER SUBSTANCES

Simultaneous: degradation products, aminophylline, furosemide, prednisolone

KEY WORDS

injections; saline

REFERENCE

Quaglia,M.G.; Bossù,E.; Desiderio,C.; Fanali,S. Use of capillary electrophoresis for testing the stability of a drugs mixture in perfusional solution, *Farmaco*, **1994**, *49*, 403–406.

SAMPLE

Matrix: solutions

Sample preparation: Prepare a 10 mg/mL solution in MeOH, inject an aliquot.

CAPILLARY ELECTROPHORESIS

Capillary: 62 cm × 53 μm fused-silica (50 cm to detector) (Polymicro Technologies)

Capillary preparation: Rinse with running buffer for 1 min between runs.

Running buffer: 50 mM pH 9.0 Phosphate borate buffer containing 50 mM dehydrocholate, 50 mM taurocholate, and 50 mM sodium dodecyl sulfate

Detector: UV 254

Migration time: 29.5

Voltage: 15 kV

Current: 56 μA

Model: Laboratory constructed

OTHER SUBSTANCES
Simultaneous: corticosterone, cortisone, cortisone acetate, deoxycorticosterone, fludrocortisone, fludrocortisone acetate, fluocinolone acetonide, hydrocortisone, hydrocortisone-21-acetate, prednisolone, prednisolone acetate, prednisone, prednisone acetate, progesterone, triamcinolone, triamcinolone acetonide

REFERENCE
Bumgarner,J.G.; Khaledi,M.G. Mixed micellar electrokinetic chromatography of corticosteroids, *Electrophoresis*, **1994**, *15*, 1260–1266.

SAMPLE
Matrix: solutions

CAPILLARY ELECTROPHORESIS
Capillary: 62 cm × 53 μm fused-silica (50 cm to detector) (Polymicro Technologies)
Capillary preparation: Rinse capillary with running buffer for 1 min between runs.
Running buffer: 50 mM pH 9.0 Phosphate/borate buffer containing 33 mM taurocholate, 33 mM glycodeoxycholate, and 70 mM butanesulfonate
Detector: UV 254
Migration time: 35
Voltage: 15 kV
Model: laboratory-constructed

OTHER SUBSTANCES
Simultaneous: corticosterone, cortisone, cortisone acetate, deoxycorticosterone, fludrocortisone, fludrocortisone acetate, fluocinolone acetonide, hydrocortisone, hydrocortisone-21-acetate, prednisolone, prednisolone acetate, prednisone, prednisone acetate, progesterone, triamcinolone, triamcinolone acetonide

REFERENCE
Bumgarner,J.G.; Khaledi,M.G. Mixed micelles of short chain alkyl surfactants and bile slats in electrokinetic chromatography: Enhanced separation of corticosteroids, *J.Chromatogr.A*, **1996**, *738*, 275–283.

SAMPLE
Matrix: solutions

CAPILLARY ELECTROPHORESIS
Capillary: 42 cm × 50 μm fused-silica packed with 3 μm CEC Hypersil ODS (packed length 30 cm, 30.1 cm to detector) (Hypersil)
Running buffer: Gradient. A was 5 mM ammonium acetate in MeCN:water 17:83. B was 5 mM ammonium acetate in MeCN:water 38:62. A:B 100:0 for 3 min, to 0:100 over 15 min, maintain at 0:100 (pumped with an HPLC pump at 0.01 mL/min for 3 min then at 0.1 mL/min).
Injection: Inject 10 μL using an HPLC injector.
Detector: UV 240
Migration time: 34
Voltage: 30 kV

OTHER SUBSTANCES
Simultaneous: adrenosterone, betamethasone, cortisone, fluocortolone, dexamethasone, hydrocortisone, prednisolone, triamcinolone, triamcinolone acetonide

KEY WORDS
electrochromatography

REFERENCE
Taylor,M.R.; Teale,P. Gradient capillary electrochromatography of drug mixtures with UV and electrospray ionisation mass spectrometric detection, *J.Chromatogr.A*, **1997**, *768*, 89–95.

Methyltestosterone

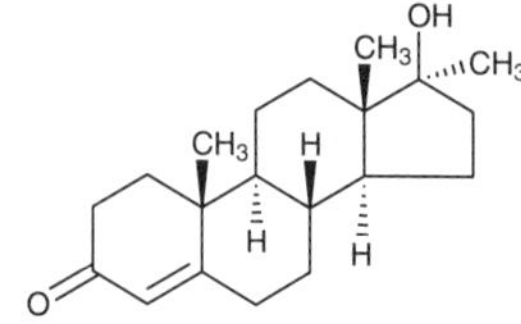

Molecular formula: $C_{20}H_{30}O_2$
Molecular weight: 302.44
CAS Registry No.: 58-18-4
Merck Index (12th ed.): 6206
Lednicer: 1 172, 4 11

SAMPLE
Matrix: formulations
Sample preparation: Tablets. Crush tablets, weigh out amount equivalent to 10 mg steroid, dissolve in 10 mL MeOH, sonicate for 15 min, filter. 1 mL Filtrate + 3 mL MeCN, make up to 10 mL with 75 mM sodium dodecyl sulfate, inject an aliquot. Suspensions. 1 mL Suspension + 50 mL MeOH, vortex for 2 min, make up to 100 mL with MeOH. Remove a 1 mL aliquot, add 3 mL MeCN, make up to 10 mL with 75 mM sodium dodecyl sulfate, inject an aliquot. Oils. 1-2 mL Oil + 50 mL MeOH, vortex for 3 min, make up to 100 mL with MeOH, let stand for several hours, filter. Remove a 1 mL aliquot of the filtrate, add 5 mL MeCN, make up to 10 mL with 75 mM sodium dodecyl sulfate, sonicate for 15 min, inject an aliquot.

CAPILLARY ELECTROPHORESIS
Capillary: 72 cm × 50 μm fused-silica (50 cm to detector) (Polymicro Technologies)
Capillary preparation: Aspirate with 1 M NaOH for 10 min, with water for 10 min, and with running buffer for 10 min.
Capillary temperature: 30
Running buffer: MeCN:buffer 40:60 (Buffer was 10 mM pH 9.0 phosphate buffer containing 10 mM borate and 75 mM sodium dodecyl sulfate.)
Injection: Inject for 1 s
Detector: UV 240
Migration time: 14.2
Voltage: 25 kV
Model: Applied Biosystems Model 270A-HT

OTHER SUBSTANCES
Simultaneous: boldenone (UV 200), clostebol acetate (UV 200), danazol (UV 280), fluoxymesterone (UV 200), methandriol (UV 200), methandrostenolone (UV 200), nandrolone (UV 200), stanolone (UV 200), stanozolol (UV 200), testosterone (UV 200)

KEY WORDS
tablets; suspensions; oils

REFERENCE
Lurie,I.S.; Sperling,A.R.; Meyers,R.P. The determination of anabolic steroids by MECC, gradient HPLC, and capillary GC, *J.Forensic Sci.*, **1994**, *39*, 74–85.

SAMPLE
Matrix: solutions

CAPILLARY ELECTROPHORESIS
Capillary: 32 cm × 100 μm fused-silica (Polymicro Technologies) packed for 24 cm with 1.5 μm Chromspher ODS (Chrompack)
Capillary preparation: Sonicate 30 mg 1.5 μm Chromspher ODS in 2 mL hexane:isopropanol 50:50 for 1 h, pump the slurry into the capillary at 600 bar using hexane:isopropanol 90:10 for 4 h, release the pressure over 1 min, wash with acetone, wash with water at 600 bar. Fabrication of frits is described in the paper.
Capillary temperature: 25
Running buffer: MeCN:water 60:40 containing 5 mM sodium dodecyl sulfate and 1.6 mM sodium tetraborate, pH 9.25
Injection: Electrokinetic injection at 10 kV.
Detector: UV (wavelength not given)

Migration time: 2.75
Voltage: 20 kV
Model: Hewlett Packard HP ^{3D}CE

OTHER SUBSTANCES
Simultaneous: hydrocortisone, progesterone, testosterone

KEY WORDS
electrochromatography

REFERENCE
Seifar,R.M.; Kok,W.T.; Kraak,J.C.; Poppe,H. Capillary electrochromatography with 1.5 μm ODS-modified non-porous silica spheres, *Chromatographia*, **1997**, *46*, 131–136.

SAMPLE
Matrix: solutions

CAPILLARY ELECTROPHORESIS
Capillary: 32 cm × 50 μm fused-silica packed for 24 cm with 1.8 μm Zorbax ODS (Polymicro Technologies) (Column preparation described in paper.)
Capillary temperature: 25
Running buffer: MeCN:0.8 mM sodium tetraborate 80:20 containing 5 mM sodium dodecyl sulfate
Injection: Electrokinetic injection at 20 kV for 1 s.
Detector: UV 254
Migration time: 3.6
Voltage: 25 kV
Model: Hewlett-Packard HP ^{3D}CE

OTHER SUBSTANCES
Simultaneous: estradiol, estriol, estrone, hydrocortisone, 4-pregnen-20α-ol-3-one, progesterone, testosterone

KEY WORDS
electrochromatography

REFERENCE
Seifar,R.M; Kraak,J.C.; Kok,W.T.; Poppe,H. Capillary electrochromatography with 1.8-μm ODS-modified porous silica particles, *J.Chromatogr.A*, **1998**, *808*, 71–77.

Methylthiouracil

Molecular formula: $C_5H_6N_2OS$
Molecular weight: 142.18
CAS Registry No.: 56-04-2
Merck Index (12th ed.): 6210
Lednicer: 1 264

SAMPLE
Matrix: urine
Sample preparation: Shake 3 mL urine with 15 mL ethyl acetate, centrifuge, repeat extraction twice more. Combine the organic layers and evaporate them to dryness under reduced pressure at 50°, reconstitute with 500 μL running buffer, inject an aliquot. (Use ethyl acetate p.a. (Merck).)

CAPILLARY ELECTROPHORESIS
Capillary: 80 cm × 100 μm fused-silica (60 cm to detector) (Polymicro Technologies)

Capillary preparation: Before use treat capillary with 1 M NaOH for 24 h, condition with
running buffer for 2 h.
Running buffer: 20 mM TES (N-tris(hydroxymethyl)methyl-2-aminoethanesulfonic acid) ad-
justed to pH 7.5 with NaOH
Injection: Hydrodynamic injection at 1 cm for 50 s.
Detector: UV 278
Migration time: 10.5
Voltage: 12 kV
Model: laboratory-constructed
Limit of detection: 0.1 ppm

OTHER SUBSTANCES
Simultaneous: propylthiouracil, thiouracil

KEY WORDS
horse

REFERENCE
Krivánková,L.; Krásensky,S.; Bocek,P. Application of capillary zone electrophoresis for analysis of thyreostatics,
Electrophoresis, **1996**, *17*, 1959–1963.

Metipranolol

Molecular formula: $C_{17}H_{27}NO_4$
Molecular weight: 309.41
CAS Registry No.: 22664-55-7, 36592-77-5 (HCl)
Merck Index (12th ed.): 6221
Lednicer: 5 17

SAMPLE
Matrix: solutions
Sample preparation: Inject an aliquot of a 100 μg/mL solution in running buffer.

CAPILLARY ELECTROPHORESIS
Capillary: 29 cm × 50 μm fused-silica (24.5 cm to detector) (Yongnian Optical Conductive Fiber
Plant, China), coated with polyacrylamide
Capillary preparation: No details of the polyacrylamide coating process are provided. However,
another paper (LC.GC 1997, 15, 40) by this group indicates that they use the procedure of
Hjertén, thus: Adjust the pH of 20 mL water to 3.5 with acetic acid, add 80 μL 3-(trimethoxysilyl)propyl methacrylate (3-methacryloxypropyltrimethoxysilane), mix, suck into capillary, let
stand at room temperature for 1 h, remove the solution, wash with water. Fill the capillary
with a deaerated 3-4% acrylamide solution containing 1 μL/mL N,N,N',N'-tetramethylethyl-
enediamine and 1 mg/mL potassium persulfate, let stand for 30 min, remove excess solution
by aspiration, rinse with water, remove water by aspiration, dry at 35° (J. Chromatogr. 1985,
347, 191).
Capillary temperature: 25
Running buffer: 100 mM NaH_2PO_4 adjusted to pH 2.5 (A) or 100 mM NaH_2PO_4 containing 45
mM hydroxypropyl-gamma-cyclodextrin, adjusted to pH 2.5 (B)
Injection: Electrokinetic injection at 15 kV for 3 s.
Detector: UV 200, UV 210
Migration time: 6.11 (A), 11.61, 12.48 (B, enantiomers)
Voltage: 15 kV
Model: Bio-Rad BioFocus 3000

OTHER SUBSTANCES
Simultaneous: albuterol, alprenolol, atenolol, baclofen, bamethan, benproperine, benserazide,
bisoprolol, bupranolol, butamirate, butethamate, carbuterol, celiprolol, clenbuterol, clobutinol,
dipivefrin, isoproterenol (isoprenaline), metaproterenol (orciprenaline), metoprolol, norfenefrine, ornidazole, oxprenolol, phenylpropanolamine, pholedrine, pirbuterol, prilocaine, procyclidine, sotalol, synephrine, terbutaline, tocainide

KEY WORDS
chiral; coated capillary

REFERENCE
Koppenhoefer,B.; Epperlein,U.; Xiaofeng,Z.; Bingcheng,L. Separation of enantiomers of drugs by capillary electrophoresis. Part 4: Hydroxypropyl-γ-cyclodextrin as chiral solvating agent, *Electrophoresis*, **1997**, *18*, 924–930.

SAMPLE
Matrix: solutions
Sample preparation: Inject an aliquot of a 100 μg/mL solution in running buffer.

CAPILLARY ELECTROPHORESIS
Capillary: 30 cm × 50 μm fused-silica (25.5 cm to detector), coated with polyacrylamide
Capillary preparation: Adjust the pH of 20 mL water to 3.5 with acetic acid, add 80 μL 3-(trimethoxysilyl)propyl methacrylate (3-methacryloxypropyltrimethoxysilane), mix, suck into capillary, let stand at room temperature for 1 h, remove the solution, wash with water. Fill the capillary with a deaerated 3-4% acrylamide solution containing 1 μL/mL N,N,N',N'-tetramethylethylenediamine and 1 mg/mL potassium persulfate, let stand for 30 min, remove excess solution by aspiration, rinse with water, remove water by aspiration, dry at 35° (J. Chromatogr. 1985, 347, 191).
Capillary temperature: 25
Running buffer: 100 mM NaH_2PO_4 containing 45 mM hydroxypropyl-gamma- cyclodextrin, adjusted to pH 2.5 with phosphoric acid
Injection: Electrokinetic injection at 15 kV for 3 s.
Detector: UV 200
Voltage: 15 kV
Model: Bio-Rad BioFocus 3000

KEY WORDS
chiral; coated capillary; comparison with the use of other cyclodextrins; this running buffer gave the greatest enantiomeric separation.; α=1.075

REFERENCE
Lin,B.; Zhu,X.; Koppenhoefer,B.; Epperlein,U. Investigation of 123 chiral drugs by cyclodextrin-modified capillary electrophoresis, *LC.GC*, **1997**, *15*, 40–46.

SAMPLE
Matrix: solutions

CAPILLARY ELECTROPHORESIS
Capillary: 29-36 cm × 50 μm fused-silica (24.5-31.5 cm to detector) (Yongnian Optical Conductive Fiber Plant, China) coated with polyacrylamide
Capillary preparation: Coat capillary as follows. Adjust the pH of 20 mL water to 3.5 with acetic acid, add 80 μL 3-(trimethoxysilyl)propyl methacrylate (3-methacryloxypropyltrimethoxysilane), mix, suck into capillary, let stand at room temperature for 1 h, remove the solution, wash with water. Fill the capillary with a deaerated 3-4% acrylamide solution containing 1 μL/mL N,N,N',N'-tetramethylethylenediamine and 1 mg/mL potassium persulfate, let stand for 30 min, remove excess solution by aspiration, rinse with water, remove water by aspiration, dry at 35° (J. Chromatogr. 1985, 347, 191).
Capillary temperature: 25
Running buffer: 100 mM pH 2.5 NaH_2PO_4 (A) or 100 mM pH 2.5 NaH_2PO_4 containing 45 mM hydroxypropyl-α-cyclodextrin (Wacker, Munich) (B)
Injection: Electromigration at 15 kV for 3 s.
Detector: UV 200; UV 210
Migration time: 6.11 (A); 11.21, 11.48 (B) (enantiomers)
Voltage: 15 kV
Model: Bio-Focus 3000

OTHER SUBSTANCES
Also analyzed: albuterol (salbutamol), alprenolol, amorolfine, atenolol, atropine, azelastine, baclofen, bamethan, benproperine, benserazide, biperiden, bisoprolol, brompheniramine, bupivacaine, bupranolol, butamirate, butethamate, carazolol, carbuterol, carteolol, carvedilol, celiprolol, chloroquine, chlorpheniramine, chlorphenoxamine, cicletanine, clenbuterol, clidinium bromide, clobutinol, dimethindene, dipivefrin, disopyramide, dobutamine, doxylamine, fendiline, flecainide, gallopamil, homatropine, ipratropium bromide, isoproterenol (isoprenaline), isothipendyl, ketamine, meclizine, mefloquine, mepindolol, mequitazine, metaclazepam, metaproterenol (orciprenaline), metoprolol, nafronyl (naftidrofuryl), nefopam, nicardipine, norfenefrine, ofloxacin, ornidazole, orphenadrine, oxomemazine, oxprenolol, oxybutynin, phenoxybenzamine, phenylpropanolamine, pholedrine, pindolol, pirbuterol, prilocaine, procyclidine, promethazine, propafenone, propranolol, reproterol, sotalol, sulpride, synephrine, talinolol, terbutaline, tetrahydrozoline (tetryzoline), theodrenaline, tioconazole, tocainide, trihexyphenidyl, trimeprazine (alimemazine), trimipramine, tropicamide, verapamil, zopiclone

KEY WORDS
coated capillary; chiral

REFERENCE
Koppenhoefer,B.; Eperlein,U.; Schlunk,R.; Zhu,X.; Lin,B. Separation of enantiomers of drugs by capillary electrophoresis. V. Hydroxypropyl-α-cyclodextrin as chiral solvating agent, *J.Chromatogr.A*, **1998**, *793*, 153–164.

Metolazone

Molecular formula: $C_{16}H_{16}ClN_3O_3S$
Molecular weight: 365.84
CAS Registry No.: 17560-51-9
Merck Index (12th ed.): 6231
Lednicer: 2 384

SAMPLE
Matrix: solutions

CAPILLARY ELECTROPHORESIS
Capillary: 40.6 cm × 50 μm fused-silica packed with 3 μm CEC Hypersil ODS (Hypersil)
Running buffer: Gradient. A was 5 mM ammonium acetate in MeCN:water 50:50. B was 5 mM ammonium acetate in MeCN:water 80:20. A:B 100:0 for 3 min, to 0:100 over 0.1 min, maintain at 0:100 (pumped with an HPLC pump at 0.01 mL/min for 3 min then at 0.1 mL/min).
Injection: Inject 5 μL using an HPLC injector.
Detector: MS, VG Biotech Platform, electrospray (details in paper)
Migration time: 27.5
Voltage: 30 kV

OTHER SUBSTANCES
Simultaneous: bendroflumethiazide, epithiazide, hydroflumethiazide, methyclothiazide

KEY WORDS
electrochromatography

REFERENCE
Taylor,M.R.; Teale,P. Gradient capillary electrochromatography of drug mixtures with UV and electrospray ionisation mass spectrometric detection, *J.Chromatogr.A*, **1997**, *768*, 89–95.

Metoprolol

Molecular formula: $C_{15}H_{25}NO_3$
Molecular weight: 267.37
CAS Registry No.: 37350-58-6, 56392-17-7 (tartrate), 119637-66-0 (fumarate), 98418-47-4 (succinate), 37350-58-6 (tartrate)
Merck Index (12th ed.): 6235
Lednicer: 2 109

SAMPLE

Matrix: blood
Sample preparation: Hydrolyze 2 mL serum with β-glucuronidase (EC 3.2.1.31, type H-1 from Helix pomatia, 416 800 U/g, Separacor) at 80° for 30 min, cool, add 900 µL MeCN, vortex for 15 min, centrifuge at 2004 g for 10 min, add ephedrine (165 µg/mL), filter (0.5 µm), inject an aliquot of the filtrate.

CAPILLARY ELECTROPHORESIS

Capillary: 58 cm × 50 µm fused-silica (50 cm to detector) (Polymicro Technologies)
Capillary preparation: Before each injection purge capillary with 5% phosphoric acid for 12 s, with water for 30 s, and with running buffer for 10 min.
Capillary temperature: 35
Running buffer: 80 mM pH 6.7 sodium phosphate buffer containing 15 mM cetyltrimethylammonium bromide
Injection: Hydrostatic injection for 20 s
Detector: UV 214
Migration time: 12.6
Internal standard: ephedrine (17.5)
Voltage: -27 kV
Model: Waters Quanta 4000

OTHER SUBSTANCES

Extracted: acebutolol, alprenolol, atenolol, labetalol, nadolol, oxprenolol, pindolol, propranolol, timolol

KEY WORDS

serum

REFERENCE

Lukkari,P.; Nyman,T.; Riekkola,M.-J. Determination of nine β-blockers in serum by micellar electrokinetic capillary chromatography, *J.Chromatogr.A*, **1994**, *674*, 241–246.

SAMPLE

Matrix: blood
Sample preparation: Filter (1 mL Amicon Centrifree Micropartition System) serum while centrifuging at 1500 g for 20 min, inject an aliquot of the ultrafiltrate.

CAPILLARY ELECTROPHORESIS

Capillary: 92 cm × 75 µm fused-silica (55 cm to detector) (Polymicro Technologies)
Capillary preparation: Rinse with running buffer for 10 min between run. At the beginning and end of each day rinse with 100 mM NaOH for 10 min and with water for 10 min.
Running buffer: Isopropanol:buffer 5:95 (Buffer was 10 mM Na_2HPO_4 containing 6 mM sodium borate and 75 mM sodium dodecyl sulfate.)
Injection: Injection via vacuum suction for 0.7 s
Detector: UV 215
Migration time: 21.5
Voltage: 25 kV
Current: 82 µA
Model: Europhor Prime Vision IV

OTHER SUBSTANCES
Extracted: atenolol, carteolol, celiprolol, levobunolol, penbutolol, pindolol, propranolol, timolol

KEY WORDS
serum; ultrafiltrate

REFERENCE
Schmutz,A.; Thormann,W. Assessment of impact of physico-chemical drug properties on monitoring drug levels by micellar electrokinetic capillary chromatography with direct serum injection, *Electrophoresis*, **1994**, *15*, 1295–1303.

SAMPLE
Matrix: blood, urine
Sample preparation: Urine. Dilute with 2 volumes of water, filter (0.5 μm), inject an aliquot. Serum. Condition a 3 mL Supelclean LC-18 SPE cartridge (Supelco) with MeOH and water. Hydrolyze 900 μL serum with β-glucuronidase (EC 3.2.1.31 type H-1 from Helix pomatia) at 60° with sonication for 30 min, add 500 μL (?) MeOH, centrifuge at 2000 g, add the supernatant to the SPE cartridge, wash with 1 mL water, dry under vacuum, elute with 2 mL MeOH:water 90:10, filter, inject an aliquot.

CAPILLARY ELECTROPHORESIS
Capillary: 68 cm × 50 μm fused-silica (60 cm to detector) (Polymicro Technologies)
Capillary preparation: Purge with running buffer for 2 min before injection.
Running buffer: pH 7.0 Phosphate buffer containing 10 mM N-cetyl-N,N,N-trimethylammonium bromide (Prepare buffer by mixing 100 mM NaH_2PO_4 and 100 mM Na_2HPO_4 to achieve a pH of 7.0.)
Injection: Hydrostatic injection for 30 s
Detector: UV 214
Migration time: 13.6
Internal standard: 2,6-dimethylphenol (only for urine) (14.5)
Voltage: -26 kV
Current: 97 μA
Model: Waters Quanta 4000

OTHER SUBSTANCES
Extracted: acebutolol, alprenolol, atenolol, labetalol, nadolol, oxprenolol, pindolol, propranolol, timolol

KEY WORDS
serum; comparison with HPLC; SPE

REFERENCE
Lukkari,P.; Sirén,H. Ion-pair chromatography and micellar electrokinetic capillary chromatography in analyzing β-adrenergic blocking agents from human biological fluids, *J.Chromatogr.A*, **1995**, *717*, 211–217.

SAMPLE
Matrix: solutions

CAPILLARY ELECTROPHORESIS
Capillary: 58 cm × 50 μm fused-silica (50 cm to detector)
Running buffer: Isopropanol:buffer 2.5:97.5 (Buffer was 80 mM pH 6.8 Phosphate buffer containing 15 mM cetyltrimethylammonium bromide.)
Injection: Hydrostatic injection for 15 s
Detector: UV 214
Migration time: 11
Voltage: -20 kV
Model: Waters Quanta 4000

OTHER SUBSTANCES
Simultaneous: acebutolol, alprenolol, atenolol, labetalol, nadolol, oxprenolol, pindolol, propranolol, sotalol, timolol

REFERENCE
Lukkari,P.; Vuorela,H.; Riekkola,M.-L. Effects of organic mobile phase modifiers on elution and separation of β-blockers in micellar electrokinetic capillary chromatography, *J.Chromatogr.A*, **1993**, *655*, 317–324.

SAMPLE
Matrix: solutions
Sample preparation: Prepare a 1 mg/mL solution in 40 mM pH 5.1 sodium phosphate buffer.

CAPILLARY ELECTROPHORESIS
Capillary: 11.5 cm × 75 μm fused silica tubing (8.5 cm to the detector) (Polymicro Technologies, Phoenix, AZ) coated on the inner surface with non-cross-linked polyacrylamide. Coat as follows. Mix 80 μL gamma-methacryloxypropyltrimethoxysilane with 20 mL water adjusted to pH 3.5 with acetic acid, suck into capillary, let stand for 1 h, remove solution, wash with water fill capillary with a 3 or 4% solution of acrylamide containing 1 mL/L N,N,N',N'-tetramethylethyl-enediamine and 1 g/L potassium persulfate, let stand for 30 min, remove excess solution, wash capillary with water, suck capillary dry, dry at 35° (J. Chromatogr. 1985, 347, 191).
Running buffer: Isopropanol:400 mM pH 5.1 sodium phosphate buffer 25:75
Injection: Fill 7 cm of tube with running buffer, fill rest of tube with 40 mg/mL cellobiohydrolase I in buffer, add a 2-3 mm plug of agarose by pressing end of tube into agarose gel, inject sample at 1000 V for 20 s. (Buffer was 400 mM pH 5.1 sodium phosphate buffer. Prepare agarose gel by boiling 20 mg agarose in 1 mL buffer, cool, allow to gel.)
Detector: UV 220
Migration time: 21, 23 (enantiomers)
Voltage: 1000 V
Model: Laboratory constructed

OTHER SUBSTANCES
Also analyzed: alprenolol, pindolol, propranolol

KEY WORDS
detector is at cathode; chiral; coated capillary

REFERENCE
Valtcheva,L.; Mohammad,J.; Pettersson,G.; Hjertén,S. Chiral separation of β-blockers by high-performance capillary electrophoresis based on non-immobilized cellulase as enantioselective protein, *J.Chromatogr.*, **1993**, *638*, 263–267.

SAMPLE
Matrix: solutions

CAPILLARY ELECTROPHORESIS
Capillary: 60 cm × 50 μm AccuSep (52.5 cm to detector) (Waters)
Capillary preparation: Before injection rinse capillary with 100 mM NaOH for 3 min and with running buffer for 3 min. Rinse new capillaries with 500 mM NaOH for 5 min
Running buffer: 50 mM pH 7.0 Na_2HPO_4 containing 25 mM (S)-N- dodecoxycarbonylvaline (Prepare (S)-N-dodecoxycarbonylvaline as follows. Prepare dodecyl chloroformate by reacting 1-dodecanol with 0.33 equivalents of triphosgene in dichloromethane solution in the presence of pyridine. Add dodecyl chloroformate dropwise to (S)-valine in 1 M NaOH solution, filter, wash with hexane, recrystallize from ether/petroleum ether.)
Injection: Hydrostatic injection 2 s
Detector: UV 214
Voltage: +12 kV
Model: Waters Quanta 4000 or 4000E

OTHER SUBSTANCES
Also analyzed: atenolol, bupivacaine, ephedrine, homatropine, ketamine, N-methylpseudo-ephedrine, norephedrine, norphenylephrine, octopamine, pindolol, terbutaline

KEY WORDS
chiral; α = 1.19

REFERENCE
Mazzeo,J.R.; Grover,E.R.; Swartz,M.E.; Petersen,J.S. Novel chiral surfactant for the separation of enantiomers by micellar electrokinetic capillary chromatography, *J.Chromatogr.A*, **1994**, *680*, 125–135.

SAMPLE
Matrix: solutions
Sample preparation: Inject an aliquot of a solution in 25 mM pH 6.8 potassium phosphate buffer.

CAPILLARY ELECTROPHORESIS
Capillary: 23.5 × 75 μm gel-filled fused-silica (16.5 cm to detector) (Polymicro Technologies)
Capillary preparation: Wash a new capillary with 100 μL water, 100 μL 1 M NaOH, 100 μL water, and 100 μL 50 mM pH 5.0 sodium citrate buffer. Pump gel mixture through capillary at 15 μL/min until it is within 1 cm of detector (detection window is filled with buffer, not gel), make sure it contains no bubbles (use a microscope), condition overnight at 2 kV (current limitation 20 μA) at reverse polarity using running buffer. (Gel mixture was 5 parts 20% bovine serum albumin in water, 21 parts 20% cellulase (from Trichoderma reesei) in 50 mM pH 5.0 sodium citrate buffer, and 4 arts 50% glutaraldehyde.)
Running buffer: Isopropanol:50 mM pH 6.8 potassium phosphate buffer 1:99
Injection: Electrokinetic injection at 2 kV for 3 s
Detector: UV 214
Migration time: 8.3 (R), 9 (S)
Voltage: 3 kV
Current: 40 μA
Model: Beckman P/ACE 2050

OTHER SUBSTANCES
Simultaneous: atenolol
Interfering: pindolol

KEY WORDS
chiral

REFERENCE
Ljungberg,H.; Nilsson,S. Protein-based capillary affinity gel electrophoresis for chiral separation of β-adrenergic blockers, *J.Liq.Chromatogr.*, **1995**, *18*, 3685–3698.

SAMPLE
Matrix: solutions
Sample preparation: Prepare a solution in MeOH/water, inject an aliquot.

CAPILLARY ELECTROPHORESIS
Capillary: 62 cm × 52 μm fused-silica (50 cm to detector) (Polymicro Technologies)
Capillary temperature: 40
Running buffer: 50 mM pH 2.50 Tetrabutylammonium phosphate
Injection: Siphon injection at 10 cm for 5 s
Detector: UV (wavelength not specified)
Migration time: 13
Voltage: 20 kV
Current: 29 μA
Model: Laboratory constructed

OTHER SUBSTANCES
Simultaneous: doxylamine, epinephrine, imidazole, isoproterenol, 2-methylphenethylamine, 1-methylphenylpropylamine, nadolol, nicotine, norepinephrine, propranolol, pseudoephedrine

REFERENCE
Quang,C.; Khaledi,M.G. Extending the scope of chiral separation of basic compounds by cyclodextrin-mediated capillary zone electrophoresis, *J.Chromatogr.A*, **1995**, *692*, 253–265.

SAMPLE
Matrix: solutions

Sample preparation: Inject an aliquot of a 300 μg/mL solution in water.

CAPILLARY ELECTROPHORESIS
Capillary: 65 cm × 50 μm fused-silica (40 cm to detector) (Isco)
Capillary preparation: Purge with running buffer for 3 min between injections. Purge daily with 1 M NaOH for 3 min, with water for 3 min, and with running buffer for 3 min.
Running buffer: Isopropanol:100 mM pH 7 phosphate buffer containing 25 mM rifamycin B 30: 70
Injection: Electrokinetic injection at 5 kV for 5 s.
Detector: UV 350
Migration time: 46.1 (first enantiomer, R_S 0.7)
Voltage: 8 kV
Model: Isco model 3850

OTHER SUBSTANCES
Simultaneous: alprenolol, epinephrine, amphetamine, norepinephrine, normetanephrine, octapamine, oxprenolol, propranolol

KEY WORDS
chiral

REFERENCE
Ward,T.J.; Dann,C.,III; Blaylock,A. Enantiomeric resolution using the macrocyclic antibiotics rifamycin B and rifamycin SV as chiral selectors for capillary electrophoresis, *J.Chromatogr.A*, **1995**, *715*, 337–344.

SAMPLE
Matrix: solutions
Sample preparation: Inject an aliquot of a solution in MeOH.

CAPILLARY ELECTROPHORESIS
Capillary: 64.5 cm × 50 μm fused-silica (Polymicro Technologies)
Capillary preparation: Between runs flush with 100 mM NaOH for 2-10 min then with running buffer for 2 min. Before use rinse with 1 M NaOH for 1 h, with 100 mM NaOH for 20 min, with water for 20 min, with 1 M acetic acid in MeOH/MeCN for 10 min, and with running buffer for 10 min.
Capillary temperature: 25
Running buffer: MeCN containing 1 M acetic acid, 200 μM Tween 20, and 30 mM sodium (+)-S-camphorsulfonate (Sodium (+)-S-camphorsulfonate can be prepared by mixing equimolar amounts of (+)-S-camphorsulfonic acid and NaOH in EtOH, collect the precipitate, wash with diethyl ether, dry.)
Injection: Pressure injection of 5 kPa for 3 s.
Detector: UV 214
Migration time: 18 (S), 19 (R)
Voltage: 30 kV
Model: Hewlett-Packard HP[3D]

KEY WORDS
chiral

REFERENCE
Bjornsdottir,I.; Hansen,S.H.; Terabe,S. Chiral separation in non-aqueous media by capillary electrophoresis using the ion-pair principle, *J.Chromatogr.A*, **1996**, *745*, 37–44.

SAMPLE
Matrix: solutions

CAPILLARY ELECTROPHORESIS
Capillary: 57 cm × 75 μm fused-silica (50 cm to detector) (Beckman)
Capillary preparation: Before each run rinse capillary with 100 mM NaOH and running buffer.
Capillary temperature: 30

Running buffer: Acetone:100 mM pH 8.1 borate buffer containing 50 mM sodium dodecyl sulfate 15:85 (A) or 100 mM phosphoric acid adjusted to pH 3.1 with triethanolamine (B)
Injection: Pressure injection for 5-10 s.
Detector: UV 200
Migration time: 12 (A), 10.7 (B)
Voltage: 25 kV
Model: Beckman P/ACE 5510

OTHER SUBSTANCES
Simultaneous: acebutolol, amiodarone, atenolol, bretylium, captopril, diltiazem, disopyramide, lidocaine, lisinopril, nicardipine, nifedipine, phenytoin, pindolol, propafenone, propranolol, quinidine, sotalol, timolol, p-toluenesulfonic acid, verapamil

KEY WORDS
only running buffer A separates all compounds; timolol interferes with running buffer A but not running buffer B

REFERENCE
Bretnall,A.E.; Clarke,G.S. Selectivity of capillary electrophoresis for the analysis of cardiovascular drugs, *J.Chromatogr.A*, **1996**, *745*, 145–154.

SAMPLE
Matrix: solutions

CAPILLARY ELECTROPHORESIS
Capillary: 43 cm × 50 μm fused-silica (36 cm to detector)
Capillary preparation: Before each injection wash with running buffer for 10 min. After each injection wash with water for 2 min. Wash new capillaries with 1 M NaOH at 60° for 50 min, with NaOH solution (?) at 60° for 10 min, and with water at 25°.
Capillary temperature: 25
Running buffer: 320 mM pH 2.0 Citrate buffer
Injection: Hydrodynamic injection at 1.5 psi for 2 s.
Detector: UV 220
Migration time: 9.4
Voltage: 15 kV
Model: Spectra-Physics Model 1000
Limit of detection: 1-18 μg/mL

OTHER SUBSTANCES
Simultaneous: acebutolol, atenolol, labetalol, levobunolol, nadolol, oxprenolol, pindolol, propranolol, timolol

REFERENCE
Lin,C.-E.; Chang,C.-C.; Lin,W.-c.; Lin,E.C. Capillary zone electrophoretic separation of β-blockers using citrate buffer at low pH, *J.Chromatogr.A*, **1996**, *753*, 133–138.

SAMPLE
Matrix: solutions

CAPILLARY ELECTROPHORESIS
Capillary: 52 cm × 75 μm fused-silica (48 cm to detector) (Polymicro Technologies)
Capillary preparation: Before each run purge with running buffer for 3 min. Every 3 runs purge with 100 mM NaOH for 5 min. Purge new capillaries with 1 M NaOH for 20 min and with 100 mM NaOH for 20 min, rinse with running buffer, equilibrate with running buffer at 12 kV for 3 h.
Running buffer: 100 mM CHES (2-(N-cyclohexylamine)ethanesulfonic acid) containing 10 mM triethylamine and 25 mM (R)-dodecoxycarbonylvaline (Waters EnantioSelect (R)-Val-1), pH adjusted to 8.8 with 1 M NaOH
Injection: Hydrostatic injection for 2 s.
Detector: UV 214
Migration time: 14, 14.1 (enantiomers)
Voltage: 12 kV

Current: ≤30 µA
Model: Waters Quanta 4000

OTHER SUBSTANCES
Simultaneous: bupivacaine, clenbuterol, disopyramide, norphenylephrine, octopamine

---

KEY WORDS
chiral

REFERENCE
Peterson,A.G.; Ahuja,E.S.; Foley,J.P. Enantiomeric separations of basic pharmaceutical drugs by micellar electrokinetic chromatography using a chiral surfactant, N-dodecoxycarbonylvaline, *J.Chromatogr.B*, **1996**, *683*, 15–28.

SAMPLE
Matrix: solutions

CAPILLARY ELECTROPHORESIS
Capillary: 55 cm × 75 µm fused-silica (40 cm to detector) (Polymicro Technologies)
Running buffer: 10 mM NaOH adjusted to pH 2.75 with 1 M phosphoric acid
Injection: Pressure injection at 20 mbar for 3.6 s.
Detector: UV 211
Migration time: 7.5
Voltage: 15.3 kV
Limit of quantitation: <19 µM

OTHER SUBSTANCES
Simultaneous: amitriptyline

KEY WORDS
paper contains details of isotachophoresis pre-concentration.

REFERENCE
Enlund,A.M.; Westerlund,D. Enhancing detectability in CE by combining an isotachophoresis preconcentration with capillary zone electrophoresis in a single capillary, *Chromatographia*, **1997**, *46*, 315–321.

SAMPLE
Matrix: solutions
Sample preparation: Inject an aliquot of a 100 µg/mL solution in running buffer.

CAPILLARY ELECTROPHORESIS
Capillary: 29 cm × 50 µm fused-silica (24.5 cm to detector) (Yongnian Optical Conductive Fiber Plant, China), coated with polyacrylamide
Capillary preparation: No details of the polyacrylamide coating process are provided. However, another paper (LC.GC 1997, 15, 40) by this group indicates that they use the procedure of Hjertén, thus: Adjust the pH of 20 mL water to 3.5 with acetic acid, add 80 µL 3-(trimethoxysilyl)propyl methacrylate (3-methacryloxypropyltrimethoxysilane), mix, suck into capillary, let stand at room temperature for 1 h, remove the solution, wash with water. Fill the capillary with a deaerated 3-4% acrylamide solution containing 1 µL/mL N,N,N',N'-tetramethylethylenediamine and 1 mg/mL potassium persulfate, let stand for 30 min, remove excess solution by aspiration, rinse with water, remove water by aspiration, dry at 35° (J. Chromatogr. 1985, 347, 191).
Capillary temperature: 25
Running buffer: 100 mM NaH$_2$PO$_4$ adjusted to pH 2.5
Injection: Electrokinetic injection at 15 kV for 3 s.
Detector: UV 200, UV 210
Migration time: 5.57
Voltage: 15 kV
Model: Bio-Rad BioFocus 3000

OTHER SUBSTANCES
Simultaneous: albuterol, alprenolol, atenolol, baclofen, bamethan, benproperine, benserazide, bisoprolol, bupranolol, butamirate, butethamate, carbuterol, celiprolol, clenbuterol, clobutinol, dipivefrin, isoproterenol (isoprenaline), metaproterenol (orciprenaline), metipranolol, norfenefrine, ornidazole, oxprenolol, phenylpropanolamine, pholedrine, pirbuterol, prilocaine, procyclidine, sotalol, synephrine, terbutaline, tocainide

KEY WORDS
coated capillary

REFERENCE
Koppenhoefer,B.; Epperlein,U.; Xiaofeng,Z.; Bingcheng,L. Separation of enantiomers of drugs by capillary electrophoresis. Part 4: Hydroxypropyl-γ-cyclodextrin as chiral solvating agent, *Electrophoresis*, **1997**, *18*, 924–930.

SAMPLE
Matrix: solutions

CAPILLARY ELECTROPHORESIS
Capillary: 67 cm $\times$ 50 μm (60 cm to detector)
Capillary preparation: Wash with running buffer for 5 min before each injection. After each injection wash with 1 M NaOH at 60° for 5 min, with 100 mM NaOH at 60° for 10 min, and with water at 25° for 5 min.
Capillary temperature: 25
Running buffer: 70 mM pH 7.0 Sodium phosphate buffer containing 15 mM cetyltriemthylammonium bromide
Injection: Hydrodynamic injection for 1 s.
Detector: UV 220
Migration time: 15
Voltage: 20 kV

OTHER SUBSTANCES
Simultaneous: acebutolol, atenolol, labetalol, levobunolol, nadolol, oxprenolol, pindolol, propranolol, timolol

REFERENCE
Lin,C.-E.; Chen,Y.-C.; Chang,C.-C.; Wang,D.-Z. Migration behavior and selectivity of β-blockers in micellar electrokinetic chromatography. Influence of micelle concentration of cationic surfactants, *J.Chromatogr.A*, **1997**, *775*, 349–357.

SAMPLE
Matrix: solutions
Sample preparation: Inject an aliquot of a 100 μM solution in water.

CAPILLARY ELECTROPHORESIS
Capillary: 65 cm $\times$ 50 μm fused-silica (56.5 cm to detector) (Polymicro Technologies)
Capillary preparation: Before each injection rinse capillary at 930 mbar with 100 mM NaOH for 1 min, with water for 2 min, and with running buffer for 4 min.
Capillary temperature: 20
Running buffer: 50 mM Phosphoric acid containing 4 mM heptakis(2,6-di-O-methyl)-β-cyclodextrin, adjusted to pH 2.5 with triethanolamine
Injection: Pressure injection at 20 mbar for 3 s.
Detector: UV 195
Migration time: 16.0 (R_S = 0.7)
Voltage: 24 kV
Model: Hewlett Packard HP[3D]

KEY WORDS
chiral

REFERENCE
Nilsson,S.; Schweitz,L.; Petersson,M. Three approaches to enantiomer separation of β-adrenergic antagonists by capillary electrochromatography, *Electrophoresis*, **1997**, *18*, 884–890.

SAMPLE
Matrix: solutions

CAPILLARY ELECTROPHORESIS
Capillary: 52-55 cm × 75 μm fused-silica (45-48 cm to detector) (Polymicro Technologies)
Capillary preparation: After each run purge with running buffer for 3 min. After every 5 runs purge with 100 mM LiOH. Purge new capillaries with 1 M LiOH for 20 min, purge with 100 mM LiOH for 20 min, rinse with running buffer, equilibrate with running buffer with voltage applied for 3 h.
Running buffer: MeCN:buffer 15:85 (Prepare a 100 mM CHES (2-[N-cyclohexylamine]ethanesulfonic acid) solution containing 25 mM (S)-N-dodecoxycarbonylvaline (Enantio-Select (S)-Val-1, Waters), adjust pH to 7 with 1 M LiOH, add triethylamine to a final concentration of 10 mM, add MeCN to a final concentration of 15%, adjust pH to 8.8 with 1 M LiOH. Preparation of (S)-N-dodecoxycarbonylvaline is as follows. Prepare dodecyl chloroformate by reacting 1-dodecanol with 0.33 equivalents of triphosgene in dichloromethane solution in the presence of pyridine. Add dodecyl chloroformate dropwise to (S)-valine in 1 M NaOH solution, filter, wash with hexane, recrystallize from ether/petroleum ether (J. Chromatogr. 1994, 680, 125).)
Injection: Hydrostatic injection for 2 s.
Detector: UV 214
Migration time: 14.8, 15 (enantiomers)
Voltage: 12 kV
Model: Waters Quanta 4000 CE

OTHER SUBSTANCES
Simultaneous: pindolol

KEY WORDS
chiral

REFERENCE
Peterson,A.G.; Foley,J.P. Influence of the inorganic counterion on the chiral micellar electrokinetic separation of basic drugs using the surfactant N-dodecoxycarbonylvaline, *J.Chromatogr.B*, **1997**, *695*, 131–145.

SAMPLE
Matrix: solutions
Sample preparation: Prepare a solution in running buffer, sonicate for 10 min, inject an aliquot.

CAPILLARY ELECTROPHORESIS
Capillary: 35 cm × 75 μm fused-silica (26.5 cm to detector) (Polymicro Technologies)
Capillary preparation: Before each run flush capillary with running buffer at 5-8 bar for several min, equilibrate until a stable current and baseline are achieved. Prepare capillary as follows. Flush capillary with 1 M NaOH for at least 30 min, flush capillary with water for at least 30 min, fill with a solution of 4 μL [(methacryloxy)propyl]trimethoxysilane in 1 mL 6 M acetic acid, let stand for at least 1 h, flush with water for several min, dry with a flow of nitrogen. Sonicate 30 mM (S)-metoprolol in sodium-dried toluene containing 3.6 g/L 2,2'-azo-bis(isobutyronitrile), 240 mM methacrylic acid, and 240 mM trimethylolpropane trimethacrylate for 10 min. Fill the capillary with this mixture, irradiate at 350 nm at -20° for 80 min, flush with several column volumes of MeCN, flush with several column volumes of running buffer. The polymer polymerizes around the imprint molecule ((S)-metoprolol). Later the imprint molecule is removed by extraction leaving a recognition site complementary in size, shape, and chemical functionality to the imprint molecule. The polymerized gel fills the whole column and is anchored to the capillary wall by the [(methacryloxy)propyl]trimethoxysilane.
Capillary temperature: 60
Running buffer: MeCN:2 M pH 3.0 ammonium acetate buffer 80:20
Injection: Electrokinetic injection at 3 kV for 3 s.
Detector: UV 195
Migration time: 6.5 (R), 7.3 (S)

Voltage: 5 kV
Model: Hewlett Packard HP 3D

KEY WORDS
electrochromatography; chiral; molecular imprinting; coated capillary

REFERENCE
Schweitz,L.; Andersson,L.I.; Nilsson,S. Capillary electrochromatography with predetermined selectivity obtained through molecular imprinting, *Anal.Chem.*, **1997**, *69*, 1179–1183.

SAMPLE
Matrix: solutions

CAPILLARY ELECTROPHORESIS
Capillary: 36 cm × 50 μm fused-silica coated with linear polyacrylamide (31.5 cm to detector) (GL Science)
Capillary preparation: At the beginning and end of each day rinse capillary with capillary wash solution (Bio-Rad Cat. No. 148-5022) at 690 kPa for more than 3 min and with water at 690 kPa for more than 3 min. Coat capillary as follows. Treat capillary with 1 M NaOH at room temperature for 1 h, rinse with water, dry by passing nitrogen gas through the capillary at 110° for 6 h. Pass thionyl chloride through the capillary using a suction pump for several min, seal capillary at both ends and heat at 70° for 6 h. Unseal the capillary and fill with 250 mM vinyl magnesium bromide in THF by suction, seal the capillary, heat at 70° for 6 h. Open the capillary and rinse it with THF for several min, rinse with distilled water, fill the capillary with polymerization solution, heat at 28 ± 2° for 1 h, condition at -100 V/cm for 30 min (Anal. Sci. 1994, 10, 1). (The polymerization solution was 5% acrylamide in water containing 49 mM Tris, 384 mM glycine, and 0.1% sodium dodecyl sulfate, degas in an ultrasonic bath. Add 40 μL 10% N,N,N',N'-tetramethylethylenediamine and 10 μL 10% ammonium persulfate to 5 mL of the degassed solution, mix thoroughly.)
Running buffer: MeOH:50 mM pH 6.0 Phosphate buffer 10:90
Injection: Before each injection rinse with water at 690 kPa for 30 s, rinse with running buffer at 690 kPa for 30 s, partially fill with separation solution (500 μM α_1-acid glycoprotein (Cohn fraction VI) (ICN) in running buffer) at 6.9 kPa for 190 s (27 cm), inject sample at 6.9 kPa for 2 s, electrophorese with running buffer (Note that α_1-acid glycoprotein from other suppliers may provide inferior results).
Detector: UV 210
Migration time: Resolution of enantiomers 1.6
Voltage: 12 kV
Model: Bio-rad BioFocus 3000

KEY WORDS
chiral; coated capillary

REFERENCE
Tanaka,Y.; Terabe,S. Separation of the enantiomers of basic drugs by affinity capillary electrophoresis using a partial filling technique and α_1-acid glycoprotein as chiral selector, *Chromatographia*, **1997**, *44*, 119–128.

SAMPLE
Matrix: solutions

CAPILLARY ELECTROPHORESIS
Capillary: 29-36 cm × 50 μm fused-silica (24.5-31.5 cm to detector) (Yongnian Optical Conductive Fiber Plant, China) coated with polyacrylamide
Capillary preparation: Coat capillary as follows. Adjust the pH of 20 mL water to 3.5 with acetic acid, add 80 μL 3-(trimethoxysilyl)propyl methacrylate (3-methacryloxypropyltrimethoxysilane), mix, suck into capillary, let stand at room temperature for 1 h, remove the solution, wash with water. Fill the capillary with a deaerated 3-4% acrylamide solution containing 1 μL/mL N,N,N',N'-tetramethylethylenediamine and 1 mg/mL potassium persulfate, let stand for 30 min, remove excess solution by aspiration, rinse with water, remove water by aspiration, dry at 35° (J. Chromatogr. 1985, 347, 191).
Capillary temperature: 25

Running buffer: 100 mM pH 2.5 NaH_2PO_4 (A) or 100 mM pH 2.5 NaH_2PO_4 containing 45 mM
hydroxypropyl-α-cyclodextrin (Wacker, Munich) (B)
Injection: Electromigration at 15 kV for 3 s.
Detector: UV 200; UV 210
Migration time: 5.57 (A); 14.41 (B) (no separation of enantiomers)
Voltage: 15 kV
Model: Bio-Focus 3000

OTHER SUBSTANCES
Also analyzed: albuterol (salbutamol), alprenolol, amorolfine, atenolol, atropine, azelastine, baclofen, bamethan, benproperine, benserazide, biperiden, bisoprolol, brompheniramine, bupivacaine, bupranolol, butamirate, butethamate, carazolol, carbuterol, carteolol, carvedilol, celiprolol, chloroquine, chlorpheniramine, chlorphenoxamine, cicletanine, clenbuterol, clidinium
bromide, clobutinol, dimethindene, dipivefrin, disopyramide, dobutamine, doxylamine, fendiline, flecainide, gallopamil, homatropine, ipratropium bromide, isoproterenol (isoprenaline),
isothipendyl, ketamine, meclizine, mefloquine, mepindolol, mequitazine, metaclazepam, metaproterenol (orciprenaline), metipranolol, nafronyl (naftidrofuryl), nefopam, nicardipine, norfenefrine, ofloxacin, ornidazole, orphenadrine, oxomemazine, oxprenolol, oxybutynin, phenox-
ybcnzamino, phenylpropanolamine, pholedrine, pindolol, pirbuterol, prilocaine, procyclidine,
promethazine, propafenone, propranolol, reproterol, sotalol, sulpride, synephrine, talinolol, terbutaline, tetrahydrozoline (tetryzoline), theodrenaline, tioconazole, tocainide, trihexyphenidyl,
trimeprazine (alimemazine), trimipramine, tropicamide, verapamil, zopiclone

---

KEY WORDS
coated capillary

REFERENCE
Koppenhoefer,B.; Eperlein,U.; Schlunk,R.; Zhu,X.; Lin,B. Separation of enantiomers of drugs by capillary electrophoresis. V. Hydroxypropyl-α-cyclodextrin as chiral solvating agent, *J.Chromatogr.A*, **1998**, *793*, 153–
164.

SAMPLE
Matrix: urine
Sample preparation: Dilute urine with 2 volumes of water, add IS, filter (0.5 μm), inject an
aliquot.

CAPILLARY ELECTROPHORESIS
Capillary: 68 cm × 50 μm fused-silica (60 cm to detector) (White Associates)
Capillary preparation: Purge capillary with running buffer for 2 min before each injection.
Running buffer: 80 mM pH 7.0 Phosphate buffer containing 10 mM N-cetyl-N,N,N-trimethyl-
ammonium bromide
Injection: Hydrostatic injection for 30 s at 10 cm.
Detector: UV 214
Migration time: 13.8
Internal standard: 2,6-dimethylphenol (14.7)
Voltage: -26 kV (at injector end)
Model: Waters Quanta 4000
Limit of detection: 10 μg/mL (S/N 3)

OTHER SUBSTANCES
Extracted: acebutolol, alprenolol, atenolol, labetalol, nadolol, oxprenolol, pindolol, propranolol,
timolol
Simultaneous: probenecid
Noninterfering: caffeine

REFERENCE
Lukkari,P.; Sirén,H.; Pantsar,M.; Riekkola,M.-L. Determination of ten β-blockers in urine by micellar electrokinetic capillary chromatography, *J.Chromatogr.*, **1993**, *632*, 143–148.

Metyrapone

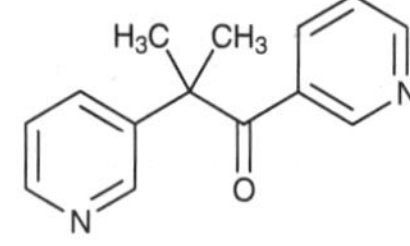

Molecular formula: $C_{14}H_{14}N_2O$
Molecular weight: 226.28
CAS Registry No.: 54-36-4, 908-35-0 (ditartrate)
Merck Index (12th ed.): 6246

SAMPLE
Matrix: blood, urine
Sample preparation: Condition a 3 mL Supelclean LC-18 SPE cartridge with 3 mL MeOH and 3 mL water. Dilute urine 1:10 with water. Precipitate proteins from serum with MeOH. Add diluted urine or protein supernatant to the SPE cartridge, wash with 3 mL water, elute with 3 mL MeOH, reconstitute to the original volume with 100 mM KOH, inject an aliquot.

CAPILLARY ELECTROPHORESIS
Capillary: 67 cm × 50 μm fused-silica (60 cm to detector) (Polymicro Technologies)
Capillary preparation: Rinse with running buffer for 2 min before run. If necessary, regenerate capillary with 100 mM NaOH for 10 min and with water for 15 min.
Capillary temperature: 20
Running buffer: 70 mM pH 4.5 Acetate buffer containing 500 mM betaine
Injection: Pressure injection for 5 s
Detector: UV 215
Migration time: 6.5
Voltage: 25 kV
Model: Beckman P/ACE 2000

OTHER SUBSTANCES
Extracted: amiloride, triamterene
Noninterfering: caffeine

KEY WORDS
serum; SPE

REFERENCE
Jumppanen,J.; Sirén,H.; Riekkola,M.-L. Screening for diuretics in urine and blood serum by capillary zone electrophoresis, *J.Chromatogr.A*, **1993**, *652*, 441–450.

Mexiletine

Molecular formula: $C_{11}H_{17}NO$
Molecular weight: 179.26
CAS Registry No.: 31828-71-4, 5370-01-4 (HCl)
Merck Index (12th ed.): 6257

SAMPLE
Matrix: solutions
Sample preparation: Inject an aliquot of a solution in running buffer.

CAPILLARY ELECTROPHORESIS
Capillary: 60 cm × 75 μm fused-silica (52.4 cm to detector)
Capillary preparation: After each run flush with 500 mM KOH for 2-3 min then with water.
Running buffer: 10 mM pH 3.8 Phosphate buffer containing 2% sulfated cyclodextrin (ds 7-10)
Injection: Hydrostatic injection.
Detector: UV 214
Migration time: 12.40, 13.08 (enantiomers)

Voltage: 15 kV
Model: Waters Quanta 4000

OTHER SUBSTANCES
Also analyzed: acebutolol, alprenolol, aminoglutethimide, brompheniramine, bupivacaine, bupropion, canadine, carbinoxamine, chloroquine, chlorpheniramine, dimethindene, disopyramide, doxylamine, hydroxychloroquine, idazoxan, isoxsuprine, ketamine, mepenzolate, mepivacaine, methoxyphenamine, midodrine, nefopam, orphenadrine, oxprenolol, oxyphencyclimine, pheniramine, phensuximide, pindolol, piperoxan, terbutaline, tetramisole, tolperisone, tranylcypromine, trihexyphenidyl, trimipramine, verapamil, warfarin

KEY WORDS
chiral; detector at anode

REFERENCE
Stalcup,A.M.; Gahm,K.H. Application of sulfated cyclodextrins to chiral separations by capillary zone electrophoresis, *Anal.Chem.*, **1996**, *68*, 1360–1368.

SAMPLE
Matrix: solutions
Sample preparation: Inject an aliquot of a 100 µg/mL solution in MeOH:water 10:90.

CAPILLARY ELECTROPHORESIS
Capillary: 37 cm × 75 µm
Capillary temperature: 15
Running buffer: 20 mM pH 2.06 Tris-phosphoric acid containing 10 mM 18-crown-6 tetracarboxylic acid (Merck, Darmstadt, Germany)
Injection: Pressure injection at 0.5 psi for 2-9 s.
Detector: UV 210, UV 235
Migration time: 6.96, 7.46 (enantiomers)
Voltage: 20 kV
Model: Beckman P/ACE 5510

OTHER SUBSTANCES
Also analyzed: aminoglutethimide, baclofen, levodopa, norephedrine, norepinephrine, octopamine, primaquine

KEY WORDS
chiral

REFERENCE
Nishi,H.; Nakamura,K.; Nakai,H.; Sato,T. Separation of enantiomers and isomers of amino compounds by capillary electrophoresis and high-performance liquid chromatography utilizing crown ethers, *J.Chromatogr.A*, **1997**, *757*, 225–235.

SAMPLE
Matrix: solutions

CAPILLARY ELECTROPHORESIS
Capillary: 36 cm × 50 µm fused-silica coated with linear polyacrylamide (31.5 cm to detector) (GL Science)
Capillary preparation: At the beginning and end of each day rinse capillary with capillary wash solution (Bio-Rad Cat. No. 148-5022) at 690 kPa for more than 3 min and with water at 690 kPa for more than 3 min. Coat capillary as follows. Treat capillary with 1 M NaOH at room temperature for 1 h, rinse with water, dry by passing nitrogen gas through the capillary at 110° for 6 h. Pass thionyl chloride through the capillary using a suction pump for several min, seal capillary at both ends and heat at 70° for 6 h. Unseal the capillary and fill with 250 mM vinyl magnesium bromide in THF by suction, seal the capillary, heat at 70° for 6 h. Open the capillary and rinse it with THF for several min, rinse with distilled water, fill the capillary with polymerization solution, heat at 28 ± 2° for 1 h, condition at -100 V/cm for 30 min (Anal. Sci. 1994, 10, 1). (The polymerization solution was 5% acrylamide in water containing 49 mM

Tris, 384 mM glycine, and 0.1% sodium dodecyl sulfate, degas in an ultrasonic bath. Add 40 μL 10% N,N,N',N'-tetramethylethylenediamine and 10 μL 10% ammonium persulfate to 5 mL of the degassed solution, mix thoroughly.)
Running buffer: Isopropanol:50 mM pH 6.0 Phosphate buffer 8:92
Injection: Before each injection rinse with water at 690 kPa for 30 s, rinse with running buffer at 690 kPa for 30 s, partially fill with separation solution (500 μM α_1-acid glycoprotein (Cohn fraction VI) (ICN) in running buffer) at 6.9 kPa for 190 s (27 cm), inject sample at 6.9 kPa for 2 s, electrophorese with running buffer (Note that α_1-acid glycoprotein from other suppliers may provide inferior results).
Detector: UV 210
Migration time: Resolution of enantiomers 1.7
Voltage: 12 kV
Model: Bio-rad BioFocus 3000

KEY WORDS
chiral; coated capillary

REFERENCE
Tanaka,Y.; Terabe,S. Separation of the enantiomers of basic drugs by affinity capillary electrophoresis using a partial filling technique and α_1-acid glycoprotein as chiral selector, *Chromatographia*, **1997**, *44*, 119–128.

SAMPLE
Matrix: solutions

CAPILLARY ELECTROPHORESIS
Capillary: 50 cm × 50 μm fused-silica (45 cm to detector)
Capillary preparation: Every other run rinse capillary with running buffer for 4 min. Before first use rinse capillary with 100 mM NaOH for 10 min, with water for 4 min, and with running buffer for 4 min.
Capillary temperature: 20
Running buffer: 40 mM Tris containing 20 mM heptakis-2,3,6-tri-O-methyl-β-cyclodextrin, adjusted to pH 2.5 with phosphoric acid
Injection: Pressure injection at 4 psi.s.
Detector: UV 210
Migration time: 11.6, 12 (enantiomers)
Voltage: 18 kV
Current: 14 μA
Model: Bio-Rad BioFocus 3000

KEY WORDS
chiral

REFERENCE
Kang,J.; Ou,Q. Chiral separation of racemic mexiletine hydrochloride using cyclodextrins as chiral additive by capillary electrophoresis, *J.Chromatogr.A*, **1998**, *795*, 394–398.

Mianserin

Molecular formula: $C_{18}H_{20}N_2$
Molecular weight: 264.37
CAS Registry No.: 24219-97-4, 21535-47-7 (HCl)
Merck Index (12th ed.): 6260
Lednicer: 2 451

SAMPLE
Matrix: blood
Sample preparation: 1 mL Plasma + 300 ng propylnorclozapine + 500 μL 1 M pH 9.4 carbonate buffer + 6 mL n-heptane:ethyl acetate 80:20, shake for 20 min, centrifuge at 8° at 3400 g for

6 min. Remove the organic layer and add it to 1.2 mL 100 mM HCl, shake for 15 min, centrifuge for 6 min. Remove the aqueous layer and add it to 1 mL 1 M pH 9.4 carbonate buffer and 150 μL toluene:isoamyl alcohol 85:15, shake for 20 min, centrifuge for 6 min. Remove the organic layer and add it to 100 μL 0.0001% diethylamine in water, evaporate under a stream of nitrogen at 40° to about 50 μL water containing diethylamine, heat at 40° (without nitrogen) for 30 min to remove traces of toluene, vortex for 15 s, inject an aliquot.

CAPILLARY ELECTROPHORESIS
Capillary: 64.5 cm × 50 μm silica (56 cm to detector) (Hewlett-Packard)
Capillary preparation: Flush capillary with 100 mM NaOH for 2 min and with running buffer for 3 min before each run. Replace running buffer in anode and cathode vials before each run.
Capillary temperature: 20
Running buffer: 75 mM Phosphoric acid containing 2 mM hydroxypropyl-β-cyclodextrin, adjusted to pH 3.0 with triethylamine
Injection: Pressure injection of water at 50 mbar for 3.7 s, electrokinetic injection of sample at 5 kV for 10 s, and pressure injection of running buffer at 50 mbar for 3.7 s.
Detector: UV 211
Migration time: 21 (S), 22.5 (R)
Internal standard: propylnorclozapine (15)
Voltage: 30 kV
Model: Hewlett-Packard
Limit of quantitation: 5 ng/mL

OTHER SUBSTANCES
Extracted: metabolites
Simultaneous: clozapine, norclozapine, thioridazine
Noninterfering: amitriptyline, citalopram, clomipramine, desipramine, fluoxetine, fluvoxamine, imipramine, maprotiline, methadone, norcitalopram, norclomipramine, norfluoxetine, norsertraline, nortrimipramine, nortriptyline, paroxetine, sertraline, trimipramine

KEY WORDS
plasma; chiral

REFERENCE
Eap,C.B.; Powell,K.; Baumann,P. Determination of the enantiomers of mianserin and its metabolites in plasma by capillary electrophoresis after liquid-liquid extraction and on-column sample preconcentration, *J.Chromatogr.Sci.*, **1997**, *35*, 315–320.

SAMPLE
Matrix: solutions
Sample preparation: Inject an aliquot of a 60 μg/mL solution.

CAPILLARY ELECTROPHORESIS
Capillary: 61 cm × 50 μm untreated fused-silica (Grom)
Capillary temperature: 21 ± 1
Running buffer: 50 mM pH 6.0 phosphate buffer containing 60 mM β-cyclodextrin
Injection: Hydrostatic injection at 10 cm for 5 s
Detector: UV 214
Migration time: 3.6 (S-(+)), 3.7 (R-(-))
Voltage: 400 V/cm
Model: Grom System 100

KEY WORDS
chiral

REFERENCE
Chankvetadze,B.; Endresz,G.; Bergenthal,D.; Blaschke,G. Enantioseparation of mianserine analogues using capillary electrophoresis with neutral and charged cyclodextrin buffer modifiers. 13C NMR study of the chiral recognition mechanism, *J.Chromatogr.A*, **1995**, *717*, 245–253.

SAMPLE
Matrix: solutions

Sample preparation: Inject an aliquot of an aqueous solution.

CAPILLARY ELECTROPHORESIS
Capillary: 49 cm × 50 μm fused-silica
Capillary preparation: Purge with running buffer for 3 min before each run. At the start of each day purge under vacuum with 500 mM NaOH then with water. Finally use an electroosmotic purge with running buffer then a vacuum purge with running buffer.
Running buffer: 50 mM L-(+)-tartrate containing 2.5% Glucidex6 maltodextrin (Roquette, Lestrem, France), adjusted to pH 3.25 with triethylamine
Injection: Gravity siphon at 10 cm for 30 s.
Detector: UV 214
Migration time: 6.2, 6.5 (enantiomers)
Voltage: 30 kV
Model: Waters Quanta 4000 CE

OTHER SUBSTANCES
Simultaneous: fluoxetine, sulpride (not chiral), sultopride (not chiral), viloxacine (not chiral)

KEY WORDS
chiral

REFERENCE
D'Hulst,A.; Verbeke,N. Chiral analysis of basic drugs by oligosaccharide-mediated capillary electrophoresis, *J.Chromatogr.A*, **1996**, *735*, 283–293.

SAMPLE
Matrix: solutions

CAPILLARY ELECTROPHORESIS
Capillary: 42 cm × 50 μm fused-silica (31 cm to detector) (Polymicro Technologies)
Capillary preparation: Condition capillary by rinsing with running buffer for 10 min, with water for 10 min, with 1 M NaOH for 10 min, with water for 10 min, and with running buffer with the voltage applied for 20 min.
Capillary temperature: 25
Running buffer: Formamide containing 150 mM citric acid, 100 mM Tris, and 100 mM β-cyclodextrin (apparent pH 5.1)
Injection: Gravity injection.
Detector: UV 254
Migration time: 8.14, 8.30 (enantiomers)
Voltage: 30 kV
Model: Beckman P/ACE 5500

OTHER SUBSTANCES
Simultaneous: chlophedianol, nefopam, propiomazine, trimeprazine, trimipramine, thioridazine

KEY WORDS
chiral

REFERENCE
Wang,F.; Khaledi,M.G. Chiral separations by nonaqueous capillary electrophoresis, *Anal.Chem.*, **1996**, *68*, 3460–3467.

SAMPLE
Matrix: solutions

CAPILLARY ELECTROPHORESIS
Capillary: 44 cm × 50 μm fused-silica
Running buffer: 10 mM pH 4.3 acetic acid/ammonium acetate buffer containing 200 μg/mL carboxymethyl ether-β-cyclodextrin (Wacker Chemie, Munich)
Injection: Hydrostatic injection at 10 cm for 5-10 s.

Detector: MS, Finnigan LCQ, electrospray, positive ion mode, ion-spray tip voltage 2.6 kV, sheath
liquid MeOH:water:acetic acid 50:49:1 at 6 μL/min, m/z 265
Migration time: 4.5, 5 (enantiomers)
Voltage: 20 kV

KEY WORDS
chiral

REFERENCE
Schulte,G.; Heitmeier,S.; Chankvetadze,B.; Blaschke,G. Chiral capillary electrophoresis-electrospray mass
spectrometry coupling with charged cyclodextrin derivatives as chiral selectors, *J.Chromatogr.A,* **1998,** *800,*
77–82.

Miconazole

Molecular formula: $C_{18}H_{14}Cl_4N_2O$
Molecular weight: 416.13
CAS Registry No.: 22916-47-8, 22832-87-7 (nitrate)
Merck Index (12th ed.): 6266
Lednicer: 2 249

SAMPLE
Matrix: solutions
Sample preparation: Prepare a 60 μg/mL solution, inject an aliquot.

CAPILLARY ELECTROPHORESIS
Capillary: 61 cm × 50 μm fused-silica (44 cm to detector) (Grom)
Capillary temperature: 21 ± 1
Running buffer: MeOH:buffer 10:90 (Buffer was 50 mM pH 3.0 (?) phosphate buffer containing
20 mM hydroxypropyl-β-cyclodextrin.)
Injection: Hydrostatic injection at 10 cm for 5 s
Detector: UV 210
Migration time: 15.46, 15.74 (enantiomers)
Voltage: 400 V/cm
Model: Grom 100

KEY WORDS
chiral

REFERENCE
Chankvetadze,B.; Endresz,G.; Blaschke,G. Enantiomeric resolution of chiral imidazole derivatives using capillary electrophoresis with cyclodextrin-type buffer modifiers, *J.Chromatogr.A,* **1995,** *700,* 43–49.

Midodrine

Molecular formula: $C_{12}H_{18}N_2O_4$
Molecular weight: 254.29
CAS Registry No.: 42794-76-3, 3092-17-9 (HCl)
Merck Index (12th ed.): 6272
Lednicer: 4 23

SAMPLE
Matrix: solutions

Sample preparation: Inject an aliquot of a solution in running buffer.

CAPILLARY ELECTROPHORESIS
Capillary: 60 cm $\times$ 75 μm fused-silica (52.4 cm to detector)
Capillary preparation: After each run flush with 500 mM KOH for 2-3 min then with water.
Running buffer: 10 mM pH 3.8 Phosphate buffer containing 2% sulfated cyclodextrin (ds 7-10)
Injection: Hydrostatic injection.
Detector: UV 214
Migration time: 14.00, 16.53 (enantiomers)
Voltage: 15 kV
Model: Waters Quanta 4000

OTHER SUBSTANCES
Also analyzed: acebutolol, alprenolol, aminoglutethimide, brompheniramine, bupivacaine, bupropion, canadine, carbinoxamine, chloroquine, chlorpheniramine, dimethindene, disopyramide, doxylamine, hydroxychloroquine, idazoxan, isoxsuprine, ketamine, mepenzolate, mepivacaine, methoxyphenamine, mexiletine, nefopam, orphenadrine, oxprenolol, oxyphencyclimine, pheniramine, phensuximide, pindolol, piperoxan, terbutaline, tetramisole, tolperisone, tranylcypromine, trihexyphenidyl, trimipramine, verapamil, warfarin

KEY WORDS
chiral; detector at anode

REFERENCE
Stalcup,A.M.; Gahm,K.H. Application of sulfated cyclodextrins to chiral separations by capillary zone electrophoresis, *Anal.Chem.*, **1996**, *68*, 1360–1368.

Minocycline

Molecular formula: $C_{23}H_{27}N_3O_7$
Molecular weight: 457.48
CAS Registry No.: 10118-90-8, 13614-98-7 (HCl)
Merck Index (12th ed.): 6289
Lednicer: 1 214

SAMPLE
Matrix: bulk

CAPILLARY ELECTROPHORESIS
Capillary: 44 cm $\times$ 50 μm fused-silica (38 cm to detector) (Polymicro Technologies)
Capillary temperature: 15
Running buffer: 25 mM pH 11.75 sodium tetraborate containing 1 mM EDTA
Injection: Hydrodynamic injection for 2 s.
Detector: UV 254
Migration time: 13
Voltage: 13 kV
Model: Spectraphoresis 500 (Thermo Separation Products)
Limit of quantitation: 200 ng/mL
Limit of detection: 100 ng/mL

OTHER SUBSTANCES
Simultaneous: impurities

REFERENCE
Li,Y.M.; Van Schepdael,A.; Roets,E.; Hoogmartens,J. Capillary zone electrophoresis of minocycline, *J.Pharm.Biomed.Anal.*, **1996**, *14*, 1095–1099.

SAMPLE
Matrix: formulations
Sample preparation: Dissolve capsule contents in 15 mM pH 7.5 sodium phosphate buffer so
  as to make a 25 mM solution, centrifuge, dilute an aliquot of the supernatant to a concentration
  of 5 mM with 15 mM pH 7.5 sodium phosphate buffer, inject an aliquot.

CAPILLARY ELECTROPHORESIS
Capillary: 111.9 cm × 75 μm fused-silica (43.4 cm to detector) (Polymicro Technologies)
Capillary preparation: Condition by rinsing with 1 M NaOH for 10 min then running buffer
  overnight.
Capillary temperature: 25
Running buffer: pH 7.5 Phosphate buffer (buffer concentration 4.3 mM, total sodium concen-
  tration 15 mM, ionic strength 18.2 mM)
Injection: Hydrodynamic injection at 2 cm for 1 min
Detector: UV 260
Migration time: 5.0
Current: 20 μA
Limit of detection: 10 μM (S/N 3)

OTHER SUBSTANCES
Simultaneous: chlortetracycline, demeclocycline, doxycycline, methacycline, oxytetracycline,
  tetracycline

KEY WORDS
capsules

REFERENCE
Tavares,M.F.M.; McGuffin,V.L. Separation and characterization of tetracycline antibiotics by capillary electrophoresis, *J.Chromatogr.A*, **1994**, *686*, 129–142.

SAMPLE
Matrix: solutions

CAPILLARY ELECTROPHORESIS
Capillary: 50 cm × 50 μm (50 cm to detector) (Polymicro Technologies)
Capillary preparation: Purge capillary with 10 mM NaOH for 2 min after each injection.
Running buffer: MeOH:buffer 40:60 (Buffer was 30 mM pH 3.0 citric acid containing 24.5 mM
  β-alanine.)
Injection: Electrokinetic injection at 10 kV for 5 s
Detector: UV 254
Migration time: 4.3
Voltage: 30 kV
Model: Perkin Elmer/Applied Biosystems Model 270A-HT

OTHER SUBSTANCES
Simultaneous: chlortetracycline, doxycycline, meclocycline, methacycline, oxytetracycline,
  tetracycline

REFERENCE
Pesek,J.J.; Matyska,M.T. Separation of tetracyclines by high-performance capillary electrophoresis and capillary electrochromatography, *J.Chromatogr.A*, **1996**, *736*, 313–320.

SAMPLE
Matrix: solutions

CAPILLARY ELECTROPHORESIS
Capillary: 43 cm × 75 μm (36 cm to the detector)
Capillary preparation: Before each run wash capillary with running buffer for 2 min. After
  each run wash capillary with water at 25° for 5 min, with 100 mM NaOH at 60 ° for 5 min,
  and with water at 25° for 5 min.
Capillary temperature: 25

Running buffer: 15 mM pH 6.5 Ammonium acetate buffer containing 20 mM sodium dodecyl
sulfate and 0.135% Brij 35
Detector: UV 265
Migration time: 9
Voltage: 15 kV

OTHER SUBSTANCES
Simultaneous: chlortetracycline, demeclocycline, doxycycline, oxytetracycline, tetracycline

REFERENCE
Chen,Y.-C.; Lin,C.-E. Migration behavior and separation of tetracycline antibiotics by micellar electrokinetic
chromatography, *J.Chromatogr.A*, **1998**, *802*, 95–105.

Moclobemide

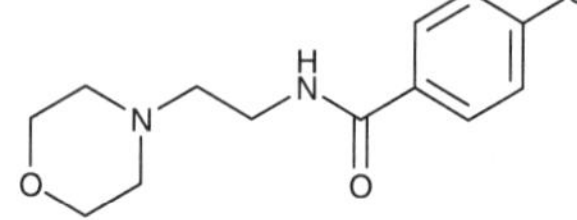

Molecular formula: $C_{13}H_{17}ClN_2O_2$
Molecular weight: 268.74
CAS Registry No.: 71320-77-9
Merck Index (12th ed.): 6309
Lednicer: 4 39

SAMPLE
Matrix: urine
Sample preparation: 10 mL Urine + 1 mL 5 M NaOH + 10 mL n-hexane, vortex for 1 min,
centrifuge at 0° at 3000 g for 5 min, repeat the extraction twice more. Combine the organic
layers and add them to 50 μL glacial acetic acid, evaporate to dryness under a stream of
nitrogen at 30°, reconstitute the residue in 50 μL 50 mM sodium taurodeoxycholate, filter (0.2
μm PTFE), inject an aliquot.

CAPILLARY ELECTROPHORESIS
Capillary: 100 cm × 50 μm fused-silica (50 cm to detector) (ISCO)
Capillary preparation: Rinse capillary with 20 μL running buffer between injections. Condition
a new capillary by filling with 1 M NaOH, let stand for 1 h, fill with 100 mM NaOH, let stand
for 1 h, rinse with buffer. Every 20 injections rinse capillary with 200 μL 1 M NaOH, 200 μL
water, and 200 μL running buffer, fill with running buffer.
Capillary temperature: 22
Running buffer: 40 mM pH 9.5 Borate buffer containing 10 mM sodium taurodeoxycholate
Injection: Load under vacuum at 7.5 kPa/s.
Detector: UV 240
Migration time: 5
Voltage: 30 kV
Model: ISCO Model 3140 electropherograph
Limit of detection: 4 ng/mL

OTHER SUBSTANCES
Extracted: acepromazine, amiodarone, amitriptyline, azaperone, chlorpromazine, cianopramine,
clomipramine, clozapine, desethylamiodarone, desipramine, diclofensine, dothiepin, doxepin,
imipramine, isocarboxazid, perphenazine, phenothiazine, pimozide, prochlorperazine, promazine, thioridazine, thiothixene, trifluoperazine, trimipramine

KEY WORDS
human; cow; pig; horse; protect from light

REFERENCE
Aumatell,A.; Wells,R.J. Determination of a cardiac antiarrhythmic, tricyclic antipsychotics and antidepressants
in human and animal urine by micellar electrokinetic capillary chromatography using a bile salt,
J.Chromatogr.B, **1995**, *669*, 331–344.

Molindone

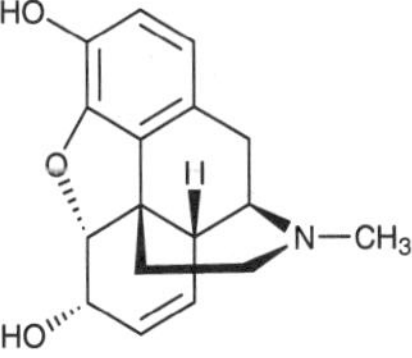

Molecular formula: $C_{16}H_{24}N_2O_2$
Molecular weight: 276.38
CAS Registry No.: 7416-34-4, 15622-65-8 (HCl)
Merck Index (12th ed.): 6315
Lednicer: 2 455

SAMPLE
Matrix: solutions

CAPILLARY ELECTROPHORESIS
Capillary: 59.6 cm × 75 μm fused-silica (52.1 cm to detector) (Polymicro Technologies)
Capillary preparation: Purge with running buffer before each run. At the beginning of each day purge using 50-60 kPa vacuum with 500 mM NaOH for 5 min, with water for 5 min, with MeCN for 5 min, and with running buffer for 5 min.
Running buffer: MeCN:MeOH:acetic acid 49:50:1 containing 20 mM ammonium acetate
Injection: Hydrostatic injection at 10 cm for 5 s.
Detector: UV 214
Migration time: 4.20
Voltage: 30 kV
Model: Waters Quanta 4000

OTHER SUBSTANCES
Simultaneous: benzquinamide, deserpidine, ethopropazine, mesoridazine, methotrimeprazine, promazine, promethazine, reserpine, thioridazine, thiothixene

REFERENCE
Leung,G.N.W.; Tang,H.P.O.; Tso,T.S.C.; Wan,T.S.M. Separation of basic drugs with non-aqueous capillary electrophoresis, *J.Chromatogr.A*, **1996**, *738*, 141–154.

Morphine

Molecular formula: $C_{17}H_{19}NO_3$
Molecular weight: 285.34
CAS Registry No.: 57-27-2, 52-26-6 (HCl), 630-81-9 (HBr), 64-13-3 (sulfate)
Merck Index (12th ed.): 6359
Lednicer: 1 286

SAMPLE
Matrix: bulk
Sample preparation: Prepare a 10 mg/mL solution in 3 mg/mL pholcodine in 10 mM HCl, dilute 5-fold with 10 mM HCl, filter (0.45 μm cellulose acetate), inject an aliquot.

CAPILLARY ELECTROPHORESIS
Capillary: 72 cm × 75 μm fused-silica (50 cm to detector) (Isco)
Capillary preparation: Flush with running buffer for 2 min between runs. Every 24 h wash capillary with 100 mM HCl for 10 min, wash with water, wash 100 mM NaOH, wash with water, fill with running buffer. Prepare new capillary by filling with 1 M NaOH, let stand for 1 h, fill with 100 mM NaOH, let stand for 1 h, wash with water, fill with running buffer.
Capillary temperature: 30
Running buffer: MeCN:buffer 10:90 (Buffer was 10 mM KH_2PO_4 containing 10 mM sodium tetraborate and 50 mM cetyltrimethylammonium bromide, pH 8.6.)
Injection: Injection by vacuum 5.0 kPa.s (4 nL)
Detector: UV 280

Migration time: 9.8
Internal standard: pholcodine (9.1)
Voltage: 15 kV
Model: Isco Model 3140

OTHER SUBSTANCES
Simultaneous: acetylcodeine, caffeine, codeine, diamorphine, ethylmorphine, noscapine, papaverine, strychnine, theophylline

REFERENCE
Trenerry,V.C.; Wells,R.J.; Robertson,J. The analysis of illicit heroin seizures by capillary zone electrophoresis, *J.Chromatogr.Sci.*, **1994**, *32*, 1–6.

SAMPLE
Matrix: bulk
Sample preparation: Dissolve 500 mg crude opium in 5 mL DMSO, make up to 50 mL with MeCN containing 25 mM ammonium acetate and 1 M acetic acid, centrifuge at 18000 g for 2 min. Dilute 1 mL of the supernatant to 20 mL with MeCN containing 25 mM ammonium acetate and 1 M acetic acid, inject an aliquot. Dilute opium tincture 200-fold with water, inject an aliquot.

CAPILLARY ELECTROPHORESIS
Capillary: 64 cm × 50 μm fused-silica (55.5 cm to detector) (Polymicro Technologies)
Capillary preparation: Flush with running buffer between runs. Before use rinse capillaries with 1 M NaOH for 1 h, with 100 mM NaOH for 20 min, and with running buffer for 10 min
Capillary temperature: 25
Running buffer: MeCN:MeOH 25:75 containing 25 mM ammonium acetate and 1 M acetic acid
Injection: Pressure injection at 5 kPa for 3 s
Detector: UV 214
Migration time: 9.3
Voltage: 25 kV
Model: Hewlett-Packard HP3D

OTHER SUBSTANCES
Simultaneous: codeine, normorphine, noscapine, papaverine, thebaine

KEY WORDS
opium; tincture

REFERENCE
Bjornsdottir,I.; Hansen,S.H. Determination of opium alkaloids in crude opium using non-aqueous capillary electrophoresis, *J.Pharm.Biomed.Anal.*, **1995**, *13*, 1473–1481.

SAMPLE
Matrix: bulk
Sample preparation: Crude preparations. Dissolve in 10 mM HCl containing 400 μg/mL pholcodine, inject an aliquot. Opium. Extract 1 g opium with 20 mL 2.5% acetic acid, make up to 100 mL with water, filter, add 20 mL of the filtrate to 60 mL water, adjust pH to 9.2 with concentrated ammonia solution, extract six times with 50 mL portions of chloroform. Dry the extracts over anhydrous sodium sulfate and evaporate to dryness under reduced pressure, reconstitute with 5 mL 50 mM HCl. Dilute 1 mL to 5 mL with water, add 2 mg pholcodine, inject an aliquot. Poppy straw. 8 g Finely-milled poppy straw + 10 mL water + 2 g calcium hydroxide, shake vigorously for 25 min, filter, dilute 20 mL filtrate with 60 mL water, adjust pH to 9.2 with 10% acetic acid, extract 5 times with 50 mL portions of chloroform:EtOH 75:5. Combine the extracts and wash them twice with 30 mL portions of water, dry over anhydrous sodium sulfate and evaporate to dryness under reduced pressure, reconstitute with 25 mL 50 mM HCl. Dilute 9 mL with 1 mL 4 mg/mL pholcodine, filter, inject an aliquot.

CAPILLARY ELECTROPHORESIS
Capillary: 70 cm × 50 μm fused-silica (45 cm to detector) (Polymicro Technologies)

Capillary preparation: Flush with running buffer for 2 min between analyses. Replace running buffer after 20 analyses. Each week wash capillary with 100 mM NaOH for 10 min and with water for 10 min then fill with running buffer.
Capillary temperature: 28
Running buffer: DMF:buffer 10:90 [MeCN:buffer 12.5:87.5 may also be used (Electrophoresis 1996, 17, 1361)] (Prepare buffer by dissolving 0.92 g cetyltrimethylammonium bromide in 50 mL 10 mM sodium tetraborate:10 mM KH_2PO_4 50:50 (pH 8.6), mix 2.5 mL DMF and 22.5 mL buffer.)
Injection: Vacuum injection 10 kPa.s (vacuum level 2)
Detector: UV 254
Migration time: 6
Internal standard: pholcodine (5.5)
Voltage: -25 kV
Model: Isco Model 3140 Electropherograph

OTHER SUBSTANCES
Simultaneous: codeine, cryptopine, narceine, noscapine, oripavine, papaverine, salutaridine, thebaine

REFERENCE
Trenerry,V.C.; Wells,R.J.; Robertson,J. Determination of morphine and related alkaloids in crude morphine, poppy straw and opium preparations by micellar electrokinetic capillary chromatography, *J.Chromatogr.A*, **1995**, *718*, 217–225.

SAMPLE
Matrix: bulk
Sample preparation: Dilute with 400 µg/mL N-propyl p-hydroxybenzoate in running buffer

CAPILLARY ELECTROPHORESIS
Capillary: 27 cm × 50 µm fused-silica (20 cm to detector)
Capillary preparation: Before each run rinse with running buffer at high pressure.
Capillary temperature: 30
Running buffer: MeCN:water 15:85 containing 40 mM sodium dodecyl sulfate, 8.5 mM sodium borate, and 8.5 mM sodium phosphate, pH 8.5
Injection: Inject at high pressure for 1 s
Detector: UV 214
Migration time: 1.2
Internal standard: N-propyl p-hydroxybenzoate (1.8)
Voltage: 20 kV
Model: Beckman P/ACE System 2100

OTHER SUBSTANCES
Simultaneous: acetaminophen, acetylcodeine, O-acetylmorphine, aspirin, caffeine, cocaine, codeine, diamorphine, diphenhydramine, hydromorphone, isoniacinamide, lidocaine, methaqualone, niacinamide, noscapine, papaverine, phenacetin, phenobarbital, phenylpropanolamine, procaine, quinine, salicylic acid, strychnine, thebaine

REFERENCE
Walker,J.A.; Krueger,S.T.; Lurie,I.S.; Marché,H.L.; Newby,N. Analysis of heroin drug seizures by micellar electrokinetic capillary chromatography (MECC), *J.Forensic Sci.*, **1995**, *40*, 6–9.

SAMPLE
Matrix: bulk
Sample preparation: Dissolve in running buffer to a concentration of 1 mg/mL, vortex for 2 min, add an equal volume of 100 µg/mL diphenhydramine in running buffer, mix, inject an aliquot.

CAPILLARY ELECTROPHORESIS
Capillary: 65 cm × 50 µm fused-silica (60 cm to detector)
Capillary preparation: Fill capillary with fresh running buffer before each run. Before use fill with 100 mM NaOH for 20 min, rinse with water, flush with running buffer

Running buffer: MeCN:50 mM 6-aminocaproic acid containing 50 mM 3-N,N-dimethylmyristylammoniopropanesulfonate (MAPS; Fluka) and 5 mM 1-heptanesulfonic acid 10:90, pH adjusted to 4.0 with 1 M phosphoric acid
Injection: Hydrodynamic injection by gravity or pressure.
Detector: UV 214
Migration time: 5.2
Internal standard: diphenhydramine (12)
Voltage: 27 kV
Current: ≤25 μA
Model: Dionex system I

OTHER SUBSTANCES
Simultaneous: acetaminophen, allobarbital, barbital, caffeine, diamorphine, niacinamide, noscapine, papaverine, phenobarbital, procaine
Interfering: codeine

REFERENCE
Naess,O.; Rasmussen,K.E. Micellar electrokinetic chromatography of charged and neutral drugs in acidic running buffers containing a zwitterionic surfactant, sulfonic acids or sodium dodecyl sulphate. Separation of heroin, basic by-products and adulterants, *J.Chromatogr.A*, **1997**, *760*, 245–251.

SAMPLE
Matrix: formulations
Sample preparation: Dilute formulations with A, inject an aliquot. Dissolve 500 mg crude opium in 5 mL DMSO, make up to 50 mL with 1 M acetic acid, centrifuge. Remove a 1 mL aliquot of the supernatant and make up to 20 mL with water, inject an aliquot.

CAPILLARY ELECTROPHORESIS
Capillary: 55 cm × 50 μm fused-silica (Polymicro Technologies)
Capillary temperature: 30
Running buffer: 50 mM pH 4.0 6-Aminocaproic acid containing 30 mM heptakis(2,6-di-O-methyl-β-cyclodextrin)
Injection: Pressure injection at 5 kPa for 3 s
Detector: UV 214
Migration time: 8.4
Voltage: 30 kV
Model: Hewlett-Packard HP³ᴰ
Limit of detection: 300 ng/mL

OTHER SUBSTANCES
Simultaneous: codeine, normorphine, noscapine, papaverine, thebaine

---

KEY WORDS
syrup; tincture; opium

REFERENCE
Bjornsdottir,I.; Hansen,S.H. Determination of opium alkaloids in opium by capillary electrophoresis, *J.Pharm.Biomed.Anal.*, **1995**, *13*, 687–693.

SAMPLE
Matrix: formulations
Sample preparation: Dilute liquid formulation with DMSO:water:1 M acetic acid 0.5:95:4.5, inject an aliquot.

CAPILLARY ELECTROPHORESIS
Capillary: 64 cm × 50 μm fused-silica (55.5 cm to detector) (Polymicro Technologies)
Capillary preparation: Flush with running buffer for 2 min between runs. Rinse new capillaries with 1 M NaOH for 1 h, with 100 mM NaOH for 20 min, with water for 20 min, and with running buffer for 10 min.
Capillary temperature: 25
Running buffer: MeCN:MeOH 25:75 containing 25 mM ammonium acetate and 1 M acetic acid
Injection: Pressure injection at 5 kPa for 3 s.

Detector: UV 214
Migration time: 7.2
Voltage: 25 kV
Current: 14 μA
Model: Hewlett-Packard HP[3D]

OTHER SUBSTANCES
Simultaneous: codeine, normorphine, noscapine, papaverine, thebaine

---

REFERENCE
Bjornsdottir,I.; Hansen,S.H. Comparison of aqueous and non-aqueous capillary electrophoresis for quantitative determination of morphine in pharmaceuticals, *J.Pharm.Biomed.Anal.*, **1997**, *15*, 1083–1089.

SAMPLE
Matrix: formulations
Sample preparation: Dilute liquid formulation with DMSO:water:1 M acetic acid 0.5:95:4.5, inject an aliquot.

CAPILLARY ELECTROPHORESIS
Capillary: 64 cm × 50 μm fused-silica (55.5 cm to detector) (Polymicro Technologies)
Capillary preparation: Flush with running buffer for 2 min between runs. Rinse new capillaries with 1 M NaOH for 1 h, with 100 mM NaOH for 20 min, with water for 20 min, and with running buffer for 10 min.
Capillary temperature: 30
Running buffer: 50 mM 6-Aminocaproic acid containing 30 mM heptakis(2,6-di-O-methyl)-β-cyclodextrin, adjusted to pH 4.0 with glacial acetic acid
Injection: Pressure injection at 5 kPa for 3 s.
Detector: UV 214
Migration time: 8.4
Voltage: 30 kV
Current: 60 μA
Model: Hewlett-Packard HP[3D]

OTHER SUBSTANCES
Simultaneous: codeine, normorphine, noscapine, papaverine, thebaine

---

REFERENCE
Bjornsdottir,I.; Hansen,S.H. Comparison of aqueous and non-aqueous capillary electrophoresis for quantitative determination of morphine in pharmaceuticals, *J.Pharm.Biomed.Anal.*, **1997**, *15*, 1083–1089.

SAMPLE
Matrix: hair
Sample preparation: Wash 25-100 mg hair with diethyl ether and 10 mM HCl, add 250 mM HCl, heat at 45° overnight, neutralize with NaOH, extract twice into the organic phase with Toxi-tubes A (Analytical Systems, Laguna Hills, CA). Remove the organic layer and evaporate it to dryness, reconstitute the residue in 20 μL running buffer:water 1:2, inject a 5-10 μL aliquot.

CAPILLARY ELECTROPHORESIS
Capillary: 40 cm × 50 μm bare silica (40 cm to detector) (Isco)
Capillary preparation: After each injection flush with 100 mM NaOH and rinse with running buffer.
Running buffer: 50 mM pH 9.2 borate buffer
Injection: Manual injection of 5-10 μL by syringe through a 1:830 splitter
Detector: UV 214
Migration time: 9.26
Internal standard: nalorphine (9.46)
Voltage: 15 kV
Model: Isco Model 3850
Limit of detection: 150 ng/g (S/N 3)

OTHER SUBSTANCES
Extracted: cocaine (UV 238)

Simultaneous: tetracaine (UV 238)
Noninterfering: acetaminophen, amitriptyline, amphetamine, atropine, benzoylecgonine, benztropine, caffeine, carbamazepine, carisoprodol, chlorpheniramine, chlorpromazine, chlorprothixene, cimetidine, cocaine, codeine, dextromethorphan, diazepam, dihydrocodeine, diphenhydramine, diphenoxylate, disopyramide, doxepin, doxylamine, emetine, erythromycin, ethylmorphine, flurazepam, glutethimide, hydrocodone, hydrocortisone, hydromorphone, hydroxyzine, imipramine, lidocaine, loxapine, meperidine, meprobamate, methadone, methamphetamine, methapyrilene, methaqualone, methocarbamol, methylphenidate, naloxone, nicotine, nordiazepam, nortriptyline, orphenadrine, oxycodone, papaverine, pentazocine, phenacetin, phencyclidine, phenmetrazine, phenolphthalein, phentermine, phenylpropanolamine, phenytoin, phetidine, prazepam, procainamide, procaine, propoxyphene, propranolol, protriptyline, pseudoephedrine, pyrilamine, quinine, salicylamide, spironolactone, strychnine, terpin hydrate, thioridazine, thiothixene, triamterene, trifluoperazine, triflupromazine, trihexyphenidyl, trimeprazine, trimethoprim, trimetobenzamide

REFERENCE

Tagliaro,F.; Poiesi,C.; Aiello,R.; Dorizzi,R.; Ghielmi,S.; Marigo,M. Capillary electrophoresis for the investigation of illicit drugs in hair: determination of cocaine and morphine, *J.Chromatogr.*, **1993**, *638*, 303–309.

SAMPLE
Matrix: solutions
Sample preparation: Inject an aliquot of a 1 mg/mL solution in MeOH.

CAPILLARY ELECTROPHORESIS
Capillary: 67 cm × 50 μm (60 cm to detector) (Composite Metal Services, Worcs., UK)
Capillary preparation: Before each run rinse with running buffer for 3 min.
Capillary temperature: 20
Running buffer: 50 mM pH 10.5 Glycine buffer containing 50 mM sodium dodecyl sulfate
Injection: Pressure injection at 30 mbar for 5 s.
Detector: UV 220
Migration time: 9.5
Voltage: 25 kV
Model: Beckman P/ACE 2050

OTHER SUBSTANCES
Simultaneous: amphetamine, caffeine, codeine, diamorphine

REFERENCE

Hyötyläinen,T.; Sirén,H.; Riekkola,M.-L. Determination of morphine analogues, caffeine and amphetamine in biological fluids by capillary electrophoresis with the marker technique, *J.Chromatogr.A*, **1996**, *735*, 439–447.

SAMPLE
Matrix: solutions
Sample preparation: Inject an aliquot of a 1-50 μg/mL solution in water.

CAPILLARY ELECTROPHORESIS
Capillary: 55 cm × 50 μm fused-silica (35 cm to detector) (J&W)
Capillary preparation: Flush with running buffer for 5 min before each run. Periodically wash with 100 mM NaOH and flush extensively with running buffer.
Running buffer: MeOH:25 mM pH 9.24 borate buffer 20:80 containing 100 mM sodium dodecyl sulfate (A) or 50 mM pH 2.35 phosphate buffer (B) or 50 mM pH 9.24 borate buffer (C)
Injection: Injection of 5 μL using a split-flow injector ratio of 1:800.
Detector: UV 200
Migration time: 11.81 (A), 10.54 (B), 8.28 (C)
Voltage: 20 kV (A, B) or 12 kV (C)
Current: <60-80 μA
Model: ISCO Model 3850

OTHER SUBSTANCES
Also analyzed: acetylcodeine, amphetamine, barbital, caffeine, codeine, cocaine, diamorphine, diazepam, flunitrazepam, lidocaine, monoacetylmorphine, nalorphine, narceine, noscapine, papaverine, pentobarbital, procaine, tetracaine, thebaine

KEY WORDS
all compounds were separated with running buffer A; some peaks overlapped with running buffers B and C.

REFERENCE
Tagliaro,F.; Smith,F.P.; Turrina,S.; Equisetto,V.; Marigo,M. Complementary use of capillary zone electrophoresis and micellar electrokinetic capillary chromatography for mutual confirmation of results in forensic drug analysis, *J.Chromatogr.A*, **1996**, *735*, 227–235.

SAMPLE
Matrix: solutions
Sample preparation: Inject an aliquot of a solution in 100 mM sodium phosphate containing 100-300 µg/mL naphazoline.

CAPILLARY ELECTROPHORESIS
Capillary: 67 cm × 50 µm fused-silica (60 cm to detector)
Capillary preparation: Rinse with running buffer for 2 min between sets of analyses.
Capillary temperature: 30
Running buffer: 200 mM pH 4.5 Phosphate buffer
Injection: Injection at high pressure for 2 s.
Detector: UV 210, UV 230
Migration time: 18.9
Internal standard: naphazoline (14)
Voltage: 20 kV
Model: Beckman P/ACE 5500

OTHER SUBSTANCES
Simultaneous: acetylcodeine, amphetamine, cocaine, codeine, diamorphine, LSD, methadone, methamphetamine, methylenedioxyamphetamine, methylenedioxymethamphetamine, PCP, psilocyn

REFERENCE
Walker,J.A.; Marché,H.L.; Newby,N.; Bechtold,E.J. A free zone capillary electrophoresis method for the quantitation of common illicit drug samples, *J.Forensic Sci.*, **1996**, *41*, 824–829.

SAMPLE
Matrix: solutions

CAPILLARY ELECTROPHORESIS
Capillary: 55 cm × 50 µm fused-silica (50 cm to detector) (Polymicro Technologies)
Capillary preparation: Between runs purge with 1 M NaOH for 2 min, with water for 2 min, and with running buffer for 3 min
Capillary temperature: 25
Running buffer: MeCN:buffer 50:50 (Buffer was 100 mM ammonium acetate adjusted to pH 3.1 with acetic acid.)
Injection: Pressure injection at 345 mbar.s.
Detector: UV 224, Finnigan MAT Model 95, electrospray (details in paper)
Migration time: 16
Voltage: 15 kV
Model: Bio-Rad BioFocus 3000

OTHER SUBSTANCES
Simultaneous: codeine, narceine, noscapine, papaverine, thebaine

REFERENCE
Unger,M.; Stöckigt,D.; Belder,D.; Stöckigt,J. General approach for the analysis of various alkaloid classes using capillary electrophoresis and capillary electrophoresis-mass spectrometry, *J.Chromatogr.A*, **1997**, *767*, 263–276.

SAMPLE
Matrix: solutions

Sample preparation: Inject an aliquot of a solution in MeOH:0.3 mM phosphoric acid 8:92.

CAPILLARY ELECTROPHORESIS
Capillary: 41 cm × 50 μm fused-silica (22 cm to detector) (Polymicro Technologies)
Capillary preparation: Rinse with running buffer for 10 min using a vacuum of 67.7 kPa.
Capillary temperature: 35
Running buffer: 40 mM pH 10.6 Sodium phosphate buffer containing 80 mM sodium dodecyl
sulfate
Injection: Vacuum injection at 16.9 kPa for 1 s.
Detector: UV 210
Migration time: 5
Voltage: 10 kV
Current: 61 μA
Model: Applied Biosystems 270A-HT

OTHER SUBSTANCES
Simultaneous: metabolites, codeine, codeine-6-glucuronide, dihydrocodeine, dihydrocodeine-6-
glucuronide, dihydromorphine, ethylmorphine, morphine-3-glucuronide, norcodeine, nordihydrocodeine, nordihydromorphine, normorphine

---

REFERENCE
Zhang,C.-X.; Thormann,W. Separation of free and glucuronidated opioids by capillary electrophoresis in aqueous, binary and micellar media, *J.Chromatogr.A*, **1997**, *764*, 157–168.

SAMPLE
Matrix: solutions

CAPILLARY ELECTROPHORESIS
Capillary: 65 cm × 50 μm polyacrylamide coated (Bio-Rad)
Capillary preparation: Between runs rinse capillary with 50 mM pH 2.5 phosphate buffer for
2 min, with 100 mM NaOH for 2 min, with water for 1 min, with washing buffer for 2 min,
and with running buffer for 2 min. (The washing buffer was the same as the running buffer
but it was contained in a separate vial to minimize contamination.)
Capillary temperature: 15
Running buffer: 50 mM pH 2.5 Phosphate buffer containing 75 mM methyl-β-cyclodextrin, 220
mM urea, and 15 mM trimethylamine
Injection: Electrokinetic injection at 5 kV for 18 s.
Detector: UV 200
Migration time: 20.3
Voltage: 25 kV
Model: Bio-Rad BioFocus 3000

OTHER SUBSTANCES
Simultaneous: meperidine, tramadol

KEY WORDS
achiral except for tramadol; coated capillary

REFERENCE
Chan,E.C.Y.; Ho,P.C. Enantiomeric separation of tramadol hydrochloride and its metabolites by cyclodextrin-
mediated capillary zone electrophoresis, *J.Chromatogr.B*, **1998**, *707*, 287–294.

SAMPLE
Matrix: solutions
Sample preparation: Inject an aliquot of an aqueous solution.

CAPILLARY ELECTROPHORESIS
Capillary: 41 cm × 50 μm fused-silica (22 cm to detector) (Polymicro Technologies)
Capillary preparation: Flush with running buffer for 5 min between runs.
Running buffer: 40 mM pH 10.7 phosphate buffer containing 70 mM sodium dodecyl sulfate
Injection: Hydrodynamic injection at 5" Hg for 1 s.

Detector: UV 210
Migration time: 4.6
Internal standard: norcodeine (2)
Voltage: 10 kV
Current: 60 μA
Model: Applied Biosystems 270A-HT

OTHER SUBSTANCES
Simultaneous: codeine, codeine-6-glucuronide, dihydrocodeine, dihydrocodeine-6-glucuronide, dihydromorphine, ethylmorphine, morphine-3-glucuronide, nordihydrocodeine, nordihydromorphine, normorphine

REFERENCE
Zhang,C.-X.; Thormann,W. Head-column field-amplified sample stacking in binary system capillary electrophoresis. 2. Optimization with a preinjection plug and application to micellar electrokinetic chromatography, *Anal.Chem.*, **1998**, *70*, 540–548.

SAMPLE
Matrix: urine
Sample preparation: Condition a Bond Elut Certify SPE cartridge with 2 mL MeOH and 2 mL water, do not allow to go dry. 5 mL Urine + 1 mL concentrated HCl, vortex, heat at 120° for 30 min, cool, adjust pH to 7 with about 1.25 mL 10 M KOH, add to the SPE cartridge, wash with 2 mL water, wash with 1 mL 100 mM pH 4 acetate buffer, wash with 2 mL MeOH, elute with 2 mL dichloromethane:isopropanol 80:20 containing 2% ammonium hydroxide. Evaporate the eluate to dryness under a stream of nitrogen at room temperature, reconstitute the residue in 100 μL running buffer, inject an aliquot.

CAPILLARY ELECTROPHORESIS
Capillary: 90 cm × 75 μm fused-silica (70 cm to detector) (Polymicro Technologies)
Capillary preparation: Before each run rinse capillary with 100 mM NaOH for 3 min and with buffer for 5 min.
Running buffer: 10 mM pH 9.1 Na_2HPO_4 containing 6 mM sodium borate and 75 mM sodium dodecyl sulfate
Injection: Gravity injection at 34 cm for 5 s
Detector: UV 195
Migration time: 17.1
Voltage: 20 kV
Current: 76-80 μA

KEY WORDS
SPE

REFERENCE
Wernly,P.; Thormann,W. Analysis of illicit drugs in human urine by micellar electrokinetic capillary chromatography with on-column fast scanning polychrome absorption detection, *Anal.Chem.*, **1991**, *63*, 2878–2882.

SAMPLE
Matrix: urine
Sample preparation: Condition a 130 mg Bond Elut Certify SPE cartridge with 2 mL MeOH and 2 mL water. 5 mL Urine + 1 mL concentrated HCl, heat at 120° for 30 min, cool, adjust pH to 7 with about 1.25 mL 10 M KOH, centrifuge at 1500 g for 3 min, add to the SPE cartridge, wash with 2 mL water, 1 mL 100 mM pH 4 acetate buffer, wash with 2 mL MeOH, dry under vacuum, elute with two 750 μL portions of MeOH containing 30% concentrated ammonium hydroxide. Evaporate the eluate to dryness under a stream of nitrogen at room temperature, reconstitute the residue in 100 μL running buffer, inject an aliquot (J. Chromatogr. B 1995, 668, 159).

CAPILLARY ELECTROPHORESIS
Capillary: 70 cm × 75 μm fused-silica (70 cm to detector)
Running buffer: 6 mM pH 9.2 Sodium Tetraborate containing 10 mM Na_2HPO_4 and 75 mM sodium dodecyl sulfate
Injection: Vacuum injection for 0.5 s.

Detector: UV 210
Migration time: 19
Voltage: 25 kV
Current: 68-78 μA
Model: Europhor Prime Vision IV

OTHER SUBSTANCES
Extracted: codeine

KEY WORDS
SPE

REFERENCE
Hufschmid,E.; Theurillat,R.; Wilder-Smith,C.H.; Thormann,W. Characterization of the genetic polymorphism
of dihydrocodeine O-demethylation in man via analysis of urinary dihydrocodeine and dihydromorphine by
micellar electrokinetic capillary chromatography, *J.Chromatogr.B*, **1996**, *678*, 43–51.

SAMPLE
Matrix: urine
Sample preparation: Condition a 500 mg Bond Elut C2 SPE cartridge with 2 mL MeOH and
2 mL 50 mM pH 7.5 Tris-HCl buffer. Condition a 100 mg Bond Elut PRS SPE cartridge with
500 μL MeOH, 500 μL water, and 500 μL 0.1% trifluoroacetic acid in water. 5 mL Urine + 5
μg nalorphine + 5 mL 50 mM pH 7.5 Tris-HCl buffer, mix, adjust pH to 7.5 with NaOH or
HCl, add to the C2 SPE cartridge, wash with 2 mL 50 mM pH 7.5 Tris-HCl buffer, elute with
2 mL MeCN:water:trifluoroacetic acid 50:50:0.1. Dilute the eluate with 8 mL 0.1% trifluoroacetic acid in water, add the mixture to the PRS SPE cartridge, wash with 500 μL 0.1% trifluoroacetic acid in water, wash with 2 mL MeOH, elute with 1 mL 3% ammonia in MeOH. Evaporate the eluate to dryness under reduced pressure, reconstitute with 50 μL MeCN:water 10:
90, inject an aliquot.

CAPILLARY ELECTROPHORESIS
Capillary: 64.5 cm × 50 μm fused-silica (56 cm to detector) (Hewlett-Packard)
Capillary preparation: Flush with running buffer for 1-4 min between runs. Flush new capil-
laries under 900 mbar pressure with 1 M NaOH for 10 min, with 100 mM NaOH for 5 min,
and with running buffer for 5 min.
Capillary temperature: 25
Running buffer: MeCN:buffer 5:95 (Buffer was 70 mM pH 9.3 borate buffer containing 70 mM
sodium dodecyl sulfate.)
Injection: Pressure injection of sample at 50 mbar for 3 s and of running buffer at 50 mbar for
1 s.
Detector: UV 195
Migration time: 10.2
Internal standard: nalorphine (12.3)
Voltage: 30 kV
Model: Hewlett-Packard HP ³ᴰCE
Limit of detection: 150 ng/mL

OTHER SUBSTANCES
Extracted: metabolites

KEY WORDS
SPE; horse

REFERENCE
Taylor,M.R.; Westwood,S.A.; Perrett,D. Determination of Phase II drug metabolites in equine urine by micellar
electrokinetic capillary chromatography, *J.Chromatogr.A*, **1996**, *745*, 155–163.

SAMPLE
Matrix: urine
Sample preparation: Condition a 300 mg Bondelut Certify SPE cartridge. 500 μL urine + 500
μL 805 ng/mL levallorphan solution, add to the SPE cartridge, wash, elute with 2 mL dichlo-

romethane:isopropanol:ammonia 80:20:2. Evaporate the eluate to dryness, reconstitute with 100 μL MeOH, dilute to 1 mL with water

CAPILLARY ELECTROPHORESIS
Capillary: 65 cm × 50 μm silica (60 cm to detector)
Capillary preparation: Before each run flush capillary with running buffer at 930 mbar for 2 min
Capillary temperature: 25
Running buffer: 100 mM pH 6 Na_2HPO_4
Injection: Electrokinetic injection at 5 kV for 10 s
Detector: UV 200
Migration time: 10.99
Internal standard: levallorphan (10.18)
Voltage: 20 kV
Model: Hewlett-Packard 3DCE
Limit of detection: 8 ng/mL

OTHER SUBSTANCES
Extracted: codeine, diamorphine, dihydrocodeine, 6-monoacetylmorphine, pholcodine

KEY WORDS
SPE

REFERENCE
Taylor,R.B.; Low,A.S.; Reid,R.G. Determination of opiates in urine by capillary electrophoresis, *J.Chromatogr.B*, **1996**, *675*, 213–223.

Nadolol

Molecular formula: $C_{17}H_{27}NO_4$
Molecular weight: 309.41
CAS Registry No.: 4200-33-9
Merck Index (12th ed.): 6431
Lednicer: 2 110

SAMPLE
Matrix: blood
Sample preparation: Hydrolyze 2 mL serum with β-glucuronidase (EC 3.2.1.31, type H-1 from Helix pomatia, 416 800 U/g, Separacor) at 80° for 30 min, cool, add 900 μL MeCN, vortex for 15 min, centrifuge at 2004 g for 10 min, add ephedrine (165 μg/mL), filter (0.5 μm), inject an aliquot of the filtrate.

CAPILLARY ELECTROPHORESIS
Capillary: 58 cm × 50 μm fused-silica (50 cm to detector) (Polymicro Technologies)
Capillary preparation: Before each injection purge capillary with 5% phosphoric acid for 12 s, with water for 30 s, and with running buffer for 10 min.
Capillary temperature: 35
Running buffer: 80 mM pH 6.7 sodium phosphate buffer containing 15 mM cetyltrimethylammonium bromide
Injection: Hydrostatic injection for 20 s
Detector: UV 214
Migration time: 11.8
Internal standard: ephedrine (17.5)
Voltage: -27 kV
Model: Waters Quanta 4000

OTHER SUBSTANCES
Extracted: acebutolol, alprenolol, atenolol, labetalol, metoprolol, oxprenolol, pindolol, propranolol, timolol

KEY WORDS
serum

REFERENCE
Lukkari,P.; Nyman,T.; Riekkola,M.-J. Determination of nine β-blockers in serum by micellar electrokinetic capillary chromatography, *J.Chromatogr.A*, **1994**, *674*, 241–246.

SAMPLE
Matrix: blood, urine
Sample preparation: Urine. Dilute with 2 volumes of water, filter (0.5 μm), inject an aliquot. Serum. Condition a 3 mL Supelclean LC-18 SPE cartridge (Supelco) with MeOH and water. Hydrolyze 900 μL serum with β-glucuronidase (EC 3.2.1.31 type H-1 from Helix pomatia) at 60° with sonication for 30 min, add 500 μL (?) MeOH, centrifuge at 2000 g, add the supernatant to the SPE cartridge, wash with 1 mL water, dry under vacuum, elute with 2 mL MeOH:water 90:10, filter, inject an aliquot.

CAPILLARY ELECTROPHORESIS
Capillary: 68 cm × 50 μm fused-silica (60 cm to detector) (Polymicro Technologies)
Capillary preparation: Purge with running buffer for 2 min before injection.
Running buffer: pH 7.0 Phosphate buffer containing 10 mM N-cetyl-N,N,N-trimethylammonium bromide (Prepare buffer by mixing 100 mM NaH_2PO_4 and 100 mM Na_2HPO_4 to achieve a pH of 7.0.)
Injection: Hydrostatic injection for 30 s
Detector: UV 214
Migration time: 12.7
Internal standard: 2,6-dimethylphenol (only for urine) (14.5)
Voltage: -26 kV
Current: 97 μA
Model: Waters Quanta 4000

OTHER SUBSTANCES
Extracted: acebutolol, alprenolol, atenolol, labetalol, metoprolol, oxprenolol, pindolol, propranolol, timolol

KEY WORDS
serum; comparison with HPLC; SPE

REFERENCE
Lukkari,P.; Sirén,H. Ion-pair chromatography and micellar electrokinetic capillary chromatography in analyzing β-adrenergic blocking agents from human biological fluids, *J.Chromatogr.A*, **1995**, *717*, 211–217.

SAMPLE
Matrix: solutions

CAPILLARY ELECTROPHORESIS
Capillary: 58 cm × 50 μm fused-silica (50 cm to detector)
Running buffer: Isopropanol:buffer 2.5:97.5 (Buffer was 80 mM pH 6.8 Phosphate buffer containing 15 mM cetyltrimethylammonium bromide.)
Injection: Hydrostatic injection for 15 s
Detector: UV 214
Migration time: 11.8
Voltage: -20 kV
Model: Waters Quanta 4000

OTHER SUBSTANCES
Simultaneous: acebutolol, alprenolol, atenolol, labetalol, oxprenolol, pindolol, propranolol, sotalol, timolol

REFERENCE
Lukkari,P.; Vuorela,H.; Riekkola,M.-L. Effects of organic mobile phase modifiers on elution and separation of β-blockers in micellar electrokinetic capillary chromatography, *J.Chromatogr.A*, **1993**, *655*, 317–324.

SAMPLE
Matrix: solutions

CAPILLARY ELECTROPHORESIS
Capillary: 82 cm × 75 μm fused-silica (Waters)
Capillary preparation: Between runs purge capillary with running buffer for 2 min. Purge a new capillary with 100 mM NaOH for 30 min, with water for 30 min, and with running buffer for 30 min.
Capillary temperature: 30
Running buffer: 30 mM pH 7.6 phosphate buffer containing 10 mM cetyltrimethylammonium bromide
Injection: Inject at high pressure for 5 s
Detector: UV 214
Migration time: 6
Voltage: 21 kV
Model: Beckman P/ACE System 2000

OTHER SUBSTANCES
Simultaneous: acebutolol, alprenolol, atenolol, oxprenolol, pindolol, propranolol, sotalol, timolol

REFERENCE
Lukkari,P.; Ennelin,A.; Sirén,H.; Riekkola,M.-L. Effect of temperature, effective capillary length, and applied voltage on the migration of nine β-blockers in micellar electrokinetic capillary chromatography, *J.Liq.Chromatogr.*, **1993**, *16*, 2069–2079.

SAMPLE
Matrix: solutions

CAPILLARY ELECTROPHORESIS
Capillary: 100 cm × 50 μm fused-silica (50 cm to detector) (Isco)
Capillary preparation: Flush capillary with 10 μL running buffer between runs. Every 40 sample injections rinse capillary with 200 μL 1 M NaOH, with 200 μL water, and 200 μL running buffer. Before use fill capillary with 1 M NaOH and allow to stand for 1 h, fill with 100 mM NaOH, allow to stand for 1 h, wash with water fill with running buffer.
Capillary temperature: 23
Running buffer: 100 mM citric acid containing 19.27 mM Na$_2$HPO$_4$ and 120 mM hydroxypropyl-β-cyclodextrin
Injection: Inject under vacuum at 4.0 kPa.s
Detector: UV 200
Migration time: 30.5 (three peaks from enantiomers and cis/trans isomers)
Voltage: 30 kV
Model: Isco Model 3140

OTHER SUBSTANCES
Simultaneous: albuterol, alprenolol, atenolol, cimaterol, clenbuterol, labetalol, oxprenolol, pindolol, pirbuterol, propranolol, terbutaline

KEY WORDS
chiral

REFERENCE
Aumatell,A.; Wells,R.J.; Wong,D.K.Y. Enantiomeric differentiation of a wide range of pharmacologically active substances by capillary electrophoresis using modified β-cyclodextrins, *J.Chromatogr.A*, **1994**, *686*, 293–307.

SAMPLE
Matrix: solutions

CAPILLARY ELECTROPHORESIS
Capillary: 62 cm × 52 μm fused-silica (50 cm to detector) (Polymicro Technologies)

Running buffer: 150 mM pH 2.5 Trimethylammonium phosphate buffer containing 20 mM hy-
droxypropyl-β-cyclodextrin
Injection: Gravity injection at 10 cm for 5 s
Migration time: 47, 48, 49 (enantiomers, isomers)
Voltage: 20 kV
Current: 82 μA
Model: Laboratory constructed

OTHER SUBSTANCES
Simultaneous: atenolol, pindolol

KEY WORDS
chiral

REFERENCE
Quang,C.; Khaledi,M.G. Direct separation of the enantiomers of β-blockers by cyclodextrin-mediated capillary
zone electrophoresis, *J.High Res.Chromatogr.*, **1994**, *17*, 99–101.

SAMPLE
Matrix: solutions
Sample preparation: Prepare a solution in MeOH/water, inject an aliquot.

CAPILLARY ELECTROPHORESIS
Capillary: 62 cm × 52 μm fused-silica (50 cm to detector) (Polymicro Technologies)
Capillary temperature: 40
Running buffer: 50 mM pH 2.50 Tetrabutylammonium phosphate
Injection: Siphon injection at 10 cm for 5 s
Detector: UV (wavelength not specified)
Migration time: 14
Voltage: 20 kV
Current: 29 μA
Model: Laboratory constructed

OTHER SUBSTANCES
Simultaneous: doxylamine, epinephrine, imidazole, isoproterenol, 2-methylphenethylamine, 1-
methylphenylpropylamine, metoprolol, nicotine, norepinephrine, propranolol, pseudoephedrine

REFERENCE
Quang,C.; Khaledi,M.G. Extending the scope of chiral separation of basic compounds by cyclodextrin-mediated
capillary zone electrophoresis, *J.Chromatogr.A*, **1995**, *692*, 253–265.

SAMPLE
Matrix: solutions

CAPILLARY ELECTROPHORESIS
Capillary: 57 cm × 75 μm fused-silica (50 cm to detector) (Beckman)
Capillary preparation: Before each run rinse capillary with 100 mM NaOH and running buffer.
Capillary temperature: 30
Running buffer: 100 mM Phosphoric acid adjusted to pH 3.1 with triethanolamine
Injection: Pressure injection for 5-10 s.
Detector: UV 200
Migration time: 11.8
Voltage: 25 kV
Model: Beckman P/ACE 5510

OTHER SUBSTANCES
Simultaneous: acebutolol, atenolol, labetalol, metoprolol, oxprenolol, pindolol, propranolol, sotalol, timolol

REFERENCE
Bretnall,A.E.; Clarke,G.S. Selectivity of capillary electrophoresis for the analysis of cardiovascular drugs,
J.Chromatogr.A, **1996**, *745*, 145–154.

SAMPLE
Matrix: solutions

CAPILLARY ELECTROPHORESIS
Capillary: 43 cm × 50 μm fused-silica (36 cm to detector)
Capillary preparation: Before each injection wash with running buffer for 10 min. After each injection wash with water for 2 min. Wash new capillaries with 1 M NaOH at 60° for 50 min, with NaOH solution (?) at 60° for 10 min, and with water at 25°.
Capillary temperature: 25
Running buffer: 320 mM pH 2.0 Citrate buffer
Injection: Hydrodynamic injection at 1.5 psi for 2 s.
Detector: UV 220
Migration time: 10
Voltage: 15 kV
Model: Spectra-Physics Model 1000
Limit of detection: 1-18 μg/mL

OTHER SUBSTANCES
Simultaneous: acebutolol, atenolol, labetalol, levobunolol, metoprolol, oxprenolol, pindolol, propranolol, timolol

REFERENCE
Lin,C.-E.; Chang,C.-C.; Lin,W.-c.; Lin,E.C. Capillary zone electrophoretic separation of β-blockers using citrate buffer at low pH, *J.Chromatogr.A*, **1996**, *753*, 133–138.

SAMPLE
Matrix: solutions

CAPILLARY ELECTROPHORESIS
Capillary: 67 cm × 50 μm (60 cm to detector)
Capillary preparation: Wash with running buffer for 5 min before each injection. After each injection wash with 1 M NaOH at 60° for 5 min, with 100 mM NaOH at 60° for 10 min, and with water at 25° for 5 min.
Capillary temperature: 25
Running buffer: 70 mM pH 7.0 Sodium phosphate buffer containing 15 mM cetyltriemthylammonium bromide
Injection: Hydrodynamic injection for 1 s.
Detector: UV 220
Migration time: 14.2
Voltage: 20 kV

OTHER SUBSTANCES
Simultaneous: acebutolol, atenolol, labetalol, levobunolol, metoprolol, oxprenolol, pindolol, propranolol, timolol

REFERENCE
Lin,C.-E.; Chen,Y.-C.; Chang,C.-C.; Wang,D.-Z. Migration behavior and selectivity of β-blockers in micellar electrokinetic chromatography. Influence of micelle concentration of cationic surfactants, *J.Chromatogr.A*, **1997**, *775*, 349–357.

SAMPLE
Matrix: urine
Sample preparation: Dilute urine with 2 volumes of water, add IS, filter (0.5 μm), inject an aliquot.

CAPILLARY ELECTROPHORESIS
Capillary: 68 cm × 50 μm fused-silica (60 cm to detector) (White Associates)
Capillary preparation: Purge capillary with running buffer for 2 min before each injection.
Running buffer: 80 mM pH 7.0 Phosphate buffer containing 10 mM N-cetyl-N,N,N-trimethyl-ammonium bromide
Injection: Hydrostatic injection for 30 s at 10 cm.

Detector: UV 214
Migration time: 12.8
Internal standard: 2,6-dimethylphenol (14.7)
Voltage: -26 kV (at injector end)
Model: Waters Quanta 4000
Limit of detection: 10 μg/mL (S/N 3)

OTHER SUBSTANCES
Extracted: acebutolol, alprenolol, atenolol, labetalol, metoprolol, oxprenolol, pindolol, propranolol, timolol
Simultaneous: probenecid
Noninterfering: caffeine

REFERENCE
Lukkari,P.; Sirén,H.; Pantsar,M.; Riekkola,M.-L. Determination of ten β-blockers in urine by micellar electrokinetic capillary chromatography, *J.Chromatogr.*, **1993**, *632*, 143–148.

SAMPLE
Matrix: urine
Sample preparation: Filter (0.2 μm), inject an aliquot of the filtrate.

CAPILLARY ELECTROPHORESIS
Capillary: 80 cm × 50 μm fused-silica (57 cm to detector)
Capillary preparation: Before each run rinse the capillary with 1 M NaOH for 3 min, with 100 mM NaOH for 3 min, with water for 3 min, and with running buffer for 10 min.
Running buffer: 110 mM Boric acid containing 56 mM NaOH and 44 mM HCl, pH 8
Injection: Vacuum injection for 1 s
Detector: UV 220
Migration time: 4.393
Voltage: 20 kV
Model: Europhor Prime Vision system IV
Limit of detection: 2.25 μM

OTHER SUBSTANCES
Extracted: acebutolol (UV 238), acetazolamide (UV 222), alprenolol (UV 220), amiloride (UV 220), atenolol (UV 228), bendroflumethiazide (UV 220), bumetanide (UV 220), chlorthalidone (UV 220), cocaine (UV 236), codeine (UV 220), ethacrynic acid (UV 220), furosemide (UV 232), hydrochlorothiazide (UV 226), methadone (UV 220), metoxiphenamine (UV 220), norcodeine (UV 220), oxprenolol (UV 220), pentazocine (UV 220), propranolol (UV 220), spironolactone (UV 244), triamterene (UV 232), xipamide (UV 234)

REFERENCE
Gonzalez,E.; Laserna,J.J. Capillary zone electrophoresis for the rapid screening of banned drugs in sport, *Electrophoresis*, **1994**, *15*, 240–243.

Nafarelin

Molecular formula: $C_{66}H_{83}N_{17}O_{13}$
Molecular weight: 1322.49
CAS Registry No.: 76932-56-4, 86220-42-0 (acetate)
Merck Index (12th ed.): 6437

SAMPLE
Matrix: solutions

Sample preparation: Prepare a 200 µg/mL solution in 10 mM pH 2.5 sodium citrate, inject an aliquot.

CAPILLARY ELECTROPHORESIS
Capillary: 100 cm × 50 µm fused-silica (78 cm to detector) (Polymicro Technologies)
Capillary preparation: Before each run flush with 100 mM NaOH for 2 min then flush with running buffer for 4 or 5 min. Before use flush with 1 M NaOH for 30 min then flush with water for 10 min.
Capillary temperature: 35
Running buffer: 50 mM pH 7.5 Tris-HCl buffer containing 20 mM borate and 20 mM cetyltrimethylammonium bromide
Injection: Injection by vacuum for 1-5 s.
Detector: UV 280
Migration time: 20.7
Voltage: -20 kV
Model: Applied Biosystems Model 270HT

OTHER SUBSTANCES
Simultaneous: buserelin, deslorelin, gonadorelin, goserelin, leuprolide

KEY WORDS
detector at anode; comparison with HPLC

REFERENCE
Corran,P.H.; Sutcliffe,N. Identification of gonadorelin (LHRH) derivatives: comparison of reversed-phase high-performance liquid chromatography and micellar electrokinetic chromatography, *J.Chromatogr.*, **1993**, *636*, 87–94.

Nafcillin

Molecular formula: $C_{21}H_{22}N_2O_5S$
Molecular weight: 414.48
CAS Registry No.: 147-52-4, 985-16-0 (Na salt), 7177-50-6 (Na salt monohydrate)
Merck Index (12th ed.): 6438
Lednicer: 1 412

SAMPLE
Matrix: solutions
Sample preparation: Prepare a 120 µg/mL solution in water, inject an aliquot.

CAPILLARY ELECTROPHORESIS
Capillary: 60 cm × 75 µm
Running buffer: 20 mM NaH_2PO_4 containing 50 mM sodium dodecyl sulfate, adjusted to pH 9.0 with sodium tetraborate
Injection: Hydrodynamic injection at 10 cm for 5 s
Detector: UV 214
Migration time: 13
Voltage: 18 kV
Model: Waters Quanta 4000

OTHER SUBSTANCES
Simultaneous: 6-aminopenicillanic acid, amoxicillin, ampicillin, cloxacillin, dicloxacillin, oxacillin, ticarcillin

REFERENCE
Swartz,M.E. Method development and selectivity control for small molecule pharmaceutical separations by capillary electrophoresis, *J.Liq.Chromatogr.*, **1991**, *14*, 923–938.

SAMPLE
Matrix: solutions

CAPILLARY ELECTROPHORESIS
Capillary: 60 cm × 50 μm fused-silica (47 cm to detector) (Polymicro Technologies)
Capillary temperature: 25
Running buffer: 20 mM pH 8.5 Sodium tetraborate containing 100 mM sodium dodecyl sulfate and 0.5 mM EDTA
Injection: Hydrodynamic injection at 50 mbar for 3.6 s (5 nL).
Detector: UV 205
Migration time: 25
Voltage: 15 kV
Model: Crystal 310 (Thermo Unicam)

OTHER SUBSTANCES
Simultaneous: amoxicillin, ampicillin, cephapirin, cloxacillin, dicloxacillin, oxacillin, penicillin G, penicillin V, phthalyl sulfathiazole, piperacillin, succinyl sulfathiazole, sulfacetamide, sulfadimethoxine, sulfaguanidine, sulfamerazine, sulfameter, sulfamethazine, sulfamethoxazole, sulfamethoxypyridazine, sulfanilamide, sulfanilic acid, sulfapyridine, sulfaquinoxaline, sulfathiazole, sulfisoxazole

REFERENCE
Hows,M.E.P.; Perrett,D.; Kay,J. Optimization of a simultaneous separation of sulphonamides, dihydrofolate reductase inhibitors and β-lactam antibiotics by capillary electrophoresis, *J.Chromatogr.A*, **1997**, *768*, 97–104.

Nafronyl

Molecular formula: $C_{24}H_{33}NO_3$
Molecular weight: 383.53
CAS Registry No.: 31329-57-4, 3200-06-4 (acid oxalate)
Merck Index (12th ed.): 6440
Lednicer: 2 213

SAMPLE
Matrix: solutions
Sample preparation: Inject an aliquot of a 100 μg/mL solution in running buffer.

CAPILLARY ELECTROPHORESIS
Capillary: 32 cm × 50 μm fused-silica (27.5 cm to detector) (Yongnian Optical Conductive Fiber Plant, China), coated with polyacrylamide
Capillary preparation: No details of the polyacrylamide coating process are provided. However, another paper (LC.GC 1997, 15, 40) by this group indicates that they use the procedure of Hjertén, thus: Adjust the pH of 20 mL water to 3.5 with acetic acid, add 80 μL 3-(trimethoxysilyl)propyl methacrylate (3-methacryloxypropyltrimethoxysilane), mix, suck into capillary, let stand at room temperature for 1 h, remove the solution, wash with water. Fill the capillary with a deaerated 3-4% acrylamide solution containing 1 μL/mL N,N,N',N'-tetramethylethylenediamine and 1 mg/mL potassium persulfate, let stand for 30 min, remove excess solution by aspiration, rinse with water, remove water by aspiration, dry at 35° (J. Chromatogr. 1985, 347, 191).
Capillary temperature: 25
Running buffer: 100 mM NaH_2PO_4 adjusted to pH 2.5 (A) or 100 mM NaH_2PO_4 containing 45 mM hydroxypropyl-gamma-cyclodextrin, adjusted to pH 2.5 (B)
Injection: Electrokinetic injection at 15 kV for 3 s.
Detector: UV 200, UV 210
Migration time: 6.15 (A), 15.61, 15.97 (B, enantiomers)
Voltage: 15 kV
Model: Bio-Rad BioFocus 3000

OTHER SUBSTANCES
Simultaneous: atropine, carazolol, cicletanine, dimethindene, fendiline, homatropine, ipratropium bromide, isothipendyl, mefloquine, metaclazepam, nefopam, nicardipine, promethazine, reproterol, tetrahydrozoline (tetryzoline), theodrenaline, tioconazole, trihexyphenidyl, trimeprazine (alimemazine), trimipramine

KEY WORDS
coated capillary; chiral

REFERENCE
Koppenhoefer,B.; Epperlein,U.; Xiaofeng,Z.; Bingcheng,L. Separation of enantiomers of drugs by capillary electrophoresis. Part 4: Hydroxypropyl-γ-cyclodextrin as chiral solvating agent, *Electrophoresis*, **1997**, *18*, 924–930.

SAMPLE
Matrix: solutions
Sample preparation: Inject an aliquot of a 100 μg/mL solution in running buffer.

CAPILLARY ELECTROPHORESIS
Capillary: 30 cm $\times$ 50 μm fused-silica (25.5 cm to detector), coated with polyacrylamide
Capillary preparation: Adjust the pH of 20 mL water to 3.5 with acetic acid, add 80 μL 3-(trimethoxysilyl)propyl methacrylate (3-methacryloxypropyltrimethoxysilane), mix, suck into capillary, let stand at room temperature for 1 h, remove the solution, wash with water. Fill the capillary with a deaerated 3-4% acrylamide solution containing 1 μL/mL N,N,N',N'-tetramethylethylenediamine and 1 mg/mL potassium persulfate, let stand for 30 min, remove excess solution by aspiration, rinse with water, remove water by aspiration, dry at 35° (J. Chromatogr. 1985, 347, 191).
Capillary temperature: 25
Running buffer: 100 mM NaH$_2$PO$_4$ containing 45 mM hydroxypropyl-gamma- cyclodextrin, adjusted to pH 2.5 with phosphoric acid
Injection: Electrokinetic injection at 15 kV for 3 s.
Detector: UV 200
Voltage: 15 kV
Model: Bio-Rad BioFocus 3000

KEY WORDS
chiral; coated capillary; comparison with the use of other cyclodextrins; this running buffer gave the greatest enantiomeric separation.; α=1.027

REFERENCE
Lin,B.; Zhu,X.; Koppenhoefer,B.; Epperlein,U. Investigation of 123 chiral drugs by cyclodextrin-modified capillary electrophoresis, *LC.GC*, **1997**, *15*, 40–46.

SAMPLE
Matrix: solutions

CAPILLARY ELECTROPHORESIS
Capillary: 29-36 cm $\times$ 50 μm fused-silica (24.5-31.5 cm to detector) (Yongnian Optical Conductive Fiber Plant, China) coated with polyacrylamide
Capillary preparation: Coat capillary as follows. Adjust the pH of 20 mL water to 3.5 with acetic acid, add 80 μL 3-(trimethoxysilyl)propyl methacrylate (3-methacryloxypropyltrimethoxysilane), mix, suck into capillary, let stand at room temperature for 1 h, remove the solution, wash with water. Fill the capillary with a deaerated 3-4% acrylamide solution containing 1 μL/mL N,N,N',N'-tetramethylethylenediamine and 1 mg/mL potassium persulfate, let stand for 30 min, remove excess solution by aspiration, rinse with water, remove water by aspiration, dry at 35° (J. Chromatogr. 1985, 347, 191).
Capillary temperature: 25
Running buffer: 100 mM pH 2.5 NaH$_2$PO$_4$ (A) or 100 mM pH 2.5 NaH$_2$PO$_4$ containing 45 mM hydroxypropyl-α-cyclodextrin (Wacker, Munich) (B)
Injection: Electromigration at 15 kV for 3 s.
Detector: UV 200; UV 210
Migration time: 6.15 (A); 10.63, 10.91 (B) (enantiomers)

Voltage: 15 kV
Model: Bio-Focus 3000

OTHER SUBSTANCES
Also analyzed: albuterol (salbutamol), alprenolol, amorolfine, atenolol, atropine, azelastine, baclofen, bamethan, benproperine, benserazide, biperiden, bisoprolol, brompheniramine, bupivacaine, bupranolol, butamirate, butethamate, carazolol, carbuterol, carteolol, carvedilol, celiprolol, chloroquine, chlorpheniramine, chlorphenoxamine, cicletanine, clenbuterol, clidinium bromide, clobutinol, dimethindene, dipivefrin, disopyramide, dobutamine, doxylamine, fendiline, flecainide, gallopamil, homatropine, ipratropium bromide, isoproterenol (isoprenaline), isothipendyl, ketamine, meclizine, mefloquine, mepindolol, mequitazine, metaclazepam, metaproterenol (orciprenaline), metipranolol, metoprolol, nefopam, nicardipine, norfenefrine, ofloxacin, ornidazole, orphenadrine, oxomemazine, oxprenolol, oxybutynin, phenoxybenzamine, phenylpropanolamine, pholedrine, pindolol, pirbuterol, prilocaine, procyclidine, promethazine, propafenone, propranolol, reproterol, sotalol, sulpride, synephrine, talinolol, terbutaline, tetrahydrozoline (tetryzoline), theodrenaline, tioconazole, tocainide, trihexyphenidyl, trimeprazine (alimemazine), trimipramine, tropicamide, verapamil, zopiclone

KEY WORDS
coated capillary; chiral

REFERENCE
Koppenhoefer,B.; Eperlein,U.; Schlunk,R.; Zhu,X.; Lin,B. Separation of enantiomers of drugs by capillary electrophoresis. V. Hydroxypropyl-α-cyclodextrin as chiral solvating agent, *J.Chromatogr.A*, **1998**, *793*, 153–164.

Nalbuphine

Molecular formula: $C_{21}H_{27}NO_4$
Molecular weight: 357.45
CAS Registry No.: 20594-83-6, 23277-43-2 (HCl)
Merck Index (12th ed.): 6444
Lednicer: 2 319

SAMPLE
Matrix: bulk
Sample preparation: Prepare a 5 mg/mL solution in water, inject an aliquot.

CAPILLARY ELECTROPHORESIS
Capillary: 72 cm × 50 μm fused-silica (50 cm to detector) (Applied Biosystems or Polymicro Technologies)
Capillary preparation: Condition a new capillary by washing with 1 M NaOH for 30 min then with water for 30 min
Capillary temperature: 30
Running buffer: 10 mM pH 7.0 NaH_2PO_4 containing 6 mM sodium borate and 50 mM sodium taurodeoxycholate
Injection: 3 nL
Detector: UV 230
Migration time: 6
Voltage: -20 kV
Model: Applied Biosystems HT 270A

OTHER SUBSTANCES
Simultaneous: impurities

REFERENCE
Williams,R.C.; Edwards,J.F.; Ainsworth,C.R. Analysis of diastereoisomer impurities in chiral pharmaceutical compounds by capillary electrophoresis, *Chromatographia*, **1994**, *38*, 441–446.

Nalidixic acid

Molecular formula: $C_{12}H_{12}N_2O_3$
Molecular weight: 232.23
CAS Registry No.: 389-08-2, 3374-05-8 (Na salt), 15769-77-4 (Na salt monohydrate)
Merck Index (12th ed.): 6446
Lednicer: 1 429

SAMPLE
Matrix: solutions
Sample preparation: Inject an aliquot of a solution in MeOH.

CAPILLARY ELECTROPHORESIS
Capillary: 59 cm × 50 µm fused-silica (43 cm to detector) (Polymicro Technologies)
Capillary preparation: Before each analysis flush with water for 3 min, with 200 mM NaOH
for 3 min, with water for 3 min, and with running buffer for 4 min. Flush new capillaries with
1 M NaOH at 2000 mbar for 10 min then with 200 mM NaOH for 10 min.
Capillary temperature: 23
Running buffer: MeCN:buffer 28:72, pH 7.3 (Buffer was 32 mM sodium borate containing 18
mM NaH_2PO_4, 39 mM sodium cholate, and 8 mM sodium heptanesulfonate.)
Injection: Hydrodynamic injection at 40 mbar for 6 s.
Detector: UV 260
Migration time: 6.3
Voltage: 30 kV
Model: Lauer Labs Prince

OTHER SUBSTANCES
Simultaneous: ciprofloxacin, enoxacin, flumequine, lomefloxacin, norfloxacin, ofloxacin, oxolinic
acid, pefloxacin, pipemidic acid, piromidic acid, rosoxacin, sparfloxacin

REFERENCE
Sun,S.-W.; Chen,L.-Y. Optimization of capillary electrophoretic separation of quinolone antibacterials using the
overlapping resolution mapping scheme, *J.Chromatogr.A*, **1997**, *766*, 215–224.

Nalorphine

Molecular formula: $C_{19}H_{21}NO_3$
Molecular weight: 311.38
CAS Registry No.: 62-67-9, 57-29-4 (HCl)
Merck Index (12th ed.): 6448

SAMPLE
Matrix: blood
Sample preparation: Condition a 1 mL 100 mg phenyl SPE cartridge (Varian) with 2 column
volumes of MeOH and 2 column volumes of water. Add 1 mL serum to the SPE cartridge, wash
with 1 mL MeOH:water 60:40, elute with four 250 µL aliquots of methanol. Filter (0.2 µm) the
eluate and evaporate it to dryness under a stream of nitrogen, reconstitute the residue in 50
µL water, inject an aliquot.

CAPILLARY ELECTROPHORESIS
Capillary: 72 cm × 50 µm fused-silica (50 cm to detector) (Polymicro Technologies)
Capillary preparation: Before each run rinse capillary with 100 mM NaOH and with running
buffer for 2 min. Condition new capillaries by rinsing with 1 M NaOH for 10 min, with water
for 10 min, and with running buffer for 10 min.
Capillary temperature: 30

Running buffer: 100 mM NaH_2PO_4 containing 15 mM methyl-β-cyclodextrin and 30 μM hexadecyltrimethylammonium bromide, pH adjusted to 2.5 with 100 mM phosphoric acid
Injection: Vacuum injection at 80 psi for 20 s.
Detector: UV 220
Migration time: 13.8
Internal standard: nalorphine
Voltage: 20 kV
Current: 45 μA
Model: Applied Biosystems ABI 270A

OTHER SUBSTANCES
Extracted: pentazocine

KEY WORDS
serum; SPE; nalorphine is IS; not a chiral assay

REFERENCE
Ameyibor,E.; Stewart,J.T. Quantitative determination of pentazocine enantiomers in human serum using derivatized β-cyclodextrin-modified capillary electrophoresis and solid phase extraction, *J.Liq.Chromatogr.Rel.Technol.*, **1998**, *21*, 953–963.

SAMPLE
Matrix: solutions
Sample preparation: Prepare a solution in running buffer:water 1:2, inject a 5-10 μL aliquot.

CAPILLARY ELECTROPHORESIS
Capillary: 40 cm × 50 μm bare silica (40 cm to detector) (Isco)
Capillary preparation: After each injection flush with 100 mM NaOH and rinse with running buffer.
Running buffer: 50 mM pH 9.2 borate buffer
Injection: Manual injection of 5-10 μL by syringe through a 1:830 splitter
Detector: UV 214
Migration time: 9.46
Voltage: 15 kV
Model: Isco Model 3850

OTHER SUBSTANCES
Simultaneous: cocaine (UV 238), morphine, tetracaine (UV 238)
Noninterfering: acetaminophen, amitriptyline, amphetamine, atropine, benzoylecgonine, benztropine, caffeine, carbamazepine, carisoprodol, chlorpheniramine, chlorpromazine, chlorprothixene, cimetidine, cocaine, codeine, dextromethorphan, diazepam, dihydrocodeine, diphenhydramine, diphenoxylate, disopyramide, doxepin, doxylamine, emetine, erythromycin, ethylmorphine, flurazepam, glutethimide, hydrocodone, hydrocortisone, hydromorphone, hydroxyzine, imipramine, lidocaine, loxapine, meperidine, meprobamate, methadone, methamphetamine, methapyrilene, methaqualone, methocarbamol, methylphenidate, naloxone, nicotine, nordiazepam, nortriptyline, orphenadrine, oxycodone, papaverine, pentazocine, phenacetin, phencyclidine, phenmetrazine, phenolphthalein, phentermine, phenylpropanolamine, phenytoin, phetidine, prazepam, procainamide, procaine, propoxyphene, propranolol, protriptyline, pseudoephedrine, pyrilamine, quinine, salicylamide, spironolactone, strychnine, terpin hydrate, thioridazine, thiothixene, triamterene, trifluoperazine, triflupromazine, trihexyphenidyl, trimeprazine, trimethoprim, trimetobenzamide

REFERENCE
Tagliaro,F.; Poiesi,C.; Aiello,R.; Dorizzi,R.; Ghielmi,S.; Marigo,M. Capillary electrophoresis for the investigation of illicit drugs in hair: determination of cocaine and morphine, *J.Chromatogr.*, **1993**, *638*, 303–309.

SAMPLE
Matrix: solutions

CAPILLARY ELECTROPHORESIS
Capillary: 59.6 cm × 75 μm fused-silica (52.1 cm to detector) (Polymicro Technologies)

Capillary preparation: Purge with running buffer before each run. At the beginning of each day purge using 50-60 kPa vacuum with 500 mM NaOH for 5 min, with water for 5 min, with MeCN for 5 min, and with running buffer for 5 min.
Running buffer: MeCN:MeOH:acetic acid 49:50:1 containing 20 mM ammonium acetate
Injection: Hydrostatic injection at 10 cm for 5 s.
Detector: UV 214
Migration time: 5.07
Voltage: 30 kV
Model: Waters Quanta 4000

OTHER SUBSTANCES
Simultaneous: diphenoxylate, ethoheptazine, fentanyl, levallorphan, meperidine, methadone, pentazocine, nikethamide

REFERENCE
Leung,C.N.W.; Tang,H.P.O.; Tso,T.S.C.; Wan,T.S.M. Separation of basic drugs with non-aqueous capillary electrophoresis, *J.Chromatogr.A*, **1996**, *738*, 141–154.

SAMPLE
Matrix: solutions
Sample preparation: Inject an aliquot of a 1-50 µg/mL solution in water.

CAPILLARY ELECTROPHORESIS
Capillary: 55 cm × 50 µm fused-silica (35 cm to detector) (J&W)
Capillary preparation: Flush with running buffer for 5 min before each run. Periodically wash with 100 mM NaOH and flush extensively with running buffer.
Running buffer: MeOH:25 mM pH 9.24 borate buffer 20:80 containing 100 mM sodium dodecyl sulfate (A) or 50 mM pH 2.35 phosphate buffer (B) or 50 mM pH 9.24 borate buffer (C)
Injection: Injection of 5 µL using a split-flow injector ratio of 1:800.
Detector: UV 200
Migration time: 22.83 (A), 11.24 (B), 8.51 (C)
Voltage: 20 kV (A, B) or 12 kV (C)
Current: <60-80 µA
Model: ISCO Model 3850

OTHER SUBSTANCES
Also analyzed: acetylcodeine, amphetamine, barbital, caffeine, codeine, cocaine, diamorphine, diazepam, flunitrazepam, lidocaine, monoacetylmorphine, morphine, narceine, noscapine, papaverine, pentobarbital, procaine, tetracaine, thebaine

KEY WORDS
all compounds were separated with running buffer A; some peaks overlapped with running buffers B and C.

REFERENCE
Tagliaro,F.; Smith,F.P.; Turrina,S.; Equisetto,V.; Marigo,M. Complementary use of capillary zone electrophoresis and micellar electrokinetic capillary chromatography for mutual confirmation of results in forensic drug analysis, *J.Chromatogr.A*, **1996**, *735*, 227–235.

SAMPLE
Matrix: urine
Sample preparation: Condition a 500 mg Bond Elut C2 SPE cartridge with 2 mL MeOH and 2 mL 50 mM pH 7.5 Tris-HCl buffer. Condition a 100 mg Bond Elut PRS SPE cartridge with 500 µL MeOH, 500 µL water, and 500 µL 0.1% trifluoroacetic acid in water. 5 mL Urine + 5 mL 50 mM pH 7.5 Tris-HCl buffer, mix, adjust pH to 7.5 with NaOH or HCl, add to the C2 SPE cartridge, wash with 2 mL 50 mM pH 7.5 Tris-HCl buffer, elute with 2 mL MeCN:water: trifluoroacetic acid 50:50:0.1. Dilute the eluate with 8 mL 0.1% trifluoroacetic acid in water, add the mixture to the PRS SPE cartridge, wash with 500 µL 0.1% trifluoroacetic acid in water, wash with 2 mL MeOH, elute with 1 mL 3% ammonia in MeOH. Evaporate the eluate to dryness under reduced pressure, reconstitute with 50 µL MeCN:water 10:90, inject an aliquot.

CAPILLARY ELECTROPHORESIS
Capillary: 64.5 cm × 50 µm fused-silica (56 cm to detector) (Hewlett-Packard)

Capillary preparation: Flush with running buffer for 1-4 min between runs. Flush new capillaries under 900 mbar pressure with 1 M NaOH for 10 min, with 100 mM NaOH for 5 min, and with running buffer for 5 min.
Capillary temperature: 25
Running buffer: MeCN:buffer 5:95 (Buffer was 70 mM pH 9.3 borate buffer containing 70 mM sodium dodecyl sulfate.)
Injection: Pressure injection of sample at 50 mbar for 3 s and of running buffer at 50 mbar for 1 s.
Detector: UV 195
Migration time: 12.3
Internal standard: nalorphine
Voltage: 30 kV
Model: Hewlett-Packard HP ^{3D}CE

OTHER SUBSTANCES
Extracted: morphine

KEY WORDS
SPE; horse; nalorphine is IS

REFERENCE
Taylor,M.R.; Westwood,S.A.; Perrett,D. Determination of Phase II drug metabolites in equine urine by micellar electrokinetic capillary chromatography, *J.Chromatogr.A*, **1996**, *745*, 155–163.

Nandrolone

Molecular formula: $C_{18}H_{26}O_2$
Molecular weight: 274.40
CAS Registry No.: 434-22-0, 360-70-3 (decanoate), 52279-57-9 (p-hydroxyphenylpropionate), 62- 90-8 (phenpropionate), 7207-92-3 (propionate), 22263-51-0 (cyclotate)
Merck Index (12th ed.): 6452
Lednicer: 1 164

SAMPLE
Matrix: blood
Sample preparation: Treat 200 μL serum with 1% sodium salicylate for 1 h, add to a 20 × 9 ASPEC pak octadecylsilane SPE cartridge (M&S Instruments Trading Inc., Osaka), wash with 20 mL water, elute with 2 mL MeOH. Evaporate the eluate to dryness under educed pressure, reconstitute with 20 μL running buffer, filter (0.5 μm), inject an aliquot.

CAPILLARY ELECTROPHORESIS
Capillary: 37.5 cm × 75 μm fused-silica
Capillary preparation: Change the running buffer for each run.
Capillary temperature: 25
Running buffer: MeCN:buffer 16:84 (Buffer was 20 mM pH 8.0 sodium tetraborate/potassium phosphate buffer containing 50 mM sodium dodecyl sulfate.)
Injection: Pressure injection at 0.05 kg/cm^2 for 2 s
Detector: UV 205
Migration time: 21
Current: 60 μA (constant current)
Model: Ohtsuka Electronic model CAPI-3000

OTHER SUBSTANCES
Extracted: hydrocortisone, hydroxypregnenolone, hydroxyprogesterone, pregnenolone

---

KEY WORDS
serum; SPE

REFERENCE
Kobayashi,Y.; Matsui,J.; Watanabe,F. Simultaneous separation of free and conjugated steroids by micellar electrokinetic chromatography and its clinical application, *Biol.Pharm.Bull.*, **1995**, *18*, 1614–1616.

SAMPLE
Matrix: bulk
Sample preparation: Dissolve compound in 75 mM (carboxymethyl)trimethylammonium chloride hydrazide (Girard's Reagent T) in EtOH:glacial acetic acid 90:10 at a concentration of 15-50 μM, let stand at room temperature for 30 min, dilute 10-fold with running buffer, inject an aliquot.

CAPILLARY ELECTROPHORESIS
Capillary: 64.5 cm $\times$ 50 μm fused-silica (56 cm to bubble cell detector) Hewlett- Packard
Capillary preparation: Rinse with 100 mM NaOH for 2 min and with running buffer for 5 min. Condition new capillaries with 1 M NaOH for 5 min, with 100 mM NaOH for 30 min, and with running buffer for 30 min.
Capillary temperature: 30
Running buffer: 20 mM pH 4.8 NaH_2PO_4
Injection: Pressure injection at 5 kPa for 5 s
Detector: UV 280
Migration time: 8.5 (nandrolone), 9.5 (nandrolone phenylpropionate)
Voltage: 25 kV
Model: Hewlett-Packard HP 3DCE

OTHER SUBSTANCES
Interfering: norethindrone (norethisterone), norethindrone enanthate

KEY WORDS
derivatization

REFERENCE
Görög,S.; Gazdag,M.; Kemenes-Bakos,P. Analysis of steroids Part 50. Derivatization of ketosteroids for their separation and determination by capillary electrophoresis, *J.Pharm.Biomed.Anal.*, **1996**, *14*, 1115–1124.

Naphazoline

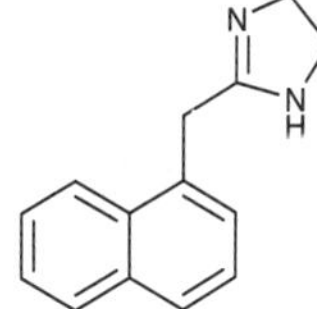

Molecular formula: $C_{14}H_{14}N_2$
Molecular weight: 210.28
CAS Registry No.: 835-31-4, 550-29-2 (HCl)
Merck Index (12th ed.): 6455
Lednicer: 1 241

SAMPLE
Matrix: solutions
Sample preparation: Inject an aliquot of a solution in 100 mM sodium phosphate containing 100-300 μg/mL naphazoline.

CAPILLARY ELECTROPHORESIS
Capillary: 67 cm $\times$ 50 μm fused-silica (60 cm to detector)
Capillary preparation: Rinse with running buffer for 2 min between sets of analyses.
Capillary temperature: 30
Running buffer: 200 mM pH 4.5 Phosphate buffer
Injection: Injection at high pressure for 2 s.
Detector: UV 210, UV 230
Migration time: 14
Voltage: 20 kV
Model: Beckman P/ACE 5500

OTHER SUBSTANCES
Simultaneous: acetylcodeine, amphetamine, cocaine, codeine, diamorphine, LSD, methadone, methamphetamine, methylenedioxyamphetamine, methylenedioxymethamphetamine, morphine, PCP, psilocyn

REFERENCE
Walker,J.A.; Marché,H.L.; Newby,N.; Bechtold,E.J. A free zone capillary electrophoresis method for the quantitation of common illicit drug samples, *J.Forensic Sci.*, **1996**, *41*, 824–829.

Naproxen

Molecular formula: $C_{14}H_{14}O_3$
Molecular weight: 230.26
CAS Registry No.: 22204-53-1, 26159-34-2 (sodium salt)
Merck Index (12th ed.): 6504
Lednicer: 1 86

SAMPLE
Matrix: blood
Sample preparation: 0.1-1 mL Plasma + 2 µg IS + 400 µL 1 M HCl + 10 mL diethyl ether, agitate mechanically for 10 min, centrifuge at 3500 g for 10 min. Remove the organic layer and evaporate it to dryness under a stream of nitrogen, reconstitute the residue in MeOH:40 mM NaH_2PO_4 50:50, inject an aliquot.

CAPILLARY ELECTROPHORESIS
Capillary: 58.7 cm × 75 µm fused-silica (50 cm to detector)
Capillary preparation: Wash with running buffer for 2 min before each injection. At the beginning of each day wash capillary with 100 mM NaOH for 10 min.
Capillary temperature: 22
Running buffer: MeOH:buffer 3:97 (Buffer was 40 mM pH 8 NaH_2PO_4 containing 104 mM sodium dodecyl sulfate.)
Injection: Hydrodynamic injection for 2 s
Detector: UV 254
Migration time: 12.7
Internal standard: benzoyl-4-phenyl-2-butyric acid (16.3)
Voltage: 20 kV
Model: Beckman P/ACE 2000
Limit of detection: 670 ng/mL (S/N 5)

OTHER SUBSTANCES
Extracted: diclofenac, diflunisal, etodolac, fenbufen, fenoprofen, flurbiprofen, ibuprofen, indomethacin, ketoprofen, niflumic acid, piroxicam, sulindac, tenoxicam, tiaprofenic acid
Noninterfering: acetaminophen, amitriptyline, caffeine, clomipramine, deoxysulindac, desipramine, diazepam, 5'-hydroxytenoxicam, imipramine, maprotiline, nortriptyline, phenobarbital, phenytoin, sulfamethoxazole, theophylline, trimipramine

KEY WORDS
plasma

REFERENCE
Maboundou,C.W.; Paintaud,G.; Bérard,M.; Bechtel,P.R. Separation of fifteen non-steroidal anti-inflammatory drugs using micellar electrokinetic capillary chromatography, *J.Chromatogr.B*, **1994**, *657*, 173–183.

SAMPLE
Matrix: blood
Sample preparation: Inject serum directly.

CAPILLARY ELECTROPHORESIS
Capillary: 105 cm × 75 μm fused-silica (68 cm to detector) (Polymicro Technologies)
Capillary preparation: Rinse with running buffer for 10 min between run. At the beginning and end of each day rinse with 100 mM NaOH for 10 min and with water for 10 min.
Running buffer: 10 mM Na_2HPO_4 containing 6 mM sodium borate and 75 mM sodium dodecyl sulfate
Injection: Injection via vacuum suction for 1 s
Detector: UV 215
Migration time: 14.2
Voltage: 23 kV
Current: 62 μA
Model: Europhor Prime Vision IV

OTHER SUBSTANCES
Extracted: acetaminophen, ethosuximide, flucytosine, primidone, sulfamethoxazole, zomepirac

KEY WORDS
serum

REFERENCE
Schmutz,A.; Thormann,W. Assessment of impact of physico-chemical drug properties on monitoring drug levels by micellar electrokinetic capillary chromatography with direct serum injection, *Electrophoresis*, **1994**, *15*, 1295–1303.

SAMPLE
Matrix: blood
Sample preparation: Centrifuge at 10000 g for 3 min, inject an aliquot.

CAPILLARY ELECTROPHORESIS
Capillary: 50 cm × 50 μm fused-silica (45 cm to detector) (Polymicro Technologies)
Capillary preparation: Between runs rinse capillary at 100 psi with 100 mM NaOH for 1 min and with running buffer for 2 min.
Capillary temperature: 20
Running buffer: 6 mM Sodium tetraborate containing 10 mM Na_2HPO_4 and 75 mM sodium dodecyl sulfate adjusted to pH 9.2 with 1 M NaOH
Injection: Pressure injection at 5 psi for 0.4 s.
Detector: UV 240
Migration time: 7.3
Voltage: 20 kV
Current: ca. 36 μA
Model: Bio-Rad BioFocus 3000

KEY WORDS
serum; direct injection

REFERENCE
Zhang,C.-X.; Thormann,W. Determination of drug levels in human serum by micellar electrokinetic capillary chromatography with direct sample injection using different quantitation strategies, *J.Capillary Electrophor.*, **1994**, *1*, 208–218.

SAMPLE
Matrix: blood
Sample preparation: 50 μL Plasma + 50 μL 0.5-1 mM sodium salicylate solution, vortex for 2 s, inject an aliquot.

CAPILLARY ELECTROPHORESIS
Capillary: 27 cm × 50 μm fused-silica (20 cm to detector) (Polymicro Technologies)
Capillary preparation: Between runs wash capillary at 5 psi with 100 mM NaOH for 3 min, with water for 3 min, and with running buffer for 3 min.
Capillary temperature: 20

Running buffer: 10 mM pH 9.2 Sodium tetraborate containing 6 mM Na_2HPO_4 and 75 mM sodium dodecyl sulfate. Replenish vials each 5-6 runs.
Injection: Pressure injection at 0.5 psi for 1 s.
Detector: F ex 325 (He-Cd laser) em 366
Migration time: 4.4
Internal standard: sodium salicylate (5.4)
Voltage: 8 kV
Current: ca. 30 μA
Model: Beckman P/ACE 5510
Limit of detection: 140 nM

OTHER SUBSTANCES
Extracted: naproxen-lysine, naproxen-protein

KEY WORDS
rat; plasma; comparison with HPLC

REFERENCE
Albrecht,C.; Reichen,J.; Visser,J.; Meijer,D.K.F.; Thormann,W. Differentiation between naproxen, naproxen-protein conjugates, and naproxen-lysine in plasma via micellar electrokinetic capillary chromatography -a new approach in the bioanalysis of drug targeting preparations, *Clin.Chem.*, **1997**, *43*, 2083–2090.

SAMPLE
Matrix: formulations
Sample preparation: Grind granules, add 70 mL MeOH, warm at 40° with occasional shaking, cool, add 20 mL 5 mg/mL methyl p-hydroxybenzoate in MeOH, make up to 100 mL with water, filter (0.45 μm), inject an aliquot.

CAPILLARY ELECTROPHORESIS
Capillary: 65 cm × 50 μm fused-silica (SGE)
Running buffer: 20 mM NaH_2PO_4 containing 100 mM sodium cholate, adjusted to pH 9.0 with 20 mM sodium tetraborate
Injection: Siphon for 10 s at 10 cm height
Detector: UV 210
Migration time: 15
Internal standard: methyl p-hydroxybenzoate (16)
Voltage: +20 kV

OTHER SUBSTANCES
Simultaneous: acetaminophen, caffeine, chlorpheniramine, dibucaine, dipyrone (sulpyrin), ethenzamide, guaifenesin, isopropylantipyrine, noscapine, phenacetin, tipepidine, trimetoquinol, triprolidine

---

KEY WORDS
granules

REFERENCE
Nishi,H.; Fukuyama,T.; Matsuo,M.; Terabe,S. Separation and determination of the ingredients of a cold medicine by micellar electrokinetic chromatography with bile salts, *J.Chromatogr.*, **1990**, *498*, 313–323.

SAMPLE
Matrix: formulations
Sample preparation: Grind tablet or powders, add 70 mL MeOH, add 20 mL 5 mg/mL methyl p-hydroxybenzoate in MeOH, sonicate for 10 min, shake for 5 min, make up to 100 mL with water, filter (0.45 μm), inject an aliquot.

CAPILLARY ELECTROPHORESIS
Capillary: 65 cm × 50 μm fused-silica (50 cm to detector) (SGE)
Running buffer: 20 mM NaH_2PO_4 containing 100 mM sodium dodecyl sulfate, adjusted to pH 9.0 with 20 mM sodium tetraborate
Injection: Siphon

Detector: UV 210
Migration time: 11
Internal standard: methyl p-hydroxybenzoate
Voltage: +20 kV

OTHER SUBSTANCES
Simultaneous: acetaminophen, caffeine, chlorpheniramine, dipyrone (sulpyrin), ethenzamide, guaifenesin, isopropylantipyrine, noscapine, phenacetin, tipepidine, trimetoquinol

KEY WORDS
tablets; powders

REFERENCE
Nishi,H.; Fukuyama,T.; Matsuo,M.; Terabe,S. Effect of surfactant structures on the separation of cold medicine ingredients by micellar electrokinetic chromatography, *J.Pharm.Sci.*, **1990**, *79*, 519–523.

SAMPLE
Matrix: formulations
Sample preparation: Sonicate tablet in 100 mL MeOH for 10 min, dilute an aliquot with running buffer to a naproxen concentration of 100 µg/mL, inject an aliquot.

CAPILLARY ELECTROPHORESIS
Capillary: 42.5 cm × 25 µm
Capillary temperature: 30
Running buffer: 20 mM pH 7.0 Phosphate containing 25 mM sodium dodecyl sulfate
Injection: Vacuum injection for 2 s
Detector: UV 230
Migration time: 4.3
Voltage: 25 kV
Current: 5 µA
Model: Applied Biosystems Model 270A

OTHER SUBSTANCES
Simultaneous: diflunisal, ibuprofen, indomethacin, sulindac, tolmetin

KEY WORDS
tablets

REFERENCE
Weinberger,R.; Albin,M. Quantitative micellar electrokinetic capillary chromatography: Linear dynamic range, *J.Liq.Chromatogr.*, **1991**, *14*, 953–972.

SAMPLE
Matrix: serum
Sample preparation: Inject serum directly.

CAPILLARY ELECTROPHORESIS
Capillary: 70 cm × 75 µm fused-silica (50 cm to detector) (Polymicro Technologies)
Capillary preparation: Rinse capillary with buffer between runs.
Running buffer: 10 mM pH 9.2 Na_2HPO_4 containing 6 mM sodium borate and 75 mM sodium dodecyl sulfate
Injection: Siphon at a height of 34 cm for 2 s
Detector: F ex 220 em 366 nm (bandpass filter)
Migration time: 7.5
Voltage: 20 kV
Current: 90 µA
Limit of detection: 200 ng/mL

OTHER SUBSTANCES
Extracted: quinidine, salicylic acid

REFERENCE
Caslavska,J.; Gassmann,E.; Thormann,W. Modification of a tunable UV-visible capillary electrophoresis detector for simultaneous absorbance and fluorescence detection: profiling of body fluids for drugs and endogenous compounds, *J.Chromatogr.A*, **1995**, *709*, 147–156.

SAMPLE
Matrix: solutions
Sample preparation: Dissolve in MeCN, dilute with water to a concentration of 25 µg/mL, inject an aliquot.

CAPILLARY ELECTROPHORESIS
Capillary: 60 cm × 75 µm fused-silica (52.5 cm to detector)
Running buffer: MeCN:30 mM pH 7.0 phosphate buffer 20:80
Injection: Hydrostatic injection at 9.8 cm for 5 s
Detector: UV 214
Migration time: 20.90
Model: Waters Quanta 4000

OTHER SUBSTANCES
Simultaneous: acemetacin, alclofenac, fenbufen, flurbiprofen, ibuprofen, indomethacin, ketoprofen, lonazolac, niflumic acid, piroxicam, tenoxicam, tiaprofenic acid, tolmetin, voltaren

REFERENCE
Donato,M.G.; Van den Eeckhout,E.; Van den Bossche,W.; Sandra,P. Capillary zone electrophoresis and micellar electrokinetic capillary chromatography of some non-steroidal antiinflammatory drugs (NSAIDs), *J.Pharm.Biomed.Anal.*, **1993**, *11*, 197–201.

SAMPLE
Matrix: solutions
Sample preparation: Prepare a solution in running buffer, inject an aliquot.

CAPILLARY ELECTROPHORESIS
Capillary: 27 cm × 25 µm fused-silica (20 cm to detector)
Capillary preparation: Rinse with 1 M HCl before each run.
Capillary temperature: 20
Running buffer: 200 mM 2-(N-morpholino)ethanesulfonic acid containing 10 mM hydroxypropyl-β-cyclodextrin and 0.4% polymeric additive (?) adjusted to pH 5.0 with tetrabutylammonium hydroxide
Injection: Inject by pressure at 0.5 psi for 3-20 s
Detector: UV 230
Migration time: 14 (R), 15.5 (S)
Internal standard: p-toluenesulfonic acid (2.5)
Voltage: 700 V/cm
Current: 3 µA
Model: Beckman P/ACE 2100
Limit of detection: 300 nM

KEY WORDS
detector at anode; chiral

REFERENCE
Guttman,A.; Cooke,N. Practical aspects in chiral separation of pharmaceuticals by capillary electrophoresis. II. Quantitative separation of naproxen enantiomers, *J.Chromatogr.A*, **1994**, *685*, 155–159.

SAMPLE
Matrix: solutions

CAPILLARY ELECTROPHORESIS
Capillary: 65.4 cm × 50 µm fused-silica (56 cm to detector) (Hewlett-Packard)
Capillary temperature: 37

Running buffer: 11.7 mM pH 8.9 Borate buffer containing 8.3 mM phosphate and 50 mM sodium dodecyl sulfate
Injection: Injection at 50 mbar.s
Detector: UV 200
Migration time: 4.7
Voltage: 370 V/cm
Model: Hewlett-Packard G1600A ^{3D}Capillary Electrophoresis system

---

OTHER SUBSTANCES
Simultaneous: acetaminophen, caffeine, guaifenesin, noscapine, phenacetin

---

REFERENCE
Heiger,D.N.; Kaltenbach,P.; Sievet,H.-J. Diode array detection in capillary electrophoresis, *Electrophoresis*, **1994**, *15*, 1234–1247.

SAMPLE
Matrix: solutions
Sample preparation: Inject an aliquot of a 50 μM solution in MeOH:water 40:60.

CAPILLARY ELECTROPHORESIS
Capillary temperature: 25
Running buffer: MeOH:water 40:60 containing 75 mM glycine and 15 mM hydroxypropyl-β-cyclodextrin adjusted to pH 9.1 with triethanolamine
Injection: Hydrodynamic injection for 2 s
Detector: UV 280
Migration time: 17.4
Voltage: 15 kV

OTHER SUBSTANCES
Simultaneous: alclofenac, bufexamac, carprofen, flufenamic acid, flurbiprofen, indomethacin, niflumic acid, piroxicam, sulindac, tiaprofenic acid
Interfering: ketoprofen

REFERENCE
Bechet,I.; Fillet,M.; Hubert,P.; Crommen,J. Improvement of achiral resolution in capillary zone electrophoresis by use of cyclodextrin additives, *Biomed.Chromatogr.*, **1995**, *9*, 267–268.

SAMPLE
Matrix: solutions
Sample preparation: Inject an aliquot of a 100 μg/mL solution.

CAPILLARY ELECTROPHORESIS
Capillary: 27 cm × 50 μm neutrally coated (20 cm to detector)
Capillary preparation: Before each run rinse capillary with 100 mM HCl for 30 s, with water for 2 min, and with running buffer for 2 min
Capillary temperature: 20
Running buffer: 25 mM pH 4.6 Sodium acetate containing 10 mM hydroxypropyl-β-cyclodextrin
Injection: Pressure injection at 0.5 psi for 5 s
Detector: UV 230
Migration time: 22.5, 27 (enantiomers)
Voltage: 500 V/cm
Current: 49 μA
Model: Beckman P/ACE system 5500

KEY WORDS
chiral; detector is at anode; coated capillary

REFERENCE
Guttman,A. Novel separation scheme for capillary electrophoresis of enantiomers, *Electrophoresis*, **1995**, *16*, 1900–1905.

SAMPLE
Matrix: solutions
Sample preparation: Inject an aliquot of an aqueous solution.

CAPILLARY ELECTROPHORESIS
Capillary: 27 cm × 25 μm fused-silica (20 cm to detector)
Capillary temperature: 20 ± 0.1
Running buffer: 200 mM 2-(N-morpholino)ethanesulfonic acid containing 10 mM hydroxypropyl-β-cyclodextrin and 0.4% poly(ethylene oxide) (MW 300 000), adjusted to pH 5.0 with tetrabutylammonium hydroxide
Injection: Pressure injection at 0.5 psi for 3 s
Detector: UV 230
Migration time: 5.3 (R), 5.4 (S)
Model: Beckman P/ACE system 2100

KEY WORDS
detector at anode; chiral; end of capillary that draws in sample must be cut square for best results

REFERENCE
Guttman,A.; Schwartz,H.E. Artifacts related to sample introduction in capillary gel electrophoresis affecting separation performance and quantitation, *Anal.Chem.*, **1995**, *67*, 2279–2283.

SAMPLE
Matrix: solutions
Sample preparation: Prepare a solution in MeOH, inject an aliquot.

CAPILLARY ELECTROPHORESIS
Capillary: 40 cm × 50 μm fused-silica (35.4 cm to detector) (Polymicro Technologies)
Capillary preparation: Between runs rinse capillary with 100 mM NaOH for 1 min and running buffer for 2 min (at 100 psi).
Capillary temperature: 20
Running buffer: THF:water 60:40 containing 40 mM phosphate, pH 2
Injection: Pressure injection at 5 psi (4 psi.s)
Detector: UV 220
Migration time: 42
Voltage: 20 kV
Current: 7 μA
Model: Bio-Rad BioFocus 3000

OTHER SUBSTANCES
Simultaneous: amiodarone, desethylamiodarone, dextromethorphan, itraconazole, ketoconazole, methadone
Noninterfering: caffeine, itraconazole, phenol, theophylline

REFERENCE
Zhang,C.-X.; von Heeren,F.; Thormann,W. Separation of hydrophobic, positively chargeable substances by capillary electrophoresis, *Anal.Chem.*, **1995**, *67*, 2070–2077.

SAMPLE
Matrix: solutions
Sample preparation: Inject an aliquot of a 100 μg/mL solution in water or running buffer.

CAPILLARY ELECTROPHORESIS
Capillary: 47 cm × 75 μm fused-silica (40 cm to detector)
Capillary preparation: Before each run rinse capillary with running buffer for 1-2 min. If peak tailing is observed fill capillary with 500 mM NaOH and let stand for 30 min, wash with water for 3 min, rinse with running buffer for 3 min.
Capillary temperature: 20
Running buffer: 20 mM pH 7.0 Phosphate buffer containing 6% dextrin (Japan Pharmacopeia grade) (Dissolve dextrin in buffer at 90° then cool to room temperature.)
Injection: Pressure injection at 0.5 psi for 2-4 s.

Detector: UV 220
Migration time: 13.537 (R), 13.739 (S)
Voltage: 20 kV
Model: Beckman P/ACE 5510

KEY WORDS
chiral

REFERENCE
Nishi,H.; Izumoto,S.; Nakamura,K.; Nakai,H.; Sato,T. Dextran and dextrin as chiral selectors in capillary zone electrophoresis, *Chromatographia*, **1996**, *42*, 617–630.

SAMPLE
Matrix: solutions

CAPILLARY ELECTROPHORESIS
Capillary: 35 cm × 50 μm coated capillary (30.5 cm to detector) (Composite Metal Services, UK)
Capillary preparation: Before each run purge at high pressure with water for 100 s and with running buffer for 120 s. If vancomycin is used, before each run purge at high pressure with water for 100 s and with running buffer not containing vancomycin for 120 s. Next purge with running buffer containing vancomycin at low pressure (175 psi.s) and inject sample. Coat capillary as follows. Adjust the pH of 20 mL water to 3.5 with acetic acid, add 80 μL 3-(trimethoxysilyl)propyl methacrylate (3-methacryloxypropyltrimethoxysilane), mix, suck into capillary, let stand at room temperature for 1 h, remove the solution, wash with water. Fill the capillary with a deaerated 3-4% acrylamide solution containing 1 μL/mL N,N,N',N'-tetramethylethylenediamine and 1 mg/mL potassium persulfate, let stand for 30 min, remove excess solution by aspiration, rinse with water, remove water by aspiration, dry at 35° (J. Chromatogr. 1985, 347, 191).
Capillary temperature: 25
Running buffer: Buffer containing 30 mM heptakis-2,3,6-tri-O-methyl-β-cyclodextrin (A), 5 mM heptamethylamino-β-cyclodextrin (B) or 5 mM vancomycin (C) (Buffer was 50 mM phosphoric acid containing 50 mM acetic acid and 50 mM boric acid, dilute with an equal volume of water, adjust pH to 5 with concentrated NaOH.)
Injection: Pressure injection at 10 psi.
Detector: UV 206
Migration time: 12.2, 13.9 (A); 10.6, 11.0 (B); 6.5, 7.0 (C) (enantiomers)
Voltage: -20 kV
Model: Bio-Rad Biofocus 3000
Limit of detection: 5 μM (A, C), 10 μM (B)

KEY WORDS
chiral; coated capillary; detector at anode

REFERENCE
Fanali,S.; Desiderio,C.; Aturki,Z. Enantiomeric resolution study by capillary electrophoresis. Selection of the appropriate chiral selector, *J.Chromatogr.A*, **1997**, *772*, 185–194.

SAMPLE
Matrix: solutions
Sample preparation: Inject an aliquot of a solution in MeOH:DMF:34 mM phosphoric acid containing 20 mM mono(6-amino-6-deoxy)-β-cyclodextrin 2:1:97.

CAPILLARY ELECTROPHORESIS
Capillary: 38.5 cm × 50 μm fused-silica (30 cm to detector) (Supelco)
Capillary temperature: 25
Running buffer: 34 mM pH 2.3 Phosphoric acid containing 20 mM mono(6-amino-6-deoxy)-β-cyclodextrin and 10 mM heptakis(2,3,6-tri-O-methyl)-β-cyclodextrin (ionic strength 24 mM)
Injection: Hydrodynamic injection at 25 mbar for 4 s.
Detector: UV 230
Migration time: 12, 13 (enantiomers)
Voltage: 20 kV
Model: Hewlett-Packard HP [3D]CE

KEY WORDS
chiral

REFERENCE
Lelièvre,F.; Gareil,P.; Bahaddi,Y.; Galons,H. Intrinsic selectivity in capillary electrophoresis for chiral separations with dual cyclodextrin systems, *Anal.Chem.*, **1997**, *69*, 393–401.

SAMPLE
Matrix: solutions

CAPILLARY ELECTROPHORESIS
Capillary: 35 cm × 50 μm polyacrylamide-coated fused-silica (30.5 cm to detector) (Composite Metal Services, UK)

Capillary preparation: Before each run rinse capillary with water for 70 s and with running buffer for 100 s. (Coat capillary as follows. Adjust the pH of 20 mL water to 3.5 with acetic acid, add 80 μL 3-(trimethoxysilyl)propyl methacrylate (3-methacryloxypropyltrimethoxysilane), mix, suck into capillary, let stand at room temperature for 1 h, remove the solution, wash with water. Fill the capillary with a deaerated 3-4% acrylamide solution containing 1 μL/mL N,N,N',N'-tetramethylethylenediamine and 1 mg/mL potassium persulfate, let stand for 30 min, remove excess solution by aspiration, rinse with water, remove water by aspiration, dry at 35° (J. Chromatogr. 1985, 347, 191).)

Capillary temperature: 25

Running buffer: Buffer containing 7.5 mM cyanoethylated-β-cyclodextrin (Cyclolab, Budapest) (Prepare buffer by adjusting the pH of 50 mM phosphoric acid containing 50 mM acetic acid and 50 mM boric acid to 5 with concentrated NaOH, add the appropriate amount of cyanoethylated-β-cyclodextrin, dilute with an equal volume of water.)

Injection: Pressure injection at 5 psi for 2 s.

Detector: UV 206

Migration time: 15.6 (second enantiomer, α = 1.060)

Voltage: 20 kV

Current: 27-40 μA

Model: Bio-Rad Biofocus 3000

KEY WORDS
detector at anode; chiral; coated capillary

REFERENCE
Aturki,Z.; Desiderio,C.; Mannina,L.; Fanali,S. Chiral separations by capillary zone electrophoresis with the use of cyanoethylated-β-cyclodextrin as chiral selector, *J.Chromatogr.A*, **1998**, *817*, 91–104.

SAMPLE
Matrix: solutions

Sample preparation: Inject an aliquot of a 50-100 μg/mL solution in 50 mM pH 4.8 ammonium acetate buffer.

CAPILLARY ELECTROPHORESIS
Capillary: 44 cm × 50 μm polyacrylamide coated

Capillary preparation: Adjust the pH of 20 mL water to 3.5 with acetic acid, add 80 μL 3-(trimethoxysilyl)propyl methacrylate (3-methacryloxypropyltrimethoxysilane), mix, suck into capillary, let stand at room temperature for 1 h, remove the solution, wash with water. Fill the capillary with a deaerated 3-4% acrylamide solution containing 1 μL/mL N,N,N',N'-tetramethylethylenediamine and 1 mg/mL potassium persulfate, let stand for 30 min, remove excess solution by aspiration, rinse with water, remove water by aspiration, dry at 35° (J. Chromatogr. 1985, 347, 191).

Running buffer: 50 mM pH 4.8 Ammonium acetate buffer containing 5 mM vancomycin

Injection: Hydrostatic injection at 10 cm for 5-10 s.

Detector: MS, Finnigan LCQ ion trap, electrospray, negative ion mode, probe tip at 2.6 kV, sheath liquid MeOH:water:ammonia 50:48:2 at 6 μL/min (Vancomycin migrates away from the detector.)

Migration time: 8.7, 9.7 (enantiomers)

Voltage: 20 kV

Model: Grom 100

OTHER SUBSTANCES
Also analyzed: carprofen, flurbiprofen, ibuprofen, ketoprofen

KEY WORDS
chiral; coated capillary

REFERENCE
Fanali,S.; Desiderio,C.; Schulte,G.; Heitmeier,S.; Strickmann,D.; Chankvetadze,B.; Blaschke,G. Chiral capillary electrophoresis-electrospray mass spectrometry coupling using vancomycin as chiral selector, *J.Chromatogr.A*, **1998**, *800*, 69–76.

SAMPLE
Matrix: tissue
Sample preparation: Homogenize (Polytron PT 3000) 1 g tissue with 5 (kidney) or 10 (liver) mL 100 mM pH 7.4 KH_2PO_4 buffer at 30000 rpm for 30 s. Remove a 100 μL aliquot and add it to 1 mL 5 M NaOH, heat at 80° for 72 h, acidify with 1.5 mL 5 M NaOH, add 100 μL 16.6 μg/mL salicylic acid, add 6 mL dichloromethane, shake vigorously for 15 min, centrifuge at 1500 g for 10 min. Remove the organic layer and evaporate it to dryness under a stream of nitrogen at 37° (ca. 30 min), reconstitute the residue in 100-120 μL running buffer, inject an aliquot.

CAPILLARY ELECTROPHORESIS
Capillary: 27 cm × 50 μm fused-silica (20 cm to detector) (Polymicro Technologies)
Capillary preparation: Between runs wash capillary at 5 psi with 100 mM NaOH for 3 min, with water for 3 min, and with running buffer for 3 min.
Capillary temperature: 20
Running buffer: 10 mM pH 9.2 Sodium tetraborate containing 6 mM Na_2HPO_4 and 75 mM sodium dodecyl sulfate
Injection: Pressure injection at 0.5 psi for 1 s.
Detector: F ex 325 (10 mW He-Cd laser) em 366 (filter)
Migration time: 4
Internal standard: salicylic acid (5)
Voltage: 8 kV
Current: ca. 30 μA
Model: Beckman P/ACE 5510
Limit of detection: 70 ng/mL

KEY WORDS
kidney; liver; rat

REFERENCE
Albrecht,C.; Thormann,W. Determination of naproxen in liver and kidney tissues by electrokinetic capillary chromatography with laser-induced fluorescence detection, *J.Chromatogr.A*, **1998**, *802*, 115–120.

Nedocromil

Molecular formula: $C_{19}H_{17}NO_7$
Molecular weight: 371.35
CAS Registry No.: 69049-73-6, 69049-74-7 (Na salt), 101626-68-0 (Ca salt)
Merck Index (12th ed.): 6524
Lednicer: 4 209

SAMPLE
Matrix: solutions

CAPILLARY ELECTROPHORESIS
Capillary: 27 cm × 75 μm

Capillary preparation: Before each run rinse capillary with 100 mM NaOH for 30 s and with running buffer for 30 s. Condition new capillaries by rinsing with 100 mM NaOH for 20 min.
Capillary temperature: 30
Running buffer: 15 mM Sodium borate
Injection: Pressure injection of sample at 25 mbar for 1 s followed by running buffer at 25 mbar for 1 s.
Detector: UV 200
Migration time: 3.96
Internal standard: aminobenzoic acid (3.5)
Voltage: 6.5 kV
Model: Beckman

OTHER SUBSTANCES
Simultaneous: aspirin, bacitracin, beclomethasone, benzoic acid, ceftizoxime, ceftriaxone, cefuroxime, cephalothin, cromolyn, embonic acid, epoprostenol, glyburide (glibenclamide), levothyroxine, nedocromil, nystatin, omeprazole, prednisolone, warfarin, zidovudine

REFERENCE
Altria,K.D.; Bryant,S.M.; Hadgett,T.A. Validated capillary electrophoresis method for the analysis of a range of acidic drugs and excipients, *J.Pharm.Biomed.Anal.*, **1997**, *15*, 1091–1101.

Nefopam

Molecular formula: $C_{17}H_{19}NO$
Molecular weight: 253.34
CAS Registry No.: 13669-70-0, 23327-57-3 (HCl)
Merck Index (12th ed.): 6529
Lednicer: 2 447

SAMPLE
Matrix: solutions
Sample preparation: Inject an aliquot of a solution in running buffer.

CAPILLARY ELECTROPHORESIS
Capillary: 60 cm $\times$ 75 μm fused-silica (52.4 cm to detector)
Capillary preparation: After each run flush with 500 mM KOH for 2-3 min then with water.
Running buffer: 10 mM pH 3.8 Phosphate buffer containing 2% sulfated cyclodextrin (ds 7-10)
Injection: Hydrostatic injection.
Detector: UV 214
Migration time: 10.75, 10.91 (enantiomers)
Voltage: 15 kV
Model: Waters Quanta 4000

OTHER SUBSTANCES
Also analyzed: acebutolol, alprenolol, aminoglutethimide, brompheniramine, bupivacaine, bupropion, canadine, carbinoxamine, chloroquine, chlorpheniramine, dimethindene, disopyramide, doxylamine, hydroxychloroquine, idazoxan, isoxsuprine, ketamine, mepenzolate, mepivacaine, methoxyphenamine, mexiletine, midodrine, orphenadrine, oxprenolol, oxyphencyclimine, pheniramine, phensuximide, pindolol, piperoxan, terbutaline, tetramisole, tolperisone, tranylcypromine, trihexyphenidyl, trimipramine, verapamil, warfarin

KEY WORDS
chiral; detector at anode

REFERENCE
Stalcup,A.M.; Gahm,K.H. Application of sulfated cyclodextrins to chiral separations by capillary zone electrophoresis, *Anal.Chem.*, **1996**, *68*, 1360–1368.

SAMPLE
Matrix: solutions

CAPILLARY ELECTROPHORESIS
Capillary: 42 cm × 50 μm fused-silica (31 cm to detector) (Polymicro Technologies)
Capillary preparation: Condition capillary by rinsing with running buffer for 10 min, with water for 10 min, with 1 M NaOH for 10 min, with water for 10 min, and with running buffer with the voltage applied for 20 min.
Capillary temperature: 25
Running buffer: Formamide containing 150 mM citric acid, 100 mM Tris, and 100 mM β-cyclodextrin (apparent pH 5.1)
Injection: Gravity injection.
Detector: UV 254
Migration time: 10.01, 10.18 (enantiomers)
Voltage: 30 kV
Model: Beckman P/ACE 5500

OTHER SUBSTANCES
Simultaneous: chlophedianol, mianserin, propiomazine, trimeprazine, trimipramine, thioridazine

KEY WORDS
chiral

REFERENCE
Wang,F.; Khaledi,M.G. Chiral separations by nonaqueous capillary electrophoresis, *Anal.Chem.*, **1996**, *68*, 3460–3467.

SAMPLE
Matrix: solutions
Sample preparation: Inject an aliquot of a 100 μg/mL solution in running buffer.

CAPILLARY ELECTROPHORESIS
Capillary: 32 cm × 50 μm fused-silica (27.5 cm to detector) (Yongnian Optical Conductive Fiber Plant, China), coated with polyacrylamide
Capillary preparation: No details of the polyacrylamide coating process are provided. However, another paper (LC.GC 1997, 15, 40) by this group indicates that they use the procedure of Hjertén, thus: Adjust the pH of 20 mL water to 3.5 with acetic acid, add 80 μL 3-(trimethoxysilyl)propyl methacrylate (3-methacryloxypropyltrimethoxysilane), mix, suck into capillary, let stand at room temperature for 1 h, remove the solution, wash with water. Fill the capillary with a deaerated 3-4% acrylamide solution containing 1 μL/mL N,N,N',N'-tetramethylethylenediamine and 1 mg/mL potassium persulfate, let stand for 30 min, remove excess solution by aspiration, rinse with water, remove water by aspiration, dry at 35° (J. Chromatogr. 1985, 347, 191).
Capillary temperature: 25
Running buffer: 100 mM NaH_2PO_4 adjusted to pH 2.5
Injection: Electrokinetic injection at 15 kV for 3 s.
Detector: UV 200, UV 210
Migration time: 4.90
Voltage: 15 kV
Model: Bio-Rad BioFocus 3000

OTHER SUBSTANCES
Simultaneous: atropine, carazolol, cicletanine, dimethindene, fendiline, homatropine, ipratropium bromide, isothipendyl, mefloquine, metaclazepam, nafronyl (naftidrofuryl), nicardipine, promethazine, reproterol, tetrahydrozoline (tetryzoline), theodrenaline, tioconazole, trihexyphenidyl, trimeprazine (alimemazine), trimipramine

KEY WORDS
coated capillary

REFERENCE

Koppenhoefer,B.; Epperlein,U.; Xiaofeng,Z.; Bingcheng,L. Separation of enantiomers of drugs by capillary electrophoresis. Part 4: Hydroxypropyl-γ-cyclodextrin as chiral solvating agent, *Electrophoresis*, **1997**, *18*, 924–930.

SAMPLE

Matrix: solutions
Sample preparation: Inject an aliquot of a 100 μg/mL solution in running buffer.

CAPILLARY ELECTROPHORESIS

Capillary: 30 cm × 50 μm fused-silica (25.5 cm to detector), coated with polyacrylamide
Capillary preparation: Adjust the pH of 20 mL water to 3.5 with acetic acid, add 80 μL 3-(trimethoxysilyl)propyl methacrylate (3-methacryloxypropyltrimethoxysilane), mix, suck into capillary, let stand at room temperature for 1 h, remove the solution, wash with water. Fill the capillary with a deaerated 3-4% acrylamide solution containing 1 μL/mL N,N,N',N'-tetramethylethylenediamine and 1 mg/mL potassium persulfate, let stand for 30 min, remove excess solution by aspiration, rinse with water, remove water by aspiration, dry at 35° (J. Chromatogr. 1985, 347, 191).
Capillary temperature: 25
Running buffer: 100 mM NaH_2PO_4 containing 45 mM hydroxypropyl-α-cyclodextrin, adjusted to pH 2.5 with phosphoric acid
Injection: Electrokinetic injection at 15 kV for 3 s.
Detector: UV 200
Voltage: 15 kV
Model: Bio-Rad BioFocus 3000

KEY WORDS

chiral; coated capillary; comparison with the use of other cyclodextrins; this running buffer gave the greatest enantiomeric separation.; α=1.028

REFERENCE

Lin,B.; Zhu,X.; Koppenhoefer,B.; Epperlein,U. Investigation of 123 chiral drugs by cyclodextrin-modified capillary electrophoresis, *LC.GC*, **1997**, *15*, 40–46.

SAMPLE

Matrix: solutions

CAPILLARY ELECTROPHORESIS

Capillary: 29-36 cm × 50 μm fused-silica (24.5-31.5 cm to detector) (Yongnian Optical Conductive Fiber Plant, China) coated with polyacrylamide
Capillary preparation: Coat capillary as follows. Adjust the pH of 20 mL water to 3.5 with acetic acid, add 80 μL 3-(trimethoxysilyl)propyl methacrylate (3-methacryloxypropyltrimethoxysilane), mix, suck into capillary, let stand at room temperature for 1 h, remove the solution, wash with water. Fill the capillary with a deaerated 3-4% acrylamide solution containing 1 μL/mL N,N,N',N'-tetramethylethylenediamine and 1 mg/mL potassium persulfate, let stand for 30 min, remove excess solution by aspiration, rinse with water, remove water by aspiration, dry at 35° (J. Chromatogr. 1985, 347, 191).
Capillary temperature: 25
Running buffer: 100 mM pH 2.5 NaH_2PO_4 (A) or 100 mM pH 2.5 NaH_2PO_4 containing 45 mM hydroxypropyl-α-cyclodextrin (Wacker, Munich) (B)
Injection: Electromigration at 15 kV for 3 s.
Detector: UV 200; UV 210
Migration time: 4.90 (A); 8.96, 9.29 (B) (enantiomers)
Voltage: 15 kV
Model: Bio-Focus 3000

OTHER SUBSTANCES

Also analyzed: albuterol (salbutamol), alprenolol, amorolfine, atenolol, atropine, azelastine, baclofen, bamethan, benproperine, benserazide, biperiden, bisoprolol, brompheniramine, bupivacaine, bupranolol, butamirate, butethamate, carazolol, carbuterol, carteolol, carvedilol, celiprolol, chloroquine, chlorpheniramine, chlorphenoxamine, cicletanine, clenbuterol, clidinium bromide, clobutinol, dimethindene, dipivefrin, disopyramide, dobutamine, doxylamine, fendi-

line, flecainide, gallopamil, homatropine, ipratropium bromide, isoproterenol (isoprenaline), isothipendyl, ketamine, meclizine, mefloquine, mepindolol, mequitazine, metaclazepam, metaproterenol (orciprenaline), metipranolol, metoprolol, nafronyl (naftidrofuryl), nicardipine, norfenefrine, ofloxacin, ornidazole, orphenadrine, oxomemazine, oxprenolol, oxybutynin, phenoxybenzamine, phenylpropanolamine, pholedrine, pindolol, pirbuterol, prilocaine, procyclidine, promethazine, propafenone, propranolol, reproterol, sotalol, sulpride, synephrine, talinolol, terbutaline, tetrahydrozoline (tetryzoline), theodrenaline, tioconazole, tocainide, trihexyphenidyl, trimeprazine (alimemazine), trimipramine, tropicamide, verapamil, zopiclone

KEY WORDS
coated capillary; chiral

REFERENCE
Koppenhoefer,B.; Eperlein,U.; Schlunk,R.; Zhu,X.; Lin,B. Separation of enantiomers of drugs by capillary electrophoresis. V. Hydroxypropyl-α-cyclodextrin as chiral solvating agent, *J.Chromatogr.A*, **1998**, *793*, 153–164.

Neomycin

Molecular formula: $C_{23}H_{46}N_6O_{13}$ (neomycin B)
Molecular weight: 614.65 (neomycin B)
CAS Registry No.: 1404-04-2, 1406-04-8 (undecylenate), 1405-10-3 (sulfate), 1405-12-5 (palmitate)
Merck Index (12th ed.): 6542

SAMPLE
Matrix: solutions
Sample preparation: Prepare a solution in 100 mM cetyltrimethylammonium bromide, inject an aliquot.

CAPILLARY ELECTROPHORESIS
Capillary: 67 cm × 50 μm fused silica capillary (60 cm to detector) (Siemens)
Running buffer: pH 5.0 Acetate buffer containing 10 mM imidazole and 50 μg/mL FC 135 (a fluorinated surfactant, Fluorad/3M)
Injection: Pressure injection for 2 s.
Detector: UV 214
Migration time: 12.5
Voltage: 12.5 kV
Model: Beckman P/ACE System 2000

OTHER SUBSTANCES
Simultaneous: amikacin, dihydrostreptomycin, gentamicin, kanamycin, lividomycin, paromomycin, sisomycin, streptomycin,
Interfering: tobramycin

KEY WORDS
indirect UV detection; detector is at anode

REFERENCE
Ackermans,M.T.; Everaerts,F.M.; Beckers,J.L. Determination of aminoglycoside antibiotics in pharmaceuticals by capillary zone electrophoresis with indirect UV detection coupled with micellar electrokinetic capillary chromatography, *J.Chromatogr.*, **1992**, *606*, 229–235.

Neostigmine bromide

Molecular formula: $C_{12}H_{19}BrN_2O_2$
Molecular weight: 303.20
CAS Registry No.: 114-80-7, 59-99-4 (neostigmine),
51-60-5 (neostigmine methylsulfate)
Merck Index (12th ed.): 6553

SAMPLE
Matrix: solutions
Sample preparation: Inject an aliquot of a solution in 0.9% NaCl.

CAPILLARY ELECTROPHORESIS
Capillary: 43 cm × 50 μm fused-silica (35 cm to detector) (Polymicro Technologies)
Capillary preparation: Before each run rinse with running buffer for 2 min. Condition new
   capillaries with 100 mM NaOH at 50° for 30 min then equilibrate with running buffer for 2 h.
Capillary temperature: 30
Running buffer: MeCN:buffer 10:100 (Dissolve 17.62 g 2-(N-cyclohexylamino)ethanesulfonic
   acid (CHES) in 100 mL 4.1 mg/mL sodium acetate in water, add 10 g MeCN, adjust pH to 7.8
   with triethylamine. Cover running buffer with a film of mineral oil to prevent MeCN
   evaporation.)
Injection: Pressure injection at 0.75 psi for 4 s.
Detector: UV 214
Migration time: 2.5
Voltage: 20 kV
Current: 32 μA
Model: SpectraPhoresis 1000 (Thermo Separation Products)

OTHER SUBSTANCES
Simultaneous: insulin

REFERENCE
Kunkel,A.; Günter,S.; Dette,C.; Wätzig,H. Quantitation of insulin by capillary electrophoresis and high-performance liquid chromatography. Method comparison and validation, *J.Chromatogr.A*, **1997**, *781*, 445–455.

SAMPLE
Matrix: solutions

CAPILLARY ELECTROPHORESIS
Capillary: 72 cm × 75 μm fused-silica Ucon-coated (57 cm to detector) (Polymicro Technologies)
Capillary preparation: Coat capillaries as follows.Rinse capillary with 1 mL dichloromethane,
   purge with nitrogen for 1 h, fill with 4 mg/mL Ucon 75 H-90,000 polymer (Alltech) in dichloromethane containing 2 mg/mL hexamethyldisilazane and 0.2 mg/mL dicumyl peroxide. Let
   stand briefly then purge with nitrogen for 30 min, seal at both ends with an oxy-acetylene
   flame without allowing air to enter the column, heat from 40° to 150° at 4°/min, maintain at
   150° for 40 min. Before use rinse with 1 mL dichloromethane, 1 mL MeOH, 1 mL water, 1 mL
   pH 2.5 phosphate buffer, 1 mL water, and 1 mL running buffer (J. Microcol. Sep. 1993, 5, 119).
Running buffer: 10 mM Triethylamine adjusted to pH 5.0 with acetic acid
Injection: 4.1 nL
Detector: UV 200
Migration time: 13
Voltage: 20 kV
Model: Model 300 Crystal CE (Thermo Bioanalysis, Franklin MA)

OTHER SUBSTANCES
Simultaneous: propantheline

KEY WORDS
coated capillary

REFERENCE
Chen,S.; Lee,M.L. Counterflow isotachopheresis-capillary zone electrophoresis on directly coupled columns of different diameters, *Anal.Chem.*, **1998**, *70*, 3777–3780.

Netilmicin

Molecular formula: $C_{21}H_{41}N_5O_7$
Molecular weight: 475.59
CAS Registry No.: 56391-56-1, 56391-57-2 (sulfate)
Merck Index (12th ed.): 6563

SAMPLE
Matrix: solutions
Sample preparation: Inject an aliquot of a 100 μM solution in buffer. (Prepare buffer by mixing equal volumes of 20 mM NaH_2PO_4 and 20 mM sodium tetraborate and adjusting the pH to 10 with 1 M NaOH.)

CAPILLARY ELECTROPHORESIS
Capillary: 110 cm × 50 μm fused-silica (100 cm to detector)
Capillary preparation: Before injection rinse capillary with running buffer at 200 mbar with no voltage for 3 min and with running buffer at 4 mbar at 20 kV for 30 s.
Capillary temperature: 40
Running buffer: 100 mM pH 10 phosphate borate buffer containing 860 μM 1-methoxycarbonylindolizine-3,5-dicarbaldehyde (Prepare 1-methoxycarbonylindolizine-3,5-dicarbaldehyde as follows. Reflux 21.4 g 2-pyridinecarboxaldehyde, 24 mL ethylene glycol, 10 g p-toluenesulfonic acid, and 300 mL benzene (Caution! Benzene is a carcinogen!) under a Dean-Stark separator for 64 h, pour into concentrated sodium carbonate solution. Remove the organic layer and extract the aqueous layer 4 times with benzene. Combine the organic layers and wash them with water, dry over anhydrous magnesium sulfate, evaporate, distil the residue to give 2-(1,3-dioxolan-2-yl)pyridine (bp 122°/4 mm Hg) (J.Org.Chem 1963, 28, 83). Reflux 15.1 g 2-(1,3-dioxolan-2-yl)pyridine and 19.5 g tert-butyl bromoacetate in 100 mL dry acetonitrile for 7 h, let stand overnight at room temperature, filter, wash the precipitate with diethyl ether to give 1-(tert-butoxycarbonylmethyl)-2-(1,3-dioxolan-2-yl)pyridinium bromide (mp 110-2° from MeCN). Suspend 51.9 g of this compound in 1. 5 L THF with stirring, add 62.1 g potassium carbonate, add 15.12 g methyl propiolate, stir at room temperature for 9 days, filter, evaporate the filtrate to dryness under reduced pressure, chromatograph the residue on silica gel with hexane:ethyl acetate 20:1-10:1, collect fractions and evaporate to dryness to give methyl 3-tert-butoxycarbonyl-5-(1,3-dioxolan-2-yl)indolizine-1-carboxylate (mp 138-9° from hexane). Reflux 20.82 g of this compound in 600 mL THF and 60 mL 10% HCl for 6 h, concentrate to one quarter of the original volume, add water, extract with chloroform. Wash the chloroform layer with water and dry it over anhydrous sodium sulfate, evaporate, chromatograph on silica gel with hexane:ethyl acetate 10:1 to give 1-methoxycarbonylindolizine-5-carbaldehyde (mp 135-7° from MeOH). Stir 12.18 g of this compound in 116 mL dry DMF at 0° under argon, add 17 mL phosphorus oxychloride, stir at room temperature for 1 h, pour into water, adjust pH to 9.0 with 5% potassium carbonate, extract with chloroform. Wash the organic layer with water and dry it over anhydrous sodium sulfate, concentrate until a precipitate forms, filter to obtain the product, concentrate the filtrate and chromatograph the residue on silica gel with hexane:ethyl acetate 5:1 to obtain more product. The product was 1-methoxycarbonylindolizine-3,5-dicarbaldehyde (mp 164-5° from methyl acetate) (J. Chromatogr. A 1996, 724, 169.))
Injection: Pressure injection at 50 mbar for 6 s (with no applied voltage).
Detector: UV 409
Migration time: 21
Voltage: 20 kV
Model: JASCO Model CE-990

OTHER SUBSTANCES
Simultaneous: amikacin, arbekacin, micromicin

KEY WORDS
derivatization; on-column derivatization

REFERENCE
Oguri,S.; Fujiyoshi,T.; Miki,Y. In-capillary derivatization with 1-methoxycarbonylindolizine-3,5-dicarbaldehyde for high-performance capillary electrophoresis, *Analyst*, **1996**, *121*, 1683–1688.

Niacin

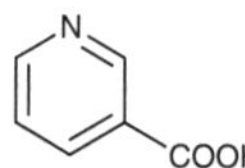

Molecular formula: $C_6H_5NO_2$
Molecular weight: 123.11
CAS Registry No.: 59-67-6
Merck Index (12th ed.): 6612

SAMPLE
Matrix: blood, hemodialysate fluid
Sample preparation: Serum. 1 mL Serum + 20 μL 100 mM phenyltriethylammonium iodide, filter (Amicon Centrifree Micropartition System, cut-off 30000) while centrifuging at 311 g for 1 h, inject an aliquot of the ultrafiltrate. Hemodialysate fluid. 9 mL Hemodialysate fluid + 1 mL 20 mM phenyltriethylammonium iodide, inject an aliquot.

CAPILLARY ELECTROPHORESIS
Capillary: 64 cm × 50 μm extended light path capillary (55.8 cm to detector) (Hewlett-Packard)
Capillary temperature: 25
Running buffer: 150 mM Boric acid adjusted to pH 9.0 with 1 M NaOH
Injection: Pressure injection at 50 mbar for 5 s
Detector: UV 210
Migration time: 11.3
Internal standard: phenyltriethylammonium iodide (3.5)
Voltage: 22 kV
Current: 32.5-34.5 μA
Model: Hewlett-Packard [3D]Capillary Electrophoresis System
Limit of detection: 2.4 μM (S/N 3)

OTHER SUBSTANCES
Extracted: allopurinol, ferulic acid, hippuric acid, 2-hydroxyhippuric acid, 4-hydroxyphenylacetic acid, hypoxanthine, indican, indole-3-acetic acid, kynurenic acid, 1-methylniacinamide, theobromine, theophylline, tryptophan, tyrosine, uracil, uric acid, xanthine

KEY WORDS
serum; ultrafiltrate

REFERENCE
Petucci,C.J.; Kantes,H.L.; Strein,T.G.; Veening,H. Capillary electrophoresis as a clinical tool. Determination of organic anions in normal and uremic serum using photodiode-array detection, *J.Chromatogr.B*, **1995**, *668*, 241–251.

SAMPLE
Matrix: formulations
Sample preparation: Grind tablet, dissolve in pH 5 citrate buffer, filter (0.2 μm), inject an aliquot.

CAPILLARY ELECTROPHORESIS
Capillary: 48.5 cm × 50 μm fused-silica (40 cm to detector)

Capillary preparation: Between runs flush with 100 mM NaOH for 2 min and with run buffer
for 3 min, inject a buffer plug. Before the first use condition capillaries with 1 M NaOH for 10
min, with 100 mM NaOH for 2 min, and with run buffer for 3 min.
Capillary temperature: 25
Running buffer: 20 mM pH 7 sodium phosphate
Injection: Injection at 40 mbar for 4.6 s (post-injection pressure (of run buffer) 40 mbar for 4 s)
Detector: UV 215
Migration time: 15.3
Voltage: 20 kV
Model: Hewlett-Packard

OTHER SUBSTANCES
Simultaneous: folinic acid, niacinamide, orotic acid, pantothenic acid, pyridoxine, thiamine, vitamin B12 (cyanocobalamin) (UV 360), ascorbic acid

KEY WORDS
tablets; buffer may interfere

REFERENCE
Jegle,U. Separation of water-soluble vitamins via high-performance capillary electrophoresis, *J.Chromatogr.A*,
1993, *652*, 495–501.

SAMPLE
Matrix: solutions
Sample preparation: Prepare a 0.5-2 mg/mL solution in water, inject an aliquot.

CAPILLARY ELECTROPHORESIS
Capillary: 65 cm × 50 μm fused-silica (50 cm to detector) (Scientific Glass Engineering)
Running buffer: 20 mM NaH_2PO_4 containing 50 mM sodium dodecyl sulfate adjusted to pH 9.0
with 20 mM sodium tetraborate
Injection: Injection by siphon at 5 cm for 5-10 s (about 1 nL)
Detector: UV 210
Migration time: 8.4
Voltage: 25 kV

OTHER SUBSTANCES
Simultaneous: niacinamide, pyridoxal, pyridoxamine, thiamine, riboflavin, pyridoxine, vitamin
B12

REFERENCE
Nishi,H.; Tsumagari,N.; Kakimoto,T.; Terabe,S. Separation of water-soluble vitamins by micellar electrokinetic
chromatography, *J.Chromatogr.*, **1989**, *465*, 331–343.

SAMPLE
Matrix: solutions
Sample preparation: Prepare a 100 μg/mL solution in MeOH:water 50:50, inject an aliquot.

CAPILLARY ELECTROPHORESIS
Capillary: 60 cm × 75 μm
Running buffer: 20 mM NaH_2PO_4 containing 50 mM sodium dodecyl sulfate, adjusted to pH 9.0
with sodium tetraborate
Injection: Hydrodynamic injection at 10 cm for 5 s
Detector: UV 214
Migration time: 10.5
Voltage: 18 kV
Model: Waters Quanta 4000

OTHER SUBSTANCES
Simultaneous: pantothenic acid, thiamine, riboflavin, pyridoxine, ascorbic acid

---

REFERENCE
Swartz,M.E. Method development and selectivity control for small molecule pharmaceutical separations by capillary electrophoresis, *J.Liq.Chromatogr.*, **1991**, *14*, 923–938.

SAMPLE
Matrix: solutions

CAPILLARY ELECTROPHORESIS
Capillary: 56.9 cm × 75 μm fused-silica (50 cm to detector)
Capillary temperature: 25
Running buffer: MeOH:60 mM pH 8.92 sodium borate buffer containing 60 mM sodium dodecyl sulfate 15:85
Detector: UV 214
Migration time: 8.5
Voltage: 30 kV
Model: Beckman PACE 2100

OTHER SUBSTANCES
Simultaneous: niacinamide, pyridoxine, thiamine, vitamin B12

REFERENCE
McLaughlin,G.M.; Nolan,J.A.; Lindahl,J.L.; Palmieri,R.H.; Anderson,K.W.; Morris,S.C.; Morrison,J.A.; Bronzert,T.J. Pharmaceutical drug separations by HPCE: Practical guidelines, *J.Liq.Chromatogr.*, **1992**, *15*, 961–1021.

SAMPLE
Matrix: solutions
Sample preparation: Prepare a 100 μg/mL solution in MeOH:water 50:50, inject an aliquot.

CAPILLARY ELECTROPHORESIS
Capillary: 47 cm × 50 μm fused-silica (40 cm to detector) (Polymicro Technologies)
Capillary temperature: 25
Running buffer: Prepare as follows. Mix n-hexane with 2 volumes of 500 mM sodium dodecyl sulfate, vortex, add 1-butanol until the mixture clears, dilute with 20 mM pH 7.0 phosphate buffer until the concentration of sodium dodecyl sulfate is 20 mM.
Injection: Inject using an overpressure of 34500000 Pa (sic) for 3 s
Detector: UV 214
Migration time: 6.3
Voltage: 10 kV
Current: about 17.5 μA
Model: Beckman P/ACE System 2100

OTHER SUBSTANCES
Simultaneous: niacinamide, pyridoxine (pyridoxol), thiamine, vitamin A, vitamin E

REFERENCE
Boso,R.L.; Bellini,M.S.; Miksík,; Deyl,Z. Microemulsion electrokinetic chromatography with different organic modifiers: separation of water- and lipid-soluble vitamins, *J.Chromatogr.A*, **1995**, *709*, 11–19.

SAMPLE
Matrix: solutions

CAPILLARY ELECTROPHORESIS
Capillary: 64.5 cm × 75 μm (56 cm to detector)
Capillary preparation: Between each run replenish inlet and outlet vials, wash capillary with fresh running buffer for 1 min, apply 30 kV for 2 min.
Capillary temperature: 25
Running buffer: 100 mM pH 8.5 Borate buffer
Injection: Pressure injection at 35 mbar for 5 s
Detector: UV 254
Migration time: 4.4

Voltage: 30 kV
Model: Hewlett-Packard HP3D CE

___

OTHER SUBSTANCES
Simultaneous: acetaminophen, aspirin, 2,4-dihydroxybenzoic acid

___

REFERENCE
Ross,G.A. Voltage pre-conditioning technique for optimization of migration-time reproducibility in capillary
electrophoresis, *J.Chromatogr.A*, **1995**, *718*, 444–447.

___

SAMPLE
Matrix: solutions
Sample preparation: Inject an aliquot of an aqueous solution.

___

CAPILLARY ELECTROPHORESIS
Capillary: 70 cm × 50 μm fused-silica (50 cm to detector) (GL Science)
Running buffer: 20 mM pH 9.1 Borate buffer containing 10 mM KOH and 150 mM sodium
dodecyl sulfate
Injection: Hydrostatic injection at 4 cm
Detector: UV 185, UV 210
Migration time: 14
Voltage: 15 kV
Current: 30 μA

___

OTHER SUBSTANCES
Simultaneous: 3-acetylpyridine, 6-aminoniacinamide, ethyl nicotinate, isoniazid, N-methylniacinamide, niacinamide, β-picoline, pyridine, pyridine-3-aldehyde, 3-pyridinemethanol, pyri-
dine-3-sulfonic acid, thioniacinamide

___

REFERENCE
Tanaka,S.; Kodama,K.; Kaneta,T.; Nakamura,H. Migration behavior of niacin derivatives in capillary electrophoresis, *J.Chromatogr.A*, **1995**, *718*, 233–237.

___

SAMPLE
Matrix: solutions

___

CAPILLARY ELECTROPHORESIS
Capillary: 37 cm × 50 μm coated (30 cm to detector) (Polymicro Technologies)
Capillary preparation: The capillary was coated with poly(N-(acryloylaminoethoxyethanol)-β-
D-glucoyranoside). Partial details are given in the referenced papers and the references quoted
therein.
Running buffer: 50 mM pH 5.5 2-(N-morpholino)ethanesulfonic acid (MES)-Tris buffer contain-
ing 1 mM copper sulfate
Injection: Electrokinetic injection for 1 s.
Detector: UV 214
Migration time: 4.5
Voltage: 15 kV
Model: Waters Quanta 4000

___

OTHER SUBSTANCES
Simultaneous: N-acryloyl-gamma-aminobutyric acid, N-acryloylglycine, N-acryloyl glycolic acid,
benzoic acid, caffeic acid, phthalic acid, p-toluenesulfonic acid

___

KEY WORDS
reversed polarity; coated capillary

___

REFERENCE
Chiari,M.; Dell'Orto,N.; Casella,L. Separation of organic acids by capillary zone electrophoresis in buffers containing divalent metal cations, *J.Chromatogr.A*, **1996**, *745*, 93–101.

SAMPLE
Matrix: solutions

CAPILLARY ELECTROPHORESIS
Capillary: 60 cm × 75 μm AccuSep (Waters) or 52.5 cm × 75 μm (45 cm to detector) (SGE)
Running buffer: 20 mM pH 7 Phosphate buffer
Injection: Hydrodynamic injection for 7 s
Detector: UV 214
Migration time: 14 (AccuSep), 15.7 (SGE)
Voltage: 15 kV
Model: Waters Quanta 4000

OTHER SUBSTANCES
Simultaneous: niacinamide, thiamine

KEY WORDS
sleeve cell increases sensitivity

REFERENCE
Djordjevic,N.M.; Ryan,K. An easy way to enhance absorbance detection sensitivity of Waters Quanta-4000 capillary electrophoresis system, *J.Liq.Chromatogr.Rel.Technol.*, **1996**, *19*, 201–206.

SAMPLE
Matrix: solutions
Sample preparation: Inject an aliquot of a solution in 1% acetic acid.

CAPILLARY ELECTROPHORESIS
Capillary: 48.5 cm × 50 μm fused-silica (40 cm to detector) (Hewlett Packard)
Capillary preparation: After each run wash with 100 mM NaOH for 7 min, with water for 1 min, and with running buffer for 3 min. At the start of each day wash with separation buffer for 10 min. Wash new capillaries with 1 M NaOH, 100 mM NaOH, water, and running buffer.
Capillary temperature: 25
Running buffer: 50 mM Borax adjusted to pH 8.5 with boric acid
Injection: Hydrodynamic injection at 5 kPa for 10 s.
Detector: UV 225
Migration time: 7.5
Voltage: 25 kV
Model: Hewlett Packard [3D]CE

OTHER SUBSTANCES
Simultaneous: adenine, biotin, niacinamide, pantothenic acid (UV 215), pyridoxine (UV 215), rutin, thiamine, riboflavin, ascorbic acid

---

REFERENCE
Fotsing,L.; Fillet,M.; Bechet,I.; Hubert,P.; Crommen,J. Determination of six water-soluble vitamins in a pharmaceutical formulation by capillary electrophoresis, *J.Pharm.Biomed.Anal.*, **1997**, *15*, 1113–1123.

SAMPLE
Matrix: solutions

CAPILLARY ELECTROPHORESIS
Capillary: 70 cm × 50 μm fused-silica (GL Science, Tokyo)
Capillary preparation: Before each run rinse capillary with running buffer at 94 kPa for 5 min before each run.
Running buffer: 50 mM pH 8.5 ammonium carbonate buffer
Injection: Pressure injection at 5 kPa (50 mbar) for 4 s.
Detector: MS, Perkin-Elmer Sciex API-300 quadrupole, electrospray (ionspray) interface, sheath liquid MeOH:running buffer 50:50 at 2.5 μL/min, ionspray voltage 5 kV, negative ion mode
Migration time: 12
Voltage: 20 kV (net voltage across capillary = 15 kV (applied voltage - electrospray voltage))
Model: Hewlett Packard 3D CE

OTHER SUBSTANCES
Simultaneous: acetaminophen, ascorbic acid, butylscopolamine bromide, caffeine, ibuprofen, ketoprofen, niacinamide, riboflavin, thiamine, vitamin B12, warfarin

REFERENCE
Tanaka,Y.; Kishimoto,Y.; Otsuka,K.; Terabe,S. Strategy for selecting separation solutions in capillary electrophoresis-mass spectrometry, *J.Chromatogr.A*, **1998**, *817*, 49–57.

Niacinamide

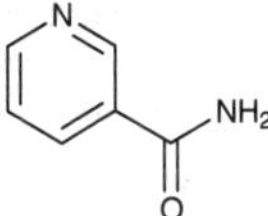

Molecular formula: $C_6H_6N_2O$
Molecular weight: 122.13
CAS Registry No.: 98-92-0
Merck Index (12th ed.): 6574

SAMPLE
Matrix: blood
Sample preparation: 0.5 mL Plasma + 10 µL 1 mg/mL barbital, mix, add 3 mL acetone:water 2:1, centrifuge. Remove the supernatant and evaporate it to dryness, reconstitute the residue in 500 µL 100 mM HCl, add 1 mL, evaporate to dryness, reconstitute in 200 µL water, inject an aliquot.

CAPILLARY ELECTROPHORESIS
Capillary: 57 cm × 50 µm unmodified fused-silica (51 cm to detector)
Capillary temperature: 25
Running buffer: 10 mM pH 9.36 sodium tetraborate
Injection: Pneumatic injection for 7 s
Detector: UV 254
Migration time: 2.97
Internal standard: barbital (4.61)
Voltage: 25 kV
Model: Beckman P/ACE 2100

OTHER SUBSTANCES
Extracted: niacin

KEY WORDS
plasma

REFERENCE
Zarzycki,P.K.; Kowalski,P.; Nowakowska,J.; Lamparczyk,H. High-performance liquid chromatographic and capillary electrophoretic determination of free nicotinic acid in human plasma and separation of its metabolites by capillary electrophoresis, *J.Chromatogr.A*, **1995**, *709*, 203–208.

SAMPLE
Matrix: bulk
Sample preparation: Dilute with 400 µg/mL n-propyl p-hydroxybenzoate in running buffer

CAPILLARY ELECTROPHORESIS
Capillary: 27 cm × 50 µm fused-silica (20 cm to detector)
Capillary preparation: Before each run rinse with running buffer at high pressure.
Capillary temperature: 30
Running buffer: MeCN:water 15:85 containing 40 mM sodium dodecyl sulfate, 8.5 mM sodium borate, and 8.5 mM sodium phosphate, pH 8.5
Injection: Inject at high pressure for 1 s
Detector: UV 214
Migration time: 0.95

Internal standard: n-propyl p-hydroxybenzoate (1.8)
Voltage: 20 kV
Model: Beckman P/ACE System 2100

OTHER SUBSTANCES
Simultaneous: acetaminophen, acetylcodeine, O-acetylmorphine, aspirin, caffeine, cocaine, codeine, diamorphine, diphenhydramine, hydromorphone, isoniacinamide, lidocaine, methaqualone, morphine, noscapine, papaverine, phenacetin, phenobarbital, phenylpropanolamine, procaine, quinine, salicylic acid, strychnine, thebaine

REFERENCE
Walker,J.A.; Krueger,S.T.; Lurie,I.S.; Marché,H.L.; Newby,N. Analysis of heroin drug seizures by micellar electrokinetic capillary chromatography (MECC), *J.Forensic Sci.*, **1995**, *40*, 6–9.

SAMPLE
Matrix: bulk
Sample preparation: Dissolve in running buffer to a concentration of 1 mg/mL, vortex for 2 min, add an equal volume of 100 μg/mL diphenhydramine in running buffer, mix, inject an aliquot.

CAPILLARY ELECTROPHORESIS
Capillary: 65 cm × 50 μm fused-silica (60 cm to detector)
Capillary preparation: Fill capillary with fresh running buffer before each run. Before use fill with 100 mM NaOH for 20 min, rinse with water, flush with running buffer
Running buffer: MeCN:50 mM 6-aminocaproic acid containing 50 mM 3-N,N-dimethylmyristylammoniopropanesulfonate (MAPS; Fluka) and 5 mM 1-heptanesulfonic acid 10:90, pH adjusted to 4.0 with 1 M phosphoric acid
Injection: Hydrodynamic injection by gravity or pressure.
Detector: UV 214
Migration time: 7.9
Internal standard: diphenhydramine (12)
Voltage: 27 kV
Current: ≤25 μA
Model: Dionex system I

OTHER SUBSTANCES
Simultaneous: acetaminophen, allobarbital, barbital, caffeine, codeine, diamorphine, morphine, noscapine, papaverine, phenobarbital, procaine

REFERENCE
Naess,O.; Rasmussen,K.E. Micellar electrokinetic chromatography of charged and neutral drugs in acidic running buffers containing a zwitterionic surfactant, sulfonic acids or sodium dodecyl sulphate. Separation of heroin, basic by-products and adulterants, *J.Chromatogr.A*, **1997**, *760*, 245–251.

SAMPLE
Matrix: formulations
Sample preparation: Grind tablet, dissolve in pH 5 citrate buffer, filter (0.2 μm), inject an aliquot.

CAPILLARY ELECTROPHORESIS
Capillary: 48.5 cm × 50 μm fused-silica (40 cm to detector)
Capillary preparation: Between runs flush with 100 mM NaOH for 2 min and with run buffer for 3 min, inject a buffer plug. Before the first use condition capillaries with 1 M NaOH for 10 min, with 100 mM NaOH for 2 min, and with run buffer for 3 min.
Capillary temperature: 25
Running buffer: 20 mM pH 7 sodium phosphate
Injection: Injection at 40 mbar for 4.6 s (post-injection pressure (of run buffer) 40 mbar for 4 s)
Detector: UV 215
Migration time: 6.2
Voltage: 20 kV
Model: Hewlett-Packard

OTHER SUBSTANCES
Simultaneous: folinic acid, niacin, orotic acid, pantothenic acid, pyridoxine, thiamine, vitamin B12 (cyanocobalamin) (UV 360), ascorbic acid

KEY WORDS
tablets

REFERENCE
Jegle,U. Separation of water-soluble vitamins via high-performance capillary electrophoresis, *J.Chromatogr.A*, **1993**, *652*, 495–501.

SAMPLE
Matrix: formulations
Sample preparation: Condition a 500 mg Bakerbond C18 SPE cartridge with 3 mL MeOH and 3 mL water:10 mM sodium heptanesulfonate in acetic acid 99:1. Heat 50-100 mg tablet and 40 mL water:10 mM sodium heptanesulfonate in acetic acid 99:1 at 55° with occasional shaking for 5 min, cool, add a 2 mL aliquot to the SPE cartridge, elute with 3.5 mL MeOH, make up the eluate to 10 mL with water, inject an aliquot.

CAPILLARY ELECTROPHORESIS
Capillary: 70 cm × 100 μm
Capillary temperature: 25
Running buffer: MeOH:buffer 10:90 (Buffer was 50 mM pH 8.0 sodium borate containing 22.5 mM sodium dodecyl sulfate.)
Injection: Inject under 3.44 kPa pressure for 5 s (30 nL)
Detector: UV 214
Migration time: 15
Voltage: 16 kV
Current: 33 μA
Model: Beckman P/ACE System 2000

OTHER SUBSTANCES
Simultaneous: pantothenic acid, pyridoxine, riboflavin, thiamine, vitamin B12, ascorbic acid

KEY WORDS
tablets; SPE

REFERENCE
Dinelli,G.; Bonetti,A. Micellar electrokinetic capillary chromatography analysis of water-soluble vitamins and multi-vitamin integrators, *Electrophoresis*, **1994**, *15*, 1147–1150.

SAMPLE
Matrix: formulations
Sample preparation: Protect from light. Grind tablet, add 50 mL 1% acetic acid, shake at 65° for 10 min, add 5 mL 10 mg/mL niacin in 1% acetic acid, cool to room temperature, make up to 100 mL with 1% acetic acid, filter (0.22 μm cellulose acetate), dilute 4-12-fold, inject an aliquot.

CAPILLARY ELECTROPHORESIS
Capillary: 48.5 cm × 50 μm fused-silica (40 cm to detector) (Hewlett Packard)
Capillary preparation: After each run wash with 100 mM NaOH for 7 min, with water for 1 min, and with running buffer for 3 min. At the start of each day wash with separation buffer for 10 min. Wash new capillaries with 1 M NaOH, 100 mM NaOH, water, and running buffer.
Capillary temperature: 25
Running buffer: 50 mM Borax adjusted to pH 8.5 with boric acid
Injection: Hydrodynamic injection at 5 kPa for 10 s.
Detector: UV 225
Migration time: 2.7
Internal standard: niacin (7.5)
Voltage: 25 kV
Model: Hewlett Packard [3D]CE

OTHER SUBSTANCES
Simultaneous: adenine, biotin, pantothenic acid (UV 215), pyridoxine (UV 215), rutin, thiamine, riboflavin, ascorbic acid

KEY WORDS
tablets

REFERENCE
Fotsing,L.; Fillet,M.; Bechet,I.; Hubert,P.; Crommen,J. Determination of six water-soluble vitamins in a pharmaceutical formulation by capillary electrophoresis, *J.Pharm.Biomed.Anal.*, **1997**, *15*, 1113–1123.

SAMPLE
Matrix: formulations
Sample preparation: Grind tablet to a homogeneous powder and dissolve in 2 mL water and 20 mL EtOH, homogenize (Ultra Turrax), centrifuge. Evaporate the supernatant to dryness under reduced pressure, reconstitute the residue with water, inject an aliquot.

CAPILLARY ELECTROPHORESIS
Capillary: 64.5 cm × 50 μm fused-silica (56 cm to detector)
Capillary preparation: Before each run flush capillary at 4 kPa with 1 M NaOH for 2 min and with buffer for 5 min. Replace buffer before each analysis.
Capillary temperature: 30
Running buffer: 1-Propanol:buffer 2:98 (Buffer was 100 mM Na_2HPO_4, 500 mM taurine, and 75 mM sodium cholate.)
Injection: Pressure injection at 4 kPa for 1 s.
Detector: UV 214
Migration time: 10.5
Voltage: 17 kV
Model: Hewlett-Packard HP[3D] CE
Limit of quantitation: 170 μM

OTHER SUBSTANCES
Simultaneous: ascorbic acid, biotin, folic acid, pyridoxamine, pyridoxine, riboflavin, thiamine, vitamin B12

KEY WORDS
tablets

REFERENCE
Buskov,S.; Moller,P.; Sorensen,H.; Sorensen,J.C.; Sorensen,S. Determination of vitamins in food based on supercritical fluid extraction prior to micellar electrokinetic capillary chromatographic analyses of individual vitamins, *J.Chromatogr.A*, **1998**, *802*, 233–241.

SAMPLE
Matrix: formulations, juice
Sample preparation: Dissolve in water, filter (0.2 μm), inject an aliquot. (Add a 50-fold excess of cysteine to solutions containing ascorbic acid.)

CAPILLARY ELECTROPHORESIS
Capillary: 60 cm × 50 μm fused-silica (51.5 cm to detector)
Capillary preparation: Condition with running buffer for 3 min before each run. Condition a new capillary 1 M NaOH for 10 min and with 100 mM NaOH for 2 min.
Capillary temperature: 25
Running buffer: 20 mM pH 8.0 Phosphate buffer
Injection: Pressure injection for 200 mbar.s
Detector: UV 200
Migration time: 3.9
Voltage: -30 kV
Model: Hewlett-Packard 3DCE
Limit of detection: 1.6 μg/mL

OTHER SUBSTANCES
Simultaneous: biotin, pyridoxine, riboflavin phosphate, thiamine, ascorbic acid

KEY WORDS
fruit

REFERENCE
Schiewe,J.; Mrestani,Y.; Neubert,R. Application and optimization of capillary zone electrophoresis in vitamin analysis, *J.Chromatogr.A*, **1995**, *717*, 255–259.

SAMPLE
Matrix: solutions
Sample preparation: Prepare a 0.5-2 mg/mL solution in water, inject an aliquot.

CAPILLARY ELECTROPHORESIS
Capillary: 65 cm × 50 μm fused-silica (50 cm to detector) (Scientific Glass Engineering)
Running buffer: 20 mM NaH_2PO_4 containing 50 mM sodium dodecyl sulfate adjusted to pH 9.0 with 20 mM sodium tetraborate
Injection: Injection by siphon at 5 cm for 5-10 s (about 1 nL)
Detector: UV 210
Migration time: 4.4
Voltage: 25 kV

OTHER SUBSTANCES
Simultaneous: niacin, pyridoxal, pyridoxamine, pyridoxine, riboflavin, thiamine, vitamin B12

REFERENCE
Nishi,H.; Tsumagari,N.; Kakimoto,T.; Terabe,S. Separation of water-soluble vitamins by micellar electrokinetic chromatography, *J.Chromatogr.*, **1989**, *465*, 331–343.

SAMPLE
Matrix: solutions

CAPILLARY ELECTROPHORESIS
Capillary: 56.9 cm × 75 μm fused-silica (50 cm to detector)
Capillary temperature: 25
Running buffer: MeOH:60 mM pH 8.92 sodium borate buffer containing 60 mM sodium dodecyl sulfate 15:85
Detector: UV 214
Migration time: 4
Voltage: 30 kV
Model: Beckman PACE 2100

OTHER SUBSTANCES
Simultaneous: niacin, pyridoxine, thiamine, vitamin B12

REFERENCE
McLaughlin,G.M.; Nolan,J.A.; Lindahl,J.L.; Palmieri,R.H.; Anderson,K.W.; Morris,S.C.; Morrison,J.A.; Bronzert,T.J. Pharmaceutical drug separations by HPCE: Practical guidelines, *J.Liq.Chromatogr.*, **1992**, *15*, 961–1021.

SAMPLE
Matrix: solutions
Sample preparation: Prepare a 100 μg/mL solution in MeOH:water 50:50, inject an aliquot.

CAPILLARY ELECTROPHORESIS
Capillary: 47 cm × 50 μm fused-silica (40 cm to detector) (Polymicro Technologies)
Capillary temperature: 25
Running buffer: Prepare as follows. Mix n-hexane with 2 volumes of 500 mM sodium dodecyl sulfate, vortex, add 1-butanol until the mixture clears, dilute with 20 mM pH 7.0 phosphate buffer until the concentration of sodium dodecyl sulfate is 20 mM.

Injection: Inject using an overpressure of 34500000 Pa (sic) for 3 s
Detector: UV 214
Migration time: 14.7
Voltage: 10 kV
Current: about 17.5 μA
Model: Beckman P/ACE System 2100

OTHER SUBSTANCES
Simultaneous: niacin, pyridoxine (pyridoxol), thiamine, vitamin A, vitamin E

REFERENCE
Boso,R.L.; Bellini,M.S.; Miksík,; Deyl,Z. Microemulsion electrokinetic chromatography with different organic
modifiers: separation of water- and lipid-soluble vitamins, *J.Chromatogr.A*, **1995**, *709*, 11–19.

SAMPLE
Matrix: solutions
Sample preparation: Inject an aliquot of an aqueous solution.

CAPILLARY ELECTROPHORESIS
Capillary: 70 cm × 50 μm fused-silica (50 cm to detector) (GL Science)
Running buffer: 20 mM pH 9.1 Borate buffer containing 10 mM KOH and 150 mM sodium
dodecyl sulfate
Injection: Hydrostatic injection at 4 cm
Detector: UV 185, UV 210
Migration time: 9
Voltage: 15 kV
Current: 30 μA

OTHER SUBSTANCES
Simultaneous: 3-acetylpyridine, 6-aminoniacinamide, ethyl nicotinate, isoniazid, N-methylniacinamide, niacin, β-picoline, pyridine, pyridine-3-aldehyde, 3-pyridinemethanol, pyridine-3-sul-
fonic acid, thioniacinamide

REFERENCE
Tanaka,S.; Kodama,K.; Kaneta,T.; Nakamura,H. Migration behavior of niacin derivatives in capillary electrophoresis, *J.Chromatogr.A*, **1995**, *718*, 233–237.

SAMPLE
Matrix: solutions

CAPILLARY ELECTROPHORESIS
Capillary: 60 cm × 75 μm AccuSep (Waters) or 52.5 cm × 75 μm (45 cm to detector) (SGE)
Running buffer: 20 mM pH 7 Phosphate buffer
Injection: Hydrodynamic injection for 7 s
Detector: UV 214
Migration time: 4.8
Voltage: 15 kV
Model: Waters Quanta 4000

OTHER SUBSTANCES
Simultaneous: niacin, thiamine

KEY WORDS
sleeve cell increases sensitivity

REFERENCE
Djordjevic,N.M.; Ryan,K. An easy way to enhance absorbance detection sensitivity of Waters Quanta-4000
capillary electrophoresis system, *J.Liq.Chromatogr.Rel.Technol.*, **1996**, *19*, 201–206.

Nicardipine

Molecular formula: $C_{26}H_{29}N_3O_6$
Molecular weight: 479.53
CAS Registry No.: 55985-32-5, 54527-84-3 (HCl)
Merck Index (12th ed.): 6579
Lednicer: 3 150

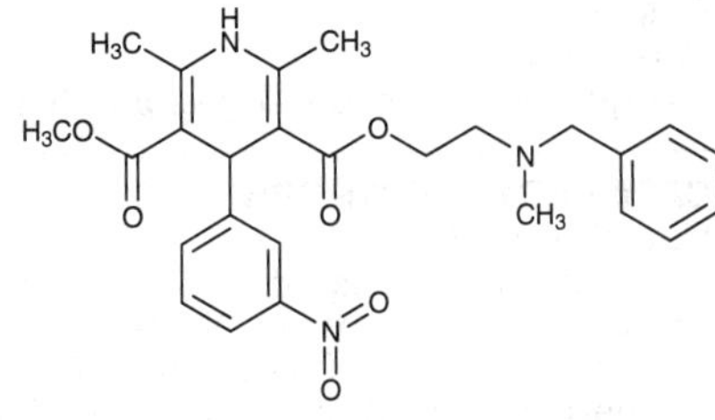

SAMPLE
Matrix: solutions
Sample preparation: Prepare a 600 μg/mL solution in MeCN:MeOH:water 25:25:50, dilute to 150 μg/mL with 100 mM pH 8.1 borate buffer containing 50 mM sodium dodecyl sulfate, inject an aliquot.

CAPILLARY ELECTROPHORESIS
Capillary: 57 cm × 75 μm fused-silica (50 cm to detector) (Beckman)
Capillary preparation: Rinse capillary with 100 mM NaOH and running buffer before each run.
Capillary temperature: 30
Running buffer: Acetone:buffer 15:85 (Buffer was 100 mM pH 8.1 borate buffer containing 50 mM sodium dodecyl sulfate.)
Injection: Pressure injection for 5 s
Detector: UV 200
Migration time: 19.5
Voltage: +25 kV
Model: Beckman P/ACE system 5510

OTHER SUBSTANCES
Simultaneous: amlodipine, atenolol, diltiazem, nifedipine, verapamil

REFERENCE
Bretnall,A.E.; Clarke,G.S. Investigation and optimisation of the use of micellar electrokinetic chromatography for the analysis of six cardiovascular drugs, *J.Chromatogr.A*, **1995**, *700*, 173–178.

SAMPLE
Matrix: solutions

CAPILLARY ELECTROPHORESIS
Capillary: 57 cm × 75 μm fused-silica (50 cm to detector) (Beckman)
Capillary preparation: Before each run rinse capillary with 100 mM NaOH and running buffer.
Capillary temperature: 30
Running buffer: Acetone:100 mM pH 8.1 borate buffer containing 50 mM sodium dodecyl sulfate 15:85
Injection: Pressure injection for 5-10 s.
Detector: UV 200
Migration time: 19
Voltage: 25 kV
Model: Beckman P/ACE 5510

OTHER SUBSTANCES
Simultaneous: acebutolol, amiodarone, atenolol, bretylium, captopril, diltiazem, disopyramide, lidocaine, lisinopril, metoprolol, nifedipine, phenytoin, pindolol, propafenone, propranolol, quinidine, sotalol, timolol, p-toluenesulfonic acid, verapamil

REFERENCE
Bretnall,A.E.; Clarke,G.S. Selectivity of capillary electrophoresis for the analysis of cardiovascular drugs, *J.Chromatogr.A*, **1996**, *745*, 145–154.

SAMPLE
Matrix: solutions
Sample preparation: Inject an aliquot of a 100 μg/mL solution in running buffer.

CAPILLARY ELECTROPHORESIS
Capillary: 32 cm × 50 μm fused-silica (27.5 cm to detector) (Yongnian Optical Conductive Fiber Plant, China), coated with polyacrylamide
Capillary preparation: No details of the polyacrylamide coating process are provided. However, another paper (LC.GC 1997, 15, 40) by this group indicates that they use the procedure of Hjertén, thus: Adjust the pH of 20 mL water to 3.5 with acetic acid, add 80 μL 3-(trimethoxysilyl)propyl methacrylate (3-methacryloxypropyltrimethoxysilane), mix, suck into capillary, let stand at room temperature for 1 h, remove the solution, wash with water. Fill the capillary with a deaerated 3-4% acrylamide solution containing 1 μL/mL N,N,N',N'-tetramethylethylenediamine and 1 mg/mL potassium persulfate, let stand for 30 min, remove excess solution by aspiration, rinse with water, remove water by aspiration, dry at 35° (J. Chromatogr. 1985, 347, 191).
Capillary temperature: 25
Running buffer: 100 mM NaH_2PO_4 adjusted to pH 2.5 (A) or 100 mM NaH_2PO_4 containing 45 mM hydroxypropyl-gamma-cyclodextrin, adjusted to pH 2.5 (B)
Injection: Electrokinetic injection at 15 kV for 3 s.
Detector: UV 200, UV 210
Migration time: 6.81 (A), 13.77, 13.93 (B, enantiomers)
Voltage: 15 kV
Model: Bio-Rad BioFocus 3000

OTHER SUBSTANCES
Simultaneous: atropine, carazolol, cicletanine, dimethindene, fendiline, homatropine, ipratropium bromide, isothipendyl, mefloquine, metaclazepam, nafronyl (naftidrofuryl), nefopam, promethazine, reproterol, tetrahydrozoline (tetryzoline), theodrenaline, tioconazole, trihexyphenidyl, trimeprazine (alimemazine), trimipramine

KEY WORDS
coated capillary; chiral

REFERENCE
Koppenhoefer,B.; Epperlein,U.; Xiaofeng,Z.; Bingcheng,L. Separation of enantiomers of drugs by capillary electrophoresis. Part 4: Hydroxypropyl-γ-cyclodextrin as chiral solvating agent, *Electrophoresis*, **1997**, *18*, 924–930.

SAMPLE
Matrix: solutions
Sample preparation: Inject an aliquot of a 100 μg/mL solution in running buffer.

CAPILLARY ELECTROPHORESIS
Capillary: 30 cm × 50 μm fused-silica (25.5 cm to detector), coated with polyacrylamide
Capillary preparation: Adjust the pH of 20 mL water to 3.5 with acetic acid, add 80 μL 3-(trimethoxysilyl)propyl methacrylate (3-methacryloxypropyltrimethoxysilane), mix, suck into capillary, let stand at room temperature for 1 h, remove the solution, wash with water. Fill the capillary with a deaerated 3-4% acrylamide solution containing 1 μL/mL N,N,N',N'-tetramethylethylenediamine and 1 mg/mL potassium persulfate, let stand for 30 min, remove excess solution by aspiration, rinse with water, remove water by aspiration, dry at 35° (J. Chromatogr. 1985, 347, 191).
Capillary temperature: 25
Running buffer: 100 mM NaH_2PO_4 containing 45 mM hydroxypropyl-α- cyclodextrin, adjusted to pH 2.5 with phosphoric acid
Injection: Electrokinetic injection at 15 kV for 3 s.
Detector: UV 200
Voltage: 15 kV
Model: Bio-Rad BioFocus 3000

KEY WORDS
chiral; coated capillary; comparison with the use of other cyclodextrins; this running buffer gave the greatest enantiomeric separation.; α=1.025

REFERENCE
Lin,B.; Zhu,X.; Koppenhoefer,B.; Epperlein,U. Investigation of 123 chiral drugs by cyclodextrin-modified capillary electrophoresis, *LC.GC*, **1997**, *15*, 40–46.

SAMPLE
Matrix: solutions

CAPILLARY ELECTROPHORESIS
Capillary: 36 cm × 50 μm fused-silica coated with linear polyacrylamide (31.5 cm to detector) (GL Science)
Capillary preparation: At the beginning and end of each day rinse capillary with capillary wash solution (Bio-Rad Cat. No. 148-5022) at 690 kPa for more than 3 min and with water at 690 kPa for more than 3 min. Coat capillary as follows. Treat capillary with 1 M NaOH at room temperature for 1 h, rinse with water, dry by passing nitrogen gas through the capillary at 110° for 6 h. Pass thionyl chloride through the capillary using a suction pump for several min, seal capillary at both ends and heat at 70° for 6 h. Unseal the capillary and fill with 250 mM vinyl magnesium bromide in THF by suction, seal the capillary, heat at 70° for 6 h. Open the capillary and rinse it with THF for several min, rinse with distilled water, fill the capillary with polymerization solution, heat at 28 ± 2° for 1 h, condition at -100 V/cm for 30 min (Anal. Sci. 1994, 10, 1). (The polymerization solution was 5% acrylamide in water containing 49 mM Tris, 384 mM glycine, and 0.1% sodium dodecyl sulfate, degas in an ultrasonic bath. Add 40 μL 10% N,N,N',N'-tetramethylethylenediamine and 10 μL 10% ammonium persulfate to 5 mL of the degassed solution, mix thoroughly.)
Running buffer: n-Propanol:50 mM pH 5.0 Phosphate buffer 10:90
Injection: Before each injection rinse with water at 690 kPa for 30 s, rinse with running buffer at 690 kPa for 30 s, partially fill with separation solution (200 μM α_1-acid glycoprotein (Cohn fraction VI) (ICN) in running buffer) at 6.9 kPa for 190 s (27 cm), inject sample at 6.9 kPa for 2 s, electrophorese with running buffer (Note that α_1-acid glycoprotein from other suppliers may provide inferior results).
Detector: UV 210
Migration time: 17, 17.4 (enantiomers)
Voltage: 12 kV
Model: Bio-rad BioFocus 3000

KEY WORDS
chiral; coated capillary

REFERENCE
Tanaka,Y.; Terabe,S. Separation of the enantiomers of basic drugs by affinity capillary electrophoresis using a partial filling technique and α_1-acid glycoprotein as chiral selector, *Chromatographia*, **1997**, *44*, 119–128.

SAMPLE
Matrix: solutions

CAPILLARY ELECTROPHORESIS
Capillary: 29-36 cm × 50 μm fused-silica (24.5-31.5 cm to detector) (Yongnian Optical Conductive Fiber Plant, China) coated with polyacrylamide
Capillary preparation: Coat capillary as follows. Adjust the pH of 20 mL water to 3.5 with acetic acid, add 80 μL 3-(trimethoxysilyl)propyl methacrylate (3-methacryloxypropyltrimethoxysilane), mix, suck into capillary, let stand at room temperature for 1 h, remove the solution, wash with water. Fill the capillary with a deaerated 3-4% acrylamide solution containing 1 μL/mL N,N,N',N'-tetramethylethylenediamine and 1 mg/mL potassium persulfate, let stand for 30 min, remove excess solution by aspiration, rinse with water, remove water by aspiration, dry at 35° (J. Chromatogr. 1985, 347, 191).
Capillary temperature: 25
Running buffer: 100 mM pH 2.5 NaH_2PO_4 (A) or 100 mM pH 2.5 NaH_2PO_4 containing 45 mM hydroxypropyl-α-cyclodextrin (Wacker, Munich) (B)
Injection: Electromigration at 15 kV for 3 s.
Detector: UV 200; UV 210
Migration time: 6.81 (A); 11.06, 11.37 (B) (enantiomers)
Voltage: 15 kV
Model: Bio-Focus 3000

OTHER SUBSTANCES
Also analyzed: albuterol (salbutamol), alprenolol, amorolfine, atenolol, atropine, azelastine, baclofen, bamethan, benproperine, benserazide, biperiden, bisoprolol, brompheniramine, bupivacaine, bupranolol, butamirate, butethamate, carazolol, carbuterol, carteolol, carvedilol, celiprolol, chloroquine, chlorpheniramine, chlorphenoxamine, cicletanine, clenbuterol, clidinium bromide, clobutinol, dimethindene, dipivefrin, disopyramide, dobutamine, doxylamine, fendiline, flecainide, gallopamil, homatropine, ipratropium bromide, isoproterenol (isoprenaline), isothipendyl, ketamine, meclizine, mefloquine, mepindolol, mequitazine, metaclazepam, metaproterenol (orciprenaline), metipranolol, metoprolol, nafronyl (naftidrofuryl), nefopam, norfenefrine, ofloxacin, ornidazole, orphenadrine, oxomemazine, oxprenolol, oxybutynin, phenoxybenzamine, phenylpropanolamine, pholedrine, pindolol, pirbuterol, prilocaine, procyclidine, promethazine, propafenone, propranolol, reproterol, sotalol, sulpride, synephrine, talinolol, terbutaline, tetrahydrozoline (tetryzoline), theodrenaline, tioconazole, tocainide, trihexyphenidyl, trimeprazine (alimemazine), trimipramine, tropicamide, verapamil, zopiclone

---

KEY WORDS
coated capillary; chiral

REFERENCE
Koppenhoefer,B.; Eperlein,U.; Schlunk,R.; Zhu,X.; Lin,B. Separation of enantiomers of drugs by capillary electrophoresis. V. Hydroxypropyl-α-cyclodextrin as chiral solvating agent, *J.Chromatogr.A*, **1998**, *793*, 153–164.

Nicotine

Molecular formula: $C_{10}H_{14}N_2$
Molecular weight: 162.23
CAS Registry No.: 54-11-5, 96055-45-7 (nicotine polacrilex)
Merck Index (12th ed.): 6611

SAMPLE
Matrix: plants
Sample preparation: 500 mg Ground tobacco + 20 mL 0.5% triethanolamine in water, sonicate for 6 min, filter (0.45 μm) an aliquot, inject an aliquot of the filtrate

---

CAPILLARY ELECTROPHORESIS
Capillary: 72 cm fused-silica (50 cm to detector)
Running buffer: 120 mM pH 2.7 Phosphate buffer
Injection: Hydrodynamic injection for 1 s
Detector: UV 260
Migration time: 8
Voltage: 18 kV
Model: Perkin-Elmer 270A-HT

OTHER SUBSTANCES
Extracted: anatabine, nornicotine

KEY WORDS
tobacco

REFERENCE
Yang,S.C.; Smetena,I. Evaluation of capillary electrophoresis for the analysis of nicotine and selected minor alkaloids from tobacco, *Chromatographia*, **1995**, *40*, 375–378.

SAMPLE
Matrix: plants

Sample preparation: Sonicate 500 mg ground tobacco with 20 mL 1% triethanolamine in water for 6 min, filter (0.45 μm), inject an aliquot of the filtrate.

CAPILLARY ELECTROPHORESIS
Capillary: 72 cm × 50 μm fused-silica (50 cm to detector)
Running buffer: 6 mM pH 9.5 Sodium phosphate buffer containing 10 mM sodium borate and 100 mM sodium dodecyl sulfate
Injection: Hydrodynamic injection for 1 s.
Detector: UV 262
Migration time: 8
Voltage: 30 kV
Model: Perkin-Elmer 270A-HT

OTHER SUBSTANCES
Simultaneous: anabasine, anatabine, myosmine, nornicotine

KEY WORDS
tobacco

REFERENCE
Yang,S.S.; Smetena,I.; Goldsmith,A.I. Evaluation of micellar electrokinetic capillary chromatography for the analysis of selected tobacco alkaloids, *J.Chromatogr.A*, **1996**, *746*, 131–136.

SAMPLE
Matrix: plants
Sample preparation: Extract 100 mg ground sample with 10 mL water with occasional shaking for 1 h, dilute extract 10-fold with water, filter (0.22 μm), inject an aliquot of the filtrate.

CAPILLARY ELECTROPHORESIS
Capillary: 44 cm × 100 μm fused-silica (Polymicro Technologies)
Capillary preparation: Before each run rinse with running buffer for 1 min. After each run rinse with 100 mM NaOH for 1 min and with water for 1 min. At the beginning of each group of runs (?) wash with 1 M NaOH at 60° for 10 min, with 100 mM NaOH at 60° for 5 min, with water at 60° for 5 min, and with running buffer at 20° for 5 min.
Capillary temperature: 20
Running buffer: 25 mM pH 2.5 Sodium phosphate buffer
Injection: Hydrodynamic injection for 2 s.
Detector: UV 260
Migration time: 6.63
Voltage: 10 kV
Model: SpectraPhoresis 500 (Thermo Separation Products)
Limit of quantitation: 1.724 μg/mL

KEY WORDS
tobacco

REFERENCE
Ralapati,S. Capillary electrophoresis as an analytical tool for monitoring nicotine in ATF regulated tobacco products, *J.Chromatogr.B*, **1997**, *695*, 117–129.

SAMPLE
Matrix: solutions

CAPILLARY ELECTROPHORESIS
Capillary: 57 cm × 50 μm fused-silica (50 cm to detector) (Polymicro Technologies)
Capillary temperature: 40
Running buffer: MeCN:buffer 10:90 (Buffer was 50 mM pH 11.0 phosphate/carbonate buffer containing 40 mM sodium dodecyl sulfate.)
Detector: UV 214
Migration time: 5
Voltage: 18 kV

Current: <60 μA
Model: Beckman P/ACE 2000

OTHER SUBSTANCES
Simultaneous: ephedrine

REFERENCE
Quang,C.; Strasters,J.K.; Khaledi,M.G. Computer-assisted modeling, prediction, and multifactor optimization in micellar electrokinetic chromatography of ionizable compounds, *Anal.Chem.*, **1994**, *66*, 1646–1653.

SAMPLE
Matrix: solutions
Sample preparation: Prepare a solution in MeOH/water, inject an aliquot.

CAPILLARY ELECTROPHORESIS
Capillary: 62 cm × 52 μm fused-silica (50 cm to detector) (Polymicro Technologies)
Capillary temperature: 40
Running buffer: 50 mM pH 2.50 Tetrabutylammonium phosphate
Injection: Siphon injection at 10 cm for 5 s
Detector: UV (wavelength not specified)
Migration time: 5.5
Voltage: 20 kV
Current: 29 μA
Model: Laboratory constructed

OTHER SUBSTANCES
Simultaneous: doxylamine, epinephrine, imidazole, isoproterenol, 2-methylphenethylamine, 1-methylphenylpropylamine, metoprolol, nadolol, norepinephrine, propranolol, pseudoephedrine

---

REFERENCE
Quang,C.; Khaledi,M.G. Extending the scope of chiral separation of basic compounds by cyclodextrin-mediated capillary zone electrophoresis, *J.Chromatogr.A*, **1995**, *692*, 253–265.

SAMPLE
Matrix: tobacco
Sample preparation: Occasionally shake 50 mg ground tobacco and 10 mL water for 1 h, dilute the extract 20-fold with water, filter (0.2 μm), inject an aliquot of the filtrate.

CAPILLARY ELECTROPHORESIS
Capillary: 44 cm × 100 μm fused-silica
Capillary preparation: Between runs rinse with 100 mM NaOH for 1 min and with water for 1 min then rinse and fill with running buffer for 1 min. Condition capillary at the start of each day by washing with 1 M NaOH at 60° for 10 min, with 100 mM NaOH at 60° for 5 min, with water at 60° for 5 min, and with running buffer at 20° for 5 min.
Capillary temperature: 20
Running buffer: 25 mM PH 2.5 NaH$_2$PO$_4$
Injection: Hydrodynamic injection for 1 s.
Detector: UV 260
Migration time: 6.2
Voltage: 10 kV
Model: SpectraPHORESIS 500 (Thermo Separation Products)
Limit of quantitation: 172 ng/mL
Limit of detection: 62 ng/mL

OTHER SUBSTANCES
Extracted: anabasine, nornicotine

REFERENCE
Lu,G.H.; Ralapati,S. Application of high-performance capillary electrophoresis to the quantitative analysis of nicotine and profiling of other alkaloids in ATF-regulated tobacco products, *Electrophoresis*, **1998**, *19*, 19–26.

Nifedipine

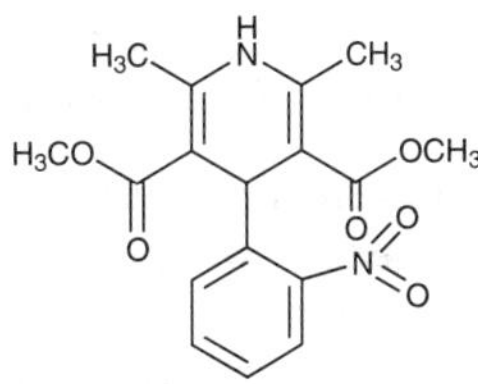

Molecular formula: $C_{17}H_{18}N_2O_6$
Molecular weight: 346.34
CAS Registry No.: 21829-25-4
Merck Index (12th ed.): 6617
Lednicer: 2 283; 4 106

SAMPLE
Matrix: solutions
Sample preparation: Prepare a 600 µg/mL solution in MeCN:MeOH:water 25:25:50, dilute to 150 µg/mL with 100 mM pH 8.1 borate buffer containing 50 mM sodium dodecyl sulfate, inject an aliquot.

CAPILLARY ELECTROPHORESIS
Capillary: 57 cm × 75 µm fused-silica (50 cm to detector) (Beckman)
Capillary preparation: Rinse capillary with 100 mM NaOH and running buffer before each run.
Capillary temperature: 30
Running buffer: Acetone:buffer 15:85 (Buffer was 100 mM pH 8.1 borate buffer containing 50 mM sodium dodecyl sulfate.)
Injection: Pressure injection for 5 s
Detector: UV 200
Migration time: 11
Voltage: +25 kV
Model: Beckman P/ACE system 5510

OTHER SUBSTANCES
Simultaneous: amlodipine, atenolol, diltiazem, nicardipine, verapamil

REFERENCE
Bretnall,A.E.; Clarke,G.S. Investigation and optimisation of the use of micellar electrokinetic chromatography for the analysis of six cardiovascular drugs, *J.Chromatogr.A*, **1995**, *700*, 173–178.

SAMPLE
Matrix: solutions

CAPILLARY ELECTROPHORESIS
Capillary: 57 cm × 75 µm fused-silica (50 cm to detector) (Beckman)
Capillary preparation: Before each run rinse capillary with 100 mM NaOH and running buffer.
Capillary temperature: 30
Running buffer: Acetone:100 mM pH 8.1 borate buffer containing 50 mM sodium dodecyl sulfate 15:85
Injection: Pressure injection for 5-10 s.
Detector: UV 200
Migration time: 7.4
Voltage: 25 kV
Model: Beckman P/ACE 5510

OTHER SUBSTANCES
Simultaneous: acebutolol, amiodarone, atenolol, bretylium, captopril, diltiazem, disopyramide, lidocaine, lisinopril, metoprolol, nicardipine, phenytoin, pindolol, propafenone, propranolol, quinidine, sotalol, timolol, p-toluenesulfonic acid, verapamil

REFERENCE
Bretnall,A.E.; Clarke,G.S. Selectivity of capillary electrophoresis for the analysis of cardiovascular drugs, *J.Chromatogr.A*, **1996**, *745*, 145–154.

Nifenalol

Molecular formula: $C_{11}H_{16}N_2O_3$
Molecular weight: 224.26
CAS Registry No.: 7413-36-7, 5704-60-9 (HCl)
Merck Index (12th ed.): 6618

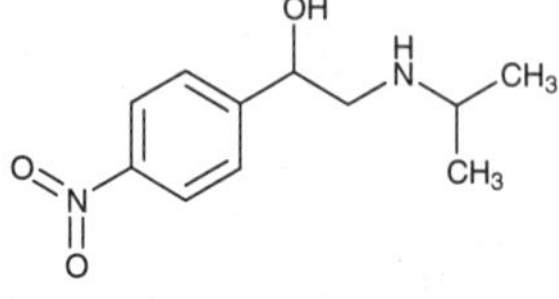

SAMPLE
Matrix: solutions

CAPILLARY ELECTROPHORESIS
Capillary: 62.1 cm × 75 μm fused-silica (44.6 cm to detector) (Polymicro Technologies)
Capillary preparation: Coat capillary as follows. Condition capillary with 1 M NaOH for 1 h, rinse with water for 30 min, fill capillary with 1% 3-(trimethoxysilyl)propyl methacrylate (methacryloxypropyltrimethoxysilane) (adjusted to pH 3.5 with acetic acid), allow to react at room temperature for 2 h, rinse with water, fill capillary with 10% 2-aminoethyl methacrylate hydrochloride (Fisher Scientific) solution containing 1-3 mg/mL N,N,N',N'-tetramethylethylenediamine and 1 mg/mL ammonium persulfate, allow to react until polymerization is complete, rinse with water, dry at 30-40° overnight.
Running buffer: 25 mM pH 4.7 Acetate buffer
Injection: Hydrodynamic injection at 5 cm for 5 s.
Detector: UV 210
Migration time: 13.20
Voltage: 15 kV
Model: laboratory constructed

OTHER SUBSTANCES
Simultaneous: acebutolol, atenolol, betaxolol, carteolol, dilevalol, pindolol, propranolol

KEY WORDS
coated capillary; detector at anode

REFERENCE
Liu,Q.; Lin,F.; Hartwick,R.A. Free solution capillary electrophoretic separation of basic proteins and drugs using cationic polymer coated capillaries, *J.Liq.Chromatogr.Rel.Technol.*, **1997**, *20*, 707–718.

SAMPLE
Matrix: solutions

CAPILLARY ELECTROPHORESIS
Capillary: 50.5 cm × 75 μm coated fused-silica (36 cm to detector) (Polymicro Technologies)
Capillary preparation: Coat capillary as follows. Condition capillary with 1 M NaOH for 1 h, rinse with water for 30 min, force 1% methacryloxypropyltrimethoxysilane (adjusted to pH 3.5 with acetic acid) through the capillary using pressure, allow to react at room temperature for 2 h, rinse with water, fill capillary with diallyldimethylammonium chloride solution containing 1 μL/mL N,N,N',N'-tetramethylethylenediamine and 1 mg/mL ammonium persulfate, allow to react until polymerization is complete, rinse with water, dry at 30-40° overnight.
Running buffer: 10 mM pH 5.0 Acetate buffer
Injection: Hydrodynamic injection at 5 cm for 5 s.
Detector: UV 210
Migration time: 12
Voltage: 12 kV
Model: laboratory-constructed

OTHER SUBSTANCES
Simultaneous: acebutolol, atenolol, dilevalol, pindolol

KEY WORDS
coated capillary

REFERENCE
Liu,Q.; Lin,F.; Hartwick,R.A. Poly(diallyldimethylammonium chloride) as a cationic coating for capillary electrophoresis, *J.Chromatogr.Sci.*, **1997**, *35*, 126–130.

SAMPLE
Matrix: solutions

CAPILLARY ELECTROPHORESIS
Capillary: 71.5 cm × 75 μm fused-silica (52.2 cm to detector) (Polymicro Technologies)
Capillary preparation: Rinse capillary with running buffer for 1 min between runs. Rinse new capillaries with 500 mM NaOH for 20 min and with water for 10 min and then condition with running buffer for 1 h.
Running buffer: 50 mM pH 3.0 containing 0.1% guaran (Guaran is the water-soluble fraction of guar gum. Prepare guaran by stirring 1 g guar gum (Sigma) in 100 mL water for 20 min, add 25 mL EtOH dropwise with stirring, let stand overnight, supercentrifuge. Add 30 mL EtOH dropwise with stirring to the centrifugate, centrifuge, collect the guaran as a precipitate, dry in air.)
Injection: Hydrodynamic injection at 5 cm for 5 s.
Detector: UV (wavelengths not given)
Migration time: 12
Voltage: 17.5 kV
Model: laboratory-constructed

OTHER SUBSTANCES
Simultaneous: acebutolol, atenolol, betaxolol, carteolol, dilevalol, pindolol, propranolol

REFERENCE
Liu,Q.; Lin,F.; Hartwick,R.A. Capillary zone electrophoretic separation of basic proteins and drugs using guaran as a buffer modifier, *Chromatographia*, **1998**, *47*, 219–224.

Nikethamide

Molecular formula: $C_{10}H_{14}N_2O$
Molecular weight: 178.23
CAS Registry No.: 59-26-7
Merck Index (12th ed.): 6635
Lednicer: 1 253

SAMPLE
Matrix: solutions

CAPILLARY ELECTROPHORESIS
Capillary: 59.6 cm × 75 μm fused-silica (52.1 cm to detector) (Polymicro Technologies)
Capillary preparation: Purge with running buffer before each run. At the beginning of each day purge using 50-60 kPa vacuum with 500 mM NaOH for 5 min, with water for 5 min, with MeCN for 5 min, and with running buffer for 5 min.
Running buffer: MeCN:MeOH:acetic acid 49:50:1 containing 20 mM ammonium acetate
Injection: Hydrostatic injection at 10 cm for 5 s.
Detector: UV 214
Migration time: 5.99
Voltage: 30 kV
Model: Waters Quanta 4000

OTHER SUBSTANCES
Simultaneous: diphenoxylate, ethoheptazine, fentanyl, levallorphan, meperidine, methadone, nalorphine, pentazocine

REFERENCE

Leung,G.N.W.; Tang,H.P.O.; Tso,T.S.C.; Wan,T.S.M. Separation of basic drugs with non-aqueous capillary electrophoresis, *J.Chromatogr.A*, **1996**, *738*, 141–154.

Nimetazepam

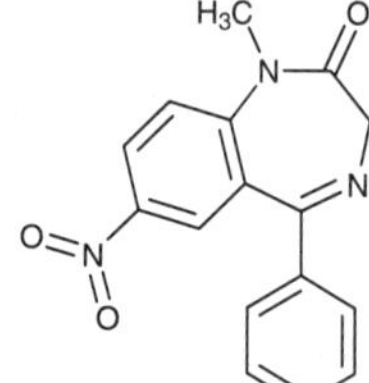

Molecular formula: $C_{16}H_{13}N_3O_3$
Molecular weight: 295.30
CAS Registry No.: 2011-67-8
Merck Index (12th ed.): 6641

SAMPLE
Matrix: solutions

CAPILLARY ELECTROPHORESIS
Capillary: 60 cm × 50 μm fused-silica (40 cm to detector), C18 coated (Supelco)
Capillary preparation: Rinse new capillaries with 250 μL 1 M NaOH, 250 μL 100 mM NaOH, water, MeOH, water, and running buffer then condition with running buffer for at least 15 min. Carry out a similar procedure between runs.
Running buffer: 100 mM pH 8.5 Borate buffer containing 25 mM sodium dodecyl sulfate and 5 M urea
Injection: Electrokinetic injection at 5 kV for 5 s.
Detector: UV 254
Migration time: 18
Voltage: 18 kV
Model: Jasco

OTHER SUBSTANCES
Simultaneous: alprazolam, amobarbital, barbital, bromazepam, clonazepam, clotiazepam, cloxazolam, diazepam, estazolam, etizolam, fludiazepam, flunitrazepam, flurazepam, haloxazolam, medazepam, mephobarbital, metharbital, nitrazepam, oxazepam, pentobarbital, phenobarbital, secobarbital, triazolam

KEY WORDS
coated capillary

REFERENCE

Jinno,K.; Han,Y.; Nakamura,M. Analysis of anxiolytic drugs by capillary electrophoresis with bare and coated capillaries, *J.Capillary Electrophor.*, **1996**, *3*, 139–145.

SAMPLE
Matrix: solutions
Sample preparation: Inject an aliquot of a 100 μg/mL solution in running buffer.

CAPILLARY ELECTROPHORESIS
Capillary: 60 cm × 50 μm acrylamide-coated fused-silica (40 cm to detector) (Supelco)
Capillary preparation: Adjust the pH of 20 mL water to 3.5 with acetic acid, add 80 μL 3-(trimethoxysilyl)propyl methacrylate (3-methacryloxypropyltrimethoxysilane), mix, suck into capillary, let stand at room temperature for 1 h, remove the solution, wash with water. Fill the capillary with a deaerated 3-4% acrylamide solution containing 1 μL/mL N,N,N',N'-tetramethylethylenediamine and 1 mg/mL ammonium persulfate, let stand for 3 h, remove excess solution by aspiration, rinse with water, remove water by aspiration, dry at 35° (cf. J. Chromatogr. 1985, 347, 191).
Running buffer: MeCN:buffer 5:95 (Buffer was 100 mM borate containing 5 M urea and 10 mM sodium dodecyl sulfate, adjusted to pH 8.5 with phosphate.)
Injection: Electrokinetic injection at 5 kV for 5 s.
Detector: UV 254
Migration time: 22.5

Voltage: 18 kV
Model: Jasco Model 870-CE

OTHER SUBSTANCES
Simultaneous: alprazolam, amobarbital, barbital, bromazepam, clonazepam, clotiazepam, cloxazolam, diazepam, estazolam, etizolam, fludiazepam, flunitrazepam, flurazepam, haloxazolam, medazepam, mephobarbital, metharbital, nitrazepam, oxazepam, pentobarbital, phenobarbital, secobarbital, triazolam

KEY WORDS
coated capillary; detector at anode

REFERENCE
Jinno,K.; Han,Y.; Sawada,H.; Taniguchi,M. Capillary electrophoretic separation of toxic drugs using a polyacrylamide-coated capillary, *Chromatographia*, **1997**, *46*, 309–314.

Nimodipine

Molecular formula: $C_{21}H_{26}N_2O_7$
Molecular weight: 418.45
CAS Registry No.: 66085-59-4
Merck Index (12th ed.): 6643
Lednicer: 3 149

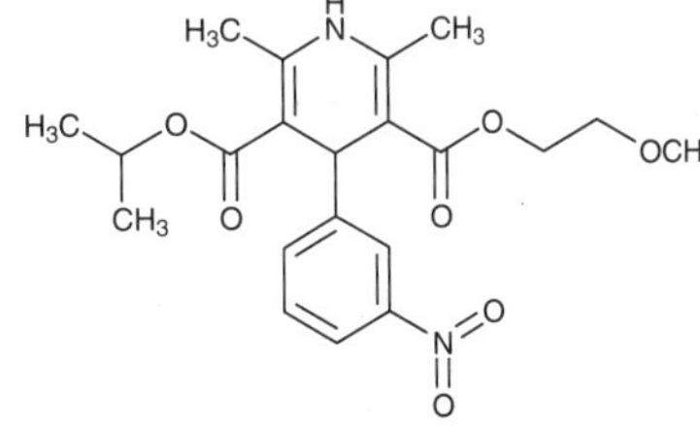

SAMPLE
Matrix: solutions

CAPILLARY ELECTROPHORESIS
Capillary: 70 cm × 75 μm fused-silica
Capillary preparation: Before each run wash with 100 mM NaOH for 5 min, with water for 5 min, and with running buffer for 5 min.
Capillary temperature: 25
Running buffer: 20 mM pH 9.4 Phosphate/borate buffer containing 1% urea and 10 mM 2-O-carboxymethyl-β-cyclodextrin
Injection: Hydrodynamic injection for 1 s.
Detector: UV 240
Migration time: 21.5, 22.5 (enantiomers)
Voltage: 20 kV
Model: SpectraPhoresis 500 (Thermo Separation Products)

OTHER SUBSTANCES
Simultaneous: amlodipine, nitrendipine

KEY WORDS
comparison with HPLC; chiral

REFERENCE
Gilar,M.; Uhrová,M.; Tesarová,E. Enantiomer separation of dihydropyridine calcium antagonists with cyclodextrins as chiral selectors: structural correlation, *J.Chromatogr.B*, **1996**, *681*, 133–141.

Nisin

Molecular formula: $C_{143}H_{230}N_{42}O_{37}S_7$
Molecular weight: 3354.12
CAS Registry No.: 1414-45-5
Merck Index (12th ed.): 6657

Abu = α-aminobutyric acid
Dha = dehydroalanine
Dhb = dehydrobutyrine

SAMPLE
Matrix: solutions
Sample preparation: Inject an aliquot of a 1 mg/mL solution in MeCN:1 mM phosphoric acid 15:85 (A) or a 375 µg/mL solution in MeCN:280 µM pH 8.05 phosphate buffer 77.5:22.5 (B)

CAPILLARY ELECTROPHORESIS
Capillary: 77 cm × 75 µm fused-silica (70 cm to detector) (Polymicro Technologies)
Running buffer: MeCN:33 mM Phosphoric acid 15:85 (A) or MeCN:12.5 mM pH 8.05 phosphate buffer 70:30 (B)
Injection: Electrokinetic injection at 8 kV for 5 s.
Detector: UV 200
Migration time: 15 (A), 7 (B)
Voltage: 15 kV (A) or 29 kV (B)
Model: Beckman P/ACE 2100

OTHER SUBSTANCES
Simultaneous: degradation products

KEY WORDS
protect from light

REFERENCE
Cruz,L.; Garden,R.W.; Kaiser,H.J.; Sweedler,J.V. Studies of the degradation products of nisin, a peptide antibiotic, using capillary electrophoresis with off-line mass spectrometry, *J.Chromatogr.A*, **1996**, *735*, 375–385.

Nisoldipine

Molecular formula: $C_{20}H_{24}N_2O_6$
Molecular weight: 388.42
CAS Registry No.: 63675-72-9
Merck Index (12th ed.): 6658

SAMPLE
Matrix: solutions

CAPILLARY ELECTROPHORESIS
Capillary: 70 cm × 75 µm fused-silica
Capillary preparation: Before each run wash with 100 mM NaOH for 5 min, with water for 5 min, and with running buffer for 5 min.
Capillary temperature: 25
Running buffer: 20 mM pH 9.4 Phosphate/borate buffer containing 1% urea, 5 mM β-cyclodextrin, and 20 mM sodium dodecyl sulfate

Injection: Hydrodynamic injection for 1 s.
Detector: UV 240
Migration time: 16, 18 (enantiomers)
Voltage: 20 kV
Model: SpectraPhoresis 500 (Thermo Separation Products)

KEY WORDS
comparison with HPLC; chiral

REFERENCE
Gilar,M.; Uhrová,M.; Tesarová,E. Enantiomer separation of dihydropyridine calcium antagonists with cyclo-dextrins as chiral selectors: structural correlation, *J.Chromatogr.B*, **1996**, *681*, 133–141.

Nitrazepam

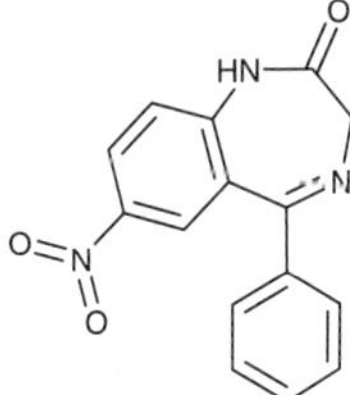

Molecular formula: $C_{15}H_{11}N_3O_3$
Molecular weight: 281.27
CAS Registry No.: 146-22-5
Merck Index (12th ed.): 6667
Lednicer: 1 366

SAMPLE
Matrix: blood
Sample preparation: Condition a Sep-Pak C18 SPE cartridge with MeOH, water, and 100 mM pH 9.0 borate buffer. 1 mL Serum + 3 mL 100 mM pH 9.0 borate buffer, mix, add to the SPE cartridge, wash with 3 mL 100 mM pH 9.0 borate buffer, wash with three 2 mL portions of water, wash with 1 mL MeCN:water 5:95, wash with 1 mL hexane, dry under full vacuum. Elute with 4 mL dichloromethane, evaporate the eluate to dryness under a stream of nitrogen, reconstitute with MeOH:water 15:85, filter (0.45 μm), inject an aliquot.

CAPILLARY ELECTROPHORESIS
Capillary: 72 cm × 50 μm fused-silica (50 cm to detector)
Capillary preparation: Before each run wash capillary with 100 mM NaOH for 4 min and with running buffer for 6 min.
Capillary temperature: 25
Running buffer: MeOH:5 mM pH 8.5 phosphate/borate buffer 15:85 containing 50 mM sodium dodecyl sulfate
Injection: Vacuum injection for 4 s.
Detector: UV 200
Migration time: 16
Internal standard: secobarbital (9)
Voltage: 25 kV
Model: Applied Biosystems Model 270A or Beckman P/ACE 5000
Limit of detection: 25 ng/mL

OTHER SUBSTANCES
Extracted: bromazepam, diazepam, estazolam, flurazepam, triazolam

KEY WORDS
SPE; serum

REFERENCE
Tomita,M.; Okuyama,T. Application of capillary electrophoresis to the simultaneous screening and quantitation of benzodiazepines, *J.Chromatogr.B*, **1996**, *678*, 331–337.

SAMPLE
Matrix: blood

Sample preparation: 500 μL Serum + 150 ng flunitrazepam + 250 μL 800 mM pH 10.0 N-cyclohexyl-2-hydroxy-3-aminopropanesulfonic acid (CAPSO) buffer + 4 mL n-pentane:ethyl acetate 75:25, vortex for 1 min, centrifuge at 25° at 1500 g for 5 min. Remove the organic layer and evaporate it to dryness at 28°, reconstitute the residue in 40 μL 50 mM pH 9.5 borate buffer containing 4.5 mM sodium dodecyl sulfate, inject an aliquot.

CAPILLARY ELECTROPHORESIS
Capillary: 47 cm × 50 μm fused-silica (40 cm to detector) (Supelco)
Capillary preparation: Before each run wash with 100 mM NaOH, water, and running buffer.
Capillary temperature: 25
Running buffer: MeCN:50 mM pH 9.5 sodium borate buffer containing 18 mM sodium dodecyl sulfate 14:86
Injection: Pressure injection at 0.5 psi for 10 s.
Detector: UV 214
Migration time: 5.5
Internal standard: flunitrazepam (5.7)
Voltage: 20 kV
Model: Beckman P/ACE 5010
Limit of quantitation: 10 ng/mL

OTHER SUBSTANCES
Extracted: clobazam, clonazepam, desmethylclobazam, desmethyldiazepam, diazepam
Noninterfering: carbamazepine, ethosuximide, phenobarbital, phenytoin, primidone, valproic acid, zonisamide

KEY WORDS
serum

REFERENCE
Imazawa,M.; Hatanaka,Y. Micellar electrokinetic capillary chromatography of benzodiazepine antiepileptics and their desmethyl metabolites in blood, *J.Pharm.Biomed.Anal.*, **1997**, *15*, 1503–1508.

SAMPLE
Matrix: solutions
Sample preparation: Inject an aliquot of a 5 μg/mL solution in running buffer.

CAPILLARY ELECTROPHORESIS
Capillary: 87 cm × 75 μm fused-silica (80 cm to detector) (Beckman)
Capillary preparation: Between runs rinse capillary with running buffer for 2 min then equilibrate for 5 min.
Capillary temperature: 35
Running buffer: 20 mM pH 7 Buffer containing 15 mM sodium cholate and 35 mM sodium deoxycholate (Buffer was 20 mM sodium borate adjusted to pH 7.0 with 20 mM NaH_2PO_4.)
Injection: Pressure injection for 2 s.
Detector: UV 214
Migration time: 16.3
Voltage: 20 kV
Model: Beckman P/ACE 2100

OTHER SUBSTANCES
Simultaneous: bromazepam, chlordiazepoxide, clobazam, clonazepam, diazepam, flunitrazepam, flurazepam, halazepam, lorazepam, lormetazepam, nordazepam, temazepam

REFERENCE
Boonkerd,S.; Detaevernier,M.R.; Vindevogel,J.; Michotte,Y. Migration behaviour of benzodiazepines in micellar electrokinetic chromatography, *J.Chromatogr.A*, **1996**, *756*, 279–286.

SAMPLE
Matrix: solutions

CAPILLARY ELECTROPHORESIS
Capillary: 60 cm × 50 μm fused-silica (40 cm to detector), C18 coated (Supelco)

Capillary preparation: Rinse new capillaries with 250 μL 1 M NaOH, 250 μL 100 mM NaOH, water, MeOH, water, and running buffer then condition with running buffer for at least 15 min. Carry out a similar procedure between runs.
Running buffer: 100 mM pH 8.5 Borate buffer containing 25 mM sodium dodecyl sulfate and 5 M urea
Injection: Electrokinetic injection at 5 kV for 5 s.
Detector: UV 254
Migration time: 16
Voltage: 18 kV
Model: Jasco

OTHER SUBSTANCES
Simultaneous: alprazolam, amobarbital, barbital, bromazepam, clonazepam, clotiazepam, cloxazolam, diazepam, estazolam, etizolam, fludiazepam, flunitrazepam, flurazepam, haloxazolam, medazepam, mephobarbital, metharbital, nimetazepam, oxazepam, pentobarbital, phenobarbital, secobarbital, triazolam

---

KEY WORDS
coated capillary

REFERENCE
Jinno,K.; Han,Y.; Nakamura,M. Analysis of anxiolytic drugs by capillary electrophoresis with bare and coated capillaries, *J.Capillary Electrophor.*, **1996**, *3*, 139–145.

SAMPLE
Matrix: solutions
Sample preparation: Inject an aliquot of a solution in MeOH

CAPILLARY ELECTROPHORESIS
Capillary: 70 cm × 50 μm fused-silica (63 cm to detector) (Composite Metal Services, Hallow, UK)
Capillary preparation: Wash with running buffer for 5 min before each injection. Purge new columns with 1 M NaOH at 60° for 5 min, with 100 mM NaOH at 60° for 5 min, and with water at 60° for 5 min. Equilibrate with run buffer for 10 min then apply voltage for 5 min.
Capillary temperature: 30
Running buffer: MeOH:20 mM pH 2.5 citric acid 15:85 (A) or MeOH:6 mM pH 8 sodium tetraborate containing 12 mM NaH_2PO_4 and 75 mM sodium dodecyl sulfate 5:95 (B)
Injection: Hydrodynamic vacuum injection at 1.5 psi for 7 s (42 nL).
Detector: UV 200
Migration time: 7.75 (A), 17.21 (B)
Voltage: 20 kV
Current: 10 μA
Model: SpectraPhoresis 1000 (ThermoSeparation Products)
Limit of detection: 1.28 μM

OTHER SUBSTANCES
Simultaneous: chlordiazepoxide, diazepam, flurazepam

KEY WORDS
comparison with HPLC, capillary GC, and polarography

REFERENCE
McGrath,G.; McClean,S.; O'Kane,E.; Smyth,W.F.; Tagliaro,F. Study of the capillary zone electrophoretic behaviour of selected drugs, and its comparison with other analytical techniques for their formulation assay, *J.Chromatogr.A*, **1996**, *735*, 237–247.

SAMPLE
Matrix: solutions

CAPILLARY ELECTROPHORESIS
Capillary: 53 cm × 50 μm fused-silica packed with 3 μm CEC Hypersil ODS (46 cm packed with ODS material) (Hypersil)

Running buffer: Gradient. A was 5 mM ammonium acetate in MeCN:water 50:50. B was 5 mM ammonium acetate in MeCN:water 80:20. A:B 100:0 for 3 min, to 0:100 over 0.1 min, maintain at 0:100 (pumped with an HPLC pump at 0.01 mL/min for 3 min then at 0.1 mL/min).
Injection: Inject 10 μL using an HPLC injector.
Detector: MS, VG Biotech Platform, electrospray (details in paper)
Migration time: 19
Voltage: 30 kV

OTHER SUBSTANCES
Simultaneous: diazepam

KEY WORDS
electrochromatography

REFERENCE
Taylor,M.R.; Teale,P. Gradient capillary electrochromatography of drug mixtures with UV and electrospray ionisation mass spectrometric detection, *J.Chromatogr.A*, **1997**, *768*, 89–95.

SAMPLE
Matrix: urine
Sample preparation: Place a silica fiber coated with an 85 μm-thick layer of polyacrylate (Supelco) in 10 mL urine, stir at 60° for 2 h, place the fiber in a capillary tube containing 20 μL MeCN, let stand at room temperature for 30 min, inject an aliquot of the solution. (Desorb a new fiber several times with MeCN before use.)

CAPILLARY ELECTROPHORESIS
Capillary: 60 cm × 50 μm acrylamide-coated fused-silica (40 cm to detector) (Supelco)
Capillary preparation: Adjust the pH of 20 mL water to 3.5 with acetic acid, add 80 μL 3-(trimethoxysilyl)propyl methacrylate (3-methacryloxypropyltrimethoxysilane), mix, suck into capillary, let stand at room temperature for 1 h, remove the solution, wash with water. Fill the capillary with a deaerated 3-4% acrylamide solution containing 1 μL/mL N,N,N',N'-tetramethylethylenediamine and 1 mg/mL ammonium persulfate, let stand for 3 h, remove excess solution by aspiration, rinse with water, remove water by aspiration, dry at 35° (cf. J. Chromatogr. 1985, 347, 191).
Running buffer: MeCN:buffer 5:95 (Buffer was 100 mM borate containing 5 M urea and 10 mM sodium dodecyl sulfate, adjusted to pH 8.5 with phosphate.)
Injection: Electrokinetic injection at 5 kV for 5 s.
Detector: UV 254
Migration time: 10.5
Voltage: 18 kV
Model: Jasco Model 870-CE
Limit of detection: <1 ppm

OTHER SUBSTANCES
Extracted: flunitrazepam, triazolam
Simultaneous: alprazolam, amobarbital, barbital, bromazepam, clonazepam, clotiazepam, cloxazolam, diazepam, estazolam, etizolam, fludiazepam, flurazepam, haloxazolam, medazepam, mephobarbital, metharbital, nimetazepam, oxazepam, pentobarbital, phenobarbital, secobarbital

KEY WORDS
coated capillary; detector at anode; SPE

REFERENCE
Jinno,K.; Han,Y.; Sawada,H.; Taniguchi,M. Capillary electrophoretic separation of toxic drugs using a polyacrylamide-coated capillary, *Chromatographia*, **1997**, *46*, 309–314.

Nitrendipine

Molecular formula: $C_{18}H_{20}N_2O_6$
Molecular weight: 360.37
CAS Registry No.: 39562-70-4
Merck Index (12th ed.): 6669

SAMPLE
Matrix: solutions

CAPILLARY ELECTROPHORESIS
Capillary: 64 cm $\times$ 75 μm (48 cm to detector) (Polymicro Technologies)
Running buffer: MeOH:buffer 0.6:99.4 (Buffer was 14 mM pH 4.5 6-aminocaproic acid buffer
    containing 8 mM adipic acid, 3% dextran (MW 39000), 0.05% polyethylene oxide (MW 300000),
    and 7.5 mM heptakis(2,6-di-O-methyl)-β-cyclodextrin
Injection: Hydrodynamic injection at 20 cm
Detector: UV 220
Migration time: 16, 16.5 (enantiomers)
Voltage: 22 kV
Model: Laboratory constructed

OTHER SUBSTANCES
Simultaneous: disopyramide, pindolol

KEY WORDS
chiral

REFERENCE
Soini,H.; Riekkola,M.-L.; Novotny,M.V. Mixed polymer networks in the direct analysis of pharmaceuticals in
    urine by capillary electrophoresis, *J.Chromatogr.A*, **1994**, *680*, 623–634.

SAMPLE
Matrix: solutions

CAPILLARY ELECTROPHORESIS
Capillary: 70 cm $\times$ 75 μm fused-silica
Capillary preparation: Before each run wash with 100 mM NaOH for 5 min, with water for 5
    min, and with running buffer for 5 min.
Capillary temperature: 25
Running buffer: 20 mM pH 9.4 Phosphate/borate buffer containing 1% urea and 10 mM 2-O-
    carboxymethyl-β-cyclodextrin
Injection: Hydrodynamic injection for 1 s.
Detector: UV 240
Migration time: 18, 18.5 (enantiomers)
Voltage: 20 kV
Model: SpectraPhoresis 500 (Thermo Separation Products)

OTHER SUBSTANCES
Simultaneous: amlodipine, nimodipine

KEY WORDS
comparison with HPLC; chiral

REFERENCE
Gilar,M.; Uhrová,M.; Tesarová,E. Enantiomer separation of dihydropyridine calcium antagonists with cyclo-
    dextrins as chiral selectors: structural correlation, *J.Chromatogr.B*, **1996**, *681*, 133–141.

Nordazepam

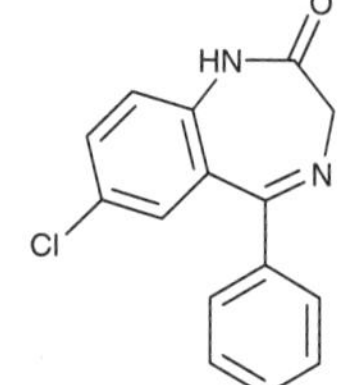

Molecular formula: $C_{15}H_{11}ClN_2O$
Molecular weight: 270.72
CAS Registry No.: 1088-11-5
Merck Index (12th ed.): 6784

SAMPLE
Matrix: solutions
Sample preparation: Inject an aliquot of a 5 µg/mL solution in running buffer.

CAPILLARY ELECTROPHORESIS
Capillary: 87 cm × 75 µm fused-silica (80 cm to detector) (Beckman)
Capillary preparation: Between runs rinse capillary with running buffer for 2 min then equilibrate for 5 min.
Capillary temperature: 35
Running buffer: 20 mM pH 7 Buffer containing 15 mM sodium cholate and 35 mM sodium deoxycholate (Buffer was 20 mM sodium borate adjusted to pH 7.0 with 20 mM NaH_2PO_4.)
Injection: Pressure injection for 2 s.
Detector: UV 214
Migration time: 23
Voltage: 20 kV
Model: Beckman P/ACE 2100

OTHER SUBSTANCES
Simultaneous: bromazepam, chlordiazepoxide, clobazam, clonazepam, diazepam, flunitrazepam, flurazepam, halazepam, lorazepam, lormetazepam, nitrazepam, temazepam

REFERENCE
Boonkerd,S.; Detaevernier,M.R.; Vindevogel,J.; Michotte,Y. Migration behaviour of benzodiazepines in micellar electrokinetic chromatography, *J.Chromatogr.A*, **1996**, *756*, 279–286.

Norepinephrine

Molecular formula: $C_8H_{11}NO_3$
Molecular weight: 169.18
CAS Registry No.: 51-41-2, 69815-49-2 (bitartrate monohydrate), 51-40-1 (bitartrate)
Merck Index (12th ed.): 6788

SAMPLE
Matrix: blood
Sample preparation: Inject an aliquot of 5-fold diluted bovine serum directly.

CAPILLARY ELECTROPHORESIS
Capillary: 50 cm × 50 µm (20 cm to detector) (GL Science)
Capillary preparation: Rinse capillary with running buffer for 5 min before each run. At the beginning of each day rinse capillary with 100 mM NaOH for 10 min and with water for 5 min.
Running buffer: 10 mM pH 2.5 Phosphate buffer containing 25 mM sodium dodecyl sulfate and 12.5 mM Tween 20
Injection: Siphon at 10 cm for 10 s.
Detector: UV 210
Migration time: 16
Voltage: 18 kV
Current: 21-22 µA

Model: Jasco CE-800

OTHER SUBSTANCES
Extracted: dopamine, epinephrine, levodopa, 3-methoxytyrosine, phenylalanine, tyrosine

KEY WORDS
serum; cow; detector is at anode

REFERENCE
Esaka,Y.; Tanaka,K.; Uno,B.; Goto,M.; Kano,K. Sodium dodecyl sulfate-Tween 20 mixed micellar electrokinetic chromatography for separation of hydrophobic cations: Application to adrenaline and its precursors, *Anal.Chem.*, **1997**, *69*, 1332–1338.

SAMPLE
Matrix: cells
Sample preparation: 5 μL Cell suspension + 200 μL water, inject an aliquot (cells lyse on contact with running buffer).

CAPILLARY ELECTROPHORESIS
Capillary: 65 × 16 μm cm fused-silica (45 cm to detector) (Polymicro Technologies)
Capillary preparation: Every few runs flush capillary with running buffer, equilibrate for 5 min.
Running buffer: 100 mM pH 2.3 Citric acid
Injection: Hydrodynamic
Detector: F ex 275 (Ar laser (or frequency doubled Kr laser at 284 nm))
Migration time: 3.6
Voltage: 30 kV
Model: Laboratory constructed

OTHER SUBSTANCES
Extracted: epinephrine

REFERENCE
Chang,H.-T.; Yeung,E.S. Determination of catecholamines in single adrenal medullary cells by capillary electrophoresis and laser-induced native fluorescence, *Anal.Chem.*, **1995**, *67*, 1079–1083.

SAMPLE
Matrix: CSF
Sample preparation: 720 μL CSF + 80 μL 100 nM dihydroxybenzylamine in 175 mM perchloric acid + 160 μL 500 mM pH 8.6 borate buffer + 80 μL 22.5 mM NaCN in water + 40 μL 5 mM naphthalene-2,3-dicarboxaldehyde in MeOH, mix, let stand at room temperature for 5 min, inject an aliquot.

CAPILLARY ELECTROPHORESIS
Capillary: 43 cm × 25 μm fused-silica (23 cm to detector) (Polymicro Technologies)
Capillary temperature: 25.5
Running buffer: 200 mM pH 7.05 ± 0.02 phosphate buffer
Injection: Hydrodynamic injection by vacuum (300 mm Hg) to give 2.5 nL sample injection followed by 60 pL 200 mM orthophosphoric acid
Detector: F ex 442 (He-Cd laser) em 490 (bandpass filter)
Migration time: 1.326
Internal standard: dihydroxybenzylamine (1.346)
Voltage: 29 kV
Model: Zeta Technology IRIS 2000
Limit of detection: 0.086 nM

OTHER SUBSTANCES
Extracted: dopamine
Noninterfering: amino acids

KEY WORDS
derivatization

REFERENCE

Bert,L.; Robert,F.; Denoroy,L.; Renaud,B. High-speed separation of subnanomolar concentrations of noradrenaline and dopamine using capillary zone electrophoresis with laser-induced fluorescence detection, *Electrophoresis*, **1996**, *17*, 523–525.

SAMPLE

Matrix: solutions
Sample preparation: Prepare a solution in running buffer, inject an aliquot.

CAPILLARY ELECTROPHORESIS

Capillary: 78.6 cm × 12.7 μm fused-silica (Polymicro Technologies)
Running buffer: 25 mM pH 5.7 2-morpholinoethanesulfonic acid buffer
Injection: Injection by electromigration at 30 kV for 2 s
Detector: E, 0.7 V vs sodium-saturated calomel electrode
Migration time: 5.2
Voltage: 30 kV

OTHER SUBSTANCES

Simultaneous: 3,4-dihydroxyphenylacetic acid, homovanillic acid, isoproterenol, 4-methylcate-
chol, serotonin, ascorbic acid

REFERENCE

Wallingford,R.A.; Ewing,A.G. Capillary zone electrophoresis with electrochemical detection in 12.7 μm diameter
columns, *Anal.Chem.*, **1988**, *60*, 1972–1975.

SAMPLE

Matrix: solutions

CAPILLARY ELECTROPHORESIS

Capillary: 87.9 cm × 26 μm fused-silica (Polymicro Technologies)
Running buffer: 20 mM pH 6.05 2-morpholinoethanesulfonic acid buffer
Injection: Inject at 5 kV for 2 s
Detector: E, 0.7 V vs sodium-saturated calomel electrode (details of detector construction in
paper)
Migration time: 8.2
Voltage: 25 kV
Current: 2 μA
Limit of detection: 0.2-0.4 fmole (S/N 3)

OTHER SUBSTANCES

Simultaneous: catechol, dopamine, epinephrine

REFERENCE

Wallingford,R.A.; Ewing,A.G. Amperometric detection of catechols in capillary zone electrophoresis with normal
and micellar solutions, *Anal.Chem.*, **1988**, *60*, 258–263.

SAMPLE

Matrix: solutions

CAPILLARY ELECTROPHORESIS

Capillary: 70 cm × 100 μm fused-silica (50 cm to detector) (Gasukuro Kogyo, Tokyo)
Running buffer: 100 mM KOH containing 200 mM boric acid
Injection: Hydrostatic injection for 10 s.
Detector: UV 217
Migration time: 16
Voltage: 9.5 kV
Current: 100 μA

OTHER SUBSTANCES

Simultaneous: dopamine, epinephrine, isoproterenol, levodopa, metanephrine, normetanephrine, vanillylmanedlic acid

REFERENCE
Tanaka,S.; Kaneta,T.; Yoshida,H. Separation of catecholamines by capillary zone electrophoresis using complexation with boric acid, *Anal.Sci.*, **1990**, *6*, 467–468.

SAMPLE
Matrix: solutions
Sample preparation: Prepare a solution in running buffer or 50 mM HCl, inject an aliquot.

CAPILLARY ELECTROPHORESIS
Capillary: 57 cm × 75 μm fused-silica (50 cm to detector) (Polymicro)
Capillary temperature: 25
Running buffer: 30 mM pH 2.5 Tris-citric acid buffer containing 20 mM hydroxypropyl-β-cyclodextrin
Injection: Hydrodynamic injection for 1 s
Detector: UV 254
Migration time: 27, 30 (enantiomers)
Voltage: 263 V/cm
Model: Beckman P/ACE 2000 or 2100

KEY WORDS
chiral

REFERENCE
Kuhn,R.; Wagner,J.; Walbroehl,Y.; Bereuter,T. Potential and limitations of an optically active crown ether for chiral separation in capillary zone electrophoresis, *Electrophoresis*, **1994**, *15*, 828–834.

SAMPLE
Matrix: solutions
Sample preparation: Inject an aliquot of a solution in running buffer.

CAPILLARY ELECTROPHORESIS
Capillary: 44 cm × 5 μm fused-silica (Polymicro Technologies)
Running buffer: Isopropanol:30 mM morpholinoethanesulfonic acid 20:80, pH adjusted to 6.0
Injection: Electromigration at 12 kV for 3 s
Detector: E, Pt electrode 0.700 V, Ag/AgCl reference electrode, details of cell in paper
Migration time: 7.3
Voltage: 25 kV

OTHER SUBSTANCES
Simultaneous: catechol, dopamine, epinephrine, isoproterenol, serotonin

REFERENCE
Chen,M.-C.; Huang,H.-J. An electrochemical cell for end-column amperometric detection in capillary electrophoresis, *Anal.Chem.*, **1995**, *67*, 4010–4014.

SAMPLE
Matrix: solutions

CAPILLARY ELECTROPHORESIS
Capillary: 56 cm × 50 μm fused-silica (36 cm to detector) (Supelco)
Running buffer: 25 mM pH 2.7 phosphate buffer containing 50 mg/mL soluble β-cyclodextrin polymer cross-linked with epichlorohydrin (MW 3000-5000) (Cyclolab, Budapest)
Injection: Hydrodynamic injection at 20 cm for 10 s
Detector: UV 206
Migration time: 13.2 (-), 13.3 (+)
Voltage: 12 kV
Model: Laboratory made

OTHER SUBSTANCES
Simultaneous: isoproterenol, propranolol

KEY WORDS
chiral

REFERENCE
Ingelse,B.A.; Everaerts,F.M.; Desiderio,C.; Fanali,S. Enantiomeric separation by capillary electrophoresis using a soluble neutral β-cyclodextrin polymer, *J.Chromatogr.A*, **1995**, *709*, 89–98.

SAMPLE
Matrix: solutions
Sample preparation: Inject an aliquot of a solution in Ringer's solution.

CAPILLARY ELECTROPHORESIS
Capillary: 65 cm × 50 μm fused-silica (Polymicro Technologies)
Capillary preparation: Before each run flush capillary with 500 mM disodium EDTA (adjusted to pH 13 with solid NaOH) for 1 min and with running buffer for 1 min.
Running buffer: 100 mM pH 4.75 Lithium acetate containing 0.5 mM EDTA
Injection: Electrokinetic injection for 3 s at 20 kV
Detector: E, Bioanalytical Systems PA-1, 33 μm carbon fiber electrode +0.85 V, Ag/AgCl reference electrode (design of capillary/detector decoupler is given in paper)
Migration time: 4.2
Voltage: 20 kV

OTHER SUBSTANCES
Simultaneous: 3,4-dihydroxybenzylamine, dopamine, isoproterenol

REFERENCE
Park,S.; Lunte,C.E. A perfluorosulfonated ionomer end-column electrical decoupler for capillary electrophoresis/electrochemical detection, *Anal.Chem.*, **1995**, *67*, 4366–4370.

SAMPLE
Matrix: solutions
Sample preparation: Prepare a solution in MeOH/water, inject an aliquot.

CAPILLARY ELECTROPHORESIS
Capillary: 62 cm × 52 μm fused-silica (50 cm to detector) (Polymicro Technologies)
Capillary temperature: 40
Running buffer: 50 mM pH 2.50 Tetrabutylammonium phosphate
Injection: Siphon injection at 10 cm for 5 s
Detector: UV (wavelength not specified)
Migration time: 10
Voltage: 20 kV
Current: 29 μA
Model: Laboratory constructed

OTHER SUBSTANCES
Simultaneous: doxylamine, epinephrine, imidazole, isoproterenol, 2-methylphenethylamine, 1-methylphenylpropylamine, metoprolol, nadolol, nicotine, propranolol, pseudoephedrine

REFERENCE
Quang,C.; Khaledi,M.G. Extending the scope of chiral separation of basic compounds by cyclodextrin-mediated capillary zone electrophoresis, *J.Chromatogr.A*, **1995**, *692*, 253–265.

SAMPLE
Matrix: solutions
Sample preparation: Inject an aliquot of a 300 μg/mL solution in water.

CAPILLARY ELECTROPHORESIS
Capillary: 65 cm × 50 μm fused-silica (40 cm to detector) (Isco)
Capillary preparation: Purge with running buffer for 3 min between injections. Purge daily with 1 M NaOH for 3 min, with water for 3 min, and with running buffer for 3 min.

Running buffer: Isopropanol:100 mM pH 7 phosphate buffer containing 25 mM rifamycin B 30:
70
Injection: Electrokinetic injection at 5 kV for 5 s.
Detector: UV 350
Migration time: 32.8 (first enantiomer, R_s 0.9)
Voltage: 8 kV
Model: Isco model 3850

OTHER SUBSTANCES
Simultaneous: alprenolol, epinephrine, amphetamine, metoprolol, normetanephrine, octapamine, oxprenolol, propranolol

KEY WORDS
chiral

REFERENCE
Ward,T.J.; Dann,C.,III; Blaylock,A. Enantiomeric resolution using the macrocyclic antibiotics rifamycin B and
rifamycin SV as chiral selectors for capillary electrophoresis, *J.Chromatogr.A,* **1995,** *715,* 337–344.

SAMPLE
Matrix: solutions

CAPILLARY ELECTROPHORESIS
Capillary: 65 cm × 50 μm coated fused-silica (65 cm to detector) (Polymicro Technologies)
Capillary preparation: Rinse with running buffer for 2 min between runs. Coat capillary as
follows. Flush capillary with 1 M NaOH for 1 h, rinse with water for 1 h, dry at 180° with a
flow of nitrogen overnight, fill the capillary with reagent using a syringe, let stand for 10 min,
force out excess reagent using pressure, dry at 90° with a flow of nitrogen overnight, wash with
acetone for 30 min, wash with MeOH for 30 min, rinse with water for 20 min, equilibrate with
running buffer for 1 h before use. (Prepare reagent by mixing 500 μL 3-aminopropyltriethox-
ysilane, 375 μL EtOH, 54 μL water, and 100 μL 1.2 m HCl, stir at room temperature for 10 h
before use.)
Capillary temperature: 22
Running buffer: 10 mM pH 3.4 Phosphate buffer
Injection: Hydrodynamic injection at 10 cm for 5 s.
Detector: UV 214
Migration time: 4.85
Voltage: -24 kV
Model: Beckman P/ACE Model 2200

OTHER SUBSTANCES
Simultaneous: epinephrine

KEY WORDS
coated capillary

REFERENCE
Guo,Y.; Imahori,G.A.; Colón,L.A. Hydrolytically stable amino-silica glass coating material for manipulation of
the electroosmotic flow in capillary electrophoresis, *J.Chromatogr.A,* **1996,** *744,* 17–29.

SAMPLE
Matrix: solutions
Sample preparation: Inject an aliquot of a 5-10 μM solution.

CAPILLARY ELECTROPHORESIS
Capillary: 70 cm × 75 μm fused-silica (55 cm to detector)
Capillary temperature: 21
Running buffer: 20 mM pH 2.7 Tris-phosphate buffer containing 0.5% hydroxypropylmethyl
cellulose and 24 mM heptakis(2,6-di-O-methyl)-β-cyclodextrin (Hydroxypropylmethyl cellulose
was from Sigma; viscosity of 2% solution = 4000 cP at 25°.)
Injection: Electrokinetic injection at 3 kV for 12 s.

Detector: UV 190
Migration time: 16.2, 17 (enantiomers)
Voltage: 21 kV
Model: Crystal 300 (ATI, Unicam)

OTHER SUBSTANCES
Simultaneous: amphetamine, epinephrine, methamphetamine, pseudoephedrine, selegiline
(deprenyl)

KEY WORDS
chiral

REFERENCE
Szökö,E.; Gyimesi,J.; Barcza,L.; Magyar,K. Determination of binding constants and the influence of methanol
on the separation of drug enantiomers in cyclodextrin modified capillary electrophoresis, *J.Chromatogr.A*,
1996, *745*, 181–187.

SAMPLE
Matrix: solutions

CAPILLARY ELECTROPHORESIS
Capillary: 100 cm × 50 μm fused-silica (Polymicro Technologies)
Running buffer: 20 mM pH 2.5 Sodium citrate
Injection: Electrokinetic injection.
Detector: E, Bioanalytical Systems BAS LC-4C, 25 μm gold wire electrode (design in paper)
+800 mV, Pt auxiliary electrode, Ag/AgCl reference electrode
Migration time: 17
Voltage: 30 kV
Limit of quantitation: 500 nM

OTHER SUBSTANCES
Simultaneous: dopamine, epinephrine

REFERENCE
Zhong,M.; Lunte,S.M. Integrated on-capillary electrochemical detector for capillary electrophoresis,
Anal.Chem., **1996**, *68*, 2488–2493.

SAMPLE
Matrix: solutions

CAPILLARY ELECTROPHORESIS
Capillary: 60 cm × 100 μm (Polymicro Technologies)
Running buffer: 10 mM pH 9.5 Sodium tetraborate buffer
Injection: Electrokinetic injection for 5 s at 15 kV.
Detector: Chemiluminescence. The running buffer passed through a porous cellulose acetate
joint (construction details in paper) and mixed with 1 mM potassium permanganate in 1 M
sulfuric acid pumped at 2.5 μL/min in an 80 cm × 460 μm PTFE tube.
Migration time: 4.60
Voltage: 15 kV
Current: 56 μA
Model: laboratory constructed
Limit of detection: 100 μM

OTHER SUBSTANCES
Simultaneous: catechol, dopamine, serotonin

REFERENCE
Lee,Y.-T.; Whang,C.-W. Off-column chemiluminescence detection in capillary electrophoresis, *J.Chromatogr.A*,
1997, *771*, 379–384.

SAMPLE
Matrix: solutions

CAPILLARY ELECTROPHORESIS
Capillary: 37 cm × 50 μm fused-silica (30 cm to detector) (Quadrex, New Haven CT)
Capillary temperature: -16
Running buffer: MeOH:water 10:90 containing 20 mM heptakis(2,6-di-O-methyl)-β-cyclodextrin, 5 M urea, and 150 mM sodium phosphate, pH 2.5
Injection: Injection for 6 s
Detector: UV 214
Migration time: 12.2, 12.5 (enantiomers)
Voltage: 30 kV
Current: 45 μA
Model: Beckman P/ACE 2210

OTHER SUBSTANCES
Simultaneous: epinephrine, β hydroxyphenethylamine, isoproterenol, octopamine

KEY WORDS
chiral

REFERENCE
Ma,S.; Horváth,C. Capillary zone electrophoresis at subzero temperatures. II: Chiral separation of biogenic amines, *Electrophoresis*, **1997**, *18*, 873–883.

SAMPLE
Matrix: solutions
Sample preparation: Inject an aliquot of a solution in running buffer.

CAPILLARY ELECTROPHORESIS
Capillary: 56 cm × 75 μm fused-silica (36 cm to detector) (Otsuka Electronics, Osaka)
Running buffer: Formamide containing 2.5 mM (+)-18-crown-6 tetracarboxylic acid (Merck, Darmstadt, Germany) and 2.5 mM tetra-n-butylammonium perchlorate
Injection: Gravity injection at 10 cm for 5 s.
Detector: UV 260-300
Migration time: 11.69, 12.13 (enantiomers)
Voltage: 20 kV
Model: Jasco CE-800

KEY WORDS
chiral

REFERENCE
Mori,Y.; Ueno,K.; Umeda,T. Enantiomeric separations of primary amino compounds by non-aqueous capillary zone electrophoresis with a chiral crown ether, *J.Chromatogr.A*, **1997**, *757*, 328–332.

SAMPLE
Matrix: solutions

CAPILLARY ELECTROPHORESIS
Capillary: 35 cm × 50 μm polyacrylamide-coated fused-silica (30.5 cm to detector) (Composite Metal Services, UK)
Capillary preparation: Before each run rinse capillary with water for 70 s and with running buffer for 100 s. (Coat capillary as follows. Adjust the pH of 20 mL water to 3.5 with acetic acid, add 80 μL 3-(trimethoxysilyl)propyl methacrylate (3-methacryloxypropyltrimethoxysilane), mix, suck into capillary, let stand at room temperature for 1 h, remove the solution, wash with water. Fill the capillary with a deaerated 3-4% acrylamide solution containing 1 μL/mL N,N,N',N'-tetramethylethylenediamine and 1 mg/mL potassium persulfate, let stand for 30 min, remove excess solution by aspiration, rinse with water, remove water by aspiration, dry at 35° (J. Chromatogr. 1985, 347, 191).)
Capillary temperature: 25

Running buffer: Buffer containing 70 mM cyanoethylated-β-cyclodextrin (Cyclolab, Budapest) (Prepare buffer by adjusting the pH of 50 mM phosphoric acid containing 50 mM acetic acid and 50 mM boric acid to 2.5 with concentrated NaOH, add the appropriate amount of cyanoethylated-β-cyclodextrin, dilute with an equal volume of water.)
Injection: Pressure injection at 5 psi for 2 s.
Detector: UV 206
Migration time: 6.3 (second enantiomer, α = 1.022)
Voltage: 20 kV
Current: 27-40 μA
Model: Bio-Rad Biofocus 3000

KEY WORDS
chiral; coated capillary

REFERENCE
Aturki,Z.; Desiderio,C.; Mannina,L.; Fanali,S. Chiral separations by capillary zone electrophoresis with the use of cyanoethylated-β-cyclodextrin as chiral selector, *J.Chromatogr.A*, **1998**, *817*, 91–104.

SAMPLE
Matrix: solutions

CAPILLARY ELECTROPHORESIS
Capillary: 85 cm × 75 μm fused-silica (Composite Metal Services, Hallow, UK)
Running buffer: pH 10 Borate buffer containing 0.5 mM EDTA (Run with a pressure of 20 mbar at the injection end.)
Injection: Hydrodynamic injection at 40 mbar for 6 s.
Detector: E, Bioanalytical Systems Unijet cell +0.8 V, 1 mm dia. glassy carbon working electrode, Ag/AgCl reference electrode, stainless steel auxiliary electrode, two 16 μm gaskets (design details in paper)
Migration time: 6.6
Model: Prince
Limit of detection: 10-20 nM

OTHER SUBSTANCES
Simultaneous: dopamine, epinephrine

REFERENCE
Durgbanshi,A.; Kok,W.T. Capillary electrophoresis and electrochemical detection with a conventional detector cell, *J.Chromatogr.A*, **1998**, *798*, 289–296.

SAMPLE
Matrix: solutions
Sample preparation: Inject an aliquot of an aqueous solution

CAPILLARY ELECTROPHORESIS
Capillary: 53 cm × 75 μm fused-silica (40.5 cm to detector) (Polymicro Technologies)
Running buffer: 5 mM pH 7.3 NaH_2PO_4 containing 5 mM sodium borate and 20 mM sodium dodecyl sulfate (A) or 10 mM pH 3.1 NaH_2PO_4 (B)
Injection: Electrokinetic injection at 17 kV for 9 s.
Detector: UV 204
Migration time: 3 (A), 8.5 (B)
Voltage: 17 kV
Current: 48 μA (A) or 26 μA (B)
Model: laboratory-constructed
Limit of detection: 740 nM

OTHER SUBSTANCES
Simultaneous: 3,4-dihydroxybenzylamine, dopamine, epinephrine

REFERENCE
Shakulashvili,N.; Finkler,C.; Engelhardt,H. Separation of catecholamines and serotonin by micellar electrokinetic chromatography with UV detection, *Chromatographia*, **1998**, *47*, 89–92.

SAMPLE
Matrix: solutrions

CAPILLARY ELECTROPHORESIS
Capillary: 50 cm × 75 μm fused-silica (36 cm to detector) (SGE)
Capillary preparation: Fill capillary with 100 mM KOH for 10 min, wash several times with running buffer.
Running buffer: 10 mM Tris adjusted to pH 6.4 with phosphoric acid
Injection: Siphon at 10 cm for 10 s
Detector: UV 206
Migration time: 3.3
Internal standard: dichloroisoproterenol (4.4)
Voltage: 16 kV
Model: Laboratory constructed

OTHER SUBSTANCES
Simultaneous: dopamine, epinephrine, isoxsuprine

REFERENCE
Fanali,S.; Cristalli,M.; Nardi,A.; Ossicini,L.; Shukla,S.K. Capillary zone electrophoresis in pharmaceutical analysis, *Farmaco*, **1990**, *45*, 693–702.

SAMPLE
Matrix: tissue
Sample preparation: Sonicate (Artix Sonic Dismembrator Model 150, power setting 30) 1-2 mg rat brain tissue and 10 μL EtOH:water 70:30 in an ice bath for 5-10 s, centrifuge at 16000 g for 10 min. 1 μL Supernatant + 1 μL 490 μM α-aminoadipic acid, mix, remove a 1 μL aliquot and add it to 5 μL 20 mM pH 9.0 sodium borate buffer, add 1.5 μL 20 mM NaCN in water, mix, add 1.5 μL 20 mM naphthalene-2,3-dicarboxaldehyde in MeCN, mix thoroughly, let stand at room temperature for 30 min (protect from light), inject an aliquot.

CAPILLARY ELECTROPHORESIS
Capillary: 115 cm × 50 μm fused-silica (95 cm to detector)
Running buffer: 20 mM pH 9.0 sodium borate buffer
Injection: Electrokinetic injection at 5 kV for 12 s
Detector: UV 420
Migration time: 13.6
Internal standard: α-aminoadipic acid (21.6)
Voltage: 30 kV

OTHER SUBSTANCES
Extracted: amino acids, gamma-aminobutyric acid, dopamine, levodopa, phosphoethanolamine, taurine

KEY WORDS
rat; brain; derivatization

REFERENCE
Weber,P.L.; O'Shea,T.J.; Lunte,S.M. Separation and quantitation of the amino acid neurotransmitters in rat brain by capillary electrophoresis, *J.Pharm.Biomed.Anal.*, **1994**, *12*, 319–324.

SAMPLE
Matrix: urine
Sample preparation: Condition a light alumina B SPE cartridge (Waters) with water. Adjust urine to pH 3 with 6 M HCl. 10 mL Acidified urine + 1 mL 200 mM EDTA + 100 μL 500 mM ascorbic acid, adjust pH to 8.5 with NaOH solution, add to the SPE cartridge, wash with two 2 mL portions of water, elute with 1 mL 100 mM HCl, inject an aliquot of the eluate.

CAPILLARY ELECTROPHORESIS
Capillary: 80 cm × 75 μm fused-silica (Composite Metal Services, Worcs., UK)
Running buffer: 30 mM pH 10 Borate buffer containing 0.5 mM EDTA and 1 mM ascorbic acid

Injection: Hydrodynamic injection at 40 mbar for 6 s.
Detector: F ex (Varian Schott glass UG11 filter) em 500 (cut-off filter) [ex 300 em 545] following post-column reaction. The end of the capillary was connected to a 11.5 cm × 75 μm fused-silica capillary by means of a porous PTFE sleeve with a narrow gap between the capillaries (described in the paper). The detector was 5.5 cm after the junction. The porous sleeve was immersed in a reagent reservoir. The reservoir was grounded and was under 30 mbar air pressure. (Reagent was 2 mM terbium(III) chloride containing 2 (?) mM EDTA and 800 mM CsCl.)
Migration time: 6.5
Voltage: 30 kV
Model: Prince
Limit of detection: 80 nM

OTHER SUBSTANCES
Extracted: 3,4-dihydroxyphenylacetic acid, 3,4-dihydroxyphenyl glycol, dopamine, epinephrine, levodopa

---

KEY WORDS
SPE; post-column reaction

REFERENCE
Zhu,R.; Kok,W.T. Determination of catecholamines and related compounds by capillary electrophoresis with post-column terbium complexation and sensitized luminescence detection, *Anal.Chem.*, **1997**, *69*, 4010–4016.

Norethindrone

Molecular formula: $C_{20}H_{26}O_2$
Molecular weight: 298.43
CAS Registry No.: 68-22-4, 51-98-9 (acetate)
Merck Index (12th ed.): 6790
Lednicer: 1 164, 165; 2 145

SAMPLE
Matrix: blood
Sample preparation: Condition a 1 mL 100 mg Bond Elut C18 SPE cartridge with 1 mL MeOH and 1 mL water. Add 500 μL plasma to the cartridge, wash with 2 mL water, wash with 2 mL MeOH:water 20:80, elute with two 500 μL portions of MeOH. Evaporate the eluate to dryness under a stream of nitrogen, reconstitute the residue in 50 μL MeCN:MeOH:20 mM pH 8 Tris-HCl buffer 22.5:22.5:45, inject an aliquot.

CAPILLARY ELECTROPHORESIS
Capillary: 50 cm × 100 μm fused-silica (Polymicro Technologies) packed with 3 μm ODS Hypersil for 20 cm (20 cm to detector) (details in paper)
Running buffer: MeCN:MeOH:20 mM pH 8 Tris-HCl buffer 37.5:37.5:25
Injection: Electrokinetic injection at 15 kV for 5 s (25 nL). (Injector modified for injection from small volume samples; details in paper.)
Detector: UV 240
Migration time: 12.5
Internal standard: norethindrone
Voltage: 25 kV (?)
Model: Isco Model 3850

OTHER SUBSTANCES
Extracted: androstenedione, 17α-hydroxyprogesterone, 20α-hydroxyprogesterone, progesterone, testosterone

---

KEY WORDS
electrochromatography; comparison with HPLC; plasma; SPE; norethindrone is IS

REFERENCE
Stead,D.A.; Reid,R.G.; Taylor,R.B. Capillary electrochromatography of steroids. Increased sensitivity by on-line concentration and comparison with high-performance liquid chromatography, *J.Chromatogr.A*, **1998**, *798*, 259–267.

SAMPLE
Matrix: bulk
Sample preparation: Dissolve compound in 75 mM (carboxymethyl)trimethylammonium chloride hydrazide (Girard's Reagent T) in EtOH:glacial acetic acid 90:10 at a concentration of 15-50 μM, let stand at room temperature for 30 min, dilute 10-fold with running buffer, inject an aliquot.

CAPILLARY ELECTROPHORESIS
Capillary: 64.5 cm × 50 μm fused-silica (56 cm to bubble cell detector) (Hewlett- Packard)
Capillary preparation: Rinse with 100 mM NaOH for 2 min and with running buffer for 5 min. Condition new capillaries with 1 M NaOH for 5 min, with 100 mM NaOH for 30 min, and with running buffer for 30 min.
Capillary temperature: 30
Running buffer: 20 mM pH 4.8 NaH_2PO_4
Injection: Pressure injection at 5 kPa for 5 s
Detector: UV 280
Migration time: 8.5 (norethindrone), 9.5 (norethindrone enanthate)
Voltage: 25 kV
Model: Hewlett-Packard HP 3DCE

OTHER SUBSTANCES
Interfering: nandrolone, nandrolone phenylpropionate

KEY WORDS
derivatization

REFERENCE
Görög,S.; Gazdag,M.; Kemencs Bakos,P. Analysis of steroids Part 50. Derivatization of ketosteroids for their separation and determination by capillary electrophoresis, *J.Pharm.Biomed.Anal.*, **1996**, *14*, 1115–1124.

SAMPLE
Matrix: solutions

CAPILLARY ELECTROPHORESIS
Capillary: 48.5 cm × 50 μm fused-silica (40 cm to detector) (Hewlett-Packard)
Capillary preparation: Before each run flush capillary with 100 mM NaOH for 2 min and with running buffer for 5 min.
Capillary temperature: 25
Running buffer: MeOH:buffer 10:90 (Buffer was 20 mM pH 8 Sodium borate/sodium phosphate buffer containing 50 mM sodium cholate.)
Injection: Pressure injection at 50 mbar for 1-2 s.
Detector: UV 210
Migration time: 7.5 (norethindrone), 10 (norethindrone acetate)
Voltage: 20 kV
Model: Hewlett-Packard [3D]CE

OTHER SUBSTANCES
Simultaneous: ethinyl estradiol, ethynodiol diacetate, mestranol, norethynodrel, norgestrel

REFERENCE
Poole,S.K.; Poole,C.F. Separation of pharmaceutically important estrogens by micellar electrokinetic chromatography, *J.Chromatogr.A*, **1996**, *749*, 247–255.

Norethynodrel

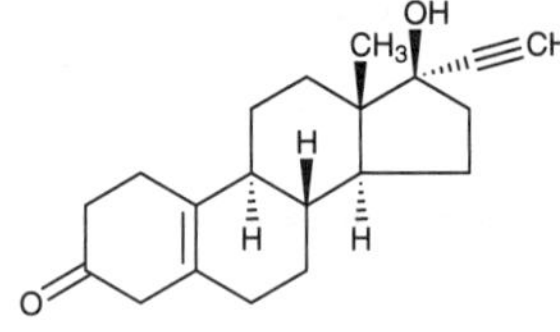

Molecular formula: $C_{20}H_{26}O_2$
Molecular weight: 298.43
CAS Registry No.: 68-23-5
Merck Index (12th ed.): 6791
Lednicer: 1 168

SAMPLE
Matrix: solutions

CAPILLARY ELECTROPHORESIS
Capillary: 48.5 cm × 50 μm fused-silica (40 cm to detector) (Hewlett-Packard)
Capillary preparation: Before each run flush capillary with 100 mM NaOH for 2 min and with running buffer for 5 min.
Capillary temperature: 25
Running buffer: MeOH:buffer 10:90 (Buffer was 20 mM pH 8 Sodium borate/sodium phosphate buffer containing 50 mM sodium cholate.)
Injection: Pressure injection at 50 mbar for 1-2 s.
Detector: UV 210
Migration time: 9
Voltage: 20 kV
Model: Hewlett-Packard ³ᴰCE

OTHER SUBSTANCES
Simultaneous: ethinyl estradiol, ethynodiol diacetate, mestranol, norethindrone, norethindrone acetate, norgestrel

REFERENCE
Poole,S.K.; Poole,C.F. Separation of pharmaceutically important estrogens by micellar electrokinetic chromatography, *J.Chromatogr.A*, **1996**, *749*, 247–255.

Norfenefrine

Molecular formula: $C_8H_{11}NO_2$
Molecular weight: 153.18
CAS Registry No.: 536-21-0, 15308-34-6 (dl-form HCl)
Merck Index (12th ed.): 6792

SAMPLE
Matrix: solutions
Sample preparation: Inject an aliquot of a 100 μg/mL solution in running buffer.

CAPILLARY ELECTROPHORESIS
Capillary: 29 cm × 50 μm fused-silica (24.5 cm to detector) (Yongnian Optical Conductive Fiber Plant, China), coated with polyacrylamide
Capillary preparation: No details of the polyacrylamide coating process are provided. However, another paper (LC.GC 1997, 15, 40) by this group indicates that they use the procedure of Hjertén, thus: Adjust the pH of 20 mL water to 3.5 with acetic acid, add 80 μL 3-(trimethoxysilyl)propyl methacrylate (3-methacryloxypropyltrimethoxysilane), mix, suck into capillary, let stand at room temperature for 1 h, remove the solution, wash with water. Fill the capillary with a deaerated 3-4% acrylamide solution containing 1 μL/mL N,N,N',N'-tetramethylethylenediamine and 1 mg/mL potassium persulfate, let stand for 30 min, remove excess solution by aspiration, rinse with water, remove water by aspiration, dry at 35° (J. Chromatogr. 1985, 347, 191).

Capillary temperature: 25
Running buffer: 100 mM NaH_2PO_4 adjusted to pH 2.5
Injection: Electrokinetic injection at 15 kV for 3 s.
Detector: UV 200, UV 210
Migration time: 3.95
Voltage: 15 kV
Model: Bio-Rad BioFocus 3000

OTHER SUBSTANCES
Simultaneous: albuterol, alprenolol, atenolol, baclofen, bamethan, benproperine, benserazide, bisoprolol, bupranolol, butamirate, butethamate, carbuterol, celiprolol, clenbuterol, clobutinol, dipivefrin, isoproterenol (isoprenaline), metaproterenol (orciprenaline), metipranolol, metoprolol, ornidazole, oxprenolol, phenylpropanolamine, pholedrine, pirbuterol, prilocaine, procyclidine, sotalol, synephrine, terbutaline, tocainide

KEY WORDS
coated capillary

REFERENCE
Koppenhoefer,B.; Epperlein,U.; Xiaofeng,Z.; Bingcheng,L. Separation of enantiomers of drugs by capillary electrophoresis. Part 4: Hydroxypropyl-γ-cyclodextrin as chiral solvating agent, *Electrophoresis*, **1997**, *18*, 924–930.

SAMPLE
Matrix: solutions
Sample preparation: Inject an aliquot of a 100 μg/mL solution in running buffer.

CAPILLARY ELECTROPHORESIS
Capillary: 30 cm × 50 μm fused-silica (25.5 cm to detector), coated with polyacrylamide
Capillary preparation: Adjust the pH of 20 mL water to 3.5 with acetic acid, add 80 μL 3-(trimethoxysilyl)propyl methacrylate (3-methacryloxypropyltrimethoxysilane), mix, suck into capillary, let stand at room temperature for 1 h, remove the solution, wash with water. Fill the capillary with a deaerated 3-4% acrylamide solution containing 1 μL/mL N,N,N',N'-tetramethylethylenediamine and 1 mg/mL potassium persulfate, let stand for 30 min, remove excess solution by aspiration, rinse with water, remove water by aspiration, dry at 35° (J. Chromatogr. 1985, 347, 191).
Capillary temperature: 25
Running buffer: 100 mM NaH_2PO_4 containing 45 mM heptakis(2,6-di-O- methyl)-β-cyclodextrin, adjusted to pH 2.5 with phosphoric acid
Injection: Electrokinetic injection at 15 kV for 3 s
Detector: UV 200
Voltage: 15 kV
Model: Bio-Rad BioFocus 3000

KEY WORDS
chiral; coated capillary; comparison with the use of other cyclodextrins; this running buffer gave the greatest enantiomeric separation.; α=1.054

REFERENCE
Lin,B.; Zhu,X.; Koppenhoefer,B.; Epperlein,U. Investigation of 123 chiral drugs by cyclodextrin-modified capillary electrophoresis, *LC.GC*, **1997**, *15*, 40–46.

SAMPLE
Matrix: solutions

CAPILLARY ELECTROPHORESIS
Capillary: 29-36 cm × 50 μm fused-silica (24.5-31.5 cm to detector) (Yongnian Optical Conductive Fiber Plant, China) coated with polyacrylamide
Capillary preparation: Coat capillary as follows. Adjust the pH of 20 mL water to 3.5 with acetic acid, add 80 μL 3-(trimethoxysilyl)propyl methacrylate (3-methacryloxypropyltrimethoxysilane), mix, suck into capillary, let stand at room temperature for 1 h, remove the solution, wash with water. Fill the capillary with a deaerated 3-4% acrylamide solution containing 1

μL/mL N,N,N',N'-tetramethylethylenediamine and 1 mg/mL potassium persulfate, let stand for 30 min, remove excess solution by aspiration, rinse with water, remove water by aspiration, dry at 35° (J. Chromatogr. 1985, 347, 191).
Capillary temperature: 25
Running buffer: 100 mM pH 2.5 NaH_2PO_4 (A) or 100 mM pH 2.5 NaH_2PO_4 containing 45 mM hydroxypropyl-α-cyclodextrin (Wacker, Munich) (B)
Injection: Electromigration at 15 kV for 3 s.
Detector: UV 200; UV 210
Migration time: 3.95 (A); 5.41, 5.48 (B) (enantiomers)
Voltage: 15 kV
Model: Bio-Focus 3000

OTHER SUBSTANCES
Also analyzed: albuterol (salbutamol), alprenolol, amorolfine, atenolol, atropine, azelastine, baclofen, bamethan, benproperine, benserazide, biperiden, bisoprolol, brompheniramine, bupivacaine, bupranolol, butamirate, butethamate, carazolol, carbuterol, carteolol, carvedilol, celiprolol, chloroquine, chlorpheniramine, chlorphenoxamine, cicletanine, clenbuterol, clidinium bromide, clobutinol, dimethindene, dipivefrin, disopyramide, dobutamine, doxylamine, fendiline, flecainide, gallopamil, homatropine, ipratropium bromide, isoproterenol (isoprenaline), isothipendyl, ketamine, meclizine, mefloquine, mepindolol, mequitazine, metaclazepam, metaproterenol (orciprenaline), metipranolol, metoprolol, nafronyl (naftidrofuryl), nefopam, nicardipine, ofloxacin, ornidazole, orphenadrine, oxomemazine, oxprenolol, oxybutynin, phenoxybenzamine, phenylpropanolamine, pholedrine, pindolol, pirbuterol, prilocaine, procyclidine, promethazine, propafenone, propranolol, reproterol, sotalol, sulpride, synephrine, talinolol, terbutaline, tetrahydrozoline (tetryzoline), theodrenaline, tioconazole, tocainide, trihexyphenidyl, trimeprazine (alimemazine), trimipramine, tropicamide, verapamil, zopiclone

---

KEY WORDS
coated capillary; chiral

REFERENCE
Koppenhoefer,B.; Eperlein,U.; Schlunk,R.; Zhu,X.; Lin,B. Separation of enantiomers of drugs by capillary electrophoresis. V. Hydroxypropyl-α-cyclodextrin as chiral solvating agent, *J.Chromatogr.A*, **1998**, *793*, 153–164.

Norfloxacin

Molecular formula: $C_{16}H_{18}FN_3O_3$
Molecular weight: 319.34
CAS Registry No.: 70458-96-7, 100587-52-8 (norfloxacin succinil)
Merck Index (12th ed.): 6793
Lednicer: 4 141, 143

SAMPLE
Matrix: solutions
Sample preparation: Inject an aliquot of a solution in MeOH.

CAPILLARY ELECTROPHORESIS
Capillary: 59 cm × 50 μm fused-silica (43 cm to detector) (Polymicro Technologies)
Capillary preparation: Before each analysis flush with water for 3 min, with 200 mM NaOH for 3 min, with water for 3 min, and with running buffer for 4 min. Flush new capillaries with 1 M NaOH at 2000 mbar for 10 min then with 200 mM NaOH for 10 min.
Capillary temperature: 23
Running buffer: MeCN:buffer 28:72, pH 7.3 (Buffer was 32 mM sodium borate containing 18 mM NaH_2PO_4, 39 mM sodium cholate, and 8 mM sodium heptanesulfonate.)
Injection: Hydrodynamic injection at 40 mbar for 6 s.
Detector: UV 260
Migration time: 4.05
Voltage: 30 kV

Model: Lauer Labs Prince

OTHER SUBSTANCES
Simultaneous: ciprofloxacin, enoxacin, flumequine, lomefloxacin, nalidixic acid, ofloxacin, oxolinic acid, pefloxacin, pipemidic acid, piromidic acid, rosoxacin, sparfloxacin

REFERENCE
Sun,S.-W.; Chen,L.-Y. Optimization of capillary electrophoretic separation of quinolone antibacterials using the overlapping resolution mapping scheme, *J.Chromatogr.A*, **1997**, *766*, 215–224.

Norgestimate

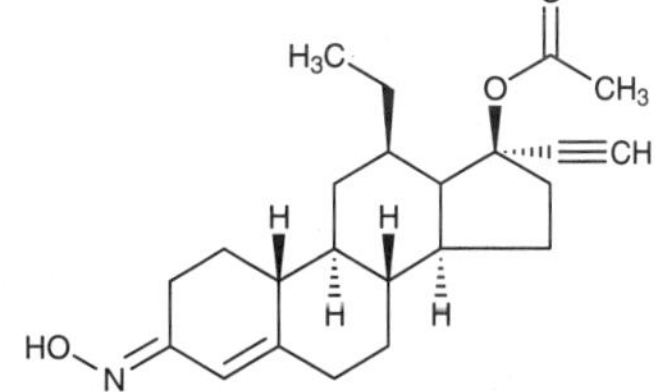

Molecular formula: $C_{23}H_{31}NO_3$
Molecular weight: 369.50
CAS Registry No.: 35189-28-7
Merck Index (12th ed.): 6796

SAMPLE
Matrix: solutions
Sample preparation: Inject an aliquot of a solution in MeOH:25 mM pH 8 Tris-HCl buffer 80: 20.

CAPILLARY ELECTROPHORESIS
Capillary: 35 cm × 100 μm fused-silica packed with 3 μm C18 (25 cm to detector) (Hewlett-Packard)
Capillary preparation: For each new capillary or buffer equilibrate with a voltage gradient of 5-10 kV for 30 min with 8 bar pressure to the inlet vial.
Capillary temperature: 25
Running buffer: MeCN:THF:25 mM pH 8 Tris-HCl buffer:water 35:20:20:25, run under 8 bar pressure on both sides.
Injection: Electromigration at 10 kV for 3 s.
Detector: UV 225
Migration time: 10 (syn), 11 (anti)
Voltage: 30 kV
Model: Hewlett-Packard 3D CE

OTHER SUBSTANCES
Simultaneous: impurities, norgestrel, norgestrel oxime, norgestrel acetate

KEY WORDS
electrochromatography

REFERENCE
Wang,J.; Schaufelberger,D.E.; Guzman,N.A. Rapid analysis of norgestimate and its potential degradation products by capillary electrophoresis, *J.Chromatogr.Sci.*, **1998**, *36*, 155–160.

Norgestrel

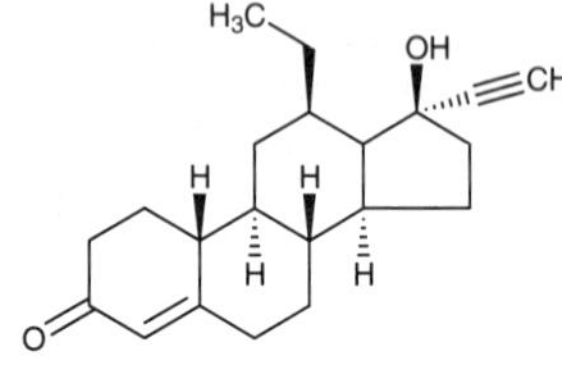

Molecular formula: $C_{21}H_{28}O_2$
Molecular weight: 312.45
CAS Registry No.: 797-63-7, 797-64-8 ((-) form), 6533-00-2
Merck Index (12th ed.): 6797
Lednicer: 1 167, 2 151, 3 84

SAMPLE
Matrix: bulk
Sample preparation: Dissolve compound in 75 mM (carboxymethyl)trimethylammonium chloride hydrazide (Girard's Reagent T) in EtOH:glacial acetic acid 90:10 at a concentration of 15-50 µM, let stand at room temperature for 30 min, dilute 10-fold with running buffer, inject an aliquot.

CAPILLARY ELECTROPHORESIS
Capillary: 64.5 cm × 50 µm fused-silica (56 cm to bubble cell detector) Hewlett- Packard
Capillary preparation: Rinse with 100 mM NaOH for 2 min and with running buffer for 5 min. Condition new capillaries with 1 M NaOH for 5 min, with 100 mM NaOH for 30 min, and with running buffer for 30 min.
Capillary temperature: 15
Running buffer: 20 mM pH 4.8 NaH_2PO_4
Injection: Pressure injection at 5 kPa for 3 s
Detector: UV 280, UV 250
Migration time: 11
Voltage: 25 kV
Model: Hewlett-Packard HP 3DCE

OTHER SUBSTANCES
Simultaneous: androstenedione (bis derivative) (UV 250), dehydroepiandrosterone (UV 250), ethisterone (UV 250), nandrolone, norethindrone (norethisterone)
Interfering: androstenedione (mono derivative) (UV 250)

KEY WORDS
derivatization

REFERENCE
Görög,S.; Gazdag,M.; Kemenes-Bakos,P. Analysis of steroids Part 50. Derivatization of ketosteroids for their separation and determination by capillary electrophoresis, *J.Pharm.Biomed.Anal.*, **1996**, *14*, 1115–1124.

SAMPLE
Matrix: solutions

CAPILLARY ELECTROPHORESIS
Capillary: 48.5 cm × 50 µm fused-silica (40 cm to detector) (Hewlett-Packard)
Capillary preparation: Before each run flush capillary with 100 mM NaOH for 2 min and with running buffer for 5 min.
Capillary temperature: 25
Running buffer: MeOH:buffer 10:90 (Buffer was 20 mM pH 8 Sodium borate/sodium phosphate buffer containing 50 mM sodium cholate.)
Injection: Pressure injection at 50 mbar for 1-2 s.
Detector: UV 210
Migration time: 8.5
Voltage: 20 kV
Model: Hewlett-Packard ^{3D}CE

OTHER SUBSTANCES
Simultaneous: ethinyl estradiol, ethynodiol diacetate, mestranol, norethindrone, norethindrone acetate, norethynodrel

REFERENCE
Poole,S.K.; Poole,C.F. Separation of pharmaceutically important estrogens by micellar electrokinetic chromatography, *J.Chromatogr.A*, **1996**, *749*, 247–255.

SAMPLE
Matrix: solutions
Sample preparation: Inject an aliquot of a solution in MeOH:25 mM pH 8 Tris-HCl buffer 80:20.

CAPILLARY ELECTROPHORESIS
Capillary: 35 cm × 100 μm fused-silica packed with 3 μm C18 (25 cm to detector) (Hewlett-Packard)
Capillary preparation: For each new capillary or buffer equilibrate with a voltage gradient of 5-10 kV for 30 min with 8 bar pressure to the inlet vial.
Capillary temperature: 25
Running buffer: MeCN:THF:25 mM pH 8 Tris-HCl buffer:water 35:20:20:25, run under 8 bar pressure on both sides.
Injection: Electromigration at 10 kV for 3 s.
Detector: UV 225
Migration time: 6
Voltage: 30 kV
Model: Hewlett-Packard 3D CE

OTHER SUBSTANCES
Simultaneous: impurities, norgestimate, norgestrel oxime, norgestrel acetate

KEY WORDS
electrochromatography

REFERENCE
Wang,J.; Schaufelberger,D.E.; Guzman,N.A. Rapid analysis of norgestimate and its potential degradation products by capillary electrophoresis, *J.Chromatogr.Sci.*, **1998**, *36*, 155–160.

Norpseudoephedrine

Molecular formula: C$_9$H$_{13}$NO
Molecular weight: 151.21
CAS Registry No.: 36393-56-3
Merck Index (12th ed.): 6811

SAMPLE
Matrix: bulk
Sample preparation: Dissolve 4 mg compound in 1 mL MeCN:water 50:50 containing 0.2% triethylamine, vortex for 30 s. Remove a 100 μL aliquot and add it to 100 μL 1.28% 2,3,4,6-tetra-O-acetyl-β-D-glucopyranosyl isothiocyanate in MeCN, vortex for 1 min, let stand for 15 min, make up to 1 mL with 10 mM pH 9.0 phosphate/borate buffer containing 100 mM sodium dodecyl sulfate, vortex for 20 s, filter (Whatman UniPrep), inject an aliquot of the filtrate.

CAPILLARY ELECTROPHORESIS
Capillary: 48 cm × 50 μm fused-silica (26 cm to detector) (Polymicro Technologies)
Capillary preparation: Condition new capillaries with 1 M NaOH for 10 min, with water for 10 min, and with running buffer for 10 min.
Capillary temperature: 30
Running buffer: MeOH:buffer 20:80 (Buffer was 10 mM pH 9.0 phosphate buffer containing 10 mM borate and 100 mM sodium dodecyl sulfate.)
Injection: Vacuum injection for 0.5 s.
Detector: UV 210
Migration time: 9.9 (+), 11.3 (-)

Voltage: 20 kV
Current: 48 μA
Model: Applied Biosystems Model 270A-HT

OTHER SUBSTANCES
Simultaneous: amphetamine, ephedrine, methamphetamine, phenylpropanolamine (norephedrine), pseudoephedrine

---

KEY WORDS
derivatization; chiral; comparison with HPLC

REFERENCE
Lurie,I.S. Micellar electrokinetic capillary chromatography of the enantiomers of amphetamine, methamphetamine and their hydroxyphenethylamine precursors, *J.Chromatogr.*, **1992**, *605*, 269–275.

SAMPLE
Matrix: solutions

CAPILLARY ELECTROPHORESIS
Capillary: 37 cm × 50 μm (30 cm to detector)
Capillary temperature: 25
Running buffer: 50 mM pH 2.5 Phosphate buffer containing 16 mM 2-hydroxy-3-trimethylammoniopropyl-β-cyclodextrin
Detector: UV 214
Migration time: 9.62 (first enantiomer; $\alpha = 1.021$; $R_S = 0.822$)
Voltage: 20 kV
Model: Beckman P/ACE 2100

OTHER SUBSTANCES
Simultaneous: cyclodrine, cyclopentolate, phendimetrazine, pholedrine, tropicamide

---

KEY WORDS
chiral

REFERENCE
Bunke,A.; Jira,T. Chiral capillary electrophoresis using a cationic cyclodextrin, *Pharmazie*, **1996**, *51*, 672–673.

Nortriptyline

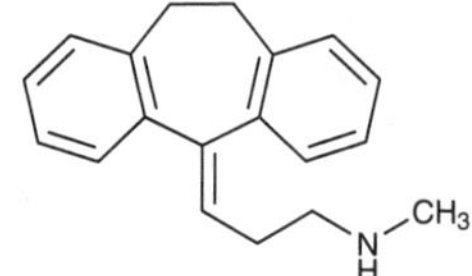

Molecular formula: C$_{19}$H$_{21}$N
Molecular weight: 263.38
CAS Registry No.: 72-69-5, 894-71-3 (HCl)
Merck Index (12th ed.): 6812
Lednicer: 1 151

SAMPLE
Matrix: blood
Sample preparation: 1 mL Serum or plasma + 50 μL 10 μL/mL trimipramine + 1 mL 2 M NaOH + 5 mL hexane:isoamyl alcohol 99:1, vortex vigorously, centrifuge at 2500 g for 5 min. Remove 4 mL of the organic layer and add it to 100 μL 100 mM HCl, extract. Remove 60 μL of the aqueous layer and evaporate it to dryness under reduced pressure, reconstitute with 20 μL water, inject an aliquot.

CAPILLARY ELECTROPHORESIS
Capillary: 71.5 cm × 50 μm fused-silica (50 cm to detector) (Polymicro Technologies)
Capillary preparation: Before each run rinse capillary with 1 M NaOH for 3 min and running buffer for 3 min (use vacuum at 508 mm Hg).

Capillary temperature: 30 or 40
Running buffer: 37.5 mM pH 8.0 Phosphate buffer containing 2 M urea and 25 mM dodecyl-trimethylammonium bromide
Injection: Vacuum injection at 508 mm Hg for 7 s
Detector: UV 254
Migration time: 10.0
Internal standard: trimipramine (11.5)
Voltage: -25 kV
Model: Applied Biosystems Model 270A
Limit of detection: 5-10 ng/mL

OTHER SUBSTANCES
Extracted: amitriptyline, desipramine, doxepin, imipramine

KEY WORDS
serum; plasma

REFERENCE
Lee,K.-J.; Lee,J.J.; Moon,D.C. Determination of tricyclic antidepressants in human plasma by micellar electro-kinetic capillary chromatography, *J.Chromatogr.*, **1993**, *616*, 135–143.

SAMPLE
Matrix: solutions
Sample preparation: Inject an aliquot of a 40 µg/mL solution in MeOH.

CAPILLARY ELECTROPHORESIS
Capillary: 64 cm × 50 µm fused silica (55.5 cm to detector) (Polymicro Technologies)
Capillary preparation: Flush with running buffer for 2 min between runs. Before use rinse with 1 M NaOH for 1 h, with 100 mM NaOH for 20 min, with water for 20 min, and with running buffer for 10 min.
Capillary temperature: 25
Running buffer: MeCN:MeOH 50:50 containing 25 mM ammonium acetate and 100 mM sodium acetate
Injection: Pressure injection at 5 kPa for 3 s
Detector: UV 214
Migration time: 6.8
Voltage: 25 kV
Model: Hewlett-Packard HP3D

OTHER SUBSTANCES
Simultaneous: amitriptyline, desmethylimipramine, didesmethylimipramine, imipramine, imipramine-N-oxide, litracene, maprotiline, methylimipramine, protriptyline

REFERENCE
Bjornsdottir,I.; Hansen,S.H. Comparison of separation selectivity in aqueous and non-aqueous capillary electrophoresis, *J.Chromatogr.A*, **1995**, *711*, 313–322.

SAMPLE
Matrix: solutions
Sample preparation: Prepare a 200 µg/mL solution in water, inject an aliquot.

CAPILLARY ELECTROPHORESIS
Capillary: 56 cm × 75 µm fused-silica (56 cm to detector) (Polymicro Technologies)
Capillary temperature: 30
Running buffer: 50 mM pH 4.0 6-Aminocaproic acid containing 25 mM 3-(N,N-dimethylmyris-tylammonium)propanesulfonate and 15 mM Tween 20
Injection: Pressure injection at 2 kPa for 3 s
Detector: UV 214
Migration time: 27
Voltage: 20 kV
Current: 62 µA
Model: Waters Quanta 4000

OTHER SUBSTANCES
Simultaneous: amitriptyline, litracene, maprotiline
Interfering: protriptyline

REFERENCE
Hansen,S.H.; Bjornsdottir,I.; Tjornelund,J. Separation of basic drug substances of very similar structure using micellar electrokinetic chromatography, *J.Pharm.Biomed.Anal.*, **1995**, *13*, 489–495.

SAMPLE
Matrix: solutions

CAPILLARY ELECTROPHORESIS
Capillary: 55.5 cm × 50 μm fused-silica (55.5 cm to detector) (Polymicro Technologies)
Capillary temperature: 25
Running buffer: MeCN:MeOH 50:50 containing 25 mM ammonium acetate and 100 mM sodium acetate
Injection: Pressure injection at 5 kPa for 3 s.
Detector: UV 214
Migration time: 6.8
Voltage: 25 kV
Current: 23 μA
Model: Hewlett Packard HP ³ᴰCE

OTHER SUBSTANCES
Simultaneous: amitriptyline, litracene, maprotiline, protriptyline

REFERENCE
Bjornsdottir,I.; Tjornelund,J.; Hansen,S.H. Nonaqueous capillary electrophoresis in pharmaceutical analysis, *J.Capillary Electrophor.*, **1996**, *3*, 83–87.

SAMPLE
Matrix: solutions

CAPILLARY ELECTROPHORESIS
Capillary: 70 cm × 50 μm coated fused-silica (55 cm to detector) (Polymicro Technologies)
Capillary preparation: Between runs rinse with water for 2 min. Coat capillaries as follows. Condition with 1 M NaOH for 15 min, rinse with water for 5 min, treat with 30 μL/mL 3-(trimethoxysilyl)propyl methacrylate in acetic acid:water 50:50 for 1 h using house vacuum, rinse with water, fill with polymerization solution, let stand for 1 h, flush, rinse with water. (Prepare polymerization solution by adding 10 μL N,N,N',N'-tetramethylethylenediamine to 10 mL of a degassed 4% solution of acrylamide in water, add 10 μL 10% ammonium persulfate in water.)
Running buffer: 10 mM pH 6.5 Citrate buffer containing 20 mM sodium dodecyl sulfate and 10 mM carboxymethyl-β-cyclodextrin (Cyclodextrin Technologies Development, Gainesville FL)
Injection: Hydrodynamic injection at 15 cm for 8-10 s.
Detector: UV 254
Migration time: 11.1
Voltage: -20 kV
Model: laboratory-constructed

OTHER SUBSTANCES
Simultaneous: amitriptyline, carbamazepine, clomipramine, desipramine, imipramine, opipramol, proptriptyline, trimipramine

KEY WORDS
coated capillary

REFERENCE
Spencer,B.J.; Zhang,W.; Purdy,W.C. Capillary electrophoretic separation of tricyclic antidepressants using charged carboxymethyl-β-cyclodextrin as a buffer additive, *Electrophoresis*, **1997**, *18*, 736–744.

SAMPLE
Matrix: solutions

CAPILLARY ELECTROPHORESIS
Capillary: 57 cm × 50 μm fused-silica (50 cm to detector) (Polymicro Technologies)
Capillary preparation: At the start of each day rinse capillary with water for 5 min, 100 mM NaOH for 30 min, with water for 10 min, and with running buffer for 15 min. Wash new capillaries with 1 M NaOH for 1 h.
Capillary temperature: 23
Running buffer: MeOH:50 mM pH 9.55 sodium phosphate buffer containing 0.06% poly(n-undecyl-α-D-glucopyranoside) (Prepare poly(n-undecyl-α-D-glucopyranoside) as follows. With exclusion of atmospheric moisture stir 40 mL alcohol-free chloroform, 5.2 g yellow mercuric oxide, 0.1 g mercuric bromide, 10 g anhydrous calcium sulfate, and 2.74 g 10-undecen-1-ol then add 10 g acetobromo-α-D-glucose (Fluka), stir at room temperature for 20 h, filter, wash the solid with chloroform. Wash the filtrate with aqueous 1 M KBr until no mercury salts are present in the aqueous layers, wash with later, concentrate. Take up the residue in benzene (Caution! Benzene is a carcinogen!) and filter it through activated silica, crystallize from n-hexane to obtain undecylenyl-2,3,4,6-tetra-O-acetyl-β-D-glucopyranoside. Add 2 mL of a 1 M solution of sodium methoxide in MeOH to a solution of 1 g of undecylenyl-2,3,4,6-tetra-O-acetyl-β-D-glucopyranoside in 50 mL dry MeOH, let stand until the reaction is complete (by TLC) to obtain undecylenyl-β-D-glucopyranoside (cf. Tenside Detergents 1978, 15, 72). Use Dowex 50W (H$^+$) resin to deionize the final product in solution. Polymerize the monomer by irradiation of an 870 μM solution in MeOH:water 20:80 with ^{60}Co gamma radiation for 48 h, purify the polymer solution by dialysis using a 1000 Da molecular mass cut-off membrane.)
Injection: Pressure injection at 0.5 psi for 5 s.
Detector: UV 214
Migration time: 11
Voltage: 22.8 kV
Model: Beckman P/ACE 5510

OTHER SUBSTANCES
Simultaneous: amitriptyline, desipramine, doxepin, imipramine, nordoxepin, protriptyline

REFERENCE
Harrell,C.W.; Dey,J.; Shamsi,S.A.; Foley,J.P.; Warner,I.M. Enhanced separation of antidepressant drugs using a polymerized nonionic surfactant as a transient capillary coating, *Electrophoresis*, **1998**, *19*, 712–718.

Noscapine

Molecular formula: C$_{22}$H$_{23}$NO$_7$
Molecular weight: 413.43
CAS Registry No.: 128-62-1, 6035-40-1 (dl-form), 912-60-7 (HCl)
Merck Index (12th ed.): 6815

SAMPLE
Matrix: bulk
Sample preparation: Mix 50 mg crude heroin with 4 mL 500 mM sulfuric acid, extract with 5 mL diethyl ether:dichloromethane 60:40. Remove the organic layer and evaporate it to dryness under a stream of nitrogen, reconstitute the residue in 750 μL MeCN:water 38:62, vortex, filter (0.45 μm), inject an aliquot of the filtrate.

CAPILLARY ELECTROPHORESIS
Capillary: 57 cm × 50 μm fused-silica (50 cm to detector) (Polymicro Technologies)
Capillary preparation: Before each use wash with 1 M NaOH for 10 min, with water for 10 min, and with running buffer for 10 min. At the start of each day wash with running buffer for 30 min. At the end of each day wash with water for 10 min.
Capillary temperature: 30

Running buffer: MeCN:buffer 10:90 (Buffer was 9 mM pH 9.0 phosphate buffer containing 9 mM borate, 45 mM sodium dodecyl sulfate, and 45 mM β-cyclodextrin sulfobutyl ether IV (CyDex L.C., Overland Park, KS)
Injection: Pressure injection for 2.5 s
Detector: UV 210 or F ex 248 (laser) em 400 ± 40 (bandpass filter)
Migration time: 30
Voltage: 21 kV
Model: Beckman Pace 5500

OTHER SUBSTANCES
Simultaneous: acetylandronornarceine, acetylnorlaudanosine, acetylnornoscapine, acetyloxyacetylanhydrodihydronornarceine, acetylthebaol, diacetylnorcodeine, dimethoxyacetyloxymethylacetamidoethylphenanthrene, meconin, methylacetamidoethylmethylenedioxymethoxyphenylacrylic acid, triacetylnormorphine

REFERENCE
Lurie,I.S.; Chan,K.C.; Spratley,T.K.; Casale,J.F.; Issaq,H.J. Separation and detection of acidic and neutral impurities in illicit heroin via capillary electrophoresis, *J.Chromatogr.B*, **1995**, *669*, 3–13.

Nylidrin

Molecular formula: $C_{19}H_{25}NO_2$
Molecular weight: 299.41
CAS Registry No.: 447-41-6, 849-55-8 (HCl)
Merck Index (12th ed.): 6830
Lednicer: 1 69

SAMPLE
Matrix: solutions

CAPILLARY ELECTROPHORESIS
Capillary: 58.2 cm × 75 μm fused-silica (50.7 cm to detector) (Polymicro Technologies)
Capillary preparation: Purge with running buffer before each run. At the beginning of each day purge using 50-60 kPa vacuum with 500 mM NaOH for 5 min, with water for 5 min, with MeCN for 5 min, and with running buffer for 5 min.
Running buffer: MeCN:MeOH:acetic acid 49:50:1 containing 20 mM ammonium acetate
Injection: Hydrostatic injection at 10 cm for 5 s.
Detector: UV 214
Migration time: 4.21
Voltage: 25 kV
Model: Waters Quanta 4000

OTHER SUBSTANCES
Simultaneous: amphetamine, benzphetamine, ephedrine, oxymetazoline, phendimetrazine, phenmetrazine, phenylephrine, phenylpropanolamine, xylometazoline

REFERENCE
Leung,G.N.W.; Tang,H.P.O.; Tso,T.S.C.; Wan,T.S.M. Separation of basic drugs with non-aqueous capillary electrophoresis, *J.Chromatogr.A*, **1996**, *738*, 141–154.

SAMPLE
Matrix: solutions

CAPILLARY ELECTROPHORESIS
Capillary: 42 cm × 75 μm acrylamide coated (36 cm to detector)
Capillary preparation: Coat column as follows. Adjust the pH of 20 mL water to 3.5 with acetic acid, add 80 μL 3-(trimethoxysilyl)propyl methacrylate (3-methacryloxypropyltrimethoxysi-

lane), mix, suck into capillary, let stand at room temperature for 1 h, remove the solution, wash with water. Fill the capillary with a deaerated 3-4% acrylamide solution containing 1 μL/mL N,N,N',N'-tetramethylethylenediamine and 1 mg/mL potassium persulfate, let stand for 30 min, remove excess solution by aspiration, rinse with water, remove water by aspiration, dry at 35° (J. Chromatogr. 1985, 347, 191).
Capillary temperature: 20
Running buffer: 100 mM pH 6 2-[N-morpholino]ethanesulfonic acid/NaOH buffer
Injection: Inject a 200 mg/mL solution of transferrin in buffer at 30 psi.s then hydrodynamically inject the sample at 10 psi.s. (Buffer was 100 mM pH 6 2-[N-morpholino]ethanesulfonic acid/ NaOH buffer. Transferrin was iron-free human serum transferrin (Behring-Werke, Marburg, Germany).)
Detector: UV 215
Migration time: 27, 28.5 (enantiomers)
Voltage: 10 kV
Model: Bio-Rad BioFocus 3000

OTHER SUBSTANCES
Simultaneous: acebutolol, chlophedianol

KEY WORDS
chiral; coated capillary

REFERENCE
Schmid,M.G.; Gübitz,G.; Kilár,F. Stereoselective interaction of drug enantiomers with human serum transferrin in capillary zone electrophoresis (II), *Electrophoresis*, **1998**, *19*, 282–287.

Nystatin

Molecular formula: $C_{47}H_{75}NO_{17}$
Molecular weight: 926.11
CAS Registry No.: 1400-61-9
Merck Index (12th ed.): 6834

SAMPLE
Matrix: solutions

CAPILLARY ELECTROPHORESIS
Capillary: 27 cm × 75 μm
Capillary preparation: Before each run rinse capillary with 100 mM NaOH for 30 s and with running buffer for 30 s. Condition new capillaries by rinsing with 100 mM NaOH for 20 min.
Capillary temperature: 30
Running buffer: 15 mM Sodium borate
Injection: Pressure injection of sample at 25 mbar for 1 s followed by running buffer at 25 mbar for 1 s.
Detector: UV 200
Migration time: 2.80
Internal standard: aminobenzoic acid (3.5)
Voltage: 6.5 kV
Model: Beckman

OTHER SUBSTANCES
Simultaneous: aspirin, bacitracin, beclomethasone, benzoic acid, ceftizoxime, ceftriaxone, cefuroxime, cephalothin, cromolyn, embonic acid, epoprostenol, glyburide (glibenclamide), levothyroxine, nedocromil, omeprazole, prednisolone, warfarin, zidovudine

REFERENCE
Altria,K.D.; Bryant,S.M.; Hadgett,T.A. Validated capillary electrophoresis method for the analysis of a range of acidic drugs and excipients, *J.Pharm.Biomed.Anal.*, **1997**, *15*, 1091–1101.

Octopamine

Molecular formula: $C_8H_{11}NO_2$
Molecular weight: 153.18
CAS Registry No.: 104-14-3, 876-04-0 (D-(-)), 770-05-8 (DL HCl)
Merck Index (12th ed.): 6856
Lednicer: 5 23

SAMPLE
Matrix: solutions

CAPILLARY ELECTROPHORESIS
Capillary: 52 cm $\times$ 75 µm fused-silica (48 cm to detector) (Polymicro Technologies)
Capillary preparation: Before each run purge with running buffer for 3 min. Every 3 runs purge with 100 mM NaOH for 5 min. Purge new capillaries with 1 M NaOH for 20 min and with 100 mM NaOH for 20 min, rinse with running buffer, equilibrate with running buffer at 12 kV for 3 h.
Running buffer: 100 mM CHES (2-(N-cyclohexylamine)ethanesulfonic acid) containing 10 mM triethylamine and 25 mM (R)-dodecoxycarbonylvaline (Waters EnantioSelect (R)-Val-1), pH adjusted to 8.8 with 1 M NaOH
Injection: Hydrostatic injection for 2 s.
Detector: UV 214
Migration time: 9, 9.2 (enantiomers)
Voltage: 12 kV
Current: $\leq$30 µA
Model: Waters Quanta 4000

OTHER SUBSTANCES
Simultaneous: bupivacaine, clenbuterol, disopyramide, metoprolol, norphenylephrine

KEY WORDS
chiral

REFERENCE
Peterson,A.G.; Ahuja,E.S.; Foley,J.P. Enantiomeric separations of basic pharmaceutical drugs by micellar electrokinetic chromatography using a chiral surfactant, N-dodecoxycarbonylvaline, *J.Chromatogr.B*, **1996**, *683*, 15–28.

SAMPLE
Matrix: solutions

CAPILLARY ELECTROPHORESIS
Capillary: 60 cm $\times$ 75 µm fused-silica (51 cm to detector) (Supelco)
Capillary preparation: At the end of each day wash capillary with 200 mM NaOH for 10 min and with water for 10 min. Condition new capillaries by washing with 200 mM NaOH for 10 min, with water for 10 min, and with running buffer for 10 min.
Running buffer: 50 mM pH 7.3 Sodium borate buffer containing 3.3% succinyl β-cyclodextrin (Wacker Chemie, Munich, Germany)
Injection: Hydrodynamic injection at 25 mbar for 6 s.
Detector: UV 208
Migration time: 41.78 (first enantiomer)
Voltage: 15 kV
Model: Prince

OTHER SUBSTANCES
Simultaneous: alprenolol, chlorthalidone, ephedrine, etilefrin, homatropine, methoxamine, norephedrine, propranolol, synephrine, trihexyphenidyl

KEY WORDS
chiral; $\alpha = 1.0835$

REFERENCE
Schmid,M.G.; Wirnsberger,K.; Gübitz,G. Chiral separation of drug enantiomers by capillary electrophoresis using succinyl-β-cyclodextrin, *Pharmazie*, **1996**, *51*, 852–854.

SAMPLE
Matrix: solutions

CAPILLARY ELECTROPHORESIS
Capillary: 37 cm × 50 μm fused-silica (30 cm to detector) (Quadrex, New Haven CT)
Capillary temperature: -16
Running buffer: MeOH:water 10:90 containing 20 mM heptakis(2,6-di-O-methyl)-β-cyclodextrin, 5 M urea, and 150 mM sodium phosphate, pH 2.5
Injection: Injection for 6 s
Detector: UV 214
Migration time: 12.8, 13 (enantiomers)
Voltage: 30 kV
Current: 45 μA
Model: Beckman P/ACE 2210

OTHER SUBSTANCES
Simultaneous: epinephrine, β-hydroxyphenethylamine, isoproterenol, norepinephrine

KEY WORDS
chiral

REFERENCE
Ma,S.; Horváth,C. Capillary zone electrophoresis at subzero temperatures. II: Chiral separation of biogenic amines, *Electrophoresis*, **1997**, *18*, 873–883.

Ofloxacin

Molecular formula: $C_{18}H_{20}FN_3O_4$
Molecular weight: 361.37
CAS Registry No.: 82419-36-1
Merck Index (12th ed.): 6865
Lednicer: 4 141-145

SAMPLE
Matrix: solutions
Sample preparation: Prepare a 200 μg/mL solution in water, inject an aliquot.

CAPILLARY ELECTROPHORESIS
Capillary: 50 cm fused-silica (20 cm to detector)
Capillary preparation: Store overnight filled with water. Before use purge with 500 mM NaOH then water.
Capillary temperature: 23
Running buffer: 20 mM pH 7.5 phosphate buffer containing 0.4% bovine serum albumin
Injection: Electrokinetic at 15 kV for 3 s
Detector: UV 300

Migration time: 9.4 (S), 10.1 (R)
Voltage: 30 kV
Current: 28 μA
Model: Jasco Model CE-800

KEY WORDS
chiral

REFERENCE
Arai,T.; Nimura,N.; Kinoshita,T. Investigation of enantioselective ofloxacin-albumin binding and displacement interactions using capillary affinity zone electrophoresis, *Biomed.Chromatogr.*, **1995**, *9*, 68–74.

SAMPLE
Matrix: solutions
Sample preparation: Inject an aliquot of a 100 μg/mL solution in water:running buffer 50:50.

CAPILLARY ELECTROPHORESIS
Capillary: 44.5 cm × 50 μm acrylamide-coated fused-silica (Bio-Rad)
Capillary temperature: 30
Running buffer: 100 mM NaH_2PO_4 containing 15 mM gamma-cyclodextrin, adjusted to pH 2.5 with phosphoric acid
Injection: Electrokinetic injection at 8 kV for 6 s.
Detector: UV 200
Migration time: 9.04
Voltage: 14 kV
Model: Bio-Rad BioFocus 3000

OTHER SUBSTANCES
Also analyzed: albuterol, alprenolol, atenolol, atropine, baclofen, bamethan, benserazide, biperiden, bisoprolol, bupivacaine, bupranolol, butetamate, carazolol, carbuterol, carvedilol, celiprolol, chloroquine, chlorpheniramine (chlorphenamine), clidinium bromide, clobutinol, disopyramide, dobutamine, flecainide, homatropine, ipratropium bromide, isoproterenol, isothipendyl, ketamine, mefloquine, mequitazine, metaproterenol (orciprenaline), metipranolol, nafronyl (naftidrofuryl), nefopam, orphenadrine, oxomemazine, oxprenolol, phenoxybenzamine, pholedrine, pindolol, pirbuterol, prilocaine, promethazine, propafenone, propranolol, sotalol, synephrine, terbutaline, tetrahydrozoline (tetryzoline), tocainide, trihexyphenidyl, trimeprazine (alimemazine), trimipramine, tropicamide, verapamil, zopiclone

KEY WORDS
coated capillary; achiral

REFERENCE
Koppenhoefer,B.; Epperlein,U.; Christian,B.; Yibing,J.; Yuying,C.; Bingcheng,L. Separation of enantiomers of drugs by capillary electrophoresis. I. γ-Cyclodextrin as chiral solvating agent, *J.Chromatogr.A*, **1995**, *717*, 181–190.

SAMPLE
Matrix: solutions
Sample preparation: Inject an aliquot of a 100 μg/mL solution in water.

CAPILLARY ELECTROPHORESIS
Capillary: 50 cm × 50 μm fused-silica (45 cm to detector)
Capillary preparation: At the start of each day purge capillary with 50 mM NaOH then water. Change reservoir buffers after each run.
Capillary temperature: 23
Running buffer: 100 mM pH 4.0 Sodium acetate buffer containing 5 mM vancomycin
Injection: Electrokinetic injection at 10 kV for 2 s.
Detector: UV 300
Migration time: 31.4 (R), 32.2 (S)
Voltage: 10 kV
Current: 20 μA
Model: Jasco CE-800

OTHER SUBSTANCES
Simultaneous: related compounds

KEY WORDS
chiral

REFERENCE
Arai,T.; Nimura,N.; Kinoshita,T. Investigation of enantioselective separation of quinolonecarboxylic acids by capillary zone electrophoresis using vancomycin as a chiral selector, *J.Chromatogr.A*, **1996**, *736*, 303–311.

SAMPLE
Matrix: solutions

CAPILLARY ELECTROPHORESIS
Capillary: 75 cm × 50 μm fused-silica (50 cm to detector) (JASCO)
Capillary preparation: Before analysis treat under pressure with 100 mM phosphoric acid, water, and running buffer for 10 min.
Running buffer: 10 mM pH 6.5 Ammonium acetate buffer containing 20 mM gamma-cyclodextrin, 10 mM zinc sulfate, and 10 mM D-phenylalanine
Injection: Siphon at 20 cm for 5 s.
Detector: UV 300
Migration time: 18 (R-(+)), 19 (S-(-))
Voltage: 10 kV
Model: JASCO CE-800

KEY WORDS
chiral

REFERENCE
Horimai,T.; Ohara,M.; Ichinose,M. Optical resolution of new quinolone drugs by capillary electrophoresis with ligand-exchange and host-guest interaction, *J.Chromatogr.A*, **1997**, *760*, 235–244.

SAMPLE
Matrix: solutions
Sample preparation: Inject an aliquot of a 100 μg/mL solution in running buffer.

CAPILLARY ELECTROPHORESIS
Capillary: 36 cm × 50 μm fused-silica (31.5 cm to detector) (Yongnian Optical Conductive Fiber Plant, China), coated with polyacrylamide
Capillary preparation: No details of the polyacrylamide coating process are provided. However, another paper (LC.GC 1997, 15, 40) by this group indicates that they use the procedure of Hjertén, thus: Adjust the pH of 20 mL water to 3.5 with acetic acid, add 80 μL 3-(trimethoxysilyl)propyl methacrylate (3-methacryloxypropyltrimethoxysilane), mix, suck into capillary, let stand at room temperature for 1 h, remove the solution, wash with water. Fill the capillary with a deaerated 3-4% acrylamide solution containing 1 μL/mL N,N,N',N'-tetramethylethylenediamine and 1 mg/mL potassium persulfate, let stand for 30 min, remove excess solution by aspiration, rinse with water, remove water by aspiration, dry at 35° (J. Chromatogr. 1985, 347, 191).
Capillary temperature: 25
Running buffer: 100 mM NaH_2PO_4 adjusted to pH 2.5 (A) or 100 mM NaH_2PO_4 containing 45 mM hydroxypropyl-gamma-cyclodextrin, adjusted to pH 2.5 (B)
Injection: Electrokinetic injection at 15 kV for 3 s.
Detector: UV 200, UV 210
Migration time: 6.42 (A), 7.46, 7.53 (B, enantiomers)
Voltage: 15 kV
Model: Bio-Rad BioFocus 3000

OTHER SUBSTANCES
Simultaneous: azelastine, biperiden, carvedilol, clidinium bromide, meclizine (meclozine), mequitazine, zopiclone

KEY WORDS
chiral; coated capillary

REFERENCE
Koppenhoefer,B.; Epperlein,U.; Xiaofeng,Z.; Bingcheng,L. Separation of enantiomers of drugs by capillary electrophoresis. Part 4: Hydroxypropyl-γ-cyclodextrin as chiral solvating agent, *Electrophoresis*, **1997**, *18*, 924–930.

SAMPLE
Matrix: solutions
Sample preparation: Inject an aliquot of a 100 μg/mL solution in running buffer.

CAPILLARY ELECTROPHORESIS
Capillary: 30 cm × 50 μm fused-silica (25.5 cm to detector), coated with polyacrylamide
Capillary preparation: Adjust the pH of 20 mL water to 3.5 with acetic acid, add 80 μL 3-(trimethoxysilyl)propyl methacrylate (3-methacryloxypropyltrimethoxysilane), mix, suck into capillary, let stand at room temperature for 1 h, remove the solution, wash with water. Fill the capillary with a deaerated 3-4% acrylamide solution containing 1 μL/mL N,N,N',N'-tetramethylethylenediamine and 1 mg/mL potassium persulfate, let stand for 30 min, remove excess solution by aspiration, rinse with water, remove water by aspiration, dry at 35° (J. Chromatogr. 1985, 347, 191).
Capillary temperature: 25
Running buffer: 100 mM NaH_2PO_4 containing 45 mM heptakis(2,6-di-O- methyl)-β-cyclodextrin, adjusted to pH 2.5 with phosphoric acid
Injection: Electrokinetic injection at 15 kV for 3 s.
Detector: UV 200
Voltage: 15 kV
Model: Bio-Rad BioFocus 3000

KEY WORDS
chiral; coated capillary; comparison with the use of other cyclodextrins; this running buffer gave the greatest enantiomeric separation.; α=1.069

REFERENCE
Lin,B.; Zhu,X.; Koppenhoefer,B.; Epperlein,U. Investigation of 123 chiral drugs by cyclodextrin-modified capillary electrophoresis, *LC.GC*, **1997**, *15*, 40–46.

SAMPLE
Matrix: solutions
Sample preparation: Inject an aliquot of a solution in MeOH.

CAPILLARY ELECTROPHORESIS
Capillary: 59 cm × 50 μm fused-silica (43 cm to detector) (Polymicro Technologies)
Capillary preparation: Before each analysis flush with water for 3 min, with 200 mM NaOH for 3 min, with water for 3 min, and with running buffer for 4 min. Flush new capillaries with 1 M NaOH at 2000 mbar for 10 min then with 200 mM NaOH for 10 min.
Capillary temperature: 23
Running buffer: MeCN:buffer 28:72, pH 7.3 (Buffer was 32 mM sodium borate containing 18 mM NaH_2PO_4, 39 mM sodium cholate, and 8 mM sodium heptanesulfonate.)
Injection: Hydrodynamic injection at 40 mbar for 6 s.
Detector: UV 260
Migration time: 4.4
Voltage: 30 kV
Model: Lauer Labs Prince

OTHER SUBSTANCES
Simultaneous: ciprofloxacin, enoxacin, flumequine, lomefloxacin, nalidixic acid, norfloxacin, oxolinic acid, pefloxacin, pipemidic acid, piromidic acid, rosoxacin, sparfloxacin

REFERENCE
Sun,S.-W.; Chen,L.-Y. Optimization of capillary electrophoretic separation of quinolone antibacterials using the overlapping resolution mapping scheme, *J.Chromatogr.A*, **1997**, *766*, 215–224.

SAMPLE
Matrix: solutions

CAPILLARY ELECTROPHORESIS
Capillary: 29-36 cm × 50 μm fused-silica (24.5-31.5 cm to detector) (Yongnian Optical Conductive Fiber Plant, China) coated with polyacrylamide
Capillary preparation: Coat capillary as follows. Adjust the pH of 20 mL water to 3.5 with acetic acid, add 80 μL 3-(trimethoxysilyl)propyl methacrylate (3-methacryloxypropyltrimethoxysilane), mix, suck into capillary, let stand at room temperature for 1 h, remove the solution, wash with water. Fill the capillary with a deaerated 3-4% acrylamide solution containing 1 μL/mL N,N,N',N'-tetramethylethylenediamine and 1 mg/mL potassium persulfate, let stand for 30 min, remove excess solution by aspiration, rinse with water, remove water by aspiration, dry at 35° (J. Chromatogr. 1985, 347, 191).
Capillary temperature: 25
Running buffer: 100 mM pH 2.5 NaH$_2$PO$_4$ (A) or 100 mM pH 2.5 NaH$_2$PO$_4$ containing 45 mM hydroxypropyl-α-cyclodextrin (Wacker, Munich) (B)
Injection: Electromigration at 15 kV for 3 s.
Detector: UV 200; UV 210
Migration time: 6.42 (A); 8.00, 8.37 (B) (enantiomers)
Voltage: 15 kV
Model: Bio-Focus 3000

OTHER SUBSTANCES
Also analyzed: albuterol (salbutamol), alprenolol, amorolfine, atenolol, atropine, azelastine, baclofen, bamethan, benproperine, benserazide, biperiden, bisoprolol, brompheniramine, bupivacaine, bupranolol, butamirate, butethamate, carazolol, carbuterol, carteolol, carvedilol, celiprolol, chloroquine, chlorpheniramine, chlorphenoxamine, cicletanine, clenbuterol, clidinium bromide, clobutinol, dimethindene, dipivefrin, disopyramide, dobutamine, doxylamine, fendiline, flecainide, gallopamil, homatropine, ipratropium bromide, isoproterenol (isoprenaline), isothipendyl, ketamine, meclizine, mefloquine, mepindolol, mequitazine, metaclazepam, metaproterenol (orciprenaline), metipranolol, metoprolol, nafronyl (naftidrofuryl), nefopam, nicardipine, norfenefrine, ornidazole, orphenadrine, oxomemazine, oxprenolol, oxybutynin, phenoxybenzamine, phenylpropanolamine, pholedrine, pindolol, pirbuterol, prilocaine, procyclidine, promethazine, propafenone, propranolol, reproterol, sotalol, sulpride, synephrine, talinolol, terbutaline, tetrahydrozoline (tetryzoline), theodrenaline, tioconazole, tocainide, trihexyphenidyl, trimeprazine (alimemazine), trimipramine, tropicamide, verapamil, zopiclone

KEY WORDS
coated capillary; chiral

REFERENCE
Koppenhoefer,B.; Eperlein,U.; Schlunk,R.; Zhu,X.; Lin,B. Separation of enantiomers of drugs by capillary electrophoresis. V. Hydroxypropyl-α-cyclodextrin as chiral solvating agent, *J.Chromatogr.A*, **1998**, *793*, 153–164.

SAMPLE
Matrix: urine
Sample preparation: Inject an aliquot directly (?).

CAPILLARY ELECTROPHORESIS
Capillary: 55.5 cm × 50 μm fused-silica (47.5 cm to detector) (Yongnian, China)
Capillary preparation: After each run rinse capillary with 100 mM NaOH for 5 min and with water for 5 min then equilibrate with running buffer for 10 min.
Capillary temperature: 24 ± 0.2
Running buffer: EtOH:40 mM borax 10:90 adjusted to pH 4.0 with phosphoric acid.
Injection: Electrokinetic at 20 ± 0.3 kV for 20 s.
Detector: UV 280
Migration time: 11
Voltage: 25 ± 0.3 kV
Model: 1229 HPCE
Limit of detection: 200 ng/mL

OTHER SUBSTANCES
Extracted: capreomycin, pasiniazide

REFERENCE
Zhang,S.S.; Liu,H.X.; Yuan,Z.B.; Yu,C.L. A reproducible, simple and sensitive high-performance capillary elec-
trophoresis method for simultaneous determination of capreomycin, ofloxacin and pasiniazide in urine,
J.Pharm.Biomed.Anal., **1998**, *17*, 617–622.

Oleandomycin

Molecular formula: $C_{35}H_{61}NO_{12}$
Molecular weight: 687.87
CAS Registry No.: 3922-90-5, 6696-47-5 (HCl), 7060-74-4
(phosphate)
Merck Index (12th ed.): 6962

SAMPLE
Matrix: solutions
Sample preparation: Inject an aliquot of a 500 μg/mL solution in MeOH:water 10:80.

CAPILLARY ELECTROPHORESIS
Capillary: 114.5 cm × 75 μm fused-silica (75 cm)
Capillary preparation: Between runs rinse capillary with running buffer for 3 min. Before use
flush capillary with 100 mM NaOH for 10 min, with water for 10 min, and equilibrate with
running buffer for 15 min. Condition a new capillary at 2000 mbar with 1 M NaOH for 30 min,
100 mM NaOH for 30 min, and water for 30 min.
Capillary temperature: 25
Running buffer: MeOH:75 mM pH 7.5 phosphate buffer 50:50
Injection: Pressure injection at 50 mbar for 5 s, ramp to operating voltage at 6 kV/s.
Detector: UV 200
Migration time: 27
Voltage: 25 kV
Model: Prince

OTHER SUBSTANCES
Simultaneous: erythromycin, josamycin

REFERENCE
Lalloo,A.K.; Chataraj,S.C.; Kanfer,I. Development of a capillary electrophoretic method for the separation of
the macrolide antibiotics, erythromycin, josamycin and oleandomycin, *J.Chromatogr.B*, **1997**, *704*, 333–341.

Omeprazole

Molecular formula: $C_{17}H_{19}N_3O_3S$
Molecular weight: 345.42
CAS Registry No.: 73590-58-6
Merck Index (12th ed.): 6977
Lednicer: 4 133

SAMPLE
Matrix: solutions

CAPILLARY ELECTROPHORESIS
Capillary: 27 cm × 75 μm
Capillary preparation: Before each run rinse capillary with 100 mM NaOH for 30 s and with running buffer for 30 s. Condition new capillaries by rinsing with 100 mM NaOH for 20 min.
Capillary temperature: 30
Running buffer: 15 mM Sodium borate
Injection: Pressure injection of sample at 25 mbar for 1 s followed by running buffer at 25 mbar for 1 s.
Detector: UV 200
Migration time: 2.52
Internal standard: aminobenzoic acid (3.5)
Voltage: 6.5 kV
Model: Beckman

OTHER SUBSTANCES
Simultaneous: aspirin, bacitracin, beclomethasone, benzoic acid, ceftizoxime, ceftriaxone, cefuroxime, cephalothin, cromolyn, embonic acid, epoprostenol, glyburide (glibenclamide), levothyroxine, nedocromil, nystatin, prednisolone, warfarin, zidovudine

REFERENCE
Altria,K.D.; Bryant,S.M.; Hadgett,T.A. Validated capillary electrophoresis method for the analysis of a range of acidic drugs and excipients, *J.Pharm.Biomed.Anal.*, **1997**, *15*, 1091–1101.

SAMPLE
Matrix: solutions
Sample preparation: Inject an aliquot of a solution in water.

CAPILLARY ELECTROPHORESIS
Capillary: 66 cm × 50 μm fused-silica (50 cm to detector)
Capillary preparation: After each analysis rinse with 100 mM NaOH for 3 min and with running buffer for 2 min.
Running buffer: n-Propanol:10 mM pH 7.4 potassium phosphate buffer containing 100 μM bovine serum albumin 7:93
Injection: Electrokinetic injection at 5-8 kV for 7 s.
Detector: UV 290
Migration time: 15.5, 16 (enantiomers)
Voltage: about 300 V/cm
Current: 35 μA (constant current)
Model: Grom CE-System 100
Limit of detection: 40 μg/mL

KEY WORDS
chiral

REFERENCE
Eberle,D.; Hummel,R.P.; Kuhn,R. Chiral resolution of pantoprazole sodium and related sulfoxides by complex formation with bovine serum albumin in capillary electrophoresis, *J.Chromatogr.A*, **1997**, *759*, 185–192.

Ondansetron

Molecular formula: $C_{18}H_{19}N_3O$

Molecular weight: 293.37

CAS Registry No.: 99614-02-5, 116002-70-1, 99614-01-4 (HCl dihydrate), 103639-04-9 (HCl dihydrate)

Merck Index (12th ed.): 6979

Lednicer: 5 164

SAMPLE
Matrix: blood
Sample preparation: Condition a 1 mL 100 mg cyanopropyl SPE cartridge (Varian) with 1 mL MeOH, 1 mL 1% ammonia in isopropanol, and 1 mL water, do not allow to dry. Mix 2 mL serum with 6.5 µL 100 µg/mL procainamide in MeOH. Add 100 µL 500 mM HCl to the SPE cartridge followed by the serum, wash with two 1 mL portions of water, allow to dry, wash with two 1 mL portions of MeCN, elute with four 500 µL portions of 1% ammonia in isopropanol. Filter the eluate and evaporate it to dryness under a stream of nitrogen, reconstitute with 90 µL water, inject an aliquot.

CAPILLARY ELECTROPHORESIS
Capillary: 72 cm × 50 µm fused-silica (50 cm to detector) (Polymicro Technologies)
Capillary preparation: Before each run rinse capillary with 100 mM NaOH for 2 min and with running buffer for 2.5 min. Condition new capillaries by rinsing with 1 M NaOH for 10 min, with water for 10 min, and with running buffer for 10 min.
Capillary temperature: 30
Running buffer: 100 mM NaH_2PO_4 containing 15 mM heptakis-(2,6-di-O-methyl)-β-cyclodextrin and 30 µM hexadecyltrimethylammonium bromide, adjusted to pH 2.5 with 100 mM phosphoric acid
Injection: Vacuum injection at 80 psi for 20 s.
Detector: UV 254
Migration time: 19.3 (S-(+)), 19.8 (R-(-))
Internal standard: procainamide (10.0)
Voltage: 20 kV
Current: 65 µA (typical)
Model: Applied Biosystems ABI 270A
Limit of quantitation: 15 ng/mL
Limit of detection: 10 ng/mL

KEY WORDS
chiral; serum; SPE

REFERENCE
Siluveru,M.; Stewart,J.T. Enantioselective determination of S-(+)- and R-(-)-ondansetron in human serum using derivatized cysclodextrin-modified capillary electrophoresis and solid-phase extraction, *J.Chromatogr.B*, **1997**, *691*, 217–222.

Opipramol

Molecular formula: $C_{23}H_{29}N_3O$
Molecular weight: 363.50
CAS Registry No.: 315-72-0, 909-39-7 (2.HCl)
Merck Index (12th ed.): 6985

SAMPLE
Matrix: solutions

CAPILLARY ELECTROPHORESIS
Capillary: 70 cm × 50 µm coated fused-silica (55 cm to detector) (Polymicro Technologies)
Capillary preparation: Between runs rinse with water for 2 min. Coat capillaries as follows. Condition with 1 M NaOH for 15 min, rinse with water for 5 min, treat with 30 µL/mL 3-(trimethoxysilyl)propyl methacrylate in acetic acid:water 50:50 for 1 h using house vacuum, rinse with water, fill with polymerization solution, let stand for 1 h, flush, rinse with water. (Prepare polymerization solution by adding 10 µL N,N,N',N'-tetramethylethylenediamine to 10 mL of a degassed 4% solution of acrylamide in water, add 10 µL 10% ammonium persulfate in water.)
Running buffer: 10 mM pH 6.5 Citrate buffer containing 20 mM sodium dodecyl sulfate and 10 mM carboxymethyl-β-cyclodextrin (Cyclodextrin Technologies Development, Gainesville FL)

Injection: Hydrodynamic injection at 15 cm for 8-10 s.
Detector: UV 254
Migration time: 10.2
Voltage: -20 kV
Model: laboratory-constructed

OTHER SUBSTANCES
Simultaneous: amitriptyline, carbamazepine, clomipramine, desipramine, imipramine, nortriptyline, proptriptyline, trimipramine

KEY WORDS
coated capillary

REFERENCE
Spencer,B.J.; Zhang,W.; Purdy,W.C. Capillary electrophoretic separation of tricyclic antidepressants using charged carboxymethyl-β-cyclodextrin as a buffer additive, *Electrophoresis*, **1997**, *18*, 736–744.

Ormetoprim

Molecular formula: $C_{14}H_{18}N_4O_2$
Molecular weight: 274.32
CAS Registry No.: 6981-18-6
Lednicer: 2 302

SAMPLE
Matrix: solutions

CAPILLARY ELECTROPHORESIS
Capillary: 90 cm × 50 μm untreated fused-silica (Polymicro Technologies)
Running buffer: Trisma buffer pH 7.2 (about 7 g Trisma HCl and 0.7 g Trisma base in 1 L water)
Injection: 3 s vacuum injection
Detector: UV 200 or MS, SCIEX API III triple quadrupole, API source, positive ion analysis, make-up flow 0.2% aqueous formic acid, nebulizing gas air, target gas argon, collision energy 20-35 eV
Migration time: 6.7
Voltage: 24.4 kV
Model: Applied Biosystems Model 270A

OTHER SUBSTANCES
Simultaneous: sulfaisomidine, sulfamethazine, trimethoprim
Interfering: erythromycin

REFERENCE
Pleasance,S.; Thibault,P.; Kelly,J. Comparison of liquid-junction and coaxial interfaces for capillary electrophoresis-mass spectrometry with application to compounds of concern to the aquaculture industry, *J.Chromatogr.*, **1992**, *591*, 325–339.

SAMPLE
Matrix: solutions
Sample preparation: Prepare a 2.5 μg/mL solution in water, filter (0.45 μm), inject an aliquot.

CAPILLARY ELECTROPHORESIS
Capillary: 71.2 cm × 50 μm fused-silica (52 cm to detector) (Applied Biosystems)
Capillary preparation: Between runs purge the capillary with 100 mM NaOH for 2 min and with running buffer for 2 min. At the start of each day purge capillary with 100 mM NaOH for 2 min, with water for 2 min, and with running buffer for 2 min

Capillary temperature: 30
Running buffer: 250 mM pH 2.1 Phosphate buffer
Injection: Vacuum injection for 10 s (40 nL)
Detector: UV 225
Migration time: 24.065
Voltage: 13 kV
Model: Applied Biosystems Model 270 A

OTHER SUBSTANCES
Simultaneous: aditoprim, brodimoprim, diaveridine, metioprim, pyrimethamine, tetroxoprim, trimethoprim

REFERENCE
Cao,J.; Cross,R.F. The separation of dihydrofolate reductase inhibitors and the determination of pKa,1 values by capillary zone electrophoresis, *J.Chromatogr.A*, **1995**, *695*, 297–308.

Ornidazole

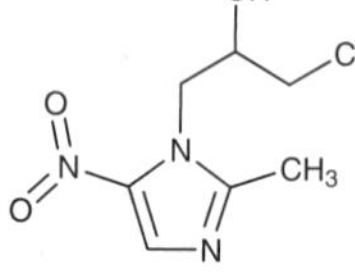

Molecular formula: $C_7H_{10}ClN_3O_3$
Molecular weight: 219.63
CAS Registry No.: 16773-42-5
Merck Index (12th ed.): 7000
Lednicer: 3 131

SAMPLE
Matrix: solutions
Sample preparation: Inject an aliquot of a 100 µg/mL solution in running buffer.

CAPILLARY ELECTROPHORESIS
Capillary: 29 cm × 50 µm fused-silica (24.5 cm to detector) (Yongnian Optical Conductive Fiber Plant, China), coated with polyacrylamide
Capillary preparation: No details of the polyacrylamide coating process are provided. However, another paper (LC.GC 1997, 15, 40) by this group indicates that they use the procedure of Hjertén, thus: Adjust the pH of 20 mL water to 3.5 with acetic acid, add 80 µL 3-(trimethoxysilyl)propyl methacrylate (3-methacryloxypropyltrimethoxysilane), mix, suck into capillary, let stand at room temperature for 1 h, remove the solution, wash with water. Fill the capillary with a deaerated 3-4% acrylamide solution containing 1 µL/mL N,N,N',N'-tetramethylethylenediamine and 1 mg/mL potassium persulfate, let stand for 30 min, remove excess solution by aspiration, rinse with water, remove water by aspiration, dry at 35° (J. Chromatogr. 1985, 347, 191).
Capillary temperature: 25
Running buffer: 100 mM NaH_2PO_4 adjusted to pH 2.5
Injection: Electrokinetic injection at 15 kV for 3 s.
Detector: UV 200, UV 210
Migration time: 13.38
Voltage: 15 kV
Model: Bio-Rad BioFocus 3000

OTHER SUBSTANCES
Simultaneous: albuterol, alprenolol, atenolol, baclofen, bamethan, benproperine, benserazide, bisoprolol, bupranolol, butamirate, butethamate, carbuterol, celiprolol, clenbuterol, clobutinol, dipivefrin, isoproterenol (isoprenaline), metaproterenol (orciprenaline), metipranolol, metoprolol, norfenefrine, oxprenolol, phenylpropanolamine, pholedrine, pirbuterol, prilocaine, procyclidine, sotalol, synephrine, terbutaline, tocainide

KEY WORDS
coated capillary

REFERENCE
Koppenhoefer,B.; Epperlein,U.; Xiaofeng,Z.; Bingcheng,L. Separation of enantiomers of drugs by capillary electrophoresis. Part 4: Hydroxypropyl-γ-cyclodextrin as chiral solvating agent, *Electrophoresis*, **1997**, *18*, 924–930.

SAMPLE
Matrix: solutions
Sample preparation: Inject an aliquot of a 100 μg/mL solution in running buffer.

CAPILLARY ELECTROPHORESIS
Capillary: 30 cm × 50 μm fused-silica (25.5 cm to detector), coated with polyacrylamide
Capillary preparation: Adjust the pH of 20 mL water to 3.5 with acetic acid, add 80 μL 3-(trimethoxysilyl)propyl methacrylate (3-methacryloxypropyltrimethoxysilane), mix, suck into capillary, let stand at room temperature for 1 h, remove the solution, wash with water. Fill the capillary with a deaerated 3-4% acrylamide solution containing 1 μL/mL N,N,N',N'-tetramethylethylenediamine and 1 mg/mL potassium persulfate, let stand for 30 min, remove excess solution by aspiration, rinse with water, remove water by aspiration, dry at 35° (J. Chromatogr. 1985, 347, 191).
Capillary temperature: 25
Running buffer: 100 mM NaH_2PO_4 containing 45 mM heptakis(2,6-di-O- methyl)-β-cyclodextrin, adjusted to pH 2.5 with phosphoric acid
Injection: Electrokinetic injection at 15 kV for 3 s.
Detector: UV 200
Voltage: 15 kV
Model: Bio-Rad BioFocus 3000

KEY WORDS
chiral; coated capillary; comparison with the use of other cyclodextrins; this running buffer gave the greatest enantiomeric separation.; α=1.025

REFERENCE
Lin,B.; Zhu,X.; Koppenhoefer,B.; Epperlein,U. Investigation of 123 chiral drugs by cyclodextrin-modified capillary electrophoresis, *LC.GC*, **1997**, *15*, 40–46.

SAMPLE
Matrix: solutions

CAPILLARY ELECTROPHORESIS
Capillary: 29-36 cm × 50 μm fused-silica (24.5-31.5 cm to detector) (Yongnian Optical Conductive Fiber Plant, China) coated with polyacrylamide
Capillary preparation: Coat capillary as follows. Adjust the pH of 20 mL water to 3.5 with acetic acid, add 80 μL 3-(trimethoxysilyl)propyl methacrylate (3-methacryloxypropyltrimethoxysilane), mix, suck into capillary, let stand at room temperature for 1 h, remove the solution, wash with water. Fill the capillary with a deaerated 3-4% acrylamide solution containing 1 μL/mL N,N,N',N'-tetramethylethylenediamine and 1 mg/mL potassium persulfate, let stand for 30 min, remove excess solution by aspiration, rinse with water, remove water by aspiration, dry at 35° (J. Chromatogr. 1985, 347, 191).
Capillary temperature: 25
Running buffer: 100 mM pH 2.5 NaH_2PO_4 (A) or 100 mM pH 2.5 NaH_2PO_4 containing 45 mM hydroxypropyl-α-cyclodextrin (Wacker, Munich) (B)
Injection: Electromigration at 15 kV for 3 s.
Detector: UV 200; UV 210
Migration time: 13.38 (A); 23.91 (B) (no separation of enantiomers)
Voltage: 15 kV
Model: Bio-Focus 3000

OTHER SUBSTANCES
Also analyzed: albuterol (salbutamol), alprenolol, amorolfine, atenolol, atropine, azelastine, baclofen, bamethan, benproperine, benserazide, biperiden, bisoprolol, brompheniramine, bupivacaine, bupranolol, butamirate, butethamate, carazolol, carbuterol, carteolol, carvedilol, celiprolol, chloroquine, chlorpheniramine, chlorphenoxamine, cicletanine, clenbuterol, clidinium bromide, clobutinol, dimethindene, dipivefrin, disopyramide, dobutamine, doxylamine, fendi-

line, flecainide, gallopamil, homatropine, ipratropium bromide, isoproterenol (isoprenaline), isothipendyl, ketamine, meclizine, mefloquine, mepindolol, mequitazine, metaclazepam, metaproterenol (orciprenaline), metipranolol, metoprolol, nafronyl (naftidrofuryl), nefopam, nicardipine, norfenefrine, ofloxacin, orphenadrine, oxomemazine, oxprenolol, oxybutynin, phenoxybenzamine, phenylpropanolamine, pholedrine, pindolol, pirbuterol, prilocaine, procyclidine, promethazine, propafenone, propranolol, reproterol, sotalol, sulpride, synephrine, talinolol, terbutaline, tetrahydrozoline (tetryzoline), theodrenaline, tioconazole, tocainide, trihexyphenidyl, trimeprazine (alimemazine), trimipramine, tropicamide, verapamil, zopiclone

KEY WORDS
coated capillary

REFERENCE
Koppenhoefer,B.; Eperlein,U.; Schlunk,R.; Zhu,X.; Lin,B. Separation of enantiomers of drugs by capillary electrophoresis. V. Hydroxypropyl-α-cyclodextrin as chiral solvating agent, *J.Chromatogr.A*, **1998**, *793*, 153–164.

Orphenadrine

Molecular formula: $C_{18}H_{23}NO$
Molecular weight: 269.39
CAS Registry No.: 83-98-7, 4682-36-4 (citrate)
Merck Index (12th ed.): 7007
Lednicer: 1 42

SAMPLE
Matrix: solutions
Sample preparation: Inject an aliquot of a 100 µg/mL solution in water:running buffer 50:50.

CAPILLARY ELECTROPHORESIS
Capillary: 44.5 cm × 50 µm acrylamide-coated fused-silica (Bio-Rad)
Capillary temperature: 30
Running buffer: 100 mM NaH_2PO_4 containing 15 mM gamma-cyclodextrin, adjusted to pH 2.5 with phosphoric acid
Injection: Electrokinetic injection at 8 kV for 6 s.
Detector: UV 200
Migration time: 13.60, 13.80 (enantiomers)
Voltage: 14 kV
Model: Bio-Rad BioFocus 3000

OTHER SUBSTANCES
Also analyzed: albuterol, alprenolol, atenolol, atropine, baclofen, bamethan, benserazide, biperiden, bisoprolol, bupivacaine, bupranolol, butetamate, carazolol, carbuterol, carvedilol, celiprolol, chloroquine, chlorpheniramine (chlorphenamine), clidinium bromide, clobutinol, disopyramide, dobutamine, flecainide, homatropine, ipratropium bromide, isoproterenol, isothipendyl, ketamine, mefloquine, mequitazine, metaproterenol (orciprenaline), metipranolol, nafronyl (naftidrofuryl), nefopam, ofloxacin, oxomemazine, oxprenolol, phenoxybenzamine, pholedrine, pindolol, pirbuterol, prilocaine, promethazine, propafenone, propranolol, sotalol, synephrine, terbutaline, tetrahydrozoline (tetryzoline), tocainide, trihexyphenidyl, trimeprazine (alimemazine), trimipramine, tropicamide, verapamil, zopiclone

KEY WORDS
chiral; coated capillary

REFERENCE
Koppenhoefer,B.; Epperlein,U.; Christian,B.; Yibing,J.; Yuying,C.; Bingcheng,L. Separation of enantiomers of drugs by capillary electrophoresis. I. γ-Cyclodextrin as chiral solvating agent, *J.Chromatogr.A*, **1995**, *717*, 181–190.

SAMPLE
Matrix: solutions
Sample preparation: Inject an aliquot of a 100 μg/mL solution in running buffer:water 50:50.

CAPILLARY ELECTROPHORESIS
Capillary: 44.5 cm × 50 μm polyacrylamide coated fused-silica (40 cm to detector) (Bio-Rad)
Running buffer: 100 mM NaH_2PO_4 containing 15 mM α-cyclodextrin, adjusted to pH 2.5 with phosphoric acid
Injection: Electromigration at 8 kV for 6 s
Detector: UV 200
Migration time: 14.75, 14.97 (enantiomers)
Voltage: 14 kV
Model: Bio-Rad Bio-Focus 3000

OTHER SUBSTANCES
Simultaneous: clidinium bromide, ketamine, oxomemazine, tetrahydrozoline, tropicamide

KEY WORDS
chiral; α = 1.015; numerous drugs were not separated into their enantiomers under these conditions; coated capillary

REFERENCE
Bingcheng,L.; Yibing,J.; Yoying,C.; Epperlein,U.; Koppenhoefer,B. Separation of drug enantiomers by capillary electrophoresis: α-Cyclodextrin as chiral solvating agent, *Chromatographia*, **1996**, *42*, 106–110.

SAMPLE
Matrix: solutions
Sample preparation: Inject an aliquot of a solution in running buffer.

CAPILLARY ELECTROPHORESIS
Capillary: 60 cm × 75 μm fused-silica (52.4 cm to detector)
Capillary preparation: After each run flush with 500 mM KOH for 2-3 min then with water.
Running buffer: 10 mM pH 3.8 Phosphate buffer containing 2% sulfated cyclodextrin (ds 7-10)
Injection: Hydrostatic injection.
Detector: UV 214
Migration time: 8.01, 8.20 (enantiomers)
Voltage: 15 kV
Model: Waters Quanta 4000

OTHER SUBSTANCES
Also analyzed: acebutolol, alprenolol, aminoglutethimide, brompheniramine, bupivacaine, bupropion, canadine, carbinoxamine, chloroquine, chlorpheniramine, dimethindene, disopyramide, doxylamine, hydroxychloroquine, idazoxan, isoxsuprine, ketamine, mepenzolate, mepivacaine, methoxyphenamine, mexiletine, midodrine, nefopam, oxprenolol, oxyphencyclimine, pheniramine, phensuximide, pindolol, piperoxan, terbutaline, tetramisole, tolperisone, tranylcypromine, trihexyphenidyl, trimipramine, verapamil, warfarin

KEY WORDS
chiral; detector at anode

REFERENCE
Stalcup,A.M.; Gahm,K.H. Application of sulfated cyclodextrins to chiral separations by capillary zone electrophoresis, *Anal.Chem.*, **1996**, *68*, 1360–1368.

SAMPLE
Matrix: solutions
Sample preparation: Inject an aliquot of a 100 μg/mL solution in running buffer.

CAPILLARY ELECTROPHORESIS
Capillary: 30 cm × 50 μm fused-silica (25.5 cm to detector) (Yongnian Optical Conductive Fiber Plant, China), coated with polyacrylamide

Capillary preparation: No details of the polyacrylamide coating process are provided. However, another paper (LC.GC 1997, 15, 40) by this group indicates that they use the procedure of Hjertén, thus: Adjust the pH of 20 mL water to 3.5 with acetic acid, add 80 µL 3-(trimethoxysilyl)propyl methacrylate (3-methacryloxypropyltrimethoxysilane), mix, suck into capillary, let stand at room temperature for 1 h, remove the solution, wash with water. Fill the capillary with a deaerated 3-4% acrylamide solution containing 1 µL/mL N,N,N',N'-tetramethylethylenediamine and 1 mg/mL potassium persulfate, let stand for 30 min, remove excess solution by aspiration, rinse with water, remove water by aspiration, dry at 35° (J. Chromatogr. 1985, 347, 191).
Capillary temperature: 25
Running buffer: 100 mM NaH_2PO_4 adjusted to pH 2.5 (A) or 100 mM NaH_2PO_4 containing 45 mM hydroxypropyl-gamma-cyclodextrin, adjusted to pH 2.5 (B)
Injection: Electrokinetic injection at 15 kV for 3 s.
Detector: UV 200, UV 210
Migration time: 4.46 (A), 12.20, 12.90 (B, enantiomers)
Voltage: 15 kV
Model: Bio-Rad BioFocus 3000

OTHER SUBSTANCES
Simultaneous: amorolfine, brompheniramine, bupivacaine, carteolol, chloroquine, chlorpheniramine, chlorphenoxamine, disopyramide, dobutamine, doxylamine, flecainide, gallopamil, ketamine, mepindolol, oxybutynin, phenoxybenzamine, pindolol, propafenone, propranolol, sulpiride, talinolol, tropicamide, verapamil

KEY WORDS
chiral; coated capillary

REFERENCE
Koppenhoefer,B.; Epperlein,U.; Xiaofeng,Z.; Bingcheng,L. Separation of enantiomers of drugs by capillary electrophoresis. Part 4: Hydroxypropyl-γ-cyclodextrin as chiral solvating agent, *Electrophoresis*, **1997**, *18*, 924–930.

SAMPLE
Matrix: solutions
Sample preparation: Inject an aliquot of a 100 µg/mL solution in running buffer.

CAPILLARY ELECTROPHORESIS
Capillary: 30 cm × 50 µm fused-silica (25.5 cm to detector), coated with polyacrylamide
Capillary preparation: Adjust the pH of 20 mL water to 3.5 with acetic acid, add 80 µL 3-(trimethoxysilyl)propyl methacrylate (3-methacryloxypropyltrimethoxysilane), mix, suck into capillary, let stand at room temperature for 1 h, remove the solution, wash with water. Fill the capillary with a deaerated 3-4% acrylamide solution containing 1 µL/mL N,N,N',N'-tetramethylethylenediamine and 1 mg/mL potassium persulfate, let stand for 30 min, remove excess solution by aspiration, rinse with water, remove water by aspiration, dry at 35° (J. Chromatogr. 1985, 347, 191).
Capillary temperature: 25
Running buffer: 100 mM NaH_2PO_4 containing 45 mM hydroxypropyl-gamma- cyclodextrin, adjusted to pH 2.5 with phosphoric acid
Injection: Electrokinetic injection at 15 kV for 3 s.
Detector: UV 200
Voltage: 15 kV
Model: Bio-Rad BioFocus 3000

KEY WORDS
chiral; coated capillary; comparison with the use of other cyclodextrins; this running buffer gave the greatest enantiomeric separation.; α=1.057

REFERENCE
Lin,B.; Zhu,X.; Koppenhoefer,B.; Epperlein,U. Investigation of 123 chiral drugs by cyclodextrin-modified capillary electrophoresis, *LC.GC*, **1997**, *15*, 40–46.

SAMPLE
Matrix: solutions

CAPILLARY ELECTROPHORESIS
Capillary: 29-36 cm × 50 μm fused-silica (24.5-31.5 cm to detector) (Yongnian Optical Conductive Fiber Plant, China) coated with polyacrylamide
Capillary preparation: Coat capillary as follows. Adjust the pH of 20 mL water to 3.5 with acetic acid, add 80 μL 3-(trimethoxysilyl)propyl methacrylate (3-methacryloxypropyltrimethoxysilane), mix, suck into capillary, let stand at room temperature for 1 h, remove the solution, wash with water. Fill the capillary with a deaerated 3-4% acrylamide solution containing 1 μL/mL N,N,N',N'-tetramethylethylenediamine and 1 mg/mL potassium persulfate, let stand for 30 min, remove excess solution by aspiration, rinse with water, remove water by aspiration, dry at 35° (J. Chromatogr. 1985, 347, 191).
Capillary temperature: 25
Running buffer: 100 mM pH 2.5 NaH_2PO_4 (A) or 100 mM pH 2.5 NaH_2PO_4 containing 45 mM hydroxypropyl-α-cyclodextrin (Wacker, Munich) (B)
Injection: Electromigration at 15 kV for 3 s.
Detector: UV 200; UV 210
Migration time: 4.46 (A); 10.43, 10.90 (B) (enantiomers)
Voltage: 15 kV
Model: Bio-Focus 3000

OTHER SUBSTANCES
Also analyzed: albuterol (salbutamol), alprenolol, amorolfine, atenolol, atropine, azelastine, baclofen, bamethan, benproperine, benserazide, biperiden, bisoprolol, brompheniramine, bupivacaine, bupranolol, butamirate, butethamate, carazolol, carbuterol, carteolol, carvedilol, celiprolol, chloroquine, chlorpheniramine, chlorphenoxamine, cicletanine, clenbuterol, clidinium bromide, clobutinol, dimethindene, dipivefrin, disopyramide, dobutamine, doxylamine, fendiline, flecainide, gallopamil, homatropine, ipratropium bromide, isoproterenol (isoprenaline), isothipendyl, ketamine, meclizine, mefloquine, mepindolol, mequitazine, metaclazepam, metaproterenol (orciprenaline), metipranolol, metoprolol, nafronyl (naftidrofuryl), nefopam, nicardipine, norfenefrine, ofloxacin, ornidazole, oxomemazine, oxprenolol, oxybutynin, phenoxybenzamine, phenylpropanolamine, pholedrine, pindolol, pirbuterol, prilocaine, procyclidine, promethazine, propafenone, propranolol, reproterol, sotalol, sulpride, synephrine, talinolol, terbutaline, tetrahydrozoline (tetryzoline), theodrenaline, tioconazole, tocainide, trihexyphenidyl, trimeprazine (alimemazine), trimipramine, tropicamide, verapamil, zopiclone

KEY WORDS
coated capillary; chiral

REFERENCE
Koppenhoefer,B.; Eperlein,U.; Schlunk,R.; Zhu,X.; Lin,B. Separation of enantiomers of drugs by capillary electrophoresis. V. Hydroxypropyl-α-cyclodextrin as chiral solvating agent, *J.Chromatogr.A*, **1998**, *793*, 153–164.

Oxacillin

Molecular formula: $C_{19}H_{19}N_3O_5S$
Molecular weight: 401.44
CAS Registry No.: 66-79-5, 7240-38-2 (Na salt monohydrate), 1173-88-2 (Na salt)
Merck Index (12th ed.): 7036
Lednicer: 1 413

SAMPLE
Matrix: solutions
Sample preparation: Prepare a 120 μg/mL solution in water, inject an aliquot.

CAPILLARY ELECTROPHORESIS
Capillary: 60 cm × 75 μm
Running buffer: 20 mM NaH_2PO_4 containing 50 mM sodium dodecyl sulfate, adjusted to pH 9.0 with sodium tetraborate

Injection: Hydrodynamic injection at 10 cm for 5 s
Detector: UV 214
Migration time: 9.5
Voltage: 18 kV
Model: Waters Quanta 4000

OTHER SUBSTANCES
Simultaneous: 6-aminopenicillanic acid, amoxicillin, ampicillin, cloxacillin, dicloxacillin, nafcillin, ticarcillin

REFERENCE
Swartz,M.E. Method development and selectivity control for small molecule pharmaceutical separations by capillary electrophoresis, *J.Liq.Chromatogr.*, **1991**, *14*, 923–938.

SAMPLE
Matrix: solutions

CAPILLARY ELECTROPHORESIS
Capillary: 60 cm × 50 μm fused-silica (47 cm to detector) (Polymicro Technologies)
Capillary temperature: 25
Running buffer: 20 mM pH 8.5 Sodium tetraborate containing 100 mM sodium dodecyl sulfate
Injection: Hydrodynamic injection at 50 mbar for 3.6 s (5 nL).
Detector: UV 205
Migration time: 9.6
Voltage: 22 kV
Model: Crystal 310 (Thermo Unicam)

OTHER SUBSTANCES
Simultaneous: amoxicillin, ampicillin, cephapirin, cloxacillin, dicloxacillin, penicillin G, penicillin V, piperacillin, pyrimethamine, sulfacetamide, sulfadimethoxine, sulfaguanidine, sulfamerazine, sulfameter, sulfamethazine, sulfanilamide, sulfanilic acid, sulfaquinoxaline, sulfisoxazole, trimethoprim
Interfering: sulfapyridine, sulfathiazole

REFERENCE
Hows,M.E.P.; Perrett,D.; Kay,J. Optimization of a simultaneous separation of sulphonamides, dihydrofolate reductase inhibitors and β-lactam antibiotics by capillary electrophoresis, *J.Chromatogr.A*, **1997**, *768*, 97–104.

SAMPLE
Matrix: solutions
Sample preparation: Inject an aliquot of a 500 μg/mL solution in 50 mM pH 7 buffer.

CAPILLARY ELECTROPHORESIS
Capillary: 48.5 cm × 50 μm fused-silica (40 cm to detector) (Hewlett Packard)
Capillary preparation: Before each injection flush capillary with 100 mM NaOH for 5 min and with running buffer for 5 min. Wash new capillaries with 1 M NaOH at 40° for 15 min, with water at 40° for 10 min, and with water at 25° for 5 min.
Capillary temperature: 25
Running buffer: 50 mM pH 7.0 Phosphate buffer (Prepare buffer by dissolving 5.29 g K_2HPO_4 and 2.61 g KH_2PO_4 in 1 L water.)
Injection: Pressure injection at 50 mbar for 9 s (18.8 nL).
Detector: UV 200
Migration time: 11.5
Voltage: 30 kV
Model: Hewlett Packard Model G1600A ^{3D}CE

OTHER SUBSTANCES
Simultaneous: ceftazidime, diclofenac, dicloxacillin, propranolol, quinine, thiamine

REFERENCE
Mrestani,Y.; Neubert,R.H.H.; Krause,A. Partition behaviour of drugs in microemulsions measured by electrokinetic chromatography, *Pharm.Res.*, **1998**, *15*, 799–801.

Oxamniquine

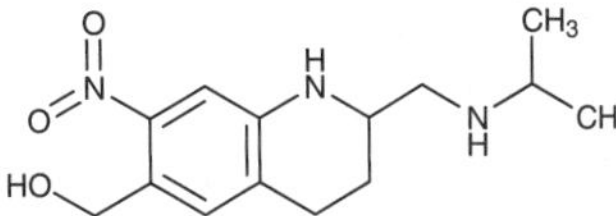

Molecular formula: $C_{14}H_{21}N_3O_3$
Molecular weight: 279.34
CAS Registry No.: 21738-42-1
Merck Index (12th ed.): 7051
Lednicer: 2 372

SAMPLE
Matrix: solutions
Sample preparation: Inject an aliquot of a solution in 10% running buffer.

CAPILLARY ELECTROPHORESIS
Capillary: 57 cm × 50 µm fused-silica (50 cm to detector) (Beckman)
Capillary preparation: Before each run wash with 100 mM NaOH for 1 min and with running buffer for 3 min. Condition new capillaries with 1 M NaOH for 1 h and with water for 20 min.
Capillary temperature: 17
Running buffer: 50 mM pH 7.0 Sodium phosphate buffer containing 15 mM sulfobutyl ether-β-cyclodextrin(4) (CyDex, Kansas) (A) or 50 mM pH 3.0 Sodium phosphate buffer containing 15 mM carboxymethyl-β-cyclodextrin (Wacker) (B)
Injection: Pressure injection.
Detector: UV 214
Migration time: 12.5, 13 (A, enantiomers), 16.5, 17 (B, enantiomers)
Voltage: 10 kV
Model: Beckman P/ACE 5510

KEY WORDS
chiral

REFERENCE
Owens,P.K.; Fell,A.F.; Coleman,M.W.; Berridge,J.C. Screening of cyclodextrins by nuclear magnetic resonance for the design of chiral capillary electrophoresis separations, *J.Chromatogr.A*, **1998**, *797*, 149–164.

Oxazepam

Molecular formula: $C_{15}H_{11}ClN_2O_2$
Molecular weight: 286.72
CAS Registry No.: 604-75-1
Merck Index (12th ed.): 7059
Lednicer: 1 366

SAMPLE
Matrix: solutions

CAPILLARY ELECTROPHORESIS
Capillary: 72 cm × 50 µm silica (50 cm to detector) (Applied Biosystems)
Capillary preparation: Before each run wash capillary with 100 mM NaOH for 3 min, wash with running buffer for 3 min, aspirate neutral marker solution (2 drops DMSO in 10 mL

water) for 1 s, place capillary end in buffer vial for 5 s, inject sample. Wash capillary at the beginning of each day by passing 1 M NaOH through for 20 min.
Capillary temperature: 30
Running buffer: MeCN:buffer 10:90 (Buffer was 30 mM pH 9.3 borate buffer containing 30 mM sodium dodecyl sulfate.)
Injection: Inject by applying vacuum for 1 s (2-3 nL)
Detector: UV 200
Migration time: 13
Voltage: +30 kV
Current: 55 μA
Model: Applied Biosystems Model 270A

OTHER SUBSTANCES
Simultaneous: chlordiazepam, desmethylchlordiazepam, desmethyldiazepam, diazepam, lorazepam, nitrazepam

REFERENCE
Evenson,M.A.; Wiktorowicz,J.E. Automated capillary electrophoresis applied to therapeutic drug monitoring, *Clin.Chem.*, **1992**, *38*, 1847–1852.

SAMPLE
Matrix: solutions
Sample preparation: Prepare a 1-50 μg/mL solution in MeOH:water 20:80, inject an aliquot.

CAPILLARY ELECTROPHORESIS
Capillary: 44 cm × 50 μm fused-silica
Capillary preparation: After each run wash capillary with water for 2 min and buffer for 3 min. At the beginning of each day equilibrate capillary with running buffer for 10 min. Condition a new capillary by washing with 1 M NaOH, 100 mM NaOH, water, and running buffer.
Capillary temperature: 25
Running buffer: MeOH:buffer 20:80 (Buffer was 75 mm pH 9.0 Glycine containing 250 mM triethanolamine and 25 mM sodium dodecyl sulfate.)
Injection: Hydrodynamic injection for 5 s
Detector: UV 235
Migration time: 9.8
Voltage: 25 kV
Current: about 15 μA
Model: Spectraphysics Spectraphoresis 1000 CE
Limit of quantitation: 700 ng/mL
Limit of detection: 200 ng/mL

OTHER SUBSTANCES
Simultaneous: bromazepam, brotizolam, clobazam, clonazepam, lorazepam, lormetazepam, nitrazepam, nordazepam, temazepam

REFERENCE
Bechet,I.; Fillet,M.; Hubert,P.; Crommen,J. Determination of benzodiazepines by micellar electrokinetic chromatography, *Electrophoresis*, **1994**, *15*, 1316–1321.

SAMPLE
Matrix: solutions

CAPILLARY ELECTROPHORESIS
Capillary: 57 cm × 75 μm fused-silica (50 cm to detector)
Capillary preparation: Before each run flush with running buffer for 2 min, equilibrate for 5 min.
Capillary temperature: 33
Running buffer: MeCN:buffer 10:90 (Buffer was 20 mM pH 10.2 borate containing 50 mM sodium cholate.)
Injection: Pressure injection for 2 s
Detector: UV 214
Migration time: 7.1

Voltage: 20 kV
Model: Beckman P/ACE 2100

OTHER SUBSTANCES
Simultaneous: lorazepam, lormetazepam, temazepam

REFERENCE
Boonkerd,S.; Detaevernier,M.R.; Michotte,Y.; Vindevogel,J. Suppression of chiral recognition of 3-hydroxy-1,4-benzodiazepines during micellar electrokinetic capillary chromatography with bile salts, *J.Chromatogr.A*, **1995**, *704*, 238–241.

SAMPLE
Matrix: solutions

CAPILLARY ELECTROPHORESIS
Capillary: 60 cm × 50 μm fused-silica (40 cm to detector), C18 coated (Supelco)
Capillary preparation: Rinse new capillaries with 250 μL 1 M NaOH, 250 μL 100 mM NaOH, water, MeOH, water, and running buffer then condition with running buffer for at least 15 min. Carry out a similar procedure between runs.
Running buffer: 100 mM pH 8.5 Borate buffer containing 25 mM sodium dodecyl sulfate and 5 M urea
Injection: Electrokinetic injection at 5 kV for 5 s.
Detector: UV 254
Migration time: 33
Voltage: 18 kV
Model: Jasco

OTHER SUBSTANCES
Simultaneous: alprazolam, amobarbital, barbital, bromazepam, clonazepam, clotiazepam, cloxazolam, diazepam, estazolam, etizolam, fludiazepam, flunitrazepam, flurazepam, haloxazolam, medazepam, mephobarbital, metharbital, nimetazepam, nitrazepam, pentobarbital, phenobarbital, secobarbital, triazolam

---

KEY WORDS
coated capillary

REFERENCE
Jinno,K.; Han,Y.; Nakamura,M. Analysis of anxiolytic drugs by capillary electrophoresis with bare and coated capillaries, *J.Capillary Electrophor.*, **1996**, *3*, 139–145.

SAMPLE
Matrix: solutions

CAPILLARY ELECTROPHORESIS
Capillary: 47 cm × 75 μm fused-silica (40 cm to detector) (Beckman)
Capillary preparation: Rinse with running buffer for 1 min before injection. Wash with 100 mM NaOH for 1 min and with water for 1 min between injections.
Capillary temperature: 30
Running buffer: THF:buffer 1:99 (Buffer was 50 mM pH 9.2 borate containing 50 mM sodium dodecyl sulfate, 2 M urea, and 20 mM gamma-cyclodextrin.)
Injection: Pressure injection at 300 kPa for 5 s.
Detector: UV 254
Migration time: 17
Voltage: 15 kV
Model: Beckman P/ACE 2000

OTHER SUBSTANCES
Simultaneous: alprazolam, bromazepam, chlordiazepoxide, clobazam, clonazepam, clorazepate, prazepam, triazolam

REFERENCE
Renou-Gonnord,M.F.; David,K. Optimized micellar electrokinetic chromatographic separation of benzodiazepines, *J.Chromatogr.A*, **1996**, *735*, 249–261.

SAMPLE
Matrix: solutions

CAPILLARY ELECTROPHORESIS
Capillary: 60 cm × 75 µm coated fused-silica (40 cm to detector) (Supelco)
Capillary preparation: Coat column as follows. Adjust the pH of 20 mL water to 3.5 with acetic
acid, add 80 µL 3-(trimethoxysilyl)propyl methacrylate (3-methacryloxypropyltrimethoxysi-
lane), mix, suck into capillary, let stand at room temperature for 1 h, remove the solution,
wash with water. Fill the capillary with a deaerated 4% acrylamide solution containing 1 mg/
mL N,N,N',N'-tetramethylethylenediamine and 1 mg/mL ammonium persulfate, let stand for
3 h, remove excess solution by aspiration, rinse with water, remove water by aspiration, dry
at 35° (cf. J. Chromatogr. 1985, 347, 191).
Running buffer: 100 mM pH 8.5 borate buffer containing 10 mM sodium dodecyl sulfate and 5
M urea
Injection: Electromigration at 5 kV for 5 s.
Detector: UV 240, UV 254
Migration time: 8.7
Voltage: 18 kV
Model: Jasco 890-CE

OTHER SUBSTANCES
Simultaneous: alprazolam, clotiazepam, cloxazolam, diazepam, estazolam, etizolam, fludiazepam, flurazepam, haloxazolam

KEY WORDS
injection at cathode; coated capillary

REFERENCE
Jinno,K.; Han,Y.; Sawada,H. Analysis of toxic drugs by capillary electrophoresis using polyacrylamide-coated
columns, *Electrophoresis*, **1997**, *18*, 284–286.

SAMPLE
Matrix: solutions
Sample preparation: Inject an aliquot of a 100 µg/mL solution in running buffer.

CAPILLARY ELECTROPHORESIS
Capillary: 60 cm × 50 µm acrylamide-coated fused-silica (40 cm to detector) (Supelco)
Capillary preparation: Adjust the pH of 20 mL water to 3.5 with acetic acid, add 80 µL 3-
(trimethoxysilyl)propyl methacrylate (3-methacryloxypropyltrimethoxysilane), mix, suck into
capillary, let stand at room temperature for 1 h, remove the solution, wash with water. Fill the
capillary with a deaerated 3-4% acrylamide solution containing 1 µL/mL N,N,N',N'-tetra-
methylethylenediamine and 1 mg/mL ammonium persulfate, let stand for 3 h, remove excess
solution by aspiration, rinse with water, remove water by aspiration, dry at 35° (cf. J. Chromatogr. 1985, 347, 191).
Running buffer: MeCN:buffer 5:95 (Buffer was 100 mM borate containing 5 M urea and 10 mM
sodium dodecyl sulfate, adjusted to pH 8.5 with phosphate.)
Injection: Electrokinetic injection at 5 kV for 5 s.
Detector: UV 254
Migration time: 11.7
Voltage: 18 kV
Model: Jasco Model 870-CE

OTHER SUBSTANCES
Simultaneous: alprazolam, amobarbital, barbital, bromazepam, clonazepam, clotiazepam, cloxazolam, diazepam, estazolam, etizolam, fludiazepam, flunitrazepam, flurazepam, haloxazolam,
medazepam, mephobarbital, metharbital, nimetazepam, nitrazepam, pentobarbital, phenobarbital, secobarbital, triazolam

KEY WORDS
coated capillary; detector at anode

REFERENCE
Jinno,K.; Han,Y.; Sawada,H.; Taniguchi,M. Capillary electrophoretic separation of toxic drugs using a polyacrylamide-coated capillary, *Chromatographia*, **1997**, *46*, 309–314.

SAMPLE
Matrix: urine
Sample preparation: Condition a 130 mg Bond Elut Certify SPE cartridge with 2 mL MeOH and 2 mL 100 mM pH 6 phosphate buffer, do not allow to go dry. Mix urine with an equal amount of 200 mM pH 5.4 buffer, add 20 μL β-glucuronidase/arylsulfatase (Helix pomatia, Boehringer Mannheim) for each 1 mL of urine, heat at 37° for 4 h. 5 (or 10) mL Urine + 2 (or 4) mL 100 mM pH 6 phosphate buffer, add to the SPE cartridge, wash with 1 mL MeOH:100 mM phosphate buffer 20:80, wash with 1 mL 1 M acetic acid, wash with 1 mL hexane, suck dry briefly, elute with 4 mL dichloromethane. Evaporate the eluate to dryness under a stream of nitrogen at room temperature, reconstitute the residue in 50-100 μL running buffer, inject an aliquot.

CAPILLARY ELECTROPHORESIS
Capillary: 105 cm $\times$ 75 μm fused-silica (68 cm to detector)
Capillary preparation: Before each run rinse capillary with 100 mM NaOH for 3.5 min and with buffer for 5 min.
Running buffer: Isopropanol:buffer 5:95 (Buffer was 10 mM pH 9.2-9.3 Na_2HPO_4 containing 6 mM sodium borate and 75 mM sodium dodecyl sulfate.)
Injection: Vacuum injection for 1 s
Detector: UV 235
Migration time: 32
Voltage: 30 kV
Current: 90 μA
Model: Europhor Prime Vision IV

OTHER SUBSTANCES
Extracted: lorazepam

KEY WORDS
SPE

REFERENCE
Schafroth,M.; Thormann,W.; Allemann,D. Micellar electrokinetic capillary chromatography of benzodiazepines in human urine, *Electrophoresis*, **1994**, *15*, 72–78.

Oxolinic acid

Molecular formula: $C_{13}H_{11}NO_5$
Molecular weight: 261.23
CAS Registry No.: 14698-29-4
Merck Index (12th ed.): 7079
Lednicer: 2 370

SAMPLE
Matrix: solutions
Sample preparation: Inject an aliquot of a solution in MeOH.

CAPILLARY ELECTROPHORESIS
Capillary: 59 cm $\times$ 50 μm fused-silica (43 cm to detector) (Polymicro Technologies)

Capillary preparation: Before each analysis flush with water for 3 min, with 200 mM NaOH for 3 min, with water for 3 min, and with running buffer for 4 min. Flush new capillaries with 1 M NaOH at 2000 mbar for 10 min then with 200 mM NaOH for 10 min.
Capillary temperature: 23
Running buffer: MeCN:buffer 28:72, pH 7.3 (Buffer was 32 mM sodium borate containing 18 mM NaH_2PO_4, 39 mM sodium cholate, and 8 mM sodium heptanesulfonate.)
Injection: Hydrodynamic injection at 40 mbar for 6 s.
Detector: UV 260
Migration time: 5.05
Voltage: 30 kV
Model: Lauer Labs Prince

OTHER SUBSTANCES
Simultaneous: ciprofloxacin, enoxacin, flumequine, lomefloxacin, nalidixic acid, norfloxacin, ofloxacin, pefloxacin, pipemidic acid, piromidic acid, rosoxacin, sparfloxacin

REFERENCE
Sun,S.-W.; Chen,L.-Y. Optimization of capillary electrophoretic separation of quinolone antibacterials using the overlapping resolution mapping scheme, *J.Chromatogr.A*, **1997**, *766*, 215–224.

Oxomemazine

Molecular formula: $C_{18}H_{22}N_2O_2S$
Molecular weight: 330.45
CAS Registry No.: 3689-50-7
Merck Index (12th ed.): 7080

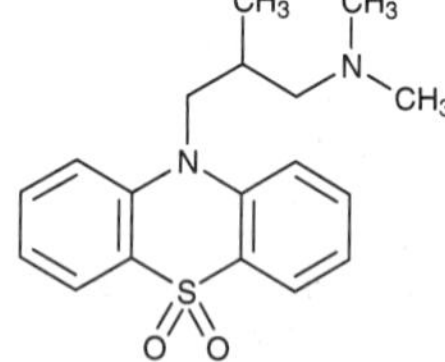

SAMPLE
Matrix: solutions
Sample preparation: Inject an aliquot of a 100 µg/mL solution in running buffer.

CAPILLARY ELECTROPHORESIS
Capillary: 30 cm × 50 µm fused-silica (25.5 cm to detector), coated with polyacrylamide
Capillary preparation: Adjust the pH of 20 mL water to 3.5 with acetic acid, add 80 µL 3-(trimethoxysilyl)propyl methacrylate (3-methacryloxypropyltrimethoxysilane), mix, suck into capillary, let stand at room temperature for 1 h, remove the solution, wash with water. Fill the capillary with a deaerated 3-4% acrylamide solution containing 1 µL/mL N,N,N',N'-tetramethylethylenediamine and 1 mg/mL potassium persulfate, let stand for 30 min, remove excess solution by aspiration, rinse with water, remove water by aspiration, dry at 35° (J. Chromatogr. 1985, 347, 191).
Capillary temperature: 25
Running buffer: 100 mM NaH_2PO_4 containing 45 mM hydroxypropyl-gamma- cyclodextrin, adjusted to pH 2.5 with phosphoric acid
Injection: Electrokinetic injection at 15 kV for 3 s.
Detector: UV 200
Voltage: 15 kV
Model: Bio-Rad BioFocus 3000

KEY WORDS
chiral; coated capillary; comparison with the use of other cyclodextrins; this running buffer gave the greatest enantiomeric separation.; α=1.113

REFERENCE
Lin,B.; Zhu,X.; Koppenhoefer,B.; Epperlein,U. Investigation of 123 chiral drugs by cyclodextrin-modified capillary electrophoresis, *LC.GC*, **1997**, *15*, 40–46.

SAMPLE
Matrix: solutions

CAPILLARY ELECTROPHORESIS
Capillary: 29-36 cm × 50 μm fused-silica (24.5-31.5 cm to detector) (Yongnian Optical Conductive Fiber Plant, China) coated with polyacrylamide
Capillary preparation: Coat capillary as follows. Adjust the pH of 20 mL water to 3.5 with acetic acid, add 80 μL 3-(trimethoxysilyl)propyl methacrylate (3-methacryloxypropyltrimethoxysilane), mix, suck into capillary, let stand at room temperature for 1 h, remove the solution, wash with water. Fill the capillary with a deaerated 3-4% acrylamide solution containing 1 μL/mL N,N,N',N'-tetramethylethylenediamine and 1 mg/mL potassium persulfate, let stand for 30 min, remove excess solution by aspiration, rinse with water, remove water by aspiration, dry at 35° (J. Chromatogr. 1985, 347, 191).
Capillary temperature: 25
Running buffer: 100 mM pH 2.5 NaH_2PO_4 (A) or 100 mM pH 2.5 NaH_2PO_4 containing 45 mM hydroxypropyl-α-cyclodextrin (Wacker, Munich) (B)
Injection: Electromigration at 15 kV for 3 s.
Detector: UV 200; UV 210
Migration time: 5.28 (A); 9.69 (B) (no separation of enantiomers)
Voltage: 15 kV
Model: Bio Focus 3000

OTHER SUBSTANCES
Also analyzed: albuterol (salbutamol), alprenolol, amorolfine, atenolol, atropine, azelastine, baclofen, bamethan, benproperine, benserazide, biperiden, bisoprolol, brompheniramine, bupivacaine, bupranolol, butamirate, butethamate, carazolol, carbuterol, carteolol, carvedilol, celiprolol, chloroquine, chlorpheniramine, chlorphenoxamine, cicletanine, clenbuterol, clidinium bromide, clobutinol, dimethindene, dipivefrin, disopyramide, dobutamine, doxylamine, fendiline, flecainide, gallopamil, homatropine, ipratropium bromide, isoproterenol (isoprenaline), isothipendyl, ketamine, meclizine, mefloquine, mepindolol, mequitazine, metaclazepam, metaproterenol (orciprenaline), metipranolol, metoprolol, nafronyl (naftidrofuryl), nefopam, nicardipine, norfenefrine, ofloxacin, ornidazole, orphenadrine, oxprenolol, oxybutynin, phenoxybenzamine, phenylpropanolamine, pholedrine, pindolol, pirbuterol, prilocaine, procyclidine, promethazine, propafenone, propranolol, reproterol, sotalol, sulpride, synephrine, talinolol, terbutaline, tetrahydrozoline (tetryzoline), theodrenaline, tioconazole, tocainide, trihexyphenidyl, trimeprazine (alimemazine), trimipramine, tropicamide, verapamil, zopiclone

KEY WORDS
coated capillary

REFERENCE
Koppenhoefer,D.; Eperlein,U.; Schlunk,R.; Zhu,X.; Lin,B. Separation of enantiomers of drugs by capillary electrophoresis. V. Hydroxypropyl-α-cyclodextrin as chiral solvating agent, *J.Chromatogr.A*, **1998**, *793*, 153–164.

Oxprenolol

Molecular formula: $C_{15}H_{23}NO_3$
Molecular weight: 265.35
CAS Registry No.: 6452-71-7, 6452-73-9 (HCl)
Merck Index (12th ed.): 7086
Lednicer: 1 117

SAMPLE
Matrix: solutions

CAPILLARY ELECTROPHORESIS
Capillary: 57 cm × 75 μm fused-silica (50 cm to detector) (Beckman)
Capillary preparation: Before each run rinse capillary with 100 mM NaOH and running buffer.
Capillary temperature: 30
Running buffer: 100 mM Phosphoric acid adjusted to pH 3.1 with triethanolamine

Injection: Pressure injection for 5-10 s.
Detector: UV 200
Migration time: 9.2
Voltage: 25 kV
Model: Beckman P/ACE 5510

OTHER SUBSTANCES
Simultaneous: acebutolol, atenolol, labetalol, metoprolol, nadolol, pindolol, propranolol, sotalol, timolol

REFERENCE
Bretnall,A.E.; Clarke,G.S. Selectivity of capillary electrophoresis for the analysis of cardiovascular drugs, *J.Chromatogr.A*, **1996**, *745*, 145–154.

SAMPLE
Matrix: solutions

CAPILLARY ELECTROPHORESIS
Capillary: 43 cm × 50 μm fused-silica (36 cm to detector)
Capillary preparation: Before each injection wash with running buffer for 10 min. After each injection wash with water for 2 min. Wash new capillaries with 1 M NaOH at 60° for 50 min, with NaOH solution (?) at 60° for 10 min, and with water at 25°.
Capillary temperature: 25
Running buffer: 320 mM pH 2.0 Citrate buffer
Injection: Hydrodynamic injection at 1.5 psi for 2 s.
Detector: UV 220
Migration time: 8.9
Voltage: 15 kV
Model: Spectra-Physics Model 1000
Limit of detection: 1-18 μg/mL

OTHER SUBSTANCES
Simultaneous: acebutolol, atenolol, labetalol, levobunolol, metoprolol, nadolol, pindolol, propranolol, timolol

REFERENCE
Lin,C.-E.; Chang,C.-C.; Lin,W.-c.; Lin,E.C. Capillary zone electrophoretic separation of β-blockers using citrate buffer at low pH, *J.Chromatogr.A*, **1996**, *753*, 133–138.

SAMPLE
Matrix: solutions

CAPILLARY ELECTROPHORESIS
Capillary: 52 cm × 75 μm fused-silica (48 cm to detector) (Polymicro Technologies)
Capillary preparation: Before each run purge with running buffer for 3 min. Every 3 runs purge with 100 mM NaOH for 5 min. Purge new capillaries with 1 M NaOH for 20 min and with 100 mM NaOH for 20 min, rinse with running buffer, equilibrate with running buffer at 12 kV for 3 h.
Running buffer: MeCN:buffer 25:75 (Buffer was 25 mM pH 8.8 sodium borate containing 10 mM triethylamine and 25 mM (R)-dodecoxycarbonylvaline (Waters EnantioSelect (R)-Val-1).)
Injection: Hydrostatic injection for 2 s.
Detector: UV 214
Migration time: 19, 19.5 (enantiomers)
Voltage: 12 kV
Current: ≤40 μA
Model: Waters Quanta 4000

OTHER SUBSTANCES
Simultaneous: alprenolol, propranolol

KEY WORDS
chiral

REFERENCE

Peterson,A.G.; Ahuja,E.S.; Foley,J.P. Enantiomeric separations of basic pharmaceutical drugs by micellar electrokinetic chromatography using a chiral surfactant, N-dodecoxycarbonylvaline, *J.Chromatogr.B*, **1996**, *683*, 15–28.

SAMPLE

Matrix: solutions
Sample preparation: Inject an aliquot of a solution in running buffer.

CAPILLARY ELECTROPHORESIS

Capillary: 60 cm × 75 μm fused-silica (52.4 cm to detector)
Capillary preparation: After each run flush with 500 mM KOH for 2-3 min then with water.
Running buffer: 10 mM pH 3.8 Phosphate buffer containing 2% sulfated cyclodextrin (ds 7-10)
Injection: Hydrostatic injection.
Detector: UV 214
Migration time: 36.92, 30.98 (enantiomers)
Voltage: 15 kV
Model: Waters Quanta 4000

OTHER SUBSTANCES

Also analyzed: acebutolol, alprenolol, aminoglutethimide, brompheniramine, bupivacaine, bupropion, canadine, carbinoxamine, chloroquine, chlorpheniramine, dimethindene, disopyramide, doxylamine, hydroxychloroquine, idazoxan, isoxsuprine, ketamine, mepenzolate, mepivacaine, methoxyphenamine, mexiletine, midodrine, nefopam, orphenadrine, oxyphencyclimine, pheniramine, phensuximide, pindolol, piperoxan, terbutaline, tetramisole, tolperisone, tranylcypromine, trihexyphenidyl, trimipramine, verapamil, warfarin

KEY WORDS

chiral; detector at anode

REFERENCE

Stalcup,A.M.; Gahm,K.H. Application of sulfated cyclodextrins to chiral separations by capillary zone electrophoresis, *Anal.Chem.*, **1996**, *68*, 1360–1368.

SAMPLE

Matrix: solutions
Sample preparation: Inject an aliquot of a 100 μg/mL solution in running buffer.

CAPILLARY ELECTROPHORESIS

Capillary: 29 cm × 50 μm fused-silica (24.5 cm to detector) (Yongnian Optical Conductive Fiber Plant, China), coated with polyacrylamide
Capillary preparation: No details of the polyacrylamide coating process are provided. However, another paper (LC.GC 1997, 15, 40) by this group indicates that they use the procedure of Hjertén, thus: Adjust the pH of 20 mL water to 3.5 with acetic acid, add 80 μL 3-(trimethoxysilyl)propyl methacrylate (3-methacryloxypropyltrimethoxysilane), mix, suck into capillary, let stand at room temperature for 1 h, remove the solution, wash with water. Fill the capillary with a deaerated 3-4% acrylamide solution containing 1 μL/mL N,N,N',N'-tetramethylethylenediamine and 1 mg/mL potassium persulfate, let stand for 30 min, remove excess solution by aspiration, rinse with water, remove water by aspiration, dry at 35° (J. Chromatogr. 1985, 347, 191).
Capillary temperature: 25
Running buffer: 100 mM NaH_2PO_4 adjusted to pH 2.5
Injection: Electrokinetic injection at 15 kV for 3 s.
Detector: UV 200, UV 210
Migration time: 5.17
Voltage: 15 kV
Model: Bio-Rad BioFocus 3000

OTHER SUBSTANCES
Simultaneous: albuterol, alprenolol, atenolol, baclofen, bamethan, benproperine, benserazide, bisoprolol, bupranolol, butamirate, butethamate, carbuterol, celiprolol, clenbuterol, clobutinol, dipivefrin, isoproterenol (isoprenaline), metaproterenol (orciprenaline), metipranolol, metoprolol, norfenefrine, ornidazole, phenylpropanolamine, pholedrine, pirbuterol, prilocaine, procyclidine, sotalol, synephrine, terbutaline, tocainide

KEY WORDS
coated capillary

REFERENCE
Koppenhoefer,B.; Epperlein,U.; Xiaofeng,Z.; Bingcheng,L. Separation of enantiomers of drugs by capillary electrophoresis. Part 4: Hydroxypropyl-γ-cyclodextrin as chiral solvating agent, *Electrophoresis*, **1997**, *18*, 924–930.

SAMPLE
Matrix: solutions
Sample preparation: Inject an aliquot of a 100 μg/mL solution in running buffer.

CAPILLARY ELECTROPHORESIS
Capillary: 30 cm × 50 μm fused-silica (25.5 cm to detector), coated with polyacrylamide
Capillary preparation: Adjust the pH of 20 mL water to 3.5 with acetic acid, add 80 μL 3-(trimethoxysilyl)propyl methacrylate (3-methacryloxypropyltrimethoxysilane), mix, suck into capillary, let stand at room temperature for 1 h, remove the solution, wash with water. Fill the capillary with a deaerated 3-4% acrylamide solution containing 1 μL/mL N,N,N',N'-tetramethylethylenediamine and 1 mg/mL potassium persulfate, let stand for 30 min, remove excess solution by aspiration, rinse with water, remove water by aspiration, dry at 35° (J. Chromatogr. 1985, 347, 191).
Capillary temperature: 25
Running buffer: 100 mM NaH_2PO_4 containing 45 mM hydroxypropyl-α- cyclodextrin, adjusted to pH 2.5 with phosphoric acid
Injection: Electrokinetic injection at 15 kV for 3 s.
Detector: UV 200
Voltage: 15 kV
Model: Bio-Rad BioFocus 3000

KEY WORDS
chiral; coated capillary; comparison with the use of other cyclodextrins; this running buffer gave the greatest enantiomeric separation.; α=1.017

REFERENCE
Lin,B.; Zhu,X.; Koppenhoefer,B.; Epperlein,U. Investigation of 123 chiral drugs by cyclodextrin-modified capillary electrophoresis, *LC.GC*, **1997**, *15*, 40–46.

SAMPLE
Matrix: solutions

CAPILLARY ELECTROPHORESIS
Capillary: 67 cm × 50 μm (60 cm to detector)
Capillary preparation: Wash with running buffer for 5 min before each injection. After each injection wash with 1 M NaOH at 60° for 5 min, with 100 mM NaOH at 60° for 10 min, and with water at 25° for 5 min.
Capillary temperature: 25
Running buffer: 70 mM pH 7.0 Sodium phosphate buffer containing 15 mM cetyltriemthylammonium bromide
Injection: Hydrodynamic injection for 1 s.
Detector: UV 220
Migration time: 16.5
Voltage: 20 kV

OTHER SUBSTANCES
Simultaneous: acebutolol, atenolol, labetalol, levobunolol, metoprolol, nadolol, pindolol, propranolol, timolol

REFERENCE
Lin,C.-E.; Chen,Y.-C.; Chang,C.-C.; Wang,D.-Z. Migration behavior and selectivity of β-blockers in micellar electrokinetic chromatography. Influence of micelle concentration of cationic surfactants, *J.Chromatogr.A*, **1997**, *775*, 349–357.

SAMPLE
Matrix: solutions

CAPILLARY ELECTROPHORESIS
Capillary: 29-36 cm × 50 μm fused-silica (24.5-31.5 cm to detector) (Yongnian Optical Conductive Fiber Plant, China) coated with polyacrylamide
Capillary preparation: Coat capillary as follows. Adjust the pH of 20 mL water to 3.5 with acetic acid, add 80 μL 3-(trimethoxysilyl)propyl methacrylate (3-methacryloxypropyltrimethoxysilane), mix, suck into capillary, let stand at room temperature for 1 h, remove the solution, wash with water. Fill the capillary with a deaerated 3-4% acrylamide solution containing 1 μL/mL N,N,N',N'-tetramethylethylenediamine and 1 mg/mL potassium persulfate, let stand for 30 min, remove excess solution by aspiration, rinse with water, remove water by aspiration, dry at 35° (J. Chromatogr. 1985, 347, 191).
Capillary temperature: 25
Running buffer: 100 mM pH 2.5 NaH_2PO_4 (A) or 100 mM pH 2.5 NaH_2PO_4 containing 45 mM hydroxypropyl-α-cyclodextrin (Wacker, Munich) (B)
Injection: Electromigration at 15 kV for 3 s.
Detector: UV 200; UV 210
Migration time: 5.17 (A); 10.77, 10.95 (B) (enantiomers)
Voltage: 15 kV
Model: Bio-Focus 3000

OTHER SUBSTANCES
Also analyzed: albuterol (salbutamol), alprenolol, amorolfine, atenolol, atropine, azelastine, baclofen, bamethan, benproperine, benserazide, biperiden, bisoprolol, brompheniramine, bupivacaine, bupranolol, butamirate, butethamate, carazolol, carbuterol, carteolol, carvedilol, celiprolol, chloroquine, chlorpheniramine, chlorphenoxamine, cicletanine, clenbuterol, clidinium bromide, clobutinol, dimethindene, dipivefrin, disopyramide, dobutamine, doxylamine, fendiline, flecainide, gallopamil, homatropine, ipratropium bromide, isoproterenol (isoprenaline), isothipendyl, ketamine, meclizine, mefloquine, mepindolol, mequitazine, metaclazepam, metaproterenol (orciprenaline), metipranolol, metoprolol, nafronyl (naftidrofuryl), nefopam, nicardipine, norfenefrine, ofloxacin, ornidazole, orphenadrine, oxomemazine, oxybutynin, phenoxybenzamine, phenylpropanolamine, pholedrine, pindolol, pirbuterol, prilocaine, procyclidine, promethazine, propafenone, propranolol, reproterol, sotalol, sulpride, synephrine, talinolol, terbutaline, tetrahydrozoline (tetryzoline), theodrenaline, tioconazole, tocainide, trihexyphenidyl, trimeprazine (alimemazine), trimipramine, tropicamide, verapamil, zopiclone

KEY WORDS
coated capillary; chiral

REFERENCE
Koppenhoefer,B.; Eperlein,U.; Schlunk,R.; Zhu,X.; Lin,B. Separation of enantiomers of drugs by capillary electrophoresis. V. Hydroxypropyl-α-cyclodextrin as chiral solvating agent, *J.Chromatogr.A*, **1998**, *793*, 153–164.

SAMPLE
Matrix: solutions
Sample preparation: Inject an aliquot of a 340 μM solution.

CAPILLARY ELECTROPHORESIS
Capillary: 58 cm × 75 μm fused-silica (50 cm to detector)
Capillary preparation: Rinse with running buffer for 2 min before each run. Before use rinse capillary at 1.36 bar with 100 mM NaOH for 5 min and with water for 10 min.

Capillary temperature: 25
Running buffer: 20 mM pH 7.0 Sodium phosphate buffer (A) or 20 mM pH 7.0 Sodium phosphate buffer containing 10 mM 1-lauroyl-2-hydroxy-sn-glycero-3-phosphocholine (Avanti Polar Lipids, Alabaster AL) (B)
Injection: Hydrodynamic injection at 34 mbar for 1 s followed by 5% MeOH for 5 s.
Detector: UV 200
Migration time: 3.97 min (A); k' 0.33(B)
Voltage: 15 kV
Current: ca. 50 μA
Model: Beckman P/ACE 2200

OTHER SUBSTANCES
Simultaneous: acebutolol, alprenolol, atenolol, metoprolol, pindolol, propranolol, timolol

REFERENCE
Masucci,J.A.; Caldwell,G.W.; Foley,J.P. Comparison of the retention behavior of β-blockers using immobilized artificial membrane chromatography and lysophospholipid micellar electrokinetic chromatography, *J.Chromatogr.A*, **1998**, *810*, 95–103.

Oxybutynin chloride

Molecular formula: $C_{22}H_{32}ClNO_3$
Molecular weight: 393.95
CAS Registry No.: 1508-65-2, 5633-20-5 (free base)
Merck Index (12th ed.): 7089
Lednicer: 1 93

SAMPLE
Matrix: solutions
Sample preparation: Inject an aliquot of a 100 μg/mL solution in running buffer.

CAPILLARY ELECTROPHORESIS
Capillary: 30 cm × 50 μm fused-silica (25.5 cm to detector) (Yongnian Optical Conductive Fiber Plant, China), coated with polyacrylamide
Capillary preparation: No details of the polyacrylamide coating process are provided. However, another paper (LC.GC 1997, 15, 40) by this group indicates that they use the procedure of Hjertén, thus: Adjust the pH of 20 mL water to 3.5 with acetic acid, add 80 μL 3-(trimethoxysilyl)propyl methacrylate (3-methacryloxypropyltrimethoxysilane), mix, suck into capillary, let stand at room temperature for 1 h, remove the solution, wash with water. Fill the capillary with a deaerated 3-4% acrylamide solution containing 1 μL/mL N,N,N',N'-tetramethylethylenediamine and 1 mg/mL potassium persulfate, let stand for 30 min, remove excess solution by aspiration, rinse with water, remove water by aspiration, dry at 35° (J. Chromatogr. 1985, 347, 191).
Capillary temperature: 25
Running buffer: 100 mM NaH_2PO_4 adjusted to pH 2.5
Injection: Electrokinetic injection at 15 kV for 3 s.
Detector: UV 200, UV 210
Migration time: 5.43
Voltage: 15 kV
Model: Bio-Rad BioFocus 3000

OTHER SUBSTANCES
Simultaneous: amorolfine, brompheniramine, bupivacaine, carteolol, chloroquine, chlorpheniramine, chlorphenoxamine, disopyramide, dobutamine, doxylamine, flecainide, gallopamil, ketamine, mepindolol, orphenadrine, phenoxybenzamine, pindolol, propafenone, propranolol, sulpiride, talinolol, tropicamide, verapamil

KEY WORDS
coated capillary

REFERENCE

Koppenhoefer,B.; Epperlein,U.; Xiaofeng,Z.; Bingcheng,L. Separation of enantiomers of drugs by capillary electrophoresis. Part 4: Hydroxypropyl-γ-cyclodextrin as chiral solvating agent, *Electrophoresis*, **1997**, *18*, 924–930.

SAMPLE
Matrix: solutions

CAPILLARY ELECTROPHORESIS
Capillary: 29-36 cm × 50 μm fused-silica (24.5-31.5 cm to detector) (Yongnian Optical Conductive Fiber Plant, China) coated with polyacrylamide
Capillary preparation: Coat capillary as follows. Adjust the pH of 20 mL water to 3.5 with acetic acid, add 80 μL 3-(trimethoxysilyl)propyl methacrylate (3-methacryloxypropyltrimethoxysilane), mix, suck into capillary, let stand at room temperature for 1 h, remove the solution, wash with water. Fill the capillary with a deaerated 3-4% acrylamide solution containing 1 μL/mL N,N,N',N'-tetramethylethylenediamine and 1 mg/mL potassium persulfate, let stand for 30 min, remove excess solution by aspiration, rinse with water, remove water by aspiration, dry at 35° (J. Chromatogr. 1985, 347, 191).
Capillary temperature: 25
Running buffer: 100 mM pH 2.5 NaH_2PO_4 (A) or 100 mM pH 2.5 NaH_2PO_4 containing 45 mM hydroxypropyl-α-cyclodextrin (Wacker, Munich) (B)
Injection: Electromigration at 15 kV for 3 s.
Detector: UV 200; UV 210
Migration time: 5.43 (A); 12.51 (B) (no separation of enantiomers)
Voltage: 15 kV
Model: Bio-Focus 3000

OTHER SUBSTANCES
Also analyzed: albuterol (salbutamol), alprenolol, amorolfine, atenolol, atropine, azelastine, baclofen, bamethan, benproperine, benserazide, biperiden, bisoprolol, brompheniramine, bupivacaine, bupranolol, butamirate, butethamate, carazolol, carbuterol, carteolol, carvedilol, celiprolol, chloroquine, chlorpheniramine, chlorphenoxamine, cicletanine, clenbuterol, clidinium bromide, clobutinol, dimethindene, dipivefrin, disopyramide, dobutamine, doxylamine, fendiline, flecainide, gallopamil, homatropine, ipratropium bromide, isoproterenol (isoprenaline), isothipendyl, ketamine, meclizine, mefloquine, mepindolol, mequitazine, metaclazepam, metaproterenol (orciprenaline), metipranolol, metoprolol, nafronyl (naftidrofuryl), nefopam, nicardipine, norfenefrine, ofloxacin, ornidazole, orphenadrine, oxomemazine, oxprenolol, phenoxybenzamine, phenylpropanolamine, pholedrine, pindolol, pirbuterol, prilocaine, procyclidine, promethazine, propafenone, propranolol, reproterol, sotalol, sulpride, synephrine, talinolol, terbutaline, tetrahydrozoline (tetryzoline), theodrenaline, tioconazole, tocainide, trihexyphenidyl, trimeprazine (alimemazine), trimipramine, tropicamide, verapamil, zopiclone

KEY WORDS
coated capillary

REFERENCE

Koppenhoefer,B.; Eperlein,U.; Schlunk,R.; Zhu,X.; Lin,B. Separation of enantiomers of drugs by capillary electrophoresis. V. Hydroxypropyl-α-cyclodextrin as chiral solvating agent, *J.Chromatogr.A*, **1998**, *793*, 153–164.

Oxymetazoline

Molecular formula: $C_{16}H_{24}N_2O$
Molecular weight: 260.38
CAS Registry No.: 1491-59-4, 2315-02-8 (HCl)
Merck Index (12th ed.): 7100
Lednicer: 1 242

SAMPLE
Matrix: solutions

CAPILLARY ELECTROPHORESIS
Capillary: 58.2 cm × 75 μm fused-silica (50.7 cm to detector) (Polymicro Technologies)
Capillary preparation: Purge with running buffer before each run. At the beginning of each day purge using 50-60 kPa vacuum with 500 mM NaOH for 5 min, with water for 5 min, with MeCN for 5 min, and with running buffer for 5 min.
Running buffer: MeCN:MeOH:acetic acid 49:50:1 containing 20 mM ammonium acetate
Injection: Hydrostatic injection at 10 cm for 5 s.
Detector: UV 214
Migration time: 3.23
Voltage: 25 kV
Model: Waters Quanta 4000

OTHER SUBSTANCES
Simultaneous: amphetamine, benzphetamine, ephedrine, nylidrin, phendimetrazine, phenmetrazine, phenylephrine, phenylpropanolamine, xylometazoline

REFERENCE
Leung,G.N.W.; Tang,H.P.O.; Tso,T.S.C.; Wan,T.S.M. Separation of basic drugs with non-aqueous capillary electrophoresis, *J.Chromatogr.A*, **1996**, *738*, 141–154.

Oxyphencyclimine

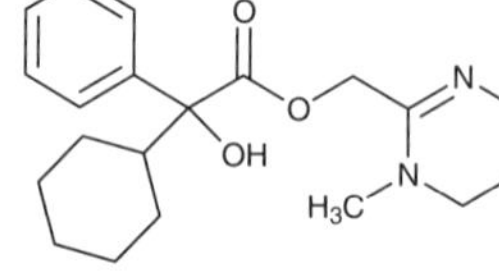

Molecular formula: $C_{20}H_{28}N_2O_3$
Molecular weight: 344.45
CAS Registry No.: 125-53-1, 125-52-0 (HCl)
Merck Index (12th ed.): 7107

SAMPLE
Matrix: solutions
Sample preparation: Inject an aliquot of a solution in running buffer.

CAPILLARY ELECTROPHORESIS
Capillary: 60 cm × 75 μm fused-silica (52.4 cm to detector)
Capillary preparation: After each run flush with 500 mM KOH for 2-3 min then with water.
Running buffer: 10 mM pH 3.8 Phosphate buffer containing 2% sulfated cyclodextrin (ds 7-10)
Injection: Hydrostatic injection.
Detector: UV 214
Migration time: 8.19, 8.26 (enantiomers)
Voltage: 15 kV
Model: Waters Quanta 4000

OTHER SUBSTANCES
Also analyzed: acebutolol, alprenolol, aminoglutethimide, brompheniramine, bupivacaine, bupropion, canadine, carbinoxamine, chloroquine, chlorpheniramine, dimethindene, disopyramide, doxylamine, hydroxychloroquine, idazoxan, isoxsuprine, ketamine, mepenzolate, mepivacaine, methoxyphenamine, mexiletine, midodrine, nefopam, orphenadrine, oxprenolol, pheniramine, phensuximide, pindolol, piperoxan, terbutaline, tetramisole, tolperisone, tranylcypromine, trihexyphenidyl, trimipramine, verapamil, warfarin

KEY WORDS
chiral; detector at anode

REFERENCE
Stalcup,A.M.; Gahm,K.H. Application of sulfated cyclodextrins to chiral separations by capillary zone electrophoresis, *Anal.Chem.*, **1996**, *68*, 1360–1368.

SAMPLE
Matrix: solutions

CAPILLARY ELECTROPHORESIS
Capillary: 36 cm × 50 μm fused-silica coated with linear polyacrylamide (31.5 cm to detector) (GL Science)
Capillary preparation: At the beginning and end of each day rinse capillary with capillary wash solution (Bio-Rad Cat. No. 148-5022) at 690 kPa for more than 3 min and with water at 690 kPa for more than 3 min. Coat capillary as follows. Treat capillary with 1 M NaOH at room temperature for 1 h, rinse with water, dry by passing nitrogen gas through the capillary at 110° for 6 h. Pass thionyl chloride through the capillary using a suction pump for several min, seal capillary at both ends and heat at 70° for 6 h. Unseal the capillary and fill with 250 mM vinyl magnesium bromide in THF by suction, seal the capillary, heat at 70° for 6 h. Open the capillary and rinse it with THF for several min, rinse with distilled water, fill the capillary with polymerization solution, heat at 28 ± 2° for 1 h, condition at -100 V/cm for 30 min (Anal. Sci. 1994, 10, 1). (The polymerization solution was 5% acrylamide in water containing 49 mM Tris, 384 mM glycine, and 0.1% sodium dodecyl sulfate, degas in an ultrasonic bath. Add 40 μL 10% N,N,N',N'-tetramethylethylenodiamine and 10 μL 10% ammonium persulfate to 5 mL of the degassed solution, mix thoroughly.)
Running buffer: Isopropanol:50 mM pH 6.0 Phosphate buffer 10:90
Injection: Before each injection rinse with water at 690 kPa for 30 s, rinse with running buffer at 690 kPa for 30 s, partially fill with separation solution (200 μM α_1-acid glycoprotein (Cohn fraction VI) (Sigma) in running buffer) at 6.9 kPa for 190 s (27 cm), inject sample at 6.9 kPa for 2 s, electrophorese with running buffer (Note that α_1-acid glycoprotein from other suppliers may provide inferior results).
Detector: UV 210
Migration time: Resolution of enantiomers 2.4
Voltage: 12 kV
Model: Bio-rad BioFocus 3000

KEY WORDS
chiral; coated capillary

REFERENCE
Tanaka,Y.; Terabe,S. Separation of the enantiomers of basic drugs by affinity capillary electrophoresis using a partial filling technique and α_1-acid glycoprotein as chiral selector, *Chromatographia*, **1997**, *44*, 119–128.

Oxytetracycline

Molecular formula: $C_{22}H_{24}N_2O_9$
Molecular weight: 460.44
CAS Registry No.: 79-57-2, 6153-64-6 (dihydate), 2058-46-0 (HCl), 15251-48-6 (Ca salt)
Merck Index (12th ed.): 7111
Lednicer: 1 212

SAMPLE
Matrix: blood, milk
Sample preparation: Condition a 3 mL 15 mg SPEC MP1 (Ansys, Irvine CA) SPE cartridge with 1 mL MeOH and 1 mL water. Mix 1 volume milk or plasma with 0.2 volumes 10% trichloroacetic acid in water, centrifuge at 1000 g for 4 min. Add 5 mL supernatant to the SPE cartridge at 2 mL/min, wash with 500 μL water, dry under 40 kPa vacuum for 5 min (more if necessary), elute with 200 μL 50 mM magnesium acetate in N-methylformamide, inject an aliquot of the eluate.

CAPILLARY ELECTROPHORESIS
Capillary: 27 cm × 75 μm fused-silica (20 cm to the detector) (Polymicro Technologies)

Capillary preparation: Flush with running buffer for 30 s before each run. Replace running
buffer after each 6 runs. Before long-term storage flush with 100 mM HCl, 100 mM NaOH,
water, MeOH, and air. Rinse new capillaries with 1 M NaOH for 1 h, with 100 mM NaOH for
20 min, with water for 20 min, with MeOH for 20 min, and with running buffer for 15 min.
Capillary temperature: 20
Running buffer: 500 mM Magnesium acetate tetrahydrate in N-methylformamide
Injection: Pressure injection at 3.5 kPa for 3 s.
Detector: UV 280, F ex 325 (He-Cd laser) em 514
Migration time: 13.7
Voltage: 15 kV
Model: Beckman P/ACE 5010
Limit of detection: 25 ng/mL

OTHER SUBSTANCES
Extracted: chlortetracycline, tetracycline

KEY WORDS
plasma; SPE

REFERENCE
Tjornelund,J.; Hansen,S.H. Use of metal complexation in non-aqueous capillary electrophoresis systems for the
separation and improved detection of tetracyclines, *J.Chromatogr.A*, **1997**, *779*, 235–243.

SAMPLE
Matrix: blood, milk, urine
Sample preparation: Prepare a column as follows. Swirl Chelating Sepharose Fast Flow resin
(Pharmacia) in its bottle, add it to a polypropylene column to give a bed volume of 1.0-1.2 mL,
wash 3 times with 2 mL portions of water, wash with 2.5 mL 10 mM copper sulfate, wash with
2 mL water. Condition a PrepSep-C18 SPE cartridge (Fisher Scientific) by allowing 3 mL 4%
dimethyldichlorosilane in toluene to drain through it by gravity, after 30 min rinse with 10 mL
EtOH, immediately before use wash with 10 mL water. Milk. Centrifuge 10 mL milk at 4° at
2000 g for 20 min, remove the lower layer and add it to 20 mL succinate buffer, centrifuge at
4° at 2000 g for 30 min, pass the supernatant through a Bond Elut SPE cartridge, wash cartridge with 2 mL succinate buffer, collect all the eluate and add it to the column. Wash with
3 mL succinate buffer, wash with 3 mL water, wash with 3 mL MeOH, wash with 3 mL water,
elute with 3 mL pH 7 buffer, when the blue band reaches the bottom of the column collect the
eluate. Add the eluate to the SPE cartridge, wash with 10 mL water, dry with air, elute with
6 mL EtOH. Evaporate the eluate to dryness under a stream of nitrogen at room temperature,
reconstitute with 20 µL running buffer adjusted to pH 7.0 with 5 M HCl, inject an aliquot.
Serum. 12 mL Serum + 12 mL PBS, filter (Amicon Centriprep-10) centrifuge at 10° at 2000 g
for 20 min, remove the filtrate, centrifuge an additional 4 times, combine the filtrates, add a
20 mL aliquot to the column. Wash with 3 mL succinate buffer, wash with 3 mL water, wash
with 3 mL MeOH, wash with 3 mL water, elute with 3 mL pH 7 buffer, when the blue band
reaches the bottom of the column collect the eluate. Add the eluate to the SPE cartridge, wash
with 10 mL water, dry with air, elute with 6 mL EtOH. Evaporate the eluate to dryness under
a stream of nitrogen at room temperature, reconstitute with 20 µL running buffer adjusted to
pH 7.0 with 5 M HCl, inject an aliquot. Urine. Centrifuge 10 mL urine at 4° at 2000 g for 15
min, add the supernatant to the column. Wash with 3 mL succinate buffer, wash with 3 mL
water, wash with 3 mL MeOH, wash with 3 mL water, elute with 3 mL pH 7 buffer, when the
blue band reaches the bottom of the column collect the eluate. Add the eluate to the SPE
cartridge, wash with 10 mL water, dry with air, elute with 6 mL EtOH. Evaporate the eluate
to dryness under a stream of nitrogen at room temperature, reconstitute with 20 µL running
buffer adjusted to pH 7.0 with 5 M HCl, inject an aliquot. (Prepare succinate buffer by dissolving 11.8 g succinic acid in 980 mL water, adjust pH to 4.0 with 10 M NaOH, make up to
1 L. Prepare PBS by dissolving 20 mG KCl, 8 g NaCl, 1.44 g Na_2HPO_4, and 24 mg KH_2PO_4 in
980 mL water, adjust pH to 7.0 with 1 M HCl, make up to 1 L. Prepare the pH 7 buffer by
dissolving 2.31 g citric acid, 7.44 g disodium EDTA dihydrate, and 5.84 g NaCl in 190 mL
water, adjust pH to 7.0 with 5 M NaOH, make up to 200 mL.)

CAPILLARY ELECTROPHORESIS
Capillary: 57 cm × 75 µm (50 cm to detector) (Beckman)
Capillary preparation: Before injection rinse with running buffer for 2 min. After each run
rinse with NaOH solution for 2 min and with water for 2 min.

Capillary temperature: 23
Running buffer: 50 mM Borate buffer containing 50 mM phosphate and 10 mM sodium dodecyl sulfate (Prepare buffer by dissolving 2.884 g sodium dodecyl sulfate, 19.07 g sodium borate decahydrate, 4.096 g Na_2HPO_4, and 2.537 g NaH_2PO_4 in 980 mL water, adjust pH to 8.5 with 5 M HCl, make up to 1 L.)
Injection: Pressure injection for 5 s (5 nL)
Detector: UV 370
Migration time: 13.6
Voltage: 15 kV
Model: Beckman P/ACE 5510
Limit of quantitation: 7.7 ppm (milk), 5.7 ppm (serum), 8.7 ppm (urine)
Limit of detection: 4.6 ppm (milk), 4.2 ppm (serum), 5.3 ppm (urine)

OTHER SUBSTANCES
Extracted: chlortetracycline, doxycycline, tetracycline

KEY WORDS
serum; cow; SPE

REFERENCE
Chen,C.-L.; Gu,X. Determination of tetracycline residues in bovine milk, serum, and urine by capillary electrophoresis, *J.AOAC Int.*, **1995**, *78*, 1369–1377.

SAMPLE
Matrix: bulk

CAPILLARY ELECTROPHORESIS
Capillary: 70 cm × 75 μm fused-silica (62 cm to detector) (Polymicro Technologies)
Capillary preparation: After each run wash capillary with 100 mM NaOH, 100 mM phosphoric acid and 20 mM EDTA.
Capillary temperature: 20
Running buffer: 20 mM pH 11.25 Sodium carbonate containing 1 mM EDTA
Injection: Hydrodynamic injection.
Detector: UV 254
Migration time: 18
Voltage: 12 kV
Model: SpectraPhoresis 500 (Thermo Separation Products)

OTHER SUBSTANCES
Simultaneous: impurities, tetracycline

KEY WORDS
comparison with HPLC

REFERENCE
Van Schepdael,A.; Van den Bergh,I.; Roets,E.; Hoogmartens,J. Purity control of oxytetracycline by capillary electrophoresis, *J.Chromatogr.A*, **1996**, *730*, 305–311.

SAMPLE
Matrix: bulk
Sample preparation: Inject an aliquot of a solution in running buffer.

CAPILLARY ELECTROPHORESIS
Capillary: 44 cm × 50 μm fused-silica (38 cm to detector) (Polymicro Technologies)
Capillary preparation: Wash capillary with 100 mM NaOH, 100 mM phosphoric acid, and 20 mM EDTA each day.
Capillary temperature: 10
Running buffer: 50 mM pH 11.0 Sodium carbonate containing 0.5% Triton X-100 and 1 mM EDTA
Injection: Hydrodynamic injection for 4 s.
Detector: UV 254

Migration time: 13.5
Voltage: 10 kV
Model: Spectraphoresis 500 (Thermo Separation Products)

OTHER SUBSTANCES
Simultaneous: degradation products, impurities, 2-acetyl-2-decarboxamidooxytetracycline, anhydrotetracycline, α-apooxytetracycline, β-apooxytetracycline, 4-epioxytetracycline, 4-epitetracycline, terrinolide, tetracycline

---

REFERENCE
Li,Y.M.; Van Schepdael,A.; Roets,E.; Hoogmartens,J. Separation of oxytetracycline and its related substances by capillary electrophoresis, *J.Liq.Chromatogr.Rel.Technol.*, **1997**, *20*, 273–282.

SAMPLE
Matrix: bulk
Sample preparation: Inject an aliquot of a solution in running buffer.

CAPILLARY ELECTROPHORESIS
Capillary: 44 cm × 50 μm fused-silica (38 cm to detector) (Polymicro Technologies)
Capillary temperature: 10
Running buffer: 50 mM pH 11.0 Sodium carbonate containing 1 mM EDTA and 0.5% Triton X-100
Injection: Hydrodynamic injection for 4 s.
Detector: UV 254
Migration time: 13
Voltage: 10 kV
Model: Spectraphoresis 500 (Thermo Separation Products)
Limit of quantitation: 1 μg/mL
Limit of detection: 500 ng/mL

OTHER SUBSTANCES
Simultaneous: impurities

REFERENCE
Li,Y.M.; Van Schepdael,A.; Roets,E.; Hoogmartens,J. Optimized methods for capillary electrophoresis of tetracyclines, *J.Pharm.Biomed.Anal.*, **1997**, *15*, 1063–1069.

SAMPLE
Matrix: formulations
Sample preparation: Dissolve capsule contents in 15 mM pH 7.5 sodium phosphate buffer so as to make a 25 mM solution, centrifuge, dilute an aliquot of the supernatant to a concentration of 5 mM with 15 mM pH 7.5 sodium phosphate buffer, inject an aliquot.

CAPILLARY ELECTROPHORESIS
Capillary: 111.9 cm × 75 μm fused-silica (43.4 cm to detector) (Polymicro Technologies)
Capillary preparation: Condition by rinsing with 1 M NaOH for 10 min then running buffer overnight.
Capillary temperature: 25
Running buffer: pH 7.5 Phosphate buffer (buffer concentration 4.3 mM, total sodium concentration 15 mM, ionic strength 18.2 mM)
Injection: Hydrodynamic injection at 2 cm for 1 min
Detector: UV 260
Migration time: 5.3
Current: 20 μA
Limit of detection: 10 μM (S/N 3)

OTHER SUBSTANCES
Simultaneous: chlortetracycline, demeclocycline, doxycycline, methacycline, minocycline, tetracycline

---

KEY WORDS
capsules

REFERENCE
Tavares,M.F.M.; McGuffin,V.L. Separation and characterization of tetracycline antibiotics by capillary electrophoresis, *J.Chromatogr.A*, **1994**, *686*, 129–142.

SAMPLE
Matrix: formulations
Sample preparation: Heat 500 mg ointment with 25 mL N-methylformamide at 61.5° for 15
min, cool in an ice bath, centrifuge, inject an aliquot of the clear supernatant.

CAPILLARY ELECTROPHORESIS
Capillary: 27 cm × 75 μm fused-silica (20 cm to detector) (Beckman or Polymicro Technologies)
Capillary preparation: Flush with running buffer for 1 min between runs. Before use condition
capillary with 1 M NaOH for 1 h, with 100 mM NaOH for 20 min, with water for 20 min, with
MeOH for 20 min, and with running buffer for 20 min.
Capillary temperature: 20
Running buffer: N-Methylformamide containing 500 mM magnesium acetate tetrahydrate
Injection: Pressure injection at 0.5 psi for 3 s.
Detector: UV 280
Migration time: 13.5
Voltage: 15 kV
Model: Beckman P/ACE 5010

OTHER SUBSTANCES
Simultaneous: degradation products, impurities, β-apooxytetracycline, 4-epianhydrotetracy-
cline, 4-epioxytetracycline, 4-epitetracycline, tetracycline
Interfering: anhydrotetracycline

KEY WORDS
ointment

REFERENCE
Tjornelund,J.; Hansen,S.H. Validation of a simple method for the determination of oxytetracycline in ointment
by non-aqueous capillary electrophoresis, *J.Pharm.Biomed.Anal.*, **1997**, *15*, 1077–1082.

SAMPLE
Matrix: solutions
Sample preparation: Inject an aliquot of a 20 μg/mL solution in water.

CAPILLARY ELECTROPHORESIS
Capillary: 20 cm × 25 μm fused-silica (Bio-Rad)
Running buffer: 200 mM pH 2.2 Sodium phosphate buffer containing 0.06% Triton X-100
Injection: Injection by electromigration at 12 kV for 10 s
Detector: UV 265
Migration time: 21
Voltage: 12 kV
Model: Bio-Rad HPE 100

OTHER SUBSTANCES
Simultaneous: chlortetracycline, quatrimycin, tetracycline

REFERENCE
Croubels,S.; Baeyens,W.; Dewaele,C.; Van Petegham,C. Capillary electrophoresis of some tetracycline antibiotics, *J.Chromatogr.A*, **1994**, *673*, 267–274.

SAMPLE
Matrix: solutions
Sample preparation: Inject an aliquot of a solution in 10 mM HCl.

CAPILLARY ELECTROPHORESIS
Capillary: 44 cm × 50 μm fused-silica (38 cm to detector) (Polymicro Technologies)

Capillary preparation: Wash capillary each day with 100 mM NaOH, 100 mM phosphoric acid, and 20 mM EDTA.
Capillary temperature: 15
Running buffer: 70 mM pH 10.50 Sodium carbonate containing 1 mM EDTA
Injection: Hydrodynamic injection for 4 s
Detector: UV 254
Migration time: 12
Voltage: 12 kV
Model: Thermo Separation Products Spectraphoresis 500
Limit of detection: 1 μg/mL

OTHER SUBSTANCES
Simultaneous: doxycycline, metacycline

REFERENCE
Van Schepdael,A.; Kibaya,R.; Roets,E.; Hoogmartens,J. Analysis of doxycycline by capillary electrophoresis, *Chromatographia*, **1995**, *41*, 367–369.

SAMPLE
Matrix: solutions

CAPILLARY ELECTROPHORESIS
Capillary: 50 cm × 50 μm (50 cm to detector) (Polymicro Technologies)
Capillary preparation: Purge capillary with 10 mM NaOH for 2 min after each injection.
Running buffer: MeOH:buffer 40:60 (Buffer was 30 mM pH 3.0 citric acid containing 24.5 mM β-alanine.)
Injection: Electrokinetic injection at 10 kV for 5 s.
Detector: UV 254
Migration time: 7.1
Voltage: 30 kV
Model: Perkin Elmer/Applied Biosystems Model 270A-HT

OTHER SUBSTANCES
Simultaneous: chlortetracycline, doxycycline, meclocycline, methacycline, minocycline, tetracycline

REFERENCE
Pesek,J.J.; Matyska,M.T. Separation of tetracyclines by high-performance capillary electrophoresis and capillary electrochromatography, *J.Chromatogr.A*, **1996**, *736*, 313–320.

SAMPLE
Matrix: solutions
Sample preparation: Inject an aliquot of a solution in MeOH.

CAPILLARY ELECTROPHORESIS
Capillary: 64 cm × 50 μm fused-silica (55.5 cm to detector) (Polymicro Technologies)
Capillary preparation: Flush capillary with running buffer for 2 min between runs. Before use rinse capillaries with 1 M NaOH for 1 h, with 100 mM NaOH for 20 min, with water for 20 min, with MeOH for 20 min, and with running buffer for 15 min.
Capillary temperature: 25
Running buffer: MeCN:MeOH:DMF 49:45:6 containing 25 mM ammonium acetate, 10 mM citric acid, and 118 mM methanesulfonic acid
Injection: Pressure injection of sample at 50 mbar for 1 s followed by pressure injection of running buffer at 50 mbar for 3 s.
Detector: UV 254
Migration time: 11.3
Voltage: 25 kV
Model: Hewlett-Packard HP[3D]

OTHER SUBSTANCES
Simultaneous: chlortetracycline, doxycycline, tetracycline

REFERENCE
Tjornelund,J.; Hansen,S.H. Determination of impurities in tetracycline hydrochloride by non-aqueous capillary electrophoresis, *J.Chromatogr.A*, **1996**, *737*, 291–300.

SAMPLE
Matrix: solutions

CAPILLARY ELECTROPHORESIS
Capillary: 43 cm × 75 µm (36 cm to the detector)
Capillary preparation: Before each run wash capillary with running buffer for 2 min. After each run wash capillary with water at 25° for 5 min, with 100 mM NaOH at 60 ° for 5 min, and with water at 25° for 5 min.
Capillary temperature: 25
Running buffer: 15 mM pH 6.5 Ammonium acetate buffer containing 20 mM sodium dodecyl sulfate and 0.135% Brij 35
Detector: UV 265
Migration time: 3.6
Voltage: 15 kV

OTHER SUBSTANCES
Simultaneous: chlortetracycline, demeclocycline, doxycycline, minocycline, tetracycline

REFERENCE
Chen,Y.-C.; Lin,C.-E. Migration behavior and separation of tetracycline antibiotics by micellar electrokinetic chromatography, *J.Chromatogr.A*, **1998**, *802*, 95–105.

Oxytocin

Molecular formula: $C_{43}H_{66}N_{12}O_{12}S_2$
Molecular weight: 1007.20
CAS Registry No.: 50-56-6
Merck Index (12th ed.): 7114

SAMPLE
Matrix: solutions

CAPILLARY ELECTROPHORESIS
Capillary: 57 cm × 75 µm fused-silica (50 cm to detector)
Capillary temperature: 20
Running buffer: 125 mM pH 2.44 NaH_2PO_4 buffer
Detector: UV 200
Migration time: 11
Voltage: 30 kV
Model: Beckman PACE 2100

OTHER SUBSTANCES
Simultaneous: angiotensin II, bombesin, bradykinin, dynorphin, leucine enkephalin, LHRH, methioine enkephalin, TRH

REFERENCE

McLaughlin,G.M.; Nolan,J.A.; Lindahl,J.L.; Palmieri,R.H.; Anderson,K.W.; Morris,S.C.; Morrison,J.A.; Bronzert,T.J. Pharmaceutical drug separations by HPCE: Practical guidelines, *J.Liq.Chromatogr.*, **1992**, *15*, 961–1021.

SAMPLE

Matrix: solutions

CAPILLARY ELECTROPHORESIS

Capillary: 50 cm $\times$ 50 μm uncoated fused-silica
Running buffer: 20 mM pH 2.2 Phosphate buffer containing 90 mM hexanesulfonic acid
Injection: Electrokinetic injection at 8 kV for 10 s
Detector: UV 200
Voltage: 9 kV
Model: Bio-Rad HPE 100

OTHER SUBSTANCES

Simultaneous: felypressin, lypressin, ornipressin

REFERENCE

Adhiambo Hagono,P.; Baeyens,W.R.G.; Van Der Weken,G. Capillary electrophoretic separation of some pharmaceutically important polypeptides, *Biomed.Chromatogr.*, **1995**, *9*, 291.

SAMPLE

Matrix: solutions
Sample preparation: Inject an aliquot of a solution in 5-fold diluted running buffer.

CAPILLARY ELECTROPHORESIS

Capillary: 35 cm $\times$ 75 μm fused-silica coated with polyacrylamide (Polymicro Technologies)
Capillary preparation: Before each run purge with 100 mM HCl for 2 min, with water for 4 min, and with running buffer for 1 min. (Coat the capillary as follows. Adjust the pH of 20 mL water to 3.5 with acetic acid, add 80 μL 3-(trimethoxysilyl)propyl methacrylate (3-methacryloxypropyltrimethoxysilane), mix, suck into capillary, let stand at room temperature for 1 h, remove the solution, wash with water. Fill the capillary with a deaerated 3-4% acrylamide solution containing 1 μL/mL N,N,N',N'-tetramethylethylenediamine and 1 mg/mL potassium persulfate, let stand for 30 min, remove excess solution by aspiration, rinse with water, remove water by aspiration, dry at 35° (J. Chromatogr. 1985, 347, 191).)
Running buffer: 100 mM pH 2.5 Sodium phosphate buffer
Injection: Pressure injection at 300 mbar.s.
Detector: UV 191
Migration time: 23
Voltage: 8 kV
Model: Hewlett-Packard HP³ᴰ CE

OTHER SUBSTANCES

Simultaneous: angiotensin II, bombesin, bradykinin, gonadorelin, leucine enkephalin, melanocyte stimulating hormone, methionine enkephalin, protirelin

KEY WORDS

coated capillary

REFERENCE

Zhang,R.; Zhang,H.X.; Eaker,D.; Hjerten,S. The effect of β-cyclodextrins as buffer additives on the (capillary) electrophoretic separation of peptides and proteins, *J.Capillary Electrophor.*, **1997**, *4*, 105–112.

Paclitaxel

Molecular formula: $C_{47}H_{51}NO_{14}$
Molecular weight: 853.92
CAS Registry No.: 33069-62-4
Merck Index (12th ed.): 7117

SAMPLE
Matrix: blood, urine
Sample preparation: 500 µL Plasma + 50 µL 18.8 µg/mL docetaxel (taxotere) in MeOH + 450 µL water + 4 mL MTBE, shake for 2 min, store at -18° until the bottom layer is frozen. Remove the organic layer and evaporate it to dryness under a stream of nitrogen at 35°, reconstitute with 50 µL running buffer, inject an aliquot. Urine. 200 µL Urine + 50 µL 18.8 µg/mL docetaxel (taxotere) in MeOH + 750 µL water + 4 mL MTBE, shake for 2 min, store at 18° until the bottom layer is frozen. Remove the organic layer and evaporate it to dryness under a stream of nitrogen at 35°, reconstitute with 50 µL running buffer, inject an aliquot.

CAPILLARY ELECTROPHORESIS
Capillary: 67 cm $\times$ 50 µm fused-silica (60 cm to detector) (Hewlett-Packard)
Capillary preparation: After each run flush capillary with 100 mM NaOH for 2 min and with running buffer for 3 min.
Running buffer: MeCN:buffer 35:65 (Buffer was 25 mM pH 8.5 Tris-phosphate buffer containing 100 mM sodium dodecyl sulfate.)
Injection: Pressure injection at 0.5 psi for 12 s.
Detector: UV 230
Migration time: 18
Internal standard: docetaxel (taxotere) (17.5)
Voltage: 28 kV
Model: Beckman P/ACE 5510
Limit of quantitation: 50 ng/mL (plasma), 125 ng/mL (urine)
Limit of detection: 20 ng/mL (plasma), 50 ng/mL (urine)

KEY WORDS
plasma

REFERENCE
Hempel,G.; Lehmkuhl,D.; Krümpelmann,S.; Blaschke,G.; Boos,J. Determination of paclitaxel in biological fluids by micellar electrokinetic chromatography, *J.Chromatogr.A*, **1996**, *745*, 173–179.

SAMPLE
Matrix: formulations
Sample preparation: Dilute 80 µL of a 480 µg/mL injection with 40 µL 100 mM sodium dodecyl sulfate and 280 µL water, inject an aliquot.

CAPILLARY ELECTROPHORESIS
Capillary: 72 cm $\times$ 50 µm fused-silica (50 cm to detector) (Polymicro Technologies)
Capillary preparation: Before each run flush capillary with water for 30 s, 200 mM NaOH for 2 min, and with water for 2 min then equilibrate with running buffer for 5 min.Wash new capillaries with 1 M NaOH for 1 h, with water for 10 min, and with 200 mM NaOH for 30 min.
Capillary temperature: 30
Running buffer: MeCN:25 mM pH 9.0 Tris-HCl 30:70 containing 40 mM sodium docecyl sulfate and 10 mM urea
Injection: Vacuum injection at 5 inches Hg for 1 s.
Detector: UV 230
Migration time: 11
Voltage: 25 kV

Model: Applied Biosystems 270A-HT

OTHER SUBSTANCES
Simultaneous: impurities, other taxanes

KEY WORDS
injections

REFERENCE
Shao,L.K.; Locke,D.C. Separation of paclitaxel and related taxanes by micellar electrokinetic capillary chromatography, *Anal.Chem.*, **1998**, *70*, 897–906.

SAMPLE
Matrix: solutions
Sample preparation: Dilute solutions in MeOH with running buffer, filter, inject an aliquot.

CAPILLARY ELECTROPHORESIS
Capillary: 87 cm × 50 μm uncoated fused-silica
Capillary preparation: Flush with running buffer after each run.
Running buffer: MeCN:buffer 25:75 (Buffer was 25 mM pH 8.5 Tris-phosphate containing 50 mM sodium dodecyl sulfate.)
Injection: Apply pressure for 3 s
Detector: UV 230
Migration time: 17
Voltage: +25 kV
Model: Beckman P/ACE 2000 or 5500

OTHER SUBSTANCES
Simultaneous: baccatin III, cephalomannine

REFERENCE
Chan,K.C.; Alvarado,A.B.; McGuire,M.T.; Muschik,G.M.; Issaq,H.J.; Snader,K.M. High-performance liquid chromatography and micellar electrokinetic chromatography of taxol and related taxanes from bark and needle extracts of *Taxus* species, *J.Chromatogr.B*, **1994**, *657*, 301–306.

SAMPLE
Matrix: solutions
Sample preparation: Prepare a 30-60 ng/mL solution in 50 mM sodium dodecyl sulfate, inject an aliquot.

CAPILLARY ELECTROPHORESIS
Capillary: 47 cm × 50 μm fused-silica
Capillary temperature: 14 kV
Running buffer: MeCN:buffer 25:75 (Buffer was 25 mM pH 8.5 Tris-phosphate containing 50 mM sodium dodecyl sulfate.)
Injection: Pressure injection for 3 s
Detector: UV 230
Migration time: 11.1
Model: Beckman P/ACE

OTHER SUBSTANCES
Simultaneous: baccatin, cephalomannine

REFERENCE
Chan,K.C.; Muschik,G.M.; Issaq,H.J.; Snader,K.M. Separation of taxol and related compounds by micellar electrokinetic chromatography, *J.High Res.Chromatogr.*, **1994**, *17*, 51–52.

Pamoic acid

Molecular formula: $C_{23}H_{16}O_6$
Molecular weight: 388.38
CAS Registry No.: 130-85-8
Merck Index (12th ed.): 7136

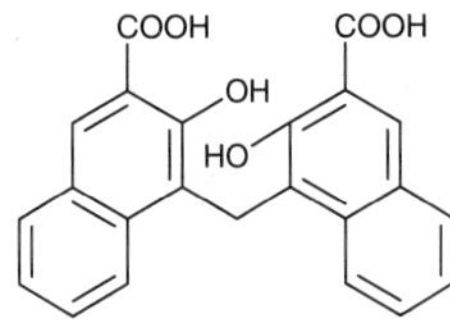

SAMPLE
Matrix: solutions

CAPILLARY ELECTROPHORESIS
Capillary: 27 cm $\times$ 75 μm
Capillary preparation: Before each run rinse capillary with 100 mM NaOH for 30 s and with running buffer for 30 s. Condition new capillaries by rinsing with 100 mM NaOH for 20 min.
Capillary temperature: 30
Running buffer: 15 mM Sodium borate
Injection: Pressure injection of sample at 25 mbar for 1 s followed by running buffer at 25 mbar for 1 s.
Detector: UV 200
Migration time: 4.10
Internal standard: aminobenzoic acid (3.5)
Voltage: 6.5 kV
Model: Beckman

OTHER SUBSTANCES
Simultaneous: aspirin, bacitracin, beclomethasone, benzoic acid, ceftizoxime, ceftriaxone, cefuroxime, cephalothin, cromolyn, epoprostenol, glyburide (glibenclamide), levothyroxine, nedocromil, nystatin, omeprazole, prednisolone, warfarin, zidovudine

REFERENCE
Altria,K.D.; Bryant,S.M.; Hadgett,T.A. Validated capillary electrophoresis method for the analysis of a range of acidic drugs and excipients, *J.Pharm.Biomed.Anal.*, **1997**, *15*, 1091–1101.

Pantoprazole

Molecular formula: $C_{16}H_{15}F_2N_3O_4S$
Molecular weight: 383.38
CAS Registry No.: 102625-70-7
Merck Index (12th ed.): 7146
Lednicer: 5 115

SAMPLE
Matrix: solutions
Sample preparation: Inject an aliquot of a solution in water.

CAPILLARY ELECTROPHORESIS
Capillary: 66 cm $\times$ 50 μm fused-silica (50 cm to detector)
Capillary preparation: After each analysis rinse with 100 mM NaOH for 3 min and with running buffer for 2 min.
Running buffer: n-Propanol:10 mM pH 7.4 potassium phosphate buffer containing 55 μM bovine serum albumin 5:95
Injection: Electrokinetic injection at 5-8 kV for 7 s.
Detector: UV 290
Migration time: 20.3, 21.5 (enantiomers)
Voltage: about 300 V/cm

Current: 35 μA (constant current)
Model: Grom CE-System 100
Limit of detection: 40 μg/mL

KEY WORDS
chiral

REFERENCE
Eberle,D.; Hummel,R.P.; Kuhn,R. Chiral resolution of pantoprazole sodium and related sulfoxides by complex formation with bovine serum albumin in capillary electrophoresis, *J.Chromatogr.A*, **1997**, *759*, 185–192.

Pantothenic acid

Molecular formula: $C_9H_{17}NO_5$
Molecular weight: 219.24
CAS Registry No.: 79-83-4 (D), 137-08-6 (Ca salt D), 6381-63-1 (Ca salt racemic), 599-54-2 (racemic)
Merck Index (12th ed.): 7147

SAMPLE
Matrix: beverages
Sample preparation: Sonicate soft drink to remove carbon dioxide, dilute 3-fold with water, inject an aliquot.

CAPILLARY ELECTROPHORESIS
Capillary: 56 cm × 75 μm fused-silica with bubble cell (Hewlett-Packard)
Capillary temperature: 15
Running buffer: MeOH:buffer 10:90 (Buffer was 60 mM pH 7.0 phosphate buffer containing 60 mM 2-hydroxypropyl-β-cyclodextrin.)
Injection: Pressure injection at 50 kPa for 4 s.
Detector: UV 200
Migration time: 21.7 (D), 22 (L)
Voltage: 20 kV
Model: Hewlett-Packard HP[3D]
Limit of detection: 3% of major enantiomer

KEY WORDS
chiral; soft drink

REFERENCE
Kodama,S.; Yamamoto,A.; Matsunaga,A. Direct chiral resolution of pantothenic acid using 2-hydroxypropyl-β-cyclodextrin in capillary electrophoresis, *J.Chromatogr.A*, **1998**, *811*, 269–273.

SAMPLE
Matrix: formulations
Sample preparation: Grind tablet, dissolve in pH 5 citrate buffer, filter (0.2 μm), inject an aliquot.

CAPILLARY ELECTROPHORESIS
Capillary: 48.5 cm × 50 μm fused-silica (40 cm to detector)
Capillary preparation: Between runs flush with 100 mM NaOH for 2 min and with run buffer for 3 min, inject a buffer plug. Before the first use condition capillaries with 1 M NaOH for 10 min, with 100 mM NaOH for 2 min, and with run buffer for 3 min.
Capillary temperature: 25
Running buffer: 20 mM pH 7 sodium phosphate
Injection: Injection at 40 mbar for 4.6 s (post-injection pressure (of run buffer) 40 mbar for 4 s)
Detector: UV 215
Migration time: 10.4

Voltage: 20 kV
Model: Hewlett-Packard

OTHER SUBSTANCES
Simultaneous: folinic acid, niacinamide, niacin, orotic acid, pyridoxine, thiamine, vitamin B12
(cyanocobalamin) (UV 360), ascorbic acid

KEY WORDS
tablets

REFERENCE
Jegle,U. Separation of water-soluble vitamins via high-performance capillary electrophoresis, *J.Chromatogr.A*,
1993, *652*, 495–501.

SAMPLE
Matrix: formulations
Sample preparation: Condition a 500 mg Bakerbond C18 SPE cartridge with 3 mL MeOH and
3 mL water:10 mM sodium heptanesulfonate in acetic acid 99:1. Heat 50-100 mg tablet and
40 mL water:10 mM sodium heptanesulfonate in acetic acid 99:1 at 55° with occasional shaking
for 5 min, cool, add a 2 mL aliquot to the SPE cartridge, elute with 3.5 mL MeOH, make up
the eluate to 10 mL with water, inject an aliquot.

CAPILLARY ELECTROPHORESIS
Capillary: 70 cm × 100 μm
Capillary temperature: 25
Running buffer: MeOH:buffer 10:90 (Buffer was 50 mM pH 8.0 sodium borate containing 22.5
mM sodium dodecyl sulfate.)
Injection: Inject under 3.44 kPa pressure for 5 s (30 nL)
Detector: UV 214
Migration time: 24
Voltage: 16 kV
Current: 33 μA
Model: Beckman P/ACE System 2000

OTHER SUBSTANCES
Simultaneous: niacinamide, pyridoxine, riboflavin, thiamine, vitamin B12, ascorbic acid

KEY WORDS
tablets; SPE

REFERENCE
Dinelli,G.; Bonetti,A. Micellar electrokinetic capillary chromatography analysis of water-soluble vitamins and
multi-vitamin integrators, *Electrophoresis*, **1994**, *15*, 1147–1150.

SAMPLE
Matrix: formulations
Sample preparation: Protect from light. Grind tablet, add 50 mL 1% acetic acid, shake at 65°
for 10 min, add 5 mL 10 mg/mL niacin in 1% acetic acid, cool to room temperature, make up
to 100 mL with 1% acetic acid, filter (0.22 μm cellulose acetate), dilute 4-12-fold, inject an
aliquot.

CAPILLARY ELECTROPHORESIS
Capillary: 48.5 cm × 50 μm fused-silica (40 cm to detector) (Hewlett Packard)
Capillary preparation: After each run wash with 100 mM NaOH for 7 min, with water for 1
min, and with running buffer for 3 min. At the start of each day wash with separation buffer
for 10 min. Wash new capillaries with 1 M NaOH, 100 mM NaOH, water, and running buffer.
Capillary temperature: 25
Running buffer: 50 mM Borax adjusted to pH 8.5 with boric acid
Injection: Hydrodynamic injection at 5 kPa for 10 s.
Detector: UV 215
Migration time: 6.5

Internal standard: niacin (7.5)
Voltage: 25 kV
Model: Hewlett Packard ^{3D}CE

OTHER SUBSTANCES
Simultaneous: adenine (UV 225), biotin (UV 225), niacinamide (UV 225), pyridoxine, rutin (UV 225), thiamine (UV 225), riboflavin (UV 225), ascorbic acid (UV 225)

KEY WORDS
tablets

REFERENCE
Fotsing,L.; Fillet,M.; Bechet,I.; Hubert,P.; Crommen,J. Determination of six water-soluble vitamins in a pharmaceutical formulation by capillary electrophoresis, *J.Pharm.Biomed.Anal.*, **1997**, *15*, 1113–1123.

SAMPLE
Matrix: solutions
Sample preparation: Prepare a 100 µg/mL solution in MeOH:water 50:50, inject an aliquot.

CAPILLARY ELECTROPHORESIS
Capillary: 60 cm × 75 µm
Running buffer: 20 mM NaH_2PO_4 containing 50 mM sodium dodecyl sulfate, adjusted to pH 9.0 with sodium tetraborate
Injection: Hydrodynamic injection at 10 cm for 5 s
Detector: UV 214
Migration time: 9
Voltage: 18 kV
Model: Waters Quanta 4000

OTHER SUBSTANCES
Simultaneous: niacin, thiamine, riboflavin, pyridoxine, ascorbic acid

---

REFERENCE
Swartz,M.E. Method development and selectivity control for small molecule pharmaceutical separations by capillary electrophoresis, *J.Liq.Chromatogr.*, **1991**, *14*, 923–938.

Papaverine

Molecular formula: $C_{20}H_{21}NO_4$
Molecular weight: 339.39
CAS Registry No.: 58-74-2, 61-25-6 (HCl)
Merck Index (12th ed.): 7151
Lednicer: 1 347

SAMPLE
Matrix: bulk
Sample preparation: Prepare a 10 mg/mL solution in 3 mg/mL pholcodine in 10 mM HCl, dilute 5-fold with 10 mM HCl, filter (0.45 µm cellulose acetate), inject an aliquot.

CAPILLARY ELECTROPHORESIS
Capillary: 72 cm × 75 µm fused-silica (50 cm to detector) (Isco)
Capillary preparation: Flush with running buffer for 2 min between runs. Every 24 h wash capillary with 100 mM HCl for 10 min, wash with water, wash 100 mM NaOH, wash with water, fill with running buffer. Prepare new capillary by filling with 1 M NaOH, let stand for 1 h, fill with 100 mM NaOH, let stand for 1 h, wash with water, fill with running buffer.
Capillary temperature: 30

Running buffer: MeCN:buffer 10:90 (Buffer was 10 mM KH_2PO_4 containing 10 mM sodium tetraborate and 50 mM cetyltrimethylammonium bromide, pH 8.6.)
Injection: Injection by vacuum 5.0 kPa.s (4 nL)
Detector: UV 280
Migration time: 14.3
Internal standard: pholcodine (9.1)
Voltage: 15 kV
Model: Isco Model 3140

OTHER SUBSTANCES
Simultaneous: acetylcodeine, caffeine, codeine, diamorphine, ethylmorphine, morphine, noscapine, strychnine, theophylline

REFERENCE
Trenerry,V.C.; Wells,R.J.; Robertson,J. The analysis of illicit heroin seizures by capillary zone electrophoresis, *J.Chromatogr.Sci.*, **1994**, *32*, 1–6.

SAMPLE
Matrix: bulk
Sample preparation: Dissolve 500 mg crude opium in 5 mL DMSO, make up to 50 mL with MeCN containing 25 mM ammonium acetate and 1 M acetic acid, centrifuge at 18000 g for 2 min. Dilute 1 mL of the supernatant to 20 mL with MeCN containing 25 mM ammonium acetate and 1 M acetic acid, inject an aliquot. Dilute opium tincture 200-fold with water, inject an aliquot.

CAPILLARY ELECTROPHORESIS
Capillary: 64 cm × 50 μm fused-silica (55.5 cm to detector) (Polymicro Technologies)
Capillary preparation: Flush with running buffer between runs. Before use rinse capillaries with 1 M NaOH for 1 h, with 100 mM NaOH for 20 min, and with running buffer for 10 min
Capillary temperature: 25
Running buffer: MeCN:MeOH 25:75 containing 25 mM ammonium acetate and 1 M acetic acid
Injection: Pressure injection at 5 kPa for 3 s
Detector: UV 214
Migration time: 11
Voltage: 25 kV
Model: Hewlett-Packard HP3D

OTHER SUBSTANCES
Simultaneous: codeine, morphine, normorphine, noscapine, thebaine

KEY WORDS
opium; tincture

REFERENCE
Bjornsdottir,I.; Hansen,S.H. Determination of opium alkaloids in crude opium using non-aqueous capillary electrophoresis, *J.Pharm.Biomed.Anal.*, **1995**, *13*, 1473–1481.

SAMPLE
Matrix: bulk
Sample preparation: Crude preparations. Dissolve in 10 mM HCl containing 400 μg/mL pholcodine, inject an aliquot. Opium. Extract 1 g opium with 20 mL 2.5% acetic acid, make up to 100 mL with water, filter, add 20 mL of the filtrate to 60 mL water, adjust pH to 9.2 with concentrated ammonia solution, extract six times with 50 mL portions of chloroform. Dry the extracts over anhydrous sodium sulfate and evaporate to dryness under reduced pressure, reconstitute with 5 mL 50 mM HCl. Dilute 1 mL to 5 mL with water, add 2 mg pholcodine, inject an aliquot. Poppy straw. 8 g Finely-milled poppy straw + 10 mL water + 2 g calcium hydroxide, shake vigorously for 25 min, filter, dilute 20 mL filtrate with 60 mL water, adjust pH to 9.2 with 10% acetic acid, extract 5 times with 50 mL portions of chloroform:EtOH 75:5. Combine the extracts and wash them twice with 30 mL portions of water, dry over anhydrous sodium sulfate and evaporate to dryness under reduced pressure, reconstitute with 25 mL 50 mM HCl. Dilute 9 mL with 1 mL 4 mg/mL pholcodine, filter, inject an aliquot.

CAPILLARY ELECTROPHORESIS
Capillary: 70 cm × 50 μm fused-silica (45 cm to detector) (Polymicro Technologies)
Capillary preparation: Flush with running buffer for 2 min between analyses. Replace running buffer after 20 analyses. Each week wash capillary with 100 mM NaOH for 10 min and with water for 10 min then fill with running buffer.
Capillary temperature: 28
Running buffer: DMF:buffer 10:90 [MeCN:buffer 12.5:87.5 may also be used (Electrophoresis 1996, 17, 1361)] (Prepare buffer by dissolving 0.92 g cetyltrimethylammonium bromide in 50 mL 10 mM sodium tetraborate:10 mM KH_2PO_4 50:50 (pH 8.6), mix 2.5 mL DMF and 22.5 mL buffer.)
Injection: Vacuum injection 10 kPa.s (vacuum level 2)
Detector: UV 254
Migration time: 7
Internal standard: pholcodine (5.5)
Voltage: -25 kV
Model: Isco Model 3140 Electropherograph

OTHER SUBSTANCES
Simultaneous: codeine, cryptopine, morphine, narceine, noscapine, oripavine, salutaridine, thebaine

REFERENCE
Trenerry,V.C.; Wells,R.J.; Robertson,J. Determination of morphine and related alkaloids in crude morphine, poppy straw and opium preparations by micellar electrokinetic capillary chromatography, *J.Chromatogr.A*, **1995**, *718*, 217–225.

SAMPLE
Matrix: bulk
Sample preparation: Dilute with 400 μg/mL n-propyl p-hydroxybenzoate in running buffer

CAPILLARY ELECTROPHORESIS
Capillary: 27 cm × 50 μm fused-silica (20 cm to detector)
Capillary preparation: Before each run rinse with running buffer at high pressure.
Capillary temperature: 30
Running buffer: MeCN:water 15:85 containing 40 mM sodium dodecyl sulfate, 8.5 mM sodium borate, and 8.5 mM sodium phosphate, pH 8.5
Injection: Inject at high pressure for 1 s
Detector: UV 214
Migration time: 2.2
Internal standard: n-propyl p-hydroxybenzoate (1.8)
Voltage: 20 kV
Model: Beckman P/ACE System 2100

OTHER SUBSTANCES
Simultaneous: acetaminophen, acetylcodeine, O-acetylmorphine, aspirin, caffeine, cocaine, codeine, diamorphine, diphenhydramine, hydromorphone, isoniacinamide, lidocaine, methaqualone, morphine, niacinamide, noscapine, phenacetin, phenobarbital, phenylpropanolamine, procaine, quinine, salicylic acid, strychnine, thebaine

REFERENCE
Walker,J.A.; Krueger,S.T.; Lurie,I.S.; Marché,H.L.; Newby,N. Analysis of heroin drug seizures by micellar electrokinetic capillary chromatography (MECC), *J.Forensic Sci.*, **1995**, *40*, 6–9.

SAMPLE
Matrix: formulations
Sample preparation: Dilute formulations with A, inject an aliquot. Dissolve 500 mg crude opium in 5 mL DMSO, make up to 50 mL with 1 M acetic acid, centrifuge. Remove a 1 mL aliquot of the supernatant and make up to 20 mL with water, inject an aliquot.

CAPILLARY ELECTROPHORESIS
Capillary: 55 cm × 50 μm fused-silica (Polymicro Technologies)
Capillary temperature: 30

Running buffer: 50 mM pH 4.0 6-Aminocaproic acid containing 30 mM heptakis(2,6-di-O-methyl-β-cyclodextrin)
Injection: Pressure injection at 5 kPa for 3 s
Detector: UV 214
Migration time: 9.3
Voltage: 30 kV
Model: Hewlett-Packard HP[3D]
Limit of detection: 200 ng/mL

OTHER SUBSTANCES
Simultaneous: codeine, morphine, normorphine, noscapine, thebaine

---

KEY WORDS
syrup; tincture; opium

REFERENCE
Bjornsdottir,I.; Hansen,S.H. Determination of opium alkaloids in opium by capillary electrophoresis, *J.Pharm.Biomed.Anal.*, **1995**, *13*, 687–693.

SAMPLE
Matrix: formulations
Sample preparation: Dilute liquid formulation with DMSO:water:1 M acetic acid 0.5:95:4.5, inject an aliquot.

CAPILLARY ELECTROPHORESIS
Capillary: 64 cm × 50 μm fused-silica (55.5 cm to detector) (Polymicro Technologies)
Capillary preparation: Flush with running buffer for 2 min between runs. Rinse new capillaries with 1 M NaOH for 1 h, with 100 mM NaOH for 20 min, with water for 20 min, and with running buffer for 10 min.
Capillary temperature: 30
Running buffer: 50 mM 6-Aminocaproic acid containing 30 mM heptakis(2,6-di-O-methyl)-β-cyclodextrin, adjusted to pH 4.0 with glacial acetic acid
Injection: Pressure injection at 5 kPa for 3 s.
Detector: UV 214
Migration time: 9.4
Voltage: 30 kV
Current: 60 μA
Model: Hewlett-Packard HP[3D]

OTHER SUBSTANCES
Simultaneous: codeine, morphine, normorphine, noscapine, thebaine

---

REFERENCE
Bjornsdottir,I.; Hansen,S.H. Comparison of aqueous and non-aqueous capillary electrophoresis for quantitative determination of morphine in pharmaceuticals, *J.Pharm.Biomed.Anal.*, **1997**, *15*, 1083–1089.

SAMPLE
Matrix: solutions
Sample preparation: Inject an aliquot of a 1-50 μg/mL solution in water.

CAPILLARY ELECTROPHORESIS
Capillary: 55 cm × 50 μm fused-silica (35 cm to detector) (J&W)
Capillary preparation: Flush with running buffer for 5 min before each run. Periodically wash with 100 mM NaOH and flush extensively with running buffer.
Running buffer: MeOH:25 mM pH 9.24 borate buffer 20:80 containing 100 mM sodium dodecyl sulfate (A) or 50 mM pH 2.35 phosphate buffer (B) or 50 mM pH 9.24 borate buffer (C)
Injection: Injection of 5 μL using a split-flow injector ratio of 1:800.
Detector: UV 200
Migration time: 36.66 (A), 11.50 (B), 7.19 (C)
Voltage: 20 kV (A, B) or 12 kV (C)
Current: <60-80 μA
Model: ISCO Model 3850

OTHER SUBSTANCES
Also analyzed: acetylcodeine, amphetamine, barbital, caffeine, codeine, cocaine, diamorphine, diazepam, flunitrazepam, lidocaine, monoacetylmorphine, morphine, nalorphine, narceine, noscapine, pentobarbital, procaine, tetracaine, thebaine

KEY WORDS
all compounds were separated with running buffer A; some peaks overlapped with running buffers B and C.

REFERENCE
Tagliaro,F.; Smith,F.P.; Turrina,S.; Equisetto,V.; Marigo,M. Complementary use of capillary zone electrophoresis and micellar electrokinetic capillary chromatography for mutual confirmation of results in forensic drug analysis, *J.Chromatogr.A*, **1996**, *735*, 227–235.

SAMPLE
Matrix: solutions

CAPILLARY ELECTROPHORESIS
Capillary: 55 cm × 50 μm fused-silica (50 cm to detector) (Polymicro Technologies)
Capillary preparation: Between runs purge with 1 M NaOH for 2 min, with water for 2 min, and with running buffer for 3 min
Capillary temperature: 25
Running buffer: MeCN:buffer 50:50 (Buffer was 100 mM ammonium acetate adjusted to pH 3.1 with acetic acid.)
Injection: Pressure injection at 345 mbar.s.
Detector: UV 224, Finnigan MAT Model 95, electrospray (details in paper)
Migration time: 15.5
Voltage: 15 kV
Model: Bio-Rad BioFocus 3000

OTHER SUBSTANCES
Simultaneous: codeine, morphine, narceine, noscapine, thebaine

REFERENCE
Unger,M.; Stöckigt,D.; Belder,D.; Stöckigt,J. General approach for the analysis of various alkaloid classes using capillary electrophoresis and capillary electrophoresis-mass spectrometry, *J.Chromatogr.A*, **1997**, *767*, 263–276.

Paromomycin

Molecular formula: $C_{23}H_{45}N_5O_{14}$
Molecular weight: 615.64
CAS Registry No.: 7542-37-2, 1263-89-4 (sulfate)
Merck Index (12th ed.): 7173

SAMPLE
Matrix: bulk
Sample preparation: Inject an aliquot of a solution in 140 mM sodium tetraborate containing 480 μg/mL sisomicin as IS.

CAPILLARY ELECTROPHORESIS
Capillary: 70 cm × 50 μm uncoated fused silica (45 cm to detector) (Polymicro Technologies or ISCO)

Capillary preparation: Rinse with running buffer for 15 min at the beginning of the day and for 4 min before each injection.
Capillary temperature: 34
Running buffer: 185 mM pH 9 sodium tetraborate
Injection: Vacuum injection at 25.0 kPa
Detector: UV 195
Migration time: 25.4
Internal standard: sisomicin (19.90)
Voltage: +15 kV
Model: ISCO Model 3140 Electropherograph

OTHER SUBSTANCES
Simultaneous: amikacin, bekanamycin, butirosin, dibekacin, dihydrostreptomycin, gentamicins, kanamycin A, ribostamycin, streptomycin, tobramycin
Noninterfering: neomycin

KEY WORDS
detector is at cathode

REFERENCE
Flurer,C.L. The analysis of aminoglycoside antibiotics by capillary electrophoresis, *J.Pharm.Biomed.Anal.*, **1995**, *13*, 809–816.

SAMPLE
Matrix: solutions
Sample preparation: Prepare a solution in 100 mM cetyltrimethylammonium bromide, inject an aliquot.

CAPILLARY ELECTROPHORESIS
Capillary: 67 cm × 50 μm fused silica capillary (60 cm to detector) (Siemens)
Running buffer: pH 5.0 Acetate buffer containing 10 mM imidazole and 50 μg/mL FC 135 (a fluorinated surfactant, Fluorad/3M)
Injection: Pressure injection for 2 s.
Detector: UV 214
Migration time: 11.5
Voltage: 12.5 kV
Model: Beckman P/ACE System 2000

OTHER SUBSTANCES
Simultaneous: amikacin, dihydrostreptomycin, gentamicin, kanamycin, lividomycin, neomycin, sisomycin, streptomycin, tobramycin

KEY WORDS
indirect UV detection; detector is at anode

REFERENCE
Ackermans,M.T.; Everaerts,F.M.; Beckers,J.L. Determination of aminoglycoside antibiotics in pharmaceuticals by capillary zone electrophoresis with indirect UV detection coupled with micellar electrokinetic capillary chromatography, *J.Chromatogr.*, **1992**, *606*, 229–235.

Pefloxacin

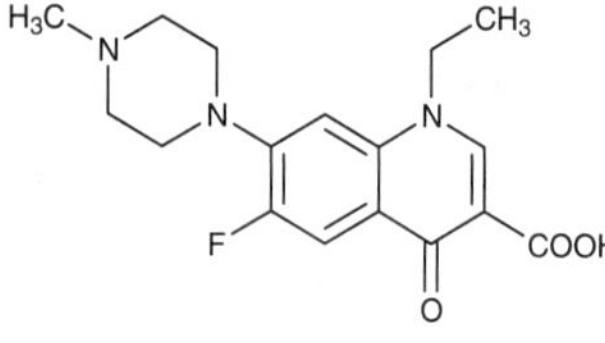

Molecular formula: $C_{17}H_{20}FN_3O_3$
Molecular weight: 333.36
CAS Registry No.: 70458-92-3, 70458-95-6 (mesylate)
Merck Index (12th ed.): 7197
Lednicer: 4 141

SAMPLE
Matrix: solutions
Sample preparation: Inject an aliquot of a solution in MeOH.

CAPILLARY ELECTROPHORESIS
Capillary: 59 cm × 50 μm fused-silica (43 cm to detector) (Polymicro Technologies)
Capillary preparation: Before each analysis flush with water for 3 min, with 200 mM NaOH
    for 3 min, with water for 3 min, and with running buffer for 4 min. Flush new capillaries with
    1 M NaOH at 2000 mbar for 10 min then with 200 mM NaOH for 10 min.
Capillary temperature: 23
Running buffer: MeCN:buffer 28:72, pH 7.3 (Buffer was 32 mM sodium borate containing 18
    mM NaH_2PO_4, 39 mM sodium cholate, and 8 mM sodium heptanesulfonate.)
Injection: Hydrodynamic injection at 40 mbar for 6 s.
Detector: UV 260
Migration time: 4.65
Voltage: 30 kV
Model: Lauer Labs Prince

OTHER SUBSTANCES
Simultaneous: ciprofloxacin, enoxacin, flumequine, lomefloxacin, nalidixic acid, norfloxacin,
    ofloxacin, oxolinic acid, pipemidic acid, piromidic acid, rosoxacin, sparfloxacin

REFERENCE
Sun,S.-W.; Chen,L.-Y. Optimization of capillary electrophoretic separation of quinolone antibacterials using the
    overlapping resolution mapping scheme, *J.Chromatogr.A*, **1997**, *766*, 215–224.

Pemoline

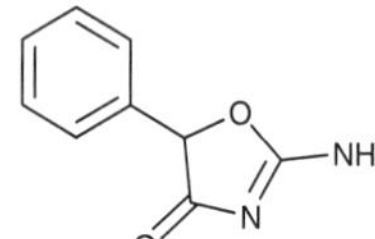

Molecular formula: $C_9H_8N_2O_2$
Molecular weight: 176.17
CAS Registry No.: 2152-34-3
Merck Index (12th ed.): 7206

SAMPLE
Matrix: solutions

CAPILLARY ELECTROPHORESIS
Capillary: 48 cm × 50 μm fused-silica (30 cm to detector) (Yongnian Optical Fiber, Hebei, China)
Capillary preparation: Between runs rinse with 100 mM NaOH for 2 min, with water for 2
    min, and with running buffer for 2 min. Rinse new capillaries with 1 M NaOH overnight,
    equilibrate with water for 30 min, and equilibrate with running buffer for 30 min.
Running buffer: 50 mM pH 3.0 Sodium phosphate buffer containing 1.0 mM sulfobutyl ether
    β-cyclodextrin (BioScience Innovations, Lawrence KS)
Injection: Hydrostatic injection at 10 cm for 5 s.
Detector: UV 214
Migration time: 16.2, 17 (enantiomers)
Voltage: 20 kV

Model: laboratory-constructed

KEY WORDS
chiral

REFERENCE
Dong,Y.; Huang,A.; Sun,Y.; Sun,Z. Chiral separation of chloroquine and pemoline by capillary zone electrophoresis with sulfobutyl ether β-cyclodextrin as buffer additive, *Chromatographia*, **1998**, *48*, 310–313.

Penbutolol

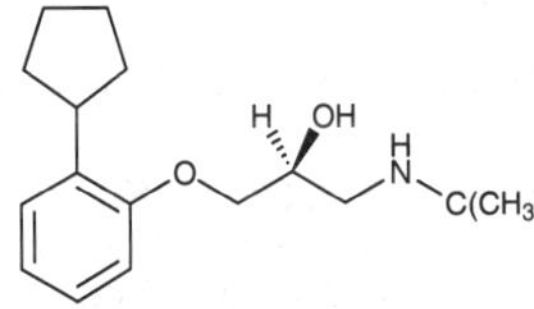

Molecular formula: $C_{18}H_{29}NO_2$
Molecular weight: 291.43
CAS Registry No.: 38363-40-5, 38363-32-5 (sulfate)
Merck Index (12th ed.): 7209

SAMPLE
Matrix: blood
Sample preparation: Filter (1 mL Amicon Centrifree Micropartition System) serum while centrifuging at 1500 g for 20 min, inject an aliquot of the ultrafiltrate.

CAPILLARY ELECTROPHORESIS
Capillary: 92 cm × 75 μm fused-silica (55 cm to detector) (Polymicro Technologies)
Capillary preparation: Rinse with running buffer for 10 min between run. At the beginning and end of each day rinse with 100 mM NaOH for 10 min and with water for 10 min.
Running buffer: Isopropanol:buffer 5:95 (Buffer was 10 mM Na_2HPO_4 containing 6 mM sodium borate and 75 mM sodium dodecyl sulfate.)
Injection: Injection via vacuum suction for 0.7 s
Detector: UV 215
Migration time: 25
Voltage: 25 kV
Current: 82 μA
Model: Europhor Prime Vision IV

OTHER SUBSTANCES
Extracted: atenolol, carteolol, celiprolol, levobunolol, metoprolol, pindolol, propranolol, timolol

KEY WORDS
serum; ultrafiltrate

REFERENCE
Schmutz,A.; Thormann,W. Assessment of impact of physico-chemical drug properties on monitoring drug levels by micellar electrokinetic capillary chromatography with direct serum injection, *Electrophoresis*, **1994**, *15*, 1295–1303.

Penciclovir

Molecular formula: $C_{10}H_{15}N_5O_3$
Molecular weight: 253.26
CAS Registry No.: 39809-25-1
Merck Index (12th ed.): 7210

SAMPLE
Matrix: solutions

Sample preparation: Inject an aliquot of a solution in running buffer.

CAPILLARY ELECTROPHORESIS
Capillary: 57 cm × 75 μm fused-silica (50 cm to detector) (Polymicro Technologies)
Capillary preparation: Rinse with running buffer for 5 min between runs. Condition a new
  capillary with 100 mM NaOH for 15 min, rinse with water for 5 min, rinse with running buffer
  for 5 min.
Capillary temperature: 20
Running buffer: 160 mM pH 9.1 Borate buffer containing 60 mM hydroxypropyl-β-cyclodextrin
Injection: Pressure injection at 3.447 kPa for 2 s.
Detector: UV 254
Migration time: 6.1
Voltage: 25 kV
Model: Beckman P/ACE 5500

OTHER SUBSTANCES
Simultaneous: adenosine, adenosine diphosphate, adenosine monophosphate, adenosine triphosphate, cytidine, cytidine diphosphate, cytidine monophosphate, cytidine triphosphate,
  guanosine, guanosine diphosphate, guanosine monophosphate, guanosine triphosphate, thymidine, thymidine diphosphate, thymidine monophosphate, thymidine triphosphate, uridine,
  uridine diphosphate, uridine monophosphate, uridine triphosphate

---

REFERENCE
Peng,X.; Chen,D.D.Y. Variance contributed by pressure induced injection in capillary electrophoresis,
  J.Chromatogr.A, **1997**, *767*, 205–216.

Penicillin G

Molecular formula: $C_{16}H_{18}N_2O_4S$
Molecular weight: 334.40
CAS Registry No.: 113-98-4 (potassium salt), 61-33-6 (free acid),
69-57-8 (Na salt), 751-84-8 (penicillin G benethamine), 1538-09-6
(penicillin G benzathine), 41372-02-5 (penicillin G benzathine tetrahydrate), 1538-11-0 (penicillin G
benzhydrylamine), 973-53-5 (Ca salt), 3344-16-9 (penicillin G hydrabamine), 6130-64-9 (penicillin G
procaine monohydrate), 54-35-3 (penicillin G procaine)
Merck Index (12th ed.): 7225

SAMPLE
Matrix: bulk
Sample preparation: Prepare a solution in pH 6 phosphate buffer, inject an aliquot.

CAPILLARY ELECTROPHORESIS
Capillary: 80 cm × 50 μm fused silica (75 cm to detector) (Polymicro Technologies or Isco)
Capillary preparation: Rinse with running buffer for 3 min between runs.
Running buffer: 100 mM pH 8 NaH_2PO_4 containing 50 mM sodium borate and 50 mM sodium
  dodecyl sulfate
Injection: Gravity injection at 10 cm for 15 s
Detector: UV 205
Migration time: 18.52
Voltage: +18 kV
Model: Dionex

OTHER SUBSTANCES
Simultaneous: degradation products, impurities, amoxicillin, ampicillin, penicillin V

REFERENCE
Flurer,C.L.; Wolnik,K.A. Chemical profiling of pharmaceuticals by capillary electrophoresis in the determination of drug origin, *J.Chromatogr.A*, **1994**, *674*, 153–163.

SAMPLE
Matrix: premix
Sample preparation: Shake 3 g premix feed and 100 mL acetone:buffer 50:50 on a wrist shaker for 30 min, centrifuge at 5000 rpm for 15 min. Remove a 3.03 mL aliquot of the supernatant and dilute it to 200 mL with buffer, dilute an aliquot 10-fold with water, filter, inject an aliquot. (Prepare buffer by dissolving 8.0 g KH_2PO_4 and 2.0 g K_2HPO_4 in 1 L water, adjust pH to 6.0 with 1 M HCl or 1 M NaOH.)

CAPILLARY ELECTROPHORESIS
Capillary: 50 cm × 50 μm fused-silica (Polymicro Technologies)
Capillary preparation: Before each run rinse with water for 80 s, with 1 M NaOH for 80 s, with water for 80 s, and with running buffer for 90 s then the ends are dipped into water
Capillary temperature: 20
Running buffer: 20 mM Sodium tetraborate containing 150 mM sodium dodecyl sulfate, pH 9.20 ± 0.05
Injection: Pressure injection at 5 psi for 6 s (42 nL).
Detector: UV 210
Migration time: 12.28
Voltage: 14 kV
Model: BioFocus 3000 (Bio-Rad)

OTHER SUBSTANCES
Simultaneous: amoxicillin, ampicillin, ceftiofur, cephapirin, cloxacillin

REFERENCE
Cutting,J.H.; Hurlbut,J.A.; Sofos,J.N. Quantitation of penicillin G in medicated premix feeds by micellar electrokinetic capillary chromatography, *J.AOAC Int.*, **1997**, *80*, 951–955.

SAMPLE
Matrix: solutions
Sample preparation: Prepare a 0.5-2 mg/mL solution in water, inject an aliquot.

CAPILLARY ELECTROPHORESIS
Capillary: 65 cm × 50 μm fused-silica (50 cm to detector) (Scientific Glass Engineering)
Running buffer: 20 mM NaH_2PO_4 containing 150 mM sodium dodecyl sulfate adjusted to pH 9.0 with 20 mM sodium tetraborate
Injection: Injection by siphon at 5 cm for 5-10 s
Detector: UV 210
Migration time: 10.3
Voltage: 20 kV

OTHER SUBSTANCES
Simultaneous: amoxicillin, ampicillin, aspoxicillin, carbenicillin, piperacillin, sulbenicillin
Also analyzed: cefmenoxime, cefminox, cefoperazone, cefotaxime, cefpimizole, cefpiramide, ceftazidime, ceftriaxone

REFERENCE
Nishi,H.; Tsumagari,N.; Kakimoto,T.; Terabe,S. Separation of β-lactam antibiotics by micellar electrokinetic chromatography, *J.Chromatogr.*, **1989**, *477*, 259–270.

SAMPLE
Matrix: solutions
Sample preparation: Inject directly.

CAPILLARY ELECTROPHORESIS
Capillary: 65 cm × 50 μm untreated fused-silica (50 cm to detector) (SGE)
Capillary preparation: Flush with running buffer every 5 runs. At the end of each day fill capillary with 100 mM KOH, let stand for 30 min, flush with water, let stand for 5 min, fill with running buffer.
Running buffer: pH 8.5 Buffer containing 100 mM sodium dodecyl sulfate (Prepare buffer by mixing 20 mM NaH_2PO_4 solution with 20 mM sodium borate solution to achieve a pH of 8.5.)

Injection: Inject by siphoning at 10 cm for 5-10 s.
Detector: UV 210
Migration time: 10.6

OTHER SUBSTANCES
Simultaneous: amoxicillin, ampicillin, aspoxicillin, carbenicillin, piperacillin, sulbenicillin

REFERENCE
Nishi,H.; Fukuyama,T.; Matsuo,M. Separation and determination of aspoxicillin in human plasma by micellar
electrokinetic chromatography with direct sample injection, *J.Chromatogr.*, **1990**, *515*, 245–255.

SAMPLE
Matrix: solutions

CAPILLARY ELECTROPHORESIS
Capillary: 60 cm × 50 μm fused-silica (47 cm to detector) (Polymicro Technologies)
Capillary temperature: 25
Running buffer: 20 mM pH 8.5 Sodium tetraborate containing 100 mM sodium dodecyl sulfate
Injection: Hydrodynamic injection at 50 mbar for 3.6 s (5 nL).
Detector: UV 205
Migration time: 10
Voltage: 22 kV
Model: Crystal 310 (Thermo Unicam)

OTHER SUBSTANCES
Simultaneous: amoxicillin, ampicillin, cephapirin, cloxacillin, dicloxacillin, oxacillin, penicillin
V, piperacillin, pyrimethamine, sulfacetamide, sulfadimethoxine, sulfaguanidine, sulfamerazine, sulfameter, sulfamethazine, sulfanilamide, sulfanilic acid, sulfapyridine, sulfaquinoxaline,
sulfathiazole, sulfisoxazole, trimethoprim

REFERENCE
Hows,M.E.P.; Perrett,D.; Kay,J. Optimization of a simultaneous separation of sulphonamides, dihydrofolate
reductase inhibitors and β-lactam antibiotics by capillary electrophoresis, *J.Chromatogr.A*, **1997**, *768*, 97–
104.

SAMPLE
Matrix: solutions

CAPILLARY ELECTROPHORESIS
Capillary: 50 cm × 50 μm fused-silica (Polymicro Technologies)
Running buffer: 10 mM pH 7.0 Phosphate buffer
Injection: Electrokinetic injection
Detector: E, Model AFDRE-5 (Pine Instrument, Grove City CA), Au working electrode, Pt aux-
iliary electrode, Ag/AgCl reference electrode (design of cell described in paper)
Migration time: 15
Voltage: 10 kV
Model: laboratory-constructed

OTHER SUBSTANCES
Simultaneous: amoxicillin, ampicillin, cloxacillin

REFERENCE
Owens,G.S.; LaCourse,W.R. Pulsed electrochemical detection of thiols and disulfides following capillary electrophoresis, *J.Chromatogr.B*, **1997**, *695*, 15–25.

Penicillin V

Molecular formula: $C_{16}H_{18}N_2O_5S$
Molecular weight: 350.40
CAS Registry No.: 87-08-1, 132-98-9 (potassium salt), 5928-84-7
(benzathine), 63690-57-3 (benzathine tetrahydrate), 6591-72-6 (hydrabamine)
Merck Index (12th ed.): 7230
Lednicer: 7230

SAMPLE
Matrix: bulk
Sample preparation: Prepare a solution in pH 6 phosphate buffer, inject an aliquot.

CAPILLARY ELECTROPHORESIS
Capillary: 80 cm × 50 μm fused silica (75 cm to detector) (Polymicro Technologies or Isco)
Capillary preparation: Rinse with running buffer for 3 min between runs.
Running buffer: 100 mM pH 8 NaH_2PO_4 containing 50 mM sodium borate and 50 mM sodium
   dodecyl sulfate
Injection: Gravity injection at 10 cm for 15 s
Detector: UV 205
Migration time: 20.91
Voltage: +18 kV
Model: Dionex

OTHER SUBSTANCES
Simultaneous: degradation products, impurities, amoxicillin, ampicillin, penicillin G

REFERENCE
Flurer,C.L.; Wolnik,K.A. Chemical profiling of pharmaceuticals by capillary electrophoresis in the determination of drug origin, *J.Chromatogr.A*, **1994**, *674*, 153–163.

SAMPLE
Matrix: solutions

CAPILLARY ELECTROPHORESIS
Capillary: 60 cm × 50 μm fused-silica (47 cm to detector) (Polymicro Technologies)
Capillary temperature: 25
Running buffer: 20 mM pH 8.5 Sodium tetraborate containing 100 mM sodium dodecyl sulfate
Injection: Hydrodynamic injection at 50 mbar for 3.6 s (5 nL).
Detector: UV 205
Migration time: 11
Voltage: 22 kV
Model: Crystal 310 (Thermo Unicam)

OTHER SUBSTANCES
Simultaneous: amoxicillin, ampicillin, cephapirin, cloxacillin, dicloxacillin, oxacillin, penicillin
   G, piperacillin, pyrimethamine, sulfacetamide, sulfadimethoxine, sulfaguanidine, sulfamerazine, sulfameter, sulfamethazine, sulfanilamide, sulfanilic acid, sulfapyridine, sulfaquinoxaline,
   sulfathiazole, sulfisoxazole, trimethoprim

REFERENCE
Hows,M.E.P.; Perrett,D.; Kay,J. Optimization of a simultaneous separation of sulphonamides, dihydrofolate
   reductase inhibitors and β-lactam antibiotics by capillary electrophoresis, *J.Chromatogr.A*, **1997**, *768*, 97–
   104.

Pentamidine

Molecular formula: $C_{19}H_{24}N_4O_2$
Molecular weight: 340.43
CAS Registry No.: 100-33-4, 6823-79-6 (dimethanesulfonate), 140-64-7 (isethionate)
Merck Index (12th ed.): 7254

SAMPLE

Matrix: blood, enzyme incubations, urine
Sample preparation: Serum. 138 μL Serum + 12 μL 500 mM phosphoric acid, mix, filter (Ultrafree-MC, regenerated cellulose, 30000 MW cut-off) while centrifuging at 5000 g for 10 min, inject an aliquot of the ultrafiltrate. Urine. Filter (Ultrafree-MC, polysulfone cellulose, 30000 MW cut-off) 120 μL urine while centrifuging at 5000 g for 10 min, discard the filtrate. Sonicate the filter with 60 μL 10 mM phosphoric acid, centrifuge, sonicate the filter with 60 μL 20 mM phosphoric acid, centrifuge, combine the two liquid fractions, inject an aliquot. Enzyme incubations. 200 μL Rat liver homogenate incubation + 50 μL MeOH, mix, centrifuge at 5000 g for 15 min, inject an aliquot of the supernatant.

CAPILLARY ELECTROPHORESIS

Capillary: 57 cm × 75 μm fused-silica (50 cm to detector) (Beckman)
Capillary preparation: Before each run rinse capillary with 100 mM NaOH for 1 min and running buffer for 2 min, fill with running buffer.
Capillary temperature: 30
Running buffer: 100 mM pH 8.35 Borate buffer containing 50 mM sodium dodecyl sulfate
Injection: Pressure injection for 5 s.
Detector: UV 214
Migration time: 9.5
Voltage: 25 kV
Model: Beckman P/ACE 2000
Limit of detection: 300 ng/mL

KEY WORDS

serum; ultrafiltrate; human; rat; liver

REFERENCE

Garzón,M.J.; Rabanal,B.; Ortiz,A.I.; Negro,A. Determination of pentamidine in serum and urine by micellar electrokinetic chromatography, *J.Chromatogr.B*, **1997**, *688*, 135–142.

Pentazocine

Molecular formula: $C_{19}H_{27}NO$
Molecular weight: 285.43
CAS Registry No.: 359-83-1, 64024-15-3 (HCl), 17146-95-1 (lactate)
Merck Index (12th ed.): 7261
Lednicer: 1 297; 2 325

SAMPLE

Matrix: blood
Sample preparation: Condition a 1 mL 100 mg phenyl SPE cartridge (Varian) with 2 column volumes of MeOH and 2 column volumes of water. 1 mL Serum + 15 μL 100 μg/mL nalorphine in MeOH, mix, add to the SPE cartridge, wash with 1 mL MeOH:water 60:40, elute with four 250 μL aliquots of methanol. Filter (0.2 μm) the eluate and evaporate it to dryness under a stream of nitrogen, reconstitute the residue in 50 μL water, inject an aliquot.

CAPILLARY ELECTROPHORESIS

Capillary: 72 cm × 50 μm fused-silica (50 cm to detector) (Polymicro Technologies)

Capillary preparation: Before each run rinse capillary with 100 mM NaOH and with running buffer for 2 min. Condition new capillaries by rinsing with 1 M NaOH for 10 min, with water for 10 min, and with running buffer for 10 min.
Capillary temperature: 30
Running buffer: 100 mM NaH_2PO_4 containing 15 mM methyl-β-cyclodextrin and 30 μM hexadecyltrimethylammonium bromide, pH adjusted to 2.5 with 100 mM phosphoric acid
Injection: Vacuum injection at 80 psi for 20 s.
Detector: UV 220
Migration time: 19.7 (-), 20.4 (+)
Internal standard: nalorphine (13.8)
Voltage: 20 kV
Current: 45 μA
Model: Applied Biosystems ABI 270A
Limit of quantitation: 35 ng/mL
Limit of detection: 25 ng/mL

KEY WORDS
serum; SPE; chiral

REFERENCE
Ameyibor,E.; Stewart,J.T. Quantitative determination of pentazocine enantiomers in human serum using derivatized β-cyclodextrin-modified capillary electrophoresis and solid phase extraction, *J.Liq.Chromatogr.Rel.Technol.*, **1998**, *21*, 953–963.

SAMPLE
Matrix: solutions

CAPILLARY ELECTROPHORESIS
Capillary: 59.6 cm × 75 μm fused-silica (52.1 cm to detector) (Polymicro Technologies)
Capillary preparation: Purge with running buffer before each run. At the beginning of each day purge using 50-60 kPa vacuum with 500 mM NaOH for 5 min, with water for 5 min, with MeCN for 5 min, and with running buffer for 5 min.
Running buffer: MeCN:MeOH:acetic acid 49:50:1 containing 20 mM ammonium acetate
Injection: Hydrostatic injection at 10 cm for 5 s.
Detector: UV 214
Migration time: 3.30
Voltage: 30 kV
Model: Waters Quanta 4000

OTHER SUBSTANCES
Simultaneous: diphenoxylate, ethoheptazine, fentanyl, levallorphan, meperidine, methadone, nalorphine, nikethamide

REFERENCE
Leung,G.N.W.; Tang,H.P.O.; Tso,T.S.C.; Wan,T.S.M. Separation of basic drugs with non-aqueous capillary electrophoresis, *J.Chromatogr.A*, **1996**, *738*, 141–154.

SAMPLE
Matrix: urine
Sample preparation: Filter (0.2 μm), inject an aliquot of the filtrate.

CAPILLARY ELECTROPHORESIS
Capillary: 80 cm × 50 μm fused-silica (57 cm to detector)
Capillary preparation: Before each run rinse the capillary with 1 M NaOH for 3 min, with 100 mM NaOH for 3 min, with water for 3 min, and with running buffer for 10 min.
Running buffer: 110 mM Boric acid containing 56 mM NaOH and 44 mM HCl, pH 8
Injection: Vacuum injection for 1 s
Detector: UV 220
Migration time: 4.497
Voltage: 20 kV
Model: Europhor Prime Vision system IV

OTHER SUBSTANCES
Extracted: acetazolamide (UV 222), alprenolol (UV 220), amiloride (UV 220), atenolol (UV 228), bendroflumethiazide (UV 220), bumetanide (UV 220), chlorthalidone (UV 220), codeine (UV 220), ethacrynic acid (UV 220), furosemide (UV 232), hydrochlorothiazide (UV 226), methadone (UV 220), metoxiphenamine (UV 220), nadolol (UV 220), oxprenolol (UV 220), propranolol (UV 220), spironolactone (UV 244), triamterene (UV 232), xipamide (UV 234)
Interfering: acebutolol (UV 238), cocaine (UV 236), norcodeine (UV 220)

REFERENCE
Gonzalez,E.; Laserna,J.J. Capillary zone electrophoresis for the rapid screening of banned drugs in sport, *Electrophoresis*, **1994**, *15*, 240–243.

Pentobarbital

Molecular formula: $C_{11}H_{18}N_2O_3$
Molecular weight: 226.28
CAS Registry No.: 76-74-4, 57-33-0 (Na salt)
Merck Index (12th ed.): 7272
Lednicer: 1 269

SAMPLE
Matrix: blood
Sample preparation: 100 µL Serum + 150 µL 80 µg/mL 3-isobutyl-1-methylxanthine in MeCN, mix for 30 s, centrifuge at 14000 g for 1 min, inject an aliquot of the supernatant.

CAPILLARY ELECTROPHORESIS
Capillary: 25 cm × 47 µm
Capillary preparation: Wash with running buffer by pressure injection for 1 min after each run. Wash daily with 50 mM phosphoric acid for 2 min, with water for 1 min, with 2 M NaOH for 2 min, with water for 1 min, and with running buffer for 2 min.
Capillary temperature: 24
Running buffer: 220 mM boric acid adjusted to pH 8.8 with 2.5 M NaOH
Injection: Pressure injection for 8 s.
Detector: UV 254
Migration time: 2.6
Internal standard: 3-isobutyl-1-methylxanthine (2.3)
Voltage: 12 kV
Model: Beckman

OTHER SUBSTANCES
Extracted: acetaminophen, caffeine, carbamazepine, iohexol, phenobarbital, phenytoin, salicylic acid, theophylline

KEY WORDS
serum

REFERENCE
Shihabi,Z.K.; Constantinescu,M.S. Iohexol in serum determined by capillary electrophoresis, *Clin.Chem.*, **1992**, *38*, 2117–2120.

SAMPLE
Matrix: blood
Sample preparation: 100 µL Serum + 150 µL 80 µg/mL IS in MeCN, vortex for 15 s, centrifuge at 15000 g for 1 min, inject an aliquot of the supernatant.

CAPILLARY ELECTROPHORESIS
Capillary: 25 cm × 50 µm

Capillary temperature: 24
Running buffer: 300 mM Boric acid adjusted to pH 8.5 with NaOH
Injection: Pressure injection for 10 s
Detector: UV 254
Migration time: 3
Internal standard: 3-isobutyl-1-methylxanthine (2.5)
Voltage: 11 kV
Model: Beckman
Limit of quantitation: 10 µg/mL

OTHER SUBSTANCES
Simultaneous: amobarbital, butabarbital, phenobarbital, secobarbital
Noninterfering: caffeine, carbamazepine, phenytoin, theophylline

KEY WORDS
serum

REFERENCE
Shihabi,Z.K. Serum pentobarbital assay by capillary electrophoresis, *J.Liq.Chromatogr.*, **1993**, *16*, 2059–2068.

SAMPLE
Matrix: blood
Sample preparation: Condition a 1 mL 100 mg Bond Elut C18 SPE cartridge (Varian) with 3 mL MeOH and 3 mL pH 9.0 phosphate buffer, do not allow to dry. 1 mL Serum + IS, mix, add to the SPE cartridge, wash with 3 mL buffer, allow to dry for 5 min, elute with 3 mL dichloromethane. Filter (nylon) the eluate and evaporate it to dryness under a stream of nitrogen, reconstitute with 1 mL MeOH:water 30:70, inject an aliquot.

CAPILLARY ELECTROPHORESIS
Capillary: 62 cm × 75 µm fused-silica (52 cm to detector) (Polymicro Technologies)
Capillary preparation: Before each run rinse capillary with 100 mM NaOH for 2 min and with running buffer for 2 min. Condition new capillaries by rinsing with 1 M NaOH for 10 min, with water for 10 min, with 100 mM HCl for 10 min, and with running buffer for 10 min.
Capillary temperature: 25
Running buffer: 50 mM NaH_2PO_4 containing 40 mM hydroxypropyl-gamma-cyclodextrin, adjusted to pH 9.0 with 100 mM NaOH
Injection: Vacuum injection at 0.5 psi for 10 s.
Detector: UV 254
Migration time: 9.86 (R-(+)), 10.10 (S-(-))
Internal standard: aprobarbital (10.86)
Voltage: 15 kV
Current: ca. 100 µA
Model: Beckman P/ACE 5000
Limit of quantitation: 1 µg/mL

OTHER SUBSTANCES
Simultaneous: clonazepam, diazepam, phenytoin

KEY WORDS
serum; SPE; chiral

REFERENCE
Srinivasan,K.; Bartlett,M.G. Capillary electrophoresis stereoselective determination of *R*-(+)- and *S*-(-)-pentobarbital from serum using hydroxypropyl-γ-cyclodextrin, solid-phase extraction and ultraviolet detection, *J.Chromatogr.B*, **1997**, *703*, 289–294.

SAMPLE
Matrix: blood, gastric contents, urine, vitreous humor
Sample preparation: Dilute gastric contents 1:10. 2 mL Serum, blood, vitreous humor, diluted gastric contents, or urine + 2.5 mL water + 200 µL mephobarbital solution, mix, add to TOXI-TUBE B (Toxi-Lab, Irvine CA), rock for 10 min, centrifuge at 2000 rpm for 5 min. Remove the organic layer and evaporate it to dryness under a stream of nitrogen at room temperature,

reconstitute the residue in 100 μL running buffer, vortex for 30 s, inject an aliquot. (Mephobarbital solution was 2 mg mephobarbital and 2 drops EtOH, made up to 20 mL with water.)

CAPILLARY ELECTROPHORESIS
Capillary: 60 cm × 75 μm AccuSep (Waters)
Capillary preparation: Purge for 1 min between samples.
Running buffer: MeCN:buffer 15:85 adjusted to pH 8.5 with 1 M phosphoric acid. (Buffer was 3.8 g sodium borate decahydrate, 1 .4 g $NaH_2PO_4.H_2O$, and 28.8 g sodium dodecyl sulfate in 1 L water.)
Injection: Hydrostatic injection for 10 s
Detector: UV 214
Migration time: 10.03
Internal standard: mephobarbital (8.88)
Voltage: 20 kV
Model: Waters Quanta 4000
Limit of detection: 100 ng/mL

OTHER SUBSTANCES
Extracted: amobarbital, butabarbital, butalbital, phenobarbital, secobarbital

KEY WORDS
serum; whole blood

REFERENCE
Ferslew,K.E.; Hagardorn,A.N.; McCormick,W.F. Application of micellar electrokinetic capillary chromatography to forensic analysis of barbiturates in biological fluids, *J.Forensic Sci.*, **1995**, *40*, 245–249.

SAMPLE
Matrix: blood, urine
Sample preparation: Serum, plasma. 200 μL Serum or plasma + 100 μL 1 M HCl + 2 mL chloroform, shake vigorously for 15 min, centrifuge at 500 g for 10 min. Remove the lower organic layer and evaporate it to dryness under a stream of nitrogen at 40°, reconstitute the residue in 200 μL running buffer, shake for 1 min, filter (0.2 μm), inject an aliquot. Urine. Condition a Bond Elut Certify SPE cartridge with 2 mL MeOH and 2 mL 100 mM pH 6 phosphate buffer, do not allow to dry. 5 mL Urine + 2 mL 100 mM pH 6 phosphate buffer, mix, add to the SPE cartridge, wash with 1 mL MeOH:100 mM phosphate buffer 20:80, dry under vacuum for 5 min, wash with 1 mL 1 M acetic acid, dry under vacuum for 10 min, wash with 1 mL hexane, elute with 4 mL dichloromethane. Evaporate the eluate to dryness under a stream of nitrogen at 40°, reconstitute the residue in 100-200 μL running buffer, filter (0.2 μm), inject an aliquot.

CAPILLARY ELECTROPHORESIS
Capillary: 90 cm × 75 μm fused-silica (70 cm to detector) (Polymicro Technologies)
Capillary preparation: Between each run rinse capillary with 100 mM NaOH for 3 min and buffer for 5 min
Running buffer: 15 mM pH 7.8 NaH_2PO_4 containing 9 mM sodium borate and 50 mM sodium dodecyl sulfate
Injection: Siphon at 34 cm for 5 s
Detector: UV 195
Migration time: 13.7
Voltage: 20 kV
Current: 60-63 μA
Model: Laboratory constructed

OTHER SUBSTANCES
Extracted: allobarbital, amobarbital, barbital, butalbital, phenobarbital, thiopental

KEY WORDS
serum; cow; human; plasma; SPE

REFERENCE
Thormann,W.; Meier,P.; Marcolli,C.; Binder,F. Analysis of barbiturates in human serum and urine by high-performance capillary electrophoresis-micellar electrokinetic capillary chromatography with on-column multi-wavelength detection, *J.Chromatogr.*, **1991**, *545*, 445–460.

SAMPLE
Matrix: solutions

CAPILLARY ELECTROPHORESIS
Capillary: 72 cm $\times$ 50 μm silica (50 cm to detector) (Applied Biosystems)
Capillary preparation: Before each run wash capillary with 100 mM NaOH for 3 min, wash with running buffer for 3 min, aspirate neutral marker solution (2 drops DMSO in 10 mL water) for 1 s, place capillary end in buffer vial for 5 s, inject sample. Wash capillary at the beginning of each day by passing 1 M NaOH through for 20 min.
Capillary temperature: 30
Running buffer: MeCN:buffer 20:80 (Buffer was 30 mM pH 9.3 borate buffer containing 30 mM sodium dodecyl sulfate.)
Injection: Inject by applying vacuum for 1 s (2-3 nL)
Detector: UV 200
Migration time: 10.3
Voltage: +30 kV
Current: 41 μA
Model: Applied Biosystems Model 270A

OTHER SUBSTANCES
Simultaneous: butabarbital, heptabarbital, hexobarbital, phenobarbital

REFERENCE
Evenson,M.A.; Wiktorowicz,J.E. Automated capillary electrophoresis applied to therapeutic drug monitoring, *Clin.Chem.*, **1992**, *38*, 1847–1852.

SAMPLE
Matrix: solutions

CAPILLARY ELECTROPHORESIS
Capillary: 60 cm $\times$ 50 μm fused-silica (40 cm to detector), bare (A) or C18 coated (B) (Supelco)
Capillary preparation: Rinse new capillaries with 250 μL 1 M NaOH, 250 μL 100 mM NaOH, water, MeOH, water, and running buffer then condition with running buffer for at least 15 min. Carry out a similar procedure between runs.
Running buffer: 100 mM pH 8.5 Borate buffer containing 25 mM sodium dodecyl sulfate and 5 M urea
Injection: Electrokinetic injection at 5 kV for 5 s.
Detector: UV 254
Migration time: 12.2 (A), 11 (B)
Voltage: 18 kV
Model: Jasco

OTHER SUBSTANCES
Simultaneous: amobarbital, barbital, mephobarbital, metharbital, phenobarbital, secobarbital

KEY WORDS
coated capillary

REFERENCE
Jinno,K.; Han,Y.; Nakamura,M. Analysis of anxiolytic drugs by capillary electrophoresis with bare and coated capillaries, *J.Capillary Electrophor.*, **1996**, *3*, 139–145.

SAMPLE
Matrix: solutions
Sample preparation: Inject an aliquot of a 1-50 μg/mL solution in water.

CAPILLARY ELECTROPHORESIS
Capillary: 55 cm × 50 μm fused-silica (35 cm to detector) (J&W)
Capillary preparation: Flush with running buffer for 5 min before each run. Periodically wash with 100 mM NaOH and flush extensively with running buffer.
Running buffer: MeOH:25 mM pH 9.24 borate buffer 20:80 containing 100 mM sodium dodecyl sulfate (A) or 50 mM pH 2.35 phosphate buffer (B) or 50 mM pH 9.24 borate buffer (C)
Injection: Injection of 5 μL using a split-flow injector ratio of 1:800.
Detector: UV 200
Migration time: 10.29 (A), 39.14 (B), 7.19 (C)
Voltage: 20 kV (A, B) or 12 kV (C)
Current: <60-80 μA
Model: ISCO Model 3850

OTHER SUBSTANCES
Also analyzed: acetylcodeine, amphetamine, barbital, caffeine, codeine, cocaine, diamorphine, diazepam, flunitrazepam, lidocaine, monoacetylmorphine, morphine, nalorphine, narceine, noscapine, papaverine, procaine, tetracaine, thebaine

KEY WORDS
all compounds were separated with running buffer A; some peaks overlapped with running buffers B and C.

REFERENCE
Tagliaro,F.; Smith,F.P.; Turrina,S.; Equisetto,V.; Marigo,M. Complementary use of capillary zone electrophoresis and micellar electrokinetic capillary chromatography for mutual confirmation of results in forensic drug analysis, *J.Chromatogr.A*, **1996**, *735*, 227–235.

SAMPLE
Matrix: solutions

CAPILLARY ELECTROPHORESIS
Capillary: 52 cm × 75 μm fused-silica
Running buffer: 50 mM pH 9.0 Phosphate buffer containing 40 mM hydroxypropyl-gamma-cyclodextrin
Detector: UV 254
Voltage: 15 kV

OTHER SUBSTANCES
Also analyzed: butalbital, secobarbital

KEY WORDS
chiral

REFERENCE
Hassan,A.; Alemayehu,B.; Anucha,T.; Chau,S.; Eradiri,O.; Ly,C.; Malaiyandi,P.; Muhuri,G.; Odidi,A.; Odidi,I.; Siwicki,J. Validation of HPLC method for assay and determination of chromatographic purity of diclofenac sodium in extended release formulations (Abstract 4173), *Pharm.Res.*, **1997**, *14*, S687.

SAMPLE
Matrix: solutions

CAPILLARY ELECTROPHORESIS
Capillary: 40 cm × 75 μm coated fused-silica (15 cm to detector) (Supelco)
Capillary preparation: Coat column as follows. Adjust the pH of 20 mL water to 3.5 with acetic acid, add 80 μL 3-(trimethoxysilyl)propyl methacrylate (3-methacryloxypropyltrimethoxysilane), mix, suck into capillary, let stand at room temperature for 1 h, remove the solution, wash with water. Fill the capillary with a deaerated 4% acrylamide solution containing 1 mg/mL N,N,N',N'-tetramethylethylenediamine and 1 mg/mL ammonium persulfate, let stand for 3 h, remove excess solution by aspiration, rinse with water, remove water by aspiration, dry at 35° (cf. J. Chromatogr. 1985, 347, 191).
Running buffer: 100 mM Tris/150 mM boric acid, pH 8.3

Injection: Electromigration at 5 kV for 5 s.
Detector: UV 240, UV 254
Migration time: 7.5
Voltage: 12 kV
Model: Jasco 890-CE

OTHER SUBSTANCES
Simultaneous: amobarbital, barbital, mephobarbital, metharbital, phenobarbital, secobarbital

KEY WORDS
injection at cathode; coated capillary

REFERENCE
Jinno,K.; Han,Y.; Sawada,H. Analysis of toxic drugs by capillary electrophoresis using polyacrylamide-coated
columns, *Electrophoresis*, **1997**, *18*, 284–286.

SAMPLE
Matrix: solutions
Sample preparation: Inject an aliquot of a 100 μg/mL solution in running buffer.

CAPILLARY ELECTROPHORESIS
Capillary: 60 cm × 50 μm acrylamide-coated fused-silica (40 cm to detector) (Supelco)
Capillary preparation: Adjust the pH of 20 mL water to 3.5 with acetic acid, add 80 μL 3-
(trimethoxysilyl)propyl methacrylate (3-methacryloxypropyltrimethoxysilane), mix, suck into
capillary, let stand at room temperature for 1 h, remove the solution, wash with water. Fill the
capillary with a deaerated 3-4% acrylamide solution containing 1 μL/mL N,N,N',N'-tetra-
methylethylenediamine and 1 mg/mL ammonium persulfate, let stand for 3 h, remove excess
solution by aspiration, rinse with water, remove water by aspiration, dry at 35° (cf. J. Chromatogr. 1985, 347, 191).
Running buffer: MeCN:buffer 5:95 (Buffer was 100 mM borate containing 5 M urea and 10 mM
sodium dodecyl sulfate, adjusted to pH 8.5 with phosphate.)
Injection: Electrokinetic injection at 5 kV for 5 s.
Detector: UV 254
Migration time: 15.8
Voltage: 18 kV
Model: Jasco Model 870-CE

OTHER SUBSTANCES
Simultaneous: alprazolam, amobarbital, barbital, bromazepam, clonazepam, clotiazepam, cloxazolam, diazepam, estazolam, etizolam, fludiazepam, flunitrazepam, flurazepam, haloxazolam,
medazepam, mephobarbital, metharbital, nimetazepam, nitrazepam, oxazepam, phenobarbital,
secobarbital, triazolam

KEY WORDS
coated capillary; detector at anode

REFERENCE
Jinno,K.; Han,Y.; Sawada,H.; Taniguchi,M. Capillary electrophoretic separation of toxic drugs using a polyacrylamide-coated capillary, *Chromatographia*, **1997**, *46*, 309–314.

SAMPLE
Matrix: solutions
Sample preparation: Inject an aliquot of a solution in 20 mM pH 5.9 ammonium acetate buffer.

CAPILLARY ELECTROPHORESIS
Capillary: 33 cm × 50 μm fused-silica (20 cm to detector) (Polymicro Technologies)
Running buffer: 10 mM pH 5.9 Ammonium acetate buffer containing 15 mM sodium dodecyl
sulfate
Injection: Electrokinetic injection at -7.4 kV for 1 s.
Detector: UV 226, MS, Finnigan MAT TSQ, negative electrospray (details in paper)
Migration time: 10.5

Voltage: -7.4 kV

OTHER SUBSTANCES
Simultaneous: amobarbital, barbital, butalbital, secobarbital

---

REFERENCE
Yang,L.; Harrata,A.K.; Lee,C.S. On-line micellar electrokinetic chromatography-electrospray ionization mass
spectrometry using anodically migrating micelles, *Anal.Chem.*, **1997**, *69*, 1820–1826.

SAMPLE
Matrix: solutions
Sample preparation: Inject an aliquot of a 20 μg/mL solution

CAPILLARY ELECTROPHORESIS
Capillary: 44 cm × 50 μm fused-silica (37 cm to detector) (Supelco)
Capillary preparation: Before each run rinse capillary with running buffer for 3 min. At the
start of each day rinse capillary with running buffer for 10 min. Treat new capillaries with 1
M NaOH, 100 mM NaOH, water, and running buffer.
Capillary temperature: 25
Running buffer: 100 mM Phosphoric acid adjusted to pH 5 with triethanolamine containing 10
mM carboxymethyl-β-cyclodextrin (Cyclolab, Budapest) and 50 mM heptakis(2,3,6-tri-O-
methyl)-β-cyclodextrin (Sigma)
Injection: Hydrodynamic injection for 5 s (13.3 nL).
Detector: UV 210
Migration time: 14.5, 17 (enantiomers)
Voltage: -25 kV
Model: Spectraphoresis 1000 CE

KEY WORDS
detector at anode; chiral; R_s = 7.7

REFERENCE
Fillet,M.; Fotsing,L.; Crommen,J. Enantioseparation of uncharged compounds by capillary electrophoresis using mixtures of anionic and neutral β-cyclodextrin derivatives, *J.Chromatogr.A*, **1998**, *817*, 113–119.

SAMPLE
Matrix: urine
Sample preparation: Place an extraction rod in 50 μL urine in a 50 mm × 1.5 mm ID length
of PTFE tubing for 30 min, place the extraction rod into 5 μL 20-40 mM pH 11.5 phosphate
buffer in a 50 mm × 1.2 mm ID length of PTFE tubing for 90 min, inject an aliquot of the
buffer. (Prepare the solid-phase extraction rod as follows. Polish a 70 × 1.1 stainless steel rod
with emery paper, clean with a Kimwipe and acetone, sonicate in EtOH, sonicate in THF, dip
in 3% PVAM in THF momentarily, hold vertically for 1 min, air dry in a hood for at least 5 h,
dip in PVC solution momentarily, hold vertically for 1 min, air dry for at least 5 h. The coating
is 3 cm long. PVAM was poly(vinyl chloride-co-vinyl acetate-co-maleic acid) consisting of 86%
vinyl chloride, 13% vinyl acetate, and 1% maleic acid. Prepare PVC solution by adding very
high molecular weight PVC slowly to THF with stirring until the concentration reaches 3.6%,
add Santicizer 141 to a concentration of 7.2%. Santicizer 141 (Monsanto) is 92% 2-ethylhexyl
diphenyl phosphate, 5% di-2-ethylhexyl phenyl phosphate, and 3% triphenyl phosphate.)

CAPILLARY ELECTROPHORESIS
Capillary: 75 cm × 75 μm fused-silica (50 cm to detector) (Polymicro Technologies)
Running buffer: 50 mM Tris adjusted to pH 7.8 with 3-[N-tris(hydroxymethyl)methylamino]-2-
hydroxypropanesulfonic acid (Tapso)
Injection: Pressure injection at 0.5 psi for 4 s.
Detector: UV 230
Migration time: 7.45
Voltage: 35 kV
Current: about 30 μA
Model: Isco 3850
Limit of detection: <1 ppm

OTHER SUBSTANCES
Extracted: amobarbital, aprobarbital, butabarbital, butalbital, mephobarbital, secobarbital, thiopental
Simultaneous: allobarbital, aspirin, phenobarbital

KEY WORDS
SPE

REFERENCE
Li,S.; Weber,S.G. Determination of barbiturates by solid-phase microextraction and capillary electrophoresis, *Anal.Chem.*, **1997**, *69*, 1217–1222.

Pentoxifylline

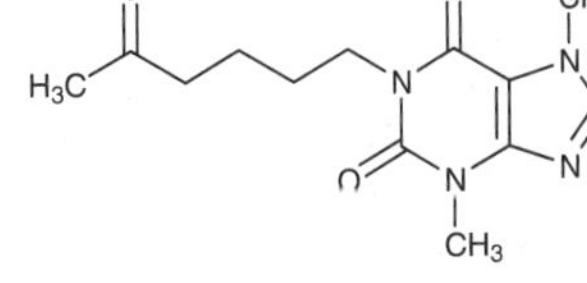

Molecular formula: $C_{13}H_{18}N_4O_3$
Molecular weight: 278.31
CAS Registry No.: 6493-05-6
Merck Index (12th ed.): 7278
Lednicer: 2 466

SAMPLE
Matrix: bulk
Sample preparation: Prepare a 20 mg/mL solution in water, inject an aliquot.

CAPILLARY ELECTROPHORESIS
Capillary: 60 cm × 75 μm fused-silica (53 cm to detector)
Capillary preparation: Rinse capillary with running buffer after each run. At the start of each day rinse with 100 mM NaOH for 5 min and with separation buffer for 10 min. Before storage rinse capillary with water and air dry.
Running buffer: 30 mM pH 9.0 Sodium borate buffer
Injection: Hydrodynamic injection at 10 cm for 5-10 s
Detector: UV 214
Migration time: 7.5
Voltage: 20 kV
Model: Waters Quanta 4000

OTHER SUBSTANCES
Simultaneous: impurities, caffeine, xanthine

REFERENCE
Korman,M.; Vindevogel,J.; Sandra,P. Application of micellar electrokinetic chromatography to the quality control of pharmaceutical formulations: the analysis of xanthine derivatives, *Electrophoresis*, **1994**, *15*, 1304–1309.

Perphenazine

Molecular formula: $C_{21}H_{26}ClN_3OS$
Molecular weight: 403.98
CAS Registry No.: 58-39-9
Merck Index (12th ed.): 7323

SAMPLE
Matrix: urine

Sample preparation: 10 mL Urine + 1 mL 5 M NaOH + 10 mL n-hexane, vortex for 1 min, centrifuge at 0° at 3000 g for 5 min. Remove the organic layer and add it to 50 μL glacial acetic acid, evaporate to dryness under a stream of nitrogen at 30°, reconstitute the residue in 50 μL 50 mM sodium taurodeoxycholate, filter (0.2 μm PTFE), inject an aliquot.

CAPILLARY ELECTROPHORESIS
Capillary: 100 cm × 50 μm fused-silica (50 cm to detector) (ISCO)
Capillary preparation: Rinse capillary with 20 μL running buffer between injections. Condition a new capillary by filling with 1 M NaOH, let stand for 1 h, fill with 100 mM NaOH, let stand for 1 h, rinse with buffer. Every 20 injections rinse capillary with 200 μL 1 M NaOH, 200 μL water, and 200 μL running buffer, fill with running buffer.
Capillary temperature: 22
Running buffer: 40 mM pH 9.5 Borate buffer containing 10 mM sodium taurodeoxycholate
Injection: Load under vacuum at 7.5 kPa/s.
Detector: UV 240
Migration time: 10.5
Voltage: 30 kV
Model: ISCO Model 3140 electropherograph
Limit of detection: 5 ng/mL

OTHER SUBSTANCES
Extracted: acepromazine, amiodarone, amitriptyline, azaperone, chlorpromazine, cianopramine, clomipramine, clozapine, desethylamiodarone, desipramine, diclofensine, dothiepin, doxepin, imipramine, isocarboxazid, moclobemide, phenothiazine, pimozide, prochlorperazine, promazine, thioridazine, thiothixene, trifluoperazine, trimipramine

KEY WORDS
human; cow; pig; horse; protect from light

REFERENCE
Aumatell,A.; Wells,R.J. Determination of a cardiac antiarrhythmic, tricyclic antipsychotics and antidepressants in human and animal urine by micellar electrokinetic capillary chromatography using a bile salt, *J.Chromatogr.B*, **1995**, *669*, 331–344.

SAMPLE
Matrix: urine
Sample preparation: Dilute 10-fold with water, inject an aliquot.

CAPILLARY ELECTROPHORESIS
Capillary: 70 cm × 50 μm fused-silica (65.4 cm to detector) (Polymicro Technologies)
Capillary temperature: 25
Running buffer: 20 mM pH 3.5 Tris-formic acid containing 50 μg/mL FC-135 (Fluorad/3M)
Injection: Pressure injection at 2 psi.
Detector: UV 240
Migration time: 8.10
Voltage: 20 kV
Model: Bio-Rad BioFocus 3000

OTHER SUBSTANCES
Extracted: prochlorperazine, promazine, thioridazine, trifluoperazine, triflupromazine

KEY WORDS
detector at anode

REFERENCE
Muijelsaar,P.G.H.M.; Claessens,H.A.; Cramers,C.A. Determination of structurally related phenothiazines by capillary zone electrophoresis and micellar electrokinetic chromatography, *J.Chromatogr.A*, **1996**, *735*, 395–402.

Phenacetin

Molecular formula: C$_{10}$H$_{13}$NO$_2$
Molecular weight: 179.22
CAS Registry No.: 62-44-2
Merck Index (12th ed.): 7344
Lednicer: 1 111

SAMPLE
Matrix: solutions
Sample preparation: Inject an aliquot of a 30-300 μg/mL solution in water.

CAPILLARY ELECTROPHORESIS
Capillary: 36 cm × 50 μm fused-silica (28 cm to detector)
Running buffer: 80 mM Reagent, 120 mM sodium dodecyl sulfate, 900 mM 1-butanol, and 10 mM borate in water where reagent is heptane (A) or octane (B) or 1-butyl chloride (C)
Injection: Hydrostatic injection.
Detector: UV 254
Migration time: 9.479 (A), 9.744 (B), 8.860 (C)
Voltage: 12 kV
Current: ca. 65 μA
Model: Waters Quanta 4000

OTHER SUBSTANCES
Simultaneous: aminopyrine, caffeine, phenobarbital

REFERENCE
Fu,X.; Lu,J.; Zhu,A. Microemulsion electrokinetic chromatographic separation of antipyretic analgesic ingredients, *J.Chromatogr.A*, **1996**, *735*, 353–356.

SAMPLE
Matrix: solutions
Sample preparation: Inject an aliquot of a solution in running buffer.

CAPILLARY ELECTROPHORESIS
Capillary: 57 cm × 50 μm fused-silica (50 cm to detector) (Polymicro Technologies)
Capillary temperature: 25
Running buffer: 25 mM NaH$_2$PO$_4$ containing 0.83% poly(sodium 10-undecenyl sulfate), adjusted to pH 7.3 with 12.5 mM sodium tetraborate (Preparation of poly(sodium 10-undecenyl sulfate is as follows. React undecylenyl alcohol (1-undecen-11-ol) with one equivalent of chlorosulfonic acid in diethyl ether under nitrogen at -5° to obtain 10-undecenyl hydrogen sulfate, react with one equivalent of NaOH to obtain sodium 10-undecenyl sulfate (J. Microcolumn Sep. 1996, 8, 115). Dissolve 20 g sodium 10-undecenyl sulfate in 40 mL water (degassed by purging with nitrogen for 24 h), add 600 mg potassium persulfate, stir at 70° under nitrogen for 50 h, cool, add 200 mL cold (0°) EtOH, filter, wash the solid with three 10 mL portions of cold EtOH, dry under vacuum at 65° overnight to obtain poly(sodium 10-undecenyl sulfate) (cf. J. Microcolumn Sep. 1992, 4, 509). Purify by dialysis using dialysis tubing with a 1000 MW cut-off (Spectrum Houston) for 24 h.)
Detector: UV 214
Migration time: 9
Voltage: 16.1 kV
Model: Beckman P/ACE 2000

OTHER SUBSTANCES
Simultaneous: acetaminophen, caffeine, guaifenesin, trimetoquinol
Interfering: ethenzamide

REFERENCE
Palmer,C.P.; Terabe,S. Micelle polymers as pseudostationary phases in MEKC: Chromatographic performance and chemical selectivity, *Anal.Chem.*, **1997**, *69*, 1852–1860.

SAMPLE
Matrix: solutions
Sample preparation: Inject an aliquot of a 10 μg/mL solution

CAPILLARY ELECTROPHORESIS
Capillary: 50 cm × 75 μm silica
Capillary temperature: 25
Running buffer: 20 mM pH 7.0 Phosphate/borate buffer containing 50 mM sodium dodecyl
  sulfate
Injection: 5 nL
Detector: UV 208
Migration time: 4
Voltage: 20 kV
Model: Otsuka CAPI-3000

OTHER SUBSTANCES
Simultaneous: acetaminophen, caffeine, ethyl paraben, guaifenesin, methyl paraben,
  salicylamide

REFERENCE
Poe,R.B.; Hayashi,Y.; Matsuda,R. Precision-optimization of wavelengths in diode-array detection in separation
  science, *Anal.Sci.*, **1997**, *13*, 951–961.

SAMPLE
Matrix: solutions

CAPILLARY ELECTROPHORESIS
Capillary: 80 cm × 50 μm fused-silica (Supelco)
Capillary preparation: Before each run rinse capillary with running buffer at 94 kPa for 5 min
  then with running buffer containing 80 mM sodium dodecyl sulfate at 94 kPa for 2.5 min before
  each run.
Running buffer: 50 mM pH 8.5 ammonium carbonate buffer
Injection: Pressure injection at 5 kPa (50 mbar) for 8 s into the capillary filled with running
  buffer containing sodium dodecyl sulfate, electrophoresis is then continued with plain running
  buffer.
Detector: MS, Perkin-Elmer Sciex API-300 quadrupole, electrospray (ionspray) interface, sheath
  liquid MeOH:running buffer 50:50 at 4 μL/min, ionspray voltage 5 kV, positive ion mode
Migration time: 32
Voltage: 25 kV (net voltage across capillary = 20 kV (applied voltage - electrospray voltage))
Model: Hewlett Packard 3D CE

OTHER SUBSTANCES
Simultaneous: acetaminophen, acetanilide, caffeine, guaifenesin, pyridoxine

REFERENCE
Tanaka,Y.; Kishimoto,Y.; Otsuka,K.; Terabe,S. Strategy for selecting separation solutions in capillary electrophoresis-mass spectrometry, *J.Chromatogr.A*, **1998**, *817*, 49–57.

Phenazopyridine

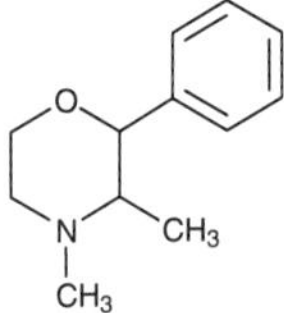

Molecular formula: $C_{11}H_{11}N_5$
Molecular weight: 213.24
CAS Registry No.: 94-78-0, 136-40-3 (HCl)
Merck Index (12th ed.): 7361
Lednicer: 1 255

SAMPLE
Matrix: formulations
Sample preparation: Grind ten tablets and weigh out an amount corresponding to 50 mg sulfisoxazole and 5 mg phenazopyridine, add 40 mL running buffer, sonicate for 30 min, make up to 50 mL, with running buffer, filter (0.45 μm). Dilute an aliquot of the filtrate 1:25 with running buffer, inject an aliquot.

CAPILLARY ELECTROPHORESIS
Capillary: 70 cm × 75 μm fused-silica
Capillary preparation: After each run refill reservoirs, rinse capillary with fresh buffer at 7 psi for 3 min.
Running buffer: 5 mM NaH_2PO_4 containing 5 mM Na_2HPO_4 and 50 mM sodium dodecyl sulfate, adjusted to pH 7.0 with NaOH
Injection: Hydrodynamic injection at 5 cm for 10 s (6.8 nL)
Detector: UV 240
Migration time: 29.3
Voltage: 15 kV
Model: Dionex Model CES I
Limit of detection: 160 ng/mL (S/N 2)

OTHER SUBSTANCES
Simultaneous: impurities, sulfanilamide, sulfisoxazole

KEY WORDS
tablets

REFERENCE
Nickerson,B.; Scypinski,S.; Sokoloff,H.; Sahota,S. Separation of sulfisoxazole, phenazopyridine, and their related impurities by micellar electrokinetic chromatography, *J.Liq.Chromatogr.*, **1995**, *18*, 3847–3875.

Phendimetrazine

Molecular formula: $C_{12}H_{17}NO$
Molecular weight: 191.27
CAS Registry No.: 634-03-7, 50-58-8 (tartrate)
Merck Index (12th ed.): 7365
Lednicer: 1 260

SAMPLE
Matrix: solutions

CAPILLARY ELECTROPHORESIS
Capillary: 37 cm × 50 μm (30 cm to detector)
Capillary temperature: 25
Running buffer: 50 mM pH 2.5 Phosphate buffer containing 16 mM 2-hydroxy-3-trimethylammoniopropyl-β-cyclodextrin
Detector: UV 214

Migration time: 11.32 (first enantiomer; α = 1.028; R_s = 0.735)
Voltage: 20 kV
Model: Beckman P/ACE 2100

OTHER SUBSTANCES
Simultaneous: cyclodrine, cyclopentolate, norpseudoephedrine, pholedrine, tropicamide

KEY WORDS
chiral

REFERENCE
Bunke,A.; Jira,T. Chiral capillary electrophoresis using a cationic cyclodextrin, *Pharmazie*, **1996**, *51*, 672–673.

SAMPLE
Matrix: solutions

CAPILLARY ELECTROPHORESIS
Capillary: 58.2 cm $\times$ 75 μm fused-silica (50.7 cm to detector) (Polymicro Technologies)
Capillary preparation: Purge with running buffer before each run. At the beginning of each
day purge using 50-60 kPa vacuum with 500 mM NaOH for 5 min, with water for 5 min, with
MeCN for 5 min, and with running buffer for 5 min.
Running buffer: MeCN:MeOH:acetic acid 49:50:1 containing 20 mM ammonium acetate
Injection: Hydrostatic injection at 10 cm for 5 s.
Detector: UV 214
Migration time: 6.04
Voltage: 25 kV
Model: Waters Quanta 4000

OTHER SUBSTANCES
Simultaneous: amphetamine, benzphetamine, ephedrine, nylidrin, oxymetazoline, phenmetrazine, phenylephrine, phenylpropanolamine, xylometazoline

REFERENCE
Leung,G.N.W.; Tang,H.P.O.; Tso,T.S.C.; Wan,T.S.M. Separation of basic drugs with non-aqueous capillary electrophoresis, *J.Chromatogr.A*, **1996**, *738*, 141–154.

Phenformin

Molecular formula: $C_{10}H_{15}N_5$
Molecular weight: 205.26
CAS Registry No.: 114-86-3, 834-28-6 (HCl)
Merck Index (12th ed.): 7376
Lednicer: 1 75

SAMPLE
Matrix: blood
Sample preparation: 100 μL Plasma + 300 μL MeCN, mix thoroughly, centrifuge at 12000 g
for 1 min. Remove the supernatant and evaporate it to dryness under reduced pressure at 70°,
reconstitute the residue in 50 μL pH 7.8 phosphate buffer, add 5 μL bromothymol blue solution,
mix, extract with 1 mL chloroform for 1 min, centrifuge at 12000 g for 2 min, repeat extraction.
Combine the organic layers and evaporate them to dryness under a stream of nitrogen at 40°,
reconstitute the residue in 100 μL 200 μM phosphoric acid, inject an aliquot. (The analyte
forms an ion pair with bromothymol blue and this analyte is extracted. Prepare bromothymol
blue solution by dissolving 62 mg bromothymol blue in 2 mL 100 mM NaOH, make up top 10
mL with water, sonicate, adjust to pH 7.8 with concentrated HCl or NaOH. Prepare the pH
7.8 phosphate buffer by dissolving 5.52 g $NaH_2PO_4.H_2O$ in 21 mL 2 M NaOH, adjust to pH 7.8
with saturated NaOH, make up to 100 mL with water.)

CAPILLARY ELECTROPHORESIS
Capillary: 40 cm × 50 μm fused-silica (32.5 cm to detector) (Polymicro Technologies)
Capillary preparation: After each run rinse capillary with 100 mM NaOH for 1 min, with water for 1 min, then with running buffer for 5 min. Condition a new capillary with 1 M NaOH at 60° for 20 min, with 100 mM NaOH at 60° for 5 min, and with water at 20° for 5 min.
Capillary temperature: 20
Running buffer: 100 mM pH 2.5 Phosphate buffer
Injection: Electrokinetic injection at 10 kV for 10 s.
Detector: UV 195
Migration time: 3
Internal standard: phenformin
Voltage: 20 kV
Current: 68 μA
Model: SpectraPhoresis 1000

OTHER SUBSTANCES
Extracted: metformin

KEY WORDS
plasma; phenformin is IS

REFERENCE
Song,J.-Z.; Chen,H.-F.; Tian,S.-J.; Sun,Z.-P. Determination of metformin in plasma by capillary electrophoresis using field-amplified sample stacking technique, *J.Chromatogr.B*, **1998**, *708*, 277–283.

Pheniramine

Molecular formula: $C_{16}H_{20}N_2$
Molecular weight: 240.35
CAS Registry No.: 86-21-5, 132-20-7 (maleate)
Merck Index (12th ed.): 7383
Lednicer: 1 77

SAMPLE
Matrix: formulations
Sample preparation: Grind tablets, extract twice with MeOH by sonicating at low temperature, centrifuge, filter (0.45 μm) the supernatant, inject an aliquot of the filtrate.

CAPILLARY ELECTROPHORESIS
Capillary: 47 cm × 50 μm (40 cm to detector) (Polymicro Technologies)
Capillary temperature: 25
Running buffer: 50 mM pH 7.5 Borate buffer containing 50 mM phosphate, 10 mM sodium dodecyl sulfate, 10 mM β-cyclodextrin, and 10 mM tetrabutylammonium hydrogen sulfate
Injection: Inject at a pressure of 0.5 psi for 1 s (5.2 nL)
Detector: UV 214
Migration time: 3.3
Voltage: 21.5 kV
Model: Beckman P/ACE 2000
Limit of detection: 53.3 pg

OTHER SUBSTANCES
Simultaneous: chlorpheniramine, cyclizine, dimenhydrinate, doxylamine, methapyrilene, promethazine, thonylamine, triprolidine

KEY WORDS
tablets

REFERENCE
Ong,C.P.; Ng,C.L.; Lee,H.K.; Li,S.F.Y. Determination of antihistamines in pharmaceuticals by capillary electrophoresis, *J.Chromatogr.*, **1991**, *588*, 335–339.

SAMPLE
Matrix: solutions
Sample preparation: Prepare a solution in running buffer, inject an aliquot.

CAPILLARY ELECTROPHORESIS
Capillary: 60 cm × 75 μm fused-silica (52.4 cm to detector)
Capillary preparation: After each run flush with 500 mM KOH for 2-3 min, flush with water, fill with running buffer
Running buffer: 10 mM Na_2HPO_4 containing 2% heparin sodium (MW 10000, 11% S, Scientific Protein Laboratories, Waunakee, WI) adjusted to pH 5 with phosphoric acid
Injection: Hydrostatic injection
Detector: UV 214
Migration time: 19.1 (first enantiomer, R_S 4.1)
Model: Waters Quanta 4000

OTHER SUBSTANCES
Also analyzed: anabasine, brompheniramine, bupivacaine, carbinoxamine, chlorcyclizine, chloroquine, chlorpheniramine, dimethindene, doxylamine, enpiroline, halofantrine, hydroxychloroquine, indapamide, mefloquine, nornicotine, primaquine, promethazine, quinacrine, tetramisole

KEY WORDS
chiral

REFERENCE
Stalcup,A.M.; Agyei,N.M. Heparin: a chiral mobile-phase additive for capillary zone electrophoresis, *Anal.Chem.*, **1994**, *66*, 3054–3059.

SAMPLE
Matrix: solutions

CAPILLARY ELECTROPHORESIS
Capillary: 62 cm × 75 μm fused-silica (54.5 cm to detector) (Polymicro Technologies)
Capillary preparation: Purge with running buffer before each run. At the beginning of each day purge using 50-60 kPa vacuum with 500 mM NaOH for 5 min, with water for 5 min, with MeCN for 5 min, and with running buffer for 5 min.
Running buffer: MeCN:MeOH:acetic acid 49:50:1 containing 20 mM ammonium acetate
Injection: Hydrostatic injection at 10 cm for 5 s.
Detector: UV 214
Migration time: 3.66
Voltage: 30 kV
Model: Waters Quanta 4000

OTHER SUBSTANCES
Simultaneous: buclizine, chlorcyclizine, chlorpheniramine, cyclizine, dimenhydrinate, methaphenilene, promethazine, pyrilamine, pyrrobutamine, tripelennamine

REFERENCE
Leung,G.N.W.; Tang,H.P.O.; Tso,T.S.C.; Wan,T.S.M. Separation of basic drugs with non-aqueous capillary electrophoresis, *J.Chromatogr.A*, **1996**, *738*, 141–154.

SAMPLE
Matrix: solutions
Sample preparation: Inject an aliquot of a solution in running buffer.

CAPILLARY ELECTROPHORESIS
Capillary: 60 cm × 75 μm fused-silica (52.4 cm to detector)

Capillary preparation: After each run flush with 500 mM KOH for 2-3 min then with water.
Running buffer: 10 mM pH 3.8 Phosphate buffer containing 2% sulfated cyclodextrin (ds 7-10)
Injection: Hydrostatic injection.
Detector: UV 214
Migration time: 8.60, 8.68 (enantiomers)
Voltage: 15 kV
Model: Waters Quanta 4000

OTHER SUBSTANCES
Also analyzed: acebutolol, alprenolol, aminoglutethimide, brompheniramine, bupivacaine, bupropion, canadine, carbinoxamine, chloroquine, chlorpheniramine, dimethindene, disopyramide, doxylamine, hydroxychloroquine, idazoxan, isoxsuprine, ketamine, mepenzolate, mepivacaine, methoxyphenamine, mexiletine, midodrine, nefopam, orphenadrine, oxprenolol, oxyphencyclimine, phensuximide, pindolol, piperoxan, terbutaline, tetramisole, tolperisone, tranylcypromine, trihexyphenidyl, trimipramine, verapamil, warfarin

KEY WORDS
chiral; detector at anode

REFERENCE
Stalcup,A.M.; Gahm,K.H. Application of sulfated cyclodextrins to chiral separations by capillary zone electrophoresis, *Anal.Chem.*, **1996**, *68*, 1360–1368.

SAMPLE
Matrix: solutions

CAPILLARY ELECTROPHORESIS
Capillary: 70 cm × 50 μm fused-silica (50 cm to detector) (Polymicro Technologies)
Capillary temperature: 30
Running buffer: 50 mM pH 4.3 Sodium phosphate containing 3.0 mM sulfobutyl ether-β-cyclodextrin (average substitution 3.9, MW 1721, Center for Drug Delivery Research, Lawrence KS)
Injection: Pressure injection at 5 inches Hg for 1.5 s.
Detector: UV 210
Migration time: 13, 13.5 (enantiomers)
Voltage: 15 kV
Current: 22 μA
Model: Perkin Elmer-Applied Biosystems Model 270

KEY WORDS
chiral

REFERENCE
Xie,G.-h.; Skanchy,D.J.; Stobaugh,J.F. Chiral separations of enantiomeric pharmaceuticals by capillary electrophoresis using sulphobutyl ether β-cyclodextrin as isomer selector, *Biomed.Chromatogr.*, **1997**, *11*, 193–199.

SAMPLE
Matrix: solutions

CAPILLARY ELECTROPHORESIS
Capillary: 57 cm × 75 μm fused-silica (34 cm to detector) (Polymicro Technologies)
Capillary preparation: Before each run rinse capillary with 100 mM NaOH for 5 min, with water for 5 min and with running buffer for 5 min. Condition new capillaries with 100 mM NaOH for 10 min, with water for 5 min and with running buffer for 5 min.
Running buffer: 10 mM pH 3.84 Phosphate buffer containing 7 mM sulfated-β-cyclodextrin
Injection: Hydrodynamic injection at 14 cm for 25 s.
Detector: UV 214
Migration time: 8, 9 (enantiomers)
Voltage: 12 kV
Model: laboratory-constructed

OTHER SUBSTANCES
Also analyzed: brompheniramine, chloroquine, chlorpheniramine, doxylamine

KEY WORDS
chiral

REFERENCE
Jin,L.J.; Li,S.F.Y. Comparison of chiral recognition capabilities of cyclodextrins for the separation of basic drugs in capillary zone electrophoresis, *J.Chromatogr.B*, **1998**, *708*, 257–266.

Phenmetrazine

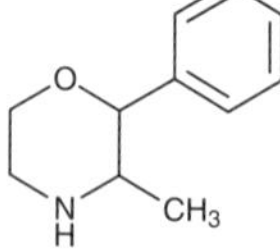

Molecular formula: $C_{11}H_{15}NO$
Molecular weight: 177.25
CAS Registry No.: 134-49-6, 1707-14-8 (HCl)
Merck Index (12th ed.): 7385
Lednicer: 1 260

SAMPLE
Matrix: blood, urine
Sample preparation: Adjust pH of 2 mL urine or plasma to 10.5 with aqueous NaOH, extract gently with chloroform:isopropanol 90:10. Remove the organic layer and evaporate it to dryness under a stream of nitrogen, reconstitute the residue in 50 (urine) or 75 (plasma) μL running buffer, filter, inject an aliquot.

CAPILLARY ELECTROPHORESIS
Capillary: 60 cm × 75 μm AccuSep uncoated silica (52.5 cm to detector) (Waters)
Capillary preparation: At the beginning of each day purge with 500 mM KOH for 5 min, with water for 5 min, and with running buffer for 10 min.
Running buffer: 50 mM NaH_2PO_4 adjusted to pH 2.35 with phosphoric acid
Injection: Hydrostatic injection at 15 cm
Detector: UV 214
Migration time: 6.01
Voltage: 22 kV
Current: 135-145 μA
Model: Waters Quanta 4000

OTHER SUBSTANCES
Extracted: acepromazine, amphetamine, benzocaine, brompheniramine, butacaine, codeine, diazepam, doxapram, lidocaine, medazepam, methamphetamine, methapyrilene, methaqualone, procaine, tetrahydrozoline
Simultaneous: meclizine

KEY WORDS
plasma

REFERENCE
Chee,G.L.; Wan,T.S.M. Reproducible and high-speed separation of basic drugs by capillary zone electrophoresis, *J.Chromatogr.*, **1993**, *612*, 172–177.

SAMPLE
Matrix: solutions

CAPILLARY ELECTROPHORESIS
Capillary: 58.2 cm × 75 μm fused-silica (50.7 cm to detector) (Polymicro Technologies)

Capillary preparation: Purge with running buffer before each run. At the beginning of each day purge using 50-60 kPa vacuum with 500 mM NaOH for 5 min, with water for 5 min, with MeCN for 5 min, and with running buffer for 5 min.
Running buffer: MeCN:MeOH:acetic acid 49:50:1 containing 20 mM ammonium acetate
Injection: Hydrostatic injection at 10 cm for 5 s.
Detector: UV 214
Migration time: 4.57
Voltage: 25 kV
Model: Waters Quanta 4000

OTHER SUBSTANCES
Simultaneous: amphetamine, benzphetamine, ephedrine, nylidrin, oxymetazoline, phendimetrazine, phenylephrine, phenylpropanolamine, xylometazoline

---

REFERENCE
Leung,G.N.W.; Tang,H.P.O.; Tso,T.S.C.; Wan,T.S.M. Separation of basic drugs with non-aqueous capillary electrophoresis, *J.Chromatogr.A*, **1996**, *738*, 141–154.

Phenobarbital

Molecular formula: $C_{12}H_{12}N_2O_3$
Molecular weight: 232.24
CAS Registry No.: 50-06-6, 57-30-7 (sodium salt)
Merck Index (12th ed.): 7386
Lednicer: 1 268

SAMPLE
Matrix: blood
Sample preparation: 50 µL Serum + 5 mL isotonic saline, add to a Sep-Pak C18 SPE cartridge, wash with 5 mL saline, wash with 5 mL water, wash with 5 mL heptane:chloroform 98:2, wash with 5 mL heptane, wash with 3 mL water, remove all water with a stream of air, elute with 3 mL MeOH (remove all MeOH with a stream of air). Evaporate the eluate to dryness under reduced pressure at 37°, reconstitute with 100 µL water, inject an aliquot.

CAPILLARY ELECTROPHORESIS
Capillary: 72 cm × 50 µm silica (50 cm to detector) (Applied Biosystems)
Capillary preparation: Before each run wash capillary with 100 mM NaOH for 3 min, wash with running buffer for 3 min, aspirate neutral marker solution (2 drops DMSO in 10 mL water) for 1 s, place capillary end in buffer vial for 5 s, inject sample. Wash capillary at the beginning of each day by passing 1 M NaOH through for 20 min.
Capillary temperature: 30
Running buffer: 30 mM pH 9.3 Borate buffer containing 30 mM sodium dodecyl sulfate
Injection: Inject by applying vacuum for 1 s (2-3 nL)
Detector: UV 200
Migration time: 5.5
Voltage: +30 kV
Current: 72 µA
Model: Applied Biosystems Model 270A

OTHER SUBSTANCES
Extracted: carbamazepine, phenytoin, primidone

KEY WORDS
serum; SPE

REFERENCE
Evenson,M.A.; Wiktorowicz,J.E. Automated capillary electrophoresis applied to therapeutic drug monitoring, *Clin.Chem.*, **1992**, *38*, 1847–1852.

SAMPLE
Matrix: blood
Sample preparation: 100 μL Serum + 150 μL 80 μg/mL 3-isobutyl-1-methylxanthine in MeCN, mix for 30 s, centrifuge at 14000 g for 1 min, inject an aliquot of the supernatant.

CAPILLARY ELECTROPHORESIS
Capillary: 25 cm × 47 μm
Capillary preparation: Wash with running buffer by pressure injection for 1 min after each run. Wash daily with 50 mM phosphoric acid for 2 min, with water for 1 min, with 2 M NaOH for 2 min, with water for 1 min, and with running buffer for 2 min.
Capillary temperature: 24
Running buffer: 220 mM boric acid adjusted to pH 8.8 with 2.5 M NaOH
Injection: Pressure injection for 8 s.
Detector: UV 254
Migration time: 2.9
Internal standard: 3-isobutyl-1-methylxanthine (2.3)
Voltage: 12 kV
Model: Beckman

OTHER SUBSTANCES
Extracted: acetaminophen, caffeine, carbamazepine, iohexol, pentobarbital, phenytoin, salicylic acid, theophylline

KEY WORDS
serum

REFERENCE
Shihabi,Z.K.; Constantinescu,M.S. Iohexol in serum determined by capillary electrophoresis, *Clin.Chem.*, **1992**, *38*, 2117–2120.

SAMPLE
Matrix: blood
Sample preparation: Vortex serum for 20 s, filter (0.45 μm), inject an aliquot.

CAPILLARY ELECTROPHORESIS
Capillary: 92 cm × 75 μm (68 cm to detector)
Capillary preparation: Before each run rinse capillary with 100 mM NaOH for 3 min and with running buffer for 5 min.
Capillary temperature: 35
Running buffer: 10 mM pH 9.2 Na_2HPO_4 containing 6 mM sodium borate and 75 mM sodium dodecyl sulfate
Injection: Vacuum suction for 1 s
Detector: UV 220, UV 245
Migration time: 13
Voltage: 20 kV
Current: 59 μA
Model: Applied Biosystems model 270A-HT

OTHER SUBSTANCES
Extracted: ethosuximide, primidone

KEY WORDS
serum

REFERENCE
Schmutz,A.; Thormann,W. Determination of phenobarbital, ethosuximide, and primidone in human serum by micellar electrokinetic capillary chromatography with direct sample injection, *Ther.Drug Monit.*, **1993**, *15*, 310–316.

SAMPLE
Matrix: blood

Sample preparation: 100 μL Serum + 150 μL MeCN, mix for 15 s, centrifuge at 15000 g for 1 min, inject an aliquot of the supernatant.

CAPILLARY ELECTROPHORESIS
Capillary: 25 cm × 50 μm
Capillary preparation: Wash with running buffer for 1 min after each run.
Capillary temperature: 24
Running buffer: 300 mM boric acid adjusted to pH 8.5 with NaOH
Injection: Pressure injection for 15 s.
Detector: UV 254
Migration time: 2.7
Voltage: 13 kV
Model: Beckman

OTHER SUBSTANCES
Extracted: iohexol, phenytoin, theophylline

KEY WORDS
serum

REFERENCE
Shihabi,Z.K. Sample matrix effects in capillary electrophoresis. II. Acetonitrile deproteinization, *J.Chromatogr.A*, **1993**, *652*, 471–475.

SAMPLE
Matrix: blood
Sample preparation: Vortex serum for 15 s, filter (0.45 μm), add sodium dodecyl sulfate to a concentration of 100 mM, let stand for 1 h, filter (Amicon Centrifree Micropartition) while centrifuging at 1500 g for 20 min, inject an aliquot of the ultrafiltrate.

CAPILLARY ELECTROPHORESIS
Capillary: 63 cm × 75 μm fused-silica (63 cm to detector)
Capillary preparation: Before each run rinse capillary with 100 mM NaOH for 3 min and with running buffer for 5 min.
Capillary temperature: 35
Running buffer: 10 mM pH 9.2 Na_2HPO_4 containing 6 mM sodium borate and 100 mM sodium dodecyl sulfate
Injection: Vacuum suction for 1 s
Detector: UV 195
Migration time: 9.4
Voltage: 25 kV
Current: 80 μA
Model: Applied Biosystems model 270A-HT

OTHER SUBSTANCES
Extracted: ethosuximide, phenytoin

KEY WORDS
serum; ultrafiltrate

REFERENCE
Schmutz,A.; Thormann,W. Factors affecting the determination of drugs and endogenous low molecular mass compounds in human serum by micellar electrokinetic capillary chromatography with direct sample injection, *Electrophoresis*, **1994**, *15*, 51–61.

SAMPLE
Matrix: blood
Sample preparation: 25 μL Serum + 25 μL IS solution, mix, inject an aliquot. (Prepare IS solution by dissolving 200 mg acetoacetanilide in 25 mL MeOH and diluting to 1 L with water.)

CAPILLARY ELECTROPHORESIS
Capillary: 50 cm × 50 μm

Capillary preparation: Between runs wash capillary with 100 mM phosphoric acid for 1 min and fill with running buffer for 1 min. Each day wash capillary with 2 M NaOH for 3 min, 100 mM phosphoric acid for 3 min, and running buffer for 3 min.
Capillary temperature: 35
Running buffer: 100 mM Boric acid containing 55.4 mM sodium dodecyl sulfate, adjusted to pH 8.4 ± 0.1 with 2 M NaOH
Injection: Pressure injection for 5 s
Detector: UV 214
Migration time: 4.4
Internal standard: acetoacetanilide (3.9)
Voltage: about 16.5 kV
Current: 38 μA (fixed)
Model: Beckman Model 2000 CE

OTHER SUBSTANCES
Extracted: felbamate
Noninterfering: acetaminophen, N-acetylprocainamide, amikacin, carbamazepine, digoxin, disopyramide, ethosuximide, lidocaine, phenytoin, primidone, procainamide, quinidine, salicylic acid, theophylline, valproic acid, vancomycin

KEY WORDS
serum; comparison with HPLC

REFERENCE
Shihabi,Z.K.; Oles,K.S. Felbamate measured in serum by two methods: HPLC and capillary electrophoresis, *Clin.Chem.*, **1994**, *40*, 1904–1908.

SAMPLE
Matrix: blood
Sample preparation: 200 μL Serum + 1 mL 5 μg/mL n-propyl p-hydroxybenzoate in ethyl acetate, vortex for 30 s, centrifuge at 13400 g for 2 min. Remove a 500 μL aliquot of the organic layer and evaporate it to dryness under a stream of nitrogen, reconstitute with 50 μL MeOH: water 5:95, inject an aliquot.

CAPILLARY ELECTROPHORESIS
Capillary: 64.5 cm × 50 μm fused-silica (56 cm to detector) (Hewlett-Packard)
Capillary preparation: Rinse capillary with running buffer at 630 mm Hg for 5 min before each run.
Capillary temperature: 30
Running buffer: 10 mM pH 8.0 Phosphate buffer containing 50 mM sodium dodecyl sulfate
Injection: Vacuum injection at 50 mm Hg for 2 s.
Detector: UV 210
Migration time: 4.1
Internal standard: n-propyl p-hydroxybenzoate (7.2)
Voltage: 30 kV
Model: Hewlett-Packard HP[3D]
Limit of detection: 2.5 μg/mL

OTHER SUBSTANCES
Extracted: carbamazepine, phenytoin, primidone, zonisamide
Noninterfering: valproic acid

KEY WORDS
serum; comparison with HPLC

REFERENCE
Makino,K.; Goto,Y.; Sueyasu,M.; Futagami,K.; Kataoka,Y.; Oishi,R. Micellar electrokinetic capillary chromatography for therapeutic drug monitoring of zonisamide, *J.Chromatogr.B*, **1997**, *695*, 417–425.

SAMPLE
Matrix: blood, gastric contents, urine, vitreous humor

Sample preparation: Dilute gastric contents 1:10. 2 mL Serum, blood, vitreous humor, diluted gastric contents, or urine + 2.5 mL water + 200 μL mephobarbital solution, mix, add to TOXI-TUBE B (Toxi-Lab, Irvine CA), rock for 10 min, centrifuge at 2000 rpm for 5 min. Remove the organic layer and evaporate it to dryness under a stream of nitrogen at room temperature, reconstitute the residue in 100 μL running buffer, vortex for 30 s, inject an aliquot. (Mephobarbital solution was 2 mg mephobarbital and 2 drops EtOH, made up to 20 mL with water.)

CAPILLARY ELECTROPHORESIS
Capillary: 60 cm × 75 μm AccuSep (Waters)
Capillary preparation: Purge for 1 min between samples.
Running buffer: MeCN:buffer 15:85 adjusted to pH 8.5 with 1 M phosphoric acid. (Buffer was 3.8 g sodium borate decahydrate, 1 .4 g $NaH_2PO_4.H_2O$, and 28.8 g sodium dodecyl sulfate in 1 L water.)
Injection: Hydrostatic injection for 10 s
Detector: UV 214
Migration time: 7.78
Internal standard: mephobarbital (8.88)
Voltage: 20 kV
Model: Waters Quanta 4000
Limit of detection: 100 ng/mL

OTHER SUBSTANCES
Extracted: amobarbital, butabarbital, butalbital, pentobarbital, secobarbital

KEY WORDS
serum; whole blood

REFERENCE
Ferslew,K.E.; Hagardorn,A.N.; McCormick,W.F. Application of micellar electrokinetic capillary chromatography to forensic analysis of barbiturates in biological fluids, *J.Forensic Sci.*, **1995**, *40*, 245–249.

SAMPLE
Matrix: blood, saliva
Sample preparation: Filter (Amicon Centrifree Micropartition system) serum or saliva at 1500 g for 20 min, inject an aliquot of the ultrafiltrate.

CAPILLARY ELECTROPHORESIS
Capillary: 90 cm × 75 μm fused-silica (70 cm to detector) (Polymicro Technologies)
Capillary preparation: Before each run rinse capillary with 100 mM NaOH for 3 min and with running buffer for 5 min
Running buffer: 10 mM pH 9.1 Na_2HPO_4 containing 6 mM sodium borate and 75 mM sodium dodecyl sulfate
Injection: Injection by siphon at 34 cm for 5 s
Detector: UV 195
Migration time: 11.5
Voltage: 20 kV
Current: about 80 μA

OTHER SUBSTANCES
Extracted: carbamazepine

KEY WORDS
serum; ultrafiltrate

REFERENCE
Thormann,W.; Lienhard,S.; Wernly,P. Strategies for the monitoring of drugs in body fluids by micellar electrokinetic capillary chromatography, *J.Chromatogr.*, **1993**, *636*, 137–148.

SAMPLE
Matrix: blood, urine

Sample preparation: Serum, plasma. 200 μL Serum or plasma + 100 μL 1 M HCl + 2 mL chloroform, shake vigorously for 15 min, centrifuge at 500 g for 10 min. Remove the lower organic layer and evaporate it to dryness under a stream of nitrogen at 40°, reconstitute the residue in 200 μL running buffer, shake for 1 min, filter (0.2 μm), inject an aliquot. Urine. Condition a Bond Elut Certify SPE cartridge with 2 mL MeOH and 2 mL 100 mM pH 6 phosphate buffer, do not allow to dry. 5 mL Urine + 2 mL 100 mM pH 6 phosphate buffer, mix, add to the SPE cartridge, wash with 1 mL MeOH:100 mM phosphate buffer 20:80, dry under vacuum for 5 min, wash with 1 mL 1 M acetic acid, dry under vacuum for 10 min, wash with 1 mL hexane, elute with 4 mL dichloromethane. Evaporate the eluate to dryness under a stream of nitrogen at 40°, reconstitute the residue in 100-200 μL running buffer, filter (0.2 μm), inject an aliquot.

CAPILLARY ELECTROPHORESIS
Capillary: 90 cm × 75 μm fused-silica (70 cm to detector) (Polymicro Technologies)
Capillary preparation: Between each run rinse capillary with 100 mM NaOH for 3 min and buffer for 5 min
Running buffer: 15 mM pH 7.8 NaH_2PO_4 containing 9 mM sodium borate and 50 mM sodium dodecyl sulfate
Injection: Siphon at 34 cm for 5 s
Detector: UV 195
Migration time: 11
Voltage: 20 kV
Current: 60-63 μA
Model: Laboratory constructed

OTHER SUBSTANCES
Extracted: allobarbital, amobarbital, barbital, butalbital, pentobarbital, thiopental

KEY WORDS
serum; cow; human; plasma; SPE

REFERENCE
Thormann,W.; Meier,P.; Marcolli,C.; Binder,F. Analysis of barbiturates in human serum and urine by high-performance capillary electrophoresis-micellar electrokinetic capillary chromatography with on-column multi-wavelength detection, *J.Chromatogr.*, **1991**, *545*, 445–460.

SAMPLE
Matrix: blood, urine
Sample preparation: Filter (Amicon Centrifree Micropartition System) at 1500 g for 20 min, inject an aliquot of the ultrafiltrate.

CAPILLARY ELECTROPHORESIS
Capillary: 90 cm × 75 μm fused-silica (70 cm to detector) (Polymicro Technologies)
Capillary preparation: Between runs rinse capillary with 100 mM NaOH for 5 min and with running buffer for 10 min.
Running buffer: 10 mM pH 9.1 Na_2HPO_4 containing 6 mM sodium borate and 75 mM sodium dodecyl sulfate
Injection: Siphon at 34 cm for 5 s
Detector: UV 195
Migration time: 11.0
Voltage: 20 kV
Current: 80 μA

OTHER SUBSTANCES
Extracted: acetaminophen, ethosuximide, primidone, salicylic acid

KEY WORDS
serum; ultrafiltrate

REFERENCE
Caslavska,J.; Lienhard,S.; Thormann,W. Comparative use of three electrokinetic capillary methods for the determination of drugs in body fluids. Prospects for rapid determination of intoxications, *J.Chromatogr.*, **1993**, *638*, 335–342.

SAMPLE
Matrix: bulk
Sample preparation: Dilute with 400 µg/mL n-propyl p-hydroxybenzoate in running buffer

CAPILLARY ELECTROPHORESIS
Capillary: 27 cm × 50 µm fused-silica (20 cm to detector)
Capillary preparation: Before each run rinse with running buffer at high pressure.
Capillary temperature: 30
Running buffer: MeCN:water 15:85 containing 40 mM sodium dodecyl sulfate, 8.5 mM sodium borate, and 8.5 mM sodium phosphate, pH 8.5
Injection: Inject at high pressure for 1 s
Detector: UV 214
Migration time: 1.6
Internal standard: n-propyl p-hydroxybenzoate (1.8)
Voltage: 20 kV
Model: Beckman P/ACE System 2100

OTHER SUBSTANCES
Simultaneous: acetaminophen, acetylcodeine, O-acetylmorphine, aspirin, caffeine, cocaine, codeine, diamorphine, diphenhydramine, hydromorphone, isoniacinamide, lidocaine, methaqualone, morphine, niacinamide, noscapine, papaverine, phenacetin, phenylpropanolamine, procaine, quinine, salicylic acid, strychnine, thebaine

REFERENCE
Walker,J.A.; Krueger,S.T.; Lurie,I.S.; Marché,H.L.; Newby,N. Analysis of heroin drug seizures by micellar electrokinetic capillary chromatography (MECC), *J.Forensic Sci.*, **1995**, *40*, 6–9.

SAMPLE
Matrix: bulk
Sample preparation: Dissolve in running buffer to a concentration of 1 mg/mL, vortex for 2 min, add an equal volume of 100 µg/mL diphenhydramine in running buffer, mix, inject an aliquot.

CAPILLARY ELECTROPHORESIS
Capillary: 65 cm × 50 µm fused-silica (60 cm to detector)
Capillary preparation: Fill capillary with fresh running buffer before each run. Before use fill with 100 mM NaOH for 20 min, rinse with water, flush with running buffer
Running buffer: MeCN:50 mM 6-aminocaproic acid containing 50 mM 3-N,N-dimethylmyristylammoniopropanesulfonate (MAPS; Fluka) and 5 mM 1-heptanesulfonic acid 10:90, pH adjusted to 4.0 with 1 M phosphoric acid
Injection: Hydrodynamic injection by gravity or pressure.
Detector: UV 214
Migration time: 16.5
Internal standard: diphenhydramine (12)
Voltage: 27 kV
Current: ≤25 µA
Model: Dionex system I

OTHER SUBSTANCES
Simultaneous: acetaminophen, allobarbital, barbital, caffeine, codeine, diamorphine, morphine, niacinamide, noscapine, papaverine, procaine

REFERENCE
Naess,O.; Rasmussen,K.E. Micellar electrokinetic chromatography of charged and neutral drugs in acidic running buffers containing a zwitterionic surfactant, sulfonic acids or sodium dodecyl sulphate. Separation of heroin, basic by-products and adulterants, *J.Chromatogr.A*, **1997**, *760*, 245–251.

SAMPLE
Matrix: perfusate
Sample preparation: 4 µL Perfusate (Ringer solution) + 2 µL 5.1 mM mesityl oxide, mix, inject an aliquot.

CAPILLARY ELECTROPHORESIS
Capillary: 44 cm $\times$ 50 μm fused-silica (34 cm to detector) (Polymicro Technologies)
Capillary preparation: Rinse with 1 M NaOH and with running buffer between injections.
Running buffer: 24 mM Sodium borate containing 50 mM NaH_2PO_4 and 50 mM sodium dodecyl
    sulfate, pH 7
Injection: Inject by vacuum for 5 s (19 nL).
Detector: UV 200
Migration time: 7.2
Internal standard: mesityl oxide (5.6)
Voltage: 19 kV
Current: about 60 μA
Model: Fast Impact automatic analyzer (Europhor Instruments)
Limit of quantitation: 2 μg/mL

KEY WORDS
rat; pharmacokinetics

REFERENCE
Tellez,S.; Forges,N.; Roussin,A.; Hernandez,L. Coupling of microdialysis with capillary electrophoresis: a new
    approach to the study of drug transfer between two compartments of the body in freely moving rats,
    J.Chromatogr., **1992**, *581*, 257–256.

SAMPLE
Matrix: solutions

CAPILLARY ELECTROPHORESIS
Capillary: 72 cm $\times$ 50 μm silica (50 cm to detector) (Applied Biosystems)
Capillary preparation: Before each run wash capillary with 100 mM NaOH for 3 min, wash
    with running buffer for 3 min, aspirate neutral marker solution (2 drops DMSO in 10 mL
    water) for 1 s, place capillary end in buffer vial for 5 s, inject sample. Wash capillary at the
    beginning of each day by passing 1 M NaOH through for 20 min.
Capillary temperature: 30
Running buffer: MeCN:buffer 20:80 (Buffer was 30 mM pH 9.3 borate buffer containing 30 mM
    sodium dodecyl sulfate.)
Injection: Inject by applying vacuum for 1 s (2-3 nL)
Detector: UV 200
Migration time: 11
Voltage: +30 kV
Current: 41 μA
Model: Applied Biosystems Model 270A

OTHER SUBSTANCES
Simultaneous: butabarbital, heptabarbital, hexobarbital, pentobarbital

REFERENCE
Evenson,M.A.; Wiktorowicz,J.E. Automated capillary electrophoresis applied to therapeutic drug monitoring,
    Clin.Chem., **1992**, *38*, 1847–1852.

SAMPLE
Matrix: solutions
Sample preparation: Inject an aliquot of a 30-300 μg/mL solution in water.

CAPILLARY ELECTROPHORESIS
Capillary: 36 cm $\times$ 50 μm fused-silica (28 cm to detector)
Running buffer: 80 mM Reagent, 120 mM sodium dodecyl sulfate, 900 mM 1-butanol, and 10
    mM borate in water where reagent is heptane (A) or octane (B) or 1-butyl chloride (C)
Injection: Hydrostatic injection.
Detector: UV 254
Migration time: 8.314 (A), 8.158 (B), 6.669 (C)
Voltage: 12 kV
Current: ca. 65 μA
Model: Waters Quanta 4000

OTHER SUBSTANCES
Simultaneous: aminopyrine, caffeine, phenacetin

REFERENCE
Fu,X.; Lu,J.; Zhu,A. Microemulsion electrokinetic chromatographic separation of antipyretic analgesic ingredients, *J.Chromatogr.A*, **1996**, *735*, 353–356.

SAMPLE
Matrix: solutions

CAPILLARY ELECTROPHORESIS
Capillary: 60 cm × 50 μm fused-silica (40 cm to detector), bare (A) or C18 coated (B) (Supelco)
Capillary preparation: Rinse new capillaries with 250 μL 1 M NaOH, 250 μL 100 mM NaOH, water, MeOH, water, and running buffer then condition with running buffer for at least 15 min. Carry out a similar procedure between runs.
Running buffer: 100 mM pH 8.5 Borate buffer containing 25 mM sodium dodecyl sulfate and 5 M urea
Injection: Electrokinetic injection at 5 kV for 5 s.
Detector: UV 254
Migration time: 10.8 (A), 9.9 (B)
Voltage: 18 kV
Model: Jasco

OTHER SUBSTANCES
Simultaneous: amobarbital, barbital, mephobarbital, metharbital, pentobarbital, secobarbital

KEY WORDS
coated capillary

REFERENCE
Jinno,K.; Han,Y.; Nakamura,M. Analysis of anxiolytic drugs by capillary electrophoresis with bare and coated capillaries, *J.Capillary Electrophor.*, **1996**, *3*, 139–145.

SAMPLE
Matrix: solutions

CAPILLARY ELECTROPHORESIS
Capillary: 57 cm × 50 μm fused-silica (50 cm to detector)
Capillary preparation: After each run rinse with 20 mM pH 9.00 borate buffer containing 90 mM sodium dodecyl sulfate for 5 min and with 100 mM NaOH for 5 min.
Running buffer: 20 mM pH 9.60 Borate buffer containing 10 mM α-cyclodextrin
Injection: Pressure injection at 3.45 kPa for 5 s.
Detector: UV 254
Migration time: 9.2
Voltage: 15 kV
Model: Beckman P/ACE 2100

OTHER SUBSTANCES
Simultaneous: cyclobarbital, thiopental

REFERENCE
Conradi,S.; Vogt,C.; Rohde,E. Separation of enantiomeric barbiturates by capillary electrophoresis using a cyclodextrin-containing run buffer, *J.Chem.Educ.*, **1997**, *74*, 1122–1125.

SAMPLE
Matrix: solutions

CAPILLARY ELECTROPHORESIS
Capillary: 60 cm × 75 μm fused-silica (Phenomenex)

Capillary preparation: Every five injections rinse capillary with 100 mM NaOH for 5 min, water for 5 min, and running buffer for 5 min.
Running buffer: 20 mM pH 8.5 Borate buffer containing 25 mM sodium dodecyl sulfate
Injection: Vacuum injection for 2 s (105 nL).
Detector: UV 200
Migration time: 8.5
Voltage: 30 kV
Model: Spectrophoresis 100 (Thermo Separation Products)

OTHER SUBSTANCES
Simultaneous: 2,3-dihydroxybenzoic acid, 2,5-dihydroxybenzoic acid, salicylic acid
Noninterfering: ascorbic acid

REFERENCE
Gökören,N.; Tunçel,M. A capillary electrophoretic method for examination of hydroxylation of salicylate ion to determine hydroxyl free radical, *Pharmazie*, **1997**, *52*, 726–727.

SAMPLE
Matrix: solutions

CAPILLARY ELECTROPHORESIS
Capillary: 40 cm × 75 μm coated fused-silica (15 cm to detector) (Supelco)
Capillary preparation: Coat column as follows. Adjust the pH of 20 mL water to 3.5 with acetic acid, add 80 μL 3-(trimethoxysilyl)propyl methacrylate (3-methacryloxypropyltrimethoxysilane), mix, suck into capillary, let stand at room temperature for 1 h, remove the solution, wash with water. Fill the capillary with a deaerated 4% acrylamide solution containing 1 mg/mL N,N,N',N'-tetramethylethylenediamine and 1 mg/mL ammonium persulfate, let stand for 3 h, remove excess solution by aspiration, rinse with water, remove water by aspiration, dry at 35° (cf. J. Chromatogr. 1985, 347, 191).
Running buffer: 100 mM Tris/150 mM boric acid, pH 8.3
Injection: Electromigration at 5 kV for 5 s.
Detector: UV 240, UV 254
Migration time: 4.5
Voltage: 12 kV
Model: Jasco 890-CE

OTHER SUBSTANCES
Simultaneous: amobarbital, barbital, mephobarbital, metharbital, pentobarbital, secobarbital

KEY WORDS
injection at cathode; coated capillary

REFERENCE
Jinno,K.; Han,Y.; Sawada,H. Analysis of toxic drugs by capillary electrophoresis using polyacrylamide-coated columns, *Electrophoresis*, **1997**, *18*, 284–286.

SAMPLE
Matrix: solutions
Sample preparation: Inject an aliquot of a 100 μg/mL solution in running buffer.

CAPILLARY ELECTROPHORESIS
Capillary: 60 cm × 50 μm acrylamide-coated fused-silica (40 cm to detector) (Supelco)
Capillary preparation: Adjust the pH of 20 mL water to 3.5 with acetic acid, add 80 μL 3-(trimethoxysilyl)propyl methacrylate (3-methacryloxypropyltrimethoxysilane), mix, suck into capillary, let stand at room temperature for 1 h, remove the solution, wash with water. Fill the capillary with a deaerated 3-4% acrylamide solution containing 1 μL/mL N,N,N',N'-tetramethylethylenediamine and 1 mg/mL ammonium persulfate, let stand for 3 h, remove excess solution by aspiration, rinse with water, remove water by aspiration, dry at 35° (cf. J. Chromatogr. 1985, 347, 191).
Running buffer: MeCN:buffer 5:95 (Buffer was 100 mM borate containing 5 M urea and 10 mM sodium dodecyl sulfate, adjusted to pH 8.5 with phosphate.)
Injection: Electrokinetic injection at 5 kV for 5 s.

Detector: UV 254
Migration time: 12.5
Voltage: 18 kV
Model: Jasco Model 870-CE

OTHER SUBSTANCES
Simultaneous: alprazolam, amobarbital, barbital, bromazepam, clonazepam, clotiazepam, cloxazolam, diazepam, estazolam, etizolam, fludiazepam, flunitrazepam, flurazepam, haloxazolam, medazepam, mephobarbital, metharbital, nimetazepam, nitrazepam, oxazepam, pentobarbital, secobarbital, triazolam

KEY WORDS
coated capillary; detector at anode

REFERENCE
Jinno,K.; Han,Y.; Sawada,H.; Taniguchi,M. Capillary electrophoretic separation of toxic drugs using a polyacrylamide-coated capillary, *Chromatographia*, **1997**, *46*, 309–314.

SAMPLE
Matrix: solutions

CAPILLARY ELECTROPHORESIS
Capillary: 57 cm × 50 μm fused-silica (50 cm to detector)
Capillary preparation: After each run rinse with 20 mM pH 9.00 borate buffer containing 90 mM sodium dodecyl sulfate for 5 min and with 100 mM NaOH for 5 min.
Running buffer: MeOH:buffer 10:90 (Buffer was 10 mM pH 8.25 borate buffer containing 10 mM phosphate and 45 mM sodium dodecyl sulfate.)
Injection: Pressure injection at 3.45 kPa for 3 s followed by a 2 s injection of 200 μM phenobarbital solution.
Detector: UV 214
Migration time: 13.3
Internal standard: phenobarbital
Voltage: 17 kV
Model: Beckman P/ACE 2100

OTHER SUBSTANCES
Simultaneous: caffeine, theobromine, theophylline

KEY WORDS
phenobarbital is IS

REFERENCE
Vogt,C.; Conradi,S.; Rohde,E. Determination of caffeine and other purine compounds in food and pharmaceuticals by micellar electrokinetic chromatography, *J.Chem.Educ.*, **1997**, *74*, 1126–1130.

Phenol

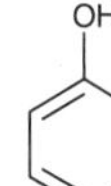

Molecular formula: C₆H₆O
Molecular weight: 94.11
CAS Registry No.: 108-95-2
Merck Index (12th ed.): 7390

SAMPLE
Matrix: microsomal incubation
Sample preparation: Analyze an aliquot directly.

CAPILLARY ELECTROPHORESIS
Capillary: 27.2 cm × 50 μm unmodified fused silica (20.2 cm to detector)

Capillary preparation: Before each injection rinse with 100 mM NaOH for 1 min and running buffer for 1 min.
Capillary temperature: 25
Running buffer: MeCN:buffer 10:90 (Buffer was 10 mM phosphate containing 7 mM borate and 50 mM sodium dodecyl sulfate
Injection: Pressure of 0.5 psi across the capillary for 3 s
Detector: UV 214
Migration time: 1.4
Voltage: raise to 15 kV over 0.2 min, maintain at 15 kV
Current: about 50 μA
Model: Beckman P/ACE System 2050

OTHER SUBSTANCES
Extracted: metabolites, microsomes, NADP+

KEY WORDS
human; liver; rat

REFERENCE
Davies,M.I.; Lunte,C.E.; Smyth,M.R. Use of micellar electrokinetic capillary chromatography in the study of in vitro metabolism of phenol by human liver microsomes, *J.Pharm.Biomed.Anal.*, **1995**, *13*, 893–897.

SAMPLE
Matrix: solutions

CAPILLARY ELECTROPHORESIS
Capillary: 65 cm × 50 μm silica (50 cm to detector) (Scientific Glass Engineering)
Running buffer: 50 mM Sodium dodecyl sulfate in buffer (Prepare buffer by mixing 25 mM sodium tetraborate solution with 50 mM NaH_2PO_4 to obtain a pH of 7.0.)
Injection: Hydrodynamic injection at 4 cm for 5-90 s.
Detector: UV 220
Migration time: 6
Voltage: about 20 kV
Current: 33 μA
Model: laboratory-constructed

OTHER SUBSTANCES
Simultaneous: p-cresol, p-ethylphenol, methanol, Sudan III, 2,6-xylenol

---

REFERENCE
Terabe,S.; Otsuka,K.; Ichikawa,K.; Tsuchiya,A.; Ando,T. Electrokinetic separations with micellar solutions and open-tubular capillaries, *Anal.Chem.*, **1984**, *56*, 111–113.

SAMPLE
Matrix: solutions
Sample preparation: Prepare a 1000 ppm solution in MeOH containing 100 ppm Sudan III, inject an aliquot.

CAPILLARY ELECTROPHORESIS
Capillary: 100 cm × 180 μm fused-silica (J&W Scientific)
Running buffer: 10 mM Borate containing 5 mM phosphate and 50 mM sodium dodecyl sulfate, pH 6.6
Injection: Gravity feed at 5 cm for 5 s
Detector: UV 254
Migration time: 36
Internal standard: Sudan III (46)
Voltage: 10 kV
Current: 130 μA
Model: Laboratory constructed

REFERENCE
Ong,C.P.; Ng,C.L.; Chong,N.C.; Lee,H.K.; Li,S.F. Retention of eleven priority phenols using micellar electrokinetic chromatography, *J.Chromatogr.*, **1990**, *516*, 263–270.

SAMPLE
Matrix: solutions
Sample preparation: Inject an aliquot of a 25 μg/mL solution by applying vacuum to the outlet for 10 s.

CAPILLARY ELECTROPHORESIS
Capillary: 100 cm × 75 μm uncoated fused silica (65 cm to detector)
Capillary preparation: At the start of each day soak capillary in 1 M NaOH for 10 min, wash with water, wash with buffer. At the end of each day wash capillary with water, wash with 100 mM NaOH, leave filled with 100 mM NaOH overnight.
Running buffer: 10 mM Na_3PO_4 containing 10 mM sodium borate, pH 9.8
Injection: Vacuum applied to outlet for 10 s.
Detector: UV 210
Migration time: 8.5
Voltage: 22.5 kV
Current: 53 μA
Model: ISCO Model 3850
Limit of detection: 380 ng/mL (S/N 3)

OTHER SUBSTANCES
Simultaneous: 4-chloro-3-methylphenol, 2-chlorophenol, 2,4-dimethylphenol, 2,4-dinitrophenol, 2-methyl-4,6-dinitrophenol, 2-nitrophenol, 4-nitrophenol, pentachlorophenol, 2,4,6-trichlorophenol

REFERENCE
Li,G.; Locke,D.C. Separation of the eleven priority pollutant phenols by capillary zone electrophoresis, *J.Chromatogr.B*, **1995**, *669*, 93–102.

SAMPLE
Matrix: solutions
Sample preparation: Inject an aliquot of a solution in running buffer.

CAPILLARY ELECTROPHORESIS
Capillary: 60 cm × 75 μm fused-silica (40 cm to detector) (Yongnian Optical Fiber Factory)
Capillary preparation: Flush with running buffer for 1 min before each injection. Before use condition capillary by flushing with 100 mM NaOH for 30 min, with water for 15 min, and with running buffer for 15 min.
Capillary temperature: 18
Running buffer: 20 mM pH 9.0 Borate buffer
Injection: Electromigration at 4 kV for 14 s
Detector: UV 220
Migration time: 2.7
Voltage: 25 kV
Model: Laboratory-constructed

OTHER SUBSTANCES
Simultaneous: amaranth, brilliant blue, indigo carmine, Ponceau 4R, sunset yellow, tartrazine

REFERENCE
Liu,H.; Zhu,T.; Zhang,Y.; Qi,S.; Huang,A.; Sun,Y. Determination of synthetic colourant food additives by capillary zone electrophoresis, *J.Chromatogr.A*, **1995**, *718*, 448–453.

SAMPLE
Matrix: solutions

CAPILLARY ELECTROPHORESIS
Capillary: 32 cm × 50 μm silica (24.5 cm to detector) (Composite Metal Services, Worcester, UK)

Capillary temperature: 26
Running buffer: 2-Butanol:ethylene glycol:MeCN:buffer:hexadimethrine bromide 25:5:10:0.001
  (Buffer was 15 mM pH 10.3 sodium phosphate buffer containing 1.25 mM sodium tetraborate.
  Hexadimethrine bromide is 1,5-dimethyl-1,5-diazaundecamethylene polymethobromide or
  polybrene.)
Injection: Hydrostatic injection at 10 cm for 8 s.
Detector: UV 214
Migration time: 2.873
Voltage: -30 kV
Current: 36 μA
Model: Waters Quanta 4000

OTHER SUBSTANCES
Simultaneous: 4-chloro-3-methylphenol, 2-chlorophenol, 2,4-dichlorophenol, 2,4-dimethylphenol,
  4,6-dinitro-2-methylphenol, 2,4-dinitrophenol, 2-nitrophenol, 4-nitrophenol, pentachlorophenol,
  2,4,6-trichlorophenol

REFERENCE
Zemann,A.; Volgger,D. Separation of priority pollutant phenols with coelectroosmotic capillary electrophoresis,
  Anal.Chem., **1997**, *69*, 3243–3250.

SAMPLE
Matrix: solutions

CAPILLARY ELECTROPHORESIS
Capillary: 57 cm $\times$ 50 μm fused-silica (57 cm to detector)
Capillary temperature: 35
Running buffer: 50 mM pH 7.0 NaH_2PO_4 containing 25 mM sodium tetraborate and 70 mM
  sodium dodecyl sulfate
Detector: UV 214
Migration time: 5.5
Voltage: 20 kV
Model: Beckman P/ACE 5010
Limit of detection: 30 μM

OTHER SUBSTANCES
Simultaneous: chlorophenols

REFERENCE
Harino,H.; Tsunoi,S.; Yoshioka,J.; Araki,T.; Masuyama,A.; Nakatsuji,Y.; Ikeda,I.; Tanaka,M. Separation of pollutant phenols by micellar electrokinetic chromatography with double-chain surfactants having two sulfo-
  nate groups, *Anal.Sci.*, **1998**, *14*, 719–724.

Phenothiazine

Molecular formula: $C_{12}H_9NS$
Molecular weight: 199.28
CAS Registry No.: 92-84-2
Merck Index (12th ed.): 7404

SAMPLE
Matrix: urine
Sample preparation: 10 mL Urine + 1 mL 5 M NaOH + 10 mL n-hexane, vortex for 1 min,
  centrifuge at 0° at 3000 g for 5 min. Remove the organic layer and add it to 50 μL glacial acetic
  acid, evaporate to dryness under a stream of nitrogen at 30°, reconstitute the residue in 50 μL
  50 mM sodium taurodeoxycholate, filter (0.2 μm PTFE), inject an aliquot.

CAPILLARY ELECTROPHORESIS
Capillary: 100 cm × 50 μm fused-silica (50 cm to detector) (ISCO)
Capillary preparation: Rinse capillary with 20 μL running buffer between injections. Condition
a new capillary by filling with 1 M NaOH, let stand for 1 h, fill with 100 mM NaOH, let stand
for 1 h, rinse with buffer. Every 20 injections rinse capillary with 200 μL 1 M NaOH, 200 μL
water, and 200 μL running buffer, fill with running buffer.
Capillary temperature: 22
Running buffer: 40 mM pH 9.5 Borate buffer containing 10 mM sodium taurodeoxycholate
Injection: Load under vacuum at 7.5 kPa/s.
Detector: UV 240
Migration time: 6.5
Voltage: 30 kV
Model: ISCO Model 3140 electropherograph
Limit of detection: 9 ng/mL

OTHER SUBSTANCES
Extracted: acepromazine, amiodarone, amitriptyline, azaperone, chlorpromazine, cianopramine,
clomipramine, clozapine, desethylamiodarone, desipramine, diclofensine, dothiepin, doxepin,
imipramine, isocarboxazid, moclobemide, perphenazine, pimozide, prochlorperazine, promazine, thioridazine, thiothixene, trifluoperazine, trimipramine

KEY WORDS
human; cow; pig; horse; protect from light

REFERENCE
Aumatell,A.; Wells,R.J. Determination of a cardiac antiarrhythmic, tricyclic antipsychotics and antidepressants
in human and animal urine by micellar electrokinetic capillary chromatography using a bile salt,
J.Chromatogr.B, **1995**, *669*, 331–344.

Phenoxybenzamine

Molcoular formula: $C_{16}H_{22}ClNO$
Molecular weight: 303.83
CAS Registry No.: 59-96-1, 63-92-3 (HCl)
Merck Index (12th ed.): 7409
Lednicer: 1 55

SAMPLE
Matrix: solutions
Sample preparation: Inject an aliquot of a 100 μg/mL solution in water:running buffer 50:50.

CAPILLARY ELECTROPHORESIS
Capillary: 44.5 cm × 50 μm acrylamide-coated fused-silica (Bio-Rad)
Capillary temperature: 30
Running buffer: 100 mM NaH_2PO_4 containing 15 mM gamma-cyclodextrin, adjusted to pH 2.5
with phosphoric acid
Injection: Electrokinetic injection at 8 kV for 6 s.
Detector: UV 200
Migration time: 11.17
Voltage: 14 kV
Model: Bio-Rad BioFocus 3000

OTHER SUBSTANCES
Also analyzed: albuterol, alprenolol, atenolol, atropine, baclofen, bamethan, benserazide, biperiden, bisoprolol, bupivacaine, bupranolol, butetamate, carazolol, carbuterol, carvedilol, celiprolol, chloroquine, chlorpheniramine (chlorphenamine), clidinium bromide, clobutinol, disopyramide, dobutamine, flecainide, homatropine, ipratropium bromide, isoproterenol, isothipendyl, ketamine, mefloquine, mequitazine, metaproterenol (orciprenaline), metipranolol, nafronyl (naftidrofuryl), nefopam, ofloxacin, orphenadrine, oxomemazine, oxprenolol, pholedrine, pindolol, pirbuterol, prilocaine, promethazine, propafenone, propranolol, sotalol, synephrine, terbutaline, tetrahydrozoline (tetryzoline), tocainide, trihexyphenidyl, trimeprazine (alimemazine), trimipramine, tropicamide, verapamil, zopiclone

KEY WORDS
coated capillary; achiral

REFERENCE
Koppenhoefer,B.; Epperlein,U.; Christian,B.; Yibing,J.; Yuying,C.; Bingcheng,L. Separation of enantiomers of drugs by capillary electrophoresis. I. γ-Cyclodextrin as chiral solvating agent, *J.Chromatogr.A*, **1995**, *717*, 181–190.

SAMPLE
Matrix: solutions
Sample preparation: Inject an aliquot of a 100 μg/mL solution in running buffer.

CAPILLARY ELECTROPHORESIS
Capillary: 30 cm × 50 μm fused-silica (25.5 cm to detector) (Yongnian Optical Conductive Fiber Plant, China), coated with polyacrylamide
Capillary preparation: No details of the polyacrylamide coating process are provided. However, another paper (LC.GC 1997, 15, 40) by this group indicates that they use the procedure of Hjertén, thus: Adjust the pH of 20 mL water to 3.5 with acetic acid, add 80 μL 3-(trimethoxysilyl)propyl methacrylate (3-methacryloxypropyltrimethoxysilane), mix, suck into capillary, let stand at room temperature for 1 h, remove the solution, wash with water. Fill the capillary with a deaerated 3-4% acrylamide solution containing 1 μL/mL N,N,N',N'-tetramethylethylenediamine and 1 mg/mL potassium persulfate, let stand for 30 min, remove excess solution by aspiration, rinse with water, remove water by aspiration, dry at 35° (J. Chromatogr. 1985, 347, 191).
Capillary temperature: 25
Running buffer: 100 mM NaH_2PO_4 adjusted to pH 2.5
Injection: Electrokinetic injection at 15 kV for 3 s.
Detector: UV 200, UV 210
Migration time: 4.63
Voltage: 15 kV
Model: Bio-Rad BioFocus 3000

OTHER SUBSTANCES
Simultaneous: amorolfine, brompheniramine, bupivacaine, carteolol, chloroquine, chlorpheniramine, chlorphenoxamine, disopyramide, dobutamine, doxylamine, flecainide, gallopamil, ketamine, mepindolol, orphenadrine, oxybutynin, pindolol, propafenone, propranolol, sulpiride, talinolol, tropicamide, verapamil

KEY WORDS
coated capillary

REFERENCE
Koppenhoefer,B.; Epperlein,U.; Xiaofeng,Z.; Bingcheng,L. Separation of enantiomers of drugs by capillary electrophoresis. Part 4: Hydroxypropyl-γ-cyclodextrin as chiral solvating agent, *Electrophoresis*, **1997**, *18*, 924–930.

SAMPLE
Matrix: solutions
Sample preparation: Inject an aliquot of a 100 μg/mL solution in running buffer.

CAPILLARY ELECTROPHORESIS
Capillary: 30 cm × 50 μm fused-silica (25.5 cm to detector), coated with polyacrylamide
Capillary preparation: Adjust the pH of 20 mL water to 3.5 with acetic acid, add 80 μL 3-(trimethoxysilyl)propyl methacrylate (3-methacryloxypropyltrimethoxysilane), mix, suck into capillary, let stand at room temperature for 1 h, remove the solution, wash with water. Fill the capillary with a deaerated 3-4% acrylamide solution containing 1 μL/mL N,N,N',N'-tetramethylethylenediamine and 1 mg/mL potassium persulfate, let stand for 30 min, remove excess solution by aspiration, rinse with water, remove water by aspiration, dry at 35° (J. Chromatogr. 1985, 347, 191).
Capillary temperature: 25
Running buffer: 100 mM NaH_2PO_4 containing 45 mM hydroxypropyl-α- cyclodextrin, adjusted to pH 2.5 with phosphoric acid
Injection: Electrokinetic injection at 15 kV for 3 s.
Detector: UV 200
Voltage: 15 kV
Model: Bio-Rad BioFocus 3000

KEY WORDS
chiral; coated capillary; comparison with the use of other cyclodextrins; this running buffer gave the greatest enantiomeric separation.; α=1.050

REFERENCE
Lin,B.; Zhu,X.; Koppenhoefer,B.; Epperlein,U. Investigation of 123 chiral drugs by cyclodextrin-modified capillary electrophoresis, *LC.GC*, **1997**, *15*, 40–46.

SAMPLE
Matrix: solutions

CAPILLARY ELECTROPHORESIS
Capillary: 29-36 cm × 50 μm fused-silica (24.5-31.5 cm to detector) (Yongnian Optical Conductive Fiber Plant, China) coated with polyacrylamide
Capillary preparation: Coat capillary as follows. Adjust the pH of 20 mL water to 3.5 with acetic acid, add 80 μL 3-(trimethoxysilyl)propyl methacrylate (3-methacryloxypropyltrimethoxysilane), mix, suck into capillary, let stand at room temperature for 1 h, remove the solution, wash with water. Fill the capillary with a deaerated 3-4% acrylamide solution containing 1 μL/mL N,N,N',N'-tetramethylethylenediamine and 1 mg/mL potassium persulfate, let stand for 30 min, remove excess solution by aspiration, rinse with water, remove water by aspiration, dry at 35° (J. Chromatogr. 1985, 347, 191).
Capillary temperature: 25
Running buffer: 100 mM pH 2.5 NaH_2PO_4 (A) or 100 mM pH 2.5 NaH_2PO_4 containing 45 mM hydroxypropyl-α-cyclodextrin (Wacker, Munich) (B)
Injection: Electromigration at 15 kV for 3 s.
Detector: UV 200; UV 210
Migration time: 4.63 (A); 9.37, 9.84 (B) (enantiomers)
Voltage: 15 kV
Model: Bio-Focus 3000

OTHER SUBSTANCES
Also analyzed: albuterol (salbutamol), alprenolol, amorolfine, atenolol, atropine, azelastine, baclofen, bamethan, benproperine, benserazide, biperiden, bisoprolol, brompheniramine, bupivacaine, bupranolol, butamirate, butethamate, carazolol, carbuterol, carteolol, carvedilol, celiprolol, chloroquine, chlorpheniramine, chlorphenoxamine, cicletanine, clenbuterol, clidinium bromide, clobutinol, dimethindene, dipivefrin, disopyramide, dobutamine, doxylamine, fendiline, flecainide, gallopamil, homatropine, ipratropium bromide, isoproterenol (isoprenaline), isothipendyl, ketamine, meclizine, mefloquine, mepindolol, mequitazine, metaclazepam, metaproterenol (orciprenaline), metipranolol, metoprolol, nafronyl (naftidrofuryl), nefopam, nicardipine, norfenefrine, ofloxacin, ornidazole, orphenadrine, oxomemazine, oxprenolol, oxybutynin, phenylpropanolamine, pholedrine, pindolol, pirbuterol, prilocaine, procyclidine, promethazine, propafenone, propranolol, reproterol, sotalol, sulpride, synephrine, talinolol, terbutaline, tetrahydrozoline (tetryzoline), theodrenaline, tioconazole, tocainide, trihexyphenidyl, trimeprazine (alimemazine), trimipramine, tropicamide, verapamil, zopiclone

KEY WORDS
coated capillary

REFERENCE
Koppenhoefer,B.; Eperlein,U.; Schlunk,R.; Zhu,X.; Lin,B. Separation of enantiomers of drugs by capillary electrophoresis. V. Hydroxypropyl-α-cyclodextrin as chiral solvating agent, *J.Chromatogr.A*, **1998**, *793*, 153–164.

Phenprocoumon

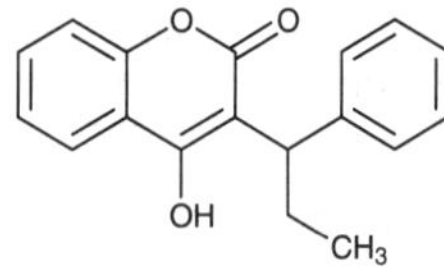

Molecular formula: $C_{18}H_{16}O_3$
Molecular weight: 280.32
CAS Registry No.: 435-97-2
Merck Index (12th ed.): 7413

SAMPLE
Matrix: solutions

CAPILLARY ELECTROPHORESIS
Capillary: 58.5 cm × 50 μm fused-silica (50 cm to detector) (Composite Metal Services, Worcs., UK)
Capillary preparation: Wash capillary with running buffer for 5 min before each run. At the start of each day wash with 100 mM NaOH for 5 min and rinse with water for 15 min.
Capillary temperature: 25
Running buffer: 40 mM pH 7 Sodium acetate buffer containing 40 mM phosphoric acid, 40 mM boric acid, and 15 mM α-cyclodextrin
Injection: Pressure injection at 150 mbar.s.
Detector: UV 202
Migration time: 5.8, 6.0 (enantiomers)
Voltage: 30 kV
Model: Hewlett-Packard [3D]CE

KEY WORDS
chiral

REFERENCE
Roos,N.; Ganzler,K.; Szemán,J.; Fanali,S. Systematic approach to cost- and time-effective method development with a starter kit for chiral separations by capillary electrophoresis, *J.Chromatogr.A*, **1997**, *782*, 257–269.

Phensuximide

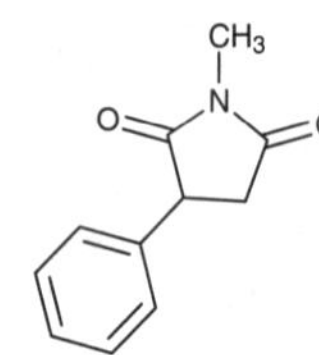

Molecular formula: $C_{11}H_{11}NO_2$
Molecular weight: 189.21
CAS Registry No.: 86-34-0
Merck Index (12th ed.): 7414
Lednicer: 1 226

SAMPLE
Matrix: solutions
Sample preparation: Prepare a 500 μg/mL solution in MeOH, inject an aliquot.

CAPILLARY ELECTROPHORESIS
Capillary: 60 cm × 75 μm fused-silica (52.4 cm to detector)

Capillary preparation: Before each run rinse capillary with 100 mM KOH for 2 min then with running buffer for 2 min.
Running buffer: 10 mM Na_2HPO_4 containing 3% sulfated β-cyclodextrin (β-cyclodextrin, sulfated) adjusted to pH 7.0 with phosphoric acid
Injection: Hydrostatic injection for 2 s
Detector: UV 214
Migration time: 19.32, 20.10 (enantiomers)
Voltage: 8 kV
Model: Waters Quanta 4000

KEY WORDS
chiral

REFERENCE
Wu,W.; Stalcup,A.M. Capillary electrophoretic chiral separations using a sulfated β-cyclodextrin-containing electrolyte, *J.Liq.Chromatogr.*, **1995**, *18*, 1289–1315.

SAMPLE
Matrix: solutions
Sample preparation: Inject an aliquot of a solution in running buffer.

CAPILLARY ELECTROPHORESIS
Capillary: 60 cm × 75 μm fused-silica (52.4 cm to detector)
Capillary preparation: After each run flush with 500 mM KOH for 2-3 min then with water.
Running buffer: 10 mM pH 3.8 Phosphate buffer containing 2% sulfated cyclodextrin (ds 7-10)
Injection: Hydrostatic injection.
Detector: UV 214
Migration time: 51.12, 60.65 (enantiomers)
Voltage: 15 kV
Model: Waters Quanta 4000

OTHER SUBSTANCES
Also analyzed: acebutolol, alprenolol, aminoglutethimide, brompheniramine, bupivacaine, bupropion, canadine, carbinoxamine, chloroquine, chlorpheniramine, dimethindene, disopyramide, doxylamine, hydroxychloroquine, idazoxan, isoxsuprine, ketamine, mepenzolate, mepivacaine, methoxyphenamine, mexiletine, midodrine, nefopam, orphenadrine, oxprenolol, oxyphencyclimine, pheniramine, pindolol, piperoxan, terbutaline, tetramisole, tolperisone, tranylcypromine, trihexyphenidyl, trimipramine, verapamil, warfarin

KEY WORDS
chiral; detector at anode

REFERENCE
Stalcup,A.M.; Gahm,K.H. Application of sulfated cyclodextrins to chiral separations by capillary zone electrophoresis, *Anal.Chem.*, **1996**, *68*, 1360–1368.

Phentermine

Molecular formula: $C_{10}H_{15}N$
Molecular weight: 149.24
CAS Registry No.: 122-09-8, 1197-21-3 (HCl)
Merck Index (12th ed.): 7415
Lednicer: 1 72

SAMPLE
Matrix: solutions
Sample preparation: Inject an aliquot of a 20 μg/mL solution in 12.5 mM pH 9.3 borate buffer.

CAPILLARY ELECTROPHORESIS
Capillary: 66.5 cm × 75 μm fused-silica (41.5 cm to detector) (Polymicro Technologies)
Running buffer: 250 mM pH 9.3 Sodium borate buffer
Injection: Pressure injection at 50 mbar for 15 s.
Detector: UV 195
Migration time: 11.2
Voltage: 8 kV
Model: Crystal 310

OTHER SUBSTANCES
Simultaneous: amphetamine, chlorpheniramine, ephedrine, MDMA, methamphetamine, phenylpropanolamine, pseudoephedrine
Noninterfering: caffeine
Interfering: MDA

REFERENCE
Lilley,K.A.; Wheat,T.E. Drug identification in biological matrices using capillary electrophoresis and chemometric software, *J.Chromatogr.B*, **1996**, *683*, 67–76.

Phenylephrine

Molecular formula: $C_9H_{13}NO_2$
Molecular weight: 167.21
CAS Registry No.: 59-42-7, 61-76-7 (HCl)
Merck Index (12th ed.): 7440
Lednicer: 1 63

SAMPLE
Matrix: solutions

CAPILLARY ELECTROPHORESIS
Capillary: 58.2 cm × 75 μm fused-silica (50.7 cm to detector) (Polymicro Technologies)
Capillary preparation: Purge with running buffer before each run. At the beginning of each day purge using 50-60 kPa vacuum with 500 mM NaOH for 5 min, with water for 5 min, with MeCN for 5 min, and with running buffer for 5 min.
Running buffer: MeCN:MeOH:acetic acid 49:50:1 containing 20 mM ammonium acetate
Injection: Hydrostatic injection at 10 cm for 5 s.
Detector: UV 214
Migration time: 4.06
Voltage: 25 kV
Model: Waters Quanta 4000

OTHER SUBSTANCES
Simultaneous: amphetamine, benzphetamine, ephedrine, nylidrin, oxymetazoline, phendimetrazine, phenmetrazine, phenylpropanolamine, xylometazoline

REFERENCE
Leung,G.N.W.; Tang,H.P.O.; Tso,T.S.C.; Wan,T.S.M. Separation of basic drugs with non-aqueous capillary electrophoresis, *J.Chromatogr.A*, **1996**, *738*, 141–154.

SAMPLE
Matrix: solutions

CAPILLARY ELECTROPHORESIS
Capillary: 102 cm × 75 μm fused-silica (80 cm to detector) (Yongnian Optical Fiber, Hebei, China)

Capillary preparation: Between injections rinse capillary with 100 mM NaOH for 2 min, with water for 2 min, and with running buffer for 5 min. Rinse new capillaries under vacuum with 500 mM NaOH for 30 min and with water for 30 min then equilibrate with running buffer for 10 min.

Running buffer: 50 mM Tris containing 12 mM β-cyclodextrin, adjusted to pH 2.3 with phosphoric acid

Injection: Electrokinetic injection.

Detector: UV 210

Migration time: 25.4, 25.7 (enantiomers)

Voltage: 26.5 kV

Model: laboratory-constructed

OTHER SUBSTANCES
Simultaneous: amphetamine

KEY WORDS
chiral

REFERENCE
Wang,Z.; Sun,Y.; Sun,Z. Enantiomeric separation of amphetamine and phenylephrine by cyclodextrin-mediated capillary zone electrophoresis, *J.Chromatogr.A*, **1996**, *735*, 295–301.

SAMPLE
Matrix: solutions

CAPILLARY ELECTROPHORESIS
Capillary: 36 cm × 50 μm fused-silica coated with linear polyacrylamide (31.5 cm to detector) (GL Science)

Capillary preparation: At the beginning and end of each day rinse capillary with capillary wash solution (Bio-Rad Cat. No. 148-5022) at 690 kPa for more than 3 min and with water at 690 kPa for more than 3 min. Coat capillary as follows. Treat capillary with 1 M NaOH at room temperature for 1 h, rinse with water, dry by passing nitrogen gas through the capillary at 110° for 6 h. Pass thionyl chloride through the capillary using a suction pump for several min, seal capillary at both ends and heat at 70° for 6 h. Unseal the capillary and fill with 250 mM vinyl magnesium bromide in THF by suction, seal the capillary, heat at 70° for 6 h. Open the capillary and rinse it with THF for several min, rinse with distilled water, fill the capillary with polymerization solution, heat at 28 ± 2° for 1 h, condition at -100 V/cm for 30 min (Anal. Sci. 1994, 10, 1). (The polymerization solution was 5% acrylamide in water containing 49 mM Tris, 384 mM glycine, and 0.1% sodium dodecyl sulfate, degas in an ultrasonic bath. Add 40 μL 10% N,N,N',N'-tetramethylethylenediamine and 10 μL 10% ammonium persulfate to 5 mL of the degassed solution, mix thoroughly.)

Running buffer: 50 mM pH 6.0 Phosphate buffer

Injection: Before each injection rinse with water at 690 kPa for 30 s, rinse with running buffer at 690 kPa for 30 s, partially fill with separation solution (1 mM α_1-acid glycoprotein (Cohn fraction VI) (ICN) in running buffer) at 6.9 kPa for 190 s (27 cm), inject sample at 6.9 kPa for 2 s, electrophorese with running buffer (Note that α_1-acid glycoprotein from other suppliers may provide inferior results).

Detector: UV 210

Migration time: Resolution of enantiomers 2.0

Voltage: 12 kV

Model: Bio-rad BioFocus 3000

KEY WORDS
chiral; coated capillary

REFERENCE
Tanaka,Y.; Terabe,S. Separation of the enantiomers of basic drugs by affinity capillary electrophoresis using a partial filling technique and α_1-acid glycoprotein as chiral selector, *Chromatographia*, **1997**, *44*, 119–128.

Phenylpropanolamine

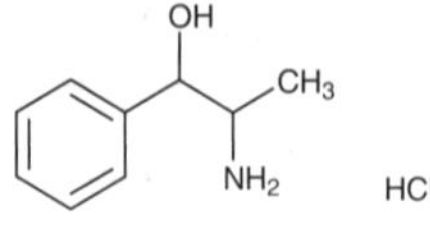

Molecular formula: $C_9H_{13}NO$
Molecular weight: 151.21
CAS Registry No.: 14838-15-4, 154-41-6 (HCl)
Merck Index (12th ed.): 7461

SAMPLE
Matrix: bulk
Sample preparation: Dissolve 4 mg compound in 1 mL MeCN:water 50:50 containing 0.2% triethylamine, vortex for 30 s. Remove a 100 μL aliquot and add it to 100 μL 1.28% 2,3,4,6-tetra-O-acetyl-β-D-glucopyranosyl isothiocyanate in MeCN, vortex for 1 min, let stand for 15 min, make up to 1 mL with 10 mM pH 9.0 phosphate/borate buffer containing 100 mM sodium dodecyl sulfate, vortex for 20 s, filter (Whatman UniPrep), inject an aliquot of the filtrate.

CAPILLARY ELECTROPHORESIS
Capillary: 48 cm × 50 μm fused-silica (26 cm to detector) (Polymicro Technologies)
Capillary preparation: Condition new capillaries with 1 M NaOH for 10 min, with water for 10 min, and with running buffer for 10 min.
Capillary temperature: 30
Running buffer: MeOH:buffer 20:80 (Buffer was 10 mM pH 9.0 phosphate buffer containing 10 mM borate and 100 mM sodium dodecyl sulfate.)
Injection: Vacuum injection for 0.5 s.
Detector: UV 210
Migration time: 12.3 (-), 15.6 (+)
Voltage: 20 kV
Current: 48 μA
Model: Applied Biosystems Model 270A-HT

OTHER SUBSTANCES
Simultaneous: amphetamine, ephedrine, methamphetamine, norpseudoephedrine, pseudoephedrine

KEY WORDS
derivatization; chiral; comparison with HPLC

REFERENCE
Lurie,I.S. Micellar electrokinetic capillary chromatography of the enantiomers of amphetamine, methamphetamine and their hydroxyphenethylamine precursors, *J.Chromatogr.*, **1992**, *605*, 269–275.

SAMPLE
Matrix: bulk
Sample preparation: Dilute with 400 μg/mL n-propyl p-hydroxybenzoate in running buffer

CAPILLARY ELECTROPHORESIS
Capillary: 27 cm × 50 μm fused-silica (20 cm to detector)
Capillary preparation: Before each run rinse with running buffer at high pressure.
Capillary temperature: 30
Running buffer: MeCN:water 15:85 containing 40 mM sodium dodecyl sulfate, 8.5 mM sodium borate, and 8.5 mM sodium phosphate, pH 8.5
Injection: Inject at high pressure for 1 s
Detector: UV 214
Migration time: 1.4
Internal standard: n-propyl p-hydroxybenzoate (1.8)
Voltage: 20 kV
Model: Beckman P/ACE System 2100

OTHER SUBSTANCES
Simultaneous: acetaminophen, acetylcodeine, O-acetylmorphine, aspirin, caffeine, cocaine, codeine, diamorphine, diphenhydramine, hydromorphone, isoniacinamide, lidocaine, methaqualone, morphine, niacinamide, noscapine, papaverine, phenacetin, phenobarbital, procaine, quinine, salicylic acid, strychnine, thebaine

REFERENCE
Walker,J.A.; Krueger,S.T.; Lurie,I.S.; Marché,H.L.; Newby,N. Analysis of heroin drug seizures by micellar electrokinetic capillary chromatography (MECC), *J.Forensic Sci.*, **1995**, *40*, 6–9.

SAMPLE
Matrix: bulk
Sample preparation: Dissolve sample in 2 mg/mL histamine diphosphate in 10 mM HCl, inject an aliquot.

CAPILLARY ELECTROPHORESIS
Capillary: 48.5 cm × 50 µm fused-silica (40 cm to detector)
Capillary preparation: Flush capillary with running buffer for 2 min before each run. Replenish buffer in vials after every 10 injections. Condition new capillaries by flushing with 1 M NaOH for 10 min, with 100 mM NaOH for 10 min, with water for 10 min, and with running buffer for 10 min.
Capillary temperature: 40
Running buffer: 75 mM NaH_2PO_4 containing 75 mM sodium borate, adjusted to pH 8.5 with 2 M NaOH
Injection: Pressure injection at 150 mbar for 3 s.
Detector: UV 230
Migration time: 5.8
Internal standard: histamine diphosphate (5)
Voltage: 10 kV
Current: 66-70 µA
Model: Hewlett Packard 3D CE
Limit of detection: 16 µg/mL (S/N = 2:1)

OTHER SUBSTANCES
Simultaneous: acetaminophen, benzocaine, cis-cinnamoylcocaine, trans-cinnamoylcocaine, cocaine, ephedrine, lidocaine, procaine, tetracaine

REFERENCE
Krawczeniuk,A.S.; Bravenec,V.A. Quantitative determination of cocaine in illicit powders by free zone capillary electrophoresis, *J.Forensic Sci.*, **1998**, *43*, 738–743.

SAMPLE
Matrix: solutions

CAPILLARY ELECTROPHORESIS
Capillary: 58.2 cm × 75 µm fused-silica (50.7 cm to detector) (Polymicro Technologies)
Capillary preparation: Purge with running buffer before each run. At the beginning of each day purge using 50-60 kPa vacuum with 500 mM NaOH for 5 min, with water for 5 min, with MeCN for 5 min, and with running buffer for 5 min.
Running buffer: MeCN:MeOH:acetic acid 49:50:1 containing 20 mM ammonium acetate
Injection: Hydrostatic injection at 10 cm for 5 s.
Detector: UV 214
Migration time: 3.74
Voltage: 25 kV
Model: Waters Quanta 4000

OTHER SUBSTANCES
Simultaneous: amphetamine, benzphetamine, ephedrine, nylidrin, oxymetazoline, phendimetrazine, phenmetrazine, phenylephrine, xylometazoline

REFERENCE
Leung,G.N.W.; Tang,H.P.O.; Tso,T.S.C.; Wan,T.S.M. Separation of basic drugs with non-aqueous capillary electrophoresis, *J.Chromatogr.A*, **1996**, *738*, 141–154.

SAMPLE
Matrix: solutions
Sample preparation: Inject an aliquot of a 20 µg/mL solution in 12.5 mM pH 9.3 borate buffer.

CAPILLARY ELECTROPHORESIS
Capillary: 66.5 cm × 75 µm fused-silica (41.5 cm to detector) (Polymicro Technologies)
Running buffer: 250 mM pH 9.3 Sodium borate buffer
Injection: Pressure injection at 50 mbar for 15 s.
Detector: UV 195
Migration time: 13.6
Voltage: 8 kV
Model: Crystal 310

OTHER SUBSTANCES
Simultaneous: amphetamine, chlorpheniramine, ephedrine, MDA, MDMA, methamphetamine, phentermine, pseudoephedrine
Noninterfering: caffeine

REFERENCE
Lilley,K.A.; Wheat,T.E. Drug identification in biological matrices using capillary electrophoresis and chemometric software, *J.Chromatogr.B*, **1996**, *683*, 67–76.

SAMPLE
Matrix: solutions

CAPILLARY ELECTROPHORESIS
Capillary: 60 cm × 75 µm fused-silica (51 cm to detector) (Supelco)
Capillary preparation: At the end of each day wash capillary with 200 mM NaOH for 10 min and with water for 10 min. Condition new capillaries by washing with 200 mM NaOH for 10 min, with water for 10 min, and with running buffer for 10 min.
Running buffer: 50 mM pH 7.3 Sodium borate buffer containing 3.3% succinyl β-cyclodextrin (Wacker Chemie, Munich, Germany)
Injection: Hydrodynamic injection at 25 mbar for 6 s.
Detector: UV 208
Migration time: 45.87, 49.95 (enantiomers)
Voltage: 15 kV
Model: Prince

OTHER SUBSTANCES
Simultaneous: alprenolol, chlorthalidone, ephedrine, etilefrin, homatropine, methoxamine, octopamine, propranolol, synephrine, trihexyphenidyl

KEY WORDS
chiral

REFERENCE
Schmid,M.G.; Wirnsberger,K.; Gübitz,G. Chiral separation of drug enantiomers by capillary electrophoresis using succinyl-β-cyclodextrin, *Pharmazie*, **1996**, *51*, 852–854.

SAMPLE
Matrix: solutions
Sample preparation: Inject an aliquot of a 100 µg/mL solution in running buffer.

CAPILLARY ELECTROPHORESIS
Capillary: 29 cm × 50 µm fused-silica (24.5 cm to detector) (Yongnian Optical Conductive Fiber Plant, China), coated with polyacrylamide

Capillary preparation: No details of the polyacrylamide coating process are provided. However, another paper (LC.GC 1997, 15, 40) by this group indicates that they use the procedure of Hjertén, thus: Adjust the pH of 20 mL water to 3.5 with acetic acid, add 80 μL 3-(trimethoxysilyl)propyl methacrylate (3-methacryloxypropyltrimethoxysilane), mix, suck into capillary, let stand at room temperature for 1 h, remove the solution, wash with water. Fill the capillary with a deaerated 3-4% acrylamide solution containing 1 μL/mL N,N,N',N'-tetramethylethylenediamine and 1 mg/mL potassium persulfate, let stand for 30 min, remove excess solution by aspiration, rinse with water, remove water by aspiration, dry at 35° (J. Chromatogr. 1985, 347, 191).
Capillary temperature: 25
Running buffer: 100 mM NaH_2PO_4 adjusted to pH 2.5
Injection: Electrokinetic injection at 15 kV for 3 s.
Detector: UV 200, UV 210
Migration time: 3.94
Voltage: 15 kV
Model: Bio-Rad BioFocus 3000

OTHER SUBSTANCES

Simultaneous: albuterol, alprenolol, atenolol, baclofen, bamethan, benproperine, benserazide, bisoprolol, bupranolol, butamirate, butethamate, carbuterol, celiprolol, clenbuterol, clobutinol, dipivefrin, isoproterenol (isoprenaline), metaproterenol (orciprenaline), metipranolol, metoprolol, norfenefrine, ornidazole, oxprenolol, pholedrine, pirbuterol, prilocaine, procyclidine, sotalol, synephrine, terbutaline, tocainide

---

KEY WORDS
coated capillary

REFERENCE
Koppenhoefer,B.; Epperlein,U.; Xiaofeng,Z.; Bingcheng,L. Separation of enantiomers of drugs by capillary electrophoresis. Part 4: Hydroxypropyl-γ-cyclodextrin as chiral solvating agent, *Electrophoresis*, **1997**, *18*, 924–930.

SAMPLE
Matrix: solutions
Sample preparation: Inject an aliquot of a 100 μg/mL solution in running buffer.

CAPILLARY ELECTROPHORESIS
Capillary: 30 cm × 50 μm fused-silica (25.5 cm to detector), coated with polyacrylamide
Capillary preparation: Adjust the pH of 20 mL water to 3.5 with acetic acid, add 80 μL 3-(trimethoxysilyl)propyl methacrylate (3-methacryloxypropyltrimethoxysilane), mix, suck into capillary, let stand at room temperature for 1 h, remove the solution, wash with water. Fill the capillary with a deaerated 3-4% acrylamide solution containing 1 μL/mL N,N,N',N'-tetramethylethylenediamine and 1 mg/mL potassium persulfate, let stand for 30 min, remove excess solution by aspiration, rinse with water, remove water by aspiration, dry at 35° (J. Chromatogr. 1985, 347, 191).
Capillary temperature: 25
Running buffer: 100 mM NaH_2PO_4 containing 45 mM hydroxypropyl-α- cyclodextrin, adjusted to pH 2.5 with phosphoric acid
Injection: Electrokinetic injection at 15 kV for 3 s.
Detector: UV 200
Voltage: 15 kV
Model: Bio-Rad BioFocus 3000

KEY WORDS
chiral; coated capillary; comparison with the use of other cyclodextrins; this running buffer gave the greatest enantiomeric separation.; α=1.031

REFERENCE
Lin,B.; Zhu,X.; Koppenhoefer,B.; Epperlein,U. Investigation of 123 chiral drugs by cyclodextrin-modified capillary electrophoresis, *LC.GC*, **1997**, *15*, 40–46.

SAMPLE
Matrix: solutions

Sample preparation: Inject an aliquot of a solution in running buffer.

CAPILLARY ELECTROPHORESIS
Capillary: 56 cm × 75 μm fused-silica (36 cm to detector) (Otsuka Electronics, Osaka)
Running buffer: Formamide containing 10 mM (+)-18-crown-6 tetracarboxylic acid (Merck, Darmstadt, Germany)
Injection: Gravity injection at 10 cm for 5 s.
Detector: UV 260-300
Migration time: 15.07, 16.56 (enantiomers)
Voltage: 20 kV
Model: Jasco CE-800

KEY WORDS
chiral

REFERENCE
Mori,Y.; Ueno,K.; Umeda,T. Enantiomeric separations of primary amino compounds by non-aqueous capillary zone electrophoresis with a chiral crown ether, *J.Chromatogr.A*, **1997**, *757*, 328–332.

SAMPLE
Matrix: solutions

CAPILLARY ELECTROPHORESIS
Capillary: 53 cm × 50 μm (44.5 cm to detector) (Polymicro Technologies)
Capillary temperature: 40
Running buffer: 70 mM Boric acid adjusted to pH 9.00 with 1 M NaOH
Injection: Pressure injection at 50 mbar for 2-5 s (2.9-7.2 nL).
Detector: UV 210
Migration time: 1.8
Voltage: 20-25 kV
Model: Hewlett-Packard [3D]CE

OTHER SUBSTANCES
Simultaneous: acetaminophen, aspirin, caffeine, salicylic acid

REFERENCE
Thompson,L.; Veening,H.; Strein,T.G. Capillary electrophoresis in the undergraduate instrumental laboratory: Determination of common analgesic formulations, *J.Chem.Educ.*, **1997**, *74*, 1117–1121.

SAMPLE
Matrix: solutions

CAPILLARY ELECTROPHORESIS
Capillary: 45 cm × 50 μm fused-silica (37.5 cm to detector) (Polymicro Technologies)
Capillary preparation: Between each run rinse with 100 mM HCl for 2 min and with running buffer for 2 min. At the start of each day condition capillary with 1 M HCl for 10 min. Before the first use rinse capillary with 1 M NaOH for 30 min and with water for 30 min.
Capillary temperature: 25
Running buffer: MeCN:30 mM pH 2.40 ethanesulfonic acid 20:80
Injection: Injection at 4 kV for 3 s.
Detector: UV 214
Migration time: 4.42
Voltage: 20 kV
Model: Waters Quanta 4000E

OTHER SUBSTANCES
Simultaneous: chloramphenicol, imipramine, lidocaine, procainamide, quinidine, trimethoprim

REFERENCE
Ding,W.; Fritz,J.S. Separation of neutral compounds and basic drugs by capillary electrophoresis in acidic solution using laurylpoly(oxyethylene) sulfate as an additive, *Anal.Chem.*, **1998**, *70*, 1859–1865.

SAMPLE
Matrix: solutions

CAPILLARY ELECTROPHORESIS
Capillary: 29-36 cm $\times$ 50 μm fused-silica (24.5-31.5 cm to detector) (Yongnian Optical Conductive Fiber Plant, China) coated with polyacrylamide

Capillary preparation: Coat capillary as follows. Adjust the pH of 20 mL water to 3.5 with acetic acid, add 80 μL 3-(trimethoxysilyl)propyl methacrylate (3-methacryloxypropyltrimethoxysilane), mix, suck into capillary, let stand at room temperature for 1 h, remove the solution, wash with water. Fill the capillary with a deaerated 3-4% acrylamide solution containing 1 μL/mL N,N,N',N'-tetramethylethylenediamine and 1 mg/mL potassium persulfate, let stand for 30 min, remove excess solution by aspiration, rinse with water, remove water by aspiration, dry at 35° (J. Chromatogr. 1985, 347, 191).

Capillary temperature: 25

Running buffer: 100 mM pH 2.5 NaH_2PO_4 (A) or 100 mM pH 2.5 NaH_2PO_4 containing 45 mM hydroxypropyl-α-cyclodextrin (Wacker, Munich) (B)

Injection: Electromigration at 15 kV for 3 s.

Detector: UV 200; UV 210

Migration time: 3.94 (A); 6.21, 6.40 (B) (enantiomers)

Voltage: 15 kV

Model: Bio-Focus 3000

OTHER SUBSTANCES
Also analyzed: albuterol (salbutamol), alprenolol, amorolfine, atenolol, atropine, azelastine, baclofen, bamethan, benproperine, benserazide, biperiden, bisoprolol, brompheniramine, bupivacaine, bupranolol, butamirate, butethamate, carazolol, carbuterol, carteolol, carvedilol, celiprolol, chloroquine, chlorpheniramine, chlorphenoxamine, cicletanine, clenbuterol, clidinium bromide, clobutinol, dimethindene, dipivefrin, disopyramide, dobutamine, doxylamine, fendiline, flecainide, gallopamil, homatropine, ipratropium bromide, isoproterenol (isoprenaline), isothipendyl, ketamine, meclizine, mefloquine, mepindolol, mequitazine, metaclazepam, metaproterenol (orciprenaline), metipranolol, metoprolol, nafronyl (naftidrofuryl), nefopam, nicardipine, norfenefrine, ofloxacin, ornidazole, orphenadrine, oxomemazine, oxprenolol, oxybutynin, phenoxybenzamine, pholedrine, pindolol, pirbuterol, prilocaine, procyclidine, promethazine, propafenone, propranolol, reproterol, sotalol, sulpride, synephrine, talinolol, terbutaline, tetrahydrozoline (tetryzoline), theodrenaline, tioconazole, tocainide, trihexyphenidyl, trimeprazine (alimemazine), trimipramine, tropicamide, verapamil, zopiclone

KEY WORDS
coated capillary; chiral

REFERENCE
Koppenhoefer,B.; Eperlein,U.; Schlunk,R.; Zhu,X.; Lin,B. Separation of enantiomers of drugs by capillary electrophoresis. V. Hydroxypropyl-α-cyclodextrin as chiral solvating agent, *J.Chromatogr.A*, **1998**, *793*, 153–164.

Phenytoin

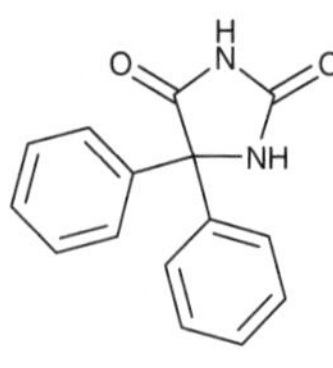

Molecular formula: $C_{15}H_{12}N_2O_2$
Molecular weight: 252.27
CAS Registry No.: 57-41-0, 630-93-3 (sodium salt)
Merck Index (12th ed.): 7475
Lednicer: 1 246

SAMPLE
Matrix: blood
Sample preparation: 50 µL Serum + 5 mL isotonic saline, add to a Sep-Pak C18 SPE cartridge, wash with 5 mL saline, wash with 5 mL water, wash with 5 mL heptane:chloroform 98:2, wash with 5 mL heptane, wash with 3 mL water, remove all water with a stream of air, elute with 3 mL MeOH (remove all MeOH with a stream of air). Evaporate the eluate to dryness under reduced pressure at 37°, reconstitute with 100 µL water, inject an aliquot.

CAPILLARY ELECTROPHORESIS
Capillary: 72 cm × 50 µm silica (50 cm to detector) (Applied Biosystems)
Capillary preparation: Before each run wash capillary with 100 mM NaOH for 3 min, wash with running buffer for 3 min, aspirate neutral marker solution (2 drops DMSO in 10 mL water) for 1 s, place capillary end in buffer vial for 5 s, inject sample. Wash capillary at the beginning of each day by passing 1 M NaOH through for 20 min.
Capillary temperature: 30
Running buffer: 30 mM pH 9.3 Borate buffer containing 30 mM sodium dodecyl sulfate
Injection: Inject by applying vacuum for 1 s (2-3 nL)
Detector: UV 200
Migration time: 6
Voltage: +30 kV
Current: 72 µA
Model: Applied Biosystems Model 270A

OTHER SUBSTANCES
Extracted: carbamazepine, phenobarbital, primidone

KEY WORDS
serum; SPE

REFERENCE
Evenson,M.A.; Wiktorowicz,J.E. Automated capillary electrophoresis applied to therapeutic drug monitoring, *Clin.Chem.*, **1992**, *38*, 1847–1852.

SAMPLE
Matrix: blood
Sample preparation: 100 µL Serum + 150 µL 80 µg/mL 3-isobutyl-1-methylxanthine in MeCN, mix for 30 s, centrifuge at 14000 g for 1 min, inject an aliquot of the supernatant.

CAPILLARY ELECTROPHORESIS
Capillary: 25 cm × 47 µm
Capillary preparation: Wash with running buffer by pressure injection for 1 min after each run. Wash daily with 50 mM phosphoric acid for 2 min, with water for 1 min, with 2 M NaOH for 2 min, with water for 1 min, and with running buffer for 2 min.
Capillary temperature: 24
Running buffer: 220 mM boric acid adjusted to pH 8.8 with 2.5 M NaOH
Injection: Pressure injection for 8 s.
Detector: UV 254
Migration time: 2.5
Internal standard: 3-isobutyl-1-methylxanthine (2.3)
Voltage: 12 kV
Model: Beckman

OTHER SUBSTANCES
Extracted: acetaminophen, caffeine, carbamazepine, iohexol, pentobarbital, phenobarbital, salicylic acid, theophylline

KEY WORDS
serum

REFERENCE
Shihabi,Z.K.; Constantinescu,M.S. Iohexol in serum determined by capillary electrophoresis, *Clin.Chem.*, **1992**, *38*, 2117–2120.

SAMPLE
Matrix: blood
Sample preparation: 100 μL Serum + 150 μL MeCN, mix for 15 s, centrifuge at 15000 g for 1 min, inject an aliquot of the supernatant.

CAPILLARY ELECTROPHORESIS
Capillary: 25 cm × 50 μm
Capillary preparation: Wash with running buffer for 1 min after each run.
Capillary temperature: 24
Running buffer: 300 mM boric acid adjusted to pH 8.5 with NaOH
Injection: Pressure injection for 15 s.
Detector: UV 254
Migration time: 2.2
Voltage: 13 kV
Model: Beckman

OTHER SUBSTANCES
Extracted: iohexol, phenobarbital, theophylline

KEY WORDS
serum

REFERENCE
Shihabi,Z.K. Sample matrix effects in capillary electrophoresis. II. Acetonitrile deproteinization, *J.Chromatogr.A*, **1993**, *652*, 471–475.

SAMPLE
Matrix: blood
Sample preparation: Vortex serum for 15 s, filter (0.45 μm), add sodium dodecyl sulfate to a concentration of 100 mM, let stand for 1 h, filter (Amicon Centrifree Micropartition) while centrifuging at 1500 g for 20 min, inject an aliquot of the ultrafiltrate.

CAPILLARY ELECTROPHORESIS
Capillary: 63 cm × 75 μm fused-silica (63 cm to detector)
Capillary preparation: Before each run rinse capillary with 100 mM NaOH for 3 min and with running buffer for 5 min.
Capillary temperature: 35
Running buffer: 10 mM pH 9.2 Na_2HPO_4 containing 6 mM sodium borate and 100 mM sodium dodecyl sulfate
Injection: Vacuum suction for 1 s
Detector: UV 195
Migration time: 14.2
Voltage: 25 kV
Current: 80 μA
Model: Applied Biosystems model 270A-HT

OTHER SUBSTANCES
Extracted: ethosuximide, phenobarbital

KEY WORDS
serum; ultrafiltrate

REFERENCE
Schmutz,A.; Thormann,W. Factors affecting the determination of drugs and endogenous low molecular mass compounds in human serum by micellar electrokinetic capillary chromatography with direct sample injection, *Electrophoresis*, **1994**, *15*, 51–61.

SAMPLE
Matrix: blood
Sample preparation: 200 μL Serum + 1 mL 5 μg/mL n-propyl p-hydroxybenzoate in ethyl acetate, vortex for 30 s, centrifuge at 13400 g for 2 min. Remove a 500 μL aliquot of the organic layer and evaporate it to dryness under a stream of nitrogen, reconstitute with 50 μL MeOH: water 5:95, inject an aliquot.

CAPILLARY ELECTROPHORESIS
Capillary: 64.5 cm × 50 μm fused-silica (56 cm to detector) (Hewlett-Packard)
Capillary preparation: Rinse capillary with running buffer at 630 mm Hg for 5 min before each run.
Capillary temperature: 30
Running buffer: 10 mM pH 8.0 Phosphate buffer containing 50 mM sodium dodecyl sulfate
Injection: Vacuum injection at 50 mm Hg for 2 s.
Detector: UV 210
Migration time: 4.2
Internal standard: n-propyl p-hydroxybenzoate (7.2)
Voltage: 30 kV
Model: Hewlett-Packard HP^{3D}
Limit of detection: 800 ng/mL

OTHER SUBSTANCES
Extracted: carbamazepine, phenobarbital, primidone, zonisamide
Noninterfering: valproic acid

KEY WORDS
serum; comparison with HPLC

REFERENCE
Makino,K.; Goto,Y.; Sueyasu,M.; Futagami,K.; Kataoka,Y.; Oishi,R. Micellar electrokinetic capillary chromatography for therapeutic drug monitoring of zonisamide, *J.Chromatogr.B*, **1997**, *695*, 417–425.

SAMPLE
Matrix: solutions
Sample preparation: Inject an aliquot of a 100 μg/mL solution in 50 mM sodium dodecyl sulfate.

CAPILLARY ELECTROPHORESIS
Capillary: 57 cm × 75 μm fused-silica (50 cm to detector) (Beckman)
Capillary preparation: Rinse capillary with 100 mM NaOH and running buffer between each run.
Capillary temperature: 30
Running buffer: Isopropanol:100 mM pH 8.1 borate buffer containing 50 mM sodium dodecyl sulfate 10:90
Injection: Pressure injection for 5 s
Detector: UV 200
Migration time: 9
Voltage: +25 kV
Model: Beckman P/ACE 5510

OTHER SUBSTANCES
Simultaneous: amiodarone, bretylium, disopyramide, lidocaine, propafenone, quinidine

REFERENCE
Bretnall,A.E.; Clarke,G.S. Investigation and optimisation of the use of organic modifiers in micellar electrokinetic chromatography, *J.Chromatogr.A*, **1995**, *716*, 49–55.

SAMPLE
Matrix: solutions

CAPILLARY ELECTROPHORESIS
Capillary: 57 cm × 75 μm fused-silica (50 cm to detector) (Beckman)
Capillary preparation: Before each run rinse capillary with 100 mM NaOH and running buffer.
Capillary temperature: 30
Running buffer: Acetone:100 mM pH 8.1 borate buffer containing 50 mM sodium dodecyl sulfate
15:85 (A) or isopropanol:100 mM pH 8.1 borate buffer containing 50 mM sodium dodecyl sulfate
10:90 (B)
Injection: Pressure injection for 5-10 s.
Detector: UV 200
Migration time: 6.5 (A), 8.5 (B)
Voltage: 25 kV
Model: Beckman P/ACE 5510

OTHER SUBSTANCES
Simultaneous: acebutolol, amiodarone, atenolol, bretylium, captopril, diltiazem, disopyramide,
lidocaine, lisinopril, metoprolol, nicardipine, nifedipine, pindolol, propafenone, propranolol,
quinidine, sotalol, timolol, p-toluenesulfonic acid, verapamil

KEY WORDS
only running buffer A separates all compounds

REFERENCE
Bretnall,A.E.; Clarke,G.S. Selectivity of capillary electrophoresis for the analysis of cardiovascular drugs,
J.Chromatogr.A, **1996**, *745*, 145–154.

Pholedrine

Molecular formula: $C_{10}H_{15}NO$
Molecular weight: 165.24
CAS Registry No.: 370-14-9, 6114-26-7 (sulfate)
Merck Index (12th ed.): 7485

SAMPLE
Matrix: solutions

CAPILLARY ELECTROPHORESIS
Capillary: 37 cm × 50 μm (30 cm to detector)
Capillary temperature: 25
Running buffer: 50 mM pH 2.5 Phosphate buffer containing 16 mM 2-hydroxy-3-trimethylam-
moniopropyl-β-cyclodextrin
Detector: UV 214
Migration time: 9.89 (first enantiomer; α = 1.009; R_s = 0.198)
Voltage: 20 kV
Model: Beckman P/ACE 2100

OTHER SUBSTANCES
Simultaneous: cyclodrine, cyclopentolate, norpseudoephedrine, phendimetrazine, tropicamide

KEY WORDS
chiral

REFERENCE
Bunke,A.; Jira,T. Chiral capillary electrophoresis using a cationic cyclodextrin, *Pharmazie*, **1996**, *51*, 672–673.

SAMPLE
Matrix: solutions
Sample preparation: Inject an aliquot of a 100 µg/mL solution in running buffer.

CAPILLARY ELECTROPHORESIS
Capillary: 29 cm × 50 µm fused-silica (24.5 cm to detector) (Yongnian Optical Conductive Fiber Plant, China), coated with polyacrylamide
Capillary preparation: No details of the polyacrylamide coating process are provided. However, another paper (LC.GC 1997, 15, 40) by this group indicates that they use the procedure of Hjertén, thus: Adjust the pH of 20 mL water to 3.5 with acetic acid, add 80 µL 3-(trimethoxysilyl)propyl methacrylate (3-methacryloxypropyltrimethoxysilane), mix, suck into capillary, let stand at room temperature for 1 h, remove the solution, wash with water. Fill the capillary with a deaerated 3-4% acrylamide solution containing 1 µL/mL N,N,N',N'-tetramethylethylenediamine and 1 mg/mL potassium persulfate, let stand for 30 min, remove excess solution by aspiration, rinse with water, remove water by aspiration, dry at 35° (J. Chromatogr. 1985, 347, 191).
Capillary temperature: 25
Running buffer: 100 mM NaH_2PO_4 adjusted to pH 2.5
Injection: Electrokinetic injection at 15 kV for 3 s.
Detector: UV 200, UV 210
Migration time: 4.10
Voltage: 15 kV
Model: Bio-Rad BioFocus 3000

OTHER SUBSTANCES
Simultaneous: albuterol, alprenolol, atenolol, baclofen, bamethan, benproperine, benserazide, bisoprolol, bupranolol, butamirate, butethamate, carbuterol, celiprolol, clenbuterol, clobutinol, dipivefrin, isoproterenol (isoprenaline), metaproterenol (orciprenaline), metipranolol, metoprolol, norfenefrine, ornidazole, oxprenolol, phenylpropanolamine, pirbuterol, prilocaine, procyclidine, sotalol, synephrine, terbutaline, tocainide

KEY WORDS
coated capillary

REFERENCE
Koppenhoefer,B.; Epperlein,U.; Xiaofeng,Z.; Bingcheng,L. Separation of enantiomers of drugs by capillary electrophoresis. Part 4: Hydroxypropyl-γ-cyclodextrin as chiral solvating agent, *Electrophoresis*, **1997**, *18*, 924–930.

SAMPLE
Matrix: solutions
Sample preparation: Inject an aliquot of a 100 µg/mL solution in running buffer.

CAPILLARY ELECTROPHORESIS
Capillary: 30 cm × 50 µm fused-silica (25.5 cm to detector), coated with polyacrylamide
Capillary preparation: Adjust the pH of 20 mL water to 3.5 with acetic acid, add 80 µL 3-(trimethoxysilyl)propyl methacrylate (3-methacryloxypropyltrimethoxysilane), mix, suck into capillary, let stand at room temperature for 1 h, remove the solution, wash with water. Fill the capillary with a deaerated 3-4% acrylamide solution containing 1 µL/mL N,N,N',N'-tetramethylethylenediamine and 1 mg/mL potassium persulfate, let stand for 30 min, remove excess solution by aspiration, rinse with water, remove water by aspiration, dry at 35° (J. Chromatogr. 1985, 347, 191).
Capillary temperature: 25
Running buffer: 100 mM NaH_2PO_4 containing 45 mM hydroxypropyl-β- cyclodextrin, adjusted to pH 2.5 with phosphoric acid
Injection: Electrokinetic injection at 15 kV for 3 s.
Detector: UV 200
Voltage: 15 kV
Model: Bio-Rad BioFocus 3000

KEY WORDS
chiral; coated capillary; comparison with the use of other cyclodextrins; this running buffer gave the greatest enantiomeric separation.; $\alpha=1.043$

REFERENCE
Lin,B.; Zhu,X.; Koppenhoefer,B.; Epperlein,U. Investigation of 123 chiral drugs by cyclodextrin-modified capillary electrophoresis, *LC.GC*, **1997**, *15*, 40–46.

SAMPLE
Matrix: solutions

CAPILLARY ELECTROPHORESIS
Capillary: 29-36 cm $\times$ 50 μm fused-silica (24.5-31.5 cm to detector) (Yongnian Optical Conductive Fiber Plant, China) coated with polyacrylamide

Capillary preparation: Coat capillary as follows. Adjust the pH of 20 mL water to 3.5 with acetic acid, add 80 μL 3-(trimethoxysilyl)propyl methacrylate (3-methacryloxypropyltrimethoxysilane), mix, suck into capillary, let stand at room temperature for 1 h, remove the solution, wash with water. Fill the capillary with a deaerated 3-4% acrylamide solution containing 1 μL/mL N,N,N',N'-tetramethylethylenediamine and 1 mg/mL potassium persulfate, let stand for 30 min, remove excess solution by aspiration, rinse with water, remove water by aspiration, dry at 35° (J. Chromatogr. 1985, 347, 191).

Capillary temperature: 25

Running buffer: 100 mM pH 2.5 NaH_2PO_4 (A) or 100 mM pH 2.5 NaH_2PO_4 containing 45 mM hydroxypropyl-α-cyclodextrin (Wacker, Munich) (B)

Injection: Electromigration at 15 kV for 3 s.

Detector: UV 200; UV 210

Migration time: 4.10 (A); 6.39 (B) (no separation of enantiomers)

Voltage: 15 kV

Model: Bio-Focus 3000

OTHER SUBSTANCES
Also analyzed: albuterol (salbutamol), alprenolol, amorolfine, atenolol, atropine, azelastine, baclofen, bamethan, benproperine, benserazide, biperiden, bisoprolol, brompheniramine, bupivacaine, bupranolol, butamirate, butethamate, carazolol, carbuterol, carteolol, carvedilol, celiprolol, chloroquine, chlorpheniramine, chlorphenoxamine, cicletanine, clenbuterol, clidinium bromide, clobutinol, dimethindene, dipivefrin, disopyramide, dobutamine, doxylamine, fendiline, flecainide, gallopamil, homatropine, ipratropium bromide, isoproterenol (isoprenaline), isothipendyl, ketamine, meclizine, mefloquine, mepindolol, mequitazine, metaclazepam, metaproterenol (orciprenaline), metipranolol, metoprolol, nafronyl (naftidrofuryl), nefopam, nicardipine, norfenofrine, ofloxacin, ornidazole, orphenadrine, oxomemazine, oxprenolol, oxybutynin, phenoxybenzamine, phenylpropanolamine, pindolol, pirbuterol, prilocaine, procyclidine, promethazine, propafenone, propranolol, reproterol, sotalol, sulpride, synephrine, talinolol, terbutaline, tetrahydrozoline (tetryzoline), theodrenaline, tioconazole, tocainide, trihexyphenidyl, trimeprazine (alimemazine), trimipramine, tropicamide, verapamil, zopiclone

KEY WORDS
coated capillary

REFERENCE
Koppenhoefer,B.; Eperlein,U.; Schlunk,R.; Zhu,X.; Lin,B. Separation of enantiomers of drugs by capillary electrophoresis. V. Hydroxypropyl-α-cyclodextrin as chiral solvating agent, *J.Chromatogr.A*, **1998**, *793*, 153–164.

Physostigmine

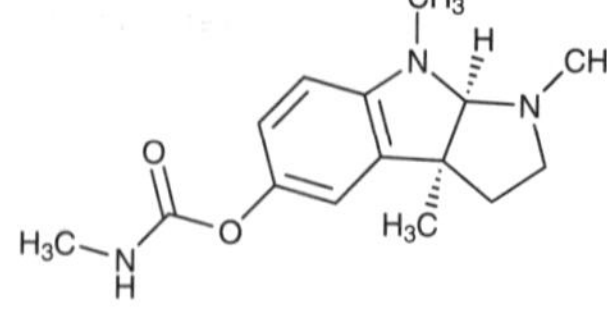

Molecular formula: $C_{15}H_{21}N_3O_2$
Molecular weight: 275.35
CAS Registry No.: 57-47-6, 57-64-7 (salicylate), 64-47-1 (sulfate)
Merck Index (12th ed.): 7540
Lednicer: 1 111

SAMPLE
Matrix: solutions
Sample preparation: Inject an aliquot of a solution in MeOH:500 µM phosphoric acid 50:50.

CAPILLARY ELECTROPHORESIS
Capillary: 80 cm × 75 µm fused-silica (70 cm to detector) (Polymicro Technologies)
Capillary preparation: Before each run wash with 4 column volumes of 100 mM NaOH, water, MeOH, and running buffer. [A complex pre-capillary extraction system is described in the paper.]
Running buffer: 100 mM pH 2.5 Phosphate buffer containing 3.9 mM dimethyl-β-cyclodextrin
Detector: UV 205
Migration time: 21
Voltage: 23 kV
Model: Prince

OTHER SUBSTANCES
Simultaneous: bambuterol

REFERENCE
Pálmarsdóttir,S.; Mathiasson,L.; Jönsson,J.Å.; Edholm,L.-E. Micro-CLC as an interface between SLM extraction and CZE for enhancement of sensitivity and selectivity in bioanalysis of drugs, *J.Capillary Electrophor.*, **1996**, *3*, 255–260.

Picumeterol

Molecular formula: $C_{21}H_{29}Cl2N_3O_2$
Molecular weight: 426.39
CAS Registry No.: 130641-36-0, 130641-37-1 (fumarate)
Lednicer: 5 16

SAMPLE
Matrix: solutions

CAPILLARY ELECTROPHORESIS
Capillary: 34 cm × 50 µm (25.5 cm (A) or 8.5 cm (B, C) to detector) (Composite Metal Services, Hallow, UK)
Running buffer: 50 mM pH 2.5 Phosphate buffer containing 25 mM dimethyl-β-cyclodextrin (A, B) or 250 mM pH 2.5 phosphate buffer containing 25 mM dimethyl-β-cyclodextrin (C)
Injection: Pressure injection at 50 mbar for 5 s.
Detector: UV 200
Migration time: 3, 3.1 (A), 1.1, 1.15 (B), 4.7, 5 (C) (enantiomers)
Voltage: 20 kV (A) or -20 kV (B) or -7.5 kV (C)
Model: Hewlett Packard 3D

OTHER SUBSTANCES
Simultaneous: clenbuterol

KEY WORDS
when the distance to the detector was 8.5 cm (B; C) injection was at the normal outlet end and voltage polarity and pressure for injection were reversed.; chiral

REFERENCE
Altria,K.D.; Kelly,M.A.; Clark,B.J. The use of a short-end injection procedure to achieve improved performance in capillary electrophoresis, *Chromatographia*, **1996**, *43*, 153–158.

Pilocarpine

Molecular formula: $C_{11}H_{16}N_2O_2$
Molecular weight: 208.26
CAS Registry No.: 92-13-7, 54-71-7 (HCl), 148-72-1 (nitrate)
Merck Index (12th ed.): 7578

SAMPLE
Matrix: formulations

CAPILLARY ELECTROPHORESIS
Capillary: 57 cm × 75 μm fused-silica (50 cm to detector) (Beckman)
Capillary preparation: Before each injection flush column with 100 mM NaOH for 2 min and running buffer for 4 min
Capillary temperature: 22
Running buffer: 20 mM pH 9.3 borate buffer containing 5% (w/v) sodium dodecyl sulfate
Injection: Pressure injection for 5 s (30 nL)
Detector: UV 214
Migration time: 15
Voltage: 20 kV
Model: Beckman P/ACE 2100
Limit of detection: 1 μg/mL

OTHER SUBSTANCES
Simultaneous: degradation products, impurities, isopilocarpic acid, isopilocarpine, pilocarpic acid

KEY WORDS
ophthalmic solutions

REFERENCE
Charman,W.N.; Humberstone,A.J.; Charman,S.A. Analysis of pilocarpine and its degradation products by micellar electrokinetic capillary chromatography, *Pharm.Res.*, **1992**, *9*, 1219–1223.

SAMPLE
Matrix: formulations
Sample preparation: Dilute a 20 mg/mL ophthalmic solution 100-fold with 50 mM HCl, mix an aliquot with an equal volume of 50 μg/mL naphazoline nitrate in 50 mM HCl, inject a sample by electromigration for 8 s.

CAPILLARY ELECTROPHORESIS
Capillary: 20 cm × 25 μm coated fused-silica, HPE Microsampler (Bio-Rad)
Running buffer: 100 mM pH 6.9 phosphate buffer containing 10 mM β-cyclodextrin
Injection: Electromigration for 8 s.
Detector: UV 217
Migration time: 4
Internal standard: naphazoline (2.3)
Voltage: 8 kV
Model: Bio-Rad HPE 100

OTHER SUBSTANCES
Simultaneous: isopilocarpine, metipranolol

KEY WORDS
ophthalmic solutions; coated capillary

REFERENCE
Baeyens,W.; Weiss,G.; Van Der Weken,G.; Van den Bossche,W. Analysis of pilocarpine and its trans epimer, isopilocarpine, by capillary electrophoresis, *J.Chromatogr.*, **1993**, *638*, 319–326.

SAMPLE
Matrix: solutions

CAPILLARY ELECTROPHORESIS
Capillary: 56 cm × 50 μm fused-silica (47.5 cm to detector) (Polymicro Technologies)
Capillary preparation: Between runs flush capillary with 100 mM NaOH for 5 min, with water for 5 min, and with running buffer for 10 min.
Capillary temperature: 20
Running buffer: pH 10.5 Sodium borate buffer (ionic strength 0.03) containing 170 mM sodium dodecyl sulfate
Injection: Pressure injection at 5 kPa for 3 s.
Detector: UV 225
Migration time: 3.8
Voltage: 30 kV
Model: Hewlett-Packard ^{3D}CE

OTHER SUBSTANCES
Simultaneous: degradation products, isopilocarpic acid, isopilocarpine, pilocarpic acid

REFERENCE
Persson,K.; Åström,O. Fractional factorial design optimization of the separation of pilocarpine and its degradation products by capillary electrophoresis, *J.Chromatogr.B*, **1997**, *697*, 207–215.

Pimozide

Molecular formula: $C_{28}H_{29}F_2N_3O$
Molecular weight: 461.55
CAS Registry No.: 2062-78-4
Merck Index (12th ed.): 7589
Lednicer: 2 290

SAMPLE
Matrix: urine
Sample preparation: 10 mL Urine + 1 mL 5 M NaOH + 10 mL n-hexane, vortex for 1 min, centrifuge at 0° at 3000 g for 5 min. Remove the organic layer and add it to 50 μL glacial acetic acid, evaporate to dryness under a stream of nitrogen at 30°, reconstitute the residue in 50 μL 50 mM sodium taurodeoxycholate, filter (0.2 μm PTFE), inject an aliquot.

CAPILLARY ELECTROPHORESIS
Capillary: 100 cm × 50 μm fused-silica (50 cm to detector) (ISCO)
Capillary preparation: Rinse capillary with 20 μL running buffer between injections. Condition a new capillary by filling with 1 M NaOH, let stand for 1 h, fill with 100 mM NaOH, let stand for 1 h, rinse with buffer. Every 20 injections rinse capillary with 200 μL 1 M NaOH, 200 μL water, and 200 μL running buffer, fill with running buffer.
Capillary temperature: 22
Running buffer: 40 mM pH 9.5 Borate buffer containing 10 mM sodium taurodeoxycholate
Injection: Load under vacuum at 7.5 kPa/s.

Detector: UV 240
Migration time: 5.5
Voltage: 30 kV
Model: ISCO Model 3140 electropherograph
Limit of detection: 6 ng/mL

OTHER SUBSTANCES
Extracted: acepromazine, amiodarone, amitriptyline, azaperone, chlorpromazine, cianopramine, clomipramine, clozapine, desethylamiodarone, desipramine, diclofensine, dothiepin, doxepin, imipramine, isocarboxazid, moclobemide, perphenazine, phenothiazine, prochlorperazine, promazine, thioridazine, thiothixene, trifluoperazine, trimipramine

KEY WORDS
human; cow; pig; horse; protect from light

REFERENCE
Aumatell,A.; Wells,R.J. Determination of a cardiac antiarrhythmic, tricyclic antipsychotics and antidepressants in human and animal urine by micellar electrokinetic capillary chromatography using a bile salt, *J.Chromatogr.B*, **1995**, *669*, 331–344.

Pindolol

Molecular formula: $C_{14}H_{20}N_2O_2$
Molecular weight: 248.33
CAS Registry No.: 13523-86-9
Merck Index (12th ed.): 7597
Lednicer: 2 342

SAMPLE
Matrix: blood
Sample preparation: Hydrolyze 2 mL serum with β-glucuronidase (EC 3.2.1.31, type H-1 from Helix pomatia, 416 800 U/g, Separacor) at 80° for 30 min, cool, add 900 μL MeCN, vortex for 15 min, centrifuge at 2004 g for 10 min, add ephedrine (165 μg/mL), filter (0.5 μm), inject an aliquot of the filtrate.

CAPILLARY ELECTROPHORESIS
Capillary: 58 cm × 50 μm fused-silica (50 cm to detector) (Polymicro Technologies)
Capillary preparation: Before each injection purge capillary with 5% phosphoric acid for 12 s, with water for 30 s, and with running buffer for 10 min.
Capillary temperature: 35
Running buffer: 80 mM pH 6.7 sodium phosphate buffer containing 15 mM cetyltrimethylammonium bromide
Injection: Hydrostatic injection for 20 s
Detector: UV 214
Migration time: 14.5
Internal standard: ephedrine (17.5)
Voltage: -27 kV
Model: Waters Quanta 4000

OTHER SUBSTANCES
Extracted: acebutolol, alprenolol, atenolol, labetalol, metoprolol, nadolol, oxprenolol, propranolol, timolol

KEY WORDS
serum

REFERENCE
Lukkari,P.; Nyman,T.; Riekkola,M.-J. Determination of nine β-blockers in serum by micellar electrokinetic capillary chromatography, *J.Chromatogr.A*, **1994**, *674*, 241–246.

SAMPLE
Matrix: blood
Sample preparation: Filter (1 mL Amicon Centrifree Micropartition System) serum while centrifuging at 1500 g for 20 min, inject an aliquot of the ultrafiltrate.

CAPILLARY ELECTROPHORESIS
Capillary: 92 cm × 75 μm fused-silica (55 cm to detector) (Polymicro Technologies)
Capillary preparation: Rinse with running buffer for 10 min between run. At the beginning and end of each day rinse with 100 mM NaOH for 10 min and with water for 10 min.
Running buffer: Isopropanol:buffer 5:95 (Buffer was 10 mM Na_2HPO_4 containing 6 mM sodium borate and 75 mM sodium dodecyl sulfate.)
Injection: Injection via vacuum suction for 0.7 s
Detector: UV 215
Migration time: 18
Voltage: 25 kV
Current: 82 μA
Model: Europhor Prime Vision IV

OTHER SUBSTANCES
Extracted: atenolol, carteolol, celiprolol, levobunolol, metoprolol, penbutolol, propranolol, timolol

KEY WORDS
serum; ultrafiltrate

REFERENCE
Schmutz,A.; Thormann,W. Assessment of impact of physico-chemical drug properties on monitoring drug levels by micellar electrokinetic capillary chromatography with direct serum injection, *Electrophoresis*, **1994**, *15*, 1295–1303.

SAMPLE
Matrix: blood, urine
Sample preparation: Urine. Dilute with 2 volumes of water, filter (0.5 μm), inject an aliquot. Serum. Condition a 3 mL Supelclean LC-18 SPE cartridge (Supelco) with MeOH and water. Hydrolyze 900 μL serum with β-glucuronidase (EC 3.2.1.31 type H-1 from Helix pomatia) at 60° with sonication for 30 min, add 500 μL (?) MeOH, centrifuge at 2000 g, add the supernatant to the SPE cartridge, wash with 1 mL water, dry under vacuum, elute with 2 mL MeOH:water 90:10, filter, inject an aliquot.

CAPILLARY ELECTROPHORESIS
Capillary: 68 cm × 50 μm fused-silica (60 cm to detector) (Polymicro Technologies)
Capillary preparation: Purge with running buffer for 2 min before injection.
Running buffer: pH 7.0 Phosphate buffer containing 10 mM N-cetyl-N,N,N-trimethylammonium bromide (Prepare buffer by mixing 100 mM NaH_2PO_4 and 100 mM Na_2HPO_4 to achieve a pH of 7.0.)
Injection: Hydrostatic injection for 30 s
Detector: UV 214
Migration time: 15.4
Internal standard: 2,6-dimethylphenol (only for urine) (14.5)
Voltage: -26 kV
Current: 97 μA
Model: Waters Quanta 4000

OTHER SUBSTANCES
Extracted: acebutolol, alprenolol, atenolol, labetalol, metoprolol, nadolol, oxprenolol, propranolol, timolol

KEY WORDS
serum; comparison with HPLC; SPE

REFERENCE
Lukkari,P.; Sirén,H. Ion-pair chromatography and micellar electrokinetic capillary chromatography in analyzing β-adrenergic blocking agents from human biological fluids, *J.Chromatogr.A*, **1995**, *717*, 211–217.

SAMPLE
Matrix: solutions

CAPILLARY ELECTROPHORESIS
Capillary: 64 cm × 50 μm uncoated silica (48 cm to detector) (Polymicro Technologies)
Running buffer: 18 mM pH 3.0 Tris buffer containing 10 mM heptakis(2,6-di-O-methyl)-β-cyclodextrin, 0.1% methylhydroxycellulose 4000, and 0.03 mM hexadecyltrimethylammonium bromide (pH adjusted with phosphoric acid)
Detector: UV 220
Migration time: 11.2, 11.6 (enantiomers)
Voltage: 24 kV
Model: Laboratory constructed

KEY WORDS
chiral

REFERENCE
Soini,H.; Riekkola,M.-L.; Novotny,M.V. Chiral separations of basic drugs and quantitation of bupivacaine enantiomers in serum by capillary electrophoresis with modified cyclodextrin buffers, *J.Chromatogr.*, **1992**, *608*, 265–274.

SAMPLE
Matrix: solutions

CAPILLARY ELECTROPHORESIS
Capillary: 82 cm × 75 μm fused-silica (Waters)
Capillary preparation: Between runs purge capillary with running buffer for 2 min. Purge a new capillary with 100 mM NaOH for 30 min, with water for 30 min, and with running buffer for 30 min.
Capillary temperature: 30
Running buffer: 30 mM pH 7.6 phosphate buffer containing 10 mM cetyltrimethylammonium bromide
Injection: Inject at high pressure for 5 s
Detector: UV 214
Migration time: 9
Voltage: 21 kV
Model: Beckman P/ACE System 2000

OTHER SUBSTANCES
Simultaneous: acebutolol, alprenolol, atenolol, nadolol, oxprenolol, propranolol, sotalol, timolol

REFERENCE
Lukkari,P.; Ennelin,A.; Sirén,H.; Riekkola,M.-L. Effect of temperature, effective capillary length, and applied voltage on the migration of nine β-blockers in micellar electrokinetic capillary chromatography, *J.Liq.Chromatogr.*, **1993**, *16*, 2069–2079.

SAMPLE
Matrix: solutions

CAPILLARY ELECTROPHORESIS
Capillary: 58 cm × 50 μm fused-silica (50 cm to detector)
Running buffer: Isopropanol:buffer 2.5:97.5 (Buffer was 80 mM pH 6.8 Phosphate buffer containing 15 mM cetyltrimethylammonium bromide.)
Injection: Hydrostatic injection for 15 s
Detector: UV 214
Migration time: 14
Voltage: -20 kV
Model: Waters Quanta 4000

OTHER SUBSTANCES
Simultaneous: acebutolol, alprenolol, atenolol, labetalol, nadolol, oxprenolol, propranolol, sotalol, timolol

REFERENCE

Lukkari,P.; Vuorela,H.; Riekkola,M.-L. Effects of organic mobile phase modifiers on elution and separation of β-blockers in micellar electrokinetic capillary chromatography, *J.Chromatogr.A*, **1993**, *655*, 317–324.

SAMPLE

Matrix: solutions
Sample preparation: Prepare a 1 mg/mL solution in 40 mM pH 5.1 sodium phosphate buffer.

CAPILLARY ELECTROPHORESIS

Capillary: 11.5 cm × 75 μm fused silica tubing (8.5 cm to the detector) (Polymicro Technologies, Phoenix, AZ) coated on the inner surface with non-cross-linked polyacrylamide. Coat as follows. Mix 80 μL gamma-methacryloxypropyltrimethoxysilane with 20 mL water adjusted to pH 3.5 with acetic acid, suck into capillary, let stand for 1 h, remove solution, wash with water fill capillary with a 3 or 4% solution of acrylamide containing 1 mL/L N,N,N',N'-tetramethylethyl-enediamine and 1 g/L potassium persulfate, let stand for 30 min, remove excess solution, wash capillary with water, suck capillary dry, dry at 35° (J. Chromatogr. 1985, 347, 191).
Running buffer: Isopropanol:400 mM pH 5.1 sodium phosphate buffer 25:75
Injection: Fill 7 cm of tube with running buffer, fill rest of tube with 40 mg/mL cellobiohydrolase I in buffer, add a 2-3 mm plug of agarose by pressing end of tube into agarose gel, inject sample at 1000 V for 20 s. (Buffer was 400 mM pH 5.1 sodium phosphate buffer. Prepare agarose gel by boiling 20 mg agarose in 1 mL buffer, cool, allow to gel.)
Detector: UV 220
Migration time: 19, 21 (enantiomers)
Voltage: 1000 V
Model: Laboratory constructed

OTHER SUBSTANCES

Also analyzed: alprenolol, metoprolol, propranolol

KEY WORDS

detector is at cathode; chiral; coated capillary

REFERENCE

Valtcheva,L.; Mohammad,J.; Pettersson,G.; Hjertén,S. Chiral separation of β-blockers by high-performance capillary electrophoresis based on non-immobilized cellulase as enantioselective protein, *J.Chromatogr.*, **1993**, *638*, 263–267.

SAMPLE

Matrix: solutions

CAPILLARY ELECTROPHORESIS

Capillary: 100 cm × 50 μm fused-silica (50 cm to detector) (Isco)
Capillary preparation: Flush capillary with 10 μL running buffer between runs. Every 40 sample injections rinse capillary with 200 μL 1 M NaOH, with 200 μL water, and 200 μL running buffer. Before use fill capillary with 1 M NaOH and allow to stand for 1 h, fill with 100 mM NaOH, allow to stand for 1 h, wash with water fill with running buffer.
Capillary temperature: 23
Running buffer: 100 mM citric acid containing 19.27 mM Na_2HPO_4 and 120 mM hydroxypropyl-β-cyclodextrin
Injection: Inject under vacuum at 4.0 kPa.s
Detector: UV 200
Migration time: 33, 33.5 (enantiomers)
Voltage: 30 kV
Model: Isco Model 3140

OTHER SUBSTANCES

Simultaneous: albuterol, alprenolol, atenolol, cimaterol, clenbuterol, labetalol, nadolol, oxpren-olol, pirbuterol, propranolol, terbutaline
Interfering: atenolol, clenbuterol

KEY WORDS

chiral

REFERENCE
Aumatell,A.; Wells,R.J.; Wong,D.K.Y. Enantiomeric differentiation of a wide range of pharmacologically active substances by capillary electrophoresis using modified β-cyclodextrins, *J.Chromatogr.A*, **1994**, *686*, 293–307.

SAMPLE
Matrix: solutions
Sample preparation: Prepare a 50 μg/mL solution in diluted running buffer, inject an aliquot.

CAPILLARY ELECTROPHORESIS
Capillary: 44 cm × 50 μm fused-silica
Capillary preparation: After each run wash capillary with water for 2 min and with running buffer for 3 min. At the beginning of each day wash capillary with water and running buffer for 5 min. Condition a new capillary with 1 M NaOH, 100 mM NaOH, water, and separation buffer.
Capillary temperature: 15
Running buffer: 100 mM Phosphoric acid containing 15 mM heptakis(2,6-di-O-methyl)-β-cyclodextrin adjusted to pH 3.0 with triethanolamine
Injection: Hydrodynamic injection for 1 s
Detector: UV 210
Voltage: 25 kV
Model: Spectra Physics Spectraphoresis 1000

KEY WORDS
chiral; enantiomer resolution 1.4

REFERENCE
Bechet,I.; Paques,P.; Fillet,M.; Hubert,P.; Crommen,J. Chiral separation of basic drugs by capillary zone electrophoresis with cyclodextrin additives, *Electrophoresis*, **1994**, *15*, 818–823.

SAMPLE
Matrix: solutions

CAPILLARY ELECTROPHORESIS
Capillary: 60 cm × 50 μm AccuSep (52.5 cm to detector) (Waters)
Capillary preparation: Before injection rinse capillary with 100 mM NaOH for 3 min and with running buffer for 3 min. Rinse new capillaries with 500 mM NaOH for 5 min
Running buffer: 50 mM pH 7.0 Na$_2$HPO$_4$ containing 25 mM (S)-N-dodecoxycarbonylvaline (Prepare (S)-N-dodecoxycarbonylvaline as follows. Prepare dodecyl chloroformate by reacting 1-dodecanol with 0.33 equivalents of triphosgene in dichloromethane solution in the presence of pyridine. Add dodecyl chloroformate dropwise to (S)-valine in 1 M NaOH solution, filter, wash with hexane, recrystallize from ether/petroleum ether.)
Injection: Hydrostatic injection 2 s
Detector: UV 214
Voltage: +12 kV
Model: Waters Quanta 4000 or 4000E

OTHER SUBSTANCES
Also analyzed: atenolol, bupivacaine, ephedrine, homatropine, ketamine, N-methylpseudoephedrine, metoprolol, norephedrine, norphenylephrine, octopamine, terbutaline

KEY WORDS
chiral; α = 1.09

REFERENCE
Mazzeo,J.R.; Grover,E.R.; Swartz,M.E.; Petersen,J.S. Novel chiral surfactant for the separation of enantiomers by micellar electrokinetic capillary chromatography, *J.Chromatogr.A*, **1994**, *680*, 125–135.

SAMPLE
Matrix: solutions

CAPILLARY ELECTROPHORESIS
Capillary: 62 cm × 52 μm fused-silica (50 cm to detector) (Polymicro Technologies)
Running buffer: 150 mM pH 2.5 Trimethylammonium phosphate buffer containing 20 mM hydroxypropyl-β-cyclodextrin
Injection: Gravity injection at 10 cm for 5 s
Migration time: 42, 43 (enantiomers)
Voltage: 20 kV
Current: 82 μA
Model: Laboratory constructed

OTHER SUBSTANCES
Simultaneous: atenolol, nadolol

KEY WORDS
chiral

REFERENCE
Quang,C.; Khaledi,M.G. Direct separation of the enantiomers of β-blockers by cyclodextrin-mediated capillary zone electrophoresis, *J.High Res.Chromatogr.*, **1994**, *17*, 99–101.

SAMPLE
Matrix: solutions

CAPILLARY ELECTROPHORESIS
Capillary: 64 cm × 75 μm (48 cm to detector) (Polymicro Technologies)
Running buffer: MeOH:buffer 0.6:99.4 (Buffer was 14 mM pH 4.5 6-aminocaproic acid buffer containing 8 mM adipic acid, 3% dextran (MW 39000), 0.05% polyethylene oxide (MW 300000), and 7.5 mM heptakis(2,6-di-O-methyl)-β-cyclodextrin
Injection: Hydrodynamic injection at 20 cm
Detector: UV 220
Migration time: 15, 15.5 (enantiomers)
Voltage: 22 kV
Model: Laboratory constructed

OTHER SUBSTANCES
Simultaneous: disopyramide, nitrendipine

KEY WORDS
chiral

REFERENCE
Soini,H.; Riekkola,M.-L.; Novotny,M.V. Mixed polymer networks in the direct analysis of pharmaceuticals in urine by capillary electrophoresis, *J.Chromatogr.A*, **1994**, *680*, 623–634.

SAMPLE
Matrix: solutions
Sample preparation: Inject an aliquot of a 100 μg/mL solution in water:running buffer 50:50.

CAPILLARY ELECTROPHORESIS
Capillary: 44.5 cm × 50 μm acrylamide-coated fused-silica (Bio-Rad)
Capillary temperature: 30
Running buffer: 100 mM NaH_2PO_4 containing 15 mM gamma-cyclodextrin, adjusted to pH 2.5 with phosphoric acid
Injection: Electrokinetic injection at 8 kV for 6 s.
Detector: UV 200
Migration time: 9.28
Voltage: 14 kV
Model: Bio-Rad BioFocus 3000

OTHER SUBSTANCES
Also analyzed: albuterol, alprenolol, atenolol, atropine, baclofen, bamethan, benserazide, biperiden, bisoprolol, bupivacaine, bupranolol, butetamate, carazolol, carbuterol, carvedilol, celiprolol, chloroquine, chlorpheniramine (chlorphenamine), clidinium bromide, clobutinol, disopyramide, dobutamine, flecainide, homatropine, ipratropium bromide, isoproterenol, isothipendyl, ketamine, mefloquine, mequitazine, metaproterenol (orciprenaline), metipranolol, nafronyl (naftidrofuryl), nefopam, ofloxacin, orphenadrine, oxomemazine, oxprenolol, phenoxybenzamine, pholedrine, pirbuterol, prilocaine, promethazine, propafenone, propranolol, sotalol, synephrine, terbutaline, tetrahydrozoline (tetryzoline), tocainide, trihexyphenidyl, trimeprazine (alimemazine), trimipramine, tropicamide, verapamil, zopiclone

KEY WORDS
achiral; coated capillary

REFERENCE
Koppenhoefer,B.; Epperlein,U.; Christian,B.; Yibing,J.; Yuying,C.; Bingcheng,L. Separation of enantiomers of drugs by capillary electrophoresis. I. γ-Cyclodextrin as chiral solvating agent, *J.Chromatogr.A*, **1995**, *717*, 181–190.

SAMPLE
Matrix: solutions
Sample preparation: Inject an aliquot of a solution in 25 mM pH 6.8 potassium phosphate buffer.

CAPILLARY ELECTROPHORESIS
Capillary: 23.5 × 75 μm gel-filled fused-silica (16.5 cm to detector) (Polymicro Technologies)
Capillary preparation: Wash a new capillary with 100 μL water, 100 μL 1 M NaOH, 100 μL water, and 100 μL 50 mM pH 5.0 sodium citrate buffer. Pump gel mixture through capillary at 15 μL/min until it is within 1 cm of detector (detection window is filled with buffer, not gel), make sure it contains no bubbles (use a microscope), condition overnight at 2 kV (current limitation 20 μA) at reverse polarity using running buffer. (Gel mixture was 5 parts 20% bovine serum albumin in water, 21 parts 20% cellulase (from Trichoderma reesei) in 50 mM pH 5.0 sodium citrate buffer, and 4 arts 50% glutaraldehyde.)
Running buffer: Isopropanol:50 mM pH 6.8 potassium phosphate buffer 1:99
Injection: Electrokinetic injection at 2 kV for 3 s
Detector: UV 214
Migration time: 8,9 (enantiomers)
Voltage: 3.5 kV
Current: 45 μA
Model: Beckman P/ACE 2050

OTHER SUBSTANCES
Simultaneous: atenolol
Interfering: metoprolol

KEY WORDS
chiral

REFERENCE
Ljungberg,H.; Nilsson,S. Protein-based capillary affinity gel electrophoresis for chiral separation of β-adrenergic blockers, *J.Liq.Chromatogr.*, **1995**, *18*, 3685–3698.

SAMPLE
Matrix: solutions
Sample preparation: Prepare a solution in MeOH/water, inject an aliquot.

CAPILLARY ELECTROPHORESIS
Capillary: 62 cm × 52 μm fused-silica (50 cm to detector) (Polymicro Technologies)
Capillary temperature: 40
Running buffer: 50 mM pH 2.50 Tetrabutylammonium phosphate containing 20 mM dimethyl-β-cyclodextrin

Injection: Siphon injection at 10 cm for 5 s
Detector: UV (wavelength not specified)
Migration time: 17.43, 17.71 (enantiomers)
Voltage: 20 kV
Model: Laboratory constructed

OTHER SUBSTANCES
Simultaneous: alprenolol, atenolol, chlorpheniramine, epinephrine, isoproterenol, labetalol, nor-epinephrine, oxprenolol, propranolol

KEY WORDS
chiral

REFERENCE
Quang,C.; Khaledi,M.G. Extending the scope of chiral separation of basic compounds by cyclodextrin-mediated capillary zone electrophoresis, *J.Chromatogr.A*, **1995**, *692*, 253–265.

SAMPLE
Matrix: solutions

CAPILLARY ELECTROPHORESIS
Capillary: 67 cm × 75 μm (60 cm to detector) (Polymicro Technologies)
Capillary preparation: Coat capillary as follows. Treat overnight with 50% 3-(trimethoxysilyl)propyl methacrylate in MeOH, fill capillary with degassed 40 g/L (acrylamide + N,N'-methylenebisacrylamide) in 100 mM pH 8.2 Tris-borate buffer containing 2 mM EDTA, 5 μL/L 10% ammonium persulfate, and 5 μL/mL N,N,N',N'-tetramethylethylenediamine, after polymerization force out gel leaving coating sticking to wall (Caution! Acrylamide is a carcinogen!), rinse with water, remove water by aspiration, dry at 35°.
Running buffer: 20 mM pH 2.5 Citric acid containing 2% carboxymethyl β-cyclodextrin
Detector: UV 214
Migration time: 12.8, 13 (enantiomers)
Voltage: 240 V/cm
Model: Beckman P/ACE 2000

OTHER SUBSTANCES
Simultaneous: arterenol, dimethindene, doxylamine, ephedrine, propranolol

KEY WORDS
chiral; coated capillary

REFERENCE
Schmitt,T.; Engelhardt,H. Optimization of enantiomeric separations in capillary electrophoresis by reversal of the migration order and using different cyclodextrins, *J.Chromatogr.A*, **1995**, *697*, 561–570.

SAMPLE
Matrix: solutions

CAPILLARY ELECTROPHORESIS
Capillary: 36 cm × 50 μm polyacrylamide-coated fused-silica (31.5 cm to detector) (Polymicro Technologies)
Capillary preparation: Before each injection rinse with water at 100 psi for 30 s, rinse with running buffer at 100 psi for 30 s, partially fill with separation solution (500 μM ovomucoid in running buffer) at 1 psi for 190 s (27 cm), inject sample, electrophorese with running buffer. Coat capillary with polyacrylamide as follows. Treat with 1 M NaOH for 1 h, rinse with water, dry at 110° by flushing with nitrogen for 6 h, pass thionyl chloride through using suction for 10 min, seal at each end, heat at 70° for 6 h, open, suck 250 mM vinylmagnesium bromide in THF into the capillary, seal at each end, heat at 70° for 6 h, open, rinse with THF for several min, rinse with water, fill by suction with a solution containing 3% acrylamide, 0.1% ammonium persulfate, and 0.1% tetramethylethylenediamine, heat at 28 ± 2° for 1 h, after 1 h wash with water (Biol.Pharm.Bull. 1993, 16, 1185).
Capillary temperature: 25

Running buffer: EtOH:50 mM pH 6.0 phosphate buffer 8:92
Injection: Inject at 1 psi for 2 s
Detector: UV 210
Migration time: 10.93, 11.36 (enantiomers)
Voltage: 12 kV
Model: Bio-Rad BioFocus 3000

KEY WORDS
chiral; partial separation zone technique; coated capillary

REFERENCE
Tanaka,Y.; Terabe,S. Partial separation zone technique for the separation of enantiomers by affinity electrokinetic chromatography with proteins as chiral pseudo-stationary phases, *J.Chromatogr.A*, **1995**, *694*, 277–
284.

SAMPLE
Matrix: solutions

CAPILLARY ELECTROPHORESIS
Capillary: 57 cm × 75 μm fused-silica (50 cm to detector) (Beckman)
Capillary preparation: Before each run rinse capillary with 100 mM NaOH and running buffer.
Capillary temperature: 30
Running buffer: Acetone:100 mM pH 8.1 borate buffer containing 50 mM sodium dodecyl sulfate
15:85 (A) or 100 mM phosphoric acid adjusted to pH 3.1 with triethanolamine (B)
Injection: Pressure injection for 5-10 s.
Detector: UV 200
Migration time: 10.5 (A), 8.8 (B)
Voltage: 25 kV
Model: Beckman P/ACE 5510

OTHER SUBSTANCES
Simultaneous: acebutolol, amiodarone, atenolol, bretylium, captopril, diltiazem, disopyramide,
lidocaine, lisinopril, metoprolol, nicardipine, nifedipine, phenytoin, propafenone, propranolol,
quinidine, sotalol, timolol, p-toluenesulfonic acid, verapamil

KEY WORDS
only running buffer A separates all compounds

REFERENCE
Bretnall,A.E.; Clarke,G.S. Selectivity of capillary electrophoresis for the analysis of cardiovascular drugs,
J.Chromatogr.A, **1996**, *745*, 145–154.

SAMPLE
Matrix: solutions

CAPILLARY ELECTROPHORESIS
Capillary: 43 cm × 50 μm fused-silica (36 cm to detector)
Capillary preparation: Before each injection wash with running buffer for 10 min. After each
injection wash with water for 2 min. Wash new capillaries with 1 M NaOH at 60° for 50 min,
with NaOH solution (?) at 60° for 10 min, and with water at 25°.
Capillary temperature: 25
Running buffer: 320 mM pH 2.0 Citrate buffer
Injection: Hydrodynamic injection at 1.5 psi for 2 s.
Detector: UV 220
Migration time: 8.6
Voltage: 15 kV
Model: Spectra-Physics Model 1000
Limit of detection: 1-18 μg/mL

OTHER SUBSTANCES
Simultaneous: acebutolol, atenolol, labetalol, levobunolol, metoprolol, nadolol, oxprenolol, propranolol, timolol

REFERENCE
Lin,C.-E.; Chang,C.-C.; Lin,W.-c.; Lin,E.C. Capillary zone electrophoretic separation of β-blockers using citrate buffer at low pH, *J.Chromatogr.A*, **1996**, *753*, 133–138.

SAMPLE
Matrix: solutions

CAPILLARY ELECTROPHORESIS
Capillary: 52 cm × 75 μm fused-silica (48 cm to detector) (Polymicro Technologies)
Capillary preparation: Before each run purge with running buffer for 3 min. Every 3 runs purge with 100 mM NaOH for 5 min. Purge new capillaries with 1 M NaOH for 20 min and with 100 mM NaOH for 20 min, rinse with running buffer, equilibrate with running buffer at 12 kV for 3 h.
Running buffer: 100 mM CHES (2-(N-cyclohexylamine)ethanesulfonic acid) containing 10 mM triethylamine and 25 mM (R)-dodecoxycarbonylvaline (Waters EnantioSelect (R)-Val-1), pH adjusted to 8.8 with 1 M NaOH
Injection: Hydrostatic injection for 2 s.
Detector: UV 214
Migration time: 11.2, 11.3 (enantiomers)
Voltage: 12 kV
Current: ≤30 μA
Model: Waters Quanta 4000

OTHER SUBSTANCES
Simultaneous: atenolol, ephedrine, methylpseudoephedrine

KEY WORDS
chiral

REFERENCE
Peterson,A.G.; Ahuja,E.S.; Foley,J.P. Enantiomeric separations of basic pharmaceutical drugs by micellar electrokinetic chromatography using a chiral surfactant, N-dodecoxycarbonylvaline, *J.Chromatogr.B*, **1996**, *683*, 15–28.

SAMPLE
Matrix: solutions
Sample preparation: Inject an aliquot of a solution in running buffer.

CAPILLARY ELECTROPHORESIS
Capillary: 60 cm × 75 μm fused-silica (52.4 cm to detector)
Capillary preparation: After each run flush with 500 mM KOH for 2-3 min then with water.
Running buffer: 10 mM pH 3.8 Phosphate buffer containing 2% sulfated cyclodextrin (ds 7-10)
Injection: Hydrostatic injection.
Detector: UV 214
Migration time: 10.87, 11.12 (enantiomers)
Voltage: 15 kV
Model: Waters Quanta 4000

OTHER SUBSTANCES
Also analyzed: acebutolol, alprenolol, aminoglutethimide, brompheniramine, bupivacaine, bupropion, canadine, carbinoxamine, chloroquine, chlorpheniramine, dimethindene, disopyramide, doxylamine, hydroxychloroquine, idazoxan, isoxsuprine, ketamine, mepenzolate, mepivacaine, methoxyphenamine, mexiletine, midodrine, nefopam, orphenadrine, oxprenolol, oxyphencyclimine, pheniramine, phensuximide, piperoxan, terbutaline, tetramisole, tolperisone, tranylcypromine, trihexyphenidyl, trimipramine, verapamil, warfarin

KEY WORDS
chiral; detector at anode

REFERENCE
Stalcup,A.M.; Gahm,K.H. Application of sulfated cyclodextrins to chiral separations by capillary zone electrophoresis, *Anal.Chem.*, **1996**, *68*, 1360–1368.

SAMPLE
Matrix: solutions
Sample preparation: Inject an aliquot of a 100 μg/mL solution in running buffer.

CAPILLARY ELECTROPHORESIS
Capillary: 30 cm × 50 μm fused-silica (25.5 cm to detector) (Yongnian Optical Conductive Fiber Plant, China), coated with polyacrylamide
Capillary preparation: No details of the polyacrylamide coating process are provided. However, another paper (LC.GC 1997, 15, 40) by this group indicates that they use the procedure of Hjertén, thus: Adjust the pH of 20 mL water to 3.5 with acetic acid, add 80 μL 3-(trimethoxysilyl)propyl methacrylate (3-methacryloxypropyltrimethoxysilane), mix, suck into capillary, let stand at room temperature for 1 h, remove the solution, wash with water. Fill the capillary with a deaerated 3-4% acrylamide solution containing 1 μL/mL N,N,N',N'-tetramethylethylenediamine and 1 mg/mL potassium persulfate, let stand for 30 min, remove excess solution by aspiration, rinse with water, remove water by aspiration, dry at 35° (J. Chromatogr. 1985, 347, 191).
Capillary temperature: 25
Running buffer: 100 mM NaH_2PO_4 adjusted to pH 2.5 (A) or 100 mM NaH_2PO_4 containing 45 mM hydroxypropyl-gamma-cyclodextrin, adjusted to pH 2.5 (B)
Injection: Electrokinetic injection at 15 kV for 3 s.
Detector: UV 200, UV 210
Migration time: 4.55 (A), 8.85, 8.94 (B, enantiomers)
Voltage: 15 kV
Model: Bio-Rad BioFocus 3000

OTHER SUBSTANCES
Simultaneous: amorolfine, brompheniramine, bupivacaine, carteolol, chloroquine, chlorpheniramine, chlorphenoxamine, disopyramide, dobutamine, doxylamine, flecainide, gallopamil, ketamine, mepindolol, orphenadrine, oxybutynin, phenoxybenzamine, propafenone, propranolol, sulpiride, talinolol, tropicamide, verapamil

KEY WORDS
chiral; coated capillary

REFERENCE
Koppenhoefer,B.; Epperlein,U.; Xiaofeng,Z.; Bingcheng,L. Separation of enantiomers of drugs by capillary electrophoresis. Part 4: Hydroxypropyl-γ-cyclodextrin as chiral solvating agent, *Electrophoresis*, **1997**, *18*, 924–930.

SAMPLE
Matrix: solutions
Sample preparation: Inject an aliquot of a 100 μg/mL solution in running buffer.

CAPILLARY ELECTROPHORESIS
Capillary: 30 cm × 50 μm fused-silica (25.5 cm to detector), coated with polyacrylamide
Capillary preparation: Adjust the pH of 20 mL water to 3.5 with acetic acid, add 80 μL 3-(trimethoxysilyl)propyl methacrylate (3-methacryloxypropyltrimethoxysilane), mix, suck into capillary, let stand at room temperature for 1 h, remove the solution, wash with water. Fill the capillary with a deaerated 3-4% acrylamide solution containing 1 μL/mL N,N,N',N'-tetramethylethylenediamine and 1 mg/mL potassium persulfate, let stand for 30 min, remove excess solution by aspiration, rinse with water, remove water by aspiration, dry at 35° (J. Chromatogr. 1985, 347, 191).
Capillary temperature: 25
Running buffer: 100 mM NaH_2PO_4 containing 45 mM hydroxypropyl-α- cyclodextrin, adjusted to pH 2.5 with phosphoric acid
Injection: Electrokinetic injection at 15 kV for 3 s.
Detector: UV 200
Voltage: 15 kV
Model: Bio-Rad BioFocus 3000

KEY WORDS
chiral; coated capillary; comparison with the use of other cyclodextrins; this running buffer gave the greatest enantiomeric separation.; α=1.022

REFERENCE
Lin,B.; Zhu,X.; Koppenhoefer,B.; Epperlein,U. Investigation of 123 chiral drugs by cyclodextrin-modified capillary electrophoresis, *LC.GC*, **1997**, *15*, 40–46.

SAMPLE
Matrix: solutions

CAPILLARY ELECTROPHORESIS
Capillary: 67 cm × 50 μm (60 cm to detector)
Capillary preparation: Wash with running buffer for 5 min before each injection. After each injection wash with 1 M NaOH at 60° for 5 min, with 100 mM NaOH at 60° for 10 min, and with water at 25° for 5 min.
Capillary temperature: 25
Running buffer: 70 mM pH 7.0 Sodium phosphate buffer containing 15 mM cetyltriemthylammonium bromide
Injection: Hydrodynamic injection for 1 s.
Detector: UV 220
Migration time: 16.8
Voltage: 20 kV

OTHER SUBSTANCES
Simultaneous: acebutolol, atenolol, labetalol, levobunolol, metoprolol, nadolol, oxprenolol, propranolol, timolol

REFERENCE
Lin,C.-E.; Chen,Y.-C.; Chang,C.-C.; Wang,D.-Z. Migration behavior and selectivity of β-blockers in micellar electrokinetic chromatography. Influence of micelle concentration of cationic surfactants, *J.Chromatogr.A*, **1997**, *775*, 349–357.

SAMPLE
Matrix: solutions

CAPILLARY ELECTROPHORESIS
Capillary: 50.5 cm × 75 μm coated fused-silica (36 cm to detector) (Polymicro Technologies)
Capillary preparation: Coat capillary as follows. Condition capillary with 1 M NaOH for 1 h, rinse with water for 30 min, force 1% methacryloxypropyltrimethoxysilane (adjusted to pH 3.5 with acetic acid) through the capillary using pressure, allow to react at room temperature for 2 h, rinse with water, fill capillary with diallyldimethylammonium chloride solution containing 1 μL/mL N,N,N',N'-tetramethylethylenediamine and 1 mg/mL ammonium persulfate, allow to react until polymerization is complete, rinse with water, dry at 30-40° overnight.
Running buffer: 10 mM pH 5.0 Acetate buffer
Injection: Hydrodynamic injection at 5 cm for 5 s.
Detector: UV 210
Migration time: 11
Voltage: 12 kV
Model: laboratory-constructed

OTHER SUBSTANCES
Simultaneous: acebutolol, atenolol, dilevalol, nifenalol

KEY WORDS
coated capillary

REFERENCE
Liu,Q.; Lin,F.; Hartwick,R.A. Poly(diallyldimethylammonium chloride) as a cationic coating for capillary electrophoresis, *J.Chromatogr.Sci.*, **1997**, *35*, 126–130.

SAMPLE
Matrix: solutions

CAPILLARY ELECTROPHORESIS
Capillary: 62.1 cm × 75 μm fused-silica (44.6 cm to detector) (Polymicro Technologies)
Capillary preparation: Coat capillary as follows. Condition capillary with 1 M NaOH for 1 h, rinse with water for 30 min, fill capillary with 1% 3-(trimethoxysilyl)propyl methacrylate (methacryloxypropyltrimethoxysilane) (adjusted to pH 3.5 with acetic acid), allow to react at room temperature for 2 h, rinse with water, fill capillary with 10% 2-aminoethyl methacrylate hydrochloride (Fisher Scientific) solution containing 1-3 mg/mL N,N,N',N'-tetramethylethylenediamine and 1 mg/mL ammonium persulfate, allow to react until polymerization is complete, rinse with water, dry at 30-40° overnight.
Running buffer: 25 mM pH 4.7 Acetate buffer
Injection: Hydrodynamic injection at 5 cm for 5 s.
Detector: UV 210
Migration time: 12.42
Voltage: 15 kV
Model: laboratory constructed

OTHER SUBSTANCES
Simultaneous: acebutolol, atenolol, betaxolol, carteolol, dilevalol, nifenalol, propranolol

KEY WORDS
coated capillary; detector at anode

REFERENCE
Liu,Q.; Lin,F.; Hartwick,R.A. Free solution capillary electrophoretic separation of basic proteins and drugs using cationic polymer coated capillaries, *J.Liq.Chromatogr.Rel.Technol.*, **1997**, *20*, 707–718.

SAMPLE
Matrix: solutions
Sample preparation: Inject an aliquot of a 100 μM solution in water.

CAPILLARY ELECTROPHORESIS
Capillary: 65 cm × 50 μm fused-silica (56.5 cm to detector) (Polymicro Technologies)
Capillary preparation: Before each injection rinse capillary at 930 mbar with 100 mM NaOH for 1 min, with water for 2 min, and with running buffer for 4 min.
Capillary temperature: 20
Running buffer: 50 mM Phosphoric acid containing 30 mM heptakis(2,6-di-O-methyl)-β-cyclodextrin, adjusted to pH 2.5 with triethanolamine
Injection: Pressure injection at 20 mbar for 3 s.
Detector: UV 195
Migration time: 18.6 (R_S = 1.7)
Voltage: 24 kV
Model: Hewlett Packard HP^{3D}

KEY WORDS
chiral

REFERENCE
Nilsson,S.; Schweitz,L.; Petersson,M. Three approaches to enantiomer separation of β-adrenergic antagonists by capillary electrochromatography, *Electrophoresis*, **1997**, *18*, 884–890.

SAMPLE
Matrix: solutions

CAPILLARY ELECTROPHORESIS
Capillary: 52-55 cm × 75 μm fused-silica (45-48 cm to detector) (Polymicro Technologies)
Capillary preparation: After each run purge with running buffer for 3 min. After every 5 runs purge with 100 mM LiOH. Purge new capillaries with 1 M LiOH for 20 min, purge with 100 mM LiOH for 20 min, rinse with running buffer, equilibrate with running buffer with voltage applied for 3 h.
Running buffer: MeCN:buffer 15:85 (Prepare a 100 mM CHES (2-[N-cyclohexylamine]ethanesulfonic acid) solution containing 25 mM (S)-N-dodecoxycarbonylvaline (EnantioSelect (S)-

Val-1, Waters), adjust pH to 7 with 1 M LiOH, add triethylamine to a final concentration of 10 mM, add MeCN to a final concentration of 15%, adjust pH to 8.8 with 1 M LiOH. Preparation of (S)-N-dodecoxycarbonylvaline is as follows. Prepare dodecyl chloroformate by reacting 1-dodecanol with 0.33 equivalents of triphosgene in dichloromethane solution in the presence of pyridine. Add dodecyl chloroformate dropwise to (S)-valine in 1 M NaOH solution, filter, wash with hexane, recrystallize from ether/petroleum ether (J. Chromatogr. 1994, 680, 125).)
Injection: Hydrostatic injection for 2 s.
Detector: UV 214
Migration time: 14, 14.3 (enantiomers)
Voltage: 12 kV
Model: Waters Quanta 4000 CE

OTHER SUBSTANCES
Simultaneous: metoprolol

KEY WORDS
chiral

REFERENCE
Peterson,A.G.; Foley,J.P. Influence of the inorganic counterion on the chiral micellar electrokinetic separation of basic drugs using the surfactant N-dodecoxycarbonylvaline, *J.Chromatogr.B*, **1997**, *695*, 131–145.

SAMPLE
Matrix: solutions

CAPILLARY ELECTROPHORESIS
Capillary: 36 cm × 50 μm fused-silica coated with linear polyacrylamide (31.5 cm to detector) (GL Science)
Capillary preparation: At the beginning and end of each day rinse capillary with capillary wash solution (Bio-Rad Cat. No. 148-5022) at 690 kPa for more than 3 min and with water at 690 kPa for more than 3 min. Coat capillary as follows. Treat capillary with 1 M NaOH at room temperature for 1 h, rinse with water, dry by passing nitrogen gas through the capillary at 110° for 6 h. Pass thionyl chloride through the capillary using a suction pump for several min, seal capillary at both ends and heat at 70° for 6 h. Unseal the capillary and fill with 250 mM vinyl magnesium bromide in THF by suction, seal the capillary, heat at 70° for 6 h. Open the capillary and rinse it with THF for several min, rinse with distilled water, fill the capillary with polymerization solution, heat at 28 ± 2° for 1 h, condition at -100 V/cm for 30 min (Anal. Sci. 1994, 10, 1). (The polymerization solution was 5% acrylamide in water containing 49 mM Tris, 384 mM glycine, and 0.1% sodium dodecyl sulfate, degas in an ultrasonic bath. Add 40 μL 10% N,N,N',N'-tetramethylethylenediamine and 10 μL 10% ammonium persulfate to 5 mL of the degassed solution, mix thoroughly.)
Running buffer: EtOH:50 mM pH 6.0 Phosphate buffer 10:90
Injection: Before each injection rinse with water at 690 kPa for 30 s, rinse with running buffer at 690 kPa for 30 s, partially fill with separation solution (100 μM α_1-acid glycoprotein (Cohn fraction VI) (Fluka) in running buffer) at 6.9 kPa for 190 s (27 cm), inject sample at 6.9 kPa for 2 s, electrophorese with running buffer (Note that α_1-acid glycoprotein from other suppliers may provide inferior results).
Detector: UV 210
Migration time: Resolution of enantiomers 7.0
Voltage: 12 kV
Model: Bio-rad BioFocus 3000

KEY WORDS
chiral; coated capillary

REFERENCE
Tanaka,Y.; Terabe,S. Separation of the enantiomers of basic drugs by affinity capillary electrophoresis using a partial filling technique and α_1-acid glycoprotein as chiral selector, *Chromatographia*, **1997**, *44*, 119–128.

SAMPLE
Matrix: solutions

CAPILLARY ELECTROPHORESIS
Capillary: 70 cm × 50 μm fused-silica (50 cm to detector) (Polymicro Technologies)
Capillary temperature: 30
Running buffer: 100 mM pH 3.9 Sodium acetate containing 5 mM sulfobutyl ether-β-cyclodextrin (average substitution 3.9, MW 1721, Center for Drug Delivery Research, Lawrence KS)
Injection: Pressure injection at 5 inches Hg for 1.5 s.
Detector: UV 210
Migration time: 18.6, 18.8 (enantiomers)
Voltage: 15 kV
Current: 26 μA
Model: Perkin Elmer-Applied Biosystems Model 270

KEY WORDS
chiral

REFERENCE
Xie,G.-h.; Skanchy,D.J.; Stobaugh,J.F. Chiral separations of enantiomeric pharmaceuticals by capillary electrophoresis using sulphobutyl ether β-cyclodextrin as isomer selector, *Biomed.Chromatogr.*, **1997**, *11*, 193–199.

SAMPLE
Matrix: solutions

CAPILLARY ELECTROPHORESIS
Capillary: 35 cm × 50 μm polyacrylamide-coated fused-silica (30.5 cm to detector) (Composite Metal Services, UK)
Capillary preparation: Before each run rinse capillary with water for 70 s and with running buffer for 100 s. (Coat capillary as follows. Adjust the pH of 20 mL water to 3.5 with acetic acid, add 80 μL 3-(trimethoxysilyl)propyl methacrylate (3-methacryloxypropyltrimethoxysilane), mix, suck into capillary, let stand at room temperature for 1 h, remove the solution, wash with water. Fill the capillary with a deaerated 3-4% acrylamide solution containing 1 μL/mL N,N,N',N'-tetramethylethylenediamine and 1 mg/mL potassium persulfate, let stand for 30 min, remove excess solution by aspiration, rinse with water, remove water by aspiration, dry at 35° (J. Chromatogr. 1985, 347, 191).)
Capillary temperature: 25
Running buffer: Buffer containing 100 mM cyanoethylated-β-cyclodextrin (Cyclolab, Budapest) (Prepare buffer by adjusting the pH of 50 mM phosphoric acid containing 50 mM acetic acid and 50 mM boric acid to 2.5 with concentrated NaOH, add the appropriate amount of cyanoethylated-β-cyclodextrin, dilute with an equal volume of water.)
Injection: Pressure injection at 5 psi for 2 s.
Detector: UV 206
Migration time: 14.9 (second enantiomer, α = 1.014)
Voltage: 20 kV
Current: 27-40 μA
Model: Bio-Rad Biofocus 3000

KEY WORDS
chiral; coated capillary

REFERENCE
Aturki,Z.; Desiderio,C.; Mannina,L.; Fanali,S. Chiral separations by capillary zone electrophoresis with the use of cyanoethylated-β-cyclodextrin as chiral selector, *J.Chromatogr.A*, **1998**, *817*, 91–104.

SAMPLE
Matrix: solutions

CAPILLARY ELECTROPHORESIS
Capillary: 29-36 cm × 50 μm fused-silica (24.5-31.5 cm to detector) (Yongnian Optical Conductive Fiber Plant, China) coated with polyacrylamide
Capillary preparation: Coat capillary as follows. Adjust the pH of 20 mL water to 3.5 with acetic acid, add 80 μL 3-(trimethoxysilyl)propyl methacrylate (3-methacryloxypropyltrimethoxysilane), mix, suck into capillary, let stand at room temperature for 1 h, remove the solution, wash with water. Fill the capillary with a deaerated 3-4% acrylamide solution containing 1

μL/mL N,N,N',N'-tetramethylethylenediamine and 1 mg/mL potassium persulfate, let stand for 30 min, remove excess solution by aspiration, rinse with water, remove water by aspiration, dry at 35° (J. Chromatogr. 1985, 347, 191).
Capillary temperature: 25
Running buffer: 100 mM pH 2.5 NaH$_2$PO$_4$ (A) or 100 mM pH 2.5 NaH$_2$PO$_4$ containing 45 mM hydroxypropyl-α-cyclodextrin (Wacker, Munich) (B)
Injection: Electromigration at 15 kV for 3 s.
Detector: UV 200; UV 210
Migration time: 4.55 (A); 7.78, 7.95 (B) (enantiomers)
Voltage: 15 kV
Model: Bio-Focus 3000

OTHER SUBSTANCES
Also analyzed: albuterol (salbutamol), alprenolol, amorolfine, atenolol, atropine, azelastine, baclofen, bamethan, benproperine, benserazide, biperiden, bisoprolol, brompheniramine, bupivacaine, bupranolol, butamirate, butethamate, carazolol, carbuterol, carteolol, carvedilol, celiprolol, chloroquine, chlorpheniramine, chlorphenoxamine, cicletanine, clenbuterol, clidinium bromide, clobutinol, dimethindene, dipivefrin, disopyramide, dobutamine, doxylamine, fendiline, flecainide, gallopamil, homatropine, ipratropium bromide, isoproterenol (isoprenaline), isothipendyl, ketamine, meclizine, mefloquine, mepindolol, mequitazine, metaclazepam, metaproterenol (orciprenaline), metipranolol, metoprolol, nafronyl (naftidrofuryl), nefopam, nicardipine, norfenefrine, ofloxacin, ornidazole, orphenadrine, oxomemazine, oxprenolol, oxybutynin, phenoxybenzamine, phenylpropanolamine, pholedrine, pirbuterol, prilocaine, procyclidine, promethazine, propafenone, propranolol, reproterol, sotalol, sulpride, synephrine, talinolol, terbutaline, tetrahydrozoline (tetryzoline), theodrenaline, tioconazole, tocainide, trihexyphenidyl, trimeprazine (alimemazine), trimipramine, tropicamide, verapamil, zopiclone

---

KEY WORDS
coated capillary

REFERENCE
Koppenhoefer,B.; Eperlein,U.; Schlunk,R.; Zhu,X.; Lin,B. Separation of enantiomers of drugs by capillary electrophoresis. V. Hydroxypropyl-α-cyclodextrin as chiral solvating agent, *J.Chromatogr.A*, **1998**, *793*, 153–164.

SAMPLE
Matrix: solutions

CAPILLARY ELECTROPHORESIS
Capillary: 71.5 cm × 75 μm fused-silica (52.2 cm to detector) (Polymicro Technologies)
Capillary preparation: Rinse capillary with running buffer for 1 min between runs. Rinse new capillaries with 500 mM NaOH for 20 min and with water for 10 min and then condition with running buffer for 1 h.
Running buffer: 50 mM pH 3.0 containing 0.1% guaran (Guaran is the water-soluble fraction of guar gum. Prepare guaran by stirring 1 g guar gum (Sigma) in 100 mL water for 20 min, add 25 mL EtOH dropwise with stirring, let stand overnight, supercentrifuge. Add 30 mL EtOH dropwise with stirring to the centrifugate, centrifuge, collect the guaran as a precipitate, dry in air.)
Injection: Hydrodynamic injection at 5 cm for 5 s.
Detector: UV (wavelengths not given)
Migration time: 12.5
Voltage: 17.5 kV
Model: laboratory-constructed

OTHER SUBSTANCES
Simultaneous: acebutolol, atenolol, betaxolol, carteolol, dilevalol, nifenalol, propranolol

---

REFERENCE
Liu,Q.; Lin,F.; Hartwick,R.A. Capillary zone electrophoretic separation of basic proteins and drugs using guaran as a buffer modifier, *Chromatographia*, **1998**, *47*, 219–224.

SAMPLE
Matrix: solutions

Sample preparation: Inject an aliquot of a 340 μM solution.

CAPILLARY ELECTROPHORESIS
Capillary: 58 cm × 75 μm fused-silica (50 cm to detector)
Capillary preparation: Rinse with running buffer for 2 min before each run. Before use rinse capillary at 1.36 bar with 100 mM NaOH for 5 min and with water for 10 min.
Capillary temperature: 25
Running buffer: 20 mM pH 7.0 Sodium phosphate buffer (A) or 20 mM pH 7.0 Sodium phosphate buffer containing 10 mM 1-lauroyl-2-hydroxy-sn-glycero-3-phosphocholine (Avanti Polar Lipids, Alabaster AL) (B)
Injection: Hydrodynamic injection at 34 mbar for 1 s followed by 5% MeOH for 5 s.
Detector: UV 200
Migration time: 3.95 min (A); k' 0.23 (B)
Voltage: 15 kV
Current: ca. 50 μA
Model: Beckman P/ACE 2200

OTHER SUBSTANCES
Simultaneous: acebutolol, alprenolol, atenolol, metoprolol, oxprenolol, propranolol, timolol

---

REFERENCE
Masucci,J.A.; Caldwell,G.W.; Foley,J.P. Comparison of the retention behavior of β-blockers using immobilized artificial membrane chromatography and lysophospholipid micellar electrokinetic chromatography, *J.Chromatogr.A*, **1998**, *810*, 95–103.

SAMPLE
Matrix: urine
Sample preparation: Dilute urine with 2 volumes of water, add IS, filter (0.5 μm), inject an aliquot.

CAPILLARY ELECTROPHORESIS
Capillary: 68 cm × 50 μm fused-silica (60 cm to detector) (White Associates)
Capillary preparation: Purge capillary with running buffer for 2 min before each injection.
Running buffer: 80 mM pH 7.0 Phosphate buffer containing 10 mM N-cetyl-N,N,N-trimethyl-ammonium bromide
Injection: Hydrostatic injection for 30 s at 10 cm.
Detector: UV 214
Migration time: 15.9
Internal standard: 2,6-dimethylphenol (14.7)
Voltage: -26 kV (at injector end)
Model: Waters Quanta 4000
Limit of detection: 10 μg/mL (S/N 3)

OTHER SUBSTANCES
Extracted: acebutolol, alprenolol, atenolol, labetalol, metoprolol, nadolol, oxprenolol, propranolol, timolol
Simultaneous: probenecid
Noninterfering: caffeine

REFERENCE
Lukkari,P.; Sirén,H.; Pantsar,M.; Riekkola,M.-L. Determination of ten β-blockers in urine by micellar electrokinetic capillary chromatography, *J.Chromatogr.*, **1993**, *632*, 143–148.

Pipemidic acid

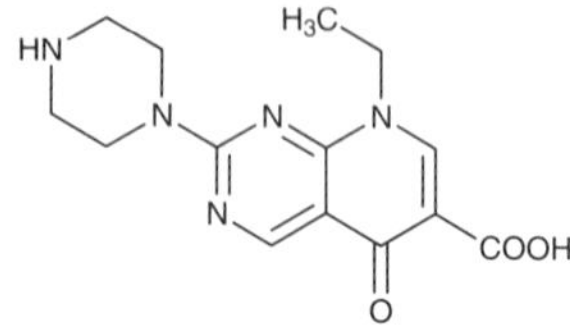

Molecular formula: $C_{14}H_{17}N_5O_3$
Molecular weight: 303.32
CAS Registry No.: 51940-44-4, 72571-82-5 (trihydrate)
Merck Index (12th ed.): 7613

SAMPLE
Matrix: solutions
Sample preparation: Inject an aliquot of a solution in MeOH.

CAPILLARY ELECTROPHORESIS
Capillary: 59 cm × 50 μm fused-silica (43 cm to detector) (Polymicro Technologies)
Capillary preparation: Before each analysis flush with water for 3 min, with 200 mM NaOH
  for 3 min, with water for 3 min, and with running buffer for 4 min. Flush new capillaries with
  1 M NaOH at 2000 mbar for 10 min then with 200 mM NaOH for 10 min.
Capillary temperature: 23
Running buffer: MeCN:buffer 28:72, pH 7.3 (Buffer was 32 mM sodium borate containing 18
  mM NaH_2PO_4, 39 mM sodium cholate, and 8 mM sodium heptanesulfonate.)
Injection: Hydrodynamic injection at 40 mbar for 6 s.
Detector: UV 260
Migration time: 4.45
Voltage: 30 kV
Model: Lauer Labs Prince

OTHER SUBSTANCES
Simultaneous: ciprofloxacin, enoxacin, flumequine, lomefloxacin, nalidixic acid, norfloxacin, ofloxacin, oxolinic acid, pefloxacin, piromidic acid, rosoxacin, sparfloxacin

REFERENCE
Sun,S.-W.; Chen,L.-Y. Optimization of capillary electrophoretic separation of quinolone antibacterials using the
  overlapping resolution mapping scheme, *J.Chromatogr.A*, **1997**, *766*, 215–224.

Piperacillin

Molecular formula: $C_{23}H_{27}N_5O_7S$
Molecular weight: 517.56
CAS Registry No.: 61477-96-1, 59703-84-3 (sodium salt)
Merck Index (12th ed.): 7616
Lednicer: 3 207; 4 179, 188

SAMPLE
Matrix: solutions
Sample preparation: Prepare a 0.5-2 mg/mL solution in water, inject an aliquot.

CAPILLARY ELECTROPHORESIS
Capillary: 65 cm × 50 μm fused-silica (50 cm to detector) (Scientific Glass Engineering)
Running buffer: 20 mM NaH_2PO_4 containing 150 mM sodium dodecyl sulfate adjusted to pH
  9.0 with 20 mM sodium tetraborate
Injection: Injection by siphon at 5 cm for 5-10 s
Detector: UV 210
Migration time: 11.4
Voltage: 20 kV

OTHER SUBSTANCES
Simultaneous: amoxicillin, ampicillin, aspoxicillin, carbenicillin, penicillin G, sulbenicillin
Also analyzed: cefmenoxime, cefminox, cefoperazone, cefotaxime, cefpimizole, cefpiramide, ceftazidime, ceftriaxone

REFERENCE
Nishi,H.; Tsumagari,N.; Kakimoto,T.; Terabe,S. Separation of β-lactam antibiotics by micellar electrokinetic chromatography, *J.Chromatogr.*, **1989**, *477*, 259–270.

SAMPLE
Matrix: solutions
Sample preparation: Inject directly.

CAPILLARY ELECTROPHORESIS
Capillary: 65 cm × 50 μm untreated fused-silica (50 cm to detector) (SGE)
Capillary preparation: Flush with running buffer every 5 runs. At the end of each day fill capillary with 100 mM KOH, let stand for 30 min, flush with water, let stand for 5 min, fill with running buffer.
Running buffer: pH 8.5 Buffer containing 100 mM sodium dodecyl sulfate (Prepare buffer by mixing 20 mM NaH_2PO_4 solution with 20 mM sodium borate solution to achieve a pH of 8.5.)
Injection: Inject by siphoning at 10 cm for 5-10 s.
Detector: UV 210
Migration time: 11.5

OTHER SUBSTANCES
Simultaneous: amoxicillin, ampicillin, aspoxicillin, carbenicillin, penicillin G, sulbenicillin

REFERENCE
Nishi,H.; Fukuyama,T.; Matsuo,M. Separation and determination of aspoxicillin in human plasma by micellar electrokinetic chromatography with direct sample injection, *J.Chromatogr.*, **1990**, *515*, 245–255.

SAMPLE
Matrix: solutions

CAPILLARY ELECTROPHORESIS
Capillary: 60 cm × 50 μm fused-silica (47 cm to detector) (Polymicro Technologies)
Capillary temperature: 25
Running buffer: 20 mM pH 8.5 Sodium tetraborate containing 100 mM sodium dodecyl sulfate
Injection: Hydrodynamic injection at 50 mbar for 3.6 s (5 nL)
Detector: UV 205
Migration time: 9.8
Voltage: 22 kV
Model: Crystal 310 (Thermo Unicam)

OTHER SUBSTANCES
Simultaneous: amoxicillin, ampicillin, cephapirin, cloxacillin, dicloxacillin, oxacillin, penicillin G, penicillin V, pyrimethamine, sulfacetamide, sulfadimethoxine, sulfaguanidine, sulfamerazine, sulfameter, sulfamethazine, sulfanilamide, sulfanilic acid, sulfapyridine, sulfaquinoxaline, sulfathiazole, sulfisoxazole, trimethoprim

REFERENCE
Hows,M.E.P.; Perrett,D.; Kay,J. Optimization of a simultaneous separation of sulphonamides, dihydrofolate reductase inhibitors and β-lactam antibiotics by capillary electrophoresis, *J.Chromatogr.A*, **1997**, *768*, 97–104.

Piperoxan

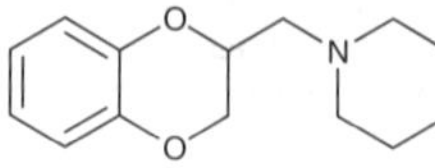

Molecular formula: $C_{14}H_{19}NO_2$
Molecular weight: 233.31
CAS Registry No.: 59-39-2
Merck Index (12th ed.): 7631
Lednicer: 1 352

SAMPLE
Matrix: solutions
Sample preparation: Inject an aliquot of a solution in running buffer.

CAPILLARY ELECTROPHORESIS
Capillary: 60 cm × 75 μm fused-silica (52.4 cm to detector)
Capillary preparation: After each run flush with 500 mM KOH for 2-3 min then with water.
Running buffer: 10 mM pH 3.8 Phosphate buffer containing 2% sulfated cyclodextrin (ds 7-10)
Injection: Hydrostatic injection.
Detector: UV 214
Migration time: 8.15, 8.61 (enantiomers)
Voltage: 15 kV
Model: Waters Quanta 4000

OTHER SUBSTANCES
Also analyzed: acebutolol, alprenolol, aminoglutethimide, brompheniramine, bupivacaine, bupropion, canadine, carbinoxamine, chloroquine, chlorpheniramine, dimethindene, disopyramide, doxylamine, hydroxychloroquine, idazoxan, isoxsuprine, ketamine, mepenzolate, mepivacaine, methoxyphenamine, mexiletine, midodrine, nefopam, orphenadrine, oxprenolol, oxyphencyclimine, pheniramine, phensuximide, pindolol, terbutaline, tetramisole, tolperisone, tranylcypromine, trihexyphenidyl, trimipramine, verapamil, warfarin

KEY WORDS
chiral; detector at anode

REFERENCE
Stalcup,A.M.; Gahm,K.H. Application of sulfated cyclodextrins to chiral separations by capillary zone electrophoresis, *Anal.Chem.*, **1996**, *68*, 1360–1368.

SAMPLE
Matrix: solutions

CAPILLARY ELECTROPHORESIS
Capillary: 17 cm × 25 μm fused-silica coated with polyacrylamide (12.4 cm to detector) (Bio-Rad)
Capillary temperature: 15
Running buffer: 54 mM pH 3 Citric acid containing 1% sulfated β-cyclodextrin (nominal 13 sulfates/cyclodextrin) (Cerestar, Hammond IN)
Injection: Hydrodynamic injection at 5 psi for 0.4 s.
Detector: UV 270
Migration time: 3.5, 5 (enantiomers)
Voltage: -3 to -8 kV
Model: Bio-Rad BioFocus 2000

KEY WORDS
detector at anode; chiral; coated capillary

REFERENCE
Sutton,R.M.C.; Gratz,S.R.; Stalcup,A.M. Use of capillary electrophoresis as a method development tool for classical gel electrophoresis, *Analyst*, **1998**, *123*, 1477–1480.

Piracetam

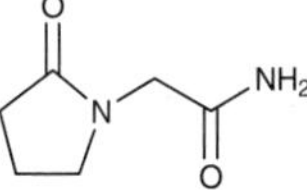

Molecular formula: $C_6H_{10}N_2O_2$
Molecular weight: 142.16
CAS Registry No.: 7491-74-9
Merck Index (12th ed.): 7641

SAMPLE
Matrix: blood
Sample preparation: 500 µL Plasma + 15 µL 1 mg/mL ephedrine hydrochloride, mix, add 3 mL MeCN, agitate on a rotary mixer for 10 min, centrifuge at 2000 g for 10 min. Remove the supernatant and evaporate it to dryness under a stream of nitrogen, reconstitute the residue in 500 µL running buffer, centrifuge at 2000 g for 54 min, inject an aliquot.

CAPILLARY ELECTROPHORESIS
Capillary: 58 cm × 50 µm fused-silica (51 cm to detector)
Capillary preparation: Between analyses flush with 100 mM HCl for 30 s, with 100 mM NaOH for 2 min, with water for 2 min, and with running buffer for 2 min. Before first use rinse capillary with 100 mM NaOH for 20 min, with water for 10 min, and with running buffer for 10 min.
Capillary temperature: 25
Running buffer: 10 mM pH 9.36 Borax containing 40 mM α-cyclodextrin
Injection: Pressure injection under argon at 0.5 psi for 7 s.
Detector: UV 200
Migration time: 7.5
Internal standard: ephedrine (4.5)
Voltage: 25 kV
Current: 16 µA
Model: Beckman P/ACE 2100
Limit of quantitation: 2 µg/mL
Limit of detection: 1 µg/mL

KEY WORDS
plasma

REFERENCE
Lamparczyk,H.; Kowalski,P.; Rajzer,D.; Nowakowska,J. Determination of piracetam in human plasma by capillary electrophoresis, *J.Chromatogr.B*, **1997**, *692*, 483–487.

Pirbuterol

Molecular formula: $C_{12}H_{20}N_2O_3$
Molecular weight: 240.30
CAS Registry No.: 38677-81-5, 29868-97-1 (di HCl), 38029-10-6 (di HCl), 65652-44-0 (acetate)
Merck Index (12th ed.): 7644
Lednicer: 2 280

SAMPLE
Matrix: solutions

CAPILLARY ELECTROPHORESIS
Capillary: 100 cm × 50 µm fused-silica (50 cm to detector) (Isco)
Capillary preparation: Flush capillary with 10 µL running buffer between runs. Every 40 sample injections rinse capillary with 200 µL 1 M NaOH, with 200 µL water, and 200 µL running

buffer. Before use fill capillary with 1 M NaOH and allow to stand for 1 h, fill with 100 mM NaOH, allow to stand for 1 h, wash with water fill with running buffer.
Capillary temperature: 23
Running buffer: 100 mM citric acid containing 19.27 mM Na_2HPO_4 and 120 mM hydroxypropyl-β-cyclodextrin
Injection: Inject under vacuum at 4.0 kPa.s
Detector: UV 200
Migration time: 16.7, 17.3 (enantiomers)
Voltage: 30 kV
Model: Isco Model 3140

OTHER SUBSTANCES
Simultaneous: albuterol, alprenolol, atenolol, cimaterol, clenbuterol, labetalol, nadolol, oxprenolol, pindolol, propranolol, terbutaline

KEY WORDS
chiral

REFERENCE
Aumatell,A.; Wells,R.J.; Wong,D.K.Y. Enantiomeric differentiation of a wide range of pharmacologically active substances by capillary electrophoresis using modified β-cyclodextrins, *J.Chromatogr.A*, **1994**, *686*, 293–307.

SAMPLE
Matrix: solutions
Sample preparation: Inject an aliquot of a 100 μg/mL solution in water:running buffer 50:50.

CAPILLARY ELECTROPHORESIS
Capillary: 44.5 cm × 50 μm acrylamide-coated fused-silica (Bio-Rad)
Capillary temperature: 30
Running buffer: 100 mM NaH_2PO_4 containing 15 mM gamma-cyclodextrin, adjusted to pH 2.5 with phosphoric acid
Injection: Electrokinetic injection at 8 kV for 6 s.
Detector: UV 200
Migration time: 6.74
Voltage: 14 kV
Model: Bio-Rad BioFocus 3000

OTHER SUBSTANCES
Also analyzed: albuterol, alprenolol, atenolol, atropine, baclofen, bamethan, benserazide, biperiden, bisoprolol, bupivacaine, bupranolol, butetamate, carazolol, carbuterol, carvedilol, celiprolol, chloroquine, chlorpheniramine (chlorphenamine), clidinium bromide, clobutinol, disopyramide, dobutamine, flecainide, homatropine, ipratropium bromide, isoproterenol, isothipendyl, ketamine, mefloquine, mequitazine, metaproterenol (orciprenaline), metipranolol, nafronyl (naftidrofuryl), nefopam, ofloxacin, orphenadrine, oxomemazine, oxprenolol, phenoxybenzamine, pholedrine, pindolol, prilocaine, promethazine, propafenone, propranolol, sotalol, synephrine, terbutaline, tetrahydrozoline (tetryzoline), tocainide, trihexyphenidyl, trimeprazine (alimemazine), trimipramine, tropicamide, verapamil, zopiclone

KEY WORDS
coated capillary; achiral

REFERENCE
Koppenhoefer,B.; Epperlein,U.; Christian,B.; Yibing,J.; Yuying,C.; Bingcheng,L. Separation of enantiomers of drugs by capillary electrophoresis. I. γ-Cyclodextrin as chiral solvating agent, *J.Chromatogr.A*, **1995**, *717*, 181–190.

SAMPLE
Matrix: solutions
Sample preparation: Inject an aliquot of a 100 μg/mL solution in running buffer.

CAPILLARY ELECTROPHORESIS
Capillary: 29 cm × 50 μm fused-silica (24.5 cm to detector) (Yongnian Optical Conductive Fiber Plant, China), coated with polyacrylamide
Capillary preparation: No details of the polyacrylamide coating process are provided. However, another paper (LC.GC 1997, 15, 40) by this group indicates that they use the procedure of Hjertén, thus: Adjust the pH of 20 mL water to 3.5 with acetic acid, add 80 μL 3-(trimethoxysilyl)propyl methacrylate (3-methacryloxypropyltrimethoxysilane), mix, suck into capillary, let stand at room temperature for 1 h, remove the solution, wash with water. Fill the capillary with a deaerated 3-4% acrylamide solution containing 1 μL/mL N,N,N',N'-tetramethylethylenediamine and 1 mg/mL potassium persulfate, let stand for 30 min, remove excess solution by aspiration, rinse with water, remove water by aspiration, dry at 35° (J. Chromatogr. 1985, 347, 191).
Capillary temperature: 25
Running buffer: 100 mM NaH_2PO_4 adjusted to pH 2.5
Injection: Electrokinetic injection at 15 kV for 3 s.
Detector: UV 200, UV 210
Migration time: 3.83
Voltage: 15 kV
Model: Bio-Rad BioFocus 3000

OTHER SUBSTANCES
Simultaneous: albuterol, alprenolol, atenolol, baclofen, bamethan, benproperine, benserazide, bisoprolol, bupranolol, butamirate, butethamate, carbuterol, celiprolol, clenbuterol, clobutinol, dipivefrin, isoproterenol (isoprenaline), metaproterenol (orciprenaline), metipranolol, metoprolol, norfenefrine, ornidazole, oxprenolol, phenylpropanolamine, pholedrine, prilocaine, procyclidine, sotalol, synephrine, terbutaline, tocainide

KEY WORDS
coated capillary

REFERENCE
Koppenhoefer,B.; Epperlein,U.; Xiaofeng,Z.; Bingcheng,L. Separation of enantiomers of drugs by capillary electrophoresis. Part 4: Hydroxypropyl-γ-cyclodextrin as chiral solvating agent, *Electrophoresis*, **1997**, *18*, 924–930.

SAMPLE
Matrix: solutions

CAPILLARY ELECTROPHORESIS
Capillary: 29-36 cm × 50 μm fused-silica (24.5-31.5 cm to detector) (Yongnian Optical Conductive Fiber Plant, China) coated with polyacrylamide
Capillary preparation: Coat capillary as follows. Adjust the pH of 20 mL water to 3.5 with acetic acid, add 80 μL 3-(trimethoxysilyl)propyl methacrylate (3-methacryloxypropyltrimethoxysilane), mix, suck into capillary, let stand at room temperature for 1 h, remove the solution, wash with water. Fill the capillary with a deaerated 3-4% acrylamide solution containing 1 μL/mL N,N,N',N'-tetramethylethylenediamine and 1 mg/mL potassium persulfate, let stand for 30 min, remove excess solution by aspiration, rinse with water, remove water by aspiration, dry at 35° (J. Chromatogr. 1985, 347, 191).
Capillary temperature: 25
Running buffer: 100 mM pH 2.5 NaH_2PO_4 (A) or 100 mM pH 2.5 NaH_2PO_4 containing 45 mM hydroxypropyl-α-cyclodextrin (Wacker, Munich) (B)
Injection: Electromigration at 15 kV for 3 s.
Detector: UV 200; UV 210
Migration time: 3.83 (A); 4.00 (B) (no separation of enantiomers)
Voltage: 15 kV
Model: Bio-Focus 3000

OTHER SUBSTANCES
Also analyzed: albuterol (salbutamol), alprenolol, amorolfine, atenolol, atropine, azelastine, baclofen, bamethan, benproperine, benserazide, biperiden, bisoprolol, brompheniramine, bupivacaine, bupranolol, butamirate, butethamate, carazolol, carbuterol, carteolol, carvedilol, celiprolol, chloroquine, chlorpheniramine, chlorphenoxamine, cicletanine, clenbuterol, clidinium bromide, clobutinol, dimethindene, dipivefrin, disopyramide, dobutamine, doxylamine, fendi-

line, flecainide, gallopamil, homatropine, ipratropium bromide, isoproterenol (isoprenaline), isothipendyl, ketamine, meclizine, mefloquine, mepindolol, mequitazine, metaclazepam, metaproterenol (orciprenaline), metipranolol, metoprolol, nafronyl (naftidrofuryl), nefopam, nicardipine, norfenefrine, ofloxacin, ornidazole, orphenadrine, oxomemazine, oxprenolol, oxybutynin, phenoxybenzamine, phenylpropanolamine, pholedrine, pindolol, prilocaine, procyclidine, promethazine, propafenone, propranolol, reproterol, sotalol, sulpride, synephrine, talinolol, terbutaline, tetrahydrozoline (tetryzoline), theodrenaline, tioconazole, tocainide, trihexyphenidyl, trimeprazine (alimemazine), trimipramine, tropicamide, verapamil, zopiclone

KEY WORDS
coated capillary

REFERENCE
Koppenhoefer,B.; Eperlein,U.; Schlunk,R.; Zhu,X.; Lin,B. Separation of enantiomers of drugs by capillary electrophoresis. V. Hydroxypropyl-α-cyclodextrin as chiral solvating agent, *J.Chromatogr.A*, **1998**, *793*, 153–164.

Piretanide

Molecular formula: $C_{17}H_{18}N_2O_5S$
Molecular weight: 362.41
CAS Registry No.: 55837-27-9
Merck Index (12th ed.): 7647
Lednicer: 3 58

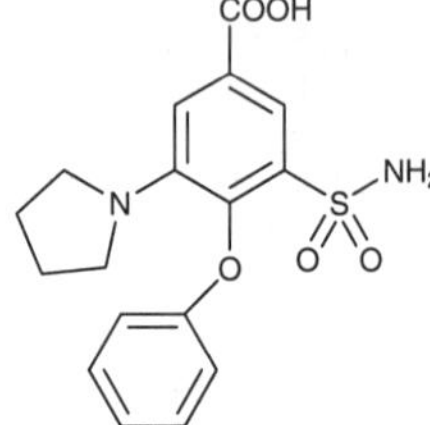

SAMPLE
Matrix: urine
Sample preparation: 5 mL Urine + 120 μL 1 M phosphoric acid + 10 mL chloroform, shake by hand for 2 min. Remove the organic layer and evaporate it to dryness under a stream of nitrogen at 45°, reconstitute with 100 μL 20 mM sodium dodecyl sulfate in water, sonicate for 1 min, inject an aliquot.

CAPILLARY ELECTROPHORESIS
Capillary: 56 cm × 50 μm fused-silica (48 cm to detector)
Capillary preparation: After each run rinse capillary with running buffer for 2 min.
Capillary temperature: 25
Running buffer: 20 mM pH 9 Sodium borate buffer containing 150 mM sodium dodecyl sulfate (Prepare by mixing 20 mM sodium tetraborate containing 150 mM sodium dodecyl sulfate with 20 mM boric acid containing 150 mM sodium dodecyl sulfate so as to achieve a pH of 9.)
Injection: Hydrodynamic injection at 50 mbar for 10 s.
Detector: UV 230
Migration time: 4.8
Voltage: 30 kV
Model: Hewlett-Packard HP³ᴰ
Limit of quantitation: 25 ng/mL

OTHER SUBSTANCES
Extracted: furosemide
Simultaneous: ethacrynic acid, torsemide

REFERENCE
Lalljie,S.P.D.; Begoña Barroso,M.; Steenackers,D.; Alonso,R.M.; Jiménez,R.M.; Sandra,P. Micellar electrokinetic chromatography as a fast screening method for the determination of the doping agents furosemide and piretanide in urine, *J.Chromatogr.B*, **1997**, *688*, 71–78.

Piromidic acid

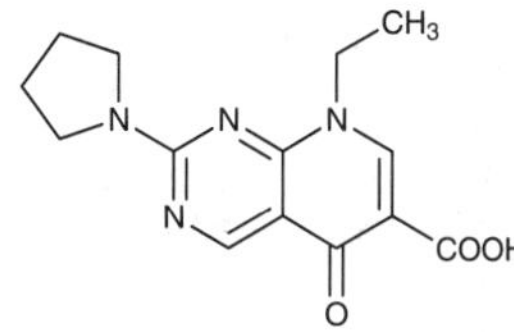

Molecular formula: $C_{14}H_{16}N_4O_3$
Molecular weight: 288.31
CAS Registry No.: 19562-30-2
Merck Index (12th ed.): 7660

SAMPLE
Matrix: solutions
Sample preparation: Inject an aliquot of a solution in MeOH.

CAPILLARY ELECTROPHORESIS
Capillary: 59 cm × 50 μm fused-silica (43 cm to detector) (Polymicro Technologies)
Capillary preparation: Before each analysis flush with water for 3 min, with 200 mM NaOH for 3 min, with water for 3 min, and with running buffer for 4 min. Flush new capillaries with 1 M NaOH at 2000 mbar for 10 min then with 200 mM NaOH for 10 min.
Capillary temperature: 23
Running buffer: MeCN:buffer 28:72, pH 7.3 (Buffer was 32 mM sodium borate containing 18 mM NaH_2PO_4, 39 mM sodium cholate, and 8 mM sodium heptanesulfonate.)
Injection: Hydrodynamic injection at 40 mbar for 6 s.
Detector: UV 260
Migration time: 5.7
Voltage: 30 kV
Model: Lauer Labs Prince

OTHER SUBSTANCES
Simultaneous: ciprofloxacin, enoxacin, flumequine, lomefloxacin, nalidixic acid, norfloxacin, ofloxacin, oxolinic acid, pefloxacin, pipemidic acid, rosoxacin, sparfloxacin

REFERENCE
Sun,S.-W.; Chen,L.-Y. Optimization of capillary electrophoretic separation of quinolone antibacterials using the overlapping resolution mapping scheme, *J.Chromatogr.A*, **1997**, *766*, 215–224.

Piroxicam

Molecular formula: $C_{15}H_{13}N_3O_4S$
Molecular weight: 331.35
CAS Registry No.: 36322-90-4, 87234-24-0 (cinnamic acid ester), 85056-47-9 (piroxicam olamine)
Merck Index (12th ed.): 7661
Lednicer: 4 173

SAMPLE
Matrix: blood
Sample preparation: 0.1-1 mL Plasma + 2 μg IS + 400 μL 1 M HCl + 10 mL diethyl ether, agitate mechanically for 10 min, centrifuge at 3500 g for 10 min. Remove the organic layer and evaporate it to dryness under a stream of nitrogen, reconstitute the residue in MeOH:40 mM NaH_2PO_4 50:50, inject an aliquot.

CAPILLARY ELECTROPHORESIS
Capillary: 58.7 cm × 75 μm fused-silica (50 cm to detector)
Capillary preparation: Wash with running buffer for 2 min before each injection. At the beginning of each day wash capillary with 100 mM NaOH for 10 min.
Capillary temperature: 22

Running buffer: MeOH:buffer 3:97 (Buffer was 40 mM pH 8 NaH_2PO_4 containing 104 mM so-
dium dodecyl sulfate.)
Injection: Hydrodynamic injection for 2 s
Detector: UV 254
Migration time: 10.7
Internal standard: benzoyl-4-phenyl-2-butyric acid (16.3)
Voltage: 20 kV
Model: Beckman P/ACE 2000
Limit of detection: 130 ng/mL (S/N 5)

OTHER SUBSTANCES

Extracted: diclofenac, diflunisal, etodolac, fenbufen, fenoprofen, flurbiprofen, ibuprofen, indomethacin, ketoprofen, naproxen, niflumic acid, sulindac, tenoxicam, tiaprofenic acid
Noninterfering: acetaminophen, amitriptyline, caffeine, clomipramine, deoxysulindac, desipramine, diazepam, 5'-hydroxytenoxicam, imipramine, maprotiline, nortriptyline, phenobarbital,
phenytoin, sulfamethoxazole, theophylline, trimipramine

KEY WORDS
plasma

REFERENCE
Maboundou,C.W.; Paintaud,G.; Bérard,M.; Bechtel,P.R. Separation of fifteen non-steroidal anti-inflammatory
drugs using micellar electrokinetic capillary chromatography, *J.Chromatogr.B*, **1994**, *657*, 173–183.

SAMPLE
Matrix: formulations
Sample preparation: Dilute 1 mL injection with running buffer, add a solution of acemetacin
in running buffer, dilute with running buffer to a final concentration of 20 µg/mL for piroxicam
and 20 µg/mL for acemetacin, inject an aliquot.

CAPILLARY ELECTROPHORESIS
Capillary: 60 cm × 75 µm (52.5 cm to detector)
Capillary preparation: Purge with running buffer for 2 min before each injection. Store capil-
lary overnight in water, rinse with 500 mM NaOH, with water, and with running buffer.
Running buffer: 30 mM pH 8.0 Phosphate buffer (Mix 0.57 mL 200 mM NaH_2PO_4 and 6.93 mL
200 mM Na_2HPO_4 tetraborate solution, make up to 50 mL with water.)
Injection: Hydrodynamic injection at 10 cm for 5 s
Detector: UV 214
Migration time: 7.4
Internal standard: acemetacin (6.8)
Voltage: 15 kV
Model: Waters Quanta 4000 CE
Limit of quantitation: 5 µg/mL

KEY WORDS
injections

REFERENCE
Donato,M.G.; Baeyens,W.; Van den Bossche,W.; Sandra,P. The determination of non-steroidal antiinflammatory
drugs in pharmaceuticals by capillary zone electrophoresis and micellar electrokinetic capillary chromatography, *J.Pharm.Biomed.Anal.*, **1994**, *12*, 21–26.

SAMPLE
Matrix: solutions
Sample preparation: Dissolve in MeCN, dilute with water to a concentration of 25 µg/mL, inject
an aliquot.

CAPILLARY ELECTROPHORESIS
Capillary: 60 cm × 75 µm fused-silica (52.5 cm to detector)
Running buffer: MeCN:30 mM pH 7.0 phosphate buffer 20:80
Injection: Hydrostatic injection at 9.8 cm for 5 s
Detector: UV 214

Migration time: 18.75
Model: Waters Quanta 4000

OTHER SUBSTANCES
Simultaneous: acemetacin, alclofenac, fenbufen, flurbiprofen, ibuprofen, indomethacin, ketoprofen, lonazolac, naproxen, niflumic acid, tenoxicam, tiaprofenic acid, tolmetin, voltaren

REFERENCE
Donato,M.G.; Van den Eeckhout,E.; Van den Bossche,W.; Sandra,P. Capillary zone electrophoresis and micellar electrokinetic capillary chromatography of some non-steroidal antiinflammatory drugs (NSAIDs), *J.Pharm.Biomed.Anal.*, **1993**, *11*, 197–201.

SAMPLE
Matrix: solutions
Sample preparation: Prepare a 20 μg/mL solution in 7.5 mM glycine adjusted to pH 8.0 with triethanolamine, inject an aliquot.

CAPILLARY ELECTROPHORESIS
Capillary: 44 cm × 50 μm fused-silica (37 cm to detector) (Supelco)
Capillary preparation: After each injection wash capillary with water for 2 min and with running buffer for 3 min. At the beginning of each day wash with running buffer for 10 min. Condition a new capillary with 1 M NaOH, 100 mM NaOH, water, and running buffer.
Capillary temperature: 25
Running buffer: 75 mM Glycine adjusted to pH 9.1 with triethanolamine
Injection: Hydrodynamic injection for 5 s
Detector: UV 280
Migration time: 5.15
Voltage: 15 kV
Model: Spectra-physics Spectraphoresis 1000 CE

OTHER SUBSTANCES
Simultaneous: alclofenac, indomethacin, sulindac, tiaprofenic acid

REFERENCE
Bechet,I.; Fillet,M.; Hubert,P.; Crommen,J. Quantitative analysis of non-steroidal anti-inflammatory drugs by capillary zone electrophoresis, *J.Pharm.Biomed.Anal.*, **1995**, *13*, 497–503.

SAMPLE
Matrix: solutions
Sample preparation: Inject an aliquot of a 50 μM solution in MeOH:water 40:60.

CAPILLARY ELECTROPHORESIS
Capillary temperature: 25
Running buffer: MeOH:water 40:60 containing 75 mM glycine and 15 mM hydroxypropyl-β-cyclodextrin adjusted to pH 9.1 with triethanolamine
Injection: Hydrodynamic injection for 2 s
Detector: UV 280
Migration time: 18.3
Voltage: 15 kV

OTHER SUBSTANCES
Simultaneous: alclofenac, bufexamac, carprofen, flufenamic acid, flurbiprofen, indomethacin, ketoprofen, naproxen, niflumic acid, sulindac, tiaprofenic acid

REFERENCE
Bechet,I.; Fillet,M.; Hubert,P.; Crommen,J. Improvement of achiral resolution in capillary zone electrophoresis by use of cyclodextrin additives, *Biomed.Chromatogr.*, **1995**, *9*, 267–268.

Polyethylene glycol

$HOCH_2(CH_2OCH_2)_nCH_2OH$

CAS Registry No.: 25322-68-3
Merck Index (12th ed.): 7729

SAMPLE
Matrix: bulk
Sample preparation: Add 1 mL 1 M phthalic anhydride in pyridine containing 30 mM imidazole to 100 mg polymer, heat at 100° for 1 h, cool, add 2 mL water, heat at 50° for 30 min. Remove a 500 μL aliquot and add it to 2 mL water, mix, inject an aliquot.

CAPILLARY ELECTROPHORESIS
Capillary: 45 cm × 75 μm μPage-3 3%T/3%C gel-filled with NO urea (40.1 cm to detector) (J&W Scientific)
Capillary preparation: Operate at -100 v/cm for 5 min, ramp to -250 V/cm over 30 min, maintain at -250 V/cm for 5 min
Running buffer: pH 8.3 Tris-borate
Injection: Electrokinetic injection at -10 kV for 1 min
Detector: UV 275
Migration time: 12-26 (depending on degree of polymerization)
Voltage: -243 V/cm
Model: Dionex CES-1

OTHER SUBSTANCES
Also analyzed: AP3S (sulfated 3 mol ethylene oxide derivative of phenol), AP7P (phosphated 7 mol ethylene oxide derivative of phenol), AP30 (30 mol ethylene oxide derivative of benzene), AP40P (phosphated 40 mol ethylene oxide derivative of phenol), surfactants

KEY WORDS
for polyethylene glycol 600; derivatization; some changes in CE conditions are needed for ionic surfactants

REFERENCE
Wallingford,R.A. Oligomeric separation of ionic and nonionic ethoxylated polymers by capillary gel electrophoresis, *Anal.Chem.*, **1996**, *68*, 2541–2548.

Pranoprofen

Molecular formula: $C_{15}H_{13}NO_3$
Molecular weight: 255.27
CAS Registry No.: 52549-17-4
Merck Index (12th ed.): 7890

SAMPLE
Matrix: solutions
Sample preparation: Inject an aliquot of a 500 μM solution in MeOH:water 50:50.

CAPILLARY ELECTROPHORESIS
Capillary: 60 cm × 50 μm fused-silica (52.5 cm to detector) (Waters)
Capillary preparation: Before each run rinse capillary with 10 mM NaOH for 1 min and with running buffer for 4 min. Condition new capillaries by flushing with 1 M NaOH for 10 min, with water for 5 min, with 6 mM NaOH containing 500 mM NaCl for 10 min, with water for 5 min, and with running buffer for 10 min.
Running buffer: 58.6 mM Formic acid containing 10 mM heptakis(trimethyl-β-cyclodextrin) adjusted to pH 4.0 with 1 M NaOH
Injection: Hydrodynamic injection at 10 cm for 10 s.

Detector: UV 254
Migration time: 30, 32 (enantiomers)
Voltage: 30 kV
Model: Waters Quanta 4000

KEY WORDS
chiral

REFERENCE
Lelièvre,F.; Gareil,P. Chiral separations of underivatized arylpropionic acids by capillary zone electrophoresis with various cyclodextrins. Acidity and inclusion constant determinations, *J.Chromatogr.A*, **1996**, *735*, 311–320.

Prazepam

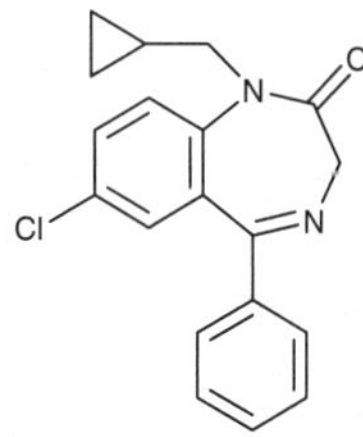

Molecular formula: $C_{19}H_{17}ClN_2O$
Molecular weight: 324.81
CAS Registry No.: 2955-38-6
Merck Index (12th ed.): 7895
Lednicer: 2 405

SAMPLE
Matrix: solutions
Sample preparation: Prepare a solution in running buffer, inject an aliquot.

CAPILLARY ELECTROPHORESIS
Capillary: 100 cm × 75 μm fused-silica (20 cm to UV detector) (Polymicro Technologies)
Capillary preparation: Between runs flush capillaries with 1.5 column volumes of 100 mM NaOH and 9 column volumes of running buffer. Flush new capillaries with 9 column volumes of 1 M NaOH, 3 column volumes of water, 3 column volumes of 100 mM HCl, 3 column volumes of water, and 9 column volumes of running buffer.
Running buffer: MeOH:water 15:85 containing 15 mM ammonium acetate, adjusted to pH 2.5 on trifluoroacetic acid
Injection: Pressure injection at 3.45 kPa (28 nL)
Detector: UV 254 or MS, Sciex TAGA 6000E triple quadrupole, API source, electrospray (ion spray) interface at 4 kV (3 kV to ion-sampling orifice), positive-ion mode, 100 μm dia sampling orifice, nitrogen gas curtain, argon collision gas, m/z 325 [M+H]+
Migration time: 6.5 (UV), 31 (MS)
Voltage: 26 kV
Model: Beckman P/ACE System 2000

OTHER SUBSTANCES
Simultaneous: chlordiazepoxide, diazepam, flurazepam

REFERENCE
Johansson,I.M.; Pavelka,R.; Henion,J.D. Determination of small drug molecules by capillary electrophoresis-atmospheric pressure ionization mass spectrometry, *J.Chromatogr.*, **1991**, *559*, 515–528.

SAMPLE
Matrix: solutions

CAPILLARY ELECTROPHORESIS
Capillary: 47 cm × 75 μm fused-silica (40 cm to detector) (Beckman)
Capillary preparation: Rinse with running buffer for 1 min before injection. Wash with 100 mM NaOH for 1 min and with water for 1 min between injections.
Capillary temperature: 30

Running buffer: THF:buffer 1:99 (Buffer was 50 mM pH 9.2 borate containing 50 mM sodium
  dodecyl sulfate, 2 M urea, and 20 mM gamma-cyclodextrin.)
Injection: Pressure injection at 300 kPa for 5 s.
Detector: UV 254
Migration time: 22
Voltage: 15 kV
Model: Beckman P/ACE 2000

OTHER SUBSTANCES
Simultaneous: alprazolam, bromazepam, chlordiazepoxide, clobazam, clonazepam, clorazepate,
  oxazepam, triazolam

REFERENCE
Renou-Gonnord,M.F.; David,K. Optimized micellar electrokinetic chromatographic separation of benzodiazepines, *J.Chromatogr.A*, **1996**, *735*, 249–261.

Praziquantel

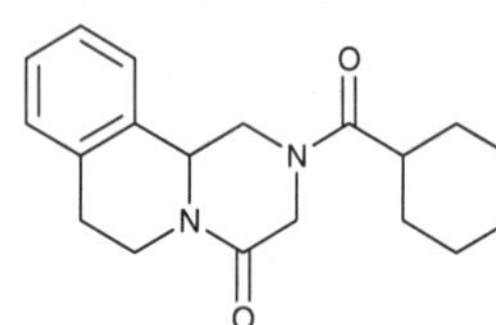

Molecular formula: $C_{19}H_{24}N_2O_2$
Molecular weight: 312.41
CAS Registry No.: 55268-74-1
Merck Index (12th ed.): 7896
Lednicer: 4 213

SAMPLE
Matrix: microsomal incubations
Sample preparation: Cool 200 μL microsomal incubation on ice, extract with 2 mL ethyl acetate, centrifuge at 2500 g for 5 min, repeat extraction. Combine the organic layers and evaporate them to dryness under a stream of nitrogen, reconstitute the residue in 35 μL 1 M urea
  solution, inject an aliquot.

CAPILLARY ELECTROPHORESIS
Capillary: 40 cm × 50 μm fused-silica (40 cm to detector) (Grom)
Capillary temperature: 25
Running buffer: 50 mM pH 7.0 Phosphate buffer containing 20 mM sodium dodecyl sulfate
Injection: Low pressure injection for 2 s.
Detector: UV 214
Migration time: 40
Voltage: 500 V/cm
Model: Beckman P/ACE 2100

OTHER SUBSTANCES
Extracted: metabolites

KEY WORDS
rat; liver; comparison with LC/MS

REFERENCE
Lerch,C.; Blaschke,G. Investigation of the stereoselective metabolism of praziquantel after incubation with rat
  liver microsomes by capillary electrophoresis and liquid chromatography-mass spectrometry,
  J.Chromatogr.B, **1998**, *708*, 267–275.

Prazosin

Molecular formula: $C_{19}H_{21}N_5O_4$
Molecular weight: 383.41
CAS Registry No.: 19216-56-9, 19237-84-4 (HCl)
Merck Index (12th ed.): 7897
Lednicer: 2 382

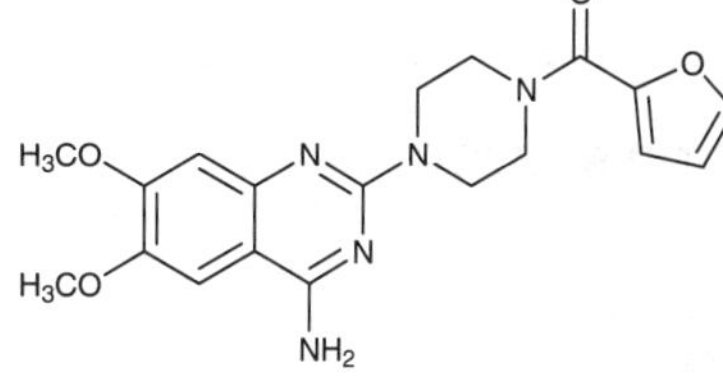

SAMPLE
Matrix: urine
Sample preparation: Centrifuge urine, dilute with 2 volumes of 32 mM pH 4.5 6-aminocaproic acid buffer containing 18 mM adipic acid, inject an aliquot.

CAPILLARY ELECTROPHORESIS
Capillary: 64 cm × 75 μm (48 cm to detector) (Polymicro Technologies)
Running buffer: MeOH:buffer 5:95 (Buffer was 32 mM pH 4.5 6-aminocaproic acid buffer containing 18 mM adipic acid, 5% dextran (MW 18300) and 0.02% polyethylene oxide (MW 300000)
Injection: Hydrodynamic injection at 20 cm
Detector: UV 220
Migration time: 14.5
Voltage: 22 kV
Model: Laboratory constructed

OTHER SUBSTANCES
Extracted: cimetidine, diltiazem, famotidine

REFERENCE
Soini,H.; Riekkola,M.-L.; Novotny,M.V. Mixed polymer networks in the direct analysis of pharmaceuticals in urine by capillary electrophoresis, *J.Chromatogr.A*, **1994**, *680*, 623–634.

Prednisolone

Molecular formula: $C_{21}H_{28}O_5$
Molecular weight: 360.45
CAS Registry No.: 50-24-8, 52438-85-4 (sesquihydrate), 52-21-1 (acetate), 2920-86-7 (hemisuccinate), 125-02-0 (disodium phosphate), 302-25-0 (dihydrogen phosphate), 1715-33-9 (sodium succinate), 2920-86-7 (hydrogen succinate), 5060-55-9 (steaglate), 7681-14-3 (tebutate), 5626-34-6 (21-diethyl-aminoacetate), 1107-99- 9(21-trimethylacetate), 630-67-1 (sodium 21-m-sulfobenzoate)
Merck Index (12th ed.): 7901
Lednicer: 1 192

SAMPLE
Matrix: formulations
Sample preparation: Add 2 mL injection to 2 mL IS solution, make up to 10 mL with water, inject an aliquot.

CAPILLARY ELECTROPHORESIS
Capillary: 50 cm × 50 μm fused-silica (Bio-Rad)
Capillary temperature: 25
Running buffer: 20 mM pH 8 phosphate buffer containing 30 mM sodium dodecyl sulfate
Injection: Injection by pressure 3 psi.sec (4.2 nL)
Detector: UV 230
Migration time: 13.9
Internal standard: hydroquinone monobenzyl ether (12)

Voltage: 17 kV
Current: 31 μA
Model: Biofocus 3000 (Bio-Rad)

OTHER SUBSTANCES
Simultaneous: degradation products, aminophylline, furosemide, methylprednisolone-21-hemi-succinate

KEY WORDS
injections; saline

REFERENCE
Quaglia,M.G.; Bossù,E.; Desiderio,C.; Fanali,S. Use of capillary electrophoresis for testing the stability of a drugs mixture in perfusional solution, *Farmaco*, **1994**, *49*, 403–406.

SAMPLE
Matrix: solutions
Sample preparation: Prepare a 10 mg/mL solution in MeOH, inject an aliquot.

CAPILLARY ELECTROPHORESIS
Capillary: 62 cm × 53 μm fused-silica (50 cm to detector) (Polymicro Technologies)
Capillary preparation: Rinse with running buffer for 1 min between runs.
Running buffer: 50 mM pH 9.0 Phosphate borate buffer containing 50 mM dehydrocholate, 50 mM taurocholate, and 50 mM sodium dodecyl sulfate
Detector: UV 254
Migration time: 22 (prednisolone), 22.5 (prednisolone acetate)
Voltage: 15 kV
Current: 56 μA
Model: Laboratory constructed

OTHER SUBSTANCES
Simultaneous: corticosterone, cortisone, cortisone acetate, deoxycorticosterone, fludrocortisone, fludrocortisone acetate, fluocinolone acetonide, hydrocortisone, hydrocortisone-21-acetate, 6α-methylprednisolone, prednisone, prednisone acetate, progesterone, triamcinolone, triamcinolone acetonide

REFERENCE
Bumgarner,J.G.; Khaledi,M.G. Mixed micellar electrokinetic chromatography of corticosteroids, *Electrophoresis*, **1994**, *15*, 1260–1266.

SAMPLE
Matrix: solutions

CAPILLARY ELECTROPHORESIS
Capillary: 62 cm × 53 μm fused-silica (50 cm to detector) (Polymicro Technologies)
Capillary preparation: Rinse capillary with running buffer for 1 min between runs.
Running buffer: 50 mM pH 9.0 Phosphate/borate buffer containing 33 mM taurocholate, 33 mM glycodeoxycholate, and 70 mM butanesulfonate
Detector: UV 254
Migration time: 21 (prednisolone), 24 (prednisolone acetate)
Voltage: 15 kV
Model: laboratory-constructed

OTHER SUBSTANCES
Simultaneous: corticosterone, cortisone, cortisone acetate, deoxycorticosterone, fludrocortisone, fludrocortisone acetate, fluocinolone acetonide, hydrocortisone, hydrocortisone-21-acetate, 6α-methylprednisolone, prednisone, prednisone acetate, progesterone, triamcinolone, triamcinolone acetonide

REFERENCE
Bumgarner,J.G.; Khaledi,M.G. Mixed micelles of short chain alkyl surfactants and bile slats in electrokinetic chromatography: Enhanced separation of corticosteroids, *J.Chromatogr.A*, **1996**, *738*, 275–283.

SAMPLE
Matrix: solutions

CAPILLARY ELECTROPHORESIS
Capillary: 27 cm × 75 μm
Capillary preparation: Before each run rinse capillary with 100 mM NaOH for 30 s and with running buffer for 30 s. Condition new capillaries by rinsing with 100 mM NaOH for 20 min.
Capillary temperature: 30
Running buffer: 15 mM Sodium borate
Injection: Pressure injection of sample at 25 mbar for 1 s followed by running buffer at 25 mbar for 1 s.
Detector: UV 200
Migration time: 3.36
Internal standard: aminobenzoic acid (3.5)
Voltage: 6.5 kV
Model: Beckman

OTHER SUBSTANCES
Simultaneous: aspirin, bacitracin, benzoic acid, ceftizoxime, ceftriaxone, cefuroxime, cephalothin, cromolyn, embonic acid, epoprostenol, glyburide (glibenclamide), levothyroxine, nedocromil, nystatin, omeprazole, warfarin, zidovudine
Interfering: beclomethasone

REFERENCE
Altria,K.D.; Bryant,S.M.; Hadgett,T.A. Validated capillary electrophoresis method for the analysis of a range of acidic drugs and excipients, *J.Pharm.Biomed.Anal.*, **1997**, *15*, 1091–1101.

SAMPLE
Matrix: solutions

CAPILLARY ELECTROPHORESIS
Capillary: 42 cm × 50 μm fused-silica packed with 3 μm CEC Hypersil ODS (packed length 30 cm, 30.1 cm to detector) (Hypersil)
Running buffer: Gradient. A was 5 mM ammonium acetate in MeCN:water 17:83. B was 5 mM ammonium acetate in MeCN:water 38:62. A:B 100:0 for 3 min, to 0:100 over 15 min, maintain at 0:100 (pumped with an HPLC pump at 0.01 mL/min for 3 min then at 0.1 mL/min).
Injection: Inject 10 μL using an HPLC injector.
Detector: UV 240
Migration time: 31
Voltage: 30 kV

OTHER SUBSTANCES
Simultaneous: adrenosterone, betamethasone, cortisone, fluocortolone, dexamethasone, methylprednisolone, triamcinolone, triamcinolone acetonide
Interfering: hydrocortisone

KEY WORDS
electrochromatography

REFERENCE
Taylor,M.R.; Teale,P. Gradient capillary electrochromatography of drug mixtures with UV and electrospray ionisation mass spectrometric detection, *J.Chromatogr.A*, **1997**, *768*, 89–95.

SAMPLE
Matrix: solutions
Sample preparation: Inject an aliquot of a 100 μg/mL solution in running buffer.

CAPILLARY ELECTROPHORESIS
Capillary: 27 cm × 50 μm fused-silica (20 cm to detector) (Composite Metal Services, Hallow, UK) packed with 3 μm Hypersil ODS for 20 cm (details in paper)
Running buffer: MeCN:2 mM pH 7.8 phosphate buffer 80:20

Injection: Electrokinetic injection at 5 kV for 5 s.
Detector: UV 214
Migration time: 6.58
Voltage: 10 kV
Model: Beckman P/ACE 2050

OTHER SUBSTANCES
Simultaneous: betamethasone, betamethasone dipropionate, betamethasone-17-valerate, clobetasol butyrate, clobetasone butyrate, fluticasone propionate, hydrocortisone

---

KEY WORDS
electrochromatography

REFERENCE
Frame,L.A.; Robinson,M.L.; Lough,W.J. Simplification of capillary electrochromatography procedures, *J.Chromatogr.A*, **1998**, *798*, 243–249.

Prednisone

Molecular formula: C$_{21}$H$_{26}$O$_5$
Molecular weight: 358.43
CAS Registry No.: 53-03-2, 125-10-0 (21-acetate)
Merck Index (12th ed.): 7904
Lednicer: 1 192

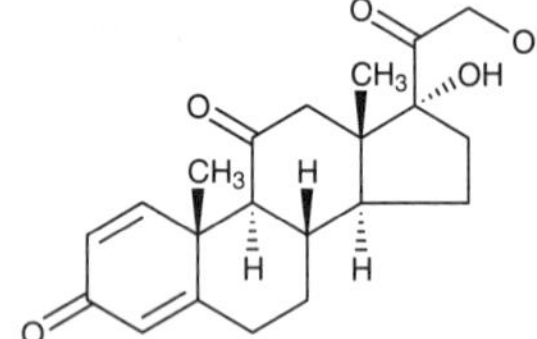

SAMPLE
Matrix: solutions
Sample preparation: Prepare a 10 mg/mL solution in MeOH, inject an aliquot.

CAPILLARY ELECTROPHORESIS
Capillary: 62 cm × 53 μm fused-silica (50 cm to detector) (Polymicro Technologies)
Capillary preparation: Rinse with running buffer for 1 min between runs.
Running buffer: 50 mM pH 9.0 Phosphate borate buffer containing 50 mM dehydrocholate, 50 mM taurocholate, and 50 mM sodium dodecyl sulfate
Detector: UV 254
Migration time: 17 (prednisone), 25 (prednisone acetate)
Voltage: 15 kV
Current: 56 μA
Model: Laboratory constructed

OTHER SUBSTANCES
Simultaneous: corticosterone, cortisone, cortisone acetate, deoxycorticosterone, fludrocortisone, fludrocortisone acetate, fluocinolone acetonide, hydrocortisone, hydrocortisone-21-acetate, 6α-methylprednisolone, prednisolone, prednisolone acetate, progesterone, triamcinolone, triamcinolone acetonide

---

REFERENCE
Bumgarner,J.G.; Khaledi,M.G. Mixed micellar electrokinetic chromatography of corticosteroids, *Electrophoresis*, **1994**, *15*, 1260–1266.

SAMPLE
Matrix: solutions

CAPILLARY ELECTROPHORESIS
Capillary: 62 cm × 53 μm fused-silica (50 cm to detector) (Polymicro Technologies)
Capillary preparation: Rinse capillary with running buffer for 1 min between runs.

Running buffer: 50 mM pH 9.0 Phosphate/borate buffer containing 33 mM taurocholate, 33 mM glycodeoxycholate, and 70 mM butanesulfonate
Detector: UV 254
Migration time: 17 (prednisone), 21.5 (prednisone acetate)
Voltage: 15 kV
Model: laboratory-constructed

OTHER SUBSTANCES
Simultaneous: corticosterone, cortisone, cortisone acetate, deoxycorticosterone, fludrocortisone, fludrocortisone acetate, fluocinolone acetonide, hydrocortisone, hydrocortisone-21-acetate, 6α-methylprednisolone, prednisolone, prednisolone acetate, progesterone, triamcinolone, triamcinolone acetonide

REFERENCE
Bumgarner,J.G.; Khaledi,M.G. Mixed micelles of short chain alkyl surfactants and bile slats in electrokinetic chromatography: Enhanced separation of corticosteroids, *J.Chromatogr.A*, **1996**, *738*, 275–283.

SAMPLE
Matrix: solutions

CAPILLARY ELECTROPHORESIS
Capillary: 33.5 cm × 100 μm fused-silica packed with 3 μm C18 for 25 cm (Innovatech, Stevenage, UK) (Polymicro Technologies)
Capillary preparation: Column packing procedure described in LC.GC 1995, 13, 800.
Capillary temperature: 20
Running buffer: MeCN:25 mM Tris HCl:water 60:20:20, run with 12 bar nitrogen pressure on each side
Injection: Electrokinetic injection at 5 kV for 3 s.
Detector: UV 200
Migration time: 3.2
Voltage: 30 kV
Model: Hewlett-Packard HP[3D]

OTHER SUBSTANCES
Simultaneous: androsterone, dehydrotestosterone, desoxycorticosterone, dexamethasone, estrone, ethinyl estradiol, hydrocortisone, pregnenolone, testosterone

KEY WORDS
electrochromatography

REFERENCE
Dittmann,M.M.; Rozing,G.P.; Ross,G.; Adam,T.; Unger,K.K. Advances in capillary electrochromatography, *J.Capillary Electrophor.*, **1997**, *4*, 201–212.

Prenalterol

Molecular formula: $C_{12}H_{19}NO_3$
Molecular weight: 225.29
CAS Registry No.: 57526-81-5, 61260-05-7 (HCl)
Merck Index (12th ed.): 7917
Lednicer: 3 30

SAMPLE
Matrix: solutions
Sample preparation: Inject an aliquot of a 100 μM solution in water.

CAPILLARY ELECTROPHORESIS
Capillary: 65 cm × 50 μm fused-silica (56.5 cm to detector) (Polymicro Technologies)

Capillary preparation: Before each injection rinse capillary at 930 mbar with 100 mM NaOH for 1 min, with water for 2 min, and with running buffer for 4 min.
Capillary temperature: 20
Running buffer: 50 mM Phosphoric acid containing 6 mM heptakis(2,6-di-O-methyl)-β-cyclodextrin, adjusted to pH 2.5 with triethanolamine
Injection: Pressure injection at 20 mbar for 3 s.
Detector: UV 195
Migration time: 15.0 (R_S = 1.6)
Voltage: 24 kV
Model: Hewlett Packard HP^{3D}

KEY WORDS
chiral

REFERENCE
Nilsson,S.; Schweitz,L.; Petersson,M. Three approaches to enantiomer separation of β-adrenergic antagonists by capillary electrochromatography, *Electrophoresis*, **1997**, *18*, 884–890.

Prilocaine

Molecular formula: $C_{13}H_{20}N_2O$
Molecular weight: 220.31
CAS Registry No.: 721-50-6, 1786-81-8 (HCl)
Merck Index (12th ed.): 7924
Lednicer: 1 17

SAMPLE
Matrix: blood
Sample preparation: Condition a 1 mL Bond-Elut C18 SPE cartridge with 1 mL 100 mM HCl, 1 mL MeOH, and 1 ml water. 1 mL Serum + 200 μL 100 mM pH 2.5 phosphate buffer + 40 μL 10 μg/mL procainamide in MeOH, vortex, add to the SPE cartridge, wash with five 500 μL portions of water (allow to dry between each wash), wash with 500 μL MeCN, allow to dry, elute with four 250 μL portions of 2% HCl in MeOH. Filter (0.2 μm) the eluate, evaporate to dryness, reconstitute the residue in 45 μL water, inject an aliquot.

CAPILLARY ELECTROPHORESIS
Capillary: 72 cm × 50 μm fused-silica (50 cm to detector) (Polymicro Technologies)
Capillary preparation: Before each analysis wash with 100 mM NaOH for 1.8 min and with running buffer for 2.2 min. Condition new capillaries with 1 M NaOH for 15 min, water for 15 min, and running buffer for 15 min.
Capillary temperature: 30
Running buffer: 100 mM NaH_2PO_4 containing 15 mM heptakis(2,6-di-O-methyl)-β-cyclodextrin and 30 μM hexadecyltrimethylammonium bromide, adjusted to pH 2.5 with 100 mM phosphoric acid
Injection: Vacuum injection at 5 inches (?) Hg for 24 s.
Detector: UV 215
Migration time: 16.0 (R), 16.4 (S)
Internal standard: procainamide (8.9)
Voltage: 25 kV
Current: 71 μA
Model: ABI Model 270A (Applied Biosystems)
Limit of quantitation: 45 ng/mL
Limit of detection: 38 ng/mL

KEY WORDS
serum; SPE; chiral

REFERENCE
Siluveru,M.; Stewart,J.T. Stereoselective determination of R-(-)- and S-(+)-prilocaine in human serum by capillary electrophoresis using a derivatized cyclodextrin and ultraviolet detection, *J.Chromatogr.B*, **1997**, *693*, 205–210.

SAMPLE
Matrix: solutions
Sample preparation: Inject an aliquot of a 100 μg/mL solution in water:running buffer 50:50.

CAPILLARY ELECTROPHORESIS
Capillary: 44.5 cm × 50 μm acrylamide-coated fused-silica (Bio-Rad)
Capillary temperature: 30
Running buffer: 100 mM NaH_2PO_4 containing 15 mM gamma-cyclodextrin, adjusted to pH 2.5 with phosphoric acid
Injection: Electrokinetic injection at 8 kV for 6 s.
Detector: UV 200
Migration time: 8.70
Voltage: 14 kV
Model: Bio-Rad BioFocus 3000

OTHER SUBSTANCES
Also analyzed: albuterol, alprenolol, atenolol, atropine, baclofen, bamethan, benserazide, biperiden, bisoprolol, bupivacaine, bupranolol, butetamate, carazolol, carbuterol, carvedilol, celiprolol, chloroquine, chlorpheniramine (chlorphenamine), clidinium bromide, clobutinol, disopyramide, dobutamine, flecainide, homatropine, ipratropium bromide, isoproterenol, isothipendyl, ketamine, mefloquine, mequitazine, metaproterenol (orciprenaline), metipranolol, nafronyl (naftidrofuryl), nefopam, ofloxacin, orphenadrine, oxomemazine, oxprenolol, phenoxybenzamine, pholedrine, pindolol, pirbuterol, promethazine, propafenone, propranolol, sotalol, synephrine, terbutaline, tetrahydrozoline (tetryzoline), tocainide, trihexyphenidyl, trimeprazine (alimemazine), trimipramine, tropicamide, verapamil, zopiclone

KEY WORDS
coated capillary; achiral

REFERENCE
Koppenhoefer,B.; Epperlein,U.; Christian,B.; Yibing,J.; Yuying,C.; Bingcheng,L. Separation of enantiomers of drugs by capillary electrophoresis. I. γ-Cyclodextrin as chiral solvating agent, *J.Chromatogr.A*, **1995**, *717*, 181–190.

SAMPLE
Matrix: solutions
Sample preparation: Inject an aliquot of a 100 μM solution in running buffer diluted 10-fold with water.

CAPILLARY ELECTROPHORESIS
Capillary: 57 cm × 25 μm fused-silica (50 cm to detector) (Polymicro Technologies)
Capillary preparation: Equilibrate with running buffer for 5 min before each analysis. Store in water overnight and rinse with 100 mM NaOH for 5 min at the start of each day and after each change of running buffer. Flush a new capillary with 100 mM HCl for 5 min, with water for 5 min, with 100 mM NaOH for 5 min, with water for 10 min, and with running buffer.
Capillary temperature: 25
Running buffer: pH 3.13 Phosphate buffer (I = 0.02) containing 14 mM taurodeoxycholate and 5.0 mM Brij-35
Injection: Pressure injection at 0.5 psi for 10 s.
Detector: UV 214
Migration time: 11.2 (R), 11.3 (S)
Voltage: 30 kV
Current: 9 μA
Model: Beckman P/ACE 2050

OTHER SUBSTANCES
Simultaneous: mepivacaine

KEY WORDS
chiral

REFERENCE
Amini,A.; Beijersten,I.; Pettersson,C.; Westerlund,D. Enantiomeric separation of local anaesthetic drugs by micellar electrokinetic capillary chromatography with taurodeoxycholate as chiral selector, *J.Chromatogr.A*, **1996**, *737*, 301–313.

SAMPLE
Matrix: solutions

CAPILLARY ELECTROPHORESIS
Capillary: 80.5 cm × 50 μm fused-silica (72 cm to detector) (Hewlett Packard)
Capillary preparation: Before each run flush with water for 1 min, with 100 mM NaOH for 4 min, with water for 1 min, and with running buffer for 4 min.
Capillary temperature: 30
Running buffer: 100 mM Phosphoric acid adjusted to pH 3.0 with triethanolamine, containing 10 mM heptakis(2,6-di-O-methyl)-β-cyclodextrin
Injection: Pressure injection at 50 mbar for 5 s (5 nL), ramp to operating voltage at 500 V/s.
Detector: UV 206
Migration time: 25, 25.8 (enantiomers)
Voltage: 30 kV
Model: Hewlett Packard HP [3D]CE

OTHER SUBSTANCES
Also analyzed: bupivacaine, mepivacaine, ropivacaine

KEY WORDS
chiral

REFERENCE
Sänger-van de Griend,C.E.; Gröningsson,K.; Westerlund,D. Chiral separation of local anaesthetics with capillary electrophoresis. Evaluation of the inclusion complex of the enantiomers with heptakis(2,6-di-O-methyl)-β-cyclodextrin, *Chromatographia*, **1996**, *42*, 263–268.

SAMPLE
Matrix: solutions

CAPILLARY ELECTROPHORESIS
Capillary: 47 cm × 50 μm fused-silica (40 cm to detector) (Polymicro Technologies)
Capillary preparation: Before each run flush with running buffer for 2 min, fill with running buffer containing 100 mM α-cyclodextrin at 0.5 psi for 4.0 min. Store capillary in water overnight, rinse with 100 mM NaOH and with water each morning. Rinse a new capillary with 100 mM HCl for 5 min, with water for 5 min, with 100 mM NaOH for 5 min, and with water for 10 min.
Capillary temperature: 17.0
Running buffer: pH 2.90 Phosphate buffer, I = 0.04
Injection: Pressure injection at 0.5 psi for 5 s, ramp to operating voltage at 1.33 kV/s.
Detector: UV 214
Migration time: 10 (S), 10.2 (R)
Voltage: 24 kV
Model: Beckman P/ACE 2100

KEY WORDS
chiral

REFERENCE
Amini,A.; Paulsen-Sörman,U. Enantioseparation of local anaesthetic drugs by capillary zone electrophoresis with cyclodextrins as chiral selectors using a partial filling technique, *Electrophoresis*, **1997**, *18*, 1019–1025.

SAMPLE
Matrix: solutions
Sample preparation: Inject an aliquot of a 100 μg/mL solution in running buffer.

CAPILLARY ELECTROPHORESIS
Capillary: 29 cm × 50 μm fused-silica (24.5 cm to detector) (Yongnian Optical Conductive Fiber Plant, China), coated with polyacrylamide
Capillary preparation: No details of the polyacrylamide coating process are provided. However, another paper (LC.GC 1997, 15, 40) by this group indicates that they use the procedure of Hjertén, thus: Adjust the pH of 20 mL water to 3.5 with acetic acid, add 80 μL 3-(trimethoxysilyl)propyl methacrylate (3-methacryloxypropyltrimethoxysilane), mix, suck into capillary, let stand at room temperature for 1 h, remove the solution, wash with water. Fill the capillary with a deaerated 3-4% acrylamide solution containing 1 μL/mL N,N,N',N'-tetramethylethylenediamine and 1 mg/mL potassium persulfate, let stand for 30 min, remove excess solution by aspiration, rinse with water, remove water by aspiration, dry at 35° (J. Chromatogr. 1985, 347, 191).
Capillary temperature: 25
Running buffer: 100 mM NaH_2PO_4 adjusted to pH 2.5
Injection: Electrokinetic injection at 15 kV for 3 s.
Detector: UV 200, UV 210
Migration time: 6.01
Voltage: 15 kV
Model: Bio-Rad BioFocus 3000

OTHER SUBSTANCES
Simultaneous: albuterol, alprenolol, atenolol, baclofen, bamethan, benproperine, benserazide, bisoprolol, bupranolol, butamirate, butethamate, carbuterol, celiprolol, clenbuterol, clobutinol, dipivefrin, isoproterenol (isoprenaline), metaproterenol (orciprenaline), metipranolol, metoprolol, norfenefrine, ornidazole, oxprenolol, phenylpropanolamine, pholedrine, pirbuterol, procyclidine, sotalol, synephrine, terbutaline, tocainide

KEY WORDS
coated capillary

REFERENCE
Koppenhoefer,B.; Epperlein,U.; Xiaofeng,Z.; Bingcheng,L. Separation of enantiomers of drugs by capillary electrophoresis. Part 4: Hydroxypropyl-γ-cyclodextrin as chiral solvating agent, *Electrophoresis*, **1997**, *18*, 924–930.

SAMPLE
Matrix: solutions
Sample preparation: Inject an aliquot of a 100 μg/mL solution in running buffer.

CAPILLARY ELECTROPHORESIS
Capillary: 30 cm × 50 μm fused-silica (25.5 cm to detector), coated with polyacrylamide
Capillary preparation: Adjust the pH of 20 mL water to 3.5 with acetic acid, add 80 μL 3-(trimethoxysilyl)propyl methacrylate (3-methacryloxypropyltrimethoxysilane), mix, suck into capillary, let stand at room temperature for 1 h, remove the solution, wash with water. Fill the capillary with a deaerated 3-4% acrylamide solution containing 1 μL/mL N,N,N',N'-tetramethylethylenediamine and 1 mg/mL potassium persulfate, let stand for 30 min, remove excess solution by aspiration, rinse with water, remove water by aspiration, dry at 35° (J. Chromatogr. 1985, 347, 191).
Capillary temperature: 25
Running buffer: 100 mM NaH_2PO_4 containing 45 mM gamma-cyclodextrin, adjusted to pH 2.5 with phosphoric acid
Injection: Electrokinetic injection at 15 kV for 3 s.
Detector: UV 200
Voltage: 15 kV
Model: Bio-Rad BioFocus 3000

KEY WORDS
chiral; coated capillary; comparison with the use of other cyclodextrins; this running buffer gave the greatest enantiomeric separation.; α=1.041

REFERENCE

Lin,B.; Zhu,X.; Koppenhoefer,B.; Epperlein,U. Investigation of 123 chiral drugs by cyclodextrin-modified capillary electrophoresis, *LC.GC*, **1997**, *15*, 40–46.

SAMPLE

Matrix: solutions
Sample preparation: Inject an aliquot of a 20 μg/mL solution in buffer with an ionic strength 10 times less than that of the running buffer.

CAPILLARY ELECTROPHORESIS

Capillary: 57 cm × 50 μm fused-silica (50 cm to detector) (Polymicro Technologies)
Capillary preparation: Between runs rinse capillary with running buffer from a different vial for 2 min, equilibrate system for 1 min before analysis. Store capillary in water overnight and rinse with 100 mM NaOH and water each morning. Rinse new capillaries with 100 mM HCl for 5 min, with water for 5 min, with 100 mM NaOH for 5 min, with water for 10 min, then with running buffer.
Capillary temperature: 15
Running buffer: pH 2.90 Phosphate buffer (I = 0.04) containing 50 mM α-cyclodextrin (Prepare buffer by mixing 19.8 mL 1 M NaOH and 23.2 mL 1 M phosphoric acid and make up to 500 mL with water.)
Injection: Pressure injection at 3.4 kPa for 5 s (6 nL).
Detector: UV 214
Migration time: 19.4 (S), 20 (R)
Voltage: 21 kV
Current: 17 μA
Model: Beckman P/ACE 2100

OTHER SUBSTANCES

Simultaneous: ketamine

KEY WORDS

chiral

REFERENCE

Amini,A.; Sörman,U.P.; Lindgren,B.H.; Westerlund,D. Enantioseparation of anaesthetic drugs by capillary zone electrophoresis using cyclodextrin-containing background electrolytes, *Electrophoresis*, **1998**, *19*, 731–737.

SAMPLE

Matrix: solutions

CAPILLARY ELECTROPHORESIS

Capillary: 29-36 cm × 50 μm fused-silica (24.5-31.5 cm to detector) (Yongnian Optical Conductive Fiber Plant, China) coated with polyacrylamide
Capillary preparation: Coat capillary as follows. Adjust the pH of 20 mL water to 3.5 with acetic acid, add 80 μL 3-(trimethoxysilyl)propyl methacrylate (3-methacryloxypropyltrimethoxysilane), mix, suck into capillary, let stand at room temperature for 1 h, remove the solution, wash with water. Fill the capillary with a deaerated 3-4% acrylamide solution containing 1 μL/mL N,N,N',N'-tetramethylethylenediamine and 1 mg/mL potassium persulfate, let stand for 30 min, remove excess solution by aspiration, rinse with water, remove water by aspiration, dry at 35° (J. Chromatogr. 1985, 347, 191).
Capillary temperature: 25
Running buffer: 100 mM pH 2.5 NaH_2PO_4 (A) or 100 mM pH 2.5 NaH_2PO_4 containing 45 mM hydroxypropyl-α-cyclodextrin (Wacker, Munich) (B)
Injection: Electromigration at 15 kV for 3 s.
Detector: UV 200; UV 210
Migration time: 6.01 (A); 8.05, 8.32 (B) (enantiomers)
Voltage: 15 kV
Model: Bio-Focus 3000

OTHER SUBSTANCES
Also analyzed: albuterol (salbutamol), alprenolol, amorolfine, atenolol, atropine, azelastine, baclofen, bamethan, benproperine, benserazide, biperiden, bisoprolol, brompheniramine, bupivacaine, bupranolol, butamirate, butethamate, carazolol, carbuterol, carteolol, carvedilol, celiprolol, chloroquine, chlorpheniramine, chlorphenoxamine, cicletanine, clenbuterol, clidinium bromide, clobutinol, dimethindene, dipivefrin, disopyramide, dobutamine, doxylamine, fendiline, flecainide, gallopamil, homatropine, ipratropium bromide, isoproterenol (isoprenaline), isothipendyl, ketamine, meclizine, mefloquine, mepindolol, mequitazine, metaclazepam, metaproterenol (orciprenaline), metipranolol, metoprolol, nafronyl (naftidrofuryl), nefopam, nicardipine, norfenefrine, ofloxacin, ornidazole, orphenadrine, oxomemazine, oxprenolol, oxybutynin, phenoxybenzamine, phenylpropanolamine, pholedrine, pindolol, pirbuterol, procyclidine, promethazine, propafenone, propranolol, reproterol, sotalol, sulpride, synephrine, talinolol, terbutaline, tetrahydrozoline (tetryzoline), theodrenaline, tioconazole, tocainide, trihexyphenidyl, trimeprazine (alimemazine), trimipramine, tropicamide, verapamil, zopiclone

KEY WORDS
coated capillary; chiral

REFERENCE
Koppenhoefer,B.; Eperlein,U.; Schlunk,R.; Zhu,X.; Lin,B. Separation of enantiomers of drugs by capillary electrophoresis. V. Hydroxypropyl-α-cyclodextrin as chiral solvating agent, *J.Chromatogr.A*, **1998**, *793*, 153–164.

Primaquine

Molecular formula: $C_{15}H_{21}N_3O$
Molecular weight: 259.35
CAS Registry No.: 90-34-6, 63-45-6 (phosphate)
Merck Index (12th ed.): 7925

SAMPLE
Matrix: solutions
Sample preparation: Prepare a 100-200 ppm solution in MeOH, inject an aliquot.

CAPILLARY ELECTROPHORESIS
Capillary: 45 cm × 50 μm fused-silica (Polymicro Technologies)
Running buffer: 50 mM pH 7.50 Phosphate buffer containing 25 mM borate and 9 mM tetrabutylammonium bromide
Injection: Hydrodynamic injection at 5 cm for 5 s (1 nL)
Detector: UV 240
Migration time: 2.9
Voltage: 15 kV
Current: 39 μA
Model: Laboratory constructed

OTHER SUBSTANCES
Simultaneous: chloroquine, dapsone, pyrimethamine, quinacrine, quinine, sulfadiazine

REFERENCE
Ng,C.L.; Toh,Y.L.; Li,S.F.Y.; Lee,H.K. Capillary electrophoresis of biologically important compounds: Optimization of separation conditions by the overlapping resolution mapping scheme, *J.Liq.Chromatogr.*, **1993**, *16*, 3653–3666.

SAMPLE
Matrix: solutions
Sample preparation: Inject an aliquot of a 15 μg/mL solution in water.

CAPILLARY ELECTROPHORESIS
Capillary: 65 cm × 50 μm silica (50 cm to detector)
Capillary preparation: Between injections wash with 100 mM NaOH, rinse with water
Capillary temperature: 20 kV
Running buffer: 25 mM Na_2HPO_4 adjusted to pH 2 with phosphoric acid
Injection: Electrokinetic injection at 5 kV for 10 s
Detector: UV 254
Migration time: 8.8
Model: Isco 3850 Capillary Electropherograph
Limit of detection: 150 ng/mL (40 ng/mL if a water plug is injected first)

OTHER SUBSTANCES
Simultaneous: 4-chlorophenylbiguanide, chloroquine, chlorproguanil, cyclochlorproguanil, cycloguanil, desethylchloroquine, mefloquine, proguanil, pyrimethamine, quinine

REFERENCE
Taylor,R.B.; Reid,R.G. Analysis of basic antimalarial drugs by CZE and MEKC. Part 1--Critical factors affecting separation, *J.Pharm.Biomed.Anal.*, **1993**, *11*, 1289–1294.

SAMPLE
Matrix: solutions
Sample preparation: Prepare a solution in running buffer, inject an aliquot.

CAPILLARY ELECTROPHORESIS
Capillary: 60 cm × 75 μm fused-silica (52.4 cm to detector)
Capillary preparation: After each run flush with 500 mM KOH for 2-3 min, flush with water, fill with running buffer
Running buffer: 10 mM Na_2HPO_4 containing 2% heparin sodium (MW 10000, 11% S, Scientific Protein Laboratories, Waunakee, WI) adjusted to pH 5 with phosphoric acid
Injection: Hydrostatic injection
Detector: UV 214
Migration time: 12.2 (first enantiomer, R_S 0.92)
Model: Waters Quanta 4000

OTHER SUBSTANCES
Also analyzed: anabasine, brompheniramine, bupivacaine, carbinoxamine, chlorcyclizine, chloroquine, chlorpheniramine, dimethindene, doxylamine, enpiroline, halofantrine, hydroxychloroquine, indapamide, mefloquine, nornicotine, pheniramine, promethazine, quinacrine, tetramisole

KEY WORDS
chiral

REFERENCE
Stalcup,A.M.; Agyei,N.M. Heparin: a chiral mobile-phase additive for capillary zone electrophoresis, *Anal.Chem.*, **1994**, *66*, 3054–3059.

SAMPLE
Matrix: solutions

CAPILLARY ELECTROPHORESIS
Capillary: 36 cm × 50 μm polyacrylamide-coated fused-silica (31.5 cm to detector) (Polymicro Technologies)
Capillary preparation: Before each injection rinse with water at 100 psi for 30 s, rinse with running buffer at 100 psi for 30 s, partially fill with separation solution (500 μM ovomucoid in running buffer) at 1 psi for 190 s (27 cm), inject sample, electrophorese with running buffer. Coat capillary with polyacrylamide as follows. Treat with 1 M NaOH for 1 h, rinse with water, dry at 110° by flushing with nitrogen for 6 h, pass thionyl chloride through using suction for 10 min, seal at each end, heat at 70° for 6 h, open, suck 250 mM vinylmagnesium bromide in THF into the capillary, seal at each end, heat at 70° for 6 h, open, rinse with THF for several min, rinse with water, fill by suction with a solution containing 3% acrylamide, 0.1% ammo-

nium persulfate, and 0.1% tetramethylethylenediamine, heat at 28 ± 2° for 1 h, after 1 h wash with water (Biol.Pharm.Bull. 1993, 16, 1185).
Capillary temperature: 25
Running buffer: 1-Propanol:50 mM pH 6.0 phosphate buffer 8:92
Injection: Inject at 1 psi for 2 s
Detector: UV 210
Migration time: 11.83, 12.24 (enantiomers)
Voltage: 12 kV
Model: Bio-Rad BioFocus 3000

OTHER SUBSTANCES
Also analyzed: chlorpheniramine, trimebutine

KEY WORDS
chiral; partial separation zone technique; coated capillary

REFERENCE
Tanaka,Y.; Terabe,S. Partial separation zone technique for the separation of enantiomers by affinity electro-kinetic chromatography with proteins as chiral pseudo-stationary phases, *J.Chromatogr.A,* **1995,** *694,* 277–284.

SAMPLE
Matrix: solutions
Sample preparation: Inject an aliquot of a 100 µg/mL solution in water or running buffer.

CAPILLARY ELECTROPHORESIS
Capillary: 47 cm × 75 µm fused-silica (40 cm to detector)
Capillary preparation: Before each run rinse capillary with running buffer for 1-2 min. If peak tailing is observed fill capillary with 500 mM NaOH and let stand for 30 min, wash with water for 3 min, rinse with running buffer for 3 min.
Capillary temperature: 25
Running buffer: 20 mM pH 2.5 Phosphate buffer containing 9% dextrin (Japan Pharmacopeia grade) (Dissolve dextrin in buffer at 90° then cool to room temperature.)
Injection: Pressure injection at 0.5 psi for 2-4 s.
Detector: UV 220
Migration time: 4.109, 4.195 (enantiomers)
Voltage: 30 kV
Model: Beckman P/ACE 5510

KEY WORDS
chiral

REFERENCE
Nishi,H.; Izumoto,S.; Nakamura,K.; Nakai,H.; Sato,T. Dextran and dextrin as chiral selectors in capillary zone electrophoresis, *Chromatographia,* **1996,** *42,* 617–630.

SAMPLE
Matrix: solutions
Sample preparation: Inject an aliquot of a 100 µg/mL solution in MeOH:water 10:90.

CAPILLARY ELECTROPHORESIS
Capillary: 47 cm × 75 µm fused-silica (40 cm to detector)
Capillary preparation: Rinse capillary with running buffer for 1 min before each run. At the end of each day wash with 100 mM NaOH and with water.
Capillary temperature: 20
Running buffer: 20 mM pH 2.9 Phosphate buffer containing 3% chondroitin sulfate C (sodium salt) (Nacalai Tesque, Kyoto)
Injection: Pressure injection at 0.5 psi for 5 s.
Detector: UV 240
Migration time: 16.88, 17.53 (enantiomers)
Voltage: 20 kV
Model: Beckman P/ACE 5510

OTHER SUBSTANCES
Also analyzed: clentiazem, diltiazem, propranolol, sulconazole, trimetoquinol, verapamil

KEY WORDS
chiral

REFERENCE
Nishi,H. Enantiomer separation of basic drugs by capillary electrophoresis using ionic and neutral polysaccharides as chiral selectors, *J.Chromatogr.A*, **1996**, *735*, 345–351.

SAMPLE
Matrix: solutions

CAPILLARY ELECTROPHORESIS
Capillary: 42 cm × 50 μm fused-silica (31 cm to detector) (Polymicro Technologies)
Capillary preparation: Condition capillary by rinsing with running buffer for 10 min, with water for 10 min, with 1 M NaOH for 10 min, with water for 10 min, and with running buffer with the voltage applied for 20 min.
Capillary temperature: 25
Running buffer: Formamide containing 150 mM citric acid, 100 mM Tris, and 100 mM methyl-β-cyclodextrin (apparent pH 5.1)
Injection: Gravity injection.
Detector: UV 254
Migration time: 8.93, 8.96 (enantiomers)
Voltage: 30 kV
Model: Beckman P/ACE 5500

OTHER SUBSTANCES
Simultaneous: trihexyphenidyl

KEY WORDS
chiral

REFERENCE
Wang,F.; Khaledi,M.G. Chiral separations by nonaqueous capillary electrophoresis, *Anal.Chem.*, **1996**, *68*, 3460–3467.

SAMPLE
Matrix: solutions
Sample preparation: Inject an aliquot of a 100 μg/mL solution in MeOH:water 10:90.

CAPILLARY ELECTROPHORESIS
Capillary: 37 cm × 75 μm
Capillary temperature: 15
Running buffer: 20 mM pH 2.06 Tris-phosphoric acid containing 10 mM 18-crown-6 tetracarboxylic acid (Merck, Darmstadt, Germany)
Injection: Pressure injection at 0.5 psi for 2-9 s.
Detector: UV 210, UV 235
Migration time: 9.60, 9.67 (enantiomers)
Voltage: 20 kV
Model: Beckman P/ACE 5510

OTHER SUBSTANCES
Also analyzed: aminoglutethimide, baclofen, levodopa, mexiletine, norephedrine, norepinephrine, octopamine

KEY WORDS
chiral; comparison with HPLC

REFERENCE
Nishi,H.; Nakamura,K.; Nakai,H.; Sato,T. Separation of enantiomers and isomers of amino compounds by capillary electrophoresis and high-performance liquid chromatography utilizing crown ethers, *J.Chromatogr.A*, **1997**, *757*, 225–235.

SAMPLE
Matrix: solutions

CAPILLARY ELECTROPHORESIS
Capillary: 36 cm × 50 μm fused-silica coated with linear polyacrylamide (31.5 cm to detector) (GL Science)
Capillary preparation: At the beginning and end of each day rinse capillary with capillary wash solution (Bio-Rad Cat. No. 148-5022) at 690 kPa for more than 3 min and with water at 690 kPa for more than 3 min. Coat capillary as follows. Treat capillary with 1 M NaOH at room temperature for 1 h, rinse with water, dry by passing nitrogen gas through the capillary at 110° for 6 h. Pass thionyl chloride through the capillary using a suction pump for several min, seal capillary at both ends and heat at 70° for 6 h. Unseal the capillary and fill with 250 mM vinyl magnesium bromide in THF by suction, seal the capillary, heat at 70° for 6 h. Open the capillary and rinse it with THF for several min, rinse with distilled water, fill the capillary with polymerization solution, heat at 28 ± 2° for 1 h, condition at -100 V/cm for 30 min (Anal. Sci. 1994, 10, 1). (The polymerization solution was 5% acrylamide in water containing 49 mM Tris, 384 mM glycine, and 0.1% sodium dodecyl sulfate, degas in an ultrasonic bath. Add 40 μL 10% N,N,N',N'-tetramethylethylenediamine and 10 μL 10% ammonium persulfate to 5 mL of the degassed solution, mix thoroughly.)
Running buffer: n-Propanol:50 mM pH 6.0 Phosphate buffer 10:90
Injection: Before each injection rinse with water at 690 kPa for 30 s, rinse with running buffer at 690 kPa for 30 s, partially fill with separation solution (500 μM α_1-acid glycoprotein (Cohn fraction VI) (Sigma) in running buffer) at 6.9 kPa for 190 s (27 cm), inject sample at 6.9 kPa for 2 s, electrophorese with running buffer (Note that α_1-acid glycoprotein from other suppliers may provide inferior results).
Detector: UV 210
Migration time: Resolution of enantiomers 7.3
Voltage: 12 kV
Model: Bio-rad BioFocus 3000

KEY WORDS
chiral; coated capillary

REFERENCE
Tanaka,Y.; Terabe,S. Separation of the enantiomers of basic drugs by affinity capillary electrophoresis using a partial filling technique and α_1-acid glycoprotein as chiral selector, *Chromatographia*, **1997**, *44*, 119–128.

Primidone

Molecular formula: $C_{12}H_{14}N_2O_2$
Molecular weight: 218.26
CAS Registry No.: 125-33-7
Merck Index (12th ed.): 7927
Lednicer: 1 276

SAMPLE
Matrix: blood
Sample preparation: 50 μL Serum + 5 mL isotonic saline, add to a Sep-Pak C18 SPE cartridge, wash with 5 mL saline, wash with 5 mL water, wash with 5 mL heptane:chloroform 98:2, wash with 5 mL heptane, wash with 3 mL water, remove all water with a stream of air, elute with 3 mL MeOH (remove all MeOH with a stream of air). Evaporate the eluate to dryness under reduced pressure at 37°, reconstitute with 100 μL water, inject an aliquot.

CAPILLARY ELECTROPHORESIS
Capillary: 72 cm × 50 μm silica (50 cm to detector) (Applied Biosystems)
Capillary preparation: Before each run wash capillary with 100 mM NaOH for 3 min, wash with running buffer for 3 min, aspirate neutral marker solution (2 drops DMSO in 10 mL water) for 1 s, place capillary end in buffer vial for 5 s, inject sample. Wash capillary at the beginning of each day by passing 1 M NaOH through for 20 min.
Capillary temperature: 30
Running buffer: 30 mM pH 9.3 Borate buffer containing 30 mM sodium dodecyl sulfate
Injection: Inject by applying vacuum for 1 s (2-3 nL)
Detector: UV 200
Migration time: 4
Voltage: +30 kV
Current: 72 μA
Model: Applied Biosystems Model 270A

OTHER SUBSTANCES
Extracted: carbamazepine, phenobarbital, phenytoin

KEY WORDS
serum; SPE

REFERENCE
Evenson,M.A.; Wiktorowicz,J.E. Automated capillary electrophoresis applied to therapeutic drug monitoring, *Clin.Chem.*, **1992**, *38*, 1847–1852.

SAMPLE
Matrix: blood
Sample preparation: Vortex serum for 20 s, filter (0.45 μm), inject an aliquot.

CAPILLARY ELECTROPHORESIS
Capillary: 92 cm × 75 μm (68 cm to detector)
Capillary preparation: Before each run rinse capillary with 100 mM NaOH for 3 min and with running buffer for 5 min.
Capillary temperature: 35
Running buffer: 10 mM pH 9.2 Na_2HPO_4 containing 6 mM sodium borate and 75 mM sodium dodecyl sulfate
Injection: Vacuum suction for 1 s
Detector: UV 220, UV 245
Migration time: 13.5
Voltage: 20 kV
Current: 59 μA
Model: Applied Biosystems model 270A-HT

OTHER SUBSTANCES
Extracted: ethosuximide, phenobarbital

KEY WORDS
serum

REFERENCE
Schmutz,A.; Thormann,W. Determination of phenobarbital, ethosuximide, and primidone in human serum by micellar electrokinetic capillary chromatography with direct sample injection, *Ther.Drug Monit.*, **1993**, *15*, 310–316.

SAMPLE
Matrix: blood
Sample preparation: Vortex serum for 15 s, filter (0.45 μm), inject an aliquot.

CAPILLARY ELECTROPHORESIS
Capillary: 63 cm × 75 μm fused-silica (63 cm to detector)

Capillary preparation: Before each run rinse capillary with 100 mM NaOH for 3 min and with running buffer for 5 min.
Capillary temperature: 35
Running buffer: 10 mM pH 9.2 Na_2HPO_4 containing 6 mM sodium borate and 200 mM sodium dodecyl sulfate
Injection: Vacuum suction for 1 s
Detector: UV 195
Migration time: 19
Voltage: 25 kV
Current: 145 µA
Model: Applied Biosystems model 270A-HT

KEY WORDS
serum; ultrafiltrate

REFERENCE
Schmutz,A.; Thormann,W. Factors affecting the determination of drugs and endogenous low molecular mass compounds in human serum by micellar electrokinetic capillary chromatography with direct sample injection, *Electrophoresis*, **1994**, *15*, 51–61.

SAMPLE
Matrix: blood
Sample preparation: Inject serum directly.

CAPILLARY ELECTROPHORESIS
Capillary: 105 cm × 75 µm fused-silica (68 cm to detector) (Polymicro Technologies)
Capillary preparation: Rinse with running buffer for 10 min between run. At the beginning and end of each day rinse with 100 mM NaOH for 10 min and with water for 10 min.
Running buffer: 10 mM Na_2HPO_4 containing 6 mM sodium borate and 75 mM sodium dodecyl sulfate
Injection: Injection via vacuum suction for 1 s
Detector: UV 215
Migration time: 13.4
Voltage: 23 kV
Current: 62 µA
Model: Europhor Prime Vision IV

OTHER SUBSTANCES
Extracted: acetaminophen, ethosuximide, flucytosine, naproxen, sulfamethoxazole, zomepirac

KEY WORDS
serum

REFERENCE
Schmutz,A.; Thormann,W. Assessment of impact of physico-chemical drug properties on monitoring drug levels by micellar electrokinetic capillary chromatography with direct serum injection, *Electrophoresis*, **1994**, *15*, 1295–1303.

SAMPLE
Matrix: blood
Sample preparation: 200 µL Serum + 1 mL 5 µg/mL n-propyl p-hydroxybenzoate in ethyl acetate, vortex for 30 s, centrifuge at 13400 g for 2 min. Remove a 500 µL aliquot of the organic layer and evaporate it to dryness under a stream of nitrogen, reconstitute with 50 µL MeOH:water 5:95, inject an aliquot.

CAPILLARY ELECTROPHORESIS
Capillary: 64.5 cm × 50 µm fused-silica (56 cm to detector) (Hewlett-Packard)
Capillary preparation: Rinse capillary with running buffer at 630 mm Hg for 5 min before each run.
Capillary temperature: 30
Running buffer: 10 mM pH 8.0 Phosphate buffer containing 50 mM sodium dodecyl sulfate
Injection: Vacuum injection at 50 mm Hg for 2 s.

Detector: UV 210
Migration time: 3.4
Internal standard: n-propyl p-hydroxybenzoate (7.2)
Voltage: 30 kV
Model: Hewlett-Packard HP[3D]

OTHER SUBSTANCES
Extracted: carbamazepine, phenobarbital, phenytoin, zonisamide
Noninterfering: valproic acid

KEY WORDS
serum; comparison with HPLC

REFERENCE
Makino,K.; Goto,Y.; Sueyasu,M.; Futagami,K.; Kataoka,Y.; Oishi,R. Micellar electrokinetic capillary chromatography for therapeutic drug monitoring of zonisamide, *J.Chromatogr.B*, **1997**, *695*, 417–425.

SAMPLE
Matrix: blood, urine
Sample preparation: Filter (Amicon Centrifree Micropartition System) at 1500 g for 20 min, inject an aliquot of the ultrafiltrate.

CAPILLARY ELECTROPHORESIS
Capillary: 90 cm × 75 μm fused-silica (70 cm to detector) (Polymicro Technologies)
Capillary preparation: Between runs rinse capillary with 100 mM NaOH for 5 min and with running buffer for 10 min.
Running buffer: 10 mM pH 9.1 Na_2HPO_4 containing 6 mM sodium borate and 75 mM sodium dodecyl sulfate
Injection: Siphon at 34 cm for 5 s
Detector: UV 195
Migration time: 10.5
Voltage: 20 kV
Current: 80 μA

OTHER SUBSTANCES
Extracted: acetaminophen, ethosuximide, phenobarbital, salicylic acid

KEY WORDS
serum; ultrafiltrate

REFERENCE
Caslavska,J.; Lienhard,S.; Thormann,W. Comparative use of three electrokinetic capillary methods for the determination of drugs in body fluids. Prospects for rapid determination of intoxications, *J.Chromatogr.*, **1993**, *638*, 335–342.

Probenecid

Molecular formula: $C_{13}H_{19}NO_4S$
Molecular weight: 285.36
CAS Registry No.: 57-66-9
Merck Index (12th ed.): 7934
Lednicer: 1 135

SAMPLE
Matrix: blood, urine
Sample preparation: Condition a 3 mL Supelclean LC-18 SPE cartridge with 3 mL MeOH and 3 mL water. Dilute urine 1:10 with water. Precipitate proteins from serum with MeOH. Add

diluted urine or protein supernatant to the SPE cartridge, wash with 3 mL water, elute with 3 mL MeOH, reconstitute to the original volume with 100 mM KOH, inject an aliquot.

CAPILLARY ELECTROPHORESIS
Capillary: 67 cm × 50 μm fused-silica (60 cm to detector) (Polymicro Technologies)
Capillary preparation: Rinse with running buffer for 2 min before run. If necessary, regenerate capillary with 100 mM NaOH for 10 min and with water for 15 min.
Capillary temperature: 20
Running buffer: 60 mM 3-(cyclohexylamino)-1-propanesulfonic acid (CAPS) adjusted to pH 10.6 with 100 mM KOH
Injection: Pressure injection for 5 s
Detector: UV 220
Migration time: 8
Voltage: 25 kV
Model: Beckman P/ACE 2000

OTHER SUBSTANCES
Extracted: acetazolamide, amiloride, bendroflumethiazide, benzthiazide, bumetanide, caffeine, chlorothiazide, chlorthalidone, clopamide, dichlorphenamide, furosemide, hydrochlorothiazide, metyrapone, triamterene, trichlormethiazide
Interfering: ethacrynic acid

KEY WORDS
serum; SPE

REFERENCE
Jumppanen,J.; Sirén,H.; Riekkola,M.-L. Screening for diuretics in urine and blood serum by capillary zone electrophoresis, *J.Chromatogr.A*, **1993**, *652*, 441–450.

Procainamide

Molecular formula: $C_{13}H_{21}N_3O$
Molecular weight: 235.33, 271.79 (HCl)
CAS Registry No.: 51-06-9, 614-39-1 (HCl)
Merck Index (12th ed.): 7936
Lednicer: 1 14

SAMPLE
Matrix: blood
Sample preparation: Condition a 1 mL Bond-Elut C18 SPE cartridge with 1 mL 100 mM HCl, 1 mL MeOH, and 1 ml water. 1 mL Serum + 200 μL 100 mM pH 2.5 phosphate buffer + 40 μL MeOH, vortex, add to the SPE cartridge, wash with five 500 μL portions of water (allow to dry between each wash), wash with 500 μL MeCN, allow to dry, elute with four 250 μL portions of 2% HCl in MeOH. Filter (0.2 μm) the eluate, evaporate to dryness, reconstitute the residue in 45 μL water, inject an aliquot.

CAPILLARY ELECTROPHORESIS
Capillary: 72 cm × 50 μm fused-silica (50 cm to detector) (Polymicro Technologies)
Capillary preparation: Before each analysis wash with 100 mM NaOH for 1.8 min and with running buffer for 2.2 min. Condition new capillaries with 1 M NaOH for 15 min, water for 15 min, and running buffer for 15 min.
Capillary temperature: 30
Running buffer: 100 mM NaH_2PO_4 containing 15 mM heptakis(2,6-di-O-methyl)-β-cyclodextrin and 30 μM hexadecyltrimethylammonium bromide, adjusted to pH 2.5 with 100 mM phosphoric acid
Injection: Vacuum injection at 5 inches (?) Hg for 24 s.
Detector: UV 215
Migration time: 8.9

Internal standard: procainamide
Voltage: 25 kV
Current: 71 μA
Model: ABI Model 270A (Applied Biosystems)

OTHER SUBSTANCES
Extracted: prilocaine

KEY WORDS
serum; SPE; procainamide is IS

REFERENCE
Siluveru,M.; Stewart,J.T. Stereoselective determination of R-(-)- and S-(+)-prilocaine in human serum by capillary electrophoresis using a derivatized cyclodextrin and ultraviolet detection, *J.Chromatogr.B*, **1997**, *693*, 205–210.

SAMPLE
Matrix: blood
Sample preparation: Condition a 1 mL 100 mg cyanopropyl SPE cartridge (Varian) with 1 mL MeOH, 1 mL 1% ammonia in isopropanol, and 1 mL water, do not allow to dry. Add 100 μL 500 mM HCl to the SPE cartridge followed by 2 mL serum, wash with two 1 mL portions of water, allow to dry, wash with two 1 mL portions of MeCN, elute with four 500 μL portions of 1% ammonia in isopropanol. Filter the eluate and evaporate it to dryness under a stream of nitrogen, reconstitute with 90 μL water, inject an aliquot.

CAPILLARY ELECTROPHORESIS
Capillary: 72 cm × 50 μm fused-silica (50 cm to detector) (Polymicro Technologies)
Capillary preparation: Before each run rinse capillary with 100 mM NaOH for 2 min and with running buffer for 2.5 min. Condition new capillaries by rinsing with 1 M NaOH for 10 min, with water for 10 min, and with running buffer for 10 min.
Capillary temperature: 30
Running buffer: 100 mM NaH_2PO_4 containing 15 mM heptakis-(2,6-di-O-methyl)-β-cyclodextrin and 30 μM hexadecyltrimethylammonium bromide, adjusted to pH 2.5 with 100 mM phosphoric acid
Injection: Vacuum injection at 80 psi for 20 s.
Detector: UV 254
Migration time: 10.0
Internal standard: procainamide
Voltage: 20 kV
Current: 65 μA (typical)
Model: Applied Biosystems ABI 270A

OTHER SUBSTANCES
Extracted: ondansetron

KEY WORDS
serum; SPE; procainamide is IS

REFERENCE
Siluveru,M.; Stewart,J.T. Enantioselective determination of S-(+)- and R-(-)-ondansetron in human serum using derivatized cysclodextrin-modified capillary electrophoresis and solid-phase extraction, *J.Chromatogr.B*, **1997**, *691*, 217–222.

SAMPLE
Matrix: solutions

CAPILLARY ELECTROPHORESIS
Capillary: 72 cm × 50 μm silica (50 cm to detector) (Applied Biosystems)
Capillary preparation: Before each run wash capillary with 100 mM NaOH for 3 min, wash with running buffer for 3 min, aspirate neutral marker solution (2 drops DMSO in 10 mL

water) for 1 s, place capillary end in buffer vial for 5 s, inject sample. Wash capillary at the beginning of each day by passing 1 M NaOH through for 20 min.
Capillary temperature: 30
Running buffer: 30 mM pH 9.3 Borate buffer containing 30 mM sodium dodecyl sulfate
Injection: Inject by applying vacuum for 1 s (2-3 nL)
Detector: UV 200
Migration time: 7.6
Voltage: +30 kV
Current: 64 μA
Model: Applied Biosystems Model 270A

OTHER SUBSTANCES
Simultaneous: N-acetylprocainamide, chlorodisopyramide, disopyramide

REFERENCE
Evenson,M.A.; Wiktorowicz,J.E. Automated capillary electrophoresis applied to therapeutic drug monitoring, *Clin.Chem.*, **1992**, *38*, 1847–1852.

SAMPLE
Matrix: solutions

CAPILLARY ELECTROPHORESIS
Capillary: 45 cm × 50 μm fused-silica (37.5 cm to detector) (Polymicro Technologies)
Capillary preparation: Between each run rinse with 100 mM HCl for 2 min and with running buffer for 2 min. At the start of each day condition capillary with 1 M HCl for 10 min. Before the first use rinse capillary with 1 M NaOH for 30 min and with water for 30 min.
Capillary temperature: 25
Running buffer: MeCN:30 mM pH 2.40 ethanesulfonic acid 20:80
Injection: Injection at 4 kV for 3 s.
Detector: UV 214
Migration time: 4.11
Voltage: 20 kV
Model: Waters Quanta 4000E

OTHER SUBSTANCES
Simultaneous: chloramphenicol, imipramine, lidocaine, phenylpropanolamine, quinidine, trimethoprim

REFERENCE
Ding,W.; Fritz,J.S. Separation of neutral compounds and basic drugs by capillary electrophoresis in acidic solution using laurylpoly(oxyethylene) sulfate as an additive, *Anal.Chem.*, **1998**, *70*, 1859–1865.

SAMPLE
Matrix: urine
Sample preparation: Urine. Dilute 100 μL urine to 1 mL with water, inject an aliquot.

CAPILLARY ELECTROPHORESIS
Capillary: 43.5 cm × 75 μm fused-silica (35.9 cm to detector)
Capillary preparation: Wash with running buffer for 2 min before each injection. At the start of each day wash the capillary with 100 mM NaOH for 10 min, with water for 10 min, and with running buffer for 10 min.
Capillary temperature: 30
Running buffer: 50 mM pH 7.7 Phosphate buffer
Injection: Pressure injection at 0.5 psi for 5 s.
Detector: UV 200
Migration time: 2.6
Voltage: 10 kV
Model: Beckman P/ACE 5500
Limit of detection: 1.235 μg/mL

OTHER SUBSTANCES
Extracted: N-acetylprocainamide

KEY WORDS
rat; human

REFERENCE
Vargas,G.; Havel,J.; Hadasová,E. Direct determination of procainamide and N-acetylprocainamide by capillary
zone electrophoresis in pharmaceutical formulations and urine, *J.Chromatogr.A*, **1997**, *772*, 271–276.

Procaine

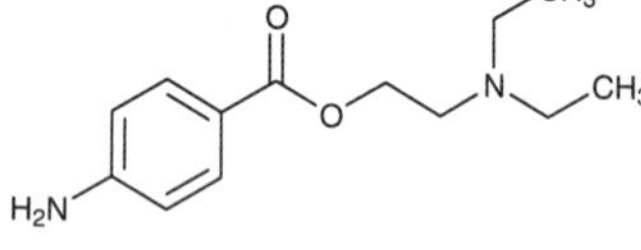

Molecular formula: $C_{13}H_{20}N_2O_2$
Molecular weight: 236.31
CAS Registry No.: 59-46-1, 51-05-8 (HCl), 149-13-3 (borate)
Merck Index (12th ed.): 7937
Lednicer: 1 9

SAMPLE
Matrix: blood, urine
Sample preparation: Adjust pH of 2 mL urine or plasma to 10.5 with aqueous NaOH, extract
gently with chloroform:isopropanol 90:10. Remove the organic layer and evaporate it to dryness
under a stream of nitrogen, reconstitute the residue in 50 (urine) or 75 (plasma) μL running
buffer, filter, inject an aliquot.

CAPILLARY ELECTROPHORESIS
Capillary: 60 cm × 75 μm AccuSep uncoated silica (52.5 cm to detector) (Waters)
Capillary preparation: At the beginning of each day purge with 500 mM KOH for 5 min, with
water for 5 min, and with running buffer for 10 min.
Running buffer: 50 mM NaH_2PO_4 adjusted to pH 2.35 with phosphoric acid
Injection: Hydrostatic injection at 15 cm
Detector: UV 214
Migration time: 5.67
Voltage: 22 kV
Current: 135-145 μA
Model: Waters Quanta 4000

OTHER SUBSTANCES
Extracted: acepromazine, amphetamine, benzocaine, brompheniramine, butacaine, codeine, diazepam, doxapram, lidocaine, medazepam, methamphetamine, methapyrilene, methaqualone,
phenmetrazine, tetrahydrozoline
Simultaneous: meclizine

KEY WORDS
plasma

REFERENCE
Chee,G.L.; Wan,T.S.M. Reproducible and high-speed separation of basic drugs by capillary zone electrophoresis,
J.Chromatogr., **1993**, *612*, 172–177.

SAMPLE
Matrix: bulk
Sample preparation: Prepare a 3 mg/mL solution in 3 mg/mL pholcodine in 10 mM HCl, make
up to 25 mL with 10 mM HCl, mix thoroughly, filter (0.45 μm cellulose acetate)

CAPILLARY ELECTROPHORESIS
Capillary: 65 cm × 75 μm fused-silica (40 cm to detector) (Isco)
Capillary preparation: Flush with running buffer for 2 min between runs. Before each batch
of samples fill capillary with 100 mM NaOH and let stand for 10 min, wash with water for 5
min, wash with 100 mM HCl for 5 min, wash with water for 5 min, fill with running buffer

(all at vacuum level 4). Clean each week by washing with 100 mM HCl for 10 min, wash with water, fill with running buffer.
Capillary temperature: 30
Running buffer: MeCN:buffer 7.5:92.5. Buffer was 10 mM KH_2PO_4 containing 10 mM sodium tetraborate and 50 mM cetyltrimethylammonium bromide, pH 8.6.)
Injection: Injection by vacuum 5.0 kPa.s (4 nL)
Detector: UV 230
Migration time: 9.1
Internal standard: pholcodine (7.4)
Voltage: -15 kV
Model: Isco Model 3140

OTHER SUBSTANCES
Simultaneous: cocaine, benzocaine, benzoylecgonine, trans-cinnamoylcocaine, lidocaine, tetracaine

REFERENCE
Trenerry,V.C.; Robertson,J.; Wells,R.J. The determination of cocaine and related substances by micellar electrokinetic capillary chromatography, *Electrophoresis*, **1994**, *15*, 103–108.

SAMPLE
Matrix: bulk
Sample preparation: Dilute with 400 μg/mL n-propyl p-hydroxybenzoate in running buffer

CAPILLARY ELECTROPHORESIS
Capillary: 27 cm × 50 μm fused-silica (20 cm to detector)
Capillary preparation: Before each run rinse with running buffer at high pressure.
Capillary temperature: 30
Running buffer: MeCN:water 15:85 containing 40 mM sodium dodecyl sulfate, 8.5 mM sodium borate, and 8.5 mM sodium phosphate, pH 8.5
Injection: Inject at high pressure for 1 s
Detector: UV 214
Migration time: 1.8
Internal standard: n-propyl p-hydroxybenzoate (1.8)
Voltage: 20 kV
Model: Beckman P/ACE System 2100

OTHER SUBSTANCES
Simultaneous: acetaminophen, acetylcodeine, O-acetylmorphine, aspirin, caffeine, cocaine, codeine, diamorphine, diphenhydramine, hydromorphone, isoniacinamide, lidocaine, methaqualone, morphine, niacinamide, noscapine, papaverine, phenacetin, phenobarbital, phenylpropanolamine, quinine, salicylic acid, strychnine, thebaine

REFERENCE
Walker,J.A.; Krueger,S.T.; Lurie,I.S.; Marché,H.L.; Newby,N. Analysis of heroin drug seizures by micellar electrokinetic capillary chromatography (MECC), *J.Forensic Sci.*, **1995**, *40*, 6–9.

SAMPLE
Matrix: bulk
Sample preparation: Dissolve in running buffer to a concentration of 1 mg/mL, vortex for 2 min, add an equal volume of 100 μg/mL diphenhydramine in running buffer, mix, inject an aliquot.

CAPILLARY ELECTROPHORESIS
Capillary: 65 cm × 50 μm fused-silica (60 cm to detector)
Capillary preparation: Fill capillary with fresh running buffer before each run. Before use fill with 100 mM NaOH for 20 min, rinse with water, flush with running buffer
Running buffer: MeCN:50 mM 6-aminocaproic acid containing 50 mM 3-N,N-dimethylmyristylammoniopropanesulfonate (MAPS; Fluka) and 5 mM 1-heptanesulfonic acid 10:90, pH adjusted to 4.0 with 1 M phosphoric acid
Injection: Hydrodynamic injection by gravity or pressure.
Detector: UV 214
Migration time: 5

Internal standard: diphenhydramine (12)
Voltage: 27 kV
Current: ≤25 μA
Model: Dionex system I

OTHER SUBSTANCES
Simultaneous: acetaminophen, allobarbital, barbital, caffeine, codeine, diamorphine, morphine, niacinamide, noscapine, papaverine, phenobarbital

REFERENCE
Naess,O.; Rasmussen,K.E. Micellar electrokinetic chromatography of charged and neutral drugs in acidic running buffers containing a zwitterionic surfactant, sulfonic acids or sodium dodecyl sulphate. Separation of heroin, basic by-products and adulterants, *J.Chromatogr.A*, **1997**, *760*, 245–251.

SAMPLE
Matrix: bulk
Sample preparation: Dissolve sample in 2 mg/mL histamine diphosphate in 10 mM HCl, inject an aliquot.

CAPILLARY ELECTROPHORESIS
Capillary: 48.5 cm × 50 μm fused-silica (40 cm to detector)
Capillary preparation: Flush capillary with running buffer for 2 min before each run. Replenish buffer in vials after every 10 injections. Condition new capillaries by flushing with 1 M NaOH for 10 min, with 100 mM NaOH for 10 min, with water for 10 min, and with running buffer for 10 min.
Capillary temperature: 40
Running buffer: 75 mM NaH_2PO_4 containing 75 mM sodium borate, adjusted to pH 8.5 with 2 M NaOH
Injection: Pressure injection at 150 mbar for 3 s.
Detector: UV 230
Migration time: 5.6
Internal standard: histamine diphosphate (5)
Voltage: 10 kV
Current: 66-70 μA
Model: Hewlett Packard 3D CE
Limit of detection: 4 μg/mL (S/N = 5:1)

OTHER SUBSTANCES
Simultaneous: acetaminophen, benzocaine, cis-cinnamoylcocaine, trans-cinnamoylcocaine, cocaine, ephedrine, lidocaine, phenylpropanolamine, tetracaine

REFERENCE
Krawczeniuk,A.S.; Bravenec,V.A. Quantitative determination of cocaine in illicit powders by free zone capillary electrophoresis, *J.Forensic Sci.*, **1998**, *43*, 738–743.

SAMPLE
Matrix: perfusate
Sample preparation: Inject an aliquot of perfusate (pH 6.5 Ringer's).

CAPILLARY ELECTROPHORESIS
Capillary: 70 cm × 50 μm untreated fused-silica (Polymicro Technologies)
Running buffer: 200 mM sodium tetraborate buffer
Injection: Injection by vacuum for 1 s
Detector: UV 200
Migration time: 6.5
Voltage: 20 kV

OTHER SUBSTANCES
Simultaneous: cocaine, lidocaine

REFERENCE

Hernandez,L.; Guzman,N.A.; Hoebel,B.G. Bidirectional microdialysis in vivo shows differential dopaminergic potency of cocaine, procaine and lidocaine in the nucleus accumbens using capillary electrophoresis for calibration of drug outward diffusion, *Psychopharmacology (Berl)*, **1991**, *105*, 264–268.

SAMPLE

Matrix: solutions
Sample preparation: Inject an aliquot of a 1-50 μg/mL solution in water.

CAPILLARY ELECTROPHORESIS

Capillary: 55 cm × 50 μm fused-silica (35 cm to detector) (J&W)
Capillary preparation: Flush with running buffer for 5 min before each run. Periodically wash with 100 mM NaOH and flush extensively with running buffer.
Running buffer: MeOH:25 mM pH 9.24 borate buffer 20:80 containing 100 mM sodium dodecyl sulfate (A) or 50 mM pH 2.35 phosphate buffer (B) or 50 mM pH 9.24 borate buffer (C)
Injection: Injection of 5 μL using a split-flow injector ratio of 1:800.
Detector: UV 200
Migration time: 27.97 (A), 8.89 (B), 6.16 (C)
Voltage: 20 kV (A, B) or 12 kV (C)
Current: <60-80 μA
Model: ISCO Model 3850

OTHER SUBSTANCES

Also analyzed: acetylcodeine, amphetamine, barbital, caffeine, codeine, cocaine, diamorphine, diazepam, flunitrazepam, lidocaine, monoacetylmorphine, morphine, nalorphine, narceine, noscapine, papaverine, pentobarbital, tetracaine, thebaine

KEY WORDS

all compounds were separated with running buffer A; some peaks overlapped with running buffers B and C.

REFERENCE

Tagliaro,F.; Smith,F.P.; Turrina,S.; Equisetto,V.; Marigo,M. Complementary use of capillary zone electrophoresis and micellar electrokinetic capillary chromatography for mutual confirmation of results in forensic drug analysis, *J.Chromatogr.A*, **1996**, *735*, 227–235.

SAMPLE

Matrix: solutions

CAPILLARY ELECTROPHORESIS

Capillary: 85 cm × 50 μm fused-silica (Polymicro Technologies)
Capillary preparation: Between runs rinse capillary with running buffer for 5 min. Before use rinse capillary with 1 M NaOH for 10-15 min and with water for 10-15 min.
Running buffer: 25 mM pH 3 Citrate buffer
Injection: Pressure injection at 30 mbar for 12 s.
Detector: MS, laboratory-constructed, electrospray 90°, sheath liquid MeOH:water:acetic acid 80:20:0.1 at 1 μL/min
Migration time: 10
Voltage: 30 kV
Current: 7.8 μA
Model: Crystal CE 300·

OTHER SUBSTANCES

Simultaneous: amphetamine, cocaine, diamorphine, methamphetamine, tetracaine

REFERENCE

Lazar,I.M.; Naisbitt,G.; Lee,M.L. Capillary electrophoresis-time-of-flight mass spectrometry of drugs of abuse, *Analyst*, **1998**, *123*, 1449–1454.

Prochlorperazine

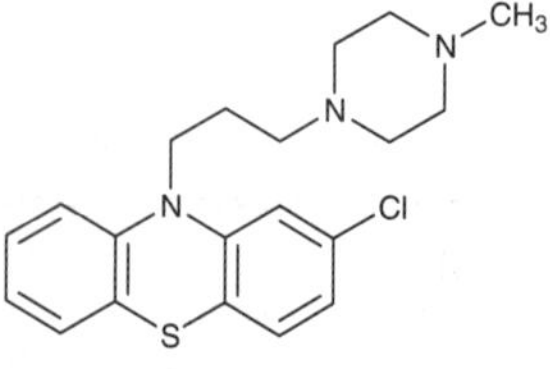

Molecular formula: $C_{20}H_{24}ClN_3S$
Molecular weight: 373.95
CAS Registry No.: 58-38-8, 84-02-6 (maleate), 1257-78-9 (edisylate)
Merck Index (12th ed.): 7942
Lednicer: 1 381

SAMPLE

Matrix: urine
Sample preparation: 10 mL Urine + 1 mL 5 M NaOH + 10 mL n-hexane, vortex for 1 min, centrifuge at 0° at 3000 g for 5 min, repeat the extraction twice more. Combine the organic layers and add them to 50 μL glacial acetic acid, evaporate to dryness under a stream of nitrogen at 30°, reconstitute the residue in 50 μL 50 mM sodium taurodeoxycholate, filter (0.2 μm PTFE), inject an aliquot.

CAPILLARY ELECTROPHORESIS

Capillary: 100 cm × 50 μm fused-silica (50 cm to detector) (ISCO)
Capillary preparation: Rinse capillary with 20 μL running buffer between injections. Condition a new capillary by filling with 1 M NaOH, let stand for 1 h, fill with 100 mM NaOH, let stand for 1 h, rinse with buffer. Every 20 injections rinse capillary with 200 μL 1 M NaOH, 200 μL water, and 200 μL running buffer, fill with running buffer.
Capillary temperature: 22
Running buffer: 40 mM pH 9.5 Borate buffer containing 10 mM sodium taurodeoxycholate
Injection: Load under vacuum at 7.5 kPa/s.
Detector: UV 240
Migration time: 14 (edisylate), 15 (dimalate)
Voltage: 30 kV
Model: ISCO Model 3140 electropherograph
Limit of detection: 5 ng/mL

OTHER SUBSTANCES

Extracted: acepromazine, amiodarone, amitriptyline, azaperone, chlorpromazine, cianopramine, clomipramine, clozapine, desethylamiodarone, desipramine, diclofensine, dothiepin, doxepin, imipramine, isocarboxazid, moclobemide, perphenazine, phenothiazine, pimozide, promazine, thioridazine, thiothixene, trifluoperazine, trimipramine

KEY WORDS

human; cow; pig; horse; protect from light

REFERENCE

Aumatell,A.; Wells,R.J. Determination of a cardiac antiarrhythmic, tricyclic antipsychotics and antidepressants in human and animal urine by micellar electrokinetic capillary chromatography using a bile salt, *J.Chromatogr.B*, **1995**, *669*, 331–344.

SAMPLE

Matrix: urine
Sample preparation: Dilute 10-fold with water, inject an aliquot.

CAPILLARY ELECTROPHORESIS

Capillary: 70 cm × 50 μm fused-silica (65.4 cm to detector) (Polymicro Technologies)
Capillary temperature: 25
Running buffer: 20 mM pH 3.5 Tris-formic acid containing 50 μg/mL FC-135 (Fluorad/3M)
Injection: Pressure injection at 2 psi.
Detector: UV 240
Migration time: 8.62
Voltage: 20 kV
Model: Bio-Rad BioFocus 3000

OTHER SUBSTANCES
Extracted: perphenazine, promazine, thioridazine, trifluoperazine, triflupromazine

KEY WORDS
detector at anode

REFERENCE
Muijelsaar,P.G.H.M.; Claessens,H.A.; Cramers,C.A. Determination of structurally related phenothiazines by capillary zone electrophoresis and micellar electrokinetic chromatography, *J.Chromatogr.A*, **1996**, *735*, 395–402.

Procyclidine

Molecular formula: $C_{19}H_{29}NO$
Molecular weight: 287.45
CAS Registry No.: 77-37-2, 1508-76-5 (HCl)
Merck Index (12th ed.): 7944
Lednicer: 1 47

SAMPLE
Matrix: solutions
Sample preparation: Inject an aliquot of a 100 µg/mL solution in running buffer.

CAPILLARY ELECTROPHORESIS
Capillary: 29 cm × 50 µm fused-silica (24.5 cm to detector) (Yongnian Optical Conductive Fiber Plant, China), coated with polyacrylamide
Capillary preparation: No details of the polyacrylamide coating process are provided. However, another paper (LC.GC 1997, 15, 40) by this group indicates that they use the procedure of Hjertén, thus: Adjust the pH of 20 mL water to 3.5 with acetic acid, add 80 µL 3-(trimethoxysilyl)propyl methacrylate (3-methacryloxypropyltrimethoxysilane), mix, suck into capillary, let stand at room temperature for 1 h, remove the solution, wash with water. Fill the capillary with a deaerated 3-4% acrylamide solution containing 1 µL/mL N,N,N',N'-tetramethylethylenediamine and 1 mg/mL potassium persulfate, let stand for 30 min, remove excess solution by aspiration, rinse with water, remove water by aspiration, dry at 35° (J. Chromatogr. 1985, 347, 191).
Capillary temperature: 25
Running buffer: 100 mM NaH_2PO_4 adjusted to pH 2.5
Injection: Electrokinetic injection at 15 kV for 3 s.
Detector: UV 200, UV 210
Migration time: 3.93
Voltage: 15 kV
Model: Bio-Rad BioFocus 3000

OTHER SUBSTANCES
Simultaneous: albuterol, alprenolol, atenolol, baclofen, bamethan, benproperine, benserazide, bisoprolol, bupranolol, butamirate, butethamate, carbuterol, celiprolol, clenbuterol, clobutinol, dipivefrin, isoproterenol (isoprenaline), metaproterenol (orciprenaline), metipranolol, metoprolol, norfenefrine, ornidazole, oxprenolol, phenylpropanolamine, pholedrine, pirbuterol, prilocaine, sotalol, synephrine, terbutaline, tocainide

KEY WORDS
coated capillary

REFERENCE
Koppenhoefer,B.; Epperlein,U.; Xiaofeng,Z.; Bingcheng,L. Separation of enantiomers of drugs by capillary electrophoresis. Part 4: Hydroxypropyl-γ-cyclodextrin as chiral solvating agent, *Electrophoresis*, **1997**, *18*, 924–930.

SAMPLE
Matrix: solutions
Sample preparation: Inject an aliquot of a 100 μg/mL solution in running buffer.

CAPILLARY ELECTROPHORESIS
Capillary: 30 cm × 50 μm fused-silica (25.5 cm to detector), coated with polyacrylamide
Capillary preparation: Adjust the pH of 20 mL water to 3.5 with acetic acid, add 80 μL 3-(trimethoxysilyl)propyl methacrylate (3-methacryloxypropyltrimethoxysilane), mix, suck into capillary, let stand at room temperature for 1 h, remove the solution, wash with water. Fill the capillary with a deaerated 3-4% acrylamide solution containing 1 μL/mL N,N,N',N'-tetramethylethylenediamine and 1 mg/mL potassium persulfate, let stand for 30 min, remove excess solution by aspiration, rinse with water, remove water by aspiration, dry at 35° (J. Chromatogr. 1985, 347, 191).
Capillary temperature: 25
Running buffer: 100 mM NaH_2PO_4 containing 45 mM hydroxypropyl-α- cyclodextrin, adjusted to pH 2.5 with phosphoric acid
Injection: Electrokinetic injection at 15 kV for 3 s.
Detector: UV 200
Voltage: 15 kV
Model: Bio-Rad BioFocus 3000

KEY WORDS
chiral; coated capillary; comparison with the use of other cyclodextrins; this running buffer gave the greatest enantiomeric separation.; α=1.020

REFERENCE
Lin,B.; Zhu,X.; Koppenhoefer,B.; Epperlein,U. Investigation of 123 chiral drugs by cyclodextrin-modified capillary electrophoresis, *LC.GC*, **1997**, *15*, 40–46.

SAMPLE
Matrix: solutions

CAPILLARY ELECTROPHORESIS
Capillary: 29-36 cm × 50 μm fused-silica (24.5-31.5 cm to detector) (Yongnian Optical Conductive Fiber Plant, China) coated with polyacrylamide
Capillary preparation: Coat capillary as follows. Adjust the pH of 20 mL water to 3.5 with acetic acid, add 80 μL 3-(trimethoxysilyl)propyl methacrylate (3-methacryloxypropyltrimethoxysilane), mix, suck into capillary, let stand at room temperature for 1 h, remove the solution, wash with water. Fill the capillary with a deaerated 3-4% acrylamide solution containing 1 μL/mL N,N,N',N'-tetramethylethylenediamine and 1 mg/mL potassium persulfate, let stand for 30 min, remove excess solution by aspiration, rinse with water, remove water by aspiration, dry at 35° (J. Chromatogr. 1985, 347, 191).
Capillary temperature: 25
Running buffer: 100 mM pH 2.5 NaH_2PO_4 (A) or 100 mM pH 2.5 NaH_2PO_4 containing 45 mM hydroxypropyl-α-cyclodextrin (Wacker, Munich) (B)
Injection: Electromigration at 15 kV for 3 s.
Detector: UV 200; UV 210
Migration time: 3.93 (A); 16.35, 16.67 (B) (enantiomers)
Voltage: 15 kV
Model: Bio-Focus 3000

OTHER SUBSTANCES
Also analyzed: albuterol (salbutamol), alprenolol, amorolfine, atenolol, atropine, azelastine, baclofen, bamethan, benproperine, benserazide, biperiden, bisoprolol, brompheniramine, bupivacaine, bupranolol, butamirate, butethamate, carazolol, carbuterol, carteolol, carvedilol, celiprolol, chloroquine, chlorpheniramine, chlorphenoxamine, cicletanine, clenbuterol, clidinium bromide, clobutinol, dimethindene, dipivefrin, disopyramide, dobutamine, doxylamine, fendiline, flecainide, gallopamil, homatropine, ipratropium bromide, isoproterenol (isoprenaline), isothipendyl, ketamine, meclizine, mefloquine, mepindolol, mequitazine, metaclazepam, metaproterenol (orciprenaline), metipranolol, metoprolol, nafronyl (naftidrofuryl), nefopam, nicardipine, norfenefrine, ofloxacin, ornidazole, orphenadrine, oxomemazine, oxprenolol, oxybutynin, phenoxybenzamine, phenylpropanolamine, pholedrine, pindolol, pirbuterol, prilocaine, pro-

methazine, propafenone, propranolol, reproterol, sotalol, sulpride, synephrine, talinolol, terbutaline, tetrahydrozoline (tetryzoline), theodrenaline, tioconazole, tocainide, trihexyphenidyl, trimeprazine (alimemazine), trimipramine, tropicamide, verapamil, zopiclone

KEY WORDS
coated capillary; chiral

REFERENCE
Koppenhoefer,B.; Eperlein,U.; Schlunk,R.; Zhu,X.; Lin,B. Separation of enantiomers of drugs by capillary electrophoresis. V. Hydroxypropyl-α-cyclodextrin as chiral solvating agent, *J.Chromatogr.A*, **1998**, *793*, 153–164.

Progesterone

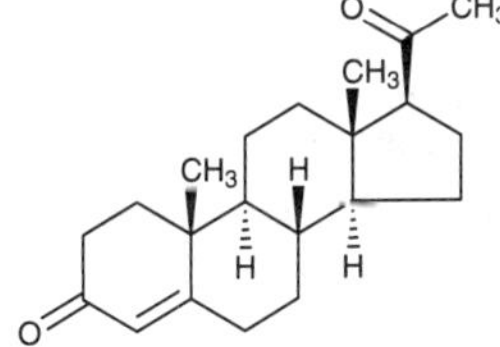

Molecular formula: $C_{21}H_{30}O_2$
Molecular weight: 314.47
CAS Registry No.: 57-83-0
Merck Index (12th ed.): 7956
Lednicer: 2 164

SAMPLE
Matrix: blood
Sample preparation: Condition a 1 mL 100 mg Bond Elut C18 SPE cartridge with 1 mL MeOH and 1 mL water. Add 500 μL plasma to the cartridge, wash with 2 mL water, wash with 2 mL MeOH:water 20:80, elute with two 500 μL portions of MeOH. Evaporate the eluate to dryness under a stream of nitrogen, reconstitute the residue in 50 μL MeCN:MeOH:20 mM pH 8 Tris-HCl buffer 22.5:22.5:45, inject an aliquot.

CAPILLARY ELECTROPHORESIS
Capillary: 50 cm × 100 μm fused-silica (Polymicro Technologies) packed with 3 μm ODS Hypersil for 20 cm (20 cm to detector) (details in paper)
Running buffer: MeCN:MeOH:20 mM pH 8 Tris-HCl buffer 37.5:37.5:25
Injection: Electrokinetic injection at 15 kV for 5 s (25 nL). (Injector modified for injection from small volume samples; details in paper.)
Detector: UV 240
Migration time: 13
Internal standard: norethindrone
Voltage: 25 kV (?)
Model: Isco Model 3850

OTHER SUBSTANCES
Extracted: androstenedione, 17α-hydroxyprogesterone, 20α-hydroxyprogesterone, testosterone

KEY WORDS
electrochromatography; comparison with HPLC; plasma; SPE

REFERENCE
Stead,D.A.; Reid,R.G.; Taylor,R.B. Capillary electrochromatography of steroids. Increased sensitivity by on-line concentration and comparison with high-performance liquid chromatography, *J.Chromatogr.A*, **1998**, *798*, 259–267.

SAMPLE
Matrix: solutions
Sample preparation: Prepare a 10 mg/mL solution in MeOH, inject an aliquot.

CAPILLARY ELECTROPHORESIS
Capillary: 62 cm × 53 μm fused-silica (50 cm to detector) (Polymicro Technologies)

Capillary preparation: Rinse with running buffer for 1 min between runs.
Running buffer: 50 mM pH 9.0 Phosphate borate buffer containing 50 mM dehydrocholate, 50 mM taurocholate, and 50 mM sodium dodecyl sulfate
Detector: UV 254
Migration time: 41
Voltage: 15 kV
Current: 56 μA
Model: Laboratory constructed

OTHER SUBSTANCES
Simultaneous: corticosterone, cortisone, cortisone acetate, deoxycorticosterone, fludrocortisone, fludrocortisone acetate, fluocinolone acetonide, hydrocortisone, hydrocortisone-21-acetate, 6α-methylprednisolone, prednisolone, prednisolone acetate, prednisone, prednisone acetate, triamcinolone, triamcinolone acetonide

REFERENCE
Bumgarner,J.G.; Khaledi,M.G. Mixed micellar electrokinetic chromatography of corticosteroids, *Electrophoresis*, **1994**, *15*, 1260–1266.

SAMPLE
Matrix: solutions

CAPILLARY ELECTROPHORESIS
Capillary: 62 cm × 53 μm fused-silica (50 cm to detector) (Polymicro Technologies)
Capillary preparation: Rinse capillary with running buffer for 1 min between runs.
Running buffer: 50 mM pH 9.0 Phosphate/borate buffer containing 33 mM taurocholate, 33 mM glycodeoxycholate, and 70 mM butanesulfonate
Detector: UV 254
Migration time: 48
Voltage: 15 kV
Model: laboratory-constructed

OTHER SUBSTANCES
Simultaneous: corticosterone, cortisone, cortisone acetate, deoxycorticosterone, fludrocortisone, fludrocortisone acetate, fluocinolone acetonide, hydrocortisone, hydrocortisone-21-acetate, 6α-methylprednisolone, prednisolone, prednisolone acetate, prednisone, prednisone acetate, triamcinolone, triamcinolone acetonide

REFERENCE
Bumgarner,J.G.; Khaledi,M.G. Mixed micelles of short chain alkyl surfactants and bile slats in electrokinetic chromatography: Enhanced separation of corticosteroids, *J.Chromatogr.A*, **1996**, *738*, 275–283.

SAMPLE
Matrix: solutions

CAPILLARY ELECTROPHORESIS
Capillary: 57 cm × 75 μm fused-silica (50 cm to detector)
Capillary preparation: Before each run rinse with running buffer for 2 min. After each run rinse capillary with 100 mM NaOH for 2 min and with water for 2 min. Condition new capillaries with 100 mM HCl for 5 min, with 100 mM NaOH for 10 min, and with water for 5 min.
Capillary temperature: 23
Running buffer: EtOH:buffer 2:98 (Buffer was 50 mM pH 9.2 borate buffer containing 75 mM sodium dodecyl sulfate.)
Injection: Low pressure injection for 5 s.
Detector: UV 214
Migration time: 64
Voltage: 20 kV
Model: Beckman P/ACE 5510

OTHER SUBSTANCES
Simultaneous: estradiol, estriol, estrone, hydrocortisone, testosterone

REFERENCE
Gu,X.; Meleka-Boules,M.; Chen,C.-L. Micellar electrokinetic capillary chromatography combined with immunoaffinity chromatography for identification and determination of dexamethasone and flumethasone in equine urine, *J.Capillary Electrophor.*, **1996**, *3*, 43–49.

SAMPLE
Matrix: solutions

CAPILLARY ELECTROPHORESIS
Capillary: 57 cm × 50 μm fused-silica (50 cm to detector) (Polymicro Technologies)
Capillary preparation: After each run rinse capillaries with 100 mM NaOH for 10 min and with water for 5 min. Fill with running buffer and apply voltage for 10 min. Rinse new capillaries with 1 M HCl for 5 min, with 100 mM NaOH for 10 min, and with water for 5 min.
Capillary temperature: 25
Running buffer: MeOH:buffer 20:80 (Buffer was 38 mM pH 7.4 phosphate buffer containing 40 mM sodium cholate.)
Injection: Pressure injection for 5 s (3.1 nL).
Detector: UV (wavelength not given)
Migration time: 16.14
Voltage: 15 kV
Model: Beckman P/ACE 5500

OTHER SUBSTANCES
Simultaneous: aldosterone, dehydroepiandrosterone, estradiol, hydrocortisone, 17α-hydroxyprogesterone

REFERENCE
Hsiao,M.-W.; Lin,S.-T. Separation of steroids by micellar electrokinetic capillary chromatography with sodium cholate, *J.Chromatogr.Sci.*, **1997**, *35*, 259–264.

SAMPLE
Matrix: solutions

CAPILLARY ELECTROPHORESIS
Capillary: 32 cm × 100 μm fused-silica (Polymicro Technologies) packed for 24 cm with 1.5 μm Chromspher ODS (Chrompack)
Capillary preparation: Sonicate 30 mg 1.5 μm Chromspher ODS in 2 mL hexane:isopropanol 50:50 for 1 h, pump the slurry into the capillary at 600 bar using hexane:isopropanol 90:10 for 4 h, release the pressure over 1 min, wash with acetone, wash with water at 600 bar. Fabrication of frits is described in the paper.
Capillary temperature: 25
Running buffer: MeCN:water 60:40 containing 5 mM sodium dodecyl sulfate and 1.6 mM sodium tetraborate, pH 9.25
Injection: Electrokinetic injection at 10 kV.
Detector: UV (wavelength not given)
Migration time: 3.2
Voltage: 20 kV
Model: Hewlett Packard HP [3D]CE

OTHER SUBSTANCES
Simultaneous: hydrocortisone, 17α-methyltestosterone, testosterone

KEY WORDS
electrochromatography

REFERENCE
Seifar,R.M.; Kok,W.T.; Kraak,J.C.; Poppe,H. Capillary electrochromatography with 1.5 μm ODS-modified nonporous silica spheres, *Chromatographia*, **1997**, *46*, 131–136.

SAMPLE
Matrix: solutions

CAPILLARY ELECTROPHORESIS
Capillary: 67 cm × 50 μm fused-silica (60 cm to detector)
Capillary preparation: Before each run equilibrate with running buffer for 3 min. After each run wash with 100 mM NaOH for 3 min and with water. At the beginning of each day rinse the capillary with 100 mM NaOH for 30 min and with water for 30 min.
Capillary temperature: 25 ± 0.1
Running buffer: 25 mM pH 9 Bicine buffer containing 10 mM sodium dodecyl sulfate and 5 mM N-dodecyl-N,N-dimethyl-3-ammonio-1-propanesulfonate (SB3-12)
Injection: High pressure injection for 5 s
Detector: UV 254
Migration time: 7.7
Voltage: 30 kV
Current: <40 μA
Model: Beckman P/ACE 5010

OTHER SUBSTANCES
Simultaneous: cortisone, 11-desoxycortisol, 21-desoxycortisol, dimethyltestosterone, hydrocortisone, 17-hydroxyprogesterone, testosterone propionate, testosterone

REFERENCE
Valbuena,G.A.; Rao,L.V.; Petersen,J.R.; Okorodudu,A.O.; Bissell,M.G.; Mohammad,A.A. Anionic-zwitterionic mixed micelles in the micellar electrokinetic separation of clinically relevant steroids on a fused-silica capillary, *J.Chromatogr.A*, **1997**, *781*, 467–474.

SAMPLE
Matrix: solutions

CAPILLARY ELECTROPHORESIS
Capillary: 32 cm × 50 μm fused-silica packed for 24 cm with 1.8 μm Zorbax ODS (Polymicro Technologies) (Column preparation described in paper.)
Capillary temperature: 25
Running buffer: MeCN:0.8 mM sodium tetraborate 80:20 containing 5 mM sodium dodecyl sulfate
Injection: Electrokinetic injection at 20 kV for 1 s.
Detector: UV 254
Migration time: 4.7
Voltage: 25 kV
Model: Hewlett-Packard HP [3D]CE

OTHER SUBSTANCES
Simultaneous: estradiol, estriol, estrone, hydrocortisone, methyltestosterone, 4-pregnen-20α-ol-3-one, testosterone

KEY WORDS
electrochromatography

REFERENCE
Seifar,R.M; Kraak,J.C.; Kok,W.T.; Poppe,H. Capillary electrochromatography with 1.8-μm ODS-modified porous silica particles, *J.Chromatogr.A*, **1998**, *808*, 71–77.

SAMPLE
Matrix: urine
Sample preparation: Vortex 5 mL urine and 2 mL hexane for 3 min, centrifuge at 3500 rpm for 3 min, repeat extraction. Combine the organic layers and evaporate them to dryness at room temperature, reconstitute the residue in 500 μL 20 mM pH 2.5 phosphate buffer containing 10 mM sodium dodecyl sulfate, inject an aliquot.

CAPILLARY ELECTROPHORESIS
Capillary: 64.5 cm × 50 μm fused-silica (56 cm to detector) (Polymicro Technologies)
Capillary preparation: Between runs flush capillary with 1 M NaOH for 1 min, with MeOH for 1 min, with 100 mM naOH for 1 min, with water for 2 min, and with running buffer for 2

min. Before use rinse capillary with 1 M NaOH for 10 min, with MeOH for 5 min, with water for 5 min, and with running buffer for 5 min.
Capillary temperature: 30
Running buffer: MeOH:100 mM pH 2 phosphate buffer 20:80 containing 50 mM sodium dodecyl sulfate
Injection: Inject a 5.8 cm water plug at 50 mbar, inject sample using voltage, when the current reaches 90% of current in a capillary filled only with running buffer replace sample with running buffer.
Detector: UV 247
Migration time: 8.5
Voltage: -25 kV
Model: Hewlett Packard 3D
Limit of detection: 10.0 ppb (S/N 3)

OTHER SUBSTANCES
Extracted: cortisone, hydrocortisone, testosterone

KEY WORDS
detector at anode

REFERENCE
Quirino,J.P.; Terabe,S. On-line concentration of neutral analytes for micellar electrokinetic chromatography. 5. Field-enhanced sample injection with reverse migrating micelles, *Anal.Chem.*, **1998**, *70*, 1893–1901.

Proglumide

Molecular formula: $C_{18}H_{26}N_2O_4$
Molecular weight: 334.42
CAS Registry No.: 6620-60-6
Merck Index (12th ed.): 7958
Lednicer: 2 93

SAMPLE
Matrix: solutions

CAPILLARY ELECTROPHORESIS
Capillary: 47 cm × 50 μm fused-silica (40 cm to detector)
Capillary preparation: Before each run rinse capillary with water for 2 min, with 100 mM HCl for 2 min, with water for 2 min, and with 500 mM KOH for 2 h
Capillary temperature: 25
Running buffer: MeOH:50 mM pH 7.0 phosphate buffer containing 15 mM d-(+)-tubocurarine chloride (Caution! Tubocurarine is highly toxic!)
Injection: Pressure injection at 0.5 psi for 1 s.
Detector: UV 254
Migration time: 15.07, 15.22 (enantiomers)
Voltage: -15 kV
Model: Beckman P/ACE 2000

OTHER SUBSTANCES
Simultaneous: ketoprofen

KEY WORDS
detector at anode; chiral

REFERENCE
Nair,U.B.; Armstrong,D.W.; Hinze,W.L. Characterization and evaluation of d-(+)-tubocurarine chloride as a chiral selector for capillary electrophoretic enantioseparations, *Anal.Chem.*, **1998**, *70*, 1059–1065.

Prolintane

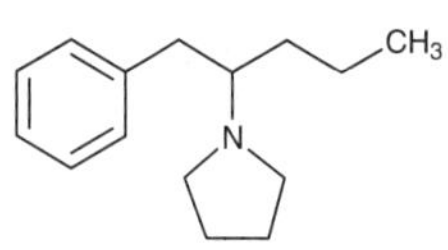

Molecular formula: C$_{15}$H$_{23}$N
Molecular weight: 217.35
CAS Registry No.: 493-92-5, 1211-28-5 (HCl)
Merck Index (12th ed.): 7964
Lednicer: 1 70

SAMPLE
Matrix: urine
Sample preparation: Mix urine with 40 mM pH 8.5 phosphate buffer in the ratio 3:10

CAPILLARY ELECTROPHORESIS
Capillary: 100 cm × 75 μm (64 cm to detector)
Capillary preparation: Before each run flush under pressure with 100 mM NaOH for 2 min, with water for 2 min, and with running buffer for 4 min.
Running buffer: 50 mM pH 8.5 Phosphate buffer
Injection: Hydrodynamic vacuum injection at 0.27 bar for 2 s.
Detector: UV 211
Migration time: 3
Voltage: 15 kV
Current: 56 μA
Model: Europhor with Prime Vision I injector
Limit of quantitation: 1 μg/mL
Limit of detection: 300 ng/mL

KEY WORDS
the metabolite, oxoprolintane, can be determined using 50 mM pH 8.5 phosphate buffer containing 4 mM sodium dodecyl sulfate and 2 mM β-cyclodextrin as running buffer.

REFERENCE
Espartero,A.G.; Pérez,J.A.; Zapardiel,A.; Bermejo,E.; Hernández,L. Direct determinationn of prolintane and its metabolite oxoprolintane in human urine by capillary zone electrophoresis and β-cyclodextrin-modified micellar electrokinetic chromatography, *J.Chromatogr.A*, **1997**, *778*, 355–361.

Promazine

Molecular formula: C$_{17}$H$_{20}$N$_2$S
Molecular weight: 284.43
CAS Registry No.: 58-40-2, 53-60-1 (HCl)
Merck Index (12th ed.): 7966
Lednicer: 1 377

SAMPLE
Matrix: solutions

CAPILLARY ELECTROPHORESIS
Capillary: 59.6 cm × 75 μm fused-silica (52.1 cm to detector) (Polymicro Technologies)
Capillary preparation: Purge with running buffer before each run. At the beginning of each day purge using 50-60 kPa vacuum with 500 mM NaOH for 5 min, with water for 5 min, with MeCN for 5 min, and with running buffer for 5 min.
Running buffer: MeCN:MeOH:acetic acid 49:50:1 containing 20 mM ammonium acetate
Injection: Hydrostatic injection at 10 cm for 5 s.
Detector: UV 214
Migration time: 3.62

Voltage: 30 kV
Model: Waters Quanta 4000

OTHER SUBSTANCES
Simultaneous: benzquinamide, deserpidine, ethopropazine, mesoridazine, methotrimeprazine, molindone, promethazine, reserpine, thioridazine, thiothixene

REFERENCE
Leung,G.N.W.; Tang,H.P.O.; Tso,T.S.C.; Wan,T.S.M. Separation of basic drugs with non-aqueous capillary electrophoresis, *J.Chromatogr.A*, **1996**, *738*, 141–154.

SAMPLE
Matrix: urine
Sample preparation: 10 mL Urine + 1 mL 5 M NaOH + 10 mL n-hexane, vortex for 1 min, centrifuge at 0° at 3000 g for 5 min, repeat the extraction twice more. Combine the organic layers and add them to 50 μL glacial acetic acid, evaporate to dryness under a stream of nitrogen at 30°, reconstitute the residue in 50 μL 50 mM sodium taurodeoxycholate, filter (0.2 μm PTFE), inject an aliquot.

CAPILLARY ELECTROPHORESIS
Capillary: 100 cm × 50 μm fused-silica (50 cm to detector) (ISCO)
Capillary preparation: Rinse capillary with 20 μL running buffer between injections. Condition a new capillary by filling with 1 M NaOH, let stand for 1 h, fill with 100 mM NaOH, let stand for 1 h, rinse with buffer. Every 20 injections rinse capillary with 200 μL 1 M NaOH, 200 μL water, and 200 μL running buffer, fill with running buffer.
Capillary temperature: 22
Running buffer: 40 mM pH 9.5 Borate buffer containing 10 mM sodium taurodeoxycholate
Injection: Load under vacuum at 7.5 kPa/s.
Detector: UV 240
Migration time: 10
Voltage: 30 kV
Model: ISCO Model 3140 electropherograph
Limit of detection: 7 ng/mL

OTHER SUBSTANCES
Extracted: acepromazine, amiodarone, amitriptyline, azaperone, chlorpromazine, cianopramine, clomipramine, clozapine, desethylamiodarone, desipramine, diclofensine, dothiepin, doxepin, imipramine, isocarboxazid, moclobemide, perphenazine, phenothiazine, pimozide, prochlorperazine, thioridazine, thiothixene, trifluoperazine, trimipramine

KEY WORDS
human; cow; pig; horse; protect from light

REFERENCE
Aumatell,A.; Wells,R.J. Determination of a cardiac antiarrhythmic, tricyclic antipsychotics and antidepressants in human and animal urine by micellar electrokinetic capillary chromatography using a bile salt, *J.Chromatogr.B*, **1995**, *669*, 331–344.

SAMPLE
Matrix: urine
Sample preparation: Dilute 10-fold with water, inject an aliquot.

CAPILLARY ELECTROPHORESIS
Capillary: 70 cm × 50 μm fused-silica (65.4 cm to detector) (Polymicro Technologies)
Capillary temperature: 25
Running buffer: 20 mM pH 3.5 Tris-formic acid containing 50 μg/mL FC-135 (Fluorad/3M)
Injection: Pressure injection at 2 psi.
Detector: UV 240
Migration time: 7.87
Voltage: 20 kV
Model: Bio-Rad BioFocus 3000

OTHER SUBSTANCES
Extracted: perphenazine, prochlorperazine, thioridazine, trifluoperazine, triflupromazine

KEY WORDS
detector at anode

REFERENCE
Muijelsaar,P.G.H.M.; Claessens,H.A.; Cramers,C.A. Determination of structurally related phenothiazines by capillary zone electrophoresis and micellar electrokinetic chromatography, *J.Chromatogr.A*, **1996**, *735*, 395–402.

Promethazine

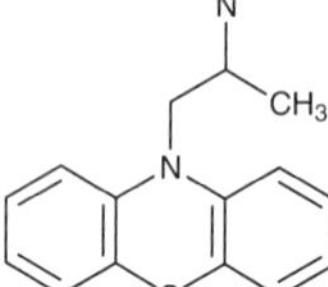

Molecular formula: $C_{17}H_{20}N_2S$
Molecular weight: 284.43
CAS Registry No.: 60-87-7, 58-33-3 (HCl), 17693-51-5 (teoclate)
Merck Index (12th ed.): 7970
Lednicer: 1 373

SAMPLE
Matrix: formulations
Sample preparation: Grind tablets, extract twice with MeOH by sonicating at low temperature, centrifuge, filter (0.45 μm) the supernatant, inject an aliquot of the filtrate.

CAPILLARY ELECTROPHORESIS
Capillary: 47 cm × 50 μm (40 cm to detector) (Polymicro Technologies)
Capillary temperature: 25
Running buffer: 50 mM pH 7.5 Borate buffer containing 50 mM phosphate, 10 mM sodium dodecyl sulfate, 10 mM β-cyclodextrin, and 10 mM tetrabutylammonium hydrogen sulfate
Injection: Inject at a pressure of 0.5 psi for 1 s (5.2 nL)
Detector: UV 214
Migration time: 4.7
Voltage: 21.5 kV
Model: Beckman P/ACE 2000
Limit of detection: 139.8 pg

OTHER SUBSTANCES
Simultaneous: chlorpheniramine, cyclizine, dimenhydrinate, doxylamine, methapyrilene, pheniramine, thonylamine, triprolidine

KEY WORDS
tablets

REFERENCE
Ong,C.P.; Ng,C.L.; Lee,H.K.; Li,S.F.Y. Determination of antihistamines in pharmaceuticals by capillary electrophoresis, *J.Chromatogr.*, **1991**, *588*, 335–339.

SAMPLE
Matrix: solutions
Sample preparation: Prepare a solution in running buffer, inject an aliquot.

CAPILLARY ELECTROPHORESIS
Capillary: 60 cm × 75 μm fused-silica (52.4 cm to detector)
Capillary preparation: After each run flush with 500 mM KOH for 2-3 min, flush with water, fill with running buffer
Running buffer: 10 mM Na_2HPO_4 containing 2% heparin sodium (MW 10000, 11% S, Scientific Protein Laboratories, Waunakee, WI) adjusted to pH 5 with phosphoric acid

Injection: Hydrostatic injection
Detector: UV 214
Migration time: 12.5
Model: Waters Quanta 4000

OTHER SUBSTANCES
Also analyzed: anabasine, brompheniramine, bupivacaine, carbinoxamine, chlorcyclizine, chloroquine, chlorpheniramine, dimethindene, doxylamine, enpiroline, halofantrine, hydroxychloroquine, indapamide, mefloquine, nornicotine, pheniramine, primaquine, quinacrine, tetramisole

REFERENCE
Stalcup,A.M.; Agyei,N.M. Heparin: a chiral mobile-phase additive for capillary zone electrophoresis, *Anal.Chem.*, **1994**, *66*, 3054–3059.

SAMPLE
Matrix: solutions

CAPILLARY ELECTROPHORESIS
Capillary: 40 cm × 50 μm fused-silica (35.5 cm to detector) (Polymicro Technologies)
Capillary preparation: Between runs purge with water for 50 s, with 100 mM NaOH for 50 s, with water for 60 s, and with running buffer for 60 s.
Running buffer: 50 mM pH 6.0 Sodium phosphate buffer containing 6 mg/mL sulfobutyl ether-β-cyclodextrin (Perkin-Elmer)
Injection: Pressure injection at 5 psi
Detector: UV 206
Migration time: 9.7, 10 (enantiomers)
Voltage: 15 kV
Model: Bio-Rad Biofocus 3000

KEY WORDS
chiral

REFERENCE
Desiderio,C.; Fanali,S. Use of negatively charged sulfobutyl ether-β-cyclodextrin for enantiomeric separation by capillary electrophoresis, *J.Chromatogr.A*, **1995**, *716*, 183–196.

SAMPLE
Matrix: solutions
Sample preparation: Inject an aliquot of a 100 μg/mL solution in water:running buffer 50:50.

CAPILLARY ELECTROPHORESIS
Capillary: 44.5 cm × 50 μm acrylamide-coated fused-silica (Bio-Rad)
Capillary temperature: 30
Running buffer: 100 mM NaH_2PO_4 containing 15 mM gamma-cyclodextrin, adjusted to pH 2.5 with phosphoric acid
Injection: Electrokinetic injection at 8 kV for 6 s.
Detector: UV 200
Migration time: 12.19, 12.44 (enantiomers)
Voltage: 14 kV
Model: Bio-Rad BioFocus 3000

OTHER SUBSTANCES
Also analyzed: albuterol, alprenolol, atenolol, atropine, baclofen, bamethan, benserazide, biperiden, bisoprolol, bupranolol, butetamate, carazolol, carbuterol, carvedilol, celiprolol, chloroquine, chlorpheniramine (chlorphenamine), clidinium bromide, clobutinol, disopyramide, dobutamine, flecainide, homatropine, ipratropium bromide, isoproterenol, isothipendyl, ketamine, mefloquine, mequitazine, metaproterenol (orciprenaline), metipranolol, nafronyl (naftidrofuryl), nefopam, ofloxacin, orphenadrine, oxomemazine, oxprenolol, phenoxybenzamine, pholedrine, pindolol, pirbuterol, prilocaine, propafenone, propranolol, sotalol, synephrine, terbutaline, tetrahydrozoline (tetryzoline), tocainide, trihexyphenidyl, trimeprazine (alimemazine), trimipramine, tropicamide, verapamil, zopiclone

KEY WORDS
chiral; coated capillary

REFERENCE
Koppenhoefer,B.; Epperlein,U.; Christian,B.; Yibing,J.; Yuying,C.; Bingcheng,L. Separation of enantiomers of drugs by capillary electrophoresis. I. γ-Cyclodextrin as chiral solvating agent, *J.Chromatogr.A*, **1995**, *717*, 181–190.

SAMPLE
Matrix: solutions

CAPILLARY ELECTROPHORESIS
Capillary: 62 cm × 75 μm fused-silica (54.5 cm to detector) (Polymicro Technologies)
Capillary preparation: Purge with running buffer before each run. At the beginning of each day purge using 50-60 kPa vacuum with 500 mM NaOH for 5 min, with water for 5 min, with MeCN for 5 min, and with running buffer for 5 min.
Running buffer: MeCN:MeOH:acetic acid 49:50:1 containing 20 mM ammonium acetate
Injection: Hydrostatic injection at 10 cm for 5 s.
Detector: UV 214
Migration time: 3.20
Voltage: 30 kV
Model: Waters Quanta 4000

OTHER SUBSTANCES
Simultaneous: buclizine, chlorcyclizine, chlorpheniramine, cyclizine, dimenhydrinate, methaphenilene, pheniramine, pyrilamine, pyrrobutamine, tripelennamine

REFERENCE
Leung,G.N.W.; Tang,H.P.O.; Tso,T.S.C.; Wan,T.S.M. Separation of basic drugs with non-aqueous capillary electrophoresis, *J.Chromatogr.A*, **1996**, *738*, 141–154.

SAMPLE
Matrix: solutions
Sample preparation: Inject an aliquot of a 100 μg/mL solution in running buffer.

CAPILLARY ELECTROPHORESIS
Capillary: 32 cm × 50 μm fused-silica (27.5 cm to detector) (Yongnian Optical Conductive Fiber Plant, China), coated with polyacrylamide
Capillary preparation: No details of the polyacrylamide coating process are provided. However, another paper (LC.GC 1997, 15, 40) by this group indicates that they use the procedure of Hjertén, thus: Adjust the pH of 20 mL water to 3.5 with acetic acid, add 80 μL 3-(trimethoxysilyl)propyl methacrylate (3-methacryloxypropyltrimethoxysilane), mix, suck into capillary, let stand at room temperature for 1 h, remove the solution, wash with water. Fill the capillary with a deaerated 3-4% acrylamide solution containing 1 μL/mL N,N,N',N'-tetramethylethylenediamine and 1 mg/mL potassium persulfate, let stand for 30 min, remove excess solution by aspiration, rinse with water, remove water by aspiration, dry at 35° (J. Chromatogr. 1985, 347, 191).
Capillary temperature: 25
Running buffer: 100 mM NaH_2PO_4 adjusted to pH 2.5 (A) or 100 mM NaH_2PO_4 containing 45 mM hydroxypropyl-gamma-cyclodextrin, adjusted to pH 2.5 (B)
Injection: Electrokinetic injection at 15 kV for 3 s.
Detector: UV 200, UV 210
Migration time: 5.06 5.06 (A), 15.35, 15.97 (B, enantiomers)
Voltage: 15 kV
Model: Bio-Rad BioFocus 3000

OTHER SUBSTANCES
Simultaneous: atropine, carazolol, cicletanine, dimethindene, fendiline, homatropine, ipratropium bromide, isothipendyl, mefloquine, metaclazepam, nafronyl (naftidrofuryl), nefopam, nicardipine, reproterol, tetrahydrozoline (tetryzoline), theodrenaline, tioconazole, trihexyphenidyl, trimeprazine (alimemazine), trimipramine

KEY WORDS
coated capillary; chiral

REFERENCE
Koppenhoefer,B.; Epperlein,U.; Xiaofeng,Z.; Bingcheng,L. Separation of enantiomers of drugs by capillary electrophoresis. Part 4: Hydroxypropyl-γ-cyclodextrin as chiral solvating agent, *Electrophoresis*, **1997**, *18*, 924–930.

SAMPLE
Matrix: solutions
Sample preparation: Inject an aliquot of a 100 μg/mL solution in running buffer.

CAPILLARY ELECTROPHORESIS
Capillary: 30 cm × 50 μm fused-silica (25.5 cm to detector), coated with polyacrylamide
Capillary preparation: Adjust the pH of 20 mL water to 3.5 with acetic acid, add 80 μL 3-(trimethoxysilyl)propyl methacrylate (3-methacryloxypropyltrimethoxysilane), mix, suck into capillary, let stand at room temperature for 1 h, remove the solution, wash with water. Fill the capillary with a deaerated 3-4% acrylamide solution containing 1 μL/mL N,N,N',N'-tetramethylethylenediamine and 1 mg/mL potassium persulfate, let stand for 30 min, remove excess solution by aspiration, rinse with water, remove water by aspiration, dry at 35° (J. Chromatogr. 1985, 347, 191).
Capillary temperature: 25
Running buffer: 100 mM NaH_2PO_4 containing 45 mM hydroxypropyl-gamma- cyclodextrin, adjusted to pH 2.5 with phosphoric acid
Injection: Electrokinetic injection at 15 kV for 3 s.
Detector: UV 200
Voltage: 15 kV
Model: Bio-Rad BioFocus 3000

KEY WORDS
chiral; coated capillary; comparison with the use of other cyclodextrins; this running buffer gave the greatest enantiomeric separation.; α=1.047

REFERENCE
Lin,B.; Zhu,X.; Koppenhoefer,B.; Epperlein,U. Investigation of 123 chiral drugs by cyclodextrin-modified capillary electrophoresis, *LC.GC*, **1997**, *15*, 40–46.

SAMPLE
Matrix: solutions

CAPILLARY ELECTROPHORESIS
Capillary: 36 cm × 50 μm fused-silica coated with linear polyacrylamide (31.5 cm to detector) (GL Science)
Capillary preparation: At the beginning and end of each day rinse capillary with capillary wash solution (Bio-Rad Cat. No. 148-5022) at 690 kPa for more than 3 min and with water at 690 kPa for more than 3 min. Coat capillary as follows. Treat capillary with 1 M NaOH at room temperature for 1 h, rinse with water, dry by passing nitrogen gas through the capillary at 110° for 6 h. Pass thionyl chloride through the capillary using a suction pump for several min, seal capillary at both ends and heat at 70° for 6 h. Unseal the capillary and fill with 250 mM vinyl magnesium bromide in THF by suction, seal the capillary, heat at 70° for 6 h. Open the capillary and rinse it with THF for several min, rinse with distilled water, fill the capillary with polymerization solution, heat at 28 ± 2° for 1 h, condition at -100 V/cm for 30 min (Anal. Sci. 1994, 10, 1). (The polymerization solution was 5% acrylamide in water containing 49 mM Tris, 384 mM glycine, and 0.1% sodium dodecyl sulfate, degas in an ultrasonic bath. Add 40 μL 10% N,N,N',N'-tetramethylethylenediamine and 10 μL 10% ammonium persulfate to 5 mL of the degassed solution, mix thoroughly.)
Running buffer: n-Propanol:50 mM pH 5.0 Phosphate buffer 10:90
Injection: Before each injection rinse with water at 690 kPa for 30 s, rinse with running buffer at 690 kPa for 30 s, partially fill with separation solution (500 μM α_1-acid glycoprotein (Cohn fraction VI) (ICN) in running buffer) at 6.9 kPa for 190 s (27 cm), inject sample at 6.9 kPa for 2 s, electrophorese with running buffer (Note that α_1-acid glycoprotein from other suppliers may provide inferior results).

Detector: UV 210
Migration time: Resolution of enantiomers 2.6
Voltage: 12 kV
Model: Bio-rad BioFocus 3000

KEY WORDS
chiral; coated capillary

REFERENCE
Tanaka,Y.; Terabe,S. Separation of the enantiomers of basic drugs by affinity capillary electrophoresis using a partial filling technique and α_1-acid glycoprotein as chiral selector, *Chromatographia*, **1997**, *44*, 119–128.

SAMPLE
Matrix: solutions

CAPILLARY ELECTROPHORESIS
Capillary: 29-36 cm × 50 μm fused-silica (24.5-31.5 cm to detector) (Yongnian Optical Conductive Fiber Plant, China) coated with polyacrylamide
Capillary preparation: Coat capillary as follows. Adjust the pH of 20 mL water to 3.5 with acetic acid, add 80 μL 3-(trimethoxysilyl)propyl methacrylate (3-methacryloxypropyltrimethoxysilane), mix, suck into capillary, let stand at room temperature for 1 h, remove the solution, wash with water. Fill the capillary with a deaerated 3-4% acrylamide solution containing 1 μL/mL N,N,N',N'-tetramethylethylenediamine and 1 mg/mL potassium persulfate, let stand for 30 min, remove excess solution by aspiration, rinse with water, remove water by aspiration, dry at 35° (J. Chromatogr. 1985, 347, 191).
Capillary temperature: 25
Running buffer: 100 mM pH 2.5 NaH_2PO_4 (A) or 100 mM pH 2.5 NaH_2PO_4 containing 45 mM hydroxypropyl-α-cyclodextrin (Wacker, Munich) (B)
Injection: Electromigration at 15 kV for 3 s.
Detector: UV 200; UV 210
Migration time: 5.06 (A); 12.29, 12.41 (B) (enantiomers)
Voltage: 15 kV
Model: Bio-Focus 3000

OTHER SUBSTANCES
Also analyzed: albuterol (salbutamol), alprenolol, amorolfine, atenolol, atropine, azelastine, baclofen, bamethan, benproperine, benserazide, biperiden, bisoprolol, brompheniramine, bupivacaine, bupranolol, butamirate, butethamate, carazolol, carbuterol, carteolol, carvedilol, celiprolol, chloroquine, chlorpheniramine, chlorphenoxamine, cicletanine, clenbuterol, clidinium bromide, clobutinol, dimethindene, dipivefrin, disopyramide, dobutamine, doxylamine, fendiline, flecainide, gallopamil, homatropine, ipratropium bromide, isoproterenol (isoprenaline), isothipendyl, ketamine, meclizine, mefloquine, mepindolol, mequitazine, metaclazepam, metaproterenol (orciprenaline), metipranolol, metoprolol, nafronyl (naftidrofuryl), nefopam, nicardipine, norfenefrine, ofloxacin, ornidazole, orphenadrine, oxomemazine, oxprenolol, oxybutynin, phenoxybenzamine, phenylpropanolamine, pholedrine, pindolol, pirbuterol, prilocaine, procyclidine, propafenone, propranolol, reproterol, sotalol, sulpride, synephrine, talinolol, terbutaline, tetrahydrozoline (tetryzoline), theodrenaline, tioconazole, tocainide, trihexyphenidyl, trimeprazine (alimemazine), trimipramine, tropicamide, verapamil, zopiclone

KEY WORDS
coated capillary; chiral

REFERENCE
Koppenhoefer,B.; Eperlein,U.; Schlunk,R.; Zhu,X.; Lin,B. Separation of enantiomers of drugs by capillary electrophoresis. V. Hydroxypropyl-α-cyclodextrin as chiral solvating agent, *J.Chromatogr.A*, **1998**, *793*, 153–164.

Propafenone

Molecular formula: $C_{21}H_{27}NO_3$
Molecular weight: 341.45
CAS Registry No.: 54063-53-5, 34183-22-7 (HCl)
Merck Index (12th ed.): 7978
Lednicer: 5 17

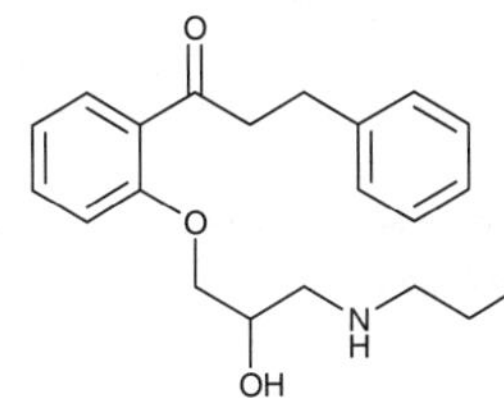

SAMPLE
Matrix: solutions
Sample preparation: Inject an aliquot of a 100 μg/mL solution in 50 mM sodium dodecyl sulfate.

CAPILLARY ELECTROPHORESIS
Capillary: 57 cm × 75 μm fused-silica (50 cm to detector) (Beckman)
Capillary preparation: Rinse capillary with 100 mM NaOH and running buffer between each run.
Capillary temperature: 30
Running buffer: Isopropanol:100 mM pH 8.1 borate buffer containing 50 mM sodium dodecyl sulfate 10:90
Injection: Pressure injection for 5 s
Detector: UV 200
Migration time: 20
Voltage: +25 kV
Model: Beckman P/ACE 5510

OTHER SUBSTANCES
Simultaneous: amiodarone, bretylium, disopyramide, lidocaine, phenytoin, quinidine

REFERENCE
Bretnall,A.E.; Clarke,G.S. Investigation and optimisation of the use of organic modifiers in micellar electrokinetic chromatography, *J.Chromatogr.A*, **1995**, *716*, 49–55.

SAMPLE
Matrix: solutions
Sample preparation: Inject an aliquot of a 100 μg/mL solution in water:running buffer 50:50.

CAPILLARY ELECTROPHORESIS
Capillary: 44.5 cm × 50 μm acrylamide-coated fused-silica (Bio-Rad)
Capillary temperature: 30
Running buffer: 100 mM NaH_2PO_4 containing 15 mM gamma-cyclodextrin, adjusted to pH 2.5 with phosphoric acid
Injection: Electrokinetic injection at 8 kV for 6 s.
Detector: UV 200
Migration time: 13.54
Voltage: 14 kV
Model: Bio-Rad BioFocus 3000

OTHER SUBSTANCES
Also analyzed: albuterol, alprenolol, atenolol, atropine, baclofen, bamethan, benserazide, biperiden, bisoprolol, bupivacaine, bupranolol, butetamate, carazolol, carbuterol, carvedilol, celiprolol, chloroquine, chlorpheniramine (chlorphenamine), clidinium bromide, clobutinol, disopyramide, dobutamine, flecainide, homatropine, ipratropium bromide, isoproterenol, isothipendyl, ketamine, mefloquine, mequitazine, metaproterenol (orciprenaline), metipranolol, nafronyl (naftidrofuryl), nefopam, ofloxacin, orphenadrine, oxomemazine, oxprenolol, phenoxybenzamine, pholedrine, pindolol, pirbuterol, prilocaine, promethazine, propranolol, sotalol, synephrine, terbutaline, tetrahydrozoline (tetryzoline), tocainide, trihexyphenidyl, trimeprazine (alimemazine), trimipramine, tropicamide, verapamil, zopiclone

KEY WORDS
coated capillary; achiral

REFERENCE
Koppenhoefer,B.; Epperlein,U.; Christian,B.; Yibing,J.; Yuying,C.; Bingcheng,L. Separation of enantiomers of drugs by capillary electrophoresis. I. γ-Cyclodextrin as chiral solvating agent, *J.Chromatogr.A*, **1995**, *717*, 181–190.

SAMPLE
Matrix: solutions

CAPILLARY ELECTROPHORESIS
Capillary: 57 cm × 75 μm fused-silica (50 cm to detector) (Beckman)
Capillary preparation: Before each run rinse capillary with 100 mM NaOH and running buffer.
Capillary temperature: 30
Running buffer: Acetone:100 mM pH 8.1 borate buffer containing 50 mM sodium dodecyl sulfate 15:85 (A) or isopropanol:100 mM pH 8.1 borate buffer containing 50 mM sodium dodecyl sulfate 10:90 (B)
Injection: Pressure injection for 5-10 s.
Detector: UV 200
Migration time: 19.5 (A), 18.3 (B)
Voltage: 25 kV
Model: Beckman P/ACE 5510

OTHER SUBSTANCES
Simultaneous: acebutolol, amiodarone, atenolol, bretylium, captopril, diltiazem, disopyramide, lidocaine, lisinopril, metoprolol, nicardipine, nifedipine, phenytoin, pindolol, propranolol, quinidine, sotalol, timolol, p-toluenesulfonic acid, verapamil

KEY WORDS
only running buffer A separates all compounds

REFERENCE
Bretnall,A.E.; Clarke,G.S. Selectivity of capillary electrophoresis for the analysis of cardiovascular drugs, *J.Chromatogr.A*, **1996**, *745*, 145–154.

SAMPLE
Matrix: solutions
Sample preparation: Inject an aliquot of a 100 μg/mL solution in running buffer.

CAPILLARY ELECTROPHORESIS
Capillary: 30 cm × 50 μm fused-silica (25.5 cm to detector) (Yongnian Optical Conductive Fiber Plant, China), coated with polyacrylamide
Capillary preparation: No details of the polyacrylamide coating process are provided. However, another paper (LC.GC 1997, 15, 40) by this group indicates that they use the procedure of Hjertén, thus: Adjust the pH of 20 mL water to 3.5 with acetic acid, add 80 μL 3-(trimethoxysilyl)propyl methacrylate (3-methacryloxypropyltrimethoxysilane), mix, suck into capillary, let stand at room temperature for 1 h, remove the solution, wash with water. Fill the capillary with a deaerated 3-4% acrylamide solution containing 1 μL/mL N,N,N',N'-tetramethylethylenediamine and 1 mg/mL potassium persulfate, let stand for 30 min, remove excess solution by aspiration, rinse with water, remove water by aspiration, dry at 35° (J. Chromatogr. 1985, 347, 191).
Capillary temperature: 25
Running buffer: 100 mM NaH$_2$PO$_4$ adjusted to pH 2.5
Injection: Electrokinetic injection at 15 kV for 3 s.
Detector: UV 200, UV 210
Migration time: 5.42
Voltage: 15 kV
Model: Bio-Rad BioFocus 3000

OTHER SUBSTANCES
Simultaneous: amorolfine, brompheniramine, bupivacaine, carteolol, chloroquine, chlorpheniramine, chlorphenoxamine, disopyramide, dobutamine, doxylamine, flecainide, gallopamil, ketamine, mepindolol, orphenadrine, oxybutynin, phenoxybenzamine, pindolol, propranolol, sulpiride, talinolol, tropicamide, verapamil

KEY WORDS
coated capillary

REFERENCE
Koppenhoefer,B.; Epperlein,U.; Xiaofeng,Z.; Bingcheng,L. Separation of enantiomers of drugs by capillary electrophoresis. Part 4: Hydroxypropyl-γ-cyclodextrin as chiral solvating agent, *Electrophoresis*, **1997**, *18*, 924–930.

SAMPLE
Matrix: solutions

CAPILLARY ELECTROPHORESIS
Capillary: 29-36 cm $\times$ 50 μm fused-silica (24.5-31.5 cm to detector) (Yongnian Optical Conductive Fiber Plant, China) coated with polyacrylamide
Capillary preparation: Coat capillary as follows. Adjust the pH of 20 mL water to 3.5 with acetic acid, add 80 μL 3-(trimethoxysilyl)propyl methacrylate (3-methacryloxypropyltrimethoxysilane), mix, suck into capillary, let stand at room temperature for 1 h, remove the solution, wash with water. Fill the capillary with a deaerated 3-4% acrylamide solution containing 1 μL/mL N,N,N',N'-tetramethylethylenediamine and 1 mg/mL potassium persulfate, let stand for 30 min, remove excess solution by aspiration, rinse with water, remove water by aspiration, dry at 35° (J. Chromatogr. 1985, 347, 191).
Capillary temperature: 25
Running buffer: 100 mM pH 2.5 NaH$_2$PO$_4$ (A) or 100 mM pH 2.5 NaH$_2$PO$_4$ containing 45 mM hydroxypropyl-α-cyclodextrin (Wacker, Munich) (B)
Injection: Electromigration at 15 kV for 3 s.
Detector: UV 200; UV 210
Migration time: 5.42 (A); 12.78 (B) (no separation of enantiomers)
Voltage: 15 kV
Model: Bio-Focus 3000

OTHER SUBSTANCES
Also analyzed: albuterol (salbutamol), alprenolol, amorolfine, atenolol, atropine, azelastine, baclofen, bamethan, benproperine, benserazide, biperiden, bisoprolol, brompheniramine, bupivacaine, bupranolol, butamirate, butethamate, carazolol, carbuterol, carteolol, carvedilol, celiprolol, chloroquine, chlorpheniramine, chlorphenoxamine, cicletanine, clenbuterol, clidinium bromide, clobutinol, dimethindene, dipivefrin, disopyramide, dobutamine, doxylamine, fendiline, flecainide, gallopamil, homatropine, ipratropium bromide, isoproterenol (isoprenaline), isothipendyl, ketamine, meclizine, mefloquine, mepindolol, mequitazine, metaclazepam, metaproterenol (orciprenaline), metipranolol, metoprolol, nafronyl (naftidrofuryl), nefopam, nicardipine, norfenofrine, ofloxacin, ornidazole, orphenadrine, oxomemazine, oxprenolol, oxybutynin, phenoxybenzamine, phenylpropanolamine, pholedrine, pindolol, pirbuterol, prilocaine, procyclidine, promethazine, propranolol, reproterol, sotalol, sulpride, synephrine, talinolol, terbutaline, tetrahydrozoline (tetryzoline), theodrenaline, tioconazole, tocainide, trihexyphenidyl, trimeprazine (alimemazine), trimipramine, tropicamide, verapamil, zopiclone

KEY WORDS
coated capillary

REFERENCE
Koppenhoefer,B.; Eperlein,U.; Schlunk,R.; Zhu,X.; Lin,B. Separation of enantiomers of drugs by capillary electrophoresis. V. Hydroxypropyl-α-cyclodextrin as chiral solvating agent, *J.Chromatogr.A*, **1998**, *793*, 153–164.

Propentofylline

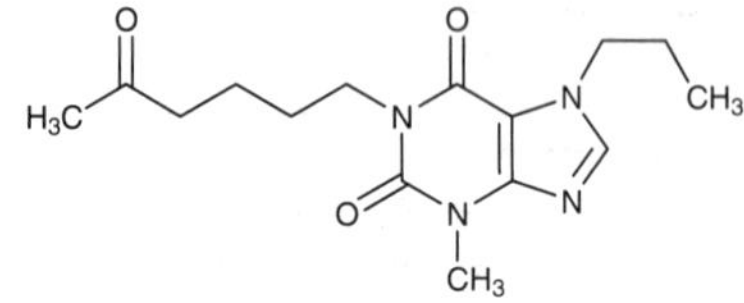

Molecular formula: $C_{15}H_{22}N_4O_3$
Molecular weight: 306.36
CAS Registry No.: 55242-55-2
Merck Index (12th ed.): 7997

SAMPLE
Matrix: solutions

CAPILLARY ELECTROPHORESIS
Capillary: 60 cm $\times$ 75 μm fused-silica (53 cm to detector)
Capillary preparation: Rinse capillary with running buffer after each run. At the start of each
    day rinse with 100 mM NaOH for 5 min and with separation buffer for 10 min. Before storage
    rinse capillary with water and air dry.
Running buffer: 30 mM pH 9.0 Sodium borate buffer containing 60 mM sodium dodecyl sulfate
Injection: Hydrodynamic injection at 10 cm for 5-10 s
Detector: UV 214
Migration time: 9.7
Voltage: 20 kV
Model: Waters Quanta 4000

OTHER SUBSTANCES
Simultaneous: caffeine, ethophylline, methylpropylxanthine, methylxanthine, pentoxifylline,
    theobromine, theophylline, xanthine

REFERENCE
Korman,M.; Vindevogel,J.; Sandra,P. Application of micellar electrokinetic chromatography to the quality con-
    trol of pharmaceutical formulations: the analysis of xanthine derivatives, *Electrophoresis*, **1994**, *15*, 1304–
    1309.

Propiomazine

Molecular formula: $C_{20}H_{24}N_2OS$
Molecular weight: 340.49
CAS Registry No.: 362-29-8, 1240-15-9 (HCl)
Merck Index (12th ed.): 8007
Lednicer: 1 376

SAMPLE
Matrix: solutions

CAPILLARY ELECTROPHORESIS
Capillary: 42 cm $\times$ 50 μm fused-silica (31 cm to detector) (Polymicro Technologies)
Capillary preparation: Condition capillary by rinsing with running buffer for 10 min, with
    water for 10 min, with 1 M NaOH for 10 min, with water for 10 min, and with running buffer
    with the voltage applied for 20 min.
Capillary temperature: 25
Running buffer: Formamide containing 150 mM citric acid, 100 mM Tris, and 100 mM β-cy-
    clodextrin (apparent pH 5.1)
Injection: Gravity injection.
Detector: UV 254
Migration time: 9.00, 9.03 (enantiomers)
Voltage: 30 kV
Model: Beckman P/ACE 5500

OTHER SUBSTANCES
Simultaneous: chlophedianol, mianserin, nefopam, trimeprazine, trimipramine, thioridazine

KEY WORDS
chiral

REFERENCE
Wang,F.; Khaledi,M.G. Chiral separations by nonaqueous capillary electrophoresis, *Anal.Chem.*, **1996**, *68*, 3460–3467.

Propoxur

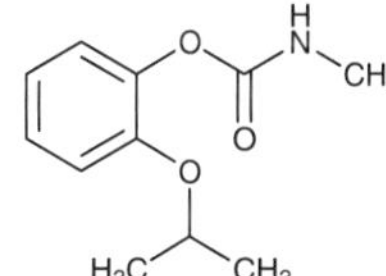

Molecular formula: $C_{11}H_{15}NO_3$
Molecular weight: 209.25
CAS Registry No.: 114-26-1
Merck Index (12th ed.): 8022

SAMPLE
Matrix: solutions
Sample preparation: Inject an aliquot of a solution in running buffer.

CAPILLARY ELECTROPHORESIS
Capillary: 60 cm × 50 μm fused-silica (53 cm to detector) (Polymicro Technologies)
Capillary preparation: Condition capillary by flushing with 100 mM NaOH for 20 min, water for 10 min, and running buffer for 10 min then equilibrate under voltage.
Running buffer: 10 mM pH 7.0 Sodium phosphate containing 60 mM sodium dodecyl sulfate
Detector: UV 210
Migration time: 15.5
Voltage: 15 kV
Model: laboratory-constructed

OTHER SUBSTANCES
Simultaneous: acephate, aldicarb, aminocarb, banol, captan, carbaryl, carbendazim, carbofuran, chlorotoluron, dichlorvos, dimethoate, fenobucarb, fensulfothion, iprodione, isofenphos, isoprocarb, malathion, mepronil, methiocarb, napropamide, paclobutrazol, parathion, pirimicarb, propyzamide, pyriodaphenthion, simazine, simetryn, thiram, triadimefon, tricyclazole

REFERENCE
Wu,Y.S.; Lee,H.K.; Li,S.F.Y. Rapid estimation of octanol-water partition coefficients of pesticides by micellar electrokinetic chromatography, *Electrophoresis*, **1998**, *19*, 1719–1727.

Propoxyphene

Molecular formula: $C_{22}H_{29}NO_2$
Molecular weight: 339.48
CAS Registry No.: 469-62-5, 1639-60-7 (HCl), 26570-10-5 (napsylate monohydrate), 55557-30-7 (l-form napsylate monohydrate), 2338-37-6 (l form), 17140-78-2 (l-form napsylate anhydrous)
Merck Index (12th ed.): 8024
Lednicer: 1 50, 298; 2 57

SAMPLE
Matrix: formulations

Sample preparation: Grind tablet, add MeCN:10 mM HCl 20:80, shake and sonicate for 15 min, centrifuge until a clear solution is obtained. Remove a 25 mL aliquot of the supernatant and add it to 10 mL 500 µg/mL propyl hydroxybenzoate, make up to 50 mL with MeCN:10 mM HCl 20:80, inject an aliquot.

CAPILLARY ELECTROPHORESIS
Capillary: 57 cm × 75 µm fused-silica (Beckman)
Capillary preparation: Pass running buffer through the capillary for at least 4 min before each injection.
Capillary temperature: 25
Running buffer: 20 mM pH 9 Borate buffer containing 25 mM sodium cholate and 50 mM sodium deoxycholate
Injection: Pressure injection for 2 s
Detector: UV 214
Migration time: 15 (dextropropoxyphene)
Internal standard: propyl hydroxybenzoate (10.5)
Voltage: 20 kV
Model: Beckman P/ACE System 2100

OTHER SUBSTANCES
Simultaneous: acetaminophen, aspirin, caffeine, chlorpheniramine, salicylic acid

KEY WORDS
tablets

REFERENCE
Boonkerd,S.; Lauwers,M.; Detaevernier,M.R.; Michotte,Y. Separation and simultaneous determination of the components in an analgesic tablet formulation by micellar electrokinetic chromatography, *J.Chromatogr.A*, **1995**, *695*, 97–102.

SAMPLE
Matrix: solutions
Sample preparation: Prepare a 50 µg/mL solution in water:buffer:MeOH 90:9:1, inject an aliquot. (Buffer was 25 mM Tris buffer adjusted to pH 2.4 with phosphoric acid.)

CAPILLARY ELECTROPHORESIS
Capillary: 82 cm × 50 µm fused-silica (60 cm to detector)
Capillary preparation: Condition by aspirating with 1 M NaOH for 10 min, with water for 10 min, and with running buffer for 10 min.
Capillary temperature: 30
Running buffer: MeOH:buffer 1.2:98.8 (Buffer was 25 mM Tris buffer adjusted to pH 2.4 with phosphoric acid containing 5 mM heptakis(2,6-di-O-methyl)-β-cyclodextrin and 0.5 mM sulfobutyl ether β-cyclodextrin (ISCO)
Injection: Vacuum injection for 0.5 s (2 nL)
Detector: UV 210
Migration time: 20.30 (+), 21.10 (-)
Voltage: 30 kV
Model: Applied Biosystems Model 270A-HT

OTHER SUBSTANCES
Simultaneous: amphetamine, cathinone, cocaine, ephedrine, methamphetamine, methcathinone, norephedrine, norpseudoephedrine, pseudoephedrine

KEY WORDS
chiral

REFERENCE
Lurie,I.S.; Klein,R.F.X.; Dal Cason,T.A.; LeBelle,M.J.; Brenneisen,R.; Weinberger,R.E. Chiral resolution of cationic drugs of forensic interest by capillary electrophoresis with mixtures of neutral and anionic cyclodextrins, *Anal.Chem.*, **1994**, *66*, 4019–4026.

Propranolol

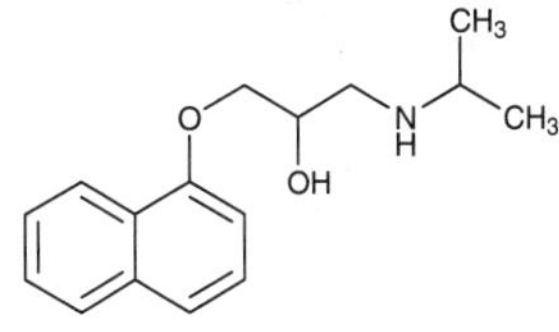

Molecular formula: $C_{16}H_{21}NO_2$
Molecular weight: 259.35
CAS Registry No.: 525-66-6, 318-98-9 (HCl)
Merck Index (12th ed.): 8025
Lednicer: 1 117; 2 105, 212

SAMPLE

Matrix: blood
Sample preparation: Hydrolyze 2 mL serum with β-glucuronidase (EC 3.2.1.31, type H-1 from Helix pomatia, 416 800 U/g, Separacor) at 80° for 30 min, cool, add 900 μL MeCN, vortex for 15 min, centrifuge at 2004 g for 10 min, add ephedrine (165 μg/mL), filter (0.5 μm), inject an aliquot of the filtrate.

CAPILLARY ELECTROPHORESIS

Capillary: 58 cm × 50 μm fused-silica (50 cm to detector) (Polymicro Technologies)
Capillary preparation: Before each injection purge capillary with 5% phosphoric acid for 12 s, with water for 30 s, and with running buffer for 10 min.
Capillary temperature: 35
Running buffer: 80 mM pH 6.7 sodium phosphate buffer containing 15 mM cetyltrimethylammonium bromide
Injection: Hydrostatic injection for 20 s
Detector: UV 214
Migration time: 16.5
Internal standard: ephedrine (17.5)
Voltage: -27 kV
Model: Waters Quanta 4000
Limit of detection: 1 μg/mL (S/N 3)

OTHER SUBSTANCES

Extracted: acebutolol, alprenolol, atenolol, metoprolol, nadolol, oxprenolol, pindolol, timolol
Interfering: labetalol

KEY WORDS

serum

REFERENCE

Lukkari,P.; Nyman,T.; Riekkola,M.-J. Determination of nine β-blockers in serum by micellar electrokinetic capillary chromatography, *J.Chromatogr.A*, **1994**, *674*, 241–246.

SAMPLE

Matrix: blood
Sample preparation: Filter (1 mL Amicon Centrifree Micropartition System) serum while centrifuging at 1500 g for 20 min, inject an aliquot of the ultrafiltrate.

CAPILLARY ELECTROPHORESIS

Capillary: 92 cm × 75 μm fused-silica (55 cm to detector) (Polymicro Technologies)
Capillary preparation: Rinse with running buffer for 10 min between run. At the beginning and end of each day rinse with 100 mM NaOH for 10 min and with water for 10 min.
Running buffer: Isopropanol:buffer 5:95 (Buffer was 10 mM Na_2HPO_4 containing 6 mM sodium borate and 75 mM sodium dodecyl sulfate.)
Injection: Injection via vacuum suction for 0.7 s
Detector: UV 215
Migration time: 24
Voltage: 25 kV
Current: 82 μA
Model: Europhor Prime Vision IV

OTHER SUBSTANCES
Extracted: atenolol, carteolol, celiprolol, levobunolol, metoprolol, penbutolol, pindolol, timolol

KEY WORDS
serum; ultrafiltrate

REFERENCE
Schmutz,A.; Thormann,W. Assessment of impact of physico-chemical drug properties on monitoring drug levels by micellar electrokinetic capillary chromatography with direct serum injection, *Electrophoresis*, **1994**, *15*, 1295–1303.

SAMPLE
Matrix: blood, urine
Sample preparation: Urine. Dilute with 2 volumes of water, filter (0.5 μm), inject an aliquot. Serum. Condition a 3 mL Supelclean LC-18 SPE cartridge (Supelco) with MeOH and water. Hydrolyze 900 μL serum with β-glucuronidase (EC 3.2.1.31 type H-1 from Helix pomatia) at 60° with sonication for 30 min, add 500 μL (?) MeOH, centrifuge at 2000 g, add the supernatant to the SPE cartridge, wash with 1 mL water, dry under vacuum, elute with 2 mL MeOH:water 90:10, filter, inject an aliquot.

CAPILLARY ELECTROPHORESIS
Capillary: 68 cm × 50 μm fused-silica (60 cm to detector) (Polymicro Technologies)
Capillary preparation: Purge with running buffer for 2 min before injection.
Running buffer: pH 7.0 Phosphate buffer containing 10 mM N-cetyl-N,N,N-trimethylammonium bromide (Prepare buffer by mixing 100 mM NaH_2PO_4 and 100 mM Na_2HPO_4 to achieve a pH of 7.0.)
Injection: Hydrostatic injection for 30 s
Detector: UV 214
Migration time: 18
Internal standard: 2,6-dimethylphenol (only for urine) (14.5)
Voltage: -26 kV
Current: 97 μA
Model: Waters Quanta 4000

OTHER SUBSTANCES
Extracted: acebutolol, alprenolol, atenolol, labetalol, metoprolol, nadolol, oxprenolol, pindolol, timolol

KEY WORDS
serum; comparison with HPLC; SPE

REFERENCE
Lukkari,P.; Sirén,H. Ion-pair chromatography and micellar electrokinetic capillary chromatography in analyzing β-adrenergic blocking agents from human biological fluids, *J.Chromatogr.A*, **1995**, *717*, 211–217.

SAMPLE
Matrix: bulk
Sample preparation: Inject an aliquot of a solution in diluted running buffer

CAPILLARY ELECTROPHORESIS
Capillary: 48.5 cm × 50 μm fused-silica (40 cm to detector)
Capillary preparation: After each run wash with water for 2 min and with running buffer for 3 min. At the beginning of each day wash capillary with running buffer for 10 min. Before use wash capillary with 1 M NaOH, 100 mM NaOH, water, and running buffer.
Capillary temperature: 15 ± 0.1
Running buffer: 100 mM Phosphoric acid containing 10 mM carboxymethyl-β-cyclodextrin (Cyclolab, Budapest), adjusted to pH 3 with triethanolamine (Alternatively hydroxypropyl-β-cyclodextrin can be used although the separation is not quite as good. The S isomer elutes at 19 min.)
Injection: Hydrodynamic injection at 5 kPa for 2-30 s
Detector: UV 210

Migration time: 21.3 (S), 22.5 (R)
Voltage: 25 kV
Model: Hewlett-Packard Model 3DCE
Limit of quantitation: 0.10% (S), 0.22% (R) (of major isomer)
Limit of detection: 0.03% (S), 0.06% (R) (of major isomer)

KEY WORDS
chiral

REFERENCE
Fillet,M.; Bechet,I.; Chiap,P.; Hubert,P.; Crommen,J. Enantiomeric purity determination of propranolol by cyclodextrin-modified capillary electrophoresis, *J.Chromatogr.A*, **1995**, *717*, 203–209.

SAMPLE
Matrix: solutions

CAPILLARY ELECTROPHORESIS
Capillary: 82 cm × 75 μm fused-silica (Waters)
Capillary preparation: Between runs purge capillary with running buffer for 2 min. Purge a new capillary with 100 mM NaOH for 30 min, with water for 30 min, and with running buffer for 30 min.
Capillary temperature: 30
Running buffer: 30 mM pH 7.6 phosphate buffer containing 10 mM cetyltrimethylammonium bromide
Injection: Inject at high pressure for 5 s
Detector: UV 214
Migration time: 10.4
Voltage: 21 kV
Model: Beckman P/ACE System 2000

OTHER SUBSTANCES
Simultaneous: acebutolol, alprenolol, atenolol, nadolol, oxprenolol, pindolol, sotalol, timolol

REFERENCE
Lukkari,P.; Ennelin,A.; Sirén,H.; Riekkola,M.-L. Effect of temperature, effective capillary length, and applied voltage on the migration of nine β-blockers in micellar electrokinetic capillary chromatography, *J.Liq.Chromatogr.*, **1993**, *16*, 2069–2079.

SAMPLE
Matrix: solutions

CAPILLARY ELECTROPHORESIS
Capillary: 58 cm × 50 μm fused-silica (50 cm to detector)
Running buffer: Isopropanol:buffer 2.5:97.5 (Buffer was 80 mM pH 6.8 Phosphate buffer containing 15 mM cetyltrimethylammonium bromide.)
Injection: Hydrostatic injection for 15 s
Detector: UV 214
Migration time: 17
Voltage: -20 kV
Model: Waters Quanta 4000

OTHER SUBSTANCES
Simultaneous: acebutolol, alprenolol, atenolol, labetalol, nadolol, oxprenolol, pindolol, sotalol, timolol

REFERENCE
Lukkari,P.; Vuorela,H.; Riekkola,M.-L. Effects of organic mobile phase modifiers on elution and separation of β-blockers in micellar electrokinetic capillary chromatography, *J.Chromatogr.A*, **1993**, *655*, 317–324.

SAMPLE
Matrix: solutions

CAPILLARY ELECTROPHORESIS
Capillary: 62 cm × 50 μm fused silica (50 cm to detector) (Polymicro Technologies)
Capillary temperature: 40
Running buffer: 150 mM pH 2.5 tetrabutylammonium phosphate buffer containing 35 mM β-cyclodextrin
Injection: Injection by gravity at 10 cm for 5 s
Detector: UV 214
Migration time: 41.6, 42.4 (enantiomers)
Voltage: 20 kV
Current: 56 μA
Model: Laboratory constructed

OTHER SUBSTANCES
Simultaneous: isoproterenol

KEY WORDS
chiral

REFERENCE
Quang,C.; Khaledi,M.G. Improved chiral separation of basic compounds in capillary electrophoresis using β-cyclodextrin and tetraalkylammonium reagents, *Anal.Chem.*, **1993**, *65*, 3354–3358.

SAMPLE
Matrix: solutions
Sample preparation: Prepare a 1 mg/mL solution in 40 mM pH 5.1 sodium phosphate buffer.

CAPILLARY ELECTROPHORESIS
Capillary: 11.5 cm × 75 μm fused silica tubing (8.5 cm to the detector) (Polymicro Technologies, Phoenix, AZ) coated on the inner surface with non-cross-linked polyacrylamide. Coat as follows. Mix 80 μL gamma-methacryloxypropyltrimethoxysilane with 20 mL water adjusted to pH 3.5 with acetic acid, suck into capillary, let stand for 1 h, remove solution, wash with water fill capillary with a 3 or 4% solution of acrylamide containing 1 mL/L N,N,N',N'-tetramethylethylenediamine and 1 g/L potassium persulfate, let stand for 30 min, remove excess solution, wash capillary with water, suck capillary dry, dry at 35° (J. Chromatogr. 1985, 347, 191).
Running buffer: Isopropanol:400 mM pH 5.1 sodium phosphate buffer 25:75
Injection: Fill 7 cm of tube with running buffer, fill rest of tube with 40 mg/mL cellobiohydrolase I in buffer, add a 2-3 mm plug of agarose by pressing end of tube into agarose gel, inject sample at 1000 V for 20 s. (Buffer was 400 mM pH 5.1 sodium phosphate buffer. Prepare agarose gel by boiling 20 mg agarose in 1 mL buffer, cool, allow to gel.)
Detector: UV 220
Migration time: 28 (R), 33 (S)
Voltage: 1000 V
Model: Laboratory constructed

OTHER SUBSTANCES
Also analyzed: alprenolol, metoprolol, pindolol

KEY WORDS
detector is at cathode; chiral; coated capillary

REFERENCE
Valtcheva,L.; Mohammad,J.; Pettersson,G.; Hjertén,S. Chiral separation of β-blockers by high-performance capillary electrophoresis based on non-immobilized cellulase as enantioselective protein, *J.Chromatogr.*, **1993**, *638*, 263–267.

SAMPLE
Matrix: solutions

CAPILLARY ELECTROPHORESIS
Capillary: 100 cm × 50 μm fused-silica (50 cm to detector) (Isco)

Capillary preparation: Flush capillary with 10 μL running buffer between runs. Every 40 sample injections rinse capillary with 200 μL 1 M NaOH, with 200 μL water, and 200 μL running buffer. Before use fill capillary with 1 M NaOH and allow to stand for 1 h, fill with 100 mM NaOH, allow to stand for 1 h, wash with water fill with running buffer.
Capillary temperature: 23
Running buffer: 100 mM citric acid containing 19.27 mM Na$_2$HPO$_4$ and 120 mM hydroxypropyl-β-cyclodextrin
Injection: Inject under vacuum at 4.0 kPa.s
Detector: UV 200
Migration time: 41 (S), 42 (R)
Voltage: 30 kV
Model: Isco Model 3140

OTHER SUBSTANCES
Simultaneous: albuterol, alprenolol, atenolol, cimaterol, clenbuterol, labetalol, nadolol, oxprenolol, pindolol, pirbuterol, terbutaline

KEY WORDS
chiral

REFERENCE
Aumatell,A.; Wells,R.J.; Wong,D.K.Y. Enantiomeric differentiation of a wide range of pharmacologically active substances by capillary electrophoresis using modified β-cyclodextrins, *J.Chromatogr.A*, **1994**, *686*, 293–307.

SAMPLE
Matrix: solutions
Sample preparation: Prepare a 50 μg/mL solution in diluted running buffer, inject an aliquot.

CAPILLARY ELECTROPHORESIS
Capillary: 44 cm × 50 μm fused-silica
Capillary preparation: After each run wash capillary with water for 2 min and with running buffer for 3 min. At the beginning of each day wash capillary with water and running buffer for 5 min. Condition a new capillary with 1 M NaOH, 100 mM NaOH, water, and separation buffer.
Capillary temperature: 15
Running buffer: 100 mM Phosphoric acid containing 15 mM heptakis(2,3,6-tri-O-methyl)-β-cyclodextrin adjusted to pH 3.0 with triethanolamine
Injection: Hydrodynamic injection for 1 s
Detector: UV 210
Voltage: 25 kV
Model: Spectra Physics Spectraphoresis 1000

KEY WORDS
chiral; enantiomer resolution 1.5

REFERENCE
Bechet,I.; Paques,P.; Fillet,M.; Hubert,P.; Crommen,J. Chiral separation of basic drugs by capillary zone electrophoresis with cyclodextrin additives, *Electrophoresis*, **1994**, *15*, 818–823.

SAMPLE
Matrix: solutions

CAPILLARY ELECTROPHORESIS
Capillary: 27 cm × 25 μm (20 cm to detector) (Polymicro Technologies)
Capillary temperature: 20
Running buffer: 200 mM 3-[(1,1-dimethyl-2-hydroxyethyl)amino]-2-hydroxypropanesulfonic acid (TAPSO) containing 10 mM hydroxypropyl-β-cyclodextrin and 0.4% hydrophilic polymeric additive (?), adjusted to pH 7.0 with tetrabutylammonium hydroxide
Injection: Pressure injection at 0.5 psi for 5 s
Detector: UV 230
Migration time: 4.4 (R), 4.6 (S)

Voltage: 700 V/cm
Model: Beckman P/ACE System 2210

KEY WORDS
chiral

REFERENCE
Guttman,A.; Cooke,N. Practical aspects of chiral separations of pharmaceuticals by capillary electrophoresis. I. Separation optimization, *J.Chromatogr.A*, **1994**, *680*, 157–162.

SAMPLE
Matrix: solutions

CAPILLARY ELECTROPHORESIS
Capillary: 60 cm × 50 μm AccuSep (52.5 cm to detector) (Waters)
Capillary preparation: Before injection rinse capillary with 100 mM NaOH for 3 min and with running buffer for 3 min. Rinse new capillaries with 500 mM NaOH for 5 min
Running buffer: MeCN:buffer 30:70 (Buffer was 25 mM Na_2HPO_4 and 25 mM sodium borate containing 25 mM (S)-N-dodecoxycarbonylvaline, pH 8.8. Prepare (S)-N-dodecoxycarbonylvaline as follows. Prepare dodecyl chloroformate by reacting 1-dodecanol with 0.33 equivalents of triphosgene in dichloromethane solution in the presence of pyridine. Add dodecyl chloroformate dropwise to (S)-valine in 1 M NaOH solution, filter, wash with hexane, recrystallize from ether/petroleum ether.)
Injection: Hydrostatic injection 2 s
Detector: UV 214
Migration time: 21, 24 (enantiomers)
Voltage: +12 kV
Model: Waters Quanta 4000 or 4000E

KEY WORDS
chiral

REFERENCE
Mazzeo,J.R.; Grover,E.R.; Swartz,M.E.; Petersen,J.S. Novel chiral surfactant for the separation of enantiomers by micellar electrokinetic capillary chromatography, *J.Chromatogr.A*, **1994**, *680*, 125–135.

SAMPLE
Matrix: solutions

CAPILLARY ELECTROPHORESIS
Capillary: 62 cm × 52 μm fused-silica (50 cm to detector) (Polymicro Technologies)
Running buffer: 70 mM pH 2.5 Trimethylammonium phosphate buffer containing 20 mM hydroxypropyl-β-cyclodextrin
Injection: Gravity injection at 10 cm for 5 s
Migration time: 42, 44 (enantiomers)
Voltage: 20 kV
Current: 43 μA
Model: Laboratory constructed

OTHER SUBSTANCES
Simultaneous: acebutolol, alprenolol, atenolol, isoproterenol, labetalol, nadolol, pindolol

KEY WORDS
chiral

REFERENCE
Quang,C.; Khaledi,M.G. Direct separation of the enantiomers of β-blockers by cyclodextrin-mediated capillary zone electrophoresis, *J.High Res.Chromatogr.*, **1994**, *17*, 99–101.

SAMPLE
Matrix: solutions

CAPILLARY ELECTROPHORESIS
Capillary: 56 cm × 50 μm fused-silica (36 cm to detector) (Supelco)
Running buffer: 25 mM pH 2.7 phosphate buffer containing 50 mg/mL soluble β-cyclodextrin
polymer cross-linked with epichlorohydrin (MW 3000-5000) (Cyclolab, Budapest)
Injection: Hydrodynamic injection at 20 cm for 10 s
Detector: UV 206
Migration time: 31.1 (-), 31.6 (+)
Voltage: 12 kV
Model: Laboratory made

OTHER SUBSTANCES
Simultaneous: isoproterenol, norepinephrine

KEY WORDS
chiral

REFERENCE
Ingelse,B.A.; Everaerts,F.M.; Desiderio,C.; Fanali,S. Enantiomeric separation by capillary electrophoresis using
a soluble neutral β-cyclodextrin polymer, *J.Chromatogr.A*, **1995**, *709*, 89–98.

SAMPLE
Matrix: solutions
Sample preparation: Inject an aliquot of a 100 μg/mL solution in water:running buffer 50:50.

CAPILLARY ELECTROPHORESIS
Capillary: 44.5 cm × 50 μm acrylamide-coated fused-silica (Bio-Rad)
Capillary temperature: 30
Running buffer: 100 mM NaH_2PO_4 containing 15 mM gamma-cyclodextrin, adjusted to pH 2.5
with phosphoric acid
Injection: Electrokinetic injection at 8 kV for 6 s.
Detector: UV 200
Migration time: 10.75
Voltage: 14 kV
Model: Bio-Rad BioFocus 3000

OTHER SUBSTANCES
Also analyzed: albuterol, alprenolol, atenolol, atropine, baclofen, bamethan, benserazide, biperiden, bisoprolol, bupivacaine, bupranolol, butetamate, carazolol, carbuterol, carvedilol, celiprolol, chloroquine, chlorpheniramine (chlorphenamine), clidinium bromide, clobutinol, disopyramide, dobutamine, flecainide, homatropine, ipratropium bromide, isoproterenol,
isothipendyl, ketamine, mefloquine, mequitazine, metaproterenol (orciprenaline), metipranolol,
nafronyl (naftidrofuryl), nefopam, ofloxacin, orphenadrine, oxomemazine, oxprenolol, phenoxybenzamine, pholedrine, pindolol, pirbuterol, prilocaine, promethazine, propafenone, sotalol,
synephrine, terbutaline, tetrahydrozoline (tetryzoline), tocainide, trihexyphenidyl, trimeprazine (alimemazine), trimipramine, tropicamide, verapamil, zopiclone

KEY WORDS
coated capillary; achiral

REFERENCE
Koppenhoefer,B.; Epperlein,U.; Christian,B.; Yibing,J.; Yuying,C.; Bingcheng,L. Separation of enantiomers of
drugs by capillary electrophoresis. I. γ-Cyclodextrin as chiral solvating agent, *J.Chromatogr.A*, **1995**, *717*,
181–190.

SAMPLE
Matrix: solutions
Sample preparation: Prepare a solution in MeOH/water, inject an aliquot.

CAPILLARY ELECTROPHORESIS
Capillary: 62 cm × 52 μm fused-silica (50 cm to detector) (Polymicro Technologies)
Capillary temperature: 40

Running buffer: 50 mM pH 2.50 Tetrabutylammonium phosphate
Injection: Siphon injection at 10 cm for 5 s
Detector: UV (wavelength not specified)
Migration time: 11.5
Voltage: 20 kV
Current: 29 μA
Model: Laboratory constructed

OTHER SUBSTANCES
Simultaneous: doxylamine, epinephrine, imidazole, isoproterenol, 2-methylphenethylamine, 1-methylphenylpropylamine, metoprolol, nadolol, nicotine, norepinephrine, pseudoephedrine

REFERENCE
Quang,C.; Khaledi,M.G. Extending the scope of chiral separation of basic compounds by cyclodextrin-mediated capillary zone electrophoresis, *J.Chromatogr.A*, **1995**, *692*, 253–265.

SAMPLE
Matrix: solutions

CAPILLARY ELECTROPHORESIS
Capillary: 67 cm × 75 μm (60 cm to detector) (Polymicro Technologies)
Capillary preparation: Coat capillary as follows. Treat overnight with 50% 3-(trimethoxysilyl)propyl methacrylate in MeOH, fill capillary with degassed 40 g/L (acrylamide + N,N'-methylenebisacrylamide) in 100 mM pH 8.2 Tris-borate buffer containing 2 mM EDTA, 5 μL/L 10% ammonium persulfate, and 5 μL/mL N,N,N',N'-tetramethylethylenediamine, after polymerization force out gel leaving coating sticking to wall (Caution! Acrylamide is a carcinogen!), rinse with water, remove water by aspiration, dry at 35°.
Running buffer: 20 mM pH 2.5 Citric acid containing 2% carboxymethyl β-cyclodextrin
Detector: UV 214
Migration time: 18.6, 19 (enantiomers)
Voltage: 240 V/cm
Model: Beckman P/ACE 2000

OTHER SUBSTANCES
Simultaneous: arterenol, dimethindene, doxylamine, ephedrine, pindolol

KEY WORDS
chiral; coated capillary

REFERENCE
Schmitt,T.; Engelhardt,H. Optimization of enantiomeric separations in capillary electrophoresis by reversal of the migration order and using different cyclodextrins, *J.Chromatogr.A*, **1995**, *697*, 561–570.

SAMPLE
Matrix: solutions

CAPILLARY ELECTROPHORESIS
Capillary: 36 cm × 50 μm polyacrylamide-coated fused-silica (31.5 cm to detector) (Polymicro Technologies)
Capillary preparation: Before each injection rinse with water at 100 psi for 30 s, rinse with running buffer at 100 psi for 30 s, partially fill with separation solution (500 μM bovine serum albumin in running buffer) at 1 psi for 190 s (27 cm), inject sample, electrophorese with running buffer. Coat capillary with polyacrylamide as follows. Treat with 1 M NaOH for 1 h, rinse with water, dry at 110° by flushing with nitrogen for 6 h, pass thionyl chloride through using suction for 10 min, seal at each end, heat at 70° for 6 h, open, suck 250 mM vinylmagnesium bromide in THF into the capillary, seal at each end, heat at 70° for 6 h, open, rinse with THF for several min, rinse with water, fill by suction with a solution containing 3% acrylamide, 0.1% ammonium persulfate, and 0.1% tetramethylethylenediamine, heat at 28 ± 2° for 1 h, after 1 h wash with water (Biol.Pharm.Bull. 1993, 16, 1185).
Capillary temperature: 25
Running buffer: 50 mM pH 6.0 Phosphate buffer
Injection: Inject at 1 psi for 2 s

Detector: UV 210
Migration time: 10.23, 10.91 (enantiomers)
Voltage: 12 kV
Model: Bio-Rad BioFocus 3000

OTHER SUBSTANCES
Simultaneous: homochlorcyclizine, oxyphencyclimine, trimebutine

KEY WORDS
chiral; partial separation zone technique; coated capillary

REFERENCE
Tanaka,Y.; Terabe,S. Partial separation zone technique for the separation of enantiomers by affinity electrokinetic chromatography with proteins as chiral pseudo-stationary phases, *J.Chromatogr.A*, **1995**, *694*, 277–284.

SAMPLE
Matrix: solutions
Sample preparation: Inject an aliquot of a 300 µg/mL solution in water.

CAPILLARY ELECTROPHORESIS
Capillary: 65 cm × 50 µm fused-silica (40 cm to detector) (Isco)
Capillary preparation: Purge with running buffer for 3 min between injections. Purge daily with 1 M NaOH for 3 min, with water for 3 min, and with running buffer for 3 min.
Running buffer: Isopropanol:100 mM pH 7 phosphate buffer containing 25 mM rifamycin B 30:70
Injection: Electrokinetic injection at 5 kV for 5 s.
Detector: UV 350
Migration time: 27.8 (first enantiomer, R_s 1.3)
Voltage: 8 kV
Model: Isco model 3850

OTHER SUBSTANCES
Simultaneous: alprenolol, epinephrine, amphetamine, metoprolol, norepinephrine, normetanephrine, octapamine, oxprenolol

KEY WORDS
chiral

REFERENCE
Ward,T.J.; Dann,C.,III; Blaylock,A. Enantiomeric resolution using the macrocyclic antibiotics rifamycin B and rifamycin SV as chiral selectors for capillary electrophoresis, *J.Chromatogr.A*, **1995**, *715*, 337–344.

SAMPLE
Matrix: solutions

CAPILLARY ELECTROPHORESIS
Capillary: 57 cm × 75 µm fused-silica (50 cm to detector) (Beckman)
Capillary preparation: Before each run rinse capillary with 100 mM NaOH and running buffer.
Capillary temperature: 30
Running buffer: Acetone:100 mM pH 8.1 borate buffer containing 50 mM sodium dodecyl sulfate 15:85 (A) or 100 mM phosphoric acid adjusted to pH 3.1 with triethanolamine (B)
Injection: Pressure injection for 5-10 s.
Detector: UV 200
Migration time: 18.3 (A), 9.4 (B)
Voltage: 25 kV
Model: Beckman P/ACE 5510

OTHER SUBSTANCES
Simultaneous: acebutolol, amiodarone, atenolol, bretylium, captopril, diltiazem, disopyramide, lidocaine, lisinopril, metoprolol, nicardipine, nifedipine, phenytoin, pindolol, propafenone, quinidine, sotalol, timolol, p-toluenesulfonic acid, verapamil

KEY WORDS
only running buffer A separates all compounds; verapamil interferes in running buffer A (interference in running buffer B has not been determined)

REFERENCE
Bretnall,A.E.; Clarke,G.S. Selectivity of capillary electrophoresis for the analysis of cardiovascular drugs, *J.Chromatogr.A*, **1996**, *745*, 145–154.

SAMPLE
Matrix: solutions
Sample preparation: Inject an aliquot of a 50-200 μM solution.

CAPILLARY ELECTROPHORESIS
Capillary: 48.5 cm × 50 μm fused-silica (44.5 cm to detector) (Polymicro Technologies)
Capillary temperature: 25
Running buffer: 100 mM Phosphoric acid containing 5 mM 6^A-methylamino-β-cyclodextrin, adjusted to pH 2.5 with tetramethylammonium hydroxide. (Synthesis of 6^A-methylamino-β-cyclodextrin was as follows. Add a solution of 3.65 g p-toluenesulfonyl chloride in 30 mL dry pyridine to 29.60 g β-cyclodextrin stirred at 5° in 300 mL dry pyridine, stir overnight at room temperature, evaporate to dryness under reduced pressure at 40°, add 700 mL diethyl ether to the residue. Collect the precipitate and recrystallize it 3 times from water to obtain mono-(6-O-p-tolylsulfonyl)-β-cyclodextrin in 31% yield (Bull. Chem. Soc. Japan 1978, 51, 3030). Heat 2 g mono-(6-O-p-tolylsulfonyl)-β-cyclodextrin with 35 mL 50% methylamine in MeOH in a sealed tube at 70° for 3 days, purify by chromatography on carboxymethylcellulose with ammonium bicarbonate solution to obtain 6^A-methylamino-β-cyclodextrin (cf. J. Am. Chem. Soc. 1980, 102, 762).
Injection: Pressure injection at 10 psi.s.
Detector: UV 206
Migration time: 44.4 (second enantiomer, R_S = 1.0)
Voltage: 18 kV
Current: 41-48 μA
Model: Biofocus 3000 (Bio-Rad)

OTHER SUBSTANCES
Simultaneous: terbutaline

KEY WORDS
chiral

REFERENCE
Fanali,S.; Camera,E. Use of methylamino-β-cyclodextrin in capillary electrophoresis. Resolution of acidic and basic enantiomers, *Chromatographia*, **1996**, *43*, 247–253.

SAMPLE
Matrix: solutions

CAPILLARY ELECTROPHORESIS
Capillary: 43 cm × 50 μm fused-silica (36 cm to detector)
Capillary preparation: Before each injection wash with running buffer for 10 min. After each injection wash with water for 2 min. Wash new capillaries with 1 M NaOH at 60° for 50 min, with NaOH solution (?) at 60° for 10 min, and with water at 25°.
Capillary temperature: 25
Running buffer: 320 mM pH 2.0 Citrate buffer
Injection: Hydrodynamic injection at 1.5 psi for 2 s.
Detector: UV 220
Migration time: 9
Voltage: 15 kV
Model: Spectra-Physics Model 1000
Limit of detection: 1-18 μg/mL

OTHER SUBSTANCES
Simultaneous: acebutolol, atenolol, labetalol, levobunolol, metoprolol, nadolol, oxprenolol, pindolol, timolol

REFERENCE
Lin,C.-E.; Chang,C.-C.; Lin,W.-c.; Lin,E.C. Capillary zone electrophoretic separation of β-blockers using citrate buffer at low pH, *J.Chromatogr.A*, **1996**, *753*, 133–138.

SAMPLE
Matrix: solutions
Sample preparation: Inject an aliquot of a 100 µg/mL solution in MeOH:water 10:90.

CAPILLARY ELECTROPHORESIS
Capillary: 47 cm × 75 µm fused-silica (40 cm to detector)
Capillary preparation: Rinse capillary with running buffer for 1 min before each run. At the end of each day wash with 100 mM NaOH and with water.
Capillary temperature: 20
Running buffer: 20 mM pH 2.9 Phosphate buffer containing 3% chondroitin sulfate C (sodium salt) (Nacalai Tesque, Kyoto)
Injection: Pressure injection at 0.5 psi for 5 s.
Detector: UV 240
Migration time: 18.40, 18.70 (enantiomers)
Voltage: 20 kV
Model: Beckman P/ACE 5510

OTHER SUBSTANCES
Also analyzed: clentiazem, diltiazem, primaquine, sulconazole, trimetoquinol, verapamil

KEY WORDS
chiral

REFERENCE
Nishi,H. Enantiomer separation of basic drugs by capillary electrophoresis using ionic and neutral polysaccharides as chiral selectors, *J.Chromatogr.A*, **1996**, *735*, 345–351.

SAMPLE
Matrix: solutions

CAPILLARY ELECTROPHORESIS
Capillary: 52 cm × 75 µm fused-silica (48 cm to detector) (Polymicro Technologies)
Capillary preparation: Before each run purge with running buffer for 3 min. Every 3 runs purge with 100 mM NaOH for 5 min. Purge new capillaries with 1 M NaOH for 20 min and with 100 mM NaOH for 20 min, rinse with running buffer, equilibrate with running buffer at 12 kV for 3 h.
Running buffer: MeCN:buffer 25:75 (Buffer was 25 mM pH 8.8 sodium borate containing 10 mM triethylamine and 25 mM (R)-dodecoxycarbonylvaline (Waters EnantioSelect (R)-Val-1).)
Injection: Hydrostatic injection for 2 s.
Detector: UV 214
Migration time: 30 (R), 30.5 (S)
Voltage: 12 kV
Current: ≤40 µA
Model: Waters Quanta 4000

OTHER SUBSTANCES
Simultaneous: alprenolol, oxprenolol

KEY WORDS
chiral

REFERENCE

Peterson,A.G.; Ahuja,E.S.; Foley,J.P. Enantiomeric separations of basic pharmaceutical drugs by micellar electrokinetic chromatography using a chiral surfactant, N-dodecoxycarbonylvaline, *J.Chromatogr.B*, **1996**, *683*, 15–28.

SAMPLE

Matrix: solutions

CAPILLARY ELECTROPHORESIS

Capillary: 60 cm × 75 μm fused-silica (51 cm to detector) (Supelco)
Capillary preparation: At the end of each day wash capillary with 200 mM NaOH for 10 min and with water for 10 min. Condition new capillaries by washing with 200 mM NaOH for 10 min, with water for 10 min, and with running buffer for 10 min.
Running buffer: 50 mM pH 7.3 Sodium borate buffer containing 3.3% succinyl β-cyclodextrin (Wacker Chemie, Munich, Germany)
Injection: Hydrodynamic injection at 25 mbar for 6 s.
Detector: UV 208
Migration time: 56.44 (first enantiomer)
Voltage: 15 kV
Model: Prince

OTHER SUBSTANCES

Simultaneous: alprenolol, chlorthalidone, ephedrine, etilefrin, homatropine, methoxamine, norephedrine, octopamine, synephrine, trihexyphenidyl

KEY WORDS

chiral; α = 1.0317

REFERENCE

Schmid,M.G.; Wirnsberger,K.; Gübitz,G. Chiral separation of drug enantiomers by capillary electrophoresis using succinyl-β-cyclodextrin, *Pharmazie*, **1996**, *51*, 852–854.

SAMPLE

Matrix: solutions

CAPILLARY ELECTROPHORESIS

Capillary: 50 cm × 50 μm fused-silica (Polymicro Technologies)
Running buffer: 50 mM NaH$_2$PO$_4$ containing 10 mM β-cyclodextrin and 40 mM t-butyl-N-hydroxycarbamate (t-BuOC(O)NHOH), pH adjusted to 2.5 with phosphoric acid
Injection: Gravity injection.
Detector: UV 214
Migration time: 16, 17 (enantiomers)
Voltage: 15 kV
Model: Dionex CES 1

KEY WORDS

chiral

REFERENCE

Billiot,E.; Wang,J.; Warner,I.M. Improved chiral separation using achiral modifiers in cyclodextrin modified capillary zone electrophoresis, *J.Chromatogr.A*, **1997**, *773*, 321–329.

SAMPLE

Matrix: solutions
Sample preparation: Inject an aliquot of a 100 μg/mL solution in running buffer.

CAPILLARY ELECTROPHORESIS

Capillary: 30 cm × 50 μm fused-silica (25.5 cm to detector) (Yongnian Optical Conductive Fiber Plant, China), coated with polyacrylamide

Capillary preparation: No details of the polyacrylamide coating process are provided. However, another paper (LC.GC 1997, 15, 40) by this group indicates that they use the procedure of Hjertén, thus: Adjust the pH of 20 mL water to 3.5 with acetic acid, add 80 μL 3-(trimethoxysilyl)propyl methacrylate (3-methacryloxypropyltrimethoxysilane), mix, suck into capillary, let stand at room temperature for 1 h, remove the solution, wash with water. Fill the capillary with a deaerated 3-4% acrylamide solution containing 1 μL/mL N,N,N',N'-tetramethylethylenediamine and 1 mg/mL potassium persulfate, let stand for 30 min, remove excess solution by aspiration, rinse with water, remove water by aspiration, dry at 35° (J. Chromatogr. 1985, 347, 191).
Capillary temperature: 25
Running buffer: 100 mM NaH_2PO_4 adjusted to pH 2.5
Injection: Electrokinetic injection at 15 kV for 3 s.
Detector: UV 200, UV 210
Migration time: 4.73
Voltage: 15 kV
Model: Bio-Rad BioFocus 3000

OTHER SUBSTANCES
Simultaneous: amorolfine, brompheniramine, bupivacaine, carteolol, chloroquine, chlorpheniramine, chlorphenoxamine, disopyramide, dobutamine, doxylamine, flecainide, gallopamil, ketamine, mepindolol, orphenadrine, oxybutynin, phenoxybenzamine, pindolol, propafenone, sulpiride, talinolol, tropicamide, verapamil

KEY WORDS
coated capillary

REFERENCE
Koppenhoefer,B.; Epperlein,U.; Xiaofeng,Z.; Bingcheng,L. Separation of enantiomers of drugs by capillary electrophoresis. Part 4: Hydroxypropyl-γ-cyclodextrin as chiral solvating agent, *Electrophoresis*, **1997**, *18*, 924–930.

SAMPLE
Matrix: solutions
Sample preparation: Inject an aliquot of a 100 μg/mL solution in running buffer.

CAPILLARY ELECTROPHORESIS
Capillary: 30 cm × 50 μm fused-silica (25.5 cm to detector), coated with polyacrylamide
Capillary preparation: Adjust the pH of 20 mL water to 3.5 with acetic acid, add 80 μL 3-(trimethoxysilyl)propyl methacrylate (3-methacryloxypropyltrimethoxysilane), mix, suck into capillary, let stand at room temperature for 1 h, remove the solution, wash with water. Fill the capillary with a deaerated 3-4% acrylamide solution containing 1 μL/mL N,N,N',N'-tetramethylethylenediamine and 1 mg/mL potassium persulfate, let stand for 30 min, remove excess solution by aspiration, rinse with water, remove water by aspiration, dry at 35° (J. Chromatogr. 1985, 347, 191).
Capillary temperature: 25
Running buffer: 100 mM NaH_2PO_4 containing 45 mM hydroxypropyl-β- cyclodextrin, adjusted to pH 2.5 with phosphoric acid
Injection: Electrokinetic injection at 15 kV for 3 s.
Detector: UV 200
Voltage: 15 kV
Model: Bio-Rad BioFocus 3000

KEY WORDS
chiral; coated capillary; comparison with the use of other cyclodextrins; this running buffer gave the greatest enantiomeric separation.; α=1.019

REFERENCE
Lin,B.; Zhu,X.; Koppenhoefer,B.; Epperlein,U. Investigation of 123 chiral drugs by cyclodextrin-modified capillary electrophoresis, *LC.GC*, **1997**, *15*, 40–46.

SAMPLE
Matrix: solutions

CAPILLARY ELECTROPHORESIS
Capillary: 67 cm × 50 μm (60 cm to detector)
Capillary preparation: Wash with running buffer for 5 min before each injection. After each injection wash with 1 M NaOH at 60° for 5 min, with 100 mM NaOH at 60° for 10 min, and with water at 25° for 5 min.
Capillary temperature: 25
Running buffer: 70 mM pH 7.0 Sodium phosphate buffer containing 15 mM cetyltriemthylammonium bromide
Injection: Hydrodynamic injection for 1 s.
Detector: UV 220
Migration time: 19.3
Voltage: 20 kV

OTHER SUBSTANCES
Simultaneous: acebutolol, atenolol, labetalol, levobunolol, metoprolol, nadolol, oxprenolol, pindolol, timolol

REFERENCE
Lin,C.-E.; Chen,Y.-C.; Chang,C.-C.; Wang,D.-Z. Migration behavior and selectivity of β-blockers in micellar electrokinetic chromatography. Influence of micelle concentration of cationic surfactants, *J.Chromatogr.A*, **1997**, *775*, 349–357.

SAMPLE
Matrix: solutions

CAPILLARY ELECTROPHORESIS
Capillary: 62.1 cm × 75 μm fused-silica (44.6 cm to detector) (Polymicro Technologies)
Capillary preparation: Coat capillary as follows. Condition capillary with 1 M NaOH for 1 h, rinse with water for 30 min, fill capillary with 1% 3-(trimethoxysilyl)propyl methacrylate (methacryloxypropyltrimethoxysilane) (adjusted to pH 3.5 with acetic acid), allow to react at room temperature for 2 h, rinse with water, fill capillary with 10% 2-aminoethyl methacrylate hydrochloride (Fisher Scientific) solution containing 1-3 mg/mL N,N,N',N'-tetramethylethylenediamine and 1 mg/mL ammonium persulfate, allow to react until polymerization is complete, rinse with water, dry at 30-40° overnight.
Running buffer: 25 mM pH 4.7 Acetate buffer
Injection: Hydrodynamic injection at 5 cm for 5 s.
Detector: UV 210
Migration time: 12.12
Voltage: 15 kV
Model: laboratory constructed

OTHER SUBSTANCES
Simultaneous: acebutolol, atenolol, betaxolol, carteolol, dilevalol, nifenalol, pindolol

KEY WORDS
coated capillary; detector at anode

REFERENCE
Liu,Q.; Lin,F.; Hartwick,R.A. Free solution capillary electrophoretic separation of basic proteins and drugs using cationic polymer coated capillaries, *J.Liq.Chromatogr.Rel.Technol.*, **1997**, *20*, 707–718.

SAMPLE
Matrix: solutions
Sample preparation: Prepare a solution in running buffer, sonicate for 10 min, inject an aliquot.

CAPILLARY ELECTROPHORESIS
Capillary: 35 cm × 75 μm fused-silica (26.5 cm to detector) (Polymicro Technologies)
Capillary preparation: Before each run flush capillary with running buffer at 5-8 bar for several min, equilibrate until a stable current and baseline are achieved. Prepare capillary as follows. Flush capillary with 1 M NaOH for at least 30 min, flush capillary with water for at least 30 min, fill with a solution of 4 μL [(methacryloxy)propyl]trimethoxysilane in 1 mL 6 M acetic

acid, let stand for at least 1 h, flush with water for several min, dry with a flow of nitrogen. Sonicate 30 mM (R)-propranolol in sodium-dried toluene containing 3.6 g/L 2,2'-azobis(isobutyronitrile), 240 mM methacrylic acid, and 240 mM trimethylolpropane trimethacrylate for 10 min. Fill the capillary with this mixture, irradiate at 350 nm at -20° for 80 min, flush with several column volumes of MeCN, flush with several column volumes of running buffer. The polymer polymerizes around the imprint molecule ((R)-propranolol). Later the imprint molecule is removed by extraction leaving a recognition site complementary in size, shape, and chemical functionality to the imprint molecule. The polymerized gel fills the whole column and is anchored to the capillary wall by the [(methacryloxy)propyl]trimethoxysilane.
Capillary temperature: 60
Running buffer: MeCN:4 M pH 3.0 ammonium acetate buffer 80:20
Injection: Electrokinetic injection at 5 kV for 5 s.
Detector: UV 214
Migration time: 2.7 (S), 2.8 (R)
Voltage: 15 kV
Model: Hewlett Packard HP 3D
Limit of detection: 1% (of S in the presence of R)

KEY WORDS
electrochromatography; chiral; molecular imprinting, coated capillary

---

REFERENCE
Schweitz,L.; Andersson,L.I.; Nilsson,S. Capillary electrochromatography with predetermined selectivity obtained through molecular imprinting, *Anal.Chem.*, **1997**, *69*, 1179–1183.

SAMPLE
Matrix: solutions

CAPILLARY ELECTROPHORESIS
Capillary: 64 cm × 50 μm fused-silica (60 cm to detector) (Polymicro Technologies)
Capillary preparation: Before each run flush capillary with 100 mM NaOH for 3 min and with running buffer for 3 min. Each day flush the capillary with 1 M NaOH for 15 min, with water for 2 min, and with running buffer for 10 min. wash new capillaries with 1 M NaOH for 1 h.
Running buffer: 50 mM pH 9.2 Sodium tetraborate buffer containing 0.5% poly(sodium N-undecylenyl-L-valinevalinate) (Prepare poly(sodium N-undecylenyl-L-valinevalinate) by analogy with the following procedure for poly(sodium N-undecylenyl-D-valinate). Add 30 mmoles undecylenic acid to 30 mmoles N-hydroxysuccinimide in 130 mL dry ethyl acetate, add 30 mmoles dicyclohexylcarbodiimide in 10 mL dry ethyl acetate, let stand at room temperature overnight, filter, recrystallize ester from EtOH. Add 1 mmole undecylenic acid N-hydroxysuccinimide ester in 10 mL THF to 1 mmole L-valine and 1 mmole sodium bicarbonate in 10 mL water, after 16 h acidify to pH 2 with 1 M HCl, remove the organic solvent under reduced pressure, add 50 mL water, filter, dry, recrystallize N-undecylenyl-L-valine from chloroform/petroleum ether (mp 91°) (J.Lipid Res. 1967, 8, 142). Prepare the sodium salt by adding an equimolar amount of sodium bicarbonate in the presence of THF. Polymerize a 100 mM aqueous solution of the sodium salt by irradiating with a ^{60}Co gamma source at 70 krad/h for 36-48 h (cf. J.Poly.Sci.: Poly.Lett.Ed. 1979, 17, 749), dialyze against water using a regenerated cellulose membrane (2000 MW cut-off), lyophilize. Dry organic solvents over Type 4A 4-8 mesh molecular sieve.)
Injection: Pressure injection for 2 s.
Detector: UV 214
Migration time: 39 (S), 40 (R)
Voltage: +20 kV
Current: 56 μA
Model: Hewlett-Packard ^{3D}CE

OTHER SUBSTANCES
Simultaneous: alprenolol

KEY WORDS
chiral

REFERENCE
Shamsi,S.A.; Macossay,J.; Warner,I.M. Improved chiral separations using a polymerized dipeptide anionic chiral surfactant in electrokinetic chromatography: Separations of basic, acidic, and neutral racemates, *Anal.Chem.*, **1997**, *69*, 2980–2987.

SAMPLE
Matrix: solutions

CAPILLARY ELECTROPHORESIS
Capillary: 90 cm × 50 μm fused-silica (50 cm to detector) (Polymicro Technologies)
Capillary temperature: 30
Running buffer: 20 mM pH 5.3 Potassium phosphate containing 20 mM tetramethylammonium phosphate and 1.0 mM sulfobutyl ether-β-cyclodextrin (average substitution 3.9, MW 1721, Center for Drug Delivery Research, Lawrence KS)
Injection: Pressure injection at 5 inches Hg for 1.5 s.
Detector: UV 210
Migration time: 21 (S), 22 (R)
Voltage: 22 kV
Current: 32 μA
Model: laboratory-constructed

KEY WORDS
chiral

REFERENCE
Xie,G.-h.; Skanchy,D.J.; Stobaugh,J.F. Chiral separations of enantiomeric pharmaceuticals by capillary electrophoresis using sulphobutyl ether β-cyclodextrin as isomer selector, *Biomed.Chromatogr.*, **1997**, *11*, 193–199.

SAMPLE
Matrix: solutions
Sample preparation: Inject an aliquot of a 100 μg/mL solution in water.

CAPILLARY ELECTROPHORESIS
Capillary: 37 cm × 50 μm eCAP polyamine coated capillary (30 cm to detector) (Beckman)
Capillary preparation: The polyamine capillary was generated by washing a neutral coated eCAP capillary with 1 M NaOH for 2 min, flushing with concentrated Polyamine Regenerator Solution (Beckman) for 2 min, and rinsing with the running buffer.
Capillary temperature: 22
Running buffer: 40 mM pH 2.4 Tris-phosphoric acid buffer containing 10 mM hydroxypropyl-β-cyclodextrin
Injection: For 2 s
Detector: UV 200
Migration time: 8.2 (R), 8.5 (S)
Voltage: 15 kV
Model: Beckman P/ACE 2210
Limit of detection: <0.5%

KEY WORDS
chiral; coated capillary

REFERENCE
Assi,K.H.; Abushoffa,A.M.; Altria,K.D.; Clark,B.J. Determination of trace enantiomeric impurities in chiral compounds by capillary electrophoresis with uncoated, cationic, anionic and neutral surface modified capillaries, *J.Chromatogr.A*, **1998**, *817*, 83–90.

SAMPLE
Matrix: solutions

CAPILLARY ELECTROPHORESIS
Capillary: 35 cm × 50 μm polyacrylamide-coated fused-silica (30.5 cm to detector) (Composite Metal Services, UK)
Capillary preparation: Before each run rinse capillary with water for 70 s and with running buffer for 100 s. (Coat capillary as follows. Adjust the pH of 20 mL water to 3.5 with acetic acid, add 80 μL 3-(trimethoxysilyl)propyl methacrylate (3-methacryloxypropyltrimethoxysilane), mix, suck into capillary, let stand at room temperature for 1 h, remove the solution, wash with water. Fill the capillary with a deaerated 3-4% acrylamide solution containing 1

μL/mL N,N,N',N'-tetramethylethylenediamine and 1 mg/mL potassium persulfate, let stand for 30 min, remove excess solution by aspiration, rinse with water, remove water by aspiration, dry at 35° (J. Chromatogr. 1985, 347, 191).)

Capillary temperature: 25
Running buffer: Buffer containing 10 mM cyanoethylated-β-cyclodextrin (Cyclolab, Budapest) (Prepare buffer by adjusting the pH of 50 mM phosphoric acid containing 50 mM acetic acid and 50 mM boric acid to 2.5 with concentrated NaOH, add the appropriate amount of cyano-ethylated-β-cyclodextrin, dilute with an equal volume of water.)
Injection: Pressure injection at 5 psi for 2 s.
Detector: UV 206
Migration time: 9.2 (second enantiomer, α = 1.009)
Voltage: 20 kV
Current: 27-40 μA
Model: Bio-Rad Biofocus 3000

KEY WORDS
chiral; coated capillary

REFERENCE
Aturki,Z.; Desiderio,C.; Mannina,L.; Fanali,S. Chiral separations by capillary zone electrophoresis with the use of cyanoethylated-β-cyclodextrin as chiral selector, *J.Chromatogr.A*, **1998**, *817*, 91–104.

SAMPLE
Matrix: solutions

CAPILLARY ELECTROPHORESIS
Capillary: 37 cm × 50 μm fused-silica (30 cm to detector) (eCAP, Beckman)
Capillary preparation: Rinse capillary with 100 mM NaOH for 2 min before each analysis.
Capillary temperature: 25
Running buffer: 50 mM pH 2.5 Phosphate buffer containing 20 mM methyl-β-cyclodextrin
Injection: Hydrodynamic injection at 0.5 psi for 3 s.
Detector: UV 280
Migration time: 14.2, 14.5 (enantiomers)
Voltage: 12 kV
Model: Beckman P/ACE 2100

KEY WORDS
chiral

REFERENCE
Jira,T.; Bunke,A.; Karbaum,A. Use of chiral and achiral ion-pairing reagents in combination with cyclodextrins in capillary electrophoresis, *J.Chromatogr.A*, **1998**, *798*, 281–288.

SAMPLE
Matrix: solutions

CAPILLARY ELECTROPHORESIS
Capillary: 29-36 cm × 50 μm fused-silica (24.5-31.5 cm to detector) (Yongnian Optical Conductive Fiber Plant, China) coated with polyacrylamide
Capillary preparation: Coat capillary as follows. Adjust the pH of 20 mL water to 3.5 with acetic acid, add 80 μL 3-(trimethoxysilyl)propyl methacrylate (3-methacryloxypropyltrimethoxysilane), mix, suck into capillary, let stand at room temperature for 1 h, remove the solution, wash with water. Fill the capillary with a deaerated 3-4% acrylamide solution containing 1 μL/mL N,N,N',N'-tetramethylethylenediamine and 1 mg/mL potassium persulfate, let stand for 30 min, remove excess solution by aspiration, rinse with water, remove water by aspiration, dry at 35° (J. Chromatogr. 1985, 347, 191).
Capillary temperature: 25
Running buffer: 100 mM pH 2.5 NaH$_2$PO$_4$ (A) or 100 mM pH 2.5 NaH$_2$PO$_4$ containing 45 mM hydroxypropyl-α-cyclodextrin (Wacker, Munich) (B)
Injection: Electromigration at 15 kV for 3 s.
Detector: UV 200; UV 210
Migration time: 4.73 (A); 10.55 (B) (no separation of enantiomers)

Voltage: 15 kV
Model: Bio-Focus 3000

OTHER SUBSTANCES
Also analyzed: albuterol (salbutamol), alprenolol, amorolfine, atenolol, atropine, azelastine, baclofen, bamethan, benproperine, benserazide, biperiden, bisoprolol, brompheniramine, bupivacaine, bupranolol, butamirate, butethamate, carazolol, carbuterol, carteolol, carvedilol, celiprolol, chloroquine, chlorpheniramine, chlorphenoxamine, cicletanine, clenbuterol, clidinium bromide, clobutinol, dimethindene, dipivefrin, disopyramide, dobutamine, doxylamine, fendiline, flecainide, gallopamil, homatropine, ipratropium bromide, isoproterenol (isoprenaline), isothipendyl, ketamine, meclizine, mefloquine, mepindolol, mequitazine, metaclazepam, metaproterenol (orciprenaline), metipranolol, metoprolol, nafronyl (naftidrofuryl), nefopam, nicardipine, norfenefrine, ofloxacin, ornidazole, orphenadrine, oxomemazine, oxprenolol, oxybutynin, phenoxybenzamine, phenylpropanolamine, pholedrine, pindolol, pirbuterol, prilocaine, procyclidine, promethazine, propafenone, reproterol, sotalol, sulpride, synephrine, talinolol, terbutaline, tetrahydrozoline (tetryzoline), theodrenaline, tioconazole, tocainide, trihexyphenidyl, trimeprazine (alimemazine), trimipramine, tropicamide, verapamil, zopiclone

KEY WORDS
coated capillary

REFERENCE
Koppenhoefer,B.; Eperlein,U.; Schlunk,R.; Zhu,X.; Lin,B. Separation of enantiomers of drugs by capillary electrophoresis. V. Hydroxypropyl-α-cyclodextrin as chiral solvating agent, *J.Chromatogr.A*, **1998**, *793*, 153–164.

SAMPLE
Matrix: solutions

CAPILLARY ELECTROPHORESIS
Capillary: 71.5 cm × 75 μm fused-silica (52.2 cm to detector) (Polymicro Technologies)
Capillary preparation: Rinse capillary with running buffer for 1 min between runs. Rinse new capillaries with 500 mM NaOH for 20 min and with water for 10 min and then condition with running buffer for 1 h.
Running buffer: 50 mM pH 3.0 containing 0.1% guaran (Guaran is the water-soluble fraction of guar gum. Prepare guaran by stirring 1 g guar gum (Sigma) in 100 mL water for 20 min, add 25 mL EtOH dropwise with stirring, let stand overnight, supercentrifuge. Add 30 mL EtOH dropwise with stirring to the centrifugate, centrifuge, collect the guaran as a precipitate, dry in air.)
Injection: Hydrodynamic injection at 5 cm for 5 s.
Detector: UV (wavelengths not given)
Migration time: 13
Voltage: 17.5 kV
Model: laboratory-constructed

OTHER SUBSTANCES
Simultaneous: acebutolol, atenolol, betaxolol, carteolol, dilevalol, nifenalol, pindolol

REFERENCE
Liu,Q.; Lin,F.; Hartwick,R.A. Capillary zone electrophoretic separation of basic proteins and drugs using guaran as a buffer modifier, *Chromatographia*, **1998**, *47*, 219–224.

SAMPLE
Matrix: solutions

CAPILLARY ELECTROPHORESIS
Capillary: 100 cm × 100 μm fused-silica (Polymicro Technologies)
Capillary preparation: Between runs wash capillary with running buffer for 10 min. Replace the electrolyte every other run. Wash new capillaries with water for 20 min and with running buffer for 20 min.
Running buffer: MeOH:water 80:20 containing 20 mM hydroxypropyl-β-cyclodextrin, 5 mM ammonium acetate and 800 mM acetic acid

Injection: Gravity injection at 15 cm for 2 s.
Detector: MS, Vestec 201, electrospray voltage 2.48 kV, sheath liquid MeOH:acetic acid 99:1
  flowing at 2.5 µL/min, m/z 260 (design of interface in J. Chromatogr. 1993, 639, 303)
Migration time: 29.35, 30.37 (enantiomers)
Voltage: 25 kV
Model: Dionex CE System I
Limit of detection: 10 µM

KEY WORDS
chiral

REFERENCE
Lu,W.; Cole,R.B. Determination of chiral pharmaceutical compounds, terbutaline, ketamine and propranolol,
  by on-line capillary electrophoresis-electrospray mass spectrometry, *J.Chromatogr.B*, **1998**, *714*, 69–75.

SAMPLE
Matrix: solutions
Sample preparation: Inject an aliquot of a 340 µM solution.

CAPILLARY ELECTROPHORESIS
Capillary: 58 cm × 75 µm fused-silica (50 cm to detector)
Capillary preparation: Rinse with running buffer for 2 min before each run. Before use rinse
  capillary at 1.36 bar with 100 mM NaOH for 5 min and with water for 10 min.
Capillary temperature: 25
Running buffer: 20 mM pH 7.0 Sodium phosphate buffer (A) or 20 mM pH 7.0 Sodium phos-
  phate buffer containing 10 mM 1-lauroyl-2-hydroxy-sn-glycero-3-phosphocholine (Avanti Polar
  Lipids, Alabaster AL) (B)
Injection: Hydrodynamic injection at 34 mbar for 1 s followed by 5% MeOH for 5 s.
Detector: UV 200
Migration time: 4.03 min (A); k' 1.37 (B)
Voltage: 15 kV
Current: ca. 50 µA
Model: Beckman P/ACE 2200

OTHER SUBSTANCES
Simultaneous: acebutolol, alprenolol, atenolol, metoprolol, oxprenolol, pindolol, timolol

REFERENCE
Masucci,J.A.; Caldwell,C.W.; Foley,J.P. Comparison of the retention behavior of β-blockers using immobilized
  artificial membrane chromatography and lysophospholipid micellar electrokinetic chromatography,
  J.Chromatogr.A, **1998**, *810*, 95–103.

SAMPLE
Matrix: solutions
Sample preparation: Inject an aliquot of a 500 µg/mL solution in 50 mM pH 7 buffer.

CAPILLARY ELECTROPHORESIS
Capillary: 48.5 cm × 50 µm fused-silica (40 cm to detector) (Hewlett Packard)
Capillary preparation: Before each injection flush capillary with 100 mM NaOH for 5 min and
  with running buffer for 5 min. Wash new capillaries with 1 M NaOH at 40° for 15 min, with
  water at 40° for 10 min, and with water at 25° for 5 min.
Capillary temperature: 25
Running buffer: 50 mM pH 7.0 Phosphate buffer (Prepare buffer by dissolving 5.29 g K_2HPO_4
  and 2.61 g KH_2PO_4 in 1 L water.)
Injection: Pressure injection at 50 mbar for 9 s (18.8 nL).
Detector: UV 200
Migration time: 8.7
Voltage: 30 kV
Model: Hewlett Packard Model G1600A [3D]CE

OTHER SUBSTANCES
Simultaneous: ceftazidime, diclofenac, dicloxacillin, oxacillin, quinine, thiamine

REFERENCE
Mrestani,Y.; Neubert,R.H.H.; Krause,A. Partition behaviour of drugs in microemulsions measured by electro-kinetic chromatography, *Pharm.Res.*, **1998**, *15*, 799–801.

SAMPLE
Matrix: solutions

CAPILLARY ELECTROPHORESIS
Capillary: 69.3 cm × 50 μm fused-silica (52.8 cm to detector) (Supelco)
Capillary preparation: Before each run rinse capillary with water and then with running buffer. At the start of each day treat capillary with 1 M NaOH, with water, and with running buffer.
Running buffer: 100 mM Phosphoric acid adjusted to pH 3.05 with triethanolamine containing 17.4 mM hydroxypropyl-β-cyclodextrin
Injection: Hydrodynamic injection at 50 mbar for 1 s.
Detector: UV 210
Migration time: 32, 32.5 (enantiomers)
Voltage: 20 kV
Model: Prince

OTHER SUBSTANCES
Simultaneous: metabolites

KEY WORDS
chiral

REFERENCE
Pak,C.; Marriott,P.J.; Carpenter,P.D.; Amiet,R.G. Enantiomeric separation of propranolol and selected metabolites by using capillary electrophoresis with hydroxypropyl-β-cyclodextrin as chiral selector, *J.Chromatogr.A*, **1998**, *793*, 357–364.

SAMPLE
Matrix: urine
Sample preparation: Dilute urine with 2 volumes of water, add IS, filter (0.5 μm), inject an aliquot.

CAPILLARY ELECTROPHORESIS
Capillary: 68 cm × 50 μm fused-silica (60 cm to detector) (White Associates)
Capillary preparation: Purge capillary with running buffer for 2 min before each injection.
Running buffer: 80 mM pH 7.0 Phosphate buffer containing 10 mM N-cetyl-N,N,N-trimethyl-ammonium bromide
Injection: Hydrostatic injection for 30 s at 10 cm.
Detector: UV 214
Migration time: 18.5
Internal standard: 2,6-dimethylphenol (14.7)
Voltage: -26 kV (at injector end)
Model: Waters Quanta 4000
Limit of detection: 10 μg/mL (S/N 3)

OTHER SUBSTANCES
Extracted: acebutolol, alprenolol, atenolol, labetalol, metoprolol, nadolol, oxprenolol, pindolol, timolol
Simultaneous: probenecid
Noninterfering: caffeine

REFERENCE
Lukkari,P.; Sirén,H.; Pantsar,M.; Riekkola,M.-L. Determination of ten β-blockers in urine by micellar electro-kinetic capillary chromatography, *J.Chromatogr.*, **1993**, *632*, 143–148.

SAMPLE
Matrix: urine
Sample preparation: Filter (0.2 μm), inject an aliquot of the filtrate.

CAPILLARY ELECTROPHORESIS
Capillary: 80 cm × 50 μm fused-silica (57 cm to detector)
Capillary preparation: Before each run rinse the capillary with 1 M NaOH for 3 min, with 100 mM NaOH for 3 min, with water for 3 min, and with running buffer for 10 min.
Running buffer: 110 mM Boric acid containing 56 mM NaOH and 44 mM HCl, pH 8
Injection: Vacuum injection for 1 s
Detector: UV 220
Migration time: 4.051
Voltage: 20 kV
Model: Europhor Prime Vision system IV

OTHER SUBSTANCES
Extracted: acebutolol (UV 238), acetazolamide (UV 222), alprenolol (UV 220), amiloride (UV 220), atenolol (UV 228), bendroflumethiazide (UV 220), bumetanide (UV 220), chlorthalidone (UV 220), cocaine (UV 236), codeine (UV 220), ethacrynic acid (UV 220), furosemide (UV 232), hydrochlorothiazide (UV 226), methadone (UV 220), metoxiphenamine (UV 220), nadolol (UV 220), norcodeine (UV 220), pentazocine (UV 220), spironolactone (UV 244), triamterene (UV 232), xipamide (UV 234)
Interfering: oxprenolol (UV 220)

REFERENCE
Gonzalez,E.; Laserna,J.J. Capillary zone electrophoresis for the rapid screening of banned drugs in sport, *Electrophoresis*, **1994**, *15*, 240–243.

Propylthiouracil

Molecular formula: $C_7H_{10}N_2OS$
Molecular weight: 170.24
CAS Registry No.: 51-52-5
Merck Index (12th ed.): 8054
Lednicer: 1 265

SAMPLE
Matrix: solutions

CAPILLARY ELECTROPHORESIS
Capillary: 41.8 cm × 100 μm fused-silica (30 cm to detector) (Polymicro Technologies)
Capillary preparation: Before use treat capillary with 1 M NaOH for 24 h, condition with running buffer for 2 h.
Running buffer: 20 mM Taurine adjusted to pH 9.5 with NaOH
Injection: Hydrodynamic injection at 1 cm for 5 s.
Detector: UV 215
Migration time: 5.9
Voltage: 6 kV
Model: laboratory-constructed

OTHER SUBSTANCES
Simultaneous: methylthiouracil, thiouracil

REFERENCE
Krivánková,L.; Krásensky,S.; Bocek,P. Application of capillary zone electrophoresis for analysis of thyreostatics, *Electrophoresis*, **1996**, *17*, 1959–1963.

Propyphenazone

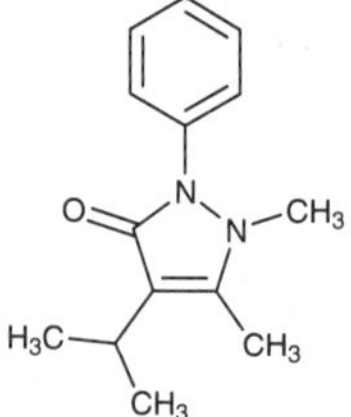

Molecular formula: $C_{14}H_{18}N_2O$
Molecular weight: 230.31
CAS Registry No.: 479-92-5
Merck Index (12th ed.): 8056

SAMPLE
Matrix: formulations
Sample preparation: Dissolve tablets in water, dilute an aliquot with water, inject an aliquot.

CAPILLARY ELECTROPHORESIS
Capillary: 57 cm × 50 μm fused-silica (50 cm to detector)
Capillary preparation: After each run rinse with 20 mM pH 9.00 borate buffer containing 90 mM sodium dodecyl sulfate for 5 min and with 100 mM NaOH for 5 min.
Running buffer: MeOH:buffer 10:90 (Buffer was 20 mM pH 8.25 borate buffer containing 45 mM sodium dodecyl sulfate.)
Injection: Pressure injection at 3.45 kPa for 3 s.
Detector: UV 214
Migration time: 17.5
Voltage: 17 kV
Model: Beckman P/ACE 2100

OTHER SUBSTANCES
Simultaneous: caffeine, acetaminophen

KEY WORDS
tablets

REFERENCE
Vogt,C.; Conradi,S.; Rohde,E. Determination of caffeine and other purine compounds in food and pharmaceuticals by micellar electrokinetic chromatography, *J.Chem.Educ.*, **1997**, *74*, 1126–1130.

Prospidium chloride

Molecular formula: $C_{18}H_{36}Cl4N_4O_2$
Molecular weight: 482.32
CAS Registry No.: 23476-83-7

SAMPLE
Matrix: tissue
Sample preparation: Homogenize tissue with an equal volume of water, add 200 μL homogenate to 600 μL MeOH, let stand at room temperature for 1 h, centrifuge. Add 150 μL supernatant to 100 μL 2 mg/mL sodium diethyldithiocarbamate in 50 mM NaOH, heat at 37° for 1.5 h, add an equal volume of MeOH, inject an aliquot.

CAPILLARY ELECTROPHORESIS
Capillary: 50 cm × 70 μm quartz (42 cm to detector)
Capillary preparation: Condition capillary with MeOH:500 mM NaOH 50:50 for 10 min, with water for 10 min, and with running buffer for 10 min.
Capillary temperature: 20
Running buffer: MeOH:20 mM pH 11.2 borate buffer 50:50
Injection: Electroosmotic flow injection for 15 s
Detector: UV 254

Migration time: 9.2
Voltage: 14 kV
Limit of detection: 1 μg/mL (S/N 3)

KEY WORDS
derivatization

REFERENCE
Okun,V.M.; Aak,O.V.; Kozlov,V.Y. Determination of the anticancer drug prospidin in human tissue by high-performance capillary electrophoresis using derivatization, *J.Chromatogr.B*, **1996**, *675*, 313–319.

Protirelin

Molecular formula: $C_{16}H_{22}N_6O_4$
Molecular weight: 362.39
CAS Registry No.: 24305-27-9
Merck Index (12th ed.): 9720

SAMPLE
Matrix: solutions
Sample preparation: Inject an aliquot of a solution in 5-fold diluted running buffer.

CAPILLARY ELECTROPHORESIS
Capillary: 35 cm × 75 μm fused-silica coated with polyacrylamide (Polymicro Technologies)
Capillary preparation: Before each run purge with 100 mM HCl for 2 min, with water for 4 min, and with running buffer for 1 min. (Coat the capillary as follows. Adjust the pH of 20 mL water to 3.5 with acetic acid, add 80 μL 3-(trimethoxysilyl)propyl methacrylate (3-methacryloxypropyltrimethoxysilane), mix, suck into capillary, let stand at room temperature for 1 h, remove the solution, wash with water. Fill the capillary with a deaerated 3-4% acrylamide solution containing 1 μL/mL N,N,N',N'-tetramethylethylenediamine and 1 mg/mL potassium persulfate, let stand for 30 min, remove excess solution by aspiration, rinse with water, remove water by aspiration, dry at 35° (J. Chromatogr. 1985, 347, 191).)
Running buffer: 100 mM pH 2.5 Sodium phosphate buffer
Injection: Pressure injection at 300 mbar.s.
Detector: UV 191
Migration time: 13
Voltage: 8 kV
Model: Hewlett-Packard HP³ᴰ CE

OTHER SUBSTANCES
Simultaneous: angiotensin II, bombesin, bradykinin, gonadorelin, leucine enkephalin, melanocyte stimulating hormone, methionine enkephalin, oxytocin

KEY WORDS
coated capillary

REFERENCE
Zhang,R.; Zhang,H.X.; Eaker,D.; Hjerten,S. The effect of β-cyclodextrins as buffer additives on the (capillary) electrophoretic separation of peptides and proteins, *J.Capillary Electrophor.*, **1997**, *4*, 105–112.

Protriptyline

Molecular formula: $C_{19}H_{21}N$
Molecular weight: 263.38
CAS Registry No.: 438-60-8, 1225-55-4 (HCl)
Merck Index (12th ed.): 8088
Lednicer: 1 152

SAMPLE
Matrix: solutions

CAPILLARY ELECTROPHORESIS
Capillary: 15 cm × 50 µm fused-silica coated with crosslinked polyacrylamide (13.5 cm to detector) (Scientific Glass Engineering)
Running buffer: 100 mM pH 6.8 Sodium phosphate buffer containing 3% G 3707 (heptaoxyethylene lauryl ether, Atlas Chemie or Sigma P 8800)
Injection: Place a 1-2 mm slug in capillary by capillary action
Detector: UV 205
Migration time: 20.3
Voltage: 3 kV
Current: 60 µA

OTHER SUBSTANCES
Simultaneous: alprenolol, benzylamine, chlorpheniramine, codeine, ephedrine, terodilin

KEY WORDS
coated capillary

REFERENCE
Hjertén,S.; Valtcheva,L.; Elenbring,K.; Eaker,D. High-performance electrophoresis of acidic and basic low-molecular-weight compounds and of proteins in the presence of polymers and neutral surfactants, *J.Liq.Chromatogr.*, **1989**, *12*, 2471–2499.

SAMPLE
Matrix: solutions
Sample preparation: Inject an aliquot of a 40 µg/mL solution in MeOH.

CAPILLARY ELECTROPHORESIS
Capillary: 64 cm × 50 µm fused silica (55.5 cm to detector) (Polymicro Technologies)
Capillary preparation: Flush with running buffer for 2 min between runs. Before use rinse with 1 M NaOH for 1 h, with 100 mM NaOH for 20 min, with water for 20 min, and with running buffer for 10 min.
Capillary temperature: 25
Running buffer: MeCN:MeOH 50:50 containing 25 mM ammonium acetate and 100 mM sodium acetate
Injection: Pressure injection at 5 kPa for 3 s
Detector: UV 214
Migration time: 6.5
Voltage: 25 kV
Model: Hewlett-Packard HP3D

OTHER SUBSTANCES
Simultaneous: amitriptyline, desmethylimipramine, didesmethylimipramine, imipramine, imipramine-N-oxide, litracene, maprotiline, methylimipramine, nortriptyline

REFERENCE
Bjornsdottir,I.; Hansen,S.H. Comparison of separation selectivity in aqueous and non-aqueous capillary electrophoresis, *J.Chromatogr.A*, **1995**, *711*, 313–322.

SAMPLE
Matrix: solutions
Sample preparation: Prepare a 200 μg/mL solution in water, inject an aliquot.

CAPILLARY ELECTROPHORESIS
Capillary: 56 cm × 75 μm fused-silica (56 cm to detector) (Polymicro Technologies)
Capillary temperature: 30
Running buffer: 50 mM pH 4.0 6-Aminocaproic acid containing 25 mM 3-(N,N-dimethylmyris-tylammonium)propanesulfonate and 15 mM Tween 20
Injection: Pressure injection at 2 kPa for 3 s
Detector: UV 214
Migration time: 27
Voltage: 20 kV
Current: 62 μA
Model: Waters Quanta 4000

OTHER SUBSTANCES
Simultaneous: amitriptyline, litracene, maprotiline
Interfering: nortriptyline

REFERENCE
Hansen,S.H.; Bjornsdottir,I.; Tjornelund,J. Separation of basic drug substances of very similar structure using micellar electrokinetic chromatography, *J.Pharm.Biomed.Anal.*, **1995**, *13*, 489–495.

SAMPLE
Matrix: solutions

CAPILLARY ELECTROPHORESIS
Capillary: 55.5 cm × 50 μm fused-silica (55.5 cm to detector) (Polymicro Technologies)
Capillary temperature: 25
Running buffer: MeCN:MeOH 50:50 containing 25 mM ammonium acetate and 100 mM sodium acetate
Injection: Pressure injection at 5 kPa for 3 s.
Detector: UV 214
Migration time: 6.5
Voltage: 25 kV
Current: 23 μA
Model: Hewlett Packard HP ^{3D}CE

OTHER SUBSTANCES
Simultaneous: amitriptyline, litracene, maprotiline, nortriptyline

REFERENCE
Bjornsdottir,I.; Tjornelund,J.; Hansen,S.H. Nonaqueous capillary electrophoresis in pharmaceutical analysis, *J.Capillary Electrophor.*, **1996**, *3*, 83–87.

SAMPLE
Matrix: solutions

CAPILLARY ELECTROPHORESIS
Capillary: 70 cm × 50 μm coated fused-silica (55 cm to detector) (Polymicro Technologies)
Capillary preparation: Between runs rinse with water for 2 min. Coat capillaries as follows. Condition with 1 M NaOH for 15 min, rinse with water for 5 min, treat with 30 μL/mL 3-(trimethoxysilyl)propyl methacrylate in acetic acid:water 50:50 for 1 h using house vacuum, rinse with water, fill with polymerization solution, let stand for 1 h, flush, rinse with water. (Prepare polymerization solution by adding 10 μL N,N,N',N'-tetramethylethylenediamine to 10 mL of a degassed 4% solution of acrylamide in water, add 10 μL 10% ammonium persulfate in water.)
Running buffer: 10 mM pH 6.5 Citrate buffer containing 20 mM sodium dodecyl sulfate and 10 mM carboxymethyl-β-cyclodextrin (Cyclodextrin Technologies Development, Gainesville FL)
Injection: Hydrodynamic injection at 15 cm for 8-10 s.

Detector: UV 254
Migration time: 8.3
Voltage: -20 kV
Model: laboratory-constructed

OTHER SUBSTANCES
Simultaneous: amitriptyline, carbamazepine, clomipramine, desipramine, imipramine, nortriptyline, opipramol, trimipramine

KEY WORDS
coated capillary

REFERENCE
Spencer,B.J.; Zhang,W.; Purdy,W.C. Capillary electrophoretic separation of tricyclic antidepressants using charged carboxymethyl-β-cyclodextrin as a buffer additive, *Electrophoresis*, **1997**, *18*, 736–744.

SAMPLE
Matrix: solutions

CAPILLARY ELECTROPHORESIS
Capillary: 57 cm × 50 μm fused-silica (50 cm to detector) (Polymicro Technologies)
Capillary preparation: At the start of each day rinse capillary with water for 5 min, 100 mM NaOH for 30 min, with water for 10 min, and with running buffer for 15 min. Wash new capillaries with 1 M NaOH for 1 h.
Capillary temperature: 23
Running buffer: MeOH:50 mM pH 9.55 sodium phosphate buffer containing 0.06% poly(n-undecyl-α-D-glucopyranoside) (Prepare poly(n-undecyl-α-D-glucopyranoside) as follows. With exclusion of atmospheric moisture stir 40 mL alcohol-free chloroform, 5.2 g yellow mercuric oxide, 0.1 g mercuric bromide, 10 g anhydrous calcium sulfate, and 2.74 g 10-undecen-1-ol then add 10 g acetobromo-α-D-glucose (Fluka), stir at room temperature for 20 h, filter, wash the solid with chloroform. Wash the filtrate with aqueous 1 M KBr until no mercury salts are present in the aqueous layers, wash with later, concentrate. Take up the residue in benzene (Caution! Benzene is a carcinogen!) and filter it through activated silica, crystallize from n-hexane to obtain undecylenyl-2,3,4,6-tetra-O-acetyl-β-D-glucopyranoside. Add 2 mL of a 1 M solution of sodium methoxide in MeOH to a solution of 1 g of undecylenyl-2,3,4,6-tetra-O-acetyl-β-D-glucopyranoside in 50 mL dry MeOH, let stand until the reaction is complete (by TLC) to obtain undecylenyl-β-D-glucopyranoside (cf. Tenside Detergents 1978, 15, 72). Use Dowex 50W (H$^+$) resin to deionize the final product in solution. Polymerize the monomer by irradiation of an 870 μM solution in MeOH:water 20:80 with ^{60}Co gamma radiation for 48 h, purify the polymer solution by dialysis using a 1000 Da molecular mass cut-off membrane.)
Injection: Pressure injection at 0.5 psi for 5 s.
Detector: UV 214
Migration time: 10
Voltage: 22.8 kV
Model: Beckman P/ACE 5510

OTHER SUBSTANCES
Simultaneous: amitriptyline, desipramine, doxepin, imipramine, nordoxepin, nortriptyline

REFERENCE
Harrell,C.W.; Dey,J.; Shamsi,S.A.; Foley,J.P.; Warner,I.M. Enhanced separation of antidepressant drugs using a polymerized nonionic surfactant as a transient capillary coating, *Electrophoresis*, **1998**, *19*, 712–718.

Pseudoephedrine

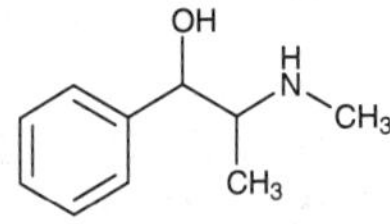

Molecular formula: $C_{10}H_{15}NO$
Molecular weight: 165.24
CAS Registry No.: 90-82-4, 345-78-8 (HCl), 7460-12-0 (sulfate)
Merck Index (12th ed.): 3645

SAMPLE

Matrix: bulk
Sample preparation: Dissolve 4 mg compound in 1 mL MeCN:water 50:50 containing 0.2% triethylamine, vortex for 30 s. Remove a 100 µL aliquot and add it to 100 µL 1.28% 2,3,4,6-tetra-O-acetyl-β-D-glucopyranosyl isothiocyanate in MeCN, vortex for 1 min, let stand for 15 min, make up to 1 mL with 10 mM pH 9.0 phosphate/borate buffer containing 100 mM sodium dodecyl sulfate, vortex for 20 s, filter (Whatman UniPrep), inject an aliquot of the filtrate.

CAPILLARY ELECTROPHORESIS

Capillary: 48 cm × 50 µm fused-silica (26 cm to detector) (Polymicro Technologies)
Capillary preparation: Condition new capillaries with 1 M NaOH for 10 min, with water for 10 min, and with running buffer for 10 min.
Capillary temperature: 30
Running buffer: MeOH:buffer 20:80 (Buffer was 10 mM pH 9.0 phosphate buffer containing 10 mM borate and 100 mM sodium dodecyl sulfate.)
Injection: Vacuum injection for 0.5 s.
Detector: UV 210
Migration time: 12.8 (+), 15 (-)
Voltage: 20 kV
Current: 48 µA
Model: Applied Biosystems Model 270A-HT

OTHER SUBSTANCES

Simultaneous: amphetamine, ephedrine, methamphetamine, norpseudoephedrine, phenylpropanolamine (norephedrine)

KEY WORDS

derivatization; chiral; comparison with HPLC

REFERENCE

Lurie,I.S. Micellar electrokinetic capillary chromatography of the enantiomers of amphetamine, methamphetamine and their hydroxyphenethylamine precursors, *J.Chromatogr.*, **1992**, *605*, 269–275.

SAMPLE

Matrix: bulk
Sample preparation: 10-50 mg Bulk compound + 1 mL 1 mg/mL caffeine in 10 mM HCl, make up to 10 mL with 10 mM HCl, sonicate for 2 min, mix thoroughly, filter (0.45 µm cellulose acetate), inject an aliquot.

CAPILLARY ELECTROPHORESIS

Capillary: 75 cm × 75 µm fused-silica (50 cm to the detector) (ISCO)
Capillary preparation: Flush with running buffer for 2 min between analyses. Condition capillary by filling with 1 M NaOH and allowing to stand for 1 h, fill with 100 mM NaOH and allow to stand for 1 h, wash with water, fill with running buffer. Each week wash capillary with 100 mM HCl for 10 min, with water, with 100 mM NaOH, and with water then fill with running buffer.
Capillary temperature: 30
Running buffer: DMSO:ethanolamine:buffer 11:1:88 (Buffer was 0.92 g cetyltrimethylammonium bromide in 100 mL 10 mM sodium tetraborate, adjust pH to 11.5 with 1 M NaOH.)
Injection: Load under vacuum, vacuum level 2, 10.0 kPa s
Detector: UV 254
Migration time: 7.5
Internal standard: caffeine (6.5)

Voltage: -15 kV
Model: ISCO Model 3140 electropherograph

OTHER SUBSTANCES
Simultaneous: amphetamine, p-aminobenzoic acid, ephedrine, methamphetamine, methylenedioxyamphetamine, methylenedioxymethamphetamine, norephedrine, pseudonorephedrine

REFERENCE
Trenerry,V.C.; Robertson,J.; Wells,R.J. Analysis of illicit amphetamine seizures by capillary electrophoresis,
J.Chromatogr.A, **1995**, *708*, 169–176.

SAMPLE
Matrix: formulations
Sample preparation: Dilute 10-fold, inject an aliquot.

CAPILLARY ELECTROPHORESIS
Capillary: 62 cm × 50 μm fused silica (50 cm to detector) (Polymicro Technologies)
Capillary temperature: 40
Running buffer: 50 mM pH 2.5 sodium phosphate buffer containing 20 mM β-cyclodextrin and
1 mM hexadecyltrimethylammonium bromide
Injection: Injection by gravity at 10 cm for 5 s
Detector: UV 214
Migration time: 9.5 (RR), 10.5 (SS)
Voltage: 20 kV
Current: 36 μA
Model: Laboratory constructed

OTHER SUBSTANCES
Simultaneous: acetaminophen, dextromethorphan, doxylamine

KEY WORDS
syrup; chiral

REFERENCE
Quang,C.; Khaledi,M.G. Improved chiral separation of basic compounds in capillary electrophoresis using β-
cyclodextrin and tetraalkylammonium reagents, *Anal.Chem.*, **1993**, *65*, 3354–3358.

SAMPLE
Matrix: formulations
Sample preparation: Add 3 mL buffer to 30 mg capsule contents, sonicate for 30 min, filter (0.2
μm Nylon), inject an aliquot of the filtrate. (Buffer was water acidified to pH 2.0 with concentrated phosphoric acid.)

CAPILLARY ELECTROPHORESIS
Capillary: 900 mm × 50 μm uncoated fused-silica (650 mm to detector)
Capillary preparation: Rinse capillary with buffer for 15 min at the start of each day and for
4 min before each analysis.
Capillary temperature: 31
Running buffer: 70 mM hydroxypropyl-β-cyclodextrin (average molar substitution 0.8, average
molecular mass 1500) containing 30 mM tetramethylammonium chloride and 10 mM sodium
dodecyl sulfate, pH 2.0
Injection: Vacuum injection at 28 kPa
Detector: UV 210
Migration time: 31.3 (-), 37.7 (+)
Voltage: +28 kV
Model: ISCO Model 3140
Limit of detection: 4 μg/mL

OTHER SUBSTANCES
Simultaneous: amphetamine, ephedrine, methamphetamine, methylephedrine, methylpseudoephedrine, norephedrine, norpseudoephedrine

KEY WORDS
capsules; chiral

REFERENCE
Flurer,C.L.; Lin,L.A.; Satzger,R.D.; Wolnik,K.A. Determination of ephedrine compounds in nutritional supplements by cyclodextrin-modified capillary electrophoresis, *J.Chromatogr.B*, **1995**, *669*, 133–139.

SAMPLE
Matrix: formulations
Sample preparation: Dilute formulation 50-fold, inject an aliquot.

CAPILLARY ELECTROPHORESIS
Capillary: 60 cm × 50 μm (52.5 cm to detector) (AccuSep)
Capillary preparation: Purge capillary with running buffer for 3 min between runs. Purge new capillaries under vacuum with 500 mM NaOH for 10 min, with water for 10 min, and with running buffer for 10 min.
Capillary temperature: 30
Running buffer: 25 mM pH 8.0 Na_2HPO_4/sodium tetraborate containing 50 mM (R)-N-dodecylcarbonylvaline (Prepare (R)-N-dodecoxycarbonylvaline as follows. Prepare dodecyl chloroformate by reacting 1-dodecanol with 0.33 equivalents of triphosgene in dichloromethane solution in the presence of pyridine. Add dodecyl chloroformate dropwise to (R)-valine in 1 M NaOH solution, filter, wash with hexane, recrystallize from ether/petroleum ether (J. Chromatogr. A 1994, 680, 125).)
Injection: Hydrostatic injection at 10 cm for 15 s.
Detector: UV 214
Migration time: 22 ((+)-enantiomer)
Voltage: 15 kV
Model: Waters Quanta 4000E

OTHER SUBSTANCES
Simultaneous: acetaminophen, dextromethorphan, doxylamine, saccharin

KEY WORDS
method would probably separate enantiomers but this has not been demonstrated

REFERENCE
Swartz,M.E.; Mazzeo,J.R.; Grover,E.R.; Brown,P.R. Validation of enantiomeric separations by micellar electrokinetic capillary chromatography using synthetic chiral surfactants, *J.Chromatogr.A*, **1996**, *735*, 303–310.

SAMPLE
Matrix: solutions

CAPILLARY ELECTROPHORESIS
Capillary: 100 cm × 50 μm fused-silica (50 cm to detector) (Isco)
Capillary preparation: Flush capillary with 10 μL running buffer between runs. Every 40 sample injections rinse capillary with 200 μL 1 M NaOH, with 200 μL water, and 200 μL running buffer. Before use fill capillary with 1 M NaOH and allow to stand for 1 h, fill with 100 mM NaOH, allow to stand for 1 h, wash with water fill with running buffer.
Capillary temperature: 23
Running buffer: 100 mM citric acid containing 19.27 mM Na_2HPO_4 and 120 mM hydroxypropyl-β-cyclodextrin
Injection: Inject under vacuum at 4.0 kPa.s
Detector: UV 200
Migration time: 27 (S), 31 (R)
Voltage: 30 kV
Model: Isco Model 3140

OTHER SUBSTANCES
Simultaneous: amphetamine, 4-bromo-2,5-dimethoxyamphetamine, ephedrine, epinephrine, methamphetamine, methyldimethoxyethylamphetamine, methyldimethoxyamphetamine, methyldimethoxymethylamphetamine

KEY WORDS
chiral

REFERENCE
Aumatell,A.; Wells,R.J.; Wong,D.K.Y. Enantiomeric differentiation of a wide range of pharmacologically active substances by capillary electrophoresis using modified β-cyclodextrins, *J.Chromatogr.A*, **1994**, *686*, 293–307.

SAMPLE
Matrix: solutions

CAPILLARY ELECTROPHORESIS
Capillary: 50 cm × 50 μm fused-silica (50 cm to detector) (Polymicro Technologies)
Capillary preparation: Rinse capillary with running buffer for 1 min between runs. Flush a new capillary with 100 mM NaOH for 30 min, flush for 20 min with running buffer, equilibrate with running buffer for 20 min at 15 kV.
Capillary temperature: 26
Running buffer: 20 mM pH 10.0 Borate buffer containing 40 mM tetrakis[6-O-(4-sulfobutyl)]-β-cyclodextrin sodium salt (Center of Drug Delivery Research, Higuchi Bioscience Center, University of Kansas, Lawrence KS)
Injection: Pressure injection at 0.5 psi for 4 s
Detector: UV 214
Migration time: 15.49 (+) (only enantiomer tested)
Voltage: 15 kV
Model: Beckman P/ACE 2000

OTHER SUBSTANCES
Simultaneous: ephedrine, methylephedrine, methylpseudoephedrine, norephedrine

KEY WORDS
chiral

REFERENCE
Dette,C.; Ebel,S.; Terabe,S. Neutral and anionic cyclodextrins in capillary zone electrophoresis: enantiomeric separation of ephedrine and related compounds, *Electrophoresis*, **1994**, *15*, 799–803.

SAMPLE
Matrix: solutions
Sample preparation: Prepare a 50 μg/mL solution in water:buffer:MeOH 90:9:1, inject an aliquot. (Buffer was 25 mM Tris buffer adjusted to pH 2.4 with phosphoric acid.)

CAPILLARY ELECTROPHORESIS
Capillary: 82 cm × 50 μm fused-silica (60 cm to detector)
Capillary preparation: Condition by aspirating with 1 M NaOH for 10 min, with water for 10 min, and with running buffer for 10 min.
Capillary temperature: 30
Running buffer: MeOH:buffer 1.2:98.8 (Buffer was 25 mM Tris buffer adjusted to pH 2.4 with phosphoric acid containing 5 mM heptakis(2,6-di-O-methyl)-β-cyclodextrin and 1 mM sulfobutyl ether β-cyclodextrin (ISCO)
Injection: Vacuum injection for 0.5 s (2 nL)
Detector: UV 210
Migration time: 11.38 (-), 12.60 (+)
Voltage: 30 kV
Model: Applied Biosystems Model 270A-HT

OTHER SUBSTANCES
Simultaneous: amphetamine, cathinone, cocaine, ephedrine, methamphetamine, methcathinone, norephedrine, norpseudoephedrine, propoxyphene

KEY WORDS
chiral

REFERENCE

Lurie,I.S.; Klein,R.F.X.; Dal Cason,T.A.; LeBelle,M.J.; Brenneisen,R.; Weinberger,R.E. Chiral resolution of cationic drugs of forensic interest by capillary electrophoresis with mixtures of neutral and anionic cyclodextrins, *Anal.Chem.*, **1994**, *66*, 4019–4026.

SAMPLE

Matrix: solutions

CAPILLARY ELECTROPHORESIS

Capillary: 60 cm × 50 μm AccuSep (52.5 cm to detector) (Waters)
Capillary preparation: Before injection rinse capillary with 100 mM NaOH for 3 min and with running buffer for 3 min. Rinse new capillaries with 500 mM NaOH for 5 min
Running buffer: 25 mM Na_2HPO_4 and 25 mM sodium borate containing 50 mM (S)-N-dodecoxycarbonylvaline, pH 8.8 (Prepare (S)-N-dodecoxycarbonylvaline as follows. Prepare dodecyl chloroformate by reacting 1-dodecanol with 0.33 equivalents of triphosgene in dichloromethane solution in the presence of pyridine. Add dodecyl chloroformate dropwise to (S)-valine in 1 M NaOH solution, filter, wash with hexane, recrystallize from ether/petroleum ether.)
Injection: Hydrostatic injection 2 s
Detector: UV 214
Migration time: 22.42, 24.55 (enantiomers)
Voltage: +12 kV
Model: Waters Quanta 4000 or 4000E

KEY WORDS

chiral

REFERENCE

Mazzeo,J.R.; Grover,E.R.; Swartz,M.E.; Petersen,J.S. Novel chiral surfactant for the separation of enantiomers by micellar electrokinetic capillary chromatography, *J.Chromatogr.A*, **1994**, *680*, 125–135.

SAMPLE

Matrix: solutions
Sample preparation: Prepare a solution in MeOH/water, inject an aliquot.

CAPILLARY ELECTROPHORESIS

Capillary: 62 cm × 52 μm fused-silica (50 cm to detector) (Polymicro Technologies)
Capillary temperature: 40
Running buffer: 50 mM pH 2.50 Tetrabutylammonium phosphate
Injection: Siphon injection at 10 cm for 5 s
Detector: UV (wavelength not specified)
Migration time: 9.5
Voltage: 20 kV
Current: 29 μA
Model: Laboratory constructed

OTHER SUBSTANCES

Simultaneous: doxylamine, epinephrine, imidazole, isoproterenol, 2-methylphenethylamine, 1-methylphenylpropylamine, metoprolol, nadolol, nicotine, norepinephrine, propranolol

REFERENCE

Quang,C.; Khaledi,M.G. Extending the scope of chiral separation of basic compounds by cyclodextrin-mediated capillary zone electrophoresis, *J.Chromatogr.A*, **1995**, *692*, 253–265.

SAMPLE

Matrix: solutions
Sample preparation: Prepare a solution in MeOH/water, inject an aliquot.

CAPILLARY ELECTROPHORESIS

Capillary: 62 cm × 52 μm fused-silica (50 cm to detector) (Polymicro Technologies)
Capillary temperature: 40

Running buffer: 50 mM pH 2.50 Tetrabutylammonium phosphate containing 20 mM hydroxy-
propyl-β-cyclodextrin
Injection: Siphon injection at 10 cm for 5 s
Detector: UV (wavelength not specified)
Migration time: 12.83, 14.32 (enantiomers)
Voltage: 20 kV
Model: Laboratory constructed

KEY WORDS
chiral

REFERENCE
Quang,C.; Khaledi,M.G. Extending the scope of chiral separation of basic compounds by cyclodextrin-mediated
capillary zone electrophoresis, *J.Chromatogr.A*, **1995**, *692*, 253–265.

SAMPLE
Matrix: solutions
Sample preparation: Inject an aliquot of a 20 μg/mL solution in 12.5 mM pH 9.3 borate buffer.

CAPILLARY ELECTROPHORESIS
Capillary: 66.5 cm × 75 μm fused-silica (41.5 cm to detector) (Polymicro Technologies)
Running buffer: 250 mM pH 9.3 Sodium borate buffer
Injection: Pressure injection at 50 mbar for 15 s.
Detector: UV 195
Migration time: 14.5
Voltage: 8 kV
Model: Crystal 310

OTHER SUBSTANCES
Simultaneous: amphetamine, chlorpheniramine, ephedrine, MDA, MDMA, methamphetamine,
phentermine, phenylpropanolamine
Noninterfering: caffeine

REFERENCE
Lilley,K.A.; Wheat,T.E. Drug identification in biological matrices using capillary electrophoresis and chemometric software, *J.Chromatogr.B*, **1996**, *683*, 67–76.

SAMPLE
Matrix: solutions

CAPILLARY ELECTROPHORESIS
Capillary: 52 cm × 75 μm fused-silica (48 cm to detector) (Polymicro Technologies)
Capillary preparation: Before each run purge with running buffer for 3 min. Every 3 runs
purge with 100 mM NaOH for 5 min. Purge new capillaries with 1 M NaOH for 20 min and
with 100 mM NaOH for 20 min, rinse with running buffer, equilibrate with running buffer at
12 kV for 3 h.
Running buffer: 100 mM CHES (2-(N-cyclohexylamine)ethanesulfonic acid) containing 10 mM
triethylamine and 25 mM (R)-dodecoxycarbonylvaline (Waters EnantioSelect (R)-Val-1), pH ad-
justed to 8.8 with 1 M NaOH
Injection: Hydrostatic injection for 2 s.
Detector: UV 214
Migration time: 11.2 (S,S), 11.8 (R.R)
Voltage: 12 kV
Current: ≤30 μA
Model: Waters Quanta 4000

OTHER SUBSTANCES
Simultaneous: salbutamol, synephrine

KEY WORDS
chiral

REFERENCE

Peterson,A.G.; Ahuja,E.S.; Foley,J.P. Enantiomeric separations of basic pharmaceutical drugs by micellar electrokinetic chromatography using a chiral surfactant, N-dodecoxycarbonylvaline, *J.Chromatogr.B*, **1996**, *683*,
15–28.

SAMPLE

Matrix: solutions
Sample preparation: Inject an aliquot of a 5-10 μM solution.

CAPILLARY ELECTROPHORESIS

Capillary: 70 cm × 75 μm fused-silica (55 cm to detector)
Capillary temperature: 21
Running buffer: 20 mM pH 2.7 Tris-phosphate buffer containing 0.5% hydroxypropylmethyl
cellulose and 24 mM heptakis(2,6-di-O-methyl)-β-cyclodextrin (Hydroxypropylmethyl cellulose
was from Sigma; viscosity of 2% solution = 4000 cP at 25°.)
Injection: Electrokinetic injection at 3 kV for 12 s.
Detector: UV 190
Migration time: 18, 19.7 (enantiomers)
Voltage: 21 kV
Model: Crystal 300 (ATI, Unicam)

OTHER SUBSTANCES

Simultaneous: methamphetamine, norepinephrine, selegiline (deprenyl)
Interfering: amphetamine, epinephrine

KEY WORDS

chiral

REFERENCE

Szökö,E.; Gyimesi,J.; Barcza,L.; Magyar,K. Determination of binding constants and the influence of methanol
on the separation of drug enantiomers in cyclodextrin modified capillary electrophoresis, *J.Chromatogr.A*,
1996, *745*, 181–187.

SAMPLE

Matrix: solutions

CAPILLARY ELECTROPHORESIS

Capillary: 72 cm × 50 μm
Capillary temperature: 30
Running buffer: 20 mM pH 2.5 Tris-phosphate buffer containing 1.5 mM sulfobutylether-β-
cyclodextrin
Detector: UV 205
Migration time: 23.5 (-), 24.5 (+)
Voltage: 30 kV
Model: Perkin-Elmer ABI 270A-HT

OTHER SUBSTANCES

Simultaneous: ephedrine

KEY WORDS

chiral

REFERENCE

Nassar,A.-E.F.; Guarco,F.J.; Gran,D.E.; Stuart,J.D.; Reuter,W.M. Optimizing a method for separating chiral
compounds by capillary electrophoresis, *J.Chromatogr.Sci.*, **1998**, *36*, 19–22.

SAMPLE

Matrix: urine
Sample preparation: Dilute 2 mL urine to 10 mL with water, inject an aliquot.

CAPILLARY ELECTROPHORESIS
Capillary: 100 cm × 75 μm (65 cm to detector) (Supelco)
Capillary preparation: Between runs rinse capillary with 100 mM NaOH for 2 min, with water for 2 min, and with running buffer for 2 min.
Running buffer: 40 mM phosphoric acid and 10 mM boric acid adjusted to pH with 9.7 with 1 M NaOH
Injection: Hydrodynamic injection for 2 s
Detector: UV 210
Migration time: 7
Voltage: 20 kV
Model: Europhor
Limit of quantitation: 1.7 μg/mL
Limit of detection: LOD 500 ng/mL (S/N 3)

OTHER SUBSTANCES
Extracted: ephedrine, epinephrine, methylephedrine, methylpseudoephedrine, norephedrine, norpseudoephedrine

REFERENCE
Chicarro,M.; Zapardiel,A.; Bermejo,E.; Pérez-López,J.A.; Hernández,L. Direct determination of ephedrine alkaloids and epinephrine in human urine by capillary electrophoresis, *J.Liq.Chromatogr.*, **1995**, *18*, 1363–1381.

Pumactant

Merck Index (12th ed.): 8126

SAMPLE
Matrix: solutions

CAPILLARY ELECTROPHORESIS
Capillary: 40 cm × 50 μm fused-silica (40 cm to detector) (Polymicro Technologies)
Capillary preparation: Between injections flush capillary with water for 2 min, with 1 M NaOH for 2 min, and with running buffer for 3 min. At the start of each day flush capillary with 1 M NaOH for 2 min, with water for 2 min, and equilibrate with running buffer for 10 min. Wash new capillaries with 1 M NaOH at 60° for 1 h and with water at 30° for 1 h.
Capillary temperature: 30
Running buffer: MeCN:MeOH:water 10:80:10 containing 100 mM boric acid and 5 mM adenosine monophosphate (monosodium salt)
Injection: Vacuum injection for 3 s.
Detector: UV 259
Migration time: 10.4 (for C16 phosphatidylglycerol)
Voltage: 30 kV
Model: Applied Biosystems 270A
Limit of detection: 3 μg/mL (for C16 phosphatidylglycerol)

OTHER SUBSTANCES
Simultaneous: cardiolipin

KEY WORDS
indirect UV detection

REFERENCE
Nowacka,J.; Oleszek,W. High performance liquid chromatography of zanhic acid glycoside in alfalfa (Medicago sativa), *Phytochem.Anal.*, **1992**, *3*, 227–230.

Pyridoxal

Molecular formula: $C_8H_9NO_3$
Molecular weight: 167.16
CAS Registry No.: 66-72-8
Merck Index (12th ed.): 8162

SAMPLE
Matrix: formulations
Sample preparation: Homogenize a multi-vitamin tablet, extract with 100 mM HCl, filter (0.45 μm).

CAPILLARY ELECTROPHORESIS
Capillary: 51 cm × 75 μm fused-silica (33 cm to detector) (Waters)
Running buffer: 20 mM pH 9.0 sodium phosphate buffer
Injection: 3 nL (5 μL with a split ratio of 1:1700)
Detector: UV 254
Migration time: 6.76
Voltage: 6.0 kV
Model: ISCO Model 3850

OTHER SUBSTANCES
Simultaneous: thiamine (vitamin B1), riboflavin (vitamin B2), pyridoxamine, pyridoxine, ascorbic acid

KEY WORDS
tablets

REFERENCE
Huopalahti,R.; Sunell,J. Use of capillary zone electrophoresis in the determination of B vitamins in pharmaceutical products, *J.Chromatogr.*, **1993**, *636*, 133–135.

SAMPLE
Matrix: solutions

CAPILLARY ELECTROPHORESIS
Capillary: 100 cm × 75 μm fused-silica (Scientific Glass Engineering)
Capillary preparation: Equilibrate with running buffer for 30 min before use.
Running buffer: 10 mM Na_2HPO_4 containing 6 mM sodium borate and 50 mM sodium dodecyl sulfate
Injection: Injection by siphon at 20 cm for 10 s or by electromigration
Detector: F ex 325 (7 mW He-Cd laser) em 430
Migration time: 16
Voltage: 30 kV

OTHER SUBSTANCES
Simultaneous: pyridoxal-5'-phosphate, pyridoxamine, pyridoxamine-5'-phosphate, pyridoxic acid, pyridoxine

REFERENCE
Swaile,D.F.; Burton,D.E.; Balchunas,A.T.; Sepaniak,M.J. Pharmaceutical analysis using micellar electrokinetic capillary chromatography, *J.Chromatogr.Sci.*, **1988**, *26*, 406–409.

SAMPLE
Matrix: solutions
Sample preparation: Prepare a 0.5-2 mg/mL solution in water, inject an aliquot.

CAPILLARY ELECTROPHORESIS
Capillary: 65 cm × 50 μm fused-silica (50 cm to detector) (Scientific Glass Engineering)
Running buffer: 20 mM NaH_2PO_4 containing 50 mM sodium dodecyl sulfate adjusted to pH 9.0
with 20 mM sodium tetraborate
Injection: Injection by siphon at 5 cm for 5-10 s (about 1 nL)
Detector: UV 210
Migration time: 6
Voltage: 25 kV

OTHER SUBSTANCES
Simultaneous: niacin, niacinamide, pyridoxamine, pyridoxine, riboflavin, thiamine, vitamin B12

REFERENCE
Nishi,H.; Tsumagari,N.; Kakimoto,T.; Terabe,S. Separation of water-soluble vitamins by micellar electrokinetic
chromatography, *J.Chromatogr.*, **1989**, *465*, 331–343.

Pyridoxamine dihydrochloride

Molecular formula: $C_8H_{14}Cl_2N_2O_2$
Molecular weight: 241.12
CAS Registry No.: 524-36-7 (di HCl)
Merck Index (12th ed.): 8164

SAMPLE
Matrix: formulations
Sample preparation: Homogenize a multi-vitamin tablet, extract with 100 mM HCl, filter (0.45
μm).

CAPILLARY ELECTROPHORESIS
Capillary: 51 cm × 75 μm fused-silica (33 cm to detector) (Waters)
Running buffer: 20 mM pH 9.0 sodium phosphate buffer
Injection: 3 nL (5 μL with a split ratio of 1:1700)
Detector: UV 254
Migration time: 5.38
Voltage: 6.0 kV
Model: ISCO Model 3850

OTHER SUBSTANCES
Simultaneous: thiamine (vitamin B1), riboflavin (vitamin B2), pyridoxal, pyridoxine, ascorbic
acid

KEY WORDS
tablets

REFERENCE
Huopalahti,R.; Sunell,J. Use of capillary zone electrophoresis in the determination of B vitamins in pharmaceutical products, *J.Chromatogr.*, **1993**, *636*, 133–135.

SAMPLE
Matrix: formulations
Sample preparation: Grind tablet to a homogeneous powder and dissolve in 2 mL water and
20 mL EtOH, homogenize (Ultra Turrax), centrifuge. Evaporate the supernatant to dryness
under reduced pressure, reconstitute the residue with water, inject an aliquot.

CAPILLARY ELECTROPHORESIS
Capillary: 64.5 cm × 50 μm fused-silica (56 cm to detector)

Capillary preparation: Before each run flush capillary at 4 kPa with 1 M NaOH for 2 min and with buffer for 5 min. Replace buffer before each analysis.
Capillary temperature: 30
Running buffer: 1-Propanol:buffer 2:98 (Buffer was 100 mM Na_2HPO_4, 500 mM taurine, and 75 mM sodium cholate.)
Injection: Pressure injection at 4 kPa for 1 s.
Detector: UV 214
Migration time: 8
Voltage: 17 kV
Model: Hewlett-Packard HP[3D] CE
Limit of quantitation: 30 μM

OTHER SUBSTANCES
Simultaneous: ascorbic acid, biotin, folic acid, niacinamide, pyridoxine, riboflavin, thiamine, vitamin B12

KEY WORDS
tablets

REFERENCE
Buskov,S.; Moller,P.; Sorensen,H.; Sorensen,J.C.; Sorensen,S. Determination of vitamins in food based on supercritical fluid extraction prior to micellar electrokinetic capillary chromatographic analyses of individual vitamins, *J.Chromatogr.A*, **1998**, *802*, 233–241.

SAMPLE
Matrix: solutions

CAPILLARY ELECTROPHORESIS
Capillary: 100 cm × 75 μm fused-silica (Scientific Glass Engineering)
Capillary preparation: Equilibrate with running buffer for 30 min before use.
Running buffer: 10 mM Na_2HPO_4 containing 6 mM sodium borate and 50 mM sodium dodecyl sulfate
Injection: Injection by siphon at 20 cm for 10 s or by electromigration
Detector: F ex 325 (7 mW He-Cd laser) em 430
Migration time: 12
Voltage: 30 kV

OTHER SUBSTANCES
Simultaneous: pyridoxal, pyridoxal-5'-phosphate, pyridoxamine-5' phosphate, pyridoxic acid, pyridoxine

REFERENCE
Swaile,D.F.; Burton,D.E.; Balchunas,A.T.; Sepaniak,M.J. Pharmaceutical analysis using micellar electrokinetic capillary chromatography, *J.Chromatogr.Sci.*, **1988**, *26*, 406–409.

SAMPLE
Matrix: solutions
Sample preparation: Prepare a 0.5-2 mg/mL solution in water, inject an aliquot.

CAPILLARY ELECTROPHORESIS
Capillary: 65 cm × 50 μm fused-silica (50 cm to detector) (Scientific Glass Engineering)
Running buffer: 20 mM NaH_2PO_4 containing 50 mM sodium dodecyl sulfate adjusted to pH 9.0 with 20 mM sodium tetraborate
Injection: Injection by siphon at 5 cm for 5-10 s (about 1 nL)
Detector: UV 210
Migration time: 4.3
Voltage: 25 kV

OTHER SUBSTANCES
Simultaneous: niacin, niacinamide, pyridoxal, pyridoxine, riboflavin, thiamine, vitamin B12

REFERENCE
Nishi,H.; Tsumagari,N.; Kakimoto,T.; Terabe,S. Separation of water-soluble vitamins by micellar electrokinetic chromatography, *J.Chromatogr.*, **1989**, *465*, 331–343.

Pyridoxine

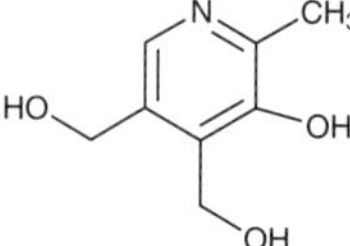

Molecular formula: $C_8H_{11}NO_3$
Molecular weight: 169.18
CAS Registry No.: 65-23-6, 58-56-0 (HCl)
Merck Index (12th ed.): 8166

SAMPLE
Matrix: formulations
Sample preparation: Homogenize a multi-vitamin tablet, extract with 100 mM HCl, filter (0.45 μm).

CAPILLARY ELECTROPHORESIS
Capillary: 51 cm × 75 μm fused-silica (33 cm to detector) (Waters)
Running buffer: 20 mM pH 9.0 sodium phosphate buffer
Injection: 3 nL (5 μL with a split ratio of 1:1700)
Detector: UV 254
Migration time: 6.45
Voltage: 6.0 kV
Model: ISCO Model 3850

OTHER SUBSTANCES
Simultaneous: thiamine (vitamin B1), riboflavin (vitamin B2), pyridoxal, pyridoxamine, ascorbic acid

KEY WORDS
tablets

REFERENCE
Huopalahti,R.; Sunell,J. Use of capillary zone electrophoresis in the determination of B vitamins in pharmaceutical products, *J.Chromatogr.*, **1993**, *636*, 133–135.

SAMPLE
Matrix: formulations
Sample preparation: Grind tablet, dissolve in pH 5 citrate buffer, filter (0.2 μm), inject an aliquot.

CAPILLARY ELECTROPHORESIS
Capillary: 48.5 cm × 50 μm fused-silica (40 cm to detector)
Capillary preparation: Between runs flush with 100 mM NaOH for 2 min and with run buffer for 3 min, inject a buffer plug. Before the first use condition capillaries with 1 M NaOH for 10 min, with 100 mM NaOH for 2 min, and with run buffer for 3 min.
Capillary temperature: 25
Running buffer: 20 mM pH 7 sodium phosphate
Injection: Injection at 40 mbar for 4.6 s (post-injection pressure (of run buffer) 40 mbar for 4 s)
Detector: UV 215
Migration time: 6.3
Voltage: 20 kV
Model: Hewlett-Packard

OTHER SUBSTANCES
Simultaneous: folinic acid, niacinamide, niacin, orotic acid, pantothenic acid, thiamine, vitamin B12 (cyanocobalamin) (UV 360), ascorbic acid

KEY WORDS
tablets

REFERENCE
Jegle,U. Separation of water-soluble vitamins via high-performance capillary electrophoresis, *J.Chromatogr.A*, **1993**, *652*, 495–501.

SAMPLE
Matrix: formulations
Sample preparation: Condition a 500 mg Bakerbond C18 SPE cartridge with 3 mL MeOH and 3 mL water:10 mM sodium heptanesulfonate in acetic acid 99:1. Heat 50-100 mg tablet and 40 mL water:10 mM sodium heptanesulfonate in acetic acid 99:1 at 55° with occasional shaking for 5 min, cool, add a 2 mL aliquot to the SPE cartridge, elute with 3.5 mL MeOH, make up the eluate to 10 mL with water, inject an aliquot.

CAPILLARY ELECTROPHORESIS
Capillary: 70 cm × 100 μm
Capillary temperature: 25
Running buffer: MeOH:buffer 10:90 (Buffer was 50 mM pH 8.0 sodium borate containing 22.5 mM sodium dodecyl sulfate.)
Injection: Inject under 3.44 kPa pressure for 5 s (30 nL)
Detector: UV 214
Migration time: 21
Voltage: 16 kV
Current: 33 μA
Model: Beckman P/ACE System 2000

OTHER SUBSTANCES
Simultaneous: niacinamide, pantothenic acid, thiamine, riboflavin, vitamin B12, ascorbic acid

KEY WORDS
tablets; SPE

REFERENCE
Dinelli,G.; Bonetti,A. Micellar electrokinetic capillary chromatography analysis of water-soluble vitamins and multi-vitamin integrators, *Electrophoresis*, **1994**, *15*, 1147–1150.

SAMPLE
Matrix: formulations
Sample preparation: Protect from light. Grind tablet, add 50 mL 1% acetic acid, shake at 65° for 10 min, add 5 mL 10 mg/mL niacin in 1% acetic acid, cool to room temperature, make up to 100 mL with 1% acetic acid, filter (0.22 μm cellulose acetate), dilute 4-12-fold, inject an aliquot.

CAPILLARY ELECTROPHORESIS
Capillary: 48.5 cm × 50 μm fused-silica (40 cm to detector) (Hewlett Packard)
Capillary preparation: After each run wash with 100 mM NaOH for 7 min, with water for 1 min, and with running buffer for 3 min. At the start of each day wash with separation buffer for 10 min. Wash new capillaries with 1 M NaOH, 100 mM NaOH, water, and running buffer.
Capillary temperature: 25
Running buffer: 50 mM Borax adjusted to pH 8.5 with boric acid
Injection: Hydrodynamic injection at 5 kPa for 10 s.
Detector: UV 215
Migration time: 4.5
Internal standard: niacin (7.5)
Voltage: 25 kV
Model: Hewlett Packard [3D]CE

OTHER SUBSTANCES
Simultaneous: adenine (UV 225), biotin (UV 225), niacinamide (UV 225), pantothenic acid, rutin (UV 225), thiamine (UV 225), riboflavin (UV 225), ascorbic acid (UV 225)

KEY WORDS
tablets

REFERENCE
Fotsing,L.; Fillet,M.; Bechet,I.; Hubert,P.; Crommen,J. Determination of six water-soluble vitamins in a pharmaceutical formulation by capillary electrophoresis, *J.Pharm.Biomed.Anal.*, **1997**, *15*, 1113–1123.

SAMPLE
Matrix: formulations
Sample preparation: Grind tablet to a homogeneous powder and dissolve in 2 mL water and 20 mL EtOH, homogenize (Ultra Turrax), centrifuge. Evaporate the supernatant to dryness under reduced pressure, reconstitute the residue with water, inject an aliquot.

CAPILLARY ELECTROPHORESIS
Capillary: 64.5 cm × 50 μm fused-silica (56 cm to detector)
Capillary preparation: Before each run flush capillary at 4 kPa with 1 M NaOH for 2 min and with buffer for 5 min. Replace buffer before each analysis.
Capillary temperature: 30
Running buffer: 1-Propanol:buffer 2:98 (Buffer was 100 mM Na_2HPO_4, 500 mM taurine, and 75 mM sodium cholate.)
Injection: Pressure injection at 4 kPa for 1 s.
Detector: UV 214
Migration time: 10.5
Voltage: 17 kV
Model: Hewlett-Packard HP[3D] CE
Limit of quantitation: 70 μM

OTHER SUBSTANCES
Simultaneous: ascorbic acid, biotin, folic acid, niacinamide, pyridoxamine, riboflavin, thiamine, vitamin B12

KEY WORDS
tablets

REFERENCE
Buskov,S.; Moller,P.; Sorensen,H.; Sorensen,J.C.; Sorensen,S. Determination of vitamins in food based on supercritical fluid extraction prior to micellar electrokinetic capillary chromatographic analyses of individual vitamins, *J.Chromatogr.A*, **1998**, *802*, 233–241.

SAMPLE
Matrix: formulations, juice
Sample preparation: Dissolve in water, filter (0.2 μm), inject an aliquot. (Add a 50-fold excess of cysteine to solutions containing ascorbic acid.)

CAPILLARY ELECTROPHORESIS
Capillary: 60 cm × 50 μm fused-silica (51.5 cm to detector)
Capillary preparation: Condition with running buffer for 3 min before each run. Condition a new capillary 1 M NaOH for 10 min and with 100 mM NaOH for 2 min.
Capillary temperature: 25
Running buffer: 20 mM pH 8.0 Phosphate buffer
Injection: Pressure injection for 200 mbar.s
Detector: UV 200
Migration time: 3.8
Voltage: -30 kV
Model: Hewlett-Packard 3DCE

OTHER SUBSTANCES
Simultaneous: biotin, niacinamide, riboflavin phosphate, thiamine, ascorbic acid

KEY WORDS
fruit

REFERENCE
Schiewe,J.; Mrestani,Y.; Neubert,R. Application and optimization of capillary zone electrophoresis in vitamin analysis, *J.Chromatogr.A*, **1995**, *717*, 255–259.

SAMPLE
Matrix: solutions

CAPILLARY ELECTROPHORESIS
Capillary: 100 cm × 75 μm fused-silica (Scientific Glass Engineering)
Capillary preparation: Equilibrate with running buffer for 30 min before use.
Running buffer: 10 mM Na_2HPO_4 containing 6 mM sodium borate and 50 mM sodium dodecyl sulfate
Injection: Injection by siphon at 20 cm for 10 s or by electromigration
Detector: F ex 325 (7 mW He-Cd laser) em 430
Migration time: 17
Voltage: 30 kV

OTHER SUBSTANCES
Simultaneous: pyridoxal, pyridoxal-5'-phosphate, pyridoxamine, pyridoxamine-5'-phosphate, pyridoxic acid

REFERENCE
Swaile,D.F.; Burton,D.E.; Balchunas,A.T.; Sepaniak,M.J. Pharmaceutical analysis using micellar electrokinetic capillary chromatography, *J.Chromatogr.Sci.*, **1988**, *26*, 406–409.

SAMPLE
Matrix: solutions
Sample preparation: Prepare a 0.5-2 mg/mL solution in water, inject an aliquot.

CAPILLARY ELECTROPHORESIS
Capillary: 65 cm × 50 μm fused-silica (50 cm to detector) (Scientific Glass Engineering)
Running buffer: 20 mM NaH_2PO_4 containing 50 mM sodium dodecyl sulfate adjusted to pH 9.0 with 20 mM sodium tetraborate
Injection: Injection by siphon at 5 cm for 5-10 s (about 1 nL)
Detector: UV 210
Migration time: 6.2
Voltage: 25 kV

OTHER SUBSTANCES
Simultaneous: niacin, niacinamide, pyridoxal, pyridoxamine, riboflavin, thiamine, vitamin B12

REFERENCE
Nishi,H.; Tsumagari,N.; Kakimoto,T.; Terabe,S. Separation of water-soluble vitamins by micellar electrokinetic chromatography, *J.Chromatogr.*, **1989**, *465*, 331–343.

SAMPLE
Matrix: solutions
Sample preparation: Prepare a 100 μg/mL solution in MeOH:water 50:50, inject an aliquot.

CAPILLARY ELECTROPHORESIS
Capillary: 60 cm × 75 μm
Running buffer: 20 mM NaH_2PO_4 containing 50 mM sodium dodecyl sulfate, adjusted to pH 9.0 with sodium tetraborate
Injection: Hydrodynamic injection at 10 cm for 5 s
Detector: UV 214
Migration time: 8
Voltage: 18 kV
Model: Waters Quanta 4000

OTHER SUBSTANCES
Simultaneous: niacin, pantothenic acid, thiamine, riboflavin, ascorbic acid

REFERENCE
Swartz,M.E. Method development and selectivity control for small molecule pharmaceutical separations by capillary electrophoresis, *J.Liq.Chromatogr.*, **1991**, *14*, 923–938.

SAMPLE
Matrix: solutions

CAPILLARY ELECTROPHORESIS
Capillary: 56.9 cm × 75 μm fused-silica (50 cm to detector)
Capillary temperature: 25
Running buffer: MeOH:60 mM pH 8.92 sodium borate buffer containing 60 mM sodium dodecyl sulfate 15:85
Detector: UV 214
Migration time: 6
Voltage: 30 kV
Model: Beckman PACE 2100

OTHER SUBSTANCES
Simultaneous: niacin, niacinamide, thiamine, vitamin B12

REFERENCE
McLaughlin,G.M.; Nolan,J.A.; Lindahl,J.L.; Palmieri,R.H.; Anderson,K.W.; Morris,S.C.; Morrison,J.A.; Bronzert,T.J. Pharmaceutical drug separations by HPCE: Practical guidelines, *J.Liq.Chromatogr.*, **1992**, *15*, 961–1021.

SAMPLE
Matrix: solutions
Sample preparation: Prepare a 100 μg/mL solution in MeOH:water 50:50, inject an aliquot.

CAPILLARY ELECTROPHORESIS
Capillary: 47 cm × 50 μm fused-silica (40 cm to detector) (Polymicro Technologies)
Capillary temperature: 25
Running buffer: Prepare as follows. Mix n-hexane with 2 volumes of 500 mM sodium dodecyl sulfate, vortex, add 1-butanol until the mixture clears, dilute with 20 mM pH 7.0 phosphate buffer until the concentration of sodium dodecyl sulfate is 20 mM.
Injection: Inject using an overpressure of 34500000 Pa (sic) for 3 s
Detector: UV 214
Migration time: 11.3
Voltage: 10 kV
Current: about 17.5 μA
Model: Beckman P/ACE System 2100

OTHER SUBSTANCES
Simultaneous: niacin, niacinamide, thiamine, vitamin A, vitamin E

REFERENCE
Boso,R.L.; Bellini,M.S.; Miksík,; Deyl,Z. Microemulsion electrokinetic chromatography with different organic modifiers: separation of water- and lipid-soluble vitamins, *J.Chromatogr.A*, **1995**, *709*, 11–19.

SAMPLE
Matrix: solutions

CAPILLARY ELECTROPHORESIS
Capillary: 80 cm × 50 μm fused-silica (Supelco)
Capillary preparation: Before each run rinse capillary with running buffer at 94 kPa for 5 min then with running buffer containing 80 mM sodium dodecyl sulfate at 94 kPa for 2.5 min before each run.
Running buffer: 50 mM pH 8.5 ammonium carbonate buffer
Injection: Pressure injection at 5 kPa (50 mbar) for 8 s into the capillary filled with running buffer containing sodium dodecyl sulfate, electrophoresis is then continued with plain running buffer.

Detector: MS, Perkin-Elmer Sciex API-300 quadrupole, electrospray (ionspray) interface, sheath liquid MeOH:running buffer 50:50 at 4 μL/min, ionspray voltage 5 kV, positive ion mode
Migration time: 12
Voltage: 25 kV (net voltage across capillary = 20 kV (applied voltage - electrospray voltage))
Model: Hewlett Packard 3D CE

OTHER SUBSTANCES
Simultaneous: acetaminophen, acetanilide, caffeine, guaifenesin, phenacetin

REFERENCE
Tanaka,Y.; Kishimoto,Y.; Otsuka,K.; Terabe,S. Strategy for selecting separation solutions in capillary electrophoresis-mass spectrometry, *J.Chromatogr.A*, **1998**, *817*, 49–57.

Pyrilamine

Molecular formula: $C_{17}H_{23}N_3O$
Molecular weight: 285.39
CAS Registry No.: 91-84-9, 59-33-6 (maleate)
Merck Index (12th ed.): 8168
Lednicer: 1 51

SAMPLE
Matrix: solutions

CAPILLARY ELECTROPHORESIS
Capillary: 62 cm × 75 μm fused-silica (54.5 cm to detector) (Polymicro Technologies)
Capillary preparation: Purge with running buffer before each run. At the beginning of each day purge using 50-60 kPa vacuum with 500 mM NaOH for 5 min, with water for 5 min, with MeCN for 5 min, and with running buffer for 5 min.
Running buffer: MeCN:MeOH:acetic acid 49:50:1 containing 20 mM ammonium acetate
Injection: Hydrostatic injection at 10 cm for 5 s.
Detector: UV 214
Migration time: 3.08
Voltage: 30 kV
Model: Waters Quanta 4000

OTHER SUBSTANCES
Simultaneous: buclizine, chlorcyclizine, chlorpheniramine, cyclizine, dimenhydrinate, methaphenilene, pheniramine, promethazine, pyrrobutamine, tripelennamine

REFERENCE
Leung,G.N.W.; Tang,H.P.O.; Tso,T.S.C.; Wan,T.S.M. Separation of basic drugs with non-aqueous capillary electrophoresis, *J.Chromatogr.A*, **1996**, *738*, 141–154.

SAMPLE
Matrix: solutions

CAPILLARY ELECTROPHORESIS
Capillary: 57 cm × 50 μm fused-silica (50 cm to detector)
Running buffer: 50 mM pH 8.0 Borate buffer containing 40 mM sodium taurodeoxycholate and 25 mM phosphatidylcholine (Prepare by adding phosphatidylcholine and stirring for 3-6 h until all cloudiness disappears. Phosphatidylcholine was 95% pure soybean lecithin, Epikuron, Lucas Meyer & Co.)
Injection: Inject a solution of the compound in the running buffer at 20 psi for 1 s, inject MeOH:water 5:95 at 20 psi for 1 s, inject a solution of halofantrine in running buffer at 20 psi for 1 s.
Detector: UV 214

Migration time: k' 15.85
Model: Beckman P/ACE 5000

OTHER SUBSTANCES
Also analyzed: acetaminophen, amoxicillin, antipyrine, aspirin, azathioprine, caffeine, captopril, carbamazepine, carprofen, chlorambucil, chlorpheniramine, chlorpromazine, cimetidine, clonidine, codeine, desipramine, diphenhydramine, ephedrine, fenoterol, flufenamic acid, flurbiprofen, haloperidol, hydroxyzine, ibuprofen, imipramine, indomethacin, ketoprofen, lidocaine, melphalan, metoprolol, nabumetone, nadolol, phenobarbital, phenol, promazine, propranolol, ranitidine, ropinirole, salicylic acid, sulfamethoxazole, testosterone, theophylline, thioridazine, tiaprofenic acid, tolfenamic acid, trifluoperazine, trimethoprim, valproic acid, verapamil

KEY WORDS
comparison with HPLC; k' = (Tr-T0)/(T0(1-Tr/Tm)) where Tr = retention time of analyte; T0 = retention time of water; and Tm = retention time of marker (halofantrine)

REFERENCE
Hanna,M.; de Biasi,V.; Bond,B.; Salter,C.; Hutt,A.J.; Camilleri,P. Estimation of the partitioning characteristics of drugs: A comparison of a large and diverse drug series utilizing chromatographic and electrophoretic methodology, *Anal.Chem.*, **1998**, *70*, 2092–2099.

Pyrimethamine

Molecular formula: $C_{12}H_{13}ClN_4$
Molecular weight: 248.71
CAS Registry No.: 58-14-0
Merck Index (12th ed.): 8169
Lednicer: 1 262

SAMPLE
Matrix: solutions
Sample preparation: Prepare a 100-200 ppm solution in MeOH, inject an aliquot.

CAPILLARY ELECTROPHORESIS
Capillary: 45 cm × 50 μm fused-silica (Polymicro Technologies)
Running buffer: 50 mM pH 7.50 Phosphate buffer containing 25 mM borate and 9 mM tetrabutylammonium bromide
Injection: Hydrodynamic injection at 5 cm for 5 s (1 nL)
Detector: UV 240
Migration time: 3.7
Voltage: 15 kV
Current: 39 μA
Model: Laboratory constructed

OTHER SUBSTANCES
Simultaneous: chloroquine, dapsone, primaquine, quinacrine, quinine, sulfadiazine

REFERENCE
Ng,C.L.; Toh,Y.L.; Li,S.F.Y.; Lee,H.K. Capillary electrophoresis of biologically important compounds: Optimization of separation conditions by the overlapping resolution mapping scheme, *J.Liq.Chromatogr.*, **1993**, *16*, 3653–3666.

SAMPLE
Matrix: solutions
Sample preparation: Inject an aliquot of a 15 μg/mL solution in water.

CAPILLARY ELECTROPHORESIS
Capillary: 65 cm × 50 μm silica (50 cm to detector)

Capillary preparation: Between injections wash with 100 mM NaOH, rinse with water
Capillary temperature: 20 kV
Running buffer: 25 mM Na_2HPO_4 adjusted to pH 2 with phosphoric acid
Injection: Electrokinetic injection at 5 kV for 10 s
Detector: UV 254
Migration time: 13.4
Model: Isco 3850 Capillary Electropherograph
Limit of detection: 150 ng/mL (40 ng/mL if a water plug is injected first)

OTHER SUBSTANCES
Simultaneous: 4-chlorophenylbiguanide, chloroquine, chlorproguanil, cyclochlorproguanil, cyclo-guanil, desethylchloroquine, mefloquine, primaquine, proguanil, quinine

REFERENCE
Taylor,R.B.; Reid,R.G. Analysis of basic antimalarial drugs by CZE and MEKC. Part 1--Critical factors affecting separation, *J.Pharm.Biomed.Anal.*, **1993**, *11*, 1289–1294.

SAMPLE
Matrix: solutions
Sample preparation: Prepare a 2.5 µg/mL solution in water, filter (0.45 µm), inject an aliquot.

CAPILLARY ELECTROPHORESIS
Capillary: 71.2 cm × 50 µm fused-silica (52 cm to detector) (Applied Biosystems)
Capillary preparation: Between runs purge the capillary with 100 mM NaOH for 2 min and with running buffer for 2 min. At the start of each day purge capillary with 100 mM NaOH for 2 min, with water for 2 min, and with running buffer for 2 min
Capillary temperature: 30
Running buffer: 250 mM pH 2.1 Phosphate buffer
Injection: Vacuum injection for 10 s (40 nL)
Detector: UV 225
Migration time: 22.44
Voltage: 13 kV
Model: Applied Biosystems Model 270 A

OTHER SUBSTANCES
Simultaneous: aditoprim, brodimoprim, diaveridine, metioprim, ormetoprim, tetroxoprim, trimethoprim

REFERENCE
Cao,J.; Cross,R.F. The separation of dihydrofolate reductase inhibitors and the determination of pKa,1 values by capillary zone electrophoresis, *J.Chromatogr.A*, **1995**, *695*, 297–308.

SAMPLE
Matrix: solutions

CAPILLARY ELECTROPHORESIS
Capillary: 60 cm × 50 µm fused-silica (47 cm to detector) (Polymicro Technologies)
Capillary temperature: 25
Running buffer: 20 mM pH 8.5 Sodium tetraborate containing 100 mM sodium dodecyl sulfate
Injection: Hydrodynamic injection at 50 mbar for 3.6 s (5 nL).
Detector: UV 205
Migration time: 24.7
Voltage: 22 kV
Model: Crystal 310 (Thermo Unicam)

OTHER SUBSTANCES
Simultaneous: amoxicillin, ampicillin, cephapirin, cloxacillin, dicloxacillin, oxacillin, penicillin G, penicillin V, piperacillin, sulfacetamide, sulfadimethoxine, sulfaguanidine, sulfamerazine, sulfameter, sulfamethazine, sulfanilamide, sulfanilic acid, sulfapyridine, sulfaquinoxaline, sulfathiazole, sulfisoxazole, trimethoprim

REFERENCE
Hows,M.E.P.; Perrett,D.; Kay,J. Optimization of a simultaneous separation of sulphonamides, dihydrofolate reductase inhibitors and β-lactam antibiotics by capillary electrophoresis, *J.Chromatogr.A*, **1997**, *768*, 97–104.

Pyrrobutamine

Molecular formula: $C_{20}H_{22}ClN$
Molecular weight: 311.85
CAS Registry No.: 91-82-7, 135-31-9 (diphosphate)
Merck Index (12th ed.): 8196
Lednicer: 1 78

SAMPLE
Matrix: solutions

CAPILLARY ELECTROPHORESIS
Capillary: 62 cm × 75 μm fused-silica (54.5 cm to detector) (Polymicro Technologies)
Capillary preparation: Purge with running buffer before each run. At the beginning of each day purge using 50-60 kPa vacuum with 500 mM NaOH for 5 min, with water for 5 min, with MeCN for 5 min, and with running buffer for 5 min.
Running buffer: MeCN:MeOH:acetic acid 49:50:1 containing 20 mM ammonium acetate
Injection: Hydrostatic injection at 10 cm for 5 s.
Detector: UV 214
Migration time: 4.36
Voltage: 30 kV
Model: Waters Quanta 4000

OTHER SUBSTANCES
Simultaneous: buclizine, chlorcyclizine, chlorpheniramine, cyclizine, dimenhydrinate, methaphenilene, pheniramine, promethazine, pyrilamine, tripelennamine

REFERENCE
Leung,G.N.W.; Tang,H.P.O.; Tso,T.S.C.; Wan,T.S.M. Separation of basic drugs with non-aqueous capillary electrophoresis, *J.Chromatogr.A*, **1996**, *738*, 141–154.

Quinacrine

Molecular formula: $C_{23}H_{30}ClN_3O$
Molecular weight: 399.96
CAS Registry No.: 83-89-6, 69-05-6 (di HCl), 6151-30-0 (di HCl dihydrate)
Merck Index (12th ed.): 8225
Lednicer: 1 396

SAMPLE
Matrix: solutions
Sample preparation: Prepare a 100-200 ppm solution in MeOH, inject an aliquot.

CAPILLARY ELECTROPHORESIS
Capillary: 45 cm × 50 μm fused-silica (Polymicro Technologies)
Running buffer: 50 mM pH 7.50 Phosphate buffer containing 25 mM borate and 9 mM tetrabutylammonium bromide

Injection: Hydrodynamic injection at 5 cm for 5 s (1 nL)
Detector: UV 240
Migration time: 3.3
Voltage: 15 kV
Current: 39 µA
Model: Laboratory constructed

OTHER SUBSTANCES
Simultaneous: chloroquine, dapsone, primaquine, pyrimethamine, quinine, sulfadiazine

REFERENCE
Ng,C.L.; Toh,Y.L.; Li,S.F.Y.; Lee,H.K. Capillary electrophoresis of biologically important compounds: Optimization of separation conditions by the overlapping resolution mapping scheme, *J.Liq.Chromatogr.*, **1993**, *16*, 3653–3666.

SAMPLE
Matrix: solutions
Sample preparation: Prepare a solution in running buffer, inject an aliquot.

CAPILLARY ELECTROPHORESIS
Capillary: 60 cm × 75 µm fused-silica (52.4 cm to detector)
Capillary preparation: After each run flush with 500 mM KOH for 2-3 min, flush with water, fill with running buffer
Running buffer: 10 mM Na_2HPO_4 containing 2% heparin sodium (MW 10000, 11% S, Scientific Protein Laboratories, Waunakee, WI) adjusted to pH 5 with phosphoric acid
Injection: Hydrostatic injection
Detector: UV 214
Migration time: 30.4
Model: Waters Quanta 4000

OTHER SUBSTANCES
Also analyzed: anabasine, brompheniramine, bupivacaine, carbinoxamine, chlorcyclizine, chloroquine, chlorpheniramine, dimethindene, doxylamine, enpiroline, halofantrine, hydroxychloroquine, indapamide, mefloquine, nornicotine, pheniramine, primaquine, promethazine, tetramisole

REFERENCE
Stalcup,A.M.; Agyei,N.M. Heparin: a chiral mobile-phase additive for capillary zone electrophoresis, *Anal.Chem.*, **1994**, *66*, 3054–3059.

Quinagolide

Molecular formula: $C_{20}H_{33}N_3O_3S$
Molecular weight: 395.57
CAS Registry No.: 87056-78-8, 94424-50-7 (HCl)
Merck Index (12th ed.): 8226

SAMPLE
Matrix: solutions

CAPILLARY ELECTROPHORESIS
Capillary: 50 cm × 75 µm fused-silica
Capillary temperature: 25
Running buffer: 50 mM pH 2.5 phosphate buffer containing 30 mM β-cyclodextrin
Injection: Hydrodynamic injection for 1 s
Detector: UV 214
Migration time: 39 (+), 40 (-)

Voltage: 15 kV
Model: Beckman P/ACE 2000

KEY WORDS
chiral

REFERENCE
Kuhn,R.; Stoecklin,F.; Erni,F. Chiral separations by host-guest complexation with cyclodextrin and crown ether in capillary zone electrophoresis, *Chromatographia*, **1992**, *33*, 32–36.

Quinidine

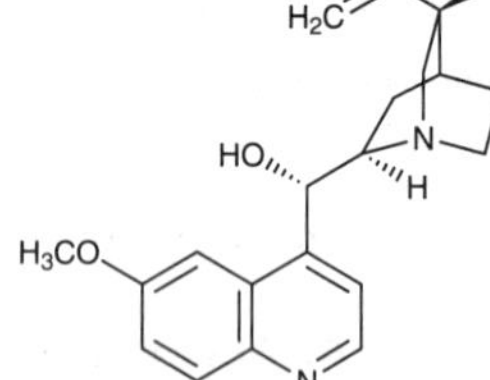

Molecular formula: $C_{20}H_{24}N_2O_2$
Molecular weight: 324.42
CAS Registry No.: 56-54-2, 7054-25-3 (gluconate), 7681-28-9 (polygalacturonate), 50-54-4 (sulfate), 6591-63-5 (sulfate dihydrate)
Merck Index (12th ed.): 8244

SAMPLE
Matrix: serum
Sample preparation: Dilute urine 2-fold, inject serum or diluted urine directly.

CAPILLARY ELECTROPHORESIS
Capillary: 70 cm × 75 μm fused-silica (50 cm to detector) (Polymicro Technologies)
Capillary preparation: Rinse capillary with buffer between runs.
Running buffer: 10 mM pH 9.2 Na_2HPO_4 containing 6 mM sodium borate and 75 mM sodium dodecyl sulfate
Injection: Siphon at a height of 34 cm for 2 s
Detector: F ex 220 em 366 nm (bandpass filter)
Migration time: 16
Voltage: 20 kV
Current: 90 μA
Limit of detection: 1.5 μg/mL

OTHER SUBSTANCES
Extracted: naproxen, salicylic acid

REFERENCE
Caslavska,J.; Gassmann,E.; Thormann,W. Modification of a tunable UV-visible capillary electrophoresis detector for simultaneous absorbance and fluorescence detection: profiling of body fluids for drugs and endogenous compounds, *J.Chromatogr.A*, **1995**, *709*, 147–156.

SAMPLE
Matrix: solutions
Sample preparation: Inject an aliquot of a 100 μg/mL solution in 50 mM sodium dodecyl sulfate.

CAPILLARY ELECTROPHORESIS
Capillary: 57 cm × 75 μm fused-silica (50 cm to detector) (Beckman)
Capillary preparation: Rinse capillary with 100 mM NaOH and running buffer between each run.
Capillary temperature: 30
Running buffer: Isopropanol:100 mM pH 8.1 borate buffer containing 50 mM sodium dodecyl sulfate 10:90
Injection: Pressure injection for 5 s
Detector: UV 200
Migration time: 12.7
Voltage: +25 kV

Model: Beckman P/ACE 5510

OTHER SUBSTANCES
Simultaneous: amiodarone, bretylium, disopyramide, lidocaine, phenytoin, propafenone

REFERENCE
Bretnall,A.E.; Clarke,G.S. Investigation and optimisation of the use of organic modifiers in micellar electrokinetic chromatography, *J.Chromatogr.A*, **1995**, *716*, 49–55.

SAMPLE
Matrix: solutions

CAPILLARY ELECTROPHORESIS
Capillary: 57 cm × 75 μm fused-silica (50 cm to detector) (Beckman)
Capillary preparation: Before each run rinse capillary with 100 mM NaOH and running buffer.
Capillary temperature: 30
Running buffer: Acetone.100 mM pH 8.1 borate buffer containing 50 mM sodium dodecyl sulfate 15:85 (A) or isopropanol:100 mM pH 8.1 borate buffer containing 50 mM sodium dodecyl sulfate 10:90 (B)
Injection: Pressure injection for 5-10 s.
Detector: UV 200
Migration time: 13.7 (A), 16 (B)
Voltage: 25 kV
Model: Beckman P/ACE 5510

OTHER SUBSTANCES
Simultaneous: acebutolol, amiodarone, atenolol, bretylium, captopril, diltiazem, disopyramide, lidocaine, lisinopril, metoprolol, nicardipine, nifedipine, phenytoin, pindolol, propafenone, propranolol, sotalol, timolol, p-toluenesulfonic acid, verapamil

KEY WORDS
only running buffer A separates all compounds

REFERENCE
Bretnall,A.E.; Clarke,G.S. Selectivity of capillary electrophoresis for the analysis of cardiovascular drugs, *J.Chromatogr.A*, **1996**, *745*, 145–154.

SAMPLE
Matrix: solutions
Sample preparation: Inject an aliquot of an aquoeus solution.

CAPILLARY ELECTROPHORESIS
Capillary: 75 cm × 75 μm fused-silica (66 cm to detector) (Polymicro Technologies)
Capillary preparation: Between runs flush capillary with running buffer for 2 min.
Capillary temperature: 28
Running buffer: MeOH:buffer 15:85 (Prepare buffer by dissolving 920 mg cetyltrimethylammonium bromide in 20 mM sodium tetraborate:20 mM KH_2PO_4 50:50, pH 8.6.)
Injection: Vacuum injection at 10 kPa/s.
Detector: UV 230
Migration time: 11.1
Internal standard: cinchonine (10.5)
Voltage: -25 kV
Model: Hewlett Packard 3D

OTHER SUBSTANCES
Simultaneous: benzoic acid, cinchonidine, quinine, saccharin, sorbic acid

KEY WORDS
comparison with HPLC

REFERENCE
Trenerry,V.C.; Ward,C.M. The separation and determination of quinine in bitter drinks by micellar electrokinetic capillary chromatography, *J.Capillary Electrophor.*, **1996**, *3*, 271–274.

SAMPLE
Matrix: solutions
Sample preparation: Inject an aliquot of a 100 μg/mL solution in MeOH:MeCN 50:50.

CAPILLARY ELECTROPHORESIS
Capillary: 64 cm × 50 μm fused-silica (55.5 cm to detector) (Polymicro Technologies)
Capillary preparation: Between runs flush capillary with running buffer for 2 min. Before use rinse capillary with 1 M NaOH for 1 h, with 100 mM NaOH for 20 min, and with running buffer for 10 min.
Capillary temperature: 25
Running buffer: MeCN:MeOH 50:50 containing 50 mM ammonium acetate and 1 M acetic acid
Injection: Pressure injection at 5 kPa for 3 s.
Detector: UV 214
Migration time: 6.5
Voltage: 30 kV
Model: Hewlett-Packard [3D]CE

OTHER SUBSTANCES
Simultaneous: quinine

REFERENCE
Hansen,S.H.; Bjornsdottir,I.; Tjornelund,J. Separation of cationic *cis--trans (Z--E)* isomers and diastereoisomers using non-aqueous capillary electrophoresis, *J.Chromatogr.A*, **1997**, *792*, 49–55.

SAMPLE
Matrix: solutions

CAPILLARY ELECTROPHORESIS
Capillary: 45 cm × 50 μm fused-silica (37.5 cm to detector) (Polymicro Technologies)
Capillary preparation: Between each run rinse with 100 mM HCl for 2 min and with running buffer for 2 min. At the start of each day condition capillary with 1 M HCl for 10 min. Before the first use rinse capillary with 1 M NaOH for 30 min and with water for 30 min.
Capillary temperature: 25
Running buffer: MeCN:30 mM pH 2.40 ethanesulfonic acid 20:80
Injection: Injection at 4 kV for 3 s.
Detector: UV 214
Migration time: 3.65
Voltage: 20 kV
Model: Waters Quanta 4000E

OTHER SUBSTANCES
Simultaneous: chloramphenicol, imipramine, lidocaine, phenylpropanolamine, procainamide, trimethoprim

REFERENCE
Ding,W.; Fritz,J.S. Separation of neutral compounds and basic drugs by capillary electrophoresis in acidic solution using laurylpoly(oxyethylene) sulfate as an additive, *Anal.Chem.*, **1998**, *70*, 1859–1865.

Quinine

Molecular formula: $C_{20}H_{24}N_2O_2$
Molecular weight: 324.42
CAS Registry No.: 130-95-0, 146-40-7 (ascorbate), 549-56-4 (bisulfate), 60-93-5 (di HCl), 83- 75-0 (ethylcarbonate), 750-90-3 (monosalicylate), 6119-70-6 (sulfate dihydrate), 804-63-7 (sulfate), 1407-83-6 (tannate), 146-06-5 (carbonate), 549-47-3 (di HBr), 130-90-5 (formate), 4325-25-1 (gluconate), 549-50-8 (HI), 549-49-5 (HBr), 130-89- 2 (HCl), 7631-46-1 (iodosulfate), 10486-11-0 (oleate), 549-52-0 (urea HCl)
Merck Index (12th ed.): 8245
Lednicer: 1 337

SAMPLE

Matrix: beverages
Sample preparation: Sonicate to degas beverage. Remove a 3 mL aliquot and add it to 1 mL 300 µg/mL cinchonine in 10 mM HCl, make up to 25 mL with water, inject an aliquot.

CAPILLARY ELECTROPHORESIS

Capillary: 75 cm × 75 µm fused-silica (66 cm to detector) (Polymicro Technologies)
Capillary preparation: Between runs flush capillary with running buffer for 2 min.
Capillary temperature: 28
Running buffer: MeOH:buffer 15:85 (Prepare buffer by dissolving 920 mg cetyltrimethylammonium bromide in 20 mM sodium tetraborate:20 mM KH_2PO_4 50:50, pH 8.6.)
Injection: Vacuum injection at 10 kPa/s.
Detector: UV 230
Migration time: 11.4
Internal standard: cinchonine (10.5)
Voltage: -25 kV
Model: Hewlett Packard 3D

OTHER SUBSTANCES

Simultaneous: benzoic acid, cinchonidine, quinidine, sorbic acid
Interfering: saccharin

KEY WORDS

comparison with HPLC

REFERENCE

Trenerry,V.C.; Ward,C.M. The separation and determination of quinine in bitter drinks by micellar electrokinetic capillary chromatography, *J.Capillary Electrophor.*, **1996**, *3*, 271–274.

SAMPLE

Matrix: bulk
Sample preparation: Dilute with 400 µg/mL n-propyl p-hydroxybenzoate in running buffer

CAPILLARY ELECTROPHORESIS

Capillary: 27 cm × 50 µm fused-silica (20 cm to detector)
Capillary preparation: Before each run rinse with running buffer at high pressure.
Capillary temperature: 30
Running buffer: MeCN:water 15:85 containing 40 mM sodium dodecyl sulfate, 8.5 mM sodium borate, and 8.5 mM sodium phosphate, pH 8.5
Injection: Inject at high pressure for 1 s
Detector: UV 214
Migration time: 3.1
Internal standard: n-propyl p-hydroxybenzoate (1.8)
Voltage: 20 kV
Model: Beckman P/ACE System 2100

OTHER SUBSTANCES
Simultaneous: acetaminophen, acetylcodeine, O-acetylmorphine, aspirin, caffeine, cocaine, codeine, diamorphine, diphenhydramine, hydromorphone, isoniacinamide, lidocaine, methaqualone, morphine, niacinamide, noscapine, papaverine, phenacetin, phenobarbital, phenylpropanolamine, procaine, salicylic acid, strychnine, thebaine

REFERENCE
Walker,J.A.; Krueger,S.T.; Lurie,I.S.; Marché,H.L.; Newby,N. Analysis of heroin drug seizures by micellar electrokinetic capillary chromatography (MECC), *J.Forensic Sci.*, **1995**, *40*, 6–9.

SAMPLE
Matrix: solutions
Sample preparation: Prepare a 100-200 ppm solution in MeOH, inject an aliquot.

CAPILLARY ELECTROPHORESIS
Capillary: 45 cm × 50 μm fused-silica (Polymicro Technologies)
Running buffer: 50 mM pH 7.50 Phosphate buffer containing 25 mM borate and 9 mM tetra-
butylammonium bromide
Injection: Hydrodynamic injection at 5 cm for 5 s (1 nL)
Detector: UV 240
Migration time: 3.4
Voltage: 15 kV
Current: 39 μA
Model: Laboratory constructed

OTHER SUBSTANCES
Simultaneous: chloroquine, dapsone, primaquine, pyrimethamine, quinacrine, sulfadiazine

REFERENCE
Ng,C.L.; Toh,Y.L.; Li,S.F.Y.; Lee,H.K. Capillary electrophoresis of biologically important compounds: Optimization of separation conditions by the overlapping resolution mapping scheme, *J.Liq.Chromatogr.*, **1993**,
16, 3653–3666.

SAMPLE
Matrix: solutions
Sample preparation: Inject an aliquot of a 15 μg/mL solution in water.

CAPILLARY ELECTROPHORESIS
Capillary: 65 cm × 50 μm silica (50 cm to detector)
Capillary preparation: Between injections wash with 100 mM NaOH, rinse with water
Capillary temperature: 20 kV
Running buffer: 25 mM Na_2HPO_4 adjusted to pH 2 with phosphoric acid
Injection: Electrokinetic injection at 5 kV for 10 s
Detector: UV 254
Migration time: 9.5
Model: Isco 3850 Capillary Electropherograph
Limit of detection: 150 ng/mL (40 ng/mL if a water plug is injected first)

OTHER SUBSTANCES
Simultaneous: 4-chlorophenylbiguanide, chloroquine, chlorproguanil, cyclochlorproguanil, cyclo-
guanil, desethylchloroquine, mefloquine, primaquine, proguanil, pyrimethamine

REFERENCE
Taylor,R.B.; Reid,R.G. Analysis of basic antimalarial drugs by CZE and MEKC. Part 1--Critical factors affecting
separation, *J.Pharm.Biomed.Anal.*, **1993**, *11*, 1289–1294.

SAMPLE
Matrix: solutions

CAPILLARY ELECTROPHORESIS
Capillary: 50 cm × 50 μm fused-silica (Tokyo Kasei)

Capillary preparation: Rinse capillary with 1 M NaOH after each run.
Running buffer: MeOH:buffer 10:90 (Buffer was 10 mM pH 7 ammonium formate containing 2% butyl acrylate-butyl methacrylate-methacrylic acid copolymer sodium salt (BBMA) (Dai-ichi Kogyo Seiyaku)
Injection: Hydrostatic injection at 15 cm for 10-30 s
Detector: MS, Hitachi M-1000 quadrupole, electrospray at 3 kV, details of interface given in paper, sheath liquid MeOH:water:formic acid 50:50:1 at 5 μL/min, positive-ion mode, drift voltage 70 V, focusing voltage 140-150 V
Migration time: 23
Voltage: 10 kV
Model: laboratory constructed

REFERENCE

Ozaki,H.; Itou,N.; Terabe,S.; Takada,Y.; Sakairi,M.; Koizumi,H. Micellar electrokinetic chromatography-mass spectrometry using a high-molecular-mass surfactant. On-line coupling with an electrospray ionization interface, *J.Chromatogr.A*, **1995**, *716*, 69–79.

SAMPLE
Matrix: solutions

CAPILLARY ELECTROPHORESIS
Capillary: 56.5 cm × 75 μm fused-silica (50 cm to detector) (CeramOptec, Bonn, Germany)
Capillary temperature: 25
Running buffer: 10 mM Disodium tetraborate containing 10 mM boric acid, 20 mM sodium dodecyl sulfate, and 3 M urea
Injection: Pressure injection for 2 s.
Detector: UV 254
Migration time: 9
Voltage: 25 kV
Model: Beckman P/ACE

OTHER SUBSTANCES
Simultaneous: anthracene, 2,4-dinitrotoluene, 2,5-dinitrotoluene, naphthalene, thiourea

REFERENCE

Bütehorn,U.; Pyell,U. Band broadening arising from stepwise alterations in mobile phase composition in micellar electrokinetic chromatography, *Chromatographia*, **1996**, *43*, 237–241.

SAMPLE
Matrix: solutions
Sample preparation: Inject an aliquot of a 100 μg/mL solution in MeOH:MeCN 50:50.

CAPILLARY ELECTROPHORESIS
Capillary: 64 cm × 50 μm fused-silica (55.5 cm to detector) (Polymicro Technologies)
Capillary preparation: Between runs flush capillary with running buffer for 2 min. Before use rinse capillary with 1 M NaOH for 1 h, with 100 mM NaOH for 20 min, and with running buffer for 10 min.
Capillary temperature: 25
Running buffer: MeCN:MeOH 50:50 containing 50 mM ammonium acetate and 1 M acetic acid
Injection: Pressure injection at 5 kPa for 3 s.
Detector: UV 214
Migration time: 6.6
Voltage: 30 kV
Model: Hewlett-Packard ³ᴰCE

OTHER SUBSTANCES
Simultaneous: quinidine

REFERENCE

Hansen,S.H.; Bjornsdottir,I.; Tjornelund,J. Separation of cationic *cis--trans (Z--E)* isomers and diastereoisomers using non-aqueous capillary electrophoresis, *J.Chromatogr.A*, **1997**, *792*, 49–55.

SAMPLE
Matrix: solutions
Sample preparation: Inject an aliquot of a 500 μg/mL solution in 50 mM pH 7 buffer.

CAPILLARY ELECTROPHORESIS
Capillary: 48.5 cm × 50 μm fused-silica (40 cm to detector) (Hewlett Packard)
Capillary preparation: Before each injection flush capillary with 100 mM NaOH for 5 min and with running buffer for 5 min. Wash new capillaries with 1 M NaOH at 40° for 15 min, with water at 40° for 10 min, and with water at 25° for 5 min.
Capillary temperature: 25
Running buffer: 50 mM pH 7.0 Phosphate buffer (Prepare buffer by dissolving 5.29 g K_2HPO_4 and 2.61 g KH_2PO_4 in 1 L water.)
Injection: Pressure injection at 50 mbar for 9 s (18.8 nL).
Detector: UV 200
Migration time: 9
Voltage: 30 kV
Model: Hewlett Packard Model G1600A [3D]CE

OTHER SUBSTANCES
Simultaneous: ceftazidime, diclofenac, dicloxacillin, oxacillin, propranolol, thiamine

REFERENCE
Mrestani,Y.; Neubert,R.H.H.; Krause,A. Partition behaviour of drugs in microemulsions measured by electrokinetic chromatography, *Pharm.Res.*, **1998**, *15*, 799–801.

SAMPLE
Matrix: solutions
Sample preparation: Mix 50 μL of a solution in 1% sodium chloride with 100 μL MeCN, vortex for 15 s, centrifuge at 14000 g for 20 s, inject an aliquot.

CAPILLARY ELECTROPHORESIS
Capillary: 40 cm × 50 μm silica
Running buffer: MeCN:buffer 50:50 (Buffer was 60 mM triethanolamine containing 50 mM tricine, pH adjusted to 8.5.)
Injection: Hydrodynamic injection at low pressure for 70 s (to fill 12% of the capillary).
Detector: UV 254
Migration time: 7.5
Voltage: 11 kV
Model: Beckman Model 2000 CE

OTHER SUBSTANCES
Simultaneous: N-acetylprocainamide, doxepin

REFERENCE
Shihabi,Z.K. Stacking of weakly cationic compounds by acetonitrile for capillary electrophoresis, *J.Chromatogr.A*, **1998**, *817*, 25–30.

Ranitidine

Molecular formula: $C_{13}H_{22}N_4O_3S$
Molecular weight: 314.41
CAS Registry No.: 66357-35-5, 66357-59-3 (HCl)
Merck Index (12th ed.): 8286
Lednicer: 3 131; 4 89, 112, 114

SAMPLE
Matrix: blood

Sample preparation: Condition a 1 mL 100 mg Supelclean LC-18 SPE cartridge with 1 mL MeOH, 1 mL water, and 1 mL 10 mM pH 7 TRIZMA buffer containing 3.2 mM acetic acid. 500 µL Serum + 500 µL 40 mM TRIZMA buffer containing 13 mM acetic acid, add to the bottom of the SPE cartridge, wash with 1 mL water from the bottom of the cartridge, elute downward with 160 µL THF:5 mM pH 7.7 TRIZMA buffer containing 1.7 mM acetic acid 50:50 while applying 150 V using platinum electrodes (negative electrode at bottom), inject an aliquot of the first 50 µL eluate.

CAPILLARY ELECTROPHORESIS
Capillary: 60 cm × 50 µm uncoated fused-silica (45 cm to detector) (Polymicro Technologies)
Capillary preparation: Rinse capillary with buffer between injections. Rinse a new capillary with water for 30 min and with 100 mM NaOH overnight, wash with water, fill with running buffer. recondition capillaries with water for 10 min, with 45% trifluoroacetic acid for 2 h, and water for 10 min.
Running buffer: 9.4 mM pH 6.4 sodium phosphate buffer containing 9.8 mM hexadecyltrimethylammonium bromide and 3.3 mM tris(hydroxymethyl)aminomethane (TRIZMA base)
Injection: Hydrodynamic injection at 13.5 cm for 10-15 s
Detector: UV 228
Migration time: 10
Internal standard: ranitidine
Voltage: -20 kV
Model: Laboratory constructed

OTHER SUBSTANCES
Extracted: cimetidine

KEY WORDS
serum; detector is at anode; SPE; ranitidine is IS

REFERENCE
Soini,H.; Tsuda,T.; Novotny,M.V. Electrochromatographic solid-phase extraction for determination of cimetidine in serum by micellar electrokinetic capillary chromatography, *J.Chromatogr.*, **1991**, *559*, 547–558.

SAMPLE
Matrix: blood
Sample preparation: Condition a 1 mL Supelclean LC-18 SPE cartridge (Supelco) with 1 mL water, 1 mL 50 mM pH 8.4 phosphate buffer, 3 mL MeOH, and 5 mL water. Add 1 mL plasma to the SPE cartridge, wash with 1 mL phosphate buffer, elute with three 1 mL portions of MeOH, evaporate the third eluate to dryness under reduced pressure, reconstitute with 50 µL 10 mM pH 2.0 NaH$_2$PO$_4$, inject an aliquot.

CAPILLARY ELECTROPHORESIS
Capillary: 24 cm × 25 µm coated capillary (Bio-Rad)
Capillary preparation: Purge capillary for a few min after each injection.
Capillary temperature: 15
Running buffer: 100 mM pH 2.0 NaH$_2$PO$_4$
Injection: Electrokinetic injection at 8 kV for 8 s, positive to negative polarity
Detector: UV 208
Migration time: 2.9
Internal standard: ranitidine
Voltage: 15 kV
Model: BioFocus 3000 (Bio-Rad)

OTHER SUBSTANCES
Extracted: cimetidine

KEY WORDS
SPE; plasma; ranitidine is IS; coated capillary

REFERENCE
Luksa,J.; Josic,D. Determination of cimetidine in human plasma by free capillary zone electrophoresis, *J.Chromatogr.B*, **1995**, *667*, 321–327.

SAMPLE
Matrix: bulk

CAPILLARY ELECTROPHORESIS
Capillary: 34 cm × 50 μm (25.5 cm (A) or 8.5 cm (B) to detector) (Composite Metal Services, Hallow, UK)
Running buffer: 10 mM pH 2.5 Triethanolamine/phosphoric acid buffer
Injection: Pressure injection at 50 mbar for 5 s.
Detector: UV 230
Migration time: 2.4 (A), 0.95 (B)
Voltage: 20 kV (A) or -7.5 kV (B)
Model: Hewlett Packard 3D

OTHER SUBSTANCES
Simultaneous: degradation products

KEY WORDS
when the distance to the detector was 8.5 cm (B) injection was at the normal outlet end and voltage polarity and pressure for injection were reversed.

REFERENCE
Altria,K.D.; Kelly,M.A.; Clark,B.J. The use of a short-end injection procedure to achieve improved performance in capillary electrophoresis, *Chromatographia*, **1996**, *43*, 153–158.

SAMPLE
Matrix: formulations
Sample preparation: Dilute injection with 3 mg/mL (+)-chlorpheniramine in water to a ranitidine concentration of 5-10 mg/mL, inject an aliquot.

CAPILLARY ELECTROPHORESIS
Capillary: 27 cm × 50 μm fused-silica (20 cm to detector) (Composite Metal Services, Hallow, UK)
Capillary preparation: Before each injection rinse with 100 mM NaOH for 1 min and with running buffer for 1 min.
Capillary temperature: 25
Running buffer: 190 mM pH 2.6 Citrate buffer (Prepare buffer by dissolving 5.52 g trisodium citrate and 21.09 g citric acid in 100 mL water.)
Injection: For 1 s.
Detector: UV 230
Migration time: 22
Internal standard: (+)-chlorpheniramine (19)
Voltage: +6 kV
Current: 90 μA
Model: Beckman P/ACE 5100 or 2050
Limit of quantitation: 0.1%
Limit of detection: 0.05%

KEY WORDS
injections; robust; validated; comparison with HPLC and TLC

REFERENCE
Kelly,M.A.; Altria,K.D.; Grace,C.; Clark,B.J. Optimisation, validation and application of a capillary electrophoresis method for the determination of ranitidine hydrochloride and related substances, *J.Chromatogr.A*, **1998**, *798*, 297–306.

SAMPLE
Matrix: solutions

CAPILLARY ELECTROPHORESIS
Capillary: 57 cm × 75 μm fused-silica (50 cm to detector) (Beckman)
Capillary temperature: 25

Running buffer: 50 mM Borax adjusted to pH 2.5 with concentrated phosphoric acid
Injection: Pressure injection for 5 s
Detector: UV 230
Migration time: 19
Voltage: +30 kV
Model: Beckman P/ACE 2000 CE

OTHER SUBSTANCES
Simultaneous: impurities

KEY WORDS
comparison with HPLC

REFERENCE
Altria,K.D.; Connolly,P.C. On-line solution stability determination of pharmaceuticals by capillary electrophoresis, *Chromatographia*, **1993**, *37*, 176–178.

SAMPLE
Matrix: solutions
Sample preparation: Dilute solution with an equal volume of 100 μg/mL imidazole solution, inject an aliquot.

CAPILLARY ELECTROPHORESIS
Capillary: 27 cm × 50 μm (20 cm to detector) (Composite Metals, Hallow, UK)
Capillary preparation: Rinse with running buffer for 0.3 min between injections.
Running buffer: 25 mM NaH_2PO_4 adjusted to pH 2.5 with concentrated phosphoric acid
Detector: UV 200
Migration time: 1.52
Internal standard: imidazole (0.97)
Model: Beckman P/ACE 5500

REFERENCE
Altria,K.D.; Traylen,E.; Turner,N. Analysis of dissolution test sample solutions by high speed capillary electrophoresis, *Chromatographia*, **1995**, *41*, 393–397.

SAMPLE
Matrix: solutions
Sample preparation: Inject an aliquot of an aqueous solution.

CAPILLARY ELECTROPHORESIS
Capillary: 37 cm × 100 μm fused-silica
Capillary preparation: Before injection rinse with 100 mM NaOH for 30 s and with running buffer for 30 s. Condition a new capillary by rinsing with 100 mM NaOH for 30 min. Store capillaries dry.
Capillary temperature: 30
Running buffer: 25 mM NaH_2PO_4 adjusted to pH 2.3 with phosphoric acid
Injection: Pressure injection for 5 s
Detector: UV 214
Migration time: 4.714 (?)
Voltage: +10 kV
Model: Beckman P/ACE 5100
Limit of detection: 50 ng/mL (S/N 3)

OTHER SUBSTANCES
Simultaneous: sumatriptan

REFERENCE
Altria,K.D.; Hadgett,T.A. An evaluation of the use of capillary electrophoresis to monitor trace drug residues following the manufacture of pharmaceuticals, *Chromatographia*, **1995**, *40*, 23–27.

SAMPLE
Matrix: solutions
Sample preparation: Inject an aliquot of an aqueous solution.

CAPILLARY ELECTROPHORESIS
Capillary: 63 cm $\times$ 50 μm fused-silica (50 cm to detector) (SGE)
Capillary preparation: Flush with 1 M NaOH at the end of each day, store overnight in water.
Etch new columns with 1 M NaOH for 1 h, flush for 15 min with each new buffer.
Running buffer: MeOH:30 mM NaH_2PO_4 10:90, pH 4.87
Injection: Hydrodynamic injection at 100 mbar for 6 s.
Detector: UV 214
Migration time: 5.4
Voltage: 23.8 kV
Model: Lauerlabs Prince

REFERENCE
Morris,V.M.; Hargreaves,C.; Overall,K.; Marriott,P.J.; Hughes,J.G. Optimization of the capillary electrophoresis separation of ranitidine and related compounds, *J.Chromatogr.A*, **1997**, *766*, 245–254.

SAMPLE
Matrix: solutions

CAPILLARY ELECTROPHORESIS
Capillary: 27 cm $\times$ 100 μm (20 cm to detector) (Composite Metal Services)
Capillary temperature: 30
Running buffer: 25 mM pH 2.3 NaH_2PO_4
Detector: UV 200
Migration time: 4.12
Internal standard: imidazole (2.46)
Voltage: 7 kV
Model: Beckman

OTHER SUBSTANCES
Simultaneous: albuterol, alosetron, lamivudine

REFERENCE
Altria,K.D.; Creasey,E.; Howells,J.S. Routine capillary electrophoresis trace level determinations of pharmaceutical and detergent residues on pharmaceutical manufacturing equipment, *J.Liq.Chromatogr.Rel.Technol.*, **1998**, *21*, 1093–1105.

SAMPLE
Matrix: solutions

CAPILLARY ELECTROPHORESIS
Capillary: 57 cm $\times$ 50 μm fused-silica (50 cm to detector)
Running buffer: 50 mM pH 8.0 Borate buffer containing 40 mM sodium taurodeoxycholate and 25 mM phosphatidylcholine (Prepare by adding phosphatidylcholine and stirring for 3-6 h until all cloudiness disappears. Phosphatidylcholine was 95% pure soybean lecithin, Epikuron, Lucas Meyer & Co.)
Injection: Inject a solution of the compound in the running buffer at 20 psi for 1 s, inject MeOH: water 5:95 at 20 psi for 1 s, inject a solution of halofantrine in running buffer at 20 psi for 1 s.
Detector: UV 214
Migration time: k' 0.71
Model: Beckman P/ACE 5000

OTHER SUBSTANCES
Also analyzed: acetaminophen, amoxicillin, antipyrine, aspirin, azathioprine, caffeine, captopril, carbamazepine, carprofen, chlorambucil, chlorpheniramine, chlorpromazine, cimetidine, clonidine, codeine, desipramine, diphenhydramine, ephedrine, fenoterol, flufenamic acid, flurbiprofen, haloperidol, hydroxyzine, ibuprofen, imipramine, indomethacin, ketoprofen, lidocaine, melphalan, metoprolol, nabumetone, nadolol, phenobarbital, phenol, promazine, propranolol, pyrilamine, ropinirole, salicylic acid, sulfamethoxazole, testosterone, theophylline, thioridazine, tiaprofenic acid, tolfenamic acid, trifluoperazine, trimethoprim, valproic acid, verapamil

KEY WORDS
comparison with HPLC; k' = (Tr-T0)/(T0(1-Tr/Tm)) where Tr = retention time of analyte; T0 = retention time of water; and Tm = retention time of marker (halofantrine)

REFERENCE
Hanna,M.; de Biasi,V.; Bond,B.; Salter,C.; Hutt,A.J.; Camilleri,P. Estimation of the partitioning characteristics of drugs: A comparison of a large and diverse drug series utilizing chromatographic and electrophoretic methodology, *Anal.Chem.*, **1998**, *70*, 2092–2099.

Remoxipride

Molecular formula: $C_{16}H_{23}BrN_2O_3$
Molecular weight: 371.27
CAS Registry No.: 80125-14-0, 82935-42-0 (HCl)
Merck Index (12th ed.): 8301
Lednicer: 4 42

SAMPLE
Matrix: solutions

CAPILLARY ELECTROPHORESIS
Capillary: 37 cm × 75 μm CElect C8 bonded capillary (30 cm to detector) (Supelco)
Capillary preparation: Flush with running buffer for 2 min, rinse with water after each set of samples.
Running buffer: pH 3.0 Phosphate buffer (I=0.05) containing 8 mM tetrapropylammonium hydroxide, 12 mM tetrabutylammonium hydroxide, and 20 mM methyl-β-cyclodextrin
Injection: Pressure injection for 2-5 s
Detector: UV 214
Migration time: 18
Voltage: 15 kV
Current: 40-55 μA
Model: Beckman P/ACE 2100
Limit of detection: 0.05%

OTHER SUBSTANCES
Simultaneous: impurities

KEY WORDS
coated capillary

REFERENCE
Stålberg,O.; Westerlund,D.; Rodby,U.-B.; Schmidt,S. Determination of impurities in remoxipride by capillary electrophoresis using UV-detection and LIF-detection; principles to handle sample overloading effects, *Chromatographia*, **1995**, *41*, 287–294.

SAMPLE
Matrix: solutions
Sample preparation: Inject an aliquot of a 100 μM solution.

CAPILLARY ELECTROPHORESIS
Capillary: 37 or 47 cm × 75 μm CElect C8 bonded capillary (effective length 30 or 40 cm) (Supelco)
Capillary preparation: Flush capillary for 2 min with running buffer before injection. Rinse capillary with water at the end of the day.
Running buffer: pH 3.0 Phosphate buffer (I = 0.05) containing 20 mM methyl-β-cyclodextrin, 8 mM tetrapropylammonium hydroxide, and 12 mM tetrabutylammonium hydroxide
Injection: Inject at 0.5 psi for 3-5 s.
Detector: UV 214
Migration time: 17.5
Voltage: 15 kV
Current: 40-55 μA
Model: Beckman P/ACE 2100

OTHER SUBSTANCES
Simultaneous: rotamers, analogs

KEY WORDS
coated capillary

REFERENCE
Stålberg,O.; Brötell,H.; Westerlund,D. Capillary electrophoretic separation of basic drugs using surface-modified C8 capillaries and derivatized cyclodextrins as structural/chiral selectors, *Chromatographia*, **1995**, *40*, 697–704.

SAMPLE
Matrix: solutions

CAPILLARY ELECTROPHORESIS
Capillary: 48.5 cm × 50 μm fused-silica polyacrylamide coated (40 cm to detector) (Polymicro Technologies)
Capillary preparation: Before each run rinse capillary with water for 3 min and with running buffer for 3 min. Store capillary in water overnight, rinse with water each morning. (Coat capillary as follows. Adjust the pH of 20 mL water to 3.5 with acetic acid, add 80 μL 3-(tri-methoxysilyl)propyl methacrylate (3-methacryloxypropyltrimethoxysilane), mix, suck into capillary, let stand at room temperature for 1 h, remove the solution, wash with water. Fill the capillary with a deaerated 3-4% acrylamide solution containing 1 μL/mL N,N,N',N'-tetra-methylethylenediamine and 1 mg/mL potassium persulfate, let stand for 30 min, remove excess solution by aspiration, rinse with water, remove water by aspiration, dry at 35° (J. Chromatogr. 1985, 347, 191).)
Capillary temperature: 25
Running buffer: 50 mM pH 4.0 Acetate buffer
Injection: Fill capillary with 200 μM α₁-acid glycoprotein (Sigma) in running buffer at 3.4 kPa for 210 s then inject sample under pressure at 3.4 kPa for 5 s (ca. 9 nL).
Detector: UV 200
Migration time: 10.2, 10.4 (enantiomers)
Voltage: 20 kV
Model: Hewlett-Packard 3D

KEY WORDS
chiral; coated capillary

REFERENCE
Amini,A.; Westerlund,D. Evaluation of association constants between drug enantiomers and human α1-acid glycoprotein by applying a partial-filling technique in affinity capillary electrophoresis, *Anal.Chem.*, **1998**, *70*, 1425–1430.

Reproterol

Molecular formula: $C_{18}H_{23}N_5O_5$
Molecular weight: 389.41
CAS Registry No.: 54063-54-6, 13055-82-8 (HCl)
Merck Index (12th ed.): 8307

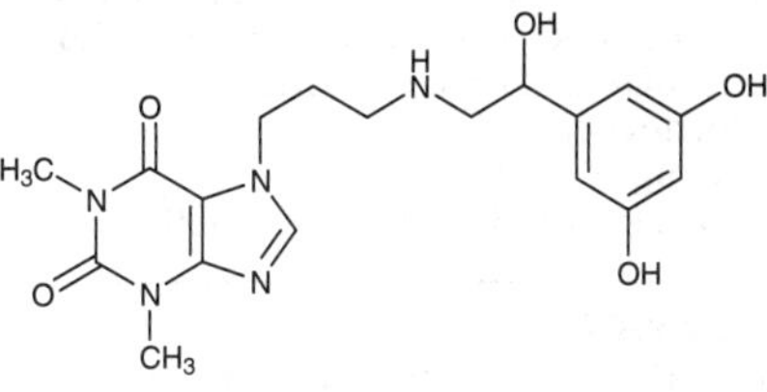

SAMPLE
Matrix: solutions
Sample preparation: Inject an aliquot of a 100 µg/mL solution in running buffer.

CAPILLARY ELECTROPHORESIS
Capillary: 32 cm × 50 µm fused-silica (27.5 cm to detector) (Yongnian Optical Conductive Fiber Plant, China), coated with polyacrylamide
Capillary preparation: No details of the polyacrylamide coating process are provided. However, another paper (LC.GC 1997, 15, 40) by this group indicates that they use the procedure of Hjertén, thus: Adjust the pH of 20 mL water to 3.5 with acetic acid, add 80 µL 3-(trimethoxysilyl)propyl methacrylate (3-methacryloxypropyltrimethoxysilane), mix, suck into capillary, let stand at room temperature for 1 h, remove the solution, wash with water. Fill the capillary with a deaerated 3-4% acrylamide solution containing 1 µL/mL N,N,N',N'-tetramethylethylenediamine and 1 mg/mL potassium persulfate, let stand for 30 min, remove excess solution by aspiration, rinse with water, remove water by aspiration, dry at 35° (J. Chromatogr. 1985, 347, 191).
Capillary temperature: 25
Running buffer: 100 mM NaH_2PO_4 adjusted to pH 2.5
Injection: Electrokinetic injection at 15 kV for 3 s.
Detector: UV 200, UV 210
Migration time: 7.21
Voltage: 15 kV
Model: Bio-Rad BioFocus 3000

OTHER SUBSTANCES
Simultaneous: atropine, carazolol, cicletanine, dimethindene, fendiline, homatropine, ipratropium bromide, isothipendyl, mefloquine, metaclazepam, nafronyl (naftidrofuryl), nefopam, nicardipine, promethazine, tetrahydrozoline (tetryzoline), theodrenaline, tioconazole, trihexyphenidyl, trimeprazine (alimemazine), trimipramine

KEY WORDS
coated capillary

REFERENCE
Koppenhoefer,B.; Epperlein,U.; Xiaofeng,Z.; Bingcheng,L. Separation of enantiomers of drugs by capillary electrophoresis. Part 4: Hydroxypropyl-γ-cyclodextrin as chiral solvating agent, *Electrophoresis*, **1997**, *18*, 924–930.

SAMPLE
Matrix: solutions
Sample preparation: Inject an aliquot of a 100 µg/mL solution in running buffer.

CAPILLARY ELECTROPHORESIS
Capillary: 30 cm × 50 µm fused-silica (25.5 cm to detector), coated with polyacrylamide
Capillary preparation: Adjust the pH of 20 mL water to 3.5 with acetic acid, add 80 µL 3-(trimethoxysilyl)propyl methacrylate (3-methacryloxypropyltrimethoxysilane), mix, suck into capillary, let stand at room temperature for 1 h, remove the solution, wash with water. Fill the capillary with a deaerated 3-4% acrylamide solution containing 1 µL/mL N,N,N',N'-tetramethylethylenediamine and 1 mg/mL potassium persulfate, let stand for 30 min, remove excess solution by aspiration, rinse with water, remove water by aspiration, dry at 35° (J. Chromatogr. 1985, 347, 191).
Capillary temperature: 25

Running buffer: 100 mM NaH_2PO_4 containing 45 mM heptakis(2,6-di-O- methyl)-β-cyclodextrin, adjusted to pH 2.5 with phosphoric acid
Injection: Electrokinetic injection at 15 kV for 3 s.
Detector: UV 200
Voltage: 15 kV
Model: Bio-Rad BioFocus 3000

KEY WORDS
chiral; coated capillary; comparison with the use of other cyclodextrins; this running buffer gave the greatest enantiomeric separation.; α=1.046

REFERENCE
Lin,B.; Zhu,X.; Koppenhoefer,B.; Epperlein,U. Investigation of 123 chiral drugs by cyclodextrin-modified capillary electrophoresis, *LC.GC*, **1997**, *15*, 40–46.

SAMPLE
Matrix: solutions

CAPILLARY ELECTROPHORESIS
Capillary: 29-36 cm × 50 μm fused-silica (24.5-31.5 cm to detector) (Yongnian Optical Conductive Fiber Plant, China) coated with polyacrylamide
Capillary preparation: Coat capillary as follows. Adjust the pH of 20 mL water to 3.5 with acetic acid, add 80 μL 3-(trimethoxysilyl)propyl methacrylate (3-methacryloxypropyltrimethoxysilane), mix, suck into capillary, let stand at room temperature for 1 h, remove the solution, wash with water. Fill the capillary with a deaerated 3-4% acrylamide solution containing 1 μL/mL N,N,N',N'-tetramethylethylenediamine and 1 mg/mL potassium persulfate, let stand for 30 min, remove excess solution by aspiration, rinse with water, remove water by aspiration, dry at 35° (J. Chromatogr. 1985, 347, 191).
Capillary temperature: 25
Running buffer: 100 mM pH 2.5 NaH_2PO_4 (A) or 100 mM pH 2.5 NaH_2PO_4 containing 45 mM hydroxypropyl-α-cyclodextrin (Wacker, Munich) (B)
Injection: Electromigration at 15 kV for 3 s.
Detector: UV 200; UV 210
Migration time: 7.21 (A); 8.94 (B) (no separation of enantiomers)
Voltage: 15 kV
Model: Bio-Focus 3000

OTHER SUBSTANCES
Also analyzed: albuterol (salbutamol), alprenolol, amorolfine, atenolol, atropine, azelastine, baclofen, bamethan, benproperine, benserazide, biperiden, bisoprolol, brompheniramine, bupivacaine, bupranolol, butamirate, butethamate, carazolol, carbuterol, carteolol, carvedilol, celiprolol, chloroquine, chlorpheniramine, chlorphenoxamine, cicletanine, clenbuterol, clidinium bromide, clobutinol, dimethindene, dipivefrin, disopyramide, dobutamine, doxylamine, fendiline, flecainide, gallopamil, homatropine, ipratropium bromide, isoproterenol (isoprenaline), isothipendyl, ketamine, meclizine, mefloquine, mepindolol, mequitazine, metaclazepam, metaproterenol (orciprenaline), metipranolol, metoprolol, nafronyl (naftidrofuryl), nefopam, nicardipine, norfenefrine, ofloxacin, ornidazole, orphenadrine, oxomemazine, oxprenolol, oxybutynin, phenoxybenzamine, phenylpropanolamine, pholedrine, pindolol, pirbuterol, prilocaine, procyclidine, promethazine, propafenone, propranolol, sotalol, sulpride, synephrine, talinolol, terbutaline, tetrahydrozoline (tetryzoline), theodrenaline, tioconazole, tocainide, trihexyphenidyl, trimeprazine (alimemazine), trimipramine, tropicamide, verapamil, zopiclone

KEY WORDS
coated capillary

REFERENCE
Koppenhoefer,B.; Eperlein,U.; Schlunk,R.; Zhu,X.; Lin,B. Separation of enantiomers of drugs by capillary electrophoresis. V. Hydroxypropyl-α-cyclodextrin as chiral solvating agent, *J.Chromatogr.A*, **1998**, *793*, 153–164.

Reserpine

Molecular formula: $C_{33}H_{40}N_2O_9$
Molecular weight: 608.69
CAS Registry No.: 50-55-5
Merck Index (12th ed.): 8314

SAMPLE
Matrix: solutions

CAPILLARY ELECTROPHORESIS
Capillary: 59.6 cm × 75 μm fused-silica (52.1 cm to detector) (Polymicro Technologies)
Capillary preparation: Purge with running buffer before each run. At the beginning of each day purge using 50-60 kPa vacuum with 500 mM NaOH for 5 min, with water for 5 min, with MeCN for 5 min, and with running buffer for 5 min.
Running buffer: MeCN:MeOH:acetic acid 49:50:1 containing 20 mM ammonium acetate
Injection: Hydrostatic injection at 10 cm for 5 s.
Detector: UV 214
Migration time: 6.02
Voltage: 30 kV
Model: Waters Quanta 4000

OTHER SUBSTANCES
Simultaneous: benzquinamide, deserpidine, ethopropazine, mesoridazine, methotrimeprazine, molindone, promazine, promethazine, thioridazine, thiothixene

REFERENCE
Leung,G.N.W.; Tang,H.P.O.; Tso,T.S.C.; Wan,T.S.M. Separation of basic drugs with non-aqueous capillary electrophoresis, *J.Chromatogr.A*, **1996**, *738*, 141–154.

SAMPLE
Matrix: solutions

CAPILLARY ELECTROPHORESIS
Capillary: 55 cm × 50 μm fused-silica (50 cm to detector) (Polymicro Technologies)
Capillary preparation: Between runs purge with 1 M NaOH for 2 min, with water for 2 min, and with running buffer for 3 min
Capillary temperature: 25
Running buffer: MeCN:buffer 50:50 (Buffer was 100 mM ammonium acetate adjusted to pH 3.1 with acetic acid.)
Injection: Pressure injection at 345 mbar.s.
Detector: UV 200, Finnigan MAT Model 95, electrospray (details in paper)
Migration time: 18.3
Voltage: 15 kV
Model: Bio-Rad BioFocus 3000

OTHER SUBSTANCES
Simultaneous: ajmaline, alstonine, corynanthine, deserpidine, gramine, β-methylajmaline, raufloridine, rescinnamine, serpentine, tabersonine, tryptamine, vinblastine, vincristine, yohimbinic acid

REFERENCE
Unger,M.; Stöckigt,D.; Belder,D.; Stöckigt,J. General approach for the analysis of various alkaloid classes using capillary electrophoresis and capillary electrophoresis-mass spectrometry, *J.Chromatogr.A*, **1997**, *767*, 263–276.

Resorcinol

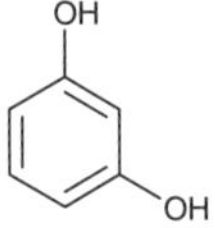

Molecular formula: $C_6H_6O_2$
Molecular weight: 110.11
CAS Registry No.: 108-46-3
Merck Index (12th ed.): 8323

SAMPLE
Matrix: solutions
Sample preparation: Inject an aliquot of a solution in running buffer.

CAPILLARY ELECTROPHORESIS
Capillary: 57 cm × 50 μm fused-silica (50 cm to detector) (Polymicro Technologies)
Capillary temperature: 25
Running buffer: 25 mM NaH_2PO_4 containing 0.83% poly(sodium 10-undecenyl sulfate), adjusted
to pH 7.3 with 12.5 mM sodium tetraborate (Preparation of poly(sodium 10-undecenyl sulfate
is as follows. React undecylenyl alcohol (1-undecen-11-ol) with one equivalent of chlorosulfonic
acid in diethyl ether under nitrogen at -5° to obtain 10-undecenyl hydrogen sulfate, react with
one equivalent of NaOH to obtain sodium 10-undecenyl sulfate (J. Microcolumn Sep. 1996, 8,
115). Dissolve 20 g sodium 10-undecenyl sulfate in 40 mL water (degassed by purging with
nitrogen for 24 h), add 600 mg potassium persulfate, stir at 70° under nitrogen for 50 h, cool,
add 200 mL cold (0°) EtOH, filter, wash the solid with three 10 mL portions of cold EtOH, dry
under vacuum at 65° overnight to obtain poly(sodium 10-undecenyl sulfate) (cf. J. Microcolumn
Sep. 1992, 4, 509). Purify by dialysis using dialysis tubing with a 1000 MW cut-off (Spectrum
Houston) for 24 h.)
Detector: UV 214
Migration time: 5.5
Voltage: 16.1 kV
Model: Beckman P/ACE 2000

OTHER SUBSTANCES
Simultaneous: 1,6-dinaphthol, 1-naphthol, 2-naphthol

REFERENCE
Palmer,C.P.; Terabe,S. Micelle polymers as pseudostationary phases in MEKC: Chromatographic performance
and chemical selectivity, *Anal.Chem.*, **1997**, *69*, 1852–1860.

Retinoic acid

Molecular formula: $C_{20}H_{28}O_2$
Molecular weight: 300.44
CAS Registry No.: 302-79-4 (tretinoin (all-trans)), 4759-48-2 (isotretinoin (13-cis)), 5300-03-8 (alitre-
tinoin (9-cis))
Merck Index (12th ed.): 8333
Lednicer: 3 12 (isotretinoin)

SAMPLE
Matrix: blood
Sample preparation: 100 μL Plasma + 1. 5 μg 13-cis-acitretin + 10 μL 5% perchloric acid, mix,
add 100 μL ethyl acetate, extract, centrifuge. Remove 20 μL of the organic layer and evaporate
it to dryness, reconstitute the residue in 100 μL MeCN:water 50:50, inject an aliquot.

CAPILLARY ELECTROPHORESIS
Capillary: 47 cm × 75 μm
Running buffer: MeCN:buffer 20:80 (Buffer was 20 mM pH 8.5 Tris-borate buffer containing 25
mM sodium dodecyl sulfate.)

Detector: UV 340
Migration time: 9.5 (isotretinoin (13-cis)), 10.3 (alitretinoin (9-cis)), 10.7 (tretinoin (all-trans))
Internal standard: 13-cis-acitretin (8.7)
Model: Beckman P/ACE

KEY WORDS
rat; plasma

REFERENCE
Chan,K.C.; Lewis,K.C.; Phang,J.M.; Issaq,H.J. Separation of retinoic acid isomers using micellar electrokinetic chromatography, *J.High Res.Chromatogr.*, **1993**, *16*, 560–561.

SAMPLE
Matrix: formulations
Sample preparation: Protect from light. Tablets. Sonicate 1 tablet with 15 mL MeOH for 15 min, centrifuge at 2000 U/min for 5 min, inject an aliquot of the supernatant. Syrup. Dilute with MeOH, inject an aliquot.

CAPILLARY ELECTROPHORESIS
Capillary: 47 cm × 50 µm fused-silica (40 cm to detector) (Polymicro Technologies)
Capillary preparation: Rinse capillary with 100 mM NaOH, water, and running
Capillary temperature: 25
Running buffer: 20 mM pH 7.9 Sodium phosphate buffer containing 3 mM Brij 35 and 75 mM sodium dodecyl sulfate
Injection: Pressure injection.
Detector: UV 330
Migration time: 15.8
Voltage: 20 kV
Model: Beckman P/ACE 5500

OTHER SUBSTANCES
Simultaneous: retinal, retinyl acetate, retinyl palmitate, vitamin A (retinol)

KEY WORDS
tablets; syrup

REFERENCE
Hsieh,Y.-Z.; Kuo,K.-L. Separation of retinoids by micellar electrokinetic capillary chromatography, *J.Chromatogr.A*, **1997**, *761*, 307–313.

SAMPLE
Matrix: solutions

CAPILLARY ELECTROPHORESIS
Capillary: 72 cm × 50 µm fused silica (50 cm to the detector)
Capillary preparation: Before each run wash with 100 mM NaOH for 3 min and with running buffer for 3 min.
Capillary temperature: 30
Running buffer: 30 mM pH 8.5 borate buffer containing 10 mM sodium dodecyl sulfate
Injection: Vacuum injection for 2.5 s
Detector: UV 345
Migration time: 5.8 (isotretinoin (13-cis)), 6.6 (tretinoin (all-trans))
Voltage: 30 kV
Model: Applied Biosystems model 270A

REFERENCE
Bempong,D.K.; Honigberg,I.L.; Meltzer,N.M. Separation of 13-*cis* and all-*trans* retinoic acid and their photo-degradation products using capillary zone electrophoresis and micellar electrokinetic chromatography (MEC), *J.Pharm.Biomed.Anal.*, **1993**, *11*, 829–833.

SAMPLE
Matrix: solutions

CAPILLARY ELECTROPHORESIS
Capillary: 50 cm × 75 μm fused-silica (50 cm to detector)
Capillary preparation: Before each run flush with 100 mM NaOH for 2 min and with water for 2 min then fill with running buffer.
Capillary temperature: 25
Running buffer: MeCN:buffer 20:80 (Buffer was 20 mM pH 9.2 sodium tetraborate containing 100 mM sodium cholate.)
Injection: Pressure injection at 0.5 psi for 1.5 s (9 nL).
Detector: UV 340
Migration time: 23
Voltage: 25 kV
Model: Beckman P/ACE 2100
Limit of detection: 10 μM

OTHER SUBSTANCES
Simultaneous: retinal, vitamin A

KEY WORDS
protect from light

REFERENCE
Profumo,A.; Profumo,V.; Vidali,G. Micellar electrokinetic capillary chromatography of three natural vitamin A derivatives, *Electrophoresis*, **1996**, *17*, 1617–1621.

Riboflavin

Molecular formula: $C_{17}H_{20}N_4O_6$
Molecular weight: 376.37
CAS Registry No.: 83-88-5, 130-40-5 (phosphate sodium)
Merck Index (12th ed.): 8367

SAMPLE
Matrix: blood
Sample preparation: Digest whole blood in 10% trichloroacetic acid at 0° for 30 min, add 4.5 M pH 6.2 sodium acetate buffer, centrifuge, mix an aliquot of the supernatant with an equal volume of running buffer, inject an aliquot.

CAPILLARY ELECTROPHORESIS
Capillary: 75 cm × 50 μm fused-silica (Scientific Glass Engineering)
Capillary preparation: Equilibrate with running buffer for 30 min before use.
Running buffer: Isopropanol:buffer 5:95 (Buffer was 10 mM Na_2HPO_4 containing 6 mM sodium borate and 50 mM sodium dodecyl sulfate.)
Injection: Injection by siphon at 20 cm for 10 s or by electromigration
Detector: F ex 488 (20 mW Ar laser) em 525
Migration time: 29
Voltage: 30 kV

OTHER SUBSTANCES
Simultaneous: flavin adenine dinucleotide, flavin mononucleotide, metabolites

KEY WORDS
whole blood

REFERENCE
Swaile,D.F.; Burton,D.E.; Balchunas,A.T.; Sepaniak,M.J. Pharmaceutical analysis using micellar electrokinetic capillary chromatography, *J.Chromatogr.Sci.*, **1988**, *26*, 406–409.

SAMPLE
Matrix: formulations
Sample preparation: Homogenize a multi-vitamin tablet, extract with 100 mM HCl, filter (0.45 μm).

CAPILLARY ELECTROPHORESIS
Capillary: 51 cm × 75 μm fused-silica (33 cm to detector) (Waters)
Running buffer: 20 mM pH 9.0 sodium phosphate buffer
Injection: 3 nL (5 μL with a split ratio of 1:1700)
Detector: UV 254
Migration time: 6.17
Voltage: 6.0 kV
Model: ISCO Model 3850

OTHER SUBSTANCES
Simultaneous: thiamine, pyridoxal, pyridoxamine, pyridoxine, ascorbic acid

KEY WORDS
tablets

REFERENCE
Huopalahti,R.; Sunell,J. Use of capillary zone electrophoresis in the determination of B vitamins in pharmaceutical products, *J.Chromatogr.*, **1993**, *636*, 133–135.

SAMPLE
Matrix: formulations
Sample preparation: Condition a 500 mg Bakerbond C18 SPE cartridge with 3 mL MeOH and 3 mL water:10 mM sodium heptanesulfonate in acetic acid 99:1. Heat 50-100 mg tablet and 40 mL water:10 mM sodium heptanesulfonate in acetic acid 99:1 at 55° with occasional shaking for 5 min, cool, add a 2 mL aliquot to the SPE cartridge, elute with 3.5 mL MeOH, make up the eluate to 10 mL with water, inject an aliquot.

CAPILLARY ELECTROPHORESIS
Capillary: 70 cm × 100 μm
Capillary temperature: 25
Running buffer: MeOH:buffer 10:90 (Buffer was 50 mM pH 8.0 sodium borate containing 22.5 mM sodium dodecyl sulfate.)
Injection: Inject under 3.44 kPa pressure for 5 s (30 nL)
Detector: UV 214
Migration time: 19
Voltage: 16 kV
Current: 33 μA
Model: Beckman P/ACE System 2000

OTHER SUBSTANCES
Simultaneous: niacinamide, pantothenic acid, pyridoxine, thiamine, vitamin B12, ascorbic acid

KEY WORDS
tablets; SPE

REFERENCE
Dinelli,G.; Bonetti,A. Micellar electrokinetic capillary chromatography analysis of water-soluble vitamins and multi-vitamin integrators, *Electrophoresis*, **1994**, *15*, 1147–1150.

SAMPLE
Matrix: formulations

Sample preparation: Protect from light. Grind tablet, add 50 mL 1% acetic acid, shake at 65° for 10 min, add 5 mL 10 mg/mL niacin in 1% acetic acid, cool to room temperature, make up to 100 mL with 1% acetic acid, filter (0.22 μm cellulose acetate), dilute 4-12-fold, inject an aliquot.

CAPILLARY ELECTROPHORESIS
Capillary: 48.5 cm × 50 μm fused-silica (40 cm to detector) (Hewlett Packard)
Capillary preparation: After each run wash with 100 mM NaOH for 7 min, with water for 1 min, and with running buffer for 3 min. At the start of each day wash with separation buffer for 10 min. Wash new capillaries with 1 M NaOH, 100 mM NaOH, water, and running buffer.
Capillary temperature: 25
Running buffer: 50 mM Borax adjusted to pH 8.5 with boric acid
Injection: Hydrodynamic injection at 5 kPa for 10 s.
Detector: UV 225
Migration time: 3.8
Internal standard: niacin (7.5)
Voltage: 25 kV
Model: Hewlett Packard [3D]CE

OTHER SUBSTANCES
Simultaneous: adenine, biotin, niacinamide, pantothenic acid (UV 215), pyridoxine (UV 215), rutin, thiamine, ascorbic acid

KEY WORDS
tablets

REFERENCE
Fotsing,L.; Fillet,M.; Bechet,I.; Hubert,P.; Crommen,J. Determination of six water-soluble vitamins in a pharmaceutical formulation by capillary electrophoresis, *J.Pharm.Biomed.Anal.*, **1997**, *15*, 1113–1123.

SAMPLE
Matrix: formulations
Sample preparation: Grind tablet to a homogeneous powder and dissolve in 2 mL water and 20 mL EtOH, homogenize (Ultra Turrax), centrifuge. Evaporate the supernatant to dryness under reduced pressure, reconstitute the residue with water, inject an aliquot.

CAPILLARY ELECTROPHORESIS
Capillary: 64.5 cm × 50 μm fused-silica (56 cm to detector)
Capillary preparation: Before each run flush capillary at 4 kPa with 1 M NaOH for 2 min and with buffer for 5 min. Replace buffer before each analysis.
Capillary temperature: 30
Running buffer: 1-Propanol:buffer 2:98 (Buffer was 100 mM Na_2HPO_4, 500 mM taurine, and 75 mM sodium cholate.)
Injection: Pressure injection at 4 kPa for 1 s.
Detector: UV 214
Migration time: 12
Voltage: 17 kV
Model: Hewlett-Packard HP[3D] CE
Limit of quantitation: 1.8 μM

OTHER SUBSTANCES
Simultaneous: ascorbic acid, biotin, folic acid, niacinamide, pyridoxamine, pyridoxine, thiamine, vitamin B12

KEY WORDS
tablets

REFERENCE
Buskov,S.; Moller,P.; Sorensen,H.; Sorensen,J.C.; Sorensen,S. Determination of vitamins in food based on supercritical fluid extraction prior to micellar electrokinetic capillary chromatographic analyses of individual vitamins, *J.Chromatogr.A*, **1998**, *802*, 233–241.

SAMPLE
Matrix: formulations, juice
Sample preparation: Dissolve in water, filter (0.2 μm), inject an aliquot. (Add a 50-fold excess of cysteine to solutions containing ascorbic acid.)

CAPILLARY ELECTROPHORESIS
Capillary: 60 cm $\times$ 50 μm fused-silica (51.5 cm to detector)
Capillary preparation: Condition with running buffer for 3 min before each run. Condition a new capillary 1 M NaOH for 10 min and with 100 mM NaOH for 2 min.
Capillary temperature: 25
Running buffer: 20 mM pH 8.0 Phosphate buffer
Injection: Pressure injection for 200 mbar.s
Detector: UV 200
Migration time: 5.9 (riboflavin phosphate)
Voltage: -30 kV
Model: Hewlett-Packard 3DCE

OTHER SUBSTANCES
Simultaneous: biotin, niacinamide, pyridoxine, thiamine, ascorbic acid

KEY WORDS
fruit

REFERENCE
Schiewe,J.; Mrestani,Y.; Neubert,R. Application and optimization of capillary zone electrophoresis in vitamin analysis, *J.Chromatogr.A*, **1995**, *717*, 255–259.

SAMPLE
Matrix: solutions
Sample preparation: Prepare a 0.5-2 mg/mL solution in water, inject an aliquot.

CAPILLARY ELECTROPHORESIS
Capillary: 65 cm $\times$ 50 μm fused-silica (50 cm to detector) (Scientific Glass Engineering)
Running buffer: 20 mM NaH_2PO_4 containing 50 mM sodium dodecyl sulfate adjusted to pH 9.0 with 20 mM sodium tetraborate
Injection: Injection by siphon at 5 cm for 5-10 s (about 1 nL)
Detector: UV 210
Migration time: 6.4
Voltage: 25 kV

OTHER SUBSTANCES
Simultaneous: niacin, niacinamide, pyridoxal, pyridoxamine, pyridoxine, thiamine, vitamin B12

REFERENCE
Nishi,H.; Tsumagari,N.; Kakimoto,T.; Terabe,S. Separation of water-soluble vitamins by micellar electrokinetic chromatography, *J.Chromatogr.*, **1989**, *465*, 331–343.

SAMPLE
Matrix: solutions
Sample preparation: Prepare a 100 μg/mL solution in MeOH:water 50:50, inject an aliquot.

CAPILLARY ELECTROPHORESIS
Capillary: 60 cm $\times$ 75 μm
Running buffer: 20 mM NaH_2PO_4 containing 50 mM sodium dodecyl sulfate, adjusted to pH 9.0 with sodium tetraborate
Injection: Hydrodynamic injection at 10 cm for 5 s
Detector: UV 214
Migration time: 10
Voltage: 18 kV
Model: Waters Quanta 4000

OTHER SUBSTANCES
Simultaneous: niacin, pantothenic acid, pyridoxine, thiamine, ascorbic acid

REFERENCE
Swartz,M.E. Method development and selectivity control for small molecule pharmaceutical separations by capillary electrophoresis, *J.Liq.Chromatogr.*, **1991**, *14*, 923–938.

SAMPLE
Matrix: solutions

CAPILLARY ELECTROPHORESIS
Capillary: 70 cm × 50 μm fused-silica (GL Science, Tokyo)
Capillary preparation: Before each run rinse capillary with running buffer at 94 kPa for 5 min before each run.
Running buffer: 50 mM pH 8.5 ammonium carbonate buffer
Injection: Pressure injection at 5 kPa (50 mbar) for 4 s.
Detector: MS, Perkin-Elmer Sciex API-300 quadrupole, electrospray (ionspray) interface, sheath liquid MeOH:running buffer 50:50 at 2.5 μL/min, ionspray voltage 5 kV, positive ion mode
Migration time: 8
Voltage: 20 kV (net voltage across capillary = 15 kV (applied voltage - electrospray voltage))
Model: Hewlett Packard 3D CE

OTHER SUBSTANCES
Simultaneous: acetaminophen, ascorbic acid, butylscopolamine bromide, caffeine, ibuprofen, ketoprofen, niacin, niacinamide, thiamine, vitamin B12, warfarin

REFERENCE
Tanaka,Y.; Kishimoto,Y.; Otsuka,K.; Terabe,S. Strategy for selecting separation solutions in capillary electrophoresis-mass spectrometry, *J.Chromatogr.A*, **1998**, *817*, 49–57.

Rimantadine

Molecular formula: $C_{12}H_{21}N$
Molecular weight: 179.31
CAS Registry No.: 13392-28-4, 1501-84-4 (HCl)
Merck Index (12th ed.): 8390
Lednicer: 2 19

SAMPLE
Matrix: solutions

CAPILLARY ELECTROPHORESIS
Capillary: 44 cm × 75 μm fused-silica (36.5 cm to detector) (Avery Dennison, MA)
Capillary preparation: Before each run wash capillary with running buffer. Before first use wash capillary with 1 M NaOH at 60° for 5 min, with 100 mM NaOH at 60° for 5 min, with water at 30° for 10 min, and with running buffer at 25° for 10 min.
Capillary temperature: 25
Running buffer: MeOH:water 20:80, pH 5.3, containing 5 mM 4-methylbenzylamine (Prepare running buffer by dissolving 130 μL 4-methylbenzylamine in 20 mL MeOH, make up to 100 mL with water, adjust pH to 5.3 with 1 M HCl.)
Injection: Hydrodynamic injection for 10 s.
Detector: UV 210
Migration time: 1.65
Voltage: 30 kV
Model: SpectraPhoresis 2000 (Thermo Bioanalysis)
Limit of detection: 1.6 ppm

OTHER SUBSTANCES
Simultaneous: amantadine

KEY WORDS
indirect UV detection; comparison with derivatization

REFERENCE
Revilla,A.L.; Hamacek,J.; Lubal,P.; Havel,J. Determination of rimantadine in pharmaceutical preparations by capillary zone electrophoresis with indirect detection or after derivatization, *Chromatographia*, **1998**, *47*, 433–439.

Ritodrine

Molecular formula: $C_{17}H_{21}NO_3$
Molecular weight: 287.36
CAS Registry No.: 26652-09-5, 23239-51-2 (HCl)
Merck Index (12th ed.): 8401
Lednicer: 2 39

SAMPLE
Matrix: solutions
Sample preparation: Inject an aliquot of a 10 μg/mL solution in water.

CAPILLARY ELECTROPHORESIS
Capillary: 68.5 × 50 μm fused-silica (60 cm to detector) (Hewlett-Packard)
Capillary preparation: After each run rinse with 100 mM NaOH for 5 min and with running buffer for 5 min. Condition a new capillary with 1 M NaOH at 40° for 20 min and with 100 mM NaOH at 40° for 10 min, rinse with water at 25° for 20 min, and rinse with running buffer.
Capillary temperature: 25
Running buffer: 50 mM pH 8.3 Tris buffer
Injection: Hydrostatic injection at 50 mbar for 2-6 s, flush with buffer for 4 s
Detector: UV 200
Migration time: 3.745
Voltage: 30 kV
Model: Hewlett-Packard HP 3DCE

OTHER SUBSTANCES
Simultaneous: albuterol, cimaterol, clenbuterol, fenoterol, isoxsuprine, metaproterenol, ractopamine, RU 42 173, terbutaline

KEY WORDS
comparison with C18 bonded capillary

REFERENCE
Chevolleau,S.; Tulliez,J. Optimization of the separation of β-agonists by capillary electrophoresis on untreated and C18 bonded silica capillaries, *J.Chromatogr.A*, **1995**, *715*, 345–354.

Ropivacaine

Molecular formula: $C_{17}H_{26}N_2O$
Molecular weight: 274.41
CAS Registry No.: 84057-95-4, 98717-15-8 (HCl), 132112-35-7 (HCl monohydrate)
Merck Index (12th ed.): 8417

SAMPLE
Matrix: bulk

CAPILLARY ELECTROPHORESIS
Capillary: 80.5 cm × 50 μm fused-silica (72 cm to detector) (Hewlett Packard)
Capillary preparation: Before each run flush with water for 1 min, with 100 mM NaOH for 4 min, with water for 1 min, and with running buffer for 4 min.
Capillary temperature: 30
Running buffer: 100 mM Phosphoric acid adjusted to pH 3.0 with triethanolamine, containing 10 mM heptakis(2,6-di-O-methyl)-β-cyclodextrin
Injection: Pressure injection at 50 mbar for 5 s (5 nL), ramp to operating voltage at 500 V/s.
Detector: UV 206
Migration time: 25 (R), 26 (S)
Voltage: 30 kV
Model: Hewlett Packard HP ³ᴰCE

OTHER SUBSTANCES
Simultaneous: bupivacaine, mepivacaine
Also analyzed: prilocaine

KEY WORDS
chiral

REFERENCE
Sänger-van de Griend,C.E.; Gröningsson,K.; Westerlund,D. Chiral separation of local anaesthetics with capillary electrophoresis. Evaluation of the inclusion complex of the enantiomers with heptakis(2,6-di-O-methyl)-β-cyclodextrin, *Chromatographia*, **1996**, *42*, 263–268.

SAMPLE
Matrix: solutions

CAPILLARY ELECTROPHORESIS
Capillary: 80.5 cm × 50 μm (72 cm to detector) (Hewlett Packard)
Capillary preparation: Before each run flush with water for 1 min, 100 mM NaOH for 4 min, with water for 1 min, and with running buffer for 4 min.
Capillary temperature: 30
Running buffer: 100 mM phosphoric acid containing 10 mM heptakis(2,6-di-O-methyl)-β-cyclodextrin, adjusted to pH 3.0 with triethanolamine
Injection: Pressure injection at 50 mbar over 5 s (5 nL), ramp to voltage at 500 V/s.
Detector: UV 206
Migration time: 24.8 (24 for R-enantiomer)
Voltage: 30 kV
Model: Hewlett Packard HP ³ᴰCE
Limit of quantitation: 0.1% (of major enantiomer)
Limit of detection: 0.05% (of major enantiomer)

OTHER SUBSTANCES
Simultaneous: bupivacaine, mepivacaine

KEY WORDS
chiral

REFERENCE
Sänger-van de Griend,C.E.; Gröningsson,K. Validation of a capillary electrophoresis method for the enantiomeric purity testing of ropivacaine, a new local anaesthetic compound, *J.Pharm.Biomed.Anal.*, **1996**, *14*,
295–304.

SAMPLE
Matrix: solutions
Sample preparation: Inject an aliquot of a 20 μg/mL solution in buffer with an ionic strength
10 times less than that of the running buffer.

CAPILLARY ELECTROPHORESIS
Capillary: 27 cm × 50 μm fused-silica (20 cm to detector) (Polymicro Technologies)
Capillary preparation: Between runs rinse capillary with running buffer from a different vial
for 2 min, equilibrate system for 1 min before analysis. Store capillary in water overnight and
rinse with 100 mM NaOH and water each morning. Rinse new capillaries with 100 mM HCl
for 5 min, with water for 5 min, with 100 mM NaOH for 5 min, with water for 10 min, then
with running buffer.
Capillary temperature: 25
Running buffer: pH 2.90 Phosphate buffer (I = 0.04) containing 100 mM α-cyclodextrin (Prepare
buffer by mixing 19.8 mL 1 M NaOH and 23.2 mL 1 M phosphoric acid and make up to 500
mL with water.)
Injection: Pressure injection at 3.4 kPa for 5 s (6 nL).
Detector: UV 214
Voltage: 10 kV
Current: 16-25 μA
Model: Beckman P/ACE 2100

KEY WORDS
chiral; α = 1.13

REFERENCE
Amini,A.; Sörman,U.P.; Lindgren,B.H.; Westerlund,D. Enantioseparation of anaesthetic drugs by capillary zone
electrophoresis using cyclodextrin-containing background electrolytes, *Electrophoresis*, **1998**, *19*, 731–737.

Rosoxacin

Molecular formula: $C_{17}H_{14}N_2O_3$
Molecular weight: 294.31
CAS Registry No.: 40034-42-2
Merck Index (12th ed.): 8426
Lednicer: 3 185

SAMPLE
Matrix: solutions
Sample preparation: Inject an aliquot of a solution in MeOH.

CAPILLARY ELECTROPHORESIS
Capillary: 59 cm × 50 μm fused-silica (43 cm to detector) (Polymicro Technologies)
Capillary preparation: Before each analysis flush with water for 3 min, with 200 mM NaOH
for 3 min, with water for 3 min, and with running buffer for 4 min. Flush new capillaries with
1 M NaOH at 2000 mbar for 10 min then with 200 mM NaOH for 10 min.
Capillary temperature: 23
Running buffer: MeCN:buffer 28:72, pH 7.3 (Buffer was 32 mM sodium borate containing 18
mM NaH_2PO_4, 39 mM sodium cholate, and 8 mM sodium heptanesulfonate.)
Injection: Hydrodynamic injection at 40 mbar for 6 s.
Detector: UV 260
Migration time: 5.45
Voltage: 30 kV

Model: Lauer Labs Prince

OTHER SUBSTANCES
Simultaneous: ciprofloxacin, enoxacin, flumequine, lomefloxacin, nalidixic acid, norfloxacin, ofloxacin, oxolinic acid, pefloxacin, pipemidic acid, piromidic acid, sparfloxacin

REFERENCE
Sun,S.-W.; Chen,L.-Y. Optimization of capillary electrophoretic separation of quinolone antibacterials using the overlapping resolution mapping scheme, *J.Chromatogr.A*, **1997**, *766*, 215–224.

Rutin

Molecular formula: $C_{27}H_{30}O_{16}$
Molecular weight: 610.53
CAS Registry No.: 153-18-4
Merck Index (12th ed.): 8456

SAMPLE
Matrix: formulations
Sample preparation: Protect from light. Grind tablet, add 50 mL 1% acetic acid, shake at 65° for 10 min, add 5 mL 10 mg/mL niacin in 1% acetic acid, cool to room temperature, make up to 100 mL with 1% acetic acid, filter (0.22 μm cellulose acetate), dilute 4-12-fold, inject an aliquot.

CAPILLARY ELECTROPHORESIS
Capillary: 48.5 cm × 50 μm fused-silica (40 cm to detector) (Hewlett Packard)
Capillary preparation: After each run wash with 100 mM NaOH for 7 min, with water for 1 min, and with running buffer for 3 min. At the start of each day wash with separation buffer for 10 min. Wash new capillaries with 1 M NaOH, 100 mM NaOH, water, and running buffer.
Capillary temperature: 25
Running buffer: 50 mM Borax adjusted to pH 8.5 with boric acid
Injection: Hydrodynamic injection at 5 kPa for 10 s.
Detector: UV 225
Migration time: 5.2
Internal standard: niacin (7.5)
Voltage: 25 kV
Model: Hewlett Packard ^{3D}CE

OTHER SUBSTANCES
Simultaneous: adenine, biotin, niacinamide, pantothenic acid (UV 215), pyridoxine (UV 215), thiamine, riboflavin, ascorbic acid

KEY WORDS
tablets

REFERENCE
Fotsing,L.; Fillet,M.; Bechet,I.; Hubert,P.; Crommen,J. Determination of six water-soluble vitamins in a pharmaceutical formulation by capillary electrophoresis, *J.Pharm.Biomed.Anal.*, **1997**, *15*, 1113–1123.

Saccharin

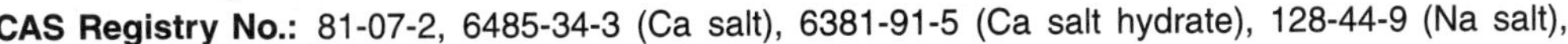

Molecular formula: $C_7H_5NO_3S$
Molecular weight: 183.19
CAS Registry No.: 81-07-2, 6485-34-3 (Ca salt), 6381-91-5 (Ca salt hydrate), 128-44-9 (Na salt), 6155-57-3 (Na salt dihydrate)
Merck Index (12th ed.): 8463

SAMPLE
Matrix: beverages, food
Sample preparation: Dilute sample with water, add dehydroacetic acid to a final concentration of 10 μg/mL, filter (0.45 μm cellulose acetate), inject an aliquot.

CAPILLARY ELECTROPHORESIS
Capillary: 75 cm × 75 μm fused-silica (50 cm to detector) (Polymicro Technologies)
Capillary preparation: Flush with running buffer for 2 min between runs. Clean each week by washing with 100 mM NaOH for 10 min and with water for 10 min, fill with running buffer.
Capillary temperature: 27
Running buffer: 10 mM Sodium borate containing 10 mM KH_2PO_4 and 50 mM sodium deoxycholate, pH 8.6
Injection: Injection by vacuum at 20 kPa.s (vacuum level 2)
Detector: UV 220
Migration time: 10.5
Internal standard: dehydroacetic acid (8.4)
Voltage: 20 kV
Model: ISCO Model 3140

OTHER SUBSTANCES
Simultaneous: acesulfame-K, alitame, aspartame, benzoic acid, caffeine, dulcin, sorbic acid

KEY WORDS
soft drink; cordial; tomato sauce; jam; sweeteners; comparison with HPLC

REFERENCE
Thompson,C.O.; Trenerry,V.C.; Kemmery,B. Micellar electrokinetic capillary chromatographic determination of artificial sweeteners in low-Joule soft drinks and other foods, *J.Chromatogr.A*, **1995**, *694*, 507–514.

SAMPLE
Matrix: formulations
Sample preparation: Dilute formulation 50-fold, inject an aliquot.

CAPILLARY ELECTROPHORESIS
Capillary: 60 cm × 50 μm (52.5 cm to detector) (AccuSep)
Capillary preparation: Purge capillary with running buffer for 3 min between runs. Purge new capillaries under vacuum with 500 mM NaOH for 10 min, with water for 10 min, and with running buffer for 10 min.
Capillary temperature: 30
Running buffer: 25 mM pH 8.0 Na_2HPO_4/sodium tetraborate containing 50 mM (R)-N-dodecyl-carbonylvaline (Prepare (R)-N-dodecoxycarbonylvaline as follows. Prepare dodecyl chloroformate by reacting 1-dodecanol with 0.33 equivalents of triphosgene in dichloromethane solution in the presence of pyridine. Add dodecyl chloroformate dropwise to (R)-valine in 1 M NaOH solution, filter, wash with hexane, recrystallize from ether/petroleum ether (J. Chromatogr. A 1994, 680, 125).)
Injection: Hydrostatic injection at 10 cm for 15 s.
Detector: UV 214
Migration time: 16.722
Voltage: 15 kV
Model: Waters Quanta 4000E

OTHER SUBSTANCES
Simultaneous: acetaminophen, dextromethorphan, doxylamine, pseudoephedrine

KEY WORDS
not a chiral compound

REFERENCE
Swartz,M.E.; Mazzeo,J.R.; Grover,E.R.; Brown,P.R. Validation of enantiomeric separations by micellar electro-kinetic capillary chromatography using synthetic chiral surfactants, *J.Chromatogr.A*, **1996**, *735*, 303–310.

Salicylamide

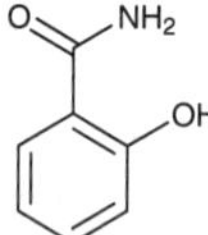

Molecular formula: $C_7H_7NO_2$
Molecular weight: 137.14
CAS Registry No.: 65-45-2
Merck Index (12th ed.): 8480
Lednicer: 1 109

SAMPLE
Matrix: bulk
Sample preparation: Prepare a 100 µg/mL solution in water, inject an aliquot.

CAPILLARY ELECTROPHORESIS
Capillary: 60 cm × 50 µm
Running buffer: 20 mM pH 11.0 sodium phosphate buffer containing 75 mM sodium dodecyl sulfate
Injection: Hydrodynamic injection at 10 cm for 10 s
Detector: UV 214
Migration time: 4.52
Voltage: 20 kV
Model: Waters Quanta 4000
Limit of detection: 16 ng/mL

OTHER SUBSTANCES
Simultaneous: impurities

REFERENCE
Swartz,M.E. Method development and selectivity control for small molecule pharmaceutical separations by capillary electrophoresis, *J.Liq.Chromatogr.*, **1991**, *14*, 923–938.

SAMPLE
Matrix: formulations
Sample preparation: Grind tablet, add 75 mL acidic MeOH, sonicate for 5 min, shake for 10 min, make up to 100 mL with acidic MeOH, centrifuge at 2500 g for 10 min. Remove a 5 mL aliquot of the supernatant and add it to 5 mL 10 mM ethyl p-aminobenzoate in MeOH, make up to 50 mL with MeOH, inject an aliquot. (Prepare acidic MeOH by adding 10 mL 100 mM HCl to 1 L MeOH.)

CAPILLARY ELECTROPHORESIS
Capillary: 80 cm × 100 µm fused-silica (50 cm to detector) (Shimadzu)
Running buffer: 20 mM pH 11 Phosphate buffer containing 50 mM sodium dodecyl sulfate
Injection: Injection by siphon at 5 cm for 5 s
Detector: UV 214
Migration time: 16
Internal standard: ethyl p-aminobenzoate (20)
Current: 100 µA (constant current)

OTHER SUBSTANCES
Simultaneous: acetaminophen, aspirin, caffeine, o-ethoxybenzamide

KEY WORDS
tablets

REFERENCE
Fujiwara,S.; Honda,S. Determination of ingredients of antipyretic analgesic preparations by micellar electro-kinetic capillary chromatography, *Anal.Chem.*, **1987**, *59*, 2773–2776.

SAMPLE
Matrix: solutions

CAPILLARY ELECTROPHORESIS
Capillary: 56.9 cm × 75 μm fused-silica (50 cm to detector)
Capillary temperature: 25
Running buffer: MeOH:60 mM pH 8.4 sodium borate buffer containing 60 mM sodium dodecyl sulfate 15:85
Detector: UV 214
Migration time: 12
Voltage: 16 kV
Model: Beckman PACE 2100

OTHER SUBSTANCES
Simultaneous: acetanilide, acetaminophen, aspirin, benzamide, caffeine, salicylic acid

REFERENCE
McLaughlin,G.M.; Nolan,J.A.; Lindahl,J.L.; Palmieri,R.H.; Anderson,K.W.; Morris,S.C.; Morrison,J.A.; Bronzert,T.J. Pharmaceutical drug separations by HPCE: Practical guidelines, *J.Liq.Chromatogr.*, **1992**, *15*, 961–1021.

SAMPLE
Matrix: solutions
Sample preparation: Inject an aliquot of a 10 μg/mL solution

CAPILLARY ELECTROPHORESIS
Capillary: 50 cm × 75 μm silica
Capillary temperature: 25
Running buffer: 20 mM pH 7.0 Phosphate/borate buffer containing 50 mM sodium dodecyl sulfate
Injection: 5 nL
Detector: UV 208
Migration time: 3
Voltage: 20 kV
Model: Otsuka CAPI-3000

OTHER SUBSTANCES
Simultaneous: acetaminophen, caffeine, ethyl paraben, guaifenesin, methyl paraben, phenacetin

REFERENCE
Poe,R.B.; Hayashi,Y.; Matsuda,R. Precision-optimization of wavelengths in diode-array detection in separation science, *Anal.Sci.*, **1997**, *13*, 951–961.

Salicylic acid

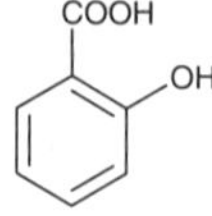

Molecular formula: $C_7H_6O_3$
Molecular weight: 138.12
CAS Registry No.: 69-72-7, 54-21-7 (Na salt)
Merck Index (12th ed.): 8484

SAMPLE

Matrix: blood
Sample preparation: 100 µL Serum + 150 µL 80 µg/mL 3-isobutyl-1-methylxanthine in MeCN, mix for 30 s, centrifuge at 14000 g for 1 min, inject an aliquot of the supernatant.

CAPILLARY ELECTROPHORESIS

Capillary: 25 cm × 47 µm
Capillary preparation: Wash with running buffer by pressure injection for 1 min after each run. Wash daily with 50 mM phosphoric acid for 2 min, with water for 1 min, with 2 M NaOH for 2 min, with water for 1 min, and with running buffer for 2 min.
Capillary temperature: 24
Running buffer: 220 mM boric acid adjusted to pH 8.8 with 2.5 M NaOH
Injection: Pressure injection for 8 s.
Detector: UV 254
Migration time: 3.9
Internal standard: 3-isobutyl-1-methylxanthine (2.3)
Voltage: 12 kV
Model: Beckman

OTHER SUBSTANCES

Extracted: acetaminophen, caffeine, carbamazepine, iohexol, pentobarbital, phenobarbital, phenytoin, theophylline

KEY WORDS

serum

REFERENCE

Shihabi,Z.K.; Constantinescu,M.S. Iohexol in serum determined by capillary electrophoresis, *Clin.Chem.*, **1992**, *38*, 2117–2120.

SAMPLE

Matrix: blood
Sample preparation: Filter (0.45 µm) plasma, inject an aliquot of the filtrate.

CAPILLARY ELECTROPHORESIS

Capillary: 43 cm × 50 µm (22 cm to detector) (Polymicro Technologies)
Capillary preparation: Rinse with running buffer at 68 kPa for 1 min between injections.
Capillary temperature: 30
Running buffer: 60 mM pH 7 Phosphate buffer containing 200 mM sodium dodecyl sulfate
Injection: Vacuum injection at 17 kPa for 1 s
Detector: UV 200
Migration time: 6
Voltage: 15 kV
Current: 140 µA
Model: Applied Biosystems 270A-HT

OTHER SUBSTANCES

Extracted: acetaminophen

KEY WORDS

plasma

REFERENCE
Wätzig,H.; Lloyd,D.K. Effect of pH and sodium dodecyl sulfate concentration on the analytical window in the direct-injection analysis of plasma samples by capillary electrophoresis, *Electrophoresis*, **1995**, *16*, 57–63.

SAMPLE
Matrix: blood
Sample preparation: Inject an aliquot of plasma directly.

CAPILLARY ELECTROPHORESIS
Capillary: 42.5 cm × 50 μm fused-silica (34.5 cm) (Polymicro Technologies)
Capillary preparation: Between runs rinse with MeCN:running buffer 50:50 for 15 min and with running buffer for 20 min. Condition new capillaries with 100 mM NaOH at 50° for 30 min and equilibrate with cathode buffer at 25° at 10 kV for 2 h. (Cathode buffer was 60 mM pH 10.0 sodium borate buffer.)
Capillary temperature: 25
Running buffer: 60 mM pH 10.0 Sodium borate buffer containing 200 mM sodium dodecyl sulfate
Injection: Vacuum injection at 0.75 psi for 3 s.
Detector: UV 200
Migration time: 16.6
Voltage: 10 kV
Current: 42-44 μA
Model: SpectraPhoresis 1000 (Thermo Separation Products)

OTHER SUBSTANCES
Extracted: acetaminophen

KEY WORDS
plasma; direct injection

REFERENCE
Kunkel,A.; Günter,S.; Wätzig,H. Quantitation of acetaminophen and salicylic acid in plasma using capillary electrophoresis without sample pretreatment. Improvement of precision, *J.Chromatogr.A*, **1997**, *768*, 125–133.

SAMPLE
Matrix: blood
Sample preparation: Filter (0.2 μm) plasma, inject an aliquot of the filtrate.

CAPILLARY ELECTROPHORESIS
Capillary: 42.5 cm × 50 μm fused-silica (34.5 cm to detector) (Polymicro Technologies)
Capillary preparation: After each run rinse capillary at 25° with MeCN:running buffer 50:50 for 15 min then re-equilibrate with running buffer for 20 min. Condition new capillaries with 100 mM NaOH at 50° for 30 min then equilibrate with cathode buffer at 25° for 2 h with voltage applied.
Capillary temperature: 25
Running buffer: 60 mM pH 10.0 Sodium borate buffer containing 200 mM sodium dodecyl sulfate (Cathode buffer was 60 mM pH 10.0 sodium borate buffer prepared by dissolving 3.170 g boric acid in 100 mL water and adjusting the pH to 10.0 with freshly-prepared 1 M NaOH.)
Injection: Vacuum injection at 0.75 psi for 3 s.
Detector: UV 200
Migration time: 15.8
Voltage: 10 kV
Current: 42 μA
Model: Spectraphoresis 1000 (Thermo Separation Products)

OTHER SUBSTANCES
Extracted: acetaminophen

KEY WORDS
plasma; direct injection

REFERENCE

Kunkel,A.; Gunter,S.; Watzig,H. Determination of pharmaceuticals in plasma by capillary electrophoresis without sample pretreatment reproducibility, limit of quantitation and limit of detection, *Electrophoresis*, **1997**, *18*, 1882–1889.

SAMPLE

Matrix: blood
Sample preparation: 100 µL Serum + 150 µL 50 µg/mL IS in MeCN, vortex, centrifuge at 13400 g for 2 min, inject an aliquot.

CAPILLARY ELECTROPHORESIS

Capillary: 64.5 cm $\times$ 75 µm fused-silica with bubble cell (56 cm to detector) (Hewlett-Packard)
Capillary preparation: Before each run rinse each capillary with 100 mM NaOH for 2 min and with running buffer for 3 min
Capillary temperature: 30
Running buffer: 100 mM Boric acid adjusted to pH 8.8 with 1 M NaOH
Injection: Vacuum injection at 50 mm Hg for 20 s (ca. 100 nL).
Detector: UV 210
Migration time: 6.5
Internal standard: 3-isobutyl-1-methylxanthine (4.3)
Voltage: 25 kV
Model: Hewlett-Packard HP^{3D} CE
Limit of quantitation: 5 µg/mL
Limit of detection: 1 µg/mL

OTHER SUBSTANCES

Extracted: aspirin, gentisic acid, salicyluric acid

KEY WORDS

serum; comparison with fluorescence polarization immunoassay

REFERENCE

Goto,Y.; Makino,K.; Kataoka,Y.; Shuto,H.; Oishi,R. Determination of salicylic acid in human serum with capillary zone electrophoresis, *J.Chromatogr.B*, **1998**, *706*, 329–335.

SAMPLE

Matrix: blood, urine
Sample preparation: Filter (Amicon Centrifree Micropartition System) at 1500 g for 20 min, inject an aliquot of the ultrafiltrate.

CAPILLARY ELECTROPHORESIS

Capillary: 90 cm $\times$ 75 µm fused-silica (70 cm to detector) (Polymicro Technologies)
Capillary preparation: Between runs rinse capillary with 100 mM NaOH for 5 min and with running buffer for 10 min.
Running buffer: 10 mM pH 9.1 Na_2HPO_4 containing 6 mM sodium borate and 75 mM sodium dodecyl sulfate
Injection: Siphon at 34 cm for 5 s
Detector: UV 195
Migration time: 14
Voltage: 20 kV
Current: 80 µA

OTHER SUBSTANCES

Extracted: acetaminophen, ethosuximide, phenobarbital, primidone

KEY WORDS

serum; ultrafiltrate

REFERENCE

Caslavska,J.; Lienhard,S.; Thormann,W. Comparative use of three electrokinetic capillary methods for the determination of drugs in body fluids. Prospects for rapid determination of intoxications, *J.Chromatogr.*, **1993**, *638*, 335–342.

SAMPLE
Matrix: blood, urine
Sample preparation: Dilute urine 2-fold, inject serum or diluted urine directly.

CAPILLARY ELECTROPHORESIS
Capillary: 70 cm × 75 μm fused-silica (50 cm to detector) (Polymicro Technologies)
Capillary preparation: Rinse capillary with buffer between runs.
Running buffer: 10 mM pH 9.2 Na_2HPO_4 containing 6 mM sodium borate and 75 mM sodium dodecyl sulfate
Injection: Siphon at a height of 34 cm for 2 s
Detector: F ex 220 em 450 nm (bandpass filter)
Migration time: 9.5
Voltage: 20 kV
Current: 90 μA

OTHER SUBSTANCES
Extracted: naproxen, quinidine

KEY WORDS
serum

REFERENCE
Caslavska,J.; Gassmann,E.; Thormann,W. Modification of a tunable UV-visible capillary electrophoresis detector for simultaneous absorbance and fluorescence detection: profiling of body fluids for drugs and endogenous compounds, *J.Chromatogr.A*, **1995**, *709*, 147–156.

SAMPLE
Matrix: blood, urine
Sample preparation: Inject an aliquot directly.

CAPILLARY ELECTROPHORESIS
Capillary: 45 cm fused-silica (45 cm to detector)
Capillary preparation: Before each run equilibrate with running buffer for 3 min. After each run wash with 100 mM NaOH then water for 3 min. Condition new capillaries by rinsing with 100 mM NaOH for 30 min and with water for 30 min.
Capillary temperature: 16 ± 0.1
Running buffer: 20 mM Sodium carbonate containing 2 mM terbium chloride and 2 mM EDTA, pH adjusted to 11 by 100 mM NaOH (The terbium absorbs the laser energy and transfers it to the analyte which then fluoresces.)
Injection: Injection at high pressure for 5 s.
Detector: F ex 325 (Omnichrome Series 39 He-Cd laser) em 547
Migration time: 8
Voltage: 20 kV
Current: <100 μA
Model: Beckman P/ACE 5500
Limit of detection: 100 nM

OTHER SUBSTANCES
Extracted: mesalamine (4-aminosalicylic acid), gentisic acid, salicyluric acid

KEY WORDS
serum; direct injection; resonance energy transfer

REFERENCE
Petersen,J.R.; Bissell,M.G.; Mohammad,A.A. Laser induced resonance energy transfer -a novel approach towards achieving high sensitivity in capillary electrophoresis. I. Clinical diagnostic application, *J.Chromatogr.A*, **1996**, *744*, 37–44.

SAMPLE
Matrix: formulations

Sample preparation: Grind tablet, add MeCN:10 mM HCl 20:80, shake and sonicate for 15 min, centrifuge until a clear solution is obtained. Remove a 25 mL aliquot of the supernatant and add it to 10 mL 500 µg/mL propyl hydroxybenzoate, make up to 50 mL with MeCN:10 mM HCl 20:80, inject an aliquot.

CAPILLARY ELECTROPHORESIS
Capillary: 57 cm × 75 µm fused-silica (Beckman)
Capillary preparation: Pass running buffer through the capillary for at least 4 min before each injection.
Capillary temperature: 25
Running buffer: 20 mM pH 9 Borate buffer containing 25 mM sodium cholate and 50 mM sodium deoxycholate
Injection: Pressure injection for 2 s
Detector: UV 214
Migration time: 13
Internal standard: propyl hydroxybenzoate (10.5)
Voltage: 20 kV
Model: Beckman P/ACE System 2100

OTHER SUBSTANCES
Simultaneous: acetaminophen, aspirin, caffeine, chlorpheniramine, dextropropoxyphene

KEY WORDS
tablets

REFERENCE
Boonkerd,S.; Lauwers,M.; Detaevernier,M.R.; Michotte,Y. Separation and simultaneous determination of the components in an analgesic tablet formulation by micellar electrokinetic chromatography, *J.Chromatogr.A*, **1995**, *695*, 97–102.

SAMPLE
Matrix: reaction mixtures

CAPILLARY ELECTROPHORESIS
Capillary: 60 cm × 75 µm fused-silica (Phenomenex)
Capillary preparation: Every five injections rinse capillary with 100 mM NaOH for 5 min, water for 5 min, and running buffer for 5 min.
Running buffer: 20 mM pH 8.5 Borate buffer containing 25 mM sodium dodecyl sulfate
Injection: Vacuum injection for 2 s (105 nL).
Detector: UV 200
Migration time: 11.5
Internal standard: phenobarbital (8.5)
Voltage: 30 kV
Model: Spectrophoresis 100 (Thermo Separation Products)
Limit of detection: 1.68 ng

OTHER SUBSTANCES
Simultaneous: 2,3-dihydroxybenzoic acid, 2,5-dihydroxybenzoic acid
Noninterfering: ascorbic acid

REFERENCE
Gökören,N.; Tunçel,M. A capillary electrophoretic method for examination of hydroxylation of salicylate ion to determine hydroxyl free radical, *Pharmazie*, **1997**, *52*, 726–727.

SAMPLE
Matrix: solutions
Sample preparation: Prepare a 200 µg/mL solution in water, inject an aliquot.

CAPILLARY ELECTROPHORESIS
Capillary: 60 cm × 50 µm

Running buffer: 15 mM pH 11.0 sodium phosphate buffer containing 25 mM sodium dodecyl sulfate
Injection: Hydrodynamic injection at 10 cm for 15 s
Detector: UV 214
Migration time: 3.7
Voltage: 30 kV
Model: Waters Quanta 4000

OTHER SUBSTANCES
Simultaneous: acetaminophen, aspirin, caffeine, salicylamide

REFERENCE
Swartz,M.E. Method development and selectivity control for small molecule pharmaceutical separations by capillary electrophoresis, *J.Liq.Chromatogr.*, **1991**, *14*, 923–938.

SAMPLE
Matrix: solutions

CAPILLARY ELECTROPHORESIS
Capillary: 56.9 cm × 75 μm fused-silica (50 cm to detector)
Capillary temperature: 25
Running buffer: MeOH:60 mM pH 8.4 sodium borate buffer containing 60 mM sodium dodecyl sulfate 15:85
Detector: UV 214
Migration time: 15
Voltage: 16 kV
Model: Beckman PACE 2100

OTHER SUBSTANCES
Simultaneous: acetanilide, acetaminophen, aspirin, benzamide, caffeine, salicylamide

REFERENCE
McLaughlin,G.M.; Nolan,J.A.; Lindahl,J.L.; Palmieri,R.H.; Anderson,K.W.; Morris,S.C.; Morrison,J.A.; Bronzert,T.J. Pharmaceutical drug separations by HPCE: Practical guidelines, *J.Liq.Chromatogr.*, **1992**, *15*, 961–1021.

SAMPLE
Matrix: solutions
Sample preparation: Inject an aliquot of a filtered (0.45 μm) aqueous solution.

CAPILLARY ELECTROPHORESIS
Capillary: 70 cm × 50 μm fused-silica (50 cm to detector)
Capillary preparation: Before each injection wash capillary with 1 M NaOH for 3 min and running buffer for 4 min
Capillary temperature: 37
Running buffer: 20 mM phosphate/borate buffer containing 100 mM sodium dodecyl sulfate
Injection: Injection by 12.5 cm Hg vacuum for 4 s (12 nL)
Detector: UV 210
Migration time: 20.5
Voltage: 20 kV
Model: Applied Biosystems Model 270A

OTHER SUBSTANCES
Simultaneous: catechol, dihydroxybenzenes

REFERENCE
Donato,M.G.; Baeyens,W.; Van den Bossche,W.; Sandra,P. The determination of non-steroidal antiinflammatory drugs in pharmaceuticals by capillary zone electrophoresis and micellar electrokinetic capillary chromatography, *J.Pharm.Biomed.Anal.*, **1994**, *12*, 21–26.

SAMPLE
Matrix: solutions
Sample preparation: Inject an aliquot of a filtered (0.45 µm) aqueous solution.

CAPILLARY ELECTROPHORESIS
Capillary: 70 cm × 50 µm fused-silica (50 cm to detector)
Capillary preparation: Before each injection wash capillary with 1 M NaOH for 3 min and running buffer for 4 min
Capillary temperature: 37
Running buffer: 20 mM phosphate/borate buffer containing 100 mM sodium dodecyl sulfate
Injection: Injection by 12.5 cm Hg vacuum for 4 s (12 nL)
Detector: UV 210
Migration time: 20.5
Voltage: 20 kV
Model: Applied Biosystems Model 270A

OTHER SUBSTANCES
Simultaneous: catechol, dihydroxybenzenes

REFERENCE
Tomita,M.; Okuyama,T.; Watanabe,S.; Watanabe,H. Quantitation of the hydroxyl radical adducts of salicylic acid by micellar electrokinetic capillary chromatography: oxidizing species formed by a Fenton reaction, *Arch.Toxicol.*, **1994**, *68*, 428–433.

SAMPLE
Matrix: solutions

CAPILLARY ELECTROPHORESIS
Capillary: 56.5 cm × 50 µm fused-silica (50 cm to detector)
Capillary preparation: Rinse with 1 M NaOH for 30 s between runs. Condition a new capillary with 1 M NaOH for 20 min.
Capillary temperature: 30
Running buffer: 30 mM 2-hydroxybutyric acid adjusted to pH 8.65 with ammonium hydroxide
Injection: Pressure injection 3 s
Detector: UV 200
Migration time: 5.5
Voltage: 25 kV
Current: 30-35 µA
Model: Beckman P/ACE Model 2100

OTHER SUBSTANCES
Simultaneous: benzoic acid

REFERENCE
Bullock,J.; Strasters,J.; Snider,J. Effect of multiple electrolyte buffers on peak symmetry, resolution, and sensitivity in capillary electrophoresis, *Anal.Chem.*, **1995**, *67*, 3246–3252.

SAMPLE
Matrix: solutions

CAPILLARY ELECTROPHORESIS
Capillary: 53 cm × 50 µm (44.5 cm to detector) (Polymicro Technologies)
Capillary temperature: 40
Running buffer: 70 mM Boric acid adjusted to pH 9.00 with 1 M NaOH
Injection: Pressure injection at 50 mbar for 2-5 s (2.9-7.2 nL).
Detector: UV 210
Migration time: 4
Voltage: 20-25 kV
Model: Hewlett-Packard ³ᴰCE

OTHER SUBSTANCES
Simultaneous: acetaminophen, aspirin, caffeine, norephedrine

REFERENCE
Thompson,L.; Veening,H.; Strein,T.G. Capillary electrophoresis in the undergraduate instrumental laboratory: Determination of common analgesic formulations, *J.Chem.Educ.*, **1997**, *74*, 1117–1121.

SAMPLE
Matrix: solutions

CAPILLARY ELECTROPHORESIS
Capillary: 57 cm × 75 μm fused-silica (50 cm to detector) (Beckman)
Capillary preparation: Rinse with running buffer for 2 min before each run. Rinse new capillaries with 1 M HCl for 5 min, water for 2 min, 1% NaOH for 10 min, water for 2 min, and running buffer for 5 min.
Capillary temperature: 22
Running buffer: 13 mM pH 9.68 ± 0.02 Sodium tetraborate buffer
Injection: Hydrodynamic injection for 10 s.
Detector: UV 214
Migration time: 8.2
Voltage: 20 kV
Model: Beckman P/ACE 5500

OTHER SUBSTANCES
Simultaneous: p-coumaric acid, ferulic acid, 4-hydroxybenzoic acid, 4-hydroxyphenylacetic acid, 3,4,5-trimethoxybenzoic acid, vanillic acid

REFERENCE
Deng,Y.; Fan,X.; Delgado,A.; Nolan,C.; Furton,K.; Zuo,Y.; Jones,R.D. Separation and determination of aromatic acids in natural water with preconcentration by capillary zone electrophoresis, *J.Chromatogr.A*, **1998**, *817*, 145–152.

SAMPLE
Matrix: solutions

CAPILLARY ELECTROPHORESIS
Capillary: 27 cm × 75 μm fused-silica (20 cm to detector) (Polymicro Technologies)
Capillary preparation: Between runs purge with running buffer.
Capillary temperature: 25
Running buffer: 16 mM Tris containing 20 mM α-hydroxyisobutyric acid and 0.005% hydroxypropylcellulose
Injection: Pressure injection at 0.5 psi for 10 s.
Detector: UV 214
Migration time: 1.7
Voltage: 20 kV
Model: Beckman P/ACE 2100

OTHER SUBSTANCES
Simultaneous: picric acid

REFERENCE
Gebauer,P.; Borecká,P.; Bocek,P. Predicting peak symmetry in capillary zone electrophoresis. Background electrolytes with two co-ions: Schizophrenic zone broadening and the role of system peaks, *Anal.Chem.*, **1998**, *70*, 3397–3406.

SAMPLE
Matrix: solutions

CAPILLARY ELECTROPHORESIS
Capillary: 57 cm × 50 μm fused-silica (50 cm to detector)

Running buffer: 50 mM pH 8.0 Borate buffer containing 40 mM sodium taurodeoxycholate and 25 mM phosphatidylcholine (Prepare by adding phosphatidylcholine and stirring for 3-6 h until all cloudiness disappears. Phosphatidylcholine was 95% pure soybean lecithin, Epikuron, Lucas Meyer & Co.)

Injection: Inject a solution of the compound in the running buffer at 20 psi for 1 s, inject MeOH: water 5:95 at 20 psi for 1 s, inject a solution of halofantrine in running buffer at 20 psi for 1 s.

Detector: UV 214

Migration time: k' 3.98

Model: Beckman P/ACE 5000

OTHER SUBSTANCES

Also analyzed: acetaminophen, amoxicillin, antipyrine, aspirin, azathioprine, caffeine, captopril, carbamazepine, carprofen, chlorambucil, chlorpheniramine, chlorpromazine, cimetidine, clonidine, codeine, desipramine, diphenhydramine, ephedrine, fenoterol, flufenamic acid, flurbiprofen, haloperidol, hydroxyzine, ibuprofen, imipramine, indomethacin, ketoprofen, lidocaine, melphalan, metoprolol, nabumetone, nadolol, phenobarbital, phenol, promazine, propranolol, pyrilamine, ranitidine, ropinirole, sulfamethoxazole, testosterone, theophylline, thioridazine, tiaprofenic acid, tolfenamic acid, trifluoperazine, trimethoprim, valproic acid, verapamil

KEY WORDS

comparison with HPLC; k' = (Tr-T0)/(T0(1-Tr/Tm)) where Tr = retention time of analyte; T0 = retention time of water; and Tm = retention time of marker (halofantrine)

REFERENCE

Hanna,M.; de Biasi,V.; Bond,B.; Salter,C.; Hutt,A.J.; Camilleri,P. Estimation of the partitioning characteristics of drugs: A comparison of a large and diverse drug series utilizing chromatographic and electrophoretic methodology, *Anal.Chem.*, **1998**, *70*, 2092–2099.

SAMPLE

Matrix: solutions

Sample preparation: Mix 100 μL of a buffer solution containing aspartame with 200 μL 22 μg/mL salicylic acid in water, inject an aliquot.

CAPILLARY ELECTROPHORESIS

Capillary: 47 cm × 50 μm fused-silica (40 cm to detector) (Beckman)

Capillary preparation: Between analyses flush capillary with 100 mM NaOH for 2 min and with running buffer for 2 min.

Capillary temperature: 20

Running buffer: 50 mM pH 9.35 Borate buffer

Injection: Pressure injection at 0.5 psi for 3 s.

Detector: UV 215

Migration time: 8

Voltage: 20 kV

Model: Beckman P/ACE 2100

OTHER SUBSTANCES

Simultaneous: aspartame

REFERENCE

Sabah,S.; Scriba,G.K.E. Determination of aspartame and its degradation and epimerization products by capillary electrophoresis, *J.Pharm.Biomed.Anal.*, **1998**, *16*, 1089–1096.

SAMPLE

Matrix: solutions

CAPILLARY ELECTROPHORESIS

Capillary: 65.5 cm × 50 μm fused-silica (57.3 cm to detector) (Yongnian Optical Factory, China)

Capillary preparation: After each run rinse capillary with 100 mM NaOH for 5 min and with water for 5 min then re-equilibrate with running buffer for 10 min.

Capillary temperature: 22 ± 0.5

Running buffer: EtOH:buffer 10:90 (Buffer was 40 mM borax containing 10 mM β-cyclodextrin, adjusted to pH 7.0 with phosphoric acid.)
Injection: Electromigration at 20 kV for 20 s.
Detector: UV 280
Migration time: 14.5
Voltage: 20 ± 0.3
Model: 1229 HPCE (Beijing)
Limit of detection: 4.86 μg/mL

OTHER SUBSTANCES
Simultaneous: p-aminophenol, benzeneazosalicylic acid, mesalamine, zinc 5-aminosalicylate

KEY WORDS
comparison with HPLC

REFERENCE
Zhang,S.S.; Liu,H.X., Yuan,Z.B. Comparison of high-performance capillary electrophoresis and liquid chromatography on analysis of zinc 5-aminosalicylate dihydrato and related materials, *J.Chromatogr.B*, **1998**, *705*, 165–170.

Scopolamine

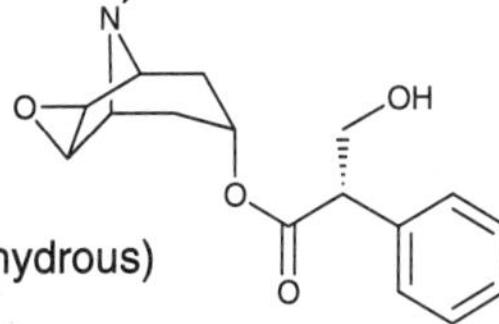

Molecular formula: $C_{17}H_{21}NO_4$
Molecular weight: 303.36
CAS Registry No.: 51-34-3, 6533-68-2 (HBr trihydrate), 114-49-8 (HBr anhydrous)
Merck Index (12th ed.): 8550

SAMPLE
Matrix: formulations
Sample preparation: Dilute formulation with water to an scopolamine concentration of about 50 μg/mL.

CAPILLARY ELECTROPHORESIS
Capillary: 64.5 cm × 50 μm fused-silica (56 cm to detector)
Capillary preparation: Between analyses wash capillary with 100 mM NaOH for 2 min, wash with water for 2 min, and equilibrate with running buffer for 3.5 min. At the start of each day flush with 1 M NaOH for 15 min and with water for 10 min.
Capillary temperature: 25
Running buffer: 100 mM pH 7 Tris-phosphate buffer
Injection: Pressure injection at 25 mbar for 20 s (12 nL), ramp voltage at 500 V/s.
Detector: UV 195
Migration time: 3.644
Voltage: 30 kV
Current: 63 μA
Model: Hewlett-Packard HP[3D]
Limit of quantitation: 3 μg/mL
Limit of detection: 1 μg/mL

OTHER SUBSTANCES
Simultaneous: atropine, homatropine

KEY WORDS
ophthalmic solutions

REFERENCE
Cherkaoui,S.; Mateus,L.; Christen,P.; Veuthey,J.-L. Development and validation of a capillary zone electrophoresis method for the determination of atropine, homatropine and scopolamine in ophthalmic solutions, *J.Chromatogr.B*, **1997**, *696*, 283–290.

SAMPLE
Matrix: formulations
Sample preparation: Dissolve tablet in 50 mL water with sonication for 5 min, centrifuge at 130 g for 10 min, filter (0.5 μm) the supernatant, inject an aliquot of the filtrate.

CAPILLARY ELECTROPHORESIS
Capillary: 37 cm × 50 μm fused-silica (30 cm to detector) (Polymicro Technologies)
Running buffer: 30 mM pH 7.00 Phosphate buffer containing 30 mM sodium dodecyl sulfate
Injection: Pressure injection for 2 s.
Detector: UV 200
Migration time: 4.5 (scopolamine), 4.8 (scopolamine N-methylbromide), 5 (scopolamine N-butylbromide)
Voltage: 24 kV
Model: Beckman P/ACE 2200

OTHER SUBSTANCES
Simultaneous: scopolamine N-oxide

KEY WORDS
tablets

REFERENCE
Wu,H.-L.; Huang,C.H.; Chen,S.-H.; Wu,S.-M. Micellar electrokinetic chromatography of scopolamine-related anticholinergics, *J.Chromatogr.A*, **1998**, *802*, 107–113.

SAMPLE
Matrix: plants
Sample preparation: Extract 50 mg lyophilized plant material with 5.0 mL 80% MeOH at 60° for 16 h, centrifuge. Remove a 500 μL aliquot of the supernatant and lyophilize it to dryness, make up to 500 μL with running buffer, inject an aliquot.

CAPILLARY ELECTROPHORESIS
Capillary: 67 cm × 75 μm fused-silica (60 cm to detector) (Isco)
Capillary preparation: Between runs rinse the capillary with 100 mM KOH for 1.5 min and with water for 1.5 min then equilibrate with running buffer for 2 min. At the start of each day purge with 100 mM KOH for 10 min, with water for 10 min, and with running buffer for 10 min.
Capillary temperature: 25
Running buffer: 40 mM pH 7.8 Phosphate buffer (Prepare buffer by mixing 40 mM Na_2HPO_4 with 40 mM KH_2PO_4 to achieve a pH of 7.8.)
Injection: Pressure injection at 0.5 psi for 4 s.
Detector: UV 214
Migration time: 5.5
Voltage: 20 kV
Model: Beckman P/ACE 2200
Limit of quantitation: 7.5 μg/mL
Limit of detection: 1.5 μg/mL

OTHER SUBSTANCES
Extracted: atropine, norscopolamine, tropic acid

REFERENCE
Eeva,M.; Salo,J.-P.; Oksman-Caldentey,K.-M. Determination of the main tropane alkaloids from transformed *Hyoscyamus muticus* plants by capillary zone electrophoresis, *J.Pharm.Biomed.Anal.*, **1998**, *16*, 717–722.

SAMPLE
Matrix: solutions
Sample preparation: Inject an aliquot of a solution in MeOH:water 10:90.

CAPILLARY ELECTROPHORESIS
Capillary: 64.5 cm × 75 μm (56 cm to detector)

Capillary preparation: Before each run flush the capillary with 100 mM NaOH for 2 min, with water for 2 min, and equilibrate with running buffer for 3.5 min. At the start of each day wash capillary with 100 mM NaOH for 15 min and with water for 10 min.
Capillary temperature: 25
Running buffer: MeCN:buffer 10:90 (Buffer was 30 mM pH 8.5 phosphate-borate buffer containing 50 mM sodium dodecyl sulfate.)
Injection: Pressure injection at 25 mbar for 10 s (ca. 30 nL), ramp to operating voltage at 500 V/s.
Detector: UV 195
Migration time: 6.5
Voltage: 30 kV
Model: Hewlett Packard HP 3D

OTHER SUBSTANCES
Simultaneous: apoatropine, homatropine, 6β-hydroxyhyoscyamine, hyoscyamine, littorine, tropic acid

REFERENCE
Cherkaoui,S.; Mateus,L.; Christen,P.; Veuthcy,J.-L. Micellar electrokinetic capillary chromatography for selected tropane alkaloid analysis in plant extract, *Chromatographia*, **1997**, *46*, 351–357.

Secobarbital

Molecular formula: $C_{12}H_{18}N_2O_3$
Molecular weight: 238.29
CAS Registry No.: 76-73-3, 309-43-3 (Na salt)
Merck Index (12th ed.): 8563
Lednicer: 1 269

SAMPLE
Matrix: blood
Sample preparation: Condition a Sep-Pak C18 SPE cartridge with MeOH, water, and 100 mM pH 9.0 borate buffer. 1 mL Serum + 3 mL 100 mM pH 9.0 borate buffer, mix, add to the SPE cartridge, wash with 3 mL 100 mM pH 9.0 borate buffer, wash with three 2 mL portions of water, wash with 1 mL MeCN:water 5:95, wash with 1 mL hexane, dry under full vacuum. Elute with 4 mL dichloromethane, evaporate the eluate to dryness under a stream of nitrogen, reconstitute with MeOH:water 15:85, filter (0.45 μm), inject an aliquot.

CAPILLARY ELECTROPHORESIS
Capillary: 72 cm × 50 μm fused-silica (50 cm to detector)
Capillary preparation: Before each run wash capillary with 100 mM NaOH for 4 min and with running buffer for 6 min.
Capillary temperature: 25
Running buffer: MeOH:5 mM pH 8.5 phosphate/borate buffer 15:85 containing 50 mM sodium dodecyl sulfate
Injection: Vacuum injection for 4 s.
Detector: UV 200
Migration time: 9
Voltage: 25 kV
Model: Applied Biosystems Model 270A or Beckman P/ACE 5000

OTHER SUBSTANCES
Extracted: bromazepam, diazepam, estazolam, flurazepam, nitrazepam, triazolam

KEY WORDS
SPE; serum

REFERENCE
Tomita,M.; Okuyama,T. Application of capillary electrophoresis to the simultaneous screening and quantitation of benzodiazepines, *J.Chromatogr.B*, **1996**, *678*, 331–337.

SAMPLE
Matrix: blood
Sample preparation: Condition a 1 mL 100 mg Bond Elut C18 SPE cartridge 3 mL MeOH and 3 mL 50 mM pH 9.0 sodium phosphate buffer, do not allow to dry. 1 mL Serum + IS, mix, add to the SPE cartridge, wash with 3 mL 50 mM pH 9.0 sodium phosphate buffer, allow to dry for 5 min, elute with 3 mL dichloromethane. Filter (0.2 μm nylon) the eluate and and evaporate it to dryness under a stream of nitrogen, reconstitute the residue in 1 mL MeOH:water 30:70, inject an aliquot.

CAPILLARY ELECTROPHORESIS
Capillary: 62 cm × 75 μm fused-silica (52 cm to detector) (Polymicro Technologies)
Capillary preparation: Before each run rinse capillary with 100 mM NaOH for 2 min and with running buffer for 2 min. Condition new capillaries by rinsing with 1 M NaOH for 10 min, with water for 10 min, with 100 mM HCl for 10 min, and with running buffer for 10 min.
Capillary temperature: 25
Running buffer: 50 mM pH 9.0 Sodium phosphate buffer containing 40 mM hydroxypropyl-gamma-cyclodextrin (Research Biochemicals International, Natick MA)
Injection: Vacuum injection at 0.5 psi for 10 s.
Detector: UV 254
Migration time: 10.86 (R), 11.17 (S)
Internal standard: aprobarbital (11.86)
Voltage: 15 kV
Current: 100 μA
Model: Beckman P/ACE 5000
Limit of quantitation: 1 μg/mL

OTHER SUBSTANCES
Simultaneous: clonazepam

KEY WORDS
chiral; serum; SPE

REFERENCE
Srinivasan,K.; Zhang,W.; Bartlett,M.G. Rapid simultaneous capillary electrophoretic determination of (*R*)- and (*S*)-secobarbital from serum and prediction of hydroxypropyl-γ-cyclodextrin-secobarbital stereoselective interaction using molecular mechanics simulation, *J.Chromatogr.Sci.*, **1998**, *36*, 85–90.

SAMPLE
Matrix: blood, gastric contents, urine, vitreous humor
Sample preparation: Dilute gastric contents 1:10. 2 mL Serum, blood, vitreous humor, diluted gastric contents, or urine + 2.5 mL water + 200 μL mephobarbital solution, mix, add to TOXI-TUBE B (Toxi-Lab, Irvine CA), rock for 10 min, centrifuge at 2000 rpm for 5 min. Remove the organic layer and evaporate it to dryness under a stream of nitrogen at room temperature, reconstitute the residue in 100 μL running buffer, vortex for 30 s, inject an aliquot. (Mephobarbital solution was 2 mg mephobarbital and 2 drops EtOH, made up to 20 mL with water.)

CAPILLARY ELECTROPHORESIS
Capillary: 60 cm × 75 μm AccuSep (Waters)
Capillary preparation: Purge for 1 min between samples.
Running buffer: MeCN:buffer 15:85 adjusted to pH 8.5 with 1 M phosphoric acid. (Buffer was 3.8 g sodium borate decahydrate, 1 .4 g $NaH_2PO_4.H_2O$, and 28.8 g sodium dodecyl sulfate in 1 L water.)
Injection: Hydrostatic injection for 10 s
Detector: UV 214
Migration time: 10.79
Internal standard: mephobarbital (8.88)
Voltage: 20 kV
Model: Waters Quanta 4000

Limit of detection: 100 ng/mL

OTHER SUBSTANCES
Extracted: amobarbital, butabarbital, butalbital, pentobarbital, phenobarbital

---

KEY WORDS
serum; whole blood

REFERENCE
Ferslew,K.E.; Hagardorn,A.N.; McCormick,W.F. Application of micellar electrokinetic capillary chromatography
to forensic analysis of barbiturates in biological fluids, *J.Forensic Sci.*, **1995**, *40*, 245–249.

SAMPLE
Matrix: solutions

CAPILLARY ELECTROPHORESIS
Capillary: 60 cm × 50 μm fused-silica (40 cm to detector), bare (A) or C18 coated (B) (Supelco)
Capillary preparation: Rinse new capillaries with 250 μL 1 M NaOH, 250 μL 100 mM NaOH,
    water, MeOH, water, and running buffer then condition with running buffer for at least 15
    min. Carry out a similar procedure between runs.
Running buffer: 100 mM pH 8.5 Borate buffer containing 25 mM sodium dodecyl sulfate and 5
    M urea
Injection: Electrokinetic injection at 5 kV for 5 s.
Detector: UV 254
Migration time: 13 (A), 11.5 (B)
Voltage: 18 kV
Model: Jasco

OTHER SUBSTANCES
Simultaneous: amobarbital, barbital, mephobarbital, metharbital, pentobarbital, phenobarbital

---

KEY WORDS
coated capillary

REFERENCE
Jinno,K.; Han,Y.; Nakamura,M. Analysis of anxiolytic drugs by capillary electrophoresis with bare and coated
    capillaries, *J.Capillary Electrophor.*, **1996**, *3*, 139–145.

SAMPLE
Matrix: solutions

CAPILLARY ELECTROPHORESIS
Capillary: 52 cm × 75 μm fused-silica
Running buffer: 50 mM pH 9.0 Phosphate buffer containing 40 mM hydroxypropyl-gamma-
    cyclodextrin
Detector: UV 254
Voltage: 15 kV

OTHER SUBSTANCES
Also analyzed: butalbital, pentobarbital

KEY WORDS
chiral

REFERENCE
Hassan,A.; Alemayehu,B.; Anucha,T.; Chau,S.; Eradiri,O.; Ly,C.; Malaiyandi,P.; Muhuri,G.; Odidi,A.; Odidi,I.;
    Siwicki,J. Validation of HPLC method for assay and determination of chromatographic purity of diclofenac
    sodium in extended release formulations (Abstract 4173), *Pharm.Res.*, **1997**, *14*, S687.

SAMPLE
Matrix: solutions

CAPILLARY ELECTROPHORESIS
Capillary: 40 cm × 75 μm coated fused-silica (15 cm to detector) (Supelco)
Capillary preparation: Coat column as follows. Adjust the pH of 20 mL water to 3.5 with acetic acid, add 80 μL 3-(trimethoxysilyl)propyl methacrylate (3-methacryloxypropyltrimethoxysilane), mix, suck into capillary, let stand at room temperature for 1 h, remove the solution, wash with water. Fill the capillary with a deaerated 4% acrylamide solution containing 1 mg/mL N,N,N',N'-tetramethylethylenediamine and 1 mg/mL ammonium persulfate, let stand for 3 h, remove excess solution by aspiration, rinse with water, remove water by aspiration, dry at 35° (cf. J. Chromatogr. 1985, 347, 191).
Running buffer: 100 mM Tris/150 mM boric acid, pH 8.3
Injection: Electromigration at 5 kV for 5 s.
Detector: UV 240, UV 254
Migration time: 7
Voltage: 12 kV
Model: Jasco 890-CE

OTHER SUBSTANCES
Simultaneous: amobarbital, barbital, mephobarbital, metharbital, pentobarbital, phenobarbital

KEY WORDS
injection at cathode; coated capillary

REFERENCE
Jinno,K.; Han,Y.; Sawada,H. Analysis of toxic drugs by capillary electrophoresis using polyacrylamide-coated columns, *Electrophoresis*, **1997**, *18*, 284–286.

SAMPLE
Matrix: solutions
Sample preparation: Inject an aliquot of a 100 μg/mL solution in running buffer.

CAPILLARY ELECTROPHORESIS
Capillary: 60 cm × 50 μm acrylamide-coated fused-silica (40 cm to detector) (Supelco)
Capillary preparation: Adjust the pH of 20 mL water to 3.5 with acetic acid, add 80 μL 3-(trimethoxysilyl)propyl methacrylate (3-methacryloxypropyltrimethoxysilane), mix, suck into capillary, let stand at room temperature for 1 h, remove the solution, wash with water. Fill the capillary with a deaerated 3-4% acrylamide solution containing 1 μL/mL N,N,N',N'-tetramethylethylenediamine and 1 mg/mL ammonium persulfate, let stand for 3 h, remove excess solution by aspiration, rinse with water, remove water by aspiration, dry at 35° (cf. J. Chromatogr. 1985, 347, 191).
Running buffer: MeCN:buffer 5:95 (Buffer was 100 mM borate containing 5 M urea and 10 mM sodium dodecyl sulfate, adjusted to pH 8.5 with phosphate.)
Injection: Electrokinetic injection at 5 kV for 5 s.
Detector: UV 254
Migration time: 16.2
Voltage: 18 kV
Model: Jasco Model 870-CE

OTHER SUBSTANCES
Simultaneous: alprazolam, amobarbital, barbital, bromazepam, clonazepam, clotiazepam, cloxazolam, diazepam, estazolam, etizolam, fludiazepam, flunitrazepam, flurazepam, haloxazolam, medazepam, mephobarbital, metharbital, nimetazepam, nitrazepam, oxazepam, pentobarbital, phenobarbital, triazolam

KEY WORDS
coated capillary; detector at anode

REFERENCE
Jinno,K.; Han,Y.; Sawada,H.; Taniguchi,M. Capillary electrophoretic separation of toxic drugs using a polyacrylamide-coated capillary, *Chromatographia*, **1997**, *46*, 309–314.

SAMPLE
Matrix: solutions
Sample preparation: Inject an aliquot of a solution in 20 mM pH 5.9 ammonium acetate buffer.

CAPILLARY ELECTROPHORESIS
Capillary: 33 cm × 50 μm fused-silica (20 cm to detector) (Polymicro Technologies)
Running buffer: 10 mM pH 5.9 Ammonium acetate buffer containing 15 mM sodium dodecyl sulfate
Injection: Electrokinetic injection at -7.4 kV for 1 s.
Detector: UV 226, MS, Finnigan MAT TSQ, negative electrospray (details in paper)
Migration time: 13
Voltage: -7.4 kV

OTHER SUBSTANCES
Simultaneous: amobarbital, barbital, butalbital, pentobarbital

REFERENCE
Yang,L.; Harrata,A.K.; Loo,C.S. On-line micellar electrokinetic chromatography-electrospray ionization mass spectrometry using anodically migrating micelles, *Anal.Chem.*, **1997**, *69*, 1820–1826.

SAMPLE
Matrix: solutions
Sample preparation: Inject an aliquot of a 20 μg/mL solution

CAPILLARY ELECTROPHORESIS
Capillary: 44 cm × 50 μm fused-silica (37 cm to detector) (Supelco)
Capillary preparation: Before each run rinse capillary with running buffer for 3 min. At the start of each day rinse capillary with running buffer for 10 min. Treat new capillaries with 1 M NaOH, 100 mM NaOH, water, and running buffer.
Capillary temperature: 25
Running buffer: 100 mM Phosphoric acid adjusted to pH 5 with triethanolamine containing 10 mM carboxymethyl-β-cyclodextrin (Cyclolab, Budapest) and 50 mM heptakis(2,3,6-tri-O-methyl)-β-cyclodextrin (Sigma)
Injection: Hydrodynamic injection for 5 s (13.3 nL).
Detector: UV 210
Voltage: -25 kV
Model: Spectraphoresis 1000 CE

KEY WORDS
detector at anode; chiral; R_s = 4.0

REFERENCE
Fillet,M.; Fotsing,L.; Crommen,J. Enantioseparation of uncharged compounds by capillary electrophoresis using mixtures of anionic and neutral β-cyclodextrin derivatives, *J.Chromatogr.A*, **1998**, *817*, 113–119.

SAMPLE
Matrix: urine
Sample preparation: Place an extraction rod in 50 μL urine in a 50 mm × 1.5 mm ID length of PTFE tubing for 30 min, place the extraction rod into 5 μL 20-40 mM pH 11.5 phosphate buffer in a 50 mm × 1.2 mm ID length of PTFE tubing for 90 min, inject an aliquot of the buffer. (Prepare the solid-phase extraction rod as follows. Polish a 70 × 1.1 stainless steel rod with emery paper, clean with a Kimwipe and acetone, sonicate in EtOH, sonicate in THF, dip in 3% PVAM in THF momentarily, hold vertically for 1 min, air dry in a hood for at least 5 h, dip in PVC solution momentarily, hold vertically for 1 min, air dry for at least 5 h. The coating is 3 cm long. PVAM was poly(vinyl chloride-co-vinyl acetate-co-maleic acid) consisting of 86% vinyl chloride, 13% vinyl acetate, and 1% maleic acid. Prepare PVC solution by adding very high molecular weight PVC slowly to THF with stirring until the concentration reaches 3.6%, add Santicizer 141 to a concentration of 7.2%. Santicizer 141 (Monsanto) is 92% 2-ethylhexyl diphenyl phosphate, 5% di-2-ethylhexyl phenyl phosphate, and 3% triphenyl phosphate.)

CAPILLARY ELECTROPHORESIS
Capillary: 75 cm × 75 μm fused-silica (50 cm to detector) (Polymicro Technologies)

Running buffer: 50 mM Tris adjusted to pH 7.8 with 3-[N-tris(hydroxymethyl)methylamino]-2-hydroxypropanesulfonic acid (Tapso)
Injection: Pressure injection at 0.5 psi for 4 s.
Detector: UV 230
Migration time: 7.6
Voltage: 35 kV
Current: about 30 μA
Model: Isco 3850
Limit of detection: <1 ppm

OTHER SUBSTANCES
Extracted: amobarbital, aprobarbital, butabarbital, butalbital, mephobarbital, pentobarbital, thiopental
Simultaneous: allobarbital, aspirin, phenobarbital

KEY WORDS
SPE

REFERENCE
Li,S.; Weber,S.G. Determination of barbiturates by solid-phase microextraction and capillary electrophoresis, *Anal.Chem.*, **1997**, *69*, 1217–1222.

Selegiline

Molecular formula: C$_{13}$H$_{17}$N
Molecular weight: 187.28
CAS Registry No.: 2323-36-6, 14611-51-9
Merck Index (12th ed.): 8569

SAMPLE
Matrix: solutions

CAPILLARY ELECTROPHORESIS
Capillary: 70 cm × 75 μm uncoated fused-silica (55 cm to detector)
Running buffer: 20 mM pH 2.8 Tris-phosphoric acid buffer containing 0.1% hydroxypropylmethylcellulose
Injection: Electrokinetic injection at 3 kV for 12 s
Detector: UV 190
Migration time: 9
Voltage: 21 kV
Current: 32 μA
Model: ATI Unicam Crystal 300

OTHER SUBSTANCES
Simultaneous: amphetamine, methamphetamine, propargylamphetamine

---

REFERENCE
Szöko,É.; Magyar,K. Chiral separation of deprenyl and its major metabolites using cyclodextrin-modified capillary zone electrophoresis, *J.Chromatogr.A*, **1995**, *709*, 157–162.

SAMPLE
Matrix: solutions

CAPILLARY ELECTROPHORESIS
Capillary: 37 cm × 50 μm fused-silica (30 cm to detector)
Capillary temperature: 20

Running buffer: 100 mM pH 2.5 Tris/phosphate buffer containing 200 mg/mL epichlorohydrin-
β-cyclodextrin polymer (MW 3000-5000, Cyclolab, Budapest)
Injection: Pressure injection at 0.5 psi for 8 s.
Detector: UV 214
Migration time: 27.5 (-), 29 (+)
Voltage: 14.8 kV
Model: Beckman P/ACE 2200

OTHER SUBSTANCES
Simultaneous: ephedrine (not chiral), methamphetamine

KEY WORDS
chiral

REFERENCE
Sevcik,J.; Stránsky,Z.; Ingelse,B.A.; Lemr,K. Capillary electrophoretic enantioseparation of selegiline, methamphetamine and ephedrine using a neutral β-cyclodextrin epichlorhydrin polymer, *J.Pharm.Biomed.Anal.*,
1996, *14*, 1089-1094.

SAMPLE
Matrix: solutions
Sample preparation: Inject an aliquot of a 5-10 µM solution.

CAPILLARY ELECTROPHORESIS
Capillary: 70 cm × 75 µm fused-silica (55 cm to detector)
Capillary temperature: 21
Running buffer: 20 mM pH 2.7 Tris-phosphate buffer containing 0.5% hydroxypropylmethyl
cellulose and 24 mM heptakis(2,6-di-O-methyl)-β-cyclodextrin (Hydroxypropylmethyl cellulose
was from Sigma; viscosity of 2% solution = 4000 cP at 25°.)
Injection: Electrokinetic injection at 3 kV for 12 s.
Detector: UV 190
Migration time: 23.8, 24.2 (enantiomers)
Voltage: 21 kV
Model: Crystal 300 (ATI, Unicam)

OTHER SUBSTANCES
Simultaneous: amphetamine, epinephrine, methamphetamine, norepinephrine, pseudoephedrine

KEY WORDS
chiral

REFERENCE
Szökö,E.; Gyimesi,J.; Barcza,L.; Magyar,K. Determination of binding constants and the influence of methanol
on the separation of drug enantiomers in cyclodextrin modified capillary electrophoresis, *J.Chromatogr.A*,
1996, *745*, 181-187.

Somatrem

Molecular formula: $C_{995}H_{1537}N_{263}O_{301}S_8$
Molecular weight: 22256.27
CAS Registry No.: 82030-87-3
Merck Index (12th ed.): 8864

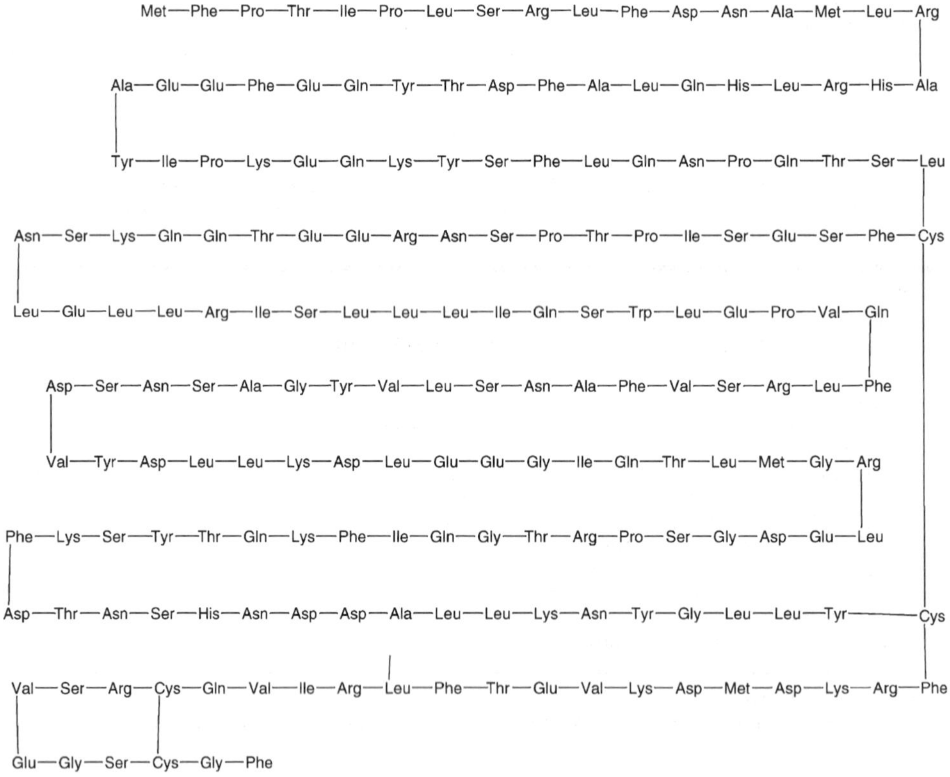

```
Met—Phe—Pro—Thr—Ile—Pro—Leu—Ser—Arg—Leu—Phe—Asp—Asn—Ala—Met—Leu—Arg
                                                                         |
      Ala—Glu—Glu—Phe—Glu—Gln—Tyr—Thr—Asp—Phe—Ala—Leu—Gln—His—Leu—Arg—His—Ala
       |
      Tyr—Ile—Pro—Lys—Glu—Gln—Lys—Tyr—Ser—Phe—Leu—Gln—Asn—Pro—Gln—Thr—Ser—Leu
                                                                             |
Asn—Ser—Lys—Gln—Gln—Thr—Glu—Glu—Arg—Asn—Ser—Pro—Thr—Pro—Ile—Ser—Glu—Ser—Phe—Cys
 |
Leu—Glu—Leu—Leu—Arg—Ile—Ser—Leu—Leu—Leu—Ile—Gln—Ser—Trp—Leu—Glu—Pro—Val—Gln
                                                                         |
    Asp—Ser—Asn—Ser—Ala—Gly—Tyr—Val—Leu—Ser—Asn—Ala—Phe—Val—Ser—Arg—Leu—Phe
     |
    Val—Tyr—Asp—Leu—Leu—Lys—Asp—Leu—Glu—Glu—Gly—Ile—Gln—Thr—Leu—Met—Gly—Arg
                                                                         |
Phe—Lys—Ser—Tyr—Thr—Gln—Lys—Phe—Ile—Gln—Gly—Thr—Arg—Pro—Ser—Gly—Asp—Glu—Leu
 |
Asp—Thr—Asn—Ser—His—Asn—Asp—Asp—Ala—Leu—Leu—Lys—Asn—Tyr—Gly—Leu—Leu—Tyr————Cys
                                     |                                       |
Val—Ser—Arg—Cys—Gln—Val—Ile—Arg—Leu—Phe—Thr—Glu—Val—Lys—Asp—Met—Asp—Lys—Arg—Phe
 |         |
Glu—Gly—Ser—Cys—Gly—Phe
```

SAMPLE
Matrix: solutions
Sample preparation: Prepare a 14 µM solution in water, inject an aliquot.

CAPILLARY ELECTROPHORESIS
Capillary: 57 cm × 50 µm uncoated fused-silica (50 cm to detector)
Capillary preparation: Before each analysis wash with 100 mM NaOH for 2 min and twice with water for 2 min, condition with running buffer for 5 min.
Capillary temperature: 23
Running buffer: 20 mM pH 2.5 Tricine containing 40 mM NaCl
Injection: Electrokinetic injection at 8 kV for 5 s
Detector: UV 200
Migration time: 14.5
Voltage: 15 kV
Model: Beckman P/ACE System 2000

OTHER SUBSTANCES
Simultaneous: somatropin

REFERENCE
Arcelloni,C.; Fermo,I.; Banfi,G.; Pontiroli,A.E.; Paroni,R. Capillary electrophoresis for protein analysis: separation of human growth hormone and human insulin molecular forms, *Anal.Biochem.*, **1993**, *212*, 160–167.

Somatropin

Molecular formula: $C_{990}H_{1529}N_{263}O_{299}S_7$
Molecular weight: 22124.09
CAS Registry No.: 9002-72-6, 12629-01-5 (human)
Merck Index (12th ed.): 8864

SAMPLE
Matrix: bulk
Sample preparation: Prepare a 12.8 mg/mL solution in 50 mM pH 8.3 ammonium carbonate buffer, inject an aliquot.

CAPILLARY ELECTROPHORESIS
Capillary: 92 cm × 50 μm untreated fused-silica (70 cm to detector) (Applied Biosystems)
Capillary preparation: Condition a new capillary by rinsing with 1 M NaOH for 20 min, washing with water for 10 min, and filling with running buffer for 20 min. At the beginning of each day purge capillary with 1 M NaOH for 10 min then water for 10 min, fill with running buffer for 20 min. Do not rinse between runs.
Capillary temperature: 30
Running buffer: 100 mM $(NH_4)_2HPO_4$ adjusted to pH 6.0 with phosphoric acid
Injection: Inject sample using vacuum (12.7 cm Hg) for 3 s then inject running buffer for 1 s.
Detector: UV 195
Migration time: 26.5
Voltage: 20 kV
Model: Applied Biosystems Model 270 A
Limit of detection: 0.3% (for impurities)

OTHER SUBSTANCES
Simultaneous: impurities

REFERENCE
Dupin,P.; Galinou,F.; Bayol,A. Analysis of recombinant human growth hormone and its related impurities by capillary electrophoresis, *J.Chromatogr.A*, **1995**, *707*, 396–400.

SAMPLE
Matrix: solutions

CAPILLARY ELECTROPHORESIS
Capillary: 50 cm × 50 μm coated with a hydrophilic polymer
Capillary preparation: Flush with buffer after each run.
Running buffer: 100 mM pH 2.56 Phosphate buffer
Injection: Electromigration at 8 kV for 5-10 s
Detector: UV 200
Migration time: 23
Voltage: 8 kV
Model: Bio-Rad HPE 100

KEY WORDS
coated capillary

REFERENCE
Frenz,J.; Wu,S.L.; Hancock,W.S. Characterization of human growth hormone by capillary electrophoresis, *J.Chromatogr.*, **1989**, *480*, 379–391.

SAMPLE
Matrix: solutions
Sample preparation: Prepare a 1 mg/mL solution in water, inject an aliquot.

CAPILLARY ELECTROPHORESIS
Capillary: 105 cm $\times$ 50 μm fused-silica (81.5 cm to detector) (Polymicro Technologies)
Capillary preparation: Replace buffer in capillary after each run.
Running buffer: 10 mM Tricine containing 5.8 mM morpholine and 20 mM NaCl, adjusted to pH 8 with 1 M NaOH
Injection: Injection by siphon at 12 cm for 15 s (2.5 nL)
Detector: UV 200
Migration time: 9.5
Voltage: 300 V/cm
Current: about 20 μA

OTHER SUBSTANCES
Simultaneous: degradation products

KEY WORDS
recombinant

REFERENCE
Nielsen,R.G.; Sittampalam,G.S.; Rickard,E.C. Capillary zone electrophoresis of insulin and growth hormone, *Anal.Biochem.*, **1989**, *177*, 20–26.

SAMPLE
Matrix: solutions
Sample preparation: Inject an aliquot of a 1 mg/mL solution.

CAPILLARY ELECTROPHORESIS
Capillary: 57 cm $\times$ 50 μm fused-silica (50 cm to UV detector) (Polymicro Technologies)
Capillary preparation: Rinse with running buffer for 5 min before each injection. Condition a new column with 100 mM NaOH for 5 min, rinse with water for 5 min, fill with running buffer for 5 min.
Capillary temperature: 25
Running buffer: MeCN:20 mM pH 9.0 ammonium acetate 20:80
Injection: Nitrogen pressure injection at 0.5 psi for 5 s (6 nL)
Detector: UV 214 or MS, Vestec Model 201A quadrupole, electrospray at 2-3 kV, block 255°, spray chamber 44°, lens 54°, spray current 240 nA, nozzle voltage 160 V, makeup solution was 3 μg/mL bradykinin (for tuning) in 1% acetic acid in MeOH:water 95:5 or 5% acetic acid in MeOH:water 50:50 pumped at 1-10 μL/min
Migration time: 5
Voltage: 20 kV
Model: Beckman P/ACE System 2050

OTHER SUBSTANCES
Simultaneous: impurities

KEY WORDS
pig; cow

REFERENCE
Tsuji,K.; Baczynskyj,L.; Bronson,G.E. Capillary electrophoresis-electrospray mass spectrometry for the analysis of recombinant bovine and porcine somatotropins, *Anal.Chem.*, **1992**, *64*, 1864–1870.

SAMPLE
Matrix: solutions

CAPILLARY ELECTROPHORESIS
Capillary: 48.5 cm $\times$ 25 μm fused-silica (40 cm to detector) (Hewlett-Packard)
Capillary preparation: After each run flush capillary with 200 mM sodium phosphate buffer containing 3 M guanidine hydrochloride for 5 min. Condition new capillaries by flushing with 100 mM nitric acid at 50° for 30 min, with 100 mM NaOH at 50° for 30 min, with 100 mM nitric acid at 50° for 30 min, with water at 50° for 15 min, and with running buffer at 50° for 1 h. Finally the capillary was heated at 50° in running buffer for up to 16 h.

Capillary temperature: 15
Running buffer: 250 mM pH 6.8 Sodium phosphate buffer containing 1% propylene glycol
Injection: Pressure injection at 50 mbar for 50 s.
Detector: UV 214
Migration time: 12
Voltage: 29 kV
Model: Hewlett-Packard HP³ᴰ
Limit of quantitation: 5 µg/mL
Limit of detection: 2 µg/mL

OTHER SUBSTANCES
Simultaneous: impurities

KEY WORDS
human; recombinant

REFERENCE
McNerney,T.M.; Watson,S.K.; Sim,J.-H.; Bridenbaugh,R.L. Separation of recombinant human growth hormone from *Escherichia coli* cell pellet extract by capillary zone electrophoresis, *J.Chromatogr.A*, **1996**, *744*, 223–229.

Sotalol

Molecular formula: $C_{12}H_{20}N_2O_3S$
Molecular weight: 272.37
CAS Registry No.: 3930-20-9, 959-24-0 (HCl)
Merck Index (12th ed.): 8876
Lednicer: 1 66; 5 23

SAMPLE
Matrix: solutions

CAPILLARY ELECTROPHORESIS
Capillary: 82 cm × 75 µm fused-silica (Waters)
Capillary preparation: Between runs purge capillary with running buffer for 2 min. Purge a new capillary with 100 mM NaOH for 30 min, with water for 30 min, and with running buffer for 30 min.
Capillary temperature: 30
Running buffer: 30 mM pH 7.6 phosphate buffer containing 10 mM cetyltrimethylammonium bromide
Injection: Inject at high pressure for 5 s
Detector: UV 214
Migration time: 7.3
Voltage: 21 kV
Model: Beckman P/ACE System 2000

OTHER SUBSTANCES
Simultaneous: acebutolol, alprenolol, atenolol, nadolol, oxprenolol, pindolol, propranolol, timolol

REFERENCE
Lukkari,P.; Ennelin,A.; Sirén,H.; Riekkola,M.-L. Effect of temperature, effective capillary length, and applied voltage on the migration of nine β-blockers in micellar electrokinetic capillary chromatography, *J.Liq.Chromatogr.*, **1993**, *16*, 2069–2079.

SAMPLE
Matrix: solutions

CAPILLARY ELECTROPHORESIS
Capillary: 58 cm × 50 μm fused-silica (50 cm to detector)
Running buffer: Isopropanol:buffer 2.5:97.5 (Buffer was 80 mM pH 6.8 Phosphate buffer containing 15 mM cetyltrimethylammonium bromide.)
Injection: Hydrostatic injection for 15 s
Detector: UV 214
Migration time: 12.2
Voltage: -20 kV
Model: Waters Quanta 4000

OTHER SUBSTANCES
Simultaneous: acebutolol, alprenolol, atenolol, labetalol, nadolol, oxprenolol, pindolol, propranolol, timolol

REFERENCE
Lukkari,P.; Vuorela,H.; Riekkola,M.-L. Effects of organic mobile phase modifiers on elution and separation of β-blockers in micellar electrokinetic capillary chromatography, *J.Chromatogr.A*, **1993**, *655*, 317–324.

SAMPLE
Matrix: solutions
Sample preparation: Prepare a 50 μg/mL solution in diluted running buffer, inject an aliquot.

CAPILLARY ELECTROPHORESIS
Capillary: 44 cm × 50 μm fused-silica
Capillary preparation: After each run wash capillary with water for 2 min and with running buffer for 3 min. At the beginning of each day wash capillary with water and running buffer for 5 min. Condition a new capillary with 1 M NaOH, 100 mM NaOH, water, and separation buffer.
Capillary temperature: 15
Running buffer: 100 mM Phosphoric acid containing 15 mM heptakis(2,6-di-O-methyl)-β-cyclodextrin adjusted to pH 3.0 with triethanolamine
Injection: Hydrodynamic injection for 1 s
Detector: UV 210
Voltage: 25 kV
Model: Spectra Physics Spectraphoresis 1000

KEY WORDS
chiral; enantiomer resolution 1.3

REFERENCE
Bechet,I.; Paques,P.; Fillet,M.; Hubert,P.; Crommen,J. Chiral separation of basic drugs by capillary zone electrophoresis with cyclodextrin additives, *Electrophoresis*, **1994**, *15*, 818–823.

SAMPLE
Matrix: solutions
Sample preparation: Inject an aliquot of a 100 μg/mL solution in water:running buffer 50:50.

CAPILLARY ELECTROPHORESIS
Capillary: 44.5 cm × 50 μm acrylamide-coated fused-silica (Bio-Rad)
Capillary temperature: 30
Running buffer: 100 mM NaH_2PO_4 containing 15 mM gamma-cyclodextrin, adjusted to pH 2.5 with phosphoric acid
Injection: Electrokinetic injection at 8 kV for 6 s.
Detector: UV 200
Migration time: 9.29
Voltage: 14 kV
Model: Bio-Rad BioFocus 3000

OTHER SUBSTANCES
Also analyzed: albuterol, alprenolol, atenolol, atropine, baclofen, bamethan, benserazide, biperiden, bisoprolol, bupivacaine, bupranolol, butetamate, carazolol, carbuterol, carvedilol, celiprolol, chloroquine, chlorpheniramine (chlorphenamine), clidinium bromide, clobutinol, disopyramide, dobutamine, flecainide, homatropine, ipratropium bromide, isoproterenol, isothipendyl, ketamine, mefloquine, mequitazine, metaproterenol (orciprenaline), metipranolol, nafronyl (naftidrofuryl), nefopam, ofloxacin, orphenadrine, oxomemazine, oxprenolol, phenoxybenzamine, pholedrine, pindolol, pirbuterol, prilocaine, promethazine, propafenone, propranolol, synephrine, terbutaline, tetrahydrozoline (tetryzoline), tocainide, trihexyphenidyl, trimeprazine (alimemazine), trimipramine, tropicamide, verapamil, zopiclone

KEY WORDS
coated capillary; achiral

REFERENCE
Koppenhoefer,B.; Epperloin,U.; Christian,B.; Yibing,J.; Yuying,C.; Bingcheng,L. Separation of enantiomers of drugs by capillary electrophoresis. I. γ-Cyclodextrin as chiral solvating agent, *J.Chromatogr.A*, **1995**, *717*, 181–190.

SAMPLE
Matrix: solutions

CAPILLARY ELECTROPHORESIS
Capillary: 57 cm × 75 μm fused-silica (50 cm to detector) (Beckman)
Capillary preparation: Before each run rinse capillary with 100 mM NaOH and running buffer.
Capillary temperature: 30
Running buffer: Acetone:100 mM pH 8.1 borate buffer containing 50 mM sodium dodecyl sulfate 15:85 (A) or 100 mM phosphoric acid adjusted to pH 3.1 with triethanolamine (B)
Injection: Pressure injection for 5-10 s.
Detector: UV 200
Migration time: 5 (A), 9.9 (B)
Voltage: 25 kV
Model: Beckman P/ACE 5510

OTHER SUBSTANCES
Simultaneous: acebutolol, amiodarone, atenolol, bretylium, captopril, diltiazem, disopyramide, lidocaine, lisinopril, metoprolol, nicardipine, nifedipine, phenytoin, pindolol, propafenone, propranolol, quinidine, timolol, p-toluenesulfonic acid, verapamil

KEY WORDS
only running buffer A separates all compounds

REFERENCE
Bretnall,A.E.; Clarke,G.S. Selectivity of capillary electrophoresis for the analysis of cardiovascular drugs, *J.Chromatogr.A*, **1996**, *745*, 145–154.

SAMPLE
Matrix: solutions
Sample preparation: Inject an aliquot of a 100 μg/mL solution in running buffer.

CAPILLARY ELECTROPHORESIS
Capillary: 29 cm × 50 μm fused-silica (24.5 cm to detector) (Yongnian Optical Conductive Fiber Plant, China), coated with polyacrylamide
Capillary preparation: No details of the polyacrylamide coating process are provided. However, another paper (LC.GC 1997, 15, 40) by this group indicates that they use the procedure of Hjertén, thus: Adjust the pH of 20 mL water to 3.5 with acetic acid, add 80 μL 3-(trimethoxysilyl)propyl methacrylate (3-methacryloxypropyltrimethoxysilane), mix, suck into capillary, let stand at room temperature for 1 h, remove the solution, wash with water. Fill the capillary with a deaerated 3-4% acrylamide solution containing 1 μL/mL N,N,N',N'-tetramethylethylenediamine and 1 mg/mL potassium persulfate, let stand for 30 min, remove excess solution

by aspiration, rinse with water, remove water by aspiration, dry at 35° (J. Chromatogr. 1985, 347, 191).
Capillary temperature: 25
Running buffer: 100 mM NaH_2PO_4 adjusted to pH 2.5
Injection: Electrokinetic injection at 15 kV for 3 s.
Detector: UV 200, UV 210
Migration time: 5.33
Voltage: 15 kV
Model: Bio-Rad BioFocus 3000

OTHER SUBSTANCES

Simultaneous: albuterol, alprenolol, atenolol, baclofen, bamethan, benproperine, benserazide, bisoprolol, bupranolol, butamirate, butethamate, carbuterol, celiprolol, clenbuterol, clobutinol, dipivefrin, isoproterenol (isoprenaline), metaproterenol (orciprenaline), metipranolol, metoprolol, norfenefrine, ornidazole, oxprenolol, phenylpropanolamine, pholedrine, pirbuterol, prilocaine, procyclidine, synephrine, terbutaline, tocainide

KEY WORDS

coated capillary

REFERENCE

Koppenhoefer,B.; Epperlein,U.; Xiaofeng,Z.; Bingcheng,L. Separation of enantiomers of drugs by capillary electrophoresis. Part 4: Hydroxypropyl-γ-cyclodextrin as chiral solvating agent, *Electrophoresis*, **1997**, *18*, 924–930.

SAMPLE

Matrix: solutions
Sample preparation: Inject an aliquot of a 100 μg/mL solution in running buffer.

CAPILLARY ELECTROPHORESIS

Capillary: 30 cm × 50 μm fused-silica (25.5 cm to detector), coated with polyacrylamide
Capillary preparation: Adjust the pH of 20 mL water to 3.5 with acetic acid, add 80 μL 3-(trimethoxysilyl)propyl methacrylate (3-methacryloxypropyltrimethoxysilane), mix, suck into capillary, let stand at room temperature for 1 h, remove the solution, wash with water. Fill the capillary with a deaerated 3-4% acrylamide solution containing 1 μL/mL N,N,N',N'-tetramethylethylenediamine and 1 mg/mL potassium persulfate, let stand for 30 min, remove excess solution by aspiration, rinse with water, remove water by aspiration, dry at 35° (J. Chromatogr. 1985, 347, 191).
Capillary temperature: 25
Running buffer: 100 mM NaH_2PO_4 containing 45 mM hydroxypropyl-β- cyclodextrin, adjusted to pH 2.5 with phosphoric acid
Injection: Electrokinetic injection at 15 kV for 3 s.
Detector: UV 200
Voltage: 15 kV
Model: Bio-Rad BioFocus 3000

KEY WORDS

chiral; coated capillary; comparison with the use of other cyclodextrins; this running buffer gave the greatest enantiomeric separation.; α=1.019

REFERENCE

Lin,B.; Zhu,X.; Koppenhoefer,B.; Epperlein,U. Investigation of 123 chiral drugs by cyclodextrin-modified capillary electrophoresis, *LC.GC*, **1997**, *15*, 40–46.

SAMPLE

Matrix: solutions

CAPILLARY ELECTROPHORESIS

Capillary: 29-36 cm × 50 μm fused-silica (24.5-31.5 cm to detector) (Yongnian Optical Conductive Fiber Plant, China) coated with polyacrylamide
Capillary preparation: Coat capillary as follows. Adjust the pH of 20 mL water to 3.5 with acetic acid, add 80 μL 3-(trimethoxysilyl)propyl methacrylate (3-methacryloxypropyltrimethox-

ysilane), mix, suck into capillary, let stand at room temperature for 1 h, remove the solution, wash with water. Fill the capillary with a deaerated 3-4% acrylamide solution containing 1 μL/mL N,N,N',N'-tetramethylethylenediamine and 1 mg/mL potassium persulfate, let stand for 30 min, remove excess solution by aspiration, rinse with water, remove water by aspiration, dry at 35° (J. Chromatogr. 1985, 347, 191).
Capillary temperature: 25
Running buffer: 100 mM pH 2.5 NaH_2PO_4 (A) or 100 mM pH 2.5 NaH_2PO_4 containing 45 mM hydroxypropyl-α-cyclodextrin (Wacker, Munich) (B)
Injection: Electromigration at 15 kV for 3 s.
Detector: UV 200; UV 210
Migration time: 5.33 (A); 6.58 (B) (no separation of enantiomers)
Voltage: 15 kV
Model: Bio-Focus 3000

OTHER SUBSTANCES
Also analyzed: albuterol (salbutamol), alprenolol, amorolfine, atenolol, atropine, azelastine, baclofen, bamcthan, benproperine, benserazide, biperiden, bisoprolol, brompheniramine, bupivacaine, bupranolol, butamirate, butcthamate, carazolol, carbuterol, carteolol, carvedilol, celiprolol, chloroquine, chlorpheniramine, chlorphenoxamine, ciclotanine, clenbuterol, clidinium bromide, clobutinol, dimethindene, dipivefrin, disopyramide, dobutamine, doxylamine, fendiline, flecainide, gallopamil, homatropine, ipratropium bromide, isoproterenol (isoprenaline), isothipendyl, ketamine, meclizine, mefloquine, mepindolol, mequitazine, metaclazepam, metaproterenol (orciprenaline), metipranolol, metoprolol, nafronyl (naftidrofuryl), nefopam, nicardipine, norfenefrine, ofloxacin, ornidazole, orphenadrine, oxomemazine, oxprenolol, oxybutynin, phenoxybenzamine, phenylpropanolamine, pholedrine, pindolol, pirbuterol, prilocaine, procyclidine, promethazine, propafenone, propranolol, reproterol, sulpride, synephrine, talinolol, terbutaline, tetrahydrozoline (tetryzoline), theodrenaline, tioconazole, tocainide, trihexyphenidyl, trimeprazine (alimemazine), trimipramine, tropicamide, verapamil, zopiclone

---

KEY WORDS
coated capillary

REFERENCE
Koppenhoefer,B.; Eperlein,U.; Schlunk,R.; Zhu,X.; Lin,B. Separation of enantiomers of drugs by capillary electrophoresis. V. Hydroxypropyl-α-cyclodextrin as chiral solvating agent, *J.Chromatogr.A*, **1998**, *793*, 153–164.

Sparfloxacin

Molecular formula: $C_{19}H_{22}F_2N_4O_3$
Molecular weight: 392.41
CAS Registry No.: 110871-86-8
Merck Index (12th ed.): 8884

SAMPLE
Matrix: solutions
Sample preparation: Inject an aliquot of a solution in MeOH.

CAPILLARY ELECTROPHORESIS
Capillary: 59 cm $\times$ 50 μm fused-silica (43 cm to detector) (Polymicro Technologies)
Capillary preparation: Before each analysis flush with water for 3 min, with 200 mM NaOH for 3 min, with water for 3 min, and with running buffer for 4 min. Flush new capillaries with 1 M NaOH at 2000 mbar for 10 min then with 200 mM NaOH for 10 min.
Capillary temperature: 23
Running buffer: MeCN:buffer 28:72, pH 7.3 (Buffer was 32 mM sodium borate containing 18 mM NaH_2PO_4, 39 mM sodium cholate, and 8 mM sodium heptanesulfonate.)
Injection: Hydrodynamic injection at 40 mbar for 6 s.
Detector: UV 260
Migration time: 3.85

Voltage: 30 kV
Model: Lauer Labs Prince

OTHER SUBSTANCES
Simultaneous: ciprofloxacin, enoxacin, flumequine, lomefloxacin, nalidixic acid, norfloxacin, ofloxacin, oxolinic acid, pefloxacin, pipemidic acid, piromidic acid, rosoxacin

REFERENCE
Sun,S.-W.; Chen,L.-Y. Optimization of capillary electrophoretic separation of quinolone antibacterials using the
overlapping resolution mapping scheme, *J.Chromatogr.A*, **1997**, *766*, 215–224.

Spirapril

Molecular formula: $C_{22}H_{30}N_2O_5S_2$
Molecular weight: 466.62
CAS Registry No.: 83647-97-6, 94841-17-5
Merck Index (12th ed.): 8905
Lednicer: 4 83

SAMPLE
Matrix: solutions

CAPILLARY ELECTROPHORESIS
Capillary: 90 cm × 50 μm
Running buffer: MeCN:10 mM sodium tetraborate 15:85, adjusted to pH 11.7 with NaOH
Detector: UV 214
Migration time: 17.5
Voltage: 25 kV
Current: 14 μA

OTHER SUBSTANCES
Simultaneous: degradation products

KEY WORDS
comparison with HPLC

REFERENCE
Steuer,W.; Grant,I.; Erni,F. Comparison of high-performance liquid chromatography, supercritical fluid chromatography and capillary zone electrophoresis in drug analysis, *J.Chromatogr.*, **1990**, *507*, 125–140.

Spironolactone

Molecular formula: $C_{24}H_{32}O_4S$
Molecular weight: 416.58
CAS Registry No.: 52-01-7
Merck Index (12th ed.): 8917
Lednicer: 1 206

SAMPLE
Matrix: urine
Sample preparation: Filter (0.2 μm), inject an aliquot of the filtrate.

CAPILLARY ELECTROPHORESIS
Capillary: 80 cm × 50 μm fused-silica (57 cm to detector)
Capillary preparation: Before each run rinse the capillary with 1 M NaOH for 3 min, with 100 mM NaOH for 3 min, with water for 3 min, and with running buffer for 10 min.
Running buffer: 110 mM Boric acid containing 56 mM NaOH and 44 mM HCl, pH 8
Injection: Vacuum injection for 1 s
Detector: UV 244
Migration time: 11.900
Voltage: 20 kV
Model: Europhor Prime Vision system IV

OTHER SUBSTANCES
Extracted: acebutolol (UV 238), acetazolamide (UV 222), alprenolol (UV 220), amiloride (UV 220), atenolol (UV 228), bendroflumethiazide (UV 220), bumetanide (UV 220), chlorthalidone (UV 220), cocaine (UV 236), codeine (UV 220), ethacrynic acid (UV 220), furosemide (UV 232), hydrochlorothiazide (UV 226), methadone (UV 220), metoxiphenamine (UV 220), nadolol (UV 220), norcodeine (UV 220), oxprenolol (UV 220), pentazocine (UV 220), propranolol (UV 220), triamterene (UV 232), xipamide (UV 234)

REFERENCE
Gonzalez,E.; Laserna,J.J. Capillary zone electrophoresis for the rapid screening of banned drugs in sport, *Electrophoresis*, **1994**, *15*, 240–243.

Stavudine

Molecular formula: $C_{10}H_{12}N_2O_4$
Molecular weight: 224.22
CAS Registry No.: 3056-17-5
Merck Index (12th ed.): 8958

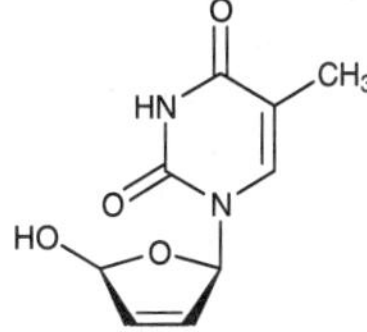

SAMPLE
Matrix: solutions

CAPILLARY ELECTROPHORESIS
Capillary: 80 cm × 75 μm fused-silica (52 cm to detector) (LC Packing, San Francisco)
Capillary temperature: 20 ± 0.5
Running buffer: 50 mM pH 6.5 Phosphate buffer containing 40 mM sodium dodecyl sulfate
Injection: Positive flow injection.
Detector: UV 260
Migration time: 7.7
Voltage: 20 kV
Current: 69 μA
Model: laboratory-constructed

OTHER SUBSTANCES
Simultaneous: didanosine, dideoxyadenosine, zalcitabine, zidovudine

REFERENCE
Singhal,R.; Xian,J.; Otim,O. Application of spherical and other polymers in capillary zone electrophoresis: separation of antiviral drugs and deoxyribonucleoside phosphates by different principles, *J.Chromatogr.A*, **1996**, *756*, 263–277.

Streptomycin

Molecular formula: $C_{21}H_{39}N_7O_{12}$
Molecular weight: 581.58
CAS Registry No.: 57-92-1, 3810-74-0 (sulfate)
Merck Index (12th ed.): 8983

SAMPLE
Matrix: bulk
Sample preparation: Inject an aliquot of an aqueous solution.

CAPILLARY ELECTROPHORESIS
Capillary: 90 cm $\times$ 50 μm uncoated fused silica (65 cm to detector) (Polymicro Technologies or ISCO)
Capillary preparation: Rinse with running buffer for 15 min at the beginning of the day and for 4 min before each injection.
Capillary temperature: 34
Running buffer: 75 mM pH 9 sodium tetraborate containing 0.5 mM myristyltrimethylammonium bromide
Injection: Vacuum injection at 25.0 kPa
Detector: UV 195
Migration time: 11
Voltage: -18 kV
Model: ISCO Model 3140 Electropherograph

OTHER SUBSTANCES
Simultaneous: dihydrostreptomycin

KEY WORDS
detector is at anode

REFERENCE
Flurer,C.L. The analysis of aminoglycoside antibiotics by capillary electrophoresis, *J.Pharm.Biomed.Anal.*, **1995**, *13*, 809–816.

SAMPLE
Matrix: bulk
Sample preparation: Inject an aliquot of an aqueous solution.

CAPILLARY ELECTROPHORESIS
Capillary: 90 cm $\times$ 50 μm uncoated fused silica (65 cm to detector) (Polymicro Technologies or ISCO)
Capillary preparation: Rinse with running buffer for 15 min at the beginning of the day and for 4 min before each injection.
Capillary temperature: 34
Running buffer: 160 mM pH 9 sodium tetraborate
Injection: Vacuum injection at 25.0 kPa
Detector: UV 195
Migration time: 16
Voltage: +18 kV
Model: ISCO Model 3140 Electropherograph

OTHER SUBSTANCES
Simultaneous: dihydrostreptomycin

KEY WORDS
detector is at cathode

REFERENCE
Flurer,C.L. The analysis of aminoglycoside antibiotics by capillary electrophoresis, *J.Pharm.Biomed.Anal.*, **1995**, *13*, 809–816.

SAMPLE
Matrix: solutions
Sample preparation: Prepare a solution in 100 mM cetyltrimethylammonium bromide, inject an aliquot.

CAPILLARY ELECTROPHORESIS
Capillary: 67 cm × 50 μm fused silica capillary (60 cm to detector) (Siemens)
Running buffer: pH 5.0 Acetate buffer containing 10 mM imidazole and 50 μg/mL FC 135 (a fluorinated surfactant, Fluorad/3M)
Injection: Pressure injection for 2 s.
Detector: UV 214
Migration time: 9.5
Voltage: 12.5 kV
Model: Beckman P/ACE System 2000

OTHER SUBSTANCES
Simultaneous: amikacin, gentamicin, kanamycin, lividomycin, neomycin, paromomycin, sisomycin, tobramycin
Interfering: dihydrostreptomycin

KEY WORDS
indirect UV detection; detector is at anode

REFERENCE
Ackermans,M.T.; Everaerts,F.M.; Beckers,J.L. Determination of aminoglycoside antibiotics in pharmaceuticals by capillary zone electrophoresis with indirect UV detection coupled with micellar electrokinetic capillary chromatography, *J.Chromatogr.*, **1992**, *606*, 229–235.

Sufentanil

Molecular formula: $C_{22}H_{30}N_2O_2S$
Molecular weight: 386.56
CAS Registry No.: 56030-54-7, 60561-17-3 (citrate)
Merck Index (12th ed.): 9056
Lednicer: 3 118

SAMPLE
Matrix: blood
Sample preparation: 1 mL Plasma + 50 μL Helix pomatia juice (Sepracor, France), mix, heat at 40° for 20 min, cool, add 2 mL n-heptane:ethyl acetate 50:50, shake at 60 cycles/min for 15 min, centrifuge at 1400 rpm for 6 min. Remove the organic layer and evaporate it to dryness under a stream of nitrogen at 40°, reconstitute with 50 μL MeOH:water 10:90, inject an aliquot.

CAPILLARY ELECTROPHORESIS
Capillary: 58 cm × 50 μm (50 cm to detector) (Hewlett-Packard)
Capillary temperature: 22

Running buffer: 100 mM MOPS adjusted to pH 7.00 with 100 mM NaOH
Injection: Hydrostatic injection at 50 mbar for 30 s.
Detector: UV 195
Migration time: 3
Voltage: 20 kV
Model: Hewlett-Packard HP 3D CE
Limit of quantitation: 8 ng/mL
Limit of detection: 3 ng/mL

KEY WORDS
plasma

REFERENCE
Rovio,S.; Sirén,H.; Riekkola,M.-L. Determination of sufentanil in human plasma by capillary electrophoresis
and gas chromatography-mass spectrometry, *J.Liq.Chromatogr.Rel.Technol.*, **1997**, *20*, 1311–1326.

Sulconazole

Molecular formula: $C_{18}H_{15}Cl_3N_2S$
Molecular weight: 397.75
CAS Registry No.: 61318-90-9, 61318-91-0 (nitrate)
Merck Index (12th ed.): 9062
Lednicer: 3 133

SAMPLE
Matrix: solutions
Sample preparation: Inject an aliquot of a 100 μg/mL solution in MeOH:water 10:90.

CAPILLARY ELECTROPHORESIS
Capillary: 47 cm × 75 μm fused-silica (40 cm to detector)
Capillary preparation: Rinse capillary with running buffer for 1 min before each run. At the
end of each day wash with 100 mM NaOH and with water.
Capillary temperature: 20
Running buffer: 20 mM pH 2.9 Phosphate buffer containing 3% chondroitin sulfate C (sodium
salt) (Nacalai Tesque, Kyoto)
Injection: Pressure injection at 0.5 psi for 5 s.
Detector: UV 240
Migration time: 18.68, 18.98 (enantiomers)
Voltage: 20 kV
Model: Beckman P/ACE 5510

OTHER SUBSTANCES
Also analyzed: clentiazem, diltiazem, primaquine, propranolol, trimetoquinol, verapamil

KEY WORDS
chiral

REFERENCE
Nishi,H. Enantiomer separation of basic drugs by capillary electrophoresis using ionic and neutral polysaccharides as chiral selectors, *J.Chromatogr.A*, **1996**, *735*, 345–351.

SAMPLE
Matrix: solutions
Sample preparation: Inject an aliquot of a 100 μg/mL solution in water or running buffer.

CAPILLARY ELECTROPHORESIS
Capillary: 47 cm × 75 μm fused-silica (40 cm to detector)

Capillary preparation: Before each run rinse capillary with running buffer for 1-2 min. If peak tailing is observed fill capillary with 500 mM NaOH and let stand for 30 min, wash with water for 3 min, rinse with running buffer for 3 min.
Capillary temperature: 25
Running buffer: 20 mM pH 2.5 Phosphate buffer containing 3% dextrin (Japan Pharmacopeia grade) (Dissolve dextrin in buffer at 90° then cool to room temperature.)
Injection: Pressure injection at 0.5 psi for 2-4 s.
Detector: UV 220
Migration time: 7.252, 7.496 (enantiomers)
Voltage: 20 kV
Model: Beckman P/ACE 5510

KEY WORDS
chiral

REFERENCE
Nishi,H.; Izumoto,S.; Nakamura,K.; Nakai,H.; Sato,T. Dextran and dextrin as chiral selectors in capillary zone electrophoresis, *Chromatographia*, **1996**, *42*, 617–630.

Sulfabenzamide

Molecular formula: $C_{13}H_{12}N_2O_3S$
Molecular weight: 276.32
CAS Registry No.: 127-71-9
Merck Index (12th ed.): 9065
Lednicer: 2 112

SAMPLE
Matrix: solutions

CAPILLARY ELECTROPHORESIS
Capillary: 47 cm × 50 μm fused-silica (40 cm to detector) (Polymicro Technologies)
Capillary preparation: Before each run rinse capillary with 100 mM ammonium hydroxide at high pressure for 2 min, with water for 2-3 min, and with running buffer for 2-3 min.
Capillary temperature: 26
Running buffer: MeCN:pH 4 acetate buffer containing 80 mM silver nitrate 15:85
Injection: Hydrodynamic injection for 3-5 s
Detector: UV 254, UV 280
Migration time: 14.7
Voltage: 20 kV
Model: Beckman P/ACE 2100

OTHER SUBSTANCES
Simultaneous: sulfacetamide, sulfadiazine, sulfadimethoxine, sulfaguanidine, sulfamerazine, sulfanilamide, sulfaquinoxaline, sulfathiazole

REFERENCE
Wright,P.B.; Dorsey,J.G. Silver(I)-mediated separations by capillary zone electrophoresis and micellar electrokinetic chromatography: Argentation electrophoresis, *Anal.Chem.*, **1996**, *68*, 415–424.

Sulfacetamide

Molecular formula: $C_8H_{10}N_2O_3S$
Molecular weight: 214.25
CAS Registry No.: 144-80-9, 127-56-0 (Na salt), 6209-17-2 (Na salt monohydrate)
Merck Index (12th ed.): 9067
Lednicer: 1 123

SAMPLE
Matrix: solutions

CAPILLARY ELECTROPHORESIS
Capillary: 47 cm $\times$ 50 μm fused-silica (40 cm to detector) (Polymicro Technologies)
Capillary preparation: Before each run rinse capillary with 100 mM ammonium hydroxide at high pressure for 2 min, with water for 2-3 min, and with running buffer for 2-3 min.
Capillary temperature: 26
Running buffer: MeCN:pH 4 acetate buffer containing 80 mM silver nitrate 15:85
Injection: Hydrodynamic injection for 3-5 s
Detector: UV 254, UV 280
Migration time: 13.4
Voltage: 20 kV
Model: Beckman P/ACE 2100

OTHER SUBSTANCES
Simultaneous: sulfabenzamide, sulfadiazine, sulfadimethoxine, sulfaguanidine, sulfamerazine, sulfanilamide, sulfaquinoxaline, sulfathiazole

REFERENCE
Wright,P.B.; Dorsey,J.G. Silver(I)-mediated separations by capillary zone electrophoresis and micellar electrokinetic chromatography: Argentation electrophoresis, *Anal.Chem.*, **1996**, *68*, 415–424.

SAMPLE
Matrix: solutions

CAPILLARY ELECTROPHORESIS
Capillary: 60 cm $\times$ 50 μm fused-silica (47 cm to detector) (Polymicro Technologies)
Capillary temperature: 25
Running buffer: 20 mM pH 8.5 Sodium tetraborate containing 100 mM sodium dodecyl sulfate
Injection: Hydrodynamic injection at 50 mbar for 3.6 s (5 nL).
Detector: UV 205
Migration time: 9.5
Voltage: 22 kV
Model: Crystal 310 (Thermo Unicam)

OTHER SUBSTANCES
Simultaneous: amoxicillin, ampicillin, cephapirin, cloxacillin, dicloxacillin, oxacillin, penicillin G, penicillin V, piperacillin, pyrimethamine, sulfadimethoxine, sulfaguanidine, sulfamerazine, sulfameter, sulfamethazine, sulfanilamide, sulfanilic acid, sulfapyridine, sulfaquinoxaline, sulfathiazole, sulfisoxazole, trimethoprim

REFERENCE
Hows,M.E.P.; Perrett,D.; Kay,J. Optimization of a simultaneous separation of sulphonamides, dihydrofolate reductase inhibitors and β-lactam antibiotics by capillary electrophoresis, *J.Chromatogr.A*, **1997**, *768*, 97–104.

Sulfachlorpyridazine

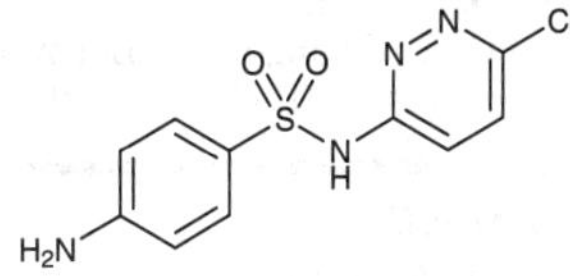

Molecular formula: $C_{10}H_9ClN_4O_2S$
Molecular weight: 284.73
CAS Registry No.: 80-32-0
Lednicer: 1 124

SAMPLE
Matrix: milk
Sample preparation: 5 mL Milk + 5 mL 3 M HCl, mix, centrifuge for 20 min. Remove the upper liquid layer and add it to 1 mL 3 M HCl, mix, centrifuge for 20 min, filter (0.45 μm) the supernatant, inject an aliquot of the filtrate.

CAPILLARY ELECTROPHORESIS
Capillary: 44 cm × 50 μm fused-silica (37 cm to detector)
Capillary preparation: Wash with running buffer for 3 min before each run.
Capillary temperature: 25
Running buffer: MeCN:buffer 9:91 (A) or buffer containing 0.5 mM β-cyclodextrin (B) (Prepare buffer by mixing 50 mM disodium tetraborate with 50 mM NaH_2PO_4 so as to produce a pH of 6.85.)
Injection: Hydrodynamic injection for 1 s.
Detector: UV 214, UV 254
Migration time: 5.2 (A); 3.5 (B)
Voltage: 20 kV
Model: Spectra-Physics Model 1000

OTHER SUBSTANCES
Simultaneous: sulfadiazine, sulfadimethoxine, sulfamerazine, sulfameter, sulfamethazine, sulfamethoxazole, sulfamethoxypyridazine, sulfamonomethoxine, sulfaquinoxaline, sulfathiazole, sulfisomidine, sulfisoxazole

KEY WORDS
cow

REFERENCE
Lin,C.-E.; Lin,W.-c.; Chiou,W.-c.; Lin,E.C.; Chang,C.-C. Migration behavior and separation of sulfonamides in capillary zone electrophoresis. I. Influence of buffer pH and electrolyte modifier, *J.Chromatogr.A*, **1996**, *755*, 261–269.

SAMPLE
Matrix: solutions

CAPILLARY ELECTROPHORESIS
Capillary: 43 cm × 50 μm fused-silica (36 cm to detector)
Capillary preparation: Wash with running buffer for 3 min before each injection and with water for 3 min after each injection. Wash a new capillary with 1 M NaOH at 60° for 50 min, with 100 mM NaOH at 60° for 10 min, and with water at 25° for 10 min.
Capillary temperature: 25
Running buffer: 30 mM pH 6.9 Citrate buffer
Injection: Hydrodynamic injection for 2 s
Detector: UV 254
Migration time: 3.55
Voltage: 20 kV
Model: Spectra-Physics Model 1000

OTHER SUBSTANCES
Simultaneous: sulfadiazine, sulfadimethoxine, sulfamerazine, sulfameter, sulfamethazine, sulfamethoxazole, sulfamethoxypyridazine, sulfamonomethoxine, sulfaquinoxaline, sulfathiazole, sulfisomidine, sulfisoxazole

REFERENCE

Lin,C.-E.; Chang,C.-C.; Lin,W.-c. Migration behavior and separation of sulfonamides in capillary zone electrophoresis. III. Citrate buffer as a background electrolyte, *J.Chromatogr.A*, **1997**, *768*, 105–112.

SAMPLE

Matrix: solutions

CAPILLARY ELECTROPHORESIS

Capillary: 44 cm × 50 μm fused-silica (37 cm to detector)

Capillary preparation: Fill with running buffer for 3 min before each injection then after each run wash with water for 2 min, with 1 M NaOH for 3 min, , and with water for 2 min. Wash new capillaries with water at 60° for 10 min, with 1 M NaOH at 60° for 1 h, and with water at 25° for 10 min.

Running buffer: MeOH:pH 6.85 50 mM phosphate/borate buffer 16.6:83.4 (A) or pH 5.8 phosphate/borate buffer containing 40 mM sodium dodecyl sulfate (B) (Prepare buffer by mixing 50 mM disodium tetraborate and 50 mM NaH_2PO_4.)

Injection: Hydrodynamic injection for 1 s.

Detector: UV 214

Migration time: 5.3 (A), 5.4 (B)

Voltage: 20 kV

Model: Spectra Physics Model 1000

OTHER SUBSTANCES

Simultaneous: sulfadiazine, sulfadimethoxine, sulfamerazine, sulfameter, sulfamethazine, sulfamethoxazole, sulfamethoxypyridazine, sulfamonomethoxine, sulfaquinoxaline, sulfathiazole, sulfisomidine, sulfisoxazole

REFERENCE

Lin,C.-E.; Lin,W.-c.; Chen,Y.-C.; Wang,S.-W. Migration behavior and selectivity of sulfonamides in capillary electrophoresis, *J.Chromatogr.A*, **1997**, *792*, 37–47.

SAMPLE

Matrix: solutions

CAPILLARY ELECTROPHORESIS

Capillary: 67 cm × 50 μm fused-silica (60 cm to detector)

Capillary preparation: Wash with running buffer for 6 min before each run.

Capillary temperature: 25

Running buffer: 500 mM pH 2.1 Sodium citrate buffer

Injection: Hydrodynamic injection for 2 s.

Detector: UV 254

Migration time: 39

Voltage: 30 kV

Current: 84 μA

Model: Spectra-Physics Model 1000

OTHER SUBSTANCES

Simultaneous: sulfadiazine, sulfadimethoxine, sulfamerazine, sulfameter, sulfamethazine, sulfamethizole, sulfamethoxazole, sulfamethoxypyridazine, sulfamonomethoxine, sulfanilamide, sulfapyridine, sulfaquinoxaline, sulfathiazole, sulfisomidine, sulfisoxazole

REFERENCE

Lin,C.-E.; Chang,C.-C.; Lin,W.-c. Migration behavior and separation of sulfonamides in capillary zone electrophoresis. II. Positively charged species at low pH, *J.Chromatogr.A*, **1997**, *759*, 203–209.

Sulfadiazine

Molecular formula: $C_{10}H_{10}N_4O_2S$
Molecular weight: 250.28
CAS Registry No.: 68-35-9, 22199-08-2 (Ag salt), 547-32-0 (Na salt)
Merck Index (12th ed.): 9071
Lednicer: 1 124

SAMPLE
Matrix: milk
Sample preparation: 5 mL Milk + 5 mL 3 M HCl, mix, centrifuge for 20 min. Remove the upper liquid layer and add it to 1 mL 3 M HCl, mix, centrifuge for 20 min, filter (0.45 μm) the supernatant, inject an aliquot of the filtrate.

CAPILLARY ELECTROPHORESIS
Capillary: 44 cm × 50 μm fused-silica (37 cm to detector)
Capillary preparation: Wash with running buffer for 3 min before each run.
Capillary temperature: 25
Running buffer: MeCN:buffer 9:91 (A) or buffer containing 0.5 mM β-cyclodextrin (B) (Prepare buffer by mixing 50 mM disodium tetraborate with 50 mM NaH_2PO_4 so as to produce a pH of 6.85.)
Injection: Hydrodynamic injection for 1 s.
Detector: UV 214, UV 254
Migration time: 4.5 (A); 3.1 (B)
Voltage: 20 kV
Model: Spectra-Physics Model 1000

OTHER SUBSTANCES
Simultaneous: sulfachlorpyridazine, sulfadimethoxine, sulfamerazine, sulfameter, sulfamethazine, sulfamethoxazole, sulfamethoxypyridazine, sulfamonomethoxine, sulfaquinoxaline, sulfathiazole, sulfisomidine, sulfisoxazole

KEY WORDS
cow

REFERENCE
Lin,C.-E.; Lin,W.-c.; Chiou,W.-c.; Lin,E.C.; Chang,C.-C. Migration behavior and separation of sulfonamides in capillary zone electrophoresis. I. Influence of buffer pH and electrolyte modifier, *J.Chromatogr.A*, **1996**, *755*, 261–269.

SAMPLE
Matrix: solutions
Sample preparation: Prepare a 100-200 ppm solution in MeOH, inject an aliquot.

CAPILLARY ELECTROPHORESIS
Capillary: 45 cm × 50 μm fused-silica (Polymicro Technologies)
Running buffer: 50 mM pH 7.50 Phosphate buffer containing 25 mM borate and 9 mM tetrabutylammonium bromide
Injection: Hydrodynamic injection at 5 cm for 5 s (1 nL)
Detector: UV 240
Migration time: 6
Voltage: 15 kV
Current: 39 μA
Model: Laboratory constructed

OTHER SUBSTANCES
Simultaneous: chloroquine, dapsone, primaquine, pyrimethamine, quinacrine, quinine

REFERENCE
Ng,C.L.; Toh,Y.L.; Li,S.F.Y.; Lee,H.K. Capillary electrophoresis of biologically important compounds: Optimization of separation conditions by the overlapping resolution mapping scheme, *J.Liq.Chromatogr.*, **1993**, *16*, 3653–3666.

SAMPLE
Matrix: solutions

CAPILLARY ELECTROPHORESIS
Capillary: 65 cm × 50 μm fused-silica (45 cm to detector)
Capillary preparation: Wash with 100 μL 100 mM NaOH for 1 min, flush with 100 μL running buffer for 1 min, allow to equilibrate under voltage for 5 min.
Running buffer: 50 mM pH 9.1 disodium tetraborate
Injection: Hydrodynamic for 5 s
Detector: UV 270
Migration time: 4.7
Voltage: 24 kV
Model: Isco Model 3850

OTHER SUBSTANCES
Simultaneous: aminobenzoic acid, thymidine, uracil

REFERENCE
Richards,R.M.E.; Xing,D.K.L. Determination of thymidine, uracil and p-aminobenzoic acid in bacteriological cultures by capillary zone electrophoresis, *J.Pharm.Biomed.Anal.*, **1994**, *12*, 1063–1068.

SAMPLE
Matrix: solutions

CAPILLARY ELECTROPHORESIS
Capillary: 50 cm × 50 μm fused-silica (Tokyo Kasei)
Capillary preparation: Rinse capillary with 1 M NaOH after each run.
Running buffer: MeOH:buffer 10:90 (Buffer was 10 mM pH 7 ammonium formate containing 2% butyl acrylate-butyl methacrylate-methacrylic acid copolymer sodium salt (BBMA) (Dai-ichi Kogyo Seiyaku)
Injection: Hydrostatic injection at 15 cm for 10-30 s
Detector: MS, Hitachi M-1000 quadrupole, electrospray at 3 kV, details of interface given in paper, sheath liquid MeOH:water:formic acid 50:50:1 at 5 μL/min, positive-ion mode, drift voltage 70 V, focusing voltage 140-150 V
Migration time: 11.3
Voltage: 10 kV
Model: laboratory constructed

OTHER SUBSTANCES
Simultaneous: sulfamethazine, sulfisomidine, sulfisoxazole

REFERENCE
Ozaki,H.; Itou,N.; Terabe,S.; Takada,Y.; Sakairi,M.; Koizumi,H. Micellar electrokinetic chromatography-mass spectrometry using a high-molecular-mass surfactant. On-line coupling with an electrospray ionization interface, *J.Chromatogr.A*, **1995**, *716*, 69–79.

SAMPLE
Matrix: solutions

CAPILLARY ELECTROPHORESIS
Capillary: 47 cm × 50 μm fused-silica (40 cm to detector) (Polymicro Technologies)
Capillary preparation: Before each run rinse capillary with 100 mM ammonium hydroxide at high pressure for 2 min, with water for 2-3 min, and with running buffer for 2-3 min.
Capillary temperature: 26
Running buffer: MeCN:pH 4 acetate buffer containing 80 mM silver nitrate 15:85
Injection: Hydrodynamic injection for 3-5 s

Detector: UV 254, UV 280
Migration time: 10.6
Voltage: 20 kV
Model: Beckman P/ACE 2100

OTHER SUBSTANCES
Simultaneous: sulfabenzamide, sulfacetamide, sulfadimethoxine, sulfaguanidine, sulfamerazine, sulfanilamide, sulfaquinoxaline, sulfathiazole

REFERENCE
Wright,P.B.; Dorsey,J.G. Silver(I)-mediated separations by capillary zone electrophoresis and micellar electrokinetic chromatography: Argentation electrophoresis, *Anal.Chem.*, **1996**, *68*, 415–424.

SAMPLE
Matrix: solutions

CAPILLARY ELECTROPHORESIS
Capillary: 67 cm × 50 μm fused-silica (60 cm to detector)
Capillary preparation: Wash with running buffer for 6 min before each run.
Capillary temperature: 25
Running buffer: 500 mM pH 2.1 Sodium citrate buffer
Injection: Hydrodynamic injection for 2 s.
Detector: UV 254
Migration time: 29
Voltage: 30 kV
Current: 84 μA
Model: Spectra-Physics Model 1000

OTHER SUBSTANCES
Simultaneous: sulfachlorpyridazine, sulfadimethoxine, sulfamerazine, sulfameter, sulfamethazine, sulfamethizole, sulfamethoxazole, sulfamethoxypyridazine, sulfamonomethoxine, sulfanilamide, sulfapyridine, sulfaquinoxaline, sulfathiazole, sulfisomidine, sulfisoxazole

REFERENCE
Lin,C.-E.; Chang,C.-C.; Lin,W.-c. Migration behavior and separation of sulfonamides in capillary zone electrophoresis. II. Positively charged species at low pH, *J.Chromatogr.A*, **1997**, *759*, 203–209.

SAMPLE
Matrix: solutions

CAPILLARY ELECTROPHORESIS
Capillary: 43 cm × 50 μm fused-silica (36 cm to detector)
Capillary preparation: Wash with running buffer for 3 min before each injection and with water for 3 min after each injection. Wash a new capillary with 1 M NaOH at 60° for 50 min, with 100 mM NaOH at 60° for 10 min, and with water at 25° for 10 min.
Capillary temperature: 25
Running buffer: 30 mM pH 6.9 Citrate buffer
Injection: Hydrodynamic injection for 2 s
Detector: UV 254
Migration time: 3.2
Voltage: 20 kV
Model: Spectra-Physics Model 1000

OTHER SUBSTANCES
Simultaneous: sulfachlorpyridazine, sulfadimethoxine, sulfamerazine, sulfameter, sulfamethazine, sulfamethoxazole, sulfamethoxypyridazine, sulfamonomethoxine, sulfaquinoxaline, sulfathiazole, sulfisomidine, sulfisoxazole

REFERENCE
Lin,C.-E.; Chang,C.-C.; Lin,W.-c. Migration behavior and separation of sulfonamides in capillary zone electrophoresis. III. Citrate buffer as a background electrolyte, *J.Chromatogr.A*, **1997**, *768*, 105–112.

SAMPLE
Matrix: solutions

CAPILLARY ELECTROPHORESIS
Capillary: 44 cm × 50 μm fused-silica (37 cm to detector)
Capillary preparation: Fill with running buffer for 3 min before each injection then after each run wash with water for 2 min, with 1 M NaOH for 3 min, , and with water for 2 min. Wash new capillaries with water at 60° for 10 min, with 1 M NaOH at 60° for 1 h, and with water at 25° for 10 min.
Running buffer: MeOH:pH 6.85 50 mM phosphate/borate buffer 16.6:83.4 (A) or pH 5.8 phosphate/borate buffer containing 40 mM sodium dodecyl sulfate (B) (Prepare buffer by mixing 50 mM disodium tetraborate and 50 mM NaH_2PO_4.)
Injection: Hydrodynamic injection for 1 s.
Detector: UV 214
Migration time: 4.5 (A), 4.0 (B)
Voltage: 20 kV
Model: Spectra Physics Model 1000

OTHER SUBSTANCES
Simultaneous: sulfachloropyridazine, sulfadimethoxine, sulfamerazine, sulfameter, sulfamethazine, sulfamethoxazole, sulfamethoxypyridazine, sulfamonomethoxine, sulfaquinoxaline, sulfathiazole, sulfisomidine, sulfisoxazole

REFERENCE
Lin,C.-E.; Lin,W.-c.; Chen,Y.-C.; Wang,S.-W. Migration behavior and selectivity of sulfonamides in capillary electrophoresis, *J.Chromatogr.A*, **1997**, *792*, 37–47.

SAMPLE
Matrix: tissue
Sample preparation: Homogenize (household food processor) 10 g tissue, extract (stomacher apparatus) with 100 mL MeCN for 5 min, centrifuge at 4000 g for 10 min, filter (0.45 μm), inject an aliquot of the filtrate.

CAPILLARY ELECTROPHORESIS
Capillary: 57 cm × 75 μm (50 cm to detector) (Beckman)
Capillary temperature: 25
Running buffer: 20 mM Borate buffer containing 20 mM phosphate, adjusted to pH 7.0 with KOH
Injection: Pressure injection for 2 s (39 nL)
Detector: UV 254
Migration time: 19
Voltage: 10 kV
Current: 26.8 μA
Model: Beckman P/ACE System 2000

OTHER SUBSTANCES
Extracted: sulfachlorpyridazine, sulfadimethoxine, sulfadoxine, sulfamerazine, sulfamethazine (sulfadimidine), sulfamethoxazole, sulfamethoxydiazine, sulfamethoxypyridazine, sulfaquinoxaline, sulfathiazole, sulfatroxazole, trimethoprim

KEY WORDS
pig

REFERENCE
Ackermans,M.T.; Beckers,J.L.; Everaerts,F.M.; Hoogland,H.; Tomassen,M.J.H. Determination of sulphonamides in pork meat extracts by capillary zone electrophoresis, *J.Chromatogr.*, **1992**, *596*, 101–109.

Sulfadimethoxine

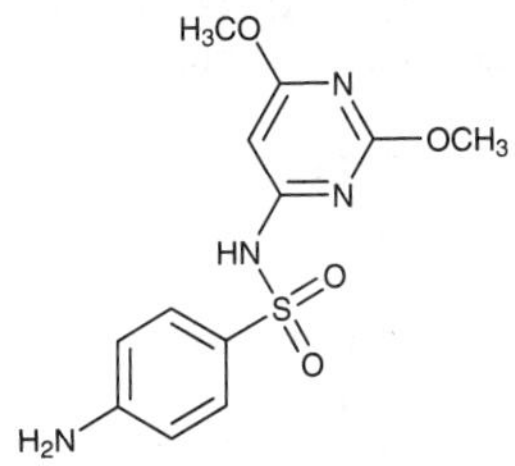

Molecular formula: $C_{12}H_{14}N_4O_4S$
Molecular weight: 310.33
CAS Registry No.: 122-11-2
Merck Index (12th ed.): 9073
Lednicer: 1 125

SAMPLE

Matrix: milk
Sample preparation: 5 mL Milk + 5 mL 3 M HCl, mix, centrifuge for 20 min. Remove the upper liquid layer and add it to 1 mL 3 M HCl, mix, centrifuge for 20 min, filter (0.45 μm) the supernatant, inject an aliquot of the filtrate.

CAPILLARY ELECTROPHORESIS

Capillary: 44 cm × 50 μm fused-silica (37 cm to detector)
Capillary preparation: Wash with running buffer for 3 min before each run.
Capillary temperature: 25
Running buffer: MeCN:buffer 9:91 (A) or buffer containing 0.5 mM β-cyclodextrin (B) (Prepare buffer by mixing 50 mM disodium tetraborate with 50 mM NaH_2PO_4 so as to produce a pH of 6.85.)
Injection: Hydrodynamic injection for 1 s.
Detector: UV 214, UV 254
Migration time: 4.6 (A); 3.25 (B)
Voltage: 20 kV
Model: Spectra-Physics Model 1000

OTHER SUBSTANCES

Simultaneous: sulfachlorpyridazine, sulfadiazine, sulfamerazine, sulfameter, sulfamethazine, sulfamethoxazole, sulfamethoxypyridazine, sulfamonomethoxine, sulfaquinoxaline, sulfathiazole, sulfisomidine, sulfisoxazole

KEY WORDS

cow

REFERENCE

Lin,C.-E.; Lin,W.-c.; Chiou,W.-c.; Lin,E.C.; Chang,C.-C. Migration behavior and separation of sulfonamides in capillary zone electrophoresis. I. Influence of buffer pH and electrolyte modifier, *J.Chromatogr.A*, **1996**, *755*, 261–269.

SAMPLE

Matrix: solutions
Sample preparation: Prepare a solution in running buffer, inject an aliquot.

CAPILLARY ELECTROPHORESIS

Capillary: 100 cm × 75 μm fused-silica (20 cm to UV detector) (Polymicro Technologies)
Capillary preparation: Between runs flush capillaries with 1.5 column volumes of 100 mM NaOH and 9 column volumes of running buffer. Flush new capillaries with 9 column volumes of 1 M NaOH, 3 column volumes of water, 3 column volumes of 100 mM HCl, 3 column volumes of water, and 9 column volumes of running buffer.
Running buffer: MeOH:water 20:80 containing 20 mM ammonium acetate, pH 6.8
Injection: Pressure injection at 3.45 kPa (5-28 nL)
Detector: UV 254 or MS, Sciex TAGA 6000E triple quadrupole, API source, electrospray (ion spray) interface at 4 kV (3 kV to ion-sampling orifice), positive-ion mode, 100 μm dia sampling orifice, nitrogen gas curtain, argon collision gas, m/z 311 [M+H]+
Migration time: 4.8 (UV), 22.5 (MS)
Voltage: 26 kV
Model: Beckman P/ACE System 2000

OTHER SUBSTANCES
Simultaneous: sulfamerazine, sulfamethazine, sulfamethoxazole, sulfanilamide, sulfathiazole

REFERENCE
Johansson,I.M.; Pavelka,R.; Henion,J.D. Determination of small drug molecules by capillary electrophoresis-atmospheric pressure ionization mass spectrometry, *J.Chromatogr.*, **1991**, *559*, 515–528.

SAMPLE
Matrix: solutions

CAPILLARY ELECTROPHORESIS
Capillary: 47 cm × 50 μm fused-silica (40 cm to detector) (Polymicro Technologies)
Capillary preparation: Before each run rinse capillary with 100 mM ammonium hydroxide at high pressure for 2 min, with water for 2-3 min, and with running buffer for 2-3 min.
Capillary temperature: 26
Running buffer: MeCN:pH 4 acetate buffer containing 80 mM silver nitrate 15:85
Injection: Hydrodynamic injection for 3-5 s
Detector: UV 254, UV 280
Migration time: 9.4
Voltage: 20 kV
Model: Beckman P/ACE 2100

OTHER SUBSTANCES
Simultaneous: sulfabenzamide, sulfacetamide, sulfadiazine, sulfaguanidine, sulfamerazine, sulfanilamide, sulfaquinoxaline, sulfathiazole

REFERENCE
Wright,P.B.; Dorsey,J.G. Silver(I)-mediated separations by capillary zone electrophoresis and micellar electrokinetic chromatography: Argentation electrophoresis, *Anal.Chem.*, **1996**, *68*, 415–424.

SAMPLE
Matrix: solutions

CAPILLARY ELECTROPHORESIS
Capillary: 60 cm × 50 μm fused-silica (47 cm to detector) (Polymicro Technologies)
Capillary temperature: 25
Running buffer: 20 mM pH 8.5 Sodium tetraborate containing 100 mM sodium dodecyl sulfate
Injection: Hydrodynamic injection at 50 mbar for 3.6 s (5 nL).
Detector: UV 205
Migration time: 9.6
Voltage: 22 kV
Model: Crystal 310 (Thermo Unicam)

OTHER SUBSTANCES
Simultaneous: amoxicillin, ampicillin, cephapirin, cloxacillin, dicloxacillin, oxacillin, penicillin G, penicillin V, piperacillin, pyrimethamine, sulfacetamide, sulfaguanidine, sulfameter, sulfamethazine, sulfanilamide, sulfanilic acid, sulfapyridine, sulfaquinoxaline, sulfathiazole, sulfisoxazole, trimethoprim
Interfering: sulfamerazine

REFERENCE
Hows,M.E.P.; Perrett,D.; Kay,J. Optimization of a simultaneous separation of sulphonamides, dihydrofolate reductase inhibitors and β-lactam antibiotics by capillary electrophoresis, *J.Chromatogr.A*, **1997**, *768*, 97–104.

SAMPLE
Matrix: solutions

CAPILLARY ELECTROPHORESIS
Capillary: 67 cm × 50 μm fused-silica (60 cm to detector)

Capillary preparation: Wash with running buffer for 6 min before each run.
Capillary temperature: 25
Running buffer: 500 mM pH 2.1 Sodium citrate buffer
Injection: Hydrodynamic injection for 2 s.
Detector: UV 254
Migration time: 43
Voltage: 30 kV
Current: 84 μA
Model: Spectra-Physics Model 1000

OTHER SUBSTANCES
Simultaneous: sulfachlorpyridazine, sulfadiazine, sulfamerazine, sulfameter, sulfamethazine, sulfamethizole, sulfamethoxazole, sulfamethoxypyridazine, sulfamonomethoxine, sulfanilamide, sulfapyridine, sulfaquinoxaline, sulfathiazole, sulfisomidine, sulfisoxazole

REFERENCE
Lin,C.-E.; Chang,C.-C.; Lin,W.-c. Migration behavior and separation of sulfonamides in capillary zone electrophoresis. II. Positively charged species at low pH, *J.Chromatogr.A*, **1997**, *759*, 203–209.

SAMPLE
Matrix: solutions

CAPILLARY ELECTROPHORESIS
Capillary: 44 cm × 50 μm fused-silica (37 cm to detector)
Capillary preparation: Fill with running buffer for 3 min before each injection then after each run wash with water for 2 min, with 1 M NaOH for 3 min, , and with water for 2 min. Wash new capillaries with water at 60° for 10 min, with 1 M NaOH at 60° for 1 h, and with water at 25° for 10 min.
Running buffer: MeOH:pH 6.85 50 mM phosphate/borate buffer 16.6:83.4 (A) or pH 5.8 phosphate/borate buffer containing 40 mM sodium dodecyl sulfate (B) (Prepare buffer by mixing 50 mM disodium tetraborate and 50 mM NaH$_2$PO$_4$.)
Injection: Hydrodynamic injection for 1 s.
Detector: UV 214
Migration time: 4.8 (A), 7.9 (B)
Voltage: 20 kV
Model: Spectra Physics Model 1000

OTHER SUBSTANCES
Simultaneous: sulfachloropyridazine, sulfadiazine, sulfamerazine, sulfameter, sulfamethazine, sulfamethoxazole, sulfamethoxypyridazine, sulfamonomethoxine, sulfaquinoxaline, sulfathiazole, sulfisomidine, sulfisoxazole

REFERENCE
Lin,C.-E.; Lin,W.-c.; Chen,Y.-C.; Wang,S.-W. Migration behavior and selectivity of sulfonamides in capillary electrophoresis, *J.Chromatogr.A*, **1997**, *792*, 37–47.

SAMPLE
Matrix: solutions

CAPILLARY ELECTROPHORESIS
Capillary: 43 cm × 50 μm fused-silica (36 cm to detector)
Capillary preparation: Wash with running buffer for 3 min before each injection and with water for 3 min after each injection. Wash a new capillary with 1 M NaOH at 60° for 50 min, with 100 mM NaOH at 60° for 10 min, and with water at 25° for 10 min.
Capillary temperature: 25
Running buffer: 30 mM pH 6.9 Citrate buffer
Injection: Hydrodynamic injection for 2 s
Detector: UV 254
Migration time: 3.25
Voltage: 20 kV
Model: Spectra-Physics Model 1000

OTHER SUBSTANCES
Simultaneous: sulfachlorpyridazine, sulfadiazine, sulfamerazine, sulfameter, sulfamethazine, sulfamethoxazole, sulfamethoxypyridazine, sulfamonomethoxine, sulfaquinoxaline, sulfathiazole, sulfisomidine, sulfisoxazole

REFERENCE
Lin,C.-E.; Chang,C.-C.; Lin,W.-c. Migration behavior and separation of sulfonamides in capillary zone electrophoresis. III. Citrate buffer as a background electrolyte, *J.Chromatogr.A*, **1997**, *768*, 105–112.

SAMPLE
Matrix: tissue
Sample preparation: Homogenize (household food processor) 10 g tissue, extract (stomacher apparatus) with 100 mL MeCN for 5 min, centrifuge at 4000 g for 10 min, filter (0.45 μm), inject an aliquot of the filtrate.

CAPILLARY ELECTROPHORESIS
Capillary: 57 cm × 75 μm (50 cm to detector) (Beckman)
Capillary temperature: 25
Running buffer: 20 mM Borate buffer containing 20 mM phosphate, adjusted to pH 7.0 with KOH
Injection: Pressure injection for 2 s (39 nL)
Detector: UV 254
Migration time: 11.3
Voltage: 10 kV
Current: 26.8 μA
Model: Beckman P/ACE System 2000

OTHER SUBSTANCES
Extracted: sulfachlorpyridazine, sulfadiazine, sulfadoxine, sulfamerazine, sulfamethazine (sulfadimidine), sulfamethoxazole, sulfamethoxydiazine, sulfamethoxypyridazine, sulfaquinoxaline, sulfathiazole, sulfatroxazole, trimethoprim

KEY WORDS
pig

REFERENCE
Ackermans,M.T.; Beckers,J.L.; Everaerts,F.M.; Hoogland,H.; Tomassen,M.J.H. Determination of sulphonamides in pork meat extracts by capillary zone electrophoresis, *J.Chromatogr.*, **1992**, *596*, 101–109.

Sulfadoxine

Molecular formula: $C_{12}H_{14}N_4O_4S$
Molecular weight: 310.33
CAS Registry No.: 2447-57-6
Merck Index (12th ed.): 9074
Lednicer: 1 125

SAMPLE
Matrix: tissue
Sample preparation: Homogenize (household food processor) 10 g tissue, extract (stomacher apparatus) with 100 mL MeCN for 5 min, centrifuge at 4000 g for 10 min, filter (0.45 μm), inject an aliquot of the filtrate.

CAPILLARY ELECTROPHORESIS
Capillary: 57 cm × 75 μm (50 cm to detector) (Beckman)
Capillary temperature: 25

Running buffer: 20 mM Borate buffer containing 20 mM phosphate, adjusted to pH 7.0 with KOH
Injection: Pressure injection for 2 s (39 nL)
Detector: UV 254
Migration time: 11.1
Voltage: 10 kV
Current: 26.8 µA
Model: Beckman P/ACE System 2000

OTHER SUBSTANCES
Extracted: sulfachlorpyridazine, sulfadiazine, sulfadimethoxine, sulfamerazine, sulfamethazine (sulfadimidine), sulfamethoxazole, sulfamethoxydiazine, sulfamethoxypyridazine, sulfaquinoxaline, sulfathiazole, sulfatroxazole, trimethoprim

KEY WORDS
pig

REFERENCE
Ackermans,M.T.; Beckers,J.L.; Everaerts,F.M.; Hoogland,H.; Tomassen,M.J.II. Determination of sulphonamides in pork meat extracts by capillary zone electrophoresis, *J.Chromatogr.*, **1992**, *596*, 101–109.

Sulfaguanidine

Molecular formula: $C_7H_{10}N_4O_2S$
Molecular weight: 214.25
CAS Registry No.: 57-67-0
Merck Index (12th ed.): 9076
Lednicer: 1 123

SAMPLE
Matrix: solutions

CAPILLARY ELECTROPHORESIS
Capillary: 60 cm × 50 µm fused-silica (47 cm to detector) (Polymicro Technologies)
Capillary temperature: 25
Running buffer: 20 mM pH 8.5 Sodium tetraborate containing 100 mM sodium dodecyl sulfate
Injection: Hydrodynamic injection at 50 mbar for 3.6 s (5 nL).
Detector: UV 205
Migration time: 7.3
Voltage: 22 kV
Model: Crystal 310 (Thermo Unicam)

OTHER SUBSTANCES
Simultaneous: amoxicillin, ampicillin, cephapirin, cloxacillin, dicloxacillin, oxacillin, penicillin G, penicillin V, piperacillin, pyrimethamine, sulfacetamide, sulfadimethoxine, sulfamerazine, sulfameter, sulfamethazine, sulfanilamide, sulfanilic acid, sulfapyridine, sulfaquinoxaline, sulfathiazole, sulfisoxazole, trimethoprim

REFERENCE
Hows,M.E.P.; Perrett,D.; Kay,J. Optimization of a simultaneous separation of sulphonamides, dihydrofolate reductase inhibitors and β-lactam antibiotics by capillary electrophoresis, *J.Chromatogr.A*, **1997**, *768*, 97–104.

Sulfalene

Molecular formula: $C_{11}H_{12}N_4O_3S$
Molecular weight: 280.31
CAS Registry No.: 152-47-6, 50933-06-7 (mixture with trimethoprim)
Merck Index (12th ed.): 9078
Lednicer: 1 125

SAMPLE
Matrix: milk
Sample preparation: 5 mL Milk + 5 mL 3 M HCl, mix, centrifuge for 20 min. Remove the upper liquid layer and add it to 1 mL 3 M HCl, mix, centrifuge for 20 min, filter (0.45 μm) the supernatant, inject an aliquot of the filtrate.

CAPILLARY ELECTROPHORESIS
Capillary: 44 cm × 50 μm fused-silica (37 cm to detector)
Capillary preparation: Wash with running buffer for 3 min before each run.
Capillary temperature: 25
Running buffer: MeCN:buffer 9:91 (A) or buffer containing 0.5 mM β-cyclodextrin (B) (Prepare buffer by mixing 50 mM disodium tetraborate with 50 mM NaH_2PO_4 so as to produce a pH of 6.85.)
Injection: Hydrodynamic injection for 1 s.
Detector: UV 214, UV 254
Migration time: 3.7 (A); 2.6 (B)
Voltage: 20 kV
Model: Spectra-Physics Model 1000

OTHER SUBSTANCES
Simultaneous: sulfachlorpyridazine, sulfadiazine, sulfadimethoxine, sulfamerazine, sulfameter, sulfamethazine, sulfamethoxazole, sulfamonomethoxine, sulfaquinoxaline, sulfathiazole, sulfisomidine, sulfisoxazole

KEY WORDS
cow

REFERENCE
Lin,C.-E.; Lin,W.-c.; Chiou,W.-c.; Lin,E.C.; Chang,C.-C. Migration behavior and separation of sulfonamides in capillary zone electrophoresis. I. Influence of buffer pH and electrolyte modifier, *J.Chromatogr.A*, **1996**, *755*, 261–269.

SAMPLE
Matrix: solutions

CAPILLARY ELECTROPHORESIS
Capillary: 44 cm × 50 μm fused-silica (37 cm to detector)
Capillary preparation: Fill with running buffer for 3 min before each injection then after each run wash with water for 2 min, with 1 M NaOH for 3 min, , and with water for 2 min. Wash new capillaries with water at 60° for 10 min, with 1 M NaOH at 60° for 1 h, and with water at 25° for 10 min.
Running buffer: MeOH:pH 6.85 50 mM phosphate/borate buffer 16.6:83.4 (A) or pH 5.8 phosphate/borate buffer containing 40 mM sodium dodecyl sulfate (B) (Prepare buffer by mixing 50 mM disodium tetraborate and 50 mM NaH_2PO_4.)
Injection: Hydrodynamic injection for 1 s.
Detector: UV 214
Migration time: 3.6 (A), 4.8 (B)
Voltage: 20 kV
Model: Spectra Physics Model 1000

OTHER SUBSTANCES
Simultaneous: sulfachloropyridazine, sulfadiazine, sulfadimethoxine, sulfamerazine, sulfameter, sulfamethazine, sulfamethoxazole, sulfamonomethoxine, sulfaquinoxaline, sulfathiazole, sulfisomidine, sulfisoxazole

REFERENCE
Lin,C.-E.; Lin,W.-c.; Chen,Y.-C.; Wang,S.-W. Migration behavior and selectivity of sulfonamides in capillary electrophoresis, *J.Chromatogr.A*, **1997**, *792*, 37–47.

SAMPLE
Matrix: solutions

CAPILLARY ELECTROPHORESIS
Capillary: 67 cm × 50 μm fused-silica (60 cm to detector)
Capillary preparation: Wash with running buffer for 6 min before each run.
Capillary temperature: 25
Running buffer: 500 mM pH 2.1 Sodium citrate buffer
Injection: Hydrodynamic injection for 2 s.
Detector: UV 254
Migration time: 28
Voltage: 30 kV
Current: 84 μA
Model: Spectra-Physics Model 1000

OTHER SUBSTANCES
Simultaneous: sulfachlorpyridazine, sulfadiazine, sulfadimethoxine, sulfamerazine, sulfameter, sulfamethazine, sulfamethizole, sulfamethoxazole, sulfamonomethoxine, sulfanilamide, sulfapyridine, sulfaquinoxaline, sulfathiazole, sulfisomidine, sulfisoxazole

REFERENCE
Lin,C.-E.; Chang,C.-C.; Lin,W.-c. Migration behavior and separation of sulfonamides in capillary zone electrophoresis. II. Positively charged species at low pH, *J.Chromatogr.A*, **1997**, *759*, 203–209.

SAMPLE
Matrix: solutions

CAPILLARY ELECTROPHORESIS
Capillary: 43 cm × 50 μm fused-silica (36 cm to detector)
Capillary preparation: Wash with running buffer for 3 min before each injection and with water for 3 min after each injection. Wash a new capillary with 1 M NaOH at 60° for 50 min, with 100 mM NaOH at 60° for 10 min, and with water at 25° for 10 min.
Capillary temperature: 25
Running buffer: 30 mM pH 6.9 Citrate buffer
Injection: Hydrodynamic injection for 2 s
Detector: UV 254
Migration time: 2.6
Voltage: 20 kV
Model: Spectra-Physics Model 1000

OTHER SUBSTANCES
Simultaneous: sulfachlorpyridazine, sulfadiazine, sulfadimethoxine, sulfamerazine, sulfameter, sulfamethazine, sulfamethoxazole, sulfamonomethoxine, sulfaquinoxaline, sulfathiazole, sulfisomidine, sulfisoxazole

REFERENCE
Lin,C.-E.; Chang,C.-C.; Lin,W.-c. Migration behavior and separation of sulfonamides in capillary zone electrophoresis. III. Citrate buffer as a background electrolyte, *J.Chromatogr.A*, **1997**, *768*, 105–112.

Sulfamerazine

Molecular formula: $C_{11}H_{12}N_4O_2S$
Molecular weight: 264.31
CAS Registry No.: 127-79-7, 127-58-2 (monosodium salt)
Merck Index (12th ed.): 9081
Lednicer: 1 124

SAMPLE
Matrix: solutions

CAPILLARY ELECTROPHORESIS
Capillary: 60 cm × 50 μm fused-silica (47 cm to detector) (Polymicro Technologies)
Capillary temperature: 25
Running buffer: 20 mM pH 8.5 Sodium tetraborate containing 100 mM sodium dodecyl sulfate
Injection: Hydrodynamic injection at 50 mbar for 3.6 s (5 nL).
Detector: UV 205
Migration time: 9.6
Voltage: 22 kV
Model: Crystal 310 (Thermo Unicam)

OTHER SUBSTANCES
Simultaneous: amoxicillin, ampicillin, cephapirin, cloxacillin, dicloxacillin, oxacillin, penicillin G, penicillin V, piperacillin, pyrimethamine, sulfacetamide, sulfaguanidine, sulfameter, sulfamethazine, sulfanilamide, sulfanilic acid, sulfapyridine, sulfaquinoxaline, sulfathiazole, sulfisoxazole, trimethoprim
Interfering: sulfadimethoxine

REFERENCE
Hows,M.E.P.; Perrett,D.; Kay,J. Optimization of a simultaneous separation of sulphonamides, dihydrofolate reductase inhibitors and β-lactam antibiotics by capillary electrophoresis, *J.Chromatogr.A*, **1997**, *768*, 97–104.

SAMPLE
Matrix: solutions

CAPILLARY ELECTROPHORESIS
Capillary: 43 cm × 50 μm fused-silica (36 cm to detector)
Capillary preparation: Wash with running buffer for 3 min before each injection and with water for 3 min after each injection. Wash a new capillary with 1 M NaOH at 60° for 50 min, with 100 mM NaOH at 60° for 10 min, and with water at 25° for 10 min.
Capillary temperature: 25
Running buffer: 30 mM pH 6.9 Citrate buffer
Injection: Hydrodynamic injection for 2 s
Detector: UV 254
Migration time: 2.7
Voltage: 20 kV
Model: Spectra-Physics Model 1000

OTHER SUBSTANCES
Simultaneous: sulfachlorpyridazine, sulfadiazine, sulfadimethoxine, sulfameter, sulfamethazine, sulfamethoxazole, sulfamethoxypyridazine, sulfamonomethoxine, sulfaquinoxaline, sulfathiazole, sulfisomidine, sulfisoxazole

REFERENCE
Lin,C.-E.; Chang,C.-C.; Lin,W.-c. Migration behavior and separation of sulfonamides in capillary zone electrophoresis. III. Citrate buffer as a background electrolyte, *J.Chromatogr.A*, **1997**, *768*, 105–112.

SAMPLE
Matrix: solutions

CAPILLARY ELECTROPHORESIS
Capillary: 67 cm × 50 μm fused-silica (60 cm to detector)
Capillary preparation: Wash with running buffer for 6 min before each run.
Capillary temperature: 25
Running buffer: 500 mM pH 2.1 Sodium citrate buffer
Injection: Hydrodynamic injection for 2 s.
Detector: UV 254
Migration time: 24.5
Voltage: 30 kV
Current: 84 μA
Model: Spectra-Physics Model 1000

OTHER SUBSTANCES
Simultaneous: sulfachlorpyridazine, sulfadiazine, sulfadimethoxine, sulfameter, sulfamethazine, sulfamethizole, sulfamethoxazole, sulfamethoxypyridazine, sulfamonomethoxine, sulfanilamide, sulfapyridine, sulfaquinoxaline, sulfathiazole, sulfisomidine, sulfisoxazole

REFERENCE
Lin,C.-E.; Chang,C.-C.; Lin,W.-c. Migration behavior and separation of sulfonamides in capillary zone electrophoresis. II. Positively charged species at low pH, *J.Chromatogr.A*, **1997**, *759*, 203–209.

SAMPLE
Matrix: solutions

CAPILLARY ELECTROPHORESIS
Capillary: 44 cm × 50 μm fused-silica (37 cm to detector)
Capillary preparation: Fill with running buffer for 3 min before each injection then after each
run wash with water for 2 min, with 1 M NaOH for 3 min, , and with water for 2 min. Wash
new capillaries with water at 60° for 10 min, with 1 M NaOH at 60° for 1 h, and with water
at 25° for 10 min.
Running buffer: MeOH:pH 6.85 50 mM phosphate/borate buffer 16.6:83.4 (A) or pH 5.8 phos-
phate/borate buffer containing 40 mM sodium dodecyl sulfate (B) (Prepare buffer by mixing 50
mM disodium tetraborate and 50 mM NaH$_2$PO$_4$.)
Injection: Hydrodynamic injection for 1 s.
Detector: UV 214
Migration time: 3.95 (A), 4.05 (B)
Voltage: 20 kV
Model: Spectra Physics Model 1000

OTHER SUBSTANCES
Simultaneous: sulfachloropyridazine, sulfadiazine, sulfadimethoxine, sulfameter, sulfamethazine, sulfamethoxazole, sulfamethoxypyridazine, sulfamonomethoxine, sulfaquinoxaline, sulfathiazole, sulfisomidine, sulfisoxazole

REFERENCE
Lin,C.-E.; Lin,W.-c.; Chen,Y.-C.; Wang,S.-W. Migration behavior and selectivity of sulfonamides in capillary
electrophoresis, *J.Chromatogr.A*, **1997**, *792*, 37–47.

Sulfameter

Molecular formula: $C_{11}H_{12}N_4O_3S$
Molecular weight: 280.31
CAS Registry No.: 651-06-9
Merck Index (12th ed.): 9082

SAMPLE
Matrix: milk
Sample preparation: 5 mL Milk + 5 mL 3 M HCl, mix, centrifuge for 20 min. Remove the upper liquid layer and add it to 1 mL 3 M HCl, mix, centrifuge for 20 min, filter (0.45 μm) the supernatant, inject an aliquot of the filtrate.

CAPILLARY ELECTROPHORESIS
Capillary: 44 cm × 50 μm fused-silica (37 cm to detector)
Capillary preparation: Wash with running buffer for 3 min before each run.
Capillary temperature: 25
Running buffer: MeCN:buffer 9:91 (A) or buffer containing 0.5 mM β-cyclodextrin (B) (Prepare buffer by mixing 50 mM disodium tetraborate with 50 mM NaH_2PO_4 so as to produce a pH of 6.85.)
Injection: Hydrodynamic injection for 1 s.
Detector: UV 214, UV 254
Migration time: 4.2 (A); 2.9 (B)
Voltage: 20 kV
Model: Spectra-Physics Model 1000

OTHER SUBSTANCES
Simultaneous: sulfachlorpyridazine, sulfadiazine, sulfadimethoxine, sulfamerazine, sulfamethazine, sulfamethoxazole, sulfamethoxypyridazine, sulfamonomethoxine, sulfaquinoxaline, sulfathiazole, sulfisomidine, sulfisoxazole

KEY WORDS
cow

REFERENCE
Lin,C.-E.; Lin,W.-c.; Chiou,W.-c.; Lin,E.C.; Chang,C.-C. Migration behavior and separation of sulfonamides in capillary zone electrophoresis. I. Influence of buffer pH and electrolyte modifier, *J.Chromatogr.A*, **1996**, *755*, 261–269.

SAMPLE
Matrix: solutions

CAPILLARY ELECTROPHORESIS
Capillary: 60 cm × 50 μm fused-silica (47 cm to detector) (Polymicro Technologies)
Capillary temperature: 25
Running buffer: 20 mM pH 8.5 Sodium tetraborate containing 100 mM sodium dodecyl sulfate
Injection: Hydrodynamic injection at 50 mbar for 3.6 s (5 nL).
Detector: UV 205
Migration time: 9.2
Voltage: 22 kV
Model: Crystal 310 (Thermo Unicam)

OTHER SUBSTANCES
Simultaneous: amoxicillin, ampicillin, cephapirin, cloxacillin, dicloxacillin, oxacillin, penicillin G, penicillin V, piperacillin, pyrimethamine, sulfacetamide, sulfadimethoxine, sulfaguanidine, sulfamerazine, sulfamethazine, sulfanilamide, sulfanilic acid, sulfapyridine, sulfaquinoxaline, sulfathiazole, sulfisoxazole, trimethoprim

REFERENCE
Hows,M.E.P.; Perrett,D.; Kay,J. Optimization of a simultaneous separation of sulphonamides, dihydrofolate reductase inhibitors and β-lactam antibiotics by capillary electrophoresis, *J.Chromatogr.A*, **1997**, *768*, 97–104.

SAMPLE
Matrix: solutions

CAPILLARY ELECTROPHORESIS
Capillary: 67 cm × 50 μm fused-silica (60 cm to detector)
Capillary preparation: Wash with running buffer for 6 min before each run.
Capillary temperature: 25
Running buffer: 500 mM pH 2.1 Sodium citrate buffer
Injection: Hydrodynamic injection for 2 s.
Detector: UV 254
Migration time: 42
Voltage: 30 kV
Current: 84 μA
Model: Spectra-Physics Model 1000

OTHER SUBSTANCES
Simultaneous: sulfachlorpyridazine, sulfadiazine, sulfadimethoxine, sulfamerazine, sulfamethazine, sulfamethizole, sulfamethoxazole, sulfamethoxypyridazine, sulfamonomethoxine, sulfanilamide, sulfapyridine, sulfaquinoxaline, sulfathiazole, sulfisomidine, sulfisoxazole

REFERENCE
Lin,C.-E.; Chang,C.-C.; Lin,W.-c. Migration behavior and separation of sulfonamides in capillary zone electrophoresis. II. Positively charged species at low pH, *J.Chromatogr.A*, **1997**, *759*, 203–209.

SAMPLE
Matrix: solutions

CAPILLARY ELECTROPHORESIS
Capillary: 43 cm × 50 μm fused-silica (36 cm to detector)
Capillary preparation: Wash with running buffer for 3 min before each injection and with water for 3 min after each injection. Wash a new capillary with 1 M NaOH at 60° for 50 min, with 100 mM NaOH at 60° for 10 min, and with water at 25° for 10 min.
Capillary temperature: 25
Running buffer: 30 mM pH 6.9 Citrate buffer
Injection: Hydrodynamic injection for 2 s
Detector: UV 254
Migration time: 2.95
Voltage: 20 kV
Model: Spectra-Physics Model 1000

OTHER SUBSTANCES
Simultaneous: sulfachlorpyridazine, sulfadiazine, sulfadimethoxine, sulfamerazine, sulfamethazine, sulfamethoxazole, sulfamethoxypyridazine, sulfamonomethoxine, sulfaquinoxaline, sulfathiazole, sulfisomidine, sulfisoxazole

REFERENCE
Lin,C.-E.; Chang,C.-C.; Lin,W.-c. Migration behavior and separation of sulfonamides in capillary zone electrophoresis. III. Citrate buffer as a background electrolyte, *J.Chromatogr.A*, **1997**, *768*, 105–112.

SAMPLE
Matrix: solutions

CAPILLARY ELECTROPHORESIS
Capillary: 44 cm × 50 μm fused-silica (37 cm to detector)
Capillary preparation: Fill with running buffer for 3 min before each injection then after each run wash with water for 2 min, with 1 M NaOH for 3 min, , and with water for 2 min. Wash

new capillaries with water at 60° for 10 min, with 1 M NaOH at 60° for 1 h, and with water at 25° for 10 min.

Running buffer: MeOH:pH 6.85 50 mM phosphate/borate buffer 16.6:83.4 (A) or pH 5.8 phosphate/borate buffer containing 40 mM sodium dodecyl sulfate (B) (Prepare buffer by mixing 50 mM disodium tetraborate and 50 mM NaH_2PO_4.)

Injection: Hydrodynamic injection for 1 s.

Detector: UV 214

Migration time: 4.05 (A), 4.6 (B)

Voltage: 20 kV

Model: Spectra Physics Model 1000

OTHER SUBSTANCES

Simultaneous: sulfachloropyridazine, sulfadiazine, sulfadimethoxine, sulfamerazine, sulfamethazine, sulfamethoxazole, sulfamethoxypyridazine, sulfamonomethoxine, sulfaquinoxaline, sulfathiazole, sulfisomidine, sulfisoxazole

REFERENCE

Lin,C.-E.; Lin,W.-c.; Chen,Y.-C.; Wang,S.-W. Migration behavior and selectivity of sulfonamides in capillary electrophoresis, *J.Chromatogr.A*, **1997**, *792*, 37–47.

Sulfamethazine

Molecular formula: $C_{12}H_{14}N_4O_2S$
Molecular weight: 278.33
CAS Registry No.: 57-68-1
Merck Index (12th ed.): 9083
Lednicer: 1 125

SAMPLE

Matrix: solutions
Sample preparation: Prepare a solution in running buffer, inject an aliquot.

CAPILLARY ELECTROPHORESIS

Capillary: 100 cm × 75 μm fused-silica (20 cm to UV detector) (Polymicro Technologies)

Capillary preparation: Between runs flush capillaries with 1.5 column volumes of 100 mM NaOH and 9 column volumes of running buffer. Flush new capillaries with 9 column volumes of 1 M NaOH, 3 column volumes of water, 3 column volumes of 100 mM HCl, 3 column volumes of water, and 9 column volumes of running buffer.

Running buffer: MeOH:water 20:80 containing 20 mM ammonium acetate, pH 6.8

Injection: Pressure injection at 3.45 kPa (5-28 nL)

Detector: UV 254 or MS, Sciex TAGA 6000E triple quadrupole, API source, electrospray (ion spray) interface at 4 kV (3 kV to ion-sampling orifice), positive-ion mode, 100 μm dia sampling orifice, nitrogen gas curtain, argon collision gas, m/z 279 [M+H]+

Migration time: 3.5 (UV), 16.6 (MS)

Voltage: 26 kV

Model: Beckman P/ACE System 2000

OTHER SUBSTANCES

Simultaneous: sulfadimethoxine, sulfamerazine, sulfamethoxazole, sulfanilamide, sulfathiazole

REFERENCE

Johansson,I.M.; Pavelka,R.; Henion,J.D. Determination of small drug molecules by capillary electrophoresis-atmospheric pressure ionization mass spectrometry, *J.Chromatogr.*, **1991**, *559*, 515–528.

SAMPLE

Matrix: solutions

CAPILLARY ELECTROPHORESIS
Capillary: 90 cm × 50 μm untreated fused-silica (Polymicro Technologies)
Running buffer: Trisma buffer pH 7.2 (about 7 g Trisma HCl and 0.7 g Trisma base in 1 L water)
Injection: 3 s vacuum injection
Detector: UV 200 or MS, SCIEX API III triple quadrupole, API source, positive ion analysis, make-up flow 0.2% aqueous formic acid, nebulizing gas air, target gas argon, collision energy 20-35 eV
Migration time: 9
Voltage: 24.4 kV
Model: Applied Biosystems Model 270A

OTHER SUBSTANCES
Simultaneous: erythromycin, ormetoprim, sulfaisomidine, trimethoprim

REFERENCE
Pleasance,S.; Thibault,P.; Kelly,J. Comparison of liquid-junction and coaxial interfaces for capillary electrophoresis-mass spectrometry with application to compounds of concern to the aquaculture industry, *J.Chromatogr.*, **1992**, *591*, 325–339.

SAMPLE
Matrix: solutions

CAPILLARY ELECTROPHORESIS
Capillary: 50 cm × 50 μm fused-silica (Tokyo Kasei)
Capillary preparation: Rinse capillary with 1 M NaOH after each run.
Running buffer: MeOH:buffer 10:90 (Buffer was 10 mM pH 7 ammonium formate containing 2% butyl acrylate-butyl methacrylate-methacrylic acid copolymer sodium salt (BBMA) (Dai-ichi Kogyo Seiyaku)
Injection: Hydrostatic injection at 15 cm for 10-30 s
Detector: MS, Hitachi M-1000 quadrupole, electrospray at 3 kV, details of interface given in paper, sheath liquid MeOH:water:formic acid 50:50:1 at 5 μL/min, positive-ion mode, drift voltage 70 V, focusing voltage 140-150 V
Migration time: 9.6
Voltage: 10 kV
Model: laboratory constructed

OTHER SUBSTANCES
Simultaneous: sulfadiazine, sulfisomidine, sulfisoxazole

REFERENCE
Ozaki,H.; Itou,N.; Terabe,S.; Takada,Y.; Sakairi,M.; Koizumi,H. Micellar electrokinetic chromatography-mass spectrometry using a high-molecular-mass surfactant. On-line coupling with an electrospray ionization interface, *J.Chromatogr.A*, **1995**, *716*, 69–79.

SAMPLE
Matrix: solutions

CAPILLARY ELECTROPHORESIS
Capillary: 60 cm × 50 μm fused-silica (47 cm to detector) (Polymicro Technologies)
Capillary temperature: 25
Running buffer: 20 mM pH 8.5 Sodium tetraborate containing 100 mM sodium dodecyl sulfate
Injection: Hydrodynamic injection at 50 mbar for 3.6 s (5 nL).
Detector: UV 205
Migration time: 11.5
Voltage: 22 kV
Model: Crystal 310 (Thermo Unicam)

OTHER SUBSTANCES
Simultaneous: amoxicillin, ampicillin, cephapirin, cloxacillin, dicloxacillin, oxacillin, penicillin G, penicillin V, piperacillin, pyrimethamine, sulfacetamide, sulfadimethoxine, sulfaguanidine, sulfamerazine, sulfameter, sulfanilamide, sulfanilic acid, sulfapyridine, sulfaquinoxaline, sulfathiazole, sulfisoxazole, trimethoprim

REFERENCE

Hows,M.E.P.; Perrett,D.; Kay,J. Optimization of a simultaneous separation of sulphonamides, dihydrofolate reductase inhibitors and β-lactam antibiotics by capillary electrophoresis, *J.Chromatogr.A*, **1997**, *768*, 97–104.

SAMPLE

Matrix: solutions

CAPILLARY ELECTROPHORESIS

Capillary: 44 cm × 50 μm fused-silica (37 cm to detector)

Capillary preparation: Fill with running buffer for 3 min before each injection then after each run wash with water for 2 min, with 1 M NaOH for 3 min, , and with water for 2 min. Wash new capillaries with water at 60° for 10 min, with 1 M NaOH at 60° for 1 h, and with water at 25° for 10 min.

Running buffer: MeOH:pH 6.85 50 mM phosphate/borate buffer 16.6:83.4 (A) or pH 5.8 phosphate/borate buffer containing 40 mM sodium dodecyl sulfate (B) (Prepare buffer by mixing 50 mM disodium tetraborate and 50 mM NaH_2PO_4.)

Injection: Hydrodynamic injection for 1 s.

Detector: UV 214

Migration time: 3.4 (A), 4.7 (B)

Voltage: 20 kV

Model: Spectra Physics Model 1000

OTHER SUBSTANCES

Simultaneous: sulfachloropyridazine, sulfadiazine, sulfadimethoxine, sulfamerazine, sulfameter, sulfamethoxazole, sulfamethoxypyridazine, sulfamonomethoxine, sulfaquinoxaline, sulfathiazole, sulfisomidine, sulfisoxazole

REFERENCE

Lin,C.-E.; Lin,W.-c.; Chen,Y.-C.; Wang,S.-W. Migration behavior and selectivity of sulfonamides in capillary electrophoresis, *J.Chromatogr.A*, **1997**, *792*, 37–47.

SAMPLE

Matrix: solutions

CAPILLARY ELECTROPHORESIS

Capillary: 43 cm × 50 μm fused-silica (36 cm to detector)

Capillary preparation: Wash with running buffer for 3 min before each injection and with water for 3 min after each injection. Wash a new capillary with 1 M NaOH at 60° for 50 min, with 100 mM NaOH at 60° for 10 min, and with water at 25° for 10 min.

Capillary temperature: 25

Running buffer: 30 mM pH 6.9 Citrate buffer

Injection: Hydrodynamic injection for 2 s

Detector: UV 254

Migration time: 2.3

Voltage: 20 kV

Model: Spectra-Physics Model 1000

OTHER SUBSTANCES

Simultaneous: sulfachlorpyridazine, sulfadiazine, sulfadimethoxine, sulfamerazine, sulfameter, sulfamethoxazole, sulfamethoxypyridazine, sulfamonomethoxine, sulfaquinoxaline, sulfathiazole, sulfisomidine, sulfisoxazole

REFERENCE

Lin,C.-E.; Chang,C.-C.; Lin,W.-c. Migration behavior and separation of sulfonamides in capillary zone electrophoresis. III. Citrate buffer as a background electrolyte, *J.Chromatogr.A*, **1997**, *768*, 105–112.

SAMPLE

Matrix: solutions

CAPILLARY ELECTROPHORESIS
Capillary: 67 cm × 50 μm fused-silica (60 cm to detector)
Capillary preparation: Wash with running buffer for 6 min before each run.
Capillary temperature: 25
Running buffer: 500 mM pH 2.1 Sodium citrate buffer
Injection: Hydrodynamic injection for 2 s.
Detector: UV 254
Migration time: 22
Voltage: 30 kV
Current: 84 μA
Model: Spectra-Physics Model 1000

OTHER SUBSTANCES
Simultaneous: sulfachlorpyridazine, sulfadiazine, sulfadimethoxine, sulfamerazine, sulfameter, sulfamethizole, sulfamethoxazole, sulfamethoxypyridazine, sulfamonomethoxine, sulfanilamide, sulfapyridine, sulfaquinoxaline, sulfathiazole, sulfisomidine, sulfisoxazole

REFERENCE
Lin,C.-E.; Chang,C.-C.; Lin,W.-c. Migration behavior and separation of sulfonamides in capillary zone electrophoresis. II. Positively charged species at low pH, *J.Chromatogr.A*, **1997**, *759*, 203–209.

SAMPLE
Matrix: tissue
Sample preparation: Homogenize (household food processor) 10 g tissue, extract (stomacher apparatus) with 100 mL MeCN for 5 min, centrifuge at 4000 g for 10 min, filter (0.45 μm), inject an aliquot of the filtrate.

CAPILLARY ELECTROPHORESIS
Capillary: 57 cm × 75 μm (50 cm to detector) (Beckman)
Capillary temperature: 25
Running buffer: 20 mM Borate buffer containing 20 mM phosphate, adjusted to pH 7.0 with KOH
Injection: Pressure injection for 2 s (39 nL)
Detector: UV 254
Migration time: 12.5
Voltage: 10 kV
Current: 26.8 μA
Model: Beckman P/ACE System 2000

OTHER SUBSTANCES
Extracted: sulfachlorpyridazine, sulfadiazine, sulfadimethoxine, sulfadoxine, sulfamerazine, sulfamethoxazole, sulfamethoxydiazine, sulfamethoxypyridazine, sulfaquinoxaline, sulfathiazole, sulfatroxazole, trimethoprim

KEY WORDS
pig

REFERENCE
Ackermans,M.T.; Beckers,J.L.; Everaerts,F.M.; Hoogland,H.; Tomassen,M.J.H. Determination of sulphonamides in pork meat extracts by capillary zone electrophoresis, *J.Chromatogr.*, **1992**, *596*, 101–109.

Sulfamethizole

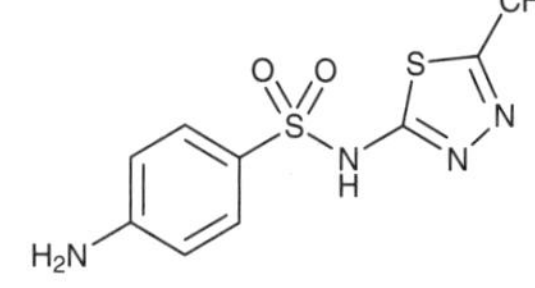

Molecular formula: $C_9H_{10}N_4O_2S_2$
Molecular weight: 270.34
CAS Registry No.: 144-82-1
Merck Index (12th ed.): 9084
Lednicer: 1 125

SAMPLE
Matrix: solutions

CAPILLARY ELECTROPHORESIS
Capillary: 67 cm × 50 μm fused-silica (60 cm to detector)
Capillary preparation: Wash with running buffer for 6 min before each run.
Capillary temperature: 25
Running buffer: 500 mM pH 2.1 Sodium citrate buffer
Injection: Hydrodynamic injection for 2 s.
Detector: UV 254
Migration time: 31.5
Voltage: 30 kV
Current: 84 μA
Model: Spectra-Physics Model 1000

OTHER SUBSTANCES
Simultaneous: sulfachlorpyridazine, sulfadiazine, sulfadimethoxine, sulfamerazine, sulfameter, sulfamethazine, sulfamethoxazole, sulfamethoxypyridazine, sulfamonomethoxine, sulfanilamide, sulfapyridine, sulfaquinoxaline, sulfathiazole, sulfisomidine, sulfisoxazole

REFERENCE
Lin,C.-E.; Chang,C.-C.; Lin,W.-c. Migration behavior and separation of sulfonamides in capillary zone electrophoresis. II. Positively charged species at low pH, *J.Chromatogr.A*, **1997**, *759*, 203–209.

Sulfamethoxazole

Molecular formula: $C_{10}H_{11}N_3O_3S$
Molecular weight: 253.28
CAS Registry No.: 723-46-6
Merck Index (12th ed.): 9086

SAMPLE
Matrix: blood
Sample preparation: Inject serum directly.

CAPILLARY ELECTROPHORESIS
Capillary: 105 cm × 75 μm fused-silica (68 cm to detector) (Polymicro Technologies)
Capillary preparation: Rinse with running buffer for 10 min between run. At the beginning and end of each day rinse with 100 mM NaOH for 10 min and with water for 10 min.
Running buffer: 10 mM Na_2HPO_4 containing 6 mM sodium borate and 75 mM sodium dodecyl sulfate
Injection: Injection via vacuum suction for 1 s
Detector: UV 215
Migration time: 13
Voltage: 23 kV
Current: 62 μA
Model: Europhor Prime Vision IV

OTHER SUBSTANCES
Extracted: acetaminophen, ethosuximide, flucytosine, naproxen, primidone, zomepirac

KEY WORDS
serum

REFERENCE
Schmutz,A.; Thormann,W. Assessment of impact of physico-chemical drug properties on monitoring drug levels by micellar electrokinetic capillary chromatography with direct serum injection, *Electrophoresis*, **1994**, *15*, 1295–1303.

SAMPLE
Matrix: blood
Sample preparation: 400 μL Plasma + 600 μL 9 g/L NaCl, mix, inject an aliquot.

CAPILLARY ELECTROPHORESIS
Capillary: 42.5 cm × 50 μm fused-silica (34.5 cm) (Polymicro Technologies)
Capillary preparation: Between runs rinse with MeCN:running buffer 50:50 for 15 min and with running buffer for 20 min. Condition new capillaries with 100 mM NaOH at 50° for 30 min and equilibrate with cathode buffer at 25° at 10 kV for 2 h. (Cathode buffer was 60 mM pH 10.0 sodium borate buffer.)
Capillary temperature: 25
Running buffer: 60 mM pH 10.0 Sodium borate buffer containing 200 mM sodium dodecyl sulfate
Injection: Vacuum injection at 0.75 psi for 3 s.
Detector: UV 220
Migration time: 11.6
Voltage: 10 kV
Current: 42-44 μA
Model: SpectraPhoresis 1000 (Thermo Separation Products)

OTHER SUBSTANCES
Extracted: salicylic acid, tolbutamide

KEY WORDS
plasma

REFERENCE
Kunkel,A.; Günter,S.; Wätzig,H. Quantitation of acetaminophen and salicylic acid in plasma using capillary electrophoresis without sample pretreatment. Improvement of precision, *J.Chromatogr.A*, **1997**, *768*, 125–133.

SAMPLE
Matrix: blood
Sample preparation: Filter (0.2 μm) plasma, inject an aliquot of the filtrate. (Note that 400 μL plasma was spiked with 600 μL of the drug solution in saline.)

CAPILLARY ELECTROPHORESIS
Capillary: 42.5 cm × 50 μm fused-silica (34.5 cm to detector) (Polymicro Technologies)
Capillary preparation: After each run rinse capillary at 25° with MeCN:running buffer 50:50 for 15 min then re-equilibrate with running buffer for 20 min. Condition new capillaries with 100 mM NaOH at 50° for 30 min then equilibrate with cathode buffer at 25° for 2 h with voltage applied.
Capillary temperature: 25
Running buffer: 60 mM pH 10.0 Sodium borate buffer containing 200 mM sodium dodecyl sulfate (Cathode buffer was 60 mM pH 10.0 sodium borate buffer prepared by dissolving 3.170 g boric acid in 100 mL water and adjusting the pH to 10.0 with freshly-prepared 1 M NaOH.)
Injection: Vacuum injection at 0.75 psi for 3 s.
Detector: UV 200
Migration time: 12
Voltage: 10 kV

Current: 42 μA
Model: Spectraphoresis 1000 (Thermo Separation Products)

OTHER SUBSTANCES
Extracted: salicylic acid, tolbutamide

KEY WORDS
plasma

REFERENCE
Kunkel,A.; Gunter,S.; Watzig,H. Determination of pharmaceuticals in plasma by capillary electrophoresis without sample pretreatment reproducibility, limit of quantitation and limit of detection, *Electrophoresis*, **1997**, *18*, 1882–1889.

SAMPLE
Matrix: milk
Sample preparation: 5 mL Milk + 5 mL 3 M HCl, mix, centrifuge for 20 min. Remove the upper liquid layer and add it to 1 mL 3 M HCl, mix, centrifuge for 20 min, filter (0.45 μm) the supernatant, inject an aliquot of the filtrate.

CAPILLARY ELECTROPHORESIS
Capillary: 44 cm × 50 μm fused-silica (37 cm to detector)
Capillary preparation: Wash with running buffer for 3 min before each run.
Capillary temperature: 25
Running buffer: MeCN:buffer 9:91 (A) or buffer containing 0.5 mM β-cyclodextrin (B) (Prepare buffer by mixing 50 mM disodium tetraborate with 50 mM NaH_2PO_4 so as to produce a pH of 6.85.)
Injection: Hydrodynamic injection for 1 s.
Detector: UV 214, UV 254
Migration time: 5.3 (A); 3.6 (B)
Voltage: 20 kV
Model: Spectra-Physics Model 1000

OTHER SUBSTANCES
Simultaneous: sulfachlorpyridazine, sulfadiazine, sulfadimethoxine, sulfamerazine, sulfameter, sulfamethazine, sulfamethoxypyridazine, sulfamonomethoxine, sulfaquinoxaline, sulfathiazole, sulfisomidine, sulfisoxazole

KEY WORDS
cow

REFERENCE
Lin,C.-E.; Lin,W.-c.; Chiou,W.-c.; Lin,E.C.; Chang,C.-C. Migration behavior and separation of sulfonamides in capillary zone electrophoresis. I. Influence of buffer pH and electrolyte modifier, *J.Chromatogr.A*, **1996**, *755*, 261–269.

SAMPLE
Matrix: solutions
Sample preparation: Prepare a solution in running buffer, inject an aliquot.

CAPILLARY ELECTROPHORESIS
Capillary: 100 cm × 75 μm fused-silica (20 cm to UV detector) (Polymicro Technologies)
Capillary preparation: Between runs flush capillaries with 1.5 column volumes of 100 mM NaOH and 9 column volumes of running buffer. Flush new capillaries with 9 column volumes of 1 M NaOH, 3 column volumes of water, 3 column volumes of 100 mM HCl, 3 column volumes of water, and 9 column volumes of running buffer.
Running buffer: MeOH:water 20:80 containing 20 mM ammonium acetate, pH 6.8
Injection: Pressure injection at 3.45 kPa (5-28 nL)
Detector: UV 254 or MS, Sciex TAGA 6000E triple quadrupole, API source, electrospray (ion spray) interface at 4 kV (3 kV to ion-sampling orifice), positive-ion mode, 100 μm dia sampling orifice, nitrogen gas curtain, argon collision gas, m/z 254 [M+H]+

Migration time: 5.3 (UV), 25.5 (MS)
Voltage: 26 kV
Model: Beckman P/ACE System 2000

OTHER SUBSTANCES
Simultaneous: sulfadimethoxine, sulfamerazine, sulfamethazine, sulfanilamide, sulfathiazole

REFERENCE
Johansson,I.M.; Pavelka,R.; Henion,J.D. Determination of small drug molecules by capillary electrophoresis-atmospheric pressure ionization mass spectrometry, *J.Chromatogr.*, **1991**, *559*, 515–528.

SAMPLE
Matrix: solutions

CAPILLARY ELECTROPHORESIS
Capillary: 67 cm × 50 μm fused-silica (60 cm to detector)
Capillary preparation: Wash with running buffer for 6 min before each run.
Capillary temperature: 25
Running buffer: 500 mM pH 2.1 Sodium citrate buffer
Injection: Hydrodynamic injection for 2 s.
Detector: UV 254
Migration time: 46
Voltage: 30 kV
Current: 84 μA
Model: Spectra-Physics Model 1000

OTHER SUBSTANCES
Simultaneous: sulfachlorpyridazine, sulfadiazine, sulfadimethoxine, sulfamerazine, sulfameter, sulfamethazine, sulfamethizole, sulfamethoxypyridazine, sulfamonomethoxine, sulfanilamide, sulfapyridine, sulfaquinoxaline, sulfathiazole, sulfisomidine, sulfisoxazole

REFERENCE
Lin,C.-E.; Chang,C.-C.; Lin,W.-c. Migration behavior and separation of sulfonamides in capillary zone electrophoresis. II. Positively charged species at low pH, *J.Chromatogr.A*, **1997**, *759*, 203–209.

SAMPLE
Matrix: solutions

CAPILLARY ELECTROPHORESIS
Capillary: 43 cm × 50 μm fused-silica (36 cm to detector)
Capillary preparation: Wash with running buffer for 3 min before each injection and with water for 3 min after each injection. Wash a new capillary with 1 M NaOH at 60° for 50 min, with 100 mM NaOH at 60° for 10 min, and with water at 25° for 10 min.
Capillary temperature: 25
Running buffer: 30 mM pH 6.9 Citrate buffer
Injection: Hydrodynamic injection for 2 s
Detector: UV 254
Migration time: 3.6
Voltage: 20 kV
Model: Spectra-Physics Model 1000

OTHER SUBSTANCES
Simultaneous: sulfachlorpyridazine, sulfadiazine, sulfadimethoxine, sulfamerazine, sulfameter, sulfamethazine, sulfamethoxypyridazine, sulfamonomethoxine, sulfaquinoxaline, sulfathiazole, sulfisomidine, sulfisoxazole

REFERENCE
Lin,C.-E.; Chang,C.-C.; Lin,W.-c. Migration behavior and separation of sulfonamides in capillary zone electrophoresis. III. Citrate buffer as a background electrolyte, *J.Chromatogr.A*, **1997**, *768*, 105–112.

SAMPLE
Matrix: solutions

CAPILLARY ELECTROPHORESIS
Capillary: 44 cm × 50 μm fused-silica (37 cm to detector)
Capillary preparation: Fill with running buffer for 3 min before each injection then after each run wash with water for 2 min, with 1 M NaOH for 3 min, , and with water for 2 min. Wash new capillaries with water at 60° for 10 min, with 1 M NaOH at 60° for 1 h, and with water at 25° for 10 min.
Running buffer: MeOH:pH 6.85 50 mM phosphate/borate buffer 16.6:83.4 (A) or pH 5.8 phosphate/borate buffer containing 40 mM sodium dodecyl sulfate (B) (Prepare buffer by mixing 50 mM disodium tetraborate and 50 mM NaH_2PO_4.)
Injection: Hydrodynamic injection for 1 s.
Detector: UV 214
Migration time: 5.4 (A), 5.3 (B)
Voltage: 20 kV
Model: Spectra Physics Model 1000

OTHER SUBSTANCES
Simultaneous: sulfachloropyridazine, sulfadiazine, sulfadimethoxine, sulfamerazine, sulfameter, sulfamethazine, sulfamethoxypyridazine, sulfamonomethoxine, sulfaquinoxaline, sulfathiazole, sulfisomidine, sulfisoxazole

REFERENCE
Lin,C.-E.; Lin,W.-c.; Chen,Y.-C.; Wang,S.-W. Migration behavior and selectivity of sulfonamides in capillary electrophoresis, *J.Chromatogr.A*, **1997**, *792*, 37–47.

SAMPLE
Matrix: solutions

CAPILLARY ELECTROPHORESIS
Capillary: 57 cm × 50 μm fused-silica (50 cm to detector)
Running buffer: 50 mM pH 8.0 Borate buffer containing 40 mM sodium taurodeoxycholate and 25 mM phosphatidylcholine (Prepare by adding phosphatidylcholine and stirring for 3-6 h until all cloudiness disappears. Phosphatidylcholine was 95% pure soybean lecithin, Epikuron, Lucas Meyer & Co.)
Injection: Inject a solution of the compound in the running buffer at 20 psi for 1 s, inject MeOH:water 5:95 at 20 psi for 1 s, inject a solution of halofantrine in running buffer at 20 psi for 1 s.
Detector: UV 214
Migration time: k' 1.58
Model: Beckman P/ACE 5000

OTHER SUBSTANCES
Also analyzed: acetaminophen, amoxicillin, antipyrine, aspirin, azathioprine, caffeine, captopril, carbamazepine, carprofen, chlorambucil, chlorpheniramine, chlorpromazine, cimetidine, clonidine, codeine, desipramine, diphenhydramine, ephedrine, fenoterol, flufenamic acid, flurbiprofen, haloperidol, hydroxyzine, ibuprofen, imipramine, indomethacin, ketoprofen, lidocaine, melphalan, metoprolol, nabumetone, nadolol, phenobarbital, phenol, promazine, propranolol, pyrilamine, ranitidine, ropinirole, salicylic acid, testosterone, theophylline, thioridazine, tiaprofenic acid, tolfenamic acid, trifluoperazine, trimethoprim, valproic acid, verapamil

KEY WORDS
comparison with HPLC; k' = (Tr-T0)/(T0(1-Tr/Tm)) where Tr = retention time of analyte; T0 = retention time of water; and Tm = retention time of marker (halofantrine)

REFERENCE
Hanna,M.; de Biasi,V.; Bond,B.; Salter,C.; Hutt,A.J.; Camilleri,P. Estimation of the partitioning characteristics of drugs: A comparison of a large and diverse drug series utilizing chromatographic and electrophoretic methodology, *Anal.Chem.*, **1998**, *70*, 2092–2099.

SAMPLE
Matrix: tissue
Sample preparation: Homogenize (household food processor) 10 g tissue, extract (stomacher apparatus) with 100 mL MeCN for 5 min, centrifuge at 4000 g for 10 min, filter (0.45 μm), inject an aliquot of the filtrate.

CAPILLARY ELECTROPHORESIS
Capillary: 57 cm × 75 μm (50 cm to detector) (Beckman)
Capillary temperature: 25
Running buffer: 20 mM Borate buffer containing 20 mM phosphate, adjusted to pH 7.0 with KOH
Injection: Pressure injection for 2 s (39 nL)
Detector: UV 254
Migration time: 22.2
Voltage: 10 kV
Current: 26.8 μA
Model: Beckman P/ACE System 2000

OTHER SUBSTANCES
Extracted: sulfachlorpyridazine, sulfadiazine, sulfadimethoxine, sulfadoxine, sulfamerazine, sulfamethazine (sulfadimidine), sulfamethoxydiazine, sulfamethoxypyridazine, sulfaquinoxaline, sulfathiazole, sulfatroxazole, trimethoprim

KEY WORDS
pig

REFERENCE
Ackermans,M.T.; Beckers,J.L.; Everaerts,F.M.; Hoogland,H.; Tomassen,M.J.H. Determination of sulphonamides in pork meat extracts by capillary zone electrophoresis, *J.Chromatogr.*, **1992**, *596*, 101–109.

SAMPLE
Matrix: urine
Sample preparation: Filter (Amicon Micron-10) urine while centrifuging at 14000 RCF for 10 min. Dilute the ultrafiltrate 5-10-fold with 5 mM pH 11.3 sodium borate buffer, inject an aliquot.

CAPILLARY ELECTROPHORESIS
Capillary: 47 cm × 50 μm fused-silica (40 cm to detector) (Polymicro Technologies)
Capillary preparation: Before each run rinse capillary with running buffer at 20 psi for 1 min. After each run rinse capillary at 20 psi with 100 mM NaOH for 30 s, with water for 30 s, and with running buffer for 1 min. Before first use rinse capillary at 20 psi with 100 mM NaOH for 20 min, with water for 20 min, and with 50 mM pH 11.3 sodium borate buffer for 20 min.
Capillary temperature: 20
Running buffer: 50 mM pH 11.3 Sodium borate buffer
Injection: Pressure injection of sample at 0.5 psi for 15 s followed by pressure injection of running buffer at 0.5 psi for 1 s.
Detector: UV 254
Migration time: 3.6
Voltage: 25 kV
Model: Beckman P/ACE 2100

OTHER SUBSTANCES
Extracted: acetaminophen, iothalamate

KEY WORDS
ultrafiltrate

REFERENCE
Bergert,J.H.; Liedtke,R.R.; Oda,R.P.; Landers,J.P.; Wilson,D.M. Development of a nonisotopic capillary electrophoresis-based method for measuring glomerular filtration rate, *Electrophoresis*, **1997**, *18*, 1827–1835.

Sulfamonomethoxine

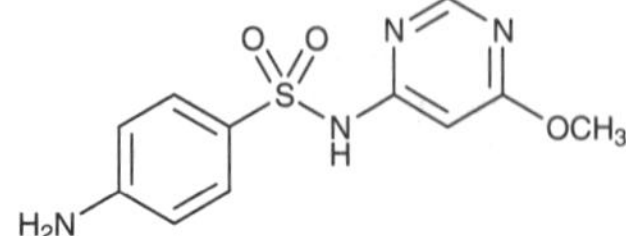

Molecular formula: $C_{11}H_{12}N_4O_3S$
Molecular weight: 280.31
CAS Registry No.: 1220-83-3

SAMPLE
Matrix: milk
Sample preparation: 5 mL Milk + 5 mL 3 M HCl, mix, centrifuge for 20 min. Remove the upper liquid layer and add it to 1 mL 3 M HCl, mix, centrifuge for 20 min, filter (0.45 μm) the supernatant, inject an aliquot of the filtrate.

CAPILLARY ELECTROPHORESIS
Capillary: 44 cm × 50 μm fused-silica (37 cm to detector)
Capillary preparation: Wash with running buffer for 3 min before each run.
Capillary temperature: 25
Running buffer: MeCN:buffer 9:91 (A) or buffer containing 0.5 mM β-cyclodextrin (B) (Prepare buffer by mixing 50 mM disodium tetraborate with 50 mM NaH_2PO_4 so as to produce a pH of 6.85.)
Injection: Hydrodynamic injection for 1 s.
Detector: UV 214, UV 254
Migration time: 4.8 (A); 3.3 (B)
Voltage: 20 kV
Model: Spectra-Physics Model 1000

OTHER SUBSTANCES
Simultaneous: sulfachlorpyridazine, sulfadiazine, sulfadimethoxine, sulfamerazine, sulfameter, sulfamethazine, sulfamethoxazole, sulfamethoxypyridazine, sulfaquinoxaline, sulfathiazole, sulfisomidine, sulfisoxazole

KEY WORDS
cow

REFERENCE
Lin,C.-E.; Lin,W.-c.; Chiou,W.-c.; Lin,E.C.; Chang,C.-C. Migration behavior and separation of sulfonamides in capillary zone electrophoresis. I. Influence of buffer pH and electrolyte modifier, *J.Chromatogr.A*, **1996**, *755*, 261–269.

SAMPLE
Matrix: solutions

CAPILLARY ELECTROPHORESIS
Capillary: 67 cm × 50 μm fused-silica (60 cm to detector)
Capillary preparation: Wash with running buffer for 6 min before each run.
Capillary temperature: 25
Running buffer: 500 mM pH 2.1 Sodium citrate buffer
Injection: Hydrodynamic injection for 2 s.
Detector: UV 254
Migration time: 33
Voltage: 30 kV
Current: 84 μA
Model: Spectra-Physics Model 1000

OTHER SUBSTANCES
Simultaneous: sulfachlorpyridazine, sulfadiazine, sulfadimethoxine, sulfamerazine, sulfameter, sulfamethazine, sulfamethizole, sulfamethoxazole, sulfamethoxypyridazine, sulfanilamide, sulfapyridine, sulfaquinoxaline, sulfathiazole, sulfisomidine, sulfisoxazole

REFERENCE
Lin,C.-E.; Chang,C.-C.; Lin,W.-c. Migration behavior and separation of sulfonamides in capillary zone electrophoresis. II. Positively charged species at low pH, *J.Chromatogr.A*, **1997**, *759*, 203–209.

SAMPLE
Matrix: solutions

CAPILLARY ELECTROPHORESIS
Capillary: 43 cm × 50 μm fused-silica (36 cm to detector)
Capillary preparation: Wash with running buffer for 3 min before each injection and with water for 3 min after each injection. Wash a new capillary with 1 M NaOH at 60° for 50 min, with 100 mM NaOH at 60° for 10 min, and with water at 25° for 10 min.
Capillary temperature: 25
Running buffer: 30 mM pH 6.9 Citrate buffer
Injection: Hydrodynamic injection for 2 s
Detector: UV 254
Migration time: 3.3
Voltage: 20 kV
Model: Spectra-Physics Model 1000

OTHER SUBSTANCES
Simultaneous: sulfachlorpyridazine, sulfadiazine, sulfadimethoxine, sulfamerazine, sulfameter, sulfamethazine, sulfamethoxazole, sulfamethoxypyridazine, sulfaquinoxaline, sulfathiazole, sulfisomidine, sulfisoxazole

REFERENCE
Lin,C.-E.; Chang,C.-C.; Lin,W.-c. Migration behavior and separation of sulfonamides in capillary zone electrophoresis. III. Citrate buffer as a background electrolyte, *J.Chromatogr.A*, **1997**, *768*, 105–112.

Sulfanilamide

Molecular formula: $C_6H_8N_2O_2S$
Molecular weight: 172.21
CAS Registry No.: 63-74-1
Merck Index (12th ed.): 9094
Lednicer: 1 121

SAMPLE
Matrix: formulations
Sample preparation: Grind ten tablets and weigh out an amount corresponding to 50 mg sulfisoxazole and 5 mg phenazopyridine, add 40 mL running buffer, sonicate for 30 min, make up to 50 mL with running buffer, filter (0.45 μm). Dilute an aliquot of the filtrate 1:25 with running buffer, inject an aliquot.

CAPILLARY ELECTROPHORESIS
Capillary: 70 cm × 75 μm fused-silica
Capillary preparation: After each run refill reservoirs, rinse capillary with fresh buffer at 7 psi for 3 min.
Running buffer: 5 mM NaH_2PO_4 containing 5 mM Na_2HPO_4 and 50 mM sodium dodecyl sulfate, adjusted to pH 7.0 with NaOH
Injection: Hydrodynamic injection at 5 cm for 10 s (6.8 nL)
Detector: UV 240
Migration time: 8.7
Voltage: 15 kV
Model: Dionex Model CES I
Limit of detection: 250 ng/mL (S/N 2)

OTHER SUBSTANCES
Simultaneous: impurities, phenazopyridine, sulfisoxazole

KEY WORDS
tablets

REFERENCE
Nickerson,B.; Scypinski,S.; Sokoloff,H.; Sahota,S. Separation of sulfisoxazole, phenazopyridine, and their related impurities by micellar electrokinetic chromatography, *J.Liq.Chromatogr.*, **1995**, *18*, 3847–3875.

SAMPLE
Matrix: solutions
Sample preparation: Prepare a solution in running buffer, inject an aliquot.

CAPILLARY ELECTROPHORESIS
Capillary: 100 cm × 75 μm fused-silica (20 cm to UV detector) (Polymicro Technologies)
Capillary preparation: Between runs flush capillaries with 1.5 column volumes of 100 mM NaOH and 9 column volumes of running buffer. Flush new capillaries with 9 column volumes of 1 M NaOH, 3 column volumes of water, 3 column volumes of 100 mM HCl, 3 column volumes of water, and 9 column volumes of running buffer.
Running buffer: MeOH:water 20:80 containing 20 mM ammonium acetate, pH 6.8
Injection: Pressure injection at 3.45 kPa (5-28 nL)
Detector: UV 254 or MS, Sciex TAGA 6000E triple quadrupole, API source, electrospray (ion spray) interface at 4 kV (3 kV to ion-sampling orifice), positive-ion mode, 100 μm dia sampling orifice, nitrogen gas curtain, argon collision gas, m/z 190 [M+NH$_4$]+
Migration time: 3.3 (UV), 16 (MS)
Voltage: 26 kV
Model: Beckman P/ACE System 2000

OTHER SUBSTANCES
Simultaneous: sulfadimethoxine, sulfamerazine, sulfamethazine, sulfamethoxazole, sulfathiazole

REFERENCE
Johansson,I.M.; Pavelka,R.; Henion,J.D. Determination of small drug molecules by capillary electrophoresis-atmospheric pressure ionization mass spectrometry, *J.Chromatogr.*, **1991**, *559*, 515–528.

SAMPLE
Matrix: solutions

CAPILLARY ELECTROPHORESIS
Capillary: 47 cm × 50 μm fused-silica (40 cm to detector) (Polymicro Technologies)
Capillary preparation: Before each run rinse capillary with 100 mM ammonium hydroxide at high pressure for 2 min, with water for 2-3 min, and with running buffer for 2-3 min.
Capillary temperature: 26
Running buffer: MeCN:pH 4 acetate buffer containing 80 mM silver nitrate 15:85
Injection: Hydrodynamic injection for 3-5 s
Detector: UV 254, UV 280
Migration time: 12.4
Voltage: 20 kV
Model: Beckman P/ACE 2100

OTHER SUBSTANCES
Simultaneous: sulfabenzamide, sulfacetamide, sulfadiazine, sulfadimethoxine, sulfaguanidine, sulfamerazine, sulfaquinoxaline, sulfathiazole

REFERENCE
Wright,P.B.; Dorsey,J.G. Silver(I)-mediated separations by capillary zone electrophoresis and micellar electrokinetic chromatography: Argentation electrophoresis, *Anal.Chem.*, **1996**, *68*, 415–424.

SAMPLE
Matrix: solutions

CAPILLARY ELECTROPHORESIS
Capillary: 60 cm × 50 μm fused-silica (47 cm to detector) (Polymicro Technologies)
Capillary temperature: 25
Running buffer: 20 mM pH 8.5 Sodium tetraborate containing 100 mM sodium dodecyl sulfate
Injection: Hydrodynamic injection at 50 mbar for 3.6 s (5 nL).
Detector: UV 205
Migration time: 6.7
Voltage: 22 kV
Model: Crystal 310 (Thermo Unicam)

OTHER SUBSTANCES
Simultaneous: amoxicillin, ampicillin, cephapirin, cloxacillin, dicloxacillin, oxacillin, penicillin G, penicillin V, piperacillin, pyrimethamine, sulfacetamide, sulfadimethoxine, sulfaguanidine, sulfamerazine, sulfameter, sulfamethazine, sulfanilic acid, sulfapyridine, sulfaquinoxaline, sulfathiazole, sulfisoxazole, trimethoprim

REFERENCE
Hows,M.E.P.; Perrett,D.; Kay,J. Optimization of a simultaneous separation of sulphonamides, dihydrofolate reductase inhibitors and β-lactam antibiotics by capillary electrophoresis, *J.Chromatogr.A*, **1997**, *768*, 97–104.

SAMPLE
Matrix: solutions

CAPILLARY ELECTROPHORESIS
Capillary: 67 cm × 50 μm fused-silica (60 cm to detector)
Capillary preparation: Wash with running buffer for 6 min before each run.
Capillary temperature: 25
Running buffer: 500 mM pH 2.1 Sodium citrate buffer
Injection: Hydrodynamic injection for 2 s.
Detector: UV 254
Migration time: 21.5
Voltage: 30 kV
Current: 84 μA
Model: Spectra-Physics Model 1000

OTHER SUBSTANCES
Simultaneous: sulfachlorpyridazine, sulfadiazine, sulfadimethoxine, sulfamerazine, sulfameter, sulfamethazine, sulfamethizole, sulfamethoxazole, sulfamethoxypyridazine, sulfamonomethoxine, sulfapyridine, sulfaquinoxaline, sulfathiazole, sulfisomidine, sulfisoxazole

REFERENCE
Lin,C.-E.; Chang,C.-C.; Lin,W.-c. Migration behavior and separation of sulfonamides in capillary zone electrophoresis. II. Positively charged species at low pH, *J.Chromatogr.A*, **1997**, *759*, 203–209.

Sulfanilic acid

Molecular formula: $C_6H_7NO_3S$
Molecular weight: 173.19
CAS Registry No.: 121-57-3, 6101-32-2 (H2O), 6106-22-5 (sodium salt 2.H2O), 31884-76-1 (zinc salt 4.H2O)
Merck Index (12th ed.): 9096

SAMPLE
Matrix: solutions

Sample preparation: Inject an aliquot of an aqueous solution.

CAPILLARY ELECTROPHORESIS
Capillary: 24 cm × 50 μm coated fused-silica (Bio-Rad)
Running buffer: 300 mM pH 8.6 Tris-boric acid containing 2 mM EDTA, 4% 6K PEG, 4% 12K PEG, 4% 20K PEG, and 4% 36K PEG (PEG is polyethylene glycol)
Injection: Pressure injection at 30 psi for 1 s
Detector: UV 254
Migration time: 3.59
Voltage: 15 kV
Model: Bio-Rad BioFocus 3000

OTHER SUBSTANCES
Simultaneous: quinobene, suramin

KEY WORDS
coated capillary

REFERENCE
Hettiarachchi,K.; Cheung,A.P. Precision in capillary electrophoresis with respect to quantitative analysis of suramin, *J.Chromatogr.A*, **1995**, *717*, 191–202.

SAMPLE
Matrix: solutions

CAPILLARY ELECTROPHORESIS
Capillary: 60 cm × 50 μm fused-silica (47 cm to detector) (Polymicro Technologies)
Capillary temperature: 25
Running buffer: 20 mM pH 8.5 Sodium tetraborate containing 100 mM sodium dodecyl sulfate
Injection: Hydrodynamic injection at 50 mbar for 3.6 s (5 nL).
Detector: UV 205
Migration time: 12.7
Voltage: 22 kV
Model: Crystal 310 (Thermo Unicam)

OTHER SUBSTANCES
Simultaneous: amoxicillin, ampicillin, cephapirin, cloxacillin, dicloxacillin, oxacillin, penicillin G, penicillin V, piperacillin, pyrimethamine, sulfacetamide, sulfadimethoxine, sulfaguanidine, sulfamerazine, sulfameter, sulfamethazine, sulfanilamide, sulfapyridine, sulfaquinoxaline, sulfathiazole, sulfisoxazole, trimethoprim

---

REFERENCE
Hows,M.E.P.; Perrett,D.; Kay,J. Optimization of a simultaneous separation of sulphonamides, dihydrofolate reductase inhibitors and β-lactam antibiotics by capillary electrophoresis, *J.Chromatogr.A*, **1997**, *768*, 97–104.

Sulfapyridine

Molecular formula: $C_{11}H_{11}N_3O_2S$
Molecular weight: 249.29
CAS Registry No.: 144-83-2, 127-57-1 (Na salt monohydrate)
Merck Index (12th ed.): 9108
Lednicer: 1 124

SAMPLE
Matrix: solutions

CAPILLARY ELECTROPHORESIS
Capillary: 60 cm × 50 µm fused-silica (47 cm to detector) (Polymicro Technologies)
Capillary temperature: 25
Running buffer: 20 mM pH 8.5 Sodium tetraborate containing 100 mM sodium dodecyl sulfate
Injection: Hydrodynamic injection at 50 mbar for 3.6 s (5 nL).
Detector: UV 205
Migration time: 9.6
Voltage: 22 kV
Model: Crystal 310 (Thermo Unicam)

OTHER SUBSTANCES
Simultaneous: amoxicillin, ampicillin, cephapirin, cloxacillin, dicloxacillin, penicillin G, penicillin V, piperacillin, pyrimethamine, sulfacetamide, sulfadimethoxine, sulfaguanidine, sulfamerazine, sulfameter, sulfamethazine, sulfanilamide, sulfanilic acid, sulfaquinoxaline, sulfisoxazole, trimethoprim
Interfering: oxacillin, sulfathiazole

REFERENCE
Howe,M.E.P.; Perrett,D.; Kay,J. Optimization of a simultaneous separation of sulphonamides, dihydrofolate reductase inhibitors and β-lactam antibiotics by capillary electrophoresis, *J.Chromatogr.A*, **1997**, *768*, 97–104.

SAMPLE
Matrix: solutions

CAPILLARY ELECTROPHORESIS
Capillary: 67 cm × 50 µm fused-silica (60 cm to detector)
Capillary preparation: Wash with running buffer for 6 min before each run.
Capillary temperature: 25
Running buffer: 500 mM pH 2.1 Sodium citrate buffer
Injection: Hydrodynamic injection for 2 s.
Detector: UV 254
Migration time: 21
Voltage: 30 kV
Current: 84 µA
Model: Spectra-Physics Model 1000

OTHER SUBSTANCES
Simultaneous: sulfachlorpyridazine, sulfadiazine, sulfadimethoxine, sulfamerazine, sulfameter, sulfamethazine, sulfamethizole, sulfamethoxazole, sulfamethoxypyridazine, sulfamonomethoxine, sulfanilamide, sulfaquinoxaline, sulfathiazole, sulfisomidine, sulfisoxazole

REFERENCE
Lin,C.-E.; Chang,C.-C.; Lin,W.-c. Migration behavior and separation of sulfonamides in capillary zone electrophoresis. II. Positively charged species at low pH, *J.Chromatogr.A*, **1997**, *759*, 203–209.

Sulfaquinoxaline

Molecular formula: $C_{14}H_{12}N_4O_2S$
Molecular weight: 300.33
CAS Registry No.: 59-40-5
Merck Index (12th ed.): 9109

SAMPLE
Matrix: milk
Sample preparation: 5 mL Milk + 5 mL 3 M HCl, mix, centrifuge for 20 min. Remove the upper liquid layer and add it to 1 mL 3 M HCl, mix, centrifuge for 20 min, filter (0.45 µm) the supernatant, inject an aliquot of the filtrate.

CAPILLARY ELECTROPHORESIS
Capillary: 44 cm × 50 μm fused-silica (37 cm to detector)
Capillary preparation: Wash with running buffer for 3 min before each run.
Capillary temperature: 25
Running buffer: MeCN:buffer 9:91 (A) or buffer containing 0.5 mM β-cyclodextrin (B) (Prepare buffer by mixing 50 mM disodium tetraborate with 50 mM NaH_2PO_4 so as to produce a pH of 6.85.)
Injection: Hydrodynamic injection for 1 s.
Detector: UV 214, UV 254
Migration time: 4.9 (A); 3.35 (B)
Voltage: 20 kV
Model: Spectra-Physics Model 1000

OTHER SUBSTANCES
Simultaneous: sulfachlorpyridazine, sulfadiazine, sulfadimethoxine, sulfamerazine, sulfameter, sulfamethazine, sulfamethoxazole, sulfamethoxypyridazine, sulfamonomethoxine, sulfathiazole, sulfisomidine, sulfisoxazole

KEY WORDS
cow

REFERENCE
Lin,C.-E.; Lin,W.-c.; Chiou,W.-c.; Lin,E.C.; Chang,C.-C. Migration behavior and separation of sulfonamides in capillary zone electrophoresis. I. Influence of buffer pH and electrolyte modifier, *J.Chromatogr.A*, **1996**, *755*, 261–269.

SAMPLE
Matrix: solutions

CAPILLARY ELECTROPHORESIS
Capillary: 47 cm × 50 μm fused-silica (40 cm to detector) (Polymicro Technologies)
Capillary preparation: Before each run rinse capillary with 100 mM ammonium hydroxide at high pressure for 2 min, with water for 2-3 min, and with running buffer for 2-3 min.
Capillary temperature: 26
Running buffer: MeCN:pH 4 acetate buffer containing 80 mM silver nitrate 15:85
Injection: Hydrodynamic injection for 3-5 s
Detector: UV 254, UV 280
Migration time: 10
Voltage: 20 kV
Model: Beckman P/ACE 2100

OTHER SUBSTANCES
Simultaneous: sulfabenzamide, sulfacetamide, sulfadiazine, sulfadimethoxine, sulfaguanidine, sulfanilamide, sulfathiazole
Interfering: sulfamerazine

REFERENCE
Wright,P.B.; Dorsey,J.G. Silver(I)-mediated separations by capillary zone electrophoresis and micellar electrokinetic chromatography: Argentation electrophoresis, *Anal.Chem.*, **1996**, *68*, 415–424.

SAMPLE
Matrix: solutions

CAPILLARY ELECTROPHORESIS
Capillary: 60 cm × 50 μm fused-silica (47 cm to detector) (Polymicro Technologies)
Capillary temperature: 25
Running buffer: 20 mM pH 8.5 Sodium tetraborate containing 100 mM sodium dodecyl sulfate
Injection: Hydrodynamic injection at 50 mbar for 3.6 s (5 nL).
Detector: UV 205
Migration time: 9.4
Voltage: 22 kV

Model: Crystal 310 (Thermo Unicam)

OTHER SUBSTANCES
Simultaneous: amoxicillin, ampicillin, cephapirin, cloxacillin, dicloxacillin, oxacillin, penicillin G, penicillin V, piperacillin, pyrimethamine, sulfacetamide, sulfadimethoxine, sulfaguanidine, sulfamerazine, sulfameter, sulfamethazine, sulfanilamide, sulfanilic acid, sulfapyridine, sulfathiazole, sulfisoxazole, trimethoprim

REFERENCE
Hows,M.E.P.; Perrett,D.; Kay,J. Optimization of a simultaneous separation of sulphonamides, dihydrofolate reductase inhibitors and β-lactam antibiotics by capillary electrophoresis, *J.Chromatogr.A*, **1997**, *768*, 97–104.

SAMPLE
Matrix: solutions

CAPILLARY ELECTROPHORESIS
Capillary: 43 cm × 50 μm fused-silica (36 cm to detector)
Capillary preparation: Wash with running buffer for 3 min before each injection and with water for 3 min after each injection. Wash a new capillary with 1 M NaOH at 60° for 50 min, with 100 mM NaOH at 60° for 10 min, and with water at 25° for 10 min.
Capillary temperature: 25
Running buffer: 30 mM pH 6.9 Citrate buffer
Injection: Hydrodynamic injection for 2 s
Detector: UV 254
Migration time: 3.35
Voltage: 20 kV
Model: Spectra-Physics Model 1000

OTHER SUBSTANCES
Simultaneous: sulfachlorpyridazine, sulfadiazine, sulfadimethoxine, sulfamerazine, sulfameter, sulfamethazine, sulfamethoxazole, sulfamethoxypyridazine, sulfamonomethoxine, sulfathiazole, sulfisomidine, sulfisoxazole

REFERENCE
Lin,C.-E.; Chang,C.-C.; Lin,W.-c. Migration behavior and separation of sulfonamides in capillary zone electrophoresis. III. Citrate buffer as a background electrolyte, *J.Chromatogr.A*, **1997**, *768*, 105–112.

SAMPLE
Matrix: solutions

CAPILLARY ELECTROPHORESIS
Capillary: 67 cm × 50 μm fused-silica (60 cm to detector)
Capillary preparation: Wash with running buffer for 6 min before each run.
Capillary temperature: 25
Running buffer: 500 mM pH 2.1 Sodium citrate buffer
Injection: Hydrodynamic injection for 2 s.
Detector: UV 254
Migration time: 44
Voltage: 30 kV
Current: 84 μA
Model: Spectra-Physics Model 1000

OTHER SUBSTANCES
Simultaneous: sulfachlorpyridazine, sulfadiazine, sulfadimethoxine, sulfamerazine, sulfameter, sulfamethazine, sulfamethizole, sulfamethoxazole, sulfamethoxypyridazine, sulfamonomethoxine, sulfanilamide, sulfapyridine, sulfathiazole, sulfisomidine, sulfisoxazole

REFERENCE
Lin,C.-E.; Chang,C.-C.; Lin,W.-c. Migration behavior and separation of sulfonamides in capillary zone electrophoresis. II. Positively charged species at low pH, *J.Chromatogr.A*, **1997**, *759*, 203–209.

SAMPLE
Matrix: solutions

CAPILLARY ELECTROPHORESIS
Capillary: 44 cm × 50 μm fused-silica (37 cm to detector)
Capillary preparation: Fill with running buffer for 3 min before each injection then after each
 run wash with water for 2 min, with 1 M NaOH for 3 min, , and with water for 2 min. Wash
 new capillaries with water at 60° for 10 min, with 1 M NaOH at 60° for 1 h, and with water
 at 25° for 10 min.
Running buffer: MeOH:pH 6.85 50 mM phosphate/borate buffer 16.6:83.4 (A) or pH 5.8 phos-
 phate/borate buffer containing 40 mM sodium dodecyl sulfate (B) (Prepare buffer by mixing 50
 mM disodium tetraborate and 50 mM NaH_2PO_4.)
Injection: Hydrodynamic injection for 1 s.
Detector: UV 214
Migration time: 5.0 (A), 10.0 (B)
Voltage: 20 kV
Model: Spectra Physics Model 1000

OTHER SUBSTANCES
Simultaneous: sulfachloropyridazine, sulfadiazine, sulfadimethoxine, sulfamerazine, sulfameter, sulfamethazine, sulfamethoxazole, sulfamethoxypyridazine, sulfamonomethoxine, sulfathiazole, sulfisomidine, sulfisoxazole

REFERENCE
Lin,C.-E.; Lin,W.-c.; Chen,Y.-C.; Wang,S.-W. Migration behavior and selectivity of sulfonamides in capillary
 electrophoresis, *J.Chromatogr.A*, **1997**, *792*, 37–47.

SAMPLE
Matrix: tissue
Sample preparation: Homogenize (household food processor) 10 g tissue, extract (stomacher
 apparatus) with 100 mL MeCN for 5 min, centrifuge at 4000 g for 10 min, filter (0.45 μm),
 inject an aliquot of the filtrate.

CAPILLARY ELECTROPHORESIS
Capillary: 57 cm × 75 μm (50 cm to detector) (Beckman)
Capillary temperature: 25
Running buffer: 20 mM Borate buffer containing 20 mM phosphate, adjusted to pH 7.0 with
 KOH
Injection: Pressure injection for 2 s (39 nL)
Detector: UV 254
Migration time: 20
Voltage: 10 kV
Current: 26.8 μA
Model: Beckman P/ACE System 2000

OTHER SUBSTANCES
Extracted: sulfachlorpyridazine, sulfadiazine, sulfadimethoxine, sulfadoxine, sulfamerazine, sulfamethazine (sulfadimidine), sulfamethoxazole, sulfamethoxydiazine, sulfamethoxypyridazine,
 sulfathiazole, sulfatroxazole, trimethoprim

KEY WORDS
pig

REFERENCE
Ackermans,M.T.; Beckers,J.L.; Everaerts,F.M.; Hoogland,H.; Tomassen,M.J.H. Determination of sulphonamides
 in pork meat extracts by capillary zone electrophoresis, *J.Chromatogr.*, **1992**, *596*, 101–109.

Sulfathiazole

Molecular formula: $C_9H_9N_3O_2S_2$
Molecular weight: 255.32
CAS Registry No.: 72-14-0, 144-74-1 (Na salt)
Merck Index (12th ed.): 9115
Lednicer: 1 124

SAMPLE

Matrix: milk
Sample preparation: 5 mL Milk + 5 mL 3 M HCl, mix, centrifuge for 20 min. Remove the upper liquid layer and add it to 1 mL 3 M HCl, mix, centrifuge for 20 min, filter (0.45 μm) the supernatant, inject an aliquot of the filtrate.

CAPILLARY ELECTROPHORESIS

Capillary: 44 cm × 50 μm fused-silica (37 cm to detector)
Capillary preparation: Wash with running buffer for 3 min before each run.
Capillary temperature: 25
Running buffer: MeCN:buffer 9:91 (A) or buffer containing 0.5 mM β-cyclodextrin (B) (Prepare buffer by mixing 50 mM disodium tetraborate with 50 mM NaH_2PO_4 so as to produce a pH of 6.85.)
Injection: Hydrodynamic injection for 1 s.
Detector: UV 214, UV 254
Migration time: 3.8 (A); 2.5 (B)
Voltage: 20 kV
Model: Spectra-Physics Model 1000

OTHER SUBSTANCES

Simultaneous: sulfachlorpyridazine, sulfadiazine, sulfadimethoxine, sulfamerazine, sulfameter, sulfamethazine, sulfamethoxazole, sulfamethoxypyridazine, sulfamonomethoxine, sulfaquinoxaline, sulfisomidine, sulfisoxazole

KEY WORDS

cow

REFERENCE

Lin,C.-E.; Lin,W.-c.; Chiou,W.-c.; Lin,E.C.; Chang,C.-C. Migration behavior and separation of sulfonamides in capillary zone electrophoresis. I. Influence of buffer pH and electrolyte modifier, *J.Chromatogr.A*, **1996**, *755*, 261–269.

SAMPLE

Matrix: solutions
Sample preparation: Prepare a solution in running buffer, inject an aliquot.

CAPILLARY ELECTROPHORESIS

Capillary: 100 cm × 75 μm fused-silica (20 cm to UV detector) (Polymicro Technologies)
Capillary preparation: Between runs flush capillaries with 1.5 column volumes of 100 mM NaOH and 9 column volumes of running buffer. Flush new capillaries with 9 column volumes of 1 M NaOH, 3 column volumes of water, 3 column volumes of 100 mM HCl, 3 column volumes of water, and 9 column volumes of running buffer.
Running buffer: MeOH:water 20:80 containing 20 mM ammonium acetate, pH 6.8
Injection: Pressure injection at 3.45 kPa (5-28 nL)
Detector: UV 254 or MS, Sciex TAGA 6000E triple quadrupole, API source, electrospray (ion spray) interface at 4 kV (3 kV to ion-sampling orifice), positive-ion mode, 100 μm dia sampling orifice, nitrogen gas curtain, argon collision gas, m/z 256 [M+H]+
Migration time: 3.8 (UV), 18.2 (MS)
Voltage: 26 kV
Model: Beckman P/ACE System 2000

OTHER SUBSTANCES
Simultaneous: sulfadimethoxine, sulfamerazine, sulfamethazine, sulfamethoxazole, sulfanilamide

REFERENCE
Johansson,I.M.; Pavelka,R.; Henion,J.D. Determination of small drug molecules by capillary electrophoresis-atmospheric pressure ionization mass spectrometry, *J.Chromatogr.*, **1991**, *559*, 515–528.

SAMPLE
Matrix: solutions

CAPILLARY ELECTROPHORESIS
Capillary: 47 cm × 50 μm fused-silica (40 cm to detector) (Polymicro Technologies)
Capillary preparation: Before each run rinse capillary with 100 mM ammonium hydroxide at high pressure for 2 min, with water for 2-3 min, and with running buffer for 2-3 min.
Capillary temperature: 26
Running buffer: MeCN:pH 4 acetate buffer containing 80 mM silver nitrate 15:85
Injection: Hydrodynamic injection for 3-5 s
Detector: UV 254, UV 280
Migration time: 10.9
Voltage: 20 kV
Model: Beckman P/ACE 2100

OTHER SUBSTANCES
Simultaneous: sulfabenzamide, sulfacetamide, sulfadiazine, sulfadimethoxine, sulfaguanidine, sulfamerazine, sulfanilamide, sulfaquinoxaline

REFERENCE
Wright,P.B.; Dorsey,J.G. Silver(I)-mediated separations by capillary zone electrophoresis and micellar electrokinetic chromatography: Argentation electrophoresis, *Anal.Chem.*, **1996**, *68*, 415–424.

SAMPLE
Matrix: solutions

CAPILLARY ELECTROPHORESIS
Capillary: 60 cm × 50 μm fused-silica (47 cm to detector) (Polymicro Technologies)
Capillary temperature: 25
Running buffer: 20 mM pH 8.5 Sodium tetraborate containing 100 mM sodium dodecyl sulfate
Injection: Hydrodynamic injection at 50 mbar for 3.6 s (5 nL).
Detector: UV 205
Migration time: 9.6
Voltage: 22 kV
Model: Crystal 310 (Thermo Unicam)

OTHER SUBSTANCES
Simultaneous: amoxicillin, ampicillin, cephapirin, cloxacillin, dicloxacillin, penicillin G, penicillin V, piperacillin, pyrimethamine, sulfacetamide, sulfadimethoxine, sulfaguanidine, sulfamerazine, sulfameter, sulfamethazine, sulfanilamide, sulfanilic acid, sulfaquinoxaline, sulfisoxazole, trimethoprim
Interfering: oxacillin, sulfapyridine

REFERENCE
Hows,M.E.P.; Perrett,D.; Kay,J. Optimization of a simultaneous separation of sulphonamides, dihydrofolate reductase inhibitors and β-lactam antibiotics by capillary electrophoresis, *J.Chromatogr.A*, **1997**, *768*, 97–104.

SAMPLE
Matrix: solutions

CAPILLARY ELECTROPHORESIS
Capillary: 67 cm × 50 μm fused-silica (60 cm to detector)

Capillary preparation: Wash with running buffer for 6 min before each run.
Capillary temperature: 25
Running buffer: 500 mM pH 2.1 Sodium citrate buffer
Injection: Hydrodynamic injection for 2 s.
Detector: UV 254
Migration time: 25
Voltage: 30 kV
Current: 84 μA
Model: Spectra-Physics Model 1000

OTHER SUBSTANCES
Simultaneous: sulfachlorpyridazine, sulfadiazine, sulfadimethoxine, sulfamerazine, sulfameter, sulfamethazine, sulfamethizole, sulfamethoxazole, sulfamethoxypyridazine, sulfamonomethoxine, sulfanilamide, sulfapyridine, sulfaquinoxaline, sulfisomidine, sulfisoxazole

REFERENCE
Lin,C.-E.; Chang,C.-C.; Lin,W.-c. Migration behavior and separation of sulfonamides in capillary zone electrophoresis. II. Positively charged species at low pH, *J.Chromatogr.A*, **1997**, *759*, 203–209.

SAMPLE
Matrix: solutions

CAPILLARY ELECTROPHORESIS
Capillary: 44 cm × 50 μm fused-silica (37 cm to detector)
Capillary preparation: Fill with running buffer for 3 min before each injection then after each run wash with water for 2 min, with 1 M NaOH for 3 min, , and with water for 2 min. Wash new capillaries with water at 60° for 10 min, with 1 M NaOH at 60° for 1 h, and with water at 25° for 10 min.
Running buffer: MeOH:pH 6.85 50 mM phosphate/borate buffer 16.6:83.4 (A) or pH 5.8 phosphate/borate buffer containing 40 mM sodium dodecyl sulfate (B) (Prepare buffer by mixing 50 mM disodium tetraborate and 50 mM NaH_2PO_4.)
Injection: Hydrodynamic injection for 1 s.
Detector: UV 214
Migration time: 3.65 (A), 4.1 (B)
Voltage: 20 kV
Model: Spectra Physics Model 1000

OTHER SUBSTANCES
Simultaneous: sulfachloropyridazine, sulfadiazine, sulfadimethoxine, sulfamerazine, sulfameter, sulfamethazine, sulfamethoxazole, sulfamethoxypyridazine, sulfamonomethoxine, sulfaquinoxaline, sulfisomidine, sulfisoxazole

REFERENCE
Lin,C.-E.; Lin,W.-c.; Chen,Y.-C.; Wang,S.-W. Migration behavior and selectivity of sulfonamides in capillary electrophoresis, *J.Chromatogr.A*, **1997**, *792*, 37–47.

SAMPLE
Matrix: solutions

CAPILLARY ELECTROPHORESIS
Capillary: 43 cm × 50 μm fused-silica (36 cm to detector)
Capillary preparation: Wash with running buffer for 3 min before each injection and with water for 3 min after each injection. Wash a new capillary with 1 M NaOH at 60° for 50 min, with 100 mM NaOH at 60° for 10 min, and with water at 25° for 10 min.
Capillary temperature: 25
Running buffer: 30 mM pH 6.9 Citrate buffer
Injection: Hydrodynamic injection for 2 s
Detector: UV 254
Migration time: 2.65
Voltage: 20 kV
Model: Spectra-Physics Model 1000

OTHER SUBSTANCES
Simultaneous: sulfachlorpyridazine, sulfadiazine, sulfadimethoxine, sulfamerazine, sulfameter, sulfamethazine, sulfamethoxazole, sulfamethoxypyridazine, sulfamonomethoxine, sulfaquinoxaline, sulfisomidine, sulfisoxazole

REFERENCE
Lin,C.-E.; Chang,C.-C.; Lin,W.-c. Migration behavior and separation of sulfonamides in capillary zone electrophoresis. III. Citrate buffer as a background electrolyte, *J.Chromatogr.A*, **1997**, *768*, 105–112.

SAMPLE
Matrix: tissue
Sample preparation: Homogenize (household food processor) 10 g tissue, extract (stomacher apparatus) with 100 mL MeCN for 5 min, centrifuge at 4000 g for 10 min, filter (0.45 μm), inject an aliquot of the filtrate.

CAPILLARY ELECTROPHORESIS
Capillary: 57 cm $\times$ 75 μm (50 cm to detector) (Beckman)
Capillary temperature: 25
Running buffer: 20 mM Borate buffer containing 20 mM phosphate, adjusted to pH 7.0 with KOH
Injection: Pressure injection for 2 s (39 nL)
Detector: UV 254
Migration time: 14
Voltage: 10 kV
Current: 26.8 μA
Model: Beckman P/ACE System 2000

OTHER SUBSTANCES
Extracted: sulfachlorpyridazine, sulfadiazine, sulfadimethoxine, sulfadoxine, sulfamerazine, sulfamethazine (sulfadimidine), sulfamethoxazole, sulfamethoxydiazine, sulfaquinoxaline, sulfatroxazole, trimethoprim
Interfering: sulfamethoxypyridazine

KEY WORDS
pig

REFERENCE
Ackermans,M.T.; Beckers,J.L.; Everaerts,F.M.; Hoogland,H.; Tomassen,M.J.H. Determination of sulphonamides in pork meat extracts by capillary zone electrophoresis, *J.Chromatogr.*, **1992**, *596*, 101–109.

Sulfisomidine

Molecular formula: $C_{12}H_{14}N_4O_2S$
Molecular weight: 278.33
CAS Registry No.: 515-64-0
Merck Index (12th ed.): 9124

SAMPLE
Matrix: milk
Sample preparation: 5 mL Milk + 5 mL 3 M HCl, mix, centrifuge for 20 min. Remove the upper liquid layer and add it to 1 mL 3 M HCl, mix, centrifuge for 20 min, filter (0.45 μm) the supernatant, inject an aliquot of the filtrate.

CAPILLARY ELECTROPHORESIS
Capillary: 44 cm $\times$ 50 μm fused-silica (37 cm to detector)
Capillary preparation: Wash with running buffer for 3 min before each run.
Capillary temperature: 25

Running buffer: MeCN:buffer 9:91 (A) or buffer containing 0.5 mM β-cyclodextrin (B) (Prepare buffer by mixing 50 mM disodium tetraborate with 50 mM NaH_2PO_4 so as to produce a pH of 6.85.)
Injection: Hydrodynamic injection for 1 s.
Detector: UV 214, UV 254
Migration time: 3.5 (A); 2.4 (B)
Voltage: 20 kV
Model: Spectra-Physics Model 1000

OTHER SUBSTANCES
Simultaneous: sulfachlorpyridazine, sulfadiazine, sulfadimethoxine, sulfamerazine, sulfameter, sulfamethazine, sulfamethoxazole, sulfamethoxypyridazine, sulfamonomethoxine, sulfaquinoxaline, sulfathiazole, sulfisoxazole

KEY WORDS
cow

REFERENCE
Lin,C.-E.; Lin,W.-c.; Chiou,W.-c.; Lin,E.C.; Chang,C.-C. Migration behavior and separation of sulfonamides in capillary zone electrophoresis. I. Influence of buffer pH and electrolyte modifier, *J.Chromatogr A*, **1996**, *755*, 261–269.

SAMPLE
Matrix: solutions

CAPILLARY ELECTROPHORESIS
Capillary: 67 cm × 50 μm fused-silica (60 cm to detector)
Capillary preparation: Wash with running buffer for 6 min before each run.
Capillary temperature: 25
Running buffer: 500 mM pH 2.1 Sodium citrate buffer
Injection: Hydrodynamic injection for 2 s.
Detector: UV 254
Migration time: 14
Voltage: 30 kV
Current: 84 μA
Model: Spectra-Physics Model 1000

OTHER SUBSTANCES
Simultaneous: sulfachlorpyridazine, sulfadiazine, sulfadimethoxine, sulfamerazine, sulfameter, sulfamethazine, sulfamethizole, sulfamethoxazole, sulfamethoxypyridazine, sulfamonomethoxine, sulfanilamide, sulfapyridine, sulfaquinoxaline, sulfathiazole, sulfisoxazole

REFERENCE
Lin,C.-E.; Chang,C.-C.; Lin,W.-c. Migration behavior and separation of sulfonamides in capillary zone electrophoresis. II. Positively charged species at low pH, *J.Chromatogr.A*, **1997**, *759*, 203–209.

SAMPLE
Matrix: solutions

CAPILLARY ELECTROPHORESIS
Capillary: 43 cm × 50 μm fused-silica (36 cm to detector)
Capillary preparation: Wash with running buffer for 3 min before each injection and with water for 3 min after each injection. Wash a new capillary with 1 M NaOH at 60° for 50 min, with 100 mM NaOH at 60° for 10 min, and with water at 25° for 10 min.
Capillary temperature: 25
Running buffer: 30 mM pH 6.9 Citrate buffer
Injection: Hydrodynamic injection for 2 s
Detector: UV 254
Migration time: 2.4
Voltage: 20 kV
Model: Spectra-Physics Model 1000

OTHER SUBSTANCES
Simultaneous: sulfachlorpyridazine, sulfadiazine, sulfadimethoxine, sulfamerazine, sulfameter, sulfamethazine, sulfamethoxazole, sulfamethoxypyridazine, sulfamonomethoxine, sulfaquinoxaline, sulfathiazole, sulfisoxazole

REFERENCE
Lin,C.-E.; Chang,C.-C.; Lin,W.-c. Migration behavior and separation of sulfonamides in capillary zone electrophoresis. III. Citrate buffer as a background electrolyte, *J.Chromatogr.A*, **1997**, *768*, 105–112.

SAMPLE
Matrix: solutions

CAPILLARY ELECTROPHORESIS
Capillary: 44 cm × 50 μm fused-silica (37 cm to detector)
Capillary preparation: Fill with running buffer for 3 min before each injection then after each run wash with water for 2 min, with 1 M NaOH for 3 min, , and with water for 2 min. Wash new capillaries with water at 60° for 10 min, with 1 M NaOH at 60° for 1 h, and with water at 25° for 10 min.
Running buffer: MeOH:pH 6.85 50 mM phosphate/borate buffer 16.6:83.4 (A) or pH 5.8 phosphate/borate buffer containing 40 mM sodium dodecyl sulfate (B) (Prepare buffer by mixing 50 mM disodium tetraborate and 50 mM NaH_2PO_4.)
Injection: Hydrodynamic injection for 1 s.
Detector: UV 214
Migration time: 3.5 (A), 3.8 (B)
Voltage: 20 kV
Model: Spectra Physics Model 1000

OTHER SUBSTANCES
Simultaneous: sulfachloropyridazine, sulfadiazine, sulfadimethoxine, sulfamerazine, sulfameter, sulfamethazine, sulfamethoxazole, sulfamethoxypyridazine, sulfamonomethoxine, sulfaquinoxaline, sulfathiazole, sulfisoxazole

REFERENCE
Lin,C.-E.; Lin,W.-c.; Chen,Y.-C.; Wang,S.-W. Migration behavior and selectivity of sulfonamides in capillary electrophoresis, *J.Chromatogr.A*, **1997**, *792*, 37–47.

Sulfisoxazole

Molecular formula: $C_{11}H_{13}N_3O_3S$
Molecular weight: 267.30
CAS Registry No.: 127-69-5, 4299-60-9 (diolamine)
Merck Index (12th ed.): 9125
Lednicer: 1 124

SAMPLE
Matrix: formulations
Sample preparation: Grind ten tablets and weigh out an amount corresponding to 50 mg sulfisoxazole and 5 mg phenazopyridine, add 40 mL running buffer, sonicate for 30 min, make up to 50 mL with running buffer, filter (0.45 μm). Dilute an aliquot of the filtrate 1:25 with running buffer, inject an aliquot.

CAPILLARY ELECTROPHORESIS
Capillary: 70 cm × 75 μm fused-silica
Capillary preparation: After each run refill reservoirs, rinse capillary with fresh buffer at 7 psi for 3 min.
Running buffer: 5 mM NaH_2PO_4 containing 5 mM Na_2HPO_4 and 50 mM sodium dodecyl sulfate, adjusted to pH 7.0 with NaOH

Injection: Hydrodynamic injection at 5 cm for 10 s (6.8 nL)
Detector: UV 240
Migration time: 12.5
Voltage: 15 kV
Model: Dionex Model CES I
Limit of detection: 320 ng/mL (S/N 2)

OTHER SUBSTANCES
Simultaneous: impurities, phenazopyridine, sulfanilamide

KEY WORDS
tablets

REFERENCE
Nickerson,B.; Scypinski,S.; Sokoloff,H.; Sahota,S. Separation of sulfisoxazole, phenazopyridine, and their related impurities by micellar electrokinetic chromatography, *J.Liq.Chromatogr.*, **1995**, *18*, 3847–3875.

SAMPLE
Matrix: milk
Sample preparation: 5 mL Milk + 5 mL 3 M HCl, mix, centrifuge for 20 min. Remove the upper liquid layer and add it to 1 mL 3 M HCl, mix, centrifuge for 20 min, filter (0.45 μm) the supernatant, inject an aliquot of the filtrate.

CAPILLARY ELECTROPHORESIS
Capillary: 44 cm × 50 μm fused-silica (37 cm to detector)
Capillary preparation: Wash with running buffer for 3 min before each run.
Capillary temperature: 25
Running buffer: MeCN:buffer 9:91 (A) or buffer containing 0.5 mM β-cyclodextrin (B) (Prepare buffer by mixing 50 mM disodium tetraborate with 50 mM NaH_2PO_4 so as to produce a pH of 6.85.)
Injection: Hydrodynamic injection for 1 s.
Detector: UV 214, UV 254
Migration time: 5.4 (A); 3.7 (B)
Voltage: 20 kV
Model: Spectra-Physics Model 1000

OTHER SUBSTANCES
Simultaneous: sulfachlorpyridazine, sulfadiazine, sulfadimethoxine, sulfamerazine, sulfameter, sulfamethazine, sulfamethoxazole, sulfamethoxypyridazine, sulfamonomethoxine, sulfaquinoxaline, sulfathiazole, sulfisomidine

KEY WORDS
cow

REFERENCE
Lin,C.-E.; Lin,W.-c.; Chiou,W.-c.; Lin,E.C.; Chang,C.-C. Migration behavior and separation of sulfonamides in capillary zone electrophoresis. I. Influence of buffer pH and electrolyte modifier, *J.Chromatogr.A*, **1996**, *755*, 261–269.

SAMPLE
Matrix: solutions

CAPILLARY ELECTROPHORESIS
Capillary: 50 cm × 50 μm fused-silica (Tokyo Kasei)
Capillary preparation: Rinse capillary with 1 M NaOH after each run.
Running buffer: MeOH:buffer 10:90 (Buffer was 10 mM pH 7 ammonium formate containing 2% butyl acrylate-butyl methacrylate-methacrylic acid copolymer sodium salt (BBMA) (Dai-ichi Kogyo Seiyaku)
Injection: Hydrostatic injection at 15 cm for 10-30 s
Detector: MS, Hitachi M-1000 quadrupole, electrospray at 3 kV, details of interface given in paper, sheath liquid MeOH:water:formic acid 50:50:1 at 5 μL/min, positive-ion mode, drift voltage 70 V, focusing voltage 140-150 V

Migration time: 11.5
Voltage: 10 kV
Model: laboratory constructed

OTHER SUBSTANCES
Simultaneous: sulfadiazine, sulfamethazine, sulfisomidine

REFERENCE
Ozaki,H.; Itou,N.; Terabe,S.; Takada,Y.; Sakairi,M.; Koizumi,H. Micellar electrokinetic chromatography-mass spectrometry using a high-molecular-mass surfactant. On-line coupling with an electrospray ionization interface, *J.Chromatogr.A*, **1995**, *716*, 69–79.

SAMPLE
Matrix: solutions

CAPILLARY ELECTROPHORESIS
Capillary: 43 cm × 50 μm fused-silica (36 cm to detector)
Capillary preparation: Wash with running buffer for 3 min before each injection and with water for 3 min after each injection. Wash a new capillary with 1 M NaOH at 60° for 50 min, with 100 mM NaOH at 60° for 10 min, and with water at 25° for 10 min.
Capillary temperature: 25
Running buffer: 30 mM pH 6.9 Citrate buffer
Injection: Hydrodynamic injection for 2 s
Detector: UV 254
Migration time: 3.7
Voltage: 20 kV
Model: Spectra-Physics Model 1000

OTHER SUBSTANCES
Simultaneous: sulfachlorpyridazine, sulfadiazine, sulfadimethoxine, sulfamerazine, sulfameter, sulfamethazine, sulfamethoxazole, sulfamethoxypyridazine, sulfamonomethoxine, sulfaquinoxaline, sulfathiazole, sulfisomidine

---

REFERENCE
Lin,C.-E.; Chang,C.-C.; Lin,W.-c. Migration behavior and separation of sulfonamides in capillary zone electrophoresis. III. Citrate buffer as a background electrolyte, *J.Chromatogr.A*, **1997**, *768*, 105–112.

SAMPLE
Matrix: solutions

CAPILLARY ELECTROPHORESIS
Capillary: 44 cm × 50 μm fused-silica (37 cm to detector)
Capillary preparation: Fill with running buffer for 3 min before each injection then after each run wash with water for 2 min, with 1 M NaOH for 3 min, , and with water for 2 min. Wash new capillaries with water at 60° for 10 min, with 1 M NaOH at 60° for 1 h, and with water at 25° for 10 min.
Running buffer: MeOH:pH 6.85 50 mM phosphate/borate buffer 16.6:83.4 (A) or pH 5.8 phosphate/borate buffer containing 40 mM sodium dodecyl sulfate (B) (Prepare buffer by mixing 50 mM disodium tetraborate and 50 mM NaH_2PO_4.)
Injection: Hydrodynamic injection for 1 s.
Detector: UV 214
Migration time: 5.5 (A), 6.0 (B)
Voltage: 20 kV
Model: Spectra Physics Model 1000

OTHER SUBSTANCES
Simultaneous: sulfachloropyridazine, sulfadiazine, sulfadimethoxine, sulfamerazine, sulfameter, sulfamethazine, sulfamethoxazole, sulfamethoxypyridazine, sulfamonomethoxine, sulfaquinoxaline, sulfathiazole, sulfisomidine

REFERENCE
Lin,C.-E.; Lin,W.-c.; Chen,Y.-C.; Wang,S.-W. Migration behavior and selectivity of sulfonamides in capillary electrophoresis, *J.Chromatogr.A*, **1997**, *792*, 37–47.

SAMPLE
Matrix: solutions

CAPILLARY ELECTROPHORESIS
Capillary: 67 cm × 50 μm fused-silica (60 cm to detector)
Capillary preparation: Wash with running buffer for 6 min before each run.
Capillary temperature: 25
Running buffer: 500 mM pH 2.1 Sodium citrate buffer
Injection: Hydrodynamic injection for 2 s.
Detector: UV 254
Migration time: 50
Voltage: 30 kV
Current: 84 μA
Model: Spectra-Physics Model 1000

OTHER SUBSTANCES
Simultaneous: sulfachlorpyridazine, sulfadiazine, sulfadimethoxine, sulfamerazine, sulfameter, sulfamethazine, sulfamethizole, sulfamethoxazole, sulfamethoxypyridazine, sulfamonomethoxine, sulfanilamide, sulfapyridine, sulfaquinoxaline, sulfathiazole, sulfisomidine

REFERENCE
Lin,C.-E.; Chang,C.-C.; Lin,W.-c. Migration behavior and separation of sulfonamides in capillary zone electrophoresis. II. Positively charged species at low pH, *J.Chromatogr.A*, **1997**, *759*, 203–209.

Sulindac

Molecular formula: $C_{20}H_{17}FO_3S$
Molecular weight: 356.42
CAS Registry No.: 38194-50-2
Merck Index (12th ed.): 9155
Lednicer: 2 210

SAMPLE
Matrix: blood
Sample preparation: 0.1-1 mL Plasma + 2 μg IS + 400 μL 1 M HCl + 10 mL diethyl ether, agitate mechanically for 10 min, centrifuge at 3500 g for 10 min. Remove the organic layer and evaporate it to dryness under a stream of nitrogen, reconstitute the residue in MeOH:40 mM NaH_2PO_4 50:50, inject an aliquot.

CAPILLARY ELECTROPHORESIS
Capillary: 58.7 cm × 75 μm fused-silica (50 cm to detector)
Capillary preparation: Wash with running buffer for 2 min before each injection. At the beginning of each day wash capillary with 100 mM NaOH for 10 min.
Capillary temperature: 22
Running buffer: MeOH:buffer 3:97 (Buffer was 40 mM pH 8 NaH_2PO_4 containing 104 mM sodium dodecyl sulfate.)
Injection: Hydrodynamic injection for 2 s
Detector: UV 254
Migration time: 21.6
Internal standard: benzoyl-4-phenyl-2-butyric acid (16.3)
Voltage: 20 kV
Model: Beckman P/ACE 2000
Limit of detection: 670 ng/mL (S/N 5)

OTHER SUBSTANCES
Extracted: diclofenac, diflunisal, etodolac, fenbufen, fenoprofen, flurbiprofen, ibuprofen, indomethacin, ketoprofen, naproxen, niflumic acid, piroxicam, tenoxicam, tiaprofenic acid
Noninterfering: acetaminophen, amitriptyline, caffeine, clomipramine, deoxysulindac, desipramine, diazepam, 5'-hydroxytenoxicam, imipramine, maprotiline, nortriptyline, phenobarbital, phenytoin, sulfamethoxazole, theophylline, trimipramine

KEY WORDS
plasma

REFERENCE
Maboundou,C.W.; Paintaud,G.; Bérard,M.; Bechtel,P.R. Separation of fifteen non-steroidal anti-inflammatory drugs using micellar electrokinetic capillary chromatography, *J.Chromatogr.B*, **1994**, *657*, 173–183.

SAMPLE
Matrix: solutions

CAPILLARY ELECTROPHORESIS
Capillary: 42.5 cm × 25 μm
Capillary temperature: 30
Running buffer: 20 mM pH 7.0 Phosphate containing 25 mM sodium dodecyl sulfate
Injection: Vacuum injection for 2 s
Detector: UV 230
Migration time: 4.5
Voltage: 25 kV
Current: 5 μA
Model: Applied Biosystems Model 270A

OTHER SUBSTANCES
Simultaneous: diflunisal, ibuprofen, indomethacin, naproxen, tolmetin

REFERENCE
Weinberger,R.; Albin,M. Quantitative micellar electrokinetic capillary chromatography: Linear dynamic range, *J.Liq.Chromatogr.*, **1991**, *14*, 953–972.

SAMPLE
Matrix: solutions
Sample preparation: Prepare a 20 μg/mL solution in 7.5 mM glycine adjusted to pH 8.0 with triethanolamine, inject an aliquot.

CAPILLARY ELECTROPHORESIS
Capillary: 44 cm × 50 μm fused-silica (37 cm to detector) (Supelco)
Capillary preparation: After each injection wash capillary with water for 2 min and with running buffer for 3 min. At the beginning of each day wash with running buffer for 10 min. Condition a new capillary with 1 M NaOH, 100 mM NaOH, water, and running buffer.
Capillary temperature: 25
Running buffer: 75 mM Glycine adjusted to pH 9.1 with triethanolamine
Injection: Hydrodynamic injection for 5 s
Detector: UV 280
Migration time: 4.95
Voltage: 15 kV
Model: Spectra-physics Spectraphoresis 1000 CE
Limit of quantitation: 2.8 μg/mL
Limit of detection: 840 ng/mL

OTHER SUBSTANCES
Simultaneous: alclofenac, indomethacin, piroxicam, tiaprofenic acid

REFERENCE
Bechet,I.; Fillet,M.; Hubert,P.; Crommen,J. Quantitative analysis of non-steroidal anti-inflammatory drugs by capillary zone electrophoresis, *J.Pharm.Biomed.Anal.*, **1995**, *13*, 497–503.

SAMPLE
Matrix: solutions
Sample preparation: Inject an aliquot of a 50 μM solution in MeOH:water 40:60.

CAPILLARY ELECTROPHORESIS
Capillary temperature: 25
Running buffer: MeOH:water 40:60 containing 75 mM glycine and 15 mM hydroxypropyl-β-cyclodextrin adjusted to pH 9.1 with triethanolamine
Injection: Hydrodynamic injection for 2 s
Detector: UV 280
Migration time: 15.8
Voltage: 15 kV

OTHER SUBSTANCES
Simultaneous: alclofenac, bufexamac, carprofen, flufenamic acid, flurbiprofen, indomethacin, ketoprofen, naproxen, niflumic acid, piroxicam, tiaprofenic acid

REFERENCE
Bechet,I.; Fillet,M.; Hubert,P.; Crommen,J. Improvement of achiral resolution in capillary zone electrophoresis by use of cyclodextrin additives, *Biomed.Chromatogr.*, **1995**, *9*, 267–268.

SAMPLE
Matrix: solutions
Sample preparation: Inject an aliquot of a solution in MeOH:water 10:90.

CAPILLARY ELECTROPHORESIS
Capillary: 44 cm × 50 μm fused-silica (37 cm to detector) (Supelco)
Capillary preparation: Wash with running buffer for 3 min after each injection. Wash with running buffer for 10 min at the end of each day.
Capillary temperature: 25
Running buffer: 100 mM Phosphoric acid containing 5 mM sulfobutyl ether-β-cyclodextrin and 10 mM heptakis(2,6 di-O-methyl)-β-cyclodextrin, adjusted to pH 3.0 with triethanolamine
Injection: Hydrodynamic injection for 5 s (13.3 nL).
Detector: UV 280
Voltage: -25 kV
Current: 60 μA
Model: SpectraPhoresis 1000 CE

KEY WORDS
chiral; detector at anode; resolution (R_s = 3.8)

REFERENCE
Fillet,M.; Hubert,P.; Crommen,J. Enantioseparation of nonsteroidal anti-inflammatory drugs by capillary electrophoresis using mixtures of anionic and uncharged β-cyclodextrins as chiral additives, *Electrophoresis*, **1997**, *18*, 1013–1018.

Sulpiride

Molecular formula: $C_{15}H_{23}N_3O_4S$
Molecular weight: 341.43
CAS Registry No.: 15676-16-1
Merck Index (12th ed.): 9163
Lednicer: 2 94

SAMPLE
Matrix: bulk

CAPILLARY ELECTROPHORESIS

Capillary: 36 cm × 50 μm fused-silica coated with linear polyacrylamide (31.5 cm to detector) (GL Science)

Capillary preparation: At the beginning and end of each day rinse capillary with capillary wash solution (Bio-Rad Cat. No. 148-5022) at 690 kPa for more than 3 min and with water at 690 kPa for more than 3 min. Coat capillary as follows. Treat capillary with 1 M NaOH at room temperature for 1 h, rinse with water, dry by passing nitrogen gas through the capillary at 110° for 6 h. Pass thionyl chloride through the capillary using a suction pump for several min, seal capillary at both ends and heat at 70° for 6 h. Unseal the capillary and fill with 250 mM vinyl magnesium bromide in THF by suction, seal the capillary, heat at 70° for 6 h. Open the capillary and rinse it with THF for several min, rinse with distilled water, fill the capillary with polymerization solution, heat at 28 ± 2° for 1 h, condition at -100 V/cm for 30 min (Anal. Sci. 1994, 10, 1). (The polymerization solution was 5% acrylamide in water containing 49 mM Tris, 384 mM glycine, and 0.1% sodium dodecyl sulfate, degas in an ultrasonic bath. Add 40 μL 10% N,N,N',N'-tetramethylethylenediamine and 10 μL 10% ammonium persulfate to 5 mL of the degassed solution, mix thoroughly.)

Running buffer: 50 mM pH 6.0 Phosphate buffer

Injection: Before each injection rinse with water at 690 kPa for 30 s, rinse with running buffer at 690 kPa for 30 s, partially fill with separation solution (1.5 mM α_1-acid glycoprotein (Cohn fraction VI) (ICN) in running buffer) at 6.9 kPa for 190 s (27 cm), inject sample at 6.9 kPa for 2 s, electrophorese with running buffer (Note that α_1-acid glycoprotein from other suppliers may provide inferior results).

Detector: UV 210

Migration time: 13.2 (S-(-)), 14 (R-(+))

Voltage: 12 kV

Model: Bio-rad BioFocus 3000

Limit of quantitation: 0.3%

Limit of detection: 0.1%

KEY WORDS

chiral; coated capillary

REFERENCE

Tanaka,Y.; Terabe,S. Separation of the enantiomers of basic drugs by affinity capillary electrophoresis using a partial filling technique and α_1-acid glycoprotein as chiral selector, *Chromatographia*, **1997**, *44*, 119–128.

SAMPLE

Matrix: solutions

Sample preparation: Inject an aliquot of a 100 μg/mL solution in running buffer.

CAPILLARY ELECTROPHORESIS

Capillary: 30 cm × 50 μm fused-silica (25.5 cm to detector) (Yongnian Optical Conductive Fiber Plant, China), coated with polyacrylamide

Capillary preparation: No details of the polyacrylamide coating process are provided. However, another paper (LC.GC 1997, 15, 40) by this group indicates that they use the procedure of Hjertén, thus: Adjust the pH of 20 mL water to 3.5 with acetic acid, add 80 μL 3-(trimethoxysilyl)propyl methacrylate (3-methacryloxypropyltrimethoxysilane), mix, suck into capillary, let stand at room temperature for 1 h, remove the solution, wash with water. Fill the capillary with a deaerated 3-4% acrylamide solution containing 1 μL/mL N,N,N',N'-tetramethylethylenediamine and 1 mg/mL potassium persulfate, let stand for 30 min, remove excess solution by aspiration, rinse with water, remove water by aspiration, dry at 35° (J. Chromatogr. 1985, 347, 191).

Capillary temperature: 25

Running buffer: 100 mM NaH_2PO_4 adjusted to pH 2.5

Injection: Electrokinetic injection at 15 kV for 3 s.

Detector: UV 200, UV 210

Migration time: 4.71

Voltage: 15 kV

Model: Bio-Rad BioFocus 3000

OTHER SUBSTANCES

Simultaneous: amorolfine, brompheniramine, bupivacaine, carteolol, chloroquine, chlorpheniramine, chlorphenoxamine, disopyramide, dobutamine, doxylamine, flecainide, gallopamil, ketamine, mepindolol, orphenadrine, oxybutynin, phenoxybenzamine, pindolol, propafenone, propranolol, talinolol, tropicamide, verapamil

KEY WORDS
coated capillary

REFERENCE
Koppenhoefer,B.; Epperlein,U.; Xiaofeng,Z.; Bingcheng,L. Separation of enantiomers of drugs by capillary electrophoresis. Part 4: Hydroxypropyl-γ-cyclodextrin as chiral solvating agent, *Electrophoresis*, **1997**, *18*, 924–930.

SAMPLE
Matrix: solutions

CAPILLARY ELECTROPHORESIS
Capillary: 29-36 cm × 50 μm fused-silica (24.5-31.5 cm to detector) (Yongnian Optical Conductive Fiber Plant, China) coated with polyacrylamide
Capillary preparation: Coat capillary as follows. Adjust the pH of 20 mL water to 3.5 with acetic acid, add 80 μL 3-(trimethoxysilyl)propyl methacrylate (3-methacryloxypropyltrimethoxysilane), mix, suck into capillary, let stand at room temperature for 1 h, remove the solution, wash with water. Fill the capillary with a deaerated 3-4% acrylamide solution containing 1 μL/mL N,N,N',N'-tetramethylethylenediamine and 1 mg/mL potassium persulfate, let stand for 30 min, remove excess solution by aspiration, rinse with water, remove water by aspiration, dry at 35° (J. Chromatogr. 1985, 347, 191).
Capillary temperature: 25
Running buffer: 100 mM pH 2.5 NaH_2PO_4 (A) or 100 mM pH 2.5 NaH_2PO_4 containing 45 mM hydroxypropyl-α-cyclodextrin (Wacker, Munich) (B)
Injection: Electromigration at 15 kV for 3 s.
Detector: UV 200; UV 210
Migration time: 4.71 (A); 6.32 (B) (no separation of enantiomers)
Voltage: 15 kV
Model: Bio-Focus 3000

OTHER SUBSTANCES
Also analyzed: albuterol (salbutamol), alprenolol, amorolfine, atenolol, atropine, azelastine, baclofen, bamethan, benproperine, benserazide, biperiden, bisoprolol, brompheniramine, bupivacaine, bupranolol, butamirate, butethamate, carazolol, carbuterol, carteolol, carvedilol, celiprolol, chloroquine, chlorpheniramine, chlorphenoxamine, cicletanine, clenbuterol, clidinium bromide, clobutinol, dimethindene, dipivefrin, disopyramide, dobutamine, doxylamine, fendiline, flecainide, gallopamil, homatropine, ipratropium bromide, isoproterenol (isoprenaline), isothipendyl, ketamine, meclizine, mefloquine, mepindolol, mequitazine, metaclazepam, metaproterenol (orciprenaline), metipranolol, metoprolol, nafronyl (naftidrofuryl), nefopam, nicardipine, norfenefrine, ofloxacin, ornidazole, orphenadrine, oxomemazine, oxprenolol, oxybutynin, phenoxybenzamine, phenylpropanolamine, pholedrine, pindolol, pirbuterol, prilocaine, procyclidine, promethazine, propafenone, propranolol, reproterol, sotalol, synephrine, talinolol, terbutaline, tetrahydrozoline (tetryzoline), theodrenaline, tioconazole, tocainide, trihexyphenidyl, trimeprazine (alimemazine), trimipramine, tropicamide, verapamil, zopiclone

KEY WORDS
coated capillary

REFERENCE
Koppenhoefer,B.; Eperlein,U.; Schlunk,R.; Zhu,X.; Lin,B. Separation of enantiomers of drugs by capillary electrophoresis. V. Hydroxypropyl-α-cyclodextrin as chiral solvating agent, *J.Chromatogr.A*, **1998**, *793*, 153–164.

Sumatriptan

Molecular formula: $C_{14}H_{21}N_3O_2S$
Molecular weight: 295.41
CAS Registry No.: 103628-46-2, 103628-48-4 (succinate)
Merck Index (12th ed.): 9172
Lednicer: 5 108

SAMPLE
Matrix: formulations

CAPILLARY ELECTROPHORESIS
Capillary: 60 cm × 75 μm fused-silica
Running buffer: 25 mM NaH_2PO_4 adjusted to pH 2.3 with concentrated phosphoric acid
Injection: Hydrodynamic injection for 20 s
Detector: UV 214
Migration time: 8
Internal standard: 2-(dimethylaminomethyl)-5-[(2-aminoethyl)sulfinylmethyl]furan (4.3)
Voltage: +20 kV
Model: Waters Quantum 4000 CE
Limit of detection: 0.1% (for impurities)

OTHER SUBSTANCES
Simultaneous: impurities

KEY WORDS
injections; saline

REFERENCE
Altria,K.D.; Filbey,S.D. Quantitative pharmaceutical analysis by capillary electrophoresis, *J.Liq.Chromatogr.*, **1993**, *16*, 2281–2292.

SAMPLE
Matrix: solutions

CAPILLARY ELECTROPHORESIS
Capillary: 57 cm × 75 μm fused-silica
Capillary preparation: Rinse with 100 mM NaOH for 1 min and rinse with running buffer for 2 min
Capillary temperature: 25
Running buffer: 100 mM pH 2.5 NaH_2PO_4
Injection: Electrokinetic injection at 10 kV for 20 s
Detector: UV 200
Migration time: 15.5
Voltage: 30 kV
Model: Beckman P/ACE 2000
Limit of detection: 2 ng/mL

OTHER SUBSTANCES
Simultaneous: albuterol

REFERENCE
Altria,K.D. Optimization and improvement of sensitivity in capillary electrophoresis for quantitation of selected pharmaceuticals, *LC.GC*, **1993**, *11*, 438–442.

SAMPLE
Matrix: solutions
Sample preparation: Inject an aliquot of an aqueous solution.

CAPILLARY ELECTROPHORESIS
Capillary: 37 cm $\times$ 100 μm fused-silica
Capillary preparation: Before injection rinse with 100 mM NaOH for 30 s and with running buffer for 30 s. Condition a new capillary by rinsing with 100 mM NaOH for 30 min. Store capillaries dry.
Capillary temperature: 30
Running buffer: 25 mM NaH_2PO_4 adjusted to pH 2.3 with phosphoric acid
Injection: Pressure injection for 5 s
Detector: UV 214
Migration time: 5.189 (?)
Voltage: +10 kV
Model: Beckman P/ACE 5100
Limit of detection: 25 ng/mL (S/N 3)

OTHER SUBSTANCES
Simultaneous: ranitidine

REFERENCE
Altria,K.D.; Hadgett,T.A. An evaluation of the use of capillary electrophoresis to monitor trace drug residues following the manufacture of pharmaceuticals, *Chromatographia*, **1995**, *40*, 23–27.

Suprofen

Molecular formula: $C_{14}H_{12}O_3S$
Molecular weight: 260.31
CAS Registry No.: 40828-46-4
Merck Index (12th ed.): 9180
Lednicer: 2 65

SAMPLE
Matrix: solutions

CAPILLARY ELECTROPHORESIS
Capillary: 60 cm $\times$ 50 μm fused-silica
Capillary preparation: Before each run purge with running buffer for 3 min. At the beginning of each day vacuum purge with 500 mM NaOH, vacuum purge with water, perform an electroosmotic purge, vacuum purge with running buffer.
Running buffer: 10 mM pH 7 Tris-phosphate buffer containing 2.5% Glucidex2 (maltodextrin preparation, Roquette, Lestrem, France)
Injection: Injection by siphon at 10 cm
Detector: UV 185
Migration time: 5.0, 5.1 (enantiomers)
Voltage: 30 kV
Model: Waters Quanta 4000 CE

OTHER SUBSTANCES
Simultaneous: fenoprofen
Interfering: ibuprofen, indoprofen

KEY WORDS
chiral

REFERENCE
D'Hulst,A.; Verbeke,N. Quantitation in chiral capillary electrophoresis: theoretical and practical considerations, *Electrophoresis*, **1994**, *15*, 854–863.

SAMPLE
Matrix: solutions

Sample preparation: Prepare a 100 µg/mL solution in 50 mM pH 7.0 phosphate buffer, inject an aliquot.

CAPILLARY ELECTROPHORESIS
Capillary: 37 cm × 50 µm (30 cm to detector)
Capillary preparation: Between runs rinse capillary with 100 mM KOH for 2 min, with water for 2 min, and with running buffer for 2 min. At the beginning of each day treat capillary with 500 mM KOH for 5 min, flush with water for 10 min, and flush with running buffer for 10 min.
Capillary temperature: 20
Running buffer: 50 mM pH 7.0 Phosphate buffer containing 2 mM vancomycin
Injection: Pressure injection at 0.5 psi for 1 s
Detector: UV 254
Migration time: 54.0, 58.4 (enantiomers)
Voltage: +5 kV
Model: Beckman P/ACE 2000

KEY WORDS
chiral

REFERENCE
Rundlett,K.L.; Armstrong,D.W. Effect of micelles and mixed micelles on efficiency and selectivity of antibiotic based capillary electrophoretic enantioseparations, *Anal.Chem.*, **1995**, *67*, 2088–2095.

SAMPLE
Matrix: solutions
Sample preparation: Inject an aliquot of a 500 µM solution in MeOH:water 50:50.

CAPILLARY ELECTROPHORESIS
Capillary: 60 cm × 50 µm fused-silica (52.5 cm to detector) (Waters)
Capillary preparation: Before each run rinse capillary with 10 mM NaOH for 1 min and with running buffer for 4 min. Condition new capillaries by flushing with 1 M NaOH for 10 min, with water for 5 min, with 6 mM NaOH containing 500 mM NaCl for 10 min, with water for 5 min, and with running buffer for 10 min.
Running buffer: Formic acid containing 10 mM heptakis(trimethyl-β-cyclodextrin) adjusted to pH 4.0 with 1 M NaOH (ionic strength, I = 75 mM)
Injection: Hydrodynamic injection at 10 cm for 15 s.
Detector: UV 254
Voltage: 30 kV
Model: Waters Quanta 4000

KEY WORDS
chiral; R_s = 1.9; electroosmotic mobility = 11.3 × 10^{-5} $cm^2V^{-1}s^{-1}$

REFERENCE
Lelièvre,F.; Gareil,P. Chiral separations of underivatized arylpropionic acids by capillary zone electrophoresis with various cyclodextrins. Acidity and inclusion constant determinations, *J.Chromatogr.A*, **1996**, *735*, 311–320.

SAMPLE
Matrix: solutions

CAPILLARY ELECTROPHORESIS
Capillary: 35 cm × 50 µm coated capillary (30.5 cm to detector) (Composite Metal Services, UK)
Capillary preparation: Before each run purge at high pressure with water for 100 s and with running buffer for 120 s. If vancomycin is used, before each run purge at high pressure with water for 100 s and with running buffer not containing vancomycin for 120 s. Next purge with running buffer containing vancomycin at low pressure (175 psi.s) and inject sample. Coat capillary as follows. Adjust the pH of 20 mL water to 3.5 with acetic acid, add 80 µL 3-(trimethoxysilyl)propyl methacrylate (3-methacryloxypropyltrimethoxysilane), mix, suck into capillary, let stand at room temperature for 1 h, remove the solution, wash with water. Fill the capillary with a deaerated 3-4% acrylamide solution containing 1 µL/mL N,N,N',N'-tetra-

methylethylenediamine and 1 mg/mL potassium persulfate, let stand for 30 min, remove excess solution by aspiration, rinse with water, remove water by aspiration, dry at 35° (J. Chromatogr. 1985, 347, 191).
Capillary temperature: 25
Running buffer: Buffer containing 30 mM heptakis-2,3,6-tri-O-methyl-β-cyclodextrin (A), 5 mM heptamethylamino-β-cyclodextrin (B) or 5 mM vancomycin (C) (Buffer was 50 mM phosphoric acid containing 50 mM acetic acid and 50 mM boric acid, dilute with an equal volume of water, adjust pH to 5 with concentrated NaOH.)
Injection: Pressure injection at 10 psi.
Detector: UV 206
Migration time: 13.9, 14.4 (A); 13.3, 14.2 (B); 7.2, 7.6 (C)
Voltage: -20 kV
Model: Bio-Rad Biofocus 3000
Limit of detection: 5 μM (A, C), 10 μM (B)

KEY WORDS
chiral; coated capillary; detector at anode

REFERENCE
Fanali,S.; Desiderio,C.; Aturki,Z. Enantiomeric resolution study by capillary electrophoresis. Selection of the appropriate chiral selector, *J.Chromatogr.A*, **1997**, *772*, 185–194.

SAMPLE
Matrix: solutions
Sample preparation: Inject an aliquot of a solution in MeOH:water 10:90.

CAPILLARY ELECTROPHORESIS
Capillary: 44 cm × 50 μm fused-silica (37 cm to detector) (Supelco)
Capillary preparation: Wash with running buffer for 3 min after each injection. Wash with running buffer for 10 min at the end of each day.
Capillary temperature: 25
Running buffer: 100 mM Phosphoric acid containing 5 mM sulfobutyl ether-β-cyclodextrin and 40 mM heptakis(2,3,6-tri-O-methyl)-β-cyclodextrin, adjusted to pH 3.0 with triethanolamine
Injection: Hydrodynamic injection for 5 s (13.3 nL).
Detector: UV 280
Voltage: -25 kV
Current: 60 μA
Model: SpectraPhoresis 1000 CE

KEY WORDS
chiral; detector at anode; resolution (R_s = 8.9)

REFERENCE
Fillet,M.; Hubert,P.; Crommen,J. Enantioseparation of nonsteroidal anti-inflammatory drugs by capillary electrophoresis using mixtures of anionic and uncharged β-cyclodextrins as chiral additives, *Electrophoresis*, **1997**, *18*, 1013–1018.

SAMPLE
Matrix: solutions
Sample preparation: Inject an aliquot of a 100 μg/mL solution in water.

CAPILLARY ELECTROPHORESIS
Capillary: 27 cm × 50 μm eCAP neutral capillary (20 cm to detector) (Beckman)
Running buffer: 100 mM pH 6.0 Sodium phosphate buffer containing 2 mM A82846B (Eli Lilly)
Injection: Fill capillary with running buffer for 5 min, apply -10 kV for 5 min, hydrodynamic injection at 0.5 psi for 2 s.
Detector: UV 254
Migration time: 5.5, 6.2 (enantiomers)
Model: Beckman P/ACE 2000

OTHER SUBSTANCES
Also analyzed: flurbiprofen, ketoprofen

KEY WORDS
chiral; detector at anode

REFERENCE
Reilly,J.; Risley,D.S. The separation of enantiomers by countercurrent capillary electrophoresis using the macrocyclic antibiotic A82846B, *LC.GC*, **1998**, *16*, 170–178.

SAMPLE
Matrix: solutions
Sample preparation: Inject an aliquot of a solution in running buffer.

CAPILLARY ELECTROPHORESIS
Capillary: 57 cm × 50 μm fused-silica (Beckman)
Capillary preparation: Rinse with 100 mM NaOH for 5 min, with water for 10 min,
Running buffer: 2-Methoxyethanol:buffer 20:80 (Buffer was 40 mM pH 6 phosphate buffer containing 500 μM Actaplanin A (Eli Lilly).)
Injection: Pressure injection for 5 s.
Detector: UV 254
Migration time: 23.1, 23.8
Model: Beckman P/ACE 2100

KEY WORDS
chiral

REFERENCE
Trelli-Seifert,L.A.; Risley,D.S. Capillary electrophoretic enantiomeric separations of nonsteroidal anti-inflammatory compounds using the macrocyclic antibiotic actaplanin A and 2-methoxyethanol, *J.Liq.Chromatogr.Rel.Technol.*, **1998**, *21*, 299–313.

SAMPLE
Matrix: solutions

CAPILLARY ELECTROPHORESIS
Capillary: 42 cm × 50 μm fused-silica (31 cm to detector) (Polymicro Technologies)
Capillary temperature: 30
Running buffer: Formamide containing 3.42% quaternary ammonium-β-cyclodextrin (American Maize Products, Hammond IN), 20 mM ammonium acetate, and 1% acetic acid (pH* = 7.1)
Detector: UV 254
Migration time: 12.76, 12.85 (enantiomers)
Voltage: -30 kV
Model: laboratory-constructed

OTHER SUBSTANCES
Also analyzed: carprofen, fenoprofen, flurbiprofen, indoprofen, ketoprofen

KEY WORDS
chiral

REFERENCE
Wang,F.; Khaledi,M.G. Nonaqueous capillary electrophoresis chiral separations with quaternary ammonium β-cyclodextrin, *J.Chromatogr.A*, **1998**, *817*, 121–128.

Suramin

Molecular formula: $C_{51}H_{40}N_6O_{23}S_6$
Molecular weight: 1297.30
CAS Registry No.: 145-63-1,
129-46-4 (hexasodium salt)
Merck Index (12th ed.): 9181

SAMPLE
Matrix: blood
Sample preparation: Mix serum with an equal volume of 510 μg/mL IS in running buffer, inject an aliquot.

CAPILLARY ELECTROPHORESIS
Capillary: 50 cm × 50 μm fused-silica (42.5 cm to detector) (Waters)
Capillary preparation: Between runs wash capillary with 100 mM NaOH for 3 min and with running buffer for 6 min. At the end of each day rinse capillary with 100 mM NaOH for 5 min and with water for 10 min. At the start of each day rinse capillary with 100 mM NaOH for 5 min and with running buffer for 10 min. Treat new capillaries with 1 M KOH for 2 min, with 100 mM NaOH for 5 min, and with water for 10 min.
Capillary temperature: 25
Running buffer: 20 mM Sodium borate containing 75 mM sodium dodecyl sulfate and 4 M urea, adjusted to pH 9.2 with 1 M HCl
Injection: Hydrostatic injection at 10 cm for 5 s.
Detector: UV 254
Migration time: 6.7
Internal standard: orange G (5)
Voltage: 25 kV
Model: Waters Quanta 4000
Limit of quantitation: 15 μg/mL
Limit of detection: 4.5 μg/mL

OTHER SUBSTANCES
Noninterfering: betamethasone, diazepam, ibuprofen, lorazepam

KEY WORDS
serum

REFERENCE
Dabas,P.C.; Vescina,M.C.; Carducci,C.N. Determination of suramin by micellar electrokinetic chromatography with direct serum injection, *J.Capillary Electrophor.*, **1997**, *4*, 253–256.

SAMPLE
Matrix: solutions
Sample preparation: Inject an aliquot of an aqueous solution.

CAPILLARY ELECTROPHORESIS
Capillary: 24 cm × 50 μm coated fused-silica (Bio-Rad)
Running buffer: 300 mM pH 8.6 Tris-boric acid containing 2 mM EDTA, 4% 6K PEG, 4% 12K PEG, 4% 20K PEG, and 4% 36K PEG (PEG is polyethylene glycol)
Injection: Pressure injection at 30 psi for 1 s
Detector: UV 254
Migration time: 9.49
Voltage: 15 kV
Model: Bio-Rad BioFocus 3000

OTHER SUBSTANCES
Simultaneous: degradation products, quinobene, sulfanilic acid

---

KEY WORDS
coated capillary

REFERENCE
Hettiarachchi,K.; Cheung,A.P. Precision in capillary electrophoresis with respect to quantitative analysis of
suramin, *J.Chromatogr.A*, **1995**, *717*, 191–202.

Synephrine

Molecular formula: $C_9H_{13}NO_2$
Molecular weight: 167.21
CAS Registry No.: 94-07-5, 5985-28-4 (HCl), 6414-49-9 (tartaric acid monoester), 16589-24-5
(tartrate)
Merck Index (12th ed.): 9189

SAMPLE
Matrix: solutions

CAPILLARY ELECTROPHORESIS
Capillary: 52 cm × 75 μm fused-silica (48 cm to detector) (Polymicro Technologies)
Capillary preparation: Before each run purge with running buffer for 3 min. Every 3 runs
purge with 100 mM NaOH for 5 min. Purge new capillaries with 1 M NaOH for 20 min and
with 100 mM NaOH for 20 min, rinse with running buffer, equilibrate with running buffer at
12 kV for 3 h.
Running buffer: 100 mM CHES (2-(N-cyclohexylamine)ethanesulfonic acid) containing 10 mM
triethylamine and 25 mM (R)-dodecoxycarbonylvaline (Waters EnantioSelect (R)-Val-1), pH ad-
justed to 8.8 with 1 M NaOH
Injection: Hydrostatic injection for 2 s.
Detector: UV 214
Migration time: 8, 8.2 (enantiomers)
Voltage: 12 kV
Current: ≤30 μA
Model: Waters Quanta 4000

OTHER SUBSTANCES
Simultaneous: albuterol, pseudoephedrine

KEY WORDS
chiral

REFERENCE
Peterson,A.G.; Ahuja,E.S.; Foley,J.P. Enantiomeric separations of basic pharmaceutical drugs by micellar electrokinetic chromatography using a chiral surfactant, N-dodecoxycarbonylvaline, *J.Chromatogr.B*, **1996**, *683*,
15–28.

SAMPLE
Matrix: solutions

CAPILLARY ELECTROPHORESIS
Capillary: 60 cm × 75 μm fused-silica (51 cm to detector) (Supelco)
Capillary preparation: At the end of each day wash capillary with 200 mM NaOH for 10 min
and with water for 10 min. Condition new capillaries by washing with 200 mM NaOH for 10
min, with water for 10 min, and with running buffer for 10 min.

Running buffer: 50 mM pH 7.3 Sodium borate buffer containing 3.3% succinyl β-cyclodextrin
(Wacker Chemie, Munich, Germany)
Injection: Hydrodynamic injection at 25 mbar for 6 s.
Detector: UV 208
Migration time: 58.18 (first enantiomer)
Voltage: 15 kV
Model: Prince

OTHER SUBSTANCES
Simultaneous: alprenolol, chlorthalidone, ephedrine, etilefrin, homatropine, methoxamine, norephedrine, octopamine, propranolol, trihexyphenidyl

KEY WORDS
chiral; α = 1.0339

REFERENCE
Schmid,M.G.; Wirnsberger,K.; Gübitz,G. Chiral separation of drug enantiomers by capillary electrophoresis
using succinyl-β-cyclodextrin, *Pharmazie*, **1996**, *51*, 852–854.

SAMPLE
Matrix: solutions
Sample preparation: Inject an aliquot of a 100 μg/mL solution in running buffer.

CAPILLARY ELECTROPHORESIS
Capillary: 29 cm × 50 μm fused-silica (24.5 cm to detector) (Yongnian Optical Conductive Fiber
Plant, China), coated with polyacrylamide
Capillary preparation: No details of the polyacrylamide coating process are provided. However,
another paper (LC.GC 1997, 15, 40) by this group indicates that they use the procedure of
Hjertén, thus: Adjust the pH of 20 mL water to 3.5 with acetic acid, add 80 μL 3-(trimethoxysilyl)propyl methacrylate (3-methacryloxypropyltrimethoxysilane), mix, suck into capillary, let
stand at room temperature for 1 h, remove the solution, wash with water. Fill the capillary
with a deaerated 3-4% acrylamide solution containing 1 μL/mL N,N,N',N'-tetramethylethyl-
enediamine and 1 mg/mL potassium persulfate, let stand for 30 min, remove excess solution
by aspiration, rinse with water, remove water by aspiration, dry at 35° (J. Chromatogr. 1985,
347, 191).
Capillary temperature: 25
Running buffer: 100 mM NaH_2PO_4 adjusted to pH 2.5
Injection: Electrokinetic injection at 15 kV for 3 s.
Detector: UV 200, UV 210
Migration time: 4.26
Voltage: 15 kV
Model: Bio-Rad BioFocus 3000

OTHER SUBSTANCES
Simultaneous: albuterol, alprenolol, atenolol, baclofen, bamethan, benproperine, benserazide,
bisoprolol, bupranolol, butamirate, butethamate, carbuterol, celiprolol, clenbuterol, clobutinol,
dipivefrin, isoproterenol (isoprenaline), metaproterenol (orciprenaline), metipranolol, metoprolol, norfenefrine, ornidazole, oxprenolol, phenylpropanolamine, pholedrine, pirbuterol, prilocaine, procyclidine, sotalol, terbutaline, tocainide

KEY WORDS
coated capillary

REFERENCE
Koppenhoefer,B.; Epperlein,U.; Xiaofeng,Z.; Bingcheng,L. Separation of enantiomers of drugs by capillary electrophoresis. Part 4: Hydroxypropyl-γ-cyclodextrin as chiral solvating agent, *Electrophoresis*, **1997**, *18*, 924–
930.

SAMPLE
Matrix: solutions
Sample preparation: Inject an aliquot of a 100 μg/mL solution in running buffer.

CAPILLARY ELECTROPHORESIS
Capillary: 30 cm × 50 μm fused-silica (25.5 cm to detector), coated with polyacrylamide
Capillary preparation: Adjust the pH of 20 mL water to 3.5 with acetic acid, add 80 μL 3-(trimethoxysilyl)propyl methacrylate (3-methacryloxypropyltrimethoxysilane), mix, suck into capillary, let stand at room temperature for 1 h, remove the solution, wash with water. Fill the capillary with a deaerated 3-4% acrylamide solution containing 1 μL/mL N,N,N',N'-tetramethylethylenediamine and 1 mg/mL potassium persulfate, let stand for 30 min, remove excess solution by aspiration, rinse with water, remove water by aspiration, dry at 35° (J. Chromatogr. 1985, 347, 191).
Capillary temperature: 25
Running buffer: 100 mM NaH_2PO_4 containing 45 mM heptakis(2,6-di-O-methyl)-β-cyclodextrin, adjusted to pH 2.5 with phosphoric acid
Injection: Electrokinetic injection at 15 kV for 3 s.
Detector: UV 200
Voltage: 15 kV
Model: Bio-Rad BioFocus 3000

KEY WORDS
chiral; coated capillary; comparison with the use of other cyclodextrins; this running buffer gave the greatest enantiomeric separation.; α=1.042

REFERENCE
Lin,B.; Zhu,X.; Koppenhoefer,B.; Epperlein,U. Investigation of 123 chiral drugs by cyclodextrin-modified capillary electrophoresis, *LC.GC*, **1997**, *15*, 40–46.

SAMPLE
Matrix: solutions

CAPILLARY ELECTROPHORESIS
Capillary: 52-55 cm × 75 μm fused-silica (45-48 cm to detector) (Polymicro Technologies)
Capillary preparation: After each run purge with running buffer for 3 min. After every 5 runs purge with 100 mM LiOH. Purge new capillaries with 1 M LiOH for 20 min, purge with 100 mM LiOH for 20 min, rinse with running buffer, equilibrate with running buffer with voltage applied for 3 h.
Running buffer: Buffer (Prepare a 100 mM CHES (2-[N-cyclohexylamine]ethanesulfonic acid) solution containing 25 mM (S)-N-dodecoxycarbonylvaline (EnantioSelect (S)-Val-1, Waters), adjust pH to 7 with 1 M LiOH, add triethylamine to a final concentration of 10 mM, adjust pH to 8.8 with 1 M LiOH. Preparation of (S)-N-dodecoxycarbonylvaline is as follows. Prepare dodecyl chloroformate by reacting 1-dodecanol with 0.33 equivalents of triphosgene in dichloromethane solution in the presence of pyridine. Add dodecyl chloroformate dropwise to (S)-valine in 1 M NaOH solution, filter, wash with hexane, recrystallize from ether/petroleum ether (J. Chromatogr. 1994, 680, 125).)
Injection: Hydrostatic injection for 2 s.
Detector: UV 214
Migration time: 8.6, 8.8 (enantiomers)
Voltage: 12 kV
Model: Waters Quanta 4000 CE

OTHER SUBSTANCES
Simultaneous: ephedrine, N-methylpseudoephedrine, metoprolol, norphenylephrine

KEY WORDS
chiral

REFERENCE
Peterson,A.G.; Foley,J.P. Influence of the inorganic counterion on the chiral micellar electrokinetic separation of basic drugs using the surfactant N-dodecoxycarbonylvaline, *J.Chromatogr.B*, **1997**, *695*, 131–145.

SAMPLE
Matrix: solutions

CAPILLARY ELECTROPHORESIS
Capillary: 58.5 cm × 50 μm fused-silica (50 cm to detector) (Composite Metal Services, Worcs., UK)
Capillary preparation: Wash capillary with running buffer for 5 min before each run. At the start of each day wash with 100 mM NaOH for 5 min and rinse with water for 15 min.
Capillary temperature: 25
Running buffer: 40 mM pH 3.0 Sodium acetate buffer containing 40 mM phosphoric acid, 40 mM boric acid, and 10 mM (2,6-di-O-methyl)-β-cyclodextrin
Injection: Pressure injection at 150 mbar.s.
Detector: UV 202
Migration time: 4.5, 4.6 (enantiomers)
Voltage: 30 kV
Model: Hewlett-Packard ^{3D}CE

KEY WORDS
chiral

REFERENCE
Roos,N.; Ganzler,K.; Szemán,J.; Fanali,S. Systematic approach to cost- and time-effective method development with a starter kit for chiral separations by capillary electrophoresis, *J.Chromatogr.A*, **1997**, *782*, 257–269.

SAMPLE
Matrix: solutions

CAPILLARY ELECTROPHORESIS
Capillary: 29-36 cm × 50 μm fused-silica (24.5-31.5 cm to detector) (Yongnian Optical Conductive Fiber Plant, China) coated with polyacrylamide
Capillary preparation: Coat capillary as follows. Adjust the pH of 20 mL water to 3.5 with acetic acid, add 80 μL 3-(trimethoxysilyl)propyl methacrylate (3-methacryloxypropyltrimethoxysilane), mix, suck into capillary, let stand at room temperature for 1 h, remove the solution, wash with water. Fill the capillary with a deaerated 3-4% acrylamide solution containing 1 μL/mL N,N,N',N'-tetramethylethylenediamine and 1 mg/mL potassium persulfate, let stand for 30 min, remove excess solution by aspiration, rinse with water, remove water by aspiration, dry at 35° (J. Chromatogr. 1985, 347, 191).
Capillary temperature: 25
Running buffer: 100 mM pH 2.5 NaH_2PO_4 (A) or 100 mM pH 2.5 NaH_2PO_4 containing 45 mM hydroxypropyl-α-cyclodextrin (Wacker, Munich) (B)
Injection: Electromigration at 15 kV for 3 s.
Detector: UV 200; UV 210
Migration time: 4.26 (A); 6.39 (B) (no separation of enantiomers)
Voltage: 15 kV
Model: Bio-Focus 3000

OTHER SUBSTANCES
Also analyzed: albuterol (salbutamol), alprenolol, amorolfine, atenolol, atropine, azelastine, baclofen, bamethan, benproperine, benserazide, biperiden, bisoprolol, brompheniramine, bupivacaine, bupranolol, butamirate, butethamate, carazolol, carbuterol, carteolol, carvedilol, celiprolol, chloroquine, chlorpheniramine, chlorphenoxamine, cicletanine, clenbuterol, clidinium bromide, clobutinol, dimethindene, dipivefrin, disopyramide, dobutamine, doxylamine, fendiline, flecainide, gallopamil, homatropine, ipratropium bromide, isoproterenol (isoprenaline), isothipendyl, ketamine, meclizine, mefloquine, mepindolol, mequitazine, metaclazepam, metaproterenol (orciprenaline), metipranolol, metoprolol, nafronyl (naftidrofuryl), nefopam, nicardipine, norfenefrine, ofloxacin, ornidazole, orphenadrine, oxomemazine, oxprenolol, oxybutynin, phenoxybenzamine, phenylpropanolamine, pholedrine, pindolol, pirbuterol, prilocaine, procyclidine, promethazine, propafenone, propranolol, reproterol, sotalol, sulpride, talinolol, terbutaline, tetrahydrozoline (tetryzoline), theodrenaline, tioconazole, tocainide, trihexyphenidyl, trimeprazine (alimemazine), trimipramine, tropicamide, verapamil, zopiclone

KEY WORDS
coated capillary

REFERENCE

Koppenhoefer,B.; Eperlein,U.; Schlunk,R.; Zhu,X.; Lin,B. Separation of enantiomers of drugs by capillary electrophoresis. V. Hydroxypropyl-α-cyclodextrin as chiral solvating agent, *J.Chromatogr.A*, **1998**, *793*, 153–164.

Tacrine

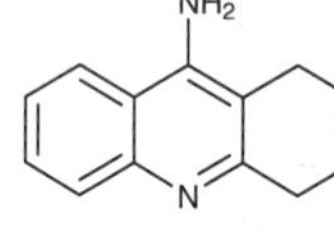

Molecular formula: $C_{13}H_{14}N_2$
Molecular weight: 198.27
CAS Registry No.: 321-64-2, 1684-40-8 (HCl)
Merck Index (12th ed.): 9199
Lednicer: 5 166

SAMPLE

Matrix: blood, urine
Sample preparation: Serum. 100 μL Serum + 150 μL MeCN, mix for 15 s, centrifuge at 15000 g for 1 min, inject an aliquot of the supernatant. Urine. Dilute 100 μL urine to 1 mL with water, inject an aliquot.

CAPILLARY ELECTROPHORESIS

Capillary: 46.6 cm × 75 μm fused-silica (39.0 cm to detector)
Capillary preparation: Before each run wash capillary with 100 mM NaOH for 2 min and with running buffer for 2 min. Replenish vials after every 5 runs. Wash capillary each day with 100 mM NaOH for 10 min, with water for 10 min, and with running buffer for 10 min.
Capillary temperature: 25
Running buffer: 300 mM pH 2.8 Phosphate buffer
Injection: Electrokinetic injection at 10 kV for 10 s.
Detector: UV 240
Migration time: 14
Voltage: 20 kV
Model: Beckman P/ACE 5500
Limit of detection: 50 ppb

OTHER SUBSTANCES

Extracted: metabolites

KEY WORDS

serum

REFERENCE

Vargas,M.G.; Havel,J.; Patocka,J. Capillary zone electrophoretic determination of some drugs against Alzheimer's disease, *J.Chromatogr.A*, **1998**, *802*, 121–128.

Talinolol

Molecular formula: $C_{20}H_{33}N_3O_3$
Molecular weight: 363.50
CAS Registry No.: 57460-41-0
Merck Index (12th ed.): 9208

SAMPLE

Matrix: solutions
Sample preparation: Inject an aliquot of a 100 μg/mL solution in running buffer.

CAPILLARY ELECTROPHORESIS
Capillary: 30 cm × 50 μm fused-silica (25.5 cm to detector) (Yongnian Optical Conductive Fiber Plant, China), coated with polyacrylamide
Capillary preparation: No details of the polyacrylamide coating process are provided. However, another paper (LC.GC 1997, 15, 40) by this group indicates that they use the procedure of Hjertén, thus: Adjust the pH of 20 mL water to 3.5 with acetic acid, add 80 μL 3-(trimethoxysilyl)propyl methacrylate (3-methacryloxypropyltrimethoxysilane), mix, suck into capillary, let stand at room temperature for 1 h, remove the solution, wash with water. Fill the capillary with a deaerated 3-4% acrylamide solution containing 1 μL/mL N,N,N',N'-tetramethylethylenediamine and 1 mg/mL potassium persulfate, let stand for 30 min, remove excess solution by aspiration, rinse with water, remove water by aspiration, dry at 35° (J. Chromatogr. 1985, 347, 191).
Capillary temperature: 25
Running buffer: 100 mM NaH_2PO_4 adjusted to pH 2.5
Injection: Electrokinetic injection at 15 kV for 3 s.
Detector: UV 200, UV 210
Migration time: 6.99
Voltage: 15 kV
Model: Bio-Rad BioFocus 3000

OTHER SUBSTANCES
Simultaneous: amorolfine, brompheniramine, bupivacaine, carteolol, chloroquine, chlorpheniramine, chlorphenoxamine, disopyramide, dobutamine, doxylamine, flecainide, gallopamil, ketamine, mepindolol, orphenadrine, oxybutynin, phenoxybenzamine, pindolol, propafenone, propranolol, sulpiride, tropicamide, verapamil

KEY WORDS
coated capillary

REFERENCE
Koppenhoefer,B.; Epperlein,U.; Xiaofeng,Z.; Bingcheng,L. Separation of enantiomers of drugs by capillary electrophoresis. Part 4: Hydroxypropyl-γ-cyclodextrin as chiral solvating agent, *Electrophoresis*, **1997**, *18*, 924–930.

SAMPLE
Matrix: solutions

CAPILLARY ELECTROPHORESIS
Capillary: 29-36 cm × 50 μm fused-silica (24.5-31.5 cm to detector) (Yongnian Optical Conductive Fiber Plant, China) coated with polyacrylamide
Capillary preparation: Coat capillary as follows. Adjust the pH of 20 mL water to 3.5 with acetic acid, add 80 μL 3-(trimethoxysilyl)propyl methacrylate (3-methacryloxypropyltrimethoxysilane), mix, suck into capillary, let stand at room temperature for 1 h, remove the solution, wash with water. Fill the capillary with a deaerated 3-4% acrylamide solution containing 1 μL/mL N,N,N',N'-tetramethylethylenediamine and 1 mg/mL potassium persulfate, let stand for 30 min, remove excess solution by aspiration, rinse with water, remove water by aspiration, dry at 35° (J. Chromatogr. 1985, 347, 191).
Capillary temperature: 25
Running buffer: 100 mM pH 2.5 NaH_2PO_4 (A) or 100 mM pH 2.5 NaH_2PO_4 containing 45 mM hydroxypropyl-α-cyclodextrin (Wacker, Munich) (B)
Injection: Electromigration at 15 kV for 3 s.
Detector: UV 200; UV 210
Migration time: 6.99 (A); 15.29 (B) (no separation of enantiomers)
Voltage: 15 kV
Model: Bio-Focus 3000

OTHER SUBSTANCES
Also analyzed: albuterol (salbutamol), alprenolol, amorolfine, atenolol, atropine, azelastine, baclofen, bamethan, benproperine, benserazide, biperiden, bisoprolol, brompheniramine, bupivacaine, bupranolol, butamirate, butethamate, carazolol, carbuterol, carteolol, carvedilol, celiprolol, chloroquine, chlorpheniramine, chlorphenoxamine, cicletanine, clenbuterol, clidinium bromide, clobutinol, dimethindene, dipivefrin, disopyramide, dobutamine, doxylamine, fendiline, flecainide, gallopamil, homatropine, ipratropium bromide, isoproterenol (isoprenaline),

isothipendyl, ketamine, meclizine, mefloquine, mepindolol, mequitazine, metaclazepam, metaproterenol (orciprenaline), metipranolol, metoprolol, nafronyl (naftidrofuryl), nefopam, nicardipine, norfenefrine, ofloxacin, ornidazole, orphenadrine, oxomemazine, oxprenolol, oxybutynin, phenoxybenzamine, phenylpropanolamine, pholedrine, pindolol, pirbuterol, prilocaine, procyclidine, promethazine, propafenone, propranolol, reproterol, sotalol, sulpride, synephrine, terbutaline, tetrahydrozoline (tetryzoline), theodrenaline, tioconazole, tocainide, trihexyphenidyl, trimeprazine (alimemazine), trimipramine, tropicamide, verapamil, zopiclone

KEY WORDS
coated capillary

REFERENCE
Koppenhoefer,B.; Eperlein,U.; Schlunk,R.; Zhu,X.; Lin,B. Separation of enantiomers of drugs by capillary electrophoresis. V. Hydroxypropyl-α-cyclodextrin as chiral solvating agent, *J.Chromatogr.A*, **1998**, *793*, 153–164.

Tamoxifen

Molecular formula: $C_{26}H_{29}NO$
Molecular weight: 371.52
CAS Registry No.: 10540-29-1, 54965-24-1 (citrate)
Merck Index (12th ed.): 9216
Lednicer: 2 127; 3 70; 4 65

SAMPLE
Matrix: blood
Sample preparation: Mix 1 part serum with 5 parts hexane:isoamyl alcohol 98:2, vortex for 0.5-1 min, centrifuge at 500 g for 5 min, freeze in dry ice/acetone. Remove the organic layer, repeat the extraction twice more, evaporate to dryness under reduced pressure, resuspend the residue in 25-50 μL 10 mM ammonium acetate in MeCN:MeOH:acetic acid 50:50:1 containing IS, inject an aliquot.

CAPILLARY ELECTROPHORESIS
Capillary: 57 cm × 50 μm fused-silica
Capillary preparation: Rinse with running buffer at 137.9 kPa for 5 min before each injection. At the start of each day wash with 100 mM NaOH at 137.9 kPa for 5 min and with water at 137.9 kPa.
Capillary temperature: 40
Running buffer: 50 mM Ammonium acetate in MeCN:MeOH:acetic acid 50:50:1
Injection: Pressure injection at 3.45 kPa for 10 s followed by an electrokinetic injection of 10 mM ammonium acetate in MeCN:MeOH:acetic acid 50:50:1 at 2 kV for 30 s.
Detector: UV 214
Migration time: 7
Internal standard: 4-dimethylaminopyridine (4.2)
Voltage: 15 kV
Current: ca. 15 μA
Model: Beckman P/ACE 5000
Limit of quantitation: 10 pg
Limit of detection: 800 amole

OTHER SUBSTANCES
Extracted: metabolites, N-desmethyltamoxifen, 4-hydroxytamoxifen

KEY WORDS
serum; cow; rat

REFERENCE
Sanders,J.M.; Burka,L.T.; Shelby,M.D.; Newbold,R.R.; Cunningham,M.L. Determination of tamoxifen and metabolites in serum by capillary electrophoresis using a nonaqueous buffer system, *J.Chromatogr.B*, **1997**, *695*, 181–185.

SAMPLE
Matrix: hepatocytes, microsomal incubations
Sample preparation: Rat liver microsomal incubations. 1 mL Microsomal incubation + 1 g NaCl, adjust pH to 9.0 with 1 M NaOH, extract with ethyl acetate, evaporate to dryness, reconstitute with 2 mL MeOH, inject an aliquot. Mouse hepatocyte incubations. Extract 1 mL hepatocyte incubation twice with 1 mL hexane:EtOH 98:2, combine the extracts and evaporate them to dryness, reconstitute with 400 µL MeOH, inject an aliquot.

CAPILLARY ELECTROPHORESIS
Capillary: 100 cm × 100 µm fused-silica (Polymicro Technologies)
Capillary preparation: Between runs wash with water for 30 min and with running buffer for 10 min. Wash new capillaries with water for 10 min, with 100 mM NaOH for 10 min, with water for 20 min, and with running buffer for 20 min.
Running buffer: MeOH containing 5 mM ammonium acetate and 100 mM acetic acid (microsomal incubations) or MeOH containing 7 mM sodium dodecyl sulfate, 2.5 mM ammonium acetate, and 50 mM acetic acid (hepatocyte incubations)
Injection: Gravity injection at 25 cm for 2 s
Detector: MS, Vestec 201, skimmer-collimator voltage difference 7 V, sheath liquid MeOH:acetic acid 99:1 flowing at 2.5 µL/min, electrospray needle 2.48 kV, flat plate counterelectrode 300 V, m/z 372
Migration time: 22.05 (microsomal incubations), 36.05 (hepatocyte incubations)
Voltage: 25 kV
Model: Dionex CE system I

OTHER SUBSTANCES
Extracted: metabolites

KEY WORDS
rat; mouse; liver

REFERENCE
Lu,W.; Poon,G.K.; Carmichael,P.L.; Cole,R.B. Analysis of tamoxifen and its metabolites by on-line capillary electrophoresis-electrospray ionization mass spectrometry employing nonaqueous media containing surfactants, *Anal.Chem.*, **1996**, *68*, 668–674.

SAMPLE
Matrix: solutions

CAPILLARY ELECTROPHORESIS
Capillary: 55 cm × 50 µm fused-silica (44 cm to detector) (Polymicro Technologies)
Running buffer: 25 mM Ammonium acetate in MeCN:MeOH 50:50 (dissolve ammonium acetate in MeOH and add MeCN)
Injection: Siphon at 5 cm for 10 s (4 nL)
Detector: UV 254
Migration time: 6.5
Voltage: 18 kV
Model: Laboratory constructed
Limit of detection: 2 ppm

OTHER SUBSTANCES
Simultaneous: metabolites

REFERENCE
Ng,C.L.; Lee,H.K.; Li,S.F.Y. Separation of tamoxifen and metabolites by capillary electrophoresis with nonaqueous buffer system, *J.Liq.Chromatogr.*, **1994**, *17*, 3847–3857.

Taurine

H₂N⌒SO₃H

Molecular formula: $C_2H_7NO_3S$
Molecular weight: 125.15
CAS Registry No.: 107-35-7
Merck Index (12th ed.): 9241

SAMPLE

Matrix: CSF
Sample preparation: 50 μL CSF + 50 μL 210 μM fluorescein isothiocyanate isomer I in acetone, mix, let stand in the dark for 2 h, dilute 100-1000 times, inject an aliquot.

CAPILLARY ELECTROPHORESIS

Capillary: 75 cm × 50 μm fused-silica (42 cm to detector) (Polymicro Technologies)
Capillary preparation: Rinse with 100 mM NaOH for 3 min, with water for 2 min, and with running buffer for 3 min.
Running buffer: 100 mM Boric acid containing 100 mM sodium dodecyl sulfate, adjusted to pH 9.3 with NaOH
Injection: Hydrodynamic injection for 2 s (15 nL)
Detector: F ex 488 (laser)
Migration time: 8.9
Voltage: +20 kV
Current: 42 μA
Model: TSP Spectraphoresis 100
Limit of detection: <0.2 nM

OTHER SUBSTANCES

Extracted: amino acids

KEY WORDS

derivatization

REFERENCE

Nouadje,G.; Rubie,H.; Chatelut,E.; Canal,P.; Nertz,M.; Puig,P.; Couderc,F. Child cerebrospinal fluid analysis by capillary electrophoresis and laser-induced fluorescence detection, *J.Chromatogr.A*, **1995**, *717*, 293–298.

SAMPLE

Matrix: dialysate
Sample preparation: Dialysate pumped at 79 nL/min mixed in a 50 nL mixing T with reagent pumped at 79 nL/min and the mixture flowed through an 8 cm × 75 μm fused-silica capillary (reaction time 2.2 min). Aliquots were automatically injected (design of injector described in paper). (Reagent was 110 mM o-phthalaldehyde and 220 mM 2-mercaptoethanol in 25 mM pH 9.5 borate buffer.)

CAPILLARY ELECTROPHORESIS

Capillary: 20 cm × 25 μm fused-silica (15 cm to detector) (Polymicro Technologies)
Capillary preparation: At the start of each day rinse with 100 mM NaOH for 10 min, with water for 10 min, and with running buffer for 10 min.
Running buffer: 175 mM 2-(N-Cyclohexylamino)ethanesulfonic acid (CHES) containing 100 mM sodium dodecyl sulfate, adjusted to pH 9.0 with 1 M NaOH
Injection: Electrokinetic injection at -100 V for 2 s.
Detector: F ex 354 (2 mW He-Cd laser, Liconix Model 4210B) em 450
Migration time: 2.5
Voltage: 400 V/cm
Limit of detection: 20-40 nM

OTHER SUBSTANCES

Extracted: alanine, aspartic acid, glutamic acid, glutamine, glycine, isoleucine, leucine, lysine, methionine, phenylalanine, serine, taurine, threonine, tyrosine

Interfering: valine

KEY WORDS
derivatization

REFERENCE
Lada,M.W.; Kennedy,R.T. Quantitative in vivo monitoring of primary amines in rat caudate nucleus using
microdialysis coupled by a flow-gated interface to capillary electrophoresis with laser-induced fluorescence
detection, *Anal.Chem.*, **1996**, *68*, 2790–2797.

SAMPLE
Matrix: tissue
Sample preparation: Sonicate (Artix Sonic Dismembrator Model 150, power setting 30) 1-2 mg
rat brain tissue and 10 µL EtOH:water 70:30 in an ice bath for 5-10 s, centrifuge at 16000 g
for 10 min. 1 µL Supernatant + 1 µL 490 µM α-aminoadipic acid, mix, remove a 1 µL aliquot
and add it to 5 µL 20 mM pH 9.0 sodium borate buffer, add 1.5 µL 20 mM NaCN in water,
mix, add 1.5 µL 20 mM naphthalene-2,3-dicarboxaldehyde in MeCN, mix thoroughly, let stand
at room temperature for 30 min (protect from light), inject an aliquot.

CAPILLARY ELECTROPHORESIS
Capillary: 115 cm × 50 µm fused-silica (95 cm to detector)
Running buffer: 20 mM pH 9.0 sodium borate buffer
Injection: Electrokinetic injection at 5 kV for 12 s
Detector: UV 420
Migration time: 15.8
Internal standard: α-aminoadipic acid (21.6)
Voltage: 30 kV
Limit of detection: 4.4 µM

OTHER SUBSTANCES
Extracted: amino acids, gamma-aminobutyric acid, dopamine, levodopa, phosphoethanolamine,
norepinephrine

KEY WORDS
rat; brain; derivatization

REFERENCE
Weber,P.L., O'Shoa,T.J.; Lunte,S.M. Separation and quantitation of the amino acid neurotransmitters in rat
brain by capillary electrophoresis, *J.Pharm.Biomed.Anal.*, **1994**, *12*, 319–324.

SAMPLE
Matrix: tissue
Sample preparation: Homogenize 2 mg rat nerve tissue in 15 mM pH 9.2 borate buffer for 20
min, acidify with concentrated perchloric acid, make up to 500 µL, centrifuge at 15000 g for
10 min. Remove a 100 µL aliquot of the supernatant and add it to 25 µL 1 mM gamma-Glu-
Glu in 15 mM pH 9.2 borate buffer and 350 µL 15 mM pH 9.2 borate buffer, add 25 µL 50 mM
fluorescein isothiocyanate, isomer I in acetone containing 0.002% pyridine, heat at 40° for 6 h,
inject an aliquot.

CAPILLARY ELECTROPHORESIS
Capillary: 51 cm × 75 µm fused-silica (46 cm to detector) (Yongnian, Hebei, China)
Capillary preparation: Before use wash capillary with 100 mM NaOH, water, and running
buffer.
Running buffer: 15 mM pH 9.2 Borate buffer
Injection: Siphon at 10 cm for 10 s.
Detector: F ex 488 (40 mW argon ion laser) CCD detector (details in paper)
Migration time: 15.3
Internal standard: gamma-Glu-Glu (17.5)
Current: 21.6 µA
Model: laboratory-constructed
Limit of detection: 3 nM

OTHER SUBSTANCES
Simultaneous: alanine, gamma-aminobutyric acid, aspartic acid, glutamic acid, glycine

KEY WORDS
rat; nerve; derivatization

REFERENCE
Zhang,L.; Chen,H.; Hu,S.; Cheng,J.; Li,Z.; Shao,M. Determination of the amino acid neurotransmitters in the dorsal root ganglion of the rat by capillary electrophoresis with a laser induced florescence-charge coupled device, *J.Chromatogr.B*, **1998**, *707*, 59–67.

Temazepam

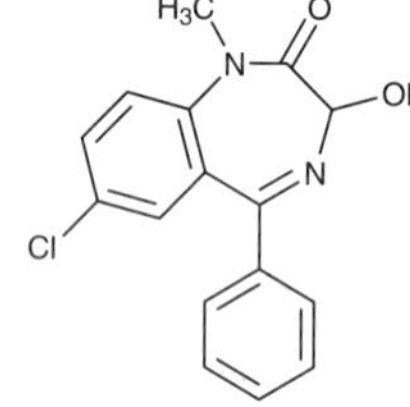

Molecular formula: $C_{16}H_{13}ClN_2O_2$
Molecular weight: 300.74
CAS Registry No.: 846-50-4
Merck Index (12th ed.): 9285
Lednicer: 2 402

SAMPLE
Matrix: solutions
Sample preparation: Prepare a 1-50 µg/mL solution in MeOH:water 20:80, inject an aliquot.

CAPILLARY ELECTROPHORESIS
Capillary: 44 cm × 50 µm fused-silica
Capillary preparation: After each run wash capillary with water for 2 min and buffer for 3 min. At the beginning of each day equilibrate capillary with running buffer for 10 min. Condition a new capillary by washing with 1 M NaOH, 100 mM NaOH, water, and running buffer.
Capillary temperature: 25
Running buffer: MeOH:buffer 20:80 (Buffer was 75 mm pH 9.0 Glycine containing 250 mM triethanolamine and 25 mM sodium dodecyl sulfate.)
Injection: Hydrodynamic injection for 5 s
Detector: UV 235
Migration time: 9.6
Voltage: 25 kV
Current: about 15 µA
Model: Spectraphysics Spectraphoresis 1000 CE
Limit of quantitation: 700 ng/mL
Limit of detection: 200 ng/mL

OTHER SUBSTANCES
Simultaneous: bromazepam, brotizolam, clobazam, clonazepam, lorazepam, lormetazepam, nitrazepam, nordazepam, oxazepam

REFERENCE
Bechet,I.; Fillet,M.; Hubert,P.; Crommen,J. Determination of benzodiazepines by micellar electrokinetic chromatography, *Electrophoresis*, **1994**, *15*, 1316–1321.

SAMPLE
Matrix: solutions

CAPILLARY ELECTROPHORESIS
Capillary: 57 cm × 75 µm fused-silica (50 cm to detector)
Capillary preparation: Before each run flush with running buffer for 2 min, equilibrate for 5 min.
Capillary temperature: 33

Running buffer: MeCN:buffer 10:90 (Buffer was 20 mM pH 10.2 borate containing 50 mM sodium cholate.)
Injection: Pressure injection for 2 s
Detector: UV 214
Migration time: 6.3
Voltage: 20 kV
Model: Beckman P/ACE 2100

OTHER SUBSTANCES
Simultaneous: lorazepam, lormetazepam, oxazepam

REFERENCE
Boonkerd,S.; Detaevernier,M.R.; Michotte,Y.; Vindevogel,J. Suppression of chiral recognition of 3-hydroxy-1,4-benzodiazepines during micellar electrokinetic capillary chromatography with bile salts, *J.Chromatogr.A*, **1995**, *704*, 238–241.

SAMPLE
Matrix: solutions
Sample preparation: Inject an aliquot of a 5 µg/mL solution in running buffer.

CAPILLARY ELECTROPHORESIS
Capillary: 87 cm × 75 µm fused-silica (80 cm to detector) (Beckman)
Capillary preparation: Between runs rinse capillary with running buffer for 2 min then equilibrate for 5 min.
Capillary temperature: 35
Running buffer: 20 mM pH 7 Buffer containing 15 mM sodium cholate and 35 mM sodium deoxycholate (Buffer was 20 mM sodium borate adjusted to pH 7.0 with 20 mM NaH_2PO_4.)
Injection: Pressure injection for 2 s.
Detector: UV 214
Migration time: 19.3
Voltage: 20 kV
Model: Beckman P/ACE 2100

OTHER SUBSTANCES
Simultaneous: bromazepam, chlordiazepoxide, clobazam, clonazepam, diazepam, flunitrazepam, flurazepam, halazepam, lorazepam, lormetazepam, nitrazepam, nordazepam

REFERENCE
Boonkerd,S.; Detaevernier,M.R.; Vindevogel,J.; Michotte,Y. Migration behaviour of benzodiazepines in micellar electrokinetic chromatography, *J.Chromatogr.A*, **1996**, *756*, 279–286.

Tenoxicam

Molecular formula: $C_{13}H_{11}N_3O_4S_2$
Molecular weight: 337.38
CAS Registry No.: 59804-37-4
Merck Index (12th ed.): 9293
Lednicer: 4 173

SAMPLE
Matrix: blood
Sample preparation: 0.1-1 mL Plasma + 2 µg IS + 400 µL 1 M HCl + 10 mL diethyl ether, agitate mechanically for 10 min, centrifuge at 3500 g for 10 min. Remove the organic layer and evaporate it to dryness under a stream of nitrogen, reconstitute the residue in MeOH:40 mM NaH_2PO_4 50:50, inject an aliquot.

CAPILLARY ELECTROPHORESIS
Capillary: 58.7 cm × 75 µm fused-silica (50 cm to detector)

Capillary preparation: Wash with running buffer for 2 min before each injection. At the beginning of each day wash capillary with 100 mM NaOH for 10 min.
Capillary temperature: 22
Running buffer: MeOH:buffer 3:97 (Buffer was 40 mM pH 8 NaH_2PO_4 containing 104 mM sodium dodecyl sulfate.)
Injection: Hydrodynamic injection for 2 s
Detector: UV 254
Migration time: 10.2
Internal standard: benzoyl-4-phenyl-2-butyric acid (16.3)
Voltage: 20 kV
Model: Beckman P/ACE 2000
Limit of detection: 130 ng/mL (S/N 5)

OTHER SUBSTANCES
Extracted: diclofenac, diflunisal, etodolac, fenbufen, fenoprofen, flurbiprofen, ibuprofen, indomethacin, ketoprofen, naproxen, niflumic acid, piroxicam, sulindac, tiaprofenic acid
Noninterfering: acetaminophen, amitriptyline, caffeine, clomipramine, deoxysulindac, desipramine, diazepam, 5'-hydroxytenoxicam, imipramine, maprotiline, nortriptyline, phenobarbital, phenytoin, sulfamethoxazole, theophylline, trimipramine

KEY WORDS
plasma

REFERENCE
Maboundou,C.W.; Paintaud,G.; Bérard,M.; Bechtel,P.R. Separation of fifteen non-steroidal anti-inflammatory drugs using micellar electrokinetic capillary chromatography, *J.Chromatogr.B*, **1994**, *657*, 173–183.

SAMPLE
Matrix: solutions
Sample preparation: Dissolve in MeCN, dilute with water to a concentration of 25 µg/mL, inject an aliquot.

CAPILLARY ELECTROPHORESIS
Capillary: 60 cm × 75 µm fused-silica (52.5 cm to detector)
Running buffer: MeCN:30 mM pH 7.0 phosphate buffer 20:80
Injection: Hydrostatic injection at 9.8 cm for 5 s
Detector: UV 214
Migration time: 19.09
Model: Waters Quanta 4000

OTHER SUBSTANCES
Simultaneous: acemetacin, alclofenac, fenbufen, flurbiprofen, ibuprofen, indomethacin, ketoprofen, lonazolac, naproxen, niflumic acid, piroxicam, tiaprofenic acid, tolmetin, voltaren

REFERENCE
Donato,M.G.; Van den Eeckhout,E.; Van den Bossche,W.; Sandra,P. Capillary zone electrophoresis and micellar electrokinetic capillary chromatography of some non-steroidal antiinflammatory drugs (NSAIDs), *J.Pharm.Biomed.Anal.*, **1993**, *11*, 197–201.

Terbutaline

Molecular formula: $C_{12}H_{19}NO_3$
Molecular weight: 225.29
CAS Registry No.: 23031-25-6, 23031-32-5 (sulfate)
Merck Index (12th ed.): 9302

SAMPLE
Matrix: solutions

CAPILLARY ELECTROPHORESIS
Capillary: 100 cm × 50 μm fused-silica (50 cm to detector) (Isco)
Capillary preparation: Flush capillary with 10 μL running buffer between runs. Every 40 sample injections rinse capillary with 200 μL 1 M NaOH, with 200 μL water, and 200 μL running buffer. Before use fill capillary with 1 M NaOH and allow to stand for 1 h, fill with 100 mM NaOH, allow to stand for 1 h, wash with water fill with running buffer.
Capillary temperature: 23
Running buffer: 100 mM citric acid containing 19.27 mM Na_2HPO_4 and 120 mM hydroxypropyl-β-cyclodextrin
Injection: Inject under vacuum at 4.0 kPa.s
Detector: UV 200
Migration time: 46, 48 (enantiomers)
Voltage: 30 kV
Model: Isco Model 3140

OTHER SUBSTANCES
Simultaneous: albuterol, alprenolol, atenolol, cimaterol, clenbuterol, labetalol, nadolol, oxprenolol, pindolol, pirbuterol, propranolol

KEY WORDS
chiral

REFERENCE
Aumatell,A.; Wells,R.J.; Wong,D.K.Y. Enantiomeric differentiation of a wide range of pharmacologically active substances by capillary electrophoresis using modified β-cyclodextrins, *J.Chromatogr.A*, **1994**, *686*, 293–307.

SAMPLE
Matrix: solutions
Sample preparation: Prepare a 50 μg/mL solution in diluted running buffer, inject an aliquot.

CAPILLARY ELECTROPHORESIS
Capillary: 44 cm × 50 μm fused-silica
Capillary preparation: After each run wash capillary with water for 2 min and with running buffer for 3 min. At the beginning of each day wash capillary with water and running buffer for 5 min. Condition a new capillary with 1 M NaOH, 100 mM NaOH, water, and separation buffer.
Capillary temperature: 15
Running buffer: 100 mM Phosphoric acid containing 15 mM hydroxypropyl-β-cyclodextrin adjusted to pH 3.0 with triethanolamine
Injection: Hydrodynamic injection for 1 s
Detector: UV 210
Voltage: 25 kV
Model: Spectra Physics Spectraphoresis 1000

KEY WORDS
chiral; enantiomer resolution 8.5

REFERENCE
Bechet,I.; Paques,P.; Fillet,M.; Hubert,P.; Crommen,J. Chiral separation of basic drugs by capillary zone electrophoresis with cyclodextrin additives, *Electrophoresis*, **1994**, *15*, 818–823.

SAMPLE
Matrix: solutions

CAPILLARY ELECTROPHORESIS
Capillary: 74 cm × 100 μm untreated fused-silica (SGE)
Running buffer: MeOH:50 mM pH 4.8 ammonium acetate 80:20
Injection: Hydrodynamic injection
Detector: MS, Finnigan MAT SSQ 710 single-quadrupole, positive-ion mode, makeup liquid was the same as running buffer (pumped at 1 μL/min), drying gas nitrogen, electrospray tip at ground potential, electrospray counter electrode at -3.8 kV, m/z 226.1

Migration time: 15.7
Voltage: +20 kV

OTHER SUBSTANCES
Simultaneous: albuterol, clenbuterol, fenoterol

REFERENCE
Lamoree,M.H.; Reinhoud,N.J.; Tjaden,U.R.; Niessen,W.M.A.; van der Greef,J. On-capillary isotachophoresis for
loadability enhancement in capillary zone electrophoresis/mass spectrometry of β-agonists, *Biol.Mass.
Spectrom.*, **1994**, *23*, 339–345.

SAMPLE
Matrix: solutions

CAPILLARY ELECTROPHORESIS
Capillary: 60 cm × 50 μm AccuSep (52.5 cm to detector) (Waters)
Capillary preparation: Before injection rinse capillary with 100 mM NaOH for 3 min and with
running buffer for 3 min. Rinse new capillaries with 500 mM NaOH for 5 min
Running buffer: 50 mM pH 7.0 Na_2HPO_4 containing 25 mM (S)-N-dodecoxycarbonylvaline (Pre-
pare (S)-N-dodecoxycarbonylvaline as follows. Prepare dodecyl chloroformate by reacting 1-
dodecanol with 0.33 equivalents of triphosgene in dichloromethane solution in the presence of
pyridine. Add dodecyl chloroformate dropwise to (S)-valine in 1 M NaOH solution, filter, wash
with hexane, recrystallize from ether/petroleum ether.)
Injection: Hydrostatic injection 2 s
Detector: UV 214
Voltage: +12 kV
Model: Waters Quanta 4000 or 4000E

OTHER SUBSTANCES
Also analyzed: atenolol, bupivacaine, ephedrine, homatropine, ketamine, N-methylpseudoephedrine, metoprolol, norephedrine, norphenylephrine, octopamine, pindolol

---

KEY WORDS
chiral; $\alpha = 1.02$

REFERENCE
Mazzeo,J.R.; Grover,E.R.; Swartz,M.E.; Petersen,J.S. Novel chiral surfactant for the separation of enantiomers
by micellar electrokinetic capillary chromatography, *J.Chromatogr.A*, **1994**, *680*, 125–135.

SAMPLE
Matrix: solutions
Sample preparation: Prepare a 200 μM solution in water, inject an aliquot.

CAPILLARY ELECTROPHORESIS
Capillary: 57 cm × 75 μm fused-silica (50 cm to detector) (Beckman)
Capillary preparation: Before each analysis wash with three column volumes of 10 mM NaOH,
three column volumes of water, and three column volumes of running buffer. At the end of each
day wash capillary with 10 column volumes of 10 mM NaOH and 10 column volumes of water.
At the beginning of each day wash capillary with 100 column volumes of 10 mM NaOH.
Capillary temperature: 25
Running buffer: 50 mM pH 2.5 Phosphate buffer containing 5 mM heptakis(2,6-di-O-methyl)-
β-cyclodextrin
Injection: Injection using pressurized nitrogen for 3 s (about 15 nL)
Detector: UV 214
Migration time: 19, 20.5 (enantiomers)
Voltage: 17 kV
Model: Beckman P/ACE 2000

KEY WORDS
chiral

REFERENCE
Pálmarsdóttir,S.; Edholm,L.-E. Capillary zone electrophoresis for separation of drug enantiomers using cyclodextrins as chiral selectors Influence of experimental parameters on separation, *J.Chromatogr.A*, **1994**, *666*,
337–350.

SAMPLE
Matrix: solutions
Sample preparation: Inject an aliquot of a 10 μg/mL solution in water.

CAPILLARY ELECTROPHORESIS
Capillary: 68.5 × 50 μm CElect-H250 C18 bonded-phase capillary (60 cm to detector) (Supelco)
Capillary preparation: After each run rinse with running buffer for 5 min. Every 10 runs wash
  with 100 mM NaOH for 1 min and with running buffer for 5 min. Condition a new capillary
  with 100 mM NaOH for 1 min, with water for 5 min, and with running buffer for 5 min. Repeat
  this cycle for 2 h.
Capillary temperature: 25
Running buffer: 50 mM pH 8.3 Tris buffer
Injection: Hydrostatic injection at 50 mbar for 2-6 s, flush with buffer for 4 s
Detector: UV 200
Migration time: 5.528
Voltage: 30 kV
Model: Hewlett-Packard HP 3DCE

OTHER SUBSTANCES
Simultaneous: albuterol, cimaterol, clenbuterol, fenoterol, isoxsuprine, metaproterenol, ractopamine, ritodrine, RU 42 173

KEY WORDS
comparison with fused-silica capillary; coated capillary

REFERENCE
Chevolleau,S.; Tulliez,J. Optimization of the separation of β-agonists by capillary electrophoresis on untreated
  and C18 bonded silica capillaries, *J.Chromatogr.A*, **1995**, *715*, 345–354.

SAMPLE
Matrix: solutions
Sample preparation: Inject an aliquot of a 10 μg/mL solution in water.

CAPILLARY ELECTROPHORESIS
Capillary: 68.5 × 50 μm fused-silica (60 cm to detector) (Hewlett-Packard)
Capillary preparation: After each run rinse with 100 mM NaOH for 5 min and with running
  buffer for 5 min. Condition a new capillary with 1 M NaOH at 40° for 20 min and with 100
  mM NaOH at 40° for 10 min, rinse with water at 25° for 20 min, and rinse with running buffer.
Capillary temperature: 25
Running buffer: 50 mM pH 8.3 Tris buffer
Injection: Hydrostatic injection at 50 mbar for 2-6 s, flush with buffer for 4 s
Detector: UV 200
Migration time: 3.539
Voltage: 30 kV
Model: Hewlett-Packard HP 3DCE

OTHER SUBSTANCES
Simultaneous: albuterol, cimaterol, clenbuterol, fenoterol, isoxsuprine, metaproterenol, ractopamine, ritodrine, RU 42 173

KEY WORDS
comparison with C18 bonded capillary

REFERENCE
Chevolleau,S.; Tulliez,J. Optimization of the separation of β-agonists by capillary electrophoresis on untreated
  and C18 bonded silica capillaries, *J.Chromatogr.A*, **1995**, *715*, 345–354.

SAMPLE
Matrix: solutions

CAPILLARY ELECTROPHORESIS
Capillary: 40 cm × 50 μm fused-silica (35.5 cm to detector) (Polymicro Technologies)
Capillary preparation: Between runs purge with water for 50 s, with 100 mM NaOH for 50 s, with water for 60 s, and with running buffer for 60 s.
Running buffer: 50 mM pH 6.0 Sodium phosphate buffer containing 1 mg/mL sulfobutyl ether-β-cyclodextrin (Perkin-Elmer)
Injection: Pressure injection at 5 psi
Detector: UV 206
Migration time: 3.2, 3.4 (enantiomers)
Voltage: 15 kV
Model: Bio-Rad Biofocus 3000

KEY WORDS
chiral

REFERENCE
Desiderio,C.; Fanali,S. Use of negatively charged sulfobutyl ether-β-cyclodextrin for enantiomeric separation by capillary electrophoresis, *J.Chromatogr.A*, **1995**, *716*, 183–196.

SAMPLE
Matrix: solutions
Sample preparation: Inject an aliquot of a 100 μg/mL solution in water:running buffer 50:50.

CAPILLARY ELECTROPHORESIS
Capillary: 44.5 cm × 50 μm acrylamide-coated fused-silica (Bio-Rad)
Capillary temperature: 30
Running buffer: 100 mM NaH_2PO_4 containing 15 mM gamma-cyclodextrin, adjusted to pH 2.5 with phosphoric acid
Injection: Electrokinetic injection at 8 kV for 6 s.
Detector: UV 200
Migration time: 9.63
Voltage: 14 kV
Model: Bio-Rad BioFocus 3000

OTHER SUBSTANCES
Also analyzed: albuterol, alprenolol, atenolol, atropine, baclofen, bamethan, benserazide, biperiden, bisoprolol, bupivacaine, bupranolol, butetamate, carazolol, carbuterol, carvedilol, celiprolol, chloroquine, chlorpheniramine (chlorphenamine), clidinium bromide, clobutinol, disopyramide, dobutamine, flecainide, homatropine, ipratropium bromide, isoproterenol, isothipendyl, ketamine, mefloquine, mequitazine, metaproterenol (orciprenaline), metipranolol, nafronyl (naftidrofuryl), nefopam, ofloxacin, orphenadrine, oxomemazine, oxprenolol, phenoxybenzamine, pholedrine, pindolol, pirbuterol, prilocaine, promethazine, propafenone, propranolol, sotalol, synephrine, tetrahydrozoline (tetryzoline), tocainide, trihexyphenidyl, trimeprazine (alimemazine), trimipramine, tropicamide, verapamil, zopiclone

KEY WORDS
coated capillary; achiral

REFERENCE
Koppenhoefer,B.; Epperlein,U.; Christian,B.; Yibing,J.; Yuying,C.; Bingcheng,L. Separation of enantiomers of drugs by capillary electrophoresis. I. γ-Cyclodextrin as chiral solvating agent, *J.Chromatogr.A*, **1995**, *717*, 181–190.

SAMPLE
Matrix: solutions
Sample preparation: Inject an aliquot of a 100 μM solution.

CAPILLARY ELECTROPHORESIS
Capillary: 37 or 47 cm × 75 μm CElect C8 bonded capillary (effective length 30 or 40 cm) (Supelco)
Capillary preparation: Flush capillary for 2 min with running buffer before injection. Rinse capillary with water at the end of the day.
Running buffer: pH 3.0 Phosphate buffer (I = 0.05) containing 60 mM hydroxypropyl-β-cyclodextrin and 20 mM tetrabutylammonium hydroxide
Injection: Inject at 0.5 psi for 3-5 s.
Detector: UV 214
Migration time: 90, 114 (enantiomers)
Voltage: 15 kV
Current: 40-55 μA
Model: Beckman P/ACE 2100

OTHER SUBSTANCES
Simultaneous: albuterol, bambuterol, clenbuterol

KEY WORDS
chiral; coated capillary

REFERENCE
Stålberg,O.; Brötell,H.; Westerlund,D. Capillary electrophoretic separation of basic drugs using surface-modified C8 capillaries and derivatized cyclodextrins as structural/chiral selectors, *Chromatographia*, **1995**, *40*, 697–704.

SAMPLE
Matrix: solutions
Sample preparation: Inject an aliquot of a solution in ethyl acetate saturated with 10 mM β-alanine adjusted to pH 5 with 3.3 M acetic acid.

CAPILLARY ELECTROPHORESIS
Capillary: 70 cm × 100 μm untreated fused-silica (SGE) coupled to 20 cm × 100 μm untreated fused-silica (BGB Analytik) with a non-conducting coupler
Capillary preparation: Condition capillary daily for 10 min with water, 250 mM KOH, water, and leading buffer
Running buffer: Leading buffer: MeOH:50 mM pH 3.3 ammonium acetate solution 80:20; terminating buffer: MeOH:12 mM β-alanine adjusted to pH 5 with 3.3 M acetic acid 15:85
Injection: Inject terminating buffer at 30 mbar for 18 s to give a 15 mm zone, inject the sample solution at 10 kV for 1 min, inject the sample solution at 10 kV for 9 min with an 8 mbar counterflow so that ethyl acetate does not enter, inject terminating buffer at 15 kV for 1 min with an 8 mbar counterflow, apply 9 kV for 6 s with a 50 mbar counterflow, apply 9 kV with a 23 mbar counterflow until the current reaches 4.0 μA, electrophorese at 21 kV with leading buffer.
Detector: MS, Finnigan MAT, TSQ-70 triple quadrupole, sheath liquid MeOH:50 mM pH 3.3 ammonium acetate solution 80:20 at 1 μL/min, electrospray needle at +3 kV, sampling capillary at ground, sampling capillary 175°, ion source 150°, repeller electrode + 30 V, m/z 226
Migration time: 14
Voltage: 9-21 kV
Model: Lauerlabs Prince
Limit of detection: 2 nM

OTHER SUBSTANCES
Simultaneous: albuterol, clenbuterol, fenoterol

REFERENCE
van der Vlis,E.; Mazereeuw,M.; Tjaden,U.R.; Irth,H.; van der Greef,J. Combined liquid-liquid electroextraction-isotachophoresis for loadability enhancement in capillary zone electrophoresis-mass spectrometry, *J.Chromatogr.A*, **1995**, *712*, 227–234.

SAMPLE
Matrix: solutions

CAPILLARY ELECTROPHORESIS
Capillary: 37 cm $\times$ 50 μm (30 cm to detector)
Capillary temperature: 25
Running buffer: 50 mM pH 5.0 Acetate buffer containing 16 mM 2-hydroxy-3-trimethylammo-
    niopropyl-β-cyclodextrin
Detector: UV 214
Migration time: 13.50 (first enantiomer; α = 1.018; R_S = 1.058)
Voltage: 20 kV
Model: Beckman P/ACE 2100

KEY WORDS
chiral

REFERENCE
Bunke,A.; Jira,T. Chiral capillary electrophoresis using a cationic cyclodextrin, *Pharmazie*, **1996**, *51*, 672–673.

SAMPLE
Matrix: solutions
Sample preparation: Inject an aliquot of a 50-200 μM solution.

CAPILLARY ELECTROPHORESIS
Capillary: 48.5 cm $\times$ 50 μm fused-silica (44.5 cm to detector) (Polymicro Technologies)
Capillary temperature: 25
Running buffer: 100 mM Phosphoric acid containing 5 mM 6^A-methylamino-β-cyclodextrin, ad-
    justed to pH 2.5 with tetramethylammonium hydroxide. (Synthesis of 6^A-methylamino-β-cy-
    clodextrin was as follows. Add a solution of 3.65 g p-toluenesulfonyl chloride in 30 mL dry
    pyridine to 29.60 g β-cyclodextrin stirred at 5° in 300 mL dry pyridine, stir overnight at room
    temperature, evaporate to dryness under reduced pressure at 40°, add 700 mL diethyl ether
    to the residue. Collect the precipitate and recrystallize it 3 times from water to obtain mono-
    (6-O-p-tolylsulfonyl)-β-cyclodextrin in 31% yield (Bull. Chem. Soc. Japan 1978, 51, 3030). Heat
    2 g mono-(6-O-p-tolylsulfonyl)-β-cyclodextrin with 35 mL 50% methylamine in MeOH in a
    sealed tube at 70° for 3 days, purify by chromatography on carboxymethylcellulose with am-
    monium bicarbonate solution to obtain 6^A-methylamino-β-cyclodextrin (cf. J. Am. Chem. Soc.
    1980, 102, 762).
Injection: Pressure injection at 10 psi.s.
Detector: UV 206
Migration time: 38.9 (second enantiomer, R_S = 2.5)
Voltage: 18 kV
Current: 41-48 μA
Model: Biofocus 3000 (Bio-Rad)

OTHER SUBSTANCES
Simultaneous: propranolol

KEY WORDS
chiral

REFERENCE
Fanali,S.; Camera,E. Use of methylamino-β-cyclodextrin in capillary electrophoresis. Resolution of acidic and
    basic enantiomers, *Chromatographia*, **1996**, *43*, 247–253.

SAMPLE
Matrix: solutions
Sample preparation: Inject an aliquot of a 50 μg/mL solution in water.

CAPILLARY ELECTROPHORESIS
Capillary: 48.5 cm $\times$ 50 μm fused-silica (40 cm to detector)
Capillary preparation: After each run wash capillary with running buffer for 3 min. At the
    start of each day wash capillary with running buffer for 10 min. Before use treat new capillaries
    with 1 M NaOH, 100 mM NaOH, water, and running buffer.
Capillary temperature: 15

Running buffer: 100 mM Phosphoric acid containing 2 mM carboxymethyl-β-cyclodextrin (Cyclolab, Budapest), adjusted to pH 3.0 with 84 mM triethanolamine
Injection: Hydrodynamic injection at 5 kPa for 2 s.
Detector: UV 210
Migration time: 16.24, 17.71 (enantiomers)
Voltage: 25 kV
Model: Hewlett Packard ³ᴰCE

OTHER SUBSTANCES
Simultaneous: chlorpheniramine, dimethindene, ephedrine, fenfluramine, isoproterenol

KEY WORDS
chiral

REFERENCE
Fillet,M.; Bechet,I.; Hubert,P.; Crommen,J. Resolution improvement by use of carboxymethyl-β-cyclodextrin as chiral additive for the enantiomeric separation of basic drugs by capillary electrophoresis, *J.Pharm.Biomed.Anal.*, **1996**, *14*, 1107–1114.

SAMPLE
Matrix: solutions
Sample preparation: Inject an aliquot of a solution in running buffer.

CAPILLARY ELECTROPHORESIS
Capillary: 60 cm × 75 μm fused-silica (52.4 cm to detector)
Capillary preparation: After each run flush with 500 mM KOH for 2-3 min then with water.
Running buffer: 10 mM pH 3.8 Phosphate buffer containing 2% sulfated cyclodextrin (ds 7-10)
Injection: Hydrostatic injection.
Detector: UV 214
Migration time: 8.37, 8.75 (enantiomers)
Voltage: 15 kV
Model: Waters Quanta 4000

OTHER SUBSTANCES
Also analyzed: acebutolol, alprenolol, aminoglutethimide, brompheniramine, bupivacaine, bupropion, canadine, carbinoxamine, chloroquine, chlorpheniramine, dimethindene, disopyramide, doxylamine, hydroxychloroquine, idazoxan, isoxsuprine, ketamine, mepenzolate, mepivacaine, methoxyphenamine, mexiletine, midodrine, nefopam, orphenadrine, oxprenolol, oxyphencyclimine, pheniramine, phensuximide, pindolol, piperoxan, tetramisole, tolperisone, tranylcypromine, trihexyphenidyl, trimipramine, verapamil, warfarin

KEY WORDS
chiral; detector at anode

REFERENCE
Stalcup,A.M.; Gahm,K.H. Application of sulfated cyclodextrins to chiral separations by capillary zone electrophoresis, *Anal.Chem.*, **1996**, *68*, 1360–1368.

SAMPLE
Matrix: solutions

CAPILLARY ELECTROPHORESIS
Capillary: 48 cm × 50 μm fused-silica (41 cm to detector)
Capillary temperature: 25.0
Running buffer: 10 mM pH 2.98 Sodium formate buffer containing 40 mM hydroxypropyl-β-cyclodextrin (degree of substitution 6.5)
Detector: UV (wavelength not given)
Migration time: 3.6, 3.8 (enantiomers)
Voltage: 30 kV
Current: <50 μA
Model: Beckman P/ACE 2200

OTHER SUBSTANCES
Simultaneous: mandelic acid

KEY WORDS
chiral

REFERENCE
Ingelse,B.A.; Sarmini,K.; Reijenga,J.C.; Kenndler,E.; Everaerts,F.M. Chiral interactions in capillary zone electrophoresis: Computer simulation and comparison with experiment, *Electrophoresis*, **1997**, *18*, 938–942.

SAMPLE
Matrix: solutions
Sample preparation: Inject an aliquot of a 100 μg/mL solution in running buffer.

CAPILLARY ELECTROPHORESIS
Capillary: 29 cm × 50 μm fused-silica (24.5 cm to detector) (Yongnian Optical Conductive Fiber Plant, China), coated with polyacrylamide
Capillary preparation: No details of the polyacrylamide coating process are provided. However, another paper (LC.GC 1997, 15, 40) by this group indicates that they use the procedure of Hjertén, thus: Adjust the pH of 20 mL water to 3.5 with acetic acid, add 80 μL 3-(trimethoxysilyl)propyl methacrylate (3-methacryloxypropyltrimethoxysilane), mix, suck into capillary, let stand at room temperature for 1 h, remove the solution, wash with water. Fill the capillary with a deaerated 3-4% acrylamide solution containing 1 μL/mL N,N,N',N'-tetramethylethylenediamine and 1 mg/mL potassium persulfate, let stand for 30 min, remove excess solution by aspiration, rinse with water, remove water by aspiration, dry at 35° (J. Chromatogr. 1985, 347, 191).
Capillary temperature: 25
Running buffer: 100 mM NaH_2PO_4 adjusted to pH 2.5
Injection: Electrokinetic injection at 15 kV for 3 s.
Detector: UV 200, UV 210
Migration time: 5.48
Voltage: 15 kV
Model: Bio-Rad BioFocus 3000

OTHER SUBSTANCES
Simultaneous: albuterol, alprenolol, atenolol, baclofen, bamethan, benproperine, benserazide, bisoprolol, bupranolol, butamirate, butethamate, carbuterol, celiprolol, clenbuterol, clobutinol, dipivefrin, isoproterenol (isoprenaline), metaproterenol (orciprenaline), metipranolol, metoprolol, norfenefrine, ornidazole, oxprenolol, phenylpropanolamine, pholedrine, pirbuterol, prilocaine, procyclidine, sotalol, synephrine, tocainide

KEY WORDS
coated capillary

REFERENCE
Koppenhoefer,B.; Epperlein,U.; Xiaofeng,Z.; Bingcheng,L. Separation of enantiomers of drugs by capillary electrophoresis. Part 4: Hydroxypropyl-γ-cyclodextrin as chiral solvating agent, *Electrophoresis*, **1997**, *18*, 924–930.

SAMPLE
Matrix: solutions
Sample preparation: Inject an aliquot of a 100 μg/mL solution in running buffer.

CAPILLARY ELECTROPHORESIS
Capillary: 30 cm × 50 μm fused-silica (25.5 cm to detector), coated with polyacrylamide
Capillary preparation: Adjust the pH of 20 mL water to 3.5 with acetic acid, add 80 μL 3-(trimethoxysilyl)propyl methacrylate (3-methacryloxypropyltrimethoxysilane), mix, suck into capillary, let stand at room temperature for 1 h, remove the solution, wash with water. Fill the capillary with a deaerated 3-4% acrylamide solution containing 1 μL/mL N,N,N',N'-tetramethylethylenediamine and 1 mg/mL potassium persulfate, let stand for 30 min, remove excess

solution by aspiration, rinse with water, remove water by aspiration, dry at 35° (J. Chromatogr. 1985, 347, 191).
Capillary temperature: 25
Running buffer: 100 mM NaH_2PO_4 containing 45 mM hydroxypropyl-β- cyclodextrin, adjusted to pH 2.5 with phosphoric acid
Injection: Electrokinetic injection at 15 kV for 3 s.
Detector: UV 200
Voltage: 15 kV
Model: Bio-Rad BioFocus 3000

KEY WORDS
chiral; coated capillary; comparison with the use of other cyclodextrins; this running buffer gave the greatest enantiomeric separation.; α=1.094

REFERENCE
Lin,B.; Zhu,X.; Koppenhoefer,B.; Epperlein,U. Investigation of 123 chiral drugs by cyclodextrin-modified capillary electrophoresis, *LC.GC*, **1997**, *15*, 40–46.

SAMPLE
Matrix: solutions
Sample preparation: Inject an aliquot of a 100 μg/mL solution in MeOH.

CAPILLARY ELECTROPHORESIS
Capillary: 58.5 cm × 50 μm fused-silica (50 cm to detector) (Composite Metal Services, Hallow UK)
Capillary temperature: 25
Running buffer: 33 mM Phosphoric acid containing 15 mM heptakis(2,6-di-O-methyl)-β-cyclodextrin (Aldrich), adjusted to pH 3 with 3 M NaOH
Injection: Pressure injection at 50 mbar for 3 s.
Detector: UV 202
Migration time: 9.9, 10.4 (enantiomers)
Voltage: 30 kV
Model: Hewlett-Packard [3D]CE

KEY WORDS
chiral

REFERENCE
Szemán,J.; Roos,N.; Csabai,K. Ruggedness of enantiomeric separation by capillary electrophoresis and high-performance liquid chromatography with methylated cyclodextrins as chiral selectors, *J.Chromatogr.A*, **1997**, *763*, 139–147.

SAMPLE
Matrix: solutions

CAPILLARY ELECTROPHORESIS
Capillary: 36 cm × 50 μm fused-silica coated with linear polyacrylamide (31.5 cm to detector) (GL Science)
Capillary preparation: At the beginning and end of each day rinse capillary with capillary wash solution (Bio-Rad Cat. No. 148-5022) at 690 kPa for more than 3 min and with water at 690 kPa for more than 3 min. Coat capillary as follows. Treat capillary with 1 M NaOH at room temperature for 1 h, rinse with water, dry by passing nitrogen gas through the capillary at 110° for 6 h. Pass thionyl chloride through the capillary using a suction pump for several min, seal capillary at both ends and heat at 70° for 6 h. Unseal the capillary and fill with 250 mM vinyl magnesium bromide in THF by suction, seal the capillary, heat at 70° for 6 h. Open the capillary and rinse it with THF for several min, rinse with distilled water, fill the capillary with polymerization solution, heat at 28 ± 2° for 1 h, condition at -100 V/cm for 30 min (Anal. Sci. 1994, 10, 1). (The polymerization solution was 5% acrylamide in water containing 49 mM Tris, 384 mM glycine, and 0.1% sodium dodecyl sulfate, degas in an ultrasonic bath. Add 40 μL 10% N,N,N',N'-tetramethylethylenediamine and 10 μL 10% ammonium persulfate to 5 mL of the degassed solution, mix thoroughly.)
Running buffer: 50 mM pH 6.0 Phosphate buffer

Injection: Before each injection rinse with water at 690 kPa for 30 s, rinse with running buffer at 690 kPa for 30 s, partially fill with separation solution (1 mM α_1-acid glycoprotein (Cohn fraction VI) (ICN) in running buffer) at 6.9 kPa for 190 s (27 cm), inject sample at 6.9 kPa for 2 s, electrophorese with running buffer (Note that α_1-acid glycoprotein from other suppliers may provide inferior results).
Detector: UV 210
Migration time: 13, 13.5 (enantiomers)
Voltage: 12 kV
Model: Bio-rad BioFocus 3000

KEY WORDS
chiral

REFERENCE
Tanaka,Y.; Terabe,S. Separation of the enantiomers of basic drugs by affinity capillary electrophoresis using a partial filling technique and α_1-acid glycoprotein as chiral selector, *Chromatographia*, **1997**, *44*, 119–128.

SAMPLE
Matrix: solutions

CAPILLARY ELECTROPHORESIS
Capillary: 35 cm × 50 μm polyacrylamide-coated fused-silica (30.5 cm to detector) (Composite Metal Services, UK)
Capillary preparation: Before each run rinse capillary with water for 70 s and with running buffer for 100 s. (Coat capillary as follows. Adjust the pH of 20 mL water to 3.5 with acetic acid, add 80 μL 3-(trimethoxysilyl)propyl methacrylate (3-methacryloxypropyltrimethoxysilane), mix, suck into capillary, let stand at room temperature for 1 h, remove the solution, wash with water. Fill the capillary with a deaerated 3-4% acrylamide solution containing 1 μL/mL N,N,N',N'-tetramethylethylenediamine and 1 mg/mL potassium persulfate, let stand for 30 min, remove excess solution by aspiration, rinse with water, remove water by aspiration, dry at 35° (J. Chromatogr. 1985, 347, 191).)
Capillary temperature: 25
Running buffer: Buffer containing 100 mM cyanoethylated-β-cyclodextrin (Cyclolab, Budapest) (Prepare buffer by adjusting the pH of 50 mM phosphoric acid containing 50 mM acetic acid and 50 mM boric acid to 2.5 with concentrated NaOH, add the appropriate amount of cyanoethylated-β-cyclodextrin, dilute with an equal volume of water.)
Injection: Pressure injection at 5 psi for 2 s.
Detector: UV 206
Migration time: 16.8 (second enantiomer, α = 1.060)
Voltage: 20 kV
Current: 27-40 μA
Model: Bio-Rad Biofocus 3000

KEY WORDS
chiral

REFERENCE
Aturki,Z.; Desiderio,C.; Mannina,L.; Fanali,S. Chiral separations by capillary zone electrophoresis with the use of cyanoethylated-β-cyclodextrin as chiral selector, *J.Chromatogr.A*, **1998**, *817*, 91–104.

SAMPLE
Matrix: solutions

CAPILLARY ELECTROPHORESIS
Capillary: 48.5 cm × 50 μm fused-silica (40 cm to detector) (Supelco)
Capillary preparation: Condition capillary with running buffer for 3 min before each injection.
Capillary temperature: 15
Running buffer: 25 mM Citric acid adjusted to pH 4.5 with Tris, containing 2% dermatan sulfate (number average 15270, mass average 22260, z average 30390, polydispersity 1.458) (Opocrin, Corlo, Italy)
Injection: Pressure injection at 5 kPa for 10 s.
Detector: UV 220

Migration time: 21, 22 (enantiomers)
Voltage: 15 kV
Model: Hewlett-Packard ^{3D}CE

OTHER SUBSTANCES
Simultaneous: albuterol
Also analyzed: metaproterenol, terbutaline

KEY WORDS
chiral

REFERENCE
Gotti,R.; Cavrini,V.; Andrisano,V.; Mascellani,G. Dermatan sulfate as useful chiral selector in capillary electrophoresis, *J.Chromatogr.A*, **1998**, *814*, 205–211.

SAMPLE
Matrix: solutions

CAPILLARY ELECTROPHORESIS
Capillary: 29-36 cm × 50 μm fused-silica (24.5-31.5 cm to detector) (Yongnian Optical Conductive Fiber Plant, China) coated with polyacrylamide
Capillary preparation: Coat capillary as follows. Adjust the pH of 20 mL water to 3.5 with acetic acid, add 80 μL 3-(trimethoxysilyl)propyl methacrylate (3-methacryloxypropyltrimethoxysilane), mix, suck into capillary, let stand at room temperature for 1 h, remove the solution, wash with water. Fill the capillary with a deaerated 3-4% acrylamide solution containing 1 μL/mL N,N,N',N'-tetramethylethylenediamine and 1 mg/mL potassium persulfate, let stand for 30 min, remove excess solution by aspiration, rinse with water, remove water by aspiration, dry at 35° (J. Chromatogr. 1985, 347, 191).
Capillary temperature: 25
Running buffer: 100 mM pH 2.5 NaH$_2$PO$_4$ (A) or 100 mM pH 2.5 NaH$_2$PO$_4$ containing 45 mM hydroxypropyl-α-cyclodextrin (Wacker, Munich) (B)
Injection: Electromigration at 15 kV for 3 s.
Detector: UV 200; UV 210
Migration time: 5.48 (A); 7.06 (B) (no separation of enantiomers)
Voltage: 15 kV
Model: Bio-Focus 3000

OTHER SUBSTANCES
Also analyzed: albuterol (salbutamol), alprenolol, amorolfine, atenolol, atropine, azelastine, baclofen, bamethan, benproperine, benserazide, biperiden, bisoprolol, brompheniramine, bupivacaine, bupranolol, butamirate, butethamate, carazolol, carbuterol, carteolol, carvedilol, celiprolol, chloroquine, chlorpheniramine, chlorphenoxamine, cicletanine, clenbuterol, clidinium bromide, clobutinol, dimethindene, dipivefrin, disopyramide, dobutamine, doxylamine, fendiline, flecainide, gallopamil, homatropine, ipratropium bromide, isoproterenol (isoprenaline), isothipendyl, ketamine, meclizine, mefloquine, mepindolol, mequitazine, metaclazepam, metaproterenol (orciprenaline), metipranolol, metoprolol, nafronyl (naftidrofuryl), nefopam, nicardipine, norfenefrine, ofloxacin, ornidazole, orphenadrine, oxomemazine, oxprenolol, oxybutynin, phenoxybenzamine, phenylpropanolamine, pholedrine, pindolol, pirbuterol, prilocaine, procyclidine, promethazine, propafenone, propranolol, reproterol, sotalol, sulpride, synephrine, talinolol, tetrahydrozoline (tetryzoline), theodrenaline, tioconazole, tocainide, trihexyphenidyl, trimeprazine (alimemazine), trimipramine, tropicamide, verapamil, zopiclone

KEY WORDS
coated capillary

REFERENCE
Koppenhoefer,B.; Eperlein,U.; Schlunk,R.; Zhu,X.; Lin,B. Separation of enantiomers of drugs by capillary electrophoresis. V. Hydroxypropyl-α-cyclodextrin as chiral solvating agent, *J.Chromatogr.A*, **1998**, *793*, 153–164.

SAMPLE
Matrix: solutions

CAPILLARY ELECTROPHORESIS
Capillary: 100 cm × 100 μm fused-silica (Polymicro Technologies)
Capillary preparation: Between runs wash capillary with running buffer for 10 min. Replace the electrolyte every other run. Wash new capillaries with water for 20 min and with running buffer for 20 min.
Running buffer: MeOH:water 80:20 containing 5 mM dimethyl-β-cyclodextrin, 5 mM ammonium acetate and 800 mM acetic acid
Injection: Gravity injection at 15 cm for 2 s.
Detector: MS, Vestec 201, electrospray voltage 2.48 kV, sheath liquid MeOH:acetic acid 99:1 flowing at 2.5 μL/min, m/z 226 (design of interface in J. Chromatogr. 1993, 639, 303)
Migration time: 29.7, 30.7 (enantiomers)
Voltage: 25 kV
Current: 13 μA
Model: Dionex CE System I
Limit of detection: 50 μM

KEY WORDS
chiral

REFERENCE
Lu,W.; Cole,R.B. Determination of chiral pharmaceutical compounds, terbutaline, ketamine and propranolol, by on-line capillary electrophoresis-electrospray mass spectrometry, *J.Chromatogr.B*, **1998**, *714*, 69–75.

SAMPLE
Matrix: urine
Sample preparation: Filter urine, inject an aliquot.

CAPILLARY ELECTROPHORESIS
Capillary: 120 cm × 75 μm fused-silica
Capillary preparation: Rinse with Bio-Rad Capillary wash Solution (20 mM phosphoric acid) and water, equilibrate with running buffer.
Running buffer: 5 mM pH 2.5 Sodium phosphate buffer containing 5 mM heptakis(2,6-di-O-methyl)-β-cyclodextrin
Injection: Electrokinetic injection at 10 kV or pressure injection for 5-10 s
Detector: MS, Sciex TAGA 6000E, API III, nebulizing gas nitrogen at 35 psi, collision gas argon, ion spray tip voltage 3.5-4.5 kV, sheath liquid 2-4 μL/min (MeOH:water 50:50 to 90:10 with or without 2 mM ammonium acetate), m/z 226 (protonated terbutaline) or 1557 (terbutaline-cyclodextrin complex) or UV 200
Migration time: 30, 31 (enantiomers)
Voltage: 30 kV
Current: 9 μA
Model: Beckman P/ACE Model 2050

KEY WORDS
chiral

REFERENCE
Sheppard,R.L.; Tong,X.; Cai,J.; Henion,J.D. Chiral separation and detection of terbutaline and ephedrine by capillary electrophoresis coupled with ion spray mass spectrometry, *Anal.Chem.*, **1995**, *67*, 2054–2058.

Testosterone

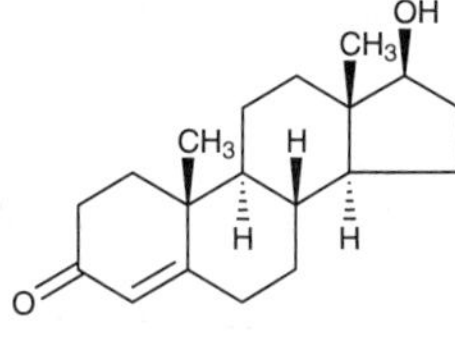

Molecular formula: $C_{19}H_{28}O_2$
Molecular weight: 288.43
CAS Registry No.: 58-22-0, 53608-96-1 (17-chloral hemiacetal), 58-20-8 (17β-cypionate), 315-37-7 (enanthate), 668-56-4 (nicotinate), 5704-03-0 (phenylacetate), 57-85-2 (propionate), 5874-98-6 (ketolaurate)
Merck Index (12th ed.): 9322
Lednicer: 1 172

SAMPLE
Matrix: blood
Sample preparation: Condition a 1 mL 100 mg Bond Elut C18 SPE cartridge with 1 mL MeOH and 1 mL water. Add 500 µL plasma to the cartridge, wash with 2 mL water, wash with 2 mL MeOH:water 20:80, elute with two 500 µL portions of MeOH. Evaporate the eluate to dryness under a stream of nitrogen, reconstitute the residue in 50 µL MeCN:MeOH:20 mM pH 8 Tris-HCl buffer 22.5:22.5:45, inject an aliquot.

CAPILLARY ELECTROPHORESIS
Capillary: 50 cm × 100 µm fused-silica (Polymicro Technologies) packed with 3 µm ODS Hypersil for 20 cm (20 cm to detector) (details in paper)
Running buffer: MeCN:MeOH:20 mM pH 8 Tris-HCl buffer 37.5:37.5:25
Injection: Electrokinetic injection at 15 kV for 5 s (25 nL). (Injector modified for injection from small volume samples; details in paper.)
Detector: UV 240
Migration time: 13
Internal standard: norethindrone
Voltage: 25 kV (?)
Model: Isco Model 3850

OTHER SUBSTANCES
Extracted: androstenedione, 17α-hydroxyprogesterone, 20α-hydroxyprogesterone, progesterone

KEY WORDS
electrochromatography; comparison with HPLC; plasma; SPE

REFERENCE
Stead,D.A.; Reid,R.G.; Taylor,R.B. Capillary electrochromatography of steroids. Increased sensitivity by on-line concentration and comparison with high-performance liquid chromatography, *J.Chromatogr.A*, **1998**, *798*, 259–267.

SAMPLE
Matrix: microsomal incubations
Sample preparation: Mix microsomal incubation with 4 µL 1 mg/mL IS in EtOH:water 40:60, add 5 mL dichloromethane, vortex for 2 min, centrifuge at 2000 g for 10 min. Remove the organic layer and evaporate it to dryness under reduced pressure, reconstitute the residue in 5 µL EtOH and 20 µL running buffer, inject an aliquot.

CAPILLARY ELECTROPHORESIS
Capillary: 72 cm × 75 µm fused-silica with Z-cell (50 cm to detector) (LC Packings International, San Francisco)
Capillary preparation: Before each run flush capillary with 100 mM NaOH for 2 min and with running buffer for 3 min.
Running buffer: EtOH:buffer 15:85 (Buffer was 50 mM pH 8.0 Tris buffer:50 mM pH 8.9 borate buffer 95:5 containing 60 mM sodium dodecyl sulfate and 15 mM β-cyclodextrin. Prepare Tris buffer by adding 50 mm HCl to 50 mm Tris and prepare borate buffer by adding 50 mm boric acid to 50 mm sodium borate.)
Injection: Vacuum injection at 17 kPa for 0.5 s.
Detector: UV 240

Migration time: 18
Internal standard: 11β-hydroxytestosterone
Voltage: 28 kV
Current: 58 μA
Model: Applied Biosystems model 270A-HT
Limit of detection: 400 nM (S/N = 3)

OTHER SUBSTANCES
Extracted: metabolites

KEY WORDS
rat; liver

REFERENCE
Fernandez,C.; Egginger,G.; Wainer,I.W.; Lloyd,D.K. Separation of testosterone metabolites in microsomal incubates using a new capillary electrophoresis assay, *J.Chromatogr.B*, **1996**, *677*, 363–368.

SAMPLE
Matrix: solutions

CAPILLARY ELECTROPHORESIS
Capillary: 57 cm × 75 μm fused-silica (50 cm to detector)
Capillary preparation: Before each run rinse with running buffer for 2 min. After each run rinse capillary with 100 mM NaOH for 2 min and with water for 2 min. Condition new capillaries with 100 mM HCl for 5 min, with 100 mM NaOH for 10 min, and with water for 5 min.
Capillary temperature: 57
Running buffer: EtOH:buffer 2:98 (Buffer was 50 mM pH 9.2 borate buffer containing 75 mM sodium dodecyl sulfate.)
Injection: Low pressure injection for 5 s.
Detector: UV 214
Migration time: 64
Voltage: 20 kV
Model: Beckman P/ACE 5510

OTHER SUBSTANCES
Simultaneous: estradiol, estriol, estrone, hydrocortisone, progesterone

REFERENCE
Gu,X.; Meleka-Boules,M.; Chen,C.-L. Micellar electrokinetic capillary chromatography combined with immunoaffinity chromatography for identification and determination of dexamethasone and flumethasone in equine urine, *J.Capillary Electrophor.*, **1996**, *3*, 43–49.

SAMPLE
Matrix: solutions

CAPILLARY ELECTROPHORESIS
Capillary: 33.5 cm × 100 μm fused-silica packed with 3 μm C18 for 25 cm (Innovatech, Stevenage, UK) (Polymicro Technologies)
Capillary preparation: Column packing procedure described in LC.GC 1995, 13, 800.
Capillary temperature: 20
Running buffer: MeCN:25 mM Tris HCl:water 60:20:20, run with 12 bar nitrogen pressure on each side
Injection: Electrokinetic injection at 5 kV for 3 s.
Detector: UV 200
Migration time: 7.5
Voltage: 30 kV
Model: Hewlett-Packard HP[3D]

OTHER SUBSTANCES
Simultaneous: androsterone, dehydrotestosterone, desoxycorticosterone, dexamethasone, estrone, ethinyl estradiol, hydrocortisone, prednisone, pregnenolone

KEY WORDS
electrochromatography

REFERENCE
Dittmann,M.M.; Rozing,G.P.; Ross,G.; Adam,T.; Unger,K.K. Advances in capillary electrochromatography, *J.Capillary Electrophor.*, **1997**, *4*, 201–212.

SAMPLE
Matrix: solutions

CAPILLARY ELECTROPHORESIS
Capillary: 17.6 cm × 50 μm fused-silica packed with 6 μm Zorbax ODS (9.6 cm to detector) (Quadrex, New Haven CT)
Capillary preparation: Sonicate 10 mg stationary phase in 110 μL acetone and 110 μL toluene for 5 min, pack in the capillary with acetone at 65 MPa for 30 min, wash with acetone at 50 MPa for 30 min, wash with water at 50 MPa, sinter frits while washing with water. Condition with running buffer at 0.15 mL/min for about 15 min.
Capillary temperature: 25
Running buffer: Gradient. A was MeCN:10 mM pH 8.0 borate buffer 65:35. B was MeCN:10 mM pH 8.0 borate buffer 85:15. A:B from 100:0 to 0:100 over 5 min, maintain at 0:100. Pump mobile phase at 0.1 mL/min.
Injection: Electrokinetic injection at 1 kV for 0.5 s.
Detector: UV 205
Migration time: 3
Voltage: 14 kV
Current: 1 μA
Model: Perkin-Elmer/Applied Biosystems ABI 270A-HT

OTHER SUBSTANCES
Simultaneous: androstan-3,17-dione, androsten-3,17-dione, corticosterone, pregnan-3,20-dione

KEY WORDS
electrochromatography

REFERENCE
Huber,C.G.; Choudhary,G.; Horváth,C. Capillary electrochromatography with gradient elution, *Anal.Chem.*, **1997**, *69*, 4429–4436.

SAMPLE
Matrix: solutions

CAPILLARY ELECTROPHORESIS
Capillary: 67 cm × 50 μm fused-silica (60 cm to detector)
Capillary preparation: Before each run equilibrate with running buffer for 3 min. After each run wash with 100 mM NaOH for 3 min and with water. At the beginning of each day rinse the capillary with 100 mM NaOH for 30 min and with water for 30 min.
Capillary temperature: 25 ± 0.1
Running buffer: 25 mM pH 9 Bicine buffer containing 10 mM sodium dodecyl sulfate and 5 mM N-dodecyl-N,N-dimethyl-3-ammonio-1-propanesulfonate (SB3-12)
Injection: High pressure injection for 5 s
Detector: UV 254
Migration time: 7.8 (testosterone), 8.3 (testosterone propionate)
Voltage: 30 kV
Current: <40 μA
Model: Beckman P/ACE 5010

OTHER SUBSTANCES
Simultaneous: cortisone, 11-desoxycortisol, 21-desoxycortisol, dimethyltestosterone, hydrocortisone, 17-hydroxyprogesterone, progesterone

---

REFERENCE

Valbuena,G.A.; Rao,L.V.; Petersen,J.R.; Okorodudu,A.O.; Bissell,M.G.; Mohammad,A.A. Anionic-zwitterionic mixed micelles in the micellar electrokinetic separation of clinically relevant steroids on a fused-silica capillary, *J.Chromatogr.A*, **1997**, *781*, 467–474.

SAMPLE

Matrix: solutions

CAPILLARY ELECTROPHORESIS

Capillary: 32 cm × 50 μm fused-silica packed for 24 cm with 1.8 μm Zorbax ODS (Polymicro Technologies) (Column preparation described in paper.)
Capillary temperature: 25
Running buffer: MeCN:0.8 mM sodium tetraborate 80:20 containing 5 mM sodium dodecyl sulfate
Injection: Electrokinetic injection at 20 kV for 1 s.
Detector: UV 254
Migration time: 3.4
Voltage: 25 kV
Model: Hewlett-Packard HP [3D]CE

OTHER SUBSTANCES

Simultaneous: estradiol, estriol, estrone, hydrocortisone, methyltestosterone, 4-pregnen-20α-ol-3-one, progesterone

---

KEY WORDS

electrochromatography

REFERENCE

Seifar,R.M; Kraak,J.C.; Kok,W.T.; Poppe,H. Capillary electrochromatography with 1.8-μm ODS-modified porous silica particles, *J.Chromatogr.A*, **1998**, *808*, 71–77.

SAMPLE

Matrix: urine
Sample preparation: Acidify 5 mL urine to pH 6, extract twice with 2 mL portions of dichloromethane. Evaporate the extracts to dryness under reduced pressure, reconstitute with 500 μL running buffer, sonicate for 5 min, inject an aliquot.

CAPILLARY ELECTROPHORESIS

Capillary: 57 cm × 50 μm fused-silica (50 cm to detector)
Capillary preparation: Before each run wash capillary with 100 mM HCl for 3 min, wash with 500 mM NaOH, wash with water, and equilibrate with running buffer for 3 min. At the beginning of each day rinse capillary with 500 mM NaOH for 30 min and with water for 30 min.
Capillary temperature: 15 ± 0.1
Running buffer: 10 mM pH 7.4 Sodium phosphate buffer containing 50 mM dodecyltrimethylammonium bromide and 5.2 mM trioctylphosphine oxide
Injection: High-pressure injection for 5 s
Detector: UV 254
Migration time: 32.5 (testosterone), 33.5 (testosterone propionate)
Voltage: 15 kV
Current: <30 μA
Model: Beckman P/ACE 5010
Limit of quantitation: 500 ng/mL

OTHER SUBSTANCES

Extracted: cortisone, 17-deoxycorticosterone, dimethyltestosterone, hydrocortisone

---

REFERENCE

Abubaker,M.A.; Petersen,J.R.; Bissell,M.G. Micellar electrokinetic capillary chromatographic separation of steroids in urine by trioctylphosphine oxide and cationic surfactant, *J.Chromatogr.B*, **1995**, *674*, 31–38.

SAMPLE

Matrix: urine

Sample preparation: Vortex 5 mL urine and 2 mL hexane for 3 min, centrifuge at 3500 rpm for 3 min, repeat extraction. Combine the organic layers and evaporate them to dryness at room temperature, reconstitute the residue in 500 μL 20 mM pH 2.5 phosphate buffer containing 10 mM sodium dodecyl sulfate, inject an aliquot.

CAPILLARY ELECTROPHORESIS
Capillary: 64.5 cm × 50 μm fused-silica (56 cm to detector) (Polymicro Technologies)
Capillary preparation: Between runs flush capillary with 1 M NaOH for 1 min, with MeOH for 1 min, with 100 mM naOH for 1 min, with water for 2 min, and with running buffer for 2 min. Before use rinse capillary with 1 M NaOH for 10 min, with MeOH for 5 min, with water for 5 min, and with running buffer for 5 min.
Capillary temperature: 30
Running buffer: MeOH:100 mM pH 2 phosphate buffer 20:80 containing 50 mM sodium dodecyl sulfate
Injection: Inject a 5.8 cm water plug at 50 mbar, inject sample using voltage, when the current reaches 90% of current in a capillary filled only with running buffer replace sample with running buffer.
Detector: UV 247
Migration time: 9
Voltage: -25 kV
Model: Hewlett Packard 3D
Limit of detection: 12.0 ppb (S/N 3)

OTHER SUBSTANCES
Extracted: cortisone, hydrocortisone, progesterone

KEY WORDS
detector at anode

REFERENCE
Quirino,J.P.; Terabe,S. On-line concentration of neutral analytes for micellar electrokinetic chromatography. 5. Field-enhanced sample injection with reverse migrating micelles, *Anal.Chem.*, **1998**, *70*, 1893–1901.

Tetracaine

Molecular formula: $C_{15}H_{24}N_2O_2$
Molecular weight: 264.37
CAS Registry No.: 94-24-6, 136-47-0 (HCl)
Merck Index (12th ed.): 9330
Lednicer: 1 110

SAMPLE
Matrix: bulk
Sample preparation: Prepare a 3 mg/mL solution in 3 mg/mL pholcodine in 10 mM HCl, make up to 25 mL with 10 mM HCl, mix thoroughly, filter (0.45 μm cellulose acetate)

CAPILLARY ELECTROPHORESIS
Capillary: 65 cm × 75 μm fused-silica (40 cm to detector) (Isco)
Capillary preparation: Flush with running buffer for 2 min between runs. Before each batch of samples fill capillary with 100 mM NaOH and let stand for 10 min, wash with water for 5 min, wash with 100 mM HCl for 5 min, wash with water for 5 min, fill with running buffer (all at vacuum level 4). Clean each week by washing with 100 mM HCl for 10 min, wash with water, fill with running buffer.
Capillary temperature: 30
Running buffer: MeCN:buffer 7.5:92.5. Buffer was 10 mM KH_2PO_4 containing 10 mM sodium tetraborate and 50 mM cetyltrimethylammonium bromide, pH 8.6.)
Injection: Injection by vacuum 5.0 kPa.s (4 nL)
Detector: UV 230

Migration time: 11.3
Internal standard: pholcodine (7.4)
Voltage: -15 kV
Model: Isco Model 3140

OTHER SUBSTANCES
Simultaneous: cocaine, benzocaine, benzoylecgonine, trans-cinnamoylcocaine, lidocaine, procaine

---

REFERENCE
Trenerry,V.C.; Robertson,J.; Wells,R.J. The determination of cocaine and related substances by micellar electrokinetic capillary chromatography, *Electrophoresis*, **1994**, *15*, 103–108.

SAMPLE
Matrix: bulk
Sample preparation: Dissolve sample in 2 mg/mL histamine diphosphate in 10 mM HCl, inject an aliquot.

CAPILLARY ELECTROPHORESIS
Capillary: 48.5 cm × 50 μm fused-silica (40 cm to detector)
Capillary preparation: Flush capillary with running buffer for 2 min before each run. Replenish buffer in vials after every 10 injections. Condition new capillaries by flushing with 1 M NaOH for 10 min, with 100 mM NaOH for 10 min, with water for 10 min, and with running buffer for 10 min.
Capillary temperature: 40
Running buffer: 75 mM NaH_2PO_4 containing 75 mM sodium borate, adjusted to pH 8.5 with 2 M NaOH
Injection: Pressure injection at 150 mbar for 3 s.
Detector: UV 230
Migration time: 6.5
Internal standard: histamine diphosphate (5)
Voltage: 10 kV
Current: 66-70 μA
Model: Hewlett Packard 3D CE
Limit of detection: 5 μg/mL (S/N = 4:1)

OTHER SUBSTANCES
Simultaneous: acetaminophen, benzocaine, cis-cinnamoylcocaine, trans-cinnamoylcocaine, cocaine, ephedrine, lidocaine, phenylpropanolamine, procaine

---

REFERENCE
Krawczeniuk,A.S.; Bravenec,V.A. Quantitative determination of cocaine in illicit powders by free zone capillary electrophoresis, *J.Forensic Sci.*, **1998**, *43*, 738–743.

SAMPLE
Matrix: solutions
Sample preparation: Prepare a solution in running buffer:water 1:2, inject a 5-10 μL aliquot.

CAPILLARY ELECTROPHORESIS
Capillary: 40 cm × 50 μm bare silica (40 cm to detector) (Isco)
Capillary preparation: After each injection flush with 100 mM NaOH and rinse with running buffer.
Running buffer: 50 mM pH 9.2 borate buffer
Injection: Manual injection of 5-10 μL by syringe through a 1:830 splitter
Detector: UV 238
Migration time: 7.69
Voltage: 15 kV
Model: Isco Model 3850

OTHER SUBSTANCES
Extracted: morphine (UV 214)
Simultaneous: cocaine, morphine (UV 214), nalorphine (UV 214)

Noninterfering: acetaminophen, amitriptyline, amphetamine, atropine, benzoylecgonine, benztropine, caffeine, carbamazepine, carisoprodol, chlorpheniramine, chlorpromazine, chlorprothixene, cimetidine, cocaine, codeine, dextromethorphan, diazepam, dihydrocodeine, diphenhydramine, diphenoxylate, disopyramide, doxepin, doxylamine, emetine, erythromycin, ethylmorphine, flurazepam, glutethimide, hydrocodone, hydrocortisone, hydromorphone, hydroxyzine, imipramine, lidocaine, loxapine, meperidine, meprobamate, methadone, methamphetamine, methapyrilene, methaqualone, methocarbamol, methylphenidate, naloxone, nicotine, nordiazepam, nortriptyline, orphenadrine, oxycodone, papaverine, pentazocine, phenacetin, phencyclidine, phenmetrazine, phenolphthalein, phentermine, phenylpropanolamine, phenytoin, phetidine, prazepam, procainamide, procaine, propoxyphene, propranolol, protriptyline, pseudoephedrine, pyrilamine, quinine, salicylamide, spironolactone, strychnine, terpin hydrate, thioridazine, thiothixene, triamterene, trifluoperazine, triflupromazine, trihexyphenidyl, trimeprazine, trimethoprim, trimetobenzamide

REFERENCE

Tagliaro,F.; Poiesi,C.; Aiello,R.; Dorizzi,R.; Ghielmi,S.; Marigo,M. Capillary electrophoresis for the investigation of illicit drugs in hair: determination of cocaine and morphine, *J.Chromatogr.*, **1993**, *638*, 303–309.

SAMPLE

Matrix: solutions
Sample preparation: Inject an aliquot of a 1-50 μg/mL solution in water.

CAPILLARY ELECTROPHORESIS

Capillary: 55 cm × 50 μm fused-silica (35 cm to detector) (J&W)
Capillary preparation: Flush with running buffer for 5 min before each run. Periodically wash with 100 mM NaOH and flush extensively with running buffer.
Running buffer: MeOH:25 mM pH 9.24 borate buffer 20:80 containing 100 mM sodium dodecyl sulfate (A) or 50 mM pH 2.35 phosphate buffer (B) or 50 mM pH 9.24 borate buffer (C)
Injection: Injection of 5 μL using a split-flow injector ratio of 1:800.
Detector: UV 200
Migration time: 46.54 (A), 9.44 (B), 9.81 (C)
Voltage: 20 kV (A, B) or 12 kV (C)
Current: <60-80 μA
Model: ISCO Model 3850

OTHER SUBSTANCES

Also analyzed: acetylcodeine, amphetamine, barbital, caffeine, codeine, cocaine, diamorphine, diazepam, flunitrazepam, lidocaine, monoacetylmorphine, morphine, nalorphine, narceine, noscapine, papaverine, pentobarbital, procaine, thebaine

KEY WORDS

all compounds were separated with running buffer A; some peaks overlapped with running buffers B and C.

REFERENCE

Tagliaro,F.; Smith,F.P.; Turrina,S.; Equisetto,V.; Marigo,M. Complementary use of capillary zone electrophoresis and micellar electrokinetic capillary chromatography for mutual confirmation of results in forensic drug analysis, *J.Chromatogr.A*, **1996**, *735*, 227–235.

SAMPLE

Matrix: solutions

CAPILLARY ELECTROPHORESIS

Capillary: 85 cm × 50 μm fused-silica (Polymicro Technologies)
Capillary preparation: Between runs rinse capillary with running buffer for 5 min. Before use rinse capillary with 1 M NaOH for 10-15 min and with water for 10-15 min.
Running buffer: 25 mM pH 3 Citrate buffer
Injection: Pressure injection at 30 mbar for 12 s.
Detector: MS, laboratory-constructed, electrospray 90°, sheath liquid MeOH:water:acetic acid 80:20:0.1 at 1 μL/min
Migration time: 10.5
Voltage: 30 kV

Current: 7.8 μA
Model: Crystal CE 300

OTHER SUBSTANCES
Simultaneous: amphetamine, cocaine, diamorphine, methamphetamine, procaine

---

REFERENCE
Lazar,I.M.; Naisbitt,G.; Lee,M.L. Capillary electrophoresis-time-of-flight mass spectrometry of drugs of abuse, *Analyst*, **1998**, *123*, 1449–1454.

Tetracycline

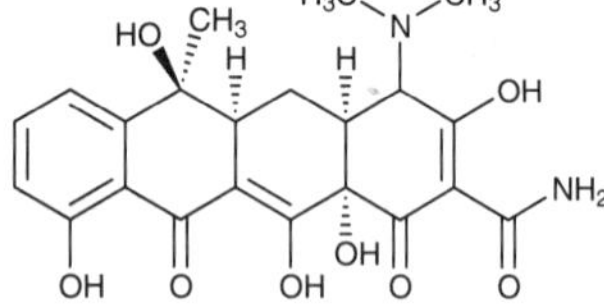

Molecular formula: $C_{22}H_{24}N_2O_8$
Molecular weight: 444.44
CAS Registry No.: 60-54-8, 6416-04-2 (trihydrate), 64-75-5 (HCl), 1336-20-5 (phosphate)
Merck Index (12th ed.): 9337
Lednicer: 1 212

SAMPLE
Matrix: blood, milk
Sample preparation: Condition a 3 mL 15 mg SPEC MP1 (Ansys, Irvine CA) SPE cartridge with 1 mL MeOH and 1 mL water. Mix 1 volume milk or plasma with 0.2 volumes 10% trichloroacetic acid in water, centrifuge at 1000 g for 4 min. Add 5 mL supernatant to the SPE cartridge at 2 mL/min, wash with 500 μL water, dry under 40 kPa vacuum for 5 min (more if necessary), elute with 200 μL 50 mM magnesium acetate in N-methylformamide, inject an aliquot of the eluate.

CAPILLARY ELECTROPHORESIS
Capillary: 27 cm × 75 μm fused-silica (20 cm to the detector) (Polymicro Technologies)
Capillary preparation: Flush with running buffer for 30 s before each run. Replace running buffer after each 6 runs. Before long-term storage flush with 100 mM HCl, 100 mM NaOH, water, MeOH, and air. Rinse new capillaries with 1 M NaOH for 1 h, with 100 mM NaOH for 20 min, with water for 20 min, with MeOH for 20 min, and with running buffer for 15 min.
Capillary temperature: 20
Running buffer: 500 mM Magnesium acetate tetrahydrate in N-methylformamide
Injection: Pressure injection at 3.5 kPa for 3 s.
Detector: UV 280, F ex 325 (He-Cd laser) em 514
Migration time: 13.3
Voltage: 15 kV
Model: Beckman P/ACE 5010
Limit of detection: 25 ng/mL

OTHER SUBSTANCES
Extracted: chlortetracycline, oxytetracycline

KEY WORDS
plasma; SPE

REFERENCE
Tjornelund,J.; Hansen,S.H. Use of metal complexation in non-aqueous capillary electrophoresis systems for the separation and improved detection of tetracyclines, *J.Chromatogr.A*, **1997**, *779*, 235–243.

SAMPLE
Matrix: blood, milk, urine
Sample preparation: Prepare a column as follows. Swirl Chelating Sepharose Fast Flow resin (Pharmacia) in its bottle, add it to a polypropylene column to give a bed volume of 1.0-1.2 mL,

wash 3 times with 2 mL portions of water, wash with 2.5 mL 10 mM copper sulfate, wash with 2 mL water. Condition a PrepSep-C18 SPE cartridge (Fisher Scientific) by allowing 3 mL 4% dimethyldichlorosilane in toluene to drain through it by gravity, after 30 min rinse with 10 mL EtOH, immediately before use wash with 10 mL water. Milk. Centrifuge 10 mL milk at 4° at 2000 g for 20 min, remove the lower layer and add it to 20 mL succinate buffer, centrifuge at 4° at 2000 g for 30 min, pass the supernatant through a Bond Elut SPE cartridge, wash cartridge with 2 mL succinate buffer, collect all the eluate and add it to the column. Wash with 3 mL succinate buffer, wash with 3 mL water, wash with 3 mL MeOH, wash with 3 mL water, elute with 3 mL pH 7 buffer, when the blue band reaches the bottom of the column collect the eluate. Add the eluate to the SPE cartridge, wash with 10 mL water, dry with air, elute with 6 mL EtOH. Evaporate the eluate to dryness under a stream of nitrogen at room temperature, reconstitute with 20 µL running buffer adjusted to pH 7.0 with 5 M HCl, inject an aliquot. Serum. 12 mL Serum + 12 mL PBS, filter (Amicon Centriprep-10) centrifuge at 10° at 2000 g for 20 min, remove the filtrate, centrifuge an additional 4 times, combine the filtrates, add a 20 mL aliquot to the column. Wash with 3 mL succinate buffer, wash with 3 mL water, wash with 3 mL MeOH, wash with 3 mL water, elute with 3 mL pH 7 buffer, when the blue band reaches the bottom of the column collect the eluate. Add the eluate to the SPE cartridge, wash with 10 mL water, dry with air, elute with 6 mL EtOH. Evaporate the eluate to dryness under a stream of nitrogen at room temperature, reconstitute with 20 µL running buffer adjusted to pH 7.0 with 5 M HCl, inject an aliquot. Urine. Centrifuge 10 mL urine at 4° at 2000 g for 15 min, add the supernatant to the column. Wash with 3 mL succinate buffer, wash with 3 mL water, wash with 3 mL MeOH, wash with 3 mL water, elute with 3 mL pH 7 buffer, when the blue band reaches the bottom of the column collect the eluate. Add the eluate to the SPE cartridge, wash with 10 mL water, dry with air, elute with 6 mL EtOH. Evaporate the eluate to dryness under a stream of nitrogen at room temperature, reconstitute with 20 µL running buffer adjusted to pH 7.0 with 5 M HCl, inject an aliquot. (Prepare succinate buffer by dissolving 11.8 g succinic acid in 980 mL water, adjust pH to 4.0 with 10 M NaOH, make up to 1 L. Prepare PBS by dissolving 20 mG KCl, 8 g NaCl, 1.44 g Na_2HPO_4, and 24 mg KH_2PO_4 in 980 mL water, adjust pH to 7.0 with 1 M HCl, make up to 1 L. Prepare the pH 7 buffer by dissolving 2.31 g citric acid, 7.44 g disodium EDTA dihydrate, and 5.84 g NaCl in 190 mL water, adjust pH to 7.0 with 5 M NaOH, make up to 200 mL.)

CAPILLARY ELECTROPHORESIS
Capillary: 57 cm × 75 µm (50 cm to detector) (Beckman)
Capillary preparation: Before injection rinse with running buffer for 2 min. After each run rinse with NaOH solution for 2 min and with water for 2 min.
Capillary temperature: 23
Running buffer: 50 mM Borate buffer containing 50 mM phosphate and 10 mM sodium dodecyl sulfate (Prepare buffer by dissolving 2.884 g sodium dodecyl sulfate, 19.07 g sodium borate decahydrate, 4.096 g Na_2HPO_4, and 2.537 g NaH_2PO_4 in 980 mL water, adjust pH to 8.5 with 5 M HCl, make up to 1 L.)
Injection: Pressure injection for 5 s (5 nL)
Detector: UV 370
Migration time: 12.9
Voltage: 15 kV
Model: Beckman P/ACE 5510
Limit of quantitation: 2.3 ppm (milk), 1.7 ppm (serum), 2.6 ppm (urine)
Limit of detection: 1.4 ppm (milk), 1.3 ppm (serum), 1.6 ppm (urine)

OTHER SUBSTANCES
Extracted: chlortetracycline, doxycycline, oxytetracycline

KEY WORDS
serum; cow; SPE

REFERENCE
Chen,C.-L.; Gu,X. Determination of tetracycline residues in bovine milk, serum, and urine by capillary electrophoresis, *J.AOAC Int.*, **1995**, *78*, 1369–1377.

SAMPLE
Matrix: bulk
Sample preparation: Inject an aliquot of a 5 mg/mL solution in MeCN:MeOH:DMF 49:45:6.

CAPILLARY ELECTROPHORESIS
Capillary: 55.5 cm × 50 μm fused-silica (55.5 cm to detector) (Polymicro Technologies)
Capillary temperature: 25
Running buffer: MeCN:MeOH:DMF 49:45:6 containing 25 mM acetic acid, 10 mM citric acid, and 118 mM methanesulfonic acid
Injection: Pressure injection at 5 kPa for 1 s.
Detector: UV 254
Migration time: 7.6
Voltage: 30 kV
Current: 65 μA
Model: Hewlett Packard HP ^{3D}CE

OTHER SUBSTANCES
Simultaneous: impurities

REFERENCE
Bjornsdottir,I.; Tjornelund,J.; Hansen,S.H. Nonaqueous capillary electrophoresis in pharmaceutical analysis, *J.Capillary Electrophor.*, **1996**, *3*, 83–87.

SAMPLE
Matrix: bulk
Sample preparation: Inject an aliquot of a solution in MeCN:MeOH:DMF 49:45:6.

CAPILLARY ELECTROPHORESIS
Capillary: 64 cm × 50 μm fused-silica (55.5 cm to detector) (Polymicro Technologies)
Capillary preparation: Flush capillary with running buffer for 2 min between runs. Before use rinse capillaries with 1 M NaOH for 1 h, with 100 mM NaOH for 20 min, with water for 20 min, with MeOH for 20 min, and with running buffer for 15 min.
Capillary temperature: 25
Running buffer: MeCN:MeOH:DMF 49:45:6 containing 25 mM ammonium acetate, 10 mM citric acid, and 118 mM methanesulfonic acid
Injection: Pressure injection of sample at 50 mbar for 1 s followed by pressure injection of running buffer at 50 mbar for 3 s.
Detector: UV 254
Migration time: 7.5
Voltage: 30 kV
Current: ca. 65 μA
Model: Hewlett-Packard HP^{3D}

OTHER SUBSTANCES
Simultaneous: impurities, chlortetracycline, doxycycline, oxytetracycline

REFERENCE
Tjornelund,J.; Hansen,S.H. Determination of impurities in tetracycline hydrochloride by non-aqueous capillary electrophoresis, *J.Chromatogr.A*, **1996**, *737*, 291–300.

SAMPLE
Matrix: bulk
Sample preparation: Inject an aliquot of a solution in running buffer.

CAPILLARY ELECTROPHORESIS
Capillary: 44 cm × 50 μm fused-silica (38 cm to detector) (Polymicro Technologies)
Capillary temperature: 15
Running buffer: 120 mM pH 10.75 Sodium carbonate containing 1 mM EDTA
Injection: Hydrodynamic injection for 4 s.
Detector: UV 270
Migration time: 15
Voltage: 12 kV
Model: Spectraphoresis 500 (Thermo Separation Products)
Limit of quantitation: 1.5 μg/mL
Limit of detection: 500 ng/mL

OTHER SUBSTANCES
Simultaneous: impurities

REFERENCE
Li,Y.M.; Van Schepdael,A.; Roets,E.; Hoogmartens,J. Optimized methods for capillary electrophoresis of tetracyclines, *J.Pharm.Biomed.Anal.*, **1997**, *15*, 1063–1069.

SAMPLE
Matrix: formulations
Sample preparation: Dissolve capsule contents in 15 mM pH 7.5 sodium phosphate buffer so as to make a 25 mM solution, centrifuge, dilute an aliquot of the supernatant to a concentration of 5 mM with 15 mM pH 7.5 sodium phosphate buffer, inject an aliquot.

CAPILLARY ELECTROPHORESIS
Capillary: 111.9 cm × 75 μm fused-silica (43.4 cm to detector) (Polymicro Technologies)
Capillary preparation: Condition by rinsing with 1 M NaOH for 10 min then running buffer overnight.
Capillary temperature: 25
Running buffer: pH 7.5 Phosphate buffer (buffer concentration 4.3 mM, total sodium concentration 15 mM, ionic strength 18.2 mM)
Injection: Hydrodynamic injection at 2 cm for 1 min
Detector: UV 260
Migration time: 5.1
Current: 20 μA
Limit of detection: 10 μM (S/N 3)

OTHER SUBSTANCES
Simultaneous: chlortetracycline, demeclocycline, doxycycline, methacycline, minocycline, oxytetracycline

KEY WORDS
capsules

REFERENCE
Tavares,M.F.M.; McGuffin,V.L. Separation and characterization of tetracycline antibiotics by capillary electrophoresis, *J.Chromatogr.A*, **1994**, *686*, 129–142.

SAMPLE
Matrix: solutions
Sample preparation: Inject an aliquot of a 20 μg/mL solution in water.

CAPILLARY ELECTROPHORESIS
Capillary: 20 cm × 25 μm fused-silica (Bio-Rad)
Running buffer: 200 mM pH 2.2 Sodium phosphate buffer containing 0.06% Triton X-100
Injection: Injection by electromigration at 12 kV for 10 s
Detector: UV 265
Migration time: 19
Voltage: 12 kV
Model: Bio-Rad HPE 100

OTHER SUBSTANCES
Simultaneous: chlortetracycline, oxytetracycline, quatrimycin

REFERENCE
Croubels,S.; Baeyens,W.; Dewaele,C.; Van Peteghem,C. Capillary electrophoresis of some tetracycline antibiotics, *J.Chromatogr.A*, **1994**, *673*, 267–274.

SAMPLE
Matrix: solutions

CAPILLARY ELECTROPHORESIS
Capillary: 50 cm × 50 μm (50 cm to detector) (Polymicro Technologies)
Capillary preparation: Purge capillary with 10 mM NaOH for 2 min after each injection.
Running buffer: MeOH:buffer 40:60 (Buffer was 30 mM pH 3.0 citric acid containing 24.5 mM β-alanine.)
Injection: Electrokinetic injection at 10 kV for 5 s.
Detector: UV 254
Migration time: 6.5
Voltage: 30 kV
Model: Perkin Elmer/Applied Biosystems Model 270A-HT

OTHER SUBSTANCES
Simultaneous: chlortetracycline, doxycycline, meclocycline, methacycline, minocycline, oxytetracycline

REFERENCE
Pesek,J.J.; Matyska,M.T. Separation of tetracyclines by high-performance capillary electrophoresis and capillary electrochromatography, *J.Chromatogr.A*, **1996**, *736*, 313–320.

SAMPLE
Matrix: solutions
Sample preparation: Inject an aliquot of a solution in running buffer.

CAPILLARY ELECTROPHORESIS
Capillary: 44 cm × 50 μm fused-silica (38 cm to detector) (Polymicro Technologies)
Capillary preparation: Wash capillary with 100 mM NaOH, 100 mM phosphoric acid, and 20 mM EDTA each day.
Capillary temperature: 10
Running buffer: 50 mM pH 11.0 Sodium carbonate containing 0.5% Triton X-100 and 1 mM EDTA
Injection: Hydrodynamic injection for 4 s.
Detector: UV 254
Migration time: 12.3
Voltage: 10 kV
Model: Spectraphoresis 500 (Thermo Separation Products)

OTHER SUBSTANCES
Simultaneous: 2-acetyl-2-decarboxamidooxytetracycline, anhydrotetracycline, α-apooxytetracycline, β-apooxytetracycline, 4-epioxytetracycline, 4-epitetracycline, oxytetracycline, terrinolide

REFERENCE
Li,Y.M.; Van Schepdael,A.; Roets,E.; Hoogmartens,J. Separation of oxytetracycline and its related substances by capillary electrophoresis, *J.Liq.Chromatogr.Rel.Technol.*, **1997**, *20*, 273–282.

SAMPLE
Matrix: solutions

CAPILLARY ELECTROPHORESIS
Capillary: 43 cm × 75 μm (36 cm to the detector)
Capillary preparation: Before each run wash capillary with running buffer for 2 min. After each run wash capillary with water at 25° for 5 min, with 100 mM NaOH at 60 ° for 5 min, and with water at 25° for 5 min.
Capillary temperature: 25
Running buffer: 15 mM pH 6.5 Ammonium acetate buffer containing 20 mM sodium dodecyl sulfate and 0.135% Brij 35
Detector: UV 265
Migration time: 4
Voltage: 15 kV

OTHER SUBSTANCES
Simultaneous: chlortetracycline, demeclocycline, doxycycline, minocycline, oxytetracycline

REFERENCE
Chen,Y.-C.; Lin,C.-E. Migration behavior and separation of tetracycline antibiotics by micellar electrokinetic chromatography, *J.Chromatogr.A*, **1998**, *802*, 95–105.

Tetrahydrozoline

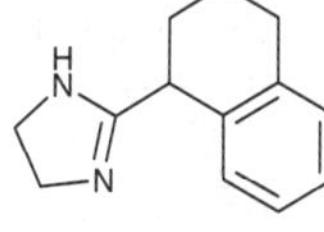

Molecular formula: $C_{13}H_{16}N_2$
Molecular weight: 200.28
CAS Registry No.: 84-22-0, 522-48-5 (HCl)
Merck Index (12th ed.): 9358
Lednicer: 1 242

SAMPLE
Matrix: blood, urine
Sample preparation: Adjust pH of 2 mL urine or plasma to 10.5 with aqueous NaOH, extract gently with chloroform:isopropanol 90:10. Remove the organic layer and evaporate it to dryness under a stream of nitrogen, reconstitute the residue in 50 (urine) or 75 (plasma) μL running buffer, filter, inject an aliquot.

CAPILLARY ELECTROPHORESIS
Capillary: 60 cm × 75 μm AccuSep uncoated silica (52.5 cm to detector) (Waters)
Capillary preparation: At the beginning of each day purge with 500 mM KOH for 5 min, with water for 5 min, and with running buffer for 10 min.
Running buffer: 50 mM NaH_2PO_4 adjusted to pH 2.35 with phosphoric acid
Injection: Hydrostatic injection at 15 cm
Detector: UV 214
Migration time: 5.90
Voltage: 22 kV
Current: 135-145 μA
Model: Waters Quanta 4000

OTHER SUBSTANCES
Extracted: acepromazine, amphetamine, benzocaine, brompheniramine, butacaine, codeine, diazepam, doxapram, lidocaine, medazepam, methamphetamine, methapyrilene, methaqualone, phenmetrazine, procaine
Simultaneous: meclizine

KEY WORDS
plasma

REFERENCE
Chee,G.L.; Wan,T.S.M. Reproducible and high-speed separation of basic drugs by capillary zone electrophoresis, *J.Chromatogr.*, **1993**, *612*, 172–177.

SAMPLE
Matrix: solutions
Sample preparation: Inject an aliquot of a 100 μg/mL solution in water:running buffer 50:50.

CAPILLARY ELECTROPHORESIS
Capillary: 44.5 cm × 50 μm acrylamide-coated fused-silica (Bio-Rad)
Capillary temperature: 30
Running buffer: 100 mM NaH_2PO_4 containing 15 mM gamma-cyclodextrin, adjusted to pH 2.5 with phosphoric acid
Injection: Electrokinetic injection at 8 kV for 6 s.
Detector: UV 200
Migration time: 8.35, 8.58 (enantiomers)
Voltage: 14 kV

Model: Bio-Rad BioFocus 3000

OTHER SUBSTANCES
Also analyzed: albuterol, alprenolol, atenolol, atropine, baclofen, bamethan, benserazide, biperiden, bisoprolol, bupivacaine, bupranolol, butetamate, carazolol, carbuterol, carvedilol, celiprolol, chloroquine, chlorpheniramine (chlorphenamine), clidinium bromide, clobutinol, disopyramide, dobutamine, flecainide, homatropine, ipratropium bromide, isoproterenol, isothipendyl, ketamine, mefloquine, mequitazine, metaproterenol (orciprenaline), metipranolol, nafronyl (naftidrofuryl), nefopam, ofloxacin, orphenadrine, oxomemazine, oxprenolol, phenoxybenzamine, pholedrine, pindolol, pirbuterol, prilocaine, promethazine, propafenone, propranolol, sotalol, synephrine, terbutaline, tocainide, trihexyphenidyl, trimeprazine (alimemazine), trimipramine, tropicamide, verapamil, zopiclone

---

KEY WORDS
chiral; coated capillary

REFERENCE
Koppenhoefer,B.; Epperlein,U.; Christian,B.; Yibing,J.; Yuying,C.; Bingcheng,L. Separation of enantiomers of drugs by capillary electrophoresis. I. γ-Cyclodextrin as chiral solvating agent, *J.Chromatogr.A*, **1995**, *717*, 181–190.

SAMPLE
Matrix: solutions
Sample preparation: Inject an aliquot of a 100 μg/mL solution in running buffer:water 50:50.

CAPILLARY ELECTROPHORESIS
Capillary: 44.5 cm × 50 μm polyacrylamide coated fused-silica (40 cm to detector) (Bio-Rad)
Running buffer: 100 mM NaH_2PO_4 containing 15 mM α-cyclodextrin, adjusted to pH 2.5 with phosphoric acid
Injection: Electromigration at 8 kV for 6 s
Detector: UV 200
Migration time: 7.18, 7.33 (enantiomers)
Voltage: 14 kV
Model: Bio-Rad Bio-Focus 3000

OTHER SUBSTANCES
Simultaneous: clidinium bromide, ketamine, orphenadrine, oxomemazine, tropicamide

---

KEY WORDS
chiral; α = 1.021; numerous drugs were not separated into their enantiomers under these conditions; coated capillary

---

REFERENCE
Bingcheng,L.; Yibing,J.; Yoying,C.; Epperlein,U.; Koppenhoefer,B. Separation of drug enantiomers by capillary electrophoresis: α-Cyclodextrin as chiral solvating agent, *Chromatographia*, **1996**, *42*, 106–110.

SAMPLE
Matrix: solutions
Sample preparation: Inject an aliquot of a 100 μg/mL solution in water:running buffer 50:50.

CAPILLARY ELECTROPHORESIS
Capillary: 44.5 cm × 50 μm polyacrylamide-coated fused-silica (40 cm to detector) (Bio-Rad)
Capillary temperature: 30
Running buffer: 100 mM NaH_2PO_4 containing 15 mM β-cyclodextrin, adjusted to pH 2.5 with phosphoric acid
Injection: Electrokinetic injection at 8 kV for 6 s.
Detector: UV 200
Migration time: 7.33, 7.50 (enantiomers)
Voltage: 14 kV
Model: Bio-Rad Bio-Focus 3000

OTHER SUBSTANCES
Simultaneous: carvedilol, chlorpheniramine, ketamine, metaproterenol (orciprenaline), tropicamide, zopiclone
Also analyzed: albuterol (not chiral), atenolol (not chiral), benserazide (not chiral), biperiden (not chiral), bupivacaine (not chiral), butethamate (not chiral), carazolol (not chiral), carbuterol (not chiral), clidinium bromide (not chiral), disopyramide (not chiral), dobutamine (not chiral), flecainide (not chiral), ipratropium (not chiral), isothipendyl (not chiral), mequitazine (not chiral), metipranolol (not chiral), nafronyl (naftidrofuryl) (not chiral), nefopam (not chiral), ofloxacin (not chiral), orphenadrine (not chiral), pindolol (not chiral), pirbuterol (not chiral), prilocaine (not chiral), propafenone (not chiral), sotalol (not chiral), tocainide (not chiral), trimipramine (not chiral)

KEY WORDS
coated capillary; chiral

REFERENCE
Koppenhoefer,B.; Epperlein,U.; Christian,R.; Lin,B.; Ji,Y.; Chen,Y. Separation of enantiomers of drugs by capillary electrophoresis. III. β-cyclodextrin as chiral solvating agent, *J.Chromatogr.A*, **1996**, *735*, 333–343.

SAMPLE
Matrix: solutions
Sample preparation: Inject an aliquot of a 100 μg/mL solution in running buffer.

CAPILLARY ELECTROPHORESIS
Capillary: 32 cm × 50 μm fused-silica (27.5 cm to detector) (Yongnian Optical Conductive Fiber Plant, China), coated with polyacrylamide
Capillary preparation: No details of the polyacrylamide coating process are provided. However, another paper (LC.GC 1997, 15, 40) by this group indicates that they use the procedure of Hjertén, thus: Adjust the pH of 20 mL water to 3.5 with acetic acid, add 80 μL 3-(trimethoxysilyl)propyl methacrylate (3-methacryloxypropyltrimethoxysilane), mix, suck into capillary, let stand at room temperature for 1 h, remove the solution, wash with water. Fill the capillary with a deaerated 3-4% acrylamide solution containing 1 μL/mL N,N,N',N'-tetramethylethylenediamine and 1 mg/mL potassium persulfate, let stand for 30 min, remove excess solution by aspiration, rinse with water, remove water by aspiration, dry at 35° (J. Chromatogr. 1985, 347, 191).
Capillary temperature: 25
Running buffer: 100 mM NaH_2PO_4 adjusted to pH 2.5 (A) or 100 mM NaH_2PO_4 containing 45 mM hydroxypropyl-gamma-cyclodextrin, adjusted to pH 2.5 (B)
Injection: Electrokinetic injection at 15 kV for 3 s.
Detector: UV 200, UV 210
Migration time: 4.24 (A), 6.71, 7.21 (B, enantiomers)
Voltage: 15 kV
Model: Bio-Rad BioFocus 3000

OTHER SUBSTANCES
Simultaneous: atropine, carazolol, cicletanine, dimethindene, fendiline, homatropine, ipratropium bromide, isothipendyl, mefloquine, metaclazepam, nafronyl (naftidrofuryl), nefopam, nicardipine, promethazine, reproterol, theodrenaline, tioconazole, trihexyphenidyl, trimeprazine (alimemazine), trimipramine

KEY WORDS
coated capillary; chiral

REFERENCE
Koppenhoefer,B.; Epperlein,U.; Xiaofeng,Z.; Bingcheng,L. Separation of enantiomers of drugs by capillary electrophoresis. Part 4: Hydroxypropyl-γ-cyclodextrin as chiral solvating agent, *Electrophoresis*, **1997**, *18*, 924–930.

SAMPLE
Matrix: solutions
Sample preparation: Inject an aliquot of a 100 μg/mL solution in running buffer.

CAPILLARY ELECTROPHORESIS
Capillary: 30 cm × 50 μm fused-silica (25.5 cm to detector), coated with polyacrylamide
Capillary preparation: Adjust the pH of 20 mL water to 3.5 with acetic acid, add 80 μL 3-(trimethoxysilyl)propyl methacrylate (3-methacryloxypropyltrimethoxysilane), mix, suck into capillary, let stand at room temperature for 1 h, remove the solution, wash with water. Fill the capillary with a deaerated 3-4% acrylamide solution containing 1 μL/mL N,N,N',N'-tetramethylethylenediamine and 1 mg/mL potassium persulfate, let stand for 30 min, remove excess solution by aspiration, rinse with water, remove water by aspiration, dry at 35° (J. Chromatogr. 1985, 347, 191).
Capillary temperature: 25
Running buffer: 100 mM NaH_2PO_4 containing 30 mM hydroxypropyl-β-cyclodextrin, adjusted to pH 2.5 with phosphoric acid
Injection: Electrokinetic injection at 15 kV for 3 s.
Detector: UV 200
Migration time: 14.7, 15.6 (enantiomers)
Voltage: 15 kV
Model: Bio-Rad BioFocus 3000

OTHER SUBSTANCES
Simultaneous: chlorpheniramine, dimethindene, homatropine, mefloquine, metaclazepam, theodrenaline

KEY WORDS
chiral; coated capillary; comparison with the use of other cyclodextrins

REFERENCE
Lin,B.; Zhu,X.; Koppenhoefer,B.; Epperlein,U. Investigation of 123 chiral drugs by cyclodextrin-modified capillary electrophoresis, *LC.GC*, **1997**, *15*, 40–46.

SAMPLE
Matrix: solutions

CAPILLARY ELECTROPHORESIS
Capillary: 29-36 cm × 50 μm fused-silica (24.5-31.5 cm to detector) (Yongnian Optical Conductive Fiber Plant, China) coated with polyacrylamide
Capillary preparation: Coat capillary as follows. Adjust the pH of 20 mL water to 3.5 with acetic acid, add 80 μL 3-(trimethoxysilyl)propyl methacrylate (3-methacryloxypropyltrimethoxysilane), mix, suck into capillary, let stand at room temperature for 1 h, remove the solution, wash with water. Fill the capillary with a deaerated 3-4% acrylamide solution containing 1 μL/mL N,N,N',N'-tetramethylethylenediamine and 1 mg/mL potassium persulfate, let stand for 30 min, remove excess solution by aspiration, rinse with water, remove water by aspiration, dry at 35° (J. Chromatogr. 1985, 347, 191).
Capillary temperature: 25
Running buffer: 100 mM pH 2.5 NaH_2PO_4 (A) or 100 mM pH 2.5 NaH_2PO_4 containing 45 mM hydroxypropyl-α-cyclodextrin (Wacker, Munich) (B)
Injection: Electromigration at 15 kV for 3 s.
Detector: UV 200; UV 210
Migration time: 4.24 (A); 6.78, 6.85 (B) (enantiomers)
Voltage: 15 kV
Model: Bio-Focus 3000

OTHER SUBSTANCES
Also analyzed: albuterol (salbutamol), alprenolol, amorolfine, atenolol, atropine, azelastine, baclofen, bamethan, benproperine, benserazide, biperiden, bisoprolol, brompheniramine, bupivacaine, bupranolol, butamirate, butethamate, carazolol, carbuterol, carteolol, carvedilol, celiprolol, chloroquine, chlorpheniramine, chlorphenoxamine, cicletanine, clenbuterol, clidinium bromide, clobutinol, dimethindene, dipivefrin, disopyramide, dobutamine, doxylamine, fendiline, flecainide, gallopamil, homatropine, ipratropium bromide, isoproterenol (isoprenaline), isothipendyl, ketamine, meclizine, mefloquine, mepindolol, mequitazine, metaclazepam, metaproterenol (orciprenaline), metipranolol, metoprolol, nafronyl (naftidrofuryl), nefopam, nicardipine, norfenefrine, ofloxacin, ornidazole, orphenadrine, oxomemazine, oxprenolol, oxybutynin, phenoxybenzamine, phenylpropanolamine, pholedrine, pindolol, pirbuterol, prilocaine, procyclidine, promethazine, propafenone, propranolol, reproterol, sotalol, sulpride, synephrine, talinolol, terbutaline, theodrenaline, tioconazole, tocainide, trihexyphenidyl, trimeprazine (alimemazine), trimipramine, tropicamide, verapamil, zopiclone

KEY WORDS
coated capillary; chiral

REFERENCE
Koppenhoefer,B.; Eperlein,U.; Schlunk,R.; Zhu,X.; Lin,B. Separation of enantiomers of drugs by capillary electrophoresis. V. Hydroxypropyl-α-cyclodextrin as chiral solvating agent, *J.Chromatogr.A*, **1998**, *793*, 153–164.

Thalidomide

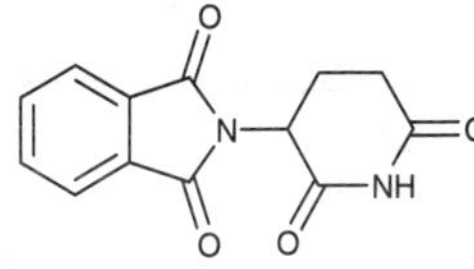

Molecular formula: $C_{13}H_{10}N_2O_4$
Molecular weight: 258.23
CAS Registry No.: 50-35-1
Merck Index (12th ed.): 9390
Lednicer: 1 257

SAMPLE
Matrix: microsomal incubations
Sample preparation: Cool 1 mL microsomal incubation to 0°, add 3 mL ethyl acetate, shake for 10 min, centrifuge at 2500 g for 10 min. Remove the organic layer and evaporate it to dryness under a stream of nitrogen, reconstitute the residue in 10 µL running buffer, inject an aliquot.

CAPILLARY ELECTROPHORESIS
Capillary: 40 cm × 50 µm fused-silica (40 cm to detector) (Grom)
Capillary preparation: Rinse capillary with 100 mM NaOH for 1 min and running buffer for 1 min before each run.
Capillary temperature: 25
Running buffer: 100 mM pH 7.0 Borate buffer containing 100 mM sodium dodecyl sulfate
Injection: Low pressure injection for 5 s (about 5 nL)
Detector: UV 214
Migration time: 16.8
Voltage: 268 V/cm
Current: 15 µA
Model: Beckman P/ACE 2100

OTHER SUBSTANCES
Extracted: metabolites

KEY WORDS
rat; liver

REFERENCE
Weinz,C.; Blaschke,G. Investigation of the in vitro biotransformation and simultaneous enantioselective separation of thalidomide and its neutral metabolites by capillary electrophoresis, *J.Chromatogr.B*, **1995**, *674*, 287–292.

SAMPLE
Matrix: solutions

CAPILLARY ELECTROPHORESIS
Capillary: 50 cm × 50 µm fused-silica (41 cm to detector) (Grom)
Capillary temperature: 21 ± 1
Running buffer: 50 mM pH 4.6 Sodium phosphate buffer containing 5 mM carboxymethyl-β-cyclodextrin (Wacker Chemie) (Anode and cathode buffers were the same but contained no cyclodextrin.)
Injection: Hydrostatic injection at 10 cm.
Detector: UV 210

Migration time: 12.95, 13.38 (enantiomers)
Voltage: 400 V/cm
Model: Grom 100

KEY WORDS
chiral

REFERENCE
Chankvetadze,B.; Endresz,G.; Blaschke,G. Capillary electrophoresis enantioseparation of noncharged and anionic chiral compouds using anionic cyclodextrin derivatives as chiral selectors, *J.Capillary Electrophor.*, **1995**, *2*, 235–240.

SAMPLE
Matrix: solutions

CAPILLARY ELECTROPHORESIS
Capillary: 60 cm × 50 μm fused-silica (Grom, Herrenberg, Germany)
Running buffer: 50 mM pH 4.5 Phosphate buffer containing 5 mM carboxymethyl-β-cyclodextrin (A) or 50 mM pH 4.5 phosphate buffer containing 7 mM spermine and 10 mM carboxymethyl-β-cyclodextrin (B)
Injection: Hydrostatic injection at 10 cm.
Detector: UV 210
Migration time: 7.2 (S (A)), 7.5 (R (A)), 28 (R (B)), 29 (S (B))
Voltage: 24 kV
Model: Grom (Herrenberg, Germany)

KEY WORDS
chiral; detector at cathode (A) and detector at anode (B)

REFERENCE
Chankvetadze,B.; Schulte,G.; Blaschke,G. Selected applications of capillaries with dynamic or permanent anodal electroosmotic flow in chiral separations by capillary electrophoresis, *J.Pharm.Biomed.Anal.*, **1997**, *15*, 1577–1584.

SAMPLE
Matrix: solutions

CAPILLARY ELECTROPHORESIS
Capillary: 60 cm × 50 μm fused-silica coated capillary (43 cm to detector) (Grom)
Capillary preparation: Adjust the pH of 20 mL water to 3.5 with acetic acid, add 80 μL 3-(trimethoxysilyl)propyl methacrylate (3-methacryloxypropyltrimethoxysilane), mix, suck into capillary, let stand at room temperature for 1 h, remove the solution, wash with water. Fill the capillary with a deaerated 3-4% acrylamide solution containing 1 μL/mL N,N,N',N'-tetramethylethylenediamine and 1 mg/mL potassium persulfate, let stand for 30 min, remove excess solution by aspiration, rinse with water, remove water by aspiration, dry at 35° (J. Chromatogr. 1985, 347, 191).
Running buffer: 50 mM pH 3.0 Phosphate buffer containing 20 mg/mL 2- hydroxypropyltrimethylamonium-β-cyclodextrin (average substitution 3.5; Wacker Chemie)
Injection: Hydrodynamic injection at 10 cm for 5 s.
Detector: UV 210
Migration time: 18, 19 (enantiomers)
Voltage: 24 kV
Model: Grom 100 (Herrenberg, Germany)

KEY WORDS
chiral; coated capillary

REFERENCE
Schulte,G.; Chankvetadze,B.; Blaschke,G. Enantioseparation in capillary electrophoresis using 2-hydroxypropyltrimethylammonium salt of β-cyclodextrin as a chiral selector, *J.Chromatogr.A*, **1997**, *771*, 259–266.

Thebaine

Molecular formula: $C_{19}H_{21}NO_3$
Molecular weight: 311.38
CAS Registry No.: 115-37-7
Merck Index (12th ed.): 9411

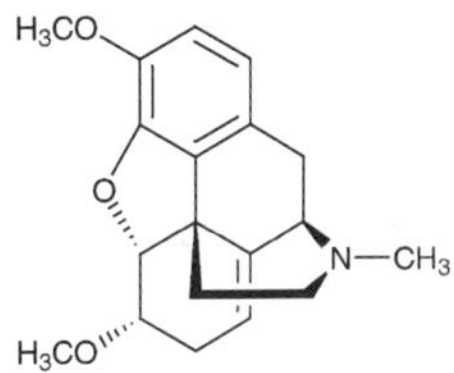

SAMPLE

Matrix: formulations
Sample preparation: Dissolve one tablet in 25 mL MeCN:40 mM pH 10 borate buffer 9:91, centrifuge, filter (glass fiber), filter (membrane), dilute filtrate 18-fold with running buffer, inject an aliquot.

CAPILLARY ELECTROPHORESIS

Capillary: 48.5 cm × 50 μm fused-silica (40 cm to detector) (Hewlett Packard)
Capillary preparation: Between runs flush with 100 mM NaOH for 5 min, with water for 2 min, and with running buffer for 5 min.
Capillary temperature: 25
Running buffer: MeCN:buffer 9:91 (Buffer was 40 mM pH 10 borate buffer containing 40 mM sodium dodecyl sulfate.)
Injection: Hydrodynamic injection at 5 kPa for 3 s.
Detector: UV 214
Migration time: 8.05
Voltage: 25 kV
Current: 44 μA
Model: Hewlett Packard ^{3D}CE

OTHER SUBSTANCES

Simultaneous: impurities, codeine, ibuprofen

KEY WORDS

tablets

REFERENCE

Persson-Stubberud,K.; Åström,O. Separation of ibuprofen, codeine phosphate, their degradation products and impurities by capillary electrophoresis. 1. Method development and optimization with fractional factorial design, *J.Chromatogr.A*, **1998**, *798*, 307–314.

SAMPLE

Matrix: solutions
Sample preparation: Inject an aliquot of a 1-50 μg/mL solution in water.

CAPILLARY ELECTROPHORESIS

Capillary: 55 cm × 50 μm fused-silica (35 cm to detector) (J&W)
Capillary preparation: Flush with running buffer for 5 min before each run. Periodically wash with 100 mM NaOH and flush extensively with running buffer.
Running buffer: MeOH:25 mM pH 9.24 borate buffer 20:80 containing 100 mM sodium dodecyl sulfate (A) or 50 mM pH 2.35 phosphate buffer (B)
Injection: Injection of 5 μL using a split-flow injector ratio of 1:800.
Detector: UV 200
Migration time: 34.47 (A), 10.54 (B)
Voltage: 20 kV
Current: <60-80 μA
Model: ISCO Model 3850

OTHER SUBSTANCES

Also analyzed: acetylcodeine, amphetamine, barbital, caffeine, codeine, cocaine, diamorphine, diazepam, flunitrazepam, lidocaine, monoacetylmorphine, morphine, nalorphine, narceine, noscapine, papaverine, pentobarbital, procaine, tetracaine

KEY WORDS
all compounds were separated with running buffer A; some peaks overlapped with running buffer B.

REFERENCE
Tagliaro,F.; Smith,F.P.; Turrina,S.; Equisetto,V.; Marigo,M. Complementary use of capillary zone electrophoresis and micellar electrokinetic capillary chromatography for mutual confirmation of results in forensic drug analysis, *J.Chromatogr.A*, **1996**, *735*, 227–235.

SAMPLE
Matrix: solutions

CAPILLARY ELECTROPHORESIS
Capillary: 20 cm × 75 μm containing 3 μm Micra bare silica (packed for 20 cm; 20 cm to detector) (Unimicro Technologies, Pleasanton CA)
Capillary preparation: Before each run condition capillary at 5 kV until a constant current is achieved. Condition new capillaries with running buffer at 500 psi.
Capillary temperature: 25
Running buffer: MeCN:10 mM pH 8.29 Tris-HCl buffer 80:20
Injection: Electrokinetic injection at 5 kV for 5 s.
Detector: UV 214
Migration time: 8
Model: Beckman P/ACE 5510

OTHER SUBSTANCES
Simultaneous: aniline, berberine, cocaine, codeine, ephedrine, jatrorrhizine

KEY WORDS
electrochromatography

REFERENCE
Wei,W.; Luo,G.A.; Hua,G.Y.; Yan,C. Capillary electrochromatographic separation of basic compounds with bare silica as stationary phase, *J.Chromatogr.A*, **1998**, *817*, 65–74.

Theodrenaline

Molecular formula: C$_{17}$H$_{21}$N$_5$O$_5$
Molecular weight: 375.38
CAS Registry No.: 13460-98-5

SAMPLE
Matrix: solutions
Sample preparation: Inject an aliquot of a 100 μg/mL solution in running buffer.

CAPILLARY ELECTROPHORESIS
Capillary: 32 cm × 50 μm fused-silica (27.5 cm to detector) (Yongnian Optical Conductive Fiber Plant, China), coated with polyacrylamide
Capillary preparation: No details of the polyacrylamide coating process are provided. However, another paper (LC.GC 1997, 15, 40) by this group indicates that they use the procedure of Hjertén, thus: Adjust the pH of 20 mL water to 3.5 with acetic acid, add 80 μL 3-(trimethoxysilyl)propyl methacrylate (3-methacryloxypropyltrimethoxysilane), mix, suck into capillary, let stand at room temperature for 1 h, remove the solution, wash with water. Fill the capillary with a deaerated 3-4% acrylamide solution containing 1 μL/mL N,N,N',N'-tetramethylethylenediamine and 1 mg/mL potassium persulfate, let stand for 30 min, remove excess solution

by aspiration, rinse with water, remove water by aspiration, dry at 35° (J. Chromatogr. 1985, 347, 191).
Capillary temperature: 25
Running buffer: 100 mM NaH_2PO_4 adjusted to pH 2.5
Injection: Electrokinetic injection at 15 kV for 3 s.
Detector: UV 200, UV 210
Migration time: 6.96
Voltage: 15 kV
Model: Bio-Rad BioFocus 3000

OTHER SUBSTANCES
Simultaneous: atropine, carazolol, cicletanine, dimethindene, fendiline, homatropine, ipratropium bromide, isothipendyl, mefloquine, metaclazepam, nafronyl (naftidrofuryl), nefopam, nicardipine, promethazine, reproterol, tetrahydrozoline (tetryzoline), tioconazole, trihexyphenidyl, trimeprazine (alimemazine), trimipramine

KEY WORDS
coated capillary

REFERENCE
Koppenhoefer,B.; Epperlein,U.; Xiaofeng,Z.; Bingcheng,L. Separation of enantiomers of drugs by capillary electrophoresis. Part 4: Hydroxypropyl-γ-cyclodextrin as chiral solvating agent, *Electrophoresis*, **1997**, *18*, 924–930.

SAMPLE
Matrix: solutions
Sample preparation: Inject an aliquot of a 100 µg/mL solution in running buffer.

CAPILLARY ELECTROPHORESIS
Capillary: 30 cm × 50 µm fused-silica (25.5 cm to detector), coated with polyacrylamide
Capillary preparation: Adjust the pH of 20 mL water to 3.5 with acetic acid, add 80 µL 3-(trimethoxysilyl)propyl methacrylate (3-methacryloxypropyltrimethoxysilane), mix, suck into capillary, let stand at room temperature for 1 h, remove the solution, wash with water. Fill the capillary with a deaerated 3-4% acrylamide solution containing 1 µL/mL N,N,N',N'-tetramethylethylenediamine and 1 mg/mL potassium persulfate, let stand for 30 min, remove excess solution by aspiration, rinse with water, remove water by aspiration, dry at 35° (J. Chromatogr. 1985, 347, 191).
Capillary temperature: 25
Running buffer: 100 mM NaH_2PO_4 containing 30 mM hydroxypropyl-β- cyclodextrin, adjusted to pH 2.5 with phosphoric acid
Injection: Electrokinetic injection at 15 kV for 3 s.
Detector: UV 200
Migration time: 19.5, 20 (enantiomers)
Voltage: 15 kV
Model: Bio-Rad BioFocus 3000

OTHER SUBSTANCES
Simultaneous: chlorpheniramine, dimethindene, homatropine, mefloquine, metaclazepam, tetrahydrozoline (tetryzoline)

KEY WORDS
chiral; coated capillary; comparison with the use of other cyclodextrins

REFERENCE
Lin,B.; Zhu,X.; Koppenhoefer,B.; Epperlein,U. Investigation of 123 chiral drugs by cyclodextrin-modified capillary electrophoresis, *LC.GC*, **1997**, *15*, 40–46.

SAMPLE
Matrix: solutions

CAPILLARY ELECTROPHORESIS
Capillary: 29-36 cm $\times$ 50 μm fused-silica (24.5-31.5 cm to detector) (Yongnian Optical Conductive Fiber Plant, China) coated with polyacrylamide
Capillary preparation: Coat capillary as follows. Adjust the pH of 20 mL water to 3.5 with acetic acid, add 80 μL 3-(trimethoxysilyl)propyl methacrylate (3-methacryloxypropyltrimethoxysilane), mix, suck into capillary, let stand at room temperature for 1 h, remove the solution, wash with water. Fill the capillary with a deaerated 3-4% acrylamide solution containing 1 μL/mL N,N,N',N'-tetramethylethylenediamine and 1 mg/mL potassium persulfate, let stand for 30 min, remove excess solution by aspiration, rinse with water, remove water by aspiration, dry at 35° (J. Chromatogr. 1985, 347, 191).
Capillary temperature: 25
Running buffer: 100 mM pH 2.5 NaH_2PO_4 (A) or 100 mM pH 2.5 NaH_2PO_4 containing 45 mM hydroxypropyl-α-cyclodextrin (Wacker, Munich) (B)
Injection: Electromigration at 15 kV for 3 s.
Detector: UV 200; UV 210
Migration time: 6.96 (A); 8.45 (B) (no separation of enantiomers)
Voltage: 15 kV
Model: Bio-Focus 3000

OTHER SUBSTANCES
Also analyzed: albuterol (salbutamol), alprenolol, amorolfine, atenolol, atropine, azelastine, baclofen, bamethan, benproperine, benserazide, biperiden, bisoprolol, brompheniramine, bupivacaine, bupranolol, butamirate, butethamate, carazolol, carbuterol, carteolol, carvedilol, celiprolol, chloroquine, chlorpheniramine, chlorphenoxamine, cicletanine, clenbuterol, clidinium bromide, clobutinol, dimethindene, dipivefrin, disopyramide, dobutamine, doxylamine, fendiline, flecainide, gallopamil, homatropine, ipratropium bromide, isoproterenol (isoprenaline), isothipendyl, ketamine, meclizine, mefloquine, mepindolol, mequitazine, metaclazepam, metaproterenol (orciprenaline), metipranolol, metoprolol, nafronyl (naftidrofuryl), nefopam, nicardipine, norfenefrine, ofloxacin, ornidazole, orphenadrine, oxomemazine, oxprenolol, oxybutynin, phenoxybenzamine, phenylpropanolamine, pholedrine, pindolol, pirbuterol, prilocaine, procyclidine, promethazine, propafenone, propranolol, reproterol, sotalol, sulpride, synephrine, talinolol, terbutaline, tetrahydrozoline (tetryzoline), tioconazole, tocainide, trihexyphenidyl, trimeprazine (alimemazine), trimipramine, tropicamide, verapamil, zopiclone

KEY WORDS
coated capillary

REFERENCE
Koppenhoefer,B.; Eperlein,U.; Schlunk,R.; Zhu,X.; Lin,B. Separation of enantiomers of drugs by capillary electrophoresis. V. Hydroxypropyl-α-cyclodextrin as chiral solvating agent, *J.Chromatogr.A*, **1998**, *793*, 153–164.

Theophylline

Molecular formula: $C_7H_8N_4O_2$
Molecular weight: 180.17
CAS Registry No.: 58-55-9, 32156-80-2 (diethanolamine), 573-41-1 (ethanolamine), 5600-19-1 (isopropanolamine), 8002-89-9 (sodium acetate), 8000-10-1 (sodium glycinate), 5967-84-0 monohydrate
Merck Index (12th ed.): 9421
Lednicer: 1 423; 2 464; 3 230; 4 165, 168, 213

SAMPLE
Matrix: beverages, blood, formulations
Sample preparation: Beverages. Degas carbonated beverages with argon, filter (0.25 μm), inject an aliquot. Serum. Filter (0.25 μm), inject an aliquot. Alternatively, vortex 600 μL serum with 900 μL ethyl acetate for 40 s, centrifuge. Remove the organic layer, add 500 μL ethyl acetate to the aqueous layer, repeat extraction. Repeat this extraction again. Combine all the organic layers and evaporate them to dryness under a stream of nitrogen, reconstitute with 20 μL

water, inject an aliquot. Tablets. Dissolve 1 tablet in 200 mL water, filter (0.25 μm), inject an aliquot.

CAPILLARY ELECTROPHORESIS
Capillary: 60 cm × 50 μm fused-silica (25 cm to detector) (Polymicro Technologies)
Capillary preparation: Before use rinse capillary hydrodynamically at 520 kPa with 500 mM pH 13 EDTA for 30 min, with water for 20 min, and with running buffer for 10 min. Fill with running buffer.
Running buffer: 20 mM pH 11.0 Sodium phosphate buffer containing 40 mM sodium dodecyl sulfate
Injection: Vacuum injection of 2 nL.
Detector: UV 274
Migration time: 1.55
Internal standard: antipyrine (1.1)
Voltage: 29 kV
Model: ISCO Model 3850
Limit of detection: 0.7 μg/mL

OTHER SUBSTANCES
Extracted: caffeine, paraxanthine, theobromine, 1,3,7-trimethyluric acid

KEY WORDS
serum; tablets; coffee; tea; soft drinks

REFERENCE
Zhao,Y.; Lunte,C.E. Determination of caffeine and its metabolites by micellar electrokinetic capillary electrophoresis, *J.Chromatogr.B*, **1997**, *688*, 265–274.

SAMPLE
Matrix: blood
Sample preparation: 100 μL Serum or plasma + 900 μL ethyl acetate, vortex for 30 s, centrifuge, remove 500 μL organic layer, repeat procedure twice more. Combine the organic layers and evaporate them to dryness under a stream of nitrogen, reconstitute the residue in 50 μL water, inject an aliquot.

CAPILLARY ELECTROPHORESIS
Capillary: 72 cm × 50 μm fused-silica capillary (50.5 cm to detector) (Applied Biosystems)
Capillary preparation: Before each run rinse capillary with 100 mM NaOH and running buffer at 0.67 bar for 3-5 min.
Capillary temperature: 26.5
Running buffer: 25 mM pH 8.0 Phosphate buffer containing 80 mM sodium dodecyl sulfate
Injection: Vacuum injection at 0.17 bar for 2 s (ca. 8.75 nL)
Detector: UV 274
Migration time: 9.5
Voltage: 21 kV
Model: Applied Biosystems Model 270A
Limit of detection: 1.6 μM

OTHER SUBSTANCES
Extracted: caffeine, paraxanthine

KEY WORDS
serum; plasma

REFERENCE
Lee,K.J.; Heo,G.S.; Kim,N.J.; Moon,D.C. Separation of theophylline and its analogues by micellar electrokinetic chromatography: application to the determination of theophylline in human plasma, *J.Chromatogr.*, **1992**, *577*, 135–141.

SAMPLE
Matrix: blood

Sample preparation: 100 μL Serum + 150 μL 80 μg/mL 3-isobutyl-1-methylxanthine in MeCN, mix for 30 s, centrifuge at 14000 g for 1 min, inject an aliquot of the supernatant.

CAPILLARY ELECTROPHORESIS
Capillary: 25 cm × 47 μm
Capillary preparation: Wash with running buffer by pressure injection for 1 min after each run. Wash daily with 50 mM phosphoric acid for 2 min, with water for 1 min, with 2 M NaOH for 2 min, with water for 1 min, and with running buffer for 2 min.
Capillary temperature: 24
Running buffer: 220 mM boric acid adjusted to pH 8.8 with 2.5 M NaOH
Injection: Pressure injection for 8 s.
Detector: UV 254
Migration time: 2.5
Internal standard: 3-isobutyl-1-methylxanthine (2.3)
Voltage: 12 kV
Model: Beckman

OTHER SUBSTANCES
Extracted: acetaminophen, caffeine, carbamazepine, iohexol, pentobarbital, phenobarbital, phenytoin, salicylic acid

KEY WORDS
serum

REFERENCE
Shihabi,Z.K.; Constantinescu,M.S. Iohexol in serum determined by capillary electrophoresis, *Clin.Chem.*, **1992**, *38*, 2117–2120.

SAMPLE
Matrix: blood
Sample preparation: 200 μL Plasma + 400 μL MeCN, shake for 2-3 min, centrifuge at 6890 g for 5 min, inject an aliquot of the supernatant.

CAPILLARY ELECTROPHORESIS
Capillary: 57 × 75 μm fused-silica (50 cm to detector) (eCAP, Beckman)
Capillary preparation: Before each injection wash capillary with 100 mM NaOH for 1 min and with running buffer for 3 min at 138 kPa. Rinse a new capillary with 1 M NaOH for 5 min, with water for 2 min, with 100 mM HCl for 2 min, with water at 138 kPa for 2 min, and with running buffer at low pressure.
Capillary temperature: 30
Running buffer: 10 mM pH 9.0 phosphate/borate buffer (Adjust pH of NaH_2PO_4 solution to pH 9.0 with sodium borate solution.)
Injection: Hydrodynamic injection at 0.5 psi (3.45 kPa) for 5 s (26 nL)
Detector: UV 280
Migration time: 7
Voltage: 12 kV
Current: 20 μA
Model: Beckman System 2100
Limit of detection: 1.8 μg/mL

KEY WORDS
plasma

REFERENCE
Johansson,I.M.; Grön-Rydberg,M.-B.; Schmekel,B. Determination of theophylline in plasma using different capillary electrophoretic systems, *J.Chromatogr.A*, **1993**, *652*, 487–493.

SAMPLE
Matrix: blood
Sample preparation: 100 μL Serum + 150 μL 25 μg/mL 3-isobutyl-1-methylxanthine in MeCN, mix for 15 s, centrifuge at 15000 g for 1 min, inject an aliquot of the supernatant.

CAPILLARY ELECTROPHORESIS
Capillary: 25 cm × 75 μm
Capillary preparation: Wash with running buffer for 1 min after each run.
Capillary temperature: 24
Running buffer: 300 mM boric acid adjusted to pH 8.5 with NaOH
Injection: Pressure injection for 6 s.
Detector: UV 280
Migration time: 2.6
Internal standard: 3-isobutyl-1-methylxanthine (2.4)
Voltage: 11 kV
Model: Beckman

OTHER SUBSTANCES
Extracted: iohexol, phenytoin,

KEY WORDS
serum

REFERENCE
Shihabi,Z.K. Sample matrix effects in capillary electrophoresis. II. Acetonitrile deproteinization, *J.Chromatogr.A*, **1993**, *652*, 471–475.

SAMPLE
Matrix: blood
Sample preparation: Centrifuge at 10000 g for 3 min, inject an aliquot.

CAPILLARY ELECTROPHORESIS
Capillary: 50 cm × 50 μm fused-silica (45 cm to detector) (Polymicro Technologies)
Capillary preparation: Between runs rinse capillary at 100 psi with 100 mM NaOH for 1 min
  and with running buffer for 2 min.
Capillary temperature: 20
Running buffer: 6 mM Sodium tetraborate containing 10 mM Na_2HPO_4 and 75 mM sodium
  dodecyl sulfate adjusted to pH 9.6 with 1 M NaOH
Injection: Pressure injection at 5 psi for 0.4 s.
Detector: UV 275
Migration time: 6.7
Voltage: 20 kV
Current: ca. 38 μA
Model: Bio-Rad BioFocus 3000

OTHER SUBSTANCES
Extracted: caffeine

KEY WORDS
serum; direct injection

REFERENCE
Zhang,C.-X.; Thormann,W. Determination of drug levels in human serum by micellar electrokinetic capillary chromatography with direct sample injection using different quantitation strategies, *J.Capillary Electrophor.*, **1994**, *1*, 208–218.

SAMPLE
Matrix: blood, hemodialysate fluid
Sample preparation: Serum. 1 mL Serum + 20 μL 100 mM phenyltriethylammonium iodide, filter (Amicon Centrifree Micropartition System, cut-off 30000) while centrifuging at 311 g for 1 h, inject an aliquot of the ultrafiltrate. Hemodialysate fluid. 9 mL Hemodialysate fluid + 1 mL 20 mM phenyltriethylammonium iodide, inject an aliquot.

CAPILLARY ELECTROPHORESIS
Capillary: 64 cm × 50 μm extended light path capillary (55.8 cm to detector) (Hewlett-Packard)
Capillary temperature: 25

Running buffer: 150 mM Boric acid adjusted to pH 9.0 with 1 M NaOH
Injection: Pressure injection at 50 mbar for 5 s
Detector: UV 210
Migration time: 7.5
Internal standard: phenyltriethylammonium iodide (3.5)
Voltage: 22 kV
Current: 32.5-34.5 µA
Model: Hewlett-Packard [3D]Capillary Electrophoresis System
Limit of detection: 1.8 µM (S/N 3)

OTHER SUBSTANCES
Extracted: allopurinol, ferulic acid, hippuric acid, 2-hydroxyhippuric acid, 4-hydroxyphenylacetic
acid, hypoxanthine, indican, indole-3-acetic acid, kynurenic acid, 1-methylniacinamide, niacin,
theobromine, tryptophan, tyrosine, uracil, uric acid, xanthine

KEY WORDS
serum; ultrafiltrate

REFERENCE
Petucci,C.J.; Kantes,H.L.; Strein,T.G.; Veening,H. Capillary electrophoresis as a clinical tool. Determination of
organic anions in normal and uremic serum using photodiode-array detection, *J.Chromatogr.B*, **1995**, *668*,
241–251.

SAMPLE
Matrix: bulk
Sample preparation: Prepare a 10 mg/mL solution in 3 mg/mL pholcodine in 10 mM HCl, dilute
5-fold with 10 mM HCl, filter (0.45 µm cellulose acetate), inject an aliquot.

CAPILLARY ELECTROPHORESIS
Capillary: 72 cm × 75 µm fused-silica (50 cm to detector) (Isco)
Capillary preparation: Flush with running buffer for 2 min between runs. Every 24 h wash
capillary with 100 mM HCl for 10 min, wash with water, wash 100 mM NaOH, wash with
water, fill with running buffer. Prepare new capillary by filling with 1 M NaOH, let stand for
1 h, fill with 100 mM NaOH, let stand for 1 h, wash with water, fill with running buffer.
Capillary temperature: 30
Running buffer: MeCN:buffer 10:90 (Buffer was 10 mM KH_2PO_4 containing 10 mM sodium
tetraborate and 50 mM cetyltrimethylammonium bromide, pH 8.6.)
Injection: Injection by vacuum 5.0 kPa.s (4 nL)
Detector: UV 280
Migration time: 6.9
Internal standard: pholcodine (9.1)
Voltage: 15 kV
Model: Isco Model 3140

OTHER SUBSTANCES
Simultaneous: acetylcodeine, caffeine, codeine, diamorphine, ethylmorphine, noscapine, morphine, papaverine, strychnine

REFERENCE
Trenerry,V.C.; Wells,R.J.; Robertson,J. The analysis of illicit heroin seizures by capillary zone electrophoresis,
J.Chromatogr.Sci., **1994**, *32*, 1–6.

SAMPLE
Matrix: formulations
Sample preparation: Dissolve tablets in water, dilute an aliquot with water, inject an aliquot.

CAPILLARY ELECTROPHORESIS
Capillary: 57 cm × 50 µm fused-silica (50 cm to detector)
Capillary preparation: After each run rinse with 20 mM pH 9.00 borate buffer containing 90
mM sodium dodecyl sulfate for 5 min and with 100 mM NaOH for 5 min.
Running buffer: MeOH:buffer 10:90 (Buffer was 10 mM pH 8.25 borate buffer containing 10
mM phosphate and 45 mM sodium dodecyl sulfate.)

Injection: Pressure injection at 3.45 kPa for 3 s followed by a 2 s injection of 200 μM phenobarbital solution.
Detector: UV 214
Migration time: 10
Internal standard: phenobarbital (13.3)
Voltage: 17 kV
Model: Beckman P/ACE 2100

OTHER SUBSTANCES
Simultaneous: caffeine, theobromine

KEY WORDS
tablets

REFERENCE
Vogt,C.; Conradi,S.; Rohde,E. Determination of caffeine and other purine compounds in food and pharmaceuticals by micellar electrokinetic chromatography, *J.Chem.Educ.*, **1997**, *74*, 1126–1130.

SAMPLE
Matrix: microsomal incubations, urine
Sample preparation: Microsomal incubations. Cool 550 μL Microsomal incubation on ice, add 50 μL 100 μg/mL 7-(β-hydroxypropyl)theophylline, centrifuge for 5 min, filter (0.2 μm) the supernatant, inject an aliquot of the filtrate. Urine. Condition a 6 mL Waters C18 SPE cartridge with 5 mL MeOH and 5 mL water. Acidify 4 mL urine to pH 4.8 with 1 M HCl, add to the SPE cartridge, wash with four 5 mL portions of 10 mM ammonium acetate, elute with three 5 mL portions of chloroform:isopropanol 95:5. Combine the eluates and evaporate them to dryness under a stream of nitrogen at 40°, reconstitute the residue in 400 μL MeOH:THF:water:acetic acid 6.5:0.75:92.7:0.05, inject an aliquot.

CAPILLARY ELECTROPHORESIS
Capillary: 67 cm × 75 μm (60 cm to detector) (Waters)
Capillary preparation: Before each injection purge capillary with running buffer for 3 min (for standards) or 6 min (for microsomal incubations, urine) at 2 kPa.
Running buffer: Buffer:200 mM sodium dodecyl sulfate 7:12, adjusted to pH 6.5 with 3 M HCl. (Buffer was 100 mM pH 8.5 NaH_2PO_4 containing 100 mM sodium borate.)
Injection: Hydrostatic injection for 20 s
Detector: UV
Migration time: 19 (microsomal incubations), 19 (urine)
Internal standard: 7-(β hydroxypropyl)theophylline (23 (microsomal incubations), 20 (urine))
Voltage: 10 kV (microsomal incubations), 9 kV (urine)
Model: Waters Quanta 4000
Limit of detection: 1 μg/mL

OTHER SUBSTANCES
Extracted: metabolites

KEY WORDS
rat; liver; SPE; human

REFERENCE
Zhang,Z.-Y.; Fasco,M.J.; Kaminsky,L.S. Determination of theophylline and its metabolites in rat liver microsomes and human urine by capillary electrophoresis, *J.Chromatogr.B*, **1995**, *665*, 201–208.

SAMPLE
Matrix: serum
Sample preparation: Mix 25 μL serum, 100 μL solution T, and 100 μL solution S, vortex for 10 s, let stand at room temperature for 10 min, inject an aliquot. (Solution T was theophylline fluorescein tracer solution and solution S was monoclonal theophylline antiserum solution from Abbott TDxFLx theophylline FPIA assay kit.)

CAPILLARY ELECTROPHORESIS
Capillary: 67 cm × 50 μm fused silica (60 cm to detector) (Polymicro Technologies)

Capillary preparation: Between runs wash capillary with 100 mM NaOH at 5 psi for 3 min, with water at 5 psi for 3 min, and with running buffer at 5 psi for 3 min.
Capillary temperature: 20
Running buffer: 6 mM Sodium tetraborate containing 10 mM Na_2HPO_4 and 75 mM sodium dodecyl sulfate, pH 9.2
Injection: Pressure injection at 0.5 psi for 1 s
Detector: F ex 488 (4 mW Ar ion laser) 520 (emission filter)
Migration time: 13.4
Voltage: 20 kV
Current: about 29 μA
Model: Beckman P/ACE System 5510
Limit of detection: 1 μg/mL

KEY WORDS
peak obtained even with no theophylline present; non-linear calibration curve; derivatization

REFERENCE
Steinmann,L.; Caslavska,J.; Thormann,W. Feasibility study of a drug immunoassay based on micellar electro-kinetic capillary chromatography with laser induced fluorescence detection: Determination of theophylline in serum, *Electrophoresis*, **1995**, *16*, 1912–1916.

SAMPLE
Matrix: solutions

CAPILLARY ELECTROPHORESIS
Capillary: 72 cm × 50 μm silica (50 cm to detector) (Applied Biosystems)
Capillary preparation: Before each run wash capillary with 100 mM NaOH for 3 min, wash with running buffer for 3 min, aspirate neutral marker solution (2 drops DMSO in 10 mL water) for 1 s, place capillary end in buffer vial for 5 s, inject sample. Wash capillary at the beginning of each day by passing 1 M NaOH through for 20 min.
Capillary temperature: 30
Running buffer: MeCN:buffer 10:90 (Buffer was 30 mM pH 9.3 borate buffer containing 30 mM sodium dodecyl sulfate.)
Injection: Inject by applying vacuum for 1 s (2-3 nL)
Detector: UV 200
Migration time: 12.7
Voltage: +30 kV
Current: 41 μA
Model: Applied Biosystems Model 270A

OTHER SUBSTANCES
Simultaneous: caffeine, dyphylline, theobromine

REFERENCE
Evenson,M.A.; Wiktorowicz,J.E. Automated capillary electrophoresis applied to therapeutic drug monitoring, *Clin.Chem.*, **1992**, *38*, 1847–1852.

SAMPLE
Matrix: solutions

CAPILLARY ELECTROPHORESIS
Capillary: 60 cm × 75 μm fused-silica (53 cm to detector)
Capillary preparation: Rinse capillary with running buffer after each run. At the start of each day rinse with 100 mM NaOH for 5 min and with separation buffer for 10 min. Before storage rinse capillary with water and air dry.
Running buffer: 30 mM pH 9.0 Sodium borate buffer containing 60 mM sodium dodecyl sulfate
Injection: Hydrodynamic injection at 10 cm for 5-10 s
Detector: UV 214
Migration time: 5.2
Voltage: 20 kV
Model: Waters Quanta 4000

OTHER SUBSTANCES
Simultaneous: caffeine, ethophylline, methylpropylxanthine, methylxanthine, pentoxifylline, propentofylline, theobromine, xanthine

REFERENCE
Korman,M.; Vindevogel,J.; Sandra,P. Application of micellar electrokinetic chromatography to the quality control of pharmaceutical formulations: the analysis of xanthine derivatives, *Electrophoresis*, **1994**, *15*, 1304–1309.

SAMPLE
Matrix: solutions

CAPILLARY ELECTROPHORESIS
Capillary: 60 cm × 50 μm
Capillary preparation: Between runs wash with 200 mM NaOH for 2 min and fill with running buffer for 2 min. At the beginning of each day wash with 200 mM NaOH for 10 min.
Capillary temperature: 35
Running buffer: 150 mM Boric acid adjusted to pH 8.8 with 1 M NaOII
Injection: Inject for 8 s
Detector: UV 280
Migration time: 7
Voltage: 300 V/cm
Model: Beckman Model 2000 CE

OTHER SUBSTANCES
Simultaneous: chlorotheophylline, isobutylmethylxanthine, xanthine

REFERENCE
Shihabi,Z.K.; Hinsdale,M.E. Some variables affecting reproducibility in capillary electrophoresis, *Electrophoresis*, **1995**, *16*, 2159-2163.

SAMPLE
Matrix: solutions

CAPILLARY ELECTROPHORESIS
Capillary: 57 cm × 50 μm fused-silica (50 cm to detector)
Running buffer: 50 mM pH 8.0 Borate buffer containing 40 mM sodium taurodeoxycholate and 25 mM phosphatidylcholine (Prepare by adding phosphatidylcholine and stirring for 3-6 h until all cloudiness disappears. Phosphatidylcholine was 95% pure soybean lecithin, Epikuron, Lucas Meyer & Co.)
Injection: Inject a solution of the compound in the running buffer at 20 psi for 1 s, inject MeOH: water 5:95 at 20 psi for 1 s, inject a solution of halofantrine in running buffer at 20 psi for 1 s.
Detector: UV 214
Migration time: k' 0.43
Model: Beckman P/ACE 5000

OTHER SUBSTANCES
Also analyzed: acetaminophen, amoxicillin, antipyrine, aspirin, azathioprine, caffeine, captopril, carbamazepine, carprofen, chlorambucil, chlorpheniramine, chlorpromazine, cimetidine, clonidine, codeine, desipramine, diphenhydramine, ephedrine, fenoterol, flufenamic acid, flurbiprofen, haloperidol, hydroxyzine, ibuprofen, imipramine, indomethacin, ketoprofen, lidocaine, melphalan, metoprolol, nabumetone, nadolol, phenobarbital, phenol, promazine, propranolol, pyrilamine, ranitidine, ropinirole, salicylic acid, sulfamethoxazole, testosterone, thioridazine, tiaprofenic acid, tolfenamic acid, trifluoperazine, trimethoprim, valproic acid, verapamil

KEY WORDS
comparison with HPLC; $k' = (Tr-T0)/(T0(1-Tr/Tm))$ where Tr = retention time of analyte; $T0$ = retention time of water; and Tm = retention time of marker (halofantrine)

REFERENCE
Hanna,M.; de Biasi,V.; Bond,B.; Salter,C.; Hutt,A.J.; Camilleri,P. Estimation of the partitioning characteristics of drugs: A comparison of a large and diverse drug series utilizing chromatographic and electrophoretic methodology, *Anal.Chem.*, **1998**, *70*, 2092–2099.

SAMPLE
Matrix: solutions
Sample preparation: Mix 50 μL of a solution in 1% sodium chloride with 100 μL MeCN, vortex for 15 s, centrifuge at 14000 g for 20 s, inject an aliquot.

CAPILLARY ELECTROPHORESIS
Capillary: 40 cm × 50 μm silica
Running buffer: MeCN:buffer 10:90 (Buffer was 160 mM triethanolamine containing 50 mM tricine, pH adjusted to 8.6.)
Injection: Hydrodynamic injection at low pressure for 70 s (to fill 12% of the capillary).
Detector: UV 254
Migration time: 11.5
Voltage: 14 kV
Model: Beckman Model 2000 CE

OTHER SUBSTANCES
Simultaneous: N-acetylprocainamide, doxepin, iothalamic acid, quinine

REFERENCE
Shihabi,Z.K. Stacking of weakly cationic compounds by acetonitrile for capillary electrophoresis, *J.Chromatogr.A*, **1998**, *817*, 25–30.

SAMPLE
Matrix: ultrafiltrate

CAPILLARY ELECTROPHORESIS
Capillary: 70 cm × 50 μm fused-silica (50 cm to detector) (Polymicro Technologies)
Capillary preparation: Before each sample rinse with 250 mM NaOH for 1 min and with running buffer for 4 min, equilibrate at 21 kV for 5 min. Every nine samples rinse for 5 min with water, for 5 min with 250 mM NaOH, and for 5 min with running buffer at 5080 mm Hg, equilibrate for 5 min.
Capillary temperature: 30
Running buffer: 25 mM pH 8.0 sodium phosphate buffer containing 80 mM sodium dodecyl sulfate
Injection: Inject using vacuum at 1270 mm Hg for 5 s (15 nL)
Detector: UV 274
Migration time: 8.5
Voltage: 21 kV
Model: Applied Biosystems 270-HT

OTHER SUBSTANCES
Extracted: metabolites

KEY WORDS
rat; pharmacokinetics

REFERENCE
Linhares,M.C.; Kissinger,P.T. Pharmacokinetic studies using micellar electrokinetic capillary chromatography with in vivo capillary ultrafiltration probes, *J.Chromatogr.*, **1993**, *615*, 327–333.

Thiamine

Molecular formula: $C_{12}H_{17}ClN_4OS$
Molecular weight: 300.81
CAS Registry No.: 59-43-8, 67-03-8 (HCl), 532-43-4 (mononitrate)
Merck Index (12th ed.): 9430

SAMPLE
Matrix: formulations
Sample preparation: Homogenize a multi-vitamin tablet, extract with 100 mM HCl, filter (0.45 μm).

CAPILLARY ELECTROPHORESIS
Capillary: 51 cm × 75 μm fused-silica (33 cm to detector) (Waters)
Running buffer: 20 mM pH 9.0 sodium phosphate buffer
Injection: 3 nL (5 μL with a split ratio of 1:1700)
Detector: UV 254
Migration time: 4.74
Voltage: 6.0 kV
Model: ISCO Model 3850

OTHER SUBSTANCES
Simultaneous: riboflavin, pyridoxal, pyridoxamine, pyridoxine, ascorbic acid

KEY WORDS
tablets

REFERENCE
Huopalahti,R.; Sunell,J. Use of capillary zone electrophoresis in the determination of B vitamins in pharmaceutical products, *J.Chromatogr.*, **1993**, *636*, 133–135.

SAMPLE
Matrix: formulations
Sample preparation: Grind tablet, dissolve in pH 5 citrate buffer, filter (0.2 μm), inject an aliquot.

CAPILLARY ELECTROPHORESIS
Capillary: 48.5 cm × 50 μm fused-silica (40 cm to detector)
Capillary preparation: Between runs flush with 100 mM NaOH for 2 min and with run buffer for 3 min, inject a buffer plug. Before the first use condition capillaries with 1 M NaOH for 10 min, with 100 mM NaOH for 2 min, and with run buffer for 3 min.
Capillary temperature: 25
Running buffer: 20 mM pH 7 sodium phosphate
Injection: Injection at 40 mbar for 4.6 s (post-injection pressure (of run buffer) 40 mbar for 4 s)
Detector: UV 215
Migration time: 4.6
Voltage: 20 kV
Model: Hewlett-Packard

OTHER SUBSTANCES
Simultaneous: folinic acid, niacinamide, niacin, orotic acid, pantothenic acid, pyridoxine, vitamin B12 (cyanocobalamin) (UV 360), ascorbic acid

KEY WORDS
tablets

REFERENCE
Jegle,U. Separation of water-soluble vitamins via high-performance capillary electrophoresis, *J.Chromatogr.A*, **1993**, *652*, 495–501.

SAMPLE
Matrix: formulations
Sample preparation: Condition a 500 mg Bakerbond C18 SPE cartridge with 3 mL MeOH and 3 mL water:10 mM sodium heptanesulfonate in acetic acid 99:1. Heat 50-100 mg tablet and 40 mL water:10 mM sodium heptanesulfonate in acetic acid 99:1 at 55° with occasional shaking for 5 min, cool, add a 2 mL aliquot to the SPE cartridge, elute with 3.5 mL MeOH, make up the eluate to 10 mL with water, inject an aliquot.

CAPILLARY ELECTROPHORESIS
Capillary: 70 cm × 100 μm
Capillary temperature: 25
Running buffer: MeOH:buffer 10:90 (Buffer was 50 mM pH 8.0 sodium borate containing 22.5 mM sodium dodecyl sulfate.)
Injection: Inject under 3.44 kPa pressure for 5 s (30 nL)
Detector: UV 214
Migration time: 29.5
Voltage: 16 kV
Current: 33 μA
Model: Beckman P/ACE System 2000

OTHER SUBSTANCES
Simultaneous: niacinamide, pantothenic acid, pyridoxine, riboflavin, vitamin B12, ascorbic acid

KEY WORDS
tablets; SPE

REFERENCE
Dinelli,G.; Bonetti,A. Micellar electrokinetic capillary chromatography analysis of water-soluble vitamins and multi-vitamin integrators, *Electrophoresis*, **1994**, *15*, 1147–1150.

SAMPLE
Matrix: formulations
Sample preparation: Protect from light. Grind tablet, add 50 mL 1% acetic acid, shake at 65° for 10 min, add 5 mL 10 mg/mL niacin in 1% acetic acid, cool to room temperature, make up to 100 mL with 1% acetic acid, filter (0.22 μm cellulose acetate), dilute 4-12-fold, inject an aliquot.

CAPILLARY ELECTROPHORESIS
Capillary: 48.5 cm × 50 μm fused-silica (40 cm to detector) (Hewlett Packard)
Capillary preparation: After each run wash with 100 mM NaOH for 7 min, with water for 1 min, and with running buffer for 3 min. At the start of each day wash with separation buffer for 10 min. Wash new capillaries with 1 M NaOH, 100 mM NaOH, water, and running buffer.
Capillary temperature: 25
Running buffer: 50 mM Borax adjusted to pH 8.5 with boric acid
Injection: Hydrodynamic injection at 5 kPa for 10 s.
Detector: UV 225
Migration time: 2
Internal standard: niacin (7.5)
Voltage: 25 kV
Model: Hewlett Packard [3D]CE

OTHER SUBSTANCES
Simultaneous: adenine, biotin, niacinamide, pantothenic acid (UV 215), pyridoxine (UV 215), rutin, riboflavin, ascorbic acid

KEY WORDS
tablets

REFERENCE
Fotsing,L.; Fillet,M.; Bechet,I.; Hubert,P.; Crommen,J. Determination of six water-soluble vitamins in a pharmaceutical formulation by capillary electrophoresis, *J.Pharm.Biomed.Anal.*, **1997**, *15*, 1113–1123.

SAMPLE
Matrix: formulations
Sample preparation: Grind tablet to a homogeneous powder and dissolve in 2 mL water and 20 mL EtOH, homogenize (Ultra Turrax), centrifuge. Evaporate the supernatant to dryness under reduced pressure, reconstitute the residue with water, inject an aliquot.

CAPILLARY ELECTROPHORESIS
Capillary: 64.5 cm × 50 μm fused-silica (56 cm to detector)
Capillary preparation: Before each run flush capillary at 4 kPa with 1 M NaOH for 2 min and with buffer for 5 min. Replace buffer before each analysis.
Capillary temperature: 30
Running buffer: 1-Propanol:buffer 2:98 (Buffer was 100 mM Na_2HPO_4, 500 mM taurine, and 75 mM sodium cholate.)
Injection: Pressure injection at 4 kPa for 1 s.
Detector: UV 214
Migration time: 7.5
Voltage: 17 kV
Model: Hewlett-Packard HP[3D] CE
Limit of quantitation: 10 μM

OTHER SUBSTANCES
Simultaneous: ascorbic acid, biotin, folic acid, niacinamide, pyridoxamine, pyridoxine, riboflavin, vitamin B12

KEY WORDS
tablets

REFERENCE
Buskov,S.; Moller,P.; Sorensen,H.; Sorensen,J.C.; Sorensen,S. Determination of vitamins in food based on supercritical fluid extraction prior to micellar electrokinetic capillary chromatographic analyses of individual vitamins, *J.Chromatogr.A*, **1998**, *802*, 233–241.

SAMPLE
Matrix: formulations, juice
Sample preparation: Dissolve in water, filter (0.2 μm), inject an aliquot. (Add a 50-fold excess of cysteine to solutions containing ascorbic acid.)

CAPILLARY ELECTROPHORESIS
Capillary: 60 cm × 50 μm fused-silica (51.5 cm to detector)
Capillary preparation: Condition with running buffer for 3 min before each run. Condition a new capillary 1 M NaOH for 10 min and with 100 mM NaOH for 2 min.
Capillary temperature: 25
Running buffer: 20 mM pH 8.0 Phosphate buffer
Injection: Pressure injection for 200 mbar.s
Detector: UV 200
Migration time: 3.0
Voltage: -30 kV
Model: Hewlett-Packard 3DCE
Limit of detection: 2.7 μg/mL

OTHER SUBSTANCES
Simultaneous: biotin, niacinamide, pyridoxine, riboflavin phosphate, ascorbic acid

KEY WORDS
fruit

REFERENCE
Schiewe,J.; Mrestani,Y.; Neubert,R. Application and optimization of capillary zone electrophoresis in vitamin analysis, *J.Chromatogr.A*, **1995**, *717*, 255–259.

SAMPLE
Matrix: solutions

Sample preparation: Prepare a 0.5-2 mg/mL solution in water, inject an aliquot.

CAPILLARY ELECTROPHORESIS
Capillary: 65 cm × 50 μm fused-silica (50 cm to detector) (Scientific Glass Engineering)
Running buffer: 20 mM NaH_2PO_4 containing 50 mM sodium dodecyl sulfate adjusted to pH 9.0 with 20 mM sodium tetraborate
Injection: Injection by siphon at 5 cm for 5-10 s (about 1 nL)
Detector: UV 210
Migration time: 11.2
Voltage: 25 kV

OTHER SUBSTANCES
Simultaneous: niacin, niacinamide, pyridoxal, pyridoxamine, pyridoxine, riboflavin, vitamin B12

REFERENCE
Nishi,H.; Tsumagari,N.; Kakimoto,T.; Terabe,S. Separation of water-soluble vitamins by micellar electrokinetic chromatography, *J.Chromatogr.*, **1989**, *465*, 331–343.

SAMPLE
Matrix: solutions
Sample preparation: Prepare a 100 μg/mL solution in MeOH:water 50:50, inject an aliquot.

CAPILLARY ELECTROPHORESIS
Capillary: 60 cm × 75 μm
Running buffer: 20 mM NaH_2PO_4 containing 50 mM sodium dodecyl sulfate, adjusted to pH 9.0 with sodium tetraborate
Injection: Hydrodynamic injection at 10 cm for 5 s
Detector: UV 214
Migration time: 17
Voltage: 18 kV
Model: Waters Quanta 4000

OTHER SUBSTANCES
Simultaneous: niacin, pantothenic acid, riboflavin, pyridoxine, ascorbic acid

REFERENCE
Swartz,M.E. Method development and selectivity control for small molecule pharmaceutical separations by capillary electrophoresis, *J.Liq.Chromatogr.*, **1991**, *14*, 923–938.

SAMPLE
Matrix: solutions

CAPILLARY ELECTROPHORESIS
Capillary: 56.9 cm × 75 μm fused-silica (50 cm to detector)
Capillary temperature: 25
Running buffer: MeOH:60 mM pH 8.92 sodium borate buffer containing 60 mM sodium dodecyl sulfate 15:85
Detector: UV 214
Migration time: 9.3
Voltage: 30 kV
Model: Beckman PACE 2100

OTHER SUBSTANCES
Simultaneous: niacin, niacinamide, pyridoxine, vitamin B12

REFERENCE
McLaughlin,G.M.; Nolan,J.A.; Lindahl,J.L.; Palmieri,R.H.; Anderson,K.W.; Morris,S.C.; Morrison,J.A.; Bronzert,T.J. Pharmaceutical drug separations by HPCE: Practical guidelines, *J.Liq.Chromatogr.*, **1992**, *15*, 961–1021.

SAMPLE
Matrix: solutions
Sample preparation: Prepare a 100 μg/mL solution in MeOH:water 50:50, inject an aliquot.

CAPILLARY ELECTROPHORESIS
Capillary: 47 cm × 50 μm fused-silica (40 cm to detector) (Polymicro Technologies)
Capillary temperature: 25
Running buffer: Prepare as follows. Mix n-hexane with 2 volumes of 500 mM sodium dodecyl sulfate, vortex, add 1-butanol until the mixture clears, dilute with 20 mM pH 7.0 phosphate buffer until the concentration of sodium dodecyl sulfate is 20 mM.
Injection: Inject using an overpressure of 34500000 Pa (sic) for 3 s
Detector: UV 214
Migration time: 6.0
Voltage: 10 kV
Current: about 17.5 μA
Model: Beckman P/ACE System 2100

OTHER SUBSTANCES
Simultaneous: niacin, niacinamide, pyridoxine (pyridoxol), vitamin A, vitamin E

REFERENCE
Boso,R.L.; Bellini,M.S.; Miksík,; Deyl,Z. Microemulsion electrokinetic chromatography with different organic modifiers: separation of water- and lipid-soluble vitamins, *J.Chromatogr.A*, **1995**, *709*, 11–19.

SAMPLE
Matrix: solutions

CAPILLARY ELECTROPHORESIS
Capillary: 60 cm × 75 μm AccuSep (Waters) or 52.5 cm × 75 μm (45 cm to detector) (SGE)
Running buffer: 20 mM pH 7 Phosphate buffer
Injection: Hydrodynamic injection for 7 s
Detector: UV 214
Migration time: 3.4
Voltage: 15 kV
Model: Waters Quanta 4000

OTHER SUBSTANCES
Simultaneous: niacinamide, niacin

KEY WORDS
sleeve cell increases sensitivity

REFERENCE
Djordjevic,N.M.; Ryan,K. An easy way to enhance absorbance detection sensitivity of Waters Quanta-4000 capillary electrophoresis system, *J.Liq.Chromatogr.Rel.Technol.*, **1996**, *19*, 201–206.

SAMPLE
Matrix: solutions
Sample preparation: Inject an aliquot of a 500 μg/mL solution in 50 mM pH 7 buffer.

CAPILLARY ELECTROPHORESIS
Capillary: 48.5 cm × 50 μm fused-silica (40 cm to detector) (Hewlett Packard)
Capillary preparation: Before each injection flush capillary with 100 mM NaOH for 5 min and with running buffer for 5 min. Wash new capillaries with 1 M NaOH at 40° for 15 min, with water at 40° for 10 min, and with water at 25° for 5 min.
Capillary temperature: 25
Running buffer: 50 mM pH 7.0 Phosphate buffer (Prepare buffer by dissolving 5.29 g K_2HPO_4 and 2.61 g KH_2PO_4 in 1 L water.)
Injection: Pressure injection at 50 mbar for 9 s (18.8 nL).
Detector: UV 200
Migration time: 7

Voltage: 30 kV
Model: Hewlett Packard Model G1600A ³ᴰCE

OTHER SUBSTANCES
Simultaneous: ceftazidime, diclofenac, dicloxacillin, oxacillin, propranolol, quinine

REFERENCE
Mrestani,Y.; Neubert,R.H.H.; Krause,A. Partition behaviour of drugs in microemulsions measured by electro-kinetic chromatography, *Pharm.Res.*, **1998**, *15*, 799–801.

SAMPLE
Matrix: solutions

CAPILLARY ELECTROPHORESIS
Capillary: 70 cm × 50 μm fused-silica (GL Science, Tokyo)
Capillary preparation: Before each run rinse capillary with running buffer at 94 kPa for 5 min before each run.
Running buffer: 40 mM pH 6.0 ammonium acetate buffer
Injection: Pressure injection at 5 kPa (50 mbar) for 4 s.
Detector: MS, Perkin-Elmer Sciex API-300 quadrupole, electrospray (ionspray) interface, sheath liquid MeOH:running buffer 50:50 at 2.5 μL/min, ionspray voltage 5 kV, positive ion mode
Migration time: 11.5
Voltage: 20 kV (net voltage across capillary = 15 kV (applied voltage - electrospray voltage))
Model: Hewlett Packard 3D CE

OTHER SUBSTANCES
Simultaneous: acetaminophen, ascorbic acid, butylscopolamine bromide, caffeine, ibuprofen, ketoprofen, niacin, niacinamide, riboflavin, vitamin B12, warfarin

REFERENCE
Tanaka,Y.; Kishimoto,Y.; Otsuka,K.; Terabe,S. Strategy for selecting separation solutions in capillary electrophoresis-mass spectrometry, *J.Chromatogr.A*, **1998**, *817*, 49–57.

Thiamphenicol

Molecular formula: $C_{12}H_{15}Cl_2NO_5S$
Molecular weight: 356.23
CAS Registry No.: 15318-45-3
Merck Index (12th ed.): 9436
Lednicer: 2 45

SAMPLE
Matrix: blood
Sample preparation: 500 μL Plasma + 20 μL 100 μg/mL chloramphenicol in water + 3-5 mg K_2HPO_4, mix thoroughly, add 5 mL ethyl acetate, rotate for 1-2 min, centrifuge at 3500 g for 5 min. Remove a 4 mL aliquot of the ethyl acetate layer and evaporate it to dryness under a stream of nitrogen at 40°, reconstitute with 100 μL water, inject an aliquot.

CAPILLARY ELECTROPHORESIS
Capillary: 70 cm × 75 μm fused-silica (62.5 cm to detector) (Yongnian Optical Fiber, Hebei, China)
Capillary preparation: Before each run wash capillary with 100 mM NaOH and water for 1 min, rinse with running buffer for 2 min. Before use condition capillary with 1 M NaOH at 60° for 20 min, with 100 mM NaOH at 60° for 5 min, and with water at 30° for 5 min.
Capillary temperature: 30
Running buffer: MeCN:buffer 10:90 (Buffer was 20 mM pH 9.2 disodium tetraborate containing 40 mM sodium dodecyl sulfate.)

Injection: Hydrodynamic injection for 10 s (40 nL).
Detector: UV 195
Migration time: 10
Internal standard: chloramphenicol (11)
Voltage: 18 kV
Current: 58-69 μA
Model: SpectraPhoresis 1000 with FOCUS detection (Thermo Separation Products)
Limit of quantitation: 100 ng/mL

KEY WORDS
plasma; pharmacokinetics

REFERENCE
Song,J.-Z.; Wu,X.-J.; Sun,Z.-P.; Tian,S.-J.; Wang,M.-L.; Wang,R.-L. Determination of thiamphenicol in human plasma by micellar electrokinetic capillary chromatography, *J.Chromatogr.B*, **1997**, *692*, 445–451.

Thiopental

Molecular formula: $C_{11}H_{17}N_2NaO_2S$
Molecular weight: 264.32
CAS Registry No.: 76-75-5, 71-73-8 (Na salt)
Merck Index (12th ed.): 9487
Lednicer: 1 274

SAMPLE
Matrix: blood
Sample preparation: 500 μL Serum or plasma + 100 μL 2 mg/mL carbamazepine in MeOH + 1 mL 70 mM pH 6.4 phosphate buffer + 5 mL n-pentane, shake vigorously for 10 min, centrifuge at 1500 g for 10 min. Remove the organic layer and evaporate it to dryness under a stream of nitrogen at 40°, reconstitute the residue in 200 μL running buffer, vortex for 1 min, inject an aliquot.

CAPILLARY ELECTROPHORESIS
Capillary: 50 cm × 50 μm fused-silica (50 cm to detector)
Capillary preparation: Before each run rinse with 100 mM NaOH for 1 min and with running buffer for 2 min
Capillary temperature: 40
Running buffer: 15 mM pH 7.8 NaH_2PO_4 containing 9 mM sodium borate and 50 mM sodium dodecyl sulfate
Injection: Inject by vacuum for 2 s
Detector: UV 290
Migration time: 4
Internal standard: carbamazepine (7.5)
Voltage: 30 kV
Current: 55-60 μA
Model: Applied Biosystems Model 270A

KEY WORDS
cow; human; serum; plasma; comparison with HPLC

REFERENCE
Meier,P.; Thormann,W. Determination of thiopental in human serum and plasma by high-performance capillary electrophoresis-micellar electrokinetic chromatography, *J.Chromatogr.*, **1991**, *559*, 505–513.

SAMPLE
Matrix: blood, urine
Sample preparation: Serum, plasma. 200 μL Serum or plasma + 100 μL 1 M HCl + 2 mL chloroform, shake vigorously for 15 min, centrifuge at 500 g for 10 min. Remove the lower

organic layer and evaporate it to dryness under a stream of nitrogen at 40°, reconstitute the residue in 200 μL running buffer, shake for 1 min, filter (0.2 μm), inject an aliquot. Urine. Condition a Bond Elut Certify SPE cartridge with 2 mL MeOH and 2 mL 100 mM pH 6 phosphate buffer, do not allow to dry. 5 mL Urine + 2 mL 100 mM pH 6 phosphate buffer, mix, add to the SPE cartridge, wash with 1 mL MeOH:100 mM phosphate buffer 20:80, dry under vacuum for 5 min, wash with 1 mL 1 M acetic acid, dry under vacuum for 10 min, wash with 1 mL hexane, elute with 4 mL dichloromethane. Evaporate the eluate to dryness under a stream of nitrogen at 40°, reconstitute the residue in 100-200 μL running buffer, filter (0.2 μm), inject an aliquot.

CAPILLARY ELECTROPHORESIS
Capillary: 90 cm × 75 μm fused-silica (70 cm to detector) (Polymicro Technologies)
Capillary preparation: Between each run rinse capillary with 100 mM NaOH for 3 min and buffer for 5 min
Running buffer: 15 mM pH 7.8 NaH_2PO_4 containing 9 mM sodium borate and 50 mM sodium dodecyl sulfate
Injection: Siphon at 34 cm for 5 s
Detector: UV 195
Migration time: 12.3
Voltage: 20 kV
Current: 60-63 μA
Model: Laboratory constructed

OTHER SUBSTANCES
Extracted: allobarbital, amobarbital, barbital, butalbital, pentobarbital, phenobarbital

KEY WORDS
serum; cow; human; plasma; SPE

REFERENCE
Thormann,W.; Meier,P.; Marcolli,C.; Binder,F. Analysis of barbiturates in human serum and urine by high-performance capillary electrophoresis-micellar electrokinetic capillary chromatography with on-column multi-wavelength detection, *J.Chromatogr.*, **1991**, *545*, 445–460.

SAMPLE
Matrix: solutions

CAPILLARY ELECTROPHORESIS
Capillary: 57 cm × 50 μm fused-silica (50 cm to detector)
Capillary preparation: After each run rinse with 20 mM pH 9.00 borate buffer containing 90 mM sodium dodecyl sulfate for 5 min and with 100 mM NaOH for 5 min.
Running buffer: 20 mM pH 9.60 Borate buffer containing 10 mM α-cyclodextrin
Injection: Pressure injection at 3.45 kPa for 5 s.
Detector: UV 254
Migration time: 7.7, 7.8 (enantiomers)
Voltage: 15 kV
Model: Beckman P/ACE 2100

OTHER SUBSTANCES
Simultaneous: cyclobarbital, phenobarbital

KEY WORDS
chiral

REFERENCE
Conradi,S.; Vogt,C.; Rohde,E. Separation of enantiomeric barbiturates by capillary electrophoresis using a cyclodextrin-containing run buffer, *J.Chem.Educ.*, **1997**, *74*, 1122–1125.

SAMPLE
Matrix: solutions
Sample preparation: Inject an aliquot of a 20 μg/mL solution

CAPILLARY ELECTROPHORESIS
Capillary: 44 cm × 50 µm fused-silica (37 cm to detector) (Supelco)
Capillary preparation: Before each run rinse capillary with running buffer for 3 min. At the start of each day rinse capillary with running buffer for 10 min. Treat new capillaries with 1 M NaOH, 100 mM NaOH, water, and running buffer.
Capillary temperature: 25
Running buffer: 100 mM Phosphoric acid adjusted to pH 5 with triethanolamine containing 10 mM carboxymethyl-β-cyclodextrin (Cyclolab, Budapest) and 50 mM heptakis(2,3,6-tri-O-methyl)-β-cyclodextrin (Sigma)
Injection: Hydrodynamic injection for 5 s (13.3 nL).
Detector: UV 210
Voltage: -25 kV
Model: Spectraphoresis 1000 CE

KEY WORDS
detector at anode; chiral; R_s = 5.2

REFERENCE
Fillet,M.; Fotsing,L.; Crommen,J. Enantioseparation of uncharged compounds by capillary electrophoresis using mixtures of anionic and neutral β-cyclodextrin derivatives, *J.Chromatogr.A*, **1998**, *817*, 113–119.

SAMPLE
Matrix: urine
Sample preparation: Place an extraction rod in 50 µL urine in a 50 mm × 1.5 mm ID length of PTFE tubing for 30 min, place the extraction rod into 5 µL 20-40 mM pH 11.5 phosphate buffer in a 50 mm × 1.2 mm ID length of PTFE tubing for 90 min, inject an aliquot of the buffer. (Prepare the solid-phase extraction rod as follows. Polish a 70 × 1.1 stainless steel rod with emery paper, clean with a Kimwipe and acetone, sonicate in EtOH, sonicate in THF, dip in 3% PVAM in THF momentarily, hold vertically for 1 min, air dry in a hood for at least 5 h, dip in PVC solution momentarily, hold vertically for 1 min, air dry for at least 5 h. The coating is 3 cm long. PVAM was poly(vinyl chloride-co-vinyl acetate-co-maleic acid) consisting of 86% vinyl chloride, 13% vinyl acetate, and 1% maleic acid. Prepare PVC solution by adding very high molecular weight PVC slowly to THF with stirring until the concentration reaches 3.6%, add Santicizer 141 to a concentration of 7.2%. Santicizer 141 (Monsanto) is 92% 2-ethylhexyl diphenyl phosphate, 5% di-2-ethylhexyl phenyl phosphate, and 3% triphenyl phosphate.)

CAPILLARY ELECTROPHORESIS
Capillary: 75 cm × 75 µm fused-silica (50 cm to detector) (Polymicro Technologies)
Running buffer: 50 mM Tris adjusted to pH 7.8 with 3-[N-tris(hydroxymethyl)methylamino]-2-hydroxypropanesulfonic acid (Tapso)
Injection: Pressure injection at 0.5 psi for 4 s.
Detector: UV 230
Migration time: 8.8
Voltage: 35 kV
Current: about 30 µA
Model: Isco 3850
Limit of detection: <1 ppm

OTHER SUBSTANCES
Extracted: amobarbital, aprobarbital, butabarbital, butalbital, mephobarbital, pentobarbital, secobarbital
Simultaneous: allobarbital, aspirin, phenobarbital

KEY WORDS
SPE

REFERENCE
Li,S.; Weber,S.G. Determination of barbiturates by solid-phase microextraction and capillary electrophoresis, *Anal.Chem.*, **1997**, *69*, 1217–1222.

Thioridazine

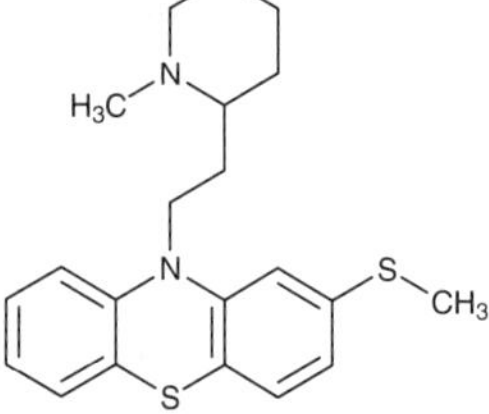

Molecular formula: $C_{21}H_{26}N_2S_2$
Molecular weight: 370.58
CAS Registry No.: 50-52-2, 130-61-0 (HCl)
Merck Index (12th ed.): 9497
Lednicer: 1 389

SAMPLE
Matrix: solutions

CAPILLARY ELECTROPHORESIS
Capillary: 59.6 cm × 75 μm fused-silica (52.1 cm to detector) (Polymicro Technologies)
Capillary preparation: Purge with running buffer before each run. At the beginning of each day purge using 50-60 kPa vacuum with 500 mM NaOH for 5 min, with water for 5 min, with MeCN for 5 min, and with running buffer for 5 min.
Running buffer: MeCN:MeOH:acetic acid 49:50:1 containing 20 mM ammonium acetate
Injection: Hydrostatic injection at 10 cm for 5 s.
Detector: UV 214
Migration time: 3.70
Voltage: 30 kV
Model: Waters Quanta 4000

OTHER SUBSTANCES
Simultaneous: benzquinamide, deserpidine, ethopropazine, mesoridazine, methotrimeprazine, molindone, promazine, promethazine, reserpine, thiothixene

REFERENCE
Leung,G.N.W.; Tang,H.P.O.; Tso,T.S.C.; Wan,T.S.M. Separation of basic drugs with non-aqueous capillary electrophoresis, *J.Chromatogr.A*, **1996**, *738*, 141–154.

SAMPLE
Matrix: solutions

CAPILLARY ELECTROPHORESIS
Capillary: 42 cm × 50 μm fused-silica (31 cm to detector) (Polymicro Technologies)
Capillary preparation: Condition capillary by rinsing with running buffer for 10 min, with water for 10 min, with 1 M NaOH for 10 min, with water for 10 min, and with running buffer with the voltage applied for 20 min.
Capillary temperature: 25
Running buffer: Formamide containing 150 mM citric acid, 100 mM Tris, and 100 mM β-cyclodextrin (apparent pH 5.1)
Injection: Gravity injection.
Detector: UV 254
Migration time: 10.61, 10.85 (enantiomers)
Voltage: 30 kV
Model: Beckman P/ACE 5500

OTHER SUBSTANCES
Simultaneous: chlophedianol, mianserin, nefopam, propiomazine, trimeprazine, trimipramine

KEY WORDS
chiral

REFERENCE
Wang,F.; Khaledi,M.G. Chiral separations by nonaqueous capillary electrophoresis, *Anal.Chem.*, **1996**, *68*, 3460–3467.

SAMPLE
Matrix: urine
Sample preparation: 10 mL Urine + 1 mL 5 M NaOH + 10 mL n-hexane, vortex for 1 min, centrifuge at 0° at 3000 g for 5 min. Remove the organic layer and add it to 50 μL glacial acetic acid, evaporate to dryness under a stream of nitrogen at 30°, reconstitute the residue in 50 μL 50 mM sodium taurodeoxycholate, filter (0.2 μm PTFE), inject an aliquot.

CAPILLARY ELECTROPHORESIS
Capillary: 100 cm × 50 μm fused-silica (50 cm to detector) (ISCO)
Capillary preparation: Rinse capillary with 20 μL running buffer between injections. Condition a new capillary by filling with 1 M NaOH, let stand for 1 h, fill with 100 mM NaOH, let stand for 1 h, rinse with buffer. Every 20 injections rinse capillary with 200 μL 1 M NaOH, 200 μL water, and 200 μL running buffer, fill with running buffer.
Capillary temperature: 22
Running buffer: 40 mM pH 9.5 Borate buffer containing 10 mM sodium taurodeoxycholate
Injection: Load under vacuum at 7.5 kPa/s.
Detector: UV 240
Migration time: 14
Voltage: 30 kV
Model: ISCO Model 3140 electropherograph
Limit of detection: 38 ng/mL

OTHER SUBSTANCES
Extracted: acepromazine, amiodarone, amitriptyline, azaperone, chlorpromazine, cianopramine, clomipramine, clozapine, desethylamiodarone, desipramine, diclofensine, dothiepin, doxepin, imipramine, isocarboxazid, moclobemide, perphenazine, phenothiazine, pimozide, prochlorperazine, promazine, thiothixene, trifluoperazine, trimipramine

KEY WORDS
human; cow; pig; horse; protect from light

REFERENCE
Aumatell,A.; Wells,R.J. Determination of a cardiac antiarrhythmic, tricyclic antipsychotics and antidepressants in human and animal urine by micellar electrokinetic capillary chromatography using a bile salt, *J.Chromatogr.B*, **1995**, *669*, 331–344.

SAMPLE
Matrix: urine
Sample preparation: Dilute 10-fold with water, inject an aliquot.

CAPILLARY ELECTROPHORESIS
Capillary: 70 cm × 50 μm fused-silica (65.4 cm to detector) (Polymicro Technologies)
Capillary temperature: 25
Running buffer: 20 mM pH 3.5 Tris-formic acid containing 50 μg/mL FC-135 (Fluorad/3M)
Injection: Pressure injection at 2 psi.
Detector: UV 240
Migration time: 7.44
Voltage: 20 kV
Model: Bio-Rad BioFocus 3000
Limit of detection: 1.36 μg/mL

OTHER SUBSTANCES
Extracted: perphenazine, prochlorperazine, promazine, trifluoperazine, triflupromazine

KEY WORDS
detector at anode

REFERENCE
Muijelsaar,P.G.H.M.; Claessens,H.A.; Cramers,C.A. Determination of structurally related phenothiazines by capillary zone electrophoresis and micellar electrokinetic chromatography, *J.Chromatogr.A*, **1996**, *735*, 395–402.

Thiothixene

Molecular formula: $C_{23}H_{29}N_3O_2S_2$
Molecular weight: 443.63
CAS Registry No.: 5591-45-7, 22189-31-7 (HCl dihydrate), 49746-09-0 (Z HCl dihydrate), 58513- 59-0 (HCl), 49746-04-5 (Z HCl)
Merck Index (12th ed.): 9503
Lednicer: 1 400

SAMPLE
Matrix: solutions

CAPILLARY ELECTROPHORESIS
Capillary: 59.6 cm × 75 μm fused-silica (52.1 cm to detector) (Polymicro Technologies)
Capillary preparation: Purge with running buffer before each run. At the beginning of each day purge using 50-60 kPa vacuum with 500 mM NaOH for 5 min, with water for 5 min, with MeCN for 5 min, and with running buffer for 5 min.
Running buffer: MeCN:MeOH:acetic acid 49:50:1 containing 20 mM ammonium acetate
Injection: Hydrostatic injection at 10 cm for 5 s.
Detector: UV 214
Migration time: 4.48
Voltage: 30 kV
Model: Waters Quanta 4000

OTHER SUBSTANCES
Simultaneous: benzquinamide, deserpidine, ethopropazine, mesoridazine, methotrimeprazine, molindone, promazine, promethazine, reserpine, thioridazine

REFERENCE
Leung,G.N.W.; Tang,H.P.O.; Tso,T.S.C.; Wan,T.S.M. Separation of basic drugs with non-aqueous capillary electrophoresis, *J.Chromatogr.A*, **1996**, *738*, 141–154.

SAMPLE
Matrix: solutions
Sample preparation: Inject an aliquot of a 100 μg/mL solution in MeOH:MeCN 50:50.

CAPILLARY ELECTROPHORESIS
Capillary: 64 cm × 50 μm fused-silica (55.5 cm to detector) (Polymicro Technologies)
Capillary preparation: Between runs flush capillary with running buffer for 2 min. Before use rinse capillary with 1 M NaOH for 1 h, with 100 mM NaOH for 20 min, and with running buffer for 10 min.
Capillary temperature: 25
Running buffer: MeCN:MeOH 50:50 containing 50 mM ammonium acetate and 1 M acetic acid
Injection: Pressure injection at 5 kPa for 3 s.
Detector: UV 214
Voltage: 30 kV
Model: Hewlett-Packard ³DCE

KEY WORDS
cis and trans isomers separated; Rs = 3.72

REFERENCE
Hansen,S.H.; Bjornsdottir,I.; Tjornelund,J. Separation of cationic *cis--trans (Z--E)* isomers and diastereoisomers using non-aqueous capillary electrophoresis, *J.Chromatogr.A*, **1997**, *792*, 49–55.

SAMPLE
Matrix: urine

Sample preparation: 10 mL Urine + 1 mL 5 M NaOH + 10 mL n-hexane, vortex for 1 min, centrifuge at 0° at 3000 g for 5 min. Remove the organic layer and add it to 50 μL glacial acetic acid, evaporate to dryness under a stream of nitrogen at 30°, reconstitute the residue in 50 μL 50 mM sodium taurodeoxycholate, filter (0.2 μm PTFE), inject an aliquot.

CAPILLARY ELECTROPHORESIS
Capillary: 100 cm × 50 μm fused-silica (50 cm to detector) (ISCO)
Capillary preparation: Rinse capillary with 20 μL running buffer between injections. Condition a new capillary by filling with 1 M NaOH, let stand for 1 h, fill with 100 mM NaOH, let stand for 1 h, rinse with buffer. Every 20 injections rinse capillary with 200 μL 1 M NaOH, 200 μL water, and 200 μL running buffer, fill with running buffer.
Capillary temperature: 22
Running buffer: 40 mM pH 9.5 Borate buffer containing 10 mM sodium taurodeoxycholate
Injection: Load under vacuum at 7.5 kPa/s.
Detector: UV 240
Migration time: 10.5
Voltage: 30 kV
Model: ISCO Model 3140 electropherograph
Limit of detection: 16 ng/mL

OTHER SUBSTANCES
Extracted: acepromazine, amiodarone, amitriptyline, azaperone, chlorpromazine, cianopramine, clomipramine, clozapine, desethylamiodarone, desipramine, diclofensine, dothiepin, doxepin, imipramine, isocarboxazid, moclobemide, perphenazine, phenothiazine, pimozide, prochlorperazine, promazine, thioridazine, trifluoperazine, trimipramine

KEY WORDS
human; cow; pig; horse; protect from light

REFERENCE
Aumatell,A.; Wells,R.J. Determination of a cardiac antiarrhythmic, tricyclic antipsychotics and antidepressants in human and animal urine by micellar electrokinetic capillary chromatography using a bile salt, *J.Chromatogr.B*, **1995**, *669*, 331–344.

Tiaprofenic acid

Molecular formula: C₁₄H₁₂O₃S
Molecular weight: 260.31
CAS Registry No.: 33005-95-7
Merck Index (12th ed.): 9562

SAMPLE
Matrix: formulations
Sample preparation: Stir 3.98 mg powdered tablet with 100 mL extraction mixture at room temperature for 30 min, filter, inject an aliquot of the filtrate. (Prepare extraction mixture by mixing 99 mL MeOH:10 mM acetate buffer 10:90 with 1 mL 1 mM IS.)

CAPILLARY ELECTROPHORESIS
Capillary: 50 cm × 50 μm fused-silica (45.4 cm to detector) (Polymicro Technologies)
Capillary preparation: Before each run was with water for 100 s and with running buffer for 150 s. Rinse a new capillary with 500 mM NaOH for 10 min and with water for 30 min.
Capillary temperature: 25
Running buffer: 50 mM pH 4.5 Acetate buffer containing 5 mM 2,3,6-tri-O-methyl-β-cyclodextrin
Injection: Pressure injection at 10 psi.s.
Detector: UV 195, UV 206
Migration time: 20.8, 21.7 (enantiomers)
Internal standard: 4-hydroxybenzoic acid (19)
Voltage: 20 kV

Model: Bio-Rad Biofocus 3000

KEY WORDS
chiral; tablets

REFERENCE
Aturki,Z.; Camera,E.; La Torre,F.; Fanali,S. Direct chiral resolution of tiaprofenic acid in pharmaceutical formulations by capillary zone electrophoresis using cyclodextrins as chiral selector, *J.Capillary Electrophor.*,
1995, *2*, 213–217.

SAMPLE
Matrix: solutions
Sample preparation: Inject an aliquot of a 50-100 μM solution.

CAPILLARY ELECTROPHORESIS
Capillary: 35 cm × 50 μm coated fused-silica (30.5 cm to detector) (Polymicro Technologies)
Capillary preparation: Coat capillary as follows. Adjust the pH of 20 mL water to 3.5 with
acetic acid, add 80 μL 3-(trimethoxysilyl)propyl methacrylate (3-methacryloxypropyltrimethox-
ysilane), mix, suck into capillary, let stand at room temperature for 1 h, remove the solution,
wash with water. Fill the capillary with a deaerated 3-4% acrylamide solution containing 1
μL/mL N,N,N',N'-tetramethylethylenediamine and 1 mg/mL potassium persulfate, let stand
for 30 min, remove excess solution by aspiration, rinse with water, remove water by aspiration,
dry at 35° (J. Chromatogr. 1985, 347, 191).
Capillary temperature: 25
Running buffer: pH 7 Buffer containing 3 mM 6^A-methylamino-β-cyclodextrin (Buffer was 50
mM phosphoric acid containing 50 mM acetic acid and 50 mM boric acid, adjusted to pH 7
with concentrated NaOH solution, diluted with an equal volume of water. Synthesis of 6^A-
methylamino-β-cyclodextrin was as follows. Add a solution of 3.65 g p-toluenesulfonyl chloride
in 30 mL dry pyridine to 29.60 g β-cyclodextrin stirred at 5° in 300 mL dry pyridine, stir
overnight at room temperature, evaporate to dryness under reduced pressure at 40°, add 700
mL diethyl ether to the residue. Collect the precipitate and recrystallize it 3 times from water
to obtain mono-(6-O-p-tolylsulfonyl)-β- cyclodextrin in 31% yield (Bull. Chem. Soc. Japan 1978,
51, 3030). Heat 2 g mono-(6-O-p-tolylsulfonyl)-β-cyclodextrin with 35 mL 50% methylamine in
MeOH in a sealed tube at 70° for 3 days, purify by chromatography on carboxymethylcellulose
with ammonium bicarbonate solution to obtain 6^A-methylamino-β-cyclodextrin (cf. J. Am.
Chem. Soc. 1980, 102, 762).)
Injection: Pressure injection at 10 psi.s.
Detector: UV 206
Migration time: 23.5, 25.5 (enantiomers)
Voltage: 15 kV
Current: 54 μA
Model: Biofocus 3000 (Bio-Rad)

OTHER SUBSTANCES
Simultaneous: acenocoumarol, phenyllactic acid, warfarin

KEY WORDS
coated capillary; chiral

REFERENCE
Fanali,S.; Camera,E. Use of methylamino-β-cyclodextrin in capillary electrophoresis. Resolution of acidic and
basic enantiomers, *Chromatographia*, **1996**, *43*, 247–253.

SAMPLE
Matrix: solutions
Sample preparation: Inject an aliquot of a solution in MeOH:water 10:90.

CAPILLARY ELECTROPHORESIS
Capillary: 44 cm × 50 μm fused-silica (37 cm to detector) (Supelco)
Capillary preparation: Wash with running buffer for 3 min after each injection. Wash with
running buffer for 10 min at the end of each day.
Capillary temperature: 25

Running buffer: 100 mM Phosphoric acid containing 5 mM sulfobutyl ether-β-cyclodextrin and 40 mM heptakis(2,3,6-tri-O-methyl)-β-cyclodextrin, adjusted to pH 3.0 with triethanolamine
Injection: Hydrodynamic injection for 5 s (13.3 nL).
Detector: UV 280
Voltage: -25 kV
Current: 60 μA
Model: SpectraPhoresis 1000 CE

KEY WORDS
chiral; detector at anode; resolution (R_s = 7.3)

REFERENCE
Fillet,M.; Hubert,P.; Crommen,J. Enantioseparation of nonsteroidal anti-inflammatory drugs by capillary electrophoresis using mixtures of anionic and uncharged β-cyclodextrins as chiral additives, *Electrophoresis*, **1997**, *18*, 1013–1018.

SAMPLE
Matrix: solutions

CAPILLARY ELECTROPHORESIS
Capillary: 57 cm × 50 μm fused-silica (50 cm to detector)
Running buffer: 50 mM pH 8.0 Borate buffer containing 40 mM sodium taurodeoxycholate and 25 mM phosphatidylcholine (Prepare by adding phosphatidylcholine and stirring for 3-6 h until all cloudiness disappears. Phosphatidylcholine was 95% pure soybean lecithin, Epikuron, Lucas Meyer & Co.)
Injection: Inject a solution of the compound in the running buffer at 20 psi for 1 s, inject MeOH: water 5:95 at 20 psi for 1 s, inject a solution of halofantrine in running buffer at 20 psi for 1 s.
Detector: UV 214
Migration time: k' 1.70
Model: Beckman P/ACE 5000

OTHER SUBSTANCES
Also analyzed: acetaminophen, amoxicillin, antipyrine, aspirin, azathioprine, caffeine, captopril, carbamazepine, carprofen, chlorambucil, chlorpheniramine, chlorpromazine, cimetidine, clonidine, codeine, desipramine, diphenhydramine, ephedrine, fenoterol, flufenamic acid, flurbiprofen, haloperidol, hydroxyzine, ibuprofen, imipramine, indomethacin, ketoprofen, lidocaine, melphalan, metoprolol, nabumetone, nadolol, phenobarbital, phenol, promazine, propranolol, pyrilamine, ranitidine, ropinirole, salicylic acid, sulfamethoxazole, testosterone, theophylline, thioridazine, tolfenamic acid, trifluoperazine, trimethoprim, valproic acid, verapamil

KEY WORDS
comparison with HPLC; k' = (Tr-T0)/(T0(1-Tr/Tm)) where Tr = retention time of analyte; T0 = retention time of water; and Tm = retention time of marker (halofantrine)

REFERENCE
Hanna,M.; de Biasi,V.; Bond,B.; Salter,C.; Hutt,A.J.; Camilleri,P. Estimation of the partitioning characteristics of drugs: A comparison of a large and diverse drug series utilizing chromatographic and electrophoretic methodology, *Anal.Chem.*, **1998**, *70*, 2092–2099.

Ticarcillin

Molecular formula: $C_{15}H_{16}N_2O_6S_2$
Molecular weight: 384.43
CAS Registry No.: 34787-01-4, 4697-14-7 (di Na salt), 74682-62-5 (Na salt)
Merck Index (12th ed.): 9568
Lednicer: 2 437

SAMPLE
Matrix: solutions
Sample preparation: Prepare a 120 μg/mL solution in water, inject an aliquot.

CAPILLARY ELECTROPHORESIS
Capillary: 60 cm × 75 μm
Running buffer: 20 mM NaH_2PO_4 containing 50 mM sodium dodecyl sulfate, adjusted to pH 9.0 with sodium tetraborate
Injection: Hydrodynamic injection at 10 cm for 5 s
Detector: UV 214
Migration time: 12
Voltage: 18 kV
Model: Waters Quanta 4000

OTHER SUBSTANCES
Simultaneous: 6-aminopenicillanic acid, amoxicillin, ampicillin, cloxacillin, dicloxacillin, nafcillin, oxacillin

REFERENCE
Swartz,M.E. Method development and selectivity control for small molecule pharmaceutical separations by capillary electrophoresis, *J.Liq.Chromatogr.*, **1991**, *14*, 923–938.

Ticlopidine

Molecular formula: $C_{14}H_{14}ClNS$
Molecular weight: 263.79
CAS Registry No.: 55142-85-3, 53885-35-1 (HCl)
Merck Index (12th ed.): 9569
Lednicer: 3 228

SAMPLE
Matrix: bulk
Sample preparation: Prepare a 1 mg/mL solution in running buffer, add norephedrine, filter (0.45 μm), inject an aliquot.

CAPILLARY ELECTROPHORESIS
Capillary: 36 cm × 50 μm uncoated fused-silica (Supelco SPE)
Capillary preparation: Before each injection wash with 100 mM NaOH for 2 min and with buffer for 3 min. At the beginning of each day wash capillary with 100 mM NaOH for 30 min.
Capillary temperature: 25
Running buffer: 60 mM pH 3 Phosphate buffer
Injection: Hydrodynamic injection for 2 s (7 nL)
Detector: UV 210, UV 210, UV 225
Migration time: 4.49
Internal standard: norephedrine (4.14)
Voltage: 20 kV
Current: 60 μA
Model: SpectraPHORESIS 1000 (Spectra Physics)

OTHER SUBSTANCES
Simultaneous: impurities

REFERENCE
Quaglia,M.G.; Farina,A.; Bossù,E.; Romolo,F. Analysis of ticlopidine and related impurities by capillary electrophoresis, *J.Pharm.Biomed.Anal.*, **1993**, *11*, 1157–1160.

Timepidium bromide

Molecular formula: $C_{17}H_{22}BrNOS_2$
Molecular weight: 400.40
CAS Registry No.: 35035-05-3
Merck Index (12th ed.): 9583

SAMPLE
Matrix: solutions
Sample preparation: Inject an aliquot of a 100 μg/mL solution in water or running buffer.

CAPILLARY ELECTROPHORESIS
Capillary: 47 cm × 75 μm fused-silica (40 cm to detector)
Capillary preparation: Before each run rinse capillary with running buffer for 1-2 min. If peak tailing is observed fill capillary with 500 mM NaOH and let stand for 30 min, wash with water for 3 min, rinse with running buffer for 3 min.
Capillary temperature: 25
Running buffer: 20 mM pH 2.5 Phosphate buffer containing 9% dextrin (Japan Pharmacopeia grade) (Dissolve dextrin in buffer at 90° then cool to room temperature.)
Injection: Pressure injection at 0.5 psi for 2-4 s.
Detector: UV 220
Migration time: 6.287, 6.440 (enantiomers)
Voltage: 30 kV
Model: Beckman P/ACE 5510

KEY WORDS
chiral

REFERENCE
Nishi,H.; Izumoto,S.; Nakamura,K.; Nakai,H.; Sato,T. Dextran and dextrin as chiral selectors in capillary zone electrophoresis, *Chromatographia*, **1996**, *42*, 617–630.

Timolol

Molecular formula: $C_{13}H_{24}N_4O_3S$
Molecular weight: 316.42
CAS Registry No.: 26839-75-8, 91524-16-2 (hemihydrate), 26921-17-5 maleate
Merck Index (12th ed.): 9585
Lednicer: 2 272

SAMPLE
Matrix: blood
Sample preparation: Hydrolyze 2 mL serum with β-glucuronidase (EC 3.2.1.31, type H-1 from Helix pomatia, 416 800 U/g, Separacor) at 80° for 30 min, cool, add 900 μL MeCN, vortex for 15 min, centrifuge at 2004 g for 10 min, add ephedrine (165 μg/mL), filter (0.5 μm), inject an aliquot of the filtrate.

CAPILLARY ELECTROPHORESIS
Capillary: 58 cm × 50 μm fused-silica (50 cm to detector) (Polymicro Technologies)

Capillary preparation: Before each injection purge capillary with 5% phosphoric acid for 12 s, with water for 30 s, and with running buffer for 10 min.
Capillary temperature: 35
Running buffer: 80 mM pH 6.7 sodium phosphate buffer containing 15 mM cetyltrimethylammonium bromide
Injection: Hydrostatic injection for 20 s
Detector: UV 214
Migration time: 12
Internal standard: ephedrine (17.5)
Voltage: -27 kV
Model: Waters Quanta 4000
Limit of detection: 50 µg/mL (S/N 3)

OTHER SUBSTANCES
Extracted: acebutolol, alprenolol, atenolol, labetalol, metoprolol, nadolol, oxprenolol, pindolol, propranolol

KEY WORDS
serum

REFERENCE
Lukkari,P.; Nyman,T.; Riekkola,M.-J. Determination of nine β-blockers in serum by micellar electrokinetic capillary chromatography, *J.Chromatogr.A*, **1994**, *674*, 241–246.

SAMPLE
Matrix: blood
Sample preparation: Filter (1 mL Amicon Centrifree Micropartition System) serum while centrifuging at 1500 g for 20 min, inject an aliquot of the ultrafiltrate.

CAPILLARY ELECTROPHORESIS
Capillary: 92 cm × 75 µm fused-silica (55 cm to detector) (Polymicro Technologies)
Capillary preparation: Rinse with running buffer for 10 min between run. At the beginning and end of each day rinse with 100 mM NaOH for 10 min and with water for 10 min.
Running buffer: Isopropanol:buffer 5:95 (Buffer was 10 mM Na_2HPO_4 containing 6 mM sodium borate and 75 mM sodium dodecyl sulfate.)
Injection: Injection via vacuum suction for 0.7 s
Detector: UV 215
Migration time: 20
Voltage: 25 kV
Current: 82 µA
Model: Europhor Prime Vision IV

OTHER SUBSTANCES
Extracted: atenolol, carteolol, celiprolol, levobunolol, metoprolol, penbutolol, pindolol, propranolol

KEY WORDS
serum; ultrafiltrate

REFERENCE
Schmutz,A.; Thormann,W. Assessment of impact of physico-chemical drug properties on monitoring drug levels by micellar electrokinetic capillary chromatography with direct serum injection, *Electrophoresis*, **1994**, *15*, 1295–1303.

SAMPLE
Matrix: blood, urine
Sample preparation: Urine. Dilute with 2 volumes of water, filter (0.5 µm), inject an aliquot. Serum. Condition a 3 mL Supelclean LC-18 SPE cartridge (Supelco) with MeOH and water. Hydrolyze 900 µL serum with β-glucuronidase (EC 3.2.1.31 type H-1 from Helix pomatia) at 60° with sonication for 30 min, add 500 µL (?) MeOH, centrifuge at 2000 g, add the supernatant to the SPE cartridge, wash with 1 mL water, dry under vacuum, elute with 2 mL MeOH:water 90:10, filter, inject an aliquot.

CAPILLARY ELECTROPHORESIS
Capillary: 68 cm × 50 μm fused-silica (60 cm to detector) (Polymicro Technologies)
Capillary preparation: Purge with running buffer for 2 min before injection.
Running buffer: pH 7.0 Phosphate buffer containing 10 mM N-cetyl-N,N,N-trimethylammonium bromide (Prepare buffer by mixing 100 mM NaH_2PO_4 and 100 mM Na_2HPO_4 to achieve a pH of 7.0.)
Injection: Hydrostatic injection for 30 s
Detector: UV 214
Migration time: 13
Internal standard: 2,6-dimethylphenol (only for urine) (14.5)
Voltage: -26 kV
Current: 97 μA
Model: Waters Quanta 4000

OTHER SUBSTANCES
Extracted: acebutolol, alprenolol, atenolol, labetalol, metoprolol, nadolol, oxprenolol, pindolol, propranolol

KEY WORDS
serum; comparison with HPLC; SPE

REFERENCE
Lukkari,P.; Sirén,H. Ion-pair chromatography and micellar electrokinetic capillary chromatography in analyzing β-adrenergic blocking agents from human biological fluids, *J.Chromatogr.A*, **1995**, *717*, 211–217.

SAMPLE
Matrix: solutions

CAPILLARY ELECTROPHORESIS
Capillary: 82 cm × 75 μm fused-silica (Waters)
Capillary preparation: Between runs purge capillary with running buffer for 2 min. Purge a new capillary with 100 mM NaOH for 30 min, with water for 30 min, and with running buffer for 30 min.
Capillary temperature: 30
Running buffer: 30 mM pH 7.6 phosphate buffer containing 10 mM cetyltrimethylammonium bromide
Injection: Inject at high pressure for 5 s
Detector: UV 214
Migration time: 6.5
Voltage: 21 kV
Model: Beckman P/ACE System 2000

OTHER SUBSTANCES
Simultaneous: acebutolol, alprenolol, atenolol, nadolol, oxprenolol, pindolol, propranolol, sotalol

REFERENCE
Lukkari,P.; Ennelin,A.; Sirén,H.; Riekkola,M.-L. Effect of temperature, effective capillary length, and applied voltage on the migration of nine β-blockers in micellar electrokinetic capillary chromatography, *J.Liq.Chromatogr.*, **1993**, *16*, 2069–2079.

SAMPLE
Matrix: solutions

CAPILLARY ELECTROPHORESIS
Capillary: 58 cm × 50 μm fused-silica (50 cm to detector)
Running buffer: Isopropanol:buffer 2.5:97.5 (Buffer was 80 mM pH 6.8 Phosphate buffer containing 15 mM cetyltrimethylammonium bromide.)
Injection: Hydrostatic injection for 15 s
Detector: UV 214
Migration time: 12
Voltage: -20 kV

Model: Waters Quanta 4000

OTHER SUBSTANCES
Simultaneous: acebutolol, alprenolol, atenolol, labetalol, nadolol, oxprenolol, pindolol, propranolol, sotalol

---

REFERENCE
Lukkari,P.; Vuorela,H.; Riekkola,M.-L. Effects of organic mobile phase modifiers on elution and separation of β-blockers in micellar electrokinetic capillary chromatography, *J.Chromatogr.A*, **1993**, *655*, 317–324.

SAMPLE
Matrix: solutions

CAPILLARY ELECTROPHORESIS
Capillary: 57 cm × 75 μm fused-silica (50 cm to detector) (Beckman)
Capillary preparation: Before each run rinse capillary with 100 mM NaOH and running buffer.
Capillary temperature: 30
Running buffer: Acetone:100 mM pH 8.1 borate buffer containing 50 mM sodium dodecyl sulfate 15:85 (A) or 100 mM phosphoric acid adjusted to pH 3.1 with triethanolamine (B)
Injection: Pressure injection for 5-10 s.
Detector: UV 200
Migration time: 12 (A), 11.3 (B)
Voltage: 25 kV
Model: Beckman P/ACE 5510

OTHER SUBSTANCES
Simultaneous: acebutolol, amiodarone, atenolol, bretylium, captopril, diltiazem, disopyramide, lidocaine, lisinopril, metoprolol, nicardipine, nifedipine, phenytoin, pindolol, propafenone, propranolol, quinidine, sotalol, p-toluenesulfonic acid, verapamil

---

KEY WORDS
only running buffer A separates all compounds; metoprolol interferes with running buffer A but not with running buffer B

REFERENCE
Bretnall,A.E.; Clarke,G.S. Selectivity of capillary electrophoresis for the analysis of cardiovascular drugs, *J.Chromatogr.A*, **1996**, *745*, 145–154.

SAMPLE
Matrix: solutions

CAPILLARY ELECTROPHORESIS
Capillary: 43 cm × 50 μm fused-silica (36 cm to detector)
Capillary preparation: Before each injection wash with running buffer for 10 min. After each injection wash with water for 2 min. Wash new capillaries with 1 M NaOH at 60° for 50 min, with NaOH solution (?) at 60° for 10 min, and with water at 25°.
Capillary temperature: 25
Running buffer: 320 mM pH 2.0 Citrate buffer
Injection: Hydrodynamic injection at 1.5 psi for 2 s.
Detector: UV 220
Migration time: 9.6
Voltage: 15 kV
Model: Spectra-Physics Model 1000
Limit of detection: 1-18 μg/mL

OTHER SUBSTANCES
Simultaneous: acebutolol, atenolol, labetalol, levobunolol, metoprolol, nadolol, oxprenolol, pindolol, propranolol

---

REFERENCE
Lin,C.-E.; Chang,C.-C.; Lin,W.-c.; Lin,E.C. Capillary zone electrophoretic separation of β-blockers using citrate buffer at low pH, *J.Chromatogr.A*, **1996**, *753*, 133–138.

SAMPLE
Matrix: solutions

CAPILLARY ELECTROPHORESIS
Capillary: 67 cm × 50 μm (60 cm to detector)
Capillary preparation: Wash with running buffer for 5 min before each injection. After each injection wash with 1 M NaOH at 60° for 5 min, with 100 mM NaOH at 60° for 10 min, and with water at 25° for 5 min.
Capillary temperature: 25
Running buffer: 70 mM pH 7.0 Sodium phosphate buffer containing 15 mM cetyltriemthylammonium bromide
Injection: Hydrodynamic injection for 1 s.
Detector: UV 220
Migration time: 14.5
Voltage: 20 kV

OTHER SUBSTANCES
Simultaneous: acebutolol, atenolol, labetalol, levobunulol, metoprolol, nadolol, oxprenolol, pindolol, propranolol

REFERENCE
Lin,C.-E.; Chen,Y.-C.; Chang,C.-C.; Wang,D.-Z. Migration behavior and selectivity of β-blockers in micellar electrokinetic chromatography. Influence of micelle concentration of cationic surfactants, *J.Chromatogr.A*, **1997**, *775*, 349–357.

SAMPLE
Matrix: solutions
Sample preparation: Inject an aliquot of a 340 μM solution.

CAPILLARY ELECTROPHORESIS
Capillary: 58 cm × 75 μm fused-silica (50 cm to detector)
Capillary preparation: Rinse with running buffer for 2 min before each run. Before use rinse capillary at 1.36 bar with 100 mM NaOH for 5 min and with water for 10 min.
Capillary temperature: 25
Running buffer: 20 mM pH 7.0 Sodium phosphate buffer (A) or 20 mM pH 7.0 Sodium phosphate buffer containing 10 mM 1-lauroyl-2-hydroxy-sn-glycero-3-phosphocholine (Avanti Polar Lipids, Alabaster AL) (B)
Injection: Hydrodynamic injection at 34 mbar for 1 s followed by 5% MeOH for 5 s.
Detector: UV 200
Migration time: 4.14 min (A); k' 0.11 (B)
Voltage: 15 kV
Current: ca. 50 μA
Model: Beckman P/ACE 2200

OTHER SUBSTANCES
Simultaneous: acebutolol, alprenolol, atenolol, metoprolol, oxprenolol, pindolol, propranolol

REFERENCE
Masucci,J.A.; Caldwell,G.W.; Foley,J.P. Comparison of the retention behavior of β-blockers using immobilized artificial membrane chromatography and lysophospholipid micellar electrokinetic chromatography, *J.Chromatogr.A*, **1998**, *810*, 95–103.

SAMPLE
Matrix: urine
Sample preparation: Dilute urine with 2 volumes of water, add IS, filter (0.5 μm), inject an aliquot.

CAPILLARY ELECTROPHORESIS
Capillary: 68 cm × 50 μm fused-silica (60 cm to detector) (White Associates)
Capillary preparation: Purge capillary with running buffer for 2 min before each injection.

Running buffer: 80 mM pH 7.0 Phosphate buffer containing 10 mM N-cetyl-N,N,N-trimethyl-ammonium bromide
Injection: Hydrostatic injection for 30 s at 10 cm.
Detector: UV 214
Migration time: 13.2
Internal standard: 2,6-dimethylphenol (14.7)
Voltage: -26 kV (at injector end)
Model: Waters Quanta 4000
Limit of detection: 20 μg/mL (S/N 3)

OTHER SUBSTANCES
Extracted: acebutolol, alprenolol, atenolol, labetalol, metoprolol, nadolol, oxprenolol, pindolol, propranolol
Noninterfering: caffeine
Interfering: probenecid

REFERENCE
Lukkari,P.; Sirén,H.; Pantsar,M.; Riekkola,M.-L. Determination of ten β-blockers in urine by micellar electrokinetic capillary chromatography, *J.Chromatogr.*, **1993**, *632*, 143–148.

Tioconazole

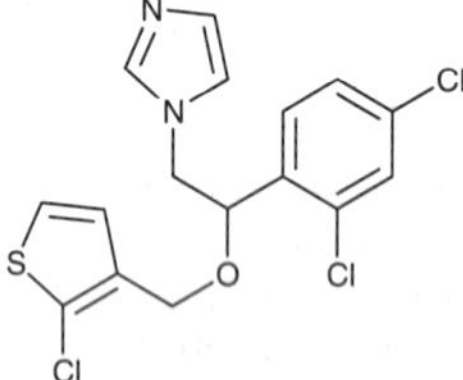

Molecular formula: $C_{16}H_{13}Cl_3N_2OS$
Molecular weight: 387.70
CAS Registry No.: 65899-73-2
Merck Index (12th ed.): 9595

SAMPLE
Matrix: solutions
Sample preparation: Prepare a 1 mM solution in running buffer.

CAPILLARY ELECTROPHORESIS
Capillary: 57 cm × 50 μm fused-silica (50 cm to detector)
Capillary temperature: 25
Running buffer: MeOH:buffer 25:75 containing 10 mM hydroxypropyl-β-cyclodextrin (degree of substitution 0.6) (Buffer was 40 mM Na_2HPO_4 and 20 mM citric acid, pH 4.3.)
Injection: Pressure for 1 s (1 nL)
Detector: UV 230
Migration time: 12.9 (-), 13.2 (+)
Model: Beckman P/ACE 2100

KEY WORDS
chiral

REFERENCE
Penn,S.G.; Goodall,D.M.; Loran,J.S. Differential binding of tioconazole enantiomers to hydroxypropyl-β-cyclodextrin studied by capillary electrophoresis, *J.Chromatogr.*, **1993**, *636*, 149–152.

SAMPLE
Matrix: solutions
Sample preparation: Prepare a 100 μM solution in the running buffer containing mesityl oxide as a neutral marker, inject an aliquot.

CAPILLARY ELECTROPHORESIS
Capillary: 57 cm × 50 μm (50 cm to detector) (Polymicro Technologies)
Capillary temperature: 25.0
Running buffer: Cyclohexanol:buffer 0.1:99.9 (Buffer was 20 mM pH 4.3 phosphate-citrate containing 5 mM β-cyclodextrin

Injection: Pressure injection for 1 s (1 nL)
Detector: UV 230
Migration time: 4.5, 5 (enantiomers)
Voltage: 30 kV
Model: Beckman PACE 2100

KEY WORDS
chiral

REFERENCE
Penn,S.G.; Bergström,E.T.; Goodall,D.M.; Loran,J.S. Capillary electrophoresis with chiral selectors: optimization of separation and determination of thermodynamic parameters for binding of tioconazole enantiomers to cyclodextrins, *Anal.Chem.*, **1994**, *66*, 2866–2873.

SAMPLE
Matrix: solutions
Sample preparation: Inject an aliquot of a 100 µg/mL solution in running buffer.

CAPILLARY ELECTROPHORESIS
Capillary: 32 cm × 50 µm fused-silica (27.5 cm to detector) (Yongnian Optical Conductive Fiber Plant, China), coated with polyacrylamide
Capillary preparation: No details of the polyacrylamide coating process are provided. However, another paper (LC.GC 1997, 15, 40) by this group indicates that they use the procedure of Hjertén, thus: Adjust the pH of 20 mL water to 3.5 with acetic acid, add 80 µL 3-(trimethoxysilyl)propyl methacrylate (3-methacryloxypropyltrimethoxysilane), mix, suck into capillary, let stand at room temperature for 1 h, remove the solution, wash with water. Fill the capillary with a deaerated 3-4% acrylamide solution containing 1 µL/mL N,N,N',N'-tetramethylethylenediamine and 1 mg/mL potassium persulfate, let stand for 30 min, remove excess solution by aspiration, rinse with water, remove water by aspiration, dry at 35° (J. Chromatogr. 1985, 347, 191).
Capillary temperature: 25
Running buffer: 100 mM NaH_2PO_4 adjusted to pH 2.5 (A) or 100 mM NaH_2PO_4 containing 45 mM hydroxypropyl gamma-cyclodextrin, adjusted to pH 2.5 (B)
Injection: Electrokinetic injection at 15 kV for 3 s.
Detector: UV 200, UV 210
Migration time: 5.48 (A), 18.81, 19.04 (B, enantiomers)
Voltage: 15 kV
Model: Bio-Rad BioFocus 3000

OTHER SUBSTANCES
Simultaneous: atropine, carazolol, cicletanine, dimethindene, fendiline, homatropine, ipratropium bromide, isothipendyl, mefloquine, metaclazepam, nafronyl (naftidrofuryl), nefopam, nicardipine, promethazine, reproterol, tetrahydrozoline (tetryzoline), theodrenaline, trihexyphenidyl, trimeprazine (alimemazine), trimipramine

KEY WORDS
coated capillary; chiral

REFERENCE
Koppenhoefer,B.; Epperlein,U.; Xiaofeng,Z.; Bingcheng,L. Separation of enantiomers of drugs by capillary electrophoresis. Part 4: Hydroxypropyl-γ-cyclodextrin as chiral solvating agent, *Electrophoresis*, **1997**, *18*, 924–930.

SAMPLE
Matrix: solutions
Sample preparation: Inject an aliquot of a 100 µg/mL solution in running buffer.

CAPILLARY ELECTROPHORESIS
Capillary: 30 cm × 50 µm fused-silica (25.5 cm to detector), coated with polyacrylamide
Capillary preparation: Adjust the pH of 20 mL water to 3.5 with acetic acid, add 80 µL 3-(trimethoxysilyl)propyl methacrylate (3-methacryloxypropyltrimethoxysilane), mix, suck into capillary, let stand at room temperature for 1 h, remove the solution, wash with water. Fill the

capillary with a deaerated 3-4% acrylamide solution containing 1 μL/mL N,N,N',N'-tetra-methylethylenediamine and 1 mg/mL potassium persulfate, let stand for 30 min, remove excess solution by aspiration, rinse with water, remove water by aspiration, dry at 35° (J. Chromatogr. 1985, 347, 191).
Capillary temperature: 25
Running buffer: 100 mM NaH_2PO_4 containing 45 mM hydroxypropyl-gamma- cyclodextrin, adjusted to pH 2.5 with phosphoric acid
Injection: Electrokinetic injection at 15 kV for 3 s.
Detector: UV 200
Voltage: 15 kV
Model: Bio-Rad BioFocus 3000

KEY WORDS
chiral; coated capillary; comparison with the use of other cyclodextrins; this running buffer gave the greatest enantiomeric separation.; α=1.015

REFERENCE
Lin,B.; Zhu,X.; Koppenhoefer,B.; Epperlein,U. Investigation of 123 chiral drugs by cyclodextrin-modified capillary electrophoresis, *LC.GC*, **1997**, *15*, 40–46.

SAMPLE
Matrix: solutions

CAPILLARY ELECTROPHORESIS
Capillary: 29-36 cm × 50 μm fused-silica (24.5-31.5 cm to detector) (Yongnian Optical Conductive Fiber Plant, China) coated with polyacrylamide
Capillary preparation: Coat capillary as follows. Adjust the pH of 20 mL water to 3.5 with acetic acid, add 80 μL 3-(trimethoxysilyl)propyl methacrylate (3-methacryloxypropyltrimethoxysilane), mix, suck into capillary, let stand at room temperature for 1 h, remove the solution, wash with water. Fill the capillary with a deaerated 3-4% acrylamide solution containing 1 μL/mL N,N,N',N'-tetramethylethylenediamine and 1 mg/mL potassium persulfate, let stand for 30 min, remove excess solution by aspiration, rinse with water, remove water by aspiration, dry at 35° (J. Chromatogr. 1985, 347, 191).
Capillary temperature: 25
Running buffer: 100 mM pH 2.5 NaH_2PO_4 (A) or 100 mM pH 2.5 NaH_2PO_4 containing 45 mM hydroxypropyl-α-cyclodextrin (Wacker, Munich) (B)
Injection: Electromigration at 15 kV for 3 s.
Detector: UV 200; UV 210
Migration time: 5.48 (A); 15.18 (B) (no separation of enantiomers)
Voltage: 15 kV
Model: Bio-Focus 3000

OTHER SUBSTANCES
Also analyzed: albuterol (salbutamol), alprenolol, amorolfine, atenolol, atropine, azelastine, baclofen, bamethan, benproperine, benserazide, biperiden, bisoprolol, brompheniramine, bupivacaine, bupranolol, butamirate, butethamate, carazolol, carbuterol, carteolol, carvedilol, celiprolol, chloroquine, chlorpheniramine, chlorphenoxamine, cicletanine, clenbuterol, clidinium bromide, clobutinol, dimethindene, dipivefrin, disopyramide, dobutamine, doxylamine, fendiline, flecainide, gallopamil, homatropine, ipratropium bromide, isoproterenol (isoprenaline), isothipendyl, ketamine, meclizine, mefloquine, mepindolol, mequitazine, metaclazepam, metaproterenol (orciprenaline), metipranolol, metoprolol, nafronyl (naftidrofuryl), nefopam, nicardipine, norfenefrine, ofloxacin, ornidazole, orphenadrine, oxomemazine, oxprenolol, oxybutynin, phenoxybenzamine, phenylpropanolamine, pholedrine, pindolol, pirbuterol, prilocaine, procyclidine, promethazine, propafenone, propranolol, reproterol, sotalol, sulpride, synephrine, talinolol, terbutaline, tetrahydrozoline (tetryzoline), theodrenaline, tocainide, trihexyphenidyl, trimeprazine (alimemazine), trimipramine, tropicamide, verapamil, zopiclone

---

KEY WORDS
coated capillary

REFERENCE
Koppenhoefer,B.; Eperlein,U.; Schlunk,R.; Zhu,X.; Lin,B. Separation of enantiomers of drugs by capillary electrophoresis. V. Hydroxypropyl-α-cyclodextrin as chiral solvating agent, *J.Chromatogr.A*, **1998**, *793*, 153–164.

Tiopronin

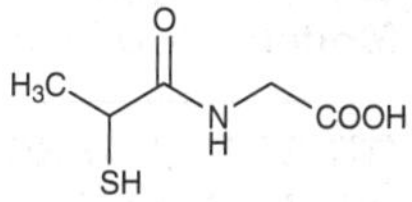

Molecular formula: $C_5H_9NO_3S$
Molecular weight: 163.20
CAS Registry No.: 1953-02-2
Merck Index (12th ed.): 9597

SAMPLE
Matrix: solutions
Sample preparation: Mix a 100 μL aliquot of a 0.1-1 mM solution in 10 mM HCl with 500 μL 100 mM pH 9.5 borate buffer, 100 μL 2.5 mM 3-hydroxy-L-tyrosine in 10 mM HCl, and 200 μL 5 mM o-phthalaldehyde in EtOH, let stand for 10 min, inject an aliquot.

CAPILLARY ELECTROPHORESIS
Capillary: 70 cm × 75 μm fused-silica
Capillary preparation: Before use wash capillary at 60° for 6 min each with water, 1 M NaOH, 100 mM NaOH, water, and running buffer.
Capillary temperature: 25
Running buffer: 100 mM pH 9.5 Borate buffer containing 10 mM β-cyclodextrin
Injection: Hydrodynamic injection for 2 s.
Detector: UV 335
Migration time: μ_a 15.3 (α = 1.23, R_s = 2.91)
Voltage: 20 kV
Model: Spectra Phoresis 1000 (Thermo Separation Products)

KEY WORDS
derivatization; chiral; comparison with HPLC; comparison with other derivatizing reagents

REFERENCE
Leroy,P.; Bellucci,L.; Nicolas,A. Chiral derivatization for separation of racemic amino and thiol drugs by liquid chromatography and capillary electrophoresis, *Chirality*, **1995**, *7*, 235–242.

Tobramycin

Molecular formula: $C_{18}H_{37}N_5O_9$
Molecular weight: 467.52
CAS Registry No.: 32986-56-4, 79645-27-5 (sulfate)
Merck Index (12th ed.): 9628

SAMPLE
Matrix: bulk
Sample preparation: Inject an aliquot of a solution in 140 mM sodium tetraborate containing 480 μg/mL sisomicin as IS.

CAPILLARY ELECTROPHORESIS
Capillary: 70 cm × 50 μm uncoated fused silica (45 cm to detector) (Polymicro Technologies or ISCO)
Capillary preparation: Rinse with running buffer for 15 min at the beginning of the day and for 4 min before each injection.
Capillary temperature: 34
Running buffer: 185 mM pH 9 sodium tetraborate
Injection: Vacuum injection at 25.0 kPa
Detector: UV 195
Migration time: 23.8
Internal standard: sisomicin (19.90)

Voltage: +15 kV
Model: ISCO Model 3140 Electropherograph

OTHER SUBSTANCES
Simultaneous: amikacin, bekanamycin, butirosin, dibekacin, dihydrostreptomycin, gentamicins, kanamycin A, paromomycin, ribostamycin, streptomycin
Noninterfering: neomycin

KEY WORDS
detector is at cathode

REFERENCE
Flurer,C.L. The analysis of aminoglycoside antibiotics by capillary electrophoresis, *J.Pharm.Biomed.Anal.*, **1995**, *13*, 809–816.

SAMPLE
Matrix: solutions
Sample preparation: Prepare a solution in 100 mM cetyltrimethylammonium bromide, inject an aliquot.

CAPILLARY ELECTROPHORESIS
Capillary: 67 cm × 50 μm fused silica capillary (60 cm to detector) (Siemens)
Running buffer: pH 5.0 Acetate buffer containing 10 mM imidazole and 50 μg/mL FC 135 (a fluorinated surfactant, Fluorad/3M)
Injection: Pressure injection for 2 s.
Detector: UV 214
Migration time: 12.5
Voltage: 12.5 kV
Model: Beckman P/ACE System 2000

OTHER SUBSTANCES
Simultaneous: amikacin, dihydrostreptomycin, gentamicin, kanamycin, lividomycin, paromomycin, sisomycin, streptomycin,
Interfering: neomycin

KEY WORDS
indirect UV detection; detector is at anode

REFERENCE
Ackermans,M.T.; Everaerts,F.M.; Beckers,J.L. Determination of aminoglycoside antibiotics in pharmaceuticals by capillary zone electrophoresis with indirect UV detection coupled with micellar electrokinetic capillary chromatography, *J.Chromatogr.*, **1992**, *606*, 229–235.

Tocainide

Molecular formula: $C_{11}H_{16}N_2O$
Molecular weight: 192.26
CAS Registry No.: 41708-72-9
Merck Index (12th ed.): 9629
Lednicer: 3 55

SAMPLE
Matrix: solutions
Sample preparation: Inject an aliquot of a 100 μg/mL solution in water:running buffer 50:50.

CAPILLARY ELECTROPHORESIS
Capillary: 44.5 cm × 50 μm acrylamide-coated fused-silica (Bio-Rad)

Capillary temperature: 30
Running buffer: 100 mM NaH$_2$PO$_4$ containing 15 mM gamma-cyclodextrin, adjusted to pH 2.5
with phosphoric acid
Injection: Electrokinetic injection at 8 kV for 6 s.
Detector: UV 200
Migration time: 8.53
Voltage: 14 kV
Model: Bio-Rad BioFocus 3000

OTHER SUBSTANCES
Also analyzed: albuterol, alprenolol, atenolol, atropine, baclofen, bamethan, benserazide, biperiden, bisoprolol, bupivacaine, bupranolol, butetamate, carazolol, carbuterol, carvedilol, celiprolol, chloroquine, chlorpheniramine (chlorphenamine), clidinium bromide, clobutinol, disopyramide, dobutamine, flecainide, homatropine, ipratropium bromide, isoproterenol, isothipendyl, ketamine, mefloquine, mequitazine, metaproterenol (orciprenaline), metipranolol, nafronyl (naftidrofuryl), nefopam, ofloxacin, orphenadrine, oxomemazine, oxprenolol, phenoxybenzamine, pholedrine, pindolol, pirbuterol, prilocaine, promethazine, propafenone, propranolol, sotalol, synephrine, tetrahydrozoline (tetryzoline), terbutaline, trihexyphenidyl, trimeprazine (alimemazine), trimipramine, tropicamide, verapamil, zopiclone

KEY WORDS
coated capillary; achiral

REFERENCE
Koppenhoefer,B.; Epperlein,U.; Christian,B.; Yibing,J.; Yuying,C.; Bingcheng,L. Separation of enantiomers of drugs by capillary electrophoresis. I. γ-Cyclodextrin as chiral solvating agent, *J.Chromatogr.A*, **1995**, *717*, 181–190.

SAMPLE
Matrix: solutions
Sample preparation: Inject an aliquot of a 100 μg/mL solution in running buffer.

CAPILLARY ELECTROPHORESIS
Capillary: 29 cm × 50 μm fused-silica (24.5 cm to detector) (Yongnian Optical Conductive Fiber Plant, China), coated with polyacrylamide
Capillary preparation: No details of the polyacrylamide coating process are provided. However, another paper (LC.GC 1997, 15, 40) by this group indicates that they use the procedure of Hjertén, thus: Adjust the pH of 20 mL water to 3.5 with acetic acid, add 80 μL 3-(trimethoxysilyl)propyl methacrylate (3-methacryloxypropyltrimethoxysilane), mix, suck into capillary, let stand at room temperature for 1 h, remove the solution, wash with water. Fill the capillary with a deaerated 3-4% acrylamide solution containing 1 μL/mL N,N,N',N'-tetramethylethylenediamine and 1 mg/mL potassium persulfate, let stand for 30 min, remove excess solution by aspiration, rinse with water, remove water by aspiration, dry at 35° (J. Chromatogr. 1985, 347, 191).
Capillary temperature: 25
Running buffer: 100 mM NaH$_2$PO$_4$ adjusted to pH 2.5
Injection: Electrokinetic injection at 15 kV for 3 s.
Detector: UV 200, UV 210
Migration time: 4.50
Voltage: 15 kV
Model: Bio-Rad BioFocus 3000

OTHER SUBSTANCES
Simultaneous: albuterol, alprenolol, atenolol, baclofen, bamethan, benproperine, benserazide, bisoprolol, bupranolol, butamirate, butethamate, carbuterol, celiprolol, clenbuterol, clobutinol, dipivefrin, isoproterenol (isoprenaline), metaproterenol (orciprenaline), metipranolol, metoprolol, norfenefrine, ornidazole, oxprenolol, phenylpropanolamine, pholedrine, pirbuterol, prilocaine, procyclidine, sotalol, synephrine, terbutaline

KEY WORDS
coated capillary

REFERENCE
Koppenhoefer,B.; Epperlein,U.; Xiaofeng,Z.; Bingcheng,L. Separation of enantiomers of drugs by capillary electrophoresis. Part 4: Hydroxypropyl-γ-cyclodextrin as chiral solvating agent, *Electrophoresis*, **1997**, *18*, 924–930.

SAMPLE
Matrix: solutions

CAPILLARY ELECTROPHORESIS
Capillary: 29-36 cm × 50 μm fused-silica (24.5-31.5 cm to detector) (Yongnian Optical Conductive Fiber Plant, China) coated with polyacrylamide
Capillary preparation: Coat capillary as follows. Adjust the pH of 20 mL water to 3.5 with acetic acid, add 80 μL 3-(trimethoxysilyl)propyl methacrylate (3-methacryloxypropyltrimethoxysilane), mix, suck into capillary, let stand at room temperature for 1 h, remove the solution, wash with water. Fill the capillary with a deaerated 3-4% acrylamide solution containing 1 μL/mL N,N,N',N'-tetramethylethylenediamine and 1 mg/mL potassium persulfate, let stand for 30 min, remove excess solution by aspiration, rinse with water, remove water by aspiration, dry at 35° (J. Chromatogr. 1985, 347, 191).
Capillary temperature: 25
Running buffer: 100 mM pH 2.5 NaH_2PO_4 (A) or 100 mM pH 2.5 NaH_2PO_4 containing 45 mM hydroxypropyl-α-cyclodextrin (Wacker, Munich) (B)
Injection: Electromigration at 15 kV for 3 s.
Detector: UV 200; UV 210
Migration time: 4.50 (A); 5.99 (B) (no separation of enantiomers)
Voltage: 15 kV
Model: Bio-Focus 3000

OTHER SUBSTANCES
Also analyzed: albuterol (salbutamol), alprenolol, amorolfine, atenolol, atropine, azelastine, baclofen, bamethan, benproperine, benserazide, biperiden, bisoprolol, brompheniramine, bupivacaine, bupranolol, butamirate, butethamate, carazolol, carbuterol, carteolol, carvedilol, celiprolol, chloroquine, chlorpheniramine, chlorphenoxamine, cicletanine, clenbuterol, clidinium bromide, clobutinol, dimethindene, dipivefrin, disopyramide, dobutamine, doxylamine, fendiline, flecainide, gallopamil, homatropine, ipratropium bromide, isoproterenol (isoprenaline), isothipendyl, ketamine, meclizine, mefloquine, mepindolol, mequitazine, metaclazepam, metaproterenol (orciprenaline), metipranolol, metoprolol, nafronyl (naftidrofuryl), nefopam, nicardipine, norfenefrine, ofloxacin, ornidazole, orphenadrine, oxomemazine, oxprenolol, oxybutynin, phenoxybenzamine, phenylpropanolamine, pholedrine, pindolol, pirbuterol, prilocaine, procyclidine, promethazine, propafenone, propranolol, reproterol, sotalol, sulpride, synephrine, talinolol, terbutaline, tetrahydrozoline (tetryzoline), theodrenaline, tioconazole, trihexyphenidyl, trimeprazine (alimemazine), trimipramine, tropicamide, verapamil, zopiclone

KEY WORDS
coated capillary

REFERENCE
Koppenhoefer,B.; Eperlein,U.; Schlunk,R.; Zhu,X.; Lin,B. Separation of enantiomers of drugs by capillary electrophoresis. V. Hydroxypropyl-α-cyclodextrin as chiral solvating agent, *J.Chromatogr.A*, **1998**, *793*, 153–164.

Tolazamide

Molecular formula: $C_{14}H_{21}N_3O_3S$
Molecular weight: 311.41
CAS Registry No.: 1156-19-0
Merck Index (12th ed.): 9644
Lednicer: 1 241

SAMPLE
Matrix: solutions
Sample preparation: Inject a dilute aqueous solution on to the tip of the capillary containing the SPE material at 138 kPa for 12 s, wash for 1 min with running buffer, desorb with elution buffer for 20 s (60 nL), inject running buffer at 3.45 kPa, electrophorese. (Elution buffer was MeCN:50 mM pH 2.5 phosphate buffer 80:20.)

CAPILLARY ELECTROPHORESIS
Capillary: 67 cm × 75 μm with a 0.5 mm long × 0.37 mm diameter column of C18 material (from a J.T. Baker SPE cartridge) at the injection end
Capillary preparation: Before each run rinse capillary with 3 column volumes of elution buffer and 10 column volumes of running buffer. After each run rinse with 500 μL elution buffer.
Capillary temperature: 20
Running buffer: 250 mM pH 8.4 Borate buffer containing 5 mM phosphate
Detector: UV 200
Migration time: 18
Voltage: 15 kV
Current: 70 μA
Model: Beckman P/ACE System 5510
Limit of detection: 5 ng/mL

OTHER SUBSTANCES
Simultaneous: acetohexamide, chlorpropamide, glipizide, glyburide, tolbutamide

KEY WORDS
SPE

REFERENCE
Strausbauch,M.A.; Xu,S.J.; Ferguson,J.E.; Nunez,M.E.; Machacek,D.; Lawson,G.M.; Wettstein,P.J.; Landers,J.P. Concentration and separation of hypoglycemic drugs using solid-phase extraction-capillary electrophoresis, *J.Chromatogr.A*, **1995**, *717*, 279–291.

SAMPLE
Matrix: solutions
Sample preparation: Inject an aliquot of a solution in MeOH.

CAPILLARY ELECTROPHORESIS
Capillary: 47 cm × 50 μm fused-silica (40 cm to detector)
Capillary preparation: Before each run rinse at 20 psi with 3 column volumes of 100 mM NaOH and 3 column volumes of running buffer. Condition new capillaries by rinsing with 20 column volumes of 1 M NaOH, water, and running buffer
Capillary temperature: 22
Running buffer: 5 mM pH 8.5 Borate buffer containing 5 mM phosphate and 75 mM sodium cholate (Prepare from 500 mM borate buffer (prepared by mixing 125 mM sodium tetraborate and 500 mM boric acid to achieve pH 8.5) and 500 mM Na$_2$HPO$_4$ solution with appropriate dilution and the addition of sodium cholate.)
Injection: Pressure injection of sample at 0.5 psi for 1-5 s followed by a pressure injection of running buffer for 1 s.
Detector: UV 200
Migration time: 3.8
Internal standard: N-acetyl-4-(2,3-dichlorophenylureido)benzenesulfonamide (5.5)
Voltage: 25 kV
Current: 61 μA
Model: Beckman P/ACE 5510

OTHER SUBSTANCES
Simultaneous: acetohexamide, chlorpropamide, glipizide, glyburide, tolbutamide

REFERENCE
Roche,M.E.; Oda,R.P.; Lawson,G.M.; Landers,J.P. Capillary electrophoretic detection of metabolites in the urine of patients receiving hypoglycemic drug therapy, *Electrophoresis*, **1997**, *18*, 1865–1874.

SAMPLE
Matrix: urine
Sample preparation: Condition a 200 μg Bond Elut C18 SPE cartridge with 1 mL MeCN. 10 mL Urine + 100 μL 10 μg/mL IS, adjust pH to 2.0 with concentrated HCl, add 15 mL dichloromethane, shake vigorously for 2 min, let stand for 5 min. Remove the organic layer and evaporate it to dryness under a stream of nitrogen at 45°, reconstitute the residue in 400 μL MeOH:water 50:50, add to the SPE cartridge, wash with 1.25 mL MeCN:water 10:90, wash with 1 mL MeCN:water 40:60, elute with 1 mL MeCN. Evaporate the eluate to dryness under a stream of nitrogen, reconstitute the residue in 20 μL MeOH, inject an aliquot.

CAPILLARY ELECTROPHORESIS
Capillary: 47 cm × 50 μm fused-silica (40 cm to detector) (Polymicro Technologies)
Capillary preparation: Before each run wash with 3 column volumes of 100 mM NaOH, dip momentarily into water, rinse with 10 column volumes of 50 mM pH 8.5 Sodium borate buffer containing 50 mM Na_2HPO_4 and 75 mM sodium cholate, rinse with 3 column volumes of running buffer. Condition a new capillary by rinsing with 20 column volumes of 100 mM NaOH, 20 column volumes of water, and 20 column volumes of running buffer.
Capillary temperature: 25
Running buffer: 5 mM pH 8.5 Sodium borate buffer containing 5 mM Na_2HPO_4 and 75 mM sodium cholate
Injection: Pressure injection at 0.5 psi for 2 s then a 1 s injection of running buffer.
Detector: UV 200
Migration time: 5.1
Internal standard: N-acetyl-4-(2,3-dichlorophenylureido)benzenesulfonamide (7.4)
Voltage: 25 kV
Current: 61 μA
Model: Beckman P/ACE System 2100 or System 5510
Limit of detection: 50 ng/mL

OTHER SUBSTANCES
Extracted: acetohexamide, chlorpropamide, glipizide, glyburide, tolbutamide

KEY WORDS
SPE

REFERENCE
Núñez,M.; Ferguson,J.E.; Machacek,D.; Jacob,G.; Oda,R.P.; Lawson,G.M.; Landers,J.P. Detection of hypoglycemic drugs in human urine using micellar electrokinetic chromatography, *Anal.Chem.*, **1995**, *67*, 3668–3675.

SAMPLE
Matrix: urine
Sample preparation: Inject an aliquot directly.

CAPILLARY ELECTROPHORESIS
Capillary: 37 cm × 50 μm fused-silica (30 cm to detector) (Polymicro Technologies)
Capillary preparation: Every 13 injections rinse capillary with 100 mM NaOH for 1 min and with running buffer for 1 min. Condition a new capillary with 20 column volumes of 100 mM NaOH, 20 column volumes of water, and 20 column volumes of running buffer.
Capillary temperature: 22
Running buffer: 5 mM pH 8.5 Borate buffer containing 5 mM phosphate and 75 mM sodium cholate
Injection: Pressure injection at 0.5 psi for 2 s.
Detector: UV 200
Migration time: 2.25
Internal standard: N-acetyl-4-(2,3-dichlorophenylureido)benzenesulfonamide (3.2)
Voltage: 28 kV
Current: 71 μA
Model: Beckman P/ACE 5510

OTHER SUBSTANCES
Extracted: acetohexamide, chlorpropamide, glipizide, glyburide, tolbutamide

KEY WORDS
direct injection

REFERENCE
Roche,M.E.; Oda,R.P.; Machacek,D.; Lawson,G.M.; Landers,J.P. Enhanced throughput with capillary electro-phoresis via continuous-sequential sample injection, *Anal.Chem.*, **1997**, *69*, 99–104.

Tolbutamide

Molecular formula: $C_{12}H_{18}N_2O_3S$
Molecular weight: 270.35
CAS Registry No.: 64-77-7, 473-41-6 (Na salt)
Merck Index (12th ed.): 9646
Lednicer: 1 136

SAMPLE
Matrix: blood
Sample preparation: 400 μL Plasma + 600 μL 9 g/L NaCl, mix, inject an aliquot.

CAPILLARY ELECTROPHORESIS
Capillary: 42.5 cm × 50 μm fused-silica (34.5 cm) (Polymicro Technologies)
Capillary preparation: Between runs rinse with MeCN:running buffer 50:50 for 15 min and with running buffer for 20 min. Condition new capillaries with 100 mM NaOH at 50° for 30 min and equilibrate with cathode buffer at 25° at 10 kV for 2 h. (Cathode buffer was 60 mM pH 10.0 sodium borate buffer.)
Capillary temperature: 25
Running buffer: 60 mM pH 10.0 Sodium borate buffer containing 200 mM sodium dodecyl sulfate
Injection: Vacuum injection at 0.75 psi for 3 s.
Detector: UV 220
Migration time: 13.7
Voltage: 10 kV
Current: 42-44 μA
Model: SpectraPhoresis 1000 (Thermo Separation Products)

OTHER SUBSTANCES
Extracted: salicylic acid, sulfamethoxazole

KEY WORDS
plasma

REFERENCE
Kunkel,A.; Günter,S.; Wätzig,H. Quantitation of acetaminophen and salicylic acid in plasma using capillary electrophoresis without sample pretreatment. Improvement of precision, *J.Chromatogr.A*, **1997**, *768*, 125–133.

SAMPLE
Matrix: blood
Sample preparation: Filter (0.2 μm) plasma, inject an aliquot of the filtrate. (Note that 400 μL plasma was spiked with 600 μL of the drug solution in saline.)

CAPILLARY ELECTROPHORESIS
Capillary: 42.5 cm × 50 μm fused-silica (34.5 cm to detector) (Polymicro Technologies)
Capillary preparation: After each run rinse capillary at 25° with MeCN:running buffer 50:50 for 15 min then re-equilibrate with running buffer for 20 min. Condition new capillaries with 100 mM NaOH at 50° for 30 min then equilibrate with cathode buffer at 25° for 2 h with voltage applied.

Capillary temperature: 25
Running buffer: 60 mM pH 10.0 Sodium borate buffer containing 200 mM sodium dodecyl sulfate (Cathode buffer was 60 mM pH 10.0 sodium borate buffer prepared by dissolving 3.170 g boric acid in 100 mL water and adjusting the pH to 10.0 with freshly-prepared 1 M NaOH.)
Injection: Vacuum injection at 0.75 psi for 3 s.
Detector: UV 200
Migration time: 14
Voltage: 10 kV
Current: 42 μA
Model: Spectraphoresis 1000 (Thermo Separation Products)

OTHER SUBSTANCES
Extracted: salicylic acid, sulfamethoxazole

KEY WORDS
plasma

REFERENCE
Kunkel,A.; Gunter,S.; Watzig,H. Determination of pharmaceuticals in plasma by capillary electrophoresis without sample pretreatment reproducibility, limit of quantitation and limit of detection, *Electrophoresis*, **1997**, *18*, 1882–1889.

SAMPLE
Matrix: solutions
Sample preparation: Inject a dilute aqueous solution on to the tip of the capillary containing the SPE material at 138 kPa for 12 s, wash for 1 min with running buffer, desorb with elution buffer for 20 s (60 nL), inject running buffer at 3.45 kPa, electrophorese. (Elution buffer was MeCN:50 mM pH 2.5 phosphate buffer 80:20.)

CAPILLARY ELECTROPHORESIS
Capillary: 67 cm × 75 μm with a 0.5 mm long × 0.37 mm diameter column of C18 material (from a J.T. Baker SPE cartridge) at the injection end
Capillary preparation: Before each run rinse capillary with 3 column volumes of elution buffer and 10 column volumes of running buffer. After each run rinse with 500 μL elution buffer.
Capillary temperature: 20
Running buffer: 250 mM pH 8.4 Borate buffer containing 5 mM phosphate
Detector: UV 200
Migration time: 18.3
Voltage: 15 kV
Current: 70 μA
Model: Beckman P/ACE System 5510
Limit of detection: 5 ng/mL

OTHER SUBSTANCES
Simultaneous: acetohexamide, chlorpropamide, glipizide, glyburide, tolazamide

KEY WORDS
SPE

REFERENCE
Strausbauch,M.A.; Xu,S.J.; Ferguson,J.E.; Nunez,M.E.; Machacek,D.; Lawson,G.M.; Wettstein,P.J.; Landers,J.P. Concentration and separation of hypoglycemic drugs using solid-phase extraction-capillary electrophoresis, *J.Chromatogr.A*, **1995**, *717*, 279–291.

SAMPLE
Matrix: solutions
Sample preparation: Inject an aliquot of a solution in MeOH.

CAPILLARY ELECTROPHORESIS
Capillary: 47 cm × 50 μm fused-silica (40 cm to detector)

Capillary preparation: Before each run rinse at 20 psi with 3 column volumes of 100 mM NaOH and 3 column volumes of running buffer. Condition new capillaries by rinsing with 20 column volumes of 1 M NaOH, water, and running buffer

Capillary temperature: 22

Running buffer: 5 mM pH 8.5 Borate buffer containing 5 mM phosphate and 75 mM sodium cholate (Prepare from 500 mM borate buffer (prepared by mixing 125 mM sodium tetraborate and 500 mM boric acid to achieve pH 8.5) and 500 mM Na_2HPO_4 solution with appropriate dilution and the addition of sodium cholate.)

Injection: Pressure injection of sample at 0.5 psi for 1-5 s followed by a pressure injection of running buffer for 1 s.

Detector: UV 200

Migration time: 4

Internal standard: N-acetyl-4-(2,3-dichlorophenylureido)benzenesulfonamide (5.5)

Voltage: 25 kV

Current: 61 μA

Model: Beckman P/ACE 5510

OTHER SUBSTANCES
Simultaneous: acetohexamide, chlorpropamide, glipizide, glyburide, tolazamide

REFERENCE
Roche,M.E.; Oda,R.P.; Lawson,G.M.; Landers,J.P. Capillary electrophoretic detection of metabolites in the urine of patients receiving hypoglycemic drug therapy, *Electrophoresis*, **1997**, *18*, 1865–1874.

SAMPLE
Matrix: urine

Sample preparation: Condition a 200 μg Bond Elut C18 SPE cartridge with 1 mL MeCN. 10 mL Urine + 100 μL 10 μg/mL IS, adjust pH to 2.0 with concentrated HCl, add 15 mL dichloromethane, shake vigorously for 2 min, let stand for 5 min. Remove the organic layer and evaporate it to dryness under a stream of nitrogen at 45°, reconstitute the residue in 400 μL MeOH:water 50:50, add to the SPE cartridge, wash with 1.25 mL MeCN:water 10:90, wash with 1 mL MeCN:water 40:60, elute with 1 mL MeCN. Evaporate the eluate to dryness under a stream of nitrogen, reconstitute the residue in 20 μL MeOH, inject an aliquot.

CAPILLARY ELECTROPHORESIS
Capillary: 47 cm × 50 μm fused-silica (40 cm to detector) (Polymicro Technologies)

Capillary preparation: Before each run wash with 3 column volumes of 100 mM NaOH, dip momentarily into water, rinse with 10 column volumes of 50 mM pH 8.5 Sodium borate buffer containing 50 mM Na_2HPO_4 and 75 mM sodium cholate, rinse with 3 column volumes of running buffer. Condition a new capillary by rinsing with 20 column volumes of 100 mM NaOH, 20 column volumes of water, and 20 column volumes of running buffer.

Capillary temperature: 25

Running buffer: 5 mM pH 8.5 Sodium borate buffer containing 5 mM Na_2HPO_4 and 75 mM sodium cholate

Injection: Pressure injection at 0.5 psi for 2 s then a 1 s injection of running buffer.

Detector: UV 200

Migration time: 5.4

Internal standard: N-acetyl-4-(2,3-dichlorophenylureido)benzenesulfonamide (7.4)

Voltage: 25 kV

Current: 61 μA

Model: Beckman P/ACE System 2100 or System 5510

Limit of detection: 50 ng/mL

OTHER SUBSTANCES
Extracted: acetohexamide, chlorpropamide, glipizide, glyburide, tolazamide

KEY WORDS
SPE

REFERENCE
Núñez,M.; Ferguson,J.E.; Machacek,D.; Jacob,G.; Oda,R.P.; Lawson,G.M.; Landers,J.P. Detection of hypoglycemic drugs in human urine using micellar electrokinetic chromatography, *Anal.Chem.*, **1995**, *67*, 3668–3675.

SAMPLE
Matrix: urine
Sample preparation: Inject an aliquot directly.

CAPILLARY ELECTROPHORESIS
Capillary: 37 cm × 50 μm fused-silica (30 cm to detector) (Polymicro Technologies)
Capillary preparation: Every 13 injections rinse capillary with 100 mM NaOH for 1 min and with running buffer for 1 min. Condition a new capillary with 20 column volumes of 100 mM NaOH, 20 column volumes of water, and 20 column volumes of running buffer.
Capillary temperature: 22
Running buffer: 5 mM pH 8.5 Borate buffer containing 5 mM phosphate and 75 mM sodium cholate
Injection: Pressure injection at 0.5 psi for 2 s.
Detector: UV 200
Migration time: 2.35
Internal standard: N-acetyl-4-(2,3-dichlorophenylureido)benzenesulfonamide (3.2)
Voltage: 28 kV
Current: 71 μA
Model: Beckman P/ACE 5510

OTHER SUBSTANCES
Extracted: acetohexamide, chlorpropamide, glipizide, glyburide, tolazamide

KEY WORDS
direct injection

REFERENCE
Roche,M.E.; Oda,R.P.; Machacek,D.; Lawson,G.M.; Landers,J.P. Enhanced throughput with capillary electrophoresis via continuous-sequential sample injection, *Anal.Chem.*, **1997**, *69*, 99–104.

Tolmetin

Molecular formula: $C_{15}H_{15}NO_3$
Molecular weight: 257.29
CAS Registry No.: 26171-23-3, 64490-92-2 (sodium salt dihydrate), 35711-34-3 (sodium salt)
Merck Index (12th ed.): 9655
Lednicer: 2 234

SAMPLE
Matrix: solutions

CAPILLARY ELECTROPHORESIS
Capillary: 42.5 cm × 25 μm
Capillary temperature: 30
Running buffer: 20 mM pH 7.0 Phosphate containing 25 mM sodium dodecyl sulfate
Injection: Vacuum injection for 2 s
Detector: UV 230
Migration time: 4.2
Voltage: 25 kV
Current: 5 μA
Model: Applied Biosystems Model 270A

OTHER SUBSTANCES
Simultaneous: diflunisal, ibuprofen, indomethacin, naproxen, sulindac

REFERENCE
Weinberger,R.; Albin,M. Quantitative micellar electrokinetic capillary chromatography: Linear dynamic range, *J.Liq.Chromatogr.*, **1991**, *14*, 953–972.

SAMPLE
Matrix: solutions
Sample preparation: Dissolve in water, dilute with water to a concentration of 25 μg/mL, inject an aliquot.

CAPILLARY ELECTROPHORESIS
Capillary: 60 cm × 75 μm fused-silica (52.5 cm to detector)
Running buffer: MeCN:30 mM pH 7.0 phosphate buffer 20:80
Injection: Hydrostatic injection at 9.8 cm for 5 s
Detector: UV 214
Migration time: 19.51
Model: Waters Quanta 4000

OTHER SUBSTANCES
Simultaneous: acemetacin, alclofenac, fenbufen, flurbiprofen, ibuprofen, indomethacin, ketoprofen, lonazolac, naproxen, niflumic acid, piroxicam, tenoxicam, tiaprofenic acid, voltaren

REFFRENCE
Donato,M.G.; Van den Eeckhout,E.; Van den Bossche,W.; Sandra,P. Capillary zone electrophoresis and micellar electrokinetic capillary chromatography of some non-steroidal antiinflammatory drugs (NSAIDs), *J.Pharm.Biomed.Anal.*, **1993**, *11*, 197–201.

Tolperisone

Molecular formula: $C_{16}H_{23}NO$
Molecular weight: 245.36
CAS Registry No.: 728-88-1, 3644-61-9 (HCl)
Merck Index (12th ed.): 9660

SAMPLE
Matrix: solutions
Sample preparation: Inject an aliquot of a solution in running buffer.

CAPILLARY ELECTROPHORESIS
Capillary: 60 cm × 75 μm fused-silica (52.4 cm to detector)
Capillary preparation: After each run flush with 500 mM KOH for 2-3 min then with water.
Running buffer: 10 mM pH 3.8 Phosphate buffer containing 2% sulfated cyclodextrin (ds 7-10)
Injection: Hydrostatic injection.
Detector: UV 214
Migration time: 9.26, 9.82 (enantiomers)
Voltage: 15 kV
Model: Waters Quanta 4000

OTHER SUBSTANCES
Also analyzed: acebutolol, alprenolol, aminoglutethimide, brompheniramine, bupivacaine, bupropion, canadine, carbinoxamine, chloroquine, chlorpheniramine, dimethindene, disopyramide, doxylamine, hydroxychloroquine, idazoxan, isoxsuprine, ketamine, mepenzolate, mepivacaine, methoxyphenamine, mexiletine, midodrine, nefopam, orphenadrine, oxprenolol, oxyphencyclimine, pheniramine, phensuximide, pindolol, piperoxan, terbutaline, tetramisole, tranylcypromine, trihexyphenidyl, trimipramine, verapamil, warfarin

KEY WORDS
chiral; detector at anode

REFERENCE
Stalcup,A.M.; Gahm,K.H. Application of sulfated cyclodextrins to chiral separations by capillary zone electrophoresis, *Anal.Chem.*, **1996**, *68*, 1360–1368.

SAMPLE
Matrix: solutions

CAPILLARY ELECTROPHORESIS
Capillary: 36 cm × 50 μm fused-silica coated with linear polyacrylamide (31.5 cm to detector) (GL Science)
Capillary preparation: At the beginning and end of each day rinse capillary with capillary wash solution (Bio-Rad Cat. No. 148-5022) at 690 kPa for more than 3 min and with water at 690 kPa for more than 3 min. Coat capillary as follows. Treat capillary with 1 M NaOH at room temperature for 1 h, rinse with water, dry by passing nitrogen gas through the capillary at 110° for 6 h. Pass thionyl chloride through the capillary using a suction pump for several min, seal capillary at both ends and heat at 70° for 6 h. Unseal the capillary and fill with 250 mM vinyl magnesium bromide in THF by suction, seal the capillary, heat at 70° for 6 h. Open the capillary and rinse it with THF for several min, rinse with distilled water, fill the capillary with polymerization solution, heat at 28 ± 2° for 1 h, condition at -100 V/cm for 30 min (Anal. Sci. 1994, 10, 1). (The polymerization solution was 5% acrylamide in water containing 49 mM Tris, 384 mM glycine, and 0.1% sodium dodecyl sulfate, degas in an ultrasonic bath. Add 40 μL 10% N,N,N',N'-tetramethylethylenediamine and 10 μL 10% ammonium persulfate to 5 mL of the degassed solution, mix thoroughly.)
Running buffer: Isopropanol:50 mM pH 6.0 Phosphate buffer 10:90
Injection: Before each injection rinse with water at 690 kPa for 30 s, rinse with running buffer at 690 kPa for 30 s, partially fill with separation solution (500 μM α_1-acid glycoprotein (Cohn fraction VI) (ICN) in running buffer) at 6.9 kPa for 190 s (27 cm), inject sample at 6.9 kPa for 2 s, electrophorese with running buffer (Note that α_1-acid glycoprotein from other suppliers may provide inferior results).
Detector: UV 210
Migration time: Resolution of enantiomers 4.1
Voltage: 12 kV
Model: Bio-rad BioFocus 3000

KEY WORDS
chiral; coated capillary

REFERENCE
Tanaka,Y.; Terabe,S. Separation of the enantiomers of basic drugs by affinity capillary electrophoresis using a partial filling technique and α_1-acid glycoprotein as chiral selector, *Chromatographia*, **1997**, *44*, 119–128.

Tramadol

Molecular formula: $C_{16}H_{25}NO_2$
Molecular weight: 263.38
CAS Registry No.: 27203-92-5, 22204-88-2 (HCl)
Merck Index (12th ed.): 9701
Lednicer: 2 17

SAMPLE
Matrix: solutions

CAPILLARY ELECTROPHORESIS
Capillary: 94 cm × 75 μm fused-silica (84 cm to detector)
Capillary temperature: 25 ± 0.10
Running buffer: 40 mM pH 9.14 sodium tetraborate
Injection: Vacuum injection for 2.5 s.
Detector: UV 215
Migration time: 13.24 (trans), 13.59 (cis)
Voltage: 17 kV
Model: SpectraPHORESIS 500 (Spectra Physics)
Limit of quantitation: 10 μg/mL

Limit of detection: 7 µg/mL

REFERENCE
Guo,W.; Zhan,Q.; Zhao,Y.; Wang,L. Determination of *cis* and *trans* isomers of tramadol hydrochloride by capillary zone electrophoresis, *Biomed.Chromatogr.*, **1998**, *12*, 13–14.

SAMPLE
Matrix: solutions

CAPILLARY ELECTROPHORESIS
Capillary: 50 cm × 50 µm polyacrylamide coated (Bio-Rad)
Capillary preparation: Between runs rinse capillary with 50 mM pH 2.5 phosphate buffer for 2 min, with 100 mM NaOH for 2 min, with water for 1 min, with washing buffer for 2 min, and with running buffer for 2 min. (The washing buffer was the same as the running buffer but it was contained in a separate vial to minimize contamination.)
Capillary temperature: 15
Running buffer: 50 mM pH 2.5 Phosphate buffer containing 75 mM methyl-β-cyclodextrin, 220 mM urea, and 0.05% 3 cps hydroxypropylmethylcellulose
Injection: Electrokinetic injection at 5 kV for 7 s.
Detector: UV 200
Migration time: 29.01 (+), 29.40 (-)
Voltage: 25 kV
Model: Bio-Rad BioFocus 3000

OTHER SUBSTANCES
Simultaneous: metabolites

KEY WORDS
chiral; coated capillary

REFERENCE
Chan,E.C.Y.; Ho,P.C. Enantiomeric separation of tramadol hydrochloride and its metabolites by cyclodextrin-mediated capillary zone electrophoresis, *J.Chromatogr.B*, **1998**, *707*, 287–294.

Tranylcypromine

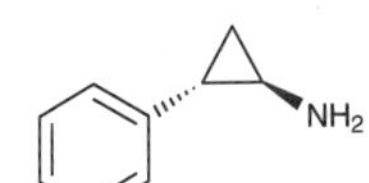

Molecular formula: $C_9H_{11}N$
Molecular weight: 133.19
CAS Registry No.: 155-09-9, 13492-01-8 (sulfate)
Merck Index (12th ed.): 9708
Lednicer: 1 73

SAMPLE
Matrix: solutions
Sample preparation: Inject an aliquot of a solution in running buffer.

CAPILLARY ELECTROPHORESIS
Capillary: 60 cm × 75 µm fused-silica (52.4 cm to detector)
Capillary preparation: After each run flush with 500 mM KOH for 2-3 min then with water.
Running buffer: 10 mM pH 3.8 Phosphate buffer containing 2% sulfated cyclodextrin (ds 7-10)
Injection: Hydrostatic injection.
Detector: UV 214
Migration time: 10.27, 10.64 (enantiomers)
Voltage: 15 kV
Model: Waters Quanta 4000

OTHER SUBSTANCES
Also analyzed: acebutolol, alprenolol, aminoglutethimide, brompheniramine, bupivacaine, bupropion, canadine, carbinoxamine, chloroquine, chlorpheniramine, dimethindene, disopyramide, doxylamine, hydroxychloroquine, idazoxan, isoxsuprine, ketamine, mepenzolate, mepivacaine, methoxyphenamine, mexiletine, midodrine, nefopam, orphenadrine, oxprenolol, oxyphencyclimine, pheniramine, phensuximide, pindolol, piperoxan, terbutaline, tetramisole, tolperisone, trihexyphenidyl, trimipramine, verapamil, warfarin

KEY WORDS
chiral; detector at anode

REFERENCE
Stalcup,A.M.; Gahm,K.H. Application of sulfated cyclodextrins to chiral separations by capillary zone electrophoresis, *Anal.Chem.*, **1996**, *68*, 1360–1368.

Tretoquinol

Molecular formula: $C_{19}H_{23}NO_5$
Molecular weight: 345.40
CAS Registry No.: 21650-42-0, 18559-63-2 (dl-form HCl)
Merck Index (12th ed.): 9719

SAMPLE
Matrix: solutions
Sample preparation: Inject an aliquot of a 100 µg/mL solution in water or running buffer.

CAPILLARY ELECTROPHORESIS
Capillary: 47 cm × 75 µm fused-silica (40 cm to detector)
Capillary preparation: Before each run rinse capillary with running buffer for 1-2 min.
Capillary temperature: 25
Running buffer: 20 mM pH 2.5 Phosphate buffer containing 15% dextran 70 (ca. 70000 Da; TCI America, 911 North Harborgate St., Portland, OR 97203; 800-423-8616, 503-283-1681, (fax) 503-283-1987; www.tciamerica.com)
Injection: Pressure injection at 0.5 psi for 2-4 s.
Detector: UV 220
Migration time: 9.254 (R), 9.408 (S)
Voltage: 30 kV
Model: Beckman P/ACE 5510

KEY WORDS
chiral

REFERENCE
Nishi,H.; Izumoto,S.; Nakamura,K.; Nakai,H.; Sato,T. Dextran and dextrin as chiral selectors in capillary zone electrophoresis, *Chromatographia*, **1996**, *42*, 617–630.

SAMPLE
Matrix: solutions
Sample preparation: Inject an aliquot of a 100 µg/mL solution in MeOH:water 10:90.

CAPILLARY ELECTROPHORESIS
Capillary: 47 cm × 75 µm fused-silica (40 cm to detector)
Capillary preparation: Rinse capillary with running buffer for 1 min before each run. At the end of each day wash with 100 mM NaOH and with water.
Capillary temperature: 20
Running buffer: 20 mM pH 2.9 Phosphate buffer containing 3% chondroitin sulfate C (sodium salt) (Nacalai Tesque, Kyoto)

Injection: Pressure injection at 0.5 psi for 5 s.
Detector: UV 240
Migration time: 18.93, 19.91 (enantiomers)
Voltage: 20 kV
Model: Beckman P/ACE 5510

OTHER SUBSTANCES
Also analyzed: clentiazem, diltiazem, primaquine, propranolol, sulconazole, verapamil

KEY WORDS
chiral

REFERENCE
Nishi,H. Enantiomer separation of basic drugs by capillary electrophoresis using ionic and neutral polysaccharides as chiral selectors, *J.Chromatogr.A*, **1996**, *735*, 345–351.

SAMPLE
Matrix: solutions

CAPILLARY ELECTROPHORESIS
Capillary: 50 cm × 75 μm fused-silica (37 cm to detector) (Beckman)
Capillary temperature: 23
Running buffer: 20 mM pH 2.5 Phosphate buffer containing 50 mM dimethyl-β-cyclodextrin
Injection: Pressure injection at 0.5 psi for 3-10 s.
Detector: UV 220
Migration time: 9.5 (S), 10 (R)
Voltage: 20 kV
Model: Beckman P/ACE 5510

OTHER SUBSTANCES
Simultaneous: denopamine

KEY WORDS
chiral

REFERENCE
Ishibuchi,K.; Izumoto,S.; Nishi,H.; Sato,T. Enantiomer separation of denopamine by capillary electrophoresis with charged and uncharged cyclodextrins, *Electrophoresis*, **1997**, *18*, 1007–1012.

SAMPLE
Matrix: solutions
Sample preparation: Inject an aliquot of a solution in running buffer.

CAPILLARY ELECTROPHORESIS
Capillary: 57 cm × 50 μm fused-silica (50 cm to detector) (Polymicro Technologies)
Capillary temperature: 25
Running buffer: 25 mM NaH_2PO_4 containing 0.83% poly(sodium 10-undecenyl sulfate), adjusted to pH 7.3 with 12.5 mM sodium tetraborate (Preparation of poly(sodium 10-undecenyl sulfate is as follows. React undecylenyl alcohol (1-undecen-11-ol) with one equivalent of chlorosulfonic acid in diethyl ether under nitrogen at -5° to obtain 10-undecenyl hydrogen sulfate, react with one equivalent of NaOH to obtain sodium 10-undecenyl sulfate (J. Microcolumn Sep. 1996, 8, 115). Dissolve 20 g sodium 10-undecenyl sulfate in 40 mL water (degassed by purging with nitrogen for 24 h), add 600 mg potassium persulfate, stir at 70° under nitrogen for 50 h, cool, add 200 mL cold (0°) EtOH, filter, wash the solid with three 10 mL portions of cold EtOH, dry under vacuum at 65° overnight to obtain poly(sodium 10-undecenyl sulfate) (cf. J. Microcolumn Sep. 1992, 4, 509). Purify by dialysis using dialysis tubing with a 1000 MW cut-off (Spectrum Houston) for 24 h.)
Detector: UV 214
Migration time: 17
Voltage: 16.1 kV

Model: Beckman P/ACE 2000

OTHER SUBSTANCES
Simultaneous: acetaminophen, caffeine, ethenzamide, guaifenesin, phenacetin

REFERENCE
Palmer,C.P.; Terabe,S. Micelle polymers as pseudostationary phases in MEKC: Chromatographic performance and chemical selectivity, *Anal.Chem.*, **1997**, *69*, 1852–1860.

SAMPLE
Matrix: solutions

CAPILLARY ELECTROPHORESIS
Capillary: 36 cm × 50 μm fused-silica coated with linear polyacrylamide (31.5 cm to detector) (GL Science)

Capillary preparation: At the beginning and end of each day rinse capillary with capillary wash solution (Bio-Rad Cat. No. 148-5022) at 690 kPa for more than 3 min and with water at 690 kPa for more than 3 min. Coat capillary as follows. Treat capillary with 1 M NaOH at room temperature for 1 h, rinse with water, dry by passing nitrogen gas through the capillary at 110° for 6 h. Pass thionyl chloride through the capillary using a suction pump for several min, seal capillary at both ends and heat at 70° for 6 h. Unseal the capillary and fill with 250 mM vinyl magnesium bromide in THF by suction, seal the capillary, heat at 70° for 6 h. Open the capillary and rinse it with THF for several min, rinse with distilled water, fill the capillary with polymerization solution, heat at 28 ± 2° for 1 h, condition at -100 V/cm for 30 min (Anal. Sci. 1994, 10, 1). (The polymerization solution was 5% acrylamide in water containing 49 mM Tris, 384 mM glycine, and 0.1% sodium dodecyl sulfate, degas in an ultrasonic bath. Add 40 μL 10% N,N,N',N'-tetramethylethylenediamine and 10 μL 10% ammonium persulfate to 5 mL of the degassed solution, mix thoroughly.)

Running buffer: MeOH:50 mM pH 6.0 Phosphate buffer 10:90

Injection: Before each injection rinse with water at 690 kPa for 30 s, rinse with running buffer at 690 kPa for 30 s, partially fill with separation solution (500 μM α_1-acid glycoprotein (Cohn fraction VI) (Sigma) in running buffer) at 6.9 kPa for 190 s (27 cm), inject sample at 6.9 kPa for 2 s, electrophorese with running buffer (Note that α_1-acid glycoprotein from other suppliers may provide inferior results).

Detector: UV 210

Migration time: Resolution of enantiomers 3.9

Voltage: 12 kV

Model: Bio-rad BioFocus 3000

KEY WORDS
chiral; coated capillary

REFERENCE
Tanaka,Y.; Terabe,S. Separation of the enantiomers of basic drugs by affinity capillary electrophoresis using a partial filling technique and α_1-acid glycoprotein as chiral selector, *Chromatographia*, **1997**, *44*, 119–128.

Triamcinolone

Molecular formula: $C_{21}H_{27}FO_6$
Molecular weight: 394.44
CAS Registry No.: 124-94-7, 76-25-5 (acetonide), 1997-15-5 (acetonide disodium phosphate), 31002-79-6 (benetonide), 5611-51-8 (hexacetonide), 67-78-7 (diacetate), 4989-94-0 (furetonide), 989-96-8 (21-(dihydrogen phosphate)
Merck Index (12th ed.): 9727
Lednicer: 1 201; 2 302

SAMPLE
Matrix: solutions

CAPILLARY ELECTROPHORESIS
Capillary: 65 cm × 50 μm fused-silica (50 cm to detector) (Scientific Glass Engineering)
Capillary preparation: Flush with running buffer for 1 min between runs. Periodically flush with water for 1 min, sweep with 100 mM KOH, let stand in 100 mM KOH for 30 min, flush with water until the effluent is neutral to pH paper, fill with running buffer, let stand for 30 min.
Running buffer: 20 mM pH 9.0 Phosphate-borate buffer containing 100 mM sodium cholate (Buffer was prepared by adjusting pH of 20 mM sodium borate to 9.0 with 20 mM NaH_2PO_4.)
Injection: Injection by siphon at 10 cm for 10 s (about 1 nL)
Detector: UV 210
Migration time: 11 (triamcinolone), 14 (triamcinolone acetonide)
Voltage: 20 kV

OTHER SUBSTANCES
Simultaneous: betamethasone, dexamethasone acetate, fluocinonide, fluocinolone acetonide, hydrocortisone acetate, hydrocortisone

REFERENCE
Nishi,H.; Fukuyama,T.; Matsuo,M.; Terabe,S. Separation and determination of lipophilic corticosteroids and benzothiazepin analogues by micellar electrokinetic chromatography using bile salts, *J.Chromatogr.*, **1990**, *513*, 279–295.

SAMPLE
Matrix: solutions
Sample preparation: Prepare a 0.2-1 mg/mL solution in MeOH, inject an aliquot.

CAPILLARY ELECTROPHORESIS
Capillary: 65 cm × 50 μm fused-silica (50 cm to detector)
Running buffer: 20 mM pH 9.0 Phosphate-borate buffer containing 50 mM sodium dodecyl sulfate, 4 M urea, and 15 mM gamma-cyclodextrin
Detector: UV 220
Migration time: 10.3 (triamcinolone acetonide)
Voltage: 20 kV

OTHER SUBSTANCES
Simultaneous: betamethasone, cortisone acetate, dexamethasone acetate, fluocinolone acetonide, fluocinonide, hydrocortisone, hydrocortisone acetate

REFERENCE
Nishi,H.; Matsuo,M. Separation of corticosteroids and aromatic hydrocarbons by cyclodextrin-modified micellar electrokinetic chromatography, *J.Liq.Chromatogr.*, **1991**, *14*, 973–986.

SAMPLE
Matrix: solutions

Sample preparation: Prepare a 10 mg/mL solution in MeOH, inject an aliquot.

CAPILLARY ELECTROPHORESIS
Capillary: 62 cm × 53 μm fused-silica (50 cm to detector) (Polymicro Technologies)
Capillary preparation: Rinse with running buffer for 1 min between runs.
Running buffer: 50 mM pH 9.0 Phosphate borate buffer containing 50 mM dehydrocholate, 50 mM taurocholate, and 50 mM sodium dodecyl sulfate
Detector: UV 254
Migration time: 14 (triamcinolone), 25.5 (triamcinolone acetonide)
Voltage: 15 kV
Current: 56 μA
Model: Laboratory constructed

OTHER SUBSTANCES
Simultaneous: corticosterone, cortisone, cortisone acetate, deoxycorticosterone, fludrocortisone, fludrocortisone acetate, fluocinolone acetonide, hydrocortisone, hydrocortisone-21-acetate, 6α-methylprednisolone, prednisolone, prednisolone acetate, prednisone, prednisone acetate, progesterone

---

REFERENCE
Bumgarner,J.G.; Khaledi,M.G. Mixed micellar electrokinetic chromatography of corticosteroids, *Electrophoresis*, **1994**, *15*, 1260–1266.

SAMPLE
Matrix: solutions

CAPILLARY ELECTROPHORESIS
Capillary: 62 cm × 53 μm fused-silica (50 cm to detector) (Polymicro Technologies)
Capillary preparation: Rinse capillary with running buffer for 1 min between runs.
Running buffer: 50 mM pH 9.0 Phosphate/borate buffer containing 33 mM taurocholate, 33 mM glycodeoxycholate, and 70 mM butanesulfonate
Detector: UV 254
Migration time: 16.5 (triamcinolone), 27.5 (triamcinolone acetonide)
Voltage: 15 kV
Model: laboratory-constructed

OTHER SUBSTANCES
Simultaneous: corticosterone, cortisone, cortisone acetate, deoxycorticosterone, fludrocortisone, fludrocortisone acetate, fluocinolone acetonide, hydrocortisone, hydrocortisone-21-acetate, 6α-methylprednisolone, prednisolone, prednisolone acetate, prednisone, prednisone acetate, progesterone

---

REFERENCE
Bumgarner,J.G.; Khaledi,M.G. Mixed micelles of short chain alkyl surfactants and bile slats in electrokinetic chromatography: Enhanced separation of corticosteroids, *J.Chromatogr.A*, **1996**, *738*, 275–283.

SAMPLE
Matrix: solutions
Sample preparation: Inject an aliquot of a solution in running buffer.

CAPILLARY ELECTROPHORESIS
Capillary: 57 cm × 75 μm fused-silica (50 cm to detector)
Capillary preparation: Before each run rinse with running buffer for 2 min. After each run rinse capillary with 100 mM NaOH for 2 min and with water for 2 min. Condition new capillaries with 100 mM HCl for 5 min, with 100 mM NaOH for 10 min, and with water for 5 min.
Capillary temperature: 23
Running buffer: MeOH:buffer 10:90 (Buffer was 40 mM pH 9.2 borate buffer containing 20 mM sodium dodecyl sulfate.)
Injection: Low pressure injection for 10 s.
Detector: UV 240
Migration time: 6
Voltage: 30 kV

Model: Beckman P/ACE 5510

OTHER SUBSTANCES
Simultaneous: betamethasone, dexamethasone, flumethasone, hydrocortisone

REFERENCE
Gu,X.; Meleka-Boules,M.; Chen,C.-L. Micellar electrokinetic capillary chromatography combined with immunoaffinity chromatography for identification and determination of dexamethasone and flumethasone in equine urine, *J.Capillary Electrophor.*, **1996**, *3*, 43–49.

SAMPLE
Matrix: solutions
Sample preparation: Inject an aliquot of a solution in running buffer.

CAPILLARY ELECTROPHORESIS
Capillary: 60 cm × 50 μm fused-silica (50 cm to detector) (Supelco)
Capillary preparation: Between runs wash capillary with 500 mM NaOH for 4 min.
Running buffer: n-Hexanol:sodium dodecyl sulfate:n-butanol:20 mM pH 10.0 phosphate buffer 0.81:3.31:6.61:89.28
Injection: Electrokinetic injection at 15 kV for 10 s.
Detector: UV 220
Migration time: 13
Voltage: 15 kV
Model: laboratory-constructed
Limit of detection: 1 pmole

OTHER SUBSTANCES
Simultaneous: aldosterone, cortexolone, corticosterone, cortisone, 11-dehydrocorticosterone, deoxycorticosterone, deoxycorticosterone acetate, dexamethasone, hydrocortisone

REFERENCE
Vomastová,L.; Miksík,I.; Deyl,Z. Microemulsion and micellar electrokinetic chromatography of steroids, *J.Chromatogr.B*, **1996**, *681*, 107–113.

SAMPLE
Matrix: solutions

CAPILLARY ELECTROPHORESIS
Capillary: 42 cm × 50 μm fused-silica packed with 3 μm CEC Hypersil ODS (packed length 30 cm, 30.1 cm to detector) (Hypersil)
Running buffer: Gradient. A was 5 mM ammonium acetate in MeCN:water 17:83. B was 5 mM ammonium acetate in MeCN:water 38:62. A:B 100:0 for 3 min, to 0:100 over 15 min, maintain at 0:100 (pumped with an HPLC pump at 0.01 mL/min for 3 min then at 0.1 mL/min).
Injection: Inject 10 μL using an HPLC injector.
Detector: UV 240
Migration time: 28 (triamcinolone), 41 (triamcinolone acetonide)
Voltage: 30 kV

OTHER SUBSTANCES
Simultaneous: adrenosterone, betamethasone, cortisone, fluocortolone, dexamethasone, methylprednisolone, prednisolone
Interfering: hydrocortisone

KEY WORDS
electrochromatography

REFERENCE
Taylor,M.R.; Teale,P. Gradient capillary electrochromatography of drug mixtures with UV and electrospray ionisation mass spectrometric detection, *J.Chromatogr.A*, **1997**, *768*, 89–95.

Triamterene

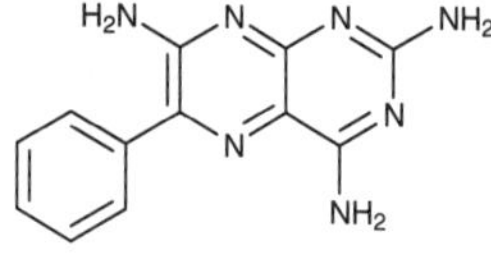

Molecular formula: $C_{12}H_{11}N_7$

Molecular weight: 253.27

CAS Registry No.: 396-01-0

Merck Index (12th ed.): 9731

Lednicer: 1 427

SAMPLE

Matrix: blood, urine

Sample preparation: Condition a 3 mL Supelclean LC-18 SPE cartridge with 3 mL MeOH and 3 mL water. Dilute urine 1:10 with water. Precipitate proteins from serum with MeOH. Add diluted urine or protein supernatant to the SPE cartridge, wash with 3 mL water, elute with 3 mL MeOH, reconstitute to the original volume with 100 mM KOH, inject an aliquot.

CAPILLARY ELECTROPHORESIS

Capillary: 67 cm × 50 μm fused-silica (60 cm to detector) (Polymicro Technologies)

Capillary preparation: Rinse with running buffer for 2 min before run. If necessary, regenerate capillary with 100 mM NaOH for 10 min and with water for 15 min.

Capillary temperature: 20

Running buffer: 70 mM pH 4.5 Acetate buffer containing 500 mM betaine

Injection: Pressure injection for 5 s

Detector: UV 215

Migration time: 5.5

Voltage: 25 kV

Model: Beckman P/ACE 2000

OTHER SUBSTANCES

Extracted: amiloride, metyrapone

Noninterfering: caffeine

KEY WORDS

serum; SPE

REFERENCE

Jumppanen,J.; Sirén,H.; Riekkola,M.-L. Screening for diuretics in urine and blood serum by capillary zone electrophoresis, *J.Chromatogr.A*, **1993**, *652*, 441–450.

SAMPLE

Matrix: urine

Sample preparation: Filter (0.2 μm), inject an aliquot of the filtrate.

CAPILLARY ELECTROPHORESIS

Capillary: 80 cm × 50 μm fused-silica (57 cm to detector)

Capillary preparation: Before each run rinse the capillary with 1 M NaOH for 3 min, with 100 mM NaOH for 3 min, with water for 3 min, and with running buffer for 10 min.

Running buffer: 110 mM Boric acid containing 56 mM NaOH and 44 mM HCl, pH 8

Injection: Vacuum injection for 1 s

Detector: UV 232

Migration time: 5.600

Voltage: 20 kV

Model: Europhor Prime Vision system IV

Limit of detection: 242 nM

OTHER SUBSTANCES
Extracted: acebutolol (UV 238), acetazolamide (UV 222), alprenolol (UV 220), amiloride (UV 220), atenolol (UV 228), bendroflumethiazide (UV 220), bumetanide (UV 220), chlorthalidone (UV 220), cocaine (UV 236), codeine (UV 220), ethacrynic acid (UV 220), furosemide (UV 232), hydrochlorothiazide (UV 226), methadone (UV 220), metoxiphenamine (UV 220), nadolol (UV 220), norcodeine (UV 220), oxprenolol (UV 220), pentazocine (UV 220), propranolol (UV 220), spironolactone (UV 244), xipamide (UV 234)

REFERENCE
Gonzalez,E.; Laserna,J.J. Capillary zone electrophoresis for the rapid screening of banned drugs in sport, *Electrophoresis*, **1994**, *15*, 240–243.

SAMPLE
Matrix: urine
Sample preparation: Centrifuge urine at 150 g for 10 min, inject an aliquot.

CAPILLARY ELECTROPHORESIS
Capillary: 100 cm × 50 μm (60 cm to detector)
Capillary preparation: Rinse capillary with running buffer for 3 min before each run. Regenerate capillary by flushing with 1 M NaOH for 10 min, with 100 mM NaOH for 10 min, and with water for 15 min.
Running buffer: pH 8 Borate buffer (concentration adjusted to keep current at 42 μA)
Injection: Electrokinetic injection at 10 kV for 10 s (ca. 6 nL).
Detector: F ex 370 em 434
Migration time: 5.2
Voltage: 30 kV
Current: 42 μA
Model: laboratory-constructed
Limit of detection: 90 nM

OTHER SUBSTANCES
Extracted: amiloride (F ex 382 em 416), bendroflumethiazide (F ex 272 em 381), bumetanide (F ex 350 em 428)

KEY WORDS
direct injection

REFERENCE
González,E.; Becerra,A.; Laserna,J.J. Direct determination of diuretic drugs in urine by capillary zone electrophoresis using fluorescence detection, *J.Chromatogr.B*, **1996**, *687*, 145 150.

Triazolam

Molecular formula: $C_{17}H_{12}Cl_2N_4$
Molecular weight: 343.21
CAS Registry No.: 28911-01-5
Merck Index (12th ed.): 9734
Lednicer: 1 368

SAMPLE
Matrix: blood
Sample preparation: Condition a Sep-Pak C18 SPE cartridge with MeOH, water, and 100 mM pH 9.0 borate buffer. 1 mL Serum + 3 mL 100 mM pH 9.0 borate buffer, mix, add to the SPE cartridge, wash with 3 mL 100 mM pH 9.0 borate buffer, wash with three 2 mL portions of water, wash with 1 mL MeCN:water 5:95, wash with 1 mL hexane, dry under full vacuum. Elute with 4 mL dichloromethane, evaporate the eluate to dryness under a stream of nitrogen, reconstitute with MeOH:water 15:85, filter (0.45 μm), inject an aliquot.

CAPILLARY ELECTROPHORESIS
Capillary: 72 cm × 50 μm fused-silica (50 cm to detector)
Capillary preparation: Before each run wash capillary with 100 mM NaOH for 4 min and with running buffer for 6 min.
Capillary temperature: 25
Running buffer: MeOH:5 mM pH 8.5 phosphate/borate buffer 15:85 containing 50 mM sodium dodecyl sulfate
Injection: Vacuum injection for 4 s.
Detector: UV 200
Migration time: 21
Internal standard: secobarbital (9)
Voltage: 25 kV
Model: Applied Biosystems Model 270A or Beckman P/ACE 5000
Limit of detection: 25 ng/mL

OTHER SUBSTANCES
Extracted: bromazepam, diazepam, estazolam, flurazepam, nitrazepam

KEY WORDS
SPE; serum

REFERENCE
Tomita,M.; Okuyama,T. Application of capillary electrophoresis to the simultaneous screening and quantitation of benzodiazepines, *J.Chromatogr.B*, **1996**, *678*, 331–337.

SAMPLE
Matrix: solutions

CAPILLARY ELECTROPHORESIS
Capillary: 60 cm × 50 μm fused-silica (40 cm to detector), C18 coated (Supelco)
Capillary preparation: Rinse new capillaries with 250 μL 1 M NaOH, 250 μL 100 mM NaOH, water, MeOH, water, and running buffer then condition with running buffer for at least 15 min. Carry out a similar procedure between runs.
Running buffer: 100 mM pH 8.5 Borate buffer containing 25 mM sodium dodecyl sulfate and 5 M urea
Injection: Electrokinetic injection at 5 kV for 5 s.
Detector: UV 254
Migration time: 42
Voltage: 18 kV
Model: Jasco

OTHER SUBSTANCES
Simultaneous: alprazolam, amobarbital, barbital, bromazepam, clonazepam, clotiazepam, cloxazolam, diazepam, estazolam, etizolam, fludiazepam, flunitrazepam, flurazepam, haloxazolam, medazepam, mephobarbital, metharbital, nimetazepam, nitrazepam, oxazepam, pentobarbital, phenobarbital, secobarbital

KEY WORDS
coated capillary

REFERENCE
Jinno,K.; Han,Y.; Nakamura,M. Analysis of anxiolytic drugs by capillary electrophoresis with bare and coated capillaries, *J.Capillary Electrophor.*, **1996**, *3*, 139–145.

SAMPLE
Matrix: solutions

CAPILLARY ELECTROPHORESIS
Capillary: 47 cm × 75 μm fused-silica (40 cm to detector) (Beckman)
Capillary preparation: Rinse with running buffer for 1 min before injection. Wash with 100 mM NaOH for 1 min and with water for 1 min between injections.

Capillary temperature: 30
Running buffer: THF:buffer 1:99 (Buffer was 50 mM pH 9.2 borate containing 50 mM sodium
  dodecyl sulfate, 2 M urea, and 20 mM gamma-cyclodextrin.)
Injection: Pressure injection at 300 kPa for 5 s.
Detector: UV 254
Migration time: 20
Voltage: 15 kV
Model: Beckman P/ACE 2000

OTHER SUBSTANCES
Simultaneous: alprazolam, bromazepam, chlordiazepoxide, clobazam, clonazepam, clorazepate,
  oxazepam, prazepam

REFERENCE
Renou-Gonnord,M.F.; David,K. Optimized micellar electrokinetic chromatographic separation of benzodiazepines, *J.Chromatogr.A*, **1996**, *735*, 249–261.

SAMPLE
Matrix: urine
Sample preparation: Place a silica fiber coated with an 85 μm-thick layer of polyacrylate (Su-
  pelco) in 10 mL urine, stir at 60° for 2 h, place the fiber in a capillary tube containing 20 μL
  MeCN, let stand at room temperature for 30 min, inject an aliquot of the solution. (Desorb a
  new fiber several times with MeCN before use.)

CAPILLARY ELECTROPHORESIS
Capillary: 60 cm × 50 μm acrylamide-coated fused-silica (40 cm to detector) (Supelco)
Capillary preparation: Adjust the pH of 20 mL water to 3.5 with acetic acid, add 80 μL 3-
  (trimethoxysilyl)propyl methacrylate (3-methacryloxypropyltrimethoxysilane), mix, suck into
  capillary, let stand at room temperature for 1 h, remove the solution, wash with water. Fill the
  capillary with a deaerated 3-4% acrylamide solution containing 1 μL/mL N,N,N',N'-tetra-
  methylethylenediamine and 1 mg/mL ammonium persulfate, let stand for 3 h, remove excess
  solution by aspiration, rinse with water, remove water by aspiration, dry at 35° (cf. J. Chromatogr. 1985, 347, 191).
Running buffer: MeCN:buffer 5:95 (Buffer was 100 mM borate containing 5 M urea and 10 mM
  sodium dodecyl sulfate, adjusted to pH 8.5 with phosphate.)
Injection: Electrokinetic injection at 5 kV for 5 s.
Detector: UV 254
Migration time: 26.5
Voltage: 18 kV
Model: Jasco Model 870 CE
Limit of detection: <1 ppm

OTHER SUBSTANCES
Extracted: flunitrazepam, nitrazepam
Simultaneous: alprazolam, amobarbital, barbital, bromazepam, clonazepam, clotiazepam, cloxazolam, diazepam, estazolam, etizolam, fludiazepam, flurazepam, haloxazolam, medazepam,
  mephobarbital, metharbital, nimetazepam, oxazepam, pentobarbital, phenobarbital, secobarbital

KEY WORDS
coated capillary; detector at anode; SPE

REFERENCE
Jinno,K.; Han,Y.; Sawada,H.; Taniguchi,M. Capillary electrophoretic separation of toxic drugs using a polyacrylamide-coated capillary, *Chromatographia*, **1997**, *46*, 309–314.

Trichlormethiazide

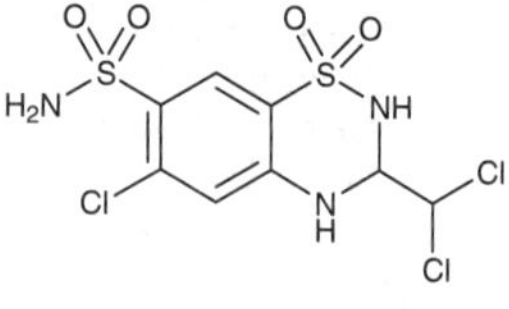

Molecular formula: $C_8H_8Cl_3N_3O_4S_2$
Molecular weight: 380.66
CAS Registry No.: 133-67-5
Merck Index (12th ed.): 9754
Lednicer: 1 359

SAMPLE
Matrix: blood, urine
Sample preparation: Condition a 3 mL Supelclean LC-18 SPE cartridge with 3 mL MeOH and 3 mL water. Dilute urine 1:10 with water. Precipitate proteins from serum with MeOH. Add diluted urine or protein supernatant to the SPE cartridge, wash with 3 mL water, elute with 3 mL MeOH, reconstitute to the original volume with 100 mM KOH, inject an aliquot.

CAPILLARY ELECTROPHORESIS
Capillary: 67 cm × 50 μm fused-silica (60 cm to detector) (Polymicro Technologies)
Capillary preparation: Rinse with running buffer for 2 min before run. If necessary, regenerate capillary with 100 mM NaOH for 10 min and with water for 15 min.
Capillary temperature: 20
Running buffer: 60 mM 3-(cyclohexylamino)-1-propanesulfonic acid (CAPS) adjusted to pH 10.6 with 100 mM KOH
Injection: Pressure injection for 5 s
Detector: UV 220
Migration time: 10.4
Voltage: 25 kV
Model: Beckman P/ACE 2000

OTHER SUBSTANCES
Extracted: acetazolamide, amiloride, bendroflumethiazide, benzthiazide, bumetanide, caffeine, chlorothiazide, chlorthalidone, clopamide, dichlorphenamide, ethacrynic acid, furosemide, hydrochlorothiazide, metyrapone, probenecid, triamterene

KEY WORDS
serum; SPE

REFERENCE
Jumppanen,J.; Sirén,H.; Riekkola,M.-L. Screening for diuretics in urine and blood serum by capillary zone electrophoresis, *J.Chromatogr.A*, **1993**, *652*, 441–450.

SAMPLE
Matrix: solutions

CAPILLARY ELECTROPHORESIS
Capillary: 72 cm × 50 μm silica (50 cm to detector) (Applied Biosystems)
Capillary preparation: Before each run wash capillary with 100 mM NaOH for 3 min, wash with running buffer for 3 min, aspirate neutral marker solution (2 drops DMSO in 10 mL water) for 1 s, place capillary end in buffer vial for 5 s, inject sample. Wash capillary at the beginning of each day by passing 1 M NaOH through for 20 min.
Capillary temperature: 30
Running buffer: MeCN:buffer 10:90 (Buffer was 30 mM pH 9.3 borate buffer containing 30 mM sodium dodecyl sulfate.)
Injection: Inject by applying vacuum for 1 s (2-3 nL)
Detector: UV 200
Migration time: 6.0
Voltage: +30 kV
Current: 56 μA
Model: Applied Biosystems Model 270A

OTHER SUBSTANCES
Simultaneous: chlorothiazide, furosemide, hydrochlorothiazide

REFERENCE
Evenson,M.A.; Wiktorowicz,J.E. Automated capillary electrophoresis applied to therapeutic drug monitoring, *Clin.Chem.*, **1992**, *38*, 1847–1852.

Trifluoperazine

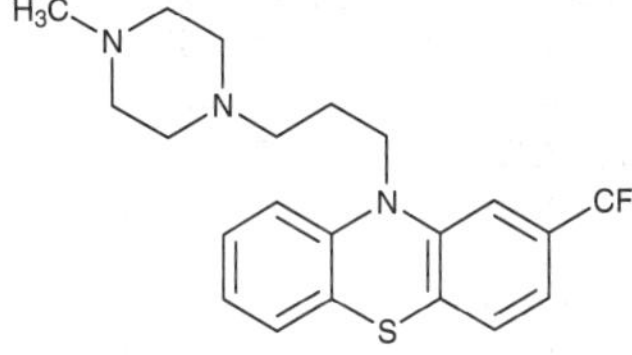

Molecular formula: $C_{21}H_{24}F_3N_3S$
Molecular weight: 407.50
CAS Registry No.: 117-89-5, 440-17-5 (di HCl)
Merck Index (12th ed.): 9811

SAMPLE
Matrix: solutions

CAPILLARY ELECTROPHORESIS
Capillary: 57 cm × 50 μm fused-silica (50 cm to detector)
Running buffer: 50 mM pH 8.0 Borate buffer containing 40 mM sodium taurodeoxycholate and 25 mM phosphatidylcholine (Prepare by adding phosphatidylcholine and stirring for 3-6 h until all cloudiness disappears. Phosphatidylcholine was 95% pure soybean lecithin, Epikuron, Lucas Meyer & Co.)
Injection: Inject a solution of the compound in the running buffer at 20 psi for 1 s, inject MeOH: water 5:95 at 20 psi for 1 s, inject a solution of halofantrine in running buffer at 20 psi for 1 s.
Detector: UV 214
Migration time: k' 151.36
Model: Beckman P/ACE 5000

OTHER SUBSTANCES
Also analyzed: acetaminophen, amoxicillin, antipyrine, aspirin, azathioprine, caffeine, captopril, carbamazepine, carprofen, chlorambucil, chlorpheniramine, chlorpromazine, cimetidine, clonidine, codeine, desipramine, diphenhydramine, ephedrine, fenoterol, flufenamic acid, flurbiprofen, haloperidol, hydroxyzine, ibuprofen, imipramine, indomethacin, ketoprofen, lidocaine, melphalan, metoprolol, nabumetone, nadolol, phenobarbital, phonol, promazine, propranolol, pyrilamine, ranitidine, ropinirole, salicylic acid, sulfamethoxazole, testosterone, theophylline, thioridazine, tiaprofenic acid, tolfenamic acid, trimethoprim, valproic acid, verapamil

KEY WORDS
comparison with HPLC; k' = (Tr-T0)/(T0(1-Tr/Tm)) where Tr = retention time of analyte; T0 = retention time of water; and Tm = retention time of marker (halofantrine)

REFERENCE
Hanna,M.; de Biasi,V.; Bond,B.; Salter,C.; Hutt,A.J.; Camilleri,P. Estimation of the partitioning characteristics of drugs: A comparison of a large and diverse drug series utilizing chromatographic and electrophoretic methodology, *Anal.Chem.*, **1998**, *70*, 2092–2099.

SAMPLE
Matrix: urine
Sample preparation: 10 mL Urine + 1 mL 5 M NaOH + 10 mL n-hexane, vortex for 1 min, centrifuge at 0° at 3000 g for 5 min. Remove the organic layer and add it to 50 μL glacial acetic acid, evaporate to dryness under a stream of nitrogen at 30°, reconstitute the residue in 50 μL 50 mM sodium taurodeoxycholate, filter (0.2 μm PTFE), inject an aliquot.

CAPILLARY ELECTROPHORESIS
Capillary: 100 cm × 50 μm fused-silica (50 cm to detector) (ISCO)

Capillary preparation: Rinse capillary with 20 µL running buffer between injections. Condition a new capillary by filling with 1 M NaOH, let stand for 1 h, fill with 100 mM NaOH, let stand for 1 h, rinse with buffer. Every 20 injections rinse capillary with 200 µL 1 M NaOH, 200 µL water, and 200 µL running buffer, fill with running buffer.
Capillary temperature: 22
Running buffer: 40 mM pH 9.5 Borate buffer containing 10 mM sodium taurodeoxycholate
Injection: Load under vacuum at 7.5 kPa/s.
Detector: UV 240
Migration time: 14
Voltage: 30 kV
Model: ISCO Model 3140 electropherograph
Limit of detection: 43 ng/mL

OTHER SUBSTANCES
Extracted: acepromazine, amiodarone, amitriptyline, azaperone, chlorpromazine, cianopramine, clomipramine, clozapine, desethylamiodarone, desipramine, diclofensine, dothiepin, doxepin, imipramine, isocarboxazid, moclobemide, perphenazine, phenothiazine, pimozide, prochlorperazine, promazine, thioridazine, thiothixene, trimipramine

KEY WORDS
human; cow; pig; horse; protect from light

REFERENCE
Aumatell,A.; Wells,R.J. Determination of a cardiac antiarrhythmic, tricyclic antipsychotics and antidepressants in human and animal urine by micellar electrokinetic capillary chromatography using a bile salt, *J.Chromatogr.B*, **1995**, *669*, 331–344.

SAMPLE
Matrix: urine
Sample preparation: Dilute 10-fold with water, inject an aliquot.

CAPILLARY ELECTROPHORESIS
Capillary: 70 cm × 50 µm fused-silica (65.4 cm to detector) (Polymicro Technologies)
Capillary temperature: 25
Running buffer: 20 mM pH 3.5 Tris-formic acid containing 50 µg/mL FC-135 (Fluorad/3M)
Injection: Pressure injection at 2 psi.
Detector: UV 240
Migration time: 8.43
Voltage: 20 kV
Model: Bio-Rad BioFocus 3000

OTHER SUBSTANCES
Extracted: perphenazine, prochlorperazine, promazine, thioridazine, triflupromazine

KEY WORDS
detector at anode

REFERENCE
Muijelsaar,P.G.H.M.; Claessens,H.A.; Cramers,C.A. Determination of structurally related phenothiazines by capillary zone electrophoresis and micellar electrokinetic chromatography, *J.Chromatogr.A*, **1996**, *735*, 395–402.

Triflupromazine

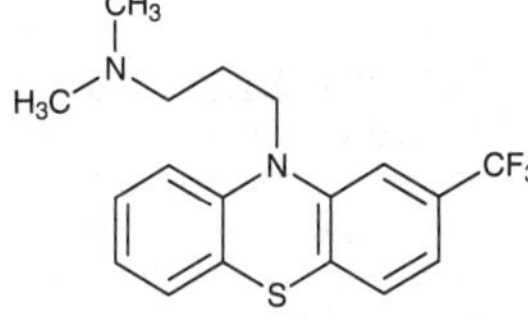

Molecular formula: $C_{18}H_{19}F_3N_2S$
Molecular weight: 352.42
CAS Registry No.: 146-54-3, 1098-60-8 (HCl)
Merck Index (12th ed.): 9814
Lednicer: 1 380

SAMPLE
Matrix: urine
Sample preparation: Dilute 10-fold with water, inject an aliquot.

CAPILLARY ELECTROPHORESIS
Capillary: 70 cm $\times$ 50 μm fused-silica (65.4 cm to detector) (Polymicro Technologies)
Capillary temperature: 25
Running buffer: 20 mM pH 3.5 Tris-formic acid containing 50 μg/mL FC-135 (Fluorad/3M)
Injection: Pressure injection at 2 psi.
Detector: UV 240
Migration time: 7.60
Voltage: 20 kV
Model: Bio-Rad BioFocus 3000
Limit of detection: 1.08 μg/mL

OTHER SUBSTANCES
Extracted: perphenazine, prochlorperazine, promazine, thioridazine, trifluoperazine

KEY WORDS
detector at anode

REFERENCE
Muijelsaar,P.G.H.M.; Claessens,H.A.; Cramers,C.A. Determination of structurally related phenothiazines by capillary zone electrophoresis and micellar electrokinetic chromatography, *J.Chromatogr.A*, **1996**, *735*, 395–402.

Trihexyphenidyl

Molecular formula: $C_{20}H_{32}ClNO$
Molecular weight: 337.93
CAS Registry No.: 144-11-6, 52-49-3 (HCl)
Merck Index (12th ed.): 9823
Lednicer: 1 47

SAMPLE
Matrix: solutions
Sample preparation: Inject an aliquot of a 100 μg/mL solution in water:running buffer 50:50.

CAPILLARY ELECTROPHORESIS
Capillary: 44.5 cm $\times$ 50 μm acrylamide-coated fused-silica (Bio-Rad)
Capillary temperature: 30
Running buffer: 100 mM NaH_2PO_4 containing 15 mM gamma-cyclodextrin, adjusted to pH 2.5 with phosphoric acid
Injection: Electrokinetic injection at 8 kV for 6 s.
Detector: UV 200
Migration time: 12.79
Voltage: 14 kV

Model: Bio-Rad BioFocus 3000

OTHER SUBSTANCES
Also analyzed: albuterol, alprenolol, atenolol, atropine, baclofen, bamethan, benserazide, biperiden, bisoprolol, bupivacaine, bupranolol, butetamate, carazolol, carbuterol, carvedilol, celiprolol, chloroquine, chlorpheniramine (chlorphenamine), clidinium bromide, clobutinol, disopyramide, dobutamine, flecainide, homatropine, ipratropium bromide, isoproterenol, isothipendyl, ketamine, mefloquine, mequitazine, metaproterenol (orciprenaline), metipranolol, nafronyl (naftidrofuryl), nefopam, ofloxacin, orphenadrine, oxomemazine, oxprenolol, phenoxybenzamine, pholedrine, pindolol, pirbuterol, prilocaine, promethazine, propafenone, propranolol, sotalol, synephrine, tetrahydrozoline (tetryzoline), terbutaline, tocainide, trimeprazine (alimemazine), trimipramine, tropicamide, verapamil, zopiclone

KEY WORDS
coated capillary; achiral

REFERENCE
Koppenhoefer,B.; Epperlein,U.; Christian,B.; Yibing,J.; Yuying,C.; Bingcheng,L. Separation of enantiomers of drugs by capillary electrophoresis. I. γ-Cyclodextrin as chiral solvating agent, *J.Chromatogr.A*, **1995**, *717*, 181–190.

SAMPLE
Matrix: solutions

CAPILLARY ELECTROPHORESIS
Capillary: 60 cm × 75 μm fused-silica (51 cm to detector) (Supelco)
Capillary preparation: At the end of each day wash capillary with 200 mM NaOH for 10 min and with water for 10 min. Condition new capillaries by washing with 200 mM NaOH for 10 min, with water for 10 min, and with running buffer for 10 min.
Running buffer: 50 mM pH 7.3 Sodium borate buffer containing 3.3% succinyl β-cyclodextrin (Wacker Chemie, Munich, Germany)
Injection: Hydrodynamic injection at 25 mbar for 6 s.
Detector: UV 208
Migration time: 78.93 (first enantiomer)
Voltage: 15 kV
Model: Prince

OTHER SUBSTANCES
Simultaneous: alprenolol, chlorthalidone, ephedrine, etilefrin, homatropine, methoxamine, norephedrine, octopamine, propranolol, synephrine

KEY WORDS
chiral; α = 1.0228

REFERENCE
Schmid,M.G.; Wirnsberger,K.; Gübitz,G. Chiral separation of drug enantiomers by capillary electrophoresis using succinyl-β-cyclodextrin, *Pharmazie*, **1996**, *51*, 852–854.

SAMPLE
Matrix: solutions
Sample preparation: Inject an aliquot of a solution in running buffer.

CAPILLARY ELECTROPHORESIS
Capillary: 60 cm × 75 μm fused-silica (52.4 cm to detector)
Capillary preparation: After each run flush with 500 mM KOH for 2-3 min then with water.
Running buffer: 10 mM pH 3.8 Phosphate buffer containing 2% sulfated cyclodextrin (ds 7-10)
Injection: Hydrostatic injection.
Detector: UV 214
Migration time: 8.25, 8.37 (enantiomers)
Voltage: 15 kV
Model: Waters Quanta 4000

OTHER SUBSTANCES
Also analyzed: acebutolol, alprenolol, aminoglutethimide, brompheniramine, bupivacaine, bupropion, canadine, carbinoxamine, chloroquine, chlorpheniramine, dimethindene, disopyramide, doxylamine, hydroxychloroquine, idazoxan, isoxsuprine, ketamine, mepenzolate, mepivacaine, methoxyphenamine, mexiletine, midodrine, nefopam, orphenadrine, oxprenolol, oxyphencyclimine, pheniramine, phensuximide, pindolol, piperoxan, terbutaline, tetramisole, tolperisone, tranylcypromine, trimipramine, verapamil, warfarin

KEY WORDS
chiral; detector at anode

REFERENCE
Stalcup,A.M.; Gahm,K.H. Application of sulfated cyclodextrins to chiral separations by capillary zone electrophoresis, *Anal.Chem.*, **1996**, *68*, 1360–1368.

SAMPLE
Matrix: solutions

CAPILLARY ELECTROPHORESIS
Capillary: 42 cm × 50 μm fused-silica (31 cm to detector) (Polymicro Technologies)
Capillary preparation: Condition capillary by rinsing with running buffer for 10 min, with water for 10 min, with 1 M NaOH for 10 min, with water for 10 min, and with running buffer with the voltage applied for 20 min.
Capillary temperature: 25
Running buffer: Formamide containing 150 mM citric acid, 100 mM Tris, and 100 mM methyl-β-cyclodextrin (apparent pH 5.1)
Injection: Gravity injection.
Detector: UV 254
Migration time: 10.86, 10.98 (enantiomers)
Voltage: 30 kV
Model: Beckman P/ACE 5500

OTHER SUBSTANCES
Simultaneous: primaquine

KEY WORDS
chiral

REFERENCE
Wang,F.; Khaledi,M.G. Chiral separations by nonaqueous capillary electrophoresis, *Anal.Chem.*, **1996**, *68*, 3460–3467.

SAMPLE
Matrix: solutions
Sample preparation: Inject an aliquot of a 100 μg/mL solution in running buffer.

CAPILLARY ELECTROPHORESIS
Capillary: 32 cm × 50 μm fused-silica (27.5 cm to detector) (Yongnian Optical Conductive Fiber Plant, China), coated with polyacrylamide
Capillary preparation: No details of the polyacrylamide coating process are provided. However, another paper (LC.GC 1997, 15, 40) by this group indicates that they use the procedure of Hjertén, thus: Adjust the pH of 20 mL water to 3.5 with acetic acid, add 80 μL 3-(trimethoxysilyl)propyl methacrylate (3-methacryloxypropyltrimethoxysilane), mix, suck into capillary, let stand at room temperature for 1 h, remove the solution, wash with water. Fill the capillary with a deaerated 3-4% acrylamide solution containing 1 μL/mL N,N,N',N'-tetramethylethylenediamine and 1 mg/mL potassium persulfate, let stand for 30 min, remove excess solution by aspiration, rinse with water, remove water by aspiration, dry at 35° (J. Chromatogr. 1985, 347, 191).
Capillary temperature: 25
Running buffer: 100 mM NaH_2PO_4 adjusted to pH 2.5
Injection: Electrokinetic injection at 15 kV for 3 s.

Detector: UV 200, UV 210
Migration time: 6.23
Voltage: 15 kV
Model: Bio-Rad BioFocus 3000

OTHER SUBSTANCES
Simultaneous: atropine, carazolol, cicletanine, dimethindene, fendiline, homatropine, ipratropium bromide, isothipendyl, mefloquine, metaclazepam, nafronyl (naftidrofuryl), nefopam, nicardipine, promethazine, reproterol, tetrahydrozoline (tetryzoline), theodrenaline, tioconazole, trimeprazine (alimemazine), trimipramine

KEY WORDS
coated capillary

REFERENCE
Koppenhoefer,B.; Epperlein,U.; Xiaofeng,Z.; Bingcheng,L. Separation of enantiomers of drugs by capillary electrophoresis. Part 4: Hydroxypropyl-γ-cyclodextrin as chiral solvating agent, *Electrophoresis*, **1997**, *18*, 924–930.

SAMPLE
Matrix: solutions
Sample preparation: Inject an aliquot of a 100 μg/mL solution in running buffer.

CAPILLARY ELECTROPHORESIS
Capillary: 30 cm × 50 μm fused-silica (25.5 cm to detector), coated with polyacrylamide
Capillary preparation: Adjust the pH of 20 mL water to 3.5 with acetic acid, add 80 μL 3-(trimethoxysilyl)propyl methacrylate (3-methacryloxypropyltrimethoxysilane), mix, suck into capillary, let stand at room temperature for 1 h, remove the solution, wash with water. Fill the capillary with a deaerated 3-4% acrylamide solution containing 1 μL/mL N,N,N',N'-tetramethylethylenediamine and 1 mg/mL potassium persulfate, let stand for 30 min, remove excess solution by aspiration, rinse with water, remove water by aspiration, dry at 35° (J. Chromatogr. 1985, 347, 191).
Capillary temperature: 25
Running buffer: 100 mM NaH_2PO_4 containing 45 mM hydroxypropyl-α-cyclodextrin, adjusted to pH 2.5 with phosphoric acid
Injection: Electrokinetic injection at 15 kV for 3 s.
Detector: UV 200
Voltage: 15 kV
Model: Bio-Rad BioFocus 3000

KEY WORDS
chiral; coated capillary; comparison with the use of other cyclodextrins; this running buffer gave the greatest enantiomeric separation.; α=1.014

REFERENCE
Lin,B.; Zhu,X.; Koppenhoefer,B.; Epperlein,U. Investigation of 123 chiral drugs by cyclodextrin-modified capillary electrophoresis, *LC.GC*, **1997**, *15*, 40–46.

SAMPLE
Matrix: solutions

CAPILLARY ELECTROPHORESIS
Capillary: 36 cm × 50 μm fused-silica coated with linear polyacrylamide (31.5 cm to detector) (GL Science)
Capillary preparation: At the beginning and end of each day rinse capillary with capillary wash solution (Bio-Rad Cat. No. 148-5022) at 690 kPa for more than 3 min and with water at 690 kPa for more than 3 min. Coat capillary as follows. Treat capillary with 1 M NaOH at room temperature for 1 h, rinse with water, dry by passing nitrogen gas through the capillary at 110° for 6 h. Pass thionyl chloride through the capillary using a suction pump for several min, seal capillary at both ends and heat at 70° for 6 h. Unseal the capillary and fill with 250 mM vinyl magnesium bromide in THF by suction, seal the capillary, heat at 70° for 6 h. Open the capillary and rinse it with THF for several min, rinse with distilled water, fill the capillary

with polymerization solution, heat at 28 ± 2° for 1 h, condition at -100 V/cm for 30 min (Anal. Sci. 1994, 10, 1). (The polymerization solution was 5% acrylamide in water containing 49 mM Tris, 384 mM glycine, and 0.1% sodium dodecyl sulfate, degas in an ultrasonic bath. Add 40 μL 10% N,N,N',N'-tetramethylethylenediamine and 10 μL 10% ammonium persulfate to 5 mL of the degassed solution, mix thoroughly.)

Running buffer: n-Propanol:50 mM pH 6.0 Phosphate buffer 10:90

Injection: Before each injection rinse with water at 690 kPa for 30 s, rinse with running buffer at 690 kPa for 30 s, partially fill with separation solution (50 μM α₁-acid glycoprotein (Cohn fraction VI) (Fluka) in running buffer) at 6.9 kPa for 190 s (27 cm), inject sample at 6.9 kPa for 2 s, electrophorese with running buffer (Note that α₁-acid glycoprotein from other suppliers may provide inferior results).

Detector: UV 210

Migration time: Resolution of enantiomers 3.6

Voltage: 12 kV

Model: Bio-rad BioFocus 3000

KEY WORDS
chiral; coated capillary

REFERENCE
Tanaka,Y.; Terabe,S. Separation of the enantiomers of basic drugs by affinity capillary electrophoresis using a partial filling technique and α₁-acid glycoprotein as chiral selector, *Chromatographia*, **1997**, *44*, 119–128.

SAMPLE
Matrix: solutions

CAPILLARY ELECTROPHORESIS
Capillary: 29-36 cm × 50 μm fused-silica (24.5-31.5 cm to detector) (Yongnian Optical Conductive Fiber Plant, China) coated with polyacrylamide

Capillary preparation: Coat capillary as follows. Adjust the pH of 20 mL water to 3.5 with acetic acid, add 80 μL 3-(trimethoxysilyl)propyl methacrylate (3-methacryloxypropyltrimethoxysilane), mix, suck into capillary, let stand at room temperature for 1 h, remove the solution, wash with water. Fill the capillary with a deaerated 3-4% acrylamide solution containing 1 μL/mL N,N,N',N'-tetramethylethylenediamine and 1 mg/mL potassium persulfate, let stand for 30 min, remove excess solution by aspiration, rinse with water, remove water by aspiration, dry at 35° (J. Chromatogr. 1985, 347, 191).

Capillary temperature: 25

Running buffer: 100 mM pH 2.5 NaH₂PO₄ (A) or 100 mM pH 2.5 NaH₂PO₄ containing 45 mM hydroxypropyl-α-cyclodextrin (Wacker, Munich) (B)

Injection: Electromigration at 15 kV for 3 s.

Detector: UV 200; UV 210

Migration time: 6.23 (A); 13.57, 13.77 (B) (enantiomers)

Voltage: 15 kV

Model: Bio-Focus 3000

OTHER SUBSTANCES
Also analyzed: albuterol (salbutamol), alprenolol, amorolfine, atenolol, atropine, azelastine, baclofen, bamethan, benproperine, benserazide, biperiden, bisoprolol, brompheniramine, bupivacaine, bupranolol, butamirate, butethamate, carazolol, carbuterol, carteolol, carvedilol, celiprolol, chloroquine, chlorpheniramine, chlorphenoxamine, cicletanine, clenbuterol, clidinium bromide, clobutinol, dimethindene, dipivefrin, disopyramide, dobutamine, doxylamine, fendiline, flecainide, gallopamil, homatropine, ipratropium bromide, isoproterenol (isoprenaline), isothipendyl, ketamine, meclizine, mefloquine, mepindolol, mequitazine, metaclazepam, metaproterenol (orciprenaline), metipranolol, metoprolol, nafronyl (naftidrofuryl), nefopam, nicardipine, norfenefrine, ofloxacin, ornidazole, orphenadrine, oxomemazine, oxprenolol, oxybutynin, phenoxybenzamine, phenylpropanolamine, pholedrine, pindolol, pirbuterol, prilocaine, procyclidine, promethazine, propafenone, propranolol, reproterol, sotalol, sulpride, synephrine, talinolol, terbutaline, tetrahydrozoline (tetryzoline), theodrenaline, tioconazole, tocainide, trimeprazine (alimemazine), trimipramine, tropicamide, verapamil, zopiclone

---

KEY WORDS
coated capillary; chiral

REFERENCE

Koppenhoefer,B.; Eperlein,U.; Schlunk,R.; Zhu,X.; Lin,B. Separation of enantiomers of drugs by capillary electrophoresis. V. Hydroxypropyl-α-cyclodextrin as chiral solvating agent, *J.Chromatogr.A*, **1998**, *793*, 153–164.

Trimebutine

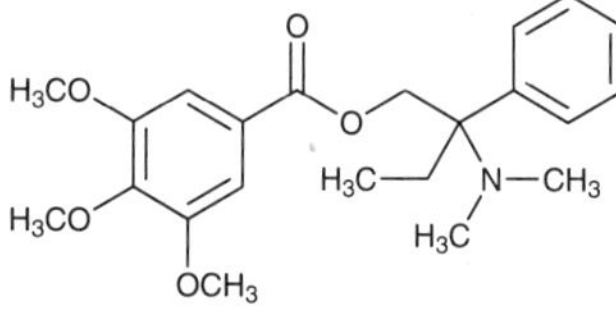

Molecular formula: $C_{22}H_{29}NO_5$
Molecular weight: 387.48
CAS Registry No.: 39133-31-8, 34140-59-5 (maleate)
Merck Index (12th ed.): 9829

SAMPLE
Matrix: solutions

CAPILLARY ELECTROPHORESIS
Capillary: 36 cm × 50 μm fused-silica coated with linear polyacrylamide (31.5 cm to detector) (GL Science)
Capillary preparation: At the beginning and end of each day rinse capillary with capillary wash solution (Bio-Rad Cat. No. 148-5022) at 690 kPa for more than 3 min and with water at 690 kPa for more than 3 min. Coat capillary as follows. Treat capillary with 1 M NaOH at room temperature for 1 h, rinse with water, dry by passing nitrogen gas through the capillary at 110° for 6 h. Pass thionyl chloride through the capillary using a suction pump for several min, seal capillary at both ends and heat at 70° for 6 h. Unseal the capillary and fill with 250 mM vinyl magnesium bromide in THF by suction, seal the capillary, heat at 70° for 6 h. Open the capillary and rinse it with THF for several min, rinse with distilled water, fill the capillary with polymerization solution, heat at 28 ± 2° for 1 h, condition at -100 V/cm for 30 min (Anal. Sci. 1994, 10, 1). (The polymerization solution was 5% acrylamide in water containing 49 mM Tris, 384 mM glycine, and 0.1% sodium dodecyl sulfate, degas in an ultrasonic bath. Add 40 μL 10% N,N,N',N'-tetramethylethylenediamine and 10 μL 10% ammonium persulfate to 5 mL of the degassed solution, mix thoroughly.)
Running buffer: n-Propanol:50 mM pH 6.0 Phosphate buffer 10:90
Injection: Before each injection rinse with water at 690 kPa for 30 s, rinse with running buffer at 690 kPa for 30 s, partially fill with separation solution (100 μM α_1-acid glycoprotein (Cohn fraction VI) (Sigma) in running buffer) at 6.9 kPa for 190 s (27 cm), inject sample at 6.9 kPa for 2 s, electrophorese with running buffer (Note that α_1-acid glycoprotein from other suppliers may provide inferior results).
Detector: UV 210
Migration time: Resolution of enantiomers 15.6
Voltage: 12 kV
Model: Bio-rad BioFocus 3000

KEY WORDS
chiral; coated capillary

REFERENCE

Tanaka,Y.; Terabe,S. Separation of the enantiomers of basic drugs by affinity capillary electrophoresis using a partial filling technique and α_1-acid glycoprotein as chiral selector, *Chromatographia*, **1997**, *44*, 119–128.

Trimeprazine

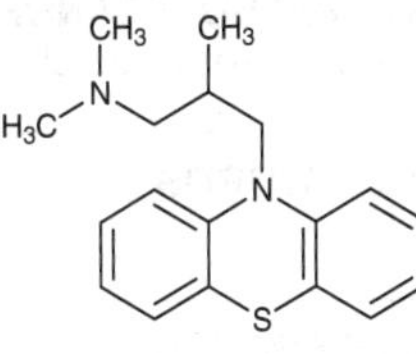

Molecular formula: $C_{18}H_{22}N_2S$
Molecular weight: 298.45
CAS Registry No.: 84-96-8, 4330-99-8 (tartrate)
Merck Index (12th ed.): 9834
Lednicer: 1 378

SAMPLE
Matrix: solutions
Sample preparation: Inject an aliquot of a 100 µg/mL solution in water:running buffer 50:50.

CAPILLARY ELECTROPHORESIS
Capillary: 44.5 cm × 50 µm acrylamide-coated fused-silica (Bio-Rad)
Capillary temperature: 30
Running buffer: 100 mM NaH_2PO_4 containing 15 mM gamma-cyclodextrin, adjusted to pH 2.5 with phosphoric acid
Injection: Electrokinetic injection at 8 kV for 6 s.
Detector: UV 200
Migration time: 14.49
Voltage: 14 kV
Model: Bio-Rad BioFocus 3000

OTHER SUBSTANCES
Also analyzed: albuterol, alprenolol, atenolol, atropine, baclofen, bamethan, benserazide, biperiden, bisoprolol, bupivacaine, bupranolol, butetamate, carazolol, carbuterol, carvedilol, celiprolol, chloroquine, chlorpheniramine (chlorphenamine), clidinium bromide, clobutinol, disopyramide, dobutamine, flecainide, homatropine, ipratropium bromide, isoproterenol, isothipendyl, ketamine, mefloquine, mequitazine, metaproterenol (orciprenaline), metipranolol, nafronyl (naftidrofuryl), nefopam, ofloxacin, orphenadrine, oxomemazine, oxprenolol, phenoxybenzamine, pholedrine, pindolol, pirbuterol, prilocaine, promethazine, propafenone, propranolol, sotalol, synephrine, terbutaline, tetrahydrozoline (tetryzoline), tocainide, trihexyphenidyl, trimipramine, tropicamide, verapamil, zopiclone

KEY WORDS
coated capillary; achiral

REFERENCE
Koppenhoefer,B.; Epperlein,U.; Christian,B.; Yibing,J.; Yuying,C.; Bingcheng,L. Separation of enantiomers of drugs by capillary electrophoresis. I. γ-Cyclodextrin as chiral solvating agent, *J.Chromatogr.A*, **1995**, *717*, 181–190.

SAMPLE
Matrix: solutions

CAPILLARY ELECTROPHORESIS
Capillary: 42 cm × 50 µm fused-silica (31 cm to detector) (Polymicro Technologies)
Capillary preparation: Condition capillary by rinsing with running buffer for 10 min, with water for 10 min, with 1 M NaOH for 10 min, with water for 10 min, and with running buffer with the voltage applied for 20 min.
Capillary temperature: 25
Running buffer: Formamide containing 150 mM citric acid, 100 mM Tris, and 100 mM β-cyclodextrin (apparent pH 5.1)
Injection: Gravity injection.
Detector: UV 254
Migration time: 9.64, 9.78 (enantiomers)
Voltage: 30 kV
Model: Beckman P/ACE 5500

OTHER SUBSTANCES
Simultaneous: chlophedianol, mianserin, nefopam, propiomazine, trimipramine, thioridazine

KEY WORDS
chiral

REFERENCE
Wang,F.; Khaledi,M.G. Chiral separations by nonaqueous capillary electrophoresis, *Anal.Chem.*, **1996**, *68*, 3460–3467.

SAMPLE
Matrix: solutions
Sample preparation: Inject an aliquot of a 100 μg/mL solution in running buffer.

CAPILLARY ELECTROPHORESIS
Capillary: 32 cm × 50 μm fused-silica (27.5 cm to detector) (Yongnian Optical Conductive Fiber Plant, China), coated with polyacrylamide
Capillary preparation: No details of the polyacrylamide coating process are provided. However, another paper (LC.GC 1997, 15, 40) by this group indicates that they use the procedure of Hjertén, thus: Adjust the pH of 20 mL water to 3.5 with acetic acid, add 80 μL 3-(trimethoxysilyl)propyl methacrylate (3-methacryloxypropyltrimethoxysilane), mix, suck into capillary, let stand at room temperature for 1 h, remove the solution, wash with water. Fill the capillary with a deaerated 3-4% acrylamide solution containing 1 μL/mL N,N,N',N'-tetramethylethylenediamine and 1 mg/mL potassium persulfate, let stand for 30 min, remove excess solution by aspiration, rinse with water, remove water by aspiration, dry at 35° (J. Chromatogr. 1985, 347, 191).
Capillary temperature: 25
Running buffer: 100 mM NaH_2PO_4 adjusted to pH 2.5 (A) or 100 mM NaH_2PO_4 containing 45 mM hydroxypropyl-gamma-cyclodextrin, adjusted to pH 2.5 (B)
Injection: Electrokinetic injection at 15 kV for 3 s.
Detector: UV 200, UV 210
Migration time: 5.18 (A), 9.70, 9.97 (B, enantiomers)
Voltage: 15 kV
Model: Bio-Rad BioFocus 3000

OTHER SUBSTANCES
Simultaneous: atropine, carazolol, cicletanine, dimethindene, fendiline, homatropine, ipratropium bromide, isothipendyl, mefloquine, metaclazepam, nafronyl (naftidrofuryl), nefopam, nicardipine, promethazine, reproterol, tetrahydrozoline (tetryzoline), theodrenaline, tioconazole, trihexyphenidyl, trimipramine

KEY WORDS
coated capillary; chiral

REFERENCE
Koppenhoefer,B.; Epperlein,U.; Xiaofeng,Z.; Bingcheng,L. Separation of enantiomers of drugs by capillary electrophoresis. Part 4: Hydroxypropyl-γ-cyclodextrin as chiral solvating agent, *Electrophoresis*, **1997**, *18*, 924–930.

SAMPLE
Matrix: solutions
Sample preparation: Inject an aliquot of a 100 μg/mL solution in running buffer.

CAPILLARY ELECTROPHORESIS
Capillary: 30 cm × 50 μm fused-silica (25.5 cm to detector), coated with polyacrylamide
Capillary preparation: Adjust the pH of 20 mL water to 3.5 with acetic acid, add 80 μL 3-(trimethoxysilyl)propyl methacrylate (3-methacryloxypropyltrimethoxysilane), mix, suck into capillary, let stand at room temperature for 1 h, remove the solution, wash with water. Fill the capillary with a deaerated 3-4% acrylamide solution containing 1 μL/mL N,N,N',N'-tetramethylethylenediamine and 1 mg/mL potassium persulfate, let stand for 30 min, remove excess

solution by aspiration, rinse with water, remove water by aspiration, dry at 35° (J. Chromatogr. 1985, 347, 191).
Capillary temperature: 25
Running buffer: 100 mM NaH_2PO_4 containing 45 mM hydroxypropyl-gamma-cyclodextrin, adjusted to pH 2.5 with phosphoric acid
Injection: Electrokinetic injection at 15 kV for 3 s.
Detector: UV 200
Voltage: 15 kV
Model: Bio-Rad BioFocus 3000

KEY WORDS
chiral; coated capillary; comparison with the use of other cyclodextrins; this running buffer gave the greatest enantiomeric separation.; α=1.021

REFERENCE
Lin,B.; Zhu,X.; Koppenhoefer,B.; Epperlein,U. Investigation of 123 chiral drugs by cyclodextrin-modified capillary electrophoresis, *LC.GC*, **1997**, *15*, 40–46.

SAMPLE
Matrix: solutions

CAPILLARY ELECTROPHORESIS
Capillary: 29-36 cm $\times$ 50 μm fused-silica (24.5-31.5 cm to detector) (Yongnian Optical Conductive Fiber Plant, China) coated with polyacrylamide
Capillary preparation: Coat capillary as follows. Adjust the pH of 20 mL water to 3.5 with acetic acid, add 80 μL 3-(trimethoxysilyl)propyl methacrylate (3-methacryloxypropyltrimethoxysilane), mix, suck into capillary, let stand at room temperature for 1 h, remove the solution, wash with water. Fill the capillary with a deaerated 3-4% acrylamide solution containing 1 μL/mL N,N,N',N'-tetramethylethylenediamine and 1 mg/mL potassium persulfate, let stand for 30 min, remove excess solution by aspiration, rinse with water, remove water by aspiration, dry at 35° (J. Chromatogr. 1985, 347, 191).
Capillary temperature: 25
Running buffer: 100 mM pH 2.5 NaH_2PO_4 (A) or 100 mM pH 2.5 NaH_2PO_4 containing 45 mM hydroxypropyl-α-cyclodextrin (Wacker, Munich) (B)
Injection: Electromigration at 15 kV for 3 s.
Detector: UV 200; UV 210
Migration time: 5.18 (A); 11.92, 12.13 (B) (enantiomers)
Voltage: 15 kV
Model: Bio-Focus 3000

OTHER SUBSTANCES
Also analyzed: albuterol (salbutamol), alprenolol, amorolfine, atenolol, atropine, azelastine, baclofen, bamethan, benproperine, benserazide, biperiden, bisoprolol, brompheniramine, bupivacaine, bupranolol, butamirate, butethamate, carazolol, carbuterol, carteolol, carvedilol, celiprolol, chloroquine, chlorpheniramine, chlorphenoxamine, cicletanine, clenbuterol, clidinium bromide, clobutinol, dimethindene, dipivefrin, disopyramide, dobutamine, doxylamine, fendiline, flecainide, gallopamil, homatropine, ipratropium bromide, isoproterenol (isoprenaline), isothipendyl, ketamine, meclizine, mefloquine, mepindolol, mequitazine, metaclazepam, metaproterenol (orciprenaline), metipranolol, metoprolol, nafronyl (naftidrofuryl), nefopam, nicardipine, norfenefrine, ofloxacin, ornidazole, orphenadrine, oxomemazine, oxprenolol, oxybutynin, phenoxybenzamine, phenylpropanolamine, pholedrine, pindolol, pirbuterol, prilocaine, procyclidine, promethazine, propafenone, propranolol, reproterol, sotalol, sulpride, synephrine, talinolol, terbutaline, tetrahydrozoline (tetryzoline), theodrenaline, tioconazole, tocainide, trihexyphenidyl, trimipramine, tropicamide, verapamil, zopiclone

KEY WORDS
coated capillary; chiral

REFERENCE
Koppenhoefer,B.; Eperlein,U.; Schlunk,R.; Zhu,X.; Lin,B. Separation of enantiomers of drugs by capillary electrophoresis. V. Hydroxypropyl-α-cyclodextrin as chiral solvating agent, *J.Chromatogr.A*, **1998**, *793*, 153–164.

SAMPLE
Matrix: urine
Sample preparation: Dilute 10-fold with water, inject an aliquot.

CAPILLARY ELECTROPHORESIS
Capillary: 70 cm × 50 μm fused-silica (65.4 cm to detector) (Polymicro Technologies)
Capillary temperature: 25
Running buffer: 20 mM pH 5.0 Tris-acetic acid containing 50 μg/mL FC-135 (Fluorad/3M) and 10 mM cetyltrimethylammonium bromide
Injection: Pressure injection at 2 psi.
Detector: UV 240
Migration time: 10.99
Voltage: 20 kV
Model: Bio-Rad BioFocus 3000

OTHER SUBSTANCES
Extracted: acepromazine, chlorpromazine, ethopropazine, fluphenazine, methotrimeprazine, perphenazine, thioridazine, triflupromazine

KEY WORDS
detector at anode

REFERENCE
Muijelsaar,P.G.H.M.; Claessens,H.A.; Cramers,C.A. Determination of structurally related phenothiazines by capillary zone electrophoresis and micellar electrokinetic chromatography, *J.Chromatogr.A*, **1996**, *735*, 395–402.

Trimethoprim

Molecular formula: $C_{14}H_{18}N_4O_3$
Molecular weight: 290.32
CAS Registry No.: 738-70-5
Merck Index (12th ed.): 9840
Lednicer: 1 262

SAMPLE
Matrix: solutions

CAPILLARY ELECTROPHORESIS
Capillary: 90 cm × 50 μm untreated fused-silica (Polymicro Technologies)
Running buffer: Trisma buffer pH 7.2 (about 7 g Trisma HCl and 0.7 g Trisma base in 1 L water)
Injection: 3 s vacuum injection
Detector: UV 200 or MS, SCIEX API III triple quadrupole, API source, positive ion analysis, make-up flow 0.2% aqueous formic acid, nebulizing gas air, target gas argon, collision energy 20-35 eV
Migration time: 7
Voltage: 24.4 kV
Model: Applied Biosystems Model 270A

OTHER SUBSTANCES
Simultaneous: erythromycin, ormetoprim, sulfamethazine, sulfaisomidine

REFERENCE
Pleasance,S.; Thibault,P.; Kelly,J. Comparison of liquid-junction and coaxial interfaces for capillary electrophoresis-mass spectrometry with application to compounds of concern to the aquaculture industry, *J.Chromatogr.*, **1992**, *591*, 325–339.

SAMPLE
Matrix: solutions

CAPILLARY ELECTROPHORESIS
Capillary: 57 cm × 50 μm eCAP polyamine-coated (50 cm to detector) (Beckman)
Capillary preparation: At the end of each run flush with 1 M NaOH for 2 min, wash with regenerator (Beckman) for 2 min, wash with running buffer for 2 min. Condition a new capillary with running buffer for 2 min and with regenerator for 2 min.
Capillary temperature: 30
Running buffer: 50 mM pH 4.2 Sodium acetate
Detector: UV 254
Migration time: 6
Voltage: 20 kV
Model: Beckman P/ACE 2210

OTHER SUBSTANCES
Simultaneous: impurities

KEY WORDS
coated capillary; comparison with uncoated capillary

REFERENCE
Assi,K.A.; Altria,K.D.; Clark,B.J. Rapid resolution of drugs and related substances with an eCAP™ polyamine coated capillary, *J.Pharm.Biomed.Anal.*, **1997**, *15*, 1041–1049.

SAMPLE
Matrix: solutions

CAPILLARY ELECTROPHORESIS
Capillary: 60 cm × 50 μm fused-silica (47 cm to detector) (Polymicro Technologies)
Capillary temperature: 25
Running buffer: 20 mM pH 8.5 Sodium tetraborate containing 100 mM sodium dodecyl sulfate
Injection: Hydrodynamic injection at 50 mbar for 3.6 s (5 nL).
Detector: UV 205
Migration time: 23.3
Voltage: 22 kV
Model: Crystal 310 (Thermo Unicam)

OTHER SUBSTANCES
Simultaneous: amoxicillin, ampicillin, cephapirin, cloxacillin, dicloxacillin, oxacillin, penicillin G, penicillin V, piperacillin, pyrimethamine, sulfacetamide, sulfadimethoxine, sulfaguanidine, sulfamerazine, sulfameter, sulfamethazine, sulfanilamide, sulfanilic acid, sulfapyridine, sulfaquinoxaline, sulfathiazole, sulfisoxazole

REFERENCE
Hows,M.E.P.; Perrett,D.; Kay,J. Optimization of a simultaneous separation of sulphonamides, dihydrofolate reductase inhibitors and β-lactam antibiotics by capillary electrophoresis, *J.Chromatogr.A*, **1997**, *768*, 97–104.

SAMPLE
Matrix: solutions

CAPILLARY ELECTROPHORESIS
Capillary: 45 cm × 50 μm fused-silica (37.5 cm to detector) (Polymicro Technologies)
Capillary preparation: Between each run rinse with 100 mM HCl for 2 min and with running buffer for 2 min. At the start of each day condition capillary with 1 M HCl for 10 min. Before the first use rinse capillary with 1 M NaOH for 30 min and with water for 30 min.
Capillary temperature: 25
Running buffer: MeCN:30 mM pH 2.40 ethanesulfonic acid 20:80
Injection: Injection at 4 kV for 3 s.
Detector: UV 214

Migration time: 5.73
Voltage: 20 kV
Model: Waters Quanta 4000E

OTHER SUBSTANCES
Simultaneous: chloramphenicol, imipramine, lidocaine, phenylpropanolamine, procainamide, quinidine

REFERENCE
Ding,W.; Fritz,J.S. Separation of neutral compounds and basic drugs by capillary electrophoresis in acidic solution using laurylpoly(oxyethylene) sulfate as an additive, *Anal.Chem.*, **1998**, *70*, 1859–1865.

SAMPLE
Matrix: solutions

CAPILLARY ELECTROPHORESIS
Capillary: 57 cm × 50 μm fused-silica (50 cm to detector)
Running buffer: 50 mM pH 8.0 Borate buffer containing 40 mM sodium taurodeoxycholate and 25 mM phosphatidylcholine (Prepare by adding phosphatidylcholine and stirring for 3-6 h until all cloudiness disappears. Phosphatidylcholine was 95% pure soybean lecithin, Epikuron, Lucas Meyer & Co.)
Injection: Inject a solution of the compound in the running buffer at 20 psi for 1 s, inject MeOH: water 5:95 at 20 psi for 1 s, inject a solution of halofantrine in running buffer at 20 psi for 1 s.
Detector: UV 214
Migration time: k' 1.29
Model: Beckman P/ACE 5000

OTHER SUBSTANCES
Also analyzed: acetaminophen, amoxicillin, antipyrine, aspirin, azathioprine, caffeine, captopril, carbamazepine, carprofen, chlorambucil, chlorpheniramine, chlorpromazine, cimetidine, clonidine, codeine, desipramine, diphenhydramine, ephedrine, fenoterol, flufenamic acid, flurbiprofen, haloperidol, hydroxyzine, ibuprofen, imipramine, indomethacin, ketoprofen, lidocaine, melphalan, metoprolol, nabumetone, nadolol, phenobarbital, phenol, promazine, propranolol, pyrilamine, ranitidine, ropinirole, salicylic acid, sulfamethoxazole, testosterone, theophylline, thioridazine, tiaprofenic acid, tolfenamic acid, trifluoperazine, valproic acid, verapamil

KEY WORDS
comparison with HPLC; k' = (Tr-T0)/(T0(1-Tr/Tm)) where Tr = retention time of analyte; T0 = retention time of water; and Tm = retention time of marker (halofantrine)

REFERENCE
Hanna,M.; de Biasi,V.; Bond,B.; Salter,C.; Hutt,A.J.; Camilleri,P. Estimation of the partitioning characteristics of drugs: A comparison of a large and diverse drug series utilizing chromatographic and electrophoretic methodology, *Anal.Chem.*, **1998**, *70*, 2092–2099.

SAMPLE
Matrix: tissue
Sample preparation: Homogenize (household food processor) 10 g tissue, extract (stomacher apparatus) with 100 mL MeCN for 5 min, centrifuge at 4000 g for 10 min, filter (0.45 μm), inject an aliquot of the filtrate.

CAPILLARY ELECTROPHORESIS
Capillary: 57 cm × 75 μm (50 cm to detector) (Beckman)
Capillary temperature: 25
Running buffer: 20 mM Borate buffer containing 20 mM phosphate, adjusted to pH 7.0 with KOH
Injection: Pressure injection for 2 s (39 nL)
Detector: UV 254
Migration time: 9.5
Voltage: 10 kV
Current: 26.8 μA

Model: Beckman P/ACE System 2000

OTHER SUBSTANCES
Extracted: sulfachlorpyridazine, sulfadiazine, sulfadimethoxine, sulfadoxine, sulfamerazine, sulfamethazine (sulfadimidine), sulfamethoxazole, sulfamethoxydiazine, sulfamethoxypyridazine, sulfaquinoxaline, sulfathiazole, sulfatroxazole

KEY WORDS
pig

REFERENCE
Ackermans,M.T.; Beckers,J.L.; Everaerts,F.M.; Hoogland,H.; Tomassen,M.J.H. Determination of sulphonamides in pork meat extracts by capillary zone electrophoresis, *J.Chromatogr.*, **1992**, *596*, 101–109.

Trimipramine

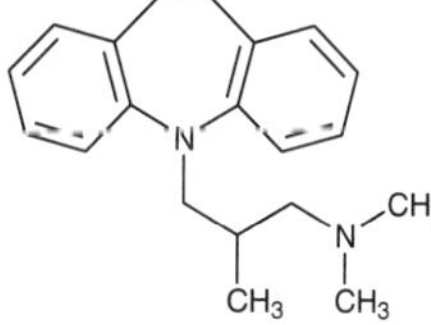

Molecular formula: $C_{20}H_{26}N_2$
Molecular weight: 294.44
CAS Registry No.: 739-71-9, 521-78-8 (maleate)
Merck Index (12th ed.): 9852

SAMPLE
Matrix: solutions
Sample preparation: Inject an aliquot of a 100 µg/mL solution in water:running buffer 50:50.

CAPILLARY ELECTROPHORESIS
Capillary: 44.5 cm × 50 µm acrylamide-coated fused-silica (Bio-Rad)
Capillary temperature: 30
Running buffer: 100 mM NaH_2PO_4 containing 15 mM gamma-cyclodextrin, adjusted to pH 2.5 with phosphoric acid
Injection: Electrokinetic injection at 8 kV for 6 s.
Detector: UV 200
Migration time: 20.13
Voltage: 14 kV
Model: Bio-Rad BioFocus 3000

OTHER SUBSTANCES
Also analyzed: albuterol, alprenolol, atenolol, atropine, baclofen, bamethan, benserazide, biperiden, bisoprolol, bupivacaine, bupranolol, butetamate, carazolol, carbuterol, carvedilol, celiprolol, chloroquine, chlorpheniramine (chlorphenamine), clidinium bromide, clobutinol, disopyramide, dobutamine, flecainide, homatropine, ipratropium bromide, isoproterenol, isothipendyl, ketamine, mefloquine, mequitazine, metaproterenol (orciprenaline), metipranolol, nafronyl (naftidrofuryl), nefopam, ofloxacin, orphenadrine, oxomemazine, oxprenolol, phenoxybenzamine, pholedrine, pindolol, pirbuterol, prilocaine, promethazine, propafenone, propranolol, sotalol, synephrine, tetrahydrozoline (tetryzoline), terbutaline, tocainide, trihexyphenidyl, trimeprazine (alimemazine), tropicamide, verapamil, zopiclone

KEY WORDS
coated capillary; achiral

REFERENCE
Koppenhoefer,B.; Epperlein,U.; Christian,B.; Yibing,J.; Yuying,C.; Bingcheng,L. Separation of enantiomers of drugs by capillary electrophoresis. I. γ-Cyclodextrin as chiral solvating agent, *J.Chromatogr.A*, **1995**, *717*, 181–190.

SAMPLE
Matrix: solutions

Sample preparation: Prepare a solution in MeOH/water, inject an aliquot.

CAPILLARY ELECTROPHORESIS
Capillary: 62 cm × 52 μm fused-silica (50 cm to detector) (Polymicro Technologies)
Capillary temperature: 40
Running buffer: 50 mM pH 2.50 Tetrabutylammonium phosphate containing 20 mM hydroxy-propyl-β-cyclodextrin
Injection: Siphon injection at 10 cm for 5 s
Detector: UV (wavelength not specified)
Migration time: 34.01, 34.37 (enantiomers)
Voltage: 20 kV
Model: Laboratory constructed

KEY WORDS
chiral

REFERENCE
Quang,C.; Khaledi,M.G. Extending the scope of chiral separation of basic compounds by cyclodextrin-mediated capillary zone electrophoresis, *J.Chromatogr.A*, **1995**, *692*, 253–265.

SAMPLE
Matrix: solutions
Sample preparation: Inject an aliquot of a solution in running buffer.

CAPILLARY ELECTROPHORESIS
Capillary: 60 cm × 75 μm fused-silica (52.4 cm to detector)
Capillary preparation: After each run flush with 500 mM KOH for 2-3 min then with water.
Running buffer: 10 mM pH 3.8 Phosphate buffer containing 2% sulfated cyclodextrin (ds 7-10)
Injection: Hydrostatic injection.
Detector: UV 214
Migration time: 8.55, 8.69 (enantiomers)
Voltage: 15 kV
Model: Waters Quanta 4000

OTHER SUBSTANCES
Also analyzed: acebutolol, alprenolol, aminoglutethimide, brompheniramine, bupivacaine, bupropion, canadine, carbinoxamine, chloroquine, chlorpheniramine, dimethindene, disopyramide, doxylamine, hydroxychloroquine, idazoxan, isoxsuprine, ketamine, mepenzolate, mepivacaine, methoxyphenamine, mexiletine, midodrine, nefopam, orphenadrine, oxprenolol, oxyphencyclimine, pheniramine, phensuximide, pindolol, piperoxan, terbutaline, tetramisole, tolperisone, tranylcypromine, trihexyphenidyl, verapamil, warfarin

KEY WORDS
chiral; detector at anode

REFERENCE
Stalcup,A.M.; Gahm,K.H. Application of sulfated cyclodextrins to chiral separations by capillary zone electrophoresis, *Anal.Chem.*, **1996**, *68*, 1360–1368.

SAMPLE
Matrix: solutions

CAPILLARY ELECTROPHORESIS
Capillary: 42 cm × 50 μm fused-silica (31 cm to detector) (Polymicro Technologies)
Capillary preparation: Condition capillary by rinsing with running buffer for 10 min, with water for 10 min, with 1 M NaOH for 10 min, with water for 10 min, and with running buffer with the voltage applied for 20 min.
Capillary temperature: 25
Running buffer: Formamide containing 150 mM citric acid, 100 mM Tris, and 100 mM β-cyclodextrin (A) or formamide containing 150 mM citric acid, 100 mM Tris, 250 mM β-cyclodextrin, and 100 mM tetrabutylammonium bromide (B) or formamide containing 150 mM citric

acid, 100 mM Tris, 250 mM β-cyclodextrin, and 100 mM tetramethylammonium bromide (C) (apparent pH 5.1 for all)
Injection: Gravity injection.
Detector: UV 254
Migration time: 9.03, 9.18 (A) or 27, 27.5 (B) or 44, 45.5 (C) (enantiomers)
Voltage: 30 kV
Model: Beckman P/ACE 5500

OTHER SUBSTANCES
Simultaneous: chlophedianol (A only), mianserin (A only), nefopam (A only), propiomazine (A only), trimeprazine (A only), thioridazine (A only)

KEY WORDS
chiral

REFERENCE
Wang,F.; Khaledi,M.G. Chiral separations by nonaqueous capillary electrophoresis, *Anal.Chem.*, **1996**, *68*, 3460–3467.

SAMPLE
Matrix: solutions
Sample preparation: Inject an aliquot of a 100 μg/mL solution in running buffer.

CAPILLARY ELECTROPHORESIS
Capillary: 32 cm × 50 μm fused-silica (27.5 cm to detector) (Yongnian Optical Conductive Fiber Plant, China), coated with polyacrylamide
Capillary preparation: No details of the polyacrylamide coating process are provided. However, another paper (LC.GC 1997, 15, 40) by this group indicates that they use the procedure of Hjertén, thus: Adjust the pH of 20 mL water to 3.5 with acetic acid, add 80 μL 3-(trimethoxysilyl)propyl methacrylate (3-methacryloxypropyltrimethoxysilane), mix, suck into capillary, let stand at room temperature for 1 h, remove the solution, wash with water. Fill the capillary with a deaerated 3-4% acrylamide solution containing 1 μL/mL N,N,N',N'-tetramethylethylenediamine and 1 mg/mL potassium persulfate, let stand for 30 min, remove excess solution by aspiration, rinse with water, remove water by aspiration, dry at 35° (J. Chromatogr. 1985, 347, 191).
Capillary temperature: 25
Running buffer: 100 mM NaH_2PO_4 adjusted to pH 2.5 (A) or 100 mM NaH_2PO_4 containing 45 mM hydroxypropyl-gamma-cyclodextrin, adjusted to pH 2.5 (B)
Injection: Electrokinetic injection at 15 kV for 3 s.
Detector: UV 200, UV 210
Migration time: 5.49 (A), 18.30, 18.54 (B, enantiomers)
Voltage: 15 kV
Model: Bio-Rad BioFocus 3000

OTHER SUBSTANCES
Simultaneous: atropine, carazolol, cicletanine, dimethindene, fendiline, homatropine, ipratropium bromide, isothipendyl, mefloquine, metaclazepam, nafronyl (naftidrofuryl), nefopam, nicardipine, promethazine, reproterol, tetrahydrozoline (tetryzoline), theodrenaline, tioconazole, trihexyphenidyl, trimeprazine (alimemazine)

KEY WORDS
coated capillary; chiral

REFERENCE
Koppenhoefer,B.; Epperlein,U.; Xiaofeng,Z.; Bingcheng,L. Separation of enantiomers of drugs by capillary electrophoresis. Part 4: Hydroxypropyl-γ-cyclodextrin as chiral solvating agent, *Electrophoresis*, **1997**, *18*, 924–930.

SAMPLE
Matrix: solutions
Sample preparation: Inject an aliquot of a 100 μg/mL solution in running buffer.

CAPILLARY ELECTROPHORESIS
Capillary: 30 cm × 50 μm fused-silica (25.5 cm to detector), coated with polyacrylamide
Capillary preparation: Adjust the pH of 20 mL water to 3.5 with acetic acid, add 80 μL 3-(trimethoxysilyl)propyl methacrylate (3-methacryloxypropyltrimethoxysilane), mix, suck into capillary, let stand at room temperature for 1 h, remove the solution, wash with water. Fill the capillary with a deaerated 3-4% acrylamide solution containing 1 μL/mL N,N,N',N'-tetramethylethylenediamine and 1 mg/mL potassium persulfate, let stand for 30 min, remove excess solution by aspiration, rinse with water, remove water by aspiration, dry at 35° (J. Chromatogr. 1985, 347, 191).
Capillary temperature: 25
Running buffer: 100 mM NaH_2PO_4 containing 45 mM gamma-cyclodextrin, adjusted to pH 2.5 with phosphoric acid
Injection: Electrokinetic injection at 15 kV for 3 s.
Detector: UV 200
Voltage: 15 kV
Model: Bio-Rad BioFocus 3000

KEY WORDS
chiral; coated capillary; comparison with the use of other cyclodextrins; this running buffer gave the greatest enantiomeric separation.; α=1.024

REFERENCE
Lin,B.; Zhu,X.; Koppenhoefer,B.; Epperlein,U. Investigation of 123 chiral drugs by cyclodextrin-modified capillary electrophoresis, *LC.GC*, **1997**, *15*, 40–46.

SAMPLE
Matrix: solutions

CAPILLARY ELECTROPHORESIS
Capillary: 70 cm × 50 μm coated fused-silica (55 cm to detector) (Polymicro Technologies)
Capillary preparation: Between runs rinse with water for 2 min. Coat capillaries as follows. Condition with 1 M NaOH for 15 min, rinse with water for 5 min, treat with 30 μL/mL 3-(trimethoxysilyl)propylmethacrylate in acetic acid:water 50:50 for 1 h using house vacuum, rinse with water, fill with polymerization solution, let stand for 1 h, flush, rinse with water. (Prepare polymerization solution by adding 10 μL N,N,N',N'-tetramethylethylenediamine to 10 mL of a degassed 4% solution of acrylamide in water, add 10 μL 10% ammonium persulfate in water.)
Running buffer: 10 mM pH 6.5 Citrate buffer containing 20 mM sodium dodecyl sulfate and 10 mM carboxymethyl-β-cyclodextrin (Cyclodextrin Technologies Development, Gainesville FL)
Injection: Hydrodynamic injection at 15 cm for 8-10 s.
Detector: UV 254
Migration time: 9.4
Voltage: -20 kV
Model: laboratory-constructed

OTHER SUBSTANCES
Simultaneous: amitriptyline, carbamazepine, clomipramine, desipramine, imipramine, nortriptyline, opipramol, proptriptyline

KEY WORDS
coated capillary

REFERENCE
Spencer,B.J.; Zhang,W.; Purdy,W.C. Capillary electrophoretic separation of tricyclic antidepressants using charged carboxymethyl-β-cyclodextrin as a buffer additive, *Electrophoresis*, **1997**, *18*, 736–744.

SAMPLE
Matrix: solutions

CAPILLARY ELECTROPHORESIS
Capillary: 36 cm × 50 μm fused-silica coated with linear polyacrylamide (31.5 cm to detector) (GL Science)

Capillary preparation: At the beginning and end of each day rinse capillary with capillary wash solution (Bio-Rad Cat. No. 148-5022) at 690 kPa for more than 3 min and with water at 690 kPa for more than 3 min. Coat capillary as follows. Treat capillary with 1 M NaOH at room temperature for 1 h, rinse with water, dry by passing nitrogen gas through the capillary at 110° for 6 h. Pass thionyl chloride through the capillary using a suction pump for several min, seal capillary at both ends and heat at 70° for 6 h. Unseal the capillary and fill with 250 mM vinyl magnesium bromide in THF by suction, seal the capillary, heat at 70° for 6 h. Open the capillary and rinse it with THF for several min, rinse with distilled water, fill the capillary with polymerization solution, heat at 28 $\pm$ 2° for 1 h, condition at -100 V/cm for 30 min (Anal. Sci. 1994, 10, 1). (The polymerization solution was 5% acrylamide in water containing 49 mM Tris, 384 mM glycine, and 0.1% sodium dodecyl sulfate, degas in an ultrasonic bath. Add 40 μL 10% N,N,N',N'-tetramethylethylenediamine and 10 μL 10% ammonium persulfate to 5 mL of the degassed solution, mix thoroughly.)

Running buffer: n-Propanol:50 mM pH 5.0 Phosphate buffer 10:90

Injection: Before each injection rinse with water at 690 kPa for 30 s, rinse with running buffer at 690 kPa for 30 s, partially fill with separation solution (120 μM α_1-acid glycoprotein (Cohn fraction VI) (Sigma) in running buffer) at 6.9 kPa for 190 s (27 cm), inject sample at 6.9 kPa for 2 s, electrophorese with running buffer (Note that α_1-acid glycoprotein from other suppliers may provide inferior results).

Detector: UV 210

Migration time: Resolution of enantiomers 9.9

Voltage: 12 kV

Model: Bio-rad BioFocus 3000

KEY WORDS
chiral; coated capillary

REFERENCE
Tanaka,Y.; Terabe,S. Separation of the enantiomers of basic drugs by affinity capillary electrophoresis using a partial filling technique and α_1-acid glycoprotein as chiral selector, *Chromatographia*, **1997**, *44*, 119–128.

SAMPLE
Matrix: solutions

CAPILLARY ELECTROPHORESIS
Capillary: 29-36 cm $\times$ 50 μm fused-silica (24.5-31.5 cm to detector) (Yongnian Optical Conductive Fiber Plant, China) coated with polyacrylamide

Capillary preparation: Coat capillary as follows. Adjust the pH of 20 mL water to 3.5 with acetic acid, add 80 μL 3-(trimethoxysilyl)propyl methacrylate (3-methacryloxypropyltrimethoxysilane), mix, suck into capillary, let stand at room temperature for 1 h, remove the solution, wash with water. Fill the capillary with a deaerated 3-4% acrylamide solution containing 1 μL/mL N,N,N',N'-tetramethylethylenediamine and 1 mg/mL potassium persulfate, let stand for 30 min, remove excess solution by aspiration, rinse with water, remove water by aspiration, dry at 35° (J. Chromatogr. 1985, 347, 191).

Capillary temperature: 25

Running buffer: 100 mM pH 2.5 NaH$_2$PO$_4$ (A) or 100 mM pH 2.5 NaH$_2$PO$_4$ containing 45 mM hydroxypropyl-α-cyclodextrin (Wacker, Munich) (B)

Injection: Electromigration at 15 kV for 3 s.

Detector: UV 200; UV 210

Migration time: 5.49 (A); 11.82 (B) (no separation of enantiomers)

Voltage: 15 kV

Model: Bio-Focus 3000

OTHER SUBSTANCES
Also analyzed: albuterol (salbutamol), alprenolol, amorolfine, atenolol, atropine, azelastine, baclofen, bamethan, benproperine, benserazide, biperiden, bisoprolol, brompheniramine, bupivacaine, bupranolol, butamirate, butethamate, carazolol, carbuterol, carteolol, carvedilol, celiprolol, chloroquine, chlorpheniramine, chlorphenoxamine, cicletanine, clenbuterol, clidinium bromide, clobutinol, dimethindene, dipivefrin, disopyramide, dobutamine, doxylamine, fendiline, flecainide, gallopamil, homatropine, ipratropium bromide, isoproterenol (isoprenaline), isothipendyl, ketamine, meclizine, mefloquine, mepindolol, mequitazine, metaclazepam, metaproterenol (orciprenaline), metipranolol, metoprolol, nafronyl (naftidrofuryl), nefopam, nicardipine, norfenefrine, ofloxacin, ornidazole, orphenadrine, oxomemazine, oxprenolol, oxybutynin,

phenoxybenzamine, phenylpropanolamine, pholedrine, pindolol, pirbuterol, prilocaine, procyclidine, promethazine, propafenone, propranolol, reproterol, sotalol, sulpride, synephrine, talinolol, terbutaline, tetrahydrozoline (tetryzoline), theodrenaline, tioconazole, tocainide, trihexyphenidyl, trimeprazine (alimemazine), tropicamide, verapamil, zopiclone

KEY WORDS
coated capillary

REFERENCE
Koppenhoefer,B.; Eperlein,U.; Schlunk,R.; Zhu,X.; Lin,B. Separation of enantiomers of drugs by capillary electrophoresis. V. Hydroxypropyl-α-cyclodextrin as chiral solvating agent, *J.Chromatogr.A*, **1998**, *793*, 153–164.

SAMPLE
Matrix: urine
Sample preparation: 10 mL Urine + 1 mL 5 M NaOH + 10 mL n-hexane, vortex for 1 min, centrifuge at 0° at 3000 g for 5 min. Remove the organic layer and add it to 50 μL glacial acetic acid, evaporate to dryness under a stream of nitrogen at 30°, reconstitute the residue in 50 μL 50 mM sodium taurodeoxycholate, filter (0.2 μm PTFE), inject an aliquot.

CAPILLARY ELECTROPHORESIS
Capillary: 100 cm × 50 μm fused-silica (50 cm to detector) (ISCO)
Capillary preparation: Rinse capillary with 20 μL running buffer between injections. Condition a new capillary by filling with 1 M NaOH, let stand for 1 h, fill with 100 mM NaOH, let stand for 1 h, rinse with buffer. Every 20 injections rinse capillary with 200 μL 1 M NaOH, 200 μL water, and 200 μL running buffer, fill with running buffer.
Capillary temperature: 22
Running buffer: 40 mM pH 9.5 Borate buffer containing 10 mM sodium taurodeoxycholate
Injection: Load under vacuum at 7.5 kPa/s.
Detector: UV 240
Migration time: 11
Voltage: 30 kV
Model: ISCO Model 3140 electropherograph
Limit of detection: 6 ng/mL

OTHER SUBSTANCES
Extracted: acepromazine, amiodarone, amitriptyline, azaperone, chlorpromazine, cianopramine, clomipramine, clozapine, desethylamiodarone, desipramine, diclofensine, dothiepin, doxepin, imipramine, isocarboxazid, moclobemide, perphenazine, phenothiazine, pimozide, prochlorperazine, promazine, thioridazine, thiothixene, trifluoperazine

KEY WORDS
human; cow; pig; horse; protect from light

REFERENCE
Aumatell,A.; Wells,R.J. Determination of a cardiac antiarrhythmic, tricyclic antipsychotics and antidepressants in human and animal urine by micellar electrokinetic capillary chromatography using a bile salt, *J.Chromatogr.B*, **1995**, *669*, 331–344.

Tripelennamine

Molecular formula: $C_{16}H_{21}N_3$
Molecular weight: 255.36
CAS Registry No.: 91-81-6, 6138-56-3 (citrate), 154-69-8 (HCl)
Merck Index (12th ed.): 9868
Lednicer: 1 51

SAMPLE
Matrix: solutions

CAPILLARY ELECTROPHORESIS
Capillary: 62 cm × 75 μm fused-silica (54.5 cm to detector) (Polymicro Technologies)
Capillary preparation: Purge with running buffer before each run. At the beginning of each day purge using 50-60 kPa vacuum with 500 mM NaOH for 5 min, with water for 5 min, with MeCN for 5 min, and with running buffer for 5 min.
Running buffer: MeCN:MeOH:acetic acid 49:50:1 containing 20 mM ammonium acetate
Injection: Hydrostatic injection at 10 cm for 5 s.
Detector: UV 214
Migration time: 2.97
Voltage: 30 kV
Model: Waters Quanta 4000

OTHER SUBSTANCES
Simultaneous: buclizine, chlorcyclizine, chlorpheniramine, cyclizine, dimenhydrinate, methaphenilene, pheniramine, promethazine, pyrilamine, pyrrobutamine

REFERENCE
Leung,G.N.W.; Tang,H.P.O.; Tso,T.S.C.; Wan,T.S.M. Separation of basic drugs with non-aqueous capillary electrophoresis, *J.Chromatogr.A*, **1996**, *738*, 141–154.

Triprolidine

Molecular formula: $C_{19}H_{22}N_2$
Molecular weight: 278.40
CAS Registry No.: 468-12-4, 6138-79-0 (HCl monohydrate), 550-70-9 (HCl)
Merck Index (12th ed.): 9877
Lednicer: 1 78

SAMPLE
Matrix: formulations
Sample preparation: Grind granules, add 70 mL MeOH, warm at 40° with occasional shaking, cool, add 20 mL 5 mg/mL methyl p-hydroxybenzoate in MeOH, make up to 100 mL with water, filter (0.45 μm), inject an aliquot.

CAPILLARY ELECTROPHORESIS
Capillary: 65 cm × 50 μm fused-silica (SGE)
Running buffer: 20 mM NaH_2PO_4 containing 100 mM sodium cholate, adjusted to pH 9.0 with 20 mM sodium tetraborate
Injection: Siphon for 10 s at 10 cm height
Detector: UV 210
Migration time: 15
Internal standard: methyl p-hydroxybenzoate (16)
Voltage: +20 kV

OTHER SUBSTANCES
Simultaneous: acetaminophen, caffeine, chlorpheniramine, dibucaine, dipyrone (sulpyrin), ethenzamide, guaifenesin, isopropylantipyrine, naproxen, noscapine, phenacetin, tipepidine, trimetoquinol

KEY WORDS
granules

REFERENCE
Nishi,H.; Fukuyama,T.; Matsuo,M.; Terabe,S. Separation and determination of the ingredients of a cold medicine by micellar electrokinetic chromatography with bile salts, *J.Chromatogr.*, **1990**, *498*, 313–323.

SAMPLE
Matrix: formulations
Sample preparation: Grind tablets, extract twice with MeOH by sonicating at low temperature, centrifuge, filter (0.45 μm) the supernatant, inject an aliquot of the filtrate.

CAPILLARY ELECTROPHORESIS
Capillary: 47 cm × 50 μm (40 cm to detector) (Polymicro Technologies)
Capillary temperature: 25
Running buffer: 50 mM pH 7.5 Borate buffer containing 50 mM phosphate, 10 mM sodium dodecyl sulfate, 10 mM β-cyclodextrin, and 10 mM tetrabutylammonium hydrogen sulfate
Injection: Inject at a pressure of 0.5 psi for 1 s (5.2 nL)
Detector: UV 214
Migration time: 4
Voltage: 21.5 kV
Model: Beckman P/ACE 2000
Limit of detection: 76.1 pg

OTHER SUBSTANCES
Simultaneous: chlorpheniramine, cyclizine, dimenhydrinate, doxylamine, methapyrilene, pheniramine, promethazine, thonylamine

KEY WORDS
tablets

REFERENCE
Ong,C.P.; Ng,C.L.; Lee,H.K.; Li,S.F.Y. Determination of antihistamines in pharmaceuticals by capillary electrophoresis, *J.Chromatogr.*, **1991**, *588*, 335–339.

Troglitazone

Molecular formula: $C_{24}H_{27}NO_5S$
Molecular weight: 441.55
CAS Registry No.: 97322-87-7
Merck Index (12th ed.): 9898

SAMPLE
Matrix: solutions
Sample preparation: Inject an aliquot of a solution in MeCN:MeOH 50:50 containing 2 mM sodium acetate.

CAPILLARY ELECTROPHORESIS
Capillary: 27 cm × 50 μm fused-silica (20 cm to detector) (Electro-Kinetic Technologies, Edinburgh, UK)
Capillary preparation: Between each injection rinse with 100 mM NaOH for 1 min, with MeOH for 1 min, and with running buffer for 1 min. Rinse new capillaries with 100 mM NaOH then with MeOH for 10 min.
Running buffer: MeCN:MeOH 50:50 containing 10 mM sodium acetate
Injection: Injection for 1 s.
Detector: UV 200
Migration time: 1.5
Voltage: 25 kV
Model: Beckman P/ACE 5000

OTHER SUBSTANCES
Simultaneous: benzoic acid, hydroxynaphthoate, naphthoxyacetic acid, penicillin (sic), warfarin

REFERENCE
Altria,K.D.; Bryant,S.M. Highly selective and efficient separations of a wide range of acidic species in capillary electrophoresis employing non-aqueous media, *Chromatographia*, **1997**, *46*, 122–122.

SAMPLE
Matrix: solutions
Sample preparation: Inject an aliquot of a solution in MeCN:water 50:50.

CAPILLARY ELECTROPHORESIS
Capillary: 27 cm × 100 μm (20 cm to detector) (Composite Metal Services)
Capillary temperature: 30
Running buffer: 15 mM Borate
Detector: UV 200
Migration time: 3
Internal standard: β-naphthoxyacetic acid (3.8)
Voltage: 6.5 kV
Model: Beckman
Limit of quantitation: 70 ng/mL
Limit of detection: 20 ng/mL

REFERENCE
Altria,K.D.; Creasey,E.; Howells,J.S. Routine capillary electrophoresis trace level determinations of pharmaceutical and detergent residues on pharmaceutical manufacturing equipment, *J.Liq.Chromatogr.Rel.Technol.*, **1998**, *21*, 1093–1105.

Troleandomycin

Molecular formula: $C_{41}H_{67}NO_{15}$
Molecular weight: 813.98
CAS Registry No.: 2751-09-9
Merck Index (12th ed.): 9899

SAMPLE
Matrix: solutions
Sample preparation: Inject an aliquot of a solution in MeCN:isopropanol:10 mM pH 7 sodium phosphate buffer 25:25:50.

CAPILLARY ELECTROPHORESIS
Capillary: 68.5 cm × 50 μm fused-silica (60 cm to detector) (Polymicro Technologies)
Capillary preparation: Rinse capillary with MeCN:water 50:50 for 2 min and with running buffer for 4 min before each run.
Capillary temperature: 15
Running buffer: MeCN:80 mM pH 6 phosphate buffer 20:80 containing 30 mM sodium cholate
Injection: Pressure injection at 50 mbar with 7.5 s for sample and 3.7 s for buffer plug
Detector: UV 205
Migration time: 15.5
Voltage: +20 kV
Model: Isco Model 3140 Electropherograph

OTHER SUBSTANCES
Simultaneous: clarithromycin, erythromycin, erythromycin estolate, erythromycin ethylsuccinate, oleandomycin, spiramycin

---

REFERENCE

Flurer,C.L. Analysis of macrolide antibiotics by capillary electrophoresis, *Electrophoresis*, **1996**, *17*, 359–366.

Tromethamine

Molecular formula: $C_4H_{11}NO_3$
Molecular weight: 121.14
CAS Registry No.: 77-86-1
Merck Index (12th ed.): 9902

SAMPLE
Matrix: formulations
Sample preparation: Dilute ophthalmic solution 100-fold with water, filter (0.45 μm nylon), inject an aliquot of the filtrate.

CAPILLARY ELECTROPHORESIS
Capillary: 50 cm × 50 μm fused-silica (Dionex)
Capillary preparation: Rinse capillary with running buffer for 2 min between injections. Condition new capillaries by flushing with 1 M NaOH for 45 min, rinsing with water for 30 min, and conditioning with running buffer for 1 h.
Running buffer: pH 6.5 Buffer (Prepare buffer by dissolving the required amount of dimethyldiphenylphosphonium hydroxide in water so as to form a 25 mM solution, heat at 80°C until dissolved, cool, pass through a Dionex anion-exchange SPE cartridge at 2 mL/min (discard first 2 mL of eluate). Dilute a 20 mL aliquot of the eluate to 100 mL, adjust the pH to 6.5 with α-hydroxyisobutyric acid, filter (0.45 μm nylon).)
Injection: Gravity injection at 10 cm for 30 s.
Detector: UV 215
Migration time: 2.4
Voltage: 20 kV
Model: Dionex CES1
Limit of detection: 2.5 μg/mL (S/N = 3)

KEY WORDS
ophthalmic solutions; indirect UV detection

REFERENCE

McArdle,F.A.; Meehan,C.J. Determination of tromethamine in an eye-care pharmaceutical by capillary electrophoresis, *Analyst*, **1998**, *123*, 1757–1760.

Tropicamide

Molecular formula: $C_{17}H_{20}N_2O_2$
Molecular weight: 284.36
CAS Registry No.: 1508-75-4
Merck Index (12th ed.): 9911

SAMPLE
Matrix: solutions

CAPILLARY ELECTROPHORESIS
Capillary: 57 cm × 75 μm (50 cm to detector)
Capillary temperature: 35
Running buffer: 150 mM pH 2.05 Na_2HPO_4 containing 3% β-cyclodextrin, 5% urea, and 40 mM (+)-10-camphorsulfonic acid

Detector: UV 254
Migration time: 21.87, 22.77 (enantiomers)
Voltage: 12 kV
Model: Beckman P/ACE 2100

OTHER SUBSTANCES
Simultaneous: cyclodrine

KEY WORDS
chiral

REFERENCE
Bunke,A.; Jira,T.; Gübitz,G. Chirale Trennung von Cyclodextrin mittels Kapillarelektrophorese, *Pharmazie*, **1995**, *50*, 570–571.

SAMPLE
Matrix: solutions
Sample preparation: Inject an aliquot of a 100 µg/mL solution in water:running buffer 50:50.

CAPILLARY ELECTROPHORESIS
Capillary: 44.5 cm × 50 µm acrylamide-coated fused-silica (Bio-Rad)
Capillary temperature: 30
Running buffer: 100 mM NaH_2PO_4 containing 15 mM gamma-cyclodextrin, adjusted to pH 2.5 with phosphoric acid
Injection: Electrokinetic injection at 8 kV for 6 s.
Detector: UV 200
Migration time: 10.46, 10.55 (enantiomers)
Voltage: 14 kV
Model: Bio-Rad BioFocus 3000

OTHER SUBSTANCES
Also analyzed: albuterol, alprenolol, atenolol, atropine, baclofen, bamethan, benserazide, biperiden, bisoprolol, bupivacaine, bupranolol, butetamate, carazolol, carbuterol, carvedilol, ccliprolol, chloroquine, chlorpheniramine (chlorphenamine), clidinium bromide, clobutinol, disopyramide, dobutamine, flecainide, homatropine, ipratropium bromide, isoproterenol, isothipendyl, ketamine, mefloquine, mequitazine, metaproterenol (orciprenaline), metipranolol, nafronyl (naftidrofuryl), nefopam, ofloxacin, orphenadrine, oxomemazine, oxprenolol, phenoxybenzamine, pholedrine, pindolol, pirbuterol, prilocaine, promethazine, propafenone, propranolol, sotalol, synephrine, terbutaline, tetrahydrozoline (tetryzoline), tocainide, trihexyphenidyl, trimeprazine (alimemazine), trimipramine, verapamil, zopiclone

KEY WORDS
chiral; coated capillary

REFERENCE
Koppenhoefer,B.; Epperlein,U.; Christian,B.; Yibing,J.; Yuying,C.; Bingcheng,L. Separation of enantiomers of drugs by capillary electrophoresis. I. γ-Cyclodextrin as chiral solvating agent, *J.Chromatogr.A*, **1995**, *717*, 181–190.

SAMPLE
Matrix: solutions
Sample preparation: Inject an aliquot of a 100 µg/mL solution in running buffer:water 50:50.

CAPILLARY ELECTROPHORESIS
Capillary: 44.5 cm × 50 µm polyacrylamide coated fused-silica (40 cm to detector) (Bio-Rad)
Running buffer: 100 mM NaH_2PO_4 containing 15 mM α-cyclodextrin, adjusted to pH 2.5 with phosphoric acid
Injection: Electromigration at 8 kV for 6 s
Detector: UV 200
Migration time: 9.63, 9.70 (enantiomers)
Voltage: 14 kV

Model: Bio-Rad Bio-Focus 3000

OTHER SUBSTANCES
Simultaneous: clidinium bromide, ketamine, orphenadrine, oxomemazine, tetrahydrozoline

KEY WORDS
chiral; α = 1.007; numerous drugs were not separated into their enantiomers under these conditions; coated capillary

REFERENCE
Bingcheng,L.; Yibing,J.; Yoying,C.; Epperlein,U.; Koppenhoefer,B. Separation of drug enantiomers by capillary electrophoresis: α-Cyclodextrin as chiral solvating agent, *Chromatographia*, **1996**, *42*, 106–110.

SAMPLE
Matrix: solutions

CAPILLARY ELECTROPHORESIS
Capillary: 37 cm $\times$ 50 μm (30 cm to detector)
Capillary temperature: 25
Running buffer: 50 mM pH 2.5 Phosphate buffer containing 16 mM 2-hydroxy-3-trimethylammoniopropyl-β-cyclodextrin
Detector: UV 214
Migration time: 18.34 (first enantiomer; α = 1.021; R_S = 0.548)
Voltage: 20 kV
Model: Beckman P/ACE 2100

OTHER SUBSTANCES
Simultaneous: cyclodrine, cyclopentolate, norpseudoephedrine, phendimetrazine, pholedrine

KEY WORDS
chiral

REFERENCE
Bunke,A.; Jira,T. Chiral capillary electrophoresis using a cationic cyclodextrin, *Pharmazie*, **1996**, *51*, 672–673.

SAMPLE
Matrix: solutions
Sample preparation: Inject an aliquot of a 100 μg/mL solution in water:running buffer 50:50.

CAPILLARY ELECTROPHORESIS
Capillary: 44.5 cm $\times$ 50 μm polyacrylamide-coated fused-silica (40 cm to detector) (Bio-Rad)
Capillary temperature: 30
Running buffer: 100 mM NaH_2PO_4 containing 15 mM β-cyclodextrin, adjusted to pH 2.5 with phosphoric acid
Injection: Electrokinetic injection at 8 kV for 6 s.
Detector: UV 200
Migration time: 10.89, 11.33 (enantiomers)
Voltage: 14 kV
Model: Bio-Rad Bio-Focus 3000

OTHER SUBSTANCES
Simultaneous: carvedilol, chlorpheniramine, ketamine, metaproterenol (orciprenaline), tetrahydrozoline, zopiclone
Also analyzed: albuterol (not chiral), atenolol (not chiral), benserazide (not chiral), biperiden (not chiral), bupivacaine (not chiral), butethamate (not chiral), carazolol (not chiral), carbuterol (not chiral), clidinium bromide (not chiral), disopyramide (not chiral), dobutamine (not chiral), flecainide (not chiral), ipratropium (not chiral), isothipendyl (not chiral), mequitazine (not chiral), metipranolol (not chiral), nafronyl (naftidrofuryl) (not chiral), nefopam (not chiral), ofloxacin (not chiral), orphenadrine (not chiral), pindolol (not chiral), pirbuterol (not chiral), prilocaine (not chiral), propafenone (not chiral), sotalol (not chiral), tocainide (not chiral), trimipramine (not chiral)

KEY WORDS
coated capillary; chiral

REFERENCE
Koppenhoefer,B.; Epperlein,U.; Christian,B.; Lin,B.; Ji,Y.; Chen,Y. Separation of enantiomers of drugs by capillary electrophoresis. III. β-cyclodextrin as chiral solvating agent, *J.Chromatogr.A*, **1996**, *735*, 333–343.

SAMPLE
Matrix: solutions
Sample preparation: Inject an aliquot of a 100 μg/mL solution in running buffer.

CAPILLARY ELECTROPHORESIS
Capillary: 30 cm × 50 μm fused-silica (25.5 cm to detector) (Yongnian Optical Conductive Fiber Plant, China), coated with polyacrylamide
Capillary preparation: No details of the polyacrylamide coating process are provided. However, another paper (LC.GC 1997, 15, 40) by this group indicates that they use the procedure of Hjertén, thus: Adjust the pH of 20 mL water to 3.5 with acetic acid, add 80 μL 3-(trimethoxysilyl)propyl methacrylate (3-methacryloxypropyltrimethoxysilane), mix, suck into capillary, let stand at room temperature for 1 h, remove the solution, wash with water. Fill the capillary with a deaerated 3-4% acrylamide solution containing 1 μL/mL N,N,N',N'-tetramethylethylenediamine and 1 mg/mL potassium persulfate, let stand for 30 min, remove excess solution by aspiration, rinse with water, remove water by aspiration, dry at 35° (J. Chromatogr. 1985, 347, 191).
Capillary temperature: 25
Running buffer: 100 mM NaH$_2$PO$_4$ adjusted to pH 2.5
Injection: Electrokinetic injection at 15 kV for 3 s.
Detector: UV 200, UV 210
Migration time: 4.73
Voltage: 15 kV
Model: Bio-Rad BioFocus 3000

OTHER SUBSTANCES
Simultaneous: amorolfine, brompheniramine, bupivacaine, carteolol, chloroquine, chlorpheniramine, chlorphenoxamine, disopyramide, dobutamine, doxylamine, flecainide, gallopamil, ketamine, mepindolol, orphenadrine, oxybutynin, phenoxybenzamine, pindolol, propafenone, propranolol, sulpiride, talinolol, verapamil

KEY WORDS
coated capillary

REFERENCE
Koppenhoefer,B.; Epperlein,U.; Xiaofeng,Z.; Bingcheng,L. Separation of enantiomers of drugs by capillary electrophoresis. Part 4: Hydroxypropyl-γ-cyclodextrin as chiral solvating agent, *Electrophoresis*, **1997**, *18*, 924–930.

SAMPLE
Matrix: solutions
Sample preparation: Inject an aliquot of a 100 μg/mL solution in running buffer.

CAPILLARY ELECTROPHORESIS
Capillary: 30 cm × 50 μm fused-silica (25.5 cm to detector), coated with polyacrylamide
Capillary preparation: Adjust the pH of 20 mL water to 3.5 with acetic acid, add 80 μL 3-(trimethoxysilyl)propyl methacrylate (3-methacryloxypropyltrimethoxysilane), mix, suck into capillary, let stand at room temperature for 1 h, remove the solution, wash with water. Fill the capillary with a deaerated 3-4% acrylamide solution containing 1 μL/mL N,N,N',N'-tetramethylethylenediamine and 1 mg/mL potassium persulfate, let stand for 30 min, remove excess solution by aspiration, rinse with water, remove water by aspiration, dry at 35° (J. Chromatogr. 1985, 347, 191).
Capillary temperature: 25
Running buffer: 100 mM NaH$_2$PO$_4$ containing 45 mM hydroxypropyl-β- cyclodextrin, adjusted to pH 2.5 with phosphoric acid
Injection: Electrokinetic injection at 15 kV for 3 s.

Detector: UV 200
Voltage: 15 kV
Model: Bio-Rad BioFocus 3000

KEY WORDS
chiral; coated capillary; comparison with the use of other cyclodextrins; this running buffer gave
the greatest enantiomeric separation.; α=1.149

REFERENCE
Lin,B.; Zhu,X.; Koppenhoefer,B.; Epperlein,U. Investigation of 123 chiral drugs by cyclodextrin-modified capillary electrophoresis, *LC.GC*, **1997**, *15*, 40–46.

SAMPLE
Matrix: solutions

CAPILLARY ELECTROPHORESIS
Capillary: 29-36 cm $\times$ 50 μm fused-silica (24.5-31.5 cm to detector) (Yongnian Optical Conductive
Fiber Plant, China) coated with polyacrylamide
Capillary preparation: Coat capillary as follows. Adjust the pH of 20 mL water to 3.5 with
acetic acid, add 80 μL 3-(trimethoxysilyl)propyl methacrylate (3-methacryloxypropyltrimethox-
ysilane), mix, suck into capillary, let stand at room temperature for 1 h, remove the solution,
wash with water. Fill the capillary with a deaerated 3-4% acrylamide solution containing 1
μL/mL N,N,N',N'-tetramethylethylenediamine and 1 mg/mL potassium persulfate, let stand
for 30 min, remove excess solution by aspiration, rinse with water, remove water by aspiration,
dry at 35° (J. Chromatogr. 1985, 347, 191).
Capillary temperature: 25
Running buffer: 100 mM pH 2.5 NaH_2PO_4 (A) or 100 mM pH 2.5 NaH_2PO_4 containing 45 mM
hydroxypropyl-α-cyclodextrin (Wacker, Munich) (B)
Injection: Electromigration at 15 kV for 3 s.
Detector: UV 200; UV 210
Migration time: 4.73 (A); 8.43 (B) (no separation of enantiomers)
Voltage: 15 kV
Model: Bio-Focus 3000

OTHER SUBSTANCES
Also analyzed: albuterol (salbutamol), alprenolol, amorolfine, atenolol, atropine, azelastine, baclofen, bamethan, benproperine, benserazide, biperiden, bisoprolol, brompheniramine, bupivacaine, bupranolol, butamirate, butethamate, carazolol, carbuterol, carteolol, carvedilol, celiprolol, chloroquine, chlorpheniramine, chlorphenoxamine, cicletanine, clenbuterol, clidinium
bromide, clobutinol, dimethindene, dipivefrin, disopyramide, dobutamine, doxylamine, fendiline, flecainide, gallopamil, homatropine, ipratropium bromide, isoproterenol (isoprenaline),
isothipendyl, ketamine, meclizine, mefloquine, mepindolol, mequitazine, metaclazepam,
metaproterenol (orciprenaline), metipranolol, metoprolol, nafronyl (naftidrofuryl), nefopam, nicardipine, norfenefrine, ofloxacin, ornidazole, orphenadrine, oxomemazine, oxprenolol, oxybutynin, phenoxybenzamine, phenylpropanolamine, pholedrine, pindolol, pirbuterol, prilocaine,
procyclidine, promethazine, propafenone, propranolol, reproterol, sotalol, sulpride, synephrine,
talinolol, terbutaline, tetrahydrozoline (tetryzoline), theodrenaline, tioconazole, tocainide, trihexyphenidyl, trimeprazine (alimemazine), trimipramine, verapamil, zopiclone

KEY WORDS
coated capillary

REFERENCE
Koppenhoefer,B.; Eperlein,U.; Schlunk,R.; Zhu,X.; Lin,B. Separation of enantiomers of drugs by capillary electrophoresis. V. Hydroxypropyl-α-cyclodextrin as chiral solvating agent, *J.Chromatogr.A*, **1998**, *793*, 153–164.

Ursodiol

Molecular formula: $C_{24}H_{40}O_4$
Molecular weight: 392.58
CAS Registry No.: 128-13-2
Merck Index (12th ed.): 10026

SAMPLE
Matrix: blood
Sample preparation: Filter (Amicon Centrifree YM) while centrifuging at 2000 rpm for 45 min, inject an aliquot of the ultrafiltrate.

CAPILLARY ELECTROPHORESIS
Capillary: fused-silica (Polymicro Technologies)
Capillary preparation: Before each run flush capillary with 100 mM NaOH for 3 min, with water for 3 min, and with running buffer for 3 min. At the start of each day flush with 1 M NaOH for 30 min, with water for 3 min, and with running buffer for 3 min. Flush new capillaries with 1 M NaOH for 1 h and with water for 10 min.
Capillary temperature: 23
Running buffer: MeOH:100 mM boric acid 75:25 containing 7 mM gamma-cyclodextrin and 5 mM adenosine 5'-monophosphate, pH 7.0
Injection: Pressure injection for 10 s.
Detector: UV 254
Migration time: 22
Voltage: 30 kV
Current: 6.8 µA
Model: Beckman P/ACE 5510
Limit of detection: 170-330 µM

OTHER SUBSTANCES
Simultaneous: chenodiol, cholic acid, deoxycholic acid, glycochenodeoxycholic acid, glycocholic acid, glycodeoxycholic acid, glycolithocholic acid, lithocholic acid, taurochenodeoxycholic acid, taurocholic acid, taurolithocholic acid, tauroursodeoxycholic acid
Interfering: glycoursodeoxycholic acid, taurodeoxycholic acid

KEY WORDS
serum; indirect UV detection; ultrafiltrate

REFERENCE
Yarabe,H.H.; Shamsi,S.A.; Warner,I.M. Capillary zone electrophoresis of bile acids with indirect photometric detection, *Anal.Chem.*, **1998**, *70*, 1412–1418.

SAMPLE
Matrix: bulk
Sample preparation: Inject an aliquot of a 1 mg/mL solution in MeOH.

CAPILLARY ELECTROPHORESIS
Capillary: 50 cm × 50 µm fused-silica (42 cm to detector)
Capillary temperature: 20
Running buffer: 100 mM pH 8.6 Tris containing 50 mM benzoic acid, 0.01% methylcellulose, and 0.5% trimethyl-β-cyclodextrin
Injection: Hydrodynamic injection for 0.5-3 s.
Detector: UV 250
Migration time: 4.2
Internal standard: cholic acid (5.0)
Voltage: 20 kV
Current: 35.4 µA
Model: Spectra PHORESIS 1000 (Thermo Separation Products)

OTHER SUBSTANCES
Simultaneous: chenodiol, deoxycholic acid, 3-hydroxy-7-chetocholanic acid, lithocholic acid, ursocholic acid

KEY WORDS
indirect UV

REFERENCE
Quaglia,M.G.; Farina,A.; Bossù,E.; Dell'Aquila,C.; Doldo,A. The indirect UV detection in the analysis of ursodeoxycholic acid and related compounds by HPCE, *J.Pharm.Biomed.Anal.*, **1997**, *16*, 281–285.

Valproic acid

Molecular formula: $C_8H_{16}O_2$
Molecular weight: 144.21
CAS Registry No.: 99-66-1, 1069-66-5 (sodium salt), 76584-70-8 (divalproex, 1:1 mixture of valproic acid and sodium valproate)
Merck Index (12th ed.): 10049

SAMPLE
Matrix: solutions

CAPILLARY ELECTROPHORESIS
Capillary: 57 cm × 50 μm fused-silica (50 cm to detector)
Running buffer: 50 mM pH 8.0 Borate buffer containing 40 mM sodium taurodeoxycholate and 25 mM phosphatidylcholine (Prepare by adding phosphatidylcholine and stirring for 3-6 h until all cloudiness disappears. Phosphatidylcholine was 95% pure soybean lecithin, Epikuron, Lucas Meyer & Co.)
Injection: Inject a solution of the compound in the running buffer at 20 psi for 1 s, inject MeOH: water 5:95 at 20 psi for 1 s, inject a solution of halofantrine in running buffer at 20 psi for 1 s.
Detector: UV 214
Migration time: k' 2.00
Model: Beckman P/ACE 5000

OTHER SUBSTANCES
Also analyzed: acetaminophen, amoxicillin, antipyrine, aspirin, azathioprine, caffeine, captopril, carbamazepine, carprofen, chlorambucil, chlorpheniramine, chlorpromazine, cimetidine, clonidine, codeine, desipramine, diphenhydramine, ephedrine, fenoterol, flufenamic acid, flurbiprofen, haloperidol, hydroxyzine, ibuprofen, imipramine, indomethacin, ketoprofen, lidocaine, melphalan, metoprolol, nabumetone, nadolol, phenobarbital, phenol, promazine, propranolol, pyrilamine, ranitidine, ropinirole, salicylic acid, sulfamethoxazole, testosterone, theophylline, thioridazine, tiaprofenic acid, tolfenamic acid, trifluoperazine, trimethoprim, verapamil

KEY WORDS
comparison with HPLC; k' = (Tr-T0)/(T0(1-Tr/Tm)) where Tr = retention time of analyte; T0 = retention time of water; and Tm = retention time of marker (halofantrine)

REFERENCE
Hanna,M.; de Biasi,V.; Bond,B.; Salter,C.; Hutt,A.J.; Camilleri,P. Estimation of the partitioning characteristics of drugs: A comparison of a large and diverse drug series utilizing chromatographic and electrophoretic methodology, *Anal.Chem.*, **1998**, *70*, 2092–2099.

Vancomycin

Molecular formula: $C_{66}H_{75}Cl_2N_9O_{24}$
Molecular weight: 1449.27
CAS Registry No.: 1404-90-6, 1404-93-9 (HCl)
Merck Index (12th ed.): 10066

SAMPLE
Matrix: solutions

CAPILLARY ELECTROPHORESIS
Capillary: 110 cm × 75 μm fused-silica (110 cm to MS, 20 cm to UV)
Running buffer: 50 mM pH 7 Ammonium acetate
Detector: MS, VG Platform, VG Biotech triaxial flow probe, electrospray, nebulizer gas nitrogen,
make-up solvent MeOH:water:formic acid 50:50:1 (pH 3), effective separation voltage 17 kV,
electrospray probe voltage 3 kV, m/z 725 or UV 214
Migration time: 4 (UV), 25 (MS)
Voltage: 20 kV
Model: Beckman P/ACE 2100

REFERENCE
Hamdan,M.; Curcuruto,O.; Di Modugno,E. Investigation of complexes between some glycopeptide antibiotics
and bacterial cell-wall analogues by electrospray- and capillary zone electrophoresis/electrospray-mass spectrometry, *Rapid.Commun.Mass.Spectrom.*, **1995**, *9*, 883–887.

Venlafaxine

Molecular formula: $C_{17}H_{27}NO_2$
Molecular weight: 277.41
CAS Registry No.: 93413-69-5, 99300-78-4 (HCl)
Merck Index (12th ed.): 10079
Lednicer: 4 26

SAMPLE
Matrix: formulations
Sample preparation: Sonicate powdered tablets with water (?) for 15 min, filter (0.45 μm),
remove a 1 mL aliquot of the filtrate, make up to 25 mL with 5 mM pH 6.5 citrate buffer
containing 100 μg/mL IS, inject an aliquot.

CAPILLARY ELECTROPHORESIS
Capillary: 64.5 cm × 50 μm fused-silica (56 cm to detector) (Hewlett Packard)
Capillary temperature: 25 (A), 15 (B)
Running buffer: 25 mM pH 6.5 Citrate buffer (A) or 100 mM pH 9.2 sodium borate buffer
containing 4 mM hydroxypropyl-β-cyclodextrin (B)

Injection: Pressure injection at 50 mbar for 8 s.
Detector: UV 200
Migration time: 7.5 (A), 9.5, 9.8 (enantiomers (B))
Internal standard: (+)-ephedrine
Voltage: 20 kV
Model: Hewlett Packard ^{3D}CE
Limit of detection: 250 ng/mL

OTHER SUBSTANCES
Simultaneous: impurities

KEY WORDS
tablets; chiral

REFERENCE
Fanali,S.; Cotichini,V.; Porra,R. Analysis of venlafaxine by capillary zone electrophoresis, *J.Capillary Electrophor.*, **1997**, *4*, 21–26.

Verapamil

Molecular formula: $C_{27}H_{38}N_2O_4$
Molecular weight: 454.61
CAS Registry No.: 52-53-9, 152-11-4 (HCl)
Merck Index (12th ed.): 10083
Lednicer: 4 34

SAMPLE
Matrix: blood
Sample preparation: 250 μL Serum + 30 μL 1 M NaOH + 500 μL ethyl acetate, shake vigorously for 2 min. Remove the organic layer and evaporate it to dryness under a stream of nitrogen, reconstitute the residue in 10 μL MeOH, inject an aliquot.

CAPILLARY ELECTROPHORESIS
Capillary: 64 cm × 50 μm uncoated silica (48 cm to detector) (Polymicro Technologies)
Running buffer: 20 mM pH 3.0 Tris buffer containing 10 mM heptakis(2,3,6-tri-O-methyl)-β-cyclodextrin, 0.1% methylhydroxycellulose 1000, and 0.05 mM hexadecyltrimethylammonium bromide
Detector: UV 215
Migration time: 10.5, 11.5 (enantiomers)
Voltage: 20 kV
Model: Laboratory constructed

KEY WORDS
serum; chiral

REFERENCE
Soini,H.; Riekkola,M.-L.; Novotny,M.V. Chiral separations of basic drugs and quantitation of bupivacaine enantiomers in serum by capillary electrophoresis with modified cyclodextrin buffers, *J.Chromatogr.*, **1992**, *608*, 265–274.

SAMPLE
Matrix: blood
Sample preparation: 1 mL Plasma + 100 μL 1 mg/mL gallopamil + 1 mL 100 mM NaOH + 5 mL hexane:isopropanol 90:10, vortex for 30 s, centrifuge at 1500 g for 10 min. Remove the organic layer and evaporate it to dryness under a stream of nitrogen, reconstitute the residue in 100 μL MeOH:water 25:75, centrifuge, inject an aliquot.

CAPILLARY ELECTROPHORESIS
Capillary: 18 cm × 75 μm fused-silica (18 cm to detector)
Capillary preparation: Before each analysis wash with 100 mM NaOH for 1 min, wash with buffer for 1 min, fill with running buffer for 2 min. (Buffer was 4 mL/L phosphoric acid in water adjusted to pH 2.5 with 1 M NaOH.)
Capillary temperature: 15
Running buffer: 60 mM pH 2.5 Phosphate buffer containing trimethyl-β-cyclodextrin
Injection: Injection by electromigration at 10 kV for 7 s
Detector: UV 200
Migration time: 8.25 (R), 8.75 (S)
Internal standard: gallopamil (8)
Voltage: 12 kV
Current: 90-100 μA
Model: Spectra-Physics Phoresis 1000
Limit of quantitation: 2.5 ng/mL

OTHER SUBSTANCES
Extracted: norverapamil

KEY WORDS
plasma; chiral

REFERENCE
Dethy,J.-M.; De Broux,S.; Lesne,M.; Longstreth,J.; Gilbert,P. Stereoselective determination of verapamil and norverapamil by capillary electrophoresis, *J.Chromatogr.B*, **1994**, *654*, 121–127.

SAMPLE
Matrix: blood
Sample preparation: 1 mL Plasma + 250 μL water + 5 mL MTBE, shake for 10 min, centrifuge at 1800 g for 10 min. Remove a 4 mL aliquot of the organic layer and add it to 40 μL 17 mM phosphoric acid, shake for 3 min, centrifuge at 2800 g for 10 min, inject an aliquot of the aqueous layer.

CAPILLARY ELECTROPHORESIS
Capillary: 64.5 cm × 50 μm fused-silica (56 cm to detector) (Hewlett-Packard)
Capillary preparation: Before each analysis purge capillary with 100 mM NaOH for 1.5 min and with running buffer for 3 min. Rinse a new capillary with 1 M NaOH for 10 min, with 100 mM NaOH for 5 min, and with running buffer for 5 min.
Capillary temperature: 25
Running buffer: 6.9 g/L pH 2.5 $NaH_2PO_4.2H_2O$
Injection: Hydrodynamic injection at 50 mbar for 40 s followed by a 50 mbar injection of running buffer for 20 s
Detector: UV 238
Migration time: 10
Internal standard: verapamil
Voltage: 30 kV
Model: Hewlett-Packard HP3D CE

OTHER SUBSTANCES
Extracted: diltiazem

KEY WORDS
plasma; verapamil is IS

REFERENCE
Coors,C.; Schulz,H.-G.; Stache,F. Development and validation of a bioanalytical method for the quantification of diltiazem and desacetyldiltiazem in plasma by capillary zone electrophoresis, *J.Chromatogr.A*, **1995**, *717*, 235–243.

SAMPLE
Matrix: solutions

Sample preparation: Prepare a 50 µg/mL solution in diluted running buffer, inject an aliquot.

CAPILLARY ELECTROPHORESIS
Capillary: 44 cm × 50 µm fused-silica
Capillary preparation: After each run wash capillary with water for 2 min and with running buffer for 3 min. At the beginning of each day wash capillary with water and running buffer for 5 min. Condition a new capillary with 1 M NaOH, 100 mM NaOH, water, and separation buffer.
Capillary temperature: 15
Running buffer: 100 mM Phosphoric acid containing 25 mM heptakis(2,3,6-tri-O-methyl)-β-cyclodextrin adjusted to pH 3.0 with triethanolamine
Injection: Hydrodynamic injection for 3 s
Detector: UV 210
Migration time: 10.1, 10.5 (enantiomers)
Voltage: 25 kV
Model: Spectra Physics Spectraphoresis 1000

OTHER SUBSTANCES
Simultaneous: norverapamil

KEY WORDS
chiral

REFERENCE
Bechet,I.; Paques,P.; Fillet,M.; Hubert,P.; Crommen,J. Chiral separation of basic drugs by capillary zone electrophoresis with cyclodextrin additives, *Electrophoresis*, **1994**, *15*, 818–823.

SAMPLE
Matrix: solutions
Sample preparation: Prepare a 600 µg/mL solution in MeCN:MeOH:water 25:25:50, dilute to 150 µg/mL with 100 mM pH 8.1 borate buffer containing 50 mM sodium dodecyl sulfate, inject an aliquot.

CAPILLARY ELECTROPHORESIS
Capillary: 57 cm × 75 µm fused-silica (50 cm to detector) (Beckman)
Capillary preparation: Rinse capillary with 100 mM NaOH and running buffer before each run.
Capillary temperature: 30
Running buffer: Acetone:buffer 15:85 (Buffer was 100 mM pH 8.1 borate buffer containing 50 mM sodium dodecyl sulfate.)
Injection: Pressure injection for 5 s
Detector: UV 200
Migration time: 18.5
Voltage: +25 kV
Model: Beckman P/ACE system 5510

OTHER SUBSTANCES
Simultaneous: amlodipine, atenolol, diltiazem, nicardipine, nifedipine

---

REFERENCE
Bretnall,A.E.; Clarke,G.S. Investigation and optimisation of the use of micellar electrokinetic chromatography for the analysis of six cardiovascular drugs, *J.Chromatogr.A*, **1995**, *700*, 173–178.

SAMPLE
Matrix: solutions
Sample preparation: Inject an aliquot of a 100 µg/mL solution in water:running buffer 50:50.

CAPILLARY ELECTROPHORESIS
Capillary: 44.5 cm × 50 µm acrylamide-coated fused-silica (Bio-Rad)
Capillary temperature: 30

Running buffer: 100 mM NaH_2PO_4 containing 15 mM gamma-cyclodextrin, adjusted to pH 2.5 with phosphoric acid
Injection: Electrokinetic injection at 8 kV for 6 s.
Detector: UV 200
Migration time: 13.76
Voltage: 14 kV
Model: Bio-Rad BioFocus 3000

OTHER SUBSTANCES
Also analyzed: albuterol, alprenolol, atenolol, atropine, baclofen, bamethan, benserazide, biperiden, bisoprolol, bupivacaine, bupranolol, butetamate, carazolol, carbuterol, carvedilol, celiprolol, chloroquine, chlorpheniramine (chlorphenamine), clidinium bromide, clobutinol, disopyramide, dobutamine, flecainide, homatropine, ipratropium bromide, isoproterenol, isothipendyl, ketamine, mefloquine, mequitazine, metaproterenol (orciprenaline), metipranolol, nafronyl (naftidrofuryl), nefopam, ofloxacin, orphenadrine, oxomemazine, oxprenolol, phenoxybenzamine, pholedrine, pindolol, pirbuterol, prilocaine, promethazine, propafenone, propranolol, sotalol, synephrine, tetrahydrozoline (tetryzoline), terbutaline, tocainide, trihexyphenidyl, trimeprazine (alimemazine), trimipramine, tropicamide, zopiclone

KEY WORDS
coated capillary; achiral

REFERENCE
Koppenhoefer,B.; Epperlein,U.; Christian,B.; Yibing,J.; Yuying,C.; Bingcheng,L. Separation of enantiomers of drugs by capillary electrophoresis. I. γ-Cyclodextrin as chiral solvating agent, *J.Chromatogr.A*, **1995**, *717*, 181–190.

SAMPLE
Matrix: solutions
Sample preparation: Prepare a solution in MeOH/water, inject an aliquot.

CAPILLARY ELECTROPHORESIS
Capillary: 62 cm × 52 μm fused-silica (50 cm to detector) (Polymicro Technologies)
Capillary temperature: 40
Running buffer: 50 mM pH 2.50 Tetrabutylammonium phosphate containing 20 mM hydroxypropyl-β-cyclodextrin
Injection: Siphon injection at 10 cm for 5 s
Detector: UV (wavelength not specified)
Migration time: 40.20, 40.60 (enantiomers)
Voltage: 20 kV
Model: Laboratory constructed

KEY WORDS
chiral

REFERENCE
Quang,C.; Khaledi,M.G. Extending the scope of chiral separation of basic compounds by cyclodextrin-mediated capillary zone electrophoresis, *J.Chromatogr.A*, **1995**, *692*, 253–265.

SAMPLE
Matrix: solutions

CAPILLARY ELECTROPHORESIS
Capillary: 36 cm × 50 μm polyacrylamide-coated fused-silica (31.5 cm to detector) (Polymicro Technologies)
Capillary preparation: Before each injection rinse with water at 100 psi for 30 s, rinse with running buffer at 100 psi for 30 s, partially fill with separation solution (500 μM ovomucoid in running buffer) at 1 psi for 190 s (27 cm), inject sample, electrophorese with running buffer. Coat capillary with polyacrylamide as follows. Treat with 1 M NaOH for 1 h, rinse with water, dry at 110° by flushing with nitrogen for 6 h, pass thionyl chloride through using suction for 10 min, seal at each end, heat at 70° for 6 h, open, suck 250 mM vinylmagnesium bromide in THF into the capillary, seal at each end, heat at 70° for 6 h, open, rinse with THF for several

min, rinse with water, fill by suction with a solution containing 3% acrylamide, 0.1% ammonium persulfate, and 0.1% tetramethylethylenediamine, heat at 28 ± 2° for 1 h, after 1 h wash with water (Biol.Pharm.Bull. 1993, 16, 1185).
Capillary temperature: 25
Running buffer: Isopropanol:50 mM pH 6.0 phosphate buffer 10:90
Injection: Inject at 1 psi for 2 s
Detector: UV 210
Migration time: 17.65, 18.02 (enantiomers)
Voltage: 12 kV
Model: Bio-Rad BioFocus 3000

KEY WORDS
chiral; partial separation zone technique; coated capillary

REFERENCE
Tanaka,Y.; Terabe,S. Partial separation zone technique for the separation of enantiomers by affinity electrokinetic chromatography with proteins as chiral pseudo-stationary phases, *J.Chromatogr.A*, **1995**, *694*, 277–284.

SAMPLE
Matrix: solutions
Sample preparation: Inject an aliquot of a 20-100 μg/mL solution in MeOH/water.

CAPILLARY ELECTROPHORESIS
Capillary: 60 cm × 75 μm fused-silica (60 cm to detector) (Polymicro Technologies)
Running buffer: 5 mM pH 9 Borate buffer containing 10 mM gamma-cyclodextrin and 0.5% poly(sodium N-undecylenyl-D-valinate) (Prepare poly(sodium N-undecylenyl-D-valinate) as follows. Add 30 mmoles undecylenic acid to 30 mmoles N-hydroxysuccinimide in 130 mL dry ethyl acetate, add 30 mmoles dicyclohexylcarbodiimide in 10 mL dry ethyl acetate, let stand at room temperature overnight, filter, recrystallize ester from EtOH. Add 1 mmole undecylenic acid N-hydroxysuccinimide ester in 10 mL THF to 1 mmole L-valine and 1 mmole sodium bicarbonate in 10 mL water, after 16 h acidify to pH 2 with 1 M HCl, remove the organic solvent under reduced pressure, add 50 mL water, filter, dry, recrystallize N-undecylenyl-L-valine from chloroform/petroleum ether (mp 91°) (J.Lipid Res. 1967, 8, 142). Prepare the sodium salt by adding an equimolar amount of NaOH in EtOH:water 1:4 (Anal.Chem. 1994, 66, 3773). Polymerize a 50 mM aqueous solution of the sodium salt by irradiating with a ^{60}Co gamma source at 0.16 Mrad/h at 21° (J.Poly.Sci.:Poly.Lett.Ed. 1979, 17, 749), lyophilize, extract with hot EtOH to remove monomer, dry under vacuum. Dry organic solvents over Type 4A 4-8 mesh molecular sieve.)
Detector: UV 280
Migration time: 13.3, 13.8 (enantiomers)
Voltage: 12 kV
Model: Dionex CES I

OTHER SUBSTANCES
Simultaneous: 1,1'-binaphthol, 1,1'-binaphthyl-2,2'-diyl hydrogen phosphate, laudanosine

KEY WORDS
chiral

REFERENCE
Wang,J.; Warner,I.M. Combined polymerized chiral micelle and γ-cyclodextrin for chiral separation in capillary electrophoresis, *J.Chromatogr.A*, **1995**, *711*, 297–304.

SAMPLE
Matrix: solutions

CAPILLARY ELECTROPHORESIS
Capillary: 57 cm × 75 μm fused-silica (50 cm to detector) (Beckman)
Capillary preparation: Before each run rinse capillary with 100 mM NaOH and running buffer.
Capillary temperature: 30

Running buffer: Acetone:100 mM pH 8.1 borate buffer containing 50 mM sodium dodecyl sulfate 15:85
Injection: Pressure injection for 5-10 s.
Detector: UV 200
Migration time: 18.5
Voltage: 25 kV
Model: Beckman P/ACE 5510

OTHER SUBSTANCES
Simultaneous: acebutolol, amiodarone, atenolol, bretylium, captopril, diltiazem, disopyramide, lidocaine, lisinopril, metoprolol, nicardipine, nifedipine, phenytoin, pindolol, propafenone, quinidine, sotalol, timolol, p-toluenesulfonic acid
Interfering: propranolol

REFERENCE
Bretnall,A.E.; Clarke,G.S. Selectivity of capillary electrophoresis for the analysis of cardiovascular drugs, *J.Chromatogr.A*, **1996**, *745*, 145–154.

SAMPLE
Matrix: solutions
Sample preparation: Inject an aliquot of a solution in MeOH.

CAPILLARY ELECTROPHORESIS
Capillary: 75 cm × 50 μm fused-silica (65 cm to detector) (Supelco)
Capillary preparation: Before each run rinse capillary with 100 mM NaOH for 2 min and with running buffer for 2 min.
Running buffer: MeOH:16 mM NaCl 25:75 containing 50 mM polyoxyethylene-4-dodecyl ether (Brij 30) (mole fraction 0.3)
Injection: Hydrodynamic injection at 15 cm for 1-6 s
Detector: UV 210
Migration time: 22.6 (-), 23.0 (+)
Voltage: 20 kV
Model: laboratory constructed

OTHER SUBSTANCES
Simultaneous: gallopamil, norverapamil

KEY WORDS
chiral

REFERENCE
Clothier,J.G.,Jr.; Tomellini,S.A. Chiral separation of verapamil and related compounds using micellar electrokinetic capillary chromatography with mixed micelles of bile salt and polyoxyethylene ethers, *J.Chromatogr.A*, **1996**, *723*, 179–187.

SAMPLE
Matrix: solutions

CAPILLARY ELECTROPHORESIS
Capillary: 47 cm × 50 μm fused-silica (40 cm to detector)
Capillary preparation: After each run rinse with 100 mM NaOH for 2-5 min, with water for 2 min, and with running buffer for 2 min
Running buffer: 80 mM Tris containing 4% dextrin 10 sulfopropyl ether, adjusted to pH 4 with 1 M phosphoric acid (Prepare dextrin 10 sulfopropyl ether as follows. Dissolve 10 g dextrin 10 in 50 mL water, stir vigorously, add 8 g 25% aqueous NaOH dropwise at room temperature over 2.5 h, at the same time add 5 g propane sultone (Caution! Propane sultone is a carcinogen!) in three portions, add 10 mL water, adjust pH to 6 with 4 M HCl, remove inorganic salts by 3-fold membrane filtration in a stirred cell (Spec, Basel; exclusion size 1000 Da; initial volume 150 mL; final volume 20 mL; cell pressure 4 bar), concentrate in a rotary evaporator, dry at 0.1 torr and 60° overnight, to gibe dextrin 10 sulfopropyl ether as a white powder (yield 5 g).)
Detector: UV 214
Migration time: 10.6, 11 (enantiomers)

Voltage: 30 kV
Model: Beckman P/ACE 5000

OTHER SUBSTANCES
Simultaneous: mianserin

KEY WORDS
chiral

REFERENCE
Jung,M.; Börnsen,K.O.; Francotte,E. Dextrin sulfopropyl ether: A novel anionic chiral buffer additive for enantiomer separation by electrokinetic chromatography, *Electrophoresis*, **1996**, *17*, 130–136.

SAMPLE
Matrix: solutions
Sample preparation: Inject an aliquot of a 100 μg/mL solution in water or running buffer.

CAPILLARY ELECTROPHORESIS
Capillary: 47 cm × 75 μm fused-silica (40 cm to detector)
Capillary preparation: Before each run rinse capillary with running buffer for 1-2 min. If peak tailing is observed fill capillary with 500 mM NaOH and let stand for 30 min, wash with water for 3 min, rinse with running buffer for 3 min.
Capillary temperature: 25
Running buffer: 20 mM pH 2.5 Phosphate buffer containing 15% dextrin (Japan Pharmacopeia grade) (Dissolve dextrin in buffer at 90° then cool to room temperature.)
Injection: Pressure injection at 0.5 psi for 2-4 s.
Detector: UV 220
Migration time: 9.784, 9.933 (enantiomers)
Voltage: 30 kV
Model: Beckman P/ACE 5510

KEY WORDS
chiral

REFERENCE
Nishi,H.; Izumoto,S.; Nakamura,K.; Nakai,H.; Sato,T. Dextran and dextrin as chiral selectors in capillary zone electrophoresis, *Chromatographia*, **1996**, *42*, 617–630.

SAMPLE
Matrix: solutions
Sample preparation: Inject an aliquot of a 100 μg/mL solution in MeOH:water 10:90.

CAPILLARY ELECTROPHORESIS
Capillary: 47 cm × 75 μm fused-silica (40 cm to detector)
Capillary preparation: Rinse capillary with running buffer for 1 min before each run. At the end of each day wash with 100 mM NaOH and with water.
Capillary temperature: 20
Running buffer: 20 mM pH 2.9 Phosphate buffer containing 3% chondroitin sulfate C (sodium salt) (Nacalai Tesque, Kyoto)
Injection: Pressure injection at 0.5 psi for 5 s.
Detector: UV 240
Migration time: 17.38, 17.62 (enantiomers)
Voltage: 20 kV
Model: Beckman P/ACE 5510

OTHER SUBSTANCES
Also analyzed: clentiazem, diltiazem, primaquine, propranolol, sulconazole, trimetoquinol

KEY WORDS
chiral

REFERENCE
Nishi,H. Enantiomer separation of basic drugs by capillary electrophoresis using ionic and neutral polysaccharides as chiral selectors, *J.Chromatogr.A*, **1996**, *735*, 345–351.

SAMPLE
Matrix: solutions
Sample preparation: Inject an aliquot of a solution in running buffer.

CAPILLARY ELECTROPHORESIS
Capillary: 60 cm $\times$ 75 μm fused-silica (52.4 cm to detector)
Capillary preparation: After each run flush with 500 mM KOH for 2-3 min then with water.
Running buffer: 10 mM pH 3.8 Phosphate buffer containing 2% sulfated cyclodextrin (ds 7-10)
Injection: Hydrostatic injection.
Detector: UV 214
Migration time: 10.45, 11.02 (enantiomers)
Voltage: 15 kV
Model: Waters Quanta 4000

OTHER SUBSTANCES
Also analyzed: acebutolol, alprenolol, aminoglutethimide, brompheniramine, bupivacaine, bupropion, canadine, carbinoxamine, chloroquine, chlorpheniramine, dimethindene, disopyramide, doxylamine, hydroxychloroquine, idazoxan, isoxsuprine, ketamine, mepenzolate, mepivacaine, methoxyphenamine, mexiletine, midodrine, nefopam, orphenadrine, oxprenolol, oxyphencyclimine, pheniramine, phensuximide, pindolol, piperoxan, terbutaline, tetramisole, tolperisone, tranylcypromine, trihexyphenidyl, trimipramine, warfarin

KEY WORDS
chiral; detector at anode

REFERENCE
Stalcup,A.M.; Gahm,K.H. Application of sulfated cyclodextrins to chiral separations by capillary zone electrophoresis, *Anal.Chem.*, **1996**, *68*, 1360–1368.

SAMPLE
Matrix: solutions
Sample preparation: Inject an aliquot of a 100 μg/mL solution in running buffer.

CAPILLARY ELECTROPHORESIS
Capillary: 30 cm $\times$ 50 μm fused-silica (25.5 cm to detector) (Yongnian Optical Conductive Fiber Plant, China), coated with polyacrylamide
Capillary preparation: No details of the polyacrylamide coating process are provided. However, another paper (LC.GC 1997, 15, 40) by this group indicates that they use the procedure of Hjertén, thus: Adjust the pH of 20 mL water to 3.5 with acetic acid, add 80 μL 3-(trimethoxysilyl)propyl methacrylate (3-methacryloxypropyltrimethoxysilane), mix, suck into capillary, let stand at room temperature for 1 h, remove the solution, wash with water. Fill the capillary with a deaerated 3-4% acrylamide solution containing 1 μL/mL N,N,N',N'-tetramethylethylenediamine and 1 mg/mL potassium persulfate, let stand for 30 min, remove excess solution by aspiration, rinse with water, remove water by aspiration, dry at 35° (J. Chromatogr. 1985, 347, 191).
Capillary temperature: 25
Running buffer: 100 mM NaH_2PO_4 adjusted to pH 2.5
Injection: Electrokinetic injection at 15 kV for 3 s.
Detector: UV 200, UV 210
Migration time: 6.74
Voltage: 15 kV
Model: Bio-Rad BioFocus 3000

OTHER SUBSTANCES
Simultaneous: amorolfine, brompheniramine, bupivacaine, carteolol, chloroquine, chlorpheniramine, chlorphenoxamine, disopyramide, dobutamine, doxylamine, flecainide, gallopamil, ketamine, mepindolol, orphenadrine, oxybutynin, phenoxybenzamine, pindolol, propafenone, propranolol, sulpiride, talinolol, tropicamide

KEY WORDS
coated capillary

REFERENCE
Koppenhoefer,B.; Epperlein,U.; Xiaofeng,Z.; Bingcheng,L. Separation of enantiomers of drugs by capillary electrophoresis. Part 4: Hydroxypropyl-γ-cyclodextrin as chiral solvating agent, *Electrophoresis*, **1997**, *18*, 924–930.

SAMPLE
Matrix: solutions

CAPILLARY ELECTROPHORESIS
Capillary: 36 cm × 50 μm fused-silica coated with linear polyacrylamide (31.5 cm to detector) (GL Science)
Capillary preparation: At the beginning and end of each day rinse capillary with capillary wash solution (Bio-Rad Cat. No. 148-5022) at 690 kPa for more than 3 min and with water at 690 kPa for more than 3 min. Coat capillary as follows. Treat capillary with 1 M NaOH at room temperature for 1 h, rinse with water, dry by passing nitrogen gas through the capillary at 110° for 6 h. Pass thionyl chloride through the capillary using a suction pump for several min, seal capillary at both ends and heat at 70° for 6 h. Unseal the capillary and fill with 250 mM vinyl magnesium bromide in THF by suction, seal the capillary, heat at 70° for 6 h. Open the capillary and rinse it with THF for several min, rinse with distilled water, fill the capillary with polymerization solution, heat at 28 ± 2° for 1 h, condition at -100 V/cm for 30 min (Anal. Sci. 1994, 10, 1). (The polymerization solution was 5% acrylamide in water containing 49 mM Tris, 384 mM glycine, and 0.1% sodium dodecyl sulfate, degas in an ultrasonic bath. Add 40 μL 10% N,N,N',N'-tetramethylethylenediamine and 10 μL 10% ammonium persulfate to 5 mL of the degassed solution, mix thoroughly.)
Running buffer: n-Propanol:50 mM pH 6.0 Phosphate buffer 10:90
Injection: Before each injection rinse with water at 690 kPa for 30 s, rinse with running buffer at 690 kPa for 30 s, partially fill with separation solution (500 μM α₁-acid glycoprotein (Cohn fraction VI) (ICN) in running buffer) at 6.9 kPa for 190 s (27 cm), inject sample at 6.9 kPa for 2 s, electrophorese with running buffer (Note that α₁-acid glycoprotein from other suppliers may provide inferior results).
Detector: UV 210
Migration time: Resolution of enantiomers 3.4
Voltage: 12 kV
Model: Bio-rad BioFocus 3000

KEY WORDS
chiral; coated capillary

REFERENCE
Tanaka,Y.; Terabe,S. Separation of the enantiomers of basic drugs by affinity capillary electrophoresis using a partial filling technique and α₁-acid glycoprotein as chiral selector, *Chromatographia*, **1997**, *44*, 119–128.

SAMPLE
Matrix: solutions

CAPILLARY ELECTROPHORESIS
Capillary: 60 cm × 50 μm fused-silica (30 cm to detector) (Polymicro Technologies)
Capillary temperature: 30
Running buffer: 40 mM pH 4.0 Sodium acetate containing 1.4 mM sulfobutyl ether-β-cyclodextrin (average substitution 3.9, MW 1721, Center for Drug Delivery Research, Lawrence KS)
Injection: Pressure injection at 5 inches Hg for 1.5 s.
Detector: UV 310
Migration time: 5.65, 5.85 (enantiomers)
Voltage: 25 kV
Current: 11 μA
Model: Perkin Elmer-Applied Biosystems Model 270

KEY WORDS
chiral

REFERENCE

Xie,G.-h.; Skanchy,D.J.; Stobaugh,J.F. Chiral separations of enantiomeric pharmaceuticals by capillary electrophoresis using sulphobutyl ether β-cyclodextrin as isomer selector, *Biomed.Chromatogr.*, **1997**, *11*, 193–199.

SAMPLE
Matrix: solutions

CAPILLARY ELECTROPHORESIS
Capillary: 29-36 cm × 50 μm fused-silica (24.5-31.5 cm to detector) (Yongnian Optical Conductive Fiber Plant, China) coated with polyacrylamide

Capillary preparation: Coat capillary as follows. Adjust the pH of 20 mL water to 3.5 with acetic acid, add 80 μL 3-(trimethoxysilyl)propyl methacrylate (3-methacryloxypropyltrimethoxysilane), mix, suck into capillary, let stand at room temperature for 1 h, remove the solution, wash with water. Fill the capillary with a deaerated 3-4% acrylamide solution containing 1 μL/mL N,N,N',N'-tetramethylethylenediamine and 1 mg/mL potassium persulfate, let stand for 30 min, remove excess solution by aspiration, rinse with water, remove water by aspiration, dry at 35° (J. Chromatogr. 1985, 347, 191).

Capillary temperature: 25

Running buffer: 100 mM pH 2.5 NaH_2PO_4 (A) or 100 mM pH 2.5 NaH_2PO_4 containing 45 mM hydroxypropyl-α-cyclodextrin (Wacker, Munich) (B)

Injection: Electromigration at 15 kV for 3 s.

Detector: UV 200; UV 210

Migration time: 6.74 (A); 12.98 (B) (no separation of enantiomers)

Voltage: 15 kV

Model: Bio-Focus 3000

OTHER SUBSTANCES
Also analyzed: albuterol (salbutamol), alprenolol, amorolfine, atenolol, atropine, azelastine, baclofen, bamethan, benproperine, benserazide, biperiden, bisoprolol, brompheniramine, bupivacaine, bupranolol, butamirate, butethamate, carazolol, carbuterol, carteolol, carvedilol, celiprolol, chloroquine, chlorpheniramine, chlorphenoxamine, cicletanine, clenbuterol, clidinium bromide, clobutinol, dimethindene, dipivefrin, disopyramide, dobutamine, doxylamine, fendiline, flecainide, gallopamil, homatropine, ipratropium bromide, isoproterenol (isoprenaline), isothipendyl, ketamine, meclizine, mefloquine, mepindolol, mequitazine, metaclazepam, metaproterenol (orciprenaline), metipranolol, metoprolol, nafronyl (naftidrofuryl), nefopam, nicardipine, norfenefrine, ofloxacin, ornidazole, orphenadrine, oxomemazine, oxprenolol, oxybutynin, phenoxybenzamine, phenylpropanolamine, pholedrine, pindolol, pirbuterol, prilocaine, procyclidine, promethazine, propafenone, propranolol, reproterol, sotalol, sulpride, synephrine, talinolol, terbutaline, tetrahydrozoline (tetryzoline), theodrenaline, tioconazole, tocainide, trihexyphenidyl, trimeprazine (alimemazine), trimipramine, tropicamide, zopiclone

KEY WORDS
coated capillary

REFERENCE

Koppenhoefer,B.; Eperlein,U.; Schlunk,R.; Zhu,X.; Lin,B. Separation of enantiomers of drugs by capillary electrophoresis. V. Hydroxypropyl-α-cyclodextrin as chiral solvating agent, *J.Chromatogr.A*, **1998**, *793*, 153–164.

Vinblastine

Molecular formula: $C_{46}H_{58}N_4O_9$
Molecular weight: 810.99
CAS Registry No.: 865-21-4, 143-67-9 (sulfate)
Merck Index (12th ed.): 10119

SAMPLE

Matrix: plants
Sample preparation: 600 mg Leaves + 4 mL isopropanol, shake on a wrist-action shaker for 15 min, filter (Acrodisc glass fiber). Evaporate the filtrate to dryness under a stream of nitrogen, reconstitute the residue in 1 mL 10 mM acetic acid, wash 3 times with 1 mL portions of cyclohexane. Evaporate the aqueous layer to dryness under a stream of nitrogen, reconstitute the residue in 200 μL 10 mM acetic acid, sonicate for 1 min, filter (glass fiber), mix with an equal volume of 100 mg/mL (sic) IS, inject an aliquot.

CAPILLARY ELECTROPHORESIS

Capillary: 72 cm × 75 μm fused-silica (57 cm to detector) (Supelco)
Capillary preparation: Wash with running buffer at 1000 mbar for 2 min before each run. After each buffer change rinse with new buffer at 1000 mbar, apply 20 kV for 10 min. At the start of each day rinse with running buffer at 1000 mbar, apply 20 kV for 10 min, repeat. Flush new capillaries at 2000 mbar with 100 mM NaOH for 5 min, water for 5 min, and running buffer for 5 min, apply 20 kV for 10 min, allow to equilibrate overnight.
Running buffer: 200 mM pH 6.2 Ammonium acetate
Detector: UV 254
Migration time: 19.2
Internal standard: palmatine (15.2)
Voltage: 10 kV
Model: Crystal CZE (ATI, Boston)
Limit of detection: 1 ppm

OTHER SUBSTANCES

Extracted: vincristine

KEY WORDS

leaves

REFERENCE

Chu,I.; Bodnar,J.A.; White,E.L.; Bowman,R.N. Quantification of vincristine and vinblastine in Catharanthus roseus plants by capillary zone electrophoresis, *J.Chromatogr.A*, **1996**, *755*, 281–288.

SAMPLE

Matrix: solutions

CAPILLARY ELECTROPHORESIS

Capillary: 55 cm × 50 μm fused-silica (50 cm to detector) (Polymicro Technologies)
Capillary preparation: Between runs purge with 1 M NaOH for 2 min, with water for 2 min, and with running buffer for 3 min
Capillary temperature: 25
Running buffer: MeCN:buffer 50:50 (Buffer was 100 mM ammonium acetate adjusted to pH 3.1 with acetic acid.)
Injection: Pressure injection at 345 mbar.s.
Detector: UV 200, Finnigan MAT Model 95, electrospray (details in paper)

Migration time: 15.8
Voltage: 15 kV
Model: Bio-Rad BioFocus 3000

OTHER SUBSTANCES
Simultaneous: ajmaline, alstonine, deserpidine, gramine, β-methylajmaline, raufloridine, re-scinnamine, reserpine, serpentine, tabersonine, tryptamine, vincristine, yohimbinic acid
Interfering: corynanthine

REFERENCE
Unger,M.; Stöckigt,D.; Belder,D.; Stöckigt,J. General approach for the analysis of various alkaloid classes using capillary electrophoresis and capillary electrophoresis-mass spectrometry, *J.Chromatogr.A*, **1997**, *767*, 263–276.

Vincristine

Molecular formula: $C_{46}H_{56}N_4O_{10}$
Molecular weight: 824.97
CAS Registry No.: 57-22-7, 2068-78-2 (sulfate)
Merck Index (12th ed.): 10124

SAMPLE
Matrix: plants
Sample preparation: 600 mg Leaves + 4 mL isopropanol, shake on a wrist-action shaker for 15 min, filter (Acrodisc glass fiber). Evaporate the filtrate to dryness under a stream of nitrogen, reconstitute the residue in 1 mL 10 mM acetic acid, wash 3 times with 1 mL portions of cyclohexane. Evaporate the aqueous layer to dryness under a stream of nitrogen, reconstitute the residue in 200 μL 10 mM acetic acid, sonicate for 1 min, filter (glass fiber), mix with an equal volume of 100 mg/mL (sic) IS, inject an aliquot.

CAPILLARY ELECTROPHORESIS
Capillary: 72 cm × 75 μm fused-silica (57 cm to detector) (Supelco)
Capillary preparation: Wash with running buffer at 1000 mbar for 2 min before each run. After each buffer change rinse with new buffer at 1000 mbar, apply 20 kV for 10 min. At the start of each day rinse with running buffer at 1000 mbar, apply 20 kV for 10 min, repeat. Flush new capillaries at 2000 mbar with 100 mM NaOH for 5 min, water for 5 min, and running buffer for 5 min, apply 20 kV for 10 min, allow to equilibrate overnight.
Running buffer: 200 mM pH 6.2 Ammonium acetate
Detector: UV 254
Migration time: 19.6
Internal standard: palmatine (15.2)
Voltage: 10 kV
Model: Crystal CZE (ATI, Boston)
Limit of detection: 1 ppm

OTHER SUBSTANCES
Extracted: vinblastine

KEY WORDS
leaves

REFERENCE
Chu,I.; Bodnar,J.A.; White,E.L.; Bowman,R.N. Quantification of vincristine and vinblastine in Catharanthus roseus plants by capillary zone electrophoresis, *J.Chromatogr.A*, **1996**, *755*, 281–288.

SAMPLE
Matrix: solutions

CAPILLARY ELECTROPHORESIS
Capillary: 55 cm × 50 μm fused-silica (50 cm to detector) (Polymicro Technologies)
Capillary preparation: Between runs purge with 1 M NaOH for 2 min, with water for 2 min, and with running buffer for 3 min
Capillary temperature: 25
Running buffer: MeCN:buffer 50:50 (Buffer was 100 mM ammonium acetate adjusted to pH 3.1 with acetic acid.)
Injection: Pressure injection at 345 mbar.s.
Detector: UV 200, Finnigan MAT Model 95, electrospray (details in paper)
Migration time: 16.1
Voltage: 15 kV
Model: Bio-Rad BioFocus 3000

OTHER SUBSTANCES
Simultaneous: ajmaline, alstonine, corynanthine, deserpidine, gramine, β-methylajmaline, raufloridine, rescinnamine, reserpine, serpentine, tabersonine, tryptamine, vinblastine, yohimbinic acid

REFERENCE
Unger,M.; Stöckigt,D.; Belder,D.; Stöckigt,J. General approach for the analysis of various alkaloid classes using capillary electrophoresis and capillary electrophoresis-mass spectrometry, *J.Chromatogr.A*, **1997**, *767*, 263–276.

Vitamin A

Molecular formula: $C_{20}H_{30}O$
Molecular weight: 286.46
CAS Registry No.: 68-26-8
Merck Index (12th ed.): 10150

SAMPLE
Matrix: formulations
Sample preparation: Protect from light. Tablets. Sonicate 1 tablet with 15 mL MeOH for 15 min, centrifuge at 2000 U/min for 5 min, inject an aliquot of the supernatant. Syrup. Dilute with MeOH, inject an aliquot.

CAPILLARY ELECTROPHORESIS
Capillary: 47 cm × 50 μm fused-silica (40 cm to detector) (Polymicro Technologies)
Capillary preparation: Rinse capillary with 100 mM NaOH, water, and running
Capillary temperature: 25
Running buffer: 20 mM pH 7.9 Sodium phosphate buffer containing 3 mM Brij 35 and 75 mM sodium dodecyl sulfate
Injection: Pressure injection.
Detector: UV 330
Migration time: 14.7
Voltage: 20 kV
Model: Beckman P/ACE 5500

OTHER SUBSTANCES
Simultaneous: retinal, retinoic acid, retinyl acetate, retinyl palmitate

KEY WORDS
tablets; syrup

REFERENCE
Hsieh,Y.-Z.; Kuo,K.-L. Separation of retinoids by micellar electrokinetic capillary chromatography, *J.Chromatogr.A*, **1997**, *761*, 307–313.

SAMPLE
Matrix: solutions
Sample preparation: Prepare a 100 μg/mL solution in MeOH:water 50:50, inject an aliquot.

CAPILLARY ELECTROPHORESIS
Capillary: 47 cm × 50 μm fused-silica (40 cm to detector) (Polymicro Technologies)
Capillary temperature: 25
Running buffer: Prepare as follows. Mix n-hexane with 2 volumes of 500 mM sodium dodecyl sulfate, vortex, add 1-butanol until the mixture clears, dilute with 20 mM pH 7.0 phosphate buffer until the concentration of sodium dodecyl sulfate is 20 mM.
Injection: Inject using an overpressure of 34500000 Pa (sic) for 3 s
Detector: UV 214
Migration time: 19.0
Voltage: 10 kV
Current: about 17.5 μA
Model: Beckman P/ACE System 2100

OTHER SUBSTANCES
Simultaneous: niacin, niacinamide, pyridoxine (pyridoxol), thiamine, vitamin E

REFERENCE
Boso,R.L.; Bellini,M.S.; Miksík,; Deyl,Z. Microemulsion electrokinetic chromatography with different organic modifiers: separation of water- and lipid-soluble vitamins, *J.Chromatogr.A*, **1995**, *709*, 11–19.

SAMPLE
Matrix: solutions

CAPILLARY ELECTROPHORESIS
Capillary: 50 cm × 75 μm fused-silica (50 cm to detector)
Capillary preparation: Before each run flush with 100 mM NaOH for 2 min and with water for 2 min then fill with running buffer.
Capillary temperature: 25
Running buffer: MeCN:buffer 20:80 (Buffer was 20 mM pH 9.2 sodium tetraborate containing 100 mM sodium cholate.)
Injection: Pressure injection at 0.5 psi for 1.5 s (9 nL).
Detector: UV 340
Migration time: 20
Voltage: 25 kV
Model: Beckman P/ACE 2100
Limit of detection: 10 μM

OTHER SUBSTANCES
Simultaneous: retinal, retinoic acid

KEY WORDS
protect from light

REFERENCE
Profumo,A.; Profumo,V.; Vidali,G. Micellar electrokinetic capillary chromatography of three natural vitamin A derivatives, *Electrophoresis*, **1996**, *17*, 1617–1621.

Vitamin B12

Molecular formula: $C_{63}H_{88}CoN_{14}O_{14}P$
Molecular weight: 1355.38
CAS Registry No.: 68-19-9
Merck Index (12th ed.): 10152

SAMPLE
Matrix: formulations
Sample preparation: Grind tablet, dissolve in pH 5 citrate buffer, filter (0.2 μm), inject an aliquot.

CAPILLARY ELECTROPHORESIS
Capillary: 48.5 cm × 50 μm fused-silica (40 cm to detector)
Capillary preparation: Between runs flush with 100 mM NaOH for 2 min and with run buffer for 3 min, inject a buffer plug. Before the first use condition capillaries with 1 M NaOH for 10 min, with 100 mM NaOH for 2 min, and with run buffer for 3 min.
Capillary temperature: 25
Running buffer: 20 mM pH 7 sodium phosphate
Injection: Injection at 40 mbar for 4.6 s (post-injection pressure (of run buffer) 40 mbar for 4 s)
Detector: UV 360
Migration time: 5.3
Voltage: 20 kV
Model: Hewlett-Packard

OTHER SUBSTANCES
Simultaneous: folinic acid (UV 215), niacinamide (UV 215), niacin (UV 215), orotic acid (UV 215), pantothenic acid (UV 215), pyridoxine (UV 215), thiamine (UV 215), ascorbic acid (UV 215)

KEY WORDS
tablets

REFERENCE
Jegle,U. Separation of water-soluble vitamins via high-performance capillary electrophoresis, *J.Chromatogr.A*, **1993**, *652*, 495–501.

SAMPLE
Matrix: formulations
Sample preparation: Condition a 500 mg Bakerbond C18 SPE cartridge with 3 mL MeOH and 3 mL water:10 mM sodium heptanesulfonate in acetic acid 99:1. Heat 50-100 mg tablet and 40 mL water:10 mM sodium heptanesulfonate in acetic acid 99:1 at 55° with occasional shaking for 5 min, cool, add a 2 mL aliquot to the SPE cartridge, elute with 3.5 mL MeOH, make up the eluate to 10 mL with water, inject an aliquot.

CAPILLARY ELECTROPHORESIS
Capillary: 70 cm $\times$ 100 μm
Capillary temperature: 25
Running buffer: MeOH:buffer 10:90 (Buffer was 50 mM pH 8.0 sodium borate containing 22.5 mM sodium dodecyl sulfate.)
Injection: Inject under 3.44 kPa pressure for 5 s (30 nL)
Detector: UV 214
Migration time: 18
Voltage: 16 kV
Current: 33 μA
Model: Beckman P/ACE System 2000

OTHER SUBSTANCES
Simultaneous: niacinamide, pantothenic acid, pyridoxine, thiamine, riboflavin, ascorbic acid

KEY WORDS
tablets; SPE

REFERENCE
Dinelli,G.; Bonetti,A. Micellar electrokinetic capillary chromatography analysis of water-soluble vitamins and multi-vitamin integrators, *Electrophoresis*, **1994**, *15*, 1147–1150.

SAMPLE
Matrix: formulations
Sample preparation: Grind tablet to a homogeneous powder and dissolve in 2 mL water and 20 mL EtOH, homogenize (Ultra Turrax), centrifuge. Evaporate the supernatant to dryness under reduced pressure, reconstitute the residue with water, inject an aliquot.

CAPILLARY ELECTROPHORESIS
Capillary: 64.5 cm $\times$ 50 μm fused-silica (56 cm to detector)
Capillary preparation: Before each run flush capillary at 4 kPa with 1 M NaOH for 2 min and with buffer for 5 min. Replace buffer before each analysis.
Capillary temperature: 30
Running buffer: 1-Propanol:buffer 2:98 (Buffer was 100 mM Na_2HPO_4, 500 mM taurine, and 75 mM sodium cholate.)
Injection: Pressure injection at 4 kPa for 1 s.
Detector: UV 214
Migration time: 11.5
Voltage: 17 kV
Model: Hewlett-Packard HP[3D] CE
Limit of quantitation: 1.5 μM

OTHER SUBSTANCES
Simultaneous: ascorbic acid, biotin, folic acid, niacinamide, pyridoxamine, pyridoxine, thiamine, riboflavin

KEY WORDS
tablets

REFERENCE
Buskov,S.; Moller,P.; Sorensen,H.; Sorensen,J.C.; Sorensen,S. Determination of vitamins in food based on supercritical fluid extraction prior to micellar electrokinetic capillary chromatographic analyses of individual vitamins, *J.Chromatogr.A*, **1998**, *802*, 233–241.

SAMPLE
Matrix: solutions
Sample preparation: Prepare a 0.5-2 mg/mL solution in water, inject an aliquot.

CAPILLARY ELECTROPHORESIS
Capillary: 65 cm $\times$ 50 μm fused-silica (50 cm to detector) (Scientific Glass Engineering)

Running buffer: 20 mM NaH_2PO_4 containing 50 mM sodium dodecyl sulfate adjusted to pH 9.0 with 20 mM sodium tetraborate
Injection: Injection by siphon at 5 cm for 5-10 s (about 1 nL)
Detector: UV 210
Migration time: 6.6
Voltage: 25 kV

OTHER SUBSTANCES
Simultaneous: niacin, niacinamide, pyridoxal, pyridoxamine, pyridoxine, riboflavin, thiamine

REFERENCE
Nishi,H.; Tsumagari,N.; Kakimoto,T.; Terabe,S. Separation of water-soluble vitamins by micellar electrokinetic chromatography, *J.Chromatogr.*, **1989**, *465*, 331–343.

SAMPLE
Matrix: solutions

CAPILLARY ELECTROPHORESIS
Capillary: 56.9 cm × 75 μm fused-silica (50 cm to detector)
Capillary temperature: 25
Running buffer: MeOH:60 mM pH 8.92 sodium borate buffer containing 60 mM sodium dodecyl sulfate 15:85
Detector: UV 214
Migration time: 5
Voltage: 30 kV
Model: Beckman PACE 2100

OTHER SUBSTANCES
Simultaneous: niacin, niacinamide, thiamine, pyridoxine

REFERENCE
McLaughlin,G.M.; Nolan,J.A.; Lindahl,J.L.; Palmieri,R.H.; Anderson,K.W.; Morris,S.C.; Morrison,J.A.; Bronzert,T.J. Pharmaceutical drug separations by HPCE: Practical guidelines, *J.Liq.Chromatogr.*, **1992**, *15*, 961–1021.

Vitamin D2

Molecular formula: $C_{28}H_{44}O$
Molecular weight: 396.66
CAS Registry No.: 50-14-6
Merck Index (12th ed.): 10156

SAMPLE
Matrix: solutions

CAPILLARY ELECTROPHORESIS
Capillary: 47 cm × 50 μm fused-silica (40 cm to detector) (Polymicro Technologies)
Capillary preparation: Before each run rinse capillary with 100 mM ammonium hydroxide at high pressure for 2 min, with water for 2-3 min, and with running buffer for 2-3 min.
Capillary temperature: 26
Running buffer: 30 mM pH 10 3-Cyclohexylamino-1-propanesulfonic acid (CAPS) buffer containing 25 mM sodium dodecyl sulfate and 25 mM silver nitrate
Injection: Hydrodynamic injection for 3-5 s

Detector: UV 254, UV 280
Migration time: 11.2
Voltage: 15 kV
Model: Beckman P/ACE 2100

OTHER SUBSTANCES
Simultaneous: cholecalciferol (vitamin D3)

REFERENCE
Wright,P.B.; Dorsey,J.G. Silver(I)-mediated separations by capillary zone electrophoresis and micellar electrokinetic chromatography: Argentation electrophoresis, *Anal.Chem.*, **1996**, *68*, 415–424.

SAMPLE
Matrix: solutions
Sample preparation: Inject an aliquot of a solution in running buffer.

CAPILLARY ELECTROPHORESIS
Capillary: 77 cm × 50 μm fused-silica (55 cm to detector) (Polymicro Technologies)
Capillary preparation: Between runs rinse capillary with 100 mM NaOH for 2 min, with water for 2 min, and with running buffer for 2 min. Condition new capillaries with 1 M NaOH for 10 min.
Running buffer: EtOH:buffer 2:98 (Buffer was 15 mM pH 7.5 phosphate buffer containing 20 mM sodium dodecyl sulfate and 40 mM heptakis(2,6-di-O-methyl)-β-cyclodextrin.)
Injection: Hydrodynamic injection at 15 cm for 8-10 s.
Detector: UV (wavelength not given)
Migration time: 25
Voltage: 15 kV

OTHER SUBSTANCES
Simultaneous: vitamin D3

KEY WORDS
comparison with HPLC

REFERENCE
Spencer,B.J.; Purdy,W.C. Comparison of the separation of fat-soluble vitamins using β-cyclodextrins in high-performance liquid chromatography and micellar electrokinetic chromatography, *J.Chromatogr.A*, **1997**, *782*, 227–235.

Vitamin D3

Molecular formula: $C_{27}H_{44}O$
Molecular weight: 384.65
CAS Registry No.: 67-97-0
Merck Index (12th ed.): 10157

SAMPLE
Matrix: solutions

CAPILLARY ELECTROPHORESIS
Capillary: 47 cm × 50 μm fused-silica (40 cm to detector) (Polymicro Technologies)
Capillary preparation: Before each run rinse capillary with 100 mM ammonium hydroxide at high pressure for 2 min, with water for 2-3 min, and with running buffer for 2-3 min.

Capillary temperature: 26
Running buffer: 30 mM pH 10 3-Cyclohexylamino-1-propanesulfonic acid (CAPS) buffer containing 25 mM sodium dodecyl sulfate and 25 mM silver nitrate
Injection: Hydrodynamic injection for 3-5 s
Detector: UV 254, UV 280
Migration time: 12
Voltage: 15 kV
Model: Beckman P/ACE 2100

OTHER SUBSTANCES
Simultaneous: ergocalciferol (vitamin D2)

REFERENCE
Wright,P.B.; Dorsey,J.G. Silver(I)-mediated separations by capillary zone electrophoresis and micellar electrokinetic chromatography: Argentation electrophoresis, *Anal.Chem.*, **1996**, *68*, 415–424.

SAMPLE
Matrix: solutions
Sample preparation: Inject an aliquot of a solution in running buffer.

CAPILLARY ELECTROPHORESIS
Capillary: 77 cm × 50 μm fused-silica (55 cm to detector) (Polymicro Technologies)
Capillary preparation: Between runs rinse capillary with 100 mM NaOH for 2 min, with water for 2 min, and with running buffer for 2 min. Condition new capillaries with 1 M NaOH for 10 min.
Running buffer: EtOH:buffer 2:98 (Buffer was 15 mM pH 7.5 phosphate buffer containing 20 mM sodium dodecyl sulfate and 40 mM heptakis(2,6-di-O-methyl)-β-cyclodextrin.)
Injection: Hydrodynamic injection at 15 cm for 8-10 s.
Detector: UV (wavelength not given)
Migration time: 24
Voltage: 15 kV

OTHER SUBSTANCES
Simultaneous: vitamin D2

KEY WORDS
comparison with HPLC

REFERENCE
Spencer,B.J.; Purdy,W.C. Comparison of the separation of fat-soluble vitamins using β-cyclodextrins in high-performance liquid chromatography and micellar electrokinetic chromatography, *J.Chromatogr.A*, **1997**, *782*, 227–235.

SAMPLE
Matrix: solutions

CAPILLARY ELECTROPHORESIS
Capillary: 67 cm × 50 μm uncoated CElect fused-silica (60 cm to detector) (Beckman)
Capillary preparation: At the start of each day rinse capillary with EtOH for 10 min, with 1 M NaOH for 10 min, with water for 25 min, and with running buffer for 3 min.
Running buffer: MeCN:water 80:20 containing 80 mM tetradecylammonium bromide
Injection: Pressure injection at 0.5 psi for 1 s.
Detector: UV 238
Migration time: 10
Voltage: 10 kV
Model: Beckman P/ACE 5510

OTHER SUBSTANCES
Simultaneous: vitamin A palmitate, vitamin E acetate

REFERENCE

Pedersen-Bjergaard,S.; Rasmussen,K.E.; Tilander,T. Separation of fat-soluble vitamins by hydrophobic interaction electrokinetic chromatography with tetradecylammonium ions as pseudostationary phase, *J.Chromatogr.A*, **1998**, *807*, 285–295.

Vitamin E

Molecular formula: $C_{29}H_{50}O_2$
Molecular weight: 430.71
CAS Registry No.: 59-02-9, 58-95-7 (acetate d-form), 52225-20-4 (acetate dl-form), 43119-47-7 (nicotinate)
Merck Index (12th ed.): 10159

SAMPLE

Matrix: solutions
Sample preparation: Prepare a 100 µg/mL solution in MeOH.water 50:50, inject an aliquot.

CAPILLARY ELECTROPHORESIS

Capillary: 47 cm × 50 µm fused-silica (40 cm to detector) (Polymicro Technologies)
Capillary temperature: 25
Running buffer: Prepare as follows. Mix n-hexane with 2 volumes of 500 mM sodium dodecyl sulfate, vortex, add 1-butanol until the mixture clears, dilute with 20 mM pH 7.0 phosphate buffer until the concentration of sodium dodecyl sulfate is 20 mM.
Injection: Inject using an overpressure of 34500000 Pa (sic) for 3 s
Detector: UV 214
Migration time: 16.0
Voltage: 10 kV
Current: about 17.5 µA
Model: Beckman P/ACE System 2100

OTHER SUBSTANCES

Simultaneous: niacin, niacinamide, pyridoxine (pyridoxol), thiamine, vitamin A

REFERENCE

Boso,R.L.; Bellini,M.S.; Miksík,; Deyl,Z. Microemulsion electrokinetic chromatography with different organic modifiers: separation of water- and lipid-soluble vitamins, *J.Chromatogr.A*, **1995**, *709*, 11–19.

SAMPLE

Matrix: solutions
Sample preparation: Inject an aliquot of a solution in running buffer.

CAPILLARY ELECTROPHORESIS

Capillary: 77 cm × 50 µm fused-silica (55 cm to detector) (Polymicro Technologies)
Capillary preparation: Between runs rinse capillary with 100 mM NaOH for 2 min, with water for 2 min, and with running buffer for 2 min. Condition new capillaries with 1 M NaOH for 10 min.
Running buffer: 20 mM pH 8.0 Phosphate buffer containing 50 mM borate, 30 mM sodium dodecyl sulfate, and 25 mM heptakis(2,6-di-O-methyl)-β-cyclodextrin.)
Injection: Hydrodynamic injection at 15 cm for 8-10 s.
Detector: UV (wavelength not given)
Migration time: 28
Voltage: 22 kV

OTHER SUBSTANCES

Simultaneous: β-tocopherol, gamma-tocopherol, δ-tocopherol

KEY WORDS

comparison with HPLC

REFERENCE

Spencer,B.J.; Purdy,W.C. Comparison of the separation of fat-soluble vitamins using β-cyclodextrins in high-performance liquid chromatography and micellar electrokinetic chromatography, *J.Chromatogr.A*, **1997**, *782*, 227–235.

Warfarin

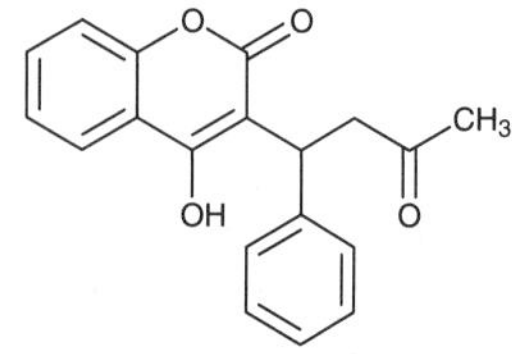

Molecular formula: $C_{19}H_{16}O_4$
Molecular weight: 308.33
CAS Registry No.: 81-81-2, 129-06-6 (sodium salt)
Merck Index (12th ed.): 10174
Lednicer: 1 131

SAMPLE

Matrix: blood
Sample preparation: 1 mL Plasma + 50 μL 25 μg/mL 5-chlorowarfarin + 100 μL 1 M HCl + 5 mL dichloromethane, stir for 10 min, centrifuge at 500 g for 5 min. Remove the organic layer and evaporate it to dryness under a stream of nitrogen at 40°, reconstitute the residue in 50 μL running buffer, inject an aliquot.

CAPILLARY ELECTROPHORESIS

Capillary: 72 cm × 50 mm fused-silica (50 cm to detector) (Applied Biosystems)
Capillary preparation: Between injections rinse with 100 mM NaOH for 1 min and with running buffer for 5 min. Condition new capillaries by flushing with 1 M NaOH for 30 min and with water for 30 min then equilibrating with running buffer for 10-30 min. Clean periodically using this procedure.
Capillary temperature: 25
Running buffer: MeOH:buffer 2:98 (Buffer was 100 mM pH 8.35 sodium phosphate buffer containing 8 mM methyl-β-cyclodextrin.)
Injection: Hydrodynamic injection at 16.9 kPa for 1 s (3.5 nL)
Detector: UV 310
Migration time: 11.2 (S-(-)), 11.4 (R-(+))
Internal standard: 5-chlorowarfarin (10)
Voltage: 20 kV
Current: 70 μA
Model: Applied Biosystems Model 270 A
Limit of detection: 200 ng/mL

KEY WORDS

chiral; plasma; cow; human

REFERENCE

Gareil,P.; Gramond,J.P.; Guyon,F. Separation and determination of warfarin enantiomers in human plasma samples by capillary zone electrophoresis using a methylated β-cyclodextrin-containing electrolyte, *J.Chromatogr.*, **1993**, *615*, 317–325.

SAMPLE

Matrix: solutions
Sample preparation: Prepare a solution in running buffer, inject an aliquot.

CAPILLARY ELECTROPHORESIS

Capillary: 50-60 cm × 50 μm fused-silica (35 cm to detector)
Capillary preparation: Flush capillary each day with 1 M KOH, EtOH, and running buffer for 5 min each
Capillary temperature: 25
Running buffer: 67 mM pH 7.4 phosphate buffer (Prepare buffer by dissolving 0.53 g KH_2PO_4 and 2.94 g $K_2HPO_4.3H_2O$ in 250 mL water.)
Injection: Electromigration at 10 kV for 2.5-60 s

Detector: UV 315
Migration time: 10
Voltage: 200 V/cm
Model: Laboratory constructed

OTHER SUBSTANCES
Simultaneous: acetone, bovine serum albumin

REFERENCE
Kraak,J.C.; Busch,S.; Poppe,H. Study of protein-drug binding using capillary zone electrophoresis, *J.Chromatogr.*, **1992**, *608*, 257–264.

SAMPLE
Matrix: solutions

CAPILLARY ELECTROPHORESIS
Capillary: 70 cm × 75 μm fused-silica (45 cm to detector) (Polymicro Technologies)
Capillary preparation: Treat with 100 mM NaOH and rinse with water before use.
Capillary temperature: 20
Running buffer: MeOH:buffer 40:60 (Buffer was 20 mM pH 6.5 sodium phosphate.)
Injection: Injection by vacuum
Detector: UV 237
Migration time: 13.2
Internal standard: coumachlor (12.8)
Voltage: 357 V/cm
Model: Isco Model 3850 or Model 3140

REFERENCE
Potter,K.J.; Allington,R.J.; Algaier,J. Separation of estrogens and rodenticides using capillary electrophoresis with aqueous-methanolic buffers, *J.Chromatogr.A*, **1993**, *652*, 427–429.

SAMPLE
Matrix: solutions

CAPILLARY ELECTROPHORESIS
Capillary: 60 cm × 50 μm fused-silica
Capillary preparation: Before each run purge with running buffer for 3 min. At the beginning of each day vacuum purge with 500 mM NaOH, vacuum purge with water, perform an electroosmotic purge, vacuum purge with running buffer.
Running buffer: 10 mM pH 7 Phosphate buffer containing 2.5% Glucidex2 (maltodextrin preparation, Roquette, Lestrem, France)
Injection: Injection by siphon at 10 cm
Detector: UV 185
Migration time: 3.65, 3.7 (enantiomers)
Voltage: 30 kV
Model: Waters Quanta 4000 CE

OTHER SUBSTANCES
Simultaneous: phenprocoumon

KEY WORDS
chiral

REFERENCE
D'Hulst,A.; Verbeke,N. Quantitation in chiral capillary electrophoresis: theoretical and practical considerations, *Electrophoresis*, **1994**, *15*, 854–863.

SAMPLE
Matrix: solutions
Sample preparation: Prepare an aqueous solution, filter (0.45 μm), inject an aliquot.

CAPILLARY ELECTROPHORESIS
Capillary: 36 cm × 50 μm (31.5 cm to detector)
Capillary preparation: Before each run rinse with water for 1 min and with running buffer for 1 min.
Capillary temperature: 25
Running buffer: THF:buffer 10:90 (Buffer was 50 mM pH 6.0 phosphate buffer containing 25 μm avidin.)
Injection: Inject at 1 psi pressure for 2 s
Detector: UV 300
Migration time: 15, 15.4 (enantiomers)
Voltage: -12 kV
Model: BioFocus 3000 (Bio-Rad)

KEY WORDS
chiral

REFERENCE
Tanaka,Y.; Matsubara,N.; Terabe,S. Separation of enantiomers by affinity electrokinetic chromatography using avidin, *Electrophoresis*, **1994**, *15*, 848–853.

SAMPLE
Matrix: solutions
Sample preparation: Inject an aliquot of a 100 μg/mL solution in MeOH:running buffer 50:50.

CAPILLARY ELECTROPHORESIS
Capillary: 37 cm × 50 μm fused-silica (30 cm to detector)
Capillary preparation: Between runs rinse capillary with 100 mM KOH for 2 min, with water for 2 min, and with running buffer for 2 min. Rinse new capillaries with 500 mM KOH for 15 min, with water for 10 min, and with running buffer for 15 min.
Capillary temperature: 25
Running buffer: 50 mM pH 7.0 Sodium phosphate buffer containing 25 mM sodium dodecyl sulfate and 2 mM vancomycin
Injection: Hydrostatic injection at 0.5 psi for 1 s
Detector: UV 254
Migration time: 20.54, 21.42 (enantiomers)
Voltage: 5 kV
Model: Beckman P/ACE 2000

OTHER SUBSTANCES
Simultaneous: bendroflumethiazide, 5-(4-hydroxyphenyl)-5-phenylhydantoin

KEY WORDS
chiral

REFERENCE
Armstrong,D.W.; Rundlett,K.L. CE resolution of neutral and anionic racemates with glycopeptide antibiotics and micelles, *J.Liq.Chromatogr.*, **1995**, *18*, 3659–3674.

SAMPLE
Matrix: solutions

CAPILLARY ELECTROPHORESIS
Capillary: 50 cm × 50 μm fused-silica (41 cm to detector) (Grom)
Capillary temperature: 21 ± 1
Running buffer: 50 mM pH 6 Sodium phosphate buffer containing 10 mM sulfobutyl ether β-cyclodextrin (degree of substitution 4.0; Center for Drug Delivery Research, University of Kansas, Lawrence KS) (Anode and cathode buffers were the same but contained no cyclodextrin.)
Injection: Hydrostatic injection at 10 cm.
Detector: UV 210
Migration time: 19.3, 20.5 (enantiomers)
Voltage: 400 V/cm

Model: Grom 100

KEY WORDS
chiral

REFERENCE
Chankvetadze,B.; Endresz,G.; Blaschke,G. Capillary electrophoresis enantioseparation of noncharged and anionic chiral compouds using anionic cyclodextrin derivatives as chiral selectors, *J.Capillary Electrophor.*, **1995**, *2*, 235–240.

SAMPLE
Matrix: solutions

CAPILLARY ELECTROPHORESIS
Capillary: 40 cm × 50 μm fused-silica (35.5 cm to detector) (Polymicro Technologies)
Capillary preparation: Between runs purge with water for 50 s, with 100 mM NaOH for 50 s, with water for 60 s, and with running buffer for 60 s.
Running buffer: 50 mM pH 6.0 Sodium phosphate buffer containing 6 mg/mL sulfobutyl ether-β-cyclodextrin (Perkin-Elmer)
Injection: Pressure injection at 5 psi
Detector: UV 206
Migration time: 12.4, 12.8 (enantiomers)
Voltage: 15 kV
Model: Bio-Rad Biofocus 3000

KEY WORDS
chiral

REFERENCE
Desiderio,C.; Fanali,S. Use of negatively charged sulfobutyl ether-β-cyclodextrin for enantiomeric separation by capillary electrophoresis, *J.Chromatogr.A*, **1995**, *716*, 183–196.

SAMPLE
Matrix: solutions
Sample preparation: Inject an aliquot of a 50-100 μM solution.

CAPILLARY ELECTROPHORESIS
Capillary: 35 cm × 50 μm coated fused-silica (30.5 cm to detector) (Polymicro Technologies)
Capillary preparation: Coat capillary as follows. Adjust the pH of 20 mL water to 3.5 with acetic acid, add 80 μL 3-(trimethoxysilyl)propyl methacrylate (3-methacryloxypropyltrimethoxysilane), mix, suck into capillary, let stand at room temperature for 1 h, remove the solution, wash with water. Fill the capillary with a deaerated 3-4% acrylamide solution containing 1 μL/mL N,N,N',N'-tetramethylethylenediamine and 1 mg/mL potassium persulfate, let stand for 30 min, remove excess solution by aspiration, rinse with water, remove water by aspiration, dry at 35° (J. Chromatogr. 1985, 347, 191).
Capillary temperature: 25
Running buffer: pH 7 Buffer containing 3 mM 6^A-methylamino-β-cyclodextrin (Buffer was 50 mM phosphoric acid containing 50 mM acetic acid and 50 mM boric acid, adjusted to pH 7 with concentrated NaOH solution, diluted with an equal volume of water. Synthesis of 6^A-methylamino-β-cyclodextrin was as follows. Add a solution of 3.65 g p-toluenesulfonyl chloride in 30 mL dry pyridine to 29.60 g β-cyclodextrin stirred at 5° in 300 mL dry pyridine, stir overnight at room temperature, evaporate to dryness under reduced pressure at 40°, add 700 mL diethyl ether to the residue. Collect the precipitate and recrystallize it 3 times from water to obtain mono-(6-O-p-tolylsulfonyl)-β- cyclodextrin in 31% yield (Bull. Chem. Soc. Japan 1978, 51, 3030). Heat 2 g mono-(6-O-p-tolylsulfonyl)-β-cyclodextrin with 35 mL 50% methylamine in MeOH in a sealed tube at 70° for 3 days, purify by chromatography on carboxymethylcellulose with ammonium bicarbonate solution to obtain 6^A-methylamino-β-cyclodextrin (cf. J. Am. Chem. Soc. 1980, 102, 762).)
Injection: Pressure injection at 10 psi.s.
Detector: UV 206
Migration time: 16 (R-(+)), 17 (S-(-))
Voltage: 15 kV

Current: 54 μA
Model: Biofocus 3000 (Bio-Rad)

OTHER SUBSTANCES
Simultaneous: acenocoumarol, phenyllactic acid, tiaprofenic acid

KEY WORDS
coated capillary; chiral

REFERENCE
Fanali,S.; Camera,E. Use of methylamino-β-cyclodextrin in capillary electrophoresis. Resolution of acidic and basic enantiomers, *Chromatographia*, **1996**, *43*, 247–253.

SAMPLE
Matrix: solutions
Sample preparation: Inject an aliquot of a 100 μg/mL solution in water or running buffer.

CAPILLARY ELECTROPHORESIS
Capillary: 47 cm × 75 μm fused-silica (40 cm to detector)
Capillary preparation: Before each run rinse capillary with running buffer for 1-2 min. If peak tailing is observed fill capillary with 500 mM NaOH and let stand for 30 min, wash with water for 3 min, rinse with running buffer for 3 min.
Capillary temperature: 20
Running buffer: 20 mM pH 7.0 Phosphate buffer containing 6% dextrin (Japan Pharmacopeia grade) (Dissolve dextrin in buffer at 90° then cool to room temperature.)
Injection: Pressure injection at 0.5 psi for 2-4 s.
Detector: UV 220
Migration time: 13.634, 13.919 (enantiomers)
Voltage: 20 kV
Model: Beckman P/ACE 5510

KEY WORDS
chiral

REFERENCE
Nishi,H.; Izumoto,S.; Nakamura,K.; Nakai,H.; Sato,T. Dextran and dextrin as chiral selectors in capillary zone electrophoresis, *Chromatographia*, **1996**, *42*, 617–630.

SAMPLE
Matrix: solutions
Sample preparation: Inject an aliquot of a solution in running buffer.

CAPILLARY ELECTROPHORESIS
Capillary: 60 cm × 75 μm fused-silica (52.4 cm to detector)
Capillary preparation: After each run flush with 500 mM KOH for 2-3 min then with water.
Running buffer: 10 mM pH 3.8 Phosphate buffer containing 2% sulfated cyclodextrin (ds 7-10)
Injection: Hydrostatic injection.
Detector: UV 214
Migration time: 25.20, 30.89 (enantiomers)
Voltage: 15 kV
Model: Waters Quanta 4000

OTHER SUBSTANCES
Also analyzed: acebutolol, alprenolol, aminoglutethimide, brompheniramine, bupivacaine, bupropion, canadine, carbinoxamine, chloroquine, chlorpheniramine, dimethindene, disopyramide, doxylamine, hydroxychloroquine, idazoxan, isoxsuprine, ketamine, mepenzolate, mepivacaine, methoxyphenamine, mexiletine, midodrine, nefopam, orphenadrine, oxprenolol, oxyphencyclimine, pheniramine, phensuximide, pindolol, piperoxan, terbutaline, tetramisole, tolperisone, tranylcypromine, trihexyphenidyl, trimipramine, verapamil

KEY WORDS
chiral; detector at anode

REFERENCE
Stalcup,A.M.; Gahm,K.H. Application of sulfated cyclodextrins to chiral separations by capillary zone electrophoresis, *Anal.Chem.*, **1996**, *68*, 1360–1368.

SAMPLE
Matrix: solutions
Sample preparation: Inject an aliquot of a 100-500 µg/mL solution in MeOH.

CAPILLARY ELECTROPHORESIS
Capillary: 55 cm × 50 µm fused-silica (50.5 cm to detector) (Polymicro Technologies)
Capillary preparation: Before each run purge capillary with water and running buffer for 2 min each.
Capillary temperature: 25
Running buffer: 25 mM pH 5.6 NaH_2PO_4 buffer containing 0.5% poly(sodium N-undecylenyl-D-valinate) (Prepare poly(sodium N-undecylenyl-D-valinate) as follows. Add 30 mmoles undecylenic acid to 30 mmoles N-hydroxysuccinimide in 130 mL dry ethyl acetate, add 30 mmoles dicyclohexylcarbodiimide in 10 mL dry ethyl acetate, let stand at room temperature overnight, filter, recrystallize ester from EtOH. Add 1 mmole undecylenic acid N-hydroxysuccinimide ester in 10 mL THF to 1 mmole L-valine and 1 mmole sodium bicarbonate in 10 mL water, after 16 h acidify to pH 2 with 1 M HCl, remove the organic solvent under reduced pressure, add 50 mL water, filter, dry, recrystallize N-undecylenyl-L-valine from chloroform/petroleum ether (mp 91°) (J.Lipid Res. 1967, 8, 142). Prepare the sodium salt by adding an equimolar amount of NaOH in EtOH:water 1:4 (Anal.Chem. 1994, 66, 3773). Polymerize a 50 mM aqueous solution of the sodium salt by irradiating with a ^{60}Co gamma source at 0.16 Mrad/h at 21° (J.Poly.Sci.: Poly.Lett.Ed. 1979, 17, 749), lyophilize, extract with hot EtOH to remove monomer, dry under vacuum. Dry organic solvents over Type 4A 4-8 mesh molecular sieve.)
Injection: Pressure injection at 2 psi.s.
Detector: UV 280
Migration time: 19.5, 20 (enantiomers)
Voltage: +20 kV
Model: Bio-Rad Biofocus 3000

OTHER SUBSTANCES
Simultaneous: coumachlor

KEY WORDS
chiral

REFERENCE
Agnew-Heard,K.A.; Sanchez Peña,M.; Shamsi,S.A.; Warner,I.M. Studies of polymerized sodium *N*-undecylenyl-L-valinate in chiral micellar electrokinetic capillary chromatography of neutral, acidic, and basic compounds, *Anal.Chem.*, **1997**, *69*, 958–964.

SAMPLE
Matrix: solutions
Sample preparation: Inject an aliquot of a solution in MeCN:MeOH 50:50 containing 2 mM sodium acetate.

CAPILLARY ELECTROPHORESIS
Capillary: 27 cm × 50 µm fused-silica (20 cm to detector) (Electro-Kinetic Technologies, Edinburgh, UK)
Capillary preparation: Between each injection rinse with 100 mM NaOH for 1 min, with MeOH for 1 min, and with running buffer for 1 min. Rinse new capillaries with 100 mM NaOH then with MeOH for 10 min.
Running buffer: MeCN:MeOH 50:50 containing 10 mM sodium acetate
Injection: Injection for 1 s.
Detector: UV 200
Migration time: 2.4
Voltage: 25 kV
Model: Beckman P/ACE 5000

OTHER SUBSTANCES
Simultaneous: benzoic acid, hydroxynaphthoate, naphthoxyacetic acid, penicillin (sic), troglitazone

---

REFERENCE
Altria,K.D.; Bryant,S.M. Highly selective and efficient separations of a wide range of acidic species in capillary electrophoresis employing non-aqueous media, *Chromatographia*, **1997**, *46*, 122–122.

SAMPLE
Matrix: solutions

CAPILLARY ELECTROPHORESIS
Capillary: 27 cm × 75 μm
Capillary preparation: Before each run rinse capillary with 100 mM NaOH for 30 s and with running buffer for 30 s. Condition new capillaries by rinsing with 100 mM NaOH for 20 min.
Capillary temperature: 30
Running buffer: 15 mM Sodium borate
Injection: Pressure injection of sample at 25 mbar for 1 s followed by running buffer at 25 mbar for 1 s.
Detector: UV 200
Migration time: 2.87
Internal standard: aminobenzoic acid (3.5)
Voltage: 6.5 kV
Model: Beckman

OTHER SUBSTANCES
Simultaneous: aspirin, bacitracin, beclomethasone, benzoic acid, ceftizoxime, ceftriaxone, cefuroxime, cephalothin, cromolyn, embonic acid, epoprostenol, glyburide (glibenclamide), levothyroxine, nedocromil, nystatin, omeprazole, prednisolone, warfarin, zidovudine

---

REFERENCE
Altria,K.D.; Bryant,S.M.; Hadgett,T.A. Validated capillary electrophoresis method for the analysis of a range of acidic drugs and excipients, *J.Pharm.Biomed.Anal.*, **1997**, *15*, 1091–1101.

SAMPLE
Matrix: solutions

CAPILLARY ELECTROPHORESIS
Capillary: 58.5 cm × 50 μm fused-silica (50 cm to detector) (Composite Metal Services, Worcs., UK)
Capillary preparation: Wash capillary with running buffer for 5 min before each run. At the start of each day wash with 100 mM NaOH for 5 min and rinse with water for 15 min.
Capillary temperature: 25
Running buffer: 40 mM pH 7 Sodium acetate buffer containing 40 mM phosphoric acid, 40 mM boric acid, and 15 mM gamma-cyclodextrin
Injection: Pressure injection at 150 mbar.s.
Detector: UV 202
Migration time: 5.1, 5.2 (enantiomers)
Voltage: 30 kV
Model: Hewlett-Packard [3D]CE

KEY WORDS
chiral

REFERENCE
Roos,N.; Ganzler,K.; Szemán,J.; Fanali,S. Systematic approach to cost- and time-effective method development with a starter kit for chiral separations by capillary electrophoresis, *J.Chromatogr.A*, **1997**, *782*, 257–269.

SAMPLE
Matrix: solutions

CAPILLARY ELECTROPHORESIS
Capillary: 60 cm × 50 μm fused-silica (30 cm to detector) (Polymicro Technologies)
Capillary temperature: 30
Running buffer: 40 mM pH 4.0 Sodium acetate containing 1.4 mM sulfobutyl ether-β-cyclo-
  dextrin (average substitution 3.9, MW 1721, Center for Drug Delivery Research, Lawrence KS)
Injection: Pressure injection at 5 inches Hg for 1.5 s.
Detector: UV 232
Migration time: 6.8, 7.5 (enantiomers)
Voltage: 25 kV
Current: 11 μA
Model: Perkin Elmer-Applied Biosystems Model 270

KEY WORDS
chiral

REFERENCE
Xie,G.-h.; Skanchy,D.J.; Stobaugh,J.F. Chiral separations of enantiomeric pharmaceuticals by capillary electrophoresis using sulphobutyl ether β-cyclodextrin as isomer selector, *Biomed.Chromatogr.*, **1997**, *11*, 193–199.

SAMPLE
Matrix: solutions
Sample preparation: Inject an aliquot of a 100 μg/mL solution in MeOH:water 45:55.

CAPILLARY ELECTROPHORESIS
Capillary: 37 cm × 75 μm eCAP neutral coated capillary (30 cm to detector) (Beckman)
Capillary temperature: 22
Running buffer: MeOH:buffer 2:98 (Buffer was 100 mM pH 8.4 NaH_2PO_4 buffer containing 8
  mM dimethyl-β-cyclodextrin.)
Injection: For 2 s
Detector: UV 200
Migration time: 13.53 (R), 14.05 (S)
Voltage: 10 kV
Model: Beckman P/ACE 2210

KEY WORDS
chiral; coated capillary

REFERENCE
Assi,K.H.; Abushoffa,A.M.; Altria,K.D.; Clark,B.J. Determination of trace enantiomeric impurities in chiral
  compounds by capillary electrophoresis with uncoated, cationic, anionic and neutral surface modified capillaries, *J.Chromatogr.A*, **1998**, *817*, 83–90.

SAMPLE
Matrix: solutions

CAPILLARY ELECTROPHORESIS
Capillary: 35 cm × 50 μm polyacrylamide-coated fused-silica (30.5 cm to detector) (Composite
  Metal Services, UK)
Capillary preparation: Before each run rinse capillary with water for 70 s and with running
  buffer for 100 s. (Coat capillary as follows. Adjust the pH of 20 mL water to 3.5 with acetic
  acid, add 80 μL 3-(trimethoxysilyl)propyl methacrylate (3-methacryloxypropyltrimethoxysi-
  lane), mix, suck into capillary, let stand at room temperature for 1 h, remove the solution,
  wash with water. Fill the capillary with a deaerated 3-4% acrylamide solution containing 1
  μL/mL N,N,N',N'-tetramethylethylenediamine and 1 mg/mL potassium persulfate, let stand
  for 30 min, remove excess solution by aspiration, rinse with water, remove water by aspiration,
  dry at 35° (J. Chromatogr. 1985, 347, 191).)
Capillary temperature: 25
Running buffer: Buffer containing 7.5 mM cyanoethylated-β-cyclodextrin (Cyclolab, Budapest)
  (Prepare buffer by adjusting the pH of 50 mM phosphoric acid containing 50 mM acetic acid
  and 50 mM boric acid to 5 with concentrated NaOH, add the appropriate amount of cyanoethylated-β-cyclodextrin, dilute with an equal volume of water.)
Injection: Pressure injection at 5 psi for 2 s.

Detector: UV 206
Migration time: 25.7 (second enantiomer, α = 1.078)
Voltage: 20 kV
Current: 27-40 μA
Model: Bio-Rad Biofocus 3000

KEY WORDS
detector at anode; chiral; coated capillary

REFERENCE
Aturki,Z.; Desiderio,C.; Mannina,L.; Fanali,S. Chiral separations by capillary zone electrophoresis with the use of cyanoethylated-β-cyclodextrin as chiral selector, *J.Chromatogr.A*, **1998**, *817*, 91–104.

SAMPLE
Matrix: solutions

CAPILLARY ELECTROPHORESIS
Capillary: 70 cm $\times$ 50 μm fused-silica (GL Science, Tokyo)
Capillary preparation: Before each run rinse capillary with running buffer at 94 kPa for 5 min before each run.
Running buffer: 50 mM pH 8.5 ammonium carbonate buffer
Injection: Pressure injection at 5 kPa (50 mbar) for 4 s.
Detector: MS, Perkin-Elmer Sciex API-300 quadrupole, electrospray (ionspray) interface, sheath liquid MeOH:running buffer 50:50 at 2.5 μL/min, ionspray voltage 5 kV, negative ion mode
Migration time: 10.2
Voltage: 20 kV (net voltage across capillary = 15 kV (applied voltage - electrospray voltage))
Model: Hewlett Packard 3D CE

OTHER SUBSTANCES
Simultaneous: acetaminophen, ascorbic acid, butylscopolamine bromide, caffeine, ibuprofen, ketoprofen, niacin, niacinamide, riboflavin, thiamine, vitamin B12

REFERENCE
Tanaka,Y.; Kishimoto,Y.; Otsuka,K.; Terabe,S. Strategy for selecting separation solutions in capillary electrophoresis-mass spectrometry, *J.Chromatogr.A*, **1998**, *817*, 49–57.

Xylometazoline

Molecular formula: $C_{16}H_{24}N_2$
Molecular weight: 244.38
CAS Registry No.: 526-36-3, 1218-35-5 (HCl)
Merck Index (12th ed.): 10219
Lednicer: 1 242

SAMPLE
Matrix: solutions

CAPILLARY ELECTROPHORESIS
Capillary: 58.2 cm $\times$ 75 μm fused-silica (50.7 cm to detector) (Polymicro Technologies)
Capillary preparation: Purge with running buffer before each run. At the beginning of each day purge using 50-60 kPa vacuum with 500 mM NaOH for 5 min, with water for 5 min, with MeCN for 5 min, and with running buffer for 5 min.
Running buffer: MeCN:MeOH:acetic acid 49:50:1 containing 20 mM ammonium acetate
Injection: Hydrostatic injection at 10 cm for 5 s.
Detector: UV 214
Migration time: 3.01
Voltage: 25 kV

Model: Waters Quanta 4000

OTHER SUBSTANCES
Simultaneous: amphetamine, benzphetamine, ephedrine, nylidrin, oxymetazoline, phendimetrazine, phenmetrazine, phenylephrine, phenylpropanolamine

REFERENCE
Leung,G.N.W.; Tang,H.P.O.; Tso,T.S.C.; Wan,T.S.M. Separation of basic drugs with non-aqueous capillary electrophoresis, *J.Chromatogr.A*, **1996**, *738*, 141–154.

Xylose

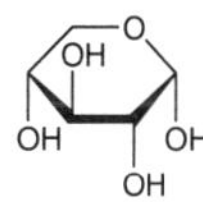

Molecular formula: $C_5H_{10}O_5$
Molecular weight: 150.13
CAS Registry No.: 58-86-6
Merck Index (12th ed.): 10220

SAMPLE
Matrix: bulk
Sample preparation: Add 50 µL 500 mM 3-methyl-1-phenyl-2-pyrazolin-5-one in MeOH and 30 µL 500 mM NaOH in water to 10-200 nmoles sugars, heat at 70° for 2 h, cool to room temperature, add 30 µL 500 mM HCl, vortex, wash 4 times with 500 µL aliquots of dichloromethane. Freeze dry the aqueous phase, reconstitute with 200-500 µL water, filter (0.2 µm nylon), inject an aliquot.

CAPILLARY ELECTROPHORESIS
Capillary: 72 cm × 50 µm fused-silica (50 cm to detector) (Polymicro Technologies)
Capillary preparation: Between runs rinse with alkaline rinse solution (NaOH/Na_3PO_4 in water) for 2 min and with running buffer for 5 min. Flush new capillaries with alkaline rinse solution for 20 min and with running buffer for 20 min.
Capillary temperature: 30
Running buffer: 30 mM Phosphoric acid containing 50 mM sodium dodecyl sulfate, adjusted to pH 7.5 with Tris (High purity sodium dodecyl sulfate (Perkin-Elmer/Applied Biosystems) should be used.)
Injection: Vacuum injection at 16.9 kPa for 3-5 s.
Detector: UV 245
Migration time: 15.9
Voltage: 20 kV
Current: ca. 29 µA
Model: Perkin-Elmer/Applied Biosystems 270A-HT
Limit of detection: 20 fmole

OTHER SUBSTANCES
Simultaneous: N-acetylmannosamine, altrose, cellobiose, 2-deoxy-D-ribose, dextrose, fucose, galactosamine, galactose, glucosamine, lactose, lyxose, maltose, mannosamine, mannose, rhamnose

KEY WORDS
derivatization

REFERENCE
Chiesa,C.; Oefner,P.J.; Zieske,L.R.; O'Neill,R.A. Micellar electrokinetic chromatography of monosaccharides derivatized with 1-phenyl-3-methyl-2-pyrazolin-5-one, *J.Capillary Electrophor.*, **1995**, *2*, 175–183.

SAMPLE
Matrix: plants
Sample preparation: Stir 40 g dried plant material with 600 mL water at 500 rpm for 24 h, filter (grid mesh 0.3 mm), centrifuge the filtrate at 17700 g for 15 min. Remove an 80 mL

aliquot of the supernatant to 400 mL EtOH, let stand at 4° for 48 h, centrifuge at 17700 g for 30 min, discard the supernatant, suspend the pellet in 100 mL ether, evaporate the ether to dryness at 40°. Add 2.5 mL 6 M trifluoroacetic acid to 20 mg of the precipitate, dissolve with gentle swirling, reflux at 110° for 4 h, lyophilize to dryness, add reagent so that the concentration of saccharide is 2 mg/mL, vortex gently, heat at 50° for 2 h, cool to room temperature, dilute 10-100-fold with MeOH, inject an aliquot. (Prepare reagent by dissolving 10 mg sodium cyanoborohydride in 1 mL MeOH containing 10% 2-aminopyridine and 10% acetic acid.)

CAPILLARY ELECTROPHORESIS
Capillary: 72 cm × 50 μm fused-silica (50 cm to detector)
Capillary preparation: Between runs wash capillary with 1 M NaOH for 3-4 min and with 1 mM NaOH for 2 min, equilibrate with running buffer for 3-6 min. Store capillaries in 1 mM NaOH overnight. Flush new capillaries with 1 M NaOH for 1 h and with 1 mM NaOH for 5 min.
Capillary temperature: 30
Running buffer: 150 mM pH 10.5 Borate buffer
Injection: Vacuum injection at 16.9 kPa for 1 s
Detector: UV 237
Migration time: 13.1
Voltage: 20 kV
Model: Applied Biosystems Model 270A

OTHER SUBSTANCES
Extracted: arabinose, 2-deoxy-D-ribose, dextrose, fucose, galactose, galacturonic acid, glucuronic acid, lyxose, maltose, maltotriose, mannuronic acid, rhamnose, ribose

KEY WORDS
derivatization

REFERENCE
Oefner,P.J.; Vorndran,A.E.; Grill,E.; Huber,C.; Bonn,G.K. Capillary zone electrophoretic analysis of carbohydrates by direct and indirect UV detection, *Chromatographia*, **1992**, *34*, 308–316.

SAMPLE
Matrix: plants
Sample preparation: Stir 40 g dried plant material with 600 mL water at 500 rpm for 24 h, filter (grid mesh 0.3 mm), centrifuge the filtrate at 17700 g for 15 min. Remove an 80 mL aliquot of the supernatant to 400 mL EtOH, let stand at 4° for 48 h, centrifuge at 17700 g for 30 min, discard the supernatant, suspend the pellet in 100 mL ether, evaporate the ether to dryness at 40°. Add 2.5 mL 6 M trifluoroacetic acid to 20 mg of the precipitate, dissolve with gentle swirling, reflux at 110° for 4 h, lyophilize to dryness, add reagent so that the concentration of saccharide is 2 mg/mL, vortex gently, heat at 50° for 2 h, cool to room temperature, dilute 10-100-fold with MeOH, inject an aliquot. (Prepare reagent by dissolving 10 mg sodium cyanoborohydride in 1 mL MeOH containing 7% p-aminobenzoic acid and 10% acetic acid.)

CAPILLARY ELECTROPHORESIS
Capillary: 72 cm × 50 μm fused-silica (50 cm to detector)
Capillary preparation: Between runs wash capillary with 1 M NaOH for 3-4 min and with 1 mM NaOH for 2 min, equilibrate with running buffer for 3-6 min. Store capillaries in 1 mM NaOH overnight. Flush new capillaries with 1 M NaOH for 1 h and with 1 mM NaOH for 5 min.
Capillary temperature: 30
Running buffer: 150 mM pH 10.0 Borate buffer
Injection: Vacuum injection at 16.9 kPa for 1 s
Detector: UV 285
Migration time: 11.5
Voltage: 28 kV
Model: Applied Biosystems Model 270A

OTHER SUBSTANCES
Extracted: arabinose, cellobiose, 2-deoxy-D-ribose, dextrose, fructose, fucose, galactose, galacturonic acid, glucuronic acid, lactose, melibiose, sorbose

KEY WORDS
derivatization

REFERENCE

Oefner,P.J.; Vorndran,A.E.; Grill,E.; Huber,C.; Bonn,G.K. Capillary zone electrophoretic analysis of carbohydrates by direct and indirect UV detection, *Chromatographia*, **1992**, *34*, 308–316.

SAMPLE

Matrix: plants
Sample preparation: Stir 40 g dried plant material with 600 mL water at 500 rpm for 24 h, filter (grid mesh 0.3 mm), centrifuge the filtrate at 17700 g for 15 min. Remove an 80 mL aliquot of the supernatant to 400 mL EtOH, let stand at 4° for 48 h, centrifuge at 17700 g for 30 min, discard the supernatant, suspend the pellet in 100 mL ether, evaporate the ether to dryness at 40°. Add 2.5 mL 6 M trifluoroacetic acid to 20 mg of the precipitate, dissolve with gentle swirling, reflux at 110° for 4 h, lyophilize to dryness, add reagent so that the concentration of saccharide is 2 mg/mL, vortex gently, heat at 50° for 2 h, cool to room temperature, dilute 40-fold with MeOH, inject an aliquot. (Prepare reagent by dissolving 10 mg sodium cyanoborohydride in 1 mL MeOH containing 10% ethyl aminobenzoate and 10% acetic acid.)

CAPILLARY ELECTROPHORESIS

Capillary: 72 cm × 50 μm fused-silica (50 cm to detector)
Capillary preparation: Between runs wash capillary with 1 M NaOH for 4 min and with 1 mM NaOH for 2 min, equilibrate with running buffer for 4 min. Flush new capillaries with 1 M NaOH for 1 h and with 1 mM NaOH for 5 min.
Capillary temperature: 30
Running buffer: 175 mM pH 10.5 Borate buffer
Injection: Vacuum injection at 16.9 kPa for 1 s
Detector: UV 305
Migration time: 9.2
Voltage: 25 kV
Model: Applied Biosystems Model 270A

OTHER SUBSTANCES

Extracted: arabinose, cellobiose, 2-deoxy-D-ribose, dextrose, fucose, galactose, galacturonic acid, glucuronic acid, lactose, maltotriose, mannuronic acid, rhamnose, ribose

KEY WORDS
derivatization

REFERENCE

Vorndran,A.E.; Grill,E.; Huber,C.; Oefner,P.J.; Bonn,G.K. Capillary zone electrophoresis of aldoses, ketoses and uronic acids derivatized with ethyl p-aminobenzoate, *Chromatographia*, **1992**, *34*, 109–114.

SAMPLE

Matrix: solutions
Sample preparation: Add 50 μL 500 mM 3-methyl-1-phenyl-2-pyrazolin-5-one in MeOH and 50 μL 300 mM NaOH to a dried sample, heat at 70° for 30 min, cool to room temperature, add 50 μL 300 mM HCl, evaporate to dryness under reduced pressure, add 200 μL water, add 200 μL chloroform, shake vigorously. Remove the aqueous layer and evaporate it to dryness, reconstitute with a small volume of MeOH, inject an aliquot.

CAPILLARY ELECTROPHORESIS

Capillary: 78 cm × 50 μm fused-silica (63 cm to detector) (Scientific Glass Engineering)
Capillary preparation: Before each run rinse with 100 mM NaOH and with running buffer. After 10 runs rinse with MeOH.
Running buffer: 200 mM pH 9.5 Borate buffer
Injection: Siphon at 5 cm for 5 s
Detector: UV 245
Migration time: 20
Internal standard: amobarbital (17.5)
Voltage: 15 kV

OTHER SUBSTANCES
Simultaneous: allose, altrose, arabinose, dextrose, galactose, gulose, idose, lyxose, mannose, oligoglucans, talose

KEY WORDS
derivatization

REFERENCE
Honda,S.; Suzuki,S.; Nose,A.; Yamamoto,K.; Kakehi,K. Capillary zone electrophoresis of reducing mono- and oligo-saccharides as the borate complexes of their 3-methyl-1-phenyl-2-pyrazolin-5-one derivatives, *Carbohydrate Res.*, **1991**, *215*, 193–198.

SAMPLE
Matrix: solutions
Sample preparation: For each 1 mg saccharides add 500 μL reagent, vortex gently, heat at 50° for 2 h, cool to room temperature, dilute 10-100 fold with MeOH, inject an aliquot. (Prepare reagent by dissolving 10 mg sodium cyanoborohydride in MeOH containing 7% p-aminobenzoic acid and 10% acetic acid.)

CAPILLARY ELECTROPHORESIS
Capillary: 72 cm × 50 μm fused-silica (50 cm to detector)
Capillary preparation: Between runs wash capillary with 1 M NaOH for 3 min, wash with 1 mM NaOH for 2 min, equilibrate with running buffer for 3 min. Store capillary in 1 mM NaOH overnight. Flush a new capillary with 1 M NaOH for 1 h then with 1 mM NaOH for 5 min.
Capillary temperature: 30
Running buffer: 150 mM Boric acid adjusted to pH 10.0 with 2 M NaOH
Injection: Vacuum injection at 16.9 kPa for 1 s
Detector: UV 285
Migration time: 11.5
Voltage: 28 kV
Current: 79 μA
Model: Applied Biosystems Model 270A
Limit of detection: 4 μM

OTHER SUBSTANCES
Simultaneous: arabinose, cellobiose, 2-deoxy-D-ribose, dextrose, fructose, fucose, galactose, galacturonic acid, glucuronic acid, lactose, melibiose, sorbose

KEY WORDS
derivatization

REFERENCE
Grill,E.; Huber,C.; Oefner,P.; Vorndran,A.; Bonn,G. Capillary zone electrophoresis of p-aminobenzoic acid derivatives of aldoses, ketoses and uronic acids, *Electrophoresis*, **1993**, *14*, 1004–1010.

SAMPLE
Matrix: solutions
Sample preparation: Add reagent to the carbohydrate solution to make a total volume of 2 mL, vortex gently, heat at 90° for 15 min, cool to room temperature, dilute 20-1000-fold with running buffer, inject an aliquot. (Just prior to use prepare reagent by dissolving 10 mg sodium cyanoborohydride in 1 mL 5% acetic acid containing 6% 4-aminobenzonitrile.)

CAPILLARY ELECTROPHORESIS
Capillary: 55 cm × 50 μm fused-silica (35 cm to detector)
Capillary temperature: 30
Running buffer: 25 mM Tris containing 100 mM sodium dodecyl sulfate, adjusted to pH 7.5 with phosphoric acid
Injection: Vacuum injection at 16.9 kPa for 1 s
Detector: UV 285
Migration time: 4.1
Voltage: 30 kV

Current: 55 μA
Model: Applied Biosystems Model 270A

OTHER SUBSTANCES
Simultaneous: arabinose, cellobiose, dextrose, fructose, galactose, lactose, lyxose, maltose, maltotriose, mannose, melibiose, ribose, sorbose

---

KEY WORDS
derivatization

REFERENCE
Schwaiger,H.; Oefner,P.J.; Huber,C.; Grill,E.; Bonn,G.K. Capillary zone electrophoresis and micellar electrokinetic chromatography of 4-aminobenzonitrile carbohydrate derivatives, *Electrophoresis*, **1994**, *15*, 941–952.

SAMPLE
Matrix: solutions
Sample preparation: Mix 2 μL of a 5 mM solution with 2 μL 100 mM 9-aminopyrene-1,4,6-trisulfonate in 4.2 M acetic acid, add 4 μL 1 M sodium cyanoborohydride in THF, centrifuge, heat at 75° for 1 h, dilute 100-to 1000-fold, inject an aliquot. (9-Aminopyrene-1,4,6-trisulfonic acid is the same as 8-aminopyrene-1,3,6-trisulfonic acid. Synthesis is as follows. Rapidly add 300 g anhydrous sodium sulfate to 1300 g concentrated sulfuric acid, cool to 58°, add 202 g finely powdered pyrene over 5 min without cooling, stir for 15 min, cool to 50-55°, add 800 g 65% oleum (with cooling with 12° water) over 20 min, stir for 5 h without cooling, dilute with ice-water, neutralize with calcium carbonate, filter, reduce the filtrate to 10 L, exchange the cation with soda, salt out with 20% sodium chloride to obtain sodium pyrenetetrasulfonate. Heat 61 g sodium pyrenetetrasulfonate with 610 mL 22% aqueous ammonia in a rotating autoclave at 200-210° for 18 h (the pressure rises to 45 atmospheres), remove the excess ammonia by distillation under vacuum, precipitate oxypyrenetrisulfonic acid with a little NaCl, precipitate 9-aminopyrene-1,4,6-trisulfonic acid by saturating with NaCl, recrystallize from dilute NaCl (Liebig's Annalen der Chemie 1939, 540, 189).)

CAPILLARY ELECTROPHORESIS
Capillary: 27 cm × 20 μm fused-silica (20 cm to detector) (Polymicro Technologies)
Capillary preparation: Between runs wash capillary with 1 M NaOH at 15 psi for 12 s, wash with water at 15 psi for 12 s, condition with running buffer for 4 min.
Running buffer: 120 mM pH 7.0 MOPS
Detector: F ex 488 (2.5 mW argon ion laser) em 520 ± 9 (narrow-band filter; notch filter at 488 nm)
Migration time: 9.2
Voltage: 25 kV
Current: 19 μΛ
Model: Beckman P/ACE 2100

OTHER SUBSTANCES
Simultaneous: N-acetylgalactosamine, N-acetylglucosamine, arabinose, dextrose, fucose, galactose, mannose, rhamnose, ribose

---

KEY WORDS
derivatization

REFERENCE
Chen,F.-T.A.; Evangelista,R.A. Analysis of mono- and oligosaccharide isomers derivatized with 9-aminopyrene-1,4,6-trisulfonate by capillary electrophoresis with laser-induced fluorescence, *Anal.Biochem.*, **1995**, *230*, 273–280.

SAMPLE
Matrix: solutions
Sample preparation: Inject an aliquot of an aqueous solution.

CAPILLARY ELECTROPHORESIS
Capillary: 100 cm × 75 μm fused-silica capillaries (Polymicro Technologies)

Capillary preparation: Etch new capillaries with 1 M NaOH for 1 h before use.
Capillary temperature: 28
Running buffer: 100 mM NaOH containing 1 mM NaCl
Injection: Inject at a pressure of 20 mbar for 12 s, ca. 15 nL
Detector: E, cuprous oxide modified carbon working electrode in a 0.5 mm i.d. PEEK tube +0.60 V, stainless steel auxiliary electrode, Ag/AgCl reference electrode. The effluent from the capillary passed through a grounded Pd cylinder and then through an 8 cm × 50 μm coupling capillary to the detector. (Prepare electrode as follows. Push copper wires in to a 4 cm length of 0.5 mm i.d. PEEK tubing so as to leave a 1 mm deep chamber at one end, seal other end with glue. Stir 300 mg conductive carbon cement (Gerhard Neubauer, Münster), 60 mg cuprous oxide (Fluka), and 300 μL acetone until a thick paste forms as the acetone evaporates. Pack paste into the chamber, allow to dry for one week. Use only electrodes with resistance less than 200 Ω. Polish with dry emery paper (grade 2/0, Oakey), 3 μm imperial micro finishing film sheet (3M), and 0.05 μm alumina particles on a Buehler pad, wash with water (Anal. Chim. Acta 1995, 300, 5).)
Migration time: 38
Voltage: 12 kV
Current: 124 μA
Model: Lauer Labs Prince programmable injector and power supply
Limit of detection: 1-2 μM

OTHER SUBSTANCES
Simultaneous: arabinose, dextrose, fucose, galactose, mannose, raffinose

---

REFERENCE
Huang,X.; Kok,W.T. Determination of sugars by capillary electrophoresis with electrochemical detection using cuprous oxide modified electrodes, *J.Chromatogr.A*, **1995**, *707*, 335–342.

SAMPLE
Matrix: solutions
Sample preparation: Evaporate 5 μL of a 1 mM solution in water to dryness under reduced pressure, add 2 μL 100 mM 8-aminopyrene-1,3,6-trisulfonic acid (Lambda Fluoreszenztechnologie, Graz, Austria), add 2 μL 1.8 M citric acid in water, add 2 μL 1 M sodium cyanoborohydride in THF, heat at 75° for 1 h, dilute to 200 μL with water, dilute 25-fold, inject an aliquot. (9-Aminopyrene-1,4,6-trisulfonic acid is the same as 8-aminopyrene-1,3,6-trisulfonic acid. Synthesis is as follows. Rapidly add 300 g anhydrous sodium sulfate to 1300 g concentrated sulfuric acid, cool to 58°, add 202 g finely powdered pyrene over 5 min without cooling, stir for 15 min, cool to 50-55°, add 800 g 65% oleum (with cooling with 12° water) over 20 min, stir for 5 h without cooling, dilute with ice-water, neutralize with calcium carbonate, filter, reduce the filtrate to 10 L, exchange the cation with soda, salt out with 20% sodium chloride to obtain sodium pyrenetetrasulfonate. Heat 61 g sodium pyrenetetrasulfonate with 610 mL 22% aqueous ammonia in a rotating autoclave at 200-210° for 18 h (the pressure rises to 45 atmospheres), remove the excess ammonia by distillation under vacuum, precipitate oxypyrenetrisulfonic acid with a little NaCl, precipitate 9-aminopyrene-1,4,6-trisulfonic acid by saturating with NaCl, recrystallize from dilute NaCl (Liebig's Annalen der Chemie 1939, 540, 189).)

CAPILLARY ELECTROPHORESIS
Capillary: 25 cm × 19 μm fused-silica
Capillary preparation: Between runs rinse capillary with 1 M NaOH at 15 psi for 12 s and with running buffer at 15 psi for 1.2 min.
Running buffer: 120 mM pH 10.2 Borate buffer
Injection: Pressure injection at 0.5 psi for 20 s
Detector: F ex 488 (laser) em 520
Migration time: 4.8
Voltage: 30 kV
Current: 26 μA
Model: Beckman P/ACE 2100

OTHER SUBSTANCES
Simultaneous: N-acetylgalactosamine, N-acetylglucosamine, dextrose, fucose, galactose, mannose

---

KEY WORDS
derivatization

REFERENCE

Evangelista,R.A.; Guttman,A.; Chen,F.-T.A. Acid-catalyzed reductive amination of aldoses with 8-aminopyrene-1,3,6-trisulfonate, *Electrophoresis*, **1996**, *17*, 347–351.

SAMPLE

Matrix: solutions
Sample preparation: 56 μmoles Dextrose + 100 μL 150 mM 8-aminonaphthalene-1,3,6-trisulfonic acid (Molecular Probes, Eugene OR) in acetic acid:water 15:85 + 100 μL 1 M sodium cyanoborohydride in DMSO, vortex, heat at 40° for 15 h, dilute 1000-fold with water, inject an aliquot.

CAPILLARY ELECTROPHORESIS

Capillary: 27 cm × 50 μm (20.5 cm to detector) (Polymicro Technologies)
Capillary preparation: Between runs flush with 100 mM NaOH for 2 min and with running buffer for 2 min.
Capillary temperature: 25
Running buffer: 150 mM Boric acid adjusted to pH 9.5 with 2 M NaOH
Injection: Hydrodynamic injection
Detector: F ex 325 (2 mW He-Cd laser) em 520 (bandpass filter)
Migration time: 12.7
Voltage: 10 kV
Model: Beckman P/ACE 2100

OTHER SUBSTANCES

Simultaneous: N-acetylgalactosamine, N-acetylglucosamine, dextrose, fucose, galactose, mannose

KEY WORDS

derivatization

REFERENCE

Klockow,A.; Amadò,R.; Widmer,H.M.; Paulus,A. The influence of buffer composition on separation efficiency and resolution on capillary electrophoresis of 8-aminonaphthalene-1,3, 6-trisulfonic acid labeled monosaccharides and complex carbohydrates, *Electrophoresis*, **1996**, *17*, 110–119.

SAMPLE

Matrix: solutions
Sample preparation: Inject an aliquot of an aqueous solution.

CAPILLARY ELECTROPHORESIS

Capillary: 57 cm × 25 μm fused-silica (34.5 cm to detector) (Polymicro Technologies)
Capillary preparation: Between injections flush capillary with 100 mM NaOH, with water, and with running buffer. Flush new capillaries with 1 M NaOH.
Running buffer: 50 mM NaOH containing 5 mM tryptophan
Injection: Vacuum injection at 0.5 psi for 1 s
Detector: UV 280
Migration time: 18.5
Voltage: 7 kV
Model: Isco 3850
Limit of detection: 30 fmole

OTHER SUBSTANCES

Simultaneous: cellobiose, dextrose, galactose, mannose, melibiose, raffinose, ribose, stachyose

KEY WORDS

indirect UV detection

REFERENCE

Lu,B.; Westerlund,D. Indirect UV detection of carbohydrates in capillary zone electrophoresis by using tryptophan as a marker, *Electrophoresis*, **1996**, *17*, 325–332.

SAMPLE
Matrix: solutions
Sample preparation: Prepare a solution containing 100 μM monosaccharide, 30 mM 6-amino-
quinoline, 5 mM sodium cyanoborohydride, and 150 mM acetic acid, heat at 80° for 45 min (40°
for 2 h for oligosaccharides), cool to room temperature, filter, inject an aliquot.

CAPILLARY ELECTROPHORESIS
Capillary: 43 cm × 30 μm fused silica (38 cm to detector) (Skandinaviska GeneTech, Kungs-
backa, Sweden)
Running buffer: 420 mM Boric acid adjusted to pH 9 with 1 M NaOH
Injection: Hydrodynamic injection at 75 m for 10 s
Detector: UV 245
Migration time: 7
Current: 55 μA (constant power 1200 mW)
Model: Dionex
Limit of detection: 1 μM

OTHER SUBSTANCES
Simultaneous: arabinose, dextrose, galactose, galacturonic acid, glucuronic acid, mannose, 4-O-
methylglucuronic acid, rhamnose

KEY WORDS
derivatization

REFERENCE
Rydlund,A.; Dahlman,O. Efficient capillary zone electrophoretic separation of wood-derived neutral and acidic
mono- and oligosaccharides, *J.Chromatogr.A*, **1996**, *738*, 129–140.

SAMPLE
Matrix: solutions
Sample preparation: Mix a 20 μL aliquot of a 5 mM solution in EtOH with 90 μL 1% sodium
cyanoborohydride in EtOH containing 1-2% acetic acid, add 10 μL benzoic hydrazide in EtOH:
water 75:25, heat at 60° for 5 h, cool to room temperature, inject an aliquot. (Adjust the concentration of benzoic hydrazide so that it is present in 20-fold excess for standards and 200-
fold excess for hydrolyzed glycoproteins.)

CAPILLARY ELECTROPHORESIS
Capillary: 70 cm × 50 μm fused-silica (52 cm to detector) (Yongnian, Hebei, China)
Capillary preparation: Flush with running buffer before each injection. Flush a new capillary
with 1 M NaOH, water, and running buffer.
Capillary temperature: 30
Running buffer: 200 mM pH 10.8 Borate buffer
Injection: Vacuum injection for 8 s.
Detector: UV 220
Migration time: 21.5
Voltage: 20 kV
Model: Spectra PHORESIS 1000 (Thermo Separation Products)
Limit of detection: 30-40 fmole

OTHER SUBSTANCES
Simultaneous: N-acetylgalactosamine, arabose, dextrose, fucose, galactose, lyxose, mannose,
rhamnose

KEY WORDS
derivatization

REFERENCE
Lin,Q.; Zhang,R.; Liu,G. Use of benzoyl hydrazine reagent for monosaccharide determination by high performance capillary electrophoresis, *J.Liq.Chromatogr.Rel.Technol.*, **1997**, *20*, 1123–1137.

Zalcitabine

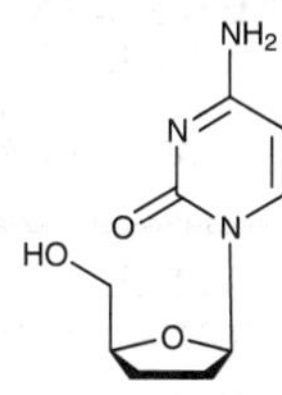

Molecular formula: $C_9H_{13}N_3O_3$
Molecular weight: 211.22
CAS Registry No.: 7481-89-2
Merck Index (12th ed.): 10242
Lednicer: 5 98

SAMPLE
Matrix: solutions

CAPILLARY ELECTROPHORESIS
Capillary: 80 cm × 75 µm fused-silica (52 cm to detector) (LC Packing, San Francisco)
Capillary temperature: 20 ± 0.5
Running buffer: 50 mM pH 6.5 Phosphate buffer containing 40 mM sodium dodecyl sulfate
Injection: Positive flow injection.
Detector: UV 260
Migration time: 8
Voltage: 20 kV
Current: 69 µA
Model: laboratory-constructed

OTHER SUBSTANCES
Simultaneous: didanosine, dideoxyadenosine, stavudine, zidovudine

REFERENCE
Singhal,R.; Xian,J.; Otim,O. Application of spherical and other polymers in capillary zone electrophoresis: separation of antiviral drugs and deoxyribonucleoside phosphates by different principles, *J.Chromatogr.A*, **1996**, *756*, 263–277.

Zidovudine

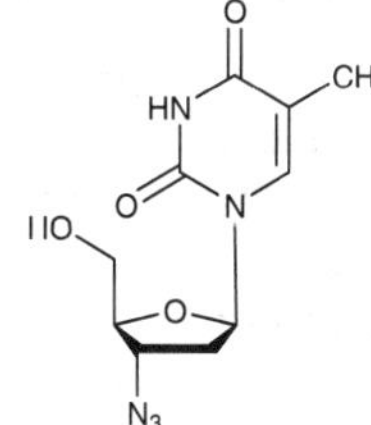

Molecular formula: $C_{10}H_{13}N_5O_4$
Molecular weight: 267.24
CAS Registry No.: 30516-87-1
Merck Index (12th ed.): 10252
Lednicer: 4 118

SAMPLE
Matrix: solutions

CAPILLARY ELECTROPHORESIS
Capillary: 80 cm × 75 µm fused-silica (52 cm to detector) (LC Packing, San Francisco)
Capillary temperature: 20 ± 0.5
Running buffer: 50 mM pH 6.5 Phosphate buffer containing 40 mM sodium dodecyl sulfate
Injection: Positive flow injection.
Detector: UV 260
Migration time: 8.8
Voltage: 20 kV
Current: 69 µA
Model: laboratory-constructed

OTHER SUBSTANCES
Simultaneous: didanosine, dideoxyadenosine, stavudine, zalcitabine

REFERENCE

Singhal,R.; Xian,J.; Otim,O. Application of spherical and other polymers in capillary zone electrophoresis: separation of antiviral drugs and deoxyribonucleoside phosphates by different principles, *J.Chromatogr.A*, **1996**, *756*, 263–277.

SAMPLE
Matrix: solutions

CAPILLARY ELECTROPHORESIS
Capillary: 27 cm × 75 µm
Capillary preparation: Before each run rinse capillary with 100 mM NaOH for 30 s and with running buffer for 30 s. Condition new capillaries by rinsing with 100 mM NaOH for 20 min.
Capillary temperature: 30
Running buffer: 15 mM Sodium borate
Injection: Pressure injection of sample at 25 mbar for 1 s followed by running buffer at 25 mbar for 1 s.
Detector: UV 200
Migration time: 2.21
Internal standard: aminobenzoic acid (3.5)
Voltage: 6.5 kV
Model: Beckman

OTHER SUBSTANCES
Simultaneous: aspirin, bacitracin, beclomethasone, benzoic acid, ceftizoxime, ceftriaxone, cefuroxime, cephalothin, cromolyn, embonic acid, epoprostenol, glyburide (glibenclamide), levothyroxine, nedocromil, nystatin, omeprazole, prednisolone, warfarin

REFERENCE

Altria,K.D.; Bryant,S.M.; Hadgett,T.A. Validated capillary electrophoresis method for the analysis of a range of acidic drugs and excipients, *J.Pharm.Biomed.Anal.*, **1997**, *15*, 1091–1101.

Zofenopril

Molecular formula: $C_{22}H_{23}NO_4S_2$
Molecular weight: 429.56
CAS Registry No.: 81872-10-8, 81938-43-4 (calcium salt)

SAMPLE
Matrix: solutions

CAPILLARY ELECTROPHORESIS
Capillary: 57 cm × 75 µm fused-silica (50 cm to detector) (Beckman)
Capillary preparation: Before each run rinse capillary with 100 mM NaOH and running buffer.
Capillary temperature: 30
Running buffer: 50 mM pH 8.2 Borate buffer containing 100 mM sodium dodecyl sulfate
Injection: Pressure injection for 5-10 s.
Detector: UV 200
Migration time: 12
Voltage: 25 kV
Model: Beckman P/ACE 5510

OTHER SUBSTANCES
Simultaneous: captopril, ceronapril, enalapril, fosinopril, lisinopril

REFERENCE

Bretnall,A.E.; Clarke,G.S. Selectivity of capillary electrophoresis for the analysis of cardiovascular drugs, *J.Chromatogr.A*, **1996**, *745*, 145–154.

Zolpidem

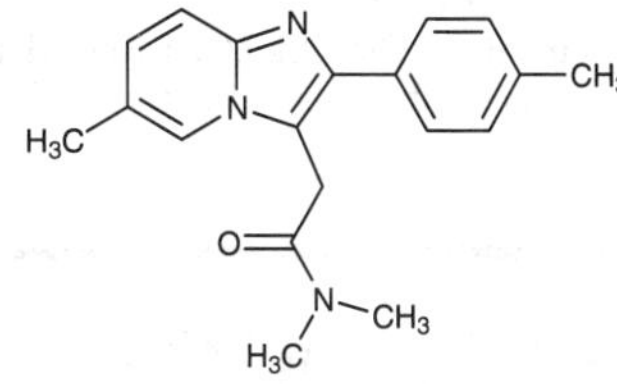

Molecular formula: $C_{19}H_{21}N_3O$
Molecular weight: 307.40
CAS Registry No.: 82626-48-0, 99294-93-6 ((+)-tartrate (2:1))
Merck Index (12th ed.): 10321
Lednicer: 4 162

SAMPLE
Matrix: solutions
Sample preparation: Inject an aliquot of a solution in MeCN:water 50:50.

CAPILLARY ELECTROPHORESIS
Capillary: 47 cm $\times$ 50 μm fused-silica (40 cm to detector)
Capillary preparation: Between runs rinse capillary with 100 mM NaOH for 1 min and with running buffer.
Capillary temperature: 20
Running buffer: 100 mM Phosphoric acid containing 16.3 mM β-cyclodextrin adjusted to pH 2.75 with triethanolamine
Injection: Pressure injection at 0.5 psi for 10 s
Detector: F ex 325 (20 mW He-Cd laser) em 450
Migration time: 12.3
Voltage: 18 kV
Model: Beckman P/ACE System 2100

OTHER SUBSTANCES
Simultaneous: zopiclone

REFERENCE
Hempel,G.; Blaschke,G. Enantioselective determination of zopiclone and its metabolites in urine by capillary electrophoresis, *J.Chromatogr.B*, **1996**, *675*, 139–146.

SAMPLE
Matrix: urine
Sample preparation: 1 mL Urine + 900 μL 100 mM pH 4.6 sodium acetate buffer containing 477 U β-glucuronidase (Helix pomatia, Sigma), cool on ice, inject an aliquot.

CAPILLARY ELECTROPHORESIS
Capillary: 67 cm $\times$ 50 μm fused-silica (60 cm to detector)
Capillary preparation: Between runs rinse capillary with 100 mM NaOH for 1 min and with running buffer.
Capillary temperature: 20
Running buffer: 50 mM pH 5.6 phosphate buffer (Prepare buffer by mixing aqueous solutions of Na_2HPO_4 and KH_2PO_4.)
Injection: Pressure injection at 0.5 psi for 10 s (10 nL)
Detector: F ex 325 (20 mW He-Cd laser) em 450
Migration time: 5
Voltage: 20 kV
Model: Beckman P/ACE System 2100
Limit of quantitation: 10 ng/mL
Limit of detection: 2 ng/mL

OTHER SUBSTANCES
Extracted: metabolites

KEY WORDS
pharmacokinetics

REFERENCE
Hempel,G.; Blaschke,G. Direct determination of zolpidem and its main metabolites in urine using capillary electrophoresis with laser-induced fluorescence detection, *J.Chromatogr.B*, **1996**, *675*, 131–137.

Zopiclone

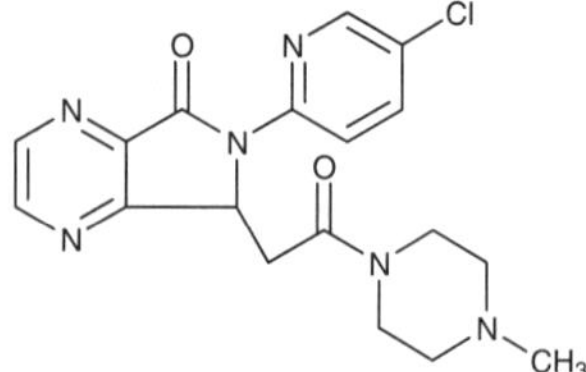

Molecular formula: $C_{17}H_{17}ClN_6O_3$
Molecular weight: 388.81
CAS Registry No.: 43200-80-2
Merck Index (12th ed.): 10324

SAMPLE
Matrix: saliva, urine
Sample preparation: 1 mL Urine or saliva + 2 mL 100 mM pH 8 borate buffer + 50 μL 1 μg/mL zolpidem in MeCN + 5 mL chloroform:isopropanol 90:10, shake for 10 min, centrifuge at 1410 g for 10 min. Remove the organic layer and evaporate it to dryness under a stream of nitrogen at 20°, reconstitute the residue in 50 μL MeCN:water 50:50, inject an aliquot.

CAPILLARY ELECTROPHORESIS
Capillary: 47 cm × 50 μm fused-silica (40 cm to detector)
Capillary preparation: Between runs rinse capillary with 100 mM NaOH for 1 min and with running buffer.
Capillary temperature: 20
Running buffer: 100 mM Phosphoric acid containing 16.3 mM β-cyclodextrin adjusted to pH 2.75 with triethanolamine
Injection: Pressure injection at 0.5 psi for 10 s
Detector: F ex 325 (20 mW He-Cd laser) em 450
Migration time: 13.4 (R-(-)), 14.0 (S-(+))
Internal standard: zolpidem (12.3)
Voltage: 18 kV
Model: Beckman P/ACE System 2100
Limit of detection: 6 ng/mL

OTHER SUBSTANCES
Extracted: metabolites

KEY WORDS
chiral; pharmacokinetics

REFERENCE
Hempel,G.; Blaschke,G. Enantioselective determination of zopiclone and its metabolites in urine by capillary electrophoresis, *J.Chromatogr.B*, **1996**, *675*, 139–146.

SAMPLE
Matrix: solutions
Sample preparation: Inject an aliquot of a 100 μg/mL solution in water:running buffer 50:50.

CAPILLARY ELECTROPHORESIS
Capillary: 44.5 cm × 50 μm acrylamide-coated fused-silica (Bio-Rad)
Capillary temperature: 30
Running buffer: 100 mM NaH_2PO_4 containing 15 mM gamma-cyclodextrin, adjusted to pH 2.5 with phosphoric acid
Injection: Electrokinetic injection at 8 kV for 6 s.
Detector: UV 200
Migration time: 10.04
Voltage: 14 kV
Model: Bio-Rad BioFocus 3000

OTHER SUBSTANCES
Also analyzed: albuterol, alprenolol, atenolol, atropine, baclofen, bamethan, benserazide, biperiden, bisoprolol, bupivacaine, bupranolol, butetamate, carazolol, carbuterol, carvedilol, celiprolol, chloroquine, chlorpheniramine (chlorphenamine), clidinium bromide, clobutinol, disopyramide, dobutamine, flecainide, homatropine, ipratropium bromide, isoproterenol, isothipendyl, ketamine, mefloquine, mequitazine, metaproterenol (orciprenaline), metipranolol, nafronyl (naftidrofuryl), nefopam, ofloxacin, orphenadrine, oxomemazine, oxprenolol, phenoxybenzamine, pholedrine, pindolol, pirbuterol, prilocaine, promethazine, propafenone, propranolol, sotalol, synephrine, tetrahydrozoline (tetryzoline), terbutaline, tocainide, trihexyphenidyl, trimeprazine (alimemazine), trimipramine, tropicamide, verapamil

KEY WORDS
coated capillary; achiral

REFERENCE
Koppenhoefer,B.; Epperlein,U.; Christian,B.; Yibing,J.; Yuying,C.; Bingcheng,L. Separation of enantiomers of drugs by capillary electrophoresis. I. γ-Cyclodextrin as chiral solvating agent, *J.Chromatogr.A*, **1995**, *717*, 181–190.

SAMPLE
Matrix: solutions
Sample preparation: Inject an aliquot of a 100 µg/mL solution in water:running buffer 50:50.

CAPILLARY ELECTROPHORESIS
Capillary: 44.5 cm × 50 µm polyacrylamide-coated fused-silica (40 cm to detector) (Bio-Rad)
Capillary temperature: 30
Running buffer: 100 mM NaH_2PO_4 containing 15 mM β-cyclodextrin, adjusted to pH 2.5 with phosphoric acid
Injection: Electrokinetic injection at 8 kV for 6 s.
Detector: UV 200
Migration time: 9.66, 10.02 (enantiomers)
Voltage: 14 kV
Model: Bio-Rad Bio-Focus 3000

OTHER SUBSTANCES
Simultaneous: carvedilol, chlorpheniramine, ketamine, metaproterenol (orciprenaline), tetrahydrozoline, tropicamide
Also analyzed: albuterol (not chiral), atenolol (not chiral), benserazide (not chiral), biperiden (not chiral), bupivacaine (not chiral), butethamate (not chiral), carazolol (not chiral), carbuterol (not chiral), clidinium bromide (not chiral), disopyramide (not chiral), dobutamine (not chiral), flecainide (not chiral), ipratropium (not chiral), isothipendyl (not chiral), mequitazine (not chiral), metipranolol (not chiral), nafronyl (naftidrofuryl) (not chiral), nefopam (not chiral), ofloxacin (not chiral), orphenadrine (not chiral), pindolol (not chiral), pirbuterol (not chiral), prilocaine (not chiral), propafenone (not chiral), sotalol (not chiral), tocainide (not chiral), trimipramine (not chiral)

KEY WORDS
coated capillary; chiral

REFERENCE
Koppenhoefer,B.; Epperlein,U.; Christian,B.; Lin,B.; Ji,Y.; Chen,Y. Separation of enantiomers of drugs by capillary electrophoresis. III. β-cyclodextrin as chiral solvating agent, *J.Chromatogr.A*, **1996**, *735*, 333–343.

SAMPLE
Matrix: solutions
Sample preparation: Inject an aliquot of a 100 µg/mL solution in running buffer.

CAPILLARY ELECTROPHORESIS
Capillary: 36 cm × 50 µm fused-silica (31.5 cm to detector) (Yongnian Optical Conductive Fiber Plant, China), coated with polyacrylamide

Capillary preparation: No details of the polyacrylamide coating process are provided. However, another paper (LC.GC 1997, 15, 40) by this group indicates that they use the procedure of Hjertén, thus: Adjust the pH of 20 mL water to 3.5 with acetic acid, add 80 µL 3-(trimethoxysilyl)propyl methacrylate (3-methacryloxypropyltrimethoxysilane), mix, suck into capillary, let stand at room temperature for 1 h, remove the solution, wash with water. Fill the capillary with a deaerated 3-4% acrylamide solution containing 1 µL/mL N,N,N',N'-tetramethylethylenediamine and 1 mg/mL potassium persulfate, let stand for 30 min, remove excess solution by aspiration, rinse with water, remove water by aspiration, dry at 35° (J. Chromatogr. 1985, 347, 191).
Capillary temperature: 25
Running buffer: 100 mM NaH_2PO_4 adjusted to pH 2.5 (A) or 100 mM NaH_2PO_4 containing 45 mM hydroxypropyl-gamma-cyclodextrin, adjusted to pH 2.5 (B)
Injection: Electrokinetic injection at 15 kV for 3 s.
Detector: UV 200, UV 210
Migration time: 6.98 (A), 8.31, 8.40 (B, enantiomers)
Voltage: 15 kV
Model: Bio-Rad BioFocus 3000

OTHER SUBSTANCES
Simultaneous: azelastine, biperiden, carvedilol, clidinium bromide, meclizine (meclozine), mequitazine, ofloxacin

---

KEY WORDS
chiral; coated capillary

REFERENCE
Koppenhoefer,B.; Epperlein,U.; Xiaofeng,Z.; Bingcheng,L. Separation of enantiomers of drugs by capillary electrophoresis. Part 4: Hydroxypropyl-γ-cyclodextrin as chiral solvating agent, *Electrophoresis*, **1997**, *18*, 924–930.

SAMPLE
Matrix: solutions
Sample preparation: Inject an aliquot of a 100 µg/mL solution in running buffer.

CAPILLARY ELECTROPHORESIS
Capillary: 30 cm × 50 µm fused-silica (25.5 cm to detector), coated with polyacrylamide
Capillary preparation: Adjust the pH of 20 mL water to 3.5 with acetic acid, add 80 µL 3-(trimethoxysilyl)propyl methacrylate (3-methacryloxypropyltrimethoxysilane), mix, suck into capillary, let stand at room temperature for 1 h, remove the solution, wash with water. Fill the capillary with a deaerated 3-4% acrylamide solution containing 1 µL/mL N,N,N',N'-tetramethylethylenediamine and 1 mg/mL potassium persulfate, let stand for 30 min, remove excess solution by aspiration, rinse with water, remove water by aspiration, dry at 35° (J. Chromatogr. 1985, 347, 191).
Capillary temperature: 25
Running buffer: 100 mM NaH_2PO_4 containing 15 mM β-cyclodextrin, adjusted to pH 2.5 with phosphoric acid
Injection: Electrokinetic injection at 15 kV for 3 s.
Detector: UV 200
Voltage: 15 kV
Model: Bio-Rad BioFocus 3000

KEY WORDS
chiral; coated capillary; comparison with the use of other cyclodextrins; this running buffer gave the greatest enantiomeric separation.; α=1.065

REFERENCE
Lin,B.; Zhu,X.; Koppenhoefer,B.; Epperlein,U. Investigation of 123 chiral drugs by cyclodextrin-modified capillary electrophoresis, *LC.GC*, **1997**, *15*, 40–46.

SAMPLE
Matrix: solutions

CAPILLARY ELECTROPHORESIS

Capillary: 29-36 cm × 50 μm fused-silica (24.5-31.5 cm to detector) (Yongnian Optical Conductive Fiber Plant, China) coated with polyacrylamide

Capillary preparation: Coat capillary as follows. Adjust the pH of 20 mL water to 3.5 with acetic acid, add 80 μL 3-(trimethoxysilyl)propyl methacrylate (3-methacryloxypropyltrimethoxysilane), mix, suck into capillary, let stand at room temperature for 1 h, remove the solution, wash with water. Fill the capillary with a deaerated 3-4% acrylamide solution containing 1 μL/mL N,N,N',N'-tetramethylethylenediamine and 1 mg/mL potassium persulfate, let stand for 30 min, remove excess solution by aspiration, rinse with water, remove water by aspiration, dry at 35° (J. Chromatogr. 1985, 347, 191).

Capillary temperature: 25

Running buffer: 100 mM pH 2.5 NaH_2PO_4 (A) or 100 mM pH 2.5 NaH_2PO_4 containing 45 mM hydroxypropyl-α-cyclodextrin (Wacker, Munich) (B)

Injection: Electromigration at 15 kV for 3 s.

Detector: UV 200; UV 210

Migration time: 6.98 (A); 10.33, 10.64 (B) (enantiomers)

Voltage: 15 kV

Model: Bio-Focus 3000

OTHER SUBSTANCES

Also analyzed: albuterol (salbutamol), alprenolol, amorolfine, atenolol, atropine, azelastine, baclofen, bamethan, benproperine, benserazide, biperiden, bisoprolol, brompheniramine, bupivacaine, bupranolol, butamirate, butethamate, carazolol, carbuterol, carteolol, carvedilol, celiprolol, chloroquine, chlorpheniramine, chlorphenoxamine, cicletanine, clenbuterol, clidinium bromide, clobutinol, dimethindene, dipivefrin, disopyramide, dobutamine, doxylamine, fendiline, flecainide, gallopamil, homatropine, ipratropium bromide, isoproterenol (isoprenaline), isothipendyl, ketamine, meclizine, mefloquine, mepindolol, mequitazine, metaclazepam, metaproterenol (orciprenaline), metipranolol, metoprolol, nafronyl (naftidrofuryl), nefopam, nicardipine, norfenefrine, ofloxacin, ornidazole, orphenadrine, oxomemazine, oxprenolol, oxybutynin, phenoxybenzamine, phenylpropanolamine, pholedrine, pindolol, pirbuterol, prilocaine, procyclidine, promethazine, propafenone, propranolol, reproterol, sotalol, sulpride, synephrine, talinolol, terbutaline, tetrahydrozoline (tetryzoline), theodrenaline, tioconazole, tocainide, trihexyphenidyl, trimeprazine (alimemazine), trimipramine, tropicamide, verapamil

KEY WORDS
coated capillary; chiral

REFERENCE

Koppenhoefer,B.; Eperlein,U.; Schlunk,R.; Zhu,X.; Lin,B. Separation of enantiomers of drugs by capillary electrophoresis. V. Hydroxypropyl-α-cyclodextrin as chiral solvating agent, *J.Chromatogr.A*, **1998**, *793*, 153–164.

BIBLIOGRAPHY OF REVIEWS OF CAPILLARY ELECTROPHORESIS

Altria, K., Kelly, T., and Clark, B., CE methods and buffer preparation. *LC.GC* **1996**, *14*, 398–404.

Altria, K. D., Clark, B. J., Filbey, S. D., Kelly, M. A., and Rudd, D. R., Application of chemometric experimental designs in capillary electrophoresis: a review. *Electrophoresis* **1995**, *16*, 2143–2148.

Altria, K. D., Determination of drug-related impurities by capillary electrophoresis. *J. Chromatogr. A* **1996**, *735*, 43–56.

Altria, K. D. and Bestford, J., Main component assay of pharmaceuticals by capillary electrophoresis: considerations regarding accuracy, and linearity data. *J.Capillary Electrophor.* **1996**, *3*, 13–23.

Altria, K. D., Elgey, J., Lockwood, P., and Moore, D., An overview of the applications of capillary electrophoresis to the analysis of pharmaceutical raw materials and excipients. *Chromatographia* **1996**, *42*, 332–342.

Altria, K. D., Bryant, S. M., Clark, B. J., and Kelly, M. A., The care and maintenance of CE capillaries. *LC.GC* **1997**, *15*, 34–38.

Altria, K. D., Smith, N. W., and Turnbull, C. H., A review of the current status of capillary electrochromatography technology and applications. *Chromatographia* **1997**, *46*, 664–674.

Altria, K. D., Smith, N. W., and Turnbull, C. H., Analysis of acidic compounds using capillary electrochromatography. *J.Chromatogr.B* **1998**, *717*, 341–353.

Armstrong, D. W. and Nair, U. B., Capillary electrophoretic enantioseparations using macrocyclic antibiotics as chiral selectors. *Electrophoresis* **1997**, *18*, 2331–2342.

Baba, Y., Prediction of the behaviour of oligonucleotides in high-performance liquid chromatography and capillary electrophoresis. *J.Chromatogr.* **1993**, *618*, 41–55.

Baba, Y., Analysis of disease-causing genes and DNA-based drugs by capillary electrophoresis. Towards DNA diagnosis and gene therapy for human diseases. *J.Chromatogr.B* **1993**, *687*, 271–302.

Banks, J. F., Recent advances in capillary electrophoresis/electrospray/mass spectrometry. *Electrophoresis* **1997**, *18*, 2255–2266.

Bao, J. J., Capillary electrophoretic immunoassays. *J.Chromatogr.B* **1997**, *699*, 463–480.

Bao, J. J., Fujima, J. M., and Danielson, N. D., Determination of minute enzymatic activities by means of capillary electrophoretic techniques. *J.Chromatogr.B* **1997**, *699*, 481–497.

Bardelmeijer, H. A., Waterval, J. C. M., Lingeman, H., van't Hof, R., Bult, A., and Underberg, W. J. M., Pre-, on- and post-column derivatization in capillary electrophoresis. *Electrophoresis* **1997**, *18*, 2214–2227.

Bardelmeijer, H. A., Lingeman, H., de Ruiter, C., and Underberg, W. J. M., Derivatization in capillary electrophoresis. *J.Chromatogr.A* **1998**, *807*, 3–26.

Bazzanella, A., Mörbel, H., Bächmann, K., Milbradt, R., Böhmer, V., and Vogt, W., Highly efficient separation of amines by electrokinetic chromatography using resorcarene-octacarboxylic acids as pseudostationary phases. *J.Chromatogr.A* **1997**, *792*, 143–149.

Beale, S. C., Capillary electrophoresis. *Anal.Chem.* **1998**, *70*, 279R-300R.

Bojarski, J. and Aboul-Enein, H. Y., Application of capillary electrophoresis for the analysis of chiral drugs in biological fluids. *Electrophoresis* **1997**, *18*, 965–969.

Bressolle, F., Audran, M., Pham, T.-N., and Vallon, J.-J., Cyclodextrins and enantiomeric separations of drugs by liquid chromatography and capillary electrophoresis: basic principles and new developments. *J.Chromatogr.B* **1996**, *687*, 303–336.

Bruchelt, G., Niethammer, D., and Schmidt, K. H., Isotachophoresis of nucleic acid constituents. *J.Chromatogr.* **1993**, *618*, 57–77.

Brumley, W. C., Techniques with potential for handling environmental samples in capillary electrophoresis. *J.Chromatogr.Sci.* **1995**, *33*, 670–685.

Brunner, L. J., DiPiro, J. T., and Feldman, S., High-performance capillary electrophoresis in the pharmaceutical sciences. *Pharmacotherapy* **1995**, *15*, 1–22.

Brunner, L. J. and DiPiro, J. T., Capillary electrophoresis for therapeutic drug monitoring. *Electrophoresis* **1998**, *19*, 2848–2855.

Cai, J. and Henion, J., Capillary electrophoresis-mass spectrometry. *J.Chromatogr.A* **1995**, *703*, 667–692.

Camilleri, P., Chiral surfactants in micellar electrokinetic capillary chromatography. *Electrophoresis* **1997**, *18*, 2322–2330.

Chang, H.-T. and Yeung, E. S., Dynamic control to improve the separation performance in capillary electrophoresis. *Electrophoresis* **1995**, *16*, 2069–2073.

Chankvetadze, B., Separation selectivity in chiral capillary electrophoresis with charged selectors. *J.Chromatogr.A* **1997**, *792*, 269–295.

Chen, F. T. and Sternberg, J. C., Characterization of proteins by capillary electrophoresis in fused-silica columns: review on serum protein analysis and application to immunoassays. *Electrophoresis* **1994**, *15*, 13–21.

Chen, S.-H. and Gallo, J. M., Use of capillary electrophoresis methods to characterize the pharmacokinetics of antisense drugs. *Electrophoresis* **1998**, *19*, 2861–2869.

Cifuentes, A. and Poppe, H., Behavior of peptides in capillary electrophoresis: Effect of peptide charge, mass and structure. *Electrophoresis* **1997**, *18*, 2362–2376.

Cikalo, M. G., Bartle, K. D., Robson, M. M., Myers, P., and Euerby, M. R., Capillary electrochromatography. Tutorial Review. *Analyst* **1993**, *123*, 87R–102R.

Colón, L. A., Guo, Y., and Fermier, A., Capillary electrochromatography. *Anal.Chem.* **1997**, *69*, 461A–467A.

Colón, L. A., Reynolds, K. J., Alicea-Maldonado, R., and Fermier, A. M., Advances in capillary electrochromatography. *Electrophoresis* **1997**, *18*, 2162–2174.

Corradini, D., Buffer additives other than the surfactant sodium dodecyl sulfate for protein separations by capillary electrophoresis. *J.Chromatogr.B* **1997**, *699*, 221–256.

Corstjens, H., Billiet, H. A. H., Frank, J., and Luyben, K. C. A. M., Optimization of selectivity in capillary electrophoresis with emphasis on micellar electrokinetic capillary chromatography. *J.Chromatogr.A* **1995**, *715*, 1–11.

Coudrec, F., Caussé, E., and Bayle, C., Drug analysis by capillary electrophoresis and laser-induced fluorescence. *Electrophoresis* **1998**, *19*, 2777–2790.

Craston, D. H. and Saeed, M., Analysis of carboxylic acids in the environment by capillary electrophoretic techniques. *J.Chromatogr.A* **1998**, *827*, 1–12.

Crego, A. L. and Marina, M. L., Capillary zone electrophoresis versus micellar electrokinetic chromatography in the separation of phenols of environmental interest. *J.Liq.Chromatogr.Rel.Technol.* **1997**, *20*, 1–20.

Cui, H., Leon, J., Reusaet, E., and Bult, A., Selective determination of peptides containing specific amino acid residues by high-performance liquid chromatography and capillary electrophoresis. *J.Chromatogr.A* **1995**, *704*, 27–36.

Dabek-Zlotorzynska, E., Capillary electrophoresis in the determination of pollutants. *Electrophoresis* **1997**, *18*, 2453–2464.

Dawson, L. A., Capillary electrophoresis and microdialysis: current technology and applications. *J.Chromatogr.B* **1997**, *697*, 89–99.

DeDionisio, L. A. and Lloyd, D. H., Capillary gel electrophoresis and antisense therapeutics. Analysis of DNA analogs. *J.Chromatogr.A* **1996**, *735*, 191–208.

Denoroy, L., Bert, L., Parrot, S., Robert, F., and Renaud, B., Assessment of pharmacodynamic and pharmacokinetic characteristics of drugs using microdialysis sampling and capillary electrophoresis. *Electrophoresis* **1998**, *19*, 2841–2847.

Denton, K. A. and Tate, S. A., Capillary electrophoresis of recombinant proteins. *J.Chromatogr.B* **1997**, *697*, 111–121.

Deyl, Z. and Struzinsky, R., Capillary zone electrophoresis: its applicability and potential in biochemical analysis. *J.Chromatogr.* **1991**, *569*, 63–122.

Deyl, Z., Tagliaro, F., and Miksik, I., Biomedical applications of capillary electrophoresis. *J.Chromatogr.B* **1994**, *656*, 3–27.

Dittman, M. M., Wienand, K., Bek, F., and Rozing, G. P., Theory and practice of capillary electrochromatography. *LC.GC* **1995**, *13*, 800–814.

Dolnik, V., Capillary zone electrophoresis of proteins. *Electrophoresis* **1997**, *18*, 2353–2361.

Dovichi, N. J., DNA sequencing by capillary electrophoresis. *Electrophoresis* **1997**, *18*, 2393–2399.

Effenhauser, C. S., Bruin, G. J. M., and Paulus, A., Integrated chip-based capillary electrophoresis. *Electrophoresis* **1997**, *18*, 2203–2213.

El Rassi, Z., Capillary electrophoresis of carbohydrates. *Adv.Chromatogr.* **1994**, *34*, 177–250.

El Rassi, Z. and Mechref, Y., Recent advances in capillary electrophoresis of carbohydrates. *Electrophoresis* **1996**, *17*, 275–301.

El Rassi, Z., Recent developments in capillary electrophoresis of carbohydrate species. *Electrophoresis* **1997**, *18*, 2400–2407.

El Rassi, Z., Capillary electrophoresis of pesticides. *Electrophoresis* **1997**, *18*, 2465–2481.

Esaka, Y. and Kano, K., Hydrophilic selectors for selectivity enhancement in capillary electrophoresis. *J.Chromatogr.A* **1997**, *792*, 445–454.

Euerby, M. R., Johnson, C. M., and Bartle, K. D., Practical experiences and applications of capillary clectrochromatography in the pharmaceutical industry. *LC.GC* **1998**, *16* 386–394.

Fanali, S., Identification of chiral drug isomers by capillary electrophoresis. *J.Chromatogr.A* **1996**, *735*, 77–121.

Fanali, S., Controlling enantioselectivity in chiral capillary electrophoresis with inclusion–complexation. *J.Chromatogr.A* **1997**, *792*, 227–268.

Fillet, M., Hubert, P., and Crommen, J., Method development strategies for the enantioseparation of drugs by capillary electrophoresis using cyclodextrins as chiral additives. *Electrophoresis* **1998**, *19*, 2834–2840.

Flurer, C. L., Analysis of antibiotics by capillary electrophoresis. *Electrophoresis* **1997**, *18*, 2427–2437.

Frantzen, F., Chromatographic and electrophoretic methods for modified hemoglobins. *J.Chromatogr.B* **1997**, *699*, 269–286.

García Campaña, A. M., Baeyens, W. R. G., and Zhao, Y., Chemiluminescence detection in capillary electrophoresis. *Anal.Chem.* **1997**, *69*, 83A–88A.

García Campaña, A. M., Baeyens, W. R. G., and Guzman, N. A., Trends toward sensitive detection in capillary electrophoresis: an overview of some recent developments. *Biomed.Chromatogr.* **1998**, *12*, 172–176.

Gas, B., Stedry, M., and Kenndler, E., Peak broadening in capillary zone electrophoresis. *Electrophoresis* **1997**, *18*, 2123–2133.

Gebauer, P. and Bocek, P., Recent application and developments of capillary isotachophoresis. *Electrophoresis* **1997**, *18*, 2154–2161.

Göttlicher, B. and Bachmann, K., Application of particles as pseudo-stationary phases in electrokinetic chromatography. *J.Chromatogr.A* **1997**, *780*, 63–73.

Grimshaw, J., Analysis of glycosaminoglycans and their oligosaccharide fragments by capillary electrophoresis. *Electrophoresis* **1997**, *18*, 2408–2414.

Guttman, A., Brunet, S., and Cooke, N., Capillary electrophoresis separation of enantiomers by cyclodextrin array chiral analysis. *LC.GC* **1996**, *14*, 32–42.

Guzman, N. A., Park, S. S., Schaufelberger, D., Hernandez, L., Paez, X., Rada, P., Tomlinson, A. J., and Naylor, S., New approaches in clinical chemistry: on-line analyte concentration and microreaction capillary electrophoresis for the determination of drugs, metabolic intermediates, and biopolymers in biological fluids. *J.Chromatogr.B* **1997**, *697*, 37–66.

Gübitz, G and Schmid, M. G., Chiral separation principles in capillary electrophoresis. *J.Chromatogr.A* **1997**, *792*, 179–225.

Haddad, P. R., Macka, M., Hilder, E. F., and Bogan, D. P., Separation of metal ions and metal-containing species by micellar electrokinetic capillary chromatography, including utilisation of metal ions in separations of other species. *J.Chromatogr.A* **1997**, *780*, 329–341.

Hage, D. S., Chiral separations in capillary electrophoresis using proteins as stereoselective binding agents. *Electrophoresis* **1997**, *18*, 2311–2321.

Hage, D. S. and Tweed, S. A., Recent advances in chromatographic and electrophoretic methods for the study of drug-protein interactions. *J.Chromatogr.B* **1997**, *699*, 499–525.

Hase, S., Precolumn derivatization for chromatographic and electrophoretic analyses of carbohydrates. *J.Chromatogr.A* **1996**, *720*, 173–182.

Heiger, D., Majors, R. E., and Lombardi, R. A., Method development strategies in capillary electrophoresis. *LC.GC* **1997**, *15*, 14–23.

Hietpas, P. B. and Ewing, A. G., On-column and post-column derivatization for capillary electrophoresis with laser-induced fluorescence for the analysis of single cells. *J.Liq.Chromatogr.* **1995**, *18*, 3557–3576.

Holland, L. A., Chetwyn, N. P., Perkins, M. D., and Lunte, S. M., Capillary electrophoresis in pharmaceutical analysis. *Pharm.Res.* **1997**, *14*, 372–387.

Honda, S., Separation of neutral carbohydrates by capillary electrophoresis. *J.Chromatogr.A* **1996**, *720*, 337–351.

Hows, M. E. P., Alfazema, L. N., and Perrett, D., Capillary electrophoresis buffers—Approaches to improving their performance. *LC.GC* **1997**, *15*, 1034–1043.

Issaq, H. J., Chan, K. C., Muschik, G. M., and Janini, G. M., Applications of capillary zone electrophoresis and micellar electrokinetic chromatography in cancer research. *J.Liq.Chromatogr.* **1995**, *18*, 1273–1288.

Issaq, H. J., Janini, G. M., Chan, K. C., and El Rassi, Z., Approaches for the optimization of experimental parameters in capillary zone electrophoresis. *Adv. Chromatogr.* **1995**, *35*, 101–169.

Issaq, H. J., Capillary electrophoresis of natural products. *Electrophoresis* **1997**, *18*, 2438–2452.

Jenkins, M. A. and Guerin, M. D., Capillary electrophoresis as a clinical tool. *J.Chromatogr.B* **1996**, *682*, 23–34.

Jenkins, M. A. and Guerin, M. D., Capillary electrophoresis procedures for serum protein analysis: comparison with established techniques. *J.Chromatogr.B* **1997**, *699*, 257–268.

Kakehi, K. and Honda, S., Analysis of glycoproteins, glycopeptides and glycoprotein-derived oligosaccharides by high-performance capillary electrophoresis. *J.Chromatogr.A* **1996**, *720*, 377–393.

Karger, B. L., Cohen, A. S., and Guttman, A., High-performance capillary electrophoresis in the biological sciences. *J.Chromatogr.* **1989**, *492*, 585–614.

Kasicka, V. and Prusik, Z., Application of capillary isotachophoresis in peptide analysis. *J.Chromatogr.* **1991**, *569*, 123–174.

Kataoka, Y., Makino, K., and Oishi, R., Capillary electrophoresis for therapeutic drug monitoring of antiepileptics. *Electrophoresis* **1998**, *19*, 2856–2860.

Katsuta, S. and Saitoh, K., Control of separation selectivity in micellar electrokinetic chromatography by modification of the micellar phase with solubilized organic compounds. *J.Chromatogr.A* **1997**, *780*, 165–178.

Khaledi, M. G., Micelles as separation media in high-performance liquid chromatography and high-performance capillary electrophoresis: overview and perspective. *J.Chromatogr.A* **1997**, *780*, 3–40.

Kitagishi, K. and Shintani, H., Analysis of compounds containing carboxyl groups in biological fluids by capillary electrophoresis. *J.Chromatogr.B* **1998**, *717*, 327–339.

Kok, S. J., Velthorst, N. H., Gooijer, C., and Brinkman, U. A. T., Analyte identification in capillary electrophoretic separation techniques. *Electrophoresis* **1998**, *19*, 2753–2776.

Krivánková, L. and Bocek, P., Synergism of capillary electrophoresis and capillary zone electrophoresis. *J.Chromatogr.B* **1997**, *689*, 13–34.

Krull, I. S. and Mazzeo, J. R., Capillary electrophoresis: the promise and the practice. *Nature* **1992**, *357*, 92–94.

Krull, I. S., Deyl, Z., and Lingeman, H., General strategies and selection of derivatization reactions for liquid chromatography and capillary electrophoresis. *J.Chromatogr.B* **1994**, *659*, 1–17.

Krull, I. S., Strong, R., Sosic, Z., Cho, B.-Y., Beale, S. C., Wang, C.-C., and Cohen, S., Labeling reactions applicable to chromatography and electrophoresis of minute amounts of proteins. *J.Chromatogr.B* **1997**, *699*, 173–208.

Kunkel, A., Degenhardt, M., Schirm, B., and Wätzig, H., Performance of instruments and aspects of methodology and validation in quantitative capillary electrophoresis. An update. *J.Chromatogr.A* **1997**, *768*, 17–27.

Landers, J. P., Capillary electrophoresis: pioneering new approaches for biomolecular analysis. *Trends Biochem.Sci.* **1993**, *18*, 409–414.

Landers, J. P., Oda, R. P., Spelsberg, T. C., Nolan, J. A., and Ulfelder, K. J., Capillary electrophoresis: a powerful microanalytical technique for biologically active molecules. *BioTechniques* **1993**, *14*, 98–111.

Landers, J. P., Clinical capillary electrophoresis. *Clin.Chem.* **1995**, *41*, 495–509.

Larsen, K. L. and Zimmermann, W., Analysis and characterisation of cyclodextrins and their inclusion complexes by affinity capillary electrophoresis. *J.Chromatogr.A* **1999**, *836*, 3–14.

Lehmann, R., Voelter, W., and Liebich, H. M., Capillary electrophoresis in clinical chemistry. *J.Chromatogr.B* **1997**, *697*, 3–35.

Levêque, D., Gallion-Renault, C., Monteil, H., and Jehl, F., Capillary electrophoresis for pharmacokinetic studies. *J.Chromatogr.B* **1997**, *697*, 67–75.

Linhardt, R. J. and Pervin, A., Separation of negatively charged carbohydrates by capillary electrophoresis. *J.Chromatogr.A* **1996**, *720*, 323–335.

Liu, X., Sosic, Z., and Krull, I. S., Capillary isoelectric focusing as a tool in the examination of antibodies, peptides and proteins of pharmaceutical interest. *J.Chromatogr.A* **1996**, *735*, 165–190.

Lloyd, D. K., Capillary electrophoretic analyses of drugs in body fluids: sample pretreatment and methods for direct injection of biofluids. *J.Chromatogr.A* **1996**, *735*, 29–42.

Lloyd, D. K., Aubry, A.-F., and De Lorenzi, E., Selectivity in capillary electrophoresis: the use of proteins. *J.Chromatogr.A* **1997**, *792*, 349–369.

Lunte, S. M. and O'Shea, T. J., Pharmaceutical and biomedical applications of capillary electrophoresis/electrochemistry. *Electrophoresis* **1994**, *15*, 79–86.

Lunte, S. M. and Lunte, C. E., Microdialysis sampling for pharmacological studies: HPLC and CE analysis. *Adv.Chromatogr.* **1996**, *36*, 383–432.

Luong, J. H. T. and Nguyen, A. L., Achiral selectivity in cyclodextrin-modified capillary electrophoresis. *J.Chromatogr.A* **1997**, *792*, 431–444.

Lurie, I. S., Application of micellar electrokinetic capillary chromatography to the analysis of illicit drug seizures. *J.Chromatogr.A* **1997**, *780*, 265–284.

Lurie, I. S., Separation selectivity in chiral and achiral capillary electrophoresis with mixed cyclodextrins. *J.Chromatogr.A* **1997**, *792*, 297–307.

Macka, M. and Haddad, P. R., Determination of metal ions by capillary electrophoresis. *Electrophoresis* **1997**, *18*, 2482–2501.

MacTaylor, C. E. and Ewing, A. G., Critical review of recent developments in fluorescence detection for capillary electrophoresis. *Electrophoresis* **1997**, *18*, 2279–2290.

Majors, R. E., The role of column diameter in chromatography and capillary electrophoresis. *LC.GC* **1996**, *14*, 10–19.

Majors, R. E., The separation of chiral compounds by GC, SFC, and CE. *LC.GC* **1997**, *15*, 412–422.

Marchi, E. and Pasacreta, R. J., Capillary electrophoresis in court: the landmark decision of The People of Tennessee versus Ware. *J.Capillary Electrophor.* **1997**, *4*, 145–156.

Matsubara, N. and Terabe, S., Micellar electrokinetic chromatography. *Methods Enzymol.* **1996**, *270*, 319–341.

Mazereeuw, M., Tjaden, U. R., and Reinhoud, N. J., Single capillary isotachophoresis-zone electrophoresis: current practice and prospects, a review. *J.Chromatogr.Sci.* **1995**, *33*, 686–697.

Mazzeo, J. R. and Krull, I. S., Coated capillaries and additives for the separation of proteins by capillary zone electrophoresis and capillary isoelectric focusing. *BioTechniques* **1991**, *10*, 638–645.

McLaughlin, G. M., Nolan, J. A., Lindahl, J. L., Palmieri, R. H., Anderson, K. W., Morris, S. C., Morrison, J. A., and Bronzert, T. J., Pharmaceutical drug separations by HPCE: practical guidelines. *J.Liq.Chromatogr.* **1992**, *15*, 961–1021.

Messana, I., Rossetti, D. V., Cassiano, L., Misiti, F., Giardina, B., and Castagnola, M., Peptide analysis by capillary (zone) electrophoresis. *J.Chromatogr.B* **1997**, *699*, 149–171.

Miyawa, J. H. and Alasandro, M. S., Capillary electrochromatography separations. *LC-GC* **1998**, *16*, 36–41.

Monnig, C. A. and Kennedy, R. T., Capillary electrophoresis. *Anal.Chem.* **1994**, *66*, 280R–314R.

Muijselaar, P. G., Retention indices in micellar electrokinetic chromatography. *J.Chromatogr.A* **1997**, *780*, 117–127.

Muijselaar, P. G., Otsuka, K., and Terabe, S., Micelles as pseudo-stationary phases in micellar electrokinetic chromatography. *J.Chromatogr.A* **1997**, *780*, 41–61.

Nguyen, N. T. and Siegler, R. W., Capillary electrophoresis of cardiovascular drugs. *J.Chromatogr.A* **1996**, *735*, 123–150.

Nishi, H. and Terabe, S., Application of micellar electrokinetic chromatography to pharmaceutical analysis. *Electrophoresis* **1990**, *11*, 691–701.

Nishi, H. and Terabe, S., Application of electrokinetic chromatography to pharmaceutical analysis. *J.Pharm.Biomed.Anal.* **1993**, *11*, 1277–1287.

Nishi, H. and Terabe, S., Optical resolution of drugs by capillary electrophoretic techniques. *J.Chromatogr.A* **1995**, *694*, 245–276.

Nishi, H., Micellar electrokinetic chromatography. Perspectives in drug analysis. *J.Chromatogr.A* **1996**, *735*, 3–27.

Nishi, H., Enantiomer separation of drugs by electrokinetic chromatography. *J.Chromatogr.A* **1996**, *735*, 57–76.

Nishi, H., Pharmaceutical applications of micelles in chromatography and electrophoresis. *J.Chromatogr.A* **1997**, *780*, 243–264.

Nishi, H., Enantioselectivity in chiral capillary electrophoresis with polysaccharides. *J.Chromatogr.A* **1997**, *792*, 327–347.

Novotny, M., Soini, H., and Stefansson, M., Chiral separation through capillary electromigration methods. *Anal.Chem.* **1994**, *66*, 646A–655A.

Novotny, M. V., Cobb, K. A., and Liu, J. P., Recent advances in capillary electrophoresis of proteins, peptides and amino acids. *Electrophoresis* **1990**, *11*, 735–749.

Novotny, M. V. and Sudor, J., High-performance capillary electrophoresis of glycoconjugates. *Electrophoresis* **1993**, *14*, 373–389.

Oefner, P. J. and Chiesa, C., Capillary electrophoresis of carbohydrates. *Glycobiology.* **1994**, *4*, 397–412.

Oravcová, J., Böhs, B., and Lindner, W., Drug-protein binding studies. New trends in analytical and experimental methodology. *J.Chromatogr.B* **1996**, *677*, 1–28.

Ossicini, L. and Fanali, S., Enantiomeric separations by electrophoretic techniques. *Chromatographia* **1997**, *45*, 428–432.

Palmer, C. P., Micelle polymers, polymer surfactants and dendrimers as pseudo-stationary phases in micellar electrokinetic chromatography. *J.Chromatogr.A* **1997**, *780*, 75–92.

Palmer, C. P. and Tanaka, M., Selectivity of polymeric and polymer-supported pseudo-stationary phases in micellar electrokinetic chromatography. *J.Chromatogr.A* **1997**, *792*, 105–124.

Paulus, A. and Klockow, A., Detection of carbohydrates in capillary electrophoresis. *J.Chromatogr.A* **1996**, *720*, 353–376.

Pesek, J. J., Matyska, M. T., Sandoval, J. E., and Williamsen, E. J., Synthesis, characterization and applications of hydride-based surface materials for HPLC, HPCE and electrochromatography. *J.Liq.Chromatogr.* **1996**, *19*, 2843–2865.

Pesek, J. J. and Matyska, M. T., Column technology in capillary electrophoresis and capillary electrochromatography. *Electrophoresis* **1997**, *18*, 2228–2238.

Poole, C. F. and Poole, S. K., Interphase model for retention and selectivity in micellar electrokinetic chromatography. *J.Chromatogr.A* **1997**, *792*, 89–104.

Pritchett, T. J., Capillary isoelectric focusing of proteins. *Electrophoresis* **1996**, *17*, 1195–1201.

Quirino, J. P. and Terabe, S., Stacking of neutral analytes in micellar electrokinetic chromatography. *J.Capillary Electrophor.* **1997**, *4*, 233–245.

Rabel, S. R. and Stobaugh, J. F., Applications of capillary electrophoresis in pharmaceutical analysis. *Pharm.Res.* **1993**, *10*, 171–186.

Rathore, A. S. and Horváth, C., Capillary electrochromatography: theories on electroosmotic flow in porous media. *J.Chromatogr.A* **1997**, *781*, 185–195.

Recio, I., Amigo, L., and López-Fandiño, R., Assessment of the quality of dairy products by capillary electrophoresis of milk proteins. *J.Chromatogr.B* **1997**, *697*, 231–242.

Reif, O. W., Lausch, R., and Freitag, R., High-performance capillary electrophoresis of human serum and plasma proteins. *Adv.Chromatogr.* **1994**, *34*, 1–56.

Riekkola, M.-L. and Jumppanen, J. H., Capillary electrophoresis of diuretics. *J.Chromatogr.A* **1996**, *735*, 151–164.

Riekkola, M.-L., Wiedmer, S. K., Valkó, I. E., and Sirén, H., Selectivity in capillary electrophoresis in the presence of micelles, chiral selectors and non-aqueous media. *J.Chromatogr.A* **1997**, *792*, 13–35.

Righetti, P. G., Recent developments in electrophoretic methods. *J.Chromatogr.* **1990**, *516*, 3–22.

Righetti, P. G. and Gelfi, C., Electrophoresis gel media: the state of the art. *J.Chromatogr.B* **1997**, *699*, 63–75.

Righetti, P. G. and Gelfi, C., Non-isocratic capillary electrophoresis for detection of DNA point mutations. *J.Chromatogr.B* **1997**, *697*, 195–205.

Righetti, P. G., Gelfi, C., Perego, M., Stoyanov, A. V., and Bossi, A., Capillary zone electrophoresis of oligonucleotides and peptides in isoelectric buffers: theory and methodology. *Electrophoresis* **1997**, *18*, 2145–2153.

Rippel, G., Corstjens, H., Billiet, H. A. H., and Frank, J., Affinity capillary electrophoresis. *Electrophoresis* **1997**, *18*, 2175–2183.

Robert, F., Bouilloux, J. P., and Denoroy, L., Capillary electrophoresis: principle and applications. *Ann.Biol.Clin.(Paris)* **1991**, *49*, 137–148.

Rodriguez-Diaz, R., Wehr, T., and Zhu, M., Capillary isoelectric focusing. *Electrophoresis* **1997**, *18*, 2134–2144.

Rundlett, K. L. and Armstrong, D. W., Methods for the estimation of binding constants by capillary electrophoresis. *Electrophoresis* **1997**, *18*, 2194–2202.

Sadecka, J., Polonsky, J., and Shintani, H., Current advancement of pharmaceutical analysis by capillary zone electrophoresis, micellar electrokinetic chromatography and isotachophoresis. *Pharmazie* **1994**, *49*, 631–641.

Sarmini, K. and Kenndler, E., Influence of organic solvents on the separation selectivity in capillary electrophoresis. *J.Chromatogr.A* **1997**, *792*, 3–11.

Schmalzing, D. and Nashabeh, W., Capillary electrophoresis based immunoassays: a critical review. *Electrophoresis* **1997**, *18*, 2184–2193.

Schweitz, L., Andersson, L. I., and Nilsson, S., Molecular imprint-based stationary phases for capillary electrochromatography. *J.Chromatogr.A* **1998**, *817*, 5–13.

Shamsi, S. A. and Warner, I. M., Monomeric and polymeric chiral surfactants as pseudo-stationary phases for chiral separations. *Electrophoresis* **1997**, *18*, 853–872.

Shelton, C. M., Koch, J. T., Desai, N., and Wheeler, J. F., Enhanced selectivity for capillary zone electrophoresis using ion-pair agents. *J.Chromatogr.A* **1997**, *792*, 455–462.

Shihabi, Z. K., Clinical applications of capillary electrophoresis. *Ann.Clin.Lab.Sci.* **1992**, *22*, 398–405.

Shihabi, Z. K., Therapeutic drug monitoring by capillary electrophoresis. *J.Chromatogr.A* **1998**, *807*, 27–36.

Smith, J. T., Developments in amino acid analysis using capillary electrophoresis. *Electrophoresis* **1997**, *18*, 2377–2392.

Smith, N. W. and Evans, M. B., Capillary zone electrophoresis in pharmaceutical and biomedical analysis. *J.Pharm.Biomed.Anal.* **1994**, *12*, 579–611.

Smyth, W. F. and McClean, S., A critical evaluation of the application of capillary electrophoresis to the detection and determination of 1,4-benzodiazepine tranquilizers in formulations and body materials. *Electrophoresis* **1998**, *19*, 2870–2882.

Song, L., Xu, Z., Kang, J., and Cheng, J., Analysis of environmental pollutants by capillary electrophoresis with emphasis on micellar electrokinetic chromatography. *J.Chromatogr.A* **1997**, *780*, 297–328.

Staller, T. D. and Sepaniak, M. J., Chemiluminescence detection in capillary electrophoresis. *Electrophoresis* **1997**, *18*, 2291–2296.

Strege, M. A. and Lagu, A. L., Micellar electrokinetic chromatography of proteins. *J.Chromatogr.A* **1997**, *780*, 285–296.

Strege, M. A. and Lagu, A. L., Capillary electrophoresis of biotechnology-derived proteins. *Electrophoresis* **1997**, *18*, 2343–2352.

Sunada, W. M. and Blanch, H. W., Polymeric separation media for capillary electrophoresis of nucleic acids. *Electrophoresis* **1997**, *18*, 2243–2254.

Sutton, R. M. C., Sutton, K. L., and Stalcup, A. M., Chiral capillary electrophoresis with noncyclic oligo- and polysaccharide chiral selectors. *Electrophoresis* **1997**, *18*, 2297–2304.

Suzuki, S. and Honda, S., A tabulated review of capillary electrophoresis of carbohydrates. *Electrophoresis* **1998**, *19*, 2539–2560.

Szökö, É., Protein and peptide analysis by capillary zone electrophoresis and micellar electrokinetic chromatography. *Electrophoresis* **1997**, *18*, 74–81.

Szulc, M. E. and Krull, I. S., Improved detection and derivatization in capillary electrophoresis. *J.Chromatogr.A* **1994**, *659*, 231–245.

Tadey, T. and Purdy, W. C., Chromatographic techniques for the isolation and purification of lipoproteins. *J.Chromatogr.B* **1995**, *671*, 237–253.

Tagliaro, F., Turrina, S., Pisi, P., Smith, F. P., and Marigo, M., Determination of illicit and/or abused drugs and compounds of forensic interest in biosamples by capillary electrophoretic/electrokinetic methods. *J.Chromatogr.B* **1998**, *713*, 27–49.

Takagi, T., Capillary electrophoresis in presence of sodium dodecyl sulfate and a sieving medium. *Electrophoresis* **1997**, *18*, 2239–2242.

Taylor, R. B., Toasaksiri, S., and Reid, R. G., A literature assessment of sample pretreatments and limits of detection for capillary electrophoresis of drugs in biological fluids and practical investigation with some antimalarials in plasma. *Electrophoresis* **1998**, *19*, 2791–2797.

Thormann, W., Lienhard, S., and Wernly, P., Strategies for the monitoring of drugs in body fluids by micellar electrokinetic capillary chromatography. *J.Chromatogr.* **1993**, *636*, 137–148.

Thormann, W., Molteni, S., Caslavska, J., and Schmutz, A., Clinical and forensic applications of capillary electrophoresis. *Electrophoresis* **1994**, *15*, 3–12.

Thormann, W., Zhang, C.-X., and Schmutz, A., Capillary electrophoreis for drug analysis in body fluids. *Ther.Drug Monit.* **1996**, *18*, 506–520.

Timerbaev, A. R., Analysis of inorganic pollutants by capillary electrophoresis. *Electrophoresis* **1997**, *18*, 185–195.

Timerbaev, A. R., Strategies for selectivity control in capillary electrophoresis of metal species. *J.Chromatogr.A* **1997**, *792*, 495–518.

Tomlinson, A. J., Benson, L. M., Guzman, N. A., and Naylor, S., Preconcentration and microreaction technology on-line with capillary electrophoresis. *J.Chromatogr.A* **1996**, *744*, 3–15.

Veltkamp, A. C., Radiochromatography in pharmaceutical and biomedical analysis. *J.Chromatogr.* **1990**, *531*, 101–129.

Ventura, R. and Segura, J., Detection of diuretic agents in doping control. *J.Chromatogr.B* **1996**, *687*, 127–144.

Verleysen, K. and Sandra, P., Separation of chiral compounds by capillary electrophoresis. *Electrophoresis* **1998**, *19*, 2798–2833.

Vespalec, R. and Bocek, P., Chiral separations by capillary electrophoresis: present state of the art. *Electrophoresis* **1994**, *15*, 755–762.

Vespalec, R. and Bocek, P., Chiral separations by capillary zone electrophoresis: Present state of the art. *Electrophoresis* **1997**, *18*, 843–852.

Vigh, G. and Sokolowski, A. D., Capillary electrophoretic separations of enantiomers using cyclodextrin containing background electrolytes. *Electrophoresis* **1997**, *18*, 2305–2310.

Voegel, P. D. and Baldwin, R. P., Electrochemical detection in capillary electrophoresis. *Electrophoresis* **1997**, *18*, 2267–2278.

von Heeren, F. and Thormann, W., Capillary electrophoresis in clinical and forensic analysis. *Electrophoresis* **1997**, *18*, 2415–2426.

Wallingford, R. A. and Ewing, A. G., Capillary electrophoresis. *Adv. Chromatogr.* **1989**, *29*, 1–76.

Ward, T. J., Chiral media for capillary electrophoresis. *Anal. Chem.* **1994**, *66*, 632A–640A.

Ward, T. J. and Oswald, T. M., Enantioselectivity in capillary electrophoresis using the macrocyclic antibiotics. *J.Chromatogr.A* **1997**, *792*, 309–325.

Watarai, H., Microemulsions in separation sciences. *J.Chromatogr.A* **1997**, *780*, 93–102.

Watzig, H. and Dette, C., Capillary electrophoresis (CE)—a review. Strategies for method development and applications related to pharmaceutical and biological sciences. *Pharmazie* **1994**, *49*, 83–96.

Wätzig, H., Degenhardt, M., and Kunkel, A., Strategies for method development and validation in CE related to pharmaceutical and biological applications. *Electrophoresis* **1998**, *19*, 2695–2752.

Werner, A., Reversed-phase and ion-pair separations of nucleotides, nucleosides and nucleobases: analysis of biological samples in health and disease. *J.Chromatogr.* **1993**, *618*, 3–14.

Wong, S. H., Advances in chromatography for clinical drug analysis: supercritical fluid chromatography, capillary electrophoresis, and selected high-performance liquid chromatography techniques. *Ther.Drug Monit.* **1993**, *15*, 576–580.

Wyss, R., Chromatographic and electrophoretic analysis of biomedically important retinoids. *J.Chromatogr.B* **1995**, *671*, 381–425.

Yang, L. and Lee, C. S., Micellar electrokinetic chromatography—mass spectrometry. *J.Chromatogr.A* **1997**, *780*, 207–218.

Yang, Q., Hidajat, K., and Li, S. F. Y., Trends in capillary electrophoresis: 1997. *J.Chromatogr.Sci.* **1997**, *35*, 358–373.